Imaging Optics

This comprehensive and self-contained text presents the fundamentals of optical imaging from the viewpoint of both ray and wave optics, within a single volume. Comprising three distinct parts, it opens with an introduction to electromagnetic theory, including electromagnetic diffraction problems and how they can be solved with the aid of standard numerical methods such as RCWA or FDTD. The second part is devoted to the basic theory of geometrical optics and the study of optical aberrations inherent in imaging systems, including large-scale telescopes and high-resolution projection lenses. A detailed overview of state-of-the-art optical system design provides readers with the necessary tools to successfully use commercial optical design software. The final part explores diffraction theory and concludes with vectorial wave propagation, image formation and image detection in high-aperture imaging systems. The wide-ranging perspective of this important book provides researchers and professionals with a comprehensive and rigorous treatise on the theoretical and applied aspects of optical imaging.

Joseph Braat is Emeritus Professor of Optics at the Delft Technical University, The Netherlands, and a Fellow of the Royal Netherlands Academy of Arts and Sciences. Previously he was based at the Philips Research Laboratories in Eindhoven where he worked on optical disc systems for video and audio recording and on high-resolution optical lithography. His further research interests are diffraction theory, astronomical imaging and optical metrology.

Peter Török is Professor of Optical Physics at the Nanyang Technological University, Singapore and at Imperial College London, UK. His research is focused on the theory of diffraction, focusing and microscopy, with particular emphasis on confocal microscopy, spectroscopic imaging and polarisation. Throughout the years he has taught vector calculus, electromagnetism, optical imaging and optical design theory.

Imaging Optics

JOSEPH BRAAT
Delft Technical University, The Netherlands
PETER TÖRÖK
Nanyang Technological University, Singapore
Imperial College of Science, Technology and Medicine, London

CAMBRIDGE
UNIVERSITY PRESS

CAMBRIDGE
UNIVERSITY PRESS

University Printing House, Cambridge CB2 8BS, United Kingdom

One Liberty Plaza, 20th Floor, New York, NY 10006, USA

477 Williamstown Road, Port Melbourne, VIC 3207, Australia

314-321, 3rd Floor, Plot 3, Splendor Forum, Jasola District Centre, New Delhi – 110025, India

79 Anson Road, #06–04/06, Singapore 079906

Cambridge University Press is part of the University of Cambridge.

It furthers the University's mission by disseminating knowledge in the pursuit of education, learning and research at the highest international levels of excellence.

www.cambridge.org
Information on this title: www.cambridge.org/9781108428088
DOI: 10.1017/9781108552264

First published 2019

Printed in the United Kingdom by TJ International Ltd, Padstow Cornwall

A catalogue record for this publication is available from the British Library.

ISBN 978-1-108-42808-8 Hardback

Additional resources for this publication at www.cambridge.org/imagingoptics.

Contents

II Geometrical Theory of Optical Imaging

III Diffraction Theory of Optical Imaging

Preface

The idea of writing a specific book on 'imaging optics' came some ten years ago and was spurred by the experience which the present authors had acquired in teaching at a university and in guiding research of M.Sc. and Ph.D. students. We noticed that the more advanced optical subjects which have to be mastered by these students to successfully accomplish their studies are rather scattered over the literature. We felt that a comprehensive, well-organised book on the theory and practice of optical imaging, using the same notation and conventions for the various subjects, was lacking. The present book, which has been conceived in the past nine years, should make it easier for students to acquire specific knowledge in the field of optical imaging.

The book comprises three parts. The first introductory part provides the physical basis of optics by means of Maxwell's equations and applies these equations to wave propagation in free space and to refraction and reflection at interfaces between media. A special topic is the propagation of fundamental and higher-order Gaussian beams. The principles needed for solving diffraction problems are explained with special attention to wave propagation and diffraction in stratified media. The rigorous coupled wave analysis and the finite-difference time-domain method are treated in some detail. The chapter on wave propagation in anisotropic media focuses on linear and circular birefringence as a preparation for polarisation aspects in imaging which are encountered in Part III of the book. Emphasis is put on the intriguing effect of conical refraction. Combined with the chapter on surface waves, the reader acquires a good overview of light diffraction and light scattering at an object surface or in an object volume of which an image has to be formed by the optical imaging system.

The second part of the book is devoted to geometrical optics, aberration theory and optical design. It provides the reader with a theoretical basis of ray optics and illustrates the limits on imaging quality based on this simplified light propagation model. Paraxial optics is treated by means of the matrix theory of refraction/reflection and ray propagation. An extensive chapter on aberration theory applied to a single surface, a single lens and to entire systems shows the practical limitations in imaging quality of an optical system. Throughout this chapter, the fundamental diffraction unsharpness in image space and the image blurring due to geometrical aberrations are jointly evaluated. In some cases the (partial) suppression of aberration in an optical system can be achieved by analytic methods. These methods are presented in some detail, together with the more widely used numerical optimisation methods. Imaging quality of an optical system can be further reduced by manufacturing errors. A statistical analysis is presented of the influence of opto-mechanical and mounting errors of lens elements and surfaces such that the expected quality of a real-world imaging system can be evaluated as well as the spread around this value. The second part ends with a longer chapter on optical design methods applied to a wide variety of low- and high-aperture optical imaging systems.

The diffraction of light is the subject of the third part of the book. Based on Maxwell's equations, the first chapter starts with an in-depth treatment of vector diffraction models which are then, step by step, reduced to the older scalar diffraction theories. Various intermediate stages of approximation between the rigorous vector model and the simplest scalar diffraction model are presented such that the reader can decide which approximation is adequate for a specific diffraction problem at hand. The point-spread function, a basic building block for the construction of the image intensity of a composite object, is discussed for an ideal and an aberrated imaging system. The classical scalar diffraction theory of Zernike and Nijboer is used for the diffraction analysis of low-aperture, aberrated imaging systems. The region of validity of this classical diffraction theory is then extended to a much larger focal volume (Extended Nijboer–Zernike theory) and provides the reader with semi-analytic results which can replace the numerical methods used for the evaluation of a general diffraction integral. The extension of this theory to ideal and aberrated point-spread functions of imaging systems with high-aperture serves to describe image formation in non-ideal high-resolution imaging systems. The chapter on point-spread functions ends with a detailed vector-based analysis of the propagation of energy, linear momentum and angular momentum in a high-aperture focused beam. Spatial-frequency analysis of the imaging by an optical system is the subject of a chapter in which the influence of the object illumination on the image is also studied. The van Cittert–Zernike coherence theorem is presented and applied to a certain number of experimental configurations. The classical two-dimensional frequency analysis is extended to the imaging of three-dimensional (volumetric) objects or object surfaces. The influence of light scattering in the optical system on the spatial frequency transfer from object to image space concludes this chapter. The final chapter of Part III discusses the systematic analysis of (vector) imaging systems. The general state of polarisation of the light radiated by the object is defined as well as the possible anisotropy of the imaging system itself, of specially inserted birefringent elements, polarisers, etc. The light propagation from object to image space is described by means of a modular, matrix-based model of wave propagation. The detection of light in image space is performed by means of a polarisation-dependent detector array in a high-aperture imaging geometry.

A number of appendices have been added and they explain, in more detail than was possible in the main body of the book, a certain number of basic definitions or analytic/numerical tools which are frequently used throughout the book. We mention the first appendix which emphasises the role of Fourier methods in modern optics. The fourth appendix provides the reader with an overview of the properties and the applications of Zernike polynomials in optics. A special appendix contains the English translation of an influential publication in the Russian language by V.S. Ignatowsky, dating back to the beginning of the twentieth century. It presents an analysis of the electromagnetic field in the focal region of a high-aperture beam focused by a lens. This publication has inspired many later researchers in this field.

The overview of subjects which is given above also shows which material associated with (classical) imaging optics is definitely missing from this book. Only incidentally and without much detail, we mention a few of the modern methods in low- and high-aperture imaging. Many acronyms circulate in the literature for special imaging methods adapted to a special type of object, illumination, state of polarisation, spectral composition of the light, interferometric detection mode, etc. Simply because of the size of this book, these methods could not be included. Another interesting topic that is missing is the (unique) retrieval of object properties from one or several recordings of an image of the object. This subject, both from the experimental and the numerical point of view, shows interesting progress, also with respect to the high-aperture imaging geometry.

The writing of a book requires continuous concentration on the subject and in this respect the first author (J. B.) was privileged because of his retired status. The absence of time-consuming managerial tasks, of proposal writing and of regular work on national or international committees permitted a permanent focus on book writing. The second author (P. T.) could not benefit from these favourable circumstances and his contribution has remained relatively small. Sections 1.2 and 1.5 and Subsection 1.8.3 of Chapter 1 and the entire Chapters 8 and 11 of the book bear his signature. The remaining part of the book and the Appendices A to F have been written by the first author.

We are confident that this textbook will be a welcome companion on the desk of a masters or graduate student in general optics and in optical imaging in particular. Part of the book can also serve as teaching material for an advanced optics course to such students. Finally, the professional in optical research and development will have at his disposal a reference book covering a wide variety of subjects in advanced optical imaging.

Acknowledgements

The first author (J. B.) would like to thank various institutions and persons who facilitated his work during the long period from the first written page up to the final wrap-up of the manuscript. During these nine years, I was able to benefit from the hospitality offered by the Department of Applied Physics of Delft University. In particular, the optics research group of the department gave me the opportunity to work in a concentrated way on this book manuscript. The 'guest' room was a very efficient environment, facing a metallic grey-coloured dead wall and giving a very restricted view of the sky. There were no perturbing sources of distraction during working-time, apart from the social contacts with the group members and students during coffee breaks. I would like to thank Professor Lucas van Vliet and Professor Paul Urbach for their efficient support by offering me this monastic room (nonetheless having worldwide internet access to scientific publications and libraries).

More directly related to the book-writing I would like to thank Jurgen Rusch of Philips Research Laboratories who offered to transport to another platform a graphics program that I had written for my own optical software. In a surprisingly short amount of time he forwarded me a pdf-based graphics program. All figures in this book related to optical systems have been produced with his new plotting routines. I would also like to thank Dr. A.J.E.M. Janssen who has been a frequent counsellor on and solver of mathematical problems related to the contents of this book. In particular, he has laid the foundation for the Extended Nijboer–Zernike diffraction theory. Especially in this field I have greatly profited from his unique analytical skills. He also kindly agreed to critically read the appendix on Zernike polynomials. Among the Ph.D. students that have graduated in the optics group of Delft University, I would like to mention three persons in particular, Arthur van de Nes, Aura Nugrowati and Sven van Haver. The long-standing collaboration with them was a great pleasure for me and, in several instances, the book content has benefited from their research results.

Two persons have contributed to the final quality of the book in an essential way. In an ideal world, the choice of expert critical readers should have been made by the author of a book before even writing the very first sentence of that book. I had the good luck to find at a later stage of the writing process two critical readers who were willing to spend much time on reviewing the manuscript. Matthew Foreman took care of most of the electromagnetic and physical optics part of the book. His sharp mind and rigorous scientific attitude, combined with his broad knowledge of the presented material, have substantially improved the level of the book and the way of presenting the material. Peter Nuyens has been an extremely strict and precise critical reader of Part II of the book. During the almost two years that we have collaborated on further improving the manuscript I have highly appreciated Peter's analytic dissection of formulas, sentences and paragraphs, his suggestions for shortening my sometimes too verbose presentation and his thorough inspection of the presented optical design examples. Peter, you used a powerful magnifying glass! I would like to apologise to both critical readers for presenting to them in several instances book material that should have been polished to a higher degree beforehand by me. Last but certainly not least, I thank my wife Anna who supported me as a sometimes absent-minded companion in life during these book-writing years.

The second author (P.T.) is grateful to a number of colleagues who contributed to developing his understanding of optics and imaging. Given that I should only have room for a contribution-proportional acknowledgement, these few lines will not be sufficient to mention everybody. Colin Sheppard played an important part in developing the consistent derivation of optical diffraction in 1998–2000 when I visited his Department in Sydney on several occasions. Peter Varga contributed to developing various focusing theories. Emmanouil Kriezis and I used to sit in the Engineering Common room in Oxford discussing how the theory of imaging extended objects could be developed. A string of amazing students inspired me and contributed to a significant degree to developing the mathematical tools and physical understanding discussed in chapters of this book. These included Peter Munro, Carlos Macías Romero and Matthew Foreman. I am truly honoured that they chose to come to work with me. As Joseph so eloquently said above, Matt has done the majority of proof reading of the text picking up typos and mistakes for which I am really grateful. I am also profoundly thankful to Joseph that he has put up with my hectic schedule and lack of progress in writing during these years. I am not sure if I could have found any other co-author willing to do what he did. Finally, I would like to give thanks to my family. Gina and Zoli are the very inspiration for my every-day existence. My wife, Janey has been the most loving, encouraging and understanding companion. I am truly blessed to have you all!

Both authors would like to acknowledge the smooth collaboration with the staff of Cambridge University Press during the publishing process, from initial manuscript transfer to copy-editing and proofreading. We especially thank Richard Smith for copy-editing the book manuscript and Roisin Munnelly for her professional guidance and assistance during the entire last year before publication.

I Electromagnetic Theory in the Optical Domain

Introduction to Part I

The first part of this book on optical imaging provides the reader with the necessary background in electromagnetic theory, relevant for solving optical problems. For a long time, optics was closely connected to mechanics, the oldest branch of science and engineering. The physical model for describing optical phenomena was largely inspired by mechanical analogues. Optical rays were represented as a stream of tiny particles, emitted by a source and propagating in a rectilinear manner, with very high speed. With respect to human vision Greek philosophers, for instance Plato, postulated the emission theory in which the eye emits beams of light that are reflected back from the environmental scene. This theory was later challenged by Euclid who wondered how one could see the very distant stars immediately after opening one's eyes during the night. It was not until the tenth century, in the work of Al-Haytham, that the eye was considered to receive independent optical rays from the outside scene, illuminated by other sources of light. The 'mechanic' nature of light has persisted through the ages, advocated among others by Descartes. A wave theory of light was put forward by Hooke and Huygens in the seventeenth century but it did not attract much attention. An eminent supporter of the particle or corpuscular theory of light was Newton. Numerous experiments on the colour of light itself and on the coloured fringes observed between two optical surfaces were performed by him in the years between 1665 and 1704 when his book *Opticks* was first published (see also **[258]**, the fourth edition of 1730) . His novel observations and experimental results were all explained in the framework of the corpuscular light theory.

The Descartes/Snell refraction law applied to Newton's mechanistic optical model requires a higher light propagation speed in glass than in air. This was made plausible by Newton by means of the attraction exerted by the glass material at the interface air/glass on the incident light corpuscles. Once inside the medium, the light corpuscles continue at the higher speed they have acquired at the transition from a less dense to a denser medium. To explain dispersion, Newton assumed that the red light particles have a different (larger) mass or shape than the blue light particles. As a consequence the red particles would experience a smaller increase of speed at the interface than the blue particles. The net effect is that refraction becomes smaller towards the red part of the spectrum. A conjecture by Newton that glasses all show the same ratio between dispersion and refraction angle was based on this assumption of a colour-dependent mass or shape of the light particles. Dispersion was thus caused by the nature of the light particles. The glass material, by means of its density, determines solely the average refraction angle. Newton's corpuscular theory was successful in explaining rectilinear propagation, refraction and reflection of light and also, to a lesser extent, the effect of diffraction (discovered by Grimaldi, published after his death in 1663 **[117]** and named 'inflection of light' by Newton).

To quantify the beam intensity of partially reflected and transmitted rays, Newton devised a theory of 'fits of easy reflection and transmission'. This property is carried by a particle from the source on, but it can be modified in the vicinity of, for instance, a glass medium. The impact of a light corpuscle on the glass interface creates a local 'wavelet' in the glass that propagates at reduced speed together with the light particle and leads to an enforcement or decrease of the total light phenomenon. An enforcement of the action of light particle and local wave leads to a 'fit of easy transmission' of the particle, the opposite to an inclination of the particle to be reflected. The distribution of the 'fits' over the corpuscles at emission from the source and their change of 'fit' at an interface were not well understood by Newton. The 'fit' property of a corpuscle was the subject of the first query (number 17) of an extended list of queries that was included by Newton in later editions of *Opticks*.

The corpuscular light theory had difficulty in explaining double refraction in a crystal of calcite, discovered by Bartholin in 1669. This strange phenomenon required at least a change in shape of the light corpuscles from spherical to flattened or rectangular. To explain the polarisation-dependent reflection and transmission coefficients at an interface between two media, the 'fits of easy reflection and transmission' of the corpuscles had to be further detailed in a rather artificial and ad hoc manner by the successors of Newton. Similar unsatisfactory assumptions about the nature of the light particles were needed to explain further experiments with polarised light by Wollaston, Malus and Brewster. In general, the corpuscular light theory was inadequate to deal with what we call today the *transverse* oscillatory nature of light.

The discovery of optical glasses with significantly different dispersion by Dollond in 1758 **[87]** was a first argument against the Newtonian light theory. Half a century later, Huygens' wave theory of light was revived by Young and Fresnel. An important extension was the notion of wavelength which immediately created the link with the colour of light. Fresnel's wave theory was very successful in accommodating the new experimental results with polarised light that were presented around 1810. Fresnel's memoir on *double refraction*, published in integral form in 1824 **[104]**, impressed the scientific community. The phenomenon of conical refraction, discovered shortly after Fresnel's untimely death, turned out to be seamlessly included in his theory. Finally, the coup de grâce for the classical corpuscular light theory was administered by the measurement of the speed of light in water, almost simultaneously by Fizeau and Foucault around 1850. It was only 75% of the speed of light in air instead of the $4 \cdot 10^8$ ms^{-1} that was required by the corpuscular light theory.

Fresnel's wave theory was a 'théorie mécanique', as stated by him in the above-mentioned memoir. Essential for the propagation of a wave is the transmission of the transverse wave motion by the molecules of the (luminiferous) aether. The all-pervading fluid of aether molecules had to be given special properties to permit the transmission of transverse wave movement into the propagation direction. Fresnel argued in his memoir that the optical polarisation experiments were so convincing that the aether fluid had to be given a mechanical property which is uncommon for a fluid, viz. a nonzero shear modulus. The existence of the aether and its relative movement with respect to moving bodies such as the planets was the subject of scientific discussion throughout the second half of the nineteenth century. The experiments by Michelson and Morley showed that no relative movement of the aether could be detected and that, most likely, it did not exist. For that reason, the original idea of Faraday that light was a high-frequency electromagnetic perturbation that could propagate in the absence of an aether medium rapidly gained ground. Since then, Maxwell's electromagnetic theory is considered to be the basis of optical wave phenomena. The twentieth century has brought further extensions of the optical theory, such as the quantum theory for black-body radiation, the quantum theory for the interaction of a photon with matter (photo-electric effect) and the quantum behaviour of the photon or assemblies of photons under the condition of low light levels.

Within the scope of this book on classical imaging optics, it is sufficient that the electromagnetic theory of light is taken as the basis for light propagation and imaging. In the first part of the book we focus on Maxwell's electromagnetic theory, applied to the domain of optical frequencies where in many instances the magnetic properties of a medium can be equated to those of vacuum. In the first chapter, after a general introduction to Maxwell's theory, we discuss the dipole source, Gaussian beam propagation and wave propagation at a perfectly smooth interface. To describe light fields emitted by a two- or three-dimensional object to be imaged, we study multilayer systems and the diffraction by periodic structures embedded in a multilayer. The second chapter of Part I is entirely devoted to wave propagation in anisotropic media, either exhibiting linear birefringence or circular birefringence. The phenomenon of conical refraction is treated in some detail. The third chapter is devoted to guided wave propagation at a planar surface. Special wave propagation properties are discussed associated with the so-called *metamaterials*. It is shown that a plane-parallel plate of an idealised metamaterial would behave as a 'perfect' imaging lens with virtually no limit on spatial resolution.

1 Electromagnetic Wave Propagation in Isotropic Media

1.1 Introduction

In the beginning of the nineteenth century, Fresnel's quantitative extension of Huygens' wave theory enabled a detailed description of light propagation in isotropic and anisotropic media, including the diffraction effects arising at sharp edges, tiny holes in a screen or at small obstructions. The wave theory of Fresnel, based on a fine-tuned mechanical aether model to produce the observed optical effects, was quite powerful in describing light wave and light energy propagation. It was not able to explain the effects of magnetic fields on light propagation or reflection, the so-called Faraday and Kerr effects.

Maxwell's electromagnetic theory was needed to establish the firm foundation of light propagation in vacuum and matter. The classical Maxwell theory can be safely used in vacuum and when the material particles involved can be considered to have macroscopic dimensions and properties of which we only need to consider the average values. It is only at very low light levels and when the light interaction with the individual atoms and molecules has to be considered, that we have to switch to the full quantum theory of propagation, transmission, reflection, absorption and scattering of light. In this chapter we use the macroscopic Maxwell's equations as the starting point for wave propagation in the optical domain with the electric and magnetic field quantities represented by three-dimensional vectors. By imposing a simplified approximate solution to Maxwell's equations, we obtain the so-called scalar wave equation and the corresponding wave solution of which the magnitude is given by a single scalar quantity, the complex amplitude of the 'light disturbance'. A further simplification of the solution to Maxwell's equations leads to the ray model of light propagation and to Fermat's principle. It is customary to speak about geometrical optics when using this model, the light energy being propagated along geometrical trajectories that in many practical cases reduce to simple straight lines. Imaging theory based on geometrical optics is subject of Part II of this book, combined with the scalar wave propagation model ('physical optics') if this is necessary to improve the accuracy of the image intensity. In this chapter we treat the parts of electromagnetic theory that, in our view, are relevant for optical imaging. An in-depth treatment of electromagnetic theory can be found in well-known textbooks like **[36]**,**[37]**,**[160]**,**[328]**.

1.2 Maxwell's Equations as Experimental Laws

It is often forgotten, especially by theoreticians, that the four equations now known as Maxwell's equations, namely Gauss' law for electric fields, Gauss' law for magnetic fields, Ampère's law and Faraday's induction law, were once separate and purely experimental laws. It was not until Maxwell realised their relationship in 1861–62, and added the displacement current to Ampère's law, that modern electromagnetism was born.

1.2.1 Electric Field, Electric Flux and Electric Potential

It is perhaps simplest to gain an understanding of Maxwell's equations by first considering the electric and magnetic fields which they govern, since these quantities directly relate to measurable forces which we are familiar with. The definition of the electric field vector **E** originates from Coulomb's experiments on the forces between charges, published in 1785. Coulomb measured the force (the so-called Coulomb force) between charges q_1 and q_2 (units Coulomb [C]) and realised that the force was proportional to both charges and inversely proportional to the square of the distance r between them. The force $\mathbf{F}_E$ (units Newton [N]), which is of course a vector quantity, is parallel to the line connecting the two point charges q_1 and q_2. Coulomb observed that like charges repel each other, and hence the force is directed away from them, whereas opposite charges attract each other, and hence the force is directed towards the charges. If the unit vector along the line connecting the two charges and pointing away from them is denoted by $\hat{\mathbf{r}}$, then $\mathbf{F}_E \propto q_1 q_2 \hat{\mathbf{r}}/r^2$. The electric field due to the charge q_1 is then defined as the force between the two charges divided by the charge q_2, $\mathbf{E} \propto q_1 \hat{\mathbf{r}}/r^2$ (units [N/C] or [V/m], i.e. the force experienced by unit charge). It is therefore clear that electric field lines[1] must start and finish on charges. By convention electric field lines point away from a positive charge and hence towards a negative charge.

The flux of the electric field through a very small *open* surface (differential flux) is defined as the projection of the electric field vector **E** onto the outward surface unit normal $\hat{\mathbf{n}}$ times the area dA of the (differential) surface element. A nonzero net charge inside a *closed* surface $\mathcal{A}$ therefore gives rise to a non-vanishing net flux of electric field **E** through the surface of the volume:

$$\oiint_{\mathcal{A}} \mathbf{E} \cdot d\mathbf{A} = \frac{1}{\epsilon} \sum_i q_i \tag{1.2.1}$$

in the case of discrete charges q_i, or

$$\oiint_{\mathcal{A}} \mathbf{E} \cdot d\mathbf{A} = \frac{1}{\epsilon} \iiint \rho dV \tag{1.2.2}$$

in the case of a distribution of charges in the volume V of volume charge density ρ (units [C/m^3]). Here $d\mathbf{A} = \hat{\mathbf{n}} dA$ is the differential surface normal and ϵ is a constant of proportionality called the permittivity, the significance of which will become clear later. The summation on the right-hand side of Eq. (1.2.1) is over all charges *inside* the closed surface, while those outside the volume do not matter. The latter can readily be explained by the fact that the electric field due to charges outside the volume has a zero *net* flux. Electric fields arising due to a set of stationary charges are also called electrostatic fields. Equations (1.2.1) and (1.2.2) mean that *electrostatic fields are due to electric charges. Field lines do not form loops; they start and end on the charges.*

Electrostatic fields are conservative, which means that if a charge is moved in a closed loop in the presence of such a field then, even though there is in general instantaneous work done along the path, the net work done for the entire path is zero. This is because along a closed loop one can resolve the electric field into two components: one parallel to it and one perpendicular. There is no work done along the component of movement perpendicular to the electric field lines. When the displacement is parallel there is work done but positive work is cancelled exactly by negative work along some other segment of the path.

When a charge is moved along an open path in the presence of other stationary charges, work is done and hence the energy of the system changes. We call this energy the *electrostatic potential energy* U (units Joules [J]) and it is defined as the work that must be done against the electrostatic field produced by a charge q_1 to bring a charge q_2 from infinity, where the electrostatic field is zero, to a distance r from q_1. An associated quantity, the *electric potential*, Φ, is defined as $\Phi = U/q_2$ (units [J/C] or [V]). As mentioned before, when a charge is moved perpendicular to the electrostatic field there is no work done and therefore the electrostatic potential energy of the system does not change. Consequently, the potential Φ does not change either. Lines and surfaces of the same potential are called *equipotential* lines and surfaces, respectively. It is clear then that the electrostatic field vector **E** is perpendicular to equipotential lines and surfaces at every point. The normal of a surface at any given point can be calculated by taking the gradient of the surface which suggests that the electrostatic field vector can be determined from the potential Φ by taking the gradient too:

$$\mathbf{E} = -\nabla\Phi\,. \tag{1.2.3}$$

Although this might first seem counterintuitive as the electric field has three independent Cartesian components whereas the potential is scalar and so it has only one, it merely underlines the fact that not all electric field vectors describe electrostatic fields and that the Cartesian components of an electrostatic field are not independent of one another.

At this juncture it is worth interjecting a mathematical note. Conservative fields have non-vanishing flux but no circulation, i.e. they are said to be irrotational. Mathematically we characterise flux density by divergence and circulation density by curl. The simultaneous knowledge of the divergence and curl uniquely represents any well-behaved vector field as follows from the fundamental theorem of vector calculus. Irrotational fields have vanishing closed loop integrals

[1] The electric field vector is tangential to electric field lines at all points.

which also means that they can be represented by a *scalar* potential function. In the case of electrostatic fields, this scalar potential function is Φ. Divergenceless fields with non-vanishing circulation can be represented by a *vector* potential function as discussed in Section 1.5.1.

1.2.2 Magnetic Flux, Ampère's Law and Maxwell's Displacement Vector

As children we all experimented with bar magnets learning from experience that they have two poles, somewhat arbitrarily called the north and south pole. Like poles repel each other whilst opposite poles attract each other. When a bar magnet is broken in half, the two halves will each have their own north and south poles, which means that it does not seem to be possible to produce a stand-alone north or south pole. The quantity used to characterise the strength and direction of the magnetic field is the vector **B** (also referred to as the 'magnetic induction vector' or 'magnetic flux density'). It is, just as the electric field vector, derived from a measurement of a force; in this case from the force the magnetic field exerts on a moving charge q. Experimental evidence suggests that the force that a moving charge experiences in a homogeneous magnetic field is mutually orthogonal to both the magnetic field and the velocity of the charge and is proportional to q and the magnitude of **v** and **B**: $\mathbf{F}_B = q\,\mathbf{v} \times \mathbf{B}$. The sum of the Coulomb force $\mathbf{F}_E$ and $\mathbf{F}_B$ is called the Lorentz force $\mathbf{F} = \mathbf{F}_E + \mathbf{F}_B = q(\mathbf{E} + \mathbf{v} \times \mathbf{B})$. It is seen that the magnetic field has units of Ns/Cm or Vs/m^2 but, more customarily, in the SI system of units, the unit of **B** is Tesla [T], though the older unit of Gauss [G] (1 G = 10^{-4} T) is still used.

If magnetic field lines are visualised by, for example, the sprinkling of iron filings on a paper placed on top of a magnet we find that they emerge from a pole of the magnet. Since poles always come in pairs and magnetic field lines also exist within magnets, it is an experimental fact that magnetic field lines always close on themselves. This should be contrasted with electric field lines which we found start and end on charges. Consequently, since it is only possible to put pairs of magnetic poles inside any closed volume, we can immediately write Gauss' law for the magnetic field as:

$$\oiint \mathbf{B} \cdot d\mathbf{A} = 0\,, \tag{1.2.4}$$

which equation simply means that *there are no magnetic monopoles. Magnetic field lines are always closed.* It is interesting to point out that the absence of magnetic monopoles has caused considerable discomfort amongst physicists as it leads to an asymmetry of Maxwell's equations as shown later. In 1931 Dirac **[85]** showed that if magnetic monopoles existed it would require all electric charges to be quantised. Therefore, since electric charges are quantised, the existence of magnetic monopoles is fully consistent with Maxwell's equations.

Jean-Baptiste Biot and Félix Savart discovered that there is a magnetic field associated with current carrying wires whose magnitude **B** is proportional to the current I (unit Ampère [A]) in the wire and inversely proportional to the distance from the wire. The magnetic field circulates around the wire forming closed loops centred on the wire as shown in Fig. 1.1. The direction of the magnetic field was found to be perpendicular to both the wire and the direction from a point on the wire to the point of observation.

In 1826 André-Marie Ampère showed experimentally on the basis of Biot and Savart's work that the closed loop integral around the wire must be proportional to the current flowing in the wire. By defining current density **J** (unit A/m^2) as the current per unit area and assigning a direction to it along the conventional current flow he was able to write

$$\oint \mathbf{B} \cdot d\mathbf{l} = \mu \iint \mathbf{J} \cdot d\mathbf{A} = \mu I\,, \tag{1.2.5}$$

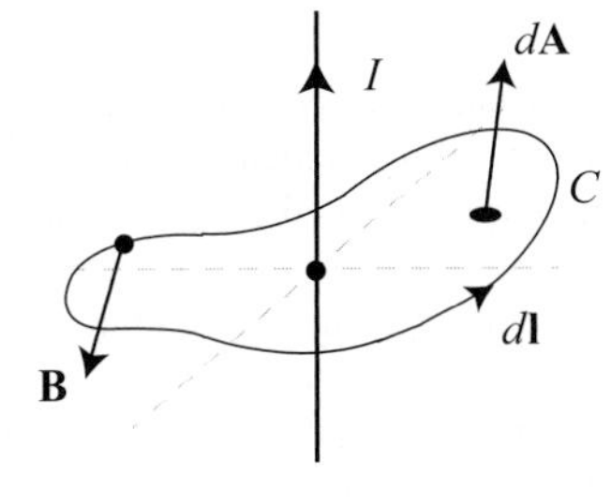

Figure 1.1: The law of Biot and Savart.

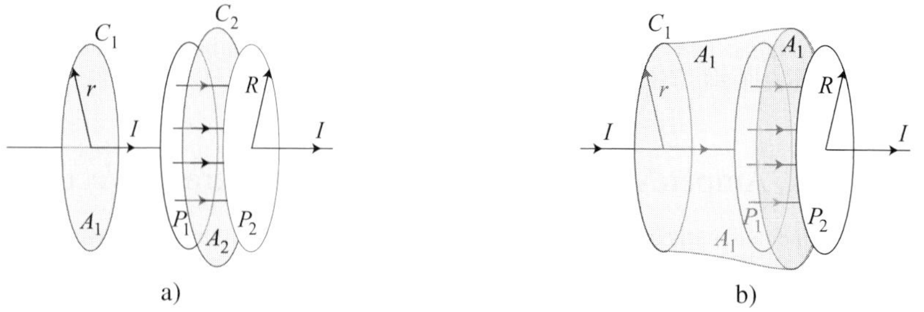

Figure 1.2: Illustrating Ampère's law and its extension by means of Maxwell's displacement field. A time varying current I charges a capacitor. P_1 and P_2 are the two capacitor plates, connected by a wire carrying a time-varying current I.
a) Ampère's law is first applied to the surface A_1 (circular planar integration curve C_1 with radius r) and then to surface A_2 (integration curve C_2).
b) Ampère's law is applied to the modified surface A_1 passing between the capacitor plates P_1 and P_2 and delimited by the circular integration curve C_1 with radius r.
The closing surfaces A_1 and A_2 have been grey-shaded in both drawings.

where μ is a constant of proportionality, called the permeability, the value of which depends on the definition of **B** and **J**, as discussed later. In Fig. 1.1 we show the integration curve C and an infinitesimal element $d\mathbf{l}$ of it. The positive direction of $d\mathbf{l}$ is connected to the direction of the outward normal of the surface element $d\mathbf{A}$ via the right-hand rule. It is important to note that the flux of the current density through the surface defined by the path along which the line integral is performed on the left-hand side must be taken into account. This surface does not have to be flat so the path does not need to be defined in a plane. This point will be further discussed below.

Maxwell used Ampère's law to calculate the magnetic field around a wire that carries a time varying current density to charge a capacitor as shown in Fig. 1.2 on the left-hand side. By arranging the first loop A_1 such that the so-defined surface is penetrated by the wire, Maxwell calculated the magnitude of the magnetic field at a distance r from the wire to be $B = \mu I/2\pi r$. He then chose the loop A_2 with surface as shown on the left-hand side in Fig. 1.2. Since the current density through the surface is zero, he concluded that the magnetic field between the electrodes must also be zero. Next, he considered the geometry shown on the right-hand side of Fig. 1.2. He again used the loop A_1 but now the associated surface was placed between the electrodes of the capacitor. Because there is no current density passing through the surface he obtained $B = 0$ again. However, this result contradicts that obtained using A_1 on the left-hand side. Therefore Maxwell asked what was so special about the volume between the electrodes of the capacitor. He inferred that, in addition to a current density, the time varying electric flux between the electrodes of the capacitor must also be responsible for generating magnetic fields. Therefore Maxwell inserted a correction term into Eq. (1.2.5) to obtain:

$$\oint \mathbf{B} \cdot d\mathbf{l} = \mu\left(\iint \mathbf{J} \cdot d\mathbf{A} + \epsilon\frac{\partial}{\partial t}\iint \mathbf{E} \cdot d\mathbf{A}\right) = \mu \iint \left(\mathbf{J} + \epsilon\frac{\partial}{\partial t}\mathbf{E}\right) \cdot d\mathbf{A}\,, \tag{1.2.6}$$

which is his extended version of Ampère's law. The line integral is performed over a closed path delimiting an open surface over which the right-hand side flux integral is evaluated. The equation states that *the circulation of magnetic field is due to the flux of a current density through a surface, whose circumference is where the circulation of the magnetic field is measured, and a time varying electric field flux through the same surface*. It is worth noting that the time varying electric field between the electrodes of the capacitor is not a conservative field and thus it is not irrotational. It is sometimes referred to as electrodynamic field.

In 1820 the Danish physicist Hans Christian Ørsted noticed that a compass deviates from its stable position if electric current flows through a wire placed in the vicinity of the compass. This was the first known experiment that connected electricity to magnetism. Michael Faraday, after seeing Ørsted's experiment, suggested that if electric current affects the compass then a magnetic field should produce a current. In order to prove this he set up two solenoids (the so-called Helmholtz coil), as shown in Fig. 1.3a. He then powered the one on the left from a battery and noticed that there was current induced in the solenoid on the right. However, he only experienced current when the switch was being flicked over. Once the switch was on, the current from the other solenoid disappeared. He hence concluded that changing (i.e. not steady) magnetic fields produce current in the other solenoid. The phenomenon is called electromagnetic induction. Heinrich Lenz later experimented to find the direction of the current that is produced by the changing magnetic field. He

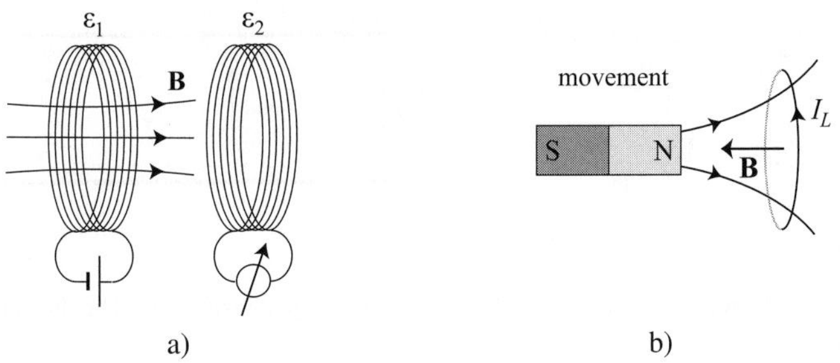

Figure 1.3: a) Faraday's experiment with solenoids.
b) Lenz's law demonstrated with a permanent magnet and a current loop.

found that the induced current in a current loop (shown with arrows in the figure above) is such that its magnetic field (denoted **B** in the figure) opposes the inducing magnetic field (see Fig. 1.3b). This is Lenz's law.

Consider now a wire loop permeated by a magnetic field of increasing magnitude. By Lenz's law, the current generated in the loop will flow such that the magnetic field it induces is opposed to the incoming magnetic field. The current must be produced by a potential difference so there has to be an electric field associated with that potential. The definition of the electromotive force (emf) [V], $\mathcal{E}_{\text{ind}}$, which is the potential in the wire is

$$\oint \mathbf{E} \cdot d\mathbf{l} = \mathcal{E}_{\text{ind}} . \tag{1.2.7}$$

The equation states what we mentioned briefly before: electrodynamic fields are not conservative, therefore they have a non-vanishing closed loop integral.

Faraday carried out a number of experiments with Helmholtz coils, as shown Fig. 1.3. He realised that the induced emf in the second coil, $\mathcal{E}_2$ is proportional to the change with time in the magnetic field produced by the first coil and also the area of the second coil. This permitted him to conclude that the quantity of importance is the change with time in the magnetic flux passing through the second coil. The magnetic flux is defined in a way similar to the flux of the electric field:

$$\Phi_B = \iint \mathbf{B} \cdot d\mathbf{A} . \tag{1.2.8}$$

Therefore

$$\mathcal{E}_2 = -\frac{\partial \Phi_B}{\partial t} = -\frac{\partial}{\partial t} \iint \mathbf{B} \cdot d\mathbf{A} , \tag{1.2.9}$$

or,

$$\oint \mathbf{E} \cdot d\mathbf{l} = -\frac{\partial}{\partial t} \iint \mathbf{B} \cdot d\mathbf{A} , \tag{1.2.10}$$

which states that *the induced emf, or circulation in the electrodynamic field, is due to time varying magnetic flux and it opposes that.* This is the fourth Maxwell's equation, Faraday's induction law.

1.2.3 Maxwell's Equations in a Material, Electric and Magnetic Polarisation

Maxwell's equations have been shown to successfully describe electromagnetic fields in vacuum and also in material media. The latter term might refer to a material that does not conduct electric current, often referred to as a dielectric. Electric fields applied to dielectrics will polarise materials. In the absence of an external electric field the atoms in dielectrics have their electron cloud evenly distributed around the nucleus, as shown in Fig. 1.4a. When an electric field **E** is applied in the direction given by Fig. 1.4b, the negative potential on the lower side gives rise to a repulsive force on the electrons and so the electron cloud will be predominantly located towards the more positive potential on the upper side of the drawing, thereby generating electric dipoles.[2] In the case of a capacitor having a dielectric material between

[2] The strength of a dipole is defined as the product of the separating distance $|\mathbf{d}|$ of the two charges with opposite sign and the magnitude q of each charge. The resulting quantity $\mathbf{p}_d = q\mathbf{d}$ is a vector and is called the dipole *moment* of the dipole. The moment vector points from the negative to the positive charge of the dipole. The strength of a dipole is expressed in units of [Cm], the net dipole moment **P** per unit volume in [Cm^{-2}]. A detailed treatment of the electromagnetic properties and the radiation pattern of an individual dipole is given in Subsection 1.6.2 of this chapter.

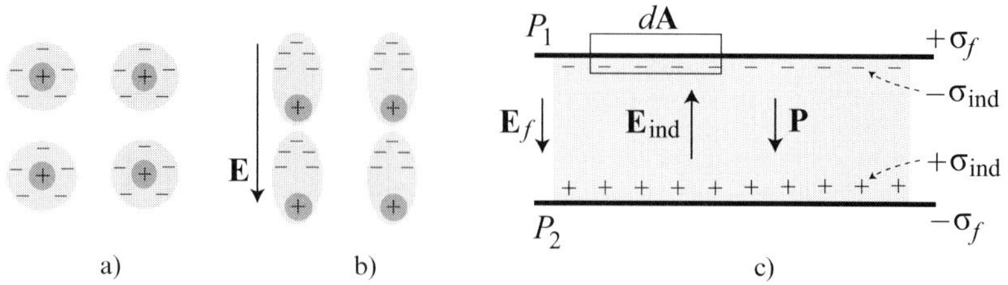

Figure 1.4: A capacitor with conducting surfaces P_1 and P_2 carrying charge densities σ_f. The dielectric material inside the capacitor has been grey-shaded.
a) Charge distribution in the unperturbed dielectric material.
b) Polarisation of the dielectric material inside the capacitor under the influence of an electric field **E**.
c) Electric fields, surface charges and polarisation in the capacitor.

the conducting plates (see Fig. 1.4c), surface charges are induced at the borders of the dielectric material, parallel to the capacitor plates P_1 and P_2. The associated surface charge density is denoted by σ_{ind} and gives rise to an induced electric field $\mathbf{E}_{ind}$ in the capacitor. The total electric field **E** is the vector sum of the electric field $\mathbf{E}_f$ in the absence of the dielectric material and the induced electric field with the dielectric material in place. The field $\mathbf{E}_f$ is produced by the surface charges $+\sigma_f$ and $-\sigma_f$ on the capacitor plates P_1 and P_2, respectively. Under the influence of the field $\mathbf{E}_f$, the dielectric material inside the capacitor produces two thin layers with surface charges $-\sigma_{ind}$ and $+\sigma_{ind}$ on the upper and lower surface of the dielectric material, respectively.

The magnitude of the induced surface charges follows from the argument that in the bulk of the dielectric material the positive and negative charges are mutually displaced; their total charge, however, remains zero on average. At the upper and lower border of the dielectric this averaging to zero of the total charge does not happen. Given the direction of the applied field $\mathbf{E}_f$, negative charges are in excess at the upper border of the dielectric material, positive charges at the lower border. The dipole moment per unit volume in the dielectric is given by **P** and is the sum of the moments of N dipoles per m^3. If the dipoles are perfectly aligned through the entire volume, the total dipole moment **P** amounts to $Nq\mathbf{d}$. The vector **P** is commonly called the electric *polarisation*. We assume that it is linearly dependent on the external field provided this is small enough. Under the influence of the electric field in the capacitor each dipole axis **d** points in the downward direction in Fig. 1.4c. The charge movement due to the dipoles leads to an induced (negative) charge dQ on the upper surface of the dielectric which is given by $-N(dA)dq$ where dA is an infinitesimally small surface element on the upper surface of the dielectric. Division by dA yields the induced surface charge density σ_{ind} of which the magnitude is thus simply given by $|\mathbf{P}|$. The corresponding electric field **E** is calculated by means of Eq. (1.2.1), applied to the shoebox in Fig. 1.4c with upper and lower surface dA. If the lateral dimensions of the capacitor are much larger than its thickness, the electric fields inside the capacitor are aligned along the vertical direction as shown in Fig. 1.4c and it is permissible to write

$$\mathbf{E} = \mathbf{E}_f + \mathbf{E}_{ind} = \frac{\sigma_f}{\epsilon_0} + \frac{\sigma_{ind}}{\epsilon_0} = \mathbf{E}_f - \frac{\mathbf{P}}{\epsilon_0} = \mathbf{E}_f - \chi\mathbf{E} ,$$

or,

$$\mathbf{E}_f = (1+\chi)\mathbf{E} = \epsilon_r \mathbf{E} . \tag{1.2.11}$$

Assuming linearity between the induced polarisation and the external field we have introduced a proportionality factor χ in Eq. (1.2.11) between $\mathbf{E}_{ind}$ and the net field **E** such that $\mathbf{P} = \epsilon_0\chi\mathbf{E}$. The dimensionless quantity χ is called the *electric susceptibility* of the dielectric material. The equally dimensionless quantity ϵ_r is called the *relative permittivity* of the material medium and ϵ_0 is called the permittivity of vacuum, though it is only a constant of proportionality depending on the system of units. In the SI system, $\epsilon_0 = 8.854 \times 10^{-12}$ Fm^{-1}. It is also not unusual to use the *displacement field* or electric flux density **D** (unit [FV/m^2], [As/m^2] or [C/m^2]) defined formally as

$$\mathbf{D} = \epsilon\mathbf{E} = \epsilon_0\epsilon_r\mathbf{E} = \epsilon_0(1+\chi)\mathbf{E} = \epsilon_0\mathbf{E} + \epsilon_0\chi\mathbf{E} = \epsilon_0\mathbf{E} + \mathbf{P} . \tag{1.2.12}$$

We note that the displacement field is not a fundamental field, meaning that it relates to a force measurement only via **E**. Note that this argument implicitly assumes that the dielectric material is linear and isotropic, meaning that the material is invariant to all rotational transformations. There are cases when the induced electric field vector is not antiparallel with the electric field inducing it. In this case χ, and consequently ϵ_r, becomes a tensor as discussed in Chapter 2. Apart from the global susceptibility χ of a material, we can also define the *polarisability* α of an elementary particle in the material, e.g. an atom or a molecule. The individual dipole moment $\mathbf{p}$ of such a particle, induced by the field $\mathbf{E}_{\text{ind}}$, equals $\mathbf{P}/N$ where N is the number of particles per unit volume. The induced dipole moment $\mathbf{p}$ of a single particle is defined as $\alpha\,\mathbf{E}$. It then follows that the polarisability is $\alpha = \epsilon_0 \chi/N$ with unit [Cm^2V^{-1}].

In a similar fashion, magnetic materials also contain magnetic dipoles due to electron currents. Depending on the type of magnetic material, when an external magnetic field, usually denoted by $\mathbf{H}$ (unit [A/m]), is applied these magnetic dipoles can orient themselves to alter the effect of the inducing magnetic field. The induced magnetic field, denoted by $\mathbf{M}$, is called *magnetisation* or *magnetic polarisation*. The magnetic field $\mathbf{H}$ and the magnetic polarisation $\mathbf{M}$ together are responsible for the overall magnetic field:

$$\mathbf{B} = \mu_0(\mathbf{H} + \mathbf{M})\,. \tag{1.2.13}$$

In diamagnetic and paramagnetic materials, the magnetisation is proportional to $\mathbf{H}$ with as constant of proportionality the *magnetic susceptibility*, χ_m, yielding $\mathbf{M} = \chi_m\mathbf{H}$ and Eq. (1.2.11) then reads

$$\mathbf{B} = \mu_0(\mathbf{H} + \chi_m\mathbf{H}) = \mu_0(1 + \chi_m)\mathbf{H} = \mu_0\mu_r\mathbf{H} = \mu\mathbf{H}\,, \tag{1.2.14}$$

where μ_r is the relative permeability (dimensionless). However, since the optical materials we are concerned with are not magnetically active in most cases, we shall restrict our discussions to $\mu_r = 1$.

1.3 Maxwell's Equations in the Optical Domain

As discussed in the previous section, the general laws governing electromagnetic phenomena are:
Coulomb's law or Gauss' law for electrostatics

$$\oiint \mathbf{D} \cdot d\mathbf{A} = \iiint \rho\, dV\,, \tag{1.3.1}$$

Gauss' law for magnetic fields

$$\oiint \mathbf{B} \cdot d\mathbf{A} = 0\,, \tag{1.3.2}$$

Ampère–Maxwell law

$$\oint \frac{\mathbf{B}}{\mu} \cdot d\mathbf{l} = \iint \left(\mathbf{J} + \frac{\partial \mathbf{D}}{\partial t}\right) \cdot d\mathbf{A}\,, \tag{1.3.3}$$

Faraday's induction law

$$\oint \mathbf{E} \cdot d\mathbf{l} = -\frac{\partial}{\partial t} \iint \mathbf{B} \cdot d\mathbf{A}\,. \tag{1.3.4}$$

In the above integrals the inner products of vector quantities and line segments or surface elements imply the evaluation of the scalar product where the line segment $d\mathbf{l}$ points in the tangential direction and the surface element vector $d\mathbf{A}$ points in the direction of the outward normal to the surface. In Eqs. (1.3.1) and (1.3.2) the volume integral is over an arbitrary volume V that is bounded by a closed surface A over which the surface integral is evaluated. In Eqs. (1.3.3) and (1.3.4), the surface integral applies to an open surface A that is bounded by a curve l along which the line integral has to be carried out (see also Fig. 1.5 for the geometrical details). In the equations above, we consider $\mathbf{E}$, the electric field and $\mathbf{B}$, the magnetic induction, as the two basic quantities that determine the electromagnetic field.

The other medium-determined quantities occurring in Eqs. (1.3.1)–(1.3.4) are the current density $\mathbf{J}$, the scalar quantity ρ, the volume charge density (unit [C/m^3]) and the dielectric displacement or electric flux density (electric induction) $\mathbf{D}$. With the aid of the electric permittivity $\epsilon = \epsilon_0\epsilon_r$, the magnetic permeability $\mu = \mu_r\mu_0$ and the specific conductivity σ, we define the following relationships between the basic field vectors $\mathbf{E}$ and $\mathbf{B}$ and the other vector quantities $\mathbf{D}$, $\mathbf{H}$ and $\mathbf{J}$ via the so-called material equations or *constitutive relations*:

$$\mathbf{D} = \epsilon\,\mathbf{E}\,, \tag{1.3.5}$$

$$\mathbf{B} = \mu\,\mathbf{H}\,, \tag{1.3.6}$$

$$\mathbf{J} = \sigma\,\mathbf{E}\,, \tag{1.3.7}$$

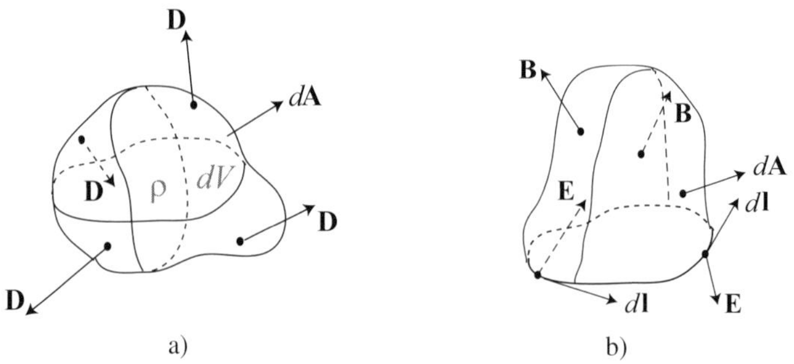

Figure 1.5: a) Illustrating the volume and closed surface integrals involved in the Coulomb/Gauss laws. The vector quantity $d\mathbf{A}$ is normal to the closed surface A. The vector $\mathbf{D}$ represents the electric displacement field vector (or electric flux density vector), ρ is the electric charge density and dV is an infinitesimal volume element.
b) The open surface integral and the line integral which appear in the Ampère and Faraday laws. The vector $d\mathbf{l}$ is tangent to the limiting curve of the open surface A. The vector $\mathbf{B}$ represents the magnetic flux density and $\mathbf{E}$ the electric field strength vector.

as follows from Eqs. (1.2.12) and (1.2.14). In homogeneous[3] and isotropic media, ϵ, μ and σ are constants. Moreover, in the optical domain, the conductivity of most dielectric materials of interest will be low or close to zero to guarantee high transmission through the medium; in many cases, it can be conveniently set to zero. In inhomogeneous isotropic media, ϵ, μ and σ are scalar functions of the position vector $\mathbf{r}$. In anisotropic media, these quantities become tensors. In the optical domain, we are generally allowed to equate the magnetic permeability of a medium to that of vacuum, μ_0. Recent developments in material engineering (metamaterials) show that this is not always necessarily the case. Some aspects of these recent material developments, like the possibility of 'perfect imaging', are treated in Chapter 3.

In order to transform the integral version of Maxwell's equations into their alternative differential form we apply the Gauss and Stokes vector integral theorems,

$$\iiint \nabla \cdot \mathbf{v} \, dV = \oiint \mathbf{v} \cdot d\mathbf{A} \,, \tag{1.3.8}$$

$$\oint \mathbf{v} \cdot d\mathbf{l} = \iint (\nabla \times \mathbf{v}) \cdot d\mathbf{A} \,, \tag{1.3.9}$$

where $\mathbf{v}$ is a general vector field, to Maxwell's equations above. On comparing Eqs. (1.3.1)–(1.3.4) with Eqs. (1.3.8) and (1.3.9), we obtain Maxwell's equations in *differential form*,

$$\nabla \cdot \mathbf{D} = \rho \,, \tag{1.3.10}$$

$$\nabla \cdot \mathbf{B} = 0 \,, \tag{1.3.11}$$

$$\nabla \times \frac{\mathbf{B}}{\mu} = \mathbf{J} + \frac{\partial \mathbf{D}}{\partial t} \,, \tag{1.3.12}$$

$$\nabla \times \mathbf{E} = -\frac{\partial \mathbf{B}}{\partial t} \,. \tag{1.3.13}$$

Note that there is a significant difference between Maxwell's equations expressed in integral and differential form. While the former are applicable to volumes of space, the latter are only applicable to points, with curl ($\nabla\times$) denoting *circulation density* and divergence ($\nabla\cdot$) denoting *flux density*. As we have seen in the introduction, it is more usual and perhaps sensible to use currents and charges with Maxwell's equations in their integral form but charge and current densities with the differential form. As noted above, because a vector function is uniquely characterised by the simultaneous knowledge of the circulation density (curl) and the divergence of the function, so the differential form of Maxwell's equations uniquely specifies both the electric and the magnetic field at a given position in space.

1.4 Electromagnetic Energy Density and Energy Transport

Electromagnetic fields create an energy density in space and can give rise to a flow of energy. In this section we establish the electric and magnetic energy density and the energy flow created by electromagnetic waves, expressed in terms of the

[3] By homogeneous we mean that ϵ and μ are not position-dependent.

basic vector quantities that characterise the electromagnetic field. The relationship is generally called Poynting's theorem after the author who first established it **[282]**.

Starting with the general Maxwell equations of Section 1.3, we perform a scalar multiplication of Eq. (1.3.12) with **E** and of Eq. (1.3.13) with **H** respectively, and subtracting the results, we obtain

$$\mathbf{E}\cdot\nabla\times\mathbf{H}-\mathbf{H}\cdot\nabla\times\mathbf{E}=-\nabla\cdot(\mathbf{E}\times\mathbf{H})=\mathbf{J}\cdot\mathbf{E}+\mathbf{E}\cdot\frac{\partial\mathbf{D}}{\partial t}+\mathbf{H}\cdot\frac{\partial\mathbf{B}}{\partial t}\,, \tag{1.4.1}$$

where we have used the property of the scalar triple product $\nabla\cdot(\mathbf{a}\times\mathbf{b})=-\mathbf{a}\cdot\nabla\times\mathbf{b}+\mathbf{b}\cdot\nabla\times\mathbf{a}$.

Assuming that the material properties described by ϵ, μ and σ are time-independent, it is permissible to write $\mathbf{E}\cdot(\partial\mathbf{D}/\partial t)=\frac{1}{2}(\partial\mathbf{E}\cdot\mathbf{D}/\partial t)$ and $\mathbf{H}\cdot(\partial\mathbf{B}/\partial t)=\frac{1}{2}(\partial\mathbf{H}\cdot\mathbf{B}/\partial t)$. Substitution of these results into Eq. (1.4.1) yields

$$\nabla\cdot(\mathbf{E}\times\mathbf{H})+\mathbf{J}\cdot\mathbf{E}+\frac{1}{2}\frac{\partial\mathbf{E}\cdot\mathbf{D}}{\partial t}+\frac{1}{2}\frac{\partial\mathbf{H}\cdot\mathbf{B}}{\partial t}=0\,. \tag{1.4.2}$$

Equation (1.4.2) establishes a quadratic relationship between the field quantities **E**, **D**, **B**, **H** and **J**, that were already linked by the linear relations of Eqs. (1.3.5)–(1.3.7). Performing the integration of Eq. (1.4.2) over a certain volume V in space, bounded by a closed surface A, we find

$$\iiint\left(\mathbf{J}\cdot\mathbf{E}+\frac{1}{2}\frac{\partial\mathbf{E}\cdot\mathbf{D}}{\partial t}+\frac{1}{2}\frac{\partial\mathbf{H}\cdot\mathbf{B}}{\partial t}\right)dV+\oiint(\mathbf{E}\times\mathbf{H})\cdot d\mathbf{A}\;=\;0\,, \tag{1.4.3}$$

where we have applied Gauss' integral theorem to the volume integral involving $\nabla\cdot(\mathbf{E}\times\mathbf{H})$.

The quantity $w_e=\mathbf{E}\cdot\mathbf{D}/2$ is commonly called the electric energy density of the electromagnetic field and $w_h=\mathbf{H}\cdot\mathbf{B}/2$ is the corresponding magnetic energy density. The integration of the electric and magnetic energy densities over the volume V yields the total electromagnetic energy w_{em} present in the volume,

$$w_{em}=\iiint(w_e+w_h)\,dV\,. \tag{1.4.4}$$

This allows us to write

$$\frac{\partial w_{em}}{\partial t}+\iiint\mathbf{J}\cdot\mathbf{E}\,dV+\oiint(\mathbf{E}\times\mathbf{H})\cdot d\mathbf{A}=0\,, \tag{1.4.5}$$

which is the conservation law for electromagnetic energy in the volume V bounded by a surface A. Equation (1.4.5) is known as the integral formulation of Poynting's theorem and it states that the time rate of change of the electromagnetic energy is equal to that lost via dissipation (the $\mathbf{J}\cdot\mathbf{E}$-term) and an outward flow through the surface A that bounds the volume. The electromagnetic quantity that determines the outward flow of energy is the vector product of **E** and **H**. This vector is called the Poynting vector $\mathbf{S}=\mathbf{E}\times\mathbf{H}$ with units of [Wm^{-2}]. Its magnitude determines the number of Watts of electromagnetic energy that flows through the closed surface. The quantity $(\mathbf{E}\times\mathbf{H})\cdot d\mathbf{A}$ in the left-hand part of Eq. (1.4.5) determines the outward flow of electromagnetic energy.

In the optical domain with typical time frequencies ω of the order of 600 THz, it is only the time-averaged absolute value of the Poynting vector which is accessible for measurement, for instance when the light intensity is detected by a photosensitive element. In the case of harmonic electric and magnetic field strength vectors with frequency ω and a phase difference of ϕ between them, the average value $\langle S\rangle$ of the Poynting vector is given by

$$\begin{aligned}\langle S\rangle&=\lim_{T\to\infty}\frac{1}{2T}\int_{-T}^{+T}E_0\cos(\omega t)\,H_0\cos(\omega t+\phi)\,dt\\&=\frac{1}{2}E_0H_0\cos\phi+\lim_{T\to\infty}\frac{1}{4T}\int_{-T}^{+T}E_0\,H_0\Big\{\cos\phi\cos(2\omega t)\;-\;\sin\phi\sin(2\omega t)\Big\}\,dt\,.\end{aligned} \tag{1.4.6}$$

The integrand of the integral on the second line contains harmonic factors with frequency 2ω and in the case that the detector integration time $T\gg 2\pi/\omega$ the integral can safely be equated to zero; the average value of the Poynting vector is then equal to $(E_0H_0\cos\phi)/2$. In low-frequency electrical circuits the 2ω-frequency can be detected, both the in-phase $\cos(2\omega t)\cos\phi$ term and the quadrature term which is proportional to $\sin(2\omega t)\sin\phi$. The net energy transport associated with the $\sin(2\omega t)$ term is zero for any integration interval $[-T,+T]$. Such oscillating energy 'propagation' with zero average is observed in, for instance, a circuit with a pure capacitor or a pure inductor as impedance (*reactive* impedance with $|\phi|=\pi/2$). For this reason, the $\sin(2\omega t)$ term of the Poynting vector is called its 'reactive' part. In Subsection 1.8.7 of this chapter we discuss transitions between optical media and it will turn out that, given the medium properties, they can behave as a capacitor or an inductor in an electric circuit and that we can attribute a reactive optical impedance to them.

Table 1.1. The various quantities transported by an electromagnetic field. The propagation speed has been denoted by v and ω is the circular frequency of the electromagnetic wave. p_C is the degree of circular polarisation.

Quantity	Density	Unit	Flow	Unit
Energy	$\frac{\mathbf{E}\cdot\mathbf{D}+\mathbf{H}\cdot\mathbf{B}}{2}$	$\mathrm{Jm^{-3}}$	$\mathbf{E}\times\mathbf{H}$	$\mathrm{Wm^{-2}}$
Linear momentum	$\mathbf{E}\times\mathbf{H}/v^2$	$\mathrm{Nsm^{-3}}$	$\mathbf{E}\times\mathbf{H}/v$	Pa
Angular momentum	$\{\mathbf{E}\times\mathbf{H}/(\omega\, v)\}\,p_C$	$\mathrm{Jsm^{-3}}$	$\{\mathbf{E}\times\mathbf{H}/\omega\}\,p_C$	$\mathrm{Jm^{-2}}$

In electromagnetic theory, the Poynting vector replaces the radiometric quantity of irradiance for a receiving surface or (radiant) exitance for an emitting surface. In the literature both denominations are often replaced, in somewhat negligent short-hand writing, by the word 'intensity'. In the frequently occurring situation of a lossless medium, Poynting's theorem reduces to

$$\frac{\partial w_{em}}{\partial t} + \oiint \mathbf{S}\cdot d\mathbf{A} = 0\,. \tag{1.4.7}$$

Equation (1.4.7) states that an increase of electromagnetic energy in a volume is due to an inward flow of energy through the bounding surface, or that a decrease of electromagnetic energy in a volume is due to the outward flow of energy through the bounding surface.

Apart from energy, the electromagnetic field also transports linear and angular momentum. Without derivation, we list the corresponding densities and flow vector strengths for linear and angular momenta in Table 1.1. The factor p_C in Table 1.1 is the degree of circular polarisation and is discussed in Section 1.6.6. In the case of a plane wave in a medium with speed of propagation v, the flow vector strengths follow from the density values by a multiplication by the propagation speed v.

Note that the Poynting vector must be applied with caution. Consider for example an electrostatic field $\mathbf{E}_s$ and a magnetostatic field $\mathbf{H}_s$ simultaneously existing in space, for example created by a set of static charges and a magnet. Clearly $\mathbf{E}_s\times\mathbf{H}_s$ can be nonzero at locations where $\mathbf{E}_s \not\parallel \mathbf{H}_s$. Non-vanishing vector product can yield a nonzero flux in Eq. (1.4.7) implying that change in energy occurs with time. Yet there cannot be any *flow* of energy because the two fields were assumed to be static. Physically there is of course no problem with Eq. (1.4.7) because it tells us that because the change in energy with time is zero there should not be any flux of electromagnetic energy, which is just what we expect. This example underlines that the use of the Poynting vector is not of general validity in electrodynamics.

In line with the conservation law for electromagnetic energy given by Eq. (1.4.7) in a lossless medium we can define conservation laws for the linear and angular momentum of an electromagnetic field. The expressions in terms of the electric and magnetic field vectors for these two properties of an electromagnetic field can be found in standard textbooks on electromagnetism, for instance those by Stratton **[328]**, Jackson **[160]** and Zangwill **[384]**. The expressions for linear and angular momentum are further examined and applied in Chapter 9 of this book, Section 9.6, where the electromagnetic energy and momentum distribution in the focal region of a converging beam with large opening angle are analysed.

To obtain a conservation law for the linear momentum, we first define the linear momentum density of the electromagnetic field and denote it by the vector $\mathbf{m}$. The flow of momentum is determined by the Maxwell stress tensor $\mathbf{T}$ whose entries T_{pq} with $p, q = (x, y, z)$ have the dimension of pressure (see Table 1.1). The value of T_{pq} equals the amount of momentum in the p-direction, transported per unit area perpendicular to the q-axis per unit time; the sign has been chosen such that the transport of momentum must be measured in the negative q-direction. In terms of force and pressure, the element T_{pq} equals the force exerted in the p-direction on a surface perpendicular to the q-direction, yielding the dimension of pressure for each tensor element. The diagonal elements of the stress tensor are called the *normal* pressure components, the off-diagonal elements are known as the *shear* pressure components. The conservation of the p-component of linear momentum is given by the expression

$$\frac{\partial m_p}{\partial t} + \frac{\partial T_{px}}{\partial x} + \frac{\partial T_{py}}{\partial y} + \frac{\partial T_{pz}}{\partial z} = 0, \tag{1.4.8}$$

where (x, y, z) are real-space Cartesian coordinates. Similarly, we denote the angular momentum density by the vector symbol $\mathbf{j}$ and the corresponding tensor by $\mathbf{M}$ with the flow components M_{pq}. The conservation of angular momentum is then expressed by

$$\frac{\partial j_p}{\partial t} + \frac{\partial M_{px}}{\partial x} + \frac{\partial M_{py}}{\partial y} + \frac{\partial M_{pz}}{\partial z} = 0, \tag{1.4.9}$$

where each tensor element M_{pq} has the dimension of Nm^{-1}.

We remark that angular momentum of the electromagnetic field has a dual origin. There is *intrinsic* angular momentum that follows from the spin of the photons. It gives rise to a certain degree of circular polarisation p_C of the optical field with either right- or left-handed sign. Another contribution is generated by a particular shape of the geometrical wavefront in the exit aperture (or exit pupil) of an imaging system. This type of angular momentum is called *orbital* angular momentum. The conservation laws of energy, linear and angular momentum will be studied in more detail in Chapter 9 in the framework of imaging pencils with a large convergence angle and a certain state of polarisation. They may also carry geometrical aberrations that give rise to orbital angular momentum.

1.5 Potential Functions and the Electromagnetic Field Vectors

Originally Maxwell wrote his famous formulae (1865–1873) in a much different format, consisting of 20 equations. The four equations we now associate with Maxwell were in fact put in their current form by Heaviside and Gibbs around 1892. Though Maxwell's initial formulation was not in terms of vectors but quaternions he used two potential functions: one that we now call the *vector potential*, denoted by $\mathbf{A}$, and the other one the electric *scalar potential*. The vector potential was a concept closely relating to Faraday's electronic state. However, Heaviside strongly criticised **[129]** the use of potentials saying: *...the fact that* $\mathbf{A}$ *often has a 'scalar potential parasite* Φ*' sometimes causes great mathematical complexity and indistinctness; and it is, for practical reasons, best to murder the whole lot, or, at any rate, merely employ them as subsidiary functions... Thus* Φ *and* $\mathbf{A}$ *are murdered, so to speak, with a great gain in definiteness and conciseness.*[4] Although in Section 1.2 we have discussed the physical significance of the scalar potential, for long it has been speculated that the vector potential function is a mere mathematical construct. However, in 1959 Aharonov and Bohm **[7]** showed in the context of quantum mechanics that *...contrary to the conclusions of classical mechanics, there exist effects of potentials on charged particles, even in the region where all the fields (and therefore the forces on the particles) vanish* revealing that the (vector) potential function had real physical meaning.

Nevertheless, the importance of potentials in classical electrodynamics is great – they are often used to solve problems that would otherwise be more difficult to solve using the traditional electric and magnetic field vectors. The main difficulty arises from the fact that, as discussed in e.g. Section 1.3, in a homogeneous and isotropic dielectric volume free of electric charges there are only two degrees of freedom in choosing components of any field quantity. Therefore, one faces considerable difficulties in specifying boundary conditions for solving the Helmholtz equation (see Section 1.6) as often it is not clear which components of the field should be specified. To a certain degree this ambiguity is removed with the use of the scalar and vector potentials because instead of specifying four vector components (say two for the electric and two for the magnetic field) one sometimes only needs to specify two components of the scalar potential to solve a problem. In addition, we shall see later that the Hertz vector, connected to both the electric and magnetic fields and the potential functions, is yet another quantity that makes solutions of electromagnetic problems less ambiguous.

1.5.1 The Vector and Scalar Potential

We have seen in previous sections that electric fields are either due to an electric charge distribution or a time varying magnetic induction. Conversely, magnetic fields are due to electric currents or time varying displacement fields. We shall now write the differential Maxwell's equations as two independent sets of equations: one describing phenomena originating from electric sources:

$$\nabla \cdot \mathbf{D}_e = \rho\,, \tag{1.5.1}$$

$$\nabla \cdot \mathbf{B}_e = 0\,, \tag{1.5.2}$$

[4] A fascinating overview of the historicity of the vector potential was written by Wu and Yang **[379]**.

$$\nabla \times \mathbf{H}_e = \mathbf{J} + \frac{\partial \mathbf{D}_e}{\partial t}, \tag{1.5.3}$$

$$\nabla \times \mathbf{E}_e = -\frac{\partial \mathbf{B}_e}{\partial t}, \tag{1.5.4}$$

and another from magnetic sources:

$$\nabla \cdot \mathbf{D}_m = 0\,, \tag{1.5.5}$$

$$\nabla \cdot \mathbf{B}_m = 0\,, \tag{1.5.6}$$

$$\nabla \times \mathbf{H}_m = \frac{\partial \mathbf{D}_m}{\partial t}, \tag{1.5.7}$$

$$\nabla \times \mathbf{E}_m = -\frac{\partial \mathbf{B}_m}{\partial t}. \tag{1.5.8}$$

The two sets of equations are coupled by the corresponding constitutive equations:

$$\mathbf{E} = \mathbf{E}_e + \mathbf{E}_m = \mathbf{E}_e + \frac{1}{\epsilon}\mathbf{D}_m\,, \tag{1.5.9}$$

$$\mathbf{D} = \mathbf{D}_e + \mathbf{D}_m\,, \tag{1.5.10}$$

$$\mathbf{H} = \mathbf{H}_e + \mathbf{H}_m = \frac{1}{\mu}\mathbf{B}_e + \mathbf{H}_m\,, \tag{1.5.11}$$

$$\mathbf{B} = \mathbf{B}_e + \mathbf{B}_m\,. \tag{1.5.12}$$

Any divergenceless vector can be represented by the curl of another vector ($\nabla \cdot \nabla \times\ . \equiv 0$) so we elect to represent $\mathbf{B}_e$ and $\mathbf{D}_m$ this way:

$$\mathbf{B}_e = \nabla \times \mathbf{A}_e\,, \qquad \mathbf{D}_m = -\nabla \times \mathbf{A}_m. \tag{1.5.13}$$

The vector fields $\mathbf{A}_e$ and $\mathbf{A}_m$ are called the *magnetic and electric vector potential*, respectively (note that the name is associated with whether the vector potential defines the magnetic induction or the electric displacement field and *not* with whichever phenomenon the vector potential is associated). Substituting the above definition for $\mathbf{B}_e$ into Eq. (1.5.4) we have

$$\nabla \times \mathbf{E}_e = -\partial_t \mathbf{B}_e = -\nabla \times (\partial_t \mathbf{A}_e) \qquad \rightarrow \qquad \nabla \times (\mathbf{E}_e + \partial_t \mathbf{A}_e) = \mathbf{0}\,,$$

and a similar substitution of the definition for $\mathbf{D}_m$ into Eq. (1.5.7) gives

$$\nabla \times (\mathbf{H}_m + \partial_t \mathbf{A}_m) = \mathbf{0}\,.$$

Bearing in mind that $\nabla \times \nabla . \equiv \mathbf{0}$ we are permitted to include the gradient of a scalar function in the antiderivative (or primitive function) of the above two curls as integration constant. Hence

$$\mathbf{E}_e = -\nabla \Phi_e - \partial_t \mathbf{A}_e\,, \tag{1.5.14}$$

$$\mathbf{H}_m = -\nabla \Phi_m - \partial_t \mathbf{A}_m\,. \tag{1.5.15}$$

In the time stationary case Eq. (1.5.14) becomes $\mathbf{E}_e = -\nabla\Phi_e$, which is seen to be identical with Eq. (1.2.3). Hence Φ_e can be identified as the *electric scalar potential*. Likewise Φ_m is referred to as the *magnetic scalar potential*. Using now Eq. (1.5.9) and (1.5.11) we may write the total electric and magnetic field as

$$\mathbf{E} = -\nabla \Phi_e - \partial_t \mathbf{A}_e - \frac{1}{\epsilon}\nabla \times \mathbf{A}_m\,, \tag{1.5.16}$$

$$\mathbf{H} = -\nabla \Phi_m - \partial_t \mathbf{A}_m + \frac{1}{\mu}\nabla \times \mathbf{A}_e\,. \tag{1.5.17}$$

Although the above equations are useful to calculate the electric and magnetic field from the scalar and vector potentials, they do not permit us to calculate the potentials themselves. Equations (1.5.16) and (1.5.17) are interesting in the sense that they tell us that scalar and vector potentials are the 'sources' of electric and magnetic fields. This, however, is not very satisfactory because we know that the actual sources of the fields must be charges and currents, which intuitively leads to the conclusion that it must be possible to express the potentials in terms of charges and currents. In doing this we take the definition of the magnetic vector potential and Eq. (1.5.14) and substitute these into Eq. (1.5.3) to obtain:

$$\frac{1}{\mu}\nabla\times\mathbf{B}_e = \mathbf{J} + \partial_t(\epsilon\mathbf{E}_e) \quad\rightarrow\quad \nabla\times\nabla\times\mathbf{A}_e = \mu\mathbf{J} + \epsilon\mu\partial_t(-\nabla\Phi_e - \partial_t\mathbf{A}_e)\,.$$

leading, via the vector identity $\nabla\times\nabla\times. = \nabla(\nabla\cdot.) - \nabla^2.$, to

$$\nabla(\nabla\cdot\mathbf{A}_e) - \nabla^2\mathbf{A}_e = \mu\mathbf{J} - \epsilon\mu\nabla\partial_t\Phi_e - \epsilon\mu\partial_t^2\mathbf{A}_e \quad\rightarrow$$

$$\nabla^2\mathbf{A}_e - \epsilon\mu\frac{\partial^2\mathbf{A}_e}{\partial t^2} = -\mu\mathbf{J} + \nabla\left(\nabla\cdot\mathbf{A}_e + \epsilon\mu\frac{\partial\Phi_e}{\partial t}\right)\,. \tag{1.5.18}$$

Alternatively, if we start from Eq. (1.5.1) and (1.5.14) we have

$$\nabla\cdot\mathbf{D}_m = \rho \quad\rightarrow\quad \nabla\cdot(-\nabla\Phi_e - \partial_t\mathbf{A}_e) = \frac{\rho}{\epsilon}\,.$$

That, after subtracting $\epsilon\mu\partial_t^2\Phi_e$ from both sides, gives

$$\nabla^2\Phi_e - \epsilon\mu\frac{\partial^2\Phi_e}{\partial t^2} = \frac{1}{\epsilon}\rho - \frac{\partial}{\partial t}\left(\nabla\cdot\mathbf{A}_e + \epsilon\mu\frac{\partial\Phi_e}{\partial t}\right). \tag{1.5.19}$$

Substituting for $\mathbf{E}_m$ and $\mathbf{H}_m$ into Eq. (1.5.8) yields, via a procedure similar to what led to Eq. (1.5.18):

$$\nabla^2\mathbf{A}_m - \epsilon\mu\frac{\partial^2\mathbf{A}_m}{\partial t^2} = \nabla\left(\nabla\cdot\mathbf{A}_m + \epsilon\mu\frac{\partial\Phi_m}{\partial t}\right). \tag{1.5.20}$$

Finally, following an identical procedure that yielded Eq. (1.5.19), starting from Eqs. (1.5.6) and (1.5.15) we obtain:

$$\nabla^2\Phi_m - \epsilon\mu\frac{\partial^2\Phi_m}{\partial t^2} = -\frac{\partial}{\partial t}\left(\nabla\cdot\mathbf{A}_m + \epsilon\mu\frac{\partial\Phi_m}{\partial t}\right). \tag{1.5.21}$$

In general, Eq. (1.5.18) is coupled with Eq. (1.5.19) and Eq. (1.5.20) is coupled with Eq. (1.5.21). In order to decouple them we may choose

$$\nabla\cdot\mathbf{A}_{e,m} + \epsilon\mu\frac{\partial\Phi_{e,m}}{\partial t} = 0 \tag{1.5.22}$$

as supplementary conditions. The above formula is known as the *Lorenz condition*. If this equation is satisfied we have from Eqs. (1.5.18)–(1.5.21):

$$\nabla^2\mathbf{A}_e - \epsilon\mu\frac{\partial^2\mathbf{A}_e}{\partial t^2} = -\mu\mathbf{J}\,, \tag{1.5.23}$$

$$\nabla^2\Phi_e - \epsilon\mu\frac{\partial^2\Phi_e}{\partial t^2} = \frac{\rho}{\epsilon}\,, \tag{1.5.24}$$

$$\nabla^2\mathbf{A}_m - \epsilon\mu\frac{\partial^2\mathbf{A}_m}{\partial t^2} = \mathbf{0}\,, \tag{1.5.25}$$

$$\nabla^2\Phi_m - \epsilon\mu\frac{\partial^2\Phi_m}{\partial t^2} = 0\,. \tag{1.5.26}$$

So we see that our intuition was correct: the (magnetic) vector potential is due to the electric current and the scalar potential is due to charges. The latter of these statements should be trivial from the historical introduction in Section 1.2.

It is important to point out that the choice of both the scalar and vector potential was quite arbitrary. To show this we shall define a set of new potentials

$$\mathbf{A}_{e,m} \rightarrow \mathbf{A}'_{e,m} = \mathbf{A}_{e,m} + \nabla f_{e,m}, \qquad \Phi \rightarrow \Phi'_{e,m} = \Phi_{e,m} - \partial_t f_{e,m}\,,$$

where the primed and unprimed quantities signify the new and old values, respectively, and $f_{e,m}(x, y, z; t)$ is an arbitrary (well-behaved) function. It is clear that when the new potentials are substituted into Eqs. (1.5.16) and (1.5.17) the electric and magnetic fields do not change. Consequently, the choice of the scalar and vector potential is not unique. However, once $f_{e,m}$ is chosen the potentials are defined uniquely. In order to see what condition the function f needs to satisfy we substitute the new values of the potentials into the Lorenz condition:[5]

$$\nabla\cdot\mathbf{A}' + \epsilon\mu\partial_t\Phi' = \nabla\cdot\mathbf{A} + \nabla\cdot\nabla f + \epsilon\mu\partial_t\Phi - \epsilon\mu\partial_t^2 f = \nabla\cdot\mathbf{A} + \epsilon\mu\partial_t\Phi + (\nabla^2 - \epsilon\mu\partial_t^2)f = 0\,,$$

[5] We omit the e, m subscripts for the following couple of equations for brevity; however, the following conclusions apply equally to e and m potentials.

meaning that

$$\nabla \cdot \mathbf{A} + \epsilon\mu\partial_t \Phi = -(\nabla^2 - \epsilon\mu\partial_t^2) f \ ,$$

so that the Lorenz condition is satisfied as long as f is a solution of the homogeneous wave equation. On account of Eq. (1.5.22), the above equation can always be solved. In Section 1.6.1 we shall be discussing the calculation of the potential functions from the source quantities ρ and $\mathbf{J}$.

An alternative way to view Eqs. (1.5.1)–(1.5.4) and Eqs. (1.5.5)–(1.5.8) is to say that the former set of equations corresponds to regions in space where there are sources of electromagnetic radiation, whilst the latter corresponds to source-free regions. The easiest way to demonstrate this is to take the curl of both sides of Eq. (1.5.4) and substitute from Eq. (1.5.3) $\nabla \times \mathbf{H}$:

$$\nabla \times \nabla \times \mathbf{E}_e = -\partial_t \nabla \times \mathbf{B}_e = -\mu\partial_t^2 \mathbf{D}_e - \mu\partial_t \mathbf{J} \ .$$

The left-hand side of this equation can be transformed to give:

$$\nabla(\nabla \cdot \mathbf{E}_e) - \nabla^2 \mathbf{E}_e = \frac{1}{\epsilon}\nabla\rho - \nabla^2 \mathbf{E}_e \ ,$$

where we have used Eq. (1.5.1). Consequently, we have

$$\nabla^2 \mathbf{E}_e - \mu\epsilon \frac{\partial^2 \mathbf{E}_e}{\partial t^2} = \frac{1}{\epsilon}\nabla\rho + \mu\frac{\partial \mathbf{J}}{\partial t} \ , \tag{1.5.27}$$

and there is a corresponding equation for the magnetic field:

$$\nabla^2 \mathbf{H}_e - \epsilon\mu \frac{\partial^2 \mathbf{H}_e}{\partial t^2} = -\nabla \times \mathbf{J} \ , \tag{1.5.28}$$

which can be obtained by taking the curl of Eq. (1.5.3) and substituting in from Eq. (1.5.4) and Eq. (1.5.2). Simultaneously, following the same procedure we can obtain the corresponding equations from Eqs. (1.5.5)–(1.5.8):

$$\nabla^2 \mathbf{E}_m - \mu\epsilon \frac{\partial^2 \mathbf{E}_m}{\partial t^2} = \mathbf{0} \tag{1.5.29}$$

and

$$\nabla^2 \mathbf{H}_m - \epsilon\mu \frac{\partial^2 \mathbf{H}_m}{\partial t^2} = \mathbf{0} \ . \tag{1.5.30}$$

It is seen that Eq. (1.5.27) and Eq. (1.5.29) form a pair of inhomogeneous and homogeneous vector wave equations, respectively. The same is true for Eq. (1.5.28) and Eq. (1.5.30).

A wealth of information is available regarding the solutions of inhomogeneous wave equations. For the present discussion we merely state that the general solution of the inhomogeneous (scalar) wave equation

$$(\nabla^2 - \epsilon\mu\partial_t^2) u(\mathbf{r}, t) = h(\mathbf{r}, t) \ , \tag{1.5.31}$$

where $\mathbf{r} = (x, y, z)$, is

$$u(\mathbf{r}, t) = u_0(\mathbf{r}, t) + \int dt \int d^3\mathbf{r}' h(\mathbf{r}', t)\, G(\mathbf{r}, \mathbf{r}'; t) \ , \tag{1.5.32}$$

where $G(\mathbf{r}, \mathbf{r}', t)$ satisfies

$$(\nabla^2 - \epsilon\mu\partial_t^2) G(\mathbf{r}, \mathbf{r}', t) = \delta(t - t')\delta(\mathbf{r} - \mathbf{r}') \tag{1.5.33}$$

and $u_0(\mathbf{r}, t)$ is a solution of the homogeneous wave equation

$$(\nabla^2 - \epsilon\mu\partial_t^2) u_0(\mathbf{r}, t) = 0 \ . \tag{1.5.34}$$

In other words, the solution to the inhomogeneous wave equation is given as the sum of the specific solution (the double integral involving h and G) and *any* solution of the homogeneous wave equation $u_0(\mathbf{r}, t)$. The function G is commonly called Green's function and it is treated in more detail in Section 1.6. Thus, when the solution (of the homogeneous wave equation) for $\mathbf{E}_m$ is added to the solution (of the inhomogeneous wave equation) for $\mathbf{E}_e$ to compute the total electric field, we merely follow the principles of solving inhomogeneous partial differential equations. Hence adding the fields $\mathbf{E}_e$ and $\mathbf{E}_m$ to yield the total field is fully physical. We note that the electric vector potential $\mathbf{A}_m$ and the magnetic scalar potential

Φ_m are often not used because the equations can be formulated such that their effect is included in the magnetic vector potential $\mathbf{A}_e$ and the electric scalar potential Φ_e.

1.5.2 The Hertz Vectors

We have seen in the previous section that knowledge of the electric and magnetic vector and scalar potentials is sufficient to determine the electric and magnetic fields uniquely. In the following section we will show that the potential functions can be determined from a knowledge of the sources. We shall now show that the knowledge of two vector potential functions is in fact sufficient to uniquely determine the electric and magnetic fields. Let us define the electric Hertz vector $\mathbf{\Pi}_e$ via

$$\nabla \cdot \mathbf{\Pi}_e = -\Phi_e \,, \tag{1.5.35}$$

and its magnetic counterpart $\mathbf{\Pi}_m$ by

$$\nabla \cdot \mathbf{\Pi}_m = -\Phi_m \,. \tag{1.5.36}$$

Then, by using the Lorenz condition Eq. (1.5.22) we have

$$\nabla \cdot (\mathbf{A}_{e,m} - \epsilon\mu\partial_t \mathbf{\Pi}_{e,m}) = 0 \,,$$

giving

$$\mathbf{A}_{e,m} = \epsilon\mu \frac{\partial \mathbf{\Pi}_{e,m}}{\partial t} \,. \tag{1.5.37}$$

With Eqs. (1.5.36) and (1.5.37) we are now equipped to express the electric and magnetic fields from the two Hertz vectors. So from Eqs. (1.5.14) and (1.5.15) we may write:

$$\mathbf{E} = \nabla(\nabla \cdot \mathbf{\Pi}_e) - \epsilon\mu \frac{\partial^2 \mathbf{\Pi}_e}{\partial t^2} - \mu \frac{\partial}{\partial t} \nabla \times \mathbf{\Pi}_m \,, \tag{1.5.38}$$

$$\mathbf{H} = \nabla(\nabla \cdot \mathbf{\Pi}_m) - \epsilon\mu \frac{\partial^2 \mathbf{\Pi}_m}{\partial t^2} + \epsilon \frac{\partial}{\partial t} \nabla \times \mathbf{\Pi}_e \,. \tag{1.5.39}$$

The electric and magnetic Hertz vectors need to satisfy the inhomogeneous and homogeneous vector wave equation respectively. To show this we start from Eq. (1.5.23) and substitute Eq. (1.5.37) into that:

$$\nabla^2 \mathbf{A}_e - \epsilon\mu\partial_t^2 \mathbf{A}_e = \epsilon\partial_t(\nabla^2 \mathbf{\Pi}_e - \epsilon\mu\partial_t^2 \mathbf{\Pi}_e) = -\mathbf{J}.$$

Integrating both sides with respect to t gives

$$\nabla^2 \mathbf{\Pi}_e - \epsilon\mu \frac{\partial^2 \mathbf{\Pi}_e}{\partial t^2} = -\frac{1}{\epsilon} \int \mathbf{J} \, dt, \tag{1.5.40}$$

and the same procedure leads to the corresponding equation for $\mathbf{\Pi}_m$:

$$\nabla^2 \mathbf{\Pi}_m - \epsilon\mu \frac{\partial^2 \mathbf{\Pi}_m}{\partial t^2} = \mathbf{0}. \tag{1.5.41}$$

The solution of these equations is discussed in Section 1.6, but it is worth pointing out that *one* of the solutions of Eq. (1.5.41) is $\mathbf{\Pi}_m = \mathbf{0}$. Indeed, this solution will be the most applicable in almost all cases within the remit of this book, given that magnetic currents do not exist. In that case we may rewrite Eqs. (1.5.38) and (1.5.39):

$$\mathbf{E} = \nabla(\nabla \cdot \mathbf{\Pi}_e) - \epsilon\mu \frac{\partial^2 \mathbf{\Pi}_e}{\partial t^2} \,, \tag{1.5.42}$$

$$\mathbf{B} = \epsilon\mu \frac{\partial}{\partial t} \nabla \times \mathbf{\Pi}_e \,. \tag{1.5.43}$$

On the other hand, for source-free regions we may set $\mathbf{J} = \mathbf{0}$. For this case, both the electric and magnetic vector wave equations read

$$\nabla^2 \mathbf{\Pi}_{e,m} - \epsilon\mu \frac{\partial^2 \mathbf{\Pi}_{e,m}}{\partial t^2} = \mathbf{0} \,, \tag{1.5.44}$$

which means that Eqs. (1.5.38) and (1.5.39) can be written as

$$\mathbf{E} = \nabla \times \nabla \times \mathbf{\Pi}_e - \mu \frac{\partial}{\partial t} \nabla \times \mathbf{\Pi}_m \,, \tag{1.5.45}$$

$$\mathbf{H} = \nabla \times \nabla \times \mathbf{\Pi}_m + \epsilon \frac{\partial}{\partial t} \nabla \times \mathbf{\Pi}_e \,. \tag{1.5.46}$$

1.6 Harmonic Solutions and the Helmholtz Equation

Given the spectral bandwidth of radiation sources most frequently used in the optical domain, such as narrow-band lasers, obtaining a time-harmonic solution of Maxwell's equations seems to be an obvious choice. One nevertheless frequently needs to deal with pulsed waves, e.g. short laser pulses, or seismic or acoustic waves that often have an impulse-like shape in the time domain. These waves, however, can be represented as a linear superposition of time-harmonic waves of different frequencies using Fourier synthesis. Here we consider electromagnetic fields with the harmonic time dependence of $\exp(-i\omega t)$ where ω is the *circular frequency* and t denotes time. For a detailed treatment of the complex amplitude notation that is frequently used for time-harmonic fields, the reader is referred to Appendix A.

The general form of a time-harmonic electric and magnetic wave is

$$\mathbf{E}_h(\mathbf{r}, t) = \mathbf{E}(\mathbf{r}) \exp(-i\omega t) \,, \tag{1.6.1}$$

$$\mathbf{B}_h(\mathbf{r}, t) = \mathbf{B}(\mathbf{r}) \exp(-i\omega t) \,, \tag{1.6.2}$$

where $\mathbf{E}(\mathbf{r})$ and $\mathbf{B}(\mathbf{r})$ are the complex amplitudes of the **E**- and **B**-fields, respectively, and $\mathbf{r} = (x, y, z)$ a general position vector in space. By substituting these expressions in Eqs. (1.3.10)–(1.3.13) the time-dependent Maxwell equations can be transformed into their time-independent counterparts:

$$\nabla \cdot \epsilon \mathbf{E} = \rho \,, \tag{1.6.3}$$

$$\nabla \cdot \mathbf{B} = 0 \,, \tag{1.6.4}$$

$$\nabla \times \frac{\mathbf{B}}{\mu} = (\sigma - i\omega\epsilon) \mathbf{E} \,, \tag{1.6.5}$$

$$\nabla \times \mathbf{E} = i\omega \mathbf{B} \,, \tag{1.6.6}$$

where we have written these equations in terms of the electric field and the magnetic induction vectors, **E** and **B**, respectively. Because the time-harmonic term cancels on both sides of all four equations, these vectors are now functions of the position vector **r** only.

General solutions of the above differential equations can be obtained in principle when the appropriate boundary conditions are given. In general, analytic and numerical methods are implemented to construct the three-dimensional solutions for the **E**- and **B**-vectors from the source to the region of interest across possible boundaries between media. Nevertheless, a simple solution can be obtained analytically when the region of interest is homogeneous and free of charges ($\rho = 0$). Note that the lack of charges does not mean a vanishing electric field **E**, it merely means that the electromagnetic disturbance is originated from outside the region of interest. In these cases one usually specifies the incident electric field on the boundary of the volume of interest and solves for the disturbance everywhere inside that volume. When $\rho = 0$, by taking the curl of Eqs. (1.6.5) and (1.6.6) and substituting $\nabla \cdot \mathbf{E} = 0$ and $\nabla \cdot \mathbf{B} = 0$ from Eqs. (1.6.3)–(1.6.4), we obtain

$$\triangle \mathbf{E} + i\omega\mu(\sigma - i\omega\epsilon)\mathbf{E} = 0 \,, \tag{1.6.7}$$

$$\triangle \mathbf{B} + i\omega\mu(\sigma - i\omega\epsilon)\mathbf{B} = 0 \,, \tag{1.6.8}$$

where $\triangle = \nabla \cdot \nabla$ denotes the Laplace operator. These time-independent Maxwell equations are commonly called the Helmholtz equations. For propagation in a lossless medium, we have the frequently used expressions

$$\triangle \mathbf{E} + k^2 \mathbf{E} = 0 \,, \tag{1.6.9}$$

$$\triangle \mathbf{B} + k^2 \mathbf{B} = 0 \,, \tag{1.6.10}$$

where $k = \omega / v$ is the radial wavenumber of the harmonic radiation field and v the speed of propagation in the medium.

1.6.1 Calculation of the Potential Functions

Before embarking upon the topic of this section, it seems a good idea to briefly review the mathematics of Green's functions that have already appeared in the previous section, Eqs. (1.5.32) and (1.5.33). Using the language of electrical engineering and linear system theory, Green's functions can be viewed as the impulse response function of an inhomogeneous differential equation. Consider for example the following inhomogeneous wave equation:

$$\nabla^2 f(\mathbf{r},t) - a\partial_t^2 f(\mathbf{r},t) = g(\mathbf{r},t)\,.$$

If we assume time-harmonic dependence of both f and g, $f(\mathbf{r},t) = f(\mathbf{r})\exp(-i\omega t)$ and $g(\mathbf{r},t) = g(\mathbf{r})\exp(-i\omega t)$, we have[6]

$$\nabla^2 f(\mathbf{r}) + a\omega^2 f(\mathbf{r}) = g(\mathbf{r})\,,$$

because the time-harmonic term can be cancelled from both sides of the equation, which can be written in a general form as

$$\mathcal{L}f(\mathbf{r}) = g(\mathbf{r})\,,$$

where $\mathcal{L}$ is a linear operator, which in our specific case is $\mathcal{L} = \nabla^2 + a\omega^2$. There exists a function $G(\mathbf{r})$ for which

$$\mathcal{L}G(\mathbf{r}) = \delta(\mathbf{r})\,,$$

or more generally

$$\mathcal{L}_{\mathbf{r}}G(\mathbf{r},\mathbf{r}_0) = \delta(\mathbf{r}-\mathbf{r}_0)\,, \tag{1.6.11}$$

where $\mathcal{L}_{\mathbf{r}}$ indicates that $\mathcal{L}$ operates on $\mathbf{r}$ whilst $\mathbf{r}_0$ is kept constant. The above equation means that $G(\mathbf{r})$ is the impulse response function of the linear operator $\mathcal{L}^{-1}$. Therefore $f(\mathbf{r})$ can be written as the linear superposition of the impulse response function and $g(\mathbf{r})$:

$$f(\mathbf{r}) = \iiint G(\mathbf{r},\mathbf{r}_0)g(\mathbf{r}_0)d\mathbf{r}_0\,, \tag{1.6.12}$$

integrating for all space, because

$$\mathcal{L}f(\mathbf{r}) = \iiint \mathcal{L}_{\mathbf{r}}G(\mathbf{r},\mathbf{r}_0)g(\mathbf{r}_0)d\mathbf{r}_0 = \iiint \delta(\mathbf{r}-\mathbf{r}_0)g(\mathbf{r}_0)d\mathbf{r}_0 = g(\mathbf{r})\,,$$

which follows from the definition of the Dirac delta-function $\iiint \delta(\mathbf{r})dV = 1$. The function $G(\mathbf{r},\mathbf{r}_0)$ is called *Green's function*, which is seen to offer a convenient way to solve inhomogeneous differential equations, such as the inhomogeneous wave equation shown above. We shall see later that in electromagnetics Green's function can be written as $G(\mathbf{r},\mathbf{r}_0) = G(\mathbf{r}-\mathbf{r}_0)$, in which case Eq. (1.6.12) is seen to become a three-dimensional convolution integral of G and g, the physical significance of which will become clear later. A possible physical interpretation of Eq. (1.6.12) is that the disturbance at any given point in space (except for $\mathbf{r}_0$) is given as the sum of contributions of a source term $G(\mathbf{r},\mathbf{r}_0)$ weighted by $g(\mathbf{r}_0)$. This interpretation is elucidated further below.

Having done with the mathematical preliminaries, in Section 1.5 we discussed the scalar and vector potential functions that allow the calculation of the electric and magnetic field vectors $\mathbf{E}$ and $\mathbf{B}$. For time-harmonic sources or currents the inhomogeneous wave equations (1.5.23) and (1.5.24) can be written in vacuum as

$$(\nabla^2 + k^2)\Phi(\mathbf{r}) = -\frac{\rho_\omega(\mathbf{r})}{\epsilon_0}\,, \qquad (\nabla^2 + k^2)\mathbf{A}(\mathbf{r}) = -\mu_0\mathbf{J}_\omega(\mathbf{r})\,. \tag{1.6.13}$$

In this equation $k = \omega/v$ is the radial wavenumber of the radiation in the medium under consideration; $\rho_\omega(\mathbf{r})$ and $\mathbf{J}_\omega(\mathbf{r})$ the harmonic components at frequency ω of the charge and current distributions, or source functions, respectively, and we have dropped the e subscript from both vector and scalar potential. As seen above the solution of the wave equations for the two potential functions requires a spatial integration over the source functions defined by the static charge distribution $\rho_\omega(\mathbf{r})$ and the current density distribution $\mathbf{J}_\omega(\mathbf{r})$. As discussed in the mathematical preliminaries, we first calculate the potential functions in the presence of a point-like source or current represented by the δ-function.

The solution of the wave equations for the scalar and vectorial potential function may show spherical symmetry if the current source is point-like and represented by the δ-function. For the solution to have spherical symmetry it is required

[6] Note that the arguments presented in this review are informative but mathematically not complete. For a fully rigorous discussion the reader is referred to Arfken **[10]**.

that there are no boundaries that could perturb the symmetry; preferably, the space in which the potential functions propagate is empty. With the δ-type source elements for charge and current, located in the point $\mathbf{r} = \mathbf{r}_0$, we have the differential equations

$$(\nabla^2 + k^2)G_\Phi = -\frac{q_\omega \delta(\mathbf{r} - \mathbf{r}_0)}{\epsilon_0}, \qquad (\nabla^2 + k^2)\mathbf{G}_A = -\mu_0 \mathbf{I}_\omega \delta(\mathbf{r} - \mathbf{r}_0). \tag{1.6.14}$$

q_ω is the localised charge at frequency ω and $\mathbf{I}_\omega$ gives the strength and the direction of the line current element, both at $\mathbf{r} = \mathbf{r}_0$. In comparison with Eq. (1.6.11) it is seen that the functions G_Φ and $\mathbf{G}_A$ are the Green's functions that give the solutions of the inhomogeneous wave equations for point-like charge or current sources. The response to more complicated source distributions is then easily obtained by integration of the point-source solutions over the source volume as discussed in conjunction with Eq. (1.6.12).

1.6.1.1 Green's Function Solution for the Static Case

To obtain the Green's function solutions we first study the special properties of the function $|\mathbf{r} - \mathbf{r}_0|^{-1}$. For the gradient of this function and for the Laplacian we have

$$\left.\begin{aligned} \nabla\left(\frac{1}{|\mathbf{r} - \mathbf{r}_0|}\right) &= -\frac{\mathbf{r} - \mathbf{r}_0}{|\mathbf{r} - \mathbf{r}_0|^3}, \\ \nabla^2\left(\frac{1}{|\mathbf{r} - \mathbf{r}_0|}\right) &= 0, \end{aligned}\right\} \quad |\mathbf{r} - \mathbf{r}_0| \neq 0. \tag{1.6.15}$$

The calculation of the value of $\nabla^2 (1/|\mathbf{r} - \mathbf{r}_0|)$ in the point $\mathbf{r} = \mathbf{r}_0$ is carried out with the aid of Fig. 1.6. We apply Gauss' theorem to the quantity $\nabla(1/|\mathbf{r} - \mathbf{r}_0|)$ and write

$$\iiint \nabla^2\left(\frac{1}{|\mathbf{r} - \mathbf{r}_0|}\right) dV = -\oiint_A \frac{(\mathbf{r} - \mathbf{r}_0)}{|\mathbf{r} - \mathbf{r}_0|^3} \cdot d\mathbf{A}, \tag{1.6.16}$$

where $d\mathbf{A}$ stands for the product of the surface area dA multiplied by the unit outward normal vector $\hat{\mathbf{v}}$ of the closed surface A. Because $|\mathbf{r} - \mathbf{r}_0|$ is the radius of the sphere and $\mathbf{r} - \mathbf{r}_0$ is parallel with the surface normal $\hat{\mathbf{v}}$, the integrand of the surface integral is constant and it therefore evaluates to 4π. Using the definition of the Dirac delta-function we arrive at the identity

$$\nabla^2\left(\frac{1}{|\mathbf{r} - \mathbf{r}_0|}\right) = -4\pi\delta(\mathbf{r}), \tag{1.6.17}$$

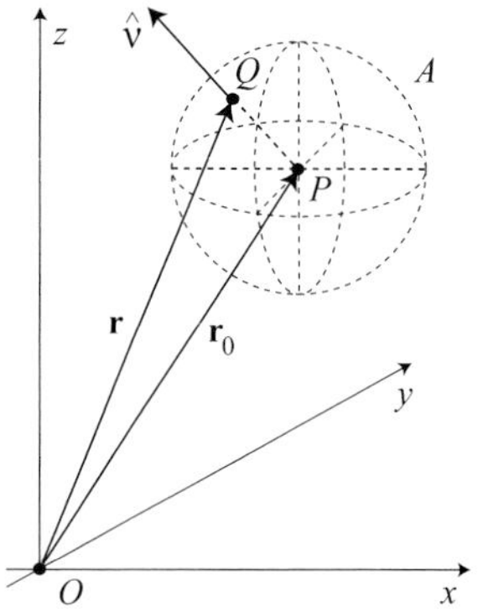

Figure 1.6: A point-like source element is found at the point P characterised by position vector $\mathbf{r}_0$. The potential function value has spherical symmetry and depends on the distance $|\mathbf{r} - \mathbf{r}_0|$ of a general point Q from the source element P.

where we remark that the three-dimensional δ-function has the dimension of m^{-3}. Expressed in terms of the one-dimensional radial δ-function, $\delta(\mathbf{r})$ is given by

$$\delta(\mathbf{r}) = \frac{\delta(r)}{2\pi r^2} , \tag{1.6.18}$$

where $\delta(r)$ has the dimension of m^{-1}.

In the static case, considering the scalar equation in Eq. (1.6.14) with $k \to 0$ and q_ω being replaced by a static charge q, we have the differential equation for the Green's function

$$\nabla^2 G(\mathbf{r}) = -\frac{q\delta(\mathbf{r})}{\epsilon_0} . \tag{1.6.19}$$

Combining this expression with the result of Eq. (1.6.17) we find,

$$G(\mathbf{r}) = \frac{q}{4\pi\epsilon_0 |\mathbf{r} - \mathbf{r}_0|} , \tag{1.6.20}$$

where $G(\mathbf{r})$ is an electric potential having the dimension of Volts.

1.6.1.2 Green's Function for Harmonically Varying Charges and Currents

In the case of a harmonically varying charge or current distribution, the left-hand differential equation of Eq. (1.6.14) for the scalar potential is written as

$$(\nabla_{\mathbf{x}}^2 + k^2) G_\Phi(r) = \left(\frac{1}{r} \frac{d^2}{dr^2} \{ r G_\Phi(r) \} + k^2 G_\Phi(r) \right) = 0 , \qquad r \neq 0 . \tag{1.6.21}$$

Here $\nabla_{\mathbf{x}}^2$ stands for the Laplacian in cartesian coordinates; we also supposed that the solution G_Φ has spherical symmetry and that the radial distance r is measured from the location where the charge is located. A solution $C\, G_\Phi(r) = \exp(ikr)/r$ satisfies the differential equation for any finite value of r and with C some arbitrary amplitude. The amplitude factor is derived following the same arguments as for the static case, with the requirement that the harmonic solution must approach the static solution for $k|\mathbf{r} - \mathbf{r}_0| \to 0$. For the constant C we find $q_\omega / 4\pi\epsilon_0$ and the general solution is then given by

$$G_\Phi(\mathbf{r}) = \frac{q}{4\pi\epsilon_0} \frac{\exp(ik|\mathbf{r} - \mathbf{r}_0|)}{|\mathbf{r} - \mathbf{r}_0|} , \qquad \mathbf{r} \neq \mathbf{r}_0 . \tag{1.6.22}$$

The solution of the second differential equation of Eq. (1.6.14) for the vector potential $\mathbf{A}_e$ is found along the same lines. To find the Green's vector function, the point-like current element $\mathbf{I}_\omega \delta(\mathbf{r} - \mathbf{r}_0)$ is introduced with the solution

$$\mathbf{G}_A(\mathbf{r}) = \frac{\mu_0 \mathbf{I}_\omega}{4\pi} \frac{\exp(ik|\mathbf{r} - \mathbf{r}_0|)}{|\mathbf{r} - \mathbf{r}_0|} , \qquad \mathbf{r} \neq \mathbf{r}_0 . \tag{1.6.23}$$

The scalar and vector potential functions for general charge and current distributions follow from the linear superposition principle (c.f. Eq. (1.6.12)):

$$\mathbf{A}_e(\mathbf{r}) = \frac{\mu_0}{4\pi} \iiint \mathbf{J}_\omega(\mathbf{r}_0) \frac{\exp(ik|\mathbf{r} - \mathbf{r}_0|)}{|\mathbf{r} - \mathbf{r}_0|} d\mathbf{r}_0 , \tag{1.6.24}$$

$$\Phi(\mathbf{r}) = \frac{1}{4\pi\epsilon_0} \iiint \rho_\omega(\mathbf{r}_0) \frac{\exp(ik|\mathbf{r} - \mathbf{r}_0|)}{|\mathbf{r} - \mathbf{r}_0|} d\mathbf{r}_0 , \tag{1.6.25}$$

where $d\mathbf{r}_0$ stands for the integration over the source and current volume, respectively.

1.6.2 Radiation Field of a Dipole Source

Most natural sources in the optical domain, radiating at frequencies in the PHz regime, are atomic or molecular dipole sources that produce a radiation pattern that depends on the radial distance and on the elevation angle θ, see Fig. 1.7. The field is invariant under a rotation around the dipole axis $\mathbf{p}$. A dipole is characterised by its moment $\mathbf{p}$. The direction of $\mathbf{p}$ is determined by the axis on which the two equal but opposite charges that make up the dipole oscillate. By agreement the

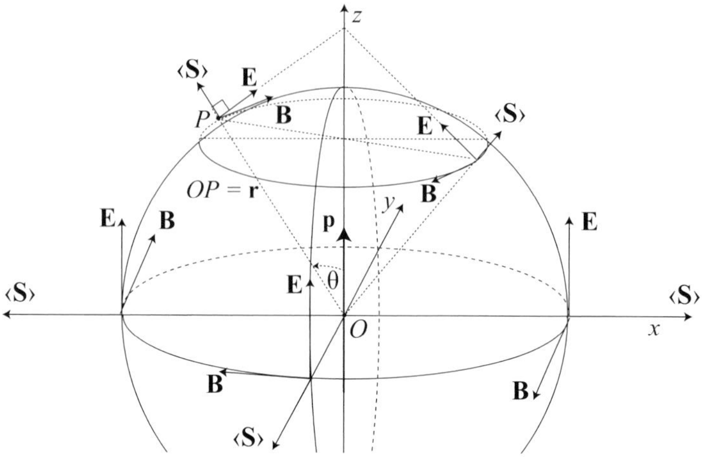

Figure 1.7: The electromagnetic field strengths produced by a typical dipole emitter $\mathbf{p}(t)$ in the optical domain. In the far field of the dipole, the electric field vector $\mathbf{E}$ lies in the plane erected by the dipole axis $\mathbf{p}$ and the position vector $\mathbf{r}$. In a small annular region perpendicular to the dipole axis ($\theta = \pi/2$), the electric and magnetic field vectors are virtually constant regarding their direction and magnitude.

dipole moment is oriented from the negative to the positive charge. The magnitude of the dipole moment is simply given by the product of the oscillation amplitude d of the two charges and the magnitude of the positive charge q. In the case of an atomic dipole the magnitudes of two charges are equal but, since the positive charge is associated with the nucleus and the negative charge with the electron, the significantly larger mass of the nucleus makes only the electron oscillate. For a harmonically oscillating dipole the moment is given in complex notation by $\mathbf{p}\exp(-i\omega t)$.

1.6.2.1 The Vector Potential Function A of a Dipole

To obtain the electric and magnetic field strengths of an oscillating dipole it is advisable to first calculate the magnetic vector potential $\mathbf{A}_e$ (we shall drop the subscript e in what follows) generated by the current distribution of a dipole. We use Eq. (1.6.24) and take the centre of the dipole oscillation as the origin by letting $\mathbf{r}_0 = \mathbf{0}$. We then apply the approximation, fully justified in the optical domain where $d \ll \lambda$, that the quantity $\exp(ikr)/r$ is practically constant when integrating over the volume of the dipole and therefore it can be taken out of the integral. The resulting equation for $\mathbf{A}$ is now given by

$$\mathbf{A}(\mathbf{r}) = \frac{\mu_0}{4\pi}\,\frac{\exp(ikr)}{r}\iiint \mathbf{J}_\omega(\mathbf{r}')d\mathbf{r}' \,, \tag{1.6.26}$$

where $\mathbf{r}'$ is a dummy variable for the integration over the volume of the dipole.

In the case of a dipole the position of the oscillating charge with magnitude q is given by

$$\rho_\omega(\mathbf{r}) = q\,\delta\{\mathbf{r} - \mathbf{d}\exp(-i\omega t)\} \,. \tag{1.6.27}$$

As before, the three-dimensional vector in the argument indicates that δ is the three-dimensional delta-function. The condition for the conservation of charge that oscillates at a frequency ω reads

$$\nabla\cdot\mathbf{J}_\omega + \frac{\partial\rho_\omega}{\partial t} = 0 \,, \tag{1.6.28}$$

which can be obtained by taking the divergence of both sides of Eq. (1.3.12) and then substituting in from Eq. (1.3.10). The evaluation of the volume integral over the current density in Eq. (1.6.26) can be carried out by integrating in parts. In three dimensions we have

$$\iiint_V (\partial_i f)g_i\, dV = \oiint f g_i\, dS - \iiint_V f(\partial_i g_i)dV \,, \tag{1.6.29}$$

where $i = x, y, z$ and dS is the surface element on the boundary surface projected onto the normal. Applying the formula to three functions g_i that form the vector function $\mathbf{g}$ and summing the resulting expressions we obtain

$$\iiint_V (\nabla f) \cdot \mathbf{g}\, dV = \oiint f\mathbf{g} \cdot \hat{\mathbf{v}} dS - \iiint_V f\, (\nabla \cdot \mathbf{g})\, dV \,, \tag{1.6.30}$$

where $\hat{\mathbf{v}}$ is the unit outward normal of S.

As the next step, we successively substitute in Eq. (1.6.30) the functions $f_1 = x$, $f_2 = y$ and $f_3 = z$ and multiply each equation by the appropriate Cartesian direction unit vector. Summing the three resulting equations yields the expression

$$\iiint_V \mathbf{g}\, dV = \oiint \mathbf{r}(\mathbf{g} \cdot \hat{\mathbf{v}}) dS - \iiint_V \mathbf{r}\, (\nabla \cdot \mathbf{g})\, dV \,. \tag{1.6.31}$$

When applied to the current distribution $\mathbf{J}_\omega$ we find

$$\iiint_V \mathbf{J}_\omega\, dV = \oiint \mathbf{r}(\mathbf{J}_\omega \cdot \hat{\mathbf{v}}) dS - \iiint_V \mathbf{r}\, (\nabla \cdot \mathbf{J}_\omega)\, dV \,, \tag{1.6.32}$$

noting that outside the source region the surface integral yields zero contribution.
We may use Eq. (1.6.28) to find the expression for $\nabla \cdot \mathbf{J}_\omega$ and the time derivative of the charge distribution is given by

$$\begin{aligned}\frac{\partial \rho_\omega}{\partial t} = i\omega q \Bigg[& d_x \frac{d\{\delta(u_x)\}}{du_x} \delta(u_y)\delta(u_z) \\ & + d_y \delta(u_x) \frac{d\{\delta(u_y)\}}{du_y} \delta(u_z) + d_z \delta(u_x)\delta(u_y) \frac{d\{\delta(u_z)\}}{du_z} \Bigg] \,,\end{aligned} \tag{1.6.33}$$

where the arguments u_x, u_y and u_z apply to the one-dimensional Cartesian δ-functions that compose the three-dimensional δ-function of Eq. (1.6.27). By using the identity

$$\int \delta(x) dx = -\int x \frac{d\delta(x)}{dx} dx = 1,$$

we may evaluate the integral of Eq. (1.6.32) to obtain

$$\iiint_V \mathbf{J}_\omega\, dV = -i\omega q \mathbf{d} = -i\omega \mathbf{p} \,, \tag{1.6.34}$$

where the dipole moment $\mathbf{p}$ is defined by $q\mathbf{d}$. Although the derivation seems rather elaborate given the relatively trivial solution for the basic dipole, the approach above can be applied equally in the case of more complicated current distributions, for instance, when oscillating three-dimensional multipole charge distributions are present.

With the result above, the vector potential function for a dipole is simply given by

$$\mathbf{A}_e(\mathbf{r}) = \frac{-i\omega\mu_0 \mathbf{p}}{4\pi} \frac{\exp(ikr)}{r} \,. \tag{1.6.35}$$

1.6.2.2 The Electromagnetic Field Vectors of a Radiating Dipole

The magnetic induction vector in vacuum can be directly calculated from the vector potential by means of Eq. (1.5.13) and we obtain for the amplitude of the harmonic component of the induction vector, with the dipole in the origin of the coordinate system,

$$\mathbf{B}_0(\mathbf{r}) = \frac{i\omega\mu_0}{4\pi}\, \mathbf{p} \times \nabla \left(\frac{\exp(ikr)}{r} \right) = -\frac{\omega\mu_0 |\mathbf{p}| \exp(ikr)}{4\pi r} \left(k + \frac{i}{r} \right) \hat{\mathbf{p}} \times \hat{\mathbf{r}} \,, \tag{1.6.36}$$

where $\hat{\mathbf{p}}$ is the unit vector parallel to the dipole moment vector and $\hat{\mathbf{r}}$ is the unit vector in the direction OP, see Fig. 1.7.

The calculation of the electric field strength in a region free of charges and currents uses the Ampère–Maxwell law of Eq. (1.3.3), which for harmonic fields reads

$$\mathbf{E}_0(\mathbf{r}) = \frac{i}{\omega\epsilon_0\mu_0} \nabla \times \mathbf{B}_0(\mathbf{r}) \,, \tag{1.6.37}$$

which yields the expression

$$\mathbf{E}_0(\mathbf{r}) = -\frac{1}{4\pi\epsilon_0} \nabla \times g(r)\mathbf{p} \times \mathbf{r} \,, \tag{1.6.38}$$

with

$$g(r) = \left(\frac{ikr - 1}{r^3}\right)\exp(ikr)\,. \tag{1.6.39}$$

We note that

$$\frac{dg(r)}{dr} = -\frac{k^2}{r^2}\exp(ikr) - 3\,\frac{g(r)}{r}\,. \tag{1.6.40}$$

With the expression for the vector triple product,

$$\nabla \times \mathbf{a} \times \mathbf{b} = \mathbf{a}(\nabla \cdot \mathbf{b}) + (\mathbf{b} \cdot \nabla)\mathbf{a} - \mathbf{b}(\nabla \cdot \mathbf{a}) - (\mathbf{a} \cdot \nabla)\mathbf{b}\,, \tag{1.6.41}$$

and putting $\mathbf{a} = g(r)\mathbf{p}$ and $\mathbf{b} = \mathbf{r}$ we obtain after some algebraic manipulation

$$\mathbf{E}_0(\mathbf{r}) = \frac{|\mathbf{p}|}{4\pi\epsilon_0}\left(\frac{\exp(ikr)}{r}\right)\left\{k^2(\hat{\mathbf{r}} \times \hat{\mathbf{p}} \times \hat{\mathbf{r}}) + \left(\frac{ik}{r} - \frac{1}{r^2}\right)[\hat{\mathbf{p}} - 3\{\hat{\mathbf{r}} \cdot \hat{\mathbf{p}}\}\hat{\mathbf{r}}]\right\}\,. \tag{1.6.42}$$

1.6.2.3 The Energy Flow from an Oscillating Dipole

The Poynting vector that determines the magnitude and direction of the energy flow in an electromagnetic field can be derived from the expressions for $\mathbf{B}_0$ and $\mathbf{E}_0$ that are given in Eqs. (1.6.36) and (1.6.42). The average value of the energy flow for a harmonic signal is given by $\langle \mathbf{S} \rangle = \Re\{\mathbf{E}_0 \times \mathbf{H}^*\}/2$. Substituting in the expressions for the electric and magnetic field strengths we obtain for the far-field contribution, proportional to $1/r^2$,

$$\begin{aligned}\langle \mathbf{S} \rangle &= \frac{1}{2}\Re\left\{-\frac{k^4 c\,|\mathbf{p}|^2}{16\pi^2\epsilon_0 r^2}(\hat{\mathbf{r}} \times \hat{\mathbf{p}} \times \hat{\mathbf{r}}) \times (\hat{\mathbf{p}} \times \hat{\mathbf{r}})\right\}\\ &= \frac{k^4 c\,|\mathbf{p}|^2}{32\pi^2\epsilon_0 r^2}\left\{1 - (\hat{\mathbf{r}} \cdot \hat{\mathbf{p}})^2\right\}\hat{\mathbf{r}}\,.\end{aligned} \tag{1.6.43}$$

The energy propagation direction is along the vector $\mathbf{r}$ and its magnitude is proportional to $(\hat{\mathbf{r}} \cdot \hat{\mathbf{p}})^2 = \sin^2\theta$. The other components of the time-averaged Poynting vector present in the near field of the dipole are proportional to r^{-3} and r^{-5}. These components oscillate at a frequency of 2ω and do not contribute to the net energy transport away from the dipole source.

The total power P_r radiated by the dipole is the flux of the time-averaged Poynting vector through the surface of a sphere centred on the dipole, as follows from Eq. (1.4.7):

$$P_r = \frac{k^4 c\,|\mathbf{p}|^2}{32\pi^2\epsilon_0}\left(2\pi\int_0^{\pi}\sin^3\theta d\theta\right) = \frac{k^4 c\,|\mathbf{p}|^2}{12\pi\epsilon_0}\,. \tag{1.6.44}$$

We can correlate the classical formula above with the result from quantum theory about the energy carried away by a photon from an atom. In the process of spontaneous emission, the atom decays from a higher to a lower energy state with a difference in energy of $\hbar ck$. We assume that the dipole moment strength shows a Gaussian profile as a function of time,

$$|\mathbf{p}(t)| = \frac{p_0}{(\pi)^{1/4}}\exp\left(-\frac{t^2}{2\tau^2}\right)\,, \tag{1.6.45}$$

where $2\tau\sqrt{2}$ is the time elapsed between the e^{-1} levels of the upcoming and decaying dipole strength. Integration over time of the squared dipole moment yields

$$\int_{-\infty}^{+\infty}\frac{p_0^2}{\sqrt{\pi}}\exp\left\{-(t/\tau)^2\right\}dt = \tau p_0^2\,. \tag{1.6.46}$$

We equate the total energy radiated by the dipole to the energy carried away by the photon and obtain

$$p_0 = \left(\frac{6\epsilon_0 h}{k^3\tau}\right)^{1/2}\,. \tag{1.6.47}$$

A photon in the visible spectrum with a photon energy of 2.2 eV yields a value for p_0 of 5.04×10^{-29} Cm ($\tau = 10$ ns). The classical dipole oscillation amplitude $|\mathbf{d}|$ then amounts to 0.31 nm.

1.6.2.4 Discussion of the Dipole Fields and Dipole Radiation

The plots a) and b) of Fig. 1.8 display $|\mathbf{E}_0|$ and $|\mathbf{H}_0|$ in the near field. Sufficiently close to the dipole, the electric field strength, see Eq. (1.6.42), is dominated by the term that is proportional to r^{-3}. The term proportional to kr^{-2} is relatively important in the region around $r \approx \lambda/2\pi$. Both terms stem from the average static charge distribution and the linear movement of the dipole charges. The oscillatory nature of the electric field is not visible in the near field because the exponential factor can be put equal to unity. The first term proportional to $1/r$ is the only term that has an appreciable amplitude at larger distances. For $r > \lambda$, the first term is the dominant one and produces the expected oscillating behaviour in the far field of the dipole. The magnetic field, see Eq. (1.6.36), is zero on the dipole axis. At a large enough lateral distance from the dipole axis, the magnetic field obeys Ampère's law in the presence of a line current. The direction of

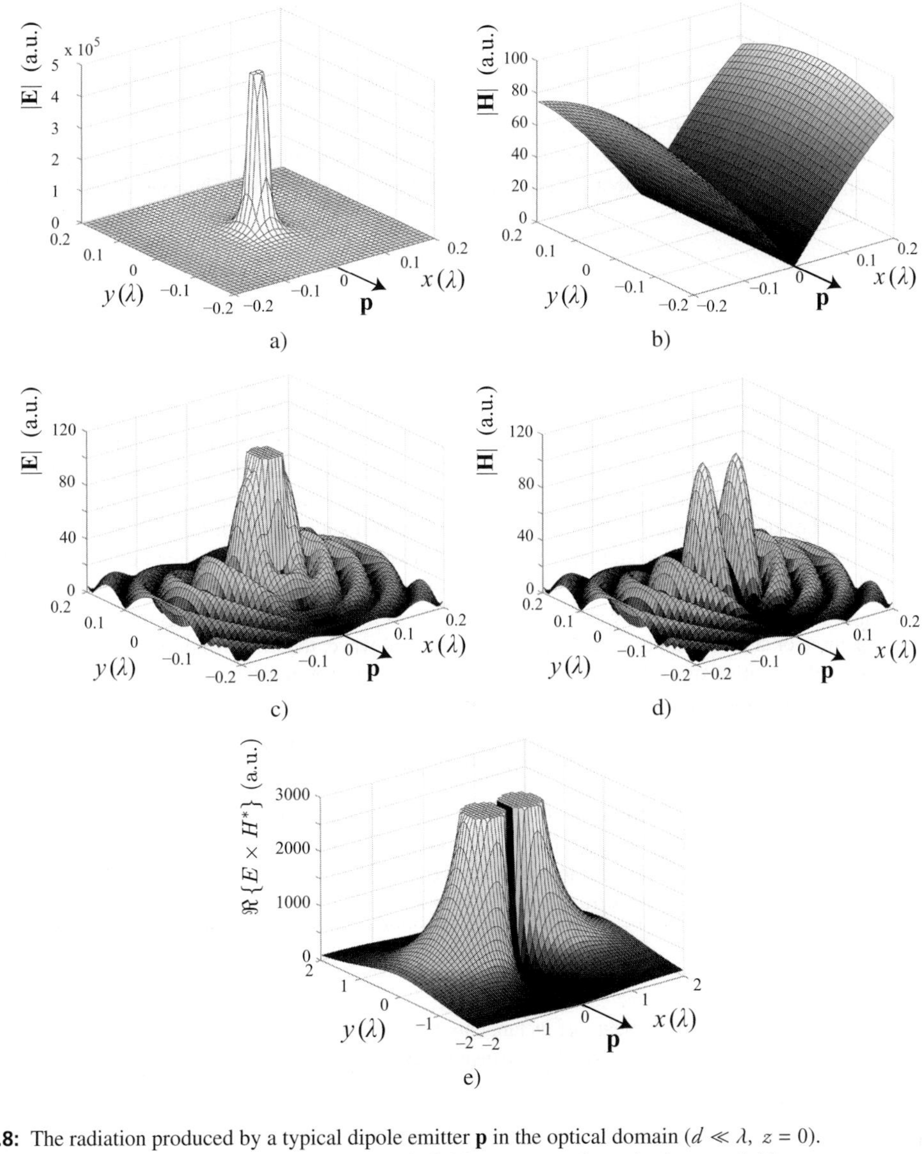

Figure 1.8: The radiation produced by a typical dipole emitter $\mathbf{p}$ in the optical domain ($d \ll \lambda$, $z = 0$).
a) and b) The absolute values of the electric and magnetic field vectors are shown in the *near field*.
c) and d) The same quantities in the relative *far field* of the dipole.
e) The magnitude of the average value of the Poynting vector.

the electric field vector in the near field does not obey a simple picture. The magnetic field is always perpendicular to the plane erected by the radius vector and the dipole axis.

In plots c) and d) of Fig. 1.8 we show the moduli of the electric and magnetic field in the relative far field of the dipole, up to $(x, y) = 2\lambda$. The singular value at the origin and the high values close to the dipole of the electric field have been truncated in graph c) to some maximum value to enhance the detailed structure of the electric field in the far-field region. The first term of the electric field in Eq. (1.6.42) is zero on the dipole axis and attains a maximum value in a direction perpendicular to this axis. The electric field is unchanged when replacing $\hat{\mathbf{r}}$ by $-\hat{\mathbf{r}}$; the magnetic field vector takes on the opposite sign for $\hat{\mathbf{r}} \rightarrow -\hat{\mathbf{r}}$. Far enough from the source, the radial periodicity with period λ becomes clearly visible. In the far field of the dipole, the vectors $\mathbf{E}_0$, $\mathbf{H}_0$ and the radius vector $\hat{\mathbf{r}}$ form a right-handed orthogonal vector set.

In Fig. 1.8e the average value of the Poynting vector has been plotted, according to Eq. (1.6.43). Extremely high values close to the dipole centre have been truncated. There is no energy flow at all along the dipole axis. For constant radial value r, the energy transport is proportional to $\sin^2\theta$, the maximum energy being radiated in a plane that contains the dipole and is perpendicular to its axis.

1.6.3 Spherical Waves

When we discussed the radiation pattern of the dipole wave in the previous sections, we pointed out that the contribution from all terms of higher than quadratic power to the far-field distribution is negligible. Therefore, from Eq. (1.6.42) after ignoring these terms we have

$$\mathbf{E} = \frac{k^2 p}{4\pi\epsilon_0}(\hat{\mathbf{r}} \times \hat{\mathbf{p}} \times \hat{\mathbf{r}})\frac{\exp(ik|\mathbf{r} - \mathbf{r}_0|)}{|\mathbf{r} - \mathbf{r}_0|}, \tag{1.6.48}$$

or in the more general time-harmonic form

$$\mathbf{E}(\mathbf{r}, t) = \mathbf{E}_a \frac{\exp[i(k|\mathbf{r} - \mathbf{r}_0| - \omega t)]}{|\mathbf{r} - \mathbf{r}_0|}, \tag{1.6.49}$$

when the source is placed at $\mathbf{r} = \mathbf{r}_0$. The above equation is the most general form of a *spherical wave* with amplitude $\mathbf{E}_a$. Note that though Eq. (1.6.48) automatically satisfies Maxwell's equations because the electric field of the dipole wave was calculated from Ampère's law, the amplitude vector $\mathbf{E}_a = (E_{a,x}, E_{a,y}, E_{a,z})$ in Eq. (1.6.49) cannot be chosen freely and one would need to make sure that a particular choice satisfies Maxwell's equations. It is also worth nothing that even if the Cartesian components are chosen such that they satisfy Maxwell's equations it does not necessarily mean that the resulting spherical wave can be practically realised. In Eq. (1.6.49) the wave vector $\mathbf{k}$ is directed away from the source or toward the sink at $\mathbf{r} = \mathbf{r}_0$. The value $E_{a,i}$ is the strength of the $i = (x, y, z)$ component of the electric field, measured at 1 m distance from the source or sink. For an inward propagating spherical wave, the minus sign in front of ω in Eq. (1.6.49) should be replaced by the positive sign. Comparable solutions can be obtained for the magnetic induction $\mathbf{B}$.

To examine the properties of the amplitude vector $\mathbf{E}_a$ we now take the divergence of the electric field $\mathbf{E}$:

$$\nabla_\mathbf{r} \cdot \mathbf{E} = \{\mathbf{E}_a \cdot (\mathbf{r} - \mathbf{r}_0)\}\left(ik - \frac{1}{r}\right)\frac{\exp[i(k|\mathbf{r} - \mathbf{r}_0|)]}{|\mathbf{r} - \mathbf{r}_0|}, \tag{1.6.50}$$

for $\mathbf{r} \neq \mathbf{r}_0$, where the subscript $\mathbf{r}$ at $\nabla_\mathbf{r}$ signifies that the divergence is calculated with respect to $\mathbf{r}$. In a medium free of unbound charges $\nabla_\mathbf{r} \cdot \mathbf{E}$ is required to be zero by Eq. (1.6.3). From this condition it follows that $\mathbf{E}_a \cdot (\mathbf{r} - \mathbf{r}_0) = 0$, that is the electric amplitude vector is perpendicular to the radius vector. A similar calculation shows that $\mathbf{B}_a \cdot (\mathbf{r} - \mathbf{r}_0) = 0$. These results mean that *the spherical wave is transversal in character*.

1.6.4 Plane Wave Solution

The surface of a sphere can be defined by means of its tangential planes. We shall borrow this idea to analyse the electric field of the spherical wave further. We invoke Weyl's expansion **[232]** which states

$$\frac{\exp(ikr)}{r} = \frac{ik}{2\pi}\iint_{-\infty}^{\infty} \exp[ik(px + qy + m|z|)]\frac{dpdq}{m}, \tag{1.6.51}$$

where $r = (x, y, z)$ and $m = (1 - p^2 - q^2)^{1/2}$ for $p^2 + q^2 \leq 1$. Electing to work in the positive half-space $z \geq 0$ it is clear that the time-independent part of Eq. (1.6.49) can now be written, without loss of generality for $\mathbf{r}_0 = \mathbf{0}$, as

$$\mathbf{E}(r) = \mathbf{E}_a \frac{\exp(ikr)}{r} = \frac{i\,\mathbf{E}_a}{2\pi} \iint\limits_{k_x^2+k_y^2 \le k} \exp(i\,\mathbf{k}\cdot\mathbf{r}) \frac{dk_x dk_y}{k_z}, \tag{1.6.52}$$

where $\mathbf{k} = (k_x, k_y, k_z) = k(p, q, m)$. The above equation represents a superposition of elementary waves of the form

$$\mathbf{E} = \mathbf{E}_b \exp(i\,\mathbf{k}\cdot\mathbf{r}), \tag{1.6.53}$$

where $\mathbf{E}_b = i\mathbf{E}_a/(2\pi k_z)$. Equation (1.6.53) can satisfy the Helmholtz equation provided $\mathbf{E}_b$ is chosen appropriately. In order to investigate how this can be done, consider the x and y Cartesian components of Eq. (1.6.53)

$$E_j(\mathbf{r}) = E_j \exp\{i(\mathbf{k}\cdot\mathbf{r} + \delta_i)\}, \tag{1.6.54}$$

where $i = x, y$ and we have introduced an arbitrary phase term δ_i for the sake of generality. Because the Helmholtz equation is already time-independent we have dropped the time-harmonic term $\exp(-i\omega t)$. It is important to emphasise that the solution that is sought for only specifies two components of the electric field. This is because in a homogeneous and isotropic medium that is free from charges and currents the electromagnetic field possesses only two degrees of freedom. Therefore, if we also specified the third component of the electric field vector, in our example the z component, the equation would become overdetermined. We thus choose to calculate the third component from Maxwell's equations directly. With the medium properties of the volume of interest that were given above and using that $\nabla \cdot \mathbf{E} = 0$ we have

$$\frac{\partial E_z}{\partial z} = -i\left(k_x E_x + k_y E_y\right), \tag{1.6.55}$$

since $\partial E_j/\partial j = jk_j E_j$ as follows from Eq. (1.6.54). We thus have that

$$E_z = \int \frac{\partial E_z}{\partial z} dz = -\frac{1}{k_z}\left(k_x E_x + k_y E_y\right), \tag{1.6.56}$$

yielding

$$\mathbf{E}(\mathbf{r}) = \mathbf{E}_a \exp\{i(\mathbf{k}\cdot\mathbf{r} + \mathbf{\Psi})\}, \tag{1.6.57}$$

where

$$\mathbf{\Psi} = \left(\phi_x, \phi_y, 0\right)^T \tag{1.6.58}$$

and

$$\mathbf{E}_a = \left(E_x, E_y, -\frac{1}{k_z}\left[k_x E_x \exp(i\phi_x) + k_y E_y \exp(i\phi_y)\right]\right)^T. \tag{1.6.59}$$

The magnetic field follows from Faraday's law, Eq. (1.6.6), yielding

$$\begin{aligned}
B_x &= -\frac{\sqrt{\epsilon\mu}}{k_z}\left[E_x k_x k_y + E_y\left(k_y^2 + k_z^2\right)\right], \\
B_y &= \frac{\sqrt{\epsilon\mu}}{k_z}\left[E_x\left(k_x^2 + k_z^2\right) + E_y k_x k_y\right], \\
B_z &= -\sqrt{\epsilon\mu}\left(E_x k_y - E_y k_x\right).
\end{aligned}$$

A general time-harmonic solution of the Helmholtz equation can thus be written in the form:

$$\mathbf{E}_h(\mathbf{r}, t) = \mathbf{E}_a \exp\{i(\mathbf{k}\cdot\mathbf{r} - \omega t)\}, \tag{1.6.60}$$

$$\mathbf{B}_h(\mathbf{r}, t) = \mathbf{B}_a \exp\{i(\mathbf{k}\cdot\mathbf{r} - \omega t)\}, \tag{1.6.61}$$

when $\delta_x = \delta_y = 0$.

An electromagnetic disturbance whose electric and magnetic fields are described by Eqs. (1.6.60) and (1.6.61) is called a *plane wave*. As follows from the above derivation, the disturbance created by a harmonically oscillating pair of opposite charges located at very large distance from the volume of interest will locally resemble very closely a plane wave. The wave vector $\mathbf{k}$ is in general complex and satisfies the so-called dispersion relation

$$\mathbf{k}\cdot\mathbf{k} = \omega^2\epsilon\mu\left(1 + i\frac{\sigma}{\epsilon\omega}\right) = \left(\frac{\omega}{v}\right)^2, \tag{1.6.62}$$

where v denotes the (complex) speed of propagation of the plane wave. In a lossless medium with $\sigma = 0$, the magnitude k of the wave vector is simply related to the wavelength λ of the harmonic wave by $k = 2\pi/\lambda$ and equals the advance in phase of the wave after propagation over the unit of length.

At this point it is useful to introduce the complex refractive index $\hat{n}$ of the medium defined by the ratio of the speed of an electromagnetic wave in vacuum and that in the specific medium at frequency ω,

$$\hat{n} = n + i\kappa = \frac{c}{v} = \left\{\frac{\epsilon\mu}{\epsilon_0\mu_0}\left(1 + i\frac{\sigma}{\epsilon\omega}\right)\right\}^{1/2} = \left\{\epsilon_r\mu_r\left(1 + i\frac{\sigma}{\epsilon_r\epsilon_0\omega}\right)\right\}^{1/2}, \tag{1.6.63}$$

where $c = (\epsilon_0\mu_0)^{-1/2}$ is the speed of light in vacuum and ϵ_r and μ_r the relative permittivity and permeability of the medium, respectively. Equation (1.6.63) yields the following relationship between the real and imaginary parts of the complex refractive index $\hat{n}$ and the medium constants ϵ_r, μ_r and σ,

$$n^2 - \kappa^2 = \epsilon_r\mu_r\,, \tag{1.6.64}$$

$$2n\kappa = \epsilon_r\mu_r\left(\frac{\sigma}{\epsilon_r\epsilon_0\omega}\right), \tag{1.6.65}$$

which leads to

$$n^2 = \frac{\epsilon_r\mu_r}{2}\left\{\sqrt{1 + \frac{\sigma^2}{\epsilon_r^2\epsilon_0^2\omega^2}} + 1\right\}, \tag{1.6.66}$$

$$\kappa^2 = \frac{\epsilon_r\mu_r}{2}\left\{\sqrt{1 + \frac{\sigma^2}{\epsilon_r^2\epsilon_0^2\omega^2}} - 1\right\}. \tag{1.6.67}$$

Using $\mathbf{k} = \hat{n}k_0\hat{\mathbf{s}} = \hat{n}(\omega/c)\hat{\mathbf{s}}$ with k_0 the wavenumber in vacuum and $\hat{\mathbf{s}}$ the unit vector in the propagation direction, we obtain for the plane wave solution

$$\mathbf{E}_h(\mathbf{r}, t) = \mathbf{E}_a \exp\left\{i\omega\left(\hat{n}\frac{\hat{\mathbf{s}}\cdot\mathbf{r}}{c} - t\right)\right\}, \tag{1.6.68}$$

$$\mathbf{B}_h(\mathbf{r}, t) = \mathbf{B}_a \exp\left\{i\omega\left(\hat{n}\frac{\hat{\mathbf{s}}\cdot\mathbf{r}}{c} - t\right)\right\}. \tag{1.6.69}$$

In the case of a conducting medium, we write for the electric field component of the plane wave

$$\mathbf{E}_h(\mathbf{r}, t) = \mathbf{E}_a \exp\left\{-\frac{\omega\kappa}{c}(\hat{\mathbf{s}}\cdot\mathbf{r})\right\}\exp\left\{i\omega\left[n\left(\frac{\hat{\mathbf{s}}\cdot\mathbf{r}}{c}\right) - t\right]\right\}, \tag{1.6.70}$$

and a corresponding expression for the $\mathbf{B}$ vector. The wave is seen to propagate with velocity $v_{ph} = c/n$, normally referred to as the *phase velocity*; κ is usually called the *extinction coefficient*. A related quantity, the so-called *absorption coefficient*, corresponds to the attenuation of the intensity of the wave during propagation and is given by $\alpha = 2\omega\kappa/c$. When waves of finite spatial extent (wave packets) are characterised, another quantity becomes important: the *group velocity* v_g. This is the velocity with which the envelope of the wave packet propagates (see Appendix B for a detailed discussion of group velocity).

In Fig. 1.9 a damped plane wave is illustrated propagating in the direction of the wave vector $\mathbf{k}_r$. For small values of conductivity $\sigma \ll \epsilon_0\omega$ the extinction coefficient κ in Eq. (1.6.67) can be approximated using the MacLaurin expansion to yield $\kappa \approx \sigma\sqrt{\mu_r/\epsilon_0}/2\epsilon_0\omega$. Equation (1.6.70) can thus be rewritten in an approximate form as

$$\mathbf{E}_h(\mathbf{r}, t) \approx \mathbf{E}_a \exp\left\{-\frac{1}{2}Z_m\sigma(\hat{\mathbf{s}}\cdot\mathbf{r})\right\}\exp\left\{i\omega\left[\epsilon_r\mu_r\left(\frac{\hat{\mathbf{s}}\cdot\mathbf{r}}{c}\right) - t\right]\right\}, \tag{1.6.71}$$

where $Z_m = (\mu/\epsilon)^{1/2}$ is defined as the impedance of the medium. The Poynting vector associated with the plane wave (often denoted inaccurately by the *intensity* of the wave) is proportional to the product of $\mathbf{E}$ and $\mathbf{B}$. This quantity thus contains the factor $\exp\{-Z_m\sigma(\hat{\mathbf{s}}\cdot\mathbf{r})\}$ and we observe that it exponentially decays during propagation due to the energy

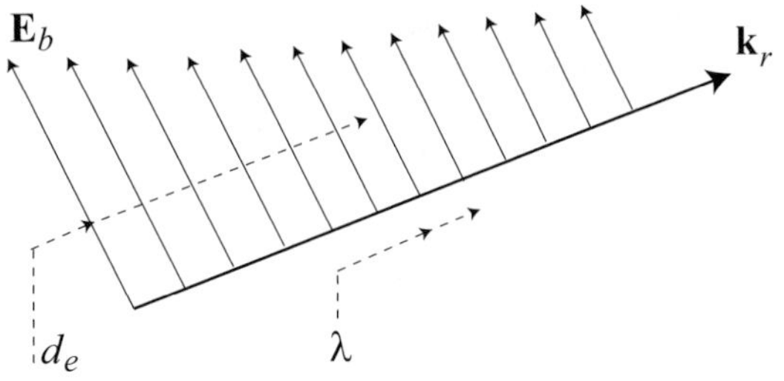

Figure 1.9: A plane wave propagating in the direction of $\mathbf{k}_r = k_r\hat{\mathbf{s}}$ where $\mathbf{k}_r$ is the real part of the wave vector and $\hat{\mathbf{s}}$ the unit propagation vector. The magnitude of the complex wave vector $\mathbf{k}$ is given by $k = (n - i\kappa)\omega/c$. The wavelength λ is the distance between two equiphase surfaces with a phase difference of 2π. The extinction distance d_e is the distance over which the wave amplitude E_a is reduced by a factor of $\exp(-\frac{1}{2})$.

dissipated in the medium. After propagation in the direction of $\hat{\mathbf{s}}$ over a distance d_e, the initial value of the Poynting vector has decreased by a factor of e, where the *extinction distance* d_e is given by

$$d_e = \frac{1}{\sigma Z_m} . \tag{1.6.72}$$

As mentioned above, the plane wave solution plays an important role in wave field analysis and synthesis. Using Fourier synthesis, a general wave field can be represented by means of a spatial spectrum of plane waves. In Appendix A we give a brief overview of the application of Fourier theory and the complex notation for waves with a harmonic time dependence. In general, it is sufficient to provide the plane wave amplitude vectors $\mathbf{E}_b$ and $\mathbf{B}_b$ as a function of the wave vector $\mathbf{k}$ to define a wave field of arbitrary nature. The spectral range of $\mathbf{k}$ is defined by, for instance, the range of its k_x and k_y components. For a general wave field that propagates almost parallel to a chosen z-axis, the k_z component of the wave vector is dominant and the k_x and k_y components will be relatively small. The range of these components will be limited and the wave representation by means of these components will be an efficient one. The k_z component of a particular plane wave vector $\mathbf{k}$ then follows from the dispersion relation of Eq. (1.6.62) that is written now as

$$k_x^2 + k_y^2 + k_z^2 = \frac{\omega^2}{c^2}\,\tilde{n}^2 , \tag{1.6.73}$$

using the complex refractive index $\tilde{n} = n + i\kappa$ to describe the properties of the medium. The value of the generally complex k_z component now follows from

$$k_z^2 = (n^2 - \kappa^2)\frac{\omega^2}{c^2} - k_x^2 - k_y^2 + 2in\kappa\frac{\omega^2}{c^2} . \tag{1.6.74}$$

In the special case of a non-conducting medium, k_z will be real unless the right-hand side of Eq. (1.6.73) is negative. We find

$$k_z = \begin{cases} \sqrt{\omega^2 n^2/c^2 - k_x^2 - k_y^2} , & \text{if} \quad n\omega/c \geq \sqrt{k_x^2 + k_y^2} , \\ +i\sqrt{k_x^2 + k_y^2 - \omega^2 n^2/c^2} , & \text{if} \quad n\omega/c < \sqrt{k_x^2 + k_y^2} . \end{cases} \tag{1.6.75}$$

The sign of the purely imaginary expression of k_z on the second line of Eq. (1.6.75) has been chosen such that the propagating wave will be damped in the $+z$-direction. These types of waves with a wave vector value k_z greater than $n\omega/c$ do not propagate over substantial distances and are called *evanescent waves*. Figure 1.10 shows the amplitude and phase behaviour of such an evanescent plane wave. Evanescent waves are usually generated at interfaces between media of different refractive indices or when light is scattered or diffracted by subwavelength structures.

It is important to point out that plane waves are transverse. Assuming an optically homogeneous volume of space with no charges inside, Eq. (1.6.3) reduces to $\nabla \cdot \mathbf{E} = 0$ as before. The substitution of the plane wave solution for the electromagnetic field vectors of Eqs. (1.6.60) and (1.6.61) into Eqs. (1.6.3) and (1.6.4) yields

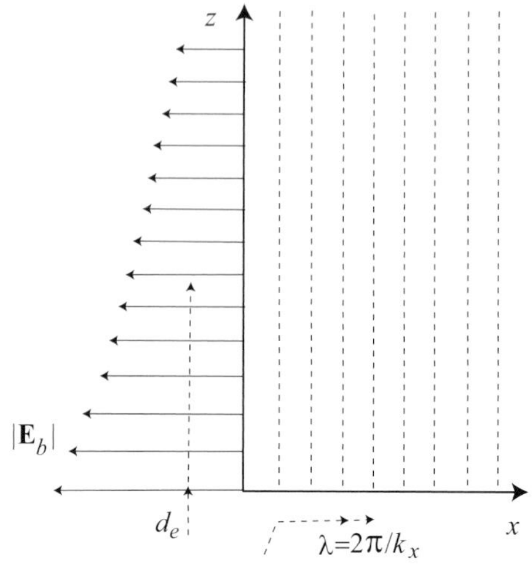

Figure 1.10: Equi-amplitude and equiphase planes of an evanescent plane wave. The damped wave, represented by Eq. (1.6.75) with $k_y = 0$, has an amplitude E_b that decreases in the z-direction. The damping distance d_e is given by $1/2|k_z|$. The planes with constant phase (dashed in the figure) are perpendicular to the x-axis. The wavelength λ is given by $2\pi/k_x$. Because the equi-amplitude and equiphase planes of this plane wave are not parallel, the wave is called an inhomogeneous wave.

$$k_x E_{b,x} + k_y E_{b,y} + k_z E_{b,z} = 0 \,, \tag{1.6.76}$$

$$k_x B_{b,x} + k_y B_{b,y} + k_z B_{b,z} = 0 \,, \tag{1.6.77}$$

from which we conclude that both the vectors $\mathbf{E}_b$ and $\mathbf{B}_b$ are orthogonal to the wave vector $\mathbf{k}$. Finally, taking the scalar product of Eq. (1.6.6) with $\mathbf{E}$ and using the result from vector analysis that the divergence of a curl is zero, we conclude that $\mathbf{E}$ and $\mathbf{B}$ are orthogonal. Consequently, for a plane wave the three vectors $\mathbf{k}$, $\mathbf{E}$ and $\mathbf{B}$ are mutually orthogonal and the wave vector $\mathbf{k}$ points in the direction of $\mathbf{E} \times \mathbf{B}$.

1.6.5 General Vector Wave Solution

In previous sections we discussed specific forms of electromagnetic radiation: dipole, spherical and plane waves. It is now time to generalise our findings and describe a general harmonic vector wave. The three Cartesian components of such a wave are given by

$$E_{b,x}(\mathbf{r},t) = A_x(\mathbf{r}) \exp\{i[S_x(\mathbf{r}) - \omega t]\} \,, \tag{1.6.78}$$

$$E_{b,y}(\mathbf{r},t) = A_y(\mathbf{r}) \exp\{i[S_y(\mathbf{r}) - \omega t]\} \,, \tag{1.6.79}$$

$$E_{b,z}(\mathbf{r},t) = A_z(\mathbf{r}) \exp\{i[S_z(\mathbf{r}) - \omega t]\} \,, \tag{1.6.80}$$

and there are a set of comparable expressions for the magnetic vector components.

In contrast to the plane wave or spherical wave solutions, the amplitudes of the field components now depend in a general way on the position vector $\mathbf{r}$. Differences between the field components may arise, for instance, at a transition between two media where the amplitudes of the field components are affected in different ways. The phase of the field components of the wave is not described by a single function S either. In the case of transmission or reflection at boundaries between media, the field components experience different phase jumps, leading to the separate phase functions S_x, S_y and S_z in Eqs. (1.6.78)–(1.6.80).

The amplitude and phase functions, A and S, respectively, are not fully independent. In a charge-free and homogeneous medium, Eq. (1.6.3) reduces to $\nabla \cdot \mathbf{E} = 0$ and this requirement imposes the following condition on the amplitude and phase functions of the general vector wave:

$$\exp(iS_x)\left\{\frac{\partial A_x}{\partial x} + iA_x\frac{\partial S_x}{\partial x}\right\} + \exp(iS_y)\left\{\frac{\partial A_y}{\partial y} + iA_y\frac{\partial S_y}{\partial y}\right\} + \exp(iS_z)\left\{\frac{\partial A_z}{\partial z} + iA_z\frac{\partial S_z}{\partial z}\right\} = 0. \tag{1.6.81}$$

Far away from the source or from the focal region of a lens we may neglect the spatial derivatives of the slowly varying amplitude function with respect to the derivatives of the quickly oscillating phase functions S_x, S_y and S_z. With this assumption, Eq. (1.6.81) reduces to

$$A_x \exp(-iS_x)\,\frac{\partial S}{\partial x} + A_y \exp(-iS_y)\,\frac{\partial S}{\partial y} + A_z \exp(-iS_z)\,\frac{\partial S}{\partial z} = \mathbf{A}' \cdot \left(\frac{\partial S_x}{\partial x}, \frac{\partial S_y}{\partial y}, \frac{\partial S_z}{\partial z}\right) = 0 \,. \tag{1.6.82}$$

Equation (1.6.82) establishes the orthogonality between the gradient vector $(\partial S_x/\partial x, \partial S_y/\partial y, \partial S_z/\partial z)$ and the complex vector $\mathbf{A}'(\mathbf{r})$ with Cartesian components $A_x(\mathbf{r})\exp\{iS_x(\mathbf{r})\}$, $A_y(\mathbf{r})\exp\{iS_y(\mathbf{r})\}$ and $A_z(\mathbf{r})\exp\{iS_z(\mathbf{r})\}$. This orthogonality relation is the general result for a vector wave that we have already encountered in a simpler form for a plane wave in Eqs. (1.6.76) and (1.6.77).

In a homogeneous and isotropic medium, the phase functions have the same dependence on the spatial coordinates $\mathbf{r} = (x, y, z)$. Their differences are given by a constant term and this yields

$$S_x(\mathbf{r}) = S_y(\mathbf{r}) + \alpha_y = S_z(\mathbf{r}) + \alpha_z = S(\mathbf{r}) \,. \tag{1.6.83}$$

The gradient vector of Eq. (1.6.82) then reduces to $\nabla S(\mathbf{r})$. Inspection of Eq. (1.6.83) shows that the condition $\mathbf{A}' \cdot \nabla S = 0$ for the complex vector $\mathbf{A}'$ and the real vector ∇ yields two equations. When the scalar function S and the constants α_y and α_z are known, we thus have two equations for the three quantities to be found, A_x, A_y, A_z, respectively.

1.6.6 State of Polarisation of a General Vector Wave

In the preceding paragraphs we have established the transverse character of an electromagnetic wave. In this subsection we study the spatio-temporal behaviour of the electric and magnetic field components in the plane perpendicular to the local propagation vector, i.e. $\nabla S(\mathbf{r})$ for a general vector wave, $\mathbf{r}$ for a spherical wave and $\mathbf{k}$ for a plane wave. Starting with the general vector wave, we define a point of observation Q in space by $\mathbf{r} = \mathbf{r}_Q$. The complex electric field components at this point are given by

$$E_{h,x}(\mathbf{r}_Q, t) = A_x(\mathbf{r}_Q)\exp\{i[S_x(\mathbf{r}_Q) - \omega t]\} \,, \tag{1.6.84}$$

$$E_{h,y}(\mathbf{r}_Q, t) = A_y(\mathbf{r}_Q)\exp\{i[S_y(\mathbf{r}_Q) - \omega t]\} \,, \tag{1.6.85}$$

$$E_{h,z}(\mathbf{r}_Q, t) = A_z(\mathbf{r}_Q)\exp\{i[S_z(\mathbf{r}_Q) - \omega t]\} \,. \tag{1.6.86}$$

Definition of the vectors $\mathbf{a}_1$ and $\mathbf{a}_2$ as

$$\mathbf{a}_1(\mathbf{r}_Q) = \left(A_x(\mathbf{r}_Q)\cos\{S_x(\mathbf{r}_Q)\}, A_y(\mathbf{r}_Q)\cos\{S_y(\mathbf{r}_Q)\}, A_z(\mathbf{r}_Q)\cos\{S_z(\mathbf{r}_Q)\}\right)^T \,, \tag{1.6.87}$$

$$\mathbf{a}_2(\mathbf{r}_Q) = \left(A_x(\mathbf{r}_Q)\sin\{S_x(\mathbf{r}_Q)\}, A_y(\mathbf{r}_Q)\sin\{S_y(\mathbf{r}_Q)\}, A_z(\mathbf{r}_Q)\sin\{S_z(\mathbf{r}_Q)\}\right)^T \,, \tag{1.6.88}$$

allows one to write the time dependence of the electric field as

$$\mathbf{E}_h(\mathbf{r}_Q, t) = \mathbf{a}_1(\mathbf{r}_Q)\cos\omega t \; + \; \mathbf{a}_2(\mathbf{r}_Q)\sin\omega t \,, \tag{1.6.89}$$

where we have switched from complex signal to analytic signal notation. Another, equivalent method of notation using the complex vector $\mathbf{a}(\mathbf{r}_Q) = \mathbf{a}_1(\mathbf{r}_Q) + i\mathbf{a}_2(\mathbf{r}_Q)$ is given by

$$\mathbf{E}_h(\mathbf{r}_Q, t) = \Re\left\{\mathbf{a}(\mathbf{r}_Q)\exp(-i\omega t)\right\} \,. \tag{1.6.90}$$

In the plane perpendicular to the local propagation vector ∇S (see Eq. (1.6.83)), the general shape of the curve described by the endpoint of the electric vector is an ellipse, the so-called *polarisation ellipse*. At each moment of time t, the electric vector is given by a linear superposition of the vectors $\mathbf{a}_1(\mathbf{r}_Q)$ and $\mathbf{a}_2(\mathbf{r}_Q)$ where the weights for the superposition are given by the instantaneous values of $\cos\omega t$ and $\sin\omega t$. In general, the two vectors are not orthogonal, as shown in Fig. 1.11.

By introducing an offset in time of τ in Eq. (1.6.89), the same ellipse can be described using another set of vectors, $\mathbf{a}_1'$ and $\mathbf{a}_2'$. By a specific choice of τ, the two vectors can be made orthogonal so that the condition $\mathbf{a}_1' \cdot \mathbf{a}_2' = 0$ is satisfied. With the time offset τ, the harmonic electric field vector is written

$$\begin{aligned}\mathbf{E}_h(\mathbf{r}_Q, t+\tau) &= \left\{\mathbf{a}_1(\mathbf{r}_Q)\cos\omega\tau + \mathbf{a}_2(\mathbf{r}_Q)\sin\omega\tau\right\}\cos\omega t \\ &\quad + \left\{-\mathbf{a}_1(\mathbf{r}_Q)\sin\omega\tau + \mathbf{a}_2(\mathbf{r}_Q)\cos\omega\tau\right\}\sin\omega t \\ &= \mathbf{a}_1'(\mathbf{r}_Q)\cos\omega t + \mathbf{a}_2'(\mathbf{r}_Q)\sin\omega t \,.\end{aligned} \tag{1.6.91}$$

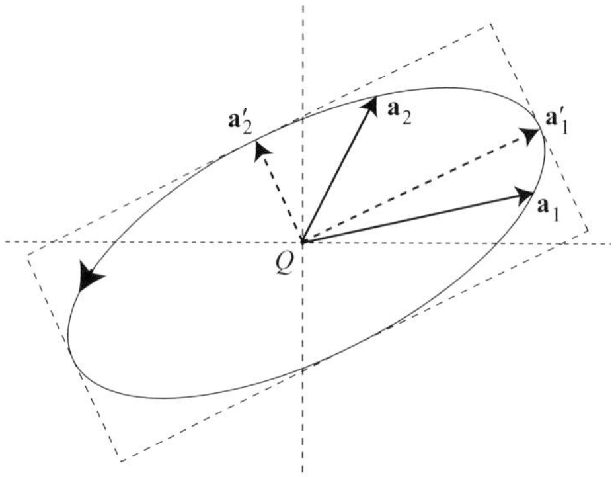

Figure 1.11: The polarisation ellipse at a general point Q for a vector wave of general nature, created by the superposition of two vectors $\mathbf{a}_1(\mathbf{r}_Q)$ and $\mathbf{a}_2(\mathbf{r}_Q)$; the same polarisation ellipse can also be represented by the superposition of two orthogonal vectors $\mathbf{a}'_1(\mathbf{r}_Q)$ and $\mathbf{a}'_2(\mathbf{r}_Q)$.

The orthogonality of the two vectors requires

$$\left\{\mathbf{a}_1(\mathbf{r}_Q)\cos\omega\tau + \mathbf{a}_2(\mathbf{r}_Q)\sin\omega\tau\right\}\cdot\left\{-\mathbf{a}_1(\mathbf{r}_Q)\sin\omega\tau + \mathbf{a}_2(\mathbf{r}_Q)\cos\omega\tau\right\} = 0\,, \tag{1.6.92}$$

or, after some rearrangement,

$$\mathbf{a}_1(\mathbf{r}_Q)\cdot\mathbf{a}_2(\mathbf{r}_Q)\cos 2\omega\tau - \frac{1}{2}\left(|\,\mathbf{a}_1(\mathbf{r}_Q)|^2 - |\,\mathbf{a}_2(\mathbf{r}_Q)|^2\right)\sin 2\omega\tau = 0\,. \tag{1.6.93}$$

Solving for τ yields the implicit result

$$\tan 2\omega\tau = \frac{2\mathbf{a}_1(\mathbf{r}_Q)\cdot\mathbf{a}_2(\mathbf{r}_Q)}{|\,\mathbf{a}_1(\mathbf{r}_Q)|^2 - |\,\mathbf{a}_2(\mathbf{r}_Q)|^2} = q_0\,. \tag{1.6.94}$$

Using basic trigonometric relations, we find

$$\tan\omega\tau = \frac{q_0}{\sqrt{q_0^2+1}+1}\,, \tag{1.6.95}$$

$$\sin\omega\tau = \frac{q_0}{\sqrt{2\left[q_0^2+1+\sqrt{q_0^2+1}\right]}}\,, \qquad \cos\omega\tau = \frac{\sqrt{q_0^2+1}+1}{\sqrt{2\left[q_0^2+1+\sqrt{q_0^2+1}\right]}}\,, \tag{1.6.96}$$

$$\sin 2\omega\tau = \frac{q_0}{\sqrt{q_0^2+1}}\,, \qquad \cos 2\omega\tau = \frac{1}{\sqrt{q_0^2+1}}\,. \tag{1.6.97}$$

The lengths of the two orthogonal vectors determine the long and short axis of the polarisation ellipse and are given by

$$|\mathbf{a}'_1|^2 = \frac{1}{2}\left\{|\mathbf{a}_1|^2 + |\mathbf{a}_2|^2 + \sqrt{(|\mathbf{a}_1|^2 - |\mathbf{a}_2|^2)^2 + 4(\mathbf{a}_1\cdot\mathbf{a}_2)^2}\right\}\,, \tag{1.6.98}$$

$$|\mathbf{a}'_2|^2 = \frac{1}{2}\left\{|\mathbf{a}_1|^2 + |\mathbf{a}_2|^2 - \sqrt{(|\mathbf{a}_1|^2 - |\mathbf{a}_2|^2)^2 + 4(\mathbf{a}_1\cdot\mathbf{a}_2)^2}\right\}\,. \tag{1.6.99}$$

The angle α_1 between the original vector $\mathbf{a}_1$ and the new vector $\mathbf{a}'_1$ can be obtained by taking the inner product of $\mathbf{a}'_1$ with itself and eliminating $\mathbf{a}_2$ using Eq. (1.6.91). We find

$$\begin{aligned}
\mathbf{a}'_1\cdot\mathbf{a}'_1 &= \left\{\mathbf{a}_1\cos\omega\tau + \frac{\mathbf{a}'_2 + \mathbf{a}_1\sin\omega\tau}{\cos\omega\tau}\sin\omega\tau\right\}\cdot\mathbf{a}'_1 \\
&= \mathbf{a}_1\cdot\mathbf{a}'_1\cos\omega\tau + \frac{\mathbf{a}_1\cdot\mathbf{a}'_1\sin^2\omega\tau}{\cos\omega\tau} \\
&= \frac{\mathbf{a}_1\cdot\mathbf{a}'_1}{\cos\omega\tau}\,.
\end{aligned} \tag{1.6.100}$$

This yields the result

$$\cos\alpha_1 = \cos\omega\tau \frac{|\mathbf{a}'_1|}{|\mathbf{a}_1|} \,. \tag{1.6.101}$$

In a comparable way, we find for the angle α_2 between $\mathbf{a}'_2$ and $\mathbf{a}_2$ that

$$\cos\alpha_2 = \cos\omega\tau \frac{|\mathbf{a}'_2|}{|\mathbf{a}_2|} \,. \tag{1.6.102}$$

Specific values of the vectors $\mathbf{a}'_1$ and $\mathbf{a}'_2$ lead to special cases of polarisation:

- *Linear polarisation*
  If $\mathbf{a}'_2$ of Eq. (1.6.99) equals zero, the polarisation ellipse reduces to a straight line. The condition to be fulfilled reads

  $$|\mathbf{a}_1|\,|\mathbf{a}_2| = \pm\mathbf{a}_1 \cdot \mathbf{a}_2 \,, \tag{1.6.103}$$

  meaning that $\mathbf{a}_1$ and $\mathbf{a}_2$ are parallel or anti-parallel.
- *Circular polarisation*
  If the two orthogonal vectors $\mathbf{a}'_1$ and $\mathbf{a}'_2$ have equal length, the ellipse reduces to a circle. The condition for circular polarisation thus reads

  $$|\mathbf{a}_1| = |\mathbf{a}_2| \,, \qquad \mathbf{a}_1 \cdot \mathbf{a}_2 = 0 \,. \tag{1.6.104}$$

  These conditions imply that the vectors $\mathbf{a}_1$ and $\mathbf{a}_2$ were already orthogonal and of equal length. The time offset τ, introduced in Eq. (1.6.91), is irrelevant for this case, which is shown by the indeterminate value of $\tan 2\omega\tau$ (Eq. (1.6.94)).
- *Elliptical polarisation*
  For the general case of elliptical polarisation the original vectors $\mathbf{a}_1$ and $\mathbf{a}_2$ were not necessarily orthogonal. The resulting vectors $\mathbf{a}'_1$ and $\mathbf{a}'_2$ have unequal length. Using complex notation, the degree of circular polarisation p_C is defined by

  $$p_C = \frac{2\Im\{A'_1(A'_2)^*\}}{|A'_1|^2 + |A'_2|^2} = \frac{2|A'_1|\,|A'_2|\,\sin(\phi'_1 - \phi'_2)}{|A'_1|^2 + |A'_2|^2} \,. \tag{1.6.105}$$

  A'_1 and A'_2 are the complex amplitudes of the orthogonal components and the vectors $\mathbf{a}'_1$, $\mathbf{a}'_2$ and the wave propagation direction form a right-handed orthogonal system. The phase difference $\phi' = \phi'_1$-ϕ'_2 equals $\pm\pi/2$ with $\phi' = +\pi/2$ yielding the right-handed elliptical state of polarisation ($p_C > 0$) and $\phi' = -\pi/2$ the left-handed state with $p_C < 0$. In the special cases of right- and left-handed *circular* polarisation of a wave that propagates in the z direction, the normalised electric field components are conveniently represented by

  $$\text{RC}: \quad p_C = +1 \,, \qquad \hat{\mathbf{E}}_0 = \frac{\exp(i\phi_0)}{\sqrt{2}}\,(1, -i, 0) \,, \tag{1.6.106}$$

  $$\text{LC}: \quad p_C = -1 \,, \qquad \hat{\mathbf{E}}_0 = \frac{\exp(i\phi_0)}{\sqrt{2}}\,(1, +i, 0) \,, \tag{1.6.107}$$

  with $\exp(i\phi_0)$ an arbitrary multiplying phase factor.

1.6.6.1 Handedness of the State of Polarisation

The literature is not consistent in how handedness is assigned to a general elliptical polarisation state. To illustrate the possible source of confusion, we first consider the convention that associates the sense of rotation in time of the electric vector in a plane perpendicular to the propagation direction with the handedness of the polarisation state. The geometry of the elliptical state of polarisation of a wave propagating in the ∇S direction is illustrated in Fig. 1.12. The dashed ellipse in the plane $z = 0$ represents the time evolution of the electric field vector. With the viewing direction being along the positive z-axis, we arrive at a counterclockwise or left-handed (*LH*) state of polarisation. A second convention assigns the sense of screw of the helical space path described by the endpoint of the electric vector to the handedness of the state of polarisation. This amounts to determining the sense of the screw of this helical curve at a fixed point in time, by observing a 'still frame' of the elliptically polarised wave. In Fig. 1.12 we have drawn two such frozen 'images' (solid and dotted space curve) of a propagating wave. The difference in time between the two curves is a quarter of a period of oscillation. To simplify the picture, the wave is planar; the screw sense of the wave is right-handed (*RH*).

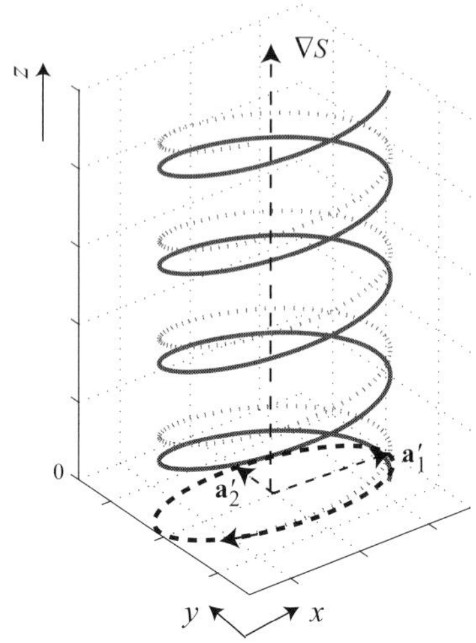

Figure 1.12: The evolution of the endpoint of the electric vector of a propagating wave with a local propagation direction defined by ∇S. The vectors $\mathbf{a}'_1$ and $\mathbf{a}'_2$ define the orthogonal axes of the polarisation ellipse; in the drawing, they have been chosen along the x and y-axes, respectively. The time evolution of the electric vector has been sketched by means of two traces of the endpoint of the electric field vector. The dotted curve is a quarter period later in time than the solid curve. The endpoints of the electric vector have been connected by a helical space curve. The handedness of this space curve is right-handed (*RH*).

The same right-handedness is detected in the figure when we observe the path of the endpoint of the electric vector in a fixed plane as a function of time ('frozen' image with respect to position), but now looking *backward* with respect to the propagation direction of the wave. The convention that is generally accepted in the optics community is the second one, such that the wave geometry of Fig. 1.12 is termed to be *RH*. The opposite handedness definition is advocated by the American Society IEEE and is frequently used in radio wave propagation. In this book we use the 'optics community' definition, the one that is commonly called the 'looking backward' or 'looking toward the source' definition. The handedness h_e according to the 'optics' convention is conveniently represented by means of the expression

$$h_e = \mathrm{sgn}\left[\phi' \,\{(\mathbf{a}'_1 \times \mathbf{a}'_2)\cdot \nabla S\}\right] \,, \tag{1.6.108}$$

where $\phi' = \phi'_1 - \phi'_2$ is defined on the interval $-\pi < \phi' \leq \pi$. A zero value of h_e corresponds to a linear state of polarisation, positive and negative values correspond to *RH* and *LH* states of polarisation, respectively.

The state of polarisation of the general vector wave was analysed in a particular point Q with position vector $\mathbf{r}_Q$. The state of polarisation in any other point, even if this point is located on the same wavefront, will generally be different so that it is not possible to speak about *the* state of polarisation of the vector wave. This is in contrast with the plane vector wave, that has a uniquely defined state of polarisation everywhere in space.

1.6.6.2 Change of Handedness upon Reflection

Special attention should be paid to the change in handedness upon reflection. If a circularly polarised wave, propagating in the z-direction, is reflected at a perfect mirror that is perpendicular to the z-axis, we observe that the two reflected linear waves have their electric field components along the x- and y-direction inverted with respect to the incident wave (see Fig. 1.13). The argument on handedness developed with the aid of the perfect mirror also applies to waves reflected from a more general transition between two media at oblique incidence. The only difference is that apart from the handedness we observe that the ellipticity of the reflected wave may also differ from that of the incident one.

The possible confusion on handedness is avoided by merely considering the orthogonal field components in a plane perpendicular to the propagation direction given by the unit wave vector $\hat{\mathbf{k}}$. For instance, for propagation in the positive z-direction, the field components $(1, -i, 0)$ identify a right-handed circularly polarised wave. At reflection, the same transverse components define a left-handed circularly polarised wave because the lateral field components of this wave and the reflected propagation vector form a left-handed coordinate system; this effect is properly accounted for by the scalar triple product of Eq. (1.6.108).

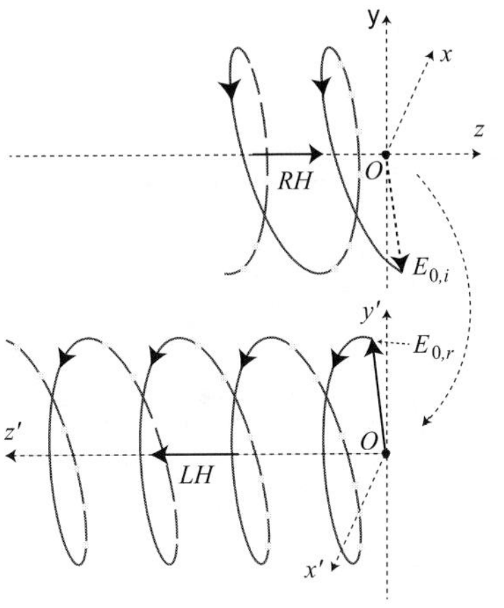

Figure 1.13: The change of handedness upon reflection of an elliptically polarised wave. The incident wave is *RH*-polarised and shown in the upper part of the graph. The coordinate system for the forward propagating wave is (x, y, z) with origin at O. The reflected wave (lower graph) is *LH*-polarised in the rotated right-handed coordinate system (x', y, z'). The reflecting surface is a perfect mirror, positioned at $z = z' = 0$. $\mathbf{E}_{0,i}$ and $\mathbf{E}_{0,r}$ are the electric field strengths of the incident and reflected wave at the point O at a certain moment in time $t = t_0$.

1.7 Gaussian Beams

In this section we study harmonic solutions of the Maxwell equations for a special type of propagating beam that is typical for optical radiation generated in an optical cavity. This new type of optical beam was observed during the first experiments with 'lasing' light sources. The properties of these beams were described immediately after the discovery of the maser and the laser in the beginning of the 1960s **[102]**, **[43]**, **[44]**, **[189]**. Due to its spatial profile, the laser beam coming out of the elongated optical cavity of, for instance, a HeNe-laser is commonly called a *Gaussian* beam for reasons that will become obvious in the following analysis. An expansion of the forward and backward propagating fields in a HeNe-laser cavity in terms of infinitely extending plane waves is not practical because of the narrow tube-like geometry of the cavity in the transverse dimension. For that reason, an approximate solution of the Helmholtz equation that adequately represents the field close to the cavity axis and shows a fast decay of field strength away from this axis of symmetry is preferable. The amplitude decay of a resonant mode of a cavity is generally not sharply delimited by a diaphragm in the cavity. The decrease of amplitude when going off-axis is caused by the fact that in the outer parts of the beam more light is lost during propagation by light diffraction and by phase mismatch. The net result is a transverse decrease in field strength that is well represented by an exponential Gaussian function of the lateral distance ρ from the axis with $\rho = (x^2 + y^2)^{1/2}$.

The representation of the phase of the field makes use of the paraxial approximation of light propagation in which a spherical wavefront is replaced by its second-order parabolic equivalent. The state of polarisation of such a 'paraxial' propagating beam is assumed to be constant over the beam cross-section since the propagation direction ∇S everywhere in the Gaussian beam has to be parallel to the z-axis for the paraxial approximation to be valid. For that reason, with a predetermined uniform state of polarisation, we are allowed to use the scalar Helmholtz equation for the calculation of the vector field strength of a propagating beam. With the assumptions above we postulate a solution for the scalar field strength U that is given by

$$\mathbf{E}(\rho, z) = \mathbf{p}\, U(\rho, z) = \mathbf{p}\, V(\rho, z) \exp(ikz) \,, \tag{1.7.1}$$

where $\mathbf{p}$ is the (complex) polarisation vector and $V(\rho, z)$ is the amplitude function of the Gaussian beam that slowly varies as a function of the propagation distance z. We substitute the scalar function $U(\rho, z)$ into the Helmholtz equation written in cylindrical coordinates yielding

$$\frac{1}{\rho}\frac{\partial}{\partial \rho}\left(\rho\,\frac{\partial V(\rho, z)}{\partial \rho}\right)\exp(ikz) + \frac{\partial^2 \{V(\rho, z)\exp(ikz)\}}{\partial z^2} + k^2\, V(\rho, z)\exp(ikz) = 0 \,. \tag{1.7.2}$$

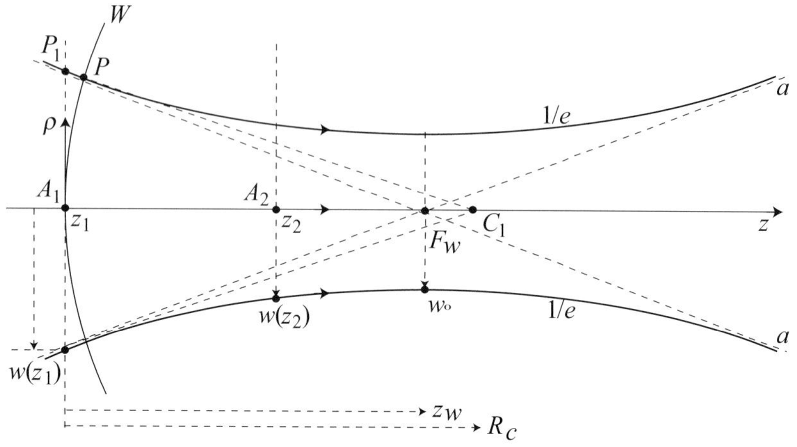

Figure 1.14: Axial cross-section of a circularly symmetric Gaussian beam, ρ is the radial coordinate. In a plane $z = z_1$ the spherical wavefront W has been drawn with its centre of curvature at C_1; the wavefront deviation is given by the optical path from P to P_1 (negative in the figure). The solid curves are the $1/e$ amplitude contours of the beam. The waist of the beam is in the plane through F_w. For large z-values, the solid $1/e$ curves tend to the dashed asymptotic lines denoted by a.

Evaluating the differential products we obtain a differential equation for the slowly varying 'envelope' function $V(\rho, z)$,

$$\frac{1}{\rho}\frac{\partial V(\rho,z)}{\partial \rho} + \frac{\partial^2 V(\rho,z)}{\partial \rho^2} + 2ik\,\frac{\partial V(\rho,z)}{\partial z} \approx 0\,. \tag{1.7.3}$$

In this expression we have neglected the second-order derivative $\partial^2 V(\rho, z)/\partial z^2$ with respect to the term $2ik\partial V(\rho, z)/\partial z$. This is allowed if the laser beam has a small beam divergence and is only slowly contracting or expanding as a function of the propagation distance z. This approximated Helmholtz equation, valid for propagating waves that are confined to a narrow lateral region along a central axis, is commonly called the *paraxial* Helmholtz equation.

We seek a solution to the differential Eq. (1.7.3) which satisfies the properties of the Gaussian type laser beams that were discussed above. As such we make the ansatz that the paraxial function $V(\rho, z)$ takes the form

$$V(\rho,z) = \frac{1}{r(z)}\,\exp\left\{-i\frac{k\rho^2}{2q(z)}\right\}\,, \tag{1.7.4}$$

which resembles the basic expression for a spherical wave emanating from a point source. When q is real, the argument of the complex exponential describes a parabolic approximation of a spherical wavefront. The argument has been given a minus sign so that a positive radius of curvature R_c (see Fig. 1.14) produces the negative pathlength $[PP_1]$, measured from the parabolic wavefront through A_1 to the reference plane perpendicular to the z-axis through the same point A_1.

When q is complex, the real part of $1/q$ equals the paraxial curvature of the wavefront and its imaginary part determines the width of the Gaussian amplitude profile in a cross-section taken perpendicular to the propagation direction. A characteristic measure for the lateral extent of the Gaussian beam is its width at the $1/e$ amplitude level relative to the peak amplitude, given by $2w(z)$. The complex beam parameter $q(z)$ is conveniently written as

$$\frac{1}{q(z)} = \frac{1}{R_c(z)} - \frac{i\lambda}{\pi w^2(z)}\,. \tag{1.7.5}$$

The leading factor $1/r(z)$ in Eq. (1.7.4), the equivalent of the factor $1/R$ for a spherical wave, takes care of the axial amplitude variation such that the total power transported by the Gaussian beam remains constant in an arbitrary cross-section. A nonzero argument of the generally complex quantity $r(z)$ accounts for a slowly varying axial phase term that is picked up by the Gaussian beam during propagation along the z-axis.

1.7.1 Solution to the Paraxial Helmholtz Equation

Having constructed a suitable form of $V(\rho, z)$ as given by Eq. (1.7.4) we now substitute into the paraxial Helmholtz equation of Eq. (1.7.3). Using the quotient differentiation rule we obtain

$$V(\rho, z)\left[-\frac{\rho^2 k^2}{q^2}\left(1+\frac{dq}{dz}\right)-2ik\left(\frac{1}{q}+\frac{1}{r}\frac{dr}{dz}\right)\right]=0\,, \tag{1.7.6}$$

where, for ease of notation, we have suppressed the explicit functional dependence of q and r on z.

A nontrivial solution for $V(\rho, z)$ that is valid for all values of ρ requires that the following two differential equations are simultaneously satisfied,

$$\frac{dq}{dz}=-1\,, \tag{1.7.7}$$

$$\frac{dr}{r}=-\frac{dz}{q}\,. \tag{1.7.8}$$

The first equation for q is independent of r and can be solved directly. With the solution for the complex parameter q, we obtain the solution for $r(z)$ from the second equation.

1.7.1.1 Curvature $1/R_c$ and Beam Width Parameter w of a Gaussian Beam

The integration of Eq. (1.7.7) from z_1 to z_2 immediately yields

$$q(z_2)=q(z_1)-(z_2-z_1)\,. \tag{1.7.9}$$

To calculate the parameters R_c and w of the Gaussian beam as a function of z we suppose that in the plane $z = z_1$ these quantities are given; we will denote them by $R_{c,1}$ and w_1, respectively. Using these initial conditions in the plane $z = z_1$, we have from Eq. (1.7.9) that $q(z_2) = q(z_1) - \Delta z$ with $\Delta z = z_2 - z_1$. We then obtain after some algebra the complex beam parameter in the plane $z = z_2$ as

$$\begin{aligned}\frac{1}{q(z_2)}&=\frac{1}{R_c(z_2)}-\frac{i\lambda}{\pi w^2(z_2)}=\frac{1}{q(z_1)-\Delta z}\\&=\frac{\pi^2 w_1^4(R_{c,1}-\Delta z)-\lambda^2 R_{c,1}^2\Delta z-i\pi w_1^2\lambda R_{c,1}^2}{\pi^2 w_1^4(R_{c,1}-\Delta z)^2+\lambda^2 R_{c,1}^2(\Delta z)^2}\,.\end{aligned} \tag{1.7.10}$$

Equating the real parts on both sides of Eq. (1.7.10) we have for the radius of curvature R_c in the plane $z = z_2$ the expression

$$\frac{1}{R_c(z_2)}=\frac{(R_{c,1}-\Delta z)-\lambda^2 R_{c,1}^2\Delta z\,/\,\pi^2 w_1^4}{(R_{c,1}-\Delta z)^2+\lambda^2 R_{c,1}^2(\Delta z)^2\,/\,\pi^2 w_1^4}\,. \tag{1.7.11}$$

A special case arises if $R_c(z_2)$ becomes infinitely large. For that z-coordinate the divergence of the beam changes sign and the wavefront at the axial position $z = z_2$ is flat. Accordingly, this position is identified as the 'best' focus of the Gaussian beam. From Eq. (1.7.11) we immediately obtain that the value of $(\Delta z)_0 = z_0 - z_1$ for finding the beam cross-section with a flat wavefront is given by

$$(\Delta z)_0=\frac{R_{c,1}}{1+[\lambda R_{c,1}/\pi w_1^2]^2}\,, \tag{1.7.12}$$

where the sign of $(\Delta_z)_0$ is the same as the sign of $R_{c,1}$. This means that the axial coordinate shift $(\Delta_z)_0$ from the plane at $z = z_1$ towards the 'best focus' at $z = z_0$ is towards the centre of curvature C_1 of the wavefront through A_1 in Fig. 1.14.

The value of the beam width parameter w in the plane $z = z_2$ is found by equating the imaginary parts of Eq. (1.7.10), yielding the expression

$$\left(\frac{w(z_2)}{w_1}\right)^2=\left(\frac{R_{c,1}-\Delta z}{R_{c,1}}\right)^2+\left(\frac{\lambda\,\Delta z}{\pi w_1^2}\right)^2\,. \tag{1.7.13}$$

We are interested in the axial position along the beam where the Gaussian beam size w has a minimum value, corresponding to the 'waist' of the Gaussian beam. By taking the derivative of the equation above, it directly follows that $(\Delta z)_w$ is given by

$$(\Delta z)_w = \frac{w_1^4 R_{c,1}}{w_1^4 + (\lambda R_{c,1}/\pi)^2} = \frac{R_{c,1}}{1 + [\lambda R_{c,1}/\pi w_1^2]^2} , \qquad (1.7.14)$$

from which it follows that the beam waist is located at the same axial position at which the beam has a flat wavefront. The value of the smallest beam width w_0 follows from insertion of the value of $(\Delta z)_w$ in Eq. (1.7.13) as

$$\left(\frac{w_0}{w_1}\right)^2 = \frac{[\lambda R_{c,1}/(\pi w_1^2)]^2}{1 + [\lambda R_{c,1}/(\pi w_1^2)]^2} . \qquad (1.7.15)$$

We observe that any shift smaller than $(\Delta z)_w$ from an initial position in the beam with $z = z_1$ in the direction of the centre of curvature C_1 produces a smaller beam width. Only in the case that the wavefront was flat at $z = z_1$ ($R_{c,1} = \infty$), do we have that $w_0 = w_1$, which confirms that the beam waist F_w is found in the cross-section with $z = z_1$. When the origin of the z-coordinate is taken at the beam waist, Eq. (1.7.15) can be used to find $1/q(z)$ by letting $R_{c,1} \rightarrow \infty$ and $w_1 = w_0$, yielding

$$\frac{1}{q(z)} = \frac{1}{R_c(z)} - \frac{i\lambda}{\pi w^2(z)} = \frac{-z - i(\pi w_0^2/\lambda)}{z^2 + \left(\pi w_0^2/\lambda\right)^2} . \qquad (1.7.16)$$

The equi-amplitude curves of a Gaussian beam are described by hyperbola branches as shown in Fig. 1.15. For the $1/e$-amplitude curve we have the equation

$$\rho^2 - \left(\frac{\lambda z}{\pi w_0}\right)^2 = w_0^2 . \qquad (1.7.17)$$

From this equation it follows that the hyperbola foci are found at the positions

$$\rho_{F_{1,2}} = \pm\, w_0 \sqrt{1 + \left(\frac{\pi w_0}{\lambda}\right)^2} \approx \frac{\pi w_0^2}{\lambda} . \qquad (1.7.18)$$

The approximate value of $\pi w_0^2/\lambda$ is justified since for all Gaussian beams it should hold that $w_0 \gg \lambda$. A general point Q on the $1/e$ curve has an azimuth θ that is given by

$$\theta = \arctan\left\{\frac{\lambda}{\pi w_0}\left[1 + \left(\frac{\pi w_0^2}{\lambda z}\right)^2\right]\right\} . \qquad (1.7.19)$$

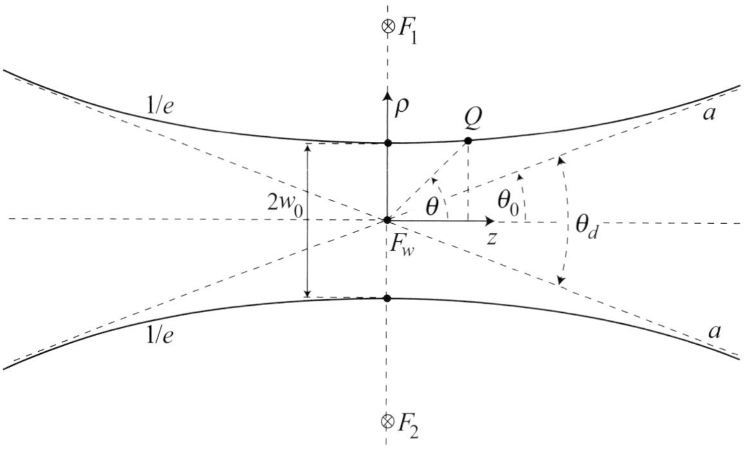

Figure 1.15: The equi-amplitude lines in a Gaussian beam. Q is a general point on the e^{-1}-amplitude curve. F_w is the centre of the beam waist; F_1 and F_2 are the foci of the hyperbola branches that produce the equi-amplitude curves of the Gaussian beam.

The angles of the asymptotes to the hyperbola branches with the z-axis are given by

$$\theta_0 = \pm \arctan\left(\frac{\lambda}{\pi w_0}\right) \approx \pm \frac{\lambda}{\pi w_0}\,, \tag{1.7.20}$$

which allows us to write $\rho_{F_{1,2}} \approx w_0/\theta_0$.

The angle θ_0 has been given the name of the *natural* divergence angle of the Gaussian beam, imposed by the ratio of the wavelength of the light and the beam waist. It immediately follows from Eq. (1.7.20) that the product $w_0\,\theta_0$ is a quantity that only depends on the wavelength of the beam. A transformation of the geometry of a Gaussian beam by a refraction or reflection at a single surface or by a lens will leave this product unaltered. The total divergence angle θ_d of the Gaussian beam is twice the value of $|\theta_0|$ and is given by $0.64\lambda/w_0$. For Gaussian beams to satisfy the paraxial Helmholtz equation, the natural divergence angle θ_0 has to remain small, typically below 10 to 20 mrad in the visible range of the spectrum. In that case, the two focal points $F_{1,2}$ of the hyperbola in Fig. 1.15 would lie far outside the drawing area.

We conclude the discussion on beam curvature and beam width with the derivation of an invariant quantity of a Gaussian beam. From Eq. (1.7.16) we find that

$$R_c(z) = -\,z\left[1+\left(\frac{\pi w_0^2}{\lambda z}\right)^2\right], \qquad \frac{w^2(z)}{w_0^2} = 1+\left(\frac{\lambda z}{\pi w_0^2}\right)^2 . \tag{1.7.21}$$

These expressions can be combined to yield specifically

$$\frac{z\,R_c(z)}{w^2(z)} = -\left(\frac{\pi w_0}{\lambda}\right)^2 = -\frac{1}{\theta_0^2}\,, \tag{1.7.22}$$

which is a conserved quantity for a Gaussian beam (the origin for the coordinate z is set at the waist F_w of the beam). The curvature of the Gaussian beam varies from zero at $|z| \rightarrow \infty$, has a maximum value for some finite $|z|$, to fall back to zero again at the beam waist. The position $z_{c,max}$ (or z_R) along the beam where the wavefront curvature is maximum is calculated with the aid of Eq. (1.7.21), together with the minimum radius of curvature $R_{c,min}$,

$$z_{c,max} = z_R = \pm\frac{\pi w_0^2}{\lambda} = \pm\frac{w_0}{\theta_0}\,, \qquad R_{c,min} = \mp\frac{2\pi w_0^2}{\lambda} . \tag{1.7.23}$$

Equation (1.7.23) shows that the centre of curvature of the wavefront through M_1 is found at M_2 and vice versa. The beam width (again defined in terms of the $1/e$ level) at M_1 and M_2 is $\sqrt{2}$ larger than the beam width w_0 at the waist F_w of the beam. In Fig. 1.16 we have shown the points M_1 and M_2 with $|z_{c,max}| = w_0/\theta_0$ where the wavefronts with maximum curvature intersect the beam axis.

The distances $F_wM_2 = M_1F_w = z_{c,max}$ are commonly denoted by z_R or the Rayleigh length of the Gaussian beam. If we multiply the Rayleigh length with the natural divergence angle θ_0 we obtain the beam width w_0. Equation (1.7.21)

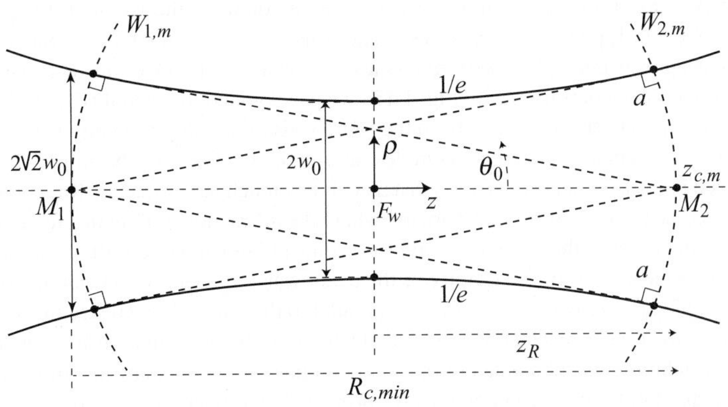

Figure 1.16: The Gaussian beam wavefronts $W_{1,m}$ and $W_{2,m}$ through the axial points M_1 and M_2, respectively, that exhibit maximum wavefront curvature. Their centres of curvature are M_2 and M_1, respectively. z_R is the Rayleigh distance of the beam, $R_{c,min} = 2z_R$ is the confocal distance of the beam.

states that the width at a plane with axial coordinate z is given by the quadratic sum of the intrinsic width at the beam waist and a geometric contribution from propagation. Beam propagation over a distance given by the Rayleigh length produces a geometric broadening of w_0, such that the total beam width is $2w_0\sqrt{2}$. The denomination 'Rayleigh length' for the Gaussian beam follows from a comparable defocusing distance that was defined by Rayleigh for a classical focused beam.

The quantity $R_{c,min} = |M_1M_2|$, equal to twice the Rayleigh distance, has been given the name 'confocal distance' of a Gaussian beam. A cavity with this length is a stable configuration for the generation of laser light with a beam divergence $\theta_d = 2\lambda/(\pi w_0)$. Finally, we find that the departure of the parabolically curved wavefront from a planar reference surface at the e^{-1} amplitude level of the Gaussian beam is given by $W = z\,\theta_0^2/2$. The parabolic wavefront deviation evaluated for the wavefronts through the Rayleigh points M_1 and M_2 yields a phase difference of ∓ 1 radian at the $1/e$ amplitude level of the beam.

1.7.1.2 The Amplitude Factor $1/r$ of a Gaussian Beam

The substitution of the result $q(z_2) = q(z_1) - z_2 + z_1$ into the second differential equation (1.7.8) yields, after integration from z_1 to z_2,

$$\ln\left[\frac{r(z_2)}{r(z_1)}\right] = \ln\left[\frac{-z_2 + q(z_1)}{-z_1 + q(z_1)}\right] ,$$

or, equivalently,

$$r(z_2) = \left[\frac{-z_2 + q(z_1)}{-z_1 + q(z_1)}\right] r(z_1) . \tag{1.7.24}$$

We take the axial position of the beam waist F_w as the origin for the z-coordinate and determine the waist value w_0 and the on-axis amplitude $1/r_0$ of the beam in the waist. Subsequently, we normalise the on-axis amplitude to unity such that $r(0) = 1$. With F_w as origin, Eq. (1.7.24) can then be written as

$$r(z) = \frac{-z + q(0)}{q(0)} = 1 + i\frac{z}{z_R} , \tag{1.7.25}$$

where we have used the fact that at the beam waist the beam parameter q follows from $1/q(0)=-i\,\lambda/(\pi w_0^2)$. The modulus and phase of the complex amplitude factor $r(z)$ are given by

$$|r(z)| = \left\{1 + \left(\frac{z}{z_R}\right)^2\right\}^{1/2} , \text{ and } \arg\{r(z)\} = \phi(z) = \arctan\left(\frac{z}{z_R}\right) = \arctan\left\{z\left(\frac{\theta_0}{w_0}\right)\right\} . \tag{1.7.26}$$

We observe that at a large axial distance from the waist F_w the modulus of the factor $r(z)$ is proportional to $|z|$. The argument of the factor $1/r(z)$ in Eq. (1.7.4) equals $-\phi(z)$ and runs from $+\pi/2$ for large negative z values to $-\pi/2$ for $z \rightarrow +\infty$. Exactly at the beam waist, the phase $\phi(0)$ of a Gaussian beam is zero. Over the distance running from $-w_0/\theta_0$ to $+w_0/\theta_0$ the phase retardation amounts to $\pi/2$. In Fig. 1.17a we have plotted the modulus $1/|r(z)|$ of the axial amplitude of a Gaussian beam as a function of the distance z from the beam waist, normalised to unity at the beam waist F_w. Graph b) of Fig. 1.17 shows the on-axis phase shift of the complex amplitude of a Gaussian beam in units of π. The total phase shift of $-\pi$ that is acquired by a Gaussian beam over a distance that is much larger than its Rayleigh distance is called the Gouy phase shift. It is discussed in more detail in Chapter 9 where the diffraction of light in a focused beam is studied. We note that with respect to a plane wave, the Gaussian beam suffers an additional phase shift of $-\pi$ radians after traversal of the beam-waist region. This means that in the vicinity of the beam waist of a Gaussian beam the phase advances slower along the z-axis than for a plane wave with its wave vector parallel to the z-axis. We will see in Chapters 8 and 9 that this phenomenon is common to focused beams that have a limited lateral extent and exhibit diffraction, as opposed to an individual plane wave that by definition is unbounded in space and propagates without any amplitude or phase perturbation.

Using the results for amplitude, phase, curvature and propagation properties from the previous subsections, the expression for an elementary Gaussian beam with symmetry of revolution can be written as

$$\mathbf{E}(\rho, z) = \frac{\mathbf{p}}{|r(z)|} \exp\left\{-\frac{\rho^2}{w^2(z)}\right\} \exp\left\{-\frac{ik\rho^2}{2R_c(z)}\right\} \exp[i\{kz - \phi(z)\}] , \tag{1.7.27}$$

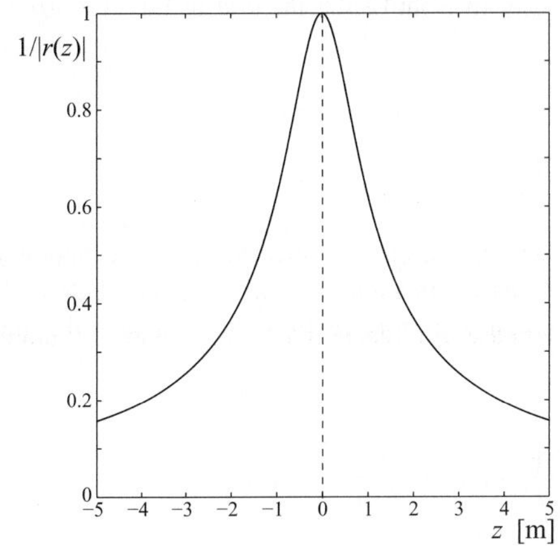

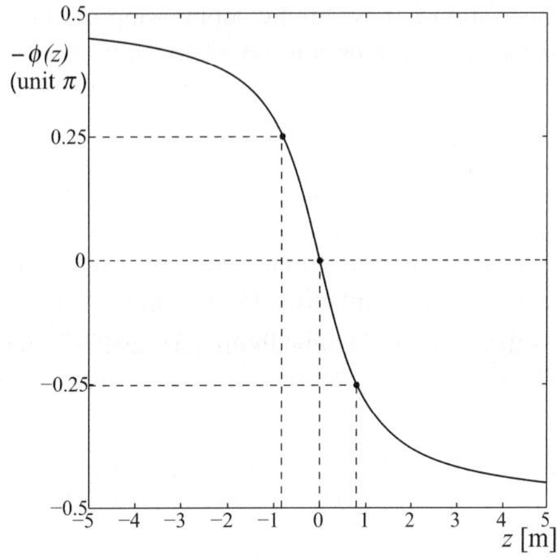

Figure 1.17: a) The modulus of the normalised on-axis amplitude $1/|r(z)|$ of a Gaussian beam.
b) The on-axis phase $-\phi(z)$ of a Gaussian beam expressed in units of π. Beam parameters are $\lambda = 633$ nm, $w_0 = 0.4$ mm yielding $\theta_0 = 0.5$ mrad. The axial distance z is expressed in units of metres; the Rayleigh distance is 0.80 m.

where $\mathbf{p}$, as before, is the light polarisation vector which is uniform and identical for any cross-section of the Gaussian beam. From this expression we can check the allowed range of beam parameter values for which the approximation $|d^2V/dz^2| \ll 2k\,|dV/dz|$, used in the derivation of the paraxial Helmholtz equation, holds. On axis at the beam waist, the derivatives of $V(\rho, z)$ are

$$\left.\frac{dV(\rho,z)}{dz}\right|_{z=0,\rho=0} = -\left.\frac{1}{r^2(z)}\frac{dr(z)}{dz}\right|_{z=0} = -i\frac{\lambda}{\pi w_0^2}\,,$$

$$\left.\frac{d^2V(\rho,z)}{dz^2}\right|_{z=0,\rho=0} = \frac{1}{r^2(z)}\left\{\frac{2}{r(z)}\left(\frac{dr(z)}{dz}\right)^2 - \frac{d^2r(z)}{dz^2}\right\}_{z=0} = \left(\frac{\lambda}{\pi w_0^2}\right)^2 . \tag{1.7.28}$$

We require that the value of the second derivative is at the most $2k\epsilon$ times the magnitude of the first derivative with ϵ a quantity $\ll 1$. This condition is satisfied if

$$w_0 \geq \frac{\lambda}{\pi}\sqrt{\frac{1}{2\epsilon}}\,. \tag{1.7.29}$$

For instance, using the conservative value of $\epsilon = 10^{-4}$, we find that $w_0 \geq 22\lambda$, or, equivalently, the natural divergence angle θ_0 of the beam should be $\leq \sqrt{2\epsilon} = 14$ mrad.

1.7.1.3 Amplitude Normalisation and Energy Transport by a Gaussian Beam

The amplitude of a Gaussian beam can be normalised in such a way that the squared modulus of $E(\rho, 0)$ (see Eq. (1.7.27)), integrated over the beam-waist cross-section, equals unity. We thus require that for $|\mathbf{p}| = 1$ it holds that

$$\frac{1}{w_0^2}\int_0^{2\pi}\int_0^{\infty} c_G^2 \exp\left(-\frac{2\rho^2}{w_0^2}\right)\rho\,d\rho = 1\,, \tag{1.7.30}$$

where, to obtain a dimensionless normalisation factor c_G, we have normalised the surface integral with the aid of the beam-waist parameter w_0. Evaluation of the integral yields the result

$$c_G = \sqrt{\frac{2}{\pi}}\,. \tag{1.7.31}$$

Another normalisation is possible by suppressing the factor $1/w_0^2$ in front of the integral in Eq. (1.7.30), yielding the normalisation factor c'_G with dimension of $[\mathrm{m}^{-1}]$,

$$c'_G = \frac{1}{w_0}\sqrt{\frac{2}{\pi}}\,. \tag{1.7.32}$$

To determine the electric energy transported by a Gaussian beam we examine the electric field distribution at the beam waist. In the paraxial approximation, the Poynting vector of a Gaussian beam is everywhere directed along the z-axis as follows from Eq. (1.7.27). The total power P_G carried by a Gaussian beam follows from the integral of the time-averaged Poynting vector over a cross-section taken at the waist, giving

$$P_G = \frac{\epsilon_0 c}{2}\,|E_0|^2 \int_0^{2\pi}\int_0^{\infty} \exp\left\{-\frac{2\rho^2}{w_0^2}\right\}\rho\, d\rho d\gamma = \frac{\pi\epsilon_0 c w_0^2}{4}\,|E_0|^2\,, \tag{1.7.33}$$

where $|E_0|$ is the modulus of the on-axis electric field strength at the waist of the beam. With the aid of the normalisation constants c_G or c'_G we find the expressions

$$P_G = \frac{\epsilon_0 c w_0^2}{2}\,|E_{0,n}|^2\,, \qquad P_G = \frac{\epsilon_0 c}{2}\,|E'_{0,n}|^2\,, \tag{1.7.34}$$

where $E_{0,n}$ is the electric field associated with the normalisation factor c_G and $E'_{0,n}$ the field related to a Gaussian beam with normalisation factor c'_G. We note that the normalisation of the Gaussian beam function with the aid of the c'_G factor yields the simplest expression for the energy transport, identical to the expression for energy transport by a plane wave.

For the non-normalised Gaussian beam we express the electric field strength E_0 as a function of P_g and w_0 using the simple arithmetic expression

$$|E_0| = 21.9\,\frac{P_G^{1/2}}{w_0}\,. \tag{1.7.35}$$

For a classical, non-focused He-Ne laser beam with a typical waist parameter $w_0 = 0.4$ mm ($\lambda = 633$ nm), we find that $|E_0| \approx 55\,000\, P_G^{1/2}$ V/m. For a typical power of 1 mW, the electric field strength in the waist amounts to 1730 Vm^{-1}. Increasing the power of such a laser beam when operating in dry air would cause spontaneous electric breakdown if the electric field strength exceeds the value of $2 \cdot 10^6$ Vm^{-1}. It is seen that the laser power to produce such an effect has to be at least 1300 W.

1.7.2 Higher-order Gaussian Beams

Depending on the geometry of a resonant cavity, the light beam which exits from it may show amplitude variations in a cross-section of the beam, different from those of the basic Gaussian beam profile that was found in the previous subsection. We assume that such a lateral amplitude pattern of a higher-order Gaussian beam is unaltered in form upon propagation and that its lateral scale varies in the same way as the beam width $w(z)$ of the basic, zero-order Gaussian beam. The assumption of an unchanged lateral profile during propagation is unphysical because it neglects the basic effect of light diffraction; however, within the approximation of the paraxial wave equation of Eq. (1.7.3), this assumption is a valid one. We will now discuss in detail Gaussian beams with a transverse amplitude pattern that can be factorised as the product of an x- and a y-dependent function, the so-called Gauss–Hermite beams. Another class of beams with an amplitude pattern that can be written as a factorised function with respect to the polar coordinates (ρ, α) in the beam cross-section is known as Gauss–Laguerre beams. The amplifying cavity of laser sources can be geometrically and optically designed in such a way that laser beams of the Gauss–Hermite or Gauss–Laguerre type are preferably produced. In practice, most sources are required to produce the basic Gaussian beam. In such a case, higher-order lasing modes of the Gauss–Hermite or Gauss–Laguerre type should be suppressed as much as possible by reducing their internal gain in the cavity.

1.7.3 Gauss–Hermite Beams

For a beam with Cartesian, as opposed to polar, symmetry we first note that $V(\rho, z)$ must be replaced by $V(x, y, z)$. For higher-order Gauss–Hermite beams we assume the form

$$V(x,y,z) = \frac{1}{r_z}\, g_x\left(\frac{x}{w_z}\right) g_y\left(\frac{y}{w_z}\right) \exp\left\{-\frac{ik(x^2+y^2)}{2q_z}\right\}, \tag{1.7.36}$$

where, for ease of notation, we have written the z-dependent functions $r(z)$, $q(z)$ and $w(z)$ as r_z, q_z and w_z, respectively. For a higher-order Gaussian beam, we use the same definitions for $q(z)$ and $w(z)$ as those given in Eqs. (1.7.16) and (1.7.21). We now substitute the trial function $V(x, y, z)$ into the Cartesian version of Eq. (1.7.3). We denote the arguments x/w_z and y/w_z of the functions g_x and g_y as $x_s = x/w_z$ and $y_s = y/w_z$, respectively. After some careful administration and ordering of derivatives of the same function and with the same spatial dependence, we obtain the equation

$$\begin{aligned}
0 &= \left[-\frac{dq_z}{dz} - 1\right]\left(x^2+y^2\right)\frac{k^2}{q_z^2} && a)\\
&+ \left[\frac{1}{w_z^2 g_x}\frac{d^2 g_x}{dx_s^2} - \frac{2ikx}{w_z}\left(\frac{1}{q_z} + \frac{1}{w_z}\frac{dw_z}{dz}\right)\frac{1}{g_x}\frac{dg_x}{dx_s}\right] && b)\\
&+ \left[\frac{1}{w_z^2 g_y}\frac{d^2 g_y}{dy_s^2} - \frac{2iky}{w_z}\left(\frac{1}{q_z} + \frac{1}{w_z}\frac{dw_z}{dz}\right)\frac{1}{g_y}\frac{dg_y}{dy_s}\right] && c)\\
&- \left[\frac{1}{q_z} + \frac{1}{r_z}\frac{dr_z}{dz}\right] 2ik\,. && d)
\end{aligned} \tag{1.7.37}$$

The differential equation for $V(x, y, z)$ has been subdivided into four parts, labelled *a*) to *d*) that together should yield zero for any point $Q(x, y, z)$ in an arbitrary beam cross-section. Part *a*) can be treated separately; it is identically zero if Eq. (1.7.9) is satisfied for the function q_z. For parts *b*) and *c*) we first calculate the common factor between round braces using Eqs. (1.7.16) and (1.7.21), with the result

$$\frac{1}{q_z} + \frac{1}{w_z}\frac{dw_z}{dz} = -\frac{i\lambda}{\pi w_z^2}\,. \tag{1.7.38}$$

We substitute the identity of Eq. (1.7.38) in parts *b*) and *c*) of Eq. (1.7.37) and introduce the change of variables

$$u = \frac{x\sqrt{2}}{w_z} = x_s\sqrt{2}\,, \qquad v = \frac{y\sqrt{2}}{w_z} = y_s\sqrt{2}\,. \tag{1.7.39}$$

The sum of parts *b*), *c*) and *d*) thus has to satisfy the equation

$$\frac{2}{w_z^2 g_x}\left[\frac{d^2 g_x}{du^2} - 2u\frac{dg_x}{du}\right] + \frac{2}{w_z^2 g_y}\left[\frac{d^2 g_y}{dv^2} - 2v\frac{dg_y}{dv}\right] - 2ik\left[\frac{1}{q_z} + \frac{1}{r_z}\frac{dr_z}{dz}\right] = 0\,. \tag{1.7.40}$$

The transformation of variables according to Eq. (1.7.39) was used to bring the two factors between square brackets in the equation above into a form which closely resembles the Hermitian differential equation,

$$\frac{d^2 g}{du^2} - 2u\frac{dg}{du} + 2m\,g = 0\,, \tag{1.7.41}$$

for which the Hermite polynomials $H_m(u)$ of degree m are solutions provided that m is a non-negative integer. Equation (1.7.40) is now written as follows,

$$\frac{2}{w_z^2 g_x}\left[\frac{d^2 g_x}{du^2} - 2u\frac{dg_x}{du} + 2m_x g_x\right] + \frac{2}{w_z^2 g_y}\left[\frac{d^2 g_y}{dv^2} - 2v\frac{dg_y}{dv} + 2m_y g_y\right] - \frac{4}{w_z^2}\left(m_x + m_y\right) - 2ik\left[\frac{1}{q_z} + \frac{1}{r_z}\frac{dr_z}{dz}\right] = 0\,. \tag{1.7.42}$$

The first two terms of Eq. (1.7.42) yields zero if the functions g_x and g_y are chosen to be Hermite polynomials of order m_x and m_y, respectively,

$$g_x\left(\frac{x}{w_z}\right)=H_{m_x}\left(\frac{x\sqrt{2}}{w_z}\right), \qquad g_y\left(\frac{y}{w_z}\right)=H_{m_y}\left(\frac{y\sqrt{2}}{w_z}\right). \tag{1.7.43}$$

The remaining two terms of Eq. (1.7.42) must also equal zero. Using the expressions for w_z and $1/q_z$ that were already found for the elementary Gaussian beam, we obtain after some manipulation the following differential equation for the axial function r_z,

$$\frac{dr_z}{r_z}=\frac{\left[i(m_x+m_y+1)+(z/z_R)\right]}{1+(z/z_R)^2}\,d(z/z_R)\,, \tag{1.7.44}$$

where, for reasons of compact notation, we have used the Rayleigh length z_R of the elementary Gaussian beam. The two terms in the numerator in the right-hand side of Eq. (1.7.44) give rise, after integration of the differential equation, to an arctangent and to a logarithmic function such that

$$\ln(r_z)=i(m_x+m_y+1)\arctan\left(\frac{z}{z_R}\right)+\frac{1}{2}\ln\left(\frac{w_z}{w_0}\right)^2. \tag{1.7.45}$$

We then find the on-axis beam amplitude to be given by

$$\frac{1}{r_H(z)}=\left[1+\left(\frac{z}{z_R}\right)^2\right]^{-1/2}\exp\left\{-i(m_x+m_y+1)\arctan\left(\frac{z}{z_R}\right)\right\}, \tag{1.7.46}$$

where we have used the expression for w_z^2/w_0^2 from Eq. (1.7.21) and the Rayleigh length z_R from Eq. (1.7.23) for ease of notation. The decrease in axial amplitude of a Gauss–Hermite beam is the same as for an elementary Gaussian beam, but the axial phase ϕ of a Gauss–Hermite beam behaves differently; it is given by

$$\phi_H(z)=-\,i(m_x+m_y+1)\arctan\left(\frac{z}{z_R}\right). \tag{1.7.47}$$

Compared with a plane wave propagating along the z-axis, the phase difference of the Gauss–Hermite beam between the axial points M_2 and M_1 with $z=z_R$ and $z=-z_R$, respectively, equals $-(m_x+m_y+1)\pi/2$.

The Gauss–Hermite beam is hence given by

$$V_H(x,y,z)\exp(ikz)=\frac{c_H(m_x,m_y)}{|r_H(z)|}H_{m_x}\left(\frac{x\sqrt{2}}{w_z}\right)H_{m_y}\left(\frac{y\sqrt{2}}{w_z}\right)$$
$$\times\exp\left\{\frac{-ik(x^2+y^2)}{2}\left(\frac{1}{R_c(z)}-\frac{i\lambda}{\pi w^2(z)}\right)\right\}\exp[i\{kz+\phi_H(z)\}], \tag{1.7.48}$$

where we have added a normalisation constant $c_H(m_x,m_y)$ such that

$$\iint_{-\infty}^{+\infty}\left|V_H(x,y,0)\right|^2 d\left(\frac{x}{w_0}\right)d\left(\frac{y}{w_0}\right)=1. \tag{1.7.49}$$

The Cartesian integration variables (x,y) in the plane of the beam waist have been made dimensionless with the aid of the beam-waist half-diameter w_0 of the corresponding elementary Gaussian beam. The normalisation of Eq. (1.7.49) has the advantage that the constant $c_H(m_x,m_y)$ itself is also dimensionless. Evaluating the normalisation integral of Eq. (1.7.49) and using the expression for the inner product of a Hermite polynomial with itself, see **[266]**, p. 439, Table 18.3.1, we obtain the normalisation factor of a Gauss–Hermite beam as

$$c_H(m_x,m_y)=\left(\pi\,2^{m_x+m_y-1}(m_x)!\,(m_y)!\right)^{-1/2}. \tag{1.7.50}$$

In a similar way as for the elementary Gaussian beam, we can define the normalisation coefficient $c'_H(m_x,m_y)=c_H(m_x,m_y)/w_0$.

The mode indices of a Gauss–Hermite beam are given by m_x and m_y. When inspecting the intensity cross-section, m_x and m_y give the number of intensity zeros along the x- and y-axis, respectively. A special case is the elementary Gaussian beam with indices $(0,0)$. Within the paraxial approximation, the state of polarisation of the Gauss–Hermite

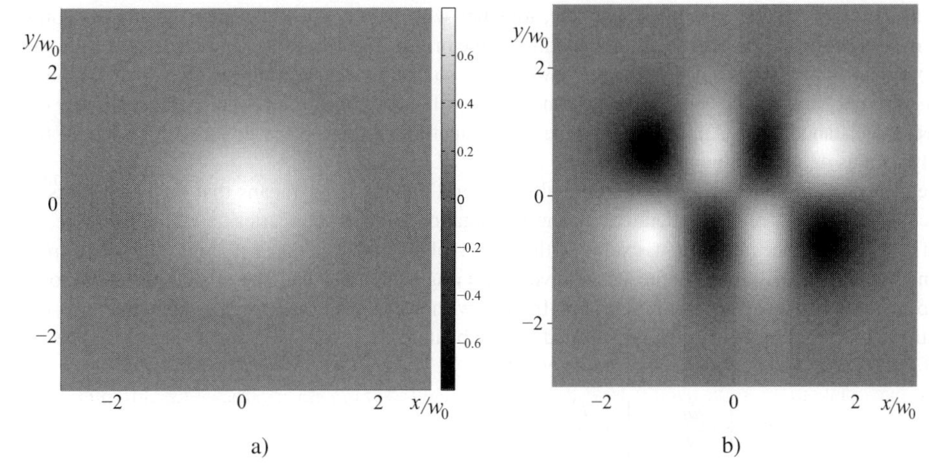

Figure 1.18: Complex amplitude (real-valued) of a Gauss–Hermite beam in the beam-waist cross-section. The Cartesian coordinates in the beam-waist cross-section are in units of the beam waist w_0.
a) $TEM_{0,0}$ beam mode.
b) $TEM_{3,1}$ beam.

beams is uniform with both the electric and magnetic field components confined to a plane perpendicular to the axis of propagation, the z-axis. For this reason, the modes are of the *TEM* (transverse electromagnetic) type which is a special class of guided mode, occurring for instance in optical waveguides and co-axial cables. In Fig. 1.18 we have shown the complex amplitude in the beam waist of a $TEM_{0,0}$ mode and of a Gauss–Hermite $TEM_{3,1}$ mode; in both cases, the complex amplitude is a real function. The grey level in both graphs of Fig. 1.18 corresponds to the zero-amplitude level. The normalisation factor according to Eq. (1.7.50) has been included in the calculation; for the special case of the $TEM_{0,0}$ mode, this factor equals $w_0\sqrt{2/\pi}$.

1.7.4 Gauss–Laguerre Beams

In the previous subsection, the 'aperture' function $g(x,y)$ was written as the product of two functions that were a function of x and y only and the Gauss–Hermite beam modes were found. Another special type of higher-order beam is found when the function g, written in polar coordinates as $g(\rho,\alpha)$, can be factorised as a product of a function of ρ and a function of α only. The scalar wave function that should satisfy the paraxial Helmholtz equation is given by $V(\rho,\alpha,z)\exp(ikz)$ and the function V should satisfy the differential equation

$$\frac{1}{\rho}\frac{\partial}{\partial\rho}\left(\rho\frac{\partial V}{\partial\rho}\right)+\frac{1}{\rho^2}\frac{\partial^2 V}{\partial\alpha^2}+2ik\frac{\partial V}{\partial z}=0\,. \tag{1.7.51}$$

The factorised trial solution for V is written as

$$V(\rho,\alpha,z)=\frac{1}{r_z}\,g_\rho\left(\frac{\rho}{w_z}\right)\exp(im_\alpha\alpha)\,\exp\left(-\frac{ik\rho^2}{2q_z}\right), \tag{1.7.52}$$

where m_α determines the azimuthal periodicity of the solution and where, as before, the subscripts z of r, w and q indicate that these quantities are functions of z only. The argument of the function g_ρ has been normalised with respect to the beam width w_z. The product function $g_\rho(\rho/w_z)\exp(im_\alpha\alpha)$ modifies the amplitude and phase in the cross-section of the Gaussian beam and should be a continuous function with finite value first and second derivatives. This continuity requirement, which is imposed to ensure a physical solution, means we can now write the function as

$$g_\rho\left(\frac{\rho}{w_z}\right)\exp(im_\alpha\alpha)=\left(\frac{\rho}{w_z}\right)^{m_\alpha} f\left\{\left(\frac{\rho}{w_z}\right)^2\right\}\,\exp(im_\alpha\alpha)\,. \tag{1.7.53}$$

The factor to the power m_α on the right-hand side, together with the exponential factor, guarantees continuity of the function g_ρ over the entire beam cross-section including at the origin. The function f can be derived from the differential Eq. (1.7.51) and introduces extra terms with even powers of ρ_s so that the degree of the product function may have powers of $\rho_s = \rho/w_z$ running from $m_\alpha, m_\alpha + 2, \cdots$, up to a maximum value, without compromising the continuity of the function $g(\rho/w_z)$; see, for instance, **[37]**, Appendix VII, where a comparable discussion is given with respect to the orthogonal Zernike polynomials defined on the unit disk.

The dependence of the function g_ρ on the normalised lateral coordinate ρ_s means that the beam cross-section, apart from the scaling factor w_z, remains identical during propagation. As for the Cartesian Gauss–Hermite solutions, this assumption is a 'risky' one because it excludes the effect of diffraction that will blur the beam cross-section on propagation over a large distance. If the beam divergence angle is small and the spatial variations of the complex amplitude over the beam cross-section remain low-frequency; however, the criterion for the correctness of the paraxial wave equation, $|\partial^2 V/\partial z^2| \ll 2k|\partial V/\partial z|$, is fulfilled. We substitute the expression for $V(\rho, \alpha, z)$ in the paraxial wave equation of Eq. (1.7.51) and obtain the differential equation

$$0 = \left[-\frac{dq_z}{dz} - 1\right]\rho^2 \frac{k^2}{q_z^2} \qquad a)$$

$$+\left[\frac{1}{w_z^2 g_{\rho_s}} \frac{d^2 g_{\rho_s}}{d\rho_s^2} - \frac{1}{\rho w_z g_{\rho_s}} \left\{2ik\rho^2\left(\frac{1}{q_z} + \frac{1}{w_z}\frac{dw_z}{dz}\right) - 1\right\} \frac{dg_{\rho_s}}{d\rho_s}\right] \qquad b)$$

$$-\frac{m_\alpha^2}{\rho^2} - 2ik\left[\frac{1}{q_z} + \frac{1}{r_z}\frac{dr_z}{dz}\right], \qquad c)$$

$$(1.7.54)$$

in which we have also used the scaled radial coordinate $\rho_s = \rho/w_z$. We equate part *a*) of this equation to zero and obtain the solution for q_z that was given previously in Eq. (1.7.9). Using the result of Eq. (1.7.38) we have the following differential equation for the function $g_{\rho_s}(\rho_s)$ from parts *b*) and *c*) of Eq. (1.7.54),

$$\left\{\frac{d^2 g_{\rho_s}}{d\rho_s^2} + \frac{1}{\rho_s}\left(-4\rho_s^2 + 1\right)\frac{dg_{\rho_s}}{d\rho_s}\right\} - \left\{\frac{m_\alpha^2}{\rho_s^2} + 2ikw_z^2\left[\frac{1}{q_z} + \frac{1}{r_z}\frac{dr_z}{dz}\right]\right\} g_{\rho_s} = 0 . \qquad (1.7.55)$$

In the next step we substitute $g_{\rho_s}(\rho_s\sqrt{2}) = g_{\rho_0}(\rho_0)$ of Eq. (1.7.53) in Eq. (1.7.55) and introduce a scaled coordinate $\rho_0 = \rho_s\sqrt{2}$ that will show its usefulness at the end of the calculation. We then obtain the differential equation,

$$\frac{d^2 g_{\rho_0}(\rho_0)}{d\rho_0^2} - \frac{1}{\rho_0}\left(2\rho_0^2 - 1\right)\frac{dg_{\rho_0}(\rho_0)}{d\rho_0} - \left\{\frac{m_\alpha^2}{\rho_0^2} + ikw_z^2\left[\frac{1}{q_z} + \frac{1}{r_z}\frac{dr_z}{dz}\right]\right\} g_{\rho_0}(\rho_0) = 0 . \qquad (1.7.56)$$

Equation (1.7.53) shows that the function g_ρ comprises a function f which has ρ^2 as argument. For this reason we introduce a change of variable $t = \rho_0^2$. The differential quotients with respect to ρ_0 in terms of the new variable t are given by

$$\frac{dg_{\rho_0}}{d\rho_0} = 2\frac{dg_t}{dt}\sqrt{t}\,, \qquad \frac{d^2 g_{\rho_0}}{d\rho_0^2} = 4t\frac{d^2 g_t}{dt^2} + 2\frac{dg_t}{dt} . \qquad (1.7.57)$$

Replacing the function g_{ρ_0} by g_t modifies the differential Eq. (1.7.56) into

$$t\frac{d^2 g_t}{dt^2} + (1-t)\frac{dg_t}{dt} - \left\{\frac{m_\alpha^2}{4t} + \frac{ikw_z^2}{4}\left[\frac{1}{q_z} + \frac{1}{r_z}\frac{dr_z}{dz}\right]\right\} g_t = 0 . \qquad (1.7.58)$$

We then evaluate the differential quotients $d^2 g_t/dt^2$ and dg_t/dt using $g_t(t) = t^{|m_\alpha|/2} f_t(t)$ and obtain, after some rearrangement, the final differential equation for the function $f_t(t)$,

$$t\frac{d^2 f_t}{dt^2} + (|m_\alpha| + 1 - t)\frac{df_t}{dt} + \left\{-\frac{|m_\alpha|}{2} - \frac{ikw_z^2}{4}\left[\frac{1}{q_z} + \frac{1}{r_z}\frac{dr_z}{dz}\right]\right\} f_t = 0 . \qquad (1.7.59)$$

Thanks to the scaling factor of $\sqrt{2}$ arising from the transformation from ρ_s to ρ_0, the differential Eq. (1.7.59) is similar to the equation

$$t\,\frac{d^2L_n^m(t)}{dt^2} + (m+1-t)\,\frac{dL_n^m(t)}{dt} + n\,L_n^m(t) = 0\,, \tag{1.7.60}$$

where $L_n^m(t)$ are the generalised Laguerre polynomials of azimuthal order m and radial order n where both n and m are non-negative integer numbers.

In terms of the original ρ-coordinate we hence have the solution

$$g_\rho(\rho) = \left(\frac{\rho\sqrt{2}}{w_z}\right)^{|m_\alpha|} L_n^{|m_\alpha|}\left(\frac{2\rho^2}{w_z^2}\right)\,. \tag{1.7.61}$$

The condition for this solution in terms of Laguerre polynomials to exist is that

$$n = -\left\{\frac{|m_\alpha|}{2} + \frac{ikw_z^2}{4}\left[\frac{1}{q_z} + \frac{1}{r_z}\frac{dr_z}{dz}\right]\right\}\,, \tag{1.7.62}$$

where the quantities q_z and w_z are those associated with an elementary Gaussian beam.

In a comparable way as was done for the Gauss–Hermite beam, we solve the differential Eq. (1.7.62) above to obtain the function r_z for a Gauss–Laguerre beam with the result

$$\frac{1}{r_L(z)} = \left\{1+\left(\frac{z}{z_R}\right)^2\right\}^{-1/2} \exp\left\{-i\,(2n+|m_\alpha|+1)\,\arctan\left(\frac{z}{z_R}\right)\right\}\,. \tag{1.7.63}$$

The decrease in axial amplitude of the Gauss–Laguerre beam is the same as for the elementary Gaussian beam. The axial phase function $\phi_L(z)$ of the Gauss–Laguerre beam is given by

$$\phi_L(z) = -\,i\,(2n+|m_\alpha|+1)\,\arctan\left(\frac{z}{z_R}\right)\,. \tag{1.7.64}$$

The Gauss–Laguerre beam is now written in full as

$$\begin{aligned} V_L(\rho,\alpha,z)\,\exp(ikz) = {} & \frac{c_L(n,|m_\alpha|)}{|r_L(z)|}\left(\frac{\rho\sqrt{2}}{w_z}\right)^{|m_\alpha|} L_n^{|m_\alpha|}\left(\frac{2\rho^2}{w_z^2}\right)\exp(im_\alpha\alpha) \\ & \times \exp\left\{\frac{-ik\rho^2}{2}\left(\frac{1}{R_c(z)} - \frac{i\lambda}{\pi w^2(z)}\right)\right\}\exp[i\{kz+\phi_L(z)\}]\,, \end{aligned} \tag{1.7.65}$$

where the normalisation constant $c_L(n,|m_\alpha|)$ is chosen such that

$$\int_0^\infty\int_0^{2\pi} \left|V_L(\rho,\alpha,0)\right|^2 \left(\frac{\rho}{w_0}\right) d\left(\frac{\rho}{w_0}\right) d\alpha = 1\,. \tag{1.7.66}$$

Carrying out this integration and using the expression for the inner product of a Laguerre polynomial $L_n^{|m_\alpha|}$ with itself (see **[266]**), we obtain the normalisation constant of a Gauss–Laguerre beam with mode numbers (n, m_α) as

$$c_L(n,|m_\alpha|) = \left(\frac{2}{\pi}\,\frac{n!}{(n+|m_\alpha|)!}\right)^{1/2}\,. \tag{1.7.67}$$

As indicated previously in this section, the amplitude normalisation constant c'_L, obtained by a spatial integration over the beam waist *without* lateral scaling of the radial coordinate by means of the beam-waist radius w_0, is simply given by $c'_L = c_L/w_0$. As was to be expected, both the Gauss–Hermite and the Gauss–Laguerre beams of lowest order $(0,0)$ yield the normalisation constant of the elementary Gaussian beam, $c(0,0) = (2/\pi)^{1/2}$.

The mode numbers of a Gauss–Laguerre beam are given by the indices n and m_α of the Laguerre polynomial $L_n^{m_\alpha}(2\rho^2/w_z^2)$. The function $g_\rho(\rho)$ is a polynomial in ρ of degree N_ρ=$2n+m_\alpha$. Although the possible number of zeros of $g_\rho(\rho)$ could thus be at maximum N_ρ, the special structure of the function $g_\rho(\rho)$ limits the number of zeros to $2n+1$ over the beam cross-section; the Laguerre polynomial with lower index n has $2n$ roots, whilst the factor ρ^{m_α} delivers an extra zero at the origin.

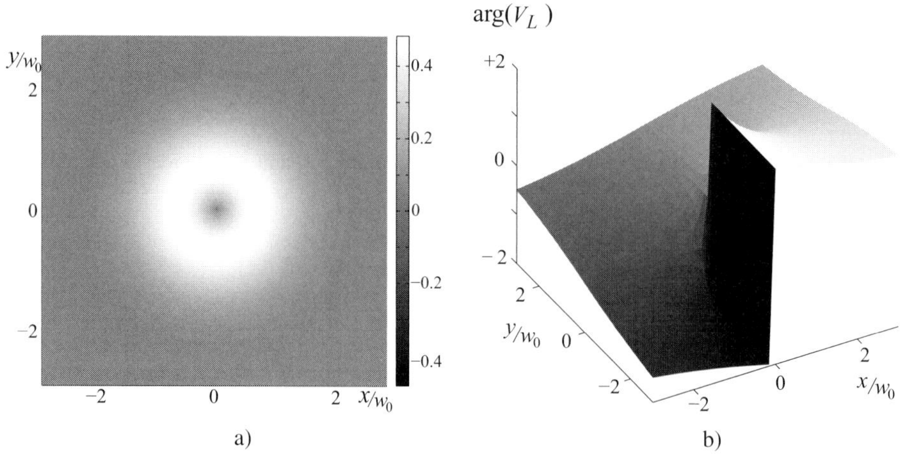

Figure 1.19: The lowest order Gauss–Laguerre beam with $n = 0$ and $m_\alpha=1$. The Cartesian coordinates in the beam-waist cross-section are in units of the beam waist w_0 of an elementary Gaussian beam.
a) Modulus of the $\mathrm{TEM}_{0,1}$ Laguerre mode.
b) Argument of the $\mathrm{TEM}_{0,1}$ mode.

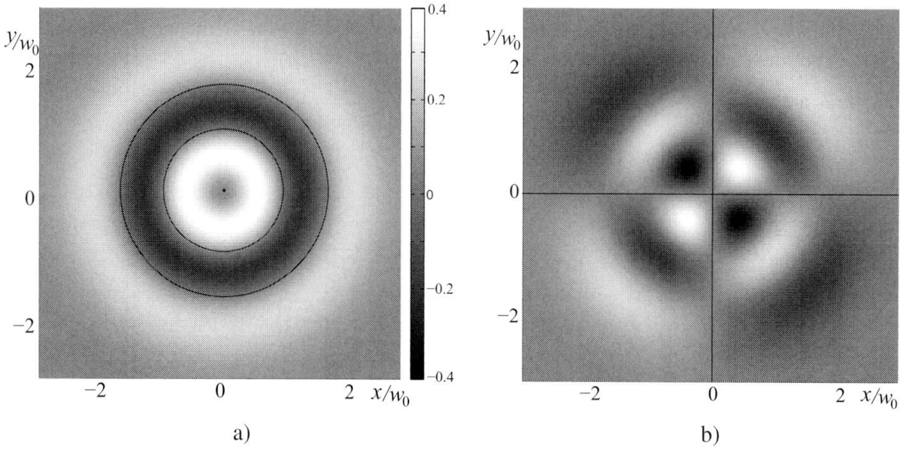

Figure 1.20: Illustrating the complex amplitude at the waist of a Gauss–Laguerre beam. The Cartesian coordinates in the beam-waist cross-section are in units of the beam-waist value w_0.
a) The function $g_\rho(\rho)$ of Eq. (1.7.61) for the beam mode $\mathrm{TEM}_{2,2}$, multiplied by the exponential waist function $\exp(-\rho^2/w_0^2)$.
b) The superposition of the complex amplitude of two Laguerre beams with identical radial mode number but opposite azimuthal number: $\mathrm{TEM}_{2,2} + \mathrm{TEM}_{2,-2}$.

Figure 1.19 shows the modulus and phase of a Gauss–Laguerre beam at the beam waist with mode numbers (0, 1). A single zero is present on axis yielding the doughnut shape of the beam cross-section. The phase of the Gauss–Laguerre beam presents a helical shape with an undetermined value at the origin where the phase assumes all values between $-m_\alpha\pi$ and $+m_\alpha\pi$. The complex amplitude is not undetermined because its modulus is exactly zero on the axis of the beam. The phase of the Gauss–Laguerre beam shows a phase vortex at the origin with a topological charge equal to the azimuthal mode number of the beam. In Chapter 9 we discuss in more detail the significance of this phase vortex for the beam properties. It suffices to say at this point that the Gauss–Laguerre beam produces a laterally integrated flux of (orbital) angular momentum that is given by $m_\alpha P_0/(kc)$ J, where P_0 is the beam power and $k = 2\pi/\lambda$. The beam profile of a higher-order Gauss–Laguerre beam is shown in Fig. 1.20. In graph a) we have shown the function $g_\rho(\rho)$ times the exponentially decreasing function $\exp(-\rho^2/w_z^2)$. The zero crossings of this product have been illustrated by the circles in the figure, together with the location of the phase vortex of this beam at the origin, with topological charge m_α=+2. In a light-generating cavity a combination of Gauss–Laguerre beams can be excited with the beams locked in phase. In graph b) we show the modulus of the complex amplitude for a combination of two beams with indices (2, 2) and (2, −2), respectively, and with a constant phase difference of π. In this case, the resulting higher-order pattern is a so-called second-order *sine*-Gauss–Laguerre beam with zero crossings along the x- and y-axes (solid lines). The summation of two beams with opposite orbital angular momentum gives a combined beam which has no net momentum. Gauss–Laguerre beams have attracted much attention, for instance in high-resolution imaging where the doughnut structure of these beams can be exploited. The orbital momentum of a Gauss–Laguerre beam can also be used as a means for information storage in optical transmission channels.

1.7.5 Gaussian Beam in an Optical Cavity

In this subsection we calculate the Gaussian beam shape that develops in an optical cavity that is part of, for instance, a laser source. After multiple roundtrips of the beam between the mirrors that delimit the cavity, a steady-state Gaussian beam results. We suppose that the wavefronts of this Gaussian beam at the location of the mirrors exactly match the shape of the (spherical) reflecting mirror surfaces. At any point on a mirror surface, the local normal to the incident wavefront and to the mirror surface are parallel and the shape of the wavefront of the reflected wave is an exact copy of that of the incident wave. In Fig. 1.21 we have sketched a cavity with two spherical reflectors and a Gaussian beam that satisfies the steady-state condition, which is normally incident on the cavity mirrors. The waist is located at W_G with an axial coordinate of $z = z_0$. The radii of curvature R_0 and R_1 of the wavefronts of the Gaussian beam at O_0 and O_1 follow from Eq. (1.7.21) as

$$R_{g,0} = z_0\left\{1 + (z_R/z_0)^2\right\}\,, \qquad R_{g,1} = -(L-z_0)\left\{1 + \left[z_r/(L-z_0)\right]^2\right\}\,. \tag{1.7.68}$$

The mirror curvatures are written in terms of a cavity parameter p defined for each mirror by

$$p_0 = 1 - c_0 L\,, \qquad p_1 = 1 + c_1 L\,, \tag{1.7.69}$$

such that p equals zero when the centre of curvature C of a mirror is located at the vertex O of the other mirror. We impose the condition $R_0 = R_{g,0}$ and $R_1 = R_{g,1}$ on the two equations (1.7.68) and find that the distances of the waist W_G to each of the mirrors and the Rayleigh distance of the Gaussian beam are given by

$$\begin{aligned} z_0 &= \frac{(1-p_0)\,p_1}{p_0 + p_1 - 2p_0p_1}\,L\,, \qquad L - z_0 = \frac{p_0\,(1-p_1)}{p_0+p_1-2p_0p_1}\,L\,, \\ z_R^2 &= \frac{p_0p_1(1-p_0p_1)}{(p_0+p_1-2p_0p_1)^2}\,L^2\,. \end{aligned} \tag{1.7.70}$$

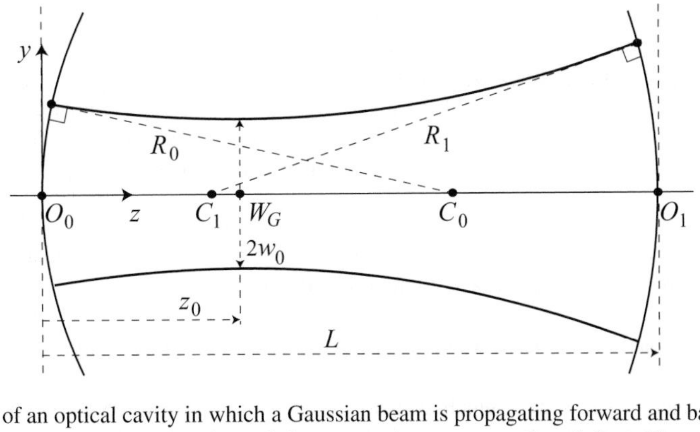

Figure 1.21: The layout of an optical cavity in which a Gaussian beam is propagating forward and backward. The two spherical reflecting surfaces through O_0 and O_1 have their centres of curvature at C_0 and C_1 and have curvatures $c_0 = 1/R_0$ and $c_1 = 1/R_1$, respectively. The cavity length is L; according to the sign convention used, R_1 is negative in the figure.

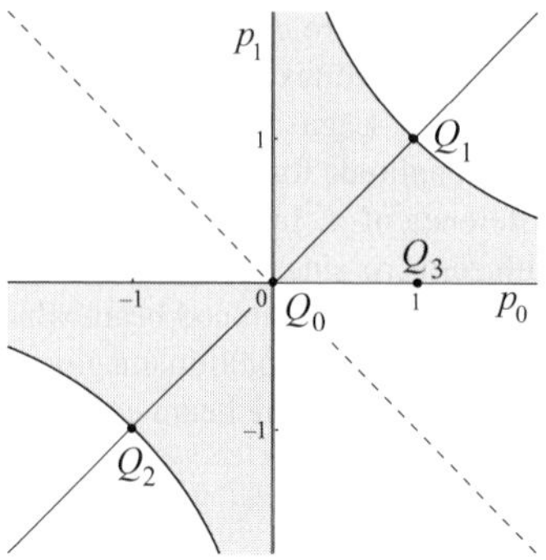

Figure 1.22: Stable region (grey-shaded in the figure) for Gaussian beam generation as a function of the geometric parameters p_0 and p_1 of the optical cavity.

The beam-waist parameter w_0 then follows from Eq. (1.7.23) as

$$w_0^2 = \frac{L\lambda}{\pi} \sqrt{\frac{p_0 p_1 (1 - p_0 p_1)}{(p_0 + p_1 - 2p_0 p_1)^2}} . \tag{1.7.71}$$

Having the position and the size of the beam-waist at our disposal, we calculate the beam parameters w_{M_0} and w_{M_1} at the mirrors with the aid of Eq. (1.7.21) and obtain after some algebra

$$w_{M_0}^2 = \frac{L\lambda}{\pi} \sqrt{\frac{p_1}{p_0 (1 - p_0 p_1)}} , \qquad w_{M_1}^2 = \frac{L\lambda}{\pi} \sqrt{\frac{p_0}{p_1 (1 - p_0 p_1)}} ,$$

or,

$$\frac{w_{M_0}^2}{w_0^2} = \frac{p_0 + p_1 - 2p_0 p_1}{p_0 (1 - p_0 p_1)} , \qquad \frac{w_{M_1}^2}{w_0^2} = \frac{p_0 + p_1 - 2p_0 p_1}{p_1 (1 - p_0 p_1)} . \tag{1.7.72}$$

For the Gaussian beam to be a physical solution it is required that the beam size on each of the mirrors is real and finite. For that reason, the mirror parameters should satisfy the condition

$$0 \le p_0 p_1 \le 1 . \tag{1.7.73}$$

If an equality in Eq. (1.7.73) applies, the beam-waist size, its axial position and other beam parameters are obtained as limiting values of Eq. (1.7.72) when the coordinate pairs (p_0, p_1) are chosen on the delimiting curves or delimiting lines $p_0 p_1 = 1$, $p_0 = 0$, $p_1 = 0$ or at the point $p_0 = p_1 = 0$. Evaluation of the limiting value of the beam waist shows that any point (p_0, p_1) on the delimiting curves leads to an unstable beam solution with $w_0 = 0$. At the point $p_0 = p_1 = 0$, however, the square root of Eq. (1.7.71) takes the finite value of 1/2.

The region within the delimiting curves of stable beam configuration is shown in Fig. 1.22 by the shaded areas, with the coordinate axis $p_0 = 0$, $p_1 = 0$ and the two branches of the hyperbola $p_1 = p_0^{-1}$ acting as delimiting curves.

Some special mirror configurations, all belonging to the symmetrical geometry with $p_0 = p_1$ are:

- Q_0: $p_0 = p_1 = 0$.
  The configuration Q_0 has $c_0 = -c_1 = 1/L$. The cavity length equals $2z_R$, the beam-waist parameter is $w_0 = \sqrt{L\lambda/2\pi}$ and the beam widths on the mirrors are a factor of $\sqrt{2}$ larger than the beam waist.
- Q_1: $p_0 = p_1 = 1$.
  The cavity mirrors are planar and the beam-waist parameter w_0 becomes infinitely large.
- Q_2: $p_0 = p_1 = -1$.
  The configuration Q_2 has $c_0 = -c_1 = 2/L$, which leads to a homocentric mirror geometry. The beam waist is zero and the beam footprint on each mirror is infinitely large. Such a configuration is very sensitive to a perturbation of the axial symmetry, for instance by a tilt of one mirror with respect to the z-axis.

It is clear that the planar and homocentric (or 'confocal') configurations Q_1 and Q_2 and their direct 'stable' neighbourhood should be avoided. The configuration Q_0 is a symmetric one with the smallest variation of beam width over the cavity length. The point Q_0 is on the border of the unstable second and fourth quadrants. An excursion in cavity geometry along the dashed line with $p_0 = -p_1$ immediately leads to a purely imaginary value of w_0^2 with the beam amplitude exponentially increasing for increasing values of ρ. To avoid such a geometric perturbation of the point Q_0 into the unstable neighbouring region, the cavity parameters p_0 and p_1 are preferably chosen on the line segment Q_0Q_1, close to the origin Q_0. This leads to a cavity geometry with the centres of curvature of the mirrors lying (just) outside the cavity.

An example of an asymmetric cavity is represented by the point Q_3 in the figure. The combination of a planar mirror and a mirror with radius equal to L gives rise to a beam with its waist at the planar mirror. The beam waist tends to zero and the beam width on the curved mirror becomes very large. A slightly modified geometry towards the interior of the stable region allows a stable configuration to be found with the beam waist still close to an exit interface of the cavity. The basic theory for Gaussian beams that was developed in this section will be further used in Chapter 4 in the framework of the geometrical optics treatment of this type of optical beams.

1.8 Wave Propagation at an Interface between Two Media

1.8.1 Introduction

The analysis of the reflection and refraction of a plane wave at a planar interface is the basic building block for the study of more complicated medium geometries in which optical waves propagate. In this chapter we derive Snell's law and the Fresnel reflection and transmission coefficients at a planar interface. We also establish the energy balance at an interface. The phenomenon of surface waves and plasmon created bounded waves at surfaces is not treated in this chapter; these special phenomena are discussed in Chapter 3 of this book. With the results obtained for a single interface, we address another important geometry in optics, the sequence of parallel interfaces between media of different refractive indices, a so-called stratified medium, which is frequently used in optical instruments and devices. The combination of thin layers, commonly called an optical coating, can be used to enhance or reduce the transmission or reflection from the base surface. The first application goes back to the 1950s when the industrial application of antireflection coatings became feasible, greatly enhancing the optical contrast in optical imaging instruments like field glasses, microscope objectives and projection lenses. The advent of the He-Ne laser required highly reflecting mirrors with reflection coefficients as large as 0.99 and with a very low absorption. Such mirror surfaces could only be obtained by depositing multilayer structures on a highly polished glass substrate.

More recent developments in the field of multilayers have been triggered by subwavelength and nanostructures and by the development of mirror structures outside the optical spectrum, for instance in the Extreme UV and the soft X-ray wavelength regime. Apart from the continuous stratified layers, micro- and nano-structured layers with one- or two-dimensional patterning have been designed to achieve new properties regarding the reflection, transmission and scattering of light at these structures. A basic understanding of the behaviour of standard multilayers is required to effectively predict and calculate the features of more complicated *structured* multilayers. In this chapter we discuss wave propagation in a multilayer and study transmission and reflection from such a structure. Expressions for the electric and magnetic fields within the layers are developed. Various matrix treatments of the wave propagation in a multilayer are discussed, with special emphasis on the stability of the various matrix representations. The theoretical treatment in this chapter is limited to isotropic and homogeneous media. An extension to anisotropic media, including chiral media (liquid crystals) is possible starting from the same framework.

1.8.2 Transmission and Reflection at a Planar Interface

In this section we study the transmission and reflection of a harmonic plane wave that is incident on a planar boundary between two semi-infinite media. The infinite extension of the two interfacing media avoids any other reflected or incident waves in the two media from interfaces at a finite distance. In order not to complicate the analysis, the media are supposed to be isotropic but they can be absorptive. The analysis will be carried out using the concept of the impedance Z of a medium. Like the common definition used for conductors, we define the impedance of a medium as the ratio between the electric potential and current in that medium. For a harmonic electromagnetic wave, we define the impedance as the ratio of the complex electric $\mathbf{E}_0$ and magnetic field vector $\mathbf{H}_0$, produced by that wave. In terms of the dielectric permittivity and the magnetic permeability, Z equals $\sqrt{\mu/\epsilon}$, see Eq. (1.6.71). The energy absorption of the medium was accounted for by the specific conductivity σ of the medium. In what follows, for the purpose of ease of notation, we extend the notion of impedance and allow a possibly complex value for $Z = \sqrt{\mu/\tilde{\epsilon}}$ with the complex permittivity now given by $\tilde{\epsilon} = \epsilon + i\sigma/\omega$. To establish this more general definition for harmonic waves, we use the differential Maxwell equations (1.6.5) and (1.6.6) and the expression for the curl of a harmonic field component to find the relationships

$$i\mathbf{k} \times \mathbf{H}_0 = -i\omega\tilde{\epsilon}\mathbf{E}_0 , \tag{1.8.1}$$

$$i\mathbf{k} \times \mathbf{E}_0 = i\omega\mu\mathbf{H}_0 . \tag{1.8.2}$$

Taking the vector product of the wave vector $\mathbf{k}$ with Eq. (1.8.1) and using the result of Eq. (1.8.2) we obtain

$$\mathbf{k} \times \mathbf{k} \times \mathbf{H}_0 = -\omega\tilde{\epsilon}\mathbf{k} \times \mathbf{E}_0 = -\omega^2\tilde{\epsilon}\mu\mathbf{H}_0 . \tag{1.8.3}$$

Because of the orthogonality of the triplet ($\mathbf{E}_0$,$\mathbf{H}_0$, $\mathbf{k}$), we readily find the relationship $\mathbf{k} \cdot \mathbf{k} = \omega^2\tilde{\epsilon}\mu$ between the wave vector, wave frequency and medium properties. We then use the fact that the vector $\mathbf{k} \times \mathbf{H}_0$ has a magnitude of kH_0 with k and H_0 the complex numbers that multiply the unit vectors along the $\mathbf{k}$ and $\mathbf{H}_0$ directions of the orthogonal triplet. The vector $\mathbf{k} \times \mathbf{H}_0$ points in the direction opposite to $\mathbf{E}_0$ and this allows us to write

$$Z = \frac{E_0}{H_0} = \frac{k}{\omega\tilde{\epsilon}} = \frac{\omega\sqrt{\tilde{\epsilon}\mu}}{\omega\tilde{\epsilon}} = \sqrt{\frac{\mu}{\tilde{\epsilon}}} . \tag{1.8.4}$$

Alternatively, by taking the vector product of Eq. (1.8.1) with $\mathbf{H}_0$ and the vector product of Eq. (1.8.2) with $\mathbf{E}_0$, we find a relationship between $\mathbf{E}_0 \cdot \mathbf{E}_0$ and $\mathbf{H}_0 \cdot \mathbf{H}_0$. The value of Z then follows from

$$Z = \sqrt{\frac{\mathbf{E}_0 \cdot \mathbf{E}_0}{\mathbf{H}_0 \cdot \mathbf{H}_0}} = \sqrt{\frac{\mu}{\tilde{\epsilon}}} . \tag{1.8.5}$$

In the case of vacuum, we find a real impedance value of 376.73 Ω. From the impedance Z of a medium we find the admittance Y of the medium from $Y = Z^{-1} = \sqrt{\tilde{\epsilon}/\mu}$. The values of the impedance and admittance as defined here are called the intrinsic values of the medium. The impedance (and admittance) are generally complex quantities. The real part is called the *resistance*, the imaginary part the *reactance* or the reactive part of the impedance.

Having defined the intrinsic impedance and admittance of a medium, we now analyse the transmission and reflection of a harmonic wave incident on the boundary between two media with permittivity and permeability values of $(\tilde{\epsilon}_1, \mu_1)$ and $(\tilde{\epsilon}_2, \mu_2)$, respectively. The complex permittivity $\tilde{\epsilon}$ is written as $\tilde{\epsilon} = \epsilon' + i\epsilon''$. Regarding the magnetic permeabilities μ_1 and μ_2, we restrict their generally complex values to be on the real axis in the optical frequency domain.

1.8.3 The Continuity of the Electric and Magnetic Field at a Boundary

We shall now discuss the transition of the electric and magnetic field vectors through a (locally) plane interface. The electric field, electric displacement, magnetic induction and magnetic field vectors below the interface will be denoted by $\mathbf{E}_i$, $\mathbf{D}_i$, $\mathbf{B}_i$ and $\mathbf{H}_i$ respectively, whereas the corresponding quantities above the interface will be denoted by $\mathbf{E}_t$, $\mathbf{D}_t$, $\mathbf{B}_t$ and $\mathbf{H}_t$. We first consider the components of the $\mathbf{D}$ and $\mathbf{B}$ vectors that are perpendicular to the planar interface. With reference to Fig. 1.23 left, we set up a Gaussian pillbox on either side of the boundary. We may then employ Gauss' law for the magnetic induction of Eq. (1.2.4) in the following way. The total flux of the magnetic field can be calculated as the flux through the top and the bottom surfaces and that through the side of the pillbox. However, in what follows we shall make $\Delta h \rightarrow 0$, causing the area of the side to approach zero. Therefore, given the finite magnetic field density, the magnetic flux through that area vanishes. Therefore the total flux is given by

$$\lim_{\Delta h \rightarrow 0} \oiint \mathbf{B} \cdot d\mathbf{A} = \iint \mathbf{B}^{(1)} \cdot \hat{\mathbf{v}}_1 \, dA + \iint \mathbf{B}^{(2)} \cdot \hat{\mathbf{v}}_2 \, dA = 0. \tag{1.8.6}$$

However, as the area of integration is very small, it is reasonable to expect the magnetic field not to vary over it so we have

$$\lim_{\Delta h \rightarrow 0} \oiint \mathbf{B} \cdot d\mathbf{A} = (\mathbf{B}^{(1)} - \mathbf{B}^{(2)}) \cdot \hat{\mathbf{v}} \, \Delta A = 0 , \tag{1.8.7}$$

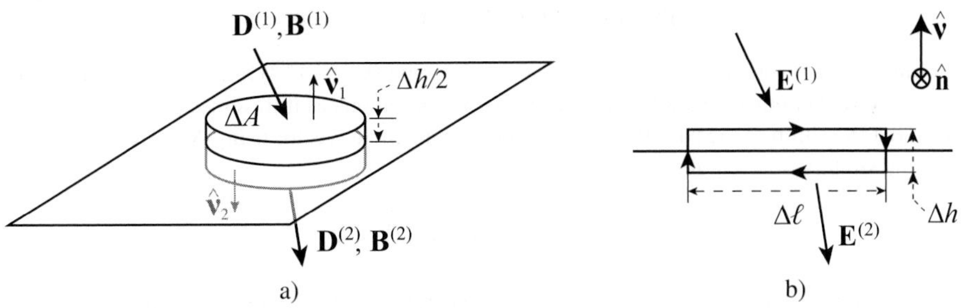

Figure 1.23: a) The surface unit normal of the top surface is denoted by $\hat{\mathbf{v}}_1$ and the corresponding unit normal for the bottom surface is denoted by $\hat{\mathbf{v}}_2 = -\hat{\mathbf{v}}_1$. The (differential) area of the top and bottom surfaces is denoted by ΔA and the height of the pillbox is Δh.
b) The integration path is chosen in a plane that is perpendicular to the interface, using the right-hand corkscrew rule, so its surface normal $\hat{\mathbf{n}}$ points away from the reader. The upper and lower part of the loop with length $\Delta\ell$ are parallel to the interface and the sides of height Δh are perpendicular to it. The loop is placed symmetrically with respect to the interface.

where we have used $\hat{\mathbf{v}} = \hat{\mathbf{v}}_1 = -\hat{\mathbf{v}}_2$, meaning that

$$(\mathbf{B}^{(1)} - \mathbf{B}^{(2)}) \cdot \hat{\mathbf{v}} = 0 \,. \tag{1.8.8}$$

In words, the component of the magnetic field vector that is normal to the surface of the plane interface is continuous through that surface.

Much the same process can be applied to the displacement field. First we define surface charge density σ, already mentioned in Section 1.2, in a more rigorous way:

$$\lim_{\Delta h \to 0} \iiint \rho \, dV = \iint \sigma \, dA \,, \tag{1.8.9}$$

where again ρ is the volume charge density. With this definition we have from Gauss' law written for $\mathbf{D}$ instead of $\mathbf{E}$:

$$\lim_{\Delta h \to 0} \oiint \mathbf{D} \cdot d\mathbf{A} = \iint \mathbf{D}^{(1)} \cdot \hat{\mathbf{v}} \, dA - \iint \mathbf{D}^{(2)} \cdot \hat{\mathbf{v}} \, dA = \iint \sigma \, dA \,. \tag{1.8.10}$$

So by stipulating that the displacement field does not vary within the area ΔA we finally have

$$(\mathbf{D}^{(1)} - \mathbf{D}^{(2)}) \cdot \hat{\mathbf{v}} = \sigma \,, \tag{1.8.11}$$

that is the component of the displacement field that is perpendicular to the plane interface changes abruptly as it traverses through that interface. The magnitude of the change is the surface charge density. In the case of dielectrics, where there are no free charges on the surface, the normal component of the displacement field, and also the electric field, is continuous as it traverses that interface.

Next, we consider the components of the electric and magnetic vectors that are parallel to the plane interface. Since we shall be using Faraday's and Ampère's laws it will be necessary to define a path for the line integrals. With reference to Fig. 1.23b, we may write Faraday's law Eq. (1.2.10), as

$$\oint \mathbf{E} \cdot d\mathbf{l} = -\frac{\partial}{\partial t} \iint \mathbf{B} \cdot d\mathbf{A} \,, \tag{1.8.12}$$

where $\mathbf{l}$ maps the path of the loop and $d\mathbf{A} = dA\,\hat{\mathbf{v}}$. If we make an assumption that the electric field does not vary within the loop we may re-write the left-hand side of the above equation as

$$\oint \mathbf{E} \cdot d\mathbf{l} = \int \mathbf{E}^{(1)} \cdot d\mathbf{l}_1 + \int \mathbf{E}^{(2)} \cdot d\mathbf{l}_2 = (\mathbf{E}^{(1)} \cdot \hat{\mathbf{l}}_1 + \mathbf{E}^{(2)} \cdot \hat{\mathbf{l}}_2)\Delta\ell \,, \tag{1.8.13}$$

where the first integral corresponds to integration along the top of the loop and the second to the bottom of the loop. Note that the component of the electric field that is perpendicular to the sides of the loop does not enter into the line integral and, since we have assumed that the electric field does not vary within the loop, those parallel to the vertical sides must cancel pairwise. Next, we shall make $\Delta h \to 0$ which means that the flux of the magnetic field on the right-hand side of Eq. (1.8.12) vanishes, resulting in

$$(\mathbf{E}^{(1)} - \mathbf{E}^{(2)}) \cdot \hat{\mathbf{l}}_1 = (\mathbf{E}^{(1)} - \mathbf{E}^{(2)}) \times \hat{\mathbf{v}} = \mathbf{0}\,, \tag{1.8.14}$$

because $\mathbf{l}_2 = -\mathbf{l}_1$ and $\hat{\mathbf{v}} \perp \hat{\mathbf{l}}_1$. Note that since $\hat{\mathbf{v}} \perp \hat{\mathbf{n}}$ we can also write $(\mathbf{E}^{(1)} - \mathbf{E}^{(2)}) \times \hat{\mathbf{n}} = \mathbf{0}$. The equation means that the component of the electric field that is parallel to the interface is continuous as it traverses the interface.

Last we look at the transition of the parallel component of the magnetic vector using Ampère's law, Eq. (1.2.6), that is reproduced here in a slightly different format:

$$\oint \mathbf{H} \cdot d\mathbf{l} = \iint \left(\mathbf{J} + \frac{\partial}{\partial t}\mathbf{D}\right) \cdot d\mathbf{A}\,. \tag{1.8.15}$$

Before we can proceed we need to define, just as for surface charge density, the surface current density. In metals most free electrons tend to arrange themselves to the surface of the conductor above the skin depth. Therefore even if the height of the loop tends to zero, $\Delta h \to 0$, the current density $\mathbf{j}$ just below the surface can be significant. In Ampère's law the current density $\mathbf{J}$ is used as a bulk quantity. Therefore,

$$\lim_{\Delta h \to 0} \iint \mathbf{J} \cdot d\mathbf{A} = \lim_{\Delta h \to 0} \iint \mathbf{J} \cdot \hat{\mathbf{n}} dA = \int \mathbf{j} \cdot \hat{\mathbf{n}}\, dl\,. \tag{1.8.16}$$

Equipped with this definition we are ready to use Ampère's law on the loop. The left-hand side of Eq. (1.8.15) gives:

$$\oint \mathbf{H} \cdot d\mathbf{l} = \int \mathbf{H}^{(1)} \cdot d\mathbf{l}_1 + \int \mathbf{H}^{(2)} \cdot d\mathbf{l}_2 = \int \left(\mathbf{H}^{(1)} - \mathbf{H}^{(2)}\right) \cdot \hat{\mathbf{l}}_1\, dl\,, \tag{1.8.17}$$

because, just as before, we assume that the magnetic field $\mathbf{H}$ does not vary within the area of the loop and so the contributions from the left-hand and right-hand sides exactly cancel. Upon reducing the height of the loop to zero the flux of the displacement field in the right-hand side of Eq. (1.8.15) yields zero and so we have

$$\int \left(\mathbf{H}^{(1)} - \mathbf{H}^{(2)}\right) \cdot \hat{\mathbf{l}}_1 dl = \int \mathbf{j} \cdot \hat{\mathbf{n}}\, dl\,, \tag{1.8.18}$$

which is only possible if

$$\left|\left(\mathbf{H}^{(1)} - \mathbf{H}^{(2)}\right) \times \hat{\mathbf{v}}\right| = |\mathbf{j} \cdot \hat{\mathbf{n}}|\,. \tag{1.8.19}$$

Equation (1.8.19) states that the component of the magnetic field that is parallel to the interface, changes abruptly upon traversing the interface with the magnitude of change being the surface current density. In dielectrics, when no surface currents are possible due to lack of unbound electrons, the parallel component of the magnetic field, and therefore the magnetic induction field $\mathbf{B}$ is continuous as it traverses the interface.

1.8.4 Global Matching of the Fields at the Boundary

In order to study the transmission and reflection of a plane wave through a plane interface it is necessary to consider the state of polarisation of that plane wave. For this, let us first define the plane of incidence as the plane containing the wave vector $\mathbf{k}_i$ and the surface normal of the boundary of the two media (Fig. 1.24). A general state of polarisation of the incident wave can be decomposed into two orthogonal components of the electric field: one that is parallel to the plane of incidence (p-polarisation or transverse magnetic (TM) polarisation, see Fig. 1.24), and one that is perpendicular to this plane (s-polarisation from the German adjective *senkrecht* or transverse electric (TE) polarisation, see Fig. 1.25). Thus, any incident wave with an arbitrary state of polarisation can be decomposed into s- and p-polarisation states and the transmitted and reflected components are then recombined to produce the complete transmitted and reflected waves. The propagation direction of the incident wave is chosen such that the scalar product $\mathbf{k}_i \cdot \hat{\mathbf{z}}$ is positive (wave progression in the $+z$-direction).

The flow of the optical energy is determined by the time-averaged value of the Poynting vector $\mathbf{S}$ which is equal to the vector product of the electric and magnetic field vectors (see Eq. (1.4.5)). In complex notation, the time-averaged value of the product of two complex quantities is given by $\Re\{\mathbf{A}_1\, \mathbf{A}_2^*\}/2$ (see Eq. (A.6.4) of Appendix A). The reference directions for the mutually perpendicular complex field vectors $\mathbf{E}_{0,i}$ and $\mathbf{H}_{0,i}$ of the incident plane wave are chosen such that the corresponding time-averaged value of the Poynting vector, $\langle \mathbf{S} \rangle$, with its power per unit surface given by $\Re\{\mathbf{E}_{0,i} \times \mathbf{H}_{0,i}^*\}/2$, points in the direction of the wave vector $\mathbf{k}_i$. A comparable choice is made for the reflected and transmitted field vectors.

The boundary conditions following from Maxwell's equations at the interface require global continuity of the components of the electric and magnetic field vectors that are tangent to the boundary surface. The complex amplitude

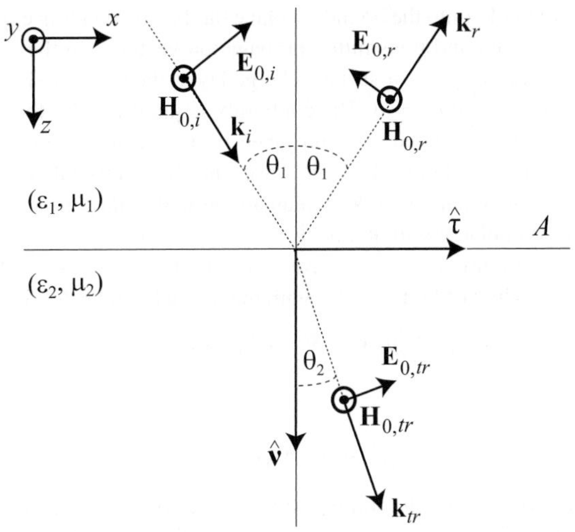

Figure 1.24: A planar interface A, perpendicular to the plane of the drawing, separates two semi-infinite media with permittivities and permeabilities (ϵ_1, μ_1) and (ϵ_2, μ_2), respectively. The unit normal vector to the plane interface is $\hat{\mathbf{v}}$. The incident plane wave propagates in the direction of $\mathbf{k}_i$, the reflected and transmitted waves have wave vectors $\mathbf{k}_r$ and $\mathbf{k}_{tr}$. The plane of the drawing is the plane of incidence defined by the vectors $\hat{\mathbf{v}}$ and $\mathbf{k}_i$. The unit vector $\hat{\boldsymbol{\tau}}$ is located in the plane of incidence and tangent to the interface. The electric and magnetic fields associated with the incident, reflected and transmitted waves are denoted by $(\mathbf{E}_{0,i}, \mathbf{H}_{0,i})$, $(\mathbf{E}_{0,r}, \mathbf{H}_{0,r})$ and $(\mathbf{E}_{0,tr}, \mathbf{H}_{0,tr})$, respectively. A special state of polarisation (p-polarisation) has been depicted in the drawing and the arrows denote the positive directions for the field vectors.

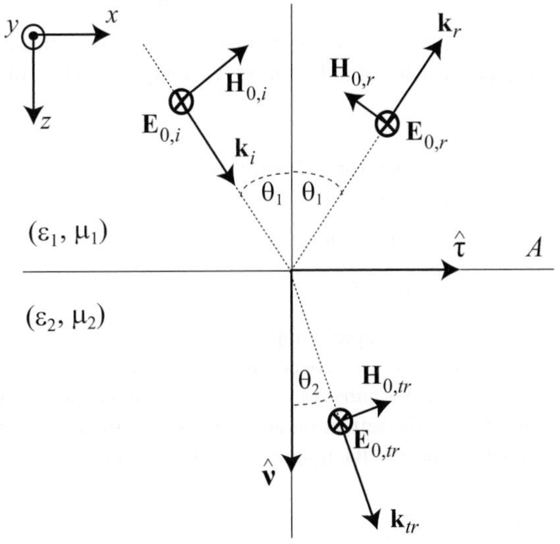

Figure 1.25: As Fig. 1.24, but now for the case of an s-polarised plane wave with the electric vectors $\mathbf{E}_{0,i}$, $\mathbf{E}_{0,r}$ and $\mathbf{E}_{0,tr}$ pointing toward the rear side of the plane of the drawing.

patterns in the boundary plane, produced by the incident, reflected and transmitted wave, have a temporal-spatial dependence that is given by

$$\exp\left\{i(\mathbf{k}_{i,t} \cdot \mathbf{r} - \omega_i t\right\} \,, \quad \exp\left\{i(\mathbf{k}_{r,t} \cdot \mathbf{r} - \omega_r t\right\} \,, \quad \exp\left\{i(\mathbf{k}_{tr,t} \cdot \mathbf{r} - \omega_t t\right\} \,,$$
$$\mathbf{k}_{i,t} = \hat{\mathbf{v}} \times \mathbf{k}_i \times \hat{\mathbf{v}} = \mathbf{k}_i - (\mathbf{k}_i \cdot \hat{\mathbf{v}})\; \hat{\mathbf{v}} \,, \qquad (1.8.20)$$

where $\mathbf{k}_{i,t}$ is the projection of the vector $\mathbf{k}_i$ onto the boundary plane (and corresponding expressions for the other tangential components $\mathbf{k}_{r,t}$ and $\mathbf{k}_{tr,t}$ of the reflected and transmitted propagation vectors). In the case of harmonic waves incident at a boundary, with no special surface cladding or surface charge layer, the transmission and reflection process is linear and the optical frequencies ω_r and ω_t equal $\omega_i = \omega$. The continuity requirement for the tangential electric and magnetic field components at the interface for any spatial position given by the vector $\mathbf{r}$ requires the identity of the projected wave vectors $\mathbf{k}_{i,t}$, $\mathbf{k}_{r,t}$ and $\mathbf{k}_{tr,t}$. From the identity $\mathbf{k}_{i,t} = \mathbf{k}_{r,t}$, we find the nontrivial solution $\mathbf{k}_r = \mathbf{k}_i - 2\,(\mathbf{k}_i \cdot \hat{\mathbf{v}})\ \hat{\mathbf{v}}$. The components of the vector $\mathbf{k}_r$ and the incident vector $\mathbf{k}_i$, which are normal to the surface, have opposite signs. The vector equation for $\mathbf{k}_r$ also shows that $\mathbf{k}_r$ is coplanar with $\mathbf{k}_i$ and $\hat{\mathbf{v}}$.

For the projected vector of the transmitted wave we have to satisfy the identity $\mathbf{k}_{i,t} = \mathbf{k}_{tr,t}$. Using the general property $\mathbf{k} \cdot \mathbf{k} = \omega^2 \epsilon\mu$ and taking the cross product of both vectors with the normal unit vector $\hat{\mathbf{v}}$ we have

$$\sqrt{\epsilon_1\mu_1}\ \hat{\mathbf{k}}_i \times \hat{\mathbf{v}} = \sqrt{\epsilon_2\mu_2}\ \hat{\mathbf{k}}_t \times \hat{\mathbf{v}}\ ,$$

or

$$\sqrt{\epsilon_1\mu_1}\ \sin\theta_1 = \sqrt{\epsilon_2\mu_2}\ \sin\theta_2\ , \tag{1.8.21}$$

where $\hat{\mathbf{k}}_i$ and $\hat{\mathbf{k}}_t$ are the unit wave vectors and θ_1 and θ_2 are the angles between the normal vector and the incident and transmitted wave vector, respectively. The quantities ϵ and μ are formally allowed to be complex; for ease of notation, we have suppressed the tilde-sign that was meant to stress their complex value. The first line of Eq. (1.8.21) states that the transmitted wave vector $\mathbf{k}_t$ is coplanar with $\mathbf{k}_i$ and $\hat{\mathbf{v}}$. In the analysis of wave transmission and reflection at an interface, we use general macroscopic medium properties that require, in essence, complex values for both ϵ and μ. This approach allows a treatment that is valid over the full electromagnetic spectrum. In the special case of negative real values, it allows the treatment of so-called 'ideal' metamaterials. For the optical regime, we restrict ourselves to the case $\mu = \mu_0$. With an absorption-free medium at optical frequencies, we then have the property $\epsilon = n^2\epsilon_0$ where n is the real refractive index of the medium. The second line of Eq. (1.8.21) then reduces to Snell's law of refraction, $n_1 \sin\theta_1 = n_2 \sin\theta_2$.

1.8.5 Amplitude Matching: Reflection and Transmission Coefficients

The local matching of the amplitudes of the electric and magnetic field components that are tangent to the interface needs a separate treatment for the two orthogonal states of polarisation. We consider each in turn now.

1.8.5.1 Electric p-polarisation (TM)

Considering the incident wave with p-polarisation (or TM wave) in Fig. 1.24 we write the continuity conditions for the electrical and magnetic tangential field components:

$$\left(\mathbf{E}_{0,i} - \mathbf{E}_{0,r}\right) \cdot \hat{\boldsymbol{\tau}} = \mathbf{E}_{0,tr} \cdot \hat{\boldsymbol{\tau}}\ , \tag{1.8.22}$$

$$Y_{m,1}\left(E_{0,i} + E_{0,r}\right) = Y_{m,2}E_{0,tr}\ , \tag{1.8.23}$$

where the italic symbols $E_{0,i}$ etc. denote the complex amplitude of the corresponding bold vector quantities. In the equation above we have used the relation $H = Y_m E$ between the amplitudes of the electric and magnetic field vectors; Y_m was defined earlier in Subsection 1.8.2 as the *intrinsic* (complex) admittance of the medium in which the fields are present. Using the angle of incidence θ_1 and the refraction angle θ_2 we find the following ratios of the reflected and the transmitted electric field amplitude with respect to the amplitude of the incident wave (p-polarisation):

$$r_p = \left.\frac{E_{0,r}}{E_{0,i}}\right|_p = \frac{\dfrac{Y_{m,2}}{\cos\theta_2} - \dfrac{Y_{m,1}}{\cos\theta_1}}{\dfrac{Y_{m,2}}{\cos\theta_2} + \dfrac{Y_{m,1}}{\cos\theta_1}} = \frac{Y_{p,2} - Y_{p,1}}{Y_{p,2} + Y_{p,1}}\ , \tag{1.8.24}$$

$$t_p = \left.\frac{E_{0,tr}}{E_{0,i}}\right|_p = \left(\frac{Y_{m,1}}{Y_{m,2}}\right)\frac{2\,\dfrac{Y_{m,2}}{\cos\theta_2}}{\dfrac{Y_{m,2}}{\cos\theta_2} + \dfrac{Y_{m,1}}{\cos\theta_1}} = \left(\frac{\cos\theta_1}{\cos\theta_2}\right)\frac{2\ Y_{p,1}}{Y_{p,2} + Y_{p,1}}\ , \tag{1.8.25}$$

where we have used the quantity $Y_p = Y_m/\cos\theta$ for the *effective* admittance of a medium in the case of p- or TM-polarisation. The reflection and transmission coefficients for the H-fields are immediately obtained by using the admittance relation $H = Y_m E$ in each medium.

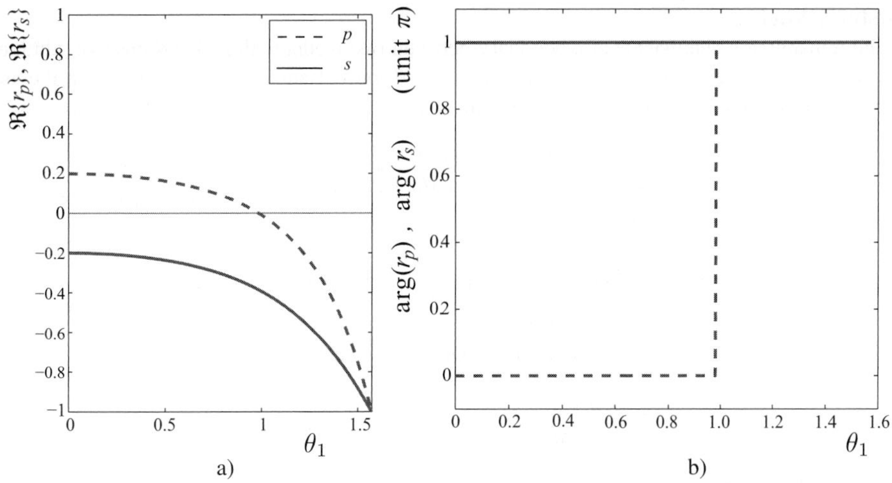

Figure 1.26: An incident wave strikes an interface with an optically denser medium. An optical contrast of $n_2/n_1 = 1.5$ has been used.
a) Real parts of the (amplitude) reflection coefficients r_p (dashed) and r_s (solid line) as a function of the angle of incidence θ_1.
b) Phase of the reflected p- and s-waves in units of π.

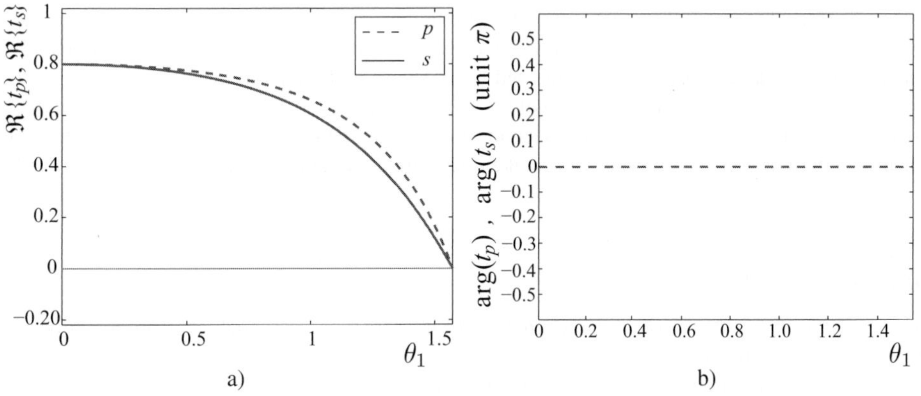

Figure 1.27: As Fig. 1.26, but now for the transmitted p- and s-polarised waves.

1.8.5.2 Electric s-polarisation (TE)

Comparable considerations regarding the continuity of the tangential field vectors can be made for the case of s- or TE-polarisation of the incident wave (see Fig. 1.25). We find the following results for the amplitude reflection and transmission coefficients:

$$r_s = \left.\frac{E_{0,r}}{E_{0,i}}\right|_s = -\frac{Y_{m,2}\cos\theta_2 - Y_{m,1}\cos\theta_1}{Y_{m,2}\cos\theta_2 + Y_{m,1}\cos\theta_1} = -\frac{Y_{s,2} - Y_{s,1}}{Y_{s,2} + Y_{s,1}} , \tag{1.8.26}$$

$$t_s = \left.\frac{E_{0,tr}}{E_{0,i}}\right|_s = \frac{2\,Y_{m,1}\cos\theta_1}{Y_{m,2}\cos\theta_2 + Y_{m,1}\cos\theta_1} = \frac{2\,Y_{s,1}}{Y_{s,2} + Y_{s,1}} , \tag{1.8.27}$$

where the effective admittance in the case of s-polarisation is given by $Y_s = Y_m \cos\theta$. In Fig. 1.26 the real values of the complex reflection coefficients r_p and r_s have been plotted, together with their arguments, for the transition towards an optically denser medium. Figure 1.27 shows the same graphs for the transmitted wave.

1.8.5.3 Brewster's Angle

The reflection and transmission coefficients at a boundary between two media with real and positive admittances (and $\mu_1 = \mu_2 = \mu_0$), can also be written in terms of the angles θ_1 and θ_2 only. Using Snell's law and with the aid of some trigonometric manipulation we have from Eqs. (1.8.24)–(1.8.27),

$$
\begin{aligned}
r_p &= \frac{\sin 2\theta_1 - \sin 2\theta_2}{\sin 2\theta_1 + \sin 2\theta_2} = \frac{\tan(\theta_1 - \theta_2)}{\tan(\theta_1 + \theta_2)} ,\\
r_s &= -\frac{\sin(\theta_1 - \theta_2)}{\sin(\theta_1 + \theta_2)} ,\\
t_p &= \frac{4\cos\theta_1 \sin\theta_2}{\sin 2\theta_1 + \sin 2\theta_2} = \frac{2\cos\theta_1 \sin\theta_2}{\sin(\theta_1 + \theta_2)\cos(\theta_1 - \theta_2)} ,\\
t_s &= \frac{2\cos\theta_1 \sin\theta_2}{\sin(\theta_1 + \theta_2)} .
\end{aligned}
\tag{1.8.28}
$$

These expressions for the amplitude reflection and transmission coefficients at an interface were first given by Fresnel and they are commonly called the Fresnel coefficients. A special case arises for the amplitude of the reflected wave with a p-polarised incident wave. The numerator of r_p in Eq. (1.8.24) becomes zero at the particular angle of incidence when $Y_{p,1} = Y_{p,2}$. From Eq. (1.8.28) we find the condition $\tan(\theta_1 + \theta_2) \rightarrow \infty$ or $\theta_1 + \theta_2 = \pi/2$. The wave vectors of the reflected and transmitted wave are perpendicular to each other. The substitution of this relation into Snell's law yields $\tan\theta_1 = n_2/n_1$.

The particular angle with zero reflection is seen in Fig. 1.26. For an optical contrast of 1.5, the angle θ_1 equals 0.313π. This angle, called Brewster's angle (θ_B) after the scholar who discovered it, allows a fully s-polarised reflected beam to be created from an incident beam of light which is unpolarised. In this respect, an *unpolarised* beam of light is a beam of which the azimuth α of the (linear) state of polarisation changes randomly over time with a uniform probability function $f(\alpha)$ in the interval $0 \le \alpha \le \pi$. The phase of the reflected beam shows a phase jump of π when the angle of incidence passes through the value θ_B. The physical explanation of the zero reflection at the Brewster angle is the fact that the dipole axis of an oscillating dipole at the boundary in the second medium points in the direction of the reflected wave vector. Since the radiation pattern of a dipole becomes zero in the direction of the dipole axis, no optical power can be radiated back into the first medium.

1.8.6 Total Reflection

Inspection of Snell's law shows that $\sin\theta_2$ is larger than unity if $n_1 \sin\theta_1/n_2 > 1$; such a situation can occur if $n_2 < n_1$. The angle of incidence corresponding to $\sin\theta_1 = n_2/n_1$ is called the critical angle. At angles of incidence larger than the critical angle, the wave penetrating in the second medium is non-propagating. To analyse the particularities of the reflected and transmitted wave beyond the critical angle, we suppose that the vector normal to the boundary points in the positive z-direction. The complex amplitude of the incident wave of Fig. 1.24 is written

$$
\mathbf{E}_i(\mathbf{r}) = \mathbf{E}_{0,i} \exp\{i(k_{i,x}x + k_{i,z}z)\} . \tag{1.8.29}
$$

We choose the plane of incidence to be parallel to the plane $y = 0$. Without loss of generality, the boundary between the two media is chosen as the plane $z = 0$. The angle of incidence θ_1 is chosen in such a way that $k_{i,x}^2 = (k_i \sin\theta_1)^2 > \epsilon_2\mu_2\omega^2$. The refraction into the second medium leaves the tangential component of the wave vector unchanged. According to the dispersion relation for a non-conducting medium (see Eq. (1.6.62)), the z-component of the transmitted wave vector $\mathbf{k}_{tr}$ is given by

$$
k_{tr,z} = (\omega^2\epsilon_2\mu_2 - k_i^2 \sin^2\theta_1)^{1/2} = +i\sqrt{k_i^2 \sin^2\theta_1 - \omega^2\epsilon_2\mu_2} . \tag{1.8.30}
$$

The amplitude of the transmitted wave is z-dependent; the choice of the sign of the square root is such that the wave amplitude decreases when the distance from the boundary increases, in our case by propagation in the positive z-direction. To calculate the amplitudes of the reflected and transmitted waves, we need the *effective* admittance values of the waves in the two media. We start with the case of parallel polarisation.

1.8.6.1 Electric p-polarisation (TM)

The effective admittance of the second medium for the transmitted wave, $Y_{p,2} = \sqrt{\epsilon_2/\mu_2} / \cos\theta_2$, needs special attention. In general we have $k_{tr,z} = k_{tr}\cos\theta_2 = \omega\sqrt{\epsilon_2\mu_2}\cos\theta_2$ and, using Eq. (1.8.30), we find

$$\cos\theta_2 = i\sqrt{\frac{\epsilon_1\mu_1\sin^2\theta_1}{\epsilon_2\mu_2} - 1}\,, \tag{1.8.31}$$

which is a purely imaginary quantity in the case of total reflection. We write the expressions for r_p and t_p of Eqs. (1.8.24) and (1.8.25) as

$$r_p = \frac{\cos\theta_1 - \sqrt{\epsilon_1\mu_2/\epsilon_2\mu_1}\,\cos\theta_2}{\cos\theta_1 + \sqrt{\epsilon_1\mu_2/\epsilon_2\mu_1}\,\cos\theta_2}\,, \tag{1.8.32}$$

$$t_p = 2\sqrt{\frac{\epsilon_1\mu_2}{\epsilon_2\mu_1}}\,\frac{\cos\theta_1}{\cos\theta_1 + \sqrt{\epsilon_1\mu_2/\epsilon_2\mu_1}\,\cos\theta_2}\,. \tag{1.8.33}$$

When the electric permittivities and magnetic permeabilities of the two media are real and positive, the reflection coefficient, r_p, takes the special form $r_p = (a - ib)/(a + ib)$ where a and b are positive numbers. The modulus of r_p is hence unity and the argument is given by $\phi_{r,p} = -2\ \arctan b/a$,

$$|r_p| = 1\,, \qquad \phi_{r,p} = -2\ \arctan\sqrt{\frac{\epsilon_1^2\sin^2\theta_1/\epsilon_2^2 - (\epsilon_1\mu_2/\epsilon_2\mu_1)}{1 - \sin^2\theta_1}}\,. \tag{1.8.34}$$

The corresponding expressions for the transmitted wave are

$$|t_p| = 2\sqrt{\frac{\epsilon_1\mu_2}{\epsilon_2\mu_1}}\sqrt{\frac{1 - \sin^2\theta_1}{1 - (\epsilon_1\mu_2/\epsilon_2\mu_1) - \sin^2\theta_1\left(\epsilon_2^2 - \epsilon_1^2\right)/\epsilon_2^2}}\,, \qquad \phi_{t,p} = \frac{1}{2}\phi_{r,p}\,. \tag{1.8.35}$$

In the optical regime of the electromagnetic spectrum where $\mu_1 = \mu_2 = \mu_0$ and for real and positive refractive indices $n_1 = (\epsilon_1)^{1/2}$ and $n_2 = (\epsilon_2)^{1/2}$, we find

$$|r_p| = 1\,, \qquad \phi_{r,p} = -2\ \arctan\left\{\frac{n_1}{n_2}\sqrt{\frac{n_1^2\sin^2\theta_1/n_2^2 - 1}{1 - \sin^2\theta_1}}\right\}, \tag{1.8.36}$$

$$|t_p| = 2\frac{n_1 n_2}{\sqrt{n_1^2 - n_2^2}}\sqrt{\frac{1 - \sin^2\theta_1}{(n_1^2 + n_2^2)\sin^2\theta_1 - n_2^2}}\,, \qquad \phi_{t,p} = \frac{1}{2}\phi_{r,p}\,. \tag{1.8.37}$$

In the regime of total internal reflection, the calculation of the field components of the transmitted wave also needs some extra attention. The tangential component of the transmitted wave is given by $k_{tr,x} = k_i\sin\theta_1$ and the z-component is given by Eq. (1.8.30). The electric field strength of the transmitted wave in the second medium is formally written as

$$\mathbf{E}_{tr}(\mathbf{r}) = \mathbf{E}_{0,tr}\ \exp\{i(k_{t,x}x + k_{tr,z}z)\}\,, \tag{1.8.38}$$

where $\mathbf{E}_{0,tr}$ is the electric field vector of the transmitted wave as shown in Fig. 1.24. Using the property $\nabla\cdot\mathbf{D} = 0$ for a source-free region, we have in the second medium

$$k_{t,x}E_{0tr,x} + k_{tr,z}E_{0tr,z} = 0\,. \tag{1.8.39}$$

This allows us to write the electric field components of the transmitted wave

$$E_{0tr,x} = \frac{k_{tr,z}}{k_{tr}}E_{0,tr}\,, \qquad E_{0tr,z} = -\frac{k_{tr,x}}{k_{tr}}E_{0,tr}\,. \tag{1.8.40}$$

$E_{0,tr}$ is the field strength of the transmitted wave and equals $t_pE_{0,i}$. The wavenumber of the transmitted wave is k_{tr} and it is given by $(\mathbf{k}_{tr}\cdot\mathbf{k}_{tr})^{1/2}$. The choice of the signs of $E_{0tr,x}$ and $E_{0tr,z}$ is such that they are in line with the field vectors depicted in Fig. 1.24, where the normal vector $\hat{\mathbf{v}}$ points in the positive z-direction.

Using the relations above, we obtain the following expression for the complex amplitude of the transmitted wave as a function of the distance z from the interface,

$$\mathbf{E}_{tr}(\mathbf{r}) = \frac{|t_p|\exp(i\phi_{t,p})E_{0,i}}{k_{tr}}\left(k_{tr,z}\,\hat{\mathbf{x}} - k_{tr,x}\,\hat{\mathbf{z}}\right)\exp\left\{-k_0\sqrt{n_1^2\sin^2\theta_1 - n_2^2}\,z\right\}\exp\{ik_{i,x}x\}\,, \quad (1.8.41)$$

where $E_{0,i}$ is the electric field strength of the incident wave and $k_0 = \omega\sqrt{\epsilon_0\mu_0}$, the wavenumber in vacuum. To calculate the magnetic field vector, we use $i\mathbf{k}_{tr} \times \mathbf{E}_{tr} = i\omega\mu\mathbf{H}_{tr}$ yielding the expression

$$\mathbf{H}_{tr}(\mathbf{r}) = \frac{|t_p|\exp(i\phi_{t,p})n_2E_{0,i}}{\mu_0 c}\,\hat{\mathbf{y}}\,\exp\left\{-k_0\sqrt{n_1^2\sin^2\theta_1 - n_2^2}\,z\right\}\exp\{ik_{i,x}x\}\,, \quad (1.8.42)$$

where we have used the fact that $(\epsilon_2/\mu_2)^{1/2} = n_2/(\mu_0 c)$ in the optical domain of the spectrum and $\hat{\mathbf{y}}$ is a unit vector in the y-direction.. The transmitted wave is exponentially decreasing as a function of the distance from the boundary. The $1/e$ distance for the wave amplitude is given by

$$|z_{1/e}| = \frac{\lambda_0}{2\pi\sqrt{n_1^2\sin^2\theta_1 - n_2^2}}\,. \quad (1.8.43)$$

The transmitted wave is an inhomogeneous wave with the equiphase surfaces and the equi-amplitude surfaces perpendicular to each other (see also Fig. 1.10). There is no net power transport by the transmitted wave into the second medium. The wave only has an appreciable amplitude over a distance of the order of the wavelength of the light, measured normal to the boundary. At larger distances from the boundary, the wave amplitude becomes negligible. The damping of the transmitted wave increases with the angle of incidence θ_1 and is maximum at grazing incidence on the boundary surface.

If a third medium with a refractive index value which is comparable to that of the denser medium is brought close to the medium boundary to which the inhomogeneous wave is bound, the attenuated wave is transmitted into the third medium as a homogeneous propagating wave. The attenuation depends on the perpendicular distance between the two (parallel) boundaries, on the state of polarisation and on the wavelength of the light. This phenomenon of light tunnelling through a low index gap between two higher index media in which normally total internal reflection would occur is commonly called *frustrated* total internal reflection. An example of light tunnelling is illustrated in Fig. 1.28d where the flat faces of two rectangular prisms are brought almost into contact. A measurement of the ratio of transmitted and reflected light allows a sub-wavelength evaluation of the gap thickness d. Alternatively, a variation of the gap thickness allows the intensity of the transmitted or reflected light to be modulated.

1.8.6.2 Electric *s*-polarisation (*TE*)

The expressions for r_s and t_s for the case of real and positive values of the electric permittivity and magnetic permeability are given by

$$|r_s| = 1\,, \qquad \phi_{r,s} = -2\arctan\sqrt{\frac{\mu_1^2\sin^2\theta_1/\mu_2^2 - (\mu_1\epsilon_2/\mu_2\epsilon_1)}{1-\sin^2\theta_1}}\,. \quad (1.8.44)$$

The expressions for the transmission coefficient and the phase of the electric vector of the transmitted wave are

$$|t_s| = 2\sqrt{\frac{1-\sin^2\theta_1}{1-(\mu_1\epsilon_2/\mu_2\epsilon_1) - \sin^2\theta_1\left(\mu_2^2-\mu_1^2\right)/\mu_2^2}}\,, \qquad \phi_{t,s} = \frac{1}{2}\phi_{r,s}\,. \quad (1.8.45)$$

Upon comparing the result for $|t_p|$ of Eq. (1.8.35) and $|t_s|$ above we note the analogous format of the two expressions in which the role of the electric permittivity and the magnetic permeability are interchanged. Identical expressions are obtained if we compare the transmission coefficient for the electric vector for *TE*-polarisation with the transmission coefficient for the magnetic vector in the case of *TM*-polarisation.

Using the ϵ and μ values for optical wavelengths that were discussed above, we have the expressions

$$|r_s| = 1\,, \qquad \phi_{r,s} = -2\arctan\left\{\frac{n_2}{n_1}\sqrt{\frac{n_1^2\sin^2\theta_1/n_2^2 - 1}{1-\sin^2\theta_1}}\right\}\,, \quad (1.8.46)$$

$$|t_s| = 2\,n_1\sqrt{\frac{1-\sin^2\theta_1}{n_1^2-n_2^2}}\,, \qquad \phi_{t,s} = \frac{1}{2}\phi_{r,s}\,. \quad (1.8.47)$$

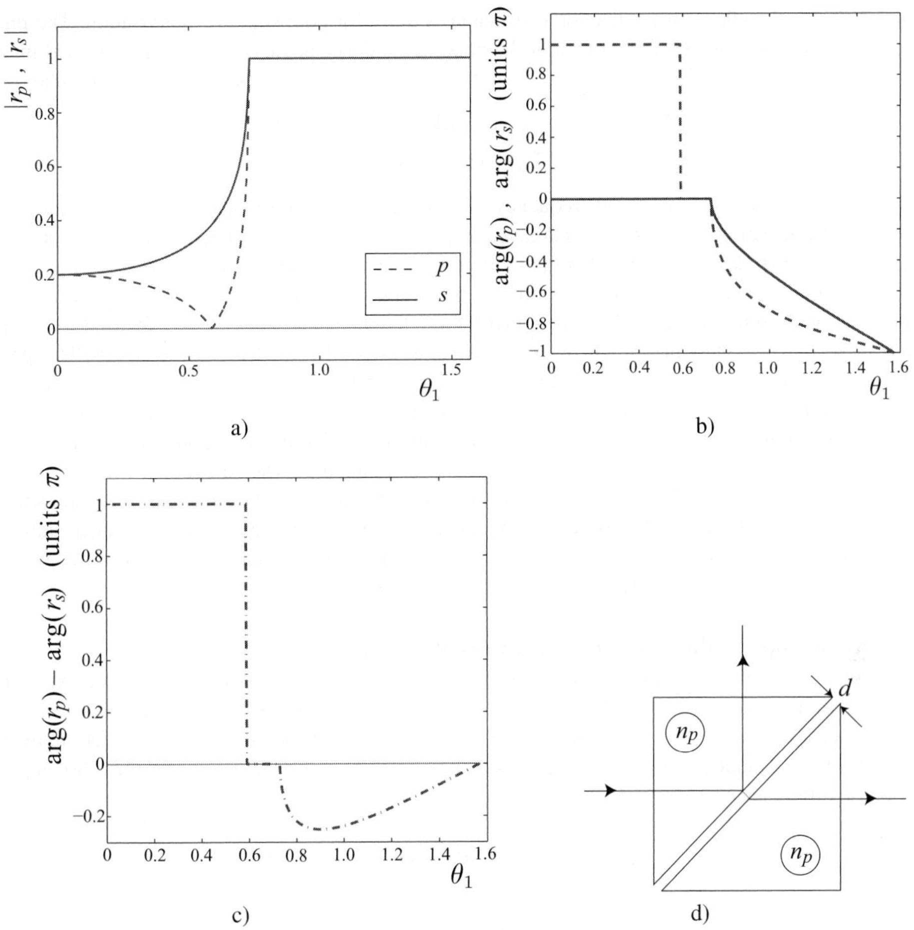

Figure 1.28: The moduli of the reflection coefficients r_p (dashed) and r_s (solid line) and their phases as a function of the angle of incidence θ_1. The incident wave strikes a boundary with an optically *less dense* medium. The optical contrast is given by $n_2/n_1 = 2/3$. The Brewster angle for the reflected p-polarised wave occurs at an angle θ_1 of 0.588 radians. Total reflection occurs at angles of incidence $\theta_1 \geq 0.73$ radians.
a) Moduli of the s- and p-reflection coefficients.
b) The phase of the reflected s- and p-waves.
c) Phase difference between an s-polarised reflected beam and a p-polarised reflected beam.
d) *Frustrated* total internal reflection with the aid of two prisms with index n_p positioned in virtual contact with a sub-wavelength air gap thickness of d.

Without further details we give the result for the electric and the magnetic field vectors of the transmitted s-polarised wave,

$$\mathbf{E}_{tr}(\mathbf{r}) = -\,|t_s| \exp(i\phi_{t,s}) E_{0,i}\, \hat{\mathbf{y}} \exp\left\{-k_0 \sqrt{n_1^2 \sin^2\theta_1 - n_2^2}\, z\right\} \exp\{ik_{i,x}x\}\,, \tag{1.8.48}$$

$$\mathbf{H}_{tr}(\mathbf{r}) = \frac{|t_s| \exp(i\phi_{t,s}) n_2 E_{0,i}}{\mu_0 c\, k_{tr}} \left(k_{tr,z}\,\hat{\mathbf{x}} - k_{tr,x}\,\hat{\mathbf{z}}\right) \times \exp\left\{-k_0 \sqrt{n_1^2 \sin^2\theta_1 - n_2^2}\, z\right\} \exp\{ik_{i,x}x\}\,. \tag{1.8.49}$$

The particularities that occur when the incident wave is incident on an optically less dense medium are shown in Fig. 1.28 for the reflected wave. In the case of p-polarisation, a reflection zero is visible at the Brewster angle; with $\tan\theta_B = n_2/n_1$, the angle is now found in the interval $0 < \theta_B < \pi/4$. The phase of the reflected wave shows a jump of

π at θ_B. Beyond the critical angle θ_c, the reflection coefficient is unity for both states of polarisation. The phase of the reflected wave gradually increases to a value of $-\pi$ at the grazing incidence angle $\theta_1 = \pi/2$. The physical interpretation of this negative phase change is the following. The reflected wave, in the case of s-polarisation, is written as

$$\begin{aligned}\mathbf{E}_r(\mathbf{r},t) &= |r_p|\,\mathbf{E}_{0,i}\,\exp\{i(\mathbf{k}_r\cdot\mathbf{r}+\phi_{r,p}-\omega t)\}\\ &= |r_p|\,\mathbf{E}_{0,i}\,\exp\{i(\mathbf{k}_r\cdot\mathbf{r}-\omega t')\}\,,\end{aligned} \tag{1.8.50}$$

where $t' = t - (\phi_{r,p}/2\pi)T$ and $1/T$ is the light frequency. The wavefront of the reflected wave passes an observer at a moment in time that, for negative values of $\phi_{r,p}$, is delayed. Equation (1.8.50) shows that the interaction at the boundary takes a finite amount of time, despite the fact that the total optical power is reflected into the first medium. An intuitive interpretation is that there is a very short dwelling time in the second medium with a very small 'penetration depth', before the optical energy is 100% reflected. For a beam with finite lateral extent, i.e. a weakly focused beam of light, the penetration into the second less dense medium gives rise to a lateral shift of the (totally) reflected beam. This phenomenon was first predicted by Goos and Hänchen in 1947 **[114]** and experimentally verified by them a few years later.

The properties of the transmitted field components assuming real admittances with $n_2 < n_1$ are shown in Fig. 1.29. With the less dense second medium, we notice a high transmitted amplitude which is larger than unity. This phenomenon is comparable with the amplitude swing observed at the end of a free-hanging rope when the rope is excited by a mechanical impulse. A high transmitted field amplitude is observed close to the critical angle. The analysis of the transmitted power at the boundary in Subsection 1.8.7 will, however, show that the net power flow away from the boundary into the second medium remains smaller than unity and is even zero beyond the critical angle. The phase of the transmitted wave increases to a limiting value of $-\pi/2$ at grazing incidence.

1.8.6.3 Phase Anisotropy in the Case of Total Internal Reflection

An intriguing effect, both for the transmitted and reflected waves, is the difference in phase as a function of the state of polarisation in the regime of total reflection. In Fig. 1.28c, the difference between the phase shifts, $\phi_{r,p} - \phi_{r,s}$, is shown for the reflected wave. The p-polarisation state yields a systematically larger phase shift at total reflection, the maximum difference being close to 0.9 radians (for $n_2/n_1 = 2/3$). Using Eqs. (1.8.36) and (1.8.46), we find the following expression describing the phase difference $\delta\phi = \phi_{r,p} - \phi_{r,s}$:

$$\tan\left(\frac{\delta\phi}{2}\right) = -\frac{\sqrt{1-\sin^2\theta_1}\,\sqrt{\sin^2\theta_1-(n_2/n_1)^2}}{\sin^2\theta_1}\,. \tag{1.8.51}$$

Differentiating this expression with respect to $\sin\theta_1$ yields an extremum for $\tan(\delta\phi/2)$ at

$$\sin\theta_{1,e} = \frac{n_2\sqrt{2}}{\sqrt{n_1^2+n_2^2}}\,. \tag{1.8.52}$$

The extremal value $\delta\phi_e$ of the phase difference is then given by

$$\tan\left(\frac{\delta\phi_e}{2}\right) = \frac{n_2^2-n_1^2}{2n_1n_2}\,. \tag{1.8.53}$$

The polarisation-dependent phase shift at total internal reflection was exploited by Fresnel to devise an effectively birefringent or optically retarding element without using crystalline materials that were commonly used for this purpose **[105]**. Figure 1.28c shows that the phase difference between a p-polarised and an s-polarised reflected beam can reach an extreme value of $-\pi/4$ for a ratio of n_2/n_1=2/3. For a minimum value of $\delta\phi_e/2$ of $-\pi/8$, we find, using $\tan(\pi/8)=\sqrt{2}-1$, the condition

$$0 < \frac{n_2}{n_1} \le 1-\sqrt{2}+\sqrt{4-2\sqrt{2}}\,, \qquad \text{or,} \qquad \frac{n_1}{n_2} \ge 1.497\,. \tag{1.8.54}$$

For a glass–air reflection, we have good-quality optical glass material with a refractive index ratio of, for instance, n_1/n_2=1.515 (standard 'crown' glass). Calculation shows that a $\pi/4$ phase shift is then feasible at angles of incidence θ_1 of 48.06 or 55.07 degrees. Two successive internal total reflections, each yielding a $\pi/4$ phase difference between two orthogonal linear states of polarisation, result in a 'quarter wave' element (see Fig. 1.30a). At the exit of the Fresnel rhomb, the phase of the s-polarised wave advances by an amount of $\pi/2$ the phase of the p-polarised wave. Such an element

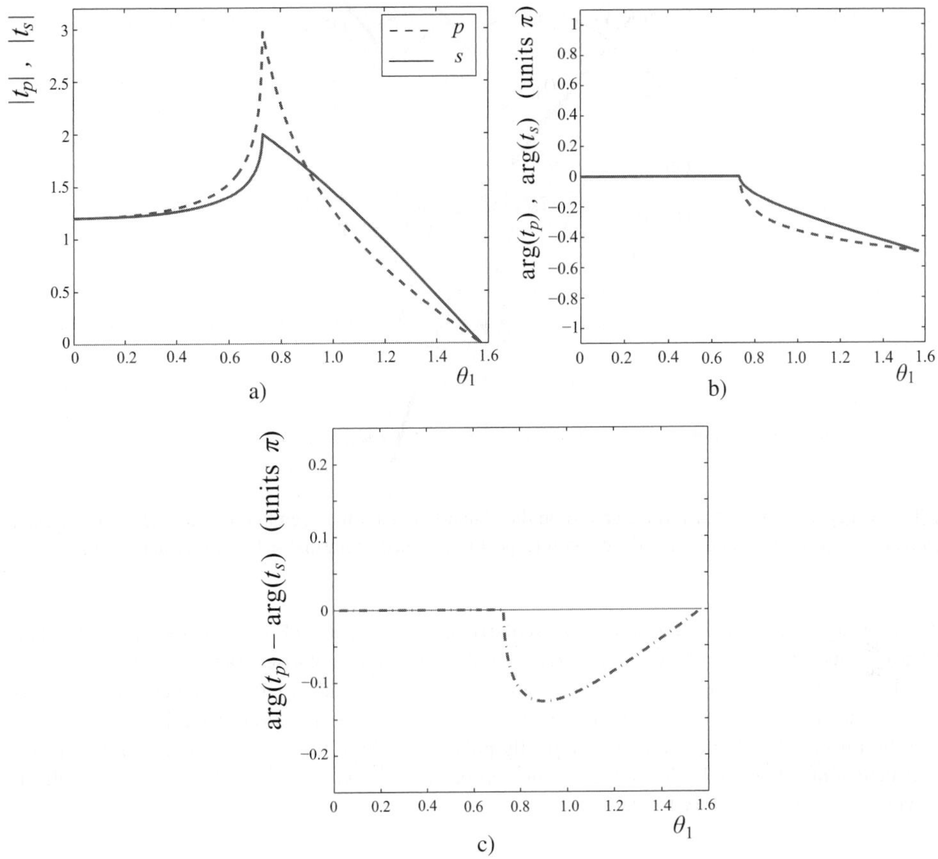

Figure 1.29: a) The moduli of the transmission coefficients t_p (dashed) and t_s (solid line) as a function of the incidence angle θ_1.
b) Arguments of the s- and p-transmission coefficient as a function of θ_1.
c) Phase difference between a transmitted p- and s-polarised wave.

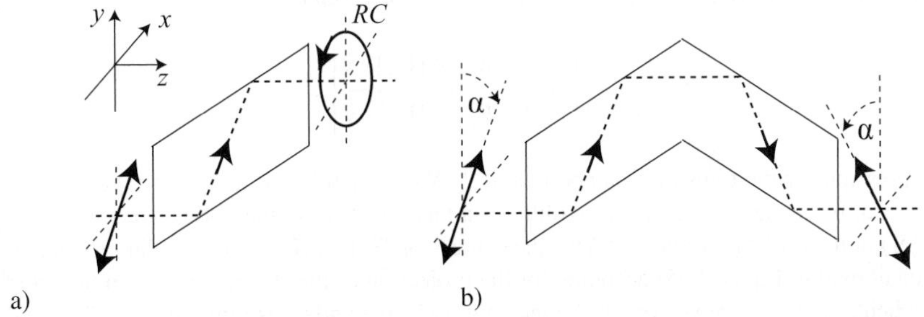

Figure 1.30: a) Fresnel rhomb with two successive internal reflections yielding a quarter-wave plate. The incident parallel beam, propagating along the positive z-axis, is linearly polarised at an angle of $\pi/4$ radians with respect to the positive x-axis. The light exiting the Fresnel rhomb is right-handed circularly polarised (RC, $\lambda/4$ wave plate action).
b) Fresnel rhomb with four internal reflections acting as a half-wave plate. The incident beam is linearly polarised at an angle α with respect to the positive y-axis. The exiting beam is linearly polarised at an angle $-\alpha$ with respect to the same axis.

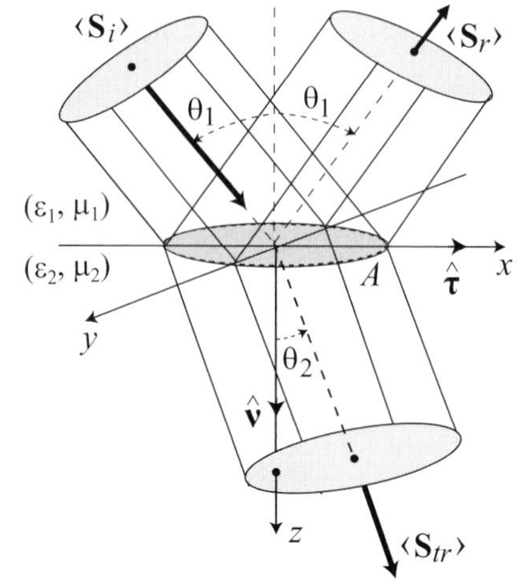

Figure 1.31: The energy flow through a surface area A on the planar boundary between two media with optical properties given by (ϵ_1, μ_1) and (ϵ_2, μ_2). The angle of incidence with respect to the surface normal is θ_1, the refraction angle is θ_2.

is capable of converting linearly polarised light into circularly polarised light. In Fig. 1.30a the plane of polarisation of the incident wave is such that the exiting wave is right-handed circularly polarised (*RC*). A rotation of the plane of polarisation of the incident wave over an angle of $\pi/2$ around the z-axis yields a left-handed circularly polarised wave (*LC*) at the exit surface of the rhomb. Any other azimuth of the plane of polarisation of the linearly polarised light at the entrance of the Fresnel rhomb gives rise to elliptically polarised light at the exit. The invention by Fresnel has the additional advantage of maintaining the desired retardation property over a wide range of the spectrum; i.e. the dispersion of the phase changes $\phi_{r,p}$ and $\phi_{r,s}$ is very limited.

1.8.7 Energy Transport at the Boundary

To calculate the energy transport through the boundary surface between two optical media characterised by their optical constants (ϵ_1, μ_1) and (ϵ_2, μ_2), we require the electric and magnetic field vectors in the incident, reflected and transmitted beam (see Fig. 1.31). The reflection and transmission coefficient associated with the electromagnetic energy flowing toward and from the boundary is defined with the aid of the time-averaged power flow (Poynting vector) in the incident, reflected and transmitted beam. For example, the reflection coefficient is given by

$$R = \left| \frac{\langle S_{r,z} \rangle}{\langle S_{i,z} \rangle} \right| = \left| \frac{\Re\left(\mathbf{E}_{0,r} \times \mathbf{H}^*_{0,r}\right) \cdot \hat{\mathbf{v}}}{\Re\left(\mathbf{E}_{0,i} \times \mathbf{H}^*_{0,i}\right) \cdot \hat{\mathbf{v}}} \right| , \qquad (1.8.55)$$

with a comparable expression for the transmission coefficient T. We are interested in the average flow of energy since this is the quantity (optical irradiance) which is accessible for a measurement in the optical frequency domain. For the time-averaged value $\langle \mathbf{S} \rangle$ of the Poynting vector we have the expression $\Re\{\mathbf{E} \times \mathbf{H}^*\}/2$. The scalar product of the Poynting vector with the surface normal in Eq. (1.8.55) accounts for the projection of the surface area A on the interface onto the planes that are perpendicular to the wave vectors of the incident, reflected and transmitted beams. The modulus is taken because we are only interested in the ratio of the flows, not in their co-propagating or counter-propagating nature. The cases of p- and s-polarisation will now be treated separately.

1.8.7.1 *p*-polarisation

Without loss of generality, we take the normal to the boundary surface in the z-direction and, using a right-handed coordinate system, take the positive x-direction along the unit vector $\hat{\tau}$ and define the plane of incidence by $y = 0$. The electric and magnetic vectors of the incident, reflected and transmitted waves are then given by

$$\begin{aligned}
\mathbf{E}_{0,i} &= (\cos\theta_1,\ 0,\ -\sin\theta_1)\ E_{0,i}\,,\\
\mathbf{E}_{0,r} &= (-r_p\cos\theta_1,\ 0,\ -r_p\sin\theta_1)\ E_{0,i}\,,\\
\mathbf{E}_{0,tr} &= (t_p\cos\theta_2,\ 0,\ -t_p\sin\theta_2)\ E_{0,i}\,;
\end{aligned} \tag{1.8.56}$$

$$\begin{aligned}
\mathbf{H}_{0,i} &= (0,\ (\epsilon_1/\mu_1)^{1/2},\ 0)\ E_{0,i}\,,\\
\mathbf{H}_{0,r} &= (0,\ r_p(\epsilon_1/\mu_1)^{1/2},\ 0)\ E_{0,i}\,,\\
\mathbf{H}_{0,tr} &= (0,\ t_p(\epsilon_2/\mu_2)^{1/2},\ 0)\ E_{0,i}\,.
\end{aligned} \tag{1.8.57}$$

Substitution of the reflected and incident fields in Eq. (1.8.55) yields the simple expression

$$R_p = |r_p|^2 = \left|\frac{Y_{p,2} - Y_{p,1}}{Y_{p,2} + Y_{p,1}}\right|^2 . \tag{1.8.58}$$

The same operation carried out for the transmitted and incident fields yields

$$T_p = |t_p|^2 \left|\frac{\Re\left\{\cos\theta_2\left[(\epsilon_2/\mu_2)^{1/2}\right]^*\right\}}{\Re\left\{\cos\theta_1\left[(\epsilon_1/\mu_1)^{1/2}\right]^*\right\}}\right| = |t_p|^2 \left|\frac{\cos\theta_2}{\cos\theta_1}\right|^2 \frac{\Re\{Y_{p,2}\}}{\Re\{Y_{p,1}\}} . \tag{1.8.59}$$

The quantities $\cos\theta_1$ and $\cos\theta_2$ become complex when the wavenumber k is larger than nk_0, the maximum allowable value for lossless wave propagation in a medium with refractive index n. Using the expression for t_p of Eq. (1.8.25) we can write

$$T_p = \left|\frac{2\,Y_{p,1}}{Y_{p,1} + Y_{p,2}}\right|^2 \frac{\Re\{Y_{p,2}\}}{\Re\{Y_{p,1}\}} . \tag{1.8.60}$$

When the admittances of both media are real the expressions become

$$R_p = \left(\frac{Y_{p,2} - Y_{p,1}}{Y_{p,2} + Y_{p,1}}\right)^2 , \qquad T_p = \frac{4\,Y_{p,1}Y_{p,2}}{Y_{p,1} + Y_{p,2}} , \tag{1.8.61}$$

and it is easily verified that in this case $R_p + T_p = 1$.

1.8.7.2 *s*-polarisation

In a comparable way, using Fig. 1.25, we calculate the energy reflection and transmission coefficients in the case of *s*-polarisation with the results

$$R_s = |r_s|^2 = \left|\frac{Y_{s,2} - Y_{s,1}}{Y_{s,2} + Y_{s,1}}\right|^2 , \tag{1.8.62}$$

$$T_s = |t_s|^2 \left|\frac{\Re\left\{\cos\theta_2(\epsilon_2/\mu_2)^{1/2}\right\}}{\Re\left\{\cos\theta_1(\epsilon_1/\mu_1)^{1/2}\right\}}\right| = \left|\frac{2\,Y_{s,1}}{Y_{s,2} + Y_{s,1}}\right|^2 \frac{\Re\{Y_{s,2}\}}{\Re\{Y_{s,1}\}} . \tag{1.8.63}$$

Again considering the case of real valued admittances we have

$$R_s = \left(\frac{Y_{s,2} - Y_{s,1}}{Y_{s,2} + Y_{s,1}}\right)^2 , \qquad T_s = \frac{4\,Y_{s,1}Y_{s,2}}{\left(Y_{s,2} + Y_{s,1}\right)^2} , \tag{1.8.64}$$

from which the energy conservation relation $R_s + T_s = 1$ follows.

1.8.7.3 Total Internal Reflection

Analysis of the power flow for the case of total internal reflection needs special treatment. It immediately follows from Eqs. (1.8.58) and (1.8.62) that the power reflection coefficient equals unity, independent of the state of polarisation of the incident wave. Restricting analysis to real and positive optical constants, we see that the transmitted power coefficients T_p and T_s both contain a $\Re(\cos\theta_2)$ factor. For total internal reflection, this quantity is zero and no net power flows into the second medium, although the field in the second medium is not zero, so as to assure the continuity of the tangential field components across the boundary between the two media. The field in the second medium has a propagating component

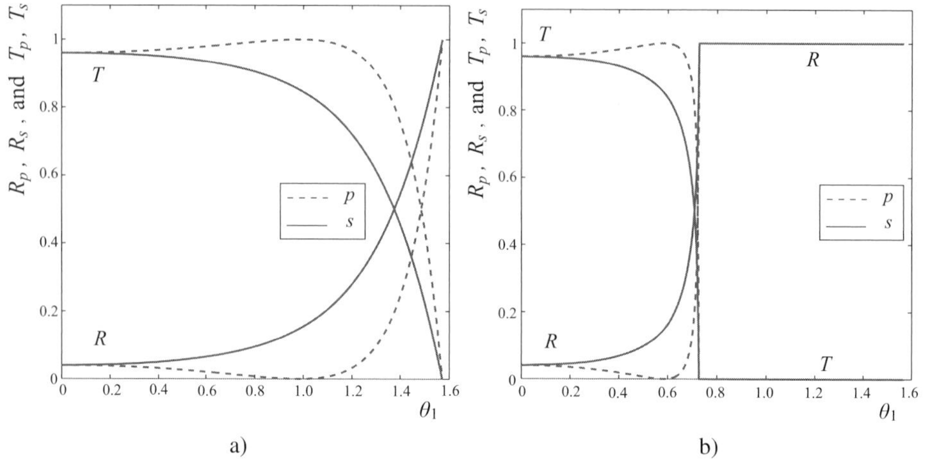

Figure 1.32: Plots of the power reflection and transmission coefficients, R_p, R_s and T_p, T_s, respectively. The dashed curves apply to p-polarisation of the incident plane wave, the solid curves to s-polarisation.
a) Transition between two media with real admittances ($n_1/n_2 = 2/3$).
b) $n_1/n_2 = 3/2$ with the phenomenon of total internal reflection beyond the critical angle θ_c.

only in the direction parallel to the boundary. This component, along the x-direction in our example, can be derived from Eqs. (1.8.56) and (1.8.57) and an intensity transmittance factor T_x in the tangential direction can be derived from it. In the case of p-polarisation we write $T_{x,p}$ as the ratio of the tangential components of the time-averaged Poynting vector $\langle \mathbf{S} \rangle$ in the plane $z = 0$, yielding

$$T_{x,p} = \left| \frac{\langle S_{tr,x} \rangle}{\langle S_{i,x} \rangle} \right|_p = \frac{\left| \Re \left(\mathbf{E}_{0,tr} \times \mathbf{H}^*_{0,tr} \right)_x \right|}{\left| \Re \left(\mathbf{E}_{0,i} \times \mathbf{H}^*_{0,i} \right)_x \right|} = \left| t_p \right|^2 \left| \frac{\Re \left\{ \sin\theta_2 \left[(\epsilon_2/\mu_2)^{1/2} \right]^* \right\}}{\Re \left\{ \sin\theta_1 \left[(\epsilon_1/\mu_1)^{1/2} \right]^* \right\}} \right| , \tag{1.8.65}$$

where the transmitted field vectors are those immediately after the boundary. The value of $T_{x,p}$ is thus associated with the value of the Poynting vector component in the second medium, immediately after the boundary.

We note that for the case of real and positive optical constants Snell's law $n_1 \sin\theta_1 = n_2 \sin\theta_2$ remains valid for the case of total internal reflection and that $\sin\theta_2$ is thus real valued. Using this property, we obtain for $T_{x,p}$ the expression

$$T_{x,p} = \left(\frac{\mu_1}{\mu_2} \right)^{1/2} \left| t_p \right|^2 . \tag{1.8.66}$$

Given the exponential decay of the transmitted wave in the second medium, the component of the Poynting vector parallel to the boundary also decays exponentially along a direction perpendicular to the boundary. We conclude by saying that the tangential power flow component $\langle S_{t,x} \rangle$ in the second medium, integrated in a plane normal to the boundary and normalised with respect to the total incident power, is zero. There is no net average flow of electromagnetic power into the second medium. The incident wave penetrates only into the second medium over an effective distance given by $z_{1/e}$ (see Eq. (1.8.43)).

In Fig. 1.32 we have plotted the reflection and transmission coefficients, R and T, respectively, for the case of a transition between two media with real admittances. In the first graph, the second medium is optically denser that the first one. In the second graph, the second medium is less dense and, at sufficiently large angles of incidence, we observe the phenomenon of total internal reflection.

1.8.7.4 Oscillatory Power Flow at the Boundary

The analysis of energy flow from the boundary given above was restricted to considerations of the time-averaged flow. In complex notation, these averages are given by $\langle \mathbf{S}_r \rangle = \frac{1}{2}\Re\{\mathbf{E}_{0,r} \times \mathbf{H}^*_{0,r}\}$ and $\langle \mathbf{S}_{tr} \rangle = \frac{1}{2}\Re\{\mathbf{E}_{0,tr} \times \mathbf{H}^*_{0,tr}\}$ (see, for instance, Table A.1 of Appendix A). To find the high-frequency component of the energy flow we must take the same vector products, however, omitting the complex conjugate sign. Using the expressions for the complex electric and magnetic field components of Eqs. (1.8.56) and (1.8.57), we find that the time-dependent power flow in the positive z-direction in the second medium is given by

$$\mathbf{S}_{tr}(\mathbf{r},t)_{2\omega} = \frac{1}{2}\Re\left\{ t_p^2\, E_{0,i}^2\, Y_{m,2}\ (\sin\theta_2\,,\,0\,,\,\cos\theta_2)\exp\{-i2\omega t\} \right.$$
$$\left. \times\exp\{i2k_{i,x}x\}\ \exp\left[-2k_0\sqrt{n_1^2\sin^2\theta_1 - n_2^2}\,z\right]\right\}. \tag{1.8.67}$$

We have considered, as an example, the case of p-polarisation and the corresponding electric and magnetic field vectors were taken from Eqs. (1.8.41) and (1.8.42). Using the complex values for $\cos\theta_2$ and t_p (see Eqs. (1.8.31) and (1.8.37)) we find, after some rearrangement, that at the interface $z = 0$:

$$T_{z,2\omega}\Big|_p = \frac{S_{tr,z}}{S_{i,z}}\Big|_{2\omega,p} = \frac{Y_{m,2}}{Y_{m,1}}\frac{\Re\{t_p^2\cos\theta_2\}}{\Re\{\cos\theta_1\}} = \frac{4n_1^2n_2^2}{n_1^2-n_2^2}\,\frac{\sqrt{1-\sin^2\theta_1}\sqrt{\sin^2\theta_1-(n_2^2/n_1^2)}}{(n_1^2+n_2^2)\sin^2\theta_1 - n_2^2}$$
$$\times\ \exp\left\{ i\left[\frac{\pi}{2} - 2\arctan\left(\frac{n_1}{n_2}\sqrt{\frac{n_1^2\sin^2\theta_1/n_2^2 - 1}{1-\sin^2\theta_1}}\right)\right]\right\}. \tag{1.8.68}$$

We thus find that exactly at the boundary between the two media there is a finite energy flux into the second medium which oscillates at frequency 2ω but averages to precisely zero. It follows from Eq. (1.8.67) that at the boundary the high-frequency power flow in the z-direction in the second medium has a phase shift of $2\phi_{t,p} + \pi/2$ with respect to the high-frequency flow in the first medium.

1.8.7.5 Reflection and Transmission for Particular Admittance Combinations

In this subsection we analyse the wave transition at a planar interface for a number of selected *effective* admittance values, Y_1 and Y_2, of the two media in contact. If we write $Y_2/Y_1 = a\exp(i\psi)$, the expressions for the reflection and transmission coefficients of an s-polarised wave become, after some trigonometric manipulation,

$$r_s = \frac{-(1-a^2) + i\,4a\sin(\psi/2)\cos(\psi/2)}{(1-a)^2 + 4a\cos^2(\psi/2)}\,,$$
$$t_s = \frac{2(1-a) + 4a\cos^2(\psi/2)\{1 - i\tan(\psi/2)\}}{(1-a)^2 + 4a\cos^2(\psi/2)}\,. \tag{1.8.69}$$

In the case that the two admittances have the same absolute value ($a = 1$) we find

$$r_{s,a=1} = i\tan(\psi/2)\,,\qquad t_{s,a=1} = 1 - i\tan(\psi/2)\,. \tag{1.8.70}$$

For the case of p-polarisation an extra multiplying factor $(\cos\theta_1/\cos\theta_2)$ must be introduced in the expression for the transmission coefficient t_p (see Eq. (1.8.25)). In Table 1.2 we have listed some combinations of admittance values.

- In the first row of the table we consider the reflection and transmission data for a transition between two media having admittances with equal argument ($\psi = 0$). Freely propagating waves in dielectric media obey this condition.

Table 1.2. Some special cases of admittance ratios at an interface between two media and the associated values of the reflection and transmission coefficients.

a	ψ	Y_2/Y_1	r_s	t_s	$\lvert r_s\rvert^2$	$\lvert t_s\rvert^2\,\Re(Y_2)/\Re(Y_1)$
a	0	a	$-\frac{1-a}{1+a}$	$\frac{2}{1+a}$	$\left(\frac{1-a}{1+a}\right)^2$	$\frac{4a}{(1+a)^2}$
1	$\pi/4$	$\frac{1+i}{\sqrt{2}}$	$i(\sqrt{2}-1)$	$1+i(1-\sqrt{2})$	$3-2\sqrt{2}$	$2(\sqrt{2}-1)$
1	$\pi/2$	$+i$	$+i$	$1-i$	1	0
1	π	-1	$+i\infty$	$-i\infty$	—	—
0	—	0	-1	2	1	0

- In the second row the two admittances have equal modulus but a difference $\psi = \pi/4$ in their argument. Such a case is encountered when a metal and a dielectric are in contact with equal modulus of the electric permittivity (we assume that the magnetic permeabilities are that of vacuum).
- In the third row we consider the special case of two media with a difference in argument of their admittances of $\pi/2$, which occurs when the electric permittivities have a difference in their argument of π. When $\psi = \pi/2$, the admittances of the two media are in phase quadrature and no light can penetrate into the second medium or vice versa. An example is found for the combination of a metal-like substance and a dielectric material, the metal having a real negative permittivity value. In the visible range of the spectrum, silver and aluminium approach the $\psi = \pi/2$ condition, when they are in contact with air or a dielectric material. A well-polished surface of these materials possesses a reflectivity close to unity. A different situation occurs if an evanescent wave is incident on such an idealised metal surface. The effective admittance of the medium for such a wave is purely imaginary and the wave can now penetrate into the metal. Due to our assumption of a negative real permittivity for the metal-like material, the associated extinction coefficient κ is zero and the wave then propagates without attenuation in this fictitious metal.
  Another example of an admittance combination with an argument difference of $\pi/2$ is encountered for the regime of total internal reflection at a transition to a less dense dielectric medium. The less dense medium exhibits a purely imaginary admittance value when the angle of incidence in the dense medium is larger than the critical angle.
- In the fourth row the difference in argument of the effective admittances is π. Such a value leads to a singular case with both r and t tending to infinity. The physical interpretation of this singular case is the excitation of a surface wave bound to the interface between the two media with all the incident power concentrated in a volume with a typical height of the order of the wavelength of the light. These surface-bound waves are associated with a metallic medium or a waveguiding structure. For the case that one of the media is metallic, the surface-bound waves are commonly called *surface plasmon* waves as they can be attributed to the excitation by the incident wave of the free electrons in the metallic medium. A further discussion of surface plasmon waves is given in Chapter 3.
- In the last row of the table we consider the example where the second admittance is zero, a special case of the general admittance ratio of a in the first row. Vacuum has a nonzero admittance of 0.00265 S and cannot meet this condition. Theoretically, using more complicated geometries with two or more interfaces combined in a stack of thin layers, the effective admittance of such a stack could be zero.

The optical properties of multiple thin layers, stacked one on top of the other and with different material properties, is the subject of the next section.

1.9 Transmission and Reflection in a Stratified Medium

In this section we study the reflection and transmission of planar waves by an assembly of thin layers with parallel planar interfaces. The global name for such a structure is 'stratified medium'; in optics, it is frequently called a multilayer structure or, briefly, a multilayer. Arbitrary values of the electric permittivity and magnetic permeability of each layer are allowed. We limit ourselves to isotropic materials and we exclude scattering of light at the interfaces or in the bulk of the thin layers. Applications of multilayers in the optical domain are anti-reflection coatings, reflection enhancement coatings for laser cavity mirrors, beam splitters, thin-layer structures for optical data storage, imaging into the thin detection layers of a solid-state detector or into the photoresist layer on top of a silicon wafer for IC-fabrication. An extensive treatment of the subject of multilayers in the optical domain is found in **[228]**. More recently, research on surface and plasmonic waves and on synthetic 'metamaterials' has further triggered the analysis of the electromagnetic properties of multilayers. These new materials allow for multilayer configurations with striking properties. As an example we cite the so-called 'perfect lens', a theoretical prediction **[356]** that is nearing an interesting practical realisation **[273]**. A discussion of the imaging capabilities of assemblies of such new materials is the subject of Chapter 3.

1.9.1 Multilayer Geometry and Transfer Matrix Algorithm

We consider the geometry of Fig. 1.33 in which a stack of thin layers is sandwiched between two media that extend from minus to plus infinity in the z-direction with indices $\tilde{n}_m$ and $\tilde{n}_s$, respectively. Without loss of generality, we choose the normals to the interfaces between the layers to be parallel to the z-axis. The wave vector is taken to be parallel

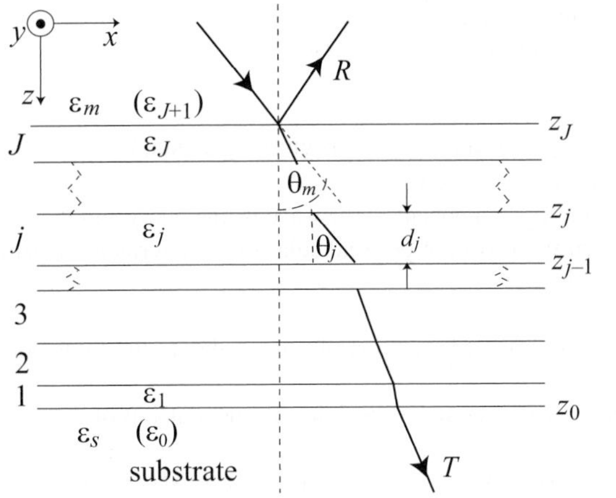

Figure 1.33: Cross-section in the $y = 0$ plane of a general multilayer stack. A plane wave is incident from the upper half-space, with angle of incidence θ_m, and is partially reflected and transmitted to the substrate below. The multiple internal reflections have not been indicated in the figure. The multilayer stack has been numbered starting at the substrate with the first layer numbered 1 and the last one J. The permittivity and permeability of the upper half-space are real (ϵ_m, μ_m), all other indices are allowed to be complex with $\epsilon_j = \epsilon_j' + i\epsilon_j''$ and $\mu_j = \mu_j' + i\mu_j''$. By extrapolation, ϵ_m and ϵ_s are numbered ϵ_{J+1} and ϵ_0, respectively, where the latter quantity should not be confused here with the electric permittivity of vacuum. The thickness of each of the layers is given by d_j. In the drawing we have only indicated the ϵ-values.

to the plane $y = 0$. Light is incident from $z = -\infty$, yielding a positive value of the incident wave vector component k_z. To ensure a finite power at the top of the multilayer, the upper half-space is loss-free with a real refractive index n_m. The indices $\tilde{n}_j$ of the various layers with thickness d_j are allowed to be complex as too is $\tilde{n}_s$, the refractive index of the lower half-space that plays the role of the substrate of the multilayer. The number of a layer is counted from 1 for the first layer on the substrate, to J for the last layer in contact with the upper half-space. The angle of incidence of the incident wave with the normal $\hat{\mathbf{v}} = (0, 0, 1)$ to the multilayer surfaces is θ_m; the corresponding wave vector components are $k_m = \omega\sqrt{\epsilon_m\mu_m}(\sin\theta_m, 0, \cos\theta_m)$.

Various calculation methods can be used to obtain the amplitude and phase of the reflected and transmitted wave. The so-called modal method considers each layer as a cavity, each with basic modes which have a cosine and sine variation as a function of z. At each interface, the mode conversion from one cavity to the next is calculated using the boundary conditions for the electric and magnetic fields at the interface. In this section we take the forward and backward propagating wave in each layer as the basic functions, each with their appropriate exponential function according to Eq. (1.8.20). We adopt the method which takes the tangential components of the electric and magnetic fields at each interface, E_t and H_t, as the quantities to be propagated through the multilayer system from the half-space, labelled m, to the substrate with label s. Their value on both sides of an interface between two layers is identical because of the Maxwell boundary conditions for the electric and magnetic fields. If we have an expression for the values of $E_{t,j+1}$ and $H_{t,j+1}$ as a function of the values $E_{t,j}$ and $H_{t,j}$ at a preceding interface, we can iterate from one interface to the next so as to determine the relationship between the tangential fields at the interface between the substrate ('layer' 0) and layer 1 and the components at the interface between layer J and the upper half-space ('layer' $J + 1$). With knowledge of the incoming and outgoing waves, at the upper and lower interface, we are then able to find the reflected and transmitted waves.

A short discussion is needed about the structure of the waves in the individual layers and the two half-spaces. The incident wave has a wave vector component $k_{m,x} = \omega\sqrt{\epsilon_m\mu_m}\sin\theta_m$. Using the real value of the $k_{m,z}$ component, we have that the propagation angle of the incident wave is $\theta_m = \arctan(k_x/k_z)$. From the electromagnetic boundary conditions, it follows that the k_x component, parallel to the interfaces, is conserved throughout the multilayer and in the substrate. The general expression for the wave vector in a layer with index j is hence given by (see also Eqs. (1.6.73)–(1.6.75)),

$$\mathbf{k}_j = \omega\left(\sqrt{\epsilon_m\mu_m}\sin\theta_m,\ 0\ ,\left\{(\epsilon_j' + i\epsilon_j'')(\mu_j' + i\mu_j'') - \epsilon_m\mu_m\sin^2\theta_m\right\}^{1/2}\right). \qquad (1.9.1)$$

Three types of waves can be present:

- *freely propagating wave*;
  $\epsilon_j'' = \mu_j'' = 0$, $\epsilon_j'\mu_j' \geq \epsilon_m\mu_m \sin^2\theta_m$;
  the $k_{z,j}$ component is real and the propagation angle θ_j is given by $\arctan(k_{x,j}/k_{z,j})$.
- *evanescent wave*; $\epsilon_j'' = \mu_j'' = 0$, $\epsilon_j'\mu_j' < \epsilon_m\mu_m \sin^2\theta_m$;
  The component $k_{x,j}$ is real but the $k_{z,j}$ component of the wave is imaginary and the wave is damped in the z-direction. The planes of constant phase of the wave are perpendicular to the interface, the surfaces of constant amplitude are planes perpendicular to the z-axis.
- *general damped wave*;
  Nonzero values of ϵ_j'' or μ_j'' cause lossy propagation of a wave due to the finite conductivity and inductance of the medium.
  The $k_{z,j}$ component is complex $= \Re(k_{z,j}) + i\Im(k_{z,j})$, and the exponential part of the resulting wave function is given by

$$\exp\left[-\Im(k_{z,j})\, z\right] \exp\left\{i(\omega\sqrt{\epsilon_m\mu_m}\, x + \Re(k_{z,j})\, z - \omega t)\right\} . \tag{1.9.2}$$

When taking the square root to obtain k_z from Eq. (1.9.1), we select the complex solution that yields a damped wave, depending on whether the wave propagates in the forward or backward direction. The damped wave in layer j with the exponential of Eq. (1.9.2) has a planar wavefront and its direction of propagation is given by

$$\theta_j = \arctan\left(\frac{\omega\sqrt{\epsilon_m\mu_m}}{\Re(k_{z,j})}\right) . \tag{1.9.3}$$

It follows from this equation that an evanescent wave can be considered as a special case of a general damped wave with the propagation angle θ_j equal to $\pi/2$.

1.9.1.1 *p*-polarisation

As was necessary for a single interface between two media, we need to discriminate between the two orthogonal states of linear polarisation. In Fig. 1.34 we have depicted the electric and magnetic field vectors in the $j-1$ and j layers for the case of a p-polarised wave, together with the tangential field components at each interface. The complex electric vector is parallel to the plane of incidence and the complex magnetic field vector points in the positive y-direction. The tangential electric vector is considered to be positive in the $+x$-direction. The positive direction for the magnetic tangential field vector is the positive y-direction. At the interface between a layer j and $j+1$, we have the tangential field components $E_{t,j}$ and $H_{t,j}$, not to be confused with components of the electromagnetic field vectors $\mathbf{E}_j^+$ and $\mathbf{H}_j^+$ of the transmitted wave. With our choice for the positive directions of the electric and magnetic vectors of the forward and backward propagating waves, the tangential components at the interface between layers $j-1$ and j, at a chosen reference position with $x = 0$, are given by

$$\begin{aligned} E_{t,j-1} &= \cos\theta_{j-1}\left(E_{j-1}^+ - E_{j-1}^-\right) = \cos\theta_j\left(E_j^+ - E_j^-\right) , \\ H_{t,j-1} &= Y_{m,j-1}\left(E_{j-1}^+ + E_{j-1}^-\right) = Y_{m,j}\left(E_j^+ + E_j^-\right) . \end{aligned} \tag{1.9.4}$$

As before, the intrinsic medium admittance $Y_{m,j}$ is defined by $(\epsilon_j/\mu_j)^{1/2}$. Without loss of generality, we have assumed that the phase of the forward and backward propagating waves, at the interface between layers j and $j-1$ $(z = z_{j-1})$, is zero. The electric field vector due to the forward and backward propagating waves in layer j is thus given by

$$\begin{aligned} \mathbf{E}_j(\mathbf{r}) = {} & (E_j^+ \cos\theta_j\, ,\, 0\, ,\, -E_j^+ \sin\theta_j)\exp\{i(k_{x,j}x + k_{z,j}z)\} \\ & -(E_j^- \cos\theta_j\, ,\, 0\, ,\, E_j^- \sin\theta_j)\exp\{i(k_{x,j}x - k_{z,j}z)\} , \end{aligned} \tag{1.9.5}$$

and a comparable expression can be established for the sum of the magnetic field vectors of the two waves. With the aid of Eq. (1.9.5) we can write the wave amplitudes at the interface between layers j and $j+1$, after a displacement in the z-direction of $\Delta z = z_j - z_{j-1} = -d_j$ (tangential reference position is $x = 0$), as

$$\begin{aligned} E_{t,j} &= \cos\theta_j\left\{E_j^+ \exp(-ik_{z,j}d_j) - E_j^- \exp(ik_{z,j}d_j)\right\} , \\ H_{t,j} &= Y_{m,j}\left\{E_j^+ \exp(-ik_{z,j}d_j) + E_j^- \exp(ik_{z,j}d_j)\right\} . \end{aligned} \tag{1.9.6}$$

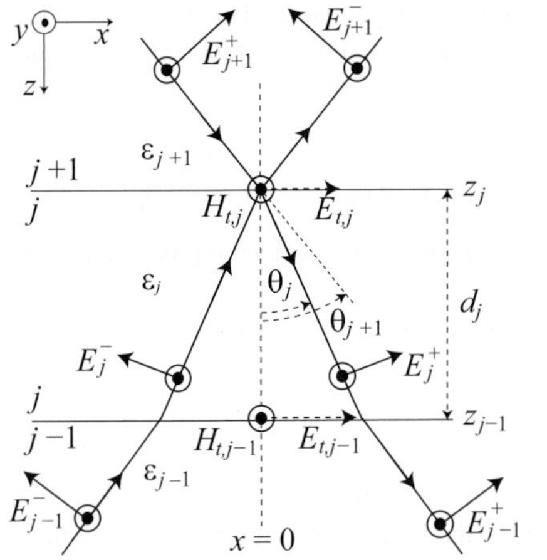

Figure 1.34: The electric and magnetic field vectors of the forward and backward propagating waves in the layers $j+1$, j and $j-1$ of a multilayer structure (electric p-polarisation). The electric field vector strength labelled E^{+} is related to a forward propagating wave and the vector strength labelled E^{-} is associated with the field vector of a backward propagating wave.

Combining Eqs. (1.9.4) and (1.9.6) yields the matrix equation

$$\begin{aligned}\begin{pmatrix} E_{t,j} \\ H_{t,j} \end{pmatrix} &= \begin{pmatrix} \cos(k_{z,j}d_j) & -i\dfrac{\cos\theta_j}{Y_{m,j}}\sin(k_{z,j}d_j) \\ -i\dfrac{Y_{m,j}}{\cos\theta_j}\sin(k_{z,j}d_j) & \cos(k_{z,j}d_j) \end{pmatrix} \begin{pmatrix} E_{t,j-1} \\ H_{t,j-1} \end{pmatrix} \\ &= \begin{pmatrix} \cos(k_{z,j}d_j) & -i\dfrac{\sin(k_{z,j}d_j)}{Y_{p,j}} \\ -iY_{p,j}\sin(k_{z,j}d_j) & \cos(k_{z,j}d_j) \end{pmatrix} \begin{pmatrix} E_{t,j-1} \\ H_{t,j-1} \end{pmatrix} \\ &= \mathbf{M}_{p,j} \begin{pmatrix} E_{t,j-1} \\ H_{t,j-1} \end{pmatrix}. \end{aligned} \tag{1.9.7}$$

The matrix $\mathbf{M}_{p,j}$ which is known as the transfer matrix for medium j, contains the admittance of the medium m for p-polarisation, given the propagation angle in the medium: $Y_{p,j} = Y_{m,j}/\cos\theta_j$.

1.9.1.2 s-polarisation

It is easily verified that the matrix $\mathbf{M}_{s,j}$ for the case of s-polarised light is given by a comparable expression as for the matrix $\mathbf{M}_{p,j}$ above. The only modification is the replacement of the admittance $Y_{p,j}$ of the medium j by the quantity $Y_{s,j} = Y_{m,j}\cos\theta_j$.

1.9.2 Matrix Properties and Composite Transfer Matrix of a Multilayer

The quantities $\cos(k_{z,j}d_j)$, $\sin(k_{z,j}d_j)$ and $\cos\theta_j$ all appear in both of the elementary transfer matrices, $\mathbf{M}_{p,j}$ and $\mathbf{M}_{s,j}$,. These terms are real when the waves in medium j are freely propagating. If the wave in medium j is evanescent or the medium is dissipative, these quantities become complex. In this case the factor $\cos\theta_j$ appearing in the medium admittance is replaced by the ratio $k_{z,j}/k_j$, that is to say the normalised z-component of the complex wave vector in the medium. For absorbing media, we choose the sign of the imaginary part of $k_{z,j}$ such that the forward propagating wave and the reflected wave decrease in amplitude during propagation. Inspection of the elementary transfer matrix shows that the determinant of $\mathbf{M}_{p,j}$ equals unity. The determinant can be considered as the Jacobian of the transformation of the tangential field vectors at interface $j-1, j$ to those at the interface $j, j+1$, according to Eq. (1.9.7). A unity determinant means that there is no scaling involved between the output and input tangential field components. Indeed, there are no physical arguments that could justify a scaling of the product of the tangential field components of a plane wave upon propagation.

The two eigenvalues $\lambda_{1,2}$ of $\mathbf{M}_{p,j}$ are $\exp(\pm ik_{z,j}d_j)$. The corresponding eigenvectors are given by

$$\begin{aligned} \lambda_1 &= \exp(+ik_{z,j}d_j)\,, \qquad & (E_{t,j}, H_{t,j}) &= (1\,,\,-Y_{p,j})\,, \\ \lambda_2 &= \exp(-ik_{z,j}d_j)\,, \qquad & (E_{t,j}, H_{t,j}) &= (1\,,\,Y_{p,j})\,. \end{aligned} \tag{1.9.8}$$

In terms of the amplitudes (E_j^+, E_j^-) of the forward and backward propagating waves we can write

$$\begin{aligned} \lambda_1 &= \exp(+ik_{z,j}d_j)\,, \qquad & (E_j^+, E_j^-) &= \frac{E_{t,j}}{\cos\theta_j}(0\,,\,-1)\,, \\ \lambda_2 &= \exp(-ik_{z,j}d_j)\,, \qquad & (E_j^+, E_j^-) &= \frac{E_{t,j}}{\cos\theta_j}(1\,,\,0)\,. \end{aligned} \tag{1.9.9}$$

The (E^+, E^-) 'eigenvectors' correspond to the particular case where in a sublayer j either the forward propagating or backward propagating wave is absent. In those cases, the multiplication by the transfer matrix is equivalent to a phase change of the tangential components given by the exponential factor $\exp(\pm ik_{z,j}d_j)$; in an absorbing medium or for an evanescent wave, the amplitude changes also because of the complex value of $k_{z,j}$.

1.9.2.1 Expression for the Composite Transfer Matrix of a Multilayer

For both p- and s-polarisation, we can determine the tangential fields $(E_{t,J}, H_{t,J})$ at the interface of the multilayer with the incident medium from the tangential fields $(E_{t,0}, H_{t,0})$ at the interface between the substrate and the first layer. It suffices to apply subsequent multiplications of the individual layer matrices $\mathbf{M}_j$ according to

$$\begin{aligned} \begin{pmatrix} E_{t,J} \\ H_{t,J} \end{pmatrix} &= \mathbf{M}_J\,\mathbf{M}_{J-1}\,\cdots\,\mathbf{M}_2\,\mathbf{M}_1 \begin{pmatrix} E_{t,0} \\ H_{t,0} \end{pmatrix} \\ &= \begin{pmatrix} m_{11} & m_{12} \\ m_{21} & m_{22} \end{pmatrix} \begin{pmatrix} E_{t,0} \\ H_{t,0} \end{pmatrix} = \mathbf{M} \begin{pmatrix} E_{t,0} \\ H_{t,0} \end{pmatrix}, \end{aligned} \tag{1.9.10}$$

where the specific matrices $\mathbf{M}_{p,j}$ or $\mathbf{M}_{s,j}$ are used, according to the state of polarisation of the incident wave. It is important to follow the correct sequence of matrix multiplication; the product in Eq. (1.9.10) has to be evaluated from right to left.

A number of general matrix properties apply to the multilayer matrix $\mathbf{M}$. For example, because $\mathbf{M}$ is the product of unimodular matrices with a determinant of unity, the matrix $\mathbf{M}$ is itself unimodular. The inverse matrix of $\mathbf{M}$ is given by

$$(\mathbf{M})^{-1} = \begin{pmatrix} m_{22} & -m_{12} \\ -m_{21} & m_{11} \end{pmatrix}, \tag{1.9.11}$$

which also has a determinant of unity. The physical explanation for the change in sign of the off-diagonal elements of $\mathbf{M}^{-1}$ as compared to that of the off-diagonal elements in $\mathbf{M}$ is the opposite value of the pathlength increment $k_{z,j}d_j$ which has to be accounted for when calculating the phase of the tangential components $(E_{t,j-1}, H_{t,j-1})$ from the values of $(E_{t,j}, H_{t,j})$ using the inverted matrix. From the individual layer matrix $M_{p,j}$ of Eq. (1.9.7) it is clear that, for the same wave propagation direction, the off-diagonal elements with their sine-dependence on the optical pathlength, must change sign when the phase is calculated in a position which is 'upstream' (matrix $\mathbf{M}$) or 'downstream' (matrix $\mathbf{M}^{-1}$) of the wave propagation direction from the upper half space labelled $J + 1$ to the substrate layer with index 0.

We finally remark that the individual layer matrices $M_{p,j}$ or $M_{s,j}$ are not Hermitian. As a consequence, the eigenvalues of these matrices are not necessarily real and their eigenvectors are generally not orthogonal. In special situations, however, the composite matrix of a multilayer may show such special properties because of a particular multilayer geometry.

1.10 Multilayer Reflection and Transmission Coefficients

To obtain the amplitude reflection and transmission coefficients of the multilayer structure, we calculate the complex amplitudes of the reflected and transmitted waves. The transfer matrix relating the tangential field components at the substrate interface to those at the interface with the upper half-space (see Eq. (1.9.10)) yields the required connection

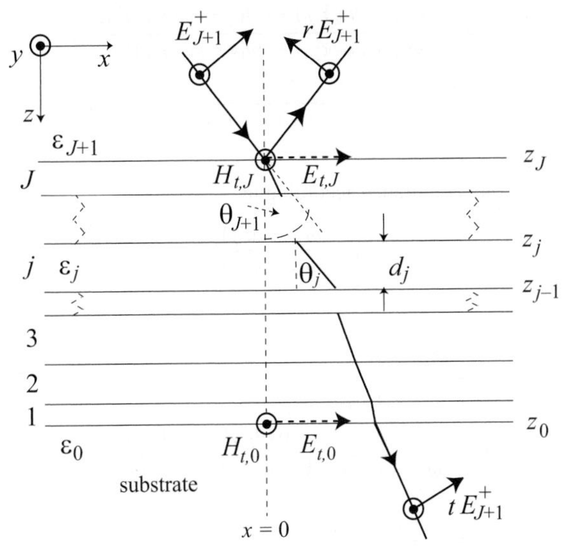

Figure 1.35: The electric and magnetic field vectors of the incident, reflected and transmitted plane waves in the upper half-space and the substrate, respectively. The drawing applies to the case of electric p-polarisation with the magnetic field vector pointing in the y-direction. At the upper interface with the incident medium and at the lower interface with the substrate, we have the tangential field components, $(E_{t,J}, H_{t,J})$ and $(E_{t,0}, H_{t,0})$. The reference directions for the tangential components E_t and H_t are the positive x- and the positive y-direction, respectively. The incident electric field strength is E^+_{J+1}. The strength of the transmitted electric field is tE^+_{J+1}, the reflected field strength is rE^+_{J+1}.

between the two media. In Fig. 1.35 we show the geometry of the multilayer and the electric and magnetic field vectors for the case of electric p-polarisation. The layer numbers are counted again from 1, for the first layer in contact with the substrate, up to J for the upper layer of the multilayer stack. The substrate and the upper half-space are numbered 0 and $J+1$, respectively. The tangential electric and magnetic field components at the upper interface $(J, J+1)$ are given by $E_{t,J}$ and $H_{t,J}$; at the interface with the substrate we have $E_{t,0}$ and $H_{t,0}$. For the magnetic field strength in each layer we have $|H_j| = Y_{m,j}|E_j|$. Given the sign conventions for the electric and magnetic field shown in Fig. 1.35, the tangential field strengths at the interfaces $(0, 1)$ and $(J, J+1)$ are

$$
\begin{aligned}
E_{t,0} &= t_p \cos\theta_0 E^+_{J+1} \, , \\
H_{t,0} &= t_p\, Y_{m,0} E^+_{J+1} \, , \\
E_{t,J} &= (1 - r_p) \cos\theta_{J+1} E^+_{J+1} \, , \\
H_{t,J} &= (1 + r_p)\, Y_{m,J+1} E^+_{J+1} \, ,
\end{aligned}
\qquad (1.10.1)
$$

where we have assumed that no wave is incident on the multilayer stack from the substrate side. To calculate the transmission and reflection coefficients r_p and t_p, we use the multilayer matrix $\mathbf{M}_p$ of Eq. (1.9.10) to eliminate, for instance, $E_{t,J}$ and $H_{t,J}$, and we obtain, after some rearrangement

$$
r_p = \frac{(m_{21} + m_{22} Y_{p,0}) - (m_{11} + m_{12} Y_{p,0}) Y_{p,J+1}}{(m_{21} + m_{22} Y_{p,0}) + (m_{11} + m_{12} Y_{p,0}) Y_{p,J+1}} \, , \qquad (1.10.2)
$$

$$
t_p = \left(\frac{\cos\theta_{J+1}}{\cos\theta_0} \right) \left(\frac{2 Y_{p,J+1}}{(m_{21} + m_{22} Y_{p,0}) + (m_{11} + m_{12} Y_{p,0}) Y_{p,J+1}} \right) . \qquad (1.10.3)
$$

As was done previously, for p-polarisation and for a propagation angle θ_j, we use the admittance value $Y_{p,j} = Y_{m,j} / \cos\theta_j$. In the case that $\mathbf{M}_p$ is the identity matrix, we effectively find the reflection and transmission coefficients that would be found in the absence of the multilayer. A comparable analysis for the case of s-polarisation yields

$$r_s = -\frac{(m_{21} + m_{22}Y_{s,0}) - (m_{11} + m_{12}Y_{s,0})Y_{s,J+1}}{(m_{21} + m_{22}Y_{s,0}) + (m_{11} + m_{12}Y_{s,0})Y_{s,J+1}} , \tag{1.10.4}$$

$$t_s = \left(\frac{2Y_{s,J+1}}{(m_{21} + m_{22}Y_{s,0}) + (m_{11} + m_{12}Y_{s,0})Y_{s,J+1}} \right) , \tag{1.10.5}$$

where $Y_{s,j}$ is defined by $Y_{m,j} \cos\theta_j$.

If we now define a modified substrate admittance according to

$$Y_0' = \frac{m_{21} + m_{22}Y_0}{m_{11} + m_{12}Y_0} , \tag{1.10.6}$$

we obtain the simplified expressions

$$\begin{aligned}
r_p &= \frac{Y_{p,0}' - Y_{p,J+1}}{Y_{p,0}' + Y_{p,J+1}} , \\
t_p &= \frac{1}{m_{11} + m_{12}Y_{p,0}} \left(\frac{\cos\theta_{J+1}}{\cos\theta_0} \right) \frac{2Y_{p,J+1}}{Y_{p,0}' + Y_{p,J+1}} , \\
r_s &= -\frac{Y_{s,0}' - Y_{s,J+1}}{Y_{s,0}' + Y_{s,J+1}} , \\
t_s &= \frac{1}{m_{11} + m_{12}Y_{s,0}} \frac{2Y_{s,J+1}}{Y_{s,0}' + Y_{s,J+1}} .
\end{aligned} \tag{1.10.7}$$

A comparison of the formulae above and the expressions of Eqs. (1.8.24)–(1.8.27) shows the strong correspondence between the resulting formulas when using the concept of a modified surface admittance Y_0' due to the presence of the multilayer.

1.10.1 Energy Transport toward and away from the Multilayer

The power reflection coefficient for a multilayer stack can be defined analogously to that for a single interface as given by Eq. (1.8.55). Following similar derivations to those presented in Subsection 1.8.7, we can find the power reflection coefficients of a multilayer stack; the expressions are again given by $R_p = |r_p|^2$ and $R_s = |r_s|^2$, depending on the state of polarisation of the incident wave. For the power transmission coefficients T_p and T_s, we have to go into some more detail. The general expression for the power transmission coefficient is

$$T = \left| \frac{\langle S_{tr,z} \rangle}{\langle S_{i,z} \rangle} \right| = \left| \frac{\Re\,(\mathbf{E}_{tr} \times \mathbf{H}_{tr}^*) \cdot \hat{\mathbf{v}}}{\Re\,(\mathbf{E}_i \times \mathbf{H}_i^*) \cdot \hat{\mathbf{v}}} \right| , \tag{1.10.8}$$

where $\mathbf{E}_{tr}$ and $\mathbf{H}_{tr}$ are the electric and magnetic field vectors of the transmitted wave in the half-space with index 0 and, likewise, $\mathbf{E}_i$ and $\mathbf{H}_i$ the incident field components in the half-space with index $J + 1$. We can write the incident and transmitted fields at the upper and lower boundaries of the multilayer as

s-polarisation

$$\begin{aligned}
\mathbf{E}_i &= (0,\ 1, 0)\ E_{J+1}^+ , \\
\mathbf{E}_{tr} &= (0, t_s, 0)\ E_{J+1}^+ , \\
\mathbf{H}_i &= (-\cos\theta_{J+1}, 0, \sin\theta_{J+1})\ (\epsilon_{J+1}/\mu_{J+1})^{1/2}\ E_{J+1}^+ , \\
\mathbf{H}_{tr} &= (-t_s \cos\theta_0, 0, t_s \sin\theta_0)\ (\epsilon_0/\mu_0)^{1/2}\ E_{J+1}^+ ,
\end{aligned} \tag{1.10.9}$$

p-polarisation

$$\begin{aligned}
\mathbf{E}_i &= (\cos\theta_{J+1}, 0, -\sin\theta_{J+1})\ E_{J+1}^+ , \\
\mathbf{E}_{tr} &= (t_p \cos\theta_0, 0, -t_p \sin\theta_0)\ E_{J+1}^+ , \\
\mathbf{H}_i &= (0,\ 1, 0)\ (\epsilon_{J+1}/\mu_{J+1})^{1/2}\ E_{J+1}^+ , \\
\mathbf{H}_{tr} &= (0, t_p, 0)\ (\epsilon_0/\mu_0)^{1/2}\ E_{J+1}^+ .
\end{aligned} \tag{1.10.10}$$

Substitution of the field components into Eq. (1.10.8) yields the expressions

$$T_s = |t_s|^2 \frac{\Re\left\{(\cos\theta_0 Y_{m,0})^*\right\}}{\Re\left\{(\cos\theta_{J+1} Y_{m,J+1})^*\right\}} = |t_s|^2 \frac{\Re\left\{Y_{s,0}\right\}}{\Re\left\{Y_{s,J+1}\right\}} ,$$

$$T_p = |t_p|^2 \frac{\Re\left\{\cos\theta_0 Y^*_{m,0}\right\}}{\Re\left\{\cos\theta_{J+1} Y^*_{m,J+1}\right\}} = |t_p|^2 \left|\frac{\cos\theta_0}{\cos\theta_{J+1}}\right|^2 \frac{\Re\left\{Y_{p,0}\right\}}{\Re\left\{Y_{p,J+1}\right\}} . \qquad (1.10.11)$$

To enhance the symmetry between the expressions for T_s and T_p, an adapted definition of t_p can be used, specifically

$$t'_p = \frac{\cos\theta_0}{\cos\theta_{J+1}} t_p . \qquad (1.10.12)$$

With this definition of the amplitude transmission coefficient t'_p, the expressions for T_s and T_p become formally identical.

The expressions for the power transmission coefficients T admit some special cases:

- $\Re\{Y_0\} = 0$.
  The admittance of the substrate medium, given the state of polarisation and the wave vector direction and magnitude, is imaginary. This case is encountered, for instance, if the transmitted wave is evanescent in the substrate medium. No net power is transported into the substrate; the amplitude in the substrate near the interface is generally not zero; a surface wave is launched.
- $\Re\{Y_{J+1}\} = 0$.
  In this case, the admittance of the incident medium is imaginary and hence the wave coming from $z = -\infty$ which creates the incident wave amplitude at the upper boundary of the multilayer is evanescent in nature. Its amplitude is, however, infinitesimally small at the entrance side (interface $J+1, J$) of the multilayer. Consequently, the amplitude at the interface $(1, 0)$ will also remain infinitely small. The power transmission factor therefore has no physical significance and we have to exclude this case. It does not mean that we cannot make statements about the wave amplitude. The quantities t (and r) can take arbitrary complex values.

Free and evanescent wave propagation in electrically absorptive media (with plasmon wave propagation as a special case) and special properties of optical waves in artificial media with more general electric and magnetic properties (metamaterials) are discussed in Chapter 3.

1.10.1.1 Energy Conservation for the Incident, Reflected and Transmitted Waves

The total power carried by a wave which is incident on a multilayer is distributed over the reflected and transmitted waves whilst part of the incident energy is absorbed in the multilayer if the refractive index of one or several layers has a nonzero imaginary part. In this subsection we analyse the distribution of incident energy over the two outgoing waves and the relative amount of optical energy absorption within the multilayer. In this framework it is interesting to compare the energy balance associated with an incident wave coming from the upper halfspace ($j = J + 1$) to that related to a second wave which is incident from the halfspace with index $j = 0$ and has identical propagation angles θ_j in each layer. In Fig. 1.36 we have depicted the two incident waves for the case of p-polarisation. Figure 1.36a shows an incident wave propagating in the positive z-direction, in Fig. 1.36b the wave propagates in the opposite direction. The amplitude reflection and transmission coefficients for the two cases are r^+, t^+ and r^-, t^-, respectively. On comparing the two cases, it is important to discriminate between the admittances of the media for the forward and backward incident wave. From Fig. 1.36b we notice that, in the case of p-polarisation, the magnetic field vector changes sign when the propagation direction of the incident wave is reversed. We include this effect in our calculations by defining a different admittance Y_p for Fig. 1.36a and b: $Y^-_{p,0} = -Y^+_{p,0}$ and $Y^-_{p,J+1} = -Y^+_{p,J+1}$.

Focusing on the intensity transmission coefficient of the multilayer stack, we first write the expression for the complex amplitude transmission coefficient for the reversed case,

$$t^-_p = \left(\frac{\cos\theta_0}{\cos\theta_{J+1}}\right)\left(\frac{2Y^-_{p,0}}{(-m_{21} + m_{11}Y^-_{p,0}) + (m_{22} - m_{12}Y^-_{p,J+1})Y^-_{p,0}}\right). \qquad (1.10.13)$$

To obtain this expression we have first used the expression for the 'downstream' tangential field components associated with a wave which propagates in the positive z-direction, using the matrix elements of the inverse matrix $\mathbf{M}^{-1}$ of Eq. (1.9.11). To obtain the tangential components which correspond to an incident wave with reversed propagation

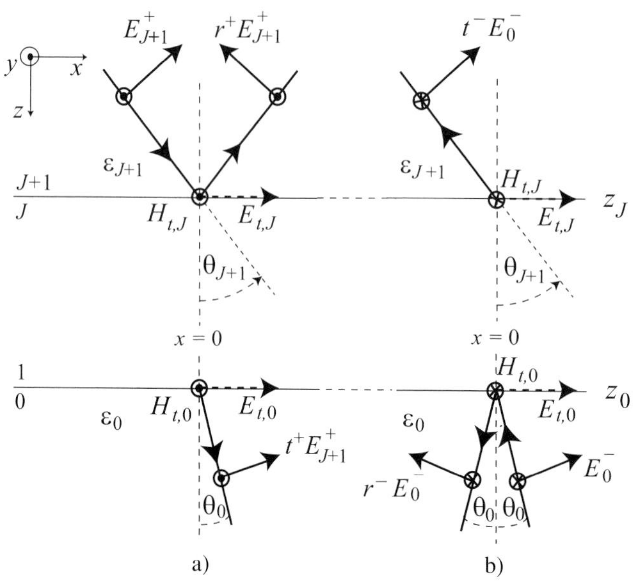

Figure 1.36: a) A wave propagating in the positive z-direction with electric field vector $\mathbf{E}^+_{J+1}$ is incident on the multilayer stack with layers 1 to J. The electric vector is p-polarised. The reflected wave has an electric field strength $r^+ E^+_{J+1}$, the transmitted wave $t^+ E^+_{J+1}$. The tangential field components at the upper and lower interfaces are $(E_{t,J}, H_{t,J})$ and $(E_{t,0}, H_{t,0})$, respectively. Note the reversal of direction of the magnetic field vector between the forward and backward propagating cases. b) The propagation direction has been reversed. The electric field strengths of the transmitted and reflected wave are $t^- E^-_0$ and $r^- E^-_0$, respectively.

direction and travelling in the $-z$-direction, we have used the material admittances primed with the minus sign. Using these primed negative admittances, the positive directions of the electric and magnetic field vectors are such that the $\mathbf{E}, \mathbf{H}, \mathbf{k}$-triplet is right-handed orthogonal for a wave propagating in the negative z-direction, as shown in Fig. 1.36b.

For the power transmission coefficients T^+_p and T^-_p we have the expressions

$$T^+_p = \left| \frac{2Y^+_{p,J+1}}{(m_{21} + m_{22}Y^+_{p,0}) + (m_{11} + m_{12}Y^+_{p,0})Y^+_{p,J+1}} \right|^2 \frac{\Re\left\{Y^+_{p,0}\right\}}{\Re\left\{Y^+_{p,J+1}\right\}},$$
$$T^-_p = \left| \frac{2Y^+_{p,0}}{(m_{21} + m_{22}Y^+_{p,0}) + (m_{11} + m_{12}Y^+_{p,0})Y^+_{p,J+1}} \right|^2 \frac{\Re\left\{-Y^+_{p,J+1}\right\}}{\Re\left\{-Y^+_{p,0}\right\}}, \tag{1.10.14}$$

where all layer admittances now apply to the case of an incident wave propagating in the positive z-direction.

The two denominators in the expressions for T^+_p and T^-_p are identical yielding

$$\frac{T^-_p}{T^+_p} = \left| \frac{Y^+_{p,0}}{Y^+_{p,J+1}} \right|^2 \left(\frac{\Re\left\{Y^+_{p,J+1}\right\}}{\Re\left\{Y^+_{p,0}\right\}} \right)^2 . \tag{1.10.15}$$

For the case where the admittances of the media $J + 1$ and 0 are real, we have $T^-_p = T^+_p$. The same result can be derived for the transmission coefficients T^-_s and T^+_s. It is not possible to derive a comparable relationship between the power reflection coefficients R^+_p and R^-_p or their s-polarised counterparts. The reason for this is that in the case of the r coefficients, the numerator for the + and − cases is, in general, not identical.

A special relationship holds for the R and T coefficients in the case of media with real admittances. In this case, each individual transfer matrix for the multilayer can be written as

$$\mathbf{M}_j = \begin{pmatrix} a & ib \\ ic & d \end{pmatrix}, \tag{1.10.16}$$

with $a, \cdots, d$ all real and $ad + bc = 1$. It is easily verified that this property is maintained when products of such matrices are formed. This means that the complete multilayer transfer matrix can be written as

$$\mathbf{M}\begin{pmatrix} m_{11} & im_{12} \\ im_{21} & m_{22} \end{pmatrix}, \tag{1.10.17}$$

with all quantities m_{ij} real and $m_{11}m_{22} + m_{12}m_{21} = 1$. In terms of these real coefficients we have for the case of an s-polarised incident wave

$$\begin{aligned} R_s = |r_s|^2 &= \left|\frac{(im_{s,21} + m_{s,22}Y_{s,0}) - (m_{s,11}Y_{s,J+1} + im_{s,12}Y_{s,0}Y_{s,J+1})}{(im_{s,21} + m_{s,22}Y_{s,0}) + (m_{s,11}Y_{s,J+1} + im_{s,12}Y_{s,0}Y_{s,J+1})}\right|^2, \\ T_s = |t_s|^2\frac{Y_{s,0}}{Y_{s,J+1}} &= \frac{4\,Y_{s,0}Y_{s,J+1}}{\left|(im_{s,21} + m_{s,22}Y_{s,0}) + (m_{s,11}Y_{s,J+1} + im_{s,12}Y_{s,0}Y_{s,J+1})\right|^2}. \end{aligned} \tag{1.10.18}$$

Evaluating the sum of R_s and T_s and using $m_{s,11}m_{s,22} + m_{s,12}m_{s,21} = 1$ yields the result $R_s + T_s = 1$. The substitution of the p-polarised quantities also gives $R_p + T_p = 1$.

The same results can be obtained for the inverted multilayer system. With our earlier result $T_s^+ = T_s^-$ and $T_p^+ = T_p^-$ for a layer system with real admittances, we conclude that in that case $R_s^+ = R_s^-$ and $R_p^+ = R_p^-$. In the more general case with absorptive layers (but with real admittances for the upper and lower half-spaces) we write

$$T^+ = T^- \; ; \qquad R^+ + A^+ = R^- + A^- , \tag{1.10.19}$$

where A^+ and A^- are the optical power ratios that are absorbed in the multilayer in the forward and backward propagation case, respectively. It is important to note that the quantities R and A, when designing a multilayer structure, can be given quite different values depending on the propagation direction through the multilayer.

1.10.2 The Electric and Magnetic Field Strengths in a Multilayer

In many applications we are interested not only in the amplitudes of the incoming and outgoing waves at a multilayer but also in the electric and magnetic field components of the two waves propagating in a particular layer. In Fig. 1.37 we show the situation for a multilayer on which two waves are incident. One comes from $z = -\infty$ with electric field strength E_{J+1}^+ and is incident at an angle θ_{J+1} onto the upper surface of the multilayer. The second wave comes from $z = +\infty$ and has a field strength E_0^- and an angle of incidence θ_0. The two angles θ_{J+1} and θ_0 have been chosen such that the wave vector component parallel to the interfaces is identical for both incident waves. Moreover, the two waves have identical frequency. At the upper and lower interface, for a fixed value of the coordinate x, the two waves have a constant phase difference of ϕ_d. The transmitted wave components of the wave incident from above and the reflected components of the wave incident from below are co-directional and show interference effects. They produce tangential electric and magnetic components at each interface in the multilayer that have the same phase and amplitude relation over the entire interface. We will use this property to find the tangential electric and magnetic field components at a general interface $(j, j+1)$ within the multilayer. From these tangential components at the interface, we can derive the electric and magnetic field strengths of the forward and backward propagating waves in layer j and layer $j+1$.

To find the field inside the multilayer, we first establish a relationship between the tangential electric and magnetic field components at the upper and lower interface of the multilayer. With the aid of the transfer matrices $\mathbf{M}$ and $\mathbf{M}^{-1}$ of the multilayer for the particular angles of incidence θ_{J+1} and θ_0 in the case of p-polarisation, we calculate the reflection and transmission coefficients of the two incident waves with field strengths E_{J+1}^+ and E_0^-, respectively. Using these coefficients, we know the field strengths of the three waves that propagate in the upper half-space and of the three waves propagating in the substrate. With reference to Fig. 1.37 we can write the tangential field components at the upper and lower interface of the multilayer as

$$\begin{aligned} E_{t,J} &= \left\{(1 - r_p^+)E_{J+1}^+ - t_p^- E_0^-\right\}\frac{k_{z,J+1}}{k_{J+1}}, \\ H_{t,J} &= \left\{(1 + r_p^+)E_{J+1}^+ + t_p^- E_0^-\right\}Y_{m,J+1}, \\ E_{t,0} &= \left\{(-1 + r_p^-)E_0^- + t_p^+ E_{J+1}^+\right\}\frac{k_{z,0}}{k_0}, \\ H_{t,0} &= \left\{(1 + r_p^-)E_0^- + t_p^+ E_{J+1}^+\right\}Y_{m,0}. \end{aligned} \tag{1.10.20}$$

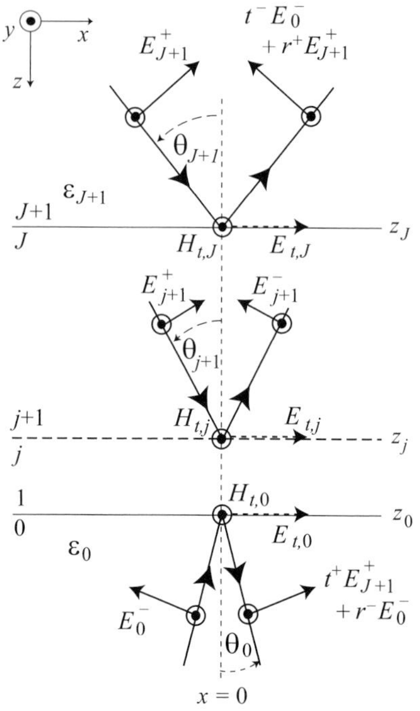

Figure 1.37: Two p-polarised waves are incident on a multilayer with electric field strengths of E^+_{J+1} and E^-_0. The incident, reflected and transmitted waves create the tangential field components $(E_{t,J}, H_{t,J})$ and $(E_{t,0}, H_{t,0})$ at the upper and lower interface of the multilayer, respectively. The tangential field components at the interface between layers j and $j+1$ are denoted by $E_{t,j}$ and $H_{t,j}$. The strengths of the electric field vectors of the forward and backward propagating wave in layer $j+1$ are E^+_{j+1} and E^-_{j+1}, respectively. The magnetic field vector is tangential to the interfaces and is positive when it points in the positive y-direction.

In the first and the third equation above, we have replaced $\cos\theta_j$ by the ratio $k_{z,j}/k_j$, with $k_j = (\mathbf{k}_j \cdot \mathbf{k}_j)^{1/2}$; this allows a uniform treatment of both freely propagating waves and evanescent or damped waves that could occur inside the multilayer or the substrate. The subscript p of the reflection and transmission coefficients refers to the state of polarisation and the superscripts $\pm$ denote the propagation direction of the incident wave from which the reflected and transmitted waves were derived.

We can find the tangential field components at the interface $(j, j+1)$ in two possible ways by using the relationships

$$\begin{pmatrix} E_{t,J} \\ H_{t,J} \end{pmatrix} = \mathbf{M}_J \cdots \mathbf{M}_{j+1} \begin{pmatrix} E_{t,j} \\ H_{t,j} \end{pmatrix},$$

or,

$$\begin{pmatrix} E_{t,j} \\ H_{t,j} \end{pmatrix} = \mathbf{M}^{-1}_{j+1} \cdots \mathbf{M}^{-1}_J \begin{pmatrix} E_{t,J} \\ H_{t,J} \end{pmatrix} = \left(\mathbf{M}_J \cdots \mathbf{M}_{j+1}\right)^{-1} \begin{pmatrix} E_{t,J} \\ H_{t,J} \end{pmatrix};$$

$$\begin{pmatrix} E_{t,j} \\ H_{t,j} \end{pmatrix} = \mathbf{M}_j \cdots \mathbf{M}_1 \begin{pmatrix} E_{t,0} \\ H_{t,0} \end{pmatrix}. \qquad (1.10.21)$$

The most frequently occurring case is the one with a single incident wave from the upper side. In that case it is sufficient to use only the equation on the second line of (1.10.21) to determine the tangential field components at the interface $(j, j+1)$. To cover the more general case with a second wave incident from the substrate, we have to add the second and third lines of Eq. (1.10.21). We obtain the result

$$\begin{pmatrix} E_{t,j} \\ H_{t,j} \end{pmatrix} = \left(\mathbf{M}_J \cdots \mathbf{M}_{j+1}\right)^{-1} \begin{pmatrix} E_{t,J} \\ H_{t,J} \end{pmatrix} + \left(\mathbf{M}_j \cdots \mathbf{M}_1\right) \begin{pmatrix} E_{t,0} \\ H_{t,0} \end{pmatrix}. \qquad (1.10.22)$$

Using the expressions of Eq. (1.10.20) the tangential components at the $(j, j+1)$ interface can then be found. When the incident fields have an arbitrary state of polarisation we have to evaluate Eq. (1.10.22) twice, once for each orthogonal state of polarisation, using the matrix formalism for p- and for s-polarisation separately. For our case where the layer interfaces are normal to the z-axis, the quantity $E_{t,j}$ then stands for a two-vector, $(E_{t,j,x}, E_{t,j,y})$; the same applies to the tangential components of the magnetic vector.

To go from the tangential components at the interface to the electric and magnetic field strengths in the medium $j+1$, we first use the property that $\nabla \cdot \mathbf{D}_0$ is zero anywhere within layer $j+1$. Applied to our harmonic plane waves in layer $j+1$ with complex electric vectors $\mathbf{E}^+$ and $\mathbf{E}^-$ (for compactness of notation the index $j+1$ has been omitted), we have for the forward and backward propagating waves in medium $j+1$,

$$k_x E_x^+ + k_y E_y^+ + k_z E_z^+ = 0 \,, \tag{1.10.23}$$

$$k_x E_x^- + k_y E_y^- - k_z E_z^- = 0 \,, \tag{1.10.24}$$

where $\mathbf{k} = (k_x, k_y, k_z)$ is the (complex) wave vector in layer $j+1$ of the forward propagating wave. The complex magnetic field vector in layer $j+1$ for each of the two propagating plane waves is derived from the relationship $\mathbf{k} \times \mathbf{E}^{\pm} = \omega\mu\mathbf{H}^{\pm}$. After some algebra we find the following expressions for the tangential field components at the interface between the media j and $j+1$,

$$E_{t,x} = E_x^+ + E_x^- \,, \tag{1.10.25}$$

$$E_{t,y} = E_y^+ + E_y^- \,, \tag{1.10.26}$$

$$H_{t,x} = \frac{Y_m}{k}\left(k_y E_z^+ - k_z E_y^+ + k_y E_z^- + k_z E_y^-\right) , \tag{1.10.27}$$

$$H_{t,y} = \frac{Y_m}{k}\left(-k_x E_z^+ + k_z E_x^+ - k_x E_z^- - k_z E_x^-\right) . \tag{1.10.28}$$

The specific admittance of medium $j+1$ is given by $Y_m = (\epsilon_{j+1}/\mu_{j+1})^{1/2}$.

The Eqs. (1.10.23)–(1.10.28) allow us to solve for the electric field components while using the tangential field components, obtained from a calculation or a measurement, as input data. After some manipulation we obtain the result

$$\begin{aligned}
E_x^+ &= \frac{1}{2}\left(E_{t,x} + \frac{k_x k_y}{k k_z Y_m} H_{t,x} + \frac{k_y^2 + k_z^2}{k k_z Y_m} H_{t,y}\right) , \\
E_y^+ &= \frac{1}{2}\left(E_{t,y} - \frac{k_x^2 + k_z^2}{k k_z Y_m} H_{t,x} - \frac{k_x k_y}{k k_z Y_m} H_{t,y}\right) , \\
E_z^+ &= \frac{1}{2}\left(-\frac{k_x E_{t,x} + k_y E_{t,y}}{k_z} + \frac{k_y}{k Y_m} H_{t,x} - \frac{k_x}{k Y_m} H_{t,y}\right) , \\
E_x^- &= \frac{1}{2}\left(E_{t,x} - \frac{k_x k_y}{k k_z Y_m} H_{t,x} - \frac{k_y^2 + k_z^2}{k k_z Y_m} H_{t,y}\right) , \\
E_y^- &= \frac{1}{2}\left(E_{t,y} + \frac{k_x^2 + k_z^2}{k k_z Y_m} H_{t,x} + \frac{k_x k_y}{k k_z Y_m} H_{t,y}\right) , \\
E_z^- &= \frac{1}{2}\left(\frac{k_x E_{t,x} + k_y E_{t,y}}{k_z} + \frac{k_y}{k Y_m} H_{t,x} - \frac{k_x}{k Y_m} H_{t,y}\right) ,
\end{aligned} \tag{1.10.29}$$

where $k = |\mathbf{k}|$. The magnetic field vectors of the forward and backward propagating waves follow from Eqs. (1.10.25)–(1.10.28) and we get

$$\begin{aligned}
H_x^+ &= \frac{1}{2}\left(H_{t,x} - \frac{Y_m\left[k_x k_y E_{t,x} + (k_y^2 + k_z^2) E_{t,y}\right]}{k k_z}\right) , \\
H_y^+ &= \frac{1}{2}\left(H_{t,y} + \frac{Y_m\left[(k_x^2 + k_z^2) E_{t,x} + k_x k_y E_{t,y}\right]}{k k_z}\right) , \\
H_z^+ &= \frac{1}{2}\left(-\frac{k_x}{k_z} H_{t,x} - \frac{k_y}{k_z} H_{t,y} - \frac{k_y Y_m}{k} E_{t,x} + \frac{k_x Y_m}{k} E_{t,y}\right) .
\end{aligned} \tag{1.10.30}$$

The expressions for the magnetic field components of the backward propagating wave follow from Eq. (1.10.30) by changing the sign of the k_z-component of the wave vector, in line with the findings for the electric field components of Eq. (1.10.29). In the frequently occurring case of a wave vector with real components only, the quantities k_z/k etc. can be replaced by their corresponding trigonometric expressions in layer $j+1$.

The cases of perpendicular and parallel polarisation give the following results where we can write $\mathbf{k}^+ = (k_x, k_y, k_z)$ and $\mathbf{k}^- = (k_x, k_y, -k_z)$ for the forward and backward propagating waves in layer $j+1$:

- *s*-polarisation
  The electric field vectors are written as

$$\mathbf{E}^+ = \frac{1}{\sqrt{k_x^2 + k_y^2}} (k_y\, , -k_x\, , 0)\, E^+ \, , \tag{1.10.31}$$

$$\mathbf{E}^- = \frac{1}{\sqrt{k_x^2 + k_y^2}} (k_y\, , -k_x\, , 0)\, E^- \tag{1.10.32}$$

where $k = (\epsilon_{j+1}\mu_{j+1}\, \omega^2)^{1/2}$. For the tangential components of the electric and magnetic field at the interface we find, with the aid of Eqs. (1.10.25)–(1.10.28),

$$\begin{aligned} E_{t,x} &= \left(\frac{k_y}{\sqrt{k_x^2 + k_y^2}} \right) (E^+ + E^-) = (E_x^+ + E_x^-)\, , \\ E_{t,y} &= -\left(\frac{k_x}{\sqrt{k_x^2 + k_y^2}} \right) (E^+ + E^-) = (E_y^+ + E_y^-)\, , \end{aligned} \tag{1.10.33}$$

$$\begin{aligned} H_{t,x} &= \frac{Y_m k_z}{k} \left(\frac{k_x}{\sqrt{k_x^2 + k_y^2}} \right) (E^+ - E^-) = \frac{Y_m k_z}{k} (-E_y^+ + E_y^-)\, , \\ H_{t,y} &= \frac{Y_m k_z}{k} \left(\frac{k_y}{\sqrt{k_x^2 + k_y^2}} \right) (E^+ - E^-) = \frac{-Y_m k_z}{k} (-E_x^+ + E_x^-)\, . \end{aligned} \tag{1.10.34}$$

- *p*-polarisation
  For the electric field vectors we have the expressions

$$\begin{aligned} \mathbf{E}^+ &= \frac{1}{k\sqrt{k_x^2 + k_y^2}} \left(k_x k_z, k_y k_z, -(k_x^2 + k_y^2) \right) E^+ \, , \\ \mathbf{E}^- &= \frac{1}{k\sqrt{k_x^2 + k_y^2}} \left(-k_x k_z, -k_y k_z, -(k_x^2 + k_y^2) \right) E^- \, . \end{aligned} \tag{1.10.35}$$

Along the same lines as we did for *s*-polarisation, we now find the following tangential components,

$$\begin{aligned} E_{t,x} &= \left(\frac{k_x k_z}{k\sqrt{k_x^2 + k_y^2}} \right) (E^+ - E^-) = (E_x^+ + E_x^-)\, , \\ E_{t,y} &= \left(\frac{k_y k_z}{k\sqrt{k_x^2 + k_y^2}} \right) (E^+ - E^-) = (E_y^+ + E_y^-)\, , \end{aligned} \tag{1.10.36}$$

$$H_{t,x} = -\frac{Y_m k_y}{\sqrt{k_x^2 + k_y^2}} (E^+ + E^-) = \frac{Y_m k}{k_z} (-E_y^+ + E_y^-)\, ,$$

$$H_{t,y} = \frac{Y_m k_x}{\sqrt{k_x^2 + k_y^2}}(E^+ + E^-) = \frac{-Y_m k}{k_z}(-E_x^+ + E_x^-) . \tag{1.10.37}$$

As is expected, the electric and magnetic tangential field vectors for the s- and p-polarisation case are mutually orthogonal. The s- and p-polarisation states are independent propagation modes of the multilayer. They can be tracked independently through the multilayer stack. In practice, one separates the incident field into these two orthogonal states of linear polarisation at the entrance surface. The amplitude reflectivity and transmittance of the multilayer are calculated for the s- and p-state. The reflected and transmitted s- and p-polarised beams are then coherently superposed to obtain the total fields that are reflected and transmitted by the stack.

In Eqs. (1.10.35) and (1.10.36), we encounter the quantities $Y_m k_z/k$ (s-polarisation) and $Y_m k/k_z$ (p-polarisation). In the two sets of equations above, we call $Y_m k_z/k$ and $Y_m k/k_z$ the *effective* admittances of the medium for the s- and p-polarisation modes, respectively. If an electric field with a certain tangential strength E_t is launched into the medium, $Y_m E_t k_z/k$ and $Y_m E_t k/k_z$ denote the resulting tangential magnetic fields at the interface for each of the two orthogonally polarised modes in the layer. With $Y_m = (\epsilon_{j+1}/\mu_{j+1})^{1/2}$ denoting the *intrinsic* admittance of the medium and assuming the wave vectors $\mathbf{k}^+$ and $\mathbf{k}^-$ are real, we find, as in previous instances, that $Y_s = Y_m \cos\theta$ and $Y_p = Y_m / \cos\theta$.

1.11 The Scattering Matrix and the Impedance Matrix Formalism

In this section we first analyse a serious shortcoming of the transfer matrix method, namely its numerical instability with respect to matrix inversion which is required when the multilayer is analysed for waves which each travel in a direction which is opposite to the original one. It turns out that numerical instability with respect to matrix inversion arises when one or several layers show a strong attenuation of the propagating waves (absorption) or when a wave is evanescent in a layer. We then briefly present the scattering or S-matrix algorithm and the impedance or R-matrix algorithm. They offer a solution to the instability of the transfer matrix method and are numerically robust in the case of evanescent wave propagation or wave propagation in absorbing media.

1.11.1 Numerical stability of the transfer matrix method

The analysis of light propagation in a stratified medium in the previous subsections was based on the transfer matrix algorithm, presented in Eq. (1.9.7). The formalism of the transfer matrix algorithm discussed in Subsection 1.9.1 and 1.10.2 uses the tangential electromagnetic components at each interface. A transfer matrix $\mathbf{M}_j$ links the tangential components at interface $(j-1, j)$ to those at the next interface between the layers j and $j+1$. The individual transfer matrices $\mathbf{M}_j$ are numerically well defined. The matrix multiplications that are needed to arrive at the total transfer matrix also yield numerically reliable results. The tangential field components at the exit of the multilayer are calculated in an exact way, only limited by the machine precision of the computer. This precision typically corresponds to that of 15 significant digits for double precision notation. The maximum or minimum number size is limited by the size of the mantissa in double precision notation and amounts to $10^{\pm 308}$. The following example, however, illustrates the numerical problems that arise when the transfer matrix method is used for an imaginary $k_{z,j}$ wave vector component with $k_{z,j} = i|k_{z,j}|$ (i.e. an evanescent wave). For the case of p-polarisation, the effective medium impedance is given by $Y_{p,j} = k_j Y_{m,j}/k_{z,j} = -ik_j Y_{m,j}/|k_{z,j}|$. The transfer matrix for the medium j is now given by

$$\mathbf{M}_j = \begin{pmatrix} \cos(i|k_{z,j|d_j}) & \dfrac{|k_{z,j}|}{k_j Y_{m,j}} \sin(i|k_{z,j}|d_j) \\ -\dfrac{k_j Y_{m,j}}{|k_{z,j}|} \sin(i|k_{z,j}|d_j) & \cos(i|k_{z,j}d_j) \end{pmatrix}$$

$$= \begin{pmatrix} \dfrac{\exp(-|k_{z,j}|d_j) + \exp(+|k_{z,j}|d_j)}{2} & \dfrac{\exp(-|k_{z,j}|d_j) - \exp(+|k_{z,j}|d_j)}{2i} \dfrac{|k_{z,j}|}{Y_{m,j}k_j} \\ -\dfrac{\exp(-|k_{z,j}|d_j) - \exp(+|k_{z,j}|d_j)}{2i} \dfrac{Y_{m,j}k_j}{|k_{z,j}|} & \dfrac{\exp -|k_{z,j}|d_j + \exp(+|k_{z,j}|d_j)}{2} \end{pmatrix} . \tag{1.11.1}$$

For long optical pathlengths $|k_{z,j}|d_j$, the exponential functions with negative exponent will be numerically obscured by the large positive exponentials, depending on the number of significant decimal places that are available in the calculations. If the exponential function with negative exponent is lost in the additions and subtractions in the matrix elements above, the matrix equation of Eq. (1.9.7) effectively becomes

$$\begin{pmatrix} E_{t,j} \\ H_{t,j} \end{pmatrix} \approx \frac{\exp(|k_{z,j}|d_j)}{2} \begin{pmatrix} 1 & +i\dfrac{|k_{z,j}|}{Y_{m,j}k_j} \\ -i\dfrac{Y_{m,j}k_j}{|k_{z,j}|} & 1 \end{pmatrix} \begin{pmatrix} E_{t,j-1} \\ H_{t,j-1} \end{pmatrix} . \tag{1.11.2}$$

As stated previously, the left-hand vector of Eq. (1.11.2) is exact up to the machine precision; however, inspection of Eq. (1.11.2) shows that the transfer matrix **M** has become singular and does *not* have an inverse. This means that with such a numerically truncated transfer matrix, it is not possible to reconstruct the original input vector $(E_{t,j-1}, H_{t,j-1})$ by evaluating the matrix product $\mathbf{M}_j\mathbf{M}_j^{-1}$. Inverse matrices are also needed to find the field inside a multilayer, see Eqs. (1.10.21) and (1.10.22), if (parts of) the incident *and* outgoing field components are specified.

As an example of the numerical problems that arise for wave propagation in an absorbing medium, we take the material aluminium with a complex refractive index at a wavelength of 500 nm of $n + i\kappa = 0.81 + i6.05$. For a layer thickness of 250 nm (half a vacuum wavelength), the transmitted field strength is given by $\exp(-\pi\kappa) = 5.6 \cdot 10^{-9}$. Using 15 significant decimal places in double precision computations, this number is negligible in additions or subtractions with respect to its inverse value of $1.8 \cdot 10^{+8}$. The transfer matrix of Eq. (1.11.1) for this relatively thin absorbing layer thus has become singular.

We conclude that the transfer matrix yields practically exact results when used in the forward direction. For strongly absorbing media, the condition number of the *inverse* transfer matrix may become of the order of $1/\delta$ with δ the relative precision of the matrix elements. The method does not allow the inverse propagation of evanescent waves in a layer of the stack if this layer has a thickness which is comparable to the vacuum wavelength of the light. The method is also inappropriate if the amplitude strongly decays in a particular layer due to absorption. The splitting of a layer into a certain number of thin sublayers does not improve the stability of the inverse matrix; the matrix product for the artificial sublayers becomes equivalent to the single matrix of the complete layer. In such cases, the transfer matrix is inadequate for an accurate description of both the forward and the backward propagating fields. Different matrix descriptions of the electric and magnetic field components on the entrance surface and the exit surface of a thin layer stack are needed and these descriptions are presented in the next subsection.

1.11.2 The Scattering Matrix S

The transfer matrix method uses as input and output vector the tangential electromagnetic components at the interface which are a superposition of the projected vectors of both the forward and the backward propagating wave in a layer. However, in the scattering matrix algorithm **[289]**, we consider the individual field strengths of the forward and backward propagating waves which, for the case of the electric field vector, are written as

$$\mathbf{E}_j(\mathbf{r})\big|_{\pm} = \mathbf{E}_j^{\pm} \exp\{i(k_{x,j}x \pm k_{z,j}z)\} . \tag{1.11.3}$$

The scattering matrix formalism uses either the electric field strengths or the magnetic ones. From the values of the electric field strengths we then obtain the magnetic values using the admittance of the medium and vice versa. A further difference between the transfer matrix algorithm and the scattering algorithm resides in the grouping of the electric (or magnetic) field strengths in the input and output vector (see Fig. 1.38). The strengths of the two incident field vectors are taken as input vector for a scattering matrix calculation. Multiplication of the input vector by the scattering matrix then yields the output vector which contains the field strengths of the two outgoing waves. The scattering matrix equation for the plane waves at an interface between medium $j + 1$ and j is thus given by

$$\begin{pmatrix} E_j^+ \\ E_{j+1}^- \end{pmatrix} = \begin{pmatrix} t_{j+1,j}^+ & r_{j+1,j}^- \\ r_{j+1,j}^+ & t_{j+1,j}^- \end{pmatrix} \begin{pmatrix} E_{j+1}^+ \\ E_j^- \end{pmatrix} , \tag{1.11.4}$$

where, for example, $r_{j+1,j}^+$ is the amplitude reflection coefficient at the interface $(j + 1, j)$ for a wave propagating in the positive z-direction and thus arriving from the medium $j + 1$. For a single set of incident and outgoing waves corresponding to one specific propagation direction, we have two entries for each E-element, namely the two orthogonally polarised electric (or magnetic) field strengths. The elements of the scattering matrix $\mathbf{S}_j$ are themselves square matrices

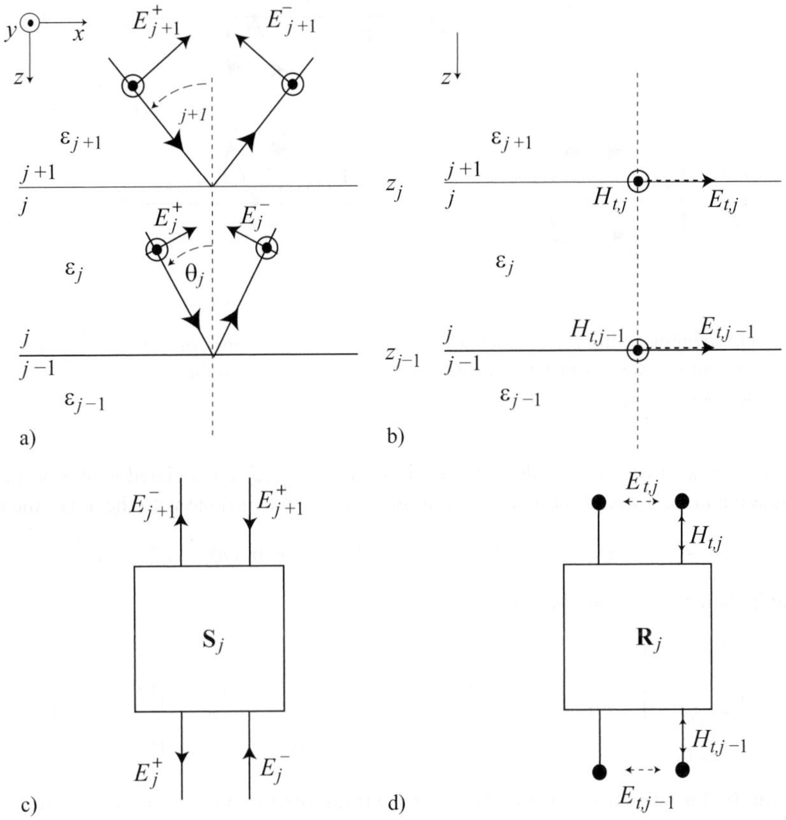

Figure 1.38: a) The input and output vectors used in the scattering matrix algorithm, and b) in the impedance matrix algorithm. The scattering matrix $\mathbf{S}_j$ applies to the field vectors entering and leaving a layer j, whilst the matrix $\mathbf{R}_j$ applies to the tangential electromagnetic vectors describing the orthogonal modes of a particular layer. c) and d) The network equivalents of the S- and R-matrix.

with dimension of 2×2. Only for the case of a single incident wave with a polarisation corresponding to the s- or p-eigenmode do the submatrices of the scattering matrix become single (complex) numbers and $\mathbf{S}_j$ reduces to a 2×2 matrix.

As an example we give the scattering matrix of an interface that is struck by incident and outgoing waves of which the electric field amplitudes have been decomposed into their s- and p-components. We obtain the matrix equation

$$\begin{pmatrix} E^+_{s,j,z_j} \\ E^+_{p,j,z_j} \\ E^-_{s,j+1,z_j} \\ E^-_{p,j+1,z_j} \end{pmatrix} = \begin{pmatrix} t^+_s & 0 & r^-_s & 0 \\ 0 & t^+_p & 0 & r^-_p \\ r^+_s & 0 & t^-_s & 0 \\ 0 & r^+_p & 0 & t^-_p \end{pmatrix} \begin{pmatrix} E^+_{s,j+1,z_j} \\ E^+_{p,j+1,z_j} \\ E^-_{s,j,z_j} \\ E^-_{p,j,z_j} \end{pmatrix}, \tag{1.11.5}$$

where, for ease of notation, we have dropped the subscript $(j, j+1)$ of the reflection and transmission coefficients. The interface is the transition from layer j to $j+1$ and the field amplitudes in both layers are evaluated, at infinitesimally small distances from the plane $z = z_j$. The propagation of the fields from the interface at $z = z_j$ to the next interface at $z = z_{j+1}$ is illustrated in Fig. 1.39b and is described by the scattering 'transport' matrix,

$$\begin{pmatrix} E^+_{s,j+1,z_j} \\ E^+_{p,j+1,z_j} \\ E^-_{s,j+1,z_{j+1}} \\ E^-_{p,j+1,z_{j+1}} \end{pmatrix} = \exp(i\phi_{j+1}) \begin{pmatrix} E^+_{s,j+1,z_{j+1}} \\ E^+_{p,j+1,z_{j+1}} \\ E^-_{s,j+1,z_j} \\ E^-_{p,j+1,z_j} \end{pmatrix}, \tag{1.11.6}$$

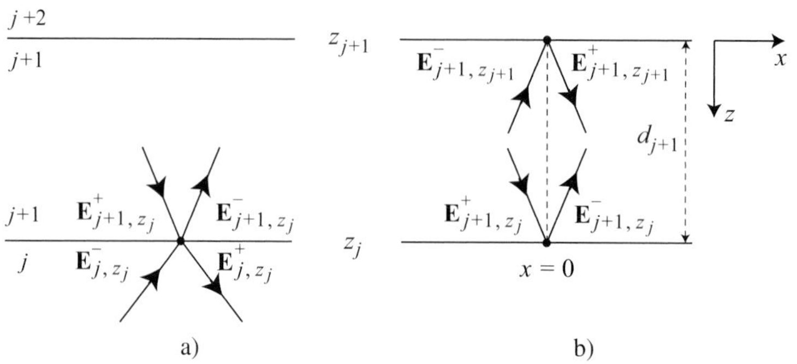

Figure 1.39: a) The field strengths of the waves associated with the scattering matrix of a single interface between medium $j+1$ to j. b) Field strengths of the waves which propagate in the layer $j+1$ between the interfaces at $z=z_j$ and $z=z_{j+1}$ and which are needed for the scattering transport matrix.

where, for compactness of notation, we have denoted the phase factor $k_{z,j+1}d_{j+1}$ associated with wave propagation in layer $j+1$ by ϕ_{j+1}. Elimination of the wave amplitudes in medium $j+1$ at $z=z_j$ is done with the aid of the relations

$$E^-_{j+1,z_j} = \exp(-i\phi_{j+1})\, E^-_{j+1,z_{j+1}}\,, \qquad E^+_{j+1,z_j} = \exp(+i\phi_{j+1})\, E^+_{j+1,z_{j+1}}\,, \tag{1.11.7}$$

and, after some manipulation, we obtain the expression

$$\begin{pmatrix} E^+_{s,j,z_j} \\ E^+_{p,j,z_j} \\ E^-_{s,j+1,z_{j+1}} \\ E^-_{p,j+1,z_{j+1}} \end{pmatrix} = \exp(i\phi_{j+1}) \begin{pmatrix} t_s^+ & 0 & r_s^-\, e^{-i\phi_{j+1}} & 0 \\ 0 & t_p^+ & 0 & r_p^-\, e^{-i\phi_{j+1}} \\ r_s^+\, e^{+i\phi_{j+1}} & 0 & t_s^- & 0 \\ 0 & r_p^+\, e^{+i\phi_{j+1}} & 0 & t_p^- \end{pmatrix} \begin{pmatrix} E^+_{s,j+1,z_{j+1}} \\ E^+_{p,j+1,z_{j+1}} \\ E^-_{s,j,z_j} \\ E^-_{p,j,z_j} \end{pmatrix}. \tag{1.11.8}$$

Equation (1.11.8) can be formulated in a different way by writing the matrix as a product of transport and interface scattering matrices, yielding

$$\begin{pmatrix} E^+_{s,j,z_j} \\ E^+_{p,j,z_j} \\ E^-_{s,j+1,z_{j+1}} \\ E^-_{p,j+1,z_{j+1}} \end{pmatrix} = \exp(i\phi_{j+1}) \begin{pmatrix} 1 & 0 & 0 & 0 \\ 0 & 1 & 0 & 0 \\ 0 & 0 & e^{+i\phi_{j+1}} & 0 \\ 0 & 0 & 0 & e^{+i\phi_{j+1}} \end{pmatrix} \begin{pmatrix} t_s^+ & 0 & r_s^- & 0 \\ 0 & t_p^+ & 0 & r_p^- \\ r_s^+ & 0 & t_s^- & 0 \\ 0 & r_p^+ & 0 & t_p^- \end{pmatrix} \begin{pmatrix} 1 & 0 & 0 & 0 \\ 0 & 1 & 0 & 0 \\ 0 & 0 & e^{-i\phi_{j+1}} & 0 \\ 0 & 0 & 0 & e^{-i\phi_{j+1}} \end{pmatrix} \begin{pmatrix} E^+_{s,j+1,z_{j+1}} \\ E^+_{p,j+1,z_{j+1}} \\ E^-_{s,j,z_j} \\ E^-_{p,j,z_j} \end{pmatrix}. \tag{1.11.9}$$

It is seen that the elements of the scattering matrix for an individual layer do not contain the differences of exponential functions, like those that occur in a sine or hyperbolic sine function of the quantity $k_{z,j+1}d_{j+1}$. The absence of such functions in the algorithm guarantees numerical stability when, for instance, absorptive materials or evanescent waves are encountered in one or several of the (thicker) layers of the stack.

The attractive feature of the transfer matrix algorithm, the concatenation of layers by a simple consecutive multiplication of matrices, is lost when using the scattering matrix approach for more than one layer. The input and output vectors in the scattering matrix formalism both contain electric field strengths at the entrance interface and at the exit interface. If two layer systems are combined and a scattering matrix of the combined layer system is required, it is necessary to eliminate the field strengths at the interface where the two layer systems are connected. A recursive scheme enables the elimination of these intermediate quantities and the elimination proceeds as follows. Having two scattering matrices defined by

$$\begin{pmatrix} E_2^+ \\ E_1^- \end{pmatrix} = \begin{pmatrix} S_{11}^{(1)} & S_{12}^{(1)} \\ S_{21}^{(1)} & S_{22}^{(1)} \end{pmatrix} \begin{pmatrix} E_1^+ \\ E_2^- \end{pmatrix}, \qquad \begin{pmatrix} E_3^+ \\ E_2^- \end{pmatrix} = \begin{pmatrix} S_{11}^{(2)} & S_{12}^{(2)} \\ S_{21}^{(2)} & S_{22}^{(2)} \end{pmatrix} \begin{pmatrix} E_2^+ \\ E_3^- \end{pmatrix}, \tag{1.11.10}$$

we require the matrix that links the input and output vectors of the combination

$$\begin{pmatrix} E_3^+ \\ E_1^- \end{pmatrix} = \begin{pmatrix} S_{11}^{(3)} & S_{12}^{(3)} \\ S_{21}^{(3)} & S_{22}^{(3)} \end{pmatrix} \begin{pmatrix} E_1^+ \\ E_3^- \end{pmatrix}. \tag{1.11.11}$$

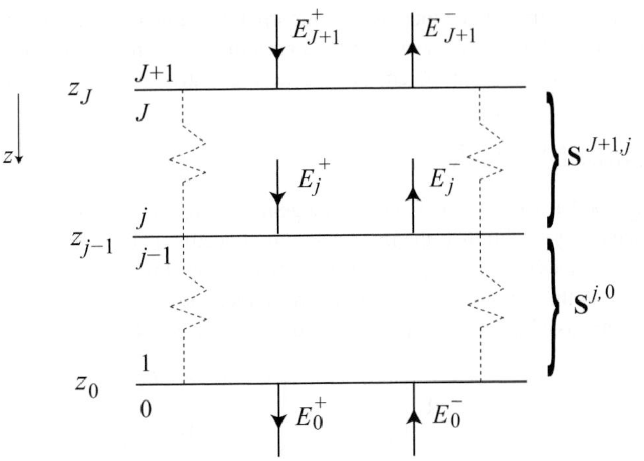

Figure 1.40: Calculation of the electric field strengths of the plane waves inside a multilayer, at the upper side of the interface $(j-1,j)$, using the scattering matrix algorithm. In general, the strength E^+_{J+1} is given of the electric field vector of a wave that is incident on the multilayer from the upper half-space.

After some manipulation we find the following expressions for the submatrices of the combination matrix $\mathbf{S}^{(3)}$,

$$S^{(3)}_{11} = S^{(2)}_{11}\left(I - S^{(1)}_{12}S^{(2)}_{21}\right)^{-1} S^{(1)}_{11} ,$$
$$S^{(3)}_{12} = S^{(2)}_{12} + S^{(2)}_{11}\left(I - S^{(1)}_{12}\,S^{(2)}_{21}\right)^{-1} S^{(1)}_{12}\,S^{(2)}_{22} ,$$
$$S^{(3)}_{21} = S^{(1)}_{21} + S^{(1)}_{22}\,S^{(2)}_{21}\left(I - S^{(1)}_{12}\,S^{(2)}_{21}\right)^{-1} S^{(1)}_{11} ,$$
$$S^{(3)}_{22} = S^{(1)}_{22}\,S^{(2)}_{21}\left(I - S^{(1)}_{12}\,S^{(2)}_{21}\right)^{-1}\left(S^{(2)}_{21}\right)^{-1} S^{(2)}_{22} = S^{(1)}_{22}\left(I - S^{(2)}_{21}\,S^{(1)}_{12}\right)^{-1} S^{(2)}_{22} , \qquad (1.11.12)$$

where I is again the identity matrix. The matrix multiplication according to the scheme above was first introduced by Redheffer **[289]** in the context of multiple input and output electrical network theory. It is usually denoted as the Redheffer 'star' product, $\mathbf{S}^{(3)} = \mathbf{S}^{(1)} \otimes \mathbf{S}^{(2)}$.

We apply the elimination procedure to a stack (total number of layers is J) in which we want to calculate the field strengths in an intermediate layer with index j, at the upper side of the interface $(j-1,j)$ (see Fig. 1.40). We assume that the field strengths of the incident waves at the outer interfaces, $(0,1)$ and $(J,J+1)$, respectively, are known and we use these values to obtain the field strengths at the interface with coordinate $z = z_{j-1}$, in the layer with index j. We proceed by applying the recursive scheme using the Redheffer star product for the field propagation from the upper half-space $J+1$ to the upper side of interface $(j-1,j)$, including the propagation part in layer j from interface $(j+1,j)$ to $(j,j-1)$. Starting the recursive scheme with the identity matrix in medium $J+1$, we obtain by recursion a scattering matrix $\mathbf{S}^{J+1,j}$. Following the same procedure and starting at the upper side of the $(j-1,j)$ interface, we find a matrix $\mathbf{S}^{j,0}$ that links the outgoing and incoming fields in layer l and the substrate. In matrix notation, we have

$$\begin{pmatrix} E^+_j \\ E^-_{J+1} \end{pmatrix} = \mathbf{S}^{J+1,j}\begin{pmatrix} E^+_{J+1} \\ E^-_j \end{pmatrix} , \qquad \begin{pmatrix} E^+_0 \\ E^-_j \end{pmatrix} = \mathbf{S}^{j,0}\begin{pmatrix} E^+_j \\ E^-_0 \end{pmatrix} . \qquad (1.11.13)$$

As was shown in Eq. (1.11.11), with knowledge about the incoming fields E^+_{J+1} and E^-_0, we can find a solution for the fields of interest E^+_j and E^-_j. In the frequently occurring case that there is no wave incident from the substrate side and the field incident from the upper half-space is known, we have the following solution in terms of the S-submatrices,

$$E^-_j = S^{j,0}_{21}\left(I - S^{J+1,j}_{12}\,S^{j,0}_{21}\right)^{-1} S^{J+1,j}_{11}\,E^+_{J+1} ,$$
$$E^+_j = \left(I - S^{J+1,j}_{12}\,S^{j,0}_{21}\right)^{-1} S^{J+1,j}_{11}\,E^+_{J+1} . \qquad (1.11.14)$$

The dimensions of the scattering matrix can be increased to incorporate many in- and outgoing waves, for instance those entering and leaving a waveguide, those being scattered by a periodic structure, etc. Because of its numerical stability, also in such more intricate wave propagation problems, the scattering matrix algorithm is widely used.

1.11.3 The Impedance Matrix R

In the case of the so-called impedance or R-matrix, we use tangential components of the electromagnetic field vectors on both sides of a layer interface or layer section as the input and output vectors in the matrix algorithm, exactly as we did in our treatment of the transfer matrices (see also Fig. 1.38 for a network-based picture of the impedance matrix algorithm). The essential difference between the transfer matrix and the impedance matrix method is the way in which the tangential electromagnetic components are assigned to the input and output vectors. The **R**-matrix equation reads as follows **[208]**,

$$\begin{pmatrix} E_{t,j} \\ \\ E_{t,j-1} \end{pmatrix} = \mathbf{R}_j \begin{pmatrix} H_{t,j} \\ \\ H_{t,j-1} \end{pmatrix} = \begin{pmatrix} R_{11}^{(j)} & R_{12}^{(j)} \\ \\ R_{21}^{(j)} & R_{22}^{(j)} \end{pmatrix} \begin{pmatrix} H_{t,j} \\ \\ H_{t,j-1} \end{pmatrix}. \tag{1.11.15}$$

The difference with the transfer matrix method is that the input and output vector each contain field values associated with opposite sides of a layer or an interface. We observe that in Eq. (1.11.15) all **R**-matrix elements have the dimension of electrical impedance. Interchanging the E- and H-vectors would lead to matrix elements which have the dimension of an electrical admittance.

The elementary matrix $\mathbf{R}_j$, describing both a transition at an interface and wave transport to the next interface, can be derived from Eq. (1.9.7) by rearranging the input and output vector. Combining the s- and p-polarised tangential components into a single matrix, we obtain

$$\begin{pmatrix} E_{t,s,j} \\ E_{t,p,j} \\ E_{t,s,j-1} \\ E_{t,p,j-1} \end{pmatrix} = \frac{i}{\sin\phi_j} \begin{pmatrix} \frac{\cos\phi_j}{Y_{s,j}} & 0 & -\frac{1}{Y_{s,j}} & 0 \\ 0 & \frac{\cos\phi_j}{Y_{p,j}} & 0 & -\frac{1}{Y_{p,j}} \\ \frac{1}{Y_{s,j}} & 0 & -\frac{\cos\phi_j}{Y_{s,j}} & 0 \\ 0 & \frac{1}{Y_{p,j}} & 0 & -\frac{\cos\phi_j}{Y_{p,j}} \end{pmatrix} \begin{pmatrix} H_{t,p,j} \\ H_{t,s,j} \\ H_{t,p,j-1} \\ H_{t,s,j-1} \end{pmatrix}, \tag{1.11.16}$$

where ϕ_j is the (complex) optical pathlength $k_{z,j}d_j$ in layer j. Note the connection between E_s and H_p for the components associated with electrical s-polarisation and between E_p and H_s for the electrically p-polarised wave.

1.11.3.1 Numerical Stability of the Impedance Matrix

For the case of, for instance, an evanescent wave with $\phi_j = ia$, the elementary $\mathbf{R}_j$-matrix for s- or p-polarisation is given by

$$\mathbf{R}_j = \frac{-1}{Y_j[e^{-a} - e^{a}]} \begin{pmatrix} e^{-a} + e^{a} & -2 \\ +2 & -e^{-a} - e^{a} \end{pmatrix},$$

$$\mathbf{R}_j^{(\infty)} = \lim_{|\phi_j|\to\infty} \mathbf{R}_j = \frac{1}{Y_j} \begin{pmatrix} 1 & 0 \\ 0 & -1 \end{pmatrix}, \tag{1.11.17}$$

where Y_j is the layer admittance for either s- or p-polarisation and $\mathbf{R}_j^{(\infty)}$ corresponds to the limiting case for large values of $|\phi_j|$. The limiting form of the matrix will also be the result of numerical calculations when truncation errors due to limited precision have occurred. Despite the loss of numerical accuracy, $\mathbf{R}_j^{(\infty)}$ has a nonzero determinant equal to $-1/Y^2$, the same as the determinant of the exact matrix $\mathbf{R}_j$. A forward and backward calculation of the tangential fields in the layer stack will return the starting values due to the existence of a stable inverse matrix.

The particular shape of matrix $\mathbf{R}_j^{(\infty)}$ is explained with the aid of Fig. 1.41. The wave attenuation in layer j is very strong because of absorption or because of the evanescent character of the wave in layer j. In the limiting case, there are no outgoing waves left and the only waves that create the tangential fields are those with amplitudes $E^{+}_{j,j+1}$ and $E^{-}_{j,j-1}$. The propagation direction of the first wave is such that $H_{t,j} = YE_{t,j}$, given the reference directions for E_t and H_t. The wave leaving the interface $(j, j-1)$, propagating in the negative z-direction, creates tangential E and H components that are linked by the relationship $H_{t,j-1} = -YE_{t,j-1}$. Alternatively, one can state that the admittance of the medium is negative for this direction of propagation. Given the strong attenuation, there is no mutual influence between the tangential field components on opposite sides of layer j, which explains the off-diagonal zeros in $\mathbf{R}_j^{(\infty)}$.

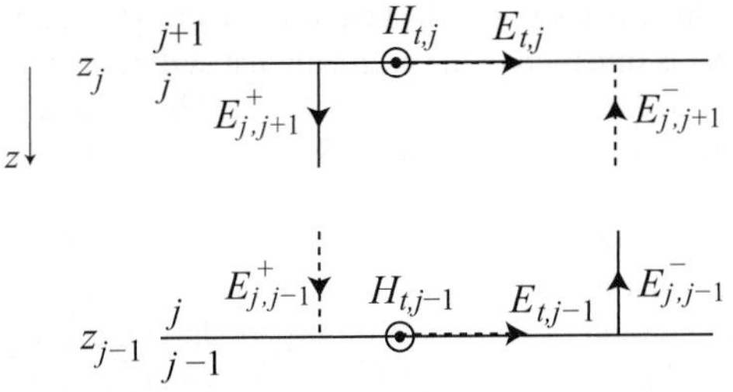

Figure 1.41: The tangential field components at both sides of a layer with index j in which strong wave attenuation takes place (p-polarisation). Only the waves in layer j that leave an interface can have appreciable amplitudes; the field strengths of the waves that arrive at an interface (dashed in the figure) have been reduced to negligibly small values.

1.11.3.2 Recursive Scheme for the Impedance Matrix

To obtain the **R**-matrix of a certain number of layers or of an entire multilayer stack, one needs a recursive scheme, as with the case of the scattering matrix algorithm. Defining two **R** matrices according to

$$\begin{pmatrix} E_{t,2} \\ E_{t,1} \end{pmatrix} = \mathbf{R}^{(1)} \begin{pmatrix} H_{t,2} \\ H_{t,1} \end{pmatrix}, \qquad \begin{pmatrix} E_{t,3} \\ E_{t,2} \end{pmatrix} = \mathbf{R}^{(2)} \begin{pmatrix} H_{t,3} \\ H_{t,2} \end{pmatrix}, \tag{1.11.18}$$

and also writing the composite matrix equation in the form

$$\begin{pmatrix} E_{t,3} \\ E_{t,1} \end{pmatrix} = \mathbf{R}^{(3)} \begin{pmatrix} H_{t,3} \\ H_{t,1} \end{pmatrix}, \tag{1.11.19}$$

we find

$$\mathbf{R}^{(3)} = \begin{pmatrix} R^{(2)}_{11} + R^{(2)}_{12}\left(R^{(1)}_{11} - R^{(2)}_{22}\right)^{-1} R^{(2)}_{21} & -R^{(2)}_{12}\left(R^{(1)}_{11} - R^{(2)}_{22}\right)^{-1} R^{(1)}_{12} \\ R^{(1)}_{21}\left(R^{(1)}_{11} - R^{(2)}_{22}\right)^{-1} R^{(2)}_{21} & R^{(1)}_{22} - R^{(1)}_{21}\left(R^{(1)}_{11} - R^{(2)}_{22}\right)^{-1} R^{(1)}_{12} \end{pmatrix}. \tag{1.11.20}$$

We conclude with a remark about the individual matrix $\mathbf{R}_j$. It becomes singular for real values of $k_{z,j}d_j$ which are equal to $n\pi$ where n is an integer. In practice, this does not need to pose a problem. Such a critical layer thickness d_j is avoided by replacing the layer with several sub-layers for which $\sin\phi$ is not zero. By applying the recursive scheme using these sub-layers, the singularity is avoided and a reliable numerical evaluation of the total **R** matrix remains feasible. The impedance matrix algorithm is frequently used when analysing light scattering by an optical structure with the aid of modal analysis. The diffraction of light by a periodic structure is also well adapted to this matrix method. The input and output vectors are the electric and magnetic field components belonging to the freely propagating and bounded diffraction orders that are created by the incident coherent wave or waves.

1.12 Stratified Medium with Laterally Modulated Periodic Sublayers

Stratified media with perfectly flat interfaces have been analysed in the previous sections with, in principle, two waves present at both the entrance face and the exit face of the total stack of thin layers. In many instances, one of the outer or internal surfaces may carry a perturbation such that a planar interface becomes deformed. It is also possible that local changes in the electric permittivity or the magnetic permeability occur. In that case it may not only be a planar surface that has deformed but a certain volume within a layer that has different optical properties. The change can be limited to a very restricted lateral area so that a virtually point-like perturbation results. Alternatively, it can be a more general perturbation that is described by some 'structure' function which depends on the lateral coordinates (x, y). Mathematically, it can be advantageous to perform a periodic Fourier decomposition of the lateral structure function and to treat the various harmonic components that were produced by the structure separately. If the object itself is not periodic, an elementary part of it can be periodically continued so that a Fourier decomposition of the structure is feasible. The simplest periodic structures are gratings, which have been used for a long time in optical spectroscopy. Such structures delimit two semi-infinite media by means of a surface that carries a periodic height variation. The height variation can be binary, triangular or have a more complicated shape that is imparted to the originally flat surface by the cutting device of a so-called ruling machine. Alternatively, analogous surface profiles can also be fabricated by using a periodically varying intensity profile to develop a photoresist layer. In all cases, such a periodic object produces multiple *diffraction orders* from a single incident plane wave.

In many instances, the lateral changes in the optical properties and shape of a particular layer in a stack form the features of a three-dimensional object that has to be imaged by an optical system and that needs to be interpreted after detection. To allow the correct interpretation of the image of a (periodic) object, the information about the amplitude, phase and state of polarisation of the diffracted orders of that object should be retrieved from the image intensity. By means of 'reverse engineering' the image data it is possible to determine the exact structure of the object. Unfortunately, some basic information is lost if the light is propagated over larger distances before it is captured by an imaging system. This means that any near-field information about the object, contained in non-propagating evanescent waves, is always lost in a 'far-field' imaging method. In this section we do not discuss the 'reverse engineering or 'object retrieval' process. We focus on the forward diffraction problem, viz. calculation of the diffraction orders, both reflected and transmitted, that are created by the periodic lateral modulation of the object structure.

In the optical domain, extensive research has been carried out on the topic of diffraction by periodic structures since the 1960s. Interesting analytical results had already been obtained before this period, for instance, by Sommerfeld **[321]**, Kottler **[192]**, **[193]** and Bouwkamp **[42]**, but the development of numerical methods to analyse more complicated diffracting geometries only took off with the advent of the modern digital computer. Optical computations could then build further on reference cases from electrical engineering (antenna, coaxial cable, waveguide). The special features of optical diffraction problems stem mostly from the wide range of electrical and magnetic material properties that are found in the optical spectrum from far infrared to deep UV wavelengths. In this section we first give an outline of light diffraction problems to be solved, mostly defined in terms of a boundary value problem. A detailed treatment of all the available methods is outside the scope of this book. A fortunate circumstance for the reader is that a number of these methods have proven to be robust and can handle a wide variety of object geometries. In practice, the diffracted fields calculated with the aid of these numerical methods serve as the electromagnetic fields emanating from a two- or three-dimensional illuminated object. The outgoing object waves propagate towards the imaging system and produce the intensity pattern in the image region.

1.12.1 Diffraction Orders Produced by a Periodic Structure

If a periodic structure is formed by a surface with a periodically repeating height profile arranged along straight lines it is called a (one-dimensional) grating. The grating pitch d is the period of the grating and it corresponds to the offset that is given to a cutting tool of the ruling machine after each passage over the surface. For the optical modelling of such a periodic surface geometry, we assume that the periodic surface delimits two semi-infinite homogeneous and isotropic media with medium properties given by the values of the electric permittivity ϵ and the magnetic permeability μ (see Fig. 1.42). In its simplest form, the height profile is binary (see insert b)). A plane wave with wave vector k_i is incident on the grating structure in the plane $y = 0$ at an angle α_1 to the positive z-axis. In the insert a) of Fig. 1.42 the three-dimensional geometry is shown with the so-called grating vector $\mathbf{v}_d$ (with $|\mathbf{v}_d| = 1/d$) parallel to the x-axis. The grating vector is by definition perpendicular to the grating lines. In the simplest geometry, the vectors k_i and $\mathbf{v}_d$ are in the plane that is perpendicular to the so-called longitudinal grating direction, i.e. parallel to the y-axis in the figure. Any translation along this direction will leave the grating diffraction geometry and the beams diffracted by the grating unchanged. A more general geometry may comprise a two-dimensional (rectangular) grating with grating periods d_x and d_y and an incident beam that is defined by the inclination angle α_i and a supplementary azimuthal angle α_a without the restriction to the special cases that α_a is 0 or π (incident beam parallel to the plane $y = 0$) or $\pi/2$ or $3\pi/2$ (incident beam parallel to the plane $x = 0$, see Fig. 1.43). For solving the diffraction problem, the one-dimensional grating geometry of Fig. 1.42 is analytically the simplest one and corresponds to the so-called classical grating mounting that is used in most spectroscopic applications. When the incident beam vector is not parallel to the plane $y = 0$, the grating geometry corresponds to a *conical* grating mounting and the analysis is less straightforward. In this subsection we first derive expressions for the wave vectors of the waves which are diffracted by a periodic structure. A simple one-dimensional periodic structure is used with an in-plane geometry for the incident wave and the result for this simple structure is then extended to a two-dimensional periodic structure.

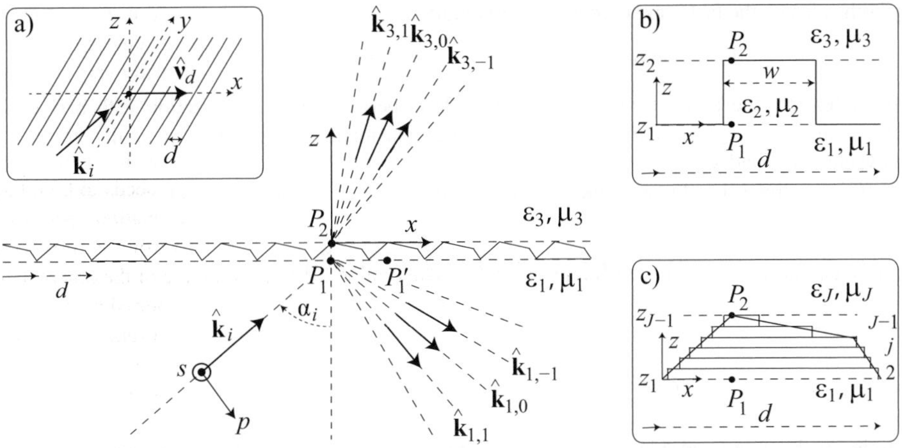

Figure 1.42: A periodic height-modulated interface (period d) between two media with permittivity and permeability given by (ϵ_1, μ_1) and (ϵ_3, μ_3), respectively. An incident plane wave in the first medium has wave vector k_i. Reflected and transmitted waves have propagation directions given by the normalised wave vectors $\hat{k}_{1,m}$ and $\hat{k}_{3,m}$, with m indicating the order number of a diffracted wave. Only some of the freely propagating plane waves in the two half-spaces have been shown. Insert a) The three-dimensional geometry (one-dimensional periodic structure.) Insert b) A binary height modulation of the surface. Insert c) Illustrating a more complicated height profile which is approximated by a certain number of sublayers, numbered $j = 2, \cdots, J-1$.

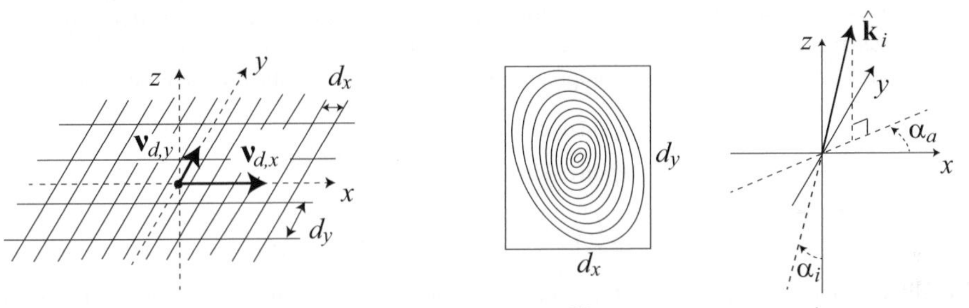

Figure 1.43: a) Two-dimensional rectangular periodic structure with periods d_x and d_y. b) Example of a height profile (contour plot) within an elementary periodic cell. c) Incident plane wave characterised by the (normalised) wave vector $\hat{k}_i$ and the inclination and azimuthal angles α_i and α_a, respectively.

1.12.1.1 Expression for the Wave Vector of a Diffraction Order

To calculate the possible directions in which light is diffracted by a periodic structure, as free propagating waves or as surface-bound evanescent waves, we use an argument that was given by Petit **[276]** and Maystre **[243]**. In Fig. 1.42 the incident normalised wave vector $\hat{k}_i$ is parallel to the plane $y = 0$. We assume that the total electric field on a vertical axis through P_1P_2 is known. To simplify the analysis, we use the transverse electric state of polarisation (*TE* or *s*-polarisation) and we denote this total field function by $E_y(x, z)$. If we shift the observation axis horizontally over a distance d, the period of the grating structure, the field function will be the same, apart from a phase difference that is caused by the shift in phase of the incident beam when moving from P_1 to P_1'. We thus have

$$E_y(x + d, z) = E_y(x, z) \exp(ik_i d \sin\alpha_i) ,$$

or ,

$$E_y(x, z) = E_y(x + d, z) \exp(-ik_i d \sin\alpha_i) . \tag{1.12.1}$$

We multiply both sides by the factor $\exp(-ik_i x \sin\alpha_i)$ to obtain

$$E_y(x,z)\exp(-ik_i x\sin\alpha_i) = E_y(x+d,z)\exp\{-ik_i(x+d)\sin\alpha_i\}\,. \tag{1.12.2}$$

This equation states that the function on the left-hand side is periodic in the x-direction with period $d = 1/\nu_d$. An analogous argument can be applied equally well to a two-dimensional grating with rectangular grating axes parallel to the grating vectors $\mathbf{v}_{d,x}$ and $\mathbf{v}_{d,y}$.

In the case of a one-dimensional grating the periodic function $E_y(x,z)$ is expanded as a Fourier series according to

$$E_y(x,z)\exp(-ik_i x\sin\alpha_i) = \sum_{m=-\infty}^{+\infty} E_{y,m}(z)\exp(i2\pi m\nu_d x)\,,$$

or,

$$E_y(x,z) = \sum_{m=-\infty}^{+\infty} E_{y,m}(z)\exp\{i(k_i\sin\alpha_i + 2\pi m\nu_d)x\}\,. \tag{1.12.3}$$

The infinite series on the right-hand side of Eq. (1.12.3) represents a set of waves with the x-component of the wave vector given by the quantity between round brackets, $k_i\sin\alpha_i + 2\pi m\nu_d$. For an electromagnetic field component $A_j(x,y,z)$ that arises in the general two-dimensional grating geometry we obtain a double periodic series of waves according to

$$A_j(x,y,z) = \sum_{m_x=-\infty}^{+\infty}\sum_{m_y=-\infty}^{+\infty} A_{j,m_x,m_y}(z)\exp\{i[(k_i\sin\alpha_i\cos\alpha_a + 2\pi m_x\nu_{d,x})x + (k_i\sin\alpha_i\sin\alpha_a + 2\pi m_y\nu_{d,y})y]\}\,, \tag{1.12.4}$$

where the x- and y-wave vector components of each wave follow from the exponential terms. We can obtain the corresponding z-component of each wave vector using the dispersion relation of Eq. (1.6.62). For non-magnetic media this relation is conveniently written as $k^2 = n_{\text{eff}}^2 k_0^2$ where n_{eff} is the (generally complex) refractive index of the inhomogeneous medium and k_0 the vacuum wavenumber of the radiation. The effective index of the medium is dependent on the lateral frequency combination (m_x, m_y). In the forthcoming subsection we provide the field solutions in each layer from which we will see that each frequency pair (m_x, m_y) can be associated with a *propagation mode* with its characteristic effective medium index n_{eff}. The wave vector components of a general term of the double Fourier series are hence

$$\begin{aligned}
k_{x,m_x} &= k_i\sin\alpha_i\cos\alpha_a + 2\pi m_x\nu_{d,x}\,,\\
k_{y,m_y} &= k_i\sin\alpha_i\sin\alpha_a + 2\pi m_y\nu_{d,y}\,,\\
k_{z,j,m_x,m_y} &= \pm\left((n_{\text{eff}}k_0)^2 - k_{x,m_x}^2 - k_{y,m_y}^2\right)^{1/2}\,.
\end{aligned} \tag{1.12.5}$$

The subscript j refers to the sublayer index of the grating structure which runs from 1 to 3 for the simplest configuration. As shown in Fig. 1.42, in this case layer 1 and 3 correspond to the incident and exit medium, respectively. Layer 2 corresponds to the modulated grating layer. The optical properties of layer 2 depend on the method of fabrication. For example, if the grating is made using a cutting machine, the permittivity of layer 2 will match that of the incident and exit medium in a position-dependent manner. A photoresist layer will, however, have a permittivity ϵ_2 which is independent of the permittivities of the entrance and exit media. More complicated geometries than a simple binary modulation scheme can be modelled by subdividing the modulated area into multiple sublayers indexed as $j = 2, \cdots, J-1$. The optical properties of sublayer j are then given by (ϵ_j, μ_j). From here on, for the optical spectral domain, we suppose that the magnetic permeability of all media is equal to that of vacuum, μ_0.

The plus or minus sign of the component k_{z,j,m_x,m_y} also depends on the medium index j. In the first medium ($j = 1$) and noting the origin for the z-coordinate is at P_1, we have that $-\infty < z \le 0$. The minus sign is used when describing propagating reflected waves in the first medium with z-components $k_{z,1,m_x,m_y}$. For evanescent waves in the first medium, we require that they tend to zero for large negative values of z. In the outgoing upper half-space, the opposite sign is chosen for k_{z,J,m_x,m_y} as compared to that of $k_{z,1,m_x,m_y}$. In that way, we only allow outgoing propagating waves and exclude any incident wave coming from z=$+\infty$, and non-propagating waves with a positive imaginary value for k_{z,J,m_x,m_y} will decay to zero for positive z-values. These choices lead to the scheme

$$z<0\,,\quad k_{z,1,m_x,m_y} = \begin{cases} -(k_1^2 - k_{x,m_x}^2 - k_{y,m_y}^2)^{1/2}\,, & k_1^2 - k_{x,m_x}^2 - k_{y,m_y}^2 \ge 0\,,\\ -i(k_{x,m_x}^2 + k_{y,m_y}^2 - k_1^2)^{1/2}\,, & k_1^2 - k_{x,m_x}^2 - k_{y,m_y}^2 < 0\,, \end{cases} \tag{1.12.6}$$

$$z > z_{J-1}\,, \qquad k_{z,J,m} = \begin{cases} (k_J^2 - k_{x,m_x}^2 - k_{y,m_y}^2)^{1/2}\,, & k_J^2 - k_{x,m_x}^2 - k_{y,m_y}^2 \geq 0\,, \\ i\,(k_{x,m_x}^2 + k_{y,m_y}^2 - k_J^2)^{1/2}\,, & k_J^2 - k_{x,m_x}^2 - k_{y,m_y}^2 < 0\,. \end{cases} \tag{1.12.7}$$

When solving a diffraction problem, a large fraction of the waves will be evanescent waves since these are needed to represent the fine structure of the electric and magnetic field at the sharp edges of the grating structure. For instance, in the example of the one-dimensional grating in insert b) of Fig. 1.42, discontinuities of the electric fields can be expected at the horizontal and vertical sections of the grating structure in the case of p- or TM-polarisation.

In the intermediate region(s) with $2 \leq j \leq J-1$, propagating and evanescent waves are possible with both signs of the z-component allowed. For each sublayer with its specific periodic permittivity function, a Fourier expansion is calculated,

$$\epsilon_j(x,y) = \sum_{m_x=-\infty}^{+\infty} \sum_{m_y=-\infty}^{+\infty} \epsilon_{j,m_x,m_y} \exp\{i2\pi(m_x \nu_{d,x} x + m_y \nu_{d,y} y)\}\,. \tag{1.12.8}$$

In practice, the infinite series are truncated to a finite number of terms, symmetrically chosen with respect to the term with $m_x = m_y = 0$. For reasons that will become clear in the following subsection, $(4M_x + 1) \times (4M_y + 1)$ terms are required in the ϵ-expansion if the Fourier expansions of the field components $E_j(x, y, z)$ and $H_j(x, y, z)$ are each limited to $(2M_x + 1) \times (2M_y + 1)$ terms.

1.12.2 The Differential Maxwell Equations in Grating Sublayers

The solution to Maxwell's equations in each sublayer is based on a Fourier expansion of the electric permittivity function (Eq. (1.12.8)). For this reason, this method of solution is called the Fourier modal method (FMM) or, alternatively, rigorous coupled wave analysis (RCWA). In this section we analyse the general case of a two-dimensional (or crossed) grating that is illuminated by a propagating plane wave coming from an arbitrary direction. The solution method for this general case was given by Moharam and Gaylord **[250]** and Noponen and Turunen **[261]**. A systematic overview of the method for various types of gratings and illumination geometries can be found in Nugrowati **[262]**. The solution of Maxwell's equations using the RCWA method which is presented in this subsection for a two-dimensional periodic structure is largely based on the analysis given in **[262]**.

In the two half-spaces labelled $j = 1$ and $j = J$ of Fig. 1.42c and in each of the modulated sublayers with indices $2 \leq j \leq J-1$, Maxwell's equations must be solved. In the absence of free charges and induced currents in the materials, we use Eqs. (1.6.5) and (1.6.6) in terms of the electric field vector E and the magnetic field vector H. For the general three-dimensional case we write,

$$\begin{aligned} &\text{a)}\quad \frac{\partial E_z}{\partial y} - \frac{\partial E_y}{\partial z} = i\omega\mu_0 H_x\,, &&\text{d)}\quad \frac{\partial H_z}{\partial y} - \frac{\partial H_y}{\partial z} = -i\omega\epsilon E_x\,, \\ &\text{b)}\quad \frac{\partial E_x}{\partial z} - \frac{\partial E_z}{\partial x} = i\omega\mu_0 H_y\,, &&\text{e)}\quad \frac{\partial H_x}{\partial z} - \frac{\partial H_z}{\partial x} = -i\omega\epsilon E_y\,, \\ &\text{c)}\quad \frac{\partial E_y}{\partial x} - \frac{\partial E_x}{\partial y} = i\omega\mu_0 H_z\,, &&\text{f)}\quad \frac{\partial H_y}{\partial x} - \frac{\partial H_x}{\partial y} = -i\omega\epsilon E_z\,. \end{aligned} \tag{1.12.9}$$

In Eq. (1.12.9) the field vector components are each obtained as the sum of a certain number of Fourier terms. An example of such a solution, with an infinite number of Fourier terms, was given in Eq. (1.12.3) for the E_y-component in the two-dimensional wave diffraction problem at a one-dimensional grating. In practice, the Fourier series are truncated with all terms with index $|m| > M$ taken to be zero (M is a non-negative integer). The total number of Fourier terms is thus limited to $2M + 1$. Differentiation of a field component with respect to x or y implies multiplication of each Fourier component with a factor $i(n_1 k_0 \sin\alpha_i \cos\alpha_a + 2\pi m_x \nu_{d,x})$ or $i(n_1 k_0 \sin\alpha_i \sin\alpha_a + 2\pi m_y \nu_{d,y})$, respectively. We write in formal vector-matrix notation

$$\frac{\partial \mathcal{E}}{\partial x} = i\,\mathbf{K}_x\,\mathcal{E}\,, \qquad \frac{\partial \mathcal{E}}{\partial y} = i\,\mathbf{K}_y\,\mathcal{E}, \tag{1.12.10}$$

where $\mathbf{K}_x$ and $\mathbf{K}_y$ are square $(2M + 1) \times (2M + 1)$ matrices with their diagonal elements $K_{x,l,l}$ and $K_{y,l,l}$ equal to $n_1 k_0 \sin\alpha_i \cos\alpha_a + 2\pi(l - M - 1)\nu_{d,x}$ and $n_1 k_0 \sin\alpha_i \sin\alpha_a + 2\pi(l - M - 1)\nu_{d,y}$, respectively, and with all off-diagonal elements equal to zero. We also note for further use that the order of multiplication of two $\mathbf{K}$-matrices is irrelevant as two diagonal matrices commute. The symbol $\mathcal{E}$ represents a column vector for which the lth element contains the $\mathcal{E}_{l-M-1}$ Fourier component of the field. We note that the Fourier coefficients are z-independent for a specific sublayer j.

Using this vector-matrix notation for the Fourier components of each of the six electromagnetic vector components in a sublayer j, we write Eqs. (1.12.9a)-(1.12.9f) as follows,

$$\begin{aligned}
&\text{a)}\quad i\mathbf{K}_y\mathcal{E}_z - \frac{\partial \mathcal{E}_y}{\partial z} = i\omega\mu_0\mathcal{H}_x\,, &&\text{d)}\quad i\mathbf{K}_y\mathcal{H}_z - \frac{\partial \mathcal{H}_y}{\partial z} = -i\omega\epsilon\mathcal{E}_x\,,\\
&\text{b)}\quad \frac{\partial \mathcal{E}_x}{\partial z} - i\mathbf{K}_x\mathcal{E}_z = i\omega\mu_0\mathcal{H}_y\,, &&\text{e)}\quad \frac{\partial \mathcal{H}_x}{\partial z} - i\mathbf{K}_x\mathcal{H}_z = -i\omega\epsilon\mathcal{E}_y\,,\\
&\text{c)}\quad i\mathbf{K}_x\mathcal{E}_y - i\mathbf{K}_y\mathcal{E}_x = i\omega\mu_0\mathcal{H}_z\,, &&f)\quad i\mathbf{K}_x\mathcal{H}_y - i\mathbf{K}_y\mathcal{H}_x = -i\omega\epsilon\mathcal{E}_z\,.
\end{aligned} \tag{1.12.11}$$

Equations (1.12.9d)–(1.12.9f) contain products of the Fourier expansion coefficients of the function $\epsilon(x,y,z)$ and of an electric field component in a sublayer. This product is written as a matrix-vector product $\mathbf{T}\mathcal{E}$ where $\mathbf{T}$ is a matrix whose elements, for the one-dimensional grating case, are composed of Fourier coefficients ϵ_m. The structure of the $\mathbf{T}$-matrix for a one-dimensional grating is given in Fig. 1.44 and follows from consideration of the general matrix equation $A = \mathbf{T}B$. The column vectors $\mathbf{A}$ and $\mathbf{B}$ each contain the Fourier coefficients of an underlying periodic function A and B. An individual Fourier term, of frequency $m\nu_d$, hence has the general form

$$A_m \exp\{i[n_1 k_0 \sin\alpha_i + 2\pi m\nu_d]x\} = \sum_{m'} \epsilon_{m'} \exp(i2\pi m'\nu_d x) \times \sum_{m''} B_{m''} \exp\{i[n_1 k_0 \sin\alpha_i + 2\pi m''\nu_d]x\}\,. \tag{1.12.12}$$

For consistency, however, both sides of Eq. (1.12.12) must have the same frequency dependence. Imposing this requirement yields the condition $m' = m - m''$, or,

$$A_m = \sum_{m''=-M}^{M} \epsilon_{m-m''}\, B_{m''}\,,$$

or, in matrix notation,

$$\mathbf{A} = \mathbf{T}\,\mathbf{B}\,. \tag{1.12.13}$$

For a one-dimensional grating, an element T_{mj} of the matrix $\mathbf{T}$ is given by the Fourier component $\epsilon_{m+M+1-j}$. The structure of the $\mathbf{T}$-matrix is that of a Toeplitz matrix with identical elements on the diagonal and on each of the off-diagonal lines parallel to the main diagonal. The total number of independent entries of a $(2M+1)\times(2M+1)$ Toeplitz matrix is $4M+1$. To cover the full range of harmonic components of the magnetic field components in the left-hand side of Eqs. (1.12.11d)–(1.12.11f), the Fourier expansion of $\epsilon_j(x)$ thus has to span the frequency range $n_1 k_0 \sin\alpha_i - 2M\nu_d \le \nu \le n_1 k_0 \sin\alpha_i + 2M\nu_d$ with a total number of $4M+1$ Fourier terms. We note that the matrix-vector product $\mathbf{T}\mathbf{B}$ for a sequence of discrete frequencies is the equivalent of a convolution integral in the continuous domain.

With the aid of this definition of the matrix $\mathbf{T}$ for a one-dimensional grating, we can easily extend the analysis to a two-dimensional grating. The electromagnetic column vector $\mathbf{B}$ for the one-dimensional case with length $2M+1$ is augmented with $2M$ extra vectors with equal length, yielding a total length of $\mathbf{B}$ of $(2M+1)\times(2M+1)$ elements. Each

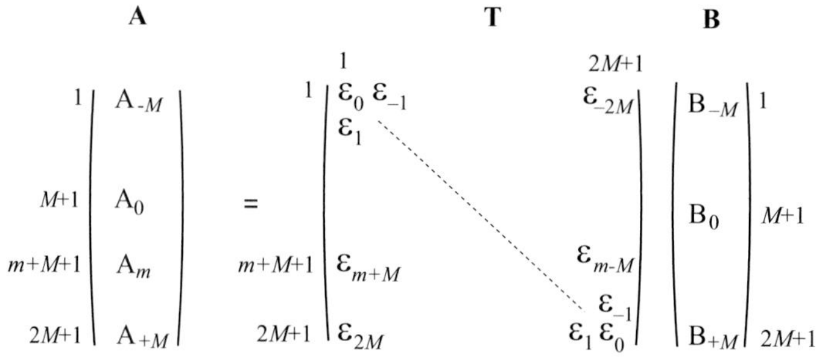

Figure 1.44: Structure of the matrix $\mathbf{T}$ in the Maxwell differential equations of Eqs. (1.12.11d)–(1.12.11f) for a one-dimensional grating.

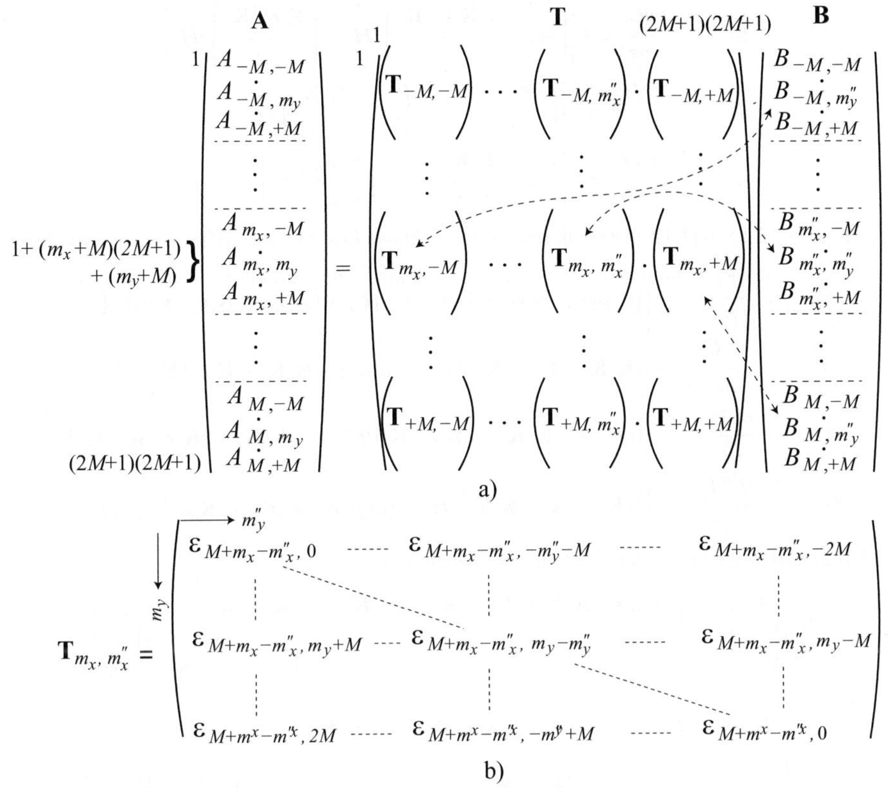

Figure 1.45: a) The structure of the **T**-matrix for the two-dimensional grating case as a block matrix of individual Toeplitz matrices $\mathbf{T}_{m_x,m''_x}$ with dimensions $(2M+1)\times(2M+1)$. The total number of elements in the matrix **T** amounts to $(2M+1)^4$. b) The structure of an individual Toeplitz matrix.

element b_{j,m_x,m_y} of vector **B** in a sublayer j follows from a Fourier expansion similar to that of Eq. (1.12.4). We adopt an ordering such that the $m_x + M + 1 + (m_y + M)(2M + 1)$ element of **B** contains the coefficient b_{j,m_x,m_y}. Similarly to the one-dimensional grating, each element of the composite vector **A** for the two-dimensional grating case is obtained from a matrix-vector multiplication given by

$$A_{m_x,m_y} = \sum_{m''_x=-M}^{M} \sum_{m''_y=-M}^{M} \epsilon_{m_x-m''_x,\, m_y-m''_y}\, B_{m''_x,\, m''_y} \,. \tag{1.12.14}$$

Similarly to Eq. (1.12.13) we write the expression of Eq. (1.12.14) in matrix notation as $\mathbf{A} = \mathbf{T}\mathbf{B}$ and the structure of the matrix **T** for the two-dimensional grating case is illustrated in Fig. 1.45, together with the structure of the composite electromagnetic column vectors. The column vectors have a length of $(2M+1)^2$. The figure shows how a subvector with elements A_{m_x,m_y}, $m_y = -M, \cdots, +M$, is calculated by the summation of the products $\mathbf{T}_{m_x,m''_x}\mathbf{B}_{m''_x,m''_y}$, $m''_y = -M, \cdots, +M$. Each of the block matrices $\mathbf{T}_{m_x,m''_x}$ is a Toeplitz matrix.

Using the Toeplitz-type matrix **T** for the two-dimensional grating case we can further manipulate Eqs. (1.12.9a)–(1.12.9f) so as to give a second-order differential equation for the x- and y-field components that is comparable to the Helmholtz equation in homogeneous space. From Eqs. (1.12.9c) and (1.12.9f) we obtain an expression for $\mathcal{E}_z$ and $\mathcal{H}_z$ in terms of the x- and y-fields. The substitution of $\mathcal{H}_z$ in Eqs. (1.12.9d) and (1.12.9e) and of $\mathcal{E}_z$ in Eqs. (1.12.9a) and (1.12.9b) yields the expressions

$$\text{a)} \qquad \frac{\partial \mathcal{E}_x}{\partial z} = i\left[\left(\frac{\mathbf{K}_x\epsilon^{-1}\mathbf{K}_y}{\omega}\right)\mathcal{H}_x + \left(\omega\mu_0 - \frac{\mathbf{K}_x\epsilon^{-1}\mathbf{K}_x}{\omega}\right)\mathcal{H}_y\right],$$

$$\text{b)} \quad \frac{\partial \mathcal{E}_y}{\partial z} = i\left[\left(-\omega\mu_0 + \frac{\mathbf{K}_y\epsilon^{-1}\mathbf{K}_y}{\omega}\right)\mathcal{H}_x - \left(\frac{\mathbf{K}_y\epsilon^{-1}\mathbf{K}_x}{\omega}\right)\mathcal{H}_y\right],$$

$$\text{c)} \quad \frac{\partial \mathcal{H}_x}{\partial z} = i\left[\left(-\frac{\mathbf{K}_x\mathbf{K}_y}{\omega\mu_0}\right)\mathcal{E}_x - \left(\omega\epsilon - \frac{\mathbf{K}_x\mathbf{K}_x}{\omega\mu_0}\right)\mathcal{E}_y\right],$$

$$\text{d)} \quad \frac{\partial \mathcal{H}_y}{\partial z} = i\left[\left(\omega\epsilon - \frac{\mathbf{K}_y\mathbf{K}_y}{\omega\mu_0}\right)\mathcal{E}_x + \left(\frac{\mathbf{K}_y\mathbf{K}_x}{\omega\mu_0}\right)\mathcal{E}_y\right]. \tag{1.12.15}$$

We differentiate Eqs. (1.12.15a)–(1.12.15d) with respect to z and use Eqs. (1.12.11a), b), d) and e) to obtain,

$$\text{a)} \quad \frac{\partial^2 \mathcal{E}_x}{\partial z^2} = -\left[\left(\omega^2\mu_0\epsilon - \mathbf{K}_y\mathbf{K}_y - \mathbf{K}_x\epsilon^{-1}\mathbf{K}_x\epsilon\right)\mathcal{E}_x + \left(\mathbf{K}_y\mathbf{K}_x - \mathbf{K}_x\epsilon^{-1}\mathbf{K}_y\epsilon\right)\mathcal{E}_y\right],$$

$$\text{b)} \quad \frac{\partial^2 \mathcal{E}_y}{\partial z^2} = -\left[\left(\mathbf{K}_x\mathbf{K}_y - \mathbf{K}_y\epsilon^{-1}\mathbf{K}_x\epsilon\right)\mathcal{E}_x + \left(\omega^2\mu_0\epsilon - \mathbf{K}_x\mathbf{K}_x - \mathbf{K}_y\epsilon^{-1}\mathbf{K}_y\epsilon\right)\mathcal{E}_y\right],$$

$$\text{c)} \quad \frac{\partial^2 \mathcal{H}_x}{\partial z^2} = -\left[\left(\omega^2\mu_0\epsilon - \mathbf{K}_x\mathbf{K}_x - \epsilon\mathbf{K}_y\epsilon^{-1}\mathbf{K}_y\right)\mathcal{H}_x + \left(-\mathbf{K}_x\mathbf{K}_y + \epsilon\mathbf{K}_y\epsilon^{-1}\mathbf{K}_x\right)\mathcal{H}_y\right],$$

$$\text{d)} \quad \frac{\partial^2 \mathcal{H}_y}{\partial z^2} = -\left[\left(\epsilon\mathbf{K}_x\epsilon^{-1}\mathbf{K}_y - \mathbf{K}_y\mathbf{K}_x\right)\mathcal{H}_x + \left(\omega^2\mu_0\epsilon - \mathbf{K}_y\mathbf{K}_y - \epsilon\mathbf{K}_x\epsilon^{-1}\mathbf{K}_x\right)\mathcal{H}_y\right]. \tag{1.12.16}$$

Equations (1.12.16a) to d) are conveniently written as two matrix equations,

$$\frac{\partial^2}{\partial z^2}\begin{pmatrix}\mathcal{E}_x \\ \mathcal{E}_y\end{pmatrix} = -\begin{pmatrix}\omega^2\mu_0\epsilon - \mathbf{K}_y\mathbf{K}_y - \mathbf{K}_x\epsilon^{-1}\mathbf{K}_x\epsilon & \mathbf{K}_y\mathbf{K}_x - \mathbf{K}_x\epsilon^{-1}\mathbf{K}_y\epsilon \\ \mathbf{K}_x\mathbf{K}_y - \mathbf{K}_y\epsilon^{-1}\mathbf{K}_x\epsilon & \omega^2\mu_0\epsilon - \mathbf{K}_x\mathbf{K}_x - \mathbf{K}_y\epsilon^{-1}\mathbf{K}_y\epsilon\end{pmatrix}\begin{pmatrix}\mathcal{E}_x \\ \mathcal{E}_y\end{pmatrix}, \tag{1.12.17}$$

$$\frac{\partial^2}{\partial z^2}\begin{pmatrix}\mathcal{H}_x \\ \mathcal{H}_y\end{pmatrix} = -\begin{pmatrix}\omega^2\mu_0\epsilon - \mathbf{K}_x\mathbf{K}_x - \epsilon\mathbf{K}_y\epsilon^{-1}\mathbf{K}_y & -\mathbf{K}_x\mathbf{K}_y + \epsilon\mathbf{K}_y\epsilon^{-1}\mathbf{K}_x \\ \epsilon\mathbf{K}_x\epsilon^{-1}\mathbf{K}_y - \mathbf{K}_y\mathbf{K}_x & \omega^2\mu_0\epsilon - \mathbf{K}_y\mathbf{K}_y - \epsilon\mathbf{K}_x\epsilon^{-1}\mathbf{K}_x\end{pmatrix}\begin{pmatrix}\mathcal{H}_x \\ \mathcal{H}_y\end{pmatrix}. \tag{1.12.18}$$

For the classical spectroscopic in-plane geometry with a one-dimensional grating ($\nu_{d,y} = 0$), the matrix equations are appreciably simplified,

$$\frac{\partial^2}{\partial z^2}\begin{pmatrix}\mathcal{E}_x \\ \mathcal{E}_y\end{pmatrix} = -\begin{pmatrix}\omega^2\mu_0\epsilon - \mathbf{K}_x\epsilon^{-1}\mathbf{K}_x\epsilon & 0 \\ 0 & \omega^2\mu_0\epsilon - \mathbf{K}_x\mathbf{K}_x\end{pmatrix}\begin{pmatrix}\mathcal{E}_x \\ \mathcal{E}_y\end{pmatrix}, \tag{1.12.19}$$

$$\frac{\partial^2}{\partial z^2}\begin{pmatrix}\mathcal{H}_x \\ \mathcal{H}_y\end{pmatrix} = -\begin{pmatrix}\omega^2\mu_0\epsilon - \mathbf{K}_x\mathbf{K}_x & 0 \\ 0 & \omega^2\mu_0\epsilon - \epsilon\mathbf{K}_x\epsilon^{-1}\mathbf{K}_x\end{pmatrix}\begin{pmatrix}\mathcal{H}_x \\ \mathcal{H}_y\end{pmatrix}, \tag{1.12.20}$$

with uncoupled equations for each of the four electromagnetic field components. We note that the formal notation of, for instance, $\mathcal{E}_x$ in Eqs. (1.12.17) and (1.12.19) refers to quite different column vectors. In Eq. (1.12.17) $\mathcal{E}_x$ is a concatenation of $(2M + 1)$ column vectors of length $2M + 1$ each. In the last equation, $\mathcal{E}_x$ is limited to a single column vector with $2M + 1$ elements.

1.12.3 Solution to Maxwell's Equations in Individual Grating Sublayers

To compute the electromagnetic fields in each sublayer j of the grating stack we solve the matrix equations of Eqs. (1.12.17) and (1.12.18) for the electric and magnetic field components. For each pair of field components a so-called *modal* solution is constructed, in analogy with the propagating modes that are encountered in electric transmission lines or optical (multi)mode fibres and waveguides (see Fig. 1.46). In the optical domain, the propagating modes of a fibre are determined by the geometry of the fibre cross-section and by the nature of its core and cladding material. The propagating modes are characterised by their propagation constant $\beta_{j,m} = n_{j,m}\,k_0$ where m is an integer which is in practice limited to a finite range $m = -M, \cdots, M$. The propagation constant determines the propagation speed $v_{j,m}=\omega/\beta_{j,m}=c/n_{j,m}$ of each of the forward and backward propagating $(2M + 1)$ fibre eigenmodes. Each mode has its specific eigenprofile that determines the spatial distribution of the electric or magnetic field in the fibre. In the case of a periodic multilayer stack, the modes are characterised by the value, $k_{z,j,m}$, of the wave vector component in the direction perpendicular to the grating interfaces (see Fig. 1.42c). The eigenmode profile, associated with the periodic grating profile, is a linear superposition

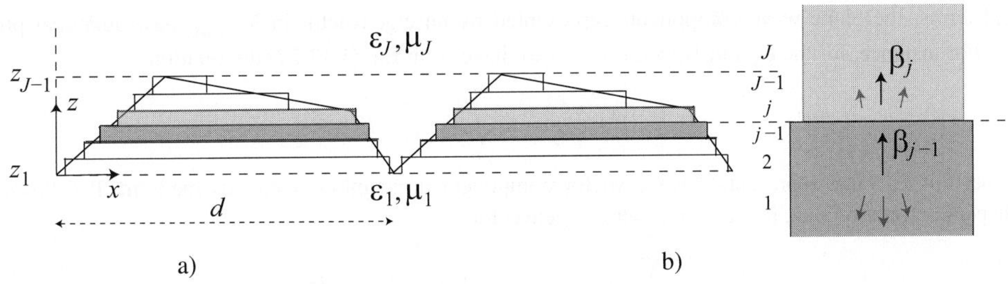

Figure 1.46: a) Wave propagation in a periodic layered grating.
b) Mode coupling and mode scattering between two waveguide or optical fibre sections with different lateral cross-sections.

of the various periodic electric or magnetic field components. In the one-dimensional case with in-plane diffraction and for the TE-polarisation case, the eigenmode field is a linear combination of periodic field components with strength $E_{y,m}$ (see the right-hand side of Eq. (1.12.3)).

1.12.3.1 The General Two-dimensional Grating Geometry

For this most general case we obtain the k_{z,j,m_x,m_y}-eigenvalues of the modes and their corresponding eigenvectors in layer j by computing the eigenvalues and eigenvectors of the square matrices $\mathbf{M}_E$ and $\mathbf{M}_H$ that are given in Eqs. (1.12.17) and (1.12.18), respectively,

$$\text{a)} \quad \frac{\partial^2}{\partial z^2}\begin{pmatrix}\mathcal{E}_{x,j}\\ \mathcal{E}_{y,j}\end{pmatrix} = \mathbf{M}_{E,j}\begin{pmatrix}\mathcal{E}_{x,j}\\ \mathcal{E}_{y,j}\end{pmatrix}, \qquad \text{b)} \quad \frac{\partial^2}{\partial z^2}\begin{pmatrix}\mathcal{H}_{x,j}\\ \mathcal{H}_{y,j}\end{pmatrix} = \mathbf{M}_{H,j}\begin{pmatrix}\mathcal{H}_{x,j}\\ \mathcal{H}_{y,j}\end{pmatrix}. \tag{1.12.21}$$

In sublayer j we obtain from each of the two matrices a total number of $2(2M+1)^2$ concatenated eigenvectors. Each eigenvector contains $(2M+1)^2$ elements of the E_x-component and the same number of E_y-components. The corresponding eigenvalue set is given by w^E_{j,m_x,m_y}. In numerical computations, relatively small eigenvalues are preferably skipped to avoid loss of precision. In that case, the total number of eigenvalues is smaller than $2(2M+1)^2$ and the eigenvalues in a certain layer j are indexed by q according to $w^E_{j,q}$. The same number of eigenvalues/eigenvectors follows from the solution of the magnetic eigenvalue problem of Eq. (1.12.21b). From Eqs. (1.12.17) and (1.12.18) we have the following expression for a particular eigenvector–eigenvalue combination,

$$\begin{pmatrix}\dfrac{\partial^2 V_{j,m_x,m_y,1}}{\partial z^2}\\ \vdots\\ \dfrac{\partial^2 V_{j,m_x,m_y,2(2M+1)^2}}{\partial z^2}\end{pmatrix} = -k^2_{z,j,m_x,m_y}\begin{pmatrix}V_{j,m_x,m_y,1}\\ \vdots\\ \vdots\\ V_{j,m_x,m_y,2(2M+1)^2}\end{pmatrix}$$

$$= -\mathbf{M}\begin{pmatrix}V_{j,m_x,m_y,1}\\ \vdots\\ \vdots\\ V_{j,m_x,m_y,2(2M+1)^2}\end{pmatrix} = -w_{j,m_x,m_y}\begin{pmatrix}V_{j,m_x,m_y,1}\\ \vdots\\ \vdots\\ V_{j,m_x,m_y,2(2M+1)^2}\end{pmatrix}. \tag{1.12.22}$$

The last subscript N of an element of the column vector $\mathbf{V}$ denotes the position of that element in the concatenated vector of eigenvectors and it is given by $N = 1 + m_x + M + (m_y + M)(2M+1)$ for the first set of eigenvectors associated with the x-polarised electric field in sublayer j and by $N = 1 + m_x + M + (1 + m_y + 3M)(2M+1)$ for the eigenvectors associated with the y-polarised electric field.

In optical terms, the plane wave components represented by an eigenvector in $\mathbf{V}_{j,m_x,m_y}$ have identical propagation coefficients. For instance, for the $\mathcal{E}_x$ and $\mathcal{E}_y$ components we have from Eq. (1.12.22) the relation,

$$k^E_{z,j,m_x,m_y} = \pm \sqrt{w^E_{j,m_x,m_y}} \,. \tag{1.12.23}$$

We recall that this k_z value represents a wave vector component in an inhomogeneous medium. It follows from the classical dispersion relation that, in the case of an E eigenvector

$$\left(k^E_{z,j,m_x,m_y}\right)^2 = w^E_{j,m_x,m_y} = \left(n^E_{\text{eff}}\right)^2 k_0^2 - k^2_{x,m_x} - k^2_{y,m_y} \,, \tag{1.12.24}$$

where we have used the effective medium constant n_{eff}. The propagation constant k_{z,j,m_x,m_y} for an E or H mode with mode numbers (m_x, m_y) equals $+(w_{j,m_x,m_y})^{1/2}$ for an upward travelling mode, whilst a downward travelling mode has the opposite sign. A general solution in layer j of the periodic stack is a linear combination of all possible propagation modes with their (x,y)-dependent amplitude profile determined by each eigenvector $\mathbf{V}_{j,m_x,m_y}$ and their propagation speed by $\pm\omega\,(w_{j,m_x,m_y})^{-1/2}$. For real eigenvalues w_{j,m_x,m_y}, the values of k_{z,j,m_x,m_y} are found on the positive real axis ($w_{j,m_x,m_y} \geq 0$) or they are purely imaginary ($w_{j,m_x,m_y} < 0$). For generally complex ϵ-values, restricted to the upper complex half-plane (passive media), the eigenvalues w_{j,m_x,m_y} are found in the same half-plane and the resulting value of k_{z,j,m_x,m_y} has positive real and imaginary parts for the upward travelling mode and the opposite sign for the downward travelling mode.

From the two sets of eigenvectors ($\mathbf{V}^{E,x}$, $\mathbf{V}^{E,y}$) and ($\mathbf{V}^{H,x}$, $\mathbf{V}^{H,y}$) that follow from the eigenvalue analysis of matrices $\mathbf{M}_{E,j}$ and $\mathbf{M}_{H,j}$, respectively, we construct the corresponding z components of the electric and magnetic field eigenvectors according to

$$\mathbf{V}^{E,z} = -\frac{1}{\omega}\left[\epsilon^{-1}\left(\mathbf{K}_x\mathbf{V}^{H,y} - \mathbf{K}_y\mathbf{V}^{H,x}\right)\right] , \qquad \mathbf{V}^{H,z} = \frac{1}{\omega\mu_0}\left(\mathbf{K}_x\mathbf{V}^{E,y} - \mathbf{K}_y\mathbf{V}^{E,x}\right) . \tag{1.12.25}$$

The complex amplitude as a function of the position in sublayer j of a certain electromagnetic component follows from the summation of all harmonic components that are contained in the eigenvectors. The strength of each eigenvector component is denoted by a^{E+}_{j,m_x,m_y} and a^{E-}_{j,m_x,m_y} for the upward and downward travelling eigenvector components, respectively, in the case of the 'electrical' solution of the eigenvalue problem of Eq. (1.12.21a). For instance, for the strength $E_{x,j}$ of the x component of the electric field in the jth layer, we find the expression,

$$\begin{aligned} E_{j,x}(x,y,z) &= \sum_{m_x=-M}^{+M}\sum_{m_y=-M}^{+M} V^{E,x}_{j,m_x,m_y,m_x+M+1+(m_y+M)(2M+1)} \\ &\times \exp\left\{i\left[(k_0 n_1 \sin\alpha_i \cos\alpha_a + 2\pi m_x \nu_{d,x})x + (k_0 n_1 \sin\alpha_i \sin\alpha_a + 2\pi m_y \nu_{d,y})y\right]\right\} \\ &\times \left[a^{E+}_{j,m_x,m_y} \exp\left\{ik^E_{z,j,m_x,m_y}(z - z_{j-1})\right\} + a^{E-}_{j,m_x,m_y} \exp\left\{-ik^E_{z,j,m_x,m_y}(z - z_j)\right\}\right] . \end{aligned} \tag{1.12.26}$$

When considering a magnetic field component, the sign in front of the second z-dependent exponential term on the third line of Eq. (1.12.26) becomes negative. The $a^{\pm}$-coefficients are still unknown quantities in each of the layers j. Their values follow from applying the appropriate boundary conditions, as discussed in detail in Subsection 1.12.4. In short, however, the first boundary condition specifies the incident wave on the grating, from the upper or lower half-space. Together with the electromagnetic boundary conditions at each grating interface it is possible to calculate the waves in each layer and the transmitted and reflected waves in both half-spaces.

1.12.3.2 The One-dimensional Grating with In-plane Geometry

For the in-plane one-dimensional case, we have the simplified, uncoupled matrices of Eqs. (1.12.19) and (1.12.20). We choose the two matrix equations for $\mathcal{E}_y$ and $\mathcal{H}_y$ because the other electric and magnetic field components will follow from their solutions in a rather straightforward manner. We first let

$$M = \begin{cases} M_E = \omega^2\mu_0\epsilon - \mathbf{K}_x\mathbf{K}_x \,, & TE-\text{mode}\,, \quad \text{a)} \\ M_H = \omega^2\mu_0\epsilon - \epsilon\mathbf{K}_x\epsilon^{-1}\mathbf{K}_x \,, & TM-\text{mode}\,. \quad \text{b)} \end{cases} \tag{1.12.27}$$

The eigenvalue analysis of these matrices yields two sets of $(2M+1)$ eigenvectors $\mathbf{V}^{E,y}_{j,m_x}$ and $\mathbf{V}^{H,y}_{j,m_x}$ with their corresponding eigenvalues $w^{E,y}_{j,m_x}$ and $w^{H,y}_{j,m_x}$. For this one-dimensional in-plane geometry, we obtain from the y eigenvector components the other Cartesian field components in sublayer j with the aid of Eqs. (1.12.9a)–(1.12.9f), whereby

$$\left(\mathbf{V}^{E,y}_{j,m_x}, k^{E,y}_{z,j,m_x}\right): \quad \mathbf{V}^{H,x}_{j,m_x} = -\frac{k^{E,y}_{z,j,m_x}\,\mathbf{V}^{E,y}_{j,m_x}}{\mu_0\omega}, \qquad \mathbf{V}^{H,z}_{j,m_x} = \frac{\mathbf{K}_x\,\mathbf{V}^{E,y}_{j,m_x}}{\mu_0\omega}. \tag{1.12.28}$$

$$\left(\mathbf{V}^{H,y}_{j,m_x}, k^{H,y}_{z,j,m_x}\right): \quad \mathbf{V}^{E,x}_{j,m_x} = \frac{\epsilon^{-1}\,k^{H,y}_{z,j,m_x}\,\mathbf{V}^{H,y}_{j,m_x}}{\omega}, \qquad \mathbf{V}^{E,z}_{j,m_x} = -\frac{\epsilon^{-1}\,\mathbf{K}_x\,\mathbf{V}^{H,y}_{j,m_x}}{\omega}. \tag{1.12.29}$$

In the one-dimensional grating example at hand, the two principal field components were E_y and H_y with the eigenvectors for the other field components provided by Eqs. (1.12.28) and (1.12.29). We could also have solved two other matrix equations, for instance the pair related to $\mathcal{E}_x$ and $\mathcal{H}_x$. This provides us with the eigenvector–eigenvalue pairs $\left(\mathbf{V}^{E,x}_{j,m_x}, w^{E,x}_{j,m_x}\right)$ and $\left(\mathbf{V}^{H,x}_{j,m_x}, w^{H,x}_{j,m_x}\right)$. With these solutions we obtain the other field components from the expressions

$$\left(\mathbf{V}^{E,x}_{j,m_x}, k^{E,x}_{z,j,m_x}\right), \quad \begin{cases} \mathbf{V}^{E,z}_{j,m_x} = k^{E,x}_{z,j,m_x}\left(\mathbf{K} - \omega^2\mu_0\,\mathbf{K}^{-1}\epsilon\right)^{-1}\mathbf{V}^{E,x}_{j,m_x}, \\ \mathbf{V}^{H,y}_{j,m_x} = -\omega\,k^{E,x}_{z,j,m_x}\,\mathbf{K}^{-1}\epsilon\left(\mathbf{K} - \omega^2\mu_0\,\mathbf{K}^{-1}\epsilon\right)^{-1}\mathbf{V}^{E,x}_{j,m_x}, \end{cases} \tag{1.12.30}$$

$$\left(\mathbf{V}^{H,x}_{j,m_x}, k^{H,x}_{z,j,m_x}\right), \quad \begin{cases} \mathbf{V}^{H,z}_{j,m_x} = k^{H,x}_{z,j,m_x}\left(\mathbf{K} - \omega^2\mu_0\,\epsilon\,\mathbf{K}^{-1}\right)^{-1}\mathbf{V}^{H,x}_{j,m_x}, \\ \mathbf{V}^{E,y}_{j,m_x} = -\omega\,\mu_0\,k^{H,x}_{z,j,m_x}\,\mathbf{K}^{-1}\left(\mathbf{K} - \omega^2\mu_0\,\epsilon\,\mathbf{K}^{-1}\right)^{-1}\mathbf{V}^{H,x}_{j,m_x}. \end{cases} \tag{1.12.31}$$

A comparison of Eqs. (1.12.28) and (1.12.29) and Eqs. (1.12.30) and (1.12.31) shows that the first eigenvalue solution based on the y components of the electric and magnetic field is computationally preferable because it avoids the extra evaluation of an inverse matrix.

For each of these field components, an eigenvector produces an x-harmonic electric or magnetic field component in layer j. For instance, the contribution of an eigenvector $\mathbf{V}^{E,y}_{j,m_x}$ to the electric y-field component $E_{j,y}$ in layer j is given by,

$$\begin{aligned} E_{j,y}(x,z) = &\sum_{m_x=-M}^{+M} V^{E,y}_{j,m_x,m_x+M+1}\,\exp\{i\,[k_0 n_1 \sin\alpha_i + 2\pi m_x \nu_d]\,x\} \\ &\times\left[a^{E+}_{j,m_x}\exp\left\{ik^{E}_{z,j,m_x}(z - z_{j-1})\right\} + a^{E-}_{j,m_x}\exp\left\{-ik^{E}_{z,j,m_x}(z - z_j)\right\}\right], \end{aligned} \tag{1.12.32}$$

where $z_{j-1} \le z \le z_j$ and $V^{E,y}_{j,m_x,m_x+M+1}$ is the m_xth element of the y component of the electric eigenvector. In the case of a magnetic field component, the sign in front of the second z-dependent exponential term on the second line of Eq. (1.12.32) becomes negative. The coefficients a^{E+}_{j,m_x} and a^{E-}_{j,m_x}, still unknown at this stage of the computation, give the strengths of the upward and downward travelling 'eigenwaves' in layer j of the stack. As denoted by the superscript E, the coefficients pertain to the 'electric' solution (TE-mode) of the eigenvalue problem of Eq. (1.12.27).

1.12.3.3 Truncation of the Fourier Series Expansion

The solution of Maxwell's equations in each sublayer of the grating structure provides an infinite number of eigenmodes. In practice, the modal expansions have to be truncated to finite ranges $-M_x < m_x < M_x$ and $-M_y < m_Y < M_y$. In practice, a single value $M = \max(|M_x|, |M_y|)$ is used for the truncation of both modal expansions. We can adopt two different strategies to determine a suitable choice of M_x and M_y, namely one based on physical arguments or, alternatively, one based on numerical arguments. With the former approach, the truncation point determines the smallest period that is included in the expansion and hence the smallest physical feature. As soon as this value approaches the atomic or molecular distances of the grating material, a further increase of M_x or M_y in the framework of macroscopic Maxwell theory is physically irrelevant. A truncation based on numerical arguments would, however, be based on the maximum ratio, r_e, between the largest and smallest relevant eigenvalue. From Eq. (1.12.24) we see that the eigenvalue w related to strongly evanescent waves tends to a value of $-4\pi^2(m_x^2\nu_{d,x}^2 + m_y^2\nu_{d,y}^2)$. On the other hand, the largest eigenvalue, with $m_x = m_y = 0$, is of the order of k_0. In the case of a high-frequency grating with $k_0 \approx 2\pi\nu_{d,x} \approx 2\pi\nu_{d,y}$ we have that $r_e = -(m_x^2 + m_y^2)$. A realistic truncation of terms can then be chosen by setting $|r_e| = 10^4$ which means that m_x and m_y should not exceed a value of typically 50 to 70.

In this subsection we have obtained the modal field solutions based on a Fourier expansion in each sublayer of the grating. In a practical implementation of this method, the number of modes included in the calculation will be limited, because of available computer memory and because of excessive computing time requirements. The solution that is found with a finite number of periodic components is not exact. Despite this fundamental inaccuracy, the solution method using the Fourier eigenmodes of the grating structure is called *rigorous* because its theoretical framework is exact.

1.12.4 Implementation of the Interface Boundary Conditions

The eigenmodes in each layer of the grating stack are coupled to those of the adjacent layers via transmission and reflection at the two interfaces and via propagation, both upward and downward within the layer. This is equivalent to the coupling of modes between various sections of a fibre with different geometric or material properties as was depicted in Fig. 1.46b. At each interface we use the continuity of the tangential electric and magnetic field components of the eigenmodes, represented by their x and y components. It goes without saying that a magnetic eigenmode, for instance obtained from Eq. (1.12.18) for a two-dimensional grating, has accompanying tangential electric field components and vice versa. These accompanying field components are provided by the six expressions in Eq. (1.12.11). With the aid of these equations we can collect the components of the four tangential components $\mathbf{E}_x$, $\mathbf{E}_y$, $\mathbf{H}_x$ and $\mathbf{H}_y$ in each sublayer and in the two semi-infinite homogeneous half-spaces. In general, in one of these half-spaces the incident plane wave has been specified. The relation between the tangential components in the jth layer at its exit surface and its entrance surface is given by

$$\begin{pmatrix} V^{E,x}_{j,m_x,m_y} \\ \vdots \\ V^{E,y}_{j,m_x,m_y} \\ \vdots \\ V^{H,x}_{j,m_x,m_y} \\ \vdots \\ V^{H,y}_{j,m_x,m_y} \end{pmatrix} = \begin{pmatrix} V^{E,x}_{j,m_x,m_y,q} & V^{E,x}_{j,m_x,m_y,q} \\ \vdots & \\ V^{E,y}_{j,m_x,m_y,q} & V^{E,y}_{j,m_x,m_y,q} \\ \vdots & \\ V^{H,x}_{j,m_x,m_y,q} & -V^{H,x}_{j,m_x,m_y,q} \\ \vdots & \\ V^{H,y}_{j,m_x,m_y,q} & -V^{H,y}_{j,m_x,m_y,q} \end{pmatrix} \begin{pmatrix} \exp\left(+ik^E_{z,j,q}z\right) & 0 \\ 0 & \exp\left(-ik^E_{z,j,q}z\right) \end{pmatrix} \begin{pmatrix} a^{E+}_{j,q} \\ a^{E-}_{j,q} \end{pmatrix}. \qquad (1.12.33)$$

In this matrix equation, the left-hand column vector has $4(2M+1)^2$ row elements. The V-matrix on the right-hand side has the same number of rows and $2q$ columns of eigenvectors, where q is the number of eigenvalues that has been kept in the solution of the eigenvalue problem with $q \le (2M+1)^2$. The matrix with the exponential terms has only nonzero-elements on its diagonal and is square with dimension $(2q \times 2q)$. The column vector with the unknown coefficients $a^{\pm}$ for the upward and downward travelling components has $2q$ elements.

To relate the field components at the interface, j, to those at the neighbouring interface, $j+1$, in a numerically stable manner, we use the scattering matrix method that was discussed in Subsection 1.11.2. Using this method, including the recursive scheme of Eqs. (1.11.13) and (1.11.14), a scattering matrix is obtained that relates the $(1,2)$ and $(J-1,J)$ interfaces according to the layer numbering in Figs. 1.42 and 1.46. This composite scattering matrix describing the entire grating stack provides two equations relating the coefficients $a^{\pm}_1$ and $a^{\pm}_J$ to the eigenvectors in the two semi-infinite half-spaces. From these equations we obtain the quantities that are important for the user of the grating. Specifically, in the half-space with the incident wave we obtain the field strengths of the propagating reflected diffraction orders of the grating and in the 'transmitting' half-space we determine the strengths of the transmitted diffraction orders. For the details of the numerical implementation of the boundary value problem we refer to Noponen and Turunen **[261]**, Li **[208]** and Nugrowati **[262]**.

1.12.5 Example of Diffraction by a One-dimensional Dielectric Grating

To demonstrate diffraction by a strongly modulated grating we consider the free-standing binary glass grating as discussed by Nugrowati in **[262]**, Section 2.7.1. Its geometry is shown in Fig. 1.47a and consists of microscopic rectangular cylinders of glass (n_g=1.5), aligned with their axes parallel to the y-axis. The cross-section of each cylinder has a width of $5\lambda_0/3$ and a height of λ_0, where λ_0 is the vacuum wavelength. The intermediate space between two glass cylinders has a width $w_x = \lambda_0/3$ and the refractive index of this open space is taken as that of air and set to unity. The period d_x of the

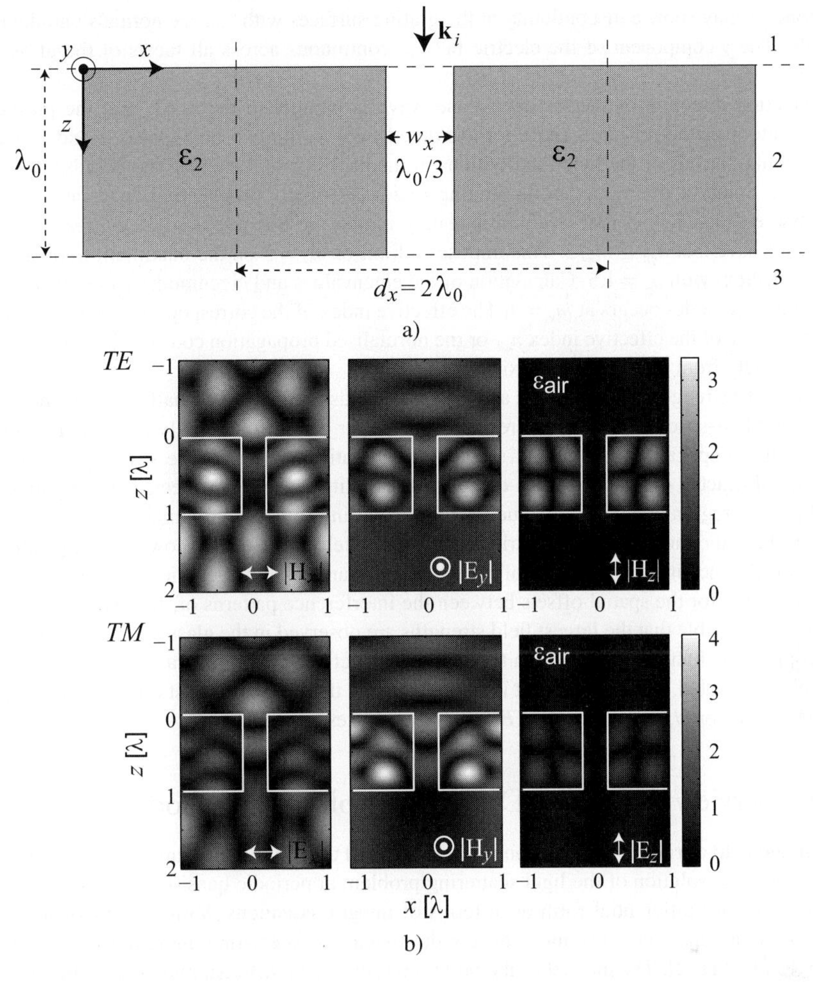

Figure 1.47: a) Geometry of the one-dimensional dielectric grating. b) The electric and magnetic field strengths in and outside the dielectric grating (reproduced from **[262]**, courtesy A.M Nugrowati). The magnetic field strength has been divided by the vacuum admittance $Y_0 = (\epsilon_0/\mu_0)^{1/2}$ so that the electric and magnetic field components have comparable numerical values in the graphs.

grating structure is $2\lambda_0$. For the particular case in which the incident plane wave is incident normally to the grating ($\mathbf{k}_i$ parallel to the z-axis), the grating has an optical contrast of $n_g/n_{air} = 1.5$ and assuming that the period and thickness of the grating are as shown in Fig. 1.47a, we can draw the following conclusions (see also Fig. 1.47b):

- the grating is strongly modulated and the approximate scalar phase difference between the optical paths through the glass and air equals π,
- the diffraction pattern is symmetric with respect to the plane $x = 0$,
- three freely propagating waves are allowed in the two half-spaces with unit refractive index,
- the reflected and transmitted orders with order number ± 2 are propagating waves with their wave vector exactly parallel to the x-axis in the two half-spaces,
- the incident wave and the freely propagating diffraction orders couple to propagating and damped waves in the modulated grating region, labelled 2.

For the magnetic field vectors, the transitions at the grating surfaces, with normal vectors parallel to the x- or the z-axis, do not constitute a change in permeability and the magnetic vectors are all continuous. For the electric field components, the x and z components may show a discontinuity at the grating surfaces with surface normals parallel to the x- and the z-axis, respectively. The y component of the electric field is continuous across all facets of the glass cylinders of the grating.

Since the propagation direction of the incident plane wave is parallel to the z-axis and the profile of the grating is symmetric, the pattern of the reflected and the transmitted field in the two half-spaces 1 and 3 is also symmetric. Moreover, the matching periods of the field distributions in media 1, 2 and 3 exclude the possibility of any odd modes in the grating region. Some of the modes in the grating region are freely propagating, whereas the higher orders are evanescent. A coarse estimate shows that freely propagating modes are possible if $n_{\text{eff}}k_0 \geq 2\pi|m_x|/d_x$. Since the period d_x is equal to $2\lambda_0$, we have that $n_{\text{eff}} \geq m_x/2$. This imposes a limit of $m_x = 3$ on the maximum frequency of propagating modes in the glass medium with $n_g = 1.5$. Calculation of the eigenvalues and eigenmodes shows that the transition from propagating to evanescent modes occurs at $m_x = 4$. The effective index of the corresponding mode is close to unity. These results show that the value of the effective index n_{eff} or the normalised propagation constant k_{z,m_x} in a spatially modulated region may differ strongly from first-order approximations.

Figure 1.47b shows the strengths of the electric and magnetic fields in the entrance half-space, in the modulated grating volume and in the exit half-space. Strong interference effects occur between the freely propagating modes in regions 1 and 3 and between the propagating and guided modes in the grating region 2. The strongly modulated interference effects show that the diffracted orders all have a comparable amplitude. This is expected as the pathlength differences in the modulated grating region amount to a value of $\lambda_0/2$ according to the approximate scalar diffraction theory. The interference patterns have the same pitch for electric and magnetic field components, however, the position of the patterns may be different. A difference in the arguments of reflection and transmission coefficients of electric and magnetic field components is responsible for the spatial offsets between the interference patterns of, for instance, E_y (*TE* case) and H_y (*TM* case). It is also remarkable that the largest field strengths are observed in the glass sections of the modulated region. Only a small fraction of the light tunnels through the narrow air sections which have a sub-wavelength width. Although the grating is deeply modulated, most of the light is transmitted via the zeroth and first diffracted orders towards the exit half-space (see the graphs for H_x and E_x for the *TE* and *TM* cases, respectively).

1.12.6 Brief Overview of Grating Diffraction Solution Methods

Pioneering work in the field of rigorous diffraction methods applied to periodic structures is described in the book edited by Petit **[276]**. It covers the solution of the light scattering problem at periodic quasi-planar structures using Maxwell's equations in their customary differential form or in terms of integral equations. Various numerical techniques like the point-matching method and the method of moments are discussed. An interesting, more recent overview of the subject has been given by Kahnert **[172]**. The interest of the optics community in different numerical solution methods depends on their respective requirements in terms of computer memory and calculation time. We limit the discussion in this subsection to the numerical methods that have the most appealing features with respect to memory and calculation time. These are the rigorous coupled wave analysis method, the finite-difference time-domain method and the finite-element method. An interesting benchmark-test of these three electromagnetic field solution methods can be found in **[200]**. An extension of the rigorous coupled wave analysis method to the propagation of ultrashort pulses through diffracting obstacles is given in **[263]**.

1.12.6.1 Rigorous Coupled Wave Analysis (RCWA)

The solution method that has been presented with some detail in the previous subsection is based on the coupling of modal expansions in each of the layers of a periodically modulated stack. A coupled wave analysis of the diffraction process in thick hologram gratings was first given by Kogelnik **[188]** and further implemented and evaluated by Knop **[187]**. The coupled wave analysis that was outlined in this section was first described by Moharam and Gaylord **[250]** and, because of the possibility to include a large number of diffracted waves, the name *rigorous* coupled wave analysis (*RCWA*) was coined. Moharam and co-authors extended the method to double-periodic structures **[249]**, **[251]**. A comprehensive overview of various grating analysis methods in two- and three-dimensional diffraction geometries has been given by Nugrowati **[262]**.

The theoretical scheme based on Eqs. (1.12.17) and (1.12.18) is robust. Maxwell's equations are solved without any approximations in each sublayer j. In practice, however, the numerical implementation of the RCWA method can give rise to convergence problems, especially for the calculation of the H_y components (*TM*-polarisation). Lalanne and Morris **[201]** discovered that the Fourier series expansion of a discontinuous permittivity function $\epsilon(x)$ gives rise to discontinuities in a product function like $\epsilon(x)E_x$. This discontinuity persists despite the inclusion of a large number of Fourier terms in the ϵ-expansion, and convergence to the correct end values of the electromagnetic field components is very slow as a function of the total number of Fourier terms $2M + 1$. The presence of an x component in the E-field only occurs in the case of TM-polarisation if an in-line two-dimensional geometry is used. As a solution to this convergence problem Lalanne and Morris used the Fourier expansion of the inverse of the permittivity function. In the core differential equations (1.12.17) and (1.12.18) and in all subsequent equations the permittivity matrix $\boldsymbol{\epsilon}$ is replaced by $(\boldsymbol{\epsilon}^{-1})^{-1}$. This relatively simple operation has put an end to a long period of research hampered by the poor convergence of the RCWA-method and other Fourier-based diffraction analysis methods which was especially problematic when modelling strongly modulated grating structures with TM-polarisation. The subtle details of the mathematical solution of this tenacious TM-convergence problem were presented by Li **[209]**.

1.12.6.2 Finite-Difference Time-Domain Method (FDTD)

The FDTD method was first used to numerically solve electrical engineering problems and dates back to a publication by Yee **[382]**. The subject was reintroduced some ten years later by Taflove and Brodwin **[334]**, with some useful rearrangements of the original finite difference formulas by Yee. A description of the present-day version of the *FDTD*-method can be found in the authoritative book by Taflove **[333]**. The FDTD method studies wave propagation in the time domain with real field values and uses Maxwell's time-dependent differential equations. The computation starts from a prescription of the initial field values of a source that is switched on at $t = 0$. To have a smooth transition from zero to finite field values, the field initialisation follows a certain time profile, for instance a Gaussian one. If there is a strong transient at $t = 0$, the temporal impulse response will be felt for an extended period of time. In optics, for spectrally narrow sources like lasers, the desired steady-state of the source is quasi-monochromatic with a well-defined oscillation frequency of the source. A Gaussian initial profile allows the steady-state to be reached in a short time. In the case of ultrashort optical pulses that are limited at most to a few source oscillations, the FDTD method can be directly applied without the need of a careful start-up procedure in the temporal domain.

- *Three-dimensional system of finite-difference Maxwell equations*

The system of finite difference equations for the electromagnetic field components uses Maxwell's time-dependent equations with the inclusion of a possible current density $\mathbf{J}$ in materials. Following **[382]**, we use Eqs. (1.3.12) and (1.3.13) to obtain

$$\begin{aligned} &a)\quad -\frac{\partial B_x}{\partial t} = \frac{\partial E_z}{\partial y} - \frac{\partial E_y}{\partial z}\,, &&d)\quad \frac{\partial D_x}{\partial t} = \frac{\partial H_z}{\partial y} - \frac{\partial H_y}{\partial z} - J_x\,,\\ &b)\quad -\frac{\partial B_y}{\partial t} = \frac{\partial E_x}{\partial z} - \frac{\partial E_z}{\partial x}\,, &&e)\quad \frac{\partial D_y}{\partial t} = \frac{\partial H_x}{\partial z} - \frac{\partial H_z}{\partial x} - J_y\,,\\ &c)\quad -\frac{\partial B_z}{\partial t} = \frac{\partial E_y}{\partial x} - \frac{\partial E_x}{\partial y}\,, &&f)\quad \frac{\partial D_z}{\partial t} = \frac{\partial H_y}{\partial x} - \frac{\partial H_x}{\partial y} - J_z\,. \end{aligned} \tag{1.12.34}$$

For the joint space-time difference scheme, the spatial domain is discretised into a grid of points which are indexed by (k, l, m) whilst points in time are indexed by n. The choice of the grid points is illustrated in Fig. 1.48. In this figure, the spatial grid increments depicted are equal although it should be mentioned that this is not an obligatory choice. In this case, the spatial coordinates of the grid points are simply given by

$$(x_k, y_l, z_m) = (k\,\delta x,\ l\,\delta y,\ m\,\delta z)\,. \tag{1.12.35}$$

The spatial and temporal dependence of the electromagnetic field components and other time–space-dependent quantities are compactly denoted by

$$f(k\,\delta x,\ l\,\delta y,\ m\,\delta z,\ n\,\delta t) = f^n(k, l, m)\,. \tag{1.12.36}$$

With reference to Fig. 1.48 in which the positions at which the electromagnetic field components are found have been indicated on an elementary cube with edges $(\delta x, \delta y, \delta z)$, we obtain the following finite difference equation for Eq. (1.12.34a),

$$\begin{aligned} &\frac{B_x^{n+1/2}(k, l+1/2, m+1/2) - B_x^{n-1/2}(k, l+1/2, m+1/2)}{\delta t}\\ &= \frac{E_y^n(k, l+1/2, m+1) - E_y^n(k, l+1/2, m)}{\delta z} - \frac{E_z^n(k, l+1, m+1/2) - E_z^n(k, l, m+1/2)}{\delta y}\,. \end{aligned} \tag{1.12.37}$$

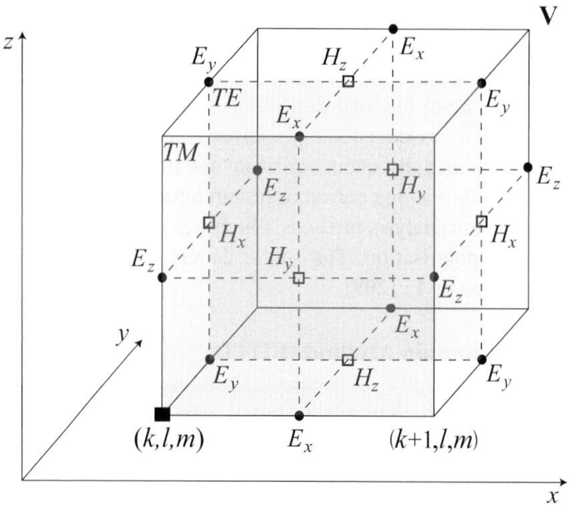

Figure 1.48: Grid points used in the finite difference scheme for solving Maxwell's time-dependent equations. The start point for an iteration is the lower-left corner with spatial indices (k, l, m) for the (x, y, z)-coordinates. The edge of the cube is equal to the spatial increments δx, δy and δz that are used to obtain the finite differences. The two-dimensional cross-sections labelled *TE* and *TM* are used if the diffraction geometry is translationally invariant along one Cartesian axis.

Comparable equations are obtained from Eqs. (1.12.34b) and (1.12.34c). The finite difference equation related to Eq. (1.12.34d) is given by,

$$\frac{D_x^n(k+1/2,l,m) - D_x^{n-1}(k+1/2,l,m)}{\delta t}$$
$$= \frac{H_z^{n-1/2}(k+1/2,l+1/2,m) - H_z^{n-1/2}(k+1/2,l-1/2,m)}{\delta y}$$
$$- \frac{H_y^{n-1/2}(k+1/2,l,m+1/2) - H_y^{n-1/2}(k+1/2,l,m-1/2)}{\delta z} + J_x^{n-1/2}(k+1/2,l,m)\,. \tag{1.12.38}$$

Following **[334]**, we rearrange Eqs. (1.12.37) and (1.12.38) as follows. The magnetic induction components B are replaced by the magnetic field components H by dividing B by the local and time-independent value of the magnetic permeability μ. In the same way we replace D by ϵE, where ϵ is the local scalar value of the permittivity. Using the constitutive relation of Eq. (1.3.7) for the current density $\mathbf{J}$ and imposing the condition that each elementary volume V is a cube with equal edges $\delta x = \delta y = \delta z = \delta$, we obtain the following six finite difference equations,

$$H_x^{n+1/2}(k,l+1/2,m+1/2) = H_x^{n-1/2}(k,l+1/2,m+1/2) + \frac{\delta t}{\mu(k,l+1/2,m+1/2)\,\delta}$$
$$\times \left\{E_y^n(k,l+1/2,m+1) - E_y^n(k,l+1/2,m) + E_z^n(k,l,m+1/2) - E_z^n(k,l+1,m+1/2)\right\}, \tag{1.12.39}$$

$$H_y^{n+1/2}(k+1/2,l,m+1/2) = H_y^{n-1/2}(k+1/2,l,m+1/2) + \frac{\delta t}{\mu(k+1/2,l,m+1/2)\,\delta}$$
$$\times \{E_z^n(k+1,l,m+1/2) - E_z^n(k,l,m+1/2) + E_x^n(k+1/2,l,m) - E_x^n(k+1/2,l,m+1)\}, \tag{1.12.40}$$

$$H_z^{n+1/2}(k+1/2,l+1/2,m) = H_z^{n-1/2}(k+1/2,l+1/2,m) + \frac{\delta t}{\mu(k+1/2,l+1/2,m)\,\delta}$$
$$\times \left\{E_x^n(k+1/2,l+1,m) - E_x^n(k+1/2,l,m) + E_y^n(k,l+1/2,m) - E_y^n(k+1,l+1/2,m)\right\}, \tag{1.12.41}$$

$$\begin{aligned}E_x^{n+1}(k+1/2,l,m) &= \left(1-\frac{\sigma(k+1/2,l,m)}{\epsilon(k+1/2,l,m)}\,\delta t\right)E_x^{n}(k+1/2,l,m)\\ &+\frac{\delta t}{\epsilon(k+1/2,l,m)\,\delta}\Big[H_z^{n+1/2}(k+1/2,l+1/2,m)-H_z^{n+1/2}(k+1/2,l-1/2,m)\\ &\qquad+H_y^{n+1/2}(k+1/2,l,m-1/2)-H_y^{n+1/2}(k+1/2,l,m+1/2)\Big]\,,\end{aligned} \tag{1.12.42}$$

$$\begin{aligned}E_y^{n+1}(k,l+1/2,m) &= \left(1-\frac{\sigma(k,l+1/2,m)}{\epsilon(k,l+1/2,m)}\,\delta t\right)E_y^{n}(k,l+1/2,m)\\ &+\frac{\delta t}{\epsilon(k,l+1/2,m)\,\delta}\Big[H_x^{n+1/2}(k,l+1/2,m+1/2)-H_x^{n+1/2}(k,l+1/2,m-1/2)\\ &\qquad+H_z^{n+1/2}(k-1/2,l+1/2,m)-H_z^{n+1/2}(k+1/2,l+1/2,m)\Big]\,,\end{aligned} \tag{1.12.43}$$

$$\begin{aligned}E_z^{n+1}(k,l,m+1/2) &= \left(1-\frac{\sigma(k,l,m+1/2)}{\epsilon(k,l,m+1/2)}\,\delta t\right)E_z^{n}(k,l,m+1/2)\\ &+\frac{\delta t}{\epsilon(k,l,m+1/2)\,\delta}\Big[H_y^{n+1/2}(k+1/2,l,m+1/2)-H_y^{n+1/2}(k-1/2,l,m+1/2)\\ &\qquad+H_x^{n+1/2}(k,l-1/2,m+1/2)-H_x^{n+1/2}(k,l+1/2,m+1/2)\Big]\,.\end{aligned} \tag{1.12.44}$$

The quantities $\delta t/\{\mu(k,l,m)\,\delta\}$ and $\delta t/\{\epsilon(k,l,m)\,\delta\}$ that appear in the equations above can be replaced by $Y(k,l,m)$ and $Z(k,l.m)$, respectively, i.e. the values of the local admittance and resistance of the medium. When the six electromagnetic components are evaluated with the aid of Eqs. (1.12.39)–(1.12.44) in this specific order, the right-hand side of each equation contains previously computed values that have already been calculated at the same lattice point of an elementary cube or at lattice points of a neighbouring elementary cube (see also Fig. 1.48).

For the majority of diffraction problems, the initial boundary condition corresponds to a plane wave, launched at $t = 0$ at a sufficiently large distance from a diffraction obstacle, such that a plane and harmonic wave is formed at the diffracting object. In other cases, a dipole source at a finite distance can be used to illuminate the object. The assembly of elementary cubes in the computational volume describes the diffracting object by means of a two- or three-dimensional spatial distribution of (ϵ,μ) values specified at each lattice point. With the progression of the time steps, the wave emanating from the source gradually pervades the volume of elementary cubes and the field components are found at the various lattice points of each cube with local origin at a general point (k,l,m). The field components are obtained for a sequence of sampling points in time given by $n\,\delta t$ for the E components and $(n+1/2)\,\delta t$ for the H components. When the maximum and minimum field values are constant as a function of the time step, the steady state has been reached and the final field has thus been found.

If the crest of the wave reaches the borders of the total simulation volume, a decision has to be made about the field values at these points. A first conjecture would be to impose zero field values on the border surface but this might cause a steep change in the field components, a phenomenon that gives rise to a reflected wave component. As a practical solution, the border surface is put sufficiently far away from the diffracting object so that a wave that is reflected at the border does not reach it before a steady-state has been established in and around this diffracting obstacle. The extra volume required to put the physical border far away leads to an extra calculation time and memory requirement. For that reason, the invention of the so-called 'perfectly matched layer' (PML) by Bérenger **[25]** was a very welcome addition to the FDTD method. Such a fictitious layer is capable of absorbing the incident radiation without a reflected component. In this respect, 'fictitious' means that the generally inhomogeneous and anisotropic layer cannot be physically realised with existing materials. With a limited number of extra elementary lattice cubes, the computational volume can be efficiently terminated with the aid of a PML layer.

For a stable iterative solution of the finite difference equations, the time increment δt should be chosen such that the corresponding change of the field components within the elementary rectangular volume V of Fig. 1.48 remains on average a small fraction of their actual values within this volume. An estimate of the maximum time increment δt as a function of the length of the edges δx, δy and δz of the elementary cell is given in **[334]** as

$$v\,\delta t \le \left(\frac{1}{(\delta x)^2}+\frac{1}{(\delta y)^2}+\frac{1}{(\delta z)^2}\right)^{-1/2}, \tag{1.12.45}$$

where v is the propagation speed in the medium. When the elementary cell is cubic we have $\delta t \leq \delta/(v\sqrt{3})$. As a rule of thumb, with a cell size $\delta = \lambda/p$, we find $\delta t \approx T/2p$ where T is the oscillation time of the (harmonic) radiation. The value of the factor p depends on the wavelength of the radiation (in the case of quasi-monochromatic radiation) and on the size of the diffracting or scattering structures. As a rule of thumb, p is typically 10. The convergence of an FDTD field solution can be studied by increasing p. For metals with strong absorption, the complex propagation speed v presents a further complication. For that reason, metallic materials with a negative real part of the permittivity lead to an instability in the calculations which can be reduced by using a much finer mesh. Note that in the standard FDTD approach all mesh distances had to be maintained over the full extent of the domain. This is a disadvantage, the most critical part of the geometry slowing down the entire calculation. Recent improvements use conforming grids and subgrids **[66]**. The typical accuracy of an FDTD calculation is 1%. Close to small perturbations, the accuracy is lower and may amount to a few percentage points. The accuracy depends on the mesh size and thus on the computational effort. Memory requirements are generally not a problem in FDTD-based computations.

Material dispersion needs a special treatment in FDTD when the spectral width of the source is large. The customary permittivity value for a specific frequency cannot be used. For a short temporal phenomenon the response function $g(t)$ of the medium to an electric field is used to obtain the time evolution of the electric flux density **D**. The response function $g(t)$ follows from the Kramers–Kronig relations, see Appendix C. In the time domain, the polarisation of the medium follows from the history of the acting electric field via a convolution integral. In the FDTD method the fact that **D** can be well approximated by means of a weighted sum of a first and second time derivative of the electric field **E** is used. In this way, **E**, **H** and **D** can be tabulated together in the FDTD scheme and the dispersion problem is incorporated in a stable way into the method.

- *Two-dimensional system of finite-difference Maxwell equations*

In the case of grating diffraction, a reduction in the dimension of the problem occurs when considering a one-dimensional grating used in the 'in-plane' mounting geometry. Figure 1.49 shows this special case for two distinct states of polarisation of the incident plane wave, namely the *TE*- and *TM*-states. Any other state of polarisation of the incident plane wave can be treated as a superposition of these two elementary states. The elementary cubic cell is reduced to a two-dimensional square, with its origin at $(k, l+1/2, m)$ for the *TE*-case and at (k, l, m) for *TM*-polarisation. The finite difference formulas simplify as follows:

TE-polarisation ($y = l + 1/2$):

$$H_x^{n+1/2}(k, m+1/2) = H_x^{n-1/2}(k, m+1/2) + Y(k, m+1/2)\left[E_y^n(k, m+1) - E_y^n(k, m)\right], \tag{1.12.46}$$

$$H_z^{n+1/2}(k+1/2, m) = H_z^{n-1/2}(k+1/2, m) + Y(k+1/2, m)\left[E_y^n(k, m) - E_y^n(k+1, m)\right], \tag{1.12.47}$$

$$\begin{aligned} E_y^{n+1}(k, m) &= \{1 - \sigma(k, m) Z(k, m)\delta\}\, E_y^n(k, m) + Z(k, m) \\ &\times\left[H_x^{n+1/2}(k, m+1/2) - H_x^{n+1/2}(k, m-1/2) + H_z^{n+1/2}(k-1/2, m) - H_z^{n+1/2}(k+1/2, m)\right]. \end{aligned} \tag{1.12.48}$$

TM-polarisation ($y = l$):

$$\begin{aligned} H_y^{n+1/2}(k+1/2, m+1/2) &= H_y^{n-1/2}(k+1/2, m+1/2) + Y(k+1/2, m+1/2) \\ &\times [E_x^n(k+1/2, m) - E_x^n(k+1/2, m+1) + E_z^n(k+1, m+1/2) - E_z^n(k, m+1/2)], \end{aligned} \tag{1.12.49}$$

$$\begin{aligned} E_x^{n+1}(k+1/2, m) &= \{1 - \sigma(k+1/2, m) Z(k+1/2, m)\delta\}\, E_x^n(k+1/2, m) \\ &+ Z(k+1/2, m)\left[H_y^{n+1/2}(k+1/2, m-1/2) - H_y^{n+1/2}(k+1/2, m+1/2)\right], \end{aligned} \tag{1.12.50}$$

$$\begin{aligned} E_z^{n+1}(k, m+1/2) &= \{1 - \sigma(k, m+1/2) Z(k, m+1/2)\delta\}\, E_z^n(k, m+1/2) \\ &+ Z(k, m+1/2)\left[H_y^{n+1/2}(k+1/2, m+1/2) - H_y^{n+1/2}(k-1/2, m+1/2)\right]. \end{aligned} \tag{1.12.51}$$

As an example of an FDTD analysis the diffraction of a plane wave by a dielectric or a perfectly conducting rod with circular cross-section is simulated. In Fig. 1.50, graphs a) and b) we show the diffracted fields produced by a dielectric and a perfectly conducting cylinder, respectively, as determined using FDTD analysis. In both graphs the incident wave is planar with its state of polarisation perpendicular to the plane of the figure (*TE*-polarisation). In Fig. 1.50a the refractive

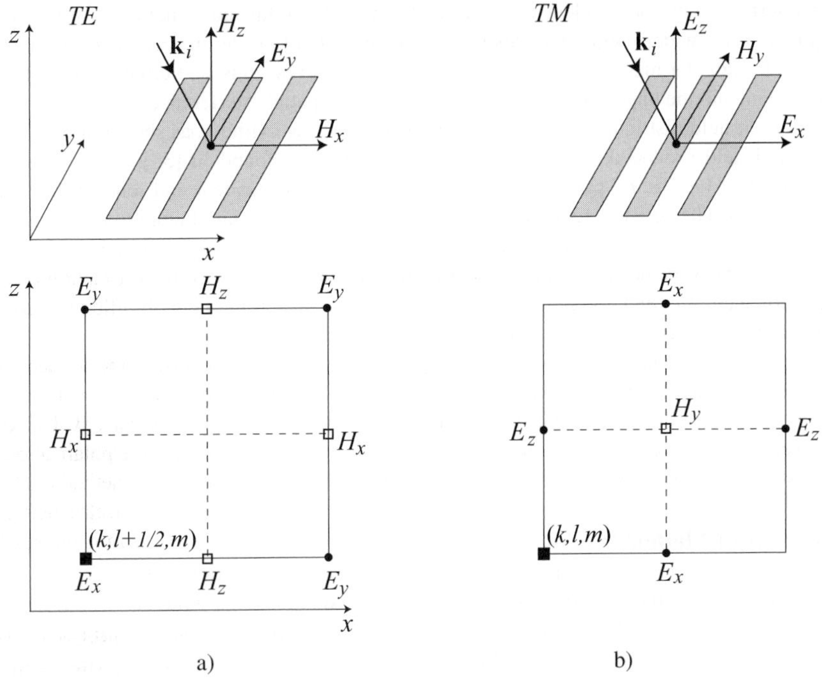

Figure 1.49: Two-dimensional finite difference grids for calculation of the diffracted field components of a one-dimensional grating (in-plane illumination with a plane wave with wave vector $\mathbf{k}_i$); a) transverse electric (*TE*), and b) transverse magnetic (*TM*) case.

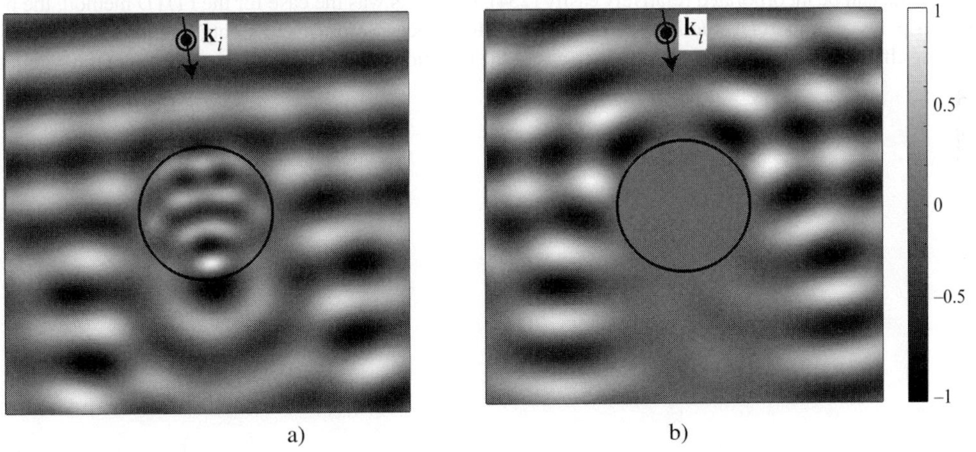

Figure 1.50: An incident harmonic plane wave with wave vector $\mathbf{k}_i$ is diffracted by a rod with a circular cross-section, diameter $2\lambda_0$ where λ_0 is the vacuum wavelength; *TE* polarisation. a) The rod material is glass with a refractive index n_g of 2. b) The rod material is highly conducting with no field penetrating into it. (courtesy Dr. O.T.A. Janssen and Dr. P.-J. Chen).

index of the glass material is relatively high, $n_g = 2$. It is seen that the radiation is refracted into the glass rod and that a diffraction focus is formed, close to the diametrically opposed surface part of the rod. This is confirmed by geometrical optics that predicts an image distance of $2R$ ($2R$ is the diameter of the circular cross-section of the rod). The light reflected from the cylinder interferes with the incident wave to yield a weakly modulated standing wave pattern. The outgoing field is seen to spread over a wider range of angles than the focused field inside, in accordance again with geometrical optics which predicts the angular magnification of the outgoing refracted beam is n_g. Further away from the cylinder, the refracted/diffracted beam and the unperturbed planar wave produce a spatial interference pattern.

A completely different field distribution is produced by the conducting cylinder, as shown in Fig. 1.50b. An analytic result for the scattered field produced by a conducting cylinder can be found in **[35]**. No field is allowed to enter the cylinder if the incident field is *TE*-polarised everywhere. The sum of incident and reflected field on the perfectly conducting surface is zero. The surface creates a reflected wave of which the initial amplitude is opposite to that of the incident wave. As a consequence, the standing wave pattern in the reflected space has a high modulation index (equal to unity in the case of normal incidence on the cylinder). The main difference with the diffraction pattern of the glass cylinder is that the conducting cylinder creates a deep shadow region which is gradually filled by diffracted light. For the case of the dielectric cylinder ($n_g = 2$), the modulation index equals $(1 - r)/(1 + r)$ where $r = 1/3$ is the reflection coefficient at normal incidence. The modulation index of the standing wave pattern close to the surface thus equals 1/2 in the case of the dielectric cylinder and this explains the resemblance of the reflected field patterns in a) and b). The weak diffracted field in the shadow of the metal rod lacks symmetry with respect to the plane defined by the axis of the rod and the incident wave vector. This effect is due to the chosen grid spacing which is not dense enough at the metal–vacuum interface. At the cost of calculation time the symmetry of the solution can be enhanced.

1.12.6.3 Finite-element Method (FEM)

The finite-element method was initially developed in the fields of elastic deformation and rheology. In the optics domain it has been implemented to solve the Laplace-type Helmholtz wave equations in a finite-element mesh that is adapted as well as possible to the permittivity and permeability function of the diffracting structure. As opposed to the mesh of grid points in FDTD that is in principle equidistant in the whole computational volume, no such requirement exists for the FEM method. The solution of the differential equations is obtained by evaluating finite differences between electric and magnetic fields at the neighbouring mesh points. The boundary conditions serve to match the fields with the permittivity and permeability functions of the medium. It is one of the basic advantages of the finite-element method that its absolute convergence to the end result is guaranteed by subsequent reductions of the mesh size in the computational volume. The practical problem in doing so is the quadratic or cubic increase in memory requirement and computation time as a function of the mesh size. The method is versatile with respect to simulating different medium properties; inhomogeneous and anisotropic media can be incorporated relatively easily **[254]**, **[351]**. As was the case for the FDTD method, the perfectly matched layer plays an important role in the FEM method. It replaces Sommerfeld's outgoing radiation condition and avoids aliasing effects in periodic structures. In both cases, the insertion of a PML allows a reduction of the computational volume.

2 Wave Propagation in Anisotropic Media

2.1 Introduction

In Chapter 1 we have discussed elementary and more general wave solutions that can exist in isotropic media. In these media, the material properties are independent of the direction of propagation of the wave and the constitutive relations take on the simple form of Eqs. (1.3.5)–(1.3.7). In several instances, the microstructure of the material in which the electromagnetic wave propagates gives rise to a directional dependence, with respect to both the speed of propagation of the wave and the rate of dissipation of the wave energy. The first example of anisotropy of a medium with respect to propagation speed was found in the crystalline material calcite, or $CaCO_3$, of which the striking optical properties were first observed by the Danish physicist Rasmus Bartholin in 1669. The large and clear $CaCO_3$ samples, imported from Iceland, beautifully demonstrated the phenomenon of double refraction, but Bartholin was not able to explain this startling phenomenon. A few years later, the Dutch physicist Huygens gave an explanation by supposing the propagation of two types of waves in such a crystal. Huygens' wave theory of light **[152]** was initially based on his so-called secondary waves with a spherical shape, the ordinary waves. By also allowing 'spheroidal' (ellipsoidal) waves in the anisotropic material, Huygens was able to qualitatively and quantitatively explain the existence and magnitude of the double refraction effect (see pp. 62–90 of the original version of his *Traité de la lumière*). He also alluded to an extra 'property' of light that discriminated between the two types of refracted waves, the ordinary ones and the new extraordinary ones and that would be known as the state of polarisation of a wave. This 'property' was needed to explain the peculiar behaviour he observed when sending light through two successive crystals of 'Iceland spar' with different orientations of their crystal axes. The intensity of the doubly refracted beams was influenced by the mutual orientation of the crystal axes and it was even possible to suppress one of them.

It was not until the discoveries by Malus, Brewster and Young that a more complete theory of light propagation in anisotropic media could be established. The main credit for the development of this theory should be given to Fresnel, who judiciously adapted the mechanical luminiferous aether model to allow for the propagation of transverse optical waves, the incorporation of polarisation properties and the description of wave propagation in anisotropic media. Fresnel's theory was able to accurately calculate the propagation of polarised waves in anisotropic media and to establish the direction of propagation and the strength of the ordinary and extraordinary rays for waves incident from an arbitrary direction with respect to the crystal axes. Another important step forward in Fresnel's theory was the consideration of propagating waves of a single frequency and wavelength, which was in contrast to Huygens' model where the light was thought to be composed of pulse-like disturbances of the aether. Of course, with the advent of Maxwell's electromagnetic theory, we will leave aside the details of Fresnel's theory. We will instead concentrate on the necessary extension of the electromagnetic theory discussed this far to deal with wave propagation in crystals or other materials that have been rendered anisotropic by some special treatment. Typical examples of such materials include plastic sheet polarisers, quarter-wave plates, stretched membranes, synthetic 'photonic crystals', etc.

2.2 Harmonic Electromagnetic Waves in an Anisotropic Medium

In the same vein as in Chapter 1, we first study the propagation of a harmonic plane wave, from which a description of more complicated wave patterns, for example in terms of their geometrical structure or spectral composition, can be constructed. The synthesis of more complicated waves from an assembly of elementary plane waves is carried out with the aid of Fourier analysis given in Appendix A. The propagation of plane waves in an anisotropic medium starts with a description of the medium properties. Initially we limit our description to linear anisotropic media, for instance calcite ($CaCO_3$); however, later in this chapter an extension is made to media that exhibit optical activity, such as crystal quartz or various types of electrically activated liquid crystals.

2.2.1 The Constitutive Equations

Up to this point we have assumed that the electric permittivity, magnetic permeability and specific conductivity could be described using a scalar value, for a given frequency ω of the electromagnetic wave, as given in Eqs. (1.3.5)–(1.3.7). For an anisotropic medium, we have to introduce a tensor relationship between the various vector quantities according to

$$\mathbf{D} = \epsilon_0 \,\overline{\overline{\epsilon}}_r\, \mathbf{E} , \tag{2.2.1}$$

$$\mathbf{B} = \mu_0 \,\overline{\overline{\mu}}_r\, \mathbf{H} , \tag{2.2.2}$$

$$\mathbf{J} = \overline{\overline{\sigma}}\, \mathbf{E} , \tag{2.2.3}$$

where the quantities $\overline{\overline{\epsilon}}_r$, $\overline{\overline{\mu}}_r$ and $\overline{\overline{\sigma}}$ are now 3×3 tensors. As a consequence of the tensor relation between, for instance, **E** and **D**, it follows that the two vectors generally are not parallel and that the ratio of their lengths is a function of the propagation direction in the anisotropic medium.

When the specific conductivity is a tensorial quantity, the medium exhibits directionally dependent absorption, typical for a crystal or a sheet polariser. In what follows, we set σ to zero, thereby restricting attention to non-dissipative or lossless media but we keep the tensor relationship for the relative permittivity and the relative permeability where we use the notation

$$\overline{\overline{\epsilon}}_r = \begin{pmatrix} \epsilon_{xx} & \epsilon_{xy} & \epsilon_{xz} \\ \epsilon_{yx} & \epsilon_{yy} & \epsilon_{yz} \\ \epsilon_{zx} & \epsilon_{zy} & \epsilon_{zz} \end{pmatrix} , \qquad \overline{\overline{\mu}}_r = \begin{pmatrix} \mu_{xx} & \mu_{xy} & \mu_{xz} \\ \mu_{yx} & \mu_{yy} & \mu_{yz} \\ \mu_{zx} & \mu_{zy} & \mu_{zz} \end{pmatrix} . \tag{2.2.4}$$

The components of the tensors $\overline{\overline{\epsilon}}_r$ and $\overline{\overline{\mu}}_r$ cannot be chosen completely independently. It was shown by Born that both tensors are symmetric[1] in a lossless medium and each of them has only six independent components. To illustrate the consequence of the symmetry property, we consider, as an example, the electric energy density in an anisotropic medium, $w_e = \frac{1}{2}\mathbf{E} \cdot \mathbf{D}$. This quantity is given by

$$w_e = \frac{\epsilon_0}{2}\left(\epsilon_{xx}E_x^2 + \epsilon_{yy}E_y^2 + \epsilon_{zz}E_z^2 + 2\epsilon_{xy}E_xE_y + 2\epsilon_{xz}E_xE_z + 2\epsilon_{yz}E_yE_z\right) . \tag{2.2.5}$$

[1] The symmetry property of, for instance, the dielectric tensor $\overline{\overline{\epsilon}}_r$ can be derived as follows **[36]**. If the medium properties are time-independent, the time derivative of the electric energy density is written as

$$\frac{\partial w_e}{\partial t} = \mathbf{E} \cdot \frac{\partial \mathbf{D}}{\partial t} = \frac{1}{2}\frac{\partial}{\partial t}(\mathbf{E} \cdot \mathbf{D}) .$$

The substitution of $\mathbf{D} = \epsilon_0 \,\overline{\overline{\epsilon}}_r\, \mathbf{E}$ in the two expressions for $\frac{\partial w_e}{\partial t}$ leads, after some arrangement, to

$$\frac{1}{2}E_j\epsilon_{jk}\frac{\partial E_k}{\partial t} = \frac{1}{2}\frac{\partial E_j}{\partial t}\epsilon_{jk}E_k ,$$

where a summation over the indices that occur twice is implied (Einstein convention). Interchanging the summation over j and k in the right-hand side of the expression above yields

$$\frac{1}{2}E_j\left(\epsilon_{jk} - \epsilon_{kj}\right)\frac{\partial E_k}{\partial t} = 0 .$$

For nonzero values of the field components, this equation can only be satisfied by requiring $\epsilon_{jk} = \epsilon_{kj}$.

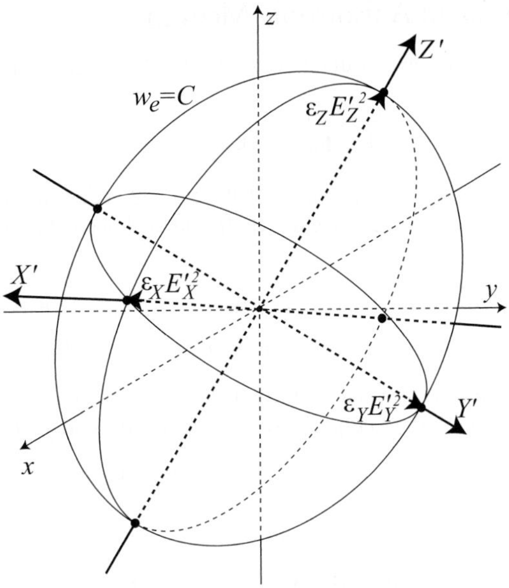

Figure 2.1: The shape of the surface with constant electric energy density in an anisotropic medium, $w_e = C$. The surface is commonly called the *index ellipsoid*. The principal axes of the surface with $w_e = C$ have been denoted by (X', Y', Z').

The expression for w_e is now modified by a rotation of the coordinate axes from the (x, y, z) system to a rotated system (X', Y', Z'), such that w_e is a function of only the squares of the electric field components measured along the new rotated coordinate axes. The required rotation of the coordinate axes can be carried out in two ways:

- Two successive Euler rotations are applied to the original coordinate system (x, y, z) and the two rotation angles are determined such that w_e in the new coordinate system is a function of squares only;
- The symmetric $\overline{\overline{\epsilon}}_r$ tensor is diagonalised. The eigenvectors that follow from this operation constitute the axes (X', Y', Z') of the rotated system and the three eigenvalues, ϵ_X, ϵ_Y and ϵ_Z, are the coefficients of the purely quadratic expression for w_e that results from this diagonalisation of $\overline{\overline{\epsilon}}_r$. The symmetry property of a real-valued tensor $\overline{\overline{\epsilon}}_r$ implies that the eigenvalues are real and that the eigenvectors are orthogonal; the rotated coordinate system (X', Y', Z') remains Cartesian.

In the rotated coordinate system the second-order function of the electric field strength components is now reduced to a sum of squares according to

$$w_e = \frac{\epsilon_0}{2}\left(\epsilon_X E_X'^2 + \epsilon_Y E_Y'^2 + \epsilon_Z E_Z'^2\right) . \tag{2.2.6}$$

The eigenvalues ϵ_X, ϵ_Y and ϵ_Z are called the principal relative permittivities belonging to the eigenvectors (principal axes) that are obtained after diagonalisation of $\overline{\overline{\epsilon}}_r$. The electric field strengths E_X', E_Y' and E_Z' are now those measured along the rotated coordinate axes in the system (X', Y', Z'). The expression for w_e can be represented by a three-dimensional surface with the contributions from the electric field strength components plotted along the principal axes of the medium. In general, the surface is an ellipsoid and it is commonly called the *index ellipsoid* (see Fig. 2.1) of the anisotropic medium. If two principal values of the permittivity tensor are identical, the surface becomes an ellipsoid of revolution. The surface reduces to a sphere in the case of an isotropic medium.

In what follows, we will also formally use the tensors that describe the inverse relationship between the vectors **E**, **D** and **H**, **B**,

$$\mathbf{E} = \epsilon_0^{-1}\left(\overline{\overline{\epsilon}}_r\right)^{-1}\mathbf{D} , \qquad \mathbf{H} = \mu_0^{-1}\left(\overline{\overline{\mu}}_r\right)^{-1}\mathbf{B} . \tag{2.2.7}$$

Once the tensor has been cast in diagonal form, the diagonal elements of the inverted tensor $\left(\overline{\overline{\epsilon}}_r\right)^{-1}$ simply equal the inverted values $1/\epsilon_X$, $1/\epsilon_Y$ and $1/\epsilon_Z$ and likewise for the diagonalised μ tensor.

2.2.2 Wave Propagation in an Anisotropic Medium

To analyse wave propagation in an anisotropic medium, we start with a harmonic plane wave with an electric field vector (in complex notation) given by,

$$\mathbf{E} = \mathbf{E}_0 \exp[i(\mathbf{k}\cdot\mathbf{r} - \omega t)]\,, \tag{2.2.8}$$

where $\mathbf{E}_0$ represents the complex amplitude components of the electric vector of the wave and $\mathbf{k}$ is the wave vector. Comparable expressions hold for the other field vectors $\mathbf{D}$, $\mathbf{H}$ and $\mathbf{B}$. Substitution of the plane wave expressions in Maxwell's equations (1.3.12) and (1.3.13) yields

$$\nabla\times[\mathbf{H}_0 \exp(i\mathbf{k}\cdot\mathbf{r})] = -i\omega\mathbf{D}_0 \exp(i\mathbf{k}\cdot\mathbf{r}) = -i\omega\epsilon_0\,\bar{\bar{\epsilon}}_r\,\mathbf{E}_0 \exp(i\mathbf{k}\cdot\mathbf{r})\,, \tag{2.2.9}$$

$$\nabla\times[\mathbf{E}_0 \exp(i\mathbf{k}\cdot\mathbf{r})] = +i\omega\mathbf{B}_0 \exp(i\mathbf{k}\cdot\mathbf{r}) = +i\omega\mu_0\,\bar{\bar{\mu}}_r\,\mathbf{H}_0 \exp(i\mathbf{k}\cdot\mathbf{r})\,. \tag{2.2.10}$$

To obtain uncoupled equations for the electric and magnetic fields, we substitute $\mathbf{H}_0$ from Eq. (2.2.10) into Eq. (2.2.9) and similarly substitute $\mathbf{E}_0$ from Eq. (2.2.9) into Eq. (2.2.10). Evaluating the curl yields

$$\nabla\times\mathbf{E} = i\mathbf{k}\times\mathbf{E}\,, \qquad \nabla\times\mathbf{H} = i\mathbf{k}\times\mathbf{H}\,, \tag{2.2.11}$$

and we obtain the result

$$\mathbf{k}\times\left[\left(\bar{\bar{\mu}}_r\right)^{-1}(\mathbf{k}\times\mathbf{E}_0)\right] + \omega^2\epsilon_0\mu_0\,\bar{\bar{\epsilon}}_r\,\mathbf{E}_0 = 0\,, \tag{2.2.12}$$

$$\mathbf{k}\times\left[\left(\bar{\bar{\epsilon}}_r\right)^{-1}(\mathbf{k}\times\mathbf{H}_0)\right] + \omega^2\epsilon_0\mu_0\,\bar{\bar{\mu}}_r\,\mathbf{H}_0 = 0\,. \tag{2.2.13}$$

The two vector equations above represent two uncoupled sets of three scalar equations, one set for the components of the electric field and the other set for the magnetic field components. Each set of equations is similar to a set of eigenvalue equations of the form

$$(\mathbf{A} - p\mathbf{I})\,\mathbf{x} = 0\,, \tag{2.2.14}$$

where $\mathbf{A}$ is a general matrix, $\mathbf{I}$ is the identity matrix, p is an associated eigenvalue and $\mathbf{x}$ represents the electric and magnetic field strength components.

There is, however, an important restriction to the solutions of Eqs. (2.2.12) and (2.2.13). Specifically, they are subject to the constraint that $\bar{\bar{\epsilon}}_r\,\mathbf{E}_0$ and $\bar{\bar{\mu}}_r\,\mathbf{H}_0$ must both be perpendicular to the wave vector $\mathbf{k}$. In terms of the dimension of the eigenvector space and the number of eigenvalues, both values are reduced from three for an unconstrained solution space to two for the particular type of eigenvalue equations (2.2.12) and (2.2.13). For instance, in the case of Eq. (2.2.12), the possible electric strengths are found in a plane and the third eigenvector, with zero eigenvalue, is directed along the wave vector $\mathbf{k}$. In the electromagnetic context, the two eigenvectors with nonzero eigenvalue are perpendicular to the Poynting vector and any nonzero electric field strength vector $\mathbf{E}$ is found by a linear superposition of the two eigenvectors in the plane perpendicular to the Poynting vector. Because of the similarities between Eqs. (2.2.12) and (2.2.13) and a classical eigenvalue equation like Eq. (2.2.14), we use the term 'eigenvalue' equation for equations of the type of Eq. (2.2.12) in the remaining part of this chapter.

With respect to the electric and magnetic eigenvalue equations, we note that the same eigenvalue k of the wave vector $\mathbf{k}$ should follow from either Eq. (2.2.12) or Eq. (2.2.13). Combining both equations in matrix notation, we can write

$$\begin{pmatrix} M_E & 0 \\ 0 & M_H \end{pmatrix}\begin{pmatrix} E_{0,x} \\ E_{0,y} \\ E_{0,z} \\ H_{0,x} \\ H_{0,y} \\ H_{0,z} \end{pmatrix} = \mathbf{0}\,, \tag{2.2.15}$$

where we have introduced two zero-padded 3×3 submatrices and two 3×3 submatrices M_E and M_H which can be constructed from Eqs. (2.2.12) and (2.2.13). Nontrivial solutions of Eq. (2.2.15) are obtained when the determinant of the matrix equals zero. We first reconsider the isotropic case for illustrative purposes, before proceeding to the anisotropic case.

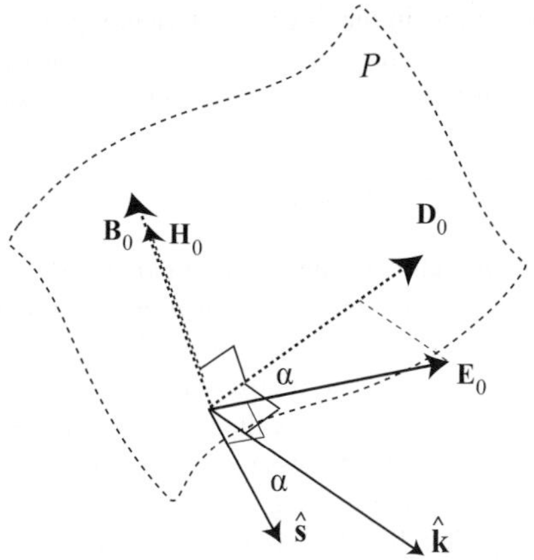

Figure 2.2: The electric and magnetic field vectors ($\mathbf{E}_0$ and $\mathbf{H}_0$) and induction vectors ($\mathbf{D}_0$ and $\mathbf{B}_0$) in an electrically anisotropic medium, together with the normalised wave vector $\hat{\mathbf{k}}$ and ray vector $\hat{\mathbf{s}}$.

2.2.2.1 Isotropic Medium

The tensors $\bar{\bar{\epsilon}}_r$ and $\bar{\bar{\mu}}_r$ are replaced by the constants ϵ_r and μ_r. The zero determinant condition leads to the following expression for the electric field components:

$$\mathbf{k} \times \mathbf{k} \times \mathbf{E}_0 + \omega^2 \epsilon_0 \mu_0 \epsilon_r \mu_r \mathbf{E}_0 = -k^2 \mathbf{E}_0 + (\mathbf{k} \cdot \mathbf{E}_0)\mathbf{k} + \omega^2 \epsilon_0 \mu_0 \epsilon_r \mu_r \mathbf{E}_0 = 0 . \tag{2.2.16}$$

It follows from Maxwell's equation (1.3.10) that the scalar product $\mathbf{k} \cdot \mathbf{E}_0$ vanishes in an isotropic medium that is free of charges. Consequently, the plane wave is of the transversal type. The quantity $\epsilon_0 \epsilon_r \mu_0 \mu_r$ is equal to $1/v^2$ with v the speed of the wave in the isotropic medium. We thus have that $\omega^2 \epsilon_0 \mu_0 \epsilon_r \mu_r = k^2$, the wavenumber squared in the medium. We also have from Eq. (1.3.11) that $\mathbf{k} \cdot \mathbf{H}_0 = 0$. Finally, Eqs. (2.2.9)–(2.2.12) show that the vectors $\mathbf{E}_0$, $\mathbf{H}_0$ and $\mathbf{k}$ form an orthogonal triplet. Of course, in an isotropic medium, the same holds for the triplet $\mathbf{D}_0$, $\mathbf{B}_0$ and $\mathbf{k}$.

2.2.2.2 Anisotropy with Respect to Permittivity (no Dissipation)

A frequently occurring case is that of dielectric anisotropy, for which the permittivity is described by a general, but symmetric $\bar{\bar{\epsilon}}_r$ tensor while the relative permeability adopts a constant scalar value of μ_r. Without loss of generality, we can introduce a rotation of the coordinate system so as to coincide with the principal axes of the dielectric tensor that is now written as

$$\bar{\bar{\epsilon}}_r = \begin{pmatrix} \epsilon_X & 0 & 0 \\ 0 & \epsilon_Y & 0 \\ 0 & 0 & \epsilon_Z \end{pmatrix} . \tag{2.2.17}$$

We use Eq. (2.2.12) and expand the vector triple product to write

$$(\mathbf{k} \cdot \mathbf{E}_0)\, \mathbf{k} - k^2 \mathbf{E}_0 + \omega^2 \mu_0 \mu_r \mathbf{D}_0 = \mathbf{0} . \tag{2.2.18}$$

Dividing the result by $k^2 = \omega^2/v^2$ we obtain

$$\mathbf{E}_0 - (\hat{\mathbf{k}} \cdot \mathbf{E}_0)\, \hat{\mathbf{k}} = \mu_0 \mu_r v^2 \mathbf{D}_0 , \tag{2.2.19}$$

where $\hat{\mathbf{k}}$ is the unit wave vector. Taking the scalar product of Eq. (2.2.19) with $\hat{\mathbf{k}}$ immediately shows that $\hat{\mathbf{k}}$ and $\mathbf{D}_0$ are orthogonal in an anisotropic medium such that

$$k_X(\epsilon_X E_{0X}) + k_Y(\epsilon_Y E_{0Y}) + k_Z(\epsilon_Z E_{0Z}) = 0 . \tag{2.2.20}$$

Figure 2.2 shows the wave vector $\hat{\mathbf{k}}$, the electric vectors $\mathbf{E}_0$ and $\mathbf{D}_0$ and the magnetic vectors $\mathbf{B}_0$ and $\mathbf{H}_0$. In an anisotropic medium, the unit propagation vector $\hat{\mathbf{k}}$ and the electric field vector $\mathbf{E}_0$ are at an angle $\pi/2 - \alpha$ with each other, where α is positive or negative depending on the values of the principal velocities of the medium and the propagation direction of the wave. The electric induction vector $\mathbf{D}_0$ is located in the plane P that is perpendicular to the propagation vector $\hat{\mathbf{k}}$. The magnetic vectors, $\mathbf{B}_0$ and $\mathbf{H}_0$, are also located in the plane P and are perpendicular to both $\mathbf{D}_0$ and $\mathbf{E}_0$ and to the propagation vector $\hat{\mathbf{k}}$. The vector $\hat{\mathbf{s}}$ points in the direction of energy propagation by the plane wave. Apart from a constant dimensional factor, the vector $\mathbf{D}_0$ is the projection of $\mathbf{E}_0$ onto the plane perpendicular to the unit propagation vector $\hat{\mathbf{k}}$. The angle α between the two vectors $\mathbf{E}_0$ and $\mathbf{D}_0$ is given by the expression

$$\cos\alpha = \frac{\mathbf{E}_0 \cdot \mathbf{D}_0}{|\mathbf{E}_0|\,|\mathbf{D}_0|} = \frac{\mathbf{E}_0^2 - (\hat{\mathbf{k}} \cdot \mathbf{E}_0)^2}{\left|\mathbf{E}_0\right|\left|\mathbf{E}_0 - (\hat{\mathbf{k}} \cdot \mathbf{E}_0)\hat{\mathbf{k}}\right|} = \left[1 - (\hat{\mathbf{k}} \cdot \hat{\mathbf{E}}_0)^2\right]^{1/2}, \qquad \sin\alpha = \hat{\mathbf{k}} \cdot \hat{\mathbf{E}}_0 , \tag{2.2.21}$$

where $\hat{\mathbf{E}}_0$ is the unit vector pointing in the direction of the electric field vector.

The unit vector $\hat{\mathbf{s}}$ that points in the direction of energy propagation follows from the vector product of the electric field strength $\mathbf{E}$ and the magnetic field vector $\mathbf{H}$ (see Eq. (1.4.5)). For a harmonic wave with complex electric and magnetic amplitude vectors $\mathbf{E}_0$ and $\mathbf{H}_0$, its direction and time-averaged strength follow from the vector $\Re\{\mathbf{E}_0 \times \mathbf{H}_0^*\}/2$. In geometrical optics, the corresponding unit vector is called the *ray vector* and the propagation velocity of the electromagnetic energy is called the *ray velocity*. We discuss these quantities and their relation with the unit wave vector $\hat{\mathbf{k}}$ and the wave propagation speed $v = \omega/|\mathbf{k}|$ in the following sections, in particular in Section 2.5.

The medium anisotropy for wave propagation, having two linear states of polarisation as eigenfunctions, is commonly called *linear birefringence*. It was originally discovered in natural crystals like calcite and quartz. They both belong to the group of uniaxial crystals with $\epsilon_X = \epsilon_Y \neq \epsilon_Z$. The more general anisotropy with $\epsilon_X \neq \epsilon_Y \neq \epsilon_Z$ is a property of the so-called biaxial materials. Examples are mica and perovskite. In the following section we discuss the particularities of plane wave propagation in such uniaxial and biaxial crystals.

2.3 Plane Wave Solutions in Uniaxial and Biaxial Media

Using the basic results for wave propagation in anisotropic media that were introduced in Section 2.1, we now study in more detail the polarisation eigenstates and the propagation speed of harmonic waves in media that possess a linear anisotropy (see Eq. (2.2.17)). It will turn out that two basic plane wave types are feasible in such an anisotropic medium, both linearly polarised and propagating at different speed. For an *ordinary* plane wave we have that the wave vector and the direction of energy propagation (Poynting vector) are parallel. The *extraordinary* plane wave does not possess this property; phase advance and energy transport during propagation follow different directions. Historically speaking the denominations 'ordinary' and 'extraordinary' wave propagation have a different origin. The first experiments with (uniaxial) anisotropic plane-parallel plates were carried out in the seventeenth century with a small pointlike source of natural light. Part of the illuminating light satisfied the combination of propagation direction and state of polarisation that gave rise to light propagation with identical speed in the crystal plate. Apparently, Snell's law was respected by this spherical wave and *ordinary* wave propagation was observed. The remaining part of the incident light had the opposite state of polarisation, seemed to violate Snell's law and thus behaved in an *extraordinary* way. The subtle differences between the anisotropic refraction and propagation of light in uniaxial or biaxial crystals could not yet be deduced from these early experiments.

In this section we will see that the basic ordinary and extraordinary *plane* waves discussed above occur as eigenstates in both uniaxial and biaxial media. The broader definition of ordinary wave propagation that is valid for an entire angular spectrum of plane waves (a spherical wave) is restricted to the class of uniaxial crystals. To find the relation between the electric field strength vector $\mathbf{E}_0$, the electric flux density vector $\mathbf{D}_0$ and the unit wave vector $\hat{\mathbf{k}}$ we use the Helmholtz equation in the form of Eq. (2.2.19). For the geometrical analysis of these vectors in uniaxial and biaxial media we write the Helmholtz equation as

$$n^2\, \hat{\mathbf{k}} \times \hat{\mathbf{k}} \times \hat{\mathbf{E}}_0 = -\,\bar{\bar{\epsilon}}_r\, \hat{\mathbf{E}}_0 = -\frac{|\mathbf{D}_0|}{\epsilon_0|\mathbf{E}_0|}\, \hat{\mathbf{D}}_0 , \tag{2.3.1}$$

where $n = c/v = k/k_0$ is the refractive index of the anisotropic medium for the plane wave with wave vector $\mathbf{k}$, which determines the phase propagation speed. To solve the Helmholtz equation we have to find vector solutions $\hat{\mathbf{E}}_0$ with both the correct direction and magnitude. We initially consider only the orientation of the unit field vectors $\hat{\mathbf{E}}_0$ and $\hat{\mathbf{D}}_0$ that occur in the Helmholtz equation above. The determination of the eigenvalues pertaining to the Helmholtz equation in terms of the refractive index n is a subsequent step of a more computational nature that is addressed in Section 2.4.

A nonzero solution of Eq. (2.3.1) requires that the vectors defined on the left- and right-hand sides are parallel to each other. The left-hand side vector is parallel to the projection of the unit electric field strength vector $\mathbf{E}_0$ on a plane perpendicular to the wave vector; this projected vector thus lies in a plane defined by the unit vectors $\hat{\mathbf{E}}_0$ and $\hat{\mathbf{k}}$. Consequently, for a nonzero solution of the Helmholtz equation, we require $\hat{\mathbf{k}}$, $\mathbf{E}_0$ and $\mathbf{D}_0$ to lie in the same plane.

The mutual orientation of these unit vectors, together with the magnetic field strength vector, is discussed for uniaxial and biaxial media in the following two subsections.

2.3.1 Wave, Ray and Field Vectors in a Uniaxial Medium

We analyse the transformation of $\mathbf{E}_0$ into $\mathbf{D}_0$ in an anisotropic medium with the aid of the permittivity tensor, $\bar{\bar{\epsilon}}_r$, expressed in the principal coordinate system. For a uniaxial medium, we have $\epsilon_X = \epsilon_Y$, such that the index ellipsoid is rotationally symmetric about Z'. In Fig. 2.3 the unit wave vector $\hat{\mathbf{k}}$ has been drawn in a general direction. It follows from the previous discussion that the endpoints of the unit vectors $\hat{\mathbf{D}}_0$ that satisfy the Helmholtz equation lie on a circle in a plane perpendicular to $\hat{\mathbf{k}}$. For a uniaxial medium with $\epsilon_X = \epsilon_Y$, one of the eigenvectors of the permittivity tensor $\bar{\bar{\epsilon}}_r$ is the unit vector $\hat{\mathbf{Z}}'$ with eigenvalue ϵ_Z. The two other identical eigenvalues $\epsilon_X = \epsilon_Y$ determine the plane $Z' = 0$ as 'eigenplane'.

If the electric field vector $\hat{\mathbf{E}}_0$ is parallel to an eigenvector of the $\bar{\bar{\epsilon}}_r$-tensor, the vectors $\hat{\mathbf{E}}_0$ and $\hat{\mathbf{D}}_0$ are also parallel. Figure 2.3 shows that this special case arises if $\hat{\mathbf{D}}_0$ intersects the 'eigenplane' $Z' = 0$. It yields the eigenvector pair $(\hat{\mathbf{D}}_2, \hat{\mathbf{E}}_2)$ that lies in the plane $\mathcal{P}_2$. A rotation of the state of polarisation over an angle of $\pi/2$ produces a state of polarisation that lies in the plane $\mathcal{P}_1$. It follows from Eq. (2.2.20) that the electric flux density vector in the plane $\mathcal{P}_1$ is given by $\hat{\mathbf{D}}_1$, perpendicular to $\hat{\mathbf{k}}$. It was shown in the previous section that the vectors $\hat{\mathbf{k}}$, $\hat{\mathbf{D}}_1$ and $\hat{\mathbf{E}}_1$ that satisfy the Helmholtz equation are coplanar; the exact value of the angle α between $\hat{\mathbf{D}}_1$ and $\hat{\mathbf{E}}_1$ follows from calculations carried out further on in this subsection. It follows from Fig. 2.4 that $\hat{\mathbf{D}}_2$ is perpendicular to $\hat{\mathbf{D}}_1$. The figure also shows that the electric field vectors $\hat{\mathbf{E}}_1$ and $\hat{\mathbf{E}}_2$ are perpendicular to each other. The unit magnetic field strength vectors $\hat{\mathbf{H}}_1$ and $\hat{\mathbf{H}}_2$ have also been drawn in Fig. 2.3; the vectors $\hat{\mathbf{k}}$, $\hat{\mathbf{D}}$ and $\hat{\mathbf{H}}$ form an orthogonal triplet. For any other state of linear polarisation of a propagating plane wave, the electric flux density vector $\mathbf{D}$ is projected onto the two orthogonal linear eigenstates of the medium, characterised by the unit vectors $\hat{\mathbf{D}}_1$ and $\hat{\mathbf{D}}_2$.

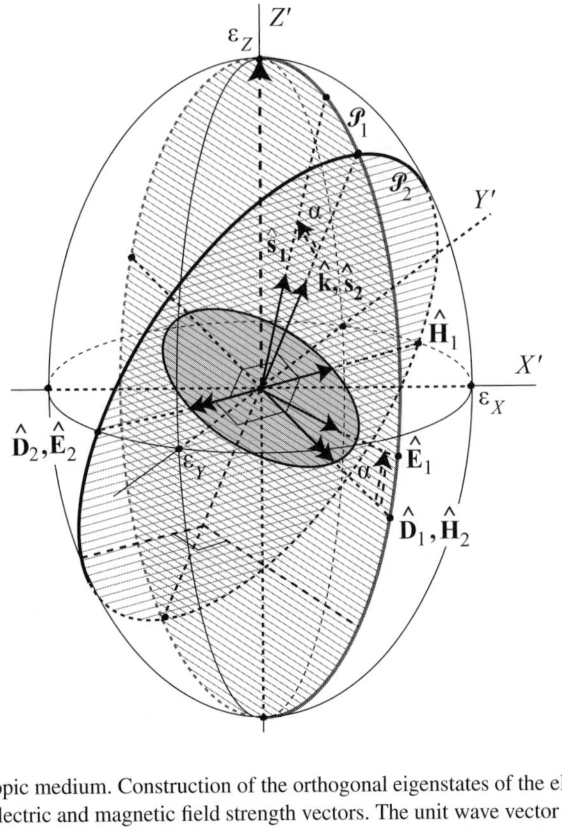

Figure 2.3: A uniaxial anisotropic medium. Construction of the orthogonal eigenstates of the electric flux density vector, $\hat{\mathbf{D}}_1$ and $\hat{\mathbf{D}}_2$, and of corresponding electric and magnetic field strength vectors. The unit wave vector is $\hat{\mathbf{k}}$; the unit vectors $\hat{\mathbf{s}}_1$ and $\hat{\mathbf{s}}_2$ point in the direction of the Poynting vectors belonging to the *extraordinary* and *ordinary* polarisation eigenstates 1 and 2, respectively.

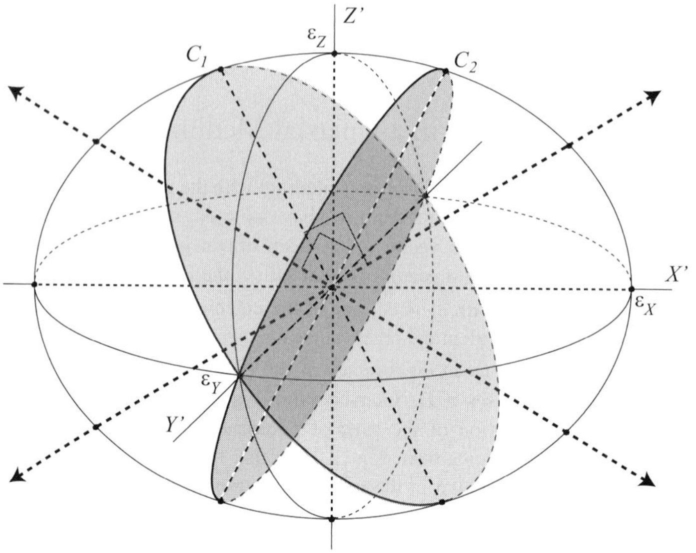

Figure 2.4: The direction of the optical axes (thick dashed lines with arrowheads in the figure), perpendicular to the two planes that produce circular intersecting curves C_1 and C_2 with the index ellipsoid. We have assumed that the principal values of the dielectric tensor satisfy $\epsilon_X > \epsilon_Y > \epsilon_Z$.

In the language of crystal optics, the energy propagation direction in an anisotropic medium is denoted by the *ray vector* $\hat{\mathbf{s}}$ and its direction follows from the Poynting vector $\Re\{\mathbf{E}_0 \times \mathbf{H}_0^*\}/2$. It is seen in Fig. 2.3 that $\hat{\mathbf{s}}_1$ is not parallel to the wave vector $\hat{\mathbf{k}}$. The vector solution $(\hat{\mathbf{k}}, \hat{\mathbf{D}}_1, \hat{\mathbf{E}}_1, \hat{\mathbf{H}}_1)$ gives rise to an angular offset between the phase propagation direction given by the wave vector $\hat{\mathbf{k}}$ and the electromagnetic energy propagation direction (the *ray* direction). A wave with this property is called an extraordinary wave. The electric vector $\hat{\mathbf{E}}_1$ lies in the plane $\mathcal{P}_1$ defined by the Z'-axis and the wave vector. The plane $\mathcal{P}_1$ is called the principal section for this particular propagation direction $\hat{\mathbf{k}}$.

For the second vector solution $(\hat{\mathbf{k}}, \hat{\mathbf{D}}_2, \hat{\mathbf{E}}_2, \hat{\mathbf{H}}_2)$ we conclude that $\hat{\mathbf{k}}$ and $\hat{\mathbf{s}}_2$ are parallel and the wave behaviour is similar to that in an isotropic medium. Consequently, $\hat{\mathbf{s}}_2$ is called the *ordinary* ray vector. The plane $\mathcal{P}_2$ containing the electric field vector is perpendicular to $\mathcal{P}_1$. Note that for the two polarisation states in a uniaxial medium we have $\hat{\mathbf{D}}_1 \cdot \hat{\mathbf{D}}_2 = 0$ and $\hat{\mathbf{E}}_1 \cdot \hat{\mathbf{E}}_2 = 0$.

When the wave vector is parallel to the Z'-axis of the anisotropic medium, the vector solutions for $(\mathbf{D}_0, \mathbf{E}_0)$ in the plane $Z' = 0$ can have an arbitrary orientation with the orientation-independent property $\mathbf{D}_0 = \epsilon_0 \epsilon_X \mathbf{E}_0$. This particular propagation direction is in the direction of the so-called *optical axis* of the medium (see the thick dashed arrow in Fig. 2.3). No special requirement on the state of polarisation of the light is imposed in this particular direction and the wave propagation is quasi-isotropic. In general, the state of polarisation of a wave propagating along the Z'-axis will be elliptical; the polarisation eigenstates for any other propagation direction are linear.

2.3.2 Wave, Ray and Field Vectors in a Biaxial Medium

The index ellipsoid for a biaxial medium, in terms of the electric flux density vector $\mathbf{D}_0$, is defined through the equation

$$w_e = \frac{\epsilon_0}{2} \left(\frac{D_{0,X}^{'2}}{\epsilon_X} + \frac{D_{0,Y}^{'2}}{\epsilon_Y} + \frac{D_{0,Z}^{'2}}{\epsilon_Z} \right), \tag{2.3.2}$$

with $\epsilon_X \neq \epsilon_Y \neq \epsilon_Z$. The surfaces of constant energy density are ellipsoidal in shape with unequal axis lengths. In this subsection we first analyse the special propagation directions that give rise to quasi-isotropic wave propagation in a biaxial medium. The next step is to find the electric and magnetic field vectors for a general propagation direction in such a medium.

2.3.2.1 Optical Axes in a Biaxial Medium

To find the wave propagation directions in which quasi-isotropic propagation is possible, we examine the collection of electric flux density vectors that are located in a plane perpendicular to an arbitrary wave vector $\hat{\mathbf{k}}$. A general unit electric flux density vector $\hat{\mathbf{D}}_0$ in this plane is written with the aid of two orthogonal base vectors $\hat{\mathbf{D}}_1$ and $\hat{\mathbf{D}}_2$,

$$\hat{\mathbf{D}}_1 = \frac{1}{\sqrt{\hat{k}_X^2 + \hat{k}_Z^2}} \begin{pmatrix} \hat{k}_Z \\ 0 \\ -\hat{k}_X \end{pmatrix} , \qquad \hat{\mathbf{D}}_2 = \frac{1}{\sqrt{\hat{k}_X^2 + \hat{k}_Z^2}} \begin{pmatrix} -\hat{k}_X \hat{k}_Y \\ 1 - \hat{k}_Y^2 \\ -\hat{k}_Y \hat{k}_Z \end{pmatrix} . \tag{2.3.3}$$

With a general unit vector $\hat{\mathbf{D}}_0$ given by $a\hat{\mathbf{D}}_1 + b\hat{\mathbf{D}}_2$ and $\sqrt{a^2 + b^2} = 1$, we calculate the electric energy density w_e and require that $\partial w_e / \partial a$ is zero. If this condition is satisfied for all a with $0 \leq a \leq 1$, the cross-section of the energy surface $w_e = C$ perpendicular to the vector $\hat{\mathbf{k}}$, where C is an arbitrary constant, is a circle. Such a circular cross-section is encountered for all propagation directions in an isotropic medium. However, for a biaxial medium it only exists for specific values of $\hat{\mathbf{k}}$. Wave propagation in these directions experiences an apparently isotropic medium. Enforcing the condition $\partial w_e / \partial a = 0$ yields the result,

$$\hat{k}_Y = 0 , \qquad \frac{\hat{k}_Z^2}{\epsilon_X} + \frac{\hat{k}_X^2}{\epsilon_Z} - \frac{(\hat{k}_X^2 + \hat{k}_Z^2)^2}{\epsilon_Y} = 0 . \tag{2.3.4}$$

From the two conditions above we find that the quasi-isotropic propagation directions in the biaxial medium are those for which

$$\hat{k}_X^2 = \frac{\epsilon_Z(\epsilon_X - \epsilon_Y)}{\epsilon_Y(\epsilon_X - \epsilon_Z)} , \qquad \hat{k}_Z^2 = \frac{\epsilon_X(\epsilon_Y - \epsilon_Z)}{\epsilon_Y(\epsilon_X - \epsilon_Z)} . \tag{2.3.5}$$

These expressions for $\hat{k}_X$ and $\hat{k}_Z$ determine propagation directions for which quasi-isotropic behaviour results. Importantly, however, the solutions of Eq. (2.3.5) come in pairs of anti-parallel vectors corresponding to forward or backward propagation along so-called *optical axes* of the medium. They are contained in the cross-section of the surface $w_e = C$ that contains the largest and the smallest principal index value of the medium.

The optical axes are shown in Fig. 2.4 for the case that $\epsilon_X > \epsilon_Y > \epsilon_Z$. In the figure we also show the circular cross-sections C_1 and C_2 of the index ellipsoids; each of the optical axes is perpendicular to such a circular cross-section. In comparison with crystals occurring in nature the maximum anisotropy $\epsilon_X - \epsilon_Z$ has been strongly exaggerated in the figure. Equation (2.3.5) yields a freely propagating solution provided that $\epsilon_Z > \epsilon_Y > \epsilon_X$. If one or both of the expressions for the squared quantities above are negative, a 'propagating' solution can possibly be found by a cyclic transformation of the solution given by $\hat{k}_Y = 0$ and Eq. (2.3.5). If one of the wave vector components is imaginary, the corresponding wave is evanescent. Such wave solutions will not be considered here. A plane wave with its wave vector aligned along an optical axis has a phase propagation speed v_p that does not depend on its state of polarisation. For detailed calculation of the speed of propagation and the field vectors in a biaxial medium we refer to Section 2.4.

2.3.2.2 The Electric and Magnetic Field Vectors in a Biaxial Medium

The calculation of the eigenvectors that satisfy Eq. (2.3.1) in a biaxial medium is more laborious than in a uniaxial medium. The requirement for a solution is the coplanarity of the three vectors $\hat{\mathbf{k}}$, $\mathbf{D}_0$ and $\mathbf{E}_0$ that is written as

$$\mathbf{D}_0 \cdot (\hat{\mathbf{k}} \times \mathbf{E}_0) = 0 . \tag{2.3.6}$$

Using the principal axes of the biaxial medium as Cartesian coordinates axes we are allowed to write

$$\mathbf{E}_0 = \frac{1}{\epsilon_0} \left(\epsilon_X^{-1} D_{0,x} , \; \epsilon_Y^{-1} D_{0,y} , \; \epsilon_Z^{-1} D_{0,z} \right) , \tag{2.3.7}$$

where, for instance, ϵ_X^{-1} represents the upper left diagonal element of the inverted permittivity tensor $\left(\overline{\overline{\epsilon}}_r\right)^{-1}$, referred to the principal axes of the anisotropic medium. We also have the condition that the wave vector and the electric flux density vector are orthogonal, leading to the two equations

$$\hat{k}_X D_{0,Y} D_{0,Z} (\epsilon_Y^{-1} - \epsilon_Z^{-1}) + \hat{k}_Y D_{0,X} D_{0,Z} (\epsilon_Z^{-1} - \epsilon_X^{-1}) + \hat{k}_Z D_{0,X} D_{0,Y} (\epsilon_X^{-1} - \epsilon_Y^{-1}) = 0 , \tag{2.3.8}$$

$$\hat{k}_X D_{0,X} + \hat{k}_Y D_{0,Y} + \hat{k}_Z D_{0,Z} = 0 . \tag{2.3.9}$$

Eliminating $D_{0,X}$ from Eqs. (2.3.8) and (2.3.9) and denoting $D_{0,Z}/D_{0,Y}$ by p, we obtain a quadratic equation in p which reads

$$\hat{k}_Y\hat{k}_Z(\epsilon_Z^{-1}-\epsilon_X^{-1})\,p^2-\left\{\hat{k}_X^2(\epsilon_Y^{-1}-\epsilon_Z^{-1})-\hat{k}_Y^2(\epsilon_Z^{-1}-\epsilon_X^{-1})\right.$$
$$\left.-\hat{k}_Z^2(\epsilon_X^{-1}-\epsilon_Y^{-1})\right\}p+\hat{k}_Y\hat{k}_Z(\epsilon_X^{-1}-\epsilon_Y^{-1})=0\,, \tag{2.3.10}$$

with the solutions

$$\begin{aligned}p_{1,2}=&\left\{\hat{k}_X^2(\epsilon_Y^{-1}-\epsilon_Z^{-1})-\hat{k}_Y^2(\epsilon_Z^{-1}-\epsilon_X^{-1})-\hat{k}_Z^2(\epsilon_X^{-1}-\epsilon_Y^{-1})\right.\\&\pm\left[\hat{k}_X^4(\epsilon_Y^{-1}-\epsilon_Z^{-1})^2+\hat{k}_Y^4(\epsilon_Z^{-1}-\epsilon_X^{-1})^2+\hat{k}_Z^4(\epsilon_X^{-1}-\epsilon_Y^{-1})^2\right.\\&-2\hat{k}_X^2\hat{k}_Y^2(\epsilon_Y^{-1}-\epsilon_Z^{-1})(\epsilon_Z^{-1}-\epsilon_X^{-1})-2\hat{k}_X^2\hat{k}_Z^2(\epsilon_Y^{-1}-\epsilon_Z^{-1})(\epsilon_X^{-1}-\epsilon_Y^{-1})\\&\left.\left.-2\hat{k}_Y^2\hat{k}_Z^2(\epsilon_Z^{-1}-\epsilon_X^{-1})(\epsilon_X^{-1}-\epsilon_Y^{-1})\right]^{1/2}\right\}\left\{2\hat{k}_Y\hat{k}_Z(\epsilon_Z^{-1}-\epsilon_X^{-1})\right\}^{-1}.\end{aligned} \tag{2.3.11}$$

Using the solution for $p_{1,2}$ and Eq. (2.3.9) we have

$$\frac{D_{0,X}}{D_{0,Y}}=-\frac{\hat{k}_Y+p_{1,2}\,\hat{k}_Z}{\hat{k}_X}\,. \tag{2.3.12}$$

Writing $p_1=a+b$ and $p_2=a-b$ for the two roots we obtain the following two solutions for the electric flux density vector,

$$\begin{aligned}\mathbf{D}_1&=\left(-\hat{k}_Y-(a+b)\hat{k}_Z\,,\ \hat{k}_X\,,\ (a+b)\hat{k}_X\right),\\\mathbf{D}_2&=\left(-\hat{k}_Y-(a-b)\hat{k}_Z\,,\ \hat{k}_X\,,\ (a-b)\hat{k}_X\right).\end{aligned} \tag{2.3.13}$$

Taking the scalar product of these two solution vectors yields the expression

$$\mathbf{D}_1\cdot\mathbf{D}_2=\hat{k}_X^2+\hat{k}_Y^2+2a\hat{k}_Y\hat{k}_Z+(a^2-b^2)(\hat{k}_X^2+\hat{k}_Z^2)\,. \tag{2.3.14}$$

From Eq. (2.3.11) it follows that

$$\begin{aligned}a^2-b^2&=\frac{\epsilon_X^{-1}-\epsilon_Y^{-1}}{\epsilon_Z^{-1}-\epsilon_X^{-1}}\,,\\2a&=\frac{\hat{k}_X^2(\epsilon_Y^{-1}-\epsilon_Z^{-1})-\hat{k}_Y^2(\epsilon_Z^{-1}-\epsilon_X^{-1})-\hat{k}_Z^2(\epsilon_X^{-1}-\epsilon_Y^{-1})}{\hat{k}_Y\hat{k}_Z(\epsilon_Z^{-1}-\epsilon_X^{-1})}\,.\end{aligned} \tag{2.3.15}$$

Substitution of these expressions into Eq. (2.3.14) yields the final result

$$\mathbf{D}_1\cdot\mathbf{D}_2=0\,. \tag{2.3.16}$$

We now use Eq. (2.3.7) to calculate the scalar product of the electric field strength vectors pertaining to the solutions $\mathbf{D}_1$ and $\mathbf{D}_2$ and obtain after some rearrangement

$$\mathbf{E}_1\cdot\mathbf{E}_2=\frac{\hat{k}_X^2}{(\epsilon_Z^{-1}-\epsilon_X^{-1})}\left\{\epsilon_X^{-2}(\epsilon_Y^{-1}-\epsilon_Z^{-1})\right.$$
$$\left.+\epsilon_Y^{-2}(\epsilon_Z^{-1}-\epsilon_X^{-1})+\epsilon_Z^{-2}(\epsilon_X^{-1}-\epsilon_Y^{-1})\right\}\neq 0\,. \tag{2.3.17}$$

The expression above assumes that the wave vector component $\hat{k}_X$ was not zero and that $\epsilon_X\neq\epsilon_Y\neq\epsilon_Z$. In the special case of a uniaxial medium with $\epsilon_X=\epsilon_Y$, we immediately see that the scalar product $\mathbf{E}_1\cdot\mathbf{E}_2$ becomes zero. This has already been demonstrated in the discussion of the solution vectors **D** and **E** for a uniaxial medium in Subsection 2.3.1.

In Fig. 2.5 we illustrate the geometry of the electric flux density and field strength vectors in a biaxial medium. The wave vector $\hat{\mathbf{k}}$ points in an arbitrary direction with respect to the principal axes of the index ellipsoid, chosen along the X'-, Y'- and Z'-axes in the figure. The solutions for the electric flux density vectors are found in a plane $\mathcal{P}_3$, tinted in Fig. 2.5, that is perpendicular to $\hat{\mathbf{k}}$. As a result of the general orientation of $\hat{\mathbf{k}}$ and hence the plane perpendicular to $\hat{\mathbf{k}}$, its intersection with the index ellipsoid is an ellipse which passes through the point Q. It can be shown that the vectors $\mathbf{D}_1$ and $\mathbf{D}_2$ of Eq. (2.3.13) that solve the Helmholtz equation coincide with the directions of the major and minor axes of the intersecting ellipse. The corresponding unit vectors have been indicated by $\hat{\mathbf{D}}_1$ and $\hat{\mathbf{D}}_2$ in Fig. 2.5. These vectors, together with the unit wave vector $\hat{\mathbf{k}}$, define two mutually orthogonal planes, $\mathcal{P}_1$ and $\mathcal{P}_2$, which are hatched in the figure. The solution vectors for the electric field strength, $\mathbf{E}_1$ are $\mathbf{E}_2$, are found in these two planes.

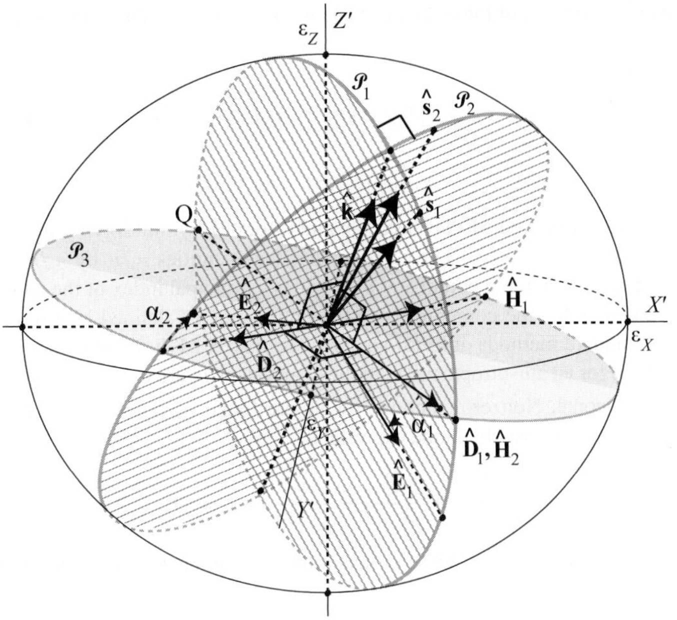

Figure 2.5: Representation of the orthogonal eigenstates of the electric flux density vector, $\hat{\mathbf{D}}_1$ and $\hat{\mathbf{D}}_2$, and of the corresponding electric and magnetic field strength vectors in a biaxial medium. The unit wave vector is $\hat{\mathbf{k}}$, the unit ray vectors $\hat{\mathbf{s}}_1$ and $\hat{\mathbf{s}}_2$ point in the direction of the Poynting vectors belonging to polarisation eigenstates 1 and 2.

Since none of the **D** and **E** solution vectors are parallel in a biaxial medium, the vectors $\mathbf{E}_1$ and $\mathbf{E}_2$ are not perpendicular in general. It is only in special cases, when the wave vector is parallel to one of the principal axes of the medium, that $\mathbf{E}_1 \cdot \mathbf{E}_2 = 0$. Orthogonality of the planes $\mathcal{P}_1$ and $\mathcal{P}_2$, however, implies that the relations $\hat{\mathbf{D}}_1 \cdot \hat{\mathbf{E}}_2 = 0$ and $\hat{\mathbf{D}}_2 \cdot \hat{\mathbf{E}}_1 = 0$ hold. In Fig. 2.5 we have also indicated the magnetic field strength unit vectors $\hat{\mathbf{H}}_1$ and $\hat{\mathbf{H}}_2$ that form orthogonal triplets with $\hat{\mathbf{k}}$ and the electric flux density vectors $\hat{\mathbf{D}}$. The unit ray vectors $\hat{\mathbf{s}}_1$ and $\hat{\mathbf{s}}_2$ form orthogonal triplets with the corresponding **E** and **H** vectors. It is seen that in the biaxial medium, for a general wave vector direction, both waves are extraordinary in the sense that the unit wave vector and the unit ray vector do not coincide.

2.4 Solution of the Helmholtz Equation in an Anisotropic Medium

In the preceding sections the Helmholtz equation was discussed qualitatively with respect to the vector solutions that can exist in an anisotropic medium. We now proceed to find the phase propagation speed v_p of the wave in the anisotropic medium and the corresponding eigenvectors for the electric flux density **D** and the field strengths **E** and **H**.

2.4.1 The Phase Propagation Speed v_p and Phase Propagation Index n_p

The starting point for the calculation of v_p is Eq. (2.2.18) with $\mathbf{D}_0$ replaced by $\epsilon_0\, \bar{\bar{\epsilon}}_r\, \mathbf{E}_0$. Introducing the principal propagation speeds according to $\epsilon_0\mu_0\mu_r\epsilon_X = 1/v_X^2$ and comparable expressions for $1/v_Y^2$ and $1/v_Z^2$, we obtain

$$\begin{pmatrix} \frac{v_p^2}{v_X^2} - \hat{k}_Y^2 - \hat{k}_Z^2 & \hat{k}_X\hat{k}_Y & \hat{k}_X\hat{k}_Z \\ \hat{k}_X\hat{k}_Y & \frac{v_p^2}{v_Y^2} - \hat{k}_X^2 - \hat{k}_Z^2 & \hat{k}_Y\hat{k}_Z \\ \hat{k}_X\hat{k}_Z & \hat{k}_Y\hat{k}_Z & \frac{v_p^2}{v_Z^2} - \hat{k}_X^2 - \hat{k}_Y^2 \end{pmatrix} \mathbf{E}_0 = \mathbf{M}\,\mathbf{E}_0 = \mathbf{0}\,. \tag{2.4.1}$$

Casting this matrix equation in the form of the eigenvalue equation (2.2.14) we obtain the expression

$$\begin{pmatrix} v_p^2 - v_X^2(1-\hat{k}_X^2) & v_X^2\hat{k}_X\hat{k}_Y & v_X^2\hat{k}_X\hat{k}_Z \\ v_Y^2\hat{k}_X\hat{k}_Y & v_p^2 - v_Y^2(1-\hat{k}_Y^2) & v_Y^2\hat{k}_Y\hat{k}_Z \\ v_Z^2\hat{k}_X\hat{k}_Z & v_Z^2\hat{k}_Y\hat{k}_Z & v_p^2 - v_Z^2(1-\hat{k}_Z^2) \end{pmatrix} \mathbf{E}_0 = (v_p^2\mathbf{I} - \mathbf{M}')\,\mathbf{E}_0 = \mathbf{0}\,, \tag{2.4.2}$$

where $\mathbf{I}$ is as usual the identity matrix and $\mathbf{M}'$ has been obtained from $\mathbf{M}$ by multiplying its first row by v_X^2, the second row by v_Y^2 and the third row by v_Z^2. In Eq. (2.4.1) we can replace the quantities v_p/v_i by n_i/n_p where n_p is the refractive index of the propagating eigenmode in the medium and $n_i = \sqrt{\epsilon_i}$ is a principal index of the anisotropic medium.

The linear system of equations for the components of the unit vector $\hat{\mathbf{E}}_0$ defined by Eq. (2.4.1) or equivalently by Eq. (2.4.2), is solved using standard methods of linear algebra. The matrix $\mathbf{M}'$ has real components and is symmetric for an isotropic medium; however, for an anisotropic medium the eigenvalues v_p^2 are not necessarily real and the eigenvectors of $\mathbf{M}'$ do not need to be orthogonal. Nonzero solutions for $\hat{\mathbf{E}}_0$ are found by putting $\det(v_p^2\mathbf{I} - \mathbf{M}') = 0$ or $\det(\mathbf{M}) = 0$. Using $p_i = v_p/v_i = n_i/n_p$ in Eq. (2.4.1) we obtain the expression

$$\frac{\hat{k}_X^2}{p_X^2 - 1} + \frac{\hat{k}_Y^2}{p_Y^2 - 1} + \frac{\hat{k}_Z^2}{p_Z^2 - 1} + 1 = 0\,. \tag{2.4.3}$$

This expression can be written in terms of the phase propagation speed v_p or in terms of the refractive index n_p. For the former we have

$$\frac{\hat{k}_X^2 v_X^2}{v_X^2 - v_p^2} + \frac{\hat{k}_Y^2 v_Y^2}{v_Y^2 - v_p^2} + \frac{\hat{k}_Z^2 v_Z^2}{v_Z^2 - v_p^2} = 1\,, \tag{2.4.4}$$

which, by substitution of $1 = \hat{k}_X^2 + \hat{k}_Y^2 + \hat{k}_Z^2$, can also be written as

$$\frac{\hat{k}_X^2}{v_X^2 - v_p^2} + \frac{\hat{k}_Y^2}{v_Y^2 - v_p^2} + \frac{\hat{k}_Z^2}{v_Z^2 - v_p^2} = 0\,. \tag{2.4.5}$$

Using the principal indices $n_i = c/v_i$ for $i = X, Y, Z$ and the phase index n_p defined by $n_p = c/v_p$, we obtain after some rearrangement the expression

$$\frac{\hat{k}_X^2}{n_p^2 - n_X^2} + \frac{\hat{k}_Y^2}{n_p^2 - n_Y^2} + \frac{\hat{k}_Z^2}{n_p^2 - n_Z^2} = \frac{1}{n_p^2}\,. \tag{2.4.6}$$

Equations (2.4.3)–(2.4.6) are various forms of the so-called Fresnel wave normal equation, which determines the propagation speed of a plane wave in a general anisotropic medium as a function of the orientation of the wave vector with respect to the principal axes of the medium. The propagation speed enters as v_p^2 in the Fresnel equation and this means that both forward and backward propagation waves are allowed for a given direction of the wave vector $|\hat{\mathbf{k}}|$.

2.4.1.1 Phase Propagation Speed v_p and Index n_p in a Biaxial Medium

For a certain wave propagation direction given by the unit vector $\hat{\mathbf{k}}$, the equations above give the propagation speed or the refractive index of the propagating eigenmode in the medium. The equations above are quadratic equations in the variables v_p^2 or n_p^2 and they are more conveniently written as

$$v_p^4 - v_p^2\left\{\hat{k}_X^2(v_Y^2 + v_Z^2) + \hat{k}_Y^2(v_X^2 + v_Z^2) + \hat{k}_Z^2(v_X^2 + v_Y^2)\right\} + \hat{k}_X^2 v_Y^2 v_Z^2 + \hat{k}_Y^2 v_X^2 v_Z^2 + \hat{k}_Z^2 v_X^2 v_Y^2 = 0, \tag{2.4.7}$$

$$n_p^4\left\{n_X^2\hat{k}_X^2 + n_Y^2\hat{k}_Y^2 + n_Z^2\hat{k}_Z^2\right\} - n_p^2\left\{\hat{k}_X^2 n_X^2(n_Y^2 + n_Z^2) + \hat{k}_Y^2 n_Y^2(n_X^2 + n_Z^2) + \hat{k}_Z^2 n_Z^2(n_X^2 + n_Y^2)\right\} + n_X^2 n_Y^2 n_Z^2 = 0. \tag{2.4.8}$$

Solutions for v_p^2 and n_p^2 are directly available using the standard expression for the roots of a quadratic equation. With the aid of a solution of $v_p(\hat{\mathbf{k}})$ we construct the vector $\mathbf{r}_p = v_p t_0 \hat{\mathbf{k}}$, where t_0 denotes time increment. The endpoint $|\mathbf{r}_p|$ = constant shows to what position the phase front of a plane wave with unit wave vector $\hat{\mathbf{k}}$ has advanced in a time t_0 from the origin O. Instead of the plane wave emitted by a single point source in an O at a large distance, we now consider an angular spectrum of plane wave components to simulate a point source at a finite distance. The planar surfaces with constant phase of each plane wave component merge into a curved equiphase surface. This surface consists of two sheets corresponding to the two solutions of the quadratic equation in v_p^2 (Eq. (2.4.8)) and is furthermore point-symmetric with respect to the

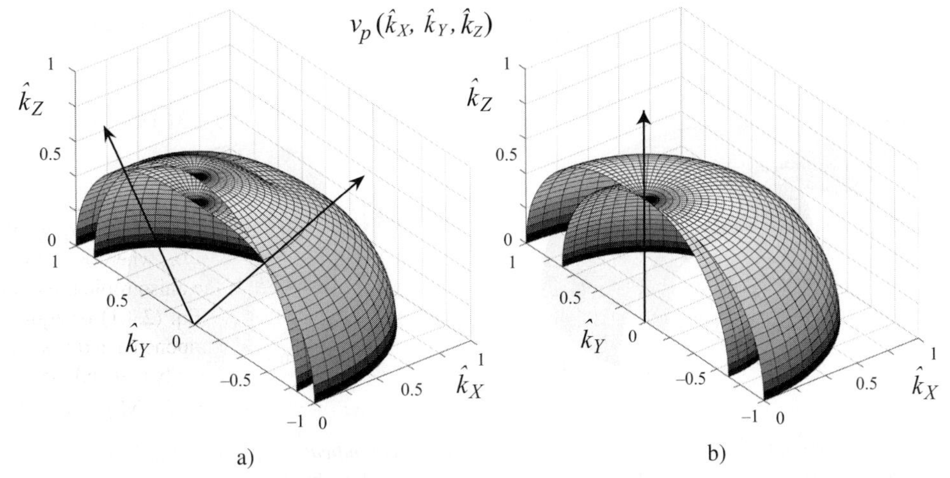

Figure 2.6: *Wave normal surface* representing the phase propagation speed v_p in an anisotropic medium as a function of the propagation direction given by the unit vector $\hat{\mathbf{k}}$.
a) Biaxial medium. The principal velocity directions have been chosen along the Cartesian X-, Y- and Z-directions. The relative values of the principal velocities are given by the values $v_X = 0.85$, $v_Y = 0.70$ and $v_Z = 1.00$. The resulting optical axes of the biaxial material are located in the plane $X = 0$, defined by the smallest and largest principal velocity directions, and have been indicated by black arrows; the propagation speed along this direction is given by v_X.
b) Wave normal surface for the special case of a uniaxial medium ($v_X = v_Y = 0.70$, $v_Z = 1.00$) with a single optical axis.

origin because the solutions for $\pm\hat{\mathbf{k}}$ yield identical values for v_p. The two sheets of the surface can be associated with the two wavefronts that arise when a point source emits a spectrum of randomly polarised plane waves ('natural' light), filling the entire 4π solid angle.

An example of such a surface is shown in Fig. 2.6a for a biaxial medium; it is commonly called the *wave normal* surface. The two sheets of the surface intersect each other at points corresponding to the optical axes of the medium. Expressions for the vector components of $\hat{\mathbf{k}}$ defining the optical axes were given in Eq. (2.3.5); however, in terms of the principal propagation speeds they read (assuming $v_X < v_Y < v_Z$)

$$\hat{k}_X = \pm\sqrt{\frac{v_Y^2 - v_X^2}{v_Z^2 - v_X^2}}, \qquad \hat{k}_Y = 0, \qquad \hat{k}_Z = \pm\sqrt{\frac{v_Z^2 - v_Y^2}{v_Z^2 - v_X^2}}\;; \qquad v_p = v_Y\,. \tag{2.4.9}$$

The resulting propagation speed is the principal value corresponding to the direction that is perpendicular to the plane defined by the two optical axes. The optical axes are found in the plane that contains the two principal directions with the largest and smallest principal propagation speed. The directions with the largest and smallest propagation speed (the Z- and Y-axes in Fig. 2.6a coincide with the bisecting planes of the optical axes. A special case arises if the wave vector is parallel to one of the principal axes. Figure 2.6a shows that in this case the two solutions for v_p or n_p are the principal values along the other two axes of the medium. In graph b) of the figure, applying to a uniaxial medium with $v_X = v_Y$, the two optical axes have merged to a single one that coincides with the Z-axis. In the example it is along this axis that the largest principal velocity is found.

We can also plot the value of the refractive index $n_p = c/v_p$ for the two plane waves with unit propagation vector $\hat{\mathbf{k}}$ by plotting the vector $\mathbf{r}_n = n_p r_0 \hat{\mathbf{k}}$. The endpoint of $\mathbf{r}_p$ indicates what value the phase of a plane wave has assumed after propagation over a distance r_0 from the origin O. This surface is commonly called the *optical indicatrix*, not to be confused with the index ellipsoid of Fig. 2.1. The relevance of the optical indicatrix will be discussed later in the framework of the electric polarisation eigenstates of an anisotropic medium. Figure 2.7 shows the wave normal and the optical indicatrix for the same biaxial medium.

The wave normal surface and the optical indicatrix can be derived from each other in the following way. Inspection of Eqs. (2.4.7) and (2.4.8) shows that, with the substitution $v_p = c/n_p$, we can write these two equations as

$$a_1 v_p^4 + a_2 v_p^2 + a_3 = 0\,, \qquad a_1 c^4 + a_2 c^2 n_p^2 + a_3 n_p^4 = 0\,, \tag{2.4.10}$$

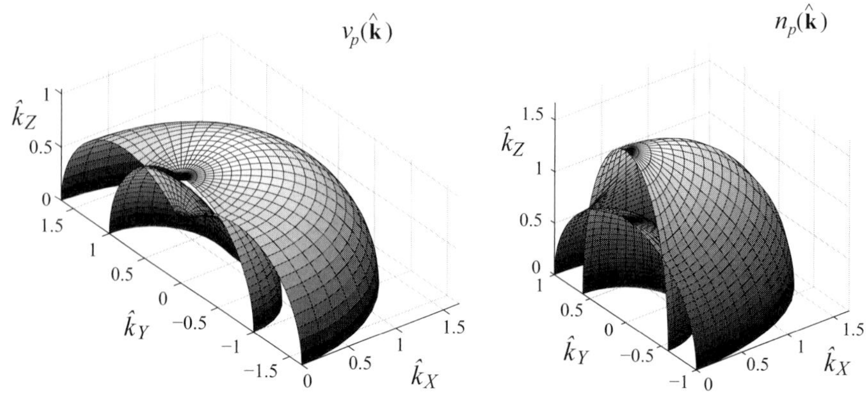

Figure 2.7: A comparison of the *wave normal surface* $v_p(\hat{\mathbf{k}})$ and the *optical indicatrix* $n_p(\hat{\mathbf{k}})$ for a biaxial medium. To visualise the differences between the shape of both surfaces, the anisotropy of the medium has been made extremely large; $v_X = 1$, $v_Y = 3/5$, $v_Z = 5/3$ (arbitrary units).

where the coefficients a_1, a_2 and a_3 follow from (2.4.7) and (2.4.8). We obtain the solutions

$$v_p^2 = \frac{-a_2 \pm (a_2^2 - 4a_1a_3)^{1/2}}{2a_1} , \qquad n_p^2 = \frac{-a_2c^2 \pm (a_2^2c^4 - 4a_1a_3c^4)^{1/2}}{2a_3} . \tag{2.4.11}$$

It then follows that

$$n_p^2 = \frac{c^2}{\hat{k}_X^2 v_Y^2 v_Z^2 + \hat{k}_Y^2 v_X^2 v_Z^2 + \hat{k}_Z^2 v_X^2 v_Y^2} \qquad v_p^2 = \frac{n_X^2 n_Y^2 n_Z^2}{\hat{k}_X^2 n_X^2 + \hat{k}_Y^2 n_Y^2 + \hat{k}_Z^2 n_Z^2} \, v_p^2 . \tag{2.4.12}$$

2.4.1.2 The Wave Normal Surface in Cylindrical Coordinates

For a further analysis of the wave normal surface with its two velocity sheets as depicted in Fig. 2.6 it is useful to represent it with the aid of cylindrical coordinates. We start with Eq. (2.4.5) and eliminate the components $\hat{k}_X$, $\hat{k}_Y$ and $\hat{k}_Z$ of the unit wave vector $\hat{\mathbf{k}}$ by writing

$$\hat{k}_X v_p = X , \qquad \hat{k}_Y v_p = Y , \qquad \hat{k}_Z v_p = Z , \tag{2.4.13}$$

where v_p is equal to $(X^2 + Y^2 + Z^2)^{1/2}$, the velocity in the direction given by the wave vector $\hat{\mathbf{k}}$. The surface $G(X, Y, Z) = 0$, described by the endpoint of the velocity vector, is then obtained from Eq. (2.4.7) by substituting in this equation the expressions of Eq. (2.4.13). In a cross-section through the Z-axis of the velocity surface $G = 0$ in which the lateral coordinate is given by $R = (X^2 + Y^2)^{1/2}$ and the azimuth of the cross-section, ϕ, is defined by $X = R\cos\phi$ and $Y = R\sin\phi$, we can write

$$\begin{aligned} G(R,\phi,Z) &= (R^2 + Z^2)(R^2 + Z^2 - v_X^2 \sin^2\phi - v_Y^2 \cos^2\phi) \\ &\quad - R^2 v_Z^2 \left(1 - \frac{v_X^2 \sin^2\phi + v_Y^2 \cos^2\phi}{R^2 + Z^2}\right) \\ &\quad - Z^2 \left(v_X^2 \cos^2\phi + v_Y^2 \sin^2\phi - \frac{v_X^2 v_Y^2}{R^2 + Z^2}\right) = 0 . \end{aligned} \tag{2.4.14}$$

For the special cases $\phi = 0$ and $\phi = \pi/2$ we have

$$\begin{aligned} G(X,0,Z) &= (X^2 + Z^2 - v_Y^2)\left\{(X^2 + Z^2)^2 - X^2 v_Z^2 - Z^2 v_X^2\right\} = 0 , \\ G(0,Y,Z) &= (Y^2 + Z^2 - v_X^2)\left\{(Y^2 + Z^2)^2 - Y^2 v_Z^2 - Z^2 v_Y^2\right\} = 0 . \end{aligned} \tag{2.4.15}$$

In the two cross-sections with $\phi = \pi/2$ and $\phi = 0$ the velocity surface consists of two curves of which one has a circular shape, giving rise to ordinary wave propagation, and the second has an oval shape corresponding to extraordinary wave

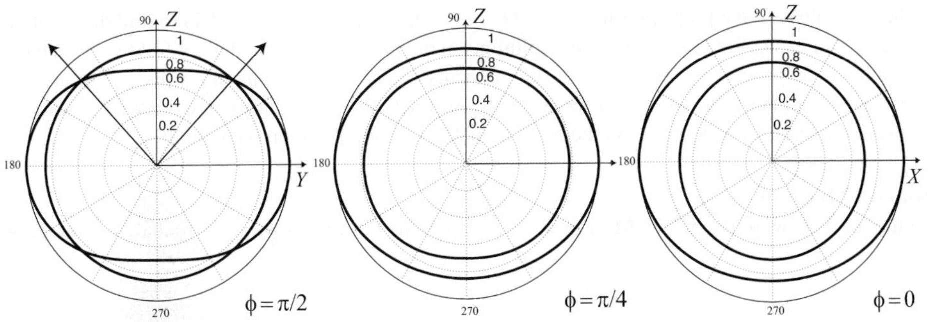

Figure 2.8: Polar plot of the propagation speed in a biaxial anisotropic medium in a cross-section comprising the medium direction with the largest principal propagation speed (Z-direction). The polar angle with the Z-direction runs from 0 to 2π. The azimuth of the cross-section plane with respect to the plane $Y = 0$ is given by the angle ϕ. The relative values of the principal velocities are the same as in Fig. 2.6, $v_X = 0.85$, $v_Y = 0.70$ and $v_Z = 1.00$. The optical axes are found in the cross-section with azimuth $\phi = \pi/2$ and have been depicted by the two arrows; the propagation speed in these two directions is given by v_X.

propagation (see Fig. 2.8). In a general cross-section with arbitrary azimuth ϕ, the velocity surface equation turns out to be given by

$$G(X,Y,Z) = G(X,0,Z)\cos^2\phi + G(0,Y,Z)\sin^2\phi = 0 . \tag{2.4.16}$$

Two curves emerge now with the property that $G(X,Y,Z) \neq G(X,0,Z)$ and $G(X,Y,Z) \neq G(0,Y,Z)$. There is no velocity curve with a spherical shape and both wavefront cross-sections correspond to extraordinary wave propagation (see Fig. 2.8, central plot for $\phi = \pi/4$).

We emphasise that in a biaxial medium the ordinary wave propagation is restricted to the two cross-sections of the wave normal surface defined by Eq. (2.4.16) by choosing $\phi = 0$ or $\phi = \pi/2$. This is in contrast to the ordinary wave propagation in a uniaxial medium. Each planar wave with a linear state of electrical polarisation that is perpendicular to the principal section defined by the wave propagation vector $\hat{\mathbf{k}}$ and the optical axis experiences ordinary light propagation in a uniaxial medium. There exists an infinity of such cross-sections, each associated with the appropriate state of linear polarisation. If the light is produced by a point source emitting natural, unpolarised radiation, half of the light intensity thus propagates in the ordinary way as a spherical wave; the other half of the light gives rise to extraordinary wave propagation. In a biaxial medium the same point source produces two wavefronts with an *aspherical* shape. Each of these wavefronts possesses one single cross-section with circular shape and these cross-sections are found in mutually orthogonal planes. For the special case of a uniaxial medium with $v_X = v_Y \neq v_Z$ we have the following solutions for the propagation speed:

$$v_1^2 = v_X^2 = v_Y^2 , \qquad v_2^2 = v_X^2 + (\hat{k}_X^2 + \hat{k}_Y^2)(v_Z^2 - v_X^2) . \tag{2.4.17}$$

Equal velocities of the ordinary and extraordinary wave are found if

$$(\hat{k}_X^2 + \hat{k}_Y^2)(v_X^2 - v_Z^2) = 0 , \tag{2.4.18}$$

and this condition is either satisfied for $\hat{k} = (0,0,1)$, corresponding to propagation along the optical axis with velocity $v_X = v_Y$, or for $v_X = v_Z$ (isotropic medium).

2.4.2 Propagation Direction and Polarisation Eigenstates in Anisotropic Media

So far, we have obtained expressions for the two possible values of the phase propagation speed v_p that correspond to a certain wave propagation vector $\mathbf{k}$. Explicitly, they are given by (see Eq. (2.4.7)),

$$\begin{aligned} v_p^2 =& \frac{1}{2}\left[v_X^2(\hat{k}_Y^2+\hat{k}_Z^2) + v_Y^2(\hat{k}_X^2+\hat{k}_Z^2) + v_Z^2(\hat{k}_X^2+\hat{k}_Y^2)\right] \\ &\pm \frac{1}{2}\left[v_X^4(\hat{k}_Y^2+\hat{k}_Z^2)^2 + v_Y^4(\hat{k}_X^2+\hat{k}_Z^2)^2 + v_Z^4(\hat{k}_X^2+\hat{k}_Y^2)^2 + 2v_X^2v_Y^2(\hat{k}_X^2\hat{k}_Y^2-\hat{k}_Z^2) + 2v_X^2v_Z^2(\hat{k}_X^2\hat{k}_Z^2-\hat{k}_Y^2) + 2v_Y^2v_Z^2(\hat{k}_Y^2\hat{k}_Z^2-\hat{k}_X^2)\right]^{1/2} . \end{aligned} \tag{2.4.19}$$

Employing the two values of the propagation speed $v_p(\hat{k}_X, \hat{k}_Y, \hat{k}_Z)$, we return to Eq. (2.4.1) to find the components of the electric field vector for each of the two plane wave solutions in the medium. It immediately follows from these equations that the electric field components are real when the components $\hat{k}_X$, $\hat{k}_Y$ and $\hat{k}_Z$ of the unit wave vector are real, implying that the state of polarisation of the associated waves is always linear. We first analyse the allowed states of polarisation in a general biaxial medium and then discuss the more tractable solutions in a uniaxial medium.

2.4.2.1 Biaxial Medium

Multiplying the second row of the matrix **M** of (2.4.1) by $\hat{k}_Z$ and the third row by $\hat{k}_Y$, we find after subtraction that

$$E_{0,Y} = \frac{\hat{k}_Y v_Y^2 (v_p^2 - v_Z^2)}{\hat{k}_Z v_Z^2 (v_p^2 - v_Y^2)} E_{0,Z} \,. \tag{2.4.20}$$

In a similar way, operating on rows 1 and 2 of Eq. (2.4.1), we obtain

$$E_{0,X} = \frac{\hat{k}_X v_X^2 (v_p^2 - v_Z^2)}{\hat{k}_Z v_Z^2 (v_p^2 - v_X^2)} E_{0,Z} \,. \tag{2.4.21}$$

The eigenvectors are therefore of the form

$$\mathbf{E}_i = \left(\frac{\hat{k}_X v_X^2}{v_{p,i}^2 - v_X^2} \,, \frac{\hat{k}_Y v_Y^2}{v_{p,i}^2 - v_Y^2} \,, \frac{\hat{k}_Z v_Z^2}{v_{p,i}^2 - v_Z^2} \right) , \tag{2.4.22}$$

for $i = 1, 2$. Alternatively, in terms of the refractive indices of the eigenstates, we have

$$\mathbf{E}_i = \left(\frac{\hat{k}_X}{n_{p,i}^2 - n_X^2} \,, \frac{\hat{k}_Y}{n_{p,i}^2 - n_Y^2} \,, \frac{\hat{k}_Z}{n_{p,i}^2 - n_Z^2} \right) . \tag{2.4.23}$$

Singular values arise when v_p or n_p takes on one of the principal values for the propagation speed or the index of the propagation mode, respectively. Two cases can be distinguished:

- $\hat{\mathbf{k}} = (1, 0, 0)$
  The solutions for the phase propagation speed follow from Eq. (2.4.19), $v_p = v_Y, \ v_Z$. Inspection of Eq. (2.4.1) then shows that the allowed polarisation states are

$$v_{p,1} = v_Y \,, \qquad \hat{\mathbf{E}}_1 = (0, 1, 0) \,,$$
$$v_{p,2} = v_Z \,, \qquad \hat{\mathbf{E}}_2 = (0, 0, 1) \,.$$

  Similarly, the cases $\hat{\mathbf{k}} = (0, 1, 0)$ and $\hat{\mathbf{k}} = (0, 0, 1)$ yield electric eigenvectors which are oriented along the two other principal axes that are perpendicular to the wave vector.
- $\hat{\mathbf{k}}$ is parallel to one of the optical axes of the biaxial medium (see Eq. (2.4.9)).
  This case needs a special treatment; it gives rise to the phenomenon of so-called *conical refraction* and is treated in Section 2.8 of this chapter.

Some other special propagation directions allow for an analytic solution of the state of polarisation of the corresponding wave. For clarity we assume that $v_Z > v_X > v_Y$. We first study the propagation of waves with the vector $\hat{\mathbf{k}}$ located in a plane perpendicular to the plane containing the two optical axes which, with reference to Fig. 2.6, we choose as the $\hat{k}_Y = 0$ plane. From Eq. (2.4.7) or (2.4.19) we obtain

$$\left\{ \begin{array}{ll} \text{a)} & v_1^2 = v_Y^2 \,, \\ \text{or ,} & \\ \text{b)} & v_2^2 = \hat{k}_X^2 v_Z^2 + \hat{k}_Z^2 v_X^2 \,. \end{array} \right. \tag{2.4.24}$$

The associated electric field eigenvectors cannot be directly derived from Eq. (2.4.22) for the case that $v_1^2 = v_Y^2$ since it leads to a singular quotient. With the aid of Eq. (2.4.2), however, we can readily determine the unit electric vectors for the two cases, which are given by

$$\left\{ \begin{array}{ll} \text{a)} & \hat{\mathbf{E}}_1 = (0, 1, 0) \,, \\ \text{b)} & \hat{\mathbf{E}}_2 = \left[\hat{k}_Z^2 v_X^4 + \hat{k}_X^2 v_Z^4 \right]^{-1/2} \left(\hat{k}_Z \, v_X^2 \,, \, 0 \,, \, -\hat{k}_X \, v_Z^2 \right) . \end{array} \right. \tag{2.4.25}$$

The first solution a) yields a wave propagation speed that is independent of the wave direction in a cross-section that bisects the optical axes. The polarisation eigenstate is perpendicular to this cross-section and thus perpendicular to **k**. Waves in this cross-section behave quasi-isotropically. The propagation speed for the second case b) shows a directional dependence and the linear polarisation eigenstate is parallel to the cross-section with $\hat{k}_Y = 0$. The vectors $\hat{\mathbf{k}}$ and **E** are not perpendicular to each other in this case.

We also analyse the behaviour for a wave propagating parallel to the plane containing the optical axes ($\hat{k}_X = 0$). We obtain from Eq. (2.4.7) with $\hat{\mathbf{k}} = (0, \hat{k}_Y, \hat{k}_Z)$:

$$\begin{cases} \text{c)} \quad v_1^2 = v_X^2 \,, \\ \text{and} \\ \text{d)} \quad v_2^2 = \hat{k}_Y^2 v_Z^2 + \hat{k}_Z^2 v_Y^2 \,. \end{cases} \tag{2.4.26}$$

The corresponding polarisation eigenstates are given by

$$\begin{cases} \text{c)} \quad \hat{\mathbf{E}}_1 = (1, 0, 0) \,, \\ \text{d)} \quad \hat{\mathbf{E}}_2 = \left[\hat{k}_Z^2\, v_Y^4 + \hat{k}_Y^2\, v_Z^4\right]^{-1/2} \left(0\,,\ \hat{k}_Z\, v_Y^2\,,\ -\hat{k}_Y\, v_Z^2\right) . \end{cases} \tag{2.4.27}$$

The solution with eigenvector $\mathbf{E}_1$ behaves in a quasi-isotropic way. The calculation of the second eigenvector $\mathbf{E}_2$ needs a special treatment if v_2 equals v_X because the expression for the X-component becomes undefined if $v_Z > v_X > v_Y$. In this case the wave vector is parallel to one of the optical axes and the phenomenon of conical refraction occurs (see Section 2.8).

2.4.2.2 Uniaxial Medium

Propagation in a uniaxial medium can be considered as a limiting case of propagation in a biaxial medium with, for instance, $v_X = v_Y = v_o \neq v_Z$. The optical axes merge to a single optical axis that, in our example, points in the Z-direction. Due to the rotational symmetry with respect to the optical axis, the solutions *a)* and *c)* for the biaxial medium are now found for any cross-section through the Z-axis. This means that quasi-isotropic wave propagation is possible for all propagation directions. A spherical wave could propagate undistorted through the medium with a propagation speed of v_o if its state of polarisation is correct. In particular, at every point the linear state of polarisation $\mathbf{E}_{0,o}$ has to be perpendicular to the plane defined by the optical axis and the local propagation vector. In a uniaxial medium this special plane, labelled $\mathcal{P}_1$ in Fig. 2.9, is called the *principal section*. A wave exhibiting isotropic behaviour is said to be an *ordinary* wave and the propagation speed is called the ordinary speed v_o (see also Fig. 2.9).

The uniaxial medium also supports the wave solutions corresponding to cases *b)* and *d)* of the biaxial medium. Here again, because of the symmetry of revolution around the optical axis, the propagation speed v_{eo} of the *extraordinary* wave only depends on the angle between the wave propagation vector and the optical axis. From Eq. (2.4.15), with $v_X = v_Y = v_0$, we find

$$v_{eo}^2 = v_Z^2 + (v_o^2 - v_Z^2)\hat{k}_Z^2 \,. \tag{2.4.28}$$

For a general wave vector $\hat{\mathbf{k}}$, using Eqs. (2.4.6)–(2.4.19), we find the following polarisation eigenstates:

- *Ordinary wave* ($v = v_o$)

$$\hat{\mathbf{E}}_{0,o} = \frac{1}{\sqrt{\hat{k}_X^2 + \hat{k}_Y^2}} (-\hat{k}_Y, \hat{k}_X, 0) \,. \tag{2.4.29}$$

The state of polarisation is perpendicular to the principal section and also perpendicular to the wave vector $\hat{\mathbf{k}}$.

- *Extraordinary wave*

The propagation speed is given by Eq. (2.4.28) and from Eq. (2.4.22) we deduce

$$\hat{\mathbf{E}}_{0,eo} = \frac{1}{\sqrt{1 - \hat{k}_Z^2 + \dfrac{(1-\hat{k}_Z^2)^2\, v_Z^4}{\hat{k}_Z^2\, v_o^4}}} \left(\hat{k}_X, \hat{k}_Y, -\frac{(1-\hat{k}_Z^2)\, v_Z^2}{\hat{k}_Z\, v_o^2}\right) . \tag{2.4.30}$$

It is seen that the vector $\hat{\mathbf{E}}_{0,eo}$ is perpendicular to the normal $(\hat{k}_Y, -\hat{k}_X, 0)$ of the principal section and hence the ordinary polarisation vector $\hat{\mathbf{E}}_{0,o}$. Consequently, the linear state of polarisation lies in the principal section. We also note that the

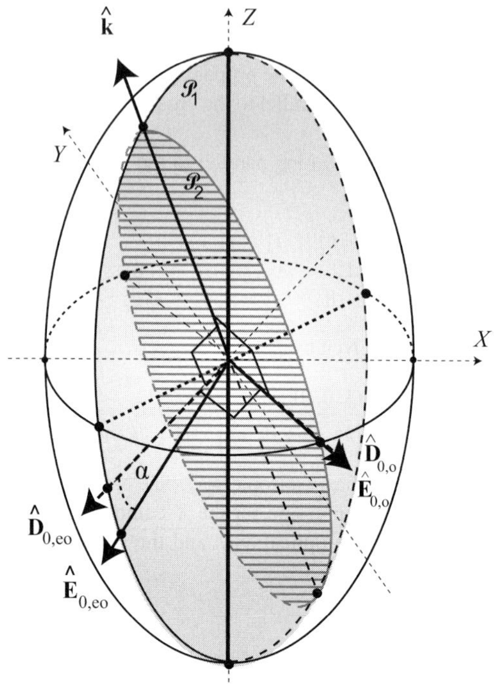

Figure 2.9: Wave propagation in a uniaxial medium for a general propagation direction $\hat{\mathbf{k}}$. The principal velocity directions have been chosen along the Cartesian X-, Y- and Z-directions and their relative magnitudes are given by $v_X = v_Y = v_o = 0.70$ and $v_Z = 1.00$. The optical axis, indicated by a thick solid line, is directed along the Z-axis. The principal section has been denoted by $\mathcal{P}_1$. The polarisation eigenstates associated with the extraordinary and ordinary waves with wave vector $\hat{\mathbf{k}}$ are given by the unit vectors $\hat{\mathbf{E}}_{0,eo}$ and $\hat{\mathbf{E}}_{0,o}$. The corresponding unit vectors in the direction of the electric flux density vector (dashed in the figure) have been denoted by $\hat{\mathbf{D}}_{0,eo}$ and $\hat{\mathbf{D}}_{0,o}$, respectively.

inner product of $\hat{\mathbf{E}}_{0,eo}$ and $\hat{\mathbf{k}}$ is not zero. The angle α between $\hat{\mathbf{D}}_{0,eo}$ and $\hat{\mathbf{E}}_{0,eo}$ follows from noting that $\sin\alpha = \hat{\mathbf{E}}_{0,eo} \cdot \hat{\mathbf{k}}$. Using Eq. (2.4.30), we find

$$\sin\alpha = \frac{(1-\hat{k}_Z^2)\left(1-\frac{v_Z^2}{v_o^2}\right)}{\sqrt{1-\hat{k}_Z^2+\frac{(1-\hat{k}_Z^2)^2\, v_Z^4}{\hat{k}_Z^2\, v_o^4}}}$$
$$= \hat{k}_Z\sqrt{1-\hat{k}_Z^2}\left(v_o^2-v_Z^2\right)\left\{v_Z^4(1-\hat{k}_Z^2)-v_o^4\hat{k}_Z^2\right\}^{-1/2} . \qquad (2.4.31)$$

Using the fact that $\epsilon_j \propto 1/v_j^2$ for the three orthogonal directions $j = X$, Y and Z, one easily verifies that the electric flux density vector $\mathbf{D}_{0,eo}$ is perpendicular to $\hat{\mathbf{k}}$.

2.5 Energy Transport in a Medium with Linear Anisotropy

In this section we first derive the relation between the energy propagation (Poynting) vectors and the wave vector $\mathbf{k}$ in a *linear* anisotropic medium. In the second part of this section we show how a modified set of eigenvalue equations describing wave propagation can be established in which the unit wave vector $\hat{\mathbf{k}}$ is replaced by the unit ray vector $\hat{\mathbf{s}}$. Critically, the vector $\hat{\mathbf{s}}$ determines the propagation direction of electromagnetic radiation in an anisotropic medium and is the quantity considered when tracing rays through a system using the simplified geometrical optics model of wave propagation.

2.5.1 Ratio of Ray and Phase Velocity

Regardless of the medium properties, the direction and magnitude of the energy transported by an electromagnetic wave are given by the vector product of $\mathbf{E}(\mathbf{r},t)$ and $\mathbf{H}(\mathbf{r},t)$ (see Eq. (1.4.5)). Using the complex notation with amplitude vectors $\mathbf{E}_0 \exp(i\mathbf{k}\cdot\mathbf{r})$ and $\mathbf{H}_0 \exp(i\mathbf{k}\cdot\mathbf{r})$ and limiting ourselves to the mean energy flow, we have the following expression for the time-averaged Poynting vector (see also Appendix A):

$$\langle \mathbf{S} \rangle = \frac{1}{2}\Re\{\mathbf{E}_0 \times \mathbf{H}_0^*\} = \frac{1}{2}|\mathbf{E}_0||\mathbf{H}_0|\cos\alpha_{eh}\,\hat{\mathbf{s}}\,, \tag{2.5.1}$$

where the unit vector $\hat{\mathbf{s}}$ points in the direction of $\Re\{\mathbf{E}_0 \times \mathbf{H}_0^*\}$ and α_{eh} is the phase offset between the electric and magnetic field strengths. In a lossless, nonmagnetic anisotropic medium with linear polarisation eigenstates, α_{eh} is zero. This result follows from the fact that the medium admittance is real in such a medium whereby the harmonic vectors $\mathbf{E}_0$ and $\mathbf{H}_0$ are in phase.

The average electromagnetic energy density of a wave can be found by taking the sum of the time-averages of the electric energy density w_e and the magnetic energy density w_m, previously defined in Eq. (1.4.3). So doing for a harmonic electromagnetic field yields

$$\frac{1}{4}\left(\mathbf{E}\cdot\mathbf{D}^* + \mathbf{H}\cdot\mathbf{B}^*\right) = \frac{1}{4}\left(|\mathbf{E}||\mathbf{D}|\cos\alpha + |\mathbf{H}||\mathbf{B}|\right)\,, \tag{2.5.2}$$

where we have assumed that the medium behaves isotropically with respect to its magnetic properties ($\bar{\bar{\mu}}_r = \mathbf{I}$, where $\mathbf{I}$ is the unit tensor). The angle α denotes the angle between the vectors $\mathbf{D}_0$ and $\mathbf{E}_0$ and is shown in Figs. 2.3 and 2.5 for wave propagation in a uniaxial and a biaxial medium, respectively.

Using Eq. (2.2.10) and Eq. (2.2.11) for harmonic waves we have

$$\mathbf{H}_0 = \frac{\mathbf{k}\times\mathbf{E}_0}{\omega_0\mu_0}\,, \qquad |\mathbf{H}_0| = \frac{|\mathbf{k}||\mathbf{E}_0|\cos\alpha}{\omega_0\mu_0}\,. \tag{2.5.3}$$

Likewise, using expression (2.2.19), we find

$$|\mathbf{D}_0| = \frac{1}{\mu_0 v^2}\left|\mathbf{E}_0 - (\hat{\mathbf{k}}\cdot\mathbf{E}_0)\hat{\mathbf{k}}\right| = \frac{|\mathbf{E}_0|}{\mu_0 v^2}\cos\alpha\,. \tag{2.5.4}$$

Now defining the value of the propagation speed of electromagnetic energy $|\mathbf{v}_{em}|$ by the ratio

$$|\mathbf{v}_{em}| = \frac{|\langle \mathbf{S}\rangle|}{\langle w_e + w_m\rangle} = \frac{\frac{1}{2}|\mathbf{E}_0 \times \mathbf{H}_0^*|}{\frac{1}{4}\left(|\mathbf{E}_0||\mathbf{D}_0|\cos\alpha + \mu_0|\mathbf{H}_0|^2\right)}\,, \tag{2.5.5}$$

where we have used $\mathbf{B}_0 = \mu_0\mathbf{H}_0$, we can substitute in the relations of Eqs. (2.5.3) and (2.5.4) to yield

$$|\mathbf{v}_{em}| = \frac{\frac{1}{2}\frac{|\mathbf{k}||\mathbf{E}_0|^2\cos\alpha}{\omega_0}}{\frac{1}{2}\frac{|\mathbf{E}_0|^2\cos^2\alpha}{v^2}} = \frac{|\mathbf{k}|v^2}{\omega_0\cos\alpha} = \frac{|v|}{\cos\alpha}\,. \tag{2.5.6}$$

The magnitude of the electromagnetic energy propagation vector $|\mathbf{v}_{em}|$ is also called the ray velocity v_r, as opposed to the quantity $v = c/n$ that is called the phase velocity. This terminology originates from the ray optics picture in which electromagnetic energy propagates along the optical rays of light.

The geometry of the electric and magnetic field vectors and the corresponding wave and ray velocity vector are shown in Fig. 2.10; the vector $\mathbf{E}_{0,p} = -\hat{\mathbf{k}}\times\hat{\mathbf{k}}\times\mathbf{E}_0$ is a vector that points in the direction of $\mathbf{D}_0$ and has a length $|\mathbf{E}_0|\cos\alpha$. Equation (2.5.6) for $|\mathbf{v}_{em}|$ states that the magnitude of the phase velocity $|\mathbf{v}| = \omega/|\mathbf{k}|$ is obtained by projecting the ray velocity vector onto the wave vector $\mathbf{k}$. The intersections of a number of planes of constant phase ϕ with the plane containing $\mathbf{v}$ and $\mathbf{E}$ are also shown in Fig. 2.10. Note that these planes are perpendicular to the wave vector $\hat{\mathbf{k}}$ but not to the ray vector $\hat{\mathbf{s}}$.

In Fig. 2.11 the velocity vectors of the ordinary and extraordinary waves propagating in a uniaxial medium are shown. The symmetry around the optical axis (dotted line along the Z-axis) means that only the cross-section containing the X and Z components of the velocity vectors needs to be considered. An expression for the phase velocity vector $\mathbf{v}_{eo}$ is given by Eq. (2.4.28). If $v_Z > v_X = v_Y = v_o$, the angle between the electric field vector and the wave vector is larger than $\pi/2$. Consequently, the extraordinary ray velocity vector $\mathbf{v}_{em}$ is rotated, relative to the phase velocity vector, by an angle α towards the Z-axis, the optical axis of the medium. The endpoints of the ordinary velocity vectors are found on the dotted circle with radius v_o; both the wave and the energy propagation vector of the ordinary wave point in the direction of $\hat{\mathbf{k}}$.

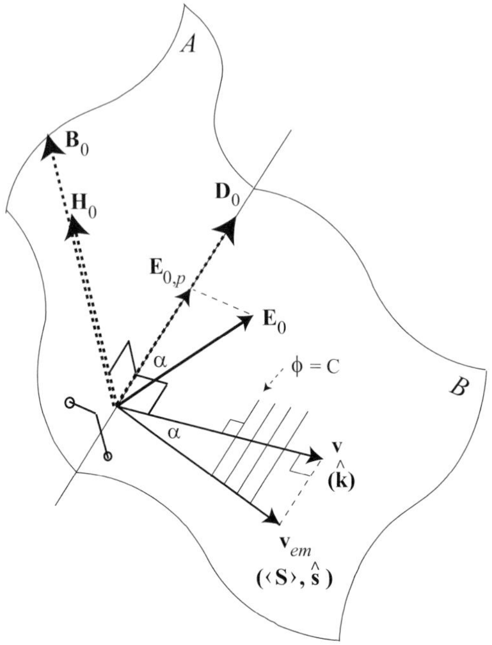

Figure 2.10: The electromagnetic field vectors $\mathbf{B}_0$, $\mathbf{H}_0$ and $\mathbf{D}_0$ are coplanar and located in the plane A. The vectors $\mathbf{D}_0$, $\mathbf{E}_0$, $\mathbf{v}$ and $\mathbf{v}_{em}$ are located in the plane B, where B is perpendicular to A. The phase velocity vector $\mathbf{v}$ points in the direction of the unit wave vector $\hat{\mathbf{k}}$, whilst the unit ray vector $\hat{\mathbf{s}}$ points in the direction of the Poynting vector $\langle \mathbf{S} \rangle$. The magnitude of $\mathbf{v}$ equals the projection of the electromagnetic energy velocity vector $\mathbf{v}_{em}$ onto $\hat{\mathbf{k}}$.

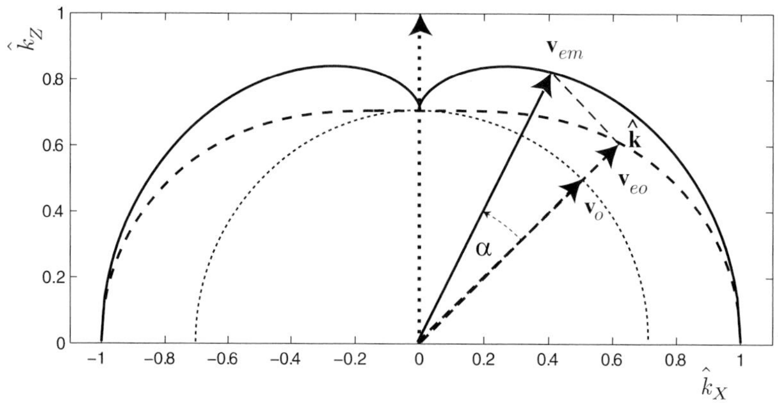

Figure 2.11: The endpoints of the extraordinary phase velocity vector $\mathbf{v}_{eo}$ (dashed curve) and the ray velocity vector $\mathbf{v}_{em}$ (solid curve) as a function of the direction of the unit wave propagation vector $\hat{\mathbf{k}}$. The medium is uniaxial and the optical axis points in the Z-direction. The angle between the energy propagation direction and the wave vector is α. This angle is a function of the wave components, $(\hat{k}_X, 0, \hat{k}_Z)$ in the example of the figure. The normalised principal propagation speeds are given by $v_Z = 1.0$ and $v_o = 0.7$.

2.5.2 Geometrical Discussion of the Eigenvalue Equations

We now focus on the eigenvalue equations of a linear anisotropic medium as given by Eqs. (2.2.12) and (2.2.13). For a magnetically isotropic medium we have the simplified expressions

$$\hat{\mathbf{k}} \times \hat{\mathbf{k}} \times \mathbf{E}_0 + (k_0^2/k^2)\, \bar{\bar{\epsilon}}_r\, \mathbf{E}_0 = \mathbf{0}\,, \tag{2.5.7}$$

$$\hat{\mathbf{k}} \times \left(\overline{\overline{\epsilon}}_r\right)^{-1} (\hat{\mathbf{k}} \times \mathbf{H}_0) + (k_0^2/k^2)\mathbf{H}_0 = \mathbf{0} , \tag{2.5.8}$$

where we have used the unit wave vector $\hat{\mathbf{k}} = \mathbf{k}/k$ and the vacuum wavenumber k_0 given by $\omega^2\epsilon_0\mu_0 = k_0^2$. Each set of equations yields the eigenvalues k and the corresponding polarisation eigenstates $\mathbf{E}_0$ or $\mathbf{H}_0$. The two sets should yield the same k-eigenvalues, although this is not immediately evident from the structure of the equations. Using the geometrical picture of the wave vector and the electromagnetic field vectors of Fig. 2.10, we can easily prove the equivalence of the eigenvalues k_E and k_H that follows from Eqs. (2.5.7) and (2.5.8). For real electromagnetic vector amplitudes $\mathbf{E}_0$, $\mathbf{H}_0$, $\mathbf{D}_0$ and $\mathbf{B}_0$, we obtain $\hat{\mathbf{k}} \times \hat{\mathbf{k}} \times \mathbf{E}_0 = -\cos\alpha|\mathbf{E}_0|\hat{\mathbf{D}}_0$, where $\hat{\mathbf{D}}_0$ is the unit vector pointing in the direction of the electric flux density vector $\mathbf{D}_0$. Using the constitutive relation $\mathbf{D}_0 = \epsilon_0\, \overline{\overline{\epsilon}}_r\, \mathbf{E}_0$ we obtain from Eq. (2.5.7) the result

$$\left(-\cos\alpha|\mathbf{E}_0| + \frac{k_0^2}{k_E^2}\frac{|\mathbf{D}_0|}{\epsilon_0}\right)\hat{\mathbf{D}}_0 = 0 , \tag{2.5.9}$$

where k_E is the modulus of the wave vector according to this eigenvalue equation. Proceeding in an analogous way with Eq. (2.5.8) we obtain

$$\left(-\frac{\epsilon_0|\mathbf{E}_0|\cos\alpha}{|\mathbf{D}_0|} + \frac{k_0^2}{k_H^2}\right)\hat{\mathbf{H}}_0 = 0 . \tag{2.5.10}$$

The two sets of equations thus yield the same eigenvalue solutions for the wave vector modulus, specifically

$$k_E^2 = k_H^2 = \frac{k_0^2|\mathbf{D}_0|}{\epsilon_0|\mathbf{E}_0|\cos\alpha} . \tag{2.5.11}$$

2.6 Eigenvalue Equations Involving the Ray Vector $\hat{\mathbf{s}}$

The eigenvalue equations defined by Eq. (2.4.1) are a set of linear equations in terms of the electric field strength components. The components of the wave vector appear as parameters in Eq. (2.4.1) and the associated lengths of the wave vector are considered as the eigenvalues of the linear system. The length of the wave vector is proportional to the effective refractive index of the medium given the propagation direction and inversely proportional to the phase propagation speed of the wave. The discussion regarding Fig. 2.10 has shown the special role of the four coplanar vectors $\mathbf{D}_0$, $\mathbf{E}_0$,$\hat{\mathbf{k}}$ and $\hat{\mathbf{s}}$ in the solution of the plane wave propagation problem. Figure 2.10, however, also suggests that a solution for the fields describing a plane wave in an anisotropic medium can be found from a linear system of equations in terms of the components of the electric flux density vector with the components of the ray vector $\hat{\mathbf{s}}$ now acting as the corresponding parameters.

2.6.1 Modification of the Equations with the Wave Vector $\hat{\mathbf{k}}$

We now modify the eigenvalue equations written in terms of the wave vector $\mathbf{k}$, Eq. (2.5.7), in such a way that the unit ray vector $\hat{\mathbf{s}}$ appears as the input direction for which the eigenvalues and eigenstates in the linear anisotropic medium are calculated. In practice, this is an important reformulation since $\hat{\mathbf{s}}$ describes the direction along which the electromagnetic energy flows in an anisotropic medium.

2.6.1.1 Relationship between the Unit Vectors $\hat{\mathbf{k}}$, $\hat{\mathbf{D}}_0$, $\hat{\mathbf{E}}_0$ and $\hat{\mathbf{s}}$

Figure 2.12 shows the unit wave vector $\hat{\mathbf{k}}$ and the possible directions of the unit electric flux vector $\hat{\mathbf{D}}_0$, the endpoints of which are located on a circle in a plane perpendicular to the wave vector. The electric flux density vector follows from the electric field strength according to $\mathbf{D}_0 = \epsilon_0\, \overline{\overline{\epsilon}}_r\, \mathbf{E}_0$. Inspection of Eq. (2.5.7) shows that a solution of this vector equation requires the vectors $\hat{\mathbf{D}}_0$ and $\hat{\mathbf{k}} \times \hat{\mathbf{k}} \times \hat{\mathbf{E}}_0$ to be parallel. This is equivalent to the requirement that $(\hat{\mathbf{k}} \times \hat{\mathbf{k}} \times \hat{\mathbf{E}}_0) \times \hat{\mathbf{D}}_0$ is the null vector, yielding

$$(\hat{\mathbf{k}} \cdot \hat{\mathbf{E}}_0)(\hat{\mathbf{k}} \times \hat{\mathbf{D}}_0) - \hat{\mathbf{E}}_0 \times \hat{\mathbf{D}}_0 = \mathbf{0} . \tag{2.6.1}$$

Solutions $(\hat{\mathbf{D}}_{0,j}, \hat{\mathbf{E}}_{0,j})$ need to satisfy the requirement that $\hat{\mathbf{k}} \times \hat{\mathbf{D}}_{0,j}$ and $\hat{\mathbf{E}}_{0,j} \times \hat{\mathbf{D}}_{0,j}$ are parallel vectors. This means that $\hat{\mathbf{k}}$, $\hat{\mathbf{D}}_{0,j}$ and $\hat{\mathbf{E}}_{0,j}$ are located in the same plane $\mathcal{P}_j$, perpendicular to $\hat{\mathbf{H}}_{0,j}$. The unit ray vector $\hat{\mathbf{s}}_j$ which is perpendicular to $\hat{\mathbf{E}}_{0,j}$ and $\hat{\mathbf{H}}_{0,j}$, is also located in the plane $\mathcal{P}_j$. The plane $\mathcal{P}_j$ is a principal section for wave propagation in the medium, related to a

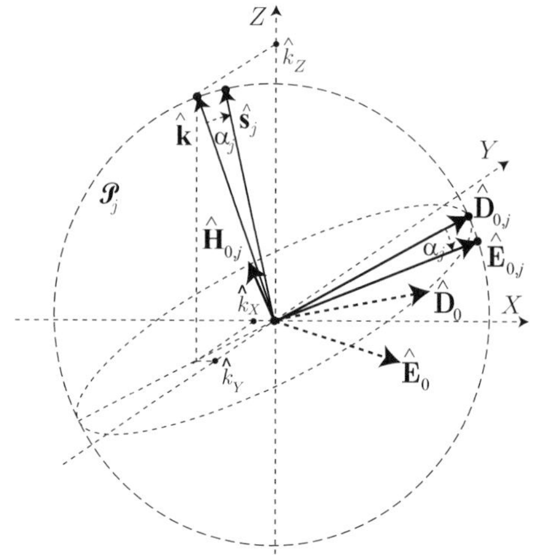

Figure 2.12: The wave vector $\hat{\mathbf{k}}$ and the electromagnetic unit vectors $\hat{\mathbf{D}}_0$ and $\hat{\mathbf{E}}_0$. Special directions of the vectors have been denoted by $\hat{\mathbf{D}}_{0,j}$, $\hat{\mathbf{E}}_{0,j}$. For these directions the vectors $\hat{\mathbf{k}}$, $\hat{\mathbf{D}}_{0,j}$ and $\hat{\mathbf{E}}_{0,j}$ lie in a single plane, the so-called principal section $\mathcal{P}_j$. The corresponding unit magnetic field vector $\hat{\mathbf{H}}_{0,j}$ is perpendicular to this plane. The resulting ray vector $\hat{\mathbf{s}}_j$ is perpendicular to $\hat{\mathbf{E}}_{0,j}$ and $\hat{\mathbf{H}}_{0,j}$ and equally located in the principal section $\mathcal{P}_j$.

(linear) eigenstate in the anisotropic medium. In general, for a particular wave propagation vector $\hat{\mathbf{k}}$, two principal sections are found. In a principal section $\mathcal{P}_j$ we find the same angular separation α_j between the vector pairs $(\hat{\mathbf{D}}_{0,j}, \hat{\mathbf{E}}_{0,j})$ and $(\hat{\mathbf{k}}, \hat{\mathbf{s}}_j)$, respectively. It then follows that $\hat{\mathbf{k}} \cdot \hat{\mathbf{E}}_{0,j} = -\hat{\mathbf{s}}_j \cdot \hat{\mathbf{D}}_{0,j}$ (see also Fig. 2.12). These geometrical properties permit us to write

$$(\hat{\mathbf{k}} \cdot \hat{\mathbf{E}}_{0,j})(\hat{\mathbf{k}} \times \hat{\mathbf{D}}_{0,j}) - \hat{\mathbf{E}}_{0,j} \times \hat{\mathbf{D}}_{0,j} = -(\hat{\mathbf{s}}_j \cdot \hat{\mathbf{D}}_{0,j})(\hat{\mathbf{s}}_j \times \hat{\mathbf{E}}_{0,j}) + \hat{\mathbf{D}}_{0,j} \times \hat{\mathbf{E}}_{0,j} = \mathbf{0}\,. \tag{2.6.2}$$

We can hence conclude, in line with Eqs. (2.5.7) and (2.6.1), that the polarisation eigenstates of the plane wave can also be found by solution of the condition

$$(\hat{\mathbf{s}} \times \hat{\mathbf{s}} \times \hat{\mathbf{D}}_0) \times \hat{\mathbf{E}}_0 = \mathbf{0}\,. \tag{2.6.3}$$

The construction of the electromagnetic eigenvectors and the associated energy propagation vectors in a linear anisotropic medium can now be further illustrated with the aid of Fig. 2.13. In part a) of the figure the endpoints of a set of unit vectors $\hat{\mathbf{E}}_0$ are plotted. They follow from the relation $\mathbf{E}_0 = \left(\bar{\bar{\epsilon}}_r\right)^{-1} \mathbf{D}_0/\epsilon_0$ in the anisotropic medium. The unit vectors $\hat{\mathbf{D}}_0$ have their endpoints on the dotted circle in the plane S that is perpendicular to the wave vector $\hat{\mathbf{k}}$. The endpoints of each vector pair $(\hat{\mathbf{D}}_0, \hat{\mathbf{E}}_0)$ are connected by a line segment in the graph. The angle between the wave vector $\hat{\mathbf{k}}$ and the Z-axis is $-\pi/6$ and the angle between the projection of $\hat{\mathbf{k}}$ on the plane $Z = 0$ and the X-axis is $\pi/3$. The medium is biaxial and the elements of the ϵ-tensor have the values $\epsilon_{XX} = 3$, $\epsilon_{YY} = 2$, $\epsilon_{ZZ} = 1$, $\epsilon_{XY} = 0.5$, $\epsilon_{XZ} = 0.2$ and $\epsilon_{YZ} = 0.3$. Figure 2.13a shows that the line segment joining the endpoints of a vector pair $(\hat{\mathbf{D}}_0, \hat{\mathbf{E}}_0)$ lies in the plane defined by $\hat{\mathbf{k}}$ and $\hat{\mathbf{D}}_0$ in only two planar cross-sections through the vector $\hat{\mathbf{k}}$. These so-called principal sections, $\mathcal{P}_1$ and $\mathcal{P}_2$, are mutually orthogonal.

The electromagnetic unit vectors and the unit wave vector in the principal sections of Fig. 2.13b satisfy Eq. (2.6.1) as the three vectors $\hat{\mathbf{k}}$, $\hat{\mathbf{D}}_0$ and $\hat{\mathbf{E}}_0$ are coplanar. Part b) shows, for a selection of vectors $\hat{\mathbf{D}}_0$ in the plane S, the vector product of the vector $\hat{\mathbf{D}}_0$ and the vector $\hat{\mathbf{k}} \times \hat{\mathbf{k}} \times \hat{\mathbf{E}}_0$. The resulting vector points in the direction of $\hat{\mathbf{k}}$ and has zero length when the two vectors composing the vector product are parallel. Part b) shows that this situation only occurs when both vectors $\hat{\mathbf{D}}_0$ and $\hat{\mathbf{E}}_0$ are located at principal sections $\mathcal{P}_j$. The energy propagation directions corresponding to each linear eigenstate are then given by the unit vectors $\hat{\mathbf{s}}_1$ and $\hat{\mathbf{s}}_2$ and they equally lie in the principal sections $\mathcal{P}_1$ and $\mathcal{P}_2$, respectively. Their direction follows from the vector product $\hat{\mathbf{E}}_{0,j} \times \hat{\mathbf{H}}_{0,j}$.

Figure 2.14 illustrates the orientation of the vectors $\hat{\mathbf{D}}_0$, $\hat{\mathbf{E}}_0$ and $\hat{\mathbf{k}}$ by means of the inner product $\hat{\mathbf{k}} \cdot \{\hat{\mathbf{D}}_0 \times (\hat{\mathbf{k}} \times \hat{\mathbf{k}} \times \hat{\mathbf{E}}_0)\}$. If $\hat{\mathbf{D}}_0$, $\hat{\mathbf{E}}_0$ and $\hat{\mathbf{k}}$ are coplanar, this quantity becomes zero. Simultaneously, we have plotted the inner product of two electric field strength vectors $\hat{\mathbf{E}}_0$ that are derived from electric flux density vectors $\mathbf{D}_0$ with a difference of $\pi/2$ in azimuth ϕ in the plane perpendicular to $\hat{\mathbf{k}}$. We denote this inner product by $\hat{\mathbf{E}}_0(\phi) \cdot \hat{\mathbf{E}}_0(\phi + \pi/2)$. Both inner products are plotted in Fig. 2.14

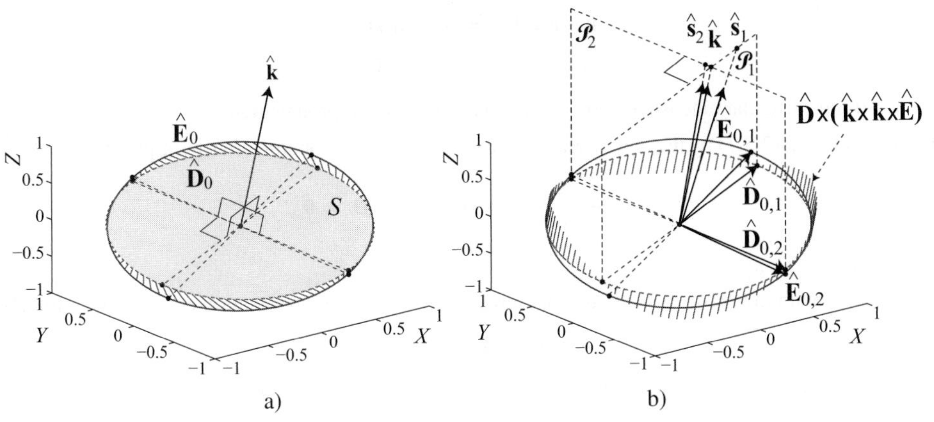

Figure 2.13: Mapping of the unit flux density vector $\hat{\mathbf{D}}_0$ onto the unit electric vector $\hat{\mathbf{E}}_0$ by means of the permittivity tensor $\bar{\bar{\epsilon}}_r$.
a) The line segments joining the endpoints of the vector $\hat{\mathbf{E}}_0$ and the vector $\hat{\mathbf{D}}_0$ are shown for endpoints of the vector $\hat{\mathbf{D}}_0$ on the dashed circle that is perpendicular to the wave vector $\hat{\mathbf{k}}$.
b) Construction of eigenvector pairs $(\hat{\mathbf{D}}_{0,j}, \hat{\mathbf{E}}_{0,j})$ and the energy propagation directions $\hat{\mathbf{s}}_j$. The plane $\mathcal{P}_j$, defined by the unit wave vector $\hat{\mathbf{k}}$ and a ray vector $\hat{\mathbf{s}}_j$, is called a principal section for the wave propagation ($j = 1, 2$).

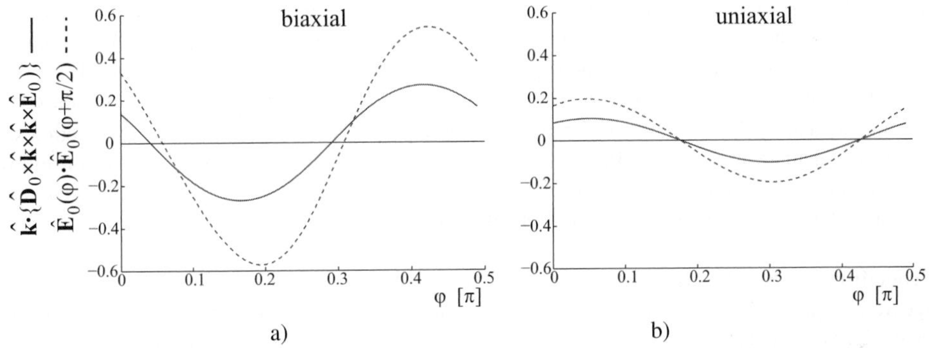

Figure 2.14: The orientation of the electric flux density unit vector $\hat{\mathbf{D}}$ and the electric field strength unit vector $\hat{\mathbf{E}}$. Solid curves: $\hat{\mathbf{k}} \cdot \{\hat{\mathbf{D}}_0 \times (\hat{\mathbf{k}} \times \hat{\mathbf{k}} \times \hat{\mathbf{E}}_0)\}$. Dashed curves: $\hat{\mathbf{E}}_0(\phi) \cdot \hat{\mathbf{E}}_0(\phi + \pi/2)$. The angle ϕ represents the azimuth of the vector $\hat{\mathbf{D}}_0$ in the plane S perpendicular to the wave vector $\hat{\mathbf{k}}$, in units of π. The direction of the wave vector $\hat{\mathbf{k}}$ is identical to that in Fig. 2.13.
a) Wave propagation in a biaxial medium with the same ϵ-tensor elements as those used in Fig. 2.13.
b) Uniaxial medium, diagonal tensor elements given by $\epsilon_{XX} = \epsilon_{YY} = 1$, $\epsilon_{ZZ} = 2$, off-diagonal elements are zero.

as a function of the azimuthal angle ϕ in the plane S (see Fig. 2.13). The product $\hat{\mathbf{k}} \cdot \{\hat{\mathbf{D}}_0 \times (\hat{\mathbf{k}} \times \hat{\mathbf{k}} \times \hat{\mathbf{E}}_0)\}$ is represented by the solid curve, the product $\hat{\mathbf{E}}_0(\phi) \cdot \hat{\mathbf{E}}_0(\phi + \pi/2)$ by the dashed curve. For the case of a biaxial medium Fig. 2.14a shows that the unit electric field vectors $\mathbf{E}_{0,1}$ and $\mathbf{E}_{0,2}$ which correspond to the two linear eigenstates of a biaxial medium are not orthogonal in general. Figure 2.14b shows that the electric eigenvectors are always orthogonal in a uniaxial medium. The magnitude of the anisotropy of the medium has been strongly exaggerated to enhance the visibility of the effects.

2.6.2 Eigenvalue Equation in Terms of the Ray Vector ŝ

Returning to Eq. (2.5.7), for a nonzero solution of $\mathbf{E}_0$ we require that the two vectors $\hat{\mathbf{k}} \times \hat{\mathbf{k}} \times \hat{\mathbf{E}}_0$ and $\bar{\bar{\epsilon}}_r\ \hat{\mathbf{E}}_0 = \hat{\mathbf{D}}_0/\epsilon_0$ are parallel. The solutions, the polarisation eigenstates of wave propagation, are thus given by coplanar vector combinations $(\mathbf{E}_{0,j}, \mathbf{D}_{0,j})$ that are located in the mutually perpendicular planes $\mathcal{P}_1$ and $\mathcal{P}_2$ (see Fig. 2.13). From Fig. 2.13 we deduce the following geometrical properties for the two polarisation eigenstates $(\mathbf{E}_{0,j}, \mathbf{D}_{0,j})$ labelled 1 and 2,

$$\hat{\mathbf{k}} \times \hat{\mathbf{k}} \times \hat{\mathbf{E}}_{0,j} = -\cos\alpha\, \hat{\mathbf{D}}_{0,j}\,, \tag{2.6.4}$$

$$\hat{\mathbf{s}} \times \hat{\mathbf{s}} \times \hat{\mathbf{D}}_{0,j} = -\cos\alpha\, \hat{\mathbf{E}}_{0,j}\,, \tag{2.6.5}$$

where $\cos\alpha = 1$ if the particular eigenstate corresponds to ordinary wave propagation.

We can now divide Eq. (2.5.7) by $|\mathbf{E}_0|$ and use Eqs. (2.6.4) and (2.6.5) to obtain, after multiplication with $\left(\bar{\bar{\epsilon}}_r\right)^{-1} |\mathbf{D}_0|$,

$$\frac{1}{n_p^2 \cos^2\alpha}\, \hat{\mathbf{s}} \times \hat{\mathbf{s}} \times \mathbf{D}_0 + \left(\bar{\bar{\epsilon}}_r\right)^{-1} \mathbf{D}_0 = \mathbf{0}\,, \tag{2.6.6}$$

where we have put $n_p^2 = k^2/k_0^2$.

The original equation with the wave vector $\mathbf{k}$ as input direction for the wave propagation reads, in a comparable format,

$$n_p^2\, \hat{\mathbf{k}} \times \hat{\mathbf{k}} \times \mathbf{E}_0 + \bar{\bar{\epsilon}}_r\, \mathbf{E}_0 = \mathbf{0}\,. \tag{2.6.7}$$

An alternative way of writing these equations in terms of the phase velocity v_p and the ray propagation speed v_r (also denoted by v_{em}) is

$$\left(\frac{v_r}{c}\right)^2 \hat{\mathbf{s}} \times \hat{\mathbf{s}} \times \mathbf{D}_0 + \left(\bar{\bar{\epsilon}}_r\right)^{-1} \mathbf{D}_0 = \mathbf{0}\,, \tag{2.6.8}$$

$$\left(\frac{c}{v_p}\right)^2 \hat{\mathbf{k}} \times \hat{\mathbf{k}} \times \mathbf{E}_0 + \bar{\bar{\epsilon}}_r\, \mathbf{E}_0 = \mathbf{0}\,. \tag{2.6.9}$$

We observe that the eigenvalue equation (2.6.6) in terms of the unit ray vector $\hat{\mathbf{s}}$ is obtained from the wave vector eigenvalue equation by replacing the combination $(\hat{\mathbf{k}}, \mathbf{E}_0)$ by $(\hat{\mathbf{s}}, \mathbf{D}_0)$, $\bar{\bar{\epsilon}}_r$ by $\bar{\bar{\epsilon}}_r^{\,-1}$, n_p by $1/(n_p \cos\alpha) = 1/n_r$, or, alternatively, v_p/c by c/v_r. The eigenvalues of Eq. (2.6.7) are $n_{p,j} = c/v_{p,j}$ with associated ray propagation vectors $\hat{\mathbf{s}}_1$ and $\hat{\mathbf{s}}_2$, corresponding to the two possible energy propagation directions. For uniaxial media and for certain propagation directions in a biaxial medium, one of the two propagation directions $\hat{\mathbf{s}}_j$ is parallel to the wave vector $\hat{\mathbf{k}}$.

The ray vector eigenvalue equation (2.6.6) yields the eigenvalues $1/n_{p,j}\cos\alpha_j = v_{r,j}/c$ $(j = 1, 2)$, and the corresponding wave eigenvectors $\hat{\mathbf{k}}_1$ and $\hat{\mathbf{k}}_2$. If $\alpha_j = 0$, we deal with an ordinary propagating wave. With the ray direction $\hat{\mathbf{s}}$ as input in Eq. (2.6.8), we can calculate an eigenstate $\mathbf{D}_0$ and obtain the corresponding eigenstate $\mathbf{E}_0$ from the constitutive relation of Eq. (2.2.1) for a linear anisotropic medium, and vice versa. The quantity $\cos\alpha = \hat{\mathbf{E}}_0 \cdot \hat{\mathbf{D}}_0$ then yields the ratio between the phase and ray velocity of the particular polarisation eigenstate.

2.6.2.1 Procedure for Solving the Wave and Ray Vector Eigenvalue Equations

The two eigenvalue equations in terms of the unit wave vector $\hat{\mathbf{k}}$ or the ray vector $\hat{\mathbf{s}}$ are solved in the following way:

Wave vector eigenvalue equation, Eq. (2.6.7):

- solve the equation $|\mathbf{M}_E| = 0$, where $\mathbf{M}_E$ is the matrix of the system of equations according to (2.6.7), to find the set of four eigenvalues n_p comprising two real positive values for forward propagating waves and two real negative ones for backward propagating solutions.
- substitute the positive values $n_{p,1}$ and $n_{p,2}$ in Eq. (2.6.7) to obtain the polarisation eigenstates of the electric field strength, respectively the unit vectors $\mathbf{E}_{0,1}$ and $\mathbf{E}_{0,2}$.
- find the corresponding electric flux densities $\mathbf{D}_{0,1}$ and $\mathbf{D}_{0,2}$ from the constitutive relation $\mathbf{D}_{0,j} = \epsilon_0\, \bar{\bar{\epsilon}}_r\, \mathbf{E}_{0,j}$ with $j = 1, 2$.
- calculate $\cos\alpha_j = \hat{\mathbf{E}}_{0,j} \cdot \hat{\mathbf{D}}_{0,j}$ for the two polarisation eigenstates.
- find the unit wave vector $\hat{\mathbf{s}}_j$ from the expression $\hat{\mathbf{s}}_j = \cos\alpha_j\, \hat{\mathbf{k}} + \sin\alpha_j\, \hat{\mathbf{D}}_{0,j}$.

Ray vector eigenvalue equation, Eq. (2.6.6):

- solve the equation $|\mathbf{M}_D| = 0$ to find the eigenvalues $(n_{p,j}\cos\alpha_j)^{-1}$.
- substitute the two positive values of $(n_{p,j}\cos\alpha_j)^{-1}$ in the ray vector equation to obtain the eigenstates $\mathbf{D}_{0,1}$ and $\mathbf{D}_{0,2}$ (real unit vectors).
- find $\mathbf{E}_{0,j}$ from the inverted constitutive relation $\mathbf{E}_{0,j} = \left(\bar{\bar{\epsilon}}\right)^{-1} \mathbf{D}_{0,j}/\epsilon_0$ and construct the corresponding unit vectors.
- calculate $\cos\alpha_j = \hat{\mathbf{E}}_{0,j} \cdot \hat{\mathbf{D}}_{0,j}$ for the two polarisation eigenstates.
- find the values of $n_{p,j}$ from the two eigenvalues $(n_{p,j}\cos\alpha_j)^{-1}$.
- find the unit wave vector $\hat{\mathbf{k}}_j$ from the expression $\hat{\mathbf{k}}_j = \cos\alpha_j\, \hat{\mathbf{s}} - \sin\alpha_j\, \hat{\mathbf{E}}_{0,j}$.
- construct the wave vectors from the expression $\mathbf{k}_j = n_{p,j} k_0 \hat{\mathbf{k}}_j$.

The eigenvectors $\mathbf{E}_{0,j}$ and $\mathbf{D}_{0,j}$ of both the $\hat{\mathbf{k}}$- and $\hat{\mathbf{s}}$-system of equations are real because the matrices $\mathbf{M}_E$ and $\mathbf{M}_D$ are real and symmetric in the case of a linear anisotropic material without losses. In the particular case that two eigenvectors $\mathbf{E}_{0,j}$ and $\mathbf{D}_{0,j}$ are parallel with $\alpha_j = 0$, the wave propagation along the corresponding wave vector $\hat{\mathbf{k}}$ is normal or *ordinary* in the restricted sense with $\hat{\mathbf{k}} \parallel \hat{\mathbf{s}}$. Ordinary wave propagation in the wide sense, where an entire spherical wave is propagating in an apparently isotropic way, is only possible in uniaxial media.

2.6.3 Relationship between Wave Vector $\hat{\mathbf{k}}$ and Ray Vector $\hat{\mathbf{s}}$

The Helmholtz equation (2.6.8) can be solved for a given wave vector $\hat{\mathbf{k}}$ and yields the eigenvalue n_p for the phase refractive index and the electric field strength vector $\mathbf{E}_0$ corresponding to an eigenmode of the medium. From these quantities, the electric flux density vector $\mathbf{D}_0$ and the ray vector $\hat{\mathbf{s}}$ are immediately obtained. In this subsection, we present a direct method to obtain the ray vectors $\hat{\mathbf{s}}$ from the wave vector $\hat{\mathbf{k}}$ without explicitly solving the Helmholtz eigenvalue equation (2.6.8). The only prior knowledge that is needed is the value of the phase refractive index n_p of the propagating wave. We refer to Fig. 2.10 for a graphical representation of the electromagnetic field vectors and the wave and ray propagation vectors.

We start with the Helmholtz equations (2.6.8) and (2.6.9) for $\mathbf{E}_0$ and $\mathbf{D}_0$, respectively, which in a slightly different notation read

$$n_p^2\, \hat{\mathbf{k}} \times \hat{\mathbf{k}} \times \mathbf{E}_0 + \bar{\bar{\epsilon}}_r\, \mathbf{E}_0 = \mathbf{0}\,, \tag{2.6.10}$$

$$(1/n_r^2)\, \hat{\mathbf{s}} \times \hat{\mathbf{s}} \times \mathbf{D}_0 + \bar{\bar{\epsilon}}_r^{\,-1}\, \mathbf{D}_0 = \mathbf{0}\,, \tag{2.6.11}$$

where n_r denotes the ray index of the anisotropic medium. The solutions of these equations have a similar appearance and are represented by the unit vectors

$$\hat{E}_{0,j} = \frac{n_p^2(\hat{\mathbf{k}} \cdot \hat{\mathbf{E}}_0)}{n_p^2 - n_j^2}\, \hat{k}_j\,, \qquad \hat{D}_{0,j} = \frac{n_j^2(\hat{\mathbf{s}} \cdot \hat{\mathbf{D}}_0)}{n_j^2 - n_r^2}\, \hat{s}_j\,, \tag{2.6.12}$$

where the index j now denotes the X-, Y- and Z- components of the vectors and n_j^2 stands for the principal values ϵ_X, ϵ_Y and ϵ_Z of the permittivity tensor $\bar{\bar{\epsilon}}_r$ of the medium. We suppose, as before, that the Cartesian axes (X, Y, Z) of the medium have been aligned with the principal axes of the anisotropic medium to facilitate the analysis.

From Fig. 2.10 we derive for the vectors in plane B that $\hat{\mathbf{k}} \cdot \hat{\mathbf{E}}_0 = -\hat{\mathbf{s}} \cdot \hat{\mathbf{D}}_0$. Substitution of this identity into Eq. (2.6.12) yields

$$\hat{s}_j = \frac{n_p^2(n_r^2 - n_j^2)}{n_j^2(n_p^2 - n_j^2)}\, \frac{\hat{D}_{0,j}}{\hat{E}_{0,j}}\, \hat{k}_j\,. \tag{2.6.13}$$

To obtain the desired expression that directly connects $\hat{\mathbf{s}}$ to $\hat{\mathbf{k}}$ without the need for previously calculated values of n_r and $\mathbf{D}_0$, we first compute the ratio $\hat{D}_{0,j}/\hat{E}_{0,j}$. With the principal axes of the permittivity tensor aligned with the Cartesian coordinate axes we have the relation

$$\hat{D}_{0,j}/\hat{E}_{0,j} = \epsilon_0 n_j^2 |\mathbf{E}_0|/|\mathbf{D}_0|\,. \tag{2.6.14}$$

To calculate the ratio $|\mathbf{E}_0|/|\mathbf{D}_0|$, we inspect Eq. (2.6.10) and note that the vector $-\hat{\mathbf{k}} \times \hat{\mathbf{k}} \times \mathbf{E}_0$ equals the projection of the vector $\mathbf{E}_0$ onto the plane A, indicated by $\mathbf{E}_{0,p}$ in Fig. 2.10. A solution of Eq. (2.6.10) requires this vector to be parallel to $\mathbf{D}_0$, subject to the condition

$$-\,n_p^2 \mathbf{E}_{0,p} + \mathbf{D}_0/\epsilon_0 = \mathbf{0}\,. \tag{2.6.15}$$

Using the property $|\mathbf{E}_{0,p}| = |\mathbf{E}_0| \cos\alpha = (n_r/n_p)|\mathbf{E}_0|$, we find $\hat{D}_{0,j}/\hat{E}_{0,j} = n_j^2/(n_p n_r)$. Substituting this result into Eq. (2.6.13) yields the expression

$$\hat{s}_j = \frac{n_p}{n_r}\left(\frac{n_r^2 - n_j^2}{n_p^2 - n_j^2}\right)\hat{k}_j\,. \tag{2.6.16}$$

The expression for the ray vector components $\hat{s}_j$ still contains the unknown ray vector velocity via the index $n_r = c/v_r$. To express n_r in terms of n_p and the components $\hat{k}_j$ of the wave vector we use the fact that $\hat{\mathbf{s}}$ and $\hat{\mathbf{k}}$ in Eq. (2.6.16) are unit vectors. We rewrite the equation as

$$n_r \hat{s}_j - n_p \hat{k}_j = n_p \left(\frac{n_r^2 - n_j^2}{n_p^2 - n_j^2} - 1 \right) \hat{k}_j = n_p (n_r^2 - n_p^2) \left(\frac{1}{n_p^2 - n_j^2} \right) \hat{k}_j , \tag{2.6.17}$$

square both sides and sum over the three components to obtain

$$n_r^2 + n_p^2 - 2n_r n_p (\hat{\mathbf{s}} \cdot \hat{\mathbf{k}}) = n_p^2 (n_p^2 - n_r^2)^2 \left\{ \frac{\hat{k}_X^2}{(n_p^2 - n_X^2)^2} + \frac{\hat{k}_Y^2}{(n_p^2 - n_Y^2)^2} + \frac{\hat{k}_Z^2}{(n_p^2 - n_Z^2)^2} \right\} . \tag{2.6.18}$$

Noting that $(\hat{\mathbf{s}} \cdot \hat{\mathbf{k}}) = \cos\alpha = n_r / n_p$, we obtain after some rearrangement,

$$\begin{aligned} \frac{n_r^2}{n_p^2} &= 1 - \left\{ \frac{\hat{k}_X^2}{(1 - n_X^2/n_p^2)^2} + \frac{\hat{k}_Y^2}{(1 - n_Y^2/n_p^2)^2} + \frac{\hat{k}_Z^2}{(1 - n_Z^2/n_p^2)^2} \right\}^{-1} \\ &= 1 - (1/F_k) . \end{aligned} \tag{2.6.19}$$

With the aid of the factor F_k, the components of the ray vector can be written as

$$\hat{s}_j = \frac{n_p}{n_r} \left\{ 1 - \frac{1}{F_k (1 - n_j^2 / n_p^2)} \right\} \hat{k}_j . \tag{2.6.20}$$

Equation (2.6.20) directly gives the ray vector components from the wave vector components. The only extra information that is required is the phase refractive index value n_p of the propagation mode under consideration. It follows that, with two independent and orthogonally polarised propagation modes, two ray vectors $\hat{\mathbf{s}}_i$ can be associated with each wave vector $\hat{\mathbf{k}}$.

If the roles of wave and ray vector are interchanged, the analysis above is similar. Each direction of energy propagation given by $\hat{\mathbf{s}}$ is associated with two wave vectors $\hat{\mathbf{k}}_i$, each with their specific state of polarisation and energy propagation speed $v_r = c/n_r$. The expression for the wave vector components related to a wave with ray index n_r reads

$$\hat{k}_j = \frac{n_r}{n_p} \left\{ 1 + \frac{1}{F_s (1 - n_j^2 / n_r^2)} \right\} \hat{s}_j ,$$

where

$$\begin{aligned} F_s &= \left\{ \frac{\hat{s}_X^2}{(1 - n_X^2/n_r^2)^2} + \frac{\hat{s}_Y^2}{(1 - n_Y^2/n_r^2)^2} + \frac{\hat{s}_Z^2}{(1 - n_Z^2/n_r^2)^2} \right\} , \\ \frac{n_p^2}{n_r^2} &= 1 + 1 / F_s . \end{aligned} \tag{2.6.21}$$

In Fig. 2.15 we show the endpoints of a set of unit wave vectors $\hat{\mathbf{k}}$ after projection of these vectors onto the plane $k_X = 0$. For each wave vector the associated projections of the ray vectors $\hat{\mathbf{s}}_1$ and $\hat{\mathbf{s}}_2$ are shown. In Fig. 2.15a the medium is uniaxial. The endpoint of each projected wave vector has been indicated by a tiny circle. The azimuthal displacement of an associated ray vector is shown by means of a line segment that joins the endpoints of the projected wave vector and the ray vector. As was demonstrated previously, one ray vector is always ordinary in a uniaxial medium and coincides with the wave vector. The graph shows that the displacement between wave vector and extraordinary ray vector is always in a plane containing the wave vector and the optical axis (the Z-axis in the graph). For $\hat{k}_Z = 0$, both orthogonally polarised waves behave as if propagating in an isotropic medium.

In Fig. 2.15b the result is shown for a biaxial medium. In general, the wave vector is at the corner point of each 'hook' in the graph. The displacements of the two ray vectors, both extraordinary, are in mutually perpendicular planes, containing the solid line segment and the dotted line segment, respectively. For the case that the wave vector $\hat{\mathbf{k}}$ lies in one of the principal planes associated with the permittivity tensor (the planes $X = 0$, $Y = 0$ and $Z = 0$), one of the ray vectors behaves 'ordinarily' and is parallel to the wave vector. For the upper and lower hemispheres of the graph, we observe a more complicated behaviour of the ray vectors, the reason for which is the small difference in azimuth between the wave vector and the optical axes. The directions of the optical axes are defined by $\hat{k}_Y = 0$, $\hat{k}_Z = \pm\sqrt{69}/12 \approx \pm 0.7$. We note that the transformation expressions above ask for a special treatment when v_r or n_r equals one of the principal values v_j or n_j, respectively, where $j = X, Y, Z$. In that case we directly derive the solutions from the $(\hat{\mathbf{k}}, \mathbf{E}_0)$ system of equations, which was discussed in Subsection 2.4.2. The special case of conical refraction which arises when $\hat{\mathbf{k}}$ is directed along one of the optical axes in a biaxial medium is discussed in Section 2.8.

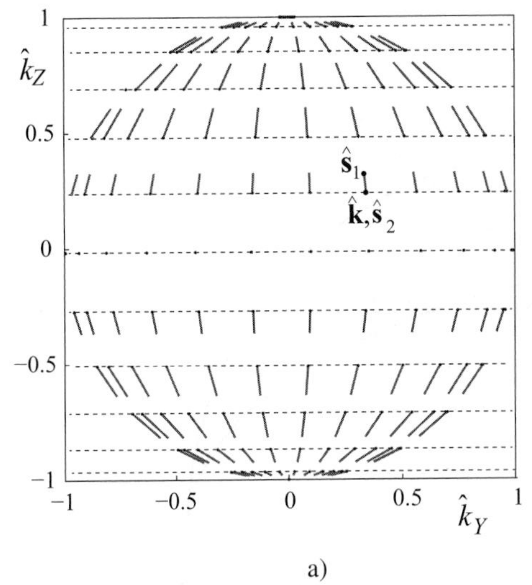

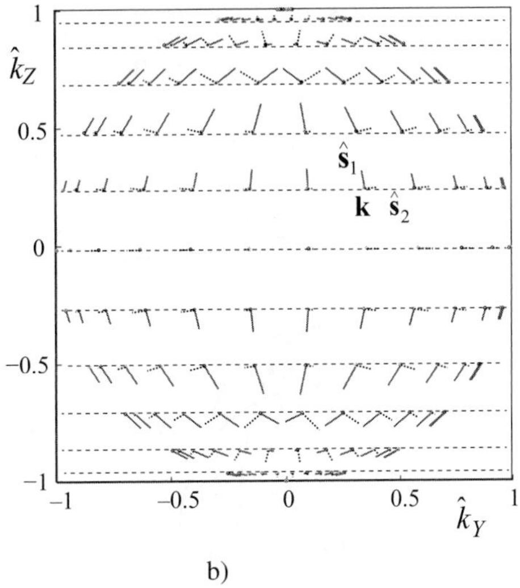

Figure 2.15: Projection onto the plane $\hat{k}_X = 0$ of the endpoints of the unit wave vector $\hat{\mathbf{k}}$, located on the dashed horizontal lines, and of the endpoints of the two corresponding unit ray vectors $\hat{\mathbf{s}}_1$ and $\hat{\mathbf{s}}_2$. The directions of the line segments that join the endpoints of the wave vector and a ray vector (solid lines: $\hat{\mathbf{k}}$, $\hat{\mathbf{s}}_1$, dotted lines: $\hat{\mathbf{k}}$, $\hat{\mathbf{s}}_2$) define the cross-sections $\mathcal{P}_1$ and $\mathcal{P}_2$ through $\hat{\mathbf{k}}$ in which the electrical polarisation eigenstates are found.
a) Uniaxial medium with $v_X = v_Y = 0.65$, $v_Z = 0.55$, arbitrary units.
b) Biaxial medium with $v_X = 0.65$, $v_Y = 0.60$, $v_Z = 0.55$.

2.7 The Functions $v_p(\hat{\mathbf{k}})$, $n_p(\hat{\mathbf{k}})$, $v_r(\hat{\mathbf{s}})$ and $n_r(\hat{\mathbf{s}})$

We have previously encountered the functions $v_p(\hat{\mathbf{k}})$ and $n_p(\hat{\mathbf{k}})$ when solving the ($\hat{\mathbf{k}}$, $\mathbf{E}_0$)-based Helmholtz equation in an anisotropic medium. The three-dimensional surfaces associated with these two functions are called the wave normal surface and the optical indicatrix, respectively. When solving Eq. (2.6.8), i.e. the ($\hat{\mathbf{s}}$,$\mathbf{D}_0$)-based Helmholtz equation in an anisotropic medium, we find in an analogous way the functions $v_r(\hat{\mathbf{s}})$ and $n_r(\hat{\mathbf{s}})$. We now define the vector $\mathbf{r}_{vr} = v_r\,\hat{\mathbf{s}}$, the endpoint of which describes a surface in three-dimensional space.

After a certain time increment δt, the electromagnetic energy radiated by a point source at the origin propagates to this surface. In an anisotropic medium, as for the surface corresponding to the phase velocity v_p, the surface defined by $\mathbf{r}_{vr}$ consists of two sheets, one for each orthogonal eigenstate of polarisation associated with the direction $\hat{\mathbf{s}}$. There is a close analogy between the functions with index p and index r. Writing Eq. (2.6.8) in matrix form in line with Eq. (2.4.1), we obtain

$$\begin{pmatrix} \frac{v_X^2}{v_r^2} - \hat{s}_Y^2 - \hat{s}_Z^2 & \hat{s}_X\hat{s}_Y & \hat{s}_X\hat{s}_Z \\ \hat{s}_X\hat{s}_Y & \frac{v_Y^2}{v_r^2} - \hat{s}_X^2 - \hat{s}_Z^2 & \hat{s}_Y\hat{s}_Z \\ \hat{s}_X\hat{s}_Z & \hat{s}_Y\hat{s}_Z & \frac{v_Z^2}{v_r^2} - \hat{s}_X^2 - \hat{s}_Y^2 \end{pmatrix} \mathbf{D}_0 = M_s\,\mathbf{D}_0 = \mathbf{0}\,. \tag{2.7.1}$$

For the ($\hat{\mathbf{s}}$, $\mathbf{D}_0$) system of equations we define a parameter $q_i = v_i/v_r = n_r/n_i$ and the eigenvalues then follow from the equation

$$\frac{\hat{s}_X^2}{q_x^2 - 1} + \frac{\hat{s}_Y^2}{q_y^2 - 1} + \frac{\hat{s}_Z^2}{q_z^2 - 1} + 1 = 0\,. \tag{2.7.2}$$

A comparison of Eq. (2.4.3) and Eq. (2.7.2) shows that the parameters p_i and q_i play identical roles in the construction of the surfaces that describe the phase and ray refractive indices or propagation speeds in an anisotropic medium. To switch from the wave to the ray domain, we replace the principal refractive indices of the medium by their inverted values. To illustrate this principle, we write down the equations for n_r and n_p,

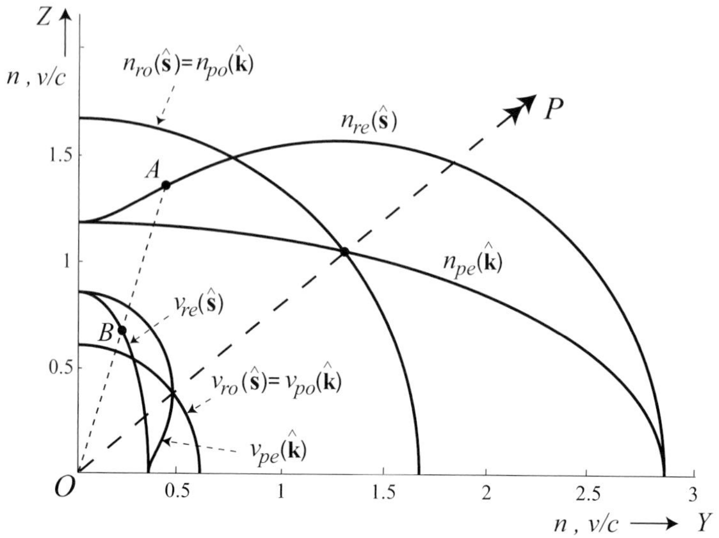

Figure 2.16: Plot of the surfaces described by the endpoints of the vectors with lengths $v_p(\hat{\mathbf{k}})$, $v_r(\hat{\mathbf{s}})$, $n_p(\hat{\mathbf{k}})$ and $n_r(\hat{\mathbf{s}})$ in a biaxial medium. The anisotropic effect has been strongly exaggerated: $v_X = 0.6c$, $v_Y = 0.85c$, $v_Z = 0.35c$. The cross-section ($X = 0$) contains the principal axes Y and Z and the optical axis OP. The phase and ray propagation speeds have been normalised with respect to c.

$$\frac{\hat{k}_X^2}{n_p^2 - n_X^2} + \frac{\hat{k}_Y^2}{n_p^2 - n_Y^2} + \frac{\hat{k}_Z^2}{n_p^2 - n_Z^2} = \frac{1}{n_p^2} \, ,$$

$$\frac{\hat{s}_X^2}{1/n_r^2 - 1/n_X^2} + \frac{\hat{s}_Y^2}{1/n_r^2 - 1/n_Y^2} + \frac{\hat{s}_Z^2}{1/n_r^2 - 1/n_Z^2} = n_r^2 \, . \tag{2.7.3}$$

In Fig. 2.16 we show polar plots for n_p, n_r, v_p and v_r in a cross-section ($X = 0$) through two principal axes of the anisotropic medium. The three-dimensional surfaces associated with the functions $v_r(\hat{\mathbf{s}})$ and $n_r(\hat{\mathbf{s}})$ are called the *ray normal surface* and the *ray optical indicatrix*, by analogy with the previously defined surfaces $v_p(\hat{\mathbf{k}})$ and $n_p(\hat{\mathbf{k}})$ that have the wave vector $\hat{\mathbf{k}}$ as parameter. The plots for v_p and v_r are normalised with respect to the speed of light in vacuum. The medium is biaxial and the optical axis OP is located in the plane of the drawing. By definition, the product of $n_r(\hat{\mathbf{s}})$ and $v_r(\hat{\mathbf{s}})/c$ is unity for all directions of propagation; $OA \cdot OB$ is unity in the figure. The same holds for the curves that represent $n_p(\hat{\mathbf{k}})$ and $v_p(\hat{\mathbf{k}})$. Curves for both the extraordinary and ordinary wave parameters are plotted and are labelled 'e' and 'o', respectively. We note that since we consider a cross-section that is perpendicular to a principal axis of the medium, the curves for the ordinary wave reduce to circles.

2.7.1 The Surfaces $\mathbf{r}_{np} = n_p(\hat{\mathbf{k}})\,\hat{\mathbf{k}}$ and $\mathbf{r}_{vr} = v_r(\hat{\mathbf{s}})\,\hat{\mathbf{s}}$

The surfaces that are swept by the endpoints of the vectors $[n_p(\hat{\mathbf{k}})]\,\hat{\mathbf{k}}$ and $[v_r(\hat{\mathbf{s}})]\,\hat{\mathbf{s}}$ have a special importance, for example when tracing rays or applying Huygens' secondary wave construction in anisotropic media. We start with the optical indicatrix surface that is produced by tracing the vector $\mathbf{r}_{np} = [n_p(\hat{\mathbf{k}})]\,\hat{\mathbf{k}}$. In the direct neighbourhood of a wave vector $\hat{\mathbf{k}}_0$, the optical indicatrix surface is swept by the endpoint of $\mathbf{r}_{np}$ and is accompanied by small changes in the electric field strength and electric flux density vectors. To calculate these changes we use Eq. (2.6.10). Defining $n_p\hat{\mathbf{k}} = \mathbf{r}_{np}$ and using the identity $\overline{\overline{\epsilon}}_r\,\mathbf{E}_0 = \mathbf{D}_0/\epsilon_0$ we have

$$\delta\left(\mathbf{r}_{np} \times \mathbf{r}_{np} \times \mathbf{E}_0 + \frac{\mathbf{D}_0}{\epsilon_0}\right) = \mathbf{0} \, . \tag{2.7.4}$$

The differential is written as

$$(2\mathbf{r}_{np}\cdot\delta\mathbf{r}_{np})\mathbf{E}_0 + \mathbf{r}_{np}^2\delta\mathbf{E}_0 - \left[\delta\mathbf{r}_{np}\cdot\mathbf{E}_0 + \mathbf{r}_{np}\cdot\delta\mathbf{E}_0\right]\mathbf{r}_{np} - (\mathbf{r}_{np}\cdot\mathbf{E}_0)\delta\mathbf{r}_{np} = \frac{\delta\mathbf{D}_0}{\epsilon_0}\,. \tag{2.7.5}$$

Scalar multiplication of Eq. (2.7.5) with $\mathbf{E}_0$ yields the expression

$$\begin{aligned}(2\mathbf{r}_{np}\cdot\delta\mathbf{r}_{np})\mathbf{E}_0^2 + \mathbf{r}_{np}^2(\delta\mathbf{E}_0\cdot\mathbf{E}_0) - (\delta\mathbf{r}_{np}\cdot\mathbf{E}_0)(\mathbf{r}_{np}\cdot\mathbf{E}_0)&\\ -(\mathbf{r}_{np}\cdot\delta\mathbf{E}_0)(\mathbf{r}_{np}\cdot\mathbf{E}_0) - (\mathbf{r}_{np}\cdot\mathbf{E}_0)(\delta\mathbf{r}_{np}\cdot\mathbf{E}_0) &= \frac{\delta\mathbf{D}_0\cdot\mathbf{E}_0}{\epsilon_0}\,.\end{aligned} \tag{2.7.6}$$

The symmetry of the permittivity tensor $\overline{\overline{\epsilon}}_r$ allows us to equate $\delta\mathbf{D}_0\cdot\mathbf{E}_0$ to $\delta\mathbf{E}_0\cdot\mathbf{D}_0$ and, after some rearrangement we obtain for the expression of Eq. (2.7.6)

$$(2\mathbf{r}_{np}\cdot\delta\mathbf{r}_{np})\mathbf{E}_0^2 - 2(\delta\mathbf{r}_{np}\cdot\mathbf{E}_0)(\mathbf{r}_{np}\cdot\mathbf{E}_0) = \delta\mathbf{E}_0\cdot\left\{\frac{\mathbf{D}_0}{\epsilon_0} - \mathbf{r}_{np}^2\mathbf{E}_0 + \mathbf{r}_{np}\cdot(\mathbf{E}_0)\mathbf{r}_{np}\right\} = 0\,. \tag{2.7.7}$$

The final expression for the differential becomes

$$2\,\delta\mathbf{r}_{np}\cdot\left(\mathbf{E}_0\times\mathbf{r}_{np}\times\mathbf{E}_0\right) = 0\,,$$

or,

$$2\,n_p\,\delta\mathbf{r}_{np}\cdot(\mathbf{E}_0\times\mathbf{k}_0\times\mathbf{E}_0) = 0\,. \tag{2.7.8}$$

The vector $\mathbf{E}_0\times\mathbf{k}_0\times\mathbf{E}_0$ is parallel to the ray vector $\hat{\mathbf{s}}$. The vector $\delta\mathbf{r}_{np}$ is tangent to the surface described by $\mathbf{r}_{np}$, the optical indicatrix. The expression of Eq. (2.7.8) therefore requires that the vector $\delta\mathbf{r}_{np}$ is perpendicular to the normalised Poynting vector $\hat{\mathbf{s}}$. Consequently, the electric field strength vector $\mathbf{E}_0$ is also tangent to the optical indicatrix.

A similar analysis can be applied to the Helmholtz equation written in terms of the vector pair $(\hat{\mathbf{s}},\mathbf{D}_0)$ (see Eq. (2.6.11)). Writing

$$v_r^2\,\hat{\mathbf{s}}\times\hat{\mathbf{s}}\times\mathbf{D}_0 + \frac{\overline{\overline{\epsilon}}_r^{\,-1}\,\mathbf{D}_0}{\epsilon_0\mu_0} = \mathbf{0}\,, \tag{2.7.9}$$

we use the surface vector $\mathbf{r}_{vr} = v_r(\hat{\mathbf{s}})\,\hat{\mathbf{s}}$ and obtain the differential equation

$$\delta\left(\mathbf{r}_{vr}\times\mathbf{r}_{vr}\times\mathbf{D}_0 + \frac{\mathbf{E}_0}{\mu_0}\right) = \mathbf{0}\,. \tag{2.7.10}$$

Proceeding along the same lines as in the derivation above, we conclude that

$$2\,v_r\,\delta\mathbf{r}_{vr}\cdot(\mathbf{D}_0\times\mathbf{s}_0\times\mathbf{D}_0) = 0\,. \tag{2.7.11}$$

The conclusion is that the vector $\mathbf{D}_0\times\mathbf{s}_0\times\mathbf{D}_0$, which is parallel to the wave vector $\hat{\mathbf{k}}$, is everywhere on the surface perpendicular to the tangent plane of the surface. This implies that the electric flux density vector $\mathbf{D}_0$ is located in the tangent plane of the surface defined by the endpoint of the vector $\mathbf{r}_{vr}$.

2.7.2 Particular Properties of Extraordinary Wave Propagation

In Fig. 2.17 we illustrate how the properties of the surfaces associated with $n_p(\hat{\mathbf{k}})$ and $v_r(\hat{\mathbf{s}})$ can be used to construct the extraordinary ray vector and the electric field vectors for a propagating wave in the extraordinary regime. The ordinary circular parts of the various surfaces have been omitted for clarity. The unit wave vector is denoted by $\hat{\mathbf{k}}_0$ in the figure. The points A and C give the phase index n_p and normalised phase velocity (v_p/c) for the propagating wave with unit wave vector $\hat{\mathbf{k}}_0$ $(OC\cdot OA = 1)$. The direction of the extraordinary ray vector $\hat{\mathbf{s}}_2$ follows from the following consideration. We construct a small pencil of waves around the direction $\hat{\mathbf{k}}_0$. The wavefronts, perpendicular to each wave vector on the curve $v_{pe}(\hat{\mathbf{k}})$ around the point C intersect each other at the point B. At this point B, the partial amplitudes of the pencil of waves constructively add and produce a maximum intensity. The point B thus defines the direction of energy flow $\hat{\mathbf{s}}_2$ in the medium for a wave vector direction given by $\hat{\mathbf{k}}_0$. The geometrical locus of the intersection points of perpendiculars dropped from each vector with its endpoint on the surface through C contracts to the (singular) point B. As we have seen in the preceding subsection, the surface through B is determined by the value of $v_{re}(\hat{\mathbf{s}})$, the energy or ray velocity. It was also seen that the electric flux density vector $\mathbf{D}_2$ is tangential to this surface (see Eq. (2.7.11)).

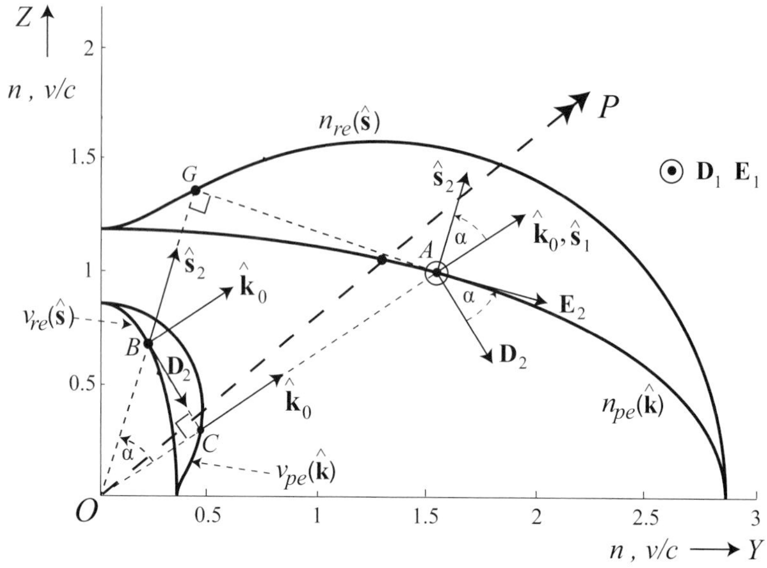

Figure 2.17: The construction of the ray vector $\hat{\mathbf{s}}_2$ and the directions of the electric vectors $\mathbf{E}_2$ and $\mathbf{D}_2$ of an extraordinary wave. $\hat{\mathbf{k}}_0$ is the unit wave vector of the plane wave. The arrow *OP* indicates the direction of an optical axis of the biaxial medium.

A more straightforward method to obtain the extraordinary ray vector follows from Eq. (2.7.11). At the intersection point *A* of the unit wave vector with the optical indicatrix, the electric field strength vector is found in the tangent plane of the indicatrix. For the special cross-section shown in Fig. 2.16, the electric vector is found in the plane containing the optical axes of the biaxial medium and the extraordinary unit ray vector $\hat{\mathbf{s}}_2$ is perpendicular to the indicatrix at the point *A*. Assembling all vectors at *A*, we obtain a complete picture upon adding the ordinary ray and field vectors, $\hat{\mathbf{s}}_1 = \hat{\mathbf{k}}_0$ and the parallel pair $(\mathbf{D}_1, \mathbf{E}_1)$. The quantity $\cos\alpha$ was calculated earlier (see Eq. (2.5.6)); however, in Fig. 2.17 we deduce $\cos\alpha = v_p/v_r$ from the ratio of the line segment lengths *OG* and *OA*. Alternatively, $\cos\alpha$ can be found from the ratio *OC*/*OB*. Using the angle α between $\hat{\mathbf{k}}_0$ and $\hat{\mathbf{s}}_2$, we can construct the surface for the ray index $n_{re}(\hat{\mathbf{s}})$ by determining the point of intersection, *G*, of the tangent plane at *A* with the vector through the origin *O*, parallel to the ray vector $\hat{\mathbf{s}}_2$. The surface corresponding to $n_{pe}(\hat{\mathbf{k}})$ is called the geometrical *pedal* surface of $n_{re}(\hat{\mathbf{s}})$. The ratio n_r/n_p of phase and ray index equals $\cos\alpha$.

In Fig. 2.17 we also show why the surface derived from $n_p(\hat{\mathbf{k}})$, the optical indicatrix, and the surface related to $v_r(\hat{\mathbf{s}})$ have a particular importance for the case of extraordinary wave propagation. The figure shows the surfaces for n_p and v_r in the principal section of a biaxial medium, in this case the one that contains the optical axes, one of which, *OP* is shown in the figure. A plane with unit wave vector $\hat{\mathbf{k}}_0$ intersects the optical indicatrix at the point *A*. The normal to the surface at this point provides the direction of the extraordinary ray vector $\hat{\mathbf{s}}_2$ (see Eq. (2.7.8)). The electric field strength vectors corresponding to the ordinary and extraordinary wave solution in the medium are located in the tangent plane to the surface at *A*. If we consider a cross-section that is parallel to one of the principal sections of the permittivity tensor, the two electric vectors are mutually perpendicular. The extraordinary one, $\mathbf{E}_2$, is located in the plane of the drawing.

As was shown previously, there is a duality in the description of wave propagation, using either the set of vectors $(\hat{\mathbf{k}}, \mathbf{E}_0)$ with eigenvalue n_p or the set $(\hat{\mathbf{s}}, \mathbf{D}_0)$ with the eigenvalue $1/n_r$. This also becomes apparent from Fig. 2.17 when considering the surfaces derived from the functions $v_p(\hat{\mathbf{k}})$ and $v_r(\hat{\mathbf{s}})$. From Eq. (2.7.11) we see that the wave vector $\hat{\mathbf{k}}_0$ is perpendicular to the tangent plane at point *B* where the ray vector intersects the extraordinary ray velocity surface. The extraordinary electric flux density vector $\mathbf{D}_2$ is found in the tangent plane at *B*. For the cross-section shown in Fig. 2.16, $\mathbf{D}_2$ is located in the plane of the drawing. The vector through *O*, perpendicular to the tangent plane in *B*, is the unit wave vector $\hat{\mathbf{k}}_0$. The intersection point *C* defines the surface associated with $v_p(\hat{\mathbf{k}})$ and $\cos\alpha$ is given by the ratio *OC*/*OB*. In general, we have for an extraordinary ray that $v_p(\hat{\mathbf{k}}).n_p(\hat{\mathbf{k}}) = v_r(\hat{\mathbf{k}}).n_r(\hat{\mathbf{k}}) = c$. This property also follows from the similarity of the triangles *OBC* and *OAG*.

In Fig. 2.18 we schematically plot in three dimensions the optical indicatrix, generated from the function $n_p(\hat{\mathbf{k}})$ for a biaxial medium. In the figure the three 'principal' cross-sections, which each contain a different pair of principal axes of the optical indicatrix, coincide with the planes $X = 0$, $Y = 0$ and $Z = 0$. The three principal velocities in the medium have

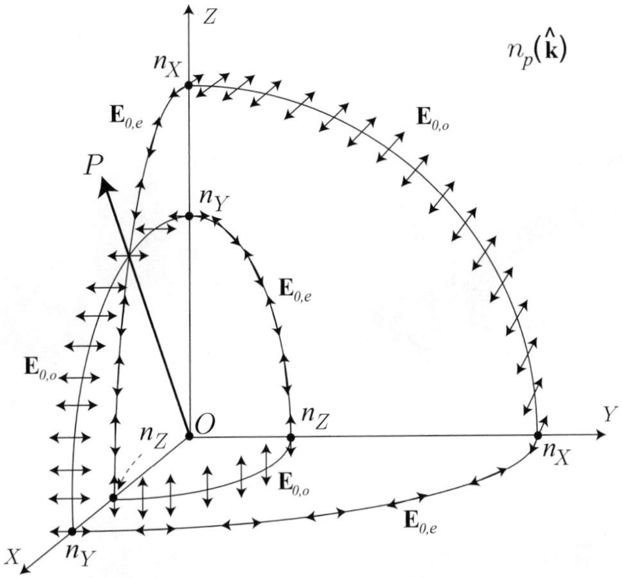

Figure 2.18: The surface n_p and the electric field strength eigenvector, tangent to this surface.

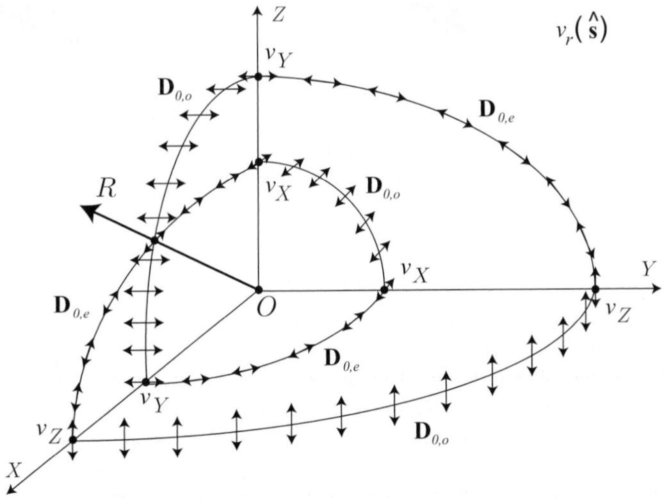

Figure 2.19: The surface v_r and the electric flux density eigenvector that is tangent to this surface.

been chosen such that $v_Z > v_Y > v_X$ and the optical axes are parallel to the plane $Y = 0$. In each of the three principal cross-sections of the indicatrix, one of the two propagating waves is ordinary. The orientation of the electric field strength vectors $\mathbf{E}_{0,o}$ and $\mathbf{E}_{0,e}$ has been indicated in the figure by means of the arrows. It is seen that for a wave vector that lies in a principal cross-section, the ordinary field strength vector is perpendicular to that principal cross-section whilst the extraordinary field vector is parallel to it. The 'principal' cross-sections in a biaxial medium play the role of the principal section, defined by the vectors $\hat{\mathbf{k}}$ and $\hat{\mathbf{E}}$, that is found in a uniaxial medium. In a biaxial medium when the wave vector $\hat{\mathbf{k}}$ does not lie in one of the three principal planes, the orthogonally polarised eigenstates are both extraordinary waves.

Figure 2.19 shows a comparable plot to Fig. 2.18 for the surface derived from $v_r(\hat{\mathbf{s}})$. For each energy propagation direction $\hat{\mathbf{s}}$, the electric flux density vector is tangent to the surface. In the three principal cross-sections we have plotted the direction of $\mathbf{D}_{0,e}$ and $\mathbf{D}_{0,o}$, associated with the extraordinary and the ordinary propagating wave. The direction given

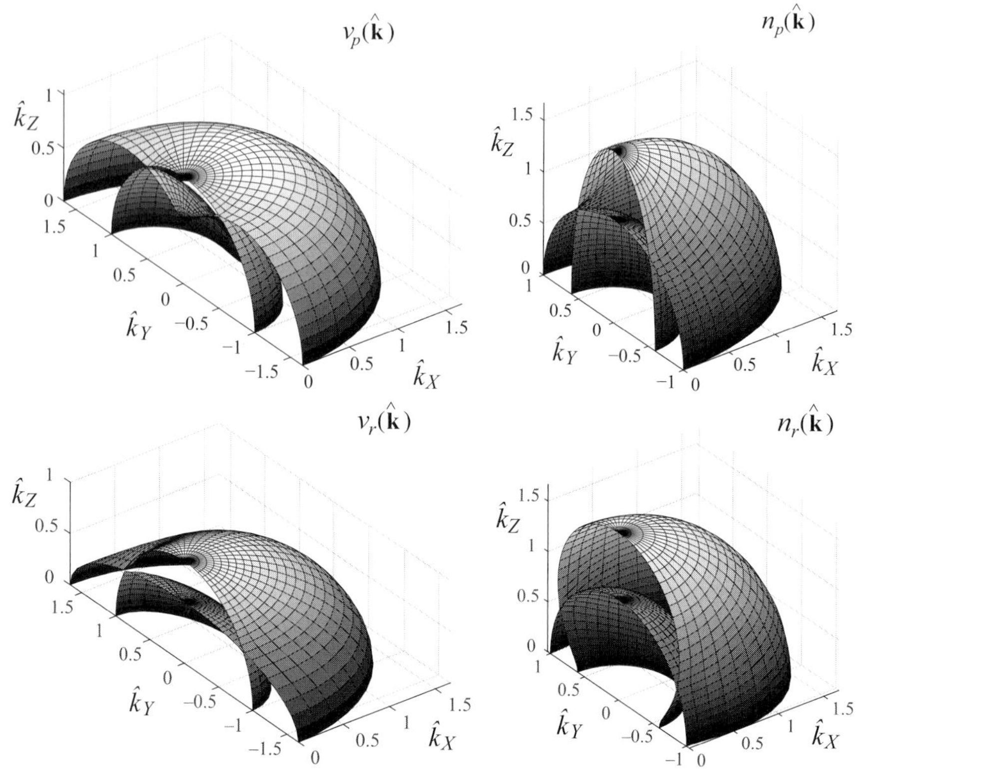

Figure 2.20: Three-dimensional plots of the four surfaces v_p, n_p, v_r and n_r for a biaxial medium (wave normal surface, optical indicatrix, ray normal surface and ray indicatrix, respectively). The anisotropy of the medium has been strongly exaggerated. The principal axes of the permittivity tensor coincide with the Cartesian coordinate axes (X, Y, Z). The principal refractive indices associated with the permittivity tensor $\bar{\bar{\epsilon}}_r$ are $n_X = n_0$, $n_Y = 5n_0/3$, $n_Z = 3n_0/5$, where n_0 is a reference index. The plotted quantities n and v have been normalised by the values n_0 and $v_0 = c/n_0$.

by the vector OR corresponds to that for which the energy propagation speeds of the two orthogonal states of polarisation are equal. For a general direction lying outside the three principal cross-sections, the propagating waves are again both extraordinary and the two orthogonal electric flux density vectors $\mathbf{D}_1$ and $\mathbf{D}_2$, both located in the tangent plane of the surface, follow from Eq. (2.3.13) or from direct solution of the Helmholtz equation in terms of $\mathbf{D}_0$ and $\hat{\mathbf{s}}$ (see Eq. (2.7.1)).

Figure 2.20 shows three-dimensional plots of the surfaces $\mathbf{r}_{vp}(\hat{\mathbf{k}})$, $\mathbf{r}_{np}(\hat{\mathbf{k}})$, $\mathbf{r}_{vr}(\hat{\mathbf{s}})$ and $\mathbf{r}_{nr}(\hat{\mathbf{s}})$. Note that in these graphs the principal velocities obey the relationship $v_Y < v_X < v_Z$. The surface derived from $n_p(\hat{\mathbf{k}})$ is the optical indicatrix. The two solutions for the electric field strength vector in the medium, $\mathbf{E}_1$ and $\mathbf{E}_2$, are tangent to the indicatrix at the point defined by the unit wave vector $\hat{\mathbf{k}}$. The electric flux density vectors $\mathbf{D}_1$ and $\mathbf{D}_2$ are tangent to the surface $v_r(\hat{\mathbf{s}})$ at the point defined by the energy propagation direction $\hat{\mathbf{s}}$.

2.8 Conical Refraction

Biaxial media show a peculiar pattern of energy propagation if the unit wave vector $\hat{\mathbf{k}}$ of a plane wave is directed along one of the optical axes of the medium. To observe this phenomenon, the angle of incidence of a parallel beam of light on the planar interface with a biaxial medium has to be such that the wave vector inside the medium is directed along the optical axis. After refraction at the interface, the so-called phenomenon of *conical* refraction is observed. Conical refraction was initially predicted by Hamilton **[124]**. Two months after the announcement, the phenomenon was experimentally confirmed by Lloyd **[215]**,**[216]**. To analyse the effect of conical refraction, we first find a solution of the eigenvalue equation for n_p or v_p for the propagation direction along an optical axis (see Eqs. (2.4.7) and (2.4.8)); the eigenvalue equals the middle value of the principal indices n_X, n_Y and n_Z of the medium. Since we consider propagation along an optical axis, the wave propagation is quasi-isotropic such that any direction of the electric flux density vector $\mathbf{D}_0$, perpendicular to $\hat{\mathbf{k}}$, is allowed. The electric field strength vectors $\mathbf{E}_i$ ($i = 1, 2$) in the anisotropic medium normally follow from Eq. (2.4.22) or (2.4.23), using the calculated eigenvalues for the refractive index or the propagation speed of the waves. It is seen that the equations for $\mathbf{E}_i$ become indeterminate when one of the eigenvalues is identical to a principal value of the medium, which for illustrative purposes we shall take as n_Y in what follows.

2.8.1 Quadratic Expansion of the Optical Indicatrix

To find the solution for the energy propagation direction $\hat{\mathbf{s}}$ and the electric field strength $\mathbf{E}$ for a plane wave with its unit wave vector $\hat{\mathbf{k}}_P$ parallel to an optical axis (see Eq. (2.3.5)) we analyse the optical indicatrix surface in the neighbourhood of the optical axis. According to Eq. (2.7.8), the energy propagation direction is perpendicular to the optical indicatrix. For the direction given by exactly $\hat{\mathbf{k}}_P$, the gradient vector of the surface is not defined. A quadratic expansion of the surface can, however, provide information about the shape of the surface in the direct vicinity of $\hat{\mathbf{k}}_P$. We start with Eq. (2.4.8) for the phase index n_p and rotate the coordinate system from (X, Y, Z) to (X', Y', Z') such that the wave vector $\hat{\mathbf{k}}'_P$ is parallel to the Z'-axis,

$$\begin{pmatrix} \hat{k}_X \\ \hat{k}_Y \\ \hat{k}_Z \end{pmatrix} = \begin{pmatrix} \hat{k}_{Z,P} & 0 & \hat{k}_{X,P} \\ 0 & 1 & 0 \\ -\hat{k}_{X,P} & 0 & \hat{k}_{Z,P} \end{pmatrix} \begin{pmatrix} \hat{k}'_X \\ \hat{k}'_Y \\ \hat{k}'_Z \end{pmatrix} . \tag{2.8.1}$$

The corresponding rotation angle θ_0 around the Y-axis, taken as positive in the clockwise direction from Z to X, is given by (see Fig. 2.21)

$$\tan\theta_0 = \frac{\hat{k}_{X,P}}{\hat{k}_{Z,P}} . \tag{2.8.2}$$

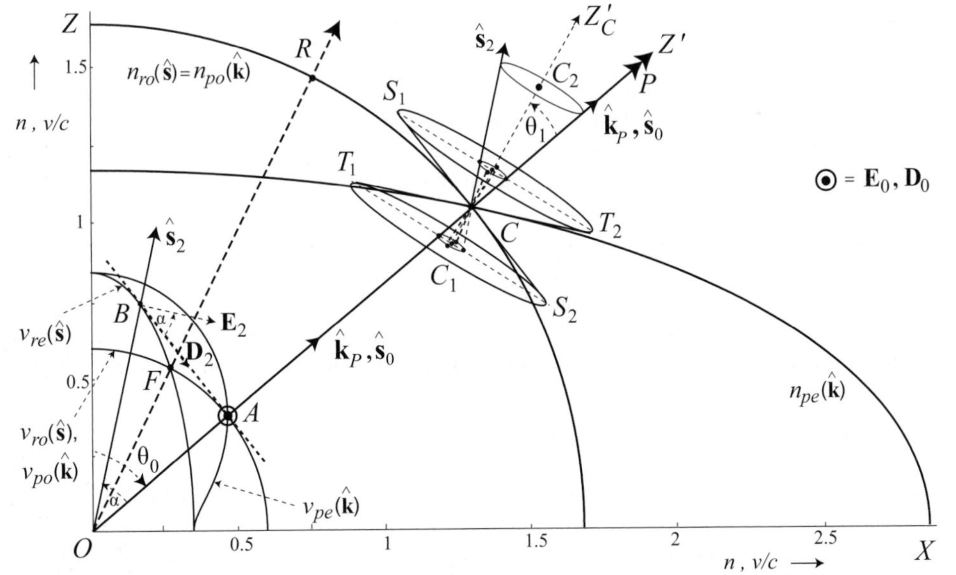

Figure 2.21: The quadratic approximation of the indicatrix surface in the neighbourhood of the (wave) optical axis OP through C for a biaxial medium; OR is the ray optical axis. For the sake of clarity, the anisotropy has been highly exaggerated with the principal values of the refractive indices given by $n_X = 1.18$, $n_Y = 1.67$, $n_Z = 2.86$. The line segment S_1S_2 is tangent to the ordinary cross-section of the indicatrix surface at the point C, whilst the line segment T_1T_2 is tangent to the extraordinary cross-section.

Equation (2.4.8) is written in the rotated coordinate system and the corresponding indicatrix surface follows by making the substitutions

$$X' = n_p \hat{k}'_X\,, \qquad Y' = n_p \hat{k}'_Y\,, \qquad Z' = n_p \hat{k}'_Z\,, \qquad X'^2 + Y'^2 + Z'^2 = n_p^2\,. \tag{2.8.3}$$

The indicatrix surface G_{np} is thus defined by the expression

$$G_{np}(X',Y',Z') = (X'^2 + Y'^2 + Z'^2)\left\{\left(\epsilon_X + \epsilon_Z - \frac{\epsilon_X\epsilon_Z}{\epsilon_Y}\right)X'^2 + \epsilon_Y Y'^2 + \frac{\epsilon_X\epsilon_Z}{\epsilon_Y}Z'^2 - 2\,\frac{\sqrt{\epsilon_X\epsilon_Z(\epsilon_Y-\epsilon_X)(\epsilon_Z-\epsilon_Y)}}{\epsilon_Y}X'Z'\right\}$$
$$-\left\{\epsilon_Y(\epsilon_X+\epsilon_Z)(X'^2+Y'^2) + 2\epsilon_X\epsilon_Z Z'^2 - 2\sqrt{\epsilon_X\epsilon_Z(\epsilon_Y-\epsilon_X)(\epsilon_Z-\epsilon_Y)}\,X'Z'\right\} + \epsilon_X\epsilon_Y\epsilon_Z = 0\,. \tag{2.8.4}$$

We seek a quadratic expansion of Eq. (2.8.4) at the point $C = (0, 0, n_Y)$ on the surface. Using the local coordinates (X'_C, Y'_C, Z'_C) where $X'_C = X'$, $Y'_C = Y'$ and $Z'_C = Z' - \sqrt{\epsilon_Y}$, the surface equation becomes

$$G_{np}(X'_C, Y'_C, Z'_C) = \frac{(\epsilon_Z-\epsilon_Y)(\epsilon_X-\epsilon_Y)}{4}\,Y_C'^2 + \epsilon_X\epsilon_Z\,Z_C'^2 - \sqrt{\epsilon_X\epsilon_Z(\epsilon_Y-\epsilon_X)(\epsilon_Z-\epsilon_Y)}\,X'_C Z'_C = 0\,. \tag{2.8.5}$$

To eliminate the cross-term with $X'_C Z'_C$ we introduce a second rotation of the coordinate system through an angle θ_1 in the plane $Y'_C = 0$ according to

$$\begin{aligned} X'_C &= X''_C \cos\theta_1 + Z''_C \sin\theta_1\,, \\ Y'_C &= Y''_C\,, \\ Z'_C &= -X''_C \sin\theta_1 + Z''_C \cos\theta_1\,. \end{aligned} \tag{2.8.6}$$

The cross-product vanishes if the rotation angle θ_1 satisfies

$$\tan 2\theta_1 = -\sqrt{\frac{(\epsilon_Y-\epsilon_X)(\epsilon_Z-\epsilon_Y)}{\epsilon_X\epsilon_Z}}\,, \tag{2.8.7}$$

whereby the resulting surface equation reads

$$X_C''^2 + \frac{Y_C''^2}{1-\tan^2\theta_1} = \frac{Z_C''^2}{\tan^2\theta_1}\,. \tag{2.8.8}$$

Equation (2.8.8) shows that the quadratically approximated surface reduces to a conical surface with its apex at C (see Fig. 2.21). The full apex angle of the conical surface is given by $\pi - 2\theta_1$. The axis C_1C_2 of the cone is found at an angle $\theta_0 + \theta_1$ with respect to the original Z-axis. In our example the sign of θ_1 was negative, which means that the cone axis is inclined from the optical axis towards the principal axis of the medium with the highest permittivity eigenvalue. A cross-section of the cone perpendicular to the cone axis has an elliptical shape, for which the ratio of the major and minor axes, a_C and b_C respectively, is found to be given by

$$\frac{b_C^2}{a_C^2} = \frac{2}{1+\sqrt{1+(\epsilon_Y-\epsilon_X)(\epsilon_Z-\epsilon_Y)/\epsilon_X\epsilon_Z}}\,. \tag{2.8.9}$$

In practical cases, biaxial anisotropy is relatively small and the angle θ_1 is of the order of a few degrees at most. Consequently, the elliptical cross-section can frequently be well approximated by a circle.

2.8.2 The Conical Assembly of Ray Vectors

To find the possible ray vectors $\hat{\mathbf{s}}_i$ associated with the wave vector $\hat{\mathbf{k}}_P$, we construct the normal vector to the cone surface at C. The gradient vectors at C to the conical surface defined by $G_{np}(X', Y', Z') = 0$ describe a second conical surface with the same axis C_1C_2 and an apex angle of $2\theta_1$ which defines the allowed/possible ray vectors $\hat{\mathbf{s}}$. It is this cone of ray vectors that was predicted by Hamilton and experimentally demonstrated by Lloyd in 1833. The experiment was carried out with the aid of a crystal of aragonite and it produced a cone of rays with an apex angle of approximately 3 degrees. Instead of two discrete solutions for energy propagation we find a continuum of directions. We shall denote a specific vector from this continuum by $\hat{\mathbf{s}}_\beta$.

A detailed picture of the geometry of the hollow cone of energy propagation vectors is given in Fig. 2.22 and in the insert a). The shape of the optical indicatrix shows a cusp in the direction of one of the optical axes of a medium, at point

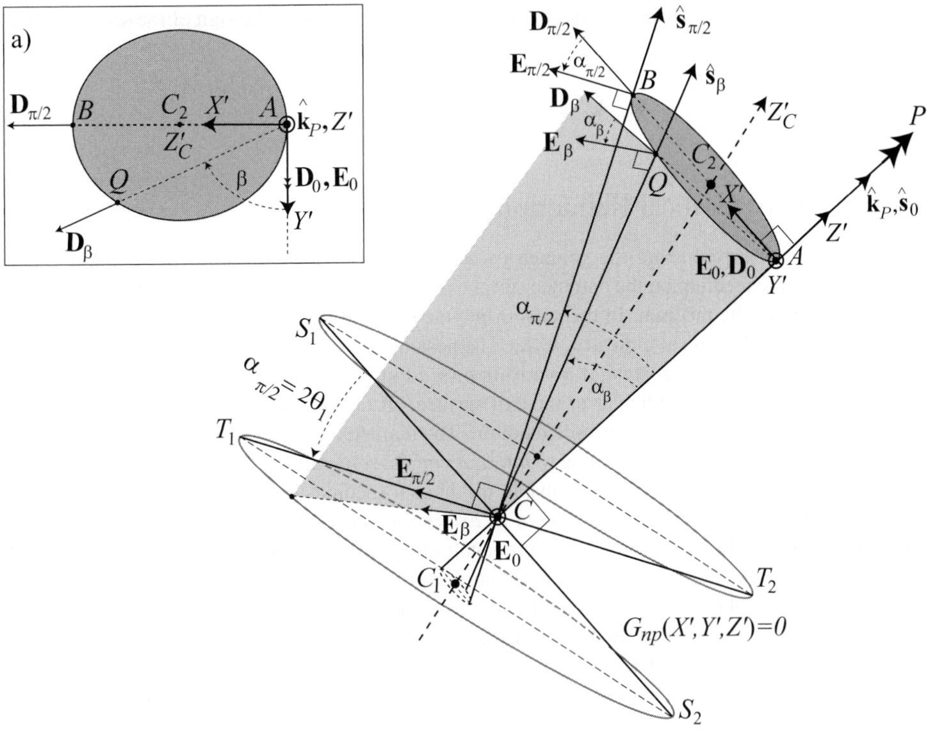

Figure 2.22: The electric field strength and flux density vectors in the case of conical refraction in a biaxial medium. At the bottom we have sketched the optical indicatrix $G_{np}(X', Y', Z') = 0$, approximated at the point C by a double-branched conical surface. The unit wave vector $\hat{\mathbf{k}}_P$ is parallel to the optical axis of the biaxial medium and perpendicular to the *ordinary* cross-section S_1S_2 of the optical indicatrix. The insert labelled a) is a view in the negative $\hat{\mathbf{k}}$-direction and shows the distribution of the electric flux density vector $\mathbf{D}_\beta$ in a plane perpendicular to the wave vector $\hat{\mathbf{k}}_P$.

C in the figure. Since in our example we assume that $\epsilon_Z > \epsilon_Y > \epsilon_X$ the plane of the drawing ($Y' = 0$) contains the optical axis which is parallel to the Z'-axis. The cross-section with $Y' = 0$ contains the straight line S_1S_2, which is tangential to the ordinary circular sheet of the indicatrix, and the line T_1T_2, which is tangential to the extraordinary sheet. The ordinarily propagating ray, denoted by $\hat{\mathbf{s}}_0$ in Fig. 2.21, is perpendicular to the ordinary sheet of the indicatrix through S_1S_2 of the conical surface and is parallel to the wave vector. The associated field vectors $\mathbf{D}_0$ and $\mathbf{E}_0$ are perpendicular to the plane of the drawing (see Fig. 2.21 at point A). In the same figure an extraordinary ray has been denoted by $\hat{\mathbf{s}}_2$ which is parallel to OB. The point B on the extraordinary sheet of the ray normal surface is the tangent point with a plane perpendicular to the wave vector $\hat{\mathbf{k}}_P$. The angle between the associated field vectors $\mathbf{D}_2$ and $\mathbf{E}_2$ is given by α.

We return to Fig. 2.22 and observe that in a plane perpendicular to $\hat{\mathbf{k}}_P$ the strength of the electric polarisation that is induced by an electromagnetic wave is independent of the azimuth of the state of polarisation and all directions of the electric flux density vector $\mathbf{D}_\beta$ are allowed where β is the angle between $\mathbf{D}_\beta$ and the Y'-axis, measured in a plane Z' = constant (see insert a)). To solve the Helmholtz equation the vectors $\hat{\mathbf{k}}_P$, $\mathbf{D}_\beta$ and $\mathbf{E}_\beta$ have to be coplanar. The grey-shaded planar cross-section AQC contains such a vector triplet. The angle α_β between $(\mathbf{D}_\beta, \mathbf{E}_\beta)$ or $(\hat{\mathbf{k}}_P, \hat{\mathbf{s}}_\beta)$ follows from the constitutive relation $\mathbf{D}_\beta = \epsilon_0 \,\bar{\bar{\epsilon}}_r\, \mathbf{E}_\beta$ with $\mathbf{D}_\beta = \epsilon_0(\epsilon_X E_{X,\beta}, \epsilon_Y E_{Y,\beta}, \epsilon_Z E_{Z,\beta})$ in a coordinate system (X, Y, Z) that is aligned along the principal axes of the medium. The coplanarity of the vectors can easily be checked by calculating the scalar product $\hat{\mathbf{k}}_P \cdot \mathbf{E}_\beta \times \mathbf{D}_\beta$, which is indeed zero independently of the azimuth β.

The unit ray vector can be constructed using the condition that $\hat{\mathbf{s}}_\beta$ should be perpendicular to the optical indicatrix, see the various vectors drawn at the cusp C of the indicatrix. Alternatively, from vector calculus, we obtain for the ray vector in a medium with linear anisotropy the expression

$$\hat{\mathbf{s}}_\beta = \hat{\mathbf{E}}_\beta \times \hat{\mathbf{k}}_P \times \hat{\mathbf{D}}_\beta = \hat{\mathbf{k}}_P \cos\alpha_\beta - (\hat{\mathbf{E}}_\beta \cdot \hat{\mathbf{k}}_P)\hat{\mathbf{D}}_\beta \,. \tag{2.8.10}$$

The sign of the angle α_β in the figure follows from the relation $\sin\alpha_\beta = -\hat{\mathbf{E}}_\beta \cdot \hat{\mathbf{k}}_P$¿ The sign of the scalar product $\hat{\mathbf{E}}_\beta \cdot \hat{\mathbf{k}}_P$ is given by

$$\mathbf{E}_\beta \cdot \hat{\mathbf{k}}_P = -\frac{(\epsilon_Y - \epsilon_X)(\epsilon_Z - \epsilon_Y)}{\epsilon_0\, \epsilon_Y^2(\epsilon_Z - \epsilon_X)} \sin\beta\,. \tag{2.8.11}$$

2.8.3 Observation of Conical Refraction

Conical refraction can be observed using the geometries depicted in Fig. 2.23. In drawing a), a single plane wave is incident from a first isotropic medium on the entrance surface of a biaxial medium. The polished surface is perpendicular to the wave-optic axis *OP* of the medium. In the anisotropic medium we have sketched the ordinary and extraordinary sheets of the cross-section of the optical indicatrix $n_p(\hat{\mathbf{k}})$. The possible ray directions in the medium are given by a hollow cone, each ray vector having a different state of polarisation (see Fig. 2.22). Since there is only a single wave vector in the medium, which is moreover perpendicular to the exit surface, all rays of the hollow cone propagate parallel to one another after transmission into the second isotropic medium. The diameter of the cone is proportional to the thickness of the biaxial plate. This geometry is said to give rise to *internal* conical refraction.

Alternatively, in Fig. 2.23b, conical refraction inside the biaxial medium is replaced by conical refraction into the second isotropic medium. This phenomenon is known as *external* conical refraction. To understand this phenomenon, it is useful to consider the duality describing wave propagation using the set $(\hat{\mathbf{k}}_0, \mathbf{E}_0, n_p)$ or $(\hat{\mathbf{s}}_0, \mathbf{D}_0, v_r/c)$. To describe external conical refraction the latter set $(\hat{\mathbf{s}}_0, \mathbf{D}_0, v_r/c)$ is more useful. To produce external conical refraction, we polish the faces of the plane-parallel plate of the biaxial material in such a way that an incident full cone of unpolarised rays is refracted towards the ray-optic axis *OR* of the biaxial medium (see also Fig. 2.19 for the definition of this ray-optic axis as opposed to the wave-optic axis *OP*). The ray-optic axis *OR* in Fig. 2.21 is parallel to a ray direction $\hat{\mathbf{s}}_a$ in the medium along which the ray propagation speed v_r is identical for an assembly of plane waves with wave vectors $\hat{\mathbf{k}}$ arranged on

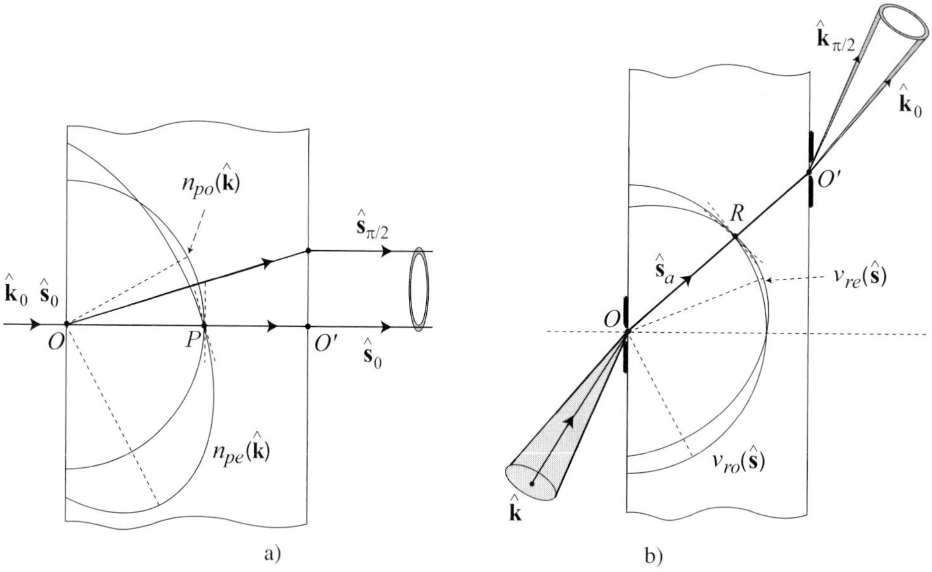

Figure 2.23: Internal and external conical refraction.
a) Internal conical refraction. A parallel beam of light (wave vector $\hat{\mathbf{k}}_0$) is normally incident from an isotropic medium onto a plane-parallel plate made of a biaxial material. An optical axis *OP* of the medium is perpendicular to the surface. The two ray vectors $\hat{\mathbf{s}}_0$ and $\hat{\mathbf{s}}_{\pi/2}$ are part of the hollow cone of rays that exits from the plane-parallel plate.
b) External conical refraction. A full cone of rays (natural light) is incident on the entrance surface. Two diaphragms at *O* and *O′* select an energy propagation direction in the crystal given by the vector $\hat{\mathbf{s}}_a$ which is parallel to the ray-optic axis *OR* of the medium. The ordinary and extraordinary ray propagation speeds are represented by the curves $v_{re}(\hat{\mathbf{s}})$ and $v_{ro}(\hat{\mathbf{s}})$. The refracted pencil of rays exiting at *O′* is a hollow cone.

the surface of a cone. The two diaphragms at O and O' select the energy propagation direction $\hat{\mathbf{s}}_a$. Only the correct plane waves with wave vector $\hat{\mathbf{k}}$ and the appropriate state of linear polarisation can couple into the energy direction $\hat{\mathbf{s}}_a$ and then pass the diaphragm at O'. The continuous set of wave vectors $\hat{\mathbf{k}}$ associated with $\hat{\mathbf{s}}_a$ are located on a hollow cone and this wave field is refracted out of the anisotropic medium as a hollow cone of light. The state of polarisation in the output cone of light varies in the same way as shown in Fig. 2.22 (see insert a)).

2.9 Optical Activity

In the uniaxial and biaxial media that were treated above, the electric field vector **E** produces an anisotropic electric flux density or dielectric displacement vector **D** under the influence of a locally anisotropic structure of the material. Calcite was the first material in which this property was observed. The discovery of anisotropy in calcite dates back to as early as the seventeenth century, because of its large degree of anisotropy and the easily visible effect of double refraction. In 1811 Arago discovered that the plane of polarisation of a linearly polarised light beam is rotated when passing through a quartz crystal. This effect of *optical rotation* was also discovered in liquids where the stochastic orientation of the molecules was expected to lead to an effectively isotropic behaviour.

A qualitative explanation of the occurrence of optical rotation was first given by Drude **[88]** who found that instead of considering only the local value of the electromagnetic field vector **E**, it is necessary to also consider the spatial derivatives of **E** and **H**. More specifically, Drude showed that to describe optical rotation in globally isotropic media like a liquid, the rotations $\nabla \times \mathbf{E}$ and $\nabla \times \mathbf{H}$ of the electric and magnetic field vector have to be included in the calculation of the electric and magnetic flux density vectors **D** and **B**. Physically speaking, the crystals or molecules in which optical rotation is observed do not only give rise to linear motion of the free or bound electrons under the influence of an external electromagnetic field, but also to helically shaped current loops. The oscillating electron transport along helical paths induces a displacement current which, according to the Ampère–Maxwell law, creates an oscillating magnetic dipole moment in the optically active structure. The magnetic dipole is aligned along the propagation direction of the light wave and the strength of the induced magnetic dipole moment is proportional to $\partial\mathbf{E}/\partial t$. Conversely, according to Faraday's law, the oscillating magnetic dipole induces an oscillating electric field that is proportional to $\partial\mathbf{H}/\partial t$. The helical motion of electric charge can exhibit a differing sense of rotation and hence the optical rotation can be right- or left-handed depending on the right- or left-handed structure of the molecule or crystal. Persistence of the bulk optical rotation of right-handed or left-handed molecules in the liquid phase follows from the fact that the molecules have an identical effect on the incident light if their orientation is inverted. The angular averaging in a liquid does, however, yield a reduction in optical activity by a factor of $1/2$ with respect to a fully aligned molecular structure in the crystalline state.

In the photon picture of light propagation in a medium, standard birefringence is related to single scattering of a photon at the lattice of a crystal or at the atomic structure of a molecule. The incident radiation has a wavelength that is very large with respect to the crystal lattice of the anisotropic medium. The phase of a scattered wave in a certain propagation direction depends on the average propagation speed in that direction, averaged over many crystal building blocks. Optical rotation arises when double scattering of an incident photon occurs. Since the probability of such multiple scattering events is much lower, the optical rotation effects are much smaller in general than the principal linear birefringence of an anisotropic crystal. A detailed analysis of optical rotation starting from the atomic and molecular level can be found in the textbook by Charney **[65]**. Higher-order anisotropic effects have been reported, relating to multiple photon scattering in isotropic crystals. The first mention of such an effect dates back to Lorentz **[221]**. In a cubic crystal, with a lattice constant a that is not negligibly small with respect to the wavelength λ of the radiation, he found that higher-order anisotropy may occur. The deviation from isotropy is proportional to $(a/\lambda)^2$.

Optical rotatory power can also be introduced in a medium by applying an external magnetic field. This surprising effect was first discovered by Faraday and carries his name **[95]**,**[96]**. Although the phenomenon is comparable to optical rotation induced by a helical three-dimensional structure of a medium, there is an important difference with respect to the reciprocity of light propagation in the two cases. Faraday rotation is discussed in greater detail in Appendix E.

2.9.1 Constitutive Relations for a Medium with Optical Rotation

Following the analysis by Drude **[88]**, Condon **[75]** and Chipman and co-workers **[244]**, the constitutive relations of an anisotropic optically active medium must be modified with respect to those given in Eqs. (2.2.1) and (2.2.2) and are instead given by

$$\mathbf{D} = \epsilon_0 \overline{\overline{\epsilon}}_r \mathbf{E} - \overline{\overline{X}} \frac{\partial \mathbf{H}}{\partial t} ,$$

$$\mathbf{B} = \mu_0 \overline{\overline{\mu}}_r \mathbf{H} + \overline{\overline{X}} \frac{\partial \mathbf{E}}{\partial t} , \tag{2.9.1}$$

where $\mathbf{E}$ and $\mathbf{H}$ are the harmonic electric and magnetic field components and the time dependence $\exp(-i\omega t)$ has been omitted. The tensor coefficients X_{kl} have the dimension of s^2m^{-1}, i.e. the inverse of an acceleration. Carrying out the differentiation with respect to time, the constitutive relations are commonly written as

$$\mathbf{D}_0 = \epsilon_0 \overline{\overline{\epsilon}}_r \mathbf{E}_0 + i \overline{\overline{G}} \mathbf{H}_0 ,$$

$$\mathbf{B}_0 = \mu_0 \overline{\overline{\mu}}_r \mathbf{H}_0 - i \overline{\overline{G}} \mathbf{E}_0 , \tag{2.9.2}$$

where the tensor elements G_{kl} are equal to ωX_{kl}. The tensor $\overline{\overline{G}}$ is called the gyrotropic tensor and its elements are real and have the dimension of sm^{-1}.

2.9.2 Eigenvalue Equation for Anisotropic and Optically Active Media

To obtain the eigenvalue equations for the electric and magnetic field components in an optically active medium, we start again from Eqs. (2.2.9) and (2.2.10), however, this time supplemented with the optical rotation effect, represented by the gyrotropic tensor $\overline{\overline{G}}$. After some rearrangement we find

$$\mathbf{0} = \mathbf{k}\times \overline{\overline{\epsilon}}_r^{-1} \mathbf{k} \times \mathbf{H}_0 + i\omega\left(\mathbf{k}\times \overline{\overline{\epsilon}}_r^{-1} \overline{\overline{G}} \mathbf{H}_0 + \overline{\overline{G}}\overline{\overline{\epsilon}}_r^{-1} \mathbf{k} \times \mathbf{H}_0\right)$$
$$+ \omega^2\left(\epsilon_0\mu_0 \overline{\overline{\mu}}_r - \overline{\overline{G}}\overline{\overline{\epsilon}}_r^{-1}\overline{\overline{G}}\right)\mathbf{H}_0 , \tag{2.9.3}$$

$$\mathbf{0} = \mathbf{k}\times \overline{\overline{\mu}}_r^{-1} \mathbf{k} \times \mathbf{E}_0 + i\omega\left(\mathbf{k}\times \overline{\overline{\mu}}_r^{-1} \overline{\overline{G}} \mathbf{E}_0 + \overline{\overline{G}}\overline{\overline{\mu}}_r^{-1} \mathbf{k} \times \mathbf{E}_0\right)$$
$$+ \omega^2\left(\epsilon_0\mu_0 \overline{\overline{\epsilon}}_r - \overline{\overline{G}}\overline{\overline{\mu}}_r^{-1}\overline{\overline{G}}\right)\mathbf{E}_0 . \tag{2.9.4}$$

A shorthand notation is possible using a pseudotensor notation for the vector product of two vectors **a** and **b** viz.

$$\mathbf{a}\times\mathbf{b} = \begin{pmatrix} 0 & -a_3 & +a_2 \\ +a_3 & 0 & -a_1 \\ -a_2 & +a_1 & 0 \end{pmatrix}\begin{pmatrix} b_1 \\ b_2 \\ b_3 \end{pmatrix} = \overline{\overline{T}}_a \mathbf{b} , \tag{2.9.5}$$

where $\overline{\overline{T}}_a$ is the pseudotensor. In matrix notation, the two eigenvalue equations above can then be written as

$$\left[\left(\overline{\overline{T}}_k + i\omega \overline{\overline{G}}\right)\overline{\overline{\mu}}_r^{-1}\left(\overline{\overline{T}}_k + i\omega \overline{\overline{G}}\right) + \omega^2\epsilon_0\mu_0 \overline{\overline{\epsilon}}_r\right]\mathbf{E}_0 = \mathbf{0} , \tag{2.9.6}$$

$$\left[\left(\overline{\overline{T}}_k + i\omega \overline{\overline{G}}\right)\overline{\overline{\epsilon}}_r^{-1}\left(\overline{\overline{T}}_k + i\omega \overline{\overline{G}}\right) + \omega^2\epsilon_0\mu_0 \overline{\overline{\mu}}_r\right]\mathbf{H}_0 = \mathbf{0} , \tag{2.9.7}$$

where $\overline{\overline{T}}_k$ is the vector product tensor associated with the vector **k**.

For nonmagnetic media, we replace $\overline{\overline{\mu}}_r$ by the identity tensor $\overline{\overline{I}}$. The eigenvalue equations for the electric field components then reduce to

$$\left[\left(\overline{\overline{T}}_k + i\omega \overline{\overline{G}}\right)^2 + \omega^2\epsilon_0\mu_0 \overline{\overline{\epsilon}}_r\right]\mathbf{E}_0 = \mathbf{M}_E \mathbf{E}_0 = \mathbf{0} . \tag{2.9.8}$$

$\mathbf{M}_E$ is equivalent to the submatrix that was already introduced in Eq. (2.2.15).

2.9.3 Polarisation Eigenstates for an Optically Active Medium

The system of linear equations for the electric field vector $\mathbf{E}_0$ allows the calculation of the electric field eigenvectors. In the general case, in which the tensors $\overline{\overline{\epsilon}}_r$ and $\overline{\overline{G}}$ do not possess any common properties deriving from crystal symmetry, the calculation of the matrix elements of $\mathbf{M}_E$ is possible but leads to intricate expressions. We instead consider the case that the linear anisotropy of the medium is absent such that $\overline{\overline{\epsilon}}_r = \overline{\overline{I}}$. A nontrivial solution of the eigenvalue equations for $\mathbf{E}_0$ requires that the determinant of the matrix $\mathbf{M}_E$ of Eq. (2.9.8) equals zero. For the linearly isotropic case where the optical activity is present in the z-direction, $\mathbf{M}_E$ is given by

$$\mathbf{M}_E = \left\{ \begin{pmatrix} 0 & -k_z & +k_y \\ +k_z & 0 & -k_x \\ -k_y & +k_x & 0 \end{pmatrix} + i\omega \begin{pmatrix} g_o & 0 & 0 \\ 0 & g_o & 0 \\ 0 & 0 & g_e \end{pmatrix} \right\}^2 + \omega^2 \epsilon_0 \mu_0 \epsilon_r \overline{\overline{I}} \,, \tag{2.9.9}$$

where ϵ_r is the scalar value of the relative permittivity of the material. We note that the gyrotropic tensor used in Eq. (2.9.9) has the same form as that of crystal quartz. The matrix elements of $\mathbf{M}_E$ are given by

$$\begin{pmatrix} \omega^2(\epsilon_0\mu_0\epsilon_r - g_o^2) - k_y^2 - k_z^2 & -i2\omega g_o k_z + k_x k_y & +i\omega(g_o+g_e)k_y + k_x k_z \\ +i2\omega g_o k_z + k_x k_y & \omega^2(\epsilon_0\mu_0\epsilon_r - g_o^2) - k_x^2 - k_z^2 & -i\omega(g_o+g_e)k_x + k_y k_z \\ -i\omega(g_o+g_e)k_y + k_x k_z & +i\omega(g_o+g_e)k_x + k_y k_z & \omega^2(\epsilon_0\mu_0\epsilon_r - g_e^2) - k_x^2 - k_y^2 \end{pmatrix} . \tag{2.9.10}$$

We divide the matrix elements by $\omega^2\epsilon_0\mu_0 = k_0^2$ and use $\mathbf{k}/k_0 = n\hat{\mathbf{k}}$ where $\hat{\mathbf{k}}$ is the unit wave vector and c/n the velocity of the propagating wave. After some rearrangement and using $\epsilon_r = n_r^2$, the matrix $\mathbf{M}_E$ is written as

$$\begin{pmatrix} n_r^2 - c^2 g_o^2 - (\hat{k}_y^2 + \hat{k}_z^2)n^2 & -i2cg_o\hat{k}_z n + \hat{k}_x\hat{k}_y n^2 & +i\omega c(g_o+g_e)\hat{k}_y n + \hat{k}_x\hat{k}_z n^2 \\ +i2cg_o\hat{k}_z n + \hat{k}_x\hat{k}_y n^2 & n_r^2 - c^2 g_o^2 - (\hat{k}_x^2 + \hat{k}_z^2)n^2 & -ic(g_o+g_e)\hat{k}_x n + \hat{k}_y\hat{k}_z n^2 \\ -ic(g_o+g_e)\hat{k}_y n + \hat{k}_x\hat{k}_z n^2 & +ic(g_o+g_e)\hat{k}_x n + \hat{k}_y\hat{k}_z n^2 & n_r^2 - c^2 g_e^2 - (\hat{k}_x^2 + \hat{k}_y^2)n^2 \end{pmatrix} . \tag{2.9.11}$$

Using the gyrotropic tensor of Eq. (2.9.9), the wave solutions should remain unchanged when a rotation around the z-axis is performed. For that reason, we can analyse the condition $|\mathbf{M}_E| = 0$ for the unit wave vector $\hat{\mathbf{k}} = (0, \sin\theta, \cos\theta)$ without loss of generality. With the substitutions

$$c = \frac{\omega g_o}{k_0}\,, \quad d = \frac{\omega g_e}{k_0}\,, \quad a = n_r^2 - \frac{\omega^2 g_o^2}{k_0^2}\,, \quad a + c^2 - d^2 = n_r^2 - \frac{\omega^2 g_e^2}{k_0^2}\,, \tag{2.9.12}$$

we obtain the following quadratic equation in the quantity n^2,

$$\begin{aligned} &n^4\left[(c^2-d^2)\cos^2\theta + (c-d)^2\sin^2\theta\cos^2\theta + a\right] \\ &-n^2\left[2a^2 + a(c^2-d^2)(1+\cos^2\theta) + 4c^2(a+c^2-d^2)\cos^2\theta + a(c+d)^2\sin^2\theta\right] \\ &+a^2(a+c^2-d^2) = 0\,. \end{aligned} \tag{2.9.13}$$

This equation yields four solutions for n, of which the positive solutions are taken for the forward propagating waves. Special cases occur for wave propagation in the z-direction, $\theta = 0$, where the optical activity of the medium is strongest, and for propagation perpendicular to the z-axis. We consider these special cases in more detail:

- $\theta = 0$

  The solution for n yields the values

$$n = n_r \pm \frac{\omega g_o}{k_0}\,. \tag{2.9.14}$$

  The corresponding polarisation eigenstates follow from Eq. (2.2.15) after substituting in the value for n. Taking the positive sign in Eq. (2.9.14), it follows that the (normalised) eigenvector is given by

$$n = n_r + \frac{\omega g_o}{k_0}\,, \qquad \hat{\mathbf{E}}_0 = \frac{1}{2}\sqrt{2}\,(1, +i, 0)\,, \tag{2.9.15}$$

corresponding to left-handed circularly polarised light. For the eigenvalue $n = n_r - \omega g_o/k_0$, the eigenvector corresponds to right-handed circularly polarised light. Since the matrix $\mathbf{M}_E$ is Hermitian, the two eigenvectors are orthogonal. When describing the general case of propagation of elliptically polarised light, we can hence decompose the light into its two orthogonal circular eigenstates with a certain phase difference between them. The different speed of propagation of the two circular components in an optically active medium leads to an increase in the phase difference as the wave progresses in the medium. A rotation of the polarisation ellipse to the left or to the right depending on the handedness of the medium hence occurs. Optically active media that exhibit two orthogonal circularly polarised eigenstates are also called *circularly birefringent*.

- $\theta = \pi/2$

  The calculation of the value of n yields

$$n^2 = \left(n_r^2 + g_o g_e \frac{\omega^2}{k_0^2}\right) \pm n_r(g_o + g_e)\frac{\omega}{k_0} \,. \tag{2.9.16}$$

If we approximate the value of n by expanding Eq. (2.9.16) up to first order in $g\omega/k_0$, the calculation of the (non-normalised) polarisation eigenstates yields the result

$$\begin{aligned} n &\approx n_r + \frac{\omega g_o}{2n_r k_0}\,, \qquad \mathbf{E}_0 \approx \left(1,\ 0,\ -i\,\frac{1 + g_o \dfrac{\omega}{n_r k_0}}{1 + \dfrac{g_o + g_e}{2}\dfrac{\omega}{n_r^2 k_0}}\right), \\ n &\approx n_r - \frac{\omega g_o}{2n_r k_0}\,, \qquad \mathbf{E}_0 \approx \left(1,\ 0,\ +i\,\frac{1 - g_o \dfrac{\omega}{n_r k_0}}{1 - \dfrac{g_o + g_e}{2}\dfrac{\omega}{n_r^2 k_0}}\right). \end{aligned} \tag{2.9.17}$$

In general, the polarisation eigenstates are left- and right-handed elliptical and mutually orthogonal. If $g_o = g_e$, the eigenstates are circularly polarised. For each polarisation eigenstate the magnetic field $\mathbf{H}_0$ follows from the substitution of the wave vector $\hat{\mathbf{k}}$ and the value of n into Eq. (2.9.7).

More general anisotropic media, with arbitrary orientations of the linear and circular birefringence, can be treated using the general expressions of Eqs. (2.9.3) and (2.9.4). Solution of the equation $|\mathbf{M}| = 0$ to find n, will in general require a numerical root finding procedure. The subject of general bi-anisotropic media, with the corresponding eigenvalue equations

$$\begin{pmatrix} M_{EE} & M_{EH} \\ M_{HE} & M_{HH} \end{pmatrix} \begin{pmatrix} \mathbf{E}_0 \\ \mathbf{H}_0 \end{pmatrix} = \mathbf{0}\,, \tag{2.9.18}$$

is treated by Mackay and Lakhtakia **[227]**.

2.10 Wave Propagation in an Anisotropic Medium Including Rotation

In the framework of geometrical optics the unit ray vector $\hat{\mathbf{s}}$ is used to describe the direction of propagation of optical energy in a medium. In an optical imaging system that contains birefringent materials, we need to trace the energy carrying rays through the homogeneous parts of the medium and study the refraction and reflection at interfaces between two media. The wave vector $\mathbf{k}$ is an auxiliary quantity that allows the progression of the phase along the optical ray to be described. In the practice of optical design and optical system analysis, it is useful to have expressions that directly yield the polarisation eigenstates and the ray and phase propagation speed from the direction of the ray vector $\hat{\mathbf{s}}$. In the preceding section, for linear anisotropic media, it was shown that an eigenvalue equation can be developed for the wave propagation in a linear anisotropic medium in terms of this unit ray vector $\hat{\mathbf{s}}$ and the electric flux density vector $\mathbf{D}_0$ of a harmonic plane wave. In this section we extend this analysis to a more general medium possessing both linear anisotropy and optical activity with an arbitrary orientation. We consider the media to be isotropic with respect to their magnetic properties.

Table 2.1. Effective vector and tensor quantities appearing in the eigenvalue equations (Eqs. (2.10.1) and (2.10.2)) for the electric field strength $\mathbf{E}_0$ and the magnetic field strength $\mathbf{H}_0$, respectively.

Eigenvalue equation	$M_{EE}\mathbf{E}_0 = \mathbf{0}$	$M_{HH}\mathbf{H}_0 = \mathbf{0}$
Propagation vector	$\overline{\overline{\mu}}_r^{-1}\,\mathbf{k}$	$\overline{\overline{\epsilon}}_r^{-1}\,\mathbf{k}$
Permittivity and permeability tensors	$\overline{\overline{\mu}}_r^{-1}\,\overline{\overline{\epsilon}}_r$	$\overline{\overline{\epsilon}}_r^{-1}\,\overline{\overline{\mu}}_r$
Gyrotropic tensor	$\overline{\overline{\mu}}_r^{-1}\,\overline{\overline{G}}$	$\overline{\overline{\epsilon}}_r^{-1}\,\overline{\overline{G}}$
Eigenstates	$\mathbf{E}_{0,j} \rightarrow \mathbf{B}_{0,j}, \mathbf{H}_{0,j}, \mathbf{D}_{0,j}$	$\mathbf{H}_{0,j} \rightarrow \mathbf{D}_{0,j}, \mathbf{E}_{0,j}, \mathbf{B}_{0,j}$
Modulus	$k_{j,E} = \omega/v_j$	$k_{j,H} = \omega/v_j$

2.10.1 Symmetry of Electromagnetic Field Equations

For the analysis of wave propagation in more general anisotropic media it is useful to examine the high degree of symmetry that exists between the eigenvalue equations for the electric and magnetic field strengths, see Eqs. (2.9.6) and (2.9.7). Multiplying the first equation with $\overline{\overline{\mu}}_r^{-1}$ and the second equation with $\overline{\overline{\epsilon}}_r^{-1}$, we obtain the following expressions:

$$\left[\left\{\overline{\overline{\mu}}_r^{-1}\left(\overline{\overline{T}}_k + i\omega\,\overline{\overline{G}}\right)\right\}^2 + \omega^2\epsilon_0\mu_0\,\overline{\overline{\mu}}_r^{-1}\,\overline{\overline{\epsilon}}_r\right]\mathbf{E}_0 = M_{EE}\,\mathbf{E}_0 = \mathbf{0}\,, \tag{2.10.1}$$

$$\left[\left\{\overline{\overline{\epsilon}}_r^{-1}\left(\overline{\overline{T}}_k + i\omega\,\overline{\overline{G}}\right)\right\}^2 + \omega^2\epsilon_0\mu_0\,\overline{\overline{\epsilon}}_r^{-1}\,\overline{\overline{\mu}}_r\right]\mathbf{H}_0 = M_{HH}\,\mathbf{H}_0 = \mathbf{0}\,. \tag{2.10.2}$$

The equations show a comparable structure regarding the way in which the electric and magnetic tensors appear. It is also seen that the gyrotropic tensor modifies the way in which the magnetic field strength originates from the electric field strength in an optically active medium. In Table 2.1 we have listed the symmetries between the associated variables and medium parameters which appear in Eqs. (2.10.1) and (2.10.2).

The eigenvalue equations require the solution of $|M_{EE}| = 0$ or $|M_{HH}| = 0$. The matrices have special properties because of the symmetry of the various tensors that compose them. As was shown in Subsection 2.2.1, the dielectric tensor and the permeability tensor have to be symmetric in lossless media; the same argument holds for the gyrotropic tensor $\overline{\overline{G}}$. Focusing now on Eq. (2.10.1) and equating $\overline{\overline{\mu}}_r$ to the identity tensor, we notice that the first term, $\overline{\overline{T}}_k + i\omega\,\overline{\overline{G}}$, can be written as

$$\begin{pmatrix} 0 & -k_z & +k_y \\ +k_z & 0 & -k_x \\ -k_y & +k_x & 0 \end{pmatrix} + i\omega \begin{pmatrix} g_{11} & g_{12} & g_{13} \\ g_{12} & g_{22} & g_{23} \\ g_{13} & g_{23} & g_{33} \end{pmatrix} = i\,\overline{\overline{M}}_h\,, \tag{2.10.3}$$

where we have used the symmetry property of $\overline{\overline{G}}$. It follows that the tensor $\overline{\overline{M}}_h$ is Hermitian with real elements on the diagonal; the same holds for the product tensor $(i\,\overline{\overline{M}}_h)^2$. When we insert this tensor in Eq. (2.10.1), we obtain

$$\left(-\,\overline{\overline{M}}_h^{\,2} + k_0^2\,\overline{\overline{\epsilon}}_r\right)\mathbf{E}_0 = \mathbf{0}\,. \tag{2.10.4}$$

For the case of a medium exhibiting optical rotation and $\overline{\overline{\epsilon}}_r = n^2\,\overline{\overline{I}}$, this equation is an eigenvalue equation for a Hermitian matrix. Consequently, the resulting wave vector eigenvalues $k_j = n_j k_0$ are real and the possibly complex polarisation eigenstates $\mathbf{E}_{0,j}$ are orthogonal. In the presence of linear birefringence, these special properties of the system of equations are maintained as the tensor $\overline{\overline{\epsilon}}_r$ is real symmetric and hence also Hermitian.

The field strengths and the flux densities are derived from the constitutive relations of Eq. (2.9.2) and from Maxwell's equations. From the eigenstates of the $\mathbf{E}_0$ system we obtain

$$\mathbf{B}_0 = \frac{\mathbf{k} \times \mathbf{E}_0}{\omega}, \tag{2.10.5}$$

$$\mathbf{H}_0 = \frac{\overline{\overline{\mu}}_r^{-1}}{\mu_0}\left(\frac{\mathbf{k} \times \mathbf{E}_0}{\omega} + i\,\overline{\overline{G}}\,\mathbf{E}_0\right), \tag{2.10.6}$$

$$\mathbf{D}_0 = \left\{\epsilon_0\,\overline{\overline{\epsilon}}_r - \frac{\overline{\overline{G}}\,\overline{\overline{\mu}}_r^{-1}\,\overline{\overline{G}}}{\mu_0}\right\}\mathbf{E}_0 + i\,\frac{\overline{\overline{G}}\,\overline{\overline{\mu}}_r^{-1}\,\mathbf{k} \times \mathbf{E}_0}{\mu_0 \omega}, \tag{2.10.7}$$

where $k = \omega / v_E$ is the wave vector eigenvalue associated with the solution of the eigenvalue equations for the electric field strength. In a comparable way, for a lossless medium for which the current density J is equal to zero, the field quantities derived from the $\mathbf{H}_0$ system are given by

$$\mathbf{D}_0 = -\frac{\mathbf{k} \times \mathbf{H}_0}{\omega}, \tag{2.10.8}$$

$$\mathbf{E}_0 = -\frac{\overline{\overline{\epsilon}}_r^{-1}}{\epsilon_0}\left(\frac{\mathbf{k} \times \mathbf{H}_0}{\omega} + i\,\overline{\overline{G}}\,\mathbf{H}_0\right), \tag{2.10.9}$$

$$\mathbf{B}_0 = \left\{\mu_0\,\overline{\overline{\mu}}_r - \frac{\overline{\overline{G}}\,\overline{\overline{\epsilon}}_r^{-1}\,\overline{\overline{G}}}{\epsilon_0}\right\}\mathbf{H}_0 + i\,\frac{\overline{\overline{G}}\,\overline{\overline{\epsilon}}_r^{-1}\,\mathbf{k} \times \mathbf{H}_0}{\epsilon_0 \omega}, \tag{2.10.10}$$

where the wave vector eigenvalue $k = \omega / v_H$ is now derived from the solution of the eigenvalue equations for $\mathbf{H}_0$.

2.10.2 Polarisation Eigenstate Calculation (Wave Vector Equation)

In this subsection we consider the expressions given in Eqs. (2.10.5)–(2.10.7) for the general case of an optically active electric/magnetic anisotropic medium. As is customary at optical frequencies, we suppose that the medium is magnetically isotropic and we use the eigenvalue equation written in terms of the electric field strength vector $\mathbf{E}_0$. Using the solutions for $\mathbf{E}_0$, the orthogonality of the solutions for the electromagnetic field vectors will be examined.

In Subsection 2.9 we used Eq. (2.9.4) to calculate the polarisation eigenstates of the electric field strength vector and the eigenvalues of the modulus of the wave vector $\mathbf{k}$ and this will again form the basis of our analysis here. From the length of the wave vector $\mathbf{k}$ we derive the wave propagation speed v using $v = \omega / |\mathbf{k}|$, and the refractive index value n for a polarisation eigenstate follows from $n = c/v$. In the case of a nonmagnetic medium with $\overline{\overline{\mu}}_r = \overline{\overline{I}}$, Eq. (2.9.4) reduces to

$$\mathbf{k} \times \mathbf{k} \times \mathbf{E}_0 + i\omega\left\{\mathbf{k} \times \overline{\overline{G}}\,\mathbf{E}_0 + \overline{\overline{G}}\,(\mathbf{k} \times \mathbf{E}_0)\right\} + \omega^2\left(\epsilon_0 \mu_0\,\overline{\overline{\epsilon}}_r - \overline{\overline{G}}\,\overline{\overline{G}}\right)\mathbf{E}_0 = \mathbf{0}. \tag{2.10.11}$$

Dividing the equation by $k_0^2 = \omega^2 \epsilon_0 \mu_0$ yields

$$n^2 \hat{\mathbf{k}} \times \hat{\mathbf{k}} \times \mathbf{E}_0 + icn\left\{\hat{\mathbf{k}} \times \overline{\overline{G}}\,\mathbf{E}_0 + \overline{\overline{G}}\,(\hat{\mathbf{k}} \times \mathbf{E}_0)\right\} + \left(\overline{\overline{\epsilon}}_r - c^2\,\overline{\overline{G}}\,\overline{\overline{G}}\right)\mathbf{E}_0 = \mathbf{0}. \tag{2.10.12}$$

Equation (2.10.12) can now be solved to find the eigenvalues n_j and the corresponding orthogonal polarisation eigenstates $\mathbf{E}_{0,j}$ where $j = 1, 2$. The magnetic induction vector then follows from Maxwell's equation (1.3.13) which, for a harmonic wave, reads

$$i\mathbf{k}_j \times \mathbf{E}_{0,j} = i\omega \mathbf{B}_{0,j},$$

or

$$n_j k_0\,\overline{\overline{T}}_{\hat{k}}\,\mathbf{E}_{0,j} = \omega \mathbf{B}_{0,j}, \tag{2.10.13}$$

where we have used the pseudotensor $\overline{\overline{T}}_{\hat{k}}$ notation for the vector product $\hat{\mathbf{k}}_j \times \mathbf{E}_{0,j}$ and the length of the wave vector $\mathbf{k}$ is given by $n_j k_0$. The other electromagnetic quantities follow from the constitutive relations of Eq. (2.9.2), yielding the following results:

$$\mathbf{B}_{0,j} = \frac{n_j \, \overline{\overline{T}}_{\hat{k}}}{c} \, \mathbf{E}_{0,j} \, ,$$

$$\mathbf{H}_{0,j} = \left(\frac{n_j \, \overline{\overline{T}}_{\hat{k}} + ic \, \overline{\overline{G}}}{c \, \mu_0} \right) \mathbf{E}_{0,j} \, ,$$

$$\mathbf{D}_{0,j} = \epsilon_0 \left(\overline{\overline{\epsilon}}_r - c^2 \, \overline{\overline{G}}\,\overline{\overline{G}} + icn_j \, \overline{\overline{G}}\,\overline{\overline{T}}_{\hat{k}} \right) \mathbf{E}_{0,j} \, . \tag{2.10.14}$$

We now return to Eq. (2.10.12) and obtain the following equation for an eigenstate of the electric field strength,

$$- n_j^2 \, \hat{\mathbf{k}} \times \hat{\mathbf{k}} \times \mathbf{E}_{0,j} - icn_j \, \hat{\mathbf{k}} \times \overline{\overline{G}} \, \mathbf{E}_{0,j} = icn_j \, \overline{\overline{G}} \, (\hat{\mathbf{k}} \times \mathbf{E}_{0,j}) + \left(\overline{\overline{\epsilon}}_r - c^2 \, \overline{\overline{G}}\,\overline{\overline{G}} \right) \mathbf{E}_{0,j} \, . \tag{2.10.15}$$

Taking the inner product of Eq. (2.10.15) and the unit wave vector $\hat{\mathbf{k}}$ yields zero for the left-hand side and we hence conclude that

$$\hat{\mathbf{k}} \cdot \epsilon_0 \left(\overline{\overline{\epsilon}}_r - c^2 \, \overline{\overline{G}}\,\overline{\overline{G}} + icn_j \, \overline{\overline{G}}\,\overline{\overline{T}}_{\hat{k}} \right) \mathbf{E}_{0,j} = 0 \, , \tag{2.10.16}$$

where we have again used the pseudotensor $\overline{\overline{T}}_{\hat{k}}$ for the vector product operation. Using the expression for $\mathbf{D}_{0,j}$ from Eq. (2.10.14), we conclude that

$$\hat{\mathbf{k}} \cdot \left(\overline{\overline{M}}_j \, (\hat{\mathbf{k}}) \, \mathbf{E}_{0,j} \right) = \hat{\mathbf{k}} \cdot \mathbf{D}_{0,j} = 0 \, . \tag{2.10.17}$$

$\overline{\overline{M}}_j \, (\hat{\mathbf{k}})$ can be considered as the complex dielectric tensor for a polarisation eigenstate j of the general anisotropic medium. This result extends the relationship $\hat{\mathbf{k}} \cdot \mathbf{D}_0$ of Eq. (2.2.20) that was valid for a medium with only linear anisotropy. A complicating factor for a medium with optical rotation is the $\hat{\mathbf{k}}$-dependence of the tensor $\overline{\overline{M}}_j$.

The tensor expressions occurring in Eq. (2.10.16) and (2.10.17) show that in the presence of an anisotropic medium with optical rotation the use of (complex) tensors is a convenient tool. The expressions for the electromagnetic field components can be written in terms of a tensor–vector product and the orthogonality of complex vectors is easily demonstrated.

2.11 Energy Propagation in a General Anisotropic Medium

In this section we first calculate the energy propagation vector $\langle \mathbf{S} \rangle$ in terms of the complex electric or magnetic eigenstates in an anisotropic medium including the effect of optical rotation. We then discuss the orthogonality properties of the electromagnetic vectors in the case of linear electric and/or magnetic anisotropy. In the final part of the section, for media with optical rotation, we establish a set of eigenvalue equations that uses the unit ray vector $\hat{\mathbf{s}} = \langle \mathbf{S} \rangle / |\langle \mathbf{S} \rangle|$ as input direction instead of the wave vector $\hat{\mathbf{k}}$.

2.11.1 Expression for the Poynting Vector $\langle \mathbf{S} \rangle$

For harmonic plane waves in an anisotropic medium, the magnitude and direction of energy propagation is given by the average value of the Poynting vector $\langle \mathbf{S} \rangle = \Re\{\mathbf{E}_0 \times \mathbf{H}_0^*\}/2$ (see Eq. (2.5.1)). The Poynting vector can be calculated from either the $\mathbf{E}_0$ system of equations or from the $\mathbf{H}_0$ system, taking each time the proper eigenvalues $k_j = \omega / v_j$ of the wave vector and the corresponding (complex) polarisation eigenstates $\mathbf{E}_{0,j}$ or $\mathbf{H}_{0,j}$. We write the complex electric and magnetic eigenstates in a medium allowing for elliptical polarisation eigenstates as

$$\mathbf{E}_{0,j} = \mathbf{E}_{0,j}^R + i\mathbf{E}_{0,j}^I \, , \qquad \mathbf{H}_{0,j} = \mathbf{H}_{0,j}^R + i\mathbf{H}_{0,j}^I \, , \tag{2.11.1}$$

where $j = 1, 2$. After some rearrangement we find that the average value of the Poynting vector is given by the two different, albeit equivalent, expressions

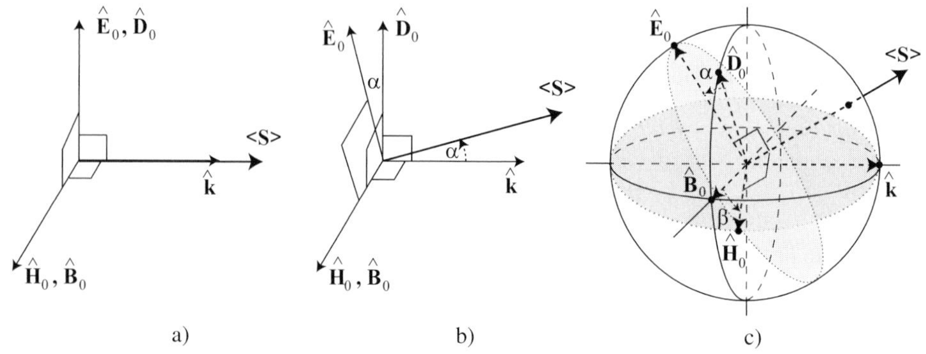

Figure 2.24: The field vectors $\mathbf{E}_0$ and $\mathbf{H}_0$, the flux density vectors $\mathbf{D}_0$ and $\mathbf{B}_0$ and the Poynting vector **S** for various optical media (unit vectors).
a) Isotropic medium, $\bar{\bar{\epsilon}}_r = n^2\bar{\bar{I}}, \bar{\bar{\mu}}_r=\bar{\bar{I}}$.
b) Linearly anisotropic, nonmagnetic medium with permittivity tensor $\bar{\bar{\epsilon}}_r$ and $\bar{\bar{\mu}}_r=\bar{\bar{I}}$.
c) Linearly anisotropic electric and magnetic medium with permittivity tensor $\bar{\bar{\epsilon}}_r$ and magnetic permeability tensor $\bar{\bar{\mu}}_r$.

$$\begin{aligned}\langle \mathbf{S}_j \rangle = {} & \frac{1}{2}\mathbf{E}^R_{0,j} \times \frac{\bar{\bar{\mu}}_r^{-1}}{\mu_0}\left[\hat{\mathbf{k}} \times \mathbf{E}^R_{0,j}/v_{E,j} - \bar{\bar{G}}\,\mathbf{E}^I_{0,j}\right] \\ & + \frac{1}{2}\mathbf{E}^I_{0,j} \times \frac{\bar{\bar{\mu}}_r^{-1}}{\mu_0}\left[\hat{\mathbf{k}} \times \mathbf{E}^I_{0,j}/v_{E,j} + \bar{\bar{G}}\,\mathbf{E}^R_{0,j}\right], \end{aligned} \tag{2.11.2}$$

$$\begin{aligned}\langle \mathbf{S}_j \rangle = {} & \frac{1}{2}\mathbf{H}^R_{0,j} \times \frac{\bar{\bar{\epsilon}}_r^{-1}}{\epsilon_0}\left[\hat{\mathbf{k}} \times \mathbf{H}^R_{0,j}/v_{H,j} - \bar{\bar{G}}\,\mathbf{H}^I_{0,j}\right] \\ & + \frac{1}{2}\mathbf{H}^I_{0,j} \times \frac{\bar{\bar{\epsilon}}_r^{-1}}{\epsilon_0}\left[\hat{\mathbf{k}} \times \mathbf{H}^I_{0,j}/v_{H,j} + \bar{\bar{G}}\,\mathbf{H}^R_{0,j}\right], \end{aligned} \tag{2.11.3}$$

where we have used the unit wave vector $\hat{\mathbf{k}} = v_j\mathbf{k}_j/\omega$ and the eigenvalue $v_j = v_{E,j}$ or $v_{H,j}$ for each system of equations.

2.11.2 Properties of the Electromagnetic Field Vectors

In Fig. 2.24 we sketch the electromagnetic field, flux density and Poynting unit vectors for various types of media. The unit wave vector $\hat{\mathbf{k}}$ that has been used when solving the eigenvalue systems for the electric and magnetic field strengths serves as a reference direction. In Fig. 2.24a the medium is both electrically and magnetically isotropic. In the isotropic case, the eigenstates for the electric and magnetic field vector are linearly polarised, perpendicular to the wave vector **k**. From Eq. (2.10.5) we conclude that $\mathbf{B}_0$ is perpendicular to **k** and $\mathbf{E}_0$. From Eq. (2.10.6) and (2.10.10) it follows that $\mathbf{H}_0$ is parallel to $\mathbf{B}_0$; the parallelism of $\mathbf{D}_0$ and $\mathbf{E}_0$ follows from Eq. (2.10.7). Finally, Eq. (2.10.8) requires that $\mathbf{D}_0$ is perpendicular to **k** and $\mathbf{H}_0$. In the isotropic case, Eq. (2.10.9) confirms the result of Eq. (2.10.7). The time-averaged Poynting vector $\langle \mathbf{S} \rangle$ is parallel to the wave vector **k**. Three orthogonal right-handed vector triplets are present in the isotropic case, ($\mathbf{E}_0$,$\mathbf{B}_0$,**k**), ($\mathbf{D}_0$,$\mathbf{H}_0$,**k**,) and ($\mathbf{E}_0$,$\mathbf{H}_0$, $\langle \mathbf{S} \rangle$). Consequently, we have that the vector sets ($\mathbf{E}_0$,$\mathbf{D}_0$,$\mathbf{H}_0$,$\mathbf{B}_0$), ($\mathbf{B}_0$,$\mathbf{H}_0$,**k**,$\langle \mathbf{S} \rangle$) and ($\mathbf{E}_0$,$\mathbf{D}_0$,**k**,$\langle \mathbf{S} \rangle$) are each coplanar.

The various unit vectors for an electrically anisotropic medium free of optical activity are represented in Fig. 2.24b. We note that the eigenstates are linear in this case. According to Eq. (2.10.7) the vectors $\mathbf{E}_0$ and $\mathbf{D}_0$ are not parallel. The scalar product of the corresponding unit vectors equals $\cos\alpha$. The vectors $\mathbf{B}_0$ and $\mathbf{H}_0$ remain parallel to each other. Equations (2.10.5) and (2.10.8) then require that the vectors ($\mathbf{D}_0$,$\mathbf{H}_0$,**k**) form a right-handed orthogonal triplet. The triplet ($\mathbf{E}_0$,$\mathbf{B}_0$,**k**) is no longer orthogonal. Evaluation of the scalar product of $\hat{\mathbf{k}}$ with the unit Poynting vector shows that the angle between $\langle \mathbf{S} \rangle$ and $\hat{\mathbf{k}}$ is α. The vectors ($\mathbf{E}_0$,$\mathbf{H}_0$,$\langle \mathbf{S} \rangle$) also form an orthogonal triplet. The vector sets ($\mathbf{E}_0$,$\mathbf{H}_0$,$\mathbf{B}_0$), ($\mathbf{D}_0$,$\mathbf{H}_0$,$\mathbf{B}_0$), ($\mathbf{B}_0$,$\mathbf{H}_0$,**k**) and ($\mathbf{E}_0$,$\mathbf{D}_0$,**k**,$\langle \mathbf{S} \rangle$) are coplanar.

The case of a medium with electric and magnetic anisotropy without optical rotation is represented in Fig. 2.24c. Equations (2.10.5) and (2.10.8) impose the orthogonality of $\mathbf{B}_0$ with respect to $\mathbf{k}$ and $\mathbf{E}_0$ and of $\mathbf{D}_0$ with respect to $\mathbf{k}$ and $\mathbf{H}_0$. In the figure $\mathbf{H}_0$ is located in the grey-shaded plane through $\mathbf{k}$ that is perpendicular to $\mathbf{D}_0$ and that has a dotted intersection curve with the unit sphere. The angle between $\mathbf{B}_0$ and $\mathbf{H}_0$ is β, while α denotes the angle between $\mathbf{D}_0$ and $\mathbf{E}_0$. The only orthogonal vector triplet for this case is formed by the vectors $(\mathbf{E}, \mathbf{H}, \hat{\mathbf{s}})$ as the Poynting vector in the medium is still defined by $\langle \mathbf{S} \rangle \sim \mathbf{E} \times \mathbf{H}$. In the combined case of linear and circular anisotropy, a geometrical representation of the $\mathbf{E}_0$ and $\mathbf{H}_0$ eigenvectors is not possible as the eigenvectors become complex. A separate treatment of their real and imaginary parts is still possible. In the absence of linear anisotropy the expressions (2.10.5)–(2.10.10) show that the effect of the circular anisotropy (optical rotation) is found exclusively in the imaginary part of the eigenvectors.

2.11.3 Eigenvalue Equations with Prescribed Unit Ray Vector $\hat{\mathbf{s}}$

To study the energy propagation in a general anisotropic medium we need to know the energy propagation speed and the electromagnetic polarisation eigenstates. In the most restricted case, it is only the energy propagation direction in the anisotropic medium that is given, for instance by means of the Poynting vector or, in terms of geometrical optics, the unit ray vector $\hat{\mathbf{s}}$. To find the two possible wave vectors $\mathbf{k}_j$ associated with the ray vector $\hat{\mathbf{s}}$ and the corresponding polarisation eigenstates, the eigenvalue equation in terms of the wave vector $\mathbf{k}$ and the electric field strength amplitude vector $\mathbf{E}_0$ needs to be transformed into an equation containing the ray vector $\hat{\mathbf{s}}$ and the electric flux density vector $\hat{\mathbf{D}}_0$. For a linear anisotropic medium, this transformation has been discussed in Subsection 2.6; however, here we extend our analysis to a medium with optical rotation for which the electric eigenstates are elliptical.

2.11.3.1 Construction of the $\hat{\mathbf{s}}$-based Eigenvalue Equations

From Eq. (2.10.3) it followed that the matrix of the eigenvalue equation for the electric field strength is Hermitian in a non-absorbing medium that is magnetically isotropic but electrically anisotropic. The Hermitian nature of the matrix implies that the eigenvalues are real and that the eigenvectors are orthogonal complex vectors. For a nonmagnetic medium the magnetic field strength vector and the magnetic flux density vector are parallel. This property links the two orthogonal vector triplets $(\hat{\mathbf{D}}_0, \hat{\mathbf{B}}_0, \hat{\mathbf{k}})$ and $(\hat{\mathbf{E}}_0, \hat{\mathbf{H}}_0, \hat{\mathbf{s}})$ via their common axis along ($\hat{\mathbf{H}}_0$ and $\hat{\mathbf{B}}_0$). The electric field strength of a resulting polarisation eigenstate is found in the plane perpendicular to the ray vector $\hat{\mathbf{s}}$, and the electric flux density vector is found in a plane perpendicular to the unit wave vector $\hat{\mathbf{k}}$. Any complex vector in a plane can be written as the sum of two orthogonal unit vectors. For instance, in the case of an eigenstate j we can write

$$\mathbf{E}_{0,j} = q_{1,j}\hat{\mathbf{E}}_1 + q_{2,j}\hat{\mathbf{E}}_2 \, , \tag{2.11.4}$$

where $\hat{\mathbf{E}}_1 \cdot \hat{\mathbf{E}}_2 = 0$ and $q_{1,j}$ and $q_{2,j}$ are complex quantities with dimension Vm^{-1}. The corresponding electric flux density vector can similarly be expressed in terms of any pair of orthogonal unit vectors in the plane perpendicular to $\hat{\mathbf{k}}'_j$. Figure 2.25 shows the orientation of the ray vector, the corresponding wave vectors and the eigenvectors $\hat{\mathbf{E}}'_j$ for the electric field strength and $\hat{\mathbf{D}}'_j$ the electric flux density. The energy propagation direction, given by the ray vector $\hat{\mathbf{s}}'$ is chosen to be in the Z'-direction as can be achieved using an appropriate rotation of the (X, Y, Z) coordinate system. All quantities in the rotated coordinate system are denoted by primed variables. $\hat{\mathbf{k}}'_1$ and $\hat{\mathbf{k}}'_2$ are the two wave vectors associated with the ray vector $\hat{\mathbf{s}}'$. A wave vector $\hat{\mathbf{k}}'_j$ is obtained from the ray vector $\hat{\mathbf{s}}' = (0, 0, 1)$ by means of a rotation in the (X', Y', Z') system through an angle α_j in the plane containing $\hat{\mathbf{s}}'$ and the eigenvector $\hat{\mathbf{E}}'_j$. ϕ_1 is the angle between the plane $\mathcal{P}_1$ containing $\hat{\mathbf{s}}'$ and $\hat{\mathbf{k}}'_1$ and the plane $Y' = 0$.

Equation (2.10.12) is the Helmholtz equation in terms of the electric field vector $\mathbf{E}_0$ for an optically active medium. The evaluation of the second term on the left-hand side of (2.10.12) requires the vector $\overset{=}{G}{}'\mathbf{E}'_0$ in the rotated system. The figure shows this vector and the vector $\hat{\mathbf{k}}'_1 \times \overset{=}{G}{}' \, \hat{\mathbf{E}}'_1$, which can be considered as a first-order perturbation term of $\mathbf{E}'_0$ due to the presence of optical activity in the medium. The component $\hat{\mathbf{k}}'_2 \times \overset{=}{G}{}' \, \hat{\mathbf{E}}'_2$ which stems from the unit vector $\hat{\mathbf{E}}'_2$ is not shown in the figure. Both $\hat{\mathbf{k}}'_1 \times \overset{=}{G}{}' \, \hat{\mathbf{E}}'_1$ and $\hat{\mathbf{k}}'_1 \times \overset{=}{G}{}' \, \hat{\mathbf{E}}'_2$ are located in the plane containing the base vectors $(\hat{\mathbf{D}}'_1, \hat{\mathbf{D}}'_2)$.

The wave propagation vector related to a polarisation eigenstate j is denoted by $\hat{\mathbf{k}}'_j$. For the $\mathbf{E}'$ and $\mathbf{D}'$ unit vectors we write

$$\begin{aligned} \hat{\mathbf{E}}'_1 &= (\cos\phi_1, \sin\phi_1, 0) \, , & \hat{\mathbf{D}}'_1 &= (\cos\alpha_1 \cos\phi_1, \cos\alpha_1 \sin\phi_1, \sin\alpha_1) \, , \\ \hat{\mathbf{E}}'_2 &= (\sin\phi_1, -\cos\phi_1, 0) \, , & \hat{\mathbf{D}}'_2 &= (\cos\alpha_2 \sin\phi_1, -\cos\alpha_2 \cos\phi_1, \sin\alpha_2) \, . \end{aligned} \tag{2.11.5}$$

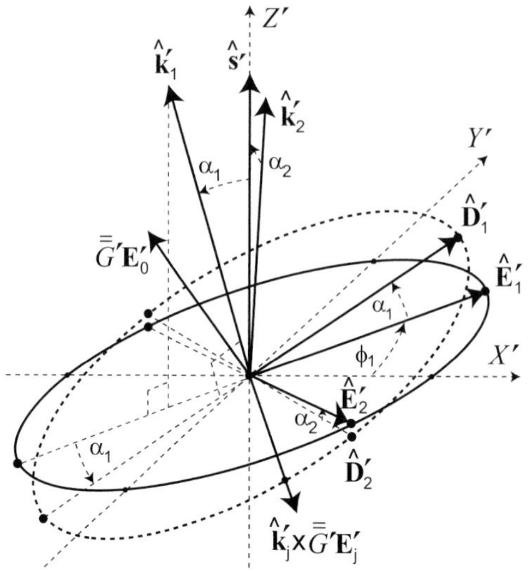

Figure 2.25: The construction of the two unit wave vectors $\hat{\mathbf{k}}'_j$ from a unit ray vector $\hat{\mathbf{s}}'$ in an anisotropic medium with optical rotation. The orthogonal unit vectors $\hat{\mathbf{E}}'_1$ and $\hat{\mathbf{E}}'_2$, perpendicular to $\hat{\mathbf{s}}'$ and located in the plane $Z' = 0$, span a complex electrical polarisation eigenstate. The orthogonal unit vectors $\hat{\mathbf{D}}'_1$ and $\hat{\mathbf{D}}'_2$ are the base vectors for describing the complex amplitude of the complex flux density eigenstate.

The rotation angles ϕ_1 and α_1, α_2 are initially unknown and their values must be found from solution of the eigenvalue problem.

To solve the eigenvalue problem for a medium exhibiting optical rotation we consider the optical rotation as a small deviation from a known solution without optical rotation. As base vectors for the electric field strength and for the electric flux density we thus provisionally use the linear eigenstates that appeared when solving the eigenvalue equations for a medium with linear anisotropy only. This amounts to equating the rotation tensor $\bar{\bar{G}}'$ to the zero tensor when solving Eq. (2.10.12). The resulting base vectors are indicated in Fig. 2.25 by the sets $(\hat{\mathbf{E}}'_1, \hat{\mathbf{E}}'_2)$ and $(\hat{\mathbf{D}}'_1, \hat{\mathbf{D}}'_2)$.

The eigenvalue equation for the general anisotropic medium is given by Eq. (2.10.12) with the rotation part $\bar{\bar{G}}' \neq \bar{\bar{0}}$ inserted in the equation. Using the expression (2.10.16) for the electric flux density $\mathbf{D}'_0$ in a medium with optical rotation, we obtain the equation

$$n^2\hat{\mathbf{k}}' \times \hat{\mathbf{k}}' \times \mathbf{E}'_0 + icn\hat{\mathbf{k}}' \times \bar{\bar{G}}'\, \mathbf{E}'_0 + \bar{\bar{M}}'\, \mathbf{E}'_0 = \mathbf{0}\,, \tag{2.11.6}$$

where $\mathbf{D}'_0 = \epsilon_0 \bar{\bar{M}}'\, \mathbf{E}'_0$ in the primed coordinate system. To eliminate the wave vector and the electric field strength vector from Eq. (2.11.6) we use the following identities:

$$\hat{\mathbf{k}}' \times \hat{\mathbf{k}}' \times \mathbf{E}'_0 = \hat{\mathbf{k}}' \times \hat{\mathbf{k}}' \times \left(q_1\hat{\mathbf{E}}'_1 + q_2\hat{\mathbf{E}}'_2\right) = -q_1 \cos\alpha\hat{\mathbf{D}}'_1 - q_2\hat{\mathbf{D}}'_2\,. \tag{2.11.7}$$

It also follows directly from Fig. 2.25 that

$$\hat{\mathbf{E}}'_1 = -\frac{\hat{\mathbf{s}}' \times \hat{\mathbf{s}}' \times \hat{\mathbf{D}}'_1}{\cos\alpha_1}\,,$$

$$\hat{\mathbf{E}}'_2 = -\,\hat{\mathbf{s}}' \times \hat{\mathbf{s}}' \times \hat{\mathbf{D}}'_2\,. \tag{2.11.8}$$

Insertion of these two equations into Eq. (2.11.6) yields the following expression:

$$n^2 q_1 \cos\alpha_1\hat{\mathbf{D}}'_1 - n^2 q_2\hat{\mathbf{D}}'_2 + \left[ic(R\,\hat{\mathbf{s}}') \times \bar{\bar{G}}' + \bar{\bar{M}}'\,(R\,\hat{\mathbf{s}}')\right] \times \left\{\hat{\mathbf{s}}' \times \hat{\mathbf{s}}' \times \left(\frac{q_1}{\cos\alpha_1}\hat{\mathbf{D}}'_1 + q_2\hat{\mathbf{D}}'_2\right)\right\} \approx \mathbf{0}\,, \tag{2.11.9}$$

where R is the rotation matrix defined by $R\hat{\mathbf{s}}' = \hat{\mathbf{k}}'$. The $\approx$ sign has been used in Eq. (2.11.9) because the vectors $\hat{\mathbf{D}}'_1$, $\hat{\mathbf{D}}'_2$, the angle α_1 and the value of n are still those that correspond to the rotation-free medium.

We denote the quantity $\left[ic(R\hat{\mathbf{s}}')\times \overline{\overline{G}}' + \overline{\overline{M}}'\,(R\hat{\mathbf{s}}')\right]$ by $[\overline{\overline{\Xi}}'\,(\hat{\mathbf{s}})]$ and multiply Eq. (2.11.9) by $[\overline{\overline{\Xi}}'\,(\hat{\mathbf{s}}')]^{-1}$. After some rearrangement, the eigenvalue equation reads

$$
\begin{aligned}
&q_{1,r}\cos\alpha_r\left(\frac{1}{n_r^2\cos^2\alpha_r}\,\hat{\mathbf{s}}'\times\hat{\mathbf{s}}'\times\hat{\mathbf{D}}'_{1,r} + \left[\overline{\overline{\Xi}}'\,(\hat{\mathbf{s}})\right]^{-1}\hat{\mathbf{D}}'_{1,r}\right)\\
&+ q_{2,r}\left(\frac{1}{n_r^2}\,\hat{\mathbf{s}}'\times\hat{\mathbf{s}}'\times\hat{\mathbf{D}}'_{2,r} + \left[\overline{\overline{\Xi}}'\,(\hat{\mathbf{s}})\right]^{-1}\hat{\mathbf{D}}'_{2,r}\right) = \mathbf{0}\,,
\end{aligned}
\qquad (2.11.10)
$$

where we have added a subscript r to all quantities that are modified because the medium is now considered to be gyrotropic.

2.11.3.2 Discussion of the Solution of the $\hat{\mathbf{s}}'$-based Eigenvalue Equations

We briefly discuss the solution of the general eigenvalue equation above; firstly for the special case of a linearly anisotropic medium and, secondly, for the general case of a medium that also exhibits optical rotation:

- *Linear anisotropy*
  For the case of a medium with linear anisotropy, we have already outlined the method by which the $\hat{\mathbf{s}}'$-based eigenvalue equations can be solved in Subsection 2.6. In Eq. (2.11.10) the case of linear anisotropy arises if the tensor $\overline{\overline{\Xi}}'$ and its inverse are real. It was shown that two orthogonal eigenvectors $\hat{\mathbf{D}}'_{1,s}$ and $\hat{\mathbf{D}}'_{2,s}$ can be found with the corresponding unit vectors $\hat{\mathbf{E}}'_{1,s}$ and $\hat{\mathbf{E}}'_{2,s}$ lying in the planes defined by the vectors $(\hat{\mathbf{s}}', \hat{\mathbf{D}}'_{1,s})$ and $(\hat{\mathbf{s}}', \hat{\mathbf{D}}'_{2,s})$, respectively. The solution vector $\hat{\mathbf{D}}'_{1,s}$ associated with the first term on the left-hand side of Eq. (2.11.10) solves the entire equation when $q_2 = 0$. This leads to a linear polarisation eigenstate for which the speed of energy propagation is characterised by the refractive index $n_s\cos\alpha$. The second eigenstate $\hat{\mathbf{D}}'_{2,s}$, obtained when $q_1 = 0$ in Eq. (2.11.10), is also a linear polarisation state, perpendicular to $\hat{\mathbf{D}}'_{1,s}$ and its propagation speed is given by the ratio c/n_s.
- *Anisotropy including optical rotation*
  Uncoupling of the first and the second term of the left-hand member of Eq. (2.11.10) is not possible in the presence of optical rotation. The action of the tensor $(\overline{\overline{\Xi}}')^{-1}$ on a vector $\hat{\mathbf{D}}'$ involves vector products of the vectors $\hat{\mathbf{k}}'$ and $\hat{\mathbf{D}}'$. Consequently, a vector $(\overline{\overline{\Xi}}')^{-1}\hat{\mathbf{D}}'$ always contains a complex-valued component that is perpendicular to $\hat{\mathbf{D}}'$. This leads to a coupled solution involving both terms in the left-hand side of Eq. (2.11.10). Such a solution comprises a linear combination of two complex orthogonal vectors, denoted here by $\hat{\mathbf{D}}'_{1,r}$ and $\hat{\mathbf{D}}'_{2,r}$. In the case that the medium has a small rotation effect, these vectors will only be slightly different from the solution that would be found for an anisotropic medium without optical rotation. The two possible orthogonal polarisation eigenstates in a general medium including optical rotation are given by the complex unit vectors

$$
\begin{aligned}
\hat{\mathbf{D}}'_{1,r} &= a\,(\cos\alpha\sin\phi, \cos\alpha\cos\phi, \sin\alpha) + i\sqrt{1-a^2}\,(\cos\phi, -\sin\phi, 0)\,,\\
\hat{\mathbf{D}}'_{2,r} &= \sqrt{1-a^2}\,(\cos\alpha\sin\phi, \cos\alpha\cos\phi, \sin\alpha) - ia\,(\cos\phi, -\sin\phi, 0)\,,
\end{aligned}
\qquad (2.11.11)
$$

  where a $(0 \le a \le 1)$ is a parameter that determines the ellipticity of the two, generally elliptical, eigenstates of the solution.

In many practical cases, the change in the polarisation eigenvectors due to the presence of optical rotation is barely noticeable. For common crystals the elements of the tensor $c\,\overline{\overline{G}}$ are several orders of magnitude smaller than the elements of the permittivity tensor. As an example, in the case of crystal quartz, the elements cg_{ij} are typically of the order of 10^{-5} and the change in the elements of the modified permittivity tensor, given by $c^2\,\overline{\overline{G}}\,\overline{\overline{G}}$, can be neglected. Consequently, when light is sent through crystal quartz in a direction with a large linear anisotropy, the polarisation eigenstates will be elliptical but with a very large eccentricity which is hardly distinguishable from the linear eigenstates that coincide with the major axes of the ellipses. It is only close to the direction of the optical axis in quartz that the typical eigenstates due to optical rotation become detectable. It was demonstrated earlier in Subsection 2.9.3 that, in the direction of the optical axis of quartz, circular polarisation eigenstates are found.

- *Numerical solution of the eigenvalue equation* (2.11.10)
The solution of the eigenvalue equation, with the ray vector $\hat{\mathbf{s}}'$ as input direction, can be carried out in two steps. In the first step, only the real part of the complex tensor $(\overline{\overline{\Xi}}'_r)^{-1}$ is included in the calculation. Equations (2.10.16), (2.10.17) and the definition of $\overline{\overline{\Xi}}'_r$ show that the real part of $(\overline{\overline{\Xi}}'_r)^{-1}$ is equal to $(\overline{\overline{\epsilon}}'_r - c^2\,\overline{\overline{G}}'\,\overline{\overline{G}}')^{-1}$, where $\overline{\overline{\epsilon}}'_r$ is the relative permittivity tensor associated with the linear anisotropy of the medium. Solving each of the equations

$$\frac{1}{n^2\cos^2\alpha}\,\hat{\mathbf{s}}'\times\hat{\mathbf{s}}'\times\hat{\mathbf{D}}'_1 + \left[\overline{\overline{\Xi}}'_r(\hat{\mathbf{s}})\right]^{-1}\hat{\mathbf{D}}'_1 = \mathbf{0}\,, \tag{2.11.12}$$

$$\frac{1}{n^2}\,\hat{\mathbf{s}}'\times\hat{\mathbf{s}}'\times\hat{\mathbf{D}}'_2 + \left[\overline{\overline{\Xi}}'_r(\hat{\mathbf{s}})\right]^{-1}\hat{\mathbf{D}}'_2 = \mathbf{0}\,, \tag{2.11.13}$$

separately gives rise to the real eigenvectors $(\hat{\mathbf{D}}'_{1,a},\hat{\mathbf{D}}'_{2,a})$ and to the approximate eigenvalues $n_a\cos\alpha_a$ and n_a. The angle ϕ_a is the rotation angle between the originally chosen base vector $\hat{\mathbf{D}}'_1$ and the approximated eigenvector $\hat{\mathbf{D}}'_{1,a}$.

In the second step we calculate the wave vector $\hat{\mathbf{k}}'_a$ from the angles ϕ_a and α_a that determine the transformation from $\hat{\mathbf{s}}'$ to $\hat{\mathbf{k}}'_a$. The approximate value for the wave vector $\hat{\mathbf{k}}'_a$ that followed from the solution of the eigenvalue equations using the real-valued tensor is then used to calculate the complex valued tensor $(\overline{\overline{\Xi}}')^{-1}$. In an iterative procedure, with a continuous update of the real eigenvalues and the complex eigenvectors, we then arrive at the eigenstates of the anisotropic medium with the effect of optical rotation included.

2.12 Reflection and Refraction at an Interface in Anisotropic Media

We conclude this chapter on wave propagation in anisotropic media with an analysis of the reflection and refraction of a plane wave at the interface between two anisotropic media, characterised by different permittivity and gyrotropic tensors, $(\overline{\overline{\epsilon}}_{r,1},\overline{\overline{G}}_1)$ and $(\overline{\overline{\epsilon}}_{r,2},\overline{\overline{G}}_2)$, respectively. The first step is to find the wave vectors of the reflected and transmitted waves in the first and in the second medium. The second step is to establish the amplitudes of the electric and magnetic field strengths in the two media and the direction and strength of the forward and backward propagating energy at the boundary between the two media.

2.12.1 Construction of the Ray Vectors at an Isotropic-to-Anisotropic Interface

Before we address the general problem of the reflection and transmission of a plane wave at an interface between two anisotropic media we discuss in this subsection a simple graphical method to find the energy propagation directions and the wavefronts in an anisotropic medium when a plane wave is (normally) incident on a planar interface that separates an isotropic from an anisotropic medium. The graphical method is illustrated in Fig. 2.26 and is based on the construction of the ray vector $\hat{\mathbf{s}}_2$ with the aid of the wave vector $\hat{\mathbf{k}}_0$ in Fig. 2.21. The surface $v_{re}(\hat{\mathbf{s}})$ defines up to what point in space the energy transported by a plane wave with ray vector $\hat{\mathbf{s}}_e$ has progressed after a certain time t_0. The surface $v_{ro}(\hat{\mathbf{s}})$ defines a comparable position in space for an ordinary plane wave with ray vector $\hat{\mathbf{s}}_o$. As the optical axes of the biaxial medium are in the plane of the drawing, we have a special case where one ray surface is spherical, corresponding to ordinarily propagating plane waves. Since the plane wave from the isotropic medium is incident normally to the interface, transmission into the anisotropic medium leaves the unit wave vector $\hat{\mathbf{k}}_0$ unaltered. To find the direction of the electric flux density vector $\mathbf{D}_{0,e}$ of the extraordinary wave we proceed as in Fig. 2.17 and construct the tangent plane to the surface $v_{re}(\hat{\mathbf{s}})$ that is normal to the wave vector $\hat{\mathbf{k}}_0$. The vector $\mathbf{D}_{0,e}$ is normal to the wave vector $\hat{\mathbf{k}}_0$. The line which joins O with the tangent point C determines the ray vector $\hat{\mathbf{s}}_e$ of the extraordinary wave. For the ordinary wave we find ray vector $\hat{\mathbf{s}}_o$ is parallel to the wave vector $\hat{\mathbf{k}}_0$. The electric field strength vector $\mathbf{E}_{0,e}$ of the extraordinary wave is tangent to the optical indicatrix (not shown in the figure). $\mathbf{E}_{0,e}$ is also perpendicular to the ray vector $\hat{\mathbf{s}}_e$ and coplanar with the vectors $\hat{\mathbf{k}}_0$ and $\mathbf{D}_{0,e}$. The ordinary ray vector $\hat{\mathbf{s}}_o$ is parallel to the wave vector $\hat{\mathbf{k}}_0$ and the two electric vectors $\mathbf{D}_{0,o}$ and $\mathbf{E}_{0,o}$ are parallel to each other and perpendicular to the principal section. In the geometry of Fig. 2.26 this is the plane of the drawing which contains the optical axis OP and the unit wave vector $\hat{\mathbf{k}}_0$. The wavefronts in the anisotropic medium have been denoted by W_e and W_o.

The geometrical construction discussed above is instructive when the anisotropic medium is uniaxial or when the refracted wave vector lies in one of the principal sections of a biaxial anisotropic medium. The general case of wave

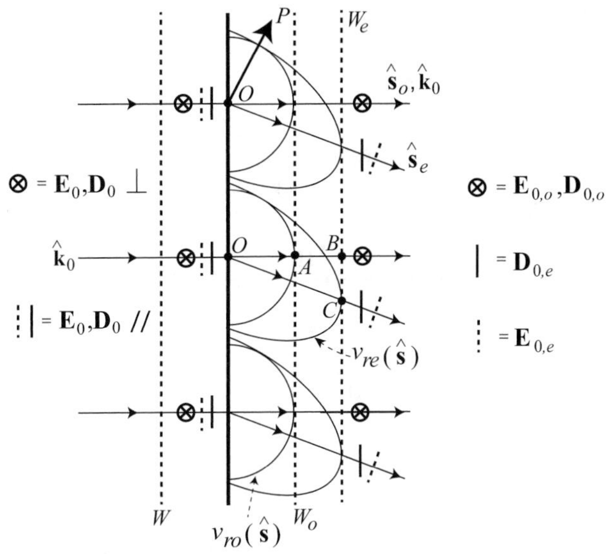

Figure 2.26: Construction of the energy flow directions (ray vectors $\hat{\mathbf{s}}_o$ and $\hat{\mathbf{s}}_e$) and wavefronts W in an anisotropic medium. The waves in the anisotropic medium are generated by the plane wave with wave vector $\hat{\mathbf{k}}_0$ which is normally incident on a surface that separates an isotropic medium from the (biaxial) anisotropic medium on the right-hand side. The optical axes are in the plane of the drawing. This plane coincides with the principal section which contains the wave vector $\hat{\mathbf{k}}_0$ and the optical axis. Two states of polarisation are considered, the extraordinary state which is parallel to the principal section and the ordinary state which is perpendicular to it. The electric field vector is represented by a dashed line, the electric flux density vector by a solid line.

propagation in a biaxial medium is more complicated. In this case, both ray vectors become extraordinary in the sense that neither of them is parallel to the wave vector. Furthermore the two ray vectors and the wave vector are no longer coplanar. In this case, the geometrical construction needs to be carried out in three dimensions (see Fig. 2.5).

2.12.2 Propagation Direction of Reflected and Transmitted Waves

In Fig. 2.27 an incident wave with wave vector $\mathbf{k}_i$ has been depicted, corresponding to a propagation eigenstate of the first medium with the associated vectors $\mathbf{E}_{0,i}$ and $\mathbf{H}_{0,i}$. The continuity of the tangential components of the electric and magnetic field strength at the boundary between the two media requires identical tangential components of the wave vectors of the reflected and of the transmitted fields. For the isotropic case these conditions have already been established in Chapter 1 (see Eq. (1.8.20)). For the general case of two anisotropic media presented in Fig. 2.27, we can write the conditions

$$\begin{aligned} \mathbf{k}_i - (\mathbf{k}_i \cdot \hat{\mathbf{v}})\,\hat{\mathbf{v}} &= \mathbf{k}_r - (\mathbf{k}_r \cdot \hat{\mathbf{v}})\,\hat{\mathbf{v}}\,, \\ \mathbf{k}_i - (\mathbf{k}_i \cdot \hat{\mathbf{v}})\,\hat{\mathbf{v}} &= \mathbf{k}_t - (\mathbf{k}_t \cdot \hat{\mathbf{v}})\,\hat{\mathbf{v}}\,. \end{aligned} \tag{2.12.1}$$

The length of the reflected or transmitted wave vector now depends on its direction and an analytic solution is only possible in some very specific cases, for instance for a linear anisotropic uniaxial medium. In general Eq. (2.12.1) can only be solved by numerical means, starting with some reasonable first guess, for instance the isotropic wave vector solution that corresponds to the average refractive index of the anisotropic medium. The first trial solution is used as the input for the set of eigenvalue equations for the electric or the magnetic field strength (see Eqs. (2.9.3) and (2.9.4)). As was shown in Section 2.3 for linear anisotropic media, four solutions arise when solving the eigenvalue equations. For a material which exhibits linear and/or circular birefringence, the matrix of the eigenvalue equations is (complex) Hermitian and, consequently, its eigenvalues k^2 are real. The four solutions correspond to forward and backward propagating waves along two different wave directions in each medium. The length of the wave vector is expressed in terms of the refractive

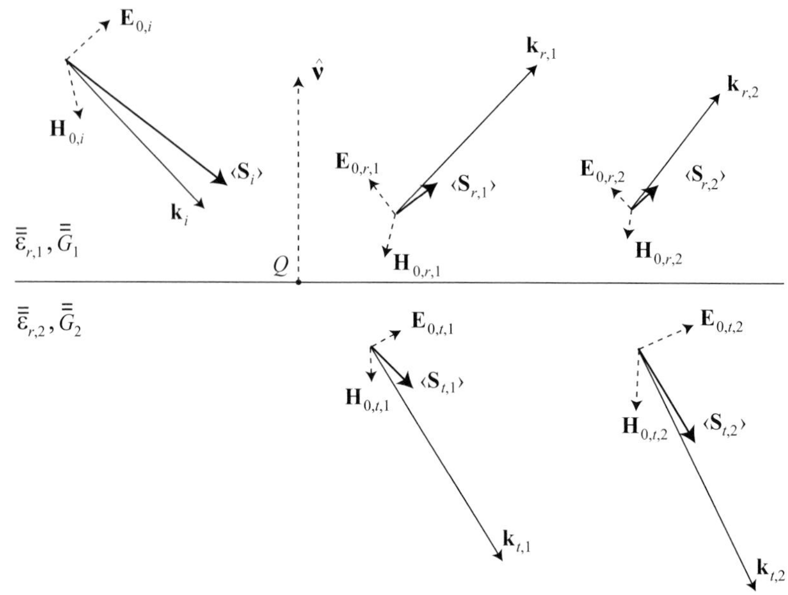

Figure 2.27: The incident, reflected and transmitted wave vectors and field strengths on both sides of an interface between two anisotropic media. The unit normal vector to the interface is $\hat{\mathbf{v}}$. The two media are characterised by their permittivity and gyrotropic tensors, $\bar{\bar{\epsilon}}_{r,1}$, $\bar{\bar{G}}_1$ for the first medium and $\bar{\bar{\epsilon}}_{r,2}$, $\bar{\bar{G}}_2$ for the second medium. The incident wave on the left-hand side corresponds to an eigenstate of the first medium, characterised by the wave vector $\mathbf{k}_i$ and the Poynting vector $\langle \mathbf{S}_i \rangle$. The transmitted power is found in two propagation directions given by $\langle \mathbf{S}_{t,1} \rangle$ and $\langle \mathbf{S}_{t,2} \rangle$ that are connected to the two solutions for the wave vector of the transmitted electromagnetic field, $\mathbf{k}_{t,1}$ and $\mathbf{k}_{t,2}$. The same considerations hold for the reflected field that also consists of two plane waves with wave vectors $\mathbf{k}_{r,1}$ and $\mathbf{k}_{r,2}$.

index $n(\hat{k}_x, \hat{k}_y, \hat{k}_z)$ with $k = nk_0$. In certain cases, one or both of the solutions for k^2 can be negative. In that case the resulting waves are evanescent as was found for the case of total internal reflection at the interface between two isotropic media. These evanescent waves should be taken into account when calculating the Fresnel coefficients of the reflected and transmitted waves; however, after an appreciable propagation distance in optical systems, their influence can be neglected. In thin layer systems with anisotropic media or in liquid crystals, their influence at a preceding or subsequent interface cannot be neglected.

When considering anisotropic media with isotropic magnetic behaviour, it is useful to solve the eigenvalue equations expressed in terms of the magnetic field strength $\mathbf{H}_0$. The resulting refractive index solutions yield the two values of the energy propagation speed v_{em}, also called the ray velocity in the language of geometrical optics. From the two solutions for the eigenstates of the magnetic field strength vector $\mathbf{H}_0$, we can find the electric field strengths $\mathbf{E}_0$ of the two polarisation eigenstates by applying Eq. (2.10.9). Constructing the vector product of the electric and magnetic field eigenstate vectors according to $\Re\{\mathbf{E} \times \mathbf{H}_0^*\}/2$ yields the direction of the Poynting vector which is used to trace the propagation of energy through a sequence of isotropic and anisotropic media.

We can define the admittance of an anisotropic medium, analogously to that of an isotropic medium, according to

$$Y_j(\hat{\mathbf{s}}_r) = \left(\frac{\mathbf{H}_{0,j}(\hat{\mathbf{s}}_r) \cdot \mathbf{H}_{0,j}(\hat{\mathbf{s}}_r)}{\mathbf{E}_{0,j}(\hat{\mathbf{s}}_r) \cdot \mathbf{E}_{0,j}(\hat{\mathbf{s}}_r)} \right)^{1/2} , \tag{2.12.2}$$

where $\mathbf{E}_{0,j}(\hat{\mathbf{s}}_r)$ and $\mathbf{H}_{0,j}(\hat{\mathbf{s}}_r)$, with $j = 1, 2$, correspond to the polarisation eigenstates that are associated with the ray velocity vector $\hat{\mathbf{s}}_r$. For the general case of elliptical polarisation whereby the eigenstates are complex, the real and imaginary magnetic and electric field components should be separately considered. The direction-dependent admittance $Y_j(\hat{\mathbf{s}}_r)$ of a general anisotropic medium is complex. For non-dissipating media, also in the presence of optical rotation, the admittance is real-valued.

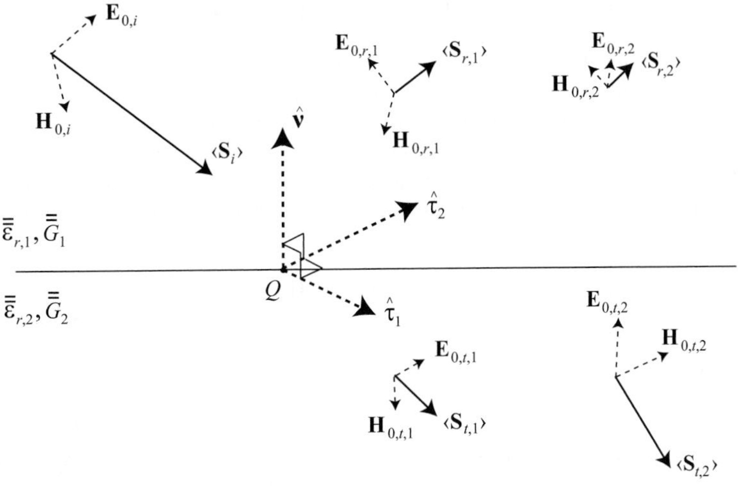

Figure 2.28: A wave, corresponding to a polarisation eigenstate $(\mathbf{E}_{0,i}, \mathbf{H}_{0,i})$ of the first medium, is incident at the interface between two anisotropic media. The two reflected propagation directions with corresponding polarisation eigenstates that match the wave vector $\mathbf{k}_{0,i}$ of the incident wave yield the electric and magnetic field strengths $(\mathbf{E}_{0,r,1}, \mathbf{H}_{0,r,1})$ and $(\mathbf{E}_{0,r,2}, \mathbf{H}_{0,r,2})$, respectively. In the second medium we have the matching polarisation eigenstates with field strengths $(\mathbf{E}_{0,t,1}, \mathbf{H}_{0,t,1})$ and $(\mathbf{E}_{0,t,2}, \mathbf{H}_{0,t,2})$. In a chosen reference point Q on the interface the normal vector $\hat{\boldsymbol{\nu}}$ and the tangential vectors $\hat{\boldsymbol{\tau}}_1$ and $\hat{\boldsymbol{\tau}}_2$ form an orthogonal, right-handed triplet. The tangential components of the incident, reflected and transmitted waves are measured along the orthogonal vector pair $(\hat{\boldsymbol{\tau}}_1, \hat{\boldsymbol{\tau}}_2)$.

For each energy propagation direction given by the unit vector $\hat{\mathbf{s}}_r$, two polarisation eigenstates can be found. The eigenstates with the same unit ray vector originate from, in general, two wave vectors that are different both in direction and length. In optical ray-tracing, it is the ray vector $\hat{\mathbf{s}}_r$ that is traced through an optical system and, in an anisotropic medium, the task is to find the corresponding wave vectors and polarisation eigenstates that belong to this energy propagation direction. The solution of this 'inverse' problem is discussed in Chapter 5, Subsection 5.11.1.

2.12.3 Amplitude and Energy Transfer at an Interface

The directions in which electromagnetic energy can be transported when an incident beam is reflected and refracted at a planar interface were derived in the previous subsection. The distribution of the incident electromagnetic energy flow between the reflected and transmitted beams is determined by the boundary conditions for the electric and magnetic field strengths at the interface. From Maxwell's equations, in the absence of surface charges or currents at the interface, we have the requirement that the tangential components of the electric and magnetic field strengths are continuous at the boundary. In the case of a transition between two isotropic media, the tangential field components are generally measured in the plane of incidence (p-polarisation) and perpendicularly to this plane (s-polarisation). These two linear states of polarisation can be considered as two orthogonal eigenstates associated with the propagation direction of the incident wave $\mathbf{k}_i$ and the unit normal vector $\hat{\boldsymbol{\nu}}$ of the surface. The general state of reflection and refraction at the interface between two anisotropic media is shown in Fig. 2.28. The polarisation eigenstates of the reflected and transmitted waves have a more general geometry and, in principle, any pair of orthogonal tangential vectors can be chosen to apply the continuity condition of the tangential field strengths at the interface. A convenient pair of tangential vectors remains $(\hat{\boldsymbol{\tau}}_1, \hat{\boldsymbol{\tau}}_2) = (\hat{\boldsymbol{\tau}}_p, \hat{\boldsymbol{\tau}}_s)$ with the indices p and s referring to the p- and s-polarisation directions; the isotropic case is now optimally included in the general anisotropic case.

2.12.3.1 Calculation of the Amplitude Reflection and Transmission Coefficients

The continuity of the tangential components of the electric field strength yields the condition

$$\left(\hat{\mathbf{E}}_{0,i} + r_1 \hat{\mathbf{E}}_{0,r,1} + r_2 \hat{\mathbf{E}}_{0,r,2}\right) \cdot \hat{\boldsymbol{\tau}}_1 = \left(t_1 \hat{\mathbf{E}}_{0,t,1} + t_2 \hat{\mathbf{E}}_{0,t,2}\right) \cdot \hat{\boldsymbol{\tau}}_1 \,, \tag{2.12.3}$$

$$\left(\hat{\mathbf{E}}_{0,i} + r_1\hat{\mathbf{E}}_{0,r,1} + r_2\hat{\mathbf{E}}_{0,r,2}\right) \cdot \hat{\boldsymbol{\tau}}_2 = \left(t_1\hat{\mathbf{E}}_{0,t,1} + t_2\hat{\mathbf{E}}_{0,t,2}\right) \cdot \hat{\boldsymbol{\tau}}_2 \,. \tag{2.12.4}$$

In these expressions, we have normalised the electric field strength of the incident wave to unity. The coefficients r_1 and r_2 are the reflection coefficients of the incident wave into the two eigenstates of reflected light in the first medium. The transmission coefficients t_1 and t_2 are defined in a comparable way for the eigenstates in the second medium. As usual, the caret sign denotes a unit vector. Note that the magnetic field strength of each plane wave, derived from the electric strength by means of Eq. (2.10.6), is generally not a unit vector. The magnetic field strengths of the incident wave and the eigenstates in the first and second medium yield the following continuity relations for the magnetic field strengths in the two media,

$$\left(\mathbf{H}_{0,i} + r_1\mathbf{H}_{0,r,1} + r_2\mathbf{H}_{0,r,2}\right) \cdot \hat{\boldsymbol{\tau}}_1 = \left(t_1\mathbf{H}_{0,t,1} + t_2\mathbf{H}_{0,t,2}\right) \cdot \hat{\boldsymbol{\tau}}_1 \,, \tag{2.12.5}$$

$$\left(\mathbf{H}_{0,i} + r_1\mathbf{H}_{0,r,1} + r_2\mathbf{H}_{0,r,2}\right) \cdot \hat{\boldsymbol{\tau}}_2 = \left(t_1\mathbf{H}_{0,t,1} + t_2\mathbf{H}_{0,t,2}\right) \cdot \hat{\boldsymbol{\tau}}_2 \,. \tag{2.12.6}$$

Equations (2.12.3)–(2.12.6) form a linear system of equations in terms of the unknown coefficients r_1, r_2, t_1 and t_2 and in matrix notation we have

$$\begin{pmatrix} \hat{\mathbf{E}}_{0,r,1}\cdot\hat{\boldsymbol{\tau}}_1 & \hat{\mathbf{E}}_{0,r,2}\cdot\hat{\boldsymbol{\tau}}_1 & -\hat{\mathbf{E}}_{0,t,1}\cdot\hat{\boldsymbol{\tau}}_1 & -\hat{\mathbf{E}}_{0,t,2}\cdot\hat{\boldsymbol{\tau}}_1 \\ \hat{\mathbf{E}}_{0,r,1}\cdot\hat{\boldsymbol{\tau}}_2 & \hat{\mathbf{E}}_{0,r,2}\cdot\hat{\boldsymbol{\tau}}_2 & -\hat{\mathbf{E}}_{0,t,1}\cdot\hat{\boldsymbol{\tau}}_2 & -\hat{\mathbf{E}}_{0,t,2}\cdot\hat{\boldsymbol{\tau}}_2 \\ \mathbf{H}_{0,r,1}\cdot\hat{\boldsymbol{\tau}}_1 & \mathbf{H}_{0,r,2}\cdot\hat{\boldsymbol{\tau}}_1 & -\mathbf{H}_{0,t,1}\cdot\hat{\boldsymbol{\tau}}_1 & -\mathbf{H}_{0,t,2}\cdot\hat{\boldsymbol{\tau}}_1 \\ \mathbf{H}_{0,r,1}\cdot\hat{\boldsymbol{\tau}}_2 & \mathbf{H}_{0,r,2}\cdot\hat{\boldsymbol{\tau}}_2 & -\mathbf{H}_{0,t,1}\cdot\hat{\boldsymbol{\tau}}_2 & -\mathbf{H}_{0,t,2}\cdot\hat{\boldsymbol{\tau}}_2 \end{pmatrix} \begin{pmatrix} r_1 \\ r_2 \\ t_1 \\ t_2 \end{pmatrix} = \mathbf{M}_a \begin{pmatrix} r_1 \\ r_2 \\ t_1 \\ t_2 \end{pmatrix} = - \begin{pmatrix} \hat{\mathbf{E}}_{0,i}\cdot\hat{\boldsymbol{\tau}}_1 \\ \hat{\mathbf{E}}_{0,i}\cdot\hat{\boldsymbol{\tau}}_2 \\ \mathbf{H}_{0,i}\cdot\hat{\boldsymbol{\tau}}_1 \\ \mathbf{H}_{0,i}\cdot\hat{\boldsymbol{\tau}}_2 \end{pmatrix} . \tag{2.12.7}$$

In general, the matrix $\mathbf{M}_a$ is complex, however, it can be split up into its real and imaginary parts, associated with the real and imaginary parts of the set of linear equations of Eq. (2.12.7). The values of the reflection and transmission coefficients for the polarisation eigenstates in the two media are given by

$$\begin{pmatrix} r_1 \\ r_2 \\ t_1 \\ t_2 \end{pmatrix} = -\left(\mathbf{M}_a\right)^{-1} \begin{pmatrix} \hat{\mathbf{E}}_{0,i}\cdot\hat{\boldsymbol{\tau}}_1 \\ \hat{\mathbf{E}}_{0,i}\cdot\hat{\boldsymbol{\tau}}_2 \\ \mathbf{H}_{0,i}\cdot\hat{\boldsymbol{\tau}}_1 \\ \mathbf{H}_{0,i}\cdot\hat{\boldsymbol{\tau}}_2 \end{pmatrix} . \tag{2.12.8}$$

In practice, numerical procedures are available to evaluate the inverse of the matrix $\mathbf{M}_a$. To avoid numerical instability, it is useful to first carry out a singular value decomposition of $\mathbf{M}_a$ so that a possibly reduced rank of the matrix can be easily identified.

2.12.3.2 Calculation of the Intensity Reflection and Transmission Coefficients

With the values of the reflection and transmission coefficients at the interface, it is a relatively small step to calculate the power that is carried away in each of the possible propagation directions in the two media. As was demonstrated in Section 1.8.7 for isotropic media, the direction and magnitude of the Poynting vector for each of the propagation directions are immediately obtained from the known electric and magnetic field strengths of each polarisation eigenstate, multiplied with the appropriate reflection or transmission coefficient. Figure 2.29 shows a general configuration at the interface of two anisotropic media with permittivity tensors $\bar{\bar{\epsilon}}_{r,1}$, $\bar{\bar{\epsilon}}_{r,2}$ and gyrotropic tensors $\bar{\bar{G}}_1$, $\bar{\bar{G}}_2$, respectively. The incident plane wave with Poynting vector $\langle \mathbf{S}_i \rangle$ is reflected and transmitted into four waves, two in each medium. In contrast to the isotropic case, the reflected and refracted Poynting vectors are not necessarily found in the plane of incidence that is formed by $\hat{\mathbf{v}}$ and $\langle \mathbf{S}_i \rangle$. The energy transfer at the interface can be expressed using the equation

$$|\langle \mathbf{S}_i \rangle| = \frac{1}{\cos\theta_i}\Big(|\langle \mathbf{S}_{r,1} \rangle|\cos\theta_{r,1} + |\langle \mathbf{S}_{r,2} \rangle|\cos\theta_{r,2} + |\langle \mathbf{S}_{t,1} \rangle|\cos\theta_{t,1} + |\langle \mathbf{S}_{t,2} \rangle|\cos\theta_{t,2}\Big) . \tag{2.12.9}$$

If we normalise the incident power to unity, the energy balance at the interface can be written in terms of the amplitude reflection and transmission coefficients that are given by Eq. (2.12.8). We use the properties that the electric and magnetic field strengths are orthogonal in a nonmagnetic medium and that the medium admittance is real in a non-dissipating medium. The latter property follows from the real value of the propagation speed for each polarisation

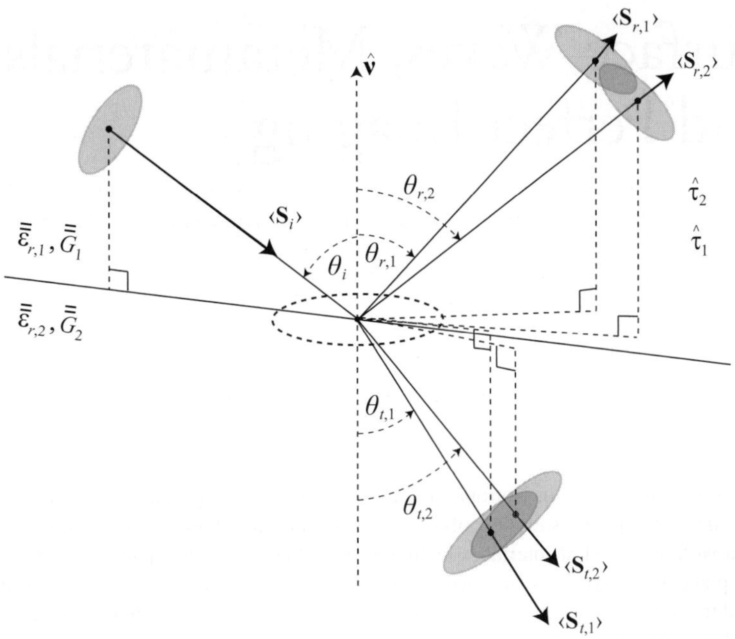

Figure 2.29: The reflected and transmitted plane waves at an interface between two anisotropic media, characterised by their permittivity and gyrotropic tensors, $\bar{\bar{\epsilon}}_r$ and $\bar{\bar{G}}$, respectively. The energy propagation direction of the incident plane wave is given by the Poynting vector $\langle \mathbf{S}_i \rangle$. The plane of incidence is defined by the unit normal vector $\hat{\mathbf{v}}$ and $\langle \mathbf{S}_i \rangle$. In general the reflected and transmitted Poynting vectors are not found in the plane of incidence.

eigenstate. The modulus of the Poynting vector of a reflected and transmitted beam is then simply given by $|S| = |E||H|/2$. Equation (2.12.9) is then replaced by the expression

$$1 = \frac{\cos\theta_{r,1}}{\cos\theta_i}|r_1|^2|S_{r,1}| + \frac{\cos\theta_{r,2}}{\cos\theta_i}|r_2|^2|S_{r,2}| + \frac{\cos\theta_{t,1}}{\cos\theta_i}|t_1|^2|S_{t,1}| + \frac{\cos\theta_{t,2}}{\cos\theta_i}|t_2|^2|S_{t,2}| , \tag{2.12.10}$$

where each modulus $S_{p,j}$ of the Poynting vectors is given by the product of the unit electric field vector and the corresponding modulus of the magnetic field strength $\mathbf{H}_{0,p,j}$ of the polarisation eigenstate of the wave.

3 Surface Waves, Metamaterials and Perfect Imaging

In Chapters 1 and 2 wave propagation in isotropic and anisotropic unbounded media was studied. More specifically, in Section 1.8, the reflection and transmission of waves at an interface between two isotropic media was treated. Particular phenomena were analysed at interfaces including the influence of the state of polarisation on the reflection and transmission of a plane wave and the occurrence of total internal reflection. Moreover, numerically stable methods to treat the reflection and transmission of light waves incident on a stratified medium with planar interfaces were described. The media considered included those with a finite value of the conductivity σ, or, equivalently, with a complex value of the electric permittivity $\epsilon = \epsilon' + i\epsilon''$. The magnetic properties of a medium were described by a real valued permeability, μ, with the tacit assumption that its value equals the vacuum value μ_0 in the optical frequency domain. The possibility of inductance of a medium, which would give rise to a complex permeability, $\mu = \mu' + i\mu''$, has been neglected so far. The reason for adopting the vacuum permeability μ_0 in the optical domain is the fact that at optical frequencies the material changes induced by external electric or magnetic fields are extremely small. The current loops that respond to these fields are created by relatively heavy ionised atoms and molecules in, for example, a crystal lattice. Their resonance frequencies are found in the GHz region, or, at the most, in the THz frequency domain. Beyond these frequencies, in the optical domain, the inertia of the charged moving bodies (other than electrons) is too large to allow any response. The material becomes magnetically 'transparent' and has a magnetic permeability identical to the vacuum value μ_0.

Tailoring of the material properties at the sub-wavelength scale, however, enables the response to electric and magnetic external effects to be modified. Such modifications are obtained in so-called artificial optical nano-crystals or, more generally, in specially manufactured electric and magnetic so-called *metamaterials*. By shifting the resonances to higher frequencies, the media can be made to react quite differently to external influences at the frequency of interest. In this way, it has been possible to adapt the electric properties of materials with specially designed sub-wavelength structures for use in the optical domain. Modification of magnetic properties has only been achieved in the THz domain and in the far infrared spectral region. To date, however, new developments may yet allow a change in the complex μ-value at optical frequencies.

In light of the growing diversity in material optical properties that are achievable through the advances mentioned above, in this chapter we discuss a particular class of waves which propagate in such media. First we consider the bound surface waves (surface plasmon polariton waves) that can propagate along the surface of certain conducting metals and that can couple to the free-electron plasmon oscillations at the metal surface. In contrast to waves with a very small wavelength that are evanescent in a transparent medium or in vacuum, the bound surface waves can propagate along a metal surface over appreciable distances. At equal wavelength, the attenuation of the surface wave can be much smaller than that of a corresponding free-space wave. Such bound waves are interesting since they transport high-frequency spatial information without suffering the large losses that are commonly experienced by evanescent waves propagating along an interface. The second topic of this chapter is the propagation of light in so-called *left-handed* metamaterials with negative values for both the electric permittivity and the magnetic permeability **[356]**. We study the peculiar properties of waves in such media with respect to energy transport and wave refraction and reflection at an interface. The third topic concerns a proposal to apply such left-handed materials to optical imaging, using the idea of a thin-layer 'perfect' lens **[273]**.

3.1 Eigenmodes of a Metal–Dielectric Interface

In this section we analyse the bound waves that can propagate along the interface of a dielectric and a metallic medium. Physically, these bound surface waves are due to a coupling between plasma oscillations of the free electrons in the metal and the electromagnetic wave that exhibits a certain spatial periodicity along the surface. This periodicity depends on the value of the wave vector component along the surface. Depending on the energy exchange between the incident plane wave and the lossy metallic medium, the surface wave shows a certain attenuation on propagation along the interface. Such a special bound surface wave, because of its physical origin, is called a surface plasmon polariton wave (SPP). The SPP waves are eigenmodes of the dielectric–metal interface.

3.1.1 Creation of Surface Plasmon Polariton (SPP) Waves

The eigenmodes of a dielectric–metal interface can be accidentally excited under the influence of some 'noisy' optical input signal. More specifically, they can be excited on purpose using a so-called Otto-configuration or a Kretschmann-configuration. The Otto-configuration uses the phenomenon of total internal reflection at one of the glass–air interfaces of a prism. A planar metal surface is brought into near-contact with the glass–air interface where the evanescent wave associated with the totally reflected beam still has appreciable amplitude. The evanescent wave then tunnels to the metal surface and produces the surface plasmon polariton wave if the resonance condition for the electron plasma oscillation in the metal is fulfilled. The closely resembling Kretschmann-configuration uses oblique reflection at a glass–metal interface. The metal material under investigation has been deposited, e.g. by evaporation, on the glass surface. The Otto-configuration is shown schematically in Fig. 3.1a.

In Fig. 3.1b we show the interface between a dielectric medium $(j+1)$ with a real-valued electric permittivity ϵ_{j+1} and an absorbing medium j with complex permittivity $\epsilon_j = \epsilon_j' + i\epsilon_j''$. For passive media, the imaginary part is taken positive such that the field strength of the wave is attenuated on propagation. The waves at the interface are two outgoing waves with field strengths denoted by A_{j+1}^- and A_j^+ and one incident wave with strength A_{j+1}^+. The dashed arrow in the figure, representing the wave with field strength A_j^- and coming from $-\infty$, has effectively zero field strength because of the absorption in the metallic half space. Applying the scattering matrix formalism to the interface (see Eq. (1.11.5)), we have, respectively for the case of s- and p-polarised waves

$$\begin{aligned}
A_{j+1,s}^- &= r_s^+ A_{j+1,s}^+ \,, \\
A_{j+1,p}^- &= r_p^+ A_{j+1,p}^+ \,, \\
A_{j,s}^+ &= t_s^+ A_{j+1,s}^+ \,, \\
A_{j,p}^+ &= t_p^+ A_{j+1,p}^+ \,.
\end{aligned} \tag{3.1.1}$$

A special case arises for the outgoing waves when the reflection and transmission coefficients become infinitely large. Although this seems to be a non-physical solution of the scattering problem, it is consistent with energy conservation if

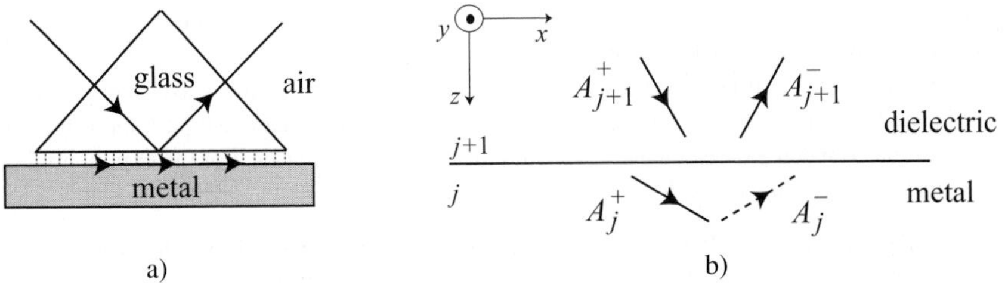

Figure 3.1: a) Otto-configuration for the creation of an SPP-wave on a metal surface by means of the evanescent wave in the air gap between a glass prism side and a metal surface (total internal reflection).
b) The ingoing and outgoing waves at the interface between two semi-infinite half spaces, of which one contains a dielectric medium and the other an absorbing medium. The wave vectors depicted in the graph are real and correspond to those of freely propagating waves. The wave vectors associated with a surface-bound wave are complex on both sides of the interface and their z-components give rise to a decreasing wave amplitude as a function of the perpendicular distance to the interface.

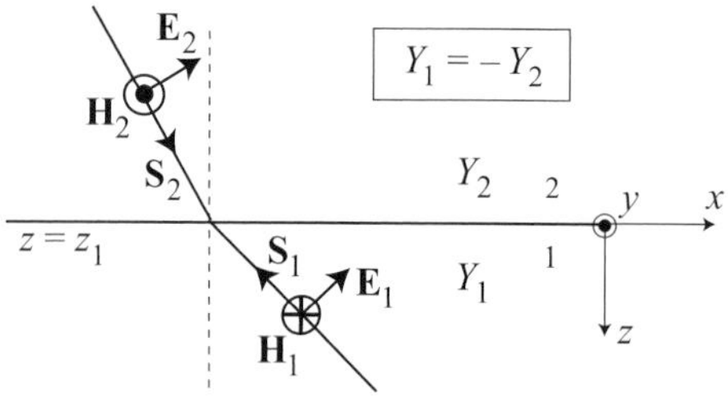

Figure 3.2: The (complex) electromagnetic field vectors of the reflected and transmitted waves in two adjacent media with opposite admittance values. The media are separated by an interface at the plane $z = z_1$. The incident wave in medium 1 has been omitted as its amplitude is negligibly small with respect to those of the (guided) reflected and transmitted waves.

the outgoing wave solutions remain bound to the surface and their normalised power flow, integrated from $z = -\infty$ to $+\infty$, is finite.

The reflection and transmission coefficients at the interface for s- and p-polarisation have a common denominator given by $Y_{j+1} + Y_j$ (see Eqs. (1.8.24)–(1.8.27)). Equating this term to zero for the p-polarised case yields

$$\frac{Y_{m,j+1}\, k_{j+1}}{k_{z,j+1}} + \frac{Y_{m,j}\, k_j}{k_{z,j}} = 0 \,, \tag{3.1.2}$$

where $Y_{m,j} = (\epsilon_j/\mu_j)^{1/2}$ is the intrinsic admittance of the medium and $k_j = \omega\sqrt{\epsilon_j \mu_j}$ is the modulus of the wave vector in medium j. Substitution of the values for Y_m and k in each medium into Eq. (3.1.2) yields the expression

$$\frac{\epsilon_{j+1}}{k_{z,j+1}} + \frac{\epsilon_j}{k_{z,j}} = 0 \,, \tag{3.1.3}$$

which is generally called the *surface plasmon* condition. An illustration of the admittance condition $Y_j = -Y_{j+1}$ with the corresponding electromagnetic vectors of the reflected and transmitted wave in the two media is shown in Fig. 3.2. We note that for the singular case under consideration the incident wave has become negligible in amplitude with respect to the infinitely large amplitudes of the reflected and transmitted waves. For a physical solution in the two semi-infinite half-spaces it is necessary that the amplitude of each of the two waves decays as a function of the perpendicular distance $|z - z_1|$ to the interface.

From the general relationship $YE = H$ we notice that when the E vectors in both media have the same direction, the corresponding H vectors have opposite directions. The Poynting vectors of the reflected and transmitted waves in the two media hence point in opposite directions. In the example of Fig. 3.2 the energy transport by the two waves is then either towards the interface or away from it. The condition for the creation of a surface-bound wave is an accumulation of optical energy on the surface and this suggests that in each medium there is a Poynting vector that points towards the interface. It will turn out that a steady-state surface wave only transports optical energy in a direction which is parallel to the surface. The z-component of the Poynting vector perpendicular to the surface is purely imaginary and, on average, does not transport energy toward or away from the surface.

In general, since we consider a surface wave solution, the k_z-components of the wave vectors of the reflected and transmitted waves in the two media should be complex with the sign of the imaginary part such that attenuation of the *outgoing* waves is guaranteed when the distance $|z - z_j|$ from the interface increases. For the exponential parts of the wave functions in the media j and $j+1$ on both sides of an interface j we write

$$j:\ \exp\{ik_{z,j}|z - z_j|\}\exp\{ik_{x,j}x\}\,, \qquad j+1:\ \exp\{ik_{z,j+1}|z - z_j|\}\exp\{ik_{x,j+1}x\}\,, \tag{3.1.4}$$

where, without loss of generality, we have limited our attention to waves with $k_y = 0$. Since the electric and magnetic field components which are parallel to the interface are continuous, the x-components of the wave vectors on both sides of the interface are identical and we will denote this component by k_x. For the surface wave to decay as a function of the distance to the interface we have that the imaginary parts of $k_{z,j}$ and $k_{z,j+1}$ are positive and the complex numbers $k_{z,j}$ and $k_{z,j+1}$ are thus found in the first or second quadrant of the complex plane.

A limiting condition for the permittivity values of the media on both sides of the interface for the guiding of a surface wave can be derived from the dispersion relation in both media if we let the modulus of the complex k_x-component of the surface-bound wave in each medium grow to very large values. The dispersion relation in medium j and $j+1$ is given by

$$j: \; k_{z,j}^2 = \epsilon_j k_0^2 - k_x^2 \,, \qquad j+1: \; k_{z,j+1}^2 = \epsilon_{j+1} k_0^2 - k_x^2 \,, \tag{3.1.5}$$

where $k_0 = \omega\sqrt{\epsilon_0\mu_0}$ is the vacuum wavenumber of the radiation. It immediately follows from these dispersion relations in the two media that $k_{z,j} = k_{z,j+1} = \pm i\, k_x$ for $k_x \rightarrow \infty$. Consequently, the surface plasmon condition of Eq. (3.1.3) imposes that $\epsilon_j = -\epsilon_{j+1}$ for the interface to be able to guide a surface wave. For the special case of a dielectric medium with ϵ_{j+1} real and positive, the required negative real value of ϵ_j can be found, for instance, in metals, not far from electric resonances in the ultraviolet spectral region. More general surface wave guiding in which finite values of k_x are involved still requires metal-like behaviour of a medium. The real negative value of ϵ_j is generalised to a complex value of ϵ_j in the second or third quadrant of the complex plane with, essentially, a negative real part of ϵ_j.

The common x-component of the wave vectors in the two media follows from the dispersion relations of Eq. (3.1.5) in these media combined with the surface plasmon condition of Eq. (3.1.3). We obtain the following expression:

$$k_x^2 = \frac{\epsilon_j^2 k_{j+1}^2 - \epsilon_{j+1}^2 k_j^2}{\epsilon_j^2 - \epsilon_{j+1}^2} = \left(\frac{\epsilon_j\,\epsilon_{j+1}}{\epsilon_j + \epsilon_{j+1}}\right) k_0^2 \,. \tag{3.1.6}$$

Equation (3.1.6) determines the dispersion of the surface wave. In general, the wave vector component k_x is complex with the imaginary part chosen such that the wave is attenuated in the propagation direction along the interface. The complex z-components of the wave vectors in the two media are given by

$$k_{z,j+1}^2 = \left(\frac{\epsilon_{j+1}^2}{\epsilon_j + \epsilon_{j+1}}\right) k_0^2 \,, \qquad k_{z,j}^2 = \left(\frac{\epsilon_j^2}{\epsilon_j + \epsilon_{j+1}}\right) k_0^2 \,. \tag{3.1.7}$$

A comparable analysis for s-polarised waves yields the eigenmode condition

$$\frac{Y_{m,j+1}\, k_{z,j+1}}{k_{j+1}} + \frac{Y_{m,j}\, k_{z,j}}{k_j} = 0 \,. \tag{3.1.8}$$

It follows from Eq. (3.1.8) that $k_{z,j+1} + k_{z,j}$ should be zero for media with identical permeability, $\mu_j = \mu_{j+1} = \mu_0$. This requirement means that the z-components of the wave vector on both sides of the interface must have opposite sign. For common materials, the requirement $k_{z,j+1} + k_{z,j} = 0$ cannot be satisfied with identical k_x-components on both sides of the interface as follows from the dispersion relation. We conclude that the bound surface waves only exist in a p-polarised state with the electric vector of the wave parallel to the plane of incidence.

3.1.2 The Electromagnetic Field Components

We have shown in the preceding subsection that the electric field vector of an SPP wave lies in the plane of incidence. In turn, this implies that the magnetic field is perpendicular to the plane of incidence. Again restricting consideration to the $k_y = 0$ case we have that the magnetic field at the interface with a magnitude H_0 points in the y-direction. This tangential component is continuous at the interface. The electric field vector, located in the plane $y = 0$, then follows from Eq. (1.8.1). We find the following expressions for the field strengths in medium j for $z > z_j$,

$$\left\{\begin{matrix} \mathbf{E}(\mathbf{r}) \\ \\ \mathbf{H}(\mathbf{r}) \end{matrix}\right\} = \left\{\begin{matrix} \dfrac{H_0}{\omega\epsilon_j}\left(k_{z,j},\; 0\,, -k_x\right) \\ \left(\,0\,,\; H_0\,,\; 0\,\right) \end{matrix}\right\} \exp\left\{i\left[k_x x + k_{z,j}\,|z - z_j|\right]\right\} \,. \tag{3.1.9}$$

The complex quantity $k_{z,j}^2$ is assumed to lie in the upper half of the complex plane as is appropriate for a standard absorbing medium. To obtain a decreasing field strength in the medium j as a function of the distance $|z - z_j|$, the value of $k_{z,j}$ is given by the positive root of $k_{z,j}^2$ and lies in the first quadrant. Since we have written the z-dependence of the exponential part of the surface wave as a function of the absolute distance to the interface at $z = z_j$, the value of $k_{z,j+1}$ is also the positive root of $k_{z,j+1}^2$, which guarantees the bound wave has zero field strength if z approaches $-\infty$.

The Otto-configuration for generating an SPP surface wave has been reproduced in Fig. 3.3 together with a schematic drawing of the exponentially decreasing electromagnetic field strengths on both sides of the interface that are typical for a surface wave. The dielectric prism, preferably with a high index of refraction, is used to operate in the regime of total internal reflection. The transmitted evanescent field in air has a wave vector component, directed along the interface that

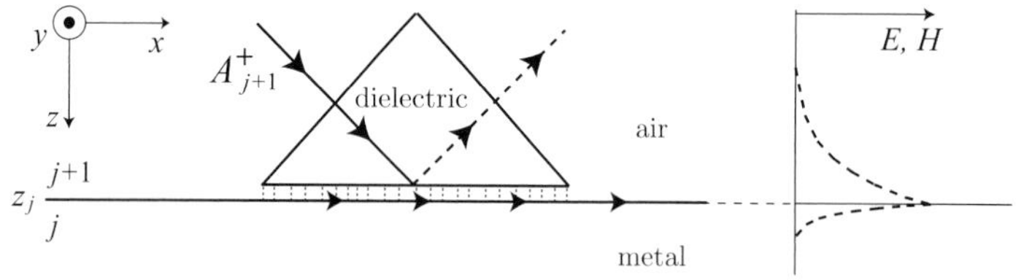

Figure 3.3: The coupling of an incident plane wave with field strength A^+_{j+1} to a metal–dielectric interface using the phenomenon of frustrated total internal reflection at the dielectric–air interface of the prism. The distance between the prism and the metal layer is much smaller than the wavelength of the light (Otto-configuration). The electric field strength E and the magnetic field strength H of a surface plasmon polariton wave, exponentially decreasing as a function of the perpendicular distance to the interface, have been schematically shown in the right-hand part of the figure. The surface wave indicated by the horizontal arrows is also damped in the propagation direction along the interface.

is higher than that of a freely propagating wave in air. This evanescent component is capable of launching an SPP wave on the interface if the distance between the exit surface of the prism and the metal surface is very small. The electric field strength profile E of the SPP wave, perpendicularly to the interface, has been sketched in a highly exaggerated way; the extent in the z-direction of a surface plasmon polariton wave is generally limited to a few wavelengths. SPP waves are highly sensitive to surface adherents or surface contamination. Simultaneously, the dispersion relation of Eq. (3.1.6) implies the coupling efficiency of surface waves has a generally strong spectral dependence. This also makes SPPs an interesting spectroscopic tool for the study of the metal medium itself or of the cladding of the surface with a thin layer of unknown thickness or composition.

3.1.3 SPP Wave and Optical Contrast at the Interface

We now study in some detail the dependence of the k_x-component of the complex wave vector of the SPP wave on the properties of the metal and dielectric. Denoting the complex permittivity of the metal medium by $\epsilon_j = \epsilon_j' + i\epsilon_j''$, we can write Eq. (3.1.6) as

$$\left(\frac{k_x}{k_0}\right)^2 = \frac{\epsilon_{j+1}}{\epsilon_0}\left(\frac{\epsilon_j'^2 + \epsilon_j''^2 + \epsilon_j'\epsilon_{j+1} + i\,\epsilon_{j+1}\epsilon_j''}{(\epsilon_j' + \epsilon_{j+1})^2 + \epsilon_j''^2}\right), \qquad (3.1.10)$$

where $k_0 = \omega(\epsilon_0\mu_0)^{1/2}$ is the vacuum wavenumber of the electromagnetic wave with frequency ω. The complex permittivity of a metal can be approximately described with the aid of the classical Drude model for the resonance behaviour of the free electrons in a metal **[37]**. The permittivity as a function of the radial frequency ω of the exciting electromagnetic wave is thus given by

$$\frac{\epsilon(\omega)}{\epsilon_0} = 1 - \frac{\omega_p^2}{\omega^2 + i\Gamma\omega} = \left(1 - \frac{1}{\omega_n^2 + \Gamma_n^2}\right) + i\,\frac{\Gamma_n}{\omega_n\left(\omega_n^2 + \Gamma_n^2\right)}, \qquad (3.1.11)$$

where ω_p is the plasma resonance frequency of the electrons, Γ is the damping coefficient of the electron oscillations, $\omega_n = \omega/\omega_p$ is the normalised frequency and $\Gamma_n = \Gamma/\omega_p$ is the normalised damping coefficient. We use the permittivity data of Ag at $\lambda_0 = 500$ nm (or $\omega = 3.77 \cdot 10^{15}$ s^{-1}) for which $\epsilon/\epsilon_0 = -8.51 + 0.76i$. From these data one can compute the values of the Drude resonance frequency in Ag, $\omega_p = 11.67 \cdot 10^{15}$ s^{-1} ($\lambda_p = 162$ nm) and the damping factor $\Gamma = 0.30 \cdot 10^{15}$ s^{-1}.

In Fig. 3.4a we show a simplified picture of the relative complex permittivity, normalised to that of vacuum, of a typical metal like silver. A single Drude resonance term has been included in the calculations which is a rather crude approximation of the real properties of silver. Measurements will reveal a more complicated picture because of the occurrence of many more than one resonance frequency in the material. According to the simple Drude model, the wavelength at which resonance occurs is 0.162 μm in the ultraviolet region of the spectrum. The visible light wavelength of 0.5 μm corresponds to the value $\omega/\omega_p = 0.323$. We notice that the resonance frequency ω_p applies to plasmon oscillations of the electrons in the bulk of the material. It will be shown that the plasmon oscillations at an interface which are relevant for the creation of an SPP wave have a lower resonance frequency. The real part of ϵ is strongly negative in the visible part of the spectrum. The admittance of the metal becomes almost purely imaginary. A freely propagating incident

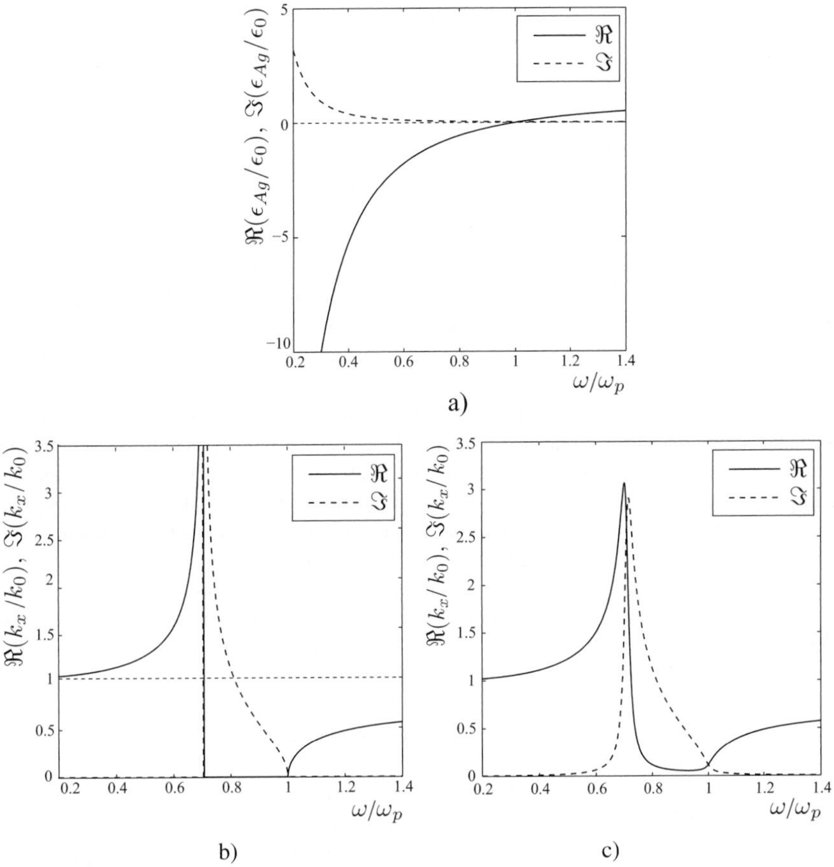

Figure 3.4: a) The complex permittivity of silver in the visible domain and in the ultraviolet spectral region (solid line: real part, dashed line: imaginary part).
b) The real part (solid line) and imaginary part (dashed) of the normalised wave vector component k_x/k_0 at the interface between air and an idealised metal with a damping coefficient Γ equal to zero.
c) The same quantities as in b), but with the Ag-permittivity data including a nonzero damping factor.

wave, coming from a medium with real admittance, will experience a high reflectivity. In the limiting case, for a purely imaginary admittance, the modulus of the reflection coefficient tends to unity, explaining the high reflectivity of a metal below the plasma frequency. The special case of $\Re(\epsilon/\epsilon_0) = 0$ follows from Eq. (3.1.11) and is found at $\omega^2 = \omega_p^2 - \Gamma^2$. Beyond the plasma frequency, in the deep ultraviolet, the permittivity gradually tends towards the vacuum value (at least in the single resonance approximation).

In Fig. 3.4b we show the real and imaginary part of the normalised wave vector component along the surface, k_x/k_0, of an SPP wave on the air–silver interface. The reference wavenumber in vacuum of a freely propagating wave is k_0 (see the horizontal dashed line at level unity in the figure). The damping in the metal has been neglected in graph b). It is seen that in the neighbourhood of the reference wavelength of $\lambda = 500$ nm, $\omega/\omega_p = 0.323$, the imaginary part of k_x/k_0 is zero. Its real part increases with frequency and tends towards infinity at a resonance frequency well below the plasma frequency ω_p. The frequency at which this singularity occurs follows from Eq. (3.1.10) by equating the denominator of the quotient in the right-hand side of this equation to zero. With $\Gamma_n = 0$, we find a singular value for k_x/k_0 if the condition $\epsilon_j' = -\epsilon_{j+1}$ is satisfied,

$$\left[1 - \frac{1}{(\omega_{SPP}/\omega_p)^2}\right] = -\frac{\epsilon_{j+1}}{\epsilon_0} ,$$

or ,

$$\omega_{SPP} = \omega_p \sqrt{\frac{1}{1 + (\epsilon_{j+1}/\epsilon_0)}} \;. \tag{3.1.12}$$

For an air–metal interface, we find the SPP resonance frequency is $\frac{1}{2}\sqrt{2}\,\omega_p$. Beyond the resonance frequency of the surface wave up to the plasmon frequency ω_p, the damping is high. For $\omega > \omega_p$, the attenuation of the surface wave is zero; but the value of $\Re\{k_x\}$ is smaller than the vacuum value.

The curves for k_x/k_0 for an air–silver interface, including damping in the metal, are given in Fig. 3.4c. The differences from the idealised damping-free metal are small, making silver a highly suitable material for guiding of surface waves. The shift of the SPP resonance frequency due to the finite damping can be calculated with the aid of Eq. (3.1.10) but for silver this shift is much smaller than the line width and hence not visible in the figure. The main differences are the finite peak values of the real and imaginary parts of k_x/k_0 and the nonzero value of the imaginary part below the resonance frequency. The value of $\Im\{k_x/k_0\}$ is small for $\omega/\omega_p < 0.6$ which makes silver an interesting material for surface wave propagation in the visible and near-infrared region for which k_x values are significantly larger than k_0. In this range, guided waves propagate along the surface with an attenuation length that is many (vacuum) wavelengths long. Such long propagation distances are useful when using surface waves to carry information from one position to another on the surface without appreciable power loss. We also remark that due to the strong dependence of the resonance frequency of the SPP wave on the permittivity ϵ_{j+1} of the dielectric, the wavelength-dependent detection of surface wave resonances is a sensitive tool for surface analysis or monitoring surface contamination.

In Fig. 3.5a we show the existence region of SPP waves in terms of the real and imaginary parts, n_j and κ_j, respectively, of the complex refractive index of the metal medium. The complex refractive index can be calculated for a certain frequency ω from the relative permittivity, permeability and conductivity values of a medium (see Eqs. (1.6.64) and (1.6.65)). For the real part n of the refractive index of a metal we have that $n \geq 0$ (a negative value of n, combined with a positive κ-value would lead to wave amplification upon propagation). To plot the three graphs of Fig. 3.5 we have calculated the components k_x, $k_{z,j}$ and $k_{z,j+1}$ of the wave vectors in media j and $j+1$. The parameters along the axes of Fig. 3.5a are the real and imaginary parts of the complex refractive index of the metal in medium j. The refractive index of the dielectric medium was taken to be 4 or equivalently $\epsilon_{j+1} = 16$. Existence of an SPP wave requires that the amplitude in the z-direction decays in both media. The large grey-shaded region in Fig. 3.5a indicates what values of (n_j, κ_j) do not satisfy this criterion. For the calculation of the wave vectors in the two media we used the expressions of Eqs. (3.1.6) and (3.1.7). For the particular case that $n_j > \kappa_j$ (which corresponds to the part of the grey-shaded area that is left of the dashed line in graph a)), it can be shown that all k_x solutions that follow from Eq. (3.1.6) have to be rejected since the value of $k_{z,j+1}$ is such that the wave amplitude in the dielectric medium shows an exponential increase as a function of the perpendicular distance to the interface.

In graphs b) and c) of Fig. 3.5 the real and imaginary parts of the wave vector component k_x have been plotted. In the region of physically relevant solutions, the real value of k_x shows, on average, an increase with n_j and κ_j in line with Eq. (3.1.10). With $\kappa_j = n_{j+1}$ and $|n_j| \rightarrow 0$, the value of k_x becomes singular (imaginary part). The damping of the surface wave in the propagation direction along the surface becomes infinitely large; this corresponds to a non-physical solution. A complex refractive index $\tilde{n}_j$ with such a particular value is not available among the common metallic materials for the optical regime. The refractive index of silver approaches zero in the blue region of the spectrum. Graph c) which depicts $\Im(k_x/k_0)$ shows that surface waves can propagate along such a metal surface with a relatively low absorption despite the medium having a large value of κ_j.

A special remark must be made on the subject of electric energy density. Applying the energy balance Equation (1.4.5) to a harmonic wave in a metal yields a negative electric energy density because of the negative value of the real part of ϵ_j ($\Re(\epsilon_j) = n_j^2 - \kappa_j^2$). This unphysical result for energy density arises from the fact that we considered a harmonic wave that is unbounded in time. In reality, we should consider a finite bandwidth signal with a starting point and end point in time. The electric energy density contribution from an infinitesimally narrow frequency band with width $\delta\omega$ in such a material is given by

$$< \delta w_e > = \frac{1}{4} \frac{\partial \Re\{\epsilon(\omega)\} |\mathbf{E}_0(\omega)|^2}{\partial \omega} \delta\omega \;, \tag{3.1.13}$$

where $\mathbf{E}_0$ is the field strength of the signal at frequency ω. Due to the strong dispersion in an absorptive material close to the resonance frequency, the energy density according to the expression above yields positive values once integrated over the temporal bandwidth of the optical signal (see Fig. 3.4a for the value of $\partial\Re\{\epsilon(\omega)\}/\partial\omega$ in the neighbourhood of an absorption line of the material). In nonlinear materials, the energy density can be negative at one frequency, if it is

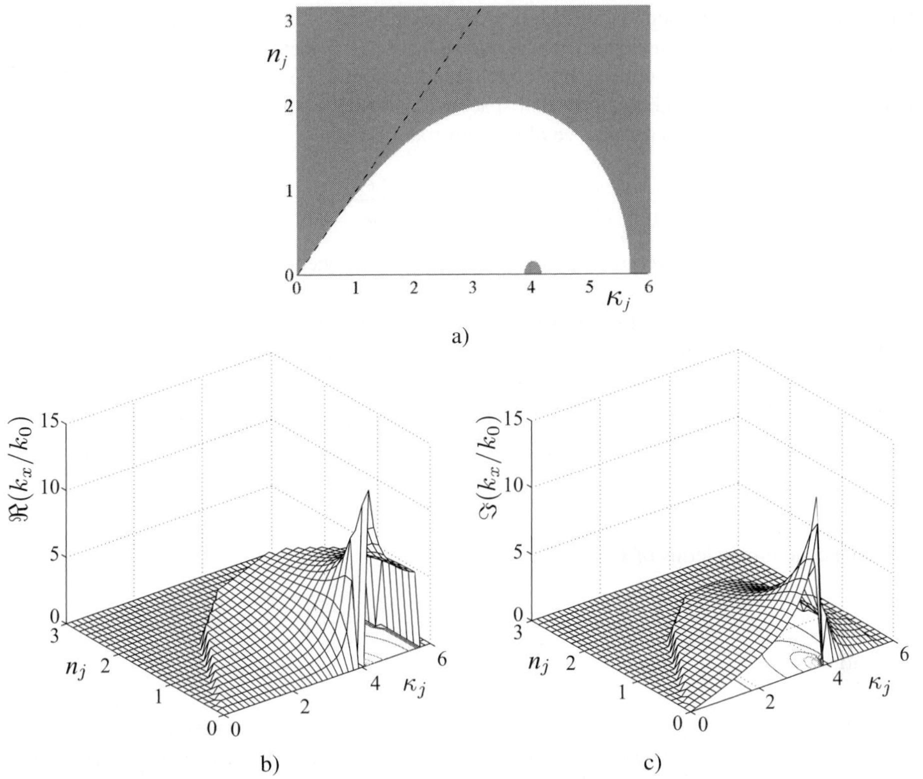

Figure 3.5: a) The region of existence of surface plasmon polariton waves, coupled to a dielectric–metal interface (white-shaded region, imaginary part of $k_x/k_0 \le 15$). The parameter values along the axes are the real and imaginary part, n_j and κ_j, respectively, of the refractive index $\tilde{n}_j = n_j + i\kappa_j$ of the metal medium.
b) and c) The real and imaginary part of the normalised wave vector component k_x/k_0 at the interface. The refractive index of the dielectric medium $j+1$ is 4.

compensated by the positive energy density at another frequency that participates in the nonlinear process. In such a case, the total energy balance of the nonlinear process needs to be taken into account.

3.1.4 Guided Surface Waves in a Multilayer Structure

In this subsection we study the condition for wave guiding along a multilayer stack that separates two semi-infinite half spaces. A wave bound to the layer structure is found by applying the guiding condition of the previous section to the multilayer. The guiding condition was equivalent to finding the poles of the reflection and transmission coefficients of the multilayer structure. Referring to Section 1.9, we first calculate the transfer matrix of the multilayer structure. Regarding the numerical stability of matrix inversion, it would be preferable to calculate the scattering matrix of the multilayer using the appropriate recursive scheme. Fortunately, in the problem at hand we do not need to invert the transfer matrix such that, for a 'forward' scattering problem, the accuracy of the transfer matrix is identical to that of the other matrix methods. We will thus use the transfer matrix method because of its mathematical simplicity.

We consider a multilayer on top of a supporting substrate with specific admittance Y_1. Following the analysis of Section 1.9 and using Eq. (1.10.6) we can consider an effective substrate admittance due to modification by the presence of the multilayer structure which for the p-polarised case is given by

$$Y'_{p,1} = \frac{m_{21,p} + m_{22,p} Y_{p,1}}{m_{11,p} + m_{12,p} Y_{p,1}} , \tag{3.1.14}$$

where m_{ij} are the elements of the transfer matrix.

The guiding condition to obtain a bound multilayer wave is given by $Y'_{p,1} + Y_{p,L+1} = 0$ where $Y_{p,L+1}$ is the admittance of the upper half space for a p-polarised incident wave. By using a numerical method, we can find solutions of the guiding condition. Either discrete values or an entire range of k_x values are found. For a single layer with index 2, sandwiched between the substrate with index 1 and the upper half space with index 3, both with equal admittances $Y_{p,3}$, we can find an explicit expression for the modified admittance of the 'substrate' layer 1,

$$Y'_{p,1} = i\,\frac{\tan(k_{z,2}d_2)}{k_{z,2}/k_2}\sqrt{\frac{\epsilon_2}{\mu_0}}\,, \tag{3.1.15}$$

where d_2 is the layer thickness of the central layer and $k_{z,2}$ is the z-component of the complex wave vector in this layer. The result of Eq. (3.1.15) was obtained with the aid of the transfer matrix elements for layer 2 (see Eq. (1.9.7)). It follows from Eq. (3.1.15) that the admittance of this layer is complex in general. To satisfy the guiding condition, the effective admittance of the outer media also has to be complex. Assuming the intrinsic admittance is real (dielectric medium), we obtain imaginary values for the effective admittance by considering evanescent incident waves with a k_x value larger than the wavenumber in the dielectric. We write the admittance of the two half-spaces as

$$Y_{p,1} = Y_{p,3} = \frac{Y_{m,3}k_3}{k_{z,3}} = -i\,\frac{1}{\sqrt{(k_x/k_3)^2 - 1}}\sqrt{\frac{\epsilon_3}{\mu_0}}\,. \tag{3.1.16}$$

Writing the guiding condition in terms of k_x we hence find

$$\frac{k_x^2}{k_3^2} = 1 + \frac{\epsilon_3}{\epsilon_2\tan^2\left(k_{z,2}d_2/2\right)}\left(1 - \frac{k_x^2}{k_2^2}\right), \tag{3.1.17}$$

where $k_{z,2}^2 = k_2^2 - k_x^2$.

We see that for real values of ϵ_1, ϵ_2 (and ϵ_3) and for $k_x < k_2$, the value of k_x will be larger than k_3; this corresponds to an evanescent incident wave, that can be coupled into the central layer by means of the prism of Fig. 3.3. If the k_x value that follows from the guiding condition is smaller than k_2, we expect guided wave propagation in the central layer with small attenuation (apart from the losses occurring at the prism coupling stage and in the adjacent media during propagation). Values of k_x that are larger than the wavenumber in the central layer give rise to strong attenuation of the guided wave. We conclude that a central layer with a high refractive index with respect to the environmental media (or cladding materials) is favourable for wave guiding with small losses over a broad range of incident wave angles. Coupling of light into the central guiding layer can also be carried out by means of a focused beam that is incident onto a specially fabricated entrance face of the central layer, perpendicular to the interfaces (known as end-butt coupling). The coupling efficiency of this method depends on the similarity of the focused field distribution at the entrance window and the electric field profile of the guided wave.

The analysis above was centred on a p-polarised incident wave. A comparable analysis can be carried out for s-polarisation and another guiding condition will be found with, in general, different solutions for the incident k_x wave vector. This is in contrast to the result that we found for surface waves on a metal–dielectric interface where only p-polarisation gave rise to wave guiding along the interface. The presence of at least two interfaces enables wave guiding under broader conditions. The general subject of two- and three-dimensional wave guiding in thin layer structures and in wave guides, active or passive, with a more general geometry is outside the scope of this book. For further study, the reader is referred to basic textbooks like **[235]** and **[151]**.

3.2 Wave Propagation in Metamaterials

In this section we address the particularities of wave propagation in media with uncommon values of electrical permittivity and magnetic permeability. As was mentioned in the introduction to this chapter, electrical resonances occur in or close to the optical frequency domain and they lead to complex values of the permittivity. Magnetic resonances occur at much lower frequencies. In the optical domain the tails of these low-frequency resonances have a barely perceivable influence on the value of the magnetic permeability and its value can be safely taken to be equal to μ_0, the vacuum value. Intensive research effort is currently being made to create artificial materials that exhibit magnetic properties at optical frequencies which differ from those of vacuum. This research effort on so-called *metamaterials* considers modification of substances at the sub-wavelength scale so that their macroscopic properties are altered without introducing appreciable light scattering from the sub-wavelength structures. At this point in time, one has to admit that most of the considerations of drastic changes of the electric and magnetic material properties beyond their classical values are still in the Gedanken-experiment phase and they do not yet correspond to a widespread physical reality.

3.2.1 Material Properties

From this stage on, we shall allow for arbitrary complex values of both the electrical permittivity ϵ and the magnetic permeability μ. We recall that the real and imaginary parts of these quantities cannot be chosen independently as a function of the frequency ω since they are connected by a Hilbert transform relationship. The connection between the real and imaginary parts of ϵ or μ is also called the dispersion relation or the Kramers–Kronig relationship (see e.g. **[10]** and Appendix C of this book). This relationship originates from the causality principle that says that a system cannot deliver a response at a time before the driving impulse was applied to the system. In our case, we consider the electromagnetic field that causes a response in the atoms and the bound or free electrons in the medium. The electric and magnetic frequency response of the medium, given by the permittivity function $\epsilon(\omega) = \epsilon'(\omega) + i\epsilon''(\omega)$ and the permeability function $\mu(\omega) = \mu'(\omega) + i\mu''(\omega)$, obeys the dispersion relations derived by Kramers and Kronig,

$$\begin{aligned}\epsilon'(\omega) - \epsilon_0 &= +\frac{2}{\pi}\int_0^{+\infty}\frac{\omega'\,\epsilon''(\omega')}{\omega'^2-\omega^2}\,d\omega',\\ \epsilon''(\omega) &= -\frac{2\omega}{\pi}\int_0^{+\infty}\frac{\epsilon'(\omega')-\epsilon_0}{\omega'^2-\omega^2}\,d\omega',\end{aligned} \tag{3.2.1}$$

with a comparable relationship for the quantities $\mu'(\omega)$ and $\mu''(\omega)$. A derivation of the Kramers–Kronig relations is given in Appendix C.

As was shown in Fig. 3.4, the real part of the electric permittivity of a metal takes on negative values in the vicinity of the plasma frequency, whilst the imaginary part remains positive. The complex quantity $\epsilon(\omega)$ is generally found in the first quadrant of the complex plane; however, close to a resonance in the material, $\epsilon(\omega)$ moves to the second quadrant. The effect of an electric resonance in a medium on the group velocity has been studied by Lamb **[202]**and Brillouin **[60]**. The magnetic permeability can also exhibit a particular behaviour close to a magnetic resonance in the material. Speculations about electromagnetic wave propagation in the case that both $\epsilon(\omega)$ and $\mu(\omega)$ lie in the second quadrant, or, still more specific, both are real and negative, were first published by Veselago **[356]**. To help us keep track of the amplitude and, especially, the phase of the generally complex permittivity and permeability of the metamaterials, we write

$$\begin{aligned}\epsilon &= |\epsilon|\exp(i\alpha_\epsilon)\,, \qquad \mu = |\mu|\exp(i\alpha_\mu)\,,\\ Y &= (\epsilon/\mu)^{1/2} = |\epsilon/\mu|^{1/2}\exp[i(\alpha_\epsilon-\alpha_\mu)/2]\,.\end{aligned} \tag{3.2.2}$$

The admittance of a medium is real if the phase angles of the permittivity and permeability are equal. Perfect metamaterials are characterised by $\alpha_\epsilon = \alpha_\mu = \pi$ and this property can be considered as a special case of a real admittance value.

3.2.2 Plane Wave Propagation Vector and Energy Flow in a Metamaterial

A harmonic plane wave in a medium with complex values of ϵ and μ is characterised by a complex propagation vector **k** which satisfies (see Eq. (1.8.3)),

$$\mathbf{k}\cdot\mathbf{k} = \omega^2\epsilon\mu\,. \tag{3.2.3}$$

We mention that in the case of a complex wave vector the quantity $\mathbf{k}\cdot\mathbf{k}$ is also complex and does not equal the modulus squared of the vector. We write the complex vector **k** as the sum of a real vector and a purely imaginary vector according to $\mathbf{k} = \mathbf{k}_R + i\mathbf{k}_I$, which allows us to write the complex amplitude of the plane wave as

$$\mathbf{E}_0(\mathbf{r}) = \mathbf{A}_0\exp\left[-\mathbf{k}_I\cdot\mathbf{r}\right]\exp\left[i\left\{\mathbf{k}_R\cdot\mathbf{r}\right\}\right]\,, \tag{3.2.4}$$

where the exponential time dependence $\exp(-i\omega t)$ has been left out according to our convention. A plane with a constant value of $\mathbf{k}_R\cdot\mathbf{r}$ is a plane of constant phase, whereas a plane with $\mathbf{k}_I\cdot\mathbf{r}$ = constant determines a plane in space with constant amplitude.

For the case that $\omega^2\epsilon\mu$ is positive and real we have

$$k_R^2 - k_I^2 = \omega^2\epsilon\mu\,, \qquad \mathbf{k}_R\cdot\mathbf{k}_I = 0\,. \tag{3.2.5}$$

The second condition implies that the two vectors $\mathbf{k}_R$ and $\mathbf{k}_I$ are orthogonal or that one of them is a zero length vector. In the first case, the planes of constant phase and those with constant amplitude are orthogonal. This situation is encountered when a wave suffers total internal reflection at the interface between two dielectrics. It applies to the evanescent wave in the lower index medium with $\mathbf{k}_R$ parallel to the interface and $\mathbf{k}_I$ perpendicular to it. For the second case, two solutions to Eq. (3.2.5) can be found, specifically

$$k_R = \pm\omega\,|\epsilon\mu|^{1/2}\exp\left\{i(\alpha_\epsilon+\alpha_\mu)/2\right\}\,, \qquad k_I = 0\,, \tag{3.2.6}$$

$$k_R = 0\,, \qquad k_I = \pm\, i\,\omega\,|\epsilon\mu|^{1/2}\exp\left\{i(\alpha_\epsilon+\alpha_\mu)/2\right\}\,. \tag{3.2.7}$$

It is seen that the solution according to Eq. (3.2.6) is basically identical to that of (3.2.7) with k_I in (3.2.7) being equal to $i\,k_R$ of (3.2.6). This means that the roles of $\mathbf{k}_R$ and $\mathbf{k}_I$ in the plane wave equation (3.2.4) are interchanged. For this reason, to avoid confusion, we consider only the solution for which $\mathbf{k}_I = \mathbf{0}$ meaning that wave propagation is free of loss (or gain).

We now proceed to consider a general plane wave with a complex wave vector $\mathbf{k}$ and electric and magnetic field vectors that are not necessarily in phase with each other. In a homogeneous and isotropic medium we have

$$\mathbf{k}\cdot\mathbf{E}_0 = 0\,, \qquad \mathbf{k}\cdot\mathbf{H}_0 = 0\,, \tag{3.2.8}$$

and from Maxwell's equations we have the relationships between electric and magnetic field vectors,

$$\mathbf{E}_0 = -\frac{1}{\omega\epsilon}\mathbf{k}\times\mathbf{H}_0\,, \qquad \mathbf{H}_0 = \frac{1}{\omega\mu}\mathbf{k}\times\mathbf{E}_0\,. \tag{3.2.9}$$

For the time-averaged value of the Poynting vector we have, as usual,

$$\langle\mathbf{S}\rangle = \frac{1}{2}\Re\left\{\mathbf{E}_0\times\mathbf{H}_0^*\right\}\,. \tag{3.2.10}$$

Using expression (3.2.9) for $\mathbf{H}_0^*$, we find

$$\mathbf{E}_0\times\mathbf{H}_0^* = \frac{\exp\{i\alpha_\mu\}}{\omega|\mu|}\left\{(\mathbf{k}_R - i\mathbf{k}_I)\left|\mathbf{E}_0\right|^2 - \left[\mathbf{E}_0\cdot(\mathbf{k}_R - i\mathbf{k}_I)\right]\mathbf{E}_0^*\right\}\,. \tag{3.2.11}$$

The second term of Eq. (3.2.11) can be written in a different form using $\mathbf{k}\cdot\mathbf{E}_0$ as given in Eq. (3.2.8). After some manipulation we obtain

$$\mathbf{E}_0\times\mathbf{H}_0^* = \frac{\exp\{i\alpha_\mu\}}{\omega|\mu|}\left\{\left|\mathbf{E}_0\right|^2\mathbf{k}_R + \mathbf{k}_I\times\Im\left(\mathbf{E}_0\times\mathbf{E}_0^*\right) - i\left[\left|\mathbf{E}_0\right|^2\mathbf{k}_I - \mathbf{k}_R\times\Im\left(\mathbf{E}_0\times\mathbf{E}_0^*\right)\right]\right\}\,, \tag{3.2.12}$$

where we have used the identity $\Im(\mathbf{a}\times\mathbf{a}^*) = -2\Re(\mathbf{a})\times\Im(\mathbf{a})$. For the time-averaged value of the energy flow (Poynting vector) we find

$$\langle\mathbf{S}\rangle = \frac{1}{2\omega|\mu|}\left\{\left|\mathbf{E}_0\right|^2\left(\cos\alpha_\mu\mathbf{k}_R + \sin\alpha_\mu\mathbf{k}_I\right) + \left(\cos\alpha_\mu\mathbf{k}_I - \sin\alpha_\mu\mathbf{k}_R\right)\times\Im\left(\mathbf{E}_0\times\mathbf{E}_0^*\right)\right\}\,. \tag{3.2.13}$$

The second term on the right-hand side of Eq. (3.2.13) is only nonzero if the vectors $\Re(\mathbf{E}_0)$ and $\Im(\mathbf{E}_0)$ are not parallel. A comparable expression can be written using the magnetic field strength $\mathbf{H}_0$ and with the phase angle α_ϵ appearing in the expression instead of α_μ.

3.2.2.1 Plane Wave Propagation and the Poynting Vector

For a plane wave in an infinitely extending medium with arbitrary values of ϵ and μ, we use the divergenceless property of the electric field to obtain

$$\mathbf{k}_R\cdot\mathbf{E}_{0,R} - \mathbf{k}_I\cdot\mathbf{E}_{0,I} = 0\,, \qquad \mathbf{k}_R\cdot\mathbf{E}_{0,I} + \mathbf{k}_I\cdot\mathbf{E}_{0,R} = 0\,, \tag{3.2.14}$$

where $\mathbf{E}_{0,R}$ and $\mathbf{E}_{0,I}$ are the real and imaginary parts of the complex electric vector. In a homogeneous and isotropic medium, far away from the source, we expect a plane wave with vectors $\mathbf{k}_R$ and $\mathbf{k}_I$ that are parallel to each other. In general, according to Eq. (3.2.3), we have

$$\begin{aligned}\mathbf{k}_R\cdot\mathbf{k}_R - \mathbf{k}_I\cdot\mathbf{k}_I &= \omega^2|\epsilon\mu|\cos(\alpha_\epsilon+\alpha_\mu)\,,\\ \mathbf{k}_R\cdot\mathbf{k}_I &= \frac{\omega^2|\epsilon\mu|}{2}\sin(\alpha_\epsilon+\alpha_\mu)\,.\end{aligned} \tag{3.2.15}$$

The solution for which $\mathbf{k}_R$ and $\mathbf{k}_I$ are parallel follows from Eq. (3.2.15) as ,

$$\mathbf{k}_R = \omega|\epsilon\mu|^{1/2}\cos[(\alpha_\epsilon + \alpha_\mu)/2]\,\hat{\mathbf{k}}\,,$$
$$\mathbf{k}_I = \omega|\epsilon\mu|^{1/2}\sin[(\alpha_\epsilon + \alpha_\mu)/2]\,\hat{\mathbf{k}}\,, \tag{3.2.16}$$

where $\hat{\mathbf{k}}$ is the unit vector pointing in the propagation direction of the plane wave. In complex notation we write $\mathbf{k} = a\hat{\mathbf{k}}$ where the complex number a is given by

$$a = \omega|\epsilon\mu|^{1/2}\exp[i(\alpha_\epsilon + \alpha_\mu)/2]\,. \tag{3.2.17}$$

Inspection of Eq. (3.2.8) for the electric field vector $\mathbf{A}_0$ in a reference plane $\hat{\mathbf{k}}\cdot\mathbf{r} = 0$ shows that $\mathbf{A}_{0,R} = \Re\{\mathbf{A}_0\}$ needs to be perpendicular to the unit vector $\hat{\mathbf{k}}$ whereas the imaginary part of $\mathbf{A}_0$ must be zero.

The magnetic field vector then follows from Eq. (3.2.9) as

$$\mathbf{H}_0(\mathbf{r}) = \left|\frac{\epsilon}{\mu}\right|^{1/2}\exp[i(\alpha_\epsilon - \alpha_\mu)/2]\,\exp\left[i\omega\,|\epsilon\mu|^{1/2}\,e^{i(\alpha_\epsilon+\alpha_\mu)/2}\,\hat{\mathbf{k}}\cdot\mathbf{r}\right]\hat{\mathbf{k}}\times\mathbf{A}_{0,R}\,. \tag{3.2.18}$$

The Poynting vector associated with this plane wave follows from Eqs. (3.2.13) and (3.2.16) and is given by

$$\langle\,\mathbf{S}\,\rangle = \frac{1}{2}\left|\frac{\epsilon}{\mu}\right|^{1/2}|\mathbf{E}_0(\mathbf{r})|^2\cos[(\alpha_\epsilon - \alpha_\mu)/2]\,\hat{\mathbf{k}} = \frac{1}{2}\Re\{Y_m\}|\mathbf{E}_0|^2\,\hat{\mathbf{k}}\,, \tag{3.2.19}$$

where we have used the expression for the medium admittance of Eq. (3.2.2).

The position dependence of the Poynting vector can be introduced more explicitly by using Eq. (3.2.4), from which we find

$$\langle\,\mathbf{S}(\mathbf{r})\,\rangle = \frac{1}{2}\Re\{Y_m\}\left(\mathbf{A}_{0,R}\right)^2\exp\left\{-2\omega|\epsilon\mu|^{1/2}\sin[(\alpha_\epsilon + \alpha_\mu)/2]\,\hat{\mathbf{k}}\cdot\mathbf{r}\right\}\,\hat{\mathbf{k}}\,. \tag{3.2.20}$$

For small values of α_ϵ and α_μ and for a small excursion $\hat{\mathbf{k}}\cdot\mathbf{r}$ from the reference position at $\mathbf{r} = \mathbf{0}$ we have the approximated value

$$\langle\,\mathbf{S}(\mathbf{r})\,\rangle \approx \frac{1}{2}\left|\frac{\epsilon}{\mu}\right|^{1/2}\left(1 - \omega|\epsilon\mu|^{1/2}(\alpha_\epsilon + \alpha_\mu)\,\hat{\mathbf{k}}\cdot\mathbf{r}\right)\mathbf{A}_{0,R}^2\,\hat{\mathbf{k}}\,. \tag{3.2.21}$$

3.2.2.2 Handedness of the Medium and of the Plane Wave

Equations (3.2.18) and (3.2.19) show that the two field vectors $\mathbf{E}_0(\mathbf{r})$, $\mathbf{H}_0(\mathbf{r})$ and the time-averaged *real* Poynting vector $\langle\,\mathbf{S}(\mathbf{r})\,\rangle$ form a right-handed orthogonal vector triplet, everywhere in the medium. According to our convention the normalised wave vector $\hat{\mathbf{k}}$ is parallel to the Poynting vector. The wave vector $\mathbf{k} = a\,\hat{\mathbf{k}}$ is generally complex and the relationship between the vectors $\mathbf{E}_0$, $\mathbf{H}_0$ and $\mathbf{k}$ is more involved now that two vectors are needed to represent the real and imaginary part of $\mathbf{k}$. In a homogeneous medium in which there is no refraction or reflection, the wave vectors $\mathbf{k}_R$ and $\mathbf{k}_I$ of a propagating but possibly attenuating plane wave are parallel.

Due to material resonances, the phase angles α_ϵ and α_μ of the complex permittivity and permeability can each show a total excursion of π as a function of frequency. Standard materials in the optical regime only exhibit electrical resonances such that the phase angle α_μ is identical to zero. In this case, the real and imaginary parts of the wave vector are both limited to the interval $[0,\ \omega|\epsilon\mu|^{1/2}]$. In other words, the endpoint of the complex wave vector is found in the first quadrant of the complex plane. Lossless wave propagation is only possible for a real valued permittivity ϵ. The introduction of a magnetic resonance with a nonzero phase angle α_μ, however, implies that the endpoint of the complex wave vector can now lie in the upper half of the complex plane. The complex factor a given by Eq. (3.2.17) lies in the range

$$-\,\omega|\epsilon\mu|^{1/2} \le \Re(a) \le \omega|\epsilon\mu|^{1/2}\,, \qquad 0 \le \Im(a) \le \omega|\epsilon\mu|^{1/2}\,. \tag{3.2.22}$$

The latter inequality for the imaginary part of a guarantees that the plane wave amplitude decreases during propagation, in line with the behaviour of a passive medium without optical amplification.

An interesting case occurs when $\alpha_\epsilon = \alpha_\mu = \pi$ (see Fig. 3.6). In this particular case, the vector $\mathbf{k}$ is real and points in the direction of $-\hat{\mathbf{k}}$. Lossless propagation is possible in such a medium for radiation with a frequency at which these special phase angles simultaneously apply. The figure shows that the vectors $\mathbf{E}_0$, $\mathbf{H}_0$ and $\mathbf{k}_R$, taken in this order, now define a *left-handed* vector triplet. For this reason, at such a particular frequency, a medium is said to be 'left-handed'. The energy propagation vector $\langle\,\mathbf{S}\,\rangle$ and the real wave vector $\mathbf{k}_R$ of a plane wave in such a medium have opposite directions.

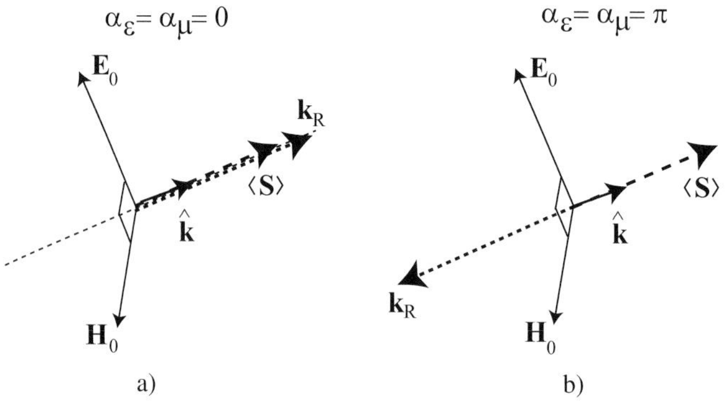

Figure 3.6: The electromagnetic field vectors $\mathbf{E}_0$, $\mathbf{H}_0$ and the time-averaged Poynting vector $\langle \mathbf{S} \rangle$ form a right-handed orthogonal vector triplet.
a) The wave vector $\mathbf{k}_R$ is parallel to the Poynting vector in a standard lossless medium at optical frequencies with $\alpha_\epsilon = \alpha_\mu = 0$.
b) The vectors in a lossless metamaterial are shown ($\alpha_\epsilon = \alpha_\mu = \pi$); $\mathbf{E}_0$, $\mathbf{H}_0$ and $\mathbf{k}_R$ form a left-handed vector triplet.

3.2.3 Reflection and Refraction of a Plane Wave

In this subsection we analyse the reflection and refraction of a plane wave at a planar interface between media having general values of the electromagnetic constants, including the special case of a left-handed medium.

3.2.3.1 Reflected Wave Vector

The reflected wave vector was calculated in Section 1.8 using the field matching conditions of Eq. (1.8.20); these conditions remain valid in the case of reflection in an absorbing medium. The complex propagation vector $\mathbf{k}_1 = \mathbf{k}_{1,R} + i\mathbf{k}_{1,I}$ of the incident wave and the complex propagation vector of the reflected wave obey the reflection law

$$\mathbf{k}_r = \mathbf{k}_1 - 2(\mathbf{k}_1 \cdot \hat{\mathbf{v}})\, \hat{\mathbf{v}} \,, \tag{3.2.23}$$

where $\hat{\mathbf{v}}$ is the unit normal vector of the planar interface between the two media. To obtain the electromagnetic field strengths of the reflected and transmitted waves, we calculate the admittances of the two media with due respect to the angle of incidence of the wave on the interface. To obtain these values, we first need to calculate the wave vector of the transmitted wave.

3.2.3.2 Transmitted Complex Wave Vector

To describe the refraction process at the interface, we need the complex tangential components of the wave vector of the incident wave. In the case of the homogeneous incident wave described in the previous subsection for which $\mathbf{k}_{1,R}$ and $\mathbf{k}_{1,I}$ are parallel, we can write the incident wave vector (assuming it lies in the xz-plane) in the form

$$\mathbf{k}_1 = \omega|\epsilon_1\mu_1|^{1/2} \exp[i(\alpha_{1,\epsilon} + \alpha_{1,\mu})/2]\, (\sin\theta_1, 0, \cos\theta_1) \,, \tag{3.2.24}$$

where θ_1 is the angle of incidence with respect to the normal of the interface that points in the positive z-direction. The tangential component of the wave vector is then proportional to $\sin\theta_1$. The dispersion relation for the wave vector components in the second medium is given by,

$$k_{x,2}^2 + k_{z,2}^2 = \omega^2|\epsilon_2\mu_2| \exp[i(\alpha_{2,\epsilon} + \alpha_{2,\mu})] \,. \tag{3.2.25}$$

Enforcing the continuity of the tangential components at the interface, we find that the squared value of the z-component of the wave vector in the second medium is given by

$$k_{z,2}^2 = \omega^2|\epsilon_2\mu_2| \exp[i(\alpha_{2,\epsilon} + \alpha_{2,\mu})] - \omega^2|\epsilon_1\mu_1| \exp[i(\alpha_{1,\epsilon} + \alpha_{1,\mu})] \sin^2\theta_1 \,. \tag{3.2.26}$$

The choice of the argument ϕ of the square root of the complex quantity $k_{z,2}^2$ needs some care. Employing a phasor representation of Eq. (3.2.26) we have

$$k_{z,2}^2 = A_2 \exp[i\phi_2] - A_1 \exp[i\phi_1] \,. \tag{3.2.27}$$

The modulus and argument of $k_{z,2}$ are denoted by $|k_{z,2}|$ and $\phi_{k_{z,2}}$, respectively, and are given by the expressions

$$|k_{z,2}| = \left\{A_2^2 + A_1^2 - 2A_1A_2\cos(\phi_2 - \phi_1)\right\}^{1/4} ,$$
$$\phi_{k_{z,2}} = \frac{1}{2}\arctan\left(\frac{A_2\sin\phi_2 - A_1\sin\phi_1}{A_2\cos\phi_2 - A_1\cos\phi_1}\right) . \qquad (3.2.28)$$

When $A_2 > A_1$ the argument $\phi_{k_{z,2}}$ lies in the interval

$$\frac{\phi_2}{2} - \frac{1}{2}\arcsin(A_1/A_2) \le \phi_{k_{z,2}} \le \frac{\phi_2}{2} + \frac{1}{2}\arcsin(A_1/A_2) . \qquad (3.2.29)$$

Regardless of the value of A_1/A_2, the argument of $k_{z,2}$ is restricted to the interval $[\phi_2/2 - \pi/4, \phi_2/2 + \pi/4]$. In a similar way, when $A_1 > A_2$, we conclude that $\phi_{k_{z,2}}$ lies in the interval

$$\frac{\phi_1}{2} - \frac{3\pi}{4} \le \phi_{k_{z,2}} \le \frac{\phi_1}{2} - \frac{\pi}{4} . \qquad (3.2.30)$$

The argument ϕ_2 is found in the interval $-2\pi < \phi_2 < 0$ (gain media) and $0 < \phi_2 < 2\pi$ (absorbing media). The special case $\phi_2 = 0$ corresponds to a dielectric medium and $\phi_2 = \pm 2\pi$ corresponds to a perfect metamaterial. The argument ϕ_1 is determined by the medium properties and by the angle of incidence of the incident wave. For the case of an evanescent incident wave, we have to replace the quantity $\sin^2\theta_1$ by the value of $(k_{x,1}/k_1)^2 > 1$.

We illustrate the findings above by means of two examples that are sketched in Fig. 3.7:

- Left-handed transmitted wave
  In Fig. 3.7a the incident wave is transmitted from a medium with absorption to a medium with gain and $A_2 > A_1$. The endpoint of $k_{z,2}^2$ lies on the dashed circle with radius A_1, centred on the endpoint of the phasor A_2. Its precise position depends on the phase angle ϕ_1. The vector component $k_{z,2}^2$ is located in the interval $[\phi_2 - \pi/2, \phi_2 + \pi/2]$, corresponding to an exponentially increasing wave amplitude in the second medium. The phasor that represents $k_{z,2}$ is in the third quadrant and has a negative real part; this indicates a 'left-handed' behaviour of the wave in the second medium. This was to be expected since the argument ϕ_2 was in the range $[-2\pi, -\pi]$, indicating a gain mechanism for both the electric and magnetic field vectors.
- Right-handed transmitted wave
  A comparable reasoning holds for the case $A_1 > A_2$. In Fig. 3.7b we show the case where the first medium is lossless and the second medium is electrically amplifying. A high-spatial frequency component $k_{x,1}$ of the wave vector along the interface yields a large $-A_1$ which dominates the value of $k_{z,2}$ along the negative real axis. If the second medium has only electrical gain, the complex quantity $A_2\exp(i\phi_2)$ with $-\pi < \phi_2 < 0$ is found, for instance, in the third quadrant. In

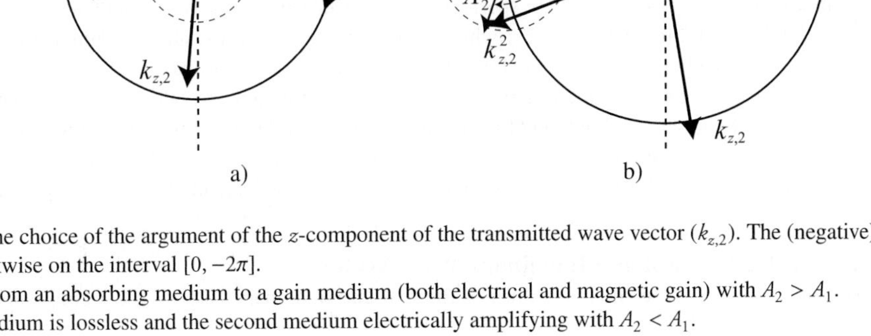

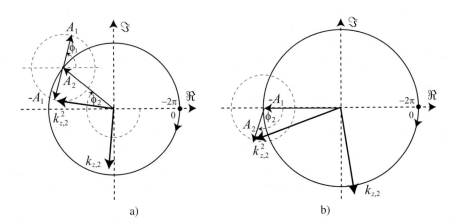

Figure 3.7: The choice of the argument of the z-component of the transmitted wave vector ($k_{z,2}$). The (negative) argument ϕ_2 is plotted clockwise on the interval $[0, -2\pi]$.
a) Transition from an absorbing medium to a gain medium (both electrical and magnetic gain) with $A_2 > A_1$.
b) The first medium is lossless and the second medium electrically amplifying with $A_2 < A_1$.

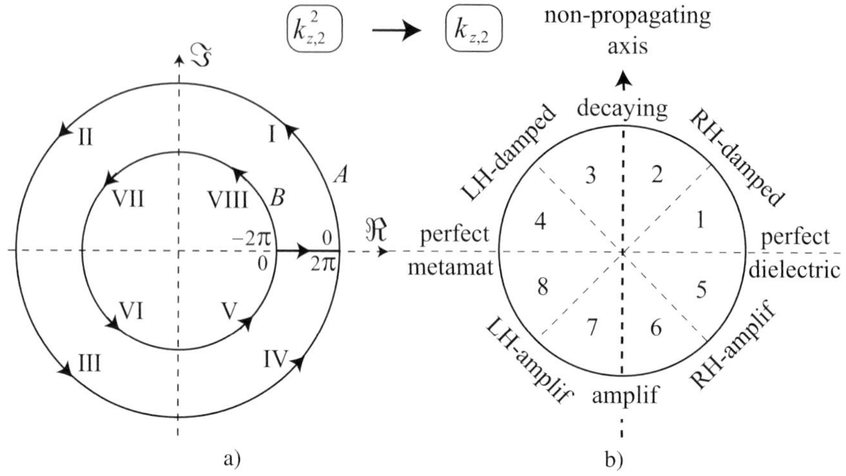

Figure 3.8: The type of plane waves occurring in media that exhibit electric and/or magnetic absorption / gain.
a) The range of complex values of $k^2_{z,2}$.
b) The complex values of $k_{z,2}$.

the drawing we then find the complex quantity $k^2_{z,2}$ in the third quadrant and the resulting argument of $k_{z,2}$ in the fourth quadrant. The z-component of the wave vector in the second medium has a positive real part and the medium behaves in a standard 'right-handed' way; this was to be expected because the absolute value of the argument of A_2 is smaller than π.

In Fig. 3.8a we show the complex quantity $k^2_{z,2}$. The argument of $k^2_{z,2}$ extends from -2π to $+2\pi$, depending on the amplifying or absorptive character of the medium. The branch cut for choosing the sign of the complex square root has been taken along the positive real axis. The two Riemann sheets for $k^2_{z,2}$ have been denoted by A and B. A corresponds to media with a lossy character (quadrants I to IV), whilst B corresponds to gain media (quadrants V to VIII).

Figure 3.8b shows the full range of solutions for $k_{z,2}$ in the complex plane. At normal incidence, with $\sin\theta_1 = 0$, we find solutions for $k_{z,2}$ in octants 1 to 4 and in 5 to 8, respectively. The choice for these regions depends on the dissipating or amplifying character of the medium. Specifically Riemann sheet A, with $0 < \alpha_{\epsilon_2} + \alpha_{\mu_2} < 2\pi$, corresponds to absorptive media, whilst Riemann sheet B, with $-2\pi < \alpha_{\epsilon_2} + \alpha_{\mu_2} < 0$, describes amplifying media. The solutions for $k_{z,2}$ in the octants 3,4, 8 and 7 correspond to *left-handed* materials. Solutions for $k_{z,2}$ along the imaginary axis yield amplified or evanescent non-propagating waves ($\Re\{k_{z,2}\} = 0$). The solutions on the negative imaginary axis can exist in a limited region of space but are not physically realisable in an infinitely extending half-space. At oblique incidence, the value of $c(\theta_1) = |\epsilon_2\mu_2| - |\epsilon_1\mu_1|\sin\theta_1^2$ determines which Riemann sheet has to be selected. When $c(\theta_1)$ is positive $k^2_{z,2}$ lies on sheet B and the wave exhibits amplification upon propagation. In the case of a negative value of $c(\theta_1)$, the solution for the wave vector is on sheet A because the amplification in the medium is counteracted by the loss introduced by evanescent wave propagation.

A special case arises when $\alpha_{2,\epsilon} + \alpha_{2,\mu}$ equals 2π (perfect metamaterial) and the wave is incident from vacuum ($\alpha_{1,\epsilon} + \alpha_{1,\mu} = 0$). For $c(\theta_1) > 0$, we find a real value for $k_{z,2}$ which is less than zero, as was already demonstrated with the aid of Eq. (3.2.16) and (3.2.17). Such a wave solution shows the uncommon feature of wave and wave energy propagation with a regressing phase. A real negative value of $c(\theta_1)$ leads to a negative imaginary value of $k_{z,2}$. The amplitude of the non-propagating wave now increases as a function of the distance in the second medium. This unphysical solution cannot exist in an unbounded region; however, in a limited region of space, for instance in a stack of thin layers, its existence is not forbidden and it will be seen that this solution is required to explain the mechanism of the 'perfect' lens in Section 3.3.

3.2.3.3 Expressions for the Real and Imaginary Wave Vectors

From Eq. (3.2.26), one can derive the squared values of the real and imaginary parts of the z-component of the wave vector. After some algebra, we find

$$k_{z,2,R}^2 = \frac{\omega^2}{2}\Big[|\epsilon_2\mu_2|\cos(\alpha_{2,\epsilon}+\alpha_{2,\mu}) - |\epsilon_1\mu_1|\sin^2\theta_1\cos(\alpha_{1,\epsilon}+\alpha_{1,\mu}) + \big\{|\epsilon_2\mu_2|^2 + |\epsilon_1\mu_1|^2\sin^4\theta_1$$
$$-2|\epsilon_1\mu_1||\epsilon_2\mu_2|\sin^2\theta_1\cos(\alpha_{2,\epsilon}+\alpha_{2,\mu}-\alpha_{1,\epsilon}-\alpha_{1,\mu})\big\}^{1/2}\Big],$$
$$k_{z,2,I}^2 = \frac{\omega^2}{2}\Big[-|\epsilon_2\mu_2|\cos(\alpha_{2,\epsilon}+\alpha_{2,\mu}) + |\epsilon_1\mu_1|\sin^2\theta_1\cos(\alpha_{1,\epsilon}+\alpha_{1,\mu}) + \big\{|\epsilon_2\mu_2|^2 + |\epsilon_1\mu_1|^2\sin^4\theta_1$$
$$-2|\epsilon_1\mu_1||\epsilon_2\mu_2|\sin^2\theta_1\cos(\alpha_{2,\epsilon}+\alpha_{2,\mu}-\alpha_{1,\epsilon}-\alpha_{1,\mu})\big\}^{1/2}\Big]. \qquad (3.2.31)$$

The expression under the square root in Eq. (3.2.31) is non-negative. Depending on the medium constants and the angle of incidence θ_1, the solutions for $k_{z,2,R}$ and $k_{z,2,I}$ can yield either real or imaginary values. Standard refraction, for instance from one dielectric to another, takes place if the conditions $\alpha_{1,\epsilon}+\alpha_{1,\mu}=0$ and $\alpha_{2,\epsilon}+\alpha_{2,\mu}=0$ are fulfilled. The refracted wave vector is given by

$$\mathbf{k}_{2,R} = \omega\left(|\epsilon_1\mu_1|^{1/2}\sin\theta_1,\ 0\ ,\ |\epsilon_2\mu_2|^{1/2}\left\{1-\frac{|\epsilon_1\mu_1|}{|\epsilon_2\mu_2|}\sin^2\theta_1\right\}^{1/2}\right),$$
$$\mathbf{k}_{2,I} = (0\ ,\ 0\ ,\ 0)\ , \qquad (3.2.32)$$

whereby propagation in the second medium is lossless.

A second solution can be found by taking the negative square root solution in Eq. (3.2.31), with the resulting refracted vectors

$$\mathbf{k}_{2,R} = (0\ ,\ 0\ ,\ 0)\ ,$$
$$\mathbf{k}_{2,I} = \omega\left(|\epsilon_1\mu_1|^{1/2}\sin\theta_1,\ 0\ ,\ |\epsilon_2\mu_2|^{1/2}\left\{\frac{|\epsilon_1\mu_1|}{|\epsilon_2\mu_2|}\sin^2\theta_1-1\right\}^{1/2}\right). \qquad (3.2.33)$$

If the argument of the square root in Eq. (3.2.33) is positive, the complex vector $\mathbf{k}_2$ corresponds to an evanescent wave with no phase progression into the second medium and a decreasing field strength away from the interface between the two media.

In general, the calculation of the real and imaginary parts of the $k_{z,2}$-component is performed in such a way that both the phase progression and the wave amplitude variation along the interface are made continuous on both sides of an interface between two media. The real part of the z-component of the wave vector in the second medium corresponds to that of the *ray* vector that follows from Snell's law for dielectric media. Even in the framework of Snell's law, in the case of total internal reflection, we must take into account the evanescent wave in the second medium by allowing an inhomogeneous wave with planes of constant amplitude and constant phase that do not coincide with each other. In the particular case of total internal reflection, these planes were at a right angle with each other. The results of this subsection have extended the concept of inhomogeneous waves in the sense that their planes of constant phase and constant amplitude can be at an arbitrary angle to each other. Such a generalisation is needed when non-dielectric media with absorption or gain, both with respect to their electric and magnetic properties, are considered. The calculation of the real and the imaginary part of the wave vector after refraction or reflection at the interface yields the orientation of the constant phase and amplitude surfaces of the waves. The medium properties and the values of $k_{z,2}^2$ and $k_{z,2}$ can assume the values that are illustrated in Fig. 3.8.

3.2.3.4 Negative Refraction into a Left-handed Medium

In the preceding pages we have already encountered the possibility of a forward progressing wave with a backward directed wave vector component $k_{z,2,I}$. Such a situation occurs when a plane wave is refracted from a dielectric medium to an amplifying medium. In this subsection we will apply our formulas to the case of a (perfect) left-handed metamaterial and come to the conclusion that the *real part* of the wave vector points in the backward direction. This inversion of the direction of phase propagation leads to the so-called effect of *negative refraction*. According to Eq. (3.2.31) the refraction from a right-handed to a left-handed medium with $\alpha_{2,\epsilon}+\alpha_{2,\mu}=2\pi$ yields the wave vectors

$$\mathbf{k}_{2,R} = \omega\left(|\epsilon_1\mu_1|^{1/2}\sin\theta_1,\ 0\ ,\ -|\epsilon_2\mu_2|^{1/2}\left\{1-\frac{|\epsilon_1\mu_1|}{|\epsilon_2\mu_2|}\sin^2\theta_1\right\}^{1/2}\right),$$
$$\mathbf{k}_{2,I} = (0\ ,\ 0\ ,\ 0)\ , \qquad (3.2.34)$$

where the z-component has an opposite sign compared to standard refraction. The regime of left-handedness starts as soon as the phase angle $\alpha_{2,\epsilon}+\alpha_{2,\mu}$ exceeds π. Exactly at the value of π, coming from a dielectric medium, no energy

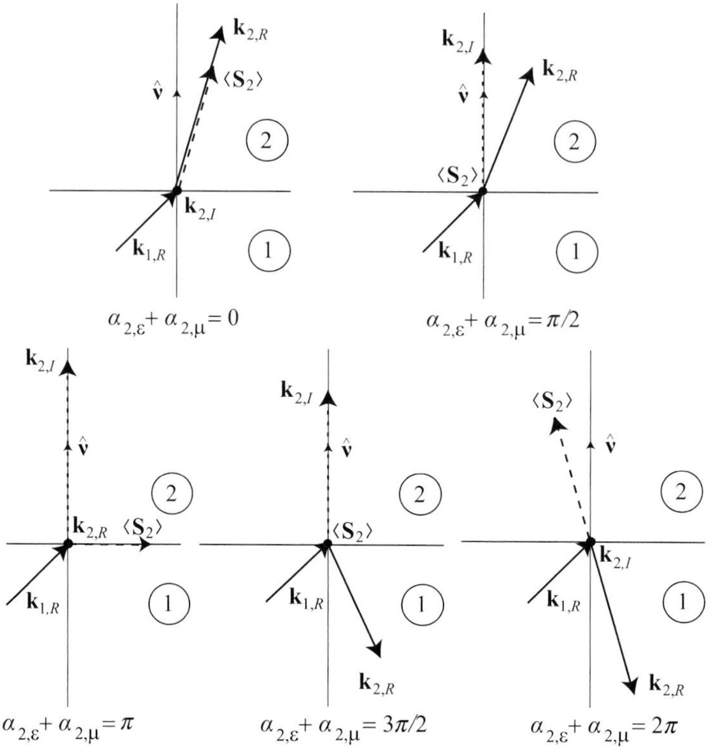

Figure 3.9: The real and imaginary parts ($\mathbf{k}_{2,R}$ and $\mathbf{k}_{2,I}$) of the transmitted wave vector. Various cases for the phase angle $\alpha_{2,\epsilon} + \alpha_{2,\mu}$ of the second medium have been depicted in a schematic way. For the first medium we took $\alpha_{1,\epsilon} + \alpha_{1,\mu} = 0$. The incident wave vector $\mathbf{k}_{1,R}$ is at an angle of $\pi/4$ with the vector $\hat{\mathbf{v}}$, the normal to the interface. The contrast ratio $|\epsilon_2\mu_2/\epsilon_1\mu_1|$ of the two media has a value of 4. A zero vector has been represented by a black dot.

can propagate into the second medium. If the phase angle π is due to an electric resonance only, the medium is called a 'perfect' metal. In the range $\pi \leq \alpha_{2,\epsilon} + \alpha_{2,\mu} \leq 2\pi$, lossy propagation occurs in the second medium. The component of the wave vector, normal to the surface ($k_{z,2,R}$), has changed sign. The vector $\mathbf{k}_{2,R}$ points backward to the first medium.

Lossless propagation in the second left-handed medium requires $\alpha_{2,\epsilon} + \alpha_{2,\mu} = 2\pi$, in which case the second medium is called a 'perfect' left-handed material. The wave vector $\mathbf{k}_2$ points backwards into the first medium; its direction is obtained by a mirror operation with respect to the plane of the interface on the wave vector $\mathbf{k}_{2,R}$ for standard refraction with $\alpha_{2,\epsilon} + \alpha_{2,\mu} = 0$ (see the first and the last drawing of Fig. 3.9). For the perfect left-handed material, the Poynting vector $\langle \mathbf{S} \rangle$ is parallel to the vector $\mathbf{k}_{2,R}$ but it points in the forward direction, into the second medium. It is because of this peculiar behaviour of the Poynting vector that one commonly speaks about *negative* refraction in such perfect left-handed media. In Section 3.3, this special feature of lossless left-handed materials is analysed in more detail.

In Fig. 3.9 we have schematically plotted the refracted vectors $\mathbf{k}_{2,R}$ and $\mathbf{k}_{2,I}$ for some typical values of the phase angle $\alpha_{2,\epsilon} + \alpha_{2,\mu}$. It is interesting to observe that the direction of the refracted vector $\mathbf{k}_{2,R}$ is influenced by the presence of absorption in the second medium (see the drawings with $\alpha_{2,\epsilon} + \alpha_{2,\mu} = \pi/2$ and $\alpha_{2,\epsilon} + \alpha_{2,\mu} = 3\pi/2$). As compared to the absorption-free case, the vector $\mathbf{k}_{2,R}$ is tilted away from the normal $\hat{\mathbf{v}}$. It can be shown that for small values of the phase angle $\alpha_{2,\epsilon} + \alpha_{2,\mu}$, this effect is a quadratic function of the magnitude of the absorption of the medium.

3.2.4 Reflection and Transmission Coefficients for s- and p-polarisation

We have not yet considered the electric and magnetic field strengths of the reflected wave and of the wave transmitted to the second medium. The derivation of the transmission and reflection coefficients is comparable to the one that led to Eqs. (1.10.33)–(1.10.36). Due care is needed now to treat the combination of complex propagation vectors and complex media constants. For the medium admittance we use the complex quantity $Y = |\epsilon/\mu|^{1/2} \exp[i(\alpha_\epsilon - \alpha_\mu)/2]$ of Eq. (3.2.2). In general, the wave vectors in the two media are also complex. We normalise such a complex vector by division by $|\mathbf{k} \cdot \mathbf{k}^*|^{1/2}$, the modulus of the complex vector. The normalised complex vector is then defined by

$$\hat{\mathbf{k}} = \frac{\mathbf{k}_R + i\mathbf{k}_I}{|\mathbf{k}|} = \cos[(\alpha_\epsilon + \alpha_\mu)/2]\, \hat{\mathbf{k}}_R + i\, \sin[(\alpha_\epsilon + \alpha_\mu)/2]\, \hat{\mathbf{k}}_I\,, \tag{3.2.35}$$

where $|\mathbf{k}|$ is given by $\omega|\epsilon\mu|^{1/2}$ in each medium.

3.2.4.1 Electric *s*-polarisation

We first consider the case when the electric field vector is *s*-polarised. Assuming the plane of incidence is the $y = 0$ plane, we thus have that the y-component of the incident field is nonzero. Defining the reflection and transmission coefficients for the electric field vector by r_E and t_E, respectively, we have $1 + r_E = t_E$ as follows from the continuity of the tangential components of the electric field at the interface.

The magnetic field components of the incident, reflected and transmitted plane waves are obtained from the general relationship $\mathbf{H}_0 = \mathbf{k} \times \mathbf{E}_0/(\omega\mu)$. For the tangential magnetic field components of the three waves we then have

$$H_i = -\frac{E_{0,i}k_{i,z}}{\omega\mu_1}\,, \qquad H_r = -r_E\frac{E_{0,i}k_{r,z}}{\omega\mu_1}\,, \qquad H_t = -t_E\frac{E_{0,i}k_{t,z}}{\omega\mu_2}\,, \tag{3.2.36}$$

where $E_{0,i}$ is the electric field strength of the incident wave. The indices i, r and t denote the field components of the incident, reflected and transmitted wave. We observe that the z-component of the reflected wave vector has the opposite sign to that of the incident wave vector and this leads, after some rearrangement, to the second continuity equation

$$(1 - r_E)\hat{k}_{i,z} = t_E\left|\frac{\epsilon_2/\mu_2}{\epsilon_1/\mu_1}\right|^{1/2} \exp[i(\alpha_{1,\mu} - \alpha_{2,\mu})]\ \hat{k}_{t,z}\,, \tag{3.2.37}$$

where $\hat{k}_{i,z} = k_{i,z}/|\mathbf{k}_i|$ and $\hat{k}_{t,z} = k_{t,z}/|\mathbf{k}_t|$. From the two continuity equations at the interface we obtain the solutions

$$\begin{aligned} r_{E,s} &= \frac{Y_1\hat{k}_{i,z}\exp[-i(\alpha_{1,\epsilon} + \alpha_{1,\mu})/2] - Y_2\hat{k}_{t,z}\exp[-i(\alpha_{2,\epsilon} + \alpha_{2,\mu})/2]}{Y_1\hat{k}_{i,z}\exp[-i(\alpha_{1,\epsilon} + \alpha_{1,\mu})/2] + Y_2\hat{k}_{t,z}\exp[-i(\alpha_{2,\epsilon} + \alpha_{2,\mu})/2]}\,, \\ t_{E,s} &= \frac{2Y_1\hat{k}_{i,z}\exp[-i(\alpha_{1,\epsilon} + \alpha_{1,\mu})/2]}{Y_1\hat{k}_{i,z}\exp[-i(\alpha_{1,\epsilon} + \alpha_{1,\mu})/2] + Y_2\hat{k}_{t,z}\exp[-i(\alpha_{2,\epsilon} + \alpha_{2,\mu})/2]}\,, \end{aligned} \tag{3.2.38}$$

where we have used the expression for the complex admittance Y given in Eq. (3.2.2).

The reflection and transmission coefficients for the magnetic field strengths are derived from the ratios $H_{0,t}/H_{0,i}$ and $H_{0,r}/H_{0,i}$, and we obtain

$$\begin{aligned} r_{H,s} &= r_{E,s}\frac{\mu_1 k_{r,z}}{\mu_1 k_{i,z}} = -r_{E,s}\,, \\ t_{H,s} &= t_{E,s}\frac{\mu_1 k_{t,z}}{\mu_2 k_{i,z}} = t_{E,s}\frac{Y_2}{Y_1}\frac{\hat{k}_{t,z}\exp[-i(\alpha_{2,\epsilon} + \alpha_{2,\mu})/2]}{\hat{k}_{i,z}\exp[-i(\alpha_{1,\epsilon} + \alpha_{1,\mu})/2]}\,. \end{aligned} \tag{3.2.39}$$

3.2.4.2 Electric *p*-polarisation

For the magnetic reflection and transmission coefficients in the case of electric *p*-polarisation we find in a comparable way

$$\begin{aligned} r_{H,p} &= \frac{Y_1/(\hat{k}_{i,z}\exp[-i(\alpha_{1,\epsilon} + \alpha_{1,\mu})/2]) - Y_2/(\hat{k}_{t,z}\exp[-i(\alpha_{2,\epsilon} + \alpha_{2,\mu})/2])}{Y_1/(\hat{k}_{i,z}\exp[-i(\alpha_{1,\epsilon} + \alpha_{1,\mu})/2]) + Y_2/(\hat{k}_{t,z}\exp[-i(\alpha_{2,\epsilon} + \alpha_{2,\mu})/2])}\,, \\ t_{H,p} &= \frac{2Y_1/(\hat{k}_{i,z}\exp[-i(\alpha_{1,\epsilon} + \alpha_{1,\mu})/2])}{Y_1/(\hat{k}_{i,z}\exp[-i(\alpha_{1,\epsilon} + \alpha_{1,\mu})/2]) + Y_2/(\hat{k}_{t,z}\exp[-i(\alpha_{2,\epsilon} + \alpha_{2,\mu})/2])}\,. \end{aligned} \tag{3.2.40}$$

The electric reflection and transmission coefficients are given by

$$
\begin{aligned}
r_{E,p} &= r_{H,p}\frac{\epsilon_1 k_{r,z}}{\epsilon_1 k_{i,z}} &&= -r_{H,p}\,, \\
t_{E,p} &= t_{H,p}\frac{\epsilon_1 k_{t,z}}{\epsilon_2 k_{i,z}} &&= t_{H,p}\frac{Y_1/(\hat{k}_{i,z}\exp[-i(\alpha_{1,\epsilon}+\alpha_{1,\mu})/2])}{Y_2/(\hat{k}_{t,z}\exp[-i(\alpha_{2,\epsilon}+\alpha_{2,\mu})/2])}\,.
\end{aligned}
\tag{3.2.41}
$$

With respect to our previous expressions for reflection and transmission in dielectric media in Section 1.8, Eqs. (1.8.24)–(1.8.27), we note that the complex admittance of a medium is still given by $Y/\hat{k}_z'$ (electric p-polarisation) and by $Y\hat{k}_z'$ (electric s-polarisation), respectively, where the prime indicates that we have to consider the effective admittances

$$
\begin{aligned}
Y_{e,s} &= \frac{k_z}{\mu} = Y_m\,\hat{k}_z' = Y_m\,\hat{k}_z\,\exp[-i(\alpha_\epsilon+\alpha_\mu)/2]\,, && \text{electric s-polarisation}\,, \\
Y_{e,p} &= \frac{\epsilon}{k_z} = \frac{Y_m}{\hat{k}_z'} = \frac{Y_m}{\hat{k}_z\exp[-i(\alpha_\epsilon+\alpha_\mu)/2]}\,, && \text{electric p-polarisation}\,.
\end{aligned}
\tag{3.2.42}
$$

Inclusion of the exponential factor in Eq. (3.2.42) allows us to easily describe transmission from one 'complex' medium to another with arbitrary electric and magnetic properties. A perfect metamaterial has an effective admittance that is equal to that of the corresponding standard medium. In general, both the medium admittances and the z-components of the wave vectors in Eqs. (3.2.38) and (3.2.41) are complex.

3.2.4.3 Non-propagating Waves in a Perfect Metamaterial

A special case arises when the wave vector of a wave entering the second medium has a component parallel to the interface whose absolute value is larger than the modulus $\omega|\epsilon_2\mu_2|^{1/2}$ of the wave vector in the second medium. To study this case further we write Eq. (3.2.34) for the complex wave vector in the second medium in a slightly different form,

$$
\begin{aligned}
\mathbf{k}_{2,R} &= \omega|\epsilon_2\mu_2|^{1/2}\left(\left|\frac{\epsilon_1\mu_1}{\epsilon_2\mu_2}\right|^{1/2}\sin\theta_1\,,\ 0\,,\ -\left\{1-\frac{|\epsilon_1\mu_1|}{|\epsilon_2\mu_2|}\sin^2\theta_1\right\}^{1/2}\right) = \omega|\epsilon_2\mu_2|^{1/2}\,\hat{\mathbf{k}}_{2,R}\,, \\
\mathbf{k}_{2,I} &= (0\,,\ 0\,,\ 0)\,.
\end{aligned}
\tag{3.2.43}
$$

The vector $\hat{\mathbf{k}}_{2,R}$ in the first line of Eq. (3.2.43) is a unit vector. The x-component of this unit vector along the interface can be larger than unity for two reasons. If the second medium has a lower 'optical contrast' than the first medium ($|\epsilon_1\mu_1| > |\epsilon_2\mu_2|$), the propagation angle θ_1 can be chosen such that the x-component of the normalised wave vector becomes larger than unity. It is also possible, independently of the optical contrast, that the wave in the first medium is evanescent and keeps this non-propagating property in the second medium. In both cases, the z-component of $\hat{\mathbf{k}}_{2,R}$ becomes purely imaginary and we can write the complex wave vector in the second medium in the form

$$
\begin{aligned}
\mathbf{k}_{2,R} &= \omega|\epsilon_2\mu_2|^{1/2}\left(\left|\frac{\epsilon_1\mu_1}{\epsilon_2\mu_2}\right|^{1/2}\sin\theta_1\,,\ 0\,,\ 0\right)\,, \\
\mathbf{k}_{2,I} &= \omega|\epsilon_2\mu_2|^{1/2}\left(0\,,\ 0\,,\ -i\left\{\frac{|\epsilon_1\mu_1|}{|\epsilon_2\mu_2|}\sin^2\theta_1 - 1\right\}^{1/2}\right)\,.
\end{aligned}
\tag{3.2.44}
$$

As usual, we have taken the value of the square root of a negative quantity to lie on the positive imaginary axis. The minus sign of the z-component in Eq. (3.2.43) brings us to the surprising conclusion that the energy density of the non-propagating wave in the second medium (metamaterial) increases as a function of the perpendicular distance to the interface. A compelling reason can be invoked for why the electromagnetic energy density, tending to infinity as a function of distance, will remain finite in practice. The nonlinear behaviour of the medium causes saturation of the energy density. A small residual dissipation of electric or magnetic energy may lead to disintegration of the medium **[137]**.

If the perpendicular distance from the interface in the metamaterial is z, the time-averaged electromagnetic energy density $\langle\, w_{em}\,\rangle$ associated with the wave has assumed a value

$$
\begin{aligned}
\langle\, w_{em}(z)\,\rangle &= \exp\left\{2\omega\left(|\epsilon_1\mu_1|\sin^2\theta_1 - |\epsilon_2\mu_2|\right)\,z\right\}\langle\, w_{em}(0)\,\rangle \\
&= \exp\left\{2\left(|\epsilon_{1,r}\,\mu_{1,r}|\sin^2\theta_1 - |\epsilon_{2,r}\,\mu_{2,r}|\right)k_0 z\right\}\langle\, w_{em}(0)\,\rangle\,,
\end{aligned}
\tag{3.2.45}
$$

where $\langle w_{em}(0) \rangle$ is the time-averaged electromagnetic energy density at the interface and where we have used the relative permittivity and permeability values in the second line of the equation. For values of z that are in the order of the vacuum wavelength λ_0 of the light, the increase in electromagnetic energy density becomes appreciable and material disintegration cannot be excluded.

3.3 The Concept of the Perfect Lens

The original paper about the peculiar properties of perfect left-handed media **[356]** mentions the 'negative' refraction effect and the inverted role of positive and negative lenses when their material is changed from right-handed to left-handed. Figure 3.10a shows the change from standard refraction to 'negative' refraction that occurs at an interface between two media with opposite handedness. Opposite handedness also alters the function of a simple lens that obeys Gauss' laws of paraxial optics (see Section 4.10 for more details on paraxial imaging with a single lens or a system of lenses). For instance, as shown in Fig. 3.10b, the classical positive plano-convex lens becomes a diverging element with negative optical power if the handedness of the lens material is opposite to that of the environmental medium. The same inversion of lens power is shown in Fig. 3.10c for a classical diverging lens that becomes a focusing element by changing the handedness of the lens material. Further consequences of the opposite handedness of two media separated by a planar interface are discussed, for instance, by Pendry in **[272]** and **[273]**. It is shown in these publications that a plane-parallel plate made of a perfect left-handed metamaterial behaves as a classical lens but with superior imaging performance.

Before discussing the special imaging properties of the plane-parallel plate of metamaterial we briefly enumerate the limitations which are associated with the imaging by a classical lens or lens system. It is well known that a standard single lens with positive power shows imaging defects when the cross-section of an incident parallel pencil of rays is made larger and the convergence angle of the focusing pencil in image space is increased. The approximate refraction law $n_1\theta_1 = n_2\theta_2$ that forms the basis of Gaussian paraxial optics is inadequate to accurately describe the direction of the focusing rays that emerge from a single lens when the maximum convergence angle is larger than, say, 0.1 radian. The sharp imaging to a point predicted by Gaussian optics cannot be achieved by imaging pencils with a larger opening angle. The focusing beam in image space does not possess a spherical wavefront but shows wavefront *aberration* and the focal spot is blurred (see Chapter 5 for a detailed treatment of imaging by pencils of rays with arbitrary convergence or divergence angles in optical imaging systems). According to Pendry, a thin plane-parallel plate of left-handed material which is positioned in a standard right-handed medium (see Fig. 3.11) does not show optical aberration and can create a perfect focus according to the ray optics model. Moreover, the imaging by the lens of Fig. 3.11 is thought to be unaffected by the so-called *diffraction limit* which says that the lateral extent of the intensity distribution in focus cannot be made substantially smaller than half the wavelength of the light. It follows that the focal spot produced by the thin plate of perfect left-handed metamaterial is free of geometrical aberration effects and, as the lens supersedes the diffraction limit, the image of a point object can have a much smaller lateral size than the wavelength of the light. A focal field distribution which is an exact copy of the emitted object field cannot be achieved as half of the emitted solid angle of radiation is lost in the imaging process by the thin plate of left-handed metamaterial.

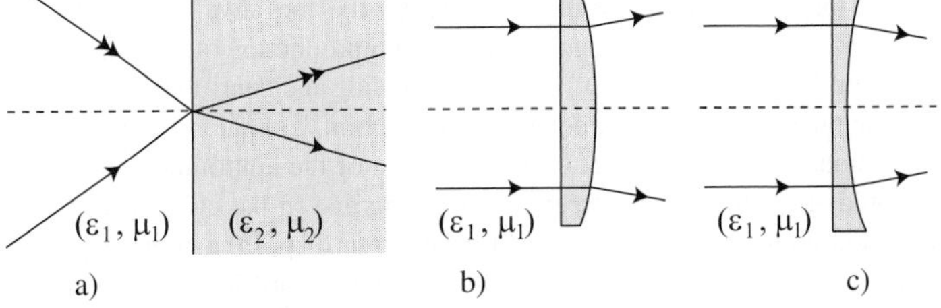

Figure 3.10: a) Negative refraction from a medium with positive real permittivity ϵ_1 and permeability μ_1 to a medium with negative real values of ϵ_2 and μ_2 (grey-shaded in the figure). The propagation direction of electromagnetic energy along an optical ray is indicated by the single or double arrows.
b) and c) The inverted focusing properties of an elementary plano-convex lens (b) and plano-concave lens (c), both made out of a left-handed material and positioned in a right-handed medium.

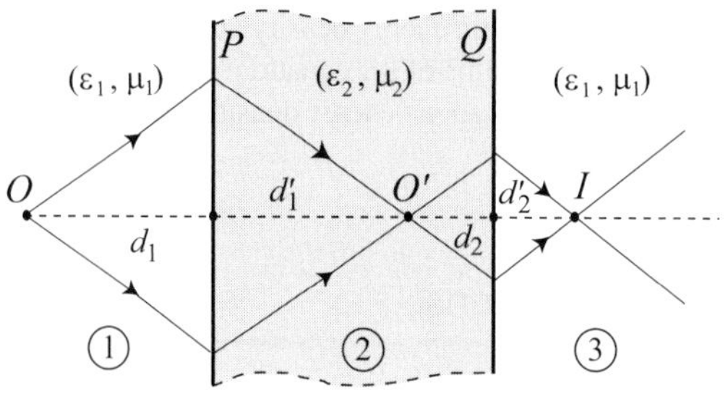

Figure 3.11: A plane-parallel plate lens made of a negative index material (grey-shaded in the figure), positioned in a positive index medium. The object point source O is imaged inside the negative index material at the point O' and then again in the positive index image space at the point I. Imaging by the plate is illustrated by means of tracing the path of some typical geometrical rays. The special case for which $|\epsilon_1| = |\epsilon_2|$ and $|\mu_1| = |\mu_2|$ is shown. The distances d_1 and d_1', as well as the distances d_2 and d_2', are equal in this case.

To study the particular imaging properties of the perfect lens, the propagation of the electromagnetic field from the source point to the image point through the plane-parallel plate of left-handed material has to be analysed. Several approaches are possible. We mention the commonly used finite-difference time-domain method (FDTD **[333]**, see also Chapter 1 of this book) which can provide the spatial distribution of an electromagnetic field throughout an inhomogeneous area of space as a function of time. Other methods for electromagnetic field modelling which are suitable for the analysis of imaging by the 'perfect' lens have been discussed in the same chapter, for instance, the Fourier modal method (FMM), the rigorous coupled wave analysis (RCWA) and the finite-element method (FEM). Alternatively, the propagation of a plane wave through the perfect lens can be analysed. The action of the lens on a spherical wave is then determined by summing the appropriate angular spectrum of plane waves that are transmitted by the lens towards the image space. In this section we shall consider the action of a negative index lens on an arbitrarily shaped incident wavefront using an angular spectrum decomposition with the aid of plane waves. The propagation of a single plane wave through such a 'perfect lens' is relatively easy to analyse because the perfect lens comprises two parallel planar interfaces, labelled P and Q in Fig. 3.11. Our earlier results of Chapter 1 for the reflection and transmission of plane waves at a planar interface can thus be used, with some minor adjustments due to the special properties of a left-handed metamaterial.

3.3.1 Plane Wave Analysis of the Perfect Lens

The plane wave analysis of the perfect lens is performed by considering freely propagating and evanescent parts of the angular spectrum emitted by the source separately. The field emitted by a single dipole source is described in Subsection 1.6.2 (see also **[160]** and **[130]**). The amplitude and phase of the corresponding vector plane wave spectrum are obtained from a Fourier transformation. For an analysis of the quality of a perfect lens, the exact shape of the emitted spectrum is left aside. We instead restrict our attention to study of the transfer of the plane wave spectrum by the plane-parallel plate 'negative' lens. Our criterion for perfect imaging by the 'negative' lens is the exact reproduction of the object-side angular spectrum of plane waves in image space. Exact reproduction means that the amplitude of an outgoing plane wave at the image point I is equal to that of the corresponding incident plane wave at the object point O. The mutual phases of the focusing plane waves, measured at the image point I, should be equal to those of the incident plane waves measured at O. We emphasise that the exact correspondence of the amplitude and phase of the plane wave field components at O and I should apply to both the freely propagating and to the evanescent part of the considered plane wave spectrum. When this condition is satisfied we would have at our disposal a genuinely 'perfect' lens that beats the performance of a standard lens made of a right-handed material. A standard lens with a focal distance which is much larger than the wavelength of the light cannot efficiently transfer an evanescent wave from the object to the image plane. Standard lens quality is thus limited by the earlier mentioned diffraction limit which means that features in the object plane which are smaller than the wavelength of the light cannot be imaged by the lens with appreciable contrast.

The simple shape of the perfect lens means that we can use the transfer matrix method for a plane wave that is incident on a sequence of planar interfaces (see e.g. Eqs. (1.9.4)–(1.9.7) for p-polarised light). The plane waves in the (right-handed) object and image space, labelled 1 and 3 (Fig. 3.11), which have equal (ϵ, μ) values, are characterised by their k_x

wave vector component and the corresponding k_z value. To simplify the analysis, we require that the admittances of the right-handed and left-handed medium are equal, which implies that $(\epsilon_1\mu_2)^{1/2} = (\epsilon_2\mu_1)^{1/2}$. We also require that the length of the wave vector in both media is equal such that the absolute values of the angle of incidence and the angle of refraction are equal at each interface. To satisfy this requirement the absolute values of the corresponding media constants on both sides of an interface should be equal ($|\epsilon_1| = |\epsilon_2|$ and $|\mu_1| = |\mu_2|$). We then have at each interface an angular magnification of the optical rays which is equal to -1 (see Fig. 3.11). Using the special real values $\epsilon_2 = -\epsilon_1$ and $\mu_2 = -\mu_1$ for the medium constants we write for the z-components of the wave vector in the media 1 and 3

$$k_{z,1} = k_{z,3} = \begin{cases} +(\omega^2\epsilon_1\mu_1 - k_{x,1}^2)^{1/2}, & k_{x,1}^2 \leq \omega^2\epsilon_1\mu_1 \\ +i(k_{x,1}^2 - \omega^2\epsilon_1\mu_1)^{1/2}, & k_{x,1}^2 > \omega^2\epsilon_1\mu_1 \end{cases} . \tag{3.3.1}$$

The z-component of the wave vector in the left-handed medium, for the special case with $\alpha_{2,\epsilon} = \alpha_{2,\mu} = \pi$, is given by

$$\begin{aligned} k_{z,2} &= (\omega^2\epsilon_1\mu_1 \exp(i2\pi) - k_{x,1}^2)^{1/2} = \exp(i\pi)\,(\omega^2\epsilon_1\mu_1 - \exp(-i2\pi)k_{x,1}^2)^{1/2} \\ &= \begin{cases} -(\omega^2\epsilon_1\mu_1 - k_{x,1}^2)^{1/2} = -k_{z,1}, & k_{x,1}^2 \leq \omega^2\epsilon_1\mu_1 \\ -i(k_{x,1}^2 - \omega^2\epsilon_1\mu_1)^{1/2} = -k_{z,1}, & k_{x,1}^2 > \omega^2\epsilon_1\mu_1 \end{cases} . \end{aligned} \tag{3.3.2}$$

To analyse the propagation from the object to the image space we must calculate the values of the reflection and transmission coefficients for s- or p-polarisation. Inspection of Eqs. (3.2.38)–(3.2.41) shows that in our special case of identical absolute values of the medium admittances ($|\epsilon_1/\mu_1| = |\epsilon_2/\mu_2|$) all transmission coefficients are unity and no light is reflected at the interfaces denoted by P and Q. To study the amplitude and phase transfer from the entrance to the exit surface, we first use Eq. (1.9.4) twice to establish the relationship between the plane wave electric field strengths A_1 in object space and A_3 in image space (see Fig. 3.12). For the tangential components of the electric and magnetic field on the interfaces P and Q we find

$$\begin{cases} E_{t,P} = \dfrac{k_{z,1}}{k_1}A_1 \\ H_{t,P} = Y_1 A_1 \end{cases}, \qquad \begin{cases} E_{t,Q} = \dfrac{k_{z,3}}{k_3}A_3 \\ H_{t,Q} = Y_3 A_3 \end{cases} . \tag{3.3.3}$$

The relative simplicity of these expressions derives from the absence of any reflected waves in object or image space. We then apply Eq. (1.9.7) to the geometry of the perfect lens with $d_1' = d_1$ and $d_2' = d_2$, whereby we have

$$\begin{pmatrix} \dfrac{k_{z,3}}{k_3}A_3 \\ Y_3 A_3 \end{pmatrix} = \begin{pmatrix} \cos\{k_{z,2}(d_1' + d_2)\} & -i\dfrac{k_{z,2}}{k_2 Y_2}\sin\{k_{z,2}(d_1' + d_2)\} \\ -i\dfrac{Y_2 k_2}{k_{z,2}}\sin\{k_{z,2}(d_1' + d_2)\} & \cos\{k_{z,2}(d_1' + d_2)\} \end{pmatrix} \begin{pmatrix} \dfrac{k_{z,1}}{k_1}A_1 \\ Y_1 A_1 \end{pmatrix} . \tag{3.3.4}$$

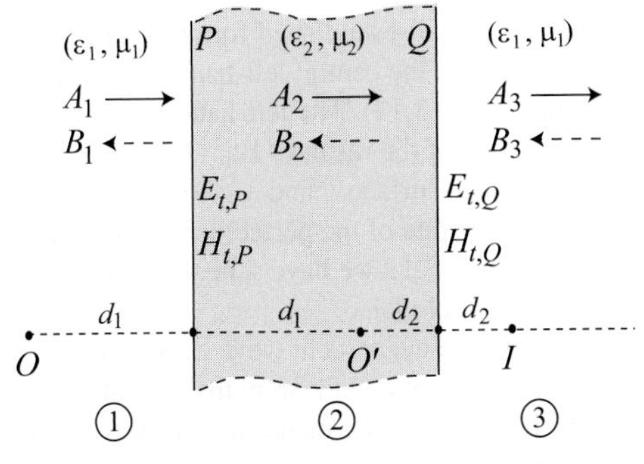

Figure 3.12: A thin plane-parallel plate lens made of a perfect left-handed material. The electric field strengths of the forward and backward propagating plane waves in the three media are given by A_j and B_j, respectively. The tangential electric and magnetic field components on the entrance and exit surfaces P and Q are denoted by $(E_{t,P}, H_{t,P})$ and $(E_{t,Q}, H_{t,Q})$. The reflected waves labelled B_1 and B_2 are zero if $|\epsilon_1/\mu_1| = |\epsilon_2/\mu_2|$.

In this equation we have replaced the original $\cos\theta_i$ function by the more general quantity $k_{z,i}/k_i$ where k_i is the length of the wave vector $\mathbf{k}_i$. With the identities $Y_3 = Y_1$ and $k_{z,3} = k_{z,1}$ and using the result of Eq. (3.3.2) we obtain

$$A_3 = -iA_1 \sin\{k_{z,1}(d_1' + d_2)\} + A_1 \cos\{k_{z,1}(d_1' + d_2)\} = A_1 \exp\{-ik_{z,1}(d_1' + d_2)\} . \tag{3.3.5}$$

For an incident evanescent wave with $k_{z,1} = +i(k_{x,1}^2 - \omega^2\epsilon_1\mu_1)^{1/2}$, we conclude that the p-polarised electromagnetic field within the left-handed material increases exponentially in the forward z-direction. An identical derivation can be performed for the case of electrical s-polarisation. In object space, for the case of an (evanescent) wave of which the field strength at O is given by A_0, the z-dependent field strength is given by $A_0 \exp(+ik_{z,1}z)$. If $k_{z,1}$ lies on the positive imaginary axis the amplitude of the wave is an exponentially decreasing function for increasing z values. The amplitude A_1 of the wave on the entrance surface P of the lens is then given by $A_1 = A_0 \exp(+ik_{z,1}d_1)$. At a distance d_1 from the first interface P, at the point O' in the left-handed medium, the amplitude attenuation of the evanescent wave in object space has been compensated for by the exponential increase in the left-handed medium. Each plane wave thus produces a field strength at O' that is equal to its original strength at the object point O. O' is called the intermediate focus of O and it possesses perfect imaging quality but is not accessible for observation. There is, however, a second image point, beyond the second interface Q of the plane-parallel plate, where the amplitude of each plane wave component emitted by the source at O has the same amplitude as at O itself. According to Eq. (3.3.5), with $d_1' = d_1$, we find this point at a distance d_2 from the second interface Q. Exactly at this point, the amplitude decrease of an evanescent plane wave in the right-handed media 1 and 3 is compensated for by the amplitude increase with the ratio A_3/A_1 in the left-handed medium. The exact reproduction at the image point I of the field strength in O is valid for both freely propagating and for evanescent waves.

For a perfect image to be formed at I the phase of a plane wave at I should be identical to the phase it has at the object point O, apart from an irrelevant phase offset that is common to all plane waves. We observe that a freely propagating plane wave travels equal distances in a right-handed and in a left-handed medium when it propagates from O to I. The net result of the phase progression in the right-handed medium and the phase regression in the left-handed medium with equal values of the k_z-components in both media gives rise to exactly the same phase of each plane wave at O and at I. The phase of the exponentially decreasing and increasing non-propagating plane waves in the two media does not change as a function of the axial position. The phase of a plane wave at I could further be influenced by the phase changes at transmission through the two interfaces P and Q. For the special case of equal absolute values of the electromagnetic constants in the right- and left-handed media these phase changes at the interfaces are zero. We thus conclude that since the amplitude and phase of the plane waves at the object point O are exactly reproduced at the image point I, the plane-parallel plate behaves as a 'perfect' lens. In theory, the lens shows unlimited resolution. In practice we expect its resolution to be far better than the classical diffraction limit of λ that constitutes the theoretical resolution limit for a classical (right-handed) lens. The practical limit to image resolution of the 'perfect' lens is the subject of the next subsection.

3.3.2 The 'Perfect Lens' Result

The derivation in the preceding subsection uses well-established formulas but does not elucidate the physical origin of the amplitude increase in the positive z-direction in the central left-handed medium of the waves that were exponentially decreasing in the right-handed media 1 and 3 (Fig. 3.12). The left-handed material in medium 2 is passive and does not possess gain properties. A physical interpretation of the result of Eq. (3.3.5) can be found by considering the appearance of two coupled surface waves, one on the entrance surface P and a second one on the exit surface Q of the plane-parallel plate. After an initial transient period, the steady-state of the perfect lens shows that much of the electromagnetic energy has been transported to the exit surface Q. In Fig. 3.13 we have schematically depicted the surface waves related to the entrance and to the exit surface of the plane-parallel plate.

As an example for wave transfer we consider an evanescent wave A, originating from a source in object space. In the case of electric p-polarisation, the electric and magnetic field strengths in media 1, 2 and 3 are obtained from Eq. (3.1.9). At this point, no special requirements are imposed on the medium constants (ϵ_1, μ_1), (ϵ_2, μ_2) and (ϵ_3, μ_3). Omitting the common exponential factor $\exp\{+ik_x x\}$ from each expression, we find

$$\mathbf{E}_1(\mathbf{r}) = H_A\left(k_{z,1}, 0, -k_x\right)\frac{1}{\omega\epsilon_1}\exp\{+ik_{z,1}z\} + H_B\left(-k_{z,1}, 0, -k_x\right)\frac{1}{\omega\epsilon_1}\exp\{-ik_{z,1}z\} ,$$

$$\mathbf{H}_1(\mathbf{r}) = H_A(0, 1, 0)\exp\{+ik_{z,1}z\} + H_B(0, 1, 0)\exp\{-ik_{z,1}z\} ,$$

$$\mathbf{E}_2(\mathbf{r}) = H_C\left(k_{z,2}, 0, -k_x\right)\frac{1}{\omega\epsilon_2}\exp\{+ik_{z,2}z\} + H_D\left(-k_{z,2}, 0, -k_x\right)\frac{1}{\omega\epsilon_2}\exp\{-ik_{z,2}z\}\,,$$
$$\mathbf{H}_2(\mathbf{r}) = H_C\,(0,1,0)\exp\{+ik_{z,2}z\} + H_D\,(0,1,0)\exp\{-ik_{z,2}z\}\,,$$
$$\mathbf{E}_3(\mathbf{r}) = H_E\left(k_{z,3}, 0, -k_x\right)\frac{1}{\omega\epsilon_3}\exp\{+ik_{z,3}(z-\varDelta)\}\,,$$
$$\mathbf{H}_3(\mathbf{r}) = H_E\,(0,1,0)\exp\{+ik_{z,3}(z-\varDelta)\}\,. \tag{3.3.6}$$

In Fig. 3.13 we suppose that the wave vector components $k_{z,l}$, $l = 1, 2, 3$, are imaginary in all three media such that the fields have an exponentially varying strength as a function of z. If medium 2 is right-handed, the field with index C decreases with increasing z and belongs to the decaying tail of the surface waves bound to the entrance surface; however, if medium 2 is left-handed, the field labelled C exponentially increases with z. In the figure we have depicted this wave profile by $C'$and it physically belongs to the inside (decaying) tail of the surface wave bound to the exit surface. The same reasoning holds for the waves D and $D'$when the medium 2 switches from right- to left-handed.

For the p-polarised coupled waves described above, we require continuity of the tangential components of **E** and **H** at the interfaces P and Q. The incident field produces a magnetic field H_A at the location $z = 0$, such that after some manipulation we can find the following expressions for the magnetic field strengths of the surface waves labelled B, C, D and E:

$$\frac{H_B}{H_A} = 2\frac{\left(1+\dfrac{k_{z,3}\epsilon_2}{k_{z,2}\epsilon_3}\right)\left(1-\dfrac{k_{z,2}\epsilon_1}{k_{z,1}\epsilon_2}\right)\exp\{-ik_{z,2}\varDelta\} + \left(1-\dfrac{k_{z,3}\epsilon_2}{k_{z,2}\epsilon_3}\right)\left(1+\dfrac{k_{z,2}\epsilon_1}{k_{z,1}\epsilon_2}\right)\exp\{+ik_{z,2}\varDelta\}}{\left(1+\dfrac{k_{z,3}\epsilon_2}{k_{z,2}\epsilon_3}\right)\left(1+\dfrac{k_{z,2}\epsilon_1}{k_{z,1}\epsilon_2}\right)\exp\{-ik_{z,2}\varDelta\} + \left(1-\dfrac{k_{z,3}\epsilon_2}{k_{z,2}\epsilon_3}\right)\left(1-\dfrac{k_{z,2}\epsilon_1}{k_{z,1}\epsilon_2}\right)\exp\{+ik_{z,2}\varDelta\}}\,, \tag{3.3.7}$$

$$\frac{H_C}{H_A} = \frac{2\left(1+\dfrac{k_{z,3}\epsilon_2}{k_{z,2}\epsilon_3}\right)\exp\{-ik_{z,2}\varDelta\}}{\left(1+\dfrac{k_{z,3}\epsilon_2}{k_{z,2}\epsilon_3}\right)\left(1+\dfrac{k_{z,2}\epsilon_1}{k_{z,1}\epsilon_2}\right)\exp\{-ik_{z,2}\varDelta\} + \left(1-\dfrac{k_{z,3}\epsilon_2}{k_{z,2}\epsilon_3}\right)\left(1-\dfrac{k_{z,2}\epsilon_1}{k_{z,1}\epsilon_2}\right)\exp\{+ik_{z,2}\varDelta\}}\,, \tag{3.3.8}$$

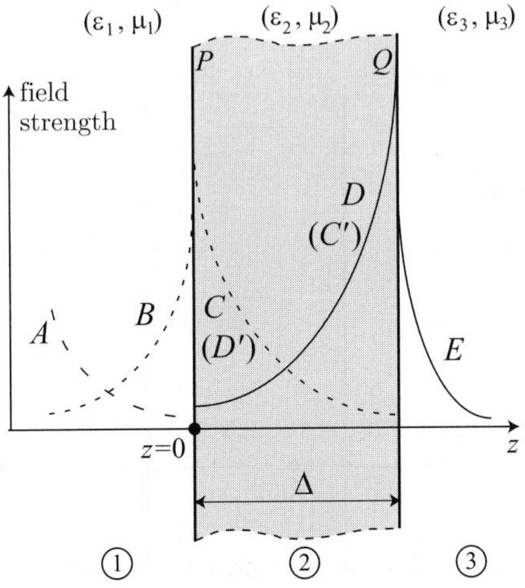

Figure 3.13: Bound surface waves on the two sides of a plane-parallel plate. The surface wave (B, C) is bound to the entrance surface P, the wave (D, E) to the exit surface Q. The thickness of the plate is $\varDelta$. In the object space we have added an incident evanescent wave, labelled A. The image space contains only the decaying plane wave component E belonging to the second surface wave (no incident wave from $z = +\infty$). The complex indices of the media are determined by (ϵ_1, μ_1), (ϵ_2, μ_2) and (ϵ_3, μ_3).

$$\frac{H_D}{H_A} = \frac{2\left(1 - \frac{k_{z,3}\epsilon_2}{k_{z,2}\epsilon_3}\right)\exp\{+ik_{z,2}\Delta\}}{\left(1 + \frac{k_{z,3}\epsilon_2}{k_{z,2}\epsilon_3}\right)\left(1 + \frac{k_{z,2}\epsilon_1}{k_{z,1}\epsilon_2}\right)\exp\{-ik_{z,2}\Delta\} + \left(1 - \frac{k_{z,3}\epsilon_2}{k_{z,2}\epsilon_3}\right)\left(1 - \frac{k_{z,2}\epsilon_1}{k_{z,1}\epsilon_2}\right)\exp\{+ik_{z,2}\Delta\}}, \quad (3.3.9)$$

$$\frac{H_E}{H_A} = \frac{4}{\left(1 + \frac{k_{z,3}\epsilon_2}{k_{z,2}\epsilon_3}\right)\left(1 + \frac{k_{z,2}\epsilon_1}{k_{z,1}\epsilon_2}\right)\exp\{-ik_{z,2}\Delta\} + \left(1 - \frac{k_{z,3}\epsilon_2}{k_{z,2}\epsilon_3}\right)\left(1 - \frac{k_{z,2}\epsilon_1}{k_{z,1}\epsilon_2}\right)\exp\{+ik_{z,2}\Delta\}}. \quad (3.3.10)$$

Comparable expressions hold for the electric field ratios of an electrically s-polarised wave.

In the general case, with arbitrary values of the electric permittivity and magnetic permeability, the two sets of surface waves have finite amplitudes. This means that a reflected evanescent wave tail B is present in the object medium as well as an exponentially increasing and an exponentially decreasing wave in medium 2.

A special case arises when medium 2 becomes left-handed and the absolute values of (ϵ, μ) in all three spaces are equal. We see that in such a case $(1 + k_{z,3}\epsilon_2/k_{z,2}\epsilon_3) = 2$ and $(1 - k_{z,3}\epsilon_2/k_{z,2}\epsilon_3)$ becomes 0. Likewise, we have $(1 + k_{z,2}\epsilon_1/k_{z,1}\epsilon_2) = 2$ and $(1 - k_{z,2}\epsilon_1/k_{z,1}\epsilon_2) = 0$. For the magnetic field strengths we then find

$$H_B = 0, \quad H_{C'} = H_A, \quad H_D = 0, \quad H_E = H_A \exp\{+ik_{z,2}\Delta\} = H_A \exp\{-ik_{z,1}\Delta\}, \quad (3.3.11)$$

with, in this special case, identical expressions for the electric fields strengths.

We note that in the presence of the left-handed medium 2 the wave vector component $k_{z,2}$ is negative imaginary. It is now the wave C' that is exponentially increasing and we attribute this wave to the surface wave that is bound to the exit surface Q. Both the wave B in object space and the exponentially decreasing wave in the medium 2 are missing, which means that there is no surface wave that is coupled to the entrance surface. All electromagnetic energy is transferred to the surface wave coupled to the exit surface. In the steady state that we study here, the field strengths $\mathbf{E}_E$ and $\mathbf{H}_E$ at the exit surface can be drastically increased with respect to the value on the entrance surface. This surprising effect should be understood as a special coupling effect of the incident wave, exclusively to the surface wave on the exit surface.

In conclusion we find that when $k_{z,2}$ is negative imaginary, only the surface wave tail C', increasing with z, is present in the left-handed medium. The decreasing wave component part is identical to zero, as well as the reflected wave tail in object space. On both sides of the interfaces, the electric and magnetic fields have equal strengths. Figure 3.14 illustrates this special case. Perfect imaging can be performed between the points O and O' within medium 2 and between O and I in the external medium 3, where O has been arbitrarily chosen in the figure. We find that at these three axial positions, the field strengths are equal regardless of which plane wave component is transmitted from O through the structure (see the

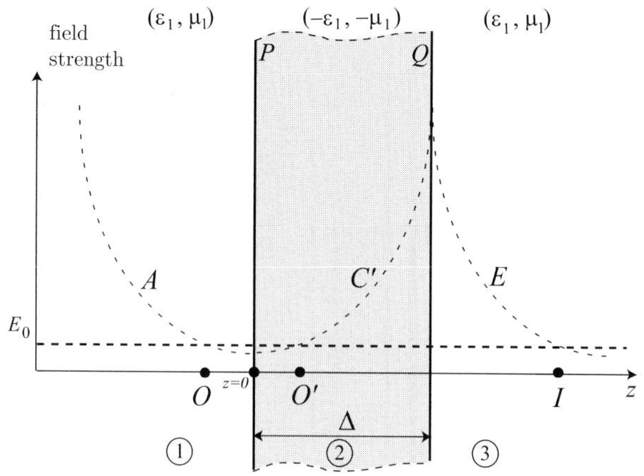

Figure 3.14: The surface-wave structure of the left-handed plane-parallel plate for the case that $|\epsilon_2| = \epsilon_1 = \epsilon_3$ and $|\mu_2| = \mu_1 = \mu_3$. The surface wave coupled to the entrance surface is absent and only the two surface wave tails bound to the exit surface Q remain. The wave C' in the left-handed medium has the same field strength as the incoming evanescent wave A on the entrance surface P. Perfect imaging is achieved between axial positions with equal field strengths, for instance between the points O, $O'$and I, which each have an electric field strength E_0.

horizontal line at level E_0 in Fig. 3.14). This holds for both freely propagating waves and for evanescent waves. Especially because of this latter property, the term 'perfect lens' has been adopted.

3.3.2.1 The Silver Lens

A practical implementation of a 'quasi' perfect lens uses a very thin planar layer of silver. With a permittivity of $\epsilon_{Ag} = -3.3 + 0.7i$ in the 400 nm vacuum wavelength range, silver approaches the ideal negative real permittivity quite closely. The permeability of the metal remains that of vacuum, $\mu_{Ag} = \mu_0$. We present an example with a source at a distance d_1 of 50 nm in front of an Ag layer of 100 nm thickness. In Fig. 3.15 we show the amplitude (left-hand graphs) and phase (right-hand graphs) of plane wave components as a function of the value of k_x of the incident wave vector, normalised with respect to k_0, the vacuum wavenumber. In the upper row the amplitude and phase are shown of the plane wave

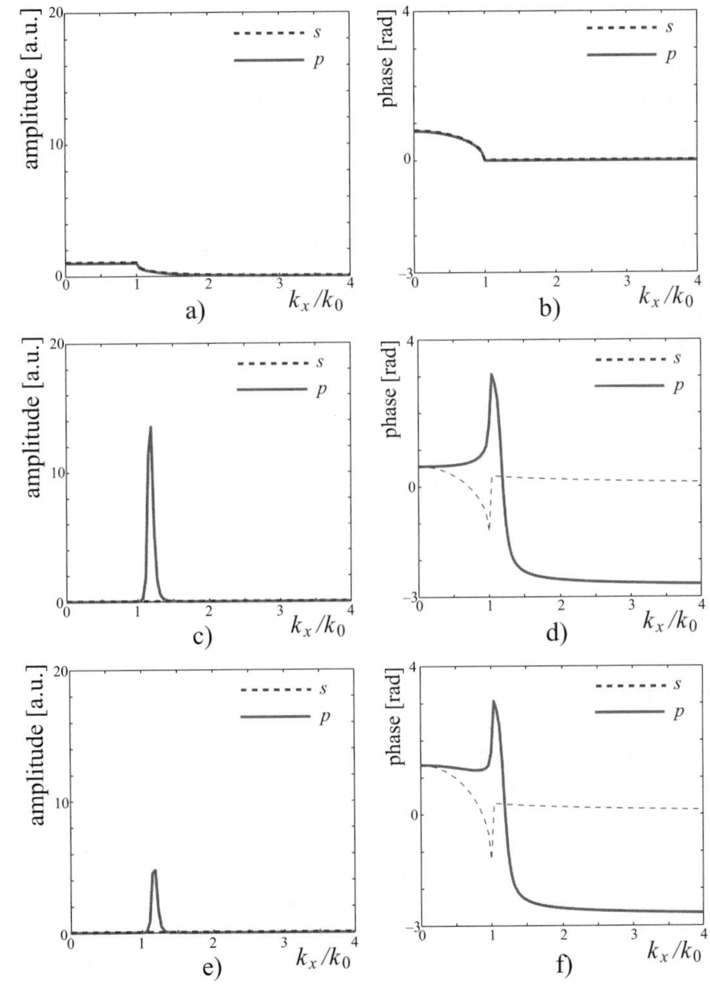

Figure 3.15: The amplitude (left-hand graphs) and phase (right-hand graphs) of plane wave components passing through a very thin plane-parallel plate of silver (thickness 100 nm), as a function of the normalised x-component (k_x/k_0) of the incident wave vector. The normal vectors to the surfaces of the silver plate point in the z-direction. The vacuum wavelength of the light is 400 nm.

a) and b) Field on the entrance surface P of a plate produced by a source at a distance of 50 nm in front of the plate.

c) and d) Field on the exit surface Q of the silver plate.

e) and f) Field at a distance of 50 nm beyond the exit surface Q.

components, measured in the axial reference point at the entrance plane of the silver plate. Up to the cutoff for free propagation, $k_x/k_0 \leq 1$, the amplitude remains unity. For $k_x/k_0 > 1$, the wave amplitude exponentially decreases as a function of k_x/k_0 because of the evanescent character of the wave. The phase of the freely propagating waves is given by $k_z d_1$ where d_1 equals 50 nm in this example. For the evanescent waves, the phase remains unchanged. Both the data for s- and p-polarisation are given in the graphs, by means of dashed and solid curves, respectively. As is expected, no polarisation dependence is present for the incident wave, propagating in vacuum.

In graphs c) and d) of Fig. 3.15 we display the amplitude and phase of the plane wave components at the exit surface of the 100 nm thick Ag-plate. The s-polarisation component is virtually zero since the admittance of Ag for such a plane wave is almost purely imaginary. The solid curve for the p-polarised component shows an enhancement in the region of the evanescent waves, close to the transition value $k_x/k_0 = 1$. This enhancement can be attributed to the appearance of p-polarised surface waves on both surfaces of the Ag plane-parallel plate. Despite the absorption in Ag, a field enhancement is present on the exit surface due to the coherent interaction between the two tails of the surface waves inside the plate. The s-polarised wave is virtually zero in the silver medium.

In the third row of Fig. 3.15, we depict the amplitude and phase at a distance of 50 nm behind the silver plate. For a perfect lens, the ideal image would appear in this plane. We observe a transfer of wave amplitude beyond the diffraction limit of $k_x/k_0 = 1$. The resonant effect that is associated with this peak of p-polarised light is accompanied by a strong variation in phase. Unfortunately, such a non-constant phase transfer is disadvantageous for the imaging quality of the plane-parallel silver plate.

3.3.3 Practical Limitations of the 'Perfect Lens' Concept

The realisation of a perfect lens suffers in practice from some intrinsic difficulties. We briefly comment on the limitations which can arise from the high energy density in the left-handed material, the manufacturing of the left-handed material and the influence of residual absorption in the left-handed material.

3.3.3.1 Energy Density

A practical choice of object distance, plate thickness and image distance is in the ten to fifty nanometre range so that near-field operation of the perfect lens is possible. In terms of the wavelength of the light, we typically have an object distance d_1 of $\lambda_0/10$, a plate thickness $d_1' + d_2$ of $\lambda_0/5$ and an image distance d_2' of $\lambda_0/10$. For reasons of simplicity, we put $|\epsilon_i| = \epsilon_0$ and $|\mu_i| = \mu_0$, i.e. the vacuum values. The transfer of a plane wave with $k_x = 10k_0$ ($k_0 = 2\pi/\lambda_0$), shows that the ratio between the electric field strengths at the exit surface Q and entrance surface P amounts to $E_E/E_A \approx \exp(4\pi)$. The corresponding factor of the energy densities is of the order of 10^{11}. This numerical example illustrates a limitation in the resolution of the perfect lens due to heating of the left-handed material in the presence of some residual absorption **[137]**.

3.3.3.2 Material Requirements

The special properties of left-handed materials discussed in this chapter are not possible in the absence of material dispersion **[356]**,**[246]**. The left-handedness induced by material properties at the atomic scale arises close to absorption lines and this induces a generally strong frequency-dependent behaviour of the refractive index and a residual absorption. Realisation of left-handed material properties over a broad frequency band is therefore technologically difficult to achieve. The simultaneous realisation of negative values of permittivity and permeability of homogeneous materials in the optical domain is also problematic (see the introductory section of this chapter). Successful operation of left-handed materials has been demonstrated in the microwave region of the spectrum by tailoring a material in the sub-wavelength domain. In the optical domain such tailoring is also possible in principle. Recent research efforts have focused on the three-dimensional structuring of a material to achieve the desired dispersion characteristics by waveguiding effects. The challenge here, however, is the realisation of left-handed behaviour over a broad range of k-vector values. In parallel, a solution has to be found for the associated problem of increased light scattering by the sub-wavelength features.

3.3.3.3 Residual Absorption and Ultimate Resolution

Residual absorption in the left-handed material leads to a frequency-dependent phase shift in the transmitted evanescent plane wave spectrum. This effect can be derived from the ratio H_E/H_A that was given in Eq. (3.3.10). A small absorption in the left-handed material is well represented by letting $\epsilon_2 = \epsilon_1 \exp\{i(\pi - \delta)\} = \epsilon_3 \exp\{i(\pi - \delta)\}$ and $\mu_2 = \mu_1 \exp\{i(\pi - \delta)\} = \mu_3 \exp\{i(\pi - \delta)\}$, where δ is a very small phase angle. In the denominator of the expression for H_E/H_A, the first product of the two factors between braces will remain close to 4, with an error that is quadratic in δ. The product in the second term of the denominator will deviate from zero with a linear dependence on δ. Omitting the details of the derivation we put

$$\left(1 - \frac{k_{z,3}\epsilon_2}{k_{z,2}\epsilon_3}\right)\left(1 - \frac{k_{z,2}\epsilon_1}{k_{z,1}\epsilon_2}\right) \approx \frac{2i\delta k_x^2}{k_x^2 - \epsilon_1\mu_1} . \tag{3.3.12}$$

The ratio E_E/E_A in this case then follows as

$$\frac{E_E}{E_A} \approx \exp\{ik_{z,2}\Delta\}\left\{1 - i\,\frac{\delta\, k_x^2 \exp\{+2ik_{z,2}\Delta\}}{2(k_x^2 - \epsilon_1\mu_1)}\right\} . \tag{3.3.13}$$

For sufficiently large values of k_x, this expression further simplifies to

$$\frac{E_E}{E_A} \approx \exp\{+k_x\Delta\}\left\{1 - i\,\frac{\delta}{2}\exp\{+2k_x\Delta\}\right\} . \tag{3.3.14}$$

Equation (3.3.14) shows that the first-order effect of the small absorption is a frequency-dependent phase shift at the exit surface of the perfect lens. To keep this phase shift below an acceptable level we require it to be smaller than, for instance, $\pi/2$. This leads to the following constraint on the absorption angle δ,

$$\delta < \pi \exp\{-2k_x\Delta\} . \tag{3.3.15}$$

Alternatively, for a given absorption angle δ, we are allowed to say that the maximum frequency that is correctly transmitted to the image space is given by

$$k_{x,\max} = -\frac{\ln(\delta/\pi)}{2\Delta} . \tag{3.3.16}$$

3.3.3.4 Axial Focal Depth

The focusing properties of a perfect lens are governed by two effects. For freely propagating plane wave components, just as for a standard lens, the resulting wavefront is exactly spherical when observed from an ideal image point I. The optical path length from object to image for different propagating components is exactly zero at this point. This is different to a standard lens where the optical path length from a real object to a real image point has a finite positive value. In the left-handed lens, the optical path length outside the lens plate is exactly cancelled by the negative phase increase inside. The axial position of the left-handed lens for correct transfer of the propagating plane waves should introduce phase shifts in the image point I that are less than, preferably, $\pi/2$ radians. This leads to a tolerance on the axial position of the image plane of the order of $\pm\lambda/4$.

With respect to the super-resolving evanescent waves, these waves do not have a z-dependent phase. The focusing of the evanescent waves is perturbed by amplitude defects on propagation. Taking the highest spatial frequency to be transferred, $k_{x,\max}$, we allow an axial shift z_f that introduces less than 20% amplitude change of this wave in a defocused plane. With the high-frequency approximation $k_z \approx k_x$, we have in image space

$$z_f \leq \frac{0.22}{k_{x,\max}} = 0.070\, q_{\text{res}} , \tag{3.3.17}$$

where the smallest resolvable detail q_{res} is defined by $\pi/k_{x,\max}$. It is seen that the focal tolerance for substantial super-resolution is of the order of 1 nm ($k_{x,\max} = 10\,k_0$, $\lambda_0 = 400$ nm).

II Geometrical Theory of Optical Imaging

Introduction to Part II

The ray-based theory of light propagation was developed and virtually brought to perfection before the advent of the electromagnetic theory of wave propagation. The original ray theory with its rectilinear propagation of light was inspired by the observation of the straight shadows cast by objects and by the observation of the directional deviation of light propagation at mirrors or lenses. A mathematical foundation of the ray-optics model was introduced by Fermat in 1662 by stating the principle of least time, or, equivalently, of least optical path. A comparable statement can already be found in the writings of Al-Haytham (AlHacen) (circa 1030), but his finding was not mathematically further developed at that time. Fermat was challenged by his fellow scientists to derive the Snell–Descartes law of refraction using his general principle. He succeeded in doing so, although it took him some five years to produce the convincing proof. With the aid of Fermat's principle of least time, the basic laws of geometrical optics can be established (see, for instance, Appendix I of **[37]**).

In this part of the book, electromagnetic wave theory is used to derive the ray-optics model of light propagation. It turns out that the geometrical ray is collinear with the Poynting vector of an electromagnetic wave and that the ray is perpendicular to the wave surfaces of constant phase in space (the *wavefronts* associated with a pencil of rays). Hamilton introduced the characteristic function associated with an optical system. With respect to reference points in object and image space, the characteristic function gives the value of the optical pathlength measured along an incident ray, through the optical system up to the image-side ray. Hamilton showed that either ray intersection points or ray directions in object and image space of an optical system can be obtained from a knowledge of the optical pathlength function of the system. The characteristic function is the equivalent of, for instance, an electric potential function generated by a distribution of charges in space. The gradient of the potential function yields the direction of the electric field lines. General properties of an optical system can be derived from Hamilton's characteristic function(s) in an elegant manner, for instance for optical systems that possess circular symmetry. A practical problem is the calculation of the multi-variable characteristic function for arbitrary directions and off-axis intersection positions of the object and image rays. Simpler expressions are available when the 'paraxial' approximation is used. In the first chapter of Part II we show how the laws of paraxial optics can be derived from the paraxially approximated characteristic functions. From a formal representation of Hamilton's characteristic functions, valid for finite ray angles and ray positions, some important imaging conditions and imaging theorems can be derived. In this respect we mention the extremely important sine-condition in optical imaging, originally stated by E. Abbe. The chapter on the basic properties of optical rays and pencils of rays is concluded by an extended section on paraxial imaging.

The second chapter of this part of the book is entirely devoted to imaging aberrations. The aberrations of a system determine the degradation of an image, within the approximated framework of ray optics. An encompassing theory of optical aberration of lowest order goes back to the 1850s. It was extremely welcome for the systematic design of the first landscape and portrait objectives for photographic cameras. In the chapter on aberration, a detailed derivation of the so-called *third-order* aberration coefficients is given, based on the wavefront aberration values of an individual refracting or reflecting surface. On purpose, aberration values in terms of the calculated wavefront deviation from the ideal spherical shape are chosen. Such a representation of optical aberration is more appropriate for well-corrected imaging systems such as microscope objectives, telescopic systems for astronomical observation and high-resolution projection lenses or mirror systems. The aberration values evaluated up to a certain maximum order of the intersection height with a surface and the aperture angle of a ray are not exact. In the practice of the design of an optical system, the analytic information on imaging aberration is supplemented by exact data on imaging defects that are gathered from so-called *finite* ray-tracing. A number of examples are given of finite ray-tracing formulas through non-standard optical surfaces. We mention aspheric surfaces and optical surfaces with a surface modification, for instance, an imprinted diffracting structure (hologram or diffractive optical element). We also address ray and wave propagation in inhomogeneous media, with certain gradient-index distributions inside the optical medium. The chapter also presents the approach to be followed when an optical ray is incident at an interface between, for instance, two anisotropic media. A total of four rays may be excited at the interface, each with its specific amplitude, propagation direction and state of polarisation.

In the modern-day practice of optical design, the digital computer plays a crucial role. The third chapter of Part II presents some analytic design methods, which require the solution of one or more (coupled) differential equations associated with the consecutive construction of ray intersection points at one or two aspheric surfaces of an optical system. The design of imaging systems of a more general nature, with larger image field angles, is carried out with the aid of a (paraxial) pre-design followed by a numerical optimisation of a chosen 'merit' function. We discuss several optimisation strategies and the various ways in which an optical merit function can be defined. In modern design, the expected mechanical tolerances during manufacturing and the tolerances on medium properties are ideally included in the optimisation process. The chapter provides the reader with an analysis of the various statistical distributions that play a role when the expected value and the variance of a set of aberration coefficients of a manufactured optical system are calculated.

The analysis and design tools that have been presented in the first three chapters of Part II are applied in the fourth chapter to a large number of optical designs. Special attention is paid to historic designs and early (photographic) objectives that clearly show how the laws of paraxial optics and the quantitative results of third-order aberration theory can be exploited in practice. The more current design examples range from relatively severely-aberrated imaging systems with large field angle to small-field microscope objectives, and from extremely large-diameter astronomical telescopes to very demanding high-resolution projection lenses and mirror systems for microlithography. It is shown that the on-paper imaging quality of the optical design and the manufacturability and stability of the final instrument should be simultaneously taken care of during the design process to achieve a successful opto-mechanical product.

4 Foundations of Geometrical Optics

4.1 Introduction

The basic carrier for light energy propagation in geometrical optics is the light ray. The assumption of rectilinear propagation of a light ray in a homogeneous medium goes back to ancient times. The change in direction of a light ray at reflection was well understood, the explicit sine law of refraction was only put forward in the beginning of the seventeenth century by Snell and Descartes. Snell based his law on experimental evidence, Descartes deduced it from his model for light propagation. A few decades later, the principle of least time was formulated by Fermat; it took him several years to deduce the sine law of refraction from his least-time principle for optical rays (1666). With these theoretical ingredients, the foundation of geometrical or ray optics had been laid and a complete theory of imaging by optical components like mirrors and lenses could be devised. Later developments in optical physics have shown the inaptness of geometrical optics to accurately describe light propagation in situations where the interference and diffraction of light become important. A striking result was put forward by Airy who calculated the energy distribution of the well-focused image of a point object by a perfect lens. Instead of the point-like image predicted by geometrical optics, he proved that the size of the image of a point object is finite. The basis for his calculation was the wave theory of light that, as of 1800, became the leading theory for light propagation due to the findings of Young and Fresnel. It followed from Airy's calculation that a lower limit on the size of a point image is set by the wavelength of the light. The discovery of the laws of electromagnetism in 1861 brought Maxwell to the conclusion that light propagation was also governed by these laws. In the next section, we follow the general practice **[224]**,**[37]**,**[186]** and derive the laws of geometrical optics from the laws of electromagnetism. Another procedure to derive the laws of geometrical optics is based on Fermat's least-time principle (see Appendix I of reference **[37]**). The advantage of starting with Maxwell's equations is that geometrical optics appears as a limiting case of Maxwell's equations and error terms of various order can be defined to bridge the gap between the result following from the exact Maxwell equations and the result based on the approximate geometrical model of light propagation **[186]**.

Why is it that geometrical optics is still widely in use in the optics community? Practical optical instruments have lateral dimensions of the order of thousands or even millions of optical wavelengths. A complete electromagnetic description at the scale of the wavelength of the light is unfeasible and not necessary for such 'large' optical components or systems. A satisfactory compromise between accuracy and calculation speed is a description of light propagation through these large components using solely the geometrical optics approximation. Mathematically, the geometrical or ray optics approximation of light propagation is obtained when the limit of Maxwell's equation is taken for $\omega \rightarrow \infty$, or, equivalently, $\lambda \rightarrow 0$. For instance, it is common practice to assume rectilinear propagation of X-rays when diagnosing the human body from a shadow projection on a planar surface. But as soon as regular structures are analysed with repeating patterns that are comparable to the X-ray wavelength, diffraction effects are clearly visible and the ray-optics approximation has to be abandoned.

More generally, when light intensity data are required in the imaging region of the instrument, the geometrical optics model is prone to yield singular results for the focused field and a more refined model needs to be put into place. The light distribution in focus is then calculated by means of the pathlength data obtained from geometrical optics for the light field

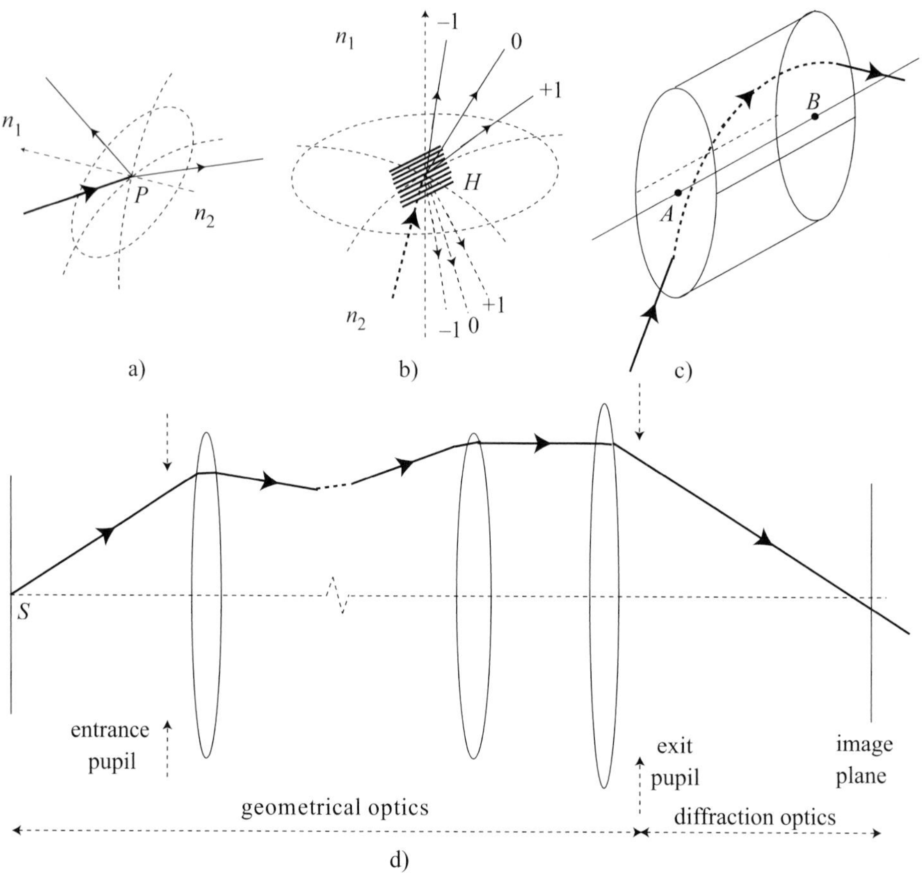

Figure 4.1: The application of the geometrical optics approximation for light propagation.

in the exit pupil of the optical system. The pathlength data enable the construction of a wavefront to which Huygens' principle is applied. In practice, this is equivalent to the calculation of a diffraction integral of a more or less refined structure. These diffraction integrals are the subject of Chapters 8 and 9 on the diffraction theory of optical imaging by perfect and aberrated optical systems.

In Fig. 4.1, the typical regions of application of geometrical optics are illustrated. In Fig. 4.1a the reflection and refraction are shown of a plane wave at a smooth, quasi-planar interface between two media with refractive indices n_1 and n_2. The incident light ray is shown thick solid, the reflected and refracted rays are thin solid. Snell's law can be safely used when the smoothness of the transition allows us to neglect light scattering. If the bending of the surface is very small, the reflected or refracted waves are assumed to remain planar waves. In Fig. 4.1b a comparable configuration is shown with a grating-like structure printed or recorded on the interface surface. Such a structure is found, for instance, on an embossed phase hologram. Extra reflected and refracted rays occur now, labelled in terms of the diffraction order number (dashed straight lines for the reflected orders, solid lines for the transmitted orders). Here too, we approximate the diffracted waves by locally flat sections of a plane wave. In Fig. 4.1c we consider wave propagation in an inhomogeneous medium. The wave vector follows a curved three-dimensional trajectory. The geometrical ray path inside the medium is indicated by the dashed line. The inhomogeneous medium inside the cylinder has an index distribution that possesses radial symmetry with respect to the medium axis AB. Snell's law is applied on the entrance and exit surface of the cylindrical inhomogeneous rod. In Fig. 4.1d we schematically represent an optical system with its entrance and exit pupil. More sophisticated optical propagation models are needed to calculate the far field of the source in the object space and to study the superposition of the diffracted, scattered and polarised fields in the image region. But from the far field of the source to the entrance pupil and through the optical system towards the exit pupil, the geometrical optics approximation is a frequently used, robust approximation. It is mainly from the exit pupil to the image region that a more sophisticated diffraction-based light propagation model is needed to calculate the detailed intensity distribution in the image plane and a close volume around it.

This two-tier approach with a sequential application of geometrical and diffraction-based light propagation is very successful in practice. It is fast in evaluating the light propagation through a complicated optical imaging system using geometrical optics as the propagation vehicle. It is sufficiently accurate because one switches to a more refined theory for detailed calculation of the light distribution in a focused image plane. The approach is less adequate if the lateral dimensions of optical components become comparable to the wavelength of the light, for instance several tens of wavelengths and less. This is the case for light-guiding structures, for structured surfaces in optics like holograms and gratings, for tiny microlenses , thin-layer coatings, etc. For several of these examples, geometrical optics is still adequate for delivering an energy propagation direction. But geometrical optics generally fails to correctly predict the amplitude or field strength associated with the ray propagation vector.

In this chapter we describe an imaging system as a succession of refracting and/or reflecting surfaces and the rays are assumed to intersect the optical surfaces in the prescribed sequence. The ray trajectories are the result of what is called *sequential* ray-tracing. The study of ray trajectories associated with (partly) reflected or scattered rays is a different subject. Multiple ray paths are combined to calculate, for instance, the irradiance in some receiving plane. Such an approach is needed in illumination systems and when studying the effects of stray light and undesired parasitic imaging in an optical system. The procedure is called *non-sequential* ray-tracing. In this chapter we limit ourselves to sequential ray-tracing in optical systems in which all centres of (paraxial) curvature of the surfaces are located on a single optical axis. Such a system is said to have rotational symmetry or symmetry of revolution.

This chapter has been organised as follows. It starts with a derivation of the laws of geometrical optics using Maxwell's equations for the electromagnetic field. An important part of the chapter is devoted to the use of the characteristic function in geometrical optics, first developed by Hamilton. With the aid of the characteristic function we introduce the laws of paraxial optics. The next subject is the derivation of some general properties of optical systems with rotational symmetry. The chapter concludes with the study of pencils of rays in an optical system, the role of diaphragm and pupils and the definition of wavefront aberration. In Section 4.10 we present the laws of paraxial optics using the so-called *matrix* theory of optical systems. The final Section 4.11 discusses the laws of optical radiometry and photometry.

4.2 Geometrical Optics Derived from Maxwell's Equations

We use the electromagnetic field equations in their differential form according to Eqs. (1.3.10)–(1.3.13). In the absence of space charges and currents, the equations for the electric and magnetic field strengths read

$$\nabla \cdot \epsilon \mathbf{E} = 0 \,, \tag{4.2.1}$$

$$\nabla \cdot \mu \mathbf{H} = 0 \,, \tag{4.2.2}$$

$$\nabla \times \mathbf{H} = \epsilon \frac{\partial \mathbf{E}}{\partial t} \,, \tag{4.2.3}$$

$$\nabla \times \mathbf{E} = -\mu \frac{\partial \mathbf{H}}{\partial t} \,, \tag{4.2.4}$$

where we assume that the material properties ϵ and μ are time-independent but vary as a function of position. For harmonic field solutions at frequency ω, a general solution in complex notation yields the electric and magnetic field expressions

$$\mathbf{E}(\mathbf{r}, t) = \mathbf{E}_0(\mathbf{r}) \exp(-i\omega t) \,, \tag{4.2.5}$$

$$\mathbf{H}(\mathbf{r}, t) = \mathbf{H}_0(\mathbf{r}) \exp(-i\omega t) \,. \tag{4.2.6}$$

The substitution of this general solution in Eqs. (4.2.1)–(4.2.4) yields

$$\nabla \cdot \epsilon \mathbf{E}_0(\mathbf{r}) = 0 \,, \tag{4.2.7}$$

$$\nabla \cdot \mu \mathbf{H}_0(\mathbf{r}) = 0 \,, \tag{4.2.8}$$

$$\nabla \times \mathbf{H}_0(\mathbf{r}) = -i\omega\epsilon \mathbf{E}_0(\mathbf{r}) \,, \tag{4.2.9}$$

$$\nabla \times \mathbf{E}_0(\mathbf{r}) = +i\omega\mu \mathbf{H}_0(\mathbf{r}) \,. \tag{4.2.10}$$

We write the planar wave solution in the following way,

$$\mathbf{E}_0(\mathbf{r}) = \mathbf{A}_p \exp\{ik(\hat{\mathbf{s}} \cdot \mathbf{r})\} , \tag{4.2.11}$$

where $\mathbf{A}_p$ is the electric field vector of the plane wave, $\hat{\mathbf{s}}$ the unit vector in the propagation direction of the plane wave and where the wavenumber k is given by the dispersion relation $\mathbf{k} \cdot \mathbf{k} = \omega^2 \epsilon\mu$. In an absorption-free medium with real permittivity and permeability, we are allowed to write $|\mathbf{k}| = k = \omega(\epsilon\mu)^{1/2}$. Using the customary assumption at optical frequencies that μ equals μ_0, the permeability of vacuum, we have

$$k = \omega \sqrt{\epsilon_r} \sqrt{\epsilon_0\mu_0} = \sqrt{\epsilon_r} \frac{\omega}{c} = n\, k_0 , \tag{4.2.12}$$

where we have used that $c = (\epsilon_0\mu_0)^{-1/2}$, the speed of light in vacuum. We further have that $(\epsilon_r)^{1/2} = n$, where ϵ_r is the relative permittivity of a medium and n its refractive index. A plane wave solution exists in a homogeneous medium when the source is at an infinitely large distance. For a source at finite distance, the solution in the far-field region in a homogeneous medium with refractive index n is well represented by a spherical wave according to

$$\mathbf{E}_0(\mathbf{r}) = \frac{\mathbf{A}_s(\mathbf{r})}{|\mathbf{r}|} \exp\{ink_0|\mathbf{r}|\} . \tag{4.2.13}$$

We define more general solutions in inhomogeneous regions of space by means of the fields

$$\mathbf{E}_0(\mathbf{r}) = \mathbf{A}_e(\mathbf{r}) \exp\{ik_0 S(\mathbf{r})\} ,$$
$$\mathbf{H}_0(\mathbf{r}) = \mathbf{A}_h(\mathbf{r}) \exp\{ik_0 S(\mathbf{r})\} , \tag{4.2.14}$$

where $\mathbf{A}_e(\mathbf{r})$ and $\mathbf{A}_h(\mathbf{r})$ are general vector functions and $S(\mathbf{r})$ is a scalar function of position which determines, as in the case of a planar or a spherical wave, the phase of the electric and magnetic field strengths of the general wave. S itself has the dimension of length. The substitution of the solutions of Eq. (4.2.14) in Eqs. (4.2.7)–(4.2.10) yields

$$\mathbf{A}_e(\mathbf{r}) \cdot \nabla S(\mathbf{r}) = \frac{i}{k_0} \left[\frac{\nabla\epsilon(\mathbf{r})}{\epsilon(\mathbf{r})} \cdot \mathbf{A}_e(\mathbf{r}) + \nabla \cdot \mathbf{A}_e(\mathbf{r}) \right] , \tag{4.2.15}$$

$$\mathbf{A}_h(\mathbf{r}) \cdot \nabla S(\mathbf{r}) = \frac{i}{k_0} \nabla \cdot \mathbf{A}_h(\mathbf{r}) , \tag{4.2.16}$$

$$\nabla S(\mathbf{r}) \times \mathbf{A}_h(\mathbf{r}) + c\epsilon(\mathbf{r})\mathbf{A}_e(\mathbf{r}) = \frac{i}{k_0} \nabla \times \mathbf{A}_h(\mathbf{r}) , \tag{4.2.17}$$

$$\nabla S(\mathbf{r}) \times \mathbf{A}_e(\mathbf{r}) - c\mu_0\mathbf{A}_h(\mathbf{r}) = \frac{i}{k_0} \nabla \times \mathbf{A}_e(\mathbf{r}) . \tag{4.2.18}$$

The equations above have been organised in such a way that the explicit dependence on the wavenumber k_0 is found in the right-hand sides. To find an expression for the scalar function $S(\mathbf{r})$, we apply the *geometrical optics approximation* by taking the limit for $1/k_0 \rightarrow 0$. In regions of space where the derivative functions in the right-hand sides of the equations remain sufficiently small, these right-hand sides all become effectively zero. Practical requirements to keep these derivative functions small are the following:

- $\nabla\epsilon(\mathbf{r})/\epsilon(\mathbf{r}) = 2\nabla n(\mathbf{r})/n(\mathbf{r}) \ll k_0$, or, $\partial n/\partial x_i \ll \pi/\lambda$, where λ is the wavelength in the medium and x_i are the Cartesian coordinates (see Eq. (4.2.15)). The spatial changes of the refractive index distribution should be small. In practice it means that the geometrical optics approximation is not valid in a region where a sudden change of the medium properties occurs at the scale of the wavelength of the light.
- We assume that in the right-hand side of Eq. (4.2.15) the quantity $\nabla \cdot \mathbf{A}_e(\mathbf{r})$ is much smaller than $k_0\mathbf{A}_e(\mathbf{r}) \cdot \nabla S(\mathbf{r})$, or, for a single Cartesian component, $(\lambda_0/2\pi)\, \partial A_{x_i,e}/\partial x_i \ll A_{x_i,e}\partial S/\partial x_i$. We write the latter expression as $(1/A_{x_i,e})\, \partial A_{x_i,e}/\, \partial x_i \ll nk_0 \approx 1$, or, $\ln(A_{x_i+\Delta x_i,e}/A_{x_i,e}) \ll k\Delta x_i$, where Δx_i is some small increment in distance along the coordinate axis with label i. This condition implies that regions in space with sudden changes in electric field strength, for instance, close to a source or a sink (absorber), should be avoided when using the geometrical optics approximation.
- Equation (4.2.16) yields a comparable condition to the above for the magnetic field strength. For the right-hand sides of Eqs. (4.2.17) and (4.2.18) to become negligibly small, we find conditions equivalent to those following from Eq. (4.2.15).

Having equated all right-hand sides to zero, we take the vector product of Eq. (4.2.17) and $\nabla S(\mathbf{r})$ and obtain

$$[\nabla S(\mathbf{r}) \cdot \mathbf{A}_h(\mathbf{r})]\, \nabla S(\mathbf{r}) - \{\nabla S(\mathbf{r}) \cdot \nabla S(\mathbf{r})\}\, \mathbf{A}_h(\mathbf{r}) + c\epsilon(\mathbf{r})\nabla S(\mathbf{r}) \times \mathbf{A}_e(\mathbf{r}) = 0 . \tag{4.2.19}$$

Using Eqs. (4.2.16) and (4.2.18) shows that Eq. (4.2.19) is satisfied, provided that

$$|\nabla S(\mathbf{r})|^2 = c^2\epsilon(\mathbf{r})\mu_0 = n^2(\mathbf{r}) \,. \tag{4.2.20}$$

Here we have used the property that the gradient of the real function S is a real vector and that we are allowed to put $\nabla S(\mathbf{r})\cdot\nabla S(\mathbf{r}) = |\nabla S(\mathbf{r})|^2$; $n(\mathbf{r})$ is the spatially varying refractive index distribution. Equation (4.2.20) is called the *eikonal equation* and the scalar function $S(\mathbf{r})$ the *eikonal* function. A surface in space with the property that $S(\mathbf{r})$ has a constant value on it is called a geometrical wavefront. The phase of the geometrical optics wave on such a surface is given by

$$\phi(\mathbf{r}) = k_0 S(\mathbf{r}) \,. \tag{4.2.21}$$

4.2.1 Energy Propagation and the Optical Ray

Using the expressions for the electric and magnetic field strength, the energy flow is given by the expression for the Poynting vector $\mathbf{S}$ (see Eq. (1.4.5)). Its time-averaged value is given by

$$\langle \mathbf{S}(\mathbf{r}) \rangle = \frac{1}{2}\,\Re\{\mathbf{A}_e(\mathbf{r}) \times \mathbf{A}_h^*(\mathbf{r})\} \,. \tag{4.2.22}$$

Taking the vector product of Eq. (4.2.18), complex conjugate version, and $\mathbf{A}_e(\mathbf{r})$, we have in the limit of $k_0 \rightarrow \infty$

$$\langle \mathbf{S}(\mathbf{r}) \rangle = \frac{1}{2c\mu_0}\,[(\mathbf{A}_e(\mathbf{r})\cdot\mathbf{A}_e^*(\mathbf{r}))\;\nabla S(\mathbf{r}) - (\mathbf{A}_e(\mathbf{r})\cdot\nabla S(\mathbf{r}))\,\mathbf{A}_e^*(\mathbf{r})] \,. \tag{4.2.23}$$

We now define the unit ray vector as

$$\hat{\mathbf{s}}(\mathbf{r}) = \frac{\nabla S(\mathbf{r})}{n(\mathbf{r})} \,. \tag{4.2.24}$$

Geometrical optics wavefronts and rays have been sketched in Fig. 4.2. Note that the rays $\hat{\mathbf{s}}$ are everywhere normal to the surfaces in space on which the eikonal function S has the same value, viz. the wavefronts in the geometrical optics approximation. From Eqs. (4.2.15) and (4.2.16) we deduce that, as for a plane wave, the electric and magnetic field vectors are perpendicular to the energy propagation vector. After some rearrangement and using that the average electric energy density $\langle w_e \rangle$ is given by $\epsilon(\mathbf{r})\mathbf{A}_e(\mathbf{r})\cdot\mathbf{A}_e^*(\mathbf{r})/2$, we obtain the expression

$$\langle \mathbf{S}(\mathbf{r}) \rangle = \frac{c}{n(\mathbf{r})}\langle w_e(\mathbf{r}) \rangle\, \hat{\mathbf{s}}(\mathbf{r}) \,. \tag{4.2.25}$$

The ray vector thus points in the direction of the time-averaged Poynting vector. The strength of the Poynting vector is proportional to the energy density and to the local propagation speed of the wave, $v(\mathbf{r}) = c/n(\mathbf{r})$. Equations (4.2.15) and (4.2.16) show that in the geometric optics approximation, sufficiently far away from obstacles, the electric and magnetic field vectors are perpendicular to the energy propagation vector, as is the case for an electromagnetic plane wave. The product of average energy density and propagation speed is called the optical excitance (of a surface element of a radiating object) or the optical irradiance at an illuminated surface element. Both quantities have the dimension Wm^{-2} and are commonly denoted by the capital letter I. More generally, we will denote by optical intensity I the flux in

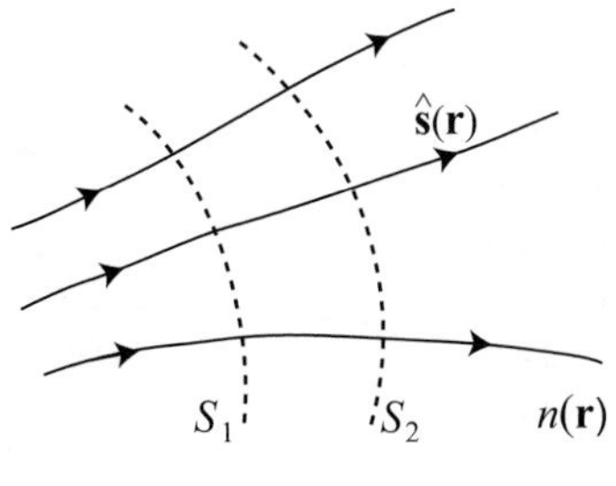

Figure 4.2: Two three-dimensional surfaces (dashed in the figure) with constant values S_1 and S_2 of the eikonal function. Three arbitrary unit optical rays $\hat{\mathbf{s}}(\mathbf{r})$ have been shown that are perpendicular to the geometrical wavefronts on which $S(\mathbf{r})$ is constant. The refractive index of the (inhomogeneous) optical medium is given by the function $n(\mathbf{r})$.

space of optical energy per unit of (projected) surface, yielding $I = dW/dS$. The unit for optical excitance or irradiance should not be confounded with the official radiometric or photometric unit for radiant/photometric intensity which has the dimension $\mathrm{W sr}^{-1}$. The latter quantity gives the radiant energy flux or luminous flux which is radiated or received by a surface element per unit solid angle. For instance, we denote by I_R the radiant intensity which plays a crucial role in radiometry (the subscript R stems from the adjective *radiant*, see also Section 4.11). In line with common (bad) practice in optics, we loosely use in this book the term *intensity* for both surface excitance and irradiance. As soon as confusion could arise between irradiance/excitance on the one hand and radiometric or photometric intensity on the other hand, we strictly adhere to the international denominations for these quantities.

4.2.2 Equation for the Propagation of Optical Intensity

In a stationary field, and in the absence of sources or sinks and without absorption or emission of radiation by the medium, we can write

$$\nabla \cdot \langle \mathbf{S}(\mathbf{r}) \rangle = \nabla \cdot (I\,\hat{\mathbf{s}}) = I\,\nabla \cdot \hat{\mathbf{s}} + \nabla I \cdot \hat{\mathbf{s}} = 0\,,$$

or,

$$\hat{\mathbf{s}} \cdot \nabla \ln I = -\nabla \cdot \hat{\mathbf{s}}\,, \tag{4.2.26}$$

where I is the power per unit surface in the optical radiation field. Using Eq. (4.2.24) for the ray vector $\hat{\mathbf{s}}$ we find the differential equation involving the eikonal function

$$\nabla \ln I(\mathbf{r}) \cdot \nabla S(\mathbf{r}) = \nabla \ln n(\mathbf{r}) \cdot \nabla S(\mathbf{r}) - \nabla^2 S(\mathbf{r})\,, \tag{4.2.27}$$

which can be numerically solved for $I(\mathbf{r})$ if the eikonal function is given in a certain spatial domain and the appropriate boundary conditions are available.

Some special cases are the following:

- $n(\mathbf{r})$ is a constant function and the light rays of the geometrical optical field are parallel, straight lines. The corresponding wavefronts are planar surfaces. The solution of Eq. (4.2.27) then requires that $\nabla \ln n(\mathbf{r}) \cdot \nabla S(\mathbf{r}) = 0$. This implies that the intensity variation in the ray field is perpendicular to the ray direction; the intensity along a ray remains unchanged.
- A collection of rays propagates from an elementary area dA_1 on the wavefront defined by $S(\mathbf{r}) = S_1$ to an elementary area dA_2 on a wavefront having the value S_2 (see Fig. 4.3). Everywhere in the volume defined by the surfaces dA_1, dA_2 and the curved surface of the energy flow tube that is defined by the outer rays, the divergence of the Poynting vector equals zero. Applying Gauss' theorem to this volume, we have $\iint I(\mathbf{r})\,\hat{\mathbf{s}} \cdot \hat{\mathbf{v}}\,dA = 0$, where the integration extends over the outer surface of the flow tube ($\hat{\mathbf{v}}$ is the outer surface normal). The contribution from the curved tube surface is zero, yielding

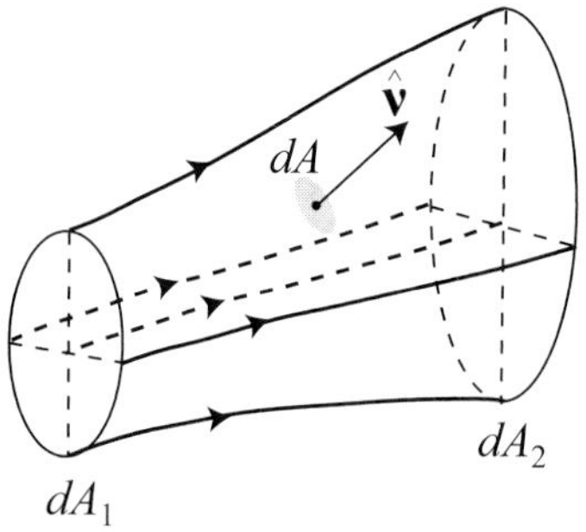

Figure 4.3: The propagation of optical energy via rays from one elementary area dA_1 on a wavefront where $S(\mathbf{r})$ has the value S_1 to another area dA_2 on the wavefront where $S(\mathbf{r})$ equals S_2. The rays are perpendicular to the wavefronts. The unit normal to the surface of the flow tube is $\hat{\mathbf{v}}$.

$$-I_1 dA_1 + I_2 dA_2 = 0 \,, \tag{4.2.28}$$

where I_1 and I_2 are the irradiance values on the two elementary wavefront areas. By tracing neighbouring (or 'differential') rays, Eq. (4.2.28) makes it possible to calculate the intensity change along the central ray of the ray tube.

- In the case of a small point-like source that emits in a homogeneous medium a power dW_0 in a solid angle $d\Omega$, the irradiance on an elementary receiving surface perpendicular to the propagation direction of the radiation is given by

$$I = \frac{1}{R^2}\frac{dW_0}{d\Omega} = \frac{I_R}{R^2} \,, \tag{4.2.29}$$

where $d\Omega$ is the elementary solid angle subtended by the receiving surface, R its distance to the source point and I_R the radiant intensity of the source. Equation (4.2.29) is called the inverse square law of geometrical optics.
- With a less stringent requirement on the high value of k_0, a geometrical optics approximation can be established for the amplitude vectors of the electric and magnetic field. This approximation brings polarisation effects within the realm of geometrical optics (see **[37]** for an introduction to this subject and **[186]** for a detailed treatment). An extension of geometrical optics to cover problems that are typically treated by using diffraction theory has been proposed in **[178]** (geometrical theory of diffraction). This extension requires the introduction of new, well-chosen geometrical rays at apertures, edges, ridges, bars, antennas, causing a ray avalanche effect in the neighbourhood of objects that are too tiny to be correctly treated by basic geometrical optics alone.

4.2.3 Optical Path and the General Ray Equation

In this subsection we define the optical path along a ray and we establish a ray equation that depends solely on the medium properties and no longer on the eikonal function $S(\mathbf{r})$. In Fig. 4.4 a point P on the wavefront with $S(\mathbf{r}_0) = S_0$ has been shown. The unit ray vector through the point P has been denoted by $\hat{\mathbf{s}}$. A point P' on this ray is at a distance ds from P and the distance from P' to the origin is given by $\mathbf{r}_0 + d\mathbf{r}$. We denote the geometrical distance along the ray from P to P' by ds. The unit ray vector is then written as

$$\hat{\mathbf{s}} = \frac{d\mathbf{r}}{ds} \,, \tag{4.2.30}$$

where the vector PP' is denoted by $d\mathbf{r}$. From Eq. (4.2.24) we thus obtain

$$\nabla S = \left(n\frac{dx}{ds}, n\frac{dy}{ds}, n\frac{dz}{ds} \right) = (nL, nM, nN) \,, \tag{4.2.31}$$

where (L, M, N) are the direction cosines of the ray. The quantities (nL, nM, nN) are commonly called the *optical* direction cosines of the ray. The value of the eikonal function in P' is given by

$$S(\mathbf{r}_0 + d\mathbf{r}) = S_0 + dS = S_0 + \nabla S \cdot d\mathbf{r} = S_0 + n ds \,, \tag{4.2.32}$$

where we have used the result of Eq. (4.2.24) and the property $d\mathbf{r} \cdot d\mathbf{r} = (ds)^2$. The quantity $dS = n ds$ is called the optical pathlength increment when going from P to P'. For a more general path from P_1 to P_2 along a generally curved ray in space we have the pathlength expression

$$S(\mathbf{r}_{P_2}) - S(\mathbf{r}_{P_1}) = \int_{P_1}^{P_2} dS = \int_{P_1}^{P_2} \nabla S(\mathbf{r}) \cdot d\mathbf{r} = \int_{P_1}^{P_2} n(\mathbf{r})\, \hat{\mathbf{s}}(\mathbf{r}) \cdot d\mathbf{r} \,, \tag{4.2.33}$$

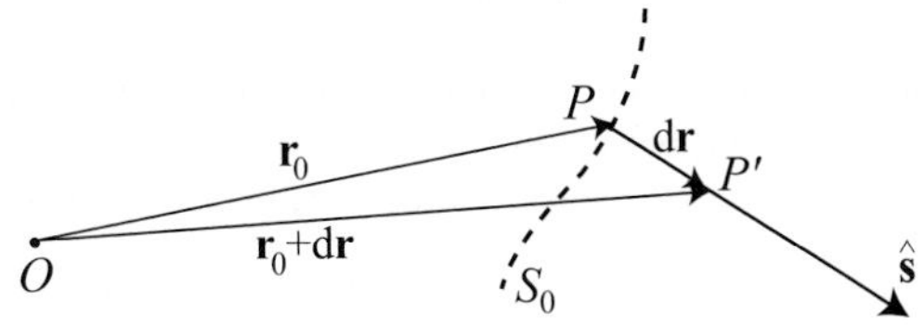

Figure 4.4: An optical wavefront, dashed in the figure, with a general point P on it. The unit geometrical ray vector $\hat{\mathbf{s}}$ through P is perpendicular to the wavefront S_0.

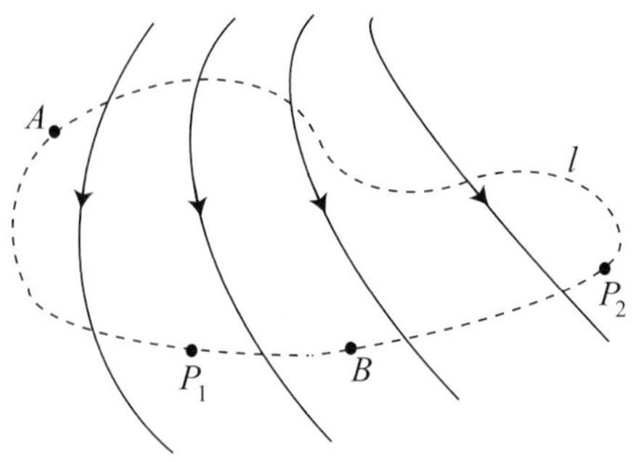

Figure 4.5: Optical pathlength from P_1 to P_2, evaluated along different trajectories P_1AP_2 and P_1BP_2 (dashed curves) in a rotation-free vector field of optical rays (solid lines with arrows); $P_1BP_2AP_1$ forms a closed path l in space.

where, by writing $n(\mathbf{r})$, we have included the possibility of a spatial variation of the refractive index. Another interesting property of optical pathlength follows from Eq. (4.2.24). Since $n\hat{\mathbf{s}}$ is the gradient of a scalar function, we have $\nabla \times (n\hat{\mathbf{s}}) = 0$. Applying Stokes' theorem to a rotation-free vector field, we find

$$\oint_l n\hat{\mathbf{s}} \cdot d\mathbf{r} = 0 , \tag{4.2.34}$$

where an arbitrary closed path in space has been denoted by l.

In Fig. 4.5 two points in space are shown, P_1 and P_2, connected by two different paths, via A and B, respectively. If Eq. (4.2.34) is used for the closed path $P_1AP_2BP_1$, we have the result

$$\int_{P_1(\rightarrow A)}^{P_2} n\hat{\mathbf{s}} \cdot d\mathbf{r} = \int_{P_1(\rightarrow B)}^{P_2} n\hat{\mathbf{s}} \cdot d\mathbf{r} , \tag{4.2.35}$$

showing that the increment in optical pathlength does not depend on the integration path in the ray field but only on the values in the endpoints. The result of Eq. (4.2.35) goes back to Lagrange and is called Lagrange's integral invariant. The optical pathlength expression requires knowledge of $\hat{\mathbf{s}}(\mathbf{r})$ in space. Equation (4.2.24) asks for a solution of $S(\mathbf{r})$ throughout space to obtain $\hat{\mathbf{s}}(\mathbf{r})$. It is desirable to obtain an expression for the calculation of $\hat{\mathbf{s}}(\mathbf{r})$ without the explicit need of the eikonal function, such that the expression only depends on the function $n(\mathbf{r})$. To this purpose we take the derivative of Eq. (4.2.24) with respect to the length parameter s along the optical ray,

$$\frac{d}{ds}\{\nabla S(\mathbf{r})\} = \frac{d}{ds}\{n(\mathbf{r})\,\hat{\mathbf{s}}(\mathbf{r})\} = \frac{d}{ds}\left(n(\mathbf{r})\,\frac{d\mathbf{r}}{ds}\right) . \tag{4.2.36}$$

The right-hand side of Eq. (4.2.36) can be further developed. We first note that the differential operator with respect to s can be written as

$$\frac{d}{ds} = \frac{\partial}{\partial x}\frac{dx}{ds} + \frac{\partial}{\partial y}\frac{dy}{ds} + \frac{\partial}{\partial z}\frac{dz}{ds} = \hat{\mathbf{s}} \cdot \nabla . \tag{4.2.37}$$

We apply this vector operator to the vector field $n(\mathbf{r})\,\hat{\mathbf{s}}(\mathbf{r})$ (see the central expression of Eq. (4.2.36)) and obtain the expression

$$\frac{d}{ds}\{n(\mathbf{r})\,\hat{\mathbf{s}}(\mathbf{r})\} = \hat{\mathbf{s}}(\mathbf{r}) \cdot \nabla\{n(\mathbf{r})\,\hat{\mathbf{s}}(\mathbf{r})\} = \nabla n(\mathbf{r}) + n(\mathbf{r})\,\hat{\mathbf{s}}(\mathbf{r}) \cdot \nabla\{\hat{\mathbf{s}}(\mathbf{r})\} . \tag{4.2.38}$$

In the last term of Eq. (4.2.38), the gradient operator acts on a vector and this needs a more detailed analysis which is given in Section A.10 of App. A. It turns out that the gradient of a vector $\mathbf{a} = (a_x, a_y, a_z)$ is a second rank tensor. The scalar product of this tensor and the vector $\mathbf{a}$ itself yields a column vector. We find that

$$(\nabla\mathbf{a}) \cdot \mathbf{a} = \frac{1}{2}\nabla(\mathbf{a} \cdot \mathbf{a}) = \frac{1}{2}\begin{pmatrix} \frac{\partial}{\partial x}(a_x^2 + a_y^2 + a_z^2) \\ \frac{\partial}{\partial y}(a_x^2 + a_y^2 + a_z^2) \\ \frac{\partial}{\partial z}(a_x^2 + a_y^2 + a_z^2) \end{pmatrix} . \tag{4.2.39}$$

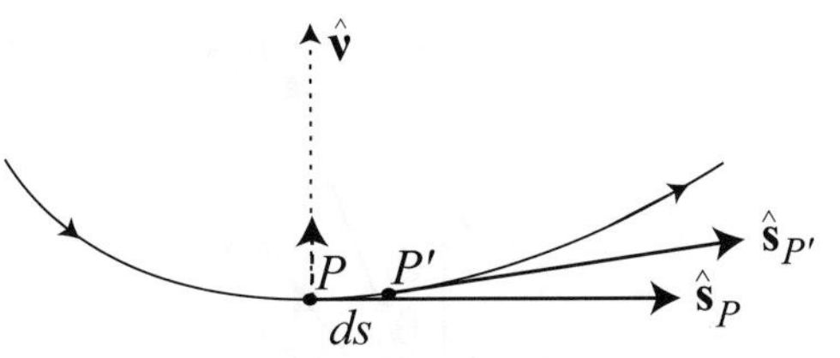

Figure 4.6: A curved ray in an inhomogeneous medium. The unit ray vector, tangent to the ray path, has been shown in two neighbouring points, P and P', separated by a distance ds along the ray path. The difference vector $\hat{\mathbf{s}}_{P'} - \hat{\mathbf{s}}_P$ is shown in P and is perpendicular to the ray path.

For the special case that $\mathbf{a}$ is a unit vector, we obtain the null vector. Equation (4.2.38) then yields the ray equation of geometrical optics,

$$\frac{d}{ds}\left\{n(\mathbf{r})\frac{d\mathbf{r}}{ds}\right\} = \nabla n(\mathbf{r}) \,. \tag{4.2.40}$$

This equation is a second-order differential equation for the position vector $\mathbf{r}$ on the ray. If initial boundary conditions for ray position and ray angle are available, the complete path of a ray can be found.

4.2.4 Ray Curvature

In Fig. 4.6 a curved ray path is shown. The change in ray vector direction has been depicted with the aid of two neighbouring points on the ray path, separated by a distance ds along the curve. For vanishingly small ds, the difference vector points in the direction of the normal vector $\hat{\mathbf{v}}$ to the curve, in a plane defined by the vectors $\hat{\mathbf{s}}_P$ and $\hat{\mathbf{s}}_{P'}$. For the differential change in ray direction we write

$$\frac{d}{ds}(\hat{\mathbf{s}}) = c_P\,\hat{\mathbf{v}} \,, \tag{4.2.41}$$

where c_P is the curvature of the ray (a positive quantity by the choice of the direction of the unit vector $\hat{\mathbf{v}}$).

An expression for c_P is obtained as follows. Starting with the general ray equation of Eq. (4.2.40), we write

$$\frac{dn}{ds}\hat{\mathbf{s}} + n\frac{d\hat{\mathbf{s}}}{ds} = \nabla n \,, \tag{4.2.42}$$

where we have omitted in the notation the $\mathbf{r}$-dependence of the vector $\hat{\mathbf{s}}$ and spatial function n. For the curvature vector we find

$$\frac{d\hat{\mathbf{s}}}{ds} = \frac{\nabla n}{n} - \frac{1}{n}\frac{dn}{ds}\hat{\mathbf{s}} \,. \tag{4.2.43}$$

This equation shows that the curvature vector is found in the plane that is defined by the gradient vector of the refractive index function and the ray vector. The scalar multiplication of both sides of Eq. (4.2.43) by the curvature vector $c_P\hat{\mathbf{v}}$ yields the result

$$c_P = \hat{\mathbf{v}} \cdot \frac{\nabla n}{n} \,. \tag{4.2.44}$$

Since we have chosen by definition that c_P is a non-negative quantity (see Eq. (4.2.41)), we conclude that the unit curvature vector always has a component in the direction of ∇n. This implies that a ray bends towards the region in space where the index of refraction is larger. If $\nabla n = 0$, the curvature becomes zero and we have $d^2\mathbf{r}/ds^2 = 0$. Integration of this second-order differential equation yields the solution $\mathbf{r} = p\,\hat{\mathbf{s}}_0 + \mathbf{r}_0$ where p is the ray parameter. The ray direction and the exact ray path follow from the initial condition $\hat{\mathbf{s}} = \hat{\mathbf{s}}_0$ in the starting point $\mathbf{r} = \mathbf{r}_0$.

4.2.5 Geometrical Optics and Snell's Law

The property of Eq. (4.2.34) can be used to derive Snell's law in the geometrical optics framework. In Fig. 4.7, a ray in a first medium with refractive index n_1 intersects an interface at the point P. The interface separates the first medium

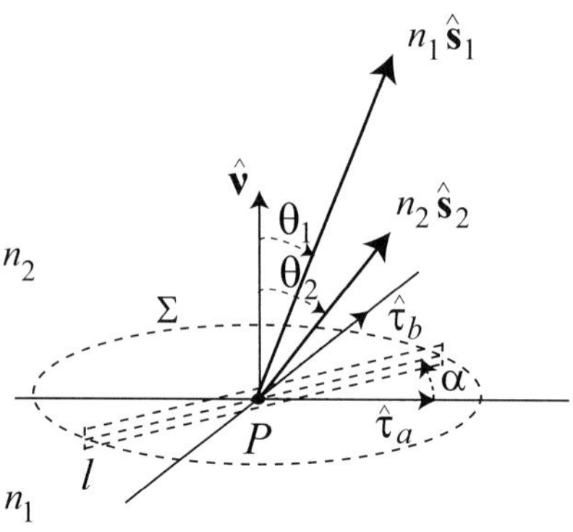

Figure 4.7: Refraction of a light ray $\hat{\mathbf{s}}_1$ from a medium with refractive index n_1 to a second medium with index n_2. P is a point on the boundary between the two media. The unit vector $\hat{\mathbf{v}}$ is normal to the boundary surface. $\hat{\boldsymbol{\tau}}_a$ and $\hat{\boldsymbol{\tau}}_b$ are mutually orthogonal unit vectors, tangent to the surface.

from a second one with refractive index n_2. The planar surface at P that is tangent to the interface between the two media has been denoted by Σ. The plane defined by the normal vector $\hat{\mathbf{v}}$ and a tangent vector $\hat{\boldsymbol{\tau}}_a$ contains the optical ray vector $n_1\hat{\mathbf{s}}_1$. A second tangent vector $\hat{\boldsymbol{\tau}}_b$ is shown which has the property that $\hat{\boldsymbol{\tau}}_a \times \hat{\boldsymbol{\tau}}_b = \hat{\mathbf{v}}$. A closed rectangular path through the ray field is chosen (dashed curve with its long side l) on the boundary between the two media. The plane containing this rectangular path is at an angle α to the tangent vector $\hat{\boldsymbol{\tau}}_a$.

4.2.5.1 Snell's Law for the Transmitted Ray

Applying the closed path integral of Eq. (4.2.34) to the rectangle with $d\mathbf{r} = (\cos\alpha\, \hat{\boldsymbol{\tau}}_a + \sin\alpha\, \hat{\boldsymbol{\tau}}_b)\, dl$ yields

$$l\,[\,\cos\alpha\,(n_1\hat{\mathbf{s}}_1 - n_2\hat{\mathbf{s}}_2)\cdot\hat{\boldsymbol{\tau}}_a - \sin\alpha\; n_2\hat{\mathbf{s}}_2\cdot\hat{\boldsymbol{\tau}}_b] = 0\,,$$

or,

$$(n_1\hat{\mathbf{s}}_1 - n_2\hat{\mathbf{s}}_2)\cdot\hat{\boldsymbol{\tau}}_a - n_2\hat{\mathbf{s}}_2\cdot\hat{\boldsymbol{\tau}}_b\;\tan\alpha = 0\,. \tag{4.2.45}$$

The result of Eq. (4.2.45) should be valid for any orientation angle α of the rectangle and this leads to the following requirements:

- $(n_1\hat{\mathbf{s}}_1 - n_2\hat{\mathbf{s}}_2)\cdot\hat{\boldsymbol{\tau}}_a = 0$, or, $n_1\sin\theta_1 = n_2\sin\theta_2$, Snell's law in restricted sense.
- $\hat{\mathbf{s}}_2\cdot\hat{\boldsymbol{\tau}}_b = 0$ which means that the vector $\hat{\mathbf{s}}_2$ is found in the plane of incidence, defined by $\hat{\mathbf{v}}$ and $\hat{\mathbf{s}}_1$, an essential complement to Snell's law. The same conclusion follows from the condition $\tan\alpha = 0$.
- the combination of both aspects of Snell's law requires that the projections of the *optical* ray vectors $n_1\hat{\mathbf{s}}_1$ and $n_2\hat{\mathbf{s}}_2$ on the plane Σ are equal in magnitude and azimuth.

Following this last item, we have across the interface continuity of the vector $n\,\hat{\mathbf{v}}\times\hat{\mathbf{s}}\times\hat{\mathbf{v}}$. It is interesting to compare this continuity condition across the boundary with Eq. (1.8.20) for the continuity on both sides of an interface of the tangential wave vector component of a plane wave in electromagnetic theory. Multiplication of the geometrical optics continuity condition by $\mathbf{k}_0$ yields the electromagnetic continuity condition for the tangential wave vector components at a boundary. The continuity of the vector $n\,\hat{\mathbf{v}}\times\hat{\mathbf{s}}\times\hat{\mathbf{v}}$ across a flat boundary leads to the vector formulation of Snell's law,

$$n_2\,\hat{\mathbf{s}}_2 = n_1\,\hat{\mathbf{s}}_1 + (n_2\cos\theta_2 - n_1\cos\theta_1)\;\hat{\mathbf{v}}\,. \tag{4.2.46}$$

Equation (4.2.46) states that the difference vector of the optical ray vectors in the two media is parallel to the normal vector $\hat{\mathbf{v}}$.

4.2.5.2 Snell's Law for the Reflected Ray

For a reflected ray, the indices of the two media are formally equal. Continuity of the vector quantity $n_1\;\hat{\mathbf{v}}\times\hat{\mathbf{s}}_1\times\hat{\mathbf{v}}$ of the incident ray is satisfied by a general ray

$$\hat{\mathbf{s}}_2 = \hat{\mathbf{s}}_1 + p\,(\hat{\mathbf{s}}_1\cdot\hat{\mathbf{v}})\,\hat{\mathbf{v}}\,. \tag{4.2.47}$$

Using the property that the reflected vector $\hat{\mathbf{s}}_2$ is a unit vector we readily find that p is either 0 or -2. The nontrivial reflected ray is found for $p = -2$. A frequently used method to obtain the reflected ray vector is the substitution of $n_2 = -n_1$ and to relate the angle θ of the reflected ray to the inverted normal vector $-\hat{\mathbf{v}}$. This approach, although effective in practice, is less rigorous than the application of Eq. (4.2.47) using $p = -2$. The assumption of both a positive and negative refractive index for the same medium is unphysical.

4.2.6 Fermat's Principle

The principle of least action or, better, of least time was proposed by Fermat in 1662. It can also be stated in terms of optical pathlength by saying that the optical path from one point in space to another is an extremum along an optical ray. In mathematical form we have that

$$[P_1P_2] = \int_{P_1}^{P_2} n ds \,, \tag{4.2.48}$$

is an extremum if the integration path from P_1 to P_2 coincides with the path of an optical ray in the medium. In the following we use bracket signs [ ] to indicate the optical path between points in space. To prove Fermat's principle, we use Lagrange's integral invariant. In Fig. 4.8 two points P_1 and P_2 are connected by a physical light ray $P_1Q_1Q_2P_2$ and by a second curved segment $P\overline{Q}_1\overline{Q}_2P_2$. The curve segment $\overline{Q}_1\overline{Q}_3$ is part of a physical ray path. Applying Lagrange's integral invariant to the closed path $\overline{Q}_1\overline{Q}_3\overline{Q}_2\overline{Q}_1$, yields

$$[nds]_{\overline{Q}_1\overline{Q}_3} + \int_{\overline{Q}_3}^{\overline{Q}_2} n\,\hat{\mathbf{s}}\cdot d\mathbf{r} - \int_{\overline{Q}_1}^{\overline{Q}_2} n\,\hat{\mathbf{s}}\cdot d\mathbf{r} = 0 \,. \tag{4.2.49}$$

The second term on the left-hand side of Eq. (4.2.49) is zero because the path is located on a wavefront and the scalar product $\hat{\mathbf{s}}\cdot d\mathbf{r}$ is rigorously zero there. Along the general path $\overline{Q}_1\overline{Q}_2$, not corresponding to an optical ray path, we have that $nds \geq nds\cos\alpha = n\hat{\mathbf{s}}\cdot d\mathbf{r}$, where α is the angle between the vector $\hat{\mathbf{s}}$ and the path vector $d\mathbf{r}$ with length ds along the trajectory $\overline{Q}_1\overline{Q}_2$. We thus have the inequality

$$[nds]_{\overline{Q}_1\overline{Q}_2} \geq \int_{\overline{Q}_1}^{\overline{Q}_2} n\,\hat{\mathbf{s}}\cdot d\mathbf{r} \,. \tag{4.2.50}$$

Using the property $[nds]_{Q_1Q_2} = [nds]_{\overline{Q}_1\overline{Q}_3}$ we deduce from Eq. (4.2.49) that

$$[nds]_{Q_1Q_2} = \int_{\overline{Q}_1}^{\overline{Q}_2} n\,\hat{\mathbf{s}}\cdot d\mathbf{r} \leq [nds]_{\overline{Q}_1\overline{Q}_2} \,. \tag{4.2.51}$$

By successive application of this inequality to the path segments on a physical ray and the segments on a general curved path, we prove, for this example, that the optical path along the physical ray is smallest.

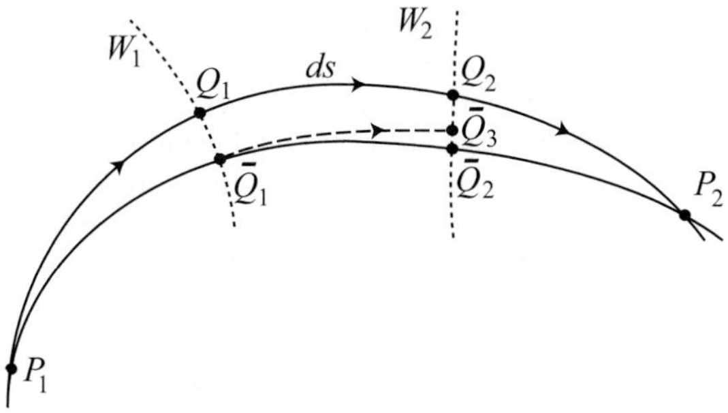

Figure 4.8: A physical light ray goes from P_1 to P_2 via the path $P_1Q_1Q_2P_2$; ds is an infinitesimally small path increment along this physical ray. A general path in the optical ray field, not corresponding to a physical light ray in the medium, is given by $P_1\overline{Q}_1\overline{Q}_2P_2$. Wavefront surfaces through Q_1 and Q_2 are dashed in the figure. Starting in $\overline{Q}_1$, a physically possible ray segment is traced from wavefront W_1 to W_2, ending in the point $\overline{Q}_3$ on W_2. The ray curve length between the points Q_1 and Q_2 on wavefronts W_1 and W_2 is ds.

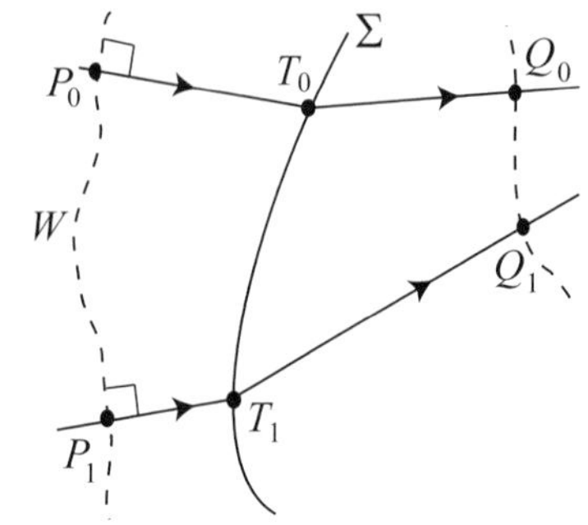

Figure 4.9: A (curved) wavefront W is incident on a surface Σ separating two media with different refractive indices. Two rays, P_0T_0 and P_1T_1 are shown in the figure, perpendicular to the wavefront W (dashed). The points Q_0 and Q_1 on the refracted rays in the second medium have equal optical pathlength from P_0 and P_1, respectively.

4.2.7 Theorem of Malus and Dupin

The orthogonality of rays and wavefronts in continuous space follows from the definition of the unit ray vector in Eq. (4.2.24) where it emerges as the gradient of the eikonal function S. The question if this orthogonality property is preserved at reflection by a curved surface was answered by Dupin in the beginning of the nineteenth century with the aid of geometric arguments, used in the same space where the incident and reflected rays and wavefronts are given. Such a geometric proof could not be given for a refracted pencil of rays because of the transition to a different medium. The proof that the refraction of a bundle of rays through a curved surface also leaves the orthogonality between rays and wavefronts intact was first given by Malus. Given their common effort on this subject, the orthogonality property of rays and their wavefront for reflected and transmitted bundles of rays is said to follow from the theorem of Malus and Dupin. In Fig. 4.9 the incident wavefront W and two rays have been shown. Applying Lagrange's integral invariant to the polygon $P_0T_0Q_0Q_1T_1P_1P_0$ yields

$$[P_0T_0Q_0] + \int_{Q_0}^{Q_1} n\,\hat{\mathbf{s}}\cdot d\mathbf{r} - [P_1T_1Q_1] + \int_{P_1}^{P_0} n\,\hat{\mathbf{s}}\cdot d\mathbf{r} = 0\,. \tag{4.2.52}$$

The fourth term on the left-hand side of Eq. (4.2.52) equals zero because of the orthogonality of rays and wavefront in the first medium. From the equality of optical pathlengths between P_0Q_0 and P_1Q_1 along the optical rays we conclude that

$$\int_{Q_0}^{Q_1} n\,\hat{\mathbf{s}}\cdot d\mathbf{r} = 0\,. \tag{4.2.53}$$

This proves that everywhere on the surface Q_0Q_1 in the second medium the optical rays are perpendicular to this surface and that the orthogonality property of Eq. (4.2.24) is maintained after a refraction. A ray bundle, originating from a point source, which has the property that a continuous surface can be constructed that is perpendicular to each ray in its intersection point with the surface is called a normal congruence of rays. This property of a ray bundle is maintained during propagation in media that are rotation-free as is generally the case in optics. In inhomogeneous media Malus' theorem is equally valid. For this case, its proof has to proceed with infinitesimal steps along the generally curved ray paths. We conclude by saying that in the case of electron optics the property $\nabla\times n\hat{\mathbf{s}} = \mathbf{0}$ is not valid when magnetic electron lenses are included. Consequently, Lagrange's integral invariant is not valid for such a lens and Malus' theorem cannot be applied to ray bundles that possess rotation.

4.3 Characteristic Function of an Optical System

The theory of geometrical optics treated in the preceding section allows the construction and analysis of rays and wavefronts if data are available about the refractive index distribution in a medium and about surfaces that separate media with a different index of refraction. The construction of ray paths and wavefronts is relatively easy in homogeneous media but tedious in inhomogeneous media. In the latter case, apart from some favoured geometries, numerical integration procedures are needed to construct the curved ray trajectories in such media. A succession of several different media with separating surfaces of, for instance, spherical shape is a very frequently occurring practical situation. Lens and mirror systems are used for imaging tasks, ranging from a simple burning glass to highly sophisticated imaging systems like earth-imaging lens systems in observation satellites or in large-field lithographic projection lenses for the semiconductor industry. The propagation of a single ray through such systems comprising a succession of homogeneous media is relatively easy. The full characterisation of such a complicated optical system requires the analysis of data gathered from a very large number of rays. For this reason, a description of the optical system by a master function which immediately provides the data on the image side when specifying the input rays on the object side is an appealing concept. The tracing of many rays through a system to characterise it is a Lagrangian approach, using the geometrical equations of motion through the system for a large number of individual light rays. The concept of a characteristic function is equivalent to developing the potential function; for instance, the electrical potential function associated with a distribution of charges in space. The calculation of a characteristic function in optics can be time-consuming but once it is available, it can almost immediately provide the user with the required data on the ray path in image space for a specified incoming ray in object space. The work on characteristic functions in optics goes back to Hamilton **[78]**. In this section we present the various propositions for a characteristic function of an optical system. As an example, we give the characteristic function of a single surface and indicate how the characteristic function of an optical system can be constructed.

4.3.1 Definition of the Characteristic Function

The definition of the characteristic function is based on the optical pathlength concept which was developed in Section 4.2. In Fig. 4.10 we show the object and image space of an optical system (not shown in the figure) with their respective coordinate origins O_0 and O_1. The coordinate axes in object and image space are parallel; the z-axes in object and image space are collinear as this is very often the case in an optical system. However, the analysis is not limited to this special choice of the coordinate systems.

4.3.1.1 Point Characteristic Function

The point characteristic function V is now defined by

$$V(x_0, y_0, z_0; x_1, y_1, z_1) = \int_{P_0}^{P_1} n \, ds \, , \tag{4.3.1}$$

where the integration is performed along the optical ray through the system that joins P_0 and P_1. Equation (4.2.33) gave the optical pathlength as the increment of the eikonal function according to

$$\int_{P_0}^{P_1} n \, ds = S(\mathbf{r}_1) - S(\mathbf{r}_0) \, . \tag{4.3.2}$$

The vectors $\mathbf{r}_0$ and $\mathbf{r}_1$ are referenced to the same origin or, if this is more practical, to the origins in object and image space, O_0 and O_1, respectively, where an implicit displacement vector has to be added in this case to the vector $\mathbf{r}_1$. The eikonal equation (4.2.24) applied to the ray vectors in object and image space yields

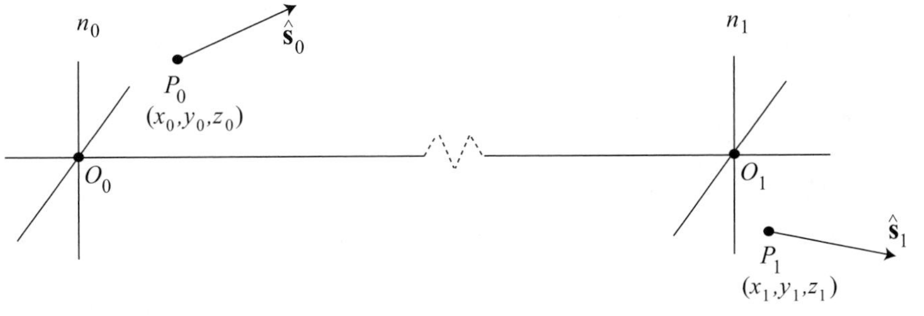

Figure 4.10: The coordinate systems, (x_0, y_0, z_0) and (x_1, y_1, z_1) and their origins, O_0 and O_1, respectively, in the object and image space of an optical system. A ray is shown that leaves a general point P_0 with unit vector $\hat{\mathbf{s}}_0$ in object space and passes through a point P_1 in image space (ray direction $\hat{\mathbf{s}}_1$). The refractive indices of object and image space are n_0 and n_1.

$$\hat{\mathbf{s}}_0 = \frac{1}{n_0}\left(\frac{\partial S}{\partial x_0}, \frac{\partial S}{\partial y_0}, \frac{\partial S}{\partial z_0}\right), \quad \hat{\mathbf{s}}_1 = \frac{1}{n_1}\left(\frac{\partial S}{\partial x_1}, \frac{\partial S}{\partial y_1}, \frac{\partial S}{\partial z_1}\right). \tag{4.3.3}$$

An infinitesimal increment nds in optical pathlength from the point P_0 in object space is also given by (see Eq. (4.3.1)),

$$nds\,|_{\text{object space}} = V(\mathbf{r}_0;\mathbf{r}_1) - V(\mathbf{r}_0 + d\mathbf{r}_0;\mathbf{r}_1) = -\nabla V(\mathbf{r}_0,\mathbf{r}_1)|_{\mathbf{r}=\mathbf{r}_0} \cdot d\mathbf{r}_0 ,$$

and, in a similar manner,

$$nds\,|_{\text{image space}} = V(\mathbf{r}_0;\mathbf{r}_1 + d\mathbf{r}_1) - V(\mathbf{r}_0;\mathbf{r}_1) = \nabla V(\mathbf{r}_0,\mathbf{r}_1)|_{\mathbf{r}=\mathbf{r}_1} \cdot d\mathbf{r}_1 . \tag{4.3.4}$$

The comparison of Eq. (4.3.3) and (4.3.4) yields for the ray vectors in object and image space,

$$\hat{\mathbf{s}}_0 = \left.\frac{\nabla S(\mathbf{r})}{n_0}\right|_{\mathbf{r}=\mathbf{r}_0} = -\left.\frac{\nabla_0 V(\mathbf{r}_0,\mathbf{r}_1)}{n_0}\right|_{\mathbf{r}=\mathbf{r}_0} , \tag{4.3.5}$$

$$\hat{\mathbf{s}}_1 = \left.\frac{\nabla S(\mathbf{r})}{n_1}\right|_{\mathbf{r}=\mathbf{r}_1} = \left.\frac{\nabla_1 V(\mathbf{r}_0,\mathbf{r}_1)}{n_1}\right|_{\mathbf{r}=\mathbf{r}_1} , \tag{4.3.6}$$

where the subscript of the nabla operator indicates with respect to which coordinates the derivatives need to be evaluated.

4.3.1.2 Angle Characteristic Function

The characteristic function $V(\mathbf{r}_0, \mathbf{r}_1)$ is called a point characteristic function because it only depends on the coordinates in object and image space. We can switch to an angle characteristic function by the following transformation (see Fig. 4.11). Choosing two reference points in object and image space, for instance the origins O_0 and O_1, we define a modified pathlength $[Q_0 Q_1]$ according to

$$[Q_0 Q_1] = [Q_0 P_0 P_1 Q_1] = [P_0 P_1] + [Q_0 P_0] + [P_1 Q_1] . \tag{4.3.7}$$

The optical paths $[P_0 Q_0]$ and $[P_1 Q_1]$ are written as $-n_0\hat{\mathbf{s}}_0 \cdot \mathbf{r_0}$ and $-n_1\hat{\mathbf{s}}_1 \cdot \mathbf{r}_1$, respectively, yielding for the total pathlength

$$[Q_0 Q_1] = V(x_0, y_0, z_0; x_1, y_1, z_1) + n_0(L_0 x_0 + M_0 y_0 + N_0 z_0) - n_1(L_1 x_1 + M_1 y_1 + N_1 z_1) . \tag{4.3.8}$$

From Eq. (4.3.8) we formally deduce that the pathlength function $[Q_0 Q_1]$ depends on the vector pairs $(\mathbf{r}_0, \mathbf{r}_1)$ and $(\hat{\mathbf{s}}_0, \hat{\mathbf{s}}_1)$. The analysis of a small variation $\delta[Q_0 Q_1]$ of $[Q_0 Q_1]$ shows that this quantity only depends on the direction cosines of the ray vectors $\hat{\mathbf{s}}_0$ and $\hat{\mathbf{s}}_1$. For that reason, the pathlength function $[Q_0 Q_1]$ is generally called an angle characteristic function and, more particularly, denominated as the *eikonal* function $E(L_0, M_0, N_0; L_1, M_1, N_1)$. Figure 4.11 also shows that a shift of the points P_0 and P_1 along the optical rays does not influence the value of $[Q_0 Q_1]$. The differential $\delta[Q_0 Q_1]$ is written as

$$\begin{aligned}&\delta E(n_0 L_0, n_0 M_0, n_0 N_0; n_1 L_1, n_1 M_1, n_1 N_1)\\ &\quad = x_0\delta(n_0 L_0) + y_0\delta(n_0 M_0) + z_0\delta(n_0 N_0) - x_1\delta(n_1 L_1) - y_1\delta(n_1 M_1) - z_1\delta(n_1 N_1) ,\end{aligned} \tag{4.3.9}$$

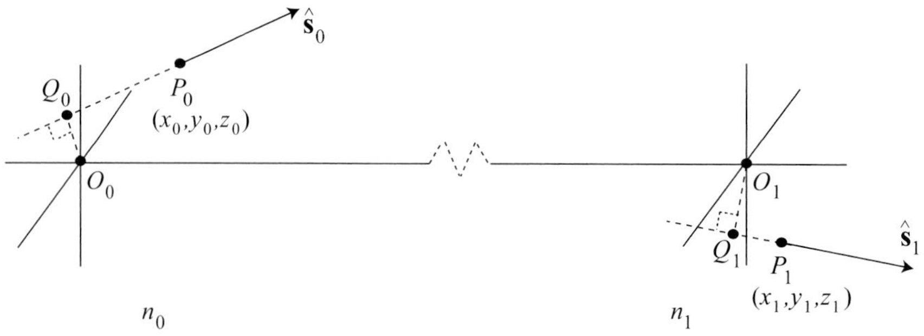

Figure 4.11: Definition of the optical pathlength $[Q_0 Q_1]$ for the angle characteristic function. As reference points in object and image space the origins O_0 and O_1 have been chosen.

and for the derivatives of E with respect to the optical direction cosines we have

$$\frac{\partial E}{\partial(n_0L_0)} = x_0 \,, \qquad \frac{\partial E}{\partial(n_1L_1)} = -x_1 \,,$$
$$\frac{\partial E}{\partial(n_0M_0)} = y_0 \,, \qquad \frac{\partial E}{\partial(n_1M_1)} = -y_1 \,,$$
$$\frac{\partial E}{\partial(n_0N_0)} = z_0 \,, \qquad \frac{\partial E}{\partial(n_1N_1)} = -z_1 \,. \tag{4.3.10}$$

The eikonal function E formally depends on six variables. However, since $\hat{\mathbf{s}}_0$ and $\hat{\mathbf{s}}_1$ are unit vectors, the number of independent variables is four. For small increments δN_0 and δN_1 we have

$$\delta N_0 = -\frac{L_0}{N_0}\delta L_0 - \frac{M_0}{N_0}\delta M_0 \,,$$
$$\delta N_1 = -\frac{L_1}{N_1}\delta L_1 - \frac{M_1}{N_1}\delta M_1 \,, \tag{4.3.11}$$

which allows us to write Eq. (4.3.9) as

$$\delta E = \left(x_0 - \frac{L_0}{N_0}z_0\right)\delta(n_0L_0) + \left(y_0 - \frac{M_0}{N_0}z_0\right)\delta(n_0M_0)$$
$$+ \left(-x_1 + \frac{L_1}{N_1}z_1\right)\delta(n_1L_1) + \left(-y_1 + \frac{M_1}{N_1}z_1\right)\delta(n_1M_1) \,. \tag{4.3.12}$$

The expression for the partial derivatives is now given by

$$\frac{\partial E}{\partial(n_0L_0)} = x_0 - \frac{L_0}{N_0}z_0 \,, \qquad \frac{\partial E}{\partial(n_1L_1)} = -x_1 + \frac{L_1}{N_1}z_1 \,,$$
$$\frac{\partial E}{\partial(n_0M_0)} = y_0 - \frac{M_0}{N_0}z_0 \,, \qquad \frac{\partial E}{\partial(n_1M_1)} = -y_1 + \frac{M_1}{N_1}z_1 \,. \tag{4.3.13}$$

Equation (4.3.13) enables the construction of two corresponding rays in object and image space. Defining an incoming and outgoing ray direction by (L_i, M_i) and (L_o, M_o), respectively, we evaluate the values of the derivatives of the function E with respect to the direction cosines for the specific values (L_i, M_i) and (L_o, M_o). The ray paths in object and image space are then given by

$$x_0 - \frac{L_i}{N_i}z_0 - \left.\frac{\partial E}{\partial(n_0L_0)}\right|_{\hat{\mathbf{s}}_i} = 0 \,, \qquad y_0 - \frac{M_i}{N_i}z_0 - \left.\frac{\partial E}{\partial(n_0M_0)}\right|_{\hat{\mathbf{s}}_i} = 0 \,,$$
$$x_1 - \frac{L_o}{N_o}z_1 + \left.\frac{\partial E}{\partial(n_1L_1)}\right|_{\hat{\mathbf{s}}_o} = 0 \,, \qquad y_1 - \frac{M_o}{N_o}z_1 + \left.\frac{\partial E}{\partial(n_1M_1)}\right|_{\hat{\mathbf{s}}_o} = 0 \,. \tag{4.3.14}$$

Equation (4.3.14) yields the ray path in object space as the intersecting line of the two planes defined by the two equations on the first line; in the same way, the ray path in image space follows from the second line of Eq. (4.3.14). In what follows, for ease of notation, we do not strictly separate the functional variables (L_0, M_0), (L_1, M_1) from the direction cosines (L_i, M_i), (L_o, M_o) of the rays under investigation of which the trajectories need to be found. The set (L_0, M_0), (L_1, M_1) is used for both the functional variables of the eikonal function and the ray direction cosines of a specific ray combination in object and image space.

4.3.1.3 Application of the Characteristic Function to Imaging

In the preceding subsections it was shown that the point characteristic function yields the ray directions for a pair of points in object and image space. The angle characteristic function requires the prescription of ray directions in object and image space and then yields the optical pathlength between two points Q_0 and Q_1 in object and image space. Each of the points Q is the projection of a chosen reference point O onto the line in space defined by the optical ray direction. The connection of the characteristic functions with optical imaging is most easily seen for the angle characteristic. In Fig. 4.11, we observe that the projecting line O_0Q_0 is tangent to an optical wavefront through Q_0, perpendicular to the ray $\hat{\mathbf{s}}_0$. This wavefront is transferred to image space. Here, the perpendicular O_1Q_1 is in the tangent plane to a wavefront through Q_1, which is perpendicular to the ray $\hat{\mathbf{s}}_1$ in image space. The angle characteristic thus delivers the pathlength

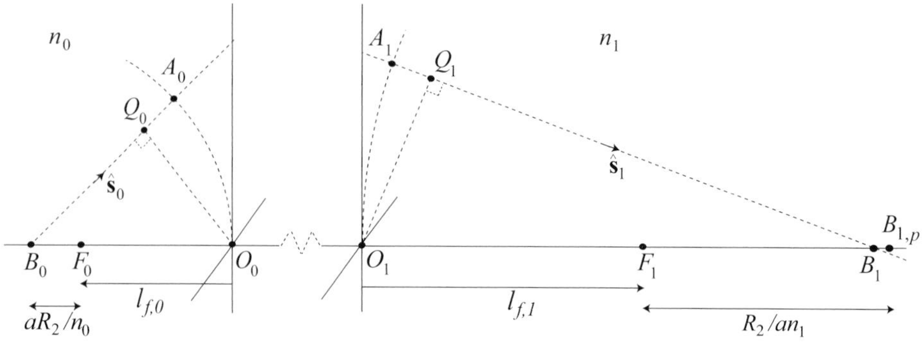

Figure 4.12: The transition of a ray from the object to the image space of an optical system. The paraxial conjugate points are B_0 and $B_{1,p}$. The object- and image-side focal points have been denoted by F_0 and F_1. The object and image space are homogeneous and have refractive indices n_0 and n_1, respectively. O_0A_0 and O_1A_1 are parts of wavefronts related to the source point B_0 and the corresponding imaging pencil with its specific ray $\hat{\mathbf{s}}_1$ intersecting the point B_1.

increment on the wavefront through Q_1 with respect to the incident wavefront through Q_0. By analysing all the ray direction combinations in object and image space with equal value of $E(L_0, M_0; L_1, M_1)$, we are able to construct a pair of connected wavefronts in object and image space that are produced by the optical system.

The following example of an eikonal function illustrates this procedure (see Fig. 4.12). A ray $\hat{\mathbf{s}}_0$ with direction cosines $\left(L_1, 0, (1-L_1^2)^{1/2}\right)$ starts in the point B_0 on the z-axis and the corresponding ray $\hat{\mathbf{s}}_1$ in image space intersects the same axis in B_1, close to the paraxial image point $B_{1,p}$. The direction cosines of $\hat{\mathbf{s}}_1$ are given by $\left(aL_1, 0, (1-a^2L_1^2)^{1/2}\right)$. For the optical pathlength $[Q_0Q_1]$ we use Lagrange's integral invariant along the closed path $O_0A_0B_0Q_0Q_1B_1A_1O_1O_0$ and equate it to zero. The distance $O_0B_0 = -z_0$ and the distance O_1B_1 from the image-side reference point to the (aberrated) point B_1 is z_1. After some rearrangement we find for the optical path $[Q_0Q_1]$ the expression

$$E(L_0, M_0; L_1, M_1)_{B_0B_1} = -n_0(1-N_0)z_0 + n_1(1-N_1)z_1 + (L_0L_1 + M_0M_1)R_2 + C\,, \tag{4.3.15}$$

where C is the constant optical path $[O_0O_1]$. The first two terms on the right-hand side of Eq. (4.3.15) can be recognised as a shift of origin along the z-axis over a distance z_0 in object space and z_1 in image space. The third term links the direction cosines of the rays in object and image space, for instance because of the presence of optical power in the imaging system. The meaning of the distance R_2 is made clear a few lines further on in this subsection. At this point in the calculation we shift the reference points to the (non-conjugate) paraxial focal points F_0 and F_1, at distances $l_{f,0}$ and $l_{f,1}$ from O_0 and O_1, respectively. Such a shift of reference points yields the modified eikonal equation

$$E(L_0, M_0; L_1, M_1)_{F_0F_1} = -n_0(1-N_0)(z_0 - l_{f,0}) + n_1(1-N_1)(z_1 - l_{f,1}) + (L_0L_1 + M_0M_1)R_2 + C\,. \tag{4.3.16}$$

We assume that the characteristic function only depends on the combinations $(1-L_0^2-M_0^2)^{1/2}$, $(1-L_1^2-M_1^2)^{1/2}$ and $(L_0L_1 + M_0M_1)$, which is typical of an optical system that shows symmetry of revolution with respect to the z-axis.

To calculate the derivatives of E with respect to the optical direction cosines in object and image space, we have to choose a certain combination of direction cosines L_0 and L_1. Without loss of generality, we put M_0 and M_1 equal to zero. We substitute $L_0 = L_i$ and $L_1 = aL_i$ and obtain the following equations for the ray directions (see Eq. (4.3.14)),

$$\begin{aligned} x_0 - L_i\left(\frac{z_0 - l_{f,0}}{(1-L_i^2)^{1/2}} + \frac{aR_2}{n_0}\right) &= 0 \qquad & y_0 &= 0\,, \\ x_1 - aL_i\left(\frac{z_1 - l_{f,1}}{(1-a^2L_i^2)^{1/2}} - \frac{R_2}{an_1}\right) &= 0 \qquad & y_1 &= 0\,. \end{aligned} \tag{4.3.17}$$

The intersection points of the ray with the z-axis in object and image space are then given by the coordinate values

$$z_0 = l_{f,0} - \frac{aR_2(1-L_i^2)^{1/2}}{n_0}\,, \qquad z_1 = l_{f,1} + \frac{R_2(1-a^2L_i^2)^{1/2}}{an_1}\,. \tag{4.3.18}$$

Equation (4.3.18) shows that the intersection points, by means of L_i, depend on the angle of incidence of the ray on the optical system. This implies that the eikonal function E of Eq. (4.3.15) does not produce perfect imaging between a pair of points in object and image space. For sufficiently small angles of incidence, we find the so-called paraxial pair of conjugate points, given by

$$z_0 = l_{f,0} - \frac{aR_2}{n_0} , \qquad z_1 = l_{f,1} + \frac{R_2}{an_1} . \tag{4.3.19}$$

With respect to the shifted reference points F_0 and F_1, we define the paraxial object and image distance by $\Delta_0 = -aR_2/n_0$ and $\Delta_1 = +R_2/(an_1)$ and we find the paraxial imaging equation,

$$\Delta_0\Delta_1 = -\frac{R_2^2}{n_0 n_1} , \tag{4.3.20}$$

as was derived by Newton. For $a = 0$ we find that the point F_0 with axial coordinate given by $z_0 = l_{f,0}$ is indeed the object-side focal point. In a similar way, for $a \to \infty$, we have that F_1 is the image-side focal point of the system. The meaning of the distance R_2 follows from the relations $f_0 = -R_2/n_0$ and $f_1 = R_2/n_1$. A further discussion of the paraxial imaging properties of an optical system is given in Section 4.10.

The characteristic functions discussed so far, the *point* characteristic function and the *angle* characteristic function, are inadequate for certain imaging cases. For instance, at a point where more than one ray passes, the point characteristic function is not able to discriminate between these rays. Especially in the focal point of a bundle of rays, the point characteristic cannot be used. The same holds for the angle characteristic function in the case of a parallel bundle of rays. The angle characteristic is unable to discriminate between these rays and cannot determine the individual ray paths of each ray of the bundle. To solve this problem, the so-called mixed characteristic has been devised that depends, for instance, on the direction cosines in object space and the image-side spatial coordinates. The number of independent coordinates of such a mixed characteristic function is five. In general, remote imaging or imaging in star space is best analysed with a point characteristic function. Finite conjugate imaging is the realm of the angle characteristic function. A serious disadvantage for the practical use of characteristic functions is the possible shift during the design stage of the conjugate object and image positions. This requires a shift from one type of characteristic function to another during the design to maintain a stable calculation scheme. A practical solution is the introduction of a fictitious perfect lens to bring the infinitely far region to the focal plane of this lens. With the aid of such a manipulation, the angle characteristic function can be used to cover all imaging conditions. This explains why the angle characteristic function is the one which is most frequently used among the four possible characteristic functions.

4.4 Angle Characteristic Function of a Single Surface

The transition of a ray through a single surface separating two homogeneous media with refractive indices n_0 and n_1 is sketched in Fig. 4.13. $P(x, y, z)$ is the intersection point of the incoming ray $\hat{\mathbf{s}}_0 = (L_0, M_0, N_0)$ with the spherical surface separating the media. The radius of curvature of this surface is R, positive in the figure, and the midpoint of the sphere, C, is located on the z-axis. The angle characteristic function belonging to the ray transition through the surface is calculated with respect to the vertex O of the surface; this point is the common reference point for rays in object and image space. Its value is given by the optical pathlength increment from object to image space,

$$\begin{aligned} E(L_0, M_0; L_1, M_1) &= [Q_0Q_1] = [Q_0P] - [Q_1P] = n_0\hat{\mathbf{s}}_0 \cdot \mathbf{r}_P - n_1\hat{\mathbf{s}}_1 \cdot \mathbf{r}_P \\ &= (n_0L_0 - n_1L_1)x_P + (n_0M_0 - n_1M_1)y_P + (n_0N_0 - n_1N_1)z_P . \end{aligned} \tag{4.4.1}$$

The expression for E still contains the spatial coordinates (x_P, y_P, z_P). These can be eliminated from Eq. (4.4.1) by using the fact that P is located on the spherical surface,

$$x_P^2 + y_P^2 + (z_P - R)^2 = R^2 , \tag{4.4.2}$$

and that the directions of an incoming and outgoing ray obey Snell's law, in vector form written as

$$(n_0\hat{\mathbf{s}}_0 - n_1\hat{\mathbf{s}}_1) \times \hat{\mathbf{v}} = \mathbf{0} , \tag{4.4.3}$$

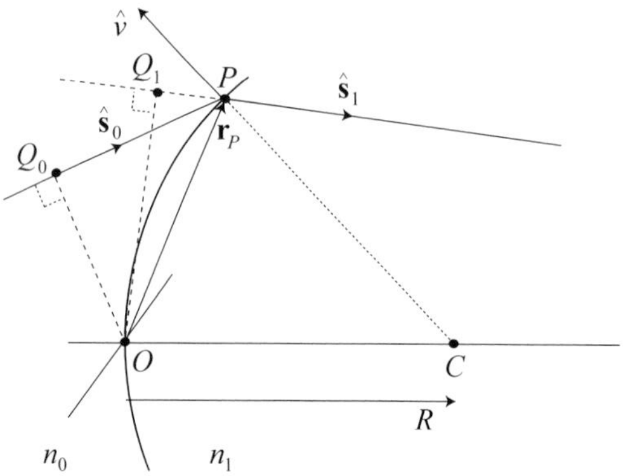

Figure 4.13: The transition of a ray through a single spherical surface, separating two media with indices n_0 and n_1. The centre of the sphere is in C, the origin is taken at O, the vertex of the sphere.

where $\hat{\mathbf{v}}$ is the (outward) normal to the spherical surface at the point P,

$$\hat{\mathbf{v}} = \frac{1}{R}(x_P, y_P, z_P - R) \ . \tag{4.4.4}$$

The substitution of the vector equation (4.4.3) into Eq. (4.4.2) yields

$$\begin{aligned}
x_P^2 &= \frac{R^2(n_0L_0 - n_1L_1)^2}{(n_0L_0 - n_1L_1)^2 + (n_0M_0 - n_1M_1)^2 + (n_0N_0 - n_1N_1)^2} \ , \\
y_P^2 &= \frac{R^2(n_0M_0 - n_1M_1)^2}{(n_0L_0 - n_1L_1)^2 + (n_0M_0 - n_1M_1)^2 + (n_0N_0 - n_1N_1)^2} \ , \\
(z_P - R)^2 &= \frac{R^2(n_0N_0 - n_1N_1)^2}{(n_0L_0 - n_1L_1)^2 + (n_0M_0 - n_1M_1)^2 + (n_0N_0 - n_1N_1)^2} \ .
\end{aligned} \tag{4.4.5}$$

The explicit expression for z_P follows from

$$z_P = R\left[1 - \frac{n_0N_0 - n_1N_1}{\sqrt{(n_0L_0 - n_1L_1)^2 + (n_0M_0 - n_1M_1)^2 + (n_0N_0 - n_1N_1)^2}}\right] , \tag{4.4.6}$$

where the minus sign in front of the fraction implies a choice for the intersection point P which is closest to the vertex of the spherical surface. Using the expressions of Eqs. (4.4.5) and (4.4.6) in Eq. (4.4.1) we obtain

$$E(L_0, M_0; L_1, M_1) = R(n_1N_1 - n_0N_0)\left[\sqrt{1 + \frac{(n_0L_0 - n_1L_1)^2 + (n_0M_0 - n_1M_1)^2}{(n_0N_0 - n_1N_1)^2}} - 1\right] . \tag{4.4.7}$$

A numerically more stable expression for the angle characteristic function is

$$E(L_0, M_0; L_1, M_1) = \frac{R}{(n_1N_1 - n_0N_0)}\left[\frac{(n_0L_0 - n_1L_1)^2 + (n_0M_0 - n_1M_1)^2}{1 + \sqrt{1 + \frac{(n_0L_0 - n_1L_1)^2 + (n_0M_0 - n_1M_1)^2}{(n_0N_0 - n_1N_1)^2}}}\right] , \tag{4.4.8}$$

which avoids the subtraction of two almost equal numbers in the case of small angles of incidence on the refracting surface.

The expression for the angle characteristic function according to Eqs. (4.4.7) or (4.4.8) still contains the z-direction cosines N_0 and N_1. These can be eliminated by using the unit vector property of $\hat{\mathbf{s}}_0$ and $\hat{\mathbf{s}}_1$.

4.4.1 First-order Angle Characteristic Function

A first impression of the imaging properties of the single surface is obtained by considering the first-order approximation of E with respect to the specific coordinate products that remain invariant when the spherical surface is rotated around the z-axis. In Fig. 4.14 we assume that the optical system possesses rotational symmetry with respect to the z-axis. The lengths of the projections of the ray vectors $\hat{\mathbf{s}}_0$ and $\hat{\mathbf{s}}_1$ onto a plane perpendicular to the z-axis, denoted by $|\mathbf{s}_{0,xy}|$ and $|\mathbf{s}_{1,xy}|$, respectively, only depend on the angles of these vectors with the z-axis and not on their azimuths with respect to the x- or y-axis. For this reason, the quantities $\mathbf{s}_{0,xy} \cdot \mathbf{s}_{0,xy}$ and $\mathbf{s}_{1,xy} \cdot \mathbf{s}_{1,xy}$ are invariant under a rotation of the z-axis. The same holds for the scalar product $\mathbf{s}_{0,xy} \cdot \mathbf{s}_{1,xy}$, which only depends on the angle between the two projected vectors. For the series expansion of the angle characteristic function E we define the rotationally invariant products of direction cosines u_{00}, u_{11} and u_{01} according to

$$u_{00} = n_0^2(L_0^2 + M_0^2)\,, \qquad u_{11} = n_1^2(L_1^2 + M_1^2)\,, \qquad u_{01} = n_0 n_1 (L_0 L_1 + M_0 M_1)\,. \tag{4.4.9}$$

The first term of the series expansion for E up to quadratic terms in the direction cosines and for $N_0 = N_1 \approx 1$ (see Eq. (4.4.7)) is then written as

$$E_2(u_{00}, u_{11}, u_{01}) = \frac{R}{2(n_1 - n_0)} (u_{00} + u_{11} - 2u_{01}) \,. \tag{4.4.10}$$

The next term in the series expansion needs the first- and second-order approximations

$$N_0 \approx 1 - \frac{L_0^2 + M_0^2}{2} = 1 - \frac{u_{00}}{2n_0^2}\,,$$
$$N_1 \approx 1 - \frac{L_1^2 + M_1^2}{2} = 1 - \frac{u_{11}}{2n_1^2}\,,$$
$$\sqrt{1 + ax} \approx 1 + \frac{a}{2}x - \frac{a^2}{8}x^2\,. \tag{4.4.11}$$

The substitution of these approximations yields the second term in the series expansion of E, of fourth order in the direction cosines,

$$E_4(u_{00}, u_{11}, u_{01}) = \frac{R}{4(n_1 - n_0)^2} (u_{00} + u_{11} - 2u_{01}) \left(\frac{u_{11}}{n_1} - \frac{u_{00}}{n_0} \right) - \frac{R}{8(n_1 - n_0)^3} (u_{00} + u_{11} - 2u_{01})^2 \,. \tag{4.4.12}$$

In the next subsection we discuss the properties of an angle characteristic function E_2 of the second order in the direction cosines with the tacit assumption that all higher-order terms E_{2n} $(n \geq 2)$ are zero. A system with such an angle characteristic function gives rise to perfect imaging. Since only the first-order term of such a characteristic function is nonzero, the optical properties are said to belong to 'first-order' optics of an optical system. In reality, the higher-order terms cannot be made zero and we are confronted with non-ideal imaging and the appearance of optical imaging aberrations. Only in some special cases, a general characteristic function can give rise to 'stigmatic' imaging. In most cases, this property is then limited to a very restricted region in space or even to a single point.

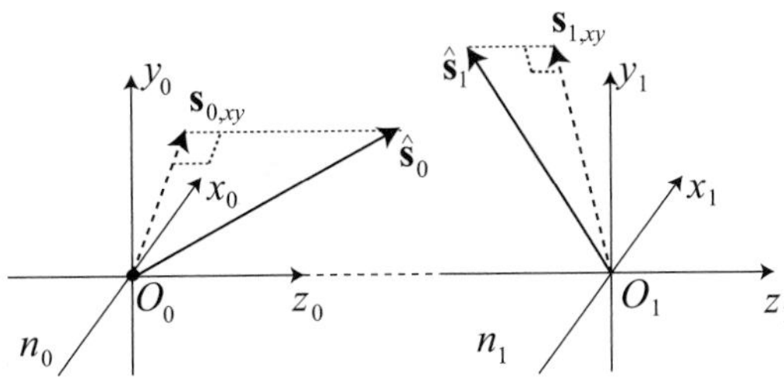

Figure 4.14: An optical system possesses rotational symmetry with respect to the z-axis. $\mathbf{s}_{0,xy}$ and $\mathbf{s}_{1,xy}$ are the projections of the unit ray vectors in object and image space onto the plane $z = 0$. Any function that describes a property of the optical system with rotational symmetry only depends on the quantities $|\mathbf{s}_{0,xy}|^2$, $|\mathbf{s}_{1,xy}|^2$ and $\mathbf{s}_{0,xy} \cdot \mathbf{s}_{1,xy}$ which are invariant under a rotation of the z-axis.

4.5 The First-order Angle Characteristic and the Paraxial Domain

In this section we consider the angle characteristic function $E_2(u_{00}, u_{11}, u_{01})$ as given by Eq. (4.4.10). The imaging is studied with the aid of Fig. 4.15. The ray paths in a system with the angle characteristic of Eq. (4.4.10) are found by applying Eq. (4.3.14). For rays in the plane $x = 0$ we have,

$$y_0 - M_0 z_0 - \frac{R}{n_1 - n_0}(n_0 M_0 - n_1 M_1) = 0 ,$$
$$y_1 - M_1 z_1 + \frac{R}{n_1 - n_0}(n_0 M_0 - n_1 M_1) = 0 . \tag{4.5.1}$$

In these equations we limit ourselves to terms that are linear in the (L, M)-direction cosines and approximate the N-direction cosines by $N_0 = N_1 = 1$. Eliminating M_0 from the two equations yields the following expression for the y-coordinate in image space:

$$y_1 = \frac{n_0 R'}{z_0 + n_0 R'} y_0 + M_1 \left(z_1 - n_1 R' + \frac{n_0 n_1 R'^2}{z_0 + n_0 R'} \right) , \tag{4.5.2}$$

where $R' = R/(n_1 - n_0)$. A comparable equation can be established for the x_1-coordinate on a ray in image space.

The condition for imaging follows from the requirement that the intersection point of an arbitrary ray $\hat{\mathbf{s}}_1$ in image space with a plane $z = z_1$ should not depend on the direction cosines (L_1, M_1) in image space. From Eq. (4.5.2) we deduce the imaging equation for the position of the so-called conjugate planes in object and image space,

$$(z_1 - n_1 R')(z_0 + n_0 R') = -n_0 n_1 R'^2 . \tag{4.5.3}$$

Two special cases arise, for either z_0 or z_1 going to infinity. We then find the object- and image-side focal points, denoted by F_0 and F_1 in Fig. 4.15, and they have the axial coordinates

$$z_{F_0} = -n_0 \frac{R}{(n_1 - n_0)} = f_0 , \qquad z_{F_1} = n_1 \frac{R}{(n_1 - n_0)} = f_1 , \tag{4.5.4}$$

where f_0 and f_1 are the object-side and image-side focal distances of the refracting surface. By referencing the object-side and image-side conjugate points to the focal points, the imaging equation (4.5.3) becomes

$$Z_1 Z_0 = f_0 f_1 , \tag{4.5.5}$$

which is the imaging equation of Newton which we have already encountered in Subsection 4.3.1.3 for a general optical system (see Eq. (4.3.20)). Another possible notation of Eq. (4.5.3) is

$$z_0 z_1 - n_1 R' z_0 + n_0 R' z_1 = 0 , \tag{4.5.6}$$

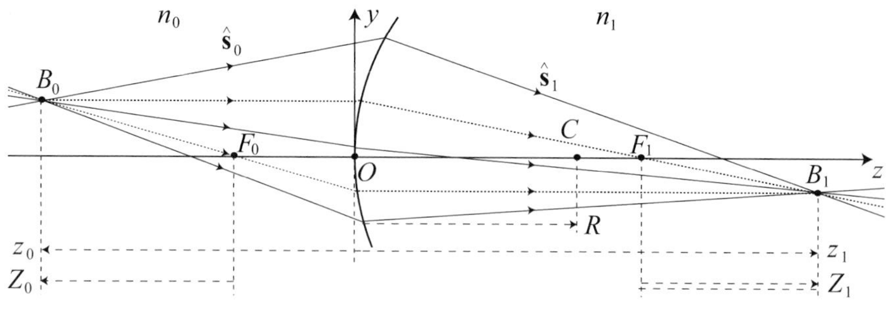

Figure 4.15: Examples of rays refracted by a spherical surface, separating two media with index n_0 and n_1. The radius of curvature of the surface is R. Imaging is observed between the points B_0 and B_1. The focal points on the object and image side are F_0 and F_1. For perfect imaging to occur for arbitrary combinations of conjugate points B_0 and B_1, the x- and y-direction cosines should remain infinitely small (paraxial optics domain).

or, after multiplication by $(n_1 - n_0)/(z_0 z_1 R)$,

$$\frac{n_1}{z_1} = \frac{n_0}{z_0} + \frac{n_1 - n_0}{R} , \tag{4.5.7}$$

the imaging equation of Gauss in terms of the distances with respect to the surface vertex O. The second quantity on the right-hand of this equation is called the optical power of the transition between the two media.

4.5.1 Transverse, Angular and Axial Magnification

The paraxial transverse magnification of the imaging between two conjugate planes follows from Eq. (4.5.2),

$$m_T = \frac{y_1}{y_0} = \frac{n_0 R'}{x_0 + n_0 R'} = \frac{-f_0}{z_0 - f_0} = -\frac{f_0}{Z_0} = -\frac{Z_1}{f_1} , \tag{4.5.8}$$

where we have used Newton's imaging equation to express the magnification with either the object-side or the image-side optical data.

The angular magnification between two conjugate planes follows from the ratio of the angles of an image ray and an object ray that intersect the axial points in the conjugate planes. With the aid of the first equation of (4.5.1), we have for the angular magnification ($y_0 = 0$),

$$m_A = \frac{M_1}{M_0} = \frac{z_0 + R' n_0}{n_1 R'} = \frac{Z_0}{f_1} = \frac{f_0}{Z_1} . \tag{4.5.9}$$

Combining the results for the transverse and the angular magnification we find

$$m_T m_A = -\frac{f_0}{f_1} = \frac{n_0}{n_1} , \tag{4.5.10}$$

the well-known paraxial result that the product of transverse and angular magnification is a constant for any pair of conjugate planes.

A less common magnification factor is the axial magnification, m_z. With the aid of Newton's imaging equation, Eq. (4.5.3), we evaluate the derivative dZ_1/dZ_0 and find

$$m_z = \frac{dZ_1}{dZ_0} = -\frac{Z_1}{Z_0} = -\frac{f_0 f_1}{Z_0^2} = \frac{n_1}{n_0} m_T^2 = \frac{n_0}{n_1} \frac{1}{m_A^2} . \tag{4.5.11}$$

We see that the axial magnification is a positive quantity for a refracting surface. In the case of an odd number of reflections in an optical system, the axial magnification is negative. The axial magnification is relevant when non-planar objects are imaged.

Other possibilities for writing the transverse and angular magnification using the conjugate distances z_0 and z_1 follow from Eq. (4.5.8) when eliminating R with the aid of Eq. (4.5.7),

$$m_T = \frac{z_1/n_1}{z_0/n_0} , \qquad m_A = \frac{z_0}{z_1} . \tag{4.5.12}$$

4.5.2 The Angle Characteristic Function Referenced to Two Conjugate Planes

The angle characteristic function E_2 of Eq. (4.4.10) is referenced to the vertex O of the refracting surface. In the case of imaging between two points B_0 and B_1 with coordinates (x_0, y_0, z_0) and (x_1, y_1, z_1) in the conjugate planes defined by $z = z_0$ and $z = z_1$, the shift of the reference point O to these conjugate points leads to a modified angle characteristic function given by

$$E_2(u_{00}, u_{11}, u_{01}) = \frac{R}{2(n_1 - n_0)}(u_{00} + u_{11} - 2u_{01}) - n_0(x_0 L_0 + y_0 M_0 + z_0 N_0) + n_1(x_1 L_1 + y_1 M_1 + z_1 N_1) . \tag{4.5.13}$$

As we have limited the expansion of E_2 to quadratic terms, we can also expand the z-direction cosines N_0 and N_1 up to this degree (see Eq. (4.4.11)). We write the positions of the conjugate planes in terms of the angular magnification m_A,

$$z_0 = (n_1 m_A - n_0)\frac{R}{n_1 - n_0} , \qquad z_1 = \left(n_1 - \frac{n_0}{m_A}\right)\frac{R}{n_1 - n_0} , \tag{4.5.14}$$

and, using the property $L_1/L_0 = m_A$, $M_1/M_0 = m_A$, we find for the resulting angle characteristic function, up to the second order in the direction cosines,

$$E_2(L_0, M_0, L_1, M_1) = \frac{R}{(n_1 - n_0)} \left[n_0^2 + n_1^2 - n_0 n_1 \left(m_A + \frac{1}{m_A} \right) \right] . \tag{4.5.15}$$

The expression for E_2 shows that the pathlength function, referenced to a pair of conjugate points, does not depend on the ray directions. The conclusion is that the property of a common intersection point in image space for all rays that originated at a single object point is equivalent to the equality of optical path along rays from the object point to the image-side *conjugate* point. Hitherto, this statement is only applicable within the domain where the quadratic approximation of the angle characteristic function E is valid. A good measure for the applicability of the quadratic approximation of E is the least-squares difference Δ^2 between the exact function and the approximated form,

$$\Delta^2 = \iint \left| E(L_0, M_0, L_1, M_1) - E_2(L_0, M_0, L_1, M_1) \right|^2 dL_1 dM_1 , \tag{4.5.16}$$

where the integration is carried out over a certain range of rays in object and image space, connected by the paraxial magnification laws. By limiting the (L_1, M_1)-integration interval and the off-axis position (x_1, y_1) of the conjugate image point, the value of Δ does not exceed a small fraction of the vacuum wavelength λ_0 of the light. In this case, the paraxial approximation is said to be valid.

4.6 Stigmatic Imaging and the Angle Characteristic Function

In the preceding sections, the perfect imaging properties in the paraxial domain were studied. We now consider the non-approximated angle characteristic function of a single surface, $E(L_0, M_0, L_1, M_1)$ as given by Eq. (4.4.7), and analyse when perfect imaging is possible outside the domain of paraxial optics. The reference points for the angle characteristic function are shifted to a pair of conjugate paraxial points on the z-axis, B_0 and B_1, with axial coordinates $z_{0,p}$ and $z_{1,p}$. We consider a pathlength difference function ΔE by taking the difference of optical pathlength between a general path along the rays $\hat{\mathbf{s}}_0$ and $\hat{\mathbf{s}}_1$ and the optical path along the z-axis. We obtain for ΔE the expression

$$\begin{aligned} \Delta E(L_0, M_0, L_1, M_1) &= E(L_0, M_0, L_1, M_1) - E(0,0,0,0) = (R - z_{0,p}) n_0 N_0 - (R - z_{1,p}) n_1 N_1 + n_0 z_{0,p} - n_1 z_{1,p} \\ &\quad - R \left[(n_0 L_0 - n_1 L_1)^2 + (n_0 M_0 - n_1 M_1)^2 + (n_0 N_0 - n_1 N_1)^2 \right]^{1/2} . \end{aligned} \tag{4.6.1}$$

Without loss of generality, because of the symmetry of revolution of the surface with respect to the z-axis and the on-axis position of the conjugate reference points, we assume that $L_0 = L_1 = 0$. Using the expressions for the axial positions of the conjugate planes (see Eq. (4.5.14)), and choosing a pair of finite M-direction cosines M_0 and M_1, where $M_1 = m_A M_0$ according to the paraxial magnification relationship which is always satisfied for sufficiently small M-values, we find

$$\begin{aligned} \Delta E(0, M_0, 0, M_1) &= \frac{n_0 n_1 R}{n_1 - n_0} \left\{ (1 - m_A) \sqrt{1 - M_0^2} - \left(\frac{1 - m_A}{m_A} \right) \sqrt{1 - m_A^2 M_0^2} + \frac{(n_1 m_A - n_0)(n_0 m_A - n_1)}{n_0 n_1 m_A} \right\} \\ &\quad - R \sqrt{n_0^2 + n_1^2 - 2 n_0 n_1 \left(m_A M_0^2 + \sqrt{(1 - M_0^2)(1 - m_A^2 M_0^2)} \right)} . \end{aligned} \tag{4.6.2}$$

Stigmatic imaging occurs if the condition $\Delta E(0, M_0, 0, m_A M_0) = 0$ is met for arbitrary values of M_0. The optical pathlength along all ray paths having $M_1 = m_A M_0$ is equal and, consequently, all rays intersect each other at a single image point on the optical axis. A wavefront at the object side has a spherical shape and so do the wavefronts in image space that have their centre of curvature at the specific image point. We emphasise that the ray mapping between object and image space is generally given by

$$M_1 = f_m(M_0) ,$$

obeying the limiting condition,

$$\lim_{M_0 \to 0} \frac{M_1}{M_0} = m_A , \tag{4.6.3}$$

where f_m is the ray mapping function. The same mapping function f_m applies to the x-direction cosines if the optical system is circularly symmetric. The choice $M_1 = m_A M_0$ which was used in Eq. (4.6.2) is a specific linear one which will turn out to be of utmost importance when aberration correction of an imaging system over a large image field is required (see Section 4.8).

The intersection point of a ray with a plane in image space can be calculated with the aid of Eq. (4.3.14). The solution to this equation requires the values of the first derivatives of the angle characteristic. In the actual case of stigmatic imaging for which $\Delta E = 0$ irrespective of the values of the direction cosines, the first derivatives are all equal to zero. For $L_0 = L_1 = 0$ and $M_1 = m_A M_0$, we find the expressions

$$y_0 = \frac{M_0}{\sqrt{1 - M_0^2}} \left(z_0 - z_{0,P}\right) , \qquad y_1 = \frac{m_A M_0}{\sqrt{1 - m_A^2 M_0^2}} \left(z_1 - z_{1,P}\right) . \tag{4.6.4}$$

We conclude that the image-side rays all intersect the z-axis in the paraxial image plane with axial coordinate $z_1 = z_{1,P}$.

4.6.1 Stigmatic Points of a Single Refracting or Reflecting Surface

We return to the angle characteristic function of a single surface, given by Eq. (4.6.2), and require it to become zero for all relevant values of the variables (L_0, M_0, L_1, M_1). The restriction 'relevant' is used because for physical reasons not all direction cosines lead to a well-defined characteristic function value. The situations of 'missing ray' (no intersection point between ray and surface) and total internal reflection have to be excluded. After some manipulation of Eq. (4.6.2), we find the special on-axis conjugate points that produce stigmatic imaging, listed in Table 4.1. The corresponding data for a spherical surface used in reflection are obtained by substituting $n_1 = -n_0$. We find two special cases, listed in the second part of Table 4.1. Both cases produce a pair of coincident conjugate points, one pair at the vertex O of the surface, a second pair is found at the centre of curvature C of the reflecting surface.

In Fig. 4.16a–c, the stigmatic imaging cases for a spherical surface, used in refraction, are presented. All three cases correspond to a combination of virtual object and real image, the first one being special in the sense that the object and image are located on the surface itself. By inverting the ray propagation direction, we find a real object and a virtual image. Although the first two imaging cases might look trivial, they can be efficiently used to reduce imaging errors or aberrations at large opening angles of the imaging pencils. The same holds for the third imaging case in Fig. 4.16c. This case was independently discovered by Roberval and Huygens. A geometrical proof by Huygens of this stigmatic imaging of the object sphere A_0 onto the image sphere A_1 is illustrated in Fig. 4.16d. By constructing two auxiliary spheres with radii $R_0 = n_1 R/n_0$ and $R_1 = n_0 R/n_1$ and having the same common centre C as the spherical refracting surface, the similarity of the triangles $\triangle PCB_1$ and $\triangle B_0 CP$ is easily proven. It then immediately follows that $n_1 \sin\theta_1 = n_0 \sin\theta_0$ and that $\hat{\mathbf{s}}_1$ is the refracted ray of the incident ray $\hat{\mathbf{s}}_0$, independently of the value of the angle of incidence θ_0 on the refracting surface. B_1 thus is a real stigmatic image of the virtual object point B_0. Since the ray refraction in any plane containing the centre C of the refracting surface is identical, the entire spherical surface A_0 is perfectly imaged onto the spherical surface A_1. By virtue of this latter property, the imaging is said to be aplanatic (from the ancient Greek $\pi\lambda\alpha\nu\tilde{\omega}\mu\alpha\iota$ = 'to

Table 4.1. Angular magnification m_A, transverse magnification m_T and axial positions z_0 and z_1 of conjugate points producing stigmatic imaging with the aid of a single refracting or reflecting surface with radius of curvature R.

	m_A	m_T	z_0	z_1
Transmission	$\frac{n_0}{n_1}$	1	0	0
	1	$\frac{n_0}{n_1}$	R	R
	$\frac{n_1}{n_0}$	$\frac{n_0^2}{n_1^2}$	$\left(\frac{n_1 + n_0}{n_0}\right) R$	$\left(\frac{n_1 + n_0}{n_1}\right) R$
Reflection	-1	$+1$	0	0
	$+1$	-1	R	R

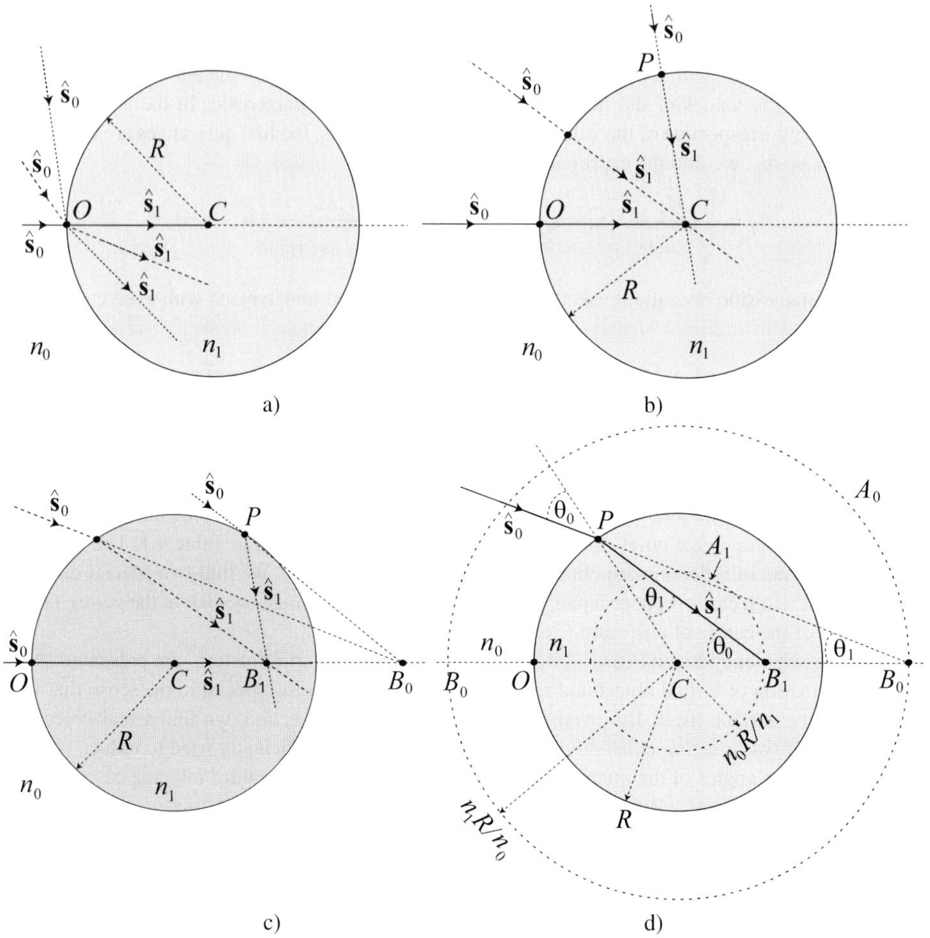

Figure 4.16: Stigmatic imaging by a spherical surface with radius of curvature R and optical contrast n_1/n_0 ($n_1 > n_0$).
a) Lateral magnification $m_T = 1$.
b) Angular magnification $m_A = 1$.
c) Lateral magnification $m_T = n_0^2/n_1^2$ (Roberval–Huygens aplanatic points).
d) Huygens' geometric construction of the aplanatic points of a spherical surface.

stray, to wander' where the *a*-prefix negates this action). The points B_0 and B_1 are said to be an aplanatic stigmatic pair of conjugate points.

4.7 Construction of the Angle Characteristic Function of a System

In this section we show the basic procedure to construct the angle characteristic function of a system from the (non-approximated) characteristic functions of the individual surfaces of the system. In Fig. 4.17 two adjacent refracting (or reflecting) surfaces with vertices O_j and O_{j+1} are shown, separated by an axial distance $d_{j+1} = O_jO_{j+1}$. A ray $\hat{\mathbf{s}}$ propagates from the surface labelled j to the surface $j+1$. We attribute the (optical) direction cosines $(n_{1,j}L_{1,j}\,,\ n_{1,j}M_{1,j})$ to this ray because it is the image ray of surface j. The ray is also the incident ray of the surface $j+1$, yielding the optical direction cosines $(n_{0,j+1}L_{0,j+1}, n_{0,j+1}M_{0,j+1})$. If the angle characteristics of the surfaces j and $j+1$ are known, the angle characteristic function $E_{j,j+1}$ of the combination is given by

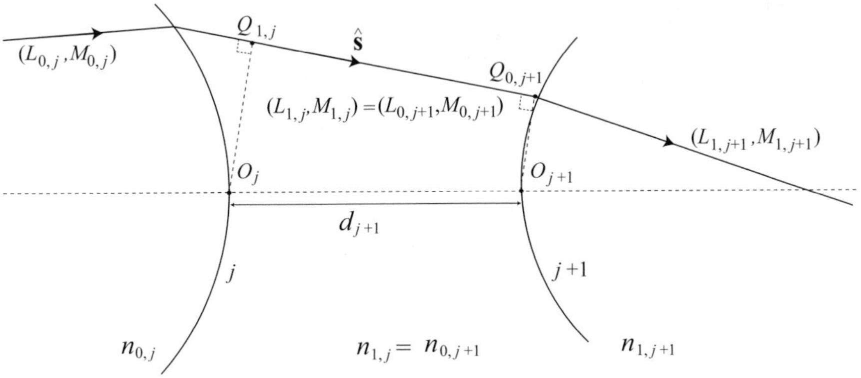

Figure 4.17: The transition of a ray $\hat{\mathbf{s}}$ from the surface j to the surface $(j+1)$ of an optical system consisting of a number of refracting and/or reflecting surfaces.

$$E_{j,j+1} = E_j(n_{0,j}L_{0,j}, n_{0,j}M_{0,j}, n_{1,j}L_{1,j}, n_{1,j}M_{1,j}) + E_{j+1}(n_{0,j+1}L_{0,j+1}, n_{0,j+1}M_{0,j+1}, n_{1,j+1}L_{1,j+1}, n_{1,j+1}M_{1,j+1}) + E_{t,j,j+1} , \qquad (4.7.1)$$

where $E_{t,j,j+1}$ is the optical pathlength increment when shifting the origin from O_j to O_{j+1}. For an axially centred system having the z-axis as axis of symmetry, the 'transport' characteristic function is written

$$E_{t,j,j+1} = n_{1,j}N_{1,j}d_{j+1} = n_{0,j+1}N_{0,j+1}d_{j+1} , \qquad (4.7.2)$$

where $N_{1,j} = N_{0,j+1}$ is the z-direction cosine of the ray vector $\hat{\mathbf{s}}$ in the intermediate space.

The total angle characteristic function of an optical system with surfaces numbered $1, \cdots, J$ is formally written as

$$E_{1,J} = \sum_{j=1}^{J} E_j(n_{0,j}L_{0,j}, n_{0,j}M_{0,j}, n_{1,j}L_{1,j}, n_{1,j}M_{1,j}) + \sum_{j=1}^{J-1} n_{0,j+1}N_{0,j+1}d_{j+1} . \qquad (4.7.3)$$

The function $E_{1,J}$ has to be written as a function of the input and output direction cosines, $n_{0,1}L_{0,1}, n_{0,1}M_{0,1}$ and $n_{1,J}L_{1,J}, n_{1,J}M_{1,J}$, respectively. Elimination of the intermediate variables is based on the argument that the change of the total characteristic function is second-order small if small changes of the ray direction in an intermediate space are introduced. This follows immediately from Fermat's principle. For a small change in $E_{1,J}$ due to a path change in the space between surfaces j and $j+1$ we can write

$$\delta E_{1,J} = \frac{\partial E_j}{\partial(n_{1,j}L_{1,j})}\delta(n_{1,j}L_{1,j}) + \frac{\partial E_{j+1}}{\partial(n_{0,j+1}L_{0,j+1})}\delta(n_{0,j+1}L_{0,j+1}) + \frac{\partial E_{t,j,j+1}}{\partial(n_{0,j+1}L_{0,j+1})}\delta(n_{0,j+1}L_{0,j+1}) = 0 , \qquad (4.7.4)$$

and a comparable expression is valid for the case of a variation of the M-direction cosines. Using the identities $n_{1,j}L_{1,j} = n_{0,j+1}L_{0,j+1}$ and $n_{1,j}M_{1,j} = n_{0,j+1}M_{0,j+1}$, we deduce from Eq. (4.7.4) the conditions

$$\frac{\partial E_j}{\partial(n_{1,j}L_{1,j})} = -\frac{\partial\left(E_{j+1} + E_{t,j,j+1}\right)}{\partial(n_{0,j+1}L_{0,j+1})} , \qquad \frac{\partial E_j}{\partial(n_{1,j}M_{1,j})} = -\frac{\partial\left(E_{j+1} + E_{t,j,j+1}\right)}{\partial(n_{0,j+1}M_{0,j+1})} . \qquad (4.7.5)$$

In this way we obtain $2(J-1)$ nonlinear equations which are sufficient to eliminate the intermediate variables. In practice, this is a task that can only be carried out by a numerical root finding procedure.

A frequently applied elimination method uses a quadratic (or paraxial) approximation of the composite angle characteristic $E_{1,J}$. Such a quadratic approximation allows for an analytic solution of the relation between the intermediate direction cosines and the entrance and exit direction cosines. In many practical cases, such an approximated elimination is satisfactory to obtain a good estimate of the imaging errors of an optical system outside the paraxial domain. The quadratic approximation method for the elimination of the intermediate variables is illustrated for the most simple case, a two-surface system, in Fig. 4.18a. The quadratic approximation of the angle characteristic function of Eq. (4.7.3) yields the expression

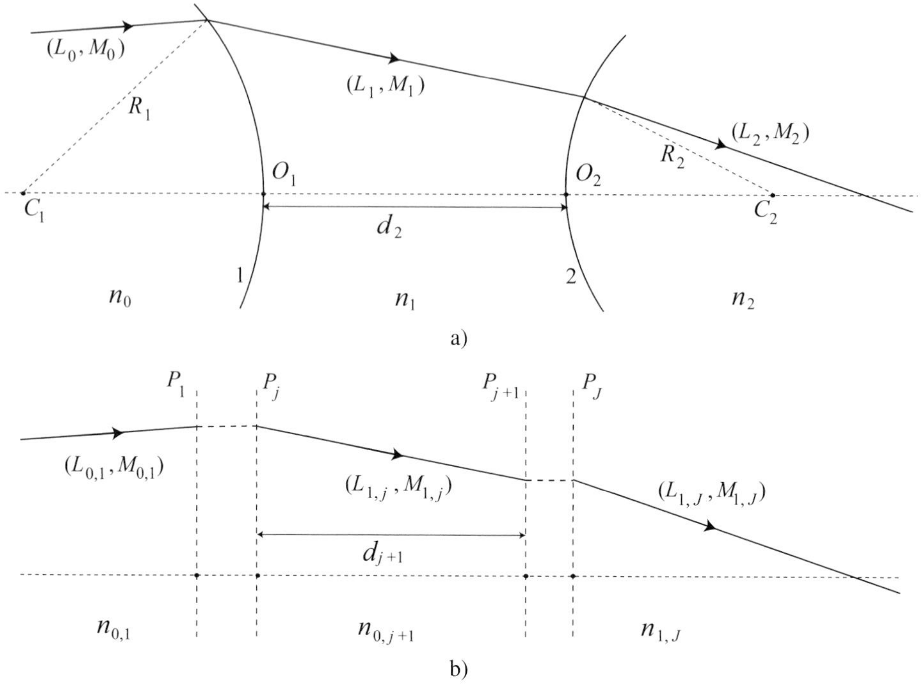

Figure 4.18: Illustrating the elimination method for the intermediate variables (L_1, M_1).
a) A quadratic approximation of the angle characteristic function for a two-surface optical system is used. The spherical surfaces have their centres of curvature in C_1 and C_2 and are separated by a distance d_2.
b) A multi-component system has been split in two parts and the components on the left- and right-hand side of an intermediate space have been replaced by their paraxial equivalents.

$$\begin{aligned}\delta E_{1,2}(n_0L_0, n_2L_2) = &\frac{R_1}{2(n_1-n_0)}\left\{n_0^2L_0^2 + n_1^2L_1^2 - 2n_0n_1L_0L_1\right\} + n_1d_2 - \frac{d_2}{2n_1}\,n_1^2L_1^2 \\ &+\frac{R_2}{2(n_2-n_1)}\left\{n_1^2L_1^2 + n_2^2L_2^2 - 2n_1n_2L_1L_2\right\},\end{aligned} \qquad (4.7.6)$$

where, for ease of notation, we have omitted the part containing the M-direction cosines.

In the case of a two-component system, Eq. (4.7.5) for the elimination of the intermediate variables reduces to $\partial E_{1,2}/\partial(n_1L_1) = 0$ and we obtain after some rearrangement

$$n_1L_1 = \frac{n_0L_0K_2 + n_2L_2K_1}{K_1 + K_2 - \frac{d_2}{n_1}K_1K_2}\,, \qquad (4.7.7)$$

where we have used the expressions for the paraxial optical power K of a surface, $K_1 = (n_1 - n_0)/R_1$ and $K_2 = (n_2 - n_1)/R_2$, respectively.

The procedure for a more complex system with many surfaces repeatedly uses the same basic step. Choosing a certain intermediate space with variables $(L_{1,j}, M_{1,j})$, the optical components on both sides are replaced by their paraxial optics approximations. Each part is represented by a certain distance between the so-called principal planes, P_1P_j and $P_{j+1}P_J$, and by its optical power (see Fig. 4.18b). The remaining distance between the principal plane P_j of the left-hand paraxial system and the principal plane P_{j+1} of the right-hand paraxial system has been denoted by d_{j+1}. Using these data, the intermediate direction cosines $(L_{1,j}, M_{1,j})$ are expressed in the input and output direction cosines, $(L_{0,1}, M_{0,1})$ and $(L_{1,J}, M_{1,J})$, by means of the paraxial data of the two neighbouring parts. This process is then repeated until all intermediate variables have been expressed in the input and output direction cosines. We repeat that the resulting angle characteristic function is not exact. However, it produces good estimates of the lowest order imaging errors; the imaging errors obtained in this way are commonly called 'pseudo-aberrations'. It can be proven that the aberrations of lowest order, the so-called third-order aberrations, are correctly obtained by means of the paraxial elimination method. For higher-order aberrations it is not allowed to neglect the influence of the aberration introduced by other surfaces when calculating the aberration contribution by a specific optical surface inside the system. When neglecting these aberration effects from the other surfaces, we miss the so-called 'induced' aberrations of a surface. The sum of *pseudo-* and *induced* aberration yields the *total* aberration of an optical surface. It is evident that the paraxial elimination method is best used for relatively short and compact optical systems, operating at small aperture angles. For such systems the difference between the 'pseudo'-aberration and the 'total' aberration remains relatively small as the heights of incidence on a surface are less perturbed by the aberration from the other surfaces than in an elongated optical system.

4.8 Isoplanatism and Aplanatism of an Optical System

The word isoplanatism is used in optical imaging when the quality of an image point is unaffected under the influence of some change in the optical system. This can be a physical change, for instance a modified distance, a change in curvature of a refracting surface, or an excursion in the object or image space. In the latter case, the optical system is called isoplanatic if its (im)perfect imaging quality is maintained over a certain limited region in image space. If the imaging quality was perfect in the reference image point and, in a first approximation, does not change in the surrounding region of this image point, the optical system is said to be aplanatic in that region. Isoplanatism and aplanatism are important features of an imaging system since in general they guarantee a finite-sized region with good or even perfect imaging properties in that region of image space. Without the isoplanatic or aplanatic property, the useful image region would be limited to a single point and its very close vicinity. In terms of imaging defects or aberrations, the isoplanatism property makes the first derivative of these defects with respect to, for instance, a coordinate change of the image point equal to zero. If not only a first derivative is made zero but also higher-order derivatives are nulled, the optical system is said to be higher-order isoplanatic or aplanatic. In what follows, we use the angle characteristic function to study the conditions for aplanatism of an optical system with respect to a spatial shift of the image point. For a system having circular symmetry with respect to the z-axis, the image shifts perpendicular and parallel to this axis are most relevant.

4.8.1 Lateral Aplanatism, the Abbe Sine Condition and the Sine Theorem

We consider an optical system with two conjugate points, O_0 and O_1, on the z-axis. For the case of perfect imaging between these points, the optical pathlength along each physical ray is an extremum according to Fermat's principle. The angle characteristic function of this optical system, with the conjugate points O_0 and O_1 as reference points, can be written as

$$E_{O_0O_1}(n_0L_0, n_0M_0, n_1L_1, n_1M_1) = C_1\, g(n_0L_0, n_0M_0, n_1L_1, n_1M_1) + C_0\,, \tag{4.8.1}$$

where C_0 and C_1 are constants with the dimension of length. The function g must have a quadratic or higher dependence on the direction cosines because of the extremum property for the optical pathlengths of the physical rays. The value of the extremum corresponds to C_0. The function g determines the mapping of the ray angles from the object to the image space of the optical system. As a consequence of the perfect imaging between O_0 and O_1, a constant optical pathlength, C_0 in Eq. (4.8.1), is found for the conjugate rays. For very small values of the direction cosines, in the paraxial domain, the ray mapping is governed by the paraxial angular magnification m_A and we have

$$L_1 = m_A L_0\,, \qquad M_1 = m_A M_0\,. \tag{4.8.2}$$

For finite values of the direction cosines, the relationship above does not necessarily hold. The mapping of finite rays from object to image space depends on the optical system. The mapping function is basically free and does not influence the perfect imaging between the axial points O_0 and O_1.

We now study the imaging between a pair of conjugate points that are laterally shifted by taking these conjugate points as new reference points for the angle characteristic function. The conjugate points according to paraxial optics should be taken since rays at small angles necessarily pass through these points. In Fig. 4.19 the points $P_0(0, dy_0, 0)$ and $P_1(0, dy_1, 0)$ are such conjugate points with $dy_1/dy_0 = m_T$, the paraxial lateral magnification. The angle characteristic function of Eq. (4.8.1), referred to the conjugate points P_0 and P_1 is now given by

$$E_{P_0P_1}(n_0L_0, n_0M_0, n_1L_1, n_1M_1) = C_1\, g(n_0L_0, n_0M_0, n_1L_1, n_1M_1) - n_0M_0dy_0 + n_1M_1dy_1 + C_0\,. \tag{4.8.3}$$

The function g correctly describes the optical pathlength of rays that are in the direct neighbourhood of the stigmatic points O_0 and O_1. This is not necessarily the case for rays outside this first-order neighbourhood.

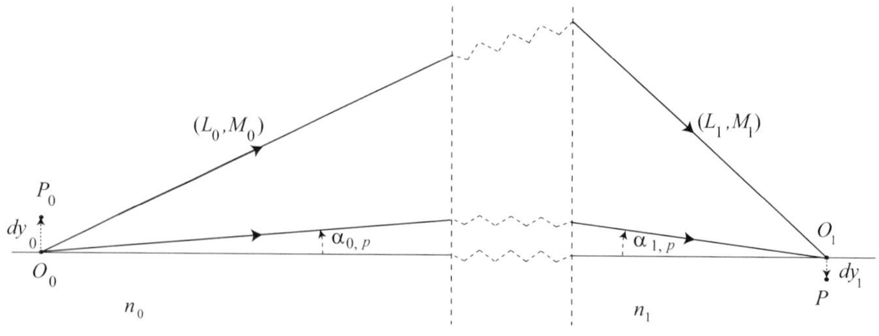

Figure 4.19: The imaging by an optical system (not shown in the figure) of an axial pair of conjugate points, O_0 and O_1, and a laterally displaced conjugate pair, P_0 and P_1. The lateral magnification dy_1/dy_0 is given by the paraxial value m_T. The imaging by rays close to the axis, with angles $\alpha_{0,p}$ and $\alpha_{1,p}$, is governed by paraxial optics. Nonparaxial rays from the axial conjugate points have x- and y-direction cosines (L_0, M_0) and (L_1, M_1).

Stigmatic imaging between P_0 and P_1 requires that for all ray pairs in object and image space the optical pathlength from P_0 to P_1 is constant and that it is an extremum with respect to neighbouring paths. From the second and third term of $E_{P_0P_1}$ we derive the condition

$$n_0 M_0 = m_T n_1 M_1 \,, \qquad \text{or} \,, \qquad \frac{M_1}{M_0} = m_A \,. \tag{4.8.4}$$

This condition imposes the ray mapping in the axial pencil from object to image space. By virtue of the axial symmetry of the optical system with respect to the z-axis, the function g is then written as

$$g(n_0 L_0, n_0 M_0, n_1 L_1, n_1 M_1) = \left(n_0 \sqrt{L_0^2 + M_0^2} - m_T n_1 \sqrt{L_1^2 + M_1^2} \right)^2 \,, \tag{4.8.5}$$

plus possible higher-order terms of even order.

The condition of Eq. (4.8.4) was derived in the context of radiation transport by Clausius **[71]**. The importance of this condition in optical imaging was pointed out by Abbe **[2]** and von Helmholtz **[133]**. Abbe especially recognised its importance for the design of optical systems and for that reason the condition is generally called the Abbe 'sine condition'. To be effective in the design of optical systems, the condition of Eq. (4.8.5) should be fulfilled for the entire pencil of rays leaving the object point O_0. If the sine condition is satisfied, a region around O_1 is imaged with the same quality as O_1 itself and the optical system shows perfect imaging over an image *field* instead of at a single point only. As a result of the general need for a large angular or spatial field in optical instruments, the Abbe sine condition has to be satisfied in the vast majority of optical imaging systems. If the sine condition is not satisfied, the quantity

$$dE_{P_0P_1} = \left(n_1 M_1 - \frac{n_0 M_0}{m_T} \right) dy_1 = \left(1 - m_A \frac{M_0}{M_1} \right) n_1 M_1 \, dy_1 \,, \tag{4.8.6}$$

is the pathlength increment for a ray pair when imaging P_0 in P_1. The term $dE_{P_0P_1}$ can be interpreted as the wavefront deviation or wavefront aberration associated with the imaging of the off-axis point. For a system with axial symmetry, this aberration is identified as comatic aberration. The sine condition fulfilment thus guarantees the absence of coma for off-axis image points in a region where higher-order dependencies on dy_1 can be neglected ('linear' coma is zero). The normalised quantity $1 - m_A M_0/M_1$ is commonly called the 'Offence against the Sine Condition' (*OSC*) and plays an important role as a target in the design of optical systems with extended field. In practice, it may occur that the Abbe sine condition can be satisfied only in a single annular region or in a few circular zones of the full pencil of rays. Even a zonal correction of the ray mapping from object to image space increases the size of the useful image field of an optical system. The Abbe sine condition of Eq. (4.8.4) becomes indeterminate when one of the two conjugates planes is at infinity, for instance when $m_T = 0$ and the image coincides with the image-side focal point (see Fig. 4.20) and the axial object point O_0 is shifted to $z_0 = -\infty$. The quantity M_0 tends to 0 and is written as $M_0 = -y_0/z_0$, where z_0 is the distance of O_0 from the entrance surface of the optical system and y_0 is the height of the point of intersection of the ray with this surface. In the case of a finite focal distance and for $z_0 \rightarrow -\infty$, we are allowed to replace z_0 by Z_0, the distance of O_0 from the object-side focal point. The Abbe sine condition now reads

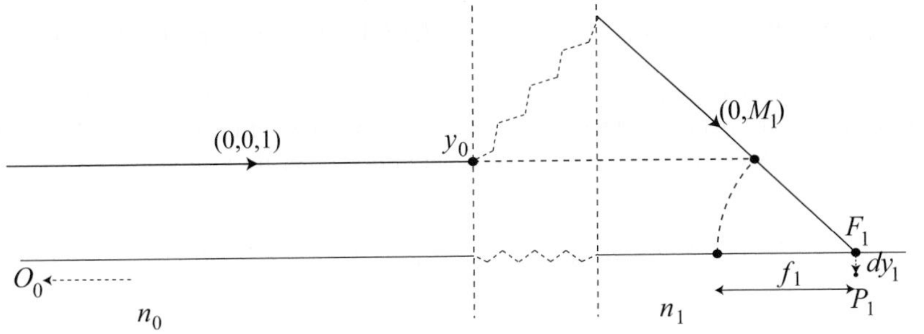

Figure 4.20: Derivation of the Abbe sine condition for the case that one of the conjugates is at infinity. The incidence height of the object ray is y_0.

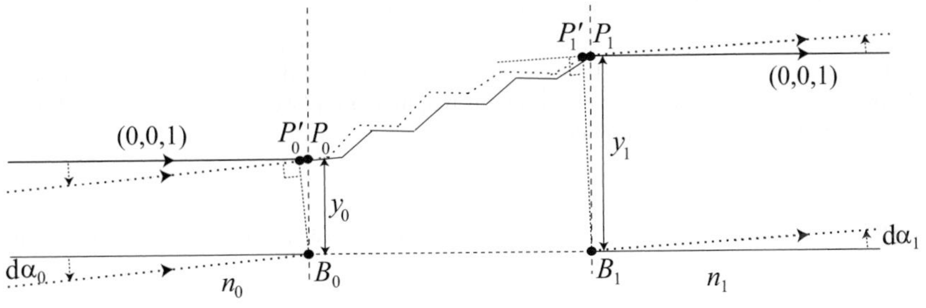

Figure 4.21: Derivation of the Abbe sine condition for the case that both conjugates are at infinity (telescopic system). B_0 and B_1 are a second pair of paraxially conjugate points at finite distances.

$$y_0 = -f_1 M_1 \,, \tag{4.8.7}$$

where we have used the relations $m_T = -f_0/Z_0$ and $n_0/f_0 = -n_1/f_1$ of Eqs. (4.5.8) and (4.5.10). From Fig. 4.20 we deduce that the lateral distance from the axis of the incoming ray should equal the height of intersection of the imaging ray with a circle of radius f_1 and centred at the focal point F_1. In this special case, the *OSC* and the wavefront aberration dW following from it are given by

$$OSC = 1 + \frac{y_0}{f_1 M_1} \,, \qquad dW = \left(1 + \frac{y_0}{M_1 f_1}\right) n_1 M_1 dy_1 \,. \tag{4.8.8}$$

The Abbe condition for a telescopic system with an optical power equal to zero reads

$$y_1 = m_T y_0 \,, \quad \text{or} \,, \qquad n_0 y_0 = n_1 m_A y_1 \,, \tag{4.8.9}$$

where y_0 and y_1 are the heights of the finite rays in object and image space and m_T and m_A are the lateral and angular magnifications of the telescopic system. A proof of this condition can also be given with the aid of Lagrange's integral invariant (see Fig. 4.21). We consider a finite object ray parallel to the z-axis with a height of incidence y_0 in the paraxial object plane through B_0 (point P_0); in the paraxial image plane through B_1, the finite ray has an intersection height y_1 (point P_1). According to paraxial optics, the ray is also parallel to the z-axis in image space. $B_0P'_0$ is part of a wavefront associated with an obliquely incident pencil at a small angle $d\alpha_0$. The paraxial ray through B_1 has an angle $d\alpha_1$ with respect to the z-axis. For the finite rays of the axial beam, in the case of perfect imaging at infinity, we have $[B_0B_1P_1P_0B_0]_a = 0$, where the subscript a refers to the rays of a beam parallel to the optical axis. The paths B_1P_1 and P_0B_0 are on the flat wavefronts associated with the axial beam and we have that $[B_0B_1]_a = [P_0P_1]_a$.

Next, we consider a parallel incident beam at a small angle $d\alpha_0$ with the optical axis. A ray of this beam leaves the telescopic system at B_1 with an inclination angle $d\alpha_1$ according to the paraxial angular magnification of the telescope. We consider the closed optical path from B_0 to P'_0, P_0, P'_1, P_1, back to P'_1, then to B_1 and finally back to B_0. We have the expression

$$[B_0P_0'P_0P_1'P_1P_1'B_1B_0]_o = [P_0'P_0] + [P_0P_1]_o - [P_1'P_1] + \int_{P_1'}^{B_1} n_1\hat{\mathbf{s}}_1 \cdot d\mathbf{r} + [B_1B_0]_o = 0\,, \tag{4.8.10}$$

where the subscript o indicates that the pathlengths along the ray portions of obliquely incident rays have to be taken. Fermat's principle states that the optical paths $[P_0P_1]_o$ and $[B_1B_0]_o$ are only different in second order from $[P_0P_1]_a$ and $[B_1B_0]_a$ and, consequently, they can be considered to be equal in first order for a sufficiently small angle of incidence $d\alpha_0$. This permits us to write for the closed path

$$[B_0P_0'P_0P_1'P_1P_1'B_1B_0]_o = [P_0'P_0] - [P_1'P_1] + \int_{P_1'}^{B_1} n_1\hat{\mathbf{s}}_1 \cdot d\mathbf{r} = 0\,,$$

or,

$$n_0y_0d\alpha_0 - n_1y_1d\alpha_1 + \int_{P_1'}^{B_1} n_1\hat{\mathbf{s}}_1 \cdot d\mathbf{r} = 0\,. \tag{4.8.11}$$

For a telescopic system that achieves perfect imaging of a small off-axis pencil of rays at an angle $d\alpha_0$, we require that the integral from P_1' to B_1 in Eq. (4.8.11) equals zero. This condition can only be satisfied if the path along the straight line $P_1'B_1$ is located on a wavefront of the outgoing beam. The outgoing oblique beam thus has a flat wavefront and the infinitely distant image produced by the telescope is perfect. For the closed-path integral of Eq. (4.8.11) to be zero we further require that $n_0y_0d\alpha_0 = n_1y_1d\alpha_1$. From paraxial optics, for small field angles $d\alpha_0$, we have that $d\alpha_1/d\alpha_0 = m_A$. Using the relation between the paraxial angular and lateral magnifications we find the Abbe condition for a telescopic system when used with infinitely distant conjugate planes,

$$y_1 = m_T y_0\,. \tag{4.8.12}$$

The offence against the sine condition, OSC, and the comatic aberration dW for an off-axis beam at infinitesimally small angle $d\alpha_1$ are given by

$$OSC = 1 - m_T\frac{y_0}{y_1}\,, \qquad dW = (OSC)\,n_1y_1d\alpha_1\,. \tag{4.8.13}$$

4.8.1.1 The Skew Ray Invariant and the Optical Sine Theorem

To prove the optical sine theorem we first discuss a property of a class of non-paraxial rays, the so-called skew rays, propagating in an optical system with rotational symmetry. In Fig. 4.22, we show a skew ray that does not intersect the axis of symmetry of the optical system, the z-axis in the figure. By virtue of Snell's law, on propagation through the system, such a ray can never intersect the optical axis, as opposed to meridional rays which propagate in a plane containing the axis of symmetry. The two rays P_0P_1 and $P_0'P_1'$ are both physical rays and we can apply the Lagrange integral invariant to them along the closed path $P_0P_0'P_1'P_1P_0$. We have

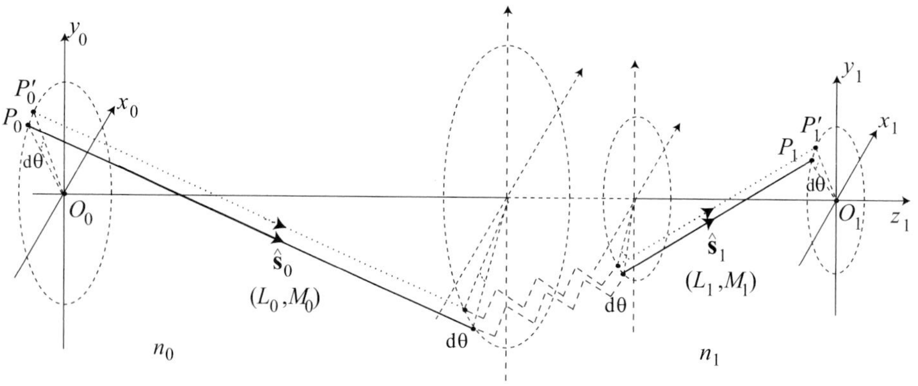

Figure 4.22: The skew ray invariant for an optical system with circular symmetry. A skew ray $\hat{\mathbf{s}}_0$ through the point $P_0(x_0, y_0)$ with direction cosines (L_0, M_0, N_0) leaves the optical system as $\hat{\mathbf{s}}_1$ having direction cosines (L_1, M_1, N_1) and the ray intersects some plane perpendicular to the z-axis at the point $P_1(x_1, y_1)$. The optical ray $P_0'P_1'$, dotted in the figure, has been obtained from the ray P_0P_1 by a minute rotation $d\theta$ around the z-axis.

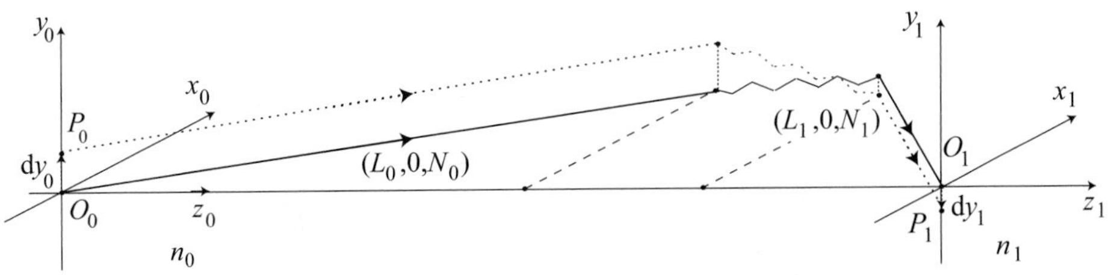

Figure 4.23: The sine theorem for an optical system with circular symmetry. O_0 and O_1 are conjugate points for the finite rays with direction cosines $(L_0, 0, N_0)$ and $(L_1, 0, N_1)$ in object and image space. P_0 is at a very small distance dy_0 from O_0 and O_0P_0 is perpendicular to the z-axis. The same holds for the point P_1, at a distance dy_1 from the z-axis.

$$[P_0P_0'] + [P_0'P_1'] + [P_1'P_1] + [P_1P_0] = 0 ,$$

or,

$$[P_0P_0'] + [P_1'P_1] = 0 . \tag{4.8.14}$$

Using $[P_0P_0'] = n_0\hat{\mathbf{s}}_0 \cdot d\mathbf{r}$, where $|d\mathbf{r}| = (x_0^2 + y_0^2)^{1/2} d\theta = P_0P_0'$, and a comparable expression for $[P_1P_1']$, we find the result

$$n_0(y_0L_0 - x_0M_0) = n_1(y_1L_1 - x_1M_1) , \tag{4.8.15}$$

or, more generally, the property that $n(yL - xM)$ is a conserved quantity in an optical system of which the z-axis is the axis of rotational symmetry. The quantity $n(yL - xM)$ is called the skew ray invariant; it is undefined for meridional rays. Two relations can be deduced from the skew ray invariant:

- We first consider an object ray through a point $P_0 = (0, y_0, z_0)$ and follow the corresponding image ray until it intersects the plane $x_1 = 0$ in the point $P_1 = (0, y_1, z_1)$. Applying the skew ray invariant to this ray yields

$$n_0y_0L_0 = n_1y_1L_1 . \tag{4.8.16}$$

  The result of Eq. (4.8.16) is generally valid, for large values of both the coordinates and the direction cosines. It is important to realise that no conjugate relation exists between the intersection points y_0 and y_1 with the plane $x = 0$.
- Another application of the skew ray invariant allows us to prove the *optical sine theorem*. We consider the imaging of a point very close to the axis by means of so-called sagittal rays. Sagittal rays are located in a plane which is perpendicular to the plane containing the optical axis and the off-axis object and image points, $P_0(0, dy_0, 0)$ and $P_1(0, dy_1, z_1)$, respectively. In Fig. 4.23, we consider the ray with direction cosines $(L_0, 0, N_0)$, propagating from an axial object point O_0 to a conjugate point O_1 on the optical axis in image space. A ray with identical direction cosines becomes a sagittal ray if we let it start at the point P_0 (see the dotted ray in the figure) which is displaced over a small distance dy_0 along the y_0 axis. The intersection point of this ray with the meridional plane $x_0 = 0$, containing the points P_0, O_0 and O_1 is P_1 and O_1P_1 is perpendicular to the z-axis. The skew ray invariant applied to the sagittal rays which intersect the points P_0 and P_1 yields the expression

$$n_0(dy_0)L_0 = n_1(dy_1)L_1 . \tag{4.8.17}$$

  Equation (4.8.17) determines the local transverse magnification dy_1/dy_0 by means of the direction cosines of the finite sagittal rays. Since the sines of the angles of the sagittal rays with the z-axis appear in the equation through the L-direction cosines, this property of sagittal rays is called the *optical sine theorem*. Confusion could arise between the sine theorem and the Abbe sine *condition*. The latter condition has to be satisfied by the meridional rays of an imaging pencil. The sine theorem for the sagittal rays of a pencil is always valid, irrespective of the mapping of the rays of an axial pencil from object to image space.

4.8.2 Axial Aplanatism and the Herschel Condition

We return to the angle characteristic function of Eq. (4.8.1) and introduce small shifts of the reference points to a new pair of paraxial conjugate points on the optical axis, P_0 and P_1. The new axial conjugate pair is found at the axial coordinates

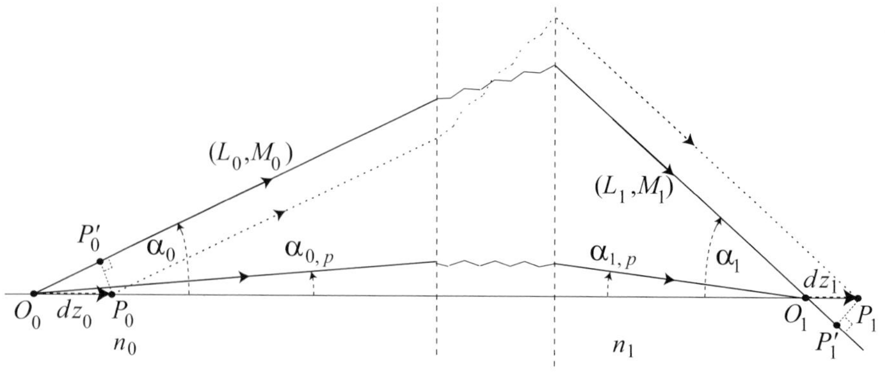

Figure 4.24: The pathlength change in the case of an axial change of conjugates. The new conjugate pair is denoted by P_0 and P_1. The optical pathlength from P_0 to P_1 is approximated by $[P_0'P_1']$.

dz_0 and dz_1, which satisfy the paraxial relation $dz_1 = n_1 m_T^2(dz_0)/n_0$. Here, we have used the expression for the axial magnification, Eq. (4.5.11), where m_T is the lateral magnification in the conjugate planes through the original reference points O_0 and O_1. In the case of infinitesimally small shifts dz_0 and dz_1, the optical pathlength between the new conjugate points is approximated by $[P_0'P_1']$ (see Fig. 4.24). The approximation is exact up to quadratic terms in the shifts dz_0 and dz_1. In terms of the angle characteristic function, having shifted the reference points to P_0 and P_1, we find

$$E_{P_0P_1}(n_0L_0, n_0M_0, n_1L_1, n_1M_1) = C_1\, g(n_0L_0, n_0M_0, n_1L_1, n_1M_1) - n_0N_0dz_0 + n_1N_1dz_1 + C_0\,, \tag{4.8.18}$$

where $C_0 = -n_1dz_1 + n_0dz_0$ to equate the pathlength increment to zero for the ray from P_0 to P_1 along the optical axis.

Perfect imaging between P_0 and P_1 requires a constant pathlength for all conjugate rays. The nulling of the sum of the second, third and fourth terms on the right-hand side of Eq. (4.8.18) leads to the condition

$$m_A^2(1 - N_0) = (1 - N_1)\,. \tag{4.8.19}$$

For $N_0 = \cos\alpha_0$ and $N_1 = \cos\alpha_1$ (see Fig. 4.24) we find the condition

$$m_A^2 \sin^2\left(\frac{\alpha_0}{2}\right) = \sin^2\left(\frac{\alpha_1}{2}\right)\,. \tag{4.8.20}$$

The requirement for stigmatic imaging in the presence of a shift of axial conjugates is due to Herschel **[134]** and is called the Herschel condition for axial aplanatism. It is relevant for optical instruments that need a large range of axial depth. The classical marine telescope is a typical example. In a similar way as for the sine condition, an expression for the limiting case $m_T \rightarrow 0$ can be derived, yielding

$$y_0^2 = 4f_1^2 \sin^2\left(\frac{\alpha_1}{2}\right)\,. \tag{4.8.21}$$

The function g of Eq. (4.8.18) should be of even order in the quantity $(m_A \sin\alpha_0/2 - \sin\alpha_1/2)$ and this yields

$$g(n_0L_0, n_0M_0, n_1L_1, n_1M_1) \approx \left(m_A\sqrt{1-\sqrt{1-L_0^2-M_0^2}} - \sqrt{1-\sqrt{1-L_1^2-M_1^2}}\right)^2, \tag{4.8.22}$$

with possible higher even orders of the same term.

As for the case of the Abbe sine condition, a deviation from the Herschel condition leads to an image defect of the shifted image point P_1. The aberration is called spherical aberration because the unsharpness figure of the aberrated image in P_1 shows circular symmetry. If the Herschel condition is not satisfied, the second and third terms of Eq. (4.8.18) yield an increment dE of the angle characteristic function according to

$$\begin{aligned} dE &= n_0(1 - N_0)dz_0 - n_1(1 - N_1)dz_1 \\ &= 2n_1(dz_1)\left[m_A^2 \sin^2\frac{\alpha_0}{2} - \sin^2\frac{\alpha_1}{2}\right]. \end{aligned} \tag{4.8.23}$$

Enforcing $dE = 0$, we obtain the following relation between conjugate plane shifts in object and image space:

$$\frac{dz_1}{dz_0} = \frac{n_0(1 - N_0)}{n_1(1 - N_1)} . \tag{4.8.24}$$

For small opening angles of the rays, we obtain the expression $n_1 m_T^2/n_0$ for the paraxial axial magnification. If we apply the direction cosines of the marginal rays of the pencil to Eq. (4.8.24), we obtain an axial magnification $dz_{1,m}/dz_0$ based on the extreme 'finite' rays with direction cosines $N_{0,m}$ and $N_{1,m}$. Such a non-paraxial value of the axial magnification serves to find an image plane shift dz_1 which adequately compensates for the spherical aberration terms that appear in dE of Eq. (4.8.23) when changing the object plane position over a distance dz_0. The residual aberration dE for rays from the pencil is then given by

$$dE = n_0 \left[(1 - N_0) - \left(\frac{1 - N_{0,m}}{1 - N_{1,m}} \right) (1 - N_1) \right] dz_0 . \tag{4.8.25}$$

As was done for the Abbe sine condition, we can define an Offence against the Herschel Condition (OHC) which is given by

$$OHC = \left(\frac{m_A \sin(\alpha_0/2)}{\sin(\alpha_1/2)} \right)^2 - 1 , \text{ or, equivalently, } \; OHC = \left(\frac{y_0}{2f_1 \sin(\alpha_1/2)} \right)^2 - 1 , \tag{4.8.26}$$

where the second expression is associated with the imaging of an infinitely distant object.

As the Abbe sine condition has, in general, the highest priority in optical design, it is interesting to see what the offence against the Herschel condition is if the Abbe sine condition is satisfied by the system. Using Eq. (4.8.4) in the expression for the OHC, we find

$$OHC = m_A^2 \left(1 - \sqrt{1 - \frac{\sin^2 \alpha_1}{m_A^2}} \right) \left(1 - \sqrt{1 - \sin^2 \alpha_1} \right)^{-1} - 1 , \quad m_T \neq 0 , \tag{4.8.27}$$

$$OHC = \left(\frac{\sin \alpha_1}{2 \sin(\alpha_1/2)} \right)^2 - 1 , \quad m_T = 0 . \tag{4.8.28}$$

The multiplication of the OHC value by the optical pathlength $2n_1(dz_1)\sin^2(\alpha_1/2)$ yields the spherical aberration of the image in P_1. Equation (4.8.27) shows that only the value $m_A = \pm 1$ allows simultaneous satisfaction of the Abbe and Herschel conditions in an optical system. With respect to the stigmatic imaging cases of Table 4.1 it is easily verified that all of them satisfy the Abbe sine condition. Three of them simultaneously satisfy the Herschel condition, provided that $|m_A| = 1$.

We note that the analysis above has been carried out independently of the exact nature of the imaging system. For sufficiently small values of dz_1 for the Herschel condition and dy_1 for the Abbe condition the results are valid. The first-order derivatives of the optical pathlength with respect to axial and lateral shifts have been made zero by satisfying the Herschel or Abbe condition. For larger excursions, the neglected higher-order terms gradually invalidate the results. It is here that the optical design influences the outcome of the approximated calculations. In general, it can be said that lateral and axial excursions of the order of 50 to 100 μm are still safe values for applying the above formulas at optical wavelengths. This is even true when $\sin \alpha$ approaches high values of the order of, for instance, 0.90.

4.9 The Definition of Transverse and Wavefront Deviation

In geometrical optics the image quality was originally judged by means of the so-called blur of a pencil of rays in image space. In Fig. 4.25 we show a ray from a pencil of rays originating in the axial object point O_0. The ray is a special one because it hits the rim of the aperture in the optical system that limits the angular extent of the pencil. It is commonly called a *marginal ray* of the pencil. The limiting aperture, circular in most cases, is called the diaphragm of the optical system. It limits the maximum elevation angles of the rays in object and image space to α_0 and α_1. The central point D of the diaphragm can be imaged towards the object space and towards the image space, yielding the paraxial images E_0 and E_1 of D by means of the parts I and II of the optical system. The point E_0 determines the axial position of the entrance pupil, the point E_1 that of the exit pupil of the optical system. According to the laws of paraxial optics, E_0 and E_1 are conjugate points and the planes through these points, perpendicular to the axis, are conjugate pupil planes. In the figure,

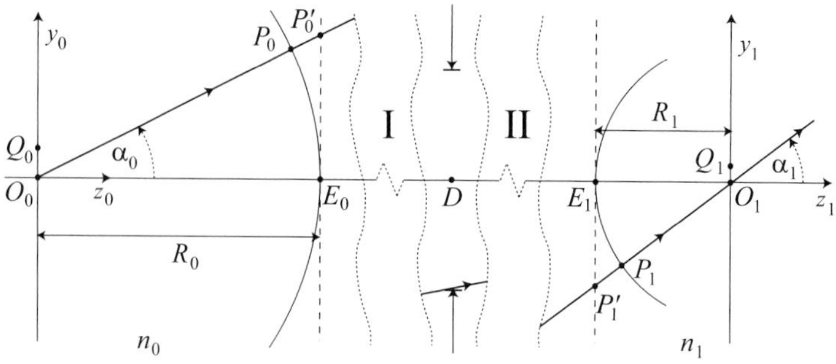

Figure 4.25: The diaphragm and pupils of an optical system. P_0 is on the spherical wavefront in the entrance pupil, centred on the object point O_0 and intersecting the optical axis in E_0. The wavefront in the exit pupil through E_1 is also spherical in the case of a perfect system. The diaphragm is found in the plane through D and it limits the maximum ray height in the ray bundle through the system. Q_0 and Q_1 are a pair of off-axis conjugate points.

the ray intersects the image plane in the axial point O_1. For an imperfect optical system, this will generally not be the case and we define the transverse aberration as

$$\delta x_t(L_0, M_0) = [x_1(L_0, M_0) - x_r] \ , \qquad \delta y_t(L_0, M_0) = [y_1(L_0, M_0) - y_r] \ , \tag{4.9.1}$$

where (x_1, y_1) are the lateral coordinates of the intersection point of the image ray with direction cosines (L_0, M_0) in object space and (x_r, y_r) the coordinates of the chosen reference point. We define the variance Δ^2 of the transverse aberration as

$$\Delta^2 = \int_{L_0}\int_{M_0} \left\{(x_1(L_0, M_0) - x_r)^2 + (y_1(L_0, M_0) - y_r)^2\right\} dL_0 dM_0 \ , \tag{4.9.2}$$

where the domain of integration is determined by the size of the diaphragm. In practice, when the system is almost perfect, the object-side direction cosines can be replaced by the image-side values, using the ray mapping function of the optical system.

In most instances, the mapping condition is dictated by the Abbe sine condition (see Eq. (4.8.4)). The small deviations between the actual direction cosines and the prescribed ones, the so-called angular aberration, should be negligible. Note that this is always the case if the exit pupil position is shifted to infinity. The value of Δ^2 is minimised by taking $x_r = \overline{x}$ and $y_r = \overline{y}$ where $\overline{x}$ and $\overline{y}$ are the average values of x_1 and y_1, respectively, after averaging over all rays of the pencil. The minimum variance is then equal to

$$\Delta^2 = \overline{x_1^2} - (\overline{x_1})^2 + \overline{y_1^2} - (\overline{y_1})^2 \ . \tag{4.9.3}$$

In the case of an imperfect optical system, Fig. 4.26 shows the wavefront deviation $P_1P'_1$ which is associated with a transverse aberration O_1A_1. We also have the angular aberration δ_1, the angle between the reference ray P_1O_1 and the aberrated ray P_1A_1. The perfect imaging in O_1 gives rise to a spherical wavefront through the centre E_1 of the exit pupil and P_1 is thus located on this sphere. The real wavefront of the imperfect system is not spherical and the aberrated ray through A_1 intersects this wavefront surface at P'_1 and the spherical wavefront at P_1. The wavefront deviation W is defined as

$$W = [O_0P_1] - [O_0E_1] = [O_0P'_1] - [O_0E_1] + [P'_1P_1] = [P'_1P_1] \ . \tag{4.9.4}$$

The sign of W formally follows from

$$W = \frac{(\mathbf{r}_{P_1} - \mathbf{r}_{P'_1}) \cdot \hat{\mathbf{s}}_{P_1}}{|\mathbf{r}_{P_1} - \mathbf{r}_{P'_1}|} \ , \tag{4.9.5}$$

where $\hat{\mathbf{s}}_{P_1}$ is the (aberrated) ray vector through the point P_1 on the reference sphere. This equation yields the correct sign of W for arbitrary propagation directions of reflected or diffracted rays. Note that the definition of W with opposite sign is frequently encountered in the literature, i.e. in **[364]**. A few textbooks use the definition of Eq. (4.9.4) where $W = n_1(P'_1P_1)$ (see, for example, **[37]**).

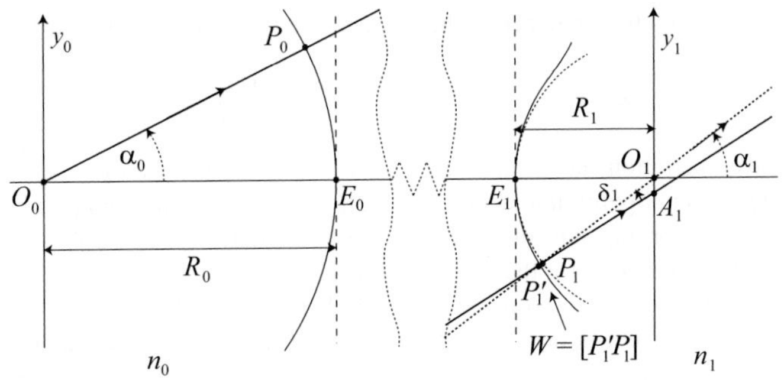

Figure 4.26: The wavefront deviation and the transverse aberration in the case of an imperfect or aberrated optical system. The aberrated ray intersects the wavefront at P_1' and the spherical reference wavefront at P_1. The wavefront deviation W equals the optical path from P_1' to P_1 (positive in the figure).

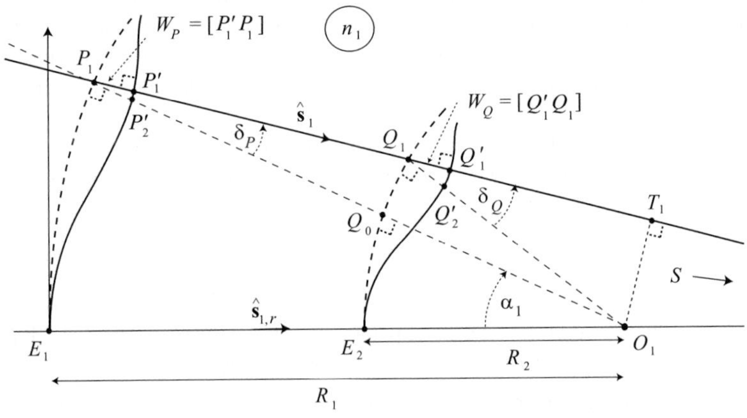

Figure 4.27: Change in the wavefront aberration W caused by a change in the radius R of the reference sphere.

4.9.1 Wavefront Aberration and a Change in Reference Sphere Radius

The wavefront aberration W defined in Fig. 4.26 is related to the reference sphere through E_1 with radius R_1 and centre of curvature O_1. The wavefront aberration W of an aberrated ray $\hat{\mathbf{s}}_1$ is shown in Fig. 4.27. The wavefront aberration is first calculated with respect to a first reference sphere through E_1 with radius of curvature R_1, centred at O_1. The reference ray is $\hat{\mathbf{s}}_{1,r}$. At a point P_1 on this reference sphere the aberration W_P is given by the optical pathlength $W_P = [P_1'P_1]$, negative in the figure. The aberrated ray propagates to a second reference sphere through E_2 having its centre of curvature also at O_1. The wavefront aberration W_Q with respect to this reference sphere is given by the optical pathlength $[Q_1'Q_1]$. From the figure we have that $P_1Q_0 = E_1E_2 = R_1 - R_2$. On the two wavefronts E_1P_1' and E_2Q_1' there are pairs of points which are joined by an orthogonal ray and we have that $P_1'Q_1' = R_1 - R_2$. In the triangle $\triangle Q_1P_1Q_0$, for sufficiently small δ_P, we have $P_1Q_0 = P_1Q_1 \cos\delta_P$. From the figure we observe that $P_1Q_1 = P_1P_1' + P_1'Q_1' + Q_1'Q_1$ and this yields the expression

$$\frac{R_1 - R_2}{\cos\delta_P} = -\frac{W_P}{n_1} + (R_1 - R_2) + \frac{W_Q}{n_1} ,$$

or ,

$$W_Q = W_P + n_1(R_1 - R_2)\left(\frac{1}{\cos\delta_P} - 1\right) \approx W_P + n_1(R_1 - R_2)\left(\frac{\delta_P^2}{2}\right) . \tag{4.9.6}$$

For the special case that $R_1 \rightarrow \infty$ and the finite perpendicular distance O_1T_1 is given, we have the limiting case

$$W_Q \approx W_P + n_1 R_2 \frac{\delta_Q^2}{2} = W_P + n_1 \frac{(O_1T_1)^2}{2R_2} . \tag{4.9.7}$$

The dependence of the wave aberration on the choice of the reference sphere can be made zero by choosing S, the intersection point of $\hat{\mathbf{s}}_1$ and $\hat{\mathbf{s}}_{1,r}$, as midpoint. In practice, for a full pencil of rays, it is not possible to satisfy the condition $\delta_P = 0$ for all rays of the pencil. For any shift of the centre of the reference sphere, Eq. (4.9.6) needs to be taken into account to correct the wavefront aberration for the new position of O_1.

If the aberrated ray and the reference ray do not intersect each other, the choice of O_1 is not straightforward. This situation frequently occurs, for instance in the case of the principal ray of a pencil and a general skew ray outside the meridional section of an oblique pencil. In this case, the midpoint of the common perpendicular to both rays can be used as centre O_1 of the reference sphere **[141]**. This midpoint will be different for each ray pair. To minimise the rms wavefront aberration over the aperture of the pencil, a midpoint O_1 is chosen that best fits all ray pairs of the pencil.

4.9.2 Relationship between Wavefront and Transverse Aberration

In Fig. 4.28 we present a diagram of an aberrated wavefront through E_1 and P_1' and the aberrated ray $P_1P_1'A_1T_1$ which intersects this wavefront at P_1'. The reference wavefront (dashed in the figure) is part of a sphere through E_1 and P_1 and the non-aberrated ray $P_1P_2'O_1$ intersects this sphere in P_2'. To discriminate between the coordinates in pupil space and the coordinates in object and image space we use capital letters for the pupil coordinates and lowercase letters for those in object and image space.

The relationship between the wavefront aberration $[P_1'P_1]$ and the transverse aberration O_1A_1 is best analysed with the aid of the point characteristic function. Using the definition of wavefront aberration according to Eq. (4.9.4) we write

$$\begin{aligned} W(X_1, Y_1) &= [O_0P_1] - [O_0E_1] = V(\mathbf{r}_0, \mathbf{r}_{P_1}) - V(\mathbf{r}_0, \mathbf{r}_{E_1}) \\ &= V(0, 0, 0, X_1, Y_1, Z_1) - V(0, 0, 0, 0, 0, Z_{E_1}) , \end{aligned} \tag{4.9.8}$$

where O_0 is the axial object point with coordinates $(x_0, y_0, z_0) = (0, 0, 0)$. The wavefront deviation is written as a function of the two independent coordinates, (X_1, Y_1), of a general point P_1 on the exit pupil sphere through E_1. The derivatives of W with respect to these coordinates X_1 and Y_1 on the reference sphere through P_1 are given by

$$\frac{\partial W(X_1, Y_1)}{\partial X_1} = \frac{\partial V(\mathbf{r}_0, \mathbf{r}_{P_1})}{\partial X_1} + \frac{\partial V(\mathbf{r}_0, \mathbf{r}_{P_1})}{\partial Z_1}\frac{\partial Z_1}{\partial X_1} , \qquad \frac{\partial W(X_1, Y_1)}{\partial Y_1} = \frac{\partial V(\mathbf{r}_0, \mathbf{r}_{P_1})}{\partial Y_1} + \frac{\partial V(\mathbf{r}_0, \mathbf{r}_{P_1})}{\partial Z_1}\frac{\partial Z_1}{\partial Y_1} , \tag{4.9.9}$$

where the dependent variable Z_1 is determined by the position of P_1 on the reference sphere with radius R_1,

$$X_1^2 + Y_1^2 + (Z_1 - R_1)^2 = R_1^2 , \tag{4.9.10}$$

where the origin for the equation of the sphere is chosen at the point E_1.

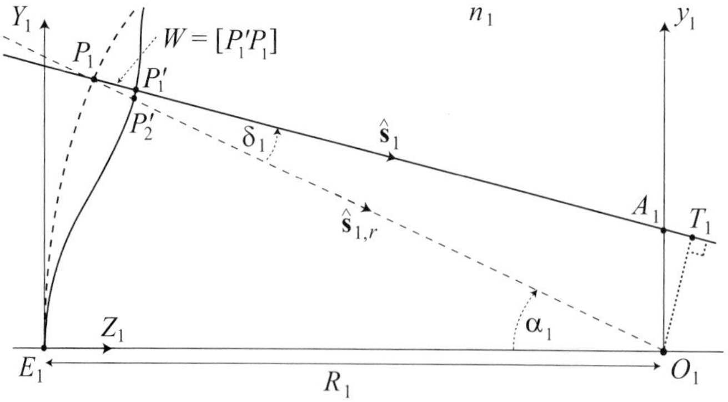

Figure 4.28: The wavefront deviation and the transverse aberration in image space for an aberrated optical system.

The differential quotients comprising Z_1 equal

$$\frac{\partial Z_1}{\partial X_1} = \frac{X_1}{R_1 - Z_1} = -\frac{L_{1,r}}{N_{1,r}} , \qquad \frac{\partial Z_1}{\partial Y_1} = \frac{Y_1}{R_1 - Z_1} = -\frac{M_{1,r}}{N_{1,r}} , \tag{4.9.11}$$

where $\hat{\mathbf{s}}_{1,r} = (L_{1,r}, M_{1,r}, N_{1,r})$ is the non-aberrated ray through P_1. The trajectory of the aberrated ray with unit vector $\hat{\mathbf{s}}_1$ is written as $\mathbf{r} = \mathbf{r}_{P_1} + p\hat{\mathbf{s}}_1$ and the intersection point A_1 with the image plane is found for the parameter value $p_{A_1} = (R_1 - Z_1)/N_1$. For the x- and y-coordinates of A_1 we find

$$x_{A_1} = X_1 + \left(\frac{R_1 - Z_1}{N_1}\right) L_1 , \qquad y_{A_1} = Y_1 + \left(\frac{R_1 - Z_1}{N_1}\right) M_1 . \tag{4.9.12}$$

With the aid of Eqs. (4.3.5) and (4.3.6) we write Eq. (4.9.9) as

$$\frac{\partial W(X_1, Y_1)}{\partial X_1} = n_1 \left\{ L_1 - L_{1,r} \left(\frac{N_1}{N_{1,r}} \right) \right\} , \qquad \frac{\partial W(X_1, Y_1)}{\partial Y_1} = n_1 \left\{ M_1 - M_{1,r} \left(\frac{N_1}{N_{1,r}} \right) \right\} . \tag{4.9.13}$$

Using the expressions of Eqs. (4.9.11) and (4.9.12) in Eq. (4.9.13) we obtain for the transverse aberration components,

$$x_{A_1} = \frac{(P_1A_1)}{n_1} \frac{\partial W(X_1, Y_1)}{\partial X_1} , \qquad y_{A_1} = \frac{(P_1A_1)}{n_1} \frac{\partial W(X_1, Y_1)}{\partial Y_1} , \tag{4.9.14}$$

where the distance P_1A_1 is equal to $(R_1 - Z_1)/N_1$. P_1A_1 depends on the z-direction cosine of the aberrated ray. However, for most optical imaging systems, the transverse aberration O_1A_1 which follows from the wave geometrical aberration distance $P_1'P_1$ is so small that P_1A_1 can be safely replaced by the radius R_1 of the reference sphere in the exit pupil.

4.9.3 Wavefront Deviation and the Angle Characteristic Function

Other definitions of wavefront aberration than that of Eq. (4.9.4) have been proposed. Referring to Fig. 4.28, the optical pathlength $[P_2'P_1]$ is also a candidate for defining the wavefront aberration. With respect to Huygens' principle, all the secondary waves are in phase on the wavefront $E_1P_2'P_1'$. By using the wavefront definition with the optical pathlength $[P_2'P_1]$, the phase of the secondary waves is projected onto the reference sphere through P_1 along the path of the non-aberrated ray $\hat{\mathbf{s}}_{1,r}$. To obtain the complex amplitude in a certain point in image space, Huygens' secondary waves on the exit pupil sphere are propagated to the desired point in image space. The amplitudes of all the secondary waves originating at the exit pupil sphere are added in that point after an adjustment of their phase by the amount of $k_0[P_2'P_1]$. In this way we find, for instance, the optical field at the reference image point O_1. Without going into detail, we just mention that the difference between the two definitions, $[P_1'P_1]$ and $[P_2'P_1]$, is of the order of $W\delta_1^2$, where δ_1 is the angular aberration of the image-side ray. In all practical cases, such a value is negligible in the optical domain.

Another definition of wavefront aberration is based on the value of the angle characteristic function. Using the reference points O_0 and O_1 in object and image space, the optical pathlength difference between an aperture ray and the axial ray is given by

$$W_H = [O_0P_1P_1'A_1T_1] - [O_0E_1O_1] , \tag{4.9.15}$$

where we have used the subscript H to refer to Hamilton's definition of the characteristic functions. We introduce the wavefront aberration expression W of Eq. (4.9.4) in Eq. (4.9.15) and this yields

$$W_H = [O_0P_1] - [O_0E_1] + [P_1T_1] - [E_1O_1] = W + n_1R_1(\cos\delta_1 - 1) \approx W - n_1R_1\frac{\delta_1^2}{2} . \tag{4.9.16}$$

Since δ_1 is of the order of the transverse blur value $O_1A_1 = \Delta$ divided by the radius R_1 of the reference sphere, it is permissible to write

$$\delta W = |W_H - W| \approx R_1 \left(\frac{\Delta}{R_1}\right)^2 , \tag{4.9.17}$$

where W_H is the less accurate quantity. The difference between the two definitions becomes zero if the exit pupil is shifted towards infinity. In principle, the preferred definitions are those related to the pathlength defects in the exit pupil region. The wavefront aberration defined by W_H is based on a prolongation of the geometrical rays beyond the exit pupil of the optical system towards the image region. In this region the diffraction unsharpness needs to be taken into account.

The choice of W_H for the calculation of wavefront aberration had an advantage in the pre-computer era. The calculation of the point of intersection of the image ray with the exit pupil sphere is not needed when using the W_H definition. Such a numerical operation was time-consuming; nowadays, such an extra effort has become negligible.

4.9.4 Wavefront Aberration and Canonical Coordinates

In this subsection we discuss a special normalised notation for image plane coordinates, pupil coordinates and wavefront aberration. This notation leads to surprisingly simple expressions for the relationship between wavefront and transverse aberration and for the Abbe sine condition. The normalising actions are the following (see Fig. 4.29):

- The wavefront aberration W is divided by λ_0, the wavelength in vacuum, yielding the primed quantity

$$W' = \frac{W}{\lambda_0} \, . \tag{4.9.18}$$

- The exit pupil coordinates (X_1, Y_1) are normalised with respect to half the transverse diameter $2\rho_1$ of the exit pupil, yielding the primed normalised coordinates

$$X_1' = \frac{X_1}{\rho_1} \, , \qquad Y_1' = \frac{Y_1}{\rho_1} \, . \tag{4.9.19}$$

- The image plane coordinates (x_1, y_1) are normalised with respect to the image unsharpness that is due to diffraction. This fundamental unsharpness is of the order of the so-called *diffraction unit* given by $\lambda_1/\sin\alpha_1$, or $\lambda_0/(n_1 \sin\alpha_1)$, where α_1 is the maximum ray angle in image space. The quantity $n_1 \sin\alpha_1$ is called the image-side numerical aperture (NA_1) or, more briefly, the numerical aperture of the imaging system. The numerical aperture is by definition a positive quantity. The diffraction unit is discussed in more detail in Chapter 9.

We now define the normalised primed coordinates in object and image space,

$$x_0' = \frac{n_0 \sin\alpha_0}{\lambda_0} \, x_0 \, , \qquad y_0' = \frac{n_0 \sin\alpha_0}{\lambda_0} \, y_0 \, , \tag{4.9.20}$$

$$x_1' = \frac{n_1 \sin\alpha_1}{\lambda_0} \, x_1 \, , \qquad y_1' = \frac{n_1 \sin\alpha_1}{\lambda_0} \, y_1 \, . \tag{4.9.21}$$

To make these normalised coordinates really *canonical* coordinates, we also want to normalise the magnification of the optical system to +1. For this reason, the sign of $\sin\alpha$ of corresponding rays in object and image space has to be taken into account. Referring to Fig. 4.29, the ray through $O_0P_0P_1O_1$ has ray angles with equal sign in object and image space and the imaging has a positive magnification. The positive direction axis for the real-space and canonical coordinates is now

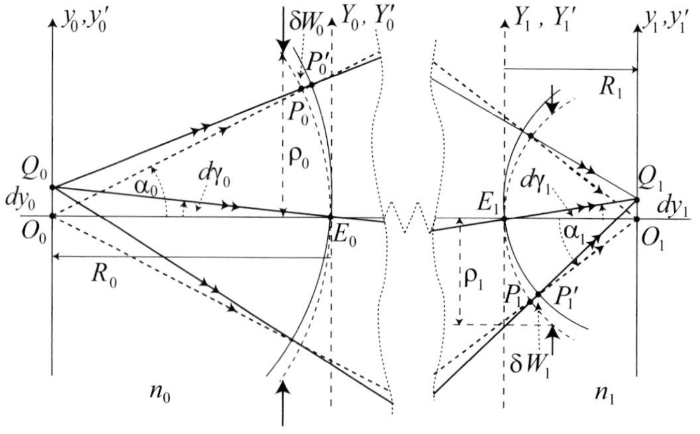

Figure 4.29: Illustrating Hopkins' canonical coordinates and off-axis imaging.

identical. However, as soon as the lateral magnification is negative, we have to take into account the opposite direction of the positive real-space and positive canonical axes in image space. The canonical coordinates were first introduced by Hopkins **[145]**. The adjective 'canonical' is inspired by the fact that some basic expressions about pupil and image plane aberration resemble Hamilton's canonical equations of motion for a mechanical system.

We now focus on Eq. (4.9.14) which yields the transverse aberration components $(\delta x_1, \delta y_1)$ (line segment O_1A_1 in Fig. 4.28) in terms of the derivative of the wavefront aberration on the reference sphere. First, the distance P_1A_1 is replaced by the fixed quantity R_1. The wavefront aberration function W is replaced by W' and the canonical coordinates (X'_1, Y'_1) are substituted for their real-space counterparts (X_1, Y_1). We obtain for the (x_1, y_1) aberration components,

$$\delta x_1 = \frac{R_1 \lambda_0}{n_1 X_{1,m}} \frac{\partial W'(X'_1, Y'_1)}{\partial X'_1} , \qquad \delta y_1 = \frac{R_1 \lambda_0}{n_1 Y_{1,m}} \frac{\partial W'(X'_1, Y'_1)}{\partial Y'_1} , \tag{4.9.22}$$

where $X_{1,m}$ and $Y_{1,m}$ are the maximum (positive) values of the pupil coordinates, determined by the size of the diaphragm of the optical system. For on-axis imaging we assume that the apertures and pupils have circular symmetry. This allows us to write

$$X_{1,m} = Y_{1,m} = -R_1 \sin\alpha_{1,m} , \tag{4.9.23}$$

where $\alpha_{1,m}$ is the maximum ray angle in image space. The minus sign in this equation stems from the sign of ray angles as a function of the value of the pupil coordinates (X_1, Y_1) and the convex or concave shape of the surface of the reference sphere in the exit pupil. The transverse aberration in terms of the canonical pupil and field coordinates can now be written in the compact form

$$\delta x'_1 = -\frac{\partial W'(X'_1, Y'_1)}{\partial X'_1} , \qquad \delta y'_1 = -\frac{\partial W'(X'_1, Y'_1)}{\partial Y'_1} . \tag{4.9.24}$$

The same canonical coordinates are now applied, with due care to the signs of optical ray angles, to the imaging of an off-axis object point Q_0 in Q_1 (see Fig. 4.29). We assume that the imaging of the axial conjugate points is perfect, implying that the dashed wavefronts in the entrance and exit pupil are spherical caps. The wavefront in the entrance pupil belonging to the off-axis object point Q_0 is spherical too (solid curve with radius R_0, negative in the figure) and has been tilted over a small angle $d\gamma_0 = dy_0/R_0$ with respect to the wavefront from O_0. At the upper rim of the entrance pupil, the point P'_0 on the tilted wavefront advances over a distance $P_0P'_0$ and the optical pathlength difference equals $n_0P_0P'_0$. The image point in image space is found at the paraxial position Q_1 where $dy_1/dy_0 = m_T$ is the paraxial magnification. The finite ray from Q_0 through P'_0 intersects the dashed wavefront in image space at the point P_1. The intersection point of the ray with the tilted sphere through E_1, having its centre at Q_1 and a radius which in first order equals R_1, is the point P'_1.

For the optical pathlength difference δW_{Q_0} between a ray from Q_0 through P'_0 to P'_1 and a paraxial ray from the same point through E_0 and E_1 we write

$$\delta W_{Q_0} = [Q_0P_0P'_0]_{Q_0} + [P_0P_1]_{Q_0} + [P_0P_1]_{Q_0} + [P_1P'_1]_{Q_0} - [Q_0E_0]_{Q_0} - [E_0E_1]_{Q_0} = -[P_0P'_0] + [P_1P'_1] , \tag{4.9.25}$$

where we have used that, in a first-order approximation for small $d\gamma_0$ and $d\gamma_1$, $[P_0P_1]_{Q_0} = [P_0P_1]_{O_0}$ and $[E_0E_1]_{Q_0} = [E_0E_1]_{O_0}$ since Fermat's principle is valid for neighbouring ray paths. Moreover, $[E_0E_1]_{O_0} = [P_0P_1]_{O_0}$, since we have assumed perfect imaging for the axial conjugate points. From the cosine rule in $\triangle Q_0O_0P_0$ we have that $[Q_0P_0] = -n_0(R_0 + \sin\alpha_0\, dy_0)$. Note that $R_0 = E_0O_0$ is a negative distance in Fig. 4.29, as well as the distance Y_1 from the axis to P_1. Using a similar result for the optical pathlength $[Q_1P_1]$ we obtain for the sum of the pathlength increments in object and image space

$$\delta W_{Q_0} \approx -n_0(dy_0)\sin\alpha_0 + n_1(dy_1)\sin\alpha_1 . \tag{4.9.26}$$

In the cross-section of Fig. 4.29 we have that $\sin\alpha_0/\sin\alpha_{0,m} = Y'_0$ and $\sin\alpha_1 / \sin\alpha_{1,m} = Y'_1$. Replacing W_{Q_0} by W' (see Eq. (4.9.18)), we can write the pathlength equation of Eq. (4.9.26) in the cross-section $X_1 = 0$ according to

$$\delta W' \approx -\left(\frac{\sin\alpha_0}{\sin\alpha_{0,m}}\right)\left(\frac{n_0\sin\alpha_{0,m}}{\lambda_0}\right)dy_0 + \left(\frac{\sin\alpha_1}{\sin\alpha_{1,m}}\right)\left(\frac{n_1\sin\alpha_{1,m}}{\lambda_0}\right)dy_1 = -Y'_0\,dy'_0 + Y'_1\,dy'_1 . \tag{4.9.27}$$

In these equations Y'_0 and Y'_1 stand for the canonical Y'-coordinates of P_0 and P_1, respectively. We then use the important property $dy'_0 = dy'_1$ of the canonical coordinates, which follows from the normalisation to the magnification value +1 for the imaging between the conjugate planes under consideration, irrespective of their real-space magnification. In two dimensions we have the following equations for the transverse ray aberration components ($\delta X_1'$, $\delta Y_1'$) of a ray that intersects the exit pupil sphere in the point with transverse coordinates (X_1', Y_1'),

$$\delta X_1' = \frac{\partial W'(X_1', Y_1')}{\partial x_1'} , \qquad \delta Y_1' = \frac{\partial W'(X_1', Y_1')}{\partial y_1'} . \tag{4.9.28}$$

In this equation $\delta X_1' = X_1' - X_0'$ and $\delta Y_1' = Y_1' - Y_0'$.

In words, we interpret Eq. (4.9.28) as follows. The difference between the normalised ray intersection heights in exit and entrance pupil determines the rate of change of the normalised wavefront aberration with respect to the normalised image distance from the axial reference point. If the transverse aberration of a ray on the exit pupil has been made zero, the aberration as a function of the lateral field coordinate is stationary. For an axial image point this means that the direct environment of this point, a small circular region, has the same imaging performance as the central point. If the condition can be satisfied for some off-axis image point, an entire annular region will have the same aberration as this specific point, provided that the imaging system possesses rotational symmetry around the optical axis. The condition of Eq. (4.9.28) is equivalent to the Abbe sine condition and it makes the optical pathlength change $dE_{P_0P_1}$ of Eq. (4.8.6) equal to zero. The offence against the sine condition (OSC) is now simply given by $\delta X_1'$ and $\delta Y_1'$ in the two orthogonal cross-sections. If we specify a maximum allowable wavefront aberration $\delta W'$, an OSC value of $\delta Y_1'$ immediately yields the maximum field excursion $\delta y_1' \approx \delta W'/\delta Y_1'$. As a rule of thumb, if $\delta W'$ is typically $1/4$ (corresponding to a *just* diffraction-limited system), the quotient $(\delta W'/\delta Y_1')^2$ is a good approximation for the number of resolvable image points with dimension $\lambda_0/(n_1 \sin\alpha_1)$.

Equations (4.9.24) and (4.9.28) are called the canonical equations. The first set of equations gives the transverse ray aberration in the image plane as a function of the wavefront slope. The second set yields the transverse displacement of a ray on the pupil sphere by means of the change of wavefront aberration as a function of the canonical image plane coordinate. The preceding analysis was given for conjugate pairs of object and image points close to the optical axis of an optical system with rotational symmetry. Hopkins' analysis **[145]** goes beyond this restriction and applies to (far) off-axis imaging pencils of rays, including the effect of a non-circular cross-section of entrance and exit pupil and the presence of an off-axis aberration such as astigmatism.

4.10 Paraxial Optics and the Matrix Analysis of Optical Systems

In this section we discuss, in some more detail than in Subsection 4.3.1.3, the properties of paraxial imaging. We first present the sign conventions which are used in this section. They are in line with those used so far in this chapter but are extended to cover more general imaging situations. We then derive anew the basic paraxial imaging properties of a single surface using the condition of pathlength equality of the rays in an imaging pencil. From the properties of single-surface imaging, we then develop the paraxial matrix theory for ray transfer and imaging by an optical system. The paraxial ray-tracing is performed given the prescribed sequence of optical surfaces (sequential ray-tracing).

4.10.1 Sign Conventions

In the literature, there is no uniformity of sign conventions for distances, curvatures, angles of incidence and wavefront deviation. We briefly state below the (sign) conventions which are used in this section on paraxial optics:

- *Distances*
  A right-handed coordinate system is used with the positive z-axis generally pointing towards the right, the assumed direction of propagation of the light which is incident on the optical system. The optical axis coincides with the z-axis. The distance AB between two points A and B is counted positive when $z_B > z_A$. The curvature of a spherical surface is counted positive when the z-coordinate of the centre of curvature is larger than the z-coordinate of the vertex or top of the surface.
- *Angles*
  Angles are counted positive in the counterclockwise direction, where the positive z-direction is taken as the reference direction. For the angle of incidence on a surface, the angle is counted positive in the counterclockwise direction with respect to the outward surface normal.

- *Refractive index*
  For formal reasons, when a reflection arises in an optical system, the refractive index of the medium after reflection is given the opposite value to the index of the medium which precedes the reflecting surface ($n_1 = -n_0$). The sign convention for distances and angles remains unchanged. Outside the paraxial domain, at large angles of incidence on a curved optical surface, it is not excluded that the reflected ray propagates in the same z-direction as the incident ray. For such cases the convention of a negative index for the 'reflection' space cannot be used.
- *Optical pathlength*
  In a homogeneous medium, the optical pathlength from A to B is defined as $[AB] = n(AB)$. When the light propagates back after an uneven number of reflections, the quantity $[AB]$ remains positive where the convention of a negative refractive index of the image space is used in the case of reflection. For a 'phase-conjugated' reflection or for propagation in a metamaterial, the index itself may take on a negative value. In such a case, if the light propagates in the positive z-direction, the change in optical pathlength becomes negative (see also Chapter 3).
- *Wavefront deviation*
  A wavefront deviation with respect to a reference surface is counted positive when the optical pathlength along a ray up to the point of intersection with the reference surface is larger than the pathlength along a reference ray to the reference surface. Note that this sign convention is also used in **[37]**, while the opposite convention is used in almost all other textbooks.

4.10.2 Defining a Paraxial System

With the aid of the position of the focal points and the position of a pair of conjugate planes having a certain lateral or angular magnification, the imaging properties of a paraxial optical system are fully defined. By prescribing four paraxial points, three distances are fixed, which is sufficient to define the paraxial system. In Fig. 4.30, the first and last surface of an optical system are found at T_0 and T_1. The points that define the paraxial imaging properties are F_0, the object-side focal point, and F_1, the image-side focal point. The extra information about the system is given by the position of so-called principal planes, P_0 and P_1, which are conjugate points with a lateral magnification of +1. By definition, the focal distance f_1 is given by the axial distance P_1F_1, since for rays propagating to image space the refracting power of the optical system seems to originate from the principal plane through P_1P_1'. This definition for the image-side focal distance of the system is plausible because an observer in image space would have to position a single lens with this focal distance at P_1 to obtain the same image ray. In the same way, as seen from object space, the focal distance f_0 equals P_0F_0, which is a negative distance in the figure.

In Fig. 4.31, the construction of the conjugate nodal points N_0 and N_1 with angular magnification +1 is illustrated, where we have used the points F_0, F_1, P_0 and P_1 as paraxial starting points. A first ray, parallel to the z-axis and originating at the off-axis point F_1' in the image-side focal plane, is traced back through the optical system and intersects the object focal point F_0 at an angle $\alpha = -y_{P_0'}/f_0$ with respect to the optical axis. A second ray from F_1' is launched back from image space with the same angle α with the z-axis. It intersects the image-side principal plane in P_1'' and, at the same height, the object-side principal plane in P_0''. From the path of this ray, we then find the object- and image-side nodal points N_0 and N_1 with angular magnification +1. From the figure we also deduce

$$F_0P_0 = N_1F_1 = -f_0 \,, \qquad F_0N_0 = P_1F_1 = f_1 \,. \tag{4.10.1}$$

In this chapter it has already been shown that $-n_0/f_0 = n_1/f_1$ and this quantity, identical in object and image space, was called the optical power of a system in units of m^{-1} (also called a *diopter*). Other definitions of a paraxial optical system

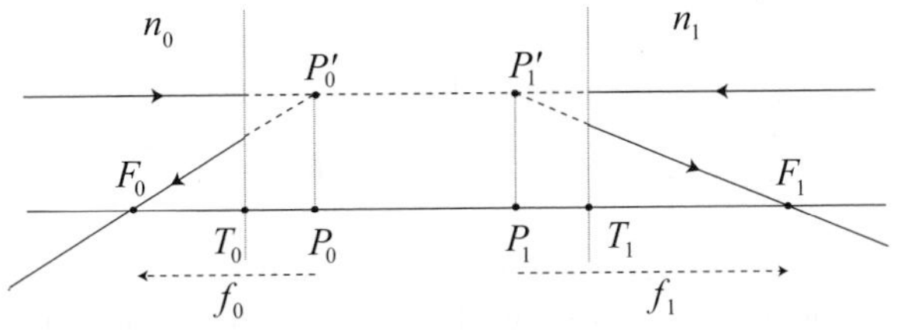

Figure 4.30: Focal points (F_0 and F_1) and principal points (P_0 and P_1) with lateral magnification +1 of an optical system.

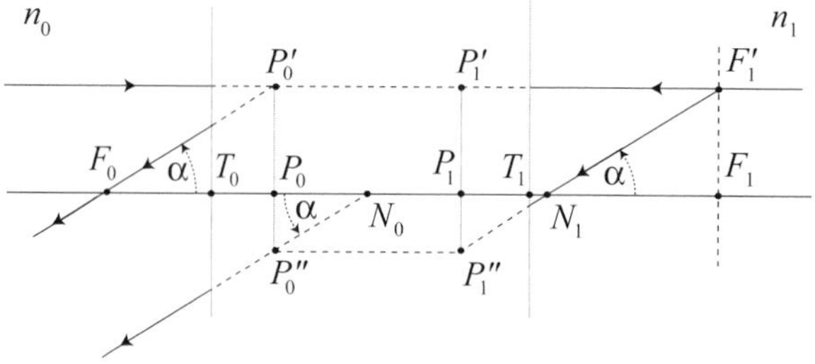

Figure 4.31: The construction of the nodal points N_0 and N_1, using the position of the focal points F_0 and F_1 and the principal points P_0 and P_1.

by means of the position of particular paraxial points are possible. In general, four of them are needed, including at least one focal point to determine the length scale of the system.

4.10.3 Paraxial Imaging by a Single Surface and by a System

The derivation of the paraxial imaging laws according to Subsection 4.3.1.3 was based on a quadratic approximation of the angle characteristic function of a single surface. If a single surface could be found with a purely quadratic characteristic function in the direction cosines (L_0, M_0) and (L_1, M_1), similar to the one in Eq. (4.4.9), such a surface would show the paraxial imaging properties for arbitrarily large values of the direction cosines. In practice, the values of the direction cosines need to be kept very small to allow the use of the quadratic approximation. Within this domain, the (L, M) direction cosines and the ray angles themselves can be interchanged. The next step in the derivation of the paraxial imaging laws is the application of the path equality property for all paraxial rays between an object and an image point. This property follows from the quadratic angle characteristic function in which the paraxial object and image points are used as reference points. Referring to Fig. 4.32, we require that the optical pathlength of a general aperture ray, $[O_0 Q_2 O_1]$, and that of the axial ray, $[O_0 E_1 O_1]$, are equal, or

$$[O_0 Q_0] + [Q_0 Q_1] + [Q_1 Q_2] + [Q_2 Q_3] + [Q_3 Q_4] + [Q_4 O_1] = [O_0 E_1] + [E_1 O_1] . \qquad (4.10.2)$$

Since Q_0 and Q_4 are located on spherical surfaces, centred on the object and on the image point, respectively, we are allowed to write

$$[Q_0 Q_1] + [Q_1 Q_2] + [Q_2 Q_3] + [Q_3 Q_4] = 0 ,$$

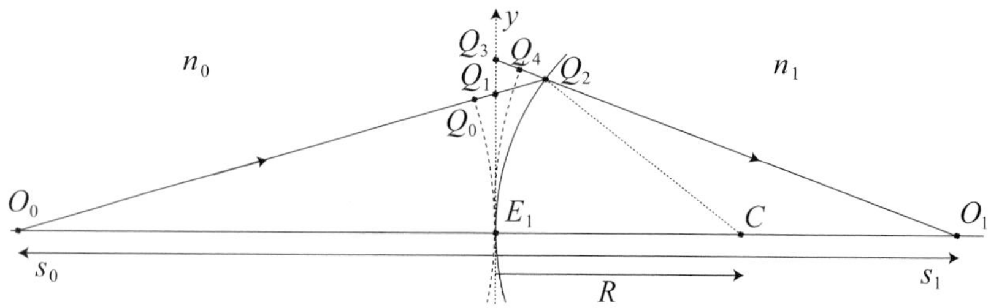

Figure 4.32: Illustrating the pathlength equality condition when imaging with the aid of a single surface. The spherically curved refracting surface, solid line in the graph, has a radius of curvature R and its centre is at C. Its point of intersection with the z-axis is E_1. The object point is at O_0, the image point at O_1. The object distance s_0 is given by $E_1 O_0$, the image distance s_1 is $E_1 O_1$.

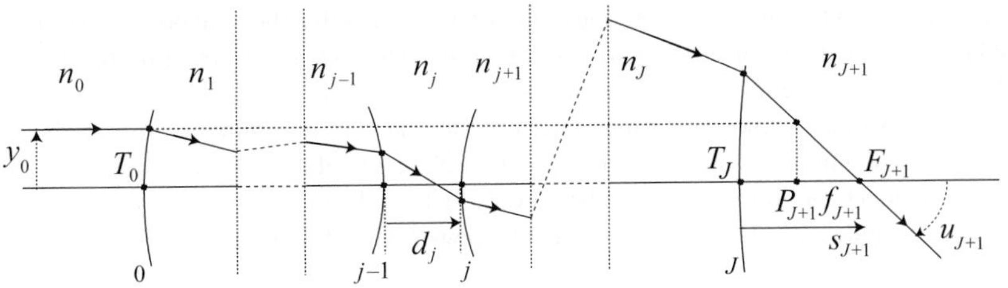

Figure 4.33: An incident parallel beam of light has paraxial intersection heights y_j on the surfaces of the optical system. The optical surfaces intersect the optical axis in the points $T_0, \cdots, T_J$. The image-side principal point and focal point of the system are P_{J+1} and F_{J+1}, respectively.

or,

$$-n_0\frac{y^2}{2s_0} + n_0\frac{y^2}{2R} - n_1\frac{y^2}{2R} + n_1\frac{y^2}{2s_1} = 0 \,. \tag{4.10.3}$$

In Eq. (4.10.3) the y-coordinates of all points Q have been given equal values. Moreover, in paraxial optics calculations, the intersection height of a ray with the optical surface is generally put equal to the y-coordinate of the paraxial ray in the reference plane through the vertex of the surface. The relative error in y that follows from this approximation is of second order and can be neglected in paraxial optics. The distance from a point on a spherical surface (the refracting surface itself or any other auxiliary surface) to the osculating plane to this spherical surface through E_1 has been put equal to $y^2/2R$, as in the case of a paraboloid. These approximations are also allowed in the case of very small values of y. We obtain the imaging equation

$$\frac{n_1}{s_1} = \frac{n_0}{s_0} + K \,, \tag{4.10.4}$$

where the quantity $K = (n_1 - n_0)/R$ denotes the optical power of the single refracting surface. Some authors introduce the notion of 'vergence' for the inverse distance with respect to a surface. The advantage of using the 'vergence' concept is that Eq. (4.10.4) and many other expressions in paraxial optics are transformed into linear relationships.

We now extend the analysis to an optical system with a sequence of optical surfaces, each characterised by its optical contrast $(n_j - n_{j-1})$, curvature $c_j = 1/R_j$ and distance d_j from the preceding surface (see Fig. 4.33). To calculate the total power of the optical system, we send in a ray, parallel to the optical axis, at a height y_0 from this axis. We can basically trace such a paraxial ray through the optical system by applying Snell's law at each surface. As a result, we obtain the heights of intersection y_j of the incident ray with each optical surface. We multiply both sides of Eq. (4.10.4) by y_j, the intersection height at the surface. Using the definition of the aperture angle $u_j = -y_j/s_j$, we obtain the following set of equations over all surfaces,

$$\begin{aligned} n_1 u_1 &= n_0 u_0 - y_0 K_0 \,, \\ n_2 u_2 &= n_1 u_1 - y_1 K_1 \,, \\ \cdots &= \cdots \qquad\qquad\quad , \\ n_{J+1} u_{J+1} &= n_J u_J - y_J K_J \,. \end{aligned} \tag{4.10.5}$$

Summing the left- and right-hand sides of all the equations above, we obtain for $u_0 = 0$,

$$n_{J+1} u_{J+1} = -\sum_{j=0}^{J} y_j K_j \,. \tag{4.10.6}$$

In Fig. 4.33, the image ray intersects the optical axis at the focal point F_{J+1}, which is at a distance f_{J+1} from the image-side principal point P_{J+1} of the system (see also Fig. 4.31 and Eq. (4.10.1) for this general property of a paraxial system). Using the expression $u_{J+1} = -y_0/f_{J+1}$, we then find

$$\frac{n_{J+1}}{f_{J+1}} = K_s = \sum_{j=0}^{J} \frac{y_j}{y_0} K_j \,, \tag{4.10.7}$$

where K_s is the optical power of the entire system. Equation (4.10.7) shows that the total power is obtained as a summation over the individual powers of the optical surfaces with a weighting factor given by the relative height of incidence y_j/y_0 on each surface of the ray that is incident from infinity. Surfaces where the incidence height becomes zero do not contribute to the total power. Such a surface coincides with an intermediate image of the infinitely distant object and it is called a 'field' surface. In the case of a thin lens at such a position, it is called a field lens of the system.

Equation (4.10.7) can also be used to determine the power of a thick lens with individual surface powers K_0 and K_1, a thickness d_1 and refractive index n_1. One obtains for the total power (Gullstrand's equation)

$$K_L = K_0 + K_1 - \frac{d_1}{n_1} K_0 K_1 \,. \tag{4.10.8}$$

The approximate expression $K_L \approx K_1 + K_2$ is known as the lens-maker's formula. The quantity d_1/n_1 is called the *reduced* thickness of the lens or its equivalent thickness in air.

4.10.4 Matrix Analysis of Paraxial Imaging

For the analysis of paraxial imaging with the aid of matrix calculus, we first derive the matrices that are related to refraction by a surface and to the ray transport from a certain surface to the next one. Using the resulting system matrix, the paraxial imaging properties of the system can be studied.

4.10.4.1 Individual Refraction and Transport Matrices and the System Matrix

To analyse the imaging properties of a system, we first transform the imaging equation of Eq. (4.10.4) into a matrix operation. The input vector has as components the optical y-direction cosine (paraxially approximated by the quantity nu) and the height of intersection with the refracting or reflecting surface. The output vector then follows from the matrix multiplication:

$$\begin{pmatrix} n_{j+1}u_{j+1} \\ y_j \end{pmatrix} = \begin{pmatrix} 1 & -K_j \\ 0 & 1 \end{pmatrix} \begin{pmatrix} n_j u_j \\ y_j \end{pmatrix} = M_{r,j} \begin{pmatrix} n_j u_j \\ y_j \end{pmatrix}, \tag{4.10.9}$$

where K_j is the optical power of the surface. The height of intersection of the refracted ray is identical to that of the incident ray. The transport of a ray from surface j to $j+1$ changes the height of intersection on the optical surfaces according to $y_{j+1} = y_j + u_{j+1} d_{j+1}$. In matrix notation the height y_{j+1} is given by

$$\begin{pmatrix} n_{j+1}u_{j+1} \\ y_{j+1} \end{pmatrix} = \begin{pmatrix} 1 & 0 \\ \dfrac{d_{j+1}}{n_{j+1}} & 1 \end{pmatrix} \begin{pmatrix} n_{j+1}u_{j+1} \\ y_j \end{pmatrix} = M_{t,j+1} \begin{pmatrix} n_{j+1}u_{j+1} \\ y_j \end{pmatrix}, \tag{4.10.10}$$

and the optical ray propagation angle u remains unchanged. In the paraxial domain, reflection is taken into account by giving the refractive index of the second medium a sign which is opposite to that of the first medium.

The matrix expression for the complete system, running from index 1 to J, is written as (see Fig. 4.33),

$$\begin{pmatrix} n_{J+1}u_{J+1} \\ y_J \end{pmatrix} = M_{r,J} M_{t,J} M_{r,J-1} \cdots M_{r,1} M_{t,1} M_{r,0} \begin{pmatrix} n_0 u_0 \\ y_0 \end{pmatrix} = M_s \begin{pmatrix} n_0 u_0 \\ y_0 \end{pmatrix}, \tag{4.10.11}$$

where the matrix product $M_s = M_{r,J} \cdots M_{r,0}$ must be evaluated from right to left. If needed, for reasons of symmetry in Eq. (4.10.11), a final transport matrix over zero distance can be added. We then end in image space with the vector data $(n_{J+1}u_{J+1})$ and y_{J+1}. The system matrix M_s is generally written as

$$M_s = \begin{pmatrix} B & -A \\ -D & C \end{pmatrix}. \tag{4.10.12}$$

The matrix definition of Eq. (4.10.12) has been described in **[131]**, **[61]**. The elements B and C are dimensionless, A has the dimension of m^{-1} and D has the dimension of length. The characters used for the matrix elements are in accordance with those that were originally given by Gauss to his paraxial system constants.

Another definition of the paraxial matrices for refraction/reflection and propagation goes back to Smith **[318]** and has been been further disseminated, among others, in **[109]**. In this definition, the order of the vector elements is inverted and the matrix elements *A*, *B*, *C* and *D* have been given different positions with the result that they are not identical with Gauss' constants,

$$\begin{pmatrix} y_J \\ n_{J+1}u_{J+1} \end{pmatrix} = \begin{pmatrix} A' & B' \\ C' & D' \end{pmatrix} \begin{pmatrix} y_0 \\ n_0 u_0 \end{pmatrix} . \tag{4.10.13}$$

The two sets of matrix elements are related as follows,

$$\begin{matrix} A & B & C & D & BC - AD = 1 \,, \\ -C' & A' & D' & -B' & A'D' - B'C' = 1 \,. \end{matrix} \tag{4.10.14}$$

Each refraction and transport matrix is unimodular and has a determinant equal to unity. As each paraxial matrix has a unit determinant, the product M_s has the same property and $BC - AD = 1$. This implies that three independent quantities are required to entirely specify a paraxial optical system, together with a reference point, for instance the position of the vertex of the first surface. This conclusion is in line with Section 4.10.2 where it was found that four positions of particular system points were needed to fully describe the imaging by a paraxial system. *A*, *B*, *C* and *D* are called the Gauss system constants because it was the mathematician Gauss who first described in a systematic way the imaging by an optical system in the paraxial approximation and identified these paraxial system constants.

In Fig. 4.34 the various intersection heights and ray angles that play a role in the calculation of the elements *ABCD* of the system matrix have been depicted. Several rays related to the imaging between the pair of conjugate planes through O_0 and O_1 are shown, originating at the object points O_0, B_0 and C_0. From these rays we select those that are needed to define the elements of the paraxial matrix related to the pair of reference planes U_0 and U_1 in object and image space. The first solid ray has double arrows, is labelled *a* and passes through Q_0 at an angle u_0 with the optical axis. It is commonly called an aperture ray. The second solid ray with one arrow, labelled *b*, passes through S_0 and has been chosen to be parallel to the optical axis. This ray is called a field ray. The rays intersect the reference plane U_1 in Q_1 and S_1, respectively. From the basic matrix equation according to Eq. (4.10.11) we obtain the following relations from Fig. 4.34,

$$\begin{aligned} &y_0 = 0 \,, && u_{1,a} = (n_0/n_1)\, B\, u_0 \,, && \\ &u_0 = 0 \,, && u_{1,b} = -A\, y_0/n_1 = -y_0/f_1 \,, && \rightarrow A = n_1/f_1 \,, \\ &y_0 = 0 \,, && y_{1,a} = -D\, n_0 u_0 \,, && \\ &u_0 = 0 \,, && y_{1,b} = C\, y_0 \,. && \end{aligned} \tag{4.10.15}$$

Given the arbitrary positions of U_0 and U_1 in the imaging configuration of Fig. 4.34, only the matrix element *A* yields a value which is independent of the position of these reference planes. Further, we observe that *B* equals the weighted ratio of the aperture ray angles in U_1 and U_0. The weighting is with respect to the refractive indices in object and image space. *C* provides the ratio of the heights of intersection of the rays with the reference planes. The quantity $n_0 D$, the 'optical' distance from U_1 to the point Q_0', determines the apparent position of the object point Q_0 after traversal of the optical system.

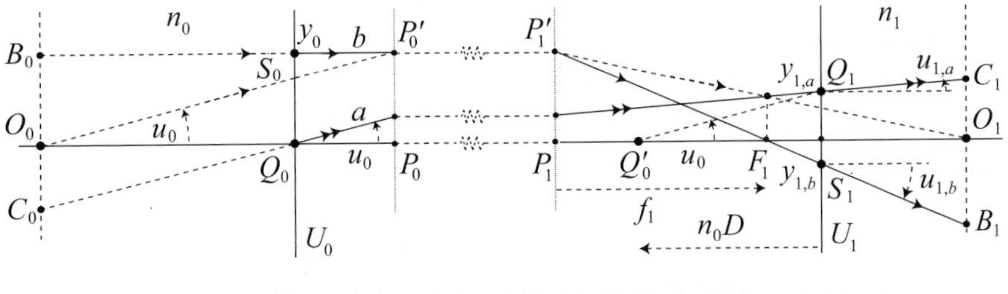

Figure 4.34: Illustrating the optical significance of the four elements of the *ABCD*-matrix. O_0 and O_1 are conjugate points, F_1 is the image-side focal distance and P_0 and P_1 are the centres of the principal planes with unit lateral magnification. The matrix elements are related to a pair of (arbitrary) reference planes through U_0 and U_1.

4.10.4.2 The Lagrange Invariant

If we evaluate the determinant of the system matrix with the aid of the matrix element values given by Eq. (4.10.15) we find that $n_1(u_{1,a}y_{1,b}-u_{1,b}y_{1,a}) = n_0u_0y_0$. If we suppress the label a of the aperture ray and provide the quantities related to the field ray b with a bar we find the customary notation of the general Lagrange invariant H applied to an aperture and a field ray in an optical system,

$$H = n_1(u_1\bar{y}_1 - \bar{u}_1y_1) = n_0(u_0\bar{y}_0 - \bar{u}_0y_0), \tag{4.10.16}$$

where the index 0 now refers to a ray pair (*aperture* and *field* ray) in object space and the index 1 to the corresponding ray pair in the image space of the optical system. The Lagrange invariant is valid for the ray quantities of such a ray pair in any transverse plane in an optical system, after any refraction or reflection of the rays under consideration. It plays an important role in aberration theory. A special case arises if we write the Lagrange invariant for a pair of conjugate planes, either the object and image plane or the conjugate pupil planes. For the object and image plane we use the angle u_0 of the marginal aperture ray and the coordinate $\bar{y}_0$ of the field point where the principal ray starts towards the central point E_0 of the entrance pupil. We take the angle γ_0 of the principal ray through E_0 and the intersection point Y_0 of the marginal aperture ray with the pupil plane and obtain for the Lagrange invariant in the conjugate object/image and pupil planes,

$$H = \underset{\text{object plane}}{n_0u_0\bar{y}_0} = \underset{\text{entrance pupil}}{-n_0\gamma_0Y_0} = \underset{\text{exit pupil}}{-n_1\gamma_1Y_1} = \underset{\text{image plane}}{n_1u_1\bar{y}_1}. \tag{4.10.17}$$

The value of H is positive or negative depending on the initial combination of aperture angle u_0 and object point coordinate $\bar{y}_0$ in the object plane.

Returning to the paraxial matrix, we introduce a shift of each of the reference planes such that they coincide with a conjugate pair of planes in object and image space and we see that the matrix elements assume the special values,

$$B = (n_1/n_0)\,m_A\,, \qquad D = 0\,, \qquad C = m_T\,, \tag{4.10.18}$$

where m_A and m_T are the angular and lateral magnification, respectively, between the conjugate planes. The quantity D is zero since in the conjugate planes the apparent position of the object point O_0 coincides with the image point O_1 and vice versa. Other special values of the matrix elements are found if the reference planes coincide with, for instance, the principal planes or the focal planes of the optical system and they are discussed in the forthcoming pages.

Another property of the system matrix of Eq. (4.10.11) is that it transforms the two-dimensional vector $(n_0u_0, \bar{y}_0)$ in a plane in object space into a vector $(n_{J+1}u_{J+1}, \bar{y}_{J+1})$ in a plane in image space, where $\bar{y}_0$ and $\bar{y}_{J+1}$ are the ray intersection heights in these planes in object and image space. We insist on using in this subsection, where needed, the barred quantities to avoid any confusion between aperture and field rays. The matrix operation leads to a scaling of the vector components in object and image space. We write for the product of the infinitesimal increments of the vector components

$$d(n_{J+1}u_{J+1})\,d\bar{y}_{J+1} = |M_s|\,d(n_0u_0)\,d\bar{y}_0\,, \tag{4.10.19}$$

where M_s plays the role of the Jacobian matrix,

$$M_s = \begin{pmatrix} \dfrac{\partial(n_{J+1}u_{J+1})}{\partial(n_0u_0)} & \dfrac{\partial(n_{J+1}u_{J+1})}{\partial\bar{y}_0} \\[2ex] \dfrac{\partial\bar{y}_{J+1}}{\partial(n_0u_0)} & \dfrac{\partial\bar{y}_{J+1}}{\partial\bar{y}_0} \end{pmatrix}, \tag{4.10.20}$$

of the coordinate transformation. Since $|M_s| = 1$, we can write

$$\iint d(n_{J+1}u_{J+1})\,d\bar{y}_{J+1} = \iint d(n_0u_0)\,d\bar{y}_0\,,$$

or, after integration,

$$n_{J+1}u_{J+1}\bar{y}_{J+1} = n_0u_0\bar{y}_0\,. \tag{4.10.21}$$

The identity of Eq. (4.10.21) is the restricted formulation of Lagrange's invariant for a pair of conjugated planes of the system which was found above in Eq. (4.10.17). With respect to Eq. (4.10.16), the intersection heights y of the aperture ray a have been equated to zero in the two conjugate planes. We then have that $nu\bar{y}$ is a conserved quantity of a paraxial optical system ($\bar{y}$ is the field coordinate in the conjugate object or image plane). Applied to conjugate planes, the invariant

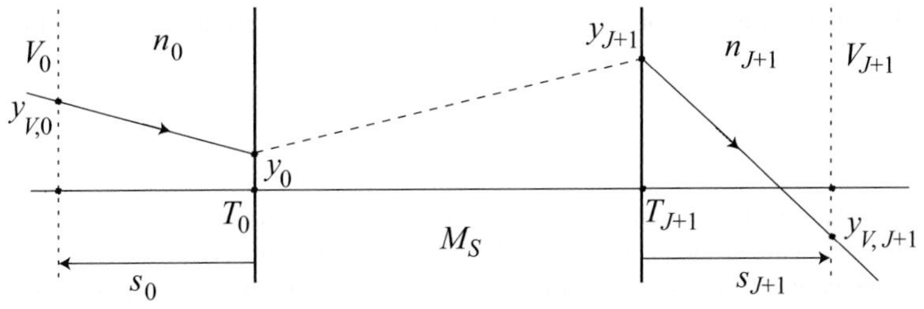

Figure 4.35: Relating the system matrix to two new reference planes in object and image space, V_0 and V_{J+1}, respectively. Note that s_0 is a negative distance in the figure. The original system matrix refers to the vertices T_0 and T_{J+1} of the optical system.

determines the relationship between transverse and angular magnification. Lagrange's invariant plays an important role in radiometry and in aberration theory.

4.10.4.3 Paraxial Imaging and the System Matrix

The system matrix of Eq. (4.10.12), including a zero-distance transport matrix $M_{t,J+1}$, is related to the two planes through the vertices T_0 and T_J of the first and last surface of the optical system. To study the ray transport between a general plane V_0 in object space and V_{J+1} in image space, we need to refer the system matrix to these planes (see Fig. 4.35). The matrix multiplications to be included are given by

$$\begin{pmatrix} n_{J+1}u_{J+1} \\ y_{V,J+1} \end{pmatrix} = \begin{pmatrix} 1 & 0 \\ \frac{s_{J+1}}{n_{J+1}} & 1 \end{pmatrix} M_s \begin{pmatrix} 1 & 0 \\ -\frac{s_0}{n_0} & 1 \end{pmatrix} \begin{pmatrix} n_0 u_0 \\ y_{V,0} \end{pmatrix} = \begin{pmatrix} B + A\frac{s_0}{n_0} & -A \\ -D + B\frac{s_{J+1}}{n_{J+1}} - C\frac{s_0}{n_0} + A\frac{s_0}{n_0}\frac{s_{J+1}}{n_{J+1}} & C - A\frac{s_{J+1}}{n_{J+1}} \end{pmatrix} \begin{pmatrix} n_0 u_0 \\ y_{V,0} \end{pmatrix}$$

$$= M_{V_0,V_{J+1}} \begin{pmatrix} n_0 u_0 \\ y_{V,0} \end{pmatrix} . \tag{4.10.22}$$

The imaging condition between the planes V_0 and V_{J+1} requires that the lower-left element of $M_{V_0,V_{J+1}}$ is zero. In this case we find an intersection height $y_{V,J+1}$ which is independent of the angles of the rays that leave the object plane at a height $y_{V,0}$. We thus have the imaging equation

$$-D + B\,\frac{s_{J+1}}{n_{J+1}} - C\,\frac{s_0}{n_0} + A\,\frac{s_0}{n_0}\,\frac{s_{J+1}}{n_{J+1}} = 0 , \tag{4.10.23}$$

and simultaneously the expression for the lateral magnification,

$$m_T = \frac{y_{V,J+1}}{y_{V,0}} = C - A\,\frac{s_{J+1}}{n_{J+1}} . \tag{4.10.24}$$

The angular magnification is obtained from a ray in the object plane leaving the axial point with $y_{V,0} = 0$. We deduce from Eq. (4.10.22) that

$$m_A = \frac{u_{J+1}}{u_0} = \frac{n_0}{n_{J+1}}\left(B + A\,\frac{s_0}{n_0}\right) . \tag{4.10.25}$$

Using the unimodular property of the matrix $M_{V_0,V_{J+1}}$, we obtain, in the case of imaging between V_0 and V_{J+1}, the relationship

$$m_T m_A = \frac{n_0}{n_{J+1}} , \quad \text{or,} \quad n_{J+1}y_{V,J+1}u_{J+1} = n_0 y_{V,0} u_0 , \tag{4.10.26}$$

a special case of the Lagrange invariant for conjugate planes. The axial magnification m_z follows from the quotient ds_{J+1}/ds_0 and we find

$$m_z = \frac{n_{J+1}}{n_0} m_T^2 = \frac{n_0}{n_{J+1}} \frac{1}{m_A^2} . \tag{4.10.27}$$

Equation (4.10.27) shows that the axial orientation of an object and that of its image are identical in object and image space for a refractive system. This property is also valid for a system with an even number of reflections. For an odd number of reflections, the axial orientation of the image is reversed. Equation (4.10.27) also shows that a shift of the object plane requires a shift in the same direction in image space to obtain anew a sharp image. It is only in the case of an odd number of reflections that the movements become opposite.

From the imaging equation (4.10.23) one obtains the positions of the object and image focal points according to

$$\begin{aligned} s_0 &\rightarrow -\infty , \qquad f_{t,J+1} = n_{J+1}\frac{C}{A} , \\ f_{t,0} &= -n_0 \frac{B}{A} , \qquad s_{J+1} \rightarrow \infty . \end{aligned} \tag{4.10.28}$$

The distances $f_{t,0}$ and $f_{t,J+1}$ are measured from the first vertex to the object-side focal point and from the last vertex to the image-side focal point, respectively. They have been given the name top focal distance or, equivalently, front and back focal length.

Special conjugate planes are those with a lateral magnification $m_T = +1$ (principal planes) and an angular magnification $m_A = +1$ (nodal planes). From Eqs. (4.10.24) and (4.10.25) we obtain for the conjugate plane positions with unit lateral magnification ($m_T = +1$),

$$s_{P,0} = -n_0 \left(\frac{B-1}{A} \right) , \qquad s_{P,J+1} = n_{J+1} \left(\frac{C-1}{A} \right) . \tag{4.10.29}$$

The distances from the principal points to the focal points in object and image space are given by

$$d_{PF,0} = f_0 = f_{t,0} - s_{P,0} = -\frac{n_0}{A} , \qquad d_{PF,J+1} = f_{J+1} = f_{t,J+1} - s_{P,J+1} = +\frac{n_{J+1}}{A} . \tag{4.10.30}$$

The absolute values of the object- and image-side focal distance become equal for $n_0 = n_{J+1}$. The quantity which is identical in object and image space is

$$A = -\frac{n_0}{f_0} = \frac{n_{J+1}}{f_{J+1}} . \tag{4.10.31}$$

This quantity is called the optical power of the system and is expressed in diopters (unit is m^{-1}). For the positions of the nodal points we find

$$s_{N,0} = \frac{-n_0 B + n_{J+1}}{A} , \qquad s_{N,J+1} = \frac{n_{J+1} C - n_0}{A} . \tag{4.10.32}$$

The distance from the nodal point to the focal point in object or image space is given by

$$d_{NF,0} = -\frac{n_{J+1}}{A} = -f_{J+1} , \qquad d_{NF,J+1} = \frac{n_0}{A} = -f_0 . \tag{4.10.33}$$

This result is in line with the positions of principal, nodal and focal points that were found in Fig. 4.31.

4.10.4.4 The Imaging Equation Related to Particular Reference Planes

The general imaging equation, with reference to the first and last vertex of the optical system (see Eq. (4.10.23)), can be made simpler by choosing new reference planes in object and image space. The following are frequently used:

Principal planes; the system matrix is given by

$$\begin{pmatrix} 1 & -A \\ 0 & 1 \end{pmatrix} . \tag{4.10.34}$$

Nodal planes; the system matrix is given by

$$\begin{pmatrix} \frac{n_{J+1}}{n_0} & -A \\ 0 & \frac{n_0}{n_{J+1}} \end{pmatrix} . \tag{4.10.35}$$

Conjugate reference planes $V_{r,0}$ and $V_{r,J+1}$ having a general value $m_{T,r}$ for the lateral magnification:
The two examples above are special cases of the more general system matrix

$$\begin{pmatrix} \frac{1}{m_{T,r}} & -A \\ 0 & m_{T,r} \end{pmatrix} . \tag{4.10.36}$$

We first treat the imaging equation for the general case of a pair of conjugated reference planes for which the modified system matrix is given by Eq. (4.10.36). The use of the principal planes or the nodal planes as reference planes then follows as a special case:

- With the aid of the matrix associated with the pair of conjugate planes with lateral magnification $m_{T,r}$, we introduce a shift of reference plane in object and image space, over distances $l_{r,0}$ and $l_{r,J+1}$, respectively, towards the planes V_0 and V_{J+1} which, for the moment, are not necessarily conjugate planes (see Fig. 4.36). The vector equation for the angle and the position of connected rays in object and image space becomes

$$\begin{pmatrix} n_{J+1}u_{J+1} \\ y_{V,J+1} \end{pmatrix} = \begin{pmatrix} \frac{1}{m_{T,r}} + A\frac{l_{r,0}}{n_0} & -A \\ \frac{l_{r,J+1}}{n_{J+1}m_{T,r}} - m_{T,r}\frac{l_{r,0}}{n_0} + A\frac{l_{r,0}}{n_0}\frac{l_{r,J+1}}{n_{J+1}} & m_{T,r} - A\frac{l_{r,J+1}}{n_{J+1}} \end{pmatrix} \begin{pmatrix} n_0 u_0 \\ y_{V,0} \end{pmatrix} . \tag{4.10.37}$$

To fulfil the imaging condition, we null the (2,1) element of the matrix, yielding the imaging equation

$$\frac{l_{r,J+1}}{n_{J+1}m_{T,r}} - m_{T,r}\frac{l_{r,0}}{n_0} + A\frac{l_{r,0}}{n_0}\frac{l_{r,J+1}}{n_{J+1}} = 0 . \tag{4.10.38}$$

We observe that the lateral magnification $m_{T,V}$ between the planes V_0 and V_{J+1}, which have now become a conjugate pair, is given by

$$m_{T,V} = m_{T,r} - A\frac{l_{r,J+1}}{n_{J+1}} , \tag{4.10.39}$$

where $l_{r,J+1}$ is the distance from the reference conjugate plane $V_{r,J+1}$ to a general conjugate plane V_{J+1} in image space. Using Eq. (4.10.38), we can write for the conjugate distances, in terms of the desired lateral magnification $m_{T,V}$ between the conjugate planes V_0 and V_{J+1},

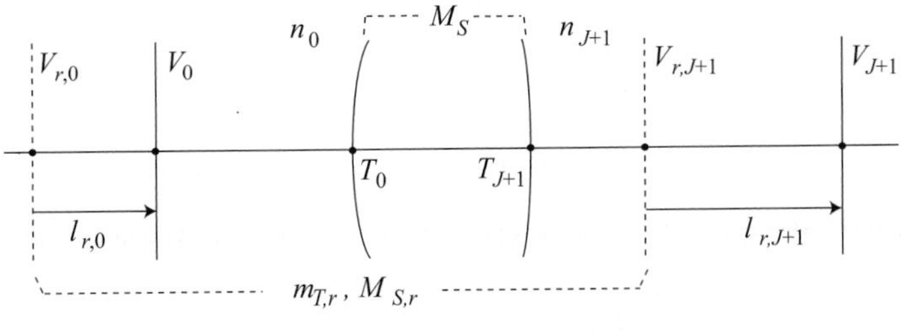

Figure 4.36: Modification of the system matrix $M_{S,r}$ by a shift of reference planes from $V_{r,0}$ to V_0 and from $V_{r,J+1}$ to V_{J+1}. $V_{r,0}$ and $V_{r,J+1}$ are conjugate planes with lateral magnification $m_{T,r}$.

$$l_{r,0} = \frac{n_0(m_{T,r} - m_{T,V})}{A\, m_{T,r} m_{T,V}} \,, \qquad l_{r,J+1} = \frac{n_{J+1}(m_{T,r} - m_{T,V})}{A} \,. \tag{4.10.40}$$

If the conjugate distances $l_{r,0}$ and $l_{r,J+1}$ are given, the lateral magnification follows from

$$m_{T,V} = \frac{1}{m_{T,r}} \frac{n_0}{n_{J+1}} \frac{l_{r,J+1}}{l_{r,0}} \,. \tag{4.10.41}$$

The axial magnification $m_{z,r}$ between a pair of conjugate planes V_0 and V_{J+1} follows from the calculation of the differential quotient $dl_{r,J+1}/dl_{r,0}$ with the aid of Eq. (4.10.38),

$$m_{z,r} = \frac{1}{m_{T,r}^2} \frac{n_0}{n_{J+1}} \frac{l_{r,J+1}^2}{l_{r,0}^2} = \frac{n_{J+1}}{n_0} m_{T,V}^2 \,. \tag{4.10.42}$$

- In the frequently used case of reference to the principal planes, we find Gauss' imaging equation and the corresponding lateral and axial magnification factors,

$$\frac{n_{J+1}}{l_{P,J+1}} = \frac{n_0}{l_{P,0}} + A \,, \qquad m_{T,P} = \frac{n_0}{n_{J+1}} \frac{l_{P,J+1}}{l_{P,0}} \,, \qquad m_{z,P} = \frac{n_{J+1}}{n_0} m_{T,P}^2 \,. \tag{4.10.43}$$

We observe that the imaging equation is identical to that for a single refracting or reflecting surface with optical power A.

- Using the nodal planes as reference planes, the paraxial imaging equation and the magnification factors are given by

$$\frac{n_0}{l_{N,J+1}} = \frac{n_{J+1}}{l_{N,0}} + A \,, \qquad m_{T,N} = \frac{l_{N,J+1}}{l_{N,0}} \,, \qquad m_{z,N} = \frac{n_{J+1}}{n_0} m_{T,N}^2 \,. \tag{4.10.44}$$

4.10.4.5 Newton's Imaging Equation

The focal planes of the optical system can also be used as reference planes for measuring object and image distance. Using the expressions for the top focal distances of Eq. (4.10.28) we obtain after some algebra the vector equation

$$\begin{pmatrix} n_{J+1}u_{J+1} \\ y_{F,J+1} \end{pmatrix} = \begin{pmatrix} 0 & -A \\ 1/A & 0 \end{pmatrix} \begin{pmatrix} n_0 u_0 \\ y_{F,0} \end{pmatrix} . \tag{4.10.45}$$

The (2,1) element of the system matrix is not zero, which means that the matrix has not been referred to a pair of conjugate planes. To analyse imaging by means of this paraxial matrix, we introduce anew a shift of reference planes by means of the distances $l_{F,0}$ and $l_{F,J+1}$. The vector equation then becomes

$$\begin{pmatrix} n_{J+1}u_{J+1} \\ y_{V,J+1} \end{pmatrix} = \begin{pmatrix} A \dfrac{l_{F,0}}{n_0} & -A \\ 1/A + A \dfrac{l_{F,0}}{n_0} \dfrac{l_{F,J+1}}{n_{J+1}} & -A \dfrac{l_{F,J+1}}{n_{J+1}} \end{pmatrix} \begin{pmatrix} n_0 u_0 \\ y_{V,0} \end{pmatrix} . \tag{4.10.46}$$

The planes V_0 and V_{J+1} are conjugate planes if the matrix element (2,1) is zero, yielding the imaging equation,

$$\frac{l_{F,J+1}}{n_{J+1}} \frac{l_{F,0}}{n_0} = -\frac{1}{A^2} \,,$$

or,

$$l_{F,0}\, l_{F,J+1} = f_0 f_{J+1} \,, \tag{4.10.47}$$

where we have used the result of Eq. (4.10.31) for the optical power of a system. The lateral and axial magnification are given by

$$m_{T,F} = -\frac{l_{F,J+1}}{f_{J+1}} = -\frac{f_0}{l_{F,0}} \,, \qquad m_{z,F} = \frac{n_{J+1}}{n_0} m_{T,F}^2 \,. \tag{4.10.48}$$

The expressions given in Eq. (4.10.47) are called Newton's imaging equations.

4.10.4.6 Afocal System

An optical system with zero power is called an afocal system. As a consequence of its most frequent application, which is to observe an infinitely distant scene, it is also called a telescopic system. The basic layout of a telescope is a compound system of two groups with a common intermediate focus. We denote the optical powers of the two groups by A_1 and A_2. The intermediate distance d_2 from the image-side principal plane of the first group to the object-side principal plane of the second group is given by $f_1' - f_2$. The system matrix M_a of an afocal system is then written as

$$M_a = \begin{pmatrix} -A_2/A_1 & 0 \\ d_2/n_2 & -A_1/A_2 \end{pmatrix} . \tag{4.10.49}$$

Note that we can replace d_2/n_2 by $(1/A_1 + 1/A_2)$. The transverse magnification of the telescope is given by

$$m_T = -\frac{A_1}{A_2} = \frac{f_2}{f_1'} . \tag{4.10.50}$$

To study imaging by the afocal system, we introduce as usual shifted reference planes V_0 and V_{J+1} in object and image space, at distances s_0 and s_{J+1} from the first and last vertex of the system. The resulting ray matrix equation is given by

$$\begin{pmatrix} n_{J+1}u_{J+1} \\ y_{V,J+1} \end{pmatrix} = \begin{pmatrix} 1/m_T & 0 \\ \dfrac{d_2}{n_2} + \dfrac{s_{J+1}}{n_{J+1}m_T} - \dfrac{s_0 m_T}{n_0} & m_T \end{pmatrix} \begin{pmatrix} n_0 u_0 \\ y_{V,0} \end{pmatrix} . \tag{4.10.51}$$

The resulting telescope imaging equation for the two-lens system (see Fig. 4.37) is given by

$$s_4 = n_4\, m_T \left(m_T \frac{s_0}{n_0} - \frac{d_2}{n_2} \right) . \tag{4.10.52}$$

Both the angular and transverse magnification of an afocal system are fixed and independent of the choice of conjugate planes. In Fig. 4.37a, we show the standard use of an afocal system, as a telescope with both object and image shifted

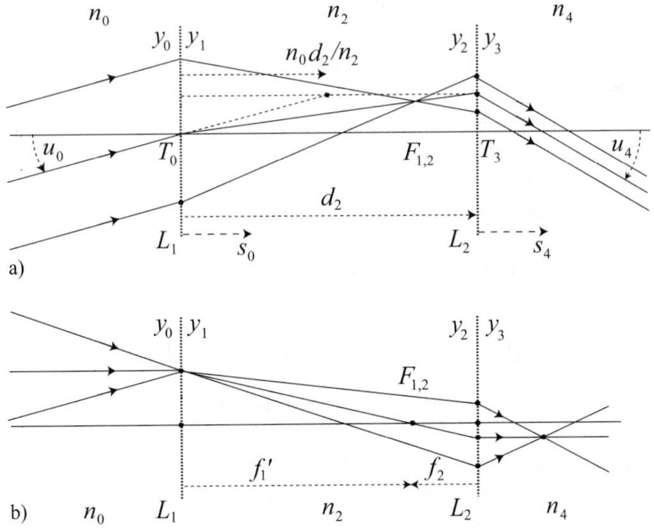

Figure 4.37: Layout of an afocal system. The positive lenses L_1 and L_2 have been schematically drawn in the figure, their common focus is in the point $F_{1,2}$. The telescope length is d_2. The refractive indices in object and image space are n_0 and n_4, that of the medium inside the telescope is n_2.
a) The afocal system is used as telescope and the conjugate planes are both at infinity.
b) The object is located at the entrance vertex of the telescopic system.

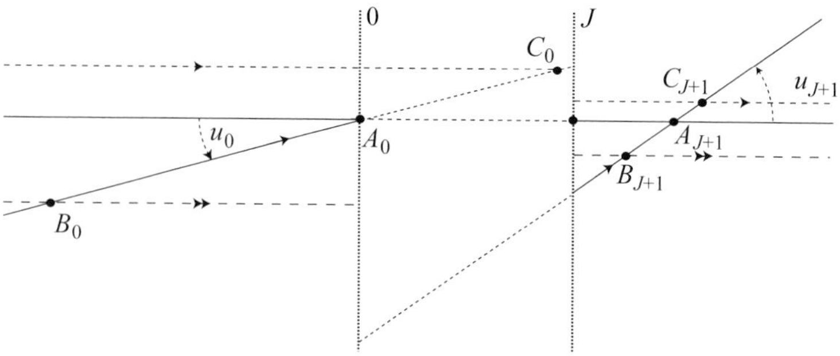

Figure 4.38: Illustrating imaging by a telescopic system. The vertices of the first and the last surface of the telescopic system have been denoted by 0 and J. The oblique ray at an angle u_0 with respect to the axis through the vertex of surface 0 is conjugate with the ray at angle u_{J+1} in image space.

to infinity. The telescope yields an angular magnification $(u_4/u_0) = n_0/(n_4 m_T)$. The 'thickness' parameter d_2/n_2 in the telescope matrix M_a plays the same role as the *reduced* lens thickness d_1/n_1 which has already been encountered in Eq. (4.10.8) for the power of a thick lens. From the system matrix M_a and for $y_0 = 0$, we derive that $u_0 = n_2 y_3/(n_0 d_2)$. For a standard telescope with two (thin) lenses in air, the refractive indices n_0 and n_2 are both equal to unity and we identify d_2 with the physical length of the telescope. If the refractive indices of object and intermediate spaces are different, $n_0 d_2/n_2$ is the equivalent thickness of the telescope as seen from the object space.

In Fig. 4.37b another pair of conjugate planes is chosen. The object plane is in virtual contact with the first lens L_1 ($s_0 = 0$). In principle, the telescope functions as a single lens, L_2 in this case, and the image is located at a distance

$$s_4 = \frac{n_4}{A_2}\left(1 + \frac{A_1}{A_2}\right) = f_2'(1 - m_T)\,. \tag{4.10.53}$$

The influence of the first lens is felt only when the object plane is moved away from its position at L_1. Regardless of the axial shift of the object plane, the lateral extent and the angular magnification of the image remain fixed. This property of an afocal system can be useful for imaging systems which have to provide image fidelity over a certain imaging depth.

Figure 4.38 shows the geometric construction of a general pair of conjugate planes of a telescopic system. Essential for the construction is the availability of an obliquely traced paraxial ray; in the figure we have shown one through the vertex 0 of the first surface, at angles u_0 and u_{J+1} in object and image space. We define an object plane by means of a general point on the oblique ray, for instance B_0 in real object space. A ray parallel to the optical axis through B_0 at a height y_0 is conjugate with an image ray parallel to the optical axis at a height y_{J+1} given by the transverse magnification m_T of the telescopic system. The point of intersection B_{J+1} of this ray with the oblique ray at angle u_{J+1} defines the conjugate real image plane. In the same way, a virtual object plane through C_0 produces a real image plane through C_{J+1}. The conjugate planes through A_0 and A_{J+1} have already been found in the second drawing of Fig. 4.37 and appear in this figure as a special case for the intersection points of the oblique ray with the two axial rays in object and image space.

4.10.5 Paraxial Optics of Diaphragm, Pupils and Windows

The imaging of an object plane onto its conjugate image plane is perfect within the paraxial approximation. Snell's law is approximated by its linearised version and optical pathlengths are approximated up to the second power in the coordinates of intersection points with optical surfaces. As soon as pencils of rays are sent through an optical system which have a lateral and angular extent such that the paraxial approximation is inadequate, imaging defects arise and the image is distorted and blurred by the effect of *optical aberrations* (see Chapter 5 for an in-depth discussion of this subject). To keep the imaging defects below a certain acceptable level, we have to limit the angular extent of the imaging pencils from the axial object point and to specify the way in which the pencils of rays from off-axis object points travel through the optical system. Moreover, we have to impose a limit on the lateral size of the object to be imaged. To achieve these goals, several limiting apertures are introduced in an optical system. Quite often, the rim of a lens is used to form the effective aperture. In this subsection we study the paraxial optics of the imaging of these limiting apertures and their role in guiding the pencils of rays through the imaging system. The aperture imaging is a second imaging process in the optical system, less important than the main imaging process from object to image plane. In the case of a self-luminous object, only the apertures in the imaging system itself need to be considered. For non-luminous objects, illuminated by means of a light source and an illumination system, the aperture imaging in the illumination system and the imaging system together have to be optimised. The goal is to convey as much radiation as possible from the source up to the final image. In general, the paraxial optics approximation suffices to simulate the light transport through the illumination and imaging system with sufficient accuracy. In high-quality systems and in wide-angle imaging systems, the defects of the aperture imaging need to be taken into account.

4.10.5.1 The Diaphragm and an Axial Pencil of Rays

The angular extent of the axial beam is generally specified by its numerical aperture $NA = n_1 \sin\alpha_1$ at the image side; n_1 is the refractive index of the image space and α_1 the maximum convergence angle of the rays of the imaging pencil. A physical stop is needed to fix the value of NA and this can be either the rim of a certain lens or a (circular) diaphragm that has been introduced on purpose. This latter option is to be preferred because internal reflections and scattering at lens edges have to be avoided in a high-quality optical system. Once the stop size and stop position have been determined (see D in Fig 4.39), we consider the diaphragm as an object which is imaged from the interior of the optical system towards both the object and the image space, via the sections I and II of the optical system, respectively. The application of the paraxial imaging laws to find the stop images in object and image space is not an exact procedure but paraxial optics yields a first approximation about position and size of the stop images. It is only in a fine-tuning stage of high-quality imaging systems that the imaging defects or aberrations of the stop images can become important. The stop image in object space is called the entrance pupil of the system, the stop image in image space is the exit pupil of the system. Both real and virtual stop images are encountered in practice. By definition, a real entrance pupil is located in front of the first vertex, a real exit pupil beyond the last vertex of the optical system. Entrance and exit pupils are both images of the diaphragm and thus paraxially conjugate. If the lateral magnification of the pupils is given by $m_{T,P} = Y_1/Y_0$ and the image magnification by $m_{T,O}$, the pupil distances R_0 and R_1 follow from Eq. (4.10.40) and are given by

$$O_0E_0 = R_0 = \frac{n_0}{A}\left(\frac{m_{T,O} - m_{T,P}}{m_{T,O}m_{T,P}}\right), \qquad O_1E_1 = R_1 = \frac{n_1}{A}\left(m_{T,O} - m_{T,P}\right). \tag{4.10.54}$$

We note that according to these expressions the conjugate distances R_0 and R_1 are measured from the conjugate reference planes with lateral magnification $m_{T,O}$, the object and image plane through O_0 and O_1. A change of sign of these distances would occur if we choose the conjugate pupil planes with lateral magnification $m_{T,P}$ as reference planes to find the object and image distances of O_0 and O_1, measured from the centres E_0 and E_1 of entrance and exit pupil. We draw attention to this sign effect because both choices can be made depending on the problem at hand. The expressions of Eq. (4.10.54)

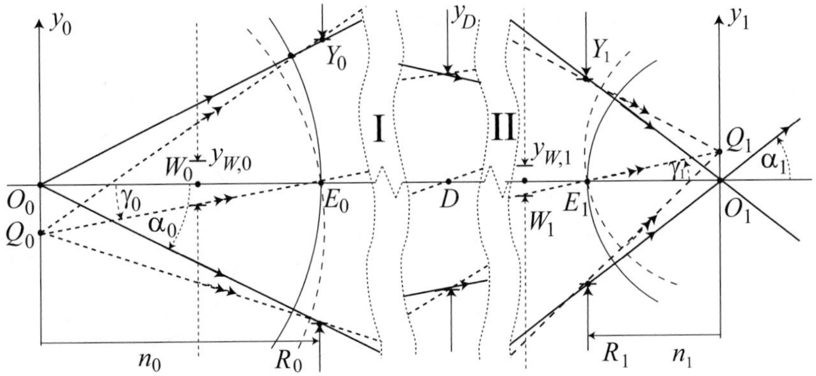

Figure 4.39: The position of diaphragm, pupils and windows in an optical system, schematically represented by its front section I and back section II. The diaphragm, located between sections I and II, intersects the optical axis in D and has a diameter $2y_D$. The entrance pupil plane intersects the optical axis in E_0, the exit pupil plane in E_1. The entrance and exit window have been denoted by W_0 and W_1, respectively.

become indeterminate if the system has zero power (afocal system). In this case, Eqs. (4.10.52) or (4.10.53) can be used to find the distances of the pupil planes to the vertices of the optical system.

4.10.5.2 Off-axis Pencils of Rays

The stop and the pupils derived from it determine the path of oblique pencils through a system. When the object and image plane are at finite distances, the central or principal ray of the pencil leaving the object point Q_0 intersects the centre of the entrance pupil at E_0 and the centre of the exit pupil at E_1, within the framework of the paraxial approximation. The Lagrange invariant H applied to the imaging pencils can be transformed into quantities related to the pupil imaging. From Fig. 4.39 it follows directly that

$$H = n_0 y_0 \alpha_0 = -n_0 \gamma_0 Y_0 = -n_1 \gamma_1 Y_1 = n_1 y_1 \alpha_1 \,, \tag{4.10.55}$$

where $y_0 = O_0Q_0$ and $y_1 = O_1Q_1$. The object field diameter is $2y_0$. The pencil of rays from the extreme object point Q_0 has a central or principal ray Q_0E_0 through the centre of the entrance pupil; the angle of the principal ray with the optical axis is γ_0. As usual, in deriving Eq. (4.10.55) the paraxial approximation has been used including the small-angle approximation $\alpha \approx \tan\alpha$. The field angles γ_0 and γ_1 in object and image space are given by the angular magnification of the pupils and they are related to the transverse pupil magnification $m_{T,P}$ by the expression $\gamma_1/\gamma_0 = n_0/(n_1 m_{T,P})$.

An extra pair of conjugate planes has been depicted in Fig. 4.40, the so-called optical 'windows'. To find the position of the window at the image side, all lens borders and mechanical obstructions of the system are imaged to image space through the intermediate optical elements that are to the right of each border. In image space, observing from the centre of the exit pupil, we determine which image of a lens border or mechanical obstruction limits the viewing angle from the exit pupil. The corresponding physical obstruction in the interior of the system is then identified as the optical window. Its images towards object space and image space are called the entrance and exit window and they are found in conjugate planes. From Fig. 4.40a one easily derives that the Lagrange invariant of the object and pupil imaging also applies to the window imaging,

$$H = n_0 y_{Q,0} \alpha_0 = -n_0 \gamma_0 Y_0 = n_0 y_{W,0} \alpha_w \,. \tag{4.10.56}$$

Comparable expressions are found for the image-side quantities. The definition of window for a single element is still formally possible. It is now a possible physical border on the object surface and the image of this border in the image plane. In this special case, the object and image window simply coincide with the object and image plane, respectively.

The concept of a window and its relation to field angle is straightforward if the lateral extent of the diaphragm tends to zero (small numerical aperture or large F-number). A pencil of rays is then well represented by its principal ray and the direct neighbourhood of this ray. The image will show a sharp low-aperture border which is determined by the window border. But for a larger pupil size, the image intensity is not sharply bounded. At large opening a gradual weakening of the image intensity is observed when crossing the low-aperture border of the image at field angle γ_1 (the image-side field angle corresponds to the object field angle γ_0 depicted in Fig. 4.40a). Outside the low-aperture border of the image, the imaging pencils of rays miss the principal ray through E_0 (see Fig. 4.40b). For this reason, a new reference ray is

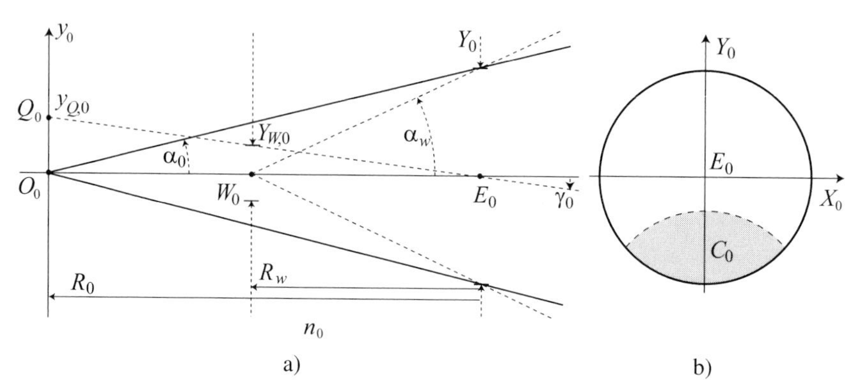

Figure 4.40: a) Illustrating the role of object, entrance pupil and entrance window in the object space of an optical system. b) The effect of pupil vignetting.

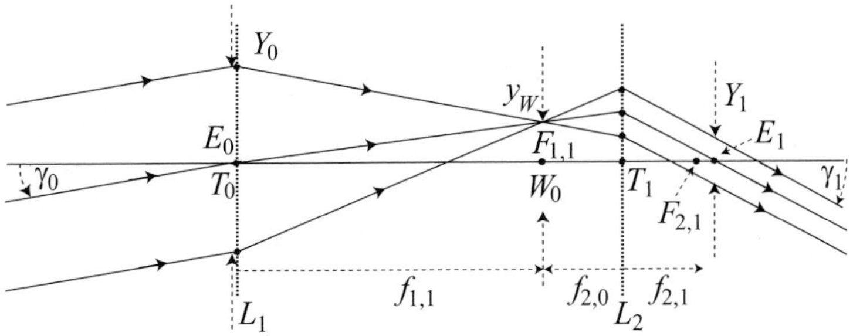

Figure 4.41: The position of object and image plane, pupils and windows in an afocal system consisting of two lenses in air, used as a telescope. The two lenses L_1 and L_2 have been schematically drawn. Their common focus is denoted by $F_{1,1}$.

defined, called the chief ray C_0 of the oblique pencil. The coordinates of the chief ray are obtained from the average X- and Y-coordinates of all rays in the pencil. If needed, in finding the coordinates of C_0, the rays of the pencil are weighted by the intensity function over the beam cross-section. The typical effect observed in Fig. 4.40b is called the *vignetting* effect of the pupils. It occurs mainly in wide-angle imaging systems and the precise intensity fall-off towards the image border is an important design specification.

To illustrate further the role of pupils and windows, we turn to an afocal system (see Fig. 4.41). The aperture diaphragm is positioned in T_0, the vertex of the first lens L_1. The exit pupil is then found at the position E_1, a distance $f_{2,1}(1 + f_{2,1}/f_{1,1})$ beyond the second lens. The physical window seen from T_0 is the aperture that has been positioned in the common focal point $F_{1,1} = F_{2,0}$ of the afocal system. The entrance and exit window are then found at infinity, coincident with the object and image of the afocal system in telescope mode of operation. The Lagrange invariant H for the object and image quantities becomes an indeterminate expression for a telescope. However, if we use that $\alpha_0 = -Y_0/R_0$ and $\gamma_0 = y_0/R_0$, we find for a remote object at distance R_0,

$$H = n_0 \alpha_0 y_0 = -n_0 \frac{Y_0}{R_0} \gamma_0 R_0 = -n_0 Y_0 \gamma_0 \, , \qquad (4.10.57)$$

showing again the conservation property of H for object and pupil imaging.

4.10.5.3 Special Positions of Entrance and Exit Pupil; Telecentric Imaging

The position of the entrance and exit pupil can be used to achieve certain effects in the paraxial imaging properties. It can also be used to avoid the transmission to image space of rays that have picked up large imaging errors (aberrations). This latter aspect is treated in more detail in the chapter on aberration theory. The first aspect is illustrated by means of the diaphragm position that forces telecentric imaging. In Fig. 4.42a, a typical imaging situation is depicted for a schematically drawn optical system S. The entrance and exit pupil are at finite distances from the object and image plane. In Fig. 4.42b, the position of the entrance pupil is at the front side focal point F_0, the exit pupil is shifted to infinity. In this case, the principal ray of an off-axis pencil is parallel to the optical axis in image space. Its intersection point with the optical axis is infinitely distant, which gives rise to the denomination of *telecentric* imaging. Within the paraxial approximation, the image formed in the nominal image plane, for instance the image Q_1 of an object point Q_0, is identical in Fig. 4.42a and b. The influence of the telecentric imaging in b) becomes visible if the receiving plane is shifted slightly away from the nominal image plane (see dashed plane in the drawings). For the case of telecentric imaging, the centre of an image point Q_1' in the shifted plane remains in the same position as measured from the axis O_0O_1; in Fig. 4.42a, the image point Q_1' moves outward when defocusing in the positive z-direction.

Telecentric imaging is required if the lateral geometry of the image should be well conserved, independent of a global defocusing, tilt or unflatness of the receiving surface. For an exit pupil at a finite distance, a shift of the receiving plane introduces an apparent change of magnification of the image. Examples of the application of telecentric imaging are found in aero-photography, optical projection lithography, but also, more generally, in microscopy. Of course, the axial shift of the image point introduces unsharpness and should be avoided. However, the slight deterioration in image quality by defocusing in the applications mentioned above is less harmful than the loss of absolute geometric fidelity of the image.

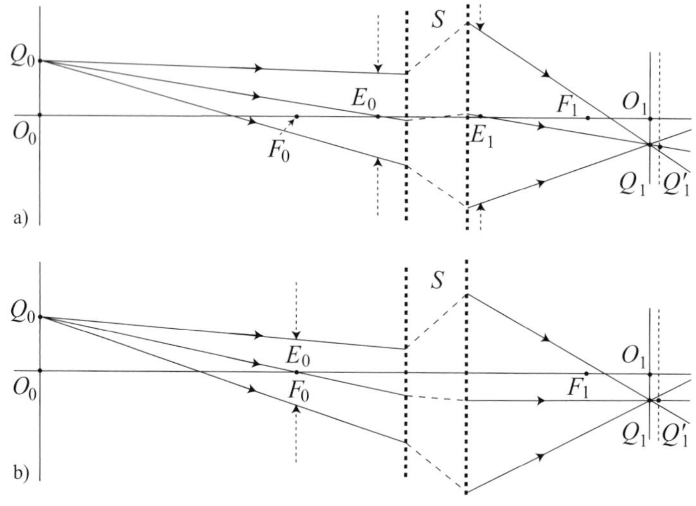

Figure 4.42: An optical system S has been schematically drawn with its object-side and image-side focal points denoted by F_0 and F_1.
a) Conjugate pupils are shown with their centres in E_0 and E_1. The pupil surfaces have finite lateral and angular magnification.
b) The entrance pupil is in the front focal point F_0, the exit pupil is shifted to infinity (telecentric imaging with angular pupil magnification equal to zero).

Double telecentric imaging where both pupils are at infinity can be achieved using an afocal system. An example of this type of imaging is found in Fig. 4.37b. The object plane is at the vertex T_0 of the afocal system. The advantage of double telecentric imaging is that the absolute image geometry is preserved independently of axial shift, tilt and unflatness of *both* object and image. This immediately follows from the fact that an afocal system has a fixed value of the lateral and angular magnification, regardless of the position of the conjugate planes.

4.10.6 Other Paraxial Imaging Properties

In this subsection we present two other interesting properties of conjugate planes which directly follow from the paraxial imaging laws. The first subject is the projection distance or the optical throw of an imaging system. The optical throw is often used as a design parameter to force the physical design into some box with prescribed length (*and* volume). For a fixed-focus system, the optical throw is an approximate target value. For an optical zoom-system with variable magnification, it is very important that the images with variable magnification are projected onto the same fixed image plane or at least onto well-defined image plane positions provided that a mechanical compensation of variations in the optical throw of the system can be allowed. The second subject is the imaging of a scene on an inclined object plane onto a similarly inclined image plane. We give the condition for the inclination angles of object and image plane to produce a sharp image and discuss the shape of the 'inclined' sharp image.

4.10.6.1 Optical Throw

A common requirement for optical zoom-systems is that their use at a different magnification should preferably not change the total distance (optical throw or projection distance) from object to image plane. For instance, when an image is formed from a remote object, the position of the focal plane should remain fixed in space for all zoom values. In this respect, it is important how the magnification of a single lens or of an entire system influences the projection distance. From Fig. 4.43 we have

$$T = -l_{F,0} - f_0 + P_0P_1 + f_1 + l_{F,1} , \tag{4.10.58}$$

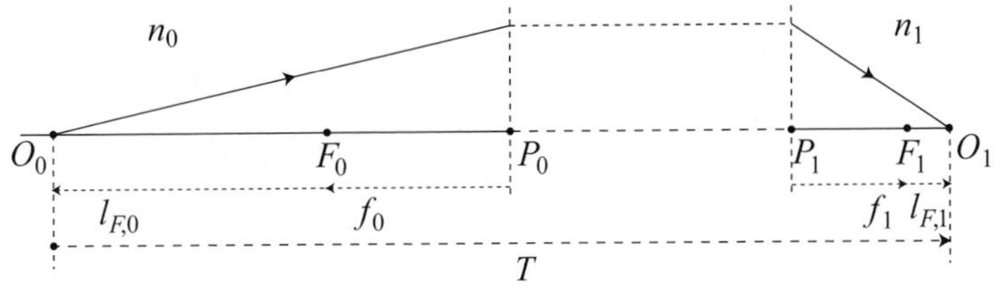

Figure 4.43: A paraxial optical system is characterised by its principal planes P_0 and P_1 and by the positions of the object-side and image-side focal points, F_0 and F_1, respectively. O_0 and O_1 are a pair of conjugate points with lateral magnification m_T. The optical throw or projection distance T is defined by the distance O_0O_1, positive in the drawing.

and, using Eqs. (4.10.47) and (4.10.48) for the object and image distances $l_{F,0}$ and $l_{F,1}$, we obtain

$$T = \left(\frac{1-m_T}{m_T}\right)(f_0 + m_T f_1) + P_0P_1 = \frac{1}{A}\left(\frac{1-m_T}{m_T}\right)(m_T n_1 - n_0) + P_0P_1 \,. \tag{4.10.59}$$

For equal refractive indices in object and image space and using that $f_0 = -f_1$, the expression reduces to

$$T = -f_1 \,\frac{(1-m_T)^2}{m_T} + P_0P_1 \,, \tag{4.10.60}$$

where f_1 is the image-side focal distance of the optical system. The optical throw possesses extreme values which follow from the condition $dT/dm_T = 0$. In the general case that $n_0 \neq n_1$ we find for the extrema,

$$\begin{aligned} m_T &= +\sqrt{\frac{n_0}{n_1}}\,, & T &= \frac{1}{A}\left(\sqrt{n_0} - \sqrt{n_1}\right)^2 + P_0P_1 \,,\\ m_T &= -\sqrt{\frac{n_0}{n_1}}\,, & T &= \frac{1}{A}\left(\sqrt{n_0} + \sqrt{n_1}\right)^2 + P_0P_1 \,. \end{aligned} \tag{4.10.61}$$

In the case of equal indices $n_0 = n_1 = n$ in object and image space, we find the extreme values

$$\begin{aligned} m_T &= +1\,, & T &= P_0P_1 \,,\\ m_T &= -1\,, & T &= 4f + P_0P_1 \,, \end{aligned} \tag{4.10.62}$$

where $f = n/A$ is the focal distance of the system.

Figure 4.44 shows two graphs of the variation of the optical throw as a function of transverse magnification. In graph a) the optical system has negative power and unequal refractive indices in object and image space. The extrema in optical throw are found at a magnification $m_T = \pm\sqrt{1/2}$. The distance ΔT between the extrema for positive and negative magnification is given by $-4\sqrt{n_0 n_1}/A = -4\sqrt{2}/A$, proportional to the geometric mean of the object-side and image-side focal distances. In graph b) the optical system has positive power and the extrema occur for $m_T = \pm 1$. The offset in optical throw is now given by $-4f_1$ (f_1 is the image-side focal distance).

4.10.6.2 Inclined Object and Image Planes

Intentionally or by accident, the object and image plane can be tilted with respect to their customary orientation, perpendicular to the optical axis. In Fig. 4.45 we have shown such a situation, with inclination angles ϕ_0 and ϕ_1 of the object and image plane, respectively. For the inclination angles we can write

$$\phi_0 = -\frac{z_0}{p_0}\,, \qquad \phi_1 = -\frac{z_1}{p_1}\,, \tag{4.10.63}$$

where $2p_0$ is the lateral extent of the (square) object in the object plane. The value z_0 is the axial shift of an object point at the rim of the object, for instance in the cross-section defined by $x = 0$. The relation between z_0 and z_1 follows from the axial magnification of the system (see Eq. (4.10.27)),

$$\frac{z_1}{z_0} = m_z = \frac{n_1}{n_0} m_T^2 \,, \tag{4.10.64}$$

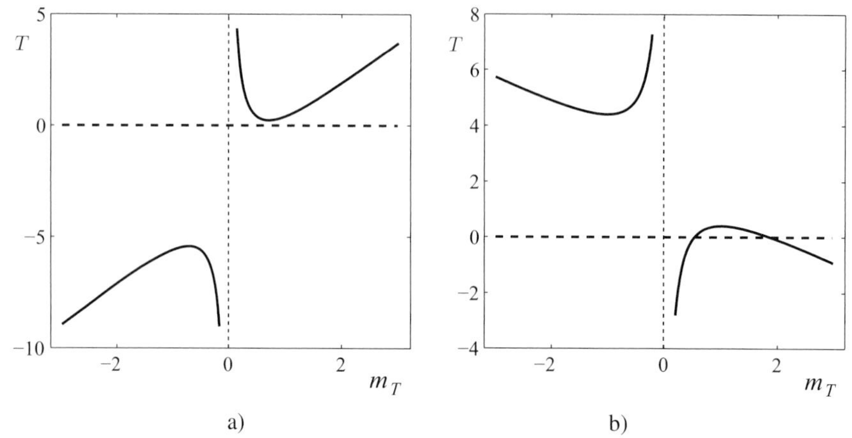

Figure 4.44: The optical throw T in units of $1/A$ as a function of the transverse magnification m_T.
a) Negative power; $A = -1\ m^{-1}$, $P_0P_1 = 0.25$ m, $n_0 = 1$, $n_1 = 2$.
b) Positive power; $A = +1\ m^{-1}$, $P_0P_1 = 0.40$ m, $n_0 = n_1 = 1$.

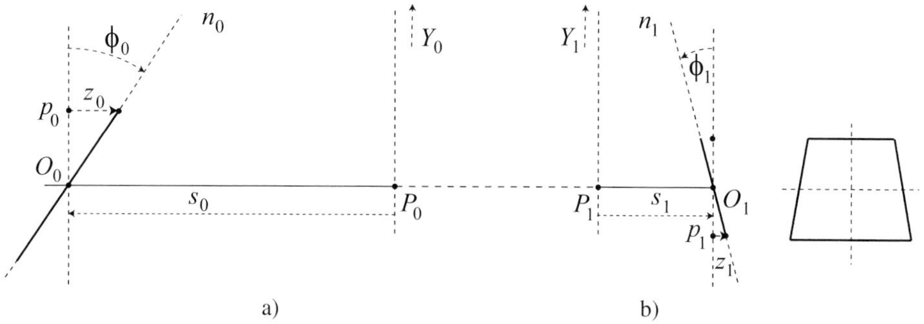

Figure 4.45: a) O_0 and O_1 are conjugate points with object and image distance given by s_0 and s_1, respectively, as measured from the principal planes P_0 and P_1. In the plane of the drawing, the intersection points of the inclined object and image planes with the principal planes are Y_0 and Y_1 (not visible in the drawing).
b) The resulting sharp image of a square object in the inclined image plane.

where $m_T = p_1/p_0 = n_0 s_1/(n_1 s_0)$ is the transverse magnification between the conjugate pair of planes through O_0 and O_1. We find the following relation between ϕ_0 and ϕ_1:

$$\frac{\phi_1}{\phi_0} = \frac{n_1}{n_0} m_T = \frac{1}{m_A} , \tag{4.10.65}$$

where m_A is the angular magnification between the conjugate planes at O_0 and O_1.

For the height of intersection of the inclined object and image plane with the principal planes through P_0 and P_1, respectively, we find

$$Y_0 = Y_1 = -\frac{s_1 p_1}{z_1} = \frac{(1 - m_T)}{\phi_1} f_1 . \tag{4.10.66}$$

With the aid of Newton's imaging equation, the image distance s_1, measured from the image-side principal plane P_1, is written as $s_1 = f_1 - m_T f_1$ (see Eq. (4.10.48)), yielding for the magnification,

$$m_T = 1 - \frac{s_1}{f_1} . \tag{4.10.67}$$

The change in magnification as a function of the axial position of the image plane is given by

$$\frac{dm_T}{ds_1} = -\frac{1}{f_1} = \frac{1}{s_1}(m_T - 1) . \quad (4.10.68)$$

For an axial shift $ds_1 = z_1$ we thus have a magnification change

$$dm_T = (m_T - 1)\frac{z_1}{s_1} = -(m_T - 1)\frac{p_1\phi_1}{s_1} . \quad (4.10.69)$$

For the relative change in magnification we then obtain

$$\frac{dm_T}{m_T} = \frac{p_1\phi_1}{m_T f_1} = -\frac{z_1}{m_T f_1} . \quad (4.10.70)$$

We conclude that the magnification in the inclined image plane is linearly dependent on the image height p_1 and does not depend on the values of the refractive indices of the media in object and image space. The shape of the image of a square object is a trapezoid (see Fig. 4.45b). In this analysis, we have assumed that the inclination angles ϕ are small. The p_0-coordinate of the rim of a square object on the tilted surface is reduced in second order because of the projection effect. In the derivation above we assumed that such a quadratic projection effect can be neglected. The maximum change in magnification over the image is given by $2z_1/(m_T f_1)$.

4.10.7 Paraxial Optics of Gaussian Beams

The basic Gaussian beam properties, continuously changing during propagation in free space, are well described by the complex q-parameter which was introduced in Section 1.7. Its real part equals the curvature of the beam wavefront and the imaginary part defines the position of the $1/e$ amplitude level in the beam cross-section. The q-parameter plays a central role in the paraxial analysis of Gaussian beam propagation through optical elements and for this reason its definition according to Eq. (1.7.5) is reproduced here,

$$\frac{1}{q(z)} = \frac{1}{R_c(z)} - \frac{i\lambda}{\pi w^2(z)} . \quad (4.10.71)$$

In this subsection on the transformation of Gaussian beams by refracting or reflecting optical elements we first introduce a single element, the thin lens, which modifies the wavefront curvature of the Gaussian beam. The thin-lens analysis is easily extended to more complicated systems by using the matrix formalism of geometrical optics which was treated in some detail in the preceding subsections. Finally, the repeated propagation forth and back of a Gaussian beam in an optical cavity is discussed and the condition for the confinement of the beam to the cavity volume is given.

4.10.7.1 Gaussian Beam Waist Imaging by a Thin Lens

In Fig. 4.46 a single thin lens L is shown and a Gaussian beam is incident on it. The half-diameter of the Gaussian beam on the entrance surface of the (thin) lens is $w_{L,0}$ and the centre of curvature of the wavefront W_0 on the same entrance surface of the lens is in O_0, at a distance R_0 (negative in the figure according to our sign convention) from the centre C of

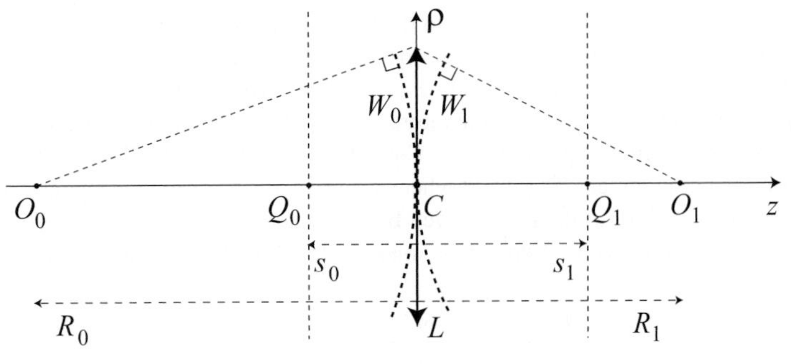

Figure 4.46: Gaussian beam transformation by a single thin lens L.

the lens. In the case of a point source, the wavefront W_0 would be produced by a source in the point O_0. The elementary lens formula would then produce an image point at a distance R_1 from the lens, at the point O_1, where O_1 is the centre of curvature of the wavefront W_1 at the exit side of the lens. We have the expression

$$\frac{1}{R_1} = \frac{1}{R_0} + \frac{1}{f} , \tag{4.10.72}$$

where f is the focal distance of the lens. The basic thin-lens equation above is used to obtain the change in q-parameter of a Gaussian beam when traversing the thin lens. For the case of equal beam half-diameters $w_{L,0} = w_{L,1}$ on the entrance and exit side of the thin lens we can write,

$$\frac{1}{q_{L,1}} = \frac{1}{q_{L,0}} + \frac{1}{f} . \tag{4.10.73}$$

The value of the q-parameter for a general position along the beam axis in object or image space follows from the propagation 'law' for the q-parameter given by Eq. (1.7.9), $q(z_2) = q(z_1) - (z_2 - z_1)$. For general positions Q_0 and Q_1 at distances s_0 and s_1 from the thin lens in object and image space, respectively, we have

$$q(s_0) = q_{L,0} - s_0 , \qquad q(s_1) = q_{L,1} - s_1 , \tag{4.10.74}$$

where the origin for the z-coordinate is at the centre C of the thin lens. The substitution in Eq. (4.10.73) of the values for $q_{L,0}$ and $q_{L,1}$ which follow from Eq. (4.10.74) yields the result,

$$q(s_1) = \frac{q(s_0)\,[1 - s_1/f] + s_0 - s_1 - s_0 s_1/f}{q(s_0)/f + [1 + s_0/f]} . \tag{4.10.75}$$

The imaging equation for the q-parameter in the case of a thin lens can be adapted to the paraxial imaging by a complete system. To this purpose we replace the optical system by its paraxially equivalent system which has two principal planes at a certain distance P_0P_1 and to which the optical power of the complete system is attributed. The expression for $q(s_1)$ that relates $q(s_0)$ in an object reference plane through Q_0 via the system refraction (or reflection) to the beam parameter $q(s_1)$ in the image-side reference plane through Q_1 is constructed with the aid of the *ABCD* constants of the paraxial matrix formalism. The paraxial matrix with respect to the principal planes is given by the expression (4.10.34) where $A = 1/f$. The reference planes are chosen at distances s_0 and s_1 in object and image space, respectively, and the paraxial matrix with respect to these planes is given by

$$\begin{pmatrix} u_1 \\ y_1 \end{pmatrix} = \begin{pmatrix} 1 & 0 \\ s_1 & 1 \end{pmatrix} \begin{pmatrix} 1 & -1/f \\ 0 & 1 \end{pmatrix} \begin{pmatrix} 1 & 0 \\ -s_0 & 1 \end{pmatrix} \begin{pmatrix} u_0 \\ y_0 \end{pmatrix} = \begin{pmatrix} 1 + s_0/f & -1/f \\ -s_0 + s_1 + s_0 s_1/f & 1 - s_1/f \end{pmatrix} \begin{pmatrix} u_0 \\ y_0 \end{pmatrix} = \begin{pmatrix} B & -A \\ -D & C \end{pmatrix} \begin{pmatrix} u_0 \\ y_0 \end{pmatrix} . \tag{4.10.76}$$

We continue with Eq. (4.10.75) which gives the value of $q(s_1)$ for a simple thin lens. In line with this equation, the value of $q(s_1)$ for a system with the general *ABCD* matrix is given by

$$q(s_1) = \frac{C\,q(s_0) + D}{A\,q(s_0) + B} , \tag{4.10.77}$$

where the distances s_0 and s_1 are measured with respect to the principal planes of the system.

With the aid of the expression for the image-side beam parameter $q(s_1)$ we calculate the position of the beam waist at the image side of the lens (see Fig. 4.47) if the position $z = s_0$ and the size w_0 of the waist in object space are given. If the beam waist itself is not given but the radius of curvature $R_c(z)$ and $w(z)$ for some general coordinate z in the beam, we obtain the beam-waist size w_0 and its position from Eqs. (1.7.14) and (1.7.15). The position of the beam waist in image space follows from the requirement $\Re(q_1) = 0$ where we use for ease of notation q_0 and q_1 for the values of the parameter q in the beam cross-sections at the points given by $z = s_0$ and $z = s_1$, respectively. If we use $q_0 = i\pi w_0^2/\lambda$ for the object-side q-parameter (beam waist), we find for the image-side value of q with the aid of Eq. (4.10.75),

$$q_1 = \frac{(1 + s_0/f)\,[s_0 - s_1(1 + s_0/f)] + (f - s_1)\left(z_{R,0}/f\right)^2 + i\,z_{R,0}}{(1 + s_0/f)^2 + \left(z_{R,0}/f\right)^2} . \tag{4.10.78}$$

Here we have used the Rayleigh distance $z_{R,0} = \pi w_0^2/\lambda$ in object space. It has been introduced in Eq. (1.7.23) and illustrated in Fig. 1.16. We impose $\Re(q_1) = 0$ and find the waist position from

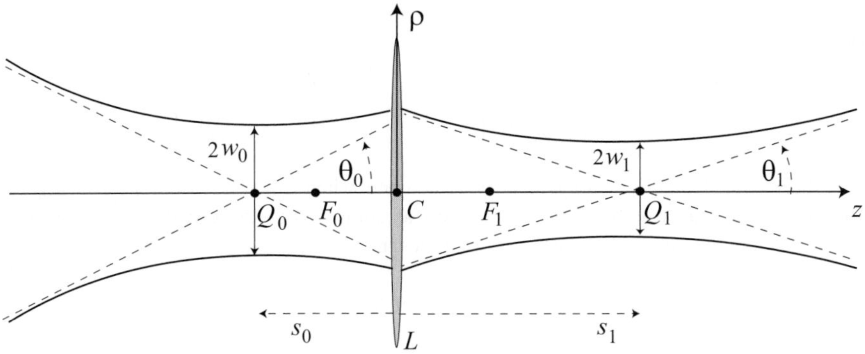

Figure 4.47: The imaging of the beam waist in the plane through Q_0 to the plane through Q_1 in image space. F_0 and F_1 are the object-side and image-side focal points of the thin lens.

$$s_1 = \frac{s_0(1+s_0/f) + f\left(z_{R,0}/f\right)^2}{(1+s_0/f)^2 + \left(z_{R,0}/f\right)^2} . \tag{4.10.79}$$

The waist size w_1 then follows from $\Im(q_1) = \pi w_1^2/\lambda$,

$$w_1^2 = \frac{w_0^2}{(1+s_0/f)^2 + \left(z_{R,0}/f\right)^2} . \tag{4.10.80}$$

From the beam waist w_1 in image space follows immediately the natural divergence angle $\theta_1 = \lambda/(\pi w_1)$ of the imaged beam. By inverse beam propagation, this divergence angle in image space yields the beam half-diameter $w_{F,0}$ in the front focal plane through F_0,

$$w_{F,0} = f\,\theta_1 = \frac{f\lambda}{\pi w_0}\left[(1+s_0/f)^2 + \left(z_{R,0}/f\right)^2\right]^{1/2} . \tag{4.10.81}$$

It is seen that for $s_0 = -f$ and having the object-side beam waist in the front focal plane, the inverse beam propagation reproduces the beam with size w_0 in the object-side focal point F_0.

Some special beam waist positions in object space are worth considering. They are compared with the results for image formation by means of a 'standard' point source emitting a spherical wave:

- $s_0 \gg f$
  The limiting value of s_1 equals f (provided $s_0 \gg f$) and the beam waist is formed at F_1. Equation (4.10.80) shows that the value of the beam waist in F_1 depends on the ratio of the object distance s_0 and the Rayleigh distance $z_{R,0}$. Two cases are considered:
  a) $z_{R,0} \ll s_0 \gg f$ (ray-optics case).
  We write $z_{R,0} = \epsilon\, s_0$ where $|\epsilon|$ is much smaller than unity. For the limiting case $\epsilon \rightarrow 0$, the substitution of $z_{R,0} = \epsilon s_0$ in Eq. (4.10.80) yields for the image-side beam waist the expression

$$w_1^2 = \left(\frac{w_0 f}{s_0}\right)^2 . \tag{4.10.82}$$

  The value of w_1/w_0 is simply given by the ratio $|f/s_0|$ of the geometrical distances s_0 and f, which means that diffraction effects can be neglected.
  b) $z_{R,0} \gg s_0 \gg f$ (diffraction case).
  The beam waist at F_1 is now given by

$$w_1 = \frac{w_0 f}{z_{R,0}} = \frac{\lambda f}{\pi w_0} . \tag{4.10.83}$$

  In practice, when optical experiments are performed with low-divergence beams over a limited propagation distance, the conditions mentioned under b) are satisfied and the beam waist formation is dominated by diffraction effects.

- $s_0 = -2f$
  We have that

$$s_1 = f\left\{\frac{2+\left(z_{R,0}/f\right)^2}{1+\left(z_{R,0}/f\right)^2}\right\}, \qquad w_1^2 = w_0^2\left\{\frac{1}{1+\left(z_{R,0}/f\right)^2}\right\}. \tag{4.10.84}$$

  It is seen that the classical image position $s_1 = 2f$ is attained when $z_{R,0} = \pi w_0^2/\lambda \ll f$. The Rayleigh distance is appreciable for Gaussian beams with a large beam waist as compared to the wavelength of the light. The smaller the beam width, the more the imaging of the waist resembles the classical imaging law for spherical waves.
- $s_0 = -f$
  The image distance s_1 is exactly f and the beam waist is given by

$$w_1^2 = f^2\left(\frac{\lambda}{\pi w_0}\right)^2 = (f\,\theta_0)^2 . \tag{4.10.85}$$

  It follows that for a Gaussian beam the geometrical optics rule applies that, in the image-side focal plane, the image size equals the focal length times the far-field object extent in angular measure; for the Gaussian beam, these quantities are evaluated for the (1/e) amplitude level of the beam.
- $s_0 = 0, \quad s_1 = f\,[(z_{R,0}/f)^2/\{1+(z_{R,0}/f)^2\}]$.
  The waist is found in the region between the lens and the image-side focal plane. Its size is given by

$$w_1^2 = \frac{w_0^2}{1+(z_{R,0}/f)^2}, \tag{4.10.86}$$

  where the extreme (s_1, w_1) combinations are given by $(0, w_0)$ for $z_{R,0} \ll f$ and $(f, \theta_0 f)$ for $z_{R,0} \gg f$.

The results above show that the imaging of the waist of a Gaussian beam follows the classical imaging laws quite closely as long as the Rayleigh distance $z_{R,0}$ of the Gaussian beam is much smaller than the focal distance f of the thin lens. For very elongated Gaussian beams with $z_{R,0} \gg f$, the waist is imaged close to the image-side focal point, independently of the waist position in object space. These effects are shown in Fig. 4.48, where the normalised image distance s_1/f of the beam waist and the relative beam waist width w_1/w_0 are plotted against the (normalised) object distance s_0/f, both for real and virtual positions of the beam waist in object space. For large absolute values of s_0/f, the waist is imaged close to the focal point F_1. The value of s_1/f assumes an extreme value for $s_0/f = -1 \pm z_{R,0}/f$. The positions of the extrema in Fig. 4.48a are given by the abscissa and ordinate pairs,

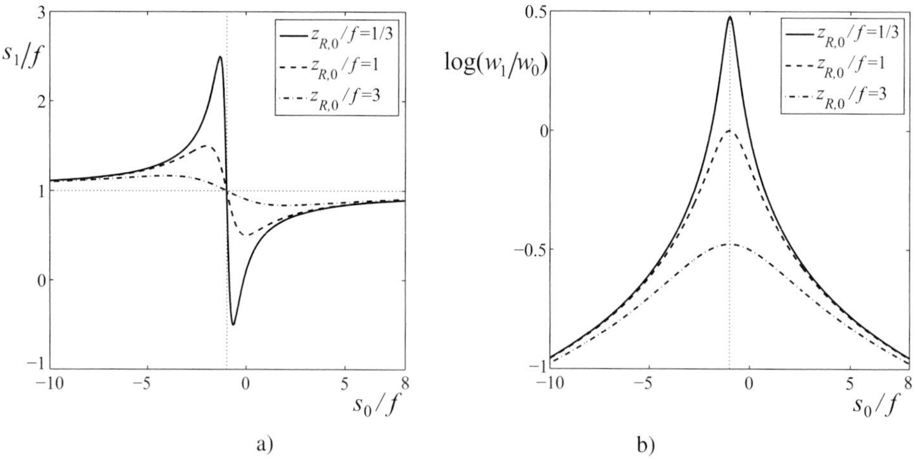

Figure 4.48: a) Normalised image distance s_1/f of the waist of a Gaussian beam as a function of the distance s_0/f of the beam waist in the object space of a thin lens with focal distance f.
b) The ratio w_1/w_0 on a logarithmic scale of the beam waists in image and object space, respectively.
Parameter in the graphs is the ratio of the Rayleigh distance $z_{R,0}$ of the Gaussian beam and the focal distance f of the lens.

$$(s_0, s_1) = \left(-f + z_{R,0}\,,\, f\left\{1 - \frac{f}{2z_{R,0}}\right\}\right), \quad (s_0, s_1) = \left(-f - z_{R,0}\,,\, f\left\{1 + \frac{f}{2z_{R,0}}\right\}\right). \tag{4.10.87}$$

The result is that the axial range where the image of the waist in image space can be found is limited to $f^2/z_{R,0}$ and that its central position corresponds to F_1.

The ratio of the beam waists in image and object space plays the role of the lateral magnification of the imaging by the lens. This ratio has a maximum if the object-side waist is in the front focal point. The maximum equals $f/z_{R,0}$. This quantity becomes infinite if $z_{R,0}$ tends to zero and such a situation corresponds to the imaging of a source point (with zero Rayleigh length) from the object focal point to infinity. A special case arises for $s_0 = -f$ and $z_{R,0} = f$; the beam waists in object and image space assume equal values.

4.10.7.2 Gaussian Beam Imaging by an Optical System

The imaging of the waist of a Gaussian beam by a general optical system is analysed in a similar way as for the thin lens. The geometry for beam waist imaging by an extended optical system is shown in Fig. 4.49. In the figure a real beam waist is found in object space at the point Q_0. If the beam waist is not explicitly given, we assume that for some beam cross-section the complex q-parameter is given and that the position of the beam waist has been derived from this parameter value. We assume that the paraxial matrix is given with respect to some arbitrary reference planes U_0 and U_1 that are not necessarily a conjugated pair. The matrix is then transformed by means of two transport matrices associated with the distances l_0 and l_1 in object and image space, respectively, and the distances are chosen such that the principal planes of the optical system are the new reference planes. Using Eqs. (4.10.29) and (4.10.34) and substituting $n_0 = n_1 = 1$ for the values of the refractive indices, we have for the distances l_0, l_1 and for the resulting matrix with respect to the principal planes through P_0 and P_1,

$$l_0 = \frac{1-B}{A}, \qquad l_1 = \frac{C-1}{A}, \qquad M_{P_0P_1} = \begin{pmatrix} 1 & -A \\ 0 & 1 \end{pmatrix}. \tag{4.10.88}$$

The next step is to propagate the Gaussian beam in object space towards the principal plane through P_0 in order to obtain the q-parameter in this plane. By definition, the beam width $w(z)$ in the two principal planes is identical, and for the change in beam parameter between the principal planes we have, in accordance with Eq. (4.10.73),

$$\frac{1}{q_{P_1}} = \frac{1}{q_{P_0}} + A\,. \tag{4.10.89}$$

In a straightforward manner, using Eqs. (4.10.76)–(4.10.78) in which the focal distance f is replaced by the inverse of the optical power A of the system, we obtain for the position s_1 and the size w_1 of the beam waist in image space,

$$s_1 = \frac{s_0(1+s_0A) + \left(z_{R,0}A\right)^2/A}{(1+s_0A)^2 + \left(z_{R,0}A\right)^2}, \qquad w_1 = \frac{w_0}{\left[(1+s_0A)^2 + \left(z_{R,0}A\right)^2\right]^{1/2}}. \tag{4.10.90}$$

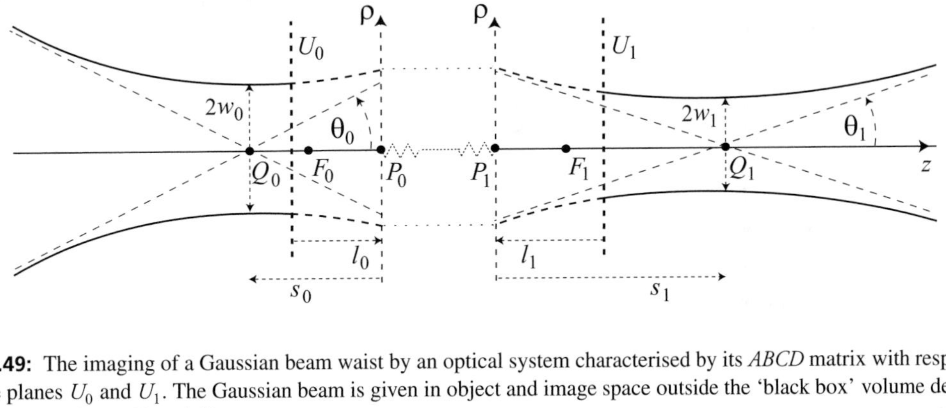

Figure 4.49: The imaging of a Gaussian beam waist by an optical system characterised by its *ABCD* matrix with respect to the reference planes U_0 and U_1. The Gaussian beam is given in object and image space outside the 'black box' volume delimited by the reference planes U_0 and U_1.

A special case of practical importance is the imaging of a Gaussian beam by a telescope or beam expander with zero optical power. Such a telescope is shown in Fig. 4.37a and, in what follows, we use the same numbering of object- and image-side quantities as in this figure. The paraxial matrix of a telescopic system is used in Eq. (4.10.51). The construction length d_2 of a classical Kepler telescope with two confocal positive lenses follows from the equation $d_2/n_2 = 1/A_1 + 1/A_2$ where A_1 and A_2 are the individual powers of the telescope lenses. For equal indices in object space, intermediate space and image space ($n_0 = n_2 = n_4 = 1$), we find for the paraxial matrix referred to two planes at distances s_0 and s_4 from the entrance and exit (thin) lenses of the telescope,

$$\begin{pmatrix} u_4 \\ y_4 \end{pmatrix} = \begin{pmatrix} 1/m_T & 0 \\ 1/A_1 + 1/A_2 + s_4/m_T - m_T s_0 & m_T \end{pmatrix} \begin{pmatrix} u_0 \\ y_0 \end{pmatrix} = \begin{pmatrix} B_r & 0 \\ -D_r & C_r \end{pmatrix} \begin{pmatrix} u_0 \\ y_0 \end{pmatrix}, \qquad (4.10.91)$$

where $m_T = -A_1/A_2$ and where we have used the short-hand B_r, C_r and D_r for the nonzero elements of the ray matrix. The equation for the beam parameter $q(s_4)$ in image space is then given by

$$q(s_4) = \frac{C_r q(s_0) + D_r}{B_r} = m_T^2 q(s_0) - s_4 + m_T^2 s_0 - m_T(1/A_1 + 1/A_2) \,. \qquad (4.10.92)$$

If s_0 corresponds to the beam-waist position in object space, we obtain for the image-side beam waist quantities, in terms of the focal distances f_1 and f_2 of the (thin) telescope lenses,

$$\frac{s_4}{f_2} = -m_T \frac{s_0}{f_1} + 1 - m_T \,, \qquad \frac{w_4}{w_0} = |m_T| \,, \qquad (4.10.93)$$

where s_4 is measured from the exit lens surface of the telescope. The telescope property of constant magnification in the entire image space is confirmed by the beam-waist size, which is constant regardless of the beam-waist position in object space.

4.10.7.3 Gaussian Beam Propagation in an Optical Cavity

Light amplification by stimulated emission of radiation is based on the repeated roundtrip of a light beam in an amplifying resonant cavity equipped with reflecting surfaces. Part of the radiation is coupled out during each roundtrip via a partly transmitting mirror surface. Although the beam propagation within the cavity should be formally described with the aid of rigorous electromagnetic field equations, the Gaussian beam paraxial approximation provides a good prediction of the outgoing field distribution. It also permits us to study the relation between the geometry of the cavity and the reproduction of the beam shape after each cavity roundtrip (see Subsection 1.7.5). In this paragraph we analyse the stability of the Gaussian beam with the aid of a ray optics stability criterion. After a very large number of cavity roundtrips, we require that the limit value of the ray intersection height with the mirrors remains finite.

In Fig. 4.50 a cavity is shown with two mirror surfaces, one intersecting the origin O_0 and one intersecting a general point O_1 at a distance L. The mirror surfaces have curvatures c_0 and c_1, respectively, the refractive index of the cavity medium is n_c. The refractive index after an even number of reflections equals $+n_c$, for the backward propagating rays in mirror space, the index equals $-n_c$. A general ray initially starts in a point Q_0 at a distance $y_{0,0}$ from the axis, in a direction defined by the angle $u_{0,0}$ where the second subscript denotes the number of completed roundtrips in the cavity. The paraxial matrix for a single cavity roundtrip is given by,

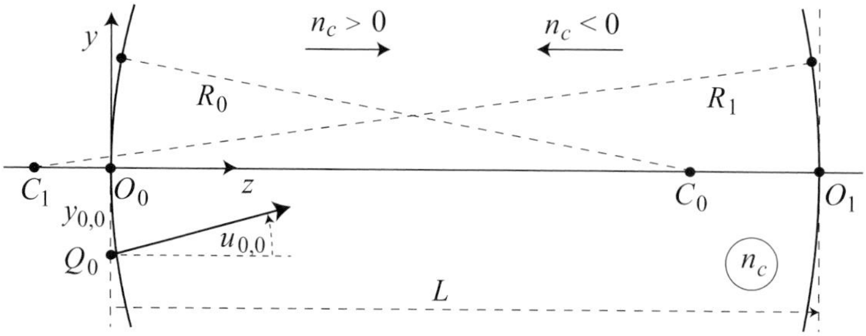

Figure 4.50: The geometry of an optical cavity equipped with spherical mirrors.

$$\begin{pmatrix} n_c u_0 \\ y_0 \end{pmatrix}_1 = \begin{pmatrix} 1 & -2n_c c_0 \\ 0 & 1 \end{pmatrix} \begin{pmatrix} 1 & 0 \\ L/n_c & 1 \end{pmatrix} \begin{pmatrix} 1 & 2n_c c_1 \\ 0 & 1 \end{pmatrix} \begin{pmatrix} 1 & 0 \\ L/n_c & 1 \end{pmatrix} \begin{pmatrix} n_c u_0 \\ y_0 \end{pmatrix}_0 = \begin{pmatrix} 1 + 2L(c_1 - 2c_0) - 4c_0 c_1 L^2 & 2n_c(c_1 - c_0 - 2c_0 c_1 L) \\ 2L(1 + c_1 L)/n_c & 1 + 2c_1 L \end{pmatrix} \begin{pmatrix} n_c u_0 \\ y_0 \end{pmatrix}_0 , \tag{4.10.94}$$

where the subscript attached to a ray vector denotes the number of roundtrips of the ray. For the ray data in the plane through O_0 after N roundtrips we have

$$\begin{pmatrix} n_c u_0 \\ y_0 \end{pmatrix}_N = \begin{pmatrix} B & -A \\ -D & C \end{pmatrix}^N \begin{pmatrix} n_c u_0 \\ y_0 \end{pmatrix}_0 , \tag{4.10.95}$$

where the matrix elements A, B, C and D are given by Eq. (4.10.94) and obey the relation $BC - AD = 1$.

To evaluate the Nth power of the paraxial matrix of the cavity, Sylvester's theorem **[332]** is used,

$$\begin{pmatrix} B & -A \\ -D & C \end{pmatrix}^N = \frac{1}{\sin\theta} \begin{pmatrix} B\sin(N\theta) - \sin[(N-1)\theta] & -A\sin(N\theta) \\ -D\sin(N\theta) & C\sin(N\theta) - \sin[(N-1)\theta] \end{pmatrix} , \tag{4.10.96}$$

where $\cos\theta = (B + C)/2$. Both the intersection height $y_{0,N}$ and the ray angle $u_{0,N}$ after N roundtrips should have finite values. This condition is only satisfied if θ has a real value or, alternatively, in terms of the matrix elements,

$$-1 \le \frac{B + C}{2} \le +1 . \tag{4.10.97}$$

From these diagonal elements of the single roundtrip matrix we immediately obtain the condition

$$0 \le (1 - c_0 L)(1 + c_1 L) \le 1 . \tag{4.10.98}$$

Another approach leading to the same end result has been illustrated in Fig. 4.51. We analyse single roundtrips in an optical cavity starting from the surface through O_0 in the forward direction using the standard *ABCD* matrix (see Fig. 4.51a). The roundtrip in the backward direction starts on reflection at the entrance surface through O_0 in the negative propagation direction and uses the inverse of the paraxial system matrix (see Fig. 4.51b). In the figure we have denoted the starting heights at the points $P_{0,f}$ and $P_{0,b}$ on the surface through O_0 by $y_{0,f}$ and $y_{0,b}$, respectively. The intersection points with the surfaces after one roundtrip are the points $P_{1,f}$ and $P_{1,b}$ with coordinates $y_{1,f}$ and $y_{1,b}$, respectively. In a comparable way we have defined the starting angles $u_{0,f}$ and $u_{0,b}$ at the points $P_{0,f}$ and $P_{0,b}$ and the final ray angles $u_{1,f}$ and $u_{1,b}$ at the points $P_{1,f}$ and $P_{1,b}$ after the second reflection in the roundtrip. In matrix notation we have the standard *forward* matrix expression associated with Fig. 4.51a,

$$\begin{pmatrix} n_c u_{1,f} \\ y_{1,f} \end{pmatrix} = \begin{pmatrix} B & -A \\ -D & C \end{pmatrix} \begin{pmatrix} n_c u_{0,f} \\ y_{0,f} \end{pmatrix} , \tag{4.10.99}$$

where n_c is the refractive index of the cavity medium. The matrix for the backward propagating rays is obtained in two steps. First we consider the rays in Fig. 4.51b to propagate in the forward direction, yielding the matrix equations

$$\begin{pmatrix} n_c u_{0,b} \\ y_{0,b} \end{pmatrix} = \begin{pmatrix} B & -A \\ -D & C \end{pmatrix} \begin{pmatrix} n_c u_{1,b} \\ y_{1,b} \end{pmatrix} , \text{ or, } \begin{pmatrix} n_c u_{1,b} \\ y_{1,b} \end{pmatrix} = \begin{pmatrix} C & A \\ D & B \end{pmatrix} \begin{pmatrix} n_c u_{0,b} \\ y_{0,b} \end{pmatrix} . \tag{4.10.100}$$

We then change the propagation direction of the rays and replace n_c by $-n_c$ as the rays effectively propagate in a mirrored space. To satisfy the second matrix equation of Eq. (4.10.100) for backward propagation rays we find the modified matrix equation

$$\begin{pmatrix} -n_c u_{1,b} \\ y_{1,b} \end{pmatrix} = \begin{pmatrix} C & -A \\ -D & B \end{pmatrix} \begin{pmatrix} -n_c u_{0,b} \\ y_{0,b} \end{pmatrix} . \tag{4.10.101}$$

A large number of rays is injected into the cavity in the forward and backward propagation directions through general points $P_{0,f}$ and $P_{0,b}$ with various heights of incidence $y_0^{(m)}$ and ray angles $u_0^{(n)}$. The integer indices m and n run from 1 to

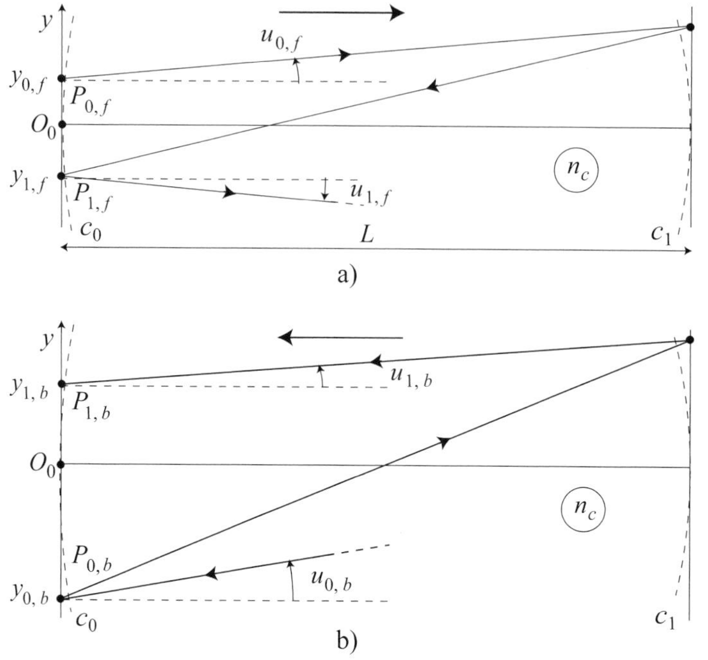

Figure 4.51: Stable cavity analysis with the aid of test rays travelling in the forward and backward direction between the mirrors of the cavity (see a) and b), respectively).

M and from 1 to N, respectively. The y-coordinates and angles u in the forward and backward system cover the relevant ranges associated with a cavity mode. The intersection heights after one roundtrip on the first mirror are given by

$$y_{1,f}^{(n,m)} = -D\,n_c u_0^{(n)} + C\,y_0^{(m)}\,, \qquad y_{1,b}^{(n,m)} = +D\,n_c u_0^{(n)} + B\,y_0^{(m)}\,. \tag{4.10.102}$$

A particular pair of rays such as the two given in Eq. (4.10.102) represents a roundtrip in the system, given a certain combination of initial ray height and ray angle for the forward system and for the backward system. For a stable cavity we require that after each roundtrip, on average, the intersection heights $y_{1,f}^{(n,m)}$ and $y_{1,b}^{(n,m)}$ are smaller than the initial values $y_0^{(m)}$ for those roundtrips. Summation of the two equations of (4.10.102) for all ray angles u_0 and ray heights y_0 on the first mirror of the cavity yields the average values of the intersection heights $\bar{y}_{1,f}$ and $\bar{y}_{1,b}$ after one roundtrip,

$$\bar{y}_{1,f} = -D\,n_c\bar{u}_0 + C\,\bar{y}_0\,, \qquad \bar{y}_{1,b} = +D\,n_c\bar{u}_0 + B\,\bar{y}_0\,. \tag{4.10.103}$$

The addition of these two equations yields an average value y_r of the ray intersection heights after many roundtrips in both directions,

$$2y_r = (B + C)\,\bar{y}_0\,. \tag{4.10.104}$$

With the aid of the stability requirement $|y_r| \le |\bar{y}_0|$ we find,

$$|B + C| \le 2\,. \tag{4.10.105}$$

Both analyses in this subsection are based on ray optics and they produce the same stability criterion as the one obtained with the aid of Gaussian beam analysis on the basis of the paraxial Helmholtz equation (see Section 1.7.5). In the wave optics analysis of Chapter 1 the requirement for a physical solution to the wave in the cavity was a finite beam width. The ray optics requirement postulates a finite ray intersection height on the mirrors after a large number of cavity roundtrips and leads to the same result. The stable cavity geometries that were found in Chapter 1 follow equally well from the ray optics analysis in this subsection.

4.10.8 Applications of the Paraxial Analysis of Imaging Systems

The laws of paraxial optics, especially those for a system with circular symmetry which were presented above, are easily implemented and computed in a systematic and efficient way using the matrix approach. The final imaging performance of an optical system is determined by its imaging defects or optical aberrations which are not included in paraxial optics. However, surprisingly, the imaging defects can be well approximated by formulas that require as input only the paraxial data. In this way, the role of paraxial optics is important when carrying out the very first design steps (prototyping) of a still unknown optical system that should fulfil a certain list of imaging requirements. The paraxial design is thus able to show the level of aberration correction and also the degree of 'smoothness' of a design or, in opposite terms, the degree of 'strain' that will most probably be present in the final design. *Smoothness* or lack of *strain* is a very important property of a design and it determines the effort that will be needed to obtain a certain state of imaging correction. It also determines the ease of manufacturing of an optical system and the measurement effort that has to be spent on individual components. If the ray properties in image space are obtained without imposing excessively large angles of incidence of individual rays on lens or mirror surfaces, the optical system will show favourable manufacturing tolerances and have a large fabrication yield in production. The paraxial design thus allows a first-order check on the angles of incidence on each optical surface of the rays in the axial beam and in off-axis beams. From this information one obtains an impression of the strain which is inherent in the optical system.

Another aspect of an optical design, the obstruction of the axial and off-axis pencils of rays by the diaphragm and the lens rims are almost immediately available from paraxial optics, with good fidelity as compared to the same data for the ultimate finite rays in the system. Such a quick estimate of the vignetting effects in an optical system is important if the practicality of a design is being assessed. This is especially the case if the system is an on-axis or off-axis mirror system. By virtue of the extremely fast calculation of the paraxial data, it is possible to numerically analyse a very large number of paraxial prototypes in a short time. This allows us to identify certain regions in design space where promising prototypes can be found which will have favourable aberration properties and show small vignetting effects. With the advent of large computing power, the subject of paraxial design prototyping has acquired much attention.

4.11 Radiometry and Photometry

In this section we describe the transport of light energy through an optical system, employing the geometrical theory of light propagation. The light emitted by a source illuminates a (planar) object. The light transmitted or reflected by the object is followed through the imaging system up to the image plane or, in the case of visual observation, up to the eye of the observer. If the light energy is measured with the aid of a detector, one uses 'objective' radiometric units (optical radiometry). For imaging systems in which, ultimately, the perception by a human eye determines the strength of the visual stimulus, we use photometric units (photometry). The intensity of the impression produced by a certain light power is now determined by both the light energy and the experimentally recorded spectral eye-sensitivity. As a result, two systems of comparable units exist side-by-side, related to radiometry and to photometry.

In this section we first define the radiometric and photometric basic units and the quantities derived from the basic units. To obtain the 'strength' of an image we first study the illumination of a plane at a certain distance from a source. The source will serve as the object for an imaging system, the plane at a certain distance plays the role of the entrance pupil of an imaging system. The following step is the computation of the illumination at the image plane. In this context we present a useful invariant of an imaging system, the so-called 'optical throughput' or 'étendue'.

4.11.1 The Radiometric and Photometric Units

In this subsection we present the radiometric and photometric units as they have been defined by the CIE (Commission Internationale de l'Éclairage). As already mentioned above, one needs the experimentally determined spectral eye-sensitivity curve to convert from radiometric to photometric energy. In Fig. 4.52 we schematically show this sensitivity curve, which is representative for the entire human population. The curve is associated with a typical *daylight* light level (*photopic* curve). At considerably lower light levels, for instance at an illumination below 3 lux, the so-called *twilight* level, the maximum eye sensitivity shifts to a shorter wavelength (λ = 510 nm, *scotopic* eye-sensitivity curve).

The radiometric and photometric quantities and their corresponding units are given in Table 4.2 With the aid of $V(\lambda)$ one could permanently employ the radiometric units, which are expressed in the basic units of the SI system. To compute

Table 4.2. Names and dimensions of radiometric and photometric quantities and units.

Quantity	Radiometry	Photometry
Energy	radiant energy J	luminous energy lm s
Energy density	radiant energy density $\mathrm{Jm^{-3}}$	luminous energy density $\mathrm{lm\,s\,m^{-3}}$
Power	radiant flux $\mathrm{Js^{-1}}$ (W)	luminous flux lm (derived unit *lumen*)
Irradiance	irradiance $\mathrm{Wm^{-2}}$	illuminance $\mathrm{lm\,m^{-2}}$ (derived unit *lux*)
Intensity	radiant intensity $\mathrm{Wsr^{-1}}$	luminous intensity $\mathrm{lm\,sr^{-1}}$ (basic unit *candela* = cd)
'Brightness'	radiance $\mathrm{Wm^{-2}sr^{-1}}$	luminance $\mathrm{lm\,m^{-2}sr^{-1} = cd\,m^{-2}}$ = nit (derived unit *nit*)

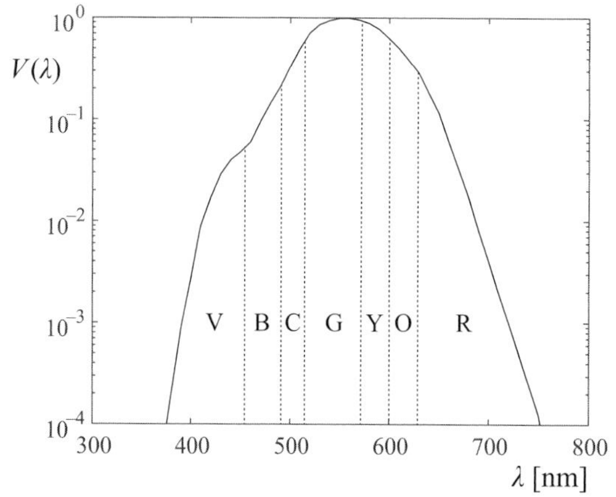

Figure 4.52: The sensitivity function $V(\lambda)$ of the human eye as a function of the wavelength in the visible spectrum (CIE 1978 curve for photopic vision). The curve has been normalised to unity at its maximum value (λ = 555 nm). The lettering in the graph denotes the colours of the visible spectrum (C is cyan).

the strength of the eye stimulus using photometric units, the spectral distribution of the light energy should be available, either from prior knowledge or by spectral measurement. The expression that converts radiant flux to luminous flux is then given by

$$\Phi_l = 683 \int_0^{\infty} V(\lambda)\, \Phi_{r,\lambda}(\lambda)\, d\lambda\,. \tag{4.11.1}$$

In this expression Φ_l is the flux in units of lm s and $\Phi_{r,\lambda}$ the spectral radiant flux in $\mathrm{W\,(nm)^{-1}}$ of the light beam (λ is expressed in nm). The factor 683 with the physical dimension of $\mathrm{lm\,W^{-1}}$, is the conversion factor from Watt to lumen for a beam of light with a wavelength of 555 nm, corresponding to the maximum of the eye-sensitivity function $V(\lambda)$. To avoid the measurement of $\Phi_{r,\lambda}$ of a light beam and the (numerical) evaluation of the integral of Eq. (4.11.1), a standardised

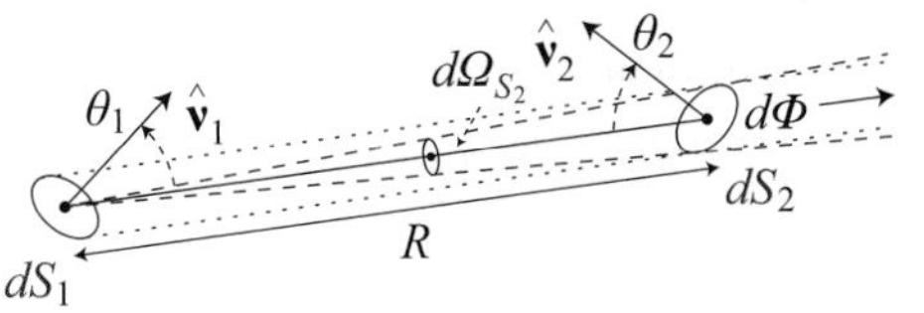

Figure 4.53: Energy propagation from an emitting surface dS_1 to a receiving surface dS_2. The solid angle subtended by dS_2 at dS_1 is given by $d\Omega$.

'white light' source has been defined. In practice it is a black-body radiator in the form of a molten platinum surface with prescribed shape and size of the light-emitting aperture. This standardised source produces by definition a luminous intensity of 1 candela (cd). A more recent definition of the candela unit specifies the candela as the luminous intensity of a source that emits monochromatic radiation with a frequency of $540 \cdot 10^{12}$ Hz ($\lambda \approx 555$ nm in standard air) and has a radiant intensity of 1/683 W sr^{-1}. In what follows we analyse energy flux in free space and in an optical (imaging) system using photometric quantities and units. It goes without saying that all results can be transferred to the radiometric domain, the only difference being the conversion from photometric to radiometric flux according to Eq. (4.11.1).

In Fig. 4.53 we illustrate how the photometric quantities illuminance, luminous intensity and luminance are associated with a light emitting source element (size dS_1) and a receiving surface element dS_2 at a distance R. Both source element and receiving element may be obliquely oriented with respect to the line that joins their centres. The illuminance E_l measured at the receiving surface dS_2 is given by the luminous flux per unit surface,

$$E_l = \frac{d\Phi_l}{dS_2} \,. \tag{4.11.2}$$

E_l is a spatially-dependent function. It is independent of the angles of incidence of the radiation since the luminous flux has been summed over all directions at the surface element dS_2. The luminous intensity of the source element dS_1 is given by the emitted luminous flux per unit solid angle,

$$I_l = \frac{d\Phi_l}{d\Omega_{S_2}} \,. \tag{4.11.3}$$

The luminous intensity I_l of an extended source is an average over the source area and is thus only a function of the viewing direction. In Fig. 4.53 we have only shown the 'elevation' angle θ. The intensity i_l can equally well depend on the azimuthal angle ϕ which is not shown in the drawing. We remark that $d\Omega_{S_2} = dS_2 \cos\theta_2 / R^2$ is the solid angle defined by the projected receiving area, viewed from the source centre. Another solid angle that will be used is $d\Omega_{S_1} = dS_1 \cos\theta_1 / R^2$ which is defined by the projected source area, as viewed from the receiving plane. For the geometry of Fig. 4.53 it follows that

$$\frac{dS_1 dS_2 \cos\theta_1 \cos\theta_2}{R^2} = d\Omega_{S_2} (dS_1 \cos\theta_1) = d\Omega_{S_1} (dS_2 \cos\theta_2) \,. \tag{4.11.4}$$

The source luminance L is defined as the luminous flux per unit solid angle and per unit projected source area (projected on a plane perpendicular to the viewing direction),

$$L = \frac{d^2\Phi_l}{d\Omega_{S_2} (dS_1 \cos\theta_1)} \,, \tag{4.11.5}$$

where we use a second-order differential for Φ_l to emphasise that L has the form of a second-order differential quotient. With the aid of Eq. (4.11.4) the quantity L can also be defined or measured using the expression

$$L = \frac{d^2\Phi_l}{d\Omega_{S_1} (dS_2 \cos\theta_2)} \,. \tag{4.11.6}$$

The luminance function L of a source depends both on the source coordinates and on the viewing direction.

The illuminance E_l and luminous intensity I_l can now be written with the aid of the source luminance function L as

$$dE_l = \frac{L\, d\Omega_{S_2} (dS_1 \cos\theta_1)}{dS_2} = \frac{L\, dS_1 \cos\theta_1 \cos\theta_2}{R^2} = L\, d\Omega_{S_1} \cos\theta_2 \,. \tag{4.11.7}$$

For the luminous intensity we obtain the expression

$$dI_l = \frac{L\, d\Omega_{S_2}\,(dS_1 \cos\theta_1)}{d\Omega_{S_2}} = L\, dS_1 \cos\theta_1 \,. \tag{4.11.8}$$

4.11.2 Special Geometries for Source Surface and Detecting Surface

We use the (differential) expressions for the illuminance and intensity to calculate the resulting quantities for an extended source:

- *Point-source.*
  The (differential) flux emitted by point source P and the receiving surface dS are shown in Fig. 4.54a. Assuming a direction-independent luminous intensity I_l, the total flux emitted by the point source follows directly from Eq. (4.11.3) and is given by

$$\Phi_l = \iint_{\text{sphere}} I_l\, d\Omega = 4\pi\, I_l \,. \tag{4.11.9}$$

- *Planar isotropic source.*
  Another frequently encountered source is a planar source that radiates uniformly as a function of the viewing angle. Such a source is called an isotropic or Lambertian source, after the physicist J. H. Lambert who first described such a source. However, the luminous intensity of such a planar source is not uniform since the projected surface diminishes as a function of the viewing angle θ according to a $\cos\theta$ function. We thus have for the intensity of the planar Lambertian source of Fig. 4.54b the expression

$$I_l(\theta_1) = I_1 \cos\theta_1 \,, \tag{4.11.10}$$

  where θ_1 is the angle between the viewing direction and the surface normal. According to Eq. (4.11.3) the differential luminous flux $d\Phi_l$ emitted by the source element dS_1 is now given by $I_l(\theta_1) d\Omega_{S_2}$. With the aid of Eqs. (4.11.5) and (4.11.10) we then find for the luminance L_1 of the source

$$L_1 = \frac{d}{dS_1}\left\{\frac{1}{\cos\theta_1}\left(\frac{d\Phi_l}{d\Omega_{S_2}}\right)\right\} = \frac{d}{dS_1} I_1 \,. \tag{4.11.11}$$

  For a uniform source we obtain the result that the luminance L_1 is given by

$$L_1 = \frac{I_1}{S_1} \,. \tag{4.11.12}$$

  The luminance of a Lambertian source is thus independent of the viewing angle. The isotropic radiation property holds well for a (perfectly) diffusing surface, for instance a glass surface that has been ground or only roughly polished. At optical wavelengths multiple scattering occurs at such an unfinished optical surface and the propagation direction of the reflected or transmitted radiation is found in a cone of scattered light. In Section 10.5, the scattering characteristics of weakly and strongly diffusing surfaces are given and their influence on the imaging quality of an optical system is studied in more detail.

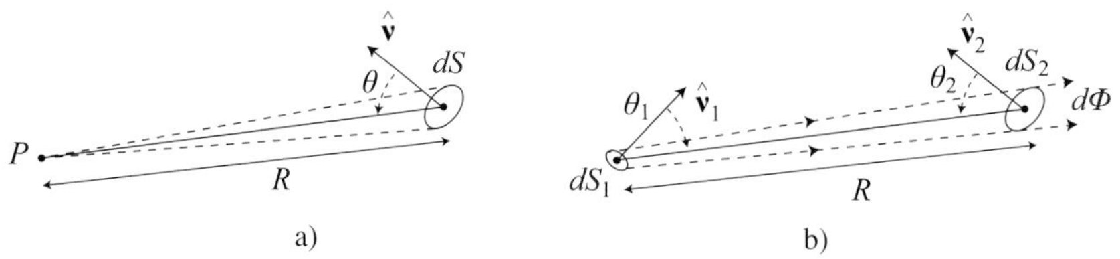

Figure 4.54: a) Radiation emitted by a point source P and received by a surface element dS in space at a distance R.
b) An isotropically diffusing surface dS_1 emits radiation towards a receiving surface dS_2 . The solid angle subtended by dS_2 at dS_1 is given by $d\Omega_{S_2}$.

- *Point source and inclined receiving plane.*
Figure 4.54a can be used to calculate the illuminance produced by a (distant) source at an inclined surface (inclination angle is θ). We use the expression for the illuminance E_l of Eq. (4.11.2) and use the expression for $d\Phi_l$ that follows from Eq. (4.11.3) to obtain

$$E_l = \frac{d\Phi_l}{dS} = \frac{I_l(\theta)d\Omega}{dS} = \frac{I\cos\theta}{R^2}. \tag{4.11.13}$$

In this equation we have replaced $I_l(\theta)$ by a constant value I, assuming that the source is sufficiently far away so that any angular variation of $I_l(\theta)$ can be neglected. Equation (4.11.13) shows the inverse square-law relationship for the source distance R and the projection effect for the receiving surface. If we replace the point source by an extended planar source as in Fig. 4.54b, we use Eq. (4.11.7) and have a (differential) contribution from the source element dS_1 according to

$$dE_l = L\, dS_1 \cos\theta_1 \left(\frac{\cos\theta_2}{R^2}\right) = dI_l \left(\frac{\cos\theta_2}{R^2}\right), \tag{4.11.14}$$

where, for the general case, the luminance function L can be a function of both the radiation direction and the source coordinates.

- *Ingoing and outgoing flux at a diffusing surface.*
In Fig. 4.55a an isotropic light emitting surface element dS has been shown and its intensity $I(\theta)$ is given by $I(\theta) = I\cos\theta$. To obtain the total flux radiated by source element dS in the forward direction, we perform an integration of the differential flux through an incremental ring-shaped cone mantle with width $d\theta$. According to Eq. (4.11.8) the source intensity dI_l at a viewing angle θ is given by $LdS\cos\theta$ and the increment in flux is thus given by

$$d^2\Phi_l = 2\pi\, L\, dS\, \cos\theta\, \sin\theta\, d\theta. \tag{4.11.15}$$

Performing the integration up to a certain maximum angle $\theta = \theta_m$ we find

$$d\Phi_{l,\theta_m} = 2\pi \int_0^{\theta_m} LdS\cos\theta\,\sin\theta\, d\theta = \pi\, L\, dS\, \sin^2\theta_m. \tag{4.11.16}$$

For the total forward-propagating flux we have

$$d\Phi_l = \pi\, L\, dS. \tag{4.11.17}$$

The expression can also be written as

$$L = \frac{1}{\pi}\frac{d\Phi_l}{dS}. \tag{4.11.18}$$

Referring to Fig. 4.55b, $d\Phi_l$ in Eq. (4.11.18) is now the flux that is focused on the diffusing surface element dS and we assume that this incident flux is isotropically scattered in the forward halfspace. The luminance L of the surface element dS is then given by Eq. (4.11.18).

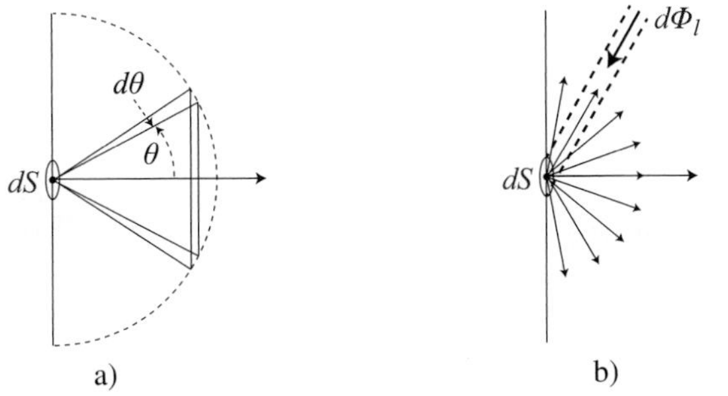

Figure 4.55: a) The 'effective' solid angle associated with a planar isotropic emitter.
b) The luminance of a diffuse surface element dS that receives an incident flux $d\Phi_l$ from an arbitrary direction.

4.11.3 Observation with the Eye: Luminance Detection

The brightness impression that is produced by an object depends on the illuminance pattern of its image on the retina. A local illuminance value on the retina determines the strength of the visual nerve impulse associated with that retina point. In Fig. 4.56 we have taken an isotropically emitting (or reflecting) surface, denoted by dS_1, as the object to be imaged by the eye. The luminous flux $d^2\Phi_l$ through the eye pupil (surface area is dS_2) is given by

$$d^2\Phi_l = L\,\frac{dS_1\cos\theta_1}{R^2}\,dS_2\,, \tag{4.11.19}$$

where we have assumed the eye pupil surface is perpendicular to the average direction of the incident luminous flux ($\theta_2 = 0$). We neglect any eye aberration that could reduce the image sharpness. In that case, the size of the sharp image dS_1' on the retina is given by the product of the solid angle subtended by the (projected) object at the eye pupil and the square of the object-side focal distance f_0 of the eye. This relation follows from the value of the angular magnification for the nodal planes (equal to +1) and the absolute distance f_0 between the image-side nodal point and the retina (see Fig. 4.31 for the special role of the image-side nodal point N'). The size of the image on the retina is thus given by

$$dS_1' = dS_1\cos\theta_1\left(\frac{f_0}{R}\right)^2 . \tag{4.11.20}$$

We note that the object-side focal distance $|f_0|$ of the average (adult) human eye equals 17 mm (focal distance in air). For the total illuminance on the retina we write

$$E_l = \tau\,\frac{d\Phi_{l,S_2}}{dS_1'} = \tau\left(\frac{S_2}{f_0^2}\right)L = \tau\,\pi\,(NA_{\text{eye}})^2\,L\,, \tag{4.11.21}$$

where the flux $d\Phi_{l,S_2}$ represents the total flux through the circular eye pupil with size S_2. The factor τ is a transmission factor that takes into account any absorption, reflections and scattering in the interior of the eye, from the outer cornea surface to the retina. The aperture NA_{eye} of the eye is measured effectively in air. For a standard pupil diameter of 4 mm, we have that $NA_{\text{eye}} = 0.12$ and the composite factor in the most right-hand side of Eq. (4.11.21) then equals $0.045\,\tau$.

It follows from Eq. (4.11.21) that the illuminance distribution on the retina surface is proportional to the spatially varying luminance function L of the observed object. The luminance function L of a radiating object is independent of the distance at which the object is observed. However, from experience we know that identical sources positioned at increasingly larger distances give rise to decreasing eye stimuli. This phenomenon is not predicted by Eq. (4.11.21) and the reason for this is the geometrical optics approximation that we used to derive the size dS_1' of the image on the retina of the object surface dS_1. In reality, the imaging on the retina is accompanied by a blurring effect due to diffraction. If the image of the object is much larger than the diffraction blur, Eq. (4.11.21) is a good approximation. However, if the geometrical image of an object becomes comparable to or even smaller than the diffraction blurring, the luminous flux to the image point on the retina is concentrated on a single cone detector and the geometrical detail about the object gets lost. In that case, we use the expression for the luminous intensity of Eq. (4.11.8), integrated over the entire object, $I_l = \int L\cos\theta_1\,dS_1$. The luminous flux into the eye according to Eq. (4.11.19) is then written as

$$d\Phi_l = I_l\,\frac{dS_2}{R^2}\,. \tag{4.11.22}$$

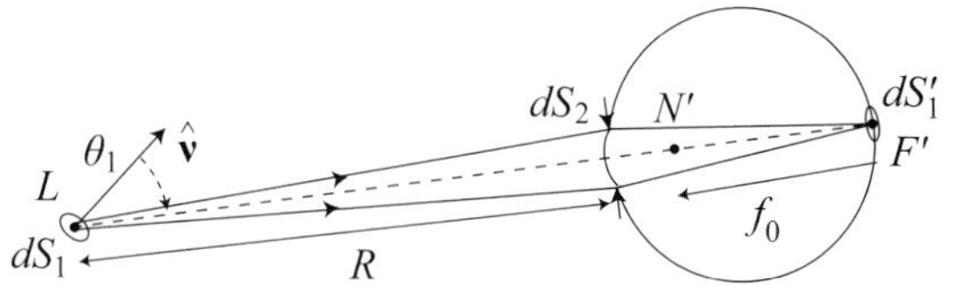

Figure 4.56: Detection of the luminous flux from a distant surface element dS_1 by the human eye. The surface of the eye pupil has been denoted by dS_2. The surface element dS_1' is the image on the retina of the object element dS_1. N' is the image-side nodal point of the eye and $F'N' = f_0$ is the object-side focal distance of the eye.

Equation (4.11.22) shows that for an unresolved object the luminous flux incident on a single cone detector is proportional to the surface-integrated intensity I_l of the object. Objects with equal luminous intensity produce stimuli that obey an inverse square law with respect to their distance R from the observer.

4.11.4 The Illuminance in the Image Plane of an Optical system

To calculate the illuminance in the image plane of an optical system we start with an isotropically radiating surface element dS_0 in the object plane (see Fig. 4.57). The object element is perpendicular to the optical axis of the system and the refractive index of the object space is n_0. We use Eq. (4.11.15) for the luminous flux through the ring-shaped zone of the spherical entrance pupil surface through E_0,

$$d^2\Phi_l = 2\pi L_0 dS_0 \cos\theta_0 \sin\theta_0 \, d\theta_0 \,, \qquad (4.11.23)$$

where θ_0 is the angle that determines the diameter of the pupil zone, dS_0 is the size of the radiating surface and L_0 the luminance of the radiating surface. It depends on the ray mapping from entrance to exit pupil where the annular zone of rays with width $d\theta_0$ at the entrance pupil is projected on the exit pupil sphere through E_1. We assume that the imaging system satisfies the Abbe sine condition (see Subsection 4.8.1). In terms of, for instance, the lateral object- and image-plane coordinate pair (y_0, y_1) we have

$$n_0 y_0 \sin\theta_0 = n_1 y_1 \sin\theta_1 \,. \qquad (4.11.24)$$

The same equation, in terms of the radial object- and image-plane coordinate yields after squaring and multiplication by π an expression in terms of circular conjugate surface areas dS_0 and dS_1 in object and image plane,

$$n_0^2 dS_0 \sin^2\theta_0 = n_1^2 dS_1 \sin^2\theta_1 \,. \qquad (4.11.25)$$

Equation (4.11.25) states that the quantity $n^2(dS)\Omega$ is conserved between object and image space of a well-corrected optical system. In this expression dS is a paraxial surface element of the object/image and Ω is the solid angle subtended by rays outside the paraxial domain. Depending on the state of correction of an imaging system, the elementary surfaces dS_0 and dS_1 can be extended to conjugate finite-size object and image areas, yielding the 'finite' conserved quantity $n^2 S\Omega$. It is in this form that the throughput conservation property of an optical system is generally presented. Equation (4.11.25) is an extension of the paraxial Helmholtz–Lagrange invariant

$$n_0 y_0 \theta_0 = n_1 y_1 \theta_1 \,, \qquad (4.11.26)$$

outside the paraxial domain. The quantity $n^2 S\Omega$ is called the 'throughput' or the 'étendue' of an optical system. The multiplication of $S\Omega$ with the luminance of the radiating object yields the luminous flux that is transported through the system.

We now return to Eq. (4.11.25) and write it in differential form,

$$2n_0^2 \, dS_0 \sin\theta_0 \cos\theta_0 \, d\theta_0 = 2n_1^2 \, dS_1 \sin\theta_1 \cos\theta_1 \, d\theta_1 \,. \qquad (4.11.27)$$

The luminous flux $d^2\Phi_1$ that leaves the optical system is given by $\tau \, d^2\Phi_0$, where τ is a system transmission factor that takes into account losses caused by reflection, absorption and scattering in the system. With the aid of Eq. (4.11.23) and Eq. (4.11.27) we write for the outgoing luminous flux

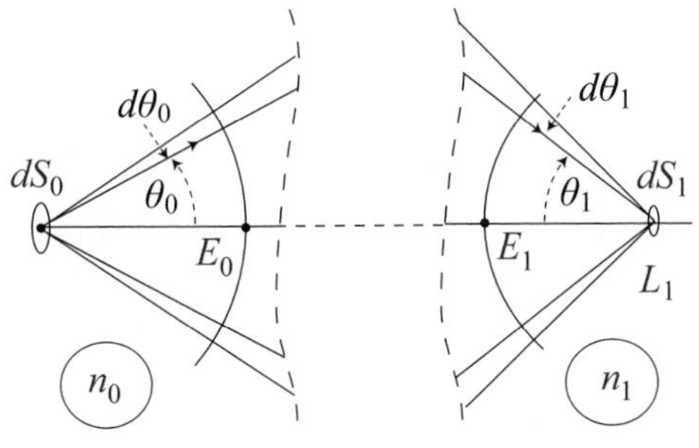

Figure 4.57: Illustrating the brilliance/luminance conservation law for an optical system that satisfies the Abbe sine condition.

$$d^2\Phi_1 = \tau \left(\frac{n_1}{n_0}\right)^2 L_0 \, dS_1 \cos\theta_1 \, 2\pi \sin\theta_1 \, d\theta_1 \, . \tag{4.11.28}$$

Inspection of Eq. (4.11.28) shows that the outgoing flux can be attributed to a surface element dS_1 with a luminance L_1 given by

$$L_1 = \tau \left(\frac{n_1}{n_0}\right)^2 L_0 \, . \tag{4.11.29}$$

Apart from the transmission factor τ, Eq. (4.11.29) is a conservation law for the quantity L/n^2 of a light-emitting or light-receiving surface in a well-corrected imaging system, obeying Abbe's sine-condition. We remark that the surface element dS_1 is not an isotropically diffusing surface since the scattering angle is limited by the image-side aperture of the optical system.

The total radiant flux received by each element of a detecting surface is determined by the irradiance. With the aid of Eq. (4.11.28), for an imaging system that obeys the Abbe sine condition, we integrate Eq. (4.11.28) over the total range of angles of incidence on the detector surface. Using a maximum angle $\theta_{1,m}$ which is determined by the aperture of the imaging system we find the expression

$$d\Phi_1 = \int_0^{\theta_{1,m}} d^2\phi_1 = \tau \left(\frac{n_1}{n_0}\right)^2 L_0 \, dS_1 \, \pi \sin^2\theta_{1,m} \, . \tag{4.11.30}$$

The illuminance is then simply given by

$$E = \frac{d\Phi_1}{dS_1} = \tau\pi \left(\frac{n_1}{n_0}\right)^2 L_0 \, \sin^2\theta_{1,m} \, . \tag{4.11.31}$$

If we use the expression for the numerical aperture in image space $n_1 \sin\theta_{1,m} = NA_1$, we can write

$$E = \tau\pi \frac{L_0}{n_0^2} (NA_1)^2 \, . \tag{4.11.32}$$

A high illuminance can be obtained by increasing the aperture NA to a high value, equal to n_1 as theoretical maximum. In practice, the maximum will be set by practical considerations, such as the increasing complexity with respect to design and manufacturing of a high-aperture system.

4.11.5 Photometry in the Presence of Diffraction Effects

When the blurring effects due to the diffraction of light cannot be neglected, the calculation of the irradiance (illuminance) is not possible using the approach given in this section since it is based on ray optics. It is still possible, however, to compare situations in which the diffraction effects are identical. We consider two cases, the observation of a distant object like a star and the photometry/radiometry associated with laser beams:

- *Observation of a distant object with a telescope.*
  Figure 4.58 shows a parallel beam of light coming from a star, entering the entrance pupil with diameter D of a telescope. The angle at which the star is seen is so small ($< 10^{-9}$ rad) that the diffraction blur λ/D strongly dominates

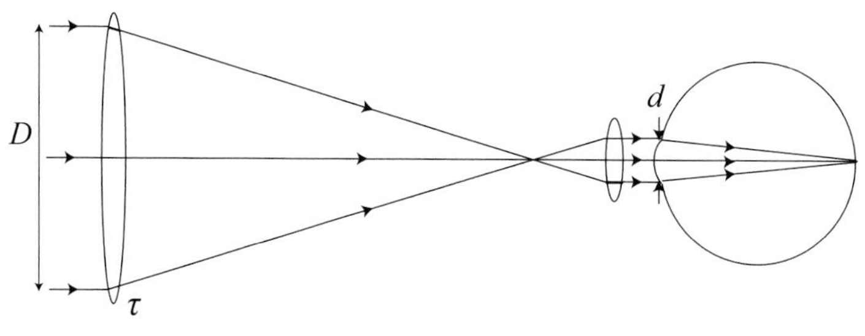

Figure 4.58: Observation of a distant object with the aid of a magnifying optical system (telescope).

the (visual) observation. The radiant flux received from a star with intensity I and collected within the diffraction image of the telescope equals

$$\Phi_{\mathrm{ol}} = \tau I \Omega = \tau I \frac{\pi D^2}{4R_t^2}, \tag{4.11.33}$$

where R_t is the distance from the star to the earth and τ is the 'energy' transmission factor of the telescope. Φ_{ol} is the flux corresponding to the 'observation limit', permitting an observer to positively decide that a stellar object is present. Observation with the naked eye of another star with the same intensity I at a distance R_e, again at the observation limit, yields the expression

$$\Phi_{\mathrm{ol}} = I \frac{\pi d^2}{4R_e^2}, \tag{4.11.34}$$

where d is the diameter of the eye pupil. It follows that observation with a telescope allows the detection of a star with identical intensity I at a larger distance given by the ratio

$$\frac{R_t}{R_e} = \frac{D}{d}\sqrt{\tau}. \tag{4.11.35}$$

- *Spatially coherent light beams.*
In the case of a Gaussian laser beam the laws of photometry are not applicable because the dimension of the beam waist of a laser beam is determined by diffraction. One can calculate that the product of the surface area × solid angle, equal to the throughput T of a laser beam, equals

$$T = S\Omega = \lambda^2. \tag{4.11.36}$$

In an analogous way as for the expression for a radiant flux,

$$\Phi = LS\Omega = LT, \tag{4.11.37}$$

one attributes a radiance value to a (monomode) laser beam by dividing the laser power P_0 by the throughput value T. In Section 1.7, the relationship between the natural divergence angle θ of a lowest order Gaussian beam and its beam waist w has been given (see Eq. (1.7.20)). For small values of θ, the solid angle Ω subtended by the Gaussian beam is given by $\pi\theta^2$. It follows from Eq. (1.7.20) that $T = S\Omega = (\pi w^2)(\pi\theta^2)$ and equals λ^2 ($2w$ is the beam waist diameter of the laser beam). It is not permissible, however, to infer from this relation the irradiance in the plane of the beam waist image when a laser beam cross-section is imaged by an optical system. To this purpose, the paraxial imaging laws for Gaussian laser beams should be used as they are given in Section 4.10.7.

5 Aberration Analysis of Optical Systems

The first steps in systematic analysis of optical imaging go back to the beginning of the sixteenth century when Kepler published the paraxial imaging equation for an optical surface or a lens. In parallel, with the advent of well-fabricated telescopes and microscopes in the seventeenth and eighteenth centuries, scientists and engineers became interested in the theoretical limitations to sharp image formation. The focus was first on the limitations in sharpness when imaging on the optical axis of an instrument, for instance when using a telescope for astronomical observation. Expressions for the on-axis image blur were developed, among others, by Huygens. Later, in microscopy, the notion of field aberration was developed and lens configurations were devised that minimise this aberration and allow a larger image field with sufficient sharpness. The advent of photography halfway through the nineteenth century gave rise to the first systematic approach to a theory of aberrations in optical imaging. The names of the mathematicians Petzval and von Seidel are associated with this breakthrough in image analysis based on ray optics. In a short period of time, some 20 to 30 years, a whole class of high-performing photographic lens systems was developed by the existing and newly founded optical industries of that time. The systematics of aberration theory were also applied in microscopy, and by the end of the nineteenth century these had already given rise to high-quality optical microscopes. These instruments have been highly beneficial for the advancement of other natural sciences.

In this chapter we first present the history of aberration analysis in optical imaging, with reference to some seventeenth century propositions for aberration correction. The next subject is a classification of aberration types. We then proceed to a study of the so-called 'third-order' Seidel aberrations which play a very important role in aberration theory. The expressions for the Seidel aberration coefficients of an optical surface are given in the domain of wavefront aberration and for a single wavelength (monochromatic aberrations). Further subjects in this chapter are the aberration coefficients of a single lens and the colour dependence of paraxial properties and aberrations. An efficient and stable representation of wavefront aberration is obtained with the aid of the Zernike circle polynomials. Finally, for aberrational analysis of an optical system at arbitrary aperture and field angles, we discuss higher-order aberration theories. We conclude the chapter with a review of ray-tracing methods for various types of optical surfaces (spherical, aspheric, diffractive optical elements) and for light propagation in inhomogeneous media and in anisotropic media.

5.1 Introduction

The paraxial optics approximation of ray propagation, refraction and reflection at a surface uses the linearised form of Snell's law, $ni = n'i'$, where i and i' are the angles of incidence and refraction at a surface (see Section 4.10). For larger angles of incidence, successive approximations of higher order could be applied, for instance, by the insertion of the third-order approximation for the sine function,

$$n\left(i - \frac{(i)^3}{6}\right) = n'\left(i' - \frac{(i')^3}{6}\right) . \tag{5.1.1}$$

If an increased accuracy in refraction angle is used as in Eq. (5.1.1), the coordinates of intersection points with an optical surface should also be calculated with increased accuracy, in this case up to the fourth order in the lateral coordinates

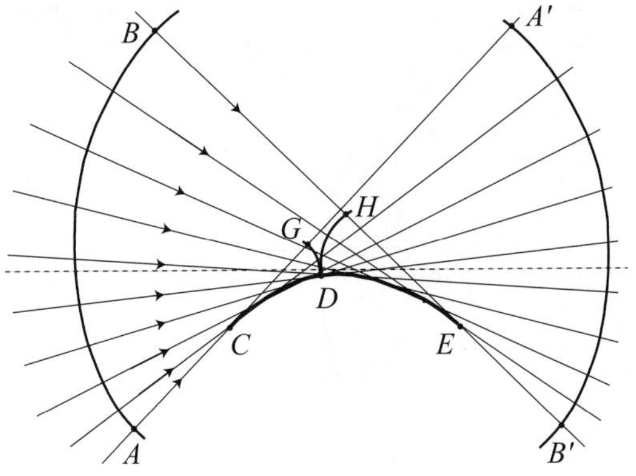

Figure 5.1: A pencil of rays traverses an aberrated focal region. The wavefront of the incident pencil is denoted by *AB*. Curve *CDE* is the evolute of curve *AB*, the envelope of its centres of curvatures or, in optical terms, the 'caustic' curve or in three dimensions the 'caustic' surface. *GDH* is a cusped wavefront in the focal region of the pencil.

x and y. Distances measured along rays, obtained from a square-root expression, should equally be developed up to the fourth order in the x- and y-coordinates of the starting point and endpoint on a ray. Aberration theories of a certain order are based on the approximation of ray angles and angles of incidence to this order and the approximation of optical distances to the same order plus one. The most widespread aberration theory is the third-order aberration theory. Of third order in the angles involved but of fourth order in terms of distances and wavefront deviation, the third-order aberration analysis of optical systems will be the main topic of this chapter.

5.1.1 Early Approach to Aberration Analysis

Before presenting the third-order theory of aberrations, dating back to the middle of the nineteenth century, we discuss some older examples of aberration analysis. In Fig. 5.1 a focused beam of light is represented by means of a set of rays and the curved wavefront through *A* and *B*, perpendicular to these rays. In the figure, the curved line *CDE* represents the locus of centres of curvature of the wavefront *AB*. Or, alternatively, *CDE* is the envelope of the normals to the wavefront. In differential geometry, the space curve *CDE* is called the evolute of the curve *AB*. Having complete knowledge of the curve *CDE*, we are able to reconstruct the wavefront *AB*. This is done by constructing the involute (or evolvent) of *CDE*. In practice, using a taut flexible rope, attached to the point *E*, touching everywhere *EDC* and having a length *EDCA*, we can trace the wavefront *AB* by unwinding the rope from *CDE*. The point *A* is obtained by having the rope exactly leave the evolute in the direction of the tangent to curve *CDE* at the point *C*. In a comparable way, all other points are obtained, with, finally, the point *B* when the rope is tangent to curve *CDE* in the extreme point *E*. The wavefront *A'B'* follows from a construction with the taut rope attached to the point *C*, giving it the length *CDEB'* and then gradually unwinding it from *CDE*. A wavefront in the intermediate region, where rays do intersect each other, needs a two-tier approach. At point *D* of the figure, the portion *GD* of the wavefront belongs to rays that diverge from each other. It is obtained by unwinding a rope section *CD* with its end attached to point *C*. The second portion of the wavefront, belonging to a still converging part of the pencil of rays, is obtained by unwinding the rope *ED*, attached at point *E*. The wavefront portions *DG* and *DH* touch each other at *D* where they form a wavefront cusp. All points on the evolute *CDE* are associated with wavefront cusps. The interest of the optics community in the wavefront evolute stems from its visibility in a practical experiment. The evolute forms the so-called caustic surface, from the Greek adjective *καυστικός* = *hot, burning*. All rays are tangent to the caustic surface and an increased optical intensity is observed on this space curve.

An interesting analysis of the creation of the caustic surface of a parallel beam which is incident on a hollow reflecting half-sphere is given in **[152]** (see Fig. 5.2a). The construction of the points of the caustic curve is further illustrated in Fig. 5.2b. A general ray from the parallel beam of light intersects the mirror at the point *D*. The angle of incidence on the mirror is $\pi/2 - \alpha$. The reflected ray is found by making the arc *DE* twice as large as the arc *BD*, yielding the point *E*, with $\angle DCE$ equal to 2α. Two neighbouring incident rays with height difference dy intersect each other after reflection at the point *G*. The point *G* divides the reflected ray segment *DE* in portions *DG* and *GE*, such that $DG/GE = 1/3$. This ratio

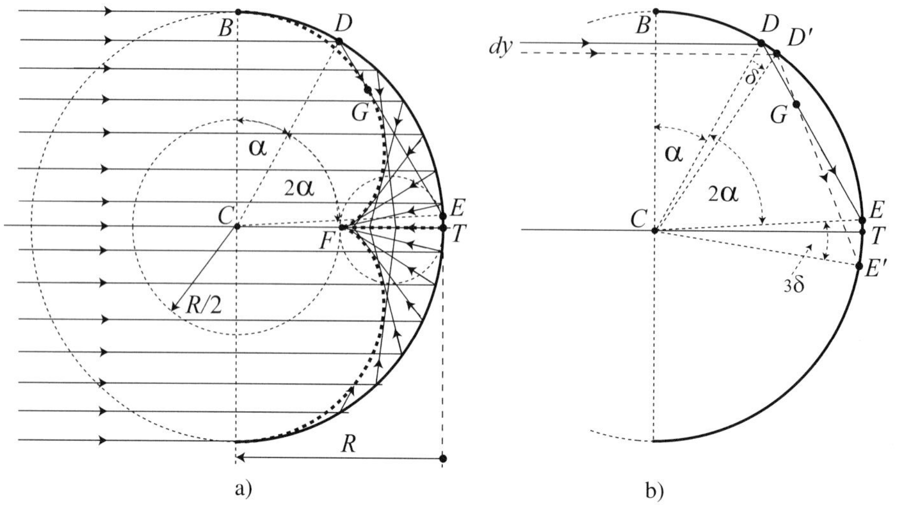

Figure 5.2: a) The reflection of a parallel beam of light by a spherical mirror and the formation of the caustic surface in the focal region (after Huygens **[152]**).
b) The construction of a point G on the caustic curve.

is deduced from Fig. 5.2b which shows that the angles $\angle DD'G$ and $\angle E'EG$ are both equal to $\pi - \alpha - 3\delta/2$. The triangles $DD'G$ and $E'EG$ are thus similar and their corresponding sides have the proportion of $DD'/E'E$. This quantity equals $1/3$ for the infinitesimally small difference in height of incidence dy of the two neighbouring rays. The set of points G that is produced by all aperture rays with $|y| \leq R$ yields the dashed curve FGB in Fig. 5.2a. It was shown by Huygens that this curve is an epicycloid, obtained by rolling the small circle through F and T with radius $|R|/4$ over the circle through F, with radius $|R|/2$ and centre of curvature C. On purpose, we use the absolute value of the radius of curvature since, according to our geometrical-optics sign conventions, the distance $R = 2TF$ has a negative value in Fig. 5.2 (F is the focal point of the spherical mirror). The total caustic curve includes the linear section FT. The complete space curve is obtained by rotating the planar graph of Fig. 5.2a with respect to the horizontal axis through C and T and this causes an infinite number of rays to intersect on the line section FT. As mentioned by Huygens, the surface that is created by rotating the curve FGB can be clearly observed when the sun is shining into a deep hollow mirror and some smoke or fine dust is blown into the reflecting cavity. Figure 5.3 illustrates this phenomenon.

In Fig. 5.4 we have drawn again the reflected pattern of rays associated with evolute curve DFB, the locus of the centres of curvature of the reflected beam. The shape of DFB being an epicycloid, differential geometry predicts that the involute or evolvent is also an epicycloid. We geometrically construct the involute using the method of Fig. 5.1. Using analytic geometry, the Cartesian coordinates of evolute and involute as a function of the parameter t are given by

$$
\begin{aligned}
x_e(t) &= (R_b + R_r)\cos(t) - R_r \cos\left\{\frac{R_b + R_r}{R_r}\, t\right\} ,\\
y_e(t) &= (R_b + R_r)\sin(t) - R_r \cos\left\{\frac{R_b + R_r}{R_r}\, t\right\} ,\\
x_i(t) &= \frac{R_b}{R_b + 2R_r}\left\{(R_b + R_r)\cos(t) + R_r \cos\left[\frac{R_b + R_r}{R_r}\, t\right]\right\} ,\\
y_i(t) &= \frac{R_b}{R_b + 2R_r}\left\{(R_b + R_r)\sin(t) + R_r \sin\left[\frac{R_b + R_r}{R_r}\, t\right]\right\} ,
\end{aligned}
\qquad (5.1.2)
$$

where (x_e, y_e) and (x_i, y_i) are the coordinates of evolute and involute, respectively. R_b is the radius of the base surface with the property $R_b = |R|/2$, and $R_r = |R|/4$ is the radius of the rolling sphere. The parameter t is the azimuthal angle of the contact point between the base sphere and the rolling sphere with radius R_r and varies from $-\pi/2$ to $+\pi/2$ for the evolute running from B via F to D. The same range of t applies to the azimuthal angle of a point on the involute curve.

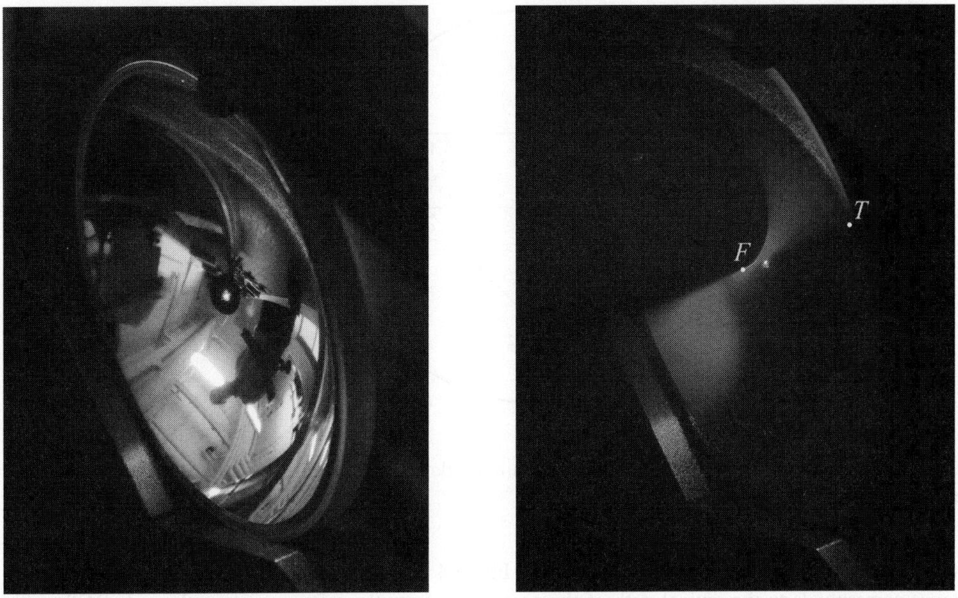

a) b)

Figure 5.3: Illustrating the caustic pattern of a spherical mirror.
a) The illuminating parallel beam (He–Ne laser) covers an upper quadrant of the mirror aperture.
b) The caustic pattern in the reflected beam has been visualised by means of light scattering in water vapour. The vertex of the mirror surface and the paraxial focal point have been indicated by T and F, respectively.

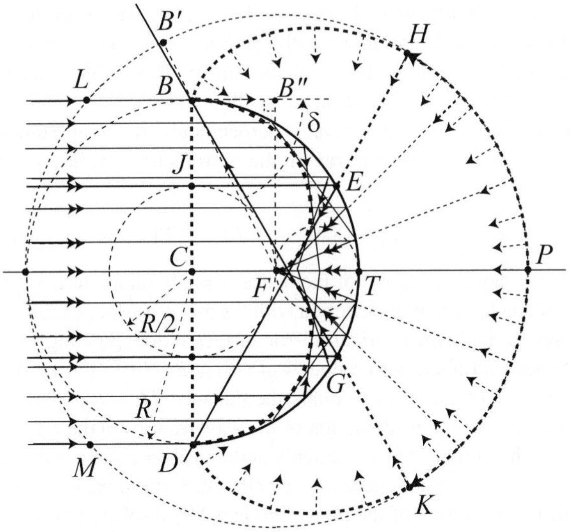

Figure 5.4: A parallel beam is incident on a spherical surface through T with radius R and having its centre of curvature at C. DFB is the evolute of the pencil of rays after a single reflection at the surface. The curve $DKPHB$ is the involute (evolvent) of the evolute DFB and represents the wavefront of the reflected beam after a single reflection. The reference wavefront is the sphere through $MPB'L$.

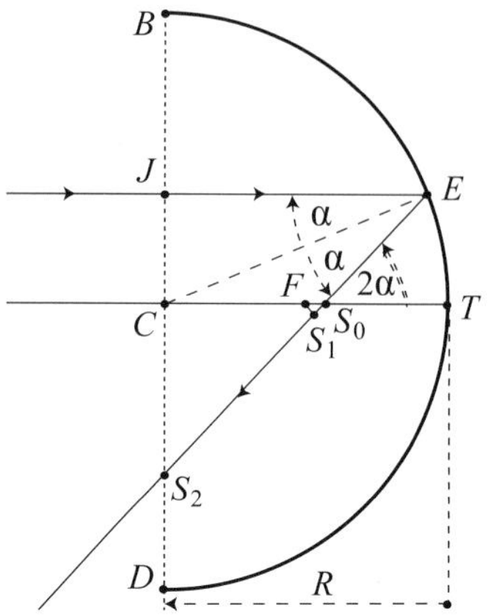

Figure 5.5: Calculation of wavefront aberration using Hamilton's definition of pathlength difference with respect to a chosen reference point (the paraxial focal point F in the figure).

The central point of the epicycloid involute curve in Fig. 5.4 is P, at a distance $PF = 3R/2$ from the focal point F. This distance corresponds to the optical path of the central ray from the reference plane through DCB to T and back to the focus F. The most extreme point of the involute is B and the total curve $DKPHB$ represents the wavefront of the reflected beam (after one single reflection). The part of the reflected wave that belongs to incident rays outside the sphere with centre of curvature C and radius $|R|/2$ will leave the reflecting hemispherical mirror after more than one reflection. We observe in Fig. 5.4 that the part of the reflected beam that leaves the mirror after a single reflection, included within the section ETG on the mirror surface, has a wavefront deviation from the reference sphere which seems to be relatively small. The wavefront aberration can be calculated with the aid of Hamilton's definition of pathlength difference between rays (see Eqs. (4.9.15)–(4.9.17)). Figure 5.5 shows a general aperture ray JES_2, which is used to calculate the wavefront aberration function of an incident parallel beam of light. As reference ray we select the axial ray through the centre of curvature C, and the reference point for calculating the pathlength difference between the two rays is the paraxial focal point F at a (negative) distance $R/2$ from the vertex T of the mirror. The pathlength difference between the aperture ray and the reference ray yields the wavefront aberration value, $W_H = [JES_0S_1] - [CTF]$. The point S_1 is the intersection point with the aperture ray JES_2 of the line through F that is perpendicular to the aperture ray. The expression for W_H as a function of the angle of incidence α of the aperture ray on the mirror surface follows from planar geometry and reads, after some rearrangement,

$$W_H(\alpha) = [JE] + [ES_0] + [S_0S_1] - [CT] - [TF] = 4R \sin^4(\alpha/2) \ . \tag{5.1.3}$$

As an example we consider the ray having an angle of incidence $\alpha = \pi/6$ on the mirror surface. The maximum aperture point D then coincides with the second point of intersection with the mirror surface of this ray and $\pi/6$ is the limiting angle of incidence for single reflection on the hemispherical mirror surface (corresponding to an off-axis height of incidence $CJ = |R|/2$). Equation (5.1.3) yields an aberration value $W_H(\pi/6) = R(7/4 - \sqrt{3})$ or $0.01795\,R = 0.0359f'$. We note that the image-side focal distance f' of the mirror equals the distance $TF = R/2$ and that it is negative in the figure according to our sign convention. In Fig. 5.4 the deviation of the wavefront from the reference sphere at H or K is hardly perceptible. However, in terms of the number of wavelengths in the visible domain, the aberration amounts to thousands of wavelengths if the mirror has a radius of curvature of typically $|R| = 100$ mm. Using the wavelength of the light as a yardstick for wavefront aberration, the quality of the image produced by a spherical mirror of an infinitely distant object point is very poor.

Figure 5.4 shows that outside the single-reflection range, for instance on the portion HB of the 'virtual' wavefront after a single reflection, the wavefront deviation from a reference sphere becomes extremely large. To further illustrate the wavefront aberration W for a general ray outside the central part of the mirror with relatively small aberration, we abandon the Hamiltonian definition of aberration and consider the wavefront aberration with respect to an exit pupil sphere intersecting the point P on the involute HPK shown in Fig. 5.4. The wavefront aberration is now defined as

$$W = [OJEH] - [OCTFP] = [JE] + [EH] - [CT] - [TF] - [FP] = JE - HE \ , \tag{5.1.4}$$

where O is the infinitely distant object point. The wavefront aberration $W = JE - HE$ equals the optical path from the planar wavefront through $BJCD$ to the sphere with its centre at F and with radius $3|R|/2$. We note that all optical distances are measured positively along the direction of propagation of the rays. This definition of W is identical to the definition of wavefront aberration in Subsection 4.9.2. An extreme situation arises when, for instance, the wavefront aberration of the outer ray along LB of the accepted pencil is considered. The value of W (after one reflection) is now given by the distance BL, a negative quantity with respect to the propagation direction of the reflected ray that, in this limiting case, propagates towards B''. The length of BL equals $|R|(\sqrt{5}-1)/2 = 0.618|R|$. There is an enormous angular aberration, $\delta = \angle B''BF$, between the reference ray BF and the 'reflected' ray BB'', the prolongation of LB at grazing incidence on the mirror. The expression for the wavefront aberration of Subsection 4.9.2 assumed that the pathlength $[BB']$ measured along the reference ray and the pathlength $[BL]$ measured along the ray itself are approximately equal. This approximation is valid up to the second order in the aberration angle δ. It turns out that this approximation is too crude for the outer incident rays. Using the definition of wavefront aberration based on Hamilton's angle characteristic function, useful reference points are C for the object pencil and F for the reflected rays of the imaging pencil. For the outer incident ray through B, the wavefront aberration is given by $[BB''] - [CTF] = -|R|$, exactly in line with the Hamiltonian expression for wavefront aberration of Eq. (5.1.3) when $\alpha = \pi/2$. This value is far different from the value of $[BB']$. The wavefront aberration based on Hamilton's characteristic function has a residual error of $(1 - \cos\delta)R_1$ (see Eq. (4.9.16)), where $\delta = \arccos(1/\sqrt{5})$ is the angular aberration of the outer ray and R_1 is the radius of the chosen reference sphere ($R_1 = 3R/2$ in our example). The numerical example shows that Hamilton's definition becomes unreliable in the presence of such an extreme aberration.

Instead of wavefront aberration, the transverse aberration in a plane through F, perpendicular to the line CT, can be measured. Another possibility, only relevant for beams originated from an axial object point, is the measurement of the *axial* aberration, the distance from the intersection point of a ray and the line segment FT to the focal point F itself. Although the calculation or measurement of the caustic surface has a physical significance by providing the observer with the enhanced-intensity surface in the focal region, its use in optical aberration analysis is very limited. More relevant is the transverse aberration. In the case of relatively high-quality optical instruments, for instance camera lenses, study of the transverse aberration immediately provides the optical designer with the geometric image blur and the imaging quality of the instrument. For very refined instruments like a microscope objective or a telescope, the lens design should yield a virtually perfect imaging quality. In that case, the wavefront aberration is the preferred quantity for judging the imaging performance of the instrument.

5.1.2 Early Approaches to Aberration Correction

Knowing the defects of optical imaging, efforts have been made to correct for these imaging defects and to compensate for the optical aberration. Stigmatic imaging by aspheric reflecting surfaces was already known in antiquity, for instance by using a paraboloid for imaging the infinitely distant object or an ellipsoidal for object and image planes at finite distances. An analysis of stigmatic imaging by a single refracting surface was first carried out by Descartes.

In Fig. 5.6 a general ray QO_0 is shown of the pencil of rays that is incident at the object point O_0, virtual in the figure. A stigmatic image should be formed at O_1. Given the distances s_0 and s_1 and the refractive indices n_0 and n_1 of object and image space, the optical pathlength equality between O_0 and O_1 for a general aperture ray through Q and the axial ray reads

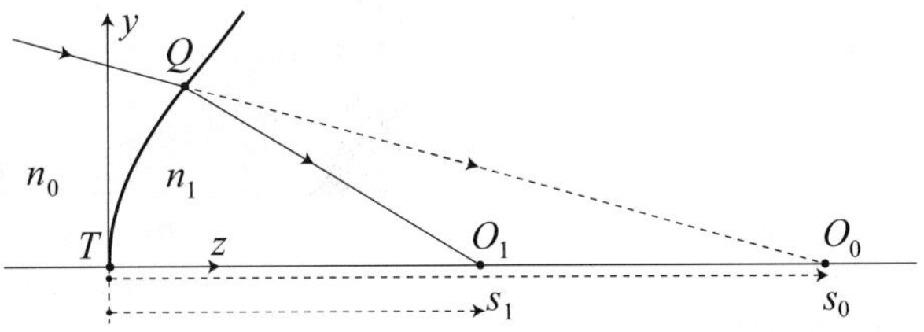

Figure 5.6: A non-paraxial incident beam with finite rays is focused at the virtual object point O_0, at a distance s_0 from the vertex T of the optical surface. The real image is formed at O_1, at a distance s_1 from T. The refractive indices of object and image space are n_0 and n_1, respectively.

$$[O_0QO_1] = [O_0TO_1]\,, \quad \text{or,} \qquad -n_0QO_0 + n_1QO_1 = -n_0TO_0 + n_1TO_1\,. \tag{5.1.5}$$

Filling in the optical parameters and the Cartesian coordinates of the general point $Q(y, z)$, we find after some lengthy algebra

$$\left[(n_1^2 - n_0^2)(y^2 + z^2) + 2(n_0^2 s_0 - n_1^2 s_1)z\right]^2$$
$$= 4n_0n_1(n_1s_1 - n_0s_0)(n_0s_1 - n_1s_0)\left\{\left(z + \frac{(n_1 - n_0)s_0s_1}{n_0s_1 - n_1s_0}\right)^2 + y^2 - \left(\frac{(n_1 - n_0)s_0s_1}{n_0s_1 - n_1s_0}\right)^2\right\}. \tag{5.1.6}$$

An expression of this type was derived by Descartes and the corresponding curves, following from this 'quartic' equation in y and z, are called Descartes' oval surfaces. As a consequence of the fourth-order dependence of the pathlength equation (5.1.6) with the general form $[f(z, y)]^2 = g(z, y)$, the solution corresponds to two curves, given by

$$f(z, y) - \sqrt{g(z, y)} = 0\,, \text{ and,} \qquad f(z, y) + \sqrt{g(z, y)} = 0\,. \tag{5.1.7}$$

Only the curves that intersect the point $T(0, 0)$ have been shown in Fig. 5.7. The first example in a) has a limited opening angle of the focusing beam due to the incident rays missing the lens body beyond a certain maximum angle. In b) the incident beam in the lens medium is produced by some other optical system and can have a larger than hemispherical solid angle.

An interesting case arises when the right-hand side $g(z, y)$ of Eq. (5.1.6) becomes zero. In that case the fourth-order dependence is lost and we simply have to solve the quadratic equation $f(z, y) = 0$. The right-hand side is made zero by choosing:

- $n_1s_1 = n_0s_0$.
  Using the definition of transverse magnification $m_T = (s_1/n_1)/(s_0/n_0)$ we find that $m_T = n_0^2/n_1^2$. The oval curve now reduces to

$$\left(z - \frac{n_0s_0}{n_0 + n_1}\right)^2 + y^2 = \left(\frac{n_0s_0}{n_0 + n_1}\right)^2, \tag{5.1.8}$$

  which is a circle with its vertex at T and having a radius of $n_0s_0/(n_0 + n_1)$. In Fig. 5.8 we show the two stigmatic imaging cases with the spherical surface. Figure 5.8a represents the stigmatic imaging case that was proposed by

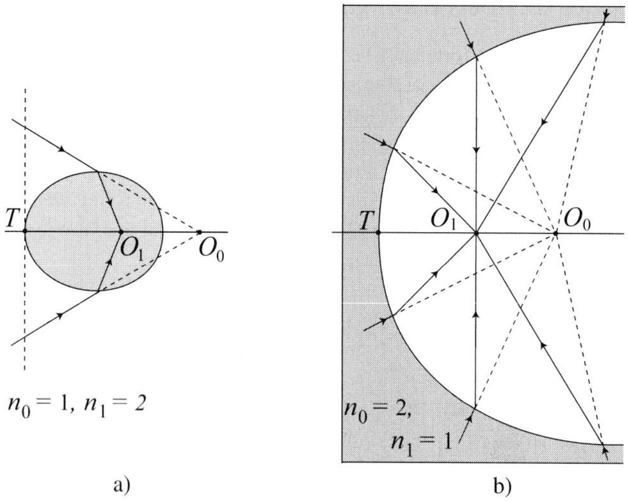

Figure 5.7: Examples of Descartes' oval surfaces.
a) $s_0 = 2$, $s_1 = 1.2$ (arbitrary units). The refractive index of the object space is unity, the lens material (grey in the figure) has an index n_1 of 2.
b) The object and image distances have the same values as in a), but now the object space has the high index value of 2 and the image space is filled with air.

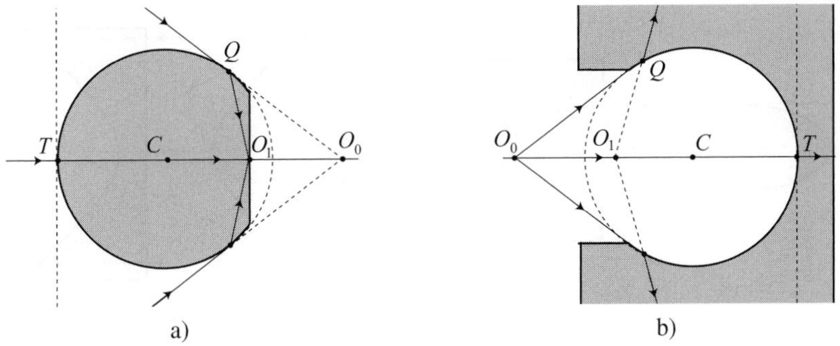

Figure 5.8: The special case of an oval surface that reduces to a sphere.
a) s_0 and s_1 are positive. A spherical cap has been taken away from the glass sphere to give access to the image point O_1.
b) s_0 and s_1 are negative. The glass volume has been shaded in the figure; the refractive index of the lens material equals 2.

de Roberval and Huygens. The geometrical construction of the so-called pair of aplanatic points, O_0 and O_1, has already been given in Section 4.6. This aplanatic imaging with the aid of a spherical surface has found a widespread application, for instance in the design of microscope objectives. Around 1840, it was Amici who first introduced the aplanatic hemispherical (or super-hemispherical) lens as the front lens of a microscope objective. The ray directions in his front lens were inverted with respect to the left-hand graph and offered a pencil of rays to the remaining part of the objective that was reduced in angular extent by a factor of $1/n_1$ where n_1 is the index of the hemispherical lens. A few years later, Amici introduced the principle of oil-immersion, obtaining a further advantage in microscope resolution. The big advantage of the spherical aplanatic lens with respect to the aspheric shape of the general Cartesian oval surface is, firstly, its relative ease of manufacturing. The second advantage is the point symmetry of a sphere and thus the compliance with the Abbe sine condition, whereas an asphere with symmetry of revolution has a symmetry *axis*. For the latter the aberration of an off-axis incident beam increases linearly with the angle of the principal ray of the beam with the axis of the aspheric surface. Equivalently, it can be shown that the Cartesian ovals and the quadratic conic sections treated in this section do not satisfy Abbe's sine condition (see Section 4.8). This makes these surfaces scientifically interesting but less apt to widespread use. An exception should be made for their use in illumination applications where the aberration requirements are much less stringent. Figure 5.8b shows another example of stigmatic imaging with the aid of a spherical surface. In this case, the object is real, the image is virtual.

- $n_0 s_1 = n_1 s_0$.
Using the paraxial imaging equation, this condition yields $s_0 = s_1 = 0$. The object and image are found in the plane through the vertex of the surface. This pair of conjugate planes, trivial at first sight, was also found in Section 4.6, using the angle characteristic function for imaging analysis. For $s_0 = s_1 = 0$, Eq. (5.1.6) leaves the shape of the oval surface indeterminate.
- Conic sections for reflecting surfaces.
Inspection of Eq. (5.1.6) shows that the fourth-order degree of the surface equation is lost when $n_1 = \pm n_0$. The non-trivial solution for $n_1 = -n_0$ applies to a reflecting surface and the resulting curve reduces to a conic section. The well-known mirror surface solutions are the parabolic mirror for imaging an infinitely distant object and the ellipsoid for imaging an object in the first focus to the second focus of the ellipsoid and vice versa. Object and image are real, located at the same side of the mirror surface. When either the object or the reflected image is virtual, one of the branches of a hyperboloid is needed for stigmatic imaging.
- Conic sections for a lens surface.
When $n_1 \neq n_0$ for the case of imaging with a lens, the quadratic form is found when either s_0 or s_1 becomes infinitely large. If $s_0 \rightarrow \infty$, Eq. (5.1.6) is written as follows,

$$\left(z - \frac{n_1 s_1}{n_1 + n_0}\right)^2 + \frac{n_1^2 y^2}{n_1^2 - n_0^2} = \left(\frac{n_1 s_1}{n_1 + n_0}\right)^2 . \qquad (5.1.9)$$

Possible lens shapes of plano-convex or plano-concave singlet lenses are shown in Fig. 5.9. In a) we have that $n_1 > n_0$ and both conic sections are ellipses. In the extreme left drawing, a solid glass ellipsoid is shown that has been sectioned perpendicularly to the axis of the ellipsoid such that the corresponding planar exit surface contains the second focal

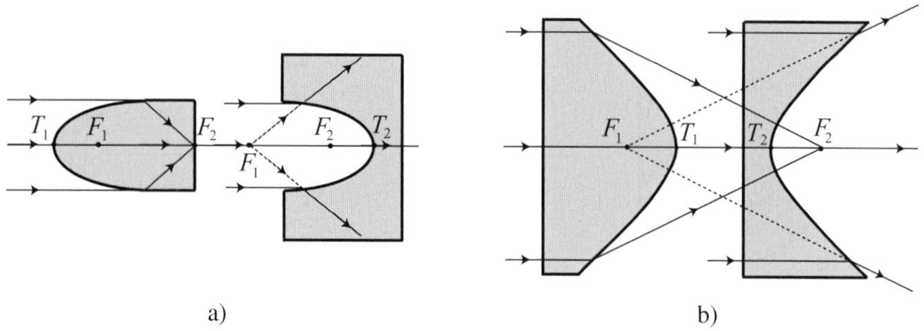

Figure 5.9: Conic sections for correcting the shape of a single lens when illuminated from infinity (positive and negative lens, index $n = 2$).
a) Elliptical cross-sections (both lenses can be shifted over an arbitrary distance).
b) Hyperbolic cross-sections (both lenses can be independently used but are shown spatially related in the figure by the two hyperbola branches and the associated focal points, F_1 and F_2).

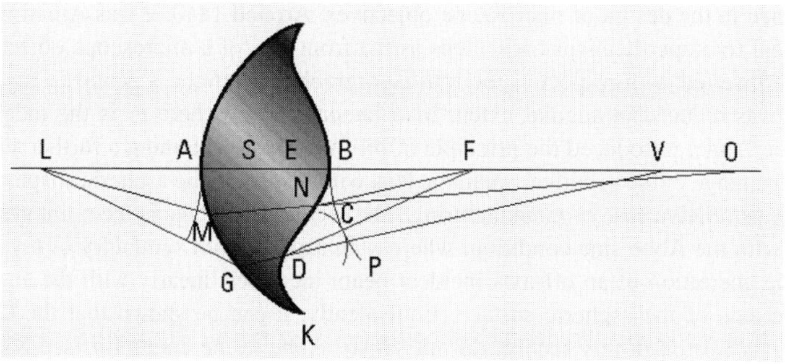

Figure 5.10: Aberration analysis and correction for a single lens according to Huygens **[152]**. The original lettering from his manuscript has been used in the drawing.

point F_2. The stigmatic image of the infinitely distant axial object point is found in F_2. The other part of the same elliptical curve has its vertex at T_2 and is used as a negative lens. The incoming parallel beam, parallel to the axis of the ellipsoid, becomes divergent in the lens medium and F_1 is the virtual image point. In Fig. 5.9b we have that $n_1 < n_0$ and the two aspheric surface cross-sections are derived from the two branches of a hyperbolic conic section. The optical power of the lenses is located on the back of each lens. The first example uses the left-hand branch of a hyperbola for imaging the parallel beam to a stigmatic point at the second focus F_2 of the hyperbola (the second negative hyperbolic lens does not play a role in this imaging step). If a negative lens is made with the aid of a hyperbolic aspheric shape, we obtain the most right-hand lens. This negative lens receives a parallel beam that leaves the negative optical surface with a stigmatic virtual image in F_1, the focus of the first branch of the hyperbola. As mentioned already for the aspheric mirror surfaces using conic sections, the imaging quality of an obliquely incident parallel beam is bad. The Abbe sine condition can generally not be satisfied by lenses having such aspheric shapes.

5.1.3 The Construction of a General Asphere According to Huygens

Huygens extensively studied the spherical aberration of a single surface. In his Traité de la Lumière **[152]**, he also discusses the spherical aberration of a lens and develops a scheme to make one surface of a biconvex lens aspheric in order to reduce the spherical aberration to zero with the lens having finite conjugate distances. He proposes asphericising the shape of the second surface of a lens (see Fig. 5.10). The source point is at L and a general ray from the source point hits the first surface at the point G. The position of the point M on this ray is such that it is on the spherical wavefront

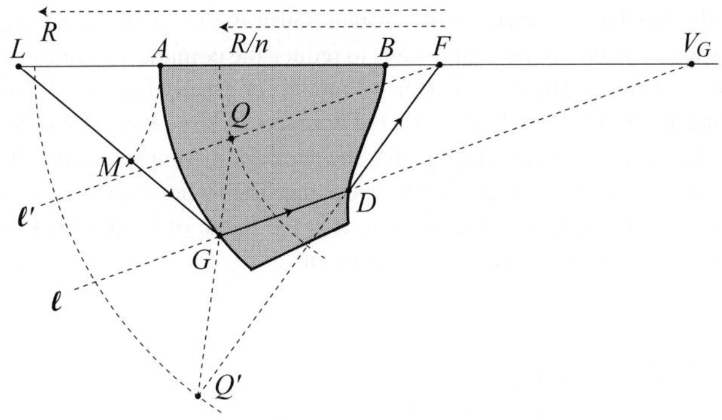

Figure 5.11: Geometrical construction of a general point D of the second lens surface through B and D to correct the spherical aberration of a single lens (this geometrical construction was brought to the attention of the first author by G. Bouwhuis).

through the vertex A of the first surface. The ray is refracted at G and intersects the optical axis, still propagating in the lens medium with index n, at the point V. The imaging in the intermediate lens space suffers from spherical aberration, demonstrated in Fig. 5.10 by the intersection point O with the optical axis of a ray through N, leaving L at a smaller angle. The aberrated beam in the intermediate lens space has to be stigmatically imaged at the point F in image space. Not surprisingly, for Huygens, the advocate of the wave theory of light, the condition of stigmatic imaging is pathlength equality between a general aperture ray and the axial ray. He states that a geometrical construction can be used, based on equal pathlength for all rays connecting L and F, to find the required positions of the points on the second lens surface.

Such a geometrical construction for a general point D on the second surface that forces the refracted ray to pass through the prescribed stigmatic image point F is illustrated in Fig. 5.11. A particular ray LMG is refracted and intersects the optical axis at the point V_G. The axial thickness of the lens, AB, and its refractive index n are given. Pathlength equality of the axial ray and the aperture ray requires

$$[AB] + [BF] = [MG] + [GD] + [DF]\ , \quad \text{or,} \quad nAB + BF = MG + nGD + DF\ . \tag{5.1.10}$$

We define the quantity $R = nAB + BF - MG$, a known optical distance for the specific ray from L through G. To find the point D on the ray GV_G, we have to satisfy the requirement

$$n\,GD + DF = R\ . \tag{5.1.11}$$

In the figure, we draw an auxiliary line l' through F, parallel to GV_G. We also draw two auxiliary circles, one with radius R and one with radius R/n, both with their centre of curvature at F. The intersection point of the small circle with l' is denoted by Q. We draw the line QG, intersecting the large circle in Q', and the line $Q'F$ that intersects the line l in D. From the similar triangles $\triangle QQ'F$ and $\triangle GQ'D$ we obtain the expression

$$GD = \frac{QF\,.\,DQ'}{FQ'} = \frac{DQ'}{n}\ . \tag{5.1.12}$$

Using $DQ' + DF = R$, it then follows that $n\,GD + DF = R$ and that D is the requested point on the second surface.

The purely geometrical construction with ruler and compass would be insufficient to attain the required positional accuracy of the points on the second surface. Analytic geometry could have produced a more accurate description of the second lens surface. But even then, the seventeenth century glass-working technology would have been inadequate for producing aspheric surfaces with the required accuracy. For some two hundred years, spherical surfaces were the only surfaces that could be manufactured with the accuracy required for use at optical wavelengths. It was only in the twentieth century that a more general use of aspheres became possible, still limited to prototyping or small series fabrication. Famous examples are the large-size aspheric telescope mirrors with sub-micron shape accuracy for stellar observation, dating back to the beginning of the twentieth century. Another breakthrough in aspheric technology was achieved at the end of that century, when mass-production of aspheric singlet and compound lenses for optical disc players was technologically mastered.

We conclude this short historic introduction by briefly mentioning some other early achievements in optical aberration analysis and aberration correction. The microscope was improved by reducing the chromatic errors of the components.

Huygens proposed and manufactured his eyepiece with constant power in the visible spectrum. Some hundred years later, the achromatic doublet lens was invented. Lister found ways to reduce the comatic aberration of the microscope objectives by adapting the radius of the cemented surface of a doublet. Fraunhofer greatly improved the manufacturing of glass and was able to design and produce high-quality colour-corrected doublets of large size. But it was not until halfway through the nineteenth century, with the advent of photography, that the need for a comprehensive theory of imaging aberrations was urgently felt. The mathematicians Petzval and Seidel were the first to develop such an aberration theory. Petzval's results were never published and were lost. For that reason, it is the name of Seidel **[301]** that is nowadays associated with the 'third-order' aberration theory of a general optical system.

5.2 Classification of Aberrations

Aberrations are best studied in the domain of transverse aberrations or in the domain of the wavefront aberration. Historically, the transverse aberrations were first to be analysed. The advent of high-quality imaging systems made the concept of wavefront aberration more relevant. In this section, we focus on the wavefront aberration of an imaging pencil. If needed, a direct switch from wavefront to transverse aberration components is possible by employing Eq. (4.9.14) for the real-space coordinates (x, y, z) or Eq. (4.9.28) for the normalised pupil and image plane (or field) coordinates. In Fig. 5.12 the reference sphere has been shown for measuring the wavefront aberration of an off-axis pencil of rays, associated with the paraxial image point A with coordinates $(x, y)_A$ (the origin is chosen at O_1). Given the paraxial magnification factors, the position of A is directly related to the position of the object point in the object plane, not shown in the figure. A general ray of the imaging pencil intersects the reference sphere at a point $P(X, Y, Z)$, where Z follows from the equation for the spherical reference surface. The origin for the (X, Y, Z) coordinate system is the centre E_1 of the reference sphere. The Z-coordinate of P is given by

$$Z = \frac{X^2 + Y^2 - 2Xx - 2Yy}{R_1\left[1 + \{1 - (X^2 + Y^2 - 2Xx - 2Yy)/R_1^2\}^{1/2}\right]}, \tag{5.2.1}$$

where $R_1 = E_1 O_1$ is positive in Fig. 5.12.

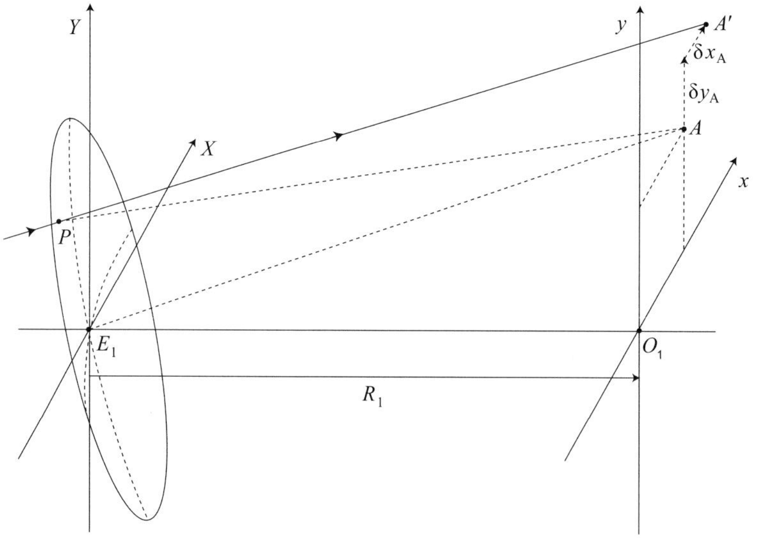

Figure 5.12: A ray intersects the reference sphere through E_1 in the exit pupil at the point P with coordinates $(X, Y, Z(X, Y))$. The reference ray through P intersects the image plane at $A(x, y)$. The aberrated ray intersects the image plane at the point A' with coordinates $(x_A + \delta x_A, y_A + \delta y_A)$. The distance $E_1 O_1$ from the centre of the exit pupil to the centre of the image plane is R_1, positive in the figure. $PA = E_1 A$ is the radius of the reference sphere for the oblique imaging pencil.

5.2.1 Number of Aberration Terms

The wavefront aberration of a general ray PA' belonging to the pencil aimed at the paraxial image point A is now written as a power series expansion according to

$$W(X,Y;x,y) = \sum c_{ijkl}\, X^i Y^j x^k y^l \,. \tag{5.2.2}$$

We are interested in the number of terms N_n^m of a certain total degree m in X, Y, x and y of this expansion,

$$i+j+k+l = m\,, \tag{5.2.3}$$

where $i, \cdots, l, m$ are all non-negative integer numbers and n, the number of independent coordinates or combinations thereof, is equal to 4 in our case. We have to solve a standard combinatorial problem with the solution

$$N_n^m = \frac{(n+m-1)!}{m!\,(n-1)!} = \binom{n+m-1}{m}\,. \tag{5.2.4}$$

The following number of combinations is found for various m values:

m	0	1	2	3	4	5	6
N_4^m	1	4	10	20	35	56	84.

(5.2.5)

Each number N_4^m comprises terms that are of order zero in the exit pupil coordinates X and Y. These terms produce a wavefront aberration that is constant over the reference sphere and do not give rise to image degradation. Only an extra constant pathlength is added to each ray of the wavefront that cannot be detected by an intensity detector. The subtraction of these constant pathlength terms leads to a reduced number of aberration terms

$$N'^m_4 = N_4^m - N_2^m = N_4^m - (m+1) = \frac{m(m+1)(m+5)}{6}\,, \tag{5.2.6}$$

where the adapted number of aberration terms is given by

m	0	1	2	3	4	5	6
N'^m_4	0	2	7	16	30	50	77

(5.2.7)

This analysis is related to an optical system without a specific symmetry property. In practice, this is the case when sequences of cylindrical lenses, toroidal optical elements, conically shaped lenses, varifocal elements, etc. are used. Diffractive optical elements are examples of optical components that may also lack any symmetry. These elements are capable of imparting a virtually arbitrary change in direction to a ray that is diffracted at its surface. However, in the vast majority of optical systems, the surfaces have a spherical shape and the centres of curvature of these surfaces are located on a single axis, the optical axis of the system. In this case, a rotation of the optical system around its optical axis leaves the imaging properties unchanged. In Subsection 4.4.1 we have already analysed which coordinate combinations remain unchanged under a rotation with respect to such an axis of symmetry. If the symmetry axis is the z-axis, we have the following second-order and higher-order products of the quantities:

$$X^2+Y^2\,, \qquad x^2+y^2\,, \qquad Xx+Yy\,, \tag{5.2.8}$$

that occur in the expression for the aberration function. Using cylindrical coordinates according to

$$X=\rho\cos\theta\,,\;\; Y=\rho\sin\theta\,, \qquad x=r\cos\phi\,,\;\; y=r\sin\phi\,, \tag{5.2.9}$$

we find the following rotationally invariant combinations:

$$\rho^2\,, \qquad r^2\,, \qquad r\rho\cos(\theta-\phi)\,. \tag{5.2.10}$$

Because of the rotational symmetry of the system, it is the azimuthal difference between the point on the exit pupil sphere and the point in the image plane (field point) that is important. In many cases, we can put $\phi=0$ without loss of generality.

For an optical system with symmetry of revolution, we then have for the wavefront expansion of Eq. (5.2.2) the modified expression

$$W\{r^2,\rho^2,r\rho\cos(\theta-\phi)\} = \sum d_{ijk}(r^2)^i(\rho^2)^j[r\rho\cos(\theta-\phi)]^k\,. \tag{5.2.11}$$

The number of aberration terms of degree m is now given by

$$i + j + k = m/2\,, \tag{5.2.12}$$

and this yields the number $N_3^{m/2}$. The number of aberration terms that only depend on the pupil coordinates is then given by

$$N'^{\,m/2}_3 = N_3^{m/2} - N_1^{m/2} = \frac{m(m+6)}{8}\,, \tag{5.2.13}$$

and we find the following aberrations of various orders m:

$$\begin{array}{lcccccc} m & 0 & 2 & 4 & 6 & 8 & 10 \\ N'^{\,m/2}_3 & 0 & 2 & 5 & 9 & 14 & 20. \\ \text{name} & \text{piston} & \text{paraxial} & \text{Seidel} & \text{Schwarzschild} & & \end{array} \tag{5.2.14}$$

The number of aberration terms is greatly reduced by the presence of an optical axis in the system. Two terms have a purely second-order dependence on the pupil and image plane coordinates, proportional to ρ^2 and $r\rho\cos(\theta-\phi)$. These terms can be classified as paraxial aberration terms and they represent a focusing error or a magnification error with respect to the paraxial reference situation. The image points are still perfectly imaged albeit in an axially or laterally shifted position. For $m \geq 4$ we have real aberration terms. The five aberration terms with $m = 4$ are the monochromatic, 'third-order' aberration terms that were first analysed by Seidel **[301]**. The terms are of *third-order* when considering the dependence of the transverse aberration components $(\delta x_A, \delta y_A)$ on the pupil and image plane (or field) coordinates, of *fourth-order* in these coordinates if the wavefront aberration function W is considered. In the case of the aberration expression in polar coordinates, the relation between transverse and wavefront aberration is

$$\begin{aligned} \delta x_A &= \frac{R_1}{n_1}\left(\cos\theta\,\frac{\partial W(\rho,\theta)}{\partial\rho} - \frac{\sin\theta}{\rho}\,\frac{\partial W(\rho,\theta)}{\partial\theta}\right)\,, \\ \delta y_A &= \frac{R_1}{n_1}\left(\sin\theta\,\frac{\partial W(\rho,\theta)}{\partial\rho} + \frac{\cos\theta}{\rho}\,\frac{\partial W(\rho,\theta)}{\partial\theta}\right)\,. \end{aligned} \tag{5.2.15}$$

Instead of the coefficients d_{ijk}, we can adopt normalised coefficients, depending on the normalised radial pupil coordinate $\rho_n = \rho/\rho_{max}$ and the normalised radial field coordinate $r_n = r/r_{max}$. Here, $2\rho_{max}$ is the diameter of the exit pupil rim and $2r_{max}$ is the diameter of the image field to be considered. After further normalisation of the aberration coefficient with respect to λ_0, we have for the normalised aberration coefficients

$$d^{(n)}_{ijk} = \frac{r_{max}^{2i+k}\,\rho_{max}^{2j+k}}{\lambda_0}\,d_{ijk} = {}_{(2i+k)}\alpha_{(2j+k,\,k)}\,. \tag{5.2.16}$$

The normalised wavefront aberration expressed with the aid of normalised aperture and field coordinates is then written as

$$W_{\lambda_0}(r_n,\rho_n,\theta) = \sum_{i,j,k} {}_{(2i+k)}\alpha_{(2j+k,\,k)}\; r_n^{2i}\,\rho_n^{2j}\,\{r_n\rho_n\cos(\theta-\phi)\}^k\,. \tag{5.2.17}$$

The α-notation with the field subscript on the left-hand side and the pupil-related subscripts $2j+k$ (power of ρ) and k (power of $\cos(\theta-\phi)$) on the right-hand side is used quite often. It shows that the power of ρ is at least equal to or larger than by an amount $2j$, $j = 0, 1, \cdots$, the power of the factor $\cos(\theta-\phi)$. This property would also follow from a continuity requirement on the wavefront surface in the origin, at $\rho = 0$. An advantage of the α-coefficient notation is that the value of the coefficient immediately yields the maximum wavefront deviation in units of the wavelength for rays of a pencil directed to the border of the image field.

With the aid of the normalised pupil coordinate ρ_n we can calculate the normalised transverse aberration components in a similar way as in Subsection 4.9.4. Normalising these transverse components by the quantity λ_0/NA, we obtain an expression for the transverse aberration in terms of the so-called canonical coordinates in pupil and field. In terms of the normalised cylindrical pupil coordinates (ρ_n, θ), we then obtain the following expressions for the transverse aberration components:

$$\begin{aligned} \delta x_{A,n} &= \cos\theta\,\frac{\partial W_{\lambda_0}(r_n,\rho_n,\theta)}{\partial\rho_n} - \frac{\sin\theta}{\rho_n}\,\frac{\partial W_{\lambda_0}(r_n,\rho_n,\theta)}{\partial\theta}\,, \\ \delta y_{A,n} &= \sin\theta\,\frac{\partial W_{\lambda_0}(r_n,\rho_n,\theta)}{\partial\rho_n} + \frac{\cos\theta}{\rho_n}\,\frac{\partial W_{\lambda_0}(r_n,\rho_n,\theta)}{\partial\theta}\,, \end{aligned} \tag{5.2.18}$$

where the wavefront aberration W_{λ_0} is expressed in units of the vacuum wavelength of the light. The transverse components are expressed in the *lateral diffraction unit* $\lambda_0/(n_1 \sin\alpha_m)$, where α_m is the opening angle of the optical system at the image side. Each aberration coefficient ${}_a\alpha_{bc}$ contained in the function $W_{\lambda_0}(r_n, \rho_n, \theta)$ should be multiplied by a factor r_n^a to obtain the aberration function for a general value of the radial coordinate r in the image field.

The general expression for the transverse aberration components due to a wavefront aberration $W_{\lambda_0,abc} = {}_a\alpha_{bc} r_n^a \rho_n^b \cos^c(\theta - \phi)$, related to an image point A with paraxial coordinates $r_n(\cos\phi, \sin\phi) r_{max}$, is given by

$$\delta x_{A,n} = {}_a\alpha_{bc} r_n^a \rho_n^{b-1} \cos^{c-1}(\theta - \phi) \left[\frac{(b+c)}{2} \cos\phi + \frac{(b-c)}{2} \cos(2\theta - \phi) \right],$$

$$\delta y_{A,n} = {}_a\alpha_{bc} r_n^a \rho_n^{b-1} \cos^{c-1}(\theta - \phi) \left[\frac{(b+c)}{2} \sin\phi + \frac{(b-c)}{2} \sin(2\theta - \phi) \right], \qquad (5.2.19)$$

where $(\delta x_{A,n}, \delta y_{A,n})$ are again expressed in terms of the lateral diffraction unit in image space.

5.2.2 Analysis of Individual Aberration Terms

In this subsection we analyse the various aberration terms up to the order $m = 4$. We present graphs of the transverse aberration vector $(\delta x_n, \delta y_n)$ for a set of rays which have specific intersection points with the exit pupil sphere (see Fig. 5.13). Contour plots are given of the aberration function W_{λ_0} as a function of the polar coordinates (ρ_n, θ) on the unit circle in the pupil. All calculations are carried out in the normalised coordinate and aberration domain with the lateral and axial coordinates in image space given by

$$r_n = \frac{r}{\lambda_0/(n_1 \sin\alpha_m)}, \qquad z_n = \frac{z}{\lambda_0/(n_1 \sin^2\alpha_m)}. \qquad (5.2.20)$$

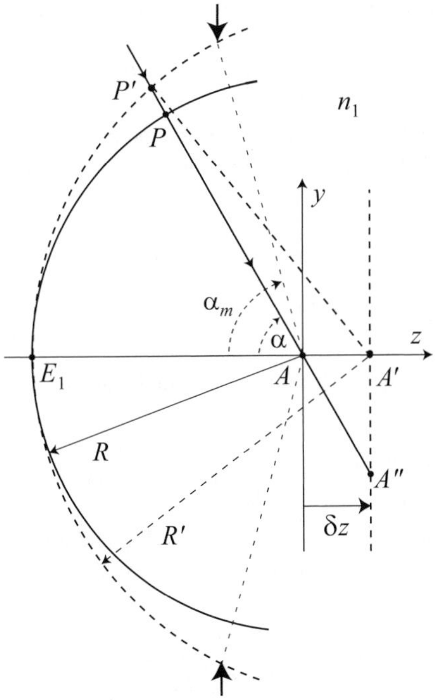

Figure 5.13: Wavefront aberration, transverse aberration and axial shift of the image point in the case of defocusing. The spherical wavefront in the exit pupil through E_1 has its midpoint at A and its radius of curvature is R. The reference sphere in the exit pupil through E_1 has its centre at A' and has a radius of curvature of R'. Both R and R' are negative in the figure; the focal shift $\delta z = AA'$ is given by $R - R'$. The maximum opening angle of the focusing beam is α_m. The refractive index of the image space is n_1.

The factor $\lambda_0/(n_1 \sin^2 \alpha_m)$ that is used to normalise the axial coordinate z is called the *axial diffraction unit*. It stems from a diffraction analysis of image formation and is discussed in Part III of this book. The coefficients of the wavefront aberration function are written as ${}_a\alpha_{bc}$ according to Eq. (5.2.16), where a is the power of the normalised radial coordinate r_n of a general point A in the image field, b is the power of the radial polar coordinate ρ_n of a point on the pupil sphere and c is the power of the azimuthal function $\cos(\theta - \phi)$.

5.2.2.1 Paraxial Defocusing Term

For $m = 2$ we have the wavefront deviation term

$$\begin{array}{cccc} \text{coefficient} & W_{\lambda_0}(\rho_n, \theta) & \delta x_n & \delta y_n \\ {}_0\alpha_{20} & \rho_n^2 & 2\rho_n \cos\theta & 2\rho_n \sin\theta\,. \end{array} \quad (5.2.21)$$

The functional dependence of the aberration function W_{λ_0} and the transverse aberration components $(\delta x_n, \delta y_n)$ of a defocused image point are given in the small table above. These functions have to be multiplied by the aberration coefficient ${}_0\alpha_{20}$, in units of the wavelength of the light. The normalised transverse aberration components are expressed in the diffraction unit in the image plane, $\lambda_0/(n_1 \sin\alpha_m)$, where α_m is the opening angle of the imaging pencils of rays. In practice, the opening angle of an oblique pencil of rays tends to decrease and its cross-section becomes non-circular. These effects, strongly dependent on the design details of an actual optical system, are not considered in what follows. The defocusing of the image plane can be varied by introducing an (extra) axial shift of the receiving plane in image space. This axial shift is given as a normalised axial distance z_n according to Eq. (5.2.20).

The exact relation between wavefront aberration and axial shift of the receiving plane is now explained with the aid of Fig. 5.13. The actual wavefront in the exit pupil is part of a sphere through E_1 and P, centred at A. The reference wavefront is related to the shifted point A' and has the radius $A'E_1 = R'$. Using our convention for the sign of wavefront aberration, the optical distance $n_1 PP'$ is the wavefront deviation W due to a shift $\delta z = AA'$ of the receiving plane from A to A'. Note that for a defocused spherical wavefront we prefer to speak about wavefront deviation instead of wavefront aberration, the latter term assuming a non-spherical shape of the wavefront. We apply the cosine rule to $\triangle P'AA'$,

$$(R')^2 = \left\{-R - \frac{W(\rho)}{n_1}\right\}^2 + (\delta z)^2 + 2\left\{-R - \frac{W(\rho)}{n_1}\right\} \delta z \cos\alpha\,, \quad (5.2.22)$$

where $\rho = R\sin\alpha$ is the (non-normalised) radial coordinate of the point P on the wavefront and α is the angle with the optical axis of a general ray PA. We solve the quadratic equation neglecting quadratic terms in δz and W to obtain

$$W(\rho) = n_1\, \delta z\, (\cos\alpha - 1)\,. \quad (5.2.23)$$

The expression for W can be expanded in powers of $\sin\alpha$ according to

$$W(\sin\alpha) = -n_1(\delta z)\left(\frac{\sin^2\alpha}{2} + \frac{\sin^4\alpha}{8} + \frac{\sin^6\alpha}{16} + \frac{5\sin^8\alpha}{128} + \cdots\right). \quad (5.2.24)$$

Using the normalised pupil coordinate $\rho_n = \sin\alpha/\sin\alpha_m$ and dividing the wavefront aberration by λ_0, the vacuum wavelength, we find

$$W_{\lambda_0}(\rho_n) = \frac{-n_1(\delta z)\sin^2\alpha_m}{2\lambda_0}\left(\rho_n^2 + \frac{\rho_n^4}{4}\sin^2\alpha_m + \frac{\rho_n^6}{8}\sin^4\alpha_m + \frac{5\rho_n^8}{64}\sin^6\alpha_m \cdots\right). \quad (5.2.25)$$

If the normalised axial shift δz_n is needed, the scaling according to Eq. (5.2.20) is used.

Equation (5.2.25) shows that the wavefront deviation from a sphere due to a defocusing is not simply described by a paraxial quadratic term with coefficient

$${}_0\alpha_{20} = -\frac{\delta z}{(2\lambda/\sin^2\alpha_m)} = -\frac{\delta z_n}{2}\,. \quad (5.2.26)$$

The quadratic term only creates a parabolic wavefront deviation in the presence of defocus. At larger values of the opening angle of the focusing beam, supplementary terms with power ρ_n^{2l} and $l > 1$ need to be included to approximate the spherical shape of the defocused wavefront. Although this suggests that the defocused beam is aberrated, this is not the case. A perfect point-like geometrical focus can be produced by shifting back the receiving plane to the focus at A over the normalised distance $-\delta z_n/2 = {}_0\alpha_{20}$. The defocus distance corresponding to the specific normalised distance $|\delta z_n| = 1/2$ is called the focal depth of the image, frequently also the *Rayleigh focal depth*. It corresponds to a maximum wavefront deviation of $\lambda/4$ at the rim of the exit pupil, where $\lambda = \lambda_0/n_1$ is the wavelength in the medium in which the image is formed. The original $\lambda/4$ criterion of Lord Rayleigh was applied by him to a pencil suffering from spherical aberration with a peak value of one wavelength. At the best-focus position, the wavefront departure of such a pencil of rays is delimited by two spheres with a $\lambda/4$ difference in radius of curvature. In the case of a defocusing of one focal depth, the Huygens secondary wavelets, originating at the reference sphere through the centre E_1 of the exit pupil, arrive in the focal plane with a maximum phase difference of $\pi/2$. For the case of an aberration-free image, $\delta z_f = \lambda/(2\sin^2\alpha_m)$ equals the real-space axial distance over which a receiving screen or detector can be displaced without serious deterioration of the image quality. The perturbing factor for image quality in the aberration-free case is the diffraction unsharpness. When the wavefront deviation in image space due to defocusing is not larger than $\lambda/4$, the image unsharpness hardly changes.

For large values of the opening angle α_m, the allowed defocusing distance should be derived from Eq. (5.2.23), yielding a (Rayleigh) focal depth δz_f of

$$\delta z_f = \frac{\lambda}{4(1-\cos\alpha_m)} \approx \frac{\lambda}{2\sin^2\alpha_m} . \tag{5.2.27}$$

The expression for the focal depth can also be written in terms of the vacuum wavelength λ_0 and the numerical aperture $NA = n_1 \sin\alpha_m$ of the focusing beam. Using these quantities we have

$$\delta z_f = \frac{\lambda_0}{4n_1(1-\cos\alpha_m)} \approx \frac{\lambda_0}{2n_1\sin^2\alpha_m} = \frac{n_1\lambda_0}{2(NA)^2} . \tag{5.2.28}$$

For strongly aberrated imaging systems, the focal depth has to be defined ad hoc as a function of the geometrical blur of the image point due to the aberration of the imaging pencil. The geometrical blur $\delta_{n,\mathrm{def}}$ due to defocusing follows from an integral over the unit circle of the quadratic (normalised) transverse aberration components δx_n and δy_n of Eq. (5.2.21) according to

$$\delta^2_{n,\mathrm{def}} = \frac{1}{\pi}\int_0^1\int_0^{2\pi} \left\{ \left(\delta x_n(\rho_n,\theta) - \overline{\delta x_n}\right)^2 + \left(\delta y_n(\rho_n,\theta) - \overline{\delta y_n}\right)^2 \right\} \rho_n \, d\rho_n d\theta , \tag{5.2.29}$$

where $\overline{\delta x_n}$ and $\overline{\delta y_n}$ are the average values of the transverse aberration components in the image plane. In the case of the defocused wavefront the average values of the transverse aberration components according to the expressions in Eq. (5.2.21) are zero. The expression for the squared transverse geometrical blur $\delta^2_{n,\mathrm{def}}$ according to Eq. (5.2.29) is then equal to $2\,({}_0\alpha_{20})^2$. The so-called geometrical focal depth δz_g in real-space is then defined as the real-space defocusing distance that produces the same root-mean-square transverse blur $\delta_{n,\mathrm{def}}$ of a perfect image as the rms blur δ_a of the aberrated point image in the nominal image plane. Figure 5.14 shows the root-mean-square lateral extent $\delta_{n,\mathrm{def}}$ of a perfectly focused pencil of rays (focus in F) when it has propagated over a distance δz_g such that the rms lateral extent equals δ_a. The defocused wave has a spherical wavefront, intersecting the z-axis at $z = \delta z_g$. The angle at which the root-mean-square extent of the pencil is found corresponds to the maximum aperture $\sin\alpha_m$ divided by $\sqrt{2}$. We then change from normalised coordinates to real-space coordinates and obtain the real-space defocusing distance from

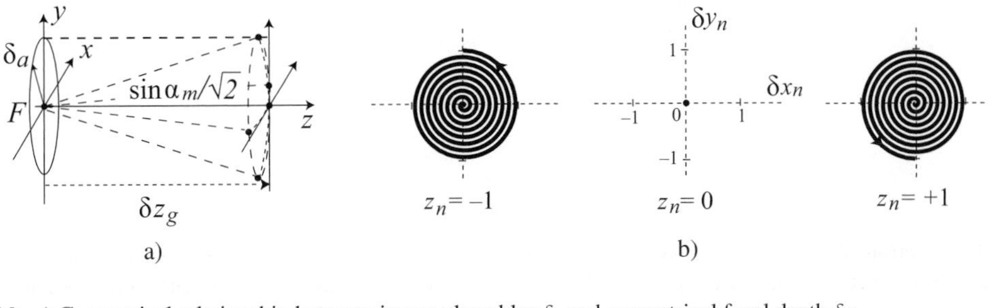

Figure 5.14: a) Geometrical relationship between image plane blur δ_a and geometrical focal depth δz_g.
b) The transverse aberration pattern in the presence of defocusing; distances are measured in diffraction units. Defocusing coefficient ${}_0\alpha_{20}$ from left to right, ${}_0\alpha_{20} = +1/2$, ${}_0\alpha_{20} = 0$, ${}_0\alpha_{20} = -1/2$.

$$\delta_{n,a} = \delta_{n,\mathrm{def}} \,, \quad \text{or,} \qquad \delta_a \frac{\sin\alpha_m}{\lambda} = \sqrt{2}\,|_0\alpha_{20}| = \left| \frac{\delta z_g \sin^2\alpha_m}{\sqrt{2}\,\lambda} \right| \,, \tag{5.2.30}$$

and the blur-defined focal depth is thus given by

$$\delta z_g = \frac{\sqrt{2}}{\sin\alpha_m}\,\delta_a \,. \tag{5.2.31}$$

Note that the geometrical focal depth is independent of the wavelength of the light, a result that was to be expected for a ray-based criterion. The quantity δz_g may depend on the position in the image field. In that case, the geometrical depth of focus of the total image is the smallest value of δz_g in the image field. In practice, a spatial weighting function can be applied. This is justified if some parts of the image have less importance than others, for instance in the case of visual observation of an image.

In Fig. 5.14b we show three plots of the transverse aberration vector in the case of defocusing. The rays in the exit pupil are selected along a linear spiral path from the centre E_1 to the rim of the pupil, with eight windings in total. The azimuthal path is clockwise when viewing in the positive z-direction. A continuous curve is plotted that joins the end points of the normalised transverse aberration vector $(\delta x_n, \delta y_n)$ in the receiving plane. In the centre plot, the wavefront deviation is rigorously zero. In the left- and right-hand plots of Fig. 5.14b the wavefront defocusing coefficient $_0\alpha_{20}$ is $+1/2$ and $-1/2$, respectively, leading to a maximum transverse aberration $\sqrt{(\delta x_n)^2 + (\delta y_n)^2}$ of $1/\sqrt{2}$ in normalised units.

5.2.2.2 Paraxial Magnification Change

The second paraxial term is given by

$$\begin{array}{cccc} \text{coefficient} & W_{\lambda_0}(r_n, \rho_n, \theta) & \delta x_n & \delta y_n \\ _1\alpha_{11} & r_n\rho_n\cos(\theta - \phi) & r_n\cos\phi & r_n\sin\phi\,. \end{array} \tag{5.2.32}$$

The relation between wavefront aberration and transverse shift of an image point follows from Fig. 5.15 by applying the cosine rule in $\triangle P'AA'$. For $\delta x = AA'$ we have for a general ray at angle α,

$$R'^2 = \left(-R + \frac{W}{n_1}\right)^2 + (\delta x)^2 - 2\left(-R + \frac{W}{n_1}\right)(\delta x)\sin\alpha \,. \tag{5.2.33}$$

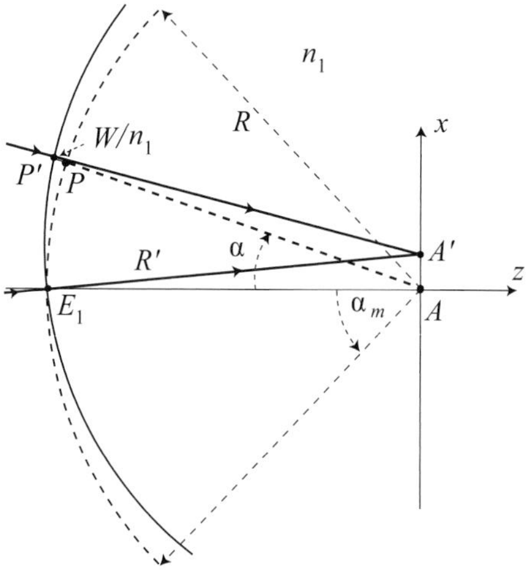

Figure 5.15: Wavefront aberration and transverse aberration in the case of a lateral shift of the image point from A to A'. The spherical reference wavefront in the exit pupil through E_1, dashed in the figure, has its centre of curvature at A and its radius of curvature is R. The physical wavefront (solid spherical section) through E_1 is focused at A' and has a radius of curvature R', both distances being negative in the figure. The wavefront deviation W is denoted by the optical distance $[P'P]$. The maximum opening angle of the focusing beam is α_m. The refractive index of the image space is n_1.

We neglect W with respect to R, use that $R \approx R'$ and find for sufficiently small values of δx the expression

$$W \sin\alpha = n_1(\delta x)\sin\alpha \,. \tag{5.2.34}$$

This expression also follows from Eq. (5.2.15), which is valid for R and $R' \to \infty$, leading to the Hamiltonian definition of wavefront aberration. We then switch to normalised coordinates and find the expression

$$W_{\lambda_0}(\rho_n) = \frac{W(\rho_n \sin\alpha_m)}{\lambda_0} = n_1(\delta x)\rho_n \frac{\sin\alpha_m}{\lambda_0} \,. \tag{5.2.35}$$

Using the expression for the wavefront aberration $W_{\lambda_0}(r_n, \rho_n, \theta) = {}_1\alpha_{11} r_n \rho_n \cos(\theta - \phi)$, we then find for the transverse shifts of a general image point A,

$$\delta x_n = {}_1\alpha_{11} r_n \cos\phi \,, \qquad \delta y_n = {}_1\alpha_{11} r_n \sin\phi \,. \tag{5.2.36}$$

Equation (5.2.36) shows that the image point displacement $\sqrt{(\delta x_n)^2 + (\delta y_n)^2}$ is given by ${}_1\alpha_{11} r_n$ with a maximum of ${}_1\alpha_{11}$ at the border of the image field. In terms of the real-space coordinates we find

$$\delta_M = \frac{\sqrt{(\delta x)^2 + (\delta y)^2}}{r_{max}} = \frac{{}_1\alpha_{11}\lambda_0}{(NA)\, r_{max}} \,, \tag{5.2.37}$$

where δ_M is the increment in magnification of the image under the influence of this paraxial term.

The two paraxial terms discussed above with an aberration degree of $m = 2$ lead to perfect imaging if an appropriate axial and lateral shift of the reference point is applied in image space. As such, these terms should not be classified as aberration terms. The terms that follow in the next subsections are real aberration terms, systematically described for the first time by L. Seidel **[301]**. In the domain of wavefront aberration, their degree m in pupil and field coordinates is 4. They either introduce a distortion of the perfect paraxial image shape or they introduce a position-dependent blur of each individual image point. The aberration terms are called field curvature, distortion, spherical aberration, coma and astigmatism.

5.2.2.3 Field Curvature

The wavefront aberration and transverse aberration components are given by

$$\begin{array}{cccc} \text{coefficient} & W_{\lambda_0}(r_n, \rho_n, \theta) & \delta x_n & \delta y_n \\ {}_2\alpha_{20} & r_n^2 \rho_n^2 & 2r_n^2 \rho_n \cos\theta & 2r_n^2 \rho_n \sin\theta \,. \end{array} \tag{5.2.38}$$

The aberration term leads to a defocusing that is quadratically dependent on the lateral distance of an image point from the centre of the image. At the border of the image field where $r = r_{max}$, we have for the normalised axial shift of the sharp image $\delta z_n = {}_2\alpha_{20}$. Taking into account the real-space axial shift $\delta z = \delta z_n (2\lambda_0)/(n_1 \sin^2\alpha_m)$, we find for the radius of curvature R_b of the curved image surface

$$R_b \approx \frac{r_{max}^2}{2\delta z} = \frac{n_1 \sin^2\alpha_m \, r_{max}^2}{4({}_2\alpha_{20})\lambda_0} \,, \tag{5.2.39}$$

where according to the fourth-order aberration theory a parabolic approximation to the shape of the curved image surface is used (see Fig. 5.16). Given the sign conventions a negative value of R_b corresponds to a concave image surface. The field angle γ in Fig. 5.16 is chosen rather large, outside the domain of applicability of the Seidel aberrations. In practical imaging systems, for such large field angles, more field curvature aberration terms of higher order could be present resulting in a more complicated shape of the surface on which the sharp image is formed.

5.2.2.4 Distortion

Wavefront aberration and transverse aberration components are given by

$$\begin{array}{cccc} \text{coefficient} & W_{\lambda_0}(r_n, \rho_n, \theta) & \delta x_n & \delta y_n \\ {}_3\alpha_{11} & r_n^3 \rho_n \cos(\theta - \phi) & r_n^3 \cos\phi & r_n^3 \sin\phi \,. \end{array} \tag{5.2.40}$$

The aberration term leads to a lateral shift of the image points. Depending on the sign of the distortion coefficient, we obtain the characteristic shapes of Fig. 5.17 for the image of a square object (the ideal square has been dashed).

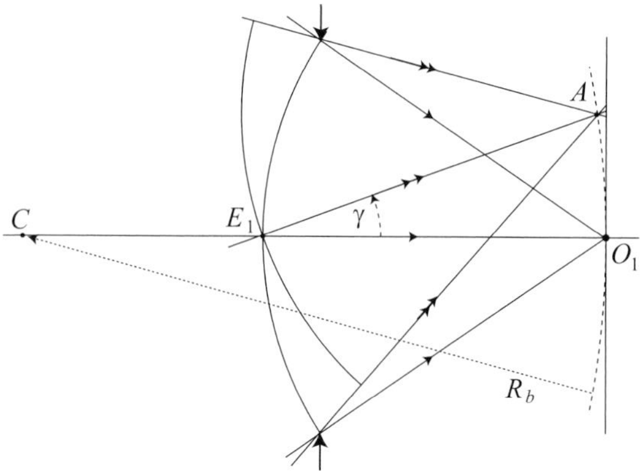

Figure 5.16: Field curvature. The wavefront in the exit pupil through E_1, generated by an on-axis object point, is focused in O_1 on the paraxial image plane. The sharp image of the complete object is formed on a curved surface, dashed in the figure, with a radius of curvature at O_1 equal to R_b and with its centre at C. The field angle of an oblique beam (double arrows) with sharp focus at A has been denoted by γ.

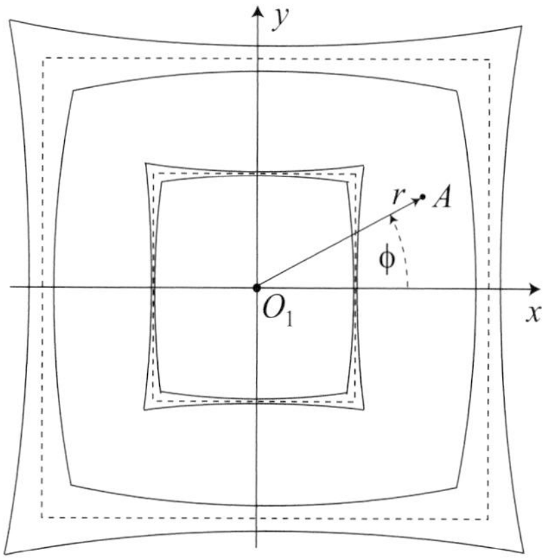

Figure 5.17: Distortion. The image of a square object is distorted. The dashed squares correspond to perfect imaging.

A pin-cushion or a barrel shape of the image arises. Each image point is shifted along the radial direction and its displacement is proportional to $(r/r_{\mathrm{max}})^3$.

5.2.2.5 Spherical Aberration

We have the following expressions for wavefront aberration and transverse aberration components,

$$
\begin{array}{cccc}
\text{coefficient} & W_{\lambda_0}(\rho_n, \theta) & \delta x_n & \delta y_n \\
{}_0\alpha_{40} & \rho_n^4 & 4\rho_n^3 \cos\theta & 4\rho_n^3 \sin\theta\,.
\end{array}
\tag{5.2.41}
$$

The wavefront aberration is only dependent on the radial coordinate ρ_n on the exit pupil sphere. Spherical aberration is the only aberration that can be present on the axis of an optical system with symmetry of revolution. Its axial value is maintained throughout the entire image field. The name spherical aberration is somewhat misleading because the resulting wavefront is *not* spherical. In the older German literature on the subject, the name Öffnungsfehler or aperture error is used. This implies, more correctly, that it is an aberration that appears only if the aperture of the optical system is increased beyond the paraxial domain. In Fig. 5.18 we have plotted rays converging to the paraxial image point O_1.

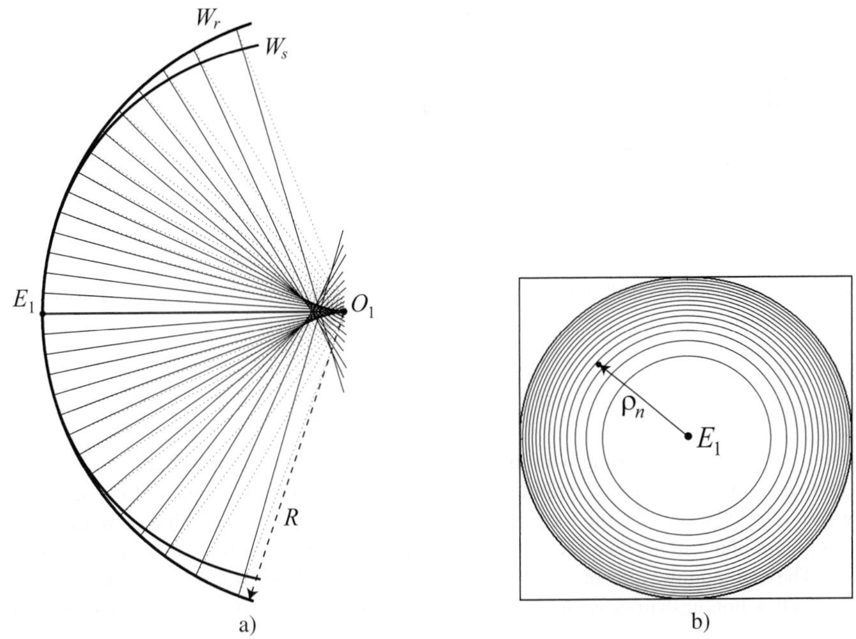

Figure 5.18: Spherical aberration.
a) The caustic figure in the focal region in the presence of a huge amount of spherical aberration. W_r is the spherical reference wavefront in the exit pupil, centred on O_1, the paraxial image point. The dotted rays are reference rays, perpendicular to the reference sphere W_r. The aberrated wavefront is W_s and the solid rays are perpendicular to this surface.
b) Contour plot of a wavefront suffering from spherical aberration ($W_s(\rho_n, \theta) \propto \rho_n^4$).

Due to the spherical aberration, a caustic figure becomes visible, comparable to the one that was sketched in Fig. 5.2. The spherical aberration is very large in this picture and unacceptable for all optical systems with some imaging quality requirement. The caustic pattern suggests that the highest ray density (and, consequently, the highest intensity according to geometrical optics) is found close to the point where the marginal rays from the edge of the pupil intersect the optical axis. More refined calculations, for instance using the minimum value of δ_a, the geometrical blur size, support such a shift of optimum focus in the direction of the marginal focus of the aberrated beam. Another feature of the caustic figure is its asymmetry with respect to some chosen axial reference plane, for instance the paraxial focal plane. The combination of circular symmetry of the aberration pattern and asymmetry in the axial direction is very helpful for detecting the presence of spherical aberration in an imaging system.

The wavefront aberration and the transverse aberration components in a defocused plane with normalised axial coordinate z_n are given by

$$W_{\lambda_0}(\rho_n, \theta) = {}_0\alpha_{40}\,\rho_n^4 + {}_0\alpha_{20}\,\rho_n^2 = {}_0\alpha_{40}\,\rho_n^4 - \frac{z_n}{2}\rho_n^2\,,$$
$$\delta x_n = \left({}_0\alpha_{40}\,\rho_n^3 - z_n\rho_n\right)\cos\theta\,, \qquad \delta y_n = \left({}_0\alpha_{40}\,\rho_n^3 - z_n\rho_n\right)\sin\theta\,. \tag{5.2.42}$$

With the aid of Eq. (5.2.29) we obtain the value of the rms blur in a general defocused plane,

$$\delta_a^2({}_0\alpha_{40}, z_n) = 4({}_0\alpha_{40})^2 - \frac{8}{3}\,{}_0\alpha_{40}\,z_n + \frac{z_n^2}{2}\,. \tag{5.2.43}$$

The minimum value of δ_a is obtained for $z_n = 8{}_0\alpha_{40}/3$ and equals $2{}_0\alpha_{40}/3$.

In the same way as for the calculation of the rms blur δ_n we can use Eq. (5.2.29) to obtain an rms wavefront aberration in a defocused plane by the substitution of the focus-dependent wavefront aberration in this equation. Using the expression for W according to Eq. (5.2.42) we readily obtain

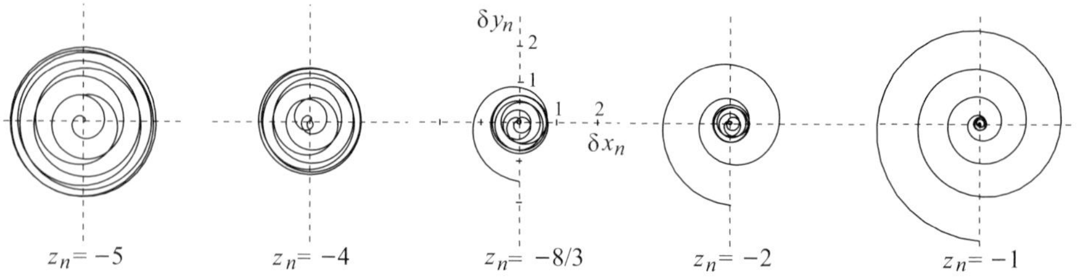

Figure 5.19: The transverse aberration pattern in the presence of spherical aberration (${}_0\alpha_{40} = -1$). Transverse aberration is measured in diffraction units.

$$\overline{W^2} - \left(\overline{W}\right)^2 = \frac{({}_0\alpha_{40})^2}{180} + \frac{({}_0\alpha_{40} - z_n/2)^2}{12} . \tag{5.2.44}$$

This expression shows that the focus-setting with minimum wavefront variance corresponds to $z_n = 2{}_0\alpha_{40}$, substantially different from the focus-setting that yields minimum geometrical blur in the image plane. In general, for all aberration types, there are different axial and lateral focus-settings that yield minimum rms blur or minimum rms wavefront aberration. The minimum-blur criterion is relevant for optical systems with a wavefront aberration that is (at least) several wavelengths large. The minimum wavefront aberration setting should be chosen for nearly perfect imaging systems that mainly suffer from diffraction unsharpness and have a wavefront aberration that is smaller than the wavelength of the light.

For spherical aberration, the asymmetry of the aberration pattern in the axial direction, already sketched in Fig. 5.18, is also present in the transverse aberration patterns of Fig. 5.19. These patterns that are traced when scanning the exit pupil outward along a spiral are shown for five axial positions in the focal region. The coefficient of spherical aberration, ${}_0\alpha_{40}$, is -1 in the figure and the maximum transverse aberration in the paraxial focus-setting ($z_n = 0$, not shown in the figure) equals 4 diffraction units. Defocusing shows the asymmetry around the 'best' focus position associated with minimum blur. This 'minimum-blur' focus-setting is associated with a minimum rms wavefront slope. The focus-setting for minimum geometric blur was derived above from Eq. (5.2.43) and is found at $z_n = 8/3{}_0\alpha_{40}$. The rms blur value δ_a then equals $2|{}_0\alpha_{40}|/3$ diffraction units and is illustrated in the central plot of Fig. 5.19. The through-focus plots show the different behaviour with respect to blur of the focus-settings $z_n = 4{}_0\alpha_{40}$ and $z_n = 2{}_0\alpha_{40}$ on opposite sides of the focus-setting with minimum blur. For the former case we have $\delta_a = 2|{}_0\alpha_{40}|\sqrt{1/3}$, whilst the latter case shows a significantly smaller blur of $\delta_a = |{}_0\alpha_{40}|\sqrt{2/3}$. This case corresponds to the focus-setting for which the wavefront aberration at the pupil rim is exactly zero, but the wavefront slope is not minimum there. Equation (5.2.44) shows that for this focus-setting the minimum wavefront variance equals $({}_0\alpha_{40})^2/180$. This particular focus-setting corresponds to what is commonly called the 'best' diffraction focus of a beam suffering from spherical aberration.

5.2.2.6 Coma

Coma is the first extra aberration that appears outside the optical axis, in the image field of the optical system. Wavefront and transverse aberration components of coma are given by

$$\begin{aligned} W_{\lambda_0}(r_n, \rho_n, \theta) &= {}_1\alpha_{31} r_n \rho_n^3 \cos(\theta - \phi) , \\ \delta x_n(r_n, \rho_n, \theta) &= {}_1\alpha_{31} r_n \rho_n^2 [2\cos\phi + \cos(2\theta - \phi)] , \\ \delta y_n(r_n, \rho_n, \theta) &= {}_1\alpha_{31} r_n \rho_n^2 [2\sin\phi + \sin(2\theta - \phi)] . \end{aligned} \tag{5.2.45}$$

The transverse aberration pattern of coma is asymmetric (see Fig. 5.20) and presents a tail. The tail-shape explains its name which comes from the Greek word κόμη, hair. The acute angle formed by the tangent lines to the in-focus aberration figure has a value of $\pi/3$. By a careful design of the optical system, such that it satisfies the Abbe sine condition (see Subsection 4.8.1), the comatic aberration is made zero. A further characteristic of the comatic transverse aberration components is their double periodicity in θ. A single rotation in the exit pupil leads to a circular transverse aberration pattern in the image plane that is completed twice. Figure 5.20 suggests that the coma pattern is symmetric with respect to defocusing. Inspection of the expression for $(\delta x_n, \delta y_n)$ shows that in the presence of defocusing we have

$$\delta x_n(r_n, \rho_n, \theta, {}_0\alpha_{20}) = \rho_n \left\{ {}_1\alpha_{31} r_n \rho_n [2\cos\phi + \cos(2\theta - \phi)] + 2{}_0\alpha_{20} \cos\theta \right\} , \tag{5.2.46}$$

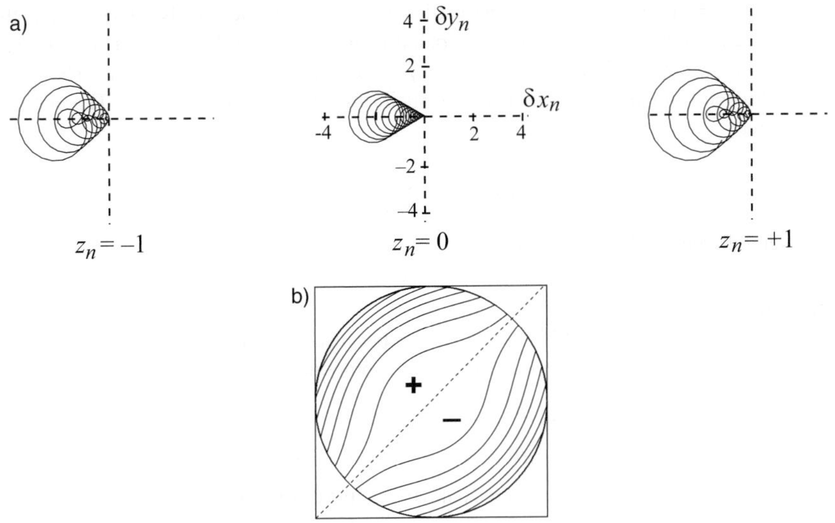

Figure 5.20: a) Transverse aberration patterns for comatic aberration with coefficient ${}_1\alpha_{31} = -1$ as a function of defocusing. b) Contour plot of the wavefront aberration $W_c(r_n, \rho_n, \theta) = {}_1\alpha_{31} r_n \rho_n^3 \cos(\theta - 3\pi/4)$ in the case of a positive value of the coefficient ${}_1\alpha_{31}$.

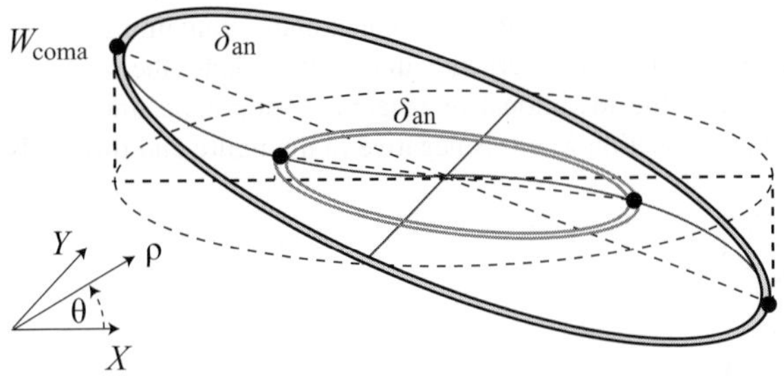

Figure 5.21: The construction of a comatic wavefront from annular sections with infinitesimal width δ_{an}, taken from planar surfaces with increasing inclination towards the rim of the wavefront.

with the property that,

$$\delta x_n(r_n, \rho_n, \theta, {}_0\alpha_{20}) = \delta x_n(r_n, \rho_n, \theta + \pi, -{}_0\alpha_{20}) \,. \tag{5.2.47}$$

An analogous expression holds for the y-transverse aberration component. The form of Eq. (5.2.47) shows that on opposite sides of the nominal focus, in two receiving planes with an equal absolute defocusing value, the same transverse aberration component is generated by aperture rays that are shifted over an azimuth of π in the exit pupil. Since coma has the property that the same transverse aberration curve is completed twice for one circular path in the pupil, the transverse aberration components belonging to pupil points with an azimuthal offset of π are identical and the aberration diagrams are thus identical in these two defocused planes.

Figure 5.20b shows that the wavefront itself is odd with respect to a dashed line that is perpendicular to the image point vector $r_n(\cos\phi, \sin\phi)$ with azimuth $\phi = 3\pi/4$ in the figure. In Fig. 5.21 we show that the continuous comatic wavefront can be constructed from annuli with infinitesimal width δ_{an} that are each time cut out of a planar surface with a certain inclination with respect to the horizontal plane in the figure. We consider the difference between the comatic wavefront of Eq. (5.2.45) and a planar surface with tilt angle a and require it to be zero,

$${}_1\alpha_{31} r_n \rho_n^3 \cos(\theta - \phi) - a\rho_n \cos(\theta - \phi) = 0 \,. \tag{5.2.48}$$

The solution is given by $a = {}_1\alpha_{31} r_n \rho_n^2$. The real-space tilt angle of the plane equals $a\lambda_0/\rho_{\max}$ and this tilt is found on the comatic wavefront in an annular region with its radial coordinate given by $\rho_n^2 = a/({}_1\alpha_{31} r_n)$. Figure 5.21 shows two examples of such annuli, for radial values ρ_n given by $\rho_n = 1/2$ and $\rho_n = 1$, respectively. The continuous comatic wavefront can be constructed from an infinite sequence of such ring-shaped planar sections. The inclination angle for a general annulus is given by

$$\gamma(\rho_n, r_n) = \frac{{}_1\alpha_{31}\lambda_0}{\rho_{\max}} r_n \rho_n^2 , \tag{5.2.49}$$

where $2\rho_{\max}$ is the lateral diameter of the exit pupil.

5.2.2.7 Astigmatism

Astigmatism is an aberration that develops at larger field angles because of its at least quadratic dependence on the lateral coordinate in image space. Wavefront and transverse aberration components are given by

$$\begin{aligned} W_{\lambda_0}(r_n, \rho_n, \theta) &= {}_2\alpha_{22} r_n^2 \rho_n^2 \cos^2(\theta - \phi) , \\ \delta x_n(r_n, \rho_n, \theta) &= 2({}_2\alpha_{22}) r_n^2 \rho_n \cos\phi \cos(\theta - \phi) , \\ \delta y_n(r_n, \rho_n, \theta) &= 2({}_2\alpha_{22}) r_n^2 \rho_n \sin\phi \cos(\theta - \phi) . \end{aligned} \tag{5.2.50}$$

The shape of an astigmatic beam in its focal region is shown in Fig. 5.22. The perfect image point of the aberration-free case is spread out in the axial direction in two mutually perpendicular orthogonal focal lines with a blurred circular image in between (circle of 'least-confusion'). In contrast with spherical aberration and coma, astigmatism does change the paraxial properties of the focusing beam. Instead of a single radius of curvature, the astigmatic beam has two principal curvatures. This is in accordance with Euler's **[94]** surface theorem which states that each surface can be locally approximated by a quadratic surface with two principal curvatures in mutually orthogonal planes. In the case of third-order astigmatism, the principal sections are the meridional plane, containing the optical axis and the principal ray E_1A to a paraxial image point A, and the plane perpendicular to this plane, the so-called sagittal plane. The curvature in the meridional plane gives rise to the meridional or tangential astigmatic line, A_tB_t in the figure. The sign of the astigmatic aberration W_a in the cross-section $X = 0$ is negative in the figure and this leads to a position of the tangential

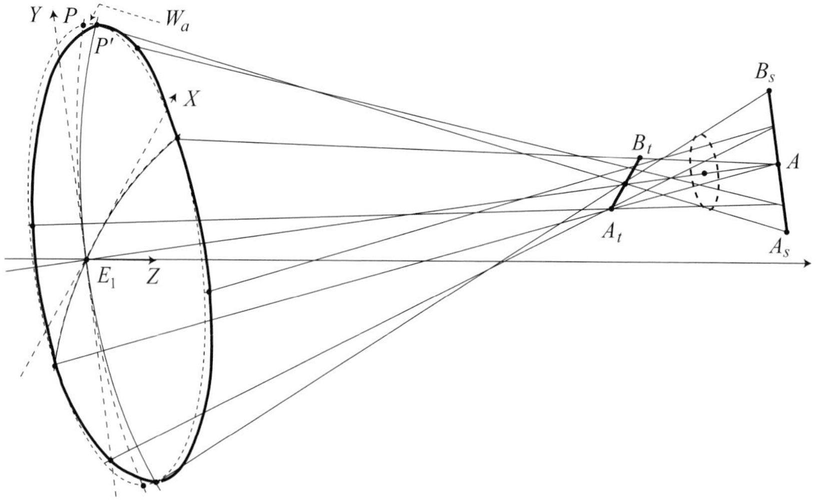

Figure 5.22: The generation of the focal lines of an astigmatic beam (${}_2\alpha_{22} < 0$, $\phi = \pi/2$). The outer contour of the spherical wavefront through a general point P and through E_1, the centre of the exit pupil, has been dashed in the figure. The contour of the astigmatic wavefront is the solid curve through the point P' on the wavefront. The maximum wavefront aberration W_a equals the optical distance $[P'P]$, negative in the figure. The tangential focal line is A_tB_t. The sagittal focal line is A_sB_s, coincident with the paraxial image plane in the drawing. The best focus has been indicated by the dashed circle halfway between the two focal lines.

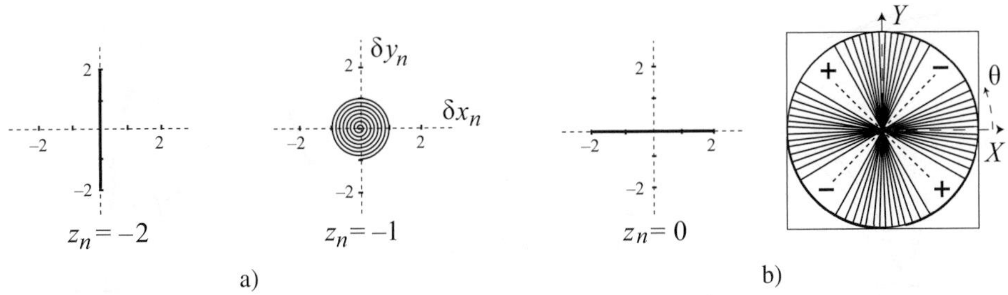

Figure 5.23: a) Transverse aberration pattern for astigmatic aberration with $W_a(r_n, \rho_n, \theta) = {}_2\alpha_{22} r_n^2 \rho_n^2 \cos^2\theta$ (${}_2\alpha_{22} = -1$). b) $\phi = \pi/4$, $W = {}_2\alpha_{22} r_n^2 \rho_n^2 (1 + \sin 2\theta)/2$. Contour plot of the θ-dependent part of the astigmatic wavefront aberration for a coefficient value ${}_2\alpha_{22} = -1$. The focal lines are at angles of $\pi/4$ and $3\pi/4$ radians with the x-axis in the image field.

astigmatic line in front of the paraxial image plane. In the other cross-section, the wavefront aberration is zero and the sagittal focal line, A_sB_s, is found in the paraxial image plane. In the figure, the sagittal astigmatic line is tilted with respect to the paraxial image plane. Although this is formally the case, the effect is generally neglected in third-order aberration theory. Between the two focal lines, a best-focus position is found with a circular cross-section; it is called the circle of least-confusion of the astigmatic pencil of rays. In the figure, some representative rays have been drawn that intersect both focal lines. With the aid of the two focal lines, it is possible to create the rays of the complete astigmatic pencil by mutually connecting all points on both astigmatic lines.

In Fig. 5.23a we show the transverse aberration pattern of the astigmatic pencil at various axial positions along the principal ray of the pencil. The transverse aberration components are given by Eq. (5.2.50). The value of ${}_2\alpha_{22}$ is -1 and the paraxial image point (with $r_n = 1$) is chosen on the x-axis with Cartesian coordinates $(1, 0)$. For a negative astigmatic coefficient, it is the tangential focal line that is in front of the sagittal line. The length of each focal line is $4\,|{}_2\alpha_{22}|$ in diffraction units. The position of the circle of least-confusion is found by requiring that $\delta_n^2 = (\delta x_n)^2 + (\delta y_n)^2$ has a constant value, independent of the azimuth θ of the aberrated ray in the exit pupil. We include a defocusing term in Eq. (5.2.50) and obtain, with ${}_0\alpha_{20} = -z_n/2$,

$$\begin{aligned}\delta x_n &= 2({}_2\alpha_{22})\, r_n^2\, \rho_n \cos\phi \cos(\theta - \phi) - \rho_n z_n \cos\theta\,,\\ \delta y_n &= 2({}_2\alpha_{22})\, r_n^2\, \rho_n \sin\phi \cos(\theta - \phi) - \rho_n z_n \sin\theta\,.\end{aligned} \tag{5.2.51}$$

For the value of δ_n^2 produced by the rays that leave the border of the exit pupil towards the extreme field point we find the expression

$$\delta_n^2 = 4\,{}_2\alpha_{22}\,({}_2\alpha_{22} - z_n)\,\cos^2(\theta - \phi) + z_n^2\,. \tag{5.2.52}$$

The circle of least-confusion is thus found in the defocused plane with normalised axial coordinate $z_n = {}_2\alpha_{22}$ and the diameter of the circle equals $2\,|z_n|$. The length of the astigmatic lines is found from this expression by putting $z_n = 0$ and $\theta = \phi$ or $\theta = \phi + \pi$ for the sagittal line and $z_n = 2\,{}_2\alpha_{22}$ and $\theta = \phi \pm \pi/2$ for the tangential focal line. In both cases, for $r_n = \rho_n = 1$, the length of the astigmatic lines at the border of the field is given by $4\,|{}_2\alpha_{22}|$. The circle of least-confusion is found on the beam axis, exactly halfway between the two astigmatic lines in the case of astigmatic aberration of the lowest order (of third order in the pupil and image plane coordinates for the transverse aberration components). The transverse rms blur $\delta_{n,\mathrm{rms}}$ of the complete astigmatic pencil as a function of defocusing is calculated with the aid of Eq. (5.2.29) with the result

$$\delta_{n,\mathrm{rms}}^2 = \frac{1}{\pi} \int_0^1 \int_0^{2\pi} \rho_n^2 \left[4\,{}_2\alpha_{22}({}_2\alpha_{22} - z_n)\cos^2(\theta - \phi) + z_n^2\right] \rho_n\, d\rho_n d\theta = {}_2\alpha_{22}\,({}_2\alpha_{22} - z_n) + \frac{z_n^2}{2}\,. \tag{5.2.53}$$

The rms blur related to the astigmatic focal lines equals $|{}_2\alpha_{22}|$. The smallest blur is found for the circle of least-confusion and has the value $\delta_{n,\mathrm{rms}} = |{}_2\alpha_{22}|/\sqrt{2}$. In Fig. 5.23b, the θ-dependent part of the astigmatic wavefront aberration is shown. For a position of the paraxial image point on a line with azimuth $\phi = \pi/4$, the total aberration function W is given by $\frac{1}{2}\,{}_2\alpha_{22} r_n^2 \rho_n^2 (1 + \sin 2\theta)$.

In Fig. 5.24 we show the location of the sagittal and tangential focal lines on curved image field surfaces. The optical system shows some field curvature. On this curved surface, the sagittal image lines are sharp. For the chosen sign of ${}_2\alpha_{22}$

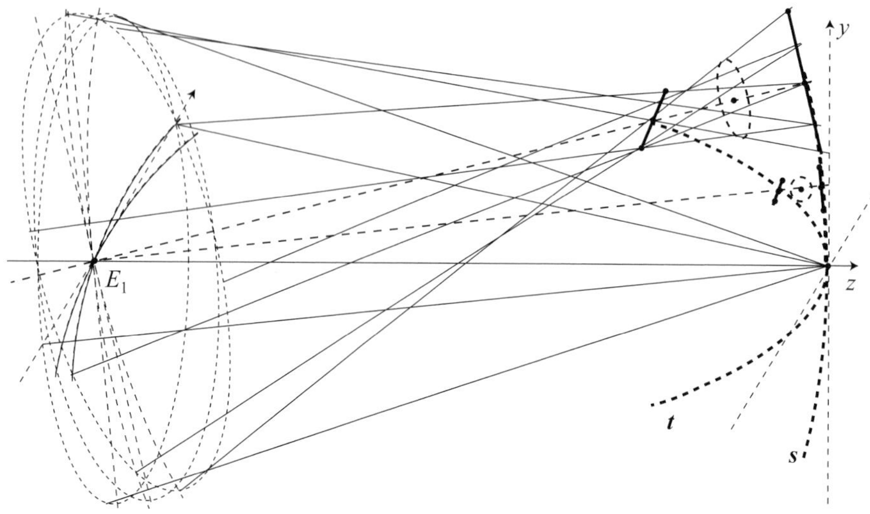

Figure 5.24: Various pencils of rays progressing to the image plane from the exit pupil with its centre at E_1. The optical system suffers from some field curvature. The sagittal focal lines are found on this curved surface labelled s. With respect to this sagittal field surface, the tangential focal lines are shifted in the z-direction over a distance that increases quadratically with the lateral field coordinate and they are found on the tangential field surface t. The aberration effects have been greatly exaggerated in the figure for reasons of clarity.

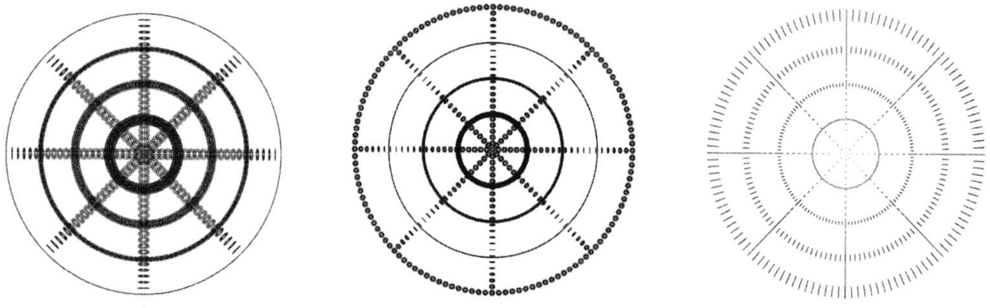

Figure 5.25: Typical image pattern in the presence of field-dependent astigmatism with coefficient ${}_2\alpha_{22} < 0$. The object is an assembly of points on straight lines through the origin and on concentric circles.
The focal settings are a) $z_n = 2\,{}_2\alpha_{22}$; b) $z_n = {}_2\alpha_{22}$; c) $z_n = 0$.

in the drawing, the tangential astigmatic lines are sharp on a tangential image surface that is more strongly bent than the sagittal image surface. In Fig. 5.25 we present the image of a circular spoked wheel in the object plane. For the case of imaging by an optical system suffering from astigmatism, typical images arise when the focus-setting is changed. In Fig. 5.25a the focus-setting is sharp for the tangential lines at the border of the field. In c) the centre of the image and the sagittal lines are sharp. The focus-setting of b) is the best compromise for image quality. The field curvature was zero in this example.

A final remark on astigmatism concerns the azimuth of the perpendicular dropped from a point on a general ray to the principal ray of an astigmatic pencil (see Fig. 5.26). A rotation is observed of this perpendicular line segment when the point on the aperture ray progresses through the focal region. Although such a rotation is present for all kinds of aberrated beams, the particularity of an astigmatic beam is that this rotation is identical for an entire planar sheet of rays with identical azimuth θ in the exit pupil. The plane in which the selected rays are initially found, in the figure at an angle

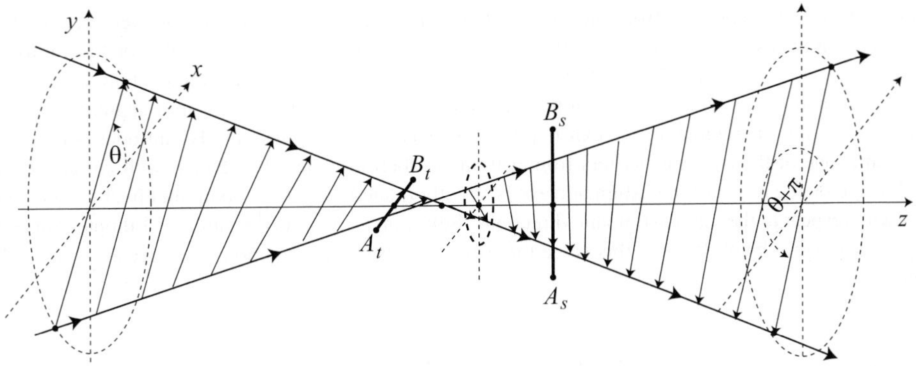

Figure 5.26: Representation of the angle θ associated with the perpendicular dropped from a general point on an aperture ray to the principal ray of an astigmatic pencil. The incident ray is part of a planar sheet of rays located in a plane with azimuth θ. In the figure the angle θ is set at an angle of approximately $\pi/4$ with respect to the astigmatic focal lines. In the circle of least-confusion, the azimuth of the rays is $-\theta$. Far away from the astigmatic focus, the perpendiculars from the rays point in the direction given by the angle $\theta + \pi$.

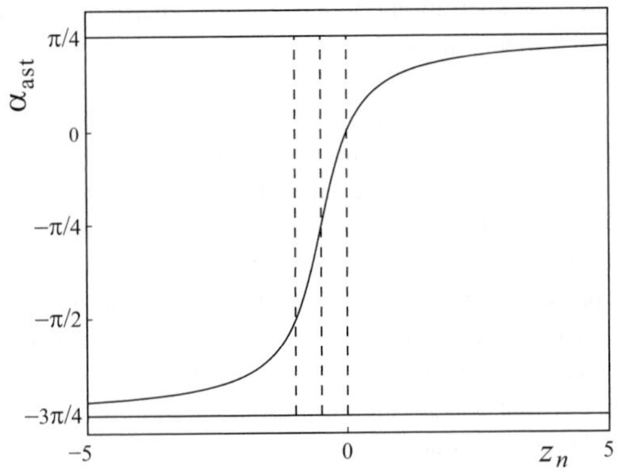

Figure 5.27: The through-focus rotation of a planar sheet of rays having a certain azimuth in the exit pupil ($-3\pi/4$ in the graph). Astigmatic coefficient: ${}_2\alpha_{22} = -1$. The left-hand and right-hand dashed lines indicate the positions of the focal lines, at $z_n = -2$ and $z_n = 0$. The central dashed line corresponds to the position of the best focus ($z_n = -1$). The exit pupil is found at $z_n \to -\infty$.

of approximately $\pi/4$ with respect to the astigmatic focal lines, rotates when passing through the astigmatic focus. Far beyond the focal region, the plane has been rotated over an angle of π. The precise azimuth α_{ast} of the rotated plane in which the rays are found when traversing the focal region can be calculated from the transverse aberration components according to Eq. (5.2.51) that have the defocus term included. The azimuth of the planar sheet of rays as a function of the focus-setting is simply given by

$$\alpha_{ast} = \arctan\left(\frac{\delta y_n}{\delta x_n}\right) . \tag{5.2.54}$$

Figure 5.27 shows the gradual variation of α_{ast} in the neighbourhood of the astigmatic focus. In this graph, for the chosen values of ${}_2\alpha_{22} r_n^2$ and $\theta - \phi$, the limiting values of α_{ast} are $-3\pi/4$ in the exit pupil and $+\pi/4$ at a sufficiently large distance beyond the focal region. A detection of the rotation angle of the planar sheet of light rays yields a measure for the axial position of the detection plane.

The rotation of the perpendiculars from the rays contained in a planar sheet of rays, or even for all rays from two opposite quadrants in the exit pupil, shows that the energy transport by such a portion of the total beam follows a half-turn screw-wise trajectory. Such a selected part of the beam thus possesses angular momentum as a consequence of its spatial structure and the astigmatic aberration. The complete astigmatic beam has zero angular momentum as the momentum from the various regions in the exit pupil is cancelled out. Astigmatism also introduces rotation effects in binocular vision if one of the two eyes suffers from astigmatic aberration and the other is well-corrected. In that case, the observed image on the retina of the astigmatic eye is slightly rotated, with the maximum effect for line features at an angle of $\pi/4$ with respect to the main astigmatic directions. Below an astigmatic deviation of 1 diopter, this effect can be tolerated by the average subject. Beyond this value the patient needs extra correction by a special aspheric shape of the spectacle glass.

5.3 Calculation of the Seidel Aberration Coefficients

In this section we calculate the Seidel aberration coefficients of an optical system in the wavefront domain. In this domain, the aberration terms depend on a total power of four with respect to the ray positions on the pupil reference sphere and in the image plane. Once these 'fourth-order' coefficients have been obtained, it is straightforward to calculate the transverse aberration components by using Eq. (5.2.15) or (5.2.18), respectively for the expressions in real-space coordinates or in terms of the canonical coordinates in pupil and image plane. The expressions for the transverse components, because of the derivations that are involved, have a total power of three in pupil and image field coordinates. Historically, because of his original derivation in the transverse aberration domain, Seidel's aberration theory is called the *third-order* theory. We use this nomenclature for Seidel's theory throughout this section, unless we explicitly talk about the fourth-order wavefront aberration theory.

For the calculation of the optical pathlengths that enter into the aberration expression, we use real-space coordinates. If needed, the final aberration expression can then be normalised to make it suitable for further processing in Eq. (5.2.18). The aberration calculation starts with analysis of the aberration that occurs when a pencil of rays is refracted or reflected by a single surface within an optical system. In such an intermediate imaging step, we need knowledge about the position of the intermediate entrance and exit pupil positions and the intermediate object and image planes. It is an important analytic advantage of the Seidel aberration theory that the aberrations of the optical system can be simply added and that for the intersection points of the rays with a surface the paraxially approximated values are adequate. This means that within the third-order Seidel theory, the aberrations that originate at a specific surface do not depend on the state of aberration of the incident beam. For that reason, the third-order aberrations are intrinsic. For higher-order aberration terms, this property does not hold. Their magnitude can be split into an *intrinsic* part and an *induced* part, that depends on the aberration incurred at preceding surfaces in the optical system. It will be clear that the contributions from induced aberrations are more important in elongated optical systems than in compact systems. It also means that the domain of validity of the Seidel aberrations becomes more restricted in such elongated systems. In that case, higher-order aberration terms can assume higher values than the generally dominating third-order ones. As a rule of thumb for the domain of validity of third-order aberration analysis, we state that a limitation of aperture and field angles to typically 0.1 radian guarantees a good estimate of the total aberration of the system by means of the third-order coefficients. In practice, the typical 0.1 radian limit of validity applies to the angles of incidence on the lens surfaces and, by means of proper *lens bending*, the aperture angle or field angle of a system may exceed the third-order limit. By a gradual build-up of convergence angles in an optical system with a substantial number of lenses, the angles of incidence at each individual surface remain within the limit of the third-order approximation. In any case, the sum of the third-order aberrations generated at each refracting or reflecting surface should be kept small in an imaging system of some quality. Knowledge of the distribution of the third-order aberrations over the surfaces of an imaging system is a prerequisite for a successful design of an optical system.

5.3.1 Aberration Change at Refraction by an Optical Surface

In Fig. 5.28 we have sketched the refraction of the principal ray $A_0Q_pA'_1$ and of a general aperture ray $A_0Q_aA'_1$ by the optical surface through T. To obtain the optical pathlength change at refraction we calculate the optical paths from the reference sphere through E_0 in the entrance pupil to the reference sphere through E_1 in the exit pupil. The difference ΔW for a general aperture ray and the principal ray through E_0 is given by

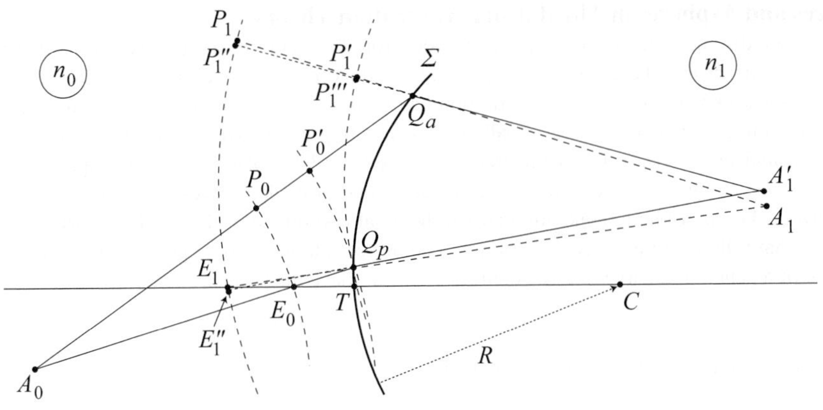

Figure 5.28: Intermediate imaging by the spherical surface Σ through T, Q_p and Q_a, with radius of curvature R (positive in the drawing) and centre at C. The object point is at A_0, the paraxial image point at A_1. The intersections of the wavefronts in entrance and exit pupil with the optical axis are E_0 and E_1. The principal ray of the pencil of rays intersects the surface in Q_p, a general aperture ray in Q_a. The spherical reference surfaces through E_0P_0 and Q_pP_0' have their centres at the object point A_0. The spherical surfaces through E_1P_1 and Q_pP_1' have their centres at the paraxial image point A_1. The refractive indices in the intermediate spaces are denoted by n_0 and n_1.

$$\Delta W = [P_0P_0'] + [P_0'Q_a] + [Q_aP_1'''] + [P_1'''P_1''] - \left([E_0Q_p] + [Q_pE_1'']\right) . \tag{5.3.1}$$

The point E_1'' lies on the reference sphere through the exit pupil, on the back-propagated, refracted principal ray. The points P_1''' and P_1' lie on the reference sphere through Q_p, on the back-propagated refracted general aperture ray. The angle $P_1Q_aP_1''$ in $\triangle Q_aP_1P_1''$ is third-order small since the transverse aberration distance A_1A_1' is of that order in the pupil and field coordinates. Consequently, the difference in length between Q_aP_1''' and Q_aP_1' is of sixth order, as well as the difference between $P_1''P_1'''$ and P_1P_1'. The same holds for the difference between the lengths of the ray segments Q_pE_1 and Q_pE_1''. This means that for the fourth-order approximation to the pathlength difference we have

$$\Delta W^{(4)} = [P_0P_0'] + [P_0'Q_a] + [Q_aP_1'] + [P_1'P_1] - \left([E_0Q_p] + [Q_pE_1]\right) = [P_0'Q_a] + [Q_aP_1'] = [P_0'Q_a] - [P_1'Q_a] . \tag{5.3.2}$$

From Fig. 5.28 we conclude that up to the fourth order we have

$$[P_1'Q_a] = [P_1'A_1] - [Q_aA_1] = [Q_pA_1] - [Q_aA_1] , \tag{5.3.3}$$

and, with the same substitution in the object space with index n_0, we find

$$\Delta W^{(4)} = n_1\left\{Q_aA_1 - Q_pA_1\right\} - n_0\left\{A_0Q_p - A_0Q_a\right\} . \tag{5.3.4}$$

Further symmetry between the pathlength differences in object and image space can be introduced by writing

$$\Delta W^{(4)} = n_1\left\{Q_aA_1 - Q_pA_1\right\} - n_0\left\{Q_aA_0 - Q_pA_0\right\} = \delta\left\{n_j(Q_aA_j - Q_pA_j)\right\} , \tag{5.3.5}$$

where the δ-symbol stands for the subtraction operation, yielding the difference of the optical pathlength expression between braces, evaluated first in image space ($j = 1$) and then in object space ($j = 0$). The advantage of the expression of Eq. (5.3.5) is that we only need the paraxial coordinates of the object and image points A_0 and A_1. It can also be proven that the (x, y) intersection heights of the two representative rays with the optical surface, viz. the ordinates of Q_p and Q_a, can be approximated by their paraxial values. The third-order error in these coordinate values, which results from this approximation, is only visible in the sixth-order terms of the pathlength expressions. For the total third-order aberration of a system, we are allowed to sum the surface contributions independently, yielding

$$W^{(4)} = \sum_l \delta\left\{n(Q_aA - Q_pA)\right\}\Big|_l , \tag{5.3.6}$$

where the index l runs over the sequence of optical surfaces in the system.

5.3.1.1 Spheres and Aspheres in Third-order Aberration Theory

The calculation of the distances that occur in the expression for $W^{(4)}$ requires an approximation that is accurate up to the fourth order. This means that the axial departure (or sag) of the optical surfaces also needs to be approximated up to that order. Optical surfaces with symmetry of revolution having a non-spherical cross-section are frequently used. The cross-section containing the optical axis is also called the generating curve. The surface that is created by rotating the generating curve around an axis is a surface with circular symmetry and the rotation axis is the optical axis of the resulting asphere. In general, one requires continuity of the surface itself and of its derivatives at the origin where the optical axis intersects the surface. For an aspheric generating curve, the axial coordinate of the resulting surface is often written as the sum of the sag that follows from the contribution of a conic section plus a more general aspheric departure from this shape. In the case of a generating curve that is a conic section, we have the surface equation

$$(1+\kappa)z^2 - 2Rz + (x^2+y^2) = 0\,, \tag{5.3.7}$$

where κ is the conic constant. The z-coordinate is given by

$$z = \frac{c(x^2+y^2)}{\left[1+\sqrt{1-(1+\kappa)c^2(x^2+y^2)}\right]}\,, \tag{5.3.8}$$

where $c = 1/R$ is the paraxial curvature of the conic section. A series expansion of the z-coordinate up to the fourth order in (x, y) yields

$$z = \frac{c}{2}(x^2+y^2) + \left\{\frac{(1+\kappa)c^3}{8} + a_4\right\}(x^2+y^2)^2 + \cdots\,, \tag{5.3.9}$$

where we have added the fourth-order coefficient a_4 associated with a more general aspheric shape of a surface. In the fourth-order approximation of the surface shape, the conic constant and the coefficient a_4 play comparable roles and are mutually redundant. For higher-order terms, the general aspheric coefficients provide more flexibility in surface shape. In practice, for reasons of numerical stability, it is better to abandon the conic-section reference and to use only the coefficients a_{2n}. In Fig. 5.29 some generating curves of aspheric surfaces have been shown that are conic sections. The curvature c is identical for all curves. The parameter in the figure is the value of the conic constant κ. The classification of the conic sections according to their κ-value is as follows:

$$\begin{array}{rl} \kappa < -1 & \text{hyperboloid} \\ \kappa = -1 & \text{paraboloid} \\ -1 < \kappa < 0 & \text{prolate ellipsoid} \\ \kappa = 0 & \text{sphere} \\ \kappa > 0 & \text{oblate ellipsoid.} \end{array} \tag{5.3.10}$$

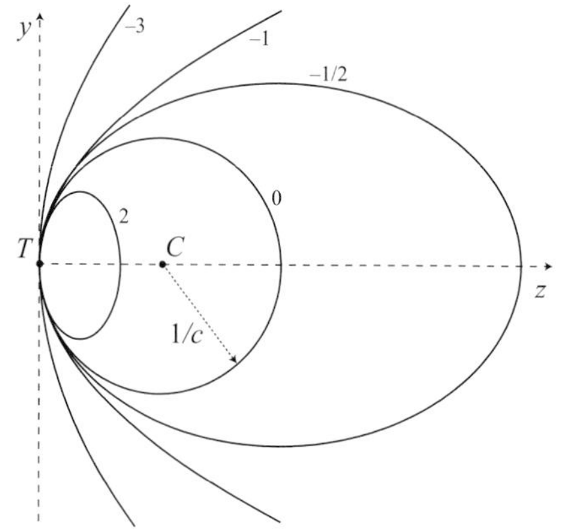

Figure 5.29: Some examples of aspheric surfaces with symmetry of revolution and a conic cross-section. The cross-section contains the vertex T of the surface and the paraxial centre of curvature C, identical for all examples. The z-axis is the axis of symmetry. The label given to each curve is the value of its conic constant κ.

We have already encountered the conic sections when discussing the subject of stigmatic imaging in Subsection 5.1.2. In that subsection, the imaging of specific pairs of conjugate points on the axis was shown to be perfect, independently of the aperture angle, when using conic sections as generating curves. In this section on third-order aberration theory, the same basic aspheric solutions are found, this time with their aspheric departure approximated up to the required fourth order.

5.3.2 Expressions for the Seidel Aberration Coefficients

The general expression of Eq. (5.3.5) allows the direct calculation of $W^{(4)}$ once a specific pair of conjugate points A_0, A_1 has been chosen and the directions of the aperture ray and the principal ray have been fixed. A more instructive approach is to use the coordinates of object and image point and the ray directions of aperture and principal ray as variables and to write the result of Eq. (5.3.5) as a function of the image field position and the intersection height of the rays in the exit pupil. With this approach we get the functional dependence of the various third-order aberration terms and their coefficients. After proper normalisation of these coefficients with respect to the wavelength of the light and the maximum pupil and field coordinates, one can easily determine their relative importance per surface in the imaging process. The most important steps to obtain such an analytic result will now be given here. The original derivation can be found in Seidel's own publication **[301]** for the transverse aberration components. A third-order aberration derivation based on the wavefront aberration approach can also be found in **[229]**,**[364]**.

Figure 5.30 shows the paraxial data (distances, angles) that are needed for the aberration calculation. The paraxial marginal ray $O_0Q_aO_1$ from an axial object point and the paraxial principal ray $A_0Q_pA_1$ from an object point at the border of the object field carry sufficient information to find the third-order aberration terms. The paraxial point of intersection with the surface of a general ray, leaving an arbitrary point in the object field, can be found by linear interpolation from the data of the marginal axial ray and the extreme principal ray. We stress again that such a simple linear interpolation is only allowed when using the third-order approximation for the aberration terms. It explains the relatively simple derivation of the third-order aberration terms and the simple summation rule for these terms when analysing an optical system.

According to Eq. (5.3.5) we need to calculate typical line segments like Q_aA_1 in a fourth-order approximation with respect to powers of the lateral field and pupil coordinates. We have

$$Q_aA_1 = \left\{x^2 + (y - p_1)^2 + (z_1 - z_{Q_a})^2\right\}^{1/2} ,$$

and

$$z_{Q_a} = \frac{c}{2}(x^2 + y^2) + \left(\frac{c^3}{8} + a_4\right)(x^2 + y^2)^2 . \qquad (5.3.11)$$

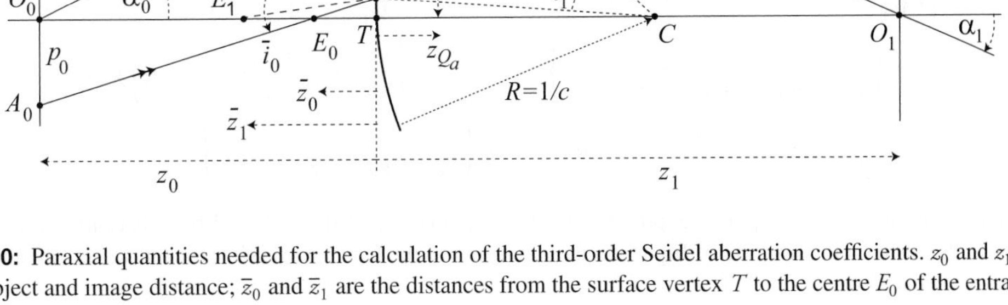

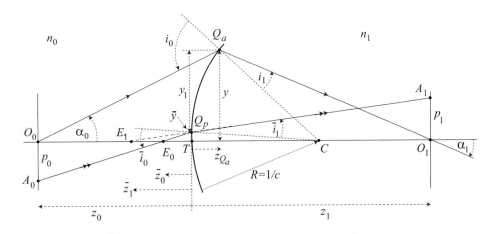

Figure 5.30: Paraxial quantities needed for the calculation of the third-order Seidel aberration coefficients. z_0 and z_1 are the paraxial object and image distance; $\bar{z}_0$ and $\bar{z}_1$ are the distances from the surface vertex T to the centre E_0 of the entrance pupil and to the centre E_1 of the exit pupil, respectively.

The quantity c is the paraxial curvature of the surface. In the case of an aspheric cross-section of the optical surface, a fourth-order surface departure has been included by means of the coefficient a_4. The expansion of the square-root expression for Q_aA_1 using the approximation of $\sqrt{1-\epsilon} \approx 1 - \epsilon/2 - \epsilon^2/8$ yields for the optical pathlength

$$[Q_aA_1] = n_1\left\{z_1 - \frac{c}{2}(x^2+y^2) - \left(\frac{c^3}{8} + a_4\right)(x^2+y^2)^2 + \left\{x^2 + (y-p_1)^2\right\}\left[\frac{1}{2z_1} + \frac{c}{4z_1^2}(x^2+y^2) - \frac{x^2+(y-p_1)^2}{8z_1^3}\right]\right\}, \quad (5.3.12)$$

and a comparable expression holds for the optical distance $[Q_aA_0]$ in object space.

The expression for the pathlength along the principal ray is easily derived from $[Q_aA_1]$ by substituting $x = 0$ and $y \rightarrow \bar{y}$, the ordinate of the point Q_p on the optical surface. We have the expression

$$[Q_pA_1] = n_1\left\{z_1 - \frac{c}{2}(\bar{y}^2) - \left(\frac{c^3}{8} + a_4\right)\bar{y}^4 + (\bar{y}-p_1)^2\left[\frac{1}{2z_1} + \frac{c}{4z_1^2}\bar{y}^2 - \frac{(\bar{y}-p_1)^2}{8z_1^3}\right]\right\}, \quad (5.3.13)$$

and a similar expression can be given for the pathlength $[Q_pA_0]$. The points A_0 and A_1 are paraxially conjugate, as well as the pupil intersection points with the optical axis, E_0 and E_1. This means that the pathlength terms up to the second order in field and pupil coordinates are zero and can be deleted from the expression for $\Delta W^{(4)}$. This reduces the pathlength expressions to their truly fourth-order terms:

$$\begin{aligned}
[Q_aA_1] &= n_1\left\{-\left(\frac{c^3}{8} + a_4\right)(x^2+y^2)^2 + \left\{x^2+(y-p_1)^2\right\}\left[\frac{c}{4z_1^2}(x^2+y^2) - \frac{x^2+(y-p_1)^2}{8z_1^3}\right]\right\},\\
[Q_aA_0] &= n_0\left\{-\left(\frac{c^3}{8} + a_4\right)(x^2+y^2)^2 + \left\{x^2+(y-p_0)^2\right\}\left[\frac{c}{4z_0^2}(x^2+y^2) - \frac{x^2+(y-p_0)^2}{8z_0^3}\right]\right\},\\
\left[Q_pA_1\right] &= n_1\left\{-\left(\frac{c^3}{8} + a_4\right)\bar{y}^4 + (\bar{y}-p_1)^2\left[\frac{c}{4z_1^2}\bar{y}^2 - \frac{(\bar{y}-p_1)^2}{8z_1^3}\right]\right\},\\
\left[Q_pA_0\right] &= n_0\left\{-\left(\frac{c^3}{8} + a_4\right)\bar{y}^4 + (\bar{y}-p_0)^2\left[\frac{c}{4z_0^2}\bar{y}^2 - \frac{(\bar{y}-p_0)^2}{8z_0^3}\right]\right\}.
\end{aligned} \quad (5.3.14)$$

The wavefront aberration is described as a function of the coordinates (x, y) with respect to the vertex T. For aberration analysis it is more instructive to write the aberration terms with the origin shifted to $Q_p(0, \bar{y})$ on the principal ray of the object- and image-side pencils. The new transverse coordinates with respect to Q_p are denoted by (x_1, y_1). Simultaneously, we eliminate the quantities p_0 and p_1 that are directly proportional to the value of $\bar{y}$. From Fig. 5.30 we deduce

$$\begin{aligned}
x = x_1, \qquad y &= y_1 + \bar{y}, \qquad p_1 = -\frac{z_1 - \bar{z}_1}{\bar{z}_1}\,\bar{y},\\
y - p_1 &= y_1 + \bar{y} - p_1 = y_1 + \frac{z_1}{\bar{z}_1}\,\bar{y},\\
x^2 + (y-p_1)^2 &= x_1^2 + y_1^2 + 2\left(\frac{z_1}{\bar{z}_1}\right)y_1\bar{y} + \left(\frac{z_1}{\bar{z}_1}\right)^2\bar{y}^2,
\end{aligned} \quad (5.3.15)$$

and similar expressions apply to the corresponding quantities in object space. Evaluation of the expression for $[Q_aA_1]$ and collection of the aberration terms with the same dependence on pupil and field coordinates yields the expression

$$\begin{aligned}
[Q_aA_1] = n_1\Bigg[&\left\{\frac{c}{4z_1^2} - \frac{1}{8z_1^3} - \frac{c^3}{8} - a_4\right\}(x_1^2+y_1^2)^2 + \left\{\frac{c}{2z_1^2} + \frac{c}{2z_1\bar{z}_1} - \frac{1}{2z_1^2\bar{z}_1} - \frac{c^3}{2} - 4a_4\right\}\bar{y}y_1(x_1^2+y_1^2)\\
&+\left\{\frac{c}{4z_1^2} + \frac{c}{4\bar{z}_1^2} - \frac{1}{4z_1\bar{z}_1^2} - \frac{c^3}{4} - 2a_4\right\}\bar{y}^2(x_1^2+y_1^2) + \left\{\frac{c}{z_1\bar{z}_1} - \frac{1}{2z_1\bar{z}_1^2} - \frac{c^3}{2} - 4a_4\right\}\bar{y}^2y_1^2\\
&+\left\{\frac{c}{2z_1\bar{z}_1} + \frac{c}{2\bar{z}_1^2} - \frac{1}{2\bar{z}_1^3} - \frac{c^3}{2} - 4a_4\right\}\bar{y}^3y_1 + \left\{\frac{c}{4\bar{z}_1^2} - \frac{z_1}{8\bar{z}_1^4} - \frac{c^3}{8} - a_4\right\}\bar{y}^4\Bigg],
\end{aligned} \quad (5.3.16)$$

and a similar expression for $[Q_aA_0]$. The expression for $[Q_pA_1]$ is obtained from $[Q_aA_1]$ by substituting $(x_1, y_1) = (0, 0)$ and reduces to the term on the last line of Eq. (5.3.16). Returning to the basic expression for the wavefront aberration of Eq. (5.3.5), we finally have

$$
\begin{aligned}
\Delta W^{(4)} =\delta \Bigg(n\Bigg[&\left\{ \frac{c}{4z^2} - \frac{1}{8z^3} - \frac{c^3}{8} - a_4 \right\} (x_1^2+y_1^2)^2 + \left\{ \frac{c}{2z^2} + \frac{c}{2z\bar{z}} - \frac{1}{2z^2\bar{z}} - \frac{c^3}{2} - 4a_4 \right\} \bar{y}y_1(x_1^2+y_1^2) \\
&+ \left\{ \frac{c}{4z^2} + \frac{c}{4\bar{z}^2} - \frac{1}{4z\bar{z}^2} - \frac{c^3}{4} - 2a_4 \right\} \bar{y}^2(x_1^2+y_1^2) + \left\{ \frac{c}{z\bar{z}} - \frac{1}{2z\bar{z}^2} - \frac{c^3}{2} - 4a_4 \right\} \bar{y}^2 y_1^2 \\
&+ \left\{ \frac{c}{2z\bar{z}} + \frac{c}{2\bar{z}^2} - \frac{1}{2\bar{z}^3} - \frac{c^3}{2} - 4a_4 \right\} \bar{y}^3 y_1 \Bigg] \Bigg) ,
\end{aligned}
\tag{5.3.17}
$$

where the δ symbol indicates that we first have to calculate the image-side expression with the index 1 inserted for the refractive index and the $z,\bar{z}$-coordinates and subtract from these the object-side quantities with everywhere the index 0. On each line of the expression for $\Delta W^{(4)}$ we recognise a specific third-order aberration term; they are spherical aberration, coma, field curvature, astigmatism and distortion, respectively. The coordinates (x_1, y_1) apply formally to the intersection point of a general ray with the optical surface. Within the paraxial approximation, these coordinates can be put equal to the paraxial intersection heights with the tangent plane to the optical surface through T. The coordinates (x_1, y_1) can also be put equal to the lateral coordinates of the point of intersection of a general ray with the reference sphere through Q_p. Figure 5.30 shows that this becomes a rather crude assumption once the opening angles, α_0 and α_1, or the field coordinates, p_0 or p_1 have appreciable values. Although the fourth-order wavefront analysis produces correct results, the total aberration value contains important sixth-order and higher-order terms that cannot be neglected in such circumstances.

The expressions of Eq. (5.3.17) are basically suitable for numerical calculations but, because of the frequently occurring fractions, they have a risk of instability. Moreover, they do not provide much analytic insight, for instance about special aberration-free imaging cases that have already been discussed in Section 4.6. In reference **[364]**, it is shown how the expressions for the various aberration terms can be streamlined by adding appropriate quantities that are identical in object and image space and disappear once the δ-subtraction operation has been carried out in Eq. (5.3.17). In Section 4.10, Eq. (4.10.4), the paraxial imaging equation was given in terms of the object and image distance to the vertex T of the surface and this equation yields the refraction-invariant quantity

$$
n_1 \left(c - \frac{1}{z_1} \right) = n_0 \left(c - \frac{1}{z_0} \right) . \tag{5.3.18}
$$

It can be multiplied by any quantity on the interface between the two media to yield other refraction-invariant quantities like

$$
A = n_0 y_1 \left(c - \frac{1}{z_0} \right) = n_0 i_0 = n_1 i_1 = n_1 y_1 \left(c - \frac{1}{z_1} \right) , \tag{5.3.19}
$$

$$
B = n_0 \bar{y} \left(c - \frac{1}{\bar{z}_0} \right) = n_0 \bar{i}_0 = n_1 \bar{i}_1 = n_1 \bar{y} \left(c - \frac{1}{\bar{z}_1} \right) , \tag{5.3.20}
$$

$$
H = n_0 \alpha_0 p_0 = n_1 \alpha_1 p_1 , \tag{5.3.21}
$$

where we have also added the Lagrange invariant, which is preserved throughout the optical system. The quantity A is a refraction invariant for the aperture ray that originates from an axial object point and intersects the optical surface and the reference sphere for that object point at a height y_1 from the axis. B is invariant at refraction by the principal ray. The Lagrange invariant H refers to the maximum aperture angles α and maximum field excursions p in the object and image plane. Another quantity that is useful in the final representation of the third-order terms but that is not invariant at refraction or reflection is

$$
\delta \left(\frac{1}{nz} \right) = - \frac{1}{r_1} \delta \left(\frac{\alpha}{n} \right) , \tag{5.3.22}
$$

where $r_1 = (x_1^2 + y_1^2)^{1/2}$ is the distance from the axis to the point of intersection of an aperture ray. The quantities A and $\delta(\alpha/n)$ are related to an aperture ray, B is associated with the principal rays of the imaging pencils. H is a mixed quantity, invariant throughout the optical system. We now take apart each of the aberration terms of Eq. (5.3.17) and adapt them to obtain expressions that are easier to interpret:

- **Spherical aberration**
  The expression for spherical aberration can be cast in a more accessible form by adding one or several terms that are invariant under a refraction or reflection. For spherical aberration it is useful to introduce a term that is proportional to

the refraction invariant A of Eq. (5.3.19). We add the quantity $n(c-1/z)\,c^2\,(x_1^2+y_1^2)^2/8$ to the first term of Eq. (5.3.17) and after some rearrangement we obtain the following expression for the spherical aberration term,

$$W_{sp} = \left\{-\frac{n_1^2}{8}\left(c-\frac{1}{z_1}\right)^2\left(\frac{1}{n_1 z_1}\right)+\frac{n_0^2}{8}\left(c-\frac{1}{z_0}\right)^2\left(\frac{1}{n_0 z_0}\right)-a_4(n_1-n_0)\right\}(x_1^2+y_1^2)^2 . \tag{5.3.23}$$

This expression can be normalised by dividing the factor $(x_1^2+y_1^2)^2$ by r_m^4, where r_m is the maximum value of the transverse coordinate $(x_1^2+y_1^2)^{1/2}$ of a marginal ray from the axial pencil. This yields the following expression for the spherical aberration term:

$$W_{sp} = \frac{1}{8}\left\{A_m^2 r_m\,\delta\left(\frac{\alpha_m}{n}\right)-8a_4 r_m^4\delta(n)\right\}\frac{(x_1^2+y_1^2)^2}{r_m^4} , \tag{5.3.24}$$

where A_m and α_m have to be evaluated for a marginal ray of the axial pencil and the δ operation was introduced earlier with the aid of Eq. (5.3.17).

- **Coma**

To eliminate the term with c^3 in the wavefront aberration expression we add the A-type refraction-invariant quantity $n(c-1/z)c^2\bar{y}y_1(x_1^2+y_1^2)/2$ of Eq. (5.3.20) to the coma term. After some rearrangement this yields the expression

$$W_{co} = \delta\left(n\left\{-\frac{1}{2z}\left(c-\frac{1}{z}\right)\left(c-\frac{1}{\bar{z}}\right)-4a_4\right\}\right)\bar{y}y_1(x_1^2+y_1^2) . \tag{5.3.25}$$

We divide the pupil coordinates by $y_{1,m}r_m^2$ and the principal ray coordinate $\bar{y}$ by its maximum value $\bar{y}_m$ for a position at the border of the image field. The expression for the comatic wavefront aberration term W_{co} is then given by

$$W_{co} = \frac{1}{2}\left\{A_m B_m y_{1,m}\,\delta\left(\frac{\alpha_m}{n}\right)-8a_4\bar{y}_m y_{1,m} r_m^2\delta(n)\right\}\frac{\bar{y}}{\bar{y}_m}\frac{y_1}{y_{1,m}}\frac{(x_1^2+y_1^2)}{r_m^2} . \tag{5.3.26}$$

The values of A_m and $\delta(\alpha_m/n)$ are obtained for a transverse coordinate value r_m. The quantity B_m is evaluated for the intersection height $\bar{y}_m$ of the principal ray of an oblique pencil towards the border of the image field.

- **Field curvature**

Proceeding in a similar way as before for spherical aberration and coma terms of the wavefront aberration we now add the refraction-invariant quantity $n(c-1/z)c^2\bar{y}^2(x_1^2+y_1^2)/4$ to the field curvature term of Eq. (5.3.17) and obtain the modified expression

$$W_{fc} = \delta\left(n\left[-\frac{c^2}{4z}-\frac{1}{4z\bar{z}^2}+\frac{c}{4z^2}+\frac{c}{4\bar{z}^2}-2a_4\right]\right)\bar{y}^2(x_1^2+y_1^2) . \tag{5.3.27}$$

We add to the terms between the brackets twice the quantity $c/(2z\bar{z})$, with opposite signs. This allows us to write the expression above as

$$W_{fc} = \delta\left(n\left[-\frac{1}{4z}\left(c-\frac{1}{\bar{z}}\right)^2+\frac{c}{4}\left(\frac{1}{z}-\frac{1}{\bar{z}}\right)^2-2a_4\right]\right)\bar{y}^2(x_1^2+y_1^2) . \tag{5.3.28}$$

The quantity $n(1/z-1/\bar{z})$ needs some further elaboration. Starting with the Lagrange invariant $H=n\alpha_m p_m$ for the largest aperture angle and field value of the image, we write

$$\alpha_m = -\frac{r_m}{z} , \quad p = -\frac{\bar{y}_m(z-\bar{z})}{\bar{z}} , \quad \rightarrow \quad n(\frac{1}{z}-\frac{1}{\bar{z}}) = -\frac{H}{r_m\bar{y}_m} . \tag{5.3.29}$$

This allows us to express the second term between the square brackets of the field curvature expression with the aid of the Lagrange invariant,

$$\frac{c}{4}\,\delta\left[n\left(\frac{1}{z}-\frac{1}{\bar{z}}\right)^2\right] = \frac{c}{4}\frac{H^2}{(r_m\bar{y}_m)^2}\,\delta\left(\frac{1}{n}\right) . \tag{5.3.30}$$

The complete expression for the field curvature is then written as

$$W_{fc} = \frac{1}{4}\left\{B_m^2 r_m\,\delta\left(\frac{\alpha_m}{n}\right)+H^2c\,\delta\left(\frac{1}{n}\right)-8a_4\bar{y}_m^2 r_m^2\delta(n)\right\}\frac{\bar{y}^2}{\bar{y}_m^2}\frac{(x_1^2+y_1^2)}{r_m^2} . \tag{5.3.31}$$

- **Astigmatism**
  To adjust the astigmatic aberration term we add the quantity $n(c - 1/z)c^2\bar{y}^2 y_1^2/2$ and, after some manipulation, we obtain the expression

$$W_{\rm as} = \frac{1}{2}\left\{\delta\left[\,n\left(c - \frac{1}{\bar{z}}\right)^2\left(-\frac{1}{z}\right)\right] - 8a_4\delta(n)\right\}\bar{y}^2 y_1^2 . \tag{5.3.32}$$

After the proper normalisation steps we have the following astigmatic term

$$W_{\rm as} = \frac{1}{2}\left\{B_m^2 r_m\,\delta\left(\frac{\alpha_m}{n}\right) - 8a_4\bar{y}_m^2 r_m^2\delta(n)\right\}\frac{\bar{y}^2}{\bar{y}_m^2}\frac{y_1^2}{r_m^2} . \tag{5.3.33}$$

- **Distortion**
  The term $n(c - 1/z)c^2\bar{y}^3 y_1/2$ is added to the distortion aberration term and we obtain the expression

$$W_{\rm di} = \frac{1}{2}\left\{\delta\left[\,n\left(c - \frac{1}{\bar{z}}\right)\left(-\frac{c}{z} + \frac{1}{\bar{z}^2}\right)\right] - 8a_4\delta(n)\right\}\bar{y}^3 y_1 . \tag{5.3.34}$$

We multiply numerator and denominator of the term between square brackets by the aperture refraction invariant $n(c - 1/z)$ and obtain, after the intentional insertion of the terms $\pm 2c/(z\bar{z})$, the expression

$$W_{\rm di} = \frac{1}{2}\left\{\frac{n(c-1/\bar{z})}{n(c-1/z)}\,\delta\left[n\left\{[-\frac{(c-1/\bar{z})^2}{z} - c\left(\frac{1}{z} - \frac{1}{\bar{z}}\right)^2\right\}\right] - 8a_4\delta(n)\right\}\bar{y}^3 y_1 . \tag{5.3.35}$$

The second term between the curly braces of the distortion expression is similar to the second term of the field curvature expression. We treat this term in the same way by introducing the Lagrange invariant H. The final expression for the distortion aberration then reads

$$W_{\rm di} = \frac{1}{2}\left\{\frac{B_m^3}{A_m}y_{1,m}\,\delta\left(\frac{\alpha_m}{n}\right) + \frac{B_m}{A_m}H^2 c\,\delta\left(\frac{1}{n}\right) - 8a_4\bar{y}_m^3 y_{1,m}\delta(n)\right\}\frac{\bar{y}^3}{\bar{y}_m^3}\frac{y_1}{y_{1,m}} , \tag{5.3.36}$$

where we have used the identity of the quadratic quantities r_m^2 and $y_{1,m}^2$. B_m has to be evaluated for the value $\bar{y}_m$ of the intersection height of the principal ray with the surface.

5.3.3 Seidel Wavefront Aberration Formula

The aberration terms derived in the previous subsection are now grouped together according to the scheme originally proposed by Seidel and summed over all optical surfaces of the optical system. In the expressions for the individual aberration terms, we replace the relative height of incidence of the principal ray, $\bar{y}/\bar{y}_m$, on each surface by the relative height p/p_m of the paraxial image point in the image plane where p_m is the maximum distance of an image point from the optical axis. The same operation is carried out for the heights of incidence of the rays on the intermediate reference spheres. The relative coordinate values $(x_1, y_1)/r_m$ on each surface are replaced by the relative coordinates $(x_E, y_E)/r_{m,E}$, where (x_E, y_E) are the coordinates of the intersection point of a ray with the exit pupil sphere of the system. The diameter of the rim of the exit pupil is $2r_{m,E}$.

5.3.3.1 Seidel Aberration Formula of an Optical System

From a paraxial ray-trace of the marginal ray from an on-axis object point and the principal ray from an outer field position, we collect the required ray data at each optical surface of the system. We then obtain the expression

$$\Delta W^{(4)}(x_E, y_E; p) = \frac{1}{8}S_1\frac{\left(x_E^2 + y_E^2\right)^2}{r_{m,E}^4} + \frac{1}{2}S_2\frac{y_E}{r_{m,E}}\frac{\left(x_E^2 + y_E^2\right)}{r_{m,E}^2}\frac{p}{p_m} + \frac{1}{2}S_3\frac{y_E^2}{r_{m,E}^2}\frac{p^2}{p_m^2} + \frac{1}{4}(S_3 + S_4)\frac{(x_E^2 + y_E^2)}{r_{m,E}^2}\frac{p^2}{p_m^2}$$
$$+ \frac{1}{2}S_5\frac{y_E}{r_{m,E}}\frac{p^3}{p_m^3} . \tag{5.3.37}$$

The proper Seidel-coefficients $S_1, \cdots, S_5$ are given by the surface sums

$$S_1 = \sum_j \left\{ A^2_{m,j} r_{m,j}\, \delta\!\left(\frac{\alpha_{m,j}}{n_j}\right) - 8a_{4,j} r^4_{m,j}\delta(n_j) \right\} ,$$

$$S_2 = \sum_j \left\{ A_{m,j} B_{m,j} r_{m,j}\, \delta\!\left(\frac{\alpha_{m,j}}{n_j}\right) - 8a_{4,j} r^3_{m,j}\bar{y}_{m,j}\delta(n_j) \right\} ,$$

$$S_3 = \sum_j \left\{ B^2_{m,j} r_{m,j}\, \delta\!\left(\frac{\alpha_{m,j}}{n_j}\right) - 8a_{4,j} r^2_{m,j}\bar{y}^2_{m,j}\delta(n_j) \right\} ,$$

$$S_4 = H^2 \sum_j c_j\, \delta\!\left(\frac{1}{n_j}\right) ,$$

$$S_5 = \sum_j \left\{ \frac{B^3_{m,j}}{A_{m,j}} r_{m,j}\, \delta\!\left(\frac{\alpha_{m,j}}{n_j}\right) + \frac{B_{m,j}}{A_{m,j}} H^2 c_j\, \delta\!\left(\frac{1}{n_j}\right) - 8a_{4,j} r_{m,j}\bar{y}^3_{m,j}\, \delta(n_j) \right\} , \tag{5.3.38}$$

where the difference operator δ subtracts the value of its argument in space j from that in space $j-1$. The Seidel coefficients S_1, S_2 and S_5 are immediately identified as those of spherical aberration, coma and distortion. The Seidel coefficient S_3 determines, alone, the astigmatism of the system. S_3 and S_4 together determine the curvature of field. In the absence of astigmatism, it is S_4 alone that is responsible for the field curvature of the system. The Seidel coefficient S_4 is a particular one. It is independent of the magnification of the system and of the position of the pupils. A surface contribution $S_{4,j}$ depends only on the curvature of that particular surface and on the inverted optical contrast, $(n_{j-1} - n_j)/n_{j-1}n_j$. Although the presence of astigmatism ($S_3 \neq 0$) in a system also influences the field curvature, a nonzero coefficient S_4 forms the basic contribution to field curvature. For that reason, S_4 is usually called the field curvature coefficient. The quantity S_4 was originally identified by Petzval. For that reason, the quantity

$$P = \sum_j c_j\, \delta\!\left(\frac{1}{n_j}\right) , \tag{5.3.39}$$

is called the Petzval sum of an optical system and, in the absence of astigmatism, it bears a direct relationship with the curvature of the image field.

The combined contribution from astigmatism and field curvature in Eq. (5.3.37) can also be written as

$$W^{(4)}(x_E, y_E; p) = \frac{1}{2} S_3 \frac{y_E^2}{r^2_{m,E}} \frac{p^2}{p^2_m} + \frac{1}{4}(S_3 + S_4) \frac{(x_E^2 + y_E^2)}{r^2_{m,E}} \frac{p^2}{p^2_m} = \left\{ \frac{1}{4}(3S_3 + S_4) \frac{y_E^2}{r^2_{m,E}} + \frac{1}{4}(S_3 + S_4) \frac{x_E^2}{r^2_{m,E}} \right\} \frac{p^2}{p^2_m} . \tag{5.3.40}$$

With reference to the meridional plane $x = 0$, the first term determines the curvature of the surface on which the tangential astigmatic focal lines are sharply imaged, the second term yields the curvature of the surface with sharply imaged sagittal focal lines on it (see Fig. 5.24, in which the astigmatic focal surfaces have been denoted by s and t). To calculate the radii of curvature R_t and R_s of the tangential and sagittal field surfaces, respectively, we need the axial distances z_t and z_s from the focal lines to the flat image surface. These distances are obtained from Eq. (5.2.24) by retaining the first term of the series, yielding

$$z = \frac{2\Delta_m}{\alpha^2_m} , \tag{5.3.41}$$

where Δ_m is the geometrical distance between wavefront and reference sphere at the rim of the pencil in the tangential or sagittal section. As before, α_m is the maximum aperture angle of the imaging pencil.

For the geometrical wavefront departure Δ_m in a medium with refractive index n_I we have W_m/n_I. To obtain the radii of curvature in the two cross-sections we use the parabolic surface approximation for the image surface,

$$R_c = \frac{p^2_m}{2z} = \frac{p^2_m \alpha^2_m}{4\Delta_m} = \frac{H^2}{4n_I^2 \Delta_m} , \tag{5.3.42}$$

where we have substituted the Lagrange invariant H of the imaging system in the expression.

The two radii of curvature are then given by

$$R_t = \frac{H^2}{n_I\,(3S_3 + S_4)} , \qquad R_s = \frac{H^2}{n_I\,(S_3 + S_4)} . \tag{5.3.43}$$

In the astigmatism-free case with $S_3 = 0$, the curvatures of both surfaces are identical. In that case we are in the presence of intrinsic field curvature with a coefficient S_4 that is exclusively given by Eq. (5.3.60). The curvature of the image field is then given by

$$c_{\mathrm{fc}} = n_I \sum_j c_j \,\delta(1/n_j) = n_I P \,, \tag{5.3.44}$$

where P is the Petzval sum of the system.

5.3.3.2 Seidel Coefficients in Terms of Image and Pupil Magnification

The Seidel coefficients for a single surface can also be expressed in terms of the lateral magnification factors of image and exit pupil, m_T and $m_{T,p}$, respectively. We employ the paraxial imaging equation and the expressions for the transverse and angular magnifications,

$$\frac{n_1}{z_1} = \frac{n_0}{z_0} + K\,, \qquad m_T = \frac{n_0 z_1}{n_1 z_0}\,, \qquad m_A = \frac{n_0}{n_1\, m_T}\,, \tag{5.3.45}$$

where $K = (n_1 - n_0)c$ is the optical power of the surface and c is, as usual, the curvature of the surface. Using comparable expressions for the imaging of the pupils, we find the expressions

$$\begin{aligned} z_1 &= \frac{n_1(1-m_T)}{K}\,, & z_0 &= \frac{n_0(1-m_T)}{m_T K}\,, \\ \bar{z}_1 &= \frac{n_1(1-m_{T,p})}{K}\,, & \bar{z}_0 &= \frac{n_0(1-m_{T,p})}{m_{T,p} K}\,. \end{aligned} \tag{5.3.46}$$

The refraction invariant A_m and the quantity $\delta(\alpha_m/n)$ are written as

$$A_m = -\frac{(n_0 - n_1 m_T)}{(n_1 - n_0)}(n_1\alpha_{1,m})\,, \qquad \delta(\alpha_m/n) = \left(\frac{1}{n_1^2} - \frac{m_T}{n_0^2}\right)(n_1\alpha_{1,m})\,, \tag{5.3.47}$$

where we have used that $y_{1,m} = -\alpha_{1,m} z_1$ ($\alpha_{1,m}$ is the maximum opening angle of the imaging pencils).

For the refraction invariant B_m of the principal ray we find the expression

$$B_m = -\frac{K(n_0 - n_1 m_{T,p})}{(n_1 - n_0)(m_{T,p} - m_T)}\, p_{1,m}\,, \tag{5.3.48}$$

where $p_{1,m}$ is the maximum off-axis distance of the image point.

We only consider the three aberration terms that introduce unsharpness: spherical aberration, coma and astigmatism. For the aberration increment at refraction or reflection at the surface we find the expressions

$$\Delta W_1 = \frac{1}{8}\left[\left\{-\frac{(1-m_T)(n_0 - n_1 m_T)^2(n_0^2 - n_1^2 m_T)}{(n_1 - n_0)^2 n_0^2 n_1^2} - 8a_4(n_1 - n_0)\frac{(1-m_T)^4}{K^3}\right\}\frac{(n_1\alpha_{1,m})^4}{K}\right]\frac{(x_1^2 + y_1^2)^2}{r_m^4}\,, \tag{5.3.49}$$

$$\begin{aligned} \Delta W_2 = \frac{1}{2}\Bigg[\Bigg\{&\frac{(1-m_T)(n_0 - n_1 m_T)(n_0^2 - n_1^2 m_T)}{(n_1 - n_0)^2 n_0^2 n_1^2}\,\frac{(n_0 - n_1 m_{T,p})}{(m_T - m_{T,p})} \\ &+ 8a_4(n_1 - n_0)\frac{(1-m_T)^3(1-m_{T,p})}{K^3(m_T - m_{T,p})}\Bigg\}p_{1,m}(n_1\alpha_{1,m})^3\Bigg]\frac{y_1}{y_{1,m}}\,\frac{(x_1^2 + y_1^2)}{r_m^2}\,\frac{p_1}{p_{1,m}}\,, \end{aligned} \tag{5.3.50}$$

$$\begin{aligned} \Delta W_3 = \frac{1}{2}\Bigg[\Bigg\{&-\frac{(1-m_T)(n_0^2 - n_1^2 m_T)}{(n_1 - n_0)^2 n_0^2 n_1^2}\left(\frac{n_0 - n_1 m_{T,p}}{m_T - m_{T,p}}\right)^2 \\ &- 8a_4(n_1 - n_0)\frac{(1-m_T)^2(1-m_{T,p})^2}{K^3(m_T - m_{T,p})^2}\Bigg\}K p_{1,m}^2(n_1\alpha_{1,m})^2\Bigg]\frac{y_1^2\, p_1^2}{r_m^2\, p_{1,m}^2}\,, \end{aligned} \tag{5.3.51}$$

where $n_1\alpha_{1,m}$ is the paraxial equivalent of the numerical aperture $n_1 \sin\alpha_{1,m}$ of the imaging pencils.

The factors between the large square brackets in Eqs. (5.3.49)–(5.3.51) are recognised as the Seidel aberration coefficients S_1, S_2 and S_3 of the specific surface in terms of the image and pupil magnification. The factor that determines the length dimension of each aberration can be written in a different way, using the height of incidence $y_{1,m}$ of the marginal ray on the surface:

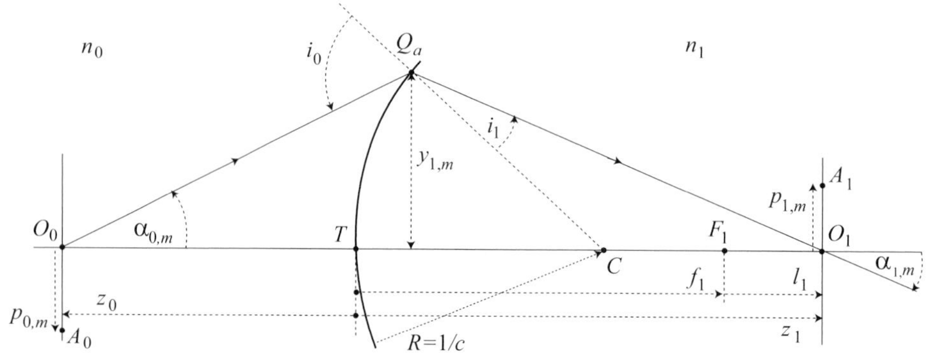

Figure 5.31: The convergence angle $\alpha_{1,m}$ and height of incidence $y_{1,m}$ on the refracting surface of a marginal ray. F_1 is the image-side focal point, at a distance $n_1 R/(n_1 - n_0)$ from the vertex T. $l_1 = F_1 O_1$ is the image distance measured from the focal point.

- *Spherical aberration*
  The factor $(n_1\alpha_{1,m})^4/K$ is modified as follows, with reference to Fig. 5.31. Using Newton's imaging equation, we have

$$z_1 = f_1 + l_1 = f_1(1 - m_T)\,, \quad \text{and,} \quad n_1\alpha_{1,m} = -\frac{n_1 y_{1,m}}{z_1} = -\frac{y_{1,m}K}{(1 - m_T)}\,.$$

  We thus obtain

$$(n_1\alpha_{1,m})^4/K = \frac{y_{1,m}^4 K^3}{(1 - m_T)^4}\,. \qquad (5.3.52)$$

- *Coma*
  Using the Lagrange invariant $H = n_1\alpha_{1,m}p_{1,m}$ we find

$$p_{1,m}(n_1\alpha_{1,m})^3 = H(n_1\alpha_{1,m})^2 = \frac{y_{1,m}^2 K^2 H}{(1 - m_T)^2}\,. \qquad (5.3.53)$$

  We note that the Lagrange invariant, in its special form $H = n\alpha p$ for an object or image plane, has a sign that depends on the field position $p_{0,m}$ in the object plane and the starting angle $\alpha_{0,m}$ of the marginal ray in object space. Once a certain initial choice has been made, the value of H remains unchanged in any cross-section of the optical system.
- *Astigmatism and field curvature*
  Both aberration terms contain the factor $Kp_{1,m}^2(n_1\alpha_{1,m})^2$ that is directly written as KH^2 or, equivalently, $K^3p_{1,m}^2y_{1,m}^2/(1 - m_T)^2$. Inspection of the aberration function shows that, at full aperture, the spherical aberration coefficient S_1 is proportional to K^{-1}, the coma term to p_1 and the astigmatic term to Kp_1^2. The optical power K can be considered as a scaling factor of the imaging geometry, proportional to the curvature of the optical surface. Low values of spherical aberration are obtained by increasing the power K of the surface. It means, not surprisingly, that a more compact imaging geometry lowers the on-axis aberration. This effect is absent for coma. Making the imaging geometry more compact would reduce the aberration but the field angle $\gamma_{1,m}$, proportional to p_1K, increases. These effects neutralise each other when considering their influence on the comatic aberration. The coefficient S_3 for astigmatism increases quadratically with the field position. Lowering the power reduces the astigmatism which implies that it is favourable to aim at a small field angle. The practical requirements on the individual aberrations determine the optimum choice for the surface curvature value.

Figures 5.32 and 5.33 show the changes of the maximum values of the three aberration terms W_1, W_2 and W_3 as a function of the lateral magnification m_T for a refracting and a reflecting optical surface. At the same value of numerical aperture, it is seen that the choice of a relatively high-index refracting material strongly reduces the spherical aberration and, to a lesser degree, also the coma and astigmatism induced at the surface. A general observation is that the aberration values are of the order of the wavelength of the light for the chosen values of numerical aperture, $n_1\alpha_{1,m} = 0.1$, and field position, $p_{1,m} = 2.5$ mm, in the second medium. At these values of aperture and field, the third-order aberration theory is

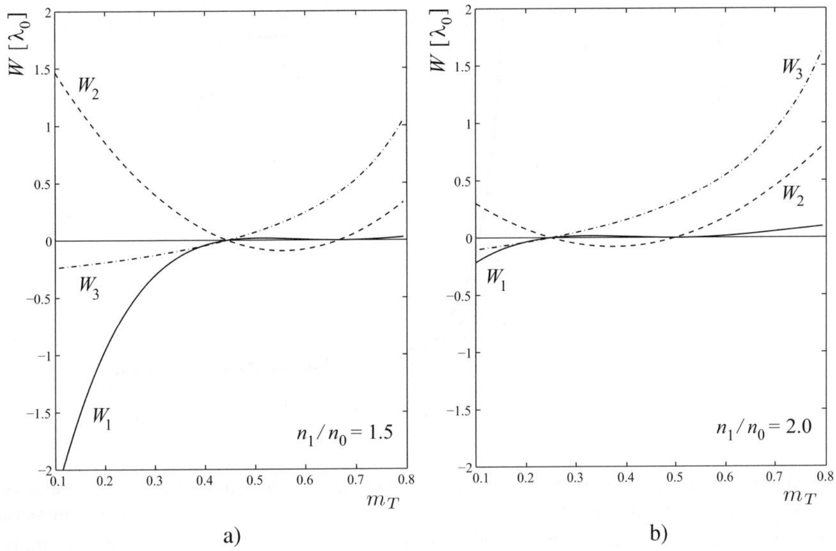

Figure 5.32: Spherical aberration, coma and astigmatism of a single refracting surface as a function of the lateral magnification m_T, in units of the vacuum wavelength λ_0 of the light. The pupil magnification $m_{T,p}$ is unity (pupils at the vertex of the surface). $K = 0.01\ \text{mm}^{-1}$, $p_{1,m} = 2.5$ mm, $n_1\alpha_{1,m} = 0.1$, $\lambda_0 = 0.5$ µm.
a) Optical contrast $n_1/n_0 = 1.5$; b) $n_1/n_0 = 2.0$.

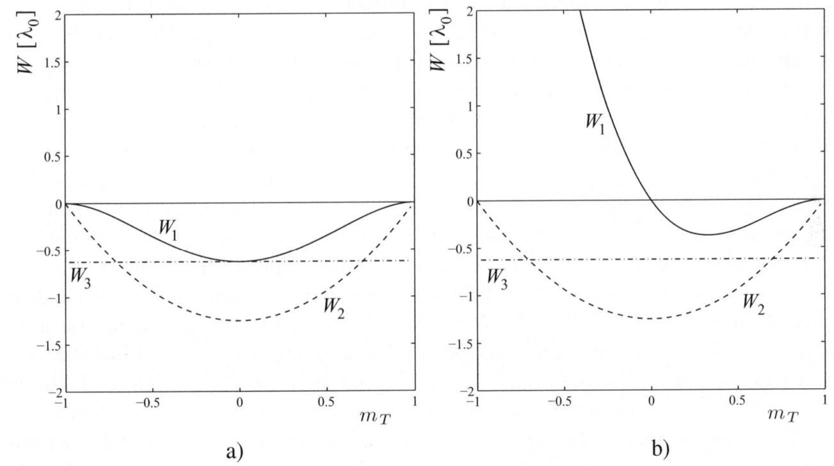

Figure 5.33: Spherical aberration, coma and astigmatism of a concave mirror as a function of the lateral magnification m_T, in units of λ_0 and for a pupil magnification $m_{T,p} = 1$. $K = 0.01\ \text{mm}^{-1}$, $p_{1,m} = 2.5$ mm, $n_1\alpha_{1,m} = 0.1$, $\lambda_0 = 0.5$ µm. The refractive index n_0 is unity.
a) Spherical mirror; b) Parabolic mirror.

a satisfactory approximation, for typical values of the focal distances such as those of the given example ($f_0 = -n_0/K$, $f_1 = n_1/K$).

Figure 5.33 shows the aberration values for a concave mirror, at equal values of numerical aperture, field position and power as in the refractive example. Spherical aberration is relatively small; astigmatism is independent of the magnification. In b), an aspheric coefficient has been introduced, $a_4 = -K^3/[8(n_1 - n_0)^3]$, turning the spherical surface

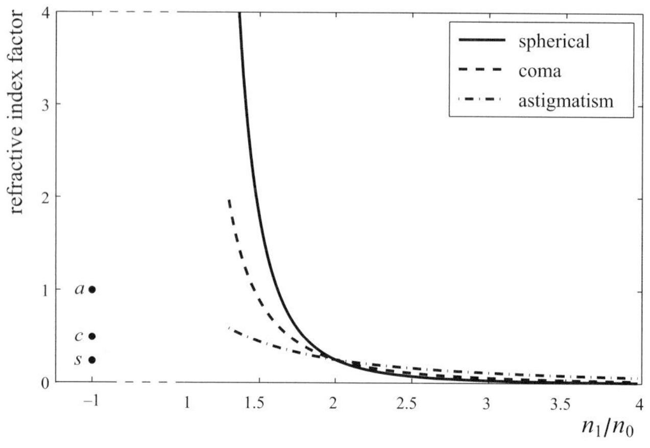

Figure 5.34: The influence of the optical contrast n_1/n_0 on the magnitude of spherical aberration, coma and astigmatism of a refracting surface. Imaging from infinity ($m_T = 0$), pupil at the surface ($m_{T,p} = +1$). The optical power, numerical aperture and field excursion are kept identical when varying the optical contrast. The index factors for a mirror surface are represented by the three dots labelled s, c and a for the value $n_1/n_0 = -1$.

into a paraboloid. It is seen that the spherical aberration is now reduced to zero for $m_T = 0$. The field aberrations are unchanged by adding the aspheric term to the surface profile. This is a consequence of the chosen pupil position: a centrally positioned pupil with lateral magnification $m_{T,p} = +1$. For this case, the fourth-order surface correction only affects the aberration of off-axis beams in sixth order and leaves the Seidel aberration terms unchanged. If the object is at infinity, the coma is relatively strong as compared to that of a refractive surface. The limited field of a (parabolic) mirror due to its coma is the principal limitation of such a surface as a mirror in a telescope. The field aberrations of the single mirror (and of the refracting surface) can be influenced by shifting the stop away from the surface ($m_{T,p} \neq +1$). The effect of a more general pupil position on the Seidel aberrations is touched upon in Subsections 5.4.3 and 5.4.4 as well as in Chapter 7, where the design and performance of various types of imaging systems are discussed.

Figure 5.34 shows the influence of the optical contrast on the magnitude of the various aberrations. The refractive index factors for the aberrations are $\{n_1 n_0(n_1 - n_0)\}^{-2}$, $(n_1 n_0)^{-2}(n_1 - n_0)^{-1}$ and $(n_1 n_0)^{-2}$, for spherical aberration, coma and astigmatism, respectively. In the figure, n_0 was put equal to unity. At equal optical power, numerical aperture and field excursion, the spherical aberration term is the most sensitive to the optical contrast. The figure shows that aberration reduction is much easier when high-index materials are chosen. Practical limits in the visible domain of the spectrum are 1.45 on the lower side and 2.0 at the upper side. In the infrared region of the spectrum, refractive indices as high as 3.5 to 4 are encountered (germanium, silicon). The physical reason for the strong aberration reduction at high-index values is the smaller curvature of the optical surface at equal power. Consequently, the angles of incidence of the rays on the surface are strongly reduced and the paraxial imaging conditions are more closely approximated. For the special values $n_1/n_0 = -1$, $m_T = 0$ and $m_{T,p} = +1$, the index factors for a reflecting surface have been plotted for the three aberrations, labelled s, c and a. If we compare for equal imaging conditions the aberrations of a mirror and of a refracting surface in the index range $1.5 < n_1/n_0 < 2$, we notice that they behave in an opposite way. Astigmatism is the relatively strongest aberration of a mirror surface, spherical aberration the smallest.

5.3.4 Special Imaging Cases

Inspection of Eq. (5.3.38) or Eqs. (5.3.49)–(5.3.51) shows that one or more Seidel coefficients of an individual surface can be made zero in the following special imaging cases. We assume that the optical surface is spherical ($a_4 = 0$).

5.3.4.1 Imaging at the Centre of Curvature

$A_m = 0$, or, $m_T = n_0/n_1$ and $z_1 = z_0 = R$ where R is the radius of curvature of the refracting or reflecting surface. In this special case, object and image are at the centre of curvature and no refraction occurs. The imaging is virtual–real or

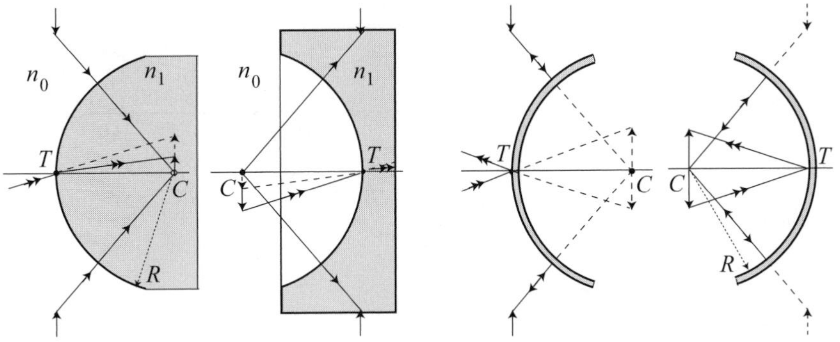

Figure 5.35: Imaging at the centre of curvature. From left to right: positive and negative refracting surface, convex and concave reflector. In all cases, the entrance and exit pupil are at the vertex T of the optical surface. Real rays and object or image heights are represented with the aid of solid lines, their virtual counterparts are dashed. The principal rays have been labelled by a double arrow.

real–virtual in the case of refraction, virtual–virtual or real–real in the case of a convex or a concave mirror (see Fig. 5.35). The imaging is free from spherical aberration and coma. As a function of magnification, the spherical aberration W_1 shows two second-order extrema. At the scale of the graphs in Fig. 5.32, these extrema are hardly visible.

5.3.4.2 Imaging at the Surface Vertex

For $r_m = 0$, or $z_1 = z_0 = 0$, we have at our disposal an (intermediate) imaging step that has zero Seidel aberration coefficients S_1, S_2 and S_3. Only field curvature and distortion are nonzero. Although this imaging step might seem a trivial one, it can be useful when only these two latter aberrations need to be changed. For instance, using a strongly negative refracting surface in such an imaging configuration permits us to compensate for the field curvature introduced by positive lenses elsewhere in the optical system. The imaging step enables a correction of the lateral and axial position of the image points without affecting the sharpness of the image, at least within the framework of the third-order aberration theory. A critical aspect is the surface quality of the optical surface. Any scratches on the surface or traces of environmental pollution, for instance dust particles, will be sharply copied onto the final image.

5.3.4.3 Imaging at Huygens' Aplanatic Points

The condition $\delta(u_j/n_j) = 0$ at refraction or, equivalently, $\delta(1/n_j z_j) = 0$, leads to a conjugate pair of points with magnification $m_T = n_0^2/n_1^2$. This pair was found in Section 4.6 when studying stigmatic imaging, namely Huygens' aplanatic points, with the axial coordinates

$$z_{0,j} = \frac{n_j + n_{j-1}}{n_{j-1}} R_j \,, \qquad z_{1,j} = \frac{n_j + n_{j-1}}{n_j} R_j \,. \tag{5.3.54}$$

In the third-order theory of aberration, we find these special points anew as they are free from spherical aberration, coma and astigmatism.

5.3.4.4 Imaging with Entrance and Exit Pupil at the Centre of Curvature

For $B_m = 0$, or $\bar{z}_1 = \bar{z}_0 = R$, the aberration coefficients for coma and astigmatism become zero. The distortion on a flat image plane is also made zero by having no refraction of the principal ray.

5.3.5 Field Curvature

Amidst the third-order aberrations, field curvature has a special status. The coefficient of field curvature is independent of the choice of conjugates, both of the object/image pair and of entrance and exit pupil. Only the paraxial quantities of curvature and optical contrast determine its magnitude. This means that the reduction of field curvature of a system is a difficult operation. A sequence of positive lenses will yield an inevitable negative value of field curvature, resulting in the

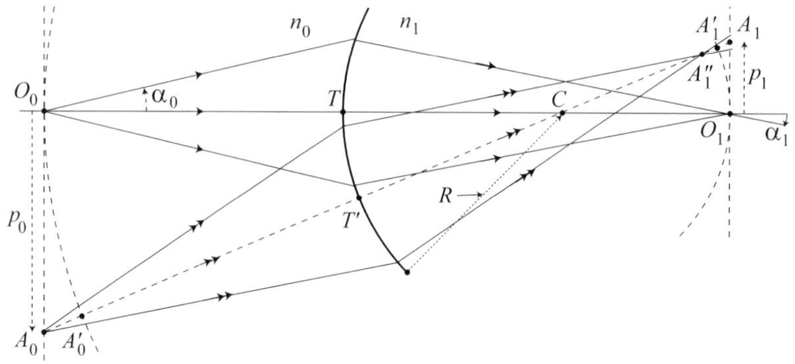

Figure 5.36: The on-axis paraxial imaging by an optical surface through T (centre of curvature at C, radius R) has been shown, for a pair of conjugate points O_0 and O_1. A_0' and A_1' are off-axis conjugate points, located on spherical surfaces through O_0 and O_1 and having their centres of curvature at C. The image point of A_0 is A_1'' on the line $A_0T'A_1$ through C.

well-known inward curving image field of a positive lens system. The reason for such an inward curving field follows from the geometry of Fig. 5.36, which depicts a single-surface transition with positive optical power. The spherical optical surface through T, with radius of curvature R and centre of curvature C, images the object point O_0 at O_1, at distances $TO_0 = z_0$ and $TO_1 = z_1$ from the vertex T. The points A_0' and A_1', both on spherical surfaces with centre of curvature C, are also conjugate points. The object point A_0 in the flat object plane through O_0 has a distance A_0A_0' from the conjugate point A_0' on the auxiliary axis $A_0A_0'T'CA_1''A_1'A_1$ through the centre of curvature C of the optical surface. The image A_1'' of A_0 follows from the shift $A_0'A_0$ and the axial magnification m_z of the system for the conjugate points A_0' and A_1'. We have, in a quadratic approximation for the distance $A_0'A_0$,

$$A_0'A_0 = -\frac{p_0^2}{2(R-z_0)}, \qquad m_z = \frac{dz_1}{dz_0} = \frac{n_0}{n_1}\frac{z_1^2}{z_0^2}. \tag{5.3.55}$$

Using this value of the paraxial axial magnification, we find for the total shift A_1A_1'' of the image point from the flat image surface through O_1 and A_1,

$$\begin{aligned} A_1A_1'' &= A_1A_1' + A_1'A_1'' \\ &= \frac{p_1^2}{2(R-z_1)} - \frac{n_0 z_1^2}{n_1 z_0^2}\,\frac{p_0^2}{2(R-z_0)}. \end{aligned} \tag{5.3.56}$$

Using the paraxial magnification between p_0 and p_1 and the paraxial imaging equation for the conjugate pair O_0 and O_1, we find after some rearrangement,

$$A_1A_1'' = \frac{p_1^2}{2}\left[\frac{1}{R-z_1} - \frac{n_1}{n_0}\frac{1}{R-z_0}\right] = -\frac{p_1^2}{2}\left\{\frac{n_1-n_0}{n_0R}\right\}. \tag{5.3.57}$$

For the radius of curvature R_{fc} of the curved image surface we thus have

$$R_{fc} = -\frac{n_0R}{n_1-n_0}. \tag{5.3.58}$$

To proceed from the geometrical radius of curvature to the wavefront aberration, we use the horizontal sag $z_{fc} = A_1A_1''$ from the flat image surface to the point A_1''. Neglecting the inclination angle of the principal ray with respect to the optical axis, we find for z_{fc} the expression given by Eq. (5.3.57). This axial shift z_{fc} of the image point corresponds to a maximum value W_{fc} of the quadratic wavefront aberration at the rim of the exit pupil,

$$W_{fc} = \frac{n_1 z_{fc}\sin^2(\alpha_1)}{2} \approx \frac{(n_1\alpha_1)^2 z_{fc}}{2n_1} = -\frac{n_1^2\alpha_1^2p_1^2(n_1-n_0)c}{4n_0n_1}, \tag{5.3.59}$$

where α_1 is the opening angle of the image-side pencil of rays.

Introducing the Lagrange invariant $H = n_1\alpha_1 p_1$, we have the final result for the Seidel coefficient S_4,

$$S_4 = cH^2\left(\frac{n_0 - n_1}{n_0 n_1}\right) = cH^2\delta\left(\frac{1}{n}\right) , \tag{5.3.60}$$

in accordance with the result of Eq. (5.3.38). The approach used above to calculate the wavefront aberration corresponding to field curvature could be extended to the other Seidel aberrations. In the case of a spherical optical surface, this method is relatively straightforward. Aspheric surfaces are less easily treated in this analysis. For that reason, we have preferred to use the more standard method of Subsection 5.3.2 for finding the Seidel aberration coefficients.

To obtain a flat image field for a system with positive power, negative lenses are used to offset the field curvature. If these negative lenses are positioned close to an intermediate image within the system, they hardly influence the total power of the system or the other aberrations but are fully effective for reducing the field curvature of the positive elements. The field curvature introduced by a mirror has, at equal power, a sign that is opposite to that of a refracting surface. In this case, the ratio of the field curvature coefficients of a mirror and of a refracting surface is given by $-n_{j-1}n_j$, when the mirror operates in a medium with index n_{j-1}. The various methods of field curvature reduction are discussed in more detail when design examples of optical systems are presented (see Chapter 7).

5.3.6 Aberration of the Pupil Imaging

The imaging quality of the exit pupil is far less critical than the quality of the image itself. Correct imaging of the entrance pupil onto the diaphragm and from the diaphragm onto the exit pupil is important for the transport of radiant energy through the optical system towards the image plane. Aberrations of the pupil imaging may degrade the efficiency of the light transport from object to image and cause field-dependent image intensity variations. For the case of a non-self-luminous object, the aberrations of the pupil imaging have to be considered together with the aberrations of the illumination system that guides the light from the source to the object. These effects are considered in more detail in Chapter 7 with design examples of optical imaging systems. In this subsection we highlight a pupil aberration that determines the through-focus behaviour of the image distortion, namely the spherical aberration of the pupil image. Since this pupil aberration is closely connected with the five Seidel image aberrations, it is frequently called the sixth Seidel aberration. A direct calculation of the spherical aberration of the pupil is possible by preparing the set of paraxial quantities that is needed for the calculation of spherical aberration. Referring to Fig. 5.30, we would need $B_{m,j}$, $\bar{y}_{m,j}$ and $\delta(\bar{\alpha}_{m,j}/n_j)$, where $\bar{\alpha}_{m,j}$ and $\bar{\alpha}_{m,j-1}$ are the maximum field angles in the intermediate image spaces. In this subsection we follow another approach that also clarifies the relation between the spherical aberration of the pupil and the aberration of the image itself.

We study the aberration that arises when the central point E_0 of the entrance pupil is imaged in the exit pupil; the paraxial image point is E_1 (see Fig. 5.37). For the imaging of this axial pair of conjugate points, we expect only a spherical aberration term. To calculate this term, we need to choose reference spheres that are centred on E_0 and E_1. In this case, we use reference spheres through O_0 and O_1, respectively. The roles of object/image and entrance/exit pupil have thus been interchanged for the object imaging and the pupil imaging. The imaging of E_0 to E_1 is studied using the pathlength data for the principal ray that were collected when imaging the object point O_0 onto O_1. The general expression for an optical pathlength Q_aA_1 was given in Eq. (5.3.16). In the last line of Eq. (5.3.16), a pathlength contribution appears which is solely a function of the intersection height $\bar{y}$ of the principal ray with the refracting surface (point Q_p). This allows us to write for the pathlength increment at refraction of the principal ray

$$[Q_pA_1] - [Q_pA_0] = \left\{ n_1\left[\frac{c}{4\bar{z}_1^2} - \frac{z_1}{8\bar{z}_1^4} - \frac{c^3}{8} - a_4\right] - n_0\left[\frac{c}{4\bar{z}_0^2} - \frac{z_0}{8\bar{z}_0^4} - \frac{c^3}{8} - a_4\right]\bar{y}^4 \right\} . \tag{5.3.61}$$

For the aberration incurred when the axial point E_0 of the entrance pupil is imaged we calculate the optical pathlength along an aperture ray and subtract the pathlength along the optical axis. Making the connection with pathlengths encountered in the imaging of O_0 to O_1, we write the fourth-order expressions as (see Fig. 5.37),

$$\Delta W^{(4)}_{E_0E_1} = [Q_pE_1] - [Q_pE_0] - \{[TE_1] - [TE_0]\} , \tag{5.3.62}$$

where the optical distances $[Q_pE_1]$ and $[Q_pE_0]$ are given by

$$\begin{aligned} \left[Q_pE_1\right] &= [Q_pA_1E_1] = [Q_pA_1] + [A_1E_1] = [Q_pA_1] + [A_1A_1''] + [A_1''E_1] , \\ \left[Q_pE_0\right] &= [Q_pA_0E_0] = [Q_pA_0] + [A_0E_0] = [Q_pA_0] + [A_0A_0'] + [A_0'E_0] . \end{aligned} \tag{5.3.63}$$

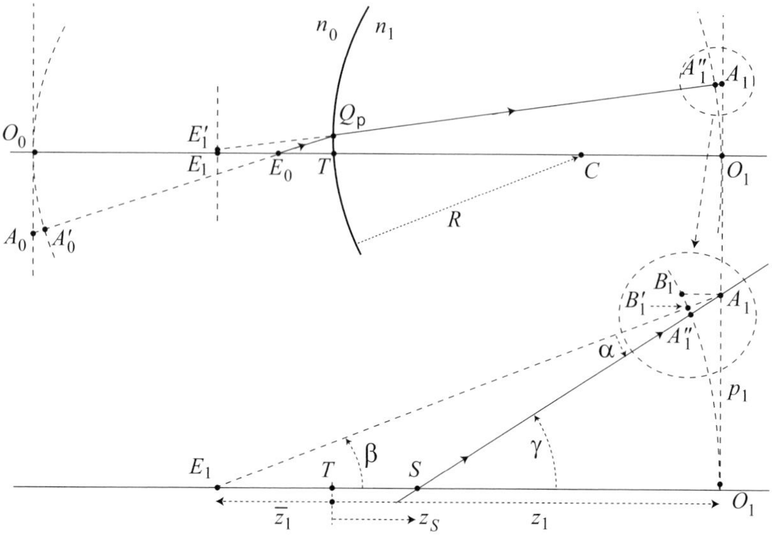

Figure 5.37: The optical pathlength increment for the imaging of the axial point E_0 of the entrance pupil onto the plane through the paraxial centre E_1 of the exit pupil. The reference spheres for the pupil imaging pass through the conjugate points O_0 and O_1 and are centred on E_0 and E_1, respectively. The intersection point of the object ray $E_0 Q_p$ with the reference sphere through O_0 has been denoted by A_0'. The intersection point of the aberrated refracted ray with the reference sphere through O_1, centred on E_1, is A_1''. The point E_1' is the intersection point of the aberrated refracted ray $Q_p A_1''$ with the plane through the paraxial image point E_1, perpendicular to the optical axis. In the lower graph, the insert shows the geometrical details of a general ray $A_1'' A_1$ which are needed to calculate the optical distance from the reference sphere through A_1'' to the paraxial image point A_1.

We obtain the final expression for the wavefront aberration

$$\Delta W^{(4)}_{E_0E_1} = [Q_pA_1] - [Q_pA_0] - \{[TO_1] - [TO_0]\} + [A_1A_1''] - [A_0A_0'] . \qquad (5.3.64)$$

The calculation of the optical pathlengths $[A_0A_0']$ and $[A_1A_1'']$ needs some extra care (see Fig. 5.37, lower graph). For the line segment A_1B_1, the surface 'sag', we have up to the fourth order in the radial coordinate p_1 of the paraxial point A_1,

$$A_1B_1 = \frac{c_r}{2} p_1^2 + \frac{c_r^3}{8} p_1^4 , \qquad (5.3.65)$$

where $R_r = 1/c_r = -z_1 + \bar{z}_1$ is the radius of the reference sphere with its centre of curvature at E_1. The line segment A_1B_1', pointing in the direction of the centre of curvature E_1 of the reference sphere, equals

$$A_1B_1' = A_1B_1 \cos\beta \approx A_1B_1 \left\{1 - \frac{c_r^2 p_1^2}{2}\right\} = \frac{c_r}{2} p_1^2 - \frac{c_r^3}{8} p_1^4 . \qquad (5.3.66)$$

For a line segment A_1A_1'' on a general ray at an angle γ, intersecting the optical axis at the point S with coordinate $z = z_S$, we find the expression, exact up to the fourth order in p_1,

$$A_1A_1'' = \frac{A_1B_1'}{\cos\alpha} \approx A_1B_1' \left\{1 + \frac{\alpha^2}{2}\right\} = \frac{c_r}{2} p_1^2 + \frac{c_r^3}{8} \left[\frac{2(z_S - \bar{z})^2}{(z_S - z_1)^2} - 1\right] p_1^4 . \qquad (5.3.67)$$

We substitute the expression for A_1A_1'' and the corresponding expression for A_0A_0' in Eq. (5.3.64) using $z_S = \bar{z}$ and, with p_1 expressed in $\bar{y}$ according to Eq. (5.3.15), we find the spherical aberration term

$$W^{(4)}_{E_0E_1} = \left[n_1 \left\{ \frac{c}{4\bar{z}_1^2} - \frac{1}{8\bar{z}_1^3} - \frac{c^3}{8} - a_4 \right\} - n_0 \left\{ \frac{c}{4\bar{z}_0^2} - \frac{1}{8\bar{z}_0^3} - \frac{c^3}{8} - a_4 \right\} \right] \bar{y}^4 . \qquad (5.3.68)$$

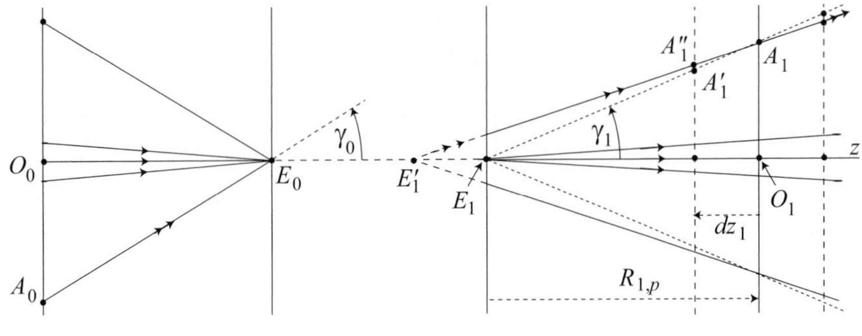

Figure 5.38: Spherical aberration of the pupil, inducing a change in distortion when the receiving plane is shifted from its nominal position in O_1. E_0 and E_1 are the paraxial positions of entrance and exit pupil. The principal ray of an off-axis pencil, originating at A_0, shows spherical aberration and intersects the optical axis at E_0 and E_1'. The dotted principal ray through E_1 and A_1 is free from pupil aberration. The distance from the paraxial exit pupil to the image plane is $R_{1,p}$.

As expected, this is exactly the result for a spherical aberration term according to Eq. (5.3.17), once the quantities for the pupil imaging have been inserted.

Figure 5.38 shows the effect of the spherical aberration of the pupil imaging. The image of the central object point O_0 is at O_1, with, as possible aberration, the spherical aberration of the image. In the image field, other aberrations may be present, among them field distortion. The presence of distortion implies that the image point A_1 is not found at the position given by the paraxial magnification. According to paraxial optics, the principal ray of the image pencil from A_0 intersects the exit pupil at its centre E_1 (see the dotted lines through E_1 in the figure). Due to spherical aberration of the pupil imaging, this intersection point is shifted, for instance, to E_1', but the principal ray still intersects the image plane at the point A_1. The influence of the pupil aberration becomes visible when the receiving plane is axially shifted. In the ideal case, with no pupil aberration, the image in the defocused plane with $z = z_1 + dz_1$ is a scaled version of the nominal image with a scaling factor $(R_{1,p} + dz_1)/R_{1,p}$. The transverse spherical aberration of the pupil image is of third order in the field angle γ_1, and the intersection point A_1'' also suffers such a third-order lateral shift from the point A_1'. Due to the spherical aberration of the pupil, an axial image shift introduces an extra distortion in the image. If the nominal image was distortion-free, the geometry of the figure leads to pincushion distortion for negative values of dz_1 and barrel distortion at the other side of the nominal focal plane. In the case of a paraxial position of the exit pupil at infinity, the linear scaling effect is absent when a defocusing is introduced (see also Subsection 4.10.5 on the subject of telecentric imaging).

5.4 Aberration of a Thin Lens

In optical design, the choice of curvatures, distances and refractive indices determines the aberrational state of the imaging system. In the case of lens elements, it is useful to have an aberration expression for the lens element itself in terms of the lens power, lens magnification, lens shape and refractive index. This avoids the summation of the aberrations of the two lens surfaces in which all paraxial data of the two intermediate imaging steps are required. For a thick lens with non-negligible axial thickness, such expressions have not been found. But in the case of zero lens thickness, tractable expressions for the Seidel aberrations have been made available.

5.4.1 Construction of an Equivalent Thin Lens

If the thickness of a lens is not negligible with respect to its radii of curvature, the assumption of zero thickness causes a serious perturbation of the ray paths through the lens. In this case, an 'equivalent thin lens' can be constructed following the procedure described in **[24]**. In Fig. 5.39 a lens with vertices T_1 and T_2 and finite thickness d is shown. The two paraxial refractions at the lens surfaces are replaced by refractions with the same angles at the positions of the paraxial object-side and image-side principal planes. The optical contrast of the lens is n; we use the index n_0 for the outer space and $n_0 n$ for the lens material itself (see Fig. 5.39). This allows us to treat, for instance, an air space with index unity between two glass media with index n_0.

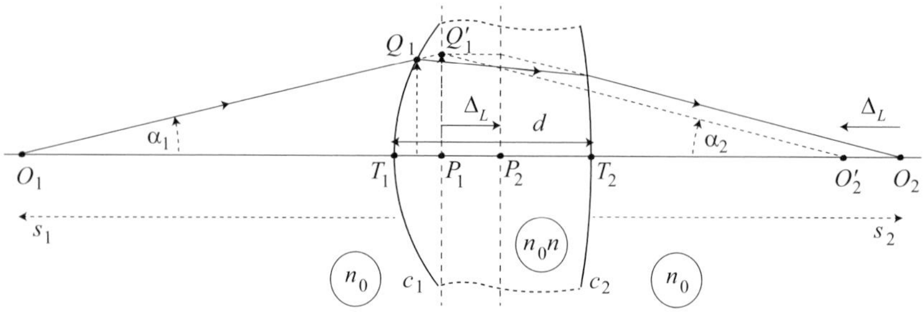

Figure 5.39: Illustrating the equivalent thin-lens concept.

To find the paraxial data of the equivalent thin lens, we first calculate the distances T_1P_1 and T_2P_2. We denote the paraxial power of the lens by K and with the aid of the paraxial lens matrix and Eq. (4.10.29) we obtain

$$K = n_0 \left\{(n-1)(c_1 - c_2) + (n-1)^2 c_1 c_2 d/n\right\} , \tag{5.4.1}$$

$$T_1P_1 = s_{P_1} = -n_0(n-1)dc_2/(nK) , \qquad T_2P_2 = s_{P_2} = -n_0(n-1)dc_1/(nK) . \tag{5.4.2}$$

The distance Δ_L, equal to the optical throw T of the lens for a transverse (or lateral) magnification $m_T = +1$, is given by

$$\Delta_L = d - s_{P_1} + s_{P_2} = \left\{1 + \frac{c_1 c_2 d}{n(c_1 - c_2) + (n-1)c_1 c_2 d}\right\}\left(\frac{n-1}{n}\right) d . \tag{5.4.3}$$

For equal refractive index n_0 of object and image space, the nodal points N_1 and N_2 coincide with the principal points P_1 and P_2, respectively (see also Eqs. (4.10.29) and (4.10.32)). Independently of the refractive indices in object and image space, the distance P_1P_2 equals the distance N_1N_2 between the nodal points. The latter distance is commonly called the *hiatus* of a thick lens or of a compound optical system. For the limiting case of a plane-parallel plate we put $c_1 = -c_2 \rightarrow 0$ and this yields the result $\Delta_L = (n-1)d/n$.

The paraxial height of incidence of an aperture ray through O_1 at an angle α_1 with the optical axis is given by y_{Q_1}. The change in direction of the ray as a consequence of the refraction at Q_1 follows from Eq. (4.10.5) and is given by

$$\Delta\alpha = -\left(\frac{n-1}{n}\right)\left(\alpha_1 + y_{Q_1} c_1\right) . \tag{5.4.4}$$

The height of incidence $y_{Q_1'}$ of the aperture ray on the principal plane through P_1 is given by $(s_1 - s_{P_1})y_{Q_1}/s_1$. For the equivalent thin lens we require that the ray directions remain unchanged at its front and exit surface. For the entrance and exit surface of the equivalent lens we derive from the equation above that their curvatures should be adjusted according to

$$c_1' = \left(\frac{s_1}{s_1 - s_{P_1}}\right) c_1 , \qquad c_2' = \left(\frac{s_2}{s_2 - s_{P_2}}\right) c_2 . \tag{5.4.5}$$

The equivalent lens is then given a thickness zero and its vertex is positioned at the principal point P_1 and the refracted ray follows the path $Q_1'O_2'$. The lens operates with new object and image distances, $s_1 - s_{P_1}$ and $s_2 - s_{P_2}$, respectively. Its new surface curvatures according to Eq. (5.4.5) have the result that the paraxial refraction and magnification of the equivalent lens are identical with that of the initial thick lens.

After aberration analysis with the aid of the equivalent thin lens, the thick lens can be reconstructed. To this end we use the ratio of the optical powers of first and second surface and, with the aid of the desired thickness d, we calculate the distances s_{P_1}, s_{P_2} and Δ_L and the new object and image distance pertaining to the finite-thickness lens. As a last step, the curvatures c_1' and c_2' are scaled back so that the ray refractions by the thin equivalent lens and by the thick lens become identical.

5.4.2 Thin Lens with Central Stop

We first examine the thin-lens aberration when the pupil magnification equals unity, meaning that the stop of the system or its image is in contact with the lens ('central' pupil position). The paraxial data that are needed in the computation of the aberrations of each imaging step have been indicated in Fig. 5.40. All data related to the first lens surface have

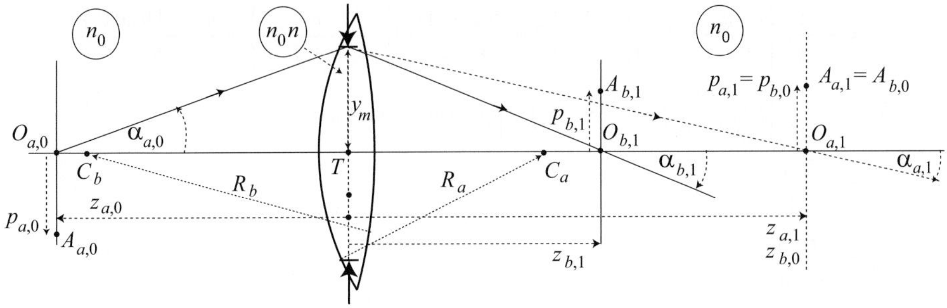

Figure 5.40: The paraxial data of a thin lens with vanishing thickness (both vertices of the lens are thought to be located at the common point T). The paraxial data of the two separate imaging steps have been given indices a and b. The diaphragm is coincident with the lens.

received the subscript a, those of the second surface the subscript b. The object-side quantities have a second subscript 0, those at the image side and those associated with both imaging steps have the extra subscript 1. The lens shape and the overall magnification m_T of the lens are expressed in terms of a shape (or bending) factor P and a magnification factor Q, given by

$$P = \frac{c_a + c_b}{c_a - c_b} \,, \qquad Q = \frac{m_T + 1}{m_T - 1} \,, \tag{5.4.6}$$

where $P = Q = 0$ corresponds to perfect paraxial symmetry between the object and image space. These substitutions are due to Coddington and they have a beneficial effect on the transparency of the final expressions for the aberrations. In these expressions (see Eqs. (5.3.49)–(5.3.51)), the partial powers K_a and K_b and the magnifications $m_{a,T}$ and $m_{b,T}$ of the imaging steps by each surface are needed. Using $K = n_0(n-1)(c_a - c_b)$ we find

$$K_a = n_0(n-1)c_a = K(P+1)/2 \,, \qquad K_b = -n_0(n-1)c_b = K(1-P)/2 \,. \tag{5.4.7}$$

The partial magnifications follow from (see Eq. (5.3.46)),

$$\begin{aligned} z_{a,0} &= \frac{n_0(1-m_{a,T})}{m_{a,T}K_a} \,, \qquad z_{a,1} = \frac{n_0 n(1-m_{a,T})}{K_a} \,, \\ z_{b,0} = z_{a,1} &= \frac{n_0 n(1-m_{b,T})}{m_{b,T}K_b} \,, \qquad z_{b,1} = \frac{n_0(1-m_{b,T})}{K_b} \,. \end{aligned} \tag{5.4.8}$$

From the thin-lens identity $z_{b,0} = z_{a,1}$ and the value m_T of the total magnification we then obtain

$$m_{a,T} = \frac{Q+1}{Q-P} \,, \qquad m_{b,T} = \frac{Q-P}{Q-1} \,. \tag{5.4.9}$$

For the calculation of the aberration terms, we also need the maximum convergence angle and the image height in the intermediate image plane at the paraxial position $z = z_{a,1}$. From the Lagrange invariant we derive $m_{b,A} m_{b,T} = n$, where $m_{b,A} = \alpha_{b,1}/\alpha_{a,1}$ is the angular magnification of the second imaging step. The following expression is obtained for the intermediate numerical aperture:

$$(n_0 n \alpha_{a,1}) = n_0 n \frac{\alpha_{b,1}}{m_{b,A}} = (n_0 \alpha_{b,1}) \frac{Q-P}{Q-1} \,, \tag{5.4.10}$$

and the intermediate image height is given by

$$p_{a,1} = \frac{p_{b,1}}{m_{b,T}} = p_{b,1} \frac{Q-1}{Q-P} \,. \tag{5.4.11}$$

Without giving the details of the rather lengthy calculation, we obtain the following results for the Seidel aberration coefficients of the thin lens (the pupil magnifications $m_{a,T,p} = m_{b,T,p}$ are both equal to unity),

$$S_1 = -\frac{1}{4n_0^2}\left\{\left(\frac{n}{n-1}\right)^2 + \frac{n+2}{n(n-1)^2}\left[P + \frac{2(n^2-1)}{n+2}Q\right]^2 - \frac{n}{n+2}Q^2\right\}\frac{[(m_T-1)n_0\alpha_{b,1}]^4}{K}, \tag{5.4.12}$$

$$S_2 = +\frac{1}{2n_0^2}\left\{\left(\frac{n+1}{n(n-1)}\right)P + \left(\frac{2n+1}{n}\right)Q\right\}p_{b,1}\,(m_T-1)^2\,(n_0\alpha_{b,1})^3, \tag{5.4.13}$$

$$S_3 = -\frac{H^2K}{n_0^2}, \tag{5.4.14}$$

$$S_4 = -\frac{H^2K}{n_0^2 n}, \tag{5.4.15}$$

$$S_5 = 0\,. \tag{5.4.16}$$

The factors containing the numerical aperture in image space can be replaced by a factor that comprises the maximum height of intersection y_m with the lens surfaces, caused by the marginal ray from the axial object point $O_{a,0}$. Using the result of Eq. (5.3.52) we have

$$(m_T-1)n_0\alpha_{b,1} = Ky_m, \text{ and, } \quad \frac{[(m_T-1)n_0\alpha_{b,1}]^4}{K} = K^3y_m^4, \tag{5.4.17}$$

$$p_{b,1}(m_T-1)^2(n_0\alpha_{b,1})^3 = HK^2y_m^2. \tag{5.4.18}$$

The wavefront aberration introduced by the thin lens then follows from Eq. (5.3.37). So far, we have not included an aspheric surface deformation of one of the lens surfaces. The aberration change due to an aspheric fourth-order deformation can be applied to the front or the back of the lens. Applying the same aspheric deformation to front or back will only change the sign of the induced aberration due to the inverted optical contrast. For the case of a central stop, the extra wavefront deformation is of the type 'spherical aberration', with an increment in the Seidel coefficient S_1 given by

$$\Delta S_1 = -8a_4n_0(n-1)y_m^4, \tag{5.4.19}$$

where a_4 is the aspheric deformation coefficient of the front surface of a lens element. The other Seidel coefficients remain unaffected by such an aspheric surface deformation because of the central position of the stop.

5.4.3 Aberration of a Thin Lens with Remote Stop

In a system of lenses, the central pupil position cannot be fulfilled for all lens elements. For the case of a remote position of the stop or its image in the space which contains the lens with index j, the footprint of an off-axis pencil of rays on the lens surface is decentred. We need to define an eccentricity parameter ϵ_j for each individual lens of the imaging system. Following the trajectory of the principal ray from the extreme off-axis object point through the system, we find the paraxial points of intersection of the principal ray with each lens. For the case of a finite-thickness lens, we approximate the height of intersection for the equivalent thin lens by the average of the two intersection heights on the thick-lens surfaces. Following this procedure, we obtain a shift parameter for each thin lens j, given by the ratio $\bar{y}_j/y_{m,j}$ where $\bar{y}_j$ is the height of intersection of the principal ray with the jth thin lens (see Fig. 5.41). In a similar way as for a single surface (see Eq. (5.3.15)), the value of $\bar{y}_j$ follows from the off-axis image coordinate $p_{1,j}$ in the intermediate image formed by lens j. We find the relation

$$\bar{y}_j = \left(\frac{m_{T,p,j}-1}{m_{T,p,j}-m_{T,j}}\right)p_{1,j}, \tag{5.4.20}$$

where $m_{T,j}$ and $m_{T,p,j}$ are the image and pupil magnification for the jth lens under consideration.

In the expression for the wavefront aberration with central pupil position according to Eq. (5.3.37), we introduce the coordinate shift $x_E = x_j$, $y_E = \bar{y}_j + y_j$ and obtain

$$\begin{aligned}\Delta W_j^{(4)}(x_j, y_j; p_{1,j}) =& \frac{1}{8}S_{1,j}\frac{\left[x_j^2+(y_j+\bar{y}_j)^2\right]^2}{y_{m,j}^4} + \frac{1}{2}S_{2,j}\frac{(y_j+\bar{y}_j)}{y_{m,j}}\frac{\left[x_j^2+(y_j+\bar{y}_j)^2\right]}{y_{m,j}^2}\frac{p_{1,j}}{p_{m,j}}\\ &+\frac{1}{2}S_{3,j}\frac{(y_j+\bar{y}_j)^2}{y_{m,j}^2}\frac{p_{1,j}^2}{p_{m,j}^2} + \frac{1}{4}(S_{3,j}+S_{4,j})\frac{x_j^2+(y_j+\bar{y}_j)^2}{y_{m,j}^2}\frac{p_{1,j}^2}{p_{m,j}^2}\,.\end{aligned} \tag{5.4.21}$$

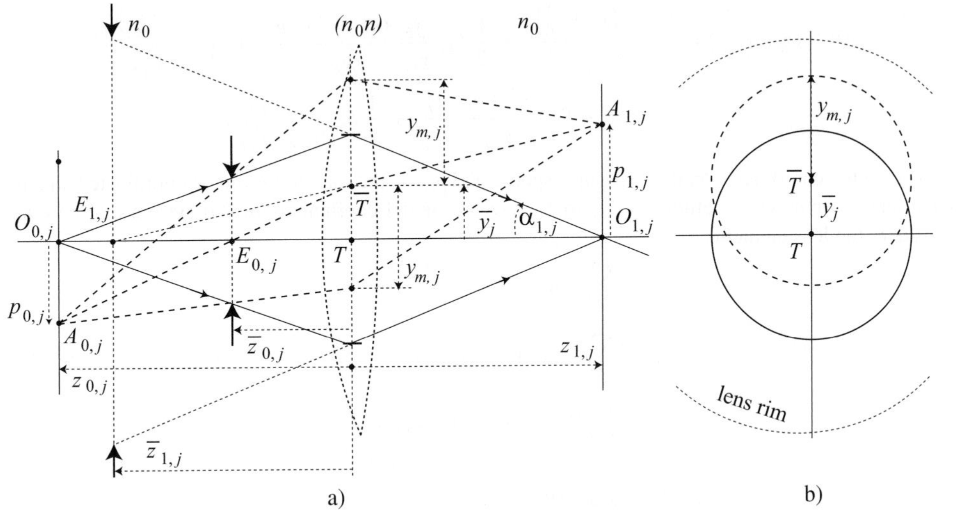

Figure 5.41: a) Imaging by a thin lens with a remote stop position (the centre of the entrance pupil is at $E_{0,j}$, that of the exit pupil at $E_{1,j}$). The principal ray of the pencil from an off-axis object point $A_{0,j}$ intersects the lens at $\overline{T}$, at an off-axis position given by $\overline{y}_j$.
b) The footprints on the lens surface of the central pencil of rays (solid circle) and of the pencil to the outer field point (dashed circle); the diameter of each footprint is $2y_{m,j}$.

We define an eccentricity parameter ϵ_j for the pencil of rays directed to the image point with coordinate $p_{1,j}$ by means of the expression

$$\frac{\overline{y}_j}{y_{m,j}} = \epsilon_j \frac{p_{1,j}}{p_{m,j}} , \tag{5.4.22}$$

where ϵ_j is given by

$$\epsilon_j = \frac{p_{m,j}}{y_{m,j}} \left(\frac{m_{T,p,j} - 1}{m_{T,p,j} - m_{T,j}} \right) , \tag{5.4.23}$$

such that for the maximum field excursion we have the eccentricity $\epsilon_j = \overline{y}_{m,j}/y_{m,j}$.

After some rearrangement, we obtain from Eq. (5.4.21) the following third-order aberration function of lens element j,

$$\begin{aligned} \Delta W_j^{(4)}(x_j, y_j; p_{1,j}) = & \frac{1}{8} S_{1,j} \frac{\left(x_j^2 + y_j^2\right)^2}{y_{m,j}^4} + \frac{1}{2}\left[S_{2,j} + \epsilon_j S_{1,j}\right] \frac{y_j}{y_{m,j}} \frac{(x_j^2 + y_j^2)}{y_{m,j}^2} \frac{p_{1,j}}{p_{m,j}} \\ & + \frac{1}{2}\left[S_{3,j} + 2\epsilon_j S_{2,j} + \epsilon_j^2 S_{1,j}\right] \frac{y_j^2}{y_{m,j}^2} \frac{p_{1,j}^2}{p_{m,j}^2} + \frac{1}{4}\left[S_{3,j} + S_{4,j} + 2\epsilon_j S_{2,j} + \epsilon_j^2 S_{1,j}\right] \frac{x_j^2 + y_j^2}{y_{m,j}^2} \frac{p_{1,j}^2}{p_{m,j}^2} \\ & + \frac{1}{2}\left[\epsilon_j (3S_{3,j} + S_{4,j}) + 3\epsilon_j^2 S_{2,j} + \epsilon_j^3 S_{1,j}\right] \frac{y_j}{y_{m,j}} \frac{p_{1,j}^3}{p_{m,j}^3} . \end{aligned} \tag{5.4.24}$$

Note that the expression above becomes indeterminate if the height of incidence $y_{m,j}$ of the marginal aperture ray becomes zero on a lens element. Simultaneously, the eccentricity factor ϵ_j for this element is undefined. Such a case is encountered when an intermediate image is formed on a lens element or when, on purpose, a so-called field lens is placed in the object or image plane. It is advisable to abandon the thin-lens formulas in such cases and to use the basic Seidel expressions per surface according to Eqs. (5.3.38).

Summing over all lens elements and replacing the intermediate normalised pupil and field coordinates by the normalised coordinates in the exit pupil and the normalised image coordinate in the image plane of the system, we find

$$\Delta W^{(4)}(x_E, y_E; p_I) = \frac{1}{8} S_{1,s} \frac{\left(x_E^2 + y_E^2\right)^2}{y_{m,E}^4} + \frac{1}{2} S_{2,s} \frac{y_E}{y_{m,E}} \frac{(x_E^2 + y_E^2)}{y_{m,E}^2} \frac{p_I}{p_{m,I}} + \frac{1}{2} S_{3,s} \frac{y_E^2}{y_{m,E}^2} \frac{p_I^2}{p_{m,I}^2}$$
$$+ \frac{1}{4} (S_{3,s} + S_{4,s}) \frac{x_E^2 + y_E^2}{y_{m,E}^2} \frac{p_I^2}{p_{m,I}^2} + \frac{1}{2} S_{5,s} \frac{y_E}{y_{m,E}} \frac{p_I^3}{p_{m,I}^3}, \qquad (5.4.25)$$

where (x_E, y_E) are the coordinates on the exit pupil sphere and $p_{m,I}$ is the maximum image coordinate value in the image plane of the lens system. The coefficients $S_{1,s}$ to $S_{5,s}$ are the Seidel coefficients of the system and they are given by summations over the lens elements,

$$S_{1,s} = \sum_j S_{1,j},$$
$$S_{2,s} = \sum_j \left\{S_{2,j} + \epsilon_j S_{1,j}\right\},$$
$$S_{3,s} = \sum_j \left\{S_{3,j} + 2\epsilon_j S_{2,j} + \epsilon_j^2 S_{1,j}\right\},$$
$$S_{4,s} = \sum_j S_{4,j},$$
$$S_{5,s} = \sum_j \left\{\epsilon_j (3S_{3,j} + S_{4,j}) + 3\epsilon_j^2 S_{2,j} + \epsilon_j^3 S_{1,j}\right\}. \qquad (5.4.26)$$

The expressions for the 'shifted' Seidel sums show that the coma, astigmatism and distortion of a lens are modified by a shift in stop position provided the lens is not free from aberration itself. For example, the astigmatism can be changed if the lens suffers from coma and spherical aberration. The relative importance of these aberration changes depends on the powers of ϵ_j. An element that is close to an intermediate image in the optical system, for instance a 'field' lens, can achieve high values of the eccentricity factor ϵ_j. On the contrary, an aplanatic lens is virtually useless if aberration balancing by means of a stop shift is aimed at. The balancing of aberrations with the aid of lenses that, on purpose, are not perfect themselves is the essence of the design of a well-corrected optical system. Very compact systems do not easily allow large values of ϵ_j; more elongated optical systems offer more flexibility in such an aberration balancing process.

In the process of aberration balancing using the effect of a remote stop position, aspheric surfaces with a nonzero a_4 coefficient can be successfully applied. Their standard use is to compensate for the spherical aberration of a system by aspherising a spherical surface that is coincident with the physical stop of the system. A well-known example is the parabolic mirror of a telescope that produces a perfect on-axis image. Aspheric surfaces that are far away from the stop and are used with a large eccentricity factor are able to modify a certain term $S_{1,j}$ in the summations of Eq. (5.4.26). In this way, using one or more optical surfaces that are close to the object or image plane or close to an intermediate image, astigmatism and distortion are influenced quite independently from the other aberrations if an eccentricity factor $\epsilon_j \gg 1$ can be achieved. An aspheric surface coincident with the stop ('central pupil') influences the off-axis aberrations only in sixth order and higher.

5.4.4 Seidel Aberrations of a Single Thin Lens

In this subsection we briefly discuss the influence of the shape or bending factor P and the magnification factor Q on the Seidel aberrations of a thin lens.

5.4.4.1 Spherical Aberration

The spherical aberration of a single lens is a quadratic function of both the shape factor P and the magnification factor Q. For each value of Q, a shape factor can be chosen that minimises the spherical aberration, according to

$$P = -\frac{2(n^2 - 1)}{n + 2} Q, \qquad (5.4.27)$$

and the minimum value of the aberration coefficient is given by

$$S_{1,\min} = -\frac{1}{4n_0^2} \left\{\frac{n^2}{(n-1)^2} - \frac{n}{n+2} Q^2\right\} y_m^4 K^3, \qquad (5.4.28)$$

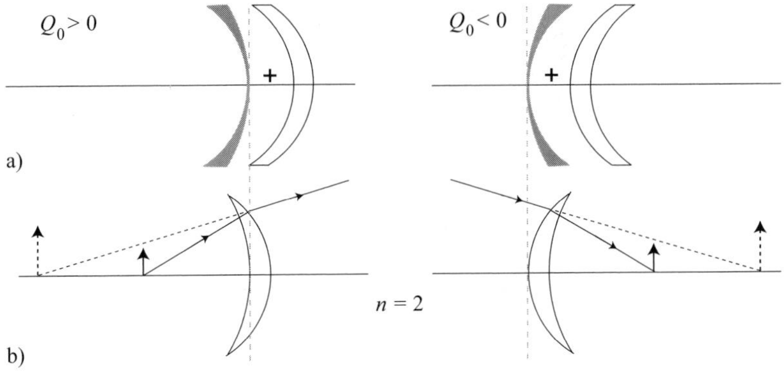

Figure 5.42: Thin lens imaging with freedom of spherical aberration.
a) The theoretical glass bodies with negative glass thickness at the lens rim, grey-shaded in the figure. Adding a meniscus lens to the lens with zero axial thickness yields the lens in b).
b) Real-world lenses with finite lens thickness on-axis. In these examples the refractive index n of the glass equals 2; $n_0 = 1$.

where y_m is the intersection height of the marginal aperture ray with the lens surfaces and K is the power of the lens. The spherical aberration is reduced to zero by choosing the magnification and shape factors equal to

$$Q_0 = \pm\frac{\sqrt{n(n+2)}}{n-1}\,, \qquad P_0 = \mp 2(n+1)\sqrt{\frac{n}{n+2}}\,. \tag{5.4.29}$$

The two solutions, with the role of object and image inverted (see Fig. 5.42), are closely related to the aplanatic lens with finite thickness which was discussed earlier in this chapter (see Fig. 5.8). In the left-hand drawing of Fig. 5.42b, we first recognise an almost refraction-free imaging step on the first surface, with object and image close to the centre of curvature of that surface. The imaging by the back surface of the lens is very similar to the imaging of the Huygens aplanatic points. The combination yields an image with a lateral magnification of approximately n, that is free from spherical aberration in the limit of zero lens thickness. The real magnification value and the conjugate positions all slightly deviate from the exact aplanatic object and image positions for the two surfaces. This is due to the fact that an ideal aplanatic lens is constructed from a zero axial thickness lens by the addition of a finite thickness meniscus lens with equal front and back curvature. The aberration of such a meniscus lens is compensated for by the slightly different imaging parameters of the lens with zero axial thickness as compared to those of the corresponding ideal aplanatic lens with zero edge thickness.

The frequently occurring imaging from infinity to the focal point requires $Q = -1$. The Seidel coefficient can then be minimised by choosing the optimum shape factor, resulting in a minimum value $S_{1,\infty}$ of the Seidel coefficient according to

$$P_\infty = \frac{2(n^2-1)}{n+2}\,; \qquad S_{1,\infty} = -\frac{1}{4n_0^2}\left(\frac{n(4n-1)}{(n+2)(n-1)^2}\right)y_m^4 K^3\,. \tag{5.4.30}$$

The value of $P_\infty = 1$ (or $c_2/c_1 = 0$) is attained if the refractive index n of the glass is 1.686. For lower index values, the lens has to be biconvex, for higher index values the convex-concave shape is optimum. In practice, a convex-plano lens is a good compromise for low spherical aberration in the range of glass index values ($1.45 \leq n \leq 2$) in the optical spectral domain (see Fig. 5.43). The same figure also shows that, at equal optical power and intersection height, a reflecting surface introduces much less spherical aberration than a lens; in the case of a refractive index of 1.5, the lens introduces 9 times more spherical aberration. For a refractive index of 2, this factor still amounts to 4.

The coefficient $S_{1,\infty}$ is always negative for a positive lens when imaging from infinity. The outer rays of the imaging pencil intersect the optical axis between the lens surface and the paraxial focus, as in the drawing of Fig. 5.18. Using a wrong lens bending factor can have a highly detrimental effect on the image quality. When creating the image of an object at infinity at the focal point of a lens in air ($Q = -1$, $n_0 = 1$), the convex-plano lens with $P = 1$ offers a relatively good correction of spherical aberration, the spherical aberration factor of Fig. 5.43 being 28/3 ($n = 3/2$). Using the same lens inverted such that P becomes -1, makes this factor 36. For a high-index lens with $n = 2$, the effect is comparable with a change from 4 to 16. The physical reason for the large influence of the bending factor is the distribution of the total refraction over the two surfaces. For $P = -1$ the flat lens surface faces the incident parallel beam and the refraction

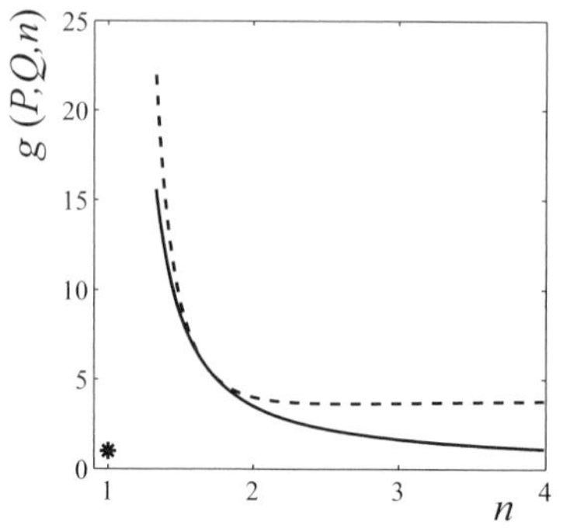

Figure 5.43: Numerical value of the factor depending on P, Q and n in the expression for the spherical aberration of a thin lens (see Eq. (5.4.12)). Lens magnification $m_T = 0$ ($Q = -1$). Solid curve: thin lens with optimum lens bending. Dashed curve: convex-plano lens shape. The asterisk for $|n| = 1$ represents the value of the same factor for a concave mirror with $n = -1$ (followed by a plane mirror in contact).

of the beam is taken care of exclusively by the curved second lens surface. Large angles of incidence are encountered on this surface leading to a more pronounced deviation from the linearised version of Snell's law.

5.4.4.2 Coma

Let us consider the magnitude and sign of the coma of a thin lens for a frequently occurring imaging situation, the imaging of a far-away scene in the focal plane of a lens with magnification $m_T = 0$. The shape of the lens is convex-plano such that the spherical aberration is small. The thin-lens parameters are then given by $P = 1$ and $Q = -1$. To better isolate the comatic aberration, a small aspheric departure is given to the first curved surface of the lens, such that the spherical aberration is made rigorously zero. The other thin-lens aberrations like astigmatism and field curvature are kept small by keeping the marginal ray angle and the field position small. In Eq. (5.4.13) we choose $n_0\alpha_{b,1} = 0.05$ and $p_{b,1} = 1$ mm; the index n_0 of the immersing medium is unity. The magnitude of the thin-lens coma is rather small for a standard refractive index like $n = 1.5$. To produce a convincing graph of the coma we make it much larger than the astigmatism by a strong reduction of the refractive index of the lens to the completely unphysical value of $n = 1.1$.

The thin lens is shown in Fig. 5.44a with three incident parallel pencils of rays at angles of −0.01, 0 and +0.01 radians. In b) we show, strongly exaggerated, the deviation between the comatic wavefront (solid curve) and the reference sphere (dashed) in the exit pupil through E_1, both related to the imaging of a parallel beam at the image point with coordinate $p_1 = -1$ mm. The sign of the coma aberration is determined by the factor between curled braces of Eq. (5.4.13) and by the second factor containing field and aperture values. For $1 < n < 1.62$ the first factor is positive. The second factor is positive for a negative field position and a negative marginal ray angle of the axial beam that corresponds to a positive intersection height in the exit pupil ($y_1 > 0$). In Eq. (5.4.18) that is often used to determine the sign of coma, it is the Lagrange invariant H which determines the sign of S_2. For $p_1 = -1$ and $\alpha_{b,1} < 0$ for positive values of the pupil coordinate y_1 we have a positive value of H, confirming the sign of the comatic wavefront in Fig 5.44b. Finally, in c) we have plotted the typical *spot diagrams* or *blur patterns* that belong to comatic aberration. They confirm the orientation of the coma tail that was predicted by the shape of the wavefront aberration in b); the coma tails are directed towards the centre of the image field.

For an incident parallel beam ($Q = -1$), the coma of a thin lens can be made zero if we abandon the simple convex-plano lens shape. The value of P that yields zero coma is given by

$$P = \frac{(2n+1)(n-1)}{n+1} \ . \tag{5.4.31}$$

For the common range of refractive indices in the optical domain ($1.45 \leq n \leq 2$), the shape is close to the convex-plano shape. The convex-plano lens shape that yields exactly zero coma follows from the following requirement on the index value:

$$n^2 - n - 1 = 0 \ , \quad \text{or,} \qquad n = \frac{1+\sqrt{5}}{2} = 1.618 \ . \tag{5.4.32}$$

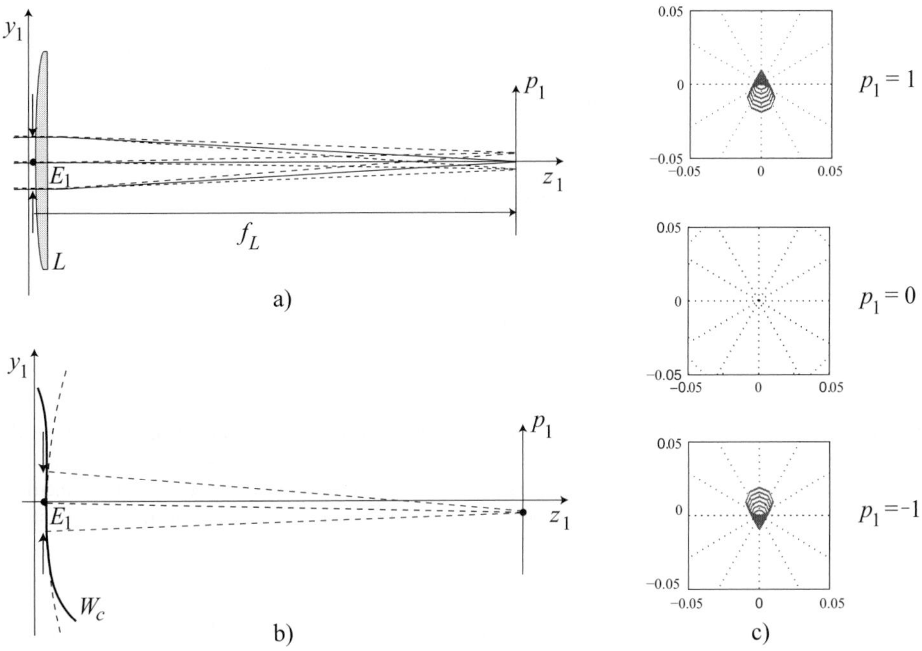

Figure 5.44: Off-axis imaging by a low-index convex-plano thin lens and sign of the comatic aberration.
a) Three parallel pencils of rays with paraxial image points at $p_1 = -1,\ 0$ and $+\,1$ mm ($f_L = 100$ mm, $NA = 0.05$).
b) The comatic wavefront W_c (solid curve) of a pencil directed towards the image point $p_1 = -1$ mm.
c) Ray intersection points in the image plane for the three image points. The coma tail is oriented towards the optical axis. The transverse aberration is in mm.

This specific value of 1.618 is a quite common index value for glasses or plastic materials in the optical domain. Using the results of the preceding paragraph, it follows that, in general, a convex-plano lens is a good choice for imaging from infinity to the focal point with small amounts of spherical aberration and coma.

A special case arises if the thin lens has a central stop and the optical imaging is point-symmetric with respect to the stop. Such a point-symmetric imaging configuration is obtained if both the bending factor P and the magnification factor Q are zero. It immediately follows from Eq. (5.4.13) that coma is zero for this case, irrespective of the refractive index of the lens. For a more general imaging geometry, it is always possible to obtain zero coma by imposing a lens bending factor P given by

$$P = -\frac{(2n+1)(n-1)}{n+1}\,Q. \tag{5.4.33}$$

Allowing a residual wavefront aberration of a quarter wavelength, we now use Eqs. (5.4.12) and (5.4.13) to see up to what aperture and field size such a simple lens can be used. For $n = 1.5$, a focal length of $f_L = 100$ mm and $\lambda = 0.55$ µm, the $\lambda/4$ criterion for spherical aberration imposes a limit of 0.05 on the aperture $\alpha_{b,1} = y_m/\{f_L(1 - m_T)\}$ of the imaging pencil. The same criterion for coma allows a field angle $\gamma_{b,1} = p_1/f_L$ of 66 mrad. From this numerical example we conclude that the choice of the optimum refractive index is important to increase the useful field of a lens. A refractive index of 1.618 would yield, according to third-order theory, an unlimited field size if coma were the only aberration. In the considerations above, we have applied the $\lambda/4$ criterion, meant for high-quality imaging, only limited by diffraction effects. For more relaxed imaging purposes, the useful aperture and field of a single convex-plano lens can become substantially larger.

5.4.4.3 Comatic Aberration in a General Point-symmetric Imaging System

It has been shown above that the special point-symmetric thin-lens imaging configuration ($P = Q = 0$) yields zero comatic aberration. We now consider the value of coma in a general imaging system with point symmetry. The analysis is not restricted to third-order aberration but is valid for any aperture and field size. A perfectly point-symmetric imaging

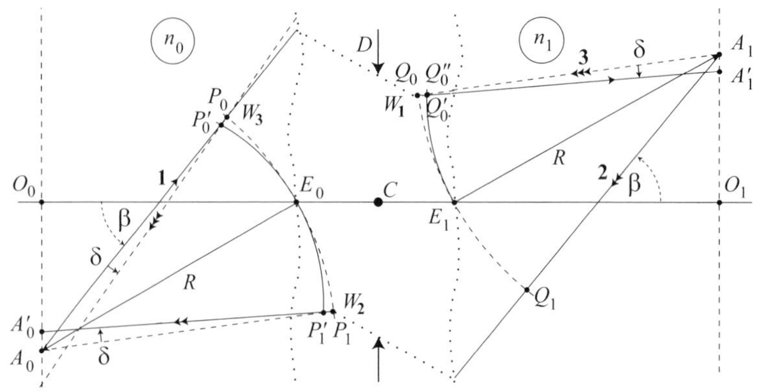

Figure 5.45: Imaging with lateral magnification −1 by an optical system that possesses point-symmetry with respect to C. The diaphragm D is located in the plane through C and the object and image spaces have equal refractive indices $n_0 = n_1 = n$. The paraxial centres of the entrance and exit pupils have been denoted by E_0 and E_1, respectively. The point-symmetric optical elements (not shown in the drawing) of the imaging system are found in the schematically drawn volume delimited by the dotted curves through E_0 and E_1.

geometry requires, apart from point-symmetric optics, a lateral magnification of −1 and equal refractive indices in object and image space. It can be shown that the asymmetric aberration coma is zero, at least in first-order approximation, if these conditions are satisfied. Figure 5.45 schematically illustrates a point-symmetric imaging geometry with respect to the central point C. The details of the real-world optical elements have been left out. To achieve symmetry for the imaging pencils, the plane of the diaphragm D should intersect the centre of symmetry C. Entrance and exit pupil have identical magnification with respect to the diaphragm and thus have a mutual magnification of unity. Consequently, the paraxial entrance and exit pupil coincide with the paraxial principal planes of the system.

To demonstrate the absence of asymmetric aberration in the imaging by a point-symmetric system, we proceed in three steps with the aid of rays that originate at an off-axis object point A_0 or its corresponding paraxial image point A_1 ($O_0A_0 = O_1A_1$). In a first instance, the reference ray for the off-axis imaging is the paraxial principal ray, $A_0E_0E_1A_1$ or its counter-propagating equivalent $A_1E_1E_0A_0$. The reference spheres in object and image space are the dashed circles $P_0E_0P_1$ and $Q_1E_1Q_0$, having their centres of curvature at A_0 and A_1, respectively:

- Ray 1 is traced from the object point A_0 and intersects the object-side reference sphere at P_0 and the image-side reference sphere at Q_0. The wavefront cross-section in image space is given by the solid curve $E_1Q_0'Q_0''$. Q_0' lies on the (solid) aberrated ray 1 through A_1', whilst Q_0'' lies on a fictitious non-aberrated dashed ray through the paraxial image point A_1. The wavefront aberration W_1 of ray 1 is defined as

$$W_1 = [P_0Q_0]_{\mathbf{1}} - [E_0E_1] = [P_0Q_0Q_0']_{\mathbf{1}} + [Q_0'Q_0]_{\mathbf{1}} - [E_0E_1] = [Q_0'Q_0]_{\mathbf{1}}\,, \tag{5.4.34}$$

  where the bold subscript 1 indicates that the distances are measured along the ray 1. The transverse aberration in the image plane is given by the line segment A_1A_1' and the angular aberration of the ray is δ. The difference between the optical paths $[Q_0'Q_0]$ and $[Q_0''Q_0]$ is second-order small in the angular aberration angle δ and is neglected in what follows.
- We exploit the point symmetry of the imaging system and trace ray 2, which starts at the image point A_1 in the backward direction and travels at the same angle β with respect to the z-axis as ray 1. Ray 2 intersects the image-side reference sphere at Q_1, such that the lateral distance of Q_1 to the principal ray segment A_1E_1 in image space is equal to that of P_0 to the object-side principal ray A_0E_0. Ray 2 intersects the object-side reference sphere at P_1 and the object-side wavefront through the point E_0 at the point P_1'. Since we have point-symmetric imaging, the angular aberration of ray 2 in object space equals δ and the transverse aberration A_0A_0' is identical to A_1A_1'. It follows that the wavefront cross-section E_0P_1' in object space is point-symmetric with respect to the image-side cross-section E_1Q_0' and that

$$W_2 = [P_1'P_1]_{\mathbf{2}} = [Q_0'Q_0]_{\mathbf{1}} = W_1\,. \tag{5.4.35}$$

- The final ray to be traced is ray 3 (dashed in the figure), which leaves the (paraxial) image point A_1 in the backward direction such that it intersects the image-side reference sphere at the point Q_0. With respect to this 'base ray' A_1Q_0, the *inverted* ray 1 is an aperture ray for the imaging of the exit pupil point Q_0 onto the corresponding entrance pupil point P_0. The pathlength difference between the two rays, measured from the common point Q_0 to the reference point P_0, is of second order in the angle δ, the angular aberration of ray 1. The intersection point of ray 3 with the object-side wavefront surface is assumed to be P_0', which is allowed if δ is small. The wavefront aberration of the dashed ray 3 is then given by

$$W_3 = [Q_0P_0]_3 - [E_1E_0] = [Q_0P_0P_0']_3 + [P_0'P_0]_3 - [E_1E_0] \approx [P_0'P_0]_3 \,. \tag{5.4.36}$$

The approximate sign has been introduced since the pathlength difference along these neighbouring rays between the points Q_0 and P_0 is second-order small in the angle δ, such that $[Q_0P_0]_3$ and $[Q_0P_0]_1$ are only equal in first order.

Equations (5.4.34)–(5.4.36) show that W_1 and W_3 are equal in first order. As a consequence, points on the wavefronts E_0P_0' and E_1Q_0' with equal lateral distance to A_0E_0 and A_1E_1, respectively, have equal wavefront aberration up to the second order in δ. The wavefront $P_1'E_0P_0'$ of a backward propagating wave originating at A_1 is thus symmetrical in first order with respect to the (paraxial) principal ray $A_1E_1E_0A_0$ and the pencil of rays from A_1 is free from coma. The same conclusion can be drawn about the symmetry of a wavefront in image space associated with a forward propagating wave from a general object point A_0.

In the discussion above, we have neglected the spherical aberration of the pupil imaging. As a consequence, the centres of the diaphragm images in object and image space are shifted along the optical axis, away from their paraxial positions, as a function of the field positions A_0 and A_1. Since the system is point-symmetric, the axial shifts in object and image space are in opposite directions but have equal length. The point-symmetry of the imaging is thus maintained and the arguments used above about symmetric wavefront cross-sections in a point-symmetric system remain valid. We conclude that, in first order, coma is absent in a point-symmetric system.

5.4.4.4 Astigmatism

Equation (5.4.14) shows that the single lens with central pupil position does not allow for any reduction of its astigmatism by influencing lens shape, magnification or refractive index. The amount of astigmatism is simply proportional to the power of the thin lens. In a compact optical system with a succession of thin lenses, all operating close to the central stop geometry, it is only the possible alternation of positive and negative power that allows a reduction of the system sum. The sign of the astigmatism of a positive lens is such that the sagittal focal lines are found on the less curved surface. With the aid of Eq. (5.3.43) it is easily verified that the radii of curvature of the sagittal and tangential image surfaces are well approximated by

$$R_s = -\frac{nf_1}{(n+1)} \,, \qquad R_t = -\frac{nf_1}{(n+3)} \,, \tag{5.4.37}$$

where the image-side focal distance f_1 is equal to n_0/K. Outside the region of validity of the third-order aberration theory, more complicated shapes of the best-image surface occur. But, close to the optical axis, the sagittal and tangential image surfaces are well approximated by second-order surfaces and their paraxial curvatures follow from Eq. (5.4.37). If we apply the aperture limitation of 0.05 that followed from the tolerance for spherical aberration of the lens example above, the $\lambda/4$ criterion for astigmatism is calculated using the value of the Lagrange invariant H that equals $n_0 p_{1,m}\alpha_{b,1}$ for the thin lens. We then obtain a maximum field angle of 33 mrad. It turns out that astigmatism will generally limit the useful field of a single lens. As mentioned before, non-diffraction-limited operation of the lens will allow a larger field.

5.4.4.5 Field Curvature

As is the case for astigmatism, the coefficient of field curvature wavefront aberration of the thin lens (with central stop) is uniquely defined by the ratio of its power and refractive index. Reduction of field curvature in a system is only possible by using both positive and negative elements. The tolerance of $\lambda/4$ applied to field curvature only yields a $\sqrt{2n}$ times larger available field than is the case for astigmatism. A further $\sqrt{2}$ in field extent is gained by allowing an equal absolute defocus at the centre and the border of the field. The in-focus region in the image is then found at a position that equals

$\sqrt{1/2}$ of the coordinate of the border of the field. In the presence of both astigmatism and field curvature, the radius of curvature R_{fit} of the best-fit curved image surface follows from Eq. (5.4.37),

$$R_{\mathrm{fit}} = -\frac{nf_L}{(n+2)} . \qquad (5.4.38)$$

5.4.5 Seidel Aberrations of a System of Thin Lenses

A system of thin lenses offers extra possibilities for aberration correction. If the system is very compact and all thin lenses are very close to the diaphragm, the aberration values of each component are very close to those found for a central stop position. Spherical aberration and coma, which have a relatively strong dependence on lens magnification, lens bending and refractive index, can be reduced by splitting a single thin lens into several thin components with the same total power. The physical reason for this aberration reduction is the increased flexibility, having several lenses, to influence the angles of incidence and the optical contrast for the aperture rays on the lens surfaces. In this way, it is relatively easy to achieve a lens system that is aplanatic, meaning free from spherical aberration and coma, at least with respect to the third-order contribution. Unfortunately, astigmatism and field curvature are independent of the lens geometry and are defined by the lens power and, to a small degree for field curvature, by the refractive index of the lens.

The expressions for an individual lens with remote stop were given in Eqs. (5.4.25) and (5.4.26). The eccentricity parameter ϵ_j on each surface plays an important role. Each aberration of a certain degree in the pupil coordinates produces aberrations of a lower degree in the pupil, that are proportional to increasing powers of the eccentricity parameter of the specific lens. The somewhat contradictory conclusion of this phenomenon is that we have to intentionally introduce defective lenses that, when used with a remote stop position, are capable of splitting-off aberrations like astigmatism, field curvature and distortion. The design with individually perfect thin lenses with the aplanatic property is useful when small-field imaging systems like a microscope objective have to be designed. But this approach is counterproductive to achieve a well-corrected system with large image field in which astigmatism, field curvature and distortion cannot be neglected. In such large-field systems it is also necessary to create sufficient space between the components so that the eccentricity factors on several elements can attain high values. An extreme example would be a lens on which the footprint of the pencil of rays is small with respect to the lens diameter and far off-centre on the lens surfaces. If such a lens has spherical aberration, it will be effective to compensate for distortion that is proportional to the third power of ϵ. The limiting case is to position a lens at an intermediate image where the footprint of the pencil of rays vanishes. Such a lens functions as a 'field' lens and serves, for example, to reduce the field curvature in the image plane.

Another method to reduce the fairly 'resisting' aberration components, astigmatism and field curvature, is to use thick elements. In third order, the aberration of a thick lens is the sum of the surface contributions and the element thickness can drastically change the stop position and the incidence height on the first and the second surface of the lens element. For this reason, thick elements have more flexibility for aberration correction than thin elements. Another favourable feature in an optical system is the possible (point) symmetry that is allowed in (part of) the system. If the imaging itself does not allow a symmetric layout ($m_T \neq \pm 1$), it can be useful to strive for a symmetric refraction with respect to the stop of at least the principal ray of each imaging pencil. The design and construction of an optical system satisfying a certain list of requirements is the subject of Chapter 7 on optical design and optical system optimisation. The various solution methods that were briefly indicated above are discussed in more detail in that chapter.

5.5 Seidel Aberrations of a Plane-parallel Plate

As part of an optical system, a window or plane-parallel plate is frequently used, for instance as a cover glass in microscopy or as a disc substrate in optical data storage when the read-out of the digital information is performed, looking through the transparent substrate. In Fig. 5.46 we have sketched a marginal ray and the principal ray that are needed for the calculation of the third-order aberration of the plate. Equations (5.3.24)–(5.3.36) are used to obtain the Seidel coefficients of the plane-parallel plate, using the auxiliary quantities A_m, B_m, y_m and $\delta(\alpha_m/n)$ on the entrance and exit surface of the window. For the refraction invariants at the two surfaces we have

$$A_{0,m} = A_{1,m} = n_0\alpha_{0,m} = n_1\alpha_{1,m} , \qquad (5.5.1)$$

$$B_{0,m} = B_{1,m} = n_0\gamma_{0,m} = n_1\gamma_{1,m} . \qquad (5.5.2)$$

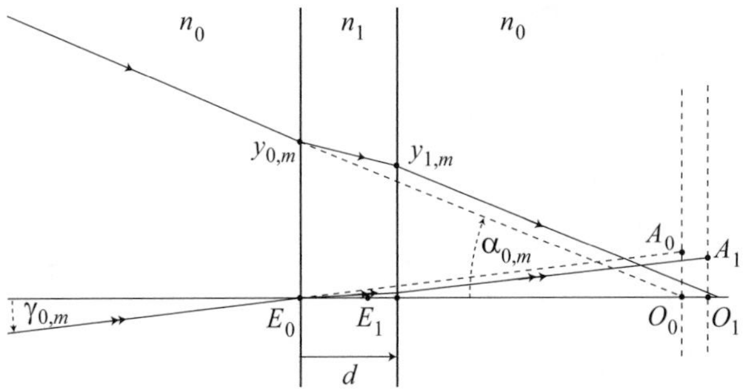

Figure 5.46: The passage of the marginal ray of an axial pencil and of the principal ray of an oblique pencil through a plane-parallel plate. The plate thickness is d, the refractive index of the environment is n_0, that of the window material is n_1 ($n_1 > n_0$ in the graph). The maximum angle of incidence of the principal ray with the outward normal to the plate is $\gamma_{0,m}$ (positive in the figure), the aperture angle of the marginal ray is $\alpha_{0,m}$ (negative in the figure).

Other quantities needed in the Seidel expressions for a plane-parallel plate are

$$\frac{\alpha_{1,m}}{n_1} - \frac{\alpha_{0,m}}{n_0} = -\frac{\alpha_{0,m}}{n_0}\left(\frac{n_1^2 - n_0^2}{n_1^2}\right), \tag{5.5.3}$$

$$y_{1,m} = y_{0,m} + \frac{n_0 \alpha_{0,m} d}{n_1}. \tag{5.5.4}$$

With the aid of Eqs. (5.5.1)–(5.5.4) we readily obtain the Seidel aberration coefficients of the plate,

$$\begin{aligned}
S_{1,p} &= n_0^2 \left(\frac{n_1^2 - n_0^2}{n_1^3} d\right) \alpha_{0,m}^4 , \\
S_{2,p} &= n_0^2 \left(\frac{n_1^2 - n_0^2}{n_1^3} d\right) \gamma_{0,m} \alpha_{0,m}^3 , \\
S_{3,p} &= n_0^2 \left(\frac{n_1^2 - n_0^2}{n_1^3} d\right) \gamma_{0,m}^2 \alpha_{0,m}^2 , \\
S_{4,p} &= 0 , \\
S_{5,p} &= n_0^2 \left(\frac{n_1^2 - n_0^2}{n_1^3} d\right) \gamma_{0,m}^3 \alpha_{0,m} .
\end{aligned} \tag{5.5.5}$$

The corresponding wavefront deviation after passage through the plane-parallel plate follows from a multiplication of each term by the appropriate relative coordinates on the pupil sphere and the relative angle of incidence of the principal ray (see also Eq. (5.4.25)),

$$\begin{aligned}
W(x_E, y_E; \gamma) = {} & \frac{1}{8} S_{1,p} \frac{\left(x_E^2 + y_E^2\right)^2}{y_{m,E}^4} + \frac{1}{2} S_{2,p} \frac{y_E}{y_{m,E}} \frac{(x_E^2 + y_E^2)}{y_{m,E}^2} \frac{\gamma_0}{\gamma_{0,m}} \\
& + \frac{1}{2} S_{3,p} \frac{y_E^2}{y_{m,E}^2} \frac{\gamma_0^2}{\gamma_{0,m}^2} + \frac{1}{4} S_{3,p} \frac{x_E^2 + y_E^2}{y_{m,E}^2} \frac{\gamma_0^2}{\gamma_{0,m}^2} + \frac{1}{2} S_{5,p} \frac{y_E}{y_{m,E}} \frac{\gamma_0^3}{\gamma_{0,m}^3} .
\end{aligned} \tag{5.5.6}$$

We note that the sign of the spherical aberration is positive and that the marginal focus of the beam is located beyond the paraxial focus. This sign is opposite to that of the spherical aberration coefficient of a standard convex-plano lens. The presence of a window in a high-aperture imaging system thus helps to reduce the spherical aberration of the system.

5.5.1 Seidel Aberration Coefficients for Large Tilt Angle γ_0

In Fig. 5.47a we have shown the refraction of a converging beam at a relatively large tilt angle γ_0. The orientation of coma and astigmatism is shown in Fig 5.47b. The sign of the astigmatism is also opposite to that of a positive lens. Along

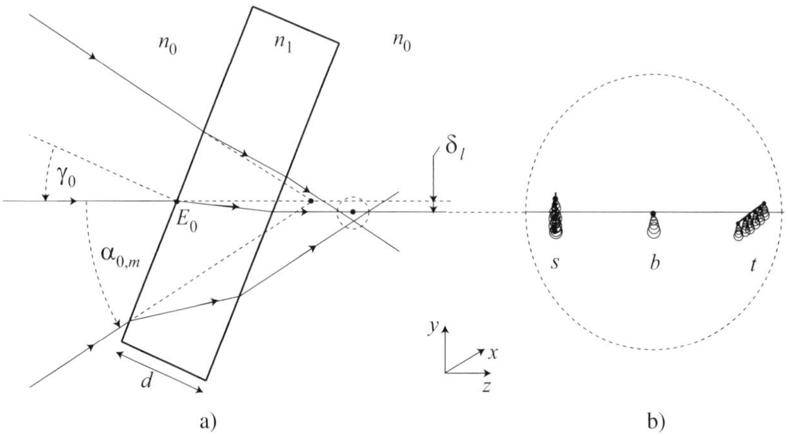

Figure 5.47: a) The aberrational state of a converging beam (aperture angle $\alpha_{0,m}$) after passage through a plane-parallel plate with refractive index n_1, environmental index is n_0. The plate has been tilted such that the principal ray of the incident beam is at an angle γ_0 (positive in the figure) with the normal to the surface.
b) Artist's impression of the comatic and astigmatic aberration in the focal region with the (aberrated) sagittal and tangential focal lines denoted by s and t and the best-focus position by b.

the propagation direction, we first observe the sagittal focal line, then the 'best focus' where the comatic aberration is best visible and, finally, we encounter the tangential focal line. The tilt angle of the window in Fig. 5.47 is already fairly large and outside the region where the third-order contributions for coma, astigmatism and distortion are well represented by the formulae of Eq. (5.5.6). Analytic formulas for the third- and higher-order aberration coefficients that better approximate the results at large tilt angle are given in **[49]** and **[91]**. A frequently occurring case is a plane-parallel plate at an angle of 45 degrees, for instance, a beam splitter. The lowest order aberration coefficients of such a plate, used at an angle $\gamma_0 = \pi/4$, are well represented by the following expressions for the Seidel coefficients of spherical aberration, coma and astigmatism ($n_0 = 1$):

$$
\begin{aligned}
S_{sp} &= \left(\frac{d\,\alpha_{0,m}^4}{(n_1^2-\sin^2\gamma_0)^{7/2}}\right)\left\{(n_1^2-\sin^2\gamma_0)^3\cos^2\gamma_0 - \frac{(n_1^2-\sin^2\gamma_0)^2}{8}(3\sin^4\gamma_0+8\cos^4\gamma_0-6\sin^2 2\gamma_0)\right.\\
&\quad\left. -\frac{3}{16}(n_1^2-\sin^2\gamma_0)(4\cos^2\gamma_0-3\sin^2\gamma_0)\,\sin^2 2\gamma_0 - \frac{15}{128}\sin^4 2\gamma_0\right\},\\
S_{co} &= \frac{(n_1^2-1)[n_1^2-(\sin^2\gamma_0)/4]\sin 2\gamma_0}{2(n_1^2-\sin^2\gamma_0)^{5/2}}\,d\,\alpha_{0,m}^3\,,\\
S_{as} &= \frac{(n_1^2-1)\sin^2\gamma_0}{(n_1^2-\sin^2\gamma_0)^{3/2}}\,d\,\alpha_{0,m}^2\,.
\end{aligned}
\qquad (5.5.7)
$$

The expressions are obtained from a power series expansion of the optical pathlength with respect to the chosen direction of the principal ray at an angle γ_0 with the normal to the plate surfaces. For a sufficiently small aperture angle $\alpha_{0,m}$, say 0.15, the formulas above represent quite well the total aberration of the refracted beam. For larger values of the aperture angle, substantial amounts of higher-order aberration components need to be included in the analysis **[49]**.

5.5.2 Geometrical Description of the Aberrated Astigmatic Focus

We conclude this subsection on the oblique plane-parallel plate in a convergent or divergent beam with some useful expressions for the lateral shift of the beam and for the (paraxial) position of the astigmatic focal lines (see Fig. 5.48). For the lateral shift δ_l between the incident and outgoing principal ray of the pencil we have the expression

$$
\delta_l = -d\,\sin\gamma_0\left\{1-\frac{n_0\cos\gamma_0}{n_1\cos\gamma_1}\right\},
$$

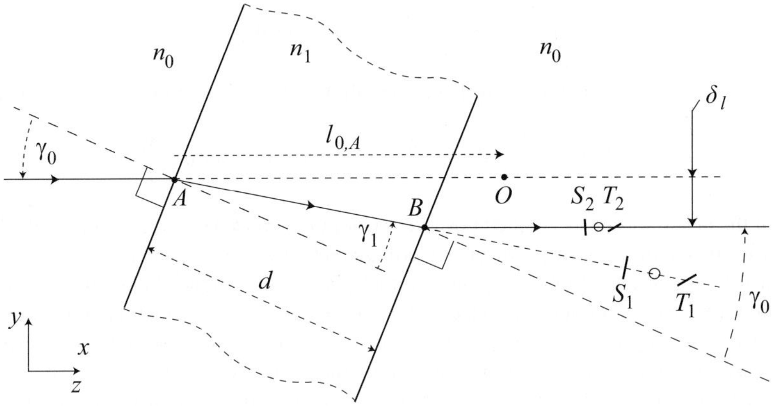

Figure 5.48: The principal ray through AO of a pencil incident on an inclined plane-parallel plate. The object point in the first medium with index n_0 is O and the object distance $AO = l_{0,A}$. The lateral shift of the principal ray is δ_l. The astigmatic foci are, in order of occurrence, S_1 and T_1 after the first refraction and S_2 and T_2 after traversal through the plate.

and, for $\gamma_0 \ll 1$,

$$\delta_l \approx -d\,\gamma_0 \left(\frac{n_1 - n_0}{n_1}\right) . \tag{5.5.8}$$

For the calculation of the axial positions of the astigmatic focal lines we apply paraxial optics analysis around the principal ray section AO outside the plate in object space, ABS_1T_1 inside the plate and BS_2T_2 in the final image space. The paraxial imaging equations with respect to an inclined axis are given by the Coddington equations which are discussed in Subsection 5.7.4 (see Eqs. (5.7.41) and (5.7.42)). For a flat surface, the Coddington imaging equations take on the simple form

$$\frac{n_1 \cos^2\gamma_1}{l_t} = \frac{n_0 \cos^2\gamma_0}{l_{0,A}} , \qquad \frac{n_1}{l_s} = \frac{n_0}{l_{0,A}} , \tag{5.5.9}$$

where l_t is the image distance of the tangential focal line and l_s the image distance of the sagittal focal line. For the imaging along the axis AB we find the image distances measured with respect to A,

$$l_{i,1,t} = \frac{n_1 \cos^2\gamma_1}{n_0 \cos^2\gamma_0}\, l_{0,A} , \qquad l_{i,1,s} = \frac{n_1}{n_0}\, l_{0,A} . \tag{5.5.10}$$

The object distances for the refraction at the point B are

$$l_{o,2,t} = l_{i,1,t} - \frac{d}{\cos\gamma_1} , \qquad l_{o,2,s} = l_{i,1,s} - \frac{d}{\cos\gamma_1} . \tag{5.5.11}$$

Applying the Coddington equations to the second surface of the plate yields the astigmatic image distances with respect to B on the principal ray BS_2T_2,

$$l_{i,2,t} = BT_2 = l_{0,A} - \frac{n_0 \cos^2\gamma_0}{n_1 \cos^3\gamma_1}\, d , \qquad l_{i,2,s} = BS_2 = l_{0,A} - \frac{n_0}{n_1 \cos\gamma_1}\, d . \tag{5.5.12}$$

The axial distance $\delta_{as} = S_2T_2 = l_{i,2,t} - l_{i,2,s}$ between the astigmatic lines is then given by

$$\delta_{\mathrm{as}} = -\frac{n_0}{n_1}\left(\frac{\cos^2\gamma_0 - \cos^2\gamma_1}{\cos^3\gamma_1}\right) d . \tag{5.5.13}$$

The normalised expression for this axial astigmatic distance is obtained by division by the axial diffraction unit $\lambda_0/(n_0 \sin^2\alpha_{0,m})$ of a focused pencil of rays with maximum opening angle $\alpha_{0,m}$. We then apply, for small tilt angles γ_0, the linearised version $n_0\gamma_0 = n_1\gamma_1$ of Snell's law and, for a maximum tilt angle $\gamma_{0,m}$ in image space, we find the expression

$$\delta_{n,\mathrm{as}} = \frac{n_0^2(n_1^2 - n_0^2)}{2n_1^3} \frac{d}{\lambda_0} \gamma_{0,m}^2 \sin^2\alpha_{0,m} \,. \tag{5.5.14}$$

From $\delta_{n,\mathrm{as}}$ we find the (normalised) astigmatic wavefront aberration of fourth order in aperture and field angle,

$$W^{(4)}_{\lambda_0,\mathrm{as}}(\rho_n, \theta, \gamma_0) = \left(\frac{n_0^2(n_1^2 - n_0^2)}{2n_1^3} \frac{d}{\lambda_0} \gamma_{0,m}^2 \alpha_{0,m}^2 \right) \rho_n^2 \sin^2\theta \left(\frac{\gamma_0}{\gamma_{0,m}} \right)^2 , \tag{5.5.15}$$

where the angle θ in the pencil of rays is counted positively from the x-axis. The expression for $W^{(4)}_{\lambda_0,\mathrm{as}}$ is in accordance with the Seidel coefficient and the astigmatic wavefront term given by Eqs. (5.5.5) and (5.5.6).

It follows that in the case of refraction by a plane-parallel plate with a higher index than that of the environment ($n_1 > n_0$), the sagittal focal line is in front of the tangential focal line. This relative position of the focal lines is independent of the convergence or the divergence of the incident pencil of rays. The paraxial 'best' astigmatic focus is exactly halfway between the astigmatic lines and has been indicated by the open circles in Fig. 5.48. We note that the obliquely used plane-parallel plate combined with a thin lens can be used to approximate a thick lens used at an off-axis field angle γ_0. The addition of the astigmatic wavefront aberration of the thin lens according to Eqs. (5.4.25) and (5.4.26) and the astigmatism of the plate according to Eq. (5.5.15) provide a good approximated value of the astigmatism of the equivalent thick lens. For the special case of normal incidence we find for the image distance $l_{2,A}$, measured from the point A on the front surface of the plate, the paraxial image shift formula once a plane-parallel plate is inserted in a focused beam,

$$l_{2,A} = d + l_{i,2} = d + l_{0,A} - \frac{n_0}{n_1} d = l_{0,A} + \frac{n_1 - n_0}{n_1} d \,. \tag{5.5.16}$$

5.6 Chromatic Aberration

So far, the calculation of wavefront aberration has been carried out for a single wavelength of the propagating light. In many applications, the spectral width of the propagating beams is not negligible. Because of the relatively strong dispersion of the optical properties of materials in the optical domain, a shift of wavelength is accompanied by a change in refractive index of the lens material. Consequently, both the paraxial properties and the aberrations of a lens system change. In practice, the change in paraxial properties is dominating the colour-dependent behaviour of an imaging system. In the literature, the change in paraxial properties of an optical system with wavelength is called chromatic aberration of the system. The term 'chromatic aberration' for this paraxial effect is not well chosen but its use is widespread in the optics community. Apart from colour effects in refraction we also have strong colour effects in diffractive optical elements for imaging, beam shaping or spectral analysis. In this section we first discuss the chromatic changes of the paraxial properties of a system and a general expression for the aberration change of finite rays. We conclude this section with a brief discussion of the chromatic effects in an optical system when using light consisting of a sequence of very short pulses.

5.6.1 Refractive Index of Air and Optical Materials

The refractive index in optics follows from the definition $n = c/v$, where v is the phase velocity in the medium. Applied to vacuum, the refractive index is exactly unity. In practice, earth-based optical systems are used in air and it is the ratio of the refractive indices of the optical material and the environmental air that determines the refraction of rays by the material.

5.6.1.1 Refractive Index of Air

The refractive index of the surrounding air is given for a particular reference composition of air at fixed temperature and pressure. Dry air is composed of nitrogen, oxygen, argon and carbon dioxide with the following volume ratios in percentages, 78.08, 20.95, 0.93 and 0.04, respectively. A recent overview of research on the refractive index of air is found in **[69]**. The refractive index as a function of wavelength can be well fitted by a dispersion formula of the Sellmeier-type **[37]**. It was first presented in **[271]** and fits well the experimental data from the ultraviolet (λ_0 = 230 nm) to the near-infrared (λ_0 = 1700 nm). For a temperature of dry air of 15 °C, a pressure of 1 atmosphere (101325 Pa) and a CO_2 content of 450 ppm (average laboratory condition) we have

$$(n-1)_r = 10^{-8} \left\{ \frac{5792105}{238.0185 - \sigma^2} + \frac{167917}{57.362 - \sigma^2} \right\} , \tag{5.6.1}$$

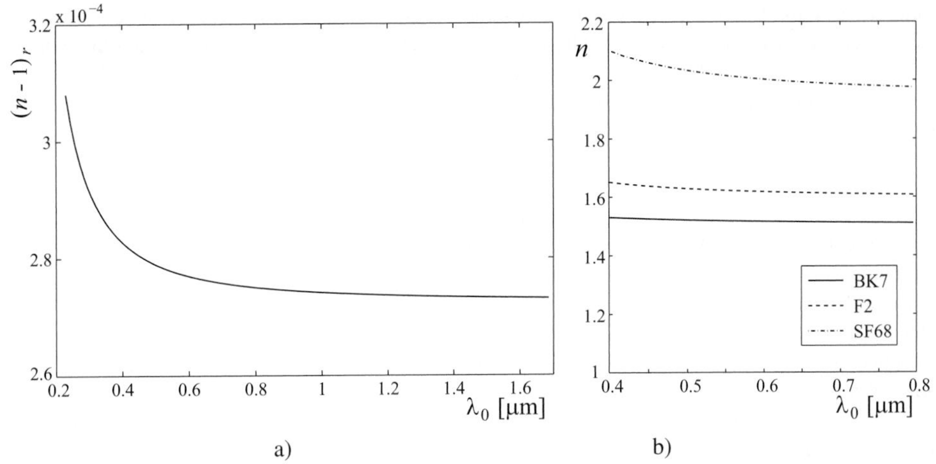

Figure 5.49: a) The refractive index function $(n-1)_r$ for standard dry air as a function of the wavelength λ_0 of the light. b) The refractive index as a function of the wavelength λ_0 for a typical low-, middle- and high-index optical glass with the relevant Sellmeier coefficients taken from the catalogue of the Schott company (visible range of the spectrum).

where $\sigma = \omega/(2\pi c)$ is the vacuum wavenumber of the light, expressed in μm^{-1}. The subscript r refers to the reference value $(n-1)$ of the optical contrast between the material and the environmental air.

A frequently used, authoritative formula for the refractive index of air including pressure, temperature and moisture influences was given in **[90]** and later improved **[27]**. For a state-of-the-art metrological overview, the reader is again referred to **[69]**. The data from the literature yield a dependence on temperature and pressure of the refractive index which is given by

$$(n-1)_{tp} = (n-1)_r \frac{p}{96095.43} \frac{\left[1 + 10^{-8} p\,(0.601 - 0.00972\, t)\right]}{(1 + 0.0036610\, t)}, \qquad (5.6.2)$$

where t is the temperature in °C, p the pressure in Pa and $(n-1)_r$ is given by Eq. (5.6.1). Further multiplicative factors can be included for the influence of water vapour and a different carbon dioxide content of the air.

Figure 5.49a shows the typical dependence of the refractive index of dry air on the wavelength of the radiation, from the ultraviolet to the near infrared. The value of the refractive index of 'standard' dry air in the centre of the visible region is of the order of 1.000278. Due to the absorption lines in the near and extreme ultraviolet (see Eq. (5.6.1)), the index strongly increases towards the shorter wavelengths. Depending on the quality requirements of the imaging system, the changes in image plane position and magnification due to pressure or temperature variations need correction. Apart from these paraxial changes, the aberrations of an imaging system also depend on the environmental conditions. In high-quality imaging, it may be necessary to work in well-stabilised temperature and pressure conditions or to introduce compensation means for the environmental changes in image errors.

5.6.1.2 Refractive Index of Optical Materials

The refractive index of optical materials is well described over the wavelength range from 0.35 μm to 1.5 μm by means of a three-term Sellmeier expression. Below the ultraviolet limit of 0.35 μm, the absorption of many visible-range materials tends to increase and makes them less suitable for high-quality optical applications. Six coefficients are sufficient to represent the refractive index with an accuracy of better than $1 \cdot 10^{-6}$ over the visible and near-infrared wavelength range. The dispersion coefficients B and C enter the dispersion equation in the following way,

$$n^2 - 1 = \frac{B_1}{1 - C_1\sigma^2} + \frac{B_2}{1 - C_2\sigma^2} + \frac{B_3}{1 - C_3\sigma^2}, \qquad (5.6.3)$$

where σ, as before, is the inverse value of the vacuum wavelength λ_0, expressed in μm^{-1}. Figure 5.49b shows the typical refractive index behaviour of a transparent optical material in the visible wavelength region.

The accurate description by means of the Sellmeier equation is often approximated by a simple linearised relation, introduced for practical reasons by E. Abbe. Each glass is represented by a six-digit code, the first three digits being the value of $(n-1)$ at a reference wavelength and the final three digits representing the value of $1/(dn/d\lambda)$, where $(dn/d\lambda)$ is the slope of the dispersion function $n(\lambda)$ of the material. The inverse of the slope is approximated with the aid of refractive index values at two reference wavelengths,

$$V_d = \frac{n_d - 1}{n_F - n_C} \,, \tag{5.6.4}$$

where V commonly denotes the Abbe-number of the optical material. The subscripts in Eq. (5.6.4) denote certain spectral lines, the He d-line (588 nm) and the hydrogen F- and C-lines at 486 and 656 nm, respectively. As an example, the very frequently used crown glass BK7 has the code 517642 with a central n-value of 1.517 at the d-line and an Abbe number of 64.2, also defined at the He d-line. Typical V-numbers of optical glasses and plastics range from 25 to 65; the refractive power of an optical transition thus changes from 1.5% ('crown' glass type) up to 4% (heavy 'flint' glass) over the visible spectral region. When an optical system is used primarily in outer parts of the optical spectrum, other dispersion numbers are defined that are more appropriate for the index behaviour in that spectral region. Historically, partial dispersion numbers have been defined according to

$$P_{\lambda_1,\lambda_2} = \frac{n_{\lambda_1} - n_{\lambda_2}}{n_F - n_C} \,, \tag{5.6.5}$$

mainly with the purpose of characterising the dispersion in the outer regions of the visible spectrum or in the near ultraviolet or infrared. Nowadays, in the practice of optical design, these special dispersion numbers are not frequently used. The general expression of Eq. (5.6.3) allows the calculation of the refractive index and its derivatives with high precision during the process of lens optimisation.

5.6.2 Change with Colour of the Paraxial Properties

In this subsection we first analyse the change in optical power of an optical transition. The next step is to study the wavefront changes that arise at a surface due to a change of the refractive indices of the adjacent media.

5.6.2.1 Optical Power

The first effect of a refractive index change is a change in optical power of a single transition between two media or of a lens. In the latter case, for a thin element, we have for the change in power between the spectral lines F and C,

$$\delta K_{F,C} = K_F - K_C = \frac{n_F - n_C}{n_d - 1}(n_d - 1)(c_1 - c_2) = \frac{K_d}{V} \,, \tag{5.6.6}$$

where K_d is the power of the thin lens at the central wavelength of the He d-line, c_1 and c_2 are the curvatures of the front and reverse side of the thin lens and V is the Abbe number of the lens material.

5.6.2.2 Paraxial Chromatic Wavefront Aberration

The chromatic effects of a single transition between two media will now be analysed by means of the pathlength changes between entrance and exit pupil that result from changes in refractive index of the two adjacent media. In Fig. 5.50 we have shown the incident wave, leaving the off-axis point A_0 and focused at the nominal image point A_1. The refracted rays Q_aA_1' and Q_pA_1' are the paraxial marginal and principal ray (dashed in the figure) for the case of an index change of the two adjacent media, δn_0 and δn_1, respectively. The changes in refractive index are the result of a change in wavelength of the light. For the nominal image situation, taking an axial object point, we have the condition that the optical paths along the marginal ray and along the axis should be equal. Denoting the object and image distance by s_0 and s_1, respectively, we have as paraxial imaging condition in the nominal situation for the on-axis points,

$$W = [O_0Q_aO_1] - [O_0TO_1] = n_0\frac{y^2}{2}\left(c - \frac{1}{s_0}\right) - n_1\frac{y^2}{2}\left(c - \frac{1}{s_1}\right) = 0 \,. \tag{5.6.7}$$

We note that this imaging condition, applied to the off-axis conjugate points A_0 and A_1 will only yield a nonzero value that is quadratic in the field coordinates p_0 or p_1 and that such a quadratic value can be neglected with respect to chromatic path length changes that are linear in p_0 and p_1.

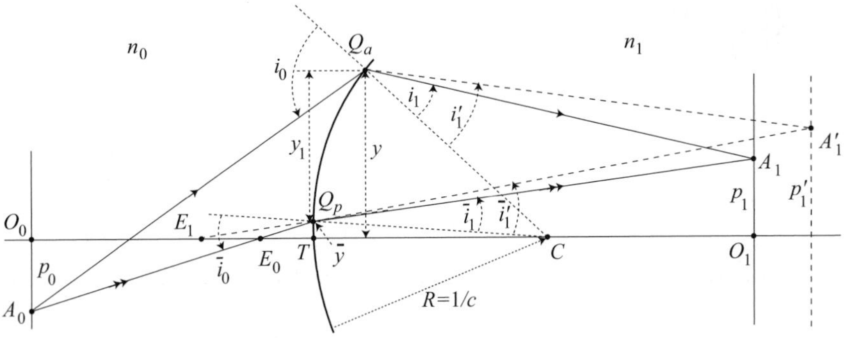

Figure 5.50: Pathlength analysis in the presence of chromatic effects. The image point A_1 is found at A_1' under the influence of the changes δn_0 and δn_1 of the refractive indices of the two media.

- *Chromatic change of wavefront aberration*

  For the axial pair of conjugate points O_0 and O_1 the change in pathlength due to changes in refractive index is given by

$$\begin{aligned}\delta W_{\mathrm{ax}}(y) &= \delta n_0 \frac{y^2}{2}\left(c - \frac{1}{s_0}\right) - \delta n_1 \frac{y^2}{2}\left(c - \frac{1}{s_1}\right) \\ &= \frac{1}{2}\frac{\delta n_0}{n_0} y\, n_0 y\left(c - \frac{1}{s_0}\right) - \frac{1}{2}\frac{\delta n_1}{n_1} y\, n_1 y\left(c - \frac{1}{s_1}\right) \\ &= -\frac{1}{2} yA\left(\frac{\delta n_1}{n_1} - \frac{\delta n_0}{n_0}\right), \end{aligned} \tag{5.6.8}$$

  where $A = n_0 i_0 = n_1 i_1$ is the refraction invariant of the marginal ray at the surface and the subscript ax of W indicates that the wavelength change corresponds to an axial shift of the image point.

  For an off-axis image point A_0, we consider the change in refraction angle of the principal ray. From Snell's law we have

$$\bar{i}_1 = \frac{n_0}{n_1}\bar{i}_0 = \frac{n_0}{n_1}\bar{y}\left(c - \frac{1}{\bar{s}_0}\right), \tag{5.6.9}$$

  where $\bar{s}_0 = TE_0$ is the distance from the top T of the surface to the centre E_0 of the entrance pupil.

  The change of the refraction angle with wavelength follows from

$$\delta\bar{i}_1 = \left(\frac{\delta n_0}{n_0 n_1} - \frac{\delta n_1}{n_1^2}\right) n_0 \bar{y}\left(c - \frac{1}{\bar{s}_0}\right). \tag{5.6.10}$$

  The change in refraction angle of the principal ray introduces an extra pathlength W_{lat} at the rim of the exit pupil with its centre of curvature at A_1. This extra pathlength is given by

$$\begin{aligned}\delta W_{\mathrm{lat}} &= n_1 y\, \delta\bar{i}_1 \\ &= -yB\left(\frac{\delta n_1}{n_1} - \frac{\delta n_0}{n_0}\right), \end{aligned} \tag{5.6.11}$$

  where $B = n_0 \bar{i}_0 = n_1 \bar{i}_1$ is the refraction invariant of the principal ray.

- *Axial chromatism*

  For a complete system, the pathlength changes are summed over all optical surfaces. The contributions δW_{ax} accumulated by a marginal ray of the axial pencil yield the quantity ΔW_{ax} that is often called the 'axial colour' of the system. By an axial shift Δz of the image plane a sharp image can be obtained for the shifted wavelength λ. The chromatic wavefront departure in the exit pupil in terms of the normalised lateral pupil coordinate ρ_n is given by

$$\Delta W_{\mathrm{ax}}(\rho_n) = \frac{1}{2} C_{\mathrm{ax}} \rho_n^2, \quad \text{where } C_{\mathrm{ax}} = -\sum_j A_{m,j}\, r_{m,j}\left(\frac{\delta n_j}{n_j} - \frac{\delta n_{j-1}}{n_{j-1}}\right). \tag{5.6.12}$$

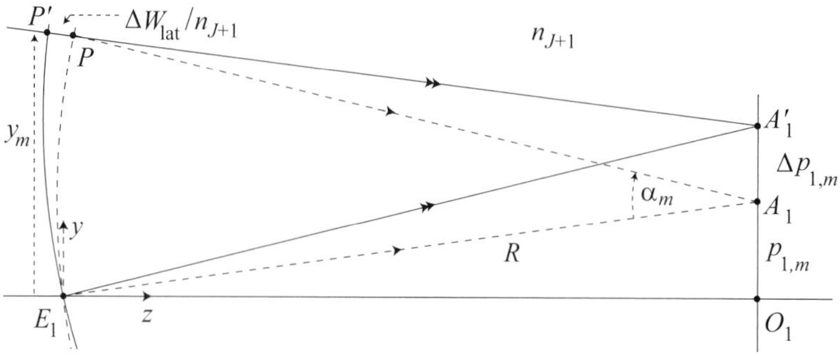

Figure 5.51: Wavefront deviation in the case of a lateral image shift Δp_1. The dashed reference sphere through E_1 has its centre at A_1. The tilted wavefront has its centre at A'_1. The wavefront deviation ΔW_{lat} for $y = y_m$ is given by $n_{J+1}(P'P)$; it is positive in the figure. The opening angle of an imaging pencil is α_m.

The maximum value C_{ax} of the quadratic wavefront departure at the rim of the exit pupil of the system determines the axial shift Δz of the image plane according to

$$\Delta z_{ax} = \frac{C_{ax}}{n_{J+1}\sin^2\alpha_m}, \qquad \Delta z_{n,ax} = \frac{C_{ax}}{\lambda_0}. \tag{5.6.13}$$

n_{J+1} is the refractive index of the image space, $\Delta z_{n,ax}$ is the normalised axial shift towards the sharp image plane for the shifted wavelength λ and α_m is the aperture angle in image space.

- *Lateral chromatism*

For the lateral shift in position of the image point A_1 we calculate the wavefront tilt ΔW_{lat} in the exit pupil of the optical system; ΔW_{lat} is currently called the 'lateral colour' of the optical system. It induces a change in magnification of the image due to chromatic effects. The value of ΔW_{lat} for an image point A_1 in the plane $x = 0$ is given by

$$\Delta W_{lat}(\rho_n, \theta, p_1) = C_{lat}\,\rho_n \sin\theta \,\frac{p_1}{p_{1,m}}, \text{ where, } C_{lat} = -\sum_j y_{m,j}\, B_{m,j}\left(\frac{\delta n_j}{n_j} - \frac{\delta n_{j-1}}{n_{j-1}}\right), \tag{5.6.14}$$

where $p_{1,m}$ is the maximum field coordinate along the y-axis.

The lateral shift of the outer image point with coordinate $p_{1,m}$ follows from Fig. 5.51. Using the approximations of paraxial optics we find the expression

$$\frac{\Delta p_{1,m}}{p_{1,m}} = \frac{C_{lat}\,R}{n_{J+1} y_m p_{1,m}} = -\frac{C_{lat}}{n_{J+1}\sin\alpha_m p_{1,m}} = -\frac{C_{lat}}{H}, \tag{5.6.15}$$

where H is the Lagrange invariant of the imaging pencil. The quantity $\Delta p_1/p_1$ is constant over the image field and equals the relative chromatic change in magnification, $\Delta m_T/m_T$, of the image.

- *Chromatic aberration of a thin lens*

The general result for the chromatic effects in a sequence of optical surfaces can be applied to the special case of a lens with its two surfaces in contact (limit of zero thickness). The axial colour introduced by a thin lens is given by

$$\Delta W_{ax} = -\frac{y^2}{2}\left(\frac{\delta n}{n-1}\right)K = -\frac{y^2}{2}\frac{K}{V}, \tag{5.6.16}$$

where we have assumed that the wavelength change δn is caused by a wavelength shift from the red spectral line C to the blue F-line.

For the lateral colour of a thin lens we find

$$\Delta W_{lat} = -y\bar{y}\left(\frac{\delta n}{n-1}\right)K = -y\bar{y}\,\frac{K}{V} = +2\frac{\bar{y}}{y}\,\Delta W_{ax}. \tag{5.6.17}$$

From the analysis above, we conclude that the colour aberrations of the thin lens are independent of the lens shape or the lens magnification. The formulas above also show, not surprisingly, that for a general surface the lateral colour is zero when the angle of incidence of the principal ray is zero ($B = 0$). For the thin lens, it is the condition of central

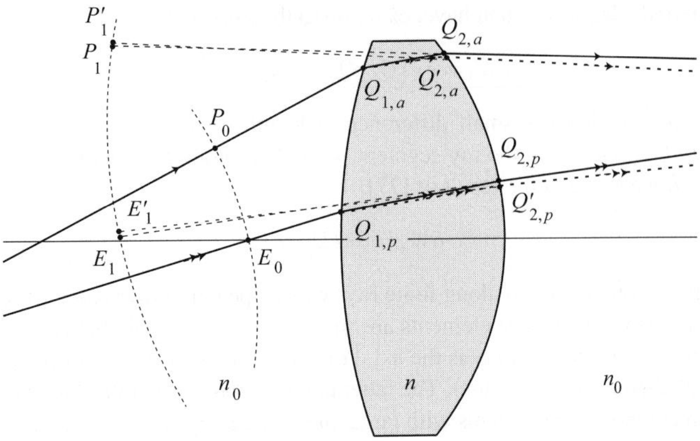

Figure 5.52: The optical pathlength deviation in the case of imaging with finite rays through a dispersive medium. The entrance pupil sphere passes through the points E_0 and P_0. The rays that have been refracted by the lens intersect the exit pupil in the space after the lens at the points E_1 and P_1 (solid rays, wavelength λ_0). The refracted rays at a wavelength $\lambda_0 + \delta\lambda$, dashed in the figure, intersect the exit pupil sphere at the points E'_1 and P'_1.

stop position that yields zero lateral colour. In this case, the principal ray is undeviated, independently of the value of the refractive index of the lens. The same property that was found for the asymmetric monochromatic aberrations coma and distortion for a complete system also holds for the lateral colour aberration. If the optical system and the imaging geometry show a point symmetry with respect to some central point C, all these aberrations are zero (see the discussion on comatic aberration and Fig. 5.45).

5.6.3 Chromatic Aberration of Finite Rays, Conrady's Formula

The chromatic pathlength changes that arise for non-paraxial, finite rays can be calculated with the aid of the following scheme **[77]**. In Fig. 5.52, the incident beam is free from aberration. Its principal ray intersects the entrance pupil at E_0, a marginal ray at P_0. At the reference wavelength λ_0, these rays are refracted by the lens via the paths $Q_{1,p}Q_{2,p}$ and $Q_{1,a}Q_{2,a}$, as shown by the solid lines in the figure. Their intersections with an exit pupil sphere, centred at an image point (not shown in the figure) are E_1 and P_1, respectively. At a different wavelength $\lambda_0 + d\lambda$, the refracted rays follow the paths $Q_{1,p}Q'_{2,p}$ and $Q_{1,a}Q'_{2,a}$ and intersect the exit pupil at E'_1 and P'_1. The pathlength difference between an aperture ray and the principal ray in the nominal situation ($\lambda = \lambda_0$) and for a shifted wavelength $\lambda_0 + \delta\lambda$ follows from the quantities

$$\begin{aligned} W_{\lambda_0} &= [P_0Q_{1,a}Q_{2,a}P_1] - [E_0Q_{1,p}Q_{2,p}E_1] \,, \\ W_{\lambda_0+\delta\lambda} &= [P_0Q_{1,a}Q'_{2,a}P'_1] - [E_0Q_{1,p}Q'_{2,p}E'_1] \,. \end{aligned} \qquad (5.6.18)$$

The pathlength difference between the situations with wavelength difference $\delta\lambda$ is thus given by

$$\begin{aligned} \delta W_\lambda = W_{\lambda_0+\delta\lambda} - W_{\lambda_0} &= (n+\delta n)(Q_{1,a}Q'_{2,a}) + n_0(Q'_{2,a}P'_1) - n(Q_{1,a}Q_{2,a}) - n_0(Q_{2,a}P_1) \\ &- (n+\delta n)(Q_{1,p}Q'_{2,p}) - n_0(Q'_{2,p}E'_1) + n(Q_{1,p}Q_{2,p}) + n_0(Q_{2,p}E_1) \,, \end{aligned} \qquad (5.6.19)$$

where δn is the change in refractive index due to the wavelength change of $\delta\lambda$.

The solid rays in the figure are real rays for the nominal wavelength λ_0. At this wavelength the dashed refracted rays are non-physical rays. For these rays, Fermat's principle for neighbouring rays can be applied, stating that the pathlength along a physical ray has an extremum value with respect to non-physical neighbouring rays. Consequently, at the wavelength λ_0, the pathlengths along the dashed non-physical rays are only in second-order different from the pathlengths along the physical rays for $\lambda = \lambda_0$. This allows us to write up to the first order in the pathlength

$$\begin{aligned} n(Q_{1,a}Q'_{2,a}) + n_0(Q'_{2,a}P'_1) &= n(Q_{1,a}Q_{2,a}) + n_0(Q_{2,a}P_1) \,, \\ n(Q_{1,p}Q'_{2,p}) + n_0(Q'_{2,p}E'_1) &= n(Q_{1,p}Q_{2,p}) + n_0(Q_{2,p}E_1) \,. \end{aligned} \qquad (5.6.20)$$

For the chromatic pathlength deviation we then have, exact up to the first order,

$$\delta W_\lambda = \delta n \left(Q_{1,a}Q'_{2,a} - Q_{1,p}Q'_{2,p} \right) . \tag{5.6.21}$$

In the next step we neglect the relatively small difference in length between the ray segment differences $Q_{1,a}Q'_{2,a} - Q_{1,p}Q'_{2,p}$ and $Q_{1,a}Q_{2,a} - Q_{1,p}Q_{2,p}$. Denoting the ray segment in the lens along the aperture ray by D and the ray segment along the principal ray by d, we arrive at the result of **[77]**,

$$\delta W_\lambda = \delta n (D - d) . \tag{5.6.22}$$

This formula yields the aberration increment along finite rays when a pencil of rays passes through a dispersive medium. The contributions from the various dispersive elements are simply summed to yield the chromatic aberration of an entire system. In the case that the chosen principal ray is the axial ray, the Conrady $(D - d)$ formula yields the change of focus *and* the change in spherical aberration of a system. The latter contribution is generally called sphero-chromatism and can be an important aberration component in systems with large aperture angles.

5.6.4 Correction of Chromatic Aberration

For a long time, the correction of chromatic errors by using different glass materials was thought to be impossible. The results from still relatively primitive experiments brought Newton to the conclusion that refractive materials all showed identical behaviour with respect to their dispersion. Colour correction, an important issue in astronomical observation, could only be achieved by using reflecting surfaces. Later measurements on the available glass types, the low-index crown glasses and the higher-index flint glasses, showed substantial differences in dispersion. Indeed, the typical V numbers of these glasses differ by a factor of almost two. Some 30 years after Newton's death, the first suggestions were put forward to combine two lenses with different glass type to eliminate their chromatic error. Unfortunately, the priority issue of this important invention gave rise to a complicated legal and moral dispute in which the names of Dollond, Chester Hall and Ramsden are mixed. In fact, as of 1760, the achromatic lens combination was available. The first cemented doublet lens seems to have been made by De Grateloup in the year 1787. Apart from the optical power, other properties of an optical system can be made achromatic. Some examples are given at the end of this subsection.

Starting with the seventeenth/eighteenth century crown and flint glasses, the glass map was extended substantially in the nineteenth and twentieth centuries. Figure 5.53 shows a glass chart provided by the Schott company, a pioneering company in the field of optical glass that was founded in the middle of the nineteenth century. The optical glasses are plotted on a map given their refractive index and dispersion. Along the horizontal axis the Abbe V number is plotted. The value of the refractive index at the spectral d-line (587.6 nm) is along the vertical axis. Some hundred optical glasses have been internationally defined. Until recently, the number of optical glasses was substantially larger. Many glasses have been discarded for economic reasons and because of environmental issues (e.g. radioactive thorium-containing glasses, glasses containing lead or arsenic compounds). The glass map can be complemented with plastic materials, like PMMA (polymethyl methacrylate), PC (polycarbonate) and a large choice of plastic compound materials. PMMA is quite close on the glass map to the very common 'crown' glass BK7. Polycarbonate resembles the flint glass F5.

5.6.4.1 Optical Power of Two Lenses in Contact, the Doublet

The power of two thin lenses in contact is simply the sum of their individual powers K_1 and K_2. For the variation of the total power K with wavelength, we have, using the Abbe numbers of the lens materials

$$\delta K_{F,C} = \frac{K_1}{V_1} + \frac{K_2}{V_2} , \tag{5.6.23}$$

where $\delta K_{F,C} = K_F - K_C$ is the power change when shifting the wavelength from the C to the F spectral line. The achromatic condition for the total power then immediately follows from $K_1/V_1 = -K_2/V_2$ where the powers K_1 and K_2 are those at the He d-line. The individual powers of the two components of the doublet lens are given by

$$K_1 = \frac{V_1}{V_1 - V_2} K , \qquad K_2 = -\frac{V_2}{V_1 - V_2} K . \tag{5.6.24}$$

For the change in power at a certain wavelength λ_1 with respect to, for instance, the blue F-line, we have in first order

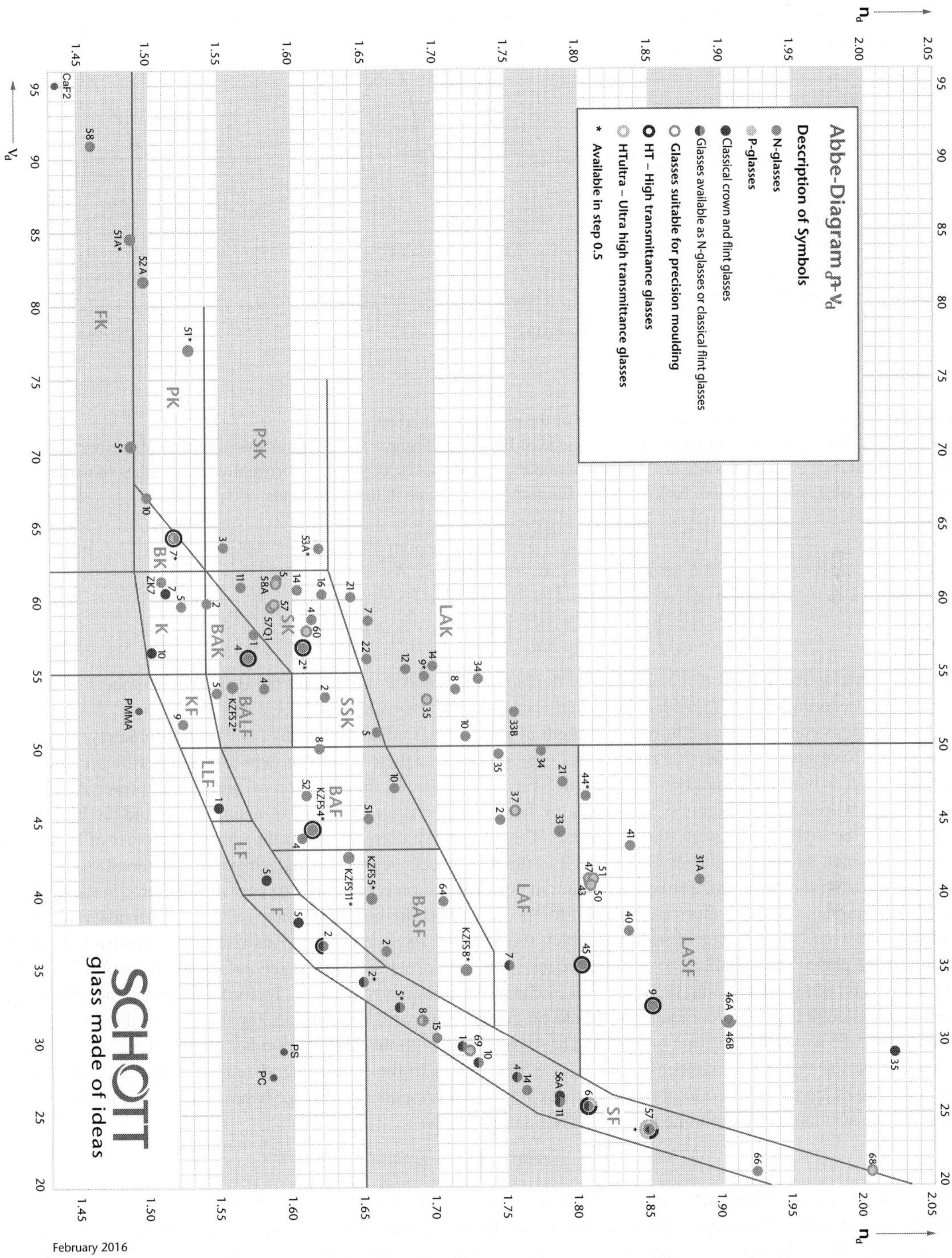

Figure 5.53: Glass chart or Abbe diagram (reproduced with the permission of Schott optical glass works, release 2016). To the glass map have been added some plastic materials like acrylic 'glass' (PMMA), polycarbonate (PC) and polystyrene (PS), and the low-dispersion material fluorite (CaF2).

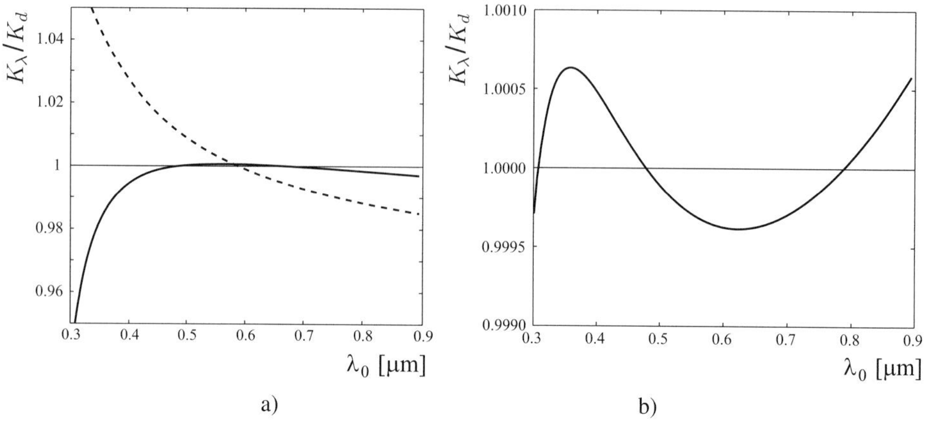

Figure 5.54: The relative variation with wavelength of the power of a doublet.
a) A single lens (BK7, dashed curve) and a doublet made of BK7 and F2 glass, with V numbers of 64 and 36, respectively.
b) The combination of calcium fluorite and BK7, V numbers 95 and 64, respectively, and virtually equal values of partial dispersion in the blue spectral region. Note the very different vertical scale in the two graphs.

$$\begin{aligned}\delta K_{\lambda_1,F} &= K_{\lambda_1} - K_F = \left(\frac{n_{\lambda_1} - n_F}{n_d - 1}\right)_1 K_1 + \left(\frac{n_{\lambda_1} - n_F}{n_d - 1}\right)_2 K_2 \\ &= \left(\frac{n_{\lambda_1} - n_F}{n_F - n_C}\right)_1 \frac{K_1}{V_1} + \left(\frac{n_{\lambda_1} - n_F}{n_F - n_C}\right)_2 \frac{K_2}{V_2} = P_1 \frac{K_1}{V_1} + P_2 \frac{K_2}{V_2} = \left(\frac{P_1 - P_2}{V_1 - V_2}\right) K_{F,C} ,\end{aligned} \tag{5.6.25}$$

where $K_{F,C}$ is the optical power of the achromatic doublet and P_1 and P_2 are the partial dispersion numbers of the two glass materials according to Eq. (5.6.5), defined in this case for the wavelength pair (λ_1, λ_F).

The doublet lens can only have a third wavelength at which it has equal power if $P_1 \approx P_2$. Unfortunately, almost all glass materials have a partial dispersion number that is proportional to their V number, which makes it difficult to achieve the condition $\delta K = 0$ at more than two wavelengths. Figure 5.54 shows the variation of normalised power K_λ/K_d as a function of the wavelength λ. Figure 5.54a shows the fairly linear change in power of a single lens and the increase in power towards the high-index region (dashed curve). The solid curve corresponds to the change in power of a standard achromatic doublet, having the glasses BK7 and F2 as the low-dispersive and the high-dispersive material, respectively. In the visible part of the spectrum, a strong reduction in power variation is obtained. At two wavelengths, in the blue and in the red spectral region, equal values are obtained for the lens power. In the intermediate region, a small quadratic power variation is observed. At the focus of such a doublet, this residual focus variation gives rise to a yellowish halo in the optimum image plane. The variations in optimum focus-setting are designated as the *secondary spectrum* of the doublet.

In the blue and ultraviolet region, the reduction in change of power is much less. To further improve the achromatic behaviour of a doublet, the two components should be given equal partial dispersion, for instance in the blue spectral region. Figure 5.55 shows the position of various glasses on a map with the Abbe V number plotted along the horizontal axis and the partial dispersion number $P_{g,F} = (n_g - n_F)/(n_F - n_C)$ in the blue spectral region along the vertical axis. 'Standard' glasses are found close to or on a reference line that represents the average behaviour of optical glasses. The equation of the standard dispersion line in the blue region is given by

$$P_{g,F} = -0.001692\, V + 0.6444 . \tag{5.6.26}$$

'Standard' dispersion curves are also available for spectral regions like the infrared and the ultraviolet.

Higher-order colour correction can be obtained by combining glasses that have almost equal values of $P_{g,F}$ and an as large as possible difference in V number. An exceptional material that violates the standard linear P/V relationship is calcium fluorite, CaF2, and it is frequently used to improve the chromatic behaviour of a lens system in the blue region. Good candidates for the components of an improved achromatic doublet are CaF2 and BK7. In such a doublet, BK7 now plays the role of the high-dispersive medium. The chromatic behaviour of this material combination is shown in Fig. 5.54b. We now observe equal power for three wavelengths and such a combination of lens materials forms an *apochromatic* doublet. The residual secondary spectrum has been strongly reduced, by typically one order of magnitude as compared to the standard achromatic doublet. A practical disadvantage of such an apochromatic doublet is its

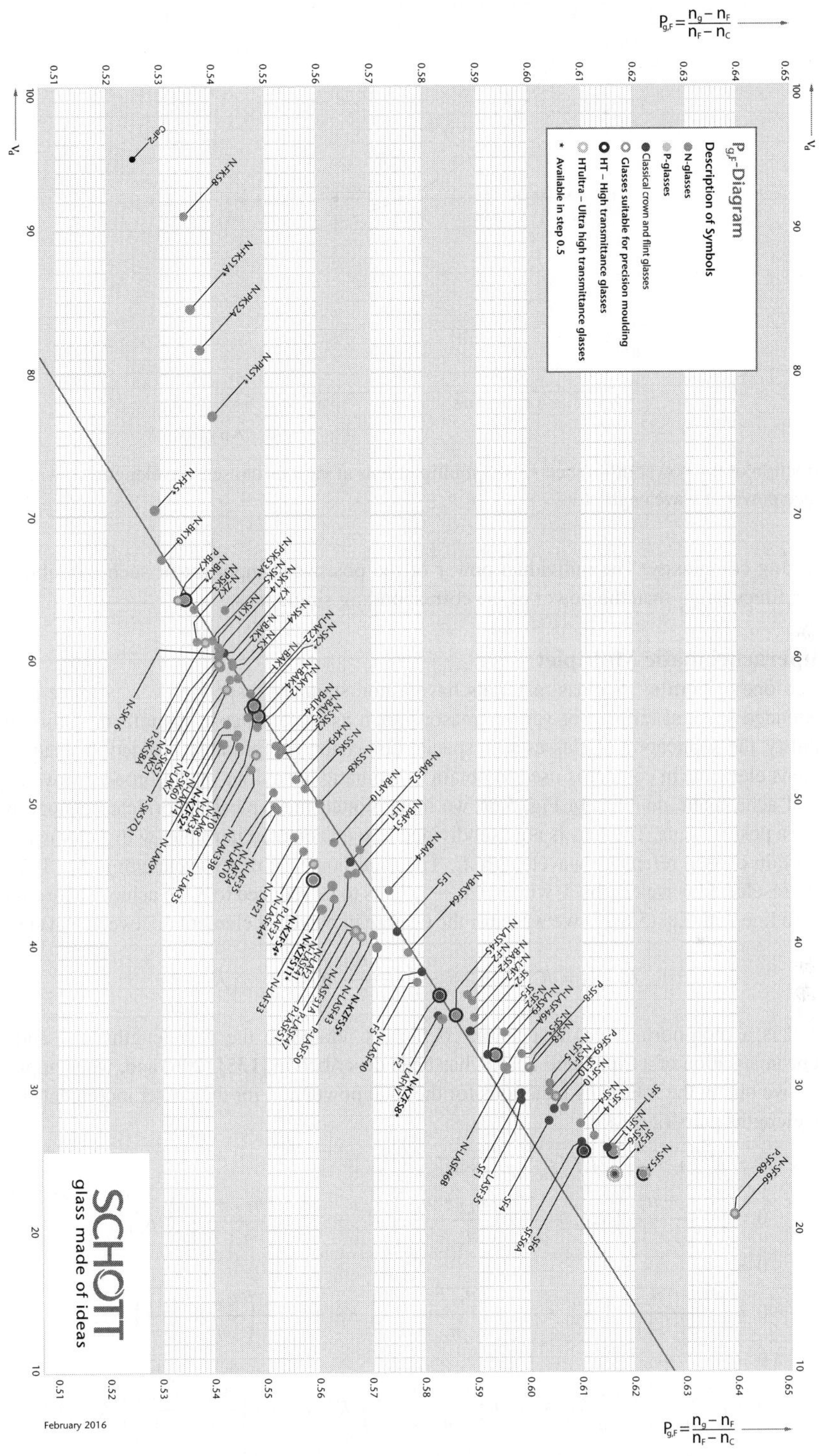

Figure 5.55: A plot of the partial dispersion number $P_{g,F}$ in the blue region of the spectrum as a function of the Abbe V number of a glass (reproduced with the permission of Schott optical glass works, release 2016). The material CaF2 (fluorite) has been added in the lower left corner of the diagram.

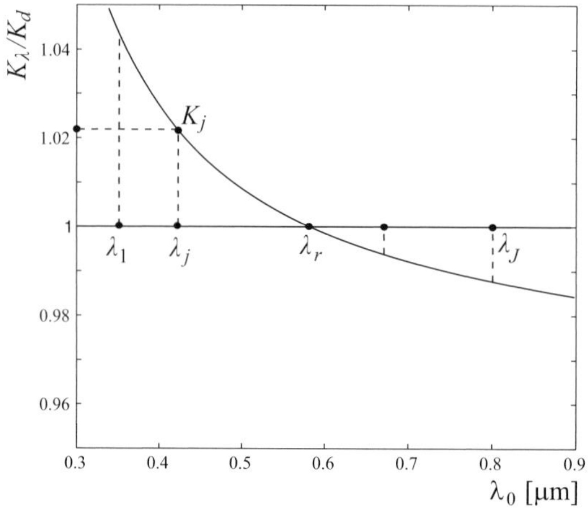

Figure 5.56: Correction of the power difference K_j of a multiplet lens at various chosen wavelengths λ_j with respect to the normalised reference power at wavelength λ_r.

relative manufacturing complexity. The individual power of the positive component of such a doublet has to be made approximately four times larger than the power of the corresponding single lens.

5.6.4.2 The Superachromatic Multiplet

A doublet of which one or both of its glass materials have a non-standard dispersion can produce a constant paraxial power over an extended spectral range. For various reasons such as material cost, manufacturability, thermal stability and chemical stability, the application of glasses with special dispersion is preferably avoided. Instead of this option, the addition of extra lens elements in contact is used to obtain achromatic behaviour over a broader wavelength region than that of the classical achromatic doublet. In Fig. 5.56 we have plotted the typical uncorrected change of optical power with wavelength of a positive lens. The lens is now subdivided into a cemented multiplet with I elements and with partial powers $K_{i,r}$ at a certain chosen reference wavelength λ_r. The total power at this wavelength is K_0. The total power at a certain number of preselected wavelengths λ_j with $j = 1, \cdots, J$ has to be reduced to K_0 to achieve achromatic behaviour at all J wavelengths. In line with Eq. (5.6.25) we write for the change in power of element i between wavelengths j and $j+1$,

$$\delta K^i_{j+1,j} = K^i_{j+1} - K^i_j = \frac{n_{i,j+1} - n_{i,j}}{n_{i,r} - 1} K_i = P^i_{\lambda_{j+1},\lambda_j} K_i \,, \tag{5.6.27}$$

where the factor P is a (non-normalised) dispersion coefficient related to the wavelengths λ_{j+1} and λ_j of the glass material i. The achromatisation of a triplet and quartet has been described in **[135]**,**[206]** and, when applied to I elements and J wavelengths, we obtain the following equations for the total power and for the power increments between the $J-1$ neighbouring wavelengths pairs $(\lambda_{j-1}, \lambda_j)$,

$$\begin{aligned}
1 &= K'_{1,r} + \quad \cdots \quad + K'_{i,r} \quad \cdots \quad + K'_{I,r} \,, \\
0 &= \left(\frac{n_{1,2} - n_{1,1}}{n_{1,r} - 1}\right) K'_{1,r} + \cdots + \left(\frac{n_{i,2} - n_{i,1}}{n_{i,r} - 1}\right) K'_{i,r} + \cdots + \left(\frac{n_{I,2} - n_{I,1}}{n_{I,r} - 1}\right) K'_{I,r} \,, \\
0 &= \quad \cdots \\
0 &= \left(\frac{n_{1,j} - n_{1,j-1}}{n_{1,r} - 1}\right) K'_{1,r} + \cdots + \left(\frac{n_{i,j} - n_{i,j-1}}{n_{i,r} - 1}\right) K'_{i,r} + \cdots + \left(\frac{n_{I,j} - n_{I,j-1}}{n_{I,r} - 1}\right) K'_{I,r} \,, \\
0 &= \quad \cdots \\
0 &= \left(\frac{n_{1,J} - n_{1,J-1}}{n_{1,r} - 1}\right) K'_{1,r} + \cdots + \left(\frac{n_{i,J} - n_{i,J-1}}{n_{i,r} - 1}\right) K'_{i,r} + \cdots + \left(\frac{n_{I,J} - n_{I,J-1}}{n_{I,r} - 1}\right) K'_{I,r} \,,
\end{aligned} \tag{5.6.28}$$

where $n_{i,j}$ is the refractive index of material i at wavelength j and $K'_{i,r}$ the power of the ith element of the multiplet, normalised with respect to the total power of the mulitplet. The total power is normalised to unity with the aid of the equation on the first line of (5.6.28).

The system of equations can be simplified further by taking a certain material i_0 as reference material and dividing each dispersion equation by the factor preceding the partial power $K'_{i_0,r}$. In matrix notation this yields the modified system

$$\begin{pmatrix} 1 & \cdots & 1 & \cdots & 1 \\ P^{1,i_0}_{\lambda_2,\lambda_1} & \cdots & 1 & \cdots & P^{I,i_0}_{\lambda_2,\lambda_1} \\ \cdots & \cdots & 1 & \cdots & \cdots \\ P^{1,i_0}_{\lambda_j,\lambda_{j-1}} & \cdots & 1 & \cdots & P^{I,i_0}_{\lambda_j,\lambda_{j-1}} \\ \cdots & \cdots & 1 & \cdots & \cdots \\ P^{1,i_0}_{\lambda_J,\lambda_{J-1}} & \cdots & 1 & \cdots & P^{I,i_0}_{\lambda_J,\lambda_{J-1}} \end{pmatrix} \begin{pmatrix} K'_{1,r} \\ \vdots \\ K'_{i_0,r} \\ \vdots \\ K'_{I,r} \end{pmatrix} = \begin{pmatrix} 1 \\ 0 \\ \vdots \\ 0 \\ \vdots \\ 0 \end{pmatrix}, \tag{5.6.29}$$

where the following definition of a generalised partial dispersion coefficient is used, referenced to the material with index i_0,

$$P^{(i,i_0)}_{(\lambda_j,\lambda_{j-1})} = \left(\frac{n_{i,j} - n_{i,j-1}}{n_{i_0,j} - n_{i_0,j-1}}\right)\left(\frac{n_{i_0,r} - 1}{n_{i,r} - 1}\right). \tag{5.6.30}$$

The system of linear equations can be analytically solved for the unknown partial powers $K'_{i,r}$ of the multiplet. A practical problem is the close resemblance of the dispersion curves of the available optical glasses. The system can easily become nearly singular and give rise to unfeasibly large partial powers. A preselection of optical glasses with sufficiently different dispersion behaviour reduces the risk of an impractical solution. The glass selection criteria are discussed in **[206]**. Each glass is represented by a point in $(J-1)$-dimensional space with coordinate axes for the $J-1$ partial dispersion coefficients $P^{(i,i_0)}_{(\lambda_j,\lambda_{j-1})}$, referenced to the material i_0. Only the glasses with a sufficiently large Cartesian distance between them in this partial dispersion space are retained. With N selected glasses, the system of equations is solved for $N!/\{I!(N-I)!\}$ possible combinations, the ordering of the selected elements in the achromatic multiplet being irrelevant. The final solution is the one that performs best with respect to the partial powers of the elements and the residual chromatic error in the $J-1$ intervals between the selected wavelengths λ_j.

5.6.4.3 Achromatic Power of an Optical System

The simplest optical refracting system consists of two (thin) lenses with powers K_1 and K_2, that are not in contact but separated by a distance d that is appreciable with respect to the focal distances of the two components. For the total power K of such a compound system in air we have

$$K = K_1 + K_2 - dK_1K_2. \tag{5.6.31}$$

Using the Abbe V number of the two lenses, we find for the variation of power with wavelength

$$\delta K = \frac{K_1}{V_1} + \frac{K_2}{V_2} - dK_1K_2\left(\frac{1}{V_1} + \frac{1}{V_2}\right). \tag{5.6.32}$$

The condition for $\delta K = 0$ reads

$$d = \frac{K_1V_2 + K_2V_1}{K_1K_2(V_1 + V_2)}. \tag{5.6.33}$$

In the particular case of identical glass materials, we find in terms of the image-side focal distances of the individual lenses,

$$d = \frac{f'_1 + f'_2}{2}. \tag{5.6.34}$$

Figure 5.57 shows a typical Huygens eyepiece that exploits the achromatic power condition of Eq. (5.6.34). The two elements are convex-plano lenses. The requirement $\delta K = 0$ has been satisfied and yields a colour-independent power of the system. It does not mean that such a two-lens eyepiece is truly achromatic. Apart from the optical power, the position of the object- and image-side principal planes should also be independent of colour. With the aid of Eq. (4.10.29), we can study the condition for $\delta s_P = 0$ for either the object- or image-side principal plane. It turns out that for the two-lens system this condition can only be met for the image-side principal plane when the power of the first lens becomes zero and vice versa for the object-side principal plane.

The constant power condition influences the axial colour of the eyepiece. For the lateral colour, the trajectory of the principal ray (dashed in Fig. 5.57) is important. No point symmetry with respect to a central diaphragm is present. Therefore, we expect an appreciable lateral colour. Both axial and lateral colour contributions from the individual lenses can be calculated with the aid of Eqs. (5.6.16) and (5.6.17). An analysis of the monochromatic aberrations shows that

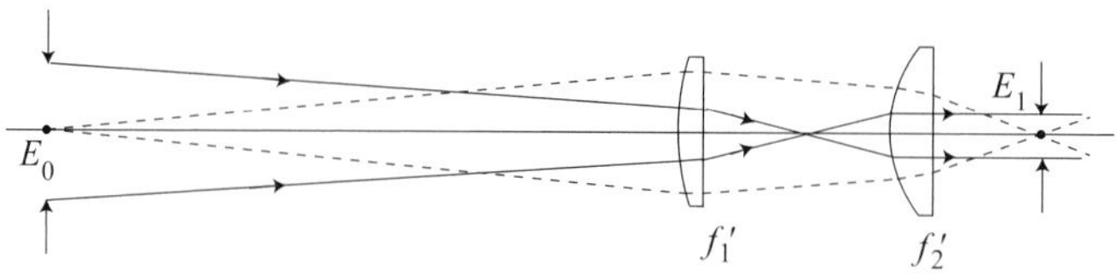

Figure 5.57: The Huygens eyepiece, consisting of two plano-convex lenses at a distance $d = (f_1' + f_2')/2$. In the drawing, $f_1' = 3f_2'$. The dashed rays are the principal rays of the extreme oblique pencils. E_0 is the exit pupil of the objective and the entrance pupil for the eyepiece. The exit pupil through E_1 is the preferred position for the iris of the eye.

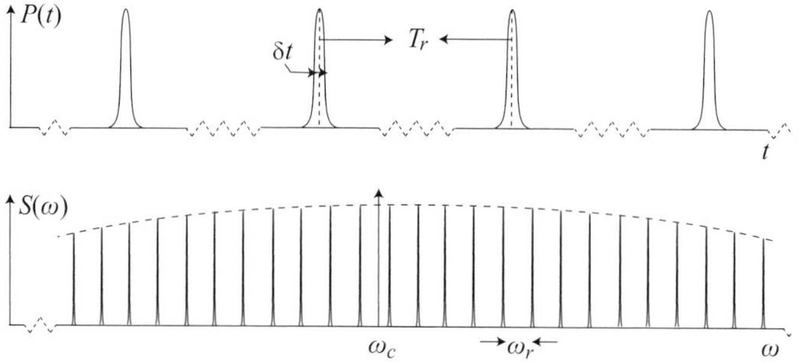

Figure 5.58: Typical time and frequency dependence of the light emitted by a pulsed frequency comb laser.

the spherical aberration is more important than the chromatic aberration. It follows that the condition of Eq. (5.6.34), although basically very interesting, is hardly of any relevance for the performance of the Huygens-type eyepiece.

5.6.5 Chromatic Correction for Pulsed-light Illumination

The analysis of chromatic effects in optical imaging was applied in the preceding paragraphs to harmonic fields. Chromatic correction aims at the equality of optical pathlength along all rays of a pencil pertaining to any frequency that is present in the light spectrum. In the aberration-free case, the pathlength equality criterion for each frequency is written

$$\Delta\mathcal{L}_\omega = \sum_j n_j(\omega)\left(l_{j,a} - l_{j,r}\right)_\omega = 0\,, \tag{5.6.35}$$

where $(l_{j,a} - l_{j,r})_\omega$ is the pathlength difference between an aperture ray and the reference ray in the jth medium, evaluated for the frequency ω. The exact value of the optical pathlength $\mathcal{L}_\omega$ from object to image point as a function of wavelength is not relevant when the image is formed by incoherent imaging over a broad spectrum with a continuous source. However, for a pulsed source, the transit time through the optical system of an individual pulse has to be equalised. Figure 5.58 shows the typical time dependence of the power $P(t)$ of a frequency comb laser. Narrow pulses with a width δt of some 20–50 cycle times of the average frequency ω_c of the source are repeated each T_r seconds (T_r is typically 10 ns). The spectral power $S(\omega)$ of such a pulsed source is shown in the lower graph. It consists of a sequence of 'spikes' with a typical width as narrow as a few Hz and a spacing ω_r between the frequency spikes that is proportional to $1/T_r$ (for instance 100 MHz). The envelope of the peak power of each spike in the frequency domain is the Fourier transform of the profile of a single time pulse.

For such a pulsed source the pathlength equality condition along a ray from object to image point is written in terms of the transit time,

$$\Delta t_\omega = \frac{1}{c}\sum_j n_j(\omega)\left(l_{j,a} - l_{j,r}\right)_\omega = 0\,. \tag{5.6.36}$$

The chromatic correction of the optical system for pulsed light requires that this condition is satisfied in the immediate neighbourhood of the central frequency and we obtain the condition

$$\frac{d(\Delta t_\omega)}{d\omega} = \frac{1}{c}\sum_j \left[\frac{dn_j(\omega)}{d\omega}\left(l_{j,a} - l_{j,r}\right)_\omega + n_j(\omega)\frac{d\left(l_{j,a} - l_{j,r}\right)_\omega}{d\omega}\right] = 0 \,. \qquad (5.6.37)$$

Using the same arguments as those of Subsection 5.6.3, we neglect the second product term between brackets since the change in pathlength of the quantity $(l_{j,a} - l_{j,r})_\omega$ is second-order small. This leads to the first-order condition to be satisfied,

$$\frac{d(\Delta t_{\omega)}}{d\omega} \approx \sum_j \frac{d}{d\omega}\left\{n_{p,j}(\omega)\right\}\left(l_{j,a} - l_{j,r}\right)_\omega = 0 \,, \qquad (5.6.38)$$

where $n_{p,j}$ is the phase refractive index of the medium j at frequency ω.

The expression above is identical to Conrady's condition and comparable arguments to those invoked for the case of harmonic radiation can now be used for the case of imaging with pulsed light. Such an imaging mode is used to enhance nonlinear effects in the imaging layer, when it is not the average image intensity but the instantaneous intensity peaks that have to be maximised. The optical system should produce equal transit times for the very short, broad-frequency wave packages that are sent through the system. The pulse transit time along all ray directions from an object point should be identical and the peak intensity in the focal region is then maximised. We write the equal transit time condition for pulses with central frequency ω_c as

$$\Delta\tau_{\omega_c} = \frac{1}{c}\sum_j n_{g,j}\left(l'_{j,a} - l'_{j,r}\right)_{\omega_c} = 0 \,, \qquad (5.6.39)$$

where τ denotes the travel time of the pulse envelope, $c/n_{g,j}$ is the group velocity in the jth medium and $l'_{j,a}$ and $l'_{j,r}$ are the geometrical distances along an aperture ray and a reference ray in a system that has been optimised for pulsed-light operation. To obtain pulse intensity enhancement in the image over a broad frequency range we require that

$$\left.\frac{d(\Delta\tau_\omega)}{d\omega}\right|_{\omega=\omega_c} = \frac{1}{c}\sum_j \frac{dn_{g,j}(\omega)}{d\omega}\left(l'_{j,a} - l'_{j,r}\right)_{\omega_c} = 0 \,. \qquad (5.6.40)$$

To establish a relation between the group velocity v_g and the phase velocity v_p at a certain frequency ω_c (see App. B), we use $\omega = v_p(\omega)k(\omega)$ and differentiate with respect to ω to obtain for $v_g = d\omega/dk$,

$$v_g = \frac{v_p}{1 - k\left(dv_p/d\omega\right)} \,. \qquad (5.6.41)$$

Changing from the frequency ω_c to the vacuum wavelength $\lambda_{c,0}$ of the radiation we obtain the following relation between the group and phase refractive indices,

$$n_g = n_p\left(1 - \frac{\lambda_{c,0}}{n_p}\frac{dn_p}{d\lambda_{c,0}}\right) \,. \qquad (5.6.42)$$

The value of n_p as a function of the vacuum wavelength $\lambda_{c,0}$ is available from measurements or calculations by means of a dispersion formula like the one given in Eq. (5.6.3). In Fig. 5.59 we have plotted the phase and group indices and their derivatives as a function of the vacuum wavelength λ_0 of a low-dispersive optical glass (BK7). The visible spectrum is a region with normal dispersion for most dielectric materials implying that $dn_p/d\lambda_0 < 0$. In this case the group velocity is always smaller than the phase velocity. The achromatic condition of Eq. (5.6.40) for imaging with pulsed light with a central frequency ω_c now reads

$$\left.\frac{d(\Delta\tau_\omega)}{d\omega}\right|_{\omega=\omega_c} = \frac{\lambda_{c,0}^3}{2\pi c^2}\sum_j \frac{d^2 n_{p,j}}{d\lambda_{c,0}^2}\left(l'_{j,a} - l'_{j,r}\right)_{\omega_c} = 0 \,. \qquad (5.6.43)$$

5.6.5.1 Correction Means Using Lenses, Prisms and Gratings

The lack of correction of the group transit time over the frequency band of a short optical pulse leads to pulse dilatation. In an optical imaging system with several lens elements, the condition of Eq. (5.6.43) can be reached by using positive and negative lenses, much as was done in the case of classical colour correction where the phase refractive index and powers of the elements were optimised. In telecommunication, the pulse width is inversely related to the maximum signal bandwidth. After propagation through an optical fibre, the dispersion of the core and cladding materials gives rise to pulse

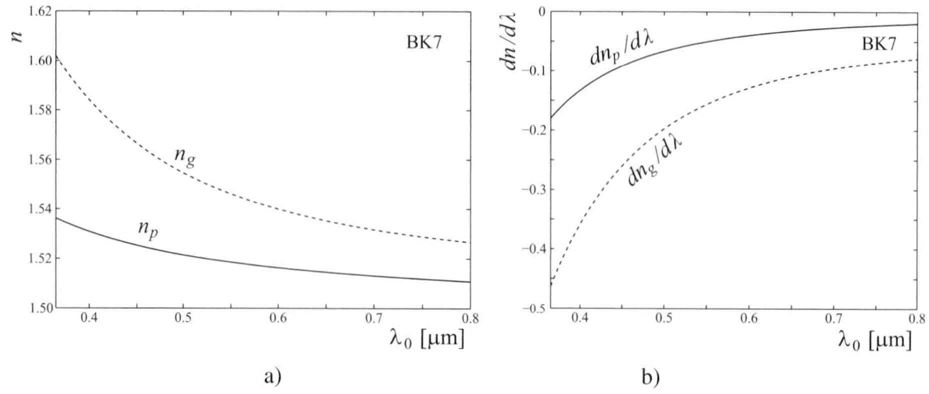

Figure 5.59: a) The phase and group index of refraction of a frequently used optical glass (BK7) in the visible part of the spectrum.
b) The derivatives of the phase index and the group index of the same glass as a function of the vacuum wavelength λ_0 (in μm).

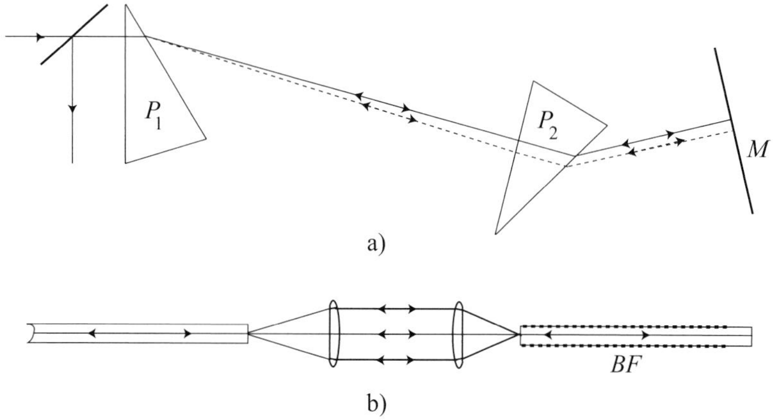

Figure 5.60: Methods for pulse compression.
a) A prism pair $P_1 - P_2$ used in reflection.
b) Wavelength-dependent dwell time in a Bragg-reflecting fibre *BF*.

broadening. An external element for reducing the pulse widening is sketched in Fig. 5.60a. With the aid of two prisms P_1 and P_2, the optical pathlength for the red and blue parts of the parallel beam is modified and the long-wavelength part accumulates a longer pathlength. The beam reflected by the mirror is coupled out with the aid of a semi-transparent mirror and has picked up the double delay. In Fig. 5.60b, a light bundle that has propagated through a fibre is coupled into a Bragg-fibre with an 'a-periodic' grating structure imprinted on the core part of this fibre. The matching of the various wavelength components with the Bragg condition ensures that the penetration depth in the fibre is made wavelength-dependent and this offers a flexible delay compensation mechanism for the pulsed-light beam.

5.7 Finite Ray-tracing

5.7.1 Role of Finite Ray-tracing in Aberration Analysis

Two basically different approaches for aberration analysis have been treated in Chapter 4 and in this chapter. Hamilton's characteristic functions provide the user with a pathlength expression for arbitrary pairs of rays in object and image space.

The pathlength is defined by the choice of reference points in object and image space or by reference directions in object and image space. A mixture of these is also allowed. The characteristic function is the equivalent of the scalar potential in a field theory in which the optical ray directions correspond to the gradient of the potential function (for instance, the force directions in an electric field). The construction of the pathlength function with sufficient precision is a very difficult task. The pathlength needs to be calculated with a precision of a fraction of the wavelength of the light, down to the nano- or picometer, and this over macroscopic distances. Within the paraxial domain it was possible to develop paraxial approximations for the characteristic function. Outside the paraxial domain, power expansions are required to find the pathlength expression with sufficient accuracy. When studying strongly converging ray fields, the characteristic function needs to be developed up to very high orders **[354]**. Although this might be do-able for a single transition or in the presence of a geometrical symmetry, the complexity of such a Hamiltonian approach becomes intractable when studying a general optical system.

Earlier in this chapter, ray aberration theory according to Seidel was presented. It brings the first refinement of ray trajectory calculation with respect to the paraxial approximation. The next expansion term, of fifth order when considering the transverse aberration terms, has been put forward by Schwarzschild (see **[37]**). A much more extensive analytic treatment of higher-order aberration theory is due to Buchdahl **[62]**. Unfortunately, the complexity of the expansions leads to almost unavoidable errors in the analysis of very high-order terms. This extreme complexity has made these higher-order theories rather unattractive and they are hardly used in the practice of optical design. A more recent approach to the calculation of high-order aberration expansion terms has been described in **[9]**,**[29]**. In principle, using a recursive numerical scheme, aberration terms of arbitrary high order can be calculated. In this approach, the *order-doubling* effect is exploited. For instance, when a distance is known with a precision up to the nth-order expansion term, the uncertainty due to the next term in a power series expansion is of order $2n$. Intermediate expansion terms of order $n, \cdots, 2n-2$ are described with sufficient precision using the ray data following from the expansion terms up to the nth order.

For the analysis of the per surface aberration build-up in an optical system the concept of *total* aberration per surface is very useful. Expressions have been derived in the past **[141]**,**[364]** that yield the wave aberration increment between two rays at refraction or reflection by a particular surface in the optical system. One ray is the reference ray, in general the principal ray of a pencil, the second is an arbitrary aperture ray of this pencil. To make the total aberration formulas useful in practice, they preferably contain only small terms of second- and higher-order. In this case, the subtraction of large optical pathlengths which carries the risk of loss of significant digits is avoided. With the total aberration data for a whole pencil of rays, after an adjustment for a single common reference sphere, one disposes of the raw material for an aberration expansion as a function of aperture per surface. The inclusion of several pencils of rays allows for an aberration expansion with respect to both aperture and field position. The optimum basis for such expansions is formed by Zernike polynomials. Such an expansion provides the system designer with precious knowledge of the aberration values in the intermediate spaces and the sensitivity of an intermediate surface for perturbations of any nature.

With the advent of ever more powerful computers in the past 50 years, the aberration expansions have lost ground with respect to direct geometrical ray-tracing. The aberrational data of individual rays are obtained, such as deviation of ray direction, shift of intersection point on a surface and optical pathlength increment from the source point. The results from tracing a large number of rays are then used to analyse the transverse aberration or wavefront aberration of an entire pencil of rays. If desirable, the data can be used to calculate the expansion coefficients of some appropriate aberration function that depends on the coordinates or ray directions in image space. Frequently applied expansion functions are the Zernike polynomials (see Appendix D). A practical limitation of the Zernike polynomials is that their favourable property of orthogonality is valid only on a disc-shaped area with circular rim. Vignetting effects in real-world optical systems limit the applicability of Zernike polynomials. With the aid of finite ray-tracing, ray data with virtually unlimited precision are obtained. The functional dependence of the aberration function on the constructional parameters of an optical system and on the imaging geometry are not available. Differential methods can be used to obtain estimates of aberration derivatives with respect to constructional data. These can then be used in optimisation schemes that are currently used in optical design. It is there that finite ray-tracing, especially in the final design stage, is extensively and almost uniquely used.

In this section we apply finite ray-tracing to various types of optical surfaces and to diffractive optical elements. A special topic is the propagation of rays in inhomogeneous media. These media are fabricated and inserted on purpose in optical systems to simplify their design and they play a comparable role for the reduction of aberration as aspheric surfaces. In optical materials like plastics, some residual inhomogeneity is frequently encountered. It results from density variations during the moulding or casting process. Other materials (crystals, moulded materials) may present anisotropy that is caused by crystal structure or induced stress in the material. In these cases, finite rays have to be traced in these media with the ray direction following from the energy flow according to the Poynting vector. Instead of a single finite ray, we observe a splitting in ordinary and extraordinary rays in such an anisotropic material. The calculation of aberration data from finite rays is the general subject of this section. The utilisation of these aberration data for the characterisation

Table 5.1. The use of finite ray data for image analysis.

	Transverse ray aberration	Wavefront aberration
Location	image plane	exit pupil reference surface
Quantities	$\delta x(X,Y)$, $\delta y(X,Y)$	$W_g(X,Y)$
Quality	δ_{rms} (rms blur)	$W_{g,\mathrm{rms}}$ (rms wavefront aberration)
Object illumination	incoherent	coherent, incoherent, partially coherent, polarised
Image amplitude / intensity	———	exit pupil diffraction integral (scalar, vectorial)
Frequency domain	geometrical optical transfer function	diffraction-based optical transfer function

of the imaging quality of an optical system is illustrated in Table 5.1. We emphasise that the geometrical wavefront aberration W_g is expressed in units of length and is defined for a general point on the exit pupil sphere with Cartesian coordinates denoted by (X, Y, Z). We use lower-case symbols (x, y, z) for the Cartesian coordinates in image space where the transverse aberration components are measured. For the normalised pupil and image space coordinates we use their primed versions.

To include wave diffraction effects, we use the aberration data on the exit pupil sphere to determine the mutual phase $\Phi(X,Y)$ with which the Huygens secondary waves travel towards the image region of the optical system and the relation $\Phi(X,Y) = kW_g(X,Y)$, where $k = n_I k_0$ is the wavenumber of the light in the image space with refractive index n_I. A complete description of the propagation of the secondary waves to the image region also requires knowledge of the amplitude $A(X,Y)$ of the secondary waves at each point on the exit pupil sphere. Amplitude and phase of the (scalar) optical field are conveniently combined in a complex function $f(X,Y)$ which is given by

$$f(X,Y) = A(X,Y)\exp\{i\Phi(X,Y)\}\,. \tag{5.7.1}$$

The so-called pupil function or lens transmission function $f(X,Y)$ plays a crucial role in the diffraction analysis of optical imaging (see Part III of this book). In the case of a scalar representation of the optical field, a single function f according to Eq. (5.7.1) is sufficient. When an electromagnetic diffraction problem has to be solved, a complex *vector* lens function is required such that at each point on the exit pupil sphere of the optical system the three components of the electromagnetic field vector are made available.

5.7.2 Pathlength Calculation and Surface Representation

The calculation of ray direction and pathlength in an optical system requires the accurate calculation of the point of intersection of a ray with a refracting, reflecting or diffracting surface. In Fig. 5.61 we have sketched a ray vector with direction cosines (L_1, M_1, N_1) which is incident on a surface with its vertex at T_1. The ray leaves the previous surface at a point P_0 with coordinates (x_0, y_0, z_0). To calculate the ray refraction at the surface through T_1, the point of intersection P_1 with this surface has to be found. The procedure is straightforward for quadratic surfaces. In this category, we find the sphere and the conic sections with rotational symmetry (paraboloid, ellipsoid, hyperboloid). First, the intersection point S_1 with the plane through the vertex T_1 of the surface is determined. Then, two intersection points with the quadratic surface are found, P_1 and P_1', by solving a quadratic equation in the distance S_1P_1. In most cases, the intersection point closest to S_1 is the required solution. The optical pathlength from P_0 to P_1 can be very large, for instance in an astronomical telescope. In such a case, it is advisable to immediately calculate the pathlength difference with the reference ray in the same space. When analysing a pencil of finite rays, the reference ray is the principal ray of this pencil. Denoting the intersection points with the surfaces of the principal ray by Q_0 and Q_1, and its direction cosines by $(L_{1,r}, M_{1,r}, N_{1,r})$, we can write

$$\begin{aligned}\Delta = [P_0P_1] - [Q_0Q_1] &= \frac{n_1}{N_1}(d_1 + z_{P_1} - z_{P_0}) - \frac{n_1}{N_{1,r}}(d_1 + z_{Q_1} - z_{Q_0})\\ &= n_1 d_1\,\frac{L_1^2 + M_1^2 - L_{1,r}^2 - M_{1,r}^2}{N_1 N_{1,r}(N_1 + N_{1,r})} + \frac{n_1}{N_1}(z_{P_1} - z_{P_0}) - \frac{n_1}{N_{1,r}}(z_{Q_1} - z_{Q_0})\,,\end{aligned} \tag{5.7.2}$$

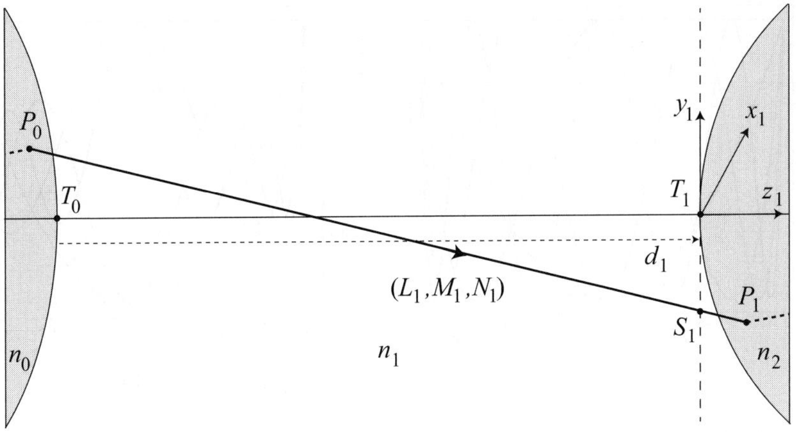

Figure 5.61: The propagation of a general ray with unit vector (L_1, M_1, N_1) in an optical system. P_1 is the point of intersection with the second surface.

where the z-coordinates are referred to the vertex of each surface. The expression for the pathlength difference Δ can suffer from loss of significant digits when the two neighbouring rays are close. Combined with a large propagation distance, the required accuracy in Δ of a fraction of the wavelength of the light is compromised. To avoid such a numerically unstable situation, the numerator of the first term in Eq. (5.7.2) is preferably written as $n_1 d_1 [(L_1 - L_{1,r})(L_1 + L_{1,r}) + \cdots]$, of second order in the (x, y)-direction cosines of the rays. The second and third terms in the expression for Δ do not risk such a numerical instability. In practice, the z-coordinates of the intersection points with the surface remain small and its value is of second order in the lateral coordinates of the points of intersection. The same argument holds for the z-direction cosines of the rays. Equation (5.7.2) is said to be of second order in the (lateral) ray variables. To construct a wavefront in the exit pupil of the optical system, the quantities Δ_j are accumulated for a set of rays from the object point to the last surface with index J of the optical system. The further ray data analysis then proceeds as indicated in Table 5.1.

5.7.2.1 General Aspheric Surface

With the advent of special surface manufacturing and moulding techniques, optical surfaces do not need to be restricted to a spherical shape. The conic sections and more general aspheric profiles with rotational symmetry have become technologically available. The classical representation of the axial coordinate z_a of a general aspheric surface with symmetry of revolution starts with the conic section part. Extra higher-order terms depending on $(x^2 + y^2)$ are added. They represent the more general aspheric departure from the base surface. The expression for the z-coordinate reads, using that $r = (x^2 + y^2)^{1/2}$,

$$z_a(r) = \frac{c\, r^2}{1 + \sqrt{1 - (1 + \kappa)\, c^2\, r^2}} + \sum_{n=2}^{n_0} a_{2n}\, r^{2n} \,, \tag{5.7.3}$$

where c is the paraxial curvature of the aspheric surface, κ the conic constant and the coefficients $a_4, a_6, \cdots a_{2n_0}$ are the higher-order aspheric coefficients of the surface.

The mixed representation of Eq. (5.7.3) is useful in surface metrology. It provides, especially in the case of a spherical base surface with $\kappa = 0$ an easily measurable reference profile. The generally small aspheric departure then follows from a differential measurement with limited range. But the representation of Eq. (5.7.3) has certain mathematical drawbacks. First, the choice of both a conic constant κ and higher-order aspheric terms introduces mutually dependent variables when measurement data have to be fitted with the aid of the formula. Secondly, the monomial terms $(x^2 + y^2)^m$ are not mutually orthogonal and this leads to a bad conditioning of the data fitting problem when using such an expansion. For these reasons, other representations of the axial coordinate z_a have been suggested. Symmetric polynomials of even order can be used. The normalised radial variable of these polynomials is $\rho = r/r_{\text{max}}$ where $2r_{\text{max}}$ is the diameter of the optically useful area on the surface. Depending on the ρ-dependent weighting function that is applied, different sets of orthogonal polynomials emerge. In Fig. 5.62 two sets of polynomials have been drawn. Using the weighting factor $(1 - x^2)^{-1/2}$, the Chebyshev polynomials $T_{2n}(x)$ are found (see Fig. 5.62a). They have the attractive property that the maximum modulation

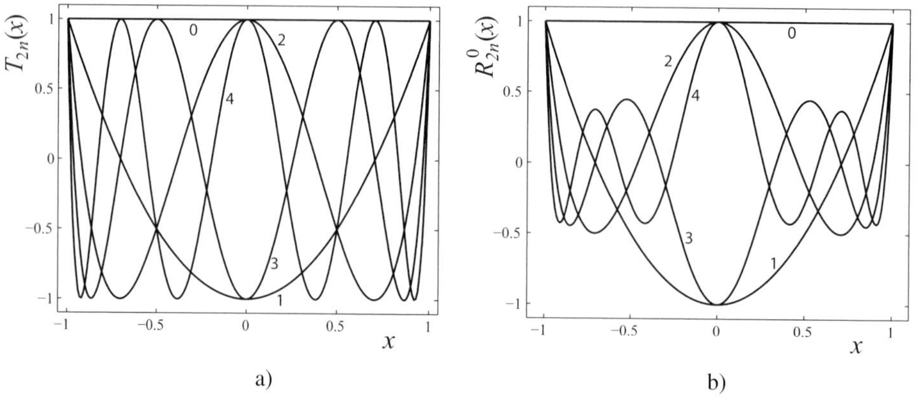

Figure 5.62: Examples of orthogonal polynomials for describing the shape of an aspheric surface or its departure with respect to a chosen reference surface. The polynomials are plotted on the interval $[-1, +1]$. The subscript $2n$ attached to a polynomial gives the maximum power of the variable x in the polynomial.
a) Chebyshev polynomials $T_{2n}(x)$.
b) Radial Zernike polynomials $R^0_{2n}(x)$.

is obtained over the full interval and that the polynomials are orthonormal over the interval $[-1, +1]$ with a fixed inner product of $\pi/2$ for $n \neq 0$. The representation of a circularly symmetric surface is given by

$$z_a(r) = g(r) + \sum_n c_{2n}\, T_{2n}(\rho)\,, \tag{5.7.4}$$

where $g(r)$ is the base surface function and the c_{2n} are the Chebyshev expansion coefficients. In practice, the interval $[-1, +1]$ for the Chebyshev polynomials is realised by assigning a measured or calculated function value at the abscissa $x = -x_0$ that is the same as that found at $x = +x_0$. The base surface $g(r)$ can also be left out and replaced by the quadratic Chebyshev polynomial $(2\rho^2 - 1)r^2_{\max}/(4R)$ together with a constant term $r^2_{\max}/(4R)$. The base surface is then effectively replaced by a paraboloid with a paraxial radius of curvature equal to R. In general, fast convergence of the Chebyshev series is obtained and the maximum degree $2n$ can be restricted to practical values of the order of 10 to 20 for the vast majority of aspheric surfaces that are used in optical systems.

In Fig. 5.62b, radial Zernike polynomials $R^0_{2n}(x)$, symmetric in the coordinate x, have been plotted. Zernike polynomials are extensively used in optical metrology and in optical design (see Appendix D for an introduction to these polynomials). The Zernike polynomials are orthogonal on the unit disc and can be used to include aspheric departures that are not necessarily circularly symmetric. An aspherically shaped surface is then given by

$$z_a(x,y) = g(x,y) + \sum_{n,m} R^m_n(\rho)\, \{b^m_{n,c} \cos m\theta + b^m_{n,s} \sin m\theta\}, \qquad \cos\theta = x/(x^2+y^2)^{1/2}, \tag{5.7.5}$$

where $g(x,y)$ is the base function and $b^m_{n,c}$ and $b^m_{n,s}$ are the cosine and sine Zernike expansion coefficients. The summation extends from $n = 0$ to a prescribed upper limit, with $n - m$ even and ≥ 0.

Other orthogonal expansions of the general aspheric term are possible. With a sphere as the base surface, the aspheric term of z_a depends on the choice of the curvature of the base surface. Possible criteria for selecting the base surface curvature are:

- the root-mean-square value of the resulting aspheric term is minimised;
- the maximum absolute value of the aspheric term is minimised;
- the root-mean-square value of the angular deviation δ between the normal to the base surface and the normal to the aspheric surface is minimised;
- the maximum absolute value of δ is minimum.

The various criteria for choosing a specific base curvature are mostly inspired by the surface metrology or the fabrication technique. According to the applied criterion, a best-fit sphere is constructed with radius R_b to be used in the aspheric surface metrology. The surface representation does not play an essential role in the ray-tracing process. For ray-tracing, it is the numerical stability of the expansion that is important. An orthogonal expansion of any type will be adequate in practice. A further discussion on the choice of orthogonal polynomial sets is found in **[101]**.

5.7.2.2 Ray Intersection Point with a General Asphere

The intersection point of a ray with an aspheric surface has to be found by an iterative procedure. The first step is to find the intersection point P_s of the incident ray with the paraxial sphere or the best-fit sphere, with radius of curvature $1/c$ or R_b, respectively. The lateral coordinate of P_s is used to find the coordinate z_a of the point $P_{a,1}$ on the aspheric surface with the same lateral coordinate. Using the surface equation of the asphere, the normal to the aspheric surface is calculated. In a next step, we calculate the point of intersection $P_{p,1}$ of the incoming ray with the plane that is tangent to the asphere at the point $P_{a,1}$. From the lateral coordinate of $P_{p,1}$, we calculate a second point on the asphere $P_{a,2}$, construct the normal and the tangent plane in that point, find the point of intersection of the ray with the new tangent plane, etc. The iteration is stopped once the z-coordinates of two subsequent points $P_{a,n}$ and $P_{p,n}$ have a difference in the value of their z-coordinate that is negligible at the scale of the wavelength of the light. This iterative procedure generally converges after just a few steps.

5.7.3 Refraction and Diffraction of a Ray

As an example we study the refraction of a ray at a surface that carries a diffracting structure. We assume that the thickness of the diffracting structure can be neglected. In the framework of geometrical optics, we only consider the possible outgoing ray directions after diffraction of the incident ray by the structure on the surface. The amount of energy that is carried by these rays is not part of the analysis; this would require the solution of an electromagnetic boundary value problem at the surface. The direction of an outgoing ray is the result of a combination of refraction at the interface and diffraction at the structure on the surface. The diffraction pattern on the optical surface can be the result of an optical interference experiment, for instance when making a holographic recording of a point source or of a more complicated object. The pattern can also be the result of the design process for a diffractive optical element (DOE). In the latter case, the design process aims at creating an outgoing beam with some angular amplitude function to modify the (uniform) amplitude distribution of the incident, optically coherent beam which has been produced by a point source. In both cases, for the hologram and the diffractive element, the phase Φ between the real (hologram) or synthetic (DOE) interfering beams is the essential input for finding the direction of the diffracted geometrical rays. The phase function Φ, defined on the optical surface, depends on the coordinates (x, y, z), where $z = z(x, y)$ follows from the surface shape, and determines the three-dimensional shape of the equal-phase lines on the surface.

5.7.3.1 Local Matching of the Tangential Field Components

To find the directions of the diffracted, transmitted and reflected rays the tangential field components of the corresponding incident, reflected and transmitted wave fields should be matched at the interface between the two media (see Eq. (1.8.20)). To determine the wave propagation directions we impose the condition that the spatial and temporal dependences of the waves on both sides of the interface are identical. For the temporal dependence, remaining in the domain of linear optics, we require equal frequencies ω for the three waves. On the interface, in the geometrical optics approximation, we approximate the general reflection and transmission functions, $r(x, y, z)$ and $t(x, y, z)$, respectively, by locally periodic ones. It is a reasonable assumption for the diffractive structure that the moduli of the complex functions r and t vary slowly along the surface. Their phase part oscillates at a high spatial frequency. The period follows from the derivative of the phase functions of r and t, represented by Φ_r and Φ_t, respectively, at the point of intersection $P_1(x_1, y_1, z_1)$ of the incident ray with the surface. For instance, in the case of the reflected wave,

$$\Phi_r\{x, y, z(x, y)\} = \Phi_r\{x_1, y_1, z_1\} + \left[\frac{\partial \Phi_r}{\partial x} + \frac{\partial \Phi_r}{\partial z}\frac{\partial z}{\partial x}\right]_{(x_1,y_1)} (x - x_1) + \left[\frac{\partial \Phi_r}{\partial y} + \frac{\partial \Phi_r}{\partial z}\frac{\partial z}{\partial y}\right]_{(x_1,y_1)} (y - y_1) . \qquad (5.7.6)$$

A comparable expression exists for the phase part of the transmitted wave in which only the constant phase term is different.

We then assume that the reflection (and transmission) functions can be expanded in the tangent plane at P_1 with the aid of Fourier series having a basic frequency $\mathbf{a}_d$, which is given by

$$\mathbf{a}_d = \frac{1}{2\pi}\nabla\Phi_{r,t}\{x, y, z(x, y)\} . \qquad (5.7.7)$$

The indices r and t are associated with the reflection and transmission case. The expansion coefficients, for instance for the reflective case, are given by

$$\Phi_r\{x, y, z(x, y)\} = \sum_m c_{m,r} \exp\{i2\pi m\, \mathbf{a}_d \cdot \mathbf{r}\} . \tag{5.7.8}$$

The quantities $c_{m,r}$ are the coefficients for the reflected wave and the zeroth-order coefficient $c_{0,r} = \Phi_r\{x_1, y_1, z_1\}$ is the phase of the reflected wave at the intersection point P_1 of the incident ray and the diffracting surface. The position vector to a general point in the tangent plane through P_1 has been denoted by $\mathbf{r}$. A comparable expansion exists for the transmitted wave with Fourier coefficients $c_{m,t}$.

The matching of the tangential field components of the three waves on the interface at P_1 requires that the periodic variations of the waves along the surface are identical. Using the geometrical-optics approximation described above, we have in the tangent plane at P_1 the following periodic exponential functions:

$$\begin{aligned} &\exp(ik_0\boldsymbol{\tau}_i \cdot \mathbf{r}) + \sum_m c_{m,r} \exp\left\{i\left(k_0\boldsymbol{\tau}_{r,m} + 2\pi m\mathbf{a}_d\right)\cdot \mathbf{r}\right\}, \quad \text{region } n_0 , \\ &\sum_m c_{m,t} \exp\left\{i\left(k_1\boldsymbol{\tau}_{t,m} + 2\pi m\mathbf{a}_d\right)\cdot \mathbf{r}\right\}, \quad \text{region } n_1 . \end{aligned} \tag{5.7.9}$$

The vectors $\boldsymbol{\tau}_i$, $\boldsymbol{\tau}_{r,m}$ and $\boldsymbol{\tau}_{t,m}$ are obtained from the projection on the interface of the unit wave vectors of the incident plane wave and the various reflected and transmitted waves that were diffracted at the interface. To construct the complete solution of the diffraction problem at the interface, the coefficients $c_{m,r}$ and $c_{m,t}$ need to be found by solving a boundary value problem. To determine only the plane wave directions in which energy is *possibly* carried away from the surface, we use Eq. (5.7.9) and require identical phases for each of the series of plane waves on both sides of the interface,

$$k_0\boldsymbol{\tau}_i = k_0\boldsymbol{\tau}_{r,m_1} - 2\pi m_1 \mathbf{a}_d = k_1\boldsymbol{\tau}_{t,m_2} - 2\pi m_2 \mathbf{a}_d , \tag{5.7.10}$$

where m_1 and m_2 are arbitrary integers associated with the plane wave expansion for the reflected and transmitted fields, respectively. In Eq. (5.7.10) we have given the terms with m_1 and m_2 a negative sign in order to comply with the common convention on *positive* and *negative* diffraction orders which has already been used in Section 1.12. Using this sign convention, the angle between the propagation direction of a positive order and that of the zero order is positive for the case of a so-called two-dimensional 'in-plane' diffraction geometry in transmission. In such a geometry, all wave vectors are located in the plane containing the surface normal and the grating vector (see Fig. 5.63).

5.7.3.2 Calculation of the Diffracted Ray Vectors

In Fig. 5.64 we show an incident ray vector $\hat{\mathbf{s}}_0$ with direction cosines (L_0, M_0, N_0) that intersects the optical surface at the point P_1. We take the origin of the coordinate system in T_1 with the positive z-axis pointing in the direction of T_1C_1. The general diffractive structure on the optical surface is locally approximated by a periodic grating-like structure with period p. The phase of the diffracting structure at P_1 is given by $\Phi(x_1, y_1, z_1)$. Within the geometrical optics approximation, we replace the curved surface at P_1 by the tangent plane at this point. If the shape of the surface is given by an expression $z = G(x, y)$, we find for the unit normal vector $\hat{\mathbf{v}}$ of the surface

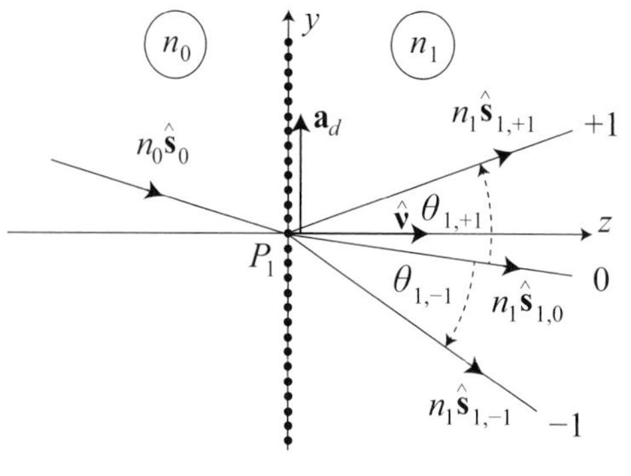

Figure 5.63: Illustrating the sign convention for diffraction orders. In-plane diffraction geometry with coplanar incident wave vector $\hat{\mathbf{s}}_0$, grating vector $\mathbf{a}_d$ and surface normal $\hat{\mathbf{v}}$.

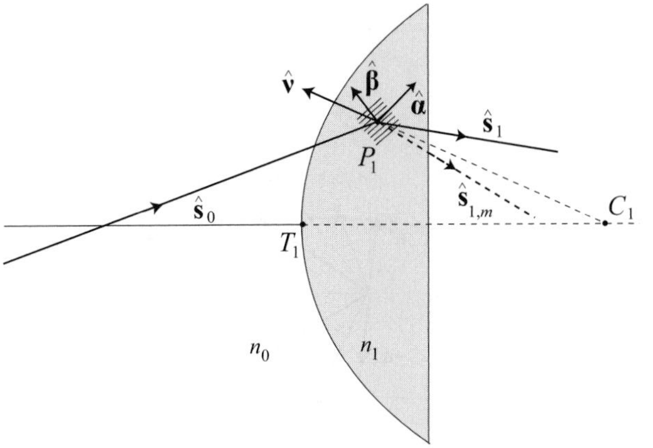

Figure 5.64: An incident ray with unit vector $\hat{\mathbf{s}}_0$ intersects the optical surface at P_1. The refracted ray is $\hat{\mathbf{s}}_1$, the refracted *and* diffracted ray of order m is $\hat{\mathbf{s}}_{1,m}$. The local period of the diffracting structure is p. $\hat{\mathbf{v}}$ is the outward normal vector, $\hat{\boldsymbol{\alpha}}$ and $\hat{\boldsymbol{\beta}}$ are unit vectors, parallel and perpendicular to the grating lines at the location of P_1. $\hat{\boldsymbol{\alpha}}$, $\hat{\boldsymbol{\beta}}$ and $\hat{\mathbf{v}}$ form a local orthogonal coordinate system with its origin at P_1. The unit vectors $\hat{\boldsymbol{\alpha}}$ and $\hat{\boldsymbol{\beta}}$ are located in the plane that is tangent to the surface at P_1.

$$\hat{\mathbf{v}} = \mathrm{sgn}(R)\left[1 + \left(\frac{\partial G}{\partial x}\right)^2 + \left(\frac{\partial G}{\partial y}\right)^2\right]^{-1/2} \left(\frac{\partial G}{\partial x}, \frac{\partial G}{\partial y}, -1\right), \tag{5.7.11}$$

where the function $\mathrm{sgn}(R) = \mathrm{sgn}(T_1C_1)$ determines the outward direction of the normal.

For the calculation of the refracted and reflected rays, we need information on the grating vector $\mathbf{a}_d = \hat{\boldsymbol{\beta}}/p$, where p is the local grating period at the point P_1. We assume that this vector quantity follows from some prescription for the diffracting structure on the optical surface, either from a holographic recording geometry or from the design data of a diffractive optical element. Using Eq. (5.7.10) for a transmitted ray with general index m, we have

$$\boldsymbol{\tau}_{t,m} = \frac{k_0}{k_1}\boldsymbol{\tau}_i + \frac{2\pi m}{k_1 p}\hat{\boldsymbol{\beta}}. \tag{5.7.12}$$

The tangent vector $\boldsymbol{\tau}_i = \hat{\mathbf{s}}_0 - (\hat{\mathbf{s}}_0 \cdot \hat{\mathbf{v}})\hat{\mathbf{v}}$ is written as a linear combination of the orthogonal vectors $\hat{\boldsymbol{\alpha}}$ and $\hat{\boldsymbol{\beta}}$,

$$\boldsymbol{\tau}_i = (\boldsymbol{\tau}_i \cdot \hat{\boldsymbol{\alpha}})\hat{\boldsymbol{\alpha}} + (\boldsymbol{\tau}_i \cdot \hat{\boldsymbol{\beta}})\hat{\boldsymbol{\beta}}. \tag{5.7.13}$$

For the tangential vector $\boldsymbol{\tau}_{t,m}$ we obtain the expression

$$\begin{aligned}\boldsymbol{\tau}_{t,m} &= \frac{k_0}{k_1}(\boldsymbol{\tau}_i \cdot \hat{\boldsymbol{\alpha}})\hat{\boldsymbol{\alpha}} + \left[\frac{k_0}{k_1}(\boldsymbol{\tau}_i \cdot \hat{\boldsymbol{\beta}}) + \frac{2\pi m}{k_1 p}\right]\hat{\boldsymbol{\beta}} \\ &= a_{t,m}\hat{\boldsymbol{\alpha}} + b_{t,m}\hat{\boldsymbol{\beta}}. \end{aligned} \tag{5.7.14}$$

As $\hat{\mathbf{s}}_{t,m}$ is a unit vector, we have two possible solutions for this vector,

$$\hat{\mathbf{s}}_{t,m} = \boldsymbol{\tau}_{t,m} \pm \sqrt{1 - a_{t,m}^2 - b_{t,m}^2}\;\hat{\mathbf{v}}. \tag{5.7.15}$$

Using $k_0/k_1 = n_0/n_1$ and the result of Eq. (5.7.14), we obtain for a transmitted diffracted vector

$$\hat{\mathbf{s}}_{t,m} = \frac{n_0}{n_1}\hat{\mathbf{s}}_0 + \frac{2\pi m}{k_1 p}\hat{\boldsymbol{\beta}} + \left\{\sqrt{1 - \frac{n_0^2}{n_1^2}\left[1 - (\hat{\mathbf{s}}_0 \cdot \hat{\mathbf{v}})^2\right] - \frac{4\pi^2 m^2}{k_1^2 p^2} - \frac{4\pi n_0 m}{n_1 p k_1}(\hat{\mathbf{s}}_0 \cdot \hat{\boldsymbol{\beta}})} - \frac{n_0}{n_1}(\hat{\mathbf{s}}_0 \cdot \hat{\mathbf{v}})\right\}\hat{\mathbf{v}}. \tag{5.7.16}$$

For a reflected ray we equate n_1 to n_0 and k_1 to k_0. We then use the minus sign in front of the square root in Eq. (5.7.15) to obtain a ray vector that propagates back in the first medium,

$$\hat{\mathbf{s}}_{r,m} = \hat{\mathbf{s}}_0 + \frac{2\pi m}{k_0 p}\hat{\boldsymbol{\beta}} - \left\{\sqrt{(\hat{\mathbf{s}}_0 \cdot \hat{\mathbf{v}})^2 - \frac{4\pi^2 m^2}{k_0^2 p^2} - \frac{4\pi m}{p k_0}(\hat{\mathbf{s}}_0 \cdot \hat{\boldsymbol{\beta}})} + (\hat{\mathbf{s}}_0 \cdot \hat{\mathbf{v}})\right\}\hat{\mathbf{v}}. \tag{5.7.17}$$

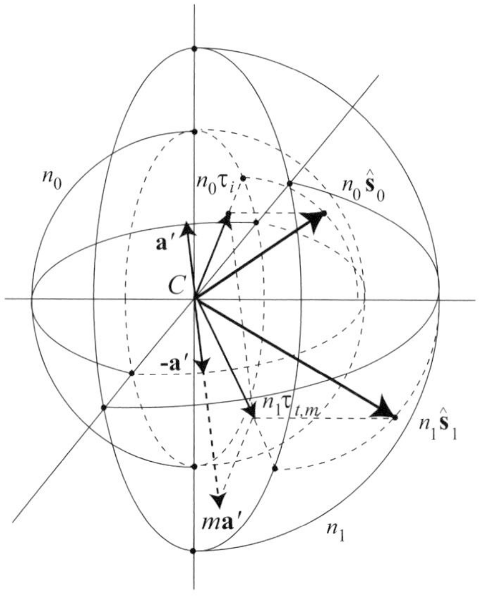

Figure 5.65: Ray diffraction at an interface. The incident ray with unit vector $\hat{\mathbf{s}}_0$ propagates in a medium with index n_0. The tangential component of the optical ray vector is given by $n_0\boldsymbol{\tau}_i$. The diffractive structure has a (local) period p and the length of the (scaled) frequency vector $\mathbf{a}'$ is $\lambda_0\mathbf{a}$ where λ_0 is the vacuum wavelength. The order number m of the diffracted wave in the figure is -3. The transmitted unit wave vector is $\hat{\mathbf{s}}_1$.

By letting p approach infinity in Eqs. (5.7.16) and (5.7.17), we recover the directions of the rays that are refracted and reflected at a non-modulated interface (see Eqs. (4.2.46) and (4.2.47)).

The diffraction process is illustrated in Fig. 5.65. The tangential wave vector component $n_1\tau_{t,m}$ of the outgoing wave is the vector sum of the tangential component of the incident wave, $n_0\tau_i$, and the scaled vector $m\mathbf{a}' = m\lambda_0\mathbf{a}$ associated with the diffracting structure. In the figure the tangential component of the outgoing vector has been constructed using a negative value of the order number m of the diffracted wave. Having constructed the tangential component of the outgoing wave vector, we find the outgoing unit ray vector $\hat{\mathbf{s}}_1$ by normalising its length to unity. In certain cases, the vector component perpendicular to the interface, $\hat{s}_{1,\perp}$, may become imaginary, producing an evanescent ray vector. Such rays can be neglected after propagation over a distance which is large with respect to their extinction length $d_e = \lambda_0/(2\pi n_1 \hat{s}_{1,\perp})$. The reader is referred to Chapter 1, Eq. (1.6.75), for a discussion on evanescent waves.

The drawing of Fig. 5.65 is related to the Ewald sphere which is used for obtaining wave propagation directions in three-dimensional diffracting or scattering structures. For the case of the planar, infinitely thin grating structure on the interface between two media with different index, two Ewald hemispheres with different radii have a common centre of curvature C at the interface. The transfer of the wave vectors through the interface is determined by the Maxwell boundary conditions at the interface. In Ewald's original publication, the outgoing wave vector in a three-dimensional scattering medium, for instance a crystal, is determined by the scattering condition

$$n_0 k_0(\hat{\mathbf{s}}_1 - \hat{\mathbf{s}}_0) = 2\pi\mathbf{a} \,, \tag{5.7.18}$$

where $\mathbf{a}$ is the three-dimensional grating vector of the periodic scattering medium.

Returning to the infinitely thin two-dimensional structure in Fig. 5.65, we find the following solutions for the zeroth diffracted order or for the case of a non-modulated interface surface,

$$n_1\hat{\mathbf{s}}_t = n_0\hat{\mathbf{s}}_0 + (n_1\cos\theta_1 - n_0\cos\theta_0)\,\hat{\mathbf{v}} \,, \tag{5.7.19}$$

$$\hat{\mathbf{s}}_r = \hat{\mathbf{s}}_0 - 2\cos\theta_0\,\hat{\mathbf{v}} \,, \tag{5.7.20}$$

where $\cos\theta_0 = (\hat{\mathbf{s}}_0\cdot\hat{\mathbf{v}})$ and $\cos\theta_1 = (\hat{\mathbf{s}}_t\cdot\hat{\mathbf{v}})$. It is easily verified that the above formulas remain valid when we use for $\hat{\mathbf{v}}$ the inward normal vector instead of the outward normal.

For the case of an optical surface with spherical shape, we write Eq. (5.7.19) as follows. The outward normal vector follows from Eq. (5.7.11),

$$\hat{\boldsymbol{\nu}} = \mathrm{sgn}(R)\left(\frac{x}{R}, \frac{y}{R}, -1 + \frac{z}{R}\right) , \tag{5.7.21}$$

and this yields the vector equation

$$n_1\hat{\mathbf{s}}_1 = n_0\hat{\mathbf{s}}_0 + \mathrm{sgn}(R)\frac{(n_1\cos\theta_1 - n_0\cos\theta_0)}{R}(x, y, -R+z) . \tag{5.7.22}$$

In analogy with the paraxial imaging law $n_1u_1 = n_0u_0 - yK$ (see Eq. (4.10.5)), we have from Eq. (5.7.22), second vector component,

$$n_1 s_{1,y} = n_0 s_{0,y} - y\,\frac{(n_1\cos\theta_1 - n_0\cos\theta_0)}{R} , \tag{5.7.23}$$

where the cosine functions are now evaluated with respect to the inward normal vector. Equation (5.7.23) shows that at angles of incidence outside the paraxial region, the optical power $(n_1 - n_0)/R$ has to be replaced by the quantity $(n_1\cos\theta_1 - n_0\cos\theta_0)/R$. This quantity is commonly called the *generalised power* of an optical transition.

5.7.3.3 Diffraction by an Interferometrically Recorded Structure

The diffracting structure on an optical surface can be made by optical means, for instance by recording the interference pattern of two point sources, P'_1 and P'_2, on the surface (see Fig. 5.66). To this purpose a photosensitive layer is deposited on the surface which is subsequently exposed to the spatial interference pattern of the two point sources. For the diffraction of an incident ray $\hat{\mathbf{s}}_0$, the unit vector $\hat{\boldsymbol{\alpha}}$, tangent to the curves of constant phase on the surface, has to be found. The vector $\hat{\boldsymbol{\alpha}}$ is perpendicular to the normal vector $\hat{\boldsymbol{\nu}}$ to the surface at the point P_1. The vector should also be perpendicular to the difference vector $\hat{\mathbf{r}}_1 - \hat{\mathbf{r}}_2$ since in that direction the pathlength difference from the points P'_1 and P'_2 is stationary. With the proper normalisation we write

$$\hat{\boldsymbol{\alpha}} = \frac{\hat{\boldsymbol{\nu}} \times (\hat{\mathbf{r}}_1 - \hat{\mathbf{r}}_2)}{\left|\hat{\boldsymbol{\nu}} \times (\hat{\mathbf{r}}_1 - \hat{\mathbf{r}}_2)\right|} . \tag{5.7.24}$$

The grating unit vector $\hat{\boldsymbol{\beta}}$ equals $\hat{\boldsymbol{\nu}} \times \hat{\boldsymbol{\alpha}}$ and the local period of the grating structure at P_1 follows from the consideration that a progression over a distance p in the direction of the grating vector $\hat{\boldsymbol{\beta}}$ produces an extra pathlength difference in the interference pattern of *one* wavelength in the recording medium,

$$p\,\hat{\boldsymbol{\beta}} \cdot (\hat{\mathbf{r}}_1 - \hat{\mathbf{r}}_2) = \frac{\lambda_R}{n_0} , \tag{5.7.25}$$

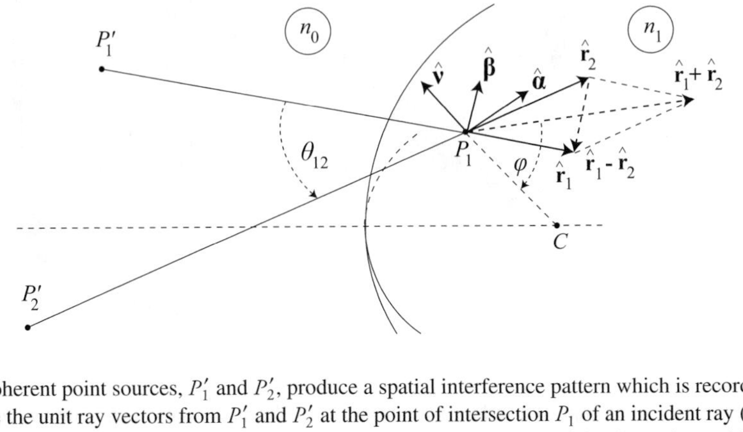

Figure 5.66: Two coherent point sources, P'_1 and P'_2, produce a spatial interference pattern which is recorded on the optical surface. $\hat{\mathbf{r}}_1$ and $\hat{\mathbf{r}}_2$ are the unit ray vectors from P'_1 and P'_2 at the point of intersection P_1 of an incident ray (not shown in the figure) with the surface. The angle between the vector $\hat{\mathbf{r}}_1 + \hat{\mathbf{r}}_2$ and the surface normal $\hat{\boldsymbol{\nu}}$ has been denoted by ϕ. C is the (local) centre of curvature of the surface. $\hat{\boldsymbol{\beta}}$ is the corresponding unit grating vector at the point P_1. The unit vectors $\hat{\boldsymbol{\alpha}}$, $\hat{\boldsymbol{\beta}}$ and the surface normal $\hat{\boldsymbol{\nu}}$ form a right-handed orthogonal coordinate system at the point P_1.

where λ_R is the vacuum recording wavelength. With the aid of Eq. (5.7.24) we obtain for the grating period p the expression

$$p = \frac{\lambda_R}{2n_0 \cos\phi \sin(\theta_{12}/2)} . \tag{5.7.26}$$

The data for the vectors $\hat{\alpha}$ and $\hat{\beta}$ and the expression for the grating period p are then used in Eqs. (5.7.16) and (5.7.17) to obtain the transmitted and reflected ray vectors, once an incident ray vector $\hat{\mathbf{s}}_0$ and the vacuum wavelength λ_0 of the radiation have been specified.

The simple interferometric recording geometry with two point sources can be extended with auxiliary optics to provide for aberrated beams that interfere on the optical surface. Such aberrations at the recording phase influence the direction and pathlength of diffracted rays and can be used to compensate for aberrations that arise in an optical system. This explains the widespread application of holographic or diffractive optical elements. It should be mentioned that the pathlength and directional changes introduced by such an element are strongly wavelength-dependent. In general, such holographic elements cannot be used over a broad spectral band.

5.7.3.4 Diffraction by a Ruled Surface

A special case of ray diffraction is encountered when the optical surface carries a mechanically engraved periodic structure. Such mechanical grating structures are used in spectroscopy and their precise manufacturing was made possible by means of a 'ruling' machine. A machine of this type is able to cut a shallow ditch-like structure in the surface of a metal substrate with the aid of a diamond tool. The tool executes a linear movement. The contact with the curved surface and the required cutting pressure are maintained by applying a spring force. A gentle curvature of the metal surface is followed by the cutting tool by means of a vertical movement. With the aid of precise metrology, the advance of the cutting tool between two subsequent cutting actions is very accurately controlled such that a well-defined periodic structure can be engraved (see Fig. 5.67). Typical periods p achieved with such ruling machines range from 10 μm down to 1 μm. The dimensions over which such metallic gratings can be made extend to several tens of centimetres. This makes them very attractive for spectroscopic applications where the spectroscopic resolution $\delta\lambda/\lambda$ is determined by the product of the number of grating periods N_p and the order number m of the diffraction order that is used. From a master grating, using a replication technique, a large number of copies can be made.

With respect to the analysis in the preceding paragraph, the mechanically ruled linear grating can be thought of as being created by means of two object points at infinity and a resulting grating period given by $p_r = \lambda_R/(2\sin\theta_{12}/2)$. λ_R is some arbitrarily chosen fictitious recording wavelength. An obvious choice is $\lambda_R = (\lambda_1 + \lambda_2)/2$, where λ_1 and λ_2 are

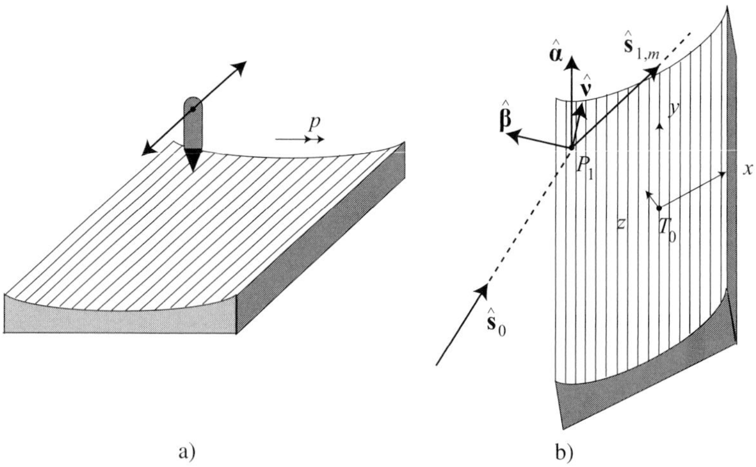

Figure 5.67: a) Manufacturing of an engraved surface with the aid of the linear movement of a cutting tool.
b) Diffraction geometry for a ruled surface. The reference point is T_0, an incident ray $\hat{\mathbf{s}}_0$ intersects the surface at P_1. The orthogonal unit vectors $\hat{\alpha}$, $\hat{\beta}$ and $\hat{\mathbf{v}}$, together with the local grating period measured in the tangent plane at P_1, determine the direction of an outgoing mth-order diffraction vector $\hat{\mathbf{s}}_{1,m}$.

the smallest and largest wavelengths at which the ruled grating will be used. θ_{12} is the angle between the wave vectors of the two fictitious interfering plane waves that give rise to the grating period p_r. An incident ray vector $\hat{\mathbf{s}}_0$ intersects the surface at P_1. When the grating lines are oriented perpendicular to the x-axis, we obtain for any point on the surface

$$\hat{\boldsymbol{\alpha}} = (0, 1, 0) , \qquad \hat{\mathbf{r}}_1 - \hat{\mathbf{r}}_2 = \left(2 \sin \frac{\theta_{12}}{2}, 0, 0\right) . \tag{5.7.27}$$

The normal vector at P_1, $\hat{\mathbf{v}} = (v_x, 0, v_z)$, follows from the surface shape. The local period of the grating, measured in the tangent plane at P_1, follows from Eq. (5.7.26),

$$p = \frac{p_r}{\cos\phi} = \frac{\lambda_R}{(2 \sin\theta_{12}/2)\, v_z} , \tag{5.7.28}$$

where ϕ is the angle between the normal vector $\hat{\mathbf{v}}$ and the vector $\hat{\mathbf{r}}_1 + \hat{\mathbf{r}}_2$ (see Fig. 5.66), which points in the z-direction in the case of the ruled surface of Fig. 5.67. The unit vector $\hat{\boldsymbol{\beta}}$ then follows from the orthogonality of the three unit vectors $\hat{\boldsymbol{\alpha}}$, $\hat{\boldsymbol{\beta}}$ and $\hat{\mathbf{v}}$ at P_1. With these data and the diffraction order number m, the reflected diffracted ray vector $\mathbf{s}_{1,m}$ can be calculated from Eq. (5.7.17).

5.7.3.5 Pathlength Calculation for a Diffracted Ray

Up to this point we have calculated the *direction* of diffracted rays, either in transmission or in reflection, created by a surface with an interference pattern recorded on it. The *pathlength* calculation along a finite ray which was given in Eq. (5.7.2) needs an extension in the case of diffracted rays. In Fig. 5.68 we have sketched the geometry which is needed for the calculation of the optical pathlength change along an incident ray and a general, out-of-plane diffracted ray. The incident ray is represented by the wave vector $\mathbf{k}_0$, with length $2\pi n_0/\lambda_0$, where λ_0 is the wavelength in vacuum. The tangent plane at the point of incidence T_0 on the surface is denoted by A in the figure, the tangential component of the incident wave vector is $\mathbf{k}_{0,t}$. The local period of the diffracting structure at T_0 produces the spatial (circular) frequency vector $2\pi\mathbf{a}_d$ in the tangent plane A. The tangential component of the diffracted vector is $\mathbf{k}_{1,t}$, in the figure sketched for a diffraction order number $m = -2$. The pathlength difference is calculated at a point Q in the tangent plane, shifted by a distance ds in the direction of the unit vector $\hat{\boldsymbol{\tau}}$. Q_0 and Q_1 are the projections of Q onto the tangential wave vector components $\mathbf{k}_{0,t}$ and $\mathbf{k}_{1,t}$, respectively. P_0 and P_1 are the projections of Q_0 and Q_1 on the incident and outgoing wave vectors $\mathbf{k}_0$ and $\mathbf{k}_1$, respectively. The diffracted ray is found by applying the continuity requirement of the tangential wave vectors in the tangent plane A. This continuity requirement reads

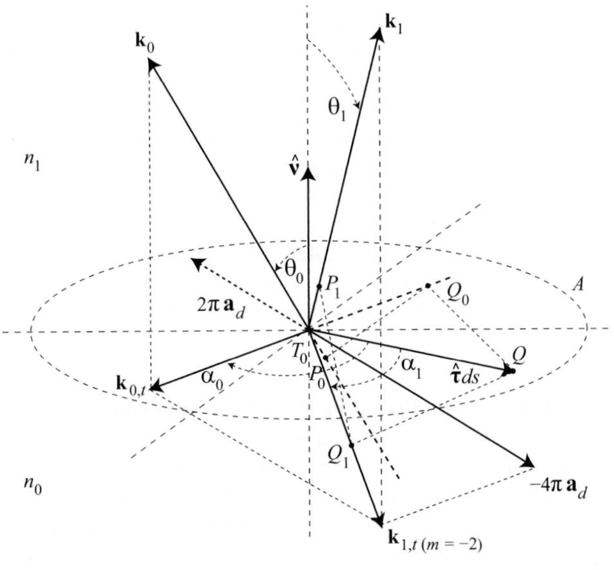

Figure 5.68: Optical pathlength calculation for a diffracted ray $\mathbf{k}_{1,t}$.

$$\mathbf{k}_{1,t} = \mathbf{k}_{0,t} + 2\pi m\,\mathbf{a}_d \,. \tag{5.7.29}$$

In the presence of a shift $\hat{\boldsymbol{\tau}}ds$ of the reference point from T_0 to a general point Q, we obtain with the aid of Fig. 5.68,

$$\mathbf{k}_{1,t} \cdot \hat{\boldsymbol{\tau}}ds = \mathbf{k}_{0,t} \cdot \hat{\boldsymbol{\tau}}ds + 2\pi m\,\mathbf{a}_d \cdot \hat{\boldsymbol{\tau}}ds \,. \tag{5.7.30}$$

Using the property $\mathbf{k}_{0,t} = \mathbf{k}_0 - (\mathbf{k}_1 \cdot \hat{\boldsymbol{\tau}})\hat{\boldsymbol{\tau}}$ and a comparable expression for $\mathbf{k}_{1,t}$, we derive from the figure

$$|\mathbf{k}_{1,t}| \sin\theta_1 \cos\alpha_1 ds = |\mathbf{k}_{0,t}| \sin\theta_0 \cos\alpha_0 ds + \frac{2\pi m}{p}\hat{\mathbf{a}}_d \cdot \hat{\boldsymbol{\tau}}ds \,, \tag{5.7.31}$$

where $\hat{\mathbf{a}}_d$ is the unit vector perpendicular to the grating lines in the tangent plane at T_0. Using the vacuum wavelength λ_0 we can write Eq. (5.7.31) as

$$n_1 \sin\theta_1 \cos\alpha_1 ds = n_0 \sin\theta_0 \cos\alpha_0 ds + m\frac{\lambda_0}{2\pi}\frac{d\Phi}{ds}ds \,, \tag{5.7.32}$$

where $\Phi(x, y, z)$ is the phase function of the diffracting structure. The grating vector $\mathbf{a}_d$ equals the gradient of $(1/2\pi)\Phi$.

From Fig. 5.68 we deduce that the infinitesimal distances $n_0 \sin\theta_0(\cos\alpha_0 ds) = \mathbf{k}_0 \cdot \hat{\mathbf{v}}\,ds$ and $n_1 \sin\theta_1(\cos\alpha_1 ds) = \mathbf{k}_1 \cdot \hat{\mathbf{v}}\,ds$ are equal to $n_0(T_0P_0)$ and $n_1(T_0P_1)$, respectively. This permits us to write Eq. (5.7.32) in its final form

$$[T_0P_1] = [T_0P_0] + m\frac{\lambda_0}{2\pi}d\Phi \,, \tag{5.7.33}$$

where, as usual, the brackets indicate that the optical pathlength should be taken.

For a finite shift of the reference point, Eq. (5.7.33) requires an integration of the phase function from the starting point T_0 to the new reference point Q. This is equivalent to the calculation of the increment in phase in the diffracting structure when going from the starting point T_0 to the endpoint Q. If the diffracting structure is given by means of its spatial frequency components, these components have to be integrated from the starting point to the end point along some path on the surface to obtain the desired phase difference Φ. If the spatial frequency on the surface is zero everywhere, the three vectors $\hat{\mathbf{v}}$, $\mathbf{k}_0$ and $\mathbf{k}_1$ become coplanar, Eq. (5.7.33) reduces to $[T_0P_1] = [T_0P_0]$ and we recover Snell's law.

5.7.4 Paraxial Ray-tracing around a Finite Ray

With the data available for a finite ray, reflected at or transmitted through an optical interface, it is possible to develop paraxial optical imaging laws with such a ray as reference system. The paraxial laws around such a ray can be developed with the aid of a method called differential ray-tracing and these laws are part of the so-called *parabasal* optics around the finite ray. The method allows very general geometries and can be applied also in inhomogeneous media. It collects the ray and pathlength information for very close neighbouring rays and extracts the wavefront departures as a function of the transverse coordinates in a plane perpendicular to the central ray. Following Euler's theorem about the orthogonal principal curvatures of any continuous surface, the wavefront can be attributed two orthogonal cross-sections with maximum difference in curvature. The rays in a narrow circular pencil around the central ray give rise to an astigmatic focus with two orthogonal lines in space, axially displaced along the central ray. The focal lines are each parallel to a principal section of the astigmatic beam. This particular refraction or reflection of a beam arises when a narrow pencil of rays is refracted at oblique incidence by a spherical surface that delimits two media with different index. In this subsection we first use basic geometry to calculate the position of the astigmatic foci of such an oblique pencil of rays. In a next step we analyse the image formation in the wavefront domain and extend the oblique pencil imaging to surfaces with holographic or periodic structures imprinted on them.

5.7.4.1 Oblique Refraction of a Narrow Pencil of Rays by a Spherical Surface

Early expressions for astigmatic refraction have been given by Barrow, Newton and Young **[383]**. The latter obtained his results in the framework of his study of the human eye. The commonly used expressions for the position of the astigmatic lines after refraction go back to Coddington **[72]**. The history of the derivation of these expressions can be found in **[182]**. Figure 5.69 shows the refraction of a (narrow) beam which is incident on a spherical surface. The principal ray of the incident beam is $\hat{\mathbf{s}}_0$ and this ray is at an angle of θ_0 with the surface normal. The refracted ray $\hat{\mathbf{s}}_1$ travels at an angle θ_1. Image formation is studied by examining the refraction of a marginal ray of the pencil at an angle $\theta_0 + d\theta_0$, refracted at an angle $\theta_1 + d\theta_1$. Without loss of generality we assume that the incident pencil of rays has a perfect focus at the point O.

- *Tangential cross-section*
We first consider the cross-section in the plane of the drawing, commonly called the meridional cross-section that contains the centre of curvature of the spherical surface. For the central ray and the marginal ray of the incident beam we have from triangles CT_0O and CT_1O the relation $d\theta_0 = d\beta - d\gamma_0$. A comparable relation holds for the refracted rays in the triangles CT_0T and CT_1T, where we assume that the tangential image is formed at the point T at a distance l_t from T_0. For the angles $d\beta$, $d\gamma_0$ and $d\theta_0$ we find

$$d\beta = c\, T_0T_1\,, \qquad d\gamma_0 = T_0T_1 \cos\theta_0/l_0\,, \qquad d\theta_0 = T_0T_1\,(c - \cos\theta_0/l_0)\,, \tag{5.7.34}$$

and comparable expressions exist for the refracted quantities. Using Snell's law in its differential form, $n_0 \cos\theta_0 d\theta_0 = n_1 \cos\theta_1 d\theta_1$, for the central and marginal rays of the narrow pencil, we obtain the classical Coddington expression for the position of the tangential focal line,

$$\frac{n_1 \cos^2\theta_1}{l_t} - \frac{n_0 \cos^2\theta_0}{l_0} = (n_1 \cos\theta_1 - n_0 \cos\theta_0)\, c\,. \tag{5.7.35}$$

- *Sagittal cross-section*
To obtain the position of the second focal line we consider the cross-section perpendicular to the plane of incidence. This cross-section is commonly called the sagittal cross-section and the rays contained in it form the so-called sagittal fan. From the rays in the sagittal fan we find the sagittal focal line. This line is perpendicular to the central refracted ray $\hat{\mathbf{s}}_1$ and located in the plane of incidence, formed by the incident ray and the axis T_0C in Fig. 5.69. In the case of a perfectly focused incident beam, the object point for imaging in the sagittal cross-section is again at O. A general consideration about the imaging by the rays in the sagittal cross-section is the following. A rotation of the spherical surface around the auxiliary axis CO leaves the direction of a refracted ray unaltered. This means that the rays in the planar sagittal cross-section, to a first approximation, all intersect the auxiliary axis at the same point. The sagittal focal line is thus found at S, the intersection point of the refracted central ray $\hat{\mathbf{s}}_1$ and CO. For the calculation of the distance l_s we use the sine rule in the triangles CT_0O and CT_0S and this yields the sagittal Coddington equation,

$$\frac{n_1}{l_s} - \frac{n_0}{l_0} = (n_1 \cos\theta_1 - n_0 \cos\theta_0)\, c\,. \tag{5.7.36}$$

In the case when the incident beam already contains astigmatism, we carry out the two calculations using the separate t- and s-focus positions and construct the auxiliary axis through the sagittal focus of the incident beam and the centre of curvature C of the surface.

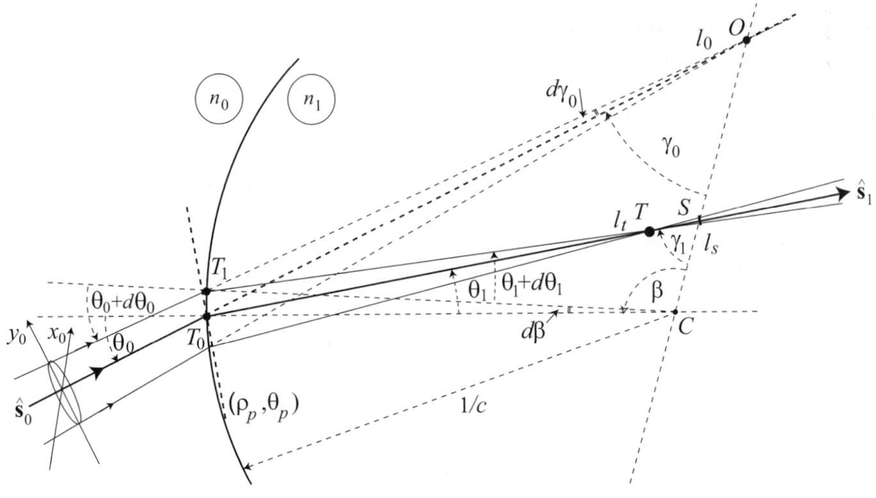

Figure 5.69: Refraction of a thin pencil of rays. The central ray $\hat{\mathbf{s}}_0$ is obliquely incident on a spherical surface with radius $R = 1/c$ and with its centre of curvature at C. The virtual object point is at O with $T_0O = l_0$. The refracted beam shows a tangential focus T and a sagittal focus S, at distances l_t and l_s from T_0, measured along the refracted principal ray $\hat{\mathbf{s}}_1$.

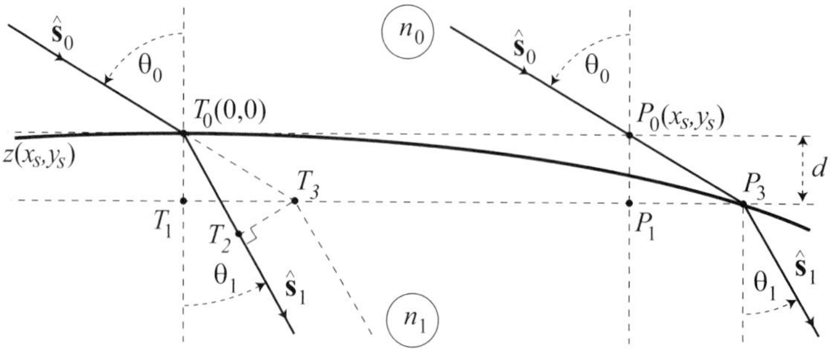

Figure 5.70: The refraction of an incident ray $\hat{\mathbf{s}}_0$ at a curved interface with profile $z(x_s, y_s) = c(x_s^2 + y_s^2)/2$; the reference zero-height level is given by the thin dashed line through T_0 and P_0.

5.7.4.2 Wavefront Analysis of Refraction at Oblique Incidence

For the wavefront analysis of the refraction of an obliquely incident beam, we first consider the pathlength difference that arises at a spherically curved surface. The geometry is shown in Fig. 5.70. A thin parallel beam of light is incident in the first medium with index n_0 on a curved surface at an angle θ_0; the refracted beam in the second medium with index n_1 propagates at an angle θ_1 with the surface normal. The curvature of the optical surface is $c = 1/R$ and, in the direct neighbourhood of the central point T_0, the surface profile is approximated by the expression $z = c\,(x_s^2 + y_s^2)/2$. The pathlength difference Δ between the central ray T_0T_2 of the pencil and a general ray through P_0P_3 is measured at the level $z = -d$ and it is given by

$$\Delta(x_s, y_s) = [P_0P_3] - [T_0T_2] = n_0P_0P_3 - n_1T_0T_2 = \frac{1}{2R}(n_0\cos\theta_0 - n_1\cos\theta_1)(x_s^2 + y_s^2)\,, \tag{5.7.37}$$

where we have taken the reference points for measuring the pathlength along the rays at P_3 and T_3, with equal z-values, both in the second medium (horizontal dashed line in the figure). Beyond this z-value, both rays propagate in the same direction with angle θ_1. In the paraxial approximation, up to quadratic terms in the lateral surface coordinates (x_s, y_s), we do not need to make a distinction between the lateral coordinates of, for instance, P_0 and P_3. Neither, for the calculation of the quadratic pathlength difference, do we need to take into account the minute difference in propagation direction in the second medium of the rays travelling through T_2 and P_3.

5.7.4.3 Wavefront Analysis and the Coddington Equations

We now apply the imaging condition along the rays $\hat{\mathbf{s}}_0$ and $\hat{\mathbf{s}}_1$ in Fig. 5.71 and require that the pathlength difference between any pair of rays in the pencil is made zero. In terms of the lateral coordinates on the incoming and outgoing beams and the coordinates on the surface we impose

$$-\,n_0\frac{(x_0^2 + y_0^2)}{2l_0} + n_1\frac{(x_1^2 + y_1^2)}{2l_1} + \frac{1}{2R}(n_0\cos\theta_0 - n_1\cos\theta_1)(x_s^2 + y_s^2) = 0\,, \tag{5.7.38}$$

where we have included the wavefront curvatures that arise when the object and image points are at finite distances, l_0 and l_1, respectively. For an incident beam at angle θ_0 with the surface normal $\hat{\mathbf{v}}$, we have the lateral coordinates

$$x_s = x_0\,, \qquad y_s = \frac{y_0}{\cos\theta_0}\,, \tag{5.7.39}$$

and a comparable expression exists for the coordinates on the surface and the lateral coordinates in the refracted beam. With the aid of these lateral coordinates we write Eq. (5.7.38) in the following way,

$$\left(\frac{n_1}{2l_1} - \frac{n_0}{2l_0}\right)x_s^2 + \left(\frac{n_1\cos^2\theta_1}{2l_1} - \frac{n_0\cos^2\theta_0}{2l_0}\right)y_s^2 = \frac{1}{2R}(n_1\cos\theta_1 - n_0\cos\theta_0)(x_s^2 + y_s^2)\,. \tag{5.7.40}$$

For a given object distance l_0, Eq. (5.7.40) can generally not be satisfied by a single image distance l_1. However, it is possible to separately reduce to zero the wavefront curvature in the plane $x_s = 0$ and in the plane $y_s = 0$. The imaging equations are given by

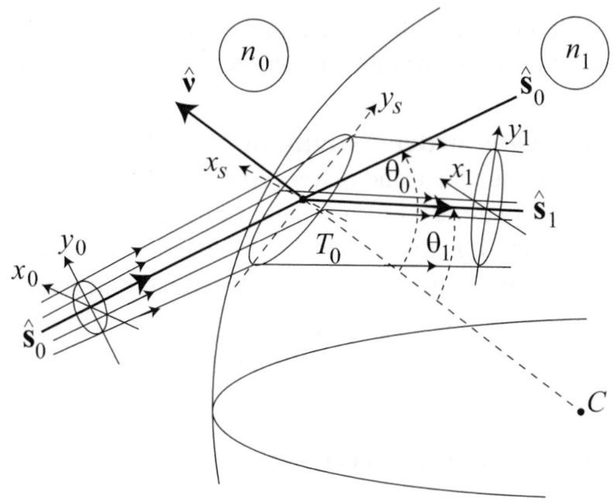

Figure 5.71: An obliquely incident narrow beam, angle of incidence θ_0 in medium with index n_0, is refracted into the second medium with index n_1 at an angle θ_1 with the normal $\hat{\mathbf{v}}$ to the spherical surface. The local surface coordinates at the point of incidence T_0 of the central ray vector $\hat{\mathbf{s}}_0$ are (x_s, y_s), the lateral coordinates on the incident and refracted beams are given by (x_0, y_0) and (x_1, y_1), respectively.

$$x_s = 0 \qquad \frac{n_1 \cos^2\theta_1}{l_t} - \frac{n_0 \cos^2\theta_0}{l_0} = \frac{(n_1\cos\theta_1 - n_0\cos\theta_0)}{R} , \tag{5.7.41}$$

$$y_s = 0 \qquad \frac{n_1}{l_s} - \frac{n_0}{l_0} = \frac{(n_1\cos\theta_1 - n_0\cos\theta_0)}{R} . \tag{5.7.42}$$

We find anew the Coddington equations with the image distances l_t and l_s for the tangential and sagittal focal lines as solutions.

In between the two focal lines, the best astigmatic focus position is found. This intermediate position is found by minimising the wavefront aberration as a function of the lateral coordinates (x_1, y_1) over the cross-section of the refracted beam. The rms wavefront aberration of the refracted beam is minimum when the departures in the $y_1 = 0$ and $x_1 = 0$ cross-sections have equal amplitude but opposite sign. For $y_0 = y_1 \cos\theta_0 / \cos\theta_1$ we obtain the best-focus image distance l_{bf} from the equation

$$\frac{n_1}{l_{bf}} = \left(\frac{\cos^2\theta_0 + \cos^2\theta_1}{2\cos^2\theta_1}\right)\frac{n_0}{l_0} + \left(\frac{1+\cos^2\theta_1}{2\cos^2\theta_1}\right)(n_1\cos\theta_1 - n_0\cos\theta_0)\, c . \tag{5.7.43}$$

It is easily verified that the inverse image distances (or the image 'vergences') obey the linear relation

$$\frac{1}{l_{bf}} = \frac{1}{2}\left(\frac{1}{l_t} + \frac{1}{l_s}\right) . \tag{5.7.44}$$

Equations (5.7.41)–(5.7.43) can also be used for the reflective case. We then have that $n_1 = -n_0$ and $\cos\theta_1 = \cos\theta_0$, yielding,

$$x_s = 0\, ; \qquad \frac{1}{l_t} = -\frac{1}{l_0} + \frac{2c}{\cos\theta_0} , \tag{5.7.45}$$

$$y_s = 0\, ; \qquad \frac{1}{l_s} = -\frac{1}{l_0} + 2c\cos\theta_0 , \tag{5.7.46}$$

$$\text{best focus}\, ; \qquad \frac{1}{l_{bf}} = -\frac{1}{l_0} + \left(\frac{1+\cos^2\theta_0}{\cos\theta_0}\right) c . \tag{5.7.47}$$

From Fig. 5.71 it is clear that the astigmatism is the result of an anamorphotic mapping of the incoming wave curvature onto the outgoing beam. Having chosen the y-coordinate direction in the plane of incidence, the stretching or compression of this coordinate at refraction causes the different curvature of the outgoing beam in the two orthogonal coordinate directions. Note that the astigmatism remains finite when the refracting optical surface is made flat. This is why a flat window used at oblique incidence will introduce astigmatism in the outgoing beam if the incident beam is converging or diverging. The astigmatism only becomes zero when the incident beam is made parallel.

5.7.4.4 Field Expansion of the Second-order Astigmatic Wavefront Aberration

The wavefront expression of Eq. (5.7.40) and the Coddington equations (5.7.41) and (5.7.42) will now be used to obtain the astigmatic wavefront aberration (second-order) as a function of the field coordinate or field angle in an imaging system. We refer to Fig. 5.69 and take the horizontal axis T_0C as the axis of symmetry of an optical system and the surface through T_0 as one of the optical transitions of the system. The wavefront aberration of the refracted beam is given as a function of the polar coordinates (ρ_p, θ_p) in the plane perpendicular to the refracted ray ($\theta_p = 0$ corresponds to the x_1-axis). As to the pupil position, this coordinate choice corresponds to the special case of a central stop. The object distance measured along a general oblique beam is identical to l_0, which means that the object is on a curved surface with radius of curvature l_0. Denoting the axial distances from T_0 to the astigmatic lines by l_t and l_s, we write for the astigmatic wavefront aberration W_a with respect to a reference sphere through T_0 and centred on the best focus between the two astigmatic lines,

$$W_a = -\frac{n_1}{4}\left(\frac{1}{l_t} - \frac{1}{l_s}\right)\rho_p^2 \cos 2\theta_p \, . \tag{5.7.48}$$

From the Coddington equations we extract the quantities $1/l_t$ and $1/l_s$ and substitute these results in Eq. (5.7.48) to obtain with the aid of Snell's law,

$$W_a = -\frac{\tan^2\theta_1}{4}\left\{\frac{n_0^2 - n_1^2}{n_0 l_0} + (n_1\cos\theta_1 - n_0\cos\theta_0)\, c\right\}\rho_p^2\cos 2\theta_p \, . \tag{5.7.49}$$

For the imaging in an intermediate image plane, perpendicular to the axis T_0C and containing the paraxial image point for $\theta_1 = 0$, we need the expression for the size of the astigmatism as a function of the transverse image plane coordinate p_1. This field coordinate is proportional to $\tan\theta_1$ and we expand W_a in powers of $\tan\theta_1$ to obtain the field behaviour of the lowest order astigmatism of the oblique pencils. We note that in the presence of field curvature the optimum focus is not located in the paraxial image plane but on a curved surface. Using the relations

$$\begin{aligned}
\cos^2\theta_1 &= 1 - \tan^2\theta_1 + \tan^4\theta_1 - \cdots \, , \\
\cos\theta_1 &= 1 - \frac{\tan^2\theta_1}{2} + \frac{3\tan^4\theta_1}{8} - \frac{5\tan^6\theta_1}{16} \cdots \, , \\
\cos\theta_0 &= 1 - \frac{n_1^2}{2n_0^2}\tan^2\theta_1 + \frac{n_1^2}{2n_0^2}\left(1 - \frac{n_1^2}{4n_0^2}\right)\tan^4\theta_1 - \frac{n_1^2}{2n_0^2}\left(1 - \frac{n_1^2}{2n_0^2} + \frac{n_1^4}{8n_0^4}\right)\tan^6\theta_1 \cdots \, ,
\end{aligned} \tag{5.7.50}$$

we obtain after some rearrangement,

$$W_a = \frac{(n_1 - n_0)\, c}{4}\left\{\left(\frac{n_0 + n_1}{n_0 l_0 c} - 1\right)\tan^2\theta_1 - \frac{n_1}{2n_0}\tan^4\theta_1 - \frac{n_1}{8n_0^3}(n_1^2 + n_1 n_0 - 3n_0^2)\tan^6\theta_1 + \cdots\right\}\rho_p^2\cos 2\theta_p \, . \tag{5.7.51}$$

The aberration term of second order in the field coordinate becomes zero if the object distance l_0 is equal to $(n_0 + n_1)R/n_0$. This object distance and the conjugate image distance $l_1 = (n_0 + n_1)R/n_1$ define the Roberval/Huygens aplanatic points. These pairs of points with perfect imaging are found on the corresponding spherical object and image surfaces. If the astigmatic wavefront aberration was calculated for points on these spheres by expanding the field aberration as a function of powers of $\sin^{2n}\theta_1$, we would have found astigmatism that is rigorously zero for all orders in the case of the Huygens points. A further important observation from Eq. (5.7.51) is the different character of the $\tan^2\theta_1$ and higher-order terms. The fourth- and higher-order terms are proportional to the optical power of the transition and to factors that are functions of the optical contrast n_1/n_0.

Instead of a curved object surface we now choose a flat object. The object distance for a general off-axis point is then given by $l_0/\cos\theta_0$. The substitution of this modified object distance in Eq. (5.7.51) yields the wavefront aberration

$$\begin{aligned}
W_a = \frac{(n_1 - n_0)\, c}{4}&\left\{\left(\frac{n_0 + n_1}{n_0 l_0 c} - 1\right)\tan^2\theta_1 - \frac{n_1\{n_0^2 l_0 c + n_1(n_1 + n_0)\}}{2n_0^3 l_0 c}\tan^4\theta_1\right. \\
&\left. - \left[\frac{n_0^2 n_1(n_1^2 + n_1 n_0 - 3n_0^2) l_0 c - n_1^2(n_1 + n_0)(4n_0^2 - n_1^2)}{8n_0^5 l_0 c}\right]\tan^6\theta_1 + \cdots\right\}\rho_p^2\cos 2\theta_p \, .
\end{aligned} \tag{5.7.52}$$

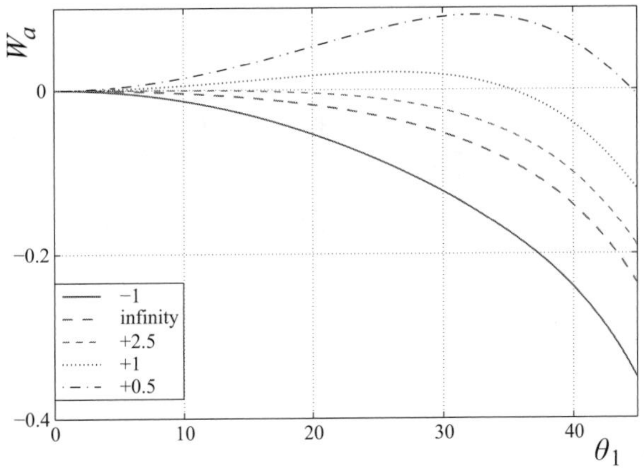

Figure 5.72: The field dependence of astigmatic wavefront aberration of second order in the pupil coordinates according to Eq. (5.7.52). The parameter in the figure is the value of the normalised object distance l_0/R. The refractive index values are $n_0 = 1$ and $n_1 = 1.5$.

Equation (5.7.52) shows the special property of the Huygens aplanatic points when considering the factor which multiplies the term with second-order field dependence, proportional to $\tan^2\theta_1$. The astigmatic wave aberration for a flat object field has been plotted in Fig. 5.72 as a function of the image field angle θ_1 for various values of the relative distance $l = l_0/R$ of the object plane to the optical transition. Going from a real object ($l = -1$) with a diverging incident pencil we proceed to a parallel beam ($l = \infty$) and, from there, to more and more converging incident beams. The diverging beam yields second- and higher-order terms with the same negative sign (tangential focal line precedes the sagittal focal line). The beam converging to the Huygens' aplanatic point ($l = 2.5$) has zero second-order astigmatism and higher-order terms with negative sign. For object planes still closer to the surface, the second-order astigmatism changes sign but the higher-order components do not. This leads to the interesting two upper curves in Fig. 5.72, which show a folding back of the initial astigmatism to zero at larger field angles. This feature can be favourably exploited in optical design for extending the useful image field. Equation (5.7.52) also illustrates the fact that, in general, it is easy to produce changes in the lowest-order aberration term by changing the magnification. The higher-order aberration terms are less easily influenced. Their reduction may require a drastic change in layout and a rearrangement of optical powers in the system.

5.7.4.5 Generalised Coddington Equations

The particular role played by the plane of incidence and the plane perpendicular to it in the preceding paragraph is lost when the outgoing beam is directed out of the plane of incidence. A possible reason for this is the presence of a diffracting structure on the optical surface with general orientation of the local grating vector. Figure 5.73 shows such a general geometry of an incident and diffracted beam with the tangential wave vector $\mathbf{k}_{t,t}$ of the transmitted wave obtained from

$$\mathbf{k}_{t,t} = \mathbf{k}_{i,t} + \frac{2\pi m}{p}\,\hat{\boldsymbol{\beta}}\,. \tag{5.7.53}$$

With the aid of the vector $\mathbf{k}_{t,t}$ we construct the outgoing wave vector which is at an angle θ_1 with respect to the normal to the tangent plane at T_0. The plane defined by the outgoing beam and the normal to the surface intersects the tangent plane along the y'-direction, at an angle ϕ with the plane of incidence through the y-axis. To express the surface coordinates in the lateral coordinates (x_0, y_0) of the incoming beam, we have as before the relations of Eq. (5.7.39). For the outgoing beam we have in the same way

$$x_1 = x_s'\,, \qquad y_1 = y_s'\cos\theta_1\,, \tag{5.7.54}$$

where the coordinates (x_s', y_s') are obtained from the coordinates (x_s, y_s) by a rotation over an angle ϕ. We then obtain for the lateral beam coordinates in the outgoing beam

$$x_1 = x_s\cos\phi + y_s\sin\phi\,, \qquad y_1 = (-x_s\sin\phi + y_s\cos\phi)\cos\theta_1\,. \tag{5.7.55}$$

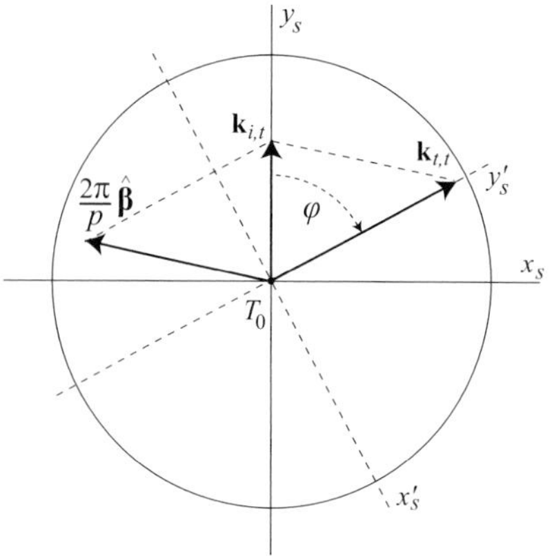

Figure 5.73: The construction of the tangential wave vector in the tangent plane through T_0. The incident wave vector produces a tangential vector $\mathbf{k}_{i,t}$, the diffractive structure has a local period p and frequency unit vector given by $\hat{\boldsymbol\beta}$. The resulting tangential vector of the refracted and diffracted beam is $\mathbf{k}_{t,t}$. The construction in the figure corresponds to a diffraction order number $m = -1$. The rotation angle ϕ is negative in the figure.

The wavefront deviation in the outgoing beam follows from Eq. (5.7.38) with the appropriate substitutions of the beam and surface coordinates. Using Eqs. (5.7.39) and (5.7.55) we obtain, after some rearrangement, for the wavefront deviation $W(x_s, y_s)$ in terms of the surface coordinates,

$$W(x_s,y_s) = \left(\frac{n_1}{2l_1}\right)\left\{x_s^2\cos^2\phi + y_s^2\sin^2\phi + 2x_sy_s\sin\phi\cos\phi + \left[x_s^2\sin^2\phi + y_s^2\cos^2\phi - 2x_sy_s\sin\phi\cos\phi\right]\cos^2\theta_1\right\}$$
$$-\left(\frac{n_0}{2l_0}\right)\left(x_s^2 + y_s^2\cos^2\theta_0\right) - \left(\frac{n_1\cos\theta_1 - n_0\cos\theta_0}{2R}\right)(x_s^2 + y_s^2)\,. \tag{5.7.56}$$

To analyse the wavefront deformation in the outgoing beam, it is preferable to express the surface coordinates in the lateral coordinates (x_1, y_1) of this beam, yielding the expression

$$W(x_1,y_1) = \frac{x_1^2}{2}\left\{\frac{n_1}{l_1} - \left(\cos^2\phi + \sin^2\phi\cos^2\theta_0\right)\frac{n_0}{l_0} - \frac{(n_1\cos\theta_1 - n_0\cos\theta_0)}{R}\right\}$$
$$+\frac{y_1^2}{2}\left\{\frac{n_1}{l_1} - \left(\frac{\sin^2\phi}{\cos^2\theta_1} + \frac{\cos^2\phi\cos^2\theta_0}{\cos^2\theta_1}\right)\frac{n_0}{l_0} - \frac{(n_1\cos\theta_1 - n_0\cos\theta_0)}{R\cos^2\theta_1}\right\}$$
$$+ x_1y_1\left(\frac{\sin\phi\cos\phi\sin^2\theta_0}{\cos\theta_1}\right)\frac{n_0}{l_0}\,. \tag{5.7.57}$$

The classical Coddington imaging equations arise as solutions of $W(x_1, y_1) = 0$ for $\phi = 0$, $x_1 = 0$ and $y_1 = 0$. A new special situation arises when ϕ takes on the special value of $\pm\pi/2$. In this case, we find the modified Coddington equations,

$$x_1 = 0\,; \qquad \frac{n_1\cos^2\theta_1}{l_t} - \frac{n_0}{l_0} = \frac{(n_1\cos\theta_1 - n_0\cos\theta_0)}{R}\,, \tag{5.7.58}$$

$$y_1 = 0\,; \qquad \frac{n_1}{l_s} - \frac{n_0\cos^2\theta_0}{l_0} = \frac{(n_1\cos\theta_1 - n_0\cos\theta_0)}{R}\,, \tag{5.7.59}$$

where l_t and l_s are the image distances of the tangential and sagittal astigmatic lines.

Further inspection of Eq. (5.7.57) shows that it is possible to have equal wavefront curvature in the principal sections $x_1 = 0$ and $y_1 = 0$ by requiring that

$$l_0 = n_0R\,\frac{\left(\cos^2\phi\cos^2\theta_1 + \sin^2\phi\cos^2\theta_0\cos^2\theta_1 - \cos^2\phi\cos^2\theta_0 - \sin^2\phi\right)}{(n_1\cos\theta_1 - n_0\cos\theta_0)\sin^2\theta_1}\,. \tag{5.7.60}$$

For an arbitrary angle $\phi \neq 0$ or $\pi/2$ we notice that the wavefront aberration cannot be made zero due to the presence of the third term on the right-hand side of Eq. (5.7.57). This term becomes zero only if the incident beam is perpendicular to the surface or the object point is at infinity. In general, an astigmatic wavefront aberration will remain that produces astigmatic lines at angles of $\pm\pi/4$ radians with the main cross-sections $x_1 = 0$ and $y_1 = 0$ of the refracted beam. This astigmatic aberration is listed in Table D.3 of Appendix D by its index $N_w = 5$ (the corresponding Zernike polynomial is commonly denoted by Z_5). This particular type of astigmatic aberration has a Zernike polynomial representation given by $\rho^2 \sin 2\theta_Z$. The polar radius ρ is given by $(x_1^2 + y_1^2)^{1/2}$ and the angle θ_Z is the angle with the x_1-axis of a ray vector to a general point (x_1, y_1) in the cross-section of the outgoing beam.

5.8 Total Aberration at a Single Surface; Formulas of Hopkins and Welford

The common procedure for the calculation of the wavefront aberration in image space from finite ray data was outlined in Subsection 5.7.1. For each intermediate space between two optical surfaces, the aberration increment Δ between a general aperture ray and a reference ray is accumulated, yielding the final wavefront aberration on the exit pupil reference sphere (see Eq. (5.7.2)). In this process, the individual surface contributions are lost unless reference spheres are constructed for each intermediate imaging step. In that case, the contribution from a specific surface can be distilled from the aberration difference on two successive reference spheres, one in the object space and one in the image space of the optical surface of interest. Although this process is quite feasible when using fast automatic computing means, it was a less attractive process in the pre-computer era. For this reason, various authors have looked for so-called *total* aberration formulas that produce numerically reliable expressions for the aberration increment between two rays at an optical surface. Using this total aberration increment of a finite ray allows for the easy construction of an aberration distribution picture for the optical system. To obtain such a map, the aberration increments for a large collection of rays of a pencil are calculated and fitted with respect to appropriate base functions, for instance, Zernike polynomials.

The aberration contribution in the pathlength change at an optical surface is of at least fourth order in the coordinates of the principal ray and a general aperture ray. For the order of the aberration, we consider the (L, M) direction cosines and lateral (x, y) surface coordinates. To include fourth-order Seidel aberrations, a total aberration expression should contain at least these fourth-order combinations of surface coordinates and direction cosines. Almost inevitably, the expression for the pathlength change contains also lower-order contributions that are present during refraction, like wavefront defocus and field-dependent wavefront tilt. Such terms are paraxial ones and generally of second order in the pupil and field coordinates. Still worse, if the pathlength change is calculated from one surface to the next one, a zero-order difference is included, namely the axial distance between the adjacent surfaces. The problem with a total aberration expression that contains terms of order lower than four is that the interesting part of it, with order four or higher in the pupil and field coordinates, can be numerically obscured by the contributions of order two (and even zero). In the case of a total aberration formula including terms of order two, it is possible to calculate the expression several times with different settings of the reference point that is used to calculate the pathlength difference between a general ray and the reference ray. The searched-for aberration term of fourth order or higher then follows from the minimum value of the total aberration as a function of the reference point position. Of course, the number of significant digits in the calculation should be such that the aberration contribution of fourth order or higher can be extracted in a numerically stable way.

For the aberration analysis of a complete pencil of rays, the aberration of each ray pair needs to be adapted for a shift towards the reference point for the *entire* pencil of rays. Such a movement of the reference point adds (small) second-order pathlength changes that are given by Eq. (4.9.6). An initially fourth-order pathlength difference formula is then reduced to second order. Since in double-precision calculations a comfortable number of significant digits (typically 15) is available, an aberration formula containing terms down to second order can be considered to be numerically robust. After subtraction of the (paraxial) second-order terms, a sufficient number of significant digits is still available to safely represent the higher-order aberration content of the pathlength change at a surface.

An elegant total aberration formula for the aberration increment between two general rays at a surface has been proposed by Hopkins **[141]**. In Fig. 5.74 an aperture ray with unit vector $\hat{\mathbf{s}}_0$ and a reference ray with vector $\hat{\mathbf{s}}_{0,r}$ are incident on an optical surface and intersect that surface at the points Q_0 and $Q_{0,r}$, respectively. The rays are not necessarily coplanar. On each ray, we can determine the point that is closest to the other ray, yielding the points A_0 and $A_{0,r}$, respectively. The point A_1 is the midpoint of the common perpendicular $A_0A_{0,r}$ of the two rays and this point is chosen as the reference point for pathlength calculation on the two rays. After refraction, a comparable construction is chosen for the refracted rays and B_1 is the midpoint of the common perpendicular of the refracted rays. The pathlength change at refraction between the aperture ray and the reference ray is then given by

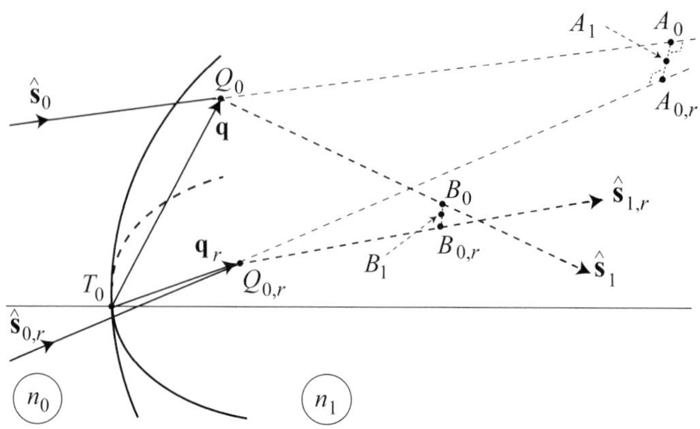

Figure 5.74: Calculation of the total aberration at an optical surface for two finite incident rays with unit vectors $\hat{\mathbf{s}}_0$ and $\hat{\mathbf{s}}_{0,r}$. The intersection points of the rays with the optical surface are Q_0 and $Q_{0,r}$. The refracted rays $\hat{\mathbf{s}}_1$ and $\hat{\mathbf{s}}_{1,r}$ are obtained by rigorous ray-tracing.

$$\Delta W_{A_1B_1} = [A_0Q_0B_0] - [A_{0,r}Q_{0,r}B_{0,r}] = n_1(Q_0B_0 - Q_{0,r}B_{0,r}) - n_0(Q_0A_0 - Q_{0,r}A_{0,r}) \,. \tag{5.8.1}$$

This definition of optical pathlength along a ray is the Hamiltonian definition and implies a reference sphere with infinitely large radius.

The expressions for the direction of the common perpendicular and its intersection points with the ray pair are obtained from vector analysis. The unit vector $\hat{\mathbf{v}}$ along the common perpendicular is given by the normalised vector product of $\hat{\mathbf{s}}_0$ and $\hat{\mathbf{s}}_{0,r}$. The shortest distance or shortest join d_v between the two rays then follows from the projection of the line segment between two general points on the rays onto this common perpendicular. Choosing Q_0 and $Q_{0,r}$ for these points we obtain,

$$\hat{\mathbf{v}} = \frac{\hat{\mathbf{s}}_0 \times \hat{\mathbf{s}}_{0,r}}{\left|\hat{\mathbf{s}}_0 \times \hat{\mathbf{s}}_{0,r}\right|} \,, \qquad d_v = \frac{(\mathbf{q} - \mathbf{q}_r)\cdot(\hat{\mathbf{s}}_0 \times \hat{\mathbf{s}}_{0,r})}{\left|\hat{\mathbf{s}}_0 \times \hat{\mathbf{s}}_{0,r}\right|} \,. \tag{5.8.2}$$

The required distances Q_0A_0 and $Q_{0,r}A_{0,r}$ follow from the property that the vector $\vec{Q_0A_0} + \vec{A_0A_{0,r}} + \vec{A_{0,r}Q_{0,r}} + \vec{Q_{0,r}Q_0} = \vec{0}$ and after some manipulation we obtain

$$Q_0A_0 = \frac{(\mathbf{q}_r - \mathbf{q})\cdot(\hat{\mathbf{s}}_{0,r} \times \hat{\mathbf{s}}_0 \times \hat{\mathbf{s}}_{0,r})}{1 - (\hat{\mathbf{s}}_0 \cdot \hat{\mathbf{s}}_{0,r})^2} \,, \qquad Q_{0,r}A_{0,r} = -\frac{(\mathbf{q}_r - \mathbf{q})\cdot(\hat{\mathbf{s}}_0 \times \hat{\mathbf{s}}_{0,r} \times \hat{\mathbf{s}}_0)}{1 - (\hat{\mathbf{s}}_0 \cdot \hat{\mathbf{s}}_{0,r})^2} \,. \tag{5.8.3}$$

The expression for ΔW then readily turns out to be

$$\Delta W_{A_1B_1} = \left\{ \frac{n_1(\hat{\mathbf{s}}_1 + \hat{\mathbf{s}}_{1,r})}{(1 + \hat{\mathbf{s}}_1 \cdot \hat{\mathbf{s}}_{1,r})} - \frac{n_0(\hat{\mathbf{s}}_0 + \hat{\mathbf{s}}_{0,r})}{(1 + \hat{\mathbf{s}}_0 \cdot \hat{\mathbf{s}}_{0,r})} \right\} \cdot (\mathbf{q}_r - \mathbf{q}) \,. \tag{5.8.4}$$

The quantity ΔW according to this equation is of second order in the direction cosines and lateral coordinates. In **[362]**, this expression has been converted into a fourth-order expression having a more favourable numerical accuracy when computing the aberration value.

The final step for obtaining an aberration map for an entire pencil of rays is to adjust the reference positions A_1 and B_1, that are specific for each pair of rays, to some global reference positions A and B. In Fig. 5.75 the reference point for the incoming pencil of rays has been shifted from the midpoint A_1 to the general point A and the pathlengths are measured to the projected points C_0 and $C_{0,r}$ on $\hat{\mathbf{s}}_0$ and $\hat{\mathbf{s}}_{0,r}$, respectively. Using analogous points B, D_0 and $D_{0,r}$ for the refracted beam we obtain for the change in pathlength ΔW_s according to the Hamiltonian definition,

$$\Delta W_s = W_{AB} - W_{A_1B_1} = n_1(B_0D_0 - B_{0,r}D_{0,r}) - n_0(A_0C_0 - A_{0,r}C_{0,r}) \,. \tag{5.8.5}$$

In object space we have $Q_0C_0 = (\mathbf{a}_0 - \mathbf{q})\cdot\hat{\mathbf{s}}_0$ and $A_0C_0 = (\mathbf{a}_0 - \mathbf{q})\cdot\hat{\mathbf{s}}_0 - Q_0A_0$. Comparable expressions for the ray in image space and for the r-indexed rays yield the result

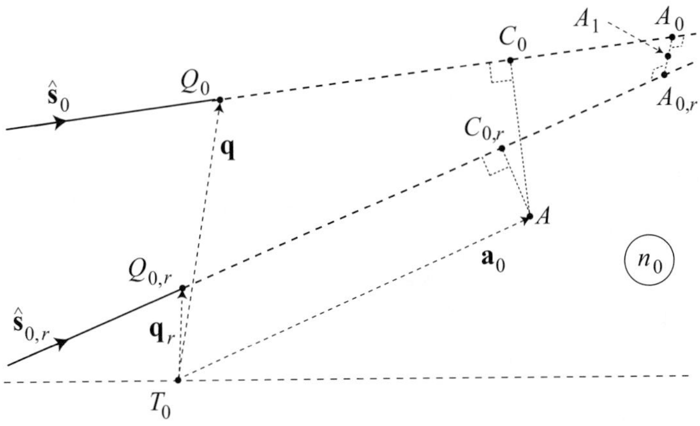

Figure 5.75: Changing the reference point from A_1 to the global point A for the incoming pencil of rays (refractive index of the medium is n_0). Q_0 and $Q_{0,r}$ are the points of intersection of the rays with the optical surface. The position of A is given by the vector $T_0A = \mathbf{a}_0$.

$$\begin{aligned}\Delta W_s = {} & n_1 \left\{(\mathbf{b}_0 - \mathbf{q}) \cdot \hat{\mathbf{s}}_1 - Q_0B_0 - (\mathbf{b}_0 - \mathbf{q}_r) \cdot \hat{\mathbf{s}}_{1,r} + Q_{0,r}B_{0,r}\right\} \\ & - n_0 \left\{(\mathbf{a}_0 - \mathbf{q}) \cdot \hat{\mathbf{s}}_0 - Q_0A_0 - (\mathbf{a}_0 - \mathbf{q}_r) \cdot \hat{\mathbf{s}}_{0,r} + Q_{0,r}A_{0,r}\right\} . \end{aligned} \quad (5.8.6)$$

Unfortunately, the aberration increment due to a change in reference point is of second order in the lateral ray quantities. The numerical advantage of converting Eq. (5.8.4) into a pure fourth-order expression is lost by the shift of reference points. In order to measure the aberration increment with respect to reference spheres with a finite radius of curvature, the wavefront correction according to Eq. (4.9.6) is needed. The line segments AC_0, $AC_{0,r}$ etc. in Fig. 5.75 play the same role as the segment O_1T_1 in Fig. 4.27.

5.9 Aperture- and Field-dependent Aberration Function of an Imaging System

The transverse aberration components and the wavefront aberration data of an optical system can be obtained from analytic power series expansions of the optical pathlength along a general ray as a function of its position in the object plane and its intersection point with the plane of the diaphragm. Alternatively, using the data from finite ray-tracing, a transverse or wavefront aberration function is obtained by fitting these ray-tracing data to appropriate basis functions of the pupil and field coordinates. Preferably, the expansion of the aberration function of the imaging system is done with the aid of orthogonal polynomials in the pupil and field coordinates.

The transverse aberration components are measured in the image plane; the wavefront aberration is generally measured on an exit pupil sphere that is centred on the paraxial image point. To illustrate the aberration calculations we have plotted in Fig. 5.76 the transverse aberration components related to an aberrated ray P_1Q_1' of an off-axis pencil and the corresponding obliquely tilted reference sphere, focused on the paraxial off-axis image point Q_1. The coordinates on the tilted sphere in the exit pupil through E_1 are denoted by ρ and θ, the coordinates of the paraxial image point Q_1 by r and ϕ. An aberrated ray $\hat{\mathbf{s}}$ through a general point P_1 on the exit pupil sphere intersects the image plane at the point Q_1' with coordinates $(x_{Q_1} + \delta x_{Q_1}, y_{Q_1} + \delta y_{Q_1})$. The polar coordinates of the general point P_1 on the exit pupil sphere are obtained as follows. The scalar product of the normal vectors to the sphere at E_1 and P_1, $\hat{\mathbf{v}}_{E_1}$ and $\hat{\mathbf{v}}_{P_1}$, respectively, is used to define the normalised radial pupil coordinate,

$$\rho = \frac{\sin\alpha}{\sin\alpha_m} = \frac{\sqrt{1-\left(\hat{\mathbf{v}}_{E_1}\cdot\hat{\mathbf{v}}_{P_1}\right)^2}}{\sin\alpha_m} , \quad (5.9.1)$$

where $\sin\alpha_m$ is derived from a point P_B on the boundary of the tilted reference sphere in the exit pupil.

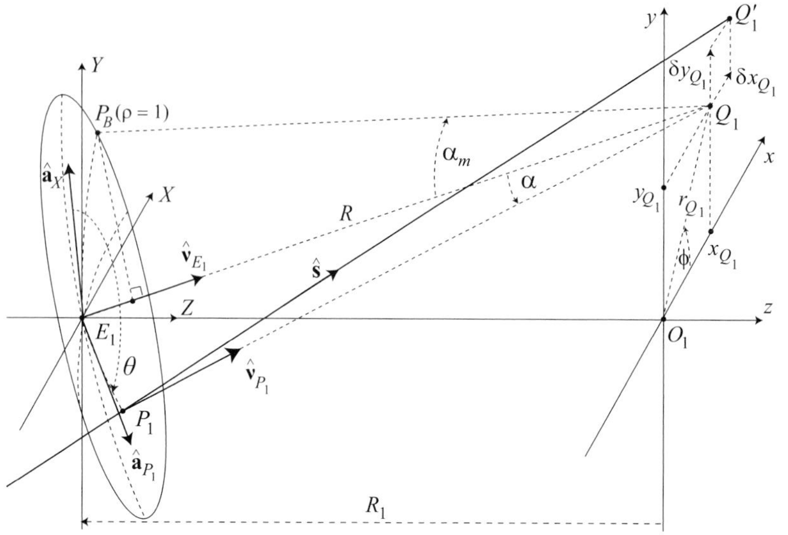

Figure 5.76: Definition of the pupil and field coordinates for a general off-axis beam. The centre of the exit pupil is E_1, the paraxial image point is Q_1. An aberrated ray $\hat{\mathbf{s}}$ through a general point P_1 of the pupil intersects the image plane at the point Q'_1. The relative pupil coordinates of P_1 are (ρ, θ), those of the paraxial image point Q_1 are (r, ϕ). The distance $E_1 O_1$ has been denoted by R_1, the distance $E_1 Q_1$ is R.

To define the azimuthal coordinate θ we first construct the unit vector $\hat{\mathbf{a}}_{P_1}$ that is perpendicular to the vector $\hat{\mathbf{v}}_{E_1}$ and parallel to the plane defined by the triangle $E_1 Q_1 P_1$. The second unit vector $\hat{\mathbf{a}}_X$ is also perpendicular to $\hat{\mathbf{v}}_{E_1}$ and is located in the plane defined by $E_1 Q_1$ and the X-axis,

$$\hat{\mathbf{a}}_{P_1} = \frac{\hat{\mathbf{v}}_{E_1} \times \hat{\mathbf{v}}_{P_1} \times \hat{\mathbf{v}}_{E_1}}{|\hat{\mathbf{v}}_{E_1} \times \hat{\mathbf{v}}_{P_1} \times \hat{\mathbf{v}}_{E_1}|} , \qquad \hat{\mathbf{a}}_X = \frac{\hat{\mathbf{v}}_{E_1} \times \hat{\mathbf{X}} \times \hat{\mathbf{v}}_{E_1}}{|\hat{\mathbf{v}}_{E_1} \times \hat{\mathbf{X}} \times \hat{\mathbf{v}}_{E_1}|} . \tag{5.9.2}$$

The azimuth θ of P_1 is then uniquely defined by the value of $\cos\theta = \hat{\mathbf{a}}_{P_1} \cdot \hat{\mathbf{a}}_X$ and the sign of the X-coordinate of P_1. In practice, pupil aberrations are present and the central ray of the pencil in the physical stop of the system does not necessarily intersect the centre E_1 of the paraxial exit pupil. In what follows, we will neglect such effects on the definition of the polar pupil coordinates (ρ, θ).

If the aberration data are obtained from ray deviation and pathlength expansions of various orders, they are expressed with the aid of 'monomials' of the coordinates on the pupil sphere and in the image plane. Such expansions have been developed by Buchdahl **[62]**. Further developments in this field use computer algebra to obtain high-order aberration terms (see Andersen **[9]** and Bociort **[29]**). But such monomial aberration expansions frequently comprise subsequent terms with alternating sign yielding poor convergence towards the end value. The expansion coefficients originate, for instance, from square-root expansions with coefficients of the binomial type that may assume large values. An expansion of the aberration function with the aid of orthogonal polynomials is preferred because of its numerical stability. Although the maximum order of the (theoretical) aberration function of an optical system can be limited to typically 10 or 12, much higher orders can be required if measured aberration data of manufactured systems need to be represented. Higher-order wavefront perturbations may arise from residual surface polishing errors and expansion orders as high as 50 or 100 may be encountered and need to be fitted by the optical metrologist.

A wavefront aberration expansion that uses Zernike polynomials to represent the aperture and field dependence of the wavefront aberration or the transverse ray components has been given in **[6]**, **[199]**. Such an expansion can be applied to an optical system with symmetry of revolution. For instance, high-quality large-field projection systems and microscope objectives are good candidates for such an expansion of the wavefront aberration function **[241]**. The coefficients of such an expansion can be used during the optimisation stage of an optical design. We also discuss an expansion of the *transverse aberration* function of a system that does not possess rotational symmetry. For example, such an expansion would be useful for a photographic objective that suffers from residual tilt and centring errors of the various elements.

These fabrication errors destroy the symmetry of revolution and introduce a much broader range of aberration terms. Equations (5.2.4)–(5.2.7) give the number of aberration terms of a certain order for such a general imaging system. A general aberration expansion is also appropriate for systems that, by nature, do not possess rotational symmetry like holographic imaging systems, free-form optical systems or cylindrical optics **[338]**.

In this section we first give a conversion scheme to modify a monomial wavefront expansion produced by classical and modern aberration theory to an expansion based on Zernike polynomials. The scheme is restricted to the wavefront expansion for an optical system with rotational symmetry with a reduced number of terms according to the classification that was given in Section 5.2. A general conversion scheme for optical systems without symmetry can be found in **[57]**. From a set of wavefront expansion coefficients we then calculate the field-averaged rms wavefront aberration. The rms wavefront aberration is discussed in Subsection 9.3.1.2 and can be used as a quality measure for an individual imaging pencil **[236]**. Its field-averaged value provides the user of an optical system with a single number to assess the performance of the system. For more severely aberrated systems we consider a polynomial expansion of the transverse aberration components as a function of pupil and field coordinates. In this context the so-called Lukosz polynomials are introduced. An expression for the field-averaged rms blur allows the assessment of a strongly aberrated system with the aid of a single quality number. The section ends with a description of some low-order aberration terms (distortion and astigmatism) which are encountered in systems without any symmetry.

5.9.1 Double Zernike Expansion for an Optical System with Rotational Symmetry

In Fig. 5.76 we have sketched an oblique pencil of rays that leaves an optical system towards a paraxial image point Q_1. For a particular ray through a point $P_1(X, Y)$ on the exit pupil sphere the transverse aberration components $(\delta x_{Q_1}, \delta y_{Q_1})$ are calculated. Simultaneously, by optical pathlength calculations along a ray or from a Hamiltonian characteristic function, we can obtain the pathlength difference of a particular ray with respect to the reference ray. The reference ray generally is the ray that intersects the centre of the diaphragm of the optical system. It does not necessarily pass through E_1, the centre of the exit pupil, since aberrations may be present in the imaging of the pupils. We neglect this effect of pupil aberration and write the aberration function as

$$W(X, Y; x, y) = [Q_0P_1] - [Q_0E_1] \,. \tag{5.9.3}$$

Q_0 is the object point for the pencil of rays (not shown in the figure) and P_1 and E_1 are the intersection points of the general aperture ray and the reference ray with the reference sphere that has its centre of curvature at the (perfect) image point Q_1. The aberration function W is defined with respect to the coordinates (X, Y) of the intersection point P_1 of the general aperture ray with the exit pupil sphere through E_1 and the image plane coordinates (x, y) of the paraxial image point Q_1. According to Eq. (5.2.2) the aberration function of a general optical system is written in terms of the Cartesian pupil and field coordinates **[140]**,

$$W(X, Y; x, y) = \sum d'_{nmlk}\, X^n Y^m x^l y^k = \sum d'_{nmlk}\, \rho_{P_1}^{n+m}\, r_{Q_1}^{l+k} \cos^n\theta \sin^m\theta \cos^l\phi \sin^k\phi \,, \tag{5.9.4}$$

where ρ_{P_1} and r_{Q_1} are the radial coordinates with respect to the origins E_1 and O_1, respectively, of the pupil and field coordinates (see Fig. 5.76).

For an optical system with an axis of rotational symmetry, the aberration function depends on coordinate combinations that are invariant with respect to a rotation around the axis O_0O_1 of the optical system. According to Eqs. (5.2.10) and (5.2.11) the rotation invariant combinations are $\boldsymbol{\rho}_{P_1} \cdot \boldsymbol{\rho}_{P_1}$, $r_{Q_1} \cdot r_{Q_1}$ and $\boldsymbol{\rho}_{P_1} \cdot r_{Q_1}$. In line with Eq. (5.2.11) the power series expansion of the aberration function is then given by

$$W\left\{\rho_{P_1}^2, r_{Q_1}^2, \rho_{P_1} r_{Q_1} \cos(\theta - \phi)\right\} = \sum d'_{nlm} \left(\rho_{P_1}^2\right)^n \left(r_{Q_1}^2\right)^l \left[r_{Q_1}\rho_{P_1} \cos(\theta - \phi)\right]^m \,, \tag{5.9.5}$$

where the expansion coefficients d' of the system with circular symmetry now have three indices nlm and do not bear an explicit relationship with the coefficients d'_{nmlk} of the general optical system. In optical aberration theory, the terms with $n = m = 0$ which only depend on the image plane coordinates are generally omitted (see Eqs. (5.2.6) and (5.2.13) and **[229]**).

To find the Zernike coefficients from a power expansion according to Eq. (5.9.5), we calculate the inner product of a general term of this expansion with coefficient d'_{nml} and a (complex) Zernike polynomial in pupil and field coordinates. The complex Zernike polynomials $Z_n^m(\rho, \theta) = R_n^{|m|}(\rho)\exp(im\theta)$ are chosen because of their much simpler manipulation as compared to the set that comprises separate cosine and sine polynomials. For a pupil function $f(\rho, \theta)$ we have the orthogonal expansion

$$f(\rho,\theta) = \sum_{n,m} c_{nm}\, R_n^{|m|}(\rho)\, \exp(im\theta)\,, \tag{5.9.6}$$

where n is a non-negative integer number, m is integer, and $n - |m|$ is even and non-negative. The coefficients c_{nm} are generally complex. The normalisation of the Zernike polynomials is such that

$$\int_0^{2\pi}\int_0^1 \left|Z_n^m(\rho,\theta)\right|^2 \rho d\rho d\theta = \frac{\pi}{n+1}\,. \tag{5.9.7}$$

Thus the c_{nm} in Eq. (5.9.6) are given by (see Appendix D),

$$c_{nm} = \frac{n+1}{\pi}\int_0^{2\pi}\int_0^1 f(\rho,\theta)\,\{Z_n^m(\rho,\theta)\}^*\, \rho d\rho d\theta\,, \tag{5.9.8}$$

where the asterisk which accompanies the Zernike polynomials denotes the complex conjugate operation. The cosine and sine coefficients of the Zernike expansion of a general complex function follow from

$$a_{c,nm} = (c_{nm} + c_{n,-m})\,, \qquad a_{s,nm} = +i(c_{nm} - c_{n,-m})\,, \qquad a_{c,n0} = c_{n0}\,, \qquad a_{s,n0} = 0\,. \tag{5.9.9}$$

In the case of a real function we have the special property $c_{n,-m} = c^*_{nm}$. Before calculating the inner product of a general term from the power series expansion and a Zernike polynomial we normalise the radial coordinates in pupil and field to unit value at the pupil rim ($\rho_{P_1} = \rho_0$) and the field border ($r_{Q_1} = r_0$) and obtain an expansion as in Eq. (5.9.5) with now unprimed expansion coefficients a_{nml}. The relationship between the primed and unprimed a-coefficients is given by

$$a_{nmlk} = a'_{nmlk}\,\rho_0^{(n+m)}\, r_0^{(l+k)}\,, \qquad\qquad a_{nlm} = a'_{nlm}\,\rho_0^{(2n+m)}\, r_0^{(2l+m)}\,. \tag{5.9.10}$$

After conversion we have the following double Zernike expansion of the aberration function,

$$W(\rho,r;\theta,\phi) = \sum_{n_1,n_2;m_1,m_2} c_{n_1 n_2 m_1 m_2}\, R_{n_1}^{|m_1|}(\rho)\, R_{n_2}^{|m_2|}(r)\, \exp\left[i(m_1\theta + m_2\phi)\right]\,, \tag{5.9.11}$$

where ρ and r are the normalised versions of the real-space radial coordinates ρ_{P_1} and r_{Q_1}.

To find the complete set of Zernike coefficients we proceed in two steps. In a first step we select a term with index nlm from a given power series expansion and calculate the inner product with a general Zernike term from the expansion of Eq. (5.9.11). The inner product is denoted by $I_{nlm}^{\,n_1 n_2 m_1 m_2}$ where $n_1 n_2 m_1 m_2$ are the indices of the Zernike polynomial. The inner product yields nonzero values for certain index combinations nlm and $n_1 n_2 m_1 m_2$, depending also on the properties of the aberration function, for instance, the presence of circular symmetry. In the second step we construct the complete Zernike expansion with the aid of a summation of the coefficients over all possible terms a_{nlm} of the power series.

For the inner product of a single term from the power series with $Z_{n_1}^{m_1}(\rho,\theta)\, Z_{n_2}^{m_2}(r,\phi)$ we write

$$I_{nlm}^{\,n_1 n_2 m_1 m_2} = \int_0^1\int_0^1\int_0^{2\pi}\int_0^{2\pi} \rho^{2n+m}\, r^{2l+m}\, \cos^m(\theta-\phi)\, R_{n_1}^{|m_1|}(\rho)\, R_{n_2}^{|m_2|}(r) \exp\left[-i(m_1\theta + m_2\phi)\right]\, \rho r d\rho dr d\theta d\phi\,. \tag{5.9.12}$$

Writing $\cos(\theta-\phi)$ as $[\exp\{i(\theta-\phi)\} + \exp\{-i(\theta-\phi)\}]/2$ and using Newton's binomial formula, we expand a power of $\cos(\theta-\phi)$ as follows,

$$\cos^m(\theta-\phi) = \frac{1}{2^m}\sum_{j=0}^{m}\binom{m}{j}\exp\left[i(m-2j)(\theta-\phi)\right]\,. \tag{5.9.13}$$

Therefore,

$$\begin{aligned}
&\int_0^{2\pi}\int_0^{2\pi} \cos^m(\theta-\phi)\exp\left[-i(m_1\theta + m_2\phi)\right]\, d\theta d\phi \\
&= \frac{1}{2^m}\sum_{j=0}^{m}\binom{m}{j}\int_0^{2\pi}\int_0^{2\pi} \exp\left[+i(m-m_1-2j)\theta\right]\, \exp\left[-i(m+m_2-2j)\phi\right]\, d\theta d\phi \\
&= \frac{4\pi^2}{2^m}\sum_{j=0}^{m}\binom{m}{j}\, \delta_{m-m_1-2j}\, \delta_{m+m_2-2j}\,,
\end{aligned} \tag{5.9.14}$$

where $\delta_n = 1$ for $n = 0$ and 0 otherwise. The right-hand side of Eq. (5.9.14) then equals

$$\begin{cases} \frac{4\pi^2}{2^m}\binom{m}{\frac{m-|m_1|}{2}} & m_1 = -m_2; \quad m - |m_{1,2}| \text{ even and non-negative,} \\ 0 & \text{otherwise.} \end{cases} \tag{5.9.15}$$

With the result of Eqs. (5.9.14) and (5.9.15), the I in Eq. (5.9.12) takes the form

$$I^{n_1 n_2 m_1 m_2}_{nlm} = \frac{4\pi^2}{2^m}\binom{m}{\frac{m-|m_1|}{2}} \int_0^1 \rho^{2n+m} R^{|m_1|}_{n_1}(\rho)\rho d\rho \int_0^1 r^{2l+m} R^{|m_1|}_{n_2}(r)\, r dr\,. \tag{5.9.16}$$

Equation (5.9.16) illustrates that for a rotationally symmetric optical system the Zernike expansion possesses three independent indices n_1, n_2 and $m_1 = m_2$ because of the coupling between the azimuthal indices of the pupil and field polynomials. We also refer to Section A.4 of Appendix A where it is shown how the Discrete Fourier Transform can be used to numerically evaluate the azimuthal part of the integral in Eq. (5.9.12).

The integrals over ρ and r in Eq. (5.9.16) have been discussed in **[52]** (see also Eq. (D.4.7) of Appendix D), and we obtain with $p_1 = (n_1 - |m_1|)/2$,

$$J^{n_1 m_1}_{nm} = \int_0^1 \rho^{2n+m} R^{|m_1|}_{n_1}(\rho)\rho\, d\rho = \frac{\frac{1}{2}\left(n + \frac{m-|m_1|}{2}\right)!\left(n + \frac{m+|m_1|}{2}\right)!}{\left(n + \frac{m-|m_1|}{2} - p_1\right)!\left(n + \frac{m+|m_1|}{2} + p_1 + 1\right)!}, \tag{5.9.17}$$

which is non-vanishing only when $n_1 = |m_1|, |m_1| + 2, \cdots, 2n + m$. For the integral over r there is a similar result, viz. $J^{n_2 m_1}_{lm}$ where $p_2 = (n_2 - |m_1|)/2$ instead of p_1. The substitution of a single power series term for $f(\rho, \theta)$ in Eq. (5.9.8) yields the following Zernike polynomial expansion

$$\rho^{2n+m} r^{2l+m} \cos^m(\theta - \phi) = \sum_{n_1=0}^{2n+m} \sum_{n_2=0}^{2l+m} \sum_{m_1=-m}^{m} b^{n_1 n_2 m_1}_{nml} R^{|m_1|}_{n_1}(\rho)\, R^{|m_1|}_{n_2}(r)\, \exp\left[i m_1(\theta - \phi)\right],$$

where

$$b^{n_1 n_2 m_1}_{nml} = \frac{4(n_1+1)(n_2+1)}{2^m}\binom{m}{\frac{m-|m_1|}{2}} J^{n_1 m_1}_{nm} J^{n_2 m_1}_{lm} = \frac{(n_1+1)(n_2+1)}{2^m}\binom{m}{\frac{m-|m_1|}{2}}$$
$$\times \frac{\left(n + \frac{m-|m_1|}{2}\right)!\left(n + \frac{m+|m_1|}{2}\right)!}{\left(n + \frac{m-|m_1|}{2} - p_1\right)!\left(n + \frac{m+|m_1|}{2} + p_1 + 1\right)!}\, \frac{\left(l + \frac{m-|m_1|}{2}\right)!\left(l + \frac{m+|m_1|}{2}\right)!}{\left(l + \frac{m-|m_1|}{2} - p_2\right)!\left(l + \frac{m+|m_1|}{2} + p_2 + 1\right)!}. \tag{5.9.18}$$

The numbers p_1 and p_2 are $(n_1 - |m_1|)/2$ and $(n_2 - |m_1|)/2$, respectively. The summation range for m_1 is restricted to the following values: $m - |m_1|$ is even and non-negative and $|m_1| \le \min(n_1, n_2)$.

For the complete power series expansion with index ranges $0 \le n \le N_p$, $0 \le l \le L_p$ and $0 \le m \le M_p$, we obtain the Zernike expansion

$$\sum_{nlm} a_{nlm} \rho^{2n+m} r^{2l+m} \cos^m(\theta - \phi) = \sum_{n_1=0}^{2N_p+M_p} \sum_{n_2=0}^{2L_p+M_p} \sum_{m_1=-M_p}^{M_p} \left\{ \sum_{nlm} a_{nlm}\, b^{n_1 n_2 m_1}_{nml} \right\} R^{|m_1|}_{n_1}(\rho)\, R^{|m_1|}_{n_2}(r)\, \exp\left[i m_1(\theta - \phi)\right]$$
$$= \sum_{n_1 n_2 m_1} c_{n_1 n_2 m_1} R^{|m_1|}_{n_1}(\rho)\, R^{|m_1|}_{n_2}(r)\, \exp\left[i m_1(\theta - \phi)\right], \tag{5.9.19}$$

where the $c_{n_1 n_2 m_1}$ are the new coefficients of the Zernike polynomial expansion and the series over m_1 is restricted in a similar way as in Eq. (5.9.18).

The expression for the b-coefficients in (5.9.18) is in closed-form with a well-defined, limited number of terms. The two expressions for the radial integral on the last line of Eq. (5.9.18) pose problems for large values of the arguments of the factorials. It is shown in Appendix A of **[57]** that the factorials can be avoided by writing the two quotients according to

$$\frac{\left(\frac{a-b}{2}\right)!\left(\frac{a+b}{2}\right)!}{\left(\frac{a-b}{2} - p\right)!\left(\frac{a+b}{2} + p + 1\right)!} = \frac{2}{a + b + 2p + 2} \prod_{j=0}^{p-1} \frac{a - b - 2j}{a + b + 2j + 2}, \tag{5.9.20}$$

where a, b and p are all integer (see also Appendix D, Eq. (D.4.8)). With the ranges of the integers for a and b that correspond to the index ranges of the Zernike polynomials it follows that each multiplication factor in the product expression is ≤ 1. The result for the azimuthal integral according to Eq. (5.9.14) can be evaluated without problems for sufficiently small m. For values of m larger than typically 30, a numerically stable method should be used, for instance the method of the discrete Fourier transform (DFT) which is discussed in Appendix A.4. Numerical exercises with some exactly prescribed pupil functions show that the maximum degree of the obtained Zernike expansion can be substantially reduced while maintaining the same level of approximation of the pupil function **[57]**. A reduction in maximum degree by a factor of typically two has been observed.

5.9.2 Field-averaged Root-mean-square Wavefront Aberration

A global quality factor for an optical imaging system is the field-averaged wavefront aberration. In the case of operation of the optical system close to the diffraction limit, such a quality factor provides a good indication of the typical spot size or of the 50% level of the modulation transfer function for spatial frequencies in the image plane. Using a complex double Zernike expansion for the field-dependent wavefront aberration function W according to Eq. (5.9.11), we have for the variance of W at a general field point with polar coordinates (r, ϕ) the expression

$$V_W(r,\phi) = \frac{1}{\pi}\int_0^1\int_0^{2\pi} |W(\rho,\theta,r,\phi)|^2 \rho d\rho d\theta - \left|\frac{1}{\pi}\int_0^1\int_0^{2\pi} W(\rho,\theta,r,\phi)\,\rho d\rho d\theta\right|^2 . \tag{5.9.21}$$

Insertion of the double Zernike expansion for W according to Eq. (5.9.11) yields the expression

$$\begin{aligned} V_W(r,\phi) =& \frac{1}{\pi}\int_0^1\int_0^{2\pi}\left(\sum_{n_1 m_1 n_2 m_2} c_{n_1 m_1 n_2 m_2} R_{n_1}^{|m_1|}(\rho)\exp(im_1\theta)\, R_{n_2}^{|m_2|}(r)\exp(im_2\phi)\right) \\ &\times\left(\sum_{n_1' m_1' n_2' m_2'} c^*_{n_1' m_1' n_2' m_2'} R_{n_1'}^{|m_1'|}(\rho)\exp(-im_1'\theta)\, R_{n_2'}^{|m_2'|}(r)\exp(-im_2'\phi)\right)\rho d\rho d\theta \\ &- \left|\frac{1}{\pi}\int_0^1\int_0^{2\pi}\sum_{n_1 m_1 n_2 m_2} c_{n_1 m_1 n_2 m_2} R_{n_1}^{|m_1|}(\rho)\exp(im_1\theta)\, R_{n_2}^{|m_2|}(r)\exp(im_2\phi)\,\rho d\rho d\theta\right|^2 , \end{aligned} \tag{5.9.22}$$

where the following conditions apply to the indices of the radial Zernike polynomials n_1 is even and non-negative, $n_1 - |m_1|$ is even, n_2 is even and non-negative, $n_2 - |m_2|$ is even, and similar conditions for the primed quantities.

Using the orthogonality properties of Zernike polynomials (see Appendix D), we find

$$\begin{aligned} V_W(r,\phi) =& \sum_{n_1 m_1}\left(\frac{1}{n_1+1}\right)\sum_{n_2 m_2 n_2' m_2'} c_{n_1 m_1 n_2 m_2} c^*_{n_1 m_1 n_2' m_2'} R_{n_2}^{|m_2|}(r)\exp(im_2\phi)\, R_{n_2'}^{|m_2'|}(r)\exp(-im_2'\phi) \\ &- \sum_{n_2 m_2 n_2' m_2'} c_{00n_2m_2} c^*_{00n_2'm_2'} R_{n_2}^{|m_2|}(r)\exp(im_2\phi)\, R_{n_2'}^{|m_2'|}(r)\exp(-im_2'\phi) . \end{aligned} \tag{5.9.23}$$

The field-averaged wavefront variance follows from Eq. (5.9.23) by integration over the unit disc in the image field, yielding

$$\begin{aligned} \overline{V_W} =& \frac{1}{\pi}\sum_{n_1 m_1;\neq(0,0)}\left(\frac{1}{n_1+1}\right)\sum_{n_2 m_2 n_2' m_2'} c_{n_1 m_1 n_2 m_2} c^*_{n_1 m_1 n_2' m_2'} \int_0^1\int_0^{2\pi} R_{n_2}^{|m_2|}(r)\, R_{n_2'}^{|m_2'|}(r)\,\exp[i(m_2-m_2')\phi]\, r dr d\phi \\ =& \sum_{n_1 m_1 n_2 m_2}\frac{|c_{n_1 m_1 n_2 m_2}|^2}{(n_1+1)(n_2+1)} - \sum_{n_2 m_2}\frac{|c_{00n_2m_2}|^2}{n_2+1} , \end{aligned} \tag{5.9.24}$$

where the ranges for the summation indices are the same as defined above in relation to Eq. (5.9.22).

The special case of an optical system with rotational symmetry and a wavefront aberration function according to the last line of Eq. (5.9.19) is treated in the same way as the general system without any symmetry property and the following result is obtained,

$$\overline{V_W} = \frac{1}{\pi} \sum_{n_1 m_1; \neq (0,0)} \left(\frac{1}{n_1 + 1} \right) \sum_{n_2 n_2'} c_{n_1 n_2 m_1} c^*_{n_1 n_2' m_1} \int_0^1 \int_0^{2\pi} R_{n_2}^{|m_1|}(r) \, R_{n_2'}^{|m_1|}(r) \, r dr d\phi$$

$$= \sum_{n_1 n_2 m_1} \frac{\left|c_{n_1 n_2 m_1}\right|^2}{(n_1 + 1)(n_2 + 1)} - \sum_{n_2} \frac{\left|c_{0 n_2 0}\right|^2}{n_2 + 1} . \tag{5.9.25}$$

We define the field-averaged root-mean-square wavefront aberration as $\overline{W}_{\mathrm{rms}} = \left(\overline{V_W}\right)^{1/2}$. In Part III of this book it is shown that the maximum relative intensity $I_{m,r}$ in the image of a point source produced by a pencil of rays leaving an optical system is, to a good approximation, proportional to

$$I_{m,r} \propto 1 - \left(\frac{2\pi W_{\mathrm{rms}}}{\lambda_0} \right)^2 , \tag{5.9.26}$$

if the value of W_{rms} of the specific pencil of rays is much smaller than the wavelength λ_0 of the light. The condition $W_{\mathrm{rms}} \ll \lambda_0$ guarantees that the blurring of the image of a point source is mainly determined by the diffraction effects at the pupil rim of the imaging system and to a lesser extent by the aberrations of the imaging system. The optical system is then said to be 'diffraction-limited' in its performance. In the range of wavefront aberration with typically $W_{\mathrm{rms}} \leq \lambda/10$, the minimisation of the field-averaged value of the root-mean-square wavefront aberration of the imaging pencils, $\overline{W}_{\mathrm{rms}}$, produces the optimum average depth of modulation in the image plane or the maximum attainable intensities of the images of point sources in the object plane.

5.9.3 Wavefront Expansion and Image Plane Transverse Aberration Components

In this subsection we discuss a wavefront aberration expansion that is suitable for optical systems with a state of correction that is far away from the diffraction limit. For such a system the aberration analysis is preferably performed in the transverse ray aberration domain and a global quality factor is the root-mean-square *blur* of an image point. The concept of image blur and the root-mean-square blur were introduced in Subsection 5.2.2, Eq. (5.2.29), and are further developed in this subsection. Figure 5.77a shows the spherical reference surface in the exit pupil with its central point E_1 on the z-axis of the optical system and its centre of curvature P_0 in the image plane with coordinates (x_0, y_0). As reference point P_0 one generally chooses the image point that would follow from an image analysis based on paraxial optics. The non-aberrated central ray would pass through the point E_1 and the paraxial image point P_0 with position vector $\boldsymbol{\delta}_r = (x_0, y_0)$. A general aberrated ray from the object point intersects the exit-pupil reference surface at the point Q_j with ray direction $\hat{\mathbf{s}}_j$.

We assume that a pencil of rays is composed of J rays in total with each ray carrying a ray intensity I_j. This ray intensity I_j is proportional to the strength of the Poynting vector of the electromagnetic field at the location of the geometrical ray. The relative weight in the aberration analysis is further determined by the relative area S_j/S_P of the reference surface in the exit pupil that is covered by ray j of the pencil. S_P is the total area of the spherical cap in the exit pupil that is covered by all the rays of the pencil. A general ray with index j intersects the image plane at the point P_j of which the position is given by the vector $\boldsymbol{\delta}_j(x_j, y_j)$. The central point $\overline{P}$ of the blurred image created by the aberrated rays is defined as

$$\overline{\boldsymbol{\delta}} = (\overline{x}, \overline{y}) = \frac{\sum_j S_j I_j \boldsymbol{\delta}_j}{\sum_j S_j I_j} , \tag{5.9.27}$$

and the transverse image shift is defined by $\overline{\boldsymbol{\delta}}_t = \overline{\boldsymbol{\delta}} - \boldsymbol{\delta}_r$. The rms blur δ of a pencil of rays with ray intensity weights I_j and relative ray areas S_j is then defined by

$$\delta = \sqrt{\frac{\sum_j (S_j I_j) \left\{ (x_j - \overline{x})^2 + (y_j - \overline{y})^2 \right\}}{\sum_j (S_j I_j)}} . \tag{5.9.28}$$

In Fig. 5.77b the various points of intersection P_0, P_j and $\overline{P}$ have been sketched and a blur circle has been drawn, centred on the point $\overline{P}$ and having a radius equal to the value δ of the rms blur.

5.9.3.1 Wavefront Aberration Function Yielding Minimum Transverse Aberration

In optical design it is convenient to continuously use the same aberrational quantity for the optimisation of an optical imaging system, independently of the status of correction of the system. In the initial phase of the design, the system will

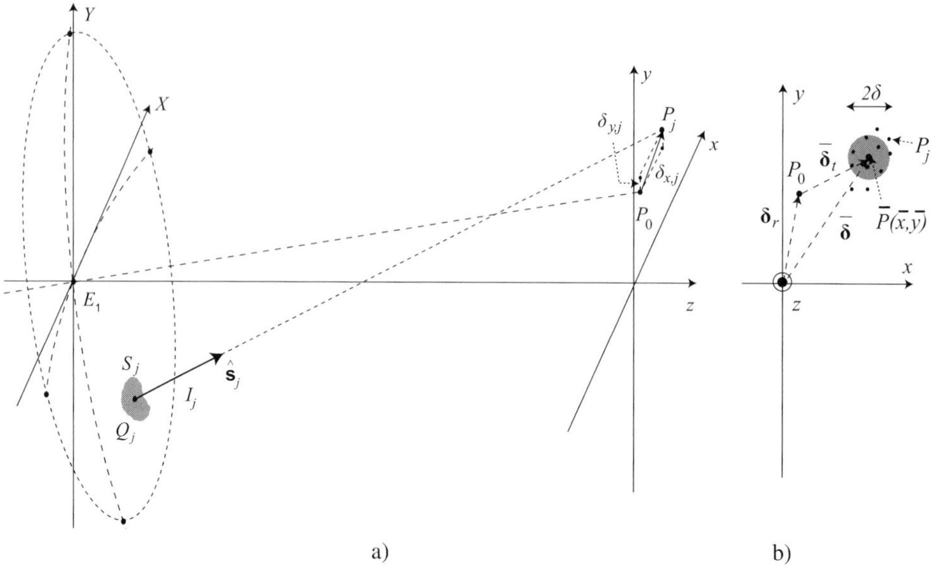

Figure 5.77: a) The definition of transverse ray aberration.
b) Root-mean-square blur δ of an intensity-weighted pencil of rays.

suffer from large aberrations and the relevant aberration is the transverse ray aberration in the image plane. In this first phase of the design, it is the wavefront slope in the exit pupil which should be minimised. In a later phase of the design, especially for highly corrected systems, the wavefront aberration itself should be minimised. To avoid an abrupt change in aberration representation during the design phase, it is recommended to use wavefront aberration throughout the process. In the early design phase with large aberrations present, the wavefront expansion is modified in such a way that each wavefront polynomial of a certain degree leads to minimum transverse aberration components in the image plane. The optimum shape of a wavefront such that it minimises the transverse aberration of a certain order in the image plane was first investigated by Lukosz **[223]**. Orthogonal expansions of the transverse aberration components were also studied by Kross **[197]**. The radial part of the wavefront polynomial of radial degree n and azimuthal order m according to Lukosz is given by

$$\begin{cases} L_n^m(\rho) = R_n^m(\rho) - R_{n-2}^m(\rho)\,, & (n \neq m)\,, \\ L_n^m(\rho) = R_n^m(\rho)\,, & (n = m)\,. \end{cases} \tag{5.9.29}$$

The proof of why a wavefront aberration term in the exit pupil corresponding to a Lukosz polynomial produces a minimum rms ray blur in the image plane can be found in **[46]**. In practice, an optical design method could first reduce the imaging aberration by means of a wavefront *slope* minimisation based on a Lukosz polynomial expansion of the wavefront in the exit pupil. If needed, the final 'diffraction-limited' performance of a system can be obtained by changing towards a wavefront expansion based on Zernike polynomials. A transition stage can be used in which the optimisation of the system is carried out with respect to both Lukosz and Zernike polynomials, each provided with an appropriate weighting factor.

Figure 5.78 shows the shape of the Lukosz polynomials L_4^0 and L_{12}^0 in comparison with the standard radial Zernike polynomials of the same order. A radial Zernike polynomial, with respect to its basic monomial expression of degree n, has been corrected by lower-order terms to flatten the wavefront over the unit disc such that the wavefront variance is minimised. The Lukosz-polynomials contain lower-order monomials with larger coefficients and, for this reason, they could be called 'overcorrected' with respect to the corresponding radial Zernike polynomial. In this context the term 'overcorrection' should not be confused with the term *overcorrection* (or undercorrection) of *spherical* aberration which is frequently used in the literature. For instance, according to this usage, a convex-plano positive lens, imaging from infinity, shows *undercorrected* spherical aberration (the marginal rays focus closer to the lens than the paraxial rays). In opposition to this linguistic usage, the 'overcorrection' of a general aberration in a Lukosz polynomial with respect to the corresponding Zernike polynomial of the same radial degree is the result of the minimisation of the geometric transverse blur in the image plane. A general wavefront expansion using the Lukosz polynomials reads

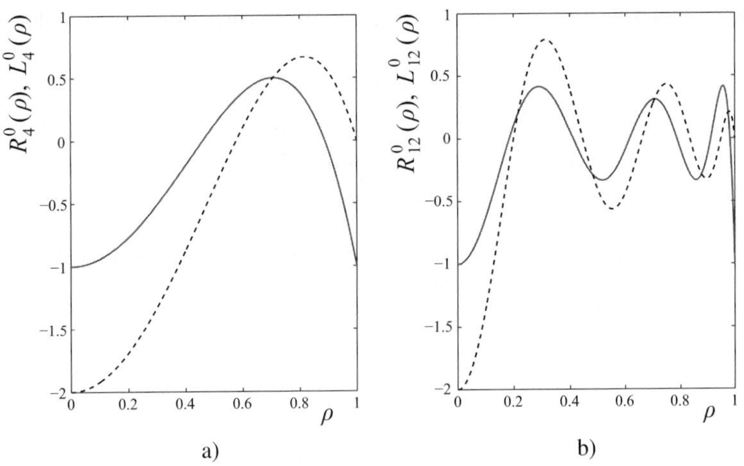

Figure 5.78: The Zernike polynomials $R_n^m(\rho)$ (solid curves) and Lukosz polynomials $L_n^m(\rho)$ (dashed curves) for minimising transverse ray aberration.
a) $R_4^0(\rho)$ and $L_4^0(\rho)$.
b) The corresponding 12th-order polynomials.

$$W(\rho,\theta) = \sum_{n=0}^{\infty}\sum_{m=0}^{\infty} L_n^m(\rho)\left\{a_{nm}\cos m\theta + b_{nm}\sin m\theta\right\} , \tag{5.9.30}$$

where $n-m$ is even and ≥ 0 and the coefficients (a_{nm}, b_{nm}) follow from the solution of a least-squares minimisation problem.

In a similar way as for the wavefront analysis in the preceding subsections (see Eq. (5.9.6)), we will adhere to the complex azimuthal notation because of its simpler manipulation. The complex expansion coefficients c_{nm} are defined by

$$W(\rho,\theta) = \sum_{n=0}^{\infty}\sum_{m=-\infty}^{m=+\infty} c_{nm} L_n^{|m|}(\rho)\exp(im\theta) . \tag{5.9.31}$$

The expressions for the transverse aberration components, expressed in diffraction units (see Eqs. (5.2.15) and (5.2.18)), are then written as

$$\begin{aligned}
\delta_x(\rho,\theta) &= \frac{1}{2}\left\{(\exp(i\theta)+\exp(-i\theta)\,\frac{\partial W}{\partial \rho} + \frac{i}{\rho}\,(\exp(i\theta)-\exp(-i\theta)\,\frac{\partial W}{\partial \theta}\right\} ,\\
\delta_y(\rho,\theta) &= \frac{1}{2}\left\{-\,i\,(\exp(i\theta)-\exp(-i\theta)\,\frac{\partial W}{\partial \rho} + \frac{1}{\rho}\,(\exp(i\theta)+\exp(-i\theta)\,\frac{\partial W}{\partial \theta}\right\} .
\end{aligned} \tag{5.9.32}$$

The substitution of the corresponding radial Zernike polynomials for each Lukosz polynomial contained in the W-derivatives above yields

$$\begin{aligned}
\delta_x(\rho,\theta) = \sum_{n,m}\frac{c_{nm}}{2}\Bigg\{ & e^{i(m+1)\theta}\left[\left(\frac{\partial R_n^{|m|}(\rho)}{\partial\rho} - \frac{m}{\rho}R_n^{|m|}(\rho)\right) - \left(\frac{\partial R_{n-2}^{|m|}(\rho)}{\partial\rho} - \frac{m}{\rho}R_{n-2}^{|m|}(\rho)\right)\right]\\
&+ e^{i(m-1)\theta}\left[\left(\frac{\partial R_n^{|m|}(\rho)}{\partial\rho} + \frac{m}{\rho}R_n^{|m|}(\rho)\right) - \left(\frac{\partial R_{n-2}^{|m|}(\rho)}{\partial\rho} + \frac{m}{\rho}R_{n-2}^{|m|}(\rho)\right)\right]\Bigg\} ,\\
\delta_y(\rho,\theta) = \sum_{n,m}\frac{ic_{nm}}{2}\Bigg\{ & -e^{i(m+1)\theta}\left[\left(\frac{\partial R_n^{|m|}(\rho)}{\partial\rho} - \frac{m}{\rho}R_n^{|m|}(\rho)\right) - \left(\frac{\partial R_{n-2}^{|m|}(\rho)}{\partial\rho} - \frac{m}{\rho}R_{n-2}^{|m|}(\rho)\right)\right]\\
&+ e^{i(m-1)\theta}\left[\left(\frac{\partial R_n^{|m|}(\rho)}{\partial\rho} + \frac{m}{\rho}R_n^{|m|}(\rho)\right) - \left(\frac{\partial R_{n-2}^{|m|}(\rho)}{\partial\rho} + \frac{m}{\rho}R_{n-2}^{|m|}(\rho)\right)\right]\Bigg\} .
\end{aligned} \tag{5.9.33}$$

The expressions between round brackets can be written as series of Zernike polynomials according to

$$\left(\frac{d}{d\rho}+\frac{m}{\rho}\right)R_n^m(\rho) = 2\sum_{k=0}^{(n-m)/2}(n-2k)\,R_{n-2k-1}^{m-1}(\rho)\,,$$
$$\left(\frac{d}{d\rho}-\frac{m}{\rho}\right)R_n^m(\rho) = 2\sum_{k=0}^{(n-m)/2-1}(n-2k)\,R_{n-2k-1}^{m+1}(\rho)\,. \qquad (5.9.34)$$

The first result was originally given by Nijboer in his thesis **[259]**, the second expression is derived in **[46]**, where in both cases it was tacitly assumed that $m \geq 0$. A more exhaustive proof, kindly communicated to the first author by Dr. A.J.E.M. Janssen, is given in Appendix D.4. It is shown there that the expressions above are equally valid for $m < 0$ by writing

$$\frac{d}{d\rho}R_n^{|m|}(\rho) \pm \frac{m}{\rho}R_n^{|m|}(\rho) = 2\sum_{k=0}^{(n-|m|)/2}(n-2k)\,R_{n-2k-1}^{|m\mp1|}(\rho)\,, \qquad (5.9.35)$$

where the common requirements on the lower and upper indices of the radial polynomials apply, viz. $n-2k-1-|m\mp1|$ even and non-negative.

We can insert the result of Eq. (5.9.35) in the expressions of Eq. (5.9.33) and obtain the final expressions,

$$\delta_x(\rho,\theta) = \sum_{n,m} nc_{nm}\left\{-e^{i\theta}R_{n-1}^{|m+1|}(\rho)+e^{-i\theta}R_{n-1}^{|m-1|}(\rho)\right\}e^{im\theta}\,,$$
$$\delta_y(\rho,\theta) = \sum_{n,m} inc_{nm}\left\{e^{i\theta}R_{n-1}^{|m+1|}(\rho)+e^{-i\theta}R_{n-1}^{|m-1|}(\rho)\right\}e^{im\theta}\,. \qquad (5.9.36)$$

In deriving these expressions for the transverse aberration components we have used the result

$$\left(\frac{\partial R_n^{|m|}(\rho)}{\partial\rho}-\frac{m}{\rho}R_n^{|m|}(\rho)\right)-\left(\frac{\partial R_{n-2}^{|m|}(\rho)}{\partial\rho}-\frac{m}{\rho}R_{n-2}^{|m|}(\rho)\right) = 2\sum_{k=0}^{(n-|m|)/2-1}(n-2k)R_{n-2k-1}^{|m+1|}(\rho)-2\sum_{k=0}^{(n-|m|)/2-2}(n-2k)R_{n-2k-3}^{|m+1|}(\rho)$$
$$= 2nR_{n-1}^{|m+1|}(\rho)\,. \qquad (5.9.37)$$

A comparable derivation is available for the Zernike polynomials with azimuthal index $|m-1|$ in Eq. (5.9.36). The proof that the variance of the transverse aberration components, $\overline{\delta^2} = \overline{\delta_x^2(\rho,\theta)+\delta_y^2(\rho,\theta)}$, is minimised by choosing the Lukosz polynomials $L_n^m(\rho)$ for the wavefront expansion can be found in **[46]**.

It is important to remember that in Eq. (5.2.17) the wavefront aberration was expressed in units of λ_0, the (vacuum) wavelength of the light. According to Eq. (5.2.18), the blur is then expressed in diffraction units λ_0/NA. To obtain the blur in real-space coordinates, the blur components of Eq. (5.9.36) have to be multiplied by λ_0/NA. If the wavefront aberration was initially expressed in μm, the real-space blur value is obtained after division by the numerical aperture *NA* of the imaging pencil.

If an imaging system operates relatively far away from the diffraction limit, the quantity of interest for judging the quality of that system is the geometrical root-mean-square blur δ in the image plane. To obtain this quantity, we separately calculate the variances of δ_x and δ_y and then take the square root of their sum. For the variance of δ_x we write

$$\overline{|\delta_x|^2} = \frac{1}{\pi}\int_0^1\int_0^{2\pi}\delta_x(\rho,\theta)\,\delta_x^*(\rho,\theta)\,\rho d\rho d\theta\,. \qquad (5.9.38)$$

Insertion of the expression for δ_x of Eq. (5.9.36) into Eq. (5.9.38) yields the expression

$$\overline{|\delta_x|^2} = \frac{1}{\pi}\sum_n\sum_{n'}\sum_m\sum_{m'}\int_0^1\int_0^{2\pi} nn'c_{nm}c_{n'm'}^*\exp[i(m-m')\theta]$$
$$\times\left[R_{n-1}^{|m+1|}(\rho)R_{n'-1}^{|m'+1|}(\rho)+\exp(2i\theta)R_{n-1}^{|m+1|}(\rho)R_{n'-1}^{|m'-1|}(\rho)\right.$$
$$\left.+R_{n-1}^{|m-1|}(\rho)R_{n'-1}^{|m'-1|}(\rho)+\exp(-2i\theta)R_{n-1}^{|m-1|}(\rho)R_{n'-1}^{|m'+1|}(\rho)\right]\rho d\rho d\theta\,. \qquad (5.9.39)$$

Using the orthogonal properties of the radial Zernike polynomials and the value of their inner product we obtain, after some rearrangement,

$$\overline{|\delta_x|^2} = \sum_n n \left\{ \sum_{0\leq|m+1|\leq n-1} c_{nm} \left(c_{nm} + c_{n,m+2}\right)^* + \sum_{0\leq|m-1|\leq n-1} c_{nm} \left(c_{nm} + c_{n,m-2}\right)^* \right\} , \tag{5.9.40}$$

where the positive and negative values of m that contribute to the two summations over m are determined by the conditions below the summation sign.

In a similar way we obtain the expression for the blur in the y-direction,

$$\overline{|\delta_y|^2} = \sum_n n \left\{ \sum_{0\leq|m+1|\leq n-1} c_{nm} \left(c_{nm} - c_{n,m+2}\right)^* + \sum_{0\leq|m-1|\leq n-1} c_{nm} \left(c_{nm} - c_{n,m-2}\right)^* \right\} . \tag{5.9.41}$$

To obtain the variance of the transverse aberration components, we use the results of Eqs. (5.9.40) and (5.9.41) and subtract the squared average values, $|\overline{\delta_x}|^2$ and $|\overline{\delta_y}|^2$, given by $|c_{1,-1} + c_{1,1}|^2$ and $|-c_{1,-1} + c_{1,1}|^2$, respectively. The expression for the blur variance is then given by

$$\begin{aligned} \delta^2 &= \overline{|\delta_x|^2} - |\overline{\delta_x}|^2 + \overline{|\delta_y|^2} - |\overline{\delta_y}|^2 \\ &= \sum_n 2n \left\{ \sum_{0\leq|m+1|\leq n-1} \left|c_{nm}\right|^2 + \sum_{0\leq|m-1|\leq n-1} \left|c_{nm}\right|^2 \right\} - 2\left\{|c_{1,-1}|^2 + |c_{1,1}|^2\right\} . \end{aligned} \tag{5.9.42}$$

The summations over m are performed for all positive and negative m-values, including $m = 0$, that satisfy the condition applying to each summation sign.

We rearrange the expression above in the following way,

$$\begin{aligned} \delta^2 &= \sum_n 2n \left\{ \sum_{m=-n}^{n-2} |c_{nm}|^2 + \sum_{m=-n+2}^{n} \left|c_{nm}\right|^2 \right\} - 2\left\{|c_{1,-1}|^2 + |c_{1,1}|^2\right\} \\ &= \sum_n 2n \left\{ \sum_{m=-n+2}^{n-2} 2|c_{nm}|^2 + |c_{n,-n}|^2 + |c_{n,n}|^2 \right\} - 2\left\{|c_{1,-1}|^2 + |c_{1,1}|^2\right\} \\ &= 2\sum_{n=2}^{N} n \left\{ \sum_{m=-n+2}^{n-2} 2|c_{nm}|^2 + |c_{n,-n}|^2 + |c_{n,n}|^2 \right\} , \end{aligned} \tag{5.9.43}$$

where N is the maximum radial degree of the wavefront expansion.

The blur variance δ^2 can also be expressed in terms of the coefficients of a cosine/sine expansion of the wavefront aberration function with coefficients a_{nm}, b_{nm}, for $m \geq 0$. From Eq. (5.9.9) we have

$$c_{nm} = \frac{a_{nm}}{2} - i\frac{b_{nm}}{2} , \quad (m > 0), \qquad c_{n,-m} = \frac{a_{nm}}{2} + i\frac{b_{nm}}{2} , \quad (m < 0), \tag{5.9.44}$$

where $c_{n0} = a_{n0}$ and $b_{n0} = 0$ for the special case $m = 0$.

It then immediately follows that even for complex-valued a_{nm} and b_{nm},

$$\delta^2 = \sum_{n=2}^{N} n \left\{ 4|a_{n0}|^2 + 2\sum_{m=1}^{n-2} \left(|a_{nm}|^2 + |b_{nm}|^2\right) + |a_{nn}|^2 + |b_{nn}|^2 \right\} , \tag{5.9.45}$$

where we have used that, for $m > 0$,

$$|c_{n,\pm m}|^2 = \frac{|a_{nm}|^2}{4} + \frac{|b_{nm}|^2}{4} \mp \frac{1}{2}\Im\{a_{nm} b^*_{nm}\} . \tag{5.9.46}$$

The result is a weighted addition of squares of the coefficients a_{nm} and b_{nm}. Equations (5.9.43) and (5.9.45) show that an expansion with Lukosz wavefront polynomials leads to an orthogonal decomposition of the transverse aberration components in the image plane. The variance of the geometrical image blur is given by a weighted sum of the squared moduli of the coefficients of the Lukosz polynomials in the wavefront expansion.

5.9.3.2 Field-averaged Geometrical Blur

A measure for the quality of an optical instrument, operating relatively far away from the diffraction limit, is the geometrical blur value that results from a quadratic averaging over the (circular) image field. To calculate such an average blur value we expand the wavefront data as a function of aperture and field coordinates in terms of Lukosz polynomials of the polar pupil coordinates (ρ, θ) and Zernike polynomials of the polar image plane coordinates (r, ϕ) according to

$$W(\rho,\theta,r,\phi) = \sum_{n_1=0}^{N_1} \sum_{m_1=-N_1}^{+N_1} \sum_{n_2=0}^{N_2} \sum_{m_2=-N_2}^{+N_2} c_{n_1 m_1 n_2 m_2}\, L_{n_1}^{|m_1|}(\rho)\, \exp(im_1\theta)\, R_{n_2}^{|m_2|}(r)\exp(im_2\phi)\,, \qquad (5.9.47)$$

where the index pairs (n_1, m_1) and (n_2, m_2) are subjected to the conditions $n \geq 0$ and $n - |m| \geq 0$ and even. We remark that the Lukosz polynomials do not form an orthogonal set on the unit disc. The inner products of the polynomials with lower indices n and $n \pm 2$ and identical upper index are nonzero. The coefficients $c_{n_1 m_1 n_2 m_2}$ of the wavefront expansion according to Eq. (5.9.47) thus cannot be obtained by the method of inner product calculation but follow from a least-squares fit of the wavefront data over the pupil area with the optimum set of coefficients c.

The variance of the transverse aberration components at a general point (r, ϕ) of the image field follows from Eq. (5.9.43) and is given by

$$\overline{\delta^2(r,\phi)}=2\sum_{n_1=2}^{N_1} n_1 \left\{ 2\left|\sum_{n_2=0}^{N_2}\sum_{m_2=-N_2}^{+N_2}\sum_{m_1=-n_1+2}^{n_1-2} c_{n_1 m_1 n_2 m_2} R_{n_2}^{|m_2|}(r)\exp(im_2\phi)\right|^2 + \left|\sum_{n_2=0}^{N_2}\sum_{m_2=-N_2}^{+N_2} c_{n_1,-n_1,n_2,m_2} \right.\right.$$

$$\left.\left. \times R_{n_2}^{|m_2|}(r)\exp(im_2\phi)\right|^2 + \left|\sum_{n_2=0}^{N_2}\sum_{m_2=-N_2}^{+N_2} c_{n_1 n_1 n_2 m_2} R_{n_2}^{|m_2|}(r)\exp(im_2\phi)\right|^2\right\}. \qquad (5.9.48)$$

The field averaged variance δ^2 over the circular image field is obtained from

$$\delta^2 = \frac{1}{\pi}\int_0^{2\pi}\int_0^1 \overline{\delta^2(r,\phi)}\; rdrd\phi\,. \qquad (5.9.49)$$

Exploiting the orthogonality properties of the Zernike polynomials, we readily obtain

$$\delta^2 = \sum_{n_2=0}^{N_2}\left(\frac{2}{n_2+1}\right)\sum_{m_2=-N_2}^{+N_2}\sum_{n_1=2}^{N_1} n_1 \left\{\sum_{m_1=-n_1+2}^{n_1-2} 2\left|c_{n_1 m_1 n_2 m_2}\right|^2 + \left|c_{n_1,-n_1,n_2,m_2}\right|^2 + \left|c_{n_1 n_1 n_2 m_2}\right|^2\right\}. \qquad (5.9.50)$$

By choosing Lukosz polynomials for the pupil wavefront expansion and Zernike polynomials for the expansion with respect to the field coordinates, the weighting functions associated with the expansion are uniform over the circular pupil and field areas. Other polynomials for the expansion with respect to the field coordinates are possible. For a square or rectangular field, Legendre polynomials would be appropriate, with uniform weighting over the image field. Chebyshev polynomials could be used for the case when the outer image field region should be weighted more heavily than the central part.

5.9.3.3 Some Examples of Low-order Field-dependent Aberration Patterns

In Fig. 5.79 we show some transverse aberration patterns which can occur in an optical imaging system without any particular symmetry property. The lack of symmetry in the optical system can be intentional, for instance in the case of optical systems that were designed using so-called 'free-form' surfaces. Unintentionally, a lack of symmetry occurs in the manufacturing phase of a lens or mirror system. Mechanical misalignments such as decentring and tilt errors of lens elements or mirror surfaces, cylindrical and higher-order figuring errors on individual optical surfaces, inhomogeneities in the refractive index distribution in a lens body, etc. destroy the as-designed rotational symmetry of the optical system. For small deviations from the rotational invariance of the optical system, symmetry with respect to a set of planes, all intersecting the former optical axis, can be observed.

In the examples which follow we focus on the low-order aberrations distortion and astigmatism, with field dependencies up to the third order. In Fig. 5.79a–i, the effect of a shift of each image point in the x-direction has been shown. Such a shift can be present in an imaging system that lacks rotational symmetry. The aperture-dependent part of the aberration function is given by $\rho\cos\theta$ and this function has been combined with the first-, second- and third-order

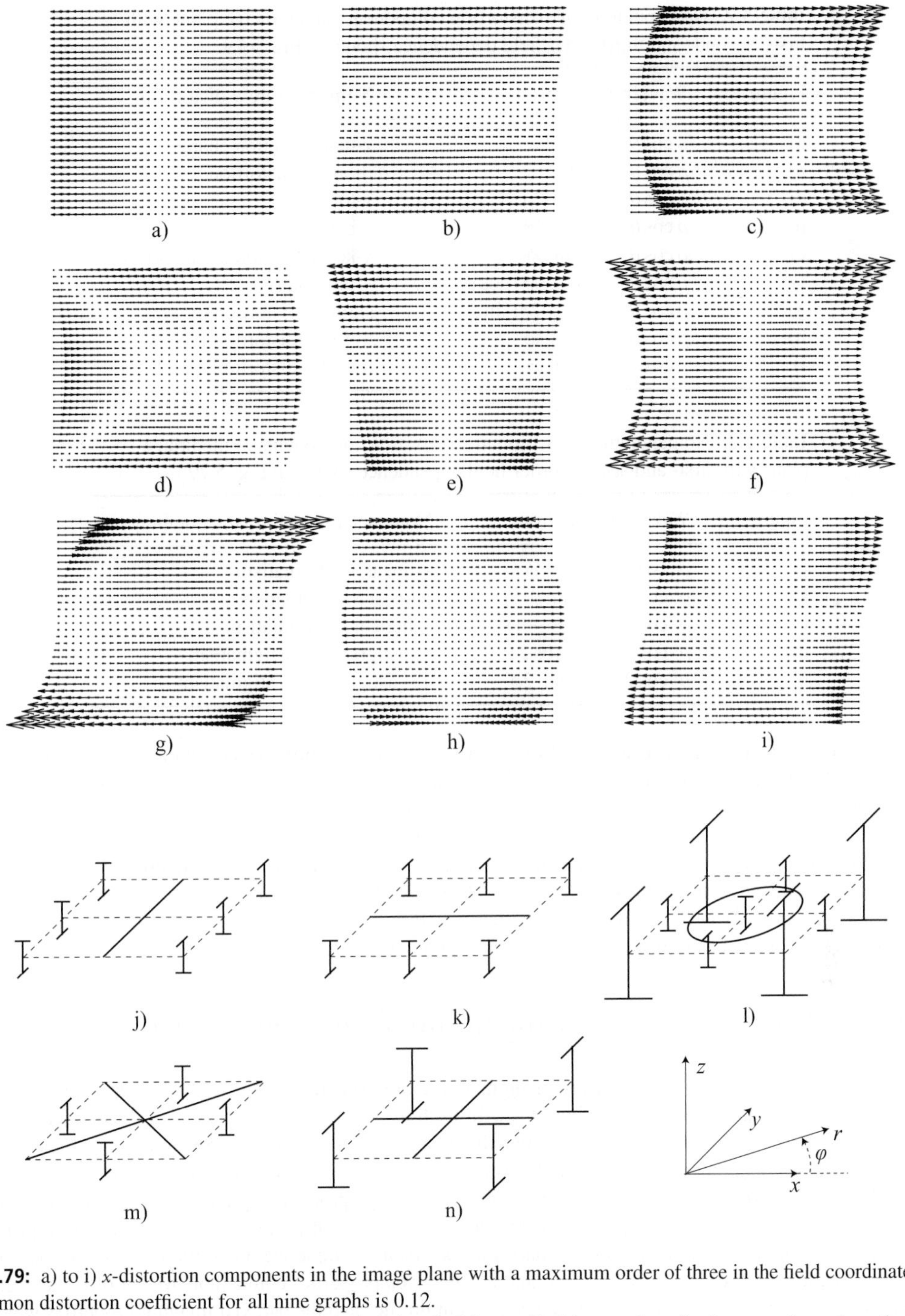

Figure 5.79: a) to i) x-distortion components in the image plane with a maximum order of three in the field coordinates (r, ϕ). The common distortion coefficient for all nine graphs is 0.12.
j) to n) Drawings of the astigmatic lines in the neighbourhood of the nominal image plane for the case of x-astigmatism with a maximum degree of two in the field coordinates (r, ϕ). For easy comparison of the five graphs, all astigmatic coefficients have been given the same (positive) value.
The image plane is a square with sides of length 2 in normalised units.

Table 5.2. Nomenclature of x-distortion components in the image plane with first-, second- and third-order field dependence shown in Fig. 5.79.

Figure	Wavefront function		Nomenclature x-distortion
a)	$\rho \cos\theta$	$r \cos\phi$	x-magnification change
b)	$\rho \cos\theta$	$r \sin\phi$	rhomb x-distortion
c)	$\rho \cos\theta$	$(2r^2 - 1)$	x-torsion
d)	$\rho \cos\theta$	$r^2 \cos 2\phi$	x-bending
e)	$\rho \cos\theta$	$r^2 \sin 2\phi$	x-keystone (wedge distortion)
f)	$\rho \cos\theta$	$(3r^3 - 2r) \cos\phi$	x-pincushion
g)	$\rho \cos\theta$	$(3r^3 - 2r) \sin\phi$	sigma x-distortion
h)	$\rho \cos\theta$	$r^3 \cos 3\phi$	trefoil x-distortion type A
i)	$\rho \cos\theta$	$r^3 \sin 3\phi$	trefoil x-distortion type B

Table 5.3. Nomenclature of astigmatic aberration components in the image plane with first- and second-order field dependence shown in Fig 5.79.

Figure	Wavefront function		Nomenclature astigmatic component
j)	$\rho^2 \cos 2\theta$	$r \cos\phi$	x-linear x-astigmatism
k)	$\rho^2 \cos 2\theta$	$r \sin\phi$	y-linear x-astigmatism
l)	$\rho^2 \cos 2\theta$	$(2r^2 - 1)$	spherical x-astigmatism
m)	$\rho^2 \cos 2\theta$	$r^2 \cos 2\phi$	0° astigmatic x-astigmatism
n)	$\rho^2 \cos 2\theta$	$r^2 \sin 2\phi$	45° astigmatic x-astigmatism

polynomials of the field coordinates. The image blur is zero but the image shift leads to characteristic distortion patterns that are described in Table 5.2. The nomenclature of distortion patterns is used to specify the tolerances on various types of residual distortion in optical systems with tight specifications on geometrical image fidelity. Examples of such systems are high-resolution cameras for imaging from space and projection systems for optical lithography.

Other examples of aperture- and field-dependent aberrations of a general nature which are present in an optical system without rotational symmetry are given in Fig. 5.79j–n. The field dependence of the wavefront aberration is given in Table 5.3 leading to the typical astigmatic image patterns of Fig. 5.79. The aperture part of the wavefront aberration relates to astigmatism and the astigmatic lines are oriented along the x- and y-axis.

5.10 Paraxial and Finite Ray-tracing in Inhomogeneous Media

A well-known example of an inhomogeneous medium is the atmosphere, in which local temperature and pressure variations give rise to changes in the refractive index. In such a medium with a gradient in the refractive index, ray propagation is not along straight lines but a ray path exhibits a (small) curvature (see Section 4.2.4). In solid materials such as optical glasses, inhomogeneity is generally considered as a defect to be avoided as much as possible. It gives rise to small angular ray deviations and leads to blurring in the image produced by an optical system. In some optical elements, index inhomogeneity with a prescribed profile is introduced on purpose, to achieve imaging properties in certain geometries that are not feasible with standard means. As an example we quote the inhomogeneous rod lens, a simple element with flat entrance and exit surface that can be given properties comparable to that of a classical positive lens. In this section we first discuss the various index gradient profiles and the paraxial imaging properties which follow from such characteristic gradient profiles. The final subject of this section is the tracing of finite rays and the calculation of optical pathlength along curved ray paths in inhomogeneous media.

5.10.1 Index Profiles

Basic refractive index profiles in inhomogeneous media are the following:

- *Axial gradient*
  The refractive index only depends on the axial coordinate z of the optical element. The index profile is represented by, for instance, a power series expansion according to

$$n(x,y,z) = n_0(1 + \alpha_1 z + \alpha_2 z^2 + \cdots)\,. \tag{5.10.1}$$

  The production of a slab of inhomogeneous material with an axial index gradient is based on the stacking of a number of homogeneous plane-parallel plates with different refractive indices. By means of a heat treatment, the diffusion tails of the individual plates into their neighbours leads to a smooth index profile in the direction perpendicular to the initial interfaces. The end product is an inhomogeneous medium in which virtually all traces of the initial index jumps have disappeared.
- *Radial gradient*
  The refractive index is a function of only the lateral coordinate $r = (x^2 + y^2)^{1/2}$ with a power series expansion

$$n(x,y,z) = n_0\left(1 - \frac{g^2}{2}r^2 + h_4 r^4 + h_6 r^6 + \cdots\right)\,. \tag{5.10.2}$$

  Equation (5.10.2) is in line with the common notation for a radial gradient material in which the g^2-coefficient with the minus sign imposes a positive lens power to a thin slab of such a material. For a more general description of a radial profile, the coefficient $-g^2$ is replaced by $2h_2$. The production of materials with a radial gradient is done with the aid of diffusion at high temperature of low-index material into glass rods with higher index that are immersed in a heated bath of such low-index glassy material. After a long diffusion process, the end product is a glass rod with a mostly parabolic lateral index profile. The advantage of this production process is that it can produce many lens rods in parallel. Its limitation is found in the maximum feasible diameter of the lens rods. The manufacturing of elements with a diameter in excess of a few centimetres becomes technologically impractical.
- *Spherical gradient*
  The refractive index is a function of the distance to some centre of symmetry C with coordinates $\mathbf{r}_0 = (x_0, y_0, z_0)$. The index profile is represented by the expression

$$n(x,y,z) = n_0\left\{1 + a_2\left|\mathbf{r} - \mathbf{r}_0\right|^2 + \cdots\right\}\,, \tag{5.10.3}$$

  where $\mathbf{r}$ is the position vector of a general point P_0 in the gradient-index medium.
  Such a profile can be obtained when glass spheres are immersed in a heated bath of a medium with different index. The geometrical centre of the glass sphere becomes the centre of symmetry of the index distribution.

5.10.2 Paraxial Properties of a Gradient-index Medium

The ray equation in an inhomogeneous medium was derived in Section 4.2.3 (see Eq. (4.2.40)),

$$\frac{d}{ds}\left(n(\mathbf{r})\frac{d\mathbf{r}}{ds}\right) = \nabla n(\mathbf{r})\,, \tag{5.10.4}$$

where the distance ds along the curved ray path is given by

$$ds = dz\sqrt{1 + \left(\frac{dx}{dz}\right)^2 + \left(\frac{dy}{dz}\right)^2}\,. \tag{5.10.5}$$

For paraxial light propagation with the z-axis as reference axis, the squares of the quantities dx/dz and dy/dz can be neglected and Eq. (4.2.40) reduces to the vector equation

$$\frac{d}{dz}\left(n(\mathbf{r})\frac{d\mathbf{r}}{dz}\right) = \nabla n(\mathbf{r})\,. \tag{5.10.6}$$

From this equation we derive two differential equations for the x- and y-coordinates of the ray path,

$$n\frac{d^2x}{dz^2} + \frac{\partial n}{\partial z}\frac{dx}{dz} - \frac{\partial n}{\partial x} = 0\,,$$

$$n\frac{d^2y}{dz^2} + \frac{\partial n}{\partial z}\frac{dy}{dz} - \frac{\partial n}{\partial y} = 0 . \tag{5.10.7}$$

The principal gradient types, axial and radial, lead to the solutions below for the curved ray path in the inhomogeneous medium.

5.10.2.1 Axial Gradient

The ray path in a medium with axial gradient has been sketched in Fig. 5.80. At the point of entry into the inhomogeneous medium, the ray angle with the normal to the surface of the inhomogeneous glass plate is θ_0 (the index of the surrounding medium is n_0). For this special case, with equal index on both sides of the surface, the ray continues without refraction in the inhomogeneous medium. The curved ray path in the medium is dashed in the figure. The exit angle with the normal to the plane-parallel plate is also θ_0 (refractive index final medium is n_0). The paraxial angles and the refractive index variations have been highly exaggerated in the figure. For the x-dependent equation we write

$$\frac{d}{dz}(nu_x) = 0 , \tag{5.10.8}$$

where we have replaced dx/dz by the paraxial ray angle u_x. The solution of Eq. (5.10.8) imposes a constant value of nu_x in the axial-gradient medium. Using the initial value for u_x at $z = 0$, we have $nu_x = n_0u_0$. The invariance of the paraxial optical ray angle nu_x follows from the argument that the axial-gradient medium can be thought of as a stack of plane-parallel plates with varying refractive indices. Snell's law requires a constant value of the product $n \sin\theta$ from layer to layer. This requirement implies that the property $n(z)u_x(z) = n_0u_0$ for a ray in an axial-gradient material is also valid outside the paraxial domain, yielding the more general property $n(z)L(z) = n_0L_0$. This result also follows directly for finite ray angles from the basic ray equation (5.10.4).

The lateral coordinate x of a paraxial ray follows from the differential equation $n(z)dx/dz = n_0u_0$ where the boundary condition is given by $x = x_0$ in the plane $z = 0$. A general solution of this equation reads

$$x = n_0u_0 \int_0^z \frac{dt}{n(t)} . \tag{5.10.9}$$

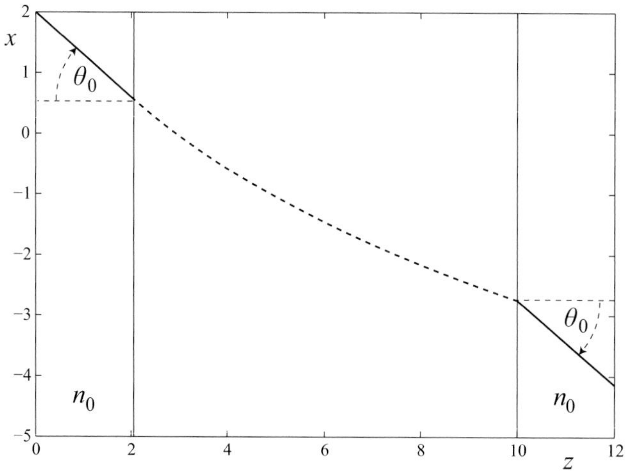

Figure 5.80: The ray path in an inhomogeneous medium with an axial gradient (z-direction in the figure) in the refractive index function.

If we retain the first two expansion terms in Eq. (5.10.1), a solution of the differential equation is given by

$$x = \frac{u_0}{\sqrt{\alpha_1^2 - 4\alpha_2}} \ln \left\{ \frac{\left(\alpha_1 + \sqrt{\alpha_1^2 - 4\alpha_2}\right) z + 2}{\left(\alpha_1 - \sqrt{\alpha_1^2 - 4\alpha_2}\right) z + 2} \right\} + x_0 \,. \tag{5.10.10}$$

If the axial gradient is sufficiently approximated by the first linear expansion term, we obtain the commonly used paraxial expression,

$$x = \frac{u_0}{\alpha_1} \ln\left(1 + \alpha_1 z\right) + x_0 \,. \tag{5.10.11}$$

Comparable solutions follow for the optical ray angle nu_y and for the y-coordinate on the ray path.

In terms of the matrix representation of a paraxial optical system (see Section 4.10.4, Eq. (4.10.12)), we have from Eq. (5.10.11) for a material thickness d,

$$\begin{pmatrix} nu \\ x \end{pmatrix} = \begin{pmatrix} B & -A \\ -D & C \end{pmatrix} \begin{pmatrix} n_0 u_0 \\ x_0 \end{pmatrix} = \begin{pmatrix} 1 & 0 \\ \dfrac{\ln(1 + \alpha_1 d)}{\alpha_1 n_0} & 1 \end{pmatrix} \begin{pmatrix} n_0 u_0 \\ x_0 \end{pmatrix} . \tag{5.10.12}$$

Expanding the logarithmic function up to the second order yields the result

$$\begin{pmatrix} nu \\ x \end{pmatrix} = \begin{pmatrix} 1 & 0 \\ \dfrac{d}{n_0}\left(1 - \dfrac{\alpha_1 d}{2}\right) & 1 \end{pmatrix} \begin{pmatrix} n_0 u_0 \\ x_0 \end{pmatrix} . \tag{5.10.13}$$

Equation (5.10.13) shows that in the paraxial approximation the axial-gradient material with thickness d can be replaced by a homogeneous material with the same thickness and a refractive index that is approximated by $n = n_0(1 + \alpha_1 d/2)$. The use of an optical material with an axial gradient might seem superfluous because there is no change in optical power. But the presence of an optical element with axial gradient can change the aberrations of the optical system. In that sense, the absence of a change in paraxial power and magnification can be considered to be an advantage because it allows for a separate treatment of the general paraxial layout of the optical system and the reduction of its aberrations.

5.10.2.2 Radial Gradient

The paraxial ray equation for the x-coordinate of a ray in a material with only a radial gradient is given by

$$\frac{d^2 x}{dz^2} + g^2 x = 0 \,. \tag{5.10.14}$$

Using the boundary conditions $u = u_{0,x}$ and $x = x_0$ in the plane $z = 0$ we find the solution

$$\begin{aligned} x(z) &= x_0 \cos(gz) + \frac{u_{0,x}}{g} \sin(gz) \,, \\ u_x(z) &= -x_0 g \sin(gz) + u_{0,x} \cos(gz) \,. \end{aligned} \tag{5.10.15}$$

Comparable equations result for the y-coordinate and the paraxial angle u_y of a ray,

$$\begin{aligned} y(z) &= y_0 \cos(gz) + \frac{u_{0,y}}{g} \sin(gz) \,, \\ u_y(z) &= -y_0 g \sin(gz) + u_{0,y} \cos(gz) \,. \end{aligned} \tag{5.10.16}$$

Combining Eqs. (5.10.15) and (5.10.16) yields the expressions

$$\begin{aligned} x(z) &= \sqrt{x_0^2 + u_{0,x}^2 g^{-2}} \cos(gz + \phi_x) \,, \\ y(z) &= \sqrt{y_0^2 + u_{0,y}^2 g^{-2}} \cos(gz + \phi_y) \,, \\ u_x(z) &= -g \sqrt{x_0^2 + u_{0,x}^2 g^{-2}} \sin(gz + \phi_x) \,, \end{aligned} \tag{5.10.17}$$

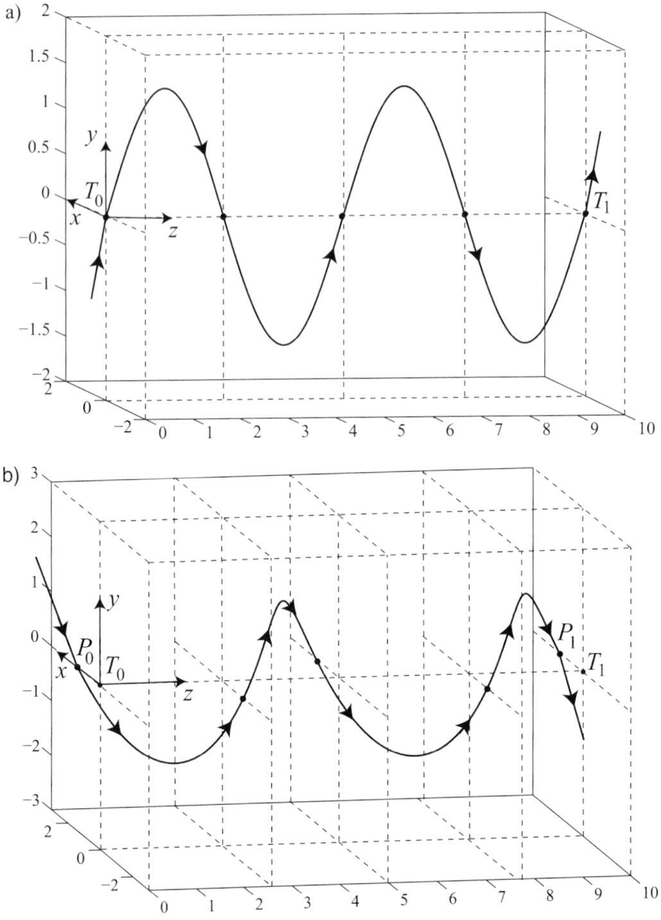

Figure 5.81: The ray path in an inhomogeneous medium with a radial gradient in the refractive index function with symmetry axis T_0T_1.
a) The ray intersects the axis of symmetry (meridional ray) at an angle $u_{0,x} = 0$ and $u_{0,y} = u_0$ and follows an oscillating path in the plane $x = 0$ with period (or pitch) $p_z = 2\pi/g$.
b) At the off-axis entry point $P_0(x_0, 0, 0)$, the ray is refracted into the inhomogeneous medium with the same paraxial angles $u_{0,x} = 0$, $u_{0,y} = u_0$. In this case, the ray path has a helical shape with a pitch $p_z = 2\pi/g$.
In both graphs, the paraxial angles have been made very large for the sake of clarity.

$$u_y(z) = -g\sqrt{y_0^2 + u_{0,y}^2 g^{-2}}\,\sin(gz + \phi_y)\,, \tag{5.10.18}$$

$$\tan(\phi_x) = -\frac{u_{0,x}}{g\,x_0}\,, \qquad \tan(\phi_y) = -\frac{u_{0,y}}{g\,y_0}\,. \tag{5.10.19}$$

In Fig. 5.81 we show ray paths in radial-gradient media. In a) the ray intersects the axis of symmetry of the radial index profile. Its paraxial angle with the y-axis is nonzero and the ray starts an oscillating path in the plane $x = 0$. The period in the z-direction of the oscillation follows from Eq. (5.10.19) and is given by $p_z = 2\pi/g$ (p_z is called the pitch of the radial-gradient material). In b), a skew ray is launched off-axis into the medium at the point $P_0(x_0, 0, 0)$ and it starts a spiralling path in the inhomogeneous medium. The paraxial ray propagates in a medium with symmetry of revolution and has to satisfy the skew ray invariant of Eq. (4.8.15), in this case given by the expression

$$n_0(y_0u_{0,x} - x_0u_{0,y}) = n_1(y_1u_{1,x} - x_1u_{1,y})\,, \tag{5.10.20}$$

where the subscripts 0 and 1 refer to two arbitrary points on the curved ray path.

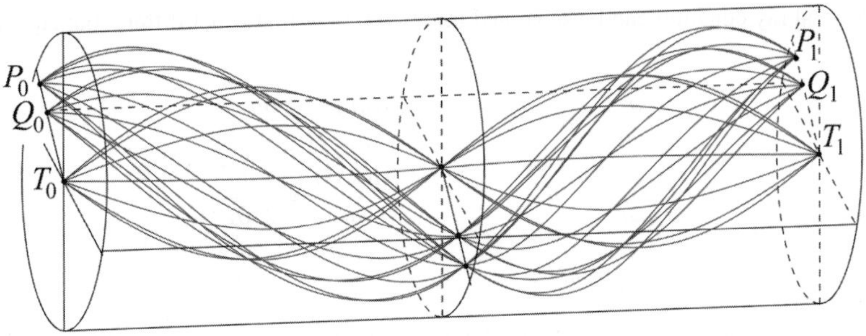

Figure 5.82: Imaging with a radial-gradient lens rod with length $T_0T_1 = 2\pi/g$ (magnification +1). At a distance $d = \pi/g$ from the entrance surface, an intermediate image is formed with a lateral magnification of –1.

The paraxial matrix formalism applied to the ray propagation in a radial-gradient medium with thickness d leads to the following expressions for the matrix elements

$$\begin{pmatrix} B & -A \\ -D & C \end{pmatrix} = \begin{pmatrix} \cos(gd) & -n_0 g \sin(gd) \\ \frac{d}{n_0} \frac{\sin(gd)}{gd} & \cos(gd) \end{pmatrix} . \tag{5.10.21}$$

All matrix elements are affected by a radial-gradient profile. The optical power coefficient shows a periodic dependence on the medium thickness as do the diagonal elements which determine the angular and lateral magnification. Some interesting special cases emerge:

- $d = \pi/(2g)$.
  The thickness corresponds to a quarter pitch ($p_z/4$) of the axial or helical path in Fig. 5.81. If a parallel beam is incident on a plane-parallel plate with this thickness, the focus is formed exactly on the back surface of the plate.
- $d = \pi/g$.
  The glass rod has zero optical power and a lateral magnification of -1. It can be used as a telescope with the entrance and exit pupil at the entrance and exit face of the glass rod.
- $d = 2\pi/g$.
  The glass rod has zero optical power and a lateral magnification of $+1$. The rod can be used as a unit-magnification telescope. Another application exploits the fact that the entrance and exit face of the rod are sharply imaged onto each other with magnification $+1$. The radial-gradient-index rod can be used as an *image relay* element. The imaging with object and image at the location of the entrance and exit faces of the lens rod is illustrated in Fig. 5.82. Three object points on a single line have been taken, the central point T_0, an outer field point P_0 and an intermediate object point Q_0 where T_0Q_0 is 0.7 times the distance T_0P_0.

5.10.3 Finite Ray-tracing in an Inhomogeneous Medium

When tracing finite rays in an optical system, the first step to be executed when encountering an inhomogeneous medium is the refraction into the medium. In the geometrical optics approximation, it is adequate to calculate the local refractive index in the inhomogeneous medium at the point of intersection of the ray and to apply Snell's law (see Eq. (5.7.19)). The further calculation of the curved ray trajectory needs the exact solution of the general ray equation (5.10.4) where s is the distance parameter along the ray. We use the substitution $ds = n dt$ yielding the following relation between the differential quotients with respect to s and t,

$$\frac{d}{ds} = \frac{1}{n}\frac{d}{dt} . \tag{5.10.22}$$

The second-order general ray equation can be separated into two sets of first-order differential equations with respect to the new parameter t,

$$\begin{aligned} \frac{dx}{dt} &= nL\,, & \frac{d(nL)}{dt} &= \frac{1}{2}\frac{\partial n^2}{\partial x}\,, \\ \frac{dy}{dt} &= nM\,, & \frac{d(nM)}{dt} &= \frac{1}{2}\frac{\partial n^2}{\partial y}\,, \\ \frac{dz}{dt} &= nN\,, & \frac{d(nN)}{dt} &= \frac{1}{2}\frac{\partial n^2}{\partial z}\,, \end{aligned} \qquad (5.10.23)$$

where (L, M, N) are the direction cosines of the tangent unit vector on the finite ray. The appropriate boundary conditions are the starting point and starting direction at a general point in the medium from where we have to perform the integration of the six differential equations. The right-hand parts of the second set of differential equations contain the function $n^2(x, y, z)$. Therefore, in the case of a medium with radial gradient, the power series expansion of the refractive index according to Eq. (5.10.2) is preferably given in the form

$$n^2(x, y, z) = n_0^2\left(1 - g^2 r^2 + h_4' r^4 + h_6' r^6 \cdots\right)\,, \qquad (5.10.24)$$

where the z-axis is the axis of symmetry and $r^2 = x^2 + y^2$. To keep track of the optical pathlength S along the curved ray, a seventh equation is conveniently added,

$$\frac{dS}{dt} = n^2\,. \qquad (5.10.25)$$

The seven simultaneous differential equations can be numerically solved with the aid of a Runge-Kutta integration scheme. Some careful interpolation is needed once the exit surface of the inhomogeneous medium is approached. When the position of the intersection point has been determined, with an accuracy of a fraction of the wavelength of the light, the refraction to the homogeneous outside region is done by applying Snell's law, using the local refractive index at the intersection point.

5.11 Polarisation Ray-tracing in Anisotropic Media

In this section we first determine the ray vectors and the corresponding polarisation eigenstates in a homogeneous part of an anisotropic medium. Secondly, we study the refraction and reflection of the geometrical rays when an interface is encountered. The interface can separate an isotropic medium from an anisotropic one or it can be the interface between two anisotropic media. In principle, both linear and circular anisotropy or a combination of both are allowed (see Section 2.3 and 2.9).

5.11.1 Polarised Rays in a General Anisotropic Medium

Ray-tracing in an anisotropic medium requires the calculation of two vectors, the wave vector and the Poynting vector. In the geometrical optics approximation, it is the direction of the Poynting vector that yields the ray vector direction in the anisotropic medium for each polarisation eigenstate. The wave vector $\mathbf{k}$ is an auxiliary quantity in the sense that its modulus provides us with the refractive index for the specific wave propagation direction via the relation $|\mathbf{k}| = k = n_p k_0$ where k_0 is the wave vector modulus of a wave with identical frequency ω, propagating in vacuum. The refractive index n_p applies to the phase propagation speed of the wave in the anisotropic medium ($v_p = c/n_p$). The wave vector modulus k in the medium determines the spatial progression of the phase of the associated plane wave and the calculation of the optical pathlength increment along a ray. Equation (2.5.6) and Fig. 2.10 showed that the progression of the phase along the ray vector only requires knowledge of the angle α between the electric field vector $\mathbf{E}_0$ and the electric flux density vector $\mathbf{D}_0$, or, equivalently, the angle between the unit wave vector $\hat{\mathbf{k}}$ and the ray vector $\hat{\mathbf{s}}$.

5.11.1.1 Calculation of the Phase Refractive Index

To find the phase refractive index n_p associated with a certain wave vector $\mathbf{k}$ in a general anisotropic medium, we use Eq. (2.9.8), valid for a nonmagnetic medium. With arbitrary orientations of the linear and circular anisotropy, a numerical solution of the equation $|M_E| = 0$ is imposed. By a root finding procedure, we obtain the possible values of the refractive index n_p. If the geometry of the medium has the special properties that are inherent to Eq. (2.9.9), an analytic solution is feasible, as given by Eq. (2.9.13). To this end, we first have to rotate the wave vector $\mathbf{k}$ of the wave under consideration in such a way that it is referenced to the symmetry axis of the anisotropic medium.

5.11.1.2 Eigenstates of the Electric Vectors E and D

Two solutions for n_p are generally found from the calculation scheme above for the forward propagating waves. They correspond to the two orthogonal polarisation eigenstates in the anisotropic medium. A complex eigenvector is associated with each polarisation eigenstate. For a linear anisotropic medium, the linear eigenstates with real eigenvectors follow from Eq. (2.4.23). A medium with general anisotropy, including optical rotation, requires a more elaborate calculation. We can use Eq. (2.10.12) of Subsection 2.10.2 which represents a linear system of equations for the electric field strength vector $\mathbf{E}_0$. For the case of a medium with only rotatory power, the matrix of the system of equations was seen to be Hermitian, implying real eigenvalues and orthogonal complex eigenvectors. For a general medium, the eigenvectors for $\mathbf{E}_0$ are generally elliptical. The corresponding electric flux density eigenvectors $\mathbf{D}_0$ follow from the constitutive relation of the anisotropic medium. In the case of the electric flux density vectors $\mathbf{D}_0$, perpendicular to the unit wave vector $\hat{\mathbf{k}}$, we form an orthogonal basis with the aid of the unit vectors:

$$\begin{aligned}
\hat{\mathbf{k}} &= (\hat{k}_X, \hat{k}_Y, \hat{k}_Z)\,, \\
\hat{\mathbf{D}}_1 &= (-\hat{k}_Y, \hat{k}_X, 0)/\sqrt{\hat{k}_X^2 + \hat{k}_Y^2}\,, \\
\hat{\mathbf{D}}_2 &= (-\hat{k}_X\hat{k}_Z, \hat{k}_Y\hat{k}_Z, \hat{k}_X^2 + \hat{k}_Y^2)/\sqrt{\hat{k}_X^2 + \hat{k}_Y^2}\,.
\end{aligned} \tag{5.11.1}$$

Any pair of orthonormal elliptical eigenstates can be written as

$$\begin{aligned}
\mathbf{D}_{e1} &= a\hat{\mathbf{D}}_1 - ib\hat{\mathbf{D}}_2\,, \qquad (RC)\,, \\
\mathbf{D}_{e2} &= b\hat{\mathbf{D}}_1 + ia\hat{\mathbf{D}}_2\,, \qquad (LC)\,,
\end{aligned} \tag{5.11.2}$$

where the nomenclature RC and LC follows from our definition in Chapter 1. For a general azimuth θ of the eigenstates with respect to the chosen basis vectors $\hat{\mathbf{D}}_1$ and $\hat{\mathbf{D}}_2$, the complex numbers are given by $a = a_1 + ia_2$ and $b = b_1 + ib_2$ where the ratio $a_2/a_1 = b_2/b_1$ equals $\tan\theta$. The scalar product $\mathbf{D}_{e1} \cdot \mathbf{D}^*_{e2}$ is zero, independently of the angle θ.

5.11.1.3 Calculation of the Ray Vectors $\hat{\mathbf{s}}_j$

Knowing the wave vector $\hat{\mathbf{k}}$ and the electric eigenvectors $\mathbf{E}_j$ we construct the unit ray vector $\hat{\mathbf{s}}_j$ by applying Eq. (2.11.2) for the Poynting vector in a general anisotropic medium and then normalising this vector. In a medium with linear anisotropy only, we can use the expressions for the eigenstates of the electric field strength and the electric flux density vectors, normalise them and apply the property of Eq. (2.8.10),

$$\hat{\mathbf{s}}_j = \hat{\mathbf{E}}_{0,j} \times \hat{\mathbf{k}} \times \hat{\mathbf{D}}_{0,j}\,. \tag{5.11.3}$$

In Fig. 5.83 we illustrate the various quantities that are needed to trace rays in an anisotropic medium. The central vector is the unit wave vector $\hat{\mathbf{k}}$. Two orthogonal (complex) vectors $\mathbf{D}_1$ and $\mathbf{D}_2$ are found in a plane that is perpendicular to $\hat{\mathbf{k}}$. They are the flux densities of the two propagation modes that are connected to the wave propagation direction $\hat{\mathbf{k}}$. Each eigenstate $\mathbf{D}_i$ gives rise to an eigenstate $\mathbf{E}_i$ for the electric field strength vector, shown on the right-hand side of the figure. An electrical eigenstate vector $\mathbf{E}_i$, together with the corresponding magnetic eigenstate vector $\mathbf{H}_i$, gives rise to a unit ray vector $\hat{\mathbf{s}}_j$ that yields the energy propagation direction. To obtain the progression of the phase along the ray vectors, planes of constant phase have been sketched in the figure (see the dashed lines perpendicular to the wave vector $\hat{\mathbf{k}}$). In a general medium with biaxial linear anisotropy, the angles α_1 and α_2 between $\hat{\mathbf{s}}_j$ and the wave vector are both finite. In some special cases, for instance in a uniaxial medium or in a biaxial medium with $\hat{\mathbf{k}}$ parallel to one of the principal sections of the permittivity tensor, one of the ray vectors is parallel to the wave vector and this gives rise to an *ordinary* eigenmode of the medium. In general, the ray vectors are found in two mutually orthogonal planes with the vector $\hat{\mathbf{k}}$ parallel to the intersection line of these two planes. Once all these quantities have been established in a first medium for a certain wave propagation direction, we are able to address the phenomena of refraction and reflection at a boundary between two media. In the process of ray-tracing, the emphasis will be on the propagation of the unit ray vectors from

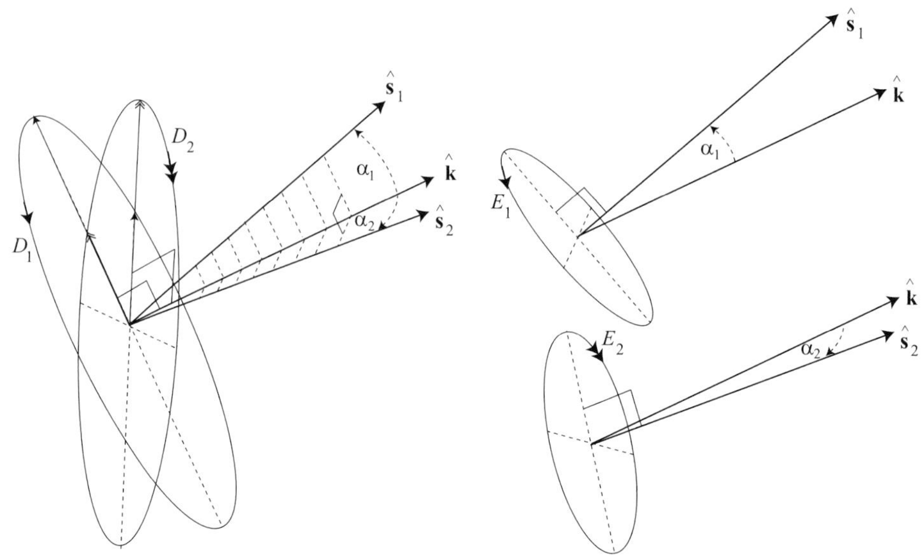

Figure 5.83: The various wave propagation and field vectors that characterise an electromagnetic plane wave in a general anisotropic medium. The electric field strength eigenvectors have been separately drawn in the right-hand part of the figure. The mutual position of the **D** and **E** eigenvectors of each (elliptical) eigenstate has not been shown in the figure since this position depends on the specific properties of the dielectric tensor of the medium under consideration.

one boundary to the next one in a sequence of anisotropic media. Although the ray vectors are the essential quantities for tracing the propagation of energy, the wave vector plays the central role in the analytical and numerical ray-tracing process.

5.11.2 Refraction and Reflection at a Planar Interface

The continuity of the tangential field components of the electric and magnetic field strength vectors has to be respected by the fields on both sides of an interface between anisotropic media. The continuity is global, both in time and in space along the interface (see Subsection 1.8.4 for a discussion on this subject). For isotropic media, with a given incident wave vector $\mathbf{k}_i$, we obtain from the continuity of the tangential wave vector components the refracted and reflected ray vectors from Eqs. (5.7.19) and (5.7.20). The unit ray and unit wave vector can be simply interchanged in an isotropic medium. For an interface between anisotropic media, we have first to define the incident wave vector. Starting with the unit wave vector of the incident plane wave, we multiply it by $n_j k_0$ to obtain the wave vector belonging to one of the two possible polarisation eigenstates in the first medium. Inserting this wave vector into the equations for the refracted and reflected wave vector, we obtain for the anisotropic case

$$n_t(\hat{\mathbf{k}}_t)\hat{\mathbf{k}}_t = n_j\hat{\mathbf{k}}_i - \left\{n_j(\hat{\mathbf{v}}\cdot\hat{\mathbf{k}}_i) - n_t(\hat{\mathbf{k}}_t)(\hat{\mathbf{v}}\cdot\hat{\mathbf{k}}_t)\right\}\hat{\mathbf{v}} , \tag{5.11.4}$$

$$n_r(\hat{\mathbf{k}}_r)\hat{\mathbf{k}}_r = n_j\hat{\mathbf{k}}_i - \left\{n_j(\hat{\mathbf{v}}\cdot\hat{\mathbf{k}}_i) - n_r(\hat{\mathbf{k}}_r)(\hat{\mathbf{v}}\cdot\hat{\mathbf{k}}_r)\right\}\hat{\mathbf{v}} . \tag{5.11.5}$$

It is interesting to see that there is not a special expression for the reflective case as was the case for isotropic media. In general, the phase refractive index of the reflected wave is different from the refractive index of the incident wave and the classical law of reflection does not apply to transitions between anisotropic media. The equations above do not provide us with explicit expressions for the quantities $(n_t, \hat{\mathbf{k}}_t)$ and $(n_r, \hat{\mathbf{k}}_r)$ that have to be found. Focusing first on Eq. (5.11.4), we denote the expression between braces in the right-hand side of the equation by Q and then square both sides of the equation. A quadratic equation in Q results,

$$Q^2 - 2n_j\left(\hat{\mathbf{v}}\cdot\hat{\mathbf{k}}_i\right)Q + n_j^2 - n_t^2 = 0 , \tag{5.11.6}$$

having as solutions

$$Q = n_j\left(\hat{\mathbf{v}}\cdot\hat{\mathbf{k}}_i\right) \pm \left[n_t^2 - n_j^2\{1-(\hat{\mathbf{v}}\cdot\hat{\mathbf{k}}_i)^2\}\right]^{1/2} . \tag{5.11.7}$$

For the refracted wave vector, we choose the Q solution with the negative sign in front of the square-root expression. Proceeding in a similar way with the equation for the reflected wave vector, we choose the square-root function with the positive sign. With these conventions, we find for the unit wave vectors of the refracted and reflected wave,

$$\hat{\mathbf{k}}_t = \frac{n_j}{n_t}\hat{\mathbf{k}}_i - \left[\frac{n_j}{n_t}(\hat{\mathbf{v}}\cdot\hat{\mathbf{k}}_i) - \left\{1-\frac{n_j^2}{n_t^2}\left[1-(\hat{\mathbf{v}}\cdot\hat{\mathbf{k}}_i)^2\right]\right\}^{1/2}\right]\hat{\mathbf{v}} , \tag{5.11.8}$$

$$\hat{\mathbf{k}}_r = \frac{n_j}{n_r}\hat{\mathbf{k}}_i - \left[\frac{n_j}{n_r}(\hat{\mathbf{v}}\cdot\hat{\mathbf{k}}_i) + \left\{1-\frac{n_j^2}{n_r^2}\left[1-(\hat{\mathbf{v}}\cdot\hat{\mathbf{k}}_i)^2\right]\right\}^{1/2}\right]\hat{\mathbf{v}} , \tag{5.11.9}$$

where we have omitted in the notation the dependence of n_t and n_r on the refracted and reflected unit wave vectors $\hat{\mathbf{k}}_t$ and $\hat{\mathbf{k}}_r$. In these expressions, we always choose the solutions with positive sign for the phase refractive index. The forward or backward propagation sense of a refracted or reflected wave follows from the wave vector itself.

To solve the implicit Eqs. (5.11.8) and (5.11.9) for the wave vectors $\hat{\mathbf{k}}_t$ and $\hat{\mathbf{k}}_r$ an iterative procedure is needed. We choose a reasonable starting point for the refractive indices, for instance, the smallest or largest refractive index value following from the permittivity tensor of the media. An index $n_t^{(l)}$ or $n_r^{(l)}$ in the (l)th iteration is found by solving the phase index eigenvalue equation of the anisotropic medium for updated values of the wave vectors $\hat{\mathbf{k}}_t^{(l)}$ and $\hat{\mathbf{k}}_r^{(l)}$. In general, a quick convergence to the solution values $n_t(\hat{\mathbf{k}}_t)$ and $n_r(\hat{\mathbf{k}}_r)$ is found. A few iteration cycles are generally enough to achieve a convergence down to 10^{-10} for the refractive index and the vector components.

The possibility exists that the square-root expression in Eq. (5.11.8) or (5.11.9) yields an imaginary value for the component of the wave vector that is normal to the interface. This leads to an evanescent refracted or reflected wave. The further processing of such a solution would yield a complex ray vector that is extinguished when propagating over a distance that is large with respect to the wavelength of the light. For ray transfer between 'distant' surfaces, such rays can be neglected. The various possibilities which arise when a plane wave is incident on an interface have been sketched in Fig. 5.84. In each sketch, the incident wave vector is the only possible one (isotropic medium) or one of the two that are associated with the polarisation eigenstates in an anisotropic medium. In the graphs, for an anisotropic medium, such an eigenstate is characterised by the phase refractive index n_j and the refraction angle $\theta_{t,j}$ or the reflection angle $\theta_{r,j}$, for $j = 1, 2$. The ray vectors associated with each incident, refracted or reflected wave have not been shown in the figure because they are not necessarily in the plane of incidence defined by the vectors $\hat{\mathbf{k}}_i$ and $\hat{\mathbf{v}}$.

5.11.3 Tracing the Ray Vectors at an Interface

For the energy transport in an optical system from one interface to the other, we need to derive the ray vectors for each refracted or reflected wave vector. The computation scheme for the ray vector components $\hat{\mathbf{s}}_j$ from the wave vector data in a linear anisotropic medium has been given in Subsection 2.6.3, Eq. (2.6.20). As initial data, the phase refractive index of the polarisation eigenstate that is considered should be given. For media that show the effect of optical rotation or circular birefringence, the eigenvalue equation that produces the refractive indices of the polarisation eigenstates requires the medium matrix that is given by Eq. (2.9.11). For the more general case of a medium with both linear and circular birefringence, we have to solve Eq. (2.10.4) with the material tensor $\overline{\overline{\mathbf{M}}}_b$ given by Eq. (2.10.3). The construction of the ray vector formally uses the electric and magnetic field strength (complex) eigenvectors that follow from the eigenvalue equations when substituting the refractive index eigenvalues in them (see Subsection 2.10.2). Equation (2.11.2) produces, with the aid of an electric polarisation eigenstate only, an expression for the Poynting vector. By normalising this vector, we obtain the ray vector in the anisotropic medium.

As was mentioned before, the *ray* vectors do not obey the coplanar property of the incident, refracted and reflected *wave* vectors at a planar interface. This 'second part' of Snell's law for ordinary ray vectors is not preserved for ray vectors belonging to an extraordinary polarisation state. It implies that for anisotropic media the transfer of linear momentum between the electromagnetic field and the medium is not limited to the plane of incidence but can have a component in the transverse direction.

The distribution of electromagnetic power over the various refracted and reflected ray vectors at an interface is normally not included in the pure ray-tracing process. The energy transfer at a planar interface between isotropic media could be

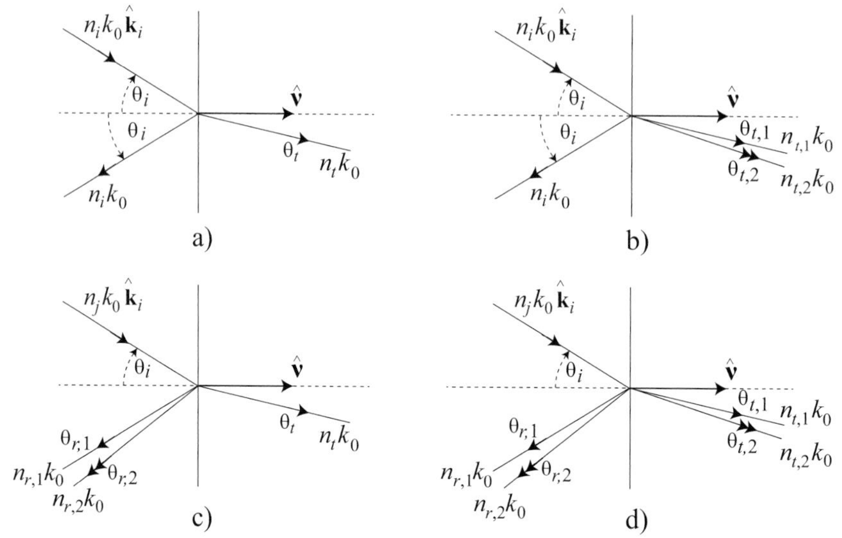

Figure 5.84: Refraction and reflection of the *wave* vector at an interface between isotropic / anisotropic media. The incident ray in an anisotropic medium has the wave vector $n_j k_0 \hat{\mathbf{k}}_i$, which is at an angle θ_i with the normal $\hat{\mathbf{v}}$ to the planar boundary between the two media. The corresponding ray vectors have not been shown in the figure.
a) Transition from isotropic to isotropic.
b) Isotropic to anisotropic transition.
c) Anisotropic to isotropic transition.
d) The general case, anisotropic to anisotropic.

analysed in terms of s- and p-eigenstates that are linear eigenstates with their preferential directions, in the plane of incidence and perpendicular to it, respectively. In the case of anisotropic media of a general nature, the eigenstates in each medium are the two (complex) eigenstates which are related to the wave vector of the refracted or reflected wave. An incident state of polarisation is projected onto the two eigenstates in each medium giving rise to four refracted and reflected ray vectors in total. The calculation scheme for obtaining the reflection and transmission coefficients has been given in Subsection 2.12.3. With the aid of the eigenstates of the incident, refracted and reflected plane waves, Eqs. (2.12.7) and (2.12.8) yield the transmission and reflection coefficients at the interface. From these coefficients and the propagation directions of the plane waves follows the energy distribution of the incident power over the four possible ray vector directions. In this calculation scheme, based on the wave amplitudes at the interface itself, the electric field components of an evanescent refracted or reflected wave should be taken into account.

6 Analytic Design and Optimisation of Optical Systems

6.1 Introduction

The history of optical system design and manufacturing is an interesting example of the mutual influence and stimulation exerted by science, technology and craftsmanship. In certain periods of time, the inventors and engineers developed observation methods or instruments that were not yet fully understood and urgently asked for a further development of the theoretical insight. The opposite also happened where theoretical possibilities appeared at the horizon which were still far away from any practical realisation. The most impressive steps forward are observed when the whole chain of theory development, measurement techniques, inventive power and technology are united in a single person or group. Great names in optical science such as Newton, Huygens, Fresnel, Fraunhofer and Abbe have enabled such huge steps forward in the more remote past. Nowadays, with increasing complexity of optical systems and instruments, it is less evident that a single name can be attached to substantial progress in a certain field. It is more a strong research group or an entire optical company that achieves the highest level in optical system design and fabrication.

If we return to the very first technical realisations in optics, we encounter descriptions of reflecting surfaces in antiquity, both in Europe and imperial China. The production of useful spectacle lenses which improved the view of visually-impaired people goes back to the twelfth century, in Italy and China. In the same regions and in the Netherlands, the first successful refracting telescopes and microscopes were manufactured around the year 1600. These achievements were supported by the paraxial imaging law for a lens which was developed by Kepler. The sine law for the refraction of light rays at finite angles, already discovered by Ibn Sahl (Baghdad) around the year 1000, was rediscovered in the western world by the publications of Snellius and Descartes. The sine law of refraction was immediately applied by Descartes to a number of still unsolved imaging problems. His new lenses required optical surfaces with an aspheric shape that, at that time, could not be made to the accuracy required for sharp optical image formation.

In the seventeenth and eighteenth century further improvements of microscope imagery were obtained, partly based on improved fabrication techniques and partly on an increased understanding of the image formation by a lens at finite opening angles. Particularly successful were the singlet lenses (magnifying glasses) produced by van Leeuwenhoek. Although resulting from a fully empirical manufacturing technique, these tiny single lenses produced surprisingly detailed images of biological specimina. For a long time, the Leeuwenhoek lenses remained superior to the compound microscopes because of the poor optical quality of the latter.

At the end of the eighteenth century, a breakthrough was reached regarding the colour correction of lenses. The statement by Newton that all transparent substances would have the same dispersion was experimentally belied and the elementary single lens could be transformed into a doublet with good colour correction (Dollond, 1761). Exhaustive work on the dispersion of glass, both experimentally and theoretically, was carried out by Joseph von Fraunhofer. His manufactured doublets with carefully selected and fabricated optical glasses were a big step forward with respect to the reduction of aberration and colour defects in lens imaging. Large-size Fraunhofer-type lenses for astronomical telescopes have been manufactured with a diameter up to 1 metre.

Systematic optical design based on aberration theory took off in the middle of the nineteenth century with the advent of photography. The design of a portrait lens by Petzval was based on his own mathematical analysis of ray propagation through a lens system. Unfortunately, his results got lost. Some ten years later, in 1856, Ludwig von Seidel published his aberration theory. This so-called 'third-order' aberration theory became the basis for the successful design of imaging systems with various applications in photography, astronomy and microscopy. Systematic design improvements, combined with better measurement techniques and reproducibility in manufacturing led to high-quality immersion microscope objectives with numerical apertures as high as 1.35 in the second half of the nineteenth century. Higher-order aberration theories were devised, finite-ray aberration formulas were used and analytic design methods were developed to eliminate the aberration of rays at large inclination angles. A famous example is the improved telescope design by Ritchey and Chrétien, which produced a substantially larger image field than the telescope designs with a primary parabolic mirror which were used before.

The collection of ray data with sufficient precision, typically up to 10 or 12 digits, was a tiresome operation relying on accurate 'human' calculators. The selection of specific ray directions for testing the performance of a design on axis and in the image field was a very sophisticated art in which much experience was invested by the optical designers. The advent of the electronic programmable computer has drastically changed the design procedures. Optimisation techniques are widely used and, provided with some initial design or prototype, the optimisation routines steer the initial design to the nearest point with optimum performance. The local optimisation routines are nowadays quite reliable and fast. But they do not offer the guarantee of finding still better optima much further away from the starting point. For this reason, as in many other fields of research, global optimisation schemes are explored which can scan the whole 'value function' or 'merit function' landscape of a design in a sufficiently short time. The optical merit function preferably contains also manufacturing tolerances, in such a way that the final design will be kept away from mechanically or materially unfeasible regions. If possible, the measurement sensitivity of a design is also incorporated in the merit function. The final design should give rise to a measurement–manufacturing feedback loop which fits to state-of-the-art measurement facilities. If the measurement accuracy is inadequate for reliably quantifying the manufacturing errors, no satisfactory end product can be made.

In the second section of this chapter we describe some analytical design techniques which permit the elimination of some basic aberrations in an optical design. The third section discusses quality criteria for optical systems and the construction of a merit function for imaging systems. In the fourth section optimisation techniques in modern optical design are discussed. In the fifth section mechanical and optical tolerancing of an optical design is introduced to assess the manufacturability of the optical system. General design considerations, design methods and the state-of-the-art of a variety of optical imaging systems are discussed in Chapter 7.

6.2 Analytic Aberration-free Design of an Optical System

It was shown earlier that aberration-free imaging is generally limited to a single point or to a small region around this point. In the first case one refers to *stigmatic* imaging and in the second case the term *aplanatic* imaging is used, see Sections 4.6 and 4.8. Higher performing optical systems which are capable of stigmatically imaging an entire volume from object to image space are generally called *absolute* systems. Still one step further, if such an optical system creates an image of which any composing curve reproduces the corresponding one on the object, the system is said to be a *perfect* system. Such systems cannot have an arbitrary magnification between object and image; the local magnification obeys the ratio n_0/n_1 where n_0 and n_1 are the refractive indices of the local regions in object and image space between which imaging is achieved **[37]**. In the case of homogeneous object and image spaces with equal indices, the only solution for perfect imaging is provided by a mirror or a system of mirrors. When the imaging requirements are loosened from three-dimensional imaging of volumes to the imaging of surfaces, we find the Huygens–Roberval aplanatic points (see Section 4.6). Because of the point symmetry of this particular imaging geometry, the aplanatic points assure the imaging of an entire spherical surface of which the lateral, in-surface magnification is given by $(n_0/n_1)^2$. This attractive property has been exploited in the design of high-numerical-aperture microscope objectives.

In the practice of optical design, the stigmatic imaging of an entire surface with arbitrary magnification is not possible. In the remainder of this section we pursue a more modest goal, the rigorous elimination of any wavefront aberration when imaging a single point in the object field of an optical system (stigmatic imaging). If this point is chosen to be on the axis of an optical system with rotational symmetry, an extension of aberration-free imaging in the direct circular neighbourhood might be feasible (aplanatic imaging). To obtain aberration-free imaging, we introduce an aspheric surface

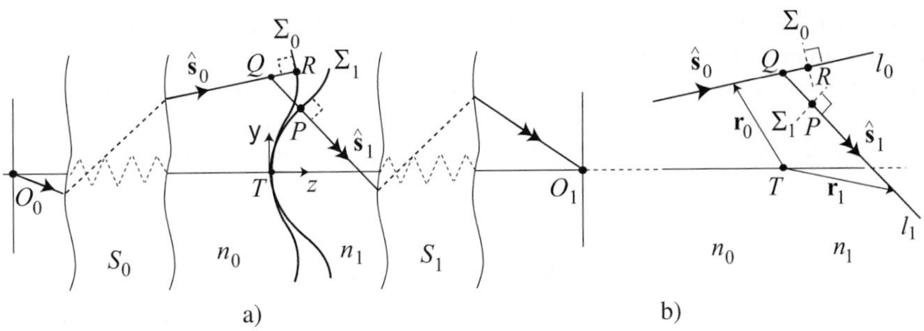

Figure 6.1: a) An axial ray and an aperture ray $\hat{\mathbf{s}}_0$ leave the object point O_0. In image space, the rays intersect at the image point O_1. In the intermediate space an aspheric surface is constructed through the point T on the optical axis; the refractive indices on both sides of the surface through T are given by n_0 and n_1. Σ_0 and Σ_1 are the wavefronts through T associated with the pencils of rays traced forward and back from O_0 and O_1 through the optical system. The local coordinate system (x, y, z) has its origin at T, the vertex of the aspheric surface.
b) The calculation of a general point Q on the aspheric surface through T.

somewhere in the optical system; as a special case, the aspheric surface could be the nearest one to the object or to the image point.

6.2.1 Stigmatic Imaging of the Axial Point

In Fig. 6.1 we have sketched the imaging of an axial object point O_0 into the point O_1 by means of a certain number of optical surfaces, schematically indicated by the parts S_0 and S_1 in the figure. After traversal of a first section S_0 of the optical system, the rays from O_0 arrive in the intermediate space which has a refractive index n_0.

6.2.1.1 Aspherisation of a Surface

Stigmatic imaging can be obtained by constructing an aspheric surface in the interior of the optical system. The first systematic description of the design of an aspheric surface in a general optical system can be found in **[370]**. In this paragraph we follow a slightly different approach. In Fig. 6.1a the surface which will be made aspheric intersects the optical axis at a point T and separates two media with refractive indices n_0 and n_1. The choice of the position of the point T is basically free. In practice it will be the front or back surface of an existing lens in the imaging system. It could also be the surface of a mirror, in which case we would apply the convention $n_1 = -n_0$. It can be profitable to locate the surface in the physical diaphragm of the optical system or in an image of it, the entrance or exit pupil. In this case, the exact pathlength correction for the axial beam will be in first order maintained for obliquely incident pencils of rays. Generally speaking, the choice of which surface should be aspherised will be determined by the best possible imaging performance in the field of the system.

The wavefront of the pencil of rays from O_0 through T is Σ_0. Each ray of the pencil is perpendicular to it. From the image point O_1 the axial ray and aperture rays are traced back towards the intermediate space with index n_1. The wavefront through T, associated with the backward-traced pencil of rays from O_1, is given by the surface Σ_1. For the construction of a general point Q of the aspheric surface through T we apply the equality-of-path condition between a general aperture ray and the axial ray,

$$[O_0Q] + [QP] + [PO_1] = [O_0T] + [TO_1] \, ,$$

or,

$$[O_0R] + [RQ] + [QP] + [PO_1] = [O_0T] + [TO_1] \, ,$$

which reduces to the basic expression

$$n_0 QR = n_1 QP \, . \tag{6.2.1}$$

This pathlength condition for the line segments QR and QP determines which ray $\hat{\mathbf{s}}_0$ from the object-side pencil will be connected with a ray $\hat{\mathbf{s}}_1$ from the image-side pencil.

The construction of the aspheric surface proceeds as follows. The wavefronts in object and image space are obtained from ray-trace data for a sufficient number of rays. By interpolation we get a smooth representation of the surface with an accuracy that is a negligible fraction of the wavelength of the light for all practical purposes. The analytic representation of the wavefronts and their normal vectors is given by

$$\begin{aligned} \Sigma_0 \;:\; z - f_0(y) = 0 \,, \qquad \hat{\mathbf{s}}_0 &= (0, M_0, N_0) = \left(\frac{0 \,,\; -df_0/dy \,,\; +1}{\sqrt{1 + (df_0/dy)^2}} \right) , \\ \Sigma_1 \;:\; z - f_1(y) = 0 \,, \qquad \hat{\mathbf{s}}_1 &= (0, M_1, N_1) = \left(\frac{0 \,,\; -df_1/dy \,,\; +1}{\sqrt{1 + (df_1/dy)^2}} \right) . \end{aligned} \tag{6.2.2}$$

The general equations for the object- and image-side ray vectors are

$$\begin{aligned} l_0 \;&:\; \qquad \mathbf{r}_0 = (0, y_R, z_R) + p_0 \hat{\mathbf{s}}_0 \,, \\ l_1 \;&:\; \qquad \mathbf{r}_1 = (0, y_P, z_P) + p_1 \hat{\mathbf{s}}_1 \,, \end{aligned} \tag{6.2.3}$$

where p_0 and p_1 are the parameters that determine the position of a point on the lines l_0 and l_1 (see Fig. 6.1b). The point Q of the aspheric surface lies on both l_0 and l_1 and should satisfy the condition $n_0 QR = n_1 QP$. From Eq. (6.2.3) we have that $\vec{QR} = \mathbf{r}_R - \mathbf{r}_Q = -p_{0,Q}\hat{\mathbf{s}}_0$ and $\vec{QP} = -p_{1,Q}\hat{\mathbf{s}}_1$. The pathlength condition then reduces to $n_0 p_{0,Q} = n_1 p_{1,Q}$; in the reflective case we have $p_{0,Q} = -p_{1,Q}$. The subtraction of $\mathbf{r}_{1,Q}$ from $\mathbf{r}_{0,Q}$ in Eq. (6.2.3) yields the expression for $p_{0,Q}$

$$p_{0,Q} \left\{ \hat{\mathbf{s}}_0 - \frac{n_0}{n_1} \hat{\mathbf{s}}_1 \right\} = (0, y_P, z_P) - (0, y_R, z_R) \,. \tag{6.2.4}$$

We eliminate $p_{0,Q}$ from this vector equation and obtain the implicit equation for the coordinate y_P,

$$y_P = y_R + \left(\frac{n_1 M_0(y_R) - n_0 M_1(y_P)}{n_1 N_0(y_R) - n_0 N_1(y_P)} \right) \{ f_1(y_P) - f_0(y_R) \} \,, \tag{6.2.5}$$

where $M_0^2 + N_0^2 = M_1^2 + N_1^2 = 1$.

Starting with a point R, the point P on the connecting ray QP is obtained by numerically solving the equation above. The point Q on the aspheric surface directly follows from Eq. (6.2.3) using the value $p_{0,Q}$ for the intersection point Q of l_0 and l_1. Monomial or, preferably, polynomial expansions of $f_0(y)$ and $f_1(y)$ are prepared from the initial ray-trace data. These expansions allow the calculation of the derivative functions and the direction cosines of the rays $\hat{\mathbf{s}}_0$ and $\hat{\mathbf{s}}_1$ for a certain height position y on the wavefronts Σ_0 and Σ_1. A numerical procedure for the construction of the aspheric surface typically requires some 100 points which are, for instance, equally distant along the y-axis. The numerical rest error in the beam focused at O_1 is then as small as 10^{-4} to 10^{-5} wavelengths. This accuracy level is largely sufficient for practical applications. Effectively, all orders of spherical aberration are suppressed by the introduction of the aspheric surface.

6.2.1.2 Asphérisation of the First or Last Surface of a System

In the special case that the surface to be aspherised is the first or the last surface of an optical system, a special construction is possible. In Fig. 6.2 the wavefront through the point T, produced by the pencil of rays issued from the object point O_0, has been denoted by Σ_0. In general, its shape will be aspheric with symmetry of revolution with respect to the optical axis O_0TO_1. The wavefront through T traced back from the image point O_1 is spherical and has a radius $\rho = O_1T$. We take a general point R on the wavefront Σ_0 and construct the ray $\hat{\mathbf{s}}_0$ which is perpendicular to Σ_0 at the point R. An auxiliary line, denoted by $\hat{\mathbf{s}}_0'$, is drawn through O_1, parallel to the ray $\hat{\mathbf{s}}_0$. An auxiliary circle with its centre at O_1 and a radius of $n_1\rho/n_0$ is constructed that intersects the line $\hat{\mathbf{s}}_0'$ at the point A. We trace the line segment AR and the point of intersection with the wavefront Σ_1 is denoted by P. The radial segment O_1P is extended and intersects the ray $\hat{\mathbf{s}}_0$ at the point Q. It follows from the construction that the triangles PQR and PAO_1 are similar and we have

$$\frac{QR}{QP} = \frac{O_1A}{O_1P} \,, \qquad \rightarrow \qquad n_0 QR = n_1 QP \,. \tag{6.2.6}$$

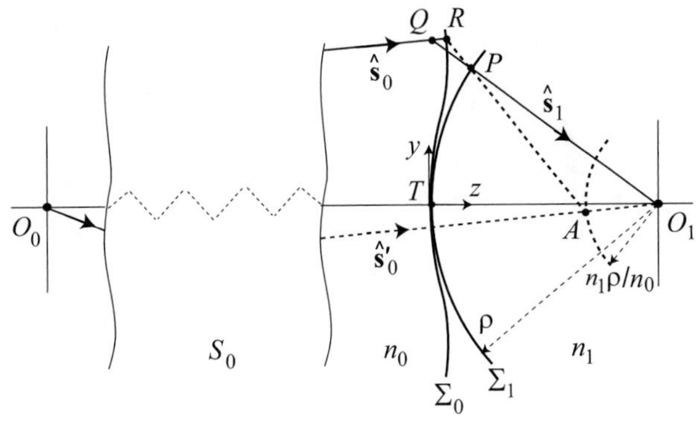

Figure 6.2: The mapping of an aspheric wavefront onto a spherical one by means of an aspheric refracting surface. Σ_0 is the wavefront resulting from a ray-trace from the object point O_0 to the penultimate medium of the system with index n_0. T is the vertex of the aspheric surface to be constructed. Σ_1 is the spherical wavefront traced back from the image point O_1 to the point T on the optical axis. Q is a general point on the aspheric surface.

This concludes the construction of a general point Q on the desired aspheric surface. The complete surface is found by applying the construction for a sufficient number of points and by applying a fitting procedure. A polynomial base in the lateral coordinate y will give an accurate representation of the aspheric surface.

So far, in the construction method of the aspheric surface, it has been assumed that the object and image side pencils of rays are stigmatic which implies that all rays originate from or converge to a single point. It goes without saying that an on-purpose deviation from the freedom of aberration of object or image pencils can be included to produce an optical system with prescribed aberration.

6.2.1.3 An Aspheric Fresnel Surface

The manufacturing of an aspheric surface can be done in various ways. Direct aspheric grinding and polishing of a glass surface are routinely used. Using aspherically shaped moulds, a glass surface can be asphericed by hot pressing with the aid of an aspheric metal mould. Alternatively, plastic lenses with aspheric surfaces can be produced by means of injection moulding. If the aspheric shape is needed on a lens or mirror substrate material that does not support the above-mentioned manufacturing methods, a thin layer can be deposited on a spherical surface and this thin layer then receives an aspheric shape with minimum thickness. In Fig. 6.3a we show, from left to right, a solid lens with one aspheric surface, a lens with a continuous aspheric cap and, in the third position, a lens with an aspheric surface showing indentations. In the case of a Fresnel-type surface with indentations, the optical pathlength condition for aperture ray and axial ray reads

$$[O_0RQPO_1] = [O_0T_1] + [T_1O_1] - [P_1P] \, ,$$

or,

$$[QP] = [QR] - [P_1P] \, , \qquad \rightarrow \qquad n_0 QR = n_1 QP - [PP_1] \, , \tag{6.2.7}$$

where the wavefronts Σ_0 and Σ_1 through T_1 are obtained, as before, from ray-trace data from the object point up to T_1 and from the image point back to T_1. For the optical pathlength defect of the Fresnel structure we take an integer number of wavelengths in the image space. The integer number K_F can be chosen arbitrarily,

$$[P_1P] = \frac{K_{F,j}\lambda_0}{n_I} \, , \tag{6.2.8}$$

where $K_{F,j}$ is related to the jth zone of the Fresnel structure, λ_0 is the vacuum wavelength and n_I the refractive index of the image space. In line with the pathlength discussion for the construction of a continuous aspheric surface, we now have a modified condition for the ray parameters $p_{0,Q}$ and $p_{1,Q}$ for the point Q on the asphere

$$n'_1 p_{1,Q} = n_0 p_{0,Q} - \frac{K_{F,j}\lambda_0}{n_I} \, , \tag{6.2.9}$$

where the geometrical distance P_1P has been schematically shown in Fig. 6.3b.

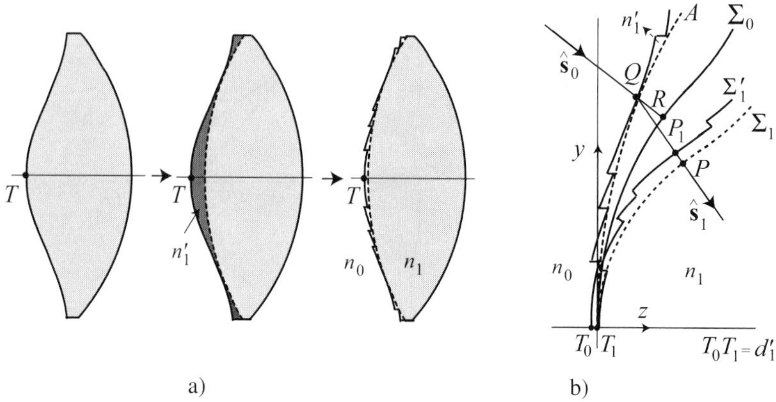

Figure 6.3: a) Various types of aspheric surfaces, two with continuous shape and one with a discontinuous aspheric cap. b) The refractive index on the left-hand side of the surface with indentations is n_0, the medium in which the surface with indentations has been machined has index n_1' and the spherical body of the lens has an index n_1. The incident beam produces a wavefront Σ_0 through T_1, the wavefront traced back from the image point is the surface Σ_1. Σ_1' is a wavefront, derived from Σ_1, with discrete pathlength differences in it, equal to an integer number of wavelengths λ_1, the wavelength in image space. The incident and refracted ray at the point Q on the dented surface are $\hat{\mathbf{s}}_0$ and $\hat{\mathbf{s}}_1$.

Solving for the common point Q on the ray segments l_0 and l_1 we obtain the following implicit equation for the y-coordinate of the point P,

$$0 = y_P - y_R - \frac{K_{F,j}\lambda_0 M_1(y_P)}{n_1' n_I} - \left(\frac{n_1' M_0(y_R) - n_0 M_1(y_P)}{n_1' N_0(y_R) - n_0 N_1(y_P)}\right)\left\{f_1(y_P) - f_0(y_R) - \frac{K_{F,j}\lambda_0 N_1(y_P)}{n_1' n_I}\right\} . \tag{6.2.10}$$

From the value of y_P we find $p_{0,Q}$ using the equation

$$p_{0,Q} = \frac{n_1' \{f_1(y_P) - f_0(y_R)\} - K_{F,j}\lambda_0 N_1(y_P)/n_I}{n_1' N_0(y_R) - n_0 N_1(y_P)} . \tag{6.2.11}$$

The coordinates of Q then follow from Eq. (6.2.3). The value of $K_{F,j}$ needs to be adapted as soon as the location of the point Q is inside the lens volume with index n_1, delimited by the surface A in the figure. The wavefront Σ_1' with indentations is adjusted for the next Fresnel zone by introducing a jump $K_{F,j+1}\lambda_0/n_I$ with respect to the preceding zone. The extent of the first central zone is determined by the axial thickness $T_0T_1 = d_1'$ of the Fresnel layer. In practical applications, the increments in pathlength from one Fresnel zone to another are chosen to be identical, as well as the extra optical path $[T_0T_1]$ on axis in the central zone. The foregoing analysis takes into account the pathlength effects due to oblique traversal of the structure and belongs to the so-called *extended* scalar analysis of a thin optical element. The predictions are reliable when a Fresnel zone has a lateral extent of many wavelengths. For transverse dimensions close to the wavelength of the light, the electromagnetic model of diffraction at the Fresnel structure should be applied. The reader is referred to **[111]** for the subject of optimising the diffracted power in the desired grating order.

6.2.1.4 Fresnel Structure and Hopkins' Pathlength Formula

The change in optical pathlength along a ray path that occurs when a ray hits a surface in a shifted position can be calculated using the drawing of Fig. 6.4. The angle of incidence θ_0 on the surface gives rise to a refracted ray $\hat{\mathbf{s}}_1$ in the nominal position of the surface. After shifting part of the surface from A_0 to A_0', the refracted ray is $\hat{\mathbf{s}}_1'$. If the shift of the surface is small with respect to the local radius of curvature R_c of the surface, the rays $\hat{\mathbf{s}}_1'$ and $\hat{\mathbf{s}}_1$ are parallel up to a rest error of the order of $|\delta|/R_c$.

The optical pathlength difference δW between the two surface positions is given by

$$\delta W = [PP_1] - [PP_2] , \tag{6.2.12}$$

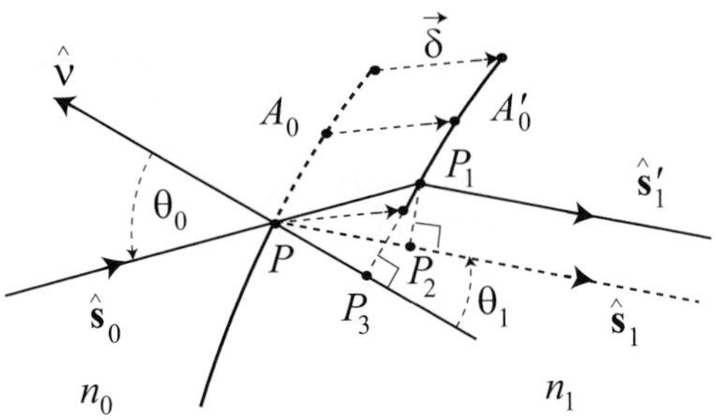

Figure 6.4: An incident ray with ray vector $\hat{\mathbf{s}}_0$ hits an optical surface at a point P at an angle of incidence θ_0 with the normal $\hat{\mathbf{v}}$ to the surface at P. Part of the surface labelled A_0 is shifted to the position characterised by the segment A_0', with a uniform displacement vector $\boldsymbol{\delta}$ for all points on the shifted surface part.

where P_1P_2 is the perpendicular from P_1 to the refracted ray $\hat{\mathbf{s}}_1$. From the drawing we also deduce that PP_3 is the projection of the surface shift vector $\boldsymbol{\delta}$ onto the outward surface normal $\hat{\mathbf{v}}$. We now obtain for the pathlength difference

$$\delta W = \frac{\boldsymbol{\delta}\cdot\hat{\mathbf{v}}}{\cos\theta_0}\left\{n_0 - n_1\cos(\theta_0-\theta_1)\right\} = \boldsymbol{\delta}\cdot\hat{\mathbf{v}}\left(n_0\cos\theta_0 - n_1\cos\theta_1\right) . \qquad (6.2.13)$$

An expression for pathlength differences along neighbouring rays is given in **[140]**, pp. 23, 24. It can be applied to the geometry of Fig. 6.4 and yields a result, identical to that of Eq. (6.2.13). The calculation of δW can be applied to find the indentation height of a Fresnel structure along the surface normal by requiring that δW equals an integer number of wavelengths of the light.

The formula for δW can also be used when surface shifts due to manufacturing errors are studied. We obtain from Eq. (6.2.13) the wavefront defect δW which arises at each 'perturbed' optical surface. The slight angular deviation between the ray $\hat{\mathbf{s}}_1'$, related to the perturbed surface and $\hat{\mathbf{s}}_1$ corresponding to the nominal state does not change the paraxial ray data and, for that reason, the fourth-order pathlength deviations $\delta W^{(4)}$ are correct for an entire optical system. For sufficiently small mechanical and optical perturbations, the lowest order aberrations, of fourth order in the ray and surface coordinates, are the only important ones.

6.2.1.5 Diffractive Optical Elements (DOE)

In Subsection 5.7.3, we discussed the role of a grating-like structure on a curved surface and the propagation of a ray that is not only refracted or reflected but also diffracted by this surface. The grating structure is characterised by the local grating vector $\mathbf{v}_d = \hat{\mathbf{v}}_d/p_d$. According to Eq. (5.7.7), a phase function is associated with the grating structure on the surface and this phase function $\Phi_Q\{x, y, z(x, y)\}$ at a general point Q on the diffracting surface determines the deviation (diffraction) from the standard reflection or refraction angle of a ray which is incident on the surface. In Fig. 6.5 we show the design procedure for a diffractive optical surface A. The pathlength difference W_Q at a general point Q on the surface A with a known shape referred to the local coordinate system (x, y, z) follows from

$$W_Q = [O_1Q] - [O_0Q] - ([O_1T] - [O_0T]) . \qquad (6.2.14)$$

The phase function Φ_Q follows from $\Phi_Q = k_1 W_Q$ where k_1 is the wavenumber in image space. The phase function Φ_Q is transformed into its truncated form $\Phi_{Q,F}$ by applying the operation

$$\Phi_{Q,F} = \Phi_Q - 2\pi m\left\lfloor\frac{\Phi_Q}{2\pi m}\right\rfloor , \qquad (6.2.15)$$

where the expression $\lfloor a\rfloor$ represents the largest integer smaller than a and m is the jump in units of 2π at each surface indentation.

Using such a diffractive optical element, a versatile design tool becomes available. If needed, the imaging at a point O_1 can be made aberrated by a known amount. In such a case, a DOE can play an important role in optical metrology. The DOE is used as a nulling element to compensate for the aberrations of the element under test. For ease of fabrication and copying, the DOE surface is often made flat. A practical drawback of diffractive optical elements is the scattering of the

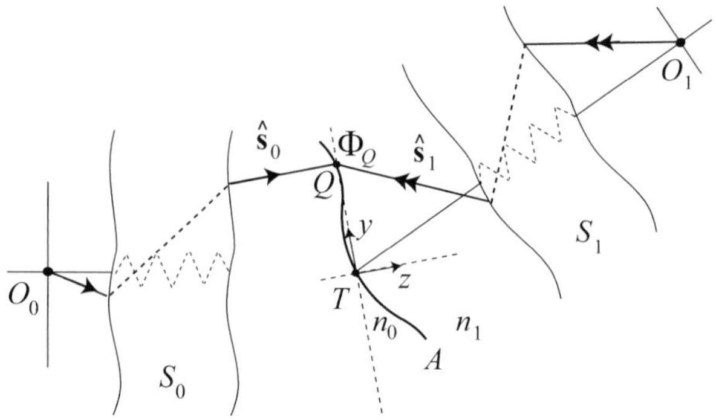

Figure 6.5: Construction of a diffractive optical element (DOE) on the surface A to achieve stigmatic imaging at O_1. The surface has its vertex at T and may have a general shape. The phase function Φ_Q at a general point Q on the surface is determined by the difference in pathlengths along the aperture rays $\hat{\mathbf{s}}_0$ and $\hat{\mathbf{s}}_1$ as compared to that between the central rays meeting at T (not shown in the drawing).

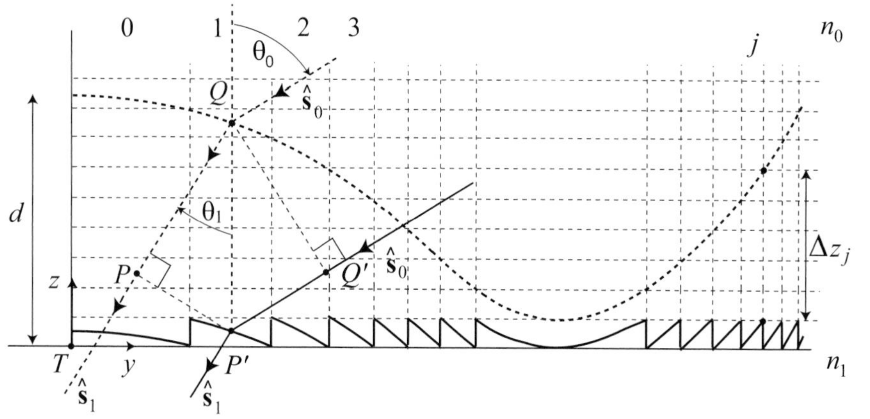

Figure 6.6: The chromatic effects of a diffractive optical element. A point Q on the continuous surface in zone 1 is shifted to Q'. The surface height reduction in a general zone numbered j is Δz_j. The incident and refracted ray are at angles θ_0 and θ_1, strongly exaggerated in the drawing, with the normal to the base surface $z = 0$. The optical contrast at the DOE surface is determined by the refractive indices n_0 and n_1.

light in unwanted directions leading to reduced contrast in the image. In general, especially for a flat diffractive element, the corrective action is limited to a very small angular spread of the incident beam with respect to the design angle.

6.2.1.6 Dispersion of a Fresnel Diffractive Optical Element

A disadvantageous property of a diffractive optical element is the strong wavelength dependence of the diffraction effect. In Fig. 6.6 we have sketched a corrective layer, for reasons of clarity of the drawing applied on a flat surface. Each time when the optical pathlength through the continuous layer (dashed curve) changes by an integer number of wavelengths in image space, an indentation is introduced with the discontinuous surface as a result. In the zone labelled 1 a ray $\hat{\mathbf{s}}_0$ is incident on the continuous layer at the point Q. A parallel ray intersects the discontinuous layer at P'. The pathlength difference between the ray refracted at Q and the ray through P', as measured at P', is given by

$$W_{P'} = [QP] - [Q'P'] = QP'\{n_1 \cos\theta_1 - n_0 \cos\theta_0\} \; . \tag{6.2.16}$$

Further pathlength differences along the rays QP and $Q'P'$ will be second-order small because these rays can be considered to be neighbouring rays and they obey Fermat's principle.

To enhance the contributions from various Fresnel zones in image space, the pathlength difference $W_{P'}$ should equal an integer number of wavelengths in the image space, yielding the condition

$$QP'_j = \frac{K_{F,j}\lambda_0}{n_I\{n_1\cos\theta_1 - n_0\cos\theta_0\}}\,, \tag{6.2.17}$$

where the integer $K_{F,j}$ multiplied by the wavelength λ_0/n_I in image space yields the pathlength jump for zone j (λ_0 is the vacuum wavelength). The designed Fresnel pattern for the specific vacuum wavelength λ_0 is then used at a different vacuum wavelength λ_1. For the optical pathlength difference $W'_{P'}$ at this modified wavelength we have

$$W'_{P'} = QP'\{n'_1\cos\theta'_1 - n'_0\cos\theta_0\} = \frac{K_{F,j}\lambda_0\{n'_1\cos\theta'_1 - n'_0\cos\theta_0\}}{n_I\{n_1\cos\theta_1 - n_0\cos\theta_0\}}\,, \tag{6.2.18}$$

where the primed angles and indices correspond to the values at wavelength λ_1. We reformulate the expression for $W'_{P'}$ as

$$\begin{aligned} W'_{P'} &= \frac{K_{F,j}\lambda_1}{n'_I}\left[\frac{n'_I}{n_I}\right]\left[\frac{n'_1\cos\theta'_1 - n'_0\cos\theta_0}{n_1\cos\theta_1 - n_0\cos\theta_0}\right]\left(\frac{\lambda_0}{\lambda_1}\right) \\ &\approx K_{F,j}\lambda_{1,I}\left(\frac{\lambda_0}{\lambda_1}\right)\,, \end{aligned} \tag{6.2.19}$$

where $\lambda_{1,I}$ is the wavelength in image space for a vacuum wavelength λ_1. The quantity $K_{F,j}\lambda_{1,I}$ equals an integer number of wavelengths during 'playback' of the diffractive optical element using the vacuum wavelength λ_1.

The approximation on the last line of the equation above is justified because the changes in refractive index and ray angle are of second order in the change in wavelength. The resulting pathlength difference in a Fresnel structure as well as in a diffractive optical element is linear in the wavelength change and is a much stronger effect than the dispersion effects observed with the refraction of rays. It is also noteworthy that the sign of the angular dispersion is opposite for the refraction and diffraction of rays. In that sense, it is possible to define an Abbe V-number for the dispersion of a focusing diffractive element or Fresnel zone plate. The power of such an element is proportional to the wavelength of the impinging light and the V-number then equals

$$V = \frac{K_d}{K_F - K_C} = \frac{\lambda_d}{\lambda_F - \lambda_C} = -3.45\,. \tag{6.2.20}$$

The extremely large dispersion number of diffractive elements, with a sign which is opposite to that of refractive elements, can be exploited for colour correction in optical systems. Details on first- and higher-order dispersion correction using a combination of diffracting and refracting surfaces can be found in **[185]**.

6.2.2 Analytic Design of an Isoplanatic System

The aspherisation of an optical surface which was demonstrated in the preceding subsection enabled the pairwise connection of a general ray from the axial input pencil of rays to some ray from the pencil that was traced back from the axial image point. In this subsection we go one step further and impose the connection of predetermined pairs of rays. A single aspheric surface is generally not capable of achieving this. The aspherisation of two surfaces in an optical system is needed to achieve such a specific ray mapping from object to image space. In Section 4.8 a ray mapping requirement was discussed which produces constancy of aberration in a small linear region around the axial point of the image field. In the case that the optical system is stigmatic on axis, the small region remains free of aberration and the ray mapping condition is the Abbe sine condition, see Eq. (4.8.4). Basically, a ray mapping condition determines in which points a ray intersects the entrance pupil sphere and the exit pupil sphere. Although the Abbe sine condition has an immense importance when imaging over an extended image field is required, some other conditions are also applied in practice. The most noticeable is the Herschel condition which also produces constancy of illumination over the entrance and exit pupil surfaces. In what follows we will focus on the ray mapping from object to image space according to the Abbe sine condition. The first design of an aspheric pair of reflecting surfaces was given by Schwarzschild **[300]**. The design obeyed the Abbe sine condition and was meant to create a telescope with an extended field of view. At the same time, the design of an aplanatic lens was proposed by Linnemann **[212]**. In both cases, the aspheric surfaces were direct neighbours and the first surfaces encountered from object and image side. The systematic design of two neighbouring aspheric surfaces in a more general optical system was given in **[359]** and later extended to the case where other reflecting or refracting surfaces are between the two aspheres to be constructed **[352]**,**[53]**.

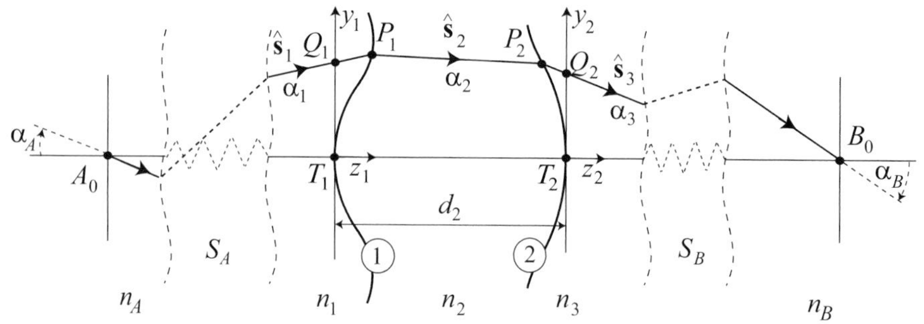

Figure 6.7: Design of an aplanatic optical system with the aid of two adjacent aspheric surfaces, labelled 1 and 2 in the figure.

6.2.2.1 Two Neighbouring Aspheric Surfaces

Figure 6.7 shows the case of an imaging system with the axial positions T_1 and T_2 where two optical surfaces are to be given an aspheric shape. The first surface with its vertex at T_1 is separated from object space by a known optical system S_A. Tracing a pencil of rays from the axial object point A_0 to the plane perpendicular to the axis through T_1 yields for each ray the point of intersection Q_1 with lateral coordinate y_1 and ray angle α_1 in the medium with index n_1. Tracing a pencil of rays backward through the known system S_B from the axial image point B_0 yields the points of intersection Q_2 of these rays with the vertex plane through T_2 and the corresponding ray angle α_3 in the medium with index n_3.

The ray $\hat{\mathbf{s}}_2 = (0, \sin\alpha_2, \cos\alpha_2)$ leaves the first aspheric surface and is directly incident on the second aspheric surface, establishing the connection between the two aspheric surfaces. The ray angles and ray intersection heights of the two pencils of rays in the planes through T_1 and T_2 are given by $M_1(\alpha_A)$ and $y_{Q,1}(\alpha_A)$ for the pencil from A_0 and $M_3(\alpha_B)$ and $y_{Q,2}(\alpha_B)$ for the backward propagated pencil from B_0. The angles α_A and α_B are the ray angles in object and image space. The ray transfer through the surfaces obeys Snell's law in vector form, $n_{j+1}\,\hat{\mathbf{v}} \times \hat{\mathbf{s}}_{j+1} \times \hat{\mathbf{v}} = n_j\,\hat{\mathbf{v}} \times \hat{\mathbf{s}}_j \times \hat{\mathbf{v}}$, where $\hat{\mathbf{v}}$ is the unit normal vector at a general point P_j on the surface with index j. A scalar multiplication by a tangent vector $\boldsymbol{\tau}_j$ in the plane containing the incident and refracted rays yields the equation

$$\left\{n_{j+1}\hat{\mathbf{s}}_{j+1} - n_j\hat{\mathbf{s}}_j\right\} \cdot \boldsymbol{\tau}_j = \mathbf{0}\,. \tag{6.2.21}$$

The tangent vector can be written as $(0, 1, dz_j/dy_j)$, yielding

$$n_{j+1}M_{j+1} - n_jM_j + \frac{dz_j}{dy_j}\left\{n_{j+1}N_{j+1} - n_jN_j\right\} = 0\,, \tag{6.2.22}$$

where M and N are the direction cosines of the ray vectors. Furthermore, we have for the coordinates of the points of intersection $P_1 = (0, y_1, z_1)$ and $P_2 = (0, y_2, z_2)$ on the two aspheric surfaces the relations

$$\begin{aligned} y_1(\alpha_A) &= y_{Q,1}(\alpha_A) + z_1(\alpha_A)\tan\alpha_1(\alpha_A)\,, \\ y_2(\alpha_B) &= y_{Q,2}(\alpha_B) + (z_2(\alpha_B) - d_2)\tan\alpha_3(\alpha_B)\,, \end{aligned} \tag{6.2.23}$$

where we note that the vertex T_1 of the first surface to be aspherised is the origin for the surface coordinate pairs (y_1, z_1) and (y_2, z_2) associated with the points P_1 and P_2 on the two surfaces. Omitting the explicit dependencies of the coordinates on α_A we can write Snell's law for the first surface as

$$\frac{dz_1}{d\alpha_A} = \frac{n_1M_1 - n_2M_2}{n_1N_1 - n_2N_2}\,\frac{dy_1}{d\alpha_A}\,. \tag{6.2.24}$$

We then take the derivative of $y_1(\alpha_A)$ in Eq. (6.2.23) with respect to α_A to obtain the differential equation

$$\frac{dz_1}{d\alpha_A} = \left(\frac{z_1}{N_1^3}\frac{dM_1}{d\alpha_A} + \frac{dy_{Q,1}}{d\alpha_A}\right) \Bigg/ \left(\frac{n_1N_1 - n_2N_2}{n_1M_1 - n_2M_2} - \frac{M_1}{N_1}\right), \tag{6.2.25}$$

where we have used that $d\alpha_j/d\alpha_A = (dM_j/d\alpha_A)/\cos\alpha_j$. In a comparable way, the application of Snell's law at the second aspheric surface yields the equation

$$\frac{dz_2}{d\alpha_A} = \frac{d\alpha_B}{d\alpha_A}\left(\frac{z_2}{N_3^3}\frac{dM_3}{d\alpha_B} + \frac{dy_{Q,2}}{d\alpha_B}\right) \Big/ \left(\frac{n_2N_2 - n_3N_3}{n_2M_2 - n_3M_3} - \frac{M_3}{N_3}\right). \tag{6.2.26}$$

Equations (6.2.25) and (6.2.26) constitute two simultaneous differential equations, linked by the Abbe aplanatic condition $n_A \sin\alpha_A = m_T n_B \sin\alpha_B$. Differentiation of the sine condition yields the expression

$$\frac{d\alpha_B}{d\alpha_A} = \frac{n_A \cos\alpha_A}{m_T n_B \cos\alpha_B}, \tag{6.2.27}$$

where m_T is the transverse magnification of the optical system. The quantities $M_1(\alpha_A)$, $M_3(\alpha_B)$, $y_{Q,1}(\alpha_A)$ and $y_{Q,2}(\alpha_B)$ are obtained from ray-tracing the object- and image-side pencils up to the vertex planes through T_1 and T_2. Stable and accurate polynomial fits of these functions are easily obtained, yielding also reliable values for the derivative functions in the equations above. M_2 follows from the coordinates of P_1 and P_2,

$$M_2 = \frac{y_2 - y_1}{\sqrt{(y_2 - y_1)^2 + (d_2 + z_2 - z_1)^2}}, \tag{6.2.28}$$

where d_2 equals the spacing T_1T_2, for instance, the thickness of the lens of which the two surfaces have to be aspherised.

Standard numerical routines are available to integrate the two differential equations, starting with the standard initial conditions $z_1 = z_2 = 0$ for $\alpha_A = \alpha_B = 0$. But it is equally possible to choose two initial points that are not on the optical axis and perform the integration towards the optical axis. It is even possible to use a single off-axis point as starting point for the two surfaces such that a lens with an acute edge will be found. In all cases, an optical system results which rigorously obeys the ray mapping condition. In the case of the Abbe sine condition, the system will be free from all orders of linear coma. The accuracy of the numerical integration scheme, for instance a four-point Runge-Kutta procedure **[283]**, can be made of the order of a tiny fraction of the wavelength of the light. A typical step in angle α_A of 0.005 radians is a good compromise between calculation speed and surface accuracy. In some cases, with awkward starting points, one or both of the surface slopes dz_j/dy_j can become very small, leading to large angles of incidence on the surface. Ray failure on a surface or total internal reflection may occur and the convergence of the integration procedure breaks down. The integration procedure can also be used in the case of a reflecting surface. The convention of negative refractive indices in mirror space has to be applied when using Eqs. (6.2.25)–(6.2.27).

6.2.2.2 Two Aspheric Surfaces with Known Optical Surfaces in Between

Figure 6.8 shows the case of an imaging system with the axial positions T_1 and T_k, separated by a number of known optical surfaces, labelled 2 to $k-1$. The surfaces 1 and k are to be given an aspheric shape. The first surface at T_1 is separated from object space by a known optical system S_A. In the same way as we did in the preceding paragraph, the tracing of a pencil of rays from the object point A_0 to the plane perpendicular to the axis through T_1 yields for each ray

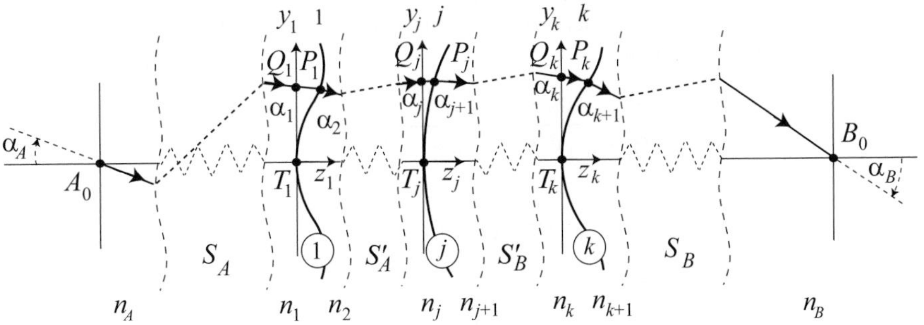

Figure 6.8: Design of an aplanatic optical system with the aid of two aspheric surfaces, labelled 1 and k in the figure. The two aspheric surfaces are separated by a number of known optical surfaces, with their labels j running from 2 to $k-1$. A general surface with label j ($2 \leq j \leq k-1$) has been explicitly represented in the drawing.

the point of intersection Q_1 with lateral coordinate y_1 and the direction cosine M_1 in the medium with index n_1. Tracing a pencil of rays backward through the known system S_B from the image point B_0 yields the points of intersection Q_k of these rays with the vertex plane through T_k and the corresponding ray direction cosine M_{k+1} in the medium with index n_{k+1}. We then fit these ray data using polynomial expansions with respect to the ray starting angle α_A in object space and the arriving angle α_B in image space.

Exactly in line with the analysis of the refraction/reflection at the surfaces 1 and 2 in the preceding subsection we obtain the differential equations for the surfaces 1 and k,

$$\frac{dz_1}{d\alpha_A} = \left(\frac{z_1}{N_1^3}\frac{dM_1}{d\alpha_A} + \frac{dy_{Q,1}}{d\alpha_A}\right) \Big/ \left(\frac{n_1N_1 - n_2N_2}{n_1M_1 - n_2M_2} - \frac{M_1}{N_1}\right) ,$$
$$\frac{dz_k}{d\alpha_A} = \frac{d\alpha_B}{d\alpha_A}\left(\frac{z_k}{N_{k+1}^3}\frac{dM_{k+1}}{d\alpha_B} + \frac{dy_{Q,k}}{d\alpha_B}\right) \Big/ \left(\frac{n_kN_k - n_{k+1}N_{k+1}}{n_kM_k - n_{k+1}M_{k+1}} - \frac{M_{k+1}}{N_{k+1}}\right) . \qquad (6.2.29)$$

The quantities M_2 and M_k in these equations are connected by the ray propagation through the intermediate surfaces but do not have a direct connection as was the case in the previous problem with two neighbouring aspheric surfaces. The refraction or reflection at each intermediate surface labelled j with $j = 2, \cdots, k-1$ is governed by Snell's law, see Eq. (6.2.22), which we formally write as

$$F_j(y_{j-1}, y_j, y_{j+1}) = 0 , \qquad (6.2.30)$$

where we have used the property that the direction cosines in Eq. (6.2.22) can be expressed in terms of the (y, z) coordinates of the surfaces $(j-1, j, j+1)$ and that each coordinate z_j follows from y_j via the surface prescription of each surface.

When performing the numerical integration of the differential equations of Eq. (6.2.29), the ray and surface data for a certain parameter increment $\Delta\alpha_A$ have to be found, starting from the solution at a certain parameter value $\alpha_{A,p}$. The index p refers to the pth step in the integration procedure, counted from the commonly used initial points $z_1 = z_k = 0$ for $\alpha_A = \alpha_B = 0$. When Snell's law is satisfied on all surfaces for the rays with parameter $\alpha_{A,0}$, the system of equations of (6.2.30) should also be satisfied for the ray parameter value $\alpha_{A,0} + \Delta\alpha_A$. In terms of the y-coordinates on the surfaces $3, \cdots, (k-2)$ we require

$$F_j(y_{j-1} + \Delta y_{j-1}, y_j + \Delta y_j, y_{j+1} + \Delta y_{j+1}) = \frac{\partial F_j}{\partial y_{j-1}}\Delta y_{j-1} + \frac{\partial F_j}{\partial y_j}\Delta y_j + \frac{\partial F_j}{\partial y_{j+1}}\Delta y_{j+1} = 0 . \qquad (6.2.31)$$

For the surfaces 2 and $k-1$ that are neighbours of the surfaces to be asphersised, we find the expressions

$$\frac{\partial F_2}{\partial y_2}\Delta y_2 + \frac{\partial F_2}{\partial y_3}\Delta y_3 = -F_2(y_1 + \Delta y_1, y_2, y_3) ,$$
$$\frac{\partial F_{k-1}}{\partial y_{k-2}}\Delta y_{k-2} + \frac{\partial F_{k-1}}{\partial y_{k-1}}\Delta y_{k-1} = -F_{k-1}(y_{k-2}, y_{k-1}, y_k + \Delta y_k) . \qquad (6.2.32)$$

The increments Δy_1 and Δy_k follow from the successive step-sizes in the Runge-Kutta integration scheme.

From the expressions for M_j and M_{j+1} as a function of the y- and z-coordinates on the surfaces the following expressions for the derivative functions above are found,

$$\frac{\partial M_j}{\partial y_j} = \frac{(d_j + z_j - z_{j-1})^2}{\{(y_j - y_{j-1})^2 + (d_j + z_j - z_{j-1})^2\}^{3/2}} = -\frac{\partial M_j}{\partial y_{j-1}} ,$$
$$\frac{\partial M_{j+1}}{\partial y_j} = -\frac{(d_{j+1} + z_{j+1} - z_j)^2}{\{(y_{j+1} - y_j)^2 + (d_{j+1} + z_{j+1} - z_j)^2\}^{3/2}} = -\frac{\partial M_{j+1}}{\partial y_{j+1}} . \qquad (6.2.33)$$

For the derivatives of the functions F_j we obtain,

$$\frac{\partial F_j}{\partial y_{j-1}} = n_j\left(1 - \frac{M_j}{N_j}\frac{dz_j}{dy_j}\right)\frac{\partial M_j}{\partial y_{j-1}} ,$$
$$\frac{\partial F_j}{\partial y_j} = n_{j+1}\left(1 - \frac{M_{j+1}}{N_{j+1}}\frac{dz_j}{dy_j}\right)\frac{\partial M_{j+1}}{\partial y_j}$$

$$- n_j \left(1 - \frac{M_j}{N_j}\frac{dz_j}{dy_j}\right)\frac{\partial M_j}{\partial y_j} + \left(n_{j+1}N_{j+1} - n_jN_j\right)\frac{d^2 z_j}{dy_j^2},$$

$$\frac{\partial F_j}{\partial y_{j+1}} = n_{j+1}\left(1 - \frac{M_{j+1}}{N_{j+1}}\frac{dz_j}{dy_j}\right)\frac{\partial M_{j+1}}{\partial y_{j+1}}. \tag{6.2.34}$$

The Eqs. (6.2.31) and (6.2.32) can be formally written as a $(k-2) \times (k-2)$ matrix equation,

$$\begin{pmatrix} \frac{\partial F_2}{\partial y_2} & \frac{\partial F_2}{\partial y_3} & 0 & \cdots & 0 & 0 & 0 \\ \cdots & & & & & & \cdots \\ 0 & \cdots & \frac{\partial F_j}{\partial y_{j-1}} & \frac{\partial F_j}{\partial y_j} & \frac{\partial F_j}{\partial y_{j+1}} & \cdots & 0 \\ \cdots & & & & & & \cdots \\ 0 & 0 & 0 & \cdots & 0 & \frac{\partial F_{k-1}}{\partial y_{k-2}} & \frac{\partial F_{k-1}}{\partial y_{k-1}} \end{pmatrix} \begin{pmatrix} \Delta y_2 \\ \cdots \\ \Delta y_j \\ \cdots \\ \Delta y_{k-1} \end{pmatrix} = \begin{pmatrix} -F_2(y_1 + \Delta y_1, y_2, y_3) \\ \cdots \\ 0 \\ \cdots \\ -F_{k-1}(y_{k-2}, y_{k-1}, y_k + \Delta y_k) \end{pmatrix}. \tag{6.2.35}$$

Each incremental step $\Delta(\alpha_A)$ of the integration routine requires a solution of the linear set of equations for the increments Δy_j of the intersection points of the ray with the intermediate surfaces. In general, the solution of the set of linear equations is a very fast and well-converging process, with typically less than ten iterations. When solving the differential equations, a good choice for the increment of the integration variable (in radians) is $10^{-3} \leq \Delta(\alpha_A) \leq 5 \cdot 10^{-3}$. This means that for a system with a high numerical aperture, the total number of surface points $p_{\max}$ on the constructed aspheres will be of the order of a hundred. The two sets of points (y_1, z_1) and (y_k, z_k) are fitted to a polynomial expansion by means of orthogonal polynomials. An optimum choice for the fitting polynomials is the Chebyshev polynomials **[4]** of the first kind and of order n, $T_n(t)$, which will show uniform convergence over the normalised t range $[0, 1]$ for the y values on the aspheres. In practice, using the property of a surface with symmetry of revolution, the interval is chosen $[-1, +1]$ and each calculated point (y_p, z_p) is doubled with its twin point $(-y_p, z_p)$. In this case the polynomial expansion will have only nonzero coefficients for the polynomials of even orders.

6.3 Merit Function of an Optical System

The analytic design methods discussed above are interesting design tools but, unfortunately, their application is limited to a relatively small class of optical systems. Basically, they produce single-point stigmatic or isoplanatic imaging systems. In the case of a system with symmetry of revolution, the design can be made free of spherical aberration and linear coma and produces a small imaging region around the axis with a virtually aberration-free imaging quality. For a system without rotational symmetry a single image point can be made aberration free and, using the Coddington equations of Eqs. (5.7.41) and (5.7.42), the paraxial astigmatism of the oblique pencil of rays can be made zero. For more general requirements, for instance good imaging quality over a large image field, third- and higher-order aberration theories can be used to evaluate the imaging defects and to improve the optical system quality (see **[62]**).

With the advent of powerful computers and elaborate numerical optimisation techniques, the focus is on the definition of a *merit function* of the optical system. This function mathematically represents the quality factors which are important for the user of the optical system. In this section we discuss the composition of such a merit function for optimisation purposes. In the next section, we present a brief overview of numerical optimisation techniques which use such merit functions to improve the imaging quality of the optical system.

A merit function can be ray-based or diffraction-based or a mixture of both. In the first case, the optical system is analysed by means of the tracing of a certain number of pencils of rays which are followed up to their intersection points with the image plane or image surface. The transverse aberration of each ray is used to evaluate the quality defects of the optical system. Such an analysis based on ray propagation according to geometrical optics is optimum when the aberration of the optical system is relatively large with respect to the 'diffraction unit' (see, for instance, Section 5.2 where this quantity was introduced). For systems that should have small aberrations, of the order of the wavelength of the light in terms of wavefront aberration, it is essential that diffraction-based quality criteria are used. In the next two subsections, the construction of a ray-based and a diffraction-based merit function is discussed.

6.3.1 Merit Function Based on Transverse Ray Aberration

The analysis of system quality with the aid of ray-tracing is mostly carried out by tracing pencils of rays originating at a chosen set of object points. In the case of a circularly symmetric system, the object points are arranged on a radial line segment in the object plane. A more systematic ray analysis uses a ray aberration expansion as a function of the aperture and image field coordinates. We first discuss the general-purpose approach which uses discrete pencils of rays and then present a ray aberration expansion which allows us to single out the coefficients of specific aberration functions from the ray-trace data.

6.3.1.1 Ray-tracing of Discrete Pencils of Rays

The transverse aberration of a pencil of rays in the image plane has been discussed in Section 4.9. In Fig. 6.9 we sketch the ray paths in an optical system of a principal ray DQ_D and a general aperture ray P_DQ_P. The rays originate from an object point Q_0. The principal ray is the ray that passes through the centre D of the physical diaphragm of the system, propagates through the image-side part of the optical system and intersects the exit pupil plane at the point E_1'. As shown in the figure, the presence of pupil aberration leads to a transverse aberration E_1E_1' of the principal ray in the exit pupil. Equation (4.9.3) gives the expression for the variance Δ^2 of the transverse aberration; Δ itself is usually called the rms blur of the imaging pencil. The choice of the reference point for calculating Δ depends on the optical application and is strongly related to the requirements on distortion and field curvature. Three possible choices for the reference point Q_1 are encountered in practice:

- *The requirements on field curvature and distortion are loose or absent.*
  The optimum reference point $Q_{1,o}(x_1, y_1, z_1)$ for the imaging pencil is derived from the transverse aberration values in a certain focal plane setting. The minimum transverse aberration is obtained by allowing a lateral shift and axial shift of the initial image-side reference point Q_1. If the receiving surface is axially shifted, each aperture ray of the pencil will intersect the surface at a new position which is laterally shifted with respect to the intersection point of the principal ray of the pencil with that surface (see Eq. (5.2.21)). At this new axial position, the smallest blur is then calculated using the squared distances between each ray intersection point (x, y) and the average values $(\bar{x}, \bar{y})$ of the intersection points of all rays in the pencil from a point Q_0 in the object plane. The average values of the lateral coordinates define the position of the 'centroid' of the pencil of rays. For an aberrated system, this position is generally different from the paraxial position. By iteration we find the optimum lateral and axial position of the reference point, $Q_{1,o}$, that produces

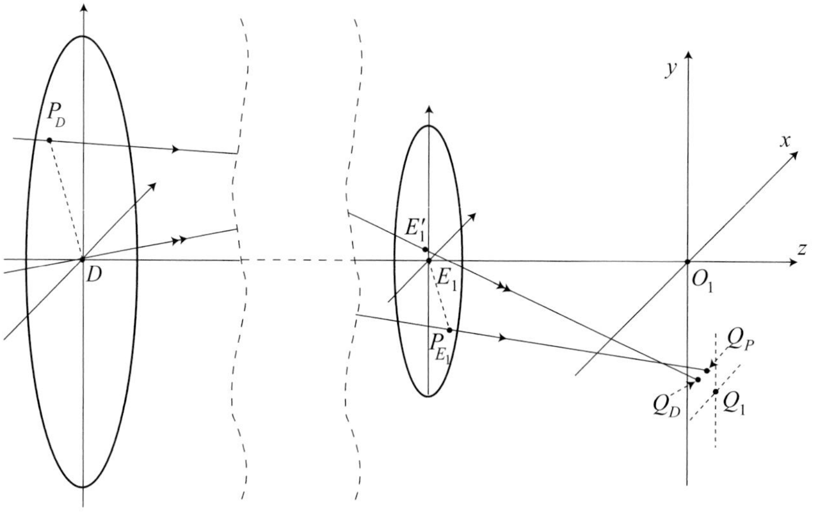

Figure 6.9: Ray propagation from the diaphragm to the exit pupil and the image region. The principal ray through D and an aperture ray through P_D have been shown in the figure. The transverse aberration of a ray is measured with respect to a reference point Q_1 with coordinates (x_1, y_1, z_1), with z_1 a defocusing of the reference point with respect to the nominal image plane. The vectors Q_1Q_D and Q_1Q_P are the transverse aberration vectors of the principal ray and of the aperture ray at the desired image position for the pencil of rays from the object point Q_0.

the minimum blur of the pencil in image space with respect to the centroid position. If the intensity distribution in the pencil of rays is not uniform, an intensity weighting factor can be given to each ray.

In general, as predicted by the third-order aberration theory, we will find a quadratically curved image surface on which all the image points Q_1 of points Q_0 in the object plane are imaged with minimum blur. Third-order theory also predicts the distortion which accompanies the imaging on such a curved image surface or on the paraxial flat image plane. The presence of higher-order aberrations at larger field angles produces a more intricate shape of the curved image surface on which the blur is minimised. There are also higher-order components of distortion present, comparable to those that were illustrated in Fig. 5.79. By systematically minimising the rms blur of a grid of image points Q_1 we thus obtain a sampled version of the curved surface on which the sharpest image is formed, together with the lateral shift of each optimised image point $Q_{1,o}$.

- *Distortion and field curvature have to satisfy certain requirements.*
  Specific values of the (x_1, y_1)- (and z_1-)coordinates of a reference point Q_1 are imposed to drive the final optimised system to these particular values.
- *Distortion and field curvature are not allowed.*
  In this case, the reference points to be chosen are found in a flat image plane and, with respect to their lateral position, they comply with the paraxial transverse magnification of the imaging system.

In an optimisation context, the quality characterisation should require a minimum of numerical effort. A limited number of imaging pencils with a small number of rays per pencil is used. In Fig. 6.10 we show a typical ray distribution over the diaphragm for an off-axis pencil. The cross-section of the pencil is approximated by four elliptical subsections with their centres at D, the axial point of the diaphragm. The weight of each ray in the optimisation is lowered in accordance with the reduction in surface area of the pencil in each quadrant. The central ray of the pencil has been denoted by C. In wide-angle systems, even more drastic vignetting may occur and, in that case, zero weight is given to the rays that were not transmitted to the image space. The distribution of the object points over the desired radial interval $[0, r_0]$ in image space can be made equidistant. A nonlinear sampling with higher density towards the edge of the field is recommended because of the quadratic and higher dependence on field position of aberrations such as distortion, field curvature and astigmatism. Quadratic or cosine-like distributions can be used according to

$$r_j = r_0 \sqrt{\frac{j - 1/2}{J}} ,$$
$$r_j = r_0 \cos\left\{\frac{\pi(2j - 1)}{4J}\right\} , \tag{6.3.1}$$

where J is the total number of field positions on the radial interval.

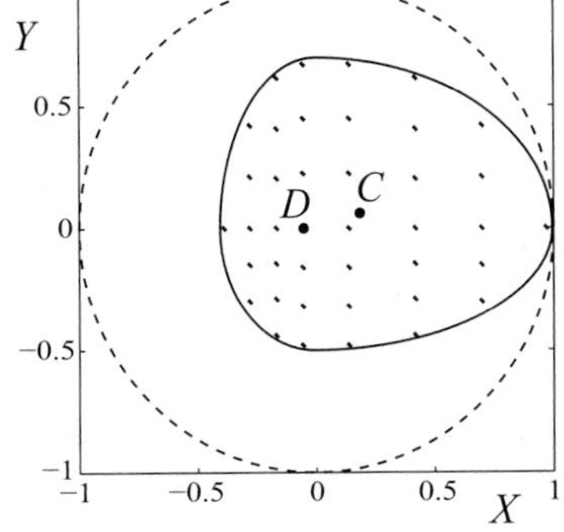

Figure 6.10: A ray grid in the physical diaphragm of the optical system in the presence of ray vignetting. The effective aperture of the pencil is approximated by four elliptical subsections with the original ray grid over the full circular diaphragm compressed in each quadrant. The principal ray intersects the diaphragm at D; the central or chief ray of the vignetted pencil has been denoted by C.

6.3.1.2 Transverse Aberration Expansion

In the case of negligible vignetting of the pencils of rays, it is possible to devise an expansion of the transverse ray aberration with respect to the aperture angle and the field position of a general ray, see Section 5.9. It is shown in this section that the reduction of the coefficients of Lukosz's wavefront polynomials can be used to minimise the transverse aberration components. The advantage of using expansion coefficients for optimisation is the possibility of isolating one single coefficient or a restricted subset of coefficients for the optimisation. In optical design problems this enables us to selectively reduce higher-order aberration coefficients that have been given a large weight while the more easily adjustable low-order aberrations are allowed to change more freely as they are given a much smaller weight in the optimisation process. In the final stage of optimisation, the low-order aberrations are also reduced by giving them a large weight. The numerical problem associated with this method is the reliable extraction of small higher-order aberration coefficients from much larger low-order terms. As soon as the evaluation of the aberration coefficients via a finite difference scheme is numerically ill-defined the optimisation becomes unstable.

6.3.2 Merit Function Based on Wavefront Aberration

Up to the 1950s the transverse aberration components were almost uniquely used for the optimisation of optical systems. The pioneering work of Hopkins **[140]** has gradually pushed the optical designers to switch from transverse to wavefront aberration as a criterion for system optimisation, at least for optical systems that operate within or close to the diffraction-limit of optical resolution. As a practical upper limit for wavefront analysis of aberration, we take an rms value of 0.5λ, measured over the exit pupil surface; the corresponding peak-to-valley aberration is $\approx 2\lambda$. The following wavefront aberration expansions are relevant if the maximum wavefront aberration is typically smaller than two to three wavelengths:

- If ray vignetting is negligible, a wavefront expansion with the aid of orthogonal Zernike polynomials can be used. Its expansion coefficients offer the same opportunity as discussed above for a transverse aberration expansion: the selective reduction of higher-order aberration coefficients. In practice, more than one image point is analysed in the required image field. Separate Zernike polynomial expansions are then used for each pencil of rays directed to a specific point in the image field.
- A mathematically more rigorous method is to use an expansion of the *global* aberration function of a circularly symmetric system with respect to both the polar pupil coordinates and the radial field coordinate. Such a 'pupil-field' expansion was discussed in Subsection 5.9.1. The expansion has coefficients $c_{n_1 n_2 m_1}$ according to Eq. (5.9.19) and offers the possibility to select for minimisation one or several wavefront aberration coefficients from the pupil-field expansion by attaching large weights to these coefficients in the merit function.
- For (initially) large wavefront aberrations, a third option is open. A weighted sum of wavefront and transverse aberration can be used for optimisation. In terms of the Zernike and Lukosz polynomials, intermediate radial polynomials $R_n^m(\rho) - p\,R_{n-2}^m(\rho)$ are used for wavefront expansion where p is some weighting factor. The transition range from wavefront to transverse aberration is covered by the interval $0 \leq p \leq 1$. More general p-values outside this interval permit enforcement of under-corrected or over-corrected wavefronts to the optimised imaging system (see Subsection 5.9.3.1).

6.3.3 Incorporation of Boundary Conditions in the Optimisation

Some boundary conditions can be adjusted without problems after optimisation. Examples of these are paraxial quantities such as the focal length of an imaging system or the optical throw from object to image plane. Such quantities are adjusted after system optimisation by a scaling operation. If the scaling factor is in the percentage range, the wavefront or transverse aberration is not substantially affected. Other quantities such as a field angle value, a telecentric pupil position or a certain distortion value for an image point have to be incorporated as active boundary conditions during optimisation. The optimisation is said to be constrained by the boundary conditions.

If a boundary condition needs to be exactly satisfied (equality constraint), it is usually incorporated by means of a Lagrange multiplier. In practice this means that a condition is imposed on the set of independent variables such that the number of independent variables is effectively reduced by one. Inequality constraints require a certain function g of the optical system to be found in a single-sided interval ($g(\mathbf{x}) > g_0$ or $g(\mathbf{x}) < g_0$) or in a two-sided finite interval ($g_0 \leq g(\mathbf{x}) \leq g_1$). An example of a single-sided boundary condition is a vignetting function that should not exceed a certain

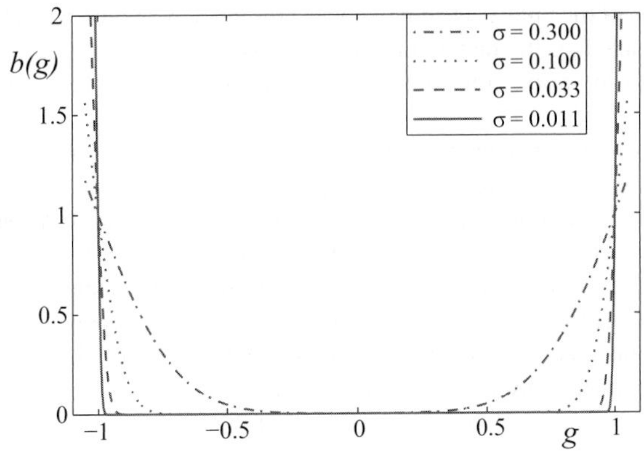

Figure 6.11: Example of a penalty function $b(g)$ which is incorporated in the optical merit function to account for a two-sided boundary inequality constraint during optimisation. The penalty function parameter is σ.

maximum value. In general, the inequality boundary conditions are appended as penalty functions to the merit function of the system. As long as a boundary function is inside the permitted interval but far away from its limits, the function is made very small. Outside the permitted interval the penalty function grows quadratically. A practical example of a penalty function $b\{g(\mathbf{x})\}$ that keeps the boundary function $g(\mathbf{x})$ confined to the two-sided (normalised) interval $[-1, +1]$ is given by

$$b\{g(\mathbf{x})\} = \begin{cases} \left(\dfrac{g+1}{2\sigma} - 1\right)^2 , & g < -1 , \\ \exp(1/2) \left[\exp\left\{-\dfrac{(g+1+\sigma)^2}{2\sigma^2}\right\} + \exp\left\{-\dfrac{(g-1-\sigma)^2}{2\sigma^2}\right\}\right] , & |g| \leq 1 , \\ \left(\dfrac{g-1}{2\sigma} + 1\right)^2 , & g > 1 . \end{cases} \tag{6.3.2}$$

The function $b(g)$ is shown in Fig. 6.11. It is seen that a small value of the penalty parameter σ, e.g. $\sigma = 0.033$, imposes a virtually zero penalty in the major part of the interval. Close to the interval limits a steep exponential increase of the penalty occurs. The function $b(g)$ itself and its first derivative are continuous in $|g(\mathbf{x})| = 1$.

6.4 Optimisation of Optical Systems

In the practice of optical design, two stages can be identified. The most important is the definition of a suitable initial design or design prototype. Real breakthroughs in optical system design require revolutionary new prototype designs. In many cases, more modest design requirements ask for an adaptation or elaboration of an existing solution to new, more demanding design specifications. The prototypes are then readily found from the existing scientific and patent literature. A wealth of existing and promising designs is available to the optical designer. The challenge is to find a fast and feasible design path to the new system with augmented imaging quality, to enhance the yield in manufacturing, to use less expensive and more durable materials, to achieve lower production cost, etc. In the past, the adaptation to such new requirements of an existing design was carried out by an experienced designer. Based on the predictions of aberration theory and with the aid of computing assistants who carried out the tedious tracing of finite rays, an improved optical system could be designed in a period of several months.

The advent of automatic computers as of the 1950s and the availability of versatile computer codes for optical design has drastically changed the design procedure. The period of months has been gradually reduced to days or less. The full analysis of an optical design with respect to issues such as manufacturing tolerances, measurement methods and cost analysis will take much more time. In this section we discuss the optimisation techniques which are used in optical design. The quantity to be optimised is a collection of performance indicators of the optical system. They can be mathematically combined into the *merit* function of the optical system that was presented in Section 6.3. The function contains first-order (paraxial) imaging properties including the maximum construction length and volume, aberration values of individual rays or of entire pencils of rays to various imaging points, values of the modulation transfer function, sensitivity to fabrication errors, etc.

The optimisation techniques can be separated into two classes. The first class is limited to the local optimisation techniques. They steer a design to best performance by a relatively limited change of the system parameters in the close environment of the starting point. The linear optimisation techniques that were originally developed for process optimisation in, for instance, the chemical process industry, have been transferred to optical system design **[97]**. Because of the rather steep changes in performance of an optical system in the presence of small parameter changes, the linear optimisation procedures have to be protected against large excursions that would move the optical system out of a feasible region.

A second class of optimisation techniques aims at global optimisation of the optical system. Mechanisms are introduced to explore regions further away from the direct local optimum in order to find more remote minima. The purpose is to obtain complete information about the 'landscape' of the optical merit function as a function of the design parameters. These techniques are still under development with the required time to find the global minimum being a key parameter for their practicality.

6.4.1 Local Optimisation of an Optical System

The local optimisation of a system starts with the definition of the optical merit function, the definition of the system parameters that can be used as optimisation variables and a description of the boundary conditions that have to be respected by the final optimised design. We first concentrate on the design variables and the optical merit function. The exact treatment of the boundary conditions is discussed later. In general, the merit function consists of a certain number of function values which represent the quality or state of correction of the optical system. The individual functions which compose the merit function can be the ray or wave aberration of judiciously chosen rays that leave a certain object point in the object field and intersect the entrance pupil sphere or the physical diaphragm at a specified point. Such raw aberration data may be replaced by expansion coefficients of a transverse ray aberration or a wave aberration function in certain basis functions, for instance Zernike polynomials; this approach was outlined in Subsection 6.3.2. The coefficients of such an expansion are representative of a certain type of aperture- and field-dependent aberration of the optical system. If this is relevant, the effect on optical quality of element fabrication and mounting errors can also be included in the merit function. In all these cases, the smaller the merit function value, the higher the quality of the system and the optimisation should lead to a minimisation of the merit function. The variables in the optimisation problem are typically the curvatures of optical surfaces, aspheric coefficients of such surfaces, mutual distances between optical elements and their thicknesses, refractive indices or glass types, coefficients that describe the geometry of the diffracting structure on a diffractive optical element, the coefficients that describe the refractive index of inhomogeneous media, etc.

Figure 6.12 shows the various cases which can be expected when optimising an optical system. In a1), the number of function values m_0 and the number of variables n_0 is equal. The system of equations can be solved exactly and the rank R of the system matrix equals $m_0 = n_0$. If for some reason the calculation of the function values suffers from numerical noise or, for instance, from inaccuracies in the series expansions of optical surfaces, it is not possible to obtain an exact solution of the system of equations. Figure 6.12b1 shows this situation. The equals sign has been replaced by the $\approx$ sign; the matrix is deficient and has a rank R which is smaller than $m_0 = n_0$.

If an optical system is optimised, the number of function values m_0 is mostly larger or much larger than the number of variables n_0 (see b1) and b2) of Fig. 6.12). The system of equations is over-determined. An approximate solution of the system is possible, in the sense that the function values reach their target values as closely as possible and the *least-squares* difference between the solution and the targeted value is minimised. The rank of the matrix equals n_0 or less when numerical instabilities influence the solution vector (see b2) of the same figure). The other cases for which $m_0 < n_0$ (see c1) and c2)) and the number of function values is small with respect to the number of variables are rather rare in optical system design. The system of equations is under-determined and the matrix rank R is $\leq m_0$. In such a case, formally, an infinite number of solutions is possible.

6.4.1.1 Linearised System of Equations

The system of equations to be solved is constructed by means of a linearisation of the merit function components $f_m(\mathbf{x})$ where $1 \leq m \leq m_0$. If we denote the values of the variables for the actual system by $\mathbf{x}_0$, the function values in the immediate vicinity of the working point $\mathbf{x} = \mathbf{x}_0$ are given by

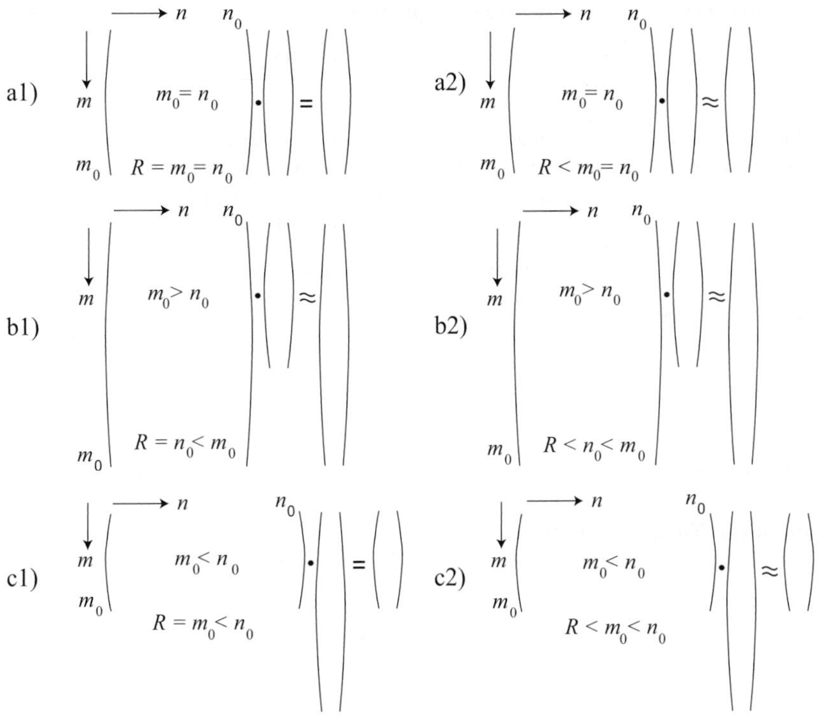

Figure 6.12: The various dimensions (m, n) of the systems of equations which are encountered when optimising an optical system and the rank R of the system matrix. In the panels a2), b2) and c2) the accuracy of the solution is lowered due to a reduced rank R of the system matrix.

$$\mathbf{f}(\mathbf{x}_0 + d\mathbf{x}) = \mathbf{f}(\mathbf{x}_0) + \begin{pmatrix} \frac{\partial f_1}{\partial x_1} & \cdots\cdots\cdots & \frac{\partial f_1}{\partial x_{n_0}} \\ \cdots\cdots & \frac{\partial f_m}{\partial x_n} & \cdots\cdots \\ \frac{\partial f_{m_0}}{\partial x_1} & \cdots\cdots\cdots & \frac{\partial f_{m_0}}{\partial x_{n_0}} \end{pmatrix} \begin{pmatrix} dx_1 \\ \cdots \\ dx_n \\ \cdots \\ dx_{n_0} \end{pmatrix} = \mathbf{f}(\mathbf{x}_0) + \mathbf{J}(d\mathbf{x}) . \tag{6.4.1}$$

A general matrix element $\partial f_m / \partial x_n$ of the matrix $\mathbf{J}$ is denoted by J_{mn}. The matrix $\mathbf{J}$ is also called the change table of the system. A Taylor expansion of each component f_m of the merit function vector $\mathbf{f}$ yields the expression

$$f_m(\mathbf{x}_0 + d\mathbf{x}) = f_m(\mathbf{x}_0) + \sum_n J_{mn} dx_n + \sum_n \sum_p H_{mnp}\, dx_n dx_p + \cdots , \tag{6.4.2}$$

where the matrix $\mathbf{J}$ is recognised as a Jacobian matrix and $\mathbf{H}$ is the Hessian tensor of the set of functions f_m. The inclusion of the Hessian tensor in the system of equations allows for a quadratic extrapolation of the functions f_m from the working point $\mathbf{x} = \mathbf{x}_0$. To find the variable changes that drive the system to the target function value $f(\mathbf{x}_0)$, the inclusion of second-order terms leads to a nonlinear system of equations in the variables x_n. Although the solution of the quadratic expression may lead to a fast and stable convergence towards the optimum value of the merit function, the initial investment in data preparation for the Hessian tensor is huge. The linear approximation according to Eq. (6.4.1) is the common approach for solving the optimisation problem.

An estimate of the linear Jacobian matrix element is obtained in practice by means of a finite difference calculation of the functions f_m, according to

$$J_{mn} = \frac{f_m(\mathbf{x}_0 + \Delta x_n/2) - f_m(\mathbf{x}_0 - \Delta x_n/2)}{\Delta x_n} . \tag{6.4.3}$$

For the element H_{mnp} of the Hessian tensor we have the expression

$$H_{mnp} = \frac{f_m(\mathbf{x}_0 + \Delta x_n + \Delta x_p) - f_m(\mathbf{x}_0 + \Delta x_n) - f_m(\mathbf{x}_0 + \Delta x_p) + f_m(\mathbf{x}_0)}{\Delta x_n \Delta x_p} . \tag{6.4.4}$$

The use of finite difference quotients instead of an exact derivative value is a source of potential instability for the solution of the system of equations, especially in the case of the second-order finite differences. This is an additional reason to omit these quotients from the system of equations. Once the Jacobian matrix elements J_{mn} have been calculated with the aid of the finite difference scheme, the linear system to be solved is given by

$$\mathbf{J}\,(\Delta\mathbf{x}) = \mathbf{f}(\mathbf{x}_0 + \Delta\mathbf{x}) - \mathbf{f}(\mathbf{x}_0) . \tag{6.4.5}$$

The function values in the vector $\mathbf{f}(\mathbf{x}_0 + \Delta\mathbf{x})$ constitute the target value vector $\mathbf{f}_t$ of the solution. In many cases, the elements of $\mathbf{f}_t$ will be simply zero, the equivalent of a perfect system. The solution vector $\Delta\mathbf{x}$ minimises the Euclidian length of the distance

$$\epsilon_d^2 = \left|\mathbf{J}(\Delta\mathbf{x}) - [\mathbf{f}_t - \mathbf{f}(\mathbf{x}_0)]\right|^2 , \tag{6.4.6}$$

where ϵ_d is the merit function residual at the optimum point in variable space.

6.4.1.2 Solution of the Normal Equations

In general, for $m_0 \neq n_0$, the matrix equation above is made square by multiplying both members by $\mathbf{J}^T$, the transpose of $\mathbf{J}$, yielding

$$(\mathbf{J}^T\mathbf{J})\,(\Delta\mathbf{x}) = \mathbf{J}^T\,[\mathbf{f}_t - \mathbf{f}(\mathbf{x}_0)] = \mathbf{J}^T(\Delta\mathbf{f}) ,$$

or,

$$\mathbf{A}\,(\Delta\mathbf{x}) = \Delta\mathbf{b} , \tag{6.4.7}$$

where the components of the vector $\Delta\mathbf{b}$ are related to the distance between the desired target values of the merit function and the actual values at the point $\mathbf{x} = \mathbf{x}_0$. The system of equations in (6.4.7) which uses the $n_0 \times n_0$ matrix $\mathbf{A}$ is called the system of normal equations for the n_0 unknowns, derived from the original rectangular system of equations with dimensions $m_0 \times n_0$. The matrix $\mathbf{A} = \mathbf{J}^T\mathbf{J}$ is symmetric and has real eigenvalues and orthogonal eigenvectors. If the matrix $\mathbf{A}$ has an inverse, the solution can formally be written as

$$\Delta\mathbf{x} = \mathbf{A}^{-1}(\Delta\mathbf{b}) . \tag{6.4.8}$$

The derivation above is formally correct but leads to unwanted numerical complications:

- The explicit calculation of the inverse of a matrix is a slow and unstable numerical process if the determinant has a small value. In practice, other numerical methods are implemented such as Cholesky factorisation **[23]** of the matrix or repeated pivoting operations of rows and columns of the matrix to obtain a more stable solution.
- The matrix $\mathbf{A}$ is the product of matrices $\mathbf{J}$ with a numerical inaccuracy given by some small quantity δ, due to the approximation of the derivative functions by finite differences. In the product matrix $\mathbf{A}$, the inaccuracies are of the order of δ^2 and to avoid loss of significant digits, the elements of matrix $\mathbf{A}$ have to be stored with doubled precision as compared to the elements of $\mathbf{J}$. Although memory storage requirements are not a real concern in optical optimisation problems, it is preferable to carry out the calculations with $\mathbf{J}$ itself.

The solution vector $\Delta\mathbf{x}$ of Eq. (6.4.8) is unsatisfactory in virtually all cases because it produces a solution that is far out of the region where the linear approximation of the system of equations according to Eq. (6.4.5) is valid. A solution with variables given by $\mathbf{x}_0 + \Delta\mathbf{x}$ may correspond to a physically unfeasible region. Rays traced through the optimised system will go beyond the allowed diameter of a lens, fail to pass the physical diaphragm or get totally reflected at a transition between two media. For this reason, a 'braking' or regularising mechanism is used when optimising a merit function with a rather nonlinear behaviour. The squared residual ϵ_d^2 of the merit function, originally defined by Eq. (6.4.6), is modified into

$$[\mathbf{J}(\Delta\mathbf{x}) - \Delta\mathbf{f}]^T[\mathbf{J}(\Delta\mathbf{x}) - \Delta\mathbf{f}] + p(\Delta\mathbf{x})^T(\Delta\mathbf{x}) . \tag{6.4.9}$$

The system of equations according to (6.4.7) is extended as follows:

$$(\mathbf{A} + p\mathbf{I})\,\Delta\mathbf{x} = \Delta\mathbf{b} . \tag{6.4.10}$$

$\mathbf{I}$ is the identity matrix and p a so-called (positive) damping factor. Using the eigenvalues λ_n and normalised eigenvectors $\mathbf{a}_n$ of matrix $\mathbf{A}$ that follow from the equation $(\mathbf{A} - \lambda_n\mathbf{I})\,\mathbf{a}_n = \mathbf{0}$, we can write

$$\Delta\mathbf{x} = \sum_n s_n\mathbf{a}_n \,, \quad \Delta\mathbf{b} = \sum_n b_n\mathbf{a}_n \,, \tag{6.4.11}$$

and Eq. (6.4.10) is then transformed into

$$\sum_n s_n(\lambda_n + p)\mathbf{a}_n = \sum_n b_n\mathbf{a}_n \,. \tag{6.4.12}$$

Since the eigenvectors $\mathbf{a}_n$ are orthogonal, we obtain the solution components

$$s_n = \frac{b_n}{\lambda_n + p} \,. \tag{6.4.13}$$

The λ_n are all real because the matrix $\mathbf{A}$ is symmetric. Moreover, with the extra property $\mathbf{A} = \mathbf{J}^T\mathbf{J}$, it follows that for any nonzero vector $\mathbf{v}$ the quadratic expression $(\mathbf{v}^T\mathbf{J}^T)(\mathbf{J}\,\mathbf{v}) = |\mathbf{J}\,\mathbf{v}|^2$ is non-negative and $\mathbf{A}$ is a positive-semidefinite matrix. All its diagonal elements are non-negative because they are each given by the sum of squares of real numbers. The eigenvalues λ_n of the positive-semidefinite matrix $\mathbf{A}$ are thus all non-negative (see **[40]**, App. C, p. 261). In consequence, when choosing a positive value for p, none of the coefficients s_n risks becoming singular. The length of the solution vector $\Delta\mathbf{x}$ is then given by

$$|\Delta\mathbf{x}|^2 = \left|\sum s_n\mathbf{a}_n\right|^2 = \sum_n \frac{b_n^2}{(\lambda_n + p)^2} \,. \tag{6.4.14}$$

Only the largest eigenvectors of matrix $\mathbf{A}$ with eigenvalues $\lambda_n > p$ substantially contribute to the solution vector. The relative contribution δ_n of an eigenvector $\mathbf{a}_n$ to the solution vector in the case of a nonzero damping factor p is given by

$$\delta_n^2 = \frac{\lambda_n^2}{(p + \lambda_n)^2} \,. \tag{6.4.15}$$

Small eigenvalues which are, in general, unreliable due to numerical inaccuracy are now ruled out from the solution and the length of the solution vector is limited. The extra term $p|\Delta\mathbf{x}|^2$ in the merit function imposes a penalty on large excursions $|\Delta\mathbf{x}|^2$. This feature, the so-called damped least-squares minimisation, was first introduced by Levenberg **[207]**. It was further refined by various authors and made available to the community via a software implementation by Marquardt **[239]**. Therefore, it is commonly referred to as the Levenberg–Marquardt damped least-squares stabilisation method.

The expression for s_n of Eq. (6.4.13) is exact within the linearised function approximation. When the solution coefficients have to be found numerically, the method described above is rather inefficient. It requires a complete analysis of the matrix $\mathbf{A}$ with respect to eigenvalues and eigenvectors and a calculation of the inner products of the vector $\Delta\mathbf{b}$ and the eigenvectors $\mathbf{a}_n$. An efficient numerical calculation uses Cholesky factorisation, writing the symmetric matrix $\mathbf{A}$ as the product of an upper and a lower triangular matrix, $\mathbf{P}$ and $\mathbf{P}^T$, respectively, **[203]**. With the aid of the triangular matrix $\mathbf{P}$ a straightforward sequential calculation of the components Δx_n of the solution vector $\Delta\mathbf{x}$ is possible.

6.4.1.3 Singular Value Decomposition of the Matrix J

A solution of the original rectangular system of equations according to Eq. (6.4.5) without using the normal equations is possible by applying a singular value decomposition to the matrix $\mathbf{J}$. An extensive description of this method can be found in **[203]**. The rectangular matrix $\mathbf{J}$ is decomposed into a matrix product,

$$\mathbf{J}_{m_0,n_0} = \mathbf{U}_{m_0,m_0}\mathbf{S}_{m_0,n_0}\mathbf{V}^T_{n_0,n_0} \,, \tag{6.4.16}$$

where the indices give the number of rows and the number of columns of each matrix. The square matrix $\mathbf{U}$ is a unitary matrix which contains as column vectors $\mathbf{u}_m$ the left-handed singular vectors that orthogonalise the space in which the functions f_m are found. The unitary matrix $\mathbf{V}$ contains the orthonormal vectors $\mathbf{v}_n$ which orthogonalise the variable space in which a solution vector $\Delta\mathbf{x}$ is found. Because of their special properties, the matrices $\mathbf{U}$ and $\mathbf{V}$ obey the relations

$$\mathbf{U}^T\mathbf{U} = \mathbf{I}_{m_0}, \qquad \mathbf{V}^T\mathbf{V} = \mathbf{I}_{n_0} \,, \qquad \mathbf{U}^T = \mathbf{U}^{-1} \,, \qquad \mathbf{V}^T = \mathbf{V}^{-1} \,, \tag{6.4.17}$$

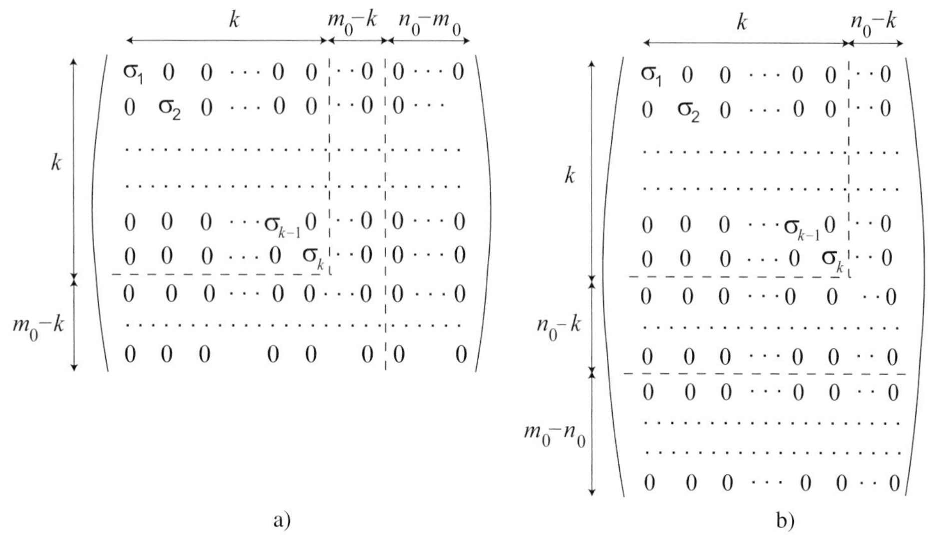

Figure 6.13: Layout of the central matrix **S** with dimensions $m_0 \times n_0$ in the singular value decomposition $\mathbf{J} = \mathbf{U}\,\mathbf{S}\,\mathbf{V}^T$.
a) $n_0 > m_0$, the system of equations is under-determined.
b) $n_0 < m_0$, an over-determined system.
The rank of the submatrix containing the eigenvalues σ is k.

where $\mathbf{I}_{m_0}$ and $\mathbf{I}_{n_0}$ are identity matrices with dimension m_0 and n_0, respectively. It can also be proven that the orthogonal vectors $\mathbf{u}_m$ are the eigenvectors of the matrix $\mathbf{J}\,\mathbf{J}^T$. In a similar way, the orthogonal set $\mathbf{v}_n$ are the eigenvectors of the matrix $\mathbf{J}^T\mathbf{J}$ and equal to the vectors $\mathbf{a}_n$ that were used when solving the system of normal equations in the preceding paragraph. The central matrix **S** collects the singular values σ_k which correspond to the matrix decomposition. The σ_k are the square roots of the eigenvalues λ_k of the matrix $\mathbf{J}^T\mathbf{J}$.

Figure 6.13 shows the layout of the central matrix **S**. In the case of an under-determined system of equations, we have the layout of Fig. 6.13a. Numerically reliable singular values σ_k have been found on the diagonal of a square submatrix with dimension $k \times k$. The remaining rows contain only zeros, as does the right-hand part of **S** in the columns $m_0 + 1$ to n_0. The more common situation of an over-determined system of equations is shown in b). Extra zero rows are found from index $n_0 + 1$ to m_0; zero columns are padded from indices $k + 1$ to n_0.

Having at our disposal the singular values σ_k in the matrix **S**, the system of equations can be solved in principle, however, the inverse of the rectangular matrix **S** is not defined. A pseudo-inverse has been proposed by Penrose **[274]**. The pseudo-inverse $\mathbf{S}^+$ of **S** is simply obtained by replacing the upper-left $k \times k$ square matrix of **S** by a matrix with the inverted values $1/\sigma_k$ on its diagonal. The dimension of $\mathbf{S}^+$ is $n_0 \times m_0$ and the remaining part of the inverse is obtained by zero padding outside the upper-left $k \times k$ square part. Equation (6.4.5) is then written as

$$\mathbf{U}\,\mathbf{S}\,\mathbf{V}^T(\Delta\mathbf{x}) = \mathbf{f}_t - \mathbf{f}(\mathbf{x}_0) = \Delta\mathbf{f}\,, \tag{6.4.18}$$

and the solution vector is given by

$$(\Delta\mathbf{x}) = \mathbf{V}_{n_0,n_0}\,\mathbf{S}^+_{n_0,m_0}\,\mathbf{U}^T_{m_0,m_0}\,\Delta\mathbf{f}\,. \tag{6.4.19}$$

A damping mechanism can be introduced in the solution of this system of equations by the minimisation of $[\mathbf{J}(\Delta\mathbf{x}) - \Delta\mathbf{f}]^T[\mathbf{J}(\Delta\mathbf{x}) - \Delta\mathbf{f}] + \sigma_d^2(\Delta\mathbf{x})^T(\Delta\mathbf{x})$, where σ_d^2 is a positive damping factor (see also Eq. (6.4.9)). Minimising the squared merit function requires the derivative of this function to become zero. For the derivative we have

$$\begin{aligned}\nabla_x\left(|\mathbf{J}(\Delta\mathbf{x}) - \Delta\mathbf{f}|^2 + \sigma_d^2\,|(\Delta\mathbf{x})|^2\right) &= \nabla_x\left\{(\Delta\mathbf{x})^T\mathbf{J}^T\mathbf{J}(\Delta\mathbf{x}) - 2(\Delta\mathbf{x})^T\mathbf{J}^T(\Delta\mathbf{f}) + (\Delta\mathbf{f})^T(\Delta\mathbf{f}) + \sigma_d^2(\Delta\mathbf{x})^T(\Delta\mathbf{x})\right\}\\ &= 2\mathbf{J}^T\mathbf{J}(\Delta\mathbf{x}) - 2\mathbf{J}^T(\Delta\mathbf{f}) + 2\sigma_d^2(\Delta\mathbf{x}) = 0\,,\end{aligned} \tag{6.4.20}$$

where the use of the ∇_x operator means that the derivatives with respect to all vector components x_n have to be taken. The solution of (6.4.20) follows from

$$\left(\mathbf{J}^T\mathbf{J} + \sigma_d^2\mathbf{I}\right)(\Delta\mathbf{x}) = \mathbf{J}^T(\Delta\mathbf{f}) \,,$$

or,

$$(\Delta\mathbf{x}) = \mathbf{J}^T\,(\mathbf{J}^T\mathbf{J} + \sigma_d^2\mathbf{I})^{-1}(\Delta\mathbf{f}) \;=\; \mathbf{J}^{+}(\Delta\mathbf{f}) \,, \tag{6.4.21}$$

where $\mathbf{J}^+$ is defined as a pseudo-inverse, identical to the $n_0 \times m_0$ rectangular matrix $\mathbf{J}^T(\mathbf{J}\mathbf{J}^T + \sigma_d^2\mathbf{I})^{-1}$. Using the decomposition $\mathbf{J} = \mathbf{U}\mathbf{S}\mathbf{V}^T$ we can write, step by step,

$$\begin{aligned}
\mathbf{J}^+ &= \mathbf{J}^T(\mathbf{J}\mathbf{J}^T + \sigma_d^2\mathbf{I})^{-1}\\
&= \mathbf{V}\mathbf{S}^T\mathbf{U}^T(\mathbf{U}\mathbf{S}\mathbf{V}^T\mathbf{V}\mathbf{S}^T\mathbf{U}^T + \sigma_d^2\mathbf{I})^{-1}\\
&= \mathbf{V}\mathbf{S}^T\mathbf{U}^T(\mathbf{U}\mathbf{S}\mathbf{S}^T\mathbf{U}^T + \sigma_d^2\mathbf{I})^{-1}\\
&= \mathbf{V}\mathbf{S}^T\mathbf{U}^T\left\{\mathbf{U}(\mathbf{S}\mathbf{S}^T + \sigma_d^2\mathbf{I})\mathbf{U}^T\right\}^{-1}\\
&= \mathbf{V}\mathbf{S}^T\mathbf{U}^T\mathbf{U}(\mathbf{S}\mathbf{S}^T + \sigma_d^2\mathbf{I})^{-1}\mathbf{U}^T\\
&= \mathbf{V}\mathbf{S}^T(\mathbf{S}\mathbf{S}^T + \sigma_d^2\mathbf{I})^{-1}\mathbf{U}^T = \mathbf{V}\mathbf{S}_\sigma^+\mathbf{U}^T \,,
\end{aligned} \tag{6.4.22}$$

where the pseudo-inverse matrix $\mathbf{S}_\sigma^+$ with dimensions n_0 x m_0 is given by

$$\mathbf{S}_\sigma^+ = \begin{pmatrix}
\sigma_1/(\sigma_1^2+\sigma_d^2) & \cdots & 0 & 0\ \cdots\ 0\\
0 & \cdots & 0 & 0\ \cdots\ 0\\
0 & \cdots\ \sigma_k/(\sigma_k^2+\sigma_d^2) & & 0\ \cdots\ 0\\
\cdots & \cdots & \cdots & \cdots\cdots\cdots\\
0 & \cdots & 0 & 0\ \cdots\ 0
\end{pmatrix} . \tag{6.4.23}$$

Consequently, in the presence of damping, each contribution to the solution following from the kth eigenvalue of $\mathbf{S}$ has a relative amplitude of $\sigma_k^2/(\sigma_k^2 + \sigma_d^2)$ with respect to the undamped solution. The influence of small and unreliable singular values, suffering from round-off errors and machine precision effects, can be effectively suppressed in the solution.

The singular value decomposition technique provides the user with useful information. It produces the orthogonal bases for the variable and function space and the relative importance of the various eigenvectors in the solution space via the corresponding singular value σ_k. From the composition of the most important singular vectors, it is possible to extract the mutual influence of the physical variables. The eigenvectors of the function space show which physical functions have comparable impact on the system merit function and which ones are possibly conflicting. These functions are then better removed from the merit function set. This extra information comes at the expense of a substantially larger number of calculation steps, albeit with half the number of significant digits.

6.4.1.4 Second-order Improvement of the Solution

An interesting improvement of the solution method described above is the introduction of data relative to the second derivative of the function values (Hessian tensor). We start with the squared merit function itself,

$$F(x_1, \cdots, x_{n_0}) = \sum_{m=1}^{m_0} f_m^2(x_1, \cdots, x_{n_0}) \,. \tag{6.4.24}$$

A weighting factor w_m can be included in each merit function component f_m and a fixed quantity $f_{t,m}$, the target value, can be subtracted from it. To find an extremum value of F, we need to find the zero value of each partial derivative $\partial F/\partial x_n$. To locate the zero-crossing position $x = x_0$ of a one-dimensional function $f(x)$ that is given at the point $x = x_P$, Newton's root finding procedure can be used,

$$x_0 - x_P = -\frac{f(x_P)}{\left(\frac{df}{dx}\right)_{x_P}} . \tag{6.4.25}$$

Using Newton's expression in the multi-dimensional space $(x_1, \cdots, x_{n_0})$, we write for the roots of the derivative functions $\partial F/\partial x_n$,

$$\nabla^2_{\mathbf{x}} F(x_1, \cdots, x_{n_0})\, \Delta\mathbf{x} = -\nabla_{\mathbf{x}} F(x_1, \cdots, x_{n_0}) . \tag{6.4.26}$$

The nabla-operator with the vector **x** as index implies that a set of equations for each x_n is considered, stored in a matrix with dimension $m_0 \times n_0$. One easily verifies that

$$\frac{\partial F}{\partial x_n} = 2\left(\frac{\partial f_1}{\partial x_n} f_1 + \cdots + \frac{\partial f_{m_0}}{\partial x_n} f_{m_0}\right),$$

or,

$$\nabla_{\mathbf{x}} F(\mathbf{x}) = 2\mathbf{J}^T\mathbf{f} . \tag{6.4.27}$$

For the second derivative matrix elements of F we find

$$\frac{\partial^2 F}{\partial x_n \partial x_{n'}} = 2\left(\frac{\partial^2 f_1}{\partial x_n \partial x_{n'}} f_1 + \frac{\partial f_1}{\partial x_n}\frac{\partial f_1}{\partial x_{n'}} + \cdots + \frac{\partial^2 f_{m_0}}{\partial x_n \partial x_{n'}} f_{m_0} + \frac{\partial f_{m_0}}{\partial x_n}\frac{\partial f_{m_0}}{\partial x_{n'}}\right),$$

or,

$$\frac{\partial^2 F}{\partial x_n \partial x_{n'}} = 2\sum_{m=1}^{m_0}\left[D_{mnn'} f_m + J^T_{nm} J_{mn'}\right] , \tag{6.4.28}$$

where $D_{mnn'}$ is the second-order derivative quotient of the function f_m with respect to the variables x_n and $x_{n'}$. The vector solution $\Delta\mathbf{x}$ for the position of the minimum of the merit function is now given by

$$\sum_{m=1}^{m_0}\left[D_{mnn'} f_m + J^T_{nm} J_{mn'}\right]\Delta\mathbf{x} = -\mathbf{J}^T\mathbf{f} .$$

Denoting $\sum_m D_{mnn'} f_m$ by the 'curvature' matrix $\mathbf{A}_c$ we write the solution $\Delta\mathbf{x}$ as

$$[\mathbf{A}_c + \mathbf{A}]\Delta\mathbf{x} = -\mathbf{J}^T\mathbf{f} , \quad \text{or} , \qquad \Delta\mathbf{x} = -[\mathbf{A}_c + \mathbf{A}]^{-1}\mathbf{J}^T\mathbf{f} , \tag{6.4.29}$$

where we have also used the notation $\mathbf{A} = \mathbf{J}^T\mathbf{J}$ of Eq. (6.4.7).

If the function space is almost flat at the working point $\mathbf{x}_0$, numerical instability is likely to occur. The presence of the curvature matrix $\mathbf{A}_c$ introduces a damping effect in the solution which is proportional to the inverse of the curvature. Note that the procedure above will drive the solution towards an extremum. To select only the nearest minimum, it is recommended to first apply the optimisation to the function of Eq. (6.4.6) and to switch to the solution of Eq. (6.4.29) once a downhill trajectory has been found. The computation of all elements of the second-order derivative matrix requires an enormous effort. A compromise is to use only the generally dominating diagonal elements of $\mathbf{A}_c$. In **[84]** a method is described which collects estimates for these diagonal elements of $\mathbf{A}_c$ from the subsequent optimisation steps towards the merit function minimum.

6.4.1.5 Numerical implementation of local optimisation

The choice of an optimisation method is a compromise between speed of convergence and reliability. In **[203]** a comparison of the required number of basic arithmetic operations is given for various optimisation methods and these data have been compiled in Table 6.1. The difference in computation time for a single or double precision multiplication or division is assumed to be given by a factor of four although this number may strongly depend on computer architecture, compiler options, memory access, etc. It turns out that the various methods do not differ significantly. Singular value decomposition seems to be average with respect to computation time but delivers, undoubtedly, the most detailed structural information on the optimisation problem at hand. It also has the least storage requirements because of the single precision possibility. Although this was an important advantage in the early computer era, it has become less important now. What is not visible from the table is the relative importance of the calculation of a single merit function value. In optical design problems, the collection of such a single function value may require extensive ray-tracing activity.

Table 6.1. Number of operations N_c (multiplication / division) in single (S) or double (D) precision for various optimisation methods. The number of functions is m_0, the number of variables n_0.

Solution method applied to $(m_0 \times n_0)$ optimisation problem	# operations N_c	Precision	N_c $m_0 = 300,\ n_0 = 100$
Singular value analysis	$2m_0n_0^2 + 4n_0^3$	S	$10 \cdot 10^6$
Calculation normal equations	$m_0n_0^2/2$	D	$6 \cdot 10^6$
Normal equations + Cholesky factorisation	$m_0n_0^2 + n_0^3/6$	D	$7.7 \cdot 10^6$
Normal equations + Gauss–Jordan method	$m_0n_0^2/2 + n_0^3/3$	D	$9.3 \cdot 10^6$
Eigenvalue analysis of normal equations	$m_0n_0^2/2 + 16n_0^3/3$	D	$11.3 \cdot 10^6$
Gram-Schmidt orthogonalisation	$m_0n_0^2$	D	$12 \cdot 10^6$

Finding the solution vector from the matrix problem is, in the majority of cases, a tiny effort as compared to the work which has to be invested in the calculation of the matrix elements themselves.

Boundary conditions which should be satisfied by the system can be incorporated in various ways. Frequently occurring boundary conditions in optical design are, for instance, a prescribed focal distance, projection length, lateral or angular magnification, etc. The deviation from the target value can be incorporated in the optical merit function and given a high weighting factor $w_{b,j}$ with respect to the components in the merit function which are associated with the imaging quality. It is also possible to reduce the dimension of the variable space and dedicate a subset of the (transformed) variable space to satisfy the boundary conditions **[97]**. The well-known method of Lagrange multipliers to treat boundary conditions is basically equivalent to the insertion of the function which describes a boundary condition in the set of merit functions; the use of the Lagrange multiplier method corresponds to the limit $w_b \rightarrow \infty$. Apart from exact boundary conditions, there are also single- or double-sided boundary conditions on a variable or on a function. These conditions can be solved by associating a (weighted) function with them which is zero within the allowed interval and shows a steep increase to unity at the borders of the interval.

In Fig. 6.14 we have plotted the iterative procedure to go from a starting point 0 to the local minimum at M. At the starting point labelled 0 the derivative data for the matrix $\mathbf{J}$ are approximated by finite difference calculations, using a typical increment for each normalised variable x_n in the range of 10^{-3} to 10^{-4}. As a result we have the numerically approximated change table $\mathbf{J}_0$. A sequence of solution vectors is calculated (six dots in the figure) each time with a smaller damping factor p. For each intermediate solution, the exact merit function is calculated with the aid of ray-tracing data from the updated system. If second-order derivative values are available, the change in damping can be steered by these data. In the example of Fig. 6.14, the damping factor for the first optimisation step from the starting point $\mathbf{x} = \mathbf{x}_0$ could have been reduced to zero. In that case, the linear extrapolation would have steered the system to the point L_1 with a real value of the merit function that is slightly smaller than the value labelled with the number 2 in the figure. In practice, the damping factor p is limited to the interval $\infty \geq p \geq p_{\text{min}}$, where p_{min} is a practical value that avoids physical problems in the ray-tracing process. Such problems are total internal reflection of a ray, the absence of a ray intersection point with an optical surface, a ray path outside the useful aperture of an optical element, etc., which make the evaluation of the merit function impossible. For that reason, using the minimum damping factor p_{min}, the solution 1 is found in the first optimisation step and a new change table $\mathbf{J}_1$ is constructed at the point $\mathbf{x} = \mathbf{x}_1$.

At the minimum labelled 2, which was found with the aid of the matrix $\mathbf{J}_1$, we calculate a new matrix $\mathbf{J}_2$ and repeat the exploration of the linear solution with the aid of solutions with decreasing damping factor. In general, with two 'change tables' available, for instance $\mathbf{J}_j$ and $\mathbf{J}_{j+1}$, we use the approximate scheme for the calculation of the second derivative matrix which was proposed in **[84]**. For the diagonal elements D_{nn} of the second-order derivative matrix we write

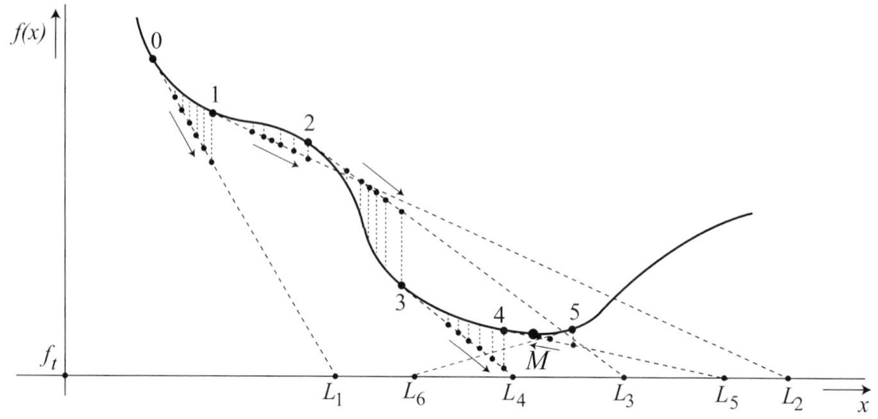

Figure 6.14: Two-dimensional illustration of the iterative optimisation steps to reach the local minimum of a function. From each new starting point j on the curve, a linear extrapolation L_{j+1} is found and the minimum point $j + 1$ which can be reached with the derivative data calculated at the point j. The target function value is f_t.

$$\left.\frac{\partial^2 f}{\partial x_n^2}\right|_{j+1} \approx \sum_{m=1}^{m_0}\left(\left.\frac{\partial f_m}{\partial x_n}\right|_{j+1} - \left.\frac{\partial f_m}{\partial x_n}\right|_{j}\right)\Big/(x_{n,j+1} - x_{n,j})$$
$$= \sum_{m=1}^{m_0}\frac{J_{j+1,mn} - J_{j,mn}}{x_{n,j+1} - x_{n,j}}, \tag{6.4.30}$$

where $x_{n,j}$ and $x_{n,j+1}$ are the values of the variable x_n at a starting point j and at the detected apparent minimum, labelled $j + 1$. Subsequent optimisation steps with, each time, updated 'change tables' will bring the solution vector finally to the real local minimum at M. The optimum choice for the damping factor is in the range 10^{-2} to 10^{-5}. Once the second derivative quotient can be included, a small damping value can be maintained to avoid very large excursions in relatively flat areas of the multi-dimensional landscape of the merit function. To avoid a possible convergence towards a local maximum with negative curvature components in multi-dimensional space, the absolute values of the second-order derivatives on the matrix diagonal are taken.

In multi-dimensional space, each local minimum can be attributed a region of starting points which will yield a solution vector in this minimum. This region is called the basin of attraction and the optimisation routine should minimise the calculation time to arrive at M. In practice, it happens that the speed of convergence is augmented at the risk of loss of numerical accuracy, for instance by using finite difference data far away from the point where they have been calculated. In some cases, loss of numerical accuracy brings the solution to a different local minimum belonging to another basin of attraction. For the user, chaotic effects appear and the reproducibility of the optimisation process is lost **[347]**.

6.4.1.6 Optical Design with the Aid of a Software Package

Basic optical design with the aid of a hired or purchased software package is quite feasible. Current design programs put at the disposal of the user a rich library of examples taken from the patent literature. These examples can be adapted to the specific needs of the designer and with the aid of automatic optimisation, an option in virtually all packages, a satisfactory solution is readily obtained. For more special design tasks, less common features of a software package need to be put into operation and this requires a careful study of the manuals or a special tutorial course on the software package. Such a special introduction into the software includes the teaching of a program-specific macro-language which allows the designer to introduce his own (non-standard) requirements during the design optimisation. A macro-language allows the user to write a personal version of the merit function, to connect variables to be optimised in a specific way, to introduce other ways of light propagation in the optical system (for instance guided-wave propagation instead of free-space propagation), to define non-standard boundary conditions, to prescribe special surface shapes, etc. Apart from knowledge of such a macro-language, the designer also needs a more profound knowledge of paraxial optics and aberration theory, of some general theorems in optical imaging, of light diffraction by periodic surfaces and of diffraction theory in general. Most of these subjects are covered in the Chapters 4 and 5 of this book; the diffraction aspects of imaging are treated in Chapters 8–11.

A good acquaintance with a specific software package and a thorough knowledge of the basic geometrical optics theory allow the optical designer to propose useful optical systems. In this context, the adjective 'useful' means feasible, manufacturable, cost-effective. A successful design needs to be tailored to the production capabilities of the optical shop or optical factory which has to produce the designed system. Moreover, from the very beginning of the design process, the available optical and mechanical measurement possibilities should be taken into account. Many of these production-related boundary conditions can be taken into account during (automatic) optimisation. But a good communication between the designers and the production departments during all phases of the design process is also important for finding *the* optimum design. For this reason, to conclude the topic of local (and global) optimisation, we treat in the last section of this chapter the subject of 'optical tolerancing' (see Section 6.5). It provides the designer with the statistical spread of the quality of a manufactured optical system given the optical and mechanical fabrication tolerances; these statistical data allow the designer to give, for example, an estimate of the yield during production of the designed optical system.

6.4.2 Methods for Global Optimisation

When we use as variables the curvatures, lens thicknesses, air distances and lens materials, the multi-dimensional landscape of the merit function of a simple optical system with only a few optical elements becomes quite complex. Many local minima of the merit function exist and the end result of an optimisation strongly depends on the starting point. The best design is the one that corresponds to the global minimum. However, there is no guarantee that a local optimisation will converge to the global minimum. In practice, with prior knowledge from the literature and from patent data, the most promising starting points can be used but the variable combination that produces the superior design corresponding to the global minimum remains unknown in principle. It is here that global optimisation techniques in optical design become interesting. Of course, any global optimisation should deliver a result in a finite amount of time. The calculation of the optical merit function of a complex optical system is a rather time-consuming operation. Certain techniques for global optimisation are, therefore, less suitable. We mention the method of finite programming supported by statistical analysis of the results, comparable to methods employed in the field of *machine learning* **[248]**. Unfortunately, the application of finite programming to an optimisation problem with a number of variables which can easily attain one hundred and a merit function with many components is not very practical. In the remaining part of this subsection we briefly consider genetic algorithm optimisation and simulated annealing as methods to detect a global minimum. A new method called saddle-point detection for global optimisation is treated in somewhat more detail.

6.4.2.1 Genetic Algorithm

Optimisation methods such as evolutionary or genetic algorithms carry out a wide search in variable space. Instead of a merit function, a fitness or viability function is attributed to the results, indicative of the reproducing power of the hypothetical biological entity that is produced from the variable space **[205]**. The entity is 'created' by combining discrete genes into a 'chromosome' and calculating the reproductive potential or evolutionary fitness of an entity possessing such a chromosome. Two evolutionary steps are considered. Mutation involves a random change of a part of the genetic code. Crossover means that subsets of two genetic codes are interchanged, comparable to what happens in the reproduction process. The crossover can be applied to randomly chosen genes or to certain fixed subsets. Parameters to be chosen in a genetic algorithm are the relative presence of the mutation and the crossover process. The combinatorial nature of a genetic algorithm makes it a less obvious choice for optical systems that are formed in a continuum variable space, especially when steep performance changes are observed for small variable changes. All variables together, given their respective ranges, are represented by binary strings and the optical performance defines the 'fitness' of the binary string in the total population. The computational effort shows an exponential increase with the string lengths and this makes genetic algorithms not very suitable for complex optical systems.

6.4.2.2 Simulated Annealing

Simulated annealing can be considered as an extension to standard optimisation in which the obligatory downward route to the (local) minimum is replaced by a stochastic acceptance of the suggested change in variable space. With the variable set x_j we calculate a new optimised function set f_m with least-squares value $f + \delta f$ at the points $x_n + \delta x_n$. The new variable set is accepted based on a probability function **[184]**, **[1]** given by

$$P(f+\delta f) = \begin{cases} 1\,, & \delta f < 0\,, \\ \exp\left(-\frac{\delta f}{a}\right)\,, & \delta f > 0\,. \end{cases} \tag{6.4.31}$$

The function P is deduced from a physical analogon and represents the probability of a particle settling itself at a position with different potential energy in a lattice when cooling down. The uphill directions correspond to temporary movements to a location with higher potential energy before the total set of particles finds its ultimate configuration with lowest potential energy. The constant a in the denominator of the exponential is proportional to the physical temperature of the assembly of particles. It is a free parameter in the simulated annealing optimisation process and its efficient choice is discussed in the references cited above.

6.4.2.3 Convex Programming

The search for a global minimum is facilitated if convex variables and convex functions can be used. The condition for a function to be convex on a certain interval $x \in [x_1, x_2]$ is given by

$$f[\alpha x_1 + (1-\alpha)x_2] \leq \alpha f(x_1) + (1-\alpha) f(x_2)\,, \tag{6.4.32}$$

where $\alpha \in [0,1]$. In terms of the second derivative of the function $f(x)$, it should satisfy the condition $d^2f/dx^2 \geq 0$. If this inequality is respected over the complete interval, the function is called strictly convex. For concave functions similar conditions are set up and each convex function can be made concave by changing its sign. Examples of convex functions are x^b with $b \geq 1$; this rather trivial convex function is shown in Fig. 6.15a, with $b = 2$. Any two points P_1 and P_2 on the convex function can be joined by a straight line P_1P_2, dotted in the figure. The vertical distance between the straight line and the convex function for a general value of the abscissa x is given by

$$P_3Q_3 = f(x_1) + \frac{x - x_1}{x_2 - x_1}\,[f(x_2) - f(x_1)] - f(x)\,, \tag{6.4.33}$$

and it is easily verified that $P_3Q_3 \geq 0$ if the condition set down in Eq. (6.4.32) holds with $(x - x_1)/(x_2 - x_1) = 1 - \alpha$. A minimisation of the convex function $f(x)$ on any general subinterval $x \in [a_1, a_2]$ will yield a minimum value. If the position of the minimum coincides with the border of the interval, the search has to be continued to an adjacent subinterval until a real minimum with $df/dx = 0$ is found. This minimum is by definition the global minimum of the convex function. The preceding reasoning for the one-dimensional function is extended in a straightforward way to a multi-dimensional variable space and boundary conditions can be included if these can also be described by convex functions. We note that, formally, a concave function is converted into a convex function by multiplication by any negative number.

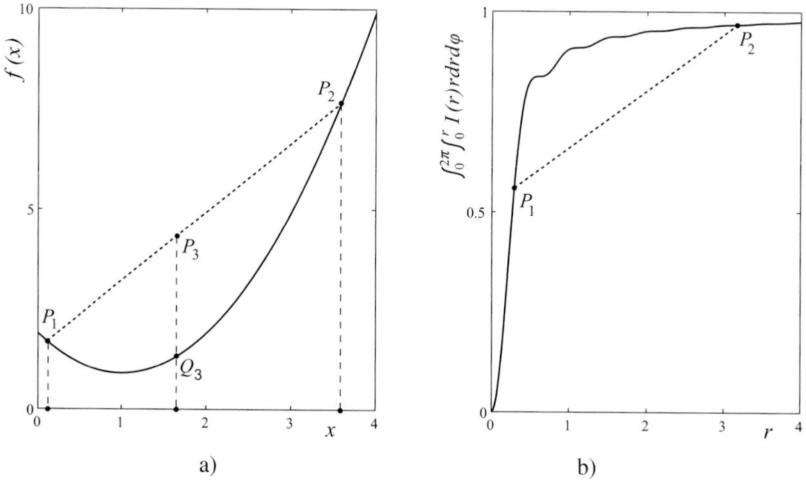

Figure 6.15: Convex functions and optimisation.
a) A simple one-dimensional convex function, $f(x) = (x-1)^2 + 1$.
b) The encircled energy in the diffraction pattern (Airy disc) of an ideal lens, a concave and increasing function for any value of the lower limit of the integral over the radial coordinate r in units of λ/NA, where NA is the numerical aperture of the lens.

The implementation of convex programming in an optical design problem suffers from the limited possibility to express the merit function components f_m and the boundary conditions in the form of convex functions. A typical optical function which is concave is the encircled energy in the diffraction pattern of a perfect lens. The function representing the encircled energy as a function of the radius of a circular region in the focal plane was first given by Rayleigh **[288]**. In Fig. 6.15b we have plotted this encircled energy flowing through a circular region in the image plane of a perfect lens. The function is concave although not in the strict sense since there are extrema (inflection points) which occur each time when a dark ring in the diffraction pattern is encountered. It is also possible to calculate the power passing through the annular region defined by $r_0 \leq r < \infty$ and to minimise this concave function. Another typical function encountered in optical design is the distance function. We require that the Euclidian distance between the merit function components and prescribed target values be made as small as possible. A distance function, for instance $|\mathbf{x} - \mathbf{a}|$, is a convex function according to the definition of Eq. (6.4.33). But the more general distance function $\sum_m |f_m(\mathbf{x}) - a_m|$ is only convex if the merit functions $f_m(\mathbf{x})$ are linear. Such a condition severely limits the application of the convex programming approach. An extensive description of convex optimisation and its applicability criteria can be found in **[45]**.

6.4.2.4 Escape Function

The stagnation of the optimisation procedure at a local minimum can be avoided by introducing an escape function. The direct environment of the local minimum is made unattractive for the optimisation routine by adding a penalty function to the merit function given by **[158]**,

$$p(\mathbf{x}) = a \exp\left\{-\frac{2}{w^2}\sum_{j=1}^{n}(x_j - x_{j,m})^2\right\} . \tag{6.4.34}$$

The values of the variables at the detected minimum are given by the vector $\mathbf{x} = \mathbf{x}_m$. The amplitude of the penalty function at the minimum itself is given by a and w is a measure for the distance in multi-dimensional space over which the penalty function has decreased to $1/e^2$ of its value. The escape function principle is illustrated in Fig. 6.16. The original function f presents two local minima, M_1 and M_2, on the interval $[-3, 4]$. The extra penalty function p_1 added to f (dashed in the figure) masks the minimum M_1. Likewise, the dashed curve $f + p_2$ obscures the second minimum. The resulting function $f + p_1 + p_2$ possesses a maximum only and the optimisation procedure continues to explore new minima outside the interval $[-3, 4]$. The optimum choice of the amplitude and the width of the extra penalty functions should be chosen such that each penalty function will make the optimisation process escape from the minimum. If the height a is not large enough, the minimum will be displaced to a nearby position. If the width parameter w is too large, the value of a minimum further away will be influenced by the penalty function in an undesirable way. A discussion on the choice of the a and w

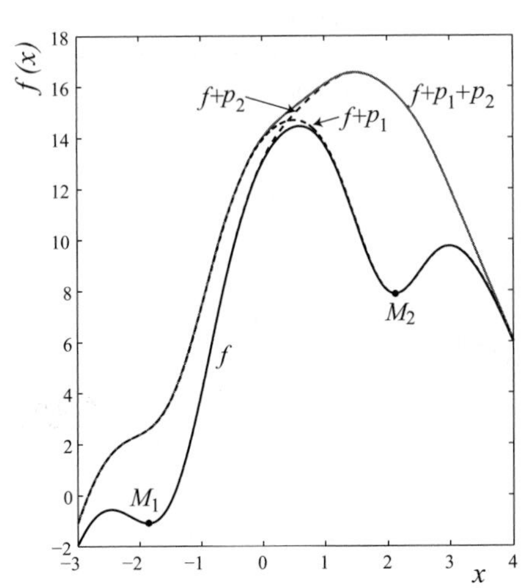

Figure 6.16: Illustrating Isshiki's escape function method for a one-dimensional function. Optimisation of the original function f with the penalty functions p_1 and p_2 added to it allows it to escape from the interval with the minima M_1 and M_2 on it.

parameters can be found in **[158]** and it is demonstrated that the choice of these parameters is not critical if the number of variables is large. Subsequent addition of an escape function at each local minimum will drive the optimisation procedure to new local minima, and, in principle, the global minimum can be selected at the end of the full search.

6.4.2.5 Saddle-point Detection

For the exploration of the complete merit function landscape, a strategy is needed to escape from the nearest local minimum to other remote minima. The genetic algorithm and the method of simulated annealing rely on statistical probabilities to escape from a local minimum. This is at the expense of a high numerical effort. A more systematic way to escape from a local minimum has been proposed by Bociort and co-workers **[30]**,**[32]**,**[31]**,**[348]**,**[147]**. The principle of the method can be explained with the aid of Fig. 6.17. In three-dimensional space we have sketched the function $f(x,y) = x^3 + y^3/2 + x^2 + y^2 - 2(x+y) + 2$ with the z-coordinate used to represent the function value. Four extrema are found at the corners of a rectangle. The different character of the extrema follows from the values of the second partial derivatives. For two of the extrema, $\partial^2 f/\partial x^2$ and $\partial^2 f/\partial y^2$ have identical sign and these points correspond to the extrema denoted by *max* and *min* in the figure. At the two other extrema, the second-order derivative coefficients have opposite signs and these are saddle points. The importance of the saddle points becomes evident when inspecting the minimum. From this point, there are two possible pathways to escape from the minimum towards a region with lower function value, either following the downward curved dashed path through S_1 or a comparable path through the second saddle point S_2. An escape route from a local minimum is thus provided by spotting the nearby saddle points in the merit function landscape and, from such a saddle point, perform a local optimisation to an adjacent local minimum. This procedure can be repeated and, basically, a closed network of minima and saddle points should be detectable. From all the local minima which have been detected, the lowest one is then declared to be the global one.

The three-dimensional intuitive picture above can be generalised in a straightforward way to n-dimensional space. For this purpose, we write the Taylor expansion of a function f_m at an extremum with the aid of Eq. (6.4.2) as

$$f_m(\mathbf{x}_0 + d\mathbf{x}) = f_m(\mathbf{x}_0) + \sum_n \sum_j H_{mnj} dx_n dx_j \,, \tag{6.4.35}$$

where the Hessian tensor $\mathbf{H}$ with elements H_{mnj} is symmetric and has real eigenvalues and orthogonal eigenvectors. The tensor can be reduced to its diagonal form. The function f_m in the direct neighbourhood of the extremum is then written as

$$f_m(\mathbf{x}_0 + d\mathbf{x}) = f_m(\mathbf{x}_0) + \sum_j \lambda_j \, (dx_j)^2 \,. \tag{6.4.36}$$

Extrema which are maxima or minima have n eigenvalues with the same sign, positive for a minimum and negative for a maximum. Saddle points arise when at least one of the eigenvalues has a sign that is different from those of the others. The Morse index of a saddle point is defined by the number of negative eigenvalues λ_j. For a minimum, the Morse index is 0, for a maximum the index amounts to n. An important category of saddle points are those with a Morse index of 1 where exactly one eigenvalue is negative. Such saddle points form the simplest connection paths between two minima.

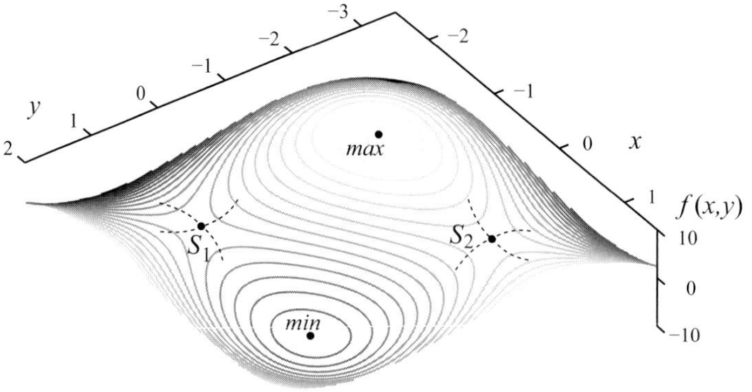

Figure 6.17: A merit function landscape with two extrema *min* and *max* and two saddle points, S_1 and S_2 which connect the extrema with other maxima or minima in the landscape.

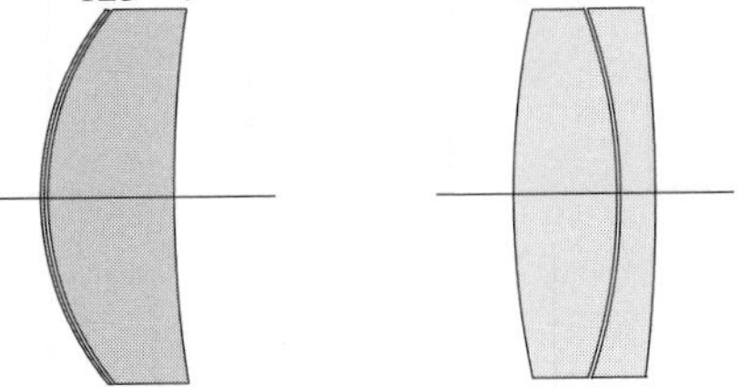

Figure 6.18: The physical creation of a saddle point by means of a thin meniscus element.
a) A glass-meniscus element with the property $c_1 = c_2 = c_3$.
b) An air-space meniscus ($c_2 = c_3$).

The physical creation of saddle points in the merit function landscape is illustrated in Fig. 6.18. The addition to the optical system of an 'infinitely thin' meniscus lens with zero power on the outside of a lens element produces a saddle point in the merit function. The condition for a saddle point is that the curvatures c_1, c_2 and c_3 in Fig. 6.18a are all equal, that the air gap has zero thickness and that the indices of the two glass elements are identical. With the aid of two extra curvature variables c_2 and c_3, the two adjacent minima of the merit function in the direct neighbourhood of the saddle point are found. Another possibility for the creation of a saddle point in the merit function is to cut a lens element into two parts by means of an infinitely thin air space and to assign some initial value to the two identical curvatures c_2 and c_3 of the new internal surfaces (see Fig. 6.18b). A systematic exploration of the extra saddle points and their associated minima, created by the meniscus-shaped glass or air 'elements', produces a merit landscape with *connecting lines* between the various saddle points and minima.

In Fig. 6.19 we show the design landscape which arises when searching for local minima of the merit function of a photographic triplet lens. The global minimum of the merit function corresponds to the triplet systems m1 and m2 in the centre of the landscape chart. The two 'best' solutions are almost perfectly mirrored versions of each other. The saddle point search has first been carried out by means of the insertion of a meniscus glass element. If it was not possible to find two associated local minima from the saddle point which was created this way, an extra search was done with the aid of an air space meniscus lens. The saddle points created with air space menisci are shown in grey boxes in the figure. The network shape and the connecting lines between minima and saddle points depend on the aperture and field angle which is used in the optimisation of the triplet lens. Some saddle points and minima may disappear when aperture and field size are changed and the connection scheme is altered. The new connecting lines are shown by the dashed lines in the figure. The merit function landscape of Fig. 6.19 shows the important role of the saddle point designs as intermediate stations or even 'hubs' when travelling from one local minimum to the other.

6.5 Optical Tolerancing

The analysis of an optical design with respect to fabrication and mounting tolerances allows us at an early stage to detect critical issues in a design and to devise the appropriate measurement and mounting techniques. More generally, optical tolerancing is capable of producing the statistical probabilities of achieving a certain optical quality of the complete instrument, for instance in terms of wavefront aberration residuals, maximum frequency transfer and Strehl intensity of the diffraction image of a source point. Such a tolerancing analysis may also include the effects of environmental factors such as homogeneous or inhomogeneous thermal heating, atmospheric pressure variations, humidity, etc. In this brief section we give a succinct calculation of the effects of deviations from the nominal design and a basic introduction to the addition of statistical variations in the constructional parameters of an optical design and the influence of these variations on the expected quality of a manufactured optical system.

6.5.1 A Finite Aberration Formula for Small Surface Perturbations

When mounting lens elements and mirror surfaces in an optical system, small geometrical perturbations occur. They can be classified as surface tilt, lateral surface displacement (decentring) and axial surface displacement. In each of these

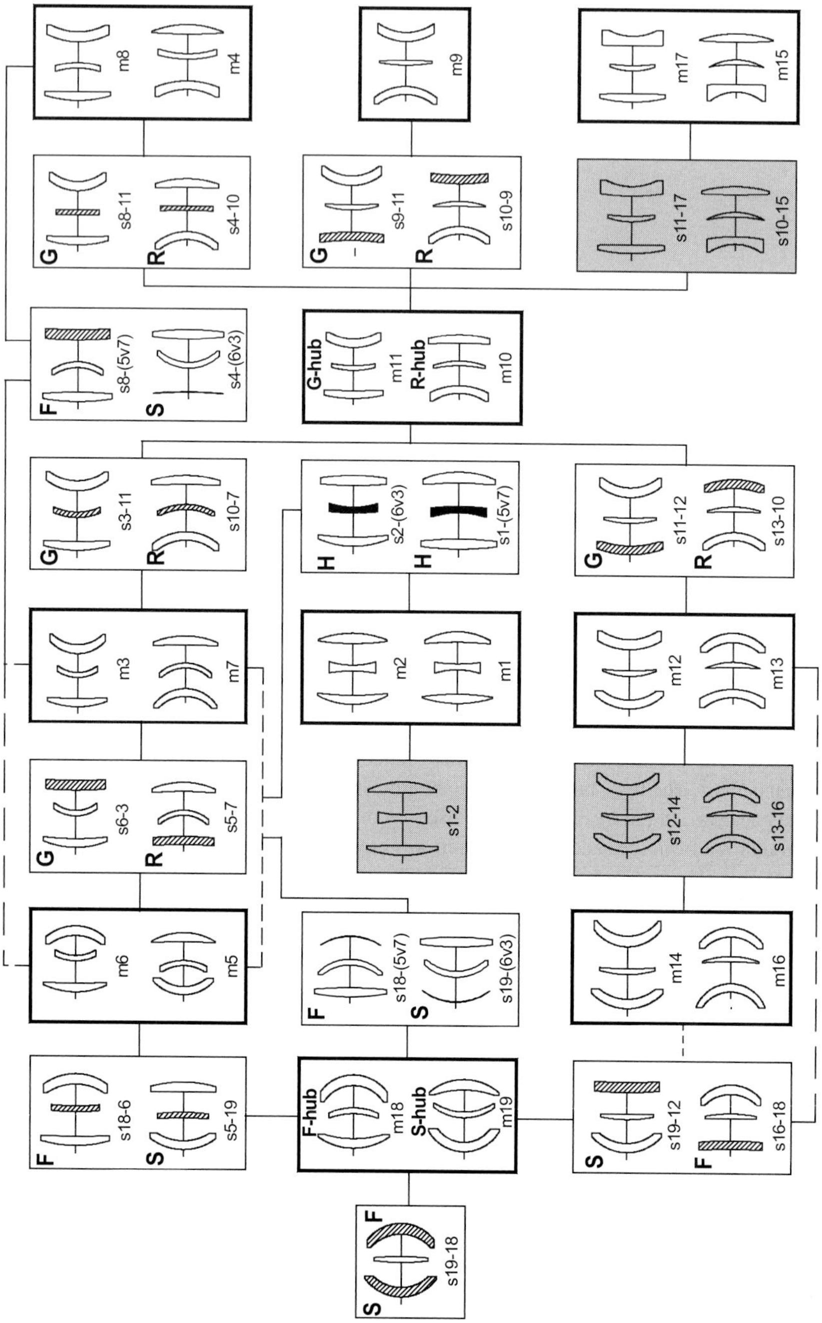

Figure 6.19: The merit function landscape of the so-called triplet lens for photography. All local minima (thick boxes) are connected by saddle points (thin boxes). An exhaustive search of all saddle points that can be reached from each local minimum reveals the complete landscape. The lens system labelled 'm1-m2' (m2 is the mirror version of m1) is the global minimum for the chosen merit function. Reproduced from **[32]**, with the permission of the learned society SPIE and the authors, F. Bociort and M. van Turnhout.

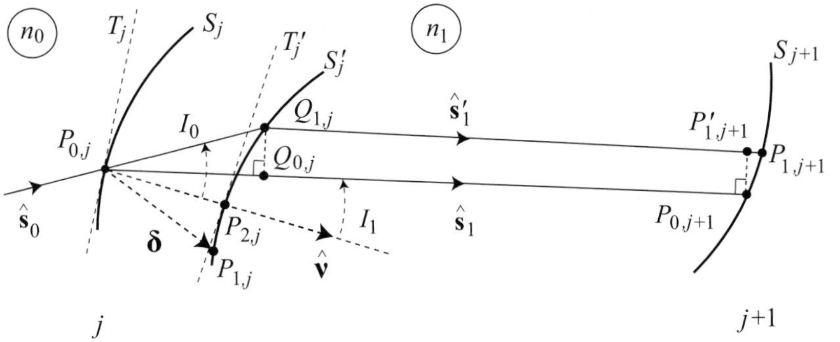

Figure 6.20: Pathlength change along a ray following a small surface displacement.

cases, the points on an optical surface follow a prescribed displacement in three dimensions, such that the entire surface is found unchanged in its new position and orientation. By monitoring the displacement of a set of points on the optical surface and calculating the corresponding perturbation of the optical pathlength at each point, we obtain a good estimate of the wavefront perturbation caused by the displaced surface in the optical system. For well-corrected imaging systems, the wavefront perturbation is preferably expressed by means of small changes in (lower order) Zernike coefficients of the wavefront propagating to a selected image point.

In Fig. 6.20 we show a refracting surface S_j of an optical system in which a ray is incident at the point $P_{0,j}$. The surface S_j separates two media with refractive indices n_0 and n_1, respectively. The unit normal vector to the surface at $P_{0,j}$ is $\hat{\mathbf{v}}$, the unit ray vector of the incident ray is $\hat{\mathbf{s}}_0$, the refracted ray has unit vector $\hat{\mathbf{s}}_1$. The surface is now displaced to a position denoted by S_j' in such a way that the point $P_{0,j}$ is found at $P_{1,j}$, where $\boldsymbol{\delta}$ is the displacement vector $P_{0,j}P_{1,j}$. The perpendicular distance between the surface and its displaced version, measured along the normal vector $\hat{\mathbf{v}}$, is given by $P_{0,j}P_{2,j} = \hat{\mathbf{v}} \cdot \boldsymbol{\delta}$. For a small enough displacement, the change in direction of the surface normal, going from $P_{1,j}$ to $P_{2,j}$, can be neglected. Moreover, we replace the curved surface by its tangent plane T_j' at $P_{2,j}$. The latter approximation influences the distances $P_{0,j}Q_{1,j}$ and $P_{0,j}Q_{0,j}$ only in the second order of $|\boldsymbol{\delta}|$.

For a specific case, namely the construction of a discontinuous Fresnel surface from a continuous aspheric surface, an expression for the change in pathlength due to a surface shift has been given, see Eq. (6.2.13) and Fig. 6.4. In what follows we calculate the pathlength change due to more general perturbations in an optical system. To separate the first-order pathlength changes from the smaller second-order ones we apply the same reasoning as was used to obtain the Conrady D–d formula for chromatic change of aberration, see Subsection 5.6.3. Optical perturbations such as a surface displacement, a surface deformation or a change in optical contrast give rise to an optical pathlength change at the surface itself which is a linear function of the (small) perturbation. But due to the surface perturbation of surface j in the system, the outgoing perturbed ray $\hat{\mathbf{s}}_1'$ has changed direction with respect to the original refracted ray $\hat{\mathbf{s}}_1$. The ray $\hat{\mathbf{s}}_1'$ intersects the next surface with index $j+1$ at a different position, denoted by $P_{1,j+1}$. We now apply Fermat's principle to the perturbed ray and the original ray and consider the perturbed ray as a non-physical neighbouring ray of the unperturbed ray. The pathlength difference between these two rays, counted along the paths $Q_{0,j}P_{0,j+1}$ and $Q_{1,j}P_{1,j+1}'$ from surface j to surface $j+1$ is only different in the second order of the angular ray deviation between $\hat{\mathbf{s}}_1'$ and $\hat{\mathbf{s}}_1$ which is caused by the surface perturbation. For the pathlength change at surface $j+1$ we then consider the refraction at the intersection point $P_{0,j+1}$ of the unperturbed ray with surface $j+1$. Therefore, using Fermat's principle, it is permissible to add independently the first-order pathlength changes of the unperturbed ray at each surface and to neglect the second-order pathlength changes between the surfaces.

Omitting the index j of the surface under consideration we write for the first-order pathlength difference δW between the two ray trajectories P_0Q_1 and P_0Q_0,

$$\delta W = n_0 P_0 Q_1 - n_1 P_0 Q_0 \, , \tag{6.5.1}$$

where the unperturbed ray path along P_0Q_0 was taken as the reference path. The point Q_0 is the intersection point of the perpendicular to $\hat{\mathbf{s}}_1$ through Q_1 and the ray $\hat{\mathbf{s}}_1$ at the second medium. Q_0Q_1 can be considered as part of a reference sphere for measuring wavefront aberration with its midpoint at infinity. The position of Q_1 is approximated by the intersection point of the incident ray $\hat{\mathbf{s}}_0$ with the tangent plane T_j' through P_2. We have

$$\vec{P_0Q_1} = \left(\frac{\boldsymbol{\delta}\cdot\hat{\mathbf{v}}}{\hat{\mathbf{s}}_0\cdot\hat{\mathbf{v}}}\right)\hat{\mathbf{s}}_0 , \qquad \vec{P_0Q_0} = \left(\frac{\boldsymbol{\delta}\cdot\hat{\mathbf{v}}}{\hat{\mathbf{s}}_0\cdot\hat{\mathbf{v}}}\right)(\hat{\mathbf{s}}_0\cdot\hat{\mathbf{s}}_1)\,\hat{\mathbf{s}}_1 , \tag{6.5.2}$$

and, using Snell's law for the refraction of rays,

$$n_1\hat{\mathbf{s}}_1 = n_0\hat{\mathbf{s}}_0 + \{n_1(\hat{\mathbf{s}}_1\cdot\hat{\mathbf{v}}) - n_0(\hat{\mathbf{s}}_0\cdot\hat{\mathbf{v}})\}\,\hat{\mathbf{v}} , \tag{6.5.3}$$

we obtain for the pathlength difference

$$\delta W = \left(\frac{\boldsymbol{\delta}\cdot\hat{\mathbf{v}}}{\hat{\mathbf{s}}_0\cdot\hat{\mathbf{v}}}\right)[n_0 - n_1\hat{\mathbf{s}}_0\cdot\hat{\mathbf{s}}_1] = -\,(\boldsymbol{\delta}\cdot\hat{\mathbf{v}})\,\{n_1(\hat{\mathbf{s}}_1\cdot\hat{\mathbf{v}}) - n_0(\hat{\mathbf{s}}_0\cdot\hat{\mathbf{v}})\} , \tag{6.5.4}$$

or, in terms of the incident ray vector only,

$$\delta W = -\,(\boldsymbol{\delta}\cdot\hat{\mathbf{v}})\,\left\{[n_1^2 - n_0^2 + n_0^2(\hat{\mathbf{s}}_0\cdot\hat{\mathbf{v}})^2]^{1/2} - n_0(\hat{\mathbf{s}}_0\cdot\hat{\mathbf{v}})\right\} . \tag{6.5.5}$$

For the reflected rays at the surfaces S and S' we find in a similar way,

$$\delta W = 2n_0\,(\boldsymbol{\delta}\cdot\hat{\mathbf{v}})\;(\hat{\mathbf{s}}_0\cdot\hat{\mathbf{v}}) , \tag{6.5.6}$$

where we have used Snell's law in vector form for the reflective case, $\hat{\mathbf{s}}_1 = \hat{\mathbf{s}}_0 - 2(\hat{\mathbf{v}}\cdot\hat{\mathbf{s}}_0)\,\hat{\mathbf{v}}$.

In general, it is advisable to avoid the reflection version of Snell's law with the common convention $n_1 = -n_0$ of paraxial optics and the corresponding definition of the sign of the angle of reflection I_1. With the aid of the formulas of Eqs. (6.5.5) and (6.5.6), unambiguous results are obtained for δW that do not depend on the choice of the normal vector (outward or inward normal) or on the sign convention for the angle of the reflected ray. Equation (6.5.5) is commonly called Hopkins' formula.

6.5.2 Efficient Calculation of Low-order Aberration Coefficients per Surface

The formula for δW can be applied when small perturbations are applied to the optical surfaces of a system with rotational symmetry. We assume that the system does not suffer from vignetting such that, in the case of off-axis beams, the beam cross-section is virtually circular. In this case, Zernike expansion of the aberration function is permitted over the entire image field without special measures to counteract the effect of a varying cross-section of the oblique beams as a function of field angle. Small amounts of axial shift, decentring and tilt of a single surface or a combination of two surfaces (lens element) adequately represent the optical and mechanical deviations from the as-designed system, see Fig. 6.21. Other

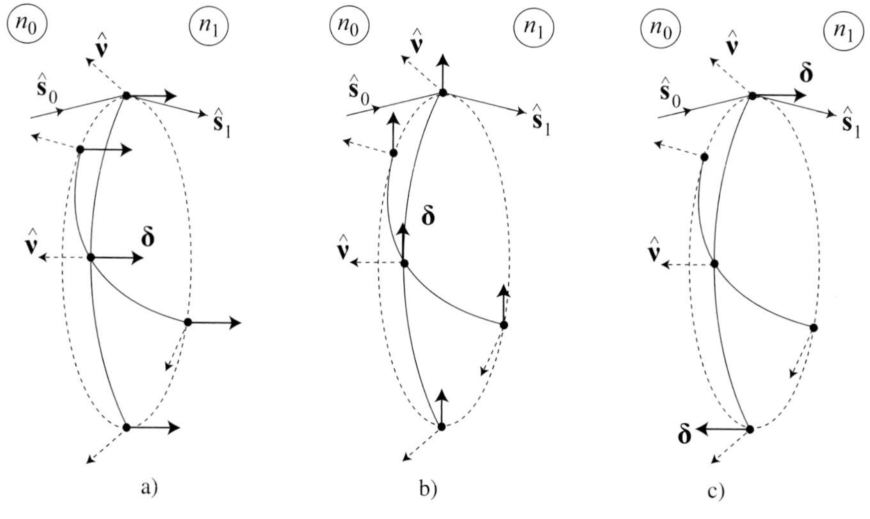

Figure 6.21: Displacements of the centre point and the four extreme points in two orthogonal cross-sections of an optical surface in the case of axial shift, decentring and tilt of the surface, depicted in a), b) and c), respectively. The unit normal vectors $\hat{\mathbf{v}}$ are dashed in the figure, the thick solid vectors $\boldsymbol{\delta}$ are the displacement vectors at each point.

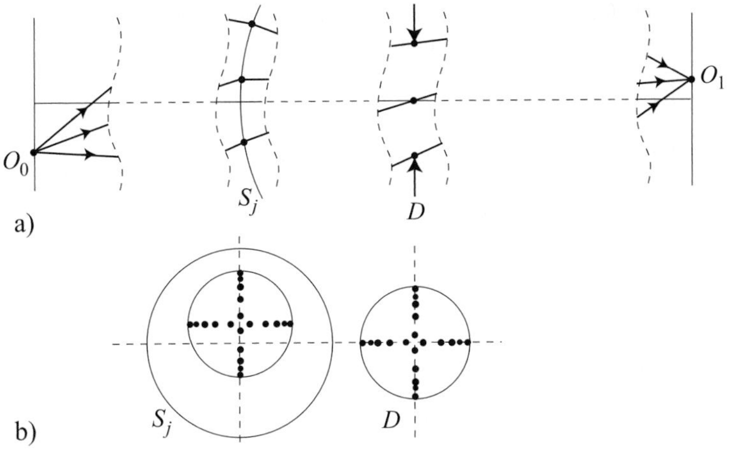

Figure 6.22: a) The selection of rays in the stop D of the optical system.
b) The footprint of the selected rays on a general surface S_j of the optical system ($K = 5$).

perturbations of an optical system can be included also; these might be a change in curvature of a surface, a cylindrical deformation of a refracting or reflecting surface, an offset in refractive index of the optical material or a difference in Abbe V number. With the aid of a judiciously chosen grid of rays in the diaphragm of the system it is possible to calculate the small values δW for each ray per individual surface. For small values of the optical and mechanical perturbations, the aberration δW is composed of the typical low-order Seidel wavefront aberrations. The wavefront deviations are expanded with respect to the Zernike polynomials Z_0^0, Z_1^1, Z_1^{-1}, Z_2^0, Z_3^1, Z_3^{-1} and Z_4^0. The astigmatic polynomials Z_2^2 and Z_2^{-2} are of second order in the surface displacements for an axial beam and could be neglected. But they can be encountered as the consequence of an imperfect optical polishing process which has left a cylindrical rest error on the optical surface. Astigmatic aberration also appears as a first-order effect of surface perturbations when considering off-axis imaging beams. For that reason, they will be included in the list of Seidel-type aberrations associated with tolerancing analysis. We also include some higher-order Zernike aberrations which are prone to occur in high-numerical-aperture systems and in systems with steep aspheric surfaces.

The efficient calculation of the coefficients of the low-order Zernike polynomials is done with the aid of a grid in the radial coordinate ρ and four azimuthal sections with $\theta = 0$, $\pi/2$, π and $3\pi/2$, defined by the meridional and sagittal section for a pencil of rays leaving an off-axis object point O_0, see Fig. 6.22a. The radial sampling of the rays is illustrated in Fig. 6.22b. A quadratic grid is used in the direction of the radial coordinate ρ with $t = \rho^2$ or a so-called cosine-grid with $\cos u = \rho$, in both cases applying equidistant sampling in the transformed coordinates,

$$t = \frac{2k-1}{2K} , \qquad \text{or,} \qquad u = \frac{\pi}{2}\left(1 - \frac{2k-1}{2K}\right) , \tag{6.5.7}$$

where $k = 1, \cdots, K$ and K is the total number of equidistant sample points on the radius.

The lower-order radial polynomials ($n \leq 4$) in terms of the t or u coordinate are shown in Table 6.2. The δW values which are calculated at each surface for the set of rays and for a specific surface perturbation are samples taken from a wavefront function $\delta W(\rho, \theta) = \sum_n \sum_m c_n^m R_n^m(\rho) \exp(im\theta)$ of which we want to compute the coefficients c_n^m. To this end we compute inner products of the sampled function δW and the Zernike polynomial associated with the coefficient (n', m'). In terms of the transformed coordinates t and u we have,

$$\begin{aligned} I_{n,n'}^{m,m'} &= \frac{1}{2\pi} \int_0^1 \int_0^{2\pi} \sum_n \sum_m c_n^m R_n^{|m|}(t) R_{n'}^{|m'|}(t) \exp[i(m-m')\,\theta]\; dt\, d\theta \\ &= \frac{1}{2\pi} \int_0^{\pi/2} \int_0^{2\pi} \sum_n \sum_m c_n^m R_n^{|m|}(u) R_{n'}^{|m'|}(u) \sin(2u)\; \exp[i(m-m')\,\theta]\; du\, d\theta \\ &= 2c_{n'}^{m'} \int_0^1 \left\{R_{n'}^{|m'|}(\rho)\right\}^2 \rho\, d\rho = \frac{c_{n'}^{m'}}{n'+1} . \end{aligned} \tag{6.5.8}$$

Table 6.2. Lower-order radial Zernike polynomials R_n^m ($n \leq 4$) using the argument substitutions $\rho = \sqrt{t}$ and $\rho = \cos u$.

$(n, \|m\|)$	$R_n^{\|m\|}(t)$	$R_n^{\|m\|}(u)$
$(0, 0)$	1	1
$(1, 1)$	$\sqrt{t}$	$\cos u$
$(2, 0)$	$2t - 1$	$\cos 2u$
$(2, 2)$	t	$(1 + \cos 2u)/2$
$(3, 1)$	$(3t - 2)\sqrt{t}$	$(\cos u + 3\cos 3u)/4$
$(4, 0)$	$6t^2 - 6t + 1$	$(1 + 3\cos 4u)/4$

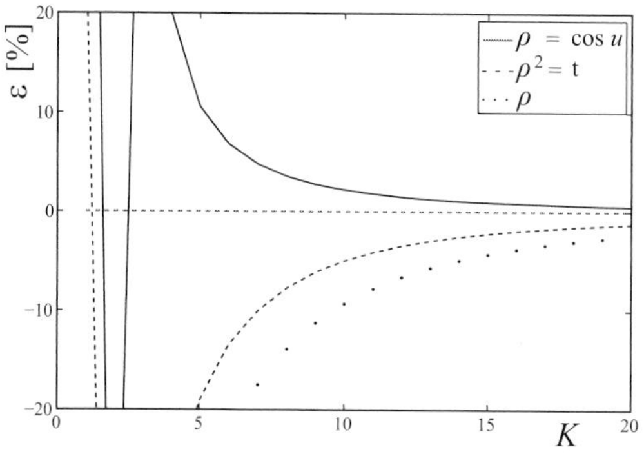

Figure 6.23: Residual error ϵ in percentages of the inner product integral of Eq. (6.5.8) for standard integration over ρ (dotted points), quadratic integration over $t = \rho^2$ (dashed curve) and using the 'cosine'-type ($\rho = \cos u$) radial sampling (solid curve), all as a function of the number of radial sample points K.

The basic midpoint quadrature rule is used to obtain the radial integrals over the ρ-, t- or u-variable. The accuracy of such an integration is shown in Fig. 6.23 as a function of the total number of sample points K. The test polynomial is Z_4^0 for Seidel spherical aberration. It follows from the figure that the cosine-sampling scheme is most advantageous **[168]** and yields sufficient convergence for tolerancing purposes when K has a value of typically 10.

The use of the finite ray aberration value δW per surface gives rise to an important saving of calculation time. From a single list of angles of incidence and surface normal directions for some 40 rays in the meridional and sagittal section we obtain the wavefront perturbations δW for each ray and each surface with a very small computational effort. The standard approach for optical tolerancing requires calculation of the wavefront in image space for each perturbation by an optical surface and calculation of the distance between this perturbed wavefront and the reference wavefront associated with the nominal system.

6.5.3 Statistical Addition of Optical and Mechanical Manufacturing Errors

In the case of well-corrected systems with circular symmetry which should operate close to the diffraction limit and in which the vignetting effects can be neglected, an error analysis based on Zernike polynomials is appropriate. As shown in the previous subsection, a quick method is available to obtain the change in Zernike coefficients by the perturbation of a specific surface or lens element in the system. By applying elementary statistical methods, one can obtain an estimate of the expected average value μ of the Zernike coefficients of a manufactured system in the presence of known manufacturing and material errors. It is also possible to calculate the expected value of the variance σ^2 of a great number of manufactured lenses.

In the case of a more general system, further away from the diffraction limit and containing generally shaped surfaces or diffractive optical elements without an axis of circular symmetry, it is advisable to simulate a great number of manufactured systems with randomly introduced perturbations in all elements simultaneously (Monte Carlo method) to create a set of statistical samples for calculating the ensemble average and ensemble variance. The evaluation of the imaging quality can be done by other means than wavefront aberration of pencils of rays. Criteria like the blur values of a set of point images or the data of the optical transfer function for various points in the image field are more informative for imaging systems which operate further away from the diffraction limit.

6.5.3.1 Tolerancing Statistics for a Well-corrected Imaging System

We assume that the nominal system possesses circular symmetry and start the tolerancing with a calculation of the 'optical' sensitivity of a mechanical or material perturbation in the optical system. In general, we can write for the wavefront perturbation δW_n^m of a certain Zernike radial and azimuthal order (n, m),

$$\delta W_n^m = \sum_{k,l} a_{n,k,l}^{m,(\alpha)} \, \epsilon_{k,l}^{\alpha} \,, \tag{6.5.9}$$

where $a_{n,k,l}^{m,(\alpha)}$ is the sensitivity coefficient of the perturbation $\epsilon_{k,l}$ of type k at surface l. The exponent α determines the dependence of the optical perturbation on the mechanical error. In most cases, α will be unity but we allow also second-order perturbations. In the case of $m = 0$, the resulting optical perturbation has circular symmetry (piston, defocusing or spherical aberration). For nonzero values of m we deal with non-circularly symmetric wavefront perturbations such as coma and astigmatism. The linear or quadratic dependence of the optical perturbation on mechanical errors is listed in Table 6.3 by means of the value of the exponent α. The value of α has been found numerically by inspection of the change in value of calculated aberration coefficients as a function of the amplitude of the surface perturbation. For the statistical addition of wavefront perturbations we assume that the mechanical and material perturbations have a normal distribution with, for each perturbation ϵ, a probability density function,

$$f(\epsilon) = \frac{1}{\sigma_\epsilon \sqrt{2\pi}} \exp\left(- \frac{\epsilon^2}{2\sigma_\epsilon^2} \right) . \tag{6.5.10}$$

The product $f(\epsilon)d\epsilon$ represents the probability of finding the random variable ϵ in the interval $[\epsilon, \epsilon + d\epsilon]$. The mean of the random variable ϵ is zero and its variance σ_ϵ^2. The statistical addition of the perturbations per Zernike aberration is treated in the following way:

Table 6.3. Value of the exponent α for various types of error-induced aberrations on the axis or in the image field of an optical system.

α	Tilt		Decentring		Axial shift		Curvature		Index	
W_n^m	axis	field	axis	field	axis	field	axis	field	axis	field
W_1^1	1	1	1	1	-	1	-	1	-	1
W_2^0	2	1	2	1	1	1	1	1	1	1
W_2^2	2	1	2	2	-	1	-	1	-	1
W_3^1	1	1	1	1	-	1	-	1	-	1
W_3^3	1	1	-	1	-	1	-	1	-	1
W_4^0	-	1	2	2	1	1	1	1	1	1
W_4^2	2	1	2	1	-	1	-	1	-	1
W_5^1	1	1	1	1	-	1	-	1	-	1
W_6^0	-	1	2	2	1	1	1	1	1	1

a) $m = 0$ and linear dependence of the Zernike aberration coefficient on the constructional errors.
The mean value of the aberration contributions δW_n^0 is zero for $\alpha = 1$ and the variance is given by addition of the variances of each individual perturbation,

$$\mu_{W_n^0} = 0 \,, \qquad \sigma^2_{W_n^0} = \sum_{k,l} \left(a^{0,(1)}_{n,k,l}\, \sigma_{k,l} \right)^2 \,, \tag{6.5.11}$$

where the coefficients α are extracted from Table 6.3 for an axial pencil of rays or for an off-axis pencil.

b) $m = 0$ and quadratic dependence of the Zernike aberration coefficient on the constructional errors.
In this case we have to calculate the statistics of a quadratic function of the constructional random variables $\epsilon_{k,l}$. The probability density function $f(W)$ of a single function $W = a\,\epsilon^2$ (a is an arbitrary coefficient) is given in **[269]**, p. 129, and reads for a normally distributed random variable ϵ,

$$\begin{aligned} f(W) &= \frac{1}{\sigma_\epsilon \sqrt{2\pi a W}} \exp\left(-\frac{W}{2a\sigma_\epsilon^2}\right) H(W) \,, && a > 0,\; W > 0 \,, \\ f(W) &= \delta(W) \,, && a = 0 \,, \\ f(W) &= \frac{1}{\sigma_\epsilon \sqrt{2\pi a W}} \exp\left(-\frac{W}{2a\sigma_\epsilon^2}\right) \{1 - H(W)\} \,, && a < 0 \,,\; W < 0 \,, \end{aligned} \tag{6.5.12}$$

where H denotes the Heaviside step-function.
The calculation of the first- and second-order stochastic moments yields the average value and the variance of the single quadratic function W with the result that

$$\mu_W = a\sigma_\epsilon^2 \,, \qquad \sigma_W^2 = 3\, a^2 \sigma_\epsilon^4 \,. \tag{6.5.13}$$

It follows that the mean value μ_W depends on the specific values of the a-coefficients and that, in general, the mean of the aberration of the perturbed system will deviate from that of the as-designed system. After summation over all surfaces of a specific Zernike aberration with indices $(n, 0)$, n is even, we have for the two stochastic moments the expressions

$$\mu_{W_n^0} = \sum_{k,l} \left(a^{0,(2)}_{n,k,l}\, \sigma^2_{k,l} \right) \,, \qquad \sigma^2_{W_n^0} = 3 \sum_{k,l} \left(a^{0,(2)}_{n,k,l}\, \sigma^2_{k,l} \right)^2 \,. \tag{6.5.14}$$

c) $m \neq 0$ and linear dependence of the Zernike aberration coefficient on the constructional errors.
The aberrations without circular symmetry are built up from contributions with stochastically varying size and a uniformly distributed azimuth θ with $0 \le \theta < 2\pi$. The stochastic sum of the aberration components is obtained from a random walk in two dimensions with stochastic variables (X, Y). To obtain the expected value for the endpoint of the random walk and the variance of the endpoint position, we proceed in two steps. We assume that the system errors that give rise to the asymmetric aberrations such as wavefront tilt, coma and astigmatism are normally distributed and produce an aberration with a size

$$\delta W_n^m = \sum_{k,l} a^{m,(\alpha)}_{n,k,l} \left(\sqrt{\epsilon^2_{k,l,x} + \epsilon^2_{k,l,y}} \right)^\alpha \,, \tag{6.5.15}$$

where the mechanical error $\epsilon_{k,l}$ should now be treated as a vector with components $(\epsilon_{k,l,x}\,,\, \epsilon_{k,l,y})$, both normally distributed with zero mean and equal variance $\sigma_{\epsilon_{k,l}}$. For the statistics of the random variable $r = \sqrt{x^2 + y^2}$ (x and y are independent Gaussian random variables) we find the Rayleigh distribution with the probability density function **[269]**,

$$f(r) = \frac{r}{\sigma^2} \exp\left(-\frac{r^2}{2\sigma^2}\right) H(r) \,, \tag{6.5.16}$$

where σ^2 is the variance of each of the independent x and y Gaussian random variables. The distribution function $F(r)$ of the random variable r_{rv} is defined as the probability that the variable assumes a value smaller than r, or, $F(r) = P\{r_{rv} \le r\}$. In the case of normal distributions for the variables x and y we have,

$$\begin{aligned} F(r) &= \int_0^{2\pi} \int_0^r f_x(r\cos\phi) f_y(r\sin\phi)\; r dr d\phi = \int_0^{2\pi} \int_0^r \frac{1}{2\pi\sigma^2} \exp\left\{-\frac{r^2}{2\sigma^2}\right\} r\, dr d\phi \\ &= 1 - \exp\left\{-\frac{r^2}{2\sigma^2}\right\} . \end{aligned} \tag{6.5.17}$$

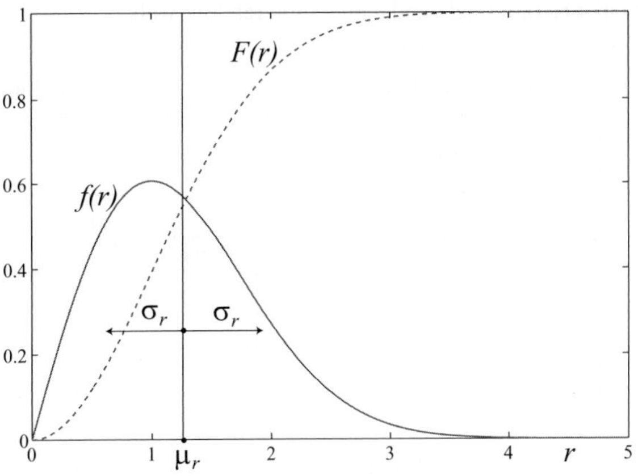

Figure 6.24: Rayleigh probability density function $f(r)$ and cumulative distribution function $F(r)$, $\sigma = 1$.

Using the general property that the derivative of the distribution function equals the probability density function $f(r)$, we immediately obtain the Rayleigh density function of Eq. (6.5.16).
The first and second statistical moments and the variance of the Rayleigh distribution are given by

$$\mu_r = \sigma\sqrt{\frac{\pi}{2}}\,, \qquad E\{r^2\} = 2\sigma^2\,, \qquad \sigma_r^2 = E\{r^2\} - \{\mu_r\}^2 = \left(2 - \frac{\pi}{2}\right)\sigma^2\,. \tag{6.5.18}$$

It follows that the probability of obtaining on average an error-free system is zero in the case of the wavefront deviations with azimuthal dependence $m\theta$. The Rayleigh probability density function has been depicted in Fig. 6.24 with a σ value of the independent x and y variables equal to unity. In the figure we have also plotted the value of the mean μ_r and the square root of the variance σ_r^2 of the statistical variable r.

d) $m \neq 0$ and quadratic dependence of the Zernike aberration coefficient on the constructional errors.
The Zernike aberration coefficients follow from the relation $t = a(x^2 + y^2)$ and this allows us to write for the distribution function $F(t') = P\{t'_r \leq t\}$ of the stochastic variable $t' = t/a = x^2 + y^2$ the expression

$$P\left\{x^2 + y^2 \leq t'\right\} = P\left\{\sqrt{x^2 + y^2} \leq \sqrt{t'}\right\} = \iint\limits_{\sqrt{x^2+y^2} \leq \sqrt{t'}} f(x)f(y)\,dxdy\,. \tag{6.5.19}$$

With equal Gaussian distributed variables x and y, we readily obtain

$$F(t') = \frac{1}{2\pi\sigma^2}\int_0^{2\pi}\int_0^{\sqrt{t'}} \exp\left\{-\frac{r^2}{2\sigma^2}\right\} r\,drd\phi = 1 - \exp\left\{-\frac{t'}{2\sigma^2}\right\}, \tag{6.5.20}$$

where the distribution function for t is obtained by replacing t' by $\mathrm{sgn}(a)\, t/|a|$ to cover both positive and negative values of the coefficient a. The probability density function follows by differentiation,

$$f(t') = \frac{1}{2\sigma^2}\exp\left\{-\frac{t'}{2\sigma^2}\right\}. \tag{6.5.21}$$

For the probability density function of the desired variable t we then find

$$f(t) = \frac{1}{2|a|\sigma^2}\exp\left\{-\mathrm{sgn}(a)\,\frac{t}{2|a|\sigma^2}\right\}. \tag{6.5.22}$$

For the statistical moments and the variance associated with $f(t)$ we have,

$$\mu_t = 2a\sigma^2\,, \qquad \overline{t^2} = 8a^2\sigma^4\,, \qquad \sigma_t = 2|a|\sigma^2\,. \tag{6.5.23}$$

The quadratic Rayleigh probability density function $f(t)$ and the cumulative distribution function $F(t)$ have been plotted in Fig. 6.25, for $\sigma = a = 1$. A comparison with the standard Rayleigh distribution graphs of Fig. 6.24 shows that the quadratic cumulative distribution function $F(t)$ has a substantially slower increase and a correspondingly larger σ_t value than the function $F(r)$. For the 95%-confidence interval, the quadratic distribution yields $t \leq 6|a|\sigma^2$. In the case of

Table 6.4. Statistics of a linear and a quadratic random walk.

	Linear random walk	Quadratic random walk
	$r = \sqrt{(ax)^2 + (ay)^2}$	$t = a(x^2 + y^2)$
$f(r),\ f(t)$	$\dfrac{r}{(a\sigma)^2} \exp\left\{-\dfrac{r^2}{2(a\sigma)^2}\right\}$	$\dfrac{1}{2\lvert a\rvert\sigma^2} \exp\left\{-\mathrm{sgn}(a)\dfrac{t}{2\lvert a\rvert\sigma^2}\right\}$
$\mu_r,\ \mu_t$	$\sqrt{\pi/2}\,\lvert a\rvert\sigma$	$2\lvert a\rvert\sigma^2$
$\overline{r^2},\ \overline{t^2}$	$2(a\sigma)^2$	$8a^2\sigma^4$
$\sigma_r^2,\ \sigma_t^2$	$\left(2 - \dfrac{\pi}{2}\right)(a\sigma)^2$	$4a^2\sigma^4$

Table 6.5. Confidence intervals in % for some statistical distributions.

	σ	2σ	3σ	6σ
Gaussian distribution	68.3	95.4	99.7	
Rayleigh (linear)	39.3	86.4	98.9	
Rayleigh (quadratic)	39.3	63.2	77.2	95

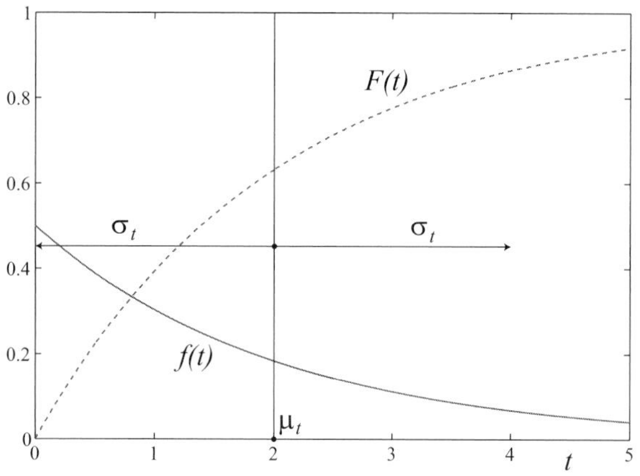

Figure 6.25: Quadratic Rayleigh density function $f(t)$ and the corresponding cumulative distribution function $F(t)$.

the standard Rayleigh distribution, 95% of the sample values are found in the interval $0 < r < 2.45a\sigma$. We note that the 'quadratic' random walk distribution is also commonly denoted as a χ-squared distribution of two independent random variables.

In Table 6.4 we have listed the results for the statistical properties of a linear and a quadratic random walk. For an easy comparison of the linear and the quadratic random walk, we have also included the proportionality factor a in the linear random walk data. The confidence intervals for the three statistical distributions that play a role in the statistical analysis of optical systems are given in Table 6.5. In the case of the quadratic Rayleigh distribution we have put $\lvert a\rvert = 1$. A short remark on the sign of the coefficient a in the quadratic expression for the stochastic variable t is needed. In the analysis, it is sufficient to take into account only the modulus of this coefficient. Any negative a-value is accounted for in the random walk related to the Zernike coefficients with upper index $m \neq 0$ by an extra angle offset of $\theta = \pi/m$.

6.5.3.2 Examples of Random Walk Statistics of an Optical System

An illustration of the two-dimensional random walk which arises for each Zernike aberration coefficient with $m \neq 0$ is illustrated in Fig. 6.26a. We show the endpoints of 100 random vectors (x, y) that obey Rayleigh statistics in the radial direction and whose azimuthal angle θ is uniformly distributed between 0 and 2π in the limit of large sample numbers. Figure 6.26b shows the corresponding walk which results when these samples are vectorially added with an azimuthal distribution θ which is uniform between 0 and 2π. The same procedure is applied in Fig. 6.26c and d for the random variable $t = x^2 + y^2$. From Table 6.4 we deduce that the ratio of the variances σ_t^2/σ_r^2 is of the order of 3^2. Despite the strong statistical variations from random walk to random walk sample, this ratio of three can be approximately observed in Fig. 6.26b and d. It confirms that for a large enough number of steps in a random walk, the expectation values for mean and variance are determined by the sum of the variances of the total number of variables. This 'large number' tendency leads to a Rayleigh probability density function for the endpoints of a random walk with the mean and variance of the random walk determined by the sum of the variances of the individual steps of the random walk, even if these steps do not have equal length **[269]**.

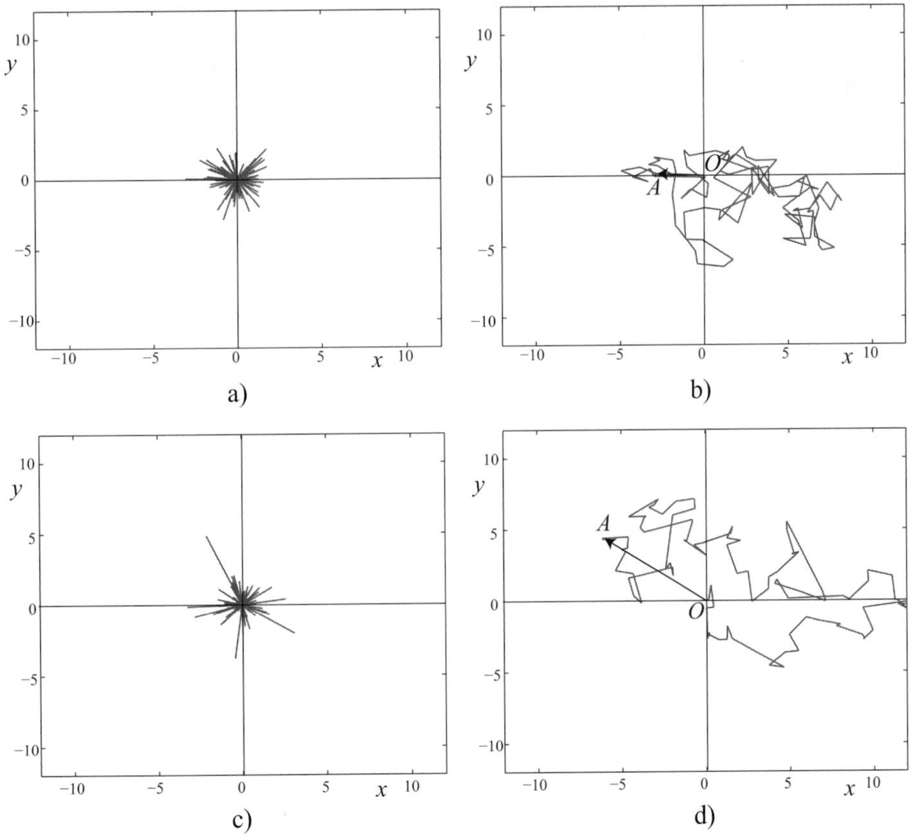

Figure 6.26: a) The length distribution of vectors with random azimuth, uniformly distributed between 0 and 2π, and with length $r = \sqrt{x^2 + y^2}$, where x and y are Gaussian random variables with zero mean and variance $\sigma_x^2 = \sigma_y^2 = 1$ (100 samples).
b) A two-dimensional random walk with the same 100 samples from the random x and y variables depicted in a).
c) and d) Same legend as for a) and b), applied to the random variable $t = x^2 + y^2$ and with the same set of x and y samples which were used in graphs a) and b).

To obtain the mean and variance of the Rayleigh distribution of a random walk, we calculate the effect of a large number of mechanical, optical and material errors by means of the sensitivity coefficients $a_{n,k,l}^{m,(\alpha)}$, each time for a perturbation value $\epsilon_{k,l}$. In practice, we apply the one σ value associated with each perturbation $\epsilon_{k,l}$. The asymmetric wavefront aberration coefficients of the optical system follow a random walk further and further away from the as-designed values, as shown in Fig. 6.26b and d. In the limit of a large number of random contributions, the mean and variance for the Zernike coefficients with indices (n, m) are given by

$$(\mu_n^m)^2 = \sum_{k,l} \left\{ \frac{\pi}{2} \left(a_{n,k,l}^{m,(1)} \, \sigma_{\epsilon_{k,l}} \right)^2 + 4 \left(a_{n,k,l}^{m,(2)} \, \sigma_{\epsilon_{k,l}}^2 \right)^2 \right\}$$

$$(\sigma_n^m)^2 = \sum_{k,l} \left\{ \left(2 - \frac{\pi}{2} \right) \left(a_{n,k,l}^{m,(1)} \sigma_{\epsilon_{k,l}} \right)^2 + 4 \left(a_{n,k,l}^{m,(2)} \sigma_{\epsilon_{k,l}}^2 \right)^2 \right\} . \qquad (6.5.24)$$

From the composite cumulative distribution function created by a large number of independent random variables such as surface or element decentring, tilt, surface deformation, distance mismatch, refractive index departure, temperature drift, aspheric surface error, etc., the statistical tolerancing provides us with an indication of the number of manufactured systems that will be found in a certain confidence interval.

An important further step in the statistical analysis, which was not discussed above, is the use of optical ‘compensators’. Compensating surfaces, elements or combinations of elements are used to reduce the as-manufactured residual aberrations to zero by introducing combinations of axial shifts, decentrings, etc. in the system. These compensating elements are effective for the residual aberrations of the Seidel type such as spherical aberration, coma, distortion and, to a lesser degree, astigmatism and field curvature. The reduction of aberrations of higher order is more difficult to achieve. In practice, this may even require polishing corrections on various optical surfaces in the system or the introduction of an adaptive optical element.

7 Design Methods for Optical Imaging Systems

7.1 Introduction

In this chapter we discuss various classes of optical imaging systems and the design methods that are used to optimise their optical performance. Historically, the telescope and microscope are the oldest imaging instruments. Systematic design of these instruments became possible from the middle of the nineteenth century when the third-order aberration theory was devised. Initially conceived by the mathematician Petzval, the new theory was made available to the scientific and industrial community by Seidel **[301]**. The existing instruments and the newly developed photographic camera profited greatly from the new theoretical insights. Simultaneously, important progress was made in chemistry (optical glass development), fine mechanics and glass-working technology. Using the new aberration theory and more advanced measurement techniques, the production quality of optical instruments could be drastically improved. For a long time optical fabrication techniques had been partly based on trial and error; from 1850 on professional mass-production became possible.

The imaging theory of the microscope developed by Abbe underpinned the ultimate role played by the diffraction of light in image formation. It was recognised that the wavelength of the light should be incorporated in the design theory. As a result of a gradual change in approach, the typically nineteenth century geometrical theory of imaging aberrations has been replaced by the wave theory of aberrations **[140]**. The last design stage, even when the initial design stage has been based on ray optics analysis, is carried out in the wavefront domain.

In this section we use both ray optics and wave optics representations of the quality of an optical instrument. A typical ray optics criterion is the blur pattern or spot diagram of an image point and the corresponding blur size, characterised by, for instance, its root-mean-square radius. For the frequency transfer of a system, the geometrical-optics-based modulation transfer function can be used ('Geo-MTF'). These geometrical approximations of the image intensity distribution and the frequency transfer function are adequate for optical systems with relative large wavefront aberrations, of the order of several wavelengths and larger. Examples are projection lenses for computer images, high-quality photographic camera lenses or objectives for surveillance.

The class of optical instruments which needs to have diffraction-limited performance or should remain close to this target requires an analysis based on wave optics. Relevant criteria for such an instrument are the wavefront shape of a pencil of rays towards a certain image point and the root-mean-square value of the wavefront aberration of such a pencil. Wavefront expansions based on Zernike polynomials (see Appendix D) are frequently used and certain targets of the values of the Zernike coefficients are set to describe the quality of the imaging system. Other criteria are the shape and the typical size of the intensity pattern of the image of a point source in the object plane. The frequency transfer by such a diffraction-limited imaging system has to be calculated by means of the diffraction-based modulation transfer function. The wavefront-based quantities require a more extensive calculation including, for example, the evaluation of a diffraction integral. These calculations were almost prohibitively time-consuming up to the early computer period and kept the optical design community for a long time in the orbit of purely geometrical optics. With the advent of fast computing capability and specially developed software, a gradual transition to wave-optics based design has occurred.

In Sections 7.2–4 we first discuss relatively simple basic imaging systems in some detail. These are the achromatic doublet, the single-element landscape lens and the Petzval portrait lens. These examples are used to introduce some basic design concepts and to illustrate the application of Seidel's aberration theory. The examples also serve to discuss various representations of optical performance which are relevant for systems having either a diffraction-limited quality or having a more coarse resolution. In Sections 7.5–9 we proceed with other imaging systems, such as photographic lenses, telescopes, microscope objectives, objectives for optical disc systems and projection lenses for high-definition computer images, selected as providing illustrative examples. A special class with extremely high imaging performance is composed of the lens and mirror systems for optical projection lithography, used in the semiconductor industry for the fabrication of microprocessors, computer logic and memory chips.

7.2 The Achromatic Doublet

The achromatic doublet was one of the first optical components that was developed once the dispersion behaviour of optical glass was better understood and measured in more detail. The British experimental opticist Dollond was the first to use different glass types to obtain colour correction in white light imaging in a microscope. Fraunhofer was the first to produce a detailed design for a doublet lens; the design principle could be used for telescope imaging, for eyepieces and for white light collimation in general. The fabrication of the cemented achromatic doublet goes back to the French opticist De Grateloup and such elements were used in early photographic lenses by Chevalier. Cemented elements were preferred because they make ghost images in the optical instrument weaker and improve the image contrast. Moreover, optical contact between two components relaxes the polishing requirements of the optical surfaces that are brought into contact via optical immersion. Another technological advantage derives from the strongly reduced centring tolerances in a cemented doublet as compared to those of a doublet with an air gap between the two elements. The immersion liquid is either oil or a resin. For a long time the preferred resin was Canada balsam, but it has a limited durability in time. Synthetic resins with a higher stability and a better resistance to environmental changes have now taken over the role of Canada balsam.

The achromatic doublet with a focal length of 100 mm and a typical lens diameter of some 40 mm can be analysed within the framework of the third-order aberration theory. If the numerical aperture is limited to 0.10, with a diaphragm diameter of 20 mm, the achromatic doublet has a performance that is close to the diffraction limit and the rms wavefront aberration of an imaging pencil can be kept below a value of approximately 0.07λ. This guarantees a 'just' diffraction-limited performance of the achromatic doublet. We analyse below the achromatic doublet using the 'thin-lens' aberration formulas as derived in Section 5.4. We first define the glass types to be used, typically a crown and a flint glass, by means of their central refractive index n_d and Abbe dispersion number V. The achromatic doublet can then be characterised by the bending factor P_1 of its first element. It is shown in the next paragraph that the important aberrations of an achromatic doublet, spherical aberration and coma, can be expressed as a function of this bending or shape factor P_1 of the first lens.

7.2.1 Thin-lens Analysis of the Achromatic Doublet

We approximate the real-world doublet with finite thicknesses of the elements by two thin lenses in contact. The second curvature of the first element and the first curvature of the second element are identical, yielding three surface curvatures (see Fig. 7.1). We have the following optical parameters:

- curvatures c_1, c_2 and c_3,
- two glass materials characterised by their refractive index at the central spectral d-line ($\lambda = 589$ nm) and their Abbe number, (n_1, V_1) and (n_2, V_2), respectively,
- the total optical power A of the doublet.

To obtain achromatic behaviour of the doublet we have for the powers of the individual elements of the doublet (see the analysis in Subsection 5.6.4),

$$A_1 = \left(\frac{V_1}{V_1 - V_2}\right) A\,, \qquad A_2 = -\left(\frac{V_2}{V_1 - V_2}\right) A\,, \qquad (7.2.1)$$

where A is the total optical power of the doublet lens. The paraxial powers of the two elements in terms of the surface curvatures are

$$A_1 = (n_1 - 1)(c_1 - c_2)\,, \qquad A_2 = (n_2 - 1)(c_2 - c_3)\,. \tag{7.2.2}$$

For the thin-lens analysis of the doublet we further need the bending factors P of elements 1 and 2 and their transverse magnification factors Q_1 and Q_2,

$$P_1 = \frac{c_1 + c_2}{c_1 - c_2}\,, \quad P_2 = \frac{c_2 + c_3}{c_2 - c_3}\,, \quad Q_1 = \frac{m_1 + 1}{m_1 - 1}\,, \quad Q_2 = \frac{m_2 + 1}{m_2 - 1}\,, \tag{7.2.3}$$

where m_1 and m_2 are given by

$$m_1 = 0\,, \qquad m_2 = \frac{V_1}{V_1 - V_2}\,. \tag{7.2.4}$$

We use P_1 as the basic variable to describe the imaging properties of the doublet with the following expressions for the quantities P_2, Q_1 and Q_2,

$$P_2 = (1 - P_1)\frac{V_1(n_2 - 1)}{V_2(n_1 - 1)} - 1\,, \qquad Q_1 = -1\,, \qquad Q_2 = \frac{2V_1 - V_2}{V_2}\,. \tag{7.2.5}$$

The evaluation of the spherical aberration and coma aberration requires the value of the intersection height Y_m with the optical surfaces of the marginal ray through B_m, belonging to the axial pencil focusing at O_2. It follows that $Y_m A = -\alpha$ (see Fig. 7.1). The Lagrange invariant H for the imaging by the two elements, operating in air, is given by $H = \alpha y$ where $y = O_2B_2$, the off-axis distance where the comatic aberration has to be evaluated. The substitution of the element data and the paraxial imaging data in Eq. (5.4.13) and the application of the relations (5.4.17) and (5.4.18) yield the following result for the comatic wavefront aberration of the doublet as a function of the bending factor P_1,

$$\begin{aligned} W_c(\rho,\theta) = \frac{1}{4}\frac{\alpha^3 y}{(V_1 - V_2)^2}\Bigg\{ &\left[\frac{V_1^2(n_1+1)}{n_1} - \frac{V_1V_2(n_2+1)}{n_2}\right]\frac{P_1}{n_1 - 1} \\ &- \frac{V_1^2(2n_1+1)}{n_1} + \frac{V_2(2V_1 - V_2)(2n_2+1)}{n_2} \\ &+ V_2\left[V_1(n_2-1) - V_2(n_1-1)\right]\frac{(n_2+1)}{n_2(n_2-1)(n_1-1)}\Bigg\}\rho^3\cos\theta\,, \end{aligned} \tag{7.2.6}$$

where $\rho = r/r_m$ is the normalised radial coordinate in the pupil and θ is the polar angle with respect to the Y-axis in a plane perpendicular to the optical axis TO_2. T determines the position of the vertices of the optical surfaces. In the case of a thin-lens approximation, all vertices coincide.

The comatic aberration, a linear function of the bending factor, can be made zero by selecting the appropriate bending of the two elements. Working out Eq. (7.2.6) we obtain the solution

$$\begin{aligned} P_1 = &\left[V_1^2(n_1-1)(2n_1+1)n_2(n_2-1) - V_1V_2n_1(n_2-1)(4n_1n_2 - 3n_2 + 2n_1 - 1)\right. \\ &\left.+ 2V_2^2n_1(n_1-1)n_2^2\right] \Big/ \left[V_1(n_2-1)\{n_2(n_1+1)V_1 - n_1(n_2+1)V_2\}\right]\,. \end{aligned} \tag{7.2.7}$$

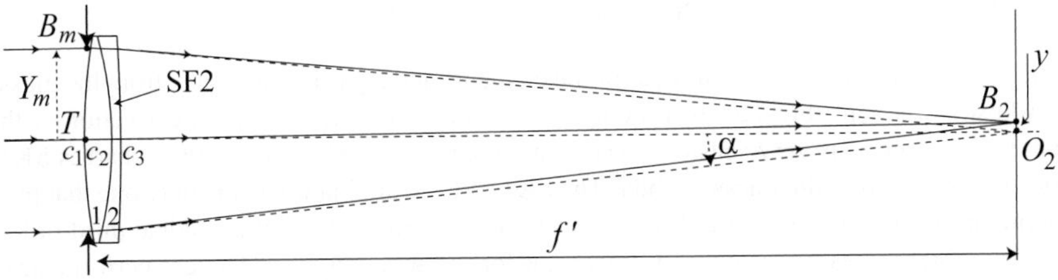

Figure 7.1: Geometrical and optical data of a cemented achromatic doublet with image-side focal distance f'. The solid rays belong to an oblique pencil, focusing at an image coordinate $y = f'/100$ from the optical axis (point B_2). The dashed rays are part of the axial pencil which is in focus at O_2. The aperture angle α of the focusing beams at the image side is 0.10 (the object plane is at infinity).

The spherical aberration of a thin lens is a quadratic function of the bending factor P (see Eq. (5.4.12)). From this equation we observe that large positive or negative values of the bending factor of a thin lens produce a large amount of negative spherical aberration. The maximum of the parabolic spherical aberration curve is found when $dS_1/dP_1 = 0$. The derivative of the spherical aberration function of the doublet with respect to the bending factor P_1 is obtained by summing the derivatives of the two elements, using expression (7.2.5) which links the bending factors of the two elements. After some algebra, the zero value of the derivative function dS_1/dP_1 of the doublet is obtained for a bending factor

$$P_1 = \Big[\, 2V_1^2 n_2(n_1^2 - 1)(n_2 - 1) - V_1 V_2 \left\{n_1(n_2 + 2)(n_2 - 1) + 4n_1(n_1 - 1)(n_2^2 - 1)\right\} + V_2^2 \left\{n_1(n_1 - 1)(n_2 + 2) + 2n_1(n_1 - 1)(n_2^2 - 1)\right\} \Big] \left[(n_2 - 1)V_1 \left\{n_2(n_1 + 2)V_1 - n_1(n_2 + 2)V_2\right\}\right]^{-1}. \quad (7.2.8)$$

7.2.2 Design Examples of Achromatic Doublets

Referring to the glass chart of Fig. 5.53, the optimum choice of the glass pair for a doublet is hampered by the limited number of optical glasses. As high-dispersive glasses, flint and heavy flint glasses are used. They can be combined with a relatively dense collection of crown glasses, on the Schott glass chart from the classes BK, K, BAK and SK. A variety of relatively standard glasses is found with a large range in refractive index, $1.51 < n_d < 1.64$.

The design of a doublet consists of the following steps:

- We first make a choice for the high-dispersion glass. In the following example we start with a certain flint glass, SF2 ($n_d = 1.648$, $V_d = 33.8$). This glass will be used for the first or the second element of the doublet.
- The flint glass is combined with a hypothetical crown glass with a V number of, for instance, 60.8. This value of the Abbe number V is the average of a group of densely packed crown glasses (SK-type) on the glass chart of Fig. 5.53.
- The analysis of the preceding subsection permits us to find the lens shape that yields zero coma and the position of the minimum absolute value of spherical aberration. By fine-tuning the refractive index n_1 of the low-index glass, the doublet shape is found with zero coma *and* zero spherical aberration.
- If needed, a fine-tuning is performed at the maximum aperture of the lens where the lens performance is still diffraction-limited according to third-order theory. The fine-tuning consists of a minimisation of the wavefront aberrations of the doublet using finite rays and a minimisation program which is able to find a local minimum.

This stepwise process can be repeated with another starting value for the Abbe number of the crown glass. An entirely new cycle is started by choosing a different flint glass, of which there are some three to four interesting candidates having a low absorption. Although they might be potentially more interesting, flint glasses with a larger dispersion show a limited transmission in the blue spectral region and have less favourable grinding and polishing properties.

The typical behaviour of the spherical and comatic wavefront aberration of an achromatic doublet as a function of its shape is shown in Fig. 7.2:

- *'crown-in-front' doublet*
  Figure 7.2a shows a doublet D_1 with virtually zero spherical and comatic wavefront aberration. In the wavefront versus bending factor graph, the maximum of the spherical aberration parabola can be made almost coincident with the intersection of the linear coma aberration function with the zero level. To this purpose, the refractive index n_1 has to be chosen as 1.5575, substantially lower than the index of the flint glass. The optimum bending factor P_1 for the first lens is –0.213 and the bending factor P_2 of the negative element is equal to 1.535.

Manufacturable solutions need finite thicknesses for the elements and a particular choice from the glasses available on the glass chart. Adjusting the lens thicknesses has only a minor effect on the aberration correction of the doublet. The choice of n_1 from the glass chart is very sensitive. The best companion of SF2 on the glass chart is SK11, both for the thin-lens solution and for the finite-thickness doublet. The SK11/SF2-combination is the only one that produces virtually no coma. The optimum bending factor P_1 for this combination is determined and the residual third-order aberrations of the doublet with finite thicknesses are calculated. Important for the wavelength-dependent performance of the doublet are the surface contributions to the (third-order) aberrations. As a rule of thumb, the chromatic aberration equals a few percent of the monochromatic aberration at each transition. For the SK11/SF2 doublet the (monochromatic) spherical aberration amounts to typically 3 wavelengths per surface. The next aberration which is important for off-axis imaging is astigmatism. For the example of the doublet, at a field angle of 0.01 radian, the astigmatism is still small, typically 0.15 wavelengths in amplitude. It was shown before that a thin lens or lens system with central diaphragm has a fixed

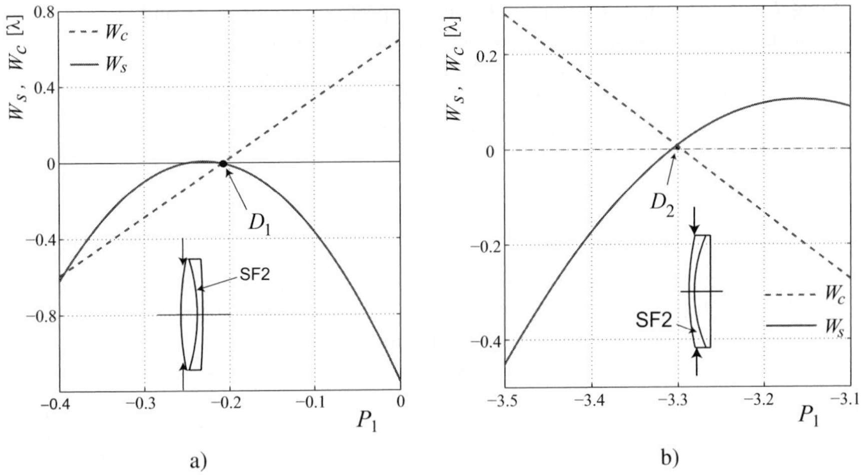

Figure 7.2: Third-order spherical wavefront aberration W_s (solid curve) and coma W_c (dashed curve) of an achromatic doublet as a function of the bending factor of the first element. Focal distance $f' = 100$ mm, $NA = 0.10$, field excursion $y = 1$ mm. Imaging from infinity.
a) 'Crown-in-front' doublet;
b) 'Flint-in-front' doublet.

amount of astigmatic aberration and field curvature, independently of lens shape or refractive index. If diffraction-limited performance is required, the useful image field of the doublet is effectively limited by astigmatism.

The manufacturing of the doublet would be facilitated if the back surface of the second element could be made flat. Unfortunately, this is not possible without compromising the aberration correction of the achromatic doublet. The maximum aperture of the achromatic doublet, typically 0.10 for diffraction-limited performance, is larger than that of a single lens. The degree of freedom that is introduced by the buried surface transition and its index contrast increases the numerical aperture by 0.03. A single lens element made of SF2 has to be stopped down to 0.07 for equal monochromatic performance. An advantage of the monochromatic singlet is that it can be given a flat exit surface as this shape virtually coincides with the absence of spherical aberration and coma.

- *'flint-in-front' doublet*
  In Fig. 7.2b the alternative solution is shown with the high-index element in front. The solution D_2 has in front a negative meniscus flint lens in contact with a positive plano-convex low-index lens. The optimum doublet shape is obtained by making the two aberration curves W_s and W_c intersect at zero level. Unfortunately, it is not possible to have the maximum of W_s coincide with the zero crossing of the W_c straight line. This makes the doublet more sensitive to index changes as a function of wavelength. A deterioration of the chromatic correction can be expected because of the effect of spherochromatism, the change of spherical aberration with wavelength. At the central wavelength, the surface contributions to the spherical aberration turn out to be twice as high as in the 'crown-in-front' option. An advantage is that the reverse side of the second element can easily be made flat without compromising its performance. The overall conclusion is that the 'flint-in-front' choice is inferior to the standard 'crown-in-front' version. Equal aberration values for D_1 and D_2 require a reduction in aperture to 0.085 of the D_2 doublet.

In Table 7.1 we list the performance data of examples D_1 and D_2 at equal aperture ($NA = 0.10$), for an off-axis pencil at an angle of 0.01 radians. The data are in terms of the optical pathlength difference over the wavefront in the exit pupil of which the rim is denoted by the arrows in Fig. 7.2. More particularly, the root-mean-square value OPD_{rms} is shown, expressed in units of $\lambda/1000$ (so-called 'milli'-wavelengths or mλ). The wavefront aberration values in the table include the effects of field curvature and distortion. The calculation of the exact total wavefront aberration with the aid of finite ray-tracing has demonstrated that, for the aperture and field value under consideration, the third-order aberration theory is a good approximation. This can be expected as the maximum paraxial angles of incidence and refraction at the surfaces are of the order of 15 degrees. At such an angle of incidence, if the optical contrast at refraction equals 1.5, the third-order approximation of Snell's law introduces an error in the refracted ray angle of approximately $6 \cdot 10^{-5}$ rad. A

Table 7.1. OPD_{rms} in units of mλ in the visible spectrum; examples D_1 and D_2, $NA = 0.10, f' = 100$ mm, $y = 1$ mm (exact ray-tracing data).

λ	486 nm	520 nm	560 nm	605 nm	656 nm
D_1	72	58	83	46	59
D_2	63	92	120	86	59

Table 7.2. OPD_{rms} in units of mλ, subdivided into various aberration contributions of third and higher order. Doublets D_1 and D_2, $NA = 0.10, f' = 100$ mm, $y = 1$ mm, $\lambda = 587.6$ nm, optimum axial position of the image plane.

	W_4^0	W_6^0	W_3^1	W_5^1	W_2^2	W_4^2	OPD_{rms}
D_1	19	4	3	1	17	0	26
D_2	50	4	13	1	17	0	54

multiplication of this value by the focal distance f' of the doublet yields the value of the lateral intersection error of a ray in the image plane ($\approx$4 μm) when third-order theory is used. Note that this value is close to that of the diffraction unsharpness $\lambda/NA = 5.9$ μm in the image plane.

An analysis of the exact wavefront data with the aid of Zernike polynomials of the fourth and higher order in the pupil coordinates leads to the wavefront data of Table 7.2. In this table, in contrast with Table 7.1, we have only included the aberration terms that lead to basic unsharpness in image space. The aberration components associated with distortion and field curvature have been omitted. A comparison of the higher-order terms obtained by finite ray-tracing with the third-order results showed that the values of the aberration terms that are of higher order than those of the Seidel aberration theory are typically less than 20% of the dominating Seidel terms.

Figure 7.3 shows the layout of a standard 'crown-in-front' doublet with the glasses SK11 and SF2. The residual chromatic deviation (secondary spectrum) is appreciable and yields a maximum value of the OPD_{rms} of 0.08 λ, just tolerable from the point of view of diffraction unsharpness of the image of a point source (see Fig. 7.3a). The solid curve yields the rms wavefront aberration in the best average focal plane, the dashed curve in the optimum focal plane for each individual colour. Figure 7.3b shows that the best axial position z for each individual colour is a quadratic function of wavelength, as predicted by the paraxial analysis of chromatic aberration in Chapter 5. The maximum deviation is of the order of 25 μm at $\lambda = 560$ nm according to the solid curve. Using the expression for the focal depth $z_f \approx \frac{1}{2}\lambda/(NA)^2 =$ 28 μm, it is seen that the maximum focal deviation due to the secondary spectrum of the doublet is just within the diffraction-limited focal range in image space. In the optimum focus position for white light we observe a lateral shift of the centre of the image point for each colour given by the dashed curve in Fig. 7.3b. The lateral shift, typically 250 nm at $\lambda = 560$ nm, is small with respect to the diffraction unit $\lambda/NA = 5.6$ μm in the image plane. For this doublet the lateral colour effects can be safely neglected.

The wavefront aberration of the SK11/SF2 doublet is shown in Fig. 7.4. For three representative wavelengths the meridional and the sagittal cross-sections of the wavefront of an oblique pencil are shown. Because of the circular symmetry of the doublet, the sagittal cross-section of the wavefront is symmetric. In the meridional cross-section an asymmetric wavefront-aberration-like coma may appear. Such an asymmetric wavefront cross-section can be partly compensated for by the subtraction or addition of a linear wavefront term. A wavefront tilt in the meridional section is obtained by a shift of the centre of the reference sphere in the y-direction. The tilt-corrected wavefront curves are solid in Fig. 7.4, the uncorrected are dashed. We observe that the meridional wavefront sections are virtually symmetric as is shown by the almost coincident solid and dashed curves. The comatic aberration of the oblique pencil is thus negligible. This aplanatic property of the doublet depends on the correct balance between the indices of the two glasses. The optimum index value following from third-order aberration analysis was 1.5575 for a V number of 60.8 ('crown-in-front' optimum). SK11 is the best candidate from the glass chart with a V number of approximately 61. The main defect of the doublet is seen to be its residual defocusing as a function of wavelength and a small amount of astigmatism. The relatively thin doublet has a quasi-fixed amount of astigmatism. In the third-order approximation it amounts to a

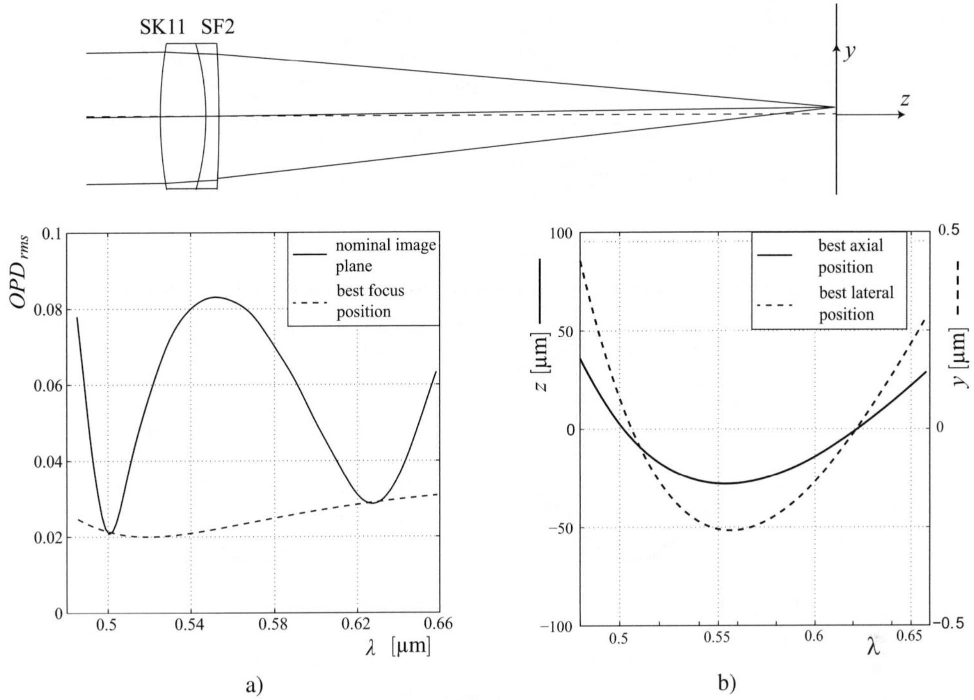

Figure 7.3: An achromatic collimator (SK11/SF2) with a slightly oblique imaging pencil (top); $NA = 0.10, f' = 100$ mm, field excursion 1 mm.
a) Optical performance in terms of OPD_{rms} (in units of the wavelength of the light) as a function of λ.
b) Best-focus position in axial (z, solid curve) and lateral (y, dashed curve) direction as a function of λ. Spectral range: 485 nm $\leq \lambda \leq$ 655 nm.

Seidel coefficient $W_{22,S} = \frac{1}{2}H^2/f'$where H is the Lagrange invariant. At a wavelength of 555 nm, this coefficient amounts to $\approx 0.09\,\lambda$. This value is confirmed by the meridional and sagittal wavefront cross-sections in Fig. 7.4 for $\lambda = 555$ nm. The difference in wavefront aberration for NA = 0.10 in these two cross-sections equals $(-0.27 + 0.18)\lambda$, showing that the third-order Seidel aberration theory is a very good approximation for the aperture and field values at which the doublet is commonly used.

A reduction of the secondary spectrum of a standard doublet requires the use of an optical glass with a non-standard dispersion curve. The flint glass KZFSN4 is such a glass and plays the role of 'high-index' glass. We choose the glass SK14 for an index match with KZFSN4 and this combination reduces the coma of the doublet. The combination produces an extended range of chromatic correction, down to a wavelength of 450 nm (see Fig. 7.5a). Figure 7.5b shows that the optimum focus curve as a function of wavelength still has a quadratic appearance and colour correction is achieved at two wavelengths only.

A substantial improvement of the colour correction is obtained by applying two glasses with non-standard dispersion. The partial dispersion rule expressed by Eq. (5.6.25) imposes two glasses with equal partial dispersion coefficients, for instance in the blue and the infrared region. It is not possible to exactly satisfy Eq. (5.6.25) because the (relative) freedom of monochromatic aberrations does not allow an arbitrary choice of the average indices of the two glass materials. A good compromise is the combination of fluorite with KZFSN4. Figure 7.6b shows the triple correction of axial colour which makes the combination an *apochromatic* doublet. The lower average indices of the glass materials as compared to the previous examples induce higher values of residual spherical aberration. To keep the performance within the just-diffraction-limited region, a reduction of the numerical aperture to the value 0.09 is needed.

In Subsection 5.6.4.2 the paraxial design of the powers of an achromatic multiplet was discussed. In principle, with N glass materials, chromatic correction of power can be obtained at N wavelengths. Unfortunately, because of the comparable dispersion of most standard glass materials, the coefficients of the system of equations are such

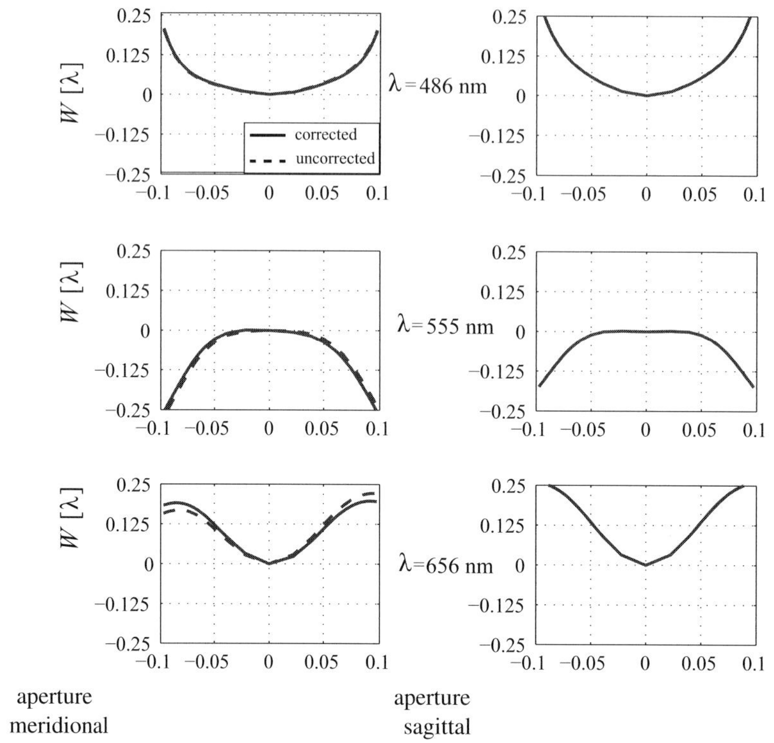

Figure 7.4: The wavefront aberration as a function of wavelength for an achromatic collimator, constructed with standard crown and flint glasses (SK11 and SF2, respectively, see the glass chart of Fig. 5.53). $NA = 0.10, f' = 100$ mm, field coordinate $y = 1$ mm.

that the system is highly dependent, yielding almost singular solutions. Using the selection rule for glass materials (see Subsection 5.6.4.2) some interesting combinations from the glass chart are possible. In Fig. 7.7 we present an apochromatic doublet which was constructed with glasses that, apart from FK51, have standard dispersion. The role of the flint glass is played by BAK2. Its index, in relation to that of FK51, enables the freedom of coma of the multiplet. The element with glass F2 is needed for fine-tuning of the chromatic correction. Figure 7.7b shows the apochromatic behaviour with axial chromatic correction at three wavelengths. The lateral colour deviation is typically 1 μm, well within the diffraction unsharpness of ≈ 5 μm. The initial paraxial choices for the powers of the elements have to be abandoned during the optimisation of the quartet with respect to monochromatic aberration. This explains why an initially expected four-wavelength chromatic correction has been reduced to the more modest *apochromatic* state of correction of the lens quartet of Fig. 7.7. It should be mentioned, however, that the wavelength range over which this apochromatic correction has been achieved with the aid of more or less standard glasses spans the near-UV, visible and near-infrared spectrum.

7.3 The Photographic Landscape Lens

The advent of photography in the nineteenth century required the development of a cheap optical lens which is capable of producing photographs of well-illuminated outside scenes. A certain minimum aperture is required for luminous efficiency; a stop number $F_{\#} = 8$ ($NA = 0.067$) is sufficient. As field angle a typical value of $2 \times 25°$ was needed. The resolution in the image had to be comparable to that of photographic printing paper which became available at the end of the nineteenth century. In this section we discuss the single lens solution for landscape photography.

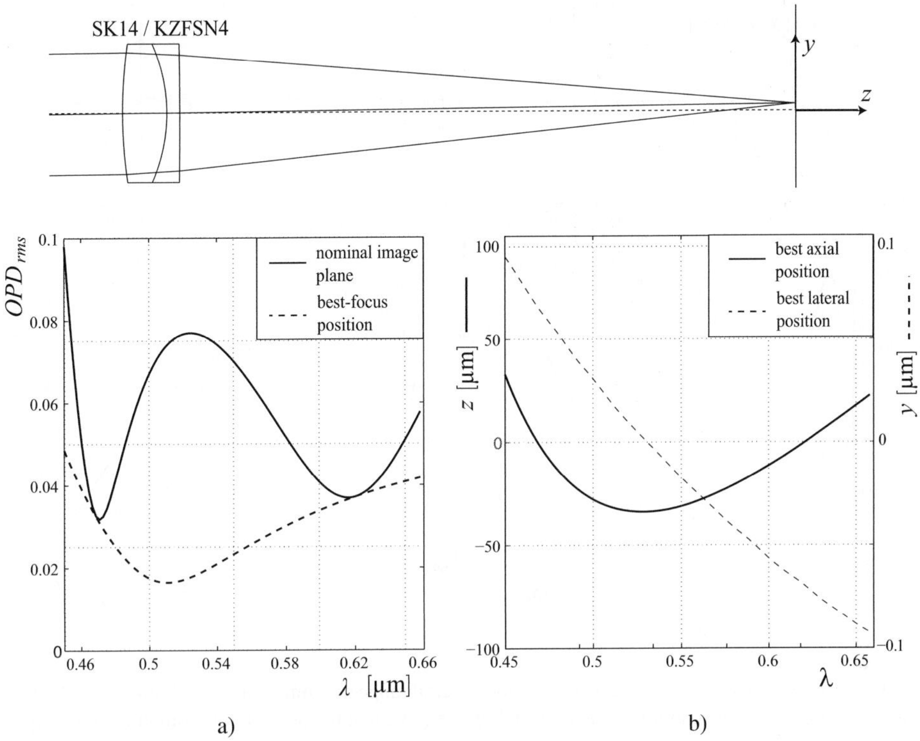

Figure 7.5: An improved achromatic collimator with achromatic behaviour up to the violet spectral region ($450 \leq \lambda \leq 660$ nm). Glass materials: SK14 (crown glass) and KZFSN4, a flint glass with a particular partial dispersion number $P_{g,F}$ in the blue spectral region; $NA = 0.10$, $f' = 100$ mm, field excursion 1 mm.
a) Optical performance in terms of OPD_{rms} (in units of the wavelength of the light) as a function of λ.
b): Best-focus position in axial (z, solid curve) and lateral (y, dashed curve) direction as a function of λ.

7.3.1 General Design Considerations

If a single lens ($f' = 50$ mm) is used as a landscape lens, a certain number of observations can be made which are based on the third-order analysis of the monochromatic and chromatic aberrations of the doublets of the previous subsection. They possess a comparable aperture and focal distance (see Fig. 7.8). The main difference is the much larger angular field of a landscape lens, typically $2 \times 23°$. The following observations with respect to the design of a landscape lens can be made:

- Third-order thin-lens analysis at a numerical aperture of 0.06 will give a reliable estimate of spherical aberration.
- Field aberrations of higher than third order are likely to play an important role at a field angle beyond 20°.
- Third-order coma of a single lens depends, among others, on the lens bending. Coma is also influenced by the stop position if the lens possesses a residual amount of spherical aberration. Zero coma can be achieved by a proper combination of lens bending and remote stop position.
- The astigmatism of a single lens with central stop is fixed. Reduction of astigmatism is possible by using the single lens with a remote stop position, provided the lens possesses residual spherical aberration and/or coma.
- Third-order field curvature is proportional to $(nf')^{-1}$. A high index is desirable.
- Distortion is absent in the case of central stop position. For the case of a remote stop position, distortion is influenced by the lens bending.
- The relative change in lens power because of dispersion in the visible part of the spectrum is given by $\pm 1/2V$, where V is the Abbe dispersion number.
- The quality of a photographic landscape lens is best described in terms of its frequency transfer, both on-axis and for various positions in the image field.

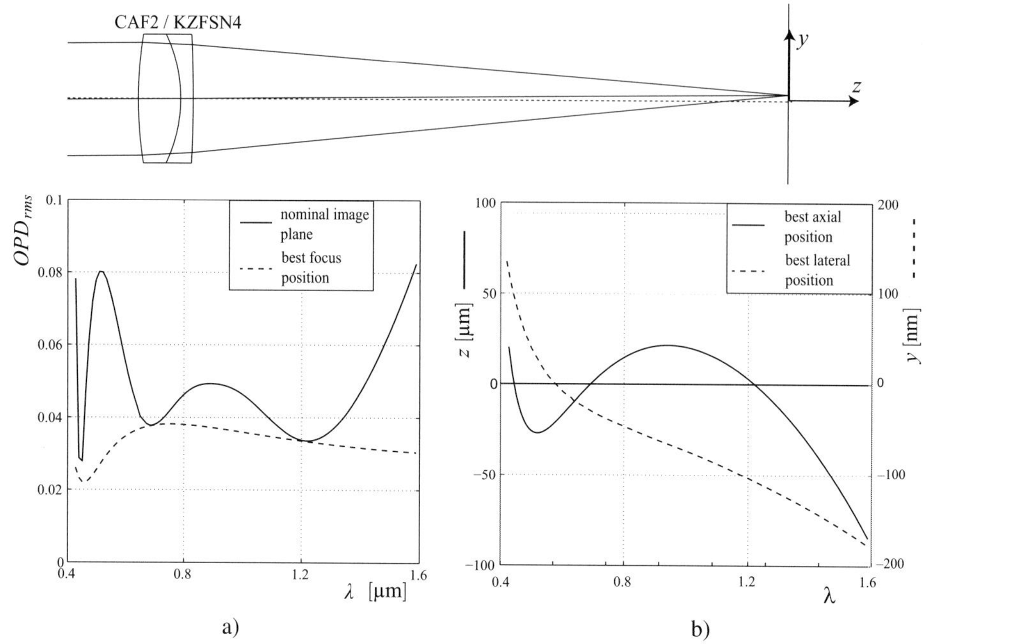

Figure 7.6: A CAF2/KZFSN4 achromatic collimator for the spectral range 405 nm $\leq \lambda \leq$ 1600 nm; NA = 0.09, f'= 100 mm, field coordinate y = 1 mm. Same legend for a) and b) as in Fig. 7.5 (note that the best lateral position y is now expressed in nm in b)).

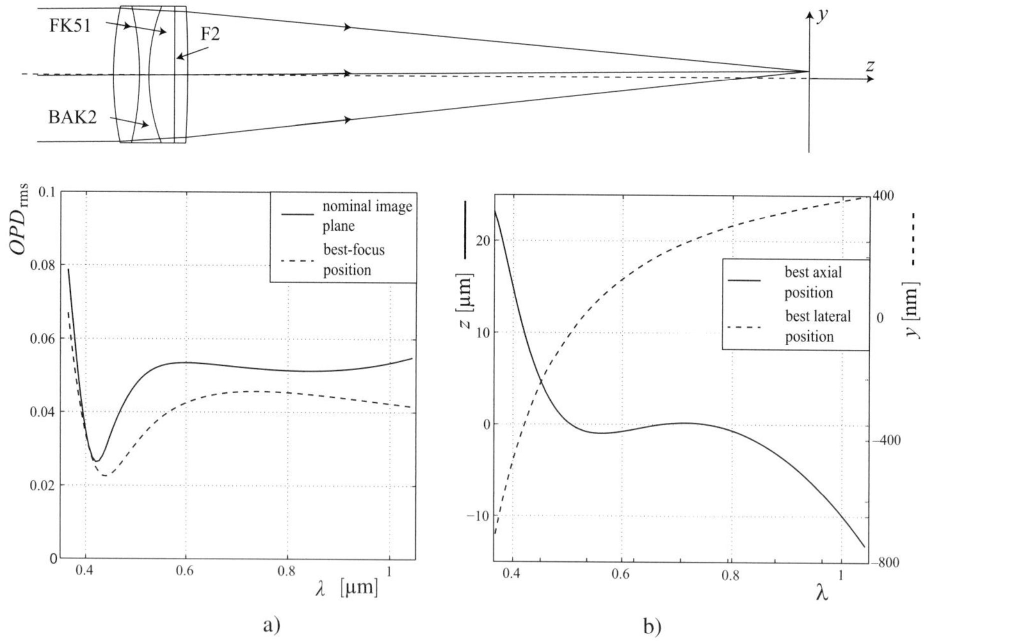

Figure 7.7: An achromatic collimator for the spectral range 365 nm $\leq \lambda \leq$ 1050 nm. Glass materials: FK51, BAK2 and F2. Apochromatic multiplet with coincident axial focus positions for three wavelengths. Same legend for a) and b) as in Fig. 7.5.

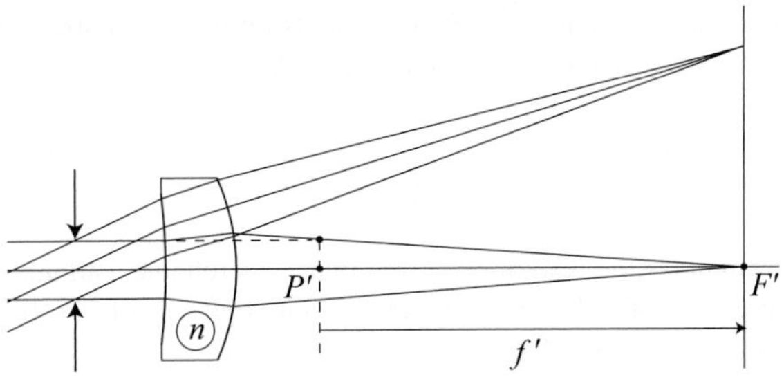

Figure 7.8: Example of a simple photographic landscape lens with remote stop. P' is the image-side principal point, F' the focal point and f' the image-side focal distance; n is the refractive index of the lens material.

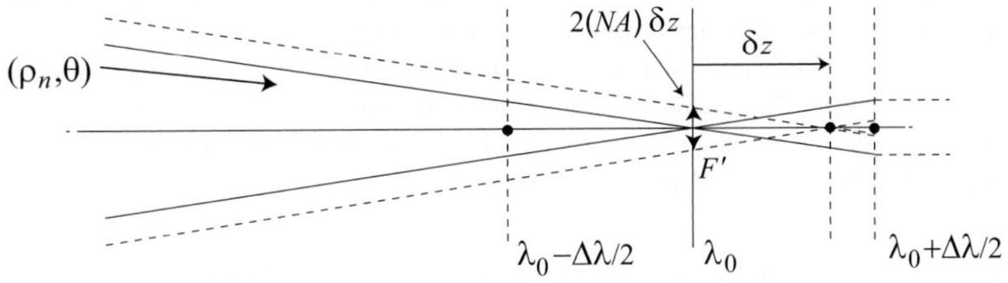

Figure 7.9: The chromatic blur in the focal plane of a single lens. F' is the position of the central focus at wavelength λ_0, the wavelength range is $\Delta\lambda$. A general ray of the imaging pencil is specified by its polar pupil coordinates (ρ_n, θ).

In the following subsections we first discuss the chromatic effects and then continue with the reduction of the monochromatic aberrations.

7.3.2 Chromatic Aberration and Frequency Transfer of a Single Lens

In the absence of chromatic correction the total focal shift over the visible spectrum is f'/V when imaging from infinity. If the monochromatic aberration of the lens is zero, an ideal point-like focus is found for the central green wavelength λ_0 in the image plane through F'. Supposing a linear dependence between wavelength and defocusing distance we can write

$$\delta z_\lambda = \left(\frac{\lambda - \lambda_0}{\Delta\lambda}\right)\frac{f'}{V} \, . \tag{7.3.1}$$

Figure 7.9 illustrates the axial chromatic effects in the focal plane through F'. Referring to Eq. (5.2.21), the transverse aberration components in the image plane through $F'$are given by

$$\left.\begin{array}{l} \delta x_\lambda(\rho_n, \theta) \\ \delta y_\lambda(\rho_n, \theta) \end{array}\right\} = (NA)\left(\frac{\lambda - \lambda_0}{\Delta\lambda}\right)\frac{f'}{V}\,\rho_n \left\{\begin{array}{l} \cos\theta \\ \sin\theta \end{array}\right. , \tag{7.3.2}$$

where (ρ_n, θ) are the polar coordinates of a ray in the exit pupil of the lens $(0 \leq \rho_n \leq 1)$.

The variance of the transverse aberration components follows from an integration over the unit circle of the quantity $(\delta x_\lambda^2 + \delta y_\lambda^2)$ and this yields

$$\delta_\lambda^2 = \frac{1}{2}\left(\frac{\lambda - \lambda_0}{\Delta\lambda}\right)^2 \left\{\frac{(NA)f'}{V}\right\}^2 . \tag{7.3.3}$$

The image blur resulting from the superposition of all wavelengths in the image plane yields

$$\delta^2 = \frac{1}{2}\left\{\frac{(NA)f'}{(\Delta\lambda)V}\right\}^2 \int_{\lambda_0-\Delta\lambda/2}^{\lambda_0+\Delta\lambda/2} w(\lambda)\,(\lambda-\lambda_0)^2 d\lambda \Bigg/ \int_{\lambda_0-\Delta\lambda/2}^{\lambda_0+\Delta\lambda/2} w(\lambda)\,d\lambda \, , \tag{7.3.4}$$

where we have introduced a weighting function $w(\lambda)$ which accounts for the power spectrum of the incident radiation. In the case of an idealised rectangular spectral profile, extending from $\lambda_0 - \Delta\lambda/2$ to $\lambda_0 + \Delta\lambda/2$ and for a uniform weighting $w(\lambda) = 1$, we obtain

$$\delta^2 = \frac{1}{24} \left\{ \frac{(NA)f'}{V} \right\}^2 . \tag{7.3.5}$$

The chromatic axial blur downgrades the resolution of the image. A reliable estimate of the spatial frequency band that can be transferred in the presence of only chromatic aberration is given in Eq. (10.2.79) and below. For the 50% modulation depth we find a spatial frequency value

$$\nu_{50} = \frac{0.16}{\delta} \approx 1.6 \, \frac{F_{\#} V}{f'} , \tag{7.3.6}$$

where we have used $NA \approx 1/2F_{\#}$ when imaging a distant scene. For a landscape lens ($f = 50$ mm) that is made of a low dispersion glass with $V = 60$, the frequency band which is transferred on-axis with at least 50% modulation depth is 15 mm^{-1} wide. For off-axis image points, the lateral chromatism of a single lens causes a further degradation in resolution, unless the stop is chosen on the lens itself (central stop position). In the case of nineteenth century grey-tone photography, the photographic emulsion was sensitive in a limited spectral region towards the blue part of the spectrum. Despite the higher dispersion in the blue, such a photographic medium enlarged the frequency pass-band in the presence of chromatic aberration. A later improvement of the spectral rendering by the emulsion (panchromatic film) limited the chromatic frequency pass-band to the typical value given by Eq. (7.3.6).

7.3.3 Third-order Monochromatic Aberrations of a Landscape Lens

To a good approximation, the third-order aberrations of a zero-thickness lens can be used to analyse the landscape lens. The change towards finite thickness will have a relatively small influence on the aberrations of the zero-thickness lens. We restrict the solution to a lens with spherically shaped surfaces and employ Eq. (5.4.24) to study the aberrations of a thin lens with, in general, a remote stop. In Subsection 5.4.4 the aberrations of the thin lens with *central* stop were discussed. Spherical aberration depends on the bending of the thin lens. The easily manufacturable convex-plano lens yields a low value, close to the minimum. Coma of a convex-plano lens can be made zero by choosing an appropriate refractive index: $n = (1 + \sqrt{5})/2$ (see Eq. (5.4.32)). It was also shown that astigmatism and field curvature only depend on the power of the lens. Distortion is zero for a central stop position. A good candidate according to third-order theory could be a lens with central stop and a refractive index n equal to 1.62, for instance the optical glass SK16 from the Schott catalogue. The performance of the lens with respect to astigmatism and field curvature needs further examination. A quick estimate is obtained from finite ray-tracing. Figure 7.10 shows the paths of the principal and marginal rays of an oblique pencil through the convex-plano lens at a large aperture of 0.20 to show the effect of the ray aberration; the

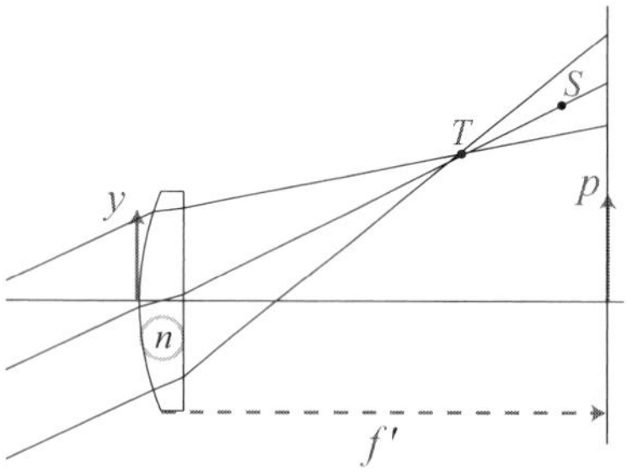

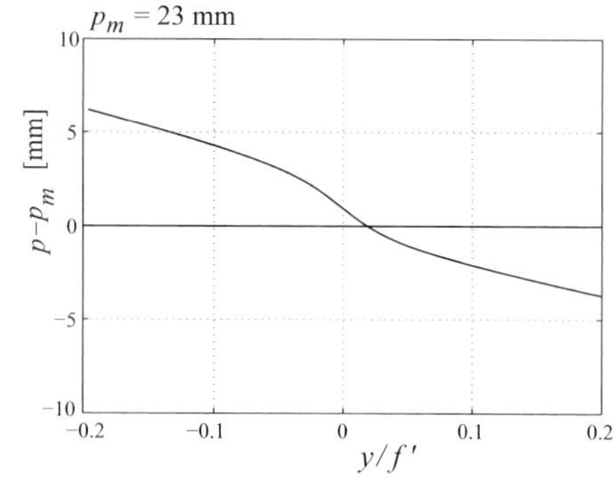

Figure 7.10: a) Finite ray picture for a convex-plano lens ($f' = 50$ mm) with central stop.
b) The transverse ray aberration $p - p_m$ in the image plane as a function of the relative intersection height y on the lens, divided by the focal distance f'. The paraxial field height p_m of the outer image point is 23 mm.

field angle is 25° corresponding to a (maximum) paraxial field point with lateral coordinate $p_m = 23$ mm. For an aperture value of 0.067, the dominating aberration in the image plane is the blur due to the tangential astigmatism, yielding a heavily defocused tangential focal point at T. The best focus would be obtained by shifting the receiving plane to an intermediate position between T and S. A quick glance at the figure shows that the resolution in such a plane would be totally insufficient.

7.3.3.1 Aberrations of a Landscape Lens with Remote Stop

Using the third-order aberration theory, we turn to other lens shapes than the $P = 1$ choice of Fig. 7.10 and study the effect of a shift of the diaphragm, away from the lens. The remote-stop thin-lens formulas for the aberrations allow us to calculate the lens shape and stop position which reduce the coma to zero. From Eqs. (5.4.12)–(5.4.17) and Eq. (5.4.24) we deduce

$$0 = S_2 + \epsilon S_1 ,$$

where

$$S_1 = -\frac{1}{4}\{f_1(P)\}(-\alpha_m)^4 f' ,$$
$$S_2 = -\frac{1}{2}\{f_2(P)\}(-\alpha_m)^3 p_m . \qquad (7.3.7)$$

f' is the image-side focal distance. The functions $f_1(P)$ and $f_2(P)$ are the expressions between braces in Eqs. (5.4.12) and (5.4.13), respectively, where the magnification factor Q was made equal to -1. α_m is the angle of the marginal ray of an axial pencil through the rim ($y = y_m$) of the lens with $\alpha_m = -y_m/f'$. All these quantities are calculated within the paraxial approximation.

The eccentricity factor ϵ is that of the oblique pencil going to a paraxial image point at p. The maximum eccentricity factor of the pencil to the outer field coordinate at $p = p_m$ follows from Eq. (5.4.20). It is written here as

$$\epsilon_m = \frac{p_m}{y_m}\frac{l_P'}{l_P' - f'} = -\frac{p_m}{y_m}\frac{l_P}{f'} . \qquad (7.3.8)$$

The quantity $2y_m$ is equal to the diameter of the lens diaphragm. The distances l_P and l_P' are the object and image distances of the entrance and exit pupil. From Eq. (7.3.7), applying the zero coma condition, we find the entrance pupil position l_P of a thin lens with bending factor P,

$$\frac{l_P}{f'} = \frac{2f_2(P)}{f_1(P)} . \qquad (7.3.9)$$

Among the remaining aberrations we first have spherical aberration, independent of the pupil position, and given by Eq. (5.4.12),

$$W_{sp}(P) = -\frac{1}{32}\{f_1(P)\}(-\alpha_m)^4 f' . \qquad (7.3.10)$$

The astigmatic wavefront aberration follows from Eq. (5.4.14). In the case of a diaphragm position corresponding to zero coma, we then have

$$W_a(P) = \frac{\alpha_m^2 p_m^2}{2f'}\left\{-1 + \frac{1}{2}\left(\frac{l_P}{f'}\right)f_2(P)\right\} = \frac{\alpha_m^2 p_m^2}{2f'}\left\{-1 + \frac{[f_2(P)]^2}{f_1(P)}\right\} . \qquad (7.3.11)$$

The field curvature follows from Eq. (5.4.15),

$$W_f(P) = \frac{\alpha_m^2 p_m^2}{4f'}\left\{-\frac{n+1}{n} + \frac{[f_2(P)]^2}{f_1(P)}\right\} . \qquad (7.3.12)$$

Finally, the distortion of the lens at the outer image point is given by

$$W_d(P) = \left(\frac{p_m^3 \alpha_m f_2(P)}{f_1(P) f'^2}\right)\left\{-\frac{3n+1}{n} + \frac{2[f_2(P)]^2}{f_1(P)}\right\} . \qquad (7.3.13)$$

More commonly used is the relative distortion $D(P)$ for the position of the outer field point, given by $\delta p_m/p_m$,

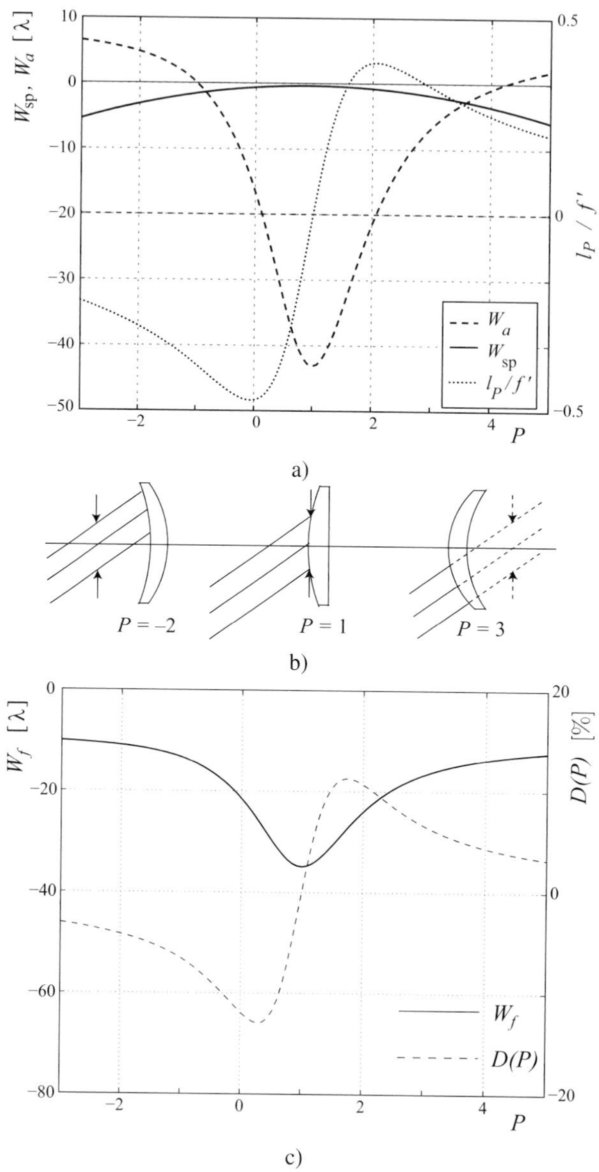

Figure 7.11: a) Axial position of the stop, l_P/f', as a function of the lens bending factor P to satisfy the zero coma condition (dotted curve). Residual spherical aberration W_{sp} (solid curve) and astigmatism W_a (dashed curve) for the zero coma lens shape and stop position.
b) Typical combinations of lens shape and stop position which yield zero coma.
c) The residual field curvature W_f (solid curve) and relative distortion D, in %, for the zero coma lens solutions. The wavefront aberration is expressed in units of λ, the wavelength of the light, and applies to an off-axis pencil to the outer image point with coordinate p_m = 23 mm (f' = 50 mm, NA = 0.067, n = 1.62, λ = 0.55 μm).

$$D(P) = -\frac{1}{p_m}\frac{W_d(P)}{\alpha_m} = -\alpha_f^2\left(\frac{f_2(P)}{f_1(P)}\right)\left\{-\frac{3n+1}{n} + \frac{2[f_2(P)]^2}{f_1(P)}\right\}, \tag{7.3.14}$$

where $\alpha_f = p_m/f'$ is the maximum field angle.

In Fig. 7.11a we have plotted (dotted curve) the axial entrance pupil position l_P, normalised to the image-side focal distance f', which yields zero coma aberration. We observe that the special value of the refractive index $n = (1 + \sqrt{5})/2$

yields zero coma for the convex-plano lens with central stop and bending factor $P = 1$. For $P < 1$ the lens shape first develops to the symmetric shape and then to the hollow meniscus shape, opened towards the object side, with the stop in front of the lens. The meniscus-shaped lenses, opened towards the image side have the virtual entrance pupil beyond the lens (see Fig. 7.11b). In a) we observe that the spherical aberration W_{sp} is relatively small and its modulus reaches a minimum close to zero for the value $P = 1$. Because of the relatively small value of the numerical aperture, spherical aberration is small with respect to the astigmatic wavefront aberration W_a. Equation (7.3.11) shows that the central stop position yields the largest absolute value of the astigmatism; for our choice of refractive index, this central stop position corresponds with the lens shape factor $P = 1$. Two lens shapes with appropriate remote stop positions yield zero astigmatism, $P \approx -1$ and $P \approx 4.3$. In Fig. 7.11c the wavefront defocusing W_f due to field curvature has been plotted and the relative distortion D. Distortion, determined by the refraction of the principal ray is only zero in the case of a central stop position.

7.3.3.2 Third-order Calculation of the Optimum Landscape Lens with Remote Stop

The amplitude of the residual wavefront aberration terms, mainly astigmatism and field curvature, is far beyond the diffraction limit of $\lambda/4$ and this allows a purely geometrical ray analysis with the rms blur as quality criterion. The optimum landscape lens is the one which produces the smallest geometrical blur in the image plane. So far, the aberration terms have been expressed in the wavefront domain. The total wavefront aberration is now expanded with the aid of Lukosz polynomials (see Subsection 5.9.3). These polynomials allow a wavefront expansion of which the coefficients, multiplied by the appropriate weighting factors (see Eq. (5.9.38)), yield the rms blur in the image plane.

The residual aberrations of the landscape lens with remote stop are spherical aberration and the combination of astigmatism and field curvature. Distortion is left out; it cannot be made zero unless a very large lens bending is applied. Using Eqs. (7.3.11)–(7.3.12), we can write the tangential and sagittal wavefront aberration components as

$$W_t(P) = \frac{\alpha_m^2 p_m^2}{4f'} \left\{ -\frac{3n+1}{n} + \frac{3[f_2(P)]^2}{f_1(P)} \right\} \left(\frac{p\,y}{p_m r_m} \right)^2 , \qquad (7.3.15)$$

$$W_s(P) = \frac{\alpha_m^2 p_m^2}{4f'} \left\{ -\frac{n+1}{n} + \frac{[f_2(P)]^2}{f_1(P)} \right\} \left(\frac{p\,x}{p_m r_m} \right)^2 , \qquad (7.3.16)$$

with (x, y) the lateral coordinates in the exit pupil and $2r_m$ the diameter of a pencil of rays.

The wavefront deviation which focuses a pencil at the 'best-focus' position between the two astigmatic lines is then given by

$$W_{bf}(P) = \frac{\alpha_m^2 p_m^2}{4f'} \left\{ -\frac{2n+1}{n} + \frac{2[f_2(P)]^2}{f_1(P)} \right\} \left(\frac{p\,r}{p_m r_m} \right)^2 . \qquad (7.3.17)$$

The maximum wavefront excursions in the tangential, sagittal and best-focus settings have been plotted in Fig. 7.12, where a) shows that interesting solutions can be found close to the two lens bending values with zero astigmatism. The subtle trade-off between the values of astigmatism and field curvature, together with the residual spherical aberration of the lens, is carried out by fitting the wavefront aberration with the aid of Lukosz polynomials. In terms of the normalised pupil coordinates (ρ, θ) we have

$$W(\rho, \theta; p) = \left\{ W_s \rho^2 + (W_t - W_s)\, \rho^2 \sin^2\theta \right\} \left(\frac{p}{p_m} \right)^2 + W_{sp}\, \rho^4 + W_{def}\, \rho^2 , \qquad (7.3.18)$$

where W_{def} is a wavefront defocusing term which corresponds to an axial shift of the image plane. In terms of the Lukosz-polynomials we write

$$\begin{aligned} W(\rho, \theta; p) = &\left\{ \left(\frac{W_s - W_t}{2} \right) \rho^2 \cos 2\theta + \left(\frac{W_t + W_s}{4} \right) (2\rho^2 - 2) \right\} \left(\frac{p}{p_m} \right)^2 \\ &+ \frac{W_{sp}}{6} \left(6\rho^4 - 8\rho^2 + 2 \right) + \left(\frac{W_{def}}{2} + \frac{2W_{sp}}{3} \right) (2\rho^2 - 2) , \end{aligned} \qquad (7.3.19)$$

where we have omitted irrelevant constant wavefront aberrations terms.

A criterion for the best lens performance over the entire image field is equal blur at the centre of the field and at the outer position with $p = p_m$. Summing the blur contributions from the various Lukosz polynomials we have,

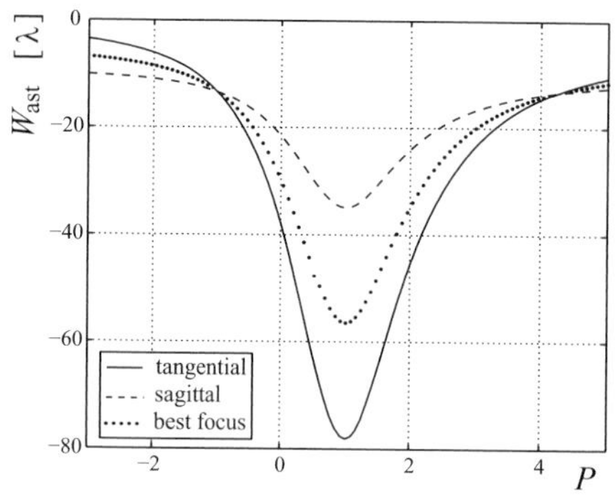

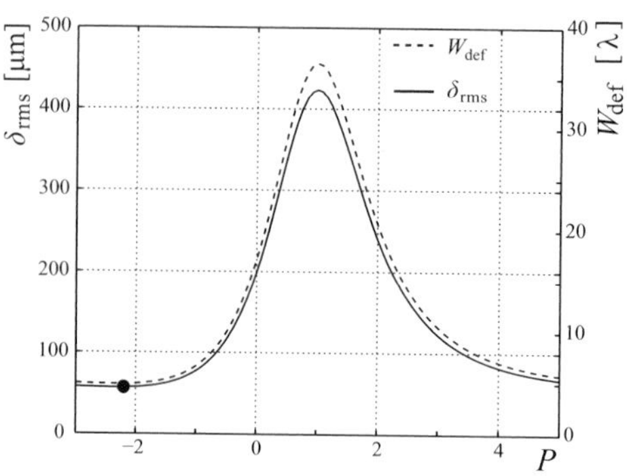

Figure 7.12: a) The maximum wavefront aberration in the tangential (solid curve) and sagittal (dashed) cross-sections and at the best-focus (dotted) position of the wavefront belonging to the imaging pencil with paraxial image coordinate $p_m = 23$ mm as a function of the bending factor P (the stop position assures the absence of third-order coma).
b) The optimum amplitude of the quadratic defocusing of the wavefront over the entire image field (W_{def}, dashed curve) and the resulting maximum geometrical blur δ_{rms} in the image field (solid curve); $f' = 50$ mm, $NA = 0.067$, $\lambda = 0.55$ μm.

$$\delta^2(0) = \frac{4W_{sp}^2}{9} + 8\left(\frac{W_{def}}{2} + \frac{2W_{sp}}{3}\right)^2 ,$$

$$\delta^2(p_m) = (W_s - W_t)^2 + \frac{4W_{sp}^2}{9} + 8\left(\frac{W_{def}}{2} + \frac{2W_{sp}}{3} + \frac{W_s + W_t}{4}\right)^2 . \tag{7.3.20}$$

Equating the two blur expressions yields the optimum defocus wavefront aberration amplitude

$$W_{def} = -\left\{\frac{3W_s^2 - 2W_sW_t + 3W_t^2}{4(W_t + W_s)}\right\} - \frac{4}{3}W_{sp} , \tag{7.3.21}$$

and the corresponding blur value is given by

$$\delta^2 = \frac{4W_{sp}^2}{9} + \frac{(3W_s^2 - 2W_sW_t + 3W_t^2)^2}{8(W_t + W_s)^2} . \tag{7.3.22}$$

Figure 7.12b shows the optimum defocus setting in terms of the Seidel coefficient W_{def} of the wavefront aberration (dashed curve). The resulting maximum blur in the image plane is found at the centre and at the edge of the image field; its rms value δ_{rms} has been plotted in b), solid curve. Larger absolute values of P give best results with a clear preference for the negative lens bending values.

A broad region with relatively small blur is found for $-2.5 < P < -2$; the typical blur value amounts to 60 μm. The black circle in Fig. 7.12b indicates where the minimum blur region is found. For this blur value we use the result of Eq. (7.3.6) which provides a rule of thumb connecting the geometrical optical blur and the spatial frequency at which 50% modulation depth is still attained; we find that for the landscape lens $\nu_{50} \approx 3$ mm^{-1}.

7.3.3.3 Finite Ray Optimisation of the Landscape Lens with Remote Stop

The third-order calculations in the preceding subsection provide a good starting point for the final optimisation of the lens. Firstly, the lens has to be given a finite thickness. Secondly, higher-order aberrations, especially depending on powers > 2 of the field coordinate can be non-negligible. The numerical aperture is small and the higher-order aperture aberrations are likely to remain small with respect to the third-order values. In Fig. 7.13 we have drawn a meniscus type lens with a bending factor P of the order of −2. The incident pencil is aimed at the paraxial image point A_1. The full aperture pencil ($NA = 0.20$ in the drawing) yields a strongly aberrated image point. By putting a remote stop into place a beam portion is selected which yields a symmetric beam with an intersection point of the rim rays which is close to the paraxial focus, both in the lateral and in the axial direction. This means that the coma and astigmatism remain small. The best image plane (solid line) is slightly in front of the paraxial image plane (dashed line). The distortion, mainly determined by the intersection height of the principal ray, remains acceptable (≤5%) for a landscape lens.

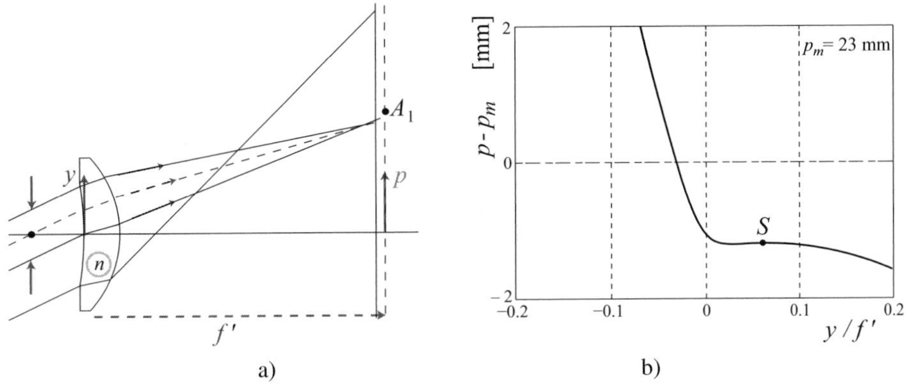

Figure 7.13: a) Finite ray picture for a meniscus-shaped lens ($f = 50$ mm) with central stop (full pencil of rays) and with remote stop of reduced size. A_1 is the paraxial image point with $p_m = 23$ mm.
b) The transverse ray aberration $p - p_m$ in the image plane as a function of the intersection height y in the pupil. S is a stationary point with respect to the transverse aberration of rays towards A_1.

In Fig. 7.13b the difference between the intersection height of a finite ray and the paraxial image height (related to a paraxial image point A_1 with $p_m = 23$ mm) has been plotted. For the lens shape of a), the quantity $p - p_m$ in b) has a relative maximum of -1.2 mm at the point S with a value of $y/f' \approx 0.06$. A pencil of rays of which the principal ray intersects the lens at this specific height ($\bar{y} = 0.06f'$) produces, in a first approximation, a stationary transverse aberration in the image plane for the rays in the cross-section given by $x = 0$. We have that y_m is equal to $(NA)f' = 3.35$ mm ($NA = 0.067$), and the eccentricity $\epsilon_m = \bar{y}/y_m$ for the outer imaging pencil then becomes 0.9. Equation (7.3.8) yields the stop position, $l_P/f' \approx -0.13$. This value is rather different from the third-order prediction of $l_P/f' \approx -0.25$ for such a lens (see Fig. 7.11a). This important deviation from third-order theory is due to the large field angle at which the landscape lens needs to be used.

Using a starting system which follows from third-order theory and from finite ray analysis of the pencil towards the extreme field point, a numerical optimisation is carried out with minimum transverse ray aberration over the image field as quality criterion. The optimised lens for a field of $2 \times 23^\circ$ and $f' = 50$ mm has been shown as an example in Fig. 7.8. The imaging performance of this landscape lens with remote stop will now be described with the aid of plots of the astigmatic wavefront aberration (*OPD*) at the pupil rim, of spot diagrams (transverse aberration patterns) and of (geometrical) modulation transfer functions. Figure 7.14a–c shows the tangential and sagittal astigmatism of various zero-thickness lenses, obtained from finite ray-tracing, in units of λ at the pupil rim. Figure 7.14b with $P = 1$ corresponds to the central stop position and shows the standard position of the astigmatic lines, where the sagittal line is closest to the paraxial image plane. The meniscus-type lenses with $P = +3.83$ and $P = -1.83$ show an inversion of the position of the astigmatic lines. It is also clear that the quadratic field dependence of third-order astigmatism is not respected for larger field angles and a flattening of the curves is observed. The solution with smallest blur for $P \approx -2$ has nonzero astigmatism but a reduced field curvature. For the lens with positive bending factor, the astigmatism is virtually zero and field curvature is the dominating imaging defect. In Fig. 7.14d we have allowed a finite lens thickness. To a first-order approximation, the finite-thickness lens is the superposition of a zero-thickness lens and a plane-parallel plate. As follows from the aberration expressions for a parallel plate, positive astigmatism and field curvature are introduced (see Eq. (5.5.5) and (5.5.6)). The minimum geometrical blur in the image plane is found by a decrease in field curvature at the expense of a slight increase of the (inverted) astigmatism of the pencils. With respect to the third-order predictions, the finite-thickness lens performs better. This is mainly due to the flattening of the astigmatism and field curvature for the outer field positions where higher-order astigmatism is introduced which slightly compensates the leading third-order terms.

In Fig. 7.15 transverse aberration patterns have been plotted, associated with various positions in the image field. The plots are obtained by using a grid of rays in the diaphragm of the lens, arranged on circles. Each circle contains 16 rays on its circumference; ten circles are present with linearly increasing radii, up to the rim of the diaphragm. The intersection points with the image plane of neighbouring rays from the same circle in the diaphragm are connected by a straight line segment. The transverse aberration analysis of the landscape lens is relevant because of its high aberration level. In the centre, we see the joint effect of defocusing and spherical aberration. The imaging quality increases in the middle field

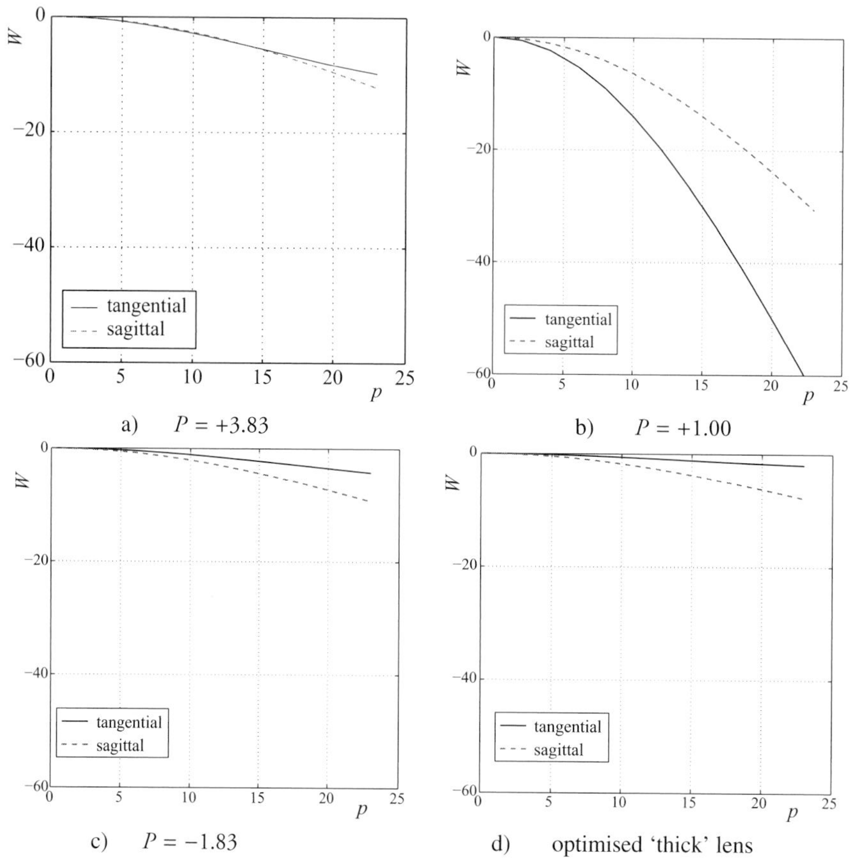

Figure 7.14: Tangential and sagittal astigmatic wavefront aberration (OPD) at the pupil rim (finite ray-tracing) as a function of the position coordinate p (in mm) in the image field for three zero-thickness lenses with various bending factors P.
a) $P = +3.83$; b) $P = +1$; c) $P = -1.83$; d) Optimised finite-thickness lens with $P = -1.90$.

area and is then dominated towards the edge by astigmatism and defocusing. Hardly any asymmetry in the y-direction is visible in the patterns with $p \neq 0$, showing the very good correction of coma. The distortion of the landscape lens is 5.8%, a relatively high value for the intended application. A reduction of the distortion is possible by a further bending of the lens.

In photography, the number of resolvable points on the diagonal or along the rectangular image frames is used for quality appraisal. Alternatively, the spatial frequency transfer by the lens to the image plane is given as a measure of imaging quality. In Fig. 7.16 the geometrical modulation transfer function has been plotted for a set of points in the image field. The four upper graphs give the modulation transfer in the image plane for the frequency interval from 0 to 10 mm^{-1}. The transfer at the centre and at the edge of the image field is lowest. As already shown in the preceding paragraph, this phenomenon is due to the balance between field curvature and astigmatism. The focus-setting of the image plane is such that the defocusing blur at the centre equals the blur due to astigmatism at the edge of the field. The graphs do not include the effect of chromatic errors. In the previous subsection on chromatic blur of a single lens, we observed that the chromatic effect would limit the frequency transfer to 50% at 15 mm^{-1} (see Eq. (7.3.6)). This value shows that the chromatic blur can be neglected with respect to the image blur coming from the monochromatic aberrations. In the lower four graphs of Fig. 7.16, the transfer at a particular frequency (5 mm^{-1} in this case) is given as a function of the axial position of the image plane ('through-focus' MTF). The balance in image blur between the centre and the edge of the field is evident from these graphs. These graphs also provide a measure for the axial positioning tolerance of the image plane.

The image size of the landscape lens has been chosen such that it allows the classical film format of 36 × 24 mm. The just-resolvable distance between two points requires a modulation depth of 20% according to the Rayleigh criterion. From

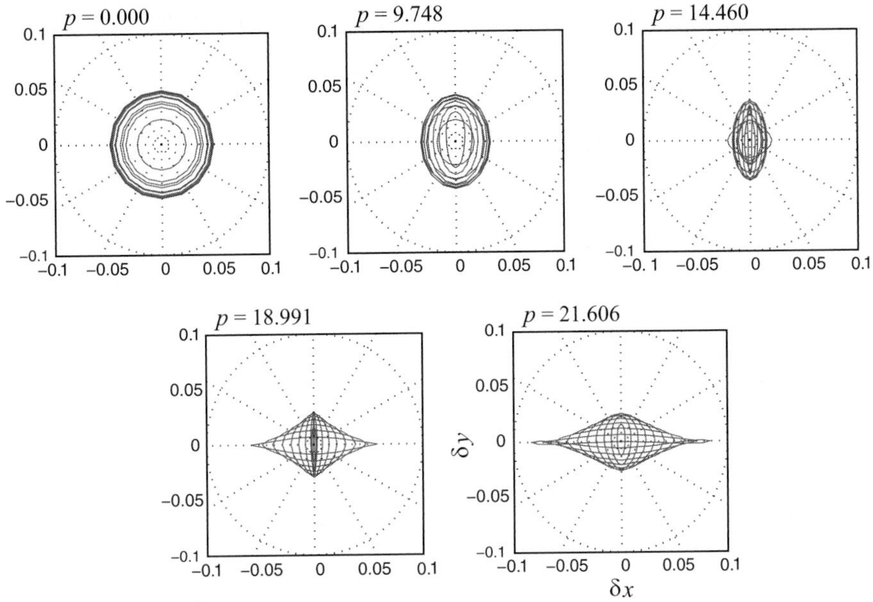

Figure 7.15: Transverse aberration plots (spot diagrams) of a simple photographic landscape lens ($f' = 50$ mm) for various paraxial positions of the image point in the field of the lens (p_{par} = 0, 10, 15, 20, 23 mm). The corresponding lateral position p of the centroid of an image point is given in the figure for each spot diagram.

the geometrical MTF-graphs we observe that this value is achieved for a spatial frequency of 8 mm^{-1} or a sequence of 16 imaged black and white pixels per mm in the image plane. The number of just-resolved pixels of an image formed by the landscape lens thus amounts to typically 600×400. This modest resolution is adequate for low-quality photography.

The lens optimum with a lens bending factor smaller than zero produces a camera with a relatively large construction depth because of the outside stop position (see Fig. 7.17). A problem of the hollow lens in a retrofit position is the vulnerability to dust and dirt and the problematic cleaning. For this reason, the opposite lens choice with a large positive P-factor has also been produced. Despite its lower optical performance, this solution was chosen in certain applications because of its mechanical and practical advantages.

7.4 The Portrait Lens

Early photography, started in 1838 with the light-sensitive material developed by Daguerre, was limited to static scenes because of the long exposure times. In the case of humans, a portrait was feasible if the subject could remain virtually immobile during the many minutes long exposure time. Petzval, a Hungarian mathematician, produced a design (1840) which enabled a much shorter exposure time. Petzval must have been the first to have had a deeper knowledge of third-order aberration theory. He has left his name to the factor which determines the field curvature of an optical system, the Petzval sum.

The main design requirement for a portrait lens is a large opening to shorten exposure time. The resolution in the centre should be high, comparable to that of the human eye (50% modulation depth at an angular frequency of 1 $mrad^{-1}$). Towards the edge of the field, a loss in imaging sharpness can easily be tolerated. The typical field angle for a portrait lens is $2 \times 15°$.

7.4.1 Design Considerations

The crucial requirement for a portrait lens in 1840 was a large opening to reduce the long exposure time for the subject. An $F_{\#}$-number smaller than 4 was needed at a magnification of typically −0.1. A large opening angle ($NA \approx 0.14$) requires

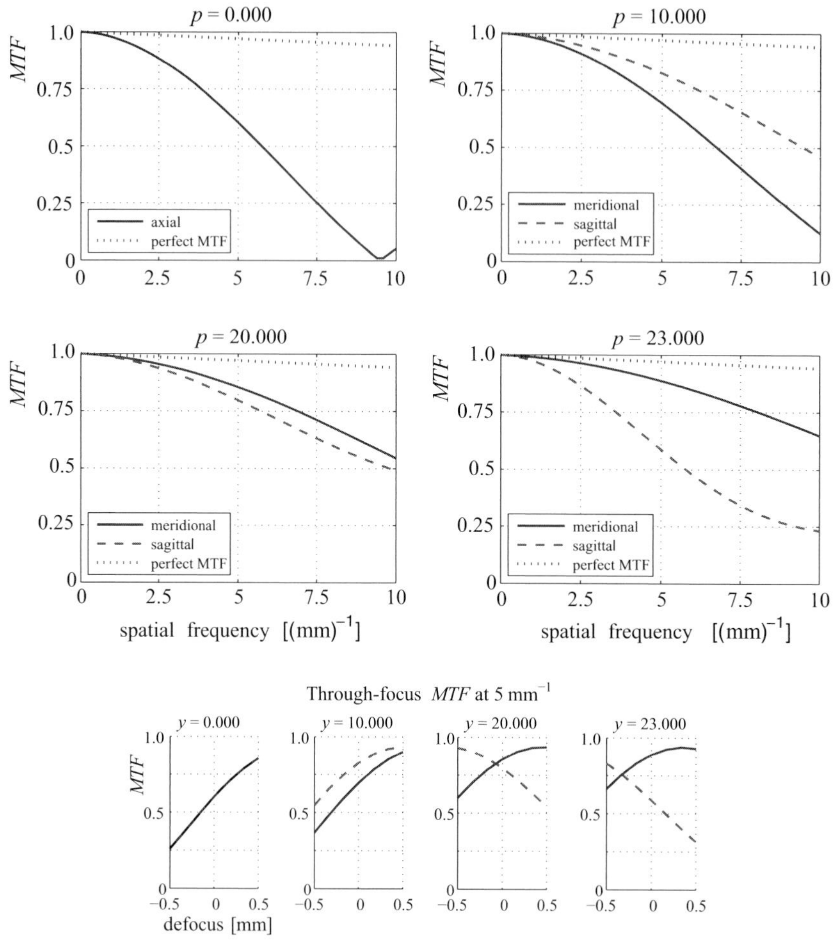

Figure 7.16: Geometrical modulation transfer function and through-focus curve at a spatial frequency of 5 mm^{-1} for various positions in the field of the single-element photographic landscape lens of Fig. 7.8 ($f = 50$ mm, $NA = 0.067$ or $F_{\#} = 8$).

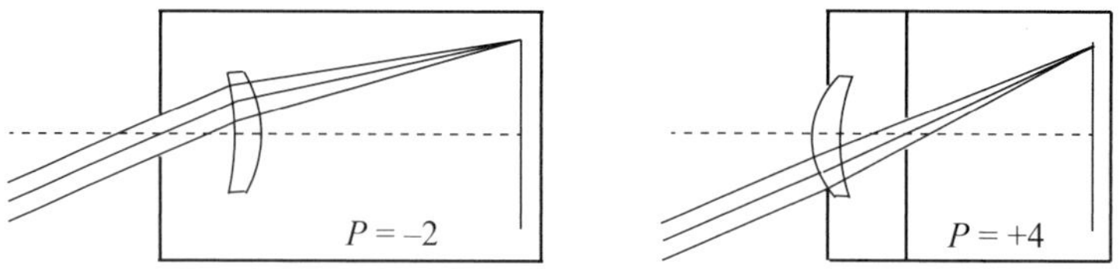

Figure 7.17: The landscape camera construction length for negative and positive bending factor P.

a good correction of the spherical aberration. The field angle of $2 \times 15^{\circ}$ asks for a good correction of coma. The remaining aberrations, astigmatism and field curvature, finally limit the lens performance. The suppression of distortion is important in portrait photography although a residual 1% can be tolerated. The high spatial resolution requirement for a portrait lens implies a good correction of the chromatic errors, both axially and laterally.

To find a good starting point we use third-order aberration theory. This approximate theory is relevant for this design problem because aperture and field angles are relatively small. A good final design must also have its third-order components well balanced. Higher-order aberration components remain small if the angles of incidence on the surfaces are also kept small (typically less then 0.2 radians) in such a low-aperture/small field system. The starting point of the design is an achromatic doublet. Figure 7.4 shows that an achromatic doublet can be well corrected for spherical aberration up to a numerical aperture of 0.10. For larger aperture, it is advisable to substantially reduce the V number difference from 27 (SK11–SF2 combination of Fig. 7.4) to a value of typically 20. From the Schott glass chart, using cheap standard glasses, we choose the combination K5–LF5 with a V number difference of 19. The advantage of the smaller V-difference is that the larger internal curvature of such a doublet produces a larger amount of positive spherical aberration at the buried surface. In this way the total spherical aberration of the doublet can be made less negative or even slightly positive. In terms of wavefront aberration, the portrait lens does not need to be of diffraction-limited quality. A practical limit on the fourth-order spherical aberration coefficient a_{sp} with $W_{sp} = a_{sp}\rho^4$ follows from the resolution criterion for the centre of the image field. The angular resolution should be at least 50% at 1 mrad^{-1}. For a focal length of $f' = 100$ mm, we require more than 50% modulation depth at 10 mm^{-1}. The rule of thumb of Eq. (10.2.79) then requires an rms blur smaller than 15 μm. The blur due to spherical aberration with wavefront coefficient a_{sp} follows from Eq. (5.9.38) with $n = 4$. We find $\delta = \sqrt{16(a_{sp}/6)^2}/NA < 15$ μm. This imposes an upper limit on a_{sp} of 3 μm or typically 5 wavelengths at $\lambda = 550$ nm. This is in contrast with the diffraction limit for spherical aberration which reads $a_{sp} \leq \lambda$.

7.4.2 Two-doublet Design

The study of the achromatic doublet in Subsection 7.2 has shown that the angular field of a single doublet is limited to a few degrees. Beyond this value, field curvature and astigmatism quickly degrade the imaging quality, far below the portrait lens requirement. An extra degree of freedom in the design is needed and this is obtained by adding a second doublet. In microscopy a very successful objective design with two cemented achromatic doublets had been devised by Lister in 1830. It doubled the available aperture of the objective; but its useful angular field remained limited to a few degrees. A two-doublet design for portrait photography has to prove that the lower resolution as compared to microscopy can be exchanged against a larger angular field.

The landscape lens design showed that the monochromatic aberrations coma and astigmatism could be influenced not only by lens bending but also by choosing a remote stop position. We can exploit this design flexibility by having the two doublets at a distance which equals some tenths of the focal distance of the system and place the physical stop midway between them. A symmetric layout with respect to the stop gives rise to an unperturbed direction of propagation of the principal ray of an oblique pencil. This property generally leads to the low values of image distortion which are needed in a portrait lens.

Third-order aberration reduction can be manually done by tabulating the Seidel aberration increments as a function of the bending of the doublets, the achromatic correction and the position of the stop. We start with a convex-plano doublet ($f_1 \approx 60$ mm) having the high V glass in front ('crown-in-front' option). The spherical aberration of this doublet is approximately zero. The second doublet, flint-in-front, is at a distance of 40 mm from the first one. It receives a bending such that the spherical aberration is also close to zero. Inspection of the third-order data shows that the comatic wavefront aberration is 5 μm but the tangential astigmatism and the sagittal astigmatism are huge (83 and 42 μm, respectively, for $\lambda_0 = 550$ nm). The astigmatism of the design must be drastically reduced. In the landscape lens design, the stop position was effectively used to adapt the astigmatism. In the portrait lens design we now increase the doublet distance to 60 mm and adapt the power of the first doublet to maintain the nominal focal distance of $f' = 100$ mm. The result is a design with the third-order wavefront aberration values shown in Table 7.3; the units are μm, measured for an imaging pencil with the (central) wavelength $\lambda_0 = 550$ nm. The coefficients a_{sp}, a_c and a_d of the Seidel spherical aberration, coma and distortion follow from Eq. (5.4.25) and equal $S_{1,s}/8$, $S_{2,s}/2$ and $S_{5,s}/2$, respectively. The coefficients of tangential and sagittal astigmatism, a_{tan} and a_{sag}, are given by $(3S_{s,3} + S_{s,4})/4$ and $(S_{s,3} + S_{s,4})/4$, respectively. The axial and lateral chromatic deviations, $\delta a_{\lambda,ax}$ and $\delta a_{\lambda,lat}$, are given by Eqs. (5.6.16) and (5.6.17). The Seidel sums can be calculated by summation over each surface taking into account the finite thicknesses of lenses. But in the pre-design phase of the portrait lens it is fully justified to use the thin-lens Seidel formulas, as the introduction of finite thicknesses will hardly influence

Table 7.3. Portrait lens with two cemented doublets.

	a_{sp}	a_c	a_{tan}	a_{sag}	a_d	$a_{\lambda,ax}$	$a_{\lambda,lat}$
$W^{(4)}$ (in units of μm)	+0.8	+21.5	−24.2	−25.8	+5	−0.39	−0.55

Table 7.4. Optimised portrait lens with two cemented doublets.

	a_{sp}	a_c	a_{tan}	a_{sag}	a_d	$a_{\lambda,ax}$	$a_{\lambda,lat}$
$W^{(4)}$ (in units of μm)	−10.9	+1.5	−19.4	−21.3	−47.3	−0.57	+4.9
ΔW (λ_0 + 100 nm)	−0.5	−0.4	−2.4	−0.8	+0.4		

the correction status of a design. If needed, the finite thickness of a lens can be taken into account by constructing the 'equivalent thin lens' **[24]** (see Subsection 5.4.1).

Coma and field curvature are the dominating aberrations. The field curvature wavefront aberration of –0.025 mm corresponds to an axial shift z_{fc} of the image at the edge of the field (p_m = 24 mm) of $-2 \times 0.025/(0.14)^2 \approx -2.5$ mm (the NA of the focusing beam is 0.14). The blur at the image border can be evaluated with the aid of Eq. (5.9.38). We find the wavefront defocusing coefficient a_2^0 of the $L_2^0(\rho)$-Lukosz polynomial for the pencil of rays to the border of the image to be $-25/2$ μm. The rms blur then equals $\sqrt{8(12.5)^2}/NA \approx 250$ μm. An image plane setting which strongly favours the resolution in the central part of the image is acceptable in portrait photography. The sharpness of the outer part of a portrait is much less important. In the early stages of photography, this part was often cut out by using an oval portrait frame.

Having accepted the appreciable amount of field curvature discussed above, we have to further reduce the coma aberration. Third-order coma increases linearly with the field coordinate. At the edge of the field the coma wavefront deviation amounts to 21.5 μm and the corresponding blur equals $\sqrt{6(21.5/3)^2}\ /\ NA \approx 125$ μm; here we have used the Lukosz polynomial $L_3^1(\rho)\cos\theta$ and Eq. (5.9.38) to calculate the geometrical blur. We see that even for an image field coordinate of $0.7p_m$, the blur value due to coma would be devastating for the portrait resolution. Several design options can be studied to reduce coma. We mention the bending of the two doublets, the inversion of the crown-flint order in each doublet, change in axial distance of the doublets, shift of the stop position. Studying the corresponding third-order aberration increments, it is not possible to find a straightforward and independent balancing of the various aberration increments such that a simultaneous reduction of spherical aberration, coma and astigmatism is achieved. The modern optical designer can use a numerical minimisation tool to reduce the aberrations of the two-doublet option. The result of such a numerical optimisation is shown in Table 7.4 where the coefficients a are expressed in μm. The coma has effectively been reduced, as well as the field curvature. Unfortunately, the spherical aberration is almost four times larger than allowed. We also notice an increase in distortion and a lateral chromatic shift which is given by $a_{\lambda,lat}/NA$ and equals 35 μm at the edge of the field. The conclusion is that this optimised design yields a better overall quality of the image but is incapable of providing the desired resolution in the centre of the field. Figure 7.18a shows a cross-section of such a portrait lens. Strongly curved inner surfaces are visible in both doublets and these make the design vulnerable in manufacturing, both from the optical and mechanical point of view ('strained' system). The sensitivity of the design to small changes is illustrated by the second line of the table where the aberration increment ΔW is given for a change in wavelength of $\Delta\lambda = +100$ nm. A substantial change in astigmatism and field curvature arises with such a change in colour.

7.4.3 Petzval Portrait Lens

The mathematician Petzval tackled the design of a portrait lens immediately after the presentation by Daguerre of his photographic chemical process Daguerrotype. Although he never communicated any details about his research, it is evident that he developed or already had at his disposal an aberration theory of optical imaging, probably very similar to the one which was published 15 years later by Seidel **[301]**. Probably, he not only relied on the application of his theory but included the analysis of his designs by means of finite ray-tracing. It is stated that Petzval had been given the computational assistance of some ten soldiers who were specialised in the calculation of cannonball trajectories.

To achieve a good portrait lens design, one or more extra degrees of freedom are needed. The addition of a third lens to the two doublets could be considered. Petzval has chosen to split the second doublet to make it an air-spaced doublet with different internal curvatures. Splitting the second doublet instead of the first is recommended because field aberrations like coma and astigmatism are more easily influenced at the location of the second doublet. The eccentricity factor ϵ is higher on the surfaces of the two split lenses than it is on the first doublet because of the reduced beam cross-section.

We use the Seidel aberration increments due to parameter changes to guide the design process. Starting configuration for the first doublet is a convex-plano shape with internal curvature such that the axial chromatism of the doublet is close to zero; the glass V numbers remain unchanged. The optical power of the doublet is approximately 60% of the total

Table 7.5. Petzval portrait lens; third-order design and reconstructed design (aberration values in units of µm).

	a_{sp}	a_c	a_{tan}	a_{sag}	a_d	$a_{\lambda,ax}$	$a_{\lambda,lat}$
Third-order design	−2.1	−5.2	−21.1	−20.9	−1.0	+0.99	−0.47
Reconstructed design	−2.9	−6.4	−25.6	−23.4	−1.5	–	–

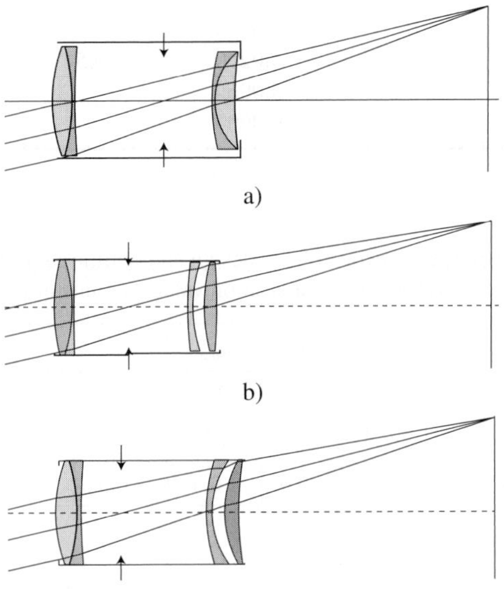

Figure 7.18: Design options for a portrait lens. Focal length f'= 100 mm, lateral magnification $m_T = -0.1$, maximum field angle γ_m = 14°.
a) Two-doublet option;
b) Petzval portrait lens;
c) Design with reduced field curvature but increased manufacturing complexity.

power. A small distance is then introduced between the two lenses which formed the second doublet (originally of the flint-in-front shape). From the bending of the first and the second lens we derive the influence on spherical aberration, coma and astigmatism. The reduction of spherical aberration, coma and astigmatism is now possible, using as parameters the bending of the first doublet and the independent bending of the two air-spaced components, which emerged from the second doublet.

7.4.3.1 A Reconstructed Petzval Design with Present-day Optical Glasses

Using the third-order aberration expressions, we execute a few iterations with adjustment of the lens thicknesses and obtain the third-order aberration coefficients which are given on the first line in Table 7.5 ('third-order design'). The third-order aberration coefficients of the Petzval portrait lens, as reconstructed by von Rohr **[295]** and Berek **[24]**, are given on the second line of Table 7.5 ('reconstructed design'). The coefficients are in units of µm for a lens system operating at λ_0 = 550 nm with $f' \approx$ 100 mm; the on-axis aperture *NA* equals 0.14. The chromatic aberration of the reconstructed Petzval lens is not given because the exact dispersion of the glasses is unknown. A comparison of Table 7.4 and Table 7.5 shows that the reduction of spherical aberration, coma and astigmatism is appreciable by splitting the second doublet. The main defect of the portrait lens is its field curvature. The layout of the Petzval portrait lens is shown in Fig. 7.18b.

A prescription of the third-order-based design with present-day optical glasses is given in Table 7.6. Applying a finite-ray optimisation to the third-order design with the glasses K5 and LF5 as crown and flint components leads to a

Table 7.6. Optical data of the third-order-based Petzval portrait lens with present-day optical glasses (K5-LF5).

Distance	Index	$V_{\#}$	Curvature	Glass	Diameter	
1075.0	1.0000		+0.0180		29.2	
5.2	1.5244	59.48	−0.0230	K5	29.2	
0.8	1.5845	40.85	+0.0020	LF5	29.2	
10.0	1.0000		0.0000		26.0	STOP
23.0	1.0000	40.85	+0.0095	LF5	24.0	
1.5	1.5845		0.0255		24.0	
3.3	1.0000	59.48	0.0200	K5	24.0	
3.6	1.5244		−0.0060		24.0	
79.5	1.0000		0.0000			

Table 7.7. Petzval portrait lens; reconstructed design and optimised design with the present-day optical glasses K5 and LF5 (aberration values in units of μm).

	a_{sp}	a_c	a_{tan}	a_{sag}	a_d	$a_{\lambda,ax}$	$a_{\lambda,lat}$
Reconstructed design	−2.9	−6.4	−25.6	−23.4	−1.5	–	–
K5-LF5 optimised design	−2.7	−4.5	−23.2	−22.7	+4.4	0.88	−0.70
ΔW (λ_0 + 100 nm)	−0.4	−0.6	−0.8	−0.3	−0.2		

Table 7.8. Petzval portrait lens: strained design with reduced field curvature.

	a_{sp}	a_c	a_{tan}	a_{sag}	a_d	$a_{\lambda,ax}$	$a_{\lambda,lat}$
$W^{(4)}$ (λ_0)	−8.1	+4.2	−9.8	−17.0	−9.9	0.83	2.09
ΔW (λ_0 + 100 nm)	−0.5	−0.9	−1.0	−0.3	0.0	–	–

slightly improved portrait lens ('K5-LF5 optimised design') with field curvature as the dominating residual aberration (see Table 7.7).

7.4.3.2 A Petzval Portrait Lens with Reduced Field Curvature

Third-order field curvature, calculated with the aid of the Petzval sum, is independent of lens shape or lens magnification. It can, however, be influenced by allowing residuals of astigmatism, coma and spherical aberration in the design (see Eq. (5.4.25)). A finite ray optimisation of the portrait lens in which the various field points are equally weighted leads to a system with nonzero astigmatism but reduced field curvature, as shown in Table 7.8. The third-order aberration coefficients of this design show that the field curvature has been substantially reduced. On the contrary, the spherical aberration has increased to a level that starts to degrade the resolution in the centre of the field. In combination with the lower field curvature, the central region of the image is on average better focused and this counteracts the resolution loss due to spherical aberration.

The performance of the original Petzval objective in terms of its modulation transfer function is shown in Fig. 7.19a. The focus-setting has been chosen such that the resolution is approximately constant up to 50% of the field radius. Beyond this value, the resolution shows a sharp degradation due to the curvature of the optimum image surface of the lens. The off-axis performance is relatively improved by the vignetting of up to 50% of the oblique pencils. The original Petzval

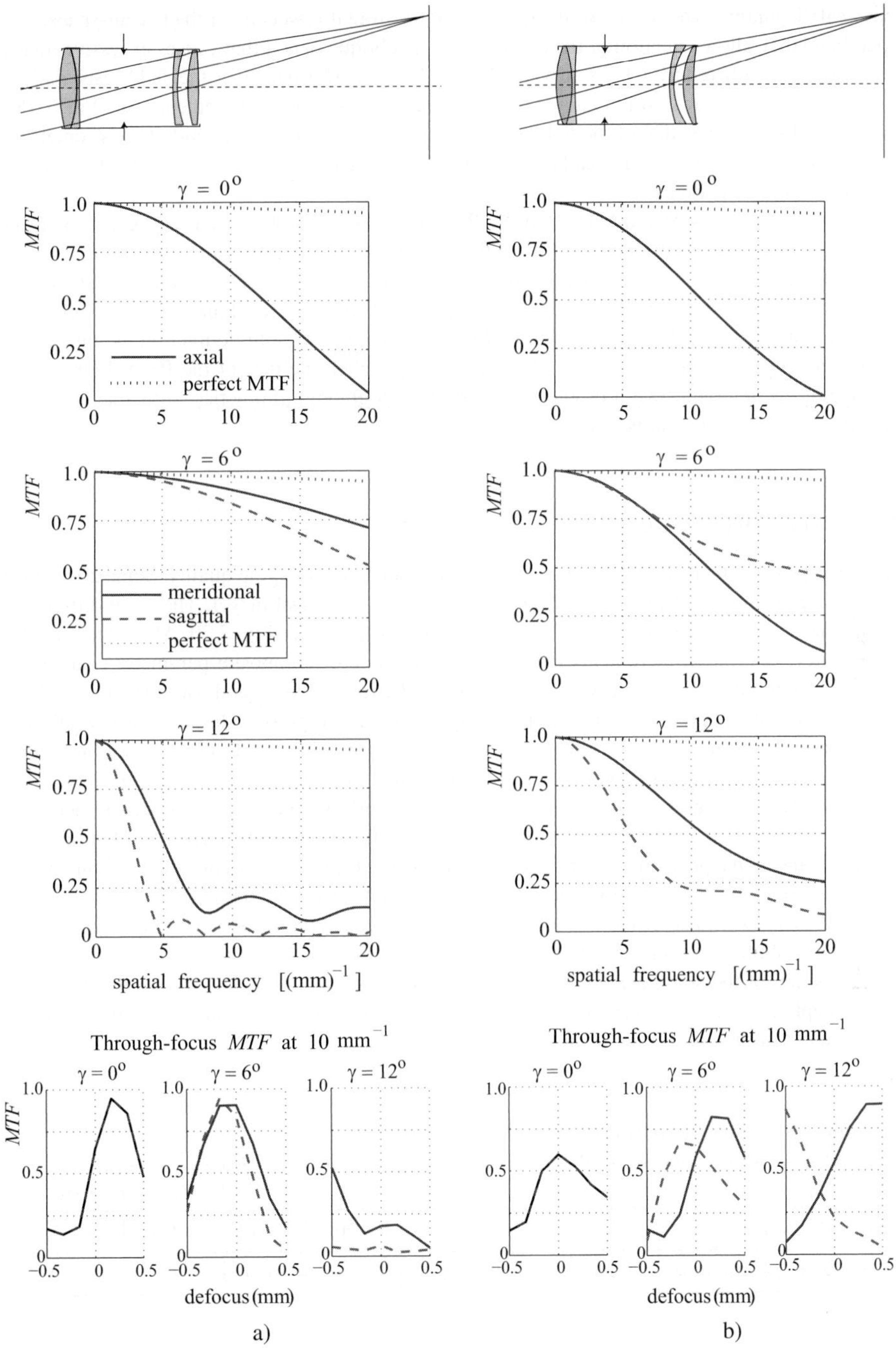

Figure 7.19: Geometrical modulation transfer function and through-focus curves at a spatial frequency of 10 mm^{-1} for three field angles γ of two Petzval portrait lens designs; $f' = 100$ mm, $NA = 0.14$ ($F_{\#} = 3.6$), $\gamma = 2 \times 14^\circ$.
a) 'K5-LF5 optimised design' of the classical Petzval lens.
b) 'Strained' version of the Petzval lens with reduced field curvature.

lens has no physical diaphragm. The mechanical mountings of the front doublet and of the two air-spaced elements define an effective pupil and they allow rejection of the outer rays of an oblique pencil that causes excessive image blur.

The 'strained' design with reduced field curvature has a better performance in the field, especially in the region $10^\circ \leq \gamma \leq 14^\circ$, at the cost of increased curvatures of the surfaces of the two split elements (see Fig. 7.18c). The better imaging performance towards the edge of the field is illustrated in Fig. 7.19b with the aid of the geometrical modulation transfer function. The penalty that has to be paid is an increased manufacturing complexity (optical surfaces and mounting tolerances) and mechanical sensitivity.

In conclusion, the portrait lens design from 1840 by Petzval (see Fig. 7.18b) was a very good compromise between imaging sharpness in the centre of the field and the fall-off of image resolution towards the border of the image field. Thanks to the application of his newly developed aberration theory, Petzval effectively eliminated spherical aberration and coma and reduced the combination of astigmatism and field curvature to a level that was acceptable for the relatively small angular field in portraiture. The key innovation of the design is the air-split doublet on the image side which adds extra degrees of freedom and, therefore, allows the substantially higher aperture of the Petzval lens. The low imaging quality at the edge of the field is less of a problem in portrait photography; in practice, oval-shaped photo frames were used to cut out the remote diagonal parts of the portrait.

7.4.4 Dallmeyer Portrait Lens

The third-order theory allows many lens solutions which yield comparable imaging quality. Some 20 years after Petzval's successful design, a comparable design was put forward and patented by Dallmeyer. The main difference between the two designs is the inversion of the glass order in the air-spaced lens group, see Fig. 7.20. Along the same lines as before, by studying third-order aberration increments as a function of single or composite parameter changes, it is relatively easy to arrive at a design which is virtually free of spherical aberration, coma and astigmatism for a full aperture of 0.14 at the image side. This approach is the original design method; the present-day approach would be based on an optimisation program. A reconstructed Dallmeyer lens was designed using third-order aberration theory with a final numerical optimisation for fine-tuning. We used K5 as the crown glass and LF5 as the flint glass. Table 7.9 shows the third-order wavefront aberration values expressed in μm. No effort needs to be invested in achieving the lowest possible aberration values. The portrait lens resolution is far from diffraction-limited. The contribution of each wavefront aberration residual to the image plane blur can be calculated using Eq. (5.9.43). From the allowed blur value of the portrait lens it is straightforward to establish the tolerances on the individual wavefront aberration coefficients. These are of the order of several wavelengths instead of the $\lambda/4$ tolerance which is needed in diffraction-limited optical instruments. The field curvature of the Dallmeyer lens is virtually identical to that of the Petzval design. The distortion can be minimised by adjusting the position of the stop between the two lens groups (as a matter of fact, the Dallmeyer lens contained an explicit physical diaphragm). For this type of lens with moderate aperture and field angles, the finite ray aberrations hardly differ from those of the third-order calculations.

Table 7.9. Dallmeyer portrait lens (reconstruction).

	a_{sp}	a_c	a_{tan}	a_{sag}	a_d	$a_{\lambda,ax}$	$a_{\lambda,lat}$
$W^{(4)}$ (λ_0)	1	0	−19	−20	−6	+1.5	+1.0
ΔW $(\lambda_0 + 100\ \text{nm})$	−0.5	−1.1	−1.3	−0.4	−0.4		

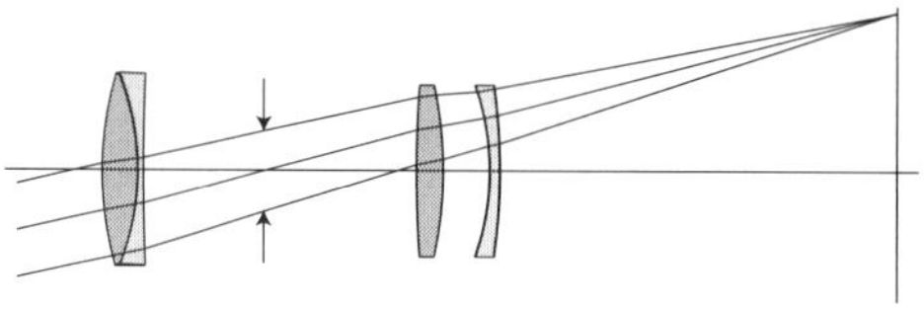

Figure 7.20: Example of a Dallmeyer portrait lens with an inverted second lens group ($f' = 100$ mm, $NA = 0.14$ on-axis).

7.5 Flat-field Imaging Systems

In many applications of optical imaging, a flat image field is a primordial requirement. Especially in the case of wide-angle imaging systems, the fulfilment of this requirement is a difficult task. In Eqs. (5.3.31) and (5.3.37) it was shown that, in the framework of third-order aberration theory, the curvature of the image field follows from a field-dependent (wavefront) defocusing term given by

$$W_{\mathrm{fc}} = \frac{1}{4}(S_3 + S_4)\left(\frac{\rho}{\rho_m}\right)^2\left(\frac{p}{p_m}\right)^2 , \tag{7.5.1}$$

where S_3 and S_4 are the coefficients of astigmatism and intrinsic field curvature; (ρ/ρ_m) and (p/p_m) are the relative aperture and field coordinates. Restricting our attention to the zero-astigmatism case, we use Eq. (5.3.44) and write

$$\frac{1}{n_I R_{\mathrm{fc}}} = P , \tag{7.5.2}$$

where R_{fc} is the radius of curvature of the image surface, n_I is the refractive index of the image space and P is the Petzval sum of the imaging system.

We showed earlier that the Petzval sum is independent of the ray trajectories through an optical surface or of the ray propagation direction. For a thin lens, Eq. (5.4.15) shows that the contribution to the Petzval sum is simply determined by the ratio of optical power K and the refractive index n of the lens. Using the thin-lens formula for field curvature, we immediately find that $R_{\mathrm{fc}} = -nf'$ where f' is the image-side focal length of the thin lens. The generally inward-curved image field of a positive lens or lens system can be flattened in the following ways (see Fig. 7.21):

- *Curved object / image field*
  In Fig. 7.21a an image is formed by a single thin lens with focal length f'. The radius of curvature of the image surface C' is $-nf'$ if the lens is placed in air. The radius of curvature is independent of the magnification of the image. Inverting the ray direction in a) yields the dashed image surface labelled C at the location of the original object plane through O. The solid object and image surfaces are conjugate surfaces as well as the dashed surfaces. If either the object or the image detecting surface can be given the appropriate curvature, the problem of (intrinsic) field curvature is solved.
- *Object- or image-side field flattener*
  In Fig. 7.21b an 'apparent' curvature of the object surface has been achieved by means of a so-called field flattening element, brought into (almost) contact with the object plane. We recall the paraxial expression for the axial shift Δz of the source point of a converging or diverging beam when it passes through a plane-parallel plate having thickness d_f and refractive index n_f,

  $$\Delta z = \left(\frac{n_f - 1}{n_f}\right) d_f . \tag{7.5.3}$$

  The (paraxial) correction of field curvature requires Δz to be equal to the distance PQ in the figure and this yields a radius of curvature R_{ff} for the hollow surface passing through O and Q_0,

  $$R_{\mathrm{ff}} = \left(\frac{n_f - 1}{n_f}\right) nf' . \tag{7.5.4}$$

  The field-flattening surface has a relatively small radius of curvature as compared to the lens focal distance. For the common refractive indices of optical materials, $R_{\mathrm{ff}} \approx f'/2$. In practice, a spherical shape is not optimum for large values of the field coordinate. Optimum flattening of the field requires an aspherically shaped field flattener. The field flattening element is best used on the short conjugate side of the lens because the lateral extension of the surface is smallest there. The use of a field flattener has serious disadvantages. The refraction of the principal ray at Q_0 produces a strong distortion effect. If the back surface of a field flattener is not in optical contact with the object surface or the image detection surface, any imperfection (scratches, dust, fingerprints) on this surface will be sharply reproduced in the image. To a lesser degree this 'scratch' problem also occurs at the hollow surface of the field flattener.
- *Layout with bulge and constriction*
  The insertion of a negative lens in a lens system which has to yield positive power can produce a reduction in field curvature. To avoid a strong reduction of positive power by the introduction of a negative element, this element is inserted in the system at a location where the height of intersection of the marginal ray of an axial pencil is relatively

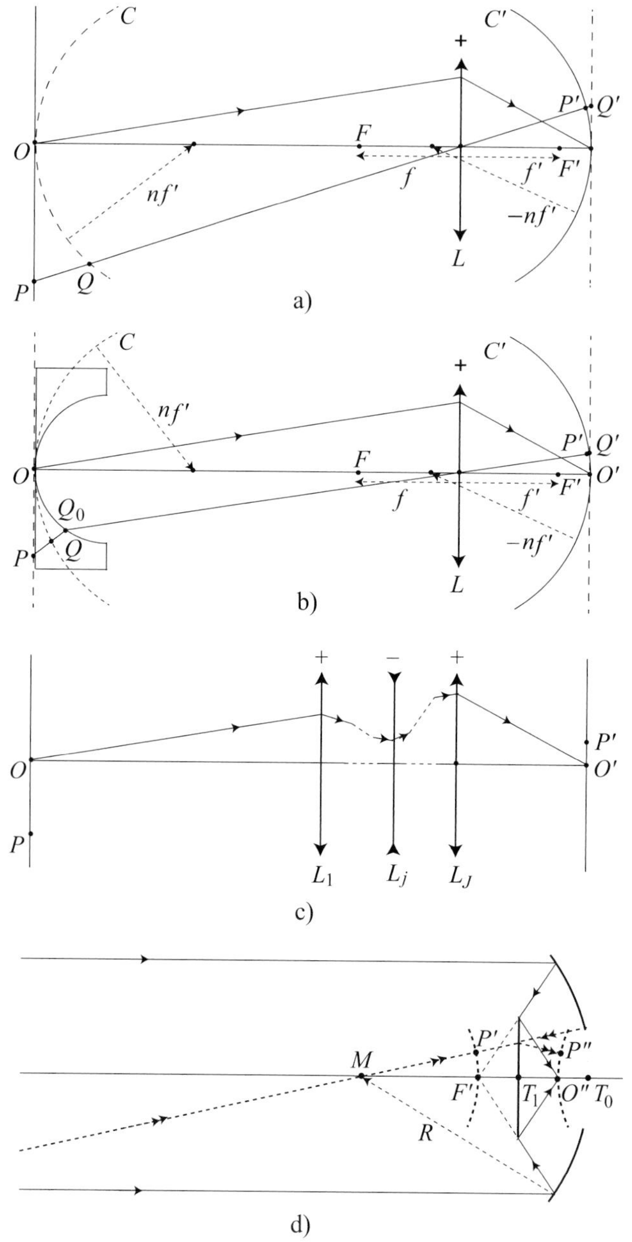

Figure 7.21: Field curvature (in the absence of astigmatism) and methods to suppress it.
a) Curved object (or image) surface.
b) Negative field-flattening lens in close contact with object or image plane.
c) Introduction of negative power in an optical system in a region with reduced intersection height of the marginal ray.
d) Exploiting the inverted field curvature produced by a reflecting surface.

low. In Subsection 4.10.3 we showed that the relative contribution to the optical power by the individual surfaces depends on the relative height of intersection of a marginal ray on that surface (see Eq. (4.10.7)). This particular equation holds for imaging from infinity. In the case of finite magnification, we consider Eq. (4.10.5) which shows that the change in ray direction at a surface is proportional to the product $y_j K_j$ where K_j is the power of the optical transition. Having a low y value for the axial pencil reduces the refraction by the lens, not only for the axial pencil but also for any oblique pencil. But the effect of the optical surface on the field curvature is unaffected by the value of y_j and only determined by its contribution to the Petzval sum of the system. In this respect, for reduction of field curvature, it is advantageous to use low-index materials for the negative elements and higher-index materials for the positive elements.

The general layout of an optical system with such a field curvature correction has been sketched in Fig. 7.21c. The axial pencil first grows in diameter, showing a first *bulge* at the location of the positive lens or lens group, is restricted in diameter at the position of the negative element (*constriction*), shows a second bulge and is then brought to focus. The bulge-and-constriction scheme of c) shows that each of the three groups operates at negative magnification such that the final magnification is also negative. If needed, this process can be repeated using several bulges and constrictions until a satisfactory compensation of field curvature has been achieved. This rather onerous procedure shows that the reduction of field curvature is an omnipresent and tenacious problem in optical design.

- *Reflecting surface*
  In Subsection 5.3.5 we emphasised that the field curvatures of a mirror and a lens with equal power have opposite signs. More precisely, for a reflecting surface in a medium with index n_I, we derive for the Petzval contribution of a mirror

$$c_{\mathrm{fc}} = (-n_I)\, P = +2c\,. \tag{7.5.5}$$

  In Fig. 7.21d a parallel beam is incident on a mirror with its vertex at T_0 and its centre of curvature at M. The image surface intersects the axis MT_0 at the focal point F' and it has a concave spherical shape with radius of curvature $R/2$. If the diaphragm is located at M, the principal ray of an oblique beam through M is focused at P'. After reflection at the plane mirror at T_1 a convex field surface through O'' and P'' is formed. The curved image surface through F' and P' can be exploited to compensate for the negative field curvature introduced by a subsequent refractive imaging system. It is important to mention that the reasoning above applies to the wavefront aberration due to the intrinsic Petzval field curvature of a mirror, in the absence of astigmatism. For a position of the diaphragm at T_0, on the mirror surface itself, the imaging of a parallel beam suffers also from astigmatism. Inspection of Eqs. (5.3.31)–(5.3.33) shows that, if the diaphragm is at the location of the mirror, identical amounts of astigmatic wavefront aberration and field curvature aberration are generated by the spherical mirror. This astigmatism has to be compensated for in the remaining part of the optical system. Fortunately, as for the field curvature, the astigmatism introduced by a concave mirror ($m_T = 0$) has the opposite sign to the astigmatism introduced by a (thin) positive lens.
- *Field curvature and astigmatism*
  The discussion of field curvature so far has not explicitly included astigmatism. In the presence of astigmatism, a flat-field requirement positions the curved tangential and sagittal image surfaces symmetrically with respect to the paraxial image plane. For a single thin lens in air with central stop this symmetric position cannot be obtained. The average tangential and sagittal wavefront aberration is given by

$$W_{\mathrm{av}} = \left(\frac{2S_3 + S_4}{4}\right) H^2 = -(2n+1)\,\frac{KH^2}{4n}\,, \tag{7.5.6}$$

  and the astigmatism enlarges by a factor of $(2n + 1)$ the field curvature that follows from the Petzval term S_4 only. An image field that is flat on average can be reached in an extended system where various elements in the system are used with a remote pupil configuration. In this way, in contrast with the single lens with central stop, the curvature of the image field due to astigmatism and the intrinsic Petzval field curvature can be uncoupled. Another phenomenon which will help to obtain a flat image field is the opposite sign of the field curvature related to the third-order terms and the curvature introduced by one or several higher-order terms. At larger field angles, an increasing higher-order term starts to counteract the basic third-order term and the sag of the image surface is reduced or even made equal to zero for a certain value of the image field coordinate p.

7.5.1 A Flat-field Photographic Triplet Lens

At the end of the nineteenth century a flat-field photographic lens was proposed by Cooke which explicitly uses a negative lens for reducing the Petzval sum of the system. Compared to the landscape lens the aperture of the triplet lens had an almost twice as large value, a less-curved image field and almost no distortion. The negative lens in the Cooke triplet design is also instrumental in the reduction of axial chromatic aberration. It is chosen from the high-dispersion glasses with the two positive lenses having lower dispersion. With the aid of Seidel's third-order theory applied to thin lenses, the system can be analytically designed; the details of this analysis can be found in **[77]**, **[181]**. The basic principle of the design was given in the preceding section. In the paraxial approximation of the imaging from infinity by the three lenses we have

$$\frac{1}{f'} = \sum_{j=1}^{3} \frac{y_j}{y_0} K_j \,, \tag{7.5.7}$$

where y_0 is the half diameter of the parallel beam coming from infinity. The quantity y_1/y_0 can be put equal to unity for the first 'thin' lens of the triplet. The other two quotients, y_2/y_0 and y_3/y_0, determine the paraxial design and influence the Petzval sum of the system. These ratios are important choices for the aberration correction of the triplet lens.

7.5.2 Third-order Optimised Triplet Lens

We first design a triplet lens with the focus entirely on the correction of the third-order aberrations. We emphasise from the very beginning that this approach to the design of a triplet lens is flawed and leads to a useless system. The main reason is that the combined requirements on aperture value and field angle introduce substantial higher-order aberrations. However, an analysis based on third-order aberrations only shows how the imaging quality of a triplet system is frustrated by the higher-order aberrations. We use an approximate third-order design by treating each lens element as a thin lens and then using the thin-lens aberration formulas. Of course, it is also possible to consider the individual surfaces and to tabulate the third-order surface contributions of the six lens surfaces. It turns out that taking into account the finite thicknesses from the beginning has a minor influence on the shape of the final third-order design.

The design starting point is governed by the choice of the positive and negative powers of the lenses such that their Petzval sum becomes zero. A solution is somewhat arbitrarily co-determined by the distances between the lenses and the corresponding intersection heights y_j/y_0 of the marginal ray of an axial beam on the three lenses. For a focal length $f'$of 50 mm, we choose the distances between the lenses typically $f'/6$ and adapt the individual lens powers such that the Petzval sum becomes zero. Although from the point of view of a zero Petzval sum it would be favourable to have a low-index glass for the negative element, we choose the glass combination SK16 and F2. This combination yields a better correction of the axial chromatic error than with a lower index glass like LF5 for the negative element. Table 7.10 lists the initial paraxial values for the elements of the triplet with a total power $K_0 = 0.02$. In the table we have used the refractive indices of SK16 (1.623) and F2 (1.624) at the central wavelength $\lambda_0 = 550$ nm; P_j is the contribution of the jth surface to the Petzval sum. The table shows that the penalty for achieving zero field curvature is heavy. High individual powers of the individual lenses are needed, especially for the central negative one.

The third-order aberration correction exploits the symmetry properties of the triplet design by locating the stop at or close to the central lens. Coma, distortion and lateral colour can be easily corrected thanks to this basic paraxial symmetry with respect to the stop. The bending of the first and the third lens is further used to null the astigmatism. The lens data of a third-order optimised design based on the paraxial starting values of Table 7.10 are given in Table 7.11, for the central wavelength $\lambda_0 = 550$ nm. With these triplet data we obtain third-order wavefront aberration coefficients as given in Table 7.12, in units of the wavelength of the light. The chromatic difference in *paraxial* focus-setting (axial chromatism) is the only 'aberration' which has not been reduced to zero. The dispersion difference between the positive and negative elements is insufficient to reach complete correction of axial colour. In practice, we should have chosen positive lenses with a substantially higher V number, for instance, close to 80. This is an unrealistic option given the

Table 7.10. Initial paraxial data for a triplet lens. P_j is the Petzval term of the jth element.

	Element 1	Element 2	Element 3
y_j/y_0	1	0.67	0.82
K_j	0.034	−0.088	0.051
$\lvert K_j/K_0 \rvert$	1.7	4.4	2.6
P_j	−0.0210	+0.0524	−0.0320

Table 7.11. Triplet design based on third-order aberration theory only.

Triplet lens	Distance	Index	Curvature		Diameter	Glass
(third-order optimum)	∞	1.000	0.04997		17.0	
	3.500	1.623	−0.00404		17.0	SK16
	8.572	1.000	−0.05744	STOP	7.2	
	2.000	1.624	0.08032		10.0	F2
	6.410	1.000	0.01823		15.0	
	3.500	1.623	−0.06586		15.0	SK16
	39.700	1.000	0.00000		36.0	

Table 7.12. Third-order aberration coefficients of a triplet lens which is optimised with respect to only third-order aberration coefficients (the data are expressed in units of the wavelength λ of the light for each wavelength).

	a_{sp}	a_c	a_{tan}	a_{sag}	a_d	$a_{\lambda,ax}$	$a_{\lambda,lat}$
$W^{(4)}$ ($\lambda = 486$ nm)	0.16	0.08	−2.56	−0.74	−0.36	−1.48	0.10
$W^{(4)}$ ($\lambda = 550$ nm)	0.15	−0.02	0.04	−0.06	−0.04	0	0
$W^{(4)}$ ($\lambda = 656$ nm)	0.16	−0.16	+4.03	+0.95	+0.84	+1.55	−0.21

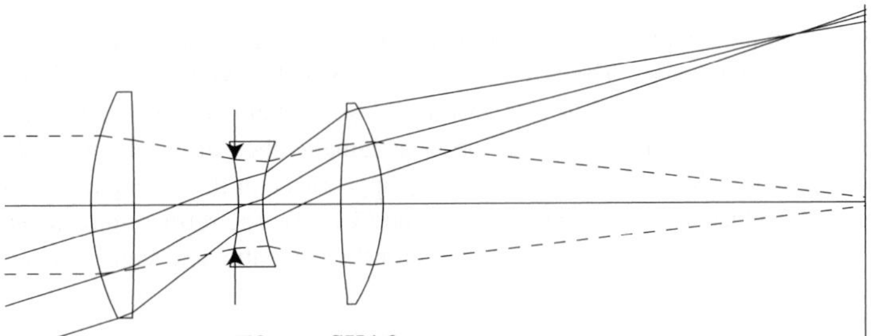

Figure 7.22: A triplet photographic lens of which all third-order aberrations are virtually zero; $f' \approx 50$ mm, $NA = 0.11$ ($F_\# = 4.5$), $\lambda_0 = 550$ nm. Wavelength range: 486 to 656 nm, $2 \times 18°$. The pencil with solid rays has the maximum field angle of 18° in object space. The dashed rays are the marginal rays of the axial pencil.

dispersion range of (standard) optical glasses. The change of astigmatism with wavelength is an upsetting feature. The reason is that at the location of especially the second lens, wavefront contributions per surface of the order of 1000 λ are detected. The balance in astigmatism at the central wavelength is substantially disturbed by the small index change when going to another part of the visible spectrum. The high powers of the individual lenses have led to a design which is very sensitive to refractive and constructional changes in the system.

The triplet layout is shown in Fig. 7.22. The drawing immediately shows that, despite the very good correction of third-order aberration, field curvature and astigmatism are present when the triplet is subjected to finite ray-tracing. Due to the high individual powers of the lenses, large angles of incidence on the surfaces occur. For a marginal ray of the axial pencil, dashed in the figure, angles of incidence of 25° occur. For the principal ray at a field angle of 25°, the angle of incidence at the back surface of the central lens amounts to approximately 50°. This means that the correction of only third-order aberrations is wholly inadequate when designing a triplet lens. In practice, the principal ray of an off-axis beam at a field angle in object space of 25° is totally reflected in the system. The maximum field angle is limited to 18° and the (higher-order) aberrations of this extreme oblique pencil are huge. At the scale of the drawing in Fig. 7.22 the blur of this pencil in the image plane is clearly visible and approaches the value of 1 mm!

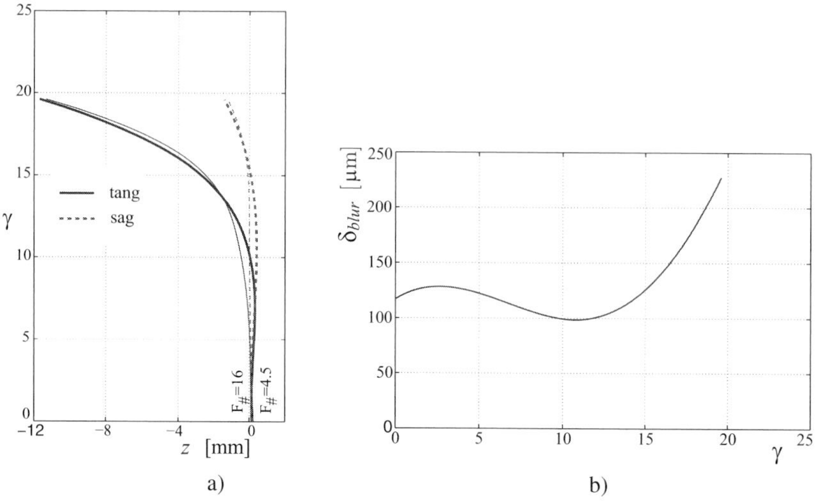

Figure 7.23: Triplet lens with virtually zero third-order (monochromatic) aberration coefficients according to Table 7.11.
a) Cross-section through the optical axis of the tangential and sagittal focal surfaces as a function of the field angle γ (in degrees). Thin solid and dashed curve: $F_\# = 16$, thick solid and dashed curve: $F_\# = 4.5$.
b) Geometrical blur δ at full aperture ($F_\# = 4.5$) of the focused spot in the optimum image plane ($z = -1.5$ mm) as a function of field angle γ.

The strong aberration of oblique imaging pencils is illustrated in Fig. 7.23, which was obtained by finite-ray analysis of this third-order optimised triplet design. The left-hand graph shows the axial position of the tangential and sagittal focal lines as a function of the field angle γ. For small values of γ, the field curvature and astigmatism are effectively zero. However, for field angles larger than 5°, a higher-order astigmatic aberration proportional to $\rho^2(p/p_m)^4$ arises and completely destroys the imaging quality of the triplet lens. The appearance of this strong astigmatism can be understood by inspection of the paraxial Coddington equations, (5.7.41) and (5.7.42). The refraction by a surface of a pencil of rays that originated at a distance l_0 yields the following expression for the tangential and sagittal image distances,

$$\frac{\cos^2\theta_1}{l_t} - \frac{1}{l_s} = -\frac{n_0}{n_1}\sin^2\theta_0\frac{1}{l_0}\,, \tag{7.5.8}$$

with θ_0 and θ_1 the angles of incidence and refraction at the surface, respectively. In the special case of an incident parallel beam we can write for the normalised astigmatic distance

$$\frac{l_t - l_s}{(l_t + l_s)/2} = -\frac{2\sin^2\theta_1}{1+\cos^2\theta_1} = -2\left(\tan^2\theta_1 - \frac{\tan^4\theta_1}{2} + \frac{\tan^6\theta_1}{4}\,\cdots\right). \tag{7.5.9}$$

This equation shows that the normalised longitudinal astigmatism of a single surface ($l_0 = \infty$) with the stop on the surface has a second and fourth-order dependence on the field coordinate $p = f'\tan\theta_1$ with opposite signs.

Although the astigmatism of a triplet lens is also strongly influenced by the astigmatic contributions of the two outer non-aplanatic lenses which have a remote stop position, the triplet lens as a whole shows the same characteristic 'folding-back' behaviour of the tangential and sagittal focal surfaces as indicated by Eq. (7.5.9). In Fig. 7.23 two sets of astigmatic focal line positions have been plotted. The thin solid and dashed curves are the astigmatic lines at small aperture of the triplet lens ($NA = 0.031$, $F_\# = 16$). Close to the axis these tangential and sagittal focal curves follow the third-order prediction of zero field curvature and zero astigmatism. The thick curves correspond to pencils of rays with full aperture, $NA = 0.11$ ($F/4.5$). Here, we notice a tiny departure from the third-order prediction and this is due to the occurrence of a higher-order term proportional to $\rho^4(p/p_m)^2$. The net effect is a slight outward curving of the best-focus astigmatic focal lines.

The quality of the imaging at the full opening of $F_\# = 4.5$ is represented in Fig. 7.23b. A best focal plane has to be selected. Although the aberration of the axial pencil is almost negligible, the search for the global optimum focus-setting introduces a large defocus at the origin ($z = -1.5$ mm), resulting in an initial blur δ of more than 100 μm. The conclusion is that this third-order optimised design is useless and would produce a 50% resolution for typically 100×100 pixels over the total field of $2 \times 20^\circ$. Compared to the huge monochromatic aberration of the design, the imperfect correction of axial chromatism as shown in Table 7.12 is of no relevance for this design.

7.5.3 Optimised Photographic Triplet Lens

The main problem of the third-order-based triplet design is the rigorous correction of the Petzval sum which led to high lens powers and large angles of incidence on the lens surfaces. A flat field surface was obtained by imposing zero Petzval sum *and* zero astigmatism in the design. A better design can be obtained by allowing a finite amount of astigmatism and requiring that the quantity W_{av} of Eq. (7.5.6) remains close to zero. We can obtain zero value by allowing a negative Petzval sum combined with inverted astigmatism. This astigmatism, inverted with respect to that of a positive lens with central stop, would produce an image with the tangential focal lines in the paraxial image plane and the sagittal focal lines beyond the paraxial plane. The astigmatism with inverted sign is introduced by the negative element. If the net astigmatism of the system is sufficiently small for the application, an effectively flat image field is obtained with $W_{av} \approx 0$.

In terms of third- and higher-order aberrations we have the following astigmatic contributions regarding their dependence on (normalised) pupil and field coordinates:

$$\begin{array}{ll}
\text{third-order} & {}_2a_{22}\,\rho_n^2\,\sin^2\theta\,(p/p_m)^2\,, \\
\text{fifth-order} & {}_4a_{22}\,\rho_n^2\,\sin^2\theta\,(p/p_m)^4\,, \\
\text{"}\quad\text{"} & {}_2a_{42}\,\rho_n^4\,\sin^2\theta\,(p/p_m)^2\,, \\
\text{seventh-order} & {}_6a_{22}\,\rho_n^2\,\sin^2\theta\,(p/p_m)^6\,, \\
\text{"}\quad\text{"} & {}_4a_{42}\,\rho_n^4\,\sin^2\theta\,(p/p_m)^4\,, \\
\text{"}\quad\text{"} & {}_2a_{62}\,\rho_n^6\,\sin^2\theta\,(p/p_m)^2\,, \\
\text{"}\quad\text{"} & {}_4a_{44}\,\rho_n^4\,\sin^4\theta\,(p/p_m)^4\,.
\end{array} \tag{7.5.10}$$

The $\sin\theta$-dependence for the astigmatic terms follows from the fact that we chose the field excursion p along the y-axis with azimuth $\phi = \pi/2$. In terms of Zernike polynomials we have, up to the fifth order, a mixture of field curvature and astigmatic contributions. For the third-order Seidel aberration we can write for a general field azimuth ϕ,

$$\begin{aligned}
{}_2W_{22}(\rho_n,\theta;r_n,\phi) &= \frac{{}_2a_{22}}{2}\,r_n^2\rho_n^2\,[1+\cos\{2(\theta-\phi)\}] \\
&= \frac{{}_2a_{22}}{8}\left\{R_2^0(r_n)+1\right\}\left\{R_2^0(\rho_n)+1\right\} \\
&\quad + \frac{{}_2a_{22}}{2}\,R_2^2(\rho_n)\cos 2\theta\,R_2^2(r_n)\cos 2\phi \\
&\quad + \frac{{}_2a_{22}}{2}\,R_2^2(\rho_n)\sin 2\theta\,R_2^2(r_n)\sin 2\phi\,,
\end{aligned} \tag{7.5.11}$$

where $r_n = p/p_m$.

Inserting $\phi = \pi/2$ in the above equation and in the comparable equations for the fifth-order aberration terms, we obtain the following Zernike representations for the Seidel (and Schwarzschild) astigmatic aberration terms,

$$\left.\begin{array}{ll}
\text{third-order} & ({}_2a_{22}/8)\left\{R_2^0(r_n)+1\right\}\left\{R_2^0(\rho_n)+1\right\} - ({}_2a_{22}/2)\,R_2^2(r_n)\,R_2^2(\rho_n)\cos 2\theta \\
\text{fifth-order} & ({}_4a_{22}/2)\left\{R_4^0(r_n)/6 + R_2^0(r_n)/2 + 1/3\right\}\left\{R_2^0(\rho_n)/2 + 1/2\right\} \\
 & \quad -\,({}_4a_{22}/2)\left[R_4^2(r_n)/4 + 3R_2^2(r_n)/4\right]\,R_2^2(\rho_n)\,\cos 2\theta\,, \\
\text{"}\quad\text{"} & ({}_2a_{42}/2)\left\{R_2^0(r_n)/2 + 1/2\right\}\left\{R_4^0(\rho_n)/6 + R_2^0(\rho_n)/2 + 1/3\right\} \\
 & \quad -\,({}_2a_{42}/2)\,R_2^2(r_n)\left[R_4^2(\rho_n)/4 + 3R_2^2(\rho_n)/4\right]\,\cos 2\theta\,.
\end{array}\right\} \tag{7.5.12}$$

In Eq. (7.5.12) for the astigmatic wavefront aberration we add the field curvature contribution of third order which directly follows from the Petzval sum of the system. We neglect constant terms and terms that depend on powers of ρ_n or r_n only. The latter terms produce a constant defocus for the entire image or a phase factor in the image that varies quadratically with r_n. These terms are irrelevant when studying the curvature and astigmatism of the image. With these restrictions we obtain the following expression regarding field curvature and astigmatism,

$$
\begin{aligned}
W(\rho_n, \theta; r_n, \pi/2) = {} & \frac{1}{8}\,[2({}_2a_{20}) + ({}_2a_{22}) + ({}_4a_{22}) + ({}_2a_{42})]\; R_2^0(r_n)\, R_2^0(\rho_n) \\
& + \frac{1}{24}\left[({}_4a_{22})R_4^0(r_n)R_2^0(\rho_n) + ({}_2a_{42})R_2^0(r_n)R_4^0(\rho_n)\right] \\
& - \frac{1}{2}\left[({}_2a_{22}) + \frac{3}{4}({}_4a_{22}) + \frac{3}{4}({}_2a_{42})\right] R_2^2(r_n)\, R_2^2(\rho_n)\cos 2\theta \\
& - \frac{1}{8}\left[({}_4a_{22})R_4^2(r_n)R_2^2(\rho_n) + ({}_2a_{42})R_2^2(r_n)R_4^2(\rho_n)\right] \cos 2\theta \,.
\end{aligned}
\tag{7.5.13}
$$

The various a-coefficients are calculated using wavefront aberration data of finite rays sampled with respect to the pupil coordinates (ρ_n, θ) and the field coordinates $(r_n, \pi/2)$. From these aberration data we obtain the a-coefficients by computing inner products with orthogonal aberration functions as given by Eq. (5.9.11). Another approach relies on the initial calculation of Seidel and Schwarzschild aberration coefficients expressed in Cartesian pupil and field coordinates. These 'Cartesian' aberration coefficients are then converted into coefficients of the orthogonal expansion of Eq. (7.5.13) with the aid of the analysis given in Subsection 5.9.1.

An improved triplet design exploits the flexibility of the third-order aberrations with respect to the lens shape factors. It allows introduction of a third-order astigmatism offset which counteracts fifth-order contributions which change much less as a function of lens shape. More precisely, from Eq. (7.5.13) we reduce the lowest order astigmatic wavefront aberration term to zero and then choose the Petzval field curvature coefficient ${}_2a_{20}$ which results in a flat image field,

$$
\begin{aligned}
{}_2a_{22} &= -3\left({}_4a_{22} + {}_2a_{42}\right)/4\,, \\
{}_2a_{20} &= ({}_2a_{22})/6\,.
\end{aligned}
\tag{7.5.14}
$$

Figure 7.24 shows the layout of a triplet lens which, in a first design stage, has been given nonzero values of the third-order a-coefficients. Such a preparatory first design stage is meant to create a starting system that is close to the global optimum. In a second and final stage, the aperture- and field-dependent aberration is further reduced with the aid of local optimisation. The first design stage, using third- *and* higher-order aberration coefficients with optimum balance, yields a 'smoother' design. This means that the heights of incidence on the second and third element, in the thin-lens approximation given by $y_2/y_0 = 0.80$ and $y_3/y_0 = 0.86$, are closer to unity than in the previous design of Subsection 7.5.2 which was exclusively based on a zero-target for the Seidel aberration coefficients. The powers of the individual lenses depend on the separations between the elements. Here, we have chosen a typical lens-to-lens distance of $f'/10$. Larger distances increase the construction length and require a higher relative power of the negative lens. We keep the low-dispersion glass SK16 for the positive elements and choose LF5 for the central negative lens. A lower refractive index for the negative lens makes it more effective in reducing the field curvature of the system. Higher refractive indices are useful for aberration reduction since the angles of incidence on the optical surfaces can be made smaller. Using the SK16–LF5

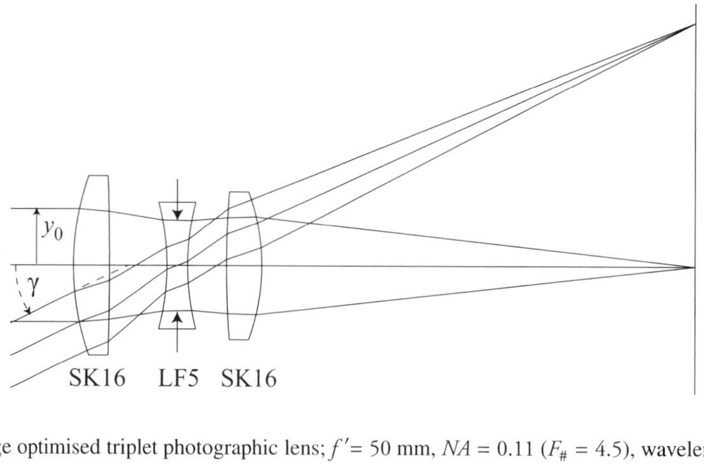

Figure 7.24: Two-stage optimised triplet photographic lens; $f' = 50$ mm, $NA = 0.11$ ($F_\# = 4.5$), wavelength range: 485 to 650 nm, $2 \times 24°$.

Table 7.13. Third-order aberration coefficients of a finite-ray optimised triplet lens (in units of μm). The central wavelength λ_0 is 550 nm.

	λ (nm)	a_{sp}	a_c	a_{tan}	a_{sag}	a_d	$a_{\lambda,ax}$	$a_{\lambda,lat}$
$\Delta W^{(4)}$	450	+0.094	−0.035	−0.397	−0.044	−0.750	−1.328	−0.112
$W^{(4)}$	550	−1.309	−1.047	−3.179	−10.243	−15.097	0	0
$\Delta W^{(4)}$	656	−0.140	+0.056	+0.683	+0.100	+1.230	+2.432	+0.167

Table 7.14. Optical data of the finite-ray optimised triplet lens.

Triplet	Distance	Index	Curvature		Diameter	Glass
(finite-ray optimum)	∞	1.0000	0.040157		16.0	
	3.500	1.6226	−0.003770		15.2	SK16
	5.713	1.0000	−0.044999	STOP	8.9	
	2.000	1.5845	0.043026		8.8	LF5
	3.778	1.0000	0.011040		12.8	
	3.500	1.6226	−0.052086		12.8	SK16
	42.17	1.0000	0.000000		50.0	.

glass choice, the paraxial powers of the lenses are approximately given by $K_1 = 1.4K$, $K_2 = -2.6K$ and $K_3 = 1.9K$ where K is the total power of the system. Using the thin-lens approximation, the Petzval sum of the system is given by

$$P = -\frac{K_1 + K_3}{n_1} - \frac{K_2}{n_2} \approx -0.42\,K, \tag{7.5.15}$$

equal to ≈70% of the Petzval curvature of a single SK16-lens with the same optical power. The maximum quadratic wavefront aberration W_{fc} due to field curvature is given by $PH^2/4$ and amounts to −13.3 μm (third-order approximation, $NA = 0.11$, maximum image field coordinate p_m is 24 mm). The reduction in field curvature looks insufficient. But the astigmatic wavefront aberration of the triplet lens amounts to +8 μm and the resulting tangential and sagittal wavefront aberration values amount to −1.4 and −9.3 μm, respectively. The average wavefront aberration due to astigmatism and field curvature at the border of the field is now given by −5.3 μm.

The complete set of third-order wavefront aberration values and their change with wavelength are given in Table 7.13. The third-order coefficients are those that result from the two-stage design procedure described above. The triplet lens system has an aperture of 0.11 in image space and an angular image field of $2 \times 25.6°$ ($p_m = 24$ mm). The paraxial estimate of the maximum angle of incidence on the second lens surfaces is limited to 0.8 radian, approximately 25% smaller than that for the third-order optimised triplet. A discussion of the aberration of the finite-ray optimised triplet lens can be limited to the effects of astigmatism and field curvature which are proportional to at least NA^2. The other aberrations, proportional to powers higher than NA^2, remain relatively small because of the small NA value of 0.11.

The design data of the two-stage optimised triplet are given in Table 7.14. The performance of the triplet lens is illustrated with the aid of the position of the astigmatic foci in the field, see Fig. 7.25a. At low aperture ($F_\# = 16$) and small field angle, the position of the astigmatic lines (thin solid and thin dashed curves) follows the third-order predictions. An increase in aperture to $NA = 0.11$ ($F_\# = 4.5$) shows the influence of the fifth-order Schwarzschild aberration term proportional to $\rho_n^4 r_n^2 \sin^2\theta$. The presence of this fifth-order aberration gives rise to blurred astigmatic lines of which the best-focus position is axially shifted with respect to that of the geometrically sharp astigmatic lines that follow from third-order aberration theory. Figure 7.25a shows that at maximum aperture the two astigmatic line curves are symmetrically arranged with respect to the flat paraxial image plane. For a field angle $\gamma > 10°$, the aberration term proportional to r_n^4 becomes important. Third- and fifth-order aberration terms are balanced in the Zernike term $R_4^2(r_n)R_2^2(\rho_n)\cos 2\theta$; the quadratic focus deviation in the field comes from the term $R_2^2(r_n)R_4^2(\rho_n)\cos 2\theta$. The latter aberration term, of fourth order in the aperture, is only significant for the astigmatic curves at maximum aperture ($F_\# = 4.5$). The higher-order terms give rise to a zero-crossing of the astigmatism, approximately at the location of the zero-crossing of the polynomial $R_4^2(r_n)$ at $r_n = \sqrt{3}/2$. The zero-crossing of the astigmatism in the image field is visible in Fig. 7.25b. The folding-back of lowest

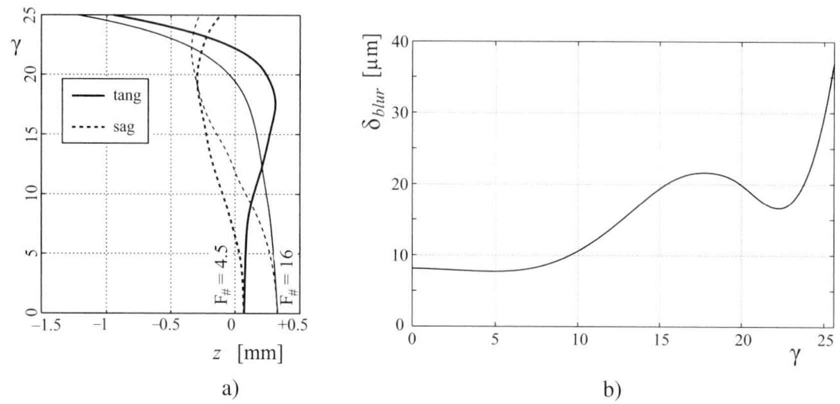

Figure 7.25: a) Axial position of the astigmatic lines of a triplet lens as a function of the field angle γ (monochromatic analysis at $\lambda = 550$ nm). Thick solid curve: tangential focal line; thick dashed curve: sagittal focal line, both for $F_\# = 4.5$. The thin curves correspond to a lens opening with $F_\# = 16$.
b) Geometrical rms blur δ of the focused spot in the image plane (full aperture, $\lambda = 550$ nm).

order astigmatism by higher-order field components has already been demonstrated and discussed in Ch. 5 for a single optical transition (see Eq. (5.7.52) and the corresponding Fig. 5.72). The image blur initially increases as a function of the field angle due to the growing astigmatism of the pencils. At approximately 90% of the field a minimum is observed with zero astigmatism. In the remaining 10% of the field, the quality severely deteriorates because of the sharp defocus of the tangential astigmatic surface.

The second stage of the design of the triplet lens, the 'fine-tuning' stage, can be numerically performed in the classical way by the reduction of the aberrations of representative rays of a set of pencils towards selected points in the image field. Analytically calculated aberration coefficients **[62]**, **[29]** present another possibility for obtaining reliable merit function data. A numerically flexible approach is to use finite ray data from a set of pencils to calculate the expansion coefficients ${}_ia_{jk}$ of a circularly symmetric system in terms of Zernike polynomials (see Section 5.9). In principle, by applying weight factors, this approach allows the reduction of aberration terms of higher orders without being perturbed by the quickly varying coefficients of lower-order aberrations, especially the third-order ones. The design of the triplet lens has shown that the reduction of aberrations terms like ${}_4a_{22}$ and ${}_2a_{42}$ determines the final quality of the triplet lens. The third-order aberrations can be given their appropriate residual values in the second design stage once the slowly varying higher-order aberration terms have been reduced in the first phase of the design.

In Fig. 7.26 we present the imaging quality of the triplet lens as a function of the field angle γ by means of the modulus of the geometrical transfer function at $\lambda = 0.55$ μm. The four *MTF* graphs show again that astigmatism is limiting the performance of the triplet lens. The meridional *MTF* curves (solid curves) show a sharp drop-off beyond a field angle of 22°. The effective field of the lens should be limited to $2 \times 24°$ because of the resolution in the meridional cross-section of the lens. The four through-focus curves for the meridional and sagittal *MTF* at a spatial frequency of 10 mm^{-1} show the axial positions of the tangential and sagittal focal lines which were given above in Fig. 7.25. The tangential and sagittal foci coincide at approximately 90% of the maximum field angle. The *MTF* curves show that a 50% modulation depth is maintained up to a field angle of 23°. If the focal distance f' equals 50 mm, the classical photographic image field of 36×24 mm can be accommodated. Expressed in the number of pixels along the horizontal direction, we then arrive at approximately 10^3 resolved pixels at 50% contrast. This number hardly changes when a spectral weighting of the imaging quality is carried out for the visible spectrum. Because of the high degree of symmetry with respect to the stop, the lateral colour of the triplet lens is negligible. If possible, the *symmetry* principle with respect to the stop is often exploited in optical design since it has the tendency to strongly reduce the asymmetrical aberrations coma and distortion. The axial colour of the triplet lens is small and reduces the spectrally weighted modulation transfer by only a few percent with respect to the monochromatic values. Because of its relative optical simplicity, decent resolution and practical field angle, the triplet solution is also used in modern digital (miniature) photography.

The design method for a triplet landscape lens using the values of fifth- and higher-order aberration coefficients to compute the necessary third-order offsets has been discussed in this subsection in some detail. This approach can equally well be used for the design of more complicated lens systems. The triplet lens itself has been exhaustively researched.

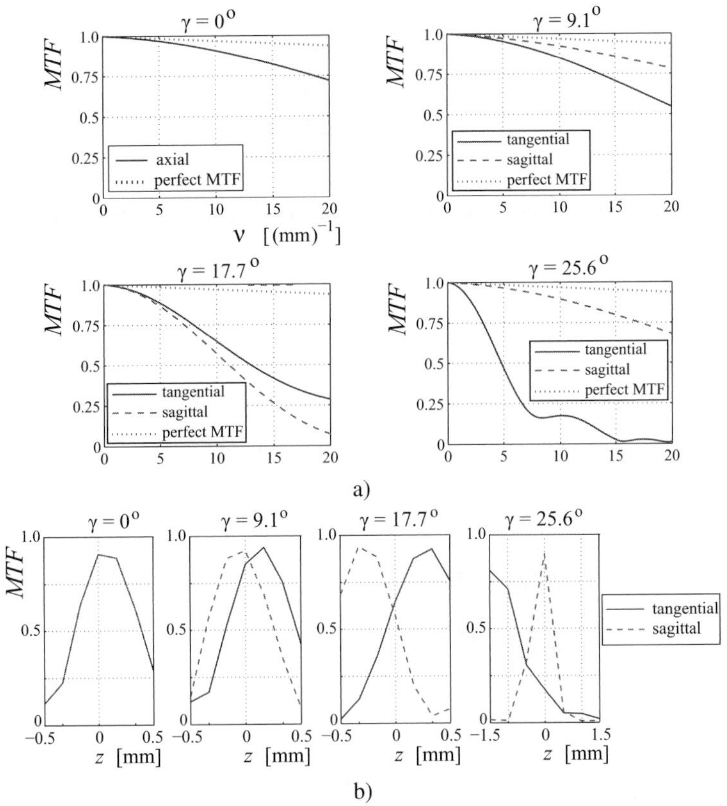

Figure 7.26: Two-stage optimised triplet lens ($f' = 50$ mm, $F_\# = 4.5$) of Fig. 7.24. a) Geometrical modulation transfer functions for the field angles $\gamma = 0, 9.1, 17.7$ and 25.6 degrees. The spatial frequency value ν is expressed in mm^{-1}. b) Through-focus *MTF* curves at a spatial frequency of 10 mm^{-1} for the same field angles γ as in a). Note the different horizontal scale for the graph corresponding to the extreme field angle $\gamma = 25.6°$.

In **[181]** a system of equations is given for the calculation of the thin-lens bending coefficients P_j of each lens such that the required offset values of the third-order aberrations coefficients are obtained. Before solving these equations a choice has to be made about the distribution of optical power over the three lens elements and the values of the third-order residuals. Each design obtained in this way needs a final local optimisation to adapt it in detail to the lens imaging requirements. This 'fine-tuning' optimisation covers aspects such as the precise vignetting of off-axis pencils, the exact chromatic correction depending on the spectrum of the light used for imaging, optimisation with respect to a mixture of wave and transverse aberration data to adapt the position of the astigmatic surfaces towards the edge of the field, etc. In all cases there remains the uncertainty as to whether the global minimum of the merit function was effectively reached. Fortunately, the saddle point method (see Subsection 6.4.2.5) has been applied to a relatively simple imaging system like the triplet lens and, consequently, the entire design landscape has become available. The triplet landscape is depicted in Fig. 6.19. The global minima of this figure (labelled m1-m2) immediately provide the user with a pre-design which can be refined as a function of the specific imaging requirements at hand. It is seen that the two-stage triplet optimum which was obtained in this subsection resembles closely the layout of the global minimum in the triplet design landscape.

7.5.4 Photographic Objective Lenses of the Double-Gauss Type

Early in the photographic era there was already a need for higher performance than that which was offered by the simple landscape lens and the Petzval portrait lens. The Cooke triplet lens offered a much better imaging quality than the landscape lens but could not satisfy the requirements of more advanced, professional photography. The extra requirements

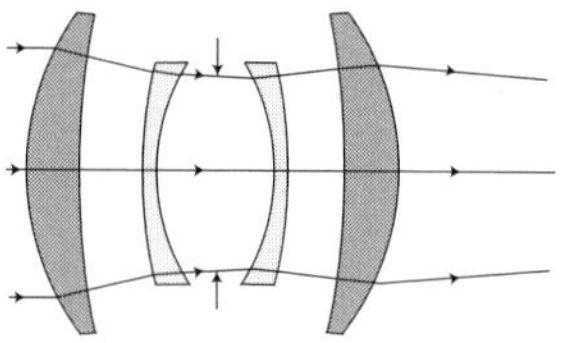

Figure 7.27: The basic Double-Gauss objective lens. Half of it forms a 'Gauss lens' which was meant to be used as an objective for an astronomical telescope.

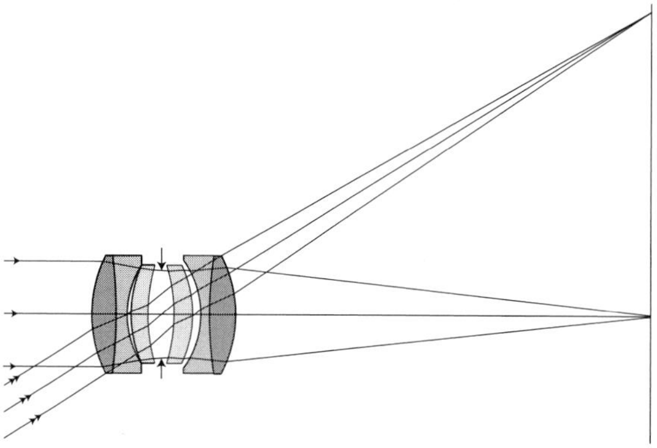

Figure 7.28: Double-Gauss objective (US patent 1,792,917, Carl Zeiss company, Jena, 1927); $f' = 100$ mm, $NA = 0.11$ ($F_\# = 4.5$), angular field $2 \times 32^\circ$.

were a smaller $F_\#$ and/or a larger angular imaging field. The basic triplet lens has been extended, for instance by making the image-side element a doublet. But a more successful layout, for a larger opening and larger angular field, turned out to be the so-called Double-Gauss type lens. The basic lens type of Fig. 7.27 was proposed at the end of the nineteenth century. The name has been derived from a proposition made by Gauss for a refracting telescope objective. He combined a principal meniscus-type positive lens with a lower power negative meniscus lens for the correction of spherical aberration. The suggestion was not put into practice because of problems with weight and mechanical tolerances. By doubling the Gauss lens combination, symmetrically with respect to a central stop, a powerful design prototype was created. Historically, the first Double-Gauss lens precedes the Cooke triplet by a few years. The designs bear a close resemblance. The Double-Gauss is obtained from the triplet by splitting the central negative lens of the triplet in two well-separated negative meniscus elements. The menisci have identical shape and face each other with the stop in between them. Positive meniscus-shaped elements on the front side and back convey the required net positive power to the design. The basic design has perfect symmetry with respect to the stop. This means that the principal ray continues undeviated. In general, the asymmetric aberrations like distortion, coma and lateral chromatism tend to be small in such a configuration. The presence of two separate negative elements yields extra degrees of freedom for the correction of spherical aberration and axial chromatism. We will see that a relatively flat image field can be obtained by allowing purposely, as in the triplet lens, an important residual of third-order astigmatism. The higher-order astigmatism at larger field angles compensates the third-order contribution yielding a zero-crossing of the astigmatic focal surfaces at some larger field angle, well outside the third-order range.

7.5.4.1 Lens Properties

As an example of a Double-Gauss objective we select a patented design dating back to 1927 (see Fig. 7.28). The constructional data of the objective are given in Table 7.15. The chosen example has a relatively small opening but a large angular field.

Table 7.15. Wide-angle Double-Gauss objective (US patent 1,792,197).

Distance	Curvature	Diameter	Index	Abbe number	Glass type
100000.000	0.038610	25.0	1.00000		
5.120	−0.010406	24.0	1.61087	55.8	≈ SK2
2.310	0.054318	19.8	1.53994	47.4	≈ LLF1
0.750	0.042052	19.8	1.00000		
3.480	0.028137	19.2	1.56064	61.1	≈ SK11
2.900	0.000000	STOP 18.3	1.00000		
2.800	−0.030193	20.0	1.00000		
3.960	−0.043975	20.0	1.56064	61.1	≈ SK11
1.570	−0.055127	20.0	1.00000		
1.880	0.012923	24.0	1.53994	47.4	≈ LLF1
5.560	−0.039417	25.0	1.61087	55.8	≈ SK2
86.508	0.000000	130.0	1.00000		

The following points are worth mentioning with respect to this Double-Gauss design:

- *Sequence of optical power.*
  In contrast to the basic layout of the Double-Gauss design in Fig. 7.27, the total power in this example is given by the sequence $- + + -$. The reason for this inversion is the wide-angle field. By means of a first negative power, the principal ray of a far off-axis pencil is bent towards the optical axis and the angles of incidence on the next surfaces are reduced by this effect.
- *Presence of cemented doublets.*
  The outer negative lenses have been made cemented doublets. Although the power sequence $- + + -$ in itself allows for a correction of the paraxial axial chromatism, the presence of the buried surfaces further facilitates this correction. Moreover, the buried transitions enable the correction of the monochromatic aberrations like spherical aberration and coma as was demonstrated in Section 7.2 for individual doublets. Figure 7.28 shows that the three refracting surfaces of the outer lenses, including the buried surfaces, are far from the stop and that they are used with an eccentricity factor ϵ which is of the order of unity. By making these outer doublet lenses non-aplanatic, they can be effectively used for the correction of astigmatism; the buried surfaces allow a reduction of the chromatic change of aberrations.
- *Petzval sum.*
  The Petzval sum of the objective equals $P = -0.18\,K$ where K is the overall power of the objective. This value is substantially smaller in absolute value than that of the optimised triplet with $P = -0.42\,K$. This enhanced reduction of Petzval field curvature in the Double-Gauss objective is needed to obtain well-flattened tangential and sagittal focal surfaces over the large angular field. The average refractive index of the negative doublets is smaller than the index of the positive lenses. At equal power of the components, this helps to reduce the absolute value of the Petzval sum of the system (see Eqs. (5.3.38) and (5.4.15)). The patent provides the index value n_d and the Abbe V number of the optical glasses. Their exact counterparts in the restricted present-day 'green' glass catalogue do not exist. The glass types mentioned in Table 7.15 are the closest possible equivalents of the original $(n_d - V)$ combinations in the patent.
- *Third-order aberration residuals.*
  With respect to the residual third-order aberration coefficients of the (optimised) triplet lens, the Double-Gauss example has substantially larger third-order residuals, especially for spherical aberration and coma. A correct comparison of Tables 7.13 and 7.16 should include a factor of two because of the difference in focal distance. A multiplication by this factor of two has been performed in the bottom row of Table 7.16. It follows from the table that the Double-Gauss lens with its 40% larger angular field needs a different third-order astigmatism offset than the triplet lens. This offset makes the third-order tangential focal surface more inward curved than the sagittal one, similarly to the astigmatic focal surfaces of a single lens.

Spherical aberration and coma of the Double-Gauss objective have larger residual values as compared to those of the triplet lens. The intermediate third-order aberration values after each surface are roughly twice as large as in the triplet lens and amount to more than 300 μm. The required third-order residuals can be obtained once reliable higher-order aberration coefficients have been calculated. In the case of the compact Double-Gauss objective, higher-order expressions like those of Schwarzschild for the *intrinsic* fifth-order aberrations can be used. Alternatively, a pupil-field wavefront expansion is constructed by means of ray-tracing which also yields the necessary higher-order aberration coefficients. From the

Table 7.16. Third-order aberration coefficients of the Double-Gauss objective of Fig. 7.28 (in units of μm). The bottom line shows the residual third-order aberration coefficients of the two-stage optimised triplet lens of Fig. 7.24, scaled to a focal length of $f' = 100$ mm.

	a_{sp}	a_c	a_{tan}	a_{sag}	a_d	$a_{\lambda,ax}$	$a_{\lambda,lat}$
$W^{(4)}$ (at λ_0)	−6.6	+8.9	−28.0	−23.4	+4.3	0	0
ΔW (λ_0 + 110 nm)	+0.04	−0.03	−0.21	−0.46	+0.12	+1.68	−1.00
$W^{(4)}$ (triplet at λ_0)	−2.6	−2.1	−6.4	−20.5	−30.2	+4.9	+0.3

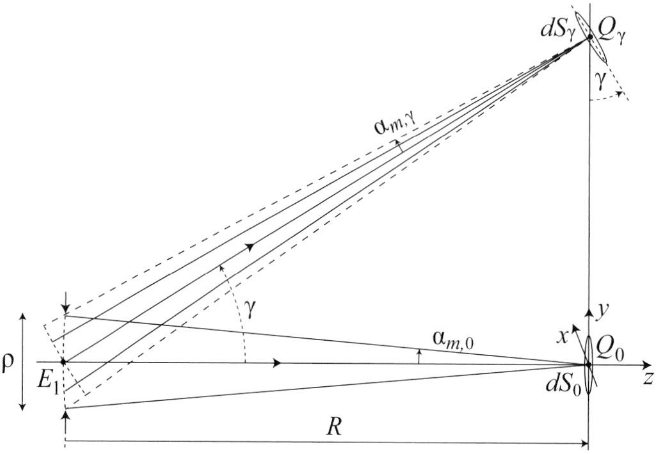

Figure 7.29: Calculation of the image plane irradiance as a function of the field angle γ.

higher-order coefficients we calculate the third-order residual values. For instance, the optimum third-order astigmatic residual and the Petzval sum follow from Eq. (7.5.14). Third-order residual values can be easily introduced in the design as they are just a small percentage of the intermediate aberration values within the lens system. When this example system was designed a century ago, an approximate third-order design with a Petzval sum suggested by experience was used as initial design. The third-order coefficients and finite ray data of carefully selected rays were used as 'merit function' data for further optimisation. By successive linear interpolation and extrapolation, the system was driven to a satisfactory final design, both with respect to imaging quality and manufacturability.

- *Vignetting and relative irradiance in the image plane.*
  Figure 7.28 clearly shows that the off-axis pencil towards the edge of the field is strongly truncated by the two outer doublets. This truncation is done on purpose to prevent strongly aberrated rays from participating in the image formation. This truncation effect can be expressed by a *vignetting* function $V(\gamma_r)$; the argument γ_r is γ/γ_m where γ_m is the maximum field angle. $V(\gamma_r)$ is the ratio of the exit pupil area where no rays pass and the total pupil area. In the case of the Double-Gauss example, the function $V(1) \approx 0.50$ and the irradiance at the edge of the field is at least reduced by a factor of $1 - V(1)$.

A more detailed calculation of the varying irradiance in the image field can now be carried out with the aid of Fig. 7.29. We assume that the light distribution in the exit pupil through E_1 of the optical system behaves as a Lambertian source having a radiance B in units of [Wm^{-2}sr^{-1}]. The energy flow d^2P_0 through an on-axis surface element dS_0 in the image plane is then given by

$$d^2P_0 = B\, d\Omega\, dS_0 \,, \tag{7.5.16}$$

where $d\Omega$ is the solid angle subtended at the point Q_0 by a surface element dS_p on the exit pupil sphere with its centre at Q_0. For the irradiance I_0 at the point Q_0, integrated over the spherical surface of the exit pupil, we use Eq. (4.11.32),

$$I_0 = \frac{dP_0}{dS_0} = B\pi \sin^2 \alpha_{m,0} = \pi B \left(\frac{\rho}{R}\right)^2 . \tag{7.5.17}$$

For an off-axis image point Q_γ a comparable calculation is performed with the projected pupil surface equal to $\pi\rho^2 \cos\gamma$ and the distance $E_1 Q_\gamma$ given by $R/\cos\gamma$. The irradiance on the inclined surface dS_γ with a size equal to dS_0 is given by

$$I_\gamma = \pi B \cos\gamma \left(\frac{\rho}{R/\cos\gamma}\right)^2 . \tag{7.5.18}$$

For the irradiance on an in-plane surface element with size dS_0 at the location of the point Q_γ we multiply I_γ by $\cos\gamma$ to take into account the surface projection effect from dS_γ to $dS_\gamma/\cos\gamma$. The irradiance on the flat image plane at the point Q_γ is then given by

$$I_{Q_\gamma} = I_0 \cos^4 \gamma . \tag{7.5.19}$$

The $\cos^4\gamma$ law for the image-plane irradiance yields a factor of 52% at the extreme corner of the image plane for the Double-Gauss example. If the vignetting function $V(1)$ is equal to 0.50 for this design, the relative irradiance is 26%. This is a low value but it is acceptable in photography. In digital photography, a signal correction can be applied, at the expense of a possible deterioration of the signal-to-noise ratio of the signal and an increased error rate.

- *Image quality.*
  The image quality of the Double-Gauss objective is illustrated in Fig. 7.30. In a) the axial shift z of the astigmatic focal surfaces has been drawn as a function of the field angle γ. According to the third-order astigmatic wavefront coefficients of Table 7.16, the focal surfaces first curve inward towards the objective. For a relative field angle $\gamma_r \approx 2/3$ the higher-order astigmatic terms give rise to an extreme value and for larger field angles the astigmatic surfaces are folded back towards the paraxial image plane. The sagittal surface even goes beyond the paraxial image setting ($z = 0$).

The best image plane setting is a compromise between the image quality in the central region and the steep deterioration at the edge of the image field. As there is an important vignetting effect for the extreme pencils of rays, the image quality deterioration at the edge of the field is reduced. The paraxial image plane setting $z = 0$ gives rise to a geometrical blur δ_{rms} as a function of γ according to Fig. 7.30b. The blur initially increases because of the average defocusing of the astigmatic surfaces and the increasing gap between them. Closer to the edge of the field, the blur slightly decreases because of the smaller focus offset of the astigmatic surfaces. At the field edge, other field aberrations like higher-order coma and astigmatism cause a further increase of the blur value.

The aberrational behaviour of the objective as a function of field angle is further illustrated in Fig. 7.30c by means of the modulation transfer functions in four field positions (paraxial image positions are $y = 0,\ 20,\ 40$ and 62 mm). The y-coordinate of the centroid of the blurred image spot is given on top of each of the *MTF* graphs. We observe that the distortion of the objective is negligibly small for photographic applications where a 1% level can easily be tolerated in the case of visual observation of the printed image. The Double-Gauss lens with distortion values of typically 0.1% is, therefore, suited for photogrammetric use. The *MTF* curves show the dominating influence of the astigmatism and field curvature on the image quality as function of the field angle. The smallest frequency transfer values are found at approximately 2/3 of the field edge; 50% modulation is still achieved at 3 mm^{-1}. With a field diagonal of 124 mm and 10 pixels per mm with 50% modulation depth, the number of 'well-resolved' pixels on the diagonal amounts to approximately 1250 for an $F_\#$ of 4.5. A value well beyond 1000 image pixels guarantees a good imaging performance for this wide-angle objective. Figure 7.30d shows the balance between meridional and sagittal resolution which has been obtained with the (paraxial) image plane choice.

Not surprisingly, the axial chromatic aberration of the objective is very good. This can be explained by the twofold correction means for axial chromatism; first by the positive/negative lens element combination and, secondly, by the introduction of a buried surface in the outer negative elements. Since the objective is highly symmetric with respect to the centre of the stop, the colour dependence of the refraction of the principal rays is compensated in the two objective halves; lateral chromatism is virtually zero!

7.5.4.2 Design Procedure for a Double-Gauss Lens

- *Paraxial design*
  The paraxial design is carried out respecting the symmetry with respect to the central stop. Because of the rather wide-angle requirement it is advisable to have the inverted power layout such that the angles of incidence of the principal ray on the outer elements can be kept small. The extra buried surface in the outer lenses is introduced for aberration and

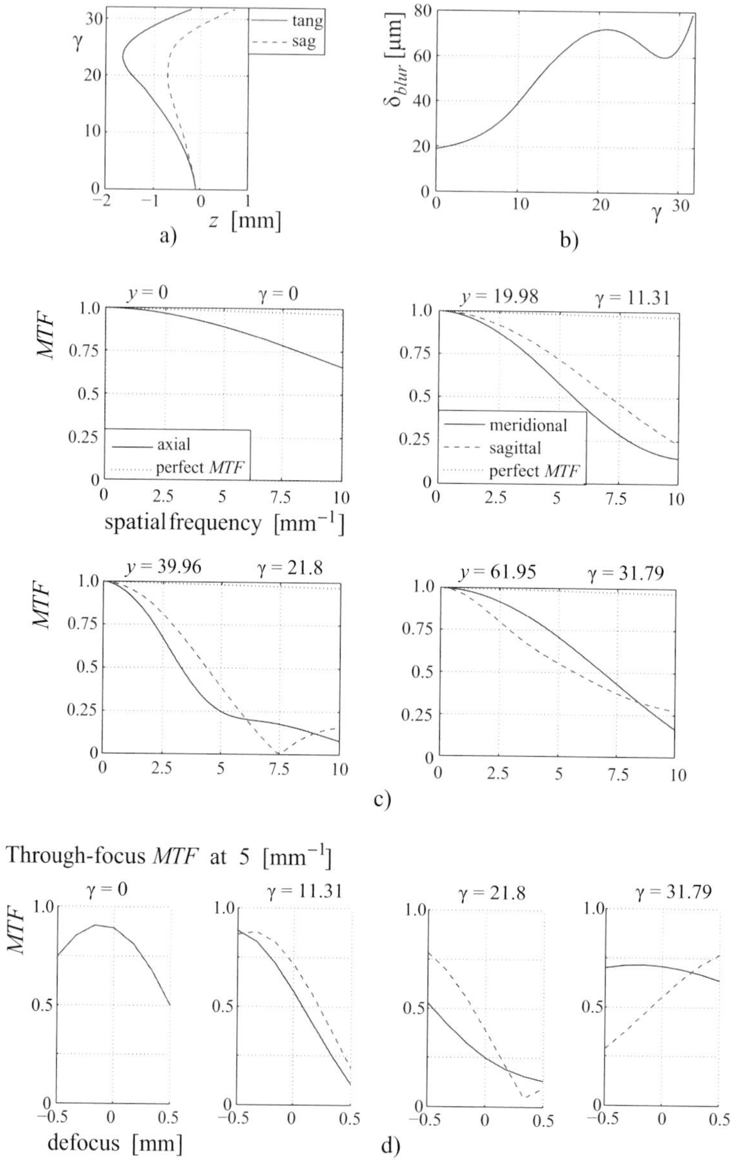

Figure 7.30: Imaging performance of the objective of Fig. 7.28.
a) Axial position of the astigmatic lines of the Double-Gauss objective as a function of the field angle γ. Solid curve: tangential astigmatic line; dashed curve: sagittal line.
b) Geometrical rms blur δ of the focused spot in the image plane.
c) Modulation transfer function *MTF*, 0 to 10 mm^{-1}, for paraxial field positions $y = 0,\ 20,\ 40,\ 62$ mm. In the figure, field positions y have been given in which the influence of distortion has been included.
d) Through-focus *MTF* curves for a spatial frequency of 5 mm^{-1}.

colour correction. The powers of the doublet elements are determined by the correction to zero of the axial chromatism. The trial paraxial design is then scaled to the desired focal length of $f' = 100$ mm.

- *Third-order design*

  It is a wrong strategy to strive for a design with zero third-order aberration coefficients. It is essential that third-order residuals are left in the design to compensate for the higher-order aberrations. In the case of a wide-angle design, the initial inward curving of the astigmatic surfaces needs to be corrected by the appearance of higher-order deformations of the astigmatic surfaces that force them back to the best-focus position in the centre of the image field. Equation (5.7.51) can be used to calculate for each surface these higher-order astigmatic effects that depend on powers like $\tan^4 \theta_1$ and $\tan^6 \theta_1$ of the image-side refraction angle θ_1 at a surface transition. Alternatively, a fifth-order aberration theory can be invoked to quantify these higher-order astigmatic effects. The third-order astigmatism and field curvature are combined in such a way that the folding-back of the astigmatic surfaces is obtained. Without this flattening of the image field the wide-angle operation of the objective would be impossible.

The same aberration balancing between third- and higher-order aberrations is needed for spherical aberration and coma. Since the aperture of the system is outside the region of validity of third-order aberration theory, there are substantial amounts of these third-order aberrations in the intermediate spaces between the lens elements. The third-order values which are encountered at each surface are typically 50 to 100 times larger than the residual value of the total system. For this reason, it is easy to create some small third-order residuals of spherical aberration and coma in image space without significant changes in the optical layout.

- *Finite-ray optimisation*

  In the pre-digital-computer era the final optimisation was carried out with judiciously chosen finite rays. Typically, a central pencil, an intermediate pencil and an extreme pencil were analysed. From each pencil the principal ray, marginal rays in orthogonal sections and a few 'zonal' rays in between were traced. Changes of the aberration values due to changes of the constructional parameters were used to obtain an improved optical system by means of (linear) interpolation. The local aberration change at a surface can be obtained from a differential aberration formula, see Eq. (6.5.4), or from a total aberration formula for a ray transition at an optical surface (see Eq. (5.8.4)). With the advent of the computing power of a modern digital computer, automatic optimisation from a paraxial/third-order pre-design to a satisfactory final design has become feasible. It is even possible to include the results of tolerancing in such an automatic optimisation process. The main problem that the optical designer faces when using an automatic optimisation program is the (implicit) enforcement of contradictory requirements. In such a case stagnation occurs and the current design, the merit function contents and the boundary conditions have to be analysed for the origin of the stagnation. Stagnation also occurs if the degrees of freedom of the design have been exhaustively used. Further improvement of the imaging quality can be achieved if extra optical transitions are made available in the design or when other surface shapes, for instance aspheres, are allowed.

Although the aberration data that were given in the previous subsections were wavefront aberration coefficients, the final optimisation of the Double-Gauss objective is better carried out using the transverse ray aberration components. The performance of such a lens system is very far from the theoretical diffraction limit and the transverse aberration components are more relevant in this case than the wavefront aberration.

7.5.5 A High-definition Flat-field Projection Lens of the Double-Gauss Type

A substantially higher resolution than that of the triplet lens or the Double-Gauss lens is needed in modern applications of projection lenses. We refer to the projection lenses that have to satisfy a high-definition standard, typically 1920 × 1080 pixels for television and computer-generated images ('optical beamer'). To satisfy the high-definition requirement, the projection lens should have a large aperture at the object side where, for instance, a liquid-crystal light modulator (LCD) or a digital micro-mirror device (DMD) is positioned. A larger numerical aperture reduces the diffraction unsharpness in the image and it increases the light collection efficiency in the case of an incoherently illuminated object panel (current pixel sizes of the order of 5 to 10 μm). A practical aperture value is $NA = 0.165$ ($F_\# = 3$). To increase the resolution of a classical Double-Gauss lens (power sequence $+ - - +$) at a larger opening angle, more lens elements are needed. Especially at the side of the object panel where the imaging pencils have a large opening angle, the refraction angles of the rays on the optical surfaces must remain moderate. To this purpose the positive lens in the Double-Gauss section between the central stop and the object panel is split in two elements, one of which is made a cemented doublet to introduce further correction of chromatic error. By making the lens closest to the object a doublet, we obtain an important lever for the lateral colour correction. The negative element of the same section, very close to the stop, is also made a doublet; it is especially effective for axial colour correction. On the small aperture side of the stop, the positive lens is also made a doublet. By means of the doublets and by the choice of the dispersion of the singlets, the axial and chromatic colour correction can both be tuned to virtually zero value. This correction is done in the paraxial domain by balancing the axial and lateral colour contributions from each lens element and from the buried surfaces of the doublets.

Table 7.17. Third-order aberration coefficients of a finite-ray optimised high-definition projection lens (central wavelength λ_0 = 550 nm; the aberration coefficients are in units of μm).

	a_{sp}	a_c	a_{tan}	a_{sag}	a_d	$a_{\lambda,ax}$	$a_{\lambda,lat}$
$W^{(4)}$ (at λ_0 = 550 nm)	−0.35	+1.56	+2.00	−3.57	+17.04	0	0
ΔW (at λ_0 + 100 nm)	+0.17	+0.54	−0.26	−0.22	+1.25	+0.59	−2.35

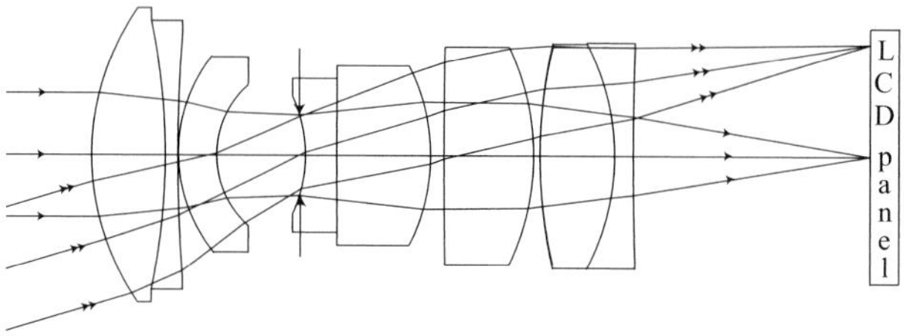

Figure 7.31: A high-resolution projection lens with an LCD-panel as the object (in this reverted layout as image); f'= 60 mm, NA = 0.165, wavelength range: 480 to 650 nm, angular field 2 × 16°.

The correction of third-order aberrations profits from the experience with the triplet and the wide-angle Double-Gauss design. In a similar way as for the triplet design, an offset of the basic astigmatism of second order in the pupil coordinates is beneficial for obtaining a larger effective flat field. Small third-order residuals for spherical aberration and coma are needed which compensate for the (small) higher-order aberrations associated with increased angles of refraction. The third-order aberration coefficients of a finite-ray optimised projection lens are given in Table 7.17 and the layout of the design is given in Fig. 7.31. The avoidance of large angles of incidence implies that the maximum third-order contributions per surface are 50 λ at the most for spherical aberration or coma. The tangential and sagittal astigmatism has final values which are the result of balancing surface contributions which amount to almost 100 λ.

As is common practice in optical design, an optical system is analysed from the far conjugate position towards the small conjugate. The reason for this is that aberrated rays will intersect the image plane in the close neighbourhood of the paraxial position. Analogously, the best image plane position will show only small axial shifts. In the case of a large conjugate distance, a change from convergence to divergence of focused beams may arise and the optimum image plane changes from real to virtual or vice versa. To avoid these undesirable effects, the analysis of the projection system is done in the opposite direction, from the low-aperture beam space to the high-aperture space with the rays propagating back from the receiving screen to the object panel. Following this sequence, the design data of the system are given in Table 7.18 with the long conjugate distance chosen such that the magnification from panel to screen is approximately −100.

The position of the astigmatic focal lines as a function of the field angle is shown in Fig. 7.32a. Thanks to the initial lowest-order astigmatism offset of the design, the best-focus position of the imaging system is bent back to the axial focus for the outer field angles. The maximum offset between tangential and sagittal focal lines is 0.1 mm at approximately 80% of the maximum field angle. Measured along the positive z-axis, the tangential field lines are found beyond the sagittal ones. This is opposite to the standard behaviour of the astigmatic focus of a positive lens. It was introduced by the intentional offset of the lowest order astigmatic wavefront component. Part b) of the same figure shows the geometrical rms blur δ of the system, projected back onto the LCD object panel. The blur is dominated by the astigmatism and the small residual field curvature and reaches its maximum at 2/3 of the field. With f'equal to 60 mm, the allowed object diagonal equals 36 mm. A 50% modulation depth is achieved by the projection system at a spatial frequency of 40 mm^{-1} on the object side. For a cell size of 12 μm we obtain a number of ≈ 3000 well-resolved pixels over the diagonal of the LCD panel or the micro-mirror device; this number of 3000 is adequate for a high-definition rendering of the object scene.

The quality of the imaging system is further illustrated in Fig. 7.33 by means of the modulation transfer function. In Fig. 7.33a, we give the diffraction-based modulation transfer function for the centre of the field and the extreme position

Table 7.18. Optical prescription of the high-definition projection lens. Finite-ray optimisation; glass material data for a wavelength of $\lambda = 550$ nm.

Distance	Index	$V_\#$	Curvature	Diameter	Glass
60000.000	1.000000		0.025644	46.0	
12.000	1.640999	55.3	−0.010055	42.0	SK18A
2.000	1.623655	36.4	0.004102	38.0	F2
0.200	1.000000		0.041739	26.0	
6.000	1.715831	53.4	0.069081	22.0	LAK8
13.347	1.000000		0.000000	13.0	STOP
1.005	1.000000		−0.047450	18.4	
5.000	1.677118	32.2	0.000742	24.0	SF5
15.000	1.640999		−0.033361	28.0	SK18A
2.200	1.000000		0.001009	34.0	
14.412	1.524987	58.5	−0.025773	34.0	B270
1.028	1.000000		0.014010	34.0	
12.000	1.524987	58.5	−0.029336	34.0	B270
3.000	1.623655	36.4	0.002058	30.0	F2
37.689	1.000000		0.000000	36.0	

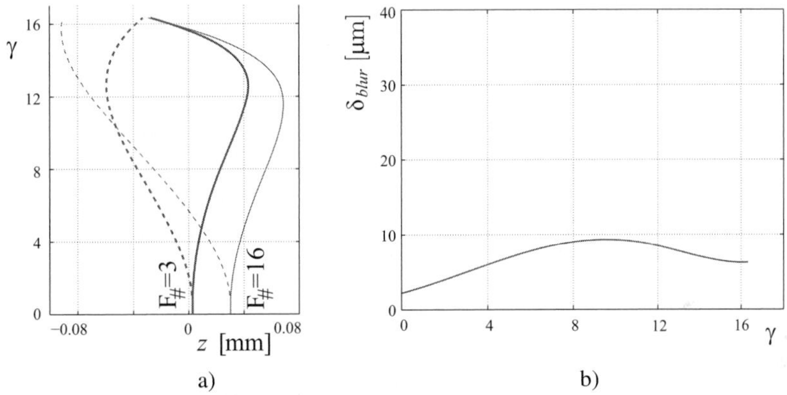

Figure 7.32: a) Axial position of the astigmatic focal lines of the high-resolution lens as a function of the field angle γ (solid curve: tangential astigmatic line, dashed curve: sagittal line).
b) Geometrical blur δ_{blur} of the focused spot in the image plane. The diameter of the stop is 13 mm ($NA = 0.165$) and the vignetting of off-axis pencils of rays is determined by the diameters of the optical surfaces as given in Table 7.18.

($\gamma = 16.5°$). The horizontal axis is in units $2NA/\lambda_0$, where $NA = 0.165$ and $\lambda_0 = 550$ nm. The projection lens is virtually perfect on axis. In the extreme corner of the image field, with a maximum vignetting of 30% of the imaging pencil, a 50% modulation level is achieved at 40 mm^{-1}; the relative irradiance is close to 60%. In Fig. 7.33b the geometrical modulation transfer function has been plotted for the same field angles, after multiplication with the aberration-free diffraction-limited transfer function. It is interesting to observe that for a highly corrected system this adapted geometrical transfer function yields results which are virtually identical to the diffraction-based transfer function in the upper row. Finally, Fig. 7.34 presents the through-focus curves of the (geometrical) modulation transfer function and illustrates the effect of the astigmatism in the oblique beams. At a field angle $\gamma = 7.5°$ the focal shift between the two astigmatic lines amounts to ±75 μm, a value which is confirmed by the axial distance between the astigmatic lines in Fig. 7.32a at the maximum aperture of $F_\# = 3$.

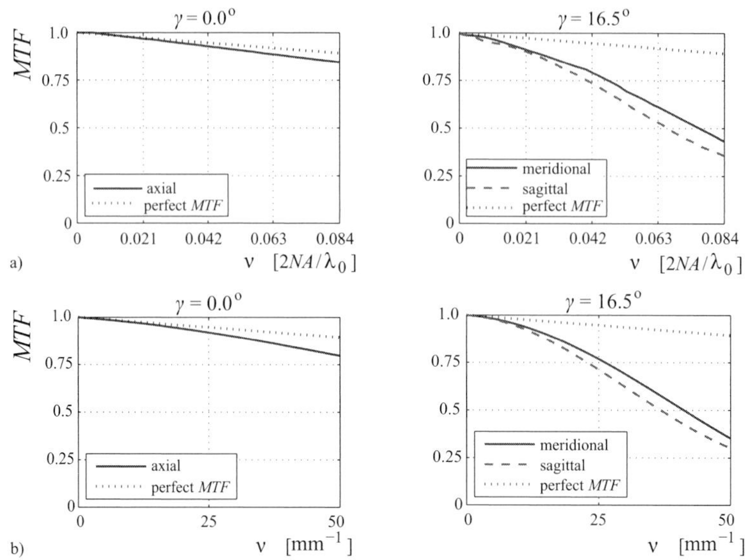

Figure 7.33: Optical performance of the high-resolution projection lens ($f' = 60$ mm) of Fig. 7.31 for various field angles γ.
a) Diffraction-based modulation transfer function.
b) Geometrical modulation transfer function.

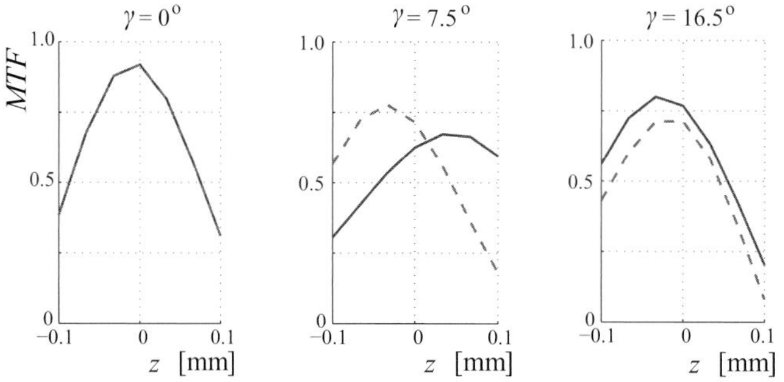

Figure 7.34: Optical performance of the high-resolution projection lens ($f' = 60$ mm) of Fig. 7.31. Through-focus modulation transfer function curves at a spatial frequency of 25 mm^{-1}.

7.5.6 A Wide-angle High-definition Aspheric Projection Lens

The projection lens of Fig. 7.35 illustrates the extra performance that can be obtained when using aspheric optical surfaces. The lens is meant to project a television image produced on the curved object surface. This surface is the inner side of the vacuum exit window of a cathode ray tube (CRT). On purpose this window has been given a curvature such that a hollow object surface is obtained, a favourable feature for compensating the natural field curvature of a positive lens system. The phosphor produces a chromatically narrow-band radiation pattern of which an as large as possible solid angle should be captured by the projection lens. The projection lens must also be compact to allow for a small television projection cabinet. The first requirement asks for a high numerical aperture at the object side and to satisfy the second requirement one needs a small focal length ($f' = 90$ mm) with respect to the object diameter of 130 mm. In b) the compact projection lens L has been placed in a projection box and the magnified image of the tube window T can be observed from the outside on a diffusing screen S.

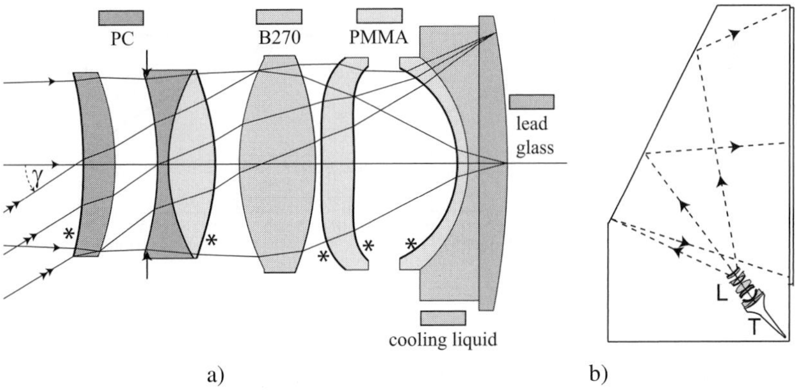

Figure 7.35: a) A wide-angle high-resolution projection lens L for a CRT-display; object diameter 130 mm, $f' = 90$ mm, $NA = 0.42$, three-colour operation, field angle $2 \times 35°$. The cathode ray tube T (not shown in this drawing) is in contact with a cooling chamber filled with a coolant liquid. The optical surfaces marked with an asterisk are aspherics.
b) The projection box with three phosphor tube/lens systems for creating a colour image on the screen S.

Regarding the design of the projection lens, the following observations can be made. Although the curved object surface facilitates the correction of field curvature, the wide-angle system needs a further correction and this is achieved by using an aspheric field flattener. Part of the material of the field flattener is the cooling liquid with a low refractive index. The aspheric shape of the field flattener is close to a paraboloid. At large field angles, a spherically shaped flattening surface would lead to an overcorrection of the field curvature of the system. The distortion that is associated with a field flattener is partly compensated by a pre-distortion of the phosphor image on the tube window. The remaining steep distortion towards the edge of the field is compensated for by the second, double-aspheric lens element with virtually zero power. The system needs a colour correction for the finite chromatic width of the phosphor radiation. This is achieved by means of the plastic doublet in which PMMA and PC play the roles of the classical crown and flint glasses for colour correction. The main power of the projection lens is provided by a lens made of a cheap optical glass. The glass element improves the thermal stability of the system. With the aid of the aspheric correction means, the maximum achievable aperture is 0.42, leading to an $F_{\#}$ of 1.06.

The quality of the projection lens is shown in Fig. 7.36 with the aid of geometrical MTF curves for various field angles. We see that the theoretical diffraction-based MTF of a perfect system is virtually unity over the range of spatial frequencies of interest, up to 10 mm^{-1}. The theoretical cut-off frequency of the lens at $\lambda_0 = 550$ nm amounts to $\approx$ 1500 mm^{-1}. The system is highly aberrated and, like all wide-angle projection lenses, it suffers from vignetting. The outer beams are truncated by 50%. Despite these high aberrations of the imaging pencils up to an rms value of $4\lambda_0$, the projection system is capable of providing a high-resolution image thanks to its large object field diameter. With a 50% modulation depth at 6 mm^{-1}, approximately 1500 pixels are well resolved over the image diagonal. Figure 7.37 shows the through-focus MTF curves at 5 mm^{-1}. It is seen that the best astigmatic focus is maintained on the image surface which gives an optimum balance between the resolution for meridional and sagittal features in the image. The through-focus curves also show the allowable focus offset at the location of the cathode ray tube. We can allow an axial deviation in position of typically 50 μm in air between the imaging part of the lens and the combination of tube and field flattener. Within this limit the high-resolution imaging quality of the projection lens is maintained. The presence of plastic materials requires an extensive mechanical and thermal tolerancing of such a lens.

7.5.7 A Wide-angle Schmidt Projection System

The German opticist Schmidt designed and produced a wide-field stellar camera in 1930. By using the camera in the opposite direction, it can also be used for the projection of enlarged images of an object which is located at the position where the stellar image is formed. The astonishingly good performance of Schmidt's wide-field camera follows from the property of isoplanatic imaging (see Fig. 7.38). The spherical mirror with its vertex at T produces an (aberrated) image from an infinitely distant object on the curved spherical surface through the focal point F. This curved surface has

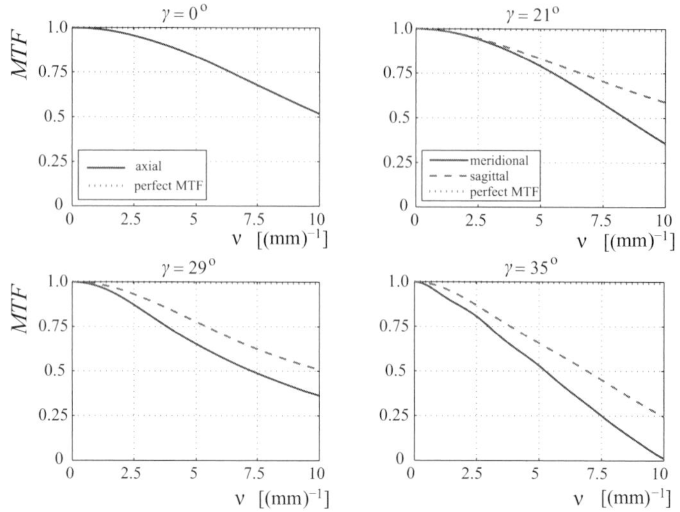

Figure 7.36: Geometrical modulation transfer function for various field angles γ of the aspheric high-definition projection lens ($f' = 90$ mm) of Fig. 7.35. All data have been traced back to the high-aperture space at the curved object surface.

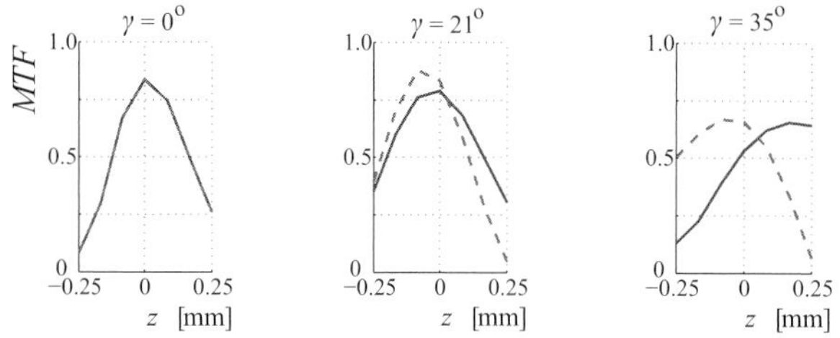

Figure 7.37: Through-focus modulation transfer function curves at a spatial frequency of 5 mm^{-1} for various field angles γ of the aspheric high-definition projection lens ($f' = 90$ mm) of Fig. 7.35.

the same centre of curvature C as the mirror surface. When the diaphragm of the system is put at the plane through C, the angle of incidence on the spherical mirror of the principal ray of an off-axis beam is zero. In terms of the refraction invariant B_m of Eq. (5.3.48), this quantity equals zero and the off-axis aberrations coma and astigmatism are absent. This interesting property is obscured by the large amount of spherical aberration which is present at the entire curved image field in focus. The invention of Schmidt is the nulling of the spherical aberration by means of an auxiliary correcting plate S located at the plane of the diaphragm. The standard way of nulling the spherical aberration is to aspherise the mirror surface, from a sphere to a parabola in the case of stellar observation. Unfortunately, the field performance of a parabola is extremely poor because it does not satisfy the Abbe sine condition. By introducing the aspheric correction at the centre of curvature C, the spherical aberration of the mirror is pre-corrected by means of the so-called 'Schmidt' plate. The limiting imaging defect of the Schmidt camera is the change of the corrective action of the plate for obliquely incident beams. This is a second-order effect which yields astigmatism at larger field angles.

The manufacturing of a large aspheric surface is a time-consuming and expensive process. The correction plate manufacturing by Schmidt used an ingenious procedure based on the deformation of a plane-parallel plate when it is used as a window of a vacuum vessel. In the presence of the appropriate residual pressure, the planar window, fixed by a ring-shaped support, is deformed according to a fourth-order profile. The outer surface of the deformed window plate is then ground and polished to obtain a flat shape. Once at atmospheric pressure, the non-machined side of the flexible plate resumes it flat shape, the other side carrying the required correction profile. Modern aspheric tooling machines avoid such a difficult, iterative process and grind and polish directly the aspheric shape in a Schmidt plate which is made of glass

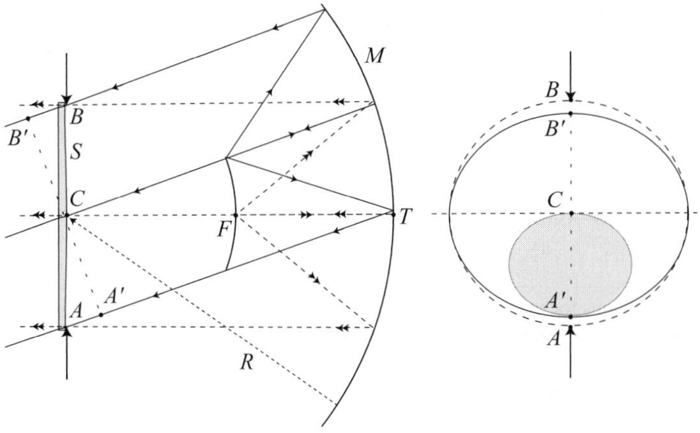

Figure 7.38: a) Concave spherical mirror with vertex at T, provided with an aspheric Schmidt plate S, for the projection of an enlarged image of the object on a curved spherical surface through F.
b) Cross-section of the outgoing beam from the extreme object point, showing the eccentric vignetting effect (grey-shaded area); $NA = 0.67$ ($F_{\#} = 0.75$).

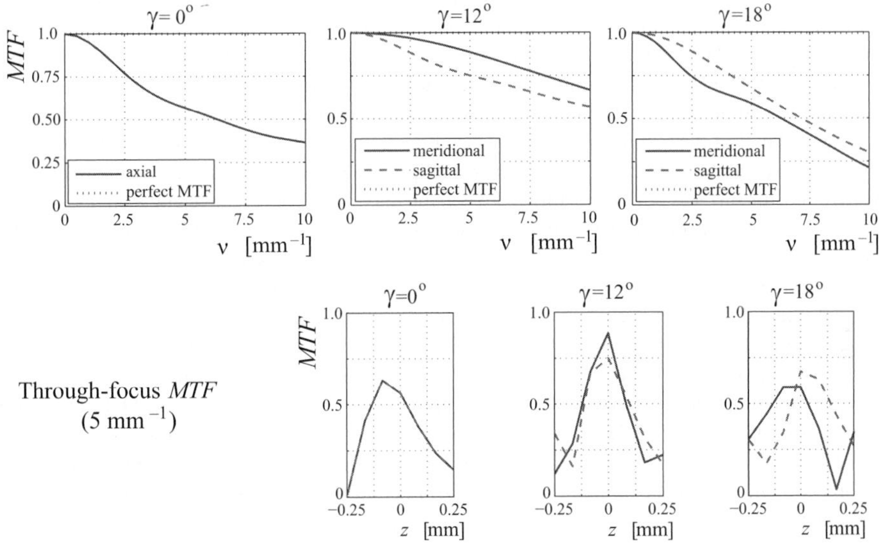

Figure 7.39: MTF curves of the Schmidt projection system for field angles of 0°, 12° and 18°, together with the through-focus curves for the same field angles. $NA = 0.65$, $f' = 100$ mm.

or silica. Plastic moulding or casting is another option to obtain the strongly aspheric correction plate for a high-aperture Schmidt system. Typical aspheric departures up to ±300 μm are found for an aperture of 0.7 and a focal distance f' of the camera equal to 100 mm.

In the case of the projection of magnified images at a large distance (see Fig. 7.38), the projection lens does not need to be of perfect quality and this means that wide-angle use with a small $F_{\#}$ is allowed. The axial beam reaches a full opening angle of almost 90°, corresponding to a numerical aperture of 0.67 ($F_{\#} \approx 0.75$) for the axial beam. Despite the central obstruction, the images produced by a Schmidt system show a superior luminosity as compared to the corresponding lens projection systems. In Fig. 7.39 we have plotted modulation transfer functions for three field angles and for beams with a numerical aperture of $NA = 0.65$ ($f' = 100$ mm).

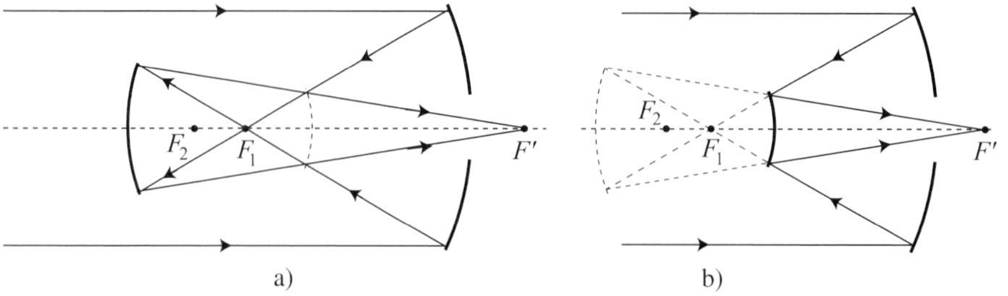

Figure 7.40: The two design options for the axially symmetric two-mirror telescope. With the aid of dashed surfaces and rays we show in each mirror system the alternative design option with equal convergence angles in image space.
a) Gregory-design, with parabolic/elliptical mirrors.
b) Cassegrain-design, with parabolic/hyperbolic mirrors.

7.6 The Astronomical Telescope

In this section we review telescopic systems for stellar observation. The main feature of a telescope is its large pupil cross-section for efficient light gathering from faint objects. The instrument should have perfect imaging quality over a broad wavelength range, leading to a mirror layout to avoid chromatic errors. The angular image field can be limited, typical values range from 1 to 100 mrad. The telescope image does not need to be formed on a flat surface and some distortion can be allowed. Separate optics after the main telescope provide imaging for different chromatic channels and for spectroscopic analysis. When designing the optics of these subchannels, initial field curvature and distortion of the main telescope can be corrected for or their influence is accounted for in the treatment of the image data. In this section we only consider the design of the main telescope. The two basic layouts are those by Gregory and by Cassegrain, both conceived in the second half of the seventeenth century (see Fig. 7.40a and 7.40b, respectively). The Gregory-design, the combination of a concave parabolic primary and a concave elliptical secondary, has an accessible intermediate image. In the case of spherical mirror surfaces which are allowed when the maximum convergence angles in the telescope remain of the order of 0.10 radians, the mirror surfaces can be individually tested with a standard Foucault test. For larger aperture angles, the real focus F_1 is tested first and then the quality of the final focus F' is measured with the secondary mirror put in place. The measuring procedure is comparable for the Cassegrain layout. In the case of a telescope with spherically shaped mirrors, the intermediate focus is aberrated; the second mirror cannot be tested independently with the aid of a point source illumination. It follows also from Fig. 7.40 that the obscuration ratio of the Gregorian telescope, at equal aperture of the final image in F', is larger than that of a Cassegrain telescope. More importantly, the total construction length of the Gregory telescope is larger and this is the reason why large-aperture astronomical telescopes are generally of the Cassegrain type. In what follows we focus on the Cassegrain telescope layout. Two examples of such telescopes using conic sections for stigmatic imaging are given; the third example is a design using general aspheric surfaces to achieve an aplanatic telescopic system. The final example is a three-mirror layout which is applied in virtually all modern telescopes with a large angular field. An exhaustive treatment of telescope optics and manufacturing issues is found in **[368]**.

7.6.1 Paraxial Optics

The paraxial design of a telescope, see Fig. 7.41, starts from some basic requirements like the pupil diameter, the construction length L and the obscuration ratio ϵ. The choice of the diameter $D = 2y_{1,m}$ of the primary mirror is primordial; the construction length L strongly influences the telescope weight and the dome size. The construction length is optically determined by the distance between the vertices $T_2 T_1$ of the primary and secondary mirror and by the back focal length $f'_b = T_2F'$. This latter distance should be larger than the optical construction length L to allow a sufficiently large clearance between the focal plane optics and the primary mirror. A very short construction length requires a large aperture in the primary image formation with high demands on mechanical stability. The aperture of the primary imaging does not exceed 0.25 in practical designs. The obscuration ratio ϵ should be kept small for two reasons. The light transmission has to be large and the telescope resolution should not be impaired by the extra ringing of the diffraction image at large obscuration values. A typical obscuration value is 0.30 or less.

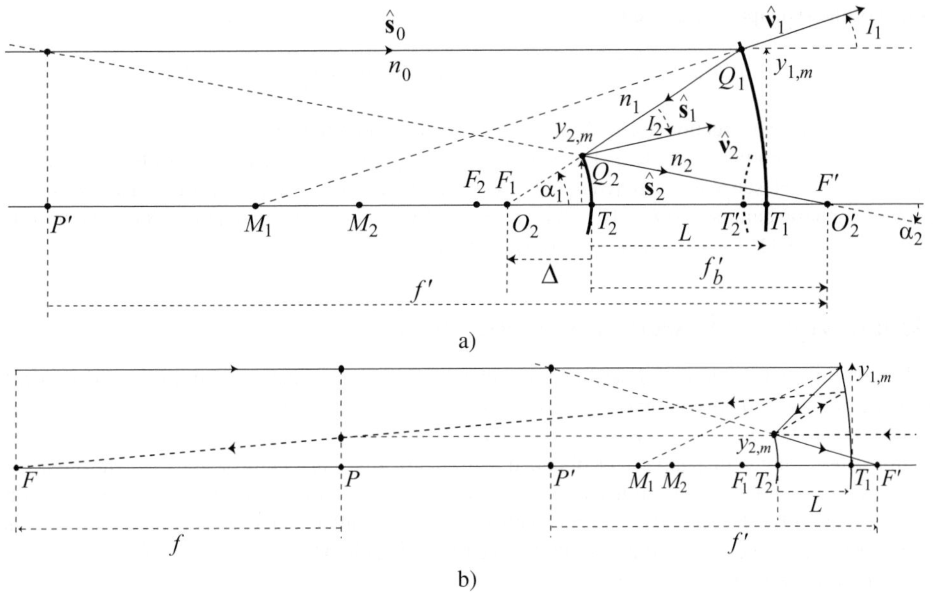

Figure 7.41: Paraxial layout of a two-mirror telescope.
a) Basic design parameters are the construction length L, the back focal length f_b' and system focal length f'. The obscuration ratio ϵ is given by $y_{2,m}/y_{1,m}$.
b) Position of the principal planes P and P' and the focal planes F and F' in the Cassegrain layout, illustrating the telephoto effect.

The paraxial curvatures and the imaging properties of the telescope are now calculated with the aid of a convenient set of telescope parameters, L, f_b' and f', together with the diameter D of the primary. The distance T_2F_1, denoted by Δ in the figure, is negative for a Cassegrain telescope and positive for a Gregory design. Δ can be expressed in the three basic parameters L, f_b' and $f'$by means of the obscuration ratio, The obscuration value ϵ is given by

$$\epsilon = \frac{-\Delta}{L-\Delta} = \frac{f_b'}{f'}, \qquad \Rightarrow \Delta = \frac{Lf_b'}{f_b'-f'}. \tag{7.6.1}$$

The power K_1 of the primary mirror is given by $K_1 = -2c_1 = -1/f_1'$. From the imaging equation for the first mirror we find

$$c_1 = \frac{1}{2(\Delta - L)} = \frac{f_b'-f'}{2Lf'}. \tag{7.6.2}$$

The curvature c_2 of the secondary mirror with power $K_2 = 2c_2$ follows from the imaging equation applied to this mirror with the object in $F_1 = O_2$ and the image in $F' = O_2'$:

$$c_2 = \frac{1}{2f_b'} + \frac{1}{2\Delta} = \frac{L+f_b'-f'}{2Lf_b'}. \tag{7.6.3}$$

The total power K of the telescope is given by

$$K = \frac{1}{f'} = K_1 + K_2 - LK_1K_2 = \frac{\Delta}{(\Delta - L)f_b'} = \frac{\epsilon}{f_b'}. \tag{7.6.4}$$

The paraxial matrix M_t of the telescope, referenced to the vertices T_1 and T_2 in object and image space, respectively, is given by

$$M_t = \begin{pmatrix} B & -A \\ -D & C \end{pmatrix} = \begin{pmatrix} 1-LK_2 & -K_1-K_2+LK_1K_2 \\ L & 1-LK_1 \end{pmatrix}. \tag{7.6.5}$$

For the position of the principal planes we find

$$s_P = T_1P = -n_0\frac{B-1}{K} = \frac{LK_2}{K}\,, \qquad s_{P'} = T_2P' = n_2\frac{C-1}{K} = -\frac{LK_1}{K}\,, \tag{7.6.6}$$

where, by convention, the construction length L of the telescope is chosen to be positive.

The distance PP' between the two principal planes is $-L(K + K_1 + K_2)/K$. The position of focal points and principal planes of the Cassegrain telescope is illustrated in Fig. 7.41b. In this section, as an example for an astronomical telescope, we use the following basic parameters: $L = 16.426$ m, $f'_b = 19.575$ m, $f' = 69.571$ m and $D = 8$ m. The mirror curvatures are then given by $c_1 = -0.0218749\ \mathrm{m}^{-1}$ and $c_2 = -0.0522021\ \mathrm{m}^{-1}$.

7.6.2 Design with Conic-section Aspheric Surfaces

The initial designs for optical telescopes used a single imaging step with a parabolic primary mirror or a two-mirror telescope (Cassegrain-type) which combined a parabolic primary and a hyperbolic secondary mirror with an intermediate stigmatic virtual image. For elongated telescopes with small opening angles of the imaging pencils and small transverse dimensions, diffraction-limited imaging can be obtained with spherically shaped mirror surfaces. The range within which spherical surfaces are permitted follows from an approximate calculation of the Seidel spherical aberration of the primary mirror. The contribution by the secondary mirror, with its ϵ times smaller diameter, is substantially less and has the opposite sign. For our coarse estimate, this contribution can be left aside. Using Eq. (5.3.38) for the primary mirror we have for its spherical wavefront aberration at the rim,

$$W_{S,1} = \left(\frac{2c_1^3 + 16a_{4,1}}{8}\right) y_{1,m}^4\,, \tag{7.6.7}$$

where $a_{4,1}$ is the fourth-order aspheric coefficient of the primary surface, c_1 the paraxial curvature of the surface and $D = 2y_{1,m}$ the diameter of the primary mirror. By choosing, for instance, $a_{4,1} = -c_1^3/8$, a parabolic shape is obtained and the spherical aberration is reduced to zero. We approximate the curvature of the primary by $c_1 \approx 1/3L$ and require that the magnitude of the third-order spherical aberration of the spherical mirror ($a_{4,1} = 0$) does not exceed one wavelength. This upper value for spherical aberration is given in Subsection 7.4.1 and guarantees that the imaging quality of a system is 'just diffraction-limited'. The one-wavelength limit for spherical aberration is confirmed by substituting the just diffraction-limited variance of a wavefront ($\lambda_0^2/180$) in Eq. (5.2.44). A spherical mirror with this level of on-axis wavefront aberration does not necessarily need a fourth-order aspheric correction term. Substitution of $NA_p = 2y_{1,m}c_1$ in Eq. (7.6.7) yields the condition

$$(NA_p)^3\, y_{1,m} \le 32\lambda_0\,, \tag{7.6.8}$$

or, in terms of D and L,

$$D^4 \le 1728\lambda_0\, L^3\,. \tag{7.6.9}$$

The white area in Fig. 7.42 defines the telescope length and diameter combination that allows spherical mirror surfaces to be used. From Eq. (7.6.8) it follows that the aperture NA_p of the primary mirror should not exceed 0.013 ($\lambda_0 = 550$ nm) in the case of an 8 metre diameter telescope. Equation (7.6.9) shows that this would lead to an extremely large construction length of at least 160 metre (see also Fig. 7.42). For small-sized telescopes with a value of NA_p equal to 0.05, the diameter of the primary should not exceed 28 cm, a typical value for an amateur astronomical telescope. The corresponding construction length amounts to typically 2 m. All intermediate choices with shorter construction length or larger diameter require aspheric mirror surfaces. The manufacturing capabilities in the seventeenth up to the nineteenth century were not adequate for producing aspheric mirrors up to the diffraction limit with figure errors below 0.1 μm. For this reason, extremely long telescope tubes and open telescopes were used.

We now briefly discuss these conic-section-based aspheric mirror designs because it turns out that they are not capable of good imaging over an extended image field. For a surface with rotational symmetry around the z-axis with as cross-section a conic section, the axial coordinate z is commonly described by

$$z_c(x,y) = \frac{c(x^2+y^2)}{1+\sqrt{1-(1+\kappa)c^2(x^2+y^2)}}\,. \tag{7.6.10}$$

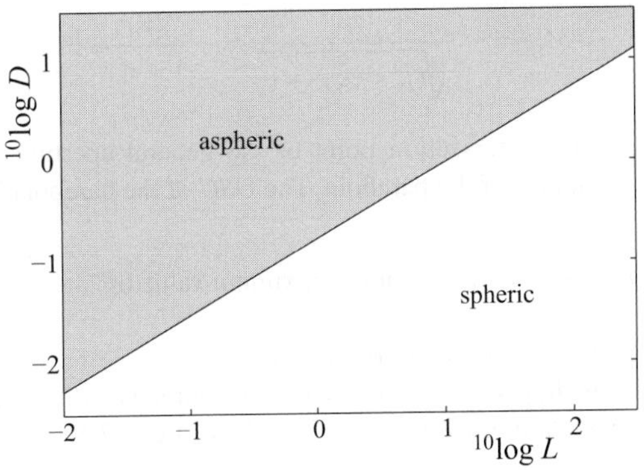

Figure 7.42: The maximum telescope diameter D as a function of the construction length L of a Cassegrain telescope with spherical mirrors having at least 'just diffraction-limited' imaging quality (white area).

Possible higher-order terms of the form $(x^2+y^2)^n$ $(n>2)$ are used to describe aspheric surfaces having a more general shape (see Eq. (5.7.3)). In the case of a parabolic cross-section, the conic constant κ equals -1. Elliptical cross-sections have $\kappa > -1$ with $\kappa = 0$ the special case of a spherical surface. For hyperbolic cross-sections, $\kappa < -1$.

For the two-mirror system in Fig. 7.41, the second mirror images the perfect virtual object O_2 to the final focus O_2'. The object distance T_2O_2 is given by Δ, negative in the figure; the image distance is f_b'. The curvature of the hyperbolic cross-section and its conic constant depend on the total distance O_2O_2' and the distance $O_2T_2 = -\Delta$ from the object point to the mirror surface. After some algebra we obtain for the z-coordinate of the hyperbolic cross-section, with its origin at T_2,

$$z_c(x,y) = \frac{-2a(x^2+y^2)/(b_2-a^2)}{1+\sqrt{1-(1-b^2/a^2)\{4a^2/(b^2-a^2)^2\}(x^2+y^2)}}\,, \tag{7.6.11}$$

where $T_2T_2' = a = f_b' + \Delta = f_b'(f_b' - f' + L)/(f_b' - f')$ and $O_2O_2' = b = f_b' - \Delta = f_b'(f_b' - f' - L)/(f_b' - f')$. O_2 and O_2' coincide with the two foci of the conic section. For the curvature and the conic constant of the hyperbolic cross-section we obtain

$$\begin{aligned} c_2 &= -\frac{2a}{b^2-a^2} = \frac{\Delta + f_b'}{2\Delta f_b'} = \frac{L+f_b'-f'}{2Lf_b'}\,,\\ \kappa_2 &= -\frac{b^2}{a^2} = -\frac{(f_b'-\Delta)^2}{(f_b'+\Delta)^2} = -\left(\frac{f'+L-f_b'}{f'-L-f_b'}\right)^2 . \end{aligned} \tag{7.6.12}$$

Part of the second branch of the hyperbolic cross-section through T_2' has been depicted by a dashed curve in Fig. 7.41. The z-coordinate of this second branch, with the origin located at T_2', is obtained from Eq. (7.6.12) by changing the sign of the paraxial curvature. The conic constant κ_2 remains unchanged. The expressions according to Eqs. (7.6.11) and (7.6.12) are also valid for other types of conic cross-sections. The elliptical cross-section is obtained by selecting positive values of Δ, making $a > b$. The circular cross-section follows from the limiting case $b = 0$. The parabola can be obtained as a limiting case of the ellipse when both a and b tend to infinity with $a - b = 2\Delta$ for $\Delta > 0$. The paraxial curvatures c_p of the two possible parabolic cross-sections are given by $c_p = \pm 1/(2\Delta)$.

7.6.3 Conic Sections and Offence against the Sine Condition (*OSC*)

The imaging with the aid of aspheric surfaces with conic cross-section suffers from a severely restricted image field diameter. In the following we briefly analyse the offence against the sine condition of such imaging steps, for instance, those using a parabolic mirror for imaging from infinity and a hyperbolic mirror for producing the accessible final image. For a parabolic mirror with its object plane at infinity, the *OSC* follows from Eq. (4.8.7) and is given by $1 + y_1/(M_{1,y}f')$, where y_1 is the height of incidence of the ray and $M_{1,y}$ the y-direction cosine of the ray after reflection at the mirror. With the aid of Fig. 7.41, using finite ray coordinates and ray directions, we have

$$M_{1,y} = -\frac{y_1}{\sqrt{(f_1' - z_1)^2 + y_1^2}} = -\frac{2y_1 c_1}{1 + c_1^2 y_1^2}, \tag{7.6.13}$$

where (y_1, z_1) are the coordinates of the intersection point of the general aperture ray with the mirror surface and $f_1' = 1/(2c_1)$ is the image-side focal distance of the parabola. The *OSC* of the parabola is then given by

$$OSC_p(y_1) = -(c_1 y_1)^2, \qquad \text{with a maximum value of} \quad -\frac{(NA_p)^2}{4}, \tag{7.6.14}$$

where $NA_p = y_{1,m}/f_1'$ denotes the aperture of the beam focused at F_1.

Using the result of Eq. (4.8.6) for the hyperbolic mirror with finite conjugate distances, we can calculate the direction cosines at the hyperbolic mirror with the aid of the distances a and b of Eqs. (7.6.11) and (7.6.12) and obtain after some algebra,

$$OSC_b(y_2) = 1 - \left[\frac{\left(\frac{1}{a+b}\right)^2 \left\{\left[b + a\sqrt{1 + \frac{4y_2^2}{b^2 - a^2}}\right]^2 + 4y_2^2\right\}}{\left(\frac{1}{b-a}\right)^2 \left\{\left[-b + a\sqrt{1 + \frac{4y_2^2}{b^2 - a^2}}\right]^2 + 4y_2^2\right\}}\right]^{1/2}, \tag{7.6.15}$$

where y_2 is the height of incidence of a general aperture ray on the hyperbolic mirror.

The contributions to the *OSC* of the parabola and the hyperbola have opposite sign in the telescope and can be balanced to obtain zero. An example of such a balancing is shown in Fig. 7.43a for an 8 metre telescope with system focal length f' equal to 69.571 m. The approximate paraxial heights of incidence on the secondary mirror were used in the calculation of the *OSC*. The compensation has been made perfect at the rim of the pupil. In the intermediate aperture zone, a residual is present which can be kept very small. In the case of the telescope example it amounts to $2.6 \cdot 10^{-5}$. If the heights of intersection y_2 of the finite rays are used in the calculation of the system *OSC* (Fig. 7.43b), an offset in *OSC* correction with respect to a) is visible. By a further slight adjustment of the conic constant κ_2 of the secondary mirror, b) can be adjusted such that balancing of the *OSC* over the telescope aperture is obtained anew.

Not visible from the *OSC* graphs is the mismatch in spherical aberration compensation that occurs when the hyperbolic shape is substantially changed. In the example of Fig. 7.43, κ_2 had to be changed from the original value −3.9, corresponding to stigmatic imaging on-axis, to −4.9 to obtain zero coma. Such a huge change in conic constant has the result that the paraxial conjugate positions do not coincide with the new foci for finite rays of the adapted hyperbola. An enormous amount of spherical aberration is the result of this hyperbola deformation.

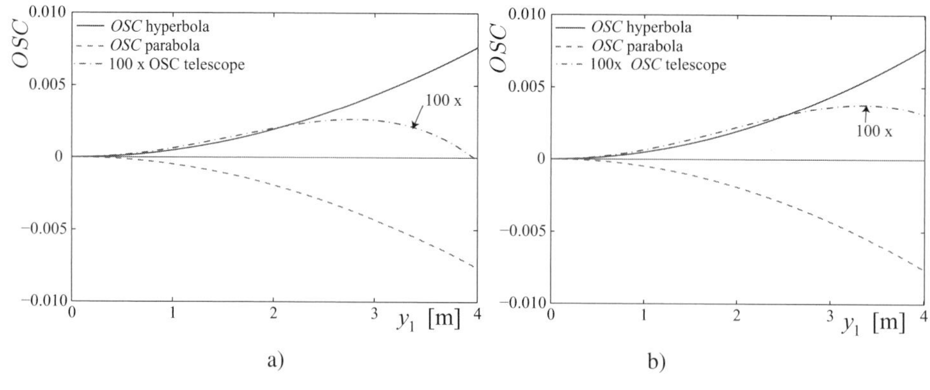

Figure 7.43: Offence against the Abbe sine condition (*OSC*); 8 metre telescope with $f' = 69.571$ m, $NA = 0.0575$.
a) *OSC* calculated with the aid of the height y_1 in the aperture of the primary parabolic mirror and the paraxial height on the secondary hyperbolic mirror. The sum of both values (100 times magnified) is given by the dot-dashed curve.
b) Same legend but now after calculation of the exact height of incidence y_2 of the rays on the secondary hyperbolic mirror.

7.6.4 The Ritchey–Chrétien Telescope Design

The only solution to simultaneously reducing spherical aberration and coma is the introduction of a new degree of freedom in the telescope layout by abandoning the parabolic shape of the primary mirror. The parabolically shaped primary mirror for large astronomical telescopes was routinely used until the 1920s. Such a shape of the primary has an important advantage in mirror metrology because of the stigmatic intermediate image. To increase the angular field of the astronomical telescope, the parabolic primary had to be replaced by a more general conic section to obtain simultaneous compensation of spherical aberration and coma in a two-mirror telescope. The theoretical analysis of such a new design yielding aplanatic imaging quality was given in a groundbreaking publication by Schwarzschild **[300]**, as early as 1905. The design and manufacturing of a real-world telescope according to Schwarzschild's analysis was carried out by the astronomer Ritchey and the optical designer Chrétien. A first telescope according to their design was constructed in the 1920s. Aplanatic telescopes are now generally called Ritchey–Chrétien telescopes; a more justified nomination would have been 'Schwarzschild–Ritchey–Chrétien' telescope.

With reference to Fig. 7.41, we calculate the third-order spherical aberration and coma of the two-mirror telescope using the third-order formulas of Eqs. (5.3.49) and (5.3.50). For the transverse magnification of the two elements we have

$$m_{T,1} = 0\,, \qquad m_{T,2} = \frac{f'_b}{n_2}\frac{n_1}{T_2F_1} = \frac{f'-f'_b}{L}\,, \tag{7.6.16}$$

where the refractive indices are given by $n_0 = -n_1 = n_2 = 1$. The diaphragm is determined by the rim of the primary and the pupil magnification values are given by

$$m_{T,p,1} = +1\,, \qquad m_{T,p,2} = \frac{s'_{p,2}}{n_2}\frac{n_1}{s_{p,2}} = \frac{1}{1-2c_2L} = \frac{f'_b}{f'-L}\,, \tag{7.6.17}$$

where $s_{p,2}$ and $s'_{p,2}$ are the pupil object and image distances for the secondary mirror. Other quantities that are encountered in the third-order formulas are

$$m_{T,2} - m_{T,p,2} = \frac{f'}{f'-L}(m_{T,2}-1), \quad 1 - m_{T,p,2} = \frac{L}{f'-L}(m_{T,2}-1), \quad 1 + m_{T,p,2} = \frac{f'+f'_b-L}{f'-L}\,. \tag{7.6.18}$$

For the apertures $n_1\alpha_1$ and $n_2\alpha_2$ of the pencils after the primary and secondary mirror, respectively, we write

$$(n_1\alpha_1) = m_{T,2}\,(n_2\alpha_2)\,. \tag{7.6.19}$$

The paraxial powers are given by $K_1 = -2c_1$ and $K_2 = 2c_2$. The axial coordinate of a conic section follows from the series expansion

$$z(x,y) = \frac{c}{2}(x^2+y^2) + \frac{(1+\kappa)c^3}{8}(x^2+y^2)^2 + \frac{(1+\kappa)^2c^5}{16}(x^2+y^2)^3 + \cdots\,, \tag{7.6.20}$$

truncated after the second term in third-order Seidel theory. The aspheric coefficients $a_{4,1}$ and $a_{4,2}$ are thus given by $\kappa_1 c_1^3/8$ and $\kappa_2 c_2^3/8$. Using these paraxial data and aspheric coefficients, we choose a positive lateral field coordinate $p_{2,m}$ in the focal plane through F' and a maximum positive ray angle $\alpha_{2,m}$. For the surface contributions, in terms of this maximum ray angle $\alpha_{2,m}$ and the outer field coordinate $p_{2,m}$, we obtain at the rim of the pupil for the spherical aberration,

$$W_{S,1} = \frac{(1+\kappa_1)\,m_{T,2}^4}{64\,c_1}\,(\alpha_{2,m})^4\,, \tag{7.6.21}$$

$$W_{S,2} = -\frac{(1-m_{T,2})^2}{64\,c_2}\left\{(1+m_{T,2})^2 + \kappa_2\,(1-m_{T,2})^2\right\}\,(\alpha_{2,m})^4\,. \tag{7.6.22}$$

For the comatic wavefront aberration we have

$$W_{C,1} = -\frac{m_{T,2}^2}{4}\,p_{2,m}(\alpha_{2,m})^3\,, \tag{7.6.23}$$

$$W_{C,2} = \frac{1}{8\,(m_{T,2}-m_{T,p,2})}\left[(1-m_{T,2})^2(1+m_{T,2})(1+m_{T,p,2}) + \kappa_2\,(1-m_{T,2})^3(1-m_{T,p,2})\right]\,p_{2,m}(\alpha_{2,m})^3\,. \tag{7.6.24}$$

If the diaphragm is coincident with the primary mirror, the coma of this surface is independent of its aspheric departure. For this reason, the aspheric departure of the secondary mirror is the only free parameter to influence the coma of the telescope. We cancel the coma contribution from the two mirrors by choosing the value of the conic constant κ_2 of the secondary mirror according to,

$$\kappa_2 = \frac{2m_{T,2}^2(m_{T,2} - m_{T,p,2}) - (1 - m_{T,2})^2(1 + m_{T,2})(1 + m_{T,p,2})}{(1 - m_{T,2})^3(1 - m_{T,2,p})} . \tag{7.6.25}$$

Once κ_2 has been chosen, the aspheric departure of the primary mirror is used to reduce to zero the spherical aberration of the system and we find,

$$\kappa_1 = -1 + \frac{c_1 \, (1 - m_{T,2})^2}{c_2 \, m_{T,2}^4} \left\{(1 + m_{T,2})^2 + \kappa_2 \, (1 - m_{T,2})^2\right\} . \tag{7.6.26}$$

We eliminate the parameters c_1, c_2 and $m_{T,p,2}$ from Eqs. (7.6.25) and (7.6.26) and express κ_1 and κ_2 in terms of f', f'_b, L and $m_{T,2}$. After some manipulation we obtain

$$\begin{aligned}\kappa_1 &= -1 - \frac{2f'_b}{L\, m_{T,2}^3} ,\\ \kappa_2 &= -\left(\frac{m_{T,2}+1}{m_{T,2}-1}\right)^2 - \frac{2f'}{L\,(m_{T,2}-1)^3} .\end{aligned} \tag{7.6.27}$$

The expressions above can be found in Schwarzschild **[300]**, Part II (On the theory of reflecting telescopes), where the eikonal version of the third-order aberration theory was applied. The formulas can also be found in **[368]**; the different sign convention of m_2 in this book should be taken into consideration. The expressions apply equally well to the Gregory and to the Cassegrain telescope. The principal difference is the change of sign of the lateral magnification of the secondary mirror (negative for the Gregory and positive for the Cassegrain telescope). The expressions for κ_1 and κ_2 can also be written in terms of the basic system parameters only,

$$\begin{aligned}\kappa_1 &= -1 - \frac{2f'_b L^2}{(f' - f'_b)^3} ,\\ \kappa_2 &= -\frac{(f' - f'_b + L)^2 (f' - f'_b - L) + 2f' L^2}{(f' - f'_b - L)^3} .\end{aligned} \tag{7.6.28}$$

To obtain the enlarged image field for the Cassegrain telescope, the primary mirror of the Ritchey–Chrétien version goes from parabolic to slightly hyperbolic. For the telescope layout of Fig. 7.41 with $m_{T,2} = +3.044$ and $m_{T,p,2} = 0.3683$, the conic constant κ_1 becomes -1.0845. The hyperbolic secondary mirror undergoes an important change. In the classical Cassegrain with parabolic primary, we have $\kappa_2 = -3.915$; the value for the Ritchey–Chrétien version equals -4.907.

For the manufacturing and measurement of the primary hyperbolic mirror of a Ritchey–Chrétien telescope, the spherical aberration at the primary focus needs to be compensated for by means of an auxiliary optic, the so-called null-corrector. The pathlength change δW_1 at the rim of the mirror ($y = y_{1,m}$) in the case of a change from parabolic to hyperbolic cross-section follows to a good approximation from Eq. (6.5.6). With the aid of Figs. 6.20 and 7.41 we obtain

$$\delta W_1 \approx -\frac{c_1^3\, y_{1,m}^4 \cos^2 I_1}{4} \left(1 + c_1^2 y_{1,m}^2\right) \delta\kappa_1 , \tag{7.6.29}$$

where $\delta\kappa_1$ is the change in conic constant and I_1 the angle of reflection (equal to I_0, the angle of incidence) at the (nominal) parabolic mirror. Using the values for the primary mirror of the Ritchey–Chrétien telescope example, $y_{1,m} = 4$, $c_1 = -0.021875$, $\delta\kappa_1 = -0.0845$ and $\sin(2I_1) = 0.175$ ($F_\# \approx 2.8$) we find a fourth-order spherical aberration term of -102.1 wavelengths and a sixth-order term of -0.8 wavelengths ($\lambda = 0.55$ μm). Figure 7.44 shows this spherical aberration (wavefront domain) at the primary focus of the Ritchey–Chrétien telescope, obtained by ray-tracing. The peak-to-valley aberration at the best intermediate focus amounts to approximately 14 μm ($\lambda = 550$ nm). From the fourth-order Zernike polynomial for spherical aberration (see App. D, Table D.1) we deduce that the best focus is found with the aid of a quadratic defocusing term $-6\rho^2$ which accompanies the fourth-order spherical aberration term given by $6\rho^4$. The peak-to-valley range of the polynomial equals 1.5, which means that this value is one fourth of the amplitude of the classical Seidel spherical aberration term with its ρ^4-dependence . It follows that the approximate calculation according to Eq. (7.6.29) is well confirmed by the result of exact ray-tracing. The rms blur at the best focus of the hyperbolic primary is 265 μm. In terms of wavefront aberration and the diffraction limit of 0.07 λ in rms value, the primary focus exceeds the diffraction

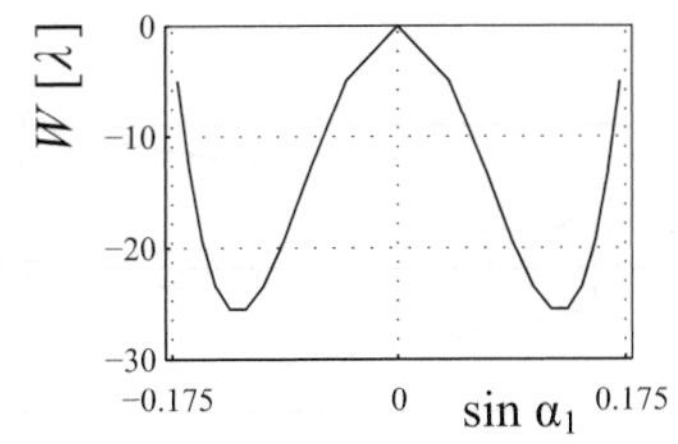

Figure 7.44: Spherical aberration at the primary focus of the RC telescope as a function of the sine of the aperture angle α_1 of a ray (result from finite ray-tracing).

limit by a factor of more than 100 in the centre of the visible wavelength range. The measurement of the surface shape of the primary mirror has to be carried out with the aid of a correcting optical system, very well calibrated, that exactly compensates for the spherical aberration introduced by the hyperbolic mirror.

Much experience has been acquired during the past 100 years to cope with this primary mirror measurement problem, the main drawback of the Ritchey–Chrétien telescope. It was in the case of the Hubble space telescope that this measurement problem was not adequately dealt with by the telescope manufacturer. One lens component of a three-component nulling optic was misplaced along the axis by more than a millimetre and this gave rise to a fourth-order spherical aberration offset at the rim of the Hubble primary mirror of 4.4 μm, on top of the spherical aberration of 0.9 μm for the nominal hyperbolic shape of the primary. The aberration offset corresponds roughly to a figure error at the edge of the primary mirror of 2.2 μm, the mirror being flatter than required. The offset in spherical aberration is eight times larger than the diffraction-limit at the centre of the visible spectrum. At the shortest wavelength of the instrument, 150 nm, the diffraction limit was even exceeded by a factor of thirty. The rms blur at the primary focus had become 20 μm instead of the 3 μm for the nominal hyperbolic primary. Unfortunately, these effects passed unnoticed before the launching of the spacecraft in 1990 and spectacular rescue and repair actions were needed before diffraction-limited operation of the telescope was finally obtained.

7.6.5 Field performance of an astronomical telescope

Having achieved (third-order) aplanatic imaging of the two-mirror telescope by the appropriate choice of the conic constants κ_1 and κ_2, the aberration that limits the useful image field is astigmatism. We assume that field curvature and distortion can be permitted in the practice of astronomical observation, for instance by using a field flattener and by calibration of the lateral image position with the aid of some test objects. In Fig. 7.45a we show the aplanatic telescope and in b) and c) the wavefront cross-sections which are obtained for a pencil of rays towards the border of the practical image field, typically 100 mm in diameter.Any asymmetry due to coma is absent in the meridional section (see Fig. 7.45b). After a correction of the axial focus-setting by 0.19 mm towards the telescope, the resulting wavefront is purely astigmatic (solid curves). The rms OPD equals 0.069 λ at 550 nm, a value which is just within the diffraction limit of aberration. The total angular field of the telescope is 1.5 mrad (or 5′).

To enlarge the angular field, the astigmatism needs to be reduced. We use again the third-order aberration formulas and obtain with the aid of Eq. (5.3.51) the astigmatic wavefront aberration contributions from the primary and secondary mirror,

$$W_{A,1} = c_1\, p_{2,m}^2 (\alpha_{2,m})^2 \,, \tag{7.6.30}$$

$$W_{A,2} = -\frac{c_2}{4}\left(\frac{1-m_{T,2}}{m_{T,2}-m_{T,p,2}}\right)^2 \left[(1+m_{T,p,2})^2 + \kappa_2\,(1-m_{T,p,2})^2\right] p_{2,m}^2 (\alpha_{2,m})^2 \,. \tag{7.6.31}$$

We use the κ_1 and κ_2 coefficients of the aplanatic telescope and impose $W_{A,1} + W_{A,2} = 0$. Eliminating c_1, c_2, $m_{T,2}$ and $m_{T,p,2}$ from the expressions, we obtain a surprisingly simple condition for zero astigmatism,

$$L = 2f' . \tag{7.6.32}$$

The credit for obtaining this zero-astigmatism condition and a corresponding telescope design is given to Couder **[79]**. It should be said that the condition can also be directly obtained from the expressions for astigmatism and field curvature

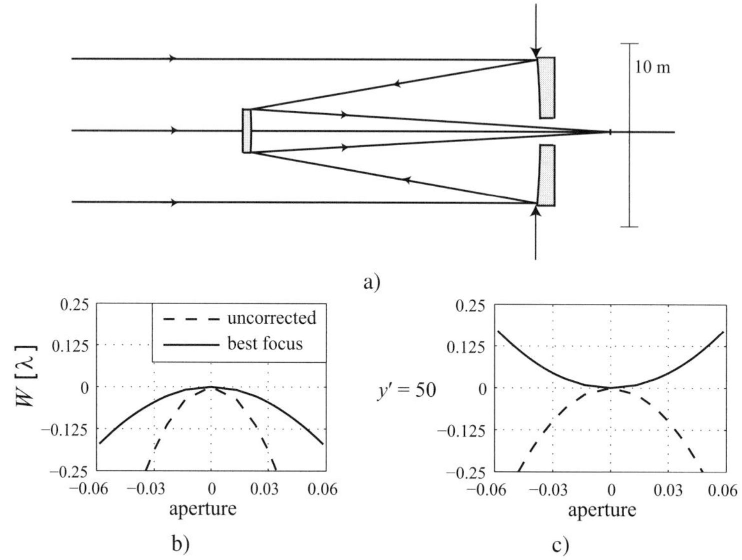

Figure 7.45: a) Optical telescope of the Ritchey–Chrétien type (f'= 69.571 m) with a pupil diameter of 8 m, NA = 0.0575, $c_1 = -0.0218749\,\mathrm{m}^{-1}$, $c_2 = -0.0522021\,\mathrm{m}^{-1}$, L = 16.426 m (telescope magnification $m_T = -10^{-5}$). Curved image field with diameter of 100 mm.
b) and c) Meridional (b) and sagittal (c) wavefront cross-sections (λ = 550 nm) for a pencil that is imaged at the border of the image field, 50 mm off-axis (or at a field angle in object space of 0.75 mrad).

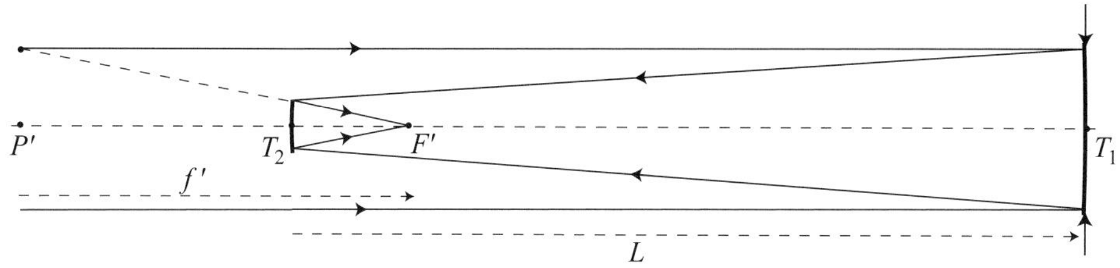

Figure 7.46: The anastigmatic telescope according to Couder with $f' = L/2$ = 20 m, f_b' = 6 m, diameter of primary 8 m, NA = 0.20. The pencil of rays is imaged 50 mm off-axis, corresponding to a field angle of 2.5 mrad in object space.

of an aplanatic telescope which were given by Schwarzschild in his basic publication on reflecting telescopes **[300]**. But the example given in this latter publication applies to an anastigmatic flat-field telescope that has a convex primary and a concave secondary, an unattractive design because of the huge dimension of the secondary mirror. For this reason the design was not accepted as a telescope design by Schwarzschild. However, the design has been successful in microscopy (Schwarzschild reflecting objective).

The layout of a Couder telescope is given in Fig. 7.46, with the same primary diameter as the Ritchey–Chrétien telescope of Fig. 7.45. For various reasons, listed below, the Couder telescope has remained rather a curiosity instead of being widely used:

- *Inconvenient tube length, bad accessibility image plane.*
  The practical inconvenience of the Couder telescope is the difficult access to the focal plane and the large telescope length. Including the extra tube length for adequate baffling, the mechanical construction length of the Couder example of the figure amounts to approximately seven times the diameter of the primary mirror. For the Ritchey–Chrétien we have a construction length that is approximately 2.5 times the primary diameter.

Table 7.19. Design parameters of a two-mirror telescope using conic-section surfaces.

	f'	L	f'_b	κ_1	κ_2	I_1	I_2	$W_{S,i,4}$	$W_{S,i,6}$
R–C	70	16	19	−1.0845246	−4.907213	0.088	0.116	102	−0.9
Couder	20	40	6	−7.99708	−0.67228	0.035	0.135	546	−4.2
C-opt				−8.20189	−0.68948				

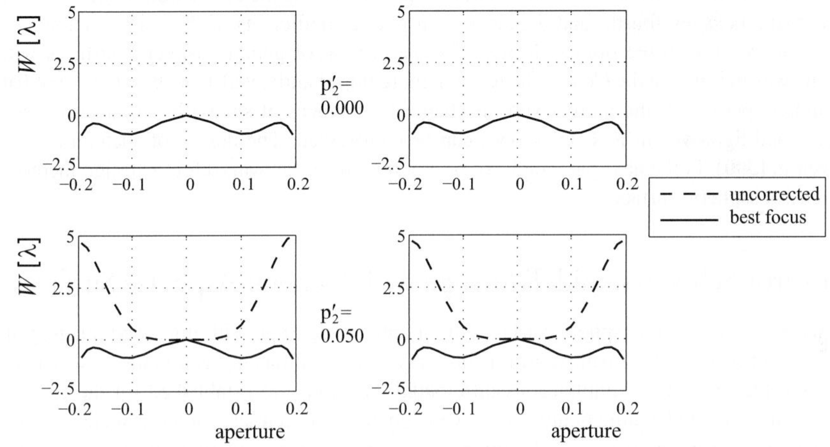

Figure 7.47: Couder telescope, primary diameter is 8 m, $c_1 = -0.00875\,\mathrm{m}^{-1}$, $c_2 = +0.054166667\,\mathrm{m}^{-1}$, $L = 40$ m. The wavefront cross-sections for the axial beam ($p'_2 = 0$) and for an off-axis pencil towards the outer field position $p'_2 = 50$ mm (telescope magnification $m_T = -10^{-5}$).

- *High-aperture imaging in the image plane at F'.*
  The secondary magnification $m_{T,2}$ of the Couder anastigmatic telescope is given by

$$m_{T,2} = \frac{1}{2} - \frac{f'_b}{2f'} . \tag{7.6.33}$$

  This value results in an angular magnification of at least 2 for the imaging on the photo-sensor and requires a small pixel size to avoid aliasing effects caused by the high-frequency content up to $2NA/\lambda$ of the projected stellar image.
- *Residual aberration of higher order.*
  At the scale of a large astronomical telescope with $D = 8$ m, the classical Couder telescope with conic-section surfaces suffers from severe higher-order aberrations as compared to the Ritchey–Chrétien (R–C) type (see Table 7.19). The reason for this is the unbalanced distribution of angles of incidence I_1 and I_2 over the two surfaces and the large amount of intermediate spherical aberration (fourth- and sixth-order terms, $W_{S,i,4}$ and $W_{S,i,6}$, in units of $\lambda = 550$ nm) in the space between the two mirrors. The Couder primary has become a strong hyperboloid, producing a highly-aberrated reflected beam. The secondary mirror, which normally would be of the hyperbolic type, has to be made elliptical to compensate for the spherical aberration of the primary, all this according to the predictions of third-order aberration theory.

The wavefront aberration of the Couder telescope with a primary diameter of 8 m, on-axis and at a field coordinate of 50 mm, is given in Fig. 7.47. The wavefront cross-sections (meridional on the left-hand side, sagittal on the right-hand side) show hardly any asymmetry or difference in focus-setting for the meridional or sagittal cross-section. We conclude that coma and astigmatism are virtually absent and that this is achieved over an angular field in stellar space of ±3 mrad (four times larger than that of the R–C telescope of the example). Unfortunately, the imaging suffers from a large amount of sixth-order spherical aberration (0.235λ rms OPD), constant over the entire image plane. We observe that the sixth-order spherical aberration has been balanced by a fourth-order component which was introduced by an offset in the

κ-values (see the third row (C-opt) of Table 7.19 where the κ-values of the optimum Couder design have received offsets of $\delta\kappa_1 = -0.2048$ and $\delta\kappa_2 = -0.0172$). Using these optimised conic constants, the amplitude of the sixth-order spherical aberration is still such that the telescope operates more than three times away from the diffraction limit. To make the Couder telescope useful for the chosen aperture in image space ($NA = 0.20$), we either have to reduce the physical size of the telescope by a factor of at least three or to improve the design. In the latter case, the restriction on the mirror surfaces to be derived from conic sections has to be abandoned. In the R–C example, the third-order aberration theory was just sufficient to describe the aberration contributions from the two mirrors. For that reason, Table 7.19 does not need a row 'R–C-opt' with an improved third-order design to cancel higher-order spherical aberration. The higher-order contribution at the aperture value of 0.0575 in image space can safely be neglected. As soon as the aperture is increased, the R–C design also suffers from higher-order residuals and starts to operate outside the diffraction limit.

The fixed relationship between fourth- and higher-order aspheric coefficients of the conic-section surfaces limits the number of degrees of freedom in the optical design. The application of general aspheric surfaces with symmetry of revolution opens the way to improve the Couder design and, more importantly, it also enables the use of Ritchey–Chrétien telescopes with higher aperture. If the more severe mechanical tolerances of such a higher aperture instrument can be dealt with, compact and light-weight R–C telescopes can be constructed. The design of such an improved two-mirror telescope was given in **[300]**. In the next paragraph we briefly reproduce Schwarzschild's design method for two-mirror telescopes with general aspheric surfaces.

7.6.6 Two-mirror Schwarzschild Telescope with General Aspheric Surfaces

To calculate the shape of the general aspheric surfaces of an optical telescope with improved imaging quality we could resort to the design method for aplanatism, valid for finite rays with arbitrary aperture angles, which was presented in Subsection 6.2.2 (see Fig. 6.7). Two simultaneous differential equations are established for the z-coordinates of each mirror surface; the equations can be numerically solved with any desired accuracy for a certain number of pairs of points on the surfaces. These two sets of points for the mirror surfaces are then converted into series expansions for the z-coordinate of each surface as a function of the lateral position y on the surface such that arbitrary rays can be traced through the mirror system.

In the case of two reflecting surfaces, with no intermediate optical elements between the object and the first mirror and between the secondary mirror and the image plane, an *analytic* calculation of the surface coordinates of the general aspheric surfaces is possible. Such an analysis was first given by Schwarzschild in **[300]** and we reproduce his treatment here in a slightly modified manner. Before this rigorous method became available in the beginning of the twentieth century, the first aplanatic telescopes were designed using Seidel third-order aberration theory (Chrétien).

7.6.6.1 Calculation of the Coordinates of General Aspheric Surfaces

In Fig. 7.48a an axial pencil of rays is brought to a perfect focus by the telescope at the focal point F'. The stigmatic imaging at F' requires equality of optical path along each ray of the pencil and this yields the equation

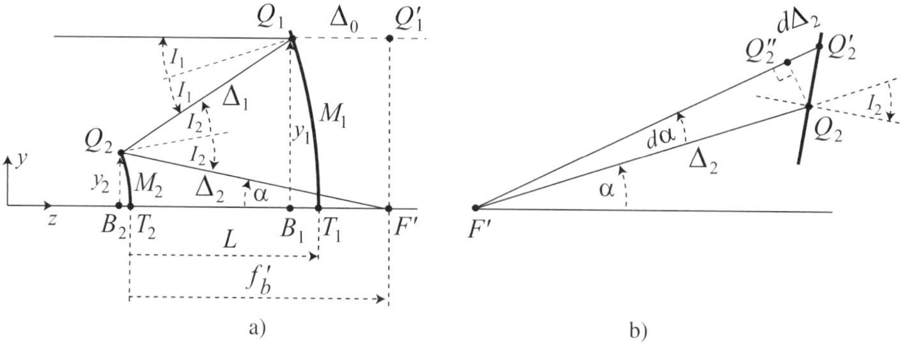

Figure 7.48: a) and b) Geometrical parameters for the derivation of the aspheric surface coordinate equations according to Schwarzschild.

$$[Q_1Q_2F'] = [B_1T_1T_2F'] = B_1F' - T_1F' + L + f'_b ,$$

or,

$$\Delta_1 + \Delta_2 = \Delta_0 + 2L . \quad (7.6.34)$$

To obtain aplanatic imaging, the Abbe sine condition has to be respected; for an infinitely distant object it reads (see Eq. (4.8.7))

$$y_1 = f' \sin\alpha . \quad (7.6.35)$$

The absence of a minus sign in this equation is explained by the definition of the sign of the angle α in Fig. 7.48a.

A differential equation is established for the distance Δ_2 from the focal point F' to the intersection point Q_2 of a ray with the second mirror M_2, with the angle α as variable. To calculate the increment $d\Delta_2$ as a function of an increment $d\alpha$ we use Fig. 7.48b. It follows that

$$\frac{1}{\Delta_2}\frac{d\Delta_2}{d\alpha} = \tan(I_2) . \quad (7.6.36)$$

To solve Eq. (7.6.36) we need an expression for $\tan(I_2)$ in terms of the telescope parameters L and f' and the aperture angle α. From Fig. 7.48a we derive

$$\begin{aligned} I_2 &= I_1 + \alpha/2 , \\ \Delta_2 \cos\alpha &= \Delta_0 + \Delta_1 \cos 2I_1 , \\ y_1 &= \Delta_2 \sin\alpha + \Delta_1 \sin 2I_1 . \end{aligned} \quad (7.6.37)$$

The elimination of Δ_0, Δ_1, y_1 and I_1 from Eqs. (7.6.34), (7.6.35) and (7.6.37) yields, after some manipulation, the differential equation for the distance Δ_2,

$$\frac{1}{\Delta_2}\frac{d\Delta_2}{d\alpha} = \tan(\alpha/2)\left\{\frac{L/f' - \Delta_2/f' + \cos^2(\alpha/2)}{L/f' - 1 + \cos^2(\alpha/2)}\right\} . \quad (7.6.38)$$

We denote L/f' by t and normalise Δ_2 according to $p = \Delta_2/f'$ and obtain the equation

$$\frac{1}{p}\frac{dp}{d\alpha} = \tan(\alpha/2)\left\{\frac{t - p + \cos^2(\alpha/2)}{t - 1 + \cos^2(\alpha/2)}\right\} . \quad (7.6.39)$$

To eliminate $\tan(\alpha/2)$ in (7.6.39), the transformation $u = \cos^{-2}(\alpha/2)$ is introduced, yielding the differential equation,

$$\frac{u}{p}\frac{dp}{du} = \left\{\frac{(t-p)u + 1}{(t-1)u + 1}\right\} . \quad (7.6.40)$$

The factor (u/p) in front of the differential quotient can be eliminated by defining a new function $v = u/p$ and, using the relation

$$\frac{dv}{du} = \frac{p - u(dp/du)}{p^2} ,$$

or ,

$$\frac{u}{p}\frac{dp}{du} = -p\frac{dv}{du} + 1 , \quad (7.6.41)$$

we obtain

$$\frac{dv}{du} = \frac{u - v}{(t-1)u + 1} ,$$

or, alternatively ,

$$0 = (v - u)du + [(t-1)u + 1]\, dv . \quad (7.6.42)$$

It follows from the equation above that the differential expression cannot be written as a total differential because $\partial(v-u)/\partial v \neq \partial[u(t-1)+1]/\partial u$. The differential equation is then multiplied by an integrating factor $g(u,v)$. To obtain a total differential we require

$$\frac{\partial}{\partial v}(v-u)g(u,v) = \frac{\partial}{\partial u}[u(t-1)+1]g(u,v)\,. \tag{7.6.43}$$

It is easily verified that the condition above can be satisfied by an integrating factor $g(u,v)=g_1(u)$ that depends only on the variable u, with the following differential equation for $g_1(u)$,

$$\frac{dg_1}{g_1} = \left[\frac{2-t}{(t-1)u+1}\right] du\,, \tag{7.6.44}$$

having the solution

$$g_1(u) = [(t-1)u+1]^{\frac{1}{t-1}-1}\,. \tag{7.6.45}$$

We multiply Eq. (7.6.42) by the integrating factor $g_1(u)$ and write

$$(v-u)g_1(u)du + [(t-1)u+1]\,g_1(u)dv \;=\; \frac{\partial F_1}{\partial u}du + \frac{\partial F_1}{\partial v}dv \;=\; dF_1 = 0. \tag{7.6.46}$$

The solution of this differential equation is $F_1(u,v)=C$, where C is an arbitrary constant.

To find the required function $v(u)$ we deduce from Eq. (7.6.46),

$$\frac{\partial F_1}{\partial v} = [(t-1)u+1]^{1/(t-1)}\,, \qquad \rightarrow \qquad F_1(u,v) = v[(t-1)u+1]^{1/(t-1)} + f(u)\,. \tag{7.6.47}$$

To find the function $f(u)$ we take $\partial F_1/\partial u$ from Eq. (7.6.46) and equate it to the same expression as follows from Eq. (7.6.47),

$$\frac{df}{du} = -u\,[(t-1)u+1]^{-1+1/(t-1)}\,. \tag{7.6.48}$$

We write u as $\{[(t-1)u+1]-1\}/(t-1)$ and obtain after integration of the equation above,

$$f(u) = -\frac{[(t-1)u+1]^{1+1/(t-1)}}{t(t-1)} + \frac{[(t-1)u+1]^{1/(t-1)}}{t-1} = \frac{1-u}{t}\,[(t-1)u+1]^{1/(t-1)}\,. \tag{7.6.49}$$

We combine Eq. (7.6.47) and Eq. (7.6.49) and, using the property that $F_1(u,v)$ equals a constant C, we obtain

$$C = \left\{v - \frac{u-1}{t}\right\}[(t-1)u+1]^{1/(t-1)}\,. \tag{7.6.50}$$

The function v is then given by

$$v(u) \;=\; C[(t-1)u+1]^{-1/(t-1)} + \frac{u-1}{t}\,. \tag{7.6.51}$$

The expression for the radius vector Δ_2 follows from $\Delta_2 = f'/[v\,\cos^2(\alpha/2)]$,

$$\frac{1}{\Delta_2} = \frac{\sin^2(\alpha/2)}{L} + \frac{A(\alpha)}{f'_b}\,, \tag{7.6.52}$$

where the function $A(\alpha)$ is given by

$$A(\alpha) \;=\; \left[\frac{t\,\cos^2(\alpha/2)}{t-\sin^2(\alpha/2)}\right]^{1/(t-1)}\,. \tag{7.6.53}$$

The value of the constant C follows from $\Delta_2 = f'_b$ for the ray along the optical axis ($\alpha = 0$). The Cartesian coordinates of a general point Q_2 on the mirror surface M_2 follow directly from the value of Δ_2.

The expression for the radius vector Δ_1 is

$$\Delta_1 = L - \Delta_2\sin^2(\alpha/2) + \frac{\sin^2(\alpha/2)\cos^2(\alpha/2)(\Delta_2 - f')^2}{L - \Delta_2\sin^2(\alpha/2)}\,. \tag{7.6.54}$$

To calculate the coordinates of a point Q_1 on the mirror M_1 we need the value of Δ_0 and, after some rearrangement, we obtain,

$$\Delta_0 = \frac{[L - 2f' \sin^2(\alpha/2)\cos^2(\alpha/2)] - [L^2 - f'^2 \sin^2(\alpha/2)\cos^2(\alpha/2)]/\Delta_2}{L/\Delta_2 - \sin^2(\alpha/2)} . \tag{7.6.55}$$

We eliminate $1/\Delta_2$ from this expression with the aid of Eq. (7.6.52) and find the expression

$$\Delta_0 = -L + \frac{f'^2 \sin^2\alpha}{4L} + f_b' \left[\frac{t - \sin^2(\alpha/2)}{t}\right]^{2+1/(t-1)} \left[\cos^2(\alpha/2)\right]^{-1/(t-1)} . \tag{7.6.56}$$

The z-coordinate of Q_1 follows directly from Δ_0, the lateral coordinate y_1 is given by the Abbe sine condition of Eq. (7.6.35).

7.6.6.2 Examples of Large-aperture Two-mirror Telescopes

In Fig. 7.49 some examples are given of large-aperture optical telescopes that require general aspheric surfaces to achieve axial stigmatism and coma-free imaging. The two graphs a) and b) have been given the same scale and the basic design parameters f', L and f_b' are found in Table 7.19. The telescope diameters have been drastically increased in the designs, with a pupil diameter of 80 m for the Cassegrain type and 40 m for the Couder telescope. For still larger apertures, limitations to the transverse dimension of the telescope are set by the physical intersection of the two mirror surfaces (Cassegrain), by the secondary mirror growing larger in size than the primary (Couder) or by the aperture in the focal plane approaching unity (Couder). In the figure, for comparison, we have also plotted the spherically-shaped mirror surfaces that follow from paraxial optics. In the two graphs, especially for the Cassegrain-type telescope, the spherical surfaces are visually different from the general aspheres. This is not the case for the conic-section surfaces as compared to the general aspheres. For identical lateral position y on the mirror surfaces, we have plotted in Fig. 7.50 the difference in axial position $z_c - z_a$ for points on the mirrors of a Cassegrain telescope, a), and a Couder telescope, b); z_c is the z-coordinate of a conic-section asphere, z_a the coordinate of a point on a generally-shaped aspheric surface. For the Cassegrain telescope, it follows that for a lateral position y up to 5 m from the axis, the coordinates z_c and z_a are comparable at the scale of the wavelength of the light. For diameters beyond 10 m, large differences arise between the two surface types, at the scale of many wavelengths. In this range of pupil diameter, the general Schwarzschild design is mandatory. For a long time, the designs shown in this paragraph were of theoretical interest only. The mechanical and thermal tolerances associated with imaging at such large convergence angles in intermediate and imaging space are alarming. At a completely different scale, in microscopy, the findings and the design method of Schwarzschild have found interesting applications. At these reduced scales, the residual astigmatism of, for instance, a Schwarzschild mirror objective is negligible. However, at the scale of a large astronomical telescope with $D = 8$ m, the astigmatism limits the diffraction-limited field diameter to 100 mm. With a focal length of 70 m, this is equivalent to a total field of view in stellar space of 1.4 mrad (or 4.8′). On top of this, there is the intrinsic curvature of field for a Cassegrain design.

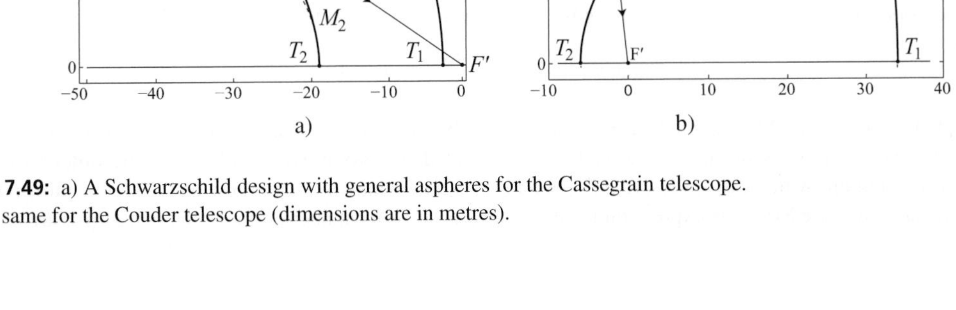

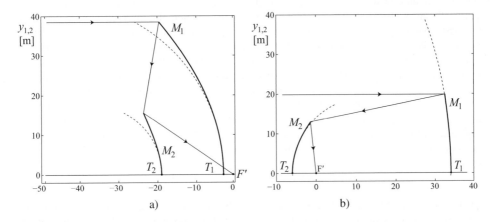

Figure 7.49: a) A Schwarzschild design with general aspheres for the Cassegrain telescope.
b) The same for the Couder telescope (dimensions are in metres).

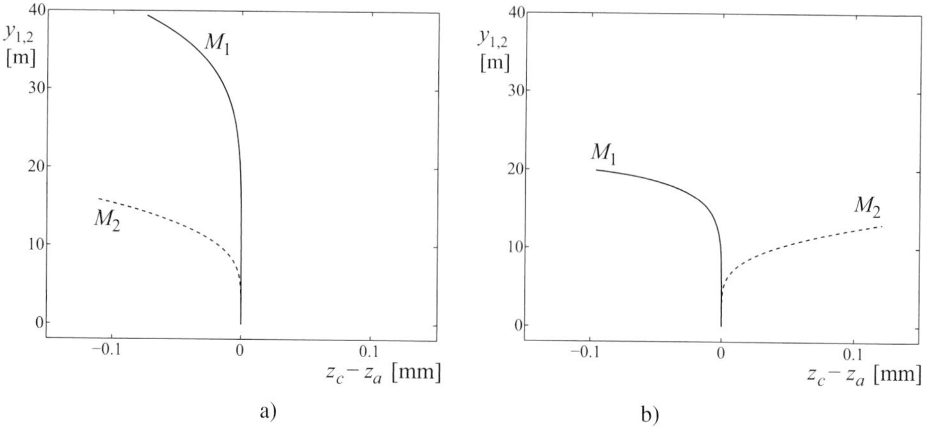

Figure 7.50: Deviation in the z-direction between the general aspheric profile according to Schwarzschild and the corresponding conic-section profile for each of the mirrors.
a) Cassegrain telescope.
b) Couder telescope.

The astigmatism in the Schwarzschild telescope with moderate aperture is of the Seidel type. According to Eqs. (7.6.30) and (7.6.31), such third-order astigmatism is given by $f_t H^2 K$, where $H = p_{2,m}\alpha_{2,m}$ is the Lagrange invariant, K the optical power of the telescope and f_t a proportionality constant depending on the telescope design. We note that the Lagrange invariant H, applied to the paraxial pupil plane, equals $D_0\gamma_0/4$ with γ_0 the total angular field in stellar space. A scaled telescope, with scaling factor S, has identical astigmatism if $S^2 D_0^2\gamma_1^2/(Sf'_0) = D_0^2\gamma_0^2/f'_0$. It follows that the angular field of the scaled telescope is given by

$$\gamma_1 = \frac{\gamma_0}{\sqrt{S}} \,. \tag{7.6.57}$$

As an example, switching from the 8 m telescope to an amateur telescope with $D = 0.2$ m, we obtain a diffraction-limited field of view which is approximately 0.5°, the angular size of the sun or the moon.

7.6.7 The Three-mirror Anastigmat

For a further improvement of telescope imaging quality the astigmatism and field curvature problem of the two-mirror system needs to be circumvented. Although the Couder telescope solved this problem in principle, its use was hampered by the impractically large tubelength. The solution comes with the addition of a third reflecting surface, leading to the three-mirror anastigmat design. It was first put forward by Paul **[270]** and further improved by Baker **[19]** by realising a flat image surface. For this reason, the three-mirror anastigmats are commonly called Paul–Baker anastigmats. Most modern large telescopes are based on the Paul–Baker anastigmat design. The three-mirror anastigmat can be considered as an extension of the two-mirror afocal telescope of Mersenne, proposed in 1636. The Mersenne telescope (see Fig. 7.51) belongs to the same class as the Cassegrain and Gregory telescopes. The only difference is that the image is relayed to infinity such that it can be used with the naked eye. The two mirror surfaces have a parabolic shape and their foci are coincident, $F_{1,2}$ in the figure. The lateral magnification follows from paraxial optics and is given by $m_T = f_2/f_1'$. The imaging quality of the Mersenne telescope is surprisingly good. Apart from the axial stigmatism provided by the parabolic mirrors, there holds for finite rays, using the conformity of the triangles $F_1Q_1T_1$ and $F_2Q_2T_2$,

$$\frac{y_2}{Q_2F_2} = \frac{y_1}{Q_1F_1}, \text{ and, with } \frac{Q_2F_2}{Q_1F_1} = \frac{T_2F_2}{T_1F_1}, \text{ we obtain } \frac{y_2}{y_1} = m_T \,. \tag{7.6.58}$$

According to Eq. (4.8.9), the Mersenne telescope satisfies the Abbe sine condition and is free from linear coma.

The astigmatism of the Mersenne telescope needs a more detailed discussion. We calculate the third-order astigmatism of a two-mirror telescope with the aid of Eq. (5.3.51) and Fig. 7.52; note that the stop is now on the second mirror surface. The relevant quantities needed in the expression for the astigmatic wavefront aberration for the first mirror are $m_{T,1} = 0$,

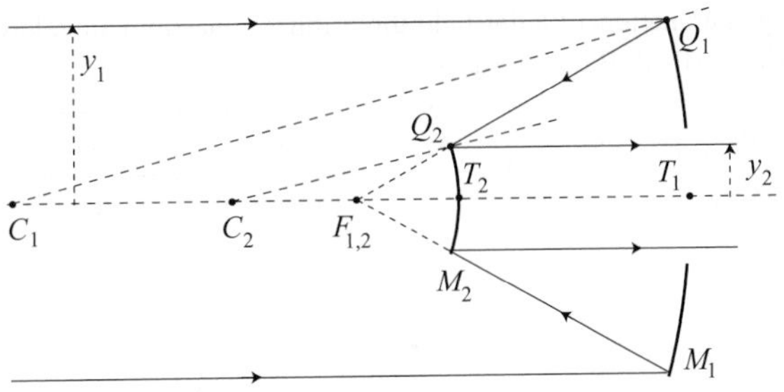

Figure 7.51: An afocal telescope according to Mersenne (version with negative power of the second mirror). The two mirrors are used in the confocal position. The common focus has been denoted $F_{1,2}$.

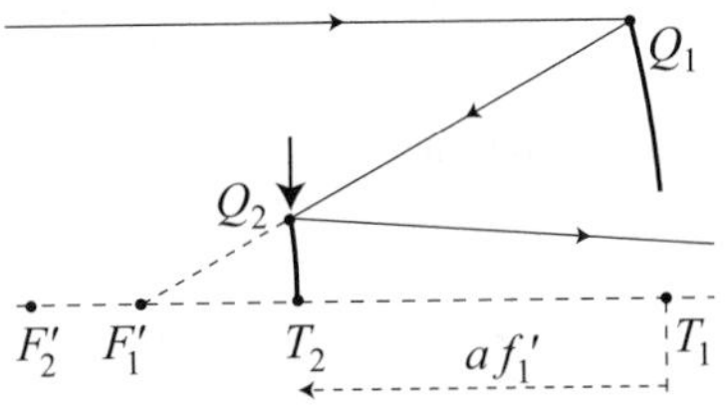

Figure 7.52: A two-mirror system with stop on the second mirror.

$K_1 = -2c_1$, $a_{4,1} = \kappa_1 c_1^3/8$, $\kappa_1 = -1$. The pupil magnification by the first mirror follows from the position of T_2, at an image distance af_1' from the first mirror. With $n_0 = -n_1 = 1$, we obtain from the imaging equation for the point T_2 by the first mirror,

$$s_{1,p} = \frac{a}{2(a-1)c_1} \,. \tag{7.6.59}$$

The lateral pupil magnification is then given by $m_{T,p,1} = 1 - a$. Filling in these data in Eq. (5.3.51) we obtain for the astigmatism $W_{A,1}$ at the rim of the first mirror

$$W_{A,1} = \frac{1}{(1-a)}\, c_1\, H^2 \,, \tag{7.6.60}$$

where H is the Lagrange invariant which equals $p_{1,m}\alpha_{1,m}$ in the image plane through F_1'.

For the second mirror with curvature c_2 and power $K_2 = 2c_2$ we first calculate the lateral object magnification with the aid of the imaging equation for the second mirror. The object distance is given by $T_2F_1' = (1-a)f_1' = (1-a)/2c_1$ and we find

$$m_{T,2} = \frac{c_1}{c_1 - (1-a)\, c_2} \,. \tag{7.6.61}$$

The pupil magnification $m_{T,p,2}$ equals unity. With $\kappa_2 = -1$, $a_{4,2} = \kappa_2 c_2^3/8$ and $n_1 = -n_2 = -1$, we obtain for the astigmatism of the second mirror

$$W_{A,2} = -\frac{K_2}{2}\, H^2 = -c_2\, H^2 , \tag{7.6.62}$$

the well-known expression for the third-order astigmatism of an optical surface that is coincident with the stop or an image of it. It follows from Eqs. (7.6.60) and (7.6.62) that the third-order astigmatism of the mirror system is zero when the condition

$$c_1 = (1-a)\, c_2 \tag{7.6.63}$$

is fulfilled. This means that the system has to be made confocal and that its optical power equals zero. Using the property that the Mersenne telescope is also free of third-order spherical aberration and coma, it follows that the third-order astigmatism is zero for any position of the stop.

Table 7.20. Design parameters of a flat-field three-mirror anastigmatic telescope using conic-section surfaces (Paul–Baker design).

c_1	κ_1	d_2	c_2	κ_2	d_3	c_3	d_4
$1/R_1$	-1	$aR_1/2$	$c_1/(1-a)$	$-1+a^3$	$(a-1)/ac_1$	$ac_1/(1-a)$	$(1-a)/2ac_1$

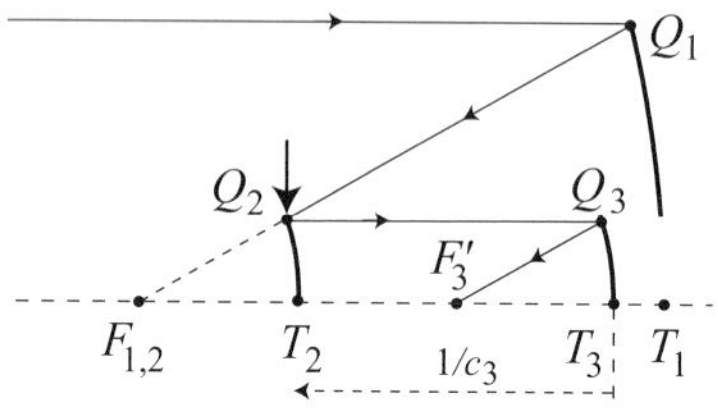

Figure 7.53: The three-mirror anastigmat according to Paul.

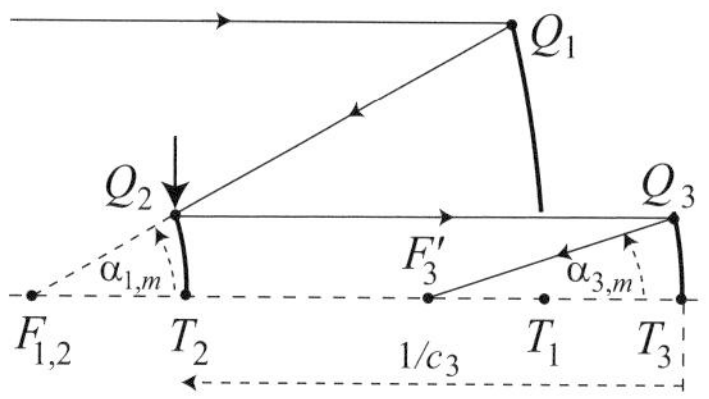

Figure 7.54: A flat-field Paul–Baker anastigmat with $a = 2/3$.

The foregoing results illustrate the special properties of the Mersenne telescope with respect to the Gregory and Cassegrain layouts which only achieve freedom of spherical aberration of all orders. We conclude by saying that the Mersenne afocal telescope is free of spherical aberration of all orders, free of coma of all orders that increase linearly with the field coordinate or field angle, and free of third-order astigmatism. Higher-order astigmatism is present as can be verified by applying the Coddington equations to the chief ray of an off-axis pencil in the system of Fig. 7.52.

The very favourable imaging properties of the Mersenne telescope were used by Paul **[270]** to produce a well-corrected image in real space by adding a third mirror, see Fig. 7.53. This third mirror is spherical and has the same curvature as the second one which is also made spherical; the centre of curvature of the third mirror coincides with T_2. The spherical aberration introduced by the second convex mirror is exactly cancelled by the spherical aberration of the third mirror. Together with the parabolic first mirror, we obtain zero spherical aberration at the image point F'_3. With respect to the third-order field aberrations, the third mirror is used with its centre of curvature coinciding with the centre of the stop at T_2. The field aberrations, proportional to $(n_2 - n_3 m_{T,p,3})$, are all zero for $m_{T,p,3} = -1$. The third mirror is used in the same manner as the spherical mirror in the Schmidt projection system. The third-order field aberrations of the Mersenne telescope remain the same when the second mirror is made spherical instead of parabolic under the condition that the stop is at this second mirror. For this special position of the stop, the aspheric coefficient $a_{4,2}$ does not influence the field aberration of the telescope. It follows that the total system is free from spherical aberration, coma and astigmatism and only suffers from field curvature.

Expressions have been derived from third-order aberration theory for the mirror surface profiles and distances between them such that the field of the three-mirror anastigmat is flattened. The basic design parameter is the obstruction ratio $1 - a$, caused by the secondary mirror. The system data for a flat field system (Fig. 7.54) are then given by the data in Table 7.20. The derivation of the values given in this table is as follows. The flat-field condition requires a Petzval sum P equal to zero, yielding,

$$P = -2[c_1 - c_2 + c_3] = 0 \, . \tag{7.6.64}$$

From the afocal Mersenne telescope we have the relation $c_1/c_2 = 1 - a$ and to obtain a zero Petzval sum we impose

$$\frac{c_3}{c_1} = -1 + \frac{c_2}{c_1} \,, \quad \text{or,} \quad \frac{c_3}{c_2} = a \,. \tag{7.6.65}$$

If $a \leq 1$ for the Mersenne telescope under consideration, we conclude that the third mirror has to be weakened in power and its vertex T_3 approaches that of the primary mirror or is even found outside the Mersenne afocal section. Keeping the parabolic shape of the primary, we have to compensate for the imbalance between the spherical aberration contributions from the spherical secondary and tertiary mirrors. The spherical aberration is made zero by allowing the secondary mirror to be aspheric with a nonzero aspheric coefficient $a_{4,2}$. We calculate the spherical aberration contributions $W_{S,2}$ and $W_{S,3}$ of these two mirrors using the expression of Eq. (5.3.49). For the second mirror we have $m_{T,2} = \infty$. We use the property that

$$\lim_{m_{T,2} \rightarrow \infty} m_{T,2}\, n_2\, \alpha_{2,m} = n_1\, \alpha_{1,m} \,, \tag{7.6.66}$$

where we have substituted $m_{T,2} = n_1/(n_2\, m_{A,2})$, $m_{A,2}$ being the angular magnification of the second mirror. We then obtain for the spherical aberration of the second aspheric mirror,

$$W_{S,2} = -\frac{(1+\kappa_2)\,(n_1\alpha_{1,m})^4}{32K_2} \,. \tag{7.6.67}$$

The spherical aberration of the third mirror with transverse magnification $m_{T,3} = 0$ and spherical profile equals

$$W_{S,2} = -\frac{(n_3\alpha_{3,m})^4}{32K_3} \,. \tag{7.6.68}$$

Zero spherical aberration of the system is obtained when

$$(1+\kappa_2)\frac{(\alpha_{1,m})^4}{K_2} + \frac{(\alpha_{3,m})^4}{K_3} = 0 \,. \tag{7.6.69}$$

Using the paraxial properties $\alpha_{3,m}/\alpha_{1,m} = a$ and $K_3/K_2 = -a$ we have the final result,

$$\kappa_2 = -1 + a^3 \,. \tag{7.6.70}$$

The second mirror has to be made elliptical. From third-order theory it follows that the coma and astigmatism of this order remain unchanged by a nonzero value of $a_{4,2}$ because the stop is found at this surface.

The composite focal length of the system is given by $f' = f_3'/m_M$ where $m_M = y_2/y_1$ is the lateral magnification of the Mersenne telescope which precedes the third mirror. Although the third-order spherical aberration, coma and astigmatism remain zero in the Paul–Baker anastigmat, the system is not free of aberration at larger apertures. For a primary mirror diameter of 8 m, the maximum aperture angle $\alpha_{3,m}$ should be limited to 0.08 rad. Beyond this value, spherical aberration starts to exceed the diffraction limit in the visual region of the spectrum. Neither does a system according to Table 7.20 rigorously satisfy the Abbe sine condition.

New-generation telescopes are planned with typical primary diameters of 40 m. To keep the telescope relatively compact, the apertures have to be substantially increased, to values of typically 0.40, approaching the F-number value of unity for the first mirror. Third-order aberration theory is entirely insufficient to describe the aberrations at these opening angles. A design strategy which can be followed is to use the paraxial layout that follows from third-order aberration theory. The extension to large numerical aperture is then carried out with the aid of the general aplanatic design method that aspherises two surfaces in an optical system to obtain axial stigmatism while respecting the Abbe sine condition (see Subsection 6.2.2). Solving the two differential equations for the aspheric surfaces, the coordinates of these surfaces can be obtained with machine accuracy. In practice, the pathlengths along the rays of an axial pencil are made identical up to a value of typically $\lambda/100000$. The offence against the sine condition according to Eq. (4.8.6) can be made as small as 10^{-10}. When applying this construction method to the mirror surfaces 1 and 2 of Fig. 7.55, a system is found that is free of spherical aberration and of linear coma. However, important residuals of astigmatism and field curvature are still present. The design is then repeated for a slightly different distance T_3F' and for a weak conic deformation of the originally spherical third mirror. By linear interpolation, one immediately finds the optimum conic constant of mirror 3 ($\kappa_3 = 0.0665$) and the change in distance T_3F', -2.1 m, which together yield a flat astigmatism-free image field. The principal field aberrations are sixth-order astigmatism, proportional to $p^2 R_4^2(\rho)\cos 2\theta$, and field spherical aberration, proportional to $p^2 R_4^0(\rho)$. To obtain minimum aberration over the image field with a diameter of 300 mm ($f' = 50$ m), the axial imaging quality is slightly deteriorated by introducing a spherical aberration offset on-axis. This is obtained by a

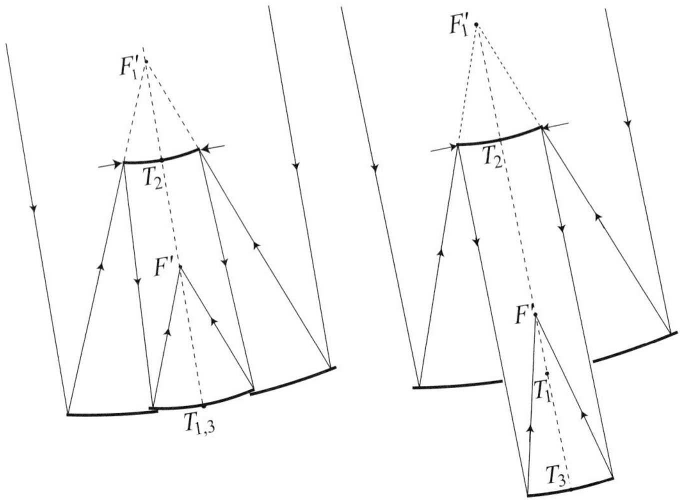

Figure 7.55: a) Three-mirror anastigmat according to Paul. The vertices T_1 and T_3 of the first and third mirror are coincident. b) Three-mirror Paul–Baker anastigmat, especially designed for a flat image field.

small offset (typically 0.1 μm) in the fourth-order aspheric coefficient of the second mirror which is coincident with the diaphragm of the system. In this way, the influence of the field spherical aberration at the border of the field is halved and the aberration balance over the image field is improved. In a final step, the system is numerically optimised with respect to aspheric coefficients of all three mirrors (up to the twelfth order) with the design result given in Table 7.21.

The z-coordinate of each of the aspheric surfaces is given by means of two representations. The first one, having coefficients $a_{2n,j}$, is a simple power series representation with even powers $2n$ (j is the index of aspheric surface). It is well known that a power series expansion, as a function of the number of considered terms, suffers from a slow convergence towards an end value with the desired accuracy. The accuracy, typically equal to $\lambda/1000$, is of the order of a nanometre in the visible spectrum. It is seen that the tenth-order term of the first asphere in Table 7.21 has not yet reached this value. Approximately three more terms would be needed to reach the nanometre region. As an alternative representation with good convergence we have given for each aspheric surface an expansion with the aid of Chebyshev polynomials $T_{2n}(r_n)$ of the first kind **[266]**. The quantity r_n is the normalised lateral coordinate according to $r_n = 2r/D$. As the modulus of the Chebyshev polynomials $T_n(r_n)$ is ≤ 1, the expansion coefficients provide the maximum contribution of each term to the end value given by the (truncated) series of terms. The coefficients $C_{2n,j}$ of a Chebyshev expansion are calculated, for instance, with the aid of the evaluation of inner products of the (r_n, z) values of a set of points on the asphere with a specific polynomial $T_{2n}(r_n)$. The Chebyshev expansion of the first asphere in Table 7.21 shows that the coefficient $C_{10,1}$ of the tenth-order polynomial has reached the required value of typically 1 nm. Without loss of numerical accuracy a power series expansion can be constructed from the corresponding Chebyshev polynomial expansion.

The same design procedure as for the Paul system is applied to the system of Fig. 7.55b, the Paul–Baker system. The procedure is easier in the sense that the Paul–Baker system has a third-order Petzval sum equal to zero. The residual field curvature after asphering mirrors 1 and 2 according to the general aplanatic design method is much less than for the system of Fig. 7.55a. A small conic constant κ_3 of 0.0052 is needed on mirror 3, together with a minute change (−0.068 m) in the distance T_3F'. To reduce the influence of the field spherical aberration, the same operation as for the first system is carried out, yielding an optimum imaging balance. This system does not need further optimisation and the maximum degree in the expansion of the aspheric surfaces is limited to eight for the first and ten for the second mirror. The imaging quality over the image field with a diameter of 480 mm ($f' = 75$ m, field angle γ is $\pm 11'$), is comparable to that of the first system, whilst the Paul–Baker design offers a 15% larger field of view. The wavefront aberration for the image field points at $p' = 0$, 80, 160 and 240 mm is given in Fig. 7.56a. We see that the lowest aberration level is found at approximately 2/3 of the field radius. The OPD_{rms} in the field is well below the diffraction limit of $\approx 0.07\,\lambda$. For an impression of the type of aberrations on-axis and in the field, Fig. 7.56b displays the spot diagrams for three field positions. The on-axis point suffers from spherical aberration. At two thirds from the field border, the correction is

Table 7.21. Design data of a Paul flat-field three-mirror anastigmatic telescope (primary diameter D is 39 m (see Fig. 7.55a). Length unit is the metre.

Paul-anastigmat, $f' = 50$ m, $\lambda = 550$ nm.				
$d_1 = \infty$	c_1	$2a_{2,1}$ (paraxial curvature)	$D/2$	19.5
	$a_{2,1}$	$-0.53191225 \cdot 10^{-02}$	$a_{4,1}$	$-0.10142149 \cdot 10^{-07}$
	$a_{6,1}$	$-0.30031333 \cdot 10^{-11}$	$a_{8,1}$	$-0.36336587 \cdot 10^{-15}$
	$a_{10,1}$	$-0.43872102 \cdot 10^{-19}$		
	$C_{0,1}$	$-0.10119018 \cdot 10^{+01}$	$C_{2,1}$	$-0.10121123 \cdot 10^{+01}$
	$C_{4,1}$	$-0.21600902 \cdot 10^{-03}$	$C_{6,1}$	$-0.56652360 \cdot 10^{-05}$
	$C_{8,1}$	$-0.66160521 \cdot 10^{-07}$	$C_{10,1}$	$-0.68118425 \cdot 10^{-09}$
$d_2 = -33.3333$	c_2	$2a_{2,2}$ (paraxial curvature)	$D/2$	5.65
	$a_{2,2}$	$-0.20487288 \cdot 10^{-01}$	$a_{4,2}$	$-0.10974822 \cdot 10^{-04}$
	$a_{6,2}$	$-0.31600387 \cdot 10^{-07}$	$a_{8,2}$	$-0.99009291 \cdot 10^{-10}$
	$a_{10,2}$	$-0.32482006 \cdot 10^{-12}$	$a_{12,2}$	$-0.24067027 \cdot 10^{-14}$
	$C_{0,2}$	$-0.33154924 \cdot 10^{+00}$	$C_{2,2}$	$-0.33312689 \cdot 10^{+00}$
	$C_{4,2}$	$-0.16163571 \cdot 10^{-02}$	$C_{6,2}$	$-0.39770277 \cdot 10^{-04}$
	$C_{8,2}$	$-0.10956403 \cdot 10^{-05}$	$C_{10,2}$	$-0.35953840 \cdot 10^{-07}$
	$C_{12,2}$	$-0.12435804 \cdot 10^{-08}$		
$d_3 = +33.3333$	c_3	$2a_{2,3}$ (paraxial curvature)	$D/2$	7.6
	$a_{2,3}$	$-0.14999999 \cdot 10^{-01}$	$a_{4,3}$	$-0.35994256 \cdot 10^{-05}$
	$a_{6,3}$	$-0.17276097 \cdot 10^{-08}$	$a_{8,3}$	$-0.10307089 \cdot 10^{-11}$
	$a_{10,3}$	$-0.78535757 \cdot 10^{-15}$		
	$C_{0,3}$	$-0.43781045 \cdot 10^{+00}$	$C_{2,3}$	$-0.43936549 \cdot 10^{+00}$
	$C_{4,3}$	$-0.15661069 \cdot 10^{-02}$	$C_{6,3}$	$-0.11164831 \cdot 10^{-04}$
	$C_{8,3}$	$-0.99487440 \cdot 10^{-07}$	$C_{10,3}$	$-0.98612825 \cdot 10^{-09}$
$d_4 = -18.7941508$				
Image quality	field angle γ (arcmin)		OPD_{rms} (units of λ)	
	0.0′		0.038	
	4.1′		0.026	
	8.2′		0.032	
	9.6′		0.056	

optimum. Higher-order astigmatism and cubic coma, proportional to the third power of the field coordinate, make their appearance. At the border of the field, these two aberrations are the dominating ones and drive the state of correction towards the just-diffraction-limited level. For comparison, in each spot diagram, the first zero-intensity ring of the Airy diffraction pattern has been drawn (dashed circle).

In Table 7.22, the design data are given for the 'Paul–Baker-anastigmat'. Other layouts are possible in which the parameter $(1 - a)$, the obstruction ratio, plays a very important role. Since general aspheric surfaces are employed, the number of degrees of freedom in the designs is made larger with respect to the original design and the necessity of keeping the first two mirrors in an afocal Mersenne geometry does not apply. The third mirror, a conic in the two presented designs, can also be given a general aspheric shape, allowing a smaller value of the obscuration ratio and a slight increase of the field angle. We note that the shape of the aspheric surfaces by means of the coefficients $a_{2n,j}$ of a power series expansion in the Tables 7.21 and 7.22 has been given without specifying the conic constant. Its influence has been absorbed in the aspheric coefficients of order four and higher of each mirror surface.

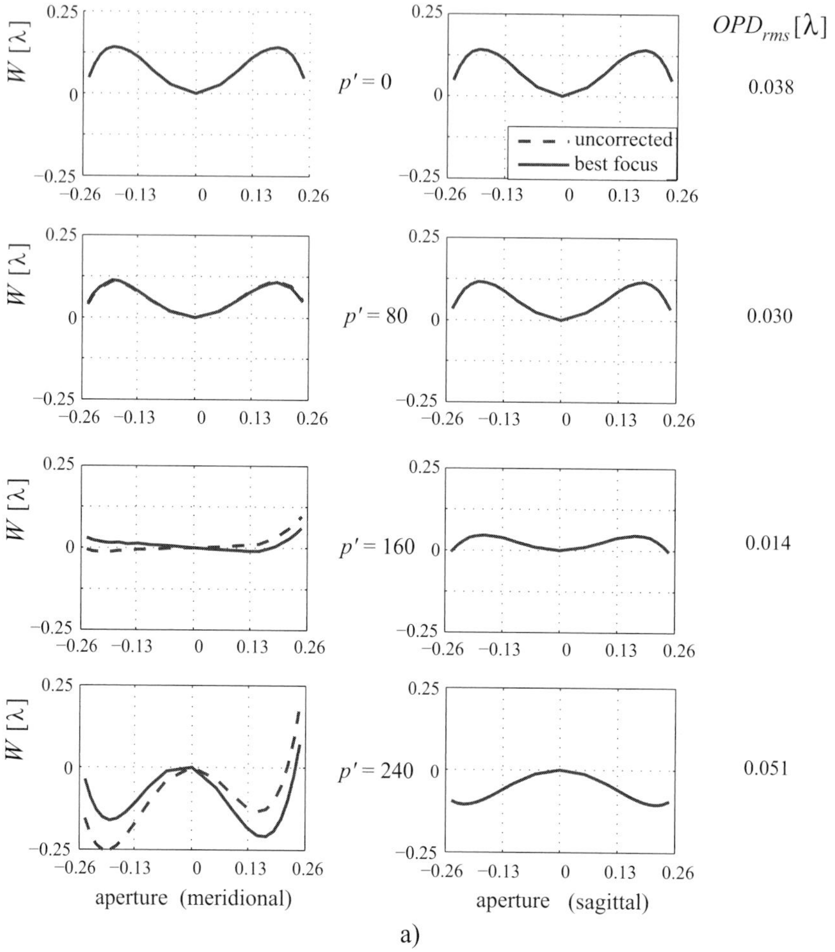

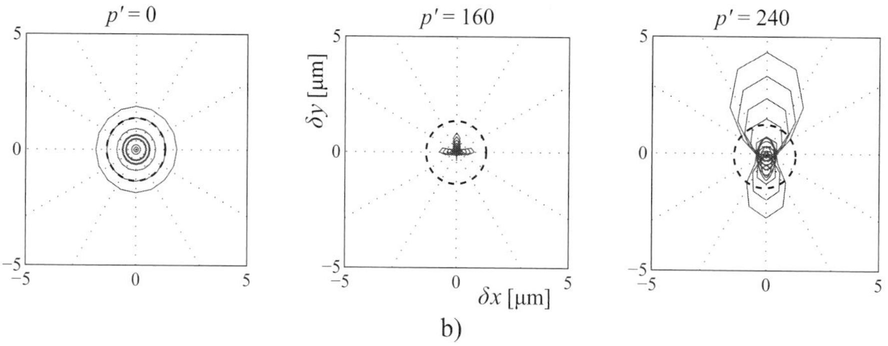

Figure 7.56: Field performance of the Paul–Baker three-mirror anastigmat of Fig. 7.55b, scaled to a focal length of $f' = 75$ m and a primary mirror diameter of 39 m ($\lambda = 550$ nm, $NA = 0.253$, telescope magnification $m_T = 0$).
a) Meridional and sagittal wavefront sections for field positions from the centre to the outer position $p' = 240$ mm.
b) Ray spot diagrams for the field positions $p' = 0$, 160 and 240 mm, illustrating the type of aberration that determines the image blur. The dashed circle corresponds to the first intensity zero of an aberration-free diffraction image.

Table 7.22. Design data of a Paul–Baker flat-field three-mirror anastigmatic telescope with a primary diameter D of 39 m (see Fig. 7.55b). Length unit is the metre.

Paul-Baker-anastigmat, $f' = 75$ m, $\lambda = 550$ nm.				
$d_1 = \infty$	c_1	$2a_{2,1}$ (paraxial curvature)	$D/2$	19.5
	$a_{2,1}$	$-0.50068051 \cdot 10^{-02}$	$a_{4,1}$	$-0.29095457 \cdot 10^{-09}$
	$a_{6,1}$	$-0.63862737 \cdot 10^{-12}$	$a_{8,1}$	$-0.45324414 \cdot 10^{-16}$
	$C_{0,1}$	$-0.95194582 \cdot 10^{+00}$	$C_{2,1}$	$-0.95195672 \cdot 10^{+00}$
	$C_{4,1}$	$-0.12049417 \cdot 10^{-04}$	$C_{6,1}$	$-0.11564719 \cdot 10^{-05}$
	$C_{8,1}$	$-0.74028528 \cdot 10^{-08}$		
$d_2 = -33.3333$	c_2	$2a_{2,2}$ (paraxial curvature)	$D/2$	6.4
	$a_{2,2}$	$-0.15088660 \cdot 10^{-01}$	$a_{4,2}$	$-0.10647662 \cdot 10^{-05}$
	$a_{6,2}$	$-0.20207273 \cdot 10^{-08}$	$a_{8,2}$	$-0.15802737 \cdot 10^{-11}$
	$a_{10,2}$	$-0.23585205 \cdot 10^{-14}$		
	$C_{0,2}$	$-0.30973033 \cdot 10^{+00}$	$C_{2,2}$	$-0.30997610 \cdot 10^{+00}$
	$C_{4,2}$	$-0.25037127 \cdot 10^{-03}$	$C_{6,2}$	$-0.46413827 \cdot 10^{-05}$
	$C_{8,2}$	$-0.40061502 \cdot 10^{-07}$		
$d_3 = +50.0000$	c_3	$2a_{2,3}$ (paraxial curvature)	$D/2$	6.7
	$a_{2,3}$	$-0.10000000 \cdot 10^{-01}$	$a_{4,3}$	$-0.10052010 \cdot 10^{-05}$
	$a_{6,3}$	$-0.20202048 \cdot 10^{-09}$	$a_{8,3}$	$-0.52424597 \cdot 10^{-13}$
	$C_{0,3}$	$-0.22521537 \cdot 10^{+00}$	$C_{2,3}$	$-0.22547146 \cdot 10^{+00}$
	$C_{4,3}$	$-0.25667211 \cdot 10^{-03}$	$C_{6,3}$	$-0.58438139 \cdot 10^{-06}$
	$C_{8,3}$	$-0.16631199 \cdot 10^{-08}$		
$d_4 = -25.067999$				
Image quality	field angle γ (arcmin)		OPD_{rms} (units of λ)	
	0.0′		0.038	
	3.7′		0.030	
	7.3′		0.014	
	10.3′		0.040	
	11.0′		0.051	

7.6.7.1 Telescope Tolerances and Atmospheric Turbulence

A classical tolerance analysis of a new-generation telescope, a focusing device with the numerical aperture of a medium-power microscope objective but a pupil diameter that is approximately 10^4 times larger, leads to exceedingly small tolerances as compared to the transverse and axial dimensions of the telescope. Taking the primary mirror with its axis of symmetry as geometrical reference, the tilt, decentring and axial shift of the secondary and tertiary mirror should remain within the limits given in Table 7.23. Within these limits, diffraction-limited operation of the telescope is guaranteed. These values cannot be maintained in a passive way. Absolute surface position stability of large mirror telescopes has been demonstrated up to diameters of 5 m. Beyond this value, for thick mirror substrates, the 24 hour period thermal drift effects of the telescope and its environment (dome and open sky) seriously degrade good observation. On top of this, the imaging quality of telescopes with pupil diameters in excess of 0.5 to 1 m is stochastically degraded by the atmospheric turbulence ('seeing'). Favourable viewing conditions with hardly any seeing effects are very rare. For this reason, diffraction-limited operation of a large telescope is not strictly required. The seeing effects introduce a lateral decoherence over the incident beam at the scale of typically 0.5 to 1 m and this limits the angular resolution to that of a telescope with such a small aperture (2 to 5 µrad, or 0.5″ to 1″).

Table 7.23. Typical mechanical tolerances of a new-generation telescope with a primary mirror diameter of 38 m.

Mirror	Perturbation	Value	Displacement at rim
2	tilt	0.7 μrad	±4 μm
3	tilt	1.5 μrad	±10 μm
2	decentring	±13 μm	
3	decentring	±70 μm	
δd_2	axial shift	±50 μm	
δd_3	axial shift	±50 mm	

This is a strong degradation with respect to the theoretical angular resolution $\delta\gamma$ of the new-generation telescope in the visible spectrum (λ = 550 nm),

$$\delta\gamma = \frac{1.22\lambda}{D} = 1.8 \text{ nrad } (0.004'') . \tag{7.6.71}$$

The mechanical tolerances can be compensated for with the aid of active optics. The telescope mirrors are not made of a single piece, but composed of a certain number of hexagonal elements, mounted on mechanically adjustable supports. A drastic reduction in mirror weight is obtained because the element thickness can be made much less than the thickness for a monolithic mirror. The shape of the mirror segments follows the as-designed profile of the mirror surface, each element being adjusted with respect to piston and tilt. The information for the mechanical adjustment of each mirror segment comes from the wavefront deviation over the cross-section of an image of the telescope pupil. Such a wavefront aberration is measured with the aid of, for instance, a Shack–Hartmann wavefront sensor **[12]**. With mechanical adjustments at a slow rate, much slower than the turbulence variations in the dome and the night sky, the overall shape and position of one or more than one of the telescope mirrors are maintained close to the nominal, as-designed values. The shape of the surfaces is preferably adjusted such that the OPD_{rms} of the imaging beams is below $\lambda/15$, requiring rms surface departures that should be definitely less than $\lambda/30$. Under such conditions, the instrument could achieve star images with a Strehl intensity as high as 80%.

In practice, the atmospheric seeing will sharply degrade the Strehl intensity. The effect of atmospheric seeing can be modelled with the aid of a stochastic phase screen, quickly varying in shape as a function of time. We define σ as the root-mean-square pathlength variation induced in the incident beam by such an atmospheric phase screen. The rms phase variation over the telescope diameter equals $2\pi\sigma$. The Strehl intensity I_S of a stellar image in the focal plane of the telescope is then given by

$$I_S = \exp\left\{-\left(\frac{2\pi\sigma}{\lambda}\right)^2\right\} , \tag{7.6.72}$$

an expression analogous to the one that results from the calculation of the specular reflection of a rough surface (see Section 10.1 where a detailed analysis of the light scattering from a stochastic surface is given). For a σ value as small as $\lambda/2$ the Strehl intensity has been effectively reduced to zero. In the presence of strong atmospheric turbulence, the ideal diffraction image of a stellar object is reduced to a blurred image with a width that is of the order of λ/L_t. The quantity L_t stands for the $1/e$ lateral correlation distance of the stochastic perturbation of the turbulent atmosphere. The resolution degradation of the telescope is given by the ratio L_t/D with D the telescope diameter. Adaptive optics seems to be the only method which is able to correct for the atmospheric turbulence in real time. By carrying out the correction with the aid of brilliant reference stars or artificial, laser-generated point sources in the upper atmosphere, even weak stellar objects can be sharply imaged. At the time of writing of this book, its successful demonstration at the scale of new-generation telescopes is planned as of 2025.

7.7 Microscope Optics

The optics of a classical wide-field microscope can be divided into three parts, the condenser optics, the objective and the eyepiece. The latter part is the detection unit comprising the human eye or a camera objective with image sensor. We briefly discuss the condenser optics and its connection with the imaging stage of the microscope. The main topic of this section is the design of a microscope objective. Especially in the case of a high-numerical-aperture objective, quite challenging design problems are encountered for which interesting solutions have been found. A special case is formed by scanning objectives such as those used in optical disc systems. In these light-weight and compact objectives it is necessary to apply aspheric surfaces to meet the stringent optical and mechanical specifications.

7.7.1 Condenser Optics of the Classical Microscope

The standard microscope illumination system goes back to Köhler **[190]** and achieves a well-separated imaging chain of source/pupil and object/image throughout the microscope. In Fig. 7.57 we show the various imaging steps from the source up to the detection plane, split up in object (O) and pupil imaging (P). The planar source S irradiates the first lens C_1 of the condenser system. The plane of C_1 is conjugate with the microscope object plane O_0 and is labelled O_{-1}. If the source S itself shows spatial emission variations, these tend to be averaged out in the plane O_{-1}. The lens at O_{-1} has been provided with an iris diaphragm which is sharply imaged onto the object plane (field diaphragm). The source S is imaged onto a plane P_0 where a second iris diaphragm has been positioned at the point F_{C2}. With the aid of the second lens C_2 of the Köhler illumination system the diaphragm P_0 is imaged at infinity. This infinitely distant pupil image in object space is focused by the microscope objective L in its back focal plane through F'_L. In this plane amplitude and phase changes of the light diffracted by the object can be introduced to enhance the imaging of the object. In the classical microscope with a prescribed tube length t, originally 160 mm, the eyepiece was inserted with its object focal point F_E coincident with the magnified intermediate image O_1 of the object O_0. In the case of visual observation, this intermediate image is then imaged with aid of the eyepiece and the eye lens onto the retina (O_2). The pupil image P_1 is imaged (demagnified) onto the iris of the eye (P_2) such that all the source light can reach the retina and produce a bright image. The practical advantages of the Köhler illumination system are the following:

- The field diaphragm allows the illumination of only a limited area of the object plane. Any light scattering from other parts of the object is avoided and the image contrast in the image region of interest is maximised.
- The aperture diaphragm allows variation of the aperture of the illuminating beams at a general point Q_0 of the object plane O_0 in an identical way. In this way, the coherence properties of the illumination are uniformly controlled over the entire object. The source image at the location of the object is at infinity and this implies telecentric imaging by the microscope objective L. Especially in microscopic metrology, telecentric imaging is a prerequisite to avoid any lateral misreading if the object is not exactly flat.

In applications where illumination uniformity over the object plane at O_0 is very important, the Köhler illumination system may show shortcomings. From Fig. 7.57 it is clear that variations in the far-field radiation pattern of the source (plane O_{-1}) will become visible in the object plane of the microscope. Such an illumination non-uniformity can be reduced by applying a multichannel condenser system which shows a similarity with the facetted eye of an insect and is commonly called a fly's eye condenser. Figure 7.58 shows a cross-section of such a composite condenser system in which the lightpaths are shown for the extreme source points through the outer facets of the facet assembly. An individual facet has been shown in the insert b) and consists, for example, of a solid glass tube with two outer spherical surfaces. The tube acts as a two-lens assembly with equal power and an intermediate distance which equals the image-side focal length of the first surface and the object-side focal length of the second surface. An incident beam focused on the entrance surface leaves the tube as a parallel beam and an incident parallel beam leaves the tube as a diverging beam with its focus on

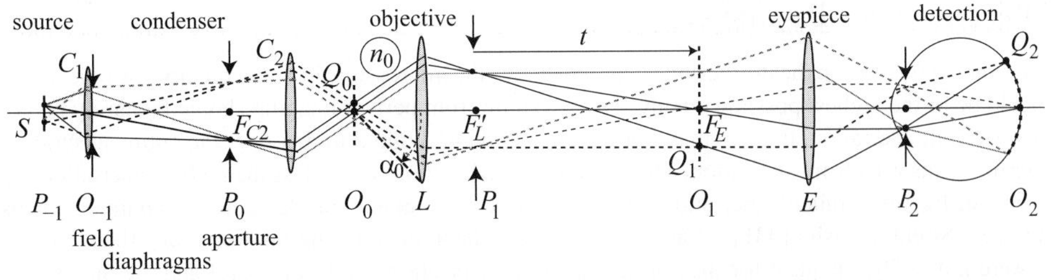

Figure 7.57: A microscope with condenser, objective, eyepiece and detection part. The layout of the condenser part is due to Köhler. A pencil with dashed rays has been shown which leaves an off-axis point P_{-1} of the source, together with a second pencil of (solid) rays leaving a diametrically opposite source point.

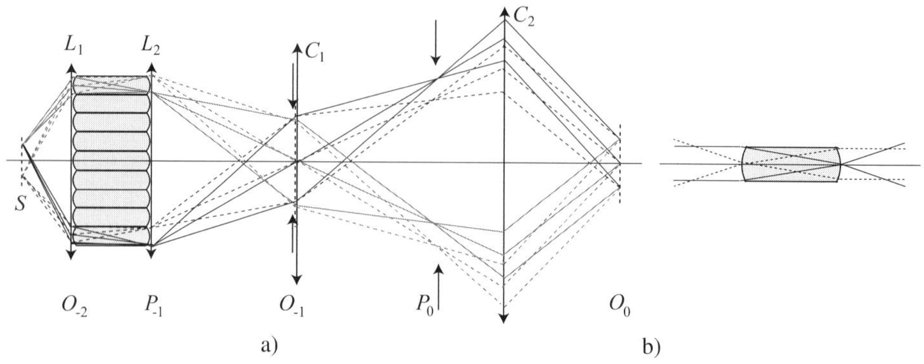

Figure 7.58: a) An improved Köhler condenser system equipped with a facetted fly's eye lens to improve the homogeneity of the illumination in the object plane.
b) Typical ray paths through an individual glass tube of the fly's eye lens.

the exit surface. The tubes are arranged in a rectangular or hexagonal array which is in close contact with two lenses L_1 and L_2. The first lens L_1 images the source S at infinity and the entrance surface of the glass tube produces a source image on its exit surface. The second lens L_2, together with the optical power of the exit surface of each tube, produces a magnified image in the plane labelled O_{-1} of the light distribution on the entrance surface of each glass tube. In each point of the plane O_{-1} we observe a superposition of rays which belong to different spatial areas of the source and to different radiation directions of the source. The net result is a very good averaging in O_{-1} of the spatial and directional intensity variations of the physical source S. Illumination variations can be kept to a level less than 1%. The lens C_1 is positioned in the plane O_{-1}, together with the field diaphragm, as is the case in the classical Köhler illumination system. The lens C_1 images the exit surfaces of the glass tubes onto the pupil diaphragm P_0 and the condenser lens C_2 projects this pupil to infinity in the object space of the microscope. The imaging in the object plane O_0 is telecentric as in the standard Köhler system. The intensity distribution in the pupil plane P_0 will show the facet structure of the fly's eye assembly. With a sufficiently large number of facets, typically 30 to 100, the influence of these periodic residual variations in the pupil illumination on the image formation remains negligible. Instead of a fly's eye condenser, hollow reflecting tubes and holographic elements have been successfully used to achieve the averaging-out of spatial and directional inhomogeneity of the source.

The optimum design of condenser lenses led to extensive scientific discussion in the late nineteenth and early twentieth century. With the advent of modern coherence theory **[70]**, **[386]**, it was concluded that the aberration correction in a condenser system is basically not relevant for the imaging quality of the microscope. The condenser should produce the desired intensity distribution in the aperture diaphragm, or, equivalently, in the far-field region of the object. The condenser optics can further be used to precorrect aberrations of the pupil imaging from P_0 to P_2 such that the light transport through the system is optimised for all off-axis imaging points Q_2 on the retina of the eye.

7.7.2 Objective Lens Parameters

In this subsection the focus is on the most critical component for the microscope imaging quality, the objective lens L. The imaging should be of the highest possible quality. The deviation from perfect imaging should be hardly perceptible and, in terms of wavefront aberration, the maximum departure should be limited to $\lambda/4$ over the entire spectral range of the source. In terms of rms wavefront aberration, a high-grade objective should have less than 0.03 λ aberration, significantly below the 'just-diffraction-limited' aberration level of 0.07 λ rms value. This leads to a maximum intensity in the diffraction image (Strehl intensity **[331]**) of at least 95% of the theoretical maximum. For a long time, the measurement techniques were not really adequate for measuring wavefront aberrations up to this level and the quality of objectives could vary substantially from one specimen to the other. Moreover, up to the 1950s, surface anti-reflection coatings were technically not yet feasible and the image contrast of high-resolution multi-element objectives was reduced because of spurious reflections. Precise interferometric wavefront measurement, both in the manufacturing stage and at the location of the user, allow the detection of any minor defect. Microscope objectives with the above-mentioned state of correction ($\leq 0.03\lambda$ aberration) are now routinely produced.

The main parameter of a microscope objective is its numerical aperture $NA = n_0 \sin\alpha_0$ where n_0 is the refractive index of the object space and α_0 half the opening angle of the cone of rays that can be captured by the objective. Further parameters to be specified are the focal distance f'_L of the objective, the object field diameter and the spectral range. To find the optimum focal distance as a function of the chosen NA value, we consider observation with the aid of the human eye. The angular separation δ_{eye} after the eyepiece of the smallest resolvable detail Δ_R in the object plane should match the angular resolution of the human eye. For the size of the smallest resolvable detail Δ_R in the object plane we invoke the Rayleigh criterion (see Section 10.1), which defines Δ_R as the minimum distance between two point sources that can be separated in image space with an intensity modulation of 20%. Following the argument related to Eq. (10.2.1) for the separation of two incoherent object points, the distance Δ_R equals 60% of the diffraction unit λ_0/NA which was introduced in Subsection 4.9.4. The angular separation δ_L after the objective of two points at a distance of Δ_R in object space is given by $\delta_L = \Delta_R/f'_L$. In the image space of the microscope, after the eyepiece E, the angular separation of the two points is multiplied by the angular magnification m_A of the eyepiece which is equal to t/f'_E, where t is the distance $F'_L F_E$, the so-called tube length of a microscope. For a long period of time, the standardised tube length of a microscope was 160 mm. The angular separation of the two object points after the eyepiece should be equal to or larger than the average angular resolution limit of the human eye of an adult person, defined by 1 minute of arc ($1' = 3\cdot10^{-4}$ rad). We thus require that

$$\frac{t}{f'_E}\frac{\Delta_R}{f'_L} = \frac{t}{f'_E}\frac{0.6\lambda_0}{(NA)f'_L} = q\,(3\cdot10^{-4})\,, \tag{7.7.1}$$

where q is a proportionality factor ≥ 1. In practice, the factor q is given a value between 2 and 4 to guarantee that the observation by the eye is free of strain and fatigue.

We are interested in the optimum value of the (lateral) magnification m_T of the compound microscope as a function of the NA value of the objective. The magnification m_T is the product of the objective and the eyepiece magnification, yielding

$$m_T = m_{T,L} m_{T,E} = \frac{t}{f'_L}\frac{l_p}{f'_E}\,, \tag{7.7.2}$$

where, as usual, the (angular) eyepiece magnification is given by the ratio of the angular extent of an object when observing this object in the front focal plane of the eyepiece and when observing it with the naked eye at a distance l_p. The distance l_p has been standardised to 250 mm since the 'average' eye of a young human being has an accommodation power of 4 diopters. We eliminate $t/f'_L f'_E$ from the Eqs. (7.7.1) and (7.7.2) and find the expression

$$m_T = 5\cdot10^{-4} q\,(NA)\,\frac{l_p}{\lambda_0}\,. \tag{7.7.3}$$

Substituting $l_p = 250$ mm and $\lambda_0 = 500$ nm, we find for the optimum magnification range of a compound microscope,

$$500\,(NA) \leq m_T \leq 1000\,(NA)\,, \tag{7.7.4}$$

where we have assumed that the value of q for relaxed observation is in the range $2 \leq q \leq 4$. The specific tube length of 160 mm, which was a standardised value for a long period of time, has been abandoned nowadays. Microscope objectives are designed for imaging at infinity and the user is free to add an auxiliary (perfect) lens with focal length f'_A to bring this infinitely distant image to an image plane at finite distance f'_A. The rule of thumb for the microscope magnification of Eq. (7.7.4) does not depend on the value of f'_A.

The lateral size Δ_r of the image on the retina of the smallest resolvable object follows from the product of the angular resolution of the eye and the focal length f_{eye} of the average human eye,

$$\Delta_r = q\,(3\cdot10^{-4}) f_{\text{eye}}\,. \tag{7.7.5}$$

The power of the eye lens is approximately 60 diopters, yielding an object-side focal length f_{eye} of 17 mm. For a value $q = 2$ we then find that the value of Δ_r is approximately given by 10 µm. This value of Δ_r is well adapted to the cone separation and cone size at the fovea (yellow spot) on the retina, which are both of the order of 3 µm. It means that several cones are used for the detection of the smallest feature in the object plane. For a value of $q < 1$, the resolution in the object plane is limited by the cone-detection mechanism.

Table 7.24. Microscope objective parameters for typical values of the numerical aperture NA ($q = 2$).

NA	0.10	0.20	0.40	0.65	0.90	1.35
f'_L (mm)	32	16	8	5	3	2
m_T ($f'_E = 25$ mm)	50	100	200	320	530	800
Field diameter (μm)	1800	900	450	275	200	130

The magnification of the microscope can be made larger by using a stronger eyepiece. It is not recommended to go beyond the value $m_T = 1000\,(NA)$. In this region of high microscope magnification, the eye is able to observe the fine structure of diffraction effects which arise at the aperture rim of the objective. The intensity variations due to diffraction effects are of no relevance for the object structure and could easily lead to erroneous conclusions about object details. Regarding the object field diameter, we choose a visual extent of the microscope image of $2 \times 10°$, equivalent to 1200 linearly resolved points of $1'$ on the retina. For $q = 2$, each resolved feature in the object field is equivalent to a distance of $0.6\lambda_0/(q\,NA)$, yielding a minimum required object diameter of $360\,\lambda_0/NA$. In Table 7.24 frequently chosen combinations of microscope parameters are listed.

The diameter Φ_{P_1} of the diaphragm in the back-focal plane (P_1) of the objective is determined by the intersection points of the marginal rays of an axial pencil and is given by $\Phi_{P_1} = 2(NA)f'_L$. This value follows from the Abbe sine condition, applied to a well-corrected objective lens with one of its conjugate planes at infinity. The absolute value of the lateral pupil magnification between the planes denoted by P_2 and P_1 is given by f'_E/t. Using the expressions for m_T of Eqs. (7.7.2) and (7.7.3) we find that the diameter Φ_{P_2} of the pupil image on the iris of the eye in the plane P_2 is given by

$$\Phi_{P_2} = 2\,\frac{(NA)f'_L f'_E}{t} = 2\,\frac{(NA)\,l_p}{m_T} = 4000\,\frac{\lambda_0}{q}\,, \tag{7.7.6}$$

where λ_0, as all other distances in this equation, has to be expressed in units of mm. If we choose $q = 2$ the value of Φ_{P_2} is exactly 1 mm. This means that the eye with its standard iris diameter of 4 mm can be allowed some freedom of movement without the risk of partly blocking the imaging pencils coming out of the microscope. If q is chosen significantly larger than 2, the beam cross-section in the eye iris becomes smaller and, consequently, the diffraction unsharpness on the retina of the image of a point object increases.

7.7.3 Design of Microscope Objectives

In this subsection we discuss the design and performance of objective lenses which are typical for the common magnification and aperture combinations in microscopy (see Table 7.24).

- $NA = 0.10$
  The simplest microscope objective with an NA value of 0.10 is an achromatic doublet, as discussed in Section 7.2. Depending on the choice of the glass materials, the wavelength range can be 'standard' achromatic (480 nm $< \lambda <$ 650 nm) or extended towards the blue/violet or infrared region (apochromatic objective). The imaging quality should be 'limited by diffraction', meaning that the wavefront aberration preferably is smaller than 0.035λ in rms value.
- $NA = 0.25$
  An objective with an aperture in the order of 0.20 to 0.25 can be made by combining two achromatic doublets. The first design of such a combination goes back to Lister **[214]**, see Fig. 7.59a. The field angle of a microscope objective being small, spherical aberration and coma are the main aberrations. In his publication Lister describes his experimental observation that the spherical aberration and coma of a doublet have equal magnitude but opposite sign when the lateral magnification is changed from $m_T = 0$ to a typical value of $m_T = 0.5$.

The change of sign of the spherical aberration was experimentally verified by observing the axial location of the focused central spot formed by either the inner rays or the outer rays of the focusing pencil. The change of sign of

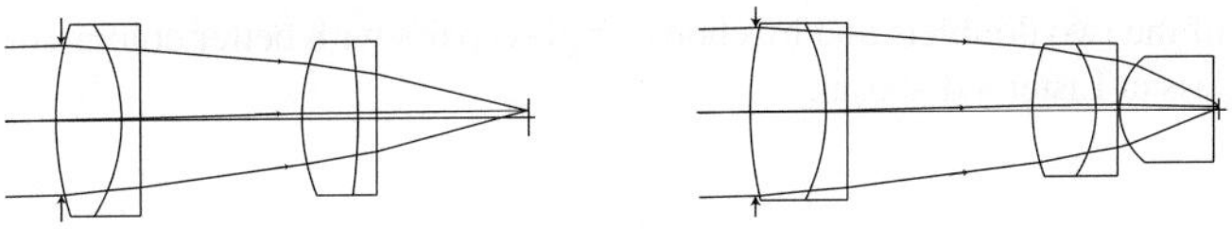

a) b)

Figure 7.59: a) A colour-corrected compound microscope objective of the Lister-type ($NA = 0.25, f'_L = 16$ mm; glasses: SK11 and F2).
b) A high-aperture objective of the Amici type ($NA = 0.65, f'_L = 5$ mm; doublets: SK11 and F2, Amici element: SK16).

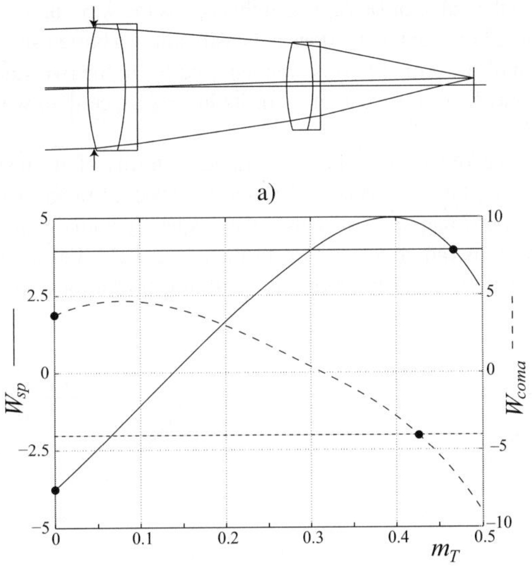

b)

Figure 7.60: a) Lister objective composed of two conformal BK7–F2 doublets; $NA = 0.22, f' = 16$ mm, field diameter = 1.0 mm;
b) Spherical aberration and coma of a convex-plano cemented doublet (BK7–F2) as a function of the lateral magnification m_T. Solid curve: spherical aberration, dashed curve: coma, both in units of λ (550 nm). $NA = 0.12, f' = 100$ mm, field angle = 0.02 .

the comatic aberration followed from the change in direction from inward to outward of the coma tail for an off-axis pencil of rays. In practice, the compensation of the aberrations in the focal plane of the composite lens system could be easily monitored when changing the distance between the two doublets. The observations of Lister are corroborated by the curves in Fig. 7.60b. They depict the spherical and comatic wavefront aberration components of a single doublet as a function of the magnification at which the doublet is used. Taking into account the scaling of the first and the second doublet which compose the Lister objective (see Fig. 7.60a), it follows that the combination of two conformal objectives at magnifications $m_T = 0$ and $m_T \approx 0.45$ produces a good compensation of the spherical and comatic aberration contributions of each doublet (see the aberration values with opposite sign represented by the thick dots in Fig. 7.60b). A typical doubling of the feasible aperture was possible using Lister's proposition. In his publication, the author illustrates the aberration compensation in the twin doublet by drawing some well-chosen rays of axial and off-axis pencils. The Lister composite objective yielded a significant increase in microscopic resolution[1] and is still used in present-day microscopy. The minor improvements which have been introduced since then were made possible by

[1] Lister's communication on this subject was presented to the Royal Society of London on 21 January 1830 and published in Vol. 120 of its Philosophical Transactions. The governors of the Society were greatly impressed by the fine experimental achievements of this amateur opticist and decided to bestow upon him, quite exceptionally, the fellowship of the Society.

allowing different shapes of the two doublets and by choosing glass pairs with better colour correction than was possible with the crown and flint glass at Lister's disposal.

- $NA = 0.40$ up to 0.65
  A further increase in numerical aperture is obtained by augmenting the Lister two-element design with a so-called Amici front lens (see Fig. 7.59b). This convex-plano front lens can be used with equal object and image distance by imaging the centre of curvature. The Amici lens is then a half sphere and the object distance is chosen to be very small to avoid significant aberration contribution from the small air space between the object plane and the flat entrance surface of the front lens. As it was shown in Table 4.1, the lateral magnification of this first imaging step from glass to air is n_1/n_0 with, generally, $n_1 \approx 1.5$ and $n_0 = 1$. The other possibility for the front lens is to exploit the conjugate distances which correspond to the Huygens (or Roberval) aplanatic points, see Fig. 4.16 and Fig. 5.8. In this case, the lateral magnification amounts to the value of $(n_1/n_0)^2$ and this opens the way for a numerical aperture as high as 0.65. The Amici lens is a strongly curved element and it does not easily allow a buried surface for colour correction. For this reason, the chromatic aberration of the Amici lens is compensated for in the two Lister-like doublets of the objective. The central doublet has now a much stronger curvature of its inner surface than what was needed in the basic Lister objective (see Fig. 7.59a and b).
  The performance of a microscope objective is best described in terms of its residual wavefront aberration, both monochromatic and polychromatic, taking into consideration the spectral range over which good imaging has to be achieved. Another representation of imaging quality uses the frequency domain and presents the modulation transfer function of the objective, for various radial positions y_i in the object field. For an $NA = 0.65$ objective of the Amici type we present the monochromatic transfer function ($\lambda = 550$ nm) for various positions in the field (see Fig. 7.61).

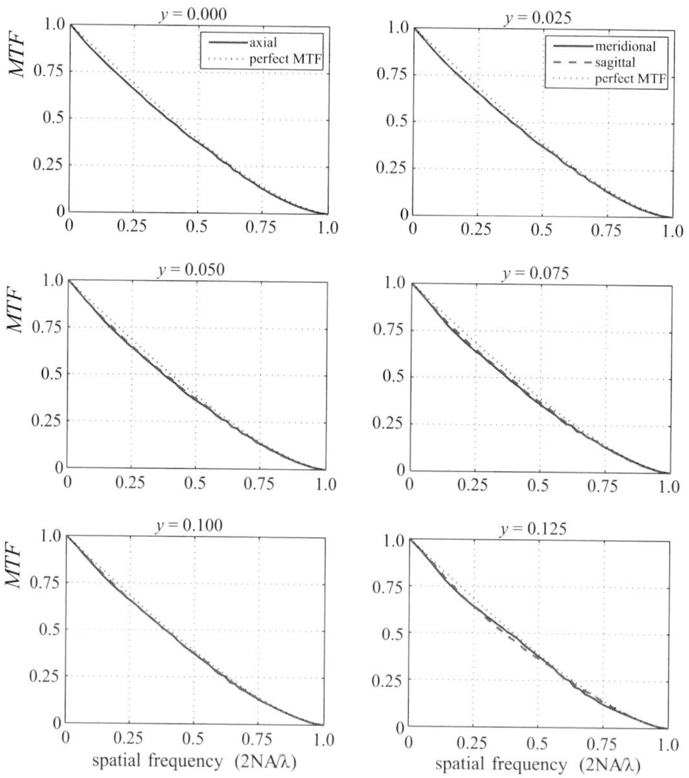

Figure 7.61: Modulation transfer function graphs of a slightly aberrated microscope objective of the Amici type (monochromatic illumination with $\lambda = 0.55$ µm, $NA = 0.65$). The radial excursion in the object plane is given by the coordinate value y.

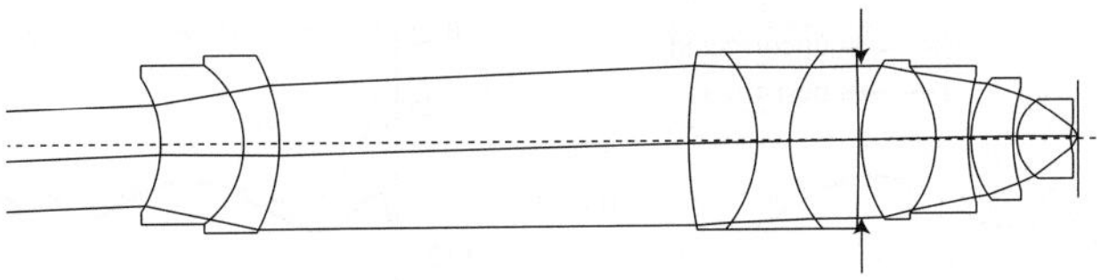

Figure 7.62: A high-numerical-aperture apochromatic objective with extended chromatic correction for a wavelength range 420 nm $\leq \lambda \leq$ 690 nm; $NA = 0.85, f' = 2$ mm.

It follows from the graphs that a well-corrected microscope objective hardly shows a deviation from its theoretical performance (dotted curves). The maximum spatial frequency that is just transmitted by this objective at a wavelength of 550 nm has a value of $\approx$ 2350 mm^{-1}. The graphs show that the 20% modulation depth level is found at a spatial frequency of approximately $4NA/3\lambda$, corresponding to periodic objects with a period of $0.75\lambda/NA$. Half a period with a typical size of 0.375 λ/NA can be just resolved using this 20% criterion for modulation depth. For the Amici objective of this example we then find the value of 0.32 μm.

- $NA \geq 0.85$ (< 1)
  Values of the numerical aperture of 0.85 and beyond ask for a second imaging step in the front part of the objective with aplanatic property. In Fig. 7.62 we have shown such an objective with a second aplanatic front lens according to Fig. 5.42, right-hand drawings. Although the example of this figure applies to a basically zero-thickness lens, a finite-thickness lens can be designed with imaging at the centre of curvature by the entrance surface and imaging at the Huygens aplanatic points by the exit surface. The net lateral magnification of this aplanatic lens with a meniscus shape amounts to the optical contrast of the lens, given by the ratio (n_1/n_0) of the lens refractive index and the index of the surrounding medium. As in the Amici front lens, it is not recommended to introduce colour correction in these elements which are used at large opening angles of the imaging pencils. In the design example of Fig. 7.62, the first colour correction is introduced in the third element, close to the stop. This element is a thick meniscus-type doublet with low power. Together with the triplet element on the other side of the stop, with virtually no net optical power but equipped with strong internal transitions for colour correction, it is possible to reduce the large axial chromatic aberration of the front elements at the high-aperture side. Further away from the stop, an extra doublet with low power has been positioned which is capable of reducing the transverse chromatic aberration of the high-NA front elements.
  The wavefront aberration of a high-NA apochromatic objective is shown in Fig. 7.63 for the extreme parts of the spectrum (λ = 420 nm and λ = 690 nm). On-axis the residual aberration is usually denoted by spherochromatism. This implies a variation of spherical aberration with wavelength; in the figure this is most pronounced in the blue part of the spectrum. The wavefront cross-sections for the outer field point (y = 70 μm) show that the comatic aberration is also wavelength-dependent. The reason for these wavelength-dependent changes in aberration is fact that the aberration changes in the high-NA front part of the objective cannot be fully compensated for in the correcting elements where the pencils have a much smaller opening angle. Optical glasses with special partial dispersion values for colour correction, sometimes home-made by the microscope manufacturer, can reduce the residual colour deviations of such high-NA objectives.
- $NA \geq 1$
  The numerical aperture can be increased above unity by filling the small volume between the object and the first lens surface of the objective with a liquid (immersion microscopy). The first immersion objectives were introduced to avoid the aberration (mainly spherical and coma) that is introduced by the customary air gap (typically 100 μm wide) in a high-NA dry objective. For practical reasons water was used as a liquid. It was not until 1879 that Abbe **[3]** drew the attention of the optical community to the resolution improvement that is obtained by object immersion and the role played by the numerical aperture $n_0 \sin\alpha_0$ in microscopic imaging. From then on, oil immersion was preferred with a typical index n_0 of 1.55. The design of an immersion objective is carried out following the same principles as those applied to the design of a dry high-NA objective. The problems to be solved are harder due to the n_0 times larger opening angles of the imaging pencils which have to be guided through the first aplanatic imaging lenses. The chromatic correction in the subsequent doublets and triplets is more onerous. Despite these design and also manufacturing challenges, it has been possible to produce well-corrected immersion lenses with a numerical aperture as high as 1.35. This corresponds to a value of $\sin\alpha_0 \approx 0.90$, the same practical limit encountered in dry objectives.

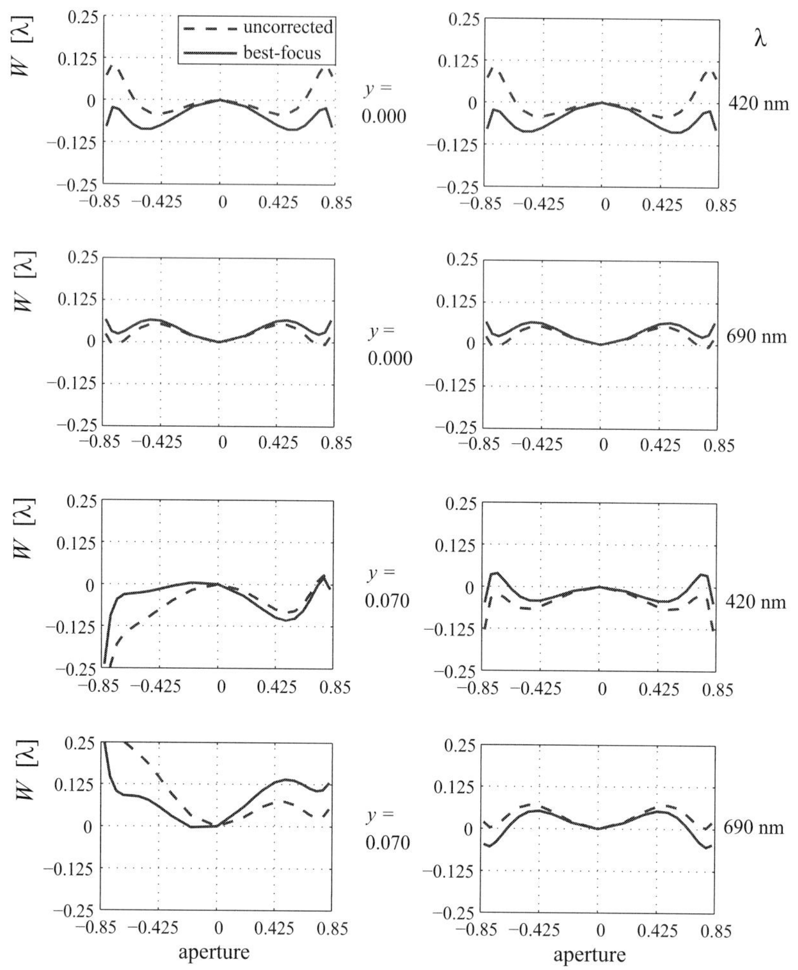

Figure 7.63: Wavefront aberration for the extreme wavelengths (420 nm and 690 nm) of a high-numerical-aperture objective with extended chromatic correction; $NA = 0.85, f' = 2$ mm. Left column: meridional cross-section of the wavefront aberration, right column: sagittal cross-section.

7.8 Aspheric Objectives for Optical Disc Systems

In scanning microscopy, especially in optical disc systems, compact and light-weight objectives are needed to satisfy their dynamical requirements. The wavefront correction of scanning objectives for optical disc read-out should be very good. The reason for this is that the scanning movements with defocusing effects and the read-out through the disc substrate layer introduce ample additional beam aberration and leave only a tiny aberration margin for the mass-produced objectives themselves. A typical measure for the residual wavefront aberration of the objective is an rms value of $0.025\ \lambda$. In terms of on-axis Strehl intensity, a value of at least 97% is required. The design requirements with respect to colour correction are alleviated because a narrow band laser source is used. When using a semiconductor laser, the central wavelength may shift substantially under the influence of its temperature. Slow shifts in temperature are possible due to the environmental conditions. Fast changes occur when the laser wavelength changes with driving current, for instance when the source is switched from the reading to the recording power level. In this section we briefly discuss the objective designs for the various optical storage systems ranging from laser disc via compact disc (CD) and the digital versatile disc (DVD) to the Blu-ray system (BD) with the highest storage capacity.

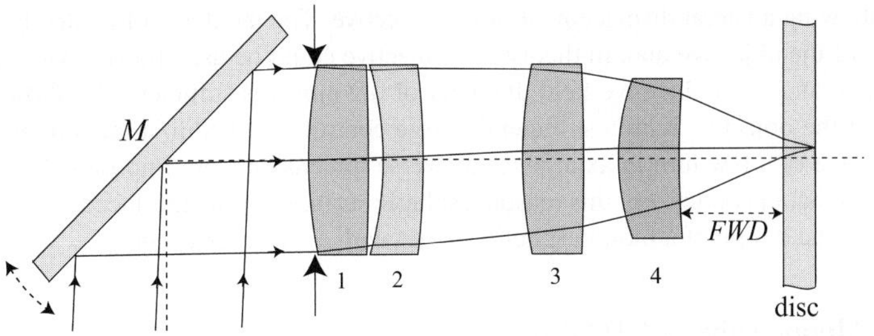

Figure 7.64: A classical microscope objective for optical disc scanning. Video Long Play System, λ = 633 nm, NA = 0.40, f'= 9.2 mm, disc thickness = 1.2 mm.

7.8.1 Video Long Play System (HeNe-laser, NA = 0.40, field diameter 800 μm)

Designs for this relatively low-aperture system are based on classical microscope designs (see Fig. 7.64). The laser beam is expanded with the aid of an afocal system and the objective is used with a parallel incident beam and lateral magnification $m_T = 0$. The objective is mounted in a vertical movement control loop to achieve active focusing on the information layer of the disc. The eccentricity of the spiral information track with respect to the axis of the rotation motor and the irregularity of the spiral itself due to the disc pressing process cause large relative movements of the track to be followed with respect to the objective axis. For this reason a radial tracking system is needed that uses a pivoting mirror to tilt the incident beam such that the focused spot can be laterally moved in the information plane. The total lateral swing for track-following amounts to 800 μm in the VLP system. The system has a large free-working-distance (FWD) between the last surface of element 4 and the top surface of the optical disc. Fast-spinning discs (30 or 25 Hz rotation frequency) with a 30 cm diameter gave rise to large vertical movements of the video disc with respect to the objective and as a mechanical safeguard the FWD-value was chosen to be 4 mm. This choice gave rise to an objective with a large focal distance f' with a value of 9.2 mm. To achieve the radial swing in the focal plane, the required angular movement of the tilting mirror is 2.5°. The elements towards the high-aperture side of the objective are used in or close to the aplanatic condition. The spherical aberration which is introduced by collimating the beam at the mirror side is compensated for by the introduction of the strongly bent meniscus-type element labelled 2 in Fig. 7.64. The large angles of incidence on the hollow side of the meniscus produce the necessary amount of spherical aberration with a sign opposite to that of the positive elements. To a minor extent, the spherical aberration introduced by the disc substrate also contributes to achieving almost zero fourth-order spherical aberration. On purpose, a small fourth-order residue is left which balances the higher-order aberration produced by the steep element 2. The stop position is chosen at the elements 1 and 2 where the spherical aberration compensation takes place. With this stop position, the residual spherical aberration does not produce field-dependent coma. The compensation mechanism for spherical aberration using the steeply curved element 2 is also effective for the reduction of coma in the field. The remaining aberration of the objective is field curvature, amounting to 6 μm at the border of the field. With a diffraction-based focal depth of typically 2 μm, the field excursion due to the tilting mirror requires a focusing action of the vertical position of the objective. The field curvature thus causes a cross-talk between the two servo control systems of the disc player. Special precautions have to be taken to avoid instability or resonances of these two control systems which are mutually influenced by the curved objective field.

In later versions of the VLP video disc system and in subsequent optical storage systems the voluminous HeNe-laser was abandoned and replaced by the compact semiconductor laser. For the CD system, the semiconductor material was an alloy composed of the elements Al, Ga and As. This type of laser operates in the near-infrared ($780 < \lambda < 820$ nm) with a typical mode spacing $\Delta\lambda$ in the wavelength domain of a few tenths of a nm. With a few competing modes in the spectrum, the laser bandwidth is of the order of 1 to 2 nm, a very substantial increase with respect to the negligible value of 0.1 pm which is typical for the bandwidth of a HeNe-laser.

7.8.2 CD System (AlGaAs-laser, NA = 0.45, field diameter 150 μm)

The compact laser source and a breakthrough in aspheric surface manufacturing led to a change in optical lightpath design. The voluminous tilting mirror for tracking the information in the radial direction was abandoned. In the presence of a much smaller 12 cm diameter disc with a better controlled shape of the information spiral, the information tracking was achieved by also allowing a lateral displacement of the objective. The incident collimated laser beam was oversized with respect to the stop of the objective and, in theory, the objective could be used for on-axis imaging only. With some mechanical mounting tolerances, an objective field diameter of 0.2 mm was sufficient. To allow the simultaneous axial and lateral movement of the objective with a sufficiently large control bandwidth, a maximum weight of the mounted objective of 1 gr and a size of a few mm in axial and transverse dimension were imposed. These requirements are not easily met with classical spherical optics. For this reason, aspheric surfaces were introduced in the design. In what follows we present in some detail the design of a mono-aspheric singlet with a flat reverse side.

7.8.2.1 Design of a Mono-aspheric CD Objective

Figure 7.65 shows a schematic drawing of such an aspheric objective, together with the optical disc with the information layer on its reverse side. The aberration analysis of such an imaging system allows for an analytical treatment in the third-order Seidel aberration domain. From the design that follows from third-order theory, the final design for finite numerical aperture is obtained. Typical design values for a CD objective are a numerical aperture of $NA = 0.45$, a central wavelength of $\lambda = 785$ nm and a field radius of 150 μm. In the case of an incident parallel beam, the lateral magnification m_T equals zero. If the entire lightpath, including the laser source, is actuated, the objective should operate at finite magnification. In the case of a CD system, with a typical laser far-field emission angle of 0.09 radians and a numerical aperture at the image side of 0.45, the angular (and lateral) magnification is given by $m_T = -0.2$.

To calculate the Seidel aberration of an aspheric singlet lens in the CD lightpath including the optical disc substrate, we use the formulas of Eqs. (5.3.37) and (5.3.38). Paraxial optics yields the various quantities $A_{m,j}$, $r_{m,j}$, $\delta(\alpha_{m,j}/n_j)$ and $B_{m,j}$ at the surfaces 1 to 3. The aberration at surface 4, the information layer itself, is zero because we force imaging at this surface. The fourth-order aspheric coefficient $a_{4,1}$ at surface 1 is used to reduce the spherical aberration of the CD imaging system to zero. For the first surface, we have the imaging distance condition

$$s_1' = \frac{n_1}{K_1}(1 - m_T), \qquad \text{with} \quad K_1 = (n_1 - n_0)c_1 \,, \tag{7.8.1}$$

where K_1 is the paraxial power of the first surface and $c_1 = 1/R_1$ the curvature of this surface. For the paraxial heights of incidence on the surfaces we find, using the conventions on signs of angles and distances of Chapter 4,

$$r_{m,1} = (-n_1\alpha_1')\left(\frac{1 - m_T}{K_1}\right),$$

$$r_{m,2} = (-n_1\alpha_1')\left(\frac{1 - m_T}{K_1}\right)\left[1 - \frac{d_1 K_1}{(1 - m_T)n_1}\right],$$

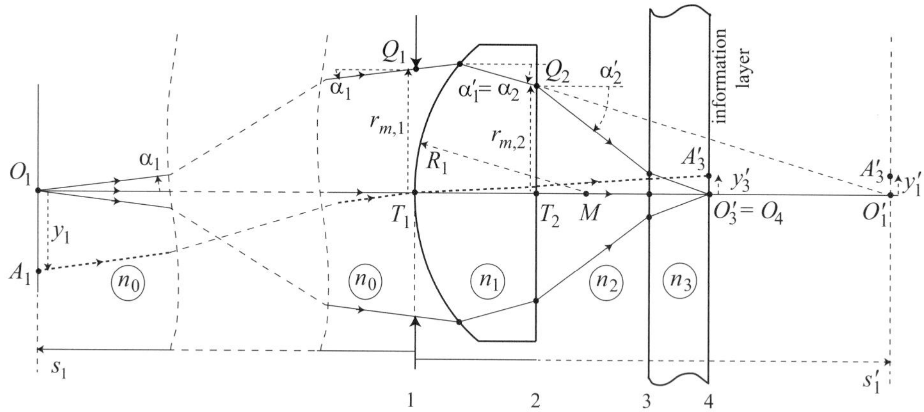

Figure 7.65: An aspheric convex-plano singlet lens for optical disc read-out. The marginal rays of an axial beam have been drawn and the principal ray of an off-axis beam from A_1 in the object field to its image point A_3' in the plane of the information layer of the optical disc with lateral coordinate $y_1' = y_3'$.

$$r_{m,3} = (-n_1\alpha_1')\left(\frac{1-m_T}{K_1}\right)\left[1 - \frac{d_1K_1}{(1-m_T)n_1} - \frac{d_2K_1}{(1-m_T)n_2}\right] ,$$

$$r_{m,4} = (-n_1\alpha_1')\left[\frac{1-m_T}{K_1} - \frac{d_1}{n_1} - \frac{d_2}{n_2} - \frac{d_3}{n_3}\right] . \tag{7.8.2}$$

The quantity $|n_1\alpha_1'|$ equals the numerical aperture of the converging beam after the first surface and its value is maintained after refraction through the subsequent planar surfaces.

The imaging condition on the back of the optical disc requires $r_{m,4}$ to be zero and this yields the imaging equation for the lens-to-disc distance d_2,

$$\frac{d_2}{n_2} = \frac{1-m_T}{K_1} - \frac{d_1}{n_1} - \frac{d_3}{n_3} , \tag{7.8.3}$$

where we have assumed that the lens thickness d_1 and the disc thickness d_3 have prescribed values. For the quantities $A_{m,j}$ we have

$$A_{m,1} = (n_1\alpha_1')\left(\frac{n_1m_T - n_0}{n_1 - n_0}\right) , \qquad A_{m,2} = A_{m,3} = n_1\alpha_1' . \tag{7.8.4}$$

The remaining factors needed for the evaluation of Eq. (5.3.38) are given by

$$\delta\left(\frac{\alpha_{m,1}}{n}\right) = (n_1\alpha_1')\left[\frac{1}{n_1^2} - \frac{m_T}{n_0^2}\right] ,$$

$$\delta\left(\frac{\alpha_{m,2}}{n}\right) = (n_1\alpha_1')\left[\frac{1}{n_2^2} - \frac{1}{n_1^2}\right] ,$$

$$\delta\left(\frac{\alpha_{m,3}}{n}\right) = (n_1\alpha_1')\left[\frac{1}{n_3^2} - \frac{1}{n_2^2}\right] ,$$

$$B_{m,1} = B_{m,2} = B_{m,3} = \frac{y_3'K_1}{1-m_T} , \tag{7.8.5}$$

where $y_3' = y_1'$ is the lateral coordinate value corresponding to the border of the required image field. Using Eq. (5.3.38), we obtain for the third-order wavefront aberration values corresponding to the marginal aperture ray α_1 and the extreme field coordinate y in Fig. 7.65,

$$W_S = -\frac{1}{8}\frac{(1-m_T)(n_1\alpha_1')^4}{K_1}\left\{\left(\frac{n_1m_T - n_0}{n_1-n_0}\right)^2\left(\frac{n_0^2 - m_Tn_1^2}{n_0^2n_1^2}\right) + 8a_{4,1}(n_1-n_0)\left(\frac{1-m_T}{K_1}\right)^3 + \left(1 - \frac{d_1K_1}{n_1(1-m_T)}\right)\left[\frac{1}{n_2^2} - \frac{1}{n_1^2}\right] + \frac{d_3K_1}{n_3(1-m_T)}\left[\frac{1}{n_3^2} - \frac{1}{n_2^2}\right]\right\} , \tag{7.8.6}$$

$$W_C = -\frac{1}{2}(n_1\alpha_1')^3y_3'\left\{\left(\frac{n_1m_T-n_0}{n_1-n_0}\right)\left(\frac{n_0^2 - m_Tn_1^2}{n_0^2n_1^2}\right) + \left(1 - \frac{d_1K_1}{n_1(1-m_T)}\right)\left[\frac{1}{n_2^2} - \frac{1}{n_1^2}\right] + \frac{d_3K_1}{n_3(1-m_T)}\left[\frac{1}{n_3^2} - \frac{1}{n_2^2}\right]\right\} , \tag{7.8.7}$$

$$W_A = -\frac{1}{2}\left(\frac{K_1}{1-m_T}\right)(n_1\alpha_1')^2(y_3')^2\left\{\left(\frac{n_0^2 - m_Tn_1^2}{n_0^2n_1^2}\right) + \left(1 - \frac{d_1K_1}{n_1(1-m_T)}\right)\left[\frac{1}{n_2^2} - \frac{1}{n_1^2}\right] + \frac{d_3K_1}{n_3(1-m_T)}\left[\frac{1}{n_3^2} - \frac{1}{n_2^2}\right]\right\} , \tag{7.8.8}$$

where W_S, W_C and W_A are the third-order spherical, coma and astigmatic aberration values of the lens system. In the expressions above, we have eliminated the distance d_2 which can be expressed in the other design parameters according to the imaging condition of Eq. (7.8.3). In the Seidel approximation, the spherical aberration W_A is reduced to zero by choosing the appropriate value of $a_{4,1}$, the fourth-order aspheric coefficient of surface 1. The comatic aberration W_C does not depend on the value of $a_{4,1}$ because the stop coincides with the first surface. It can be made zero by choosing a particular value d_1 as a function of the refractive index n_1 of the lens material,

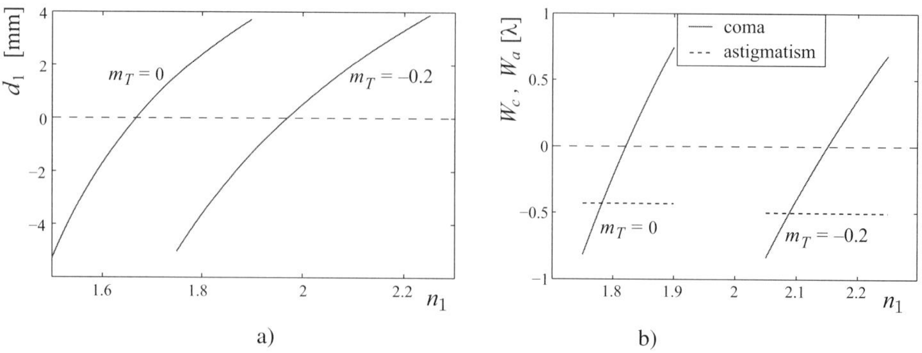

Figure 7.66: a) The optimum thickness d_1 of the aspheric singlet as a function of its refractive index n_1 according to the Seidel aberration theory ($m_T = 0$ and $m_T = -0.2$).
b) The third-order residual coma and astigmatism of the aspheric singlet ($m_T = 0$ and $m_T - 0.2$, respectively, $y_3' = 150$ μm). CD system, $\lambda = 785$ nm, $NA = 0.45, f' = 4.5$ mm, disc thickness $d_3 = 1.2$ mm.

$$d_1 = \frac{n_1^3(n_2^2 - n_3^2)}{n_3^3(n_1^2 - n_2^2)} + \frac{n_1(1 - m_T)}{K_1}\left(1 + \frac{n_2^2(n_1 m_T - n_0)(n_0^2 - m_T n_1^2)}{n_0^2(n_1^2 - n_2^2)(n_1 - n_0)}\right). \tag{7.8.9}$$

In Fig. 7.66a the thickness d_1 has been plotted as a function of the refractive index n_1 of the lens material to achieve zero third-order coma. In the case of a parallel incident beam, positive values of d_1 arise for $n_1 > 1.67$. For a practical lens thickness ($d_1/f_1' \geq 0.6$), we see from the graph that a high index of at least 1.80 is needed. This high value can be explained by the following argument. In the thin-lens approximation, with $m_T = 0$, the zero coma condition for a lens in air was fulfilled for an index $n_1 = (1 + \sqrt{5})/2 = 1.618$ (see Eq. (5.4.32)). In the case of the CD objective, the focusing distance f_1' in air is partly replaced by the finite lens thickness d_1 and the thickness of the optical disc substrate d_3. The coma introduced by these extra media is cancelled by giving the lens a higher index than the one required in the thin-lens approximation.

In Fig. 7.66b we have chosen a realistic finite thickness $d_1 = 2.7$ mm of the lens. For this fixed value of d_1, the comatic and astigmatic aberration is plotted as a function of the refractive index n_1. The coma is made zero for $n_1 \approx 1.82$ ($m_T = 0$) and for $n_1 \approx 2.14$ ($m_T = -0.2$). The first refractive index value is close to that of an available optical glass (e.g. LaSFN9), the second index asks for the more exotic optical material zircone oxide (ZrO_2). The figure also shows that the astigmatism of both systems hardly changes as a function of refractive index. The first lens surface with optical power $K = 1/f'$ yields an astigmatic aberration which is independent of the lens refractive index. This astigmatism is then slightly reduced by the astigmatism introduced at the planar transitions. This reduction only weakly depends on the indices of lens and disc material.

The design for full aperture is carried out by applying the method for the attainment of axial stigmatism, described in Subsection 6.2 and illustrated in Fig. 6.1. The spherical aberration of the aspheric lens is made rigorously zero and the coma is minimised by applying the third-order predictions for the lens thickness d_1 and the air distance d_2. Because of the appearance of higher-order comatic aberration, a small third-order residual is needed. For a prescribed value d_1, we find a slightly smaller value of the optimum index n_1 than the one that followed from third-order Seidel theory. Using the value $d_1 K_1 = 0.6$, the index n_1 amounts to 1.774 instead of the third-order prediction of 1.821. The aspheric objective lens can be made by glass moulding, for instance using the LaF33 glass from the Schott catalogue. In Fig. 7.67a we observe that the cross-section of the aspheric surface is close to that of an ellipse with its second focal point at F_2. This shape would be obtained if the lens medium with index n_1 were to extend up to F_2. The eccentricity of the ellipse, defined as the ratio of the horizontal and vertical axes of such an elliptical lens surface follows from Eq. (5.1.9) and equals $[1 - (n_0/n_1)^2]^{-1/2}$. The introduction of the media with indices n_2 and n_3 gives a slight deviation from the basic elliptical shape of surface 1. The total amount of aspheric correction that is needed at an aperture of 0.45 equals approximately 25 wavelengths.

An alternative solution, adopted earlier when mould making and glass moulding technology were less advanced, is the application of a plastic aspheric correction layer. The plastic material is, for instance, a UV hardening lacquer. A few drops of lacquer are applied on the spherical lens surface and an aspheric transparent mould, aligned with the plano-spherical lens body, is then brought into contact. The exposure to UV light through the transparent mould hardens the

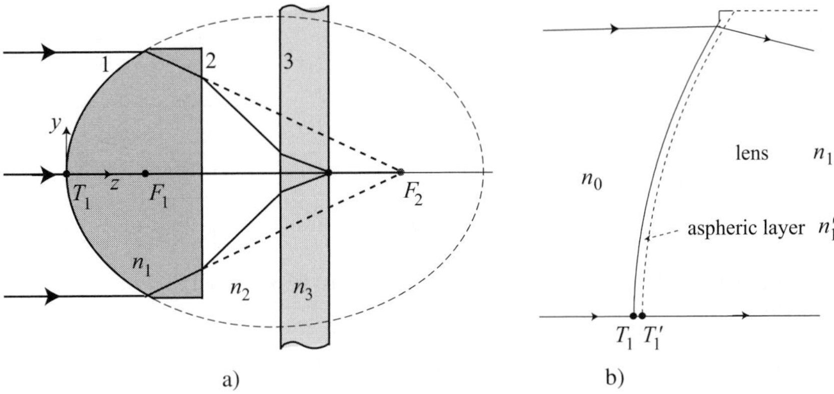

Figure 7.67: a) A single-element aspheric objective lens with planar reverse side.
b) The wavefront correction is carried out by applying an aspheric plastic cap (refractive index n_1') to the front surface. Cross-section (not to scale) of the aspheric correcting plastic layer on the spherical lens body.
CD system, λ = 785 nm, NA = 0.45, f'= 4.5 mm, disc thickness d_3 is 1.2 mm.

lacquer and, with the appropriate correction of the mould shape for shrinkage during hardening, good aspheric objectives are manufactured. In Fig. 7.67b, the aspheric profile of such a correcting plastic cap has been depicted. The thickness of the layer varies from 15 μm on-axis to a minimum of practically zero to again 15 μm at the outside. The lateral extent of the layer also determines the effective diameter of the lens stop. The field diameter of such an objective is limited to 150 μm, mostly by the inevitable residual astigmatism. Although the field diameter of a CD objective could be very small because of the absence of a scanning mirror in the lightpath, some minimum field size is needed for alignment tolerances. In terms of the angular field, we have a tolerance of $\pm 1^o$ and this is sufficient in the mass-production of CD lightpaths.

7.8.2.2 Design of a Bi-aspheric CD Objective

The design of a mono-aspheric objective with convex-plano lens body has the advantage of easy alignment of the objective in the lightpath and with respect to the disc substrate. But the high-index lens body must be made of optical glass which poses a disadvantage in cheap mass-production because of price. The use of two aspheric surfaces in an optical system allows for the simultaneous correction of all orders of spherical aberration and of all coma components that have a linear dependence on the field coordinate. Such an aplanatic design with two aspheric surfaces allows the use of a substantially lower refractive index, for instance, in the region where optical plastics are available. The first plastic bi-aspheric prototype for optical disc systems was machined on a precision lathe in PMMA (polymethylmethacrylate) with n_1 = 1.49 **[123]**. The analytic design method for achieving such an aplanatic imaging system, with arbitrary location of the two aspheric surfaces in the optical system, was described in Subsection 6.2.2. For the CD and DVD system, the plastic ‘bi-asphere’ has become the standard solution for cheap mass-production of high-quality scanning lenses (see Fig. 7.68). The residual aberration in the field is astigmatism. For relatively thin lenses, Seidel’s theory predicts that the astigmatism is proportional to the power of the objective and the square of the field coordinate. The total value of the astigmatism of the CD or DVD imaging system is reduced by the astigmatism with opposite sign which is introduced by the disc substrate. Using the third-order theory and the thin-lens approximation for the objective, we can establish the ratio between the astigmatism originated by the disc substrate and by the thin lens. Using Eqs. (5.4.14) and (5.5.5) for a thin lens in air and a disc substrate with thickness d_3 and index n_3, we obtain,

$$\frac{S_{3,s}}{S_{3,l}} = -\frac{n_3^2-1}{n_3^3}\, d_3 K\,, \tag{7.8.10}$$

where $S_{3,s}$ and $S_{3,l}$ are the Seidel astigmatic coefficients of substrate and lens, d_3 is the substrate thickness and K the power of the objective. The reduction in astigmatism is of the order of 10%, both for the CD and DVD system.

A further reduction in astigmatism (and field curvature) can be obtained by changing the distribution of paraxial powers over the front and exit surface of the objective. The basic bi-convex lens shape is converted into a lens shape with small power either on the front surface or on the back surface. This leads to lens shapes that are, in accordance with camera lens

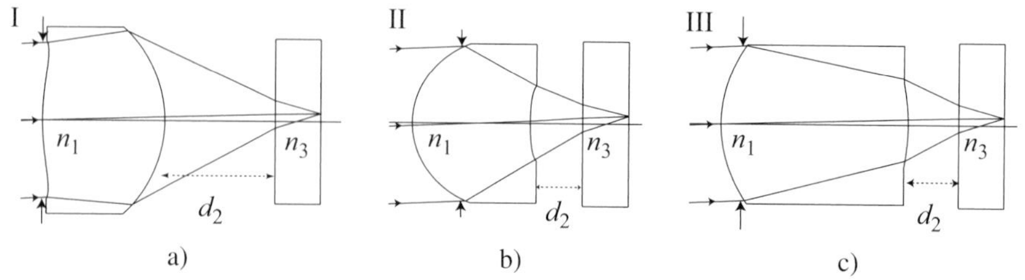

Figure 7.68: a) A bi-aspheric CD objective, inverted telephoto (retrofocus) layout (type I).
b) A bi-aspheric objective, telephoto layout (type II).
c) A standard bi-convex CD objective with two aspheric surfaces (type III).
$NA = 0.45, f' = 4.5$ mm, $\lambda = 785$ nm. Substrate thickness (polycarbonate) 1.2 mm. The stop is on the first lens surface.

design types, of the inverted-telephoto (retrofocus) or of the telephoto type. In the first case, the power of the first surface is close to zero or negative (type I), in the second case we have a small power on the second surface of the lens (type II). In Fig. 7.68 we have depicted the 'exotic' bi-aspheric designs of Type I and Type II, together with the more conventional design of a biconvex aspheric lens, denoted by Type III. The imaging performance of these three objectives is illustrated in Fig. 7.69. Two curves are shown in each graph. The uncorrected wavefront aberration graph has been referred to the paraxial image point. The corrected wavefront cross-sections have been obtained by introducing an axial and lateral shift of the reference point such that the rms wavefront aberration of the full pencil is minimised. In each graph, the horizontal coordinate 'aperture' denotes the sine of the angle of a particular ray in the imaging pencil with respect to the principal ray of the pencil.

The type I objective has a first surface which is positioned close to the paraxial centre of curvature of the second surface. The stop is located on the first surface. Such a geometry fulfils the isoplanatic imaging condition, which can be made aplanatic by giving the first surface an aspheric shape such that the spherical aberration of the lens is reduced to zero. The rigorous concentric position of the stop surface and the second surface can be abandoned when the second surface is made aspheric too. The aspheric departure of the first surface is very large, ±60 μm with respect to the best-fit sphere.[2] The second surface is hardly aspheric (±3 μm) because a spherical shape would, in principle, have been sufficient to achieve zero spherical aberration of the image. By using the lens thickness d_1 as a free parameter, the astigmatism of the image can be reduced. The first two rows of Fig. 7.69 show that the main defect of this bi-aspheric lens is inward field curvature and that the astigmatism is very small. As expected, the on-axis aberration is zero (not shown in the figure) and the coma is negligible. The type II objective has a positive/negative sequence of surface powers; this is advantageous for reducing the field curvature. The aspheric departures of both surfaces are of the order of ±20 μm with respect to the best-fit spherical surface. The third objective example shows a gradual change of convergence of the imaging pencil with two positive power surfaces. For this reason, the aspheric departures are small, ±7 μm and ±3 μm respectively. The residual field curvature and astigmatism of design III are appreciable.

The final choice of the optimum design, to be produced in extremely large quantities, has to include tolerances during manufacturing. Certain geometric demands have also to be taken into account, for instance, the free-working-distance FWD, d_2 in Fig. 7.68, and the field curvature, expressed in the corresponding axial shift in μm for each field point. In Table 7.25 we have listed these quantities which are related to the imaging quality and to the manufacturability of the three types of objectives. The imaging quality is expressed in rms OPD (units of λ) with respect to the optimum reference point of which we give the axial shift in μm. The manufacturing tolerances are given by the deviation from the nominal value which would give rise to an rms OPD (on-axis) of 0.015λ (equivalent to a loss in Strehl intensity of 0.8% with respect to the theoretical maximum of 100%). The design selection on the basis of field diameter favours Type I. As soon as we include manufacturability in the design evaluation, we see that both Type I and Type II have to be abandoned. The lateral decentring tolerance δy_1 or δy_2 of surface 1 with respect to surface 2 is extremely small. With a tolerance of 0.015λ rms OPD for an individual fabrication error, the decentring tolerance amounts to 0.2 μm for the Type I objective and a mere 0.3 μm for the Type II. On top of this, we also observe stringent values for the lens thickness error δd_2 and the tilt errors α_1 and α_2 of the surfaces of the objectives I and II with respect to the optical axis. These high sensitivities are

[2] The aspheric departure of an aspheric surface is measured along the surface normal to a best-fit reference sphere with a certain radius. The best-fit sphere has a radius such that the largest positive and largest negative departure have equal absolute value.

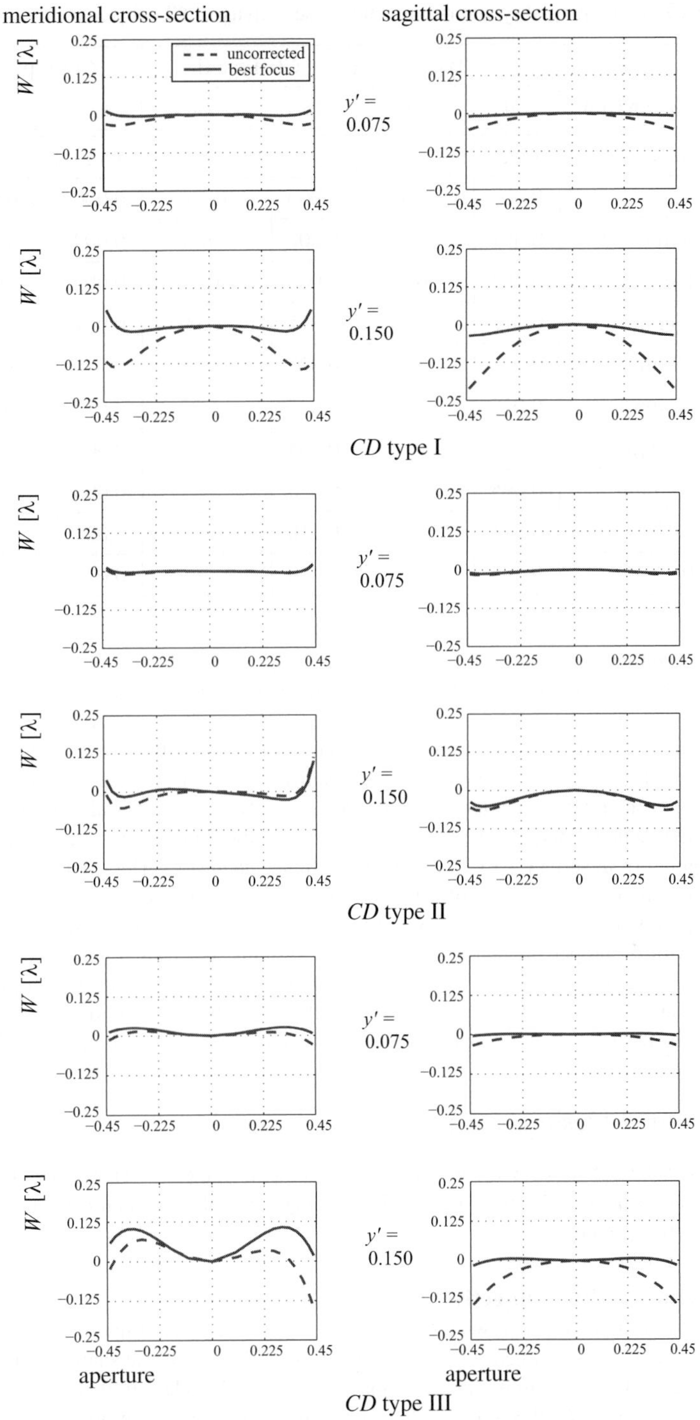

Figure 7.69: Wavefront aberration curves at two field positions ($y' = 75$ and 150 μm off-axis) of the CD objectives of type I, II and III ($\lambda = 785$ nm). On the left-hand side, the meridional cross-section of a wavefront is shown, on the right-hand side the sagittal cross-section.

Table 7.25. Imaging quality and manufacturing sensitivity of CD objectives.

	CD Type I		*CD* Type II		*CD* Type III	
Aspheric sag (μm)	±60	±3	±18	±20	±7	±3
Perturbation	*OPD*	$\delta z_3'$	*OPD*	$\delta z_3'$	*OPD*	$\delta z_3'$
$y_3' = 0$ μm	0.000	0.00	0.000	0.00	0.000	0.00
$y_3' = 40$ μm	0.001	−0.10	0.002	−0.01	0.002	−0.07
$y_3' = 80$ μm	0.004	−0.39	0.006	−0.04	0.009	−0.27
$y_3' = 120$ μm	0.010	−0.87	0.015	−0.09	0.021	−0.61
$y_3' = 160$ μm	0.018	−1.54	0.027	−0.15	0.036	−1.09
$y_3' = 200$ μm	0.028	−2.40	0.045	−0.22	0.057	−1.73
$\delta(d_2) = 10$ μm	0.029		0.164		0.002	
$\delta(y_2) = 10$ μm	0.665		0.420		0.003	
$\alpha_1 = 10^{-3}$	0.011		0.118		0.023	
$\alpha_2 = 10^{-3}$	0.208		0.010		0.025	

due to the strongly aberrated intermediate pencil of rays propagating from surface 1 to 2. Each bi-aspheric design will satisfy by definition the Abbe sine condition for the imaging from object to image space. The intermediate imaging steps at surface 1 or surface 2 do not necessarily obey the isoplanatic condition (for insensitivity to decentring or tilt) or the Herschel condition for an axial shift of conjugate position (insensitivity for lens thickness change).

Although we see a smaller field for objective III accompanied by a considerable field curvature, the fabrication tolerances of this design with 'relaxed' angles of incidence on the two surfaces make it the only feasible type for mass-production. The intermediate imaging steps are close to satisfying the isoplanatic condition and the gentle steps in convergence angles going from surface 1 to 2 avoid a large offence against the Herschel condition. Ample tolerances for geometric lens perturbations are the result of this relaxed design. The field curvature can be tolerated because of the active axial focusing system. The cross-talk between the radial track-following and the axial focusing due to field curvature can be dealt with by proper precautions in the servo systems. For a final design, other requirements like a certain minimum free-working-distance d_2 may introduce a slight deviation from the design III presented in Table 7.25. A decrease in decentring and thickness tolerance will occur but these quantities remain of the order of at least 10 μm, an acceptable value for mass-produced objectives using injection moulding or glass moulding technology.

7.8.3 Unconventional Designs for the CD-Lightpath

We conclude the section on the CD system with two examples of compact lightpaths. These so-called integrated lightpaths incorporate both the semiconductor laser substrate and the imaging optics in the actuator system for axial focusing and radial tracking of the information layer.

7.8.3.1 Reflective Schwarzschild Objective

In Fig. 7.70a we show a reflective objective which is based on the Schwarzschild microscope objective design with a small convex mirror M_1 followed by a concave focusing mirror M_2 with an aspheric shape. Both reflecting surfaces are of the Mangin type using internal reflection within the glass material (BK7 glass in this case). The stigmatic imaging is obtained by asphierising mirror M_2; this surface also functions as the stop of the objective; mirror M_1 can be kept spherical. The system is a combination of a strong negative power and a weaker positive power, typical for the *inverted telephoto* or *retrofocus* layout of an optical system. Therefore, the effective focal distance f' is much smaller than that of a lens with comparable diameter and in the example of the figure we have $f' = 0.83$ mm. The reflective objective also leads to a large free-working-distance d_4 (typically 1.6 mm) as compared to its focal distance. The beam reflected by the information layer follows the same way back. A holographic structure H has been recorded on the convex mirror M_1

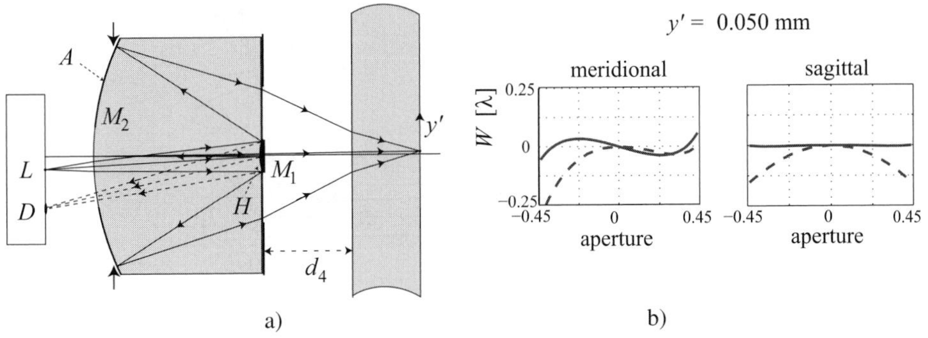

Figure 7.70: a) A reflective Schwarzschild objective with one aspheric surface, denoted by A.
b) The meridional and sagittal wavefront cross-sections of the imaging pencil towards an off-axis image point with lateral coordinate $y' = 50$ μm in the plane of the optical disc.

and part of the reflected beam is imaged off-axis (dashed pencil of rays) at the location of the detector array D which is positioned on the laser chip next to the source L. Figure 7.70b shows the wavefront aberration in the meridional and the sagittal cross-section of the off-axis beam which is shown in panel a). The residual aberration is coma and field curvature. The coma residual could be reduced to zero by asphericising both mirror surfaces. However, the angular field of the objective is quite large due to its small focal distance and there is no real need to asphericise both M_1 and M_2. The reflecting objective has not been put in mass-production. The manufacturing of the reflecting surfaces would have to be done with a sixfold more severe tolerance than is customary for a plastic lens. Neither could the reflecting CD objective design be easily extended to future generations with higher numerical aperture.

7.8.3.2 A Waveguiding Optical Lightpath with Focusing-grating-coupler (FGC)

An interesting read-out principle which has not been experimentally realised is the CD lightpath using guided wave propagation **[350]** (see Fig. 7.71a). The light from a semiconductor laser L is coupled into a planar waveguide. Preferably, the coupling is done in physical contact between the laser chip and the waveguide to avoid supplementary focusing elements. When propagating through the planar waveguide, the beam spreads out laterally and hits a region with a modulated top surface. This structure, a focusing-grating-coupler FGC with its optical centre at T_1, focuses the light in F on the information layer of the CD disc through the disc substrate D. For ease of alignment, the central ray of the outcoupled focused wave is generally perpendicular to the waveguiding substrate.

In Fig. 7.71a some of the grating lines of the focusing-grating-coupler have been shown (dashed lines). The calculation of the grating line pattern is illustrated with the aid of Fig. 7.71b. A general ray $\hat{\mathbf{s}}_0$ emitted by the laser source L and a central ray LT_0 are traced through the waveguide. The mode propagation index of the waveguide for the vacuum wavelength λ_0 of the source yields the effective refractive index n_g of the waveguide. We assume that a planar wave with a cylindrical wavefront, centred at L, propagates in the guide. At the beginning of the modulated area of the FGC, on a line through T_0 that is parallel to the x-axis, we directly have an expression for the x-direction cosine $L(x)$ of a general ray $\hat{\mathbf{s}}_0$ through P_0. At the exit of the FGC, we require a generally aberrated pencil of rays which is brought to a perfect focus after traversal of the optical system S with circular symmetry. The refractive index of the space adjacent to the waveguide is n_s. The ray direction of a general ray $\hat{\mathbf{s}}_1$ of the aberrated pencil of rays that exits from the optical system S towards the focusing-grating-coupler is obtained by tracing rays back from the image point F through the system S. Since the system S possesses circular symmetry it is sufficient to know the ray component perpendicular to the axis T_1F of the optical system S. We denote this component by $D(\rho)$. It is found, for instance, as the x-direction cosine in the plane FT_1Q_1 and depends solely on the radial coordinate ρ in the circle C through T_1, perpendicular to FT_1. The direction cosine function $D(\rho)$ follows from the ray data $\hat{\mathbf{s}}_1$ when tracing rays back from the focal point F towards the circle C.

The equal pathlength condition for any general aperture ray LPF with respect to the reference ray LT_1F is assured by taking into account the phase function Φ that arises when an optical field is diffracted by a grating structure (see Eq. (5.7.33)). The extra diffraction phase Φ which is needed to match the guided field from L to the freely propagating field to be focused in F follows from a pathlength analysis of the FGC **[59]**. The pathlength analysis of two rays $\hat{\mathbf{s}}_0$ and $\hat{\mathbf{s}}_1$, to be connected at the point P, is applied to the closed loop $T_0P_0PP_1T_1T_0$. Applying Lagrange's integral theorem to this closed loop we have for the general point $P(x, z)$ on the grating coupler the pathlength function $\mathcal{L}$ given by

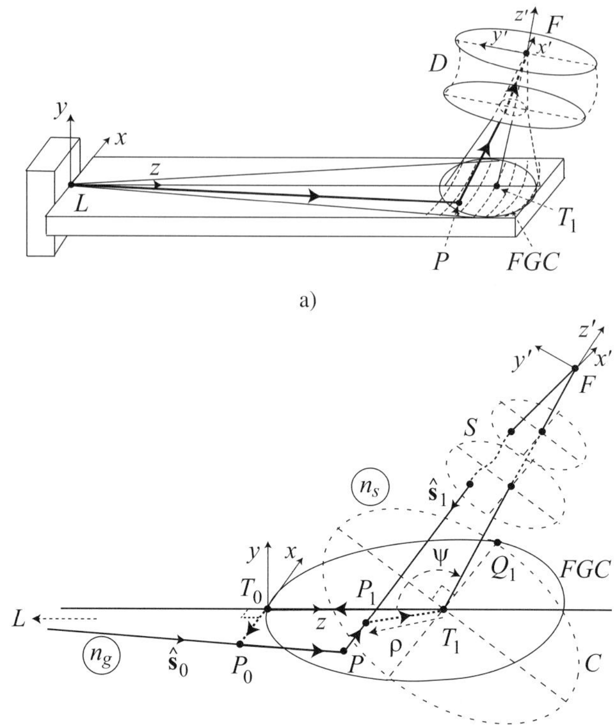

Figure 7.71: a) An integrated lightpath for CD based on waveguiding. The laser chip L is butt-coupled to the planar waveguiding section with a schematically drawn focusing-grating-coupler (FGC) on top of it (centre at T_1). The focus F is on the back of the optical disc.
b) Calculation of the optical pathlength increments along a closed path $T_0P_0PP_1T_1T_0$ and the application of Lagrange's integral invariant to the focusing-grating-coupler.

$$\begin{aligned}\mathcal{L}(x,z) &= [T_0P_0] + [P_0P] + [PP_1] + [P_1T_1] + [T_1T_0] \\ &= \int_{T_0}^{P_0} n_g L(x)dx + n_g P_0P + n_s PP_1 + \int_{P_1}^{T_1} n_s L(\rho)d\rho - n_g T_0T_1 \neq 0 \,, \end{aligned} \tag{7.8.11}$$

where n_g is the effective (mode) index of the light propagating in the grating coupler and n_s the refractive index of free space. To obtain a smooth and continuous wavefront converging to F, we need to include a grating structure that produces an extra phase $\Phi(x,z)$ when diffraction occurs at the point P such that $\mathcal{L}(x,z)$ will be made zero for a general point P on the grating-coupler. This yields the condition

$$\mathcal{L}(x,z) + \frac{\Phi(x,z)}{2\pi}\lambda_0 = 0 \,, \tag{7.8.12}$$

where the planar curves $\Phi(x,z) = j(2\pi)$ define the grating lines for integer values of j.

The phase function Φ can be tabulated on a grid in the plane $y = 0$. The spatial frequency components then follow from

$$\nu_x(x,z) = \frac{1}{2\pi}\frac{\partial\Phi(x,z)}{\partial x} \,, \qquad \nu_z(x,z) = \frac{1}{2\pi}\frac{\partial\Phi(x,z)}{\partial z} \,. \tag{7.8.13}$$

Using the values of the spatial frequency vector components $\hat{\mathbf{v}} = (\nu_x, 0, \nu_z)$ in the modulated area of the grating coupler, a general diffracted ray $\hat{\mathbf{s}}_1$ through P can be calculated (see Eq. (5.7.16)).

The calculation of the intensity distribution in the focal region around F requires an electromagnetic analysis of the diffraction phenomenon in the waveguide that carries a corrugated structure on its top surface. The surface modulation along the propagation direction T_0T_1 has to be tuned in such a way that the diffracted field converging towards F has equal amplitude for all rays. The build-up of the radiation field from the guided wave in each point of the focusing-grating-coupler and the formation of the three-dimensional focus in the neighbourhood of F is treated in **[39]** and **[38]**.

The FGC lightpath has not been used for the read-out of CD discs. The main problems of the device are its small imaging field and the change of aberration in the presence of a wavelength drift of the source. Figure 7.72 illustrates these effects. The aberration graphs in this figure refer to a FGC geometry with the value $\psi = \pi/2$ (perpendicular diffraction of the central ray). The distance from the source point L to the centre T_1 of the grating-coupler is 50 mm. As effective index of the waveguide mode we have chosen $n_g = 1.5$. The design wavelength for the diffracting structure of the FGC is 800 nm. The distance from the grating surface to the back of the optical disc is 2.2 mm, the disc thickness has the standardised CD value of 1.2 mm. Figure 7.72a shows the distorted spot diagrams at the best-focus position for the case of a lateral shift of the laser source point over a distance $\Delta x = 10$ μm. The resulting aberration is primordially coma and, to a lesser degree, astigmatism, both of third-order nature. The coma aberration is due to the fact that the FGC, being a planar diffracting device, cannot obey the Abbe sine condition. It necessarily satisfies a tangent relationship between the lateral coordinate of a diffracted ray on the FGC and its ray angle with respect to the z'-axis in image space (see Fig. 7.71). The total rms OPD due to coma and astigmatism is 85 mλ. The maximum source shift should thus be limited to a few microns which is a very severe requirement for a component in a mass-produced consumer device. We observe in Fig. 7.72a that the geometrical blur of the point F is of the order of 3 μm (the rms value amounts to 1.3 μm). Figure 7.72b shows that the tolerance on the wavelength drift of the source is also severe. A shift towards a shorter wavelength has as principal effects a lateral shift of the (paraxial) focal point in the positive y'-direction, a defocusing beyond the nominal focus F and a comatic aberration. If the wavelength shift is a slowly varying phenomenon, the active focusing and lateral tracking mechanisms in an optical disc system are able to compensate for the lateral and axial offsets. The remaining comatic aberration at the best-focus position, however, has an rms wavefront aberration of almost 100 mλ ($\delta\lambda = -2$ nm) and this means that the wavelength shift should not exceed a typical value of ±0.4 nm. In Fig. 7.72b wavefront cross-sections are shown and we observe that the peak-to-valley wavefront aberration amounts to typically 0.3 λ for the case of a source shift of $\Delta x = 0.01$ mm. The wavefront aberration in the bottom row ($\delta\lambda = -2$ nm) has not yet been corrected for the axial focus offset of the focal point and amounts to more than one wavelength P-V aberration. At the best-focus position, there remain coma and astigmatism which drive the imaging system away from the 'diffraction-limited'-quality level of 0.07 λ rms aberration. The tight tolerance on wavelength shift or spectral width of the source is a critical issue in an optical disc player. Single-mode operation of the semiconductor laser is needed and the aberrational effect of its natural wavelength drift with temperature of 0.25 nmK^{-1} should be compensated for in the lightpath.

7.8.4 DVD and Blu-ray System

The designs for CD objectives can be extended to the higher density systems like DVD and Blu-ray, the main difference being the higher numerical aperture and the shorter wavelength of the laser source. Because of tighter tolerances and reduced performance in the image field, some CD designs can no longer be used when the aperture needs to be increased.

7.8.4.1 DVD Objective

Important parameters of a read-out objective for the DVD system are the wavelength λ of the light source (semiconductor laser of the AlGaInP-type, typical wavelength range of 650–670 nm), the aperture of the objective ($NA = 0.60$), the focal distance ($f' = 3.3$ mm) and the required objective field diameter (150 μm). The mono-aspheric designs for CD can be extrapolated to the higher numerical aperture of DVD as shown in Fig. 7.73b, but they suffer from a severe reduction in size of the useful image field. The wavefront cross-sections in Fig. 7.74a show a higher-order comatic residual. Together with the inevitable astigmatism, the useful field is limited to a diameter of 80 μm. To determine this diameter we apply the criterion of $OPD_{rms} \leq 0.03\ \lambda$ for an imaging pencil to the border of the field. The conclusion is that the mono-aspheric design cannot fulfil the field diameter requirement for the DVD system.

A bi-aspheric DVD objective is shown in Fig. 7.73c. Its field performance allows a diameter which is twice as large as that of the mono-aspheric design of Fig. 7.73b. The refractive index n_1 of the objective has been chosen relatively low such that a plastic material can be used. This allows mass-production with the injection moulding process. The throughput of such a fabrication process is enhanced by applying several aspheric surface pairs, typically eight, in parallel in a single moulding step (multi-cavity mould). The use of plastic induces an important change of refractive index with temperature. A typical value for the index temperature coefficient is 10^{-4}K^{-1}, leading to a refractive index change of $5 \cdot 10^{-3}$ for a typical temperature change during operation of the optical lightpath. The use of larger indices is favourable for the lens

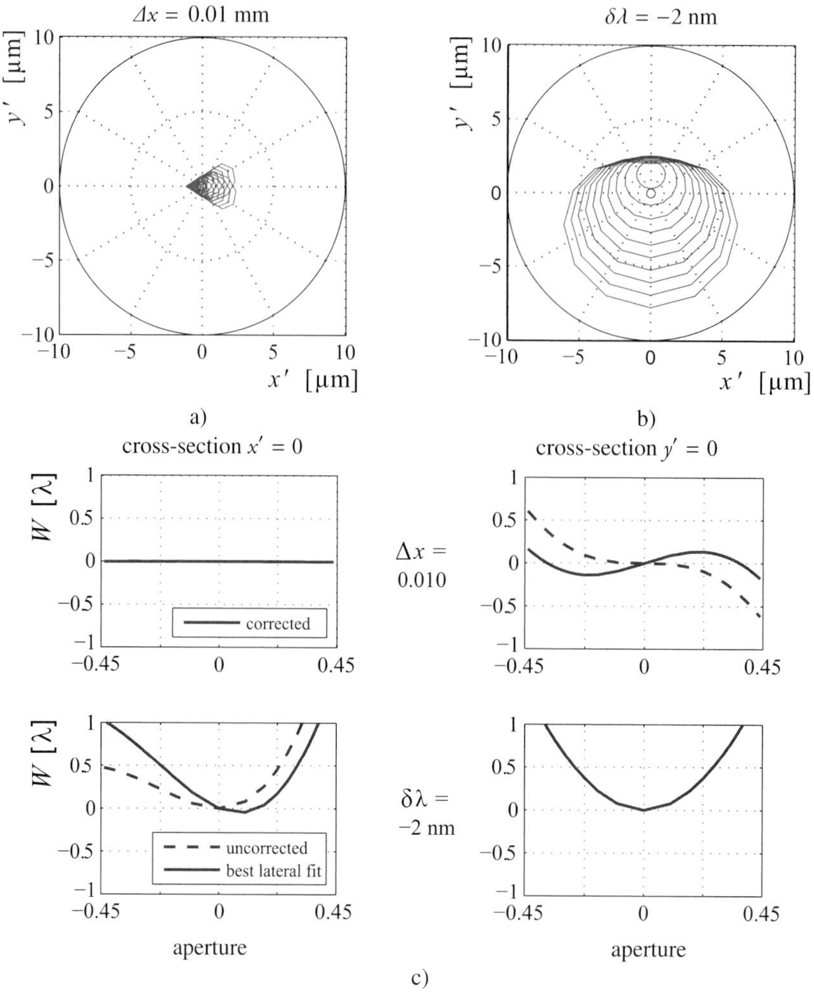

Figure 7.72: Geometrical spot diagrams produced by a FGC ($\psi = \pi/2$).
a) Transverse shift ($\Delta x = 0.01$ mm) of the source L with respect to the waveguide axis LT_1.
b) Wavelength shift $\delta\lambda = -2$ nm of the source.
c) Wavefront cross-sections of the wave leaving the FGC. Upper row: misalignment $\Delta x = 0.01$ mm of the laser source. Lower row: wavelength shift $\delta\lambda = -2$ nm of the source L (left-hand graphs: cross-section $x' = 0$, right-hand graphs: cross-section $y' = 0$).

performance and the manufacturing tolerances. In this case, glass moulding needs to be applied. As a positive side-effect of using glass, the temperature effects on the optical quality are greatly reduced (see Table 7.26).

7.8.4.2 Blu-ray Objective

The objective is used with a blue laser source (GaN-laser, $\lambda = 405$ nm), its NA is 0.85, $f' = 1.75$ mm and the field diameter should be at least 80 μm. The λ/NA ratio of the Blu-ray system allows for a more than fivefold capacity increase of the optical disc with respect to DVD. As objective designs, we show an extension to high aperture of the single bi-aspheric lens and two examples of two-element designs. In the latter case, the surface facing the disc substrate is flat. The first element has an aspheric shape A_1 on the entrance surface. For better performance, a second aspheric profile A_2 is imparted to the second element. The design examples are shown in Fig. 7.75a and b, respectively. The focal distance of the objectives is fixed at 1.75 mm. The field diameter of the objective then determines the tilt angle at which the objective

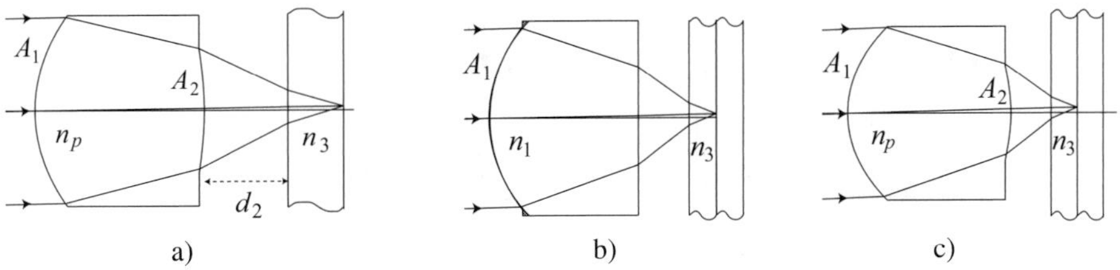

Figure 7.73: a) A plastic bi-aspheric objective for the CD system; $NA = 0.45, f' = 4.5$ mm, $\lambda = 785$ nm, field diameter is 250 µm. Lens material index $n_p = 1.53$. Disc cover layer thickness 1.2 mm.
b) Mono-aspheric objective for the DVD system (drawn to scale with the CD objective); $NA = 0.60, f' = 3.3$ mm, $\lambda = 650$ nm, field diameter is 80 µm. Lens material: LaSFN9 glass with $n_1 = 1.84$. Disc thickness 0.6 mm. Aspheric profile on the first surface (A_1).
c) Plastic bi-aspheric DVD objective; field diameter 160 µm, $n_p = 1.53$.

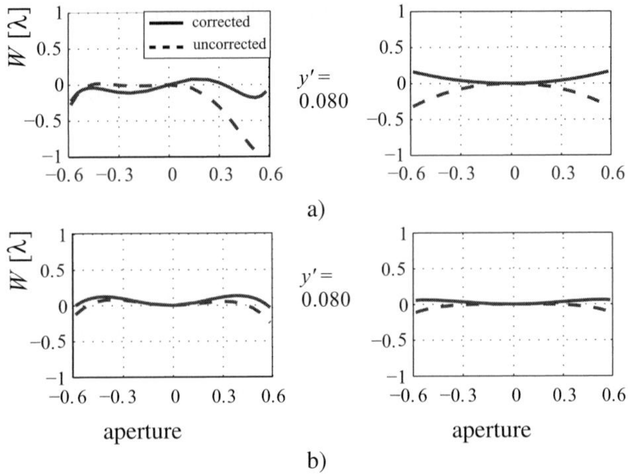

Figure 7.74: a) Meridional (left) and sagittal (right) wavefront cross-section of a pencil of rays focused towards an image point at an off-axis distance of $y' = 80$ µm (mono-aspheric DVD objective of Fig. 7.73b).
b) The same for the bi-aspheric objective of Fig. 7.73c. The dashed curves are the uncorrected wavefront cross-sections, the solid curves apply to the cross-sections referenced to the best axial and lateral focus position.

can still be used with sufficient imaging quality. With an angular tolerance of ±1.3° for the objective orientation during read-out, we require a useful field diameter of 80 µm. The definition of 'useful' implies that the increase in OPD_{rms} value does not exceed 0.030λ with respect to the on-axis value. This limit is approximately half the 'just' diffraction-limited OPD value of $\lambda/14$ with Strehl intensity of 80%. The OPD requirement of $\leq 0.030\,\lambda$ leads to a Strehl intensity of $\geq 97\%$.

In Fig. 7.76 we have shown the meridional and sagittal cross-sections of the wavefront belonging to the pencil imaged at the border of the field. Figure 7.76a is associated with a bi-aspheric single lens with an index $n_1 = 1.68$. The rms OPD of this pencil, focused approximately 40 µm from the optical axis in the best imaging plane (solid line), is ≈ 100 mλ. We observe that the aplanatism of the design leads to an almost perfectly symmetric meridional cross-section. The dominating aberrations are field spherical aberration and astigmatism of second and fourth order in the radial pupil coordinate. The acceptable field radius is just 22 µm. The choice of a higher index yields a slightly larger field, up to a radius of 25 µm for $n_1 = 1.85$. The conclusion is that a single lens is not attractive for the Blu-ray system.

In Fig. 7.76b the same wavefront cross-sections are shown for a two-element objective with one aspheric cap A_1 on the first convex surface. The pencil is focused at an image point with $y' = 40$ µm. The objective is not rigorously aplanatic and higher-order coma is present in the meridional cross-section. At the best-focus position (solid lines), the rms OPD amounts to $0.03\,\lambda$; the design just satisfies the image field criterion ($y' = 40$ µm) that was given above. By making a

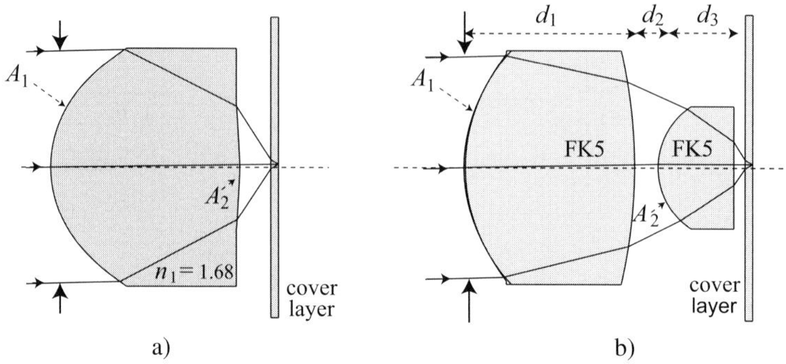

Figure 7.75: a) A bi-aspheric objective for Blu-ray, with aspheric surface profiles denoted by A_1 and A_2. $NA = 0.85$, $f' = 1.75$ mm, $\lambda = 405$ nm, refractive index $n_1 = 1.68$ (glass moulding). Disc cover layer thickness is 100 μm. Air gap thickness is 400 μm.
b) A two-element solution with a convex–convex and a convex–plano lens body made of Schott FK5 glass ($n = 1.499$). Either one (A_1) or both elements (A_1 and A_2) are provided with aspheric plastic caps. Air gap thickness is 150 μm.

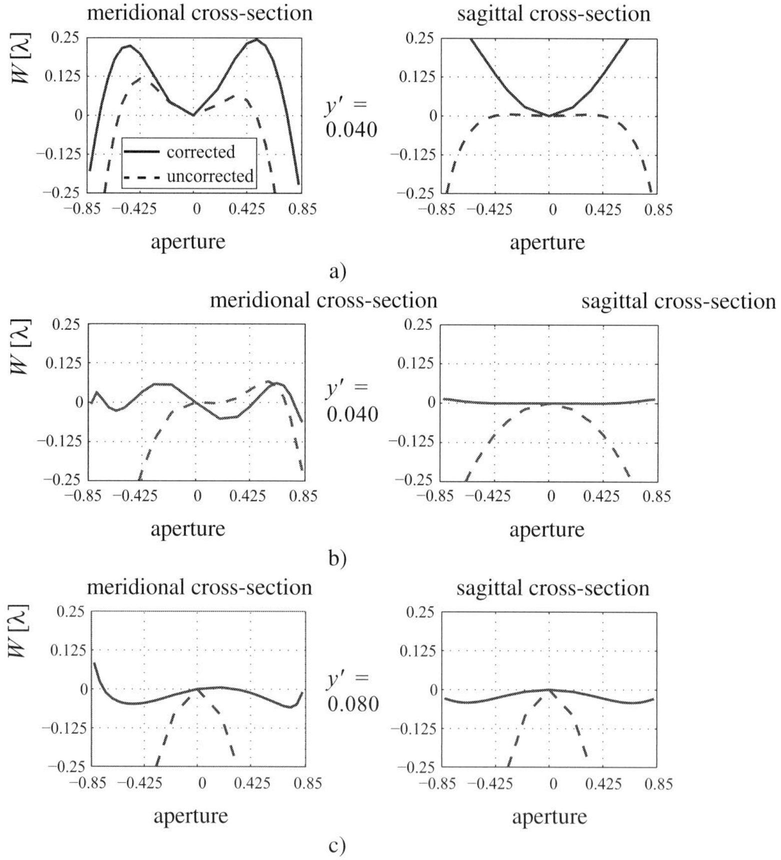

Figure 7.76: a) Bi-aspheric Blu-ray objective of Fig. 7.75a: meridional and sagittal wavefront cross-sections for a pencil of rays to an image point at 40 μm from the optical axis.
b) The same for a two-element objective with a single aspheric cap A_1 on the first lens element of Fig. 7.75b.
c) Two-element objective according to Fig. 7.75b with aspheric caps A_1 and A_2 on both elements.

Table 7.26. Comparison of imaging quality and manufacturing sensitivity of objectives for the CD, DVD and Blu-ray optical storage systems.

System	CD	DVD	DVD	Blu-ray	Blu-ray	Blu-ray
Type	bi-asph	one asph	bi-asph	bi-asph	2 elem.	2 elem.
					one asph.	two asph.
Index n_1	1.53	1.84	1.53	1.68	1.499 (FK5)	1.499 (FK5)
Asphericity (μm)	±7 / ±3	±13	±19 / ±7	±33 / ±6	±13	±13 / ±3
Field diameter (μm)	320	80	160	45	80	190
Field angle	±2.0°	±0.8°	±1.3°	±0.7°	±1.4°	±2.9°
Tolerances						
δd_1 (μm)	100	15	20	1.1	-	-
δy_2 (μm)	45	-	12	1.1	-	-
α_1 (mrad)	0.7	0.3	0.3	0.1	-	-
α_2 (mrad)	0.6	0.3	0.2	0.1	-	-
δn_1	$7 \cdot 10^{-3}$	$1.5 \cdot 10^{-3}$	$2.7 \cdot 10^{-3}$	$8 \cdot 10^{-4}$	-	-
		plastic cap				
α_1 (elem. 1, mrad)	-	-	-	-	3	0.8
α_2 (elem. 2, mrad)	-	-	-	-	0.5	0.5
δd_1 (μm)	-	-	-	-	3	4
δd_2 (μm)	-	-	-	-	1.8	3
δd_3 (μm)	-	-	-	-	1.5	1.5
δy_2 (elem. 2, μm)	-	-	-	-	1.7	5
δn (elem. 1)	-	-	-	-	$3 \cdot 10^{-3}$	$7 \cdot 10^{-4}$
δn (elem. 2)	-	-	-	-	$1.3 \cdot 10^{-3}$	$1.8 \cdot 10^{-3}$

second surface aspheric, the coma can be strongly reduced and the rms OPD limit of 0.03 λ is now reached at a field position $y' = 95$ μm. In Fig. 7.76c. we show the wavefront aberration of this objective for $y' = 80$ μm with an rms value in best focus of 22 mλ.

Since the numerical aperture is large and the wavelength is small, the manufacturing tolerances of a Blu-ray objective are very tight for a mass-production process. The design optimum is strongly influenced by the critical tolerances. In the case of the bi-aspheric single lens, the critical tolerances are on lens thickness (δd_1), lens surface decentring (δy_2), lens surface tilts (α_1 and α_2) and refractive index drift (δn_1). In the case of the two-element design, we have lens and mounting thickness tolerances δd_1, δd_2 and δd_3, tilt angles α_1 and α_2 of the lens elements and a mutual decentring δy_2 of the two elements. In Table 7.26 we have listed the field coordinate y' that produces an OPD_{rms} equal to 0.03 λ and the optical or mechanical deviations from the as-designed values that, each individually, give rise to an OPD_{rms} of 15 mλ. Since the temperature range in which the objective should operate is large, the influence of refractive index changes has also been listed. It follows from the table that the bi-aspheric lens, originally designed for CD, still has acceptable tolerances for the DVD system. Some tolerances like those on lens thickness and surface decentring are strongly reduced, but they are still within the acceptable range for mass-production. Also the tolerance on refractive index variation (plastic components under the influence of temperature variation) remains acceptable. The other option from the CD lightpath, the mono-aspheric singlet with glass body and plastic aspheric cap, becomes questionable for DVD. The main problem is the reduced angular field of less than 1°. The alignment of the objective, mounted in a two-dimensional mechanical actuator, with respect to the incident collimated laser beam is difficult to achieve with such a small angular tolerance. A second tight tolerance here is put on the refractive index change with temperature of the plastic cap.

The application of the bi-aspheric singlet to the Blu-ray system (see Fig. 7.76a), leads to unacceptably tight tolerances. If the lens material has a low index in the range of optical plastics, no satisfactory optical design is feasible. In the example of Table 7.26 the index has been raised to $n_1 = 1.68$. Glass moulding is the appropriate technology for mass-production when such an index is required. The accompanying tolerances of lens thickness, mould centring and mould tilt during the moulding process are prohibitive for mass-production. The generally accepted design solution for the high-*NA* Blu-ray objective is a two-element objective (see Fig. 7.76b). The advantage of the two-element solution is that low refractive indices can be used again. In the example the optical glass FK5 has been used; it can be optically exchanged with a standard plastic like PMMA (polymethylmethacrylate). The first element can be made a mono-aspheric lens and the second element forms a spherical element which operates, together with the air gap and the disc cover layer, close to the aplanatic condition. The performance and the tolerances of the two-element design with one aspheric surface A_1 on the first surface have been shown in the penultimate column of Table 7.26. Its tolerance on field angle is acceptable. A doubling of the field angle is obtained by adding a second aspheric surface A_2 on the second element. The axial distances and decentring tolerances are tight in any Blu-ray design. It is possible to add a third aspheric surface to slightly relax the axial or decentring tolerances. The most critical tolerance is the one on refractive index variation. In the presence of a ±25K temperature change, the indices of plastics will exceed the tolerance values listed in the table. For this reason, a mechanism for the compensation of spherical aberration is generally incorporated in the optical lightpath. A possible approach is the introduction of a change of conjugate planes in the objective by means of a defocusing of the laser source in the collimator section of the lightpath.

7.8.4.3 Objective Compatibility for CD, DVD and Blu-ray Systems

An optical disc player capable of accepting all three basic formats should be equipped with three lasers emitting at wavelengths of 785, 650 and 405 nm. The lightpath can be given two objectives, one for the red and infrared discs and one for the blue wavelength disc. The ultimate goal is a single-element objective that is compatible with each of the three wavelength, aperture and disc-thickness combinations. The reference system is Blu-ray, the most demanding one regarding manufacturing and material tolerances. The transition to, for example, the CD reading conditions needs several changes in the lightpath, each of them producing spherical aberration for the axial image point. First, we need to introduce a change in conjugate plane positions in air. In this way the extra disc substrate thickness, equal to $(1.1/n_d)$ mm in air, can be accommodated (n_d is the refractive index of the disc material at $\lambda = 785$ nm). The next step is the replacement of this equivalent air space by the plastic substrate material. Finally, the objective is used at $NA = 0.45$ and with $\lambda = 785$ nm, causing a substantial change in refractive index of the lens materials due to dispersion. When the objective is used in CD mode, the total spherical aberration of the axial image point comprises the following contributions:

- *Shift of conjugate planes*
  The spherical aberration change following a shift of conjugate planes is given by the Offence against the Herschel Condition (*OHC*) which was discussed in Chapter 4. We will denote this aberration change by $\delta W_{S,H}$. The nominal design of the objective satisfies the Abbe condition. For this special case, Eq. (4.8.28) says to what extent the Herschel condition is violated (*OHC*). Multiplication of this quantity by the index-weighted conjugate shift in image space yields the expression

$$\delta W_{S,H} = \frac{\left\{\sin^2\alpha - 4\sin^2(\alpha/2)\right\}}{2}\, n_d\,(\delta d) = -2\sin^4(\alpha/2)\, n_d\,(\delta d)\,, \tag{7.8.14}$$

  where $n_d \sin\alpha$ is the numerical aperture of the CD read-out system, δd the extra substrate thickness of 1.1 mm and $n_d = 1.62$ the refractive index of polycarbonate at $\lambda = 405$ nm.
- *Change of disc substrate thickness*
  The spherical aberration increment $\delta W_{S,d}$ caused by the extra substrate thickness δd follows from Eqs. (5.5.5) and (5.5.6) and is given by

$$\delta W_{S,d} = \frac{n_d^2 - 1}{8n_d^3}\sin^4\alpha\,(\delta d)\,, \tag{7.8.15}$$

  a positive quantity when the disc is used in air.

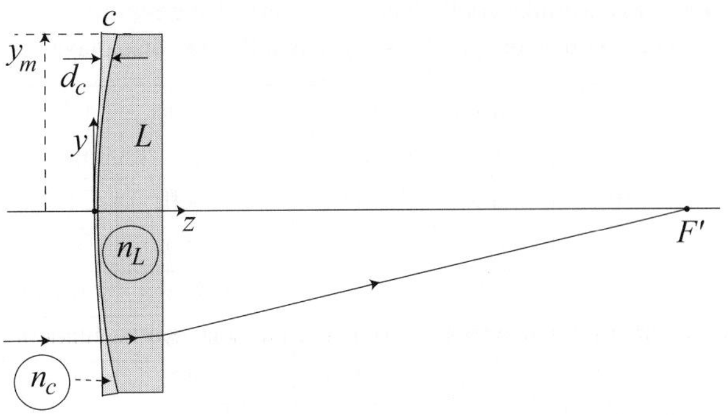

Figure 7.77: A thin lens L with an aspheric correction layer (cap c). The refractive index of the lens material is n_L and that of the aspheric cap is n_c. The wavelength shift of the light is from $\lambda = 405$ nm to $\lambda = 785$ nm (Blu-ray to CD read-out).

- *Dispersion effect*
  The change in refractive index when going from one extreme of the visible spectrum to the other is very noticeable. For the highly dispersive polycarbonate disc substrate the change is approximately 0.05. For PMMA or a low-dispersion glass such as FK5, the change is of the order of 0.02. In the case of a single thin lens, Eq. (5.4.12) yields the third-order spherical aberration coefficient. If we select the first dominating term, the spherical wavefront aberration $W_{S,L}$ (thin-lens approximation) is given by

$$W_{S,L} = -\frac{n_L^2}{32(n_L-1)^2} K_L^3 y_m^4 = -C\left(\frac{n_L}{n_L-1}\right)^2 , \qquad (7.8.16)$$

where K_L is the optical power of the lens, y_m the height of incidence of a marginal ray of the axial pencil and C is a positive quantity if the power K_L of the lens is positive. The lens magnification m_L is zero. A (thin) lens with aspheric correction layer c is shown in Fig. 7.77. Correction of the spherical aberration $W_{S,L}$ of the lens is provided by the aspheric cap with coordinates (y, z_c) such that the total spherical aberration $W_{S,t}$ of the element is zero. The spherical aberration introduced by the capping layer c is given by $W_{S,c} = (1 - n_c)\, z_c$. The requirement for zero total aberration is thus given by

$$W_{S,c} = C\left(\frac{n_L}{n_L-1}\right)^2 . \qquad (7.8.17)$$

For the change of spherical aberration as a function of a wavelength change we find

$$\delta W_{S,t} = \delta W_{S,L} + \delta W_{S,c} = \left\{\frac{2\,\delta n_L}{n_L-1} + \frac{n_L\,\delta n_c}{n_c-1}\right\}\frac{n_L C}{(n_L-1)^2} . \qquad (7.8.18)$$

We note that for a wavelength shift from blue to infrared and for $K_L > 0$, the quantity $\delta W_{S,t}$ is negative. The relative change $\delta W_{S,t}/W_{S,L}$ of spherical aberration with wavelength is given by

$$\frac{\delta W_{S,t}}{W_{S,L}} = -\frac{2}{n_L}\left(\frac{\delta n_L}{n_L-1}\right) - \left(\frac{\delta n_c}{n_c-1}\right) . \qquad (7.8.19)$$

The result of the analysis is that the change in conjugate position gives rise to a negative offset of spherical aberration. The dispersion effect of the lens/cap material also yields a negative increment of the spherical aberration when going from the blue to the infrared side of the spectrum. The increase in substrate layer thickness for CD read-out gives rise to a positive aberration increment. The approximate analysis above applies to first-order changes in optical geometry and material properties. The linear perturbation produces an increase in third-order spherical aberration of $-5.3\,\lambda$ when using the CD read-out conditions. This is equivalent to a Zernike coefficient $a_4^0 = -0.88\,\lambda$; the corresponding $\mathrm{OPD_{rms}}$ value would amount to $0.88/\sqrt{5} = 0.39\,\lambda$. The important optical changes when switching from Blu-ray to DVD or CD read-out conditions also give rise to higher-order perturbation terms. Moreover, the dispersion effect has been approximated by using the thin-lens model. As a matter of fact, a finite-ray analysis shows

Table 7.27. Zernike coefficients for spherical aberration associated with a change of disc type in a Blu-ray disc player.

	Blu-ray → DVD (λ = 650 nm)		Blu-ray → CD (λ = 785 nm)	
Z-coeff.	value	OPD_{rms}	Z-coeff	OPD_{rms}
a_4^0	−1.998	0.893	−2.659	1.189
a_6^0	−0.129	0.049	−0.285	0.108
a_8^0	−0.010	0.003	−0.028	0.009
a_{10}^0	−0.001	0.000	−0.002	0.001
Total		0.895		1.194
P-V value	3.10 λ		4.25 λ	

substantially different aberration values for both the transition to DVD and to CD. We have listed the spherical aberration values of third and higher order in Table 7.27. The combined effects of a change in wavelength and aperture and a change in disc substrate thickness are included in the table; all values are in units of λ of the corresponding laser source for each system. A substantially higher amount of spherical aberration has to be compensated for than the one predicted from small amplitude perturbation analysis according to Eqs. (7.8.14)–(7.8.18). We notice that at large angles of incidence in a high-aperture objective, the predictions of the Seidel theory, even if these are applied to small perturbations, are insufficient. With Seidel-based corrections only, we are left with residuals of non-negligible higher-order aberrations. After correction of the third-order aberration these residuals give rise to a final wavefront that is *not* diffraction-limited.

The compensation of spherical aberration can be carried out by means of a discrete level diffractive-optical-element (DOE), used in transmission. It should produce a virtually unperturbed zeroth order at 405 nm and strong first orders at the DVD and CD wavelengths. This latter condition can be fulfilled by exploiting the dispersion of the DOE material and the step height between adjacent circular annuli of the diffractive element. Instead of having a step height on the DOE that yields a single wavelength jump in the wavefront, one allows step heights of $N_1\lambda_1$ at the blue reference wavelength λ_1, where N_1 is an integer number that can be different for each annulus. The optical pathlength changes for the wavelengths λ_2 (DVD) and λ_3 (CD) are chosen such that the wavefront shape of the first-order beams at λ_2 and λ_3 carry the required correction for the aberration of the objective at these wavelengths. Some 60 to 70 annuli with discrete heights are needed over the beam cross-section. The positions of the rims of each annulus and its height follow from the solution of a general optimisation problem with respect to the wavefront aberration of the zeroth order at 405 nm and the first diffracted wavefronts at the 650 and 785 nm wavelengths. The zeroth order Blu-ray efficiency of the DOE can be brought close to 100%, the first-order efficiencies for the DVD and CD read-out are above 80%. A drawing of such a diffractive-optical-element for achieving compatibility of read-out for the three disc formats is shown in Fig. 7.78. The drawing indeed suggests that the on-axis imaging can be made perfect. What is not visible from the graph is the fact that the imaging beams for DVD and CD wavelengths no longer obey the Abbe sine condition. Consequently, the image field is very limited, typically less than 4 μm in diameter. Since a high amount of wavefront correction is introduced at the back of the DOE, the centring of the element with respect to the optical axis is very critical. For the CD beam, Table 7.27 gave a peak-to-valley excursion of the wavefront correction by the DOE of 4.25 λ, mainly of fourth-order nature. This fourth-order component is given by $W_4^0(x,y) = a_4^0[6(x^2+y^2)^2 - 6(x^2+y^2)+1]$ where (x,y) are the normalised lateral coordinates on the wavefront. In the presence of a decentring δy, the comatic wavefront component is given by $\delta W_C = 24a_4^0(\delta y)[x^2+y^2]y$. With a tolerance of 0.2 λ for the coefficient $24a_4^0(\delta y)$ of δW_C, we obtain the condition $\delta y < 0.002$ in normalised coordinate value. With a lateral dimension of the CD beam of approximately 1.6 mm, the decentring tolerance becomes ±1.5 μm. With the tiny image field diameter and the tight centring tolerance, the DOE solution for read-out compatibility is unacceptable for a mass-produced item. In practice, the two-objective solution has been implemented with one objective designed for Blu-ray and a second one for the DVD/CD combination.

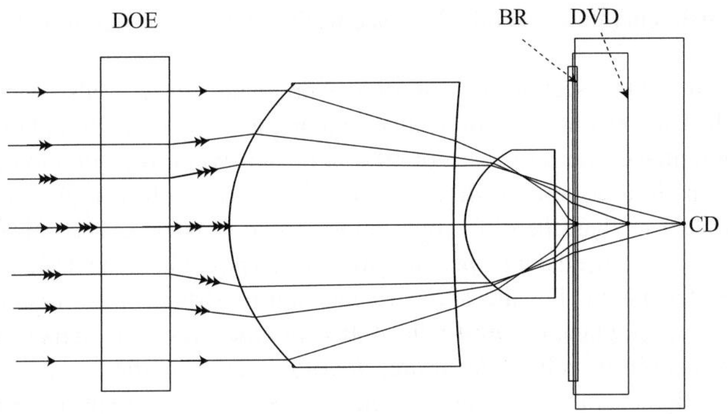

Figure 7.78: Diffractive-optical-element (DOE) in front of a high-*NA* Blu-ray objective to achieve compatible read-out of lower-density DVD and CD discs. The central ray and the marginal rays of the pencils for Blu-ray (→), DVD (→→) and CD (→→→) have been shown.

7.9 Large-field Projection Systems with Diffraction-limited Quality

In this section we concentrate on diffraction-limited, flat-field projection systems with zero distortion or with a user-prescribed distortion pattern. These projection systems are applied in laser-scanning systems, in high-resolution systems for cartography, in imaging systems for earth observation operating in a space orbit and in microlithography for microprocessors and memory chips. In this latter field, one also encounters mixed lens–mirror 'catadioptric' systems and pure mirror systems. In most cases, because of the desire to have imaging properties that are constant over the field, the exit pupil should be located at infinity such that telecentric imaging is obtained. The design history of these systems builds on high-performance photographic lenses which were already available halfway through the twentieth century and used as a basic building block the flat-field Double-Gauss system (see Fig. 7.27 and reference **[378]**). The flat field of the Double-Gauss system is the consequence of the introduction of negative elements at a position where the beam diameter is relatively small. The contribution to the total power of these elements is reduced by a constriction factor $y_j/y_0 < 1$ which applies to the surfaces of these elements, in line with Eq. (4.10.7). In the absence of astigmatism and aspheric surfaces, the wavefront deviation associated with field curvature is determined by the Petzval sum, in accordance with Eq. (5.3.31),

$$\delta W_{\mathrm{fc}} = \frac{H^2}{4}\sum_j P_j = \frac{H^2}{4}\sum_j c_j\,\delta\left(\frac{1}{n_j}\right), \tag{7.9.1}$$

where H is the Lagrange invariant of the imaging system and P_j the contribution to the Petzval sum of the system by the jth transition from a medium with index n_j to the next one with index n_{j+1}. Without special precautions, an imaging system with positive total power will have an inward curved image surface and a negative Petzval sum. For a single lens with refractive index n, the ratio P/K of Petzval sum and optical power is simply $-1/n$. To reduce the Petzval sum to zero by means of optical surfaces with negative power K_j, it is advantageous to make the ratio $P_j/K_j = -1/(n_j n_{j+1})$ as large as possible in absolute value. This is achieved by choosing relatively low refractive indices for the negative lens elements in the system. In this respect we should mention the advantage of using a combination of mirrors and lenses. The quantity P_j/K_j equals $+1/n_j^2$ and this yields a positive Petzval contribution for a positive power mirror. If possible, the introduction of a mirror surface (catadioptric system) is very interesting for achieving a flat image plane because it avoids the need for introducing a large amount of negative power in the imaging system solely to satisfy the flat field requirement.

In the presence of residual astigmatism, the optimum focal plane is found halfway between the focal surfaces for the tangential and sagittal focal lines. An example of such an optimum focus-setting for a non-diffraction-limited optical system was given in Fig. 7.32. A maximum third-order astigmatism amplitude is reached at approximately 3/4 of the field diameter. The onset of higher-order astigmatism with opposite sign then reduces the total astigmatism, leaving a small residual field curvature effect at the border of the field. In the example of the projection lens of Fig. 7.32, the focal lines were approximately ten focal depths apart. For a high-quality projection lens, these balancing effects between astigmatic line position and field curvature should take place well within the Rayleigh focal depth of $\pm\lambda/(2NA^2)$ of the projection lens.

As mentioned in the beginning of this section, the distortion of projection lenses is also tightly specified. Distortion and field curvature, in terms of their wavefront aberration values, require a severe correction. For instance, a maximum shift of the centre of the diffraction image over a distance of 10% of its half-width is a common requirement in lithographic projection lenses. This amounts to a wavefront tilt at the rim of the exit pupil of typically $\lambda/20$ and a corresponding OPD_{rms} of 0.025 λ, three times less than the diffraction-limited value of 0.07 λ. The correction of distortion in these highly qualified projection lenses is carried out by pursuing as much as possible in the design a symmetry of the principal ray path with respect to the diaphragm. For the ultimate reduction to virtually zero distortion, the addition of some elements close to the object or image plane is useful. The footprint of the imaging pencils on the optical surfaces of such elements is small and the introduced wavefront departures are of low order in the pupil coordinates, mainly wavefront tilt, wavefront curvature and astigmatism. As indicated by Eqs. (5.3.31)–(5.3.36), aspheric departures on optical surfaces close to object or image plane can also be exploited to preferentially influence these lower-order aberrations because the eccentricity factor $(\bar{y}_m/y_m)$ is large on these surfaces.

In the following we give some examples of these diffraction-limited imaging systems with field angles or field sizes which greatly exceed those of another typical diffraction-limited system that we have already discussed, the microscope objective. The first one is a laser-scanning objective, the others are examples of microlithographic imaging systems with, in the course of time, an ever increasing complexity. A basic difference between these large-field systems and the microscope objective is the limited spectral range at which these systems have to operate. Without this very restricted spectral width (laser illumination or narrow-bandwidth filtering of a classical source), the imaging performance in terms of aperture and field would be much more modest.

7.9.1 Laser-scanning Objective of the $f-\theta$ Type

The angular scanning of a modulated focused laser beam (wavelength $\lambda_0 = 785$ nm) can be used to write linear tracks of information bits on a flat, tape-like recording medium. The length l of a recorded information track is $2f'\theta_m$ where θ_m is the maximum field angle which can be imaged by the objective. The $f-\theta$ relationship is used to ascertain a constant scanning or recording speed on the medium in the focal plane at F'. For equal recording conditions over the image field, telecentric imaging is imposed. With respect to classical distortion-free imaging according to $y' = f' \tan\theta$, a barrel-type distortion is introduced when applying the $f-\theta$ relationship for the field position of an image point. For fast angular scanning of the laser beam, a rotating polygon is used. In the example, a maximum off-axis angle θ_m of 22° is used. With a focal length of $f' = -1.25$ mm, the total length l of a recorded information track is 0.96 mm.

In Fig. 7.79, we show a schematic layout of the polygon/objective combination. A reflecting facet of the rotating mirror polygon is positioned at the diaphragm D, with the polygon facet at the maximum deflection angle θ_m. The diaphragm is re-imaged at $D'$and then imaged to infinity (telecentric imaging). Information is stored along lines with a length $l = 2f'\theta_m$ in the y-direction, for each angular scan by a facet of the polygon. A sequence of parallel tracks is recorded on the moving tape medium. Apart from the planar glass window W, the mechanical and optical details of the housing of the reflecting polygon mirror have been omitted in the drawing. The object- and image-side principal planes have been denoted by P and P', the focal points by F and F', respectively. The paraxial Lagrange invariant H, expressed in the pupil data of the imaging system, is given by the product of the laser beam half diameter y_0 (0.6 mm) and the maximum scanning angle θ_m of the polygon. In the image plane, the Lagrange invariant is well approximated by $f'\theta(NA)$, the product of the half-width of the field and the numerical aperture $NA = 0.42$ of the imaging beam.

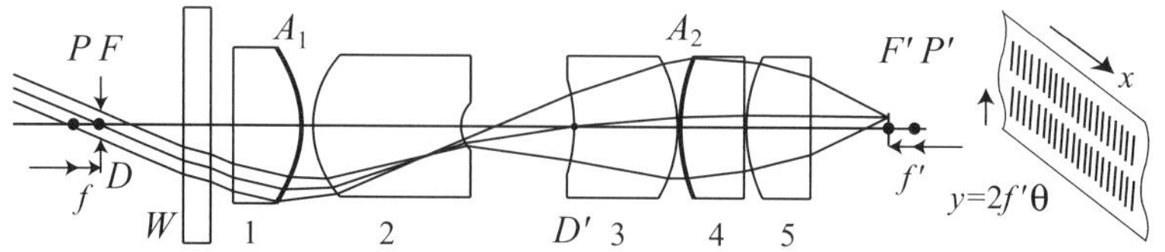

Figure 7.79: An $f-\theta$ scanning objective with a rotating polygon positioned at the effective stop D of the system, behind the exit window W of the polygon housing. The optical information is stored in tracks on a moving tape medium in the focal plane of the objective at F'.

The first stage of the objective consists of elements 1 and 2 and has a focal distance of approximately 2.1 mm. This stage is meant to re-image the diaphragm D onto the aperture D' of the second focusing part of the objective (elements 3, 4 and 5). The lateral magnification of this second stage is -1.7, yielding a compound focal distance f' of -1.25 mm. Because of the large field angle of 22°, strong field curvature reduction means are needed. In this particular case, an intermediate focus of the incident parallel beam is created close to the strongly negative surface at the exit of lens element 2. With a constriction factor of 0.05 with respect to the maximum diameter in the second stage, the negative power of this surface hardly contributes to the total optical power. The system has a Petzval sum P equal to -0.024. For the ratio P/K of Petzval sum and optical power of the system we find +0.03. This is a strong reduction with respect to the value $-K/n$ which would amount to 0.45 when the refractive index value n of the lenses used in this objective equals 1.785.

An aspheric surface A_1 (thick solid curve) has been added in the first stage to be able to influence the distortion of the system. The image D' of the diaphragm D is imaged to infinity by the second imaging stage. This imaging stage has been given three elements to allow a relatively smooth transition from the initial divergence of 0.22 at the location of D' to the final convergence with $NA = 0.42$ in image space. The layout of this second stage is comparable to that of a classical medium power microscope objective as shown in Fig. 7.64. Despite the larger change in convergence as encountered in a more standard microscope objective, the number of elements could be limited to three in the second stage of the $f-\theta$ objective. This could be achieved by the insertion of an aspheric surface A_2 on element 4 where the diameter of the imaging pencils is largest. Where possible, flat-convex elements have been used because of ease of manufacturing and alignment. The aspheric surfaces are obtained by applying plastic caps with an aspheric profile on top of a spherical surface. The imaging is virtually monochromatic and colour correction is not needed. A high-index glass (Schott SF6, $n = 1.785$) could be used with the advantage of a lower aberration level at equal power of the elements as compared to imaging with low-index glasses with lower dispersion.

In Fig. 7.80 we show graphs of the meridional and sagittal wavefront cross-sections of pencils propagating to the image points at $y' = 0$, 199, 390 and 479 μm. The aberration curves according to the nominal design are well within the diffraction limit as demonstrated by the OPD_{rms} values. The residual aberrations are field spherical aberration and astigmatism. The field spherical aberration is compensated by a small residual with opposite sign on-axis. At a position in the image field half-way to its edge, a small asymmetry in the meridional section shows the existence of comatic aberration. This coma is compensated by a cubic field-dependent term at the border of the field.

The tolerance analysis of the $f-\theta$ projection lens according to the scheme of Subsection 6.5.3 allows the calculation of the statistically expected values μ_Z of Zernike aberration coefficients and of their associated σ_Z values. The latter values give the spread around the expected average values of aberration coefficients for an as-manufactured system. In Table 7.28 we give the values for the pencil towards the border of the field. All OPD and σ values are in units of milli-wavelengths (mλ). With respect to the nominal system, certain perturbations are introduced. The increase in rms value of the optical path difference OPD with respect to its as-designed value has been denoted by $\Delta\text{OPD}_{\text{rms}}$ in the table for each perturbation. It is seen that the chosen surface or lens decentring ($\sigma_\epsilon = 5$ μm) and surface or lens tilt ($\sigma_\epsilon = 0.5$ mrad) values have the largest influence on the deterioration of the imaging quality. The first three perturbations are less critical for the as-manufactured imaging quality. The table shows that the deterioration of the system is almost negligible if the curvature deviations of the spherical surfaces are maintained within a margin of ± three interference rings.[3]

A special requirement, not included in Table 7.28, is to describe the fine structure of the two aspheric surfaces. The typical (stochastic) rms surface ripple with value σ_a of a machined or moulded aspheric surface gives rise to angular scattering. The result is a loss of specular transmittance of the aspheric surfaces and the appearance of a continuous background level of scattered light in the image plane. In Chapter 10 the subject of light scattering from rough surfaces is treated and Eq. (10.5.10) gives the expression for the specular transmission T_d of such a 'rough' surface. If we require that $T_d \geq 0.98$ for an optical transition from a medium with index $n_1 = 1.60$ to a medium with index $n_2 = 1$, we obtain at normal incidence a value $\sigma_a/\lambda \leq 0.04$. For a source wavelength $\lambda = 785$ nm, we find $\sigma_d \leq 30$ nm. The polishing of spherical surfaces is fully adequate to meet such a requirement. For mould surfaces which are produced mechanically, for instance by diamond turning, a σ_d value of 30 nm needs special care.

An inspection of the calculated sensitivity coefficients $a_{n,k,l}^{m,(\alpha)}$ of an optical system informs the designer and manufacturer about the critical surfaces or elements in the fabrication (see Subsection 6.5.3) . Depending on the manufacturing process, special care can be given to the most sensitive surfaces. In small-quantity production, an aberration measurement of the finished system can be carried out and a new position or shape of one or several elements is determined such that the aberration is reduced. In large-volume production such individual adjustments are impractical and a statistical tolerancing

[3] One interference ring, measured in reflection at normal incidence with light of a wavelength of 546 nm, corresponds to a surface height deviation at the rim of the surface of 273 nm.

Table 7.28. Tolerancing results $f-\theta$ lens ($NA = 0.42$).

	Perturbation(s)	ΔOPD_{rms}	OPD_{rms}	σ_{OPD}
Nominal system ($y' = 479$ µm)	—	0	33	0
Axial distance/thickness	5 µm	5	34	5
Surface curvature	3 rings	6	34	7
Refractive index	0.0005	1	34	6
Decentring	5 µm	23	40	28
Tilt	0.0005	17	37	11
As-manufactured ($y' = 479$ µm)	accumulated	30	45	32
Nominal system ($y' = 0$)	—	0	15	0
As-manufactured ($y' = 0$)	accumulated	12	19	6

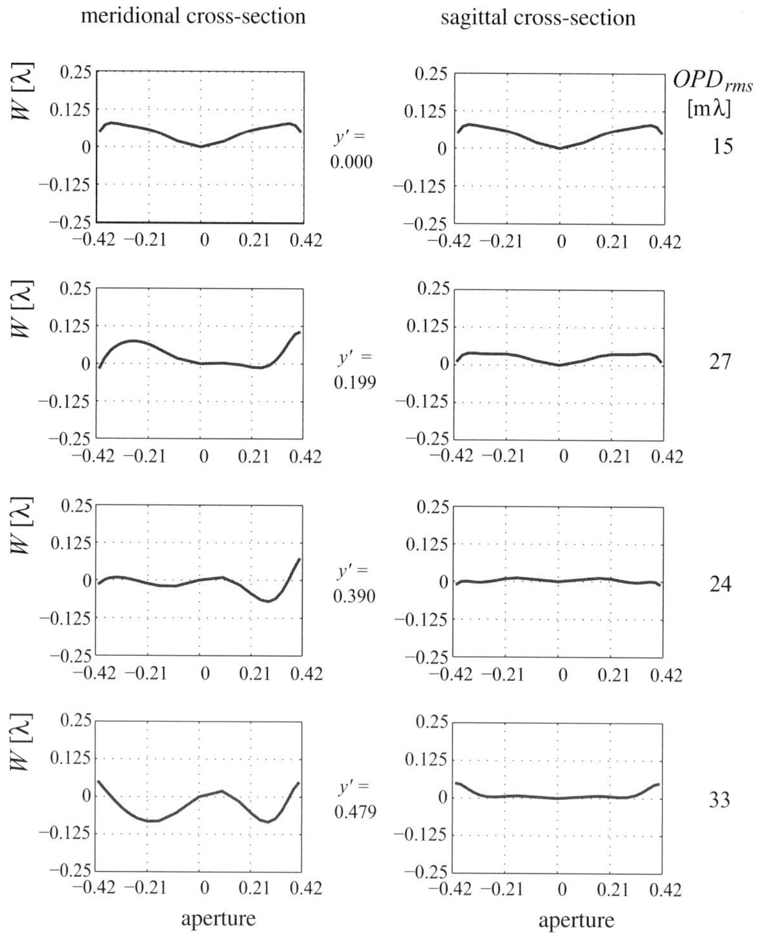

Figure 7.80: Meridional and sagittal wavefront cross-sections and OPD_{rms} for several positions in the image field of the $f-\theta$ scanning objective with $f' = -1.25$ mm of Fig. 7.79. $NA = 0.42$, $\lambda_0 = 785$ nm; $y' = 0$, 199, 390 and 479 µm.

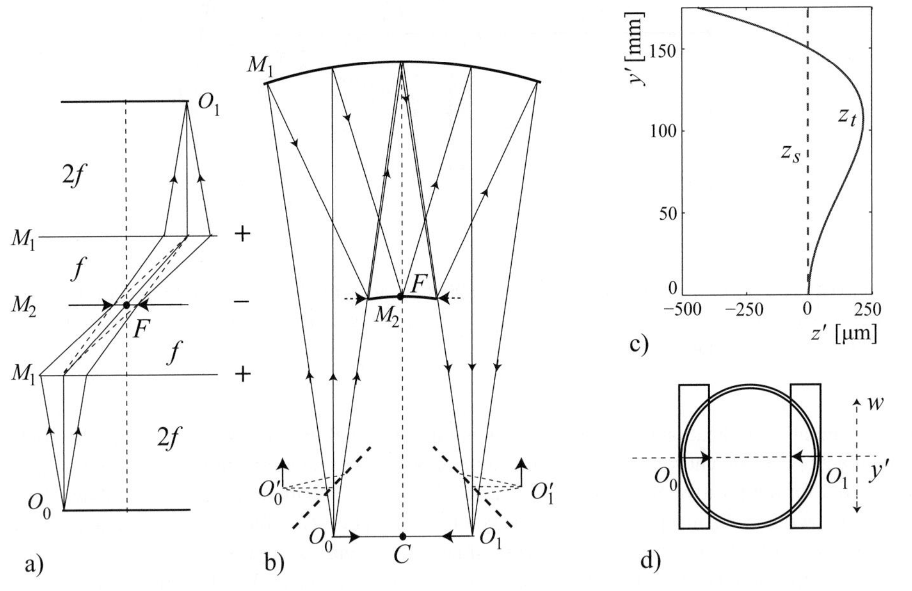

Figure 7.81: An Offner 1:1 projection system with a small well-corrected annular field at O_0 and O_1 ($NA = 0.15$, $\lambda = 365$ nm, annulus diameter 150 mm, annulus width 1 mm, radius R_1 of the chief mirror is -1000 mm).
a) Lens-equivalent scheme.
b) Mirror layout.
c) Axial position of the tangential and sagittal focal lines (z_t and z_s) as a function of the lateral distance $y'$of an image from the origin at C.
d) Useful annular image field.

table will give the percentage of manufactured systems that are found within a prescribed specification. In the case of the scanning lens with Table 7.28, for manufacturing without individual tests, the decentring and tilt manufacturing tolerances should be tightened to typically 3 μm and 0.3 mrad, respectively. In this case a value for the $\text{OPD}_{\text{rms}}+2\sigma_{\text{OPD}}$ can be obtained which is below the just-diffraction-limit of 70 mλ. Certain perturbations can be left out in practice. For instance, the refractive index tolerance of 0.0005 has a small effect and optical glasses are commonly delivered within this tolerance interval of ±0.0005.

7.9.2 All-mirror Microlithographic System with Unit Magnification (Offner)

Projection lenses for microlithography have been produced since the 1960s. The initial systems operated at unit magnification and were meant to print mask features with a typical size of 5 μm on a semiconductor substrate. The required field size was of the order of 10×10 cm^2. Back-to-back photographic projectors have been used as well as unit magnification catadioptric or mixed lens-mirror systems of the Dyson and Wynne type **[381]**. Unit magnification is an interesting option for microlithographic projection lenses since these systems can easily be made free of distortion. Absence of distortion down to the 10^{-6} level is a key requirement in projection microlithography. However, the reflective unit-magnification systems had to be abandoned when smaller printed features were required and the numerical aperture of the projection systems had to be increased beyond, say, 0.30. More versatile refractive systems and mixed refractive–catadioptric systems were designed with a typical absolute value of the magnification equal to 1/5 to 1/4. With the advent of extremely short wavelengths in extreme UV lithography, pure (aspheric) mirror systems are being used again.

In this subsection we first discuss an all-mirror system which was frequently used in the earlier days of microlithography when a feature size of typically 1 to 2 μm was required. Such a system was proposed in 1975 by Offner **[264]**. Figure 7.81a shows the paraxial layout of the unit-magnification mirror system, sketched in transmission mode. The system forms a zero-power telescope with two confocal, equal power positive elements. A negative element is positioned at the common focus F with twice the (absolute) power of the positive elements such that the Petzval sum of the system

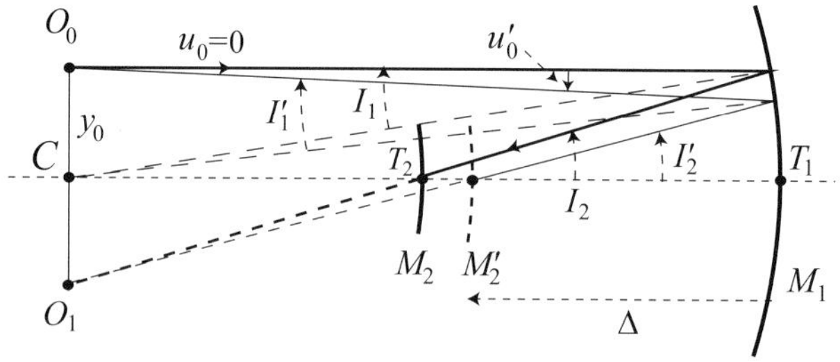

Figure 7.82: Paraxial ray-trace of the principal ray in a unit magnification Offner system. An axial shift of the second convex mirror from the nominal position M_2 to M_2' is executed and the radius of curvature c_2 is adjusted such that the centre of curvature of M_2' is again found at the point C.

equals zero. The lateral magnification of the telescope is minus unity and the stop is positioned at the common focus F. The object and image plane are chosen such that each element operates at magnification -1, with conjugates at $-2f_i'$ and $+2f_i'$, where $i = 1, 3$ for the positive elements and $i = 2$ for the negative element. Figure 7.81b shows the layout of the mirror system with the object O_0 in a plane through the centre of curvature C of the first mirror. The first mirror produces a real paraxial image at O_1, the second mirror a virtual image at O_0 and the first mirror picks up this virtual image to produce the final real image in O_1. Because of the presence of the second mirror M_2, the object and image points have to be chosen sufficiently far away from the common centre of curvature C to avoid obstruction of the imaging beams. In b) we have chosen radii of curvature for M_1 and M_2 of -1000 mm and -500 mm, respectively, and, for a lateral position O_0 which is 150 mm off-axis, the imaging beams pass unobstructed if their numerical aperture is limited to 0.15. Regarding the field curvature, the choice of the radii of the mirrors with double passage at mirror 1 yields a Petzval sum of zero. Up to the Seidel approximation, a flat image field can be expected.

The imaging quality is perfect on-axis. Moreover, since the magnification equals -1 and there is symmetry with respect to the stop, the system is, in first order, free from all odd aberrations such as coma and distortion. Especially the latter property is important in a microlithographic system. Unfortunately, the imaging quality deteriorates severely when going off-axis due to astigmatism, a prominent aberration of a mirror when imaging a plane through the centre of curvature. For the off-axis position of $y' = 150$ mm ($R_1 = 1/c_1 = -1000$ mm), the astigmatism is prohibitively large when the two radii of curvature c_1 and c_2 are chosen in a ratio of 1/2. The astigmatism as a function of the field position can be studied with the aid of Eq. (5.7.40) which yields the pathlength changes around a finite ray at refraction or reflection at a surface.

In Fig. 7.82 the two mirror surfaces M_1 and M_2 are shown and a shifted version of M_2, labelled M_2', with an adjusted radius of curvature such that a new concentric system is created with curvatures c_1 and c_2' given by

$$\frac{c_2'}{c_1} = \frac{1}{1 - c_1\Delta} = 2(1+\epsilon) \,, \tag{7.9.2}$$

where ϵ is a small relative change in c_2 with respect to the nominal value $c_2 = 2c_1$ for the basic Offner system. We calculate the paraxial angle u_0' with respect to the optical axis of a ray leaving the point O_0 and going through the centre of the second mirror, M_2', at an axial distance Δ from the first mirror M_1. In a later stage of the design, this ray path, using finite ray-tracing, becomes that of the principal ray in the Offner system and we require that the angle of incidence u_0 of the finite ray on the first mirror is zero to achieve telecentric imaging. But at this stage of the optical design we are only concerned with paraxial rays and paraxial angles of incidence. After some manipulation we find for the paraxial angle u_0' and the angles of incidence I_1' and I_2' on the two mirrors,

$$\begin{aligned} u_0' &= \left(\frac{2c_1\Delta - 1}{c_1\Delta - 1}\right) c_1 y_0 \approx -2\epsilon c_1 y_0 \,, \\ I_1' &= y_0 c_1 \,, \qquad I_2' = \frac{c_1 y_0}{c_1\Delta - 1} = -2(1+\epsilon) c_1 y_0 \,. \end{aligned} \tag{7.9.3}$$

It turns out that the paraxial angle of incidence I_1' on the first mirror is independent of the angle u_0'.

We now apply Eq. (5.7.40) for the calculation of the wavefront change at reflection around a finite ray. We use the approximate paraxial values for the angles of incidence I_1' and I_2' on the mirror surfaces M_1 and M_2' and neglect the spherical aberration of the principal ray. With $n_0 = -n_1 = 1$ for mirror M_1, we have for the wavefront change at the first mirror, in terms of the outgoing transverse beam coordinates (x_1, y_1),

$$
\begin{aligned}
\delta W_1 &= -\frac{x_1^2+y_1^2}{2}\left(\frac{1}{l_0}+\frac{1}{l_1}\right)+c_1\cos I_1'\left(x_1^2+\frac{y_1^2}{\cos^2 I_1'}\right)\\
&= \frac{x_1^2+y_1^2}{2}\left\{-\frac{1}{l_0}+\frac{1}{l_1}+\frac{c_1}{2}\left[\cos I_1'+\frac{1}{\cos I_1'}\right]\right\}\\
&\quad+(x_1^2-y_1^2)\,\frac{c_1}{2}\left[\cos I_1'-\frac{1}{\cos I_1'}\right].
\end{aligned}
\tag{7.9.4}
$$

The second term on the right-hand side of Eq. (7.9.4) yields the astigmatic wavefront aberration, which is written as

$$
\delta W_{a,1} = -\frac{1}{2}c_1\,\frac{\sin^2 I_1'}{\cos I_1'}\,\rho_1^2\cos 2\theta\,, \tag{7.9.5}
$$

where we have used the polar transverse beam coordinates (ρ_1, θ_1) instead of the Cartesian coordinates (x_1, y_1). In a similar way we obtain for mirror M_2 ($n_0 = -n_1 = -1$) the expression

$$
\delta W_{a,2} = \frac{1}{2}c_2\,\frac{\sin^2 I_2'}{\cos I_2'}\,\rho_2^2\cos 2\theta\,. \tag{7.9.6}
$$

We find the relative beam diameter $2\rho_2/(2\rho_1)$ at the second mirror with the aid of the ratio $\rho_1^2/\rho_2^2 = (c_2/c_1)^2$ and this quantity equals $4(1+\epsilon)^2$. For a movement of M_2 towards M_1 with $\epsilon < 0$, the ratio ρ_1/ρ_2 is reduced. Using the angles of incidence I_1' and I_2' according to Eq. (7.9.3), we obtain for the total astigmatism of the three times reflected beam,

$$
\delta W_a(\rho,\theta) = -c_1\,\rho_1^2\left[\frac{\sin^2 I_1'}{\cos I_1'}-\left(\frac{1}{4(1+\epsilon)}\right)\frac{\sin^2 I_2'}{\cos I_2'}\right]\rho^2\cos 2\theta\,, \tag{7.9.7}
$$

where $2\rho_1$ is the diameter of the (paraxial) beam cross-section on the first mirror and ρ is the normalised radial coordinate in the beam cross-section on mirror M_1 $(0 \le \rho \le 1)$. For the peak-to-valley astigmatic deviation we have two times the amplitude of the aberration function $\delta W_a(\rho,\theta)$.

The preceding paraxial analysis has been verified with the aid of finite ray-tracing ($NA = 0.15$) for a system with a primary mirror radius R_1 of -1000 mm. The peak-to-valley astigmatic wavefront aberration has been plotted in Fig. 7.83 in units of the wavelength of the light (365 nm). It is clear from the figure that a good correspondence exists between the exact ray-tracing result and the astigmatism calculation based on the Coddington equations. For a design with zero astigmatism at an off-axis position y_0 of 150 mm, we have $\epsilon = -0.0113823$ according to the paraxial analysis. The finite

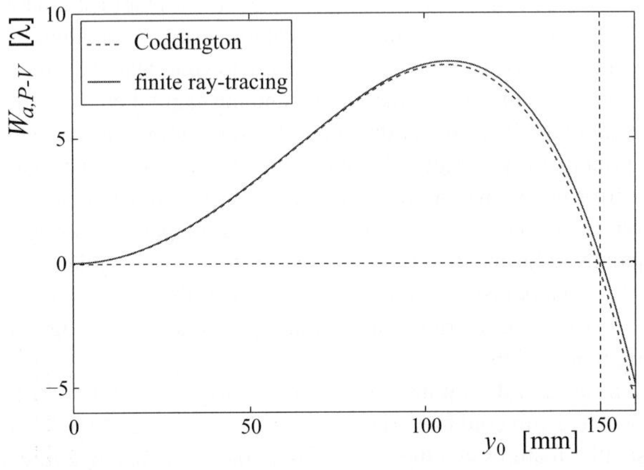

Figure 7.83: The astigmatic wavefront aberration (peak-to-valley, in units of the wavelength $\lambda = 365$ nm) as a function of the off-axis distance y_0 in mm. Unit-magnification Offner system, $NA = 0.15$, $c_1 = -0.001$ mm^{-1}, $\epsilon = -0.01124$. Solid curve: finite ray-tracing results; dashed curve: paraxial analysis using the Coddington equations.

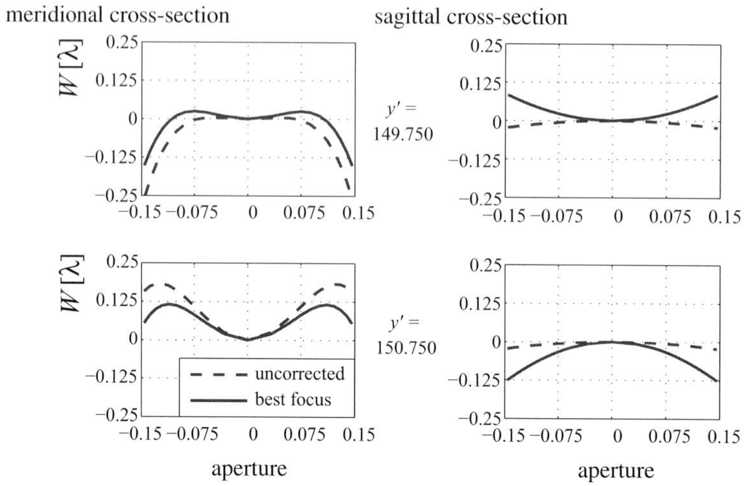

Figure 7.84: Meridional and sagittal wavefront cross-sections of an Offner 1:1 projection system at the border positions of the annular field ($NA = 0.15$, $\lambda = 365$ nm, annulus running from $y' = 149.75 \rightarrow 150.75$).

ray-tracing yields $\epsilon = -0.01124$ with the mirror M_2 shifted over a distance of 5.684 mm towards M_1. Since the region of nearly zero astigmatism is rather small, the diffraction-limited imaging quality of the mirror system is limited to a narrow circular slit (see Fig. 7.81d). The central object and image points O_0 and O_1 are in the middle of the annulus at radially opposite positions. The annulus within the rectangular region centred on O_0 is illuminated by the source and produces an annular image at O_1. By scanning a mask with the lithographic features in the direction of the arrow at O_0 and simultaneously scanning the image receiving wafer at O_1 in the opposite direction, an extended rectangular image field with width w is obtained. To avoid scanning in opposite directions, two planar mirrors are positioned at 45° close to the object and image field (see Fig. 7.81b). With such an extension, the lithographic mask at O_0' and the silicon wafer at O_1' can be mounted on a common support and translated in parallel with high accuracy.

The meridional and sagittal wavefront cross-sections of the imaging pencils towards the inner and outer border of the small annular field (width 1 mm) with sufficiently small astigmatism have been depicted in Fig. 7.84. A small tilt of ±3.5 μm of the optimum image field is observed over the width of the circular slit. It shows up in Fig. 7.84 as a different best-focus setting for the two extreme field excursions. The tilt of the best-focus surface in the image slit is within the Rayleigh focal depth (±8.1 μm) of the imaging pencils. The Seidel astigmatism changes sign in the radial direction across the annulus. Virtually constant astigmatic aberration terms in the annulus are those associated with the Zernike polynomials $Z_4^2(\rho,\theta)$ and $Z_4^4(\rho,\theta)$. Together with a small field spherical aberration term $Z_4^0(\rho)$, the $Z_4^2(\rho,\theta)$ term yields an optimum wavefront correction in both the meridional and the sagittal cross-section of the imaging pencils. The rms value of the OPD over the annular slit is smaller than 0.06 λ, with a minimum value at the centre of 0.025 λ.

The trajectory of the finite principal ray, meant to be telecentric on the object and image side, normally suffers from the spherical aberration of a mirror when imaging rays from infinity. A fortunate coincidence is the fact that we have a final solution that deviates from the initial $c_2 = 2c_1$ design. A finite principal ray through the centre of M_2' in Fig. 7.82 with $M_2M_2' \approx 5.7$ mm corresponds to an incident finite ray on the first mirror which is virtually parallel to the optical axis ($u_0' = 0.035$ mrad). Its angle of incidence I_1' on the first mirror is 0.1509 rad (the paraxial prediction is exactly 0.1500 rad). Finite ray-tracing shows that the telecentricity of the finite principal ray through O_0 at $y_0 = 150.25$ mm is found within a very comfortable margin of 0.05 mrad.

The Offner system has been a successful printing tool for microlithography. With a further reduction in wavelength to 250 nm, lines and spacings down to 1 μm could be printed, at the cost of a slightly reduced width of the annular field. A practical drawback with still smaller features was the upper limit to the numerical aperture and the unit magnification. An increase of imaging aperture has been obtained by adding concentric refracting shells close to the object and image plane, together with a larger radius of the annular field. Finally, it has been the unit magnification restriction with expensive lithographic mask manufacturing which has led to different system choices, of the more versatile catadioptric or purely refractive type.

7.9.3 Microlithographic Projection Lens Systems

In this subsection we discuss two lens designs, one with a modest numerical aperture and meant for the printing of features as small as 0.4 μm, typical for the state-of-the-art in microlithography in 1985, and a second with a numerical aperture beyond unity which is capable of printing dense features down to 50 nm.

7.9.3.1 *NA* = 0.38 (field diameter 20 mm, $\lambda = 405 \pm 4$ nm, $m_T = -0.2$)

Projection systems with such specifications have been described in **[110]** and an example with design details can be found in **[47]**. As already pointed out in the introductory section on large flat-field projection systems, the correction of field curvature is the dominant requirement and it asks for a basic layout with at least one 'constriction' and two 'bulges' in the design. When the numerical aperture is increased from a typical value of 0.20 to 0.40, the attainment of a small Petzval sum with the aid of a constriction has to be carried out more gradually, using more surfaces per positive and per negative group. This more cautious procedure will create less aberration change per surface and less accumulation of aberration in the intermediate spaces between the elements. The importance of a relaxed design has been emphasised in **[110]**. Relaxed constructional tolerances are crucial in the manufacturing process. A less strained system will also behave more favourably when undergoing environmental changes like temperature variations, air pressure changes or radiation load of lens elements. A relaxed design will use more space of the total construction length (optical throw) and contain more elements than a compact but strained imaging system. With respect to a classical system of the Double-Gauss type, the design will become much more elongated and an extra bulge and constriction can be added, pushing the number of positive and negative groups to four or five in total.

In Fig. 7.85 we show a design that complies with such a relaxed design approach. It is meant to operate in the violet spectral region at a central wavelength $\lambda = 405$ nm with a total bandwidth of 8 nm (mercury source, *h*-line). The required field diameter is 20 mm and, with the feature size to be printed, down to 0.4 μm, a numerical aperture of approximately 0.40 is desired. This means that, in the case of dense lines and spacings, the spatial frequency to be imaged by the projection lens is approximately half the incoherent cut-off frequency of $2NA/\lambda$. Aberration correction well within the diffraction limit of 0.07 λ for the OPD_{rms} is required over the entire image field. An as-designed OPD target value of 0.035 λ is desirable.

Starting at the higher *NA* side of the projection lens, we first see the section, labelled 5, that closely resembles a microscope objective of the retrofocus type (see for instance Fig. 7.64). Such a section is successfully used for imaging over a small image field at a modest aperture of 0.40. With respect to the monochromatic microscope objective of Fig. 7.64, a buried surface for colour correction has been added in section 5. For field curvature correction, two constrictions have been created in the complete system, in sections 4 and 2. For the Petzval sum we find a value $P = -0.00027$ mm^{-1}. This means that the third-order field curvature corresponds to an inward curving field with a deviation at the field border (10 mm) of −13.5 μm. With a focal length f' of 77.4 mm, the ratio P/K has been reduced to −0.02. The nonzero Petzval sum is required to counteract higher-order non-Seidel field curvature and the field curving effects when a small amount of residual astigmatism is present. The negative section 4 has a second colour correcting doublet. Going further back towards the object plane, we encounter the positive section 3 that produces the object-side constriction in section 2 with negative elements. Finally, section 1 has rather eccentric footprints of the pencils of rays on the elements and is also close to the object plane, such that it is useful to influence astigmatism and distortion. The position of the diaphragm D of the system has been chosen in section 4 where the (finite) principal ray from the field point Q_1 intersects the optical axis. Figure 7.85 also shows the paraxial focal and principal points F, F', P and P' of

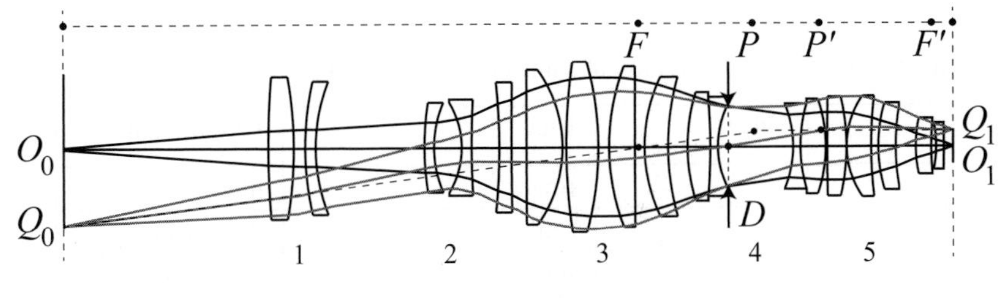

Figure 7.85: Design of a lithographic projection system with relatively modest specifications ($NA = 0.38$, $\lambda = 405 \pm 4$ nm, image field diameter 20 mm), optical throw $T = 602$ mm. Lateral image magnification $m_T = -0.2$.

the system. It is seen that the principal ray from Q_1 (dashed ray) intersects the paraxial object-side focal point F such that telecentric imaging is attained according to paraxial optics. There will be some spherical aberration involved when imaging the stop to infinity, but the deviations from parallelism of the principal rays towards the image field are within a range of ±1.5 mrad.

With respect to colour correction, it is sufficient to have a first-order correction of the paraxial colour effects. Axial colour is corrected by means of the two cemented doublets in section 4/5 and by choosing more dispersive glass materials for the negative elements in section 2. The correction of paraxial transverse colour is relatively easy in a multi-element system with a quite well-balanced number of elements on both sides of the stop. The finite ray optimisation of the projection lens has been carried out with the aid of a Zernike–Chebyshev aberration expansion with respect to the polar aperture coordinates (Zernike) and the radial field coordinate (Chebyshev). The field aberrations that finally limit the performance are field spherical aberration, denoted by ${}_2W_4^0$, and higher-order astigmatism ${}_2W_4^2$; the numerical aperture has to be limited to 0.38. Field curvature can be kept to within ±0.15 μm, with the Rayleigh depth of focus equal to 1.35 μm. Distortion amounts to ±0.01 μm, to be compared with 0.50 μm, the half-width of the Airy disc. The non-negligible chromatic effects in the presence of the bandwidth of 8 nm are visible in the colour effects on astigmatism and distortion. Especially the latter effect, equal to ±0.09 μm over the spectral bandwidth, affects the resolution of the projection lens at the border of the field, with an increase in half-width of 5% for the point-spread function.

An important external factor is the atmospheric pressure and its influence on the optical contrast for the lens elements. Projection systems can be used at quite different average atmospheric pressure, for instance at just 78 kPa at a height of 2000 m above sea level. To the average atmospheric pressure should be added the weather influence which easily gives rise to variations of ±5 kPa. Following Eq. (5.6.2), the refractive index of (dry) air changes by $+1.2 \cdot 10^{-5}$ when the air pressure rises by 5 kPa. There is an almost negligible influence on the aberrational correction of the lens but a strong effect on defocusing, amounting to 5.6 μm. This defocusing would completely spoil the imaging quality and for this reason an air pressure related focus-setting is needed. Simultaneously, a magnification change of the order of 10^{-5} is found and, to a first approximation, this effect can be compensated for by an axial shift of the object plane. Operation of the lens at a substantially different average atmospheric pressure requires an aberrational adjustment of the lens.

Statistical analysis of the projection lens has been performed using typical mounting, shape and material errors. The resulting expected values for the OPD_{rms} and the confidence intervals in production are given in Table 7.29. The typical σ values for the perturbations are given in Table 7.30. With the perturbation values from this table, systems could be built with an imaging performance that satisfies the 'just-diffraction' limit of 0.07 λ OPD_{rms} over the entire square image field of 14 × 14 mm^2. For lithographic imaging, including external errors like defocusing, silicon wafer unflatness, wafer tilt and optical contrast change due to air pressure variation, it is advisable to maintain some margin for these errors

Table 7.29. Tolerancing results for a lithographic lens ($NA = 0.38$) .

Field position	OPD_{rms} design value	OPD_{rms} expected value	OPD_{rms} 95% interval
[mm]	[mλ]	[mλ]	[mλ]
0.000	0.045	0.050	0.055
7.000	0.030	0.039	0.052
10.000	0.052	0.060	0.072

Table 7.30. Standard deviation of some opto-mechanical perturbations.

	Value	Unit	Perturbation
σ_1	25	μrad	tilt
σ_2	1	μm	decentring
σ_3	2	μm	axial shift
σ_4	1	interference ring	curvature change
σ_5	10^{-5}	—	refractive index

by reducing the OPD_{rms} of the projection lens itself. For this reason, a manufacturing process that merely relies on the statistical predictions of Table 7.29 is not applied in practice. A measurement of imaging quality is used to adjust certain elements or combination of elements such that the finished product approaches as nearly as possible the as-designed aberration values. The statistical analysis is then relevant for the possible aberration drift of the system during its lifetime under the influence of opto-mechanical changes.

7.9.3.2 NA = 1.10 (field diameter 28 mm, λ_0 =193 nm ± 0.4 pm, m_T = −0.25)

The second example is a much more advanced system which reflects the state-of-the-art of microlithographic projection lenses about the year 2010, using a deep UV wavelength ($\lambda_0 = 193$ nm is the vacuum wavelength) with water immersion at the image side. The deep UV radiation is produced by a small-bandwidth laser source. The example appears as embodiment 6 in a US patent **[86]** with principal inventor A. Dodoc. The diameter of the image field is 28 mm such that a rectangular field of 10×26 mm^2 can be used in a scanning mode to produce illuminated wafer regions up to a size of 50×26 mm^2. The lateral magnification m_T is $-1/4$. Despite the large image dimensions, the mask size can be kept below 20 cm in linear dimension.

The objective is very well corrected with a residual aberration over the entire field of less than 0.01 λ_0 for the OPD_{rms} of the imaging pencils. With such a very good state of correction, the printed lithographic features (lines or spacings) can be made as small as $0.35\lambda_0/NA \approx 50$ nm. The distortion of such a lens should not exceed a few nm and the unflatness of the image field should be much smaller than the Rayleigh focal depth z_R. For an imaging pencil with high aperture this value is given by

$$z_R = \frac{\lambda_0}{4n\left(1-\cos\alpha_m\right)}, \qquad (7.9.8)$$

where α_m equals half the opening angle of the imaging pencils and where the 'geometrical' aperture is given by $\sin\alpha_m$ (see Eq. (5.2.27)). The refractive index of water at $\lambda_0 = 193$ nm equals 1.435. This yields a geometrical aperture of the immersion lens equal to 0.77 and z_R equals 94 nm. The as-designed lens shows a maximum deviation of the tangential and the sagittal focal surfaces from the optimum image plane which is of the order of 5 nm.

The evolution from the previous lens types, several decades ago, to the design generation represented by Fig. 7.86 is discussed in **[349]** and has been made possible by the following design and technological breakthroughs:

- water-immersion at the high-*NA* side, enabling the $NA > 1$ regime,
- quasi-monochromatic laser source with $\delta\lambda_0/\lambda_0 \approx 2\cdot10^{-6}$, eliminating the need for colour correction,
- fused quartz material ($n = 1.560284$ at $\lambda_0 = 193.3$ nm) with a very high degree of homogeneity and purity,
- extensive use of aspheric surfaces,
- a sub-nanometer metrology of individual surface shapes, element positions, together with new opto-mechanical solutions.

The layout of the projection lens is governed by the flat-field requirement. It contains two negative groups, one close to the object plane (1) and one at the centre (3). There are two strong bulges numbered 2 and 4, finally converging to the imaging section 5. This latter section consists mainly of elements with an aplanatic meniscus shape; the final two elements can be characterised as a single concentric element with an aspheric air space inside. It should be noted that the imaging system is not used in standard air but in a nitrogen environment with a slightly larger index of refraction ($3\cdot10^{-6}$) than that of standard air. The sections 1 and 3 have a contribution to the Petzval sum P of $508\cdot10^{-5}$ and $1006\cdot10^{-5}$, respectively. For the groups 2, 4 and 5 we have $-476\cdot10^{-5}$, $-361\cdot10^{-5}$ and $-679\cdot10^{-5}$, yielding a total Petzval sum P of $-2\cdot10^{-5}$. In the immersion medium, this yields a curvature c_{fc} of the field equal to $n_I\, P = -0.000029$. From this value of the field curvature we derive that the inward movement of the focal surface at the border of the field equals 2.84 μm (parabolic approximation). This is a large value as compared to the Rayleigh focal depth of 94 nm; however, such a third-order 'residual' is required to compensate for higher-order field curvature effects.

To compensate for the on-axis and field-dependent aberrations, an extensive use of aspheric surfaces has been made. The abundance of aspheric surfaces allows for a more compact design with a substantial reduction in the maximum diameter and construction length of the system **[349]**. Consequently, the volume and weight are diminished, with a substantial cost advantage. Another advantage is the reduction of the internal aberrations in the system. The use of aspheric surfaces has made it possible to have maximum values below 5000 λ_0 and this influences in a positive way the opto-mechanical tolerances in the system. The magnitude of the asphericity is generally expressed in terms of the absolute value Δ_{asp} of the maximum positive or negative radial departure from the best-fit sphere. In the design, large

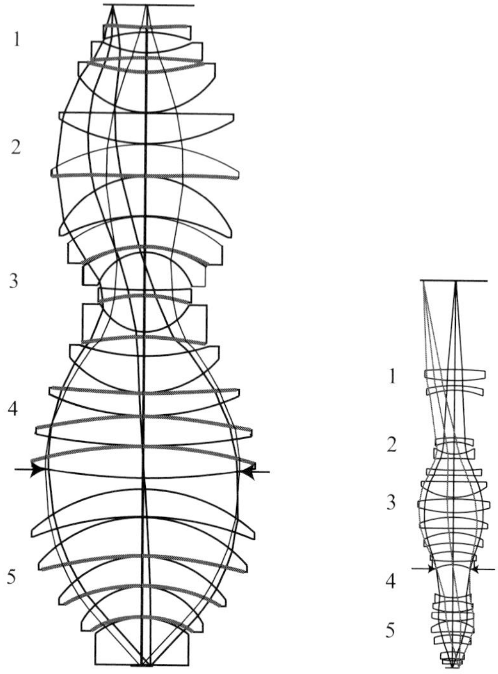

Figure 7.86: A lithographic projection system for water immersion; $NA = 1.10$, $\lambda = 193.3$ nm, image field diameter 28 mm, optical track-length $T \approx 1030$ mm. Transverse magnification $m_T = -0.25$, $f' = 158.38$ mm. At the same scale, the low-aperture projection lens of Fig. 7.85 with $f' = 77.43$ mm is shown on the right.

values of Δ_{asp} are encountered, up to 1.1 mm (lens #5). In the case of an interferometric measurement with respect to the best-fit sphere, this value would amount to approximately 3000 interference lines at a typical measurement wavelength of 633 nm, an impracticably large number. Null tests with the aid of, for example, diffractive optical elements, are applied to test these huge aspheric surface departures.

The imaging performance in the field has been depicted in Fig. 7.87. Two important aspects are the residual aberration in the image field and the distortion. The OPD_{rms} of the imaging pencils, expressed in the vacuum wavelength of the light, does not exceed 0.01 λ_0, an extremely small number. In terms of the Strehl intensity of the projected diffracted image, the as-designed value is as high as 99.5%. The second graph of Fig. 7.87 shows the distortion in nm; it has a maximum value of ±5 nm.

For the manufacturing and maintenance of the lens, it is interesting to know the optical and mechanical tolerances to be respected for high-quality imaging. In Table 7.31 we give the increments ΔOPD_{rms} due to a perturbation of the optical system, together with the expected value of OPD_{rms} (corrected for defocus and distortion) and the spread σ_{OPD} around this OPD value. The values have been collected for the imaging on-axis and at the border of the image field ($y' = 14$ mm). The sensitivity to circularly symmetric perturbations (axial shift of surfaces or elements, surface curvature change or refractive index deviation) is extremely large. However, the effects of the asymmetric perturbations of surfaces or elements (decentring and tilt) are the most impressive. On-axis and in the field, a tilt value of 1 μrad (typically ±0.1 μm displacement at the rim of a lens element) causes mainly coma with an expected OPD_{rms} value of approximately 30 mλ. The same aberration is found for a decentring of 0.05 μm which produces an expected value of 6 to 7 mλ coma. According to the last line of Table 7.31, both effects increase the aberration of a manufactured system to an expected value of 50% of the diffraction limit of 0.07 λ OPD_{rms}. The lens hardly shows a difference in opto-mechanical sensitivity for imaging pencils towards the on-axis point or towards the rim of the field. This effect can be understood by inspection of the corresponding ray paths in Fig. 7.86. The second bulge of the immersion system has the largest optical diameters of the elements and it is here that the sensitivity to opto-mechanical perturbations is largest. The drawing shows that the ray paths of the axial pencil and of the pencil to the outer field position are almost identical in this part of the system.

Table 7.31. Tolerancing results for an immersion projection lens (NA = 1.10).

	Perturbation(s)	ΔOPD_{rms}	OPD_{rms}	σ_{OPD}
Nominal system (y'= 0 mm)	—	0	7	0
Axial distance/thickness	0.2 μm	0	7	10
Surface curvature	0.25 rings	0	7	6
Refractive index	$1 \cdot 10^{-6}$	0	7	5
Decentring	0.05 μm	7	10	3
Tilt	1 μrad	28	29	15
As-manufactured (y'= 0 mm)	accumulated	29	30	21
Nominal system (y'= 14 mm)	—	0	7	0
Axial distance/thickness	0.2 μm	10	12	12
Surface curvature	0.25 rings	10	12	9
Refractive index	$1 \cdot 10^{-6}$	5	9	6
Decentring	0.05 μm	6	9	6
Tilt	1 μrad	30	31	16
As-manufactured (y'= 14 mm)	accumulated	34	35	23

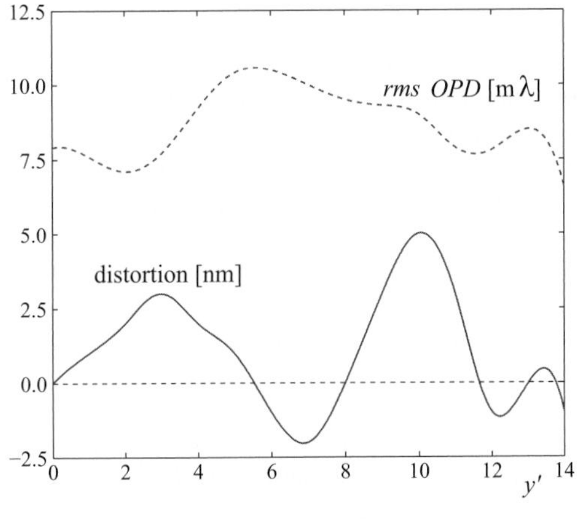

Figure 7.87: High-NA aspheric immersion lens; OPD_{rms} (in units of milli-wavelengths of 193 nm) and distortion (in nm) as a function of the position in the field.

The 'blind' fabrication of lens systems in which such tolerances of the order of 100 nm have to be respected is impossible. Compensating schemes must be devised to correct for the imaging errors (aberration including distortion and field curvature) which are measured when an optical system has been assembled. Table 7.31 and the associated sensitivity coefficients of each surface/element are useful as a source of information for the stabilisation of the finished product. During the lifetime of the projection lens, such mechanical compensation schemes are also used to compensate for the measured image deterioration due to opto-mechanical drift. Using the information on environmental conditions like average temperature, temperature gradients, atmospheric pressure and wavelength shift of the source, the lens is subjected to quasi-permanent adjustments which follow from imaging quality measurements during operation. The opto-mechanical sensitivity is so high that the system cannot be used in a stand-alone, passive way.

Another important aspect for imaging quality by such a complex multi-element system is the image contrast at low spatial frequencies due to surface scattering. This low-frequency scattering can be attributed to polishing irregularity such that the surface becomes very weakly diffusing, mainly at small angles. Equation (10.5.10) provides us with the intensity transmission per surface. A worst case estimate at normal incidence for N_S surfaces is given by the product of the intensity transmittance factors at each transition, yielding the specular intensity transmission T_s for the system

$$T_s = \exp\left\{-N_S\,[k_0(n_1 - n_2)\sigma]^2\right\}\,, \tag{7.9.9}$$

where $|n_1 - n_2| = 0.56$ is the optical contrast between air and fused quartz at a wavelength of 193.3 nm and σ is the root-mean-square surface roughness. It is permissible to use the intensity transmittance factors provided that the spatial coherence area in the object plane has an extent of only a few Airy disc diameters. If we allow 3% of intensity scattering for the total system, Eq. (7.9.9) imposes a maximum roughness value σ of 1.5 nm. This is a severe requirement, at the limits of what classical polishing on curved surfaces can achieve. With the aid of ion-beam polishing, this value can be attained even on aspheric surfaces. Another aspect regarding the light efficiency of the projection lens and the image contrast in the image plane is the light reflection at each transition. Energy is lost for the image formation and spurious light can reach the image plane after two reflections. The total intensity transmission T_r after N_S transitions is given by $(T_j)^{N_S}$ with T_j the intensity transmittance of an individual surface. With the aid of anti-reflective coatings with an intensity reflectance of 0.2% at most, it is possible to achieve a total specular transmittance, including the scattering effects above, of approximately 90%.

7.9.4 Microlithographic Mirror Systems for the Extreme UV Spectral Region

In the search for finer image detail, the immersion lens operating in the deep UV spectral region (vacuum wavelength λ_0 = 193 nm, ArF laser source) represents the end of a development. Slightly larger *NA* values have been used, but the progress in resolution remains marginal. A smaller wavelength system with an F_2 (fluorine) laser source radiating at 157 nm is not adequate for microlithography because of the minute but non-negligible second-order birefringence effect which is present in the lens material fluorite (CaF_2). This material is the only high-grade optical material that exhibits good transmission in this wavelength domain. Its structure is cubic but the dielectric tensor comprises second-order terms, whose magnitudes are proportional to the square of the ratio of the typical lattice distance a in the crystal and the wavelength λ_0 of the light. At λ_0 = 157 nm, these second-order effects influence the refractive index of the material as a function of the ray propagation direction up to a level of 10^{-6}. Birefringence effects due to 'spatial dispersion' in cubic crystals were originally predicted by Lorentz **[221]** and studied in more detail in **[302]** and **[303]**. In the ray optics model, the minute double refraction at each interface leads to ray doubling and a total of 2^J rays after propagation through J lens elements. A ray blurring arises, sufficiently large to compromise the gain in resolution which would follow from the reduction in wavelength from 193 to 157 nm.

Searching for a still shorter wavelength and materials with high transmission or reflection, a problematic spectral region is first encountered, ranging from 150 nm down to 20 nm. A material like silicon carbide (SiC) is capable of a reflectivity at normal incidence of only 40% at 60 nm wavelength **[179]**. Appreciable transmission through materials in this wavelength region is limited to thicknesses of less than one wavelength of the light. A breakthrough in normal incidence reflectivity was presented by Spiller **[324]** who opened the extreme UV wavelength region below 15 nm for (near) normal incidence reflective optics by applying resonant, extremely thin $\lambda/4$ layers. Later research showed that for a central wavelength of 13.4 nm, with the aid of a stack consisting of 40 bilayers of the materials Mo and Si a theoretical near-normal reflectivity of 72% can be obtained **[169]**. Practical values as high as 68% have been reported in the literature **[222]**. Other material combinations like Be/Mo can achieve still higher theoretical reflection coefficients, for example 77% to 80% at λ_0 = 11.3 nm, but this option has been abandoned because of the extremely toxic nature of the metal beryllium. Fine-tuning of the construction of the multilayer (varying layer thickness, introduction of a third material like ruthenium) and optimisation of the final protective capping layer have led to an optimised reflected power after nine mirrors that reaches 5.0% instead of the non-optimised value of 3.1% following from 0.68^9 **[316]**. For the Be/Mo multilayers, these values were 10.4% and, non-optimised, 5.9% (= 0.73^9). Unfortunately, these numbers show that the radiant power efficiency of a six-mirror projection system (including three normal-incidence surfaces in the illumination part) is low. A very powerful extreme UV (EUV) source is needed to create an irradiance level of at least 100 W in the image plane of the projection system. In the remaining part of this subsection we discuss the typical layout of non-obstructing EUV mirror systems, their imaging quality and fabrication requirements regarding surface accuracy and positioning tolerances.

7.9.4.1 Paraxial Layout

To obtain the highest possible resolution corresponding to a certain imaging aperture, a classical mirror system with central obstruction is not allowed for microlithography. The diffraction effects that are caused by the inner rim of the system diaphragm create extra 'ringing' effects of the point-spread function. For this reason, an off-axis mirror system is required, for instance like the Offner system which was discussed earlier in this section. A reduction ratio of the order of 4 to 5 is needed to keep the manufacturing of the lithographic masks feasible (electron beam writing). The design challenge is to find unobstructed off-axis lightpaths through a multi-mirror system with pencil diameters which correspond to the required imaging aperture, typically 0.25 to 0.30. A systematic analysis of such unobstructed off-axis mirror systems with a 'ring-field' has been carried out in **[20]**, for systems with four, six and eight mirrors. The analysis is based on paraxial optics for the pencil leaving an off-axis object point. The lateral extent of such a pencil is determined by the required aperture of the pencil in image space and the system magnification (see Fig. 7.88). The propagation between each sequential pair of mirrors with indices j and $j+1$ is analysed in the axial interval given by the smallest and the largest value of the z-coordinates of the intersection points $P_{j,0}$, $P_{j,1}$, $P_{j+1,0}$ and $P_{j+1,1}$ of the extreme meridional rays of the pencil. The possible intersection points of these two extreme rays are determined with the section of an arbitrary mirror surface, labelled i, which is used by the paraxial beam elsewhere in the sequential system with a lateral beam footprint given by the y-coordinates of the points $P_{i,0}$ and $P_{i,1}$. In the figure beam obstruction occurs since one endpoint, $P_{i,1}$ is found in the interval determined by the intersection points $P'_{i,0}$ and $P'_{i,1}$. Unobstructed paraxial layouts are obtained in certain geometric classes. A useful classification method relies on the sign of the angle of incidence of the principal ray on each mirror surface. This sign follows from the quantity $(s_z v_y - s_y v_z)/|s_z v_y - s_y v_z|$ where $\hat{\mathbf{s}} = (0, s_y, s_z)$ is the paraxial ray vector and $\hat{\mathbf{v}} = (0, v_y, v_z)$ is the mirror normal vector satisfying $\hat{\mathbf{v}} \cdot \hat{\mathbf{s}} < 0$. A geometric class is then defined by the sequence of signs for each mirror surface. In **[20]** a more practical definition is introduced. A binary 'one' is attached to a negative sign for the angle of incidence and a binary 'zero' to a positive sign. Alternatively, starting with a positive y-coordinate of the object point O_0, we follow the direction of the principal ray of the paraxial pencil and attribute a binary 1 to a reflected ray that is deflected to the right and a binary 0 for a ray reflected towards the left. The sequence of binary digits, written down from left to right according to the mirror sequence in the imaging system, forms a binary number. This binary number is then replaced by its decimal version and used to indicate the mirror class. In each class we have a subdivision for positive and negative lateral magnification. Because of the extreme speed with which paraxial system analysis can be carried out, it is possible to determine with exhaustive computer calculations which geometric classes yield unobstructed imaging systems for four, six, eight and even ten mirror systems.

The capability of each class to produce useful imaging systems has to be checked with respect to some extra design parameters. First of all, the third-order Petzval sum is made equal to zero, a common requirement for lithographic systems and even more important for an off-axis system. Although higher-order field curvature will require a slight offset, it is good practice to require initially a zero Petzval sum. At the image side, a large free-working-distance of 20 to 30 mm is needed such that the mirror substrates can be made thick enough for mechanical stability. There should also be room available to allow measurement beams to pass through this free volume above the scanning lithographic wafer stage. A very important design parameter is related to the performance of the multilayers on a surface. Each multilayer can be tuned to the average angle of incidence of the rays of an imaging pencil. Close to normal incidence, the allowed spread in angle of incidence is of the order of ±7°; for larger angles of incidence, up to 20°, the angular spread over the beam cross-section is reduced to only a few degrees. On top of this, appreciable differences between the amplitude and phase of *TE* and *TM* polarised light start to appear. A solution for this problem is the introduction of a variable thickness of the multilayer as a function of the position on the mirror surface. In each point, the layer thicknesses are adjusted such that resonant reflection occurs for the average angle of incidence at that region of the mirror. To accommodate the large aperture of a pencil in image space, the conjugate planes for the last imaging step should be close to the centre of (paraxial)

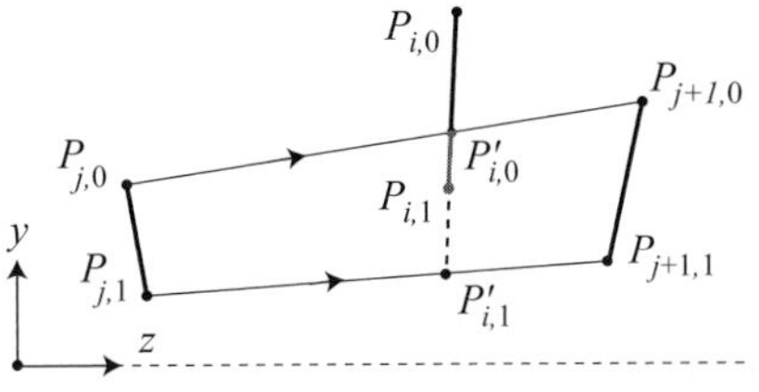

Figure 7.88: Detection of (partial) beam obstruction.

curvature of the last reflecting surface. These extra conditions on the paraxially unobstructed designs reduce the large number of potential classes to only a few attractive ones which deserve further exploration. It was immediately discovered that imaging systems with an aperture larger than 0.10 require the extra degrees of freedom provided by aspheric surface profiles such that the systems can be given diffraction-limited imaging quality at a wavelength of 13 nm. As with the 1:1 Offner-system, the optical system has circular symmetry and a ring-field image is projected. With a scanning mask and wafer, moving with speeds of v and $m_T v$, respectively, a rectangular exposed field is produced on the wafer.

7.9.4.2 System I

NA = 0.20, field annulus radius 36.5 ± 1 mm, λ_0 = 13.4 nm, m_T = −0.25

This system **[50]** belongs to the paraxial class 26− and comprises a first two-mirror system at the object side with a large focal distance (see Fig. 7.89a). The object plane with an off-axis object point O_0 is close to the front focal plane of this first section and produces an almost parallel beam ($m_T \approx +2$), propagating towards the imaging section consisting of four mirrors. The first two constitute a positive group operating at negative magnification ($m_T \approx -1/3$) and the last two mirrors have the typical shape of a Schwarzschild microscope objective, used off-axis ($m_T \approx +1/3$). The physical stop is at mirror 5 in the front focal plane of mirror 6 and is projected to infinity (telecentric imaging). In Fig. 7.89a we have schematically drawn the axial beam in a paraxial lens system with the same sequence of positive and negative powers as in the off-axis mirror system.

7.9.4.3 System II

NA = 0.25, field annulus radius 30.0 ± 1 mm, λ_0 = 13.4 nm, m_T = +0.25

The second system **[234]** (see Fig. 7.89c) belongs to geometry class 41+ and can be divided into a first four-mirror system that produces an intermediate image with a paraxial magnification $m_T \approx -0.8$ at the entrance of a second two-mirror group. The presence of an intermediate image plane helps to avoid obstruction in the system. On the other hand,

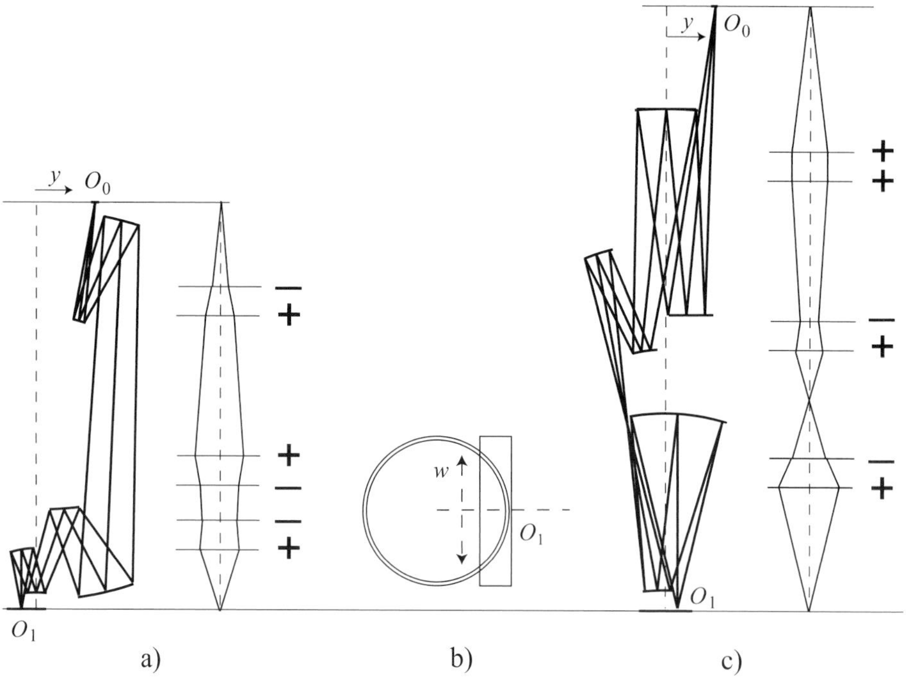

Figure 7.89: Mirror imaging systems for extreme UV radiation (λ_0 = 13.4 nm).
a) System I, NA = 0.20, m_T = −0.25, optical throw T = 1 m.
b) The well-corrected ring-field, producing a rectangular image field with width w in scanning imaging mode.
c) System II, NA = 0.25, m_T = +0.25, T = 1.5 m.

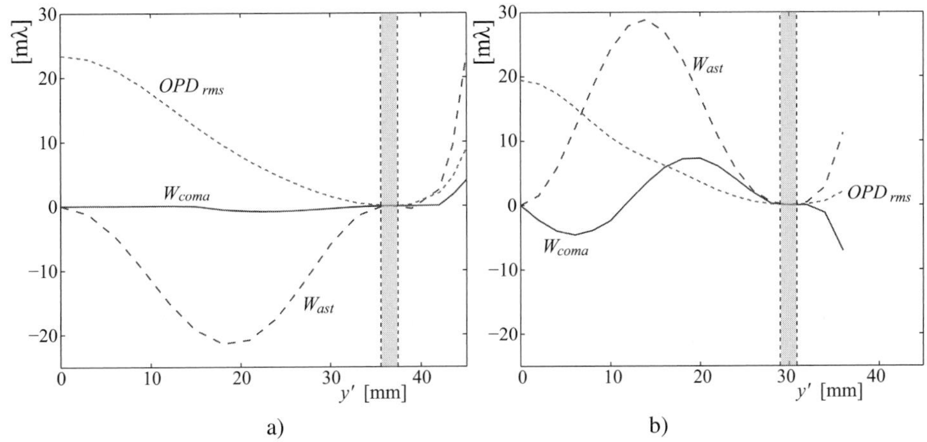

Figure 7.90: Field-dependent coma and astigmatism in an extreme UV mirror projection system.
a) Mirror system I, *NA* = 0.20.
b) Mirror system II, *NA* = 0.25.

with respect to aberration correction, it requires anew the change from small divergence (0.08) of the pencils to the beam convergence (0.25) in image space. Extra strain is likely to exist in such a final two-mirror section with a total swing in aperture of 0.33. The physical stop is at surface 2. An equivalent refractive system shows the successive converging and diverging steps in this system. With respect to system I, the second system has a 50% larger construction length T which helps to attain good imaging over an annulus with the same width w = 2 mm but at a 25% larger aperture value.

The way in which the well-corrected annular image zone is formed has been shown in Fig. 7.90. As a function of the off-axis distance y' the main wavefront aberration components, astigmatism and coma, have been calculated, with disregard to the beam obstruction that occurs when image field positions are chosen for which the mirror dimensions have not been optimised. The dominant aberration is astigmatism, with a Seidel term with opposite sign for the two systems. This third-order aberration term is reduced by a fifth-order aberration component with a fourth-order field dependence. Close to the well-corrected annular region, seventh- and ninth-order terms lead to an inflection point in system I, at zero aberration level. In system II, the astigmatism shows a relative minimum at zero level in the annular region by a cancelling of the third-, fifth- and seventh-order astigmatic terms. Beyond the annulus diameter, the balance between the Seidel and higher-order astigmatic terms is lost. This behaviour is comparable to what was observed and calculated for the 1:1 Offner mirror system (see Fig. 7.83) for the third- and fifth-order astigmatism of this spherical system. The calculations which could be carried out for the Offner system (see Eq. (7.9.7)) are less transparent for the aspheric six-mirror systems. Using aspheric surfaces, the same compensation mechanism between higher-order astigmatic terms can be used by applying a general aspheric shape to each mirror surface. The compensation of coma proceeds along the same lines, with field-dependent terms of first up to ninth order carrying an appreciable aberration amplitude.

The distortion over the annular field width should be very well corrected, down to the 1 nm level. In the off-axis systems, the paraxial magnification defined for the region close to the optical axis is irrelevant. The local magnification in the field annulus should be stationary and have the exact nominal value to allow the machine-to-machine exchange of exposed silicon wafers. In Fig. 7.91 the normalised deviation $\delta m/m$ from the desired magnification m has been shown. This quantity is defined as $y'/(my_0) - 1$, where y_0 is the object point coordinate and $y'$the image field coordinate that yields a minimum rms value of the wavefront aberration when evaluated with respect to this image point. The axial position of the image plane is fixed when calculating $\delta m/m$ and corresponds to the optimum focus-setting for the chosen annular image field. For system I we see a maximum at the desired magnification of −0.25, the system II shows an inflection point. Within the shaded region corresponding to the annular field of each system, the theoretical distortion is less than 1 nm. In Figs. 7.92 and 7.93 graphs are presented of the meridional and sagittal cross-sections of the wavefronts of pencils of rays propagating to three positions in the well-corrected ring fields of systems I and II. In both systems, small residuals of field curvature, (higher-order) astigmatism and coma are observed. The rms OPD values within the ring fields are all well within the just-diffraction-limited value of 0.07 λ_0. In line with the findings about the field-dependent astigmatism in Fig. 7.90, system I with the inflection point at y'= 36.5 mm shows a change of sign of the residual

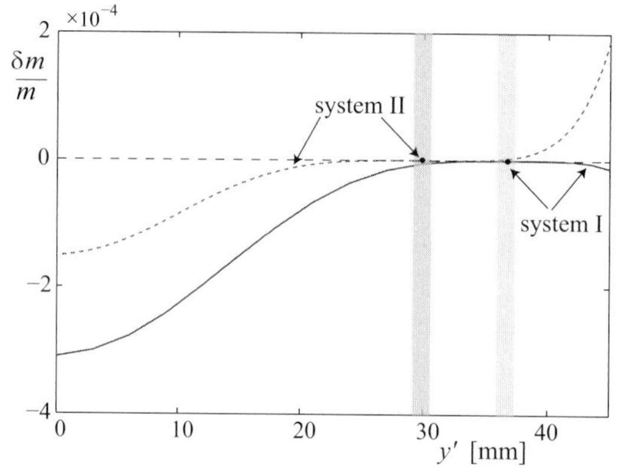

Figure 7.91: The normalised magnification deviation $\delta m/m$ as a function of the field position y'.

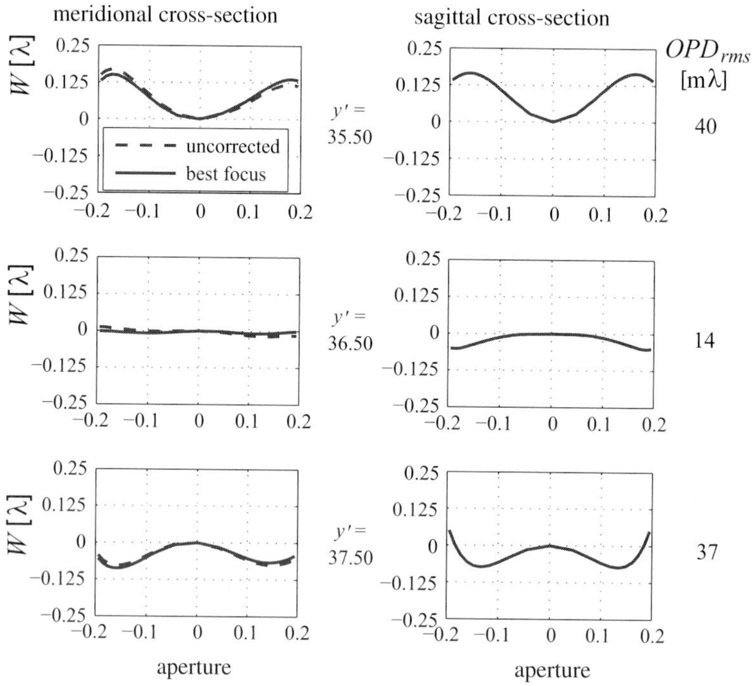

Figure 7.92: Meridional and sagittal wavefront cross-sections for three field positions in the annular field of a six-mirror imaging system for extreme UV radiation; $\lambda_0 = 13.4$ nm, $NA = 0.20$, $m_T = -0.25$, optical throw $T = 1$ m (system I).

astigmatism over the ring width; system II has virtually equal astigmatism at the inner and outer border of the annular field. For successful imaging with EUV mirror systems, the light diffusion from each mirror surface should not exceed a certain value. Referring to Eq. (7.9.9), we require, somewhat arbitrarily, that the total light scattering does not exceed 5% for a six-mirror system. Because of the high optical contrast of two of the mirror surfaces at near normal incidence, the surface roughness value σ of each mirror surface should not exceed 0.2 nm. Polishing of aspheric surfaces down to this roughness level is extremely challenging but has been demonstrated in practical systems; ion-beam polishing is the preferred technique to achieve these very low roughness levels. Depending on the roughness spectrum of aspheric surfaces, part of the light scattering will be at large angles and never reach the image plane. This scattering contribution

Table 7.32. Tolerancing results for six-mirror systems I and II; OPD values in units of milli-wavelengths (mλ).

	Perturbation(s)	ΔOPD_{rms}	OPD_{rms}	σ_{OPD}
System I (NA = 0.20, y'= 36.5 mm)	—	0	14	0
Axial distance/thickness	1 μm	25	31	14
Decentring	0.1 μm	33	37	13
Tilt	1 μrad	47	52	25
As-manufactured (y'= 36.5 mm)	accumulated	81	83	28
System II (NA = 0.25, y'= 30 mm)	—	0	17	0
Axial distance/thickness	1 μm	18	20	10
Decentring	0.1 μm	50	53	20
Tilt	1 μrad	68	69	36
As-manufactured (y'= 30 mm)	accumulated	119	119	42

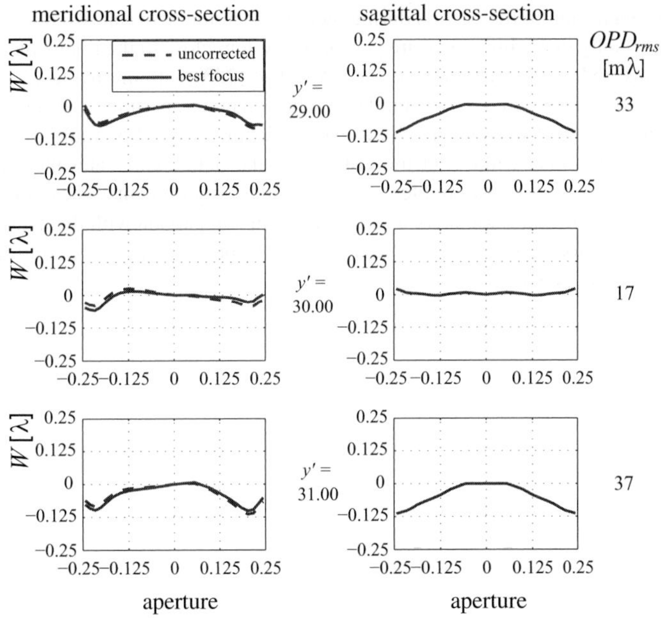

Figure 7.93: Same legend as for Fig. 7.92, but now for NA = 0.25, m_T = +0.25 and optical throw T = 1.5 m (system II).

lowers the specular transmission of the system. At smaller scattering angles, the diffused light reaches the image plane and will give rise to contrast loss in the image due to this background light. Very low spatial frequencies in the roughness spectrum of a mirror surface are equivalent to higher-order aberrations of a random nature and lead to a resolution loss in the image.

The mirror systems to be used at the short EUV wavelength have very small constructional tolerances. Statistical analysis carried out for these systems yields the values listed in Table 7.32. The values in the table show that extremely tight tolerances are encountered, especially with respect to decentring and tilt of the mirror surfaces. We conclude that the mirror system should be stabilised to within 0.1 μm in the lateral direction. In the axial direction, the tolerance on movements is more relaxed. Moreover, there is the required accuracy of the shape of the reflecting surfaces themselves.

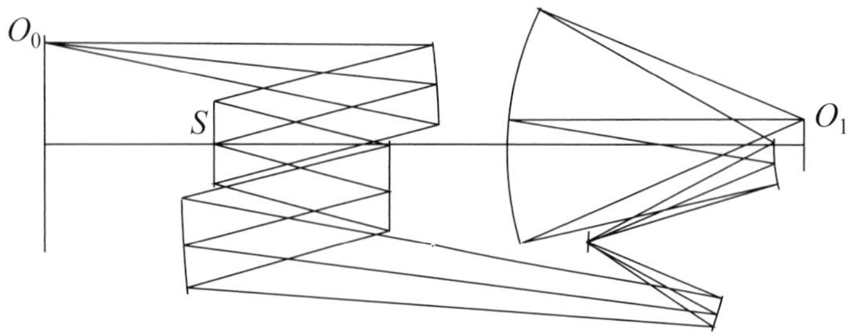

Figure 7.94: A compact eight-mirror projection system with $NA = 0.40$ and an annular field width w of 1 mm ($26.5 \leq y' \leq 27.5$); $m_T = +0.25$, optical throw $T = 838$ mm.

It is shown by Eq. (10.5.48) that the statistical addition of surface errors on each mirror requires a σ value of 0.04 nm. This means that the peak-to-valley error on a surface should be limited to 0.25 nm. Such mechanical requirements cannot be guaranteed during the lifetime of a mirror system. As was the case for the high-resolution immersion lens, the mirror system has to be continuously monitored regarding imaging quality. Based on the measurements of residual aberration, distortion, focus offset and field curvature over the ring width w, the position *and* the shape of the mirror surfaces have to be actively adjusted in three dimensions. The extreme requirements on the mutual position of the optical surfaces can be understood by the large aberration of the pencils during their propagation between two mirrors. Values as large as 10^4 wavelengths are found, especially in the high-aperture part of the system close to the image plane.

Mirror projection systems with eight mirrors have been designed **[233]** with an increased aperture as large as 0.40. Figure 7.94 shows an embodiment of such a system. The higher imaging aperture is important for further resolution improvement but also brings some disadvantages. Two extra reflections by EUV multilayer stacks reduce the light efficiency by a factor of two. The small width w (1 mm) of the annular field complicates the exposure uniformity on the silicon wafer. The higher aperture also imposes tighter opto-mechanical requirements. The tolerable values for decentring and tilt of the mirrors are reduced by approximately a factor of two in comparison with the data for the six-mirror system with $NA = 0.25$ in Table 7.32. More powerful EUV sources need to be developed to permit the introduction of these promising eight-mirror projection systems.

III Diffraction Theory of Optical Imaging

Introduction to Part III

Part III of this book is entirely devoted to the diffraction of light and its 'degrading' effect on image formation. The first observation of the spreading-out of a light beam after passage through a small opening or around an obstruction goes back to Grimaldi, in the 1650s. In his major work **[117]**, published after his death in 1663, he describes two new phenomena, mostly qualitatively. A beam of sunlight passes through a (not too) small hole in a curtain. The transmitted beam hits a central obstruction at a few metres distance and Grimaldi describes the light intensity distribution at a few metres beyond the obstruction. The first observation was a suffusion of the sunlight into the central geometrical shadow region of the illuminating beam. A more detailed second observation in the geometrical shadow region revealed a first white intensity maximum and two others which had a coloured appearance. In both coloured regions, the colour closest to the geometrical shadow was blueish, on the other side the light intensity was rubicund. Grimaldi assumes that colours appear by the breaking-up of the white light into its composing colours, in the same way as a flowing water mass is split by the pillar of a bridge. The analogy between the observed optical phenomenon and fluid transport made him choose the Latin verb *diffringere* (= to break-up, to shatter) for this *fourth manner* in which light propagates. In his publication Grimaldi states that the novel diffraction-based light propagation has the same importance as the rectilinear propagation, refraction and reflection of light. The questions why and how colours do appear in the diffraction process could not be satisfactorily answered by Grimaldi.

It took a long time before detailed observations and measurements of the diffraction of light became possible. The diffraction of light by the limited aperture of, for instance, a telescopic lens system produces a characteristic diffraction figure in the focal region. However, the details of the light diffraction were obscured by the colour defects of lenses and their manufacturing imperfections, such as air inclusions in the volume of glass material, polishing defects of the optical surfaces and light scattering. It was not until the end of the eighteenth century that really outstanding colour-corrected optical instruments could be manufactured by Joseph von Fraunhofer. Earlier observations of light diffraction by Fraunhofer were performed at a very large distance from the diffracting aperture. In the focus of his lenses with superior optical quality, the typical diffraction pattern of a circular aperture could also be observed in the image plane, in white light or in monochromatic light. The mathematical analysis of light diffraction using his newly developed wave theory of light was described by Fresnel with the aid of his diffraction integrals which permitted the calculation of the optical field close to the geometrical shadow of an obstacle or in the focal region of a lens. When the integral was applied to exactly the focal plane of a lens or, an equivalent situation, to a very distant plane in the absence of a lens, it produced the so-called Fraunhofer diffraction pattern. Fraunhofer also linked his name to the diffraction grating to be used in a spectrometer. He produced a wire-grating with 300 wires over a distance of 25 mm and a ruled grating in a gold foil with a pitch as small as 3 μm, both to serve as dispersive elements in his spectrometer.

The statement in Grimaldi's book that the diffraction-based way of light propagation was in no way a subject to be neglected with respect to the better-known ways of light propagation was a prophetic one. With respect to research in optics, the diffraction of light and the propagation of diffracted waves have attracted the interest of a great number of outstanding scientists. We mention Kirchhoff, Rayleigh, Sommerfeld, Debye, Zernike, Wolf, etc. In this third part of the book the subject of light-diffraction is first treated in its full extent, including the vector character of the electromagnetic field components of the diffracted optical field. In this general manner, the vector versions of the Kirchhoff diffraction integral and of the Rayleigh–Sommerfeld integral are derived. These vector-based diffraction integrals are then subjected to the appropriate boundary-value problem to enable a unique solution of the diffracted vector field. Simplified versions of the above-mentioned diffraction integrals are treated, among others, by the Debye–Wolf integral, which yields a valid approximation of the diffracted field if the number of Fresnel zones in the diffracting aperture is large. It is a relatively straightforward step to derive the frequently used scalar diffraction integrals from their vector equivalents. This yields, in a final approximating step, the well-known Fresnel and Fraunhofer diffraction integrals. The chapter concludes with a discussion on the regions of validity for the various vector and scalar diffraction integral formulas and, in more detail, for the widely used Debye–Wolf diffraction integral.

The second chapter of Part III continues the study of diffraction integrals in the presence of small or larger ($\gg \lambda/4$) aberrations in the imaging system. The diffraction theory developed by Zernike and Nijboer for weakly aberrated imaging systems is reviewed. The efficient representation of aberrations in terms of an orthogonal expansion with the aid of the Zernike circle polynomials is discussed. With the aid of the coefficients of a Zernike polynomial expansion of the optical aberration function, the computation of the (three-dimensional) shape of the scalar point-spread function of an imaging system can be carried out in a mathematically elegant way. The Nijboer–Zernike diffraction theory is then further extended to allow for more severely aberrated imaging systems and for larger excursions from the nominal image plane, whilst conserving the semi-analytic character of the original Nijboer–Zernike theory. The Extended Nijboer–Zernike aberration theory is then applied to the 'vector' Debye diffraction integral such that the electromagnetic components in the focal region can be calculated and the (aberrated) point-spread function of high-numerical-aperture imaging systems is obtained. The chapter on aberrated point-spread functions is concluded by a study of the flow of optical energy, linear momentum and angular momentum in the focal region of a high-numerical-aperture imaging system.

The subject of the third chapter of Part III is the frequency analysis of optical imaging systems. As in the preceding chapter on the optical point-spread function, the influence of imaging aberrations is included in the calculation of the frequency transfer by an optical system. The same holds for the state of coherence of the light that illuminates the object to be imaged. In this respect, the classical theory of partial coherence (van Cittert–Zernike theorem) is presented, in the framework of optical Fourier analysis in the spatial domain. Using these building blocks for frequency analysis, the imaging of periodic objects by an optical system is studied. Linearised versions of the imaging equations are given for the case of weakly-modulated objects. Both the classical static imaging systems are subjected to frequency transfer analysis in this chapter, as well as the more recently developed scanning imaging systems. The chapter also presents a short introduction to the statistical analysis of light scattering in an imaging system and its influence on the frequency transfer towards the image space. An important part of the chapter on frequency transfer has been devoted to the three-dimensional transfer of spatial frequencies by an imaging system. The extra dimension is relevant when an extended *volumetric* object has to be imaged. This type of (dynamic) object is frequently encountered when living specimina have to be imaged, for instance biological cells.

The fourth chapter of Part III presents experimental methods for high-resolution imaging using polarised light. An introduction to vector ray-tracing is given in which the complex orthogonal field components of a single ray are tracked during propagation through a high-aperture imaging system containing birefringent and polarising optical elements. The vector amplitude of the field components is monitored by means of generalised Jones matrices. In a next step, an assembly of rays associated with an entire imaging pencil is pursued from object to image plane to enable the fully vectorial treatment of an object pencil in a high-aperture imaging system. This approach is applied to both transmission microscopes and to reflecting microscopes of the confocal type. The potential of the vector-imaging model is illustrated by means of the imaging of fluorescent molecules embedded in a multilayer.

8 Vectorial and Scalar Theory of Diffraction and Focusing

This chapter is concerned with the foundation of vectorial and scalar diffraction theories used in this book. The definition of the term 'diffraction' is given below so it should suffice to say that in this chapter we are discussing how the propagation of, usually plane, electromagnetic waves is perturbed by apertures and how we can construct a mathematical handle for the treatment of focusing by lenses from a wave optics point of view. Even after many years of practice most optical physicists and engineers are left yearning for deeper understanding of the assumptions individual diffraction theories impose. Furthermore, it is not unusual to find erroneous statements in the literature regarding the interrelation of these theories. Therefore, this chapter presents a consistent formulation of major diffraction theories in an attempt to shed light on this sometimes confusing part of physical optics.

Starting from the source-free version of Maxwell's equations, we present a consistent formulation of diffraction theories in the sense that we use the Stratton–Chu integral theorem as the starting point and all subsequent derivations assume only knowledge of this formula. In Section 8.1 we derive the Stratton–Chu integral theorem, the integral theorem of Kirchhoff, the e- and *m*-theories and the (first) vectorial Rayleigh–Sommerfeld integral. In Section 8.2 we discuss the problems associated with choosing the boundary values when solving a diffraction problem and, in particular, the removal of the so-called back radiation from the diffraction integrals. We also discuss briefly problems associated with the discontinuous aperture rim upon which diffraction occurs. This section is followed by a thorough discussion of the Debye–Wolf and related diffraction theories in Section 8.3. This is of particular importance as the Debye–Wolf theory is perhaps the most convenient way to mathematically formulate the problem of vectorial focusing. Scalar diffraction theories are derived in Section 8.4. Finally, a general discussion on the asymptotic boundary condition is presented in Section 8.5.

Historical review of the diffraction theories is omitted in this chapter. This we do since the most significant theories were derived before or during the first half of the twentieth century and many reviews have been published on the subject since. These are most notably the individual papers by Bouwkamp **[42]** and Kottler **[192, 193]** or books by Born and Wolf **[37]**, Stamnes **[325]** and Mandel and Wolf **[232]**. Nevertheless, during our derivation, we frequently refer to the original papers or books of various authors and these publications are readily available should such need arise from the reader.

8.1 Foundation of Vector Diffraction

Consider a volume V with a closed boundary S for which the *outward* surface normal is denoted by $\hat{\mathbf{v}}$. Let $\mathbf{f}(\mathbf{r} = (x, y, z))$ and $\mathbf{g}(\mathbf{r})$ be two vector functions which are, together with their first derivatives, continuous on S and V, i.e. well behaved. A direct consequence of this assumption is that diffraction integrals derived below can be directly applied to closed, continuous volumes. There are, however, certain mathematical techniques that can be used to avoid this restriction. These will be discussed later.

From the divergence theorem of Gauss, Eq. (1.3.8), upon making the substitution $\mathbf{v} = \mathbf{f} \times (\nabla \times \mathbf{g})$ one obtains:

$$\iiint_V \nabla \cdot [\mathbf{f} \times (\nabla \times \mathbf{g})]\, dV = \oiint_S [\mathbf{f} \times (\nabla \times \mathbf{g})] \cdot \hat{\mathbf{v}}\, dA. \tag{8.1.1}$$

After expanding the $\nabla\cdot$ operator by using the identity (see Appendix A, Section A.10):

$$\nabla\cdot(\mathbf{v}\times\mathbf{w}) = \mathbf{w}\cdot(\nabla\times\mathbf{v}) - \mathbf{v}\cdot(\nabla\times\mathbf{w}), \tag{8.1.2}$$

Eq. (8.1.1) yields the vector analogue:

$$\iiint_V \{(\nabla\times\mathbf{f})\cdot(\nabla\times\mathbf{g}) - \mathbf{f}\cdot[\nabla\times(\nabla\times\mathbf{g})]\}\,dV = \oiint_S [\mathbf{f}\times(\nabla\times\mathbf{g})]\cdot\hat{\mathbf{v}}dA, \tag{8.1.3}$$

of Green's first identity:

$$\iiint_V \left(\nabla f\cdot\nabla g + f\,\nabla^2 g\right) dV = \oiint_S (f\,\nabla g)\cdot\hat{\mathbf{v}}dA, \tag{8.1.4}$$

which was obtained after the $\mathbf{v} = f\,\nabla g$ substitution to the same equation.

If we now interchange $\mathbf{f}$ and $\mathbf{g}$ in Eq. (8.1.3) and subtract the resulting equation from Eq. (8.1.3) we obtain:

$$\iiint_V [\mathbf{g}\cdot\nabla\times(\nabla\times\mathbf{f}) - \mathbf{f}\cdot\nabla\times(\nabla\times\mathbf{g})]\,dV = \oiint_S [\mathbf{f}\times(\nabla\times\mathbf{g}) - \mathbf{g}\times(\nabla\times\mathbf{f})]\cdot\hat{\mathbf{v}}dA, \tag{8.1.5}$$

which is the vector analogue of Green's second identity. These two integral theorems will be the basis of our further derivations.

8.1.1 The Stratton–Chu Theory

In Section 1.6 we derived the vector wave equations for the electric and magnetic fields. We pointed out that although a very useful formula, the Helmholtz equation cannot be used to describe electromagnetic problems because each vector equation must be separated into three independent scalar wave equations, one for each Cartesian co-ordinate component of the electric and the magnetic field. The resulting six equations must then be solved independently. This is not satisfactory because the Cartesian components of the electric and magnetic fields are interdependent via the divergence equations. In addition, the time-dependent electric and magnetic fields are coupled via Faraday's and Ampère's laws. It is possible however to derive a more generally valid expression, which we do in this section.

8.1.1.1 The Stratton–Chu Integral Theorem

First, we shall only be concerned with the source-free, time-independent Maxwell's equations, which can be written from Eqs. (1.6.3)–(1.6.6) as

$$\nabla\cdot\mathbf{E} = 0, \tag{8.1.6}$$

$$\nabla\cdot\mathbf{H} = 0, \tag{8.1.7}$$

$$\nabla\times\mathbf{H} = -i\omega\epsilon\mathbf{E}, \tag{8.1.8}$$

$$\nabla\times\mathbf{E} = i\omega\mu\mathbf{H}, \tag{8.1.9}$$

by taking the charge density to zero, $\rho = 0$ which also results in zero current density. This means that our ensuing derivation will only be valid in volumes of space absent of free charges and therefore currents. Note that the above equations also assume that the material constants ϵ and μ are position and direction independent, i.e. that the medium of propagation is homogeneous and isotropic.

We now return to Eq. (8.1.5) and set $\mathbf{f}(\mathbf{r}) = \mathbf{E}(\mathbf{r})$ and $\mathbf{g}(\mathbf{r}) = G(\mathbf{r})\hat{\mathbf{a}}$ **[328]** where $\hat{\mathbf{a}}$ is an arbitrary constant unit vector and where for the moment we only require that $G(\mathbf{r})$ satisfies the scalar time-independent wave equation $(\nabla^2 + k^2)G = 0$ on and within S. Substituting for $\mathbf{f}$ and $\mathbf{g}$ into the first term of the kernel on the left side of Eq. (8.1.5) gives:

$$G\hat{\mathbf{a}}\cdot[\nabla\times(\nabla\times\mathbf{E})] = G\hat{\mathbf{a}}\cdot\left[\nabla(\nabla\cdot\mathbf{E}) - \nabla^2\mathbf{E}\right] = -G\hat{\mathbf{a}}\cdot\nabla^2\mathbf{E} = k^2G(\hat{\mathbf{a}}\cdot\mathbf{E}), \tag{8.1.10}$$

while the second term on the left-hand side gives

$$\begin{aligned} -\mathbf{E}\cdot\{\nabla\times[\nabla\times(G\hat{\mathbf{a}})]\} &= -\mathbf{E}\cdot\left\{\nabla[\nabla\cdot(G\hat{\mathbf{a}})] - \nabla^2(G\hat{\mathbf{a}})\right\} = -\mathbf{E}\cdot\nabla(\hat{\mathbf{a}}\cdot\nabla G) + \mathbf{E}\cdot\hat{\mathbf{a}}\nabla^2 G \\ &= -\mathbf{E}\cdot[\nabla(\hat{\mathbf{a}}\cdot\nabla G)] - (\mathbf{E}\cdot\hat{\mathbf{a}})k^2 G, \end{aligned} \tag{8.1.11}$$

where we have used Eq. (8.1.6). Clearly the term $k^2 G(\hat{\mathbf{a}}\cdot\mathbf{E})$ in Eq. (8.1.10) exactly cancels the second term in Eq. (8.1.11), therefore the integrand on the left-hand side of Eq. (8.1.5) becomes $-\mathbf{E}\cdot[\nabla(\hat{\mathbf{a}}\cdot\nabla G)]$. If we now use the identity

$$\nabla \diamond (f\mathbf{v}) = (\nabla f) \diamond \mathbf{v} + (\nabla \diamond \mathbf{v}) f, \tag{8.1.12}$$

where the symbol $\diamond$ stands for either $\cdot$ or $\times$, we have

$$-\mathbf{E}\cdot[\nabla(\hat{\mathbf{a}}\cdot\nabla G)] = -\nabla\cdot[(\hat{\mathbf{a}}\cdot\nabla G)\,\mathbf{E}] + (\hat{\mathbf{a}}\cdot\nabla G)(\nabla\cdot\mathbf{E}) = -\nabla\cdot[(\hat{\mathbf{a}}\cdot\nabla G)\,\mathbf{E}]. \tag{8.1.13}$$

Hence the left-hand side of Eq. (8.1.5) now reads:

$$-\iiint_V \nabla\cdot[(\hat{\mathbf{a}}\cdot\nabla G)\,\mathbf{E}]\,dV = -\oiint_S [(\hat{\mathbf{a}}\cdot\nabla G)\,\mathbf{E}]\cdot\hat{\mathbf{v}}dA = -\hat{\mathbf{a}}\cdot \oiint_S (\mathbf{E}\cdot\hat{\mathbf{v}})\,\nabla G dA, \tag{8.1.14}$$

which follows from the divergence theorem Eq. (1.3.8). Now let us consider the right-hand side of Eq. (8.1.5). The first term in the integrand gives:

$$[\mathbf{f}\times(\nabla\times\mathbf{g})]\cdot\hat{\mathbf{v}} = [\mathbf{E}\times(\nabla G\times\hat{\mathbf{a}})]\cdot\hat{\mathbf{v}} = \hat{\mathbf{a}}\cdot[\nabla G\times(\mathbf{E}\times\hat{\mathbf{v}})] = \hat{\mathbf{a}}\cdot[(\hat{\mathbf{v}}\times\mathbf{E})\times\nabla G], \tag{8.1.15}$$

where we have used the identity

$$[\mathbf{t}\times(\mathbf{u}\times\mathbf{v})]\cdot\mathbf{w} = [\mathbf{u}\times(\mathbf{t}\times\mathbf{w})]\cdot\mathbf{v}. \tag{8.1.16}$$

The second term in right-hand side, the kernel of Eq. (8.1.5), gives:

$$[\mathbf{g}\times(\nabla\times\mathbf{f})]\cdot\hat{\mathbf{v}} = [G\hat{\mathbf{a}}\times(\nabla\times\mathbf{E})]\cdot\hat{\mathbf{v}} = \hat{\mathbf{a}}\cdot[(\nabla\times\mathbf{E})\times\hat{\mathbf{v}}]\,G = -i\omega\mu\hat{\mathbf{a}}\cdot(\hat{\mathbf{v}}\times\mathbf{H})\,G, \tag{8.1.17}$$

where we have used Eq. (8.1.9) and the identity $\mathbf{t}\cdot(\mathbf{u}\times\mathbf{v}) = \mathbf{u}\cdot(\mathbf{v}\times\mathbf{t})$. It is seen that $\hat{\mathbf{a}}$ is a factor common to all terms in Eqs. (8.1.14), (8.1.15) and (8.1.17). Hence, gathering the results from these expressions and substituting them into Eq. (8.1.5), noting that $\hat{\mathbf{a}}$ is arbitrary, it is a straightforward matter to write

$$\oiint_S [i\omega\mu(\hat{\mathbf{v}}\times\mathbf{H})\,G + (\hat{\mathbf{v}}\times\mathbf{E})\times\nabla G + (\hat{\mathbf{v}}\cdot\mathbf{E})\,\nabla G]\,dA = \mathbf{0}. \tag{8.1.18}$$

Equation (8.1.18) is usually referred to as the *Stratton–Chu integral theorem.*

Hitherto we introduced the function G and required that it should be a solution of the scalar wave equation. We now give more physical meaning to this function. As discussed in Section 1.6.1, the scalar Green's function G can be interpreted as the scalar potential at position (x, y, z) due to a charge located at (x_p, y_p, z_p). Indeed, following Eq. (1.6.22) we may set

$$G = \frac{\exp(ikr_-)}{r_-} = \frac{\exp(ik|\mathbf{r}-\mathbf{r}_p|)}{|\mathbf{r}-\mathbf{r}_p|}, \tag{8.1.19}$$

where

$$r_- = |\mathbf{r}-\mathbf{r}_p| = \sqrt{(x-x_p)^2 + (y-y_p)^2 + (z-z_p)^2}, \tag{8.1.20}$$

where $\mathbf{r} = (x, y, z)$ is the position vector of an infinitesimal surface element dA on S and $\mathbf{r}_p = (x_p, y_p, z_p)$ is a (fixed) observation point inside S (see Fig. 8.1).

It is worth stopping here for a while to appreciate the significance of Eq. (8.1.18). As shown by Stratton **[328]**, this equation states that were the closed surface S constructed from an infinitely thin conducting sheet, the integrand would consist of an electric current density (first terms in Eq. (8.1.18)), a magnetic current density (second term) and a surface electric charge density (third term), which together integrated for S would give zero. In the case of diffraction calculations this interpretation leads to erroneous conclusions as shown in Section 8.1.3. Nevertheless, Eq. (8.1.18) clearly permits the calculation of the electric field in the closed interior of S if the electromagnetic field on the surface S is known, as we shall now show.

8.1.1.2 Charges Placed Inside S

Our derivation of Eq. (8.1.5) assumed that both **f** and **g** are continuous and the volume bounded by S is homogeneous. As clearly shown by Eq. (8.1.19), we encounter a singularity in G for $r_- = 0$, i.e. when $\mathbf{r} = \mathbf{r}_p$ or, in other words, when the observation point is on S. Furthermore, there must be sources of electromagnetic radiation located somewhere in space otherwise there would not be diffraction. By virtue of Eqs. (8.1.6)–(8.1.9) we have assumed that all free charges must be

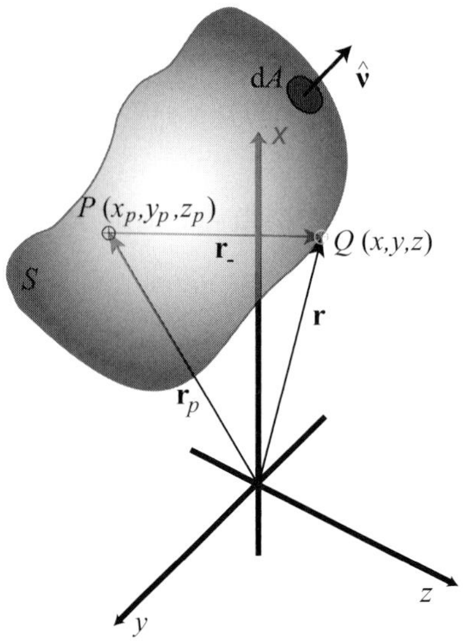

Figure 8.1: The closed surface S and the observation point P inside S.

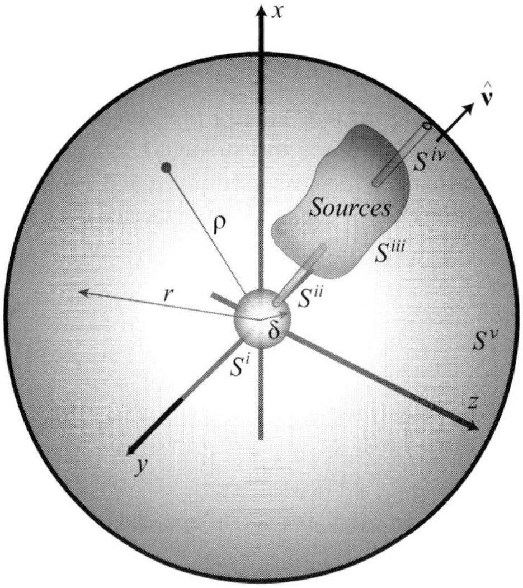

Figure 8.2: The closed surface of integration. Reproduced from **[344]** with the permission of the Optical Society of America.

exterior to S. However, in many cases of practical importance it is more convenient to assume that all free charges are located within a closed surface. The field is then known on the surface of that volume and the task is to determine the electric and magnetic fields anywhere else in space. For this, we need to redefine the surface S as follows.

Let us first shift the origin of the Cartesian co-ordinate system to P and circumscribe the observation point $P = \mathbf{r}_p$ by a small sphere of surface S^i and radius δ, as shown in Fig. 8.2. Note that due to the linearity of Maxwell's equations the choice of the origin of the co-ordinate system is arbitrary. We also circumscribe *all* free charges in space by a surface S^{iii}. Let us now construct a large spherical surface S^v, centred on P, of radius ρ and connect S^i and S^{iii} by a small diameter tube of surface S^{ii}, and S^{iii} and S^v by another small diameter tube of surface S^{iv}. The original surface S is now made up by elementary surfaces $S^i \ldots S^v$ resulting in a piecewise, smooth, orientable and simply connected surface. The originally *outward* normal $\hat{\mathbf{v}}$ of S is now an *inward* normal for $S^i \ldots S^{iv}$ and an outward normal for S^v. Consequently we have from Eq. (8.1.18):

$$\iint_{S^i} [\ldots]\, dA^i + \iint_{S^{ii}} [\ldots]\, dA^{ii} + \iint_{S^{iii}} [\ldots]\, dA^{iii} + \iint_{S^{iv}} [\ldots]\, dA^{iv} + \iint_{S^v} [\ldots]\, dA^v = \mathbf{0}. \tag{8.1.21}$$

If the diameters of tubes S^{ii} and S^{iv} are made infinitesimally small then the contribution from these surfaces becomes vanishingly small, hence we may write:

$$-\oiint_{S^i} [\ldots]\, dA^i - \oiint_{S^{iii}} [\ldots]\, dA^{iii} + \oiint_{S^v} [\ldots]\, dA^v = \mathbf{0}, \tag{8.1.22}$$

noting that now the integrals are again closed because S^i, S^{iii} and S^v are separate closed surfaces. Also note that the signs of the first and second terms have been changed so that for all surfaces $\hat{\mathbf{v}}$ is an *outward* normal. We first concentrate on the integral performed over S^i. The gradient ∇G, calculated for (x, y, z) is given by:

$$\nabla G(r) = G(r)\left(ik - \frac{1}{r}\right)\hat{\mathbf{r}}, \tag{8.1.23}$$

where $\hat{\mathbf{r}} = \mathbf{r}/|\mathbf{r}| = \mathbf{r}/r$. Referring to Fig. 8.1 and Fig. 8.2, it is clear that since the sphere was circumscribed about P, $\hat{\mathbf{r}} = \hat{\mathbf{v}}$. Whence we have:

$$-\oiint_{S^i} [\ldots]\, dA^i = -\oiint_{S^i} \left[i\omega\mu\,(\hat{\mathbf{v}} \times \mathbf{H})\, G + G\left(ik - \frac{1}{r}\right)(\hat{\mathbf{v}} \times \mathbf{E}) \times \hat{\mathbf{v}} + G\left(ik - \frac{1}{r}\right)(\hat{\mathbf{v}}\cdot\mathbf{E})\,\hat{\mathbf{v}} \right] dA^i. \tag{8.1.24}$$

The second and third terms in Eq. (8.1.24) easily transform to give

$$G\left(ik - \frac{1}{r}\right)[-\hat{\mathbf{v}} \times (\hat{\mathbf{v}} \times \mathbf{E}) + (\hat{\mathbf{v}}\cdot\mathbf{E})\,\hat{\mathbf{v}}] = \mathbf{E}\frac{\exp(ikr)}{r}\left(ik - \frac{1}{r}\right). \tag{8.1.25}$$

We clearly would like the volume S^i to be as small as possible because the singularity only occurs at P. Therefore, on the surface S^i, $dA^i = \delta^2 d\Omega$, where Ω is the associated solid angle and δ is the radius of the sphere, we might as well choose the radius of the sphere to be vanishingly small:

$$-\lim_{\delta\to 0} \iint_{\Omega} \left[i\omega\mu\,(\hat{\mathbf{v}} \times \mathbf{H})\, \frac{\exp(ikr)}{r} + \mathbf{E}\frac{\exp(ikr)}{r}\left(ik - \frac{1}{r}\right)\right]_{r=\delta} \delta^2 d\Omega = 4\pi\mathbf{E}(\mathbf{r}_p = \mathbf{0}). \tag{8.1.26}$$

We now turn to discuss the integral over S^v. In a manner similar to before, we have:

$$\oiint_{S^v} \left[i\omega\mu\,(\hat{\mathbf{v}} \times \mathbf{H})\, \frac{\exp(ikr)}{r} + \mathbf{E}\frac{\exp(ikr)}{r}\left(ik - \frac{1}{r}\right)\right]_{r=\rho} dA^v = \oiint_{S^v} \left\{ \left[(\hat{\mathbf{v}} \times \nabla \times \mathbf{E} + ik\mathbf{E})\frac{1}{r} - \mathbf{E}\frac{1}{r^2}\right] \exp(ikr) \right\}_{r=\rho} dA^v, \tag{8.1.27}$$

which, after noting that $dA^v = \rho^2 d\Omega$, can be written as an integral over the solid angle Ω as

$$\iint_{\Omega} \{[(\hat{\mathbf{v}} \times \nabla \times \mathbf{E} + ik\mathbf{E})\, r - \mathbf{E}] \exp(ikr)\}_{r=\rho}\, d\Omega = \iint_{\Omega} \{[(\mathbf{r} \times \nabla \times \mathbf{E} + ikr\mathbf{E}] - \mathbf{E}\} \exp(ikr)\}_{r=\rho}\, d\Omega, \tag{8.1.28}$$

where we have used that $\mathbf{r} = r\hat{\mathbf{v}} = \rho\hat{\mathbf{v}}$ on S^v.

Because the surface S^v limits the volume within which our calculation of the electromagnetic field is valid, we would like to extend this volume such that it encompasses the entire space, that is $\rho \to \infty$. Since S^{iii} is a surface encompassing all sources of electromagnetic radiation and the surface S^v is merely a mathematical construction it is not physically tenable to expect its presence to perturb the field at P. Consequently, as $\rho \to \infty$ the contribution of Eq. (8.1.28) to Eq. (8.1.22) should be vanishingly small. In order for this to happen we must have:

$$|\rho\mathbf{E}| < K, \tag{8.1.29}$$

$$\rho\,[\hat{\mathbf{v}} \times (\nabla \times \mathbf{E}) + ik\mathbf{E}] \to 0, \tag{8.1.30}$$

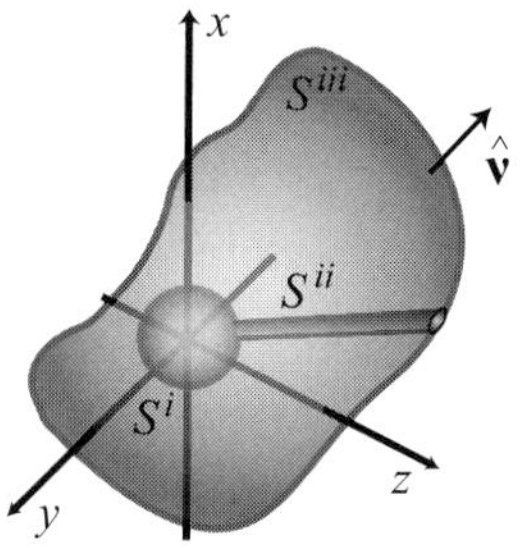

Figure 8.3: The closed surface of integration. Reproduced from **[344]** with the permission of the Optical Society of America.

as $\rho \to \infty$ where K is an arbitrary finite constant. While it is difficult to give a general physical interpretation to Eq. (8.1.30), Eq. (8.1.29) makes good physical sense because the electric field due to a finite volume of charges must decay with $1/r$ so the equation merely keeps the contribution to S_v within finite bounds. The conditions given by Eqs. (8.1.29) and (8.1.30) shall be termed the *vectorial radiation conditions*[1] in analogy to the *scalar* radiation conditions of Sommerfeld **[18]**. It is important to point out that these conditions arise because there cannot be a contribution to the field at P from a physically non-existent boundary surface and therefore they do not contradict the existence of a steady-state solution. In contrast, the argument in Born and Wolf **[37]** (Section 8.3.2) seems to imply that in the case of a steady-state solution there is a contribution from the boundary surface at infinity. These authors stipulate that the radiation only existed since $t = t_0$, therefore removing the boundary surface at a distance of $c(t - t_0)$ would not affect the field at P. However, clearly at the steady-state limit $t \to \infty$ this assumption does not hold and, more importantly a mathematical boundary surface must not contribute to the physical solution of the problem.

The only integral in Eq. (8.1.18) that is left to discuss is the one over the surface enclosing all free charges, S^{iii}. After substituting Eq. (8.1.26) into Eq. (8.1.22), the electric field on the *open exterior* volume of S^{iii} is given from Eq. (8.1.18) by:

$$\mathbf{E}(\mathbf{r}) = \frac{1}{4\pi} \oiint_{S^{iii}} \left[i\omega\mu (\hat{\mathbf{v}} \times \mathbf{H})\, G + (\hat{\mathbf{v}} \times \mathbf{E}) \times \nabla G + (\hat{\mathbf{v}} \cdot \mathbf{E}) \nabla G \right] dA^{iii}. \tag{8.1.31}$$

Due to the symmetry of Maxwell's Eqs. (8.1.6)–(8.1.9) the substitution $\mathbf{E} \to \mu\mathbf{H}$ and $\mathbf{H} \to -\epsilon\mathbf{E}$ yields for the magnetic field:

$$\mathbf{H}(\mathbf{r}) = -\frac{1}{4\pi} \oiint_{S^{iii}} \left[i\omega\epsilon (\hat{\mathbf{v}} \times \mathbf{E})\, G - (\hat{\mathbf{v}} \times \mathbf{H}) \times \nabla G - (\hat{\mathbf{v}} \cdot \mathbf{H}) \nabla G \right] dA^{iii}. \tag{8.1.32}$$

Note that this formula is different from that published in the literature, including the original paper of Stratton and Chu **[329]** and Stratton's monograph **[328]**. These workers defined the problem slightly differently, as we shall now discuss.

Consider the volume shown in Fig. 8.3. The surface S is now constructed from the sphere surrounding the point of observation whose surface is denoted by S^i; the small diameter tube of surface S^{ii} and finally the surface of the outer shell S^{iii}. Since the volume where the integral Eq. (8.1.18) is valid is the exterior of S^i and S^{ii} and the interior of S^{iii}, all free charges must be located *outside* this surface. The contribution from surface S^{ii} vanishes as the diameter of the tube is made zero and the contribution from S^i is just $4\pi\mathbf{E}$ as before. Consequently, the electric field is given, within the *closed interior* of S^{iii} by:

$$\mathbf{E}(\mathbf{r}) = -\frac{1}{4\pi} \oiint_{S^{iii}} \left[i\omega\mu (\hat{\mathbf{v}} \times \mathbf{H})\, G + (\hat{\mathbf{v}} \times \mathbf{E}) \times \nabla G + (\hat{\mathbf{v}} \cdot \mathbf{E}) \nabla G \right] dA^{iii}, \tag{8.1.33}$$

and the expression for the magnetic field is given by:

$$\mathbf{H}(\mathbf{r}) = \frac{1}{4\pi} \oiint_{S^{iii}} \left[i\omega\epsilon (\hat{\mathbf{v}} \times \mathbf{E})\, G - (\hat{\mathbf{v}} \times \mathbf{H}) \times \nabla G - (\hat{\mathbf{v}} \cdot \mathbf{H}) \nabla G \right] dA^{iii}. \tag{8.1.34}$$

On comparing Eqs. (8.1.31)–(8.1.34) we initially note a sign difference for both the electric and magnetic fields. On closer inspection we notice that there is also a difference in the surface over which the integration occurs. In Eq. (8.1.33)

[1] Note that the second condition is identical to that of Nédélec **[255]**.

the surface encompasses a source-free volume of space and $\hat{\mathbf{v}}$ is the outward surface normal. The point of observation P is inside the volume. In the case of Eq. (8.1.31) the surface S^{iii} encompasses all sources in space and $\hat{\mathbf{v}}$ is again an outward normal. The point of observation is outside the enclosed volume. Consequently we recognise that Eqs. (8.1.31) and (8.1.33) are *solutions of two different problems*. Of course, in certain cases the two formalisms may be used to describe the same problem that must result in the same electric field. In these cases, we need to keep in mind that for Eq. (8.1.31) $\hat{\mathbf{v}}$ points *away* from sources and in Eq. (8.1.33) $\hat{\mathbf{v}}$ points, mostly but not always, *towards* the sources. Nevertheless, we shall keep both of these solutions because it depends on the actual application which one we wish to use.

8.1.1.3 The Stratton–Chu Formula Satisfies Maxwell's Equations

Now we show that Eq. (8.1.33) is a rigorous solution of Maxwell's equation. It will trivially follow that Eq. (8.1.31) is also a solution of Maxwell's equations. First we introduce the notation ∇_p to distinguish the vector differential operator taken at the observation point from the one, ∇, taken at a point on the surface S and use S to denote the closed surface of integration for simplicity. We have to show that i) $\nabla_p\cdot\mathbf{E} = 0$ and ii) $\nabla_p \times \mathbf{H} = -i\omega\epsilon\mathbf{E}$. Let us start with i):

$$4\pi\nabla_p\cdot\mathbf{E} = -i\omega\mu \oiint_S \nabla_p\cdot[(\hat{\mathbf{v}}\times\mathbf{H})\,G]\,dA - \oiint_S \nabla_p\cdot[(\hat{\mathbf{v}}\times\mathbf{E})\times\nabla G]\,dA - \oiint_S \nabla_p\cdot[(\hat{\mathbf{v}}\cdot\mathbf{E})\,\nabla G]\,dA. \tag{8.1.35}$$

The first integral can be re-written with the use of Eq. (8.1.12) to give:

$$-i\omega\mu \oiint_S \nabla_p\cdot[(\hat{\mathbf{v}}\times\mathbf{H})\,G]\,dA = -i\omega\mu \oiint_S \left[\nabla_p\cdot(\hat{\mathbf{v}}\times\mathbf{H})\right] GdA - i\omega\mu \oiint_S (\hat{\mathbf{v}}\times\mathbf{H})\cdot\nabla_p GdA = i\omega\mu \oiint_S (\hat{\mathbf{v}}\times\mathbf{H})\cdot\nabla GdA, \tag{8.1.36}$$

because $(\hat{\mathbf{v}}\times\mathbf{H})$ is only a function of (x, y, z) and $\nabla_p G = -\nabla G$ (see Eq. (8.1.19)). The second integral in Eq. (8.1.35) yields identically zero after using the rule of triple scalar products on the kernel and by noticing that $\nabla\times\nabla G \equiv \mathbf{0}$. The third term is a bit more complicated to transform. We first write:

$$-\oiint_S \nabla_p\cdot[(\hat{\mathbf{v}}\cdot\mathbf{E})\,\nabla G]\,dA = \oiint_S (\hat{\mathbf{v}}\cdot\mathbf{E})\,\nabla^2 GdA = -\omega^2\mu\epsilon \oiint_S (\hat{\mathbf{v}}\cdot\mathbf{E})\,GdA, \tag{8.1.37}$$

because G satisfies the scalar wave equation. With the use of Eq. (8.1.8) we can write:

$$-\omega^2\mu\epsilon \oiint_S (\hat{\mathbf{v}}\cdot\mathbf{E})\,GdA = -i\omega\mu \oiint_S [G\,(\nabla\times\mathbf{H})]\cdot\hat{\mathbf{v}}dA. \tag{8.1.38}$$

We know from Eq. (8.1.12) that

$$G(\nabla\times\mathbf{H}) = \nabla\times(G\mathbf{H}) - (\nabla G\times\mathbf{H}), \tag{8.1.39}$$

so now Eq. (8.1.38) reads:

$$\begin{aligned}-i\omega\mu \oiint_S [G\,(\nabla\times\mathbf{H})]\cdot\hat{\mathbf{v}}dA &= -\,i\omega\mu \oiint_S [\nabla\times(G\mathbf{H})]\cdot\hat{\mathbf{v}}dA + i\omega\mu \oiint_S (\nabla G\times\mathbf{H})\cdot\hat{\mathbf{v}}dA \\ &= -\,i\omega\mu \oiint_S [\nabla\times(G\mathbf{H})]\cdot\hat{\mathbf{v}}dA - i\omega\mu \oiint_S (\hat{\mathbf{v}}\times\mathbf{H})\cdot\nabla GdA.\end{aligned} \tag{8.1.40}$$

Equation (1.3.9) reads with our usual notation:

$$\iint_S (\nabla\times\mathbf{v})\cdot\hat{\mathbf{v}}dA = \oint_C \mathbf{v}\cdot d\mathbf{s}, \tag{8.1.41}$$

where C is a closed contour around an open surface S. It is easy to see that if the surface is closed, the value of the line integral vanishes. Hence, the first term in the right-hand side of Eq. (8.1.40) also vanishes so we are left with:

$$-\oiint_S \nabla_p\cdot[(\hat{\mathbf{v}}\cdot\mathbf{E})\,\nabla G]\,dA = -i\omega\mu \oiint_S (\hat{\mathbf{v}}\times\mathbf{H})\cdot\nabla GdA. \tag{8.1.42}$$

When substituting back from Eqs. (8.1.36) and (8.1.42) into Eq. (8.1.35) we obtain:

$$\nabla_p\cdot\mathbf{E} = 0, \tag{8.1.43}$$

which completes the first part of the proof. We now prove that $\nabla_p\times\mathbf{H} = -i\omega\epsilon\mathbf{E}$. Let us consider:

$$4\pi\nabla_p\times\mathbf{H} = i\omega\epsilon \oiint_S \nabla_p\times[(\hat{\mathbf{v}}\times\mathbf{E})G]\,dA - \oiint_S \nabla_p\times[(\hat{\mathbf{v}}\times\mathbf{H})\times\nabla G]\,dA, \tag{8.1.44}$$

because, again, $\nabla \times \nabla G \equiv \mathbf{0}$. A procedure similar to that described above for the previous proof yields:

$$\nabla_p \times \mathbf{H} = \frac{i\omega\epsilon}{4\pi} \oiint_S [i\omega\mu(\hat{\mathbf{v}} \times \mathbf{E})G + (\hat{\mathbf{v}} \times \mathbf{E}) \times \nabla G + (\hat{\mathbf{v}}\cdot\mathbf{E})\nabla G]\, dA = -i\omega\epsilon\mathbf{E}. \tag{8.1.45}$$

With this equation we have completed the proof.

In conclusion, *Eqs. (8.1.33) and (8.1.34) are unique,*[2] *self-consistent and rigorous solutions of Maxwell's equations for the electric and magnetic field, respectively, at a point within any closed surface S bounding a charge and current free, homogeneous and isotropic volume*. Also, *Eqs. (8.1.31) and (8.1.32) are unique, self-consistent and rigorous solutions of Maxwell's equations for the electric and magnetic field, respectively, at any exterior point of a closed surface bounding all sources*. Interestingly, although the Stratton–Chu integral theorem has been known for almost 80 years, it has not been very widely used in physical optics. However, we show that this integral can be the basis for deriving other, well known diffraction theories.

8.1.2 The Vectorial Integral Theorem of Kirchhoff

The most frequently used diffraction integral of physical optics is the scalar integral theorem of Kirchhoff. We now derive the vectorial integral theorem of Kirchhoff, which will serve as the basis to obtain its scalar version later on in this chapter. The derivation of this formula is quite dry – essentially we need to perform a succession of transformations on the kernel of the integral to arrive at the desired form. We may rewrite Eq. (8.1.18) in the following form:[3]

$$\begin{aligned}\mathbf{0} &= \oiint_S [i\omega\mu\,(\hat{\mathbf{v}} \times \mathbf{H})\, G + 2\,(\hat{\mathbf{v}}\cdot\nabla G)\,\mathbf{E} - (\mathbf{E}\cdot\nabla G)\,\hat{\mathbf{v}} + (\hat{\mathbf{v}}\cdot\mathbf{E})\,\nabla G - (\hat{\mathbf{v}}\cdot\nabla G)\,\mathbf{E}]\, dA \\ &= \oiint_S [i\omega\mu\,(\hat{\mathbf{v}} \times \mathbf{H})\, G + 2\,(\hat{\mathbf{v}}\cdot\nabla G)\,\mathbf{E} - (\mathbf{E}\cdot\nabla G)\,\hat{\mathbf{v}} + \hat{\mathbf{v}} \times (\nabla G \times \mathbf{E})]\, dA,\end{aligned} \tag{8.1.46}$$

which leads to

$$\begin{aligned}\mathbf{0} &= \oiint_S \{[\hat{\mathbf{v}} \times (\nabla \times \mathbf{E})]\, G + 2\,(\hat{\mathbf{v}}\cdot\nabla G)\,\mathbf{E} - \hat{\mathbf{v}}\,[\nabla\cdot(G\mathbf{E})] + \hat{\mathbf{v}} \times (\nabla G \times \mathbf{E})\}\, dA \\ &= \oiint_S \{2\,(\hat{\mathbf{v}}\cdot\nabla G)\,\mathbf{E} - \hat{\mathbf{v}}\,[\nabla\cdot(G\mathbf{E})] + \hat{\mathbf{v}} \times [\nabla \times (G\mathbf{E})]\}\, dA,\end{aligned} \tag{8.1.47}$$

where we have used Eqs. (8.1.6) and (8.1.9) and Eq. (8.1.40). The use of two variants of the divergence theorem,

$$\oiint_S \psi\hat{\mathbf{v}}\, dA = \iiint_V \nabla\psi\, dV, \tag{8.1.48}$$

$$\oiint_S (\hat{\mathbf{v}} \times \mathbf{v})\, dA = \iiint_V \nabla \times \mathbf{v}\, dV, \tag{8.1.49}$$

leads to

$$\begin{aligned}\mathbf{0} &= \oiint_S \{2\,(\hat{\mathbf{v}}\cdot\nabla G)\,\mathbf{E} - \hat{\mathbf{v}}\,[\nabla\cdot(G\mathbf{E})] + \hat{\mathbf{v}} \times [\nabla \times (G\mathbf{E})]\}\, dA \\ &= \oiint_S [2\,(\hat{\mathbf{v}}\cdot\nabla G)\,\mathbf{E}]\, dA - \iiint_V \{\nabla\,[\nabla\cdot(G\mathbf{E})] - \nabla \times [\nabla \times (G\mathbf{E})]\}\, dV.\end{aligned} \tag{8.1.50}$$

We can convert the kernel of the volume integral in Eq. (8.1.50) with use of the identity $\nabla^2\mathbf{v} = \nabla(\nabla\cdot\mathbf{v}) - \nabla \times (\nabla \times \mathbf{v})$ to give

$$\mathbf{0} = \oiint_S [2\,(\hat{\mathbf{v}}\cdot\nabla G)\,\mathbf{E}]\, dA - \iiint_V \left[\nabla^2\,(G\mathbf{E})\right] dV. \tag{8.1.51}$$

Because the notation in the kernel of the volume integral can lead to some misunderstanding, we continue by discussing this term alone. It is convenient to write

$$\iiint_V \left[\nabla^2\,(G\mathbf{E})\right] dV = \sum_{n=x,y,z} \mathbf{1}_n \iiint_V \left[\nabla^2\,(GE_n)\right] dV, \tag{8.1.52}$$

[2] Note that the uniqueness follows from the fact that Maxwell's equations lead to a unique solution for a given set of boundary conditions, which is a consequence of the Uniqueness Theorem **[328]**.

[3] This derivation chiefly follows Jackson **[160]**.

with $\mathbf{l}_x = \hat{\mathbf{x}}, \hat{\mathbf{y}}, \hat{\mathbf{z}}$ and $n = x, y, z$. Next we use Green's theorem,

$$\iiint_V \nabla^2 \psi dV = \oint\!\!\oint_S (\nabla\psi)\cdot\hat{\mathbf{v}} dA \triangleq \oint\!\!\oint_S \frac{\partial \psi}{\partial v} dA, \tag{8.1.53}$$

where ψ is a well-behaved scalar function, to obtain:

$$\iiint_V \left[\nabla^2 (G\mathbf{E})\right] dV = \sum_{n=x,y,z} \mathbf{l}_n \oint\!\!\oint_S [\nabla (GE_n)]\cdot\hat{\mathbf{v}} dA. \tag{8.1.54}$$

After performing the differentiation we obtain the final result:

$$\mathbf{0} = \oint\!\!\oint_S \left\{\mathbf{E}(\hat{\mathbf{v}}\cdot\nabla G) - [G(\hat{\mathbf{v}}\cdot\nabla)\mathbf{E}]^T\right\} dA. \tag{8.1.55}$$

The gradient of a vector has the dimensions of a 3×3 tensor that, after the scalar product with a vector, reduces to a row vector (see Appendix A, Section A.10). The transpose of the second term in the above integrand is needed to convert the row vector to a column vector. In order to keep the notation simple, we omit the transpose sign from the following equations. The surface S limits the volume of validity of the above expression. Depending on the definition of this surface we can solve for any point of a closed interior within the source-free homogeneous medium,

$$\mathbf{E}(\mathbf{r}) = -\frac{1}{4\pi} \oint\!\!\oint_S [\mathbf{E}(\hat{\mathbf{v}}\cdot\nabla G) - G(\hat{\mathbf{v}}\cdot\nabla)\mathbf{E}] dA \tag{8.1.56}$$

or

$$\mathbf{E}(\mathbf{r}) = -\frac{1}{4\pi} \oint\!\!\oint_S \left[\mathbf{E}\frac{\partial G}{\partial v} - G\frac{\partial \mathbf{E}}{\partial v}\right] dA. \tag{8.1.57}$$

Alternatively, if we wish to calculate the field at any point lying in the open exterior of a surface encompassing all sources in space we obtain:

$$\mathbf{E}(\mathbf{r}) = \frac{1}{4\pi} \oint\!\!\oint_S [\mathbf{E}(\hat{\mathbf{v}}\cdot\nabla G) - G(\hat{\mathbf{v}}\cdot\nabla)\mathbf{E}] dA \tag{8.1.58}$$

or

$$\mathbf{E}(\mathbf{r}) = \frac{1}{4\pi} \oint\!\!\oint_S \left[\mathbf{E}\frac{\partial G}{\partial v} - G\frac{\partial \mathbf{E}}{\partial v}\right] dA. \tag{8.1.59}$$

Any one of Eqs. (8.1.56)–(8.1.59) is usually referred to as the *vectorial integral theorem of Kirchhoff*. We note that the shorthand in the above equations means:

$$(\hat{\mathbf{v}}\cdot\nabla)\mathbf{E} = \hat{\mathbf{x}}(\nabla E_x\cdot\hat{\mathbf{v}}) + \hat{\mathbf{y}}\left(\nabla E_y\cdot\hat{\mathbf{v}}\right) + \hat{\mathbf{z}}(\nabla E_z\cdot\hat{\mathbf{v}}). \tag{8.1.60}$$

The transition from vectorial to scalar theories of diffraction can be difficult to justify. The usual argument from authors is that in certain problems the strength of one particular vector component might be significantly greater than the other two and so the weak components can be ignored. However, this reasoning is in our opinion not really acceptable because at any given moment it is always possible to find a local coordinate system in which the electric field has a single component. Therefore we prefer to approach scalar diffraction by regarding it as an approximation to a rigorous (vector) solution of a problem that might provide a description that is more accurate than geometrical optics and, under certain conditions, is quite similar to what a rigorous solution would yield. In these cases, we may elect to use a single vector component of the electric vector given by Eqs. (8.1.56)–(8.1.59) to give the scalar diffraction integral:

$$E(\mathbf{r}) = -\frac{1}{4\pi} \oint\!\!\oint_S \left[E\frac{\partial G}{\partial v} - G\frac{\partial E}{\partial v}\right] dA. \tag{8.1.61}$$

The above equation is seen to be formally identical to equation §8.3 (8) in **[37]**, which is usually referred to as the *scalar integral theorem of Helmholtz and Kirchhoff*. Equations (8.1.31) and (8.1.61) serve as the basis of our further derivations of various vectorial and scalar diffraction theories, respectively. It is noted that it is also possible to derive Eq. (8.1.61) directly as the scalar wave equation applies to any of the individual Cartesian component of the electric vector **E**.

8.1.3 The *m* and *e* Theories

In order to be able to solve the problem of diffraction, certain components of the electric and magnetic fields need to be known. Equation (8.1.31), for example, shows that for calculation of the electric field a knowledge of the tangential components of both **E** and **H** is needed, as well as the normal component of **E**, which means that five field components need to be known to obtain a solution. One might argue that, since **E** and **H** are connected via Maxwell's equations, Eq. (8.1.31) overspecifies the problem. Let us assume that we know the tangential components of **E**, then from Eq. (8.1.6) we can obtain the normal component. By knowing **E** we can readily determine the corresponding **H** from, for example, Eq. (8.1.8). This argument implies that, provided there are no free charges in the volume, the components of the electric field vector **E** tangential to the surface of the volume are sufficient to determine the resulting field.[4] However, Maxwell's equations comprise four separate equations whereas in the case of the diffraction formulas we use a single equation. This is the real reason why some of the theories of both vectorial and scalar diffraction require overspecified incident fields.

8.1.3.1 Diffraction and the Hertz Vectors

We start our derivation from Eqs. (1.5.42) and (1.5.43) that are now written for the time-harmonic case:

$$\mathbf{E} = \nabla \times \nabla \times \mathbf{\Pi}_e + i\omega\mu\nabla \times \mathbf{\Pi}_m, \tag{8.1.62}$$

$$\mathbf{H} = \nabla \times \nabla \times \mathbf{\Pi}_m - i\omega\epsilon\nabla \times \mathbf{\Pi}_e. \tag{8.1.63}$$

Considering first Eq. (8.1.34) we have

$$\mathbf{H}(\mathbf{r}_p) = \frac{i\omega\epsilon}{4\pi}\oiint[\hat{\mathbf{v}}(\mathbf{r}) \times \mathbf{E}(\mathbf{r})]G(\mathbf{r},\mathbf{r}_p)d\mathbf{r} - \frac{1}{4\pi}\oiint[\hat{\mathbf{v}}(\mathbf{r}) \times \mathbf{H}(\mathbf{r})] \times \nabla G(\mathbf{r},\mathbf{r}_p)d\mathbf{r} - \frac{1}{4\pi}\oiint[\hat{\mathbf{v}}(\mathbf{r})\cdot\mathbf{H}(\mathbf{r})]\nabla G(\mathbf{r},\mathbf{r}_p)d\mathbf{r}. \tag{8.1.64}$$

Note that we have carefully shown the functional dependencies of various quantities for reasons that will become obvious in a moment. The values of the surface normal $\hat{\mathbf{v}}(\mathbf{r})$ and the electric and magnetic fields, $\mathbf{E}(\mathbf{r})$ and $\mathbf{H}(\mathbf{r})$, respectively are taken at the surface over which the integration is carried out (hence the rather unorthodox notation $d\mathbf{r}$). The Green's function depends on both the point on the surface and that of the observation, $\mathbf{r}$ and $\mathbf{r}_p$, respectively. In order to turn the equation for the magnetic field to one for the electric field, let us take the curl $\nabla_p\times$, operating on the observation point, of both sides of the above equation:

$$\begin{aligned}\nabla_p \times \mathbf{H}(\mathbf{r}_p) = {} & \frac{i\omega\epsilon}{4\pi}\nabla_p \times \oiint[\hat{\mathbf{v}}(\mathbf{r}) \times \mathbf{E}(\mathbf{r})]G(\mathbf{r},\mathbf{r}_p)d\mathbf{r} \\ & - \frac{1}{4\pi}\nabla_p \times \nabla_p \times \oiint[\hat{\mathbf{v}}(\mathbf{r}) \times \mathbf{H}(\mathbf{r})]G(\mathbf{r},\mathbf{r}_p)d\mathbf{r} \\ & + \frac{1}{4\pi}\nabla_p \times \nabla_p\left\{\oiint[\hat{\mathbf{v}}(\mathbf{r})\cdot\mathbf{H}(\mathbf{r})]G(\mathbf{r},\mathbf{r}_p)d\mathbf{r}\right\},\end{aligned} \tag{8.1.65}$$

where we have used $\nabla G(\mathbf{r},\mathbf{r}_p) = -\nabla_p G(\mathbf{r},\mathbf{r}_p)$ and $(\hat{\mathbf{v}} \times \mathbf{H}) \times \nabla_p G = -\nabla_p \times [(\hat{\mathbf{v}} \times \mathbf{H})G]$. The third term in the above equation vanishes because the curl of a gradient is identically zero. Using Eq. (8.1.8) we have

$$\mathbf{E}(\mathbf{r}_p) = -\frac{1}{4\pi}\nabla_p \times \oiint[\hat{\mathbf{v}}(\mathbf{r}) \times \mathbf{E}(\mathbf{r})]G(\mathbf{r},\mathbf{r}_p)d\mathbf{r} + \frac{1}{4\pi}\frac{1}{i\omega\epsilon}\nabla_p \times \nabla_p \times \oiint[\hat{\mathbf{v}}(\mathbf{r}) \times \mathbf{H}(\mathbf{r})]G(\mathbf{r},\mathbf{r}_p)d\mathbf{r}. \tag{8.1.66}$$

A similar line of argument starting from Eq. (8.1.31) leads to

$$\mathbf{H}(\mathbf{r}_p) = -\frac{1}{4\pi}\nabla_p \times \oiint[\hat{\mathbf{v}}(\mathbf{r}) \times \mathbf{H}(\mathbf{r})]G(\mathbf{r},\mathbf{r}_p)d\mathbf{r} - \frac{1}{4\pi}\frac{1}{i\omega\mu}\nabla_p \times \nabla_p \times \oiint[\hat{\mathbf{v}}(\mathbf{r}) \times \mathbf{E}(\mathbf{r})]G(\mathbf{r},\mathbf{r}_p)d\,\mathbf{r}. \tag{8.1.67}$$

On comparing Eqs. (8.1.66) and (8.1.67) with Eqs. (8.1.62) and (8.1.63) it follows that

$$\mathbf{\Pi}_e(\mathbf{r}_p) = \frac{1}{4\pi}\frac{1}{i\omega\epsilon}\oiint[\hat{\mathbf{v}}(\mathbf{r}) \times \mathbf{H}(\mathbf{r})]G(\mathbf{r},\mathbf{r}_p)d\mathbf{r}, \tag{8.1.68}$$

$$\mathbf{\Pi}_m(\mathbf{r}_p) = -\frac{1}{4\pi}\frac{1}{i\omega\mu}\oiint[\hat{\mathbf{v}}(\mathbf{r}) \times \mathbf{E}(\mathbf{r})]G(\mathbf{r},\mathbf{r}_p)d\mathbf{r}. \tag{8.1.69}$$

[4] The uniqueness theorem of Stratton in fact proves that the tangential components of either **E** or **H** are sufficient for a unique solution. As the proof of uniqueness is outside the scope of the present work, the reader is referred to the monograph of Stratton **[328]**, p. 486.

Let's look at the above definition of the Hertz vectors. As discussed in Section 1.8.3 the component of $\mathbf{H}$ that is parallel to an interface separating two media is discontinuous: $\mathbf{H}^{(1)} \times \hat{\mathbf{v}} = \mathbf{H}^{(2)} \times \hat{\mathbf{v}} + \mathbf{j}\cdot\hat{\mathbf{n}}$ where $\mathbf{j}$ is the surface current density and $\mathbf{j}\cdot\hat{\mathbf{n}}$ is the flux of that current density in the $\hat{\mathbf{n}}$ direction. Dimensionally $\mathbf{H} \times \hat{\mathbf{v}}$ is a flux of current density. Likewise an argument could be made that $\mathbf{E} \times \hat{\mathbf{v}}$ is a flux of current density of magnetic charges, although this statement does not follow from Eq. (1.8.14). The reason for this is that we have elected to base our discussions firmly on experimental facts which provide no evidence of magnetic monopoles. However, it is not unusual to find statements in the literature declaring that $\mathbf{\Pi}_e$ is due to currents of magnetic monopoles and $\mathbf{\Pi}_m$ is then due to currents of free electric charges. The only problem with these arguments is that we elected to consider a physical problem where free electric charges are located in a closed volume and we seek to calculate the electric and magnetic fields at some distance from these, outside the closed volume. The surface that the Stratton–Chu integral is performed over is not real and the space is required to be homogeneous and isotropic. Therefore it is not reasonable to assert that the Stratton–Chu integral is a sum of surface electric currents, surface magnetic currents and free surface charges because in our description the surface is completely fictitious. However, for completeness it needs to be pointed out that the Stratton–Chu integral is a rigorous formulation of Love's equivalence principle when the effects of sources of electromagnetic radiation leave an 'impressed' distribution of fictitious electric and magnetic surface currents and charges on the mathematical surface **[290]**.

If we consider the Stratton–Chu integral written for the case of enclosed charges, Eqs. (8.1.31) and (8.1.32), we have

$$\mathbf{E}(\mathbf{r}_p) = \frac{1}{4\pi}\nabla_p \times \oiint [\hat{\mathbf{v}}(\mathbf{r}) \times \mathbf{E}(\mathbf{r})] G(\mathbf{r}, \mathbf{r}_p) d\mathbf{r} - \frac{1}{4\pi}\frac{1}{i\omega\epsilon}\nabla_p \times \nabla_p \times \oiint [\hat{\mathbf{v}}(\mathbf{r}) \times \mathbf{H}(\mathbf{r})] G(\mathbf{r}, \mathbf{r}_p) d\mathbf{r}\,, \tag{8.1.70}$$

$$\mathbf{H}(\mathbf{r}_p) = \frac{1}{4\pi}\nabla_p \times \oiint [\hat{\mathbf{v}}(\mathbf{r}) \times \mathbf{H}(\mathbf{r})] G(\mathbf{r}, \mathbf{r}_p) d\mathbf{r} + \frac{1}{4\pi}\frac{1}{i\omega\mu}\nabla_p \times \nabla_p \times \oiint [\hat{\mathbf{v}}(\mathbf{r}) \times \mathbf{E}(\mathbf{r})] G(\mathbf{r}, \mathbf{r}_p) d\mathbf{r}\,. \tag{8.1.71}$$

Evidently for this case

$$\mathbf{\Pi}_e(\mathbf{r}_p) = -\frac{1}{4\pi}\frac{1}{i\omega\epsilon}\oiint [\hat{\mathbf{v}}(\mathbf{r}) \times \mathbf{H}(\mathbf{r})] G(\mathbf{r}, \mathbf{r}_p)\, d\mathbf{r}, \tag{8.1.72}$$

$$\mathbf{\Pi}_m(\mathbf{r}_p) = \frac{1}{4\pi}\frac{1}{i\omega\mu}\oiint [\hat{\mathbf{v}}(\mathbf{r}) \times \mathbf{E}(\mathbf{r})] G(\mathbf{r}, \mathbf{r}_p)\, d\mathbf{r}\,. \tag{8.1.73}$$

Returning now to the discussion of the e and m theories, consider Eq. (8.1.70) in a slightly different format:

$$4\pi\nabla_p \times \mathbf{H}(\mathbf{r}_p) = \nabla_p \times \oiint \{\hat{\mathbf{v}}(\mathbf{r}) \times [\nabla \times \mathbf{H}(r)]\}\, G(\mathbf{r}, \mathbf{r}_p) d\mathbf{r} + \nabla_p \times \nabla_p \times \oiint [\hat{\mathbf{v}}(\mathbf{r}) \times \mathbf{H}(\mathbf{r})] G(\mathbf{r}, \mathbf{r}_p) d\mathbf{r}, \tag{8.1.74}$$

where we have used Eq. (8.1.8). Note that if we had started from Eq. (8.1.67) instead of Eq. (8.1.66) we would have ended up with a formally identical equation for $\mathbf{E}$. Therefore by introducing $\mathbf{C} = \mathbf{E}, \mathbf{H}$ we can write a general expression as

$$4\pi\nabla_p \times \mathbf{C}(\mathbf{r}_p) = \nabla_p \times \oiint \{\hat{\mathbf{v}}(\mathbf{r}) \times [\nabla \times \mathbf{C}(r)]\}\, G(\mathbf{r}, \mathbf{r}_p) d\mathbf{r} + \nabla_p \times \nabla_p \times \oiint [\hat{\mathbf{v}}(\mathbf{r}) \times \mathbf{C}(\mathbf{r})] G(\mathbf{r}, \mathbf{r}_p) d\mathbf{r}\,. \tag{8.1.75}$$

Let us now introduce two auxiliary vector functions $\mathbf{G}(\mathbf{r}, \mathbf{r}_p) = \mathbf{a}G(\mathbf{r}, \mathbf{r}_p)$ where $\mathbf{a}$ is an arbitrary vector, and $\mathbf{v}(\mathbf{r})$ as an arbitrary vector function of $\mathbf{r}$ only, i.e. independent of $\mathbf{r}_p$. The following identities can be readily proven using the chain rule for differentiation and the rule of triple scalar products:

$$\mathbf{a}\cdot\left\{\nabla_p \times [\mathbf{v}G]\right\} = -\mathbf{v}\cdot\left\{\nabla_p \times \mathbf{G}\right\} = \mathbf{v}\cdot\{\nabla \times \mathbf{G}\} \tag{8.1.76}$$

and

$$\hat{\mathbf{a}}\cdot\left\{\nabla_p \times \left[\nabla_p \times (\mathbf{v}G)\right]\right\} = \mathbf{v}\cdot\left[\nabla_p \times \left(\nabla_p \times \mathbf{G}\right)\right] = \mathbf{v}\cdot[\nabla \times (\nabla \times \mathbf{G})]\,. \tag{8.1.77}$$

Hence, after taking the inner product of Eq. (8.1.75) and $\hat{\mathbf{a}}$, and re-arranging the resulting equation we have:

$$\begin{aligned}
&4\pi\hat{\mathbf{a}}\cdot[\nabla_p \times \mathbf{C}(\mathbf{r}_p)] \\
&= -\oiint_S [\hat{\mathbf{v}} \times (\nabla \times \mathbf{C})]\cdot\left(\nabla_p \times \mathbf{G}\right) d\mathbf{r} + \oiint_S (\hat{\mathbf{v}} \times \mathbf{C})\cdot\left[\nabla_p \times \left(\nabla_p \times \mathbf{G}\right)\right] d\mathbf{r} \\
&= \oiint_S [\hat{\mathbf{v}} \times (\nabla \times \mathbf{C})]\cdot(\nabla \times \mathbf{G})\, d\mathbf{r} + \oiint_S (\hat{\mathbf{v}} \times \mathbf{C})\cdot[\nabla \times (\nabla \times \mathbf{G})]\, d\mathbf{r}.
\end{aligned} \tag{8.1.78}$$

The substitution $\mathbf{K}(\mathbf{r},\mathbf{r}_p) = \nabla \times \mathbf{G}$ leads to:

$$4\pi\hat{\mathbf{a}}\cdot[\nabla_p \times \mathbf{C}(\mathbf{r}_p)] = \oiint_S [\hat{\mathbf{v}} \times (\nabla \times \mathbf{C})]\cdot\mathbf{K}\,d\mathbf{r} + \oiint_S (\hat{\mathbf{v}} \times \mathbf{C})\cdot(\nabla \times \mathbf{K})\,d\mathbf{r}$$
$$= -\oiint_S (\hat{\mathbf{v}} \times \mathbf{K})\cdot(\nabla \times \mathbf{C})d\mathbf{r} - \oiint_S [\hat{\mathbf{v}} \times (\nabla \times \mathbf{K})]\cdot\mathbf{C}d\mathbf{r}. \qquad (8.1.79)$$

8.1.3.2 Solution for a Plane Interface

In order to use symmetry properties of some of the functions in this derivation, it is necessary to make part of the surface S planar which we then denote by S_1. The rest of the surface S shall be referred to as S_2. Therefore the integration that was previously carried out over S now needs to be performed over S_1 and S_2:

$$\oiint_S \cdots = \iint_{S_1} \cdots + \iint_{S_2} \cdots$$

When the surface S_2 is taken to infinity, as required by the radiation condition, the contribution from the second integral vanishes and so the integration over the closed surface S can be replaced by integration over the open surface of S_1.

In what follows we assume that the outward normal on S_1 is given by $\hat{\mathbf{v}} = (0,0,1)^T$. We first take another look at Eq. (8.1.79) under this assumption. The equation states that the curl of either the **E** or the **H** field vectors at the point P may be obtained from knowledge of the tangential component of curl of **E** *or* **H** (first term in the right-hand side) *and* from the transverse component of **E** *or* **H**. However, we have said before that in general, if the boundary at which the field values are specified is known, then any two components of either **E** or **H** would be sufficient to uniquely determine the field at P. This implies that either the first or the second term of Eq. (8.1.79) should uniquely specify the field at P. In an attempt to find the appropriate formulas we first construct the mirror image $P'(x_p, y_p, -z_p) = P'(\mathbf{r}'_p)$ of the point of observation $P(x_p, y_p, z_p)$ about S_1 (Fig. 8.4). Note that we do not have to bother with the new singularity introduced at P' because P' is inside S. We now specify **[304]** the vector functions $\mathbf{K}^{(1)}(\mathbf{r};\mathbf{r}_p,\mathbf{r}'_p)$ and $\mathbf{K}^{(2)}(\mathbf{r};\mathbf{r}_p,\mathbf{r}'_p)$ which satisfy:

$$\hat{\mathbf{v}} \times \mathbf{K}^{(1)} = \mathbf{0}, \qquad (8.1.80)$$
$$\hat{\mathbf{v}} \times (\nabla \times \mathbf{K}^{(2)}) = \mathbf{0}, \qquad (8.1.81)$$

at the surface $z = 0$. A possible set of $\mathbf{K}^{(1)}$ and $\mathbf{K}^{(2)}$ that satisfies Eqs. (8.1.80) and (8.1.81) is:

$$\mathbf{K}^{(1)} = \nabla \times \left[\hat{\mathbf{a}}G(\mathbf{r},\mathbf{r}_p) + \hat{\mathbf{a}}'G(\mathbf{r},\mathbf{r}'_p)\right], \qquad (8.1.82)$$
$$\mathbf{K}^{(2)} = \nabla \times \left[\hat{\mathbf{a}}G(\mathbf{r},\mathbf{r}_p) - \hat{\mathbf{a}}'G(\mathbf{r},\mathbf{r}'_p)\right], \qquad (8.1.83)$$

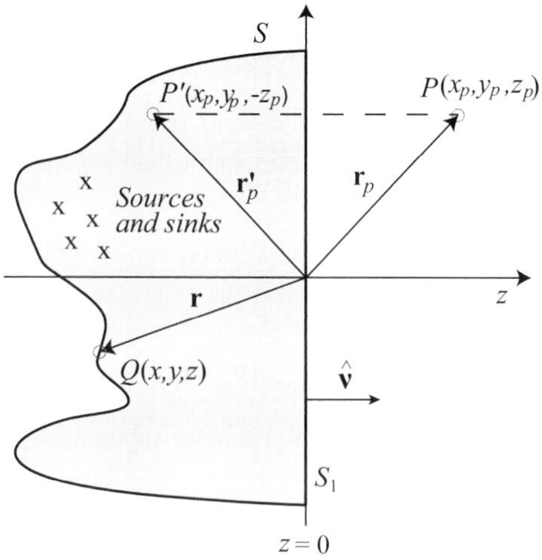

Figure 8.4: For the derivation of the *e* and *m* theories.

where

$$\hat{\mathbf{a}} = (a_x, a_y, a_z), \qquad \hat{\mathbf{a}}' = (a_x, a_y, -a_z), \tag{8.1.84}$$

are constant vectors. Vector algebra helps us to prove the following two identities:

$$[\hat{\mathbf{v}} \times (\nabla \times \mathbf{K}^{(1)})]\cdot\mathbf{C} = -2\hat{\mathbf{a}}\cdot\nabla_p \times \left\{\nabla_p \times [(\hat{\mathbf{v}} \times \mathbf{C})\, G]\right\} \tag{8.1.85}$$

and

$$(\hat{\mathbf{v}} \times \mathbf{K}^{(2)})\cdot(\nabla \times \mathbf{C}) = -2\hat{\mathbf{a}}\cdot\nabla_p \times [\hat{\mathbf{v}} \times (\nabla \times \mathbf{C})G] \tag{8.1.86}$$

for $z = 0$. It is now a straightforward matter to substitute Eqs. (8.1.85) and (8.1.86) into Eq. (8.1.79) to obtain, for the above two cases, the relations:

$$2\pi\nabla_p \times \mathbf{C} = \nabla_p \times \nabla_p \times \iint_{S_1} (\hat{\mathbf{v}} \times \mathbf{C})GdA, \tag{8.1.87}$$

$$2\pi\nabla_p \times \mathbf{C} = \nabla_p \times \iint_{S_1} [\hat{\mathbf{v}} \times (\nabla \times \mathbf{C})]GdA, \tag{8.1.88}$$

where we note that the integration is now performed over the surface S_1 since it was necessary to assume that $\hat{\mathbf{v}} = (0, 0, 1)^T$ in order to obtain these expressions. We shall discuss this step in Section 8.2 in more detail. By substituting either **E** or **H** for **C** in Eqs. (8.1.87) and (8.1.88), we find,

$$2\pi i\omega\epsilon\mathbf{E}^{(e)} = -\nabla_p \times \nabla_p \times \iint_{S_1} (\hat{\mathbf{v}} \times \mathbf{H})GdA, \tag{8.1.89}$$

$$2\pi\mathbf{H}^{(e)} = \nabla_p \times \iint_{S_1} (\hat{\mathbf{v}} \times \mathbf{H})GdA, \tag{8.1.90}$$

and

$$2\pi\mathbf{E}^{(m)} = \nabla_p \times \iint_{S_1} (\hat{\mathbf{v}} \times \mathbf{E})GdA, \tag{8.1.91}$$

$$2\pi i\omega\mu\mathbf{H}^{(m)} = \nabla_p \times \nabla_p \times \iint_{S_1} (\hat{\mathbf{v}} \times \mathbf{E})GdA. \tag{8.1.92}$$

Alternatively we can write from Eq. (8.1.72)

$$\mathbf{E}^{(e)} = 2\nabla_p \times \nabla_p \times \mathbf{\Pi}_e, \tag{8.1.93}$$

$$\mathbf{H}^{(e)} = -2i\omega\epsilon\nabla_p \times \mathbf{\Pi}_e, \tag{8.1.94}$$

and

$$\mathbf{E}^{(m)} = 2i\omega\mu\nabla_p \times \mathbf{\Pi}_m, \tag{8.1.95}$$

$$\mathbf{H}^{(m)} = 2\nabla_p \times \nabla_p \times \mathbf{\Pi}_m, \tag{8.1.96}$$

where the integration in $\mathbf{\Pi}_e$ and $\mathbf{\Pi}_m$ is of course now performed over S_1. The name *e* and *m* theory was coined to refer to Eqs. (8.1.93) and (8.1.94) and Eqs. (8.1.95) and (8.1.96) respectively by Karczewski and Wolf **[176]**, because $\mathbf{\Pi}_e$ is the electric and $\mathbf{\Pi}_m$ is the magnetic Hertz vector.

We note that because of the equivalence of Eqs. (8.1.18) and (8.1.79), the electric and magnetic fields **E** and **H** given by the Stratton–Chu integral are the arithmetic average of $\mathbf{E}^{(e)}$ and $\mathbf{E}^{(m)}$, and $\mathbf{H}^{(e)}$ and $\mathbf{H}^{(m)}$, respectively (c.f. Eqs. (8.1.62) and (8.1.63)):

$$\mathbf{E} = \frac{1}{2}(\mathbf{E}^{(e)} + \mathbf{E}^{(m)}), \qquad \mathbf{H} = \frac{1}{2}(\mathbf{H}^{(e)} + \mathbf{H}^{(m)}). \tag{8.1.97}$$

What is fascinating about Eqs. (8.1.93)–(8.1.96) is that they show that knowledge of the tangential component of *either* **E** or **H**, specified over a plane surface, is sufficient to uniquely determine the electric and magnetic fields everywhere on the right-hand side of the flat surface (screen). Therefore it seems that the various versions of the Stratton–Chu integrals are redundant in terms of requiring overspecificaiton of the incident electromagnetic field. We, however, should remember that imposing a plane surface as a part of the boundary surface restricts the problem such that the *e* and *m* theory is not equivalent to the Stratton–Chu theory. The fact that we have two different equations expressing what appears to be the same electric and magnetic fields is a much discussed problem. For further reading on this subject the reader is referred to the monograph of Stamnes **[325]**.

8.1.4 The Vectorial Rayleigh–Sommerfeld theory

It now remains to derive the vectorial Rayleigh–Sommerfeld formulas, which use all the three coordinate components of $\mathbf{E}$ at the surface $z = 0$ to determine the value of the electric field anywhere on the right-hand side of the surface. The geometry we consider is identical to that shown in Fig. 8.4.

Let the Green's function be defined by

$$G^{\pm}(\mathbf{r};\mathbf{r}_p,\mathbf{r}'_p) = G(\mathbf{r};\mathbf{r}_p) \pm G(\mathbf{r};\mathbf{r}'_p) = G^{\pm}(r_-,r_+) = G(r_-) \pm G(r_+), \tag{8.1.98}$$

with

$$r_- = \sqrt{(x-x_p)^2 + (y-y_p)^2 + (z-z_p)^2}, \tag{8.1.99}$$

$$r_+ = \sqrt{(x-x_p)^2 + (y-y_p)^2 + (z+z_p)^2}. \tag{8.1.100}$$

Let us now use the new definition of the Green's function in the vectorial integral theorem of Kirchhoff, Eq. (8.1.58):

$$\mathbf{E}^{\pm}(\mathbf{r}_p) = \frac{1}{4\pi}\iint_{S_1} \left[\mathbf{E}\,(\hat{\mathbf{v}}\cdot\nabla G^{\pm}) - G^{\pm}\,(\hat{\mathbf{v}}\cdot\nabla)\,\mathbf{E}\right] dA, \tag{8.1.101}$$

where, just as in the previous section we have taken S_2 to infinity so that it has zero contribution to the integral. On S_1 we have both $z = 0$ and $\hat{\mathbf{v}} = \hat{\mathbf{k}}$ such that

$$\nabla G^{\pm}\Big|_{z=0}\cdot\hat{\mathbf{v}} = \nabla G^{\pm}\Big|_{z=0}\cdot\hat{\mathbf{k}} = \frac{\partial G^{\pm}}{\partial z}\bigg|_{z=0}. \tag{8.1.102}$$

The derivative can be written as

$$\frac{\partial G^{\pm}}{\partial z} = \frac{dG(r_-)}{dr_-}\frac{\partial r_-}{\partial z} \pm \frac{dG(r_+)}{dr_+}\frac{\partial r_+}{\partial z}. \tag{8.1.103}$$

However, at the screen $z = 0$ such that

$$\frac{dG(r_-)}{dr_-}\bigg|_{z=0} = \frac{dG(r_+)}{dr_+}\bigg|_{z=0}, \tag{8.1.104}$$

which means that

$$\frac{\partial G^{\pm}}{\partial z}\bigg|_{z=0} = \frac{dG(r_-)}{dr_-}\bigg|_{z=0}\left(\frac{\partial r_-}{\partial z} \pm \frac{\partial r_+}{\partial z}\right)\bigg|_{z=0}. \tag{8.1.105}$$

Hence, for $z = 0$ we have $r = r'$ and thus

$$G^{+}(r_-,r_+) = 2G(r_-), \tag{8.1.106}$$

$$G^{-}(r_-,r_+) = 0, \tag{8.1.107}$$

$$\nabla G^{+}(r_-,r_+)\cdot\hat{\mathbf{v}} = 0, \tag{8.1.108}$$

$$\nabla G^{-}(r_-,r_+)\cdot\hat{\mathbf{v}} = 2\nabla G(r_-)\cdot\hat{\mathbf{v}} = 2\frac{\partial G(r_-)}{\partial v}. \tag{8.1.109}$$

As a result of substituting either G^+ or G^- into Eq. (8.1.101) we obtain two separate equations:

$$\mathbf{E}^{-}(\mathbf{r}_p) = \frac{1}{2\pi}\iint_{S_1} \mathbf{E}(\nabla G\cdot\hat{\mathbf{v}})dA = \frac{1}{2\pi}\iint_{S_1} \mathbf{E}\frac{\partial G}{\partial v}dA, \tag{8.1.110}$$

$$\mathbf{E}^{+}(\mathbf{r}_p) = -\frac{1}{2\pi}\iint_{S_1} G\,(\hat{\mathbf{v}}\cdot\nabla)\,\mathbf{E}dA = -\frac{1}{2\pi}\iint_{S_1} G\frac{\partial \mathbf{E}}{\partial v}dA, \tag{8.1.111}$$

which are the first and second *vectorial Rayleigh–Sommerfeld integrals*, respectively. We see that their arithmetic average gives the vectorial integral theorem of Kirchhoff, Eq. (8.1.58):

$$\mathbf{E}(\mathbf{r}_p) = \frac{1}{2}\left(\mathbf{E}^{+}(\mathbf{r}_p) + \mathbf{E}^{-}(\mathbf{r}_p)\right). \tag{8.1.112}$$

8.2 Boundary Value Problems in Diffraction

In the section above we derived two versions of two integral theorems (Stratton–Chu and Kirchhoff), both of which are only applicable to closed surfaces. Such surfaces, however, are quite rarely useful in practical physical optics. In the following sections we shall frequently use a set of particular conditions. When this set of conditions is used to solve equations of diffraction such that initial values (e.g. of the field) at a point or surface of space are specified then this set of conditions is referred to as the boundary value associated with a particular solution.

It should be clearly understood that we speak about diffraction when light waves interact with solids, as is usually the case, and as a result of this process the phase and/or the amplitude of the light wave is perturbed. The diffraction pattern is given by a superposition of the resulting waves. If, for example, a spherical wave propagates without being perturbed then we do not encounter diffraction. The most commonly discussed case of diffraction is that light is perturbed by a circular aperture situated in a physical screen (which is not necessarily plane). For the time being, however, we are not concerned with the aperture opening. Instead we concentrate on the problems associated with the surface S being closed.

We may now understand the difficulties associated with applying any of the integral theorems of the preceding section: the surface S must be closed over the aperture and also over either the image or the object space, depending on which theorem we use. For the case when all charges are within the volume limited by S the point of observation may be anywhere in space apart from the interior of S. During the derivation of the e and m theories, and that of the vectorial Rayleigh–Sommerfeld formulas, we assumed that a portion of S is a flat surface at $z = 0$. We denoted this surface by S_1 and in what follows we shall refer to it as a screen. If we were to open an aperture, $\mathcal{A}$, on S_1 so that light may pass through diffraction would result. It would be necessary to specify that the rest of S_1 and also the remaining surface of S are *perfectly black*. In this case there is no contribution from this surface portion and so the integration needs only to be performed over $\mathcal{A}$.

8.2.1 Applying the Radiation Condition

The situation is slightly more complicated when considering the case when P is valid everywhere in the interior of S and all sources are outside the volume V. The screen, containing the aperture opening, is considered to be a part of the surface S_1. In order to have a closed surface we have to add another surface, S_2. S_1 and S_2 form a closed surface and they enclose the volume V. We choose both S_1 and S_2 such that any sources of electromagnetic radiation that are present in the system are excluded from V because of the condition specifying that V cannot include charges. When eventually we open a portion of the screen and electromagnetic radiation enters the volume it may be perturbed by the presence of S_2. A possibility to avoid this is to take S_2 very far from the screen. However, in classical optics we are usually concerned with stationary systems, i.e. those we have left undisturbed for a long period of time such that all transient processes have already decayed. Hence, back radiation from S_2, even if it is very far from the screen, could still influence the stationary state at a given point of the volume V. In the physical sense, of course, there should not be such a second surface and we do not expect such a hypothetical surface to produce back radiation either. Therefore, there will be no contradiction with our arguments based on *physical* reasoning if we find the *mathematical* conditions under which S_2 produces a vanishing contribution, i.e. is said to be free of back-radiation. For this it will be advantageous to take S_2 to be a hemispherical shell centred on the origin of a Cartesian coordinate system (see Fig. 8.5). Let us denote the radius of the hemispherical shell S_2 by r_h. The vector $\mathbf{r} = (x, y, z)$ points as usual to the surface S_2 and the vector $\mathbf{r}_p = (x_p, y_p, z_p)$ points to the observation point. The outward surface normal of S_2 is denoted by $\hat{\mathbf{v}}$ as usual,

$$\hat{\mathbf{v}} = \frac{\mathbf{r}}{|\mathbf{r}|} = \left(\frac{x}{r}, \frac{y}{r}, \frac{z}{r}\right). \tag{8.2.1}$$

We now use the Green's function

$$G\left(\mathbf{r}, \mathbf{r}_p\right) = \frac{\exp\left(ik\left|\mathbf{r} - \mathbf{r}_p\right|\right)}{\left|\mathbf{r} - \mathbf{r}_p\right|} = \frac{\exp(ik|\mathbf{r}\_|)}{|\mathbf{r}\_|}, \tag{8.2.2}$$

with

$$\begin{aligned} r_- = |\mathbf{r}_-| = \left|\mathbf{r} - \mathbf{r}_p\right| &= \sqrt{\left(x - x_p\right)^2 + \left(y - y_p\right)^2 + \left(z - z_p\right)^2} \\ &= \left[x^2 + y^2 + z^2 - 2\left(xx_p + yy_p + zz_p\right) + x_p^2 + y_p^2 + z_p^2\right]^{1/2}. \end{aligned} \tag{8.2.3}$$

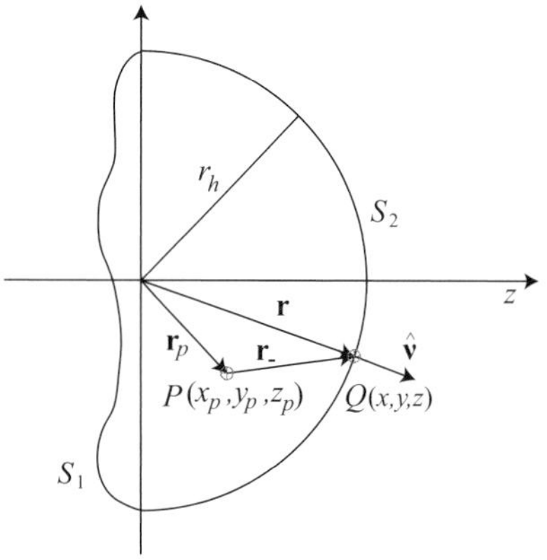

Figure 8.5: Illustrating the geometry of derivation.

If we assume that $|\mathbf{r}_p| \ll |\mathbf{r}|$, i.e. the observation point is very far from the surface S_2, then we may use the Taylor expansion to obtain $|\mathbf{r} - \mathbf{r}_p| \approx r - \hat{\mathbf{v}}\cdot\mathbf{r}_p$. Hence, when r_h is taken to infinity we may write

$$\lim_{r_h\to\infty} G\left(\mathbf{r}, \mathbf{r}_p\right) = \frac{\exp(ikr)}{r} \exp\left(-ik\hat{\mathbf{v}}\cdot\mathbf{r}_p\right) = g(\mathbf{r}, \mathbf{r}_p). \tag{8.2.4}$$

We note that $r_h \to \infty$ actually corresponds to the case where S_1 becomes an infinite plane surface. Since in Eqs. (8.1.33) and (8.1.34) both the Green's function and its gradient are present we also need to calculate the gradient. It is given as

$$\nabla g(\mathbf{r}, \mathbf{r}_p) = g(\mathbf{r}, \mathbf{r}_p)\left[\left(ik - \frac{1}{r}\right)\hat{\mathbf{v}} - ik\left(\frac{\mathbf{r}_p}{r} - \hat{\mathbf{v}}\frac{\mathbf{r}_-\cdot\mathbf{r}_p}{r^2}\right)\right] = ik\hat{\mathbf{v}}g(\mathbf{r}, \mathbf{r}_p) + O\left(\frac{1}{r}\right), \tag{8.2.5}$$

where we took the gradient at $\mathbf{r}$. Hence

$$\lim_{r_h\to\infty} \nabla G\left(\mathbf{r}, \mathbf{r}_p\right) = ik\hat{\mathbf{v}}g(\mathbf{r}, \mathbf{r}_p). \tag{8.2.6}$$

We now examine what value the Stratton–Chu integral gives when calculated on the surface S_2, with the approximate Green's function. It is clear that we may write Eq. (8.1.33) in the following form:

$$\mathbf{E} = -\frac{1}{4\pi}\iint_{S_1}(\ldots)\,dA - \frac{1}{4\pi}\iint_{S_2}(\ldots)\,dA. \tag{8.2.7}$$

We now consider only the integral over S_2 on the right-hand side of Eq. (8.2.7). From Eqs. (8.2.4)–(8.2.6) and Eq. (8.2.7) we have:

$$\frac{i\omega\mu}{4\pi}\iint_{S_2}\left\{(\hat{\mathbf{v}}\times\mathbf{H}) + \sqrt{\frac{\epsilon}{\mu}}\left[(\hat{\mathbf{v}}\times\mathbf{E})\times\hat{\mathbf{v}} + (\hat{\mathbf{v}}\cdot\mathbf{E})\hat{\mathbf{v}}\right]\right\}g(\mathbf{r}, \mathbf{r}_p)dA = \frac{i\omega\mu}{4\pi}\iint_{S_2}\left[(\hat{\mathbf{v}}\times\mathbf{H}) + \sqrt{\frac{\epsilon}{\mu}}\mathbf{E}\right]g(\mathbf{r}, \mathbf{r}_p)dA. \tag{8.2.8}$$

However, since the hemispherical shell has been taken to infinity it is plausible to assume that on the surface S_2 the electric $\mathbf{E}$ and the magnetic $\mathbf{H}$ field vectors and the outward surface normal $\hat{\mathbf{v}}$ are mutually orthogonal. On the other hand, the electric and magnetic vectors must satisfy Maxwell's equations on the surface S_2. Hence we may write (c.f. Sections 1.6.3 and 1.6.4):

$$\mathbf{E} = \sqrt{\frac{\mu}{\epsilon}}\mathbf{H}\times\hat{\mathbf{v}}, \tag{8.2.9}$$

so the total contribution from the surface S_2 vanishes.[5] Hence we are left with:

$$\mathbf{E}\left(\mathbf{r}_p\right) = -\frac{1}{4\pi}\iint_{S_1}\left[i\omega\mu\,(\hat{\mathbf{v}}\times\mathbf{H})\,G + (\hat{\mathbf{v}}\times\mathbf{E})\times\nabla G + (\hat{\mathbf{v}}\cdot\mathbf{E})\,\nabla G\right]dA, \tag{8.2.10}$$

[5] Because of Eqs. (8.2.4) and (8.2.6) the error is $O(1/r)$. Also note that the vanishing contribution from S_2 could have been directly imposed by remembering that the Stratton–Chu integral was derived with the assumption of the vectorial radiation condition.

and similarly for the magnetic field:

$$\mathbf{H}\left(\mathbf{r}_p\right) = \frac{1}{4\pi}\iint_{S_1} [i\omega\epsilon\,(\hat{\mathbf{v}}\times\mathbf{E})\,G - (\hat{\mathbf{v}}\times\mathbf{H})\times\nabla G - (\hat{\mathbf{v}}\cdot\mathbf{H})\,\nabla G]\,dA. \tag{8.2.11}$$

It is interesting to compare Eqs. (8.2.10) and (8.2.11) to Eqs. (8.1.33) and (8.1.34). We see that the only difference between the corresponding formulas is the region of integration. Furthermore, because of the equivalence of Eqs. (8.1.33) and (8.1.56), exactly the same procedure would yield for the vectorial integral theorem of Kirchhoff, Eq. (8.1.56), the following formula:

$$\mathbf{E}(\mathbf{r}_p) = -\frac{1}{4\pi}\iint_{S_1} [\mathbf{E}\,(\hat{\mathbf{v}}\cdot\nabla G) - G\,(\hat{\mathbf{v}}\cdot\nabla)\,\mathbf{E}]\,dA = -\frac{1}{4\pi}\iint_{S_1}\left[\mathbf{E}\frac{\partial G}{\partial v} - G\frac{\partial \mathbf{E}}{\partial v}\right] dA, \tag{8.2.12}$$

or, if starting from Eq. (8.1.61):

$$\mathbf{E}(\mathbf{r}_p) = \frac{1}{4\pi}\iint_{S_1} [\mathbf{E}\,(\hat{\mathbf{v}}\cdot\nabla G) - G\,(\hat{\mathbf{v}}\cdot\nabla)\,\mathbf{E}]\,dA = \frac{1}{4\pi}\iint_{S_1}\left[\mathbf{E}\frac{\partial G}{\partial v} - G\frac{\partial \mathbf{E}}{\partial v}\right] dA, \tag{8.2.13}$$

where again, after matching the surface normal the same expression results.

It should be clearly understood that the mathematical conditions given by Eqs. (8.2.4) and (8.2.6) follow from a physical phenomenon, namely that there is no contribution to the total field from S_2. This assumption is very similar to the one made when we derived the vectorial radiation conditions, Eqs. (8.1.29) and (8.1.30). These conditions also possess an important physical significance, as shown by Eq. (8.2.9). This equation states, since $\hat{\mathbf{v}}$ is the outward surface normal, that on the surface S_2 the field is outgoing, i.e. propagating away from the observation point P.

In setting our conditions above we have made no assumption concerning the surface element S_1. In order to study diffraction problems associated with an opaque screen with an aperture, it is customary to consider part of S_1 to be fully transparent and another part of S_1 to be opaque, as discussed above. It was advantageous in the preceding derivation to take S_2 to be a hemispherical shell. The reason for this is that we used the fact that the surface normal $\hat{\mathbf{v}}$ could be expressed in terms of the surface position vector $\mathbf{r}$. Then the electric and magnetic field vectors and the surface normal are mutually orthogonal. We have seen in the previous section that it can be advantageous to take S_1 to be a plane surface. The reason for this is that we chose a Green's function with special properties on a plane.

8.2.2 Kottler's Diffraction Integral

A solution to the problem associated with an open aperture was suggested by Kottler **[193]**. He divided the surface S_1 into two regions, the screen $\mathcal{B}$ and the opening $\mathcal{A}$. If the screen is assumed perfectly black, the transverse electric and transverse magnetic fields on it are zero. Hence the integral over $\mathcal{B}$ is zero. The two regions are separated from each other by a continuous contour C. Both $\mathbf{E}$ and $\mathbf{H}$ and their first derivatives are continuous and they are solutions of Maxwell's equations on $\mathcal{A}$ *and* $\mathcal{B}$. When, however, they pass through C their tangential components suffer an abrupt change.

Equations (8.1.33) and (8.1.34) consist of three terms which respectively arise from electric and magnetic currents and an electric charge density on S. The discontinuity in the tangential components of $\mathbf{E}$ and $\mathbf{H}$ implies an abrupt change in surface current density. Kottler suggested that along C there should be an accumulation of charges in such a case. It is well outside the scope of this work to go into details regarding this problem. However, for example, Stratton **[328]** showed that when there is an aperture opening the electric field may be expressed in the form:

$$\mathbf{E}\left(\mathbf{r}_p\right) = -\frac{1}{i\omega\epsilon}\frac{1}{4\pi}\oint_C \nabla G\mathbf{H}\cdot d\mathbf{l} - \frac{1}{4\pi}\iint_{\mathcal{A}} [i\omega\mu\,(\hat{\mathbf{v}}\times\mathbf{H})\,G + (\hat{\mathbf{v}}\times\mathbf{E})\times\nabla G + (\hat{\mathbf{v}}\cdot\mathbf{E})\,\nabla G]\,dA. \tag{8.2.14}$$

This formula, known as *Kottler's diffraction integral*, should be contrasted to Eq. (8.1.33). On comparing these equations we see that in Eq. (8.2.14) the only addition is a contour integral around C. When, however, the contour integral is evaluated under usual conditions of light optics its value may be seen to be 10 to 15 orders of magnitude *smaller* than that of the second term in Eq. (8.2.14). In addition, Asvestas **[15]** in a little-known paper showed that Kottler's solution actually *does not satisfy Maxwell's equations for points exactly on the rim of the aperture.* Unfortunately, Asvestas did not suggest a solution that would uniformly satisfy Maxwell's equations. We conclude this line of argument by noting that Eqs. (8.2.10) and (8.2.11) shall be regarded as *sufficiently good approximations* to the electric and magnetic fields in optical systems of practical importance.

Before we move on we stop for a moment to mention the problem associated with the substance of the screen. Perhaps the most complete discussion of this problem is due to Bouwkamp **[42]**. The role of an aperture opening was not discussed fully during the derivation of the e and m theories, Eqs. (8.1.93)–(8.1.96); we merely remarked at certain places that the aperture should be considered to be 'black'. However, there is an equivalence of these formulas and Eq. (8.1.97) with Kottler's diffraction integral, Eq. (8.2.14). Hence, it is possible to write Eq. (8.1.93) for a screen with an aperture opening. The resulting formulas require the material properties of the screen to be specific. For example, Eq. (8.1.93) assumes that the tangential component of the electric field $\mathbf{E}$ is vanishing on the screen, such that the screen corresponds to a perfectly conducting material. On the other hand, Eq. (8.1.95) requires the tangential component of the magnetic field to vanish and hence the screen may be regarded as having infinite inductivity. Interestingly enough, the derivation of Eq. (8.2.14) requires a perfectly black screen, meaning that on the opaque portion of the screen both $\mathbf{E}$ and $\mathbf{H}$ should vanish. A more thorough discussion of these problems is outside of the scope of the current work and so here we simply note that this matter is strongly connected to the inconsistencies encountered in scalar diffraction with regards to the Kirchhoff boundary conditions.

8.3 The Debye–Wolf and Related Diffraction Theories

8.3.1 The Debye–Wolf Integral for Finite Fresnel Numbers

This section is devoted to the Debye–Wolf integral, which is probably the best suited diffraction theory to describe practical optical focusing systems. To obtain the electric field we take Eq. (8.2.10) as our starting point.[6] The notation of the arrangement is shown in Fig. 8.6. The position vector $\mathbf{r}$ points to Q on $\mathcal{A}$. The point of observation P is characterised, as usual, by $\mathbf{r}_p$. The vector $\mathbf{r}_-$ is given as $\mathbf{r}_- = \mathbf{r} - \mathbf{r}_p$. The electric and magnetic field values in the kernel of the Stratton–Chu formula are those over $\mathcal{A}$ and the gradient ∇G is calculated at Q. The outward surface normal of $\mathcal{A}$ is denoted by $\hat{\mathbf{v}}$ and the unit vector along $\mathbf{r}_-$ is denoted by $\hat{\mathbf{n}}$. By way of this figure we introduce ϑ, measured from the $-z$ direction, as the supplementary angle of θ and note that the angle α, to be used later, is measured from the $-z$ direction. With reference to the figure we make the following assumptions:[7]

i) A circular aperture opening will be cut in the screen.
ii) The field incident on the aperture opening is a perfect, convergent spherical wave originating very far from the screen which converges towards the centre of the co-ordinate system. The distance from any point on the rim of the aperture to the centre of convergence is denoted by f and is referred to as the focal length. The angle at which the rim of the aperture is seen from the focus is denoted by α and is usually referred to as the convergence angle. The surface of

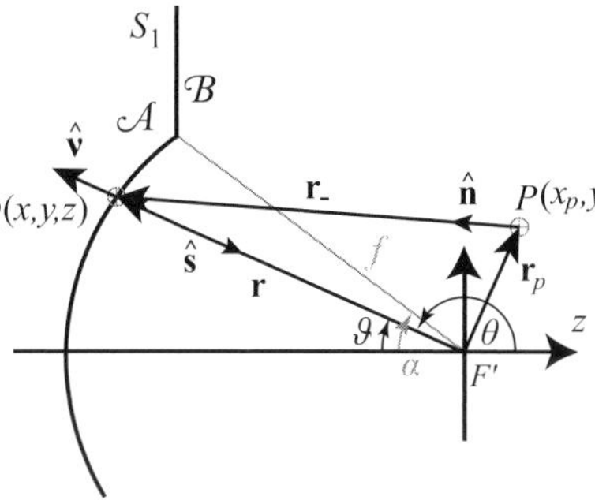

Figure 8.6: Notation for the derivation of the Debye–Wolf integral. Only part of the geometry is shown. A perfect spherical spherical wave $\mathcal{B}$ converges from a circular aperture on a screen $\mathcal{A}$ towards the back focal point F'. A point in the exit pupil on the Gaussian sphere is denoted by $Q(x, y, z)$ and the electric field needs to be computed at point $P(x_p, y_p, z_p)$. The unit vector from Q to P along a typical non-skew ray is denoted by $\hat{\mathbf{s}}$. The usual azimuthal angle θ is measured from the $+z$ direction and its supplementary angle ϑ, measured clockwise from the $-z$ direction, is introduced for historical reasons. The angle of the marginal ray is α.

[6] This derivation follows the work of Hsu and Barakat **[148]** and Török **[340]**. The latter paper contains some typographical errors that are corrected here.
[7] These assumptions are necessary to simplify the Stratton–Chu integral. A different set of assumptions could also be chosen.

integration is taken to be composed of a plane screen $\mathcal{B}$, and a spherical surface $\mathcal{A}$, which coincides with a phase front of the converging incident wave. The relationship

$$\mathbf{E} = -\sqrt{\frac{\mu}{\epsilon}}\mathbf{H} \times \hat{\mathbf{v}} \tag{8.3.1}$$

on the surface of the spherical wavefront then holds due to the locally transverse nature of spherical waves (c.f. Section 1.6.3). Note that the sign of this expression is opposite to Eq. (8.2.9) because now the surface normal $\hat{\mathbf{v}}$ is pointing opposite to the direction of propagation.

iii) The electric and magnetic fields on $\mathcal{A}$, $\mathbf{E}^a, \mathbf{H}^a$, are assumed to be equal to the field $\mathbf{E}^i, \mathbf{H}^i$ very far from the aperture, and zero on the screen:

$$\mathbf{E}^a, \mathbf{H}^a = \begin{cases} \mathbf{E}^i, \mathbf{H}^i & \text{on } \mathcal{A} \\ 0 & \text{on } \mathcal{B}, \end{cases} \tag{8.3.2}$$

that is to say the presence of the aperture does not perturb the field.

iv) The observation point is not too close[8] to $\mathcal{A}$, so that $|\mathbf{r}| \gg |\mathbf{r}_p|$. Under this condition

$$\hat{\mathbf{v}} \approx \hat{\mathbf{n}}. \tag{8.3.3}$$

The Green's function and its gradient are given by:

$$G = \frac{\exp\left(ik\left|\mathbf{r} - \mathbf{r}_p\right|\right)}{\left|\mathbf{r} - \mathbf{r}_p\right|} = \frac{\exp(ikr_-)}{r_-}, \tag{8.3.4}$$

$$\nabla G = \hat{\mathbf{n}}\frac{\exp(ikr_-)}{r_-}\left(ik - \frac{1}{r_-}\right) \approx ik\hat{\mathbf{n}}G + O\left(\frac{1}{r_-}\right). \tag{8.3.5}$$

Considering assumptions ii)–iv) and Eq. (8.3.4), (8.3.5), and (8.2.9), the Stratton–Chu integral Eq. (8.2.10) now reads:

$$\mathbf{E}\left(\mathbf{r}_p\right) = -\frac{ik}{4\pi}\iint_{\mathcal{A}} [\mathbf{E} - (\hat{\mathbf{v}} \times \mathbf{E}) \times \hat{\mathbf{n}}]\, G dA. \tag{8.3.6}$$

From Eq. (8.3.3) and assumption ii) Eq. (8.3.6) simplifies to

$$\mathbf{E}\left(\mathbf{r}_p\right) = -\frac{i}{\lambda}\iint_{\mathcal{A}} \mathbf{E} G dA. \tag{8.3.7}$$

Note that the corresponding expression for the magnetic field is

$$\mathbf{H}\left(\mathbf{r}_p\right) = -\frac{i}{\lambda}\iint_{\mathcal{A}} \mathbf{H} G dA. \tag{8.3.8}$$

It now remains to define a spherical and a cylindrical co-ordinate system. Then

$$Q = f(\sin\vartheta\cos\phi, \sin\vartheta\sin\phi, -\cos\vartheta), \tag{8.3.9}$$

$$P = (r_p\cos\gamma, r_p\sin\gamma, z_p), \tag{8.3.10}$$

where $f = |\mathbf{r}|$ and the distance r_- in Eq. (8.3.4) is given by:

$$r_- = \left[(f + z_p)^2 + r_p^2 - 2fz_p(1 + \cos\vartheta) - 2fr_p\sin\vartheta\cos(\gamma - \phi)\right]^{1/2}. \tag{8.3.11}$$

Under condition iii), Eq. (8.3.11) can be written by using the Maclaurin expansion as

$$r_- \approx f + z_p + \frac{1}{2(f + z_p)}\left[r_p^2 - 2fz_p(1 + \cos\vartheta) - 2fr_p\sin\vartheta\cos(\gamma - \phi)\right] = \Phi_0 + \Phi_1, \tag{8.3.12}$$

where

$$\Phi_0 = f + z_p + \frac{r_p^2 - 2fz_p}{2(f + z_p)}, \tag{8.3.13}$$

[8] As it turns out after solving the problem this condition is not too restrictive, as when $|\mathbf{r}_p| \gg \lambda$ is satisfied the energy has decayed so much that in practice it may be already below the detection limit.

$$\Phi_1 = \frac{f}{f+z_p} z_p \cos\vartheta - \frac{f}{f+z_p} r_p \sin\vartheta \cos(\gamma - \phi). \tag{8.3.14}$$

A number of papers have been published on the expansion of Eq. (8.3.11). For a review of these the reader is referred to **[311]**. The key issue is to recognise that whereas in the case of the usual Fresnel or Fraunhofer approximations the approximate distance is obtained as a power series expansion about either a point on the aperture or the observation point, the expansion of Eq. (8.3.11) involves an approximation of the form $\sqrt{1-\mu}$ where μ is a non-physical quantity in the sense that it depends on all system parameters. Clearly, just by taking μ to be small compared to unity, no direct assumptions are made regarding the physical size of the aperture or the location of the observation point.

It is relatively simple to analyse the error that is implicit in the approximations in Eqs. (8.3.12) and (8.3.14), but it is more difficult to establish a threshold condition describing when it is still acceptable to approximate r_- because the electric field $\mathbf{E}(P)$ at the observation point depends on all possible positions of Q and is hence a complicated function of r_-. Of course the only rigorous measure of error is to calculate the field using the approximate r_- and then compare the result to the exact formula. There are special cases when we can compute the field from more exact formulas than Eq. (8.3.7). Details of this work are not shown here so let it suffice to state that when the comparison is done, one can identify a tolerance level beyond which the approximate formula cannot be used.

If we now set

$$\mathcal{R} = \frac{f}{f+z_p} r_p, \tag{8.3.15}$$

$$\mathcal{Z} = \frac{f}{f+z_p} z_p, \tag{8.3.16}$$

the point of observation P is

$$\mathbf{r}' = (\mathcal{R}\cos\gamma, \mathcal{R}\sin\gamma, \mathcal{Z}), \tag{8.3.17}$$

represented by the vector $\mathbf{r}'$ that is the position vector of point P in a transformed co-ordinate system. We define a unit vector $\hat{\mathbf{s}}$ of a typical ray direction:

$$\hat{\mathbf{s}} = (\hat{s}_x, \hat{s}_y, \hat{s}_z) = (-\sin\vartheta\cos\phi, -\sin\vartheta\sin\phi, \cos\vartheta). \tag{8.3.18}$$

The surface element dA can be expressed as

$$dA = f^2 d\Omega = f^2 \frac{d\hat{s}_x\, d\hat{s}_y}{\hat{s}_z} = f^2 \sin\vartheta d\vartheta d\phi. \tag{8.3.19}$$

The Green's function in Eq. (8.3.7) can be approximated as

$$\frac{\exp(ikr)}{r} = \frac{\exp(ik\Phi_0)}{f+z_p} \exp(ik\Phi_1). \tag{8.3.20}$$

The reason why the more relaxed approximation can be made for the denominator of G is that it contributes to the value of the integrand comparatively little when $f \gg \lambda$. Finally, Eq. (8.3.7) gives:

$$\mathbf{E}\left(\mathbf{r}_p\right) = -\frac{if^2 \exp(ik\Phi_0)}{\lambda(f+z_p)} \iint_{\hat{s}_x^2+\hat{s}_y^2 \le 1} \frac{\mathcal{E}(\hat{s}_x, \hat{s}_y)}{\hat{s}_z} \exp(ik\hat{\mathbf{s}}\cdot\mathbf{r}') d\hat{s}_x\, d\hat{s}_y. \tag{8.3.21}$$

A similar procedure yields for the magnetic field

$$\mathbf{H}\left(\mathbf{r}_p\right) = -\frac{if^2 \exp(ik\Phi_0)}{\lambda(f+z_p)} \iint_{\hat{s}_x^2+\hat{s}_y^2 \le 1} \frac{\mathcal{H}(\hat{s}_x, \hat{s}_y)}{\hat{s}_z} \exp(ik\hat{\mathbf{s}}\cdot\mathbf{r}') d\hat{s}_x\, d\hat{s}_y, \tag{8.3.22}$$

where we have introduced $\mathcal{E}$ and $\mathcal{H}$ to denote the electric and magnetic vectors in the kernel of the integral. The significance of this notation is made clear later on in this section. The result we have just obtained is rather significant. Equations (8.3.21) and (8.3.22) are powerful formulas that can be used to calculate the fields in a range of diffraction problems. They can also be used to describe high-aperture focusing in almost all situations one encounters in physical optics including systems of low Fresnel number N, which is defined as:

$$N = \left(\frac{a}{f}\right)^2 \frac{f}{\lambda}, \tag{8.3.23}$$

where $a = f\sin(\alpha)$ is the radius of the aperture. We term the expression given by Eqs. (8.3.21) and (8.3.22) the *scaled Debye–Wolf integral* as the high angle vectorial analogue of the scalar Li and Wolf theory. The scaled Debye–Wolf integral is applicable to optical focusing systems of finite Fresnel number and arbitrary convergence angle. The coordinate transformation given by Eq. (8.3.16) is known as the Li and Wolf scaling.

8.3.2 The Debye–Wolf Integral for Infinite Fresnel Numbers

We now specialise the above formulation to systems of large Fresnel numbers. Let us keep the factor in Eq. (8.3.23)

$$\left(\frac{a}{f}\right)^2 = \text{const}, \tag{8.3.24}$$

so that by increasing f/λ in Eq. (8.3.23) the Fresnel number increases. Under the condition that $f \gg z_p$ we have from Eq. (8.3.16) $\mathcal{R} = r_p$ and $\mathcal{Z} = z_p$ so Eqs. (8.3.21) and (8.3.22) read:

$$\mathbf{E}\left(\mathbf{r}_p\right) = -\frac{if}{\lambda} \iint\limits_{\hat{s}_x^2+\hat{s}_y^2\leq 1} \frac{\mathcal{E}(\hat{s}_x,\hat{s}_y)}{\hat{s}_z} \exp(ik\hat{\mathbf{s}}\cdot\mathbf{r}_p) d\hat{s}_x d\hat{s}_y, \tag{8.3.25}$$

$$\mathbf{H}\left(\mathbf{r}_p\right) = -\frac{if}{\lambda} \iint\limits_{\hat{s}_x^2+\hat{s}_y^2\leq 1} \frac{\mathcal{H}(\hat{s}_x,\hat{s}_y)}{\hat{s}_z} \exp(ik\hat{\mathbf{s}}\cdot\mathbf{r}_p) d\hat{s}_x d\hat{s}_y. \tag{8.3.26}$$

These formulas are usually referred to as the *Debye–Wolf integral* or the Richards and Wolf integral. They were derived in a different manner by Ignatowsky **[155]**,**[153]**,**[154]** (also see Appendix G, Luneburg **[225]**, Richards and Wolf **[293]** and Stamnes **[325]**). In the above derivation we have not explicitly specified the incident electric and magnetic field vectors $\mathcal{E}$ and $\mathcal{H}$, respectively, but have instead only assumed that Eqs. (8.3.1) and (8.3.2) hold. In order to have a clear understanding how the incident field vectors should be specified it is advantageous to discuss the derivation of the Debye–Wolf integral as given by Luneburg **[225]**. We note that this derivation does not apply to the scaled Debye–Wolf integral, Eqs. (8.3.21) and (8.3.22).

8.3.2.1 Luneburg's Derivation

Consider an optical system of rotational symmetry. Let the point P_1 in the image space be a perfect conjugate pair of P_0 in the object space (see Fig. 8.7). In this arrangement a perfect spherical wave will diverge from P_0 and after refraction by the elements of the optical system (which can also introduce aberrations), an *imperfect* spherical wave will converge towards P_1. Suppose for a minute that the spherical waves did converge to P_1. In this case, with reference to Fig. 8.7b the equation of a particular convergent wavefront is of the form:

$$\psi(x, y, z) = C - R, \tag{8.3.27}$$

where C is the optical distance between P_0 and P_1 and is given from the Fermat principle:

$$C = [P_0P_1] = \int_{P_0}^{P_1} n d\ell = \text{const} \tag{8.3.28}$$

and

$$R = \sqrt{x^2 + y^2 + z^2}. \tag{8.3.29}$$

Consider now the case when the spherical waves do not converge to P_1. Figure 8.7c shows an aberrated ray corresponding to an aberrated wavefront (both indicated by grey lines) and the unaberrated ray corresponding to an

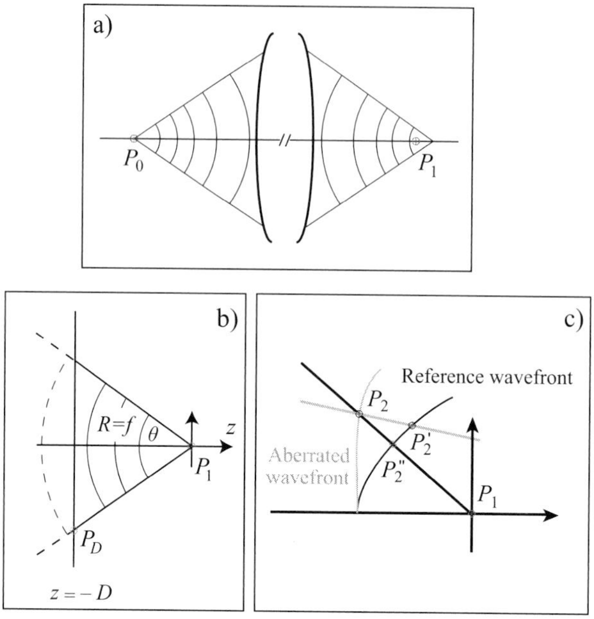

Figure 8.7: The notation for the Luneburg integral.

unaberrated wavefront (indicated by black lines). The optical path difference, measured along the aberrated ray, is $[P_2P_2']$. Consequently, one would have

$$\psi(x, y, z) = C - R - [P_2P_2']. \tag{8.3.30}$$

For most practical purposes it is a good approximation to assume the angle $P_2'P_2P_2''$ is small in which case $[P_2P_2'] \approx [P_2P_2''] = W$, hence,

$$\psi(x, y, z) = C - R - W, \tag{8.3.31}$$

where $W(x, y, z)$ is the wavefront aberration function (see Chapter 5).

Let the electric and magnetic vectors corresponding to a given ray be denoted by $\mathcal{E}_0$ and $\mathcal{H}_0$, respectively. Thus these vectors are known on a particular wavefront, say for $R = f$, i.e. when $\psi(x, y, z) = C - f - W$. On an arbitrary wavefront the electric and magnetic vectors are given by:

$$\mathcal{E} = \frac{f}{R}\mathcal{E}_0(x, y), \tag{8.3.32}$$

$$\mathcal{H} = \frac{f}{R}\mathcal{H}_0(x, y). \tag{8.3.33}$$

Let us denote a solution of Maxwell's equations by **E** and **H**. These solutions are then approximated by taking $\lambda \to 0$ to yield $\mathbf{E}_0$ and $\mathbf{H}_0$ which are, respectively, the electric and magnetic field vectors of the geometric optics approximation. $\mathbf{E}_0$ and $\mathbf{H}_0$, can be written from Eq. (8.3.30) and Eqs. (8.3.32) and (8.3.33) by:

$$\mathbf{E}_0 = \frac{f}{R}\mathcal{E}_0(x, y)\exp[ik(C - R - W)], \tag{8.3.34}$$

$$\mathbf{H}_0 = \frac{f}{R}\mathcal{H}_0(x, y)\exp[ik(C - R - W)]. \tag{8.3.35}$$

The solution for the problem of an imperfect spherical wave converging towards P_1 can be obtained in the following manner: we find the solutions of Maxwell's equations which have the same boundary values at infinity as the solution of geometrical optics, i.e.

$$\lim_{R\to\infty} R(\mathbf{E} - \mathbf{E}_0) = \mathbf{0}, \tag{8.3.36}$$

$$\lim_{R\to\infty} R(\mathbf{H} - \mathbf{H}_0) = \mathbf{0}. \tag{8.3.37}$$

The boundary values of the problem are the simplest to be considered on a plane whose equation is $z = -D$. Any point on the plane is characterised by its Cartesian co-ordinates $(x, y, -D)$. The geometric optics field on the plane is given by substituting $R = \sqrt{x^2 + y^2 + D^2}$ in the phase term of Eqs. (8.3.34) and (8.3.35). A solution of Maxwell's equations $\mathbf{E}$ is given, for example, by the first vectorial Rayleigh–Sommerfeld integral Eq. (8.1.110):

$$\mathbf{E} = \frac{1}{2\pi}\iint_{S_1} \mathbf{E}_0 \left.\frac{\partial G}{\partial z}\right|_{z=-D} dA, \tag{8.3.38}$$

with free space Green's function $G = \exp(ikr_-)/r_-$ and r_- is the distance between the point of observation P and a point on the plane S_1 as usual. We need the normal derivative of the Green's function, which is given by:

$$\left.\frac{\partial G}{\partial z}\right|_{z=-D} = -\exp(ikr_-)(D + z_p)\left(\frac{ik}{r_-^2} - \frac{1}{r_-^3}\right). \tag{8.3.39}$$

Consequently Eq. (8.3.38) now gives:

$$\mathbf{E} = -\frac{f}{2\pi}(D + z_p)\exp(ikC)\iint_{S_1} \mathbf{E}_0(x,y)\exp[ik(r_- - R)]\exp(-ikW)\left(\frac{ik}{r_-^2} - \frac{1}{r_-^3}\right)\frac{dxdy}{R}, \tag{8.3.40}$$

where

$$r_- = \sqrt{(x - x_p)^2 + (y - y_p)^2 + (D + z_p)^2}. \tag{8.3.41}$$

Since

$$R = \frac{D}{\cos\vartheta} \tag{8.3.42}$$

and

$$x = R\sin\vartheta\cos\phi = D\tan\vartheta\cos\phi, \tag{8.3.43}$$

$$y = R\sin\vartheta\sin\phi = D\tan\vartheta\sin\phi, \tag{8.3.44}$$

we may write from Eq. (8.3.41):[9]

$$\begin{aligned} r_- &= \left(\frac{D^2}{\cos^2\vartheta} + 2z_pD - 2D\tan\vartheta(x_p\cos\phi + y_p\sin\phi) + x_p^2 + y_p^2 + z_p^2\right)^{1/2} \\ &= \frac{D}{\cos\vartheta}\left(1 + \frac{2z_p\cos^2\vartheta}{D} - \frac{2\sin\vartheta\cos\vartheta}{D}(x_p\cos\phi + y_p\sin\phi)\right. \\ &\quad \left. + \cos^2\vartheta\frac{x_p^2 + y_p^2 + z_p^2}{D^2}\right)^{1/2}. \end{aligned} \tag{8.3.45}$$

We may use the Maclaurin expansion to approximate the square root in the above expression because in what follows we shall take D to infinity. The Maclaurin expansion of Eq. (8.3.45) gives:

$$\begin{aligned} r_- &= \frac{D}{\cos\vartheta} + z_p\cos\vartheta - \sin\vartheta(x_p\cos\phi + y_p\sin\phi) + \ldots \\ &\approx R + z_p\cos\vartheta - \sin\vartheta(x_p\cos\phi + y_p\sin\phi). \end{aligned} \tag{8.3.46}$$

where we have dropped terms of order $1/D^2$ and higher. We now take the axial location of the plane to infinity so from Eqs. (8.3.42)–(8.3.46) we have

$$\lim_{D\to\infty}\frac{D}{r_-} = \cos\vartheta, \tag{8.3.47}$$

$$\lim_{D\to\infty} r_- - R = z_p\cos\vartheta - \sin\vartheta(x_p\cos\phi + y_p\sin\phi). \tag{8.3.48}$$

9 There is a small typographical error in the corresponding equation of **[225]**. Our Eq. (8.3.45) corrects this misprint.

Furthermore, since

$$\frac{D}{R}dxdy = \frac{D^2}{\cos^2\vartheta}\sin\vartheta d\vartheta d\phi,$$

the term proportional to $1/r_- \to 0$ as $D \to \infty$, so it follows from Eqs. (8.3.40) and (8.3.41) and Eqs. (8.3.47) and (8.3.48) that, with reference to Fig. 8.6

$$\mathbf{E}(\mathbf{r}_p) = -\frac{if}{\lambda}\exp(ikC)\int_0^{\alpha}\int_0^{2\pi}\mathcal{E}_0(\vartheta,\phi)\exp(-ikW)\exp\{-ik[\sin\vartheta(x_p\cos\phi + y_p\sin\phi) - z_p\cos\vartheta]\}\sin\vartheta d\vartheta d\phi. \tag{8.3.49}$$

This formula maybe re-written in the form

$$\mathbf{E}(\mathbf{r}_p) = -\frac{if}{\lambda}\exp(ikC)\iint_{\hat{s}_x^2+\hat{s}_y^2\le 1}\frac{\mathcal{E}_0(\hat{s}_x,\hat{s}_y)}{\hat{s}_z}\exp(-ikW)\exp(ik\hat{\mathbf{s}}\cdot\mathbf{r}_p)d\hat{s}_x d\hat{s}_y, \tag{8.3.50}$$

where $\hat{\mathbf{s}}$ is given by Eq. (8.3.18). If we assume that the optical system introduces no aberration, then we may set $W = 0$ which leads to

$$\mathbf{E}(\mathbf{r}_p) = -\frac{if}{\lambda}\exp(ikC)\iint_{\hat{s}_x^2+\hat{s}_y^2\le 1}\frac{\mathcal{E}_0(\hat{s}_x,\hat{s}_y)}{\hat{s}_z}\exp(ik\hat{\mathbf{s}}\cdot\mathbf{r}_p)d\hat{s}_x d\hat{s}_y \tag{8.3.51}$$

and

$$\mathbf{H}(\mathbf{r}_p) = -\frac{if}{\lambda}\exp(ikC)\iint_{\hat{s}_x^2+\hat{s}_y^2\le 1}\frac{\mathcal{H}_0(\hat{s}_x,\hat{s}_y)}{\hat{s}_z}\exp(ik\hat{\mathbf{s}}\cdot\mathbf{r}_p)d\hat{s}_x d\hat{s}_y. \tag{8.3.52}$$

Note that Eq. (8.3.51) is identical, apart from a constant phase term, to Eq. (8.3.25).

The formulation of Eq. (8.3.51) reveals why the integral is such a powerful formula of physical optics. Practically all high-numerical-aperture lenses are constructed in such a way that the aperture stop is placed in the front focal plane of the lens system. Consequently, such a lens system is telecentric from the image side or, in other words, the aperture stop appears at infinity if viewed from the image side focus. If we say that the plane on which we specified the electric vector of the geometrical optics approximation corresponds to the plane of the aperture we immediately see the relevance and potential use of this diffraction integral.

The following general statement is therefore justified: *Let us specify the electric vector of the geometrical optics approximation on the surface of a sphere centred on the Gaussian focal point in image space. Our results show that, providing the Fresnel number of the lens is sufficiently high, the electric field in the focal region of this lens is given by the integral of Eq. (8.3.51).* The way in which we may specify the electric vector of the geometrical optics approximation on the spherical surface is described in Chapter 11 and is referred to as the generalised Jones matrix formalism.

8.4 Scalar Diffraction Theories

In the previous sections we have characterised the field by a vector quantity. Traditionally and indeed in many practically important cases it is sufficient to treat the field as a scalar quantity. We then speak of scalar theories of diffraction. The criteria which determine when vectorial diffraction theories should be used are discussed in Chapter 11. The aim of the present section is to derive the scalar diffraction theories for electromagnetic waves from the electromagnetic theories derived previously. These scalar theories are also applicable for diffraction of scalar waves and can alternatively be derived directly from the scalar wave equation.

8.4.1 The Fresnel–Kirchhoff Integral

We start by recalling Eq. (8.1.61), which is formally the scalar integral theorem of Helmholtz and Kirchhoff, but applies to one of the Cartesian coordinate components of the field, now denoted by U. The integration in this equation is still performed over a closed surface. Equation (8.2.12), which is formally identical to Eq. (8.1.61) and is derived from

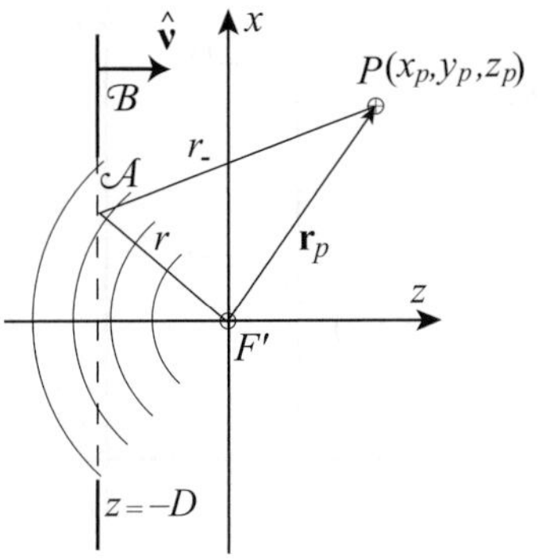

Figure 8.8: The derivation of the Kirchhoff integral.

Eq. (8.1.56), excludes the surface S_2 from the integration. It is therefore of a suitable form from which to start our present derivation. Before we start we make a change in notation here due to historical reasons: we shall use the *inward* surface normal instead of the outer one used in previous sections. One component U of the electric field $\mathbf{E}$ is therefore given from Eq. (8.2.12) as

$$U(\mathbf{r}_p) = \frac{1}{4\pi}\iint_{S_1}\left[U\frac{\partial G}{\partial v} - G\frac{\partial U}{\partial v}\right] dA, \tag{8.4.1}$$

where we note that U also has the dimension of V/m. In order to be able to solve Eq. (8.4.1) we have to impose a set of boundary conditions. We shall assume that, in the presence of the screen, the field and its normal derivative do not appreciably differ from that when the screen is absent:

$$\begin{aligned} U &= U^{(i)}; \ \partial_v U = \partial_v U^{(i)} \quad \text{on } \mathcal{A}, \\ U &= 0; \quad \partial_v U = 0 \qquad \text{on } \mathcal{B}, \end{aligned} \tag{8.4.2}$$

where $\mathcal{A}$ denotes the aperture opening, $\mathcal{B}$ is the opaque portion of the screen and $U^{(i)}$ denotes the incident field in the absence of the screen. The formulas given by Eq. (8.4.2) are usually called the Kirchhoff boundary conditions. With the assumptions of Eq. (8.4.2), Eq. (8.4.1) now reads

$$U(\mathbf{r}_p) = \frac{1}{4\pi}\iint_{\mathcal{A}}\left[U\frac{\partial G}{\partial v} - G\frac{\partial U}{\partial v}\right] dA. \tag{8.4.3}$$

Let us consider a spherical wave, converging towards a point C, incident on a plane screen at $z = -D$ (see Fig. 8.8). The Gaussian image point, which coincides with the back focal point F', is taken as the origin of a Cartesian coordinate system. Hence the incident field is given by

$$U = U_0 f\frac{\exp(-ikr)}{r}, \tag{8.4.4}$$

where[10]

$$r = \sqrt{x^2 + y^2 + z^2} \tag{8.4.5}$$

is the distance from F' and an arbitrary point (x, y, z) in the image space, $z \geq 0$, and U_0 is the strength of the wave at a distance f from F'. The Green's function is given by

$$G = \frac{\exp(ikr_-)}{r_-}, \tag{8.4.6}$$

[10] Clearly in the above formula U_0 must have field dimensions (V/m).

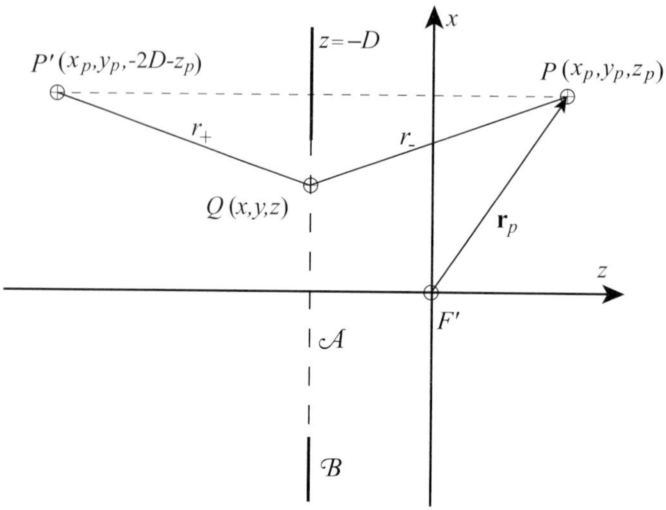

Figure 8.9: The derivation of the Rayleigh–Sommerfeld integrals.

where

$$r_- = \sqrt{(x - x_p)^2 + (y - y_p)^2 + (z - z_p)^2} \tag{8.4.7}$$

is the distance between the observation point $P = (x_p, y_p, z_p)$ and an arbitrary point (x, y, z) in the image space. To determine the field we shall need to calculate

$$\left.\frac{\partial U}{\partial v}\right|_{z=-D} = \left.\frac{\partial U}{\partial z}\right|_{z=-D} = \frac{\exp(-ikr)}{r}\frac{D}{r}\left(ik + \frac{1}{r}\right)U_0 f\,, \tag{8.4.8}$$

$$\left.\frac{\partial G}{\partial v}\right|_{z=-D} = \left.\frac{\partial G}{\partial z}\right|_{z=-D} = -\frac{\exp(ikr_-)}{r_-}\frac{D + z_p}{r_-}\left(ik - \frac{1}{r_-}\right). \tag{8.4.9}$$

In the above equation $\partial_v f = \partial_z f$, follows since the aperture is located on a plane perpendicular to the z axis. In usual optical systems the physical dimensions of the aperture and distances are much greater than the wavelength so we can safely ignore the $1/r$ and $1/r_-$ terms in Eqs. (8.4.8) and (8.4.9). Hence we have the following result:

$$U(\mathbf{r}_p) = -\frac{if}{2\lambda}\iint_{\mathcal{A}} U_0 \frac{\exp[ik(r_- - r)]}{rr_-}\left(\frac{D}{r} + \frac{D + z_p}{r_-}\right) dA\,. \tag{8.4.10}$$

Equation (8.4.10) is usually referred to as the *Fresnel–Kirchhoff integral.*

8.4.2 The Rayleigh–Sommerfeld Integrals

Equation (8.4.10), though experimentally verified for small convergence angles, leads to mathematical inconsistencies[11] in the Kirchhoff boundary conditions (Eq. (8.4.2)), which arise from the fact that it is not possible to specify arbitrarily both the field and its normal derivative on the surface, as discussed by Sommerfeld **[322]**. These inconsistencies can be reconciled, just as we did in Section 8.1.4, by defining two different Green's functions (see Fig. 8.9):

$$G^{\pm} = \frac{\exp(ikr_-)}{r_-} \pm \frac{\exp(ikr_+)}{r_+}\,, \tag{8.4.11}$$

where

$$r_+ = \sqrt{(x - x_p)^2 + (y - y_p)^2 + (z + z_p + 2D)^2}\,. \tag{8.4.12}$$

[11] Note that by using a different set of boundary conditions, as shown by Mandel and Wolf **[232]**, it is possible to make the Kirchhoff integral consistent.

Just as before we take the point of observation P and its mirror image P' on the plane screen. The singularity at P, i.e. for $r_- = 0$, has already been removed when deriving Eq. (8.1.56). The singularity at P', i.e. for $r_+ = 0$, is irrelevant because it is on the left-hand side of the screen, and hence P' is excluded from the integration.

It now remains to calculate the normal derivatives of the Green's functions:

$$\left.\frac{\partial G^+}{\partial z}\right|_{z=-D} = 0\,, \tag{8.4.13}$$

$$\left.\frac{\partial G^-}{\partial z}\right|_{z=-D} = -2\frac{\exp(ikr_-)}{r_-}\frac{D+z_p}{r_-}\left.\left(ik - \frac{1}{r_-}\right)\right|_{z=-D}\,. \tag{8.4.14}$$

The significance of the choice of the two Green's functions Eq. (8.4.11) should now be clear. While G^+ has a finite value on the screen and aperture, its normal derivative vanishes at $z = -D$. G^- has a vanishing value on the screen and aperture and a finite normal derivative at $z = -D$. Thus for G^-, the second term of Eq. (8.4.1) vanishes on the screen, and if the amplitude on the screen can be assumed to be zero then the contribution from $\mathcal{B}$ is zero. This corresponds to the physical case of a perfectly hard screen. Alternatively, for G^+, the first term of Eq. (8.4.1) vanishes on the screen, and the contribution from $\mathcal{B}$ is zero if the normal derivative of the field can be assumed zero on the screen. This is called a perfectly soft screen. Hence we have the resulting expressions for the field:

$$U^-(\mathbf{r}_p) = -\frac{if}{\lambda}\iint_{\mathcal{A}} U_0\frac{\exp[ik(r_- - r)]}{rr_-}\frac{D+z_p}{r_-}dA\,, \tag{8.4.15}$$

$$U^+(\mathbf{r}_p) = -\frac{if}{\lambda}\iint_{\mathcal{A}} U_0\frac{\exp[ik(r_- - r)]}{rr_-}\frac{D}{r}dA\,, \tag{8.4.16}$$

where we have again ignored the $1/r_-$ term *in comparison with* k. Equations (8.4.15) and (8.4.16) are usually referred to as the *first and second scalar Rayleigh–Sommerfeld integrals*, respectively.

We note that the arithmetic mean of the first and second Rayleigh–Sommerfeld integrals is the Fresnel–Kirchhoff integral:

$$U = \frac{1}{2}(U^+ + U^-)\,. \tag{8.4.17}$$

8.4.3 Approximations of the Exact Formula – the Fresnel and Fraunhofer Integrals

In Eqs. (8.4.15) and (8.4.16) the value of either $(D + z_p)/r_-$ or D/r, frequently referred to as the *inclination term* or *obliquity factor*, often does not change significantly, because the lateral coordinates of both P and F' are usually much smaller than their respective z coordinates, so we may then write:

$$U(\mathbf{r}_p) = U^+(\mathbf{r}_p) = U^-(\mathbf{r}_p) = -\frac{if}{\lambda}\iint_{\mathcal{A}} U_0\frac{\exp[ik(r_- - r)]}{rr_-}dA\,. \tag{8.4.18}$$

It is possible to obtain a more general form of Eq. (8.4.18), called Collins' integral **[73]**, by allowing for an arbitrary incident field. We now shift origin of the coordinate system to the aperture plane, a step which is not strictly necessary but will simplify notation without compromising generality. Considering Eq. (8.4.4) we therefore have

$$U(\mathbf{r}_p) = -\frac{i}{\lambda}\iint_{\mathcal{A}} U(\mathbf{r})\frac{\exp(ikr_-)}{r_-}dA\,. \tag{8.4.19}$$

The above formula is often used for numerical computations of the field. In order to gauge the level of difficulty one faces let us first consider the phase term $\exp(ikr_-) = \cos(kr_-) + i\sin(kr_-)$. The problem with this term is that since k is typically on the order of 10^7 m^{-1} even small changes in the value of r_- will introduce oscillations in the function value, hence rendering the integral almost impossible to numerically compute. However, if we use the Taylor expansion of r_- we might be able to simplify the phase term:

$$\begin{aligned} r_- &= z_p\sqrt{1 + \frac{(x-x_p)^2 + (y-y_p)^2}{z_p^2}} \approx z\left(1 + \frac{(x-x_p)^2 + (y-y_p)^2}{2z_p^2}\right) \\ &= z_p + \frac{(x-x_p)^2 + (y-y_p)^2}{2z_p} = z_p + \frac{x_p^2 + y_p^2}{2z_p} + \frac{x^2+y^2}{2z_p} - \frac{xx_p + yy_p}{z_p}\,. \end{aligned} \tag{8.4.20}$$

Now, in terms of the integration that is carried out over (x, y), the first two terms are constants. The third and the last terms, however, need to be included in the integral. A further approximation is also admissible which is in fact implicit in making the inclination factor constant: if $D \gg \lambda$ we may replace the amplitude term r_-^{-1} in the integrand by z_p^{-1} which then becomes:

$$U(\mathbf{r}_p) = -\frac{i\exp(ikz_p)}{z_p\lambda}\exp\left(ik\frac{x_p^2+y_p^2}{2z_p}\right)\iint_{-\infty}^{\infty} U(x,y)\exp\left(ik\frac{x^2+y^2}{2z_p}\right)\exp\left(-ik\frac{xx_p+yy_p}{z_p}\right)dxdy, \tag{8.4.21}$$

where we have replaced the integration on the screen $\mathcal{A}$ with one for the entire plane by noting that the limitation in the spatial extent of $\mathcal{A}$ is now lumped into the definition of $U(x, y)$. Equation (8.4.21) is known as the *Fresnel diffraction integral*. An alternative way of writing this equation is:

$$U(\mathbf{r}_p) = -\frac{i\exp(ikz_p)}{z_p\lambda}\exp\left(ik\frac{x_p^2+y_p^2}{2z_p}\right)\tilde{U}(m,n)\Big|_{m=\frac{x_p}{z_p\lambda},\,n=\frac{y_p}{z_p\lambda}}, \tag{8.4.22}$$

where

$$\tilde{U}(m,n) = \mathcal{F}\left\{U(x,y)\exp\left[\frac{ik}{2z}\left(x^2+y^2\right)\right]\right\},$$

and $\mathcal{F}$ denotes the two-dimensional Fourier transform. In certain cases when the integration is confined to a small area of the $z = 0$ plane, it is reasonable to assume that the term $x^2 + y^2$ will fall off faster than the $xx_p + yy_p$ term so the first exponential in Eq. (8.4.21) may be ignored in comparison with the second exponential to give

$$U(\mathbf{r}_p) = -\frac{i\exp(ikz_p)}{z_p\lambda}\exp\left(ik\frac{x_p^2+y_p^2}{2z_p}\right)\iint_{-\infty}^{\infty} U(x,y)\exp\left[-\frac{ik}{z_p}(xx_p+yy_p)\right]dxdy. \tag{8.4.23}$$

The approximation that allowed us to ignore the quadratic phase term is called the *Fraunhofer approximation* and hence Eq. (8.4.23) is called the *Fraunhofer diffraction integral*.

An alternative form of the Fraunhofer diffraction integral is the following:

$$\begin{aligned} U(\mathbf{r}_p) &= -\frac{i\exp(ikz_p)}{z_p\lambda}\exp\left(ik\frac{x_p^2+y_p^2}{2z_p}\right)\iint_{-\infty}^{\infty} U(x,y)\exp\left[-2\pi i\left(\frac{x_p}{z_p\lambda}x+\frac{y_p}{z_p\lambda}y\right)\right]dxdy \\ &= -\frac{i\exp(ikz_p)}{z_p\lambda}\exp\left(ik\frac{x_p^2+y_p^2}{2z_p}\right)\tilde{U}(m,n)\Big|_{m=\frac{x_p}{z_p\lambda},\,n=\frac{y_p}{z_p\lambda}}, \end{aligned} \tag{8.4.24}$$

where $\tilde{U}(m, n)$ is the Fourier transform of $U(x, y)$. The result shows that the Fraunhofer distribution is essentially the scaled Fourier transform of the initial field distribution at the aperture plane $z = 0$ with the scaling factor being $1/z_p\lambda$. There is a pre-multiplier factor which is a quadratic phase term and it is also noted that the field decays in a linear manner as the distance of the observation plane is increased.

It is worth inspecting how the approximation to the phase term changed the integral of Eq. (8.4.22) The original wavefront was spherical ($\exp(ikr)/r$) as shown in Fig. 8.10. The Fresnel approximation is equivalent to taking the first two terms of the Taylor expansion, which results in a quadratic phase function as also shown in the figure. The Fraunhofer approximation, however, only keeps the first term of the Taylor expansion and hence it is approximating the original spherical wave by a single plane wave. The quality of the approximation (i.e. how close the approximate wavefront is to the original spherical wavefront) depends on at what angle the wavefronts are observed and how far the observation point is from the optic axis. The error associated with the former case is depicted in Fig. 8.10 where Δ_{Fres} and Δ_{Fraun} indicate the angle-dependent wavefront error for the Fresnel and Fraunhofer approximations respectively.

It is interesting to re-write the Fresnel diffraction integral Eq. (8.4.23) in a slightly different format to give:

$$U(\mathbf{r}_p) = -\frac{i\exp(ikz_p)}{z_p\lambda}\iint_{-\infty}^{\infty} U(x,y)\exp\left\{\frac{ik}{2z_p}\left[(x_p-x)^2+(y_p-y)^2\right]\right\}dxdy, \tag{8.4.25}$$

or if we let

$$h(\alpha,\beta) = \exp\left[\frac{ik}{2z}(\alpha^2+\beta^2)\right],$$

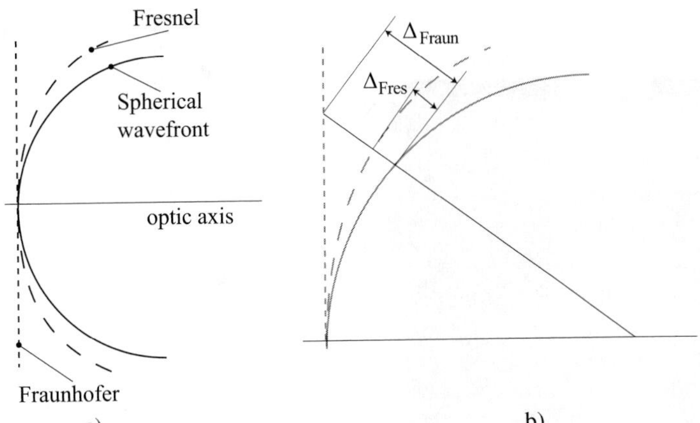

a) b)

Figure 8.10: a) The Fresnel and Fraunhofer approximations.
b) The accuracy of each of the approximations.

then we have

$$U(\mathbf{r}_p) = -\frac{i\exp(ikz_p)}{z_p\lambda}\iint_{-\infty}^{\infty} U(x,y)h(x_p - x, y_p - y)\,dxdy\,. \tag{8.4.26}$$

Equation (8.4.26) shows that the Fresnel diffraction integral may be written as a convolution of the field in the $z = 0$ plane with the quadratic wavefront represented by the function h.

An empirical formula for the distance z_F at which the Fraunhofer approximation applies is given by:

$$z_F > \frac{a^2}{2\lambda},$$

where a is the linear dimension of the aperture. In practice a can be on the order of tens of millimetres in optics with a mean visible light wavelength of 500 nm. This gives a typical distance of 100 m!

In reality both the Fresnel and Fraunhofer diffraction integrals are approximations which are valid at different distances from the aperture. Figure 8.11 shows the evolution of the field after diffraction by a slit and the different regions. It is apparent from the figure that there are no hard boundaries separating the individual regions so the distance from the screen at which a given approximation is applicable is somewhat arbitrary.

8.4.4 The Scalar Debye Integral

Returning now to the Fresnel–Kirchhoff integral, Eq. (8.4.10), the integration on the plane screen $\mathcal{A}$ may be replaced by integration over a spherical surface. This can be accomplished by considering that on the surface of the sphere that is centred on the origin of the coordinate system, the phase due to an incident convergent and spherical wavefront is constant. Hence, we need to replace the

$$\left.\frac{\exp(-ikr)}{r}\right|_{z=0}$$

term by

$$\frac{\exp(-ikf)}{f},$$

where f is the distance from the origin to the rim of the aperture. Hence we obtain from Eq. (8.4.18):

$$U(\mathbf{r}_p) = -\frac{i\exp(-ikf)}{\lambda}\iint_{\mathcal{A}} U_0\frac{\exp(ikr_-)}{r_-}dA\,. \tag{8.4.27}$$

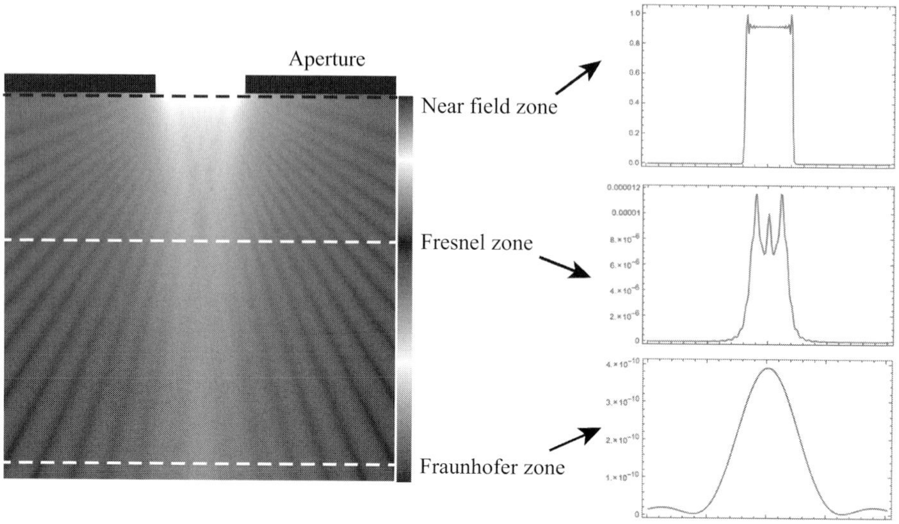

Figure 8.11: Diffraction of a plane wave by a slit aperture with the Fresnel and Fraunhofer regions indicated. Logarithmic plot of the intensity. The line scans (plotted on a linear scale) are taken from the grey-scale distribution at the locations indicated by the dashed lines. The figures are not quantitative, they are only for illustrative purposes.

Points P and Q are characterised by their cylindrical and spherical polar coordinates, respectively, as given by Eq. (8.3.9). A procedure, identical to that presented in Section 8.3, yields

$$U(\mathbf{r}_p) = -\frac{if^2 \exp(ik\Phi_0)}{\lambda(f+z_p)} \iint_{\hat{s}_x^2+\hat{s}_y^2 \leq 1} U_0 \exp(ik\,\hat{\mathbf{s}}\cdot\mathbf{r}') \frac{d\hat{s}_x\, d\hat{s}_y}{\hat{s}_z}, \tag{8.4.28}$$

where the definitions for Φ_0, $\hat{\mathbf{s}}$ and $\mathbf{r}'$ are given by Eqs. (8.3.12), (8.3.18) and (8.3.17) respectively. We term Eq. (8.4.28) the *scaled Debye integral*. Note that the validity of this formula is identical to that of Eq. (8.3.21).

If we restrict our treatment of the scaled Debye integral to large Fresnel numbers then following exactly the same procedure that resulted in Eq. (8.3.25) from Eq. (8.3.21) yields:

$$U(\mathbf{r}_p) = -\frac{if}{\lambda} \iint_{\hat{s}_x^2+\hat{s}_y^2 \leq 1} U_0 \exp(ik\,\hat{\mathbf{s}}\cdot\mathbf{r}_p) \frac{d\hat{s}_x\, d\hat{s}_y}{\hat{s}_z}. \tag{8.4.29}$$

Equation (8.4.29) is of basic importance in physical optics and is usually referred to as the *scalar Debye integral*.

We now continue by obtaining an integral expression for an optical system of low convergence angle, but arbitrary Fresnel number. For this we begin with Eq. (8.4.28) which, with reference to Fig. 8.6, can be written in the form:

$$U(\mathbf{r}_p) = -\frac{if^2 \exp(ik\Phi_0)}{\lambda(f+z_p)} \times \int_0^{\alpha}\int_0^{2\pi} U_0 \exp\left[-ik\frac{f}{f+z_p} r_p \sin\vartheta \cos(\phi-\gamma)\right] \exp\left[ik\frac{f}{f+z_p} z_p \cos\vartheta\right] \sin\vartheta\, d\vartheta\, d\phi, \tag{8.4.30}$$

where (r_p, γ, z_p) are the cylindrical coordinates of the observation point (the distance measured from the Gaussian focus). Let us now assume that the marginal ray angle α is small. Putting

$$\rho_Q = f \sin\vartheta, \tag{8.4.31}$$

we then have

$$\cos\vartheta \approx 1 - \frac{\rho_Q^2}{2f^2}, \tag{8.4.32}$$

and

$$d\rho_Q \approx f d\vartheta, \tag{8.4.33}$$

so if the radius of the aperture is a, Eq. (8.4.30) gives

$$U(\mathbf{r}_p) = -\frac{i\exp(ik\Phi_0')}{\lambda(f+z_p)}\int_0^a\int_0^{2\pi} U_0 \exp\left[\frac{-ik}{f}\frac{f}{f+z_p}r_p\rho_Q\cos(\phi-\gamma)\right]\exp\left[-\frac{ik}{2f^2}\frac{f}{f+z_p}z_p\rho_Q^2\right]d\phi\,\rho_Q d\rho_Q, \tag{8.4.34}$$

where

$$\Phi_0' = f + z_p + \frac{r_p^2}{2(f+z_p)}. \tag{8.4.35}$$

This expression is equivalent to the Li and Wolf formula **[210]** which was derived using an expression of Erkkila and Rogers **[93]**, who applied a parabolic approximation in their expressions for the spherical wavefront. If we now assume that $z_p \ll f$ we obtain:

$$U(\mathbf{r}_p) = -\frac{i}{f\lambda}\exp(ikz_p)\exp\left(\frac{ikr_p^2}{2f}\right)\int_0^a\int_0^{2\pi} U_0\exp\left[-\frac{ik}{f}r_p\rho_Q\cos(\phi-\gamma)\right]\exp\left(-\frac{ik}{2f^2}z_p\rho_Q^2\right)d\phi\,\rho_Q d\rho_Q, \tag{8.4.36}$$

where we have omitted the distance f in the phase term Φ_0' by taking the nominal focal point as the new phase reference point. Taking U_0 as constant and using the identity **[360]**

$$\int_0^{2\pi}\exp(in\phi)\exp\left[\pm i\rho\cos(\phi-\gamma)\right]d\phi = 2\pi(\pm i)^n J_n(\rho)\exp(in\gamma), \tag{8.4.37}$$

where J_n is the Bessel function of the first kind, order n, we have

$$U(\mathbf{r}_p) = -\frac{ikU_0}{f}\exp(ikz_p)\exp\left(\frac{ikr_p^2}{2f}\right)\int_0^a J_0\left(\frac{kr_p\rho_Q}{f}\right)\exp\left[-\frac{ik}{2f^2}z_p\rho_Q^2\right]\rho_Q d\rho_Q. \tag{8.4.38}$$

Equation (8.4.38) is known as the *diffraction integral of Lommel*. It was derived assuming an optical system of low numerical aperture and for an observation point sufficiently close to the focal plane. The condition $z_p \ll f$ is equivalent to the condition

$$2\pi N \gg kz_p\sin^2\alpha, \tag{8.4.39}$$

that is it is satisfied if the Fresnel number is large enough. The Lommel diffraction integral is thus applicable to optical systems with low numerical aperture and high Fresnel number. Since this integral is of practical importance and serves as a basis for paraxial imaging theory, we shall now look at two special cases.

8.4.4.1 The On-axis Case ($r_p = 0$)

When $r_p = 0$, $J_0(0) = 1$ and $\exp(0) = 1$, which permits us to re-write Eq. (8.4.38) as

$$U_f(0, z_p) = -\frac{iU_0k\exp(ikz_p)}{f}\int_0^a \exp\left(-\frac{ikz_p}{2f^2}\rho_Q^2\right)\rho_Q d\rho_Q. \tag{8.4.40}$$

Perhaps the simplest way to integrate the above expression is to substitute

$$\rho_Q^2 = x, \quad \rightarrow \quad \rho_Q = \sqrt{x}, \quad d\rho_Q = \frac{1}{2}\frac{dx}{\sqrt{x}}, \quad \rho_Q d\rho_Q = \frac{1}{2}dx,$$

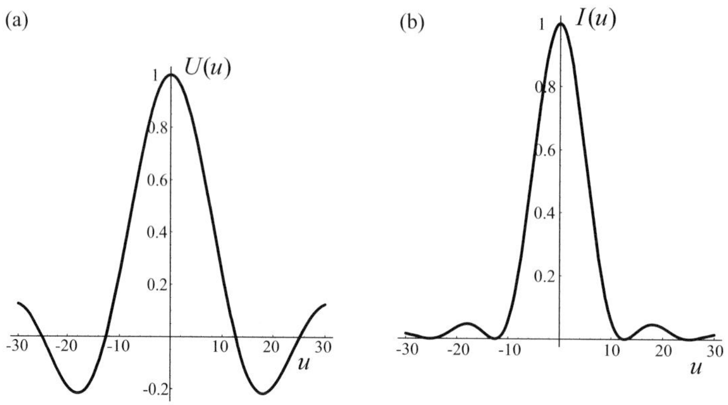

Figure 8.12: The normalised axial field a) and intensity b) distribution in the vicinity of the focus of a thin lens.

so that the field distribution on the optic axis is given by:

$$U(0, z_p) = -\frac{iU_0 k \exp(ikz_p)}{2f} \int_0^{a^2} \exp\left(-\frac{ik}{2f^2} x z_p\right) dx = \frac{U_0 f}{z_p} \exp(ikz_p) \left[\exp\left(-i\frac{kz_p}{2f^2}a^2\right) - 1\right].$$

The intensity distribution is thus given by

$$I(0, z_p) = |U(0, z_p)|^2 = \frac{I_0 f^2}{z_p^2} \left|\exp\left(-i\frac{kz_p}{2f^2}a^2\right) - 1\right|^2 ,$$

where $I_0 = |U_0|^2$. The former equation can be re-written using Euler's identify $\exp(ia) = \cos(a) + i\sin(a)$ as

$$I(0, z_p) = I_0 \left(\frac{a^2 k}{2f}\right)^2 \left[\frac{\sin(u/4)}{u/4}\right]^2 = I_0 \left(\frac{a^2 k}{2f}\right)^2 \operatorname{sinc}^2\left(\frac{u}{4}\right), \tag{8.4.41}$$

where we have introduced the *normalised axial optical co-ordinate*

$$u = k\left(\frac{a}{f}\right)^2 z_p = k(NA)^2 z_p , \tag{8.4.42}$$

where $NA = a/f = \tan\alpha \approx \sin\alpha$ is the (paraxial) *numerical aperture* (*NA*) of the lens. The axial field and intensity distributions are shown in Fig. 8.12.

The *positive* zeros of this distribution occur at

$$u/4 = m\pi , \qquad m = 1, 2, \ldots ,$$

which, by using Eq. (8.4.42), can be solved for z_p to give

$$z_p = \frac{2m\lambda}{(NA)^2} , \qquad m = 1, 2, \ldots$$

In particular, for the first zero $m = 1$ and so

$$z_p = \frac{2\lambda}{(NA)^2} .$$

This is of course just the axial position of the first positive zero. If we wish to estimate the *depth of field* of the lens (i.e. the width of the central region within which the intensity does not fall to zero) we need to take into account the distances before and after the focus, hence we need to double this length to give

$$z_{\text{DOF}} = \frac{4\lambda}{(NA)^2} . \tag{8.4.43}$$

This expression is worth remembering by heart as it serves as an important estimation rule in practical optics.

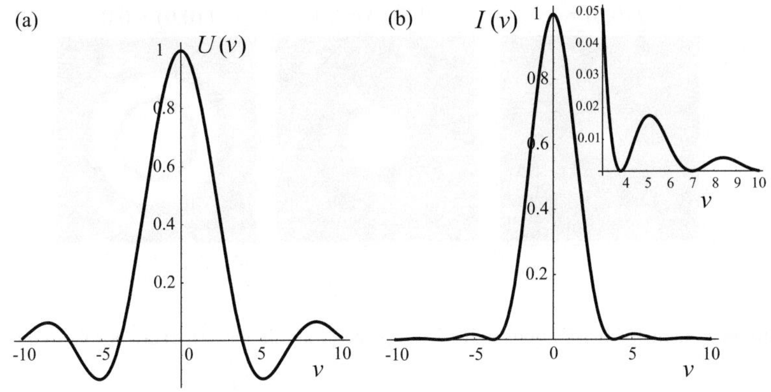

Figure 8.13: The normalised transverse field a) and intensity b) distributions in the focal plane of a thin lens. Insets shows details of sidelobes.

We remind the reader of the other quantity δz_f that was introduced in Chapter 5, Eq. (5.2.28), the *Rayleigh focal depth*. The distance δz_f is a small shift from the nominal image plane such that the maximum intensity of the point-spread function of an ideal imaging system hardly changes (20%). It is seen that for a perfect imaging system the DOF value is larger by a factor of 8 than the Rayleigh focal depth of the same system.

8.4.4.2 The In-focus Case ($z_p = 0$)

When $z_p = 0$ we can directly write the field distribution in the focal plane from Eq. (8.4.38),

$$\begin{aligned} U(r_p, 0) &= -\frac{ikU_0}{f}\exp\left(ik\frac{r_p^2}{2f}\right)\int_0^a J_0\left(\frac{kr_p\rho_Q}{f}\right)\rho_Q d\rho_Q \\ &= -\frac{ikU_0}{f}\exp\left(ik\frac{r_p^2}{2f}\right)a^2\frac{J_1(kr_pa/f)}{kr_pa/f} \\ &= -\frac{ikU_0}{2f}\exp\left(ik\frac{r_p^2}{2f}\right)a^2\left(\frac{2J_1(v)}{v}\right), \end{aligned} \tag{8.4.44}$$

where we have introduced the *transverse normalised optical co-ordinate*, v:

$$v = k\left(\frac{a}{f}\right)r_p = k\,NA\,r_p\,. \tag{8.4.45}$$

The intensity distribution is now a simple matter to express as

$$I(r_p, 0) = |U(r_p, 0)|^2 = I_0\left(\frac{ka^2}{2f}\right)^2\left(\frac{2J_1(v)}{v}\right)^2. \tag{8.4.46}$$

Line traces of the transverse (or lateral) field and intensity distributions are shown in Fig. 8.13 with the insert showing the side-lobe structure of the intensity. Figure 8.14 shows the intensity distributions in the focal plane with varying degree of saturation. In a) the shading scale is normalised to 1, in b) it is normalised to 0.1 and in c) to 0.01. The saturation reveals the fringe structure around the central main peak.

In order to work out the width of the central peak of the distribution we start by stating that the first zero of the distribution occurs at $v = 3.832$. We can solve Eq. (8.4.45) for r_p by writing:

$$r_p = \frac{3.832}{2\pi}\frac{\lambda}{NA} \approx 0.61\frac{\lambda}{NA}\,. \tag{8.4.47}$$

Therefore the width of the centre intensity distribution is given by

$$r_p = 1.22\frac{\lambda}{NA}\,. \tag{8.4.48}$$

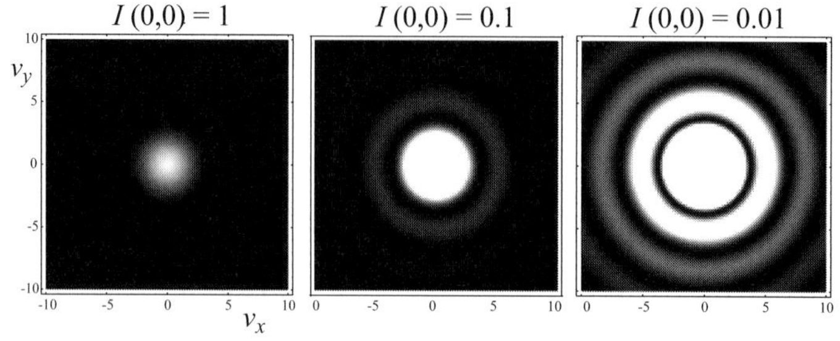

Figure 8.14: The transverse intensity distributions in the focal plane of a thin lens. a)–c) are normalised to 1, 0.1 and 0.01 in the centre, respectively.

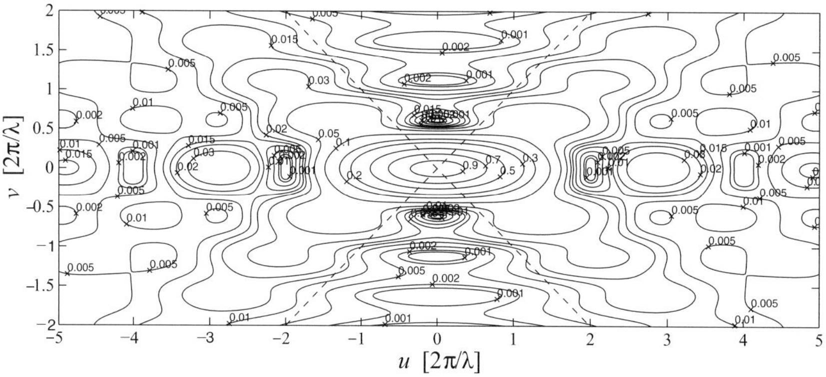

Figure 8.15: The intensity distribution in the x–z plane in the focal region of a thin lens.

This formula should also be remembered as one of the basic rules of thumb of physical optics. For sake of completeness, we also show the distribution of the intensity in the x–z plane in Fig. 8.15

The important conclusion from the above two special cases is that by using normalised optical co-ordinates v and u it is possible to write the Lommel integral as

$$U(v, u) = U_0 C \int_0^1 J_0(vr) \exp\left(-\frac{iu}{2} r^2\right) r \, dr , \tag{8.4.49}$$

where $r = \rho / a$ and $C = -ika^2 \exp(ikr_p^2/2f)/f$. This means that the field distribution does not change in shape when the NA is changed; the distribution merely scales with the numerical aperture. This is the same behaviour we have already observed in connection with the Fraunhofer diffraction integral.

8.4.5 Interrelation of Optical Diffraction Theories

Before proceeding to discuss certain details associated with results of the previous sections it is worthwhile to stop and summarise what has been found so far. The derivations given in Sections 8.1–8.4 present a logically consistent formulation of the most commonly used diffraction theories of physical optics. An overview of these theories is given in Table 8.1.

As shown in the table our starting point was the Stratton–Chu integral theorem that was obtained, without imposing approximations, from Maxwell's equations. From this, by considering either vector or scalar quantities, we obtained the vectorial integral theorem of Kirchhoff or the integral theorem of Helmholtz and Kirchhoff. All of these theorems are valid over any closed surface. For an aperture in a plane screen, we argued that the contribution from the infinite hemisphere needed to close the surface can be neglected. Starting from the Stratton–Chu theory and assuming an infinite plane screen we derived the e and m theories, and the first and second vectorial Rayleigh–Sommerfeld integrals.

Table 8.1. Interrelation of various focusing theories in physical optics.

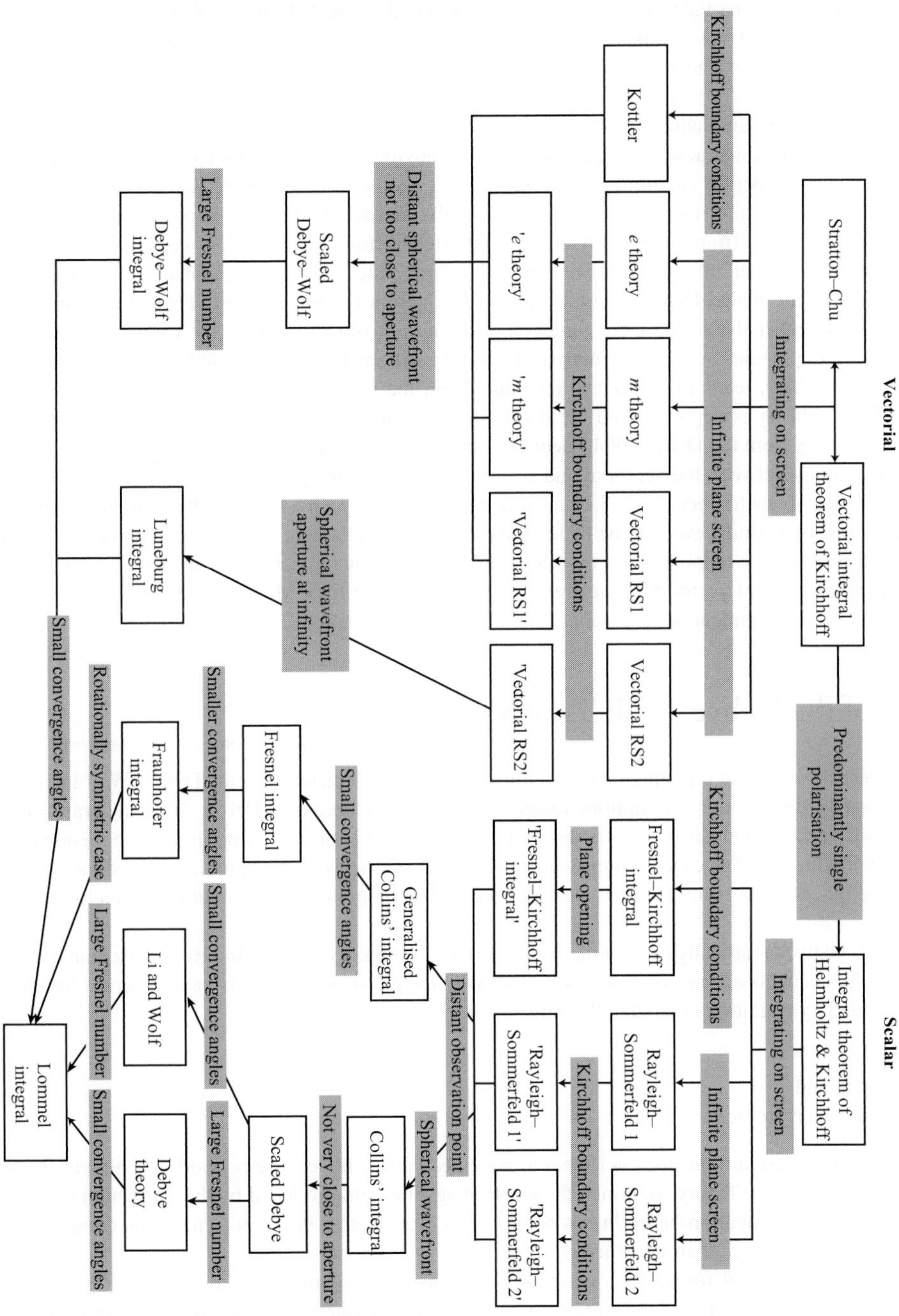

Using these theories we derived alternative expressions after imposing different boundary conditions. The expressions obtained were identical to their respective original theories with the difference that now the range of integration was confined to the aperture opening, but in some cases with an additional contour integral around the aperture edge. In physical optics the numerical value of the contour integral is very small and hence the contour integral can usually be ignored in comparison with the other terms. From the first vectorial Rayleigh–Sommerfeld theory we obtained Luneburg's integral.

From the Kottler integral, after ignoring the contour integral and assuming a spherical wavefront incident from infinity and restricting the observation point to be far from the aperture, we obtained the scaled Debye–Wolf theory. From the scaled Debye–Wolf theory we obtained, after imposing the large Fresnel number condition, the Debye–Wolf integral which was shown to be identical to the Luneburg integral. Both of these theories yield the Lommel integral for small angles of convergence.

The scalar theories were derived in an analogous manner. The Helmholtz–Kirchhoff integral theorem results in the first and second Rayleigh–Sommerfeld diffraction integrals, after first removing the contribution from the hemispherical portion of the screen and then considering an infinite plane screen and aperture. The Fresnel–Kirchhoff integral for a plane screen is obtained from the scalar integral theorem of Helmholtz and Kirchhoff after imposing the Kirchhoff boundary conditions. From either this or from the first Rayleigh–Sommerfeld diffraction integral, by considering a distant observation point and a convergent spherical wavefront we obtained Collins' integral. An approximate form of Collins' integral was obtained and called the scaled Debye integral. This representation is valid for finite Fresnel numbers and high-numerical apertures. From either the scaled Debye integral or from the Collins' integral, the Li and Wolf theory can be obtained, by imposing the condition of small numerical aperture. Alternatively, the scaled Debye theorem, on imposing very high Fresnel numbers, leads to the Debye integral. When the Li and Wolf theory is restricted to high Fresnel numbers, or the Debye theory is approximated for small numerical apertures, we obtain the Lommel integral. This may also be obtained directly from the Debye–Wolf theory on setting the numerical aperture small.

Finally we note that an analogy between the vectorial and scalar theories is quite apparent from our derivation: it is possible to relate certain theories to each other. Table 8.1 has been constructed in such a way that the complexity of the different theories is represented by various levels in the table. The most obvious example is the scaled Debye–Wolf and Debye theories: the scalar theory may be obtained from the vectorial counterpart by replacing the incident vector field by a scalar quantity. Whilst for the incident vector field we are usually required to satisfy Maxwell's equations, the scalar quantity U must satisfy the time-independent homogeneous (Helmholtz) wave equation.

8.5 The Validity of the Debye–Wolf Theory

There has been much interest shown in the literature on the validity conditions associated with the Debye–Wolf integral. For this reason we now briefly discuss the assumptions under which the Luneburg and Debye–Wolf integrals were derived. However, before we start our analysis it is worth discussing certain aspects of the vectorial focusing theories.[12]

A diffraction theory can provide either an exact or approximate solution to a given physical problem. The same theory, on the other hand and quite independent from this, may provide an exact or an approximate solution to a mathematical problem.

A theory is usually said to be mathematically exact if the solution of a scalar theory satisfies the Helmholtz equation or in the vectorial case Maxwell's equations. The condition associated with the physical problem is more complex because the task is to specify the boundary conditions such that they are:

A1) physically appropriate,
A2) mathematically correct, and
A3) do not overspecify the given problem.

Simultaneously satisfying all these conditions does not seem to have been possible because no work has hitherto been published on the high-aperture vectorial theory for which the electromagnetic field produced by the diffraction integral would be an exact solution of the physical problem. In this sense one is faced with a decision as to which condition(s) should be considered to be more important than the other(s).

Historically, the first full mathematical formulation of the scalar focusing problem is due to Kirchhoff, who used Eq. (8.1.61) with the boundary conditions, Eq. (8.4.2), to obtain the field in the focal region, Eq. (8.4.3). As pointed out by Sommerfeld **[322]** the Kirchhoff boundary conditions, Eq. (8.4.2), lead to certain inconsistencies because, according to Riemann's theorem of functions, if a harmonic function with its normal derivative vanishes on a finite portion of a surface then the function will vanish over the entire plane **[191]**. Furthermore the Fresnel–Kirchhoff integral may be written as half of the sum of the first and second Rayleigh–Sommerfeld integral. Hence a new problem arises because the first Rayleigh–Sommerfeld integral, Eq. (8.4.15) requires a knowledge of the field in the aperture plane and the second Rayleigh–Sommerfeld integral, Eq. (8.4.16) requires knowledge of the normal derivative of the field. This means that the

[12] Some of the discussion presented in this section is based on a private communication from Professor Emil Wolf.

Fresnel–Kirchhoff integral requires knowledge of both the field and its normal derivative, that is to say it overspecifies the boundary values. The last problem of the Fresnel–Kirchhoff integral is that it fails to reproduce the field in the aperture plane as specified by the Kirchhoff boundary conditions so it is, in this sense, mathematically inconsistent. According to what has been said above the Fresnel–Kirchhoff integral violates both conditions 2) and 3). It would, however, be quite incorrect to state that for these reasons the Fresnel–Kirchhoff integral should not be applied to problems of physical optics. Experimental evidence suggests that the Fresnel–Kirchhoff integral gives a fairly accurate description of the focused field, even under extreme conditions, such as low Fresnel number lenses. The success of the Fresnel–Kirchhoff integral lies in the specification of physically appropriate boundary conditions, which, even though mathematically inconsistent, result in a reasonably accurate description.

Having discussed problems associated with the Fresnel–Kirchhoff integral we now turn to those of the Debye–Wolf theory. We have shown that the Debye–Wolf integral Eq. (8.3.25) is an exact solution of Maxwell's equations. This statement also follows from visual inspection of Eq. (8.3.25) by noting that the equation expresses superposition of plane waves.

Having established that the Debye–Wolf theory is mathematically rigorous we now turn to discuss the physical conditions associated with this integral. As stated in Section 8.3 there are three derivations of the formula available. It is not surprising that it turns out, although at least two of the three derivations are formally different, that the boundary conditions specified by the three different solutions are equivalent. To show this we start by analysing the solution given by Luneburg, Eqs. (8.3.27)–(8.3.51). Luneburg's derivation is based on the first vectorial Rayleigh–Sommerfeld integral which not only results in a solution that is a mathematically exact solution of Maxwell's equations, but also the boundary conditions are mathematically consistent, physically admissible and they do not overspecify the problem.

It was shown clearly in the derivation that the first assumption made by Luneburg is removing the plane upon which the boundary value is specified to infinity. This assumption is equivalent to specifying an asymptotic boundary value. Furthermore, Luneburg in his original resulting formula specifies the integration over the surface of a sphere. He, however, in subsequent derivations limits the integration to a solid angle to coincide with the convergence angle of a lens that produces the perfect spherical wave. We may now recognise an important difference between the Fresnel–Kirchhoff (or indeed the Rayleigh–Sommerfeld) and the Luneburg and Debye theories. Whilst the integration for the first group of theories occurs over the plane of aperture opening of the screen, the integration for the Luneburg theory is performed over the angle subtended at the geometric shadow of the aperture, i.e. the angle of the outermost geometric optics rays. It has been argued before **[377, 358, 357]** and experimental evidence shows **[171]** that, far from the aperture opening and the geometric shadow region, the focused field produced by the Debye, and hence the Luneburg, integral looks like a cut-off portion of a spherical wave. It should be clearly recognised that limiting the angle of integration in the Luneburg integral eliminates the possible contribution of the diffracted waves propagating along or close to the geometric optics shadow region to the resulting field. Also, by specifying a large (asymptotic) distance between the aperture opening and the observation point any possible contribution in the diffraction pattern from evanescent waves[13] is eliminated. We may, therefore conclude that Luneburg's derivation consists of three assumptions:

L1) The contribution from evanescent waves is ignored;
L2) The contribution to diffraction from the aperture edge is ignored;
L3) An asymptotic boundary condition is specified (infinite Fresnel number lenses).

It is interesting that Luneburg did not seem to recognise the existence of assumptions L 1) and L 2), but he was not alone. As noted by Wolf and Li **[377]** and stated by Sommerfeld **[322]** (p. 319) in connection with the Debye theory: *He [Debye] ... obtained a solution of the differential equation of optics which is valid in the whole space and exactly describes ... the diffraction patterns in the vicinity of the focal point...* Sommerfeld goes on to state: *Debye's method is not limited to Kirchhoff's approximation but is based on the fundamentals of wave optics. His solution can claim the same degree of exactness as, for instance, our treatment of the problem of the straight edge...* It has been argued before **[377]** that this statement is not only misleading but also incorrect because of the above assumptions associated with Luneburg's derivation.

Since the mathematical forms of the Debye–Wolf, Eq. (8.3.25) and Luneburg integral, Eq. (8.3.51) are identical, the validity of the Debye–Wolf integral is the same as that of the Luneburg integral. We may therefore conclude that even though the Debye–Wolf theory provides an exact solution to a mathematical problem, it is only an approximate solution of the physical problem of diffraction. Furthermore, the Debye–Wolf solution is only valid, but then is an excellent approximation, if the Fresnel number of the optical system is much greater than unity.

[13] For a thorough discussion on evanescent waves the reader is referred to the work of Mandel and Wolf **[232]**.

9 The Aberrated Scalar and Vector Point-spread Function

9.1 Introduction

For a long time the finite lateral size of the focal 'point' in the image plane of an optical instrument was known to the observer, as it set a limit to the possibility of separating two stellar objects in the sky or two spectral lines in a spectrograph. During the seventeenth and eighteenth century, through a continuous improvement of the manufacturing techniques for lens and mirror systems, it became apparent that there is a fundamental minimum size for the image of a point-like object like a star. Building further on the work of Fraunhofer, it was the English Astronomer Royal, Airy, who gave the first description of the focal distribution of a telescope **[8]**. In this important publication Airy gave the integral that has to be evaluated to obtain the amplitude and intensity distribution in the focal plane. The 'Airy disc' result was given as a table of numerically calculated values of this integral expression. The concise solution of this integral $J_1(2\pi r)/(\pi r)$ in terms of a Bessel function of the first kind was not known to Airy, simply because of the fact that a systematic overview of functions of this type was only submitted for publication in 1824 **[26]** and published two years later. The Bessel functions were not common knowledge in the optics community at that time.

An important extension of Airy's result was the calculation of the complex amplitude of the diffraction point-spread function produced by a perfect imaging system outside the nominal focal plane. The integral to be evaluated is given by Eq. (8.4.38), and a semi-analytic result for an ideal imaging system was given by Lommel **[218]** in terms of (infinite) series of higher-order Bessel functions. In Fig. 9.1 we reproduce the through-focus photographs of the diffraction point-spread function that are printed in the *Atlas of Optical Phenomena* by Cagnet *et al.* **[63]**. These pictures illustrate the symmetry around focus that applies to imaging systems with a large Fresnel number (see Eq. (8.3.23)). The spatial extent of each cross-section of a defocused point-spread function largely corresponds to the geometrical optics prediction; however, a fine structure arising from wave diffraction can also be seen superimposed on the average intensity. A striking effect is visible in the cross-section taken two axial diffraction units $\lambda/(\sin\alpha_1)^2$ away from the focus F'. Specifically, the central intensity shows a strong minimum which is effectively zero. It corresponds to a defocusing for which the number of Fresnel zones that can be constructed between the spherical wavefront and the defocused plane is exactly two. The phase differences between all optical secondary waves arriving at the axial point A' span the full interval from 0 to 2π and the resulting amplitude for a uniform beam is zero. Lommel's expansion is able to account for all these diffraction phenomena in the focal region and produces the in-focus Airy disc as a special case.

The through-focus analysis by Lommel does not cover practical optical systems that fail to produce focusing beams with a perfectly spherical wavefront, but instead carry a phase defect. If the phase defect is smaller than $\pi/2$ or the root-mean-square (rms) wavefront aberration is smaller than $\lambda/14$ or, defined in a still different way, if the peak-to-valley (PV) wavefront excursion is smaller than $\lambda/4$, the system is said to be weakly aberrated. For aberrations up to several wavelengths, diffraction analysis of the image formation is still needed. Beyond an rms aberration value of the order of *one* wavelength of the light in the medium under consideration, the much simpler ray optics approach is adequate (see Subsections 10.2.4.4–10.2.4.6 for arguments about the choice of this transition point). Apart from phase defects, the amplitude of the complex pupil function or lens transmission function can be non-uniform (see Eq. (5.7.1)), for

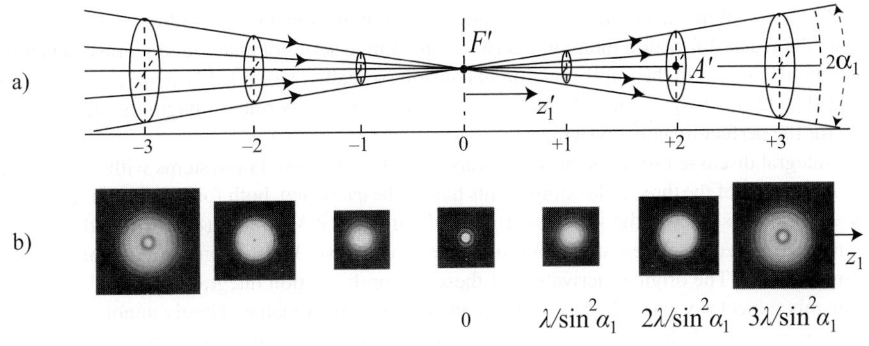

Figure 9.1: a) A focused beam (focal point is F').
b) The intensity point-spread function produced by a perfect lens in various defocused planes towards and away from the focal plane at F'.
The normalised axial coordinate z_1' is the real-space coordinate z_1 divided by the axial diffraction unit $\lambda/(\sin\alpha_1)^2$ where $NA = n_1 \sin\alpha_1$ is the image-side numerical aperture of the imaging system. The pictures in b) have been reproduced from plate 23 of the *Atlas optischer Erscheinungen* **[63]**, with the permission of Springer Verlag.

instance if the incident beam is partly obstructed or has a Gaussian laser profile. In all these cases, a general complex lens transmission function f_L' has to be taken into account and inserted into the scalar through-focus diffraction integral of Eq. (8.4.38). Before addressing this general problem, various authors studied the image formation by imperfect optical systems with an emphasis on the phase defect of the system, i.e. the amplitude part of the lens function was assumed to have a constant value. A quality criterion was developed by Strehl **[330]** in terms of the achievable on-axis intensity as compared to the theoretically 100% value for a perfect system. The denomination 'Strehl definition' is commonly used and a value of at least 0.80 is considered to indicate a close to perfect performance of the imaging system. A system with a Strehl ratio equal to 0.80 is called 'just diffraction-limited'. Various authors have studied the scalar diffraction integral in the presence of small aberration of a special type. In this respect we mention Conrady **[76]**, Steward **[326]**,**[327]** and Richter **[294]** who give specific results for systems suffering from third- and fifth-order spherical aberration and Picht **[278]**,**[279]** for systems with comatic and astigmatic defects.

In general, much effort has been put into the quest for the optimal lateral and axial focus setting yielding a maximum Strehl ratio given a certain collection of aberration terms of various type and order. This subject of 'aberration balancing' was important for achieving high-quality designs with a desired field shape and distortion properties. In some cases, for instance in photogrammetric or cartographic systems, a rigorous flat-field and distortion-free imaging close to the diffraction limit are required. For such systems, a purely ray optics design neglecting wave diffraction effects would yield erroneous results. The description of wavefront aberrations with the aid of an expansion in terms of orthogonal polynomials that are defined on a unit disc was an important milestone. They greatly facilitated the incorporation of wavefront aberration into the diffraction integral. The orthogonal *circle* polynomials were introduced in optics by Zernike **[385]** whilst carrying out his research on phase-contrast microscopy. In Appendix D we give a detailed description of these polynomials. They were successfully used by Zernike and his pupil Nijboer **[259]** to obtain expressions for the complex amplitude in the focal region of a weakly aberrated optical system. One of the side effects of these polynomials is that because of their orthogonality they each give a well-defined contribution to the variance of the wavefront aberration. This contribution is independent of the presence of other polynomials in the wavefront expansion. In practice, Seidel and higher-order aberration terms of the same aberration type are combined in an optical design in such a way that their magnitudes are proportional to the coefficients of the various power terms in a Zernike polynomial of the same aberration type. By such optimum balancing of aberrations of various orders, the wavefront variance is minimised (see also Section D.2). These aberration balancing schemes facilitated by the Zernike polynomials have gradually replaced the personal rules of thumb that were used by individual designers. With the advent of automatic computing, the coefficients of a Zernike polynomial expansion became variables of an optimisation scheme.

A general relationship between wavefront variance and Strehl intensity was established by Nijboer **[259]** and Maréchal **[237]**. The so-called Maréchal criterion (see Eq. (9.3.22) of this chapter) is independent of the type of aberration and states that a wavefront variance of $\lambda^2/(20\pi^2)$ or an rms value of $\approx \lambda/14$ produces a Strehl ratio of 80%, corresponding to a 'just diffraction-limited' system. When a Zernike representation of the pupil function according to Eq. (5.7.1) is used, it is possible to evaluate the scalar diffraction integral in terms of an infinite series of analytic functions depending on each of the Zernike polynomials present in the pupil function expansion (so-called Extended Nijboer Zernike (ENZ) diffraction theory, see **[162]**,**[51]**). In this way, a general solution is obtained for an aberrated imaging system that reduces to the Lommel solution for the perfect imaging system.

The diffraction integral discussed so far is the scalar version of Eq. (8.4.38). For systems with a high angular aperture, diffraction integrals for each of the three field components have to be evaluated, both for the electric field and the magnetic field vectors. In what follows we use the diffraction integrals of Eqs. (8.3.25) and (8.3.26) for the components of the electric and magnetic field vectors. They are commonly called the Debye–Wolf integrals and apply to imaging systems with a large Fresnel number. The original derivation of these vector diffraction integrals goes back to Ignatowsky **[153]**. The publication on this subject was published in the Russian language and remained largely unnoticed by the international community. We present an English translation of this important publication in Appendix G. In line with the results for the scalar case, the evaluation of aberrated vector diffraction integrals yields an infinite series of analytic vector ENZ functions that again depend on the specific Zernike term under consideration in the pupil function of the imaging system **[52]**.

In the next section of this chapter we give an overview of various representation methods for the pupil function of an imaging system suffering from amplitude and phase imperfections. In the third section, the scalar Debye integral is evaluated using the original Nijboer–Zernike diffraction theory. The next section is devoted to the Extended Nijboer–Zernike diffraction theory (ENZ), in turn allowing calculation of a heavily aberrated point-spread function in a strongly defocused plane. The extension of an aberrated point-spread function to the vector diffraction case is the subject of the fifth section. In the last section of this chapter we discuss energy density, energy flow and momentum flow in the focal region of a high-numerical-aperture system.

9.2 Pupil Function Expansion Using Zernike Polynomials

For calculation of a defocused point-spread function of a non-perfect imaging system we need the aberration function and the amplitude transmission or reflection function of the imaging system. As discussed in Chapter 4, these functions depend on the field angle at which the imaging beam is sent through the system. We suppose that the wavefront aberration function W of the imaging beam is known, e.g. from aberration theory or from a measurement, as well as its intensity cross-section including the vignetting of off-axis beams. From these two quantities we construct the lens function in the exit pupil with Cartesian coordinates (X_1, Y_1):

$$f_L(X_1, Y_1) = A(X_1, Y_1) \exp\{ik_0 W(X_1, Y_1)\} , \tag{9.2.1}$$

where $W(X_1, Y_1)$ is the *optical* pathlength along a ray from the wavefront to the reference sphere and k_0 the wavenumber in vacuum. This definition of the lens function f_L is equivalent to the previously given definition according to Eq. (5.7.1) in Chapter 5 on geometrical aberration theory. The phase function Φ in Eq. (5.7.1) is obtained from the geometrical aberration W_g by multiplying W_g by the wavenumber k_1 in the medium with index n_1, where we have that $k_1 = n_1 k_0$ (k_0 is the vacuum wavenumber). The aberration function $W(X_1, Y_1)$ which is used in this chapter on diffraction imaging is given in terms of the *optical* pathlength $[P_1'P_1] = n_1 P_1'P_1$ between a point P_1' on the wavefront and the corresponding point P_1 on the exit pupil sphere (see Fig. 9.2). The resulting phase function $\Phi(X_1, Y_1)$ on the exit pupil sphere is then given by $k_0 W(X_1, Y_1)$. The advantage associated with the use of the optical pathlength function $W(X_1, Y_1)$ in diffraction problems is that a division of W by the vacuum wavelength λ_0 immediately yields the phase difference function Φ, regardless of the refractive index of the medium in which the light is propagating. The function $W'(X_1, Y_1) = W(X_1, Y_1)/\lambda_0$ is commonly called the normalised (optical) aberration function and the phase function is then given by $\Phi(X_1, Y_1) = 2\pi W'(X_1, Y_1)$.

In this chapter we denote the various quantities in object space with the subscript 0 and the corresponding quantities in image space with the subscript 1. Using this convention, Fig. 4.26 shows the geometry for an on-axis imaging beam. The real-space exit pupil coordinates are (X_1, Y_1) with their origin at the centre, E_1, of the exit pupil. The amplitude function $A(X_1, Y_1)$ is the square root of the measured or calculated intensity on the exit pupil sphere through E_1 with radius R_1. This amplitude is expressed in arbitrary units, and the (optical) aberration function $W(X_1, Y_1)$ is given in units of length. According to our convention the wavefront aberration W of an off-axis beam at a certain point on the exit pupil sphere through E_1 is given by the optical path $[P_1'P_1]$ shown in Fig. 9.2 and its magnitude and sign are obtained from Eqs. (4.9.4) and (4.9.5). If wavefront information from other textbooks or from software or measurement programs has to be used, a

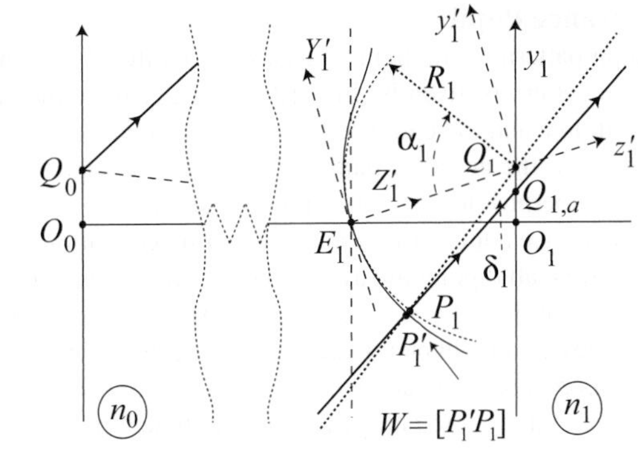

Figure 9.2: Definition of the coordinate systems in the exit pupil and the image plane used to calculate an off-axis point-spread function in the region around the paraxial image point Q_1 of the point object Q_0. $P_1'P_1Q_{1,a}$ is an aberrated ray with an angular aberration δ_1 with respect to a hypothetical perfect ray through P_1Q_1. The lateral aberration is $Q_1Q_{1,a}$ and the wavefront aberration W equals the optical path $[P_1'P_1]$. The figure shows a cross-section of the three-dimensional wavefront geometry with the plane $X_1' = 0$.

careful check is needed of how wavefront aberration is defined, especially in the case of reflecting or diffracting imaging systems. For oblique imaging beams, the coordinate axes are chosen along the principal ray and in a plane through E_1 perpendicular to this ray. The radius of the reference sphere R_1 is set equal to the new distance E_1Q_1 where Q_1 is an off-axis image point in the image plane through O_1.

In what follows, we switch to the normalised amplitude function A' through division by the on-axis amplitude A_0 and to the normalised optical aberration function W' that is obtained from the optical pathlength function W through division by the *vacuum* wavelength λ_0. The lateral exit pupil coordinates (X_1, Y_1) are normalised to unity at the pupil rim through division by ρ_0 where $2\rho_0$ is the diameter of the exit pupil; these normalised Cartesian coordinates are denoted by (X_1', Y_1') (see Fig. 9.2). The image plane coordinates (x_1, y_1) are normalised by the diffraction unit $\lambda/\sin\alpha_1 = \lambda_0/(n_1 \sin\alpha_1)$ in image space and denoted by (x_1', y_1'). Both the pupil and image plane coordinate systems have their z-axes aligned along the paraxial principal ray through E_1Q_1. In a final step we switch from Cartesian to polar coordinates in pupil and image space to obtain the normalised sets of coordinates (ρ, θ, Z_1') in pupil space and (r, γ, z_1') in image space. The latter normalised axial coordinate is obtained by division by the axial diffraction unit $\lambda/(\sin\alpha_1)^2 = n_1\lambda_0/(n_1 \sin\alpha_1)^2$ in the medium. In terms of these normalised coordinates and functions we write

$$f_L'(\rho, \theta) = A'(\rho, \theta) \exp\{i2\pi W'(\rho, \theta)\} . \tag{9.2.2}$$

The point-spread function (PSF) of an off-axis object point Q_0 is calculated in the tilted coordinate systems of Fig. 9.2. To obtain the intensity in the image plane through O_1, which is perpendicular to the optical axis, a simple projection is formally not adequate. The formation of the PSF in an oblique plane has been discussed by McCutchen **[245]**. A modified pupil function is constructed for the oblique beam which then serves as the input for an 'in-plane' diffraction integral. In this chapter, we assume that the field angle for an off-axis image point is small, in which case the intensity projection step is a satisfactory approximation.

9.2.1 The Amplitude Function in the Exit Pupil

The amplitude distribution on the exit pupil sphere is determined by several factors. We first mention the amplitude distribution in the entrance pupil that is the result of the (coherent) light emanating from the object. Secondly, amplitude changes occur during propagation through the optical system from the entrance to the exit pupil. Finally, the specific design of an optical system determines how a pencil of rays in object space is connected to a pencil of rays in image space and the choice of the designer influences the amplitude distribution on the exit pupil sphere. In the following three subsections we address these points with some emphasis on the influence of the optical system design which is especially important in the case of high-aperture imaging systems.

9.2.1.1 Optical Field in the Entrance Pupil

We first consider the angular radiation pattern of the light emanating from the object under study. The simplest object is a dipole source. Its electric far-field pattern is given by Eq. (1.6.42). The part of the far-field pattern that fits into the solid angle determined by the object dipole and the entrance pupil surface is transmitted through the optical system and gives rise to amplitude non-uniformity on the entrance pupil surface. More complicated objects which are coherently illuminated can be considered as the superposition of coherently radiating dipoles with varying electric strength and orientation. The coherent field in the entrance pupil is then obtained by a superposition of all the far-fields of the coherent object dipoles. In the scalar approximation, all dipoles are assumed to be aligned and only a small central part of their far field is collected. In that case, the picture of the spherical wave emitted from a point source replaces the physical dipole. Other radiation patterns that were discussed in Chapter 1 are those associated with various types of Gaussian beams. When imaging systems with a large acceptance angle are considered the variation of the state of polarisation of the light in the entrance pupil should also be taken into account and a vector amplitude function is needed to describe the incident source field.

9.2.1.2 Amplitude Changes during Propagation through the Optical System

The optical field in the entrance pupil propagates towards the exit pupil with abrupt changes in amplitude. The amplitude changes stem from transmission and/or reflection losses at optical interfaces, e.g. the surfaces of lens elements or mirrors. The amplitude changes are calculated with the aid of the Fresnel formulas for reflection and transmission at an optical interface as given by Eqs. (1.8.24)–(1.8.27) or by using the general formulas of Eqs. (1.10.2)–(1.10.5) for a multi-layer covered transition between two media. The amplitude changes introduced at an interface depend on the state of polarisation of the incident field. For sufficiently small angles of incidence, it is permissible to take the average of the values that are associated with s- and p-polarisation. A sufficiently large number of rays is traced from the entrance pupil to the exit pupil of the optical system and the local transmission or reflection factors at each surface are tabulated. The amplitude attenuation factors associated with a single ray are multiplied and yield the amplitude attenuation at the intersection point of the ray with the exit pupil sphere. The amplitude on the exit pupil sphere is further influenced by a possible vignetting effect. A strong vignetting effect is found in, e.g., wide-angle imaging systems. In more conventional imaging systems a small vignetting of imaging pencils is introduced on purpose by the designer to eliminate strongly aberrated rays which would give rise to an excessive blurring of the image. In general, phase changes are also introduced when light is transmitted or reflected at an interface inside the optical system. The phase changes at each interface along a ray are tabulated and the cumulative phase change is included in the phase function on the exit pupil of the optical system.

9.2.1.3 A Design Aspect: Ray Mapping between the Object and Image Space Pencils

A final amplitude effect which is relevant in optical systems with a high aperture on either the object or the image side or on both of them originates from the specific ray mapping between entrance and exit pupil that has been adopted for the optical system. The effect of ray mapping on the amplitude in the exit pupil and the global amplitude change that follows from the lateral (paraxial) magnification between exit and entrance pupil are discussed in more detail in the remainder of this subsection.

If we consider an optical system which is able to produce a perfect image O_1 of a specific axial object point O_0, there is no privileged way in which the rays that originate at the axial object point are connected to the rays of the homocentric pencil in image space which is focused in O_1. When perfect imaging is required of an object point Q_0 which is laterally or axially displaced from O_0 over a small distance, the ray mapping of the axial pencil which connects O_0 to O_1 has to obey a certain condition. It was shown in Chapter 4, Section 4.8, that the ray mapping which is used in most lens designs is the ray mapping that respects the Abbe sine condition since it produces optimal imaging in a plane through O_1 which is perpendicular to the optical axis. In Fig. 9.3, using non-normalised real-space coordinates in the object, image and pupil planes, we show a general mapping of rays from object to image space. We now analyse the energy flow through annular cross-sections on the two pupil spheres, at angles α_0 and α_1, respectively. We denote the average electric energy densities on the entrance and exit pupil annuli by $\epsilon_0 n_0^2 |\mathbf{E}_{0,p}|^2/2$ and $\epsilon_0 n_1^2 |\mathbf{E}_{1,p}|^2/2$, where $\mathbf{E}_{0,p}$ and $\mathbf{E}_{1,p}$ are the electric field vectors on the entrance and exit pupil surfaces, respectively. To determine the energy flow we multiply these quantities by the propagation speeds in object and image space, given by $v_0 = c/n_0$ and $v_1 = c/n_1$, respectively. In the absence of absorption and reflection losses, the energy flow through the differential annuli on the pupil surfaces is identical and we have

$$\pi \epsilon_0 c n_0 |\mathbf{E}_{0,p}|^2 R_0^2 \sin \alpha_0 d\alpha_0 = \pi \epsilon_0 c n_1 |\mathbf{E}_{1,p}|^2 R_1^2 \sin \alpha_1 d\alpha_1 \, ,$$

or,

$$\frac{n_1|\mathbf{E}_{1,p}|^2}{n_0|\mathbf{E}_{0,p}|^2} = \frac{R_0^2 \sin\alpha_0 \, d\alpha_0}{R_1^2 \sin\alpha_1 \, d\alpha_1} \, . \tag{9.2.3}$$

The latter expression equals the ratio of the Poynting vectors on the exit and the entrance pupil.

At this point it is useful to introduce the paraxial lateral pupil magnification $m_{T,p}$. We use the expression for $m_{T,p}$ that is given by Eq. (4.10.41). The pupil imaging is referred to the object and image planes through O_0 and O_1, respectively. The lateral paraxial magnification between these conjugate planes is m_T. The object and image distances of the pupils with respect to the conjugate planes through O_0 and O_1 are given by the distances R_0 and R_1, respectively. We then obtain from Eq. (4.10.41) the paraxial pupil magnification as

$$m_{T,p} = \frac{n_0 R_1}{n_1 R_0} \frac{1}{m_T} \, , \tag{9.2.4}$$

and the ratio of the Poynting vectors on exit and entrance pupils is hence given by

$$\frac{n_1|\mathbf{E}_{1,p}|^2}{n_0|\mathbf{E}_{0,p}|^2} = \frac{n_0^2}{n_1^2} \frac{1}{m_T^2 \, m_{T,p}^2} \frac{\sin\alpha_0 \, d\alpha_0}{\sin\alpha_1 \, d\alpha_1} \, . \tag{9.2.5}$$

Equation (9.2.5) takes different forms, depending on the angular mapping between the entrance and exit pupil.

We now consider two ray mapping options, namely the Herschel and the Abbe condition (see Section 4.8). As mentioned before, the Abbe condition is most frequently used since it is essential to obtain an extended image *plane* with good imaging quality. The Herschel condition guarantees good imaging on a line segment only, namely the optical axis.

- *Herschel condition*, $n_0 \sin(\alpha_0/2) = m_T n_1 \sin(\alpha_1/2)$.
  For the ratio of the differential aperture angles we have

$$\frac{d\alpha_0}{d\alpha_1} = \frac{m_T n_1 \cos(\alpha_1/2)}{n_0 \cos(\alpha_0/2)} \, , \tag{9.2.6}$$

  and the Poynting vector ratio of Eq. (9.2.5) is then given by

$$\frac{n_1|\mathbf{E}_{1,p}|^2}{n_0|\mathbf{E}_{0,p}|^2} = \frac{1}{m_{T,p}^2} \, . \tag{9.2.7}$$

  Given $m_{T,p}^2$ is equal to the paraxial pupil surface ratio $dS_{p,1}/dS_{p,0}$, we deduce from Eq. (9.2.7) that the ratio $n_1|\mathbf{E}_{1,p}|^2/n_0|\mathbf{E}_{0,p}|^2$ for a finite aperture angle is equal to the paraxial ratio $dS_{p,0}/dS_{p,1}$ and is uniform over the full aperture. The Herschel condition was originally meant to provide constancy of aberration along the optical axis, but the side-effect with respect to energy flow through the pupils makes this particular ray mapping condition also interesting when uniform illumination on a spherical surface is required.

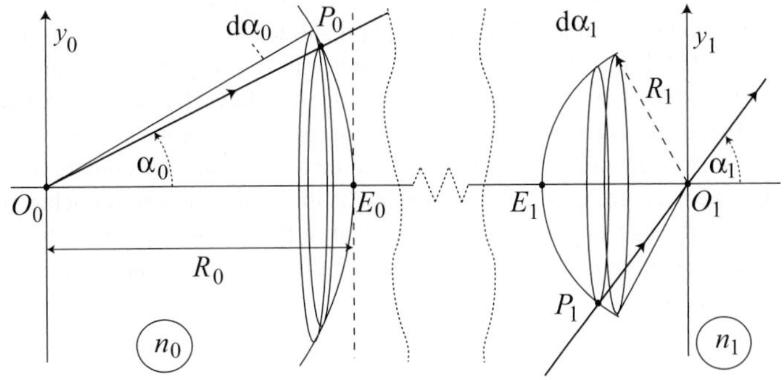

Figure 9.3: The energy flow through annular areas of the entrance and exit pupil and the ray mapping from object to image space.

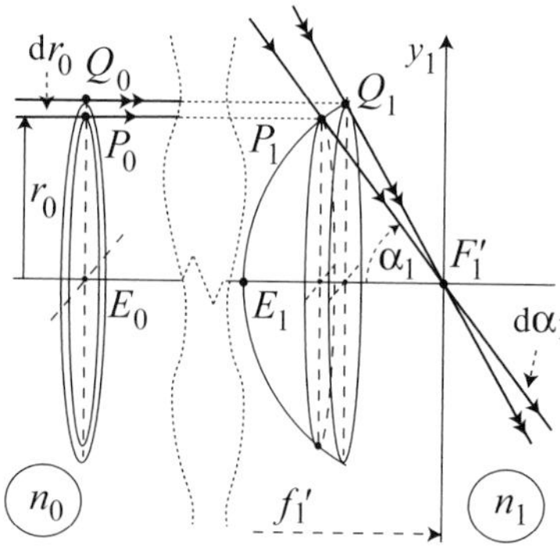

Figure 9.4: The radiometric effect.

- *Abbe sine condition,* $n_0 \sin\alpha_0 = m_T n_1 \sin\alpha_1$.
  Along the same lines as for the Herschel condition, we readily obtain the result

$$\frac{n_1|\mathbf{E}_{1,p}|^2}{n_0|\mathbf{E}_{0,p}|^2} = \frac{1}{m_{T,p}^2}\frac{\cos\alpha_1}{\cos\alpha_0}. \tag{9.2.8}$$

  Using this ratio for the Poynting vectors at non-paraxial aperture angles we find an electric field strength or *amplitude* ratio that is proportional to $(\cos\alpha_1/\cos\alpha_0)^{1/2}$. This amplitude effect is called the 'radiometric' effect (also called 'apodisation function') that is associated with the Abbe sine condition for ray mapping. The radiometric effect is generally neglected in the scalar diffraction integral where the apertures in object and image space should be limited to modest values. An optical system that images from infinity to its focal plane in air with a numerical aperture of 0.50 will have a radiometric effect at the rim of the aperture of 0.93. The radiometric effect becomes significant for apertures as high as 0.90 where vector diffraction integrals should be applied to obtain the point-spread function.

The radiometric effect for an imaging system that satisfies the Abbe sine condition and has a lateral magnification m_T equal to zero is illustrated in Fig. 9.4. The paraxial entrance and exit pupil are located in the principal planes and the image is formed at the focal point F_1' at a distance f_1' from the centre of the exit pupil E_1. For this special case, with the Abbe sine condition written as $r_0 = -f_1' \sin\alpha_1$ and the pupil magnification equal to unity, the ratio of the Poynting vectors is given by

$$\frac{n_1|\mathbf{E}_{1,p}|^2}{n_0|\mathbf{E}_{0,p}|^2} = \cos\alpha_1 . \tag{9.2.9}$$

Figure 9.4 shows that the $\cos\alpha_1$ factor for the Abbe sine condition is caused by the projection effect of the plane entrance pupil onto the spherically curved exit pupil surface. We have a direct comparison in size between the flat annular area in object space and the projected annulus on the spherical exit pupil surface in image space. For large aperture angles, this latter surface is strongly expanded. Figure 9.5 shows the difference in ray mapping between a system that satisfies the Herschel condition and a system that obeys the Abbe sine condition. It follows that at the maximum possible aperture the relative pupil diameter of a Herschel system is $\sqrt{2}$ times larger than that of an Abbe system.

Unless stated otherwise, we shall consider an imaging system that obeys the sine condition since the vast majority of all imaging systems are designed according to this prescription as this is a necessity for achieving a large lateral image field. The relative amplitude in the exit pupil as compared to that in the entrance pupil is then given by Eq. (9.2.8). In view of the application of Zernike polynomials defined on the unit disc, we use the normalised radial coordinate ρ in the entrance and exit pupils as given by

$$\rho = \frac{\sin\alpha_0}{\sin\alpha_{0,m}} = \frac{\sin\alpha_1}{\sin\alpha_{1,m}}, \tag{9.2.10}$$

where $\sin\alpha_{0,m} = s_0$ and $\sin\alpha_{1,m} = s_1$ are the sines of the maximum aperture angles in object and image space. These sines are related by the Abbe sine condition and we thus obtain for the radiometric effect on the exit pupil sphere,

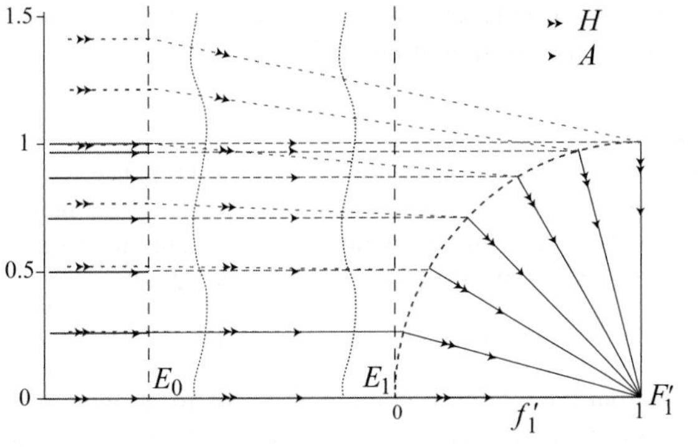

Figure 9.5: The ray mapping from object to image space for a system satisfying the Herschel (H) or the Abbe condition (A); lateral magnification $m_T = 0$. A full imaging pencil up to 90° convergence angle has been shown. The dashed (H) and solid (A) ray paths from the planar entrance pupil through the point E_0 to the spherical exit pupil through E_1 have been schematically shown; the focal distance f_1' has been normalised to unity.

$$\frac{|\mathbf{E}_{1,p}|}{|\mathbf{E}_{0,p}|} = \frac{1}{m_{T,p}} \sqrt{\frac{n_0}{n_1}} \left(\frac{(1 - s_1^2 \rho^2)}{1 - (n_1 m_T / n_0)^2 s_1^2 \rho^2} \right)^{1/4} . \tag{9.2.11}$$

In many instances, when the absolute value of the image plane irradiance is not required but just a relative 'intensity', the two constant factors in the expression for $|\mathbf{E}_{1,p}/\mathbf{E}_{0,p}|$ are omitted.

9.2.2 The Phase Function Φ in the Exit Pupil

The phase function in the exit pupil is composed of two parts. The diffraction of a coherent wave by the object produces a certain phase distribution in the entrance pupil. The propagation through the optical system to the exit pupil imparts local phase changes to the wave because of aberrations and phase jumps at optical interfaces. In most cases, we study the aberrations of the imaging system separately and suppose that the wavefront in the entrance pupil is spherical or flat. For the image formation in image space, we have to include an extra phase factor that is associated with the choice of the reference point for image formation in image space. For a lateral shift of the reference point, a linear wavefront term has to be included. In the case of a defocusing with respect to the paraxial image plane, the pathlength difference along an imaging ray between two spherical wavefronts also has to be taken into account.

9.2.2.1 The Effect of Aberrations

In the scalar approximation the wave emitted by a point source in the object plane has a uniform phase on the spherical entrance pupil surface of an optical system. For a more general object structure both amplitude and phase are functions of the polar coordinates (ρ, θ) on the entrance pupil surface. For a point source, the propagation from entrance to exit pupil through the optical system modifies the originally uniform phase to an aberrated phase function $2\pi W' = \Phi$ that determines the exponential part of the pupil function $f_L'(\rho, \theta)$ of Eq. (9.2.2). The phase function $\Phi(\rho, \theta)$ can be written as an expansion with the aid of Zernike polynomials, see Appendix D, viz.

$$\Phi(\rho, \theta) = \sum_{n=0}^{\infty} \sum_{m=0}^{\infty} R_n^m(\rho) \left\{ \alpha_{n,c}^m \cos m\theta + \alpha_{n,s}^m \sin m\theta \right\} , \tag{9.2.12}$$

with the integer indices $n, m \geq 0$, subject to $n - m \geq 0$ and even.

We strongly recommend use of complex exponentials instead of the basic trigonometric functions because this allows a much more elegant description of the properties of Zernike polynomials, for instance when taking the inner product of two polynomials, multiplication of polynomials, derivative operation, etc. In this case we have the expression

$$\Phi(\rho, \theta) = \sum_{n=0}^{\infty} \sum_{m=-\infty}^{+\infty} \alpha_n^m R_n^{|m|}(\rho) \exp(im\theta) , \tag{9.2.13}$$

with integer n and m, $n \geq 0$ and $n - |m| \geq 0$ and even. The α-coefficients can be complex if terms with a $\sin\theta$ dependence are present in the original phase function expansion of Eq. (9.2.12).

9.2.2.2 The Defocusing Factor

The optical pathlength difference W between an aperture ray and the reference ray in a focusing pencil of rays was illustrated in Fig. 5.13. In this figure the focus of the physical wavefront is at the point A. We choose the origin of the coordinate system in the point A' of the figure, corresponding to $z_1 = 0$ in our image space. The point A' is also the centre of the reference sphere (dashed in Fig. 5.13). We then have from Eq. (5.2.23) the expression

$$W(\rho) \approx n_1\, z_1\, (\cos\alpha_1 - 1)\,, \tag{9.2.14}$$

where $z_1 = A'A$ is the axial shift from the reference point at A' to the focal point at A (the defocusing distance z_1 is negative in the figure) and $\cos\alpha_1$ is expressed in terms of ρ with the aid of Eq. (9.2.10). It was shown in Subsection 5.2.2 that Eq. (9.2.14) is a good approximation when the defocusing distance z_1 can be neglected with respect to the radius of the exit pupil sphere. This is always the case when the exit pupil has been shifted to infinity (telecentric imaging) whereby Eq. (9.2.14) holds exactly.

Letting $k_{n_1} = 2\pi n_1/\lambda_0$ be the wavenumber in image space and $s_1 = \sin\alpha_{1,m}$, the exponential defocusing function is written as

$$\exp\{i\Phi_d(\rho,\theta)\} = \exp\left\{-ik_{n_1} z_1\left(1 - \sqrt{1 - s_1^2\rho^2}\right)\right\} = \exp\left\{-if'_v\left(\frac{1 - \sqrt{1 - s_1^2\rho^2}}{1 - \sqrt{1 - s_1^2}}\right)\right\}\,, \tag{9.2.15}$$

where f'_v (with the lower index v denoting the 'vectorial' diffraction case) is the high-aperture defocusing phase factor, given by $k_{n_1} z_1\{1 - (1 - s_1^2)^{1/2}\}$. For $f'_v = \pi$, the phase departure at the exit pupil rim due to defocusing equals π and we have that the pupil area exactly contains one Fresnel zone. Using this value we define the 'high-aperture' axial diffraction unit $z_{1,d} = \lambda_{n_1}/\{2(1 - (1 - s_1^2)^{1/2})\} = \lambda_{n_1}/\{2(1 - \cos\alpha_1)\}$, with λ_{n_1}, the wavelength in image space, given by the vacuum wavelength λ_0 divided by n_1. For small aperture angles, these expressions reduce to

$$\exp\{i\Phi_d(\rho,\theta)\} = \exp\left\{-\frac{i\pi z_1 s_1^2\rho^2}{\lambda_{n_1}}\right\} = \exp(-if'_d\rho^2)\,, \tag{9.2.16}$$

where $f'_d = \pi s_1^2 z_1/\lambda_{n_1}$ and the axial diffraction unit $z_{1,d}$ is given by λ_{n_1}/s_1^2. The small-aperture approximated quantities are commonly used in the scalar diffraction integrals, whereas the high-aperture equivalents must be used in the vector diffraction integrals.

9.2.2.3 The Pupil Function Including Amplitude and Phase

The coefficients $\alpha_{n,c}^m$ and $\alpha_{n,s}^m$ in Eq. (9.2.12) are real if the function to be expanded is real itself. It was proposed by Kintner and Sillitto **[183]** to generalise the Zernike expansion by allowing a complex function to be expanded, in this case the complex pupil function $f'_L(\rho,\theta)$ of Eq. (9.2.2) including its modulus *and* phase part. The defocusing phase term is not included in this complex function as it is meant to be a free parameter when evaluating a diffraction integral. The cosine and sine expansion coefficients, $\alpha_{n,c}^m$ and $\alpha_{n,s}^m$, are now associated with the complex function f'_L and become complex. Once more switching to complex exponentials we can write the pupil function in the form

$$f'_L(\rho,\theta) = \sum_{n=0}^{\infty}\sum_{m=-\infty}^{+\infty} \beta_n^m\, R_n^{|m|}(\rho)\, \exp(im\theta)\,, \tag{9.2.17}$$

where β_n^m are complex Zernike coefficients and there are the same restrictions on the indices n and m as those that were previously given for the α_n^m coefficients, just below Eq. (9.2.13). We recall that the possibly complex α coefficients are related to the complex exponential expansion of the argument Φ of the complex function f'_L. The complex β coefficients apply to the complex exponential expansion of the function f'_L itself.

The relation between the complex coefficients $\beta_{n,c}^m$ and $\beta_{n,s}^m$ of a cosine/sine expansion of the complex function f'_L and those of the new complex exponential expansion in terms of β_n^m coefficients follows from

$$m \neq 0,\quad \begin{cases} \Re\left(\beta_{n,c}^m\right) = \Re\left(\beta_n^m + \beta_n^{-m}\right)\,, & \Im\left(\beta_{n,c}^m\right) = \Im\left(\beta_n^m + \beta_n^{-m}\right)\,, \\ \Re\left(\beta_{n,s}^m\right) = -\Im\left(\beta_n^m - \beta_n^{-m}\right)\,, & \Im\left(\beta_{n,s}^m\right) = \Re\left(\beta_n^m - \beta_n^{-m}\right)\,, \end{cases}$$

$$m = 0, \quad \begin{cases} \Re\left(\beta^0_{n,c}\right) = \Re\left(\beta^0_n\right), & \Im\left(\beta^0_{n,c}\right) = \Im\left(\beta^0_n\right), \\ \Re\left(\beta^0_{n,s}\right) = 0, & \Im\left(\beta^0_{n,s}\right) = 0. \end{cases} \tag{9.2.18}$$

From this point on, we use the pupil function expansion with the complex coefficients β^m_n. Such a 'forward' expansion of the pupil function can be made arbitrarily accurate by including a sufficiently large number of terms in the expansion. The inverse process, reconstruction of the complex pupil function f'_L from a set of complex coefficients β^m_n suffers from a phase ambiguity of multiples of 2π. A reliable reconstruction of f'_L needs extra information, for instance about the continuity of the function itself or of one or more of its derivatives. The reconstruction of a complex function with a modulo 2π uncertainty into a smooth complex function is done with the aid of a phase-*unwrapping* method, comparable to those that are used in surface reconstruction from interferometric measurement data **[126]**.

With the advent of laser sources, Gaussian-type amplitude and phase functions are introduced for the so-called Gauss–Hermite (GH) and Gauss–Laguerre (GL) beams with their accompanying Hermite and Laguerre polynomials, see Section 1.7.2. Such wave functions **[315]** are not a solution of the rigorous wave equation but only satisfy the paraxial approximation of the Helmholtz equation. The GH and GL laser modes are not well adapted to a typical lens pupil with its sharp rim. But the Gauss–Laguerre wave functions have attracted special attention because they can be used to represent optical beams with orbital angular momentum (helical phase profile). Modifications of the Gauss–Laguerre laser modes have been proposed **[21]** to make them satisfy the rigorous Helmholtz equation. We note that Zernike polynomials can also be used to simulate a helical phase profile. To this purpose we restrict the expansion of Eq. (9.2.17) to a single nonzero coefficient β^m_n. With $n = m = 1$ we obtain the elementary helical profile with the phase running from 0 to 2π, combined with a transmission function running from zero in the centre of the pupil to unity at the rim.

9.3 The Point-spread Function and the Nijboer–Zernike Diffraction Theory

It was shown in Chapter 8 that the scalar Debye integral given in Eq. (8.4.29) is an approximated version of the scalar Rayleigh diffraction integrals given by Eqs. (8.4.15) and (8.4.16). If the Fresnel number $N_F = (\sin\alpha_{1,m})^2 R_1/\lambda_{n_1}$ of the imaging geometry is sufficiently large, i.e. $> 10^3$, the Debye approximation is a valid one. In practice, for most visual optics imaging devices like a microscope objective or a collimating lens, the Debye integral is a fully justified approximation. The approximation should not be used when micro-optic elements ($R_1/\lambda_{n_1} < 100$) are used or when very low-aperture focusing beams ($\sin\alpha_1 < 0.01$) are involved. To illustrate this effect we show in Fig. 9.6 the through-focus axial intensity that is obtained with a perfect imaging system. The Debye approximation for the diffraction integral leads to an axial intensity function that is proportional to $\mathrm{sinc}^2(\pi z'_1/2)$ where z'_1 is the normalised axial coordinate. This function is symmetrical with respect to the paraxial focal plane as is clearly evident from the dashed curves in the four graphs. The intensity I_n in each graph has been normalised with respect to the maximum intensity calculated with the aid of the scalar Rayleigh diffraction integral.

The Rayleigh-integral values for the axial intensity show a very different behaviour from the Debye approximation when the Fresnel number N_F is small. In graph a) we have that $N_F = 2$ and the maximum axial intensity is strongly shifted towards the lens aperture. The value $z_1 \approx -2 \cdot 10^4\lambda$ in the figure corresponds to the plane of the exit pupil. The Rayleigh diffraction integral should theoretically reproduce the axial value of the integrand function at this particular point. The occurrence of a singular value in the kernel of the Rayleigh integral does not permit positioning the receiving screen too close to the exit pupil plane of the imaging system. Figure 9.6b corresponds to $N = 5$ and the Rayleigh and the Debye results resemble each other more closely; however, an oscillatory intensity variation is still present between the exit pupil and the paraxial focus that is not correctly reproduced by the Debye result. The focal shift between the Rayleigh maximum and the Debye maximum is much less than in a) and their maximum values differ by only 5%. In Fig. 9.6c, corresponding to $N_F = 20$, the axial shift between the maxima of the Rayleigh and Debye diffraction integrals is hardly perceptible, the main difference is found in the position and height of the intensity sidelobes. Figure 9.6d with $N_F = 100$ produces virtually equal curves for both integrals, including an almost perfect symmetry with respect to the paraxial focus point. A medium-power microscope objective with $NA = 0.25$ and image-side focal distance $f'_1 = 10$ mm has a Fresnel number of 1250 when $\lambda_{n_1} = 0.5$ μm; the Debye integral can hence be used without problems if such imaging conditions apply.

The aberrated point-spread functions that are calculated in the remainder of this section use the Debye diffraction integral of Eq. (8.4.36). Figure 9.7 shows the exit pupil/image space geometry as a reminder of the notation used in this section. In pupil space we define a general point Q by its real-space Cartesian coordinates (X_1, Y_1, Z_1) and the

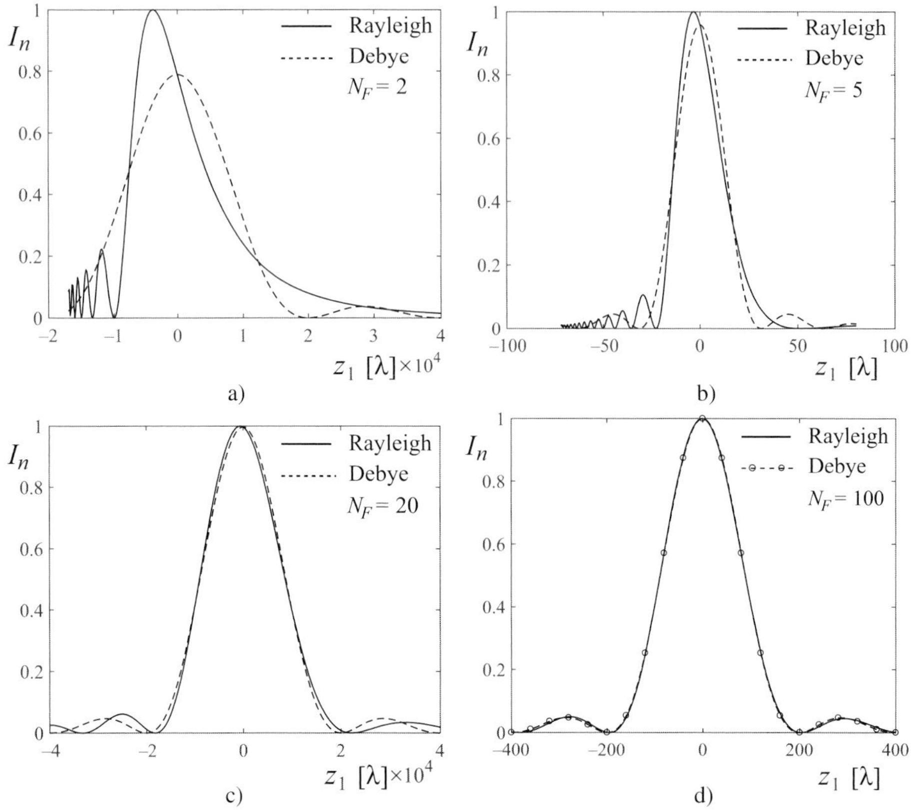

Figure 9.6: Normalised axial intensity distribution I_n in the focal region of an imaging system, calculated with the aid of the Rayleigh diffraction integral (solid curves) and the Debye integral (dashed curves). The defocusing distance z_1 is expressed in units of the wavelength $\lambda = 0.5$ µm in image space ($n_1 = 1$). The range of z_1 is limited to the interval $-2\lambda_{n_1}/(NA)^2 \leq z_1 \leq +2\lambda_{n_1}/(NA)^2$, or to a smaller negative value if the proximity of the diffracting imaging system imposes this as in graphs a) and b).
a) $N_F = 2$. $f = R_1 = 10$ mm, $a = 0.1$ mm ($NA = 0.01$).
b) $N_F = 5$. $f = 0.04$ mm, $R_1 = 0.0387$ mm, $a = 0.01$ mm ($NA = 0.25$).
c) $N_F = 20$. $f = R_1 = 100$ mm, $a = 1$ mm ($NA = 0.01$).
d) $N_F = 100$. $f = 5$ mm, $R_1 = 4.97$ mm, $a = 0.5$ mm ($NA = 0.1$).

polar coordinates (ρ_Q, θ_Q, Z_1). In this chapter we preferably use the normalised equivalents, i.e. the primed Cartesian coordinates (X_1', Y_1', Z_1') and their cylindrical equivalents (ρ, θ, Z_1'). The reason why the normalised radial coordinate ρ has not been provided with a prime is to avoid a conflict with the widespread use of (normalised) polar coordinates (ρ, θ) in diffraction problems. In particular, when Zernike polynomials are used as orthogonal polynomials on the unit circle for the expansion of a general pupil function $f_L'(\rho, \theta)$, the (ρ, θ) notation for the polar coordinates on the unit circle is largely adopted in the literature. The real space radial coordinate ρ_Q and the normalised radial coordinate ρ are connected through the relation $\rho_Q = \rho_0 \rho$ where $2\rho_0$ is the diameter of the circular rim of the exit pupil. The azimuthal coordinate θ of the normalised polar coordinate set (ρ, θ) is equal to the azimuth θ_Q of the original polar coordinates (ρ_Q, θ_Q).

In image space we use the real-space coordinates (x_1, y_1, z_1) that have their normalised equivalents (x_1', y_1', z_1'). In Section 8.4, where various types of (scalar) diffraction integrals were discussed, the image-space coordinates have been denoted by (x_p, y_p, z_p). However, since in this chapter we also discuss imaging systems with finite conjugate distances and thus need to include object-space coordinates, we replace the p-subscript of the coordinates in image-space by the subscript 1 and attribute the subscript 0 to the corresponding object-space coordinates. In both spaces, the lateral coordinates are scaled by the diffraction unit, for instance in image space by λ_{n_1}/s_1 where $s_1 = \sin\alpha_{1,m}$. In a similar way, the axial coordinates in object and image space are scaled by means of the distance $\lambda/\sin^2(\alpha_m)$, for instance by λ_{n1}/s_1^2

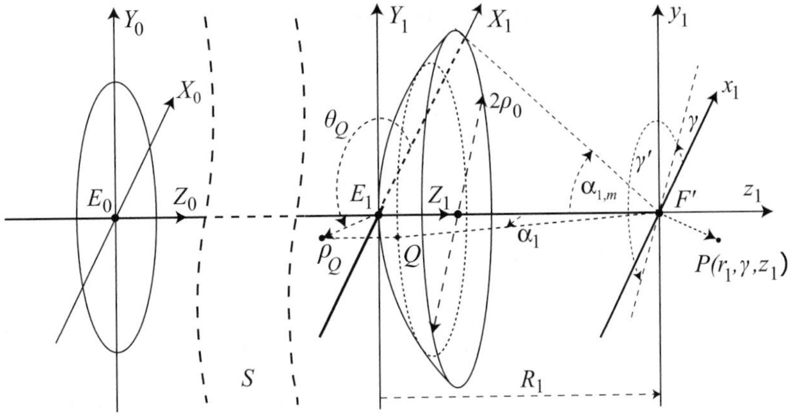

Figure 9.7: Exit pupil and image space coordinates used for evaluation of the scalar Debye diffraction integral. The maximum aperture $\sin\alpha_{1,m}$ of the focusing beam is given by $s_1 = \rho_0/|R_1|$. S denotes the schematically drawn optical system. The planar entrance pupil intersects the axial point E_0, the centre of the spherical exit pupil intersects E_1.

in image space. From these normalised Cartesian coordinates we derive the normalised cylindrical coordinates (r_1', γ, z_1') where the azimuthal coordinate γ is unchanged. The radius of the exit pupil is denoted by R_1. With these coordinate choices and with $k_{n_1} = 2\pi n_1/\lambda_0$, Eq. (8.4.36) is written as

$$\begin{aligned} U_p(r_1,\gamma,z_1) &= -\frac{i}{\lambda_{n_1} R_1}\exp(ik_{n_1} z_1)\exp\left(\frac{ik_{n_1} r_1^2}{2R_1}\right) \\ &\times \int_0^{s_1 R_1}\int_0^{2\pi} U_0(\rho_Q,\theta_Q)\exp\left[-\frac{ik_{n_1}}{|R_1|} r_1\rho_Q\cos(\theta_Q-\gamma)\right]\exp\left(-\frac{ik_{n_1}}{2R_1^2} z_1\rho_Q^2\right)\rho_Q\, d\rho_Q d\theta_Q \,, \end{aligned} \tag{9.3.1}$$

where $U_0(\rho_Q,\theta_Q)$ is the complex amplitude of the focusing wave on the exit pupil sphere at a distance R_1 from the paraxial image plane.

To obtain Eq. (9.3.1) we have replaced the integration variables (s_x, s_y) in Eqs. (8.3.51) and (8.3.52) by the normalised polar coordinates (ρ,θ) on the exit pupil sphere. In Fig. 9.7 we observe that positive lateral components (s_x, s_y) of a unit wave vector **s** give rise to negative Cartesian coordinate values (X_1, Y_1) of the intersection point of this unit wave vector **s** with the exit pupil sphere. This sign relation between the integration variables (s_x, s_y) in Eq. (8.3.51) and the real space pupil coordinates (X_1, Y_1) applies to a positive distance $R_1 = E_1F'$ from the convex exit pupil sphere to the image plane through F'. To be able to cover a general geometry in image space which includes also the case of a concave pupil sphere with $R_1 < 0$ we write the vector **s** as

$$\mathbf{s} = \left(-\frac{R_1}{|R_1|}|\sin\alpha_1|\cos\theta,\ -\frac{R_1}{|R_1|}|\sin\alpha_1|\sin\theta,\ \cos\alpha_1\right), \tag{9.3.2}$$

where the quantity $R_1/|R_1|$ determines the sign of the pupil distance from the image plane. The ‘lateral’ pathlength exponential factor in the Debye integral is then given by

$$\begin{aligned} \exp[ik_{n_1}(s_x r_{x,1} + s_y r_{y,1})] &= \exp\left[ik_{n_1}\left(\frac{-R_1}{|R_1|}\right)|\sin\alpha_1|\, r_p\cos(\theta-\gamma)\right] \\ &= \exp\left[ik_{n_1}\left(\frac{-R_1}{|R_1|}\right)\frac{\rho_Q}{|R_1|}\, r_1\cos(\theta-\gamma)\right], \end{aligned} \tag{9.3.3}$$

where r_1 is the radial distance to the z_1-axis of a general point in image space. When $R_1 > 0$ (as depicted in Fig. 9.7) the exponential carries a minus sign, as in Eq. (9.3.1). Alternatively, this minus sign can be incorporated in the argument of the cosine-function by writing $\cos[\theta-(\gamma-\pi)]$. In our convention the azimuthal reference angle θ_0 is equal to zero such that the polar axis $(\rho_Q, 0)$ coincides with the Cartesian X_1-axis in pupil space.

We now change to the normalised pupil and image space coordinates and obtain the expression, for positive values of R_1,

$$U_p(r_1', \gamma, z_1') = -i s_1^2 R_1' \exp\left(\frac{i2\pi z_1'}{s_1^2}\right) \exp\left(\frac{i\pi r_1'^2}{R_1'}\right)$$
$$\times \int_0^1 \int_0^{2\pi} f_L'(\rho, \theta) \exp(-i f_d' \rho^2) \exp\{-i2\pi r_1' \rho \cos(\theta - \gamma)\} \rho \, d\rho d\theta , \qquad (9.3.4)$$

where we have used the normalised radius $R_1' = R_1 s_1^2 / \lambda_{n_1}$ of the exit pupil sphere. The paraxial defocusing parameter f_d' was defined in Eq. (9.2.16) and equals $\pi z_1'$. We have dropped the index Q of the azimuthal polar coordinate now that we use the normalised version $\rho = \rho_Q / \rho_0$ of the radial polar coordinate. The complex amplitude function $U_0(\rho, \theta)$ has also been scaled to this normalised radial coordinate and is hence equal to the normalised pupil function $f_L'(\rho, \theta)$ that was discussed in the preceding section.

For the particular case that the defocusing factor f_d' is zero, the integral at the right-hand side of Eq. (9.3.4) can be identified with a *two-dimensional Fourier transform*. It suffices to reintroduce Cartesian coordinates in the exit pupil and the image plane instead of using polar coordinates. It is important to note that, according to our sign convention for spatial Fourier transforms in Appendix A, the Cartesian diffraction integral is equivalent to an (inverse) Fourier transformation of the *far-field* pupil function to the *near-field* point-spread function. Since the sign of the second exponential function in the kernel of the integral of Eq. (9.3.4) is negative, the inverse Fourier transformation is with respect to the *negative* Cartesian coordinate pair $(-x_1', -y_1')$ (or with respect to the azimuth $\gamma' = \gamma - \pi$ when using polar coordinates). This sign subtlety is of no concern when the pupil function is circularly symmetric and, for that reason, it is not always correctly taken into account in textbooks or publications. In Fig. A.8 of Appendix A we have sketched the forward and backward propagation of focused waves in an optical system. The corresponding sign of the (X_1, Y_1)-dependent part of the exponential pathlength function is shown in this figure, together with the sign of the Cartesian arguments of the transformed far-field and near-field functions.

Excluding the multiplicative factors in front of the integral, the complex amplitude distributions on the exit pupil sphere and in the nominal image plane are thus each other's Fourier transform. For $z_1' = 0$ and $R_1' \rightarrow \infty$, the multiplicative factor in front of the integral is real. This special case is found when the exit pupil of the optical system is at infinity and the corresponding entrance pupil is located in the front focal plane of the system. As a consequence, if we place an object with a complex transmittance function $t(X_0, Y_0)$ in the front focal plane of a perfect lens, the amplitude distribution in the focal plane at F' is the Fourier transform of the complex function t, evaluated for the negative coordinates in the focal plane. For any other axial position of the transmittance function t, a supplementary quadratic phase factor $\exp(i\pi r_1'^2 / R_1')$ is present.

If the pupil function f is circularly symmetric and the defocusing factor f_d' is zero, the integral at the right-hand side of Eq. (9.3.4) can be further simplified into

$$U_p(r_1') = -i2\pi s_1^2 R_1' \exp\left(\frac{i\pi r_1'^2}{R_1^1}\right) \int_0^1 f_L'(\rho) J_0(2\pi r_1' \rho) \rho \, d\rho , \qquad (9.3.5)$$

where we equated $\gamma - \pi$ to zero as a consequence of the circular symmetry of f_L' and further used that $\int_0^{2\pi} \exp(ia\cos\theta) \, d\theta = 2\pi J_0(a)$. The transformation of a radially symmetric function according to the integral expression of Eq. (9.3.5) is commonly called a *Hankel* transform of order zero. The upper limit of the Hankel transform integral is formally ∞. In Eq. (9.3.5) the upper integral limit is unity since the function $f_L'(\rho)$ is zero for $\rho > 1$. In the particular case of a uniform pupil function f_L' with circular rim, we encounter the Fourier transform pair

$$\text{circ}[(X_1'^2 + Y_1'^2)^{1/2}] \quad \Leftrightarrow \quad \frac{J_1[2\pi(x_1'^2 + y_1'^2)^{1/2}]}{\pi(x_1'^2 + y_1'^2)^{1/2}} . \qquad (9.3.6)$$

A final remark must be made about the distance R_1' appearing in Eq. (9.3.4) in the leading factor of the expression at the right-hand side. This quantity can be combined with the modulus $|f_L'(\rho, \theta)|$ of the exit pupil amplitude function. The product $R_1'|f_L'|$ is an invariant quantity along an optical ray of the focusing wave field and remains constant when the position of the exit pupil is changed. For this reason, the quantity $R_1'|f_L'|$ is called the ray-invariant amplitude. In electromagnetic units, it would have the dimension of volt. For the purpose of complete normalisation, the ray-invariant amplitude $R_1'|f_L'|$ can be used inside the integrand of the Debye integral instead of the modulus part $|f_L'|$.

9.3.1 Nijboer–Zernike Diffraction Theory for Weakly Aberrated Systems

The use of the Zernike circle polynomials $Z_n^m(\rho,\theta)$ for the representation of the wavefront aberration in the exit pupil of an optical system was proposed in **[385]**,**[390]**. Simultaneously, a basic result for the radial part $R_n^m(\rho)$ of the circle polynomials was derived in these publications, namely

$$\int_0^1 R_n^{|m|}(\rho) J_m(q\rho)\rho d\rho = (-1)^{\frac{n-m}{2}} \frac{J_{n+1}(q)}{q} , \tag{9.3.7}$$

where J_n is the Bessel function of the first kind and order n. The original result was intended to be used for $m \geq 0$, but it is equally valid for negative integer values of m. This relationship was exploited by Zernike in his study of the diffraction of a focused light wave by a sharp edge, for instance the knife-edge in the Foucault mirror test. In **[259]** this result was further applied to the analytic evaluation of the Debye diffraction integral for an imaging system that suffers from small amounts of aberration or defocusing. It plays a crucial role in obtaining the complex amplitude of the point-spread function in image space and can be considered as a higher-order version of the basic Airy diffraction pattern. In what follows we refer to the expression of Eq. (9.3.7) as the 'basic result' of the Nijboer–Zernike diffraction theory.

9.3.1.1 Point-spread Function in the Presence of Wavefront Aberration

The pupil function f'_L with uniform transmission and weak aberration is written in a first approximation as

$$f'_L(\rho,\theta) = \exp\{i\Phi(\rho,\theta)\} \approx 1 + i\,\Phi(\rho,\theta) = 1 + i\sum_{n=0}^{\infty}\sum_{m=-\infty}^{\infty} \alpha_n^m R_n^{|m|}(\rho) \exp(im\theta) . \tag{9.3.8}$$

With this expression for the pupil function f'_L the coefficients α_n^m are the coefficients of the Zernike expansion of the *phase* aberration function $\Phi(\rho,\theta)$. The coefficient α_0^0 yields the constant phase part of Φ. In most imaging applications this constant phase is irrelevant and can be put equal to zero; consequently, the term with $n = m = 0$ can be omitted from the double summation in Eq. (9.3.8). We insert this approximated pupil function into the Debye diffraction integral of Eq. (9.3.4) with $f'_d = 0$. For the image plane complex amplitude we obtain,

$$\begin{aligned} U_p(r'_1,\gamma',0) \;=\; &- is_1^2 R'_1 \exp\left(\frac{i\pi r_1'^2}{R'_1}\right) \int_0^1\int_0^{2\pi}\left[1 + i\sum_{n=1}^{\infty}\sum_{m=-\infty}^{\infty} \alpha_n^m R_n^{|m|}(\rho)\exp(im\theta)\right] \\ &\times \exp\{i2\pi r'_1\rho\cos(\theta-\gamma')\}\,\rho\,d\rho d\theta , \end{aligned} \tag{9.3.9}$$

where, as before, $\gamma' = \gamma - \pi$ for positive values of R_1. The integration with respect to θ is carried out using the following expression for the Bessel function J_m of the first kind and of order m (see **[266]**, §10.9.2),

$$J_m(a) = \frac{i^{-m}}{2\pi}\int_0^{2\pi} \exp\{i(a\cos\theta + m\theta)\}\,d\theta = \frac{i^{-m}}{2\pi}\int_0^{2\pi} \cos m\theta \exp\{ia\cos\theta\}\,d\theta , \tag{9.3.10}$$

where the integral with $\sin m\theta$ in the integrand yields zero.

The substitution of this result for a single term with coefficient α_n^m into Eq. (9.3.9) yields the integral expression

$$\begin{aligned} &i\alpha_n^m \exp(im\gamma')\int_0^1\int_0^{2\pi} R_n^{|m|}(\rho)\exp\{i[2\pi r'_1\rho\cos(\theta-\gamma') + m(\theta-\gamma')]\}\,\rho\,d\rho d\theta \\ &= 2\pi i^{m+1}\alpha_n^m \exp(im\gamma')\int_0^1 R_n^{|m|}(\rho) J_m(2\pi r'_1\rho)\,\rho\,d\rho . \end{aligned} \tag{9.3.11}$$

The integral over ρ is evaluated with the aid of Eq. (9.3.7) and we obtain for the image plane amplitude the expression

$$\begin{aligned} U_p(r'_1,\gamma',0) = &- i\pi s_1^2 R'_1 \exp\left(\frac{i\pi r_1'^2}{R'_1}\right) \\ &\times\left[\left(\frac{2J_1(2\pi r'_1)}{2\pi r'_1}\right) + i\sum_{n=1}^{\infty}\sum_{m=-\infty}^{\infty} i^n\,\alpha_n^m \left(\frac{2J_{n+1}(2\pi r'_1)}{2\pi r'_1}\right)\exp(im\gamma')\right] . \end{aligned} \tag{9.3.12}$$

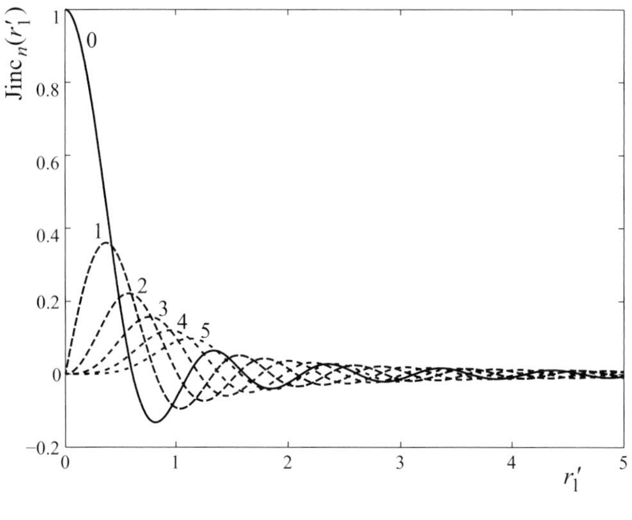

Figure 9.8: The Jinc_n functions $2J_{n+1}(2\pi r'_1)/(2\pi r'_1)$ plotted for the parameter values $n = 0, \cdots, 5$.

This shows that any wavefront perturbation in the exit pupil contributes to the complex amplitude outside the centre of the diffraction image. The first term between brackets in Eq. (9.3.12), the Airy disc function, is the only term that has a nonzero value at the origin. All other aberration terms with $n \neq 0$ are zero at the centre of the diffraction image and yield their maximum contribution to the complex amplitude on annuli that are further and further away from the centre with increasing order n. The functions $2J_{n+1}(x)/x$ are called the Jinc_n functions of a circular aperture, analogous to the sinc function. This latter arises when considering a square aperture whilst the Jinc function (with index $n = 0$) corresponds to the ideal Airy diffraction pattern from a circular aperture. The blurring effect due to aberration terms has been illustrated in Fig. 9.8 where the functions $2J_{n+1}(x)/x$ have been plotted for the argument $r'_1 = x/2\pi$ with r'_1 running from 0 to 5. The position of the first maximum of each Jinc_n function follows from the solution x_{max} closest to the origin of the equation

$$(n+2)J_{n+2}(x_{\mathrm{max}}) = nJ_n(x_{\mathrm{max}}) \, . \tag{9.3.13}$$

The value $r'_1 = x_{\mathrm{max}}/2\pi$, expressed in diffraction units, is a good measure for the blurring of the point spread function in the image plane due to a specific wavefront aberration of order n in the exit pupil.

The first-order approximation of the pupil function f'_L has been refined in **[259]** by including higher-order expansion terms. As an example, we consider a single aberration term of the phase function Φ with azimuthal cosine dependence and coefficient $\alpha^m_{n,c}$. The pupil function is then given by

$$\begin{aligned} f'(\rho,\theta) &= \exp\{i\Phi(\rho,\theta)\} \approx 1 + i\,\Phi - \frac{\Phi^2}{2} - i\,\frac{\Phi^3}{6} + \cdots \\ &= 1 + i\frac{\alpha^m_{n,c}}{2} R^{|m|}_n(\rho)\,[\exp\{im(\theta-\theta_0)\} + \exp\{-im(\theta-\theta_0)\} \\ &\quad - \frac{1}{8}(\alpha^m_{n,c})^2\,[R^{|m|}_n(\rho)]^2\,\{2 + \exp[i2m(\theta-\theta_0)] + \exp[-i2m(\theta-\theta_0)]\} + \cdots \end{aligned} \tag{9.3.14}$$

The challenge in determining these extra terms so as to obtain a better approximation of U_p, arises in the ad hoc calculation of powers and (multiple) products of Zernike polynomials. For a complete expansion of the aberration function up to a high order, this requires a rather cumbersome expansion of the (multiple) product terms in basic polynomials with their proper coefficients and a subsequent addition of coefficients associated with the same polynomial. As an example we give the second-order approximation for U_p in the case of the comatic wavefront aberration term of lowest order with cosine dependence, $\Phi = \alpha^1_{3,c}\,R^1_3(\rho)\cos(\theta-\theta_0)$, where $R^1_3(\rho) = 3\rho^3 - 2\rho$. For the exponential function $\exp[i\Phi(\rho,\theta)]$ we have the approximate second-order expression:

$$\exp\{i\Phi(\rho,\theta)\} \approx 1 - \frac{(\alpha_{3,c}^1)^2}{16} + i\,\frac{\alpha_{3,c}^1}{2}\,R_3^1(\rho)\,\{\exp[i(\theta-\theta_0)] + \exp[-i(\theta-\theta_0)]\}$$
$$- \frac{(\alpha_{3,c}^1)^2}{4}\left\{\frac{9}{20}R_6^0(\rho) + \frac{1}{4}R_4^0(\rho) + \frac{1}{20}R_2^0(\rho) + \frac{1}{10}\left[3R_6^2(\rho) + 2R_2^2(\rho)\right]\right.$$
$$\left.\times\left[\exp\{i2(\theta-\theta_0)\} + \exp\{-i2(\theta-\theta_0)\}\right]\right\}. \tag{9.3.15}$$

With the aid of this expression and the basic result of Eq. (9.3.7) we obtain the value of U_p as a sum of Jinc functions of various orders. Using the definition $\mathrm{Jinc}_n(x) = 2J_{n+1}(x)/x$ we obtain the (normalised) expression,

$$U_p(r_1', \gamma', 0) = \left[1 - \frac{(\alpha_{3,c}^1)^2}{16}\right]\mathrm{Jinc}_0(2\pi r_1') + \alpha_{3,c}^1\,\mathrm{Jinc}_3(2\pi r_1')\cos(\gamma'-\theta_0)$$
$$- \frac{(\alpha_{3,c}^1)^2}{4}\left\{-\frac{9}{20}\mathrm{Jinc}_6(2\pi r_1') + \frac{1}{4}\mathrm{Jinc}_4(2\pi r_1') - \frac{1}{20}\mathrm{Jinc}_2(2\pi r_1')\right.$$
$$\left. - \left[\frac{3}{5}\mathrm{Jinc}_6(2\pi r_1') + \frac{2}{5}\mathrm{Jinc}_2(2\pi r_1')\right]\cos[2(\gamma'-\theta_0)]\right\}. \tag{9.3.16}$$

Figure 9.9 shows the properties of two (normalised) point-spread functions: in the left-hand column the PSF of a perfect imaging system and on the right-hand side the PSF for an imaging system suffering from lowest-order coma with Zernike coefficients $\alpha_3^1 = \alpha_3^{-1} = \alpha_{3,c}^1/2 = 0.207\,\pi$ for the polynomials Z_3^1 and Z_3^{-1}, respectively. These values have been chosen such that the central normalised intensity of the PSF amounts to ≈ 0.80, corresponding to the 'just diffraction-limited' imaging quality. In Eq. (9.3.16), the first term in the right-hand side is the only one that contributes to the on-axis amplitude. To obtain an 80% axial intensity, we require that the square of the factor that multiplies the Jinc_0 function equals 0.80 and this yields $\alpha_{3,c}^1 \approx 1.300$. Since U_p in Eq. (9.3.16) is a second-order approximation, the squared factor $[1-(\alpha_{3,c}^1)^2/16]^2$ is also reliable up to second order and is then given by $1-(\alpha_{3,c}^1)^2/8$. In terms of the wavefront aberration W, we have two exponential Zernike polynomials with coefficients of $0.1035\lambda_0$ each. The proper third-order coma term, being the sum of these two terms and proportional to $3\rho^3$, has a peak-to-valley amplitude of $0.621\,\lambda_0$, measured on the rim of the exit pupil. It is easily verified that, with these two Zernike coefficients, the comatic wavefront aberration surface lies between two spherical shells that are separated in the exit pupil by a distance of approximately $0.25\,\lambda_{n_1}$.

For the example of the comatic aberration, the phase variance due to the wavefront aberration is the sum of the variances of the two exponential components of the coma aberration and is given by

$$\overline{\Phi^2} = 2\,\frac{1}{(n+1)}\left(\frac{\alpha_{n,c}^m}{2}\right)^2 = 0.211\,, \tag{9.3.17}$$

where we have used the expression for the inner product of a Zernike polynomial as given in Appendix D, Eq. (D.2.23). With $n=3$, $m=1$, $\alpha_{3,c}^1/2=0.650$, this yields the value $\overline{\Phi^2} = 0.211$. The Strehl intensity then amounts to 78.9%. The difference from the originally intended on-axis intensity value of 80% stems from the expansion to second order only of the phase function Φ and the associated image plane amplitude U_p of Eq. (9.3.16).

A comparison of the second-order PSF with comatic aberration and the intensity distribution obtained by numerical integration of Eq. (9.3.4) shows the intensity deviations in the image plane that are due to the insufficient number of terms that have been included in the expansion of the pupil phase function Φ in Eqs. (9.3.15) and (9.3.16). For instance, Fig. 9.9f shows the approximately correct axial intensity (78.8% instead of 80%) but the peak intensity of the first asymmetric diffraction ring is 50% larger (0.12) than the correct value of 0.08. The convergence of the series expansion for Φ, when using more terms, is slow since each subsequent term essentially yields an amplitude contribution that is smaller by a factor of

$$\left|\frac{\Phi^{n+1}}{\Phi^n}\right| = \frac{n}{n+1}\,\frac{\alpha_n^m}{n+1}\,, \tag{9.3.18}$$

where we have assumed that each subsequent term has a Zernike expansion coefficient that is of the order of or smaller than unity and that the maximum amplitude contribution by a $\mathrm{Jinc}_n(x)$ function is proportional to $1/n$. A grosso modo $1/n$ convergence results and this means that a substantially larger number of terms is needed in the case of α_n^m values that are close to or larger than unity.

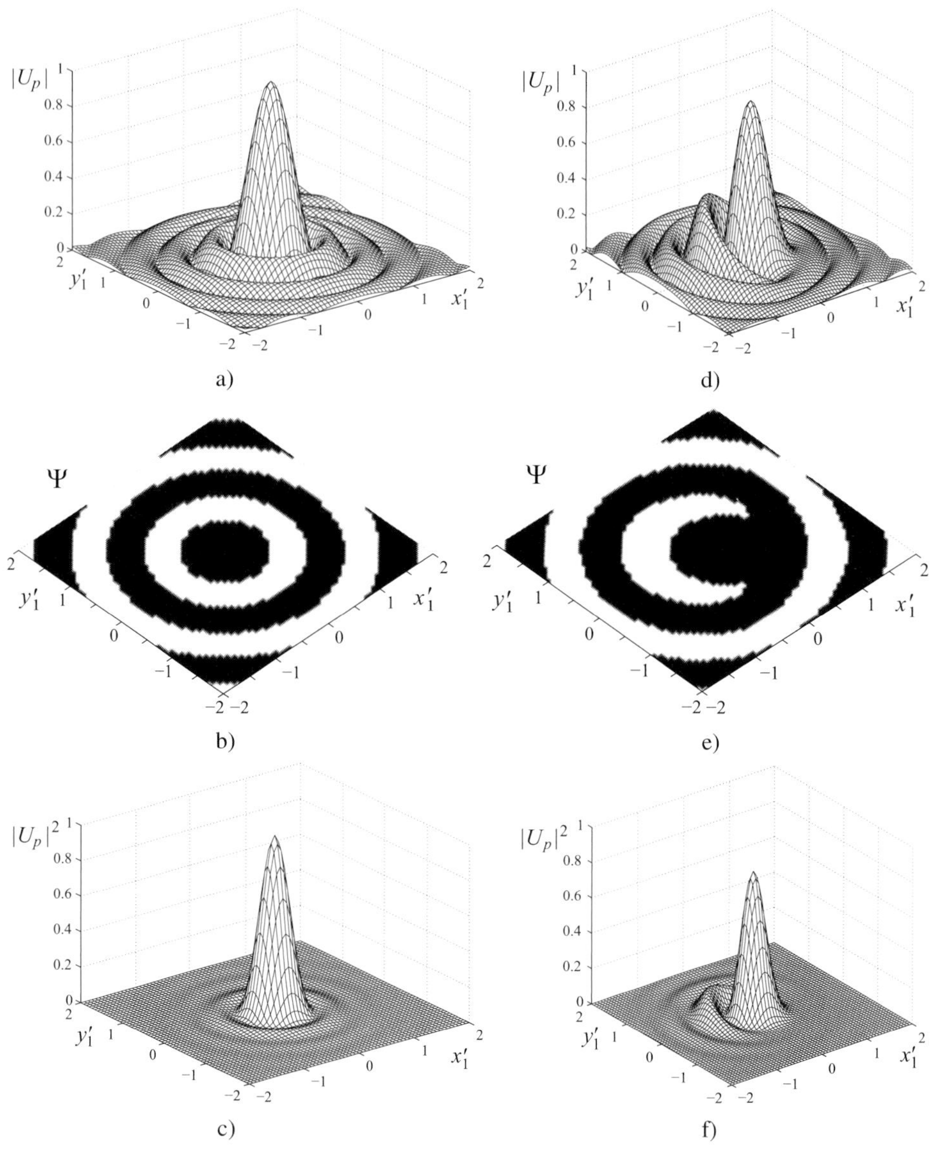

Figure 9.9: Calculation of the amplitude, phase and intensity of the point-spread function PSF (Nijboer–Zernike diffraction integral with expansion of the aberration function Φ up to the second order). a) and d) Modulus $|U_p|$ of the complex amplitude of the PSF. b) and e) Binary phase Ψ of the PSF (dark shade is zero, white is π). c) and f) Intensity ($|U_p|^2$) of the PSF.
a), b) and c) Perfect imaging system.
d), e) and f) Comatic aberration, Zernike coefficients $\alpha_3^1 = \alpha_3^{-1} = \alpha_{3,c}^1/2 = 0.207\pi$, $\theta_0 = 3\pi/4$.

In Appendix D, Section D.4, a method is given for the systematic calculation of the expansion coefficients of the product of two functions that are, each of them, given by a Zernike polynomial expansion. Triple and still higher multiple products can be obtained by a repeated application of this product rule (that goes back to Tango **[336]**) and is the result of a group-theoretical treatment of the two-dimensional Zernike disc polynomials.

An alternative approach, presented in Eq. (9.2.17), is to construct a Zernike expansion of the complex amplitude function on the exit pupil sphere and to use the generally complex expansion coefficients β_n^m of this expansion directly in the diffraction integral. In what follows, we adhere to this latter method. It requires a single-time least-squares fit of the complex pupil function on a Zernike polynomial basis to obtain the complex expansion coefficients β_n^m. From that point on, the analytic results of the Nijboer–Zernike diffraction theory are directly accessible.

9.3.1.2 Strehl Intensity and Maréchal's Criterion for Weak Aberration

At this point it is useful to discuss the relationship between the on-axis intensity (Strehl intensity) of the diffraction image of a point source and the combined influence of the amplitude *and* phase variations of the exit pupil function f'_L of the imaging system; f'_L is the normalised pupil function according to Eq. (9.2.2) with the wavefront aberration W' expressed in units of the vacuum wavelength λ_0. According to Huygens' principle, the complex amplitude in the focus of an imaging system is the superposition of all secondary wave disturbances coming from the exit pupil sphere. In the case of an aberrated imaging system with a wavefront aberration $W'(\rho,\theta)$, measured with respect to the exit pupil sphere, we find the normalised intensity at the focal point, denoted by $r'_1 = 0$, is given by

$$I(0,0) = \left|\iint_S A'(\rho,\theta)\exp\{i2\pi W'(\rho,\theta)\}dS\right|^2 \Bigg/ \left|\iint_S A'(\rho,\theta)dS\right|^2 , \tag{9.3.19}$$

where S is the area of the exit pupil and A' is the modulus of the complex amplitude on the exit pupil sphere. The normalised axial intensity in the presence of aberration according to Eq. (9.3.19) is called the Strehl intensity **[330]**. For weak aberrations the expression for $I(0,0)$ can be approximated by an expansion up to the second order of the exponential function in the numerator of Eq. (9.3.19), which yields

$$I(0,0) = 1 - 4\pi^2\left(\frac{\overline{A'}\,(\overline{A'W'^2}) - \left(\overline{A'W'}\right)^2}{\left(\overline{A'}\right)^2}\right), \tag{9.3.20}$$

where the averaging bars denote the integration of the various terms of the second-order expansion over the circular area of the exit pupil.

For the case of a uniform light amplitude A' in the exit pupil this expression simplifies to the original version for the axial intensity given in **[259]**, i.e.

$$I(0,0) = 1 - \Delta I(0,0) = 1 - 4\pi^2\left\{\overline{W'^2} - \left(\overline{W'}\right)^2\right\} = 1 - 4\pi^2\, V_{W'} = 1 - 4\pi^2\,(W'_{\text{rms}})^2 , \tag{9.3.21}$$

where $V_{W'}$ denotes the variance of the wavefront aberration over the exit pupil cross-section and W'_{rms} is the corresponding standard deviation, all expressed in units of the vacuum wavelength λ_0. We note that the quantity $2\pi W'$ is equal to the phase aberration Φ on the exit pupil sphere and write

$$I(0,0) = 1 - (\Phi_{\text{rms}})^2 . \tag{9.3.22}$$

The basic expression of Eq. (9.3.21) goes back to Nijboer **[259]** who assumed that $W' \ll \lambda_0$. The result became better known as Maréchal's criterion after an independent, but later, publication by this author **[236]**.

By convention, an optical system is said to be 'just diffraction-limited' if the reduction in peak intensity is 20% and the (axial) Strehl intensity thus amounts to 80%. Equation (9.3.21) shows that this is the case if the root-mean-square wavefront aberration, commonly called the rms OPD (Optical Path Difference), equals $0.071\,\lambda_0$. Previously, separate criteria for wavefront aberration were given for each type of aberration. The elegance of the Nijboer–Maréchal criterion is that it can be applied to all types of aberration and equally well to other system perturbations like, for instance, defocusing.

9.3.1.3 Point-spread Function in the Presence of Defocusing; the Phase Anomaly

In the case of a uniform amplitude in the exit pupil and in the absence of aberration, the complex amplitude in a defocused plane follows from Eq. (9.3.4) and is given by

$$U_p(r'_1,\gamma,z') = -i s_1^2 R'_1 \exp\left(\frac{i2\pi z'_1}{s_1^2}\right)\exp\left(\frac{i\pi r_1'^2}{R'_1}\right)$$
$$\times \int_0^1\int_0^{2\pi} \exp(-i\pi z'_1\rho^2)\exp\{i2\pi r'_1\rho\cos(\theta-\gamma)\}\,\rho\, d\rho d\theta . \tag{9.3.23}$$

In **[259]**, the first exponential factor in the integrand is written as an infinite series using Bauer's formula **[4]**, §10.1.47,

$$\exp(iz\cos\theta) = \sum_{l=0}^{\infty}(2l+1)\, i^l\, j_l(z)\, P_l(\cos\theta)\,, \tag{9.3.24}$$

where $j_l(x)$ is the spherical Bessel function of the first kind and of order l. The substitution $\cos\theta = 2\rho^2 - 1$ is made and a property is used that links the symmetrical Zernike polynomials $R_{2l}^0(\rho)$ to the Legendre polynomials, $P_l(2\rho^2 - 1) = R_{2l}^0(\rho)$ (see Eq. (D.2.12)), with the result that

$$\exp(-i\pi z_1'\rho^2) = \exp\left(-i\frac{\pi z_1'}{2}\right)\sum_{l=0}^{\infty}(-i)^l\,(2l+1)\, j_l(\pi z_1'/2)\, R_{2l}^0(\rho)\,. \tag{9.3.25}$$

Bauer's formula according to Eq. (9.3.25) can be inserted into Eq. (9.3.23) and the integration over θ is performed with the aid of Eq. (9.3.10). With the basic result of the Nijboer–Zernike theory of Eq. (9.3.7) we then obtain

$$\begin{aligned} U_p(r_1', \gamma, z_1') = &-i\pi s_1^2 R_1' \exp\left(\frac{i2\pi z_1'}{s_1^2}\right)\exp\left\{i\pi\left(\frac{r_1'^2}{R_1'} - \frac{z_1'}{2}\right)\right\} \\ &\times \sum_{l=0}^{\infty} i^l\,(2l+1)\, j_l(\pi z_1'/2)\left[\frac{2J_{2l+1}(2\pi r_1')}{2\pi r_1'}\right]. \end{aligned} \tag{9.3.26}$$

The on-axis amplitude has a particularly simple form because the only nonzero term in the summation series is that with $l = 0$, whereby

$$U_p(0,0,z_1') = -i\pi s_1^2 R_1' \exp\left\{\frac{i2\pi z_1'}{s_1^2}\left(1 - \frac{s_1^2}{4}\right)\right\}\operatorname{sinc}\left(\frac{\pi z_1'}{2}\right). \tag{9.3.27}$$

The phase of the diffracted amplitude U_p in the focal region shows a particular behaviour. For the on-axis amplitude an analytic expression of the phase can be found by inspection of the argument of $U_p(0,0,z_1')$. The particular phase development of the focused wave with respect to the phase of a hypothetical spherical wave is found by removing the phase term $\exp(i2\pi z_1'/s_1^2)$ from Eq. (9.3.27), which corresponds to the phase along the optical axis of a spherical wave. The remaining phase δ_f of the focusing diffracted wave is then given by

$$\delta_f(z_1') = \arg\{-i\exp[-i\pi z_1'/2]\operatorname{sinc}(\pi z_1'/2)\} = -\frac{\pi}{2}\left\{z_1' + \operatorname{sgn}\left[\operatorname{sinc}\left(\frac{\pi z_1'}{2}\right)\right]\right\}. \tag{9.3.28}$$

The phase factor δ_f is commonly called the anomalous phase factor of the focused wave. The phase at the focal point equals $-\pi/2$. For any pair of points that are symmetrically arranged with respect to the geometrical focus we have that

$$\delta_f(z_1') + \delta_f(-z_1') = -\pi\operatorname{sgn}\left[\operatorname{sinc}\left(\frac{\pi z_1'}{2}\right)\right] = -\pi\,, \tag{9.3.29}$$

where we project the phase values modulo 2π on the interval $[-3\pi/2\,,\,\pi/2]$. The anomalous phase factor δ_f has been plotted in Fig. 9.10 as a function of z_1' and for $r_1' = 0$. The graph shows the slower phase progression of the focused wave along the axis, accumulating a total phase lag of π or an optical pathlength of $\lambda_{n_1}/2$ when moving through the central lobe of the axial amplitude function $\operatorname{sinc}(\pi z_1'/2)$.

The phase anomaly was first discovered experimentally by Gouy **[116]** when performing interferometric experiments. A special interferometer for the testing of lenses or lens surfaces that was designed by Linnik **[213]** and rediscovered by Smartt and Strong **[317]**, is very appropriate to demonstrate the effect (see Fig. 9.11). The focused beam and the virtually spherical wave produced by the small pinhole in the absorbing layer at P are brought to interfere on a distant screen. Circular interference fringes are visible. In Fig. 9.11a the axial fringe is dark because of the extra phase difference of $-\pi$ that is imparted to the focused wave after the creation of the spherical reference beam. In b) both beams have been given the phase difference occurring in the focal region and a bright fringe is visible on the axis.

The anomalous phase δ_f was easily calculated for the axial point using Eq. (9.3.28). More generally, for a propagation direction at an angle α_1 with the optical axis, the phase difference with respect to a spherical reference beam follows from the general expression for U_p given in Eq. (9.3.26). The phase anomaly follows from the argument of the complex amplitude U_p minus the phase of the spherical wave measured from the focal point F' to a general point $P(r_1', z_1')$ on a

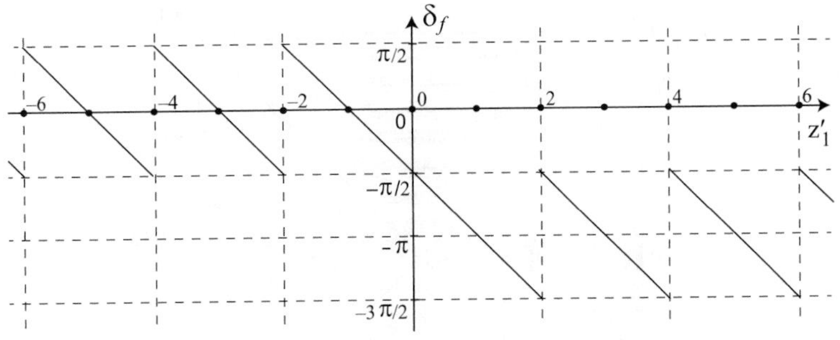

Figure 9.10: The anomalous phase $\delta_f(z_1')$ of the amplitude $U_p(0, 0, z_1')$ measured along the central axis ($r_1' = 0$) of a focused beam.

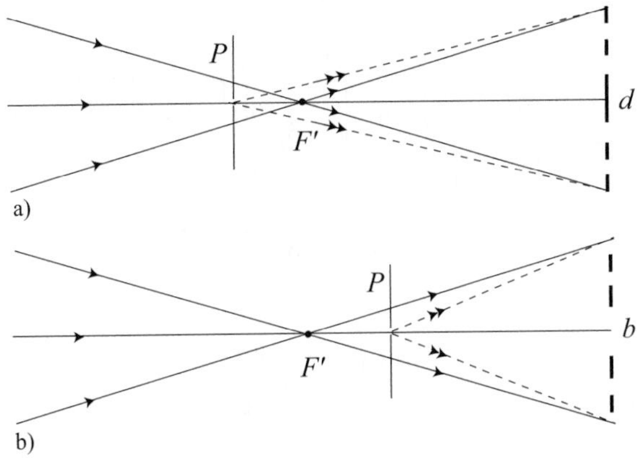

Figure 9.11: Measurement of the phase anomaly around focus with the aid of a Linnik interferometer. The plate P carries an absorbing coating with a small hole in it that has a diameter that is smaller than the point-spread function produced by the focusing beam (focus at F'). The dark axial fringe has been denoted by d (a), the bright fringe by b (b).

ray through F' at an angle α_1 with the axis. The first contribution γ_1 to δ_f comes from the two exponential factors in Eq. (9.3.26) and from the pathlength $F'P$ of the spherical reference wave, according to

$$\gamma_1 = \exp\left(\frac{i2\pi z_1'}{s_1^2}\right) \exp\left(-\frac{i\pi z_1'}{2}\right) \exp\left(-\frac{i2\pi z_1'}{s_1^2}\sqrt{1 + \frac{r_1'^2 s_1^2}{z_1'^2}}\right), \tag{9.3.30}$$

where we have expressed the pathlength $F'P$ in terms of the normalised axial and transverse radial coordinate and where we have assumed that $r_1'^2 \ll R_1'$. The radial normalised coordinate r_1' and the axial normalised coordinate z_1' on a ray at angle α_1 are related through $r_1'/z_1' = \tan(\alpha_1)/s_1 \approx \alpha_1/\alpha_{1,m}$, where the latter approximation is valid when using the small-angle scalar diffraction integral. Denoting this ratio by p, we have that the marginal ray of the focused beam corresponds to the value $p = 1$. We obtain the following expression for the phase difference γ_1,

$$\gamma_1 = \exp\left[-\frac{i\pi z_1'}{2}\left(1 + \frac{4p^2}{1 + \sqrt{1 + s_1^2 p^2}}\right)\right]. \tag{9.3.31}$$

With the aid of Eqs. (9.3.26) and (9.3.31) we can numerically calculate the anomalous through-focus phase δ_f for a general elevation angle α_1 in the focused beam; the extra phase terms come from the leading factor $-i$ and from the argument of the summation term in Eq. (9.3.26).

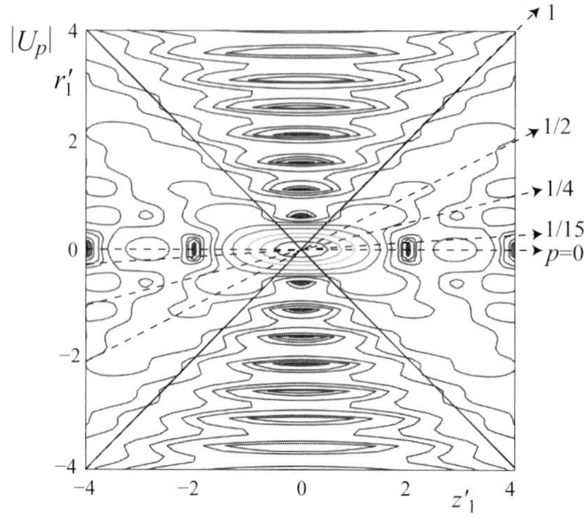

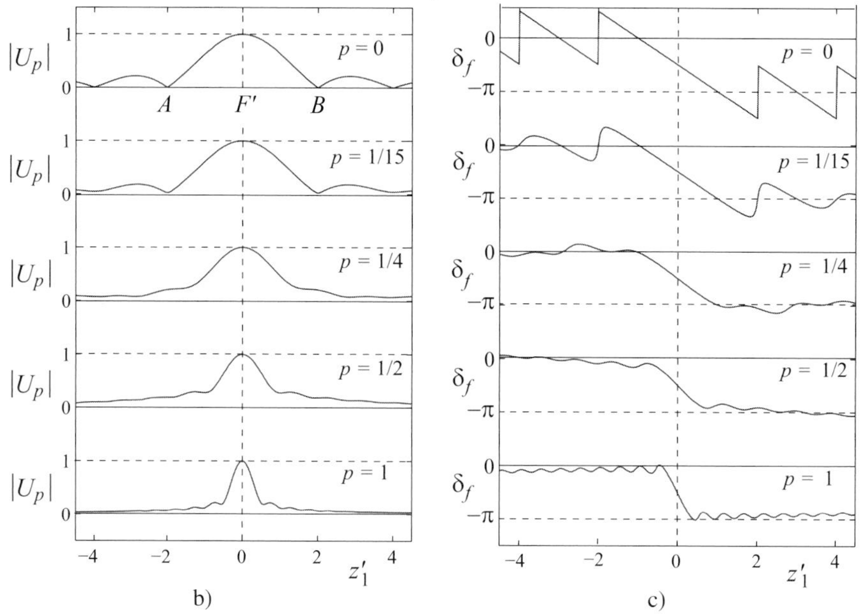

Figure 9.12: a) A contour plot of $|U_p|$ with the traces (dashed lines) for various p-values that are used in graphs b) and c).
b) The modulus of the (normalised) complex amplitude $|U_p|$ measured along a certain propagation direction $p = \alpha_1/s_1$ in the focal region; horizontal coordinate is z'_1.
c) The phase in the focal region measured along the same propagation directions (s_1=0.30 or $F_\# \approx 1.7$).

The amplitude and phase of U_p are shown in Fig. 9.12a–c, for five angles α_1 that are given by means of the parameter $p = \alpha_1/s_1$. The value $p = 1$ corresponds to the transition towards the geometrical shadow of the focused beam. The amplitude on-axis is described by the sinc($\pi z'_1/2$) function. For an oblique propagation direction the half-width of the central part of the amplitude function becomes narrower with respect to the axial coordinate z'_1. Regarding the phase of U_p, this quantity equals $-\pi/2$ in the geometric focus F'. The phase lag of π upon propagation through focus is accumulated across the full range between the first axial intensity zeros A and B on both sides of the geometrical focus F' if $p \ll 1$. The distance AB equals eight Rayleigh focal depths (one focal depth corresponds to $z'_1 = 1/2$) of the focusing beam. For steeper propagation directions, the axial range over which the phase lag of π occurs is reduced. For the marginal ray, the anomalous phase change of π is accumulated over an axial distance of approximately 1.5 focal depths. At large distances from the geometrical focus, the limit values of δ_f equal 0 and $-\pi$ for $z_1' < 0$ and > 0, respectively.

9.3.1.4 Fresnel Zone Analysis of the Phase Anomaly around Focus

The phase anomaly around focus can be explained in an approximate way using the concept of the Fresnel zone decomposition of a wavefront **[99]**,**[100]**,**[173]**. In Fig. 9.13a we have indicated the size of the first Fresnel zone Z_1 on a spherical wavefront (focus at F') with the reference point for the zone construction at the axial point D_f. Because of the axial symmetry, the wavefront normal at the origin O of the wavefront points towards D_f and this makes O the centre of the first zone. The sphere S_0 with radius OD_f through O is an osculating sphere of the spherical focusing wavefront at O. The outer limit of the mth Fresnel zone is found by finding the intersection curve on the wavefront in the exit pupil with a second spherical surface S_m with its centre of curvature at D_f and with a radius that is equal to $OD_f \pm m\lambda_{n_1}/2$. In Fig. 9.13b we have shown the two cross-sections of the spheres S_0 and S_1 in the plane of the graph. The point L is on the intersection curve of S_1 and the wavefront surface W; for the case of a spherical wavefront, the intersection curve is a circle. For such a spherical wavefront W, we consider the phase at the point D_f when many Fresnel zones cover the exit pupil surface. In this limiting case, the complex amplitude at D_f equals half the complex amplitude that is carried by the first Fresnel zone. From Fig. 9.13b we can deduce that the phase at D_f is $\pi/2$ plus $k_0[OD_f]$; the extra phase of $\pi/2$ is the phase averaged over the first zone. With respect to a fictitious spherical wave, the extra phase at D_f thus amounts to $+\pi/2$. If we consider a point that is far beyond the focal point F' of the beam, the same reasoning will yield a phase difference of $-\pi/2$.

Using the Fresnel zone construction in the exit pupil for the spherical wave W, we thus obtain an extra phase change of $-\pi$ as we move the point D_f through the focal region. This means that co-phasal surfaces with some chosen phase increment in the focal region are separated by a larger distance than in the far-field zone of the focused beam. Some authors have stated that the wave propagation speed in the focal region is increased or that the wavelength of the light has become larger but these statements are physically incorrect since the medium properties are unchanged. The anomalous phase behaviour in the focal region has to be attributed to a combined effect of diffraction and interference that follows from the truncation of the focusing beam.

The qualitative results obtained with the aid of the construction of Fresnel zones are adequate for explaining the phase jump of $-\pi$ through focus, however, the exact value of $-\pi/2$ at the focal point itself does not follow from the Fresnel zone construction. To obtain this phase lag in focus, we have to invoke the phase of the Huygens secondary waves which arise on the wavefront through the exit pupil sphere. In his memoir of 1818 on the diffraction of light (**[103]**, pp. 39–42) Fresnel made the heuristic assumption that the secondary waves have a phase lag of $\pi/2$ with respect to the primary wave. This assumption was mathematically confirmed by Kirchhoff since a multiplicative factor $-i$ appears in front of his diffraction

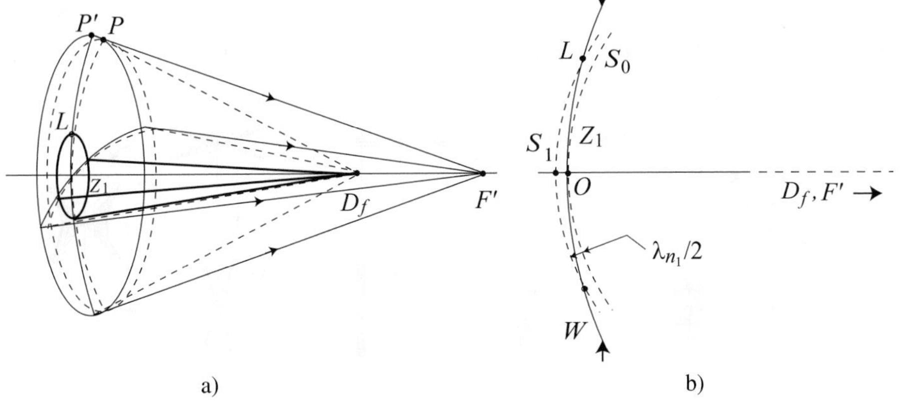

Figure 9.13: a) Construction of the first Fresnel zone Z_1 on a spherical focused wave with its focus at F'. The zone is constructed with respect to a defocused axial reference point D_f, that is between the exit pupil and the focus F'.
b) An enlarged view of the two limiting surfaces of the first Fresnel zone, separated by a distance $\lambda_{n_1}/2$ where λ_{n_1} is the wavelength in the medium with index n_1.

integral (see Eqs. (8.4.15)–(8.4.17) in Chapter 8 on scalar and vector diffraction theory). The phase retardation of $\pi/2$ of the secondary waves on the exit pupil sphere thus accounts for the absolute phase of the converging wave in the focal point F'.

9.3.1.5 Fresnel Zone Analysis of the Phase in an Astigmatic Focused Beam

Gouy first gave the theoretical analysis of the anomalous phase for focused acoustic waves. At a later stage, he showed that optical waves should exhibit a similar behaviour and he also gave a description of the anomalous phase progression in an astigmatic beam with two principal curvatures. The anomalous phase in the focal region of an astigmatic beam can also be treated with the aid of Fresnel zones. The only extension is that the Fresnel zones on the astigmatic wavefront are delimited by circles but their borders are defined by conic sections **[100]**. Far away from the focal region, the borders of the zones are elliptical; in between the focal lines they become hyperbolic sections and their asymptotes. In what follows we discuss in somewhat more detail these special shapes of the first Fresnel zone(s) in the case of an astigmatic beam and obtain the phase behaviour of the focused wave on the axis of such a beam.

We start with a reference point D_f relatively far away from the astigmatic focus, at a distance d from O (see Fig. 9.14). Using a quadratic approximation for the z-coordinates of the Fresnel surfaces S_0 and S_m in image space and of the wavefront surface W in Fig. 9.13 we can write

$$z_{S,0} = \frac{1}{2d}(X_1^2 + Y_1^2)\,, \quad z_{S,m} = \frac{X_1^2 + Y_1^2}{2(d + m\lambda_{n_1}/2)} - \frac{m\lambda_{n_1}}{2}\,, \quad z_W = \frac{X_1^2}{2R_1} + \frac{Y_1^2}{2R_2}\,. \tag{9.3.32}$$

In the expression for $z_{S,m}$, we can neglect in most practical cases the quantity $m\lambda_{n_1}/2$ with respect to d. The intersection curve of the wavefront surface W and the sphere S_m is thus defined by the equation,

$$\frac{X_1^2(R_1 - d)}{R_1 d} + \frac{Y_1^2(R_2 - d)}{R_2 d} = m\lambda_{n_1}\,. \tag{9.3.33}$$

We now define the 'best focus' distance $f_b = (R_1 + R_2)/2$ and the distance $z_a = R_2 - R_1$, the distance between the two orthogonal astigmatic lines. Denoting the defocus distance measured from F' by $\Delta = d - f_b$, the intersection curve on the wavefront W of the outer rim of the mth Fresnel zone can be written

$$X_1^2 \Bigg/ \left(\frac{m f_b^2 \lambda_{n_1}}{-z_a/2 - \Delta}\right) + Y_1^2 \Bigg/ \left(\frac{m f_b^2 \lambda_{n_1}}{z_a/2 - \Delta}\right) = \pm\frac{X_1^2}{a^2} \pm \frac{Y_1^2}{b^2} = 1\,, \tag{9.3.34}$$

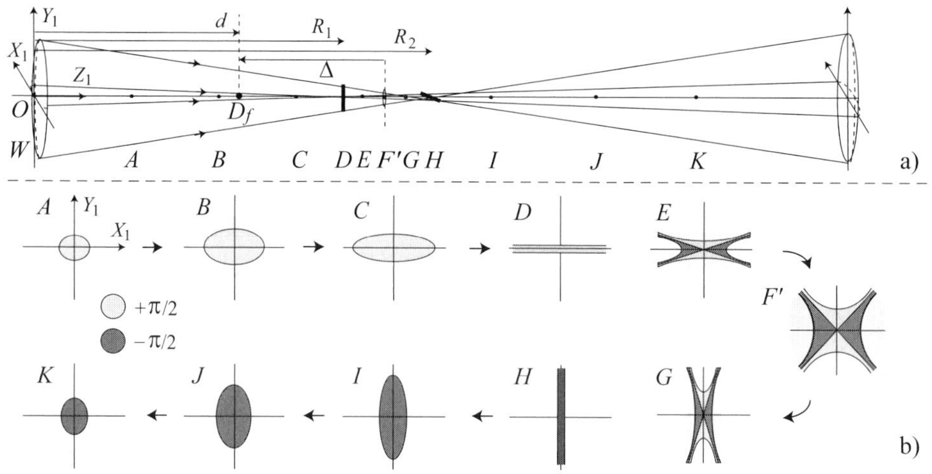

Figure 9.14: a) An astigmatic focusing beam with its incident astigmatic wavefront W through O. The two line foci are found at D and H at distances R_1 and R_2 from O, respectively. The best focus is found at F'.
b) The shape of the first Fresnel zone(s) projected onto the wavefront W through O as a function of the distance d of the reference point from the best-focus point F' (the various positions of the reference point are given by the characters $A, \cdots, K$). The average phase of the Fresnel zones or sections thereof is $+\pi/2$ (light-shaded) or $-\pi/2$ (dark-shaded).

where the case with two minus signs is excluded so as to ensure the coordinates of the intersection curve are real. Depending on these signs, the equation represents an ellipse or a hyperbola.

The following cases arise when considering the first Fresnel zone (with its origin at O) as a function of the defocusing distance Δ:

- $\Delta < -z_a/2$
  Both denominators in Eq. (9.3.34) are positive for $m = +1$. The shape of the first Fresnel zone is hence an ellipse and, with $m = +1$, its phase varies from 0 at O to $+\pi$ at the rim, such that the average value over the first zone is $\pi/2$. In Fig. 9.14b this phase value has been indicated by a light shading of the ellipse in sub-graph A. For very large values of Δ, the zone becomes virtually circular with a radius given by $f_b\sqrt{\lambda_{n_1}/\Delta}$. Closer to the first vertical focal line in C, the zone has become an ellipse that is elongated in the X_1-direction.
- $\Delta = -z_a/2$
  The value $m = +1$ is needed to obtain a real solution for the intersection curve and the first zone degenerates into a horizontal stripe with a width of $2Y_1 = 2f_b\sqrt{\lambda_{n_1}/z_a}$. The average phase contributed from this zone equals $+\pi/2$.
- $-z_a/2 < \Delta < 0$ Two solutions for the first zone are feasible in this case ($a^2 > b^2$). For $m = +1$, we have a hyperbolic equation when the sign of the X_1^2 term is negative. The zone is delimited by the asymptote axes of the hyperbola and the hyperbolic sections that intersect the Y_1-axis; its average phase is $+\pi/2$. Another first zone is found for $m = -1$ and it is delimited by the same asymptotes and by the hyperbolic sections that intersect the X_1-axis. Its average phase is $-\pi/2$. For this position in the focal region, it is seen that the zone with phase $+\pi/2$ covers the largest area on the wavefront surface and the total net phase from both zones is positive.
- $\Delta = 0$
  This particular case for the best focus position at F' leads to hyperbolas with orthogonal asymptotes that bisect the four quadrants of the (X_1, Y_1) plane. With $a^2 = b^2$ and $m = +1$, we have $-X_1^2 + Y_1^2 = a^2$, which leads to a zone area delimited by the hyperbola asymptotes and the two hyperbolic sections that intersect the Y_1-axis. For $m = -1$, 90° rotated zone sections are obtained with an average phase of $-\pi/2$. The total phase from both equal-sized first zones is exactly zero.
- $\Delta > 0$
  Analogous considerations for positive Δ values lead to the graphs G to K in Fig. 9.14b. They have a phase that tends towards an average value of $-\pi/2$ for large positive values of Δ, well away from the focal region.

The analysis with the aid of Fresnel zones shows that the phase anomaly for an astigmatic beam consists of a two-stage phase delay. From a large defocusing value to the best focus we observe a first delay of $\pi/2$. Further progression along the optical axis through the second focal line imparts a further delay of $\pi/2$. The total phase change thus amounts to $-\pi$. The astigmatic beam is the basic geometry of a focused beam. The spherical focused beam is a special case and shows the phase shift of $-\pi$ in a single step when passing through the central axial amplitude lobe in the focal region.

9.3.2 Aberrated and Defocused Point-spread Function (Strong Aberration)

A defocusing of the receiving plane in image space can be included in the Nijboer–Zernike diffraction theory. We assume that the pupil function f'_L in Eq. (9.3.4) has a modulus of unity and is determined solely by the aberration of the imaging system,

$$f'_L(\rho, \theta - \pi) = \exp\{i\Phi(\rho,\theta)\} \ . \tag{9.3.35}$$

In what follows, we assume that a single aberration term is present so that $\Phi(\rho,\theta) = \alpha_n^m \, R_n^{|m|}(\rho)\exp(im\theta)$. To expand the exponential defocusing factor $\exp(-if'_d\rho^2)$ in the integrand of the integral in Eq. (9.3.4) we use Bauer's formula of Eq. (9.3.25). For ease of notation, we collect the factors outside the integral in Eq. (9.3.4) into a single factor $g_n(z'_1, r'_1)$ and obtain the expression

$$\begin{aligned} U_p(r'_1,\gamma',z'_1) &= g_n(z'_1,r'_1)\int_0^1\int_0^{2\pi} \exp\left\{i\left[-\pi z'_1\rho^2 + 2\pi r'_1\rho\cos(\theta-\gamma') + \Phi(\rho,\theta)\right]\right\}\rho\, d\rho d\theta \\ &= g_n(z'_1,r'_1)\exp\left(-i\frac{\pi z'_1}{2}\right)\sum_{l=0}^{\infty}(-i)^l\,(2l+1)j_l\left(\frac{\pi z'_1}{2}\right)\int_0^1\int_0^{2\pi} R_{2l}^0(\rho) \end{aligned}$$

$$\times\left[1 + i\alpha_n^m R_n^{|m|}(\rho)\exp(im\theta) - \frac{1}{2}(\alpha_n^m)^2\left\{R_n^{|m|}(\rho)\right\}^2\exp(i2m\theta) + \cdots\right]$$
$$\times\exp\{i2\pi r_1'\rho\cos(\theta-\gamma')\}\,\rho\,d\rho d\theta\,. \tag{9.3.36}$$

The integration with respect to θ yields the result

$$U_p(r_1',\gamma',z_1') = \pi g_n(z_1',r_1')\exp\left(-i\frac{\pi z_1'}{2}\right)\sum_{l=0}^{\infty}(-i)^l(2l+1)j_l\left(\frac{\pi z_1'}{2}\right)\int_0^1 R_{2l}^0(\rho)\Big[2 + 2i^{m+1}\alpha_n^m R_n^{|m|}(\rho)J_m(2\pi r_1'\rho)\exp(im\gamma')$$
$$+\, i^{2m+2}(\alpha_n^m)^2\left\{R_n^{|m|}(\rho)\right\}^2 J_{2m}(2\pi r_1'\rho)\exp(i2m\gamma') + \cdots\Big]\rho\,d\rho\,. \tag{9.3.37}$$

The integral in Eq. (9.3.37) can be reduced to the standard result of Eq. (9.3.7) if, for instance, the second term between brackets in the integrand, containing the product of two Zernike polynomials $R_{2n}^0(\rho)\,R_n^{|m'|}(\rho)$ can be written as a series of individual polynomials $R_k^{|l|}$, with the usual conditions on the indices (k,l) of each polynomial. In **[259]** the following general expression for the product of two Zernike polynomials is given,

$$R_k^l(\rho)\,R_{k'}^{l'}(\rho) = \sum_{j=0}^{(k+k'-l-l'+2s)/2} C_j\,R_{k+k'-2j}^{l+l'-2s}(\rho)\,. \tag{9.3.38}$$

In the special case above, this expression reduces to

$$R_{2l}^0(\rho)\,R_n^{|m|}(\rho) = \sum_{j=0}^{(2l+n-|m|+2s)/2} C_j\,R_{2l+n-2j}^{|m|-2s}(\rho)\,. \tag{9.3.39}$$

The coefficient C_j can be found by evaluating the inner product of the left-hand term in Eq. (9.3.39) with the Zernike polynomial with the appropriate lower index $2l+n-2j$ in the right-hand side of this equation.

A second approach relies on recurrence relations for the product of a power of ρ and a Zernike polynomial, with the linear combination of the resulting polynomials having higher or lower azimuthal order than the original one (see **[259]**). As an example we give two linearisation expressions for the product of a first and a second power in ρ with a Zernike polynomial $R_n^m(\rho)$,

$$\rho R_n^m(\rho) = \frac{n-m+2}{2(n+1)}R_{n+1}^{m-1}(\rho) + \frac{n+m}{2(n+1)}R_{n-1}^{m-1}(\rho)\,, \tag{9.3.40}$$

$$\rho^2 R_n^m(\rho) = \frac{(n-m)(n-m+2)}{4(n+1)(n+2)}R_{n+2}^{m-2}(\rho) + \frac{(n-m+2)(n+m)}{2n(n+2)}R_n^{m-2}(\rho)$$
$$+\frac{(n+m)(n+m-2)}{4n(n+1)}R_{n-2}^{m-2}(\rho)\,. \tag{9.3.41}$$

Comparable formulas have been devised with increased upper index in the right-hand side of the expressions,

$$\rho R_n^m(\rho) = \frac{n+m+2}{2(n+1)}R_{n+1}^{m+1}(\rho) + \frac{n-m}{2(n+1)}R_{n-1}^{m+1}(\rho)\,, \tag{9.3.42}$$

$$\rho^2 R_n^m(\rho) = \frac{(n+m+2)(n+m+4)}{4(n+1)(n+2)}R_{n+2}^{m+2}(\rho) + \frac{(n-m)(n+m+2)}{2n(n+2)}R_n^{m+2}(\rho)$$
$$+\frac{(n-m)(n-m-2)}{4n(n+1)}R_{n-2}^{m+2}(\rho)\,. \tag{9.3.43}$$

Multiplying Eq. (9.3.40) by ρ and then using Eq. (9.3.42) twice leads to an expression containing radial polynomials that all have equal upper index,

$$\rho^2 R_n^m(\rho) = \frac{(n-m+2)(n+m+2)}{4(n+1)(n+2)}R_{n+2}^m(\rho) + \left\{\frac{(n+m+2)^2}{4(n+1)(n+2)} + \frac{(n-m)^2}{4n(n+1)}\right\}R_n^m(\rho)$$
$$+\frac{(n-m)(n+m)}{4n(n+1)}R_{n-2}^m(\rho)\,. \tag{9.3.44}$$

The repeated application of such relationships makes it possible to force the simple or multiple products of Zernike polynomials in Eq. (9.3.37) into a shape that allows the application of the basic integral result of the Nijboer–Zernike diffraction theory according to Eq. (9.3.7).

Instead of the rather tentative and laborious procedure given above, a straightforward method to handle these products was developed by Tango **[336]**. We choose the desired upper index m_3 for the polynomials that are used for the linearisation and obtain

$$Z_{n_1}^{m_1}(\rho,\theta)\, Z_{n_2}^{m_2}(\rho,\theta) = \sum_{n_3,m_3} C_{n_1,n_2,n_3}^{m_1,m_2,m_3}\, Z_{n_3}^{m_3}(\rho,\theta)\,, \tag{9.3.45}$$

where the C coefficients are now given by the Clebsch–Gordan coefficients (see Appendix D, Eqs. (D.4.9) and (D.4.10)). In this way, the integrals of the various terms in the integrand of Eq. (9.3.37) can be cast into a form for which the upper index of the Zernike polynomial and the order of the Bessel function are equal. The basic Nijboer–Zernike result for the integration over ρ can then be applied.

9.3.2.1 Symmetry Properties of the Intensity in the Focal Region

To study the structure of the intensity distribution of the point-spread function in the focal region we consider the expression for the complex amplitude U_p on the first line of Eq. (9.3.36) and insert a single aberration (cosine) term for the phase function Φ into the integrand,

$$\exp\{i\Phi(\rho,\theta)\} = \exp[i\alpha_{n,c}^m R_n^m(\rho)\cos m\theta] = \sum_{p=0}^{\infty} \frac{\left(i\alpha_{n,c}^m\right)^p}{p!} \{R_n^m(\rho)\}^p \cos^p m\theta\,, \tag{9.3.46}$$

where $\alpha_{n,c}^m$ is the cosine coefficient pertaining to the Zernike aberration term with indices (n, m). In accordance with **[259]**, p. 46, we write the powers of $\cos\theta$ as as series of cosines with arguments that are multiples of θ according to

$$\cos^m\theta = 2^{-m} \sum_{k=0}^{\lfloor m/2\rfloor} \binom{m}{k} \left(2-\delta_{m-2k,0}\right) \cos\{(m-2k)\theta\}\,, \tag{9.3.47}$$

where $\lfloor a \rfloor$ denotes the largest integer $< a$ and δ_{kl} is the Kronecker delta. Performing the integration with respect to θ in the expression for U_p of Eq. (9.3.36) yields the result

$$\begin{aligned}
U_p(r_1',\gamma',z_1') = \pi\, g_n(z_1',r_1') \Bigg\{ & 2\int_0^1 \exp(-i\pi z_1'\rho^2) J_0(2\pi r_1'\rho)\,\rho\, d\rho \\
& + 2i\alpha_{n,c}^m\, i^m \cos m\gamma' \int_0^1 \exp(-i\pi z_1'\rho^2) R_n^m(\rho) J_m(2\pi r_1'\rho)\,\rho\, d\rho \\
& + \frac{\left(i\alpha_{n,c}^m\right)^2}{2!} \left[\int_0^1 \exp(-i\pi z_1'\rho^2)\, \{R_n^m(\rho)\}^2\, J_0(2\pi r_1'\rho)\,\rho\, d\rho \right.\\
& \left. + i^{2m} \cos 2m\gamma' \int_0^1 \exp(-i\pi z_1'\rho^2)\, \{R_n^m(\rho)\}^2\, J_{2m}(2\pi r_1'\rho)\,\rho\, d\rho \right] \\
& + \frac{\left(i\alpha_{n,c}^m\right)^3}{3!}\, \frac{1}{2} \left[3\, i^m \cos m\gamma' \int_0^1 \exp(-i\pi z_1'\rho^2)\, \{R_n^m(\rho)\}^3\, J_m(2\pi r_1'\rho)\,\rho\, d\rho \right.\\
& \left. + i^{3m} \cos 3m\gamma' \int_0^1 \exp(-i\pi z_1'\rho^2)\, \{R_n^m(\rho)\}^3\, J_{3m}(2\pi r_1'\rho)\,\rho\, d\rho \right] + \cdots \Bigg\}\,.
\end{aligned} \tag{9.3.48}$$

The expression above can be further developed analytically once the powers of the specific radial polynomial have been decomposed into a linearised sum of radial polynomials with the appropriate upper indices m, $2m$, $3m$, etc. Independent of such further analytic developments, we can draw some conclusions from Eq. (9.3.48) about the symmetry of the amplitude and intensity distribution in image space:

- *Symmetry with respect to the z_1'-axis.*
  The lowest order azimuthal dependence that occurs in the expression is given by $\cos m\gamma'$. This means that the z'-axis is an m-fold axis of symmetry for the amplitude and intensity pattern in the focal region; a rotation around the z_1'-axis over an angle of $2\pi/m$ leaves U_p unchanged. It also means that each plane that makes an angle of $l\pi/m$ $(l = 0, 1, \cdots, m-1)$

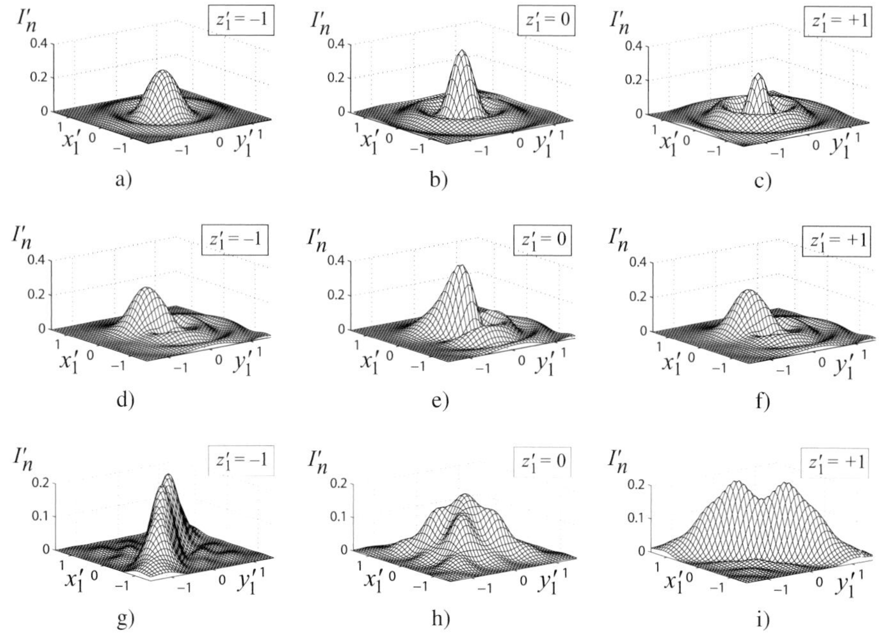

Figure 9.15: Plots of the normalised through-focus intensity I'_n of the PSF in the presence of aberration. The axial position of the image plane is given by the normalised coordinate z'_1 which, from left to right, is equal to -1, 0 and $+1$ in the graphs. The units along the image plane axes are $\lambda_{n_1} / \sin\alpha_1$. The intensity has been normalised with respect to the on-axis value of an aberration-free imaging system.
a), b) and c) Spherical aberration ($\alpha_4^0 = 2\pi/3{=}2.094$).
d), e) and f) Comatic aberration ($\alpha_3^{-1} = \alpha_3^1 = 0.210 \times 2\pi{=}1.32$, $\theta_3^{|1|} = 3\pi/4$).
g), h) and i) Astigmatic focused beam ($\alpha_2^{-2} = \alpha_2^2 = \pi/2$, $\theta_2^{|2|} = 3\pi/4$).

with the plane $x'_1 = 0$ is a plane of symmetry. Figure 9.15 illustrates these symmetry properties for the case of coma, graphs d)–f), and astigmatism, graphs g)–i). Graphs a)–c) for spherical aberration show rotational symmetry around the z'-axis.

- *Symmetry with respect to the focal plane* ($z'_1 = 0$, $m \neq 0$).
  To study the axial symmetry properties, we first observe that the integrals in Eq. (9.3.48) are each other's complex conjugates when evaluated for two values of z'_1 with equal modulus but with opposite sign. The same holds for the prefactor $g(z'_1, r'_1)$. The various factors that multiply each integral in Eq. (9.3.48) are real when m is odd. This means that the complex amplitudes U_p at equal axial distance from the focal plane are also each other's complex conjugates and that the intensities are equal at any pair of points $(x'_1, y'_1, \pm z'_1)$. This phenomenon is illustrated in Fig. 9.15d–f. For positive nonzero and even values of m, the factors that contain odd powers of $\alpha^m_{n,c}$ are purely imaginary. Symmetry in intensity would be obtained if the imaginary factor were to have opposite sign on the other side of the focal plane. This can be achieved by a rotation of the angle γ' to a new azimuth γ'' where $\gamma'' = \gamma' + \pi/m$. This means that for even and nonzero m, equal intensity is found in two points of which the second one is obtained from the first by a reflection at the plane z'_1 and a subsequent rotation around the z'_1-axis through an angle of π/m. This phenomenon can be checked for the case of astigmatism with $m = 2$ that is illustrated in Fig. 9.15g–i, where the required rotation angle is $\pi/2$.

A special symmetry is found in the focal plane itself. For m even and $z'_1 = 0$, Eq. (9.3.48) shows that the terms with odd powers of $\alpha^m_{n,c}$ are purely imaginary whilst the terms with even powers of $\alpha^m_{n,c}$ are all real. By a rotation of an angle π/m in the focal plane, the imaginary terms change sign. The intensity in the focal plane along a line through the z'_1-axis that is rotated through an angle π/m with respect to the reference line for which $x'_1 = 0$ is unchanged. This phenomenon is also clearly visible in Fig. 9.15h.

- *On-axis symmetry with respect to the focal plane for $m = 0$.*
  In general, as was shown for spherical aberration with rotational symmetry, there is no symmetry with respect to the focal plane; however, along the axis of the focusing beam (the optical axis or, for an oblique beam, the axis defined by the principal ray) symmetry in intensity with respect to the focal point on this axis may occur. The on-axis complex amplitude follows from the first line of Eq. (9.3.36) with $r_1' = 0$. When the aberration phase $\Phi(\rho, \theta)$ is given by a single spherical aberration term we obtain

$$U_p(r_1', \gamma', z_1') = 2\pi\, g_n(z_1', r_1') \int_0^1 \exp\left[-i\pi z_1' \rho^2 + i\alpha_{2n}^m R_{2n}^0(\rho)\right] \rho\, d\rho \,. \tag{9.3.49}$$

In the integral of Eq. (9.3.49) we use the relationship between the radial Zernike polynomials with upper index zero and the Legendre polynomials, $R_{2n}^0(\rho) = P_n(2\rho^2 - 1)$ (see Eq. (D.2.12)) and then perform a power series expansion of the exponential term describing the spherical aberration term. With the aid of the substitution $u = 2\rho^2 - 1$ we obtain

$$U_p(0, z_1') = \frac{\pi}{2}\, g_n(z_1', r_1') \exp\left(-i\frac{\pi z_1'}{2}\right) \sum_{l=0}^{\infty} \frac{\left(i\alpha_{2n}^0\right)^l}{l!} \int_{-1}^{1} \exp\left[-i\frac{\pi z_1' u}{2}\right] \{P_n(u)\}^l \, du \,. \tag{9.3.50}$$

From this expression for the on-axis complex amplitude, we conclude that the on-axis intensity remains unchanged when the sign of z_1' is changed, on the condition that the order n of the Legendre polynomial is even. In that case the amplitude at points of equal distance from the focus is related through a complex conjugate operation and equal intensities are found at these points. In Fig. 9.15a–c this phenomenon is clearly visible. The degree $2n$ of the spherical aberration term equals 4 and identical peak intensity is observed in these two graphs despite the other disparities between the intensity patterns in the two planes perpendicular to the optical axis.

9.4 The Extended Nijboer–Zernike Diffraction Theory

The analysis by Zernike and Nijboer of the scalar Debye diffraction integral in the presence of aberration was carried out in the years 1930–1950 when the possibility of numerical computation was still very limited. Their approximate analytic solutions of the diffraction integral were a big step forward. Some practical drawbacks of their approach, which we have already discussed in the previous section, are the restricted through-focus range of their solutions and the practical limit on the size of the aberration. To proceed to larger defocus and aberration values, the higher-order terms in the expansions of the defocus and aberration terms require the evaluation of multiple products of Zernike polynomials. A systematic approach to the linearisation of these products of polynomials in terms of new polynomials with prescribed upper index was not available to these authors. A further limitation for the practical application of the Nijboer–Zernike (NZ) diffraction theory is the fact that many optical systems do not satisfy the low-aperture scalar imaging condition. High-numerical-aperture vector imaging and imaging involving the transition from one medium to another via a planar interface **[346]** or a sequence of planar interfaces (multilayer stack) cannot be handled with the basic NZ theory. For that reason, numerical methods became the dominant tool for solving diffraction problems, enabled by the fast development of computational power since the 1960s.

Two research results have allowed further development of the NZ diffraction theory to what is now commonly called the Extended Nijboer–Zernike diffraction theory (*ENZ*). An interesting result regarding the linearisation of products of Zernike polynomials was published by Tango **[336]**. It allows a systematic solution of the linearisation problem and goes back to the theory of angular momentum in quantum mechanics as developed by Wigner **[367]**. This result remained relatively unnoticed by the optics community. A breakthrough with respect to the range of defocusing and the size of aberration that could be treated was achieved by Janssen **[162]** who found semi-analytic expressions for the diffraction integral that remain stable for large axial and lateral excursions from the nominal focal point. The strength of the aberration that can be handled can be increased by using a complex Zernike expansion of the pupil function describing both its modulus (amplitude) and argument (phase aberration) simultaneously **[183]** (see Eqs. (9.2.17) and (9.2.18)).

For the case in which large aberrations or abrupt amplitude changes in the pupil function are present, Zernike polynomials of high order in radial and azimuthal direction are needed to represent the complex pupil function with high fidelity. A (numerical) fit of the complex pupil function $f_L'(\rho, \theta)$ to the orthogonal Zernike basis needs to be carried out a single time. The calculation of the complex expansion coefficients β_n^m of Eq. (9.2.17) is performed in a mathematically straightforward way by the repeated evaluation of the integral over the unit circle of the product of the function $f_L'(\rho, \theta)$

and a (complex conjugate) Zernike polynomial $R_n^m(\rho)\exp(-im\theta)$. Including normalisation of the integral with respect to the area of the unit circle, we find the following expression for a β_n^m coefficient of the Zernike expansion of the pupil function,

$$\frac{1}{\pi}\int_0^{2\pi}\int_0^1 f'_L(\rho,\theta)\,R_n^m(\rho)\exp(-im\theta)\,\rho d\rho d\theta = \frac{1}{\pi}\int_0^{2\pi}\int_0^1 \sum_{n'}\sum_{m'}\beta_{n'}^{m'}\,R_{n'}^{|m'|}(\rho)\exp(im'\theta)\,R_n^{|m|}(\rho)\exp(-im\theta)\,\rho d\rho d\theta = \frac{\beta_n^m}{n+1}\,, \quad (9.4.1)$$

where we have used Eq. (D.2.22) for the evaluation of the double integral in Eq. (9.4.1). The quadratic fit error between the function $f'_L(\rho,\theta)$ and its Zernike polynomial expansion is monitored as a function of the largest radial and azimuthal order of the expansion. A rest error of typically 10^{-3} in amplitude, 10^{-6} in intensity is acceptable for most practical applications.

The calculation of expansion coefficients β_n^m by means of inner product evaluation according to Eq. (9.4.1) is attractive because of the transparency of the method, but its numerical efficiency is not optimum. Direct evaluation of a Zernike polynomial using the basic expressions of Eq. (D.2.9) and (D.2.10) suffers a loss of precision when factorial functions with arguments of the order of 30 and higher have to be computed. Various methods have been devised to improve the accuracy of the evaluation of Zernike polynomials of high order, i.e. by using recursive schemes **[68]**. Despite these improvements, a least-squares minimisation approach to the evaluation of Zernike coefficients **[231]** is generally more efficient with respect to convergence and computation time. With the aid of the computed set of complex Zernike coefficients β_n^m and the semi-analytic expressions for the image space amplitude for each Zernike term, the total complex amplitude at essentially any point in image space can be calculated with a high accuracy. In this section we present the scalar version of the ENZ diffraction theory.

9.4.1 Scalar Version of the *ENZ* Diffraction Theory

In the scalar diffraction integral of Eq. (9.3.4) we replace the complex pupil function $f'_L(\rho,\theta)$ by its complex Zernike expansion according to Eq. (9.2.17). We temporarily omit the factors in front of the integral and divide the remaining expression by π, the area of the unit disc, with the purpose of normalising the on-axis and in-focus amplitude of an ideal imaging system to unity. We restrict ourselves to a convex exit pupil for which the distance $R_1 > 0$. Denoting the normalised complex amplitude by $U'_p(r'_1,\gamma',z'_1)$ where $\gamma' = \gamma - \pi$, we obtain the expression

$$U'_p(r'_1,\gamma',z'_1) = \frac{1}{\pi}\sum_{n=0}^{\infty}\sum_{m=-\infty}^{+\infty}\beta_n^m\int_0^{2\pi}\int_0^1 \exp(-if'_d\rho^2)\,R_n^{|m|}(\rho)\,\exp(im\theta)$$
$$\times\exp\{i2\pi r'_1\rho\cos(\theta-\gamma')\}\,\rho\,d\rho d\theta = 2\sum_{n=0}^{\infty}\sum_{m=-\infty}^{+\infty} i^m\beta_n^m\,V_n^m(r'_1,f'_d)\,\exp(im\gamma')\,, \quad (9.4.2)$$

where we have introduced the defocusing parameter $f'_d = \pi z'_1$ and used the result of Eq. (9.3.10) when evaluating the integral over θ. The V_n^m functions are given by

$$V_n^m(r'_1,f'_d) = \int_0^1 \exp(-if'_d\rho^2)\,R_n^{|m|}(\rho)\,J_m(2\pi r'_1\rho)\,\rho\,d\rho\,. \quad (9.4.3)$$

It was shown by Janssen **[162]** that, upon performing a power series expansion of the defocus exponential, the V integral becomes

$$V_n^m(r'_1,f'_d) = \exp(-if'_d)\sum_{l=1}^{\infty}\left(2if'_d\right)^{l-1}\sum_{j=0}^{p} v_{lj}\,\frac{J_{|m|+l+2j}(2\pi r'_1)}{l\,(2\pi r'_1)^l}\,. \quad (9.4.4)$$

Letting $p = (n-|m|)/2$ and $q = (n+|m|)/2$, the coefficients v_{lj} can be written

$$v_{lj} = (-1)^{(n-m)/2}\,(|m|+l+2j)\binom{|m|+j+l-1}{l-1}\binom{j+l-1}{l-1}\binom{l-1}{p-j}\Big/\binom{q+l+j}{l}, \quad (9.4.5)$$

for $j = 0, 1, \cdots, p$ and $l = 1, 2, \cdots$ We note that the binomial coefficient $\binom{n}{m}$ is zero if $n < m$.

An alternative method to evaluate the V_n^m integral of Eq. (9.4.3) is based on Bauer's formula for the expansion of the defocus exponential (see Eq. (9.3.25)). This formula was previously used by Nijboer for exploring, among other things, the on-axis diffracted amplitude. The insertion of Bauer's formula leads to products of Zernike polynomials of the form $R_{2l}^0(\rho)\,R_n^{|m|}(\rho)$ for which linearisation schemes, based on the Clebsch–Gordan coefficients of Appendix D, are available. The expression for the V integral is then given by

$$V_n^m(r_1',f_d') = \epsilon_m \exp(-if_d'/2)\sum_{l=0}^{\infty}(2l+1)\,i^l\,j_l(-f_d'/2)\int_0^1 R_{2l}^0(\rho)\,R_n^{|m|}(\rho)\,J_m(2\pi r_1'\rho)\,\rho\,d\rho\,, \tag{9.4.6}$$

where $\epsilon_m = -1$ for odd $m<0$ and $\epsilon_m = +1$ otherwise. Linearisation of the Zernike polynomial product yields the expression

$$R_{2l}^0(\rho)\,R_{|m|+2p}^{|m|}(\rho) = \sum_{j=0}^{l+p} c_{lj}(n,m)\,R_{|m|+2j}^{|m|}(\rho)\,, \tag{9.4.7}$$

where the values $c_{lj}(n,m)$ follow from the Clebsch–Gordan coefficients. Using the basic NZ integral result for evaluating the integral in Eq. (9.4.6) we obtain

$$V_n^m(r_1',f_d') = \epsilon_m \exp(-if_d'/2)\sum_{l=0}^{\infty}(2l+1)\,i^l\,j_l(-f_d'/2)\sum_{j=0}^{l+p}(-1)^j\,c_{lj}(n,m)\,\frac{J_{|m|+2j+1}(2\pi r_1')}{2\pi r_1'}\,, \tag{9.4.8}$$

where, as before, $p=(n-|m|)/2$. Expressions for the coefficients $c_{lk}(n,m)$ are given in **[55]** and in **[127]**.

The semi-analytic expressions in Eq. (9.4.4) and (9.4.8) for the functions $V_n^m(r_1',f_d')$ are called the power–Bessel and the Bessel–Bessel expansion, respectively. Although formally yielding the same result, their numerical implementation with a limited number of available significant digits shows quite different behaviour. The power series expansion with respect to f_d' typically needs $3f_d'+5$ terms in the expansion to obtain convergence toward the final result. In the case of double precision calculations with 15 significant digits, the largest term in the expansion is larger by more than a factor of 10^9 with respect to the initial one if $|f_d'|\geq 25$. For $f_d'=25$ the disparity in the value of the terms limits the accuracy of the end result to typically 10^{-6}. A much more favourable numerical performance is obtained when the Bessel–Bessel expansion of Eq. (9.4.8) is used. The typical number of terms that are required for convergence equals $2\pi r_1'|f_d'|$ and virtually no loss-of-digits is incurred because of the favourable decay properties of the spherical Bessel functions j_l. For that reason, although the coefficients c_{lk} cannot be given in fully analytic form (unlike the coefficients v_{lj} appearing in Eq. (9.4.5)), the Bessel–Bessel expansion is preferred when numerical implementation with a limited number of significant digits is needed.

The result for the V function can be compared to previous analytic results for the aberration-free or aberrated case. We consider the following examples:

- *Aberrated in-focus amplitude distribution according to Nijboer*
  We consider the power–Bessel expression, $f_d'=0$ and limiting the summation over l in Eq. (9.4.4) to the $l=1$ term. The only nonzero v_{1j} coefficient in the summation over j is the one with $j=p$ and we have that $v_{1,(n-|m|)/2}=(-1)^{(n-m)/2}$. Assuming the coefficients $\beta_0^0=1$ and $\beta_n^m=\beta_n^{-m}=i\alpha_n^m/2$, we can represent the weak phase aberration Φ according to Eq. (9.3.8) with a $\cos m\theta$ dependence in the exit pupil. After some rearrangement, we obtain the complex amplitude $U_p'(r_1',\gamma',0)$ according to Eq. (9.4.2) in the form

$$U_p'(r_1',\gamma',0) = \frac{2J_1(2\pi r_1')}{2\pi r_1'} + i\sum_{n=0}^{\infty}\sum_{m=0}^{\infty} i^n\,\alpha_n^m\,\frac{2J_{n+1}(2\pi r_1')}{2\pi r_1'}\cos(m\gamma')\,, \tag{9.4.9}$$

  where the term with $n=m=0$ should be excluded from the double summation. This result is in agreement with Nijboer's original result, see Eq. (9.3.12).
- *The through-focus result according to Lommel*
  The through-focus complex amplitude distribution for an aberration-free imaging system was originally given by Lommel **[218]**, **[37]**. The amplitude U_p' is given by the function $2V_0^0(r_1',f_d')$ and its relation with the Lommel functions follows from

$$U_p'(r_1',f_d') = 2V_0^0(r_1',f_d') = 2\int_0^1 \exp(-if_d'\rho^2)\,J_0(2\pi r_1'\rho)\,\rho\,d\rho = C(r_1',-2f_d') + iS(r_1',-2f_d')\,, \tag{9.4.10}$$

where the Lommel functions C and S are given by (see **[37]**)

$$C(r_1',f_d') = \frac{\cos(f_d'/2)}{f_d'/2}\, G_1(r_1',f_d') + \frac{\sin(f_d'/2)}{f_d'/2}\, G_2(r_1',f_d') \,,$$

$$S(r_1',f_d') = \frac{\sin(f_d'/2)}{f_d'/2}\, G_1(r_1',f_d') - \frac{\cos(f_d'/2)}{f_d'/2}\, G_2(r_1',f_d') \,,$$

$$G_k(r_1',f_d') = \sum_{j=0}^{\infty} (-i)^{2j} \left(\frac{f_d'}{2\pi r_1'}\right)^{k+2j} J_{k+2j}(2\pi r_1') \,. \tag{9.4.11}$$

It is readily verified that the sum of two Lommel functions according to $C(r_1',-2f_d') + iS(r_1',-2f_d')$ gives

$$U_p'(r_1',f_d') = \exp(-if_d') \sum_{l=1}^{\infty} \left(2if_d'\right)^{l-1} \frac{2J_l(2\pi r_1'}{(2\pi r_1')^l} \,. \tag{9.4.12}$$

This result also follows from Eq. (9.4.4) when $n = m = 0$ is substituted. The only nonzero coefficient v_{lj} in the second summation is $v_{l0} = l$ and, with $U_p'(r_1',f_d') = 2V_0^0(r_1',f_d')$, we obtain the same result as Lommel's above, written in compact complex notation.

- *The on-axis intensity function*

  For the on-axis intensity function we use Eq. (9.4.2) with $r_1' = 0$. For the case of spherical aberration, the Zernike expansion of the pupil function comprises the polynomials $R_{2n}^0(\rho)$. We therefore require the functions $V_{2n}^0(0,f_d')$ for $n = 1, 2, \cdots$ which are given by

$$V_{2n}^0(0,f_d') = \int_0^1 \exp(-if_d'\rho^2)\, R_{2n}^0(\rho)\,\rho\, d\rho \,. \tag{9.4.13}$$

  Applying Bauer's formula to expand the defocus exponential we obtain

$$\begin{aligned} V_{2n}^0(0,f_d') &= \exp(-if_d'/2) \sum_{l=0}^{\infty}(-i)^l\,(2l+1)\, j_l(f_d'/2) \int_0^1 R_{2n}^0(\rho)\, R_{2l}^0(\rho)\,\rho\, d\rho \\ &= \frac{1}{2}\, \exp(-if_d'/2)\,(-i)^n\, j_n(f_d'/2) \,. \end{aligned} \tag{9.4.14}$$

  The complex amplitude $U_p'(0,f_d')$ of the aberrated beam with Zernike coefficients β_{2n}^0 is then given by

$$U_p'(0,f_d') = \exp(-if_d'/2) \left\{ \beta_0^0\, j_0(f_d'/2) + \sum_{n=1}^{\infty} (-i)^n\, \beta_{2n}^0\, j_n(f_d'/2) \right\} \,, \tag{9.4.15}$$

  where we have used the expression for the inner product of two radial Zernike polynomials (see Eq. (D.2.25)). For the aberration-free case, the expression for U_p' comprises only the spherical Bessel function of zeroth order. In terms of the normalised axial defocusing distance z_1' we then have

$$U_p'(0,f_d') = \exp(-if_d'/2)\, \mathrm{sinc}\left(\frac{\pi z_1'}{2}\right) \,, \tag{9.4.16}$$

  a result that was given in Eq. (9.3.27).

If spherical aberration is present in the focusing beam, we observe that the on-axis intensity is symmetric with respect to the focal point for even values of n. The complex on-axis amplitudes have complex conjugate values at equal but opposite distances from the focus. For odd values of n, this symmetry is absent; a trivial example is $n = 1$ which is equivalent to an axial shift of the geometrical focal plane. The symmetry properties can, however, also be demonstrated for fourth-order and sixth-order spherical aberration ($n = 2$ and 3, respectively) as follows. To obtain the β_{2n}^0 coefficients for these two cases, we must expand the pupil function $\exp[i\Phi(\rho)]$ where $\Phi(\rho) = \alpha_{2n}^0\, R_{2n}^0(\rho)$, up to second order. For the cases $n = 2$ and 3 we evaluate $\left\{R_{2n}^0(\rho)\right\}^2$ using the Clebsch–Gordan coefficients of Eqs. (D.4.9)–(D.4.13) in Appendix D for the case that $m_1 = m_2 = m_3 = 0$ and we obtain,

$$\begin{aligned} \left\{R_4^0(\rho)\right\}^2 &= \frac{18}{35} R_8^0(\rho) + \frac{2}{7} R_4^0(\rho) + \frac{1}{5} \,, \\ \left\{R_6^0(\rho)\right\}^2 &= \frac{100}{231} R_{12}^0(\rho) + \frac{18}{77} R_8^0(\rho) + \frac{4}{21} R_4^0(\rho) + \frac{1}{7} \,. \end{aligned} \tag{9.4.17}$$

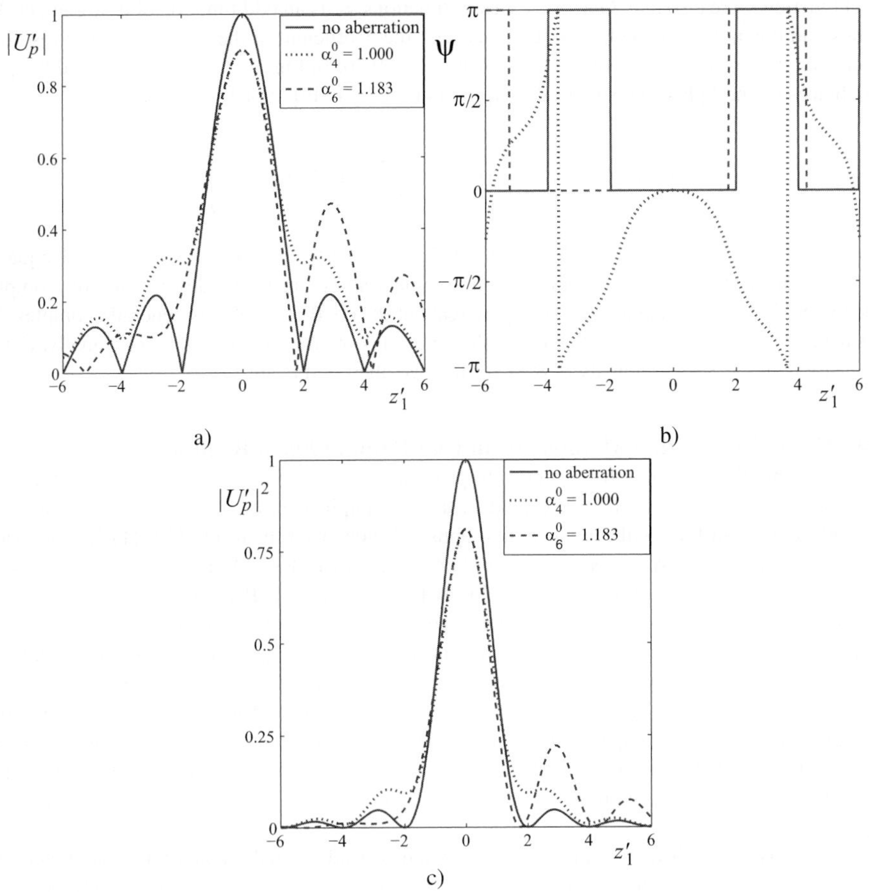

Figure 9.16: On-axis amplitude, phase and intensity of a focused beam in the focal region.
a) The modulus of the axial amplitude ($|U'_p(0, z'_1)|$) of a perfect beam and two other beams suffering from spherical aberration. Solid line: perfect imaging beam, dotted line: fourth-order spherical aberration, dashed line: sixth-order spherical aberration.
b) The axial phase function Ψ associated with the same beams.
c) The axial intensity function $|U'_p(0, z'_1)|^2$ of the three beams.

We can now calculate $f'_L(\rho) \approx 1 + i\Phi(\rho) - \Phi^2(\rho)/2$ and, upon grouping the terms of equal order, we obtain the following nonzero β-coefficients,

$$n = 2\,, \qquad \beta_0^0 = 1 - \frac{\left(\alpha_4^0\right)^2}{10}\,, \qquad \beta_4^0 = i\alpha_4^0 - \frac{\left(\alpha_4^0\right)^2}{7}\,, \qquad \beta_8^0 = -\frac{9\left(\alpha_4^0\right)^2}{35}\,, \tag{9.4.18}$$

$$n = 3\,, \qquad \beta_0^0 = 1 - \frac{\left(\alpha_6^0\right)^2}{14}\,, \qquad \beta_4^0 = -\frac{2\left(\alpha_6^0\right)^2}{21}\,, \qquad \beta_6^0 = i\alpha_6^0\,,$$
$$\beta_8^0 = -\frac{9\left(\alpha_6^0\right)^2}{77}\,, \qquad \beta_{12}^0 = -\frac{50\left(\alpha_6^0\right)^2}{231}\,. \tag{9.4.19}$$

In Fig. 9.16 we have plotted the axial amplitude, phase and intensity of an aberration-free beam and the same quantities for two others beams with fourth- and sixth-order spherical aberration, respectively. For the aberrated cases we choose the value of α_{2n}^0, such that the variance of the in-focus wavefront equals the 'just-diffraction-limited' value with a Strehl intensity of 80%. The complex amplitude of the aberration-free beam is proportional to the $\sin(\pi z'_1/2)$ function and is real everywhere on the axis. For the sixth-order spherical aberration case with $n = 3$, all coefficients $(-i)^n \beta_{2n}^0$ that multiply the spherical Bessel function j_n are also real. One of the Bessel functions, j_3, is an odd function of its argument. Moreover, the coefficient of this odd polynomial is the largest, together with the coefficient of the j_0 function. The resulting amplitude is thus not symmetric with respect to the origin, as shown in Fig. 9.16. In Fig. 9.16b we have plotted the phase Ψ of the quantity within braces on the right-hand side of the expression for U'_p in Eq. (9.4.15),

$$\Psi(0, z'_1) = \arg\left\{\beta_0^0 j_0(f'_d/2) + \sum_{n=1}^{\infty}(-i)^n \beta_{2n}^0 j_n(f'_d/2)\right\}, \tag{9.4.20}$$

where, as before, $f'_d = \pi z'_1$. Because the complex amplitude is real everywhere for $n = 3$, the angle Ψ equals zero or π.

For the $n = 2$ case, all spherical Bessel functions in the summation have an even index and thus the complex amplitude is an even function of z'_1. The coefficients of j_0 and j_4 are real but the coefficient of j_2 is generally complex. This explains the continuous behaviour of the argument Ψ of U'_p as shown in the dotted curve of Fig. 9.16b. Finally, Fig. 9.16c shows the on-axis intensity with the Strehl intensity of 80% for the two aberrated beams.

9.4.1.1 Other Wave Propagation Methods from Exit Pupil to Focal Region

For completeness, we briefly mention other methods to propagate waves from the exit pupil to the focal region. These methods provide series expansions of the wave field in the exit pupil in terms of functions that can be analytically integrated to yield the field in the focal region. A first approach goes back to Kant **[174]**,**[175]**. The evaluated vector diffraction integral is written as a series expansion of products of spherical Bessel functions and Gegenbauer polynomials without the need of numerical integration. A second, closely related method **[310]** writes the field in the exit pupil as an expansion of physical multipole waves. They are represented as the product of a spherical Bessel function for the radial dependence and a spherical harmonic function to represent the elevation angle and the azimuthal angle in the three-dimensional wave field. The spherical Bessel functions are the radial solutions of the Helmholtz time-independent wave equation when spherical coordinates (ρ, θ, γ) are employed. The propagation from the exit pupil to the focal region of the various components of the wave expansion is analytic and avoids the onerous evaluation of a diffraction integral. The incorporation of wave aberration in the exit pupil of an optical system is possible if the aberration has circular symmetry. No solutions have been found so far to represent asymmetrical wave forms (coma) with the aid of this multipole expansion.

Eigenfunction analysis of the diffraction integral has shown that band-limited Slepian-type eigenfunctions exist. They form, apart from a scaling factor, an identical orthonormal basis on the exit pupil sphere and in the image plane. With the aid of such eigenfunctions, wave propagation from exit pupil to image plane is reduced to finding the new set of expansion coefficients in the image plane from the known set of coefficients of identical eigenfunctions in the exit pupil. The eigenfunction analysis has been performed for both scalar and high-numerical-aperture vector imaging systems **[312]**, **[161]**.

9.5 Vector Point-spread Function and the ENZ Diffraction Theory

In the scalar diffraction integral that was considered in the previous section the electric and magnetic components of the field in the focal region have been tacitly assumed to be perpendicular to the average propagation direction of the focused field. A scalar treatment of the diffraction integral also assumes that the state of polarisation in the focal region is identical to that in the exit pupil. For focusing beams with a large angular aperture, these approximations are too coarse. The image space intensity distribution is proportional to $|\mathbf{E}|^2$ which contains non-negligible contributions from electric field components that were absent in the field in the pupil. For instance, in the case of an x-polarised pupil field, there will be y- and z-field components in the focal region in image space. As a rule of thumb, the difference between the 'scalar' intensity and the intensity according to the exact solution exceeds the one-percent level for an aperture angle α_1 in image space for which $\sin\alpha_1 > 0.50$. When determining the exact solution, the three orthogonal vector components of the optical field on the exit pupil sphere are established and separately propagated from there to the image region by evaluation of the diffraction integral associated with each of the orthogonal components. It is a matter of choice whether Cartesian or cylindrical orthogonal components are used.

In this chapter we neglect any polarisation-dependent light absorption and phase changes that occur during propagation from entrance to exit pupil. In reality these effects arise at transmission through the lens surfaces or reflections at mirror surfaces because of the differences between light transmission and reflection for s- and p-polarised light (polarisation-dependent Fresnel coefficients). It is also possible that birefringent effects cause differences between orthogonally polarised components of the light. These effects may be caused, for instance, by residual birefringence or stress in glass or plastic optical components or by polarisation-sensitive elements that are part of the imaging system (e.g. a polarising beam-splitter, Wollaston prism, quarter-wave plate, etc.). The systematic treatment of these effects is postponed to Chapter 11 where, among other things, two- and three-dimensional Jones matrices will be exploited to accurately describe the polarisation-dependent transfer of the electromagnetic field components through an optical system from the entrance to the exit pupil. In this chapter we only consider the polarisation-independent projection effects that determine the change in the Cartesian electric field components when going from object to image space in a high-numerical-aperture system with a lateral magnification of m_T and that satisfies the Abbe sine condition.

In the case of the Debye diffraction integral, the expressions for the vector components in the focal region of an aberration-free imaging system follow from Eqs. (8.3.25) and (8.3.26). To treat the point-spread function of a high-numerical-aperture aberrated imaging system, we introduce the lens function $f'_L(\rho, \theta)$ in the integrand of these equations. The modulus of this function is a measure of the transmission of the light when it propagates from entrance to exit pupil. The argument of $f'_L(\rho, \theta)$ follows from the phase aberration function $\Phi(\rho, \theta)$ in the exit pupil.

To obtain the modulus of the 'lens function' f'_L in the case of high-aperture imaging, one has to take into account the particular mapping of rays from the entrance to exit pupil performed by the imaging system beyond the paraxial domain. Moreover, for the defocused point-spread function, one has to refine the quadratic scalar defocus exponential $\exp(-if'_d\rho^2)$ such that the parabolic wavefront approximation for defocusing is replaced by the exact spherical wavefront deviation. We briefly discuss these three corrections to the scalar approximation and give the complex values of the field components on the exit pupil sphere. These values are the input for the three diffraction integrals that yield the three orthogonal electric and magnetic field components in the focal region of the high-*NA* imaging system. Throughout this chapter, we use the Cartesian orthogonal electromagnetic field components in the object and receiving plane. The modulus of the amplitude of the object field on the entrance pupil of the imaging system is assumed to be uniform. For real sources, for instance an electric or magnetic dipole source, this uniform amplitude has to be replaced by the typical far-field distribution of such a dipole.

9.5.1 The Electromagnetic Field Components on the Exit Pupil Sphere

To obtain the Cartesian electromagnetic field components on the exit pupil sphere, we have to study the change in strength when the field is propagated through the imaging system from the entrance to exit pupil. Depending on the 'vergence' of the imaging beam in object and image space, the transverse electromagnetic components are projected in a different way onto the Cartesian (x, y, z)-axes in object and image space. In many textbooks, the case of an incident parallel beam ($m_T = 0$) with unit pupil magnification $m_{T,p}$=+1 and equal refractive indices in object and image space is taken. In this subsection we consider the general case of an arbitrary object and pupil magnification; we also allow the refractive indices in object and image space to be different to accommodate, for instance, a water- or oil-immersed object or image. As a special case we consider the point-spread function in a particular thin layer of a multilayer stack in the focal region. Such a situation is often encountered, for instance, when an image is produced in the sensitive layer of an image sensor or in a photoresist layer on top of a substrate. With the projection rules for the Cartesian field vectors in object and image space at our disposal, we further include the additional changes in field strength and phase that can be incurred by the electromagnetic field upon propagation through the imaging system. These changes are due to the following three effects:

- *Aberration function*
  To represent the phase part of the complex pupil function $f'_L(\rho, \theta)$ in the exit pupil we use the Zernike expansion of the function $\exp\{i\Phi(\rho, \theta)\}$ according to Eq. (9.2.17). The complex coefficients β_n^m are numerically calculated, for instance by means of a least squares minimisation of the differences between the fitted function and the phase aberration values in each sample point on the exit pupil.
- *Defocusing exponential*
  To allow both for large defocusing values and for high-numerical-aperture imaging, it is necessary to replace the quadratic defocusing exponential by the exact formula for $\exp\{i\Phi_d(\rho)\}$ as was given in Eq. (9.2.15).
- *Radiometric effect*
  We adopt the Abbe sine condition for ray mapping from entrance to exit pupil as this condition holds for the vast majority of optical imaging systems. The corresponding factor in the integrand due to the radiometric effect is then given by Eq. (9.2.11).

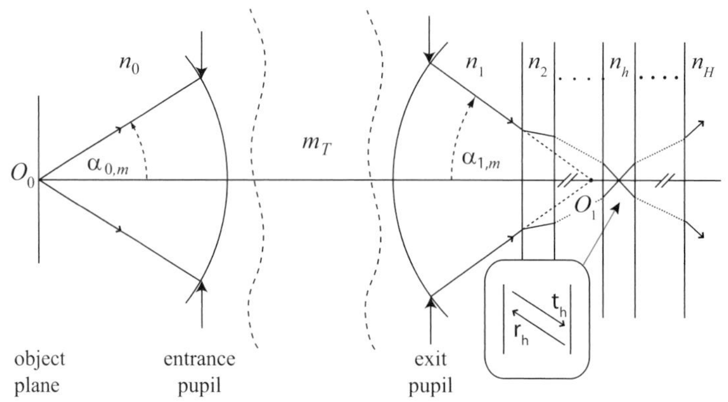

Figure 9.17: A high-aperture imaging system (lateral magnification is m_T) with a thin layer stack added in image space. The imaging properties are studied in a particular layer with index n_h.

- *Reflection and transmission at interfaces in the focal region*
  In many instances, the three-dimensional image is formed in a region where transitions between various media are present. One can think of the anti-reflection and absorbing layers on a photodetecting layer or of the impedance-matching and recording layers (photoresist) that cover the substrate of a semiconductor surface on which the nanometre-scale features of an electronic microcircuit have to be printed. In these cases, the image is formed by the interplay of forward and backward propagating waves in the particular layer in the imaging region. We thus have to consider the diffraction of both forward and backward propagating plane wave spectra (see Fig. 9.17).

To calculate the Cartesian electric field vectors on the entrance and exit pupil spheres, we start with a hypothetical point source that produces known field components on the entrance pupil sphere. In the case of a very distant source, these field components can be considered as constant over the entrance pupil surface. If the source is a dipole and its distance to the entrance pupil is small, the field components will be those of the dipole far-field pattern that fits in the solid angle subtended by the entrance pupil. In what follows we consider a general distribution of electric field components in the entrance pupil. At each point on the entrance pupil surface, the electric field components are assumed to be transverse to the normal $\hat{\mathbf{n}}$ to the pupil surface; the same is assumed to hold for the field components on the exit pupil sphere. These assumptions are exact for a source at a large distance and for an imaging system with small aberrations. It is reasonable to suppose that for a well-corrected imaging system the local wave vector deviates less than 10^{-3} radians from the pupil surface normal. The same deviations can also occur for the directions of the transverse field components of the propagating wave, and we neglect these small angular deviations. For the rotation of the field components from the $(\hat{\mathbf{X}}_0, \hat{\mathbf{Y}}_0, \hat{\mathbf{Z}}_0)$ to the $(\hat{\mathbf{p}}_0, \hat{\mathbf{s}}_0, \hat{\mathbf{k}}_0)$ coordinate system at a particular point Q_0 on the entrance pupil sphere, we refer to Fig. 9.18. The coordinate transformation from the unit vector basis $(\hat{\mathbf{X}}_0, \hat{\mathbf{Y}}_0, \hat{\mathbf{Z}}_0)$ to the basis $(\hat{\mathbf{p}}_0, \hat{\mathbf{s}}_0, \hat{\mathbf{k}}_0)$ follows from a rotation over an angle θ around the $\hat{Z}_0$-axis and, subsequently, over an angle $-\alpha_0$ around the $\hat{\mathbf{s}}_0$-axis, illustrated in Fig. 9.18b and 9.18c. To accommodate both concave and convex pupil spheres we use, e.g. for the entrance pupil sphere, the expression $\sin\alpha_0 = (R_0/|R_0|)\sqrt{1-\cos^2\alpha_0}$. In the case of the entrance pupil configuration of Fig. 9.18, R_0 is the radius of a concave pupil sphere and is taken to be negative. The expressions for the unit vectors $\hat{\mathbf{p}}_0$, $\hat{\mathbf{s}}_0$ and $\hat{\mathbf{k}}_0$ are then given by

$$
\begin{aligned}
\hat{\mathbf{p}}_0 &= \cos\alpha_0\left(\cos\theta\,\hat{\mathbf{X}}_0 + \sin\theta\,\hat{\mathbf{Y}}_0\right) + \frac{R_0}{|R_0|}\,(1-\cos^2\alpha_0)^{1/2}\,\hat{\mathbf{Z}}_0\,,\\
\hat{\mathbf{s}}_0 &= -\sin\theta\,\hat{\mathbf{X}}_0 + \cos\theta\,\hat{\mathbf{Y}}_0\,,\\
\hat{\mathbf{k}}_0 &= -\frac{R_0}{|R_0|}\,(1-\cos^2\alpha_0)^{1/2}\left(\cos\theta\,\hat{\mathbf{X}}_0 + \sin\theta\,\hat{\mathbf{Y}}_0\right) + \cos\alpha_0\,\hat{\mathbf{Z}}_0\,.
\end{aligned}
\qquad (9.5.1)
$$

The transverse s- and p-components of the field strength on the entrance pupil sphere follow from the Cartesian components of the field strength $\mathbf{E}_0$ at the point Q_0. We assume that these components are known given the properties of the source under consideration.

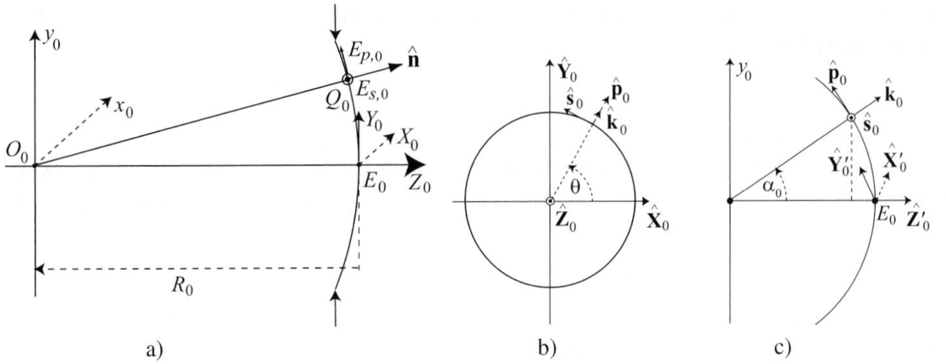

Figure 9.18: Definition of the Cartesian coordinate system (X_0, Y_0, Z_0) and the local right-handed orthogonal $(\hat{\mathbf{p}}_0, \hat{\mathbf{s}}_0, \hat{\mathbf{k}}_0)$-system on the (concave) entrance pupil sphere at Q_0 located at an elevation angle of α_0.
a) The source point is at O_0 and the transverse electric field components $E_{s,0}$ and $E_{p,0}$ at Q_0 are tangent to the entrance pupil sphere at the point Q_0; the entrance pupil is concave and the distance R_0 is negative in the figure.
b) Cross-section through the plane $Z_0 = 0$.
c) The $(\hat{\mathbf{p}}_0, \hat{\mathbf{s}}_0, \hat{\mathbf{k}}_0)$ coordinate-system for an arbitrary azimuthal angle θ on the entrance pupil sphere.
The cross-sections of the diagrams b) and c) illustrate the geometry of the rotated coordinate basis for the s and p field components.

A common source is the (electric) line dipole of which the radiation pattern is given by Eq. (9.6.42) We will treat the X- and Y-components of the electric field as variables. Sufficiently far away from the source the Z-component follows from the orthogonality of the transverse field components with respect to the unit propagation vector $\hat{\mathbf{k}}_0$ at the point Q_0, yielding

$$\mathbf{E}_0 \approx \left(E_{x,0}\,,\; E_{y,0}\,,\; \frac{R_0}{|R_0|}\,(E_{x,0}\cos\theta + E_{y,0}\sin\theta)\,\frac{(1-\cos^2\alpha_0)^{1/2}}{\cos\alpha_0} \right) . \tag{9.5.2}$$

With the aid of Eq. (9.5.1) we can obtain the field components in the coordinate system $(\hat{\mathbf{p}}_0, \hat{\mathbf{s}}_0, \hat{\mathbf{k}}_0)$ as

$$\mathbf{E}_0 \approx \left(E_{p,0}\,,\; E_{s,0}\,,\; E_{k,0}\right) = \left(\frac{E_{x,0}\cos\theta + E_{y,0}\sin\theta}{\cos\alpha_0}\,,\; -E_{x,0}\sin\theta + E_{y,0}\cos\theta\,,\; 0 \right) . \tag{9.5.3}$$

It follows from this equation that the convex or concave shape of the pupil sphere does not influence the strength of the p- and s-components. The shape only influences the sign of the z-component of the $\hat{\mathbf{p}}$ unit vector and thus the sign of the Cartesian component E_z on the entrance or exit pupil. The approximate sign in the two equations above applies to the near field of a dipole source where the value of $E_{k,0}$ is not rigorously zero. After propagation from the source point O_0 over a distance of a few wavelengths, the $E_{k,0}$-component of the radiation field can be safely put equal to zero.

9.5.1.1 Propagation of the Field Components from Entrance to Exit Pupil

The propagation of the field on the entrance pupil sphere to the exit pupil surface is treated using the laws of geometrical optics. Along each propagation direction defined by a geometrical optical ray, we suppose that the amplitude and phase of the electric field are multiplied by the complex lens transmission function along that propagation direction. For sufficiently small aberrations, we are allowed to use the value of the lens transmission function along the aberration-free propagation direction. We also suppose that the imaging system behaves in an isotropic way such that the s- and p-components of the field strength are transmitted through the optical system from the entrance to exit pupil with equal changes in strength and phase. The ray mapping from object to image space obeys the Abbe sine condition. The various quantities in object and image spaces are distinguished using the subscripts 0 and 1, respectively. We consider imaging of the source object into a particular layer of a multilayer stack placed in the focal region. Quantities within this specific layer are distinguished using the subscript h.

Using the Abbe sine condition, we have the following relationships between the transverse unit wave vector components in object and image space,

$$n_1 m_T \hat{k}_{x,1} = n_0 \hat{k}_{x,0} \,, \qquad n_1 m_T \hat{k}_{y,1} = n_0 \hat{k}_{y,0} \,, \tag{9.5.4}$$

where m_T is the paraxial lateral magnification of the imaging system and n_0 and n_1 are the refractive indices of object and image space, respectively. The points of intersection of a ray with the entrance and exit pupil, Q_0 and Q_1, respectively, follow from the unit wave vector components in object and image space; for instance, for a point Q_1 on the exit pupil sphere,

$$X_1 = -R_1 \hat{k}_{x,1} \,, \qquad Y_1 = -R_1 \hat{k}_{y,1} \,. \tag{9.5.5}$$

The sign of the radius of curvature R of the pupil surfaces is as indicated in Fig. 9.18. We also use the normalised radial coordinate ρ on the pupil spheres. Applying the Abbe sine condition, its value is given by $\sin\alpha_0/s_0$ and $\sin\alpha_1/s_1$, respectively, where $s_0 = \sin\alpha_{0,m}$ and $s_1 = \sin\alpha_{1,m}$ are equal to the sines of the maximum aperture angles in object and image space. Other frequently used relations pertaining to the positions in the entrance and exit pupil are

$$\begin{aligned} s_0 &= |n_1 m_T s_1/n_0| \,, & s_j &= |n_1 s_1/n_j| \,, \\ |\sin\alpha_0| &= \rho s_0 \,, & \cos\alpha_0 &= \left(1 - s_0^2\rho^2\right)^{1/2} \,, \\ |\sin\alpha_1| &= \rho s_1 \,, & \cos\alpha_1 &= \left(1 - s_1^2\rho^2\right)^{1/2} \,, \\ |\sin\alpha_j| &= \rho s_j \,, & \cos\alpha_j &= \left(1 - s_j^2\rho^2\right)^{1/2} \,, \end{aligned} \tag{9.5.6}$$

where n_j $(j = 1, 2, \cdots, h, \cdots, J)$ is the refractive index of a possibly present planar layer in the focal region, perpendicular to the z-axis. The layer is either an isolated one or it is part of a multilayer stack; s_j is the absolute value of the sine of the maximum aperture angle in such a layer, see Fig. 9.17. The index h is preferably used for the layer of the multilayer stack in which the geometrical focus is formed.

The directions of the transverse electric field components on the exit pupil sphere follow from the geometrical-optics ray mapping from object to image space. In our approximation where reflection or transmission losses, birefringent components and residual birefringence in the optical components are excluded, the s- and p-components on the exit pupil sphere are proportional to those on the entrance pupil sphere (for a complete treatment of wave propagation from entrance to exit pupil the reader is referred to Chapter 11). The relative strengths of s- and p-components follow from a consideration of the energy flow through corresponding annular regions on the entrance and exit pupil sphere. This subject was treated in Subsection 9.2.1 for the scalar diffraction case. Here, we briefly reproduce this analysis, now in terms of the wave vector components, the exit pupil position and the image side focal distance of the optical system.

For the energy flow dP_S through a differential region on a pupil sphere we write

$$dP_S = \epsilon\, v\, |E|^2\, dS \,, \tag{9.5.7}$$

where $\epsilon = n^2$ is the electric permittivity of the medium, $v = c/n$ is the electromagnetic propagation speed and $|E|$ the field strength of either an s- or a p-electric field component that is tangent to the pupil sphere. For two corresponding cross-sections on the entrance and exit pupil sphere we then have, in the absence of absorption, scattering and transmission or reflection losses,

$$n_0\, |E_0|^2\, dS_0 = n_1\, |E_1|^2\, dS_1 \,. \tag{9.5.8}$$

The expression for the differential elementary area dS on a pupil sphere is $dS = R^2 d\hat{k}_x d\hat{k}_y/\hat{k}_z$ where R is the radius of the pupil sphere. We thus find that the relation between the electric field strengths is

$$n_0\, R_0^2\, \frac{d\hat{k}_{x,0}\, d\hat{k}_{y,0}}{\hat{k}_{z,0}} |E_0|^2 = n_1\, R_1^2\, \frac{d\hat{k}_{x,1}\, d\hat{k}_{y,1}}{\hat{k}_{z,1}} |E_1|^2 \,, \tag{9.5.9}$$

and the value of E_1 is given by

$$|E_1| = \left|\frac{R_0}{R_1}\right| \sqrt{\frac{n_0\, \hat{k}_{z,1}}{n_1\, \hat{k}_{z,0}}} \sqrt{\frac{d\hat{k}_{x,0}\, d\hat{k}_{y,0}}{d\hat{k}_{x,1}\, d\hat{k}_{y,1}}}\, |E_0| \,. \tag{9.5.10}$$

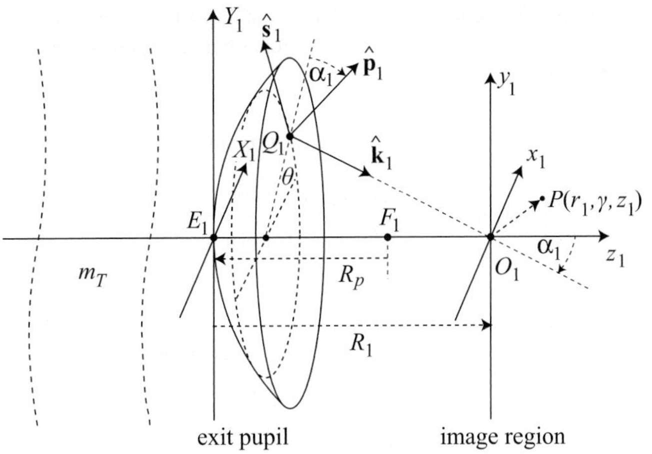

Figure 9.19: The local orthogonal coordinate system $(\hat{\mathbf{p}}_1, \hat{\mathbf{s}}_1, \hat{\mathbf{k}}_1)$ on the exit pupil sphere at a particular point Q_1. R_1 is the radius of the (convex) exit pupil and by convention denotes a positive distance in the figure.

The second square root in this equation is evaluated using the Abbe sine condition according to Eq. (9.5.4), yielding the result

$$|E_1| = \left|\frac{m_T R_0}{R_1}\right| \sqrt{\frac{n_1}{n_0}} \sqrt{\frac{\hat{k}_{z,1}}{\hat{k}_{z,0}}} |E_0|$$

$$= \left|\frac{m_T R_0}{R_1}\right| \sqrt{\frac{n_1}{n_0}} \frac{\left(1 - s_1^2\rho^2\right)^{1/4}}{\left(1 - s_0^2\rho^2\right)^{1/4}} |E_0| , \qquad (9.5.11)$$

where $(1 - s_1^2\rho^2)^{1/2} = \cos\alpha_1$ and $(1 - s_0^2\rho^2)^{1/2} = \left\{1 - (n_1^2/n_0^2)m_T^2 s_1^2\rho^2\right\}^{1/2} = \cos\alpha_0$.

We can eliminate the pupil object distances R_0 by using the expression for the pupil magnification, $R_1/R_0 = (n_1/n_0)m_T\, m_{T,p}$ and further using Newton's imaging equation (4.10.48) to write the pupil magnification as $m_{T,p} = -R_p/f_1$ where R_p is the distance F_1E_1 from the image side focal point to the centre of the exit pupil and f_1 the image side focal distance (see Fig. 9.19). The strength of the field in a particular annular region on the exit pupil sphere (normalised radius ρ) is then given by

$$|E_1| = \left|\frac{f_1}{R_p}\right| \sqrt{\frac{n_0}{n_1}} \frac{(1 - s_1^2\rho^2)^{1/4}}{\left\{1 - (n_1^2/n_0^2)m_T^2 s_1^2\rho^2\right\}^{1/4}} |E_0|$$

$$= \left|\frac{f_1}{R_p}\right| \sqrt{\frac{n_0}{n_1}}\, T_R(\rho)\, |E_0| , \qquad (9.5.12)$$

where the transmission factor

$$T_R(\rho) = \sqrt{\frac{\cos\alpha_1}{\cos\alpha_0}} = \frac{(1 - s_1^2\rho^2)^{1/4}}{\left\{1 - (n_1^2/n_0^2)m_T^2 s_1^2\rho^2\right\}^{1/4}} \qquad (9.5.13)$$

originates from the radiometric effect in an imaging system that obeys the Abbe sine condition.

9.5.1.2 The Electric Field Components on the Exit Pupil Sphere

The field on the exit pupil sphere is further defined by the complex transmission function of the imaging system. As stated earlier, we neglect any polarisation effects in the imaging system and only consider the wavefront deformation and amplitude attenuation (or amplification) that occurs when light propagates from the entrance to the exit pupil. The transmission function is then adequately represented by $f'_L(\rho, \theta)$ of Eq. (9.2.2) and it affects the orthogonal s-and p-components on the entrance pupil sphere equally. For the orthogonal field components on the exit pupil sphere we then obtain, with the aid of Eq. (9.5.3) and Fig. 9.19:

$$E_{s,1} = \left|\frac{f_1}{R_p}\right| T_R(\rho) f'_L(\rho, \theta) \sqrt{\frac{n_0}{n_1}} \left(-E_{x,0}(\rho, \theta) \sin\theta + E_{y,0}(\rho, \theta)\cos\theta\right) , \tag{9.5.14}$$

$$E_{p,1} = \left|\frac{f_1}{R_p}\right| T_R(\rho) f'_L(\rho, \theta) \sqrt{\frac{n_0}{n_1}} \frac{E_{x,0}(\rho, \theta) \cos\theta + E_{y,0}(\rho, \theta)\sin\theta}{\left\{1 - n_1^2 m_T^2 s_1^2 \rho^2 / n_0^2\right\}^{1/2}} . \tag{9.5.15}$$

The Cartesian field components in image space are then given by

$$\begin{aligned} E_{x,1} &= E_{p,1}(\rho, \theta) \cos\alpha_1 \cos\theta - E_{s,1}(\rho, \theta) \sin\theta , \\ E_{y,1} &= E_{p,1}(\rho, \theta) \cos\alpha_1 \sin\theta + E_{s,1}(\rho, \theta) \cos\theta , \\ E_{z,1} &= \frac{R_1}{|R_1|} E_{p,1}(\rho, \theta)(1 - \cos^2\alpha_1)^{1/2} . \end{aligned} \tag{9.5.16}$$

With the aid of Eq. (9.5.15) we obtain an expression for the Cartesian field components in the exit pupil as a function of the Cartesian components on the entrance pupil sphere,

$$\begin{aligned} E_{x,1} = \left|\frac{f_1}{R_p}\right| T_R(\rho) f'_L(\rho, \theta) \sqrt{\frac{n_0}{n_1}} \Bigg\{ & \frac{(1 - s_1^2\rho^2)^{1/2}}{\left\{1 - (n_1^2/n_0^2) m_T^2 s_1^2 \rho^2\right\}^{1/2}} \Big[E_{x,0}(\rho, \theta)\cos^2\theta \\ & + E_{y,0}(\rho, \theta)\cos\theta\sin\theta\Big] + E_{x,0}(\rho, \theta)\sin^2\theta - E_{y,0}(\rho, \theta)\cos\theta\sin\theta \Bigg\} , \\ E_{y,1} = \left|\frac{f_1}{R_p}\right| T_R(\rho) f'_L(\rho, \theta) \sqrt{\frac{n_0}{n_1}} \Bigg\{ & \frac{(1 - s_1^2\rho^2)^{1/2}}{\left\{1 - (n_1^2/n_0^2) m_T^2 s_1^2 \rho^2\right\}^{1/2}} \Big[E_{x,0}(\rho, \theta)\cos\theta\sin\theta \\ & + E_{y,0}(\rho, \theta)\sin^2\theta\Big] - E_{x,0}(\rho, \theta)\cos\theta\sin\theta + E_{y,0}(\rho, \theta)\cos^2\theta \Bigg\} , \\ E_{z,1} = \left|\frac{f_1}{R_p}\right| & \frac{T_R(\rho) f'_L(\rho, \theta)}{\left\{1 - (n_1^2/n_0^2) m_T^2 s_1^2 \rho^2\right\}^{1/2}} \sqrt{\frac{n_0}{n_1}} \left(\frac{R_1}{|R_1|}\right) s_1 \rho \left[E_{x,0}(\rho, \theta)\cos\theta + E_{y,0}(\rho, \theta)\sin\theta\right] . \end{aligned} \tag{9.5.17}$$

For the case of a source at a finite distance, the incident field components, $E_{x,0}(\rho, \theta)$ and $E_{y,0}(\rho, \theta)$ are quantities that vary over the cross-section of the entrance pupil. The same holds for the lens function $f'_L(\rho, \theta)$ in the case of, for instance, an aberrated imaging system.

We can represent the product of the incident field components and the lens function by means of two Zernike expansions, specifically we write

$$E_{x,0}(\rho, \theta) f'_L(\rho, \theta) = \sum_{n,m} \beta^m_{n,x} R^{|m|}_n(\rho) \exp(im\theta) , \tag{9.5.18}$$

$$E_{y,0}(\rho, \theta) f'_L(\rho, \theta) = \sum_{n,m} \beta^m_{n,y} R^{|m|}_n(\rho) \exp(im\theta) . \tag{9.5.19}$$

The substitution of these two expansions into Eq. (9.5.17) yields the following expressions for the Cartesian field components on the exit pupil sphere,

$$\begin{aligned} E_{x,1}(\rho, \theta) = \left|\frac{f_1}{2R_p}\right| \frac{T_R(\rho)}{\cos\alpha_0} \sqrt{\frac{n_0}{n_1}} & \sum_{n,m} R^{|m|}_n(\rho) \exp(im\theta) \\ \times \Bigg\{ \beta^m_{n,x} & \left[\cos\alpha_1 + \cos\alpha_0 - \frac{(\sin^2\alpha_1 - \sin^2\alpha_0)\cos 2\theta}{\cos\alpha_1 + \cos\alpha_0}\right] \\ & - \beta^m_{n,y}\left[\frac{(\sin^2\alpha_1 - \sin^2\alpha_0)\sin 2\theta}{\cos\alpha_1 + \cos\alpha_0}\right]\Bigg\} , \end{aligned} \tag{9.5.20}$$

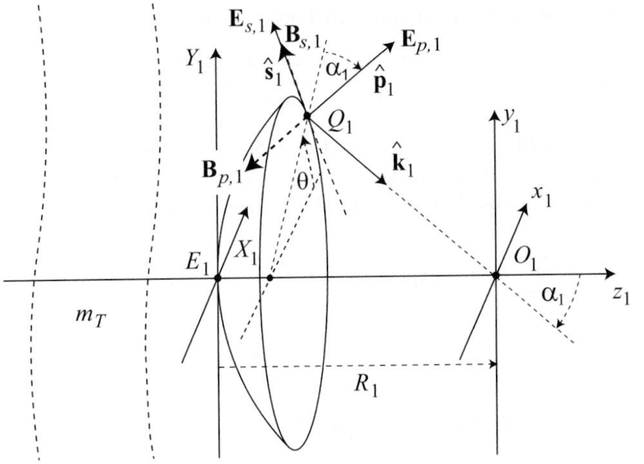

Figure 9.20: The convention for the s-and p- components of the electric field vector $\mathbf{E}_1$ and the magnetic induction vector $\mathbf{B}_1$; unit vectors $\hat{\mathbf{p}}_1$ and $\hat{\mathbf{k}}_1$ are parallel to the meridional plane defined by the points E_1, Q_1 and O_1.

$$E_{y,1}(\rho,\theta) = \left|\frac{f_1}{2R_p}\right| \frac{T_R(\rho)}{\cos\alpha_0}\sqrt{\frac{n_0}{n_1}} \sum_{n,m} R_n^{|m|}(\rho)\exp(im\theta)$$
$$\times\left\{\beta^m_{n,y}\left[\cos\alpha_1 + \cos\alpha_0 + \frac{(\sin^2\alpha_1 - \sin^2\alpha_0)\cos 2\theta}{\cos\alpha_1 + \cos\alpha_0}\right]\right.$$
$$\left. - \beta^m_{n,x}\left[\frac{(\sin^2\alpha_1 - \sin^2\alpha_0)\sin 2\theta}{\cos\alpha_1 + \cos\alpha_0}\right]\right\}, \qquad (9.5.21)$$

$$E_{z,1}(\rho,\theta) = \left|\frac{f_1}{R_p}\right| \frac{T_R(\rho)}{\cos\alpha_0}\sqrt{\frac{n_0}{n_1}}\left(\frac{R_1}{|R_1|}\right) s_1 \sum_{n,m}\rho\, R_n^{|m|}(\rho)\exp(im\theta)\left\{\beta^m_{n,x}\cos\theta + \beta^m_{n,y}\sin\theta\right\}, \qquad (9.5.22)$$

where we have used that the length of the z-components is proportional to $(1-\cos^2\alpha_1)^{1/2} = |\sin\alpha_1| = \rho s_1$. Note the sign of the square root factor $(1-\cos^2\alpha_1)^{1/2}$ follows from the value of $R_1/|R_1|$ in image space.

9.5.1.3 The Magnetic Field Components on the Exit Pupil Sphere

For calculation of a second-order electromagnetic quantity such as the Poynting vector in the image region, we need both the electric and the magnetic induction components on the exit pupil sphere. The directions of the magnetic induction components follow from the orthogonality and transversality of the unit vectors along the electric, magnetic and wave propagation vectors, $\hat{\mathbf{e}}$, $\hat{\mathbf{b}}$ and $\hat{\mathbf{k}}$, respectively, in a homogeneous and isotropic space. In the far field of a source, the magnitude of the magnetic induction vector $|\mathbf{B}|$ equals $n|\mathbf{E}|/c$ where n is the refractive index of the medium. On the exit pupil sphere (see Fig. 9.20) we thus have that $\mathbf{B}_1 = n_1\,\hat{\mathbf{k}}\times\mathbf{E}_1/c$. We could apply this expression for the magnetic induction vector to the Cartesian electric components, such that we immediately obtain the Cartesian components of $\mathbf{B}$ that are needed in the Debye-integral. We prefer, however, to separately treat the s- and p-electric components to obtain expressions for the corresponding p- and s-components of the magnetic induction vector. We have for the p-component,

$$\mathbf{B}_{p,1}(\rho,\theta) = \left|\frac{n_1 f_1}{cR_p}\right| T(\rho) f'_L(\rho,\theta)\,(-E_{x,0}\sin\theta + E_{y,0}\cos\theta)\,\hat{\mathbf{k}}_1\times\hat{\mathbf{s}}_1$$
$$= -\left|\frac{n_1 f_1}{cR_p}\right| T(\rho) f'_L(\rho,\theta)\,(-E_{x,0}\sin\theta + E_{y,0}\cos\theta)\,\hat{\mathbf{p}}_1$$
$$= -\left|\frac{n_1 f_1}{cR_p}\right| T(\rho) f'_L(\rho,\theta)\,(-E_{x,0}\sin\theta + E_{y,0}\cos\theta)\left(\cos\alpha_1\cos\theta,\ \cos\alpha_1\sin\theta,\ \frac{R_1}{|R_1|}|\sin\alpha_1|\right), \qquad (9.5.23)$$

and for the s-component of the magnetic induction it readily follows that

$$\mathbf{B}_{s,1}(\rho,\theta) = + \frac{n_1 f_1\, T(\rho) f_L'(\rho,\theta)}{cR_p} \left(\frac{(E_{x,0}\cos\theta + E_{y,0}\sin\theta)}{\cos\alpha_0} \right) (-\sin\theta\,,\ \cos\theta\,,\ 0)\ , \tag{9.5.24}$$

where we have used in the last line of Eqs. (9.5.23) and in Eq. (9.5.24) the expressions for the Cartesian components of the $\hat{\mathbf{p}}_1$ and $\hat{\mathbf{s}}_1$ unit vectors, respectively, defined analogously to Eq. (9.5.1). The insertion of the expansions for the electric field components $E_{x,0}$ and $E_{y,0}$ given by Eqs. (9.5.18) and (9.5.19) then yields

$$\begin{aligned} B_{x,1}(\rho,\theta) = \left|\frac{f_1}{2cR_p}\right| \frac{T_R(\rho)}{\cos\alpha_0} \sqrt{n_0\, n_1} \sum_{n,m} R_n^{|m|}(\rho) \exp(im\theta) \Big\{ &-\beta_{n,x}^m \big(1-\cos\alpha_0\cos\alpha_1\big)\sin 2\theta \\ &+ \beta_{n,y}^m \Big[-(1+\cos\alpha_0\cos\alpha_1) + (1-\cos\alpha_0\cos\alpha_1)\cos 2\theta \Big] \Big\}\ , \end{aligned} \tag{9.5.25}$$

$$\begin{aligned} B_{y,1}(\rho,\theta) = \left|\frac{f_1}{2cR_p}\right| \frac{T_R(\rho)}{\cos\alpha_0} \sqrt{n_0\, n_1} \sum_{n,m} R_n^{|m|}(\rho) \exp(im\theta) \Big\{ &\beta_{n,x}^m \Big[(1+\cos\alpha_0\cos\alpha_1) + (1-\cos\alpha_0\cos\alpha_1)\cos 2\theta \Big] \\ &+ \beta_{n,y}^m (1-\cos\alpha_0\cos\alpha_1)\sin 2\theta \Big\}\ , \end{aligned} \tag{9.5.26}$$

$$B_{z,1}(\rho,\theta) = \left|\frac{f_1}{cR_p}\right| T_R(\rho)\, s_1 \sqrt{n_0\, n_1} \left(\frac{-R_1}{|R_1|}\right) \sum_{n,m} \rho\, R_n^{|m|}(\rho) \exp(im\theta) \Big\{ -\beta_{n,x}^m \sin\theta + \beta_{n,y}^m \cos\theta \Big\}\,. \tag{9.5.27}$$

9.5.2 The Debye Integral and the Electromagnetic Field in the Image Region

The three Cartesian field components on the exit pupil sphere given by Eqs. (9.5.20)–(9.5.22) and Eqs. (9.5.25)–(9.5.27) are used in the following analysis as integrand functions in the Debye diffraction integral, to obtain the electric field and the magnetic induction components in the image region. The diffraction of such a vector field on the exit pupil sphere was first treated by Ignatowsky **[153]**,**[154]**,**[155]** and was reconsidered in the framework of a plane wave expansion in **[293]**,**[374]**. The vector field in the entrance pupil of the optical system was assumed to be generated by a distant point source (object at infinity) but the analysis can equally well be applied to a more complicated field distribution in the entrance pupil, for instance the field created by the coherent diffraction from an assembly of point sources or from a more general object structure. Here, we allow the source to be at a finite distance and the position of the centre of the diffracting aperture or of its paraxial image is determined by the distance R_p from the image side focal point F_1. The vector form of the Debye integral given by Eq. (8.3.25) is used to obtain the electric field components whereas the comparable Eq. (8.3.26) can be used to obtain the magnetic components. As discussed before for the scalar case, these formulas can be safely used when the Fresnel number N_F of the imaging configuration is large, typically > 1000 (see Fig. 9.6).

9.5.2.1 The Electric Field Components in the Image Region

The Debye integral for the electric field vector $\mathbf{E}_i$ in the image space is given by Eq. (8.3.25),

$$\mathbf{E}_i\left(\mathbf{r}_p\right) = -\frac{i|f|}{\lambda} \iint_{s_x^2+s_y^2\leq 1} \frac{\mathbf{E}(s_x,s_y)}{s_z} \exp(ik\mathbf{s}\cdot\mathbf{r}_\mathbf{p})\, ds_x ds_y\ , \tag{9.5.28}$$

where the ratio $|f|/\lambda$ is the amplitude concentration factor when the light propagates from the exit pupil at a distance f to the focal plane. We use the absolute value $|f|$ in this integral to allow for the possibility of a negative focal distance. We slightly adapt this equation here to accommodate an imaging system with a finite magnification and an arbitrarily positioned exit pupil. We also introduce the polar coordinates (ρ,θ) in the exit pupil that are customarily adopted when using Zernike polynomials to represent the lens transmission function $f_L'(\rho,\theta)$. It is important to remark at this stage that the transition to polar coordinates implies a possible azimuthal shift of π depending on the sign of f in the original Debye integral above. With the general distance R_1 measured from the exit pupil to the image plane, we define the polar coordinates as

$$s_x = s_1\rho\cos(\theta-\theta_0)\,,\quad s_y = s_1\rho\sin(\theta-\theta_0)\,,\quad s_z = (1-s_1^2\rho^2)^{1/2}\ ,\quad \theta_0 = \frac{\pi}{2}(1+R_1/|R_1|)\ , \tag{9.5.29}$$

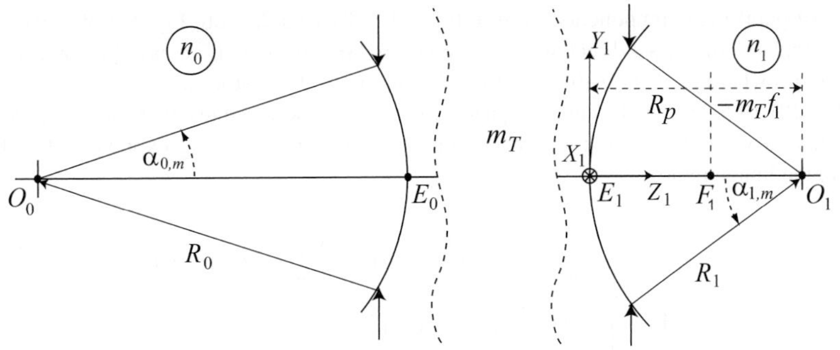

Figure 9.21: Exit pupil with its centre at E_1, image side focal point F_1 and image plane through O_1. The polar angle θ in the (X_1, Y_1) pupil plane is counted positive in the counter-clockwise direction with respect to the X_1-axis when viewed in the negative Z_1-direction. $F_1E_1 = R_p$, $F_1O_1 = -m_T f_1$. In the figure, according to our convention, the radii of curvature of the pupils, R_0 and R_1, are negative and positive, respectively.

where $s_1 = |\sin\alpha_{1,m}|$ is the sine of the maximum aperture angle in image space. We can thus express the Debye integral in polar coordinates as

$$\mathbf{E}_i(r'_1, \gamma, f'_v) = -is_1^2 \frac{|R_1|}{\lambda_{n_1}} \exp\left(\frac{if'_v}{u_1}\right) \iint_{C_1} \frac{\mathbf{E}_1(\rho,\theta)}{(1-s_1^2\rho^2)^{1/2}} \exp\left\{-i\frac{f'_v}{u_1}\left[1-(1-s_1^2\rho^2)^{1/2}\right]\right\}$$
$$\times \exp\{-i(R_1/|R_1|)\,2\pi r'_1\rho\cos(\theta-\gamma)\}\,\rho d\rho d\theta\,. \qquad (9.5.30)$$

To arrive at Eq. (9.5.30) from Eq. (9.5.28), the following must be considered:

- r'_1 is the normalised lateral radial coordinate.
- The quantity f'_v, not to be confused with the complex lens function $f'_L(\rho,\theta)$, was defined in Eq. (9.2.15) and is the high-aperture defocusing parameter. The normalised defocusing parameter f'_v for the high-aperture case is thus given by

$$f'_v = \frac{2\pi n_1 u_1 z_1}{\lambda_0} \approx \frac{\pi s_1^2}{\lambda_{n_1}} z_1 = \pi z'_1 = f'_d\,, \qquad (9.5.31)$$

  where $\lambda_{n_1} = \lambda_0/n_1$ and $u_1 = 1-(1-s_1^2)^{1/2}$. The axial diffraction unit $z_{1,v}$ for the high-aperture case follows from $f'_v = \pi$ and is given by $z_{1,v} = \lambda_{n_1}/(2u_1)$. The quantity f'_d is the previously defined low-aperture defocusing parameter for the scalar case (see Eq. (9.4.2)).
- The exponential function in the integrand of Eq. (9.5.28) has been split into a defocusing part and a transverse part, in line with the separation that was already made for the scalar case, see Eq. (9.3.4).
- C_1 is the unit circle associated with the normalised radial coordinate ρ on the exit pupil sphere. In the normalised polar coordinate system (ρ,θ) defined by Eq. (9.5.29), C_1 determines the rim of the solid angle that is subtended by the exit pupil rim with respect to the image point.

The insertion of the exit pupil field components in the Debye integral of Eq. (9.5.30) gives rise to a field concentration factor $|R_1 f_1/R_p|$. From Newton's equation we have that $F_1O_1 = -m_T f_1$ and this yields $R_1 f_1/R_p = f_1/[1 + m_T f_1/R_1]$ (see Fig. 9.21). A special case arises when the exit pupil is at infinity (telecentric imaging). In that case the field concentration factor equals $|f_1|$. An entrance pupil at infinity poses a problem because the field components $E_{p,0}$ and $E_{s,0}$ on the entrance pupil tend to zero since the finite energy emitted by the source is spread out over an infinitely large surface. The Zernike coefficients $\beta^m_{n,x}$ and $\beta^m_{n,x}$ are then undefined. In this case, the entrance pupil sphere is fictitiously shifted to a large finite distance R_0 and the field is calculated on this sphere using the finite energy flow from the source. From the entrance pupil field components we obtain the field on the corresponding exit pupil sphere with radius R_1 and this enables the calculation of the quantity $R_1\mathbf{E}_1$ that is invariant along a ray when going from exit pupil to image plane. The entrance pupil is then shifted back to infinity and we obtain the ray invariant value $f_1\mathbf{E}_{F_1}$ on the exit pupil sphere through the image side focal point F_1.

Insertion of the exit pupil field components given in Eqs. (9.5.20)–(9.5.22) into Eq. (9.5.30) gives rise to an integral over the azimuthal variable θ that has a direct analytical solution. We have that $\int_0^{2\pi} \exp(im\theta)\exp[i2\pi r_1'\rho\cos(\theta-\gamma')]d\theta = 2\pi i^m J_m(2\pi r_1'\rho)\exp(im\gamma')$ where J_m is the mth-order Bessel function of the first kind. We then express the trigonometric functions in Eqs. (9.5.20)–(9.5.22) that depend on α_0 and α_1 in terms of the variable ρ using the expressions of Eq. (9.5.6). After some manipulation we obtain, using vector column notation, the following expression for the Cartesian field components of $\mathbf{E}_i$ in the image region,

$$\mathbf{E}_i(r_1',\gamma,f_v') = \frac{-i\pi n_1 s_1^2}{\lambda_0}\left|\frac{f_1}{1+m_T f_1/R_1}\right|\sqrt{\frac{n_0}{n_1}}\exp\left(\frac{if_v'}{u_1}\right)\sum_{n,m}[-(R_1/|R_1|)\,i]^m\exp[im\gamma]$$
$$\times\left[\beta_{n,x}^m\begin{pmatrix} V_{n,0}^{m,E}-\frac{1}{2}\left\{V_{n,+2}^{m,E}\exp[+2i\gamma]+V_{n,-2}^{m,E}\exp[-2i\gamma]\right\} \\ +\frac{i}{2}\left\{V_{n,+2}^{m,E}\exp[+2i\gamma]-V_{n,-2}^{m,E}\exp[-2i\gamma]\right\} \\ -i\left\{V_{n,+1}^{m,E}\exp[+i\gamma]-V_{n,-1}^{m,E}\exp[-i\gamma]\right\}\end{pmatrix}\right.$$
$$\left.+\beta_{n,y}^m\begin{pmatrix} +\frac{i}{2}\left\{V_{n,+2}^{m,E}\exp[+2i\gamma]-V_{n,-2}^{m,E}\exp[-2i\gamma]\right\} \\ V_{n,0}^{m,E}+\frac{1}{2}\left\{V_{n,+2}^{m,E}\exp[+2i\gamma]+V_{n,-2}^{m,E}\exp[-2i\gamma]\right\} \\ -\left\{V_{n,+1}^{m,E}\exp[+i\gamma]+V_{n,-1}^{m,E}\exp[-i\gamma]\right\}\end{pmatrix}\right], \tag{9.5.32}$$

where, for integer $j=-2,\cdots,+2$, we have

$$V_{n,j}^{m,E}(r_1',f_v') = \int_0^1 \rho^{|j|} f_{|j|}^E(\rho)\exp\left[-\frac{if_v'}{u_1}\left(1-\sqrt{1-s_1^2\rho^2}\right)\right]R_n^{|m|}(\rho)J_{m+j}(2\pi r_1'\rho)\rho d\rho\,, \tag{9.5.33}$$

where the functions $f_{|j|}^E(\rho)$ for $j=0,1,2$ are given by

$$f_0^E(\rho) = \frac{\left\{\left(1-s_0^2\rho^2\right)^{1/2}+\left(1-s_1^2\rho^2\right)^{1/2}\right\}}{\left(1-s_1^2\rho^2\right)^{1/4}\left(1-s_0^2\rho^2\right)^{3/4}}\,, \tag{9.5.34}$$

$$f_1^E(\rho) = \frac{s_1}{\left(1-s_1^2\rho^2\right)^{1/4}\left(1-s_0^2\rho^2\right)^{3/4}}\,, \tag{9.5.35}$$

$$f_2^E(\rho) = \frac{s_0^2-s_1^2}{\left\{\left(1-s_1^2\rho^2\right)^{1/2}+\left(1-s_0^2\rho^2\right)^{1/2}\right\}\left(1-s_1^2\rho^2\right)^{1/4}\left(1-s_0^2\rho^2\right)^{3/4}}\,. \tag{9.5.36}$$

Equation (9.5.32), complemented by Eqs. (9.5.33)–(9.5.36), shows a structure in which the influence of the source (coefficients $\beta_{n,x}^m$ and $\beta_{n,y}^m$) and the imaging system ($V_{n,j}^{m,E}$-functions) are clearly separated. The Zernike coefficients represent modular expansion coefficients of the field on the exit pupil sphere and the V functions are the typical modular 'response' functions in the diffracted field in the image region. The evaluation of the V functions must be performed only once for a preset grid of points in the (r_1',f_v') domain. Once these data are available for a specific image-side aperture value s_1 and lateral magnification m_T, the calculation of the electric field strength $\mathbf{E}_i$ follows the simple scheme of Eq. (9.5.32).

Calculation of $\mathbf{B}_i$ is carried out along the same path. After some manipulation we obtain the following expressions,

$$\mathbf{B}_i(r_1',\gamma,f_v') = \frac{-i\pi n_1 s_1^2}{c\lambda_0}\left|\frac{f_1}{1+m_T f_1/R_1}\right|\sqrt{n_0\,n_1}\exp\left(\frac{if_v'}{u_1}\right)\sum_{n,m}[-(R_1/|R_1|)\,i]^m\exp[im\gamma]$$
$$\times\left[\beta_{n,x}^m\begin{pmatrix} +\frac{i}{2}\left\{V_{n,+2}^{m,B}\exp[+2i\gamma]-V_{n,-2}^{m,B}\exp[-2i\gamma]\right\} \\ V_{n,0}^{m,B}+\frac{1}{2}\left\{V_{n,+2}^{m,B}\exp[+2i\gamma]+V_{n,-2}^{m,B}\exp[-2i\gamma]\right\} \\ -\left\{V_{n,+1}^{m,B}\exp[+i\gamma]+V_{n,-1}^{m,B}\exp[-i\gamma]\right\}\end{pmatrix}\right.$$
$$\left.+\beta_{n,y}^m\begin{pmatrix} -V_{n,0}^{m,B}+\frac{1}{2}\left\{V_{n,+2}^{m,B}\exp[+2i\gamma]+V_{n,-2}^{m,B}\exp[-2i\gamma]\right\} \\ -\frac{i}{2}\left\{V_{n,+2}^{m,B}\exp[+2i\gamma]-V_{n,-2}^{m,B}\exp[-2i\gamma]\right\} \\ +i\left\{V_{n,+1}^{m,B}\exp[+i\gamma]-V_{n,-1}^{m,B}\exp[-i\gamma]\right\}\end{pmatrix}\right]. \tag{9.5.37}$$

The V functions for the magnetic case are given by $(j = -2, \cdots, +2)$,

$$V_{n,j}^{m,B}(r_1', f_v') = \int_0^1 \rho^{|j|} f_{|j|}^B(\rho) \exp\left[-\frac{if_v'}{u_1}\left(1 - \sqrt{1 - s_1^2\rho^2}\right)\right] R_n^{|m|}(\rho) J_{m+j}(2\pi r_1'\rho)\, \rho d\rho\,, \tag{9.5.38}$$

where the functions $f_{|j|}^B(\rho)$ for $j = 0, 1, 2$ are now given by

$$f_0^B(\rho) = \frac{1 + \left(1 - s_0^2\rho^2\right)^{1/2}\left(1 - s_1^2\rho^2\right)^{1/2}}{\left(1 - s_1^2\rho^2\right)^{1/4}\left(1 - s_0^2\rho^2\right)^{3/4}}\,, \tag{9.5.39}$$

$$f_1^B(\rho) = \frac{s_1}{\left(1 - s_1^2\rho^2\right)^{1/4}\left(1 - s_0^2\rho^2\right)^{1/4}}\,, \tag{9.5.40}$$

$$f_2^B(\rho) = \frac{-s_0^2(1 - s_1^2\rho^2)^{1/2} - s_1^2(1 - s_0^2\rho^2)^{1/2}}{\left\{\left(1 - s_0^2\rho^2\right)^{1/2} + \left(1 - s_1^2\rho^2\right)^{1/2}\right\}\left(1 - s_1^2\rho^2\right)^{1/4}\left(1 - s_0^2\rho^2\right)^{3/4}}\,. \tag{9.5.41}$$

We note that, for $j = 0, 1, 2$, the f_j^E and f_j^B functions are identical if $m_T = s_0 = 0$ (imaging from infinity). It is also worth noting that the functions f_j^E and f_j^B are of jth order in the quantities s_1 and/or s_0. This means that for decreasing aperture we can restrict the result to the leading terms comprising the V functions with index $j = 0$.

9.5.2.2 Numerical and Semi-analytic Evaluation of the V Integrals

Before we address the particularities of the evaluation of the V integrals we consider a frequently used numerical integration method, the fast Fourier transform (FFT) method. Since a complex exponential is present in the basic expression for the Debye integral of Eq. (9.5.30), the FFT method can be applied when this expression is rewritten in terms of Cartesian coordinates. The integrand functions associated with the three electric field components $E_{x,1}$, $E_{y,1}$ and $E_{z,1}$ on the exit pupil sphere follow by the substitution of Eqs. (9.5.20)–(9.5.22) into the Debye integral.

For a comparison of the FFT method with the V integral method, a single Zernike polynomial $R_n^{|m|}(\rho)\exp(im\theta)$ is selected from the sum over (n, m) in Eqs. (9.5.20)–(9.5.22). The integrand function is evaluated at a number of (equidistant) sample points that are arranged on a Cartesian grid. Performing the FFT yields the value of the Debye integral simultaneously in a grid of points which are located in a (defocused) plane in image space. The computation time for a single FFT evaluation of the Debye integral is proportional to $N(^2\log N)$ where N is the number of sample grid points. The basic time for each operation is put equal to T_g which is comparable to the time required for the multiplication of two complex numbers or of a complex number and a complex exponential function. Simple additions or multiplications take a much smaller time and are omitted from the computational budget. To evaluate the numerical effort we have to consider separately the preparation of the integrand values in the sample points and the FFT integration step.

- FFT evaluation of the Debye integral for a single Zernike term

  For the FFT method the integrand values are needed at a square grid of points and the effort per sample point can be estimated with the aid of the expressions for the Cartesian components of the electric field vector $\mathbf{E}_1$, in the integrand of Eq. (9.5.30). An estimate for the computation time of the values of the $\mathbf{E}_1$-components at the grid points is given by $N_t^2(15T_g + T_Z)$ where T_g is the time needed for a trigonometric function evaluation and T_Z the time for the (recursive) evaluation of a radial Zernike polynomial (of higher order). The FFT integration step requires $2N_t^2(^2\log N_t)$ complex multiplications with duration T_C. If we assume that $T_C \approx T_g$ we obtain for the total number of operations $N_{D,\mathrm{FFT}}$ with typical duration T_g to evaluate the Debye integral for a single Zernike polynomial,

$$N_{D,\mathrm{FFT}} = 3\,N_t^2\,[\{15 + 2\,^2\log N_t\}T_g + T_Z/3]/T_g \approx 3\,N_t^2\,\{18 + 2\,^2\log N_t\}\,, \tag{9.5.42}$$

  where the calculation time for a radial Zernike polynomial using the recursive method according to Eq. (D.3.4) in Appendix D was put equal to 10 units of T_g and is counted in the calculations as $10\,T_g/3$ since the polynomials are identical for the three field components on the exit pupil. We use the number $N_{D,\mathrm{FFT}}$ as a reference number for a single ‘brute force’ calculation of the electric field components in a certain (defocused) image plane. We emphasise that this electric field distribution in a defocused image plane is associated with a field distribution on the exit pupil sphere which corresponds to a single Zernike polynomial.

  The ‘modular’ evaluation of the Debye integral yields the expressions of Eqs. (9.5.32) and (9.5.37)) in which the total electric field is written as the sum of separate evaluations associated with a specific Zernike polynomial with indices (n, m) which is present in the x- and y-polarised components of $\mathbf{E}_1$ on the exit pupil sphere. To obtain the total field, repeated evaluations are required of five V integrals which depend on the radial and axial coordinates in image space. The integrals $V_{n,j}^{m,E}$ (and $V_{n,j}^{m,B}$) have to be evaluated at a discrete set of radial and axial points (expressed in terms of the normalised r_1' coordinate and the f_v' parameter which accounts for the axial position of a defocused plane). The integrals depend on a number of imaging parameters such as the image side aperture s_1, the lateral magnification m_T of the imaging system where $|s_0| = |m_T s_1|$, and, of course, on the Zernike aberration term under consideration (with radial and azimuthal indices n and m, respectively). With the aid of pre-calculated (one-dimensional) V integrals (see Eqs. (9.5.33) and (9.5.38)), the electric and magnetic fields are computed in a straightforward way with the aid of Eq. (9.5.32) and (9.5.37). As shown previously in Subsection 9.3.2.1, certain symmetry properties of the electric field on the exit pupil sphere can strongly reduce the total number of required integral evaluations.

We will first discuss the computational effort that is required for a *numerical* evaluation of the V integrals for the high-aperture case. To compute the one-dimensional integral over ρ, several numerical approaches are possible. We mention the trapezoidal rule, Simpson's rule, Gauss quadrature, Romberg's method, etc. To avoid at this point a discussion on numerical quadrature methods, we adopt the simplest approach, the trapezoidal rule, for the evaluation of the V integrals associated with each Zernike polynomial with coefficients $\beta_{n,x}^m$ and $\beta_{n,y}^m$. In a somewhat arbitrary manner, we adopt the same number of sample points, N_t, on the integration interval $0 \leq \rho \leq 1$ as we used before in the FFT computation. The relative error associated with the trapezoidal rule is proportional to $1/(12N_t^2)$ and the value $N_t = 512$ will generally be sufficient to achieve an error which is less than 10^{-3}.

- Numerical evaluation of the V integrals

  For the evaluation of the V integral we can reduce the number of lateral sample points N_r along the r_1'-axis to typically 100 and arrange them on an interval where the point-spread function has a nonzero value; in practice an interval for r_1' up to the normalised value of 10 is sufficient. This reduction is allowed because the point-spread function is generally a smoother function in the radial direction than the pupil function f_L' when the latter contains a high-order Zernike polynomial which quickly oscillates near the pupil rim. The calculation of an integrand sample point along the ρ-axis is comparable in computational effort to the effort that was needed for the calculation of an E_1 component for the FFT case, apart from the extra evaluation of the Bessel function J_{m+j} to which we attribute a time T_B. The one-dimensional V integration step itself (trapezoidal rule) is almost negligible in time. The next step comprises the complex multiplications of V integral values and exponential trigonometric functions which appear in Eq. (9.5.32). Each image point requires eight complex multiplications and we assume that there are N_r^2 image points with equal radial and azimuthal subdivisions of the r_1' and γ intervals, respectively. With these assumptions we arrive at the total number of operations $N_{D,V}$ with typical duration T_g,

$$N_{D,V} = N_r[5N_t(15T_g + T_Z/5 + T_B) + 8N_r T_g]/T_g = N_r[135N_t + 8N_r] , \tag{9.5.43}$$

  where we have assumed that the (recursive) calculation of a radial Zernike polynomial and a Bessel function of average order (typically 10) both require a time of 10 T_g.

  The result of the comparison is illustrated in the first two columns of Table 9.1, in units of a typical computation time T_g. The table shows that it is advantageous to use the analytical refinement which led to the solutions for $\mathbf{E}_i$ comprising the V integrals. An advantage of a factor of almost four in calculation time is obtained with respect to a standard FFT computation scheme.

- Semi-analytic evaluation of the V-integrals

  A further acceleration of the ENZ computations has been devised by the development of a semi-analytic scheme for the various V integrals such that the numerical integration is replaced by series expressions with analytic coefficients and known analytic functions. For the low-aperture scalar case the double series expansion for the V integral with an infinite number of terms for the summation associated with the scalar defocusing term f_d' are given by Eqs. (9.4.4)–(9.4.8). A truncation rule has been established that limits the number of terms when a certain final accuracy is prescribed. Such semi-analytic expressions for the field strength components can be computed substantially faster as compared to the calculation employing a numerical quadrature rule for the V integral.

The truncation rule for the first summation over the index l in Eq. (9.4.4) limits the number of terms in the summation to $l_{\max} = 3f_d' + 5$. We use a typical average value of π for the defocusing parameter f_d', which yields a normalised defocusing coordinate $z_1' = 1$. The second summation with respect to j depends on the indices of the Zernike polynomial via the parameter $p = (n - |m|)/2$. Here we adopt an average value of p equal to 5. With respect to calculation time, the crucial factor is the Bessel function of order $|m| + l + 2j$. The number of Bessel functions which have to be computed then amounts to $(n + m)/2 + 3f_d' + 5 \approx 20$. The recursive evaluation of binomial coefficients is almost negligible in

Table 9.1. Numerical effort for Debye-integral evaluation (single Zernike term).

Method	FFT	V integrals (trap. rule)	V integrals (semi-analytic)
	Eq. (9.5.30)	Eq. (9.5.32)	Eqs. (9.4.4)–(9.4.5)
	($N_t = 512$)	($N_t = 512$, $N_r = 100$)	($N_r = 100$)
N_D (units T_g)	$26 \cdot 10^6$	$7 \cdot 10^6$	$6 \cdot 10^5$

comparison with the Bessel function. Since extra Zernike polynomials arise in the high-aperture case and Clebsch–Gordan coefficients need to be calculated, experience has shown that, as a crude estimate, it is advisable to multiply the total number of operations by a factor of four. With these assumptions the total number of operations $N_{D,Vsa}$ for the semi-analytic evaluation of the V integral is approximatelygiven by

$$N_{D,Vsa} = N_r[5.4\{20(T_B + 2T_g)\} + 8N_r T_g]/T_g = N_r[4800 + 8N_r] \,, \tag{9.5.44}$$

where the typical duration of each operation is T_g.

The specific value when using $N_r = 100$ is shown in the last column of Table 9.1. The semi-analytic evaluation of the V integral yields on average a ten times shorter calculation time as compared to a numerical integration of the same integral. Still shorter calculation times are possible when the maximum order of the Zernike expansion on the exit pupil is limited to the basic primary aberrations. For much higher orders of the Zernike expansion or for a large defocusing parameter, the numerical effort associated with the semi-analytic solution increases substantially.

The important reduction in computing time by using a (truncated) semi-analytic expression of the V integral is especially appreciated when these integrals have to be used in recursive schemes with repeated evaluations; for instance, when solving an aberration retrieval problem for a lens system based on intensity measurements in various defocused planes in the image region. In the context of repeated evaluations of the diffraction integrals, an efficient use of the FFT method requires the storage in memory of the computational results that are obtained for each Zernike aberration term. The same holds for the V integral computation. If we compare the number of data points to be kept in memory for the FFT and for the V integral approach, we find a storage requirement that is at least two orders of magnitude less when using the V integral-based approach.

9.5.2.3 Brief Outline of the Semi-analytic Evaluation of a V Integral

Depending on the imaging geometry and the sophistication of the imaging model, various versions of the V integrals exist. Historically, the first V_n^m integral was derived within the context of a scalar diffraction problem with the object plane at infinity The corresponding semi-analytic expression has been given by Janssen **[162]**. The V integrals for the vector diffraction problem with the object at infinity were developed later. They are discussed in **[52]** and each V integral is specified by three indices, $V_{n,j}^m$. In the aberration-free case, the absolute value of the extra index j with $|j| = 0, 1, 2$ refers to the three diffraction integrals that were first analysed by Ignatowsky **[153]** and later by Richards and Wolf **[293]**,**[374]**. So far, the form of the V integrals used to compute the $\mathbf{E}_i$ and $\mathbf{B}_i$ fields has been identical. The V integrals that are treated in this section apply to the vector imaging case with finite magnification. These integrals have a different integrand depending on whether the electric or the magnetic field strength in image space is considered. A further extension of the imaging geometry which discussed later is the introduction of a thin layer stack in the direct neighbourhood of the image plane. In that case, different V integrals are needed for the forward and backward propagating waves in a particular thin layer.

For the scalar V integrals it was possible to construct a series expression with analytic coefficients that comprise binomial factors **[162]**. These binomials are readily calculated by means of the constituent factorials provided that these remain small, typically smaller than 30!. Alternative series expressions have been devised to deal with a defocusing exponential function with a large argument, the so-called Bessel–Bessel expansions where the power series expansion of f'_d or f'_v is replaced by a better converging series expression in terms of spherical Bessel functions **[166]**. In the vector diffraction case, the integrands of the $V_{n,j}^{m,E}$ and $V_{n,j}^{m,B}$ integrals of Eqs. (9.5.33) and (9.5.38) assume a more complicated form than in the scalar case, both in relation to the radial weighting factors $f_j^{E,B}(\rho)$ and the exponential defocusing factor. In both vectorial V integrals, the integrand can be split up into two main factors, for instance,

$$V_{n,j}^{m,E}(r'_1, f'_v) = \int_0^1 g_{|j|}^E(\rho) \left\{ \rho^{|j|} R_n^{|m|}(\rho) J_{m+j}(2\pi r'_1 \rho) \right\} \rho d\rho \,, \tag{9.5.45}$$

where the function $g^E_{|j|}(\rho)$ denotes the product of $f^E_{|j|}$ and the defocusing exponential function. The functions $f^E_{|j|}$ have been intentionally defined in such a way that the factor between brackets in the integrand contains the power $\rho^{|j|}$. The product $\rho^{|j|}\,R^{|m|}_n(\rho)$ can then subsequently be transformed into a series of Zernike polynomials with upper index $|m+j|$, identical to the order of the Bessel function J_{m+j} in the expression in curled brackets. This transformation is carried out with the aid of Eqs. (9.3.40)–(9.3.43) and allows the expression in brackets to be cast into a form such that the integral can be analytically evaluated using the basic result of the Zernike diffraction theory given in Eq. (9.3.7). Once the upper index of the Zernike polynomial and the order of the Bessel function have been made equal, it is required to incorporate the function $g^E_{|j|}(\rho)$ into $R^{|m+j|}_n$ by the construction of a series of radial Zernike polynomials with varying lower index n.

An example of the incorporation of a factor of ρ^2 into a radial polynomial $R^m_n(\rho)$ was given in Eq. (9.3.44), yielding polynomials with the same upper index and with augmented and decreased lower indices $n \pm 2$. Comparable expressions can be found for the product of two radial Zernike polynomials with different indices using the more general relationships given in the Appendix on Zernike polynomials, Section D.4.2. The basic challenge is to construct a power series expansion or a Zernike polynomial series or products of these for the complex product function $g^E_{|j|}(\rho)$. Such series expressions, multiplied with the radial polynomial $R^{|m+j|}_n(\rho)$ that has replaced the initial product $\rho^{|j|}R^{|m|}_n(\rho)$, allow the application of the basic analytic result of Eq. (9.3.7). The complexity of the construction of series expressions and the evaluation of products of them increases when going from the scalar to the vector diffraction case and from the imaging of an infinitely distant object to the imaging of an object with finite magnification factor. The details of the series expansions of the functions $g_{|j|}(\rho)$ for the electric or magnetic case are outside the scope of this chapter. An exhaustive overview of the computation of the coefficients and functions that appear in the various $V^{m,E}_{n,j}$ integrals is found in **[127]**. The corresponding truncation rules for the number of terms to be used in the infinite series expressions are discussed in **[128]**.

9.5.3 Examples of Perfect and Aberrated Vector Point-spread Functions

In this subsection we apply the semi-analytic results obtained for the various V integrals to a practical diffraction problem. We limit ourselves to an object at infinity and use Eqs. (9.5.32)–(9.5.36) to calculate the electric field vector in a (defocused) image plane. Both a perfect imaging system and an aberrated one are considered. We also study the influence of a central obstruction on the diffraction image.

For the case of an infinitely distant source we assume that the resulting field on the entrance pupil is uniform. We inject linear X-polarised light and suppose that the imaging system is aberration free and uniformly transmitting. Polarisation effects in the interior of the optical system are neglected with the result that the only nonzero Zernike coefficient is $\beta^0_{0,x} \equiv 1$. With $s_0 = m_T = 0$, $R_1 \to \infty$ and $n_0 = n_1 = 1$, the electric and magnetic fields in the focal region of a telecentric imaging system with positive power are given by

$$\mathbf{E}_i(r_1', \gamma, f_v') = -\,i\pi s_1^2\,\frac{|f_1|}{\lambda_0}\,\exp\left(\frac{if_v'}{u_1}\right)\begin{pmatrix} V^{0,E}_{0,0} - V^{0,E}_{0,2}\cos 2\gamma \\ -V^{0,E}_{0,2}\sin 2\gamma \\ -\,2iV^{0,E}_{0,1}\cos\gamma \end{pmatrix}, \tag{9.5.46}$$

$$\mathbf{B}_i(r_1', \gamma, f_v') = -\,i\pi s_1^2\,\frac{|f_1|}{c\lambda_0}\,\exp\left(\frac{if_v'}{u_1}\right)\begin{pmatrix} -V^{0,E}_{0,2}\sin 2\gamma \\ V^{0,E}_{0,0} + V^{0,E}_{0,2}\cos 2\gamma \\ -\,2iV^{0,E}_{0,1}\sin\gamma \end{pmatrix}. \tag{9.5.47}$$

The results of Eqs. (9.5.46) and (9.5.47) for $f_v' = 0$ can be directly compared with the expressions in the original publication by Ignatowsky **[153]**, see Appendix G, Eqs. (G.4.12)–(G.4.14) for the electric field components and Eqs. (G.4.18)–(G.4.20) for the I_1, I_2 and I_3 integrals. For the $V^{0,E}_{0,j}$ integrals with $|j| = 0, 1, 2$ we have that I_1 corresponds to $s_1^2 V^{0,E}_{0,1}$, I_2 to $s_1^2 V^{0,E}_{0,0}$ and I_3 to $-s_1^2 V^{0,E}_{0,2}$. These properties follow from the substitutions $|\sin\alpha| = \rho s_1$, $s_0 = 0$ and $\delta = \pi/2$ in Eqs. (G.4.18)–(G.4.20). The sign difference between the two representations with respect to the terms that have a 2γ-dependence derives from the relation $I_3 = -s_1^2 V^{0,E}_{0,2}$. There is also a sign difference in both representations between the imaginary constant factor $\pi B = i\pi|f_1|/\lambda_0$, (see Eq. (G.2.10) with $A_0 \equiv 1$), and the factor $-i\pi s_1^2|f_1|/\lambda_0$ in Eq. (9.5.46). The same holds for the imaginary z-vector components; these are e_1 and h_1 in Eqs. (G.4.12)–(G.4.15) on the one hand and $E_{z,i}$ and $B_{z,i}$ in Eqs. (9.5.46)–(9.5.47) on the other hand. Such an occurrence of a sign difference in an imaginary factor or term is due to the difference in definition of the phase of an electromagnetic propagating wave. In the original publications by Ignatowsky, the complex signal convention $\exp\{+i\omega(t - d/c)\}$ is used where d is the optical path and c the speed of light in vacuum, whereas throughout this book the opposite convention $\exp\{-i\omega(t - d/c)\}$ is used. Equations (A.8.1) and

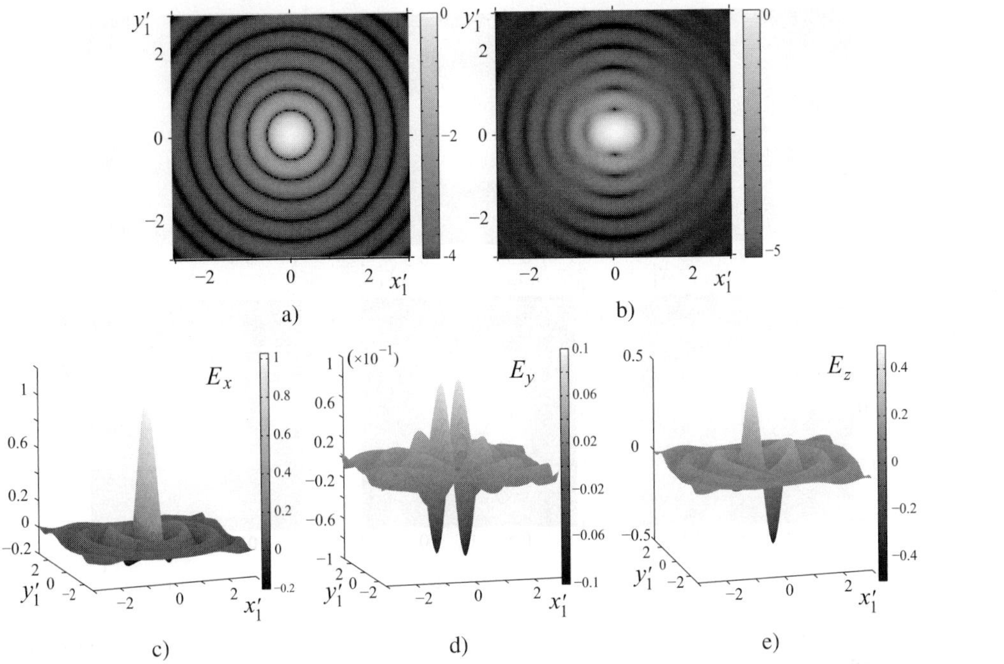

Figure 9.22: Scalar and vector point-spread function in the image plane. The coordinates (x'_1, y'_1) are in units of λ_0/s_1.
a) The logarithm (base 10) of the intensity of the scalar diffraction point-spread function ($s_1 \rightarrow 0$).
b) The logarithm of the electric energy density $|\mathbf{E}_i|^2$ (aperture value s_1 is 0.95).
c) to e) The Cartesian components of the electric field strength. Aberration-free case, $s_1 = 0.95$. Linearly polarised light in the entrance pupil ($E_{x,0} = 1$, $E_{y,0} = 0$).

(A.8.2) of Appendix A show that the complex amplitudes which follow from each convention are each other's complex conjugates. The final space-time-dependent solutions are necessarily identical.

The difference between the scalar diffraction image of a single source point in the nominal image plane and the high-aperture vectorial equivalent is illustrated in Fig. 9.22 for the nominal image plane. The main difference between the scalar and vector in-focus point-spread functions is the broadening of the central lobe of the latter in the direction of linear polarisation of the incident light (horizontal axis in the figure). The circular shape of the first set of diffraction rings is also perturbed in the high-aperture case. Figure 9.22c shows the modulus of the individual electric field strength components. The broadening of the energy density in the x'_1 direction is partly caused by E_x itself but the main contribution stems from the superposition of the E_z component, proportional to s_1. The E_y component is zero on the x- and y-axes and is generally weaker because it is proportional to s_1^2 and to the relatively small $V^0_{0,2}$-function.

In Fig. 9.23a–c the through-focus behaviour of the high-aperture energy density e_w has been illustrated in three image planes corresponding to defocus parameters of $f'_v = -\pi$, 0 and $+\pi$. In the high-aperture case, a value $|f'_v| = \pi/2$ is equivalent to a defocus of one Rayleigh focal depth of $\lambda/(2s_1^2)$ in the low-aperture case. The upper row corresponds to the case where the incident parallel beam ($s_0 = 0$) is linearly polarised (electric field) along the X_0-axis, i.e. $\mathbf{E}_0 = E_{x,0}\,(1, 0, 0)$. The prominent feature of the high-aperture in-focus ($f'_v = 0$) cross-section, b), is its lack of circular symmetry with respect to the scalar approximation. For the aperture considered, $s_1 = 0.95$, the ratio between the half-widths measured along the two orthogonal axes x'_1 and y'_1 is found to be 1.35. It is seen from the expressions for the electric field $\mathbf{E}_1$ on the exit pupil sphere that for $\theta = 0$ and π a relatively strong z-component, proportional to s_1, is present when $\beta^0_{0,y} = 0$. There is no z-component in the azimuthal cross-sections corresponding to $\theta = \pi/2$ and $3\pi/2$. For a high-aperture value like $s_1 = 0.95$ we thus have a significant z-component in the image plane electric field vector $\mathbf{E}_i$, with a $\cos\gamma$ azimuthal dependence. This z-component is absent in the cross-section through $x'_1 = 0$. However, it causes a strong broadening of the function $w_e = \epsilon|\mathbf{E}|^2/2$ in the cross-section through $y'_1=0$ in the image plane. Further away from focus, this polarisation broadening gets obscured by the general broadening of the diffraction pattern due to the defocusing of the image plane.

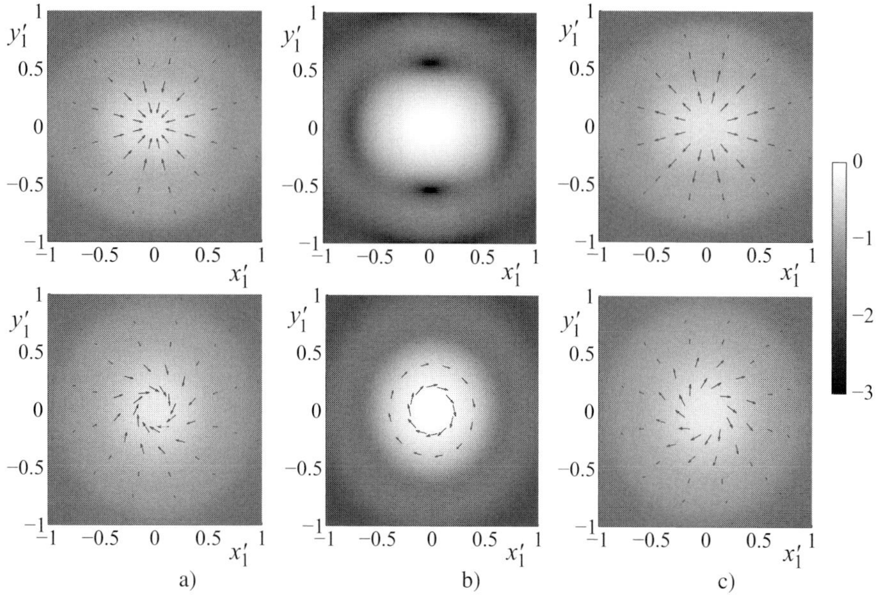

Figure 9.23: Cross-sections of the (normalised) electric energy density w_e (logarithmic scale). The axial image plane positions are given by $f'_v = -\pi$, 0 and $+\pi$ in columns a), b) and c), respectively. Imaging from infinity, $s_1 = 0.95$, aberration-free system.
Upper row: Linear state of polarisation, $(\beta^0_{0,x}, \beta^0_{0,y}) = (1, 0)$.
Lower row: right-circular polarisation, $(\beta^0_{0,x}, \beta^0_{0,y}) = (1, -i)/\sqrt{2}$.

Supplementary information is provided in the graphs by means of the arrows. Figure 9.23a and c show the lateral energy flow component in the electromagnetic field in the image volume. Exactly in focus, the energy flow is exclusively in the axial direction.

In the second row of Fig. 9.23 the energy density has been shown for a right-handed circularly polarised wave with $\mathbf{E}_0 = E_0(1, -i, 0)/\sqrt{2}$. Necessarily, as there is no preferential direction in the entrance pupil field, the electric energy density distribution e_w shows symmetry of revolution in image space. The central lobe of the electric density is slightly broadened with respect to the Airy disc, typically by some 10%. The presence of the circular polarisation becomes visible from the way in which the electric energy is transported through the image volume. The arrows in the graphs, representing the lateral energy flow components when viewed in the positive z-direction, clearly show the right-handed helical movement of the energy flow through focus. The effects of a linear state of polarisation have been further illustrated in Fig. 9.24 (see also **[52]**). In the aberration-free case and in the nominal focal plane ($f'_v = 0$), Eq. (9.5.46) yields the following expression for the electrical energy density,

$$
\begin{aligned}
\left|\mathbf{E}_i\right|^2 \propto & \left[V^{0,E}_{0,0}(r'_1, 0)\right]^2 + \left[V^{0,E}_{0,2}(r'_1, 0)\right]^2 + 2\left[V^{0,E}_{0,1}(r'_1, 0)\right]^2 \\
& + 2\left\{\left[V^{0,E}_{0,1}(r'_1, 0)\right]^2 + V^{0,E}_{0,0}(r'_1, 0)\, V^{0,E}_{0,2}(r'_1, 0)\right] \cos 2\gamma\,, \qquad (9.5.48)
\end{aligned}
$$

where all V functions are real in the nominal image plane. For $\gamma = 0$, we find that $\left|\mathbf{E}_i\right|^2 \propto (V_{0,0} + V_{0,2})^2 + 4(V^0_{0,1})^2$; for $\gamma = \pi/2$ we have $(V^0_{0,0} - V^0_{0,2})^2$. It follows that for real V, the cross-section with $\gamma = 0$ is broadened with respect to the one with $\gamma = \pi/2$. For $\gamma = 0$ the energy density is the sum of two squared quantities of which the zero values do not coincide. The cross-section with $\gamma = \pi/2$ yields zero value if $V^0_{0,0}(r'_1, 0) = V^0_{0,20}(r'_1, 0)$; the resulting value of r'_1 is smaller than the first zero of $V^0_{0,0}$, which, for small s_1 values, equals the radius of the first dark Airy ring. It follows that the function $w_e(r'_1, \pi/2, 0)$ has a smaller halfwidth than the classical Airy disc. The reason for this phenomenon is that, with respect to the Airy disc, the integrand of the high-*NA* Debye integral carries an extra factor $(\cos\alpha_1)^{-1/4}$. For $s_1 = 0.95$, this quantity amounts to 1.338 at the rim of the pupil and has the same effect on the diffraction pattern as an annular pupil, namely a narrowing of the central lobe. This explains the $\approx$10% reduction in halfwidth of the cross-section of w_e with $\gamma = \pi/2$. The curve with $\gamma = \pi/2$ is produced by a single real electric component, $E_{x,i}$, that assumes zero values for certain r'_1 values.

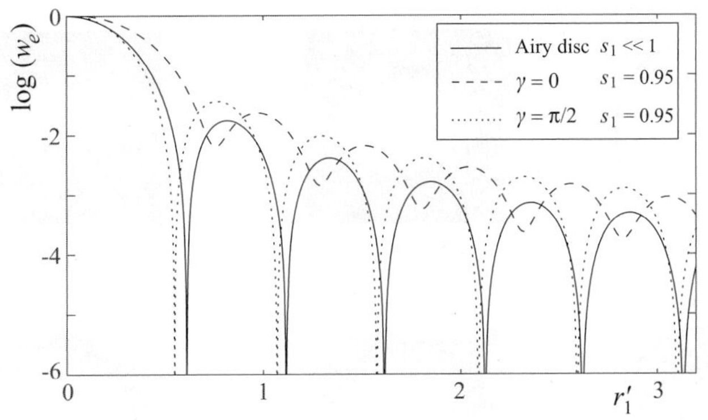

Figure 9.24: Radial cross-sections of the normalised electric energy density w_e of the in-focus aberration-free point-spread function ($s_0 = 0$, $s_1 = 0.95$, X-polarisation in the entrance pupil); w_e for azimuths $\gamma = 0$ and $\gamma = \pi/2$, as compared to the intensity of the scalar Airy disc, is shown. r'_1 is the normalised lateral radial coordinate.

In Fig. 9.25 a comparison is made between high-aperture imaging, imaging at a reduced aperture and high-aperture imaging with an annular pupil. The values of the energy density have all been calculated using the vector point-spread function formalism. In b) with an aperture value of 0.25, we observe almost perfect circular symmetry as we would expect in the case of the classical scalar point-spread function (Airy disc). The defocused graph at low aperture is almost identical to the in-focus graph because the defocusing value f'_v expressed with the aid of the low-aperture value of 0.25 is only 0.046π. The two graphs in row c) show phenomena that can already be derived from the scalar diffraction integral applied to an annular pupil. To compute the scalar result, the annular pupil $f'_{L,a}(\rho, \theta)$ is written as the difference of two uniform and aberration-free pupil functions, one with a full radial extent and one with a reduced extent (obscuration factor a),

$$f'_{L,a}(\rho) = \text{circ}(\rho) - \text{circ}(\rho/a) \,. \tag{9.5.49}$$

According to Eq. (9.3.12) the (in-focus) intensity of the scalar diffraction image is then proportional to

$$\left|U_p(r'_1, 0, 0)\right|^2 \propto \left\{ \left(\frac{2J_1(2\pi r'_1)}{2\pi r'_1} \right) - a^2 \left(\frac{2J_1(2\pi a r'_1)}{2\pi a r'_1} \right) \right\}^2 . \tag{9.5.50}$$

The presence of the obstructing central region with relative radius a reduces the maximum intensity in the image by a factor of $(1 - a^2)^2$, decreases the halfwidth of the central lobe of the diffraction image and enhances the intensity of the diffraction rings. On top of this, one observes a larger focal depth. This is easily verified by applying Eq. (9.3.27) twice, once for the full aperture and once for the reduced aperture with $s_{1,a} = a\, s_1$ and $z'_{1,a} = a^2 z'_1$. After subtraction of the expression for the reduced aperture, the normalised on-axis intensity turns out to be,

$$\begin{aligned} \left|U_p(0, 0, z'_1)\right|^2 &= \text{sinc}^2\left(\frac{\pi z'_1}{2}\right) + a^4\, \text{sinc}^2\left(\frac{\pi a^2 z'_1}{2}\right) \\ &\quad - 2a^2\, \text{sinc}\left(\frac{\pi z'_1}{2}\right) \text{sinc}\left(\frac{\pi a^2 z'_1}{2}\right) \cos\left[\frac{\pi(1 - a^2) z'_1}{2}\right] . \end{aligned} \tag{9.5.51}$$

The scalar results for the intensity of the point-spread function for an annular aperture are summarised in Fig. 9.26. Graphs a) and b) show the amplitude and intensity as a function of the normalised radial coordinate r'_1 at full aperture. Graph c) shows the axial intensity with a pronounced increase of the focal depth once the value a of the central obstruction is increased. These phenomena are also observed in the Fig. 9.25c, however the typical vector effects of a linearly polarised field at the entrance pupil also become clearly evident. Note that the shading has been adapted in each row to the maximum value of the energy density of the in-focus point-spread function.

At this point a remark should be made about the relevance of the electric energy density w_e or the Poynting vector **S** when light is detected, for instance with the aid of a pixelated photosensor or by using a photoresist layer to record a spatially varying focused light pattern. In an absorbing layer, the electromagnetic power balance measured in a certain

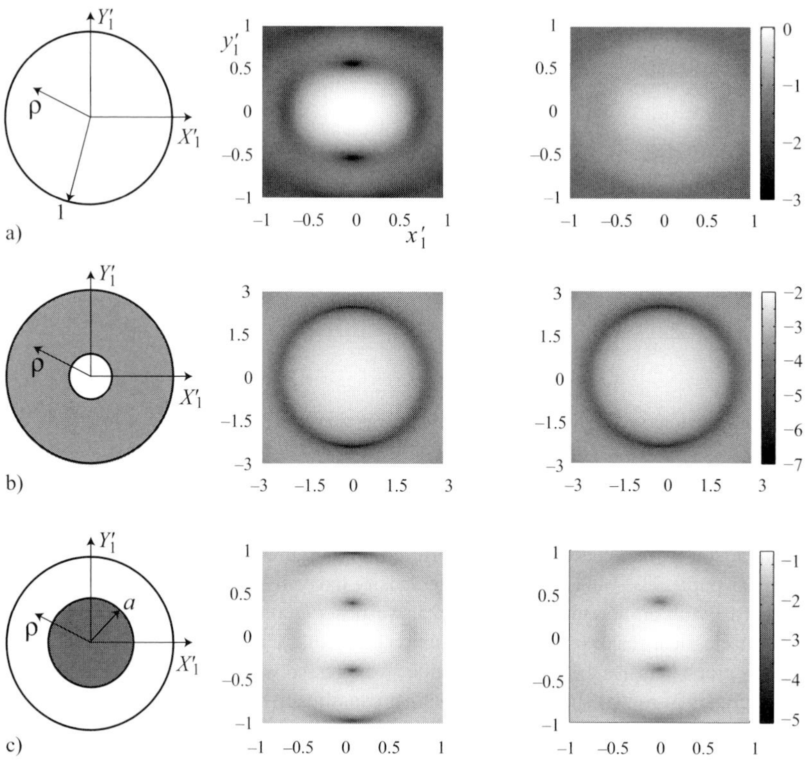

Figure 9.25: A comparison of high-aperture imaging.
a) $s_1 = 0.95$, linear X-polarisation in the entrance pupil, imaging with a reduced aperture.
b) $s_1 = 0.25$ and imaging with an annular aperture.
c) $s_1 = 0.95$, inner radius annulus at $\sin\alpha_1 = 0.475$.
The energy density w_e has been plotted (shading). The lateral distances in all graphs are in normalised coordinates λ_0/s_1 with $s_1 = 0.95$. Central column: in-focus values with $f'_v = 0$, right-hand column: $f'_v = +\pi$, evaluated at $s_1 = 0.95$.

volume V that is enclosed by a surface A is given by Eq. (1.4.5). For the stationary case, the first term in this equation is zero and we are left with a simplified balance equation for the optical power of the electromagnetic field,

$$\iiint \mathbf{J}\cdot\mathbf{E}\, dV = -\oiint (\mathbf{E}\times\mathbf{H})\cdot d\mathbf{A}\,. \tag{9.5.52}$$

The term on the left-hand side represents the absorbed energy in the layer volume with $\mathbf{J}$ given by $\sigma\mathbf{E}$. The value of σ is connected with the complex refractive index of the absorbing medium by Eq. (1.6.63). In complex notation the absorbed energy is thus proportional to $|\mathbf{E}|^2$ and this quantity needs to be evaluated over the entire layer volume. The right-hand side of Eq. (9.5.52) requires calculation of $\mathbf{E}$ and $\mathbf{H}$ vectors over the surface A and evaluation of the surface integral of the resulting Poynting vector $\mathbf{S}$.

In general, evaluation of the two-dimensional surface integral of the Poynting vector would be preferable as it is faster than the evaluation of the volume integral of a quantity proportional to the energy density. However, for weak absorption in a thin layer, the volume integral of $(\sigma/\epsilon)w_e$ can be replaced by the product of this quantity at the centre of the layer and the volume dV. This approximate value can be numerically evaluated in a shorter time than calculation of the difference in $\mathbf{S}$ over the layer thickness times the surface dA of the thin layer. For a single thin planar layer, the surface integral of $\mathbf{S}$ requires calculation of the electromagnetic fields in two distinct axial planes or the surface gradient of $\mathbf{S}$ and this computational effort is larger than the evaluation of w_e in the thin layer itself.

An example of a high-aperture aberrated point-spread function is shown in Fig. 9.27. The aberration considered is that of astigmatism whereby the phase aberration in the exit pupil, in terms of Zernike aberration coefficients, is given by

$$\Phi(\rho,\theta) = \alpha_2^2 R_2^2(\rho)\exp(2i\theta) + \alpha_2^{-2} R_2^2(\rho)\exp(-2i\theta)\,. \tag{9.5.53}$$

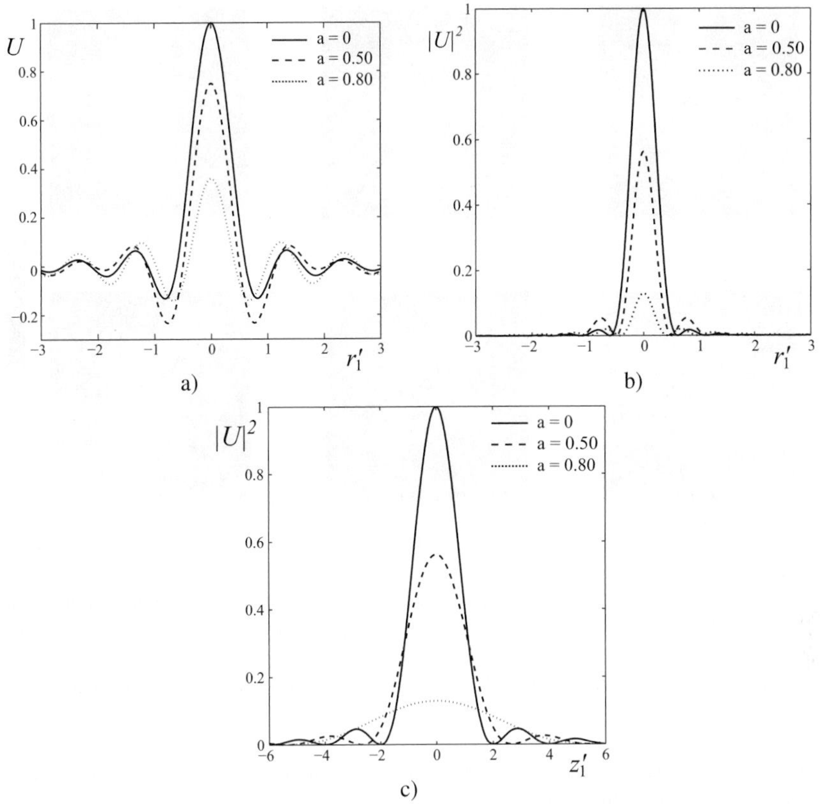

Figure 9.26: Amplitude U, a), and intensity $|U|^2$, b), of a scalar, in-focus diffraction image with annular pupil (central obstruction up to $\rho = a$) as a function of the lateral normalised coordinate r'_1.
c) Through-focus axial intensity as a function of the size a of the central obstruction, scalar approximation; z'_1 is the axial normalised coordinate for the full aperture.

The value of the α-coefficients is such that the peak-to-valley wavefront aberration equals $\lambda_0/2$. According to geometrical optics, the astigmatic focal lines are then found at focus positions given by a (low-aperture) defocusing factor of $f'_d = \pm\pi$. From the two α-coefficients, a set of β-coefficients is calculated with the aid of a power series expansion of the exit pupil lens function $f'_L(\rho, \theta) = \exp[i\Phi(\rho, \theta)]$. Figure 9.27 shows the energy density function in three image planes with f'_v equal to $-\pi$, 0 and $+\pi$, a), b) and c), respectively. The rotation of the astigmatic focal lines is visible, associated with a broadening of the astigmatic line parallel to the y_1-axis in image space (c). The horizontal focal line in front of the best focus (upper a) is approximately 35% narrower than the vertical line beyond the best focus (upper c). This disparity between the two focal lines is reduced in the lower graphs where none of the principal astigmatic axes is parallel to the linear polarisation direction. The arrows in both sets of graphs indicate the lateral energy flow in the focal volume. As was the case for circularly polarised light, the astigmatic focus also shows a helical flow pattern in the focal volume between the two astigmatic lines. It already follows from geometrical optics that the ray rotation between the two focal lines amounts to $\pi/2$ (see Figs. 5.26 and 5.27 and Eq. (5.2.54)). The total rotation from exit pupil through focus to the far field amounts to π.

9.5.4 Vector Point-spread Function in a Multilayer Structure

In many practical situations, the image produced by an optical system is captured by a sensor or a recording medium. In a first approximation, this sensor or recording medium can be assumed to be a medium with a planar interface and infinite extent. Only a reflection at the medium transition, for instance from n_1, the index of the image space, to n_b, the index of the sensor or recording medium half-space, has to be taken into account. A more realistic picture is to consider a (thin)

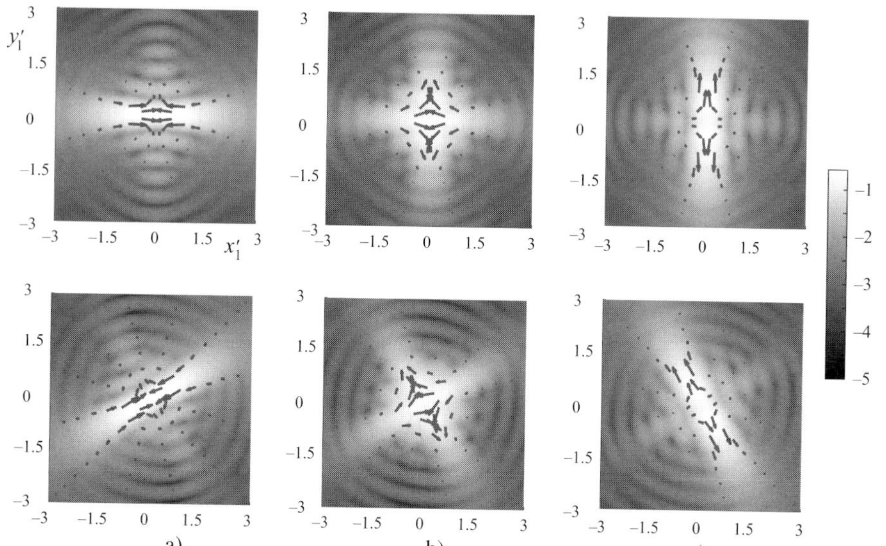

Figure 9.27: The electric energy density w_e in three image planes of an astigmatic high-aperture imaging system with $s_1 = 0.95$, $\alpha_2^2 = \alpha_2^{-2} = \pi/4$. The focus settings are given by $f'_v = -\pi$, 0 and $+\pi$ in a), b) and c), respectively. Linear polarisation in the entrance pupil with azimuth $\theta_0 = 0$ was assumed (X_0-direction).
Upper row: astigmatism at zero degrees with the X_1-axis in the exit pupil. Lower row: astigmatism at an azimuth of $\pi/6$ with the X_1-axis.

multilayer in which one of the layers is the sensing or recording medium. The layers in front of the sensing layer are needed, for instance, to minimise the light reflection back to the imaging system so that the sensing or recording device has an optimum efficiency. These layers can also be used to create a spectral filter for colour sensing purposes. At the reverse side of the multilayer, a substrate is generally present. Any layers between the sensing layer and the substrate can be used to enhance the substrate reflectivity, again for increasing the sensing or recording efficiency of the structure. To accurately describe a realistic image detection device, we thus need the ability to model the presence of such thin layers in front of or beyond the crucial sensing or recording layer itself. The general layout of an imaging and detection layer stack was given in Fig. 9.17 with a schematic representation of the forward *and* backward propagating plane waves in a given layer h. Figure 9.28 shows the forward and backward travelling waves in a particular thin layer of a composite stack in more detail. The figure also shows the definition of the parallel and perpendicular electric field components, denoted by p and s, respectively. In the layer with index n_h, we consider plane wave components with opposite propagation vectors that correspond to initial wave vectors in the exit pupil with azimuths θ and $\theta + \pi$ that are point-symmetric with respect to the centre E_1 of the exit pupil.

Computation of the Cartesian field components in the hth layer of the stack in the image region follows from a diffraction integral in which the electric fields of the forward and backward propagating waves in layer h are projected back to the exit pupil. To this purpose, the initial p- and s-components at a general point Q_1 on the exit pupil sphere are multiplied with the transmission factors $t_{p,h}(\rho)$ and $t_{s,h}(\rho)$, respectively, when considering the forward propagating wave in layer h and with $r_{p,h}(\rho)$ and $r_{s,h}(\rho)$, respectively, for the backward propagating wave. This operation produces two p- and two s-fields on the exit pupil sphere that are then transformed into Cartesian components that can be used in the vector Debye integral. The lateral offset δ between the fictitious forward propagating field component and backward propagating component that follows from Fig. 9.28 is neglected; this is allowed when the distance R_1 from the image region to the exit pupil is very large as compared to this lateral shift in layer h. The values of the complex t and r factors for the transmitted and reflected waves in layer h follow, for instance, from a thin layer analysis as presented in Chapter 1 of this book. The reference point for the amplitude and phase in a particular layer is chosen preferably at the entrance point C_s of the particular layer h in the stack. The amplitude and phase of an incoming plane wave are measured with respect to this reference point; the same holds for the reflected wave that is created in this layer. To obtain the complex electric and magnetic field components in layer h of an isotropic multilayer stack, the admittance matrix analysis of reflection and transmission at the interfaces of a stratified medium can be used (see Subsection 1.9.1).

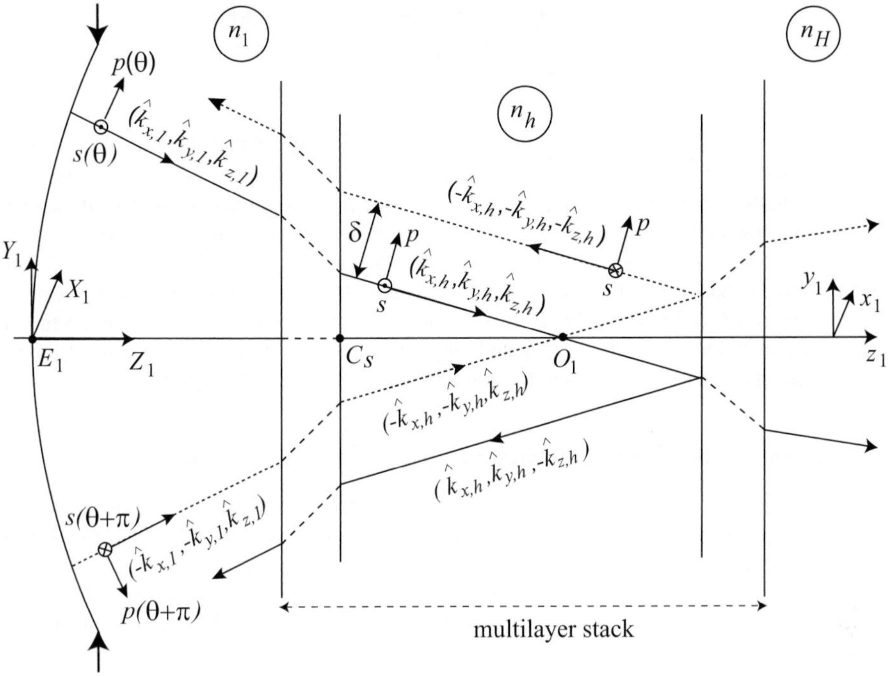

Figure 9.28: Definition of the s- and p-polarisation directions in the azimuthal planes defined by the angles θ and $\theta + \pi$. The transmitted wave vector $\hat{\mathbf{k}}_{h,t} = (\hat{k}_{x,h}, \hat{k}_{y,h}, \hat{k}_{z,h})$ in the thin layer with refractive index n_h is derived from a wave vector $\hat{\mathbf{k}}_1 = (\hat{k}_{x,1}, \hat{k}_{y,1}, \hat{k}_{z,1})$ in image space with index n_1 (paraxial focus at O_1). The s- and p-components of the oppositely directed reflected wave with unit vector $\hat{\mathbf{k}}_{h,r} = -\hat{\mathbf{k}}_{h,t}$ are obtained from the diametrically opposed direction of incidence with unit wave vector $(-\hat{k}_{x,1}, -\hat{k}_{y,1}, \hat{k}_{z,1})$ and azimuth $\theta + \pi$. The p-directions of the transmitted and reflected wave in layer h in the upper part of the diagram are chosen to be parallel, their s-components have opposite directions.

Using the same notation as previously for a homogeneous image space, we can write the p and s components in layer h in the form

$$\begin{aligned} E_{s,h}(\rho,\theta) &= \left|\frac{f_1}{R_p}\right| T_R(\rho) f'_L(\rho,\theta) \sqrt{\frac{n_0}{n_1}} \left\{ t_{s,h}(\rho) E_{s,0}(\rho,\theta) - r_{s,h}(\rho) E_{s,0}(\rho,\theta+\pi) \right\} \\ &= \left|\frac{f_1}{R_p}\right| T_R(\rho) f'_L(\rho,\theta) \sqrt{\frac{n_0}{n_1}} \left\{ t_{s,h}(\rho) \left[-E_{x,0}(\rho,\theta)\sin\theta + E_{y,0}(\rho,\theta)\cos\theta \right] \right. \\ &\qquad \left. - r_{s,h}(\rho) \left[E_{x,0}(\rho,\theta+\pi)\sin\theta - E_{y,0}(\rho,\theta+\pi)\cos\theta \right] \right\}, \end{aligned} \tag{9.5.54}$$

$$\begin{aligned} E_{p,h}(\rho,\theta) &= \left|\frac{f_1}{R_p}\right| T_R(\rho) f'_L(\rho,\theta) \sqrt{\frac{n_0}{n_1}} \left\{ t_{p,h}(\rho) E_{p,0}(\rho,\theta) + r_{p,h}(\rho) E_{p,0}(\rho,\theta+\pi) \right\} \\ &= \left|\frac{f_1}{R_p}\right| \frac{T_R(\rho) f'_L(\rho,\theta)}{\cos\alpha_0} \sqrt{\frac{n_0}{n_1}} \left\{ t_{p,h}(\rho) \left[E_{x,0}(\rho,\theta)\cos\theta + E_{y,0}(\rho,\theta)\sin\theta \right] \right. \\ &\qquad \left. - r_{p,h}(\rho) \left[E_{x,0}(\rho,\theta+\pi)\cos\theta + E_{y,0}(\rho,\theta+\pi)\sin\theta \right] \right\}. \end{aligned} \tag{9.5.55}$$

The minus sign in front of $r_{s,h}$ in Eq. (9.5.54) follows from the opposite sign convention for s-polarisation of the transmitted wave and the reflected wave (see Fig. 9.28). The Cartesian components in the hth layer are given by

$$E_{x,h}(\rho,\theta) = E_{p,h}(\rho,\theta)\cos(\alpha_h)\cos\theta - E_{s,h}(\rho,\theta)\sin\theta\,, \tag{9.5.56}$$

$$E_{y,h}(\rho,\theta) = E_{p,h}(\rho,\theta)\cos(\alpha_h)\sin\theta + E_{s,h}(\rho,\theta)\cos\theta\,, \tag{9.5.57}$$

$$E_{z,h}(\rho,\theta) = E_{p,h}(\rho,\theta)\left(\frac{R_1}{|R_1|}\right)|\sin(\alpha_h)|\,, \tag{9.5.58}$$

where $\cos(\alpha_h) = \sqrt{1 - n_1^2 s_1^2 \rho^2 / n_h^2}$ and $\sin(\alpha_h) = n_1 s_1 \rho / n_h$. With the aid of the expressions for the s and p components given in Eqs. (9.5.54) and (9.5.55) at a general point Q_1 on the exit pupil sphere we obtain the Cartesian components of the electric field strength according to Eqs. (9.5.56)–(9.5.58). The more detailed derivation of these expressions and their subsequent expansion with the aid of Zernike polynomials is omitted from this section and deferred to Appendix F. Equations (F.2.3) and (F.2.4) of this Appendix yield the electric field strength vectors $\mathbf{E}_i^t(r_1', \gamma, f'_{v,h})$ and $\mathbf{E}_i^r(r_1', \gamma, f'_{v,h})$ of, respectively, the forward and backward propagating waves in layer h of the multilayer stack. In these expressions, $f'_{v,h}$ is the defocusing factor in layer h as defined by Eq. (F.2.2). The basic V integrals to be evaluated are given in Eqs. (F.2.5)–(F.2.8). These integrals, as before, need to be evaluated only once, given the refractive index n_h of the thin layer, the maximum aperture s_1 in image space and the lateral magnification m_T of the imaging system.

9.5.5 Examples of Vector Point-spread Functions in a Thin Layer

The formation of the point-spread function in a particular thin layer of a multilayer stack is illustrated by means of two examples **[56]**. The first is related to image formation in microlithography where an image has to be created in a thin photoresist layer that produces, after development, a positive or negative surface profile corresponding to the features to be printed on a microcircuit. The second example illustrates image formation in a microscope when the incident light tunnels through an air gap of subwavelength thickness to the object surface and the image is obtained in reflection. Such an imaging method is referred to as 'solid immersion' and the corresponding microscope objective is called a solid immersion lens (SIL). As for the case of liquid immersion, the solid immersion imaging method allows imaging of finer lateral and axial detail in the object. For both examples study of the three-dimensional shape of the point-spread function is indicative of the finest detail that can be achieved in a more complicated image.

9.5.5.1 Immersion Point-spread Function in a Photoresist Layer

To decrease the size of the finest detail that can be imaged in a photo-lithographic process, the imaging is carried out in water-immersion, with an increase in resolution that is comparable to that of classical oil-immersion microscopy. The various configurations for water-immersion lithography are shown in Fig. 9.29. In a) a single transition from water to a semi-infinite medium consisting of photoresist is shown. The multilayer is reduced in this case to the single transition from water to resist. In b) the photoresist has been given a typical finite thickness of two recording wavelengths, approximately 400 nm in the case of deep UV lithography for which a wavelength of 193 nm is commonly used. The substrate is assumed to be silicon with a complex index of refraction of $n_s = 0.78 + 2.46\,i$. The modulus of the amplitude reflection coefficient of the transition from photoresist ($n_r = 1.76$) to silicon is rather large, specifically 75% at normal

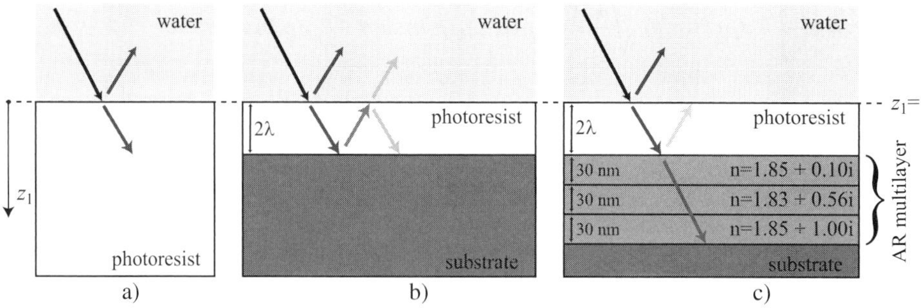

Figure 9.29: Point-spread function creation in a photoresist layer using water-immersion; $NA = 1.36$, refractive indices of water and photoresist at $\lambda_0 = 193$ nm are $n_w = 1.44$ and $n_r = 1.76$, respectively. Reproduced from **[56]**.
a) Single transition.
b) The substrate is included in the model.
c) The substrate carries an anti-reflection coating.

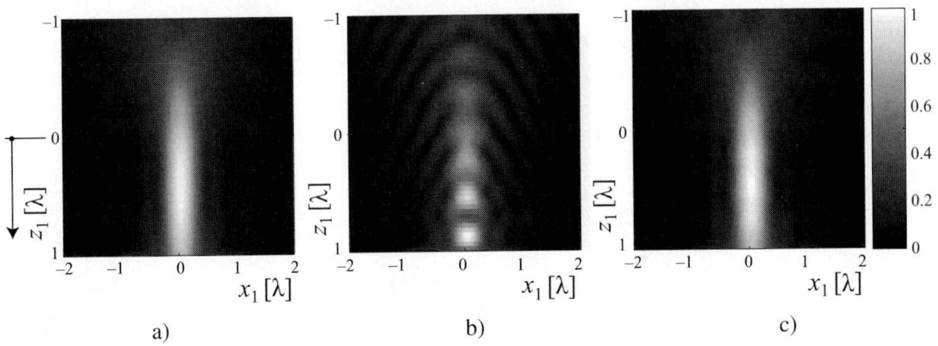

Figure 9.30: Electric energy density w_e of the vector point-spread function in a microlithographic imaging system using water immersion (reproduced from **[56]**). The x–z cross-sections of w_e in a), b) and c) correspond to the geometries of a), b) and c) in Fig. 9.29. $NA = 1.36$, $\lambda_0 = 193$ nm, aberration-free imaging system, circular polarisation in the entrance pupil.

incidence. Such a high substrate reflectivity gives rise to a standing wave pattern in the photoresist layer with appreciable modulation depth and such a pattern plays a detrimental role during the development process of the photoresist material. To strongly reduce the standing wave effect, an anti-reflection coating (AR) is inserted between the photoresist layer and the silicon substrate. Figure 9.29c shows such a substrate with an AR-coating consisting of three layers. This coating gives rise to an impedance matching between the resist layer and the silicon substrate.

In Fig. 9.30 cross-sections of the electric energy density in the photoresist layer are shown for the three geometries that are depicted in Fig. 9.29. The incident wave, propagating in the positive z_1-direction, was initially focused in the water environment in the plane $z_1 = 0$, the same plane where the interface between the water and the photoresist medium is located. It is seen that the best focus is shifted in the positive z_1-direction. This effect can be attributed to (positive) spherical aberration that occurs when the light enters the higher-index resist medium; the marginal rays are focused beyond the paraxial focus. Although some light is reflected at the interface back into the immersion medium, the field in the resist layer is due to forward propagating waves only. The diffraction pattern shows the typical through-focus cylindrical shape (the incident state of polarisation was assumed to be circular). With the geometrical aperture that was used in the water medium ($\sin\alpha_{b,\max} = 0.95$, $NA = 1.36$), the lateral halfwidth of the electric energy density is approximately half a wavelength (95 nm) and the axial halfwidth amounts to 175 nm. In Fig. 9.30b the cross-section is modulated by a strong standing wave pattern. The period of a standing wave pattern produced by two plane waves at an angle α equals $\lambda_0/[2n_r(\sin\alpha/2)]$; n_r is the real part of the refractive index of the (weakly absorbing) resist medium. For oppositely travelling waves with $\alpha = \pi$, the period is 55 nm. The graph shows a standing wave period p_s with an axial period of approximately 65 nm. This value corresponds to an averaged value of the standing wave periods that pertain to the spectrum of plane waves that is contained within the two cones of oppositely travelling waves in the thin layer. The first maximum of the standing wave pattern is found at a distance of $p_s/4$ from the interface of the photoresist layer and the substrate in the case of a perfectly conducting metal substrate. In this case a phase offset of π occurs at the interface between the incident and reflected waves. The positions of the maxima in b) show that this phase, averaged over all angles of incidence, is approximately 1.35π for the resist–silicon interface. In c) the reflectivity at the silicon boundary has been strongly reduced by the anti-reflection coating (the indices and thicknesses are given in Fig. 9.29). As a result we find almost exactly the electric density pattern of a) for the infinitely extending photoresist medium.

9.5.5.2 Point-spread Function of a Solid-immersion Lens (*SIL*)

Instead of using liquid immersion for producing finer details in an image one can also use the principle of 'solid' immersion. In this case the transition of the light from the high index exit medium of the imaging system to the high-index recording medium or to the imaging medium is obtained by tunnelling through a lower index medium, for instance air. The principle of frustrated total internal reflection is used, as was discussed in Section 1.8.6 (see also Fig. 1.28). Such a system is used in a scanning-mode for super high-density optical disc reading or recording. The substrate, with index n_s, is the optical disc in which the information is stored. The two drawings on the left-hand side of Fig. 9.31 schematically show the focusing optics and a magnified view of the lower part of the SIL-lens, the air gap of height d_g and the recording medium with the geometrical focal point F in the plane $z_1 = 0$. Note that the positive z_1-axis points downward.

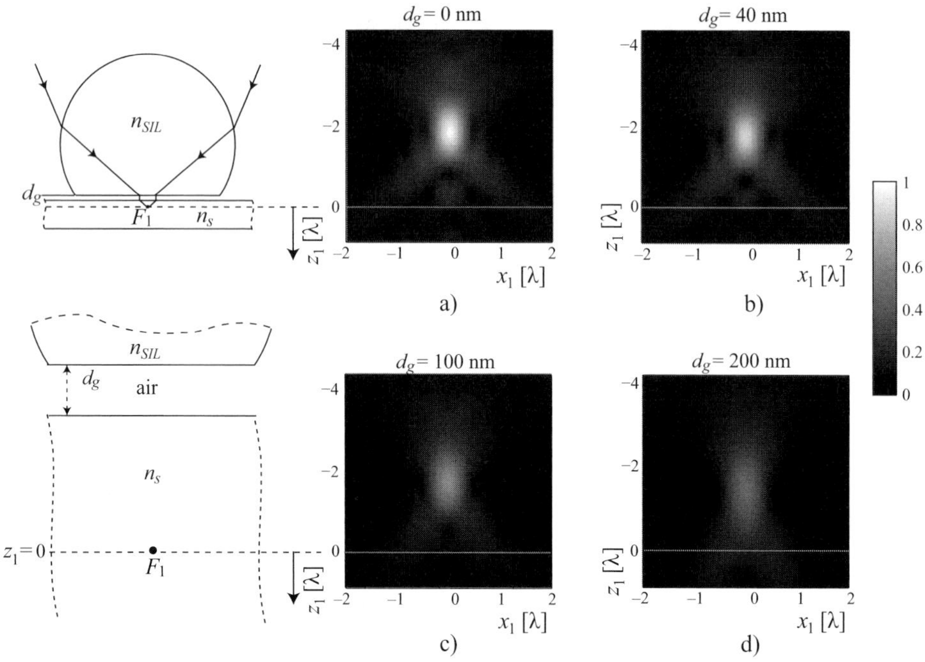

Figure 9.31: Point-spread function produced by a solid-immersion lens as a function of the air gap thickness d_g. $NA = 1.45$, $\lambda_0 = 405$ nm, $n_{\mathrm{SIL}} = 2.086$, $n_s = 1.62$; aberration-free SIL lens (reproduced from **[56]**).

The electric energy density of the point-spread function of a solid-immersion lens is shown in Fig. 9.31 for various values of the air gap width d_g. In the case that the SIL (with index n_{SIL}) and the disc medium are in physical contact, refraction takes place at the interface between these two media. The geometrical aperture in the SIL lens equals $1.45/n_{\mathrm{SIL}} = 0.70$. In the disc medium this produces a geometrical aperture of $\sin\alpha_{s,max} = 1.45/1.62 = 0.90$. This shows that all plane wave components contained in the focusing cone of light in the SIL lens can be transferred to the disc medium as propagating plane waves with a maximum aperture angle given by arcsin(0.90). The x_1-F_1z_1 cross-section of the energy density of the point-spread function for $d_g = 0$ is given in Fig. 9.31a. The maximum energy density has been normalised to unity. The geometrical focus F_1 in the disc medium is found at a prefixed coordinate $z_1 = 0$, a small distance below the top surface of the disc which is determined by the thickness of the thin layer that protects the information on the optical disc. It is seen that the maximum energy density of the point-spread function is found in a position closer to the top surface of the disc. This phenomenon can be ascribed to the spherical aberration caused by the transition from the high-index SIL material to the lower index disc medium. The lower index disc medium produces spherical aberration with a negative sign according to our convention (the marginal rays focus closer to the lens). For a certain focus position within the disc medium, this spherical aberration should be compensated for in the design of the auxiliary lens that accompanies the SIL lens.

When the air gap thickness d_g is increased to $\lambda_0/10$, see Fig. 9.31b, the changes in the lateral and axial dimensions of the energy density cross-section are hardly perceptible, it is only the decrease in maximum energy density that is clearly visible. This phenomenon becomes more pronounced in c) where $d_g = \lambda_0/4$, together with a slight increase in lateral and axial width of the energy density cross-section. Finally, in d) with a gap width of $\lambda_0/2$, virtually all non-propagating plane wave components in the air gap have been so strongly attenuated at the location of the top surface of the disc that they are effectively lost and hence do not contribute to the formation of the light spot in the disc medium. The *numerical* aperture in the disc medium thus equals unity at the most and the *geometrical* aperture has been reduced to $1/1.62 = 0.62$, to be compared with the maximum value of 0.90 that was achieved for zero gap width. The annular aperture in the SIL defined by the lower value of 0.62 and the maximum value of 0.90 is filled entirely with light which has tunnelled through the air gap as evanescent wave and has suffered an amplitude loss after passage through the air gap. When the air gap is too large, this annular part of the focused cone of light is effectively dark and the lateral size of the point-spread function is increased. The increase in lateral and axial dimensions by a factor of approximately 1.5 is clearly visible in Fig. 9.31d. In practice, to obtain a sufficiently high reading power in an optical disc system, the optimum gap was found to be 25 nm ($\lambda/15$) **[204]**. A still smaller gap thickness is not interesting. The optical reading power is only slightly increased while the constancy of the gap thickness and the micro-roughness of the top surface of the disc are subjected to impractically severe constraints.

9.5.6 The Electromagnetic Field Components in an Absorbing Thin Layer

The analysis of wave propagation in a multilayer stack allows for the presence of absorbing layers or layers in which evanescent waves appear. The matrix formalism is adequate for dealing with waves with a complex wave vector exponential since the multilayer stack, or part of it, can have a complex admittance. The Debye integral, however, can only be applied to a thin layer with a real index n_h and to the case where all plane waves coming from the exit pupil are freely propagating. To determine the nature of a plane wave present in a thin layer with general complex index n_h, we inspect the associated wave vector of a particular wave which also pertains to a certain point $Q_1(\rho, \theta)$ on the exit pupil sphere. In accordance with Eq. (9.5.1) we write,

$$\hat{\mathbf{k}}_h(\rho,\theta) = \left(-\frac{R_1}{|R_1|}\sin\alpha_h\cos\theta\,,\ -\frac{R_1}{|R_1|}\sin\alpha_h\sin\theta\,,\ \cos\alpha_h\right)$$
$$= \left(-\frac{R_1}{|R_1|}\rho\left\{\frac{n_1 s_1}{n_h}\right\}\cos\theta\,,\ -\frac{R_1}{|R_1|}\rho\left\{\frac{n_1 s_1}{n_h}\right\}\sin\theta\,,\ \left\{1-(\rho\, n_1 s_1/n_h)^2\right\}^{1/2}\right). \tag{9.5.59}$$

Two special cases arise for the components of the wave vector $\hat{k}_h$:

- n_h is real; $\sin\alpha_h = \rho\, s_h = \rho\,(n_1 s_1/n_h) > 1$.
  The value of $\cos\alpha_h$ becomes purely imaginary.
- n_h is complex (dissipative or amplifying medium); $n_h = n_{R,h} + i\, n_{I,h}$.

In the case of a dissipating medium, the refractive index of the medium has a positive imaginary part and a wave is attenuated on propagation through the medium.

In a multilayer stack with planar interfaces, the continuity of the tangential electric and magnetic field components requires that the lateral components of the wave vector, parallel to each interface, are continuous throughout the stack, such that $k_t^2 = k_x^2 + k_y^2$ is a constant for the plane wave. Using this value of k_t^2 we obtain from the plane wave analysis in Chapter 1 with the aid of Eqs. (1.6.73)–(1.6.75),

$$k_{z,h}^2 = n_h^2 k_0^2 - k_t^2 = k_0^2\left(n_{R,h}^2 - n_{I,h}^2 - n_1^2 s_1^2 \rho^2 + i\, 2 n_{R,h} n_{I,h}\right) = k_0^2\,(a+ib)\ , \tag{9.5.60}$$

where k_0 is the (circular) wavenumber in vacuum. For an absorbing medium the quantity $k_{z,h}^2$ is found in the upper half of the complex plane. We have

$$k_{z,h} = k_0\left(a^2+b^2\right)^{1/4}\exp(i\gamma/2)\ , \tag{9.5.61}$$

with $\tan(\gamma) = b/a$.

The solution $k_{z,h}$ is generally found in the first quadrant of the complex plane. For $n_{I,h} > 0$, the amplitude of the plane wave decays and its phase advances when the wave propagates in the positive z-direction. In the special case with $n_{I,h} = 0$ and $n_1 s_1 \rho > n_{R,h}$, the component $k_{z,h}$ is purely imaginary and we have an evanescent wave solution. Note that in each layer of the stack a wave can also have its counter-propagating equivalent for which the wave amplitude decreases when propagating in the negative z-direction. Formally, gain media can also be included when we assume that $n_{I,h} < 0$.

9.5.6.1 Modification of the V Integral for Decaying Plane Waves in a Thin Layer

When a given layer, h, in a multilayer is formed from a dissipative medium, the $\cos\alpha_h$ and $|\sin\alpha_h|$ factors found in the expressions for the Cartesian electric field components (Eqs. (9.5.56)–(9.5.58)) must be modified. The quantity $\cos(\alpha_h)$ must be replaced by an expression containing the z-component of the wave vector according to

$$n_h\cos(\alpha_h) = \frac{k_{z,h}}{k_0} = \left(a^2+b^2\right)^{1/4}\exp(i\gamma/2)\ , \tag{9.5.62}$$

where a and b follow from Eq. (9.5.60) and γ from Eq. (9.5.61).

The calculation of the electric field components derived in Appendix F, Eqs. (F.1.16)–(F.1.21), remains valid. In turn this brings us in a straightforward manner to the Debye integral of Eq. (F.2.1) and the associated solutions for the forward and backward propagating waves given by Eqs. (F.2.3) and (F.2.4). Combining the $f'_{v,b}$-dependent exponential outside the integral and the ρ- and $f'_{v,b}$-dependent exponential inside the integrand yields a new factor in the integrand of the first Debye integral which takes the form

$$\exp\left\{\frac{if'_{v,b}(1-s_b^2\rho^2)^{1/2}}{u_b}\right\}=\exp\left\{\frac{if'_{v,b}k_{z,b}}{n_b k_0 u_b}\right\} . \tag{9.5.63}$$

Using Eq. (F.2.2) we obtain for the exponential in the first integral,

$$\exp\left\{\frac{if'_{v,b}(1-s_b^2\rho^2)^{1/2}}{u_b}\right\}=\exp(ik_{z,b}z_1) , \tag{9.5.64}$$

where we have used $k_0 = 2\pi/\lambda_0$. This finally brings us to the modified V functions of Eq. (F.2.5) which read

$$V^m_{n,j,t_\pm}(r'_1,f'_{v,b})=\int_0^1 \rho^{|j|}\frac{\left\{[k_{z,b}/(n_b k_0)]\pm\left(1-s_0^2\rho^2\right)^{\frac{1}{2}}\right\}^{-|j|+1}}{\left(1-s_1^2\rho^2\right)^{\frac{1}{4}}\left(1-s_0^2\rho^2\right)^{\frac{3}{4}}}\exp(ik_{z,b}z_1)\, R_n^{|m|}(\rho)J_{m+j}(2\pi r'_1\rho)\,\rho d\rho . \tag{9.5.65}$$

9.5.6.2 Modification of the V Integral for Evanescent Plane Waves in a Thin Layer

For $n_{I,b} = 0$ only a small change in the expression for the V integral is needed. $\cos\alpha_b$ is replaced by

$$\cos(\alpha_b)=\begin{cases}\left(1-s_b\rho^2\right)^{1/2} , & n_1 s_1\rho \le n_b ,\\[2ex] +i\left(s_b\rho^2-1\right)^{1/2} , & n_1 s_1 \rho > n_b .\end{cases} \tag{9.5.66}$$

The $f'_{v,b}$ exponential multiplied by the factor $\exp(if'_{v,b}/u_1)$ found in front of the Debye integral is thus given by

$$\exp\left\{\frac{if'_{v,b}(1-s_b^2\rho^2)^{1/2}}{u_b}\right\}=\begin{cases}\exp\left\{+i\,n_b k_0\left(1-s_b^2\rho^2\right)^{1/2}z_1\right\} & n_1 s_1\rho \le n_b ,\\[2ex] \exp\left\{-\,n_b k_0\left(s_b^2\rho^2-1\right)^{1/2}z_1\right\} & n_1 s_1\rho > n_b .\end{cases} \tag{9.5.67}$$

The Debye diffraction integral for the backward propagating waves follows the same scheme; however, the opposite sign is taken in the $f'_{v,b}$ exponential.

9.6 Energy and Momentum Density and Their Flow Components

In this section we analyse the energy and momentum density and their flow components in the focal volume of an imaging system. The role of the electric energy density for optical detection and optical recording of information was discussed earlier in this section with the aid of Eq. (9.5.52). The linear and angular momentum of an optical (focused) beam can be used to manipulate particles that can freely move or float in a gas or in a liquid. With the aid of one or several optical beams, the two- or three-dimensional position and orientation of such a particle can be varied using the linear and angular momentum of the impinging beam(s). By dynamically adjusting the optical power and position of an optical beam, the position or azimuth of a particle can be frozen in time with the aid of a position-sensing system and a feedback loop. A requirement is that the optical contrast of the particle with its environment (gas or liquid) is large. Such 'optical trap' or 'tweezer' systems are extensively used for the manipulation of nano- and micro-particles **[13]**,**[14]**. The initial analysis of optical traps used a low-aperture diffraction approximation. For high-resolution optical tweezers, the linear and angular momentum distribution in a high-aperture focused spot is needed so as to be able to design the optimum tweezer configuration and adjust the beam power levels. This section provides the reader with such a high-*NA* analysis of the energy and momentum density and flow in a focal volume.

To study the three-dimensional energy and momentum flow we position, without loss of generality, a point source of light in an infinitely distant plane on the optical axis of the imaging system. In this case, the wave that is incident on the optical system is a plane wave of which the state of polarisation over the flat entrance pupil surface is constant and the formulas that provide the electromagnetic field components in the focal volume are somewhat simplified. For an incident parallel beam, the state of polarisation is characterised by two complex numbers a and b, which denote the complex amplitudes of the X- and Y-component of the electric field strength, respectively. The possible transmission non-uniformity and aberration of the optical system lead to a complex lens function $f'_L(\rho,\theta)$ in the exit pupil. The product of the incident X and Y field components and the lens function $f'_L(\rho,\theta)$ which was initially given by Eq. (9.5.18) and (9.5.19) is now simplified to

$$\begin{Bmatrix} E_{x,0}(\rho,\theta) \\ \\ E_{y,0}(\rho,\theta) \end{Bmatrix} f'_L(\rho,\theta) = \begin{Bmatrix} a \\ \\ b \end{Bmatrix} \sum_{n,m} \beta_n^m R_n^{|m|}(\rho) \exp(im\theta) . \tag{9.6.1}$$

The expressions for the components of the field vectors $\mathbf{E}_i$ and $\mathbf{B}_i$ in image space according to Eqs. (9.5.32) and (9.5.37) are hence given by

$$\begin{aligned} \mathbf{E}_i(r'_1,\gamma,f'_v) = &-i\pi s_1^2 \sqrt{n_0 n_1}\,\frac{|f_1|}{\lambda_0} \exp\left(\frac{if'_v}{u_1}\right) \sum_{n,m} [-(R_1/|R_1|)i]^m\, \beta_n^m \exp[im\gamma] \\ &\times \begin{pmatrix} aV_{n,0}^m - \frac{1}{2}\left\{(a-ib)V_{n,2}^m \exp[2i\gamma] + (a+ib)V_{n,-2}^m \exp[-2i\gamma]\right\} \\ bV_{n,0}^m + i\frac{1}{2}\left\{(a-ib)V_{n,2}^m \exp[2i\gamma] - (a+ib)V_{n,-2}^m \exp[-2i\gamma]\right\} \\ -i\left\{(a-ib)V_{n,1}^m \exp[i\gamma] - (a+ib)V_{n,-1}^m \exp[-i\gamma]\right\} \end{pmatrix} , \end{aligned} \tag{9.6.2}$$

$$\begin{aligned} \mathbf{B}_i(r'_1,\gamma,f'_v) = &-i\pi s_1^2 \sqrt{n_0 n_1}\,\frac{n_1|f_1|}{c\lambda_0} \exp\left(\frac{if'_v}{u_1}\right) \sum_{n,m} [-(R_1/|R_1|)i]^m\, \beta_n^m \exp[im\gamma] \\ &\times \begin{pmatrix} -bV_{n,0}^m + i\frac{1}{2}\left\{(a-ib)V_{n,2}^m \exp[2i\gamma] - (a+ib)V_{n,-2}^m \exp[-2i\gamma]\right\} \\ aV_{n,0}^m + \frac{1}{2}\left\{(a-ib)V_{n,2}^m \exp[2i\gamma] + (a+ib)V_{n,-2}^m \exp[-2i\gamma]\right\} \\ -\left\{(a-ib)V_{n,1}^m \exp[i\gamma] + (a+ib)V_{n,-1}^m \exp[-i\gamma]\right\} \end{pmatrix} . \end{aligned} \tag{9.6.3}$$

The V functions for the electric and magnetic case are identical for the case of zero lateral magnification and, for that reason, the upper index E or B of each V integral has been suppressed.

The calculation of the Poynting vector and the linear and angular momentum density and flow components requires (vector) products of the electric field strength and the magnetic induction vectors. To remain aligned with the general notation in this chapter, we use normalised lateral and axial coordinates in image space. The conservation laws for energy and momentum density which were given in Eqs. (1.4.7)–(1.4.9) for real space coordinates are transformed into

$$\frac{\partial w_{em}}{\partial t} + \frac{n_1 s_1}{\lambda_0}\left\{\frac{\partial S_{x'_1}}{\partial x'_1} + \frac{\partial S_{y'_1}}{\partial y'_1}\right\} + \frac{2\pi n_1 u_1}{\lambda_0}\frac{\partial S_{f'_v}}{\partial f'_v} = 0 , \tag{9.6.4}$$

$$\frac{\partial m_p}{\partial t} + \frac{n_1 s_1}{\lambda_0}\left\{\frac{\partial T_{px'_1}}{\partial x'_1} + \frac{\partial T_{py'_1}}{\partial y'_1}\right\} + \frac{2\pi n_1 u_1}{\lambda_0}\frac{\partial T_{pf'_v}}{\partial f'_v} = 0 , \tag{9.6.5}$$

$$\frac{\partial j_p}{\partial t} + \frac{n_1 s_1}{\lambda_0}\left\{\frac{\partial M_{px'_1}}{\partial x'_1} + \frac{\partial M_{py'_1}}{\partial y'_1}\right\} + \frac{2\pi n_1 u_1}{\lambda_0}\frac{\partial M_{pf'_v}}{\partial f'_v} = 0 , \tag{9.6.6}$$

where the index $p = x, y, z$ denotes the Cartesian components of the linear and angular momentum density vectors. The real-space coordinates (x_1, y_1, z_1) have been replaced by (x'_1, y'_1), the normalised lateral coordinates in image space, and by f'_v, the normalised axial coordinate associated with a high-aperture imaging system (see Eq. (9.2.15)). n_1 is the image space refractive index and $u_1 = 1 - (1 - s_1^2)^{1/2}$.

The momentum density components m_p and j_p in the Eqs. (9.6.4)–(9.6.6) are elements of the vectors **m** and **j**, respectively, whereas the quantities T_{pq} and M_{pq}, $p, q = (x, y, z)$, are elements of the tensors **T** and **M**, respectively. The flow of linear momentum is given by the Maxwell stress tensor **T** whose entries T_{pq} have the dimension of pressure. The value of T_{pq} yields the flow per unit surface of the pth component of the linear momentum vector in the direction q. The diagonal elements of the stress tensor are called the *normal* pressure components, the off-diagonal elements the *shear* pressure components.

With the aid of the expressions for $\mathbf{E}_i$ and $\mathbf{B}_i$ in image space we construct semi-analytic expressions for the energy and momentum density and for their respective flow components. A detailed derivation of these expressions can be found in **[54]**. In this section we mainly reproduce the results of the calculations and draw conclusions about particular density functions and flow patterns in the focal volume. Parameters of the calculated results are the state of polarisation in the entrance pupil and the specific aberration of the imaging system.

9.6.1 Analytic Expressions for the Components of the Poynting Vector

The energy flow in the focal region of a converging beam requires a description of the components of the electromagnetic Poynting vector $\mathbf{S}$ in a Cartesian or, preferably, cylindrical coordinate system. In a medium with refractive index n we have for the time-averaged value of this vector the expression

$$\langle \mathbf{S} \rangle = \frac{1}{2} \Re \{\mathbf{E}_i \times \mathbf{H}_i^*\} = \frac{\epsilon v^2}{2} \Re \{\mathbf{E}_i \times \mathbf{B}_i^*\} , \tag{9.6.7}$$

where $v = c/n$ is the propagation speed in image space and $\epsilon = n_1^2$ is the permittivity of image space. When evaluating the product of a component of $\mathbf{E}_i$ and a component of $\mathbf{B}_i$, we encounter typical terms of the form

$$\begin{aligned} G_{k,l}(\beta) &= \sum_{n,m} [-(R_1/|R_1|)\, i]^m \exp[im\gamma]\; \beta_n^m\, V_{n,k}^m(r_1', f_v') \exp[ik\gamma] \sum_{n',m'} [(R_1/|R_1|)\, i]^{m'} \exp[-im'\gamma] \beta_{n'}^{m'\,*}\, V_{n',l}^{m'\,*}(r_1', f_v') \exp[-il\gamma] \\ &= \sum_{n,m,n',m'} \exp[-i(m-m')(R_1/|R_1|)\pi/2] \exp[i(m-m'+k-l)\gamma] \beta_n^m \beta_{n'}^{m'\,*}\, V_{n,k}^m(r_1', f_v')\, V_{n',l}^{m'\,*}(r_1', f_v') \,. \end{aligned} \tag{9.6.8}$$

For the indices k and l we have that $k, l = -2, \cdots, +2$, so that the total number of $G_{k,l}$ functions is 25. We also have that $G_{k,l} = G_{l,k}^*$ and that $G_{k,k}$ is real. There are hence only 15 independent G functions in the general case with an arbitrary pupil function. In the aberration-free case, the additional relationship $G_{-k,-k} = G_{k,k}$ holds, thus reducing the number of independent G functions to 13.

With the aid of the $G_{k,l}$ functions we find, after some manipulation, the following expressions for the three Cartesian components of the Poynting vector,

$$\begin{aligned} S_x = \frac{\epsilon_0 c\pi^2 s_1^4 n_0 n_1^2 |f_1|^2}{2\lambda_0^2} \Big\{ &-2\Im(ab^*)\Im\left(G_{0,1} + G_{0,-1}\right) + (|a|^2 + |b|^2)\Im\left(G_{0,1} - G_{0,-1}\right) \\ &- \Big[2\Im(ab^*)\Im\left(G_{2,1} + G_{-2,-1}\right) - (|a|^2 + |b|^2)\Im\left(G_{2,1} - G_{-2,-1}\right)\Big]\Big\}, \end{aligned} \tag{9.6.9}$$

$$\begin{aligned} S_y = \frac{\epsilon_0 c\pi^2 s_1^4 n_0 n_1^2 |f_1|^2}{2\lambda_0^2} \Big\{ &-2\Im(ab^*)\Re\left(G_{0,1} - G_{0,-1}\right) + (|a|^2 + |b^2)\Re\left(G_{0,1} + G_{0,-1}\right) \\ &+ \Big[2\Im(ab^*)\Re\left(G_{2,1} - G_{-2,-1}\right) - (|a|^2 + |b|^2)\Re\left(G_{2,1} + G_{-2,-1}\right)\Big]\Big\}, \end{aligned} \tag{9.6.10}$$

$$S_z = \frac{\epsilon_0 c\pi^2 s_1^4 n_0 n_1^2 |f_1|^2}{2\lambda_0^2} \left\{ (|a|^2 + |b|^2)\left[G_{0,0} - \frac{1}{2}\left(G_{2,2} - G_{-2,-2}\right)\right] + \Im(ab^*)\left(G_{2,2} - G_{-2,-2}\right)\right\}. \tag{9.6.11}$$

9.6.2 The Aberration-free System as a Special Case

For the case of an aberration-free imaging system, the general expressions for the Poynting vector components are greatly simplified. All β_n^m coefficients are zero except β_0^0 which is equal to 1. The G functions therefore reduce to

$$G_{k,l}(\beta) = \exp\{i(k-l)\gamma\}\, V_{0,k}^0(r_1', f_v') V_{0,l}^{0*}(r_1', f_v') \,. \tag{9.6.12}$$

The Cartesian components of the Poynting vector are then given by

$$\begin{aligned} S_x = \frac{\epsilon_0 c\pi^2 s_1^4 n_0 n_1^2 |f_1|^2}{\lambda_0^2} \Big\{ &\Big[2\Im(ab^*)\Re\Big\{V_{0,0}^0(r_1', f_v')V_{0,1}^{0*}(r_1', f_v')\Big\}\sin\gamma \\ &+ (|a|^2 + |b|^2)\Im\Big\{V_{0,0}^0(r_1', f_v')V_{0,1}^{0*}(r_1', f_v')\Big\}\cos\gamma\Big] \\ &- \Big[2\Im(ab^*)\Re\Big\{V_{0,2}^0(r_1', f_v')V_{0,1}^{0*}(r, f)\Big\}\sin\gamma \end{aligned}$$

$$- (|a|^2 + |b|^2)\Im\left\{V^0_{0,2}(r'_1,f'_v)V^{0*}_{0,1}(r'_1,f'_v)\right\}\cos\gamma\Big]\Big\}, \tag{9.6.13}$$

$$\begin{aligned} S_y = \frac{\epsilon_0 c\pi^2 s_1^4 n_0 n_1^2 |f_1|^2}{\lambda_0^2} \Big\{\Big[&- 2\Im(ab^*)\Re\left\{V^0_{0,0}(r'_1,f'_v)V^{0*}_{0,1}(r'_1,f'_v)\right\}\cos\gamma \\ &+ (|a|^2 + |b|^2)\Im\left\{V^0_{0,0}(r'_1,f'_v)V^{0*}_{0,1}(r'_1,f'_v)\right\}\sin\gamma\Big] \\ &+ \Big[2\Im(ab^*)\Re\left\{V^0_{0,2}(r'_1,f'_v)V^{0*}_{0,1}(r'_1,f'_v)\right\}\cos\gamma \\ &+ (|a|^2 + |b|^2)\Im\left\{V^0_{0,2}(r'_1,f'_v)V^{0*}_{0,1}(r'_1,f'_v)\right\}\sin\gamma\Big]\Big\}, \end{aligned} \tag{9.6.14}$$

$$S_z = \frac{\epsilon_0 c\pi^2 s_1^4 n_0 n_1^2 |f_1|^2}{2\lambda_0^2}\left(|a|^2 + |b|^2\right)\left[\left|V^0_{0,0}(r'_1,f'_v)\right|^2 - \left|V^0_{0,2}(r'_1,f'_v)\right|^2\right]. \tag{9.6.15}$$

The leading factor in the expression for the z-component of the Poynting vector can be given a physical explanation. Specifically, it describes the increase of the irradiance (power flow per unit area) when going from the exit pupil to the centre of the point-spread function in the image plane. We can regroup this factor as follows:

$$\frac{\epsilon_0 c\pi^2 s_1^4 n_0 n_1^2 |f_1|^2}{2\lambda_0^2}\left(|a|^2 + |b|^2\right) = \frac{n_1\epsilon_0 c\left(|a_1|^2 + |b_1|^2\right)}{2}\,\frac{\pi^2\rho_0^2}{(\lambda_{n_1}/s_1)^2} = |S_z|_{ex}\,\frac{\pi\rho_0^2}{\pi(\lambda_{n_1}/\pi s_1)^2}. \tag{9.6.16}$$

The z-component of the Poynting vector of the incident parallel beam in the entrance pupil, denoted by $S_{z,0}$, equals $n_0\epsilon_0 c\left(|a|^2 + |b|^2\right)/2$. For the case of unit pupil magnification we have that the irradiance $|S_z|_{ex}$ through the centre of the exit pupil in image space is equal to the corresponding quantity $S_{z,0}$ in the entrance pupil. The expression for the exit pupil irradiance, $|S_z|_{ex}$, is thus given by the quantity $n_1\epsilon_0 c\left(|a_1|^2 + |b_1|^2\right)/2$ where a_1 and b_1 are the orthogonal electric fields in the centre of exit pupil. The second factor in the right-hand expression of Eq. (9.6.16) is the ratio of the projected area of the exit pupil ($2\rho_0 = 2s_1|f_1|$ is the exit pupil diameter in real-space coordinates) and a fictitious circular area in the image plane with a radius of $\lambda_{n_1}/(\pi s_1)$ (the diffraction unit in image space equals λ_{n_1}/s_1). In this simplified picture that neglects diffraction effects, the power flow through the circular exit pupil passes through this fictitious circular area in the image plane with uniform power flow yielding an irradiance enhancement factor $C_I = (\pi\rho_0 s_1/\lambda_{n_1})^2$. Equation (9.6.15) shows that exactly at the centre of the diffraction image the enhancement factor equals $C_I\left|V^0_{0,0}(0,0)\right|^2$. The function $V^0_{0,0}(0,0)$ is unity for very small imaging apertures. Furthermore, $V^0_{0,0}$ remains close to unity for larger apertures; for example, it amounts to 1.03 for an aperture of 0.95.

The modulus of the Poynting vector and its direction in three dimensions have been depicted in Fig. 9.32 for an aberration-free imaging system. The propagation direction of the optical energy is upward in both graphs. The lower horizontal cross-section in each graph shows the energy density distribution in the nominal focal plane. The two other horizontal cross-sections correspond to defocused image planes with normalised axial coordinates $z'_1 = +1.1$ and $z'_1 = +2.2$, respectively. In a) the initial state of polarisation in the entrance pupil is linear, in b) it is right-circular. The arrows represent the flow direction of the energy in the focal region. The sharply defined cone in Fig. 9.32a represents the limitation of the geometrical pencil of rays. Figure 9.32b for RC-polarised light in the entrance pupil shows that the energy transport in the focused beam has a left-handed rotation sense with respect to the right-handed coordinate system (x'_1, y'_1, z'_1). With respect to the definition of the handedness of circularly polarised light according to Section 1.6.6, we observe that it is consistent with the right-handedness of the electric field vector movement in a horizontal plane when viewed *against* the central propagation direction of the optical field.

The handedness of the energy transport in the propagating optical field in the case of elliptical polarisation is more easily seen when the Poynting vector components are written in a cylindrical coordinate system in which we define $S_r = S_x\cos\gamma + S_y\sin\gamma$ and $S_\gamma = -S_x\sin\gamma + S_y\cos\gamma$, such that

$$S_r = \frac{\epsilon_0 c\pi^2 s_1^4 n_0 n_1^2 |f_1|^2}{\lambda_0^2}(|a|^2 + |b|^2)\,\Im\left[\left(V^0_{0,0}(r'_1,f'_v) + V^0_{0,2}(r'_1,f'_v)\right)V^{0*}_{0,1}(r'_1,f'_v)\right], \tag{9.6.17}$$

$$S_\gamma = \frac{-2\,\epsilon_0 c\pi^2 s_1^4 n_0 n_1^2 |f_1|^2}{\lambda_0^2}\,\Im(ab^*)\,\Re\left[\left(V^0_{0,0}(r'_1,f'_v) - V^0_{0,2}(r'_1,f'_v)\right)V^{0*}_{0,1}(r'_1,f'_v)\right]. \tag{9.6.18}$$

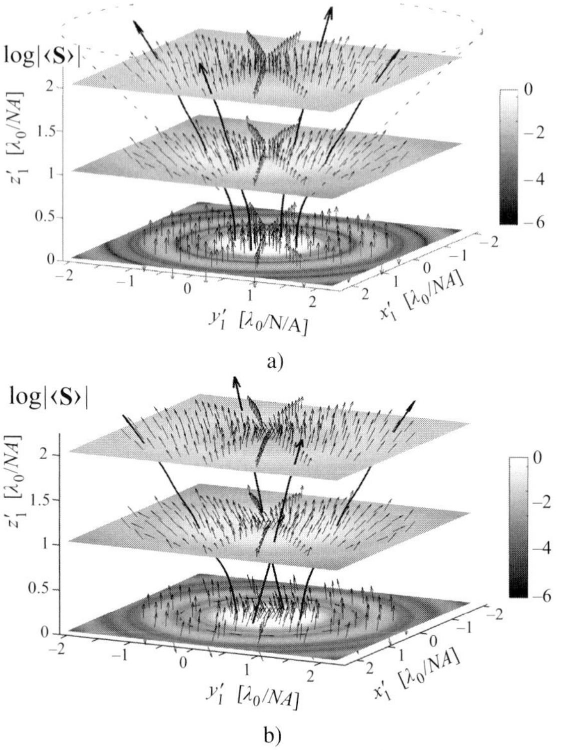

Figure 9.32: The logarithm of the absolute value $\log|\langle \mathbf{S} \rangle|$ of the average energy flow in the focal region of a high-numerical-aperture aberration-free imaging system with $NA = 0.85$ (reproduced from **[52]** with the permission of the Optical Society of America). The tiny arrows in both pictures show the direction of flow of the energy. The state of polarisation of the incident plane wave is:
a) Linear X-polarisation.
b) Right-circularly polarised light.

The expressions for S_r and S_γ permit us to draw some general conclusions about the Poynting vector components in the focal region of an aberration-free system. Using the property that $V^0_{0,k}(r'_1, -f'_v)\, V^{0*}_{0,l}(r'_1, -f'_v) = V^{0*}_{0,k}(r'_1, f'_v) V^0_{0,l}(r'_1, f'_v)$, we observe the following:

- For an aberration-free focused beam, the on-axis power flow ($r'_1 = 0$) is in the z-direction of the beam everywhere.
- The power flow component S_r in the radial direction has opposite sign in the converging part and the diverging part of the focused beam. The azimuthal component maintains the same sign on both sides of the optimum focus. Exactly in focus, the power flow is directed in the (positive or negative) z-direction over the total beam cross-section.
- The axial component S_z and the radial component S_r do not depend on the state of polarisation of the incident beam.
- The azimuthal component of an aberration-free focused beam is zero if the incident radiation is linearly polarised ($\arg(a) = \arg(b)$). As soon as this condition is not satisfied, angular momentum is present in the focused beam.
- The azimuthal component is maximum for the case of circularly polarised light as this state of polarisation maximises the absolute value of $\Im(ab^*)$.
- For large s_1 and in the neighbourhood of the focal plane, the three components S_z, S_r and S_γ can locally change sign with respect to their average values. Since the prefactors in the expressions for S_z, S_r and S_γ are position-independent, a change in sign is only possible if one of the factors between square brackets in Eqs. (9.6.15), (9.6.17) or (9.6.18) becomes zero since they contain (r'_1, f'_v)-dependent V functions. A change in sign of e.g. S_z gives rise to a (local) energy flow in the negative z-direction. A change in sign of the azimuthal component may cause a vortex in the energy flow pattern.
- If the aperture s_1 approaches zero, the lateral components S_x and S_y of the Poynting vector become negligibly small and the flow is uniquely directed in the axial direction.

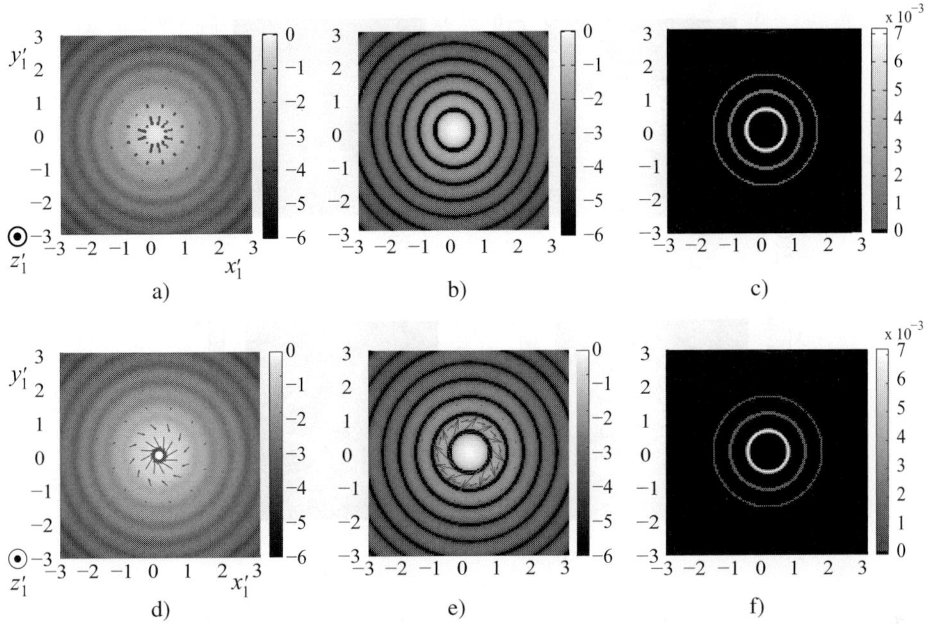

Figure 9.33: Poynting vector components in the image region of a high-*NA* lens system (*NA*= 0.95, *X*-polarisation).
a) Defocus distance $z_1' = -1$. The pattern corresponds to the logarithm of the axial component S_z of the Poynting vector. The lateral components of the Poynting vector have been represented by means of the inwardly directed arrows. Aberration-free case.
b) Same legend as a), now for $z_1' = 0$. Negative values of S_z have been given the value −6 on the logarithmic scale.
c) The value (linear scale) of the modulus of S_z for the case that $S_z < 0$.
d), e) and f) Same legend as for a), b) and c), but now for *RC*-polarised light.

The properties of the Poynting vector components in the focal region are illustrated in Fig. 9.33. Figures a) to c) are related to an incident beam in the entrance pupil with linear polarisation in the x-direction. For e) to f) the state of polarisation is right-circular ($a = 1/\sqrt{2}$, $b = -i/\sqrt{2}$). a) and d) correspond to the field in a defocused plane ($z_1' = -1$), b) and e) to the nominal image plane ($z_1' = 0$). For the linearly polarised case and for an aberration-free system, the Poynting vector shows circular symmetry. This is in contrast with the electric or magnetic energy density for the case of linear polarisation, which both show a pattern with a striking 2γ-periodicity. Since the Poynting vector is composed of products of the electric and magnetic field strength components, this lack of circular symmetry of the field components is averaged out. The circular symmetry directly follows from the azimuthal component of the Poynting vector (see Eq. (9.6.18)), since it is zero in the case of linear polarisation in the entrance pupil. In c) and f) we have plotted the regions in the nominal focal plane where the axial component S_z assumes a negative value. These regions correspond to a reverse flow of optical energy and are associated with locations where a flow singularity occurs, for instance, a flow vortex. These regions with an anomalous power flow are discussed in more detail with the aid of Fig. 9.35.

The difference in energy flow between the upper and lower rows of Fig. 9.33 is entirely due to the presence of an azimuthal energy flow component in the lower figures. The *RC*-polarised beam possesses angular momentum. e) shows the in-focus right-handed flow components for the case of *RC*-polarised light if the energy flow in the beam is viewed in the negative z-direction. Exactly in focus, the radial flow component is rigorously zero. The out-of-focus graphs for $z_1' = -1$ show an extra inward radial flow component. We note that the plots c) and f) of the negative values of the axial component S_z should be formally identical as the axial component of the Poynting vector depends only on the power of the incident beam and not on the state of polarisation. The differences between the two plots stem from the slightly different grey scale setting and the loss of numerical precision for the low values of S_z in these two graphs.

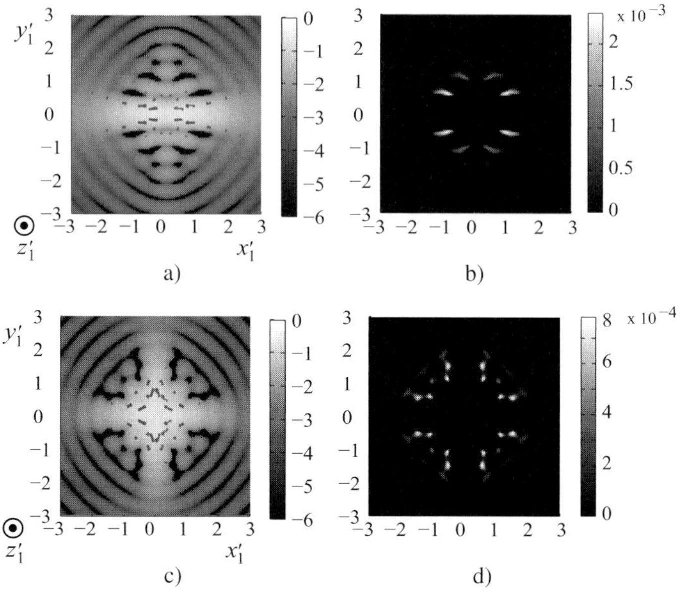

Figure 9.34: Poynting vector components in the image region of a high-*NA* lens system affected by astigmatic aberration ($\alpha_2^2 = \alpha_2^{-2} = \pi/2$, *NA*= 0.95, *X*-polarisation).
a) Defocus distance $z'_1 = -1$. The pattern corresponds to the logarithm of the axial component S_z of the Poynting vector. The lateral components of the Poynting vector have been represented by means of the inwardly directed arrows.
b) Plot of the modulus of S_z for $S_z < 0$ (linear scale).
c) and d) Same legend as a) and b), but now for the in-focus case ($z'_1 = 0$).

In Fig. 9.34 we show the logarithm of S_z in the focal region if astigmatic aberration is present. In comparison with Fig. 9.27 where the electric energy density w_e of the astigmatic line width was influenced by the angle between the linear polarisation azimuth and the astigmatic lines, no such effect is seen in the defocused and in-focus graphs of the Poynting vector. As the product of the electric and magnetic field strength components must be evaluated to determine the components of the Poynting vector, the azimuthal dependence of the electric and magnetic energy density distributions is averaged out.

Finally, in Fig. 9.35 we show examples of characteristic energy flow patterns in a high-aperture beam close to the ring-shaped zeros of the scalar intensity distribution (Airy disc). Figure 9.35a shows a (z'_1, x'_1) cross-section of the flow pattern in the neighbourhood of the first radial zero in the nominal focal plane ($z'_1 = 0$). The state of polarisation of the optical field in the entrance pupil is linear, parallel to the Y-axis. A vortex (V) in the energy flow pattern arises, for instance, when the electric or magnetic energy density exhibits a zero value. At such a point the phase of the electric (or magnetic) field is undetermined and all values between 0 and 2π are allowed in the direct vicinity of such a point in space. In Fig. 9.35a a standard circular vortex is found at the point $(x'_1, z'_1) = (0.55, 0)$ in the image plane, where the electric field strength is zero (see Fig. 9.24). The vortex is accompanied by a saddle-point (S) in the flow pattern at $x'_1 \approx 0.68$. The next vortex is found at $(x'_1, z'_1) \approx (1.05, 0)$ and is outside Fig. 9.35a.

In the vicinity of the vortices and saddle points, negative axial flow components are visible with energy transport in a direction that is opposite to the average flow direction in the focal region. In b) we have plotted the flow pattern over an extended radial interval ($0 \leq x'_1 \leq 1.5$), to show the appearance of repeating vortex/saddle point pairs in the neighbourhood of the zero intensity rings of the scalar diffraction distribution. To complete the energy flow picture, Fig. 9.35c shows the 'scalar' flow pattern which comprises the first two radial zeros of the Airy disc intensity. The energy flow is parallel to the z'_1-axis everywhere. The energy density zeros follow from the zeros of the Bessel function of the first order and of the first kind and are given by $r'_1 = 0.6098$ and 1.1166. The vertical flow line through a zero-intensity ring of the Airy pattern decreases to zero at the focal plane $z'_1 = 0$ and increases to finite values for positive z'_1 values. For a steady-state electromagnetic field, without sources or sinks, such an energy flow geometry violates the conservation law of electromagnetic energy given by Eq. (1.4.7).

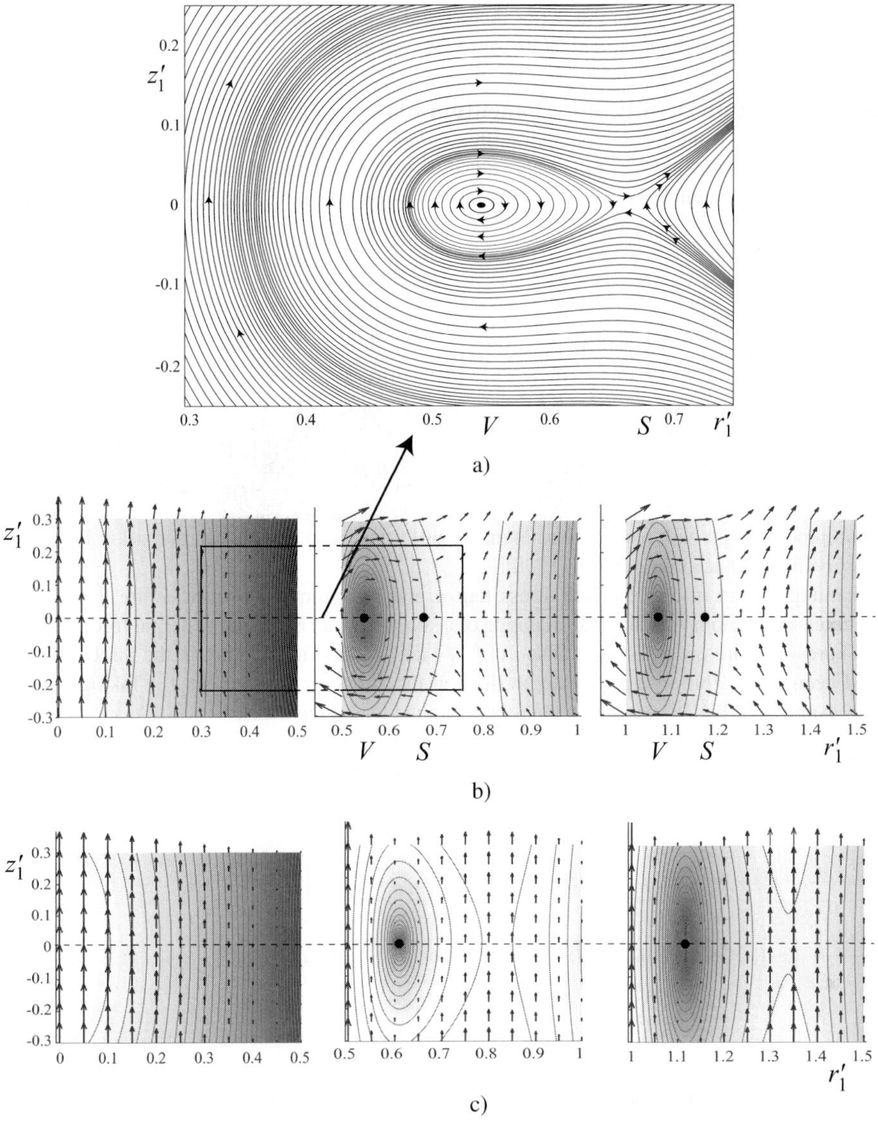

Figure 9.35: Axial cross-sections (r'_1, z'_1-plane) of the energy flow pattern (Poynting vector). The incident light in the entrance pupil has unit strength and is linearly polarised along the Y_0-axis.
a) First off-axis vortex (V, $x'_1 = 0.55$) and saddle-point (S, $x'_1 = 0.68$) in the plane $z'_1 = 0$ ($NA = 0.95$, aberration-free system).
b) Further singular flow points in the plane $z'_1 = 0$ ($NA = 0.95$, aberration-free system). The flow vectors are superimposed on contour lines of the energy density w_{em} of the electromagnetic field in the focal region. The variation of the electric energy density has been further emphasised by means of grey-shading. Three adjacent parts of a radial cross-section are shown, running from the centre of the focused field to the radial position $r'_1 = 1.5$ in normalised units of the lateral coordinate. The rectangular window corresponds to that of the flow diagram in the upper row.
c) $NA \ll 1$. The scalar diffraction case.

9.6.3 Linear and Angular Momentum Flux in a Focused Light Beam

Having seen the particularities of the energy transport in a focused beam we now study the linear and angular momentum density and their flux components in a focused beam of light. These quantities were briefly discussed in Section 1.4 and they are related to the impulse and spin of the photons (linear and spin momentum) and to the geometric structure of the beam (orbital angular momentum). With the aid of the general expressions of Chaptere 1, the momentum density and momentum flow are derived for a high-numerical-aperture focused beam.

9.6.3.1 Linear Momentum Density and Momentum Flow

The linear momentum density and the corresponding flux components are given by the vector $\epsilon\,\mathbf{E}\times\mathbf{B}$ and a tensor $\mathbf{T}$, respectively. The role of the elements T_{pq} of the Maxwell stress tensor has been discussed in Section 1.4. The expressions for the elements T_{pq} can be found, for example, in the textbooks by Stratton **[328]**, Jackson **[160]** and Zangwill **[384]**. Their time-averaged values are given in terms of the electromagnetic energy density and products of field components by means of the matrix expression

$$\mathbf{T}=\begin{pmatrix} \frac{1}{2}\left(\epsilon|\mathbf{E}|^2+\frac{1}{\mu_0}|\mathbf{B}|^2\right) -\epsilon|E_x|^2-\frac{1}{\mu_0}|B_x|^2 & -\epsilon E_xE_y^*-\frac{1}{\mu_0}B_xB_y^* & -\epsilon E_xE_z^*-\frac{1}{\mu_0}B_xB_z^* \\ -\epsilon E_yE_x^*-\frac{1}{\mu_0}B_yB_x^* & \frac{1}{2}\left(\epsilon|\mathbf{E}|^2+\frac{1}{\mu_0}|\mathbf{B}|^2\right) -\epsilon|E_y|^2-\frac{1}{\mu_0}|B_y|^2 & -\epsilon E_yE_z^*-\frac{1}{\mu_0}B_yB_z^* \\ -\epsilon E_zE_x^*-\frac{1}{\mu_0}B_zB_x^* & -\epsilon E_zE_y^*-\frac{1}{\mu_0}B_zB_y^* & \frac{1}{2}\left(\epsilon|\mathbf{E}|^2+\frac{1}{\mu_0}|\mathbf{B}|^2\right) -\epsilon|E_z|^2-\frac{1}{\mu_0}|B_z|^2 \end{pmatrix}. \tag{9.6.19}$$

Note that the difference in sign with respect to reference **[160]** arises from the chosen definition for the **T**-tensor in the momentum conservation law according to Eq. (1.4.8).

The physical interpretation of the Maxwell stress tensor and its symmetry properties are discussed in **[328]** with reference to the three-dimensional mechanical analogue of a solid material body with external forces exerted on it. In this subsection the expressions are given for the transport of x, y and z linear momentum in the three orthogonal directions inside a focused beam with arbitrary aperture by evaluating the stress tensor element T_{ij}. Taking the real parts of the elements of $\mathbf{T}$ in expression (9.6.19) we obtain the time-averaged tensor elements,

$$T_{xx}=\frac{\epsilon_0\pi^2 s_1^4 n_0 n_1^3|f_1|^2}{\lambda_0^2}\Big\{(|a|^2+|b|^2)[G_{1,1}+G_{-1,-1}+\Re(G_{2,0}+G_{-2,0})]-2\Im(ab^*)[(G_{1,1}-G_{-1,-1})+\Re(G_{2,0}-G_{-2,0})]\Big\}, \tag{9.6.20}$$

$$T_{xy}=\frac{\epsilon_0\pi^2 s_1^4 n_0 n_1^3|f_1|^2}{\lambda_0^2}\Big\{(|a|^2+|b|^2)\Im(G_{2,0}-G_{-2,0})-2\Im(ab^*)\Im(G_{2,0}+G_{-2,0})\Big\}, \tag{9.6.21}$$

$$T_{xz}=\frac{\epsilon_0\pi^2 s_1^4 n_0 n_1^3|f_1|^2}{\lambda_0^2}\Big\{-(|a|^2+|b|^2)\Im[G_{1,0}-G_{-1,0}-(G_{1,2}-G_{-1,-2})]+2\Im(ab^*)\Im[(G_{1,0}+G_{-1,0})-(G_{1,2}+G_{-1,-2})]\Big\}, \tag{9.6.22}$$

$$T_{yx}=T_{xy}\,, \tag{9.6.23}$$

$$T_{yy}=\frac{\epsilon_0\pi^2 s_1^4 n_0 n_1^3|f_1|^2}{\lambda_0^2}\Big\{(|a|^2+|b|^2)[G_{1,1}+G_{-1,-1}-\Re(G_{2,0}+G_{-2,0})]-2\Im(ab^*)[(G_{1,1}-G_{-1,-1})-\Re(G_{2,0}-G_{-2,0})]\Big\}, \tag{9.6.24}$$

$$T_{yz}=\frac{\epsilon_0\pi^2 s_1^4 n_0 n_1^3|f_1|^2}{\lambda_0^2}\Big\{(|a|^2+|b|^2)\Re[G_{1,0}+G_{-1,0}+(G_{1,2}+G_{-1,-2})]-2\Im(ab^*)\Re[(G_{1,0}-G_{-1,0})+(G_{1,2}-G_{-1,-2})]\Big\}, \tag{9.6.25}$$

$$T_{zx}=T_{xz}\,, \tag{9.6.26}$$

$$T_{zy}=T_{yz}\,, \tag{9.6.27}$$

$$T_{zz}=\frac{\epsilon_0\pi^2 s_1^4 n_0 n_1^3|f_1|^2}{\lambda_0^2}\Big\{(|a|^2+|b|^2)[G_{0,0}-(G_{1,1}+G_{-1,-1})+\frac{1}{2}(G_{2,2}+G_{-2,-2})]+2\Im(ab^*)[(G_{1,1}-G_{-1,-1})-\frac{1}{2}(G_{2,2}-G_{-2,-2})]\Big\}. \tag{9.6.28}$$

For an ideal, aberration-free optical system the tensor elements T_{zq} are generally the largest ones and they are given by

$$T_{zx}=\frac{-2\epsilon_0\pi^2 s_1^4 n_0 n_1^3|f_1|^2}{\lambda_0^2}\Big\{(|a|^2+|b|^2)\Im[V_{0,1}^0(V_{0,0}^0-V_{0,2}^0)^*]\cos\gamma-2\Im(ab^*)\Re[V_{0,1}^0(V_{0,0}^0-V_{0,2}^0)^*]\sin\gamma\Big\}, \tag{9.6.29}$$

$$T_{zy} = \frac{-2\epsilon_0 \pi^2 s_1^4 n_0 n_1^3 |f_1|^2}{\lambda_0^2} \left\{ (|a|^2 + |b|^2) \Im[V^0_{0,1}(V^0_{0,0} + V^0_{0,2})^*] \sin\gamma + 2\Im(ab^*) \Re[V^0_{0,1}(V^0_{0,0} + V^0_{0,2})^*] \cos\gamma \right\}, \qquad (9.6.30)$$

$$T_{zz} = \frac{\epsilon_0 \pi^2 s_1^4 n_0 n_1^3 |f_1|^2}{\lambda_0^2} (|a|^2 + |b|^2) \left\{ \left|V^0_{0,0}\right|^2 - 2\left|V^0_{0,1}\right|^2 + \left|V^0_{0,2}\right|^2 \right\}. \qquad (9.6.31)$$

The T_{zz} tensor element that determines the magnitude of the flow of axial linear momentum in the z-direction is the dominant element when the aperture of the beam that creates the point-spread function is small. In the aberration-free case, T_{zz} is proportional to the average electromagnetic energy density w_{em} in the entrance pupil which is given by $w_{em} = 2w_e = \epsilon_0 n_0^2(|a|^2 + |b|^2)$, provided the medium in the exit pupil and image space is a dielectric. The on-axis axial momentum in the nominal image plane is then given by $2C_I(n_1/n_0)w_e$ where C_I is the irradiance enhancement factor that was defined in the previous subsection. In the high-aperture case, this quantity needs to be multiplied by $|V^0_{0,0}(0,0)|^2$.

9.6.3.2 Angular Momentum Flow

The angular momentum in a beam of light can have a double origin. It can be caused by the spin of the photons, or, in terms of classical optics, by a nonlinear state of polarisation of the light. For the elementary light wave, a plane wave, this is the only possible source of angular momentum. For a more complicated geometrical wave pattern, produced for instance by the superposition of a spectrum of plane waves, the spatially varying phase of the resulting wave may have a helical phase pattern with a phase discontinuity along a radial line equal to a multiple of 2π (*orbital* angular momentum). Such phase discontinuities are present in higher-order Gaussian beams of the Gauss–Laguerre type and they were discussed in Section 1.7 (see also Fig. 1.19 where the 'helical' phase pattern in a cross-section of such a beam is shown and the corresponding vortex point in the centre of the beam). According to Eq. (1.7.65), a Gauss–Laguerre beam presents an azimuthal phase described by means of the exponential function $\exp(im\theta)$ where θ is the azimuthal angle in the beam cross-section and m the azimuthal order of the beam. Using the expression $\exp\{i(kz - \omega t)\}$ for a wave that propagates in the positive z-direction, we have that the helical shape of the wavefront corresponds to that of a left-handed screw for positive values of m. Instead of using (small-aperture) Gauss–Laguerre beams, we use in this subsection high-aperture beams of which the phase pattern is described by means of Zernike polynomials. For a focused beam, these polynomials permit us to represent more general wavefront departures from the spherical shape. For instance, the phase jump can be made to correspond to a value other than a multiple of 2π as for the Gauss–Laguerre beams. In such a case, the representation by means of Zernike polynomials will be onerous because of the presence of a line phase-discontinuity in the wavefront.

The angular momentum density vector with components j_p, $p = x, y, z$, follows from the Poynting vector components by means of the expression

$$\mathbf{j} = \epsilon\, \mathbf{r}_1 \times (\mathbf{E} \times \mathbf{B}^*). \qquad (9.6.32)$$

Similarly, the elements M_{pq} of the angular momentum tensor $\mathbf{M}$ are derived in a straightforward manner from the Maxwell stress tensor elements T_{pq} by means of the vector product

$$\mathbf{M} = \mathbf{r}_1 \times \mathbf{T}. \qquad (9.6.33)$$

The elements of $\mathbf{M}$ are evaluated using $M_{pq} = \sum_{mn} \epsilon_{pmn} r_{1,m} T_{nq}$ where ϵ_{pmn} is the Levi–Civita symbol and where, as before, the elements T_{nq} are the time-averaged values of the Maxwell stress-tensor elements.

Let us consider now the component $M_{zz}(r'_1, f'_v, \gamma)$ which yields the axial angular momentum flux density per unit time through a surface element with unit surface size perpendicular to the z_1-axis. In many instances, this tensor element will be the dominating one. The element M_{zz} is given by

$$M_{zz} = \frac{\lambda_0 r'_1}{n_1 s_1} \left[T_{yz} \cos\gamma - T_{xz} \sin\gamma \right]. \qquad (9.6.34)$$

In deriving this expression we have used that $r'_1 = r_1 n_1 s_1/\lambda_0$, connecting the real-space distance r_1 to the normalised dimensionless distance r'_1. We also note that $\lambda = 2\pi c/(n_1\omega)$. For a general imaging system, including aberrations, the substitution of all the tensor elements T_{pq} in Eq. (9.6.34) yields, after some rearrangement, for the M_{zz}-component,

$$
\begin{aligned}
M_{zz} = \frac{\epsilon_0 \pi^2 s_1^3 n_0 n_1^2 |f_1|^2}{\lambda_0} r_1' \Bigg[&(|a|^2 + |b|^2) \Big\{ \Re \left[G_{1,0} + G_{-1,0} + (G_{1,2} + G_{-1,-2}) \right] \cos\gamma \\
&+ \Im \left[(G_{1,0} - G_{-1,0}) - (G_{1,2} - G_{-1,-2}) \right] \sin\gamma \Big\} \\
&- 2\Im(ab^*) \Big\{ \Re \left[G_{1,0} - G_{-1,0} + (G_{1,2} - G_{-1,-2}) \right] \cos\gamma \\
&+ \Im \left[(G_{1,0} + G_{-1,0}) - (G_{1,2} + G_{-1,-2}) \right] \sin\gamma \Big\} \Bigg] .
\end{aligned}
\tag{9.6.35}
$$

The focusing of an aberration-free, high-aperture converging beam in the nominal image plane (all functions $V^m_{n,k}$ are real, $\beta^0_0 = 1$, all other $\beta^m_n = 0$) leads to a simplified expression,

$$
M_{zz} = \frac{-4\epsilon_0 \pi^2 s_1^3 n_0 n_1^2 |f_1|^2}{\lambda_0} r_1' \, \Im(ab^*) \left[V^0_{0,1} (V^0_{0,0} + V^0_{0,2}) \right] . \tag{9.6.36}
$$

In the absence of any particular beam geometry, a nonzero value of M_{zz} is necessarily produced by the spin momentum of the beam and its value is proportional to $\Im(ab^*)$, where $p_C = 2\Im(ab^*)/(|a|^2 + |b|^2)$ is the degree of circular polarisation as defined by Eq. (1.6.105).

The introduction of angular momentum in a focused beam was discussed at the beginning of this subsection. A helical phase structure in the beam cross-section produces (orbital) angular momentum in the beam. A basic discontinuous phase-structure with a phase jump of 2π is well described by the lowest order Zernike polynomial $\rho \exp\{i\theta\}$. In the example below we approximate the exit pupil function by taking $\beta^1_1 \equiv 1$ and equating all other β coefficients to zero. In the nominal focal plane we then obtain the expression

$$
\begin{aligned}
M_{zz} = \frac{\epsilon_0 \pi^2 s_1^3 n_0 n_1^2 |f_1|^2}{\lambda_0} (|a|^2 + |b|^2) r_1' \Bigg\{ &\left[(V^1_{1,1} + V^1_{1,-1}) V^1_{1,0} + (V^1_{1,1} V^1_{1,2} + V^1_{1,-1} V^1_{1,-2}) \right] \\
&- \frac{2\Im(ab^*)}{(|a|^2 + |b|^2)} \left[(V^1_{1,1} - V^1_{1,-1}) V^1_{1,0} + (V^1_{1,1} V^1_{1,2} - V^1_{1,-1} V^1_{1,-2}) \right] \Bigg\} .
\end{aligned}
\tag{9.6.37}
$$

We observe that the element M_{zz} comprises two terms. The first term between braces is proportional to the beam power $(|a|^2 + |b|^2)$ and its origin must be the geometrical deformation of the spherical wavefront by the helical phase function (orbital angular momentum). The second main term of Eq. (9.6.37) is proportional to the degree of circular polarisation, p_C, of the focused beam. We note that the V-dependent factor between square brackets that multiplies this second term is substantially smaller than the V-dependent factor of the orbital angular momentum term. When focusing a beam with small aperture, the relative importance of the 'photon spin' term becomes negligible.

9.6.3.3 Examples of Linear and Angular Momentum in High-aperture Focused Beams

In this paragraph we examine the flux components of linear and angular momentum in the image region of a high-aperture focused beam. For reasons of convenience, the electric and magnetic fields in the entrance pupil have been scaled such that the on-axis value of the T_{zz} component is equal to one in units of Nm^{-2} (aberration-free imaging system). The values of the **M**-tensor elements in units of Jm^{-2} and the Poynting vector components in Wm^{-2} are computed using the same scaled input fields. The results of the calculations are shown in a series of graphs. Each shows a cross-section, perpendicular to the z_1'-axis, of a linear momentum or an angular momentum flux component in the nominal image plane, with the normalised lateral coordinates x_1' and y_1' plotted along the horizontal and vertical axis of each graph. The three top rows of each figure show the linear momentum tensor elements, the three bottom rows show the angular momentum tensor elements. We recall that the momentum density, both linear and angular, is proportional to electromagnetic cross-terms that are sums of quadratic expressions of the electric *and* magnetic field strengths. As opposed to the electric and magnetic energy density separately which do not show circular symmetry, the momentum densities are circularly symmetric in the focal plane of a perfect imaging system. The individual tensor elements do not necessarily show this circular symmetry.

In Fig. 9.36 we show the values of the linear and angular momentum tensor elements as a function of the (normalised) image plane coordinates (x_1', y_1') for a high-aperture focused beam. The focused beam is generated by an incident plane wave with linear polarisation along the X_0-direction in the entrance pupil; $(a, b) = (1, 0)$, $n_0 = n_1 = 1$ and $s_1 = 0.95$. In the upper three rows of the figure we notice that the dominant linear momentum element is T_{zz}, displaying circular symmetry with respect to the z-axis. In the case of a small aperture this is the only non-negligible element. At high-aperture value,

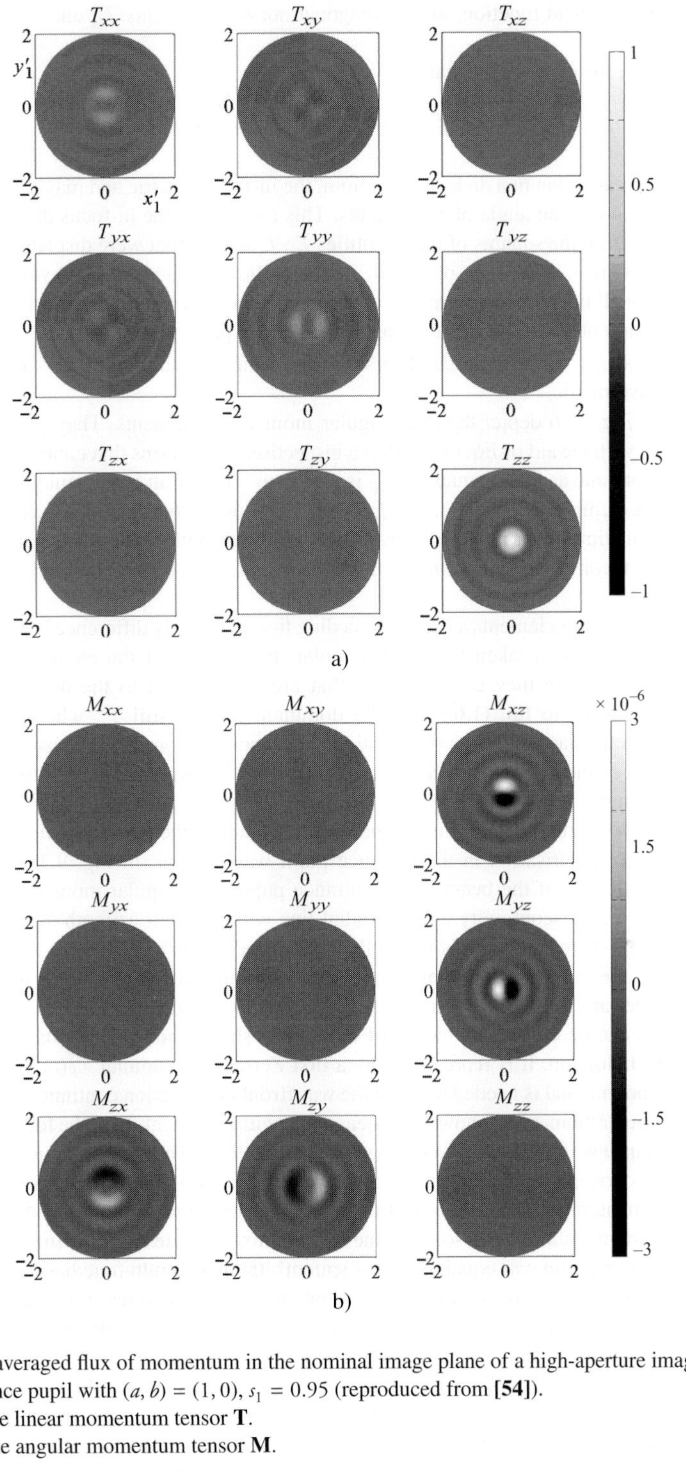

Figure 9.36: The time-averaged flux of momentum in the nominal image plane of a high-aperture imaging system. Linear polarisation in the entrance pupil with $(a, b) = (1, 0)$, $s_1 = 0.95$ (reproduced from **[54]**).
a) The components of the linear momentum tensor **T**.
b) The components of the angular momentum tensor **M**.

outside the centre of the point-spread function, we see diagonal nonzero elements T_{xx} and T_{yy} and weaker off-diagonal elements $T_{xy} = T_{yx}$.

As an example we consider the element T_{xx} that is given by

$$T_{xx} = \frac{\epsilon}{2}\left\{-|E_x|^2 + |E_y|^2 + |E_z|^2\right\} + \frac{1}{\mu_0}\left\{-|B_x|^2 + |B_y|^2 + |B_z|^2\right\} , \tag{9.6.38}$$

where, in the case of linear polarisation in a dielectric medium, the in-focus electric and magnetic energy density patterns are equal apart from a rotation over an angle of $\pi/2$ radians. This means that the in-focus diagonal patterns, like that of $T_{xx}(x_1', y_1')$, can be composed from the squares of the quantities E_x, E_y and E_z that were displayed in Fig. 9.22c–e. For the off-diagonal elements we note that, in the case of T_{xy}, B_x can be replaced by a 90° rotated version of E_y/c and B_y by the rotated version of E_x/c. Since all field strengths are real in the focal plane, the resulting pattern is that given by the sum of $\epsilon E_x E_y$ and a 90° rotated version of it. This yields the 'quadrupole'-type pattern for T_{xy} and T_{yx}. Finally, the dominant T_{zz} pattern is composed of $|E_x|^2 + |E_y|^2$ and a 90°-rotated version from which are subtracted $|E_z|^2$ and its 90°-rotated version; the end result is circularly symmetric.

The three lower rows of Fig. 9.36 depict the nine angular momentum elements. They are obtained from the linear momentum tensor elements with the aid of Eq. (9.6.33). An inspection of the terms that contribute to M_{zz} shows that both T_{xz} and T_{yz} are rigorously zero and as a consequence M_{zz} is also zero. This result is as would be expected for an incident linearly polarised beam. The elements $M_{xz} = M_{zx}$ and $M_{yz} = M_{zy}$ show nonzero values. This means that a spatial selection of part of the point-spread function would produce a new 'distorted' beam with nonzero angular momentum. However, as expected from the momentum conservation law, the total angular momentum over the entire beam cross-section remains zero.

Figure 9.37 shows the same tensor elements as in the preceding figure, the only difference being the state of polarisation in the entrance pupil which has been taken to be left-circular. It is seen that the elements $T_{xz} = T_{zx}$ and $T_{yz} = T_{zy}$ now assume nonzero values because they contain terms that are proportional to the degree of circular polarisation $p_c = 2\Im(ab^*)/(|a|^2 + |b|^2)$ according to Eq. (1.6.105). The dominant term is still T_{zz} which again shows a circularly symmetric pattern as seen in the case of linear polarisation. The angular momentum tensor elements in rows four to six are dominated by the M_{zz} element, which is zero on axis and shows a doughnut shape with annular regions around it with alternating momentum sign.

In Fig. 9.37 the origin of the angular momentum of the focusing beam is the nonzero spin momentum of the photons ($\Im(ab^*) \neq 0$) since the state of polarisation in the entrance pupil was circular. In Fig. 9.38 the origin of the angular momentum is the special geometry of the beam in the entrance pupil. The angular momentum is now produced by a wavefront deformation with a 2π discontinuity along a radial line with constant azimuth θ_0 in the entrance pupil. This wavefront deformation with a sharp transition is faithfully reproduced in the exit pupil. A small blurring effect of the sharp wavefront transition is present at the exit pupil due to the aberration associated with the imaging of the pupils. This unsharpness, however, can be safely neglected with respect to the lateral extent $2\rho_1$ of the exit pupil. Such a discontinuity in the wavefront of a beam can be created by means of a helically shaped height deformation on a substrate or by a specially designed (digital) hologram. It is represented by a first Zernike polynomial $Z_1^1(\rho, \theta) = \rho \exp(i\theta)$. The linear ρ dependence in the Zernike polynomial is needed to make the wavefront deformation continuous at the centre of the pupil.

Figure 9.38 on orbital angular momentum shows the linear and angular momentum in the focal plane of a focused beam which bear a close resemblance with the corresponding graphs of Fig. 9.37. The angular momentum tensor elements are very similar in form to the corresponding elements of a focused beam with circular polarisation. This means that the introduction of angular momentum by means of wavefront tailoring (orbital angular momentum) is virtually equivalent to the effect of spin angular momentum if the absolute value of the azimuthal index l is one. In our example the momentum state of the orbital angular momentum was equal to that of spin orbital momentum (one basic unit of $\hbar$ imparted to each photon). The main difference between the spin and the orbital momentum pictures is found in the linear momentum element T_{zz}. Its pattern has a doughnut shape with a negative value on axis. The effect of the phase discontinuity in the wavefront also becomes visible in the energy transport. The three components of the Poynting vector are shown in Fig. 9.38c. It is seen that the z-component of the Poynting vector also has a doughnut shape and shows a central zero.

The effect of a wavefront aberration on the linear and angular momentum is shown in Fig. 9.39 for the case of astigmatism. The incident state of polarisation is assumed to be linear. The linear momentum tensor elements are comparable to those of Fig. 9.37, indicating that some correspondence between circular polarisation and astigmatic aberration is present with respect to momentum transport in the focused beam. In the lower part of the figure, the similarity with a beam having angular momentum is further supported by inspection of the M_{zz} element. The pattern in focus contains quadrants with a strong axial momentum value; the signs are such that the total momentum remains

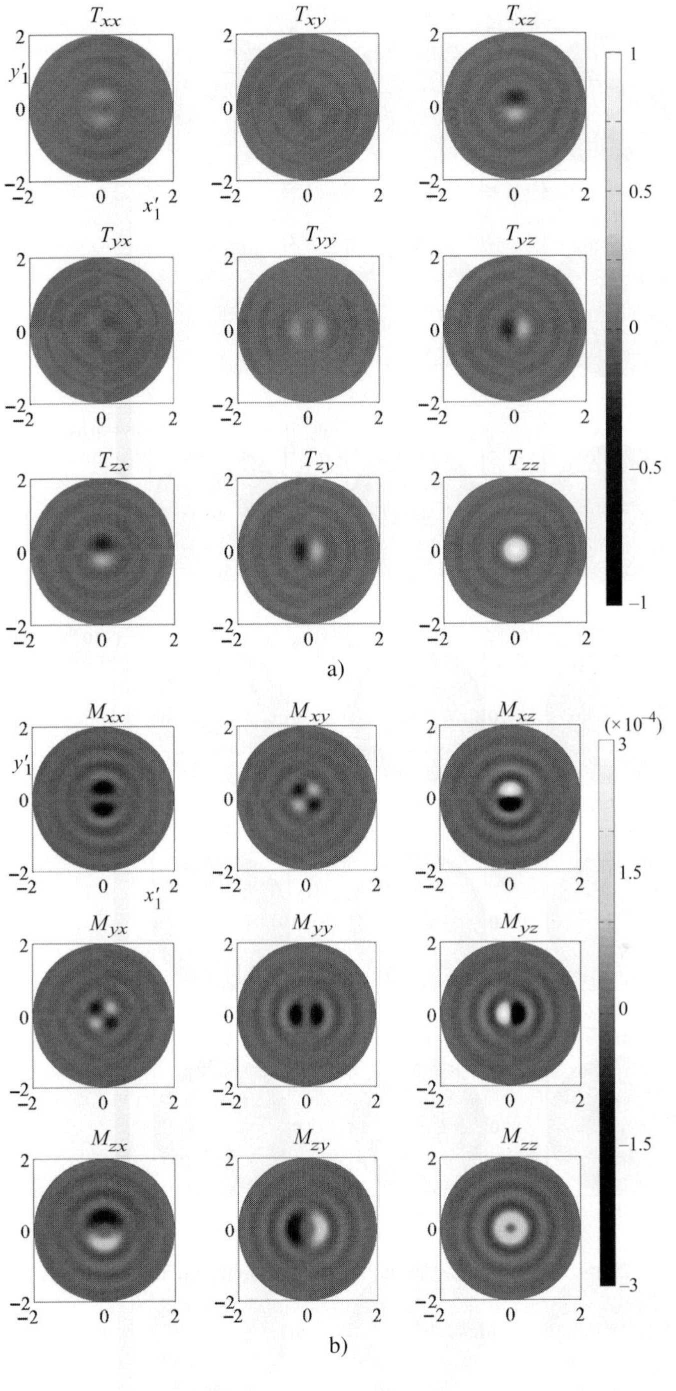

Figure 9.37: a) The time-averaged **T** and b) **M** tensor elements in the nominal focal plane of a high-aperture imaging system. Beam parameters: $s_1 = 0.95$, LC-polarisation in the entrance pupil ($(a, b) = (i, -1)/\sqrt{2}$). Reproduced from **[54]**.

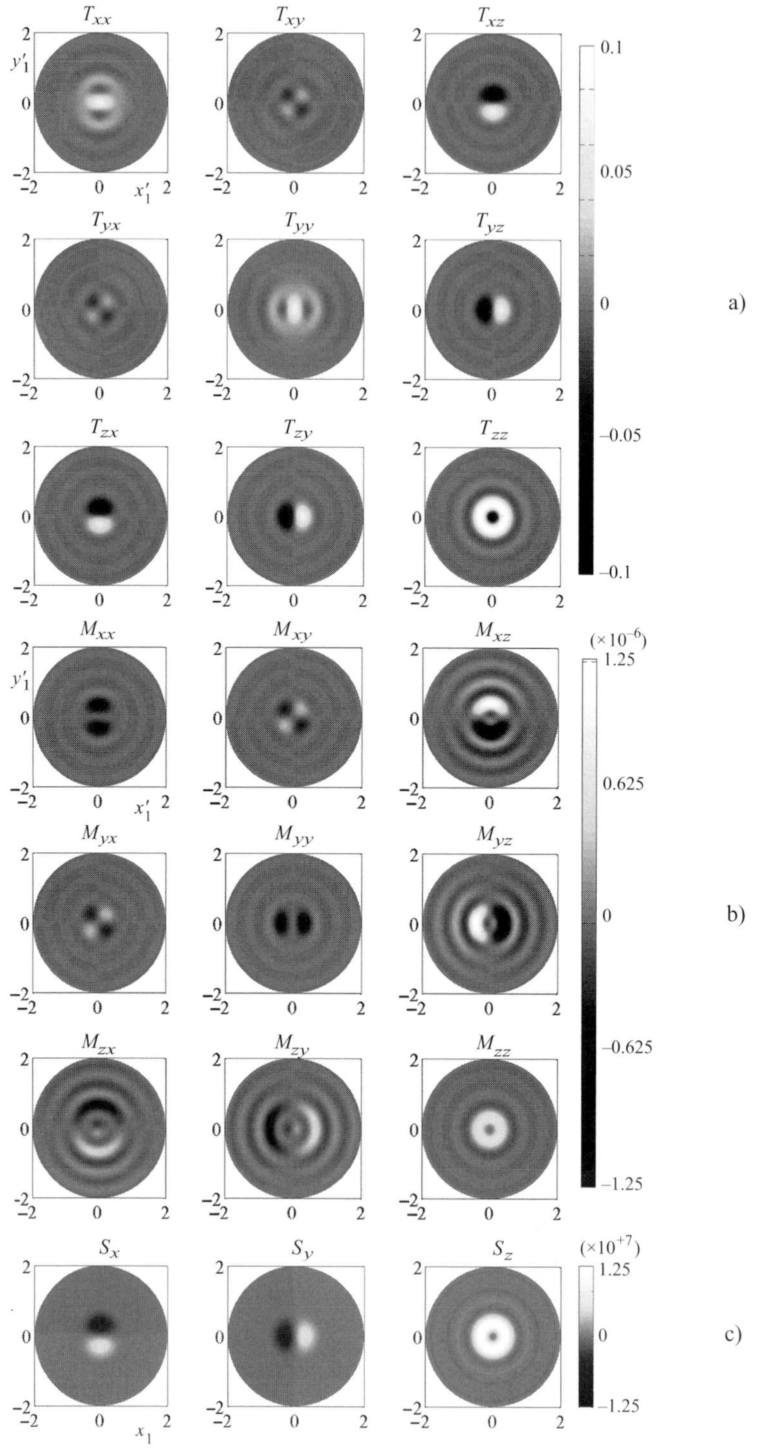

Figure 9.38: a) and b) The time-averaged **T** and **M** tensor elements and c) the Poynting vector components in the nominal focal plane. Beam parameters: $s_1 = 0.95$, linear polarisation in the entrance pupil ($a = 1,\ b = 0$), wavefront with helical phase profile ($\beta_1^1 = 1$). Reproduced from **[54]**.

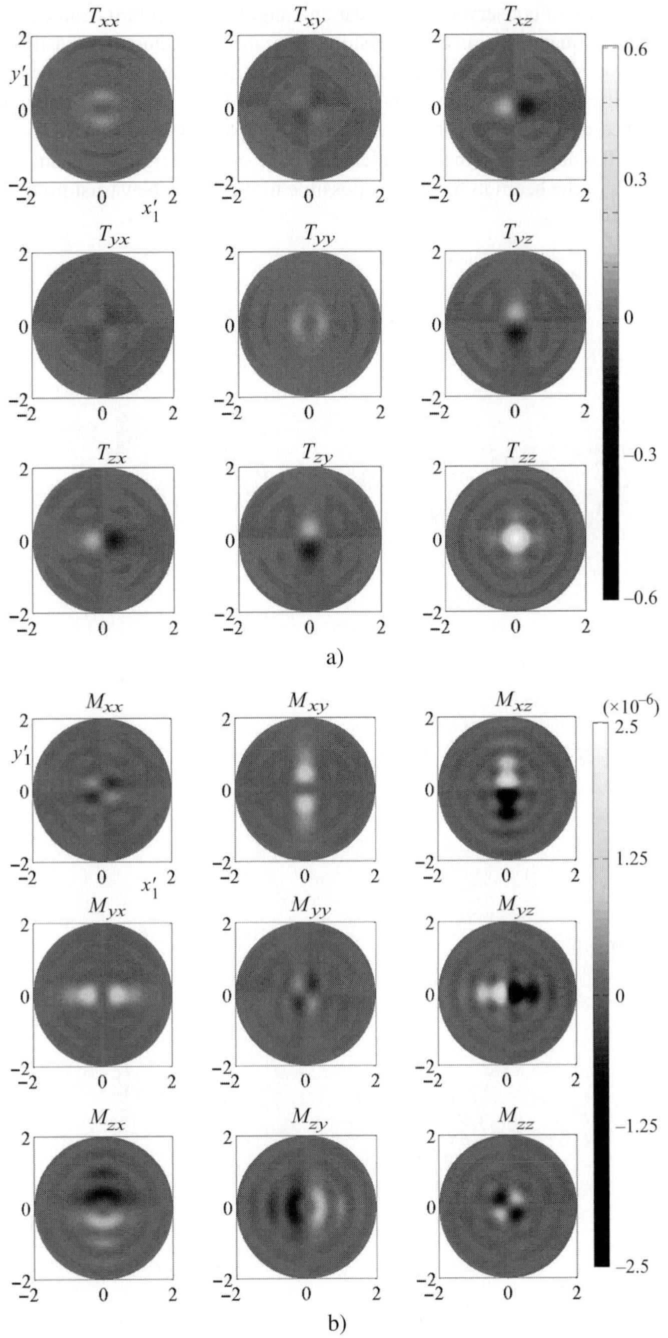

Figure 9.39: a) The time-averaged **T** and b) **M** tensor elements in the nominal focal plane of a high-aperture imaging system. Beam parameters: $s_1 = 0.95$, linear polarisation in the entrance pupil, $(a = 1,\ b = 0)$, astigmatic wavefront with Zernike coefficients $\alpha_2^2 = \alpha_2^{-2} = i\pi/4$. Reproduced from **[54]**.

zero as should be the case because of conservation of linear and angular momentum from entrance pupil to image plane. The angular z-momentum in the quadrants can be understood from the ray picture of an astigmatic focus (see Figs. 5.26 and 5.27). A geometric test ray outside the principal sections of the astigmatic beam shows a changing azimuth when progressing from one focal line to the other. In the best intermediate focal plane, the azimuthal change of the test ray is largest when the ray started at an azimuth at 45° with the principal planes of the beam. This ray optics picture is reproduced in the pattern of the angular momentum element, M_{zz}, of the astigmatic beam. By spatially selecting two diagonally opposite quadrants of the beam in focus, it is possible to produce a new (distorted) beam with a large amount of angular momentum.

10 Frequency Analysis of Optical Imaging

10.1 Introduction

The traditional method to calculate, analyse and assess the quality of a composite image formed by an optical instrument is based on the image formation of an isolated point object. Much attention has been paid in the past to the most simple composite object, two point sources or two line sources separated by a small distance. In astronomy, the observer has to decide whether the image of the celestial object is that of an isolated star or that it is created, for instance, by two individual neighbouring stars. Criteria have been developed to objectively discriminate between the image of a single point-like object and that of a dual source. These criteria are based on the superposition of intensity[1] patterns of point source objects, for instance on the retina of the human eye or on the sensitive layer of a photographic plate. A frequently used criterion for distinguishableness was put forward by Rayleigh **[287]** in the framework of his work on spectroscopy. The prism-dispersed image of the entrance slit in a spectrograph contains discrete line images if the light source spectrum is composed of isolated spectral lines. According to Rayleigh, two spectral components can be discriminated if their separation in the image plane is such that the first zero of the diffraction image of one spectral line is coincident with the maximum of the image of the neighbouring spectral line. Figure 10.1 shows the intensity pattern which arises when the intensities of the diffraction images associated with two adjacent spectral lines are superimposed. The normalised intensity pattern of the diffraction image of a very narrow line source in the focus of a lens is given by

$$I_l(x) = \left(\frac{\sin(2\pi x)}{2\pi x}\right)^2 . \qquad (10.1.1)$$

The coordinate x is the coordinate in the image plane of the lens, in the direction perpendicular to the slit, expressed in diffraction units of λ_0/NA. The image-side numerical aperture of the lens is given by NA and λ_0 is the vacuum wavelength of the radiation. The separation Δx_l of two spectral maxima according to the Rayleigh criterion should be at least $0.5\lambda_0/NA$ to allow the visual resolution of the two images. The relative minimum of the accumulated intensity pattern at $x = 0$ is then given by $8/\pi^2 = 0.81$. The presence of the relative minimum and its value with respect to unity are indicative of the ease with which, in the presence of noise, the existence of the two separate spectral lines with equal strength can be detected.

In the case of two adjacent images of a point source we have to consider the Airy disc intensity pattern, which is given by (expressed in units of the diffraction unit)

$$I_p(x) = \left(\frac{2\,J_1(2\pi x)}{2\pi x}\right)^2 . \qquad (10.1.2)$$

1 We apologise for using in this chapter for reasons of convenience the word *intensity* for both object excitance and image irradiance (unit $\mathrm{W m^{-2}}$). Formally, according to the SI system of units, radiometric or photometric *intensity* (unit $\mathrm{W\,sr^{-1}}$ and $\mathrm{lm\,sr^{-1}}$, respectively) should only be used for the optical energy flow from a source or towards a receiving surface per unit of solid angle (basic photometric SI unit is the *candela*, see also Section 4.11).

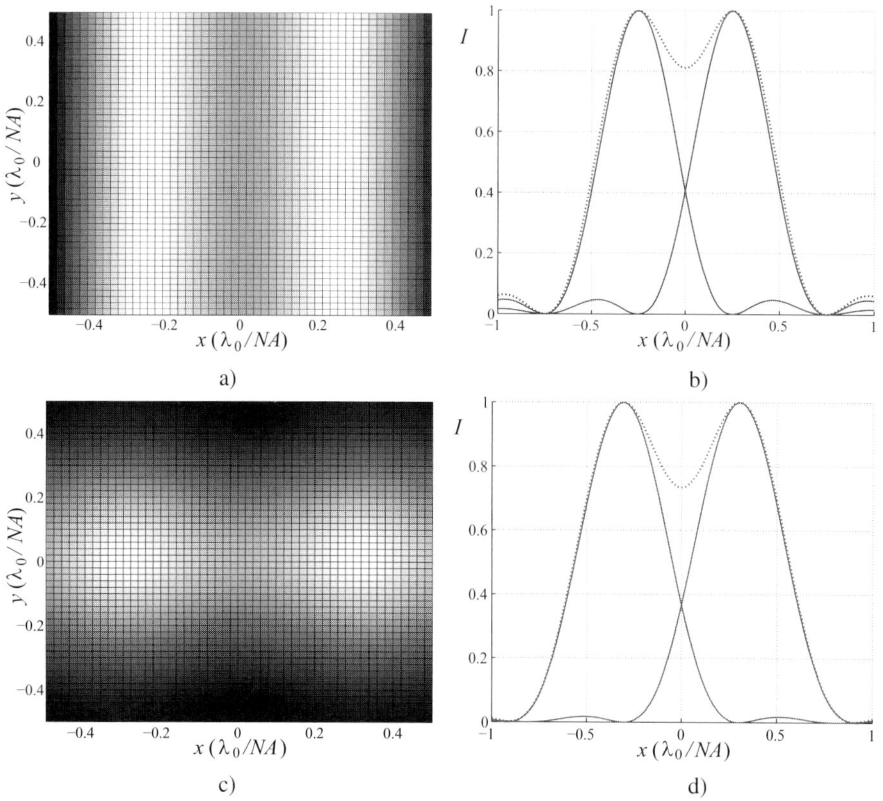

Figure 10.1: Resolution criteria for the separation of two line objects or two point objects at an infinitely large distance or in the focal plane of a lens.
a) and b) Two line sources with vanishing widths, separated by a distance of 0.5 λ_0/NA; a) two-dimensional cumulative intensity profile, b) intensity cross-sections, perpendicular to the two line sources, of the individual sources (solid curves) and of the sum of the two sources (dashed curve).
c) and d) Two point sources separated by a distance 0.61 λ_0/NA; c) two-dimensional cumulative intensity profile in the image plane, d) intensity cross-sections ($y = 0$) of the images of the two individual point sources (solid curves) and of their cumulative intensity (dashed curve).

If we now apply the Rayleigh criterion for the separation of the images of two point sources, namely the maximum of the first image coincides with the first minimum of the second, we obtain a required distance of $\Delta_p = 0.6098\ \lambda_0/NA$. Figure 10.1c and d show the accumulated intensity of two images and its cross-section in the plane $y = 0$ if, indeed, the maximum of one image coincides with a position in the first dark ring of the second image. The intensity at the relative minimum is now 0.735. This value demonstrates that the Rayleigh criterion provides us with a good indication of the separability of two sources but that its quantitative application is problematic.

Another criterion for the separability of two sources is defined by means of the separating distance that produces the onset of a relative minimum in the cumulative intensity pattern of the two images. The change of sign of the second derivative of the intensity pattern, halfway along the line joining the centres of the two source images, would produce the onset of the relative minimum. This criterion has been put forward by Sparrow **[323]** and the distance which produces a zero value of the second derivative is commonly called the Sparrow distance. The second derivatives of the intensity patterns of a line source image and of a point source image are given by,

$$\frac{d^2 I_l(x)}{dx^2} = 2\,\frac{\cos(4\pi x)}{x^2} - 2\,\frac{\sin(4\pi x)}{\pi x^3} + 3\frac{\sin^2(2\pi x)}{2\pi^2 x^4}\,, \qquad (10.1.3)$$

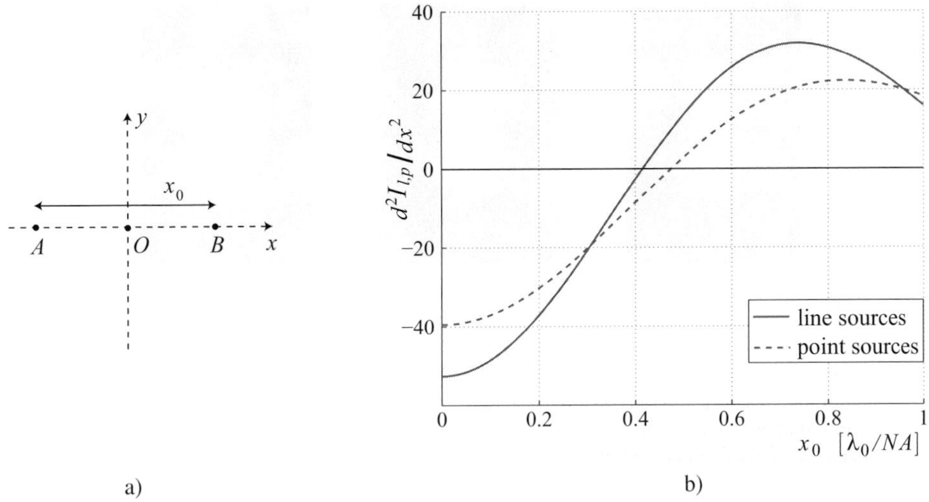

Figure 10.2: a) A and B are on the centre lines of the images of two line sources or, equivalently, the centres of two point-source images, symmetrically positioned with respect to the origin O. The separating distance AB is x_0 (in units of λ_0/NA).
b) The value of the second derivative in the origin O of an intensity pattern due to two diffraction images in A and B. The midpoints of the images have coordinates $\pm x_0/2$. Solid curve: image of two line sources, dashed curve: the equivalent for two point sources.

$$\frac{d^2 I_p(x)}{dx^2} = \frac{2\{J_0(2\pi x) - J_2(2\pi x)\}^2 + 2J_1(2\pi x)\{J_3(2\pi x) - 3J_1(2\pi x)\}}{x^2} - 8\,\frac{J_1(2\pi x)\{J_0(2\pi x) - J_2(2\pi x)\}}{\pi x^3} + 6\,\frac{J_1^2(2\pi x)}{\pi^2 x^4}\,, \tag{10.1.4}$$

where the values of the limit at $x = 0$ for the line image are given by $d^2 I_l(x)/dx^2\big|_{x=0} = -(2/3)(2\pi)^2$ and for the point image by $d^2 I_p(x)/dx^2\big|_{x=0} = -(1/2)(2\pi)^2$.

To study the separability of the images of two line or point sources we take two equal intensity objects, located symmetrically with respect to an origin O (see Fig. 10.2). The intensity in the image plane is given by (line source)

$$I(x) = \left(\frac{\sin[2\pi(x - x_0/2)]}{2\pi(x - x_0/2)}\right)^2 + \left(\frac{\sin[2\pi(x + x_0/2)]}{2\pi(x + x_0/2)}\right)^2 . \tag{10.1.5}$$

Starting from zero separation, the first zero of the second derivative indicates that a plateau appears in the intensity distribution. The detection of such a plateau, although very sensitive to perturbations and noise, confirms the presence of more than a single source. For positive values of the second derivative of $I(x)$, a relative minimum appears in O and the high-intensity region of the pattern is provided with a weakly modulated intensity oscillation with an approximate period of x_0. As was the case for the Rayleigh criterion, the zero second derivative criterion yields different separation values for a line source image or a point source image (see Fig. 10.2). The two curves show that separation distances of $0.42\lambda_0/NA$ and $0.46\lambda_0/NA$ can be resolved in the presence of a line source or a point source, respectively. An experimental comparison of two-point resolution according to the Rayleigh and Sparrow criterion is shown in Fig. 10.3 as a function of the separation x_0 in the image plane of the centres of the image points in units of λ_0/NA.

A complicating factor for the Rayleigh and Sparrow two-source 'resolution' criteria is the possible coherence between the two diffraction images. Intensity summation of the individual diffraction images was carried out in the previous calculations. For celestial objects this is correct because of the incoherence of stellar objects. In the case of objects that are illuminated by a common source, partial coherence can be expected between their images. Interference effects in the intensity pattern produced by two adjacent sources will complicate the measurement of the just resolvable distance by

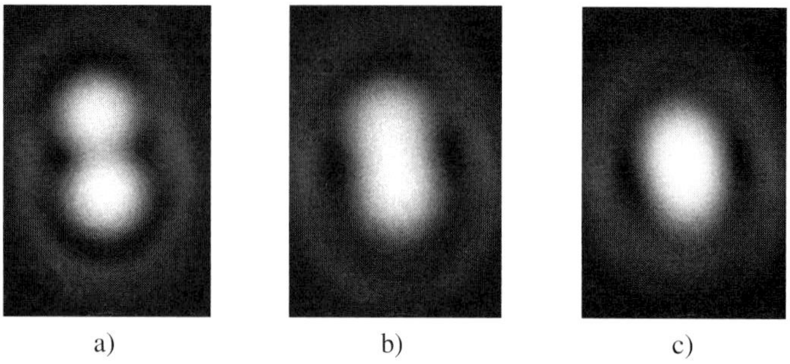

Figure 10.3: Experimental illustration of the separability of diffraction-limited images of two incoherent point sources.
a) $x_0 \approx 0.85$, well-separated image points,
b) $x_0 \approx 0.60$, separation limit according to the Rayleigh criterion,
c) $x_0 \approx 0.45$, separation limit according to the Sparrow criterion.
The separation distance x_0 is expressed in the diffraction unit λ_0/NA.

an instrument. We conclude that a resolution measurement in the spatial domain is ill-defined and influenced by other factors like type of object, illumination method and state of polarisation of the light.

The classical Rayleigh and Sparrow criteria are applied to, for instance, an intensity measurement as a function of the position of the detection slit in the focal plane of a spectrograph. In the case of measured images where the intensity information is collected using a pixelated image sensor, the classical criteria are of less interest. The two-dimensional image data are subjected to an imaging model for two shifted sources with unknown positions, intensity, degree of coherence and polarisation. The measured data are then matched as well as possible to the theoretical imaging model by adjustment of the unknown parameters of the sources. The system and detector noise are also included in the imaging process. The outcome of such a parameter estimation yields the values of the unknown parameters in terms of their expectation values and standard deviations. Considerations on imaging resolution in this more general sense can be found in a survey of this subject by den Dekker and van den Bos **[81]**.

A quite different treatment of image resolution is found in the early theory of imaging by a microscope, proposed by Abbe in 1873 **[2]**. Imaging is described as the result of interference between partial waves that are diffracted by a *periodic* test object. The image of a general object is obtained by the summation of the images of the periodic components into which the general object can be decomposed. The illumination of the object has to be coherent in Abbe's microscope. The principle of analysing a general object by means of periodic samples has been further developed by Duffieux **[89]** who explicitly discussed the transfer of spatial frequency components in an object by the optical imaging system. The application of the imaging of a general object in terms of the transfer of its spatial frequencies was further developed by Hopkins and Barham **[146]** and Hopkins **[142]** under general conditions of object illumination. The frequency analysis of optical imaging and the action of an optical system as a filter of spatial frequencies has been described in several textbooks (see Maréchal and Françon **[238]** and Goodman **[113]**). In this chapter we first reproduce the basic theory of spatial frequency analysis of optical imaging. In the second section we study the two-dimensional frequency transfer which is appropriate for planar images. This analysis is based on scalar diffraction theory. The third section is devoted to the three-dimensional transfer of spatial frequencies for the case of volumetric objects or object structures on a three-dimensional surface. The last section considers the change in (two-dimensional) frequency transfer if sources of light scattering are present in an optical system, for instance, optical surfaces with a residual surface roughness.

10.2 Optical Transfer Function of a Classical Wide-field Imaging System

In this section we study the transfer of spatial frequencies through an optical system using the scalar approximation of the optical vector field. Such an approximation is allowed if the convergence angles of the imaging beams remain small and the state of polarisation is unchanged during propagation through the optical system. In the object plane, perpendicular to the z-axis, we consider a general object of which the complex amplitude transmission function is decomposed into a spectrum of spatial frequencies by means of a Fourier transformation. The definition of the spatial Fourier transformation that is used in this chapter is given in Appendix A, Eqs. (A.8.5) and (A.8.6). Each component of the spatial spectrum represents a plane wave with a propagation direction determined by the wave vector components k_x and k_y. The (complex) 'spectral' value associated with such a plane wave determines the amplitude and phase of the plane wave. The adjective 'spectral' is associated here with the spatial spectrum of the object transmission or reflection function, as opposed to the more common meaning of this word in optics which designates a temporal frequency (or colour) of the light disturbance itself.

The transfer of a spatial frequency by an optical system is calculated by establishing the ratio of the intensity modulation related to a specific frequency in the intensity distribution in the image plane and the original intensity modulation in the object plane for this frequency. A phase shift of the periodic pattern associated with a certain frequency in the image plane is taken care of by the argument of this complex-valued ratio. The data for all the transmitted frequency components are collected in the *optical transfer function* (*OTF*). This generally complex function has a modulus part that is called the *modulation transfer function* (MTF) and a phase part with the name *phase transfer function* (*PTF*). Depending on the magnitude of the system aberrations, two schemes are employed for the calculation of the optical transfer function. For sufficiently small aberrations, below the diffraction limit or just slightly beyond this value, the image plane intensity is calculated using, for instance, the Debye diffraction integral formalism with the complex amplitude of the field on the exit pupil sphere as input. For large wavefront aberration values, typically several wavelengths or more, it is allowable to use a ray-optics model of light propagation. The rays emanate from a single point in object space and the ray intersection points in the image plane are used to obtain the image blur and image shift. From the data obtained with this ray-based model of wave propagation, a good estimate of the complex-valued optical transfer function is obtained for moderately to heavily aberrated imaging systems. The advantage of doing so is the much shorter calculation time for the ray-based *OTF*.

10.2.1 The Spatial Frequency Concept

In this subsection we briefly discuss the construction of an object from its constituent spatial frequency components. The object is assumed to be an infinitely thin transparency to which at each point a generally complex transmission factor is attributed. This transparent object is illuminated by a single plane wave or a cone of plane waves in a way that was discussed in Subsection 7.7. The connection of the ingoing and outgoing optical field by means of a single complex number associated with a single point in space is an approximation that is only allowed for an infinitely thin object structure. In general, the object will have a finite thickness and diffraction will take place within the object volume. By solving an electromagnetic boundary value problem for an incident plane wave as stipulated in Chapter 1, we obtain the electromagnetic field components in a plane beyond the modulated object volume where they propagate anew in a homogeneous medium. In the thin-object approximation, the finite volume of the three-dimensional object is neglected as well as the influence of the state of polarisation of the incident light wave on the local transmission factor. Therefore, this approximation is called the scalar thin-object approximation. For an object that influences only weakly the amplitude and phase of the incident wave and whose transmission, in the plane of the object, varies slowly with respect to the wavelength of the light, the scalar thin-object approximation is justified. This approximation is also satisfactory for the case of imaging in reflection.

With respect to notation we employ the vector quantity $\mathbf{v} = (\nu_x, \nu_y, \nu_z)$ to define a spatial frequency (unit m^{-1}) in three-dimensional space. If the spatial frequency is delimited to a plane, for instance a plane perpendicular to the (average) propagation direction of light along the z-axis of an optical system, the vector $\mathbf{v}$ is reduced to a two-dimensional vector in the (x, y) plane. Parallel to the vector $\mathbf{v}$ we use the spatial frequency vector $\boldsymbol{\omega}$ which denotes the angular spatial frequency in units of rad m^{-1} with the relation $\boldsymbol{\omega} = 2\pi\mathbf{v}$. The reason for using a second symbol to denote the same physical entity is a practical one; the use of $\boldsymbol{\omega}$ instead of $\mathbf{v}$ may save numerous factors of 2π in mathematical equations. A specific spatial frequency, for instance a grating structure in the object plane of an imaging system, is denoted by the vector $\mathbf{u} = (u_x, u_y, u_z)$. In accordance with previous calculations on the diffraction of an optical ray by a grating structure on an optical surface (see Subsection 5.7.3), we denote a grating unit vector not only by $\hat{\mathbf{u}}$ but also by $\hat{\boldsymbol{\beta}}$. A possible confusion between the spatial frequency vector $\mathbf{v}$ and the unit normal vector $\hat{\mathbf{v}}$ to a three-dimensional surface will be avoided as much as possible in this chapter. If the risk of confusion is real, the surface normal vector is exceptionally denoted by $\hat{\mathbf{n}}$.

Having defined the notation conventions for spatial frequency vectors and their components, we are now in the position to apply Fourier transform theory to object structures to be imaged. To this purpose we perform a two-dimensional Fourier transformation with respect to the spatial coordinates (x, y) of the transmission function $f(x, y)$ of a thin object structure, located in the object plane ($z = 0$) of an imaging system,

$$F(\nu_x, \nu_y) = \iint_{-\infty}^{+\infty} f(x, y) \exp[-2\pi i(\nu_x x + \nu_y y)]\, dxdy\,. \tag{10.2.1}$$

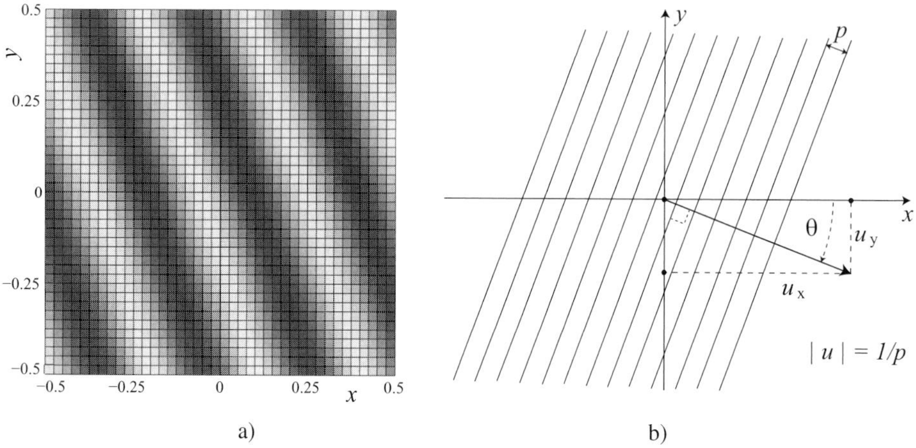

Figure 10.4: a) A grey-scale plot in the plane $z = 0$ of the real part of the periodic function $f(x, y) = a + b\exp\{i[2\pi(u_x x + u_y y) - \phi]\}$. The spatial frequency vector **u** is (3.3,1.3,0), a = 0.6, b = 0.3, $\phi = -\pi/3$. b) Definition of the length and the azimuth of a grating vector **u** (two-dimensional example).

We have used in this expression the spatial frequency components (ν_x, ν_y) as the Fourier transformed coordinates with dimension m^{-1}. The sign of the argument of the complex exponential function in the integrand of Eq. (10.2.1) has been chosen in line with the sign convention of Eqs. (A.8.5) and (A.8.6) in Appendix A. The complex function F is commonly called the spatial spectrum of the object function f and $|F|^2$ the spatial power spectrum of the function f. In Fig. 10.4a we have plotted an elementary periodic function with a specific frequency (u_x, u_y). The definition of the two-dimensional part of the grating vector **u** is illustrated in Fig. 10.4b; the length of the vector is proportional to $1/p$ and its azimuth follows from $\arctan\theta = u_y/u_x$. Equation (10.2.1) shows that a *forward* Fourier transformation yields the spatial spectrum $F(\nu_x, \nu_y)$. An inverse or *backward* transformation according to

$$f(x, y) = \iint_{-\infty}^{+\infty} F(\nu_x, \nu_y) \exp[+2\pi i(\nu_x x + \nu_y y)]\, d\nu_x d\nu_y \tag{10.2.2}$$

yields the original spatial object function $f(x, y)$ which shows that any general object function can be represented by a spectrum of elementary periodic functions (plane waves in this context).

A general periodic component in a thin object, positioned in the plane $z = 0$ and superimposed on a constant background, is given by

$$f(x, y) = a + b\exp[2\pi i(u_x x + u_y y) - i\phi] . \tag{10.2.3}$$

The function $f(x, y)$ applies to the amplitude transmission or reflection function of the object. Its intensity transmission function is given by $|f(x, y)|^2$. The Fourier transform of $f(x, y)$ is given by

$$\begin{aligned} F(\nu_x, \nu_y) &= \iint_{-\infty}^{+\infty} \left\{a + b\exp[2\pi i(u_x x + u_y y) - i\phi]\right\} \exp[-2\pi i(\nu_x x + \nu_y y)]\, dxdy \\ &= a\,\delta(\nu_x, \nu_y) + b\exp(-i\phi)\,\delta(\nu_x - u_x, \nu_y - u_y) , \end{aligned} \tag{10.2.4}$$

where we have used the sifting property of the delta-function and its scaling property which is given by Eq. (A.2.5).

10.2.1.1 Spatial Frequency Components and Plane Wave Propagation

We consider a plane wave which is incident on a thin planar object, perpendicular to the z-axis. The object exhibits a periodic spatial modulation with a spatial frequency having the components $(u_x, u_y, 0)$. The incident plane wave is split up in diffracted plane waves with order numbers m. The plane wave with order number $m = 0$ is the basic refracted wave obeying Snell's law. When the object is a thin free-standing structure in a medium with a certain refractive index, the zero-order wave maintains the propagation direction of the incident wave. To find the propagation directions of the diffracted

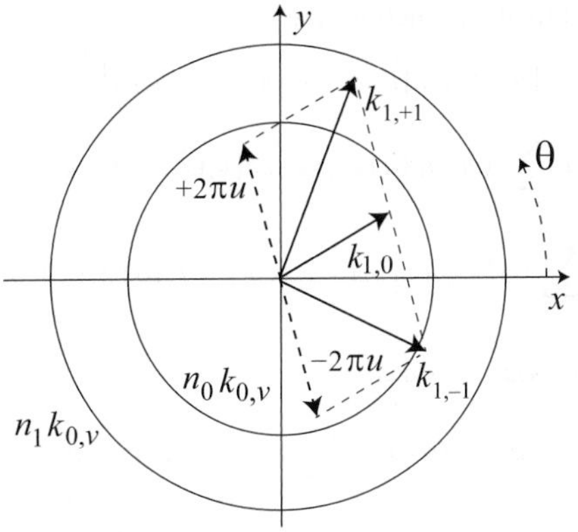

Figure 10.5: The (x, y) components of the plane wave vectors $\mathbf{k}_{1,0}$, $\mathbf{k}_{1,+1}$ and $\mathbf{k}_{1,-1}$, related to the zeroth and first diffracted orders.

plane waves with $m \neq 0$ we use the analysis of Subsection 5.7.3. Equation (5.7.12) gives the relationship between the in-plane components of the wave vectors on both sides of the interface. As the vector $\hat{\boldsymbol{\beta}}/p$ is parallel to the plane $z = 0$, the vector quantity $\hat{\boldsymbol{\beta}}/p$ has the components $(u_x, u_y, 0)$. In Eq. (5.7.12) we multiply the expressions for the components of the tangential unit vector on both sides of the modulated surface by k_1, the radial wavenumber in the second medium, and obtain

$$(k_{1,x}, k_{1,y}, 0) = (k_{0,x}, k_{0,y}, 0) + 2\pi m(u_x, u_y, 0) , \tag{10.2.5}$$

where m is the order number of the diffracted wave, $k_1 = n_1 k_{0,v}$, $k_0 = n_0 k_{0,v}$ and $k_{0,v}$ is the vacuum wavenumber of the light. Using polar coordinates in the plane of the object with azimuthal angle θ, we have the alternative expression

$$n_1(\cos\theta_1, \sin\theta_1) = n_0(\cos\theta_0, \sin\theta_0) + m\frac{\lambda_0}{p}(\cos\theta_u, \sin\theta_u) , \tag{10.2.6}$$

where λ_0 is the vacuum wavelength of the light and $p = 1/|u|$ is the period of the structure. Equation (10.2.5) yields the (x, y) components of the wave vector in the second medium. The z components follow directly from the dispersion relation of Eq. (1.6.62) which, for a real wave vector $\mathbf{k}$, reads $k_1^2 = n_1^2 k_{0,v}^2$. For the case that $k_{1,x}^2 + k_{1,y}^2 > n_1^2 k_{0,v}^2$, the z component of $\mathbf{k}_1$ becomes imaginary and its sign is chosen such that the plane wave amplitude decreases in the positive z-direction in the second medium.

In Fig. 10.5 we show wave vector projections of the incident wave and two diffracted waves in the plane of the object that contains the spatial frequency component with period $p = 1/u$. The projections in the object plane of two first-order wave vectors with indices $(1, +1)$ and $(1, -1)$ in the second medium (index n_1) have been shown. Both vector solutions in the object plane have a length that is smaller than $n_1 k_{0,v}$, the length of the wave vector in the second medium. This means that the two solutions correspond to freely propagating plane waves.

10.2.1.2 Periodic Amplitude and Periodic Phase Object

Special examples of a periodic object are the amplitude-only and the phase-only object. The amplitude-only object influences the modulus part of the complex amplitude of the transmitted (or reflected) light. The light transmitted by a phase-only object has a modified phase function. The complex amplitude transmission function $f_0(x, y)$ of a periodic amplitude object having a shift (x_s, y_s) with respect to the origin and its Fourier transform $F_0(v_x, v_y)$ are given by

$$\begin{aligned} f_0(x, y) &= a + b\cos\{2\pi[u_x(x - x_s) + u_y(y - y_s)]\} , \\ F_0(v_x, v_y) &= a\,\delta(v_x, v_y) + \frac{b}{2}\exp\{2\pi i(u_x x_s + u_y y_s)\}\,\delta(v_x + u_x, v_y + u_y) \\ &\quad + \frac{b}{2}\exp\{-2\pi i(u_x x_s + u_y y_s)\}\,\delta(v_x - u_x, v_y - u_y) . \end{aligned} \tag{10.2.7}$$

For a phase-only object the corresponding dual functions are

$$f_0(x,y) = a\exp\left[i\alpha\cos\{2\pi[u_x(x-x_s)+u_y(y-y_s)]\}\right]$$
$$= a\sum_{m=-\infty}^{+\infty}(i)^m J_m(\alpha)\exp\{im2\pi[u_x(x-x_s)+u_y(y-y_s)]\},$$
$$F_0(\nu_x,\nu_y) = a\sum_{m=-\infty}^{+\infty}(i)^m \exp\{2\pi im(u_x x_s+u_y y_s)\}J_m(\alpha)\,\delta(\nu_x+mu_x,\nu_y+mu_y), \tag{10.2.8}$$

where we have used the identity

$$\exp(ia\cos\theta) = \sum_{m=-\infty}^{+\infty}(i)^m J_m(a)\,\exp(im\theta)\,. \tag{10.2.9}$$

A frequently occurring case is the weak phase-only object with $\alpha \ll 1$ and in this case we use the approximating expression

$$f_0(x,y) = a\left[1+i\alpha\cos\{2\pi[u_x(x-x_s)+u_y(y-y_s)]\}\right],$$
$$F_0(\nu_x,\nu_y) = a\left\{\delta(\nu_x,\nu_y)+i\frac{\alpha}{2}\left[\exp\{2\pi i(u_x x_s+u_y y_s)\}\,\delta(\nu_x+u_x,\nu_y+u_y)+\exp\{-2\pi i(u_x x_s+u_y y_s)\}\,\delta(\nu_x-u_x,\nu_y-u_y)\right]\right\}. \tag{10.2.10}$$

The Fourier transforms of the pure amplitude object and the weak phase-only object differ with respect to the phase offset between the zero-frequency component and the first-order components. The first and zeroth order of the pure amplitude object are in phase or in phase opposition, depending on the sign of b/a. The first and zeroth orders of the weak phase object are in phase quadrature. A 'phasor' representation of the scalar optical disturbance illustrates these differences between pure amplitude and pure phase modulation (see Fig. 10.6). The modulus (amplitude) and phase of the optical disturbance are represented by the length and azimuth of a vector, commonly called a 'phasor'. The optical time frequency ω determines the rotation frequency of the phasor. In the case of temporally coherent light, all phasors rotate at the same frequency. In the upper row, the influence of an amplitude modulation (AM) of the incident light has been shown with a picture of the phasor for each displacement in the object plane over a quarter period, $p/4$, in a direction parallel to the spatial frequency vector **u**. The change in complex amplitude, given by the difference between the transmitted light amplitude and the unperturbed amplitude is in phase or in phase opposition with the incident light (vertical phasors). For phase modulation of the light (PM), the phasor azimuth is modulated and the difference vector, for a small enough value α of the phase modulation, is now perpendicular to the incident light phasors. It turns out that different imaging methods are needed for the optimum visualisation of amplitude-only or (weak) phase-only objects. We briefly mention Zernike's 'phase-contrast' imaging method for weak phase objects, as explained by him in his Nobel prize address in 1953 **[389]**. The essence of his method is the selective change of the phase of the non-diffracted zeroth-order light of an object by

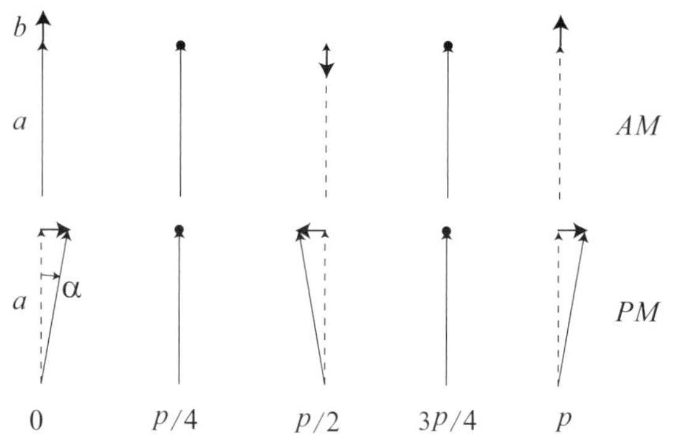

Figure 10.6: Amplitude-only (AM) and phase-only (PM) modulation of the incident light by an object. The amplitude and phase of the diffracted light with respect to the unperturbed light (vertical phasors) are represented by the thick arrows.

an amount of $\pm\pi/2$, for instance by the introduction of a phase-shifting plate **[387],[388]** in the central region (centred illumination source) or in an annular region (ring-shaped source) of the exit pupil (plane P_1 in Fig. 7.57) of a microscope objective. In comparison with the lower row of Fig. 10.6, the zeroth-order phasors of the weak phase object are rotated by an angle of 90 degrees. The interference of zeroth and first diffracted orders of the weak phase object is now identical to the interferences produced by a weak amplitude object and a (weak) intensity modulation is present in the (nominal) image plane. A refinement of the method consists of a reduction of the intensity of the zeroth-order light by means of a light-absorbing phase plate such that, at a lower average intensity level, the modulation depth of the intensity pattern in the image plane is increased.

10.2.2 Diffraction-based Optical Transfer Function

To obtain the value of the optical transfer function (*OTF*) we need to calculate the intensity modulation in the image of a periodic object. The *OTF* value then follows from the ratio of this intensity modulation in the image plane and the measured or calculated value of the intensity modulation of the light in a plane immediately beyond the object plane. To avoid confusion between the amplitude and intensity modulation of an object or an image produced by this object, we assume weakly modulated structures of the 'amplitude' type. The definition of a frequency transfer function based on the imaging of a weakly modulated phase object is not practical. It would require ad hoc definitions which depend on the type of phase object and on the special means employed in the optical system to visualise the particular phase object. In what follows we define the optical transfer function with respect to the imaging of weakly modulated amplitude objects. In the presence of such weakly modulated objects it is permissible to neglect the very small contributions from second and higher-order diffracted light. In consequence, the amplitude modulation and intensity modulation in the image plane simply differ by a factor of two.

10.2.2.1 Position of Diffraction Orders in Entrance and Exit Pupil

For the imaging process from source to object and to the entrance pupil, and then further through the optical system to the exit pupil and the image plane, we refer to Figs. 10.7 and 10.8. The illuminating beam originating at an infinitesimally small source element dS at (X_S, Y_S) is imaged by the condenser lens C onto the spherical entrance pupil surface, intersecting the point E_0 on the axis of the lens system $\mathcal{L}$. The centre of curvature of the entrance pupil sphere associated with the imaging of the point Q_0 is the point Q_0 itself. We select a small area in the object at Q_0, for instance by means of a field diaphragm, and position a periodic object at Q_0 with lateral spatial frequency components $(u_{x,0}, u_{y,0})$. The zeroth-order beam, labelled 0 in Figs. 10.7 and 10.8, has a unit wave vector with direction cosines $(L_{0,0}, M_{0,0}, N_{0,0})$. For the direction cosines of a diffracted beam which is created by the periodic object in Q_0, we have from Eq. (10.2.6),

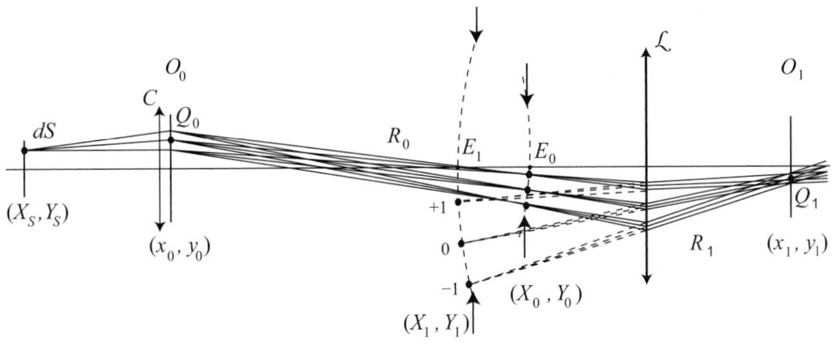

Figure 10.7: A source element dS illuminates the object plane. At the point Q_0 a periodic structure is present. The far-field pattern of the object area around Q_0 is projected onto the spherical entrance pupil through the axial point E_0 of the lens $\mathcal{L}$. From the entrance pupil the light propagates towards the spherical exit pupil which intersects the lens axis in E_1. Aberrations of the lens $\mathcal{L}$ and amplitude changes on propagation through the lens elements are included in the (complex) lens transmission function $f(X_1, Y_1)$. The far-field diffraction pattern on the exit pupil sphere gives rise to a periodic intensity pattern in the image region in the direct vicinity of the image point Q_1; $R_0 = E_0Q_0$, $R_1 = E_1Q_1$.

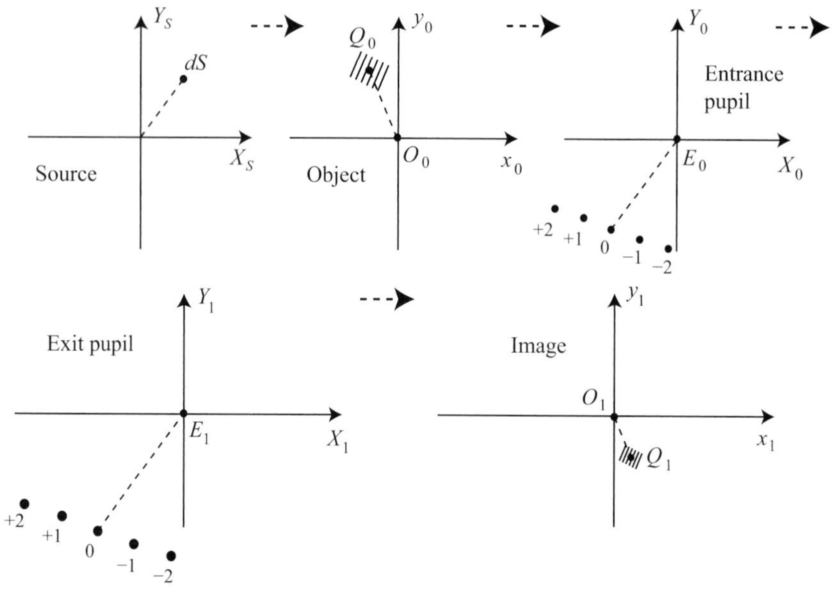

Figure 10.8: Propagation of light from the source plane to the object plane and then to the spherical entrance pupil in E_0. A periodic object is present in the immediate neighbouring region of the point Q_0. The diffracted orders are shown up to the second order. The exit pupil shows a comparable distribution of diffraction orders, the only difference being a scaling factor caused by the lateral pupil magnification $m_{T,p}$. The periodic image, demagnified in the picture, is formed in the vicinity of Q_1.

$$L_{0,m} = L_{0,0} + m\lambda_0 u_{x,0} \,, \qquad M_{0,m} = M_{0,0} + m\lambda_0 u_{y,0} \,, \tag{10.2.11}$$

where we have assumed that the refractive indices on both sides of the object plane are equal to n_0. The $N_{0,m}$ direction cosine follows from the dispersion relation in a medium with index n_0. For the case of a weakly modulated object at Q_0 we have that $|m| \leq 1$.

The point where the central (or principal) rays of the diffracted beams intersect the entrance pupil surface can be derived from Eq. (10.2.11). To this purpose we determine the points of intersection with a reference sphere which intersects the centre E_0 of the entrance pupil and is centred at the object point Q_0. The radius of this sphere is $E_0Q_0 = R_0$ and is negative in the chosen geometry of Fig. 10.7. In terms of the coordinates of the intersection points with the entrance pupil sphere we write Eq. (10.2.11) as follows,

$$-\left(\frac{X_{0,m} - x_0}{R_0}\right) + \left(\frac{X_{0,0} - x_0}{R_0}\right) = m\lambda_0 u_{x,0} \,,$$

or,

$$X_{0,m} = X_{0,0} - m\lambda_0 R_0 u_{x,0} \,,$$

and, similarly for the Y-coordinate,

$$Y_{0,m} = Y_{0,0} - m\lambda_0 R_0 u_{y,0} \,, \tag{10.2.12}$$

where the second subscript of an entrance pupil coordinate denotes the order number of the diffracted central ray. The coordinates $Z_{0,m}$ follow from the condition that each point of intersection is located on the entrance-pupil sphere, with its midpoint in the object point Q_0.

To find the exact intersection points of the central rays of the diffraction orders with the exit pupil sphere we use the Abbe sine condition which applies to any (well-corrected) lens system with a finite image field size. The Abbe sine condition (see also the detailed discussion of this condition in Subsection 4.8.1) states that points close to the optical axis are imaged with the same aberration as the on-axis point if the following condition holds,

$$\frac{X_0}{R_0} = m_T \frac{X_1}{R_1} , \tag{10.2.13}$$

where X_0 and X_1 are the intersection points of the marginal ray of a pencil of rays originating at the axial object point with the entrance and exit pupil spheres. We derive from Eq. (4.10.41) that the ratio $R_1/(m_T R_0)$ is the magnification of two conjugate planes that have distances R_0 and R_1 with respect to two conjugate planes with magnification m_T. Applied to the distances of entrance and exit pupil, measured from the conjugate planes O_0 and O_1, respectively, we conclude that $(-R_1)/(-m_T R_0)$ equals $m_{T,p}$, the transverse pupil magnification. Multiplying Eq. (10.2.12) by this transverse pupil magnification factor, we find

$$X_{1,m} = X_{1,0} - m\lambda_0 R_0 u_{x,0} \frac{R_1}{m_T R_0} = X_{1,0} - m\lambda_0 \frac{|m_T|}{m_T} R_1 \frac{u_{x,0}}{|m_T|} . \tag{10.2.14}$$

We introduce the spatial frequency components $(u_{x,1}, u_{y,1})$ in the image plane, equal to $(u_{x,0}, u_{y,0})/|m_T|$. We also define the quantity

$$R_1^{(s)} = -\frac{|m_T|}{m_T} R_1 = -\text{sgn}(m_T) R_1 , \tag{10.2.15}$$

where the sgn-function is the signum function. The definition of $R_1^{(s)}$ allows us to smoothly treat various object and pupil imaging geometries. We note that for the common imaging system with real object and image distances and pupil positions at or close to the principal points of the imaging system we find a positive value for $R_1^{(s)}$. Using the quantity $R_1^{(s)}$ in Eq. (10.2.14) we find the expressions

$$\begin{aligned} X_{1,m} &= X_{1,0} + m\lambda_0 R_1^{(s)} u_{x,1} , \\ Y_{1,m} &= Y_{1,0} + m\lambda_0 R_1^{(s)} u_{y,1} . \end{aligned} \tag{10.2.16}$$

10.2.2.2 Scalar Image Formation by the Diffracted Orders in the Exit Pupil

The amplitude distribution in the image plane through O_1 is obtained by the superposition of the three diffracted beams in the image region around Q_1, the paraxial image point of Q_0. Because of the small extent of the beams in the transverse direction we approximate them locally by plane waves. The relative phase of these beams is influenced by three factors:

- During the diffraction process at the grating in the object plane a phase difference ϕ_m may arise between the mth diffraction order and the zeroth order (we equate the relative phase of the zeroth order to zero). The relative strength of the diffracted beams with respect to the zeroth-order beam is also important. Combining the strength and the phase difference of a diffracted beam we define the complex amplitude of the transmitted mth-order beam as

$$t_m = a_m \exp(i\phi_m) . \tag{10.2.17}$$

- During their traversal of the optical system the diffracted beams follow different paths. If the system shows aberrations when an image is formed of the general point Q_0, the phase change of an mth diffracted beam follows from the wave-front aberration in the exit pupil according to

$$\Phi(X_{1,m}, Y_{1,m}) = k_{0,v} W(X_{1,m}, Y_{1,m}) , \tag{10.2.18}$$

  where $k_{0,v}$ is the vacuum wavenumber and $W(X_1, Y_1)$ is the wavefront aberration of a general point on the exit pupil sphere through E_1. Of course, the function W also depends on the coordinates (x_0, y_0) of the object point Q_0.
- It is possible, particularly in the case of oblique beams, that certain parts of a beam are cut off by the diaphragm or by other limiting apertures of the imaging system (vignetting).

We take all these effects into account by means of a complex exit pupil function $f(X_1, Y_1)$ which is given by

$$f(X_1, Y_1) = A(X_1, Y_1) \exp\{i\,\Phi(X_1, Y_1)\} . \tag{10.2.19}$$

For the complex amplitude of a diffracted wave with order number m in a point (x_1, y_1, z_1) at the nominal image plane image ($z_1 = 0$) or in a close neighbourhood of it we find

$$U_m(x_1, y_1, z_1) = t_m f(X_{1,m}, Y_{1,m}) \exp[i\,k_{0,v}\mathbf{r_1} \cdot \hat{\mathbf{s}}_{1,m}] , \tag{10.2.20}$$

where $\mathbf{r}_1$ is the displacement vector from the intersection point $(X_{1,m}, Y_{1,m}, Z_{1,m})$ of the mth diffracted order with the exit pupil sphere to the point (x_1, y_1, z_1) in image space and $\hat{\mathbf{s}}_{1,m}$ is the unit propagation vector of the central ray of the mth-order beam in image space. Using the local plane wave approximation for the (spherical) diffracted wave in the direct neighbourhood of the point Q_1, we find for the complex amplitude in the image plane

$$U_m(x_1,y_1) = t_m f(X_{1,m}, Y_{1,m}) \exp\Big[i\, k_{0,v} \Big\{L_{1,m}\big[x_1 - x_{1,Q_1} + (x_{1,Q_1} - X_{1,m})\big] + M_{1,m}\big[y_1 - y_{1,Q_1} + (y_{1,Q_1} - Y_{1,m})\big] + N_{1,m}(R_1^{(s)} - Z_{1,m})\Big\}\Big] , \tag{10.2.21}$$

where $L_{1,m}$, $M_{1,m}$ and $N_{1,m}$ are the direction cosines of the central ray of the mth-order beam. We also have that

$$\exp\Big\{i\, k_{0,v}\big[L_{1,m}(x_{1,Q_1} - X_{1,m}) + M_{1,m}(y_{1,Q_1} - Y_{1,m}) + N_{1,m}[R_1^{(s)} - Z_{1,m}]\big]\Big\} = \exp(ik_{0,v}R_1^{(s)}) , \tag{10.2.22}$$

and this yields the following expression for Eq. (10.2.21),

$$U_m(\bar{x}_1, \bar{y}_1) = t_m f(X_{1,m}, Y_{1,m}) \exp\Big\{i\, k_{0,v}\big(L_{1,m}\bar{x}_1 + M_{1,m}\bar{y}_1\big)\Big\} , \tag{10.2.23}$$

where the constant phase factor of Eq. (10.2.22) has been omitted and the shifted coordinates $(\bar{x}_1, \bar{y}_1)$ have been introduced according to

$$\bar{x}_1 = x_1 - x_{1,Q_1} , \qquad \bar{y}_1 = y_1 - y_{1,Q_1} . \tag{10.2.24}$$

The total amplitude in a general point with coordinates $(\bar{x}_1, \bar{y}_1)$ is obtained by summing over all (relevant) diffracted beams. The intensity I in a general point is then given by,

$$I(\bar{x}_1, \bar{y}_1) = \left| \sum_m U_m(\bar{x}_1, \bar{y}_1) \right|^2 . \tag{10.2.25}$$

The expression for I in Eq. (10.2.25) is substantially simplified if we assume that the relative strength of the first orders is small with respect to that of the zeroth order and that all other diffracted amplitude terms can be omitted in the summation. Equation (10.2.25) then reduces to

$$I(\bar{x}_1, \bar{y}_1) = \left| U_0(\bar{x}_1, \bar{y}_1) + U_{+1}(\bar{x}_1, \bar{y}_1) + U_{-1}(\bar{x}_1, \bar{y}_1) \right|^2 , \tag{10.2.26}$$

where the products of first orders can also be neglected. After this linearisation operation the intensity distribution in the image plane is given by

$$\begin{aligned} I(\bar{x}_1, \bar{y}_1; u_{x,1}, u_{y,1}) =& t_0^2 \left|f(X_{1,0}, Y_{1,0})\right|^2 + 2\Re\Big\{t_0\, t_{+1}^* f(X_{1,0}, Y_{1,0}) f^*(X_{1,0} + \lambda_0 u_{x,1} R_1^{(s)}, Y_{1,0} + \lambda_0 u_{y,1} R_1^{(s)}) \exp\big[2\pi i(u_{x,1}\bar{x}_1 + u_{y,1}\bar{y}_1)\big]\Big\} \\ &+ 2\Re\Big\{t_0\, t_{-1}^* f(X_{1,0}, Y_{1,0}) f^*(X_{1,0} - \lambda_0 u_{x,1} R_1^{(s)}, Y_{1,0} - \lambda_0 u_{y,1} R_1^{(s)}) \exp\big[-2\pi i(u_{x,1}\bar{x}_1 + u_{y,1}\bar{y}_1)\big]\Big\} . \end{aligned} \tag{10.2.27}$$

Equation (10.2.27) consists of a constant term and of two terms which show a periodicity in the image plane. The spatial frequency vector is given by $(u_x, u_y)/(|m_T|)$. From this property it follows that the period p_1 of the intensity distribution in the image plane is given by the period p_0 in the object plane multiplied by the magnification of the system. The azimuth of lines with constant intensity is identical in object and image plane. The 'phase' of the grating lines, which causes a shift of the pattern with respect to the reference point Q_1, is determined by the arguments of the complex factor t_m and the pupil function f.

Figure 10.8 shows where the satellite source images, which appear because of the presence of the periodic structure in the object plane at Q_0, are found in the entrance pupil. Taking into account the pupil magnification $m_{T,p}$, the relative positions of the source images (diffraction orders) in the exit pupil are identical to those in the entrance pupil. By mutual interference, the orders in the exit pupil generate an intensity pattern that, apart from the image magnification factor m_T, is a copy of the grating pattern in the object plane. The phase relation between the diffraction orders which existed after diffraction of the incident wave by the grating in the object plane changes during the transit through the optical system if it suffers from optical aberration. In addition, diffraction orders can be blocked by the physical stop (diaphragm) of the imaging system so that they cannot contribute to the image formation. These two effects cause a spatial shift and a reduction in modulation depth of the imaged periodic intensity pattern with respect to its original in the object plane.

When deriving expression (10.2.27) we limited ourselves to the interference effects between the zeroth and first orders. This approximation is only justified when we are dealing with so-called weakly modulated objects. It is essential that the first orders are relatively small and that the higher orders can be neglected. From Fig. 10.8 one can, for example, conclude that the interference pattern produced by the orders +1 and −1 will have a period which is half that of the basic pattern, which is produced by interference of the order pairs (0, +1) and (0, −1) (see Eq. (10.2.27)). In principle, the interference contribution (+1,−1) is always present and causes the image produced by an optical system to be nonlinear. If the amplitude of the first orders is sufficiently small, this term can be neglected and the intensity function in the image plane is considered to be a linear function of the amplitude modulation factor b/a of the object grating, defined by Eq. (10.2.4).

10.2.2.3 Influence of the Source Size

The expression (10.2.27) for the image intensity would suffice if the source indeed had very small dimensions. In practice, however, one often prefers to increase the source dimension, among others to obtain a sufficiently large irradiance in the image plane. For the case of, for instance, an incandescent lamp or a gas discharge tube, arbitrarily small parts of the radiating surface radiate in a fully uncoordinated manner. As there is no phase correlation whatsoever between elementary areas dS of the source, the *intensity* patterns in the image plane associated with each source element must be added. The summation of intensities implies that differences in optical path length of the various source elements to the point Q_0 are not important, as they only influence the phase in the point Q_0. Brightness variations in the source do have an effect. They give rise to a spatial source intensity weighting function when summing the individual intensity patterns related to each source element dS.

Using the magnification factors from the source to the entrance pupil ($m_{T,S}$) and from the entrance pupil to the exit pupil ($m_{T,p}$) we can write for the coordinates of the zeroth diffraction order on the exit pupil,

$$X_{1,0} = m_{T,S} m_{T,p} X_S \,, \qquad Y_{1,0} = m_{T,S} m_{T,p} Y_S \,. \tag{10.2.28}$$

The integration of expression (10.2.27) over an extended source then yields,

$$\begin{aligned} I(\bar{x}_1, \bar{y}_1; u_{x,1}, u_{y,1}) = & \, t_0^2 \iint_S |f(X_{1,0}, Y_{1,0})|^2 \, dX_{1,0} \, dY_{1,0} \\ & + 2\Re\Big\{ t_0 \, t_{+1}^* \exp\left[2\pi i(u_{x,1}\bar{x}_1 + u_{y,1}\bar{y}_1)\right] \\ & \times \iint_S f(X_{1,0}, Y_{1,0}) f^*(X_{1,0} + \lambda_0 u_{x,1} R_1^{(s)}, Y_{1,0} + \lambda_0 u_{y,1} R_1^{(s)}) \, dX_{1,0} \, dY_{1,0} \Big\} \\ & + 2\Re\Big\{ t_0 \, t_{-1}^* \exp\left[-2\pi i(u_{x,1}\bar{x}_1 + u_{y,1}\bar{y}_1)\right] \\ & \times \iint_S f(X_{1,0}, Y_{1,0}) f^*(X_{1,0} - \lambda_0 u_{x,1} R_1^{(s)}, Y_{1,0} - \lambda_0 u_{y,1} R_1^{(s)}) \, dX_{1,0} \, dY_{1,0} \Big\} , \end{aligned} \tag{10.2.29}$$

where the two-dimensional integration area S is determined by the dimensions of the projection of the source S onto the exit pupil, the magnification factor of which is given by Eq. (10.2.28). As mentioned previously, optical path differences in the space between the source and the object do not influence the shape of the intensity pattern in the image plane. This means that, in principle, the condenser lens may show large aberrations. The aberrations of the condenser lens can, however, influence the shape of the source imaged in the entrance pupil, and thus modify the effective integration range in Eq. (10.2.29) and the weighting function for the source intensity. We note that these latter effects have been omitted in Eq. (10.2.29).

10.2.3 Calculation of the Optical Transfer Function (*OTF*)

Before we calculate the optical transfer function it is necessary to point out what type of periodic object we are going to use, because this influences the amplitude and phase of the diffraction orders through the complex values of t_0, t_{+1} and t_{-1}. In addition, it is useful to have a graphic representation of the behaviour of the complex exponential functions which appear in the general expression (10.2.29) for the intensity in the image plane.

10.2.3.1 Amplitude and Phase of the Diffraction Orders

In our analysis the imaging process should be linear in intensity which requires a periodic object with a weak amplitude modulation. The spatial amplitude transmission function in the object plane is then given by

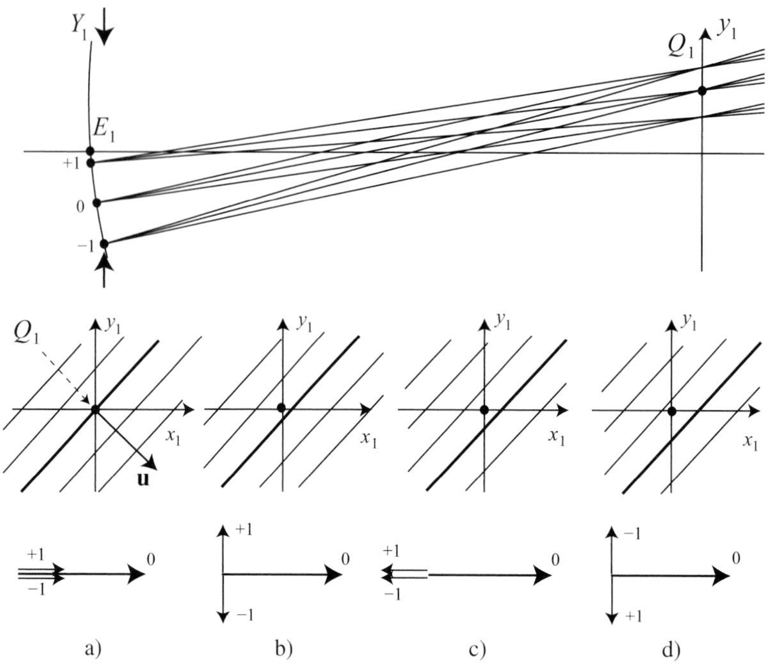

Figure 10.9: The image formation at Q_1 by the diffraction orders on the exit pupil sphere through E_1 (upper graph). Shifts of the periodic object at Q_0 are introduced, each time over a distance of $p/4$, parallel to the grating vector $(u_{x,0}, u_{y,0})$. The shifts in the image plane, multiplied by the paraxial magnification factor m_T, are shown in the middle row. In the lower row, four phasor graphs are shown, labelled a) to d), with the relative phase of the zeroth and first orders in the exit pupil each time as a function of the object shift at Q_0.

$$f_0(x_0, y_0) = 1 + \frac{b}{2}\cos\left[2\pi\left\{u_x(x_0 - x_{0,s}) + u_y(y_0 - y_{0,s})\right\}\right], \tag{10.2.30}$$

where b is real and $\ll 1$. The amplitude transmission function with modulation depth $b/2$ yields an intensity transmission with a two times larger modulation depth, equal to b. A translation of the periodic structure with respect to the origin O_0 in the object plane is taken into account by the vector $(x_{0,s}, y_{0,s})$. The complex amplitudes in the image plane of the diffraction orders associated with the periodic object function at Q_0 are given by Eq. (10.2.23). For an aberration-free optical system we put $f(X_1, Y_1) = \text{circ}\left\{(X_1^2 + Y_1^2)^{1/2}/\rho_1\right\}$ where $2\rho_1$ is the lateral diameter of the exit pupil. The function circ(r) is defined for $r \geq 0$ and equals 1 for $r \leq 1$ and zero elsewhere. For analysis of the interfering diffraction orders we consider the following three plane waves, interfering in the region around Q_1 (see Fig. 10.9). Their complex amplitudes are given by,

$$\begin{aligned} U_0(x_1, y_1, z_1) &= \exp\{ik_{0,v}N_{1,0}z_1\}, \\ U_{+1}(x_1, y_1, z_1) &= \frac{b}{4}\exp\{2\pi i[u_{x,1}(x_1 - x_{1,s}) + u_{y,1}(y_1 - y_{1,s})]\}\ \exp\{ik_{0,v}N_{1,+1}z_1\}, \\ U_{-1}(x_1, y_1, z_1) &= \frac{b}{4}\exp\{-2\pi i[u_{x,1}(x_1 - x_{1,s}) + u_{y,1}(y_1 - y_{1,s})]\}\ \exp\{ik_{0,v}N_{1,-1}z_1\}, \end{aligned} \tag{10.2.31}$$

where $x_{1,s} = m_T x_{0,s}$ and $y_{1,s} = m_T y_{0,s}$. Note that we have omitted from these expressions for the complex amplitudes U_m a phase factor $\exp\{i\,k_{0,v}(L_{1,0}x_1 + M_{1,0}y_1)\}$ which is common to all three diffracted waves and, consequently, does not affect the irradiance pattern in the image region. The z-direction cosines in Eq. (10.2.31) follow from Eq. (10.2.14),

$$N_{1,m}^2 = 1 - (L_{1,0} + m\lambda_0 u_{x,1})^2 - (M_{1,0} + m\lambda_0 u_{y,1})^2, \tag{10.2.32}$$

where $L_{1,0} = (x_1 - X_{1,0})/R_1^{(s)}$, $M_{1,0} = (y_1 - Y_{1,0})/R_1^{(s)}$ and $m = -1, 0, +1$.

Figure 10.9 schematically shows the position of the zeroth and first diffraction orders on the exit pupil sphere with radius of curvature R_1 and centre Q_1. According to Eq. (10.2.31) the three waves which originate from the periodic object are in phase at Q_1 if the object shift $(x_{0,s}, y_{0,s})$ is zero. In the case of a quarter period shift, the phase of the + 1st order has increased by $\pi/2$, while the phase of the −1st order has decreased by the same value. For each shifted object position we add the three complex amplitudes of the diffraction orders at Q_1 and obtain the irradiance in the image plane by calculating the squared modulus of this sum.

10.2.3.2 Definition of the Optical Transfer Function

Using the periodic object of Eq. (10.2.30) the values of t_0, t_{+1} and t_{-1} are real and $t_{+1} = t_{-1}$. In the general expression (10.2.29) for the image intensity the product terms $t_0 t_1^*$ and $t_0 t_{-1}^*$ can thus be placed in front of the $\Re$ sign. The strength of a transferred frequency component, after normalisation by the zero frequency component, is then given by

$$OTF(u_{x,1}, u_{y,1}) = \frac{\iint_S f(X_1, Y_1) f^*(X_1 + \lambda_0 u_{x,1} R_1^{(s)}, Y_1 + \lambda_0 u_{y,1} R_1^{(s)})\, dX_1 dY_1}{\iint_S |f(X_1, Y_1)|^2\, dX_1 dY_1}, \qquad (10.2.33)$$

where we have omitted, for ease of notation, the second subscript of the coordinates $(X_{1,0}, Y_{1,0})$ of the position of the zeroth order on the exit pupil sphere.

The name *OTF* for Optical Transfer Function is historically restricted to the transfer function that applies to the imaging of an incoherently illuminated object. The expression for the *OTF* of Eq. (10.2.33) is associated with the diffracted light with order number $m = +1$. We also need to take into account the transfer of spatial frequencies via the minus first-order diffracted wave, following from the third term in the right-hand side of Eq. (10.2.27). The complex amplitudes of both diffraction orders are real and equal in the case of the test amplitude grating of Eq. (10.2.30). The complex function $OTF(u_{x,1}, u_{y,1})$ can be written as

$$OTF(u_{x,1}, u_{y,1}) = MTF(u_{x,1}, u_{y,1}) \exp\left\{i\, PTF(u_{x,1}, u_{y,1})\right\}, \qquad (10.2.34)$$

where the *MTF* and *PTF* functions have the following properties,

$$MTF(u_{x,1}, u_{y,1}) = MTF(-u_{x,1}, -u_{y,1}), \quad PTF(u_{x,1}, u_{y,1}) = -PTF(-u_{x,1}, -u_{y,1}). \qquad (10.2.35)$$

In the case of a pupil function with circular symmetry, both with respect to its modulus and phase, the function *PTF* is identically zero. In the absence of circular symmetry, the intensity distribution in the image plane is in general given by

$$I(\bar{x}_1, \bar{y}_1; u_{x,1}, u_{y,1}) = 1 + b\, MTF(u_{x,1}, u_{y,1}) \cos\left\{2\pi(u_{x,1}\bar{x}_1 + u_{y,1}\bar{y}_1) + PTF(u_{x,1}, u_{y,1})\right\}. \qquad (10.2.36)$$

$MTF(u_{x,1}, u_{y,1})$ is generally called the modulation transfer function and $PTF(u_{x,1}, u_{y,1})$ the phase transfer function. In the limiting case, for low spatial frequencies, the modulation depth of the intensity distribution in the image plane is equal to b. This modulation depth is also expected on the basis of the assumed amplitude transmission function of the object with modulation depth $b/2$. The phase transfer function approaches zero at low spatial frequencies.

10.2.3.3 The Coherent Transfer Function

When the source is small and well approximated by a point source, the resulting illumination in the object plane can be referred to as spatially coherent. The phase of the light is not necessarily uniform in the object plane, but any phase differences between points in the object plane will be constant in time. If the object area at Q_0 is sufficiently small, the phase can be considered to be uniform. Using the values for t_0 and t_1 which follow from Eq. (10.2.31) we can determine the so-called coherent transfer function with the aid of Eq. (10.2.33). The abbreviation in the literature for this function is commonly *CTF*. We assume that the pupil function f is uniform and free of aberration, such that

$$\begin{cases} f(X_1, Y_1) = 1, & X_1^2 + Y_1^2 \le \rho_1^2 \\ f(X_1, Y_1) = 0, & \text{elsewhere} \end{cases}, \qquad (10.2.37)$$

where ρ_1 represents the radius of the exit pupil in the lateral dimension. After imaging by the condenser lens C and the imaging system $\mathcal{L}$ the point-like source is positioned at the point S with coordinates $(X_{1,S}, Y_{1,S})$ in the exit pupil (see Fig. 10.10a). The relation between the coordinate pairs in the source plane and the exit pupil are given by Eq. (10.2.28). The optical transfer function *CTF* for the case of the point source is then given by,

$$CTF(u_{x,1}, u_{y,1}) = f(X_{1,S}, Y_{1,S})\, f^*(X_{1,S} + \lambda_0 u_{x,1} R_1^{(s)}, Y_{1,S} + \lambda_0 u_{y,1} R_1^{(s)}). \qquad (10.2.38)$$

The *CTF* equals 1 as long as the imaged source point and the diffraction order in the exit pupil are found inside the exit pupil area, delimited by the pupil rim $(X_1^2 + Y_1^2) \le \rho_1^2$). The circular rim of the exit pupil and the position of the source

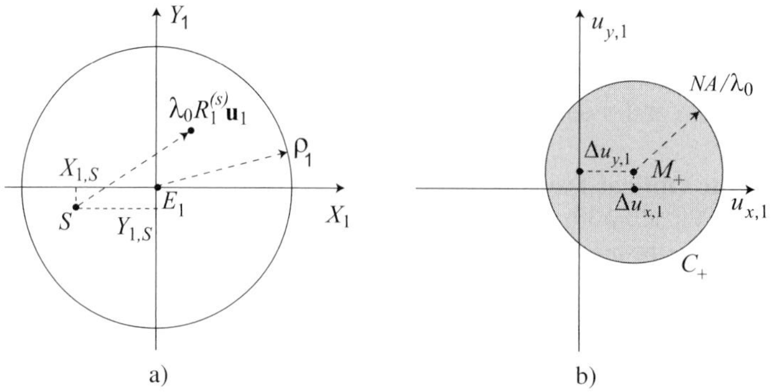

Figure 10.10: a) Diffracted orders in the exit pupil (point-source image at S).
b) The region of nonzero frequency transfer for the *CTF* in the spatial frequency plane. The central frequency of the circular region has been denoted by $(\Delta u_{x,1}, \Delta u_{y,1})$. The source coordinates $(X_{1,S}, Y_{1,S})$ in the exit pupil are both negative in the figure.

image are shown in Fig 10.10a. The vector $\lambda_0 R_1^{(s)} \mathbf{u}_1$ represents the shift of the diffraction order with $m = +1$. The zeroth order and the positive first diffracted order are transmitted both through the lens system if

$$(X_{1,S} + \lambda_0 u_{x,1} R_1^{(s)})^2 + (Y_{1,S} + \lambda_0 u_{y,1} R_1^{(s)})^2 \leq \rho_1^2 \,, \qquad X_{1,S}^2 + Y_{1,S}^2 \leq \rho_1^2 \,. \tag{10.2.39}$$

For the transmitted spatial frequencies with the aid of this plus first order we obtain, after division of Eq. (10.2.39) by $(\lambda_0 R_1^{(s)})^2$,

$$\left(u_{x,1} + \frac{X_{1,S}}{\lambda_0 R_1} \right)^2 + \left(u_{y,1} + \frac{Y_{1,S}}{\lambda_0 R_1} \right)^2 \leq \left(\frac{(NA)}{\lambda_0} \right)^2 , \tag{10.2.40}$$

with comparable equations for the frequencies that can be transmitted by the combination of the zeroth and the minus first diffraction order. The range of spatial frequencies that can be transmitted through the optical system by means of the zeroth and the +1 diffraction order has been grey-shaded in Fig. 10.10b. The range of spatial frequencies is represented by a disc-shaped area with its rim denoted by C_+, its centre point M_+ by $(\Delta u_{x,1}, \Delta u_{y,1}) = -(X_{1,S}, Y_{1,S}) \,/\, \lambda_0 R_1^{(s)}$, whilst the disc radius is given by NA/λ_0 where NA is the numerical aperture of the optical system at the image side.

Taking into account both the positive and negative diffraction orders, the intensity distribution in the image plane is given by Eq. (10.2.27). Using the object function $f_0(x_0, y_0)$ of Eq. (10.2.30) we find

$$\begin{aligned} I(\bar{x}_1, \bar{y}_1; , u_{x,1}, u_{y,1}) = 1 + \frac{b}{4} f(X_{1,S}, Y_{1,S}) \big[& f^*(X_{1,S} + \lambda_0 u_{x,1} R_1^{(s)}, Y_{1,S} + \lambda_0 u_{y,1} R_1^{(s)}) \exp\{2\pi i (u_{x,1}\bar{x}_1 + u_{y,1}\bar{y}_1)\} \\ & + f^*(X_{1,S} - \lambda_0 u_{x,1} R_1^{(s)}, Y_{1,S} - \lambda_0 u_{y,1} R_1^{(s)}) \exp\{-2\pi i (u_{x,1}\bar{x}_1 + u_{y,1}\bar{y}_1)\} \big] . \end{aligned} \tag{10.2.41}$$

The intensity function in the image plane is the superposition of the interference pattern of the zeroth and plus first order and the interference pattern of the zeroth and minus first order. The modulation depth of the first pattern is given by the *CTF* associated with the plus first diffraction order and that of the second pattern by the *CTF* related to the minus first diffraction order. For the case of an aberration-free imaging system the superposition of both patterns yields a (non-shifted) cosine function for the image intensity.

In Fig. 10.11a, for the case of a non-centred source, the area has been sketched where the summed transfer function is not equal to zero. It is composed of two circular regions C_+ and C_-, with their centres in M_+ and M_-, respectively. In the overlap area of the two circular regions the transfer function is 2 as both the +1st and the −1st order contribute constructively to the image formation. In the remaining area the transfer function equals 1. In this region *single-side-band imaging* occurs because only one diffraction order contributes to the image.

In Fig. 10.11b we have plotted the modulation depth in the image plane along the azimuthal line line *A-B* for the case of a decentred source. The maximum frequency is given by

$$u_{max} = \frac{NA}{\lambda_0} + \frac{\sqrt{X_{1,S}^2 + Y_{1,S}^2}}{\lambda_0 R_1^{(s)}} = \frac{NA + \sin \alpha_S}{\lambda_0} \,, \tag{10.2.42}$$

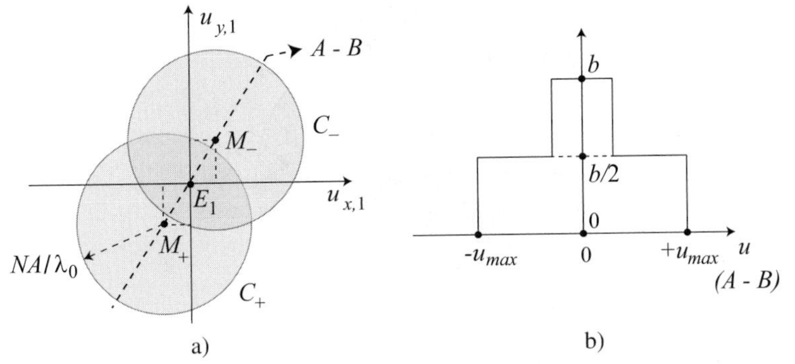

Figure 10.11: a) Spatial frequency transfer by the +1st and −1st diffraction orders in the case of off-axis illumination by a point source (the source coordinates $X_{1,S}$ and $Y_{1,S}$ are both positive in the figure).
b) The modulation depth of the imaged periodic pattern as a function of the spatial frequency u.

where we have introduced $\sin\alpha_S$, the sine of the offset angle α_S of the decentred point source. The maximum frequency which can be transferred is found in the azimuthal direction defined by the source decentring vector $E_1S = (X_{1,S}, Y_{1,S})$. A maximum value $|\mathbf{u}_{1,\max}| = 2NA/\lambda_0$ is possible if the source is also maximally decentred, exactly at the rim of the exit pupil. In this limiting case, the rims of the discs C_+ and C_- in Fig. 10.11a touch each other in the origin at E_1 and only single-sideband imaging is possible. Moreover, a periodic pattern with a spatial frequency vector that is normal to the line M_+M_- cannot be imaged when using such an extreme source position. In the reference situation with a centred source and an aberration-free system, the maximum transferable frequency equals NA/λ_0 and the modulation depth is independent of the orientation of the periodic object.

10.2.3.4 The Partially Coherent Transfer Function

For the case of a finite size of the illuminating source, the region around Q_0 in the object plane over which a constant phase relationship between the optical disturbances is maintained will be reduced in size. Outside this reduced area, the phase relation between an optical disturbance and the field in Q_0 has become random and we can effectively speak about incoherent illumination outside the restricted coherent area around Q_0. Because of the remaining coherency inside the restricted area around Q_0, this type of illumination is called *partially coherent* and we denote the corresponding transfer function by OTF_p. The classical van Cittert–Zernike theorem **[70]**, **[386]** about limited spatial coherence in a plane illuminated by an extended source is discussed in Subsection 10.2.5.

For a source with finite size, the image intensity is given by the general expression of Eq. (10.2.29), which, with respect to the $(\bar{x}_1, \bar{y}_1)$ dependent part, consists of two contributions. One contribution comes from the positive frequencies, and originates from the interference of a zeroth-order source point and its +1st-order satellite. The second contribution is associated with the negative frequencies. In Fig. 10.12 we illustrate how the interference contributions between zeroth-order source points associated with an extended source and their first-order satellites can be pictured as originating from the overlap region of two exit pupil discs, corresponding to the darker shaded area in the figure. Each zero-order source point in this overlap area contributes to the imaging of the periodic object by mutual interference in the image plane with its first-order satellite that is contained within the exit pupil rim. In the case of a +1st diffraction order, the distance between a (zeroth-order) source point and its plus first-order satellite is given by the vector $+\lambda_0 R_1^{(s)}\mathbf{u}$. In the figure two examples have been shown of a zero order and its +1st satellite. The zeroth-order A_0 and the plus first-order A_{+1} both pass without any obstruction through the exit pupil. For the couple B_0 and B_{+1}, the +1st order is located on the rim of the exit pupil and it constitutes a limiting case. All first-order satellite source points on the rim of the exit pupil originate from zero-order source points on the rim of the shifted disc D_{+1} with centre M_{+1}, defined by the displacement vector $M_{+1}E_1 = \lambda_0 R_1^{(s)}\mathbf{u}$. The other delimitation for the zero-order source points is the rim of the exit pupil itself. The useful area is thus given by the overlap area of the exit pupil function $f(X_1, Y_1)$ and its +1st shifted version, viz. the disc-shaped function $f(X_1 + \lambda_0 R_1^{(s)} u_{x,1}, Y_1 + \lambda_0 R_1^{(s)} u_{y,1})$ with its centre in $M_{+1} = -\lambda_0 R_1^{(s)}(u_{x,1}, u_{y,1})$. The argument about the overlap area is valid if the source image in the exit pupil covers the entire exit pupil. For the optical transfer function one then finds the expression of Eq. (10.2.33) in which the source integration area can be put equal to the disc given by the nonzero

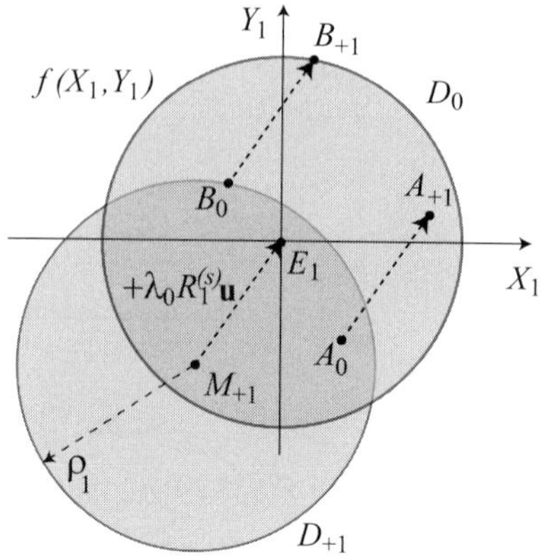

Figure 10.12: The pupil overlap region with nonzero contribution to the image formation by the zeroth and +1st order. The centre of the exit pupil is the point E_1.

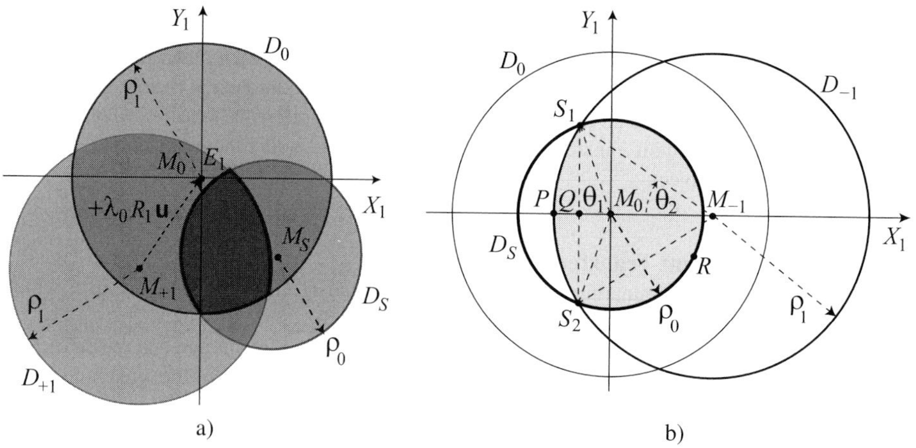

Figure 10.13: a) Exit pupil geometry with a decentred disc-shaped illuminating source with its centre in M_S.
b) Illustrating the calculation of the partially coherent transfer function OTF_p for a perfect imaging system with a centred source, which is proportional to the surface area of the shaded region.

region of the exit pupil function $f(X_1, Y_1)$. The (normalised) transfer function value which follows from the geometry of Fig. 10.12 is said to be the 'incoherent' one.

For a more general source geometry and in the presence of an aberration-free pupil function one can identify the transfer function at one specific spatial frequency $(u_{x,1}, u_{y,1})$ with the dark-shaded area of Fig. 10.13a. The discs associated with the pupil functions $f(X_1, Y_1)$ and $f(X_1 + \lambda_0 R_1^{(s)} u_{x,1}, Y_1 + \lambda_0 R_1^{(s)} u_{y,1})$ as they appear in the integrand of Eq. (10.2.33) have been denoted by D_0 and D_{+1}, respectively. The disc D_S represents the finite source projected onto the exit pupil. The transfer function is now given by the dark-shaded area in Fig. 10.13a, equal to the overlap area of the exit pupil disc and its shifted version, insofar as this common area is covered by the area of the projected source. For normalisation purposes, the area is normalised by the surface area of the disc D_S that is contained within the disc D_0.

In Fig. 10.13b a frequently occurring situation is schematically shown with a centred source (diameter $2\rho_0$) which is smaller than the diameter $2\rho_1$ of the exit pupil. The common area of the zeroth-order disc D_0 and the −1st-order disc D_{-1}, which is covered by the source image (disc D_S with radius ρ_0), is given by the shaded area, delimited by the points S_1, P, S_2 and R. For the coordinates of the intersection points S_1 and S_2 of the rims of the discs D_S and D_{-1} we find,

$$x_{S_1} = x_{S_2} = \frac{(\lambda_0 R_1^{(s)} u_1)^2 + \rho_0^2 - \rho_1^2}{2\lambda_0 R_1^{(s)} u_1} ,$$

$$y_{S_1} = -y_{S_2} = \sqrt{\rho_0^2 - x_{S_1}^2} , \qquad (10.2.43)$$

where $(\lambda_0 R_1^{(s)} u_1, 0)$ are the coordinates of the centre M_{-1} of disc D_{-1}. The disc segment S_1PS_2Q is given by $S_1PS_2M_{-1} - S_1QS_2M_{-1}$ and equals

$$S_1PS_2Q = \theta_2\rho_1^2 - (\lambda_0 R_1^{(s)} u_1 - x_{S_1})y_{S_1} , \qquad (10.2.44)$$

where $\theta_2 = \arccos[(\lambda_0 R_1^{(s)} u_1 - x_{S_1})/\rho_1]$. For the area S_1QS_2R on the disc D_S we find

$$S_1QS_2R = (\pi - \theta_1)\rho_0^2 - x_{S_1}y_{S_1} , \qquad (10.2.45)$$

where the angle θ_1 is given by $\arccos(-x_{S_1}/\rho_0)$. The total overlap area S_1PS_2R is then given by

$$\begin{aligned} S_1PS_2R &= \theta_2\rho_1^2 + (\pi - \theta_1)\rho_0^2 - \lambda_0 R_1^{(s)} u_1 y_{S_1} , && \rho_1 - \rho_0 \leq \lambda_0 R_1^{(s)} u_1 \leq \rho_1 + \rho_0 , \\ &= \pi\rho_0^2 , && 0 \leq \lambda_0 R_1^{(s)} u_1 < \rho_1 - \rho_0 , \\ &= 0 , && \lambda_0 R_1^{(s)} u_1 > \rho_1 + \rho_0 , \end{aligned} \qquad (10.2.46)$$

where the ranges for u_1 are determined by the existence of the intersection points S_1 and S_2. The OTF_p value for the aberration-free case is obtained by dividing the expressions of Eq. (10.2.46) by $\pi\rho_0^2$, the area of the source. For the particular case that the source fills the entire exit pupil ($\rho_0 = \rho_1$), we have $x_{S_1} = \lambda_0 R_1^{(s)} u_1/2$ and $\theta_1 = \pi - \theta_2$, yielding,

$$S_1PS_2R = 2\theta_2\rho_1^2 - \lambda_0 R_1^{(s)} u_1 \sqrt{\rho_1^2 - (\lambda_0 R_1^{(s)} u_1/2)^2} , \qquad (10.2.47)$$

where $\theta_2 = \arccos[\lambda_0 R_1^{(s)} u_1/(2\rho_1)]$. To obtain the OTF value, the quantity S_1PS_2R is divided by $\pi\rho_1^2$, the area of the exit pupil.

It is convenient to normalise the spatial frequency u_1 with respect to NA/λ_0 yielding the quantity $u_1' = u_1\lambda_0/NA = u_1\lambda_0 R_1^{(s)}/\rho_1$. In terms of this normalised spatial frequency, the optical transfer function for a partially coherent illuminated object ($\rho_0 < \rho_1$) is written as

$$\begin{aligned} OTF_p(u_1', 0) &= 1 , && 0 \leq u_1' < 1 - s , \\ &= \frac{1}{\pi}\left\{\frac{1}{s^2}\arccos\left(\frac{u_1'}{2} + \frac{1-s^2}{2u_1'}\right) + \arccos\left(\frac{1}{s}\left[\frac{u_1'}{2} - \frac{1-s^2}{2u_1'}\right]\right)\right. \\ &\quad \left. - \frac{u_1'}{s^2}\sqrt{\frac{1+s^2}{2} - \left(\frac{u_1'}{2}\right)^2 - \left(\frac{1-s^2}{2u_1'}\right)^2}\right\} , && 1 - s \leq u_1' \leq 1 + s , \\ &= 0 , && u_1' > 1 + s , \end{aligned} \qquad (10.2.48)$$

where we have introduced the (in)coherency factor $s = \rho_0/\rho_1$. The limiting case of incoherent illumination ($s \geq 1$) yields the expression

$$OTF(u_1', 0) = \frac{2}{\pi}\left\{\arccos\left(\frac{|u_1'|}{2}\right) - \frac{|u_1'|}{2}\sqrt{1 - \left(\frac{u_1'}{2}\right)^2}\right\} , \qquad -2 \leq u_1' \leq +2 . \qquad (10.2.49)$$

If the normalised spatial frequency $(u_1', 0)$ increases, the centre of the disc D_{-1} shifts outward along the line M_0M_{-1} and the shaded area decreases. The shaded area disappears and the OTF_p becomes zero if

$$u_1' > 1 + s , \quad \text{or} , \quad u_1 > \frac{NA + NA_S}{\lambda_0} , \qquad (10.2.50)$$

where s is the ratio of the source aperture NA_S and the image-side aperture NA. NA_S is the sine of the (half) aperture angle of the source image in the exit pupil as seen from the centre of the image plane.

For the transfer of high spatial frequencies it is therefore useful to increase the source aperture. The upper limit of the transmitted frequencies is given by

$$u_{1,\max} = \frac{2NA}{\lambda_0} , \tag{10.2.51}$$

corresponding to $NA_S \geq NA$ since in this range of source sizes the dimension of the source does not influence the size of the shaded area anymore. The common area of the discs D_0 and $D_{\pm 1}$ thus determines the value of OTF_p. If the dimensions of the projected source exceed those of the exit pupil, the value of OTF_p does not depend on the source size and one employs the name *incoherent* transfer function.

10.2.3.5 The Relation between the *OTF* and the Point-spread Function

According to Eq. (10.2.33) the *OTF* is the normalised autocorrelation function of the exit pupil function f. From the Fourier theory for linear systems it follows that the Fourier transform of the frequency response function (*OTF*) equals the impulse response of the system. For the case of the incoherent *OTF* we find the intensity impulse response

$$\mathcal{F}\left\{OTF(u_{x,1}, u_{y,1})\right\} = \frac{\mathcal{F}\left\{R_{ff}\left|_{-\lambda_0 R_1^{(s)}(u_{x,1}, u_{y,1})}\right.\right\}}{\mathcal{F}\left\{R_{ff}\left|_{(0,0)}\right.\right\}} = \frac{\left|F\left(-\frac{x_1}{\lambda_0 R_1^{(s)}}, -\frac{y_1}{\lambda_0 R_1^{(s)}}\right)\right|^2}{|F(0,0)|^2} . \tag{10.2.52}$$

The function R_{ff} denotes the autocorrelation function (positive shift) of the pupil function $f(X_1, Y_1)$, as defined in Appendix A, Table A.2. The symbol $\mathcal{F}$ stands for the Fourier transform operation and $F(x_1, y_1)$ represents the amplitude of the point-spread function. We note that the minus sign in the argument of the function F in Eq. (10.2.52) stems from the definition of the forward Fourier transform in Appendix A, Section A.9, and the sign of the argument of the complex exponential function in the kernel of the diffraction integral for light propagation from exit pupil to image plane (see also the discussion on this subject in Section 9.3). The expression above applies to the real-space coordinates (X_1, Y_1) and (x_1, y_1) on the exit pupil sphere and in the image plane, respectively. The same expression in terms of the normalised coordinates (X_1', Y_1') on the exit pupil and (x_1', y_1') in the image plane, where

$$(X_1', Y_1') = \frac{(X_1, Y_1)}{\rho_1} , \qquad (x_1', y_1') = \frac{(x_1, y_1)}{\lambda_0/NA} , \tag{10.2.53}$$

is given by

$$\mathcal{F}\left\{OTF(u_{x,1}', u_{y,1}')\right\} = \frac{\mathcal{F}\left\{R_{ff}\left|_{(-u_{x,1}', -u_{y,1}')}\right.\right\}}{\mathcal{F}\left\{R_{ff}\left|_{(0,0)}\right.\right\}} = \frac{|F(-x_1', -y_1')|^2}{|F(0,0)|^2} . \tag{10.2.54}$$

This expression was obtained by using the pupil function $f(X_1', Y_1')$ with the scaled, normalised arguments (X_1', Y_1') and the image plane amplitude function $F(x_1', y_1')$ with scaled arguments (x_1', y_1'). The normalised spatial frequency components $(u_{x,1}', u_{y,1}')$ are given by $(u_{x,1}, u_{y,1})\lambda_0 R_1^{(s)}/\rho_1$, where $\rho_1/R_1^{(s)}$ is the sine of the angle $\alpha_{1,\max}$ between the marginal ray and the principal ray of an imaging pencil, in accordance with Eq. (10.2.53). The numerical aperture is then given by $NA = n_1 \sin\alpha_{1,\max}$ where n_1 is the refractive index of the medium in image space. In general, the Fourier transform of the pupil function f equals the amplitude distribution F of the diffraction image of the optical system (see also Section 9.3). The impulse response of Eq. (10.2.54) thus represents the (normalised) intensity distribution of the diffraction image. We observe that for sufficiently large source dimensions ($NA_S \geq NA$) the imaging system can be considered to be linear in intensity, provided that the modulation depth of the object is sufficiently small. If this is not the case, linearity in intensity formally only occurs if the projected source is 'infinitely' large with respect to the exit pupil area.

In Fig. 10.14 the transfer function and the image plane point-spread function are shown for the two extreme cases of coherent or incoherent illumination of the object. The figure illustrates the inverse relationship that exists between the width of the spatial spectrum (left-hand graphs) and the lateral extent of the impulse response function (right-hand graphs). Figure 10.14b shows the strong oscillations of the coherent impulse response which are a consequence of the sharp boundary at $|u_1'| = 1|$ in the top-hat function of Fig. 10.14a. This ringing effect in the impulse response is much less in d) for the incoherent case. The corresponding *OTF* function with the appearance of a 'Chinese hat' approaches zero at $u_1' = 2$ in a smooth way, without a discontinuity in the function itself or in its first derivative.

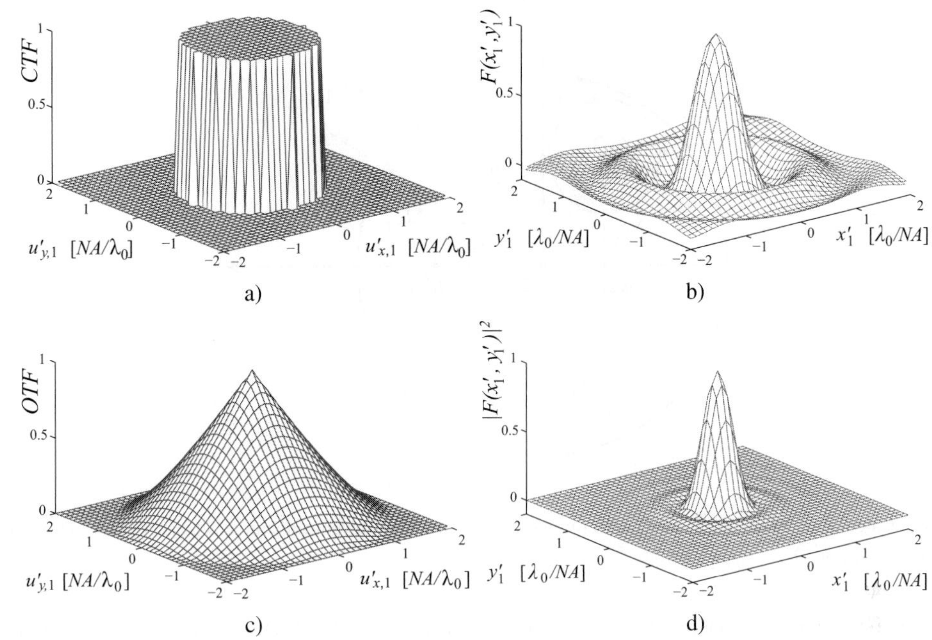

Figure 10.14: a) and b) The transfer function *CTF* for coherent object illumination and the corresponding amplitude impulse response or point-spread function in the image plane.
c) and d) The transfer function *OTF* for incoherent illumination of the object and the intensity point-spread function $|F|^2$.

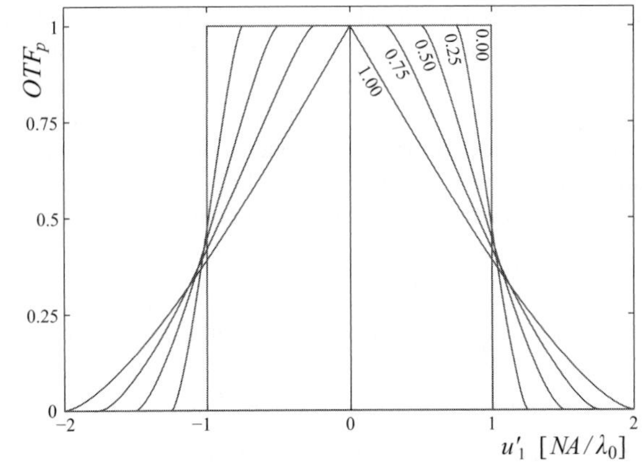

Figure 10.15: Various examples of partially coherent *OTF* functions as a function of the spatial frequency u'_1 in normalised units NA/λ_0 (centred source). The parameter for each curve is s, the ratio of the projected source diameter $2\rho_0$ in the exit pupil and the lateral size $2\rho_1$ of the exit pupil.

10.2.3.6 Examples of Partially Coherent Transfer Functions

An important parameter for the value of OTF_p is the position of the source in the exit pupil and its relative size, given by the parameter s. The source size modifies the height and width of the OTF_p. Figure 10.15 illustrates the influence of the size s of a *centred* source on the OTF_p for an otherwise perfect imaging system. A number of transfer functions are shown in which the source size s varies from 0 to 1 in steps of 0.25. The coherently illuminated object with $s = 0$

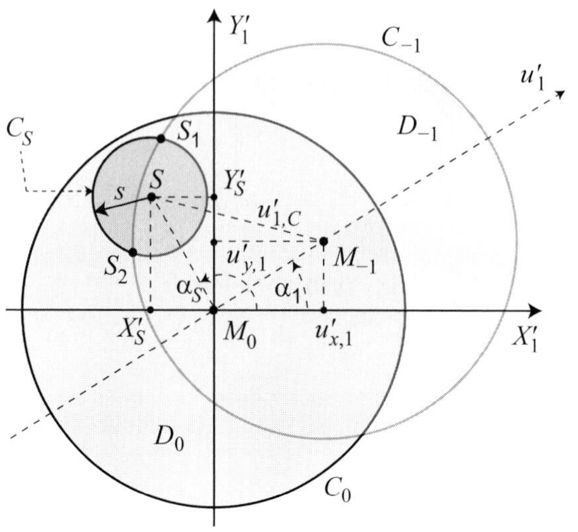

Figure 10.16: Calculation of the OTF_p of a perfect imaging system with a decentred source S (relative size s) for a general frequency vector $(u'_{x,1}, u'_{y,1})$.

yields the limiting case of the coherent CTF of a centred source which was discussed in the previous subsection. The transfer is maximum but limited to a range of $\pm NA/\lambda_0$. A larger source size increases the maximum frequency that can be transferred but reduces the strength of the transferred frequencies below NA/λ_0. For the limiting case where the source has an equal or larger size than the exit pupil we find the so-called *incoherent OTF* with a cut-off frequency of $2NA/\lambda_0$. This maximum frequency is a basic limitation for any linear optical system. It corresponds to the maximum diffraction angle in the imaging system. In this case, a zeroth-order source point touches the exit pupil rim and its first-order diffracted satellite touches the pupil rim at a diametrically opposite position.

The OTF_p can be further modified by using a decentred source, of which the projected area is entirely contained within the exit pupil disc. The OTF_p function of a decentred source is derived from that of a centred source with the aid of Fig. 10.16. The source with diameter $2s$ has its midpoint in S with (normalised) coordinates (X'_S, Y'_S). The azimuth of the source is given by the angle α_S. The frequency axis along which the transfer has to be obtained is the axis M_0M_{-1}, at an angle α_1 with the X'_1-axis. A general frequency vector has components $(u'_{x,1}, u'_{y,1})$ and a length denoted by u'_1. Four special cases arise for the frequency transfer, each time the circles C_{-1} and C_S touch each other as a function of the position of M_{-1} on the u'_1-axis. The condition for an internal or external contact point of both circles reads,

$$(u'_{x,1} - X'_S)^2 + (u'_{y,1} - Y'_S)^2 = (1 \pm s)^2 \, . \tag{10.2.55}$$

Using the angles α_1, α_S and the length $r'_S = \sqrt{X'^2_S + Y'^2_S}$ we have the following solutions for the four frequencies,

$$u'_1 = r'_S \cos(\alpha_1 - \alpha_S) \pm \sqrt{(1 \pm s)^2 - r'^2_S \sin^2(\alpha_1 - \alpha_S)} \, . \tag{10.2.56}$$

To obtain the frequency transfer we define the following four frequencies,

$$\begin{aligned}
u'^{(1)}_1 &= -\sqrt{(1+s)^2 - r'^2_S \sin^2(\alpha_1 - \alpha_S)} + r'_S \cos(\alpha_1 - \alpha_S) \, , \\
u'^{(2)}_1 &= -\sqrt{(1-s)^2 - r'^2_S \sin^2(\alpha_1 - \alpha_S)} + r'_S \cos(\alpha_1 - \alpha_S) \, , \\
u'^{(3)}_1 &= \sqrt{(1-s)^2 - r'^2_S \sin^2(\alpha_1 - \alpha_S)} + r'_S \cos(\alpha_1 - \alpha_S) \, , \\
u'^{(4)}_1 &= \sqrt{(1+s)^2 - r'^2_S \sin^2(\alpha_1 - \alpha_S)} + r'_S \cos(\alpha_1 - \alpha_S) \, .
\end{aligned} \tag{10.2.57}$$

These four frequencies are symmetrically positioned with respect to zero frequency if the frequency axis u'_1 and the line segment M_0S are orthogonal. In that case, for a perfect optical system, the OTF_p is symmetric with respect to zero frequency. In all other cases, an off-axis source position gives rise to an asymmetric OTF_p function.

For the calculation of the OTF_p for an arbitrary value of u'_1 we use the distance $M_{-1}S$ and call it the spatial frequency $u'_{1,C}$, defined by

$$u'^{\,2}_{1,C} = u'^{\,2}_1 + r'^{\,2}_S - 2u'_1 r'_S \cos(\alpha_1 - \alpha_S) \,. \qquad (10.2.58)$$

We observe that the geometry of the source disc S and the shifted pupil disc D_{-1} is identical with that of Fig. 10.13b and that it is thus permitted to use Eq. (10.2.48), which is associated with a centred source. For the OTF_p we then obtain the following expressions

$$\begin{aligned}
OTF_p(u'_{x,1}, u'_{y,1}) &= 1 \,, \qquad u'^{(2)}_1 \leq u'_1 \leq u'^{(3)}_1 , \\
OTF_p(u'_{x,1}, u'_{y,1}) &= \frac{1}{\pi}\left\{\frac{1}{s^2}\arccos\left(\frac{u'_{1,C}}{2} + \frac{1-s^2}{2u'_{1,C}}\right) + \arccos\left(\frac{1}{s}\left[\frac{u'_{1,C}}{2} - \frac{1-s^2}{2u'_{1,C}}\right]\right)\right. \\
&\quad \left. - \frac{u'_{1,C}}{s^2}\sqrt{\frac{1+s^2}{2} - \left(\frac{u'_{1,C}}{2}\right)^2 - \left(\frac{1-s^2}{2u'_{1,C}}\right)^2}\right\} , \\
&\qquad u'^{(1)}_1 < u'_1 < u'^{(2)}_1 , \quad u'^{(3)}_1 < u'_1 < u'^{(4)}_1 , \\
OTF_p(u'_{x,1}, u'_{y,1}) &= 0 \,, \qquad u'_1 \leq u'^{(1)}_1 , \quad u'_1 \geq u'^{(4)}_1 .
\end{aligned} \qquad (10.2.59)$$

These expressions are now applied to one or several decentred source images in the exit pupil of the imaging system. In Fig. 10.17a we have plotted four source images S_1 to S_4, arranged on the quadrant diagonals in the entrance and exit pupil. This illumination geometry is called *quadrupole* illumination. Figure 10.17c shows the frequency transfer for spatial frequencies along the X'_1-axis ($\alpha_1 = 0$) and for a frequency axis at an angle of 45° with respect to the X'_1-axis.

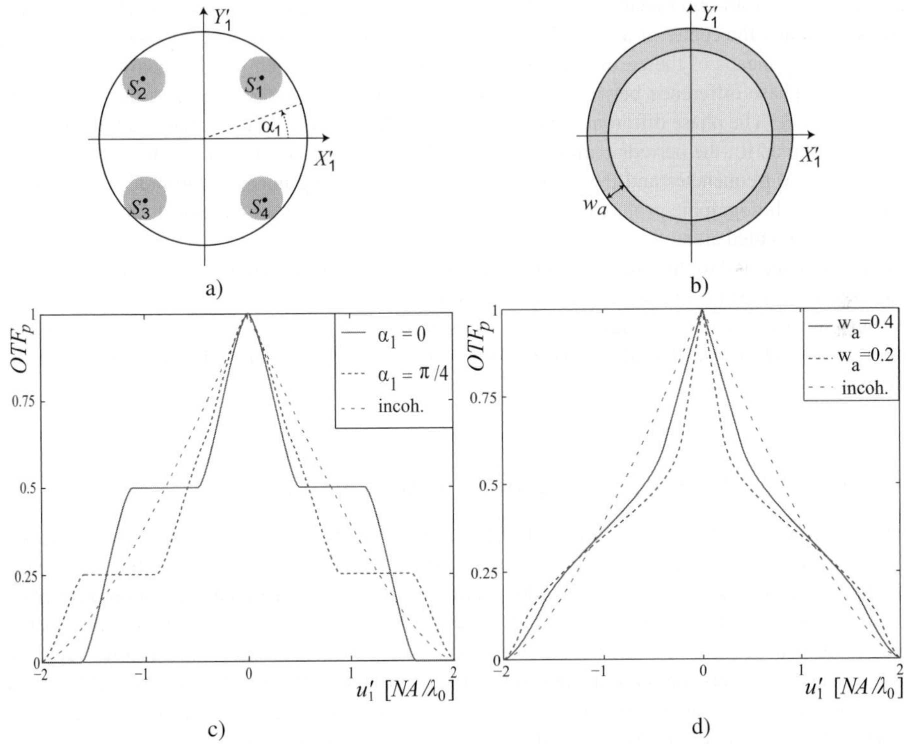

Figure 10.17: a) The four source elements (diameter $2s = 0.40$) of the quadrupole illumination.
b) Annular illumination with an annulus width $w_a = 0.2$.
c) 'Quadrupole' illumination ($s = 0.20$) and the corresponding OTF_p for the frequency azimuths $\alpha_1 = 0$ and $\pi/4$.
d) OTF_p for illumination with a ring-shaped source with annulus width w_a. The incoherent OTF has been plotted for comparison.

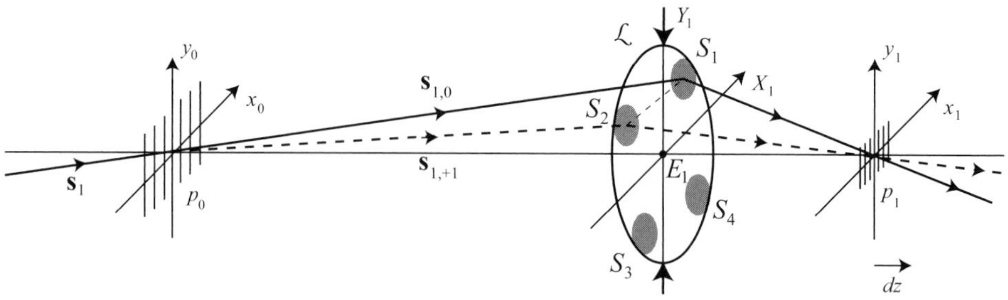

Figure 10.18: The central ray $\mathbf{s}_{1,0}$ of the zeroth-order wave (solid line) and $\mathbf{s}_{1,+1}$ of the first-order diffracted wave (dashed line) of a periodic structure in the object plane. The central ray of the incident beam $\mathbf{s}_1$ is directed towards the centre of a projected source element S_1 in the exit pupil (quadrupole illumination system). In the drawing, the imaging lens $\mathcal{L}$, the entrance pupil and the exit pupil are coincident and each intersects the optical axis at the point E_1.

The advantage of the quadrupole illumination with a typical value of $s = 0.2$ is the enhancement of the OTF_p for the medium and higher spatial frequencies, especially in the interval $1.0 < u_1' < 1.5$, both in the X_1'- and Y_1'-direction. For object structures with predominant features along these two axes, the quadrupole illumination offers a serious advantage as compared to classical incoherent illumination. Another advantage, not directly visible from the OTF_p graphs, is the relative insensitivity to defocusing of the image plane. In Fig. 10.18 we show an incident illumination beam directed to source S_1 of a quadrupole illumination system. A periodic structure with grating vector $1/p_0$, parallel to the x_0-axis, is imaged by means of the zeroth-order beam $s_{1,0}$ and its first-order satellite beam $s_{1,+1}$. The period p_0 is such that the first-order beam passes through the centre of a second source image S_2. In image space, these two beams create the periodic image with period p_1 by means of the zeroth- and first-order beam that are each other's mirror images with respect to the plane $x_1 = 0$. The phase difference between these two beams remains unchanged when the image plane is axially shifted over a distance dz. The phase difference changes slightly when the period p_0 is changed but a reduced sensitivity to defocus blur is observed for the periods p that are close to the optimum period p_0. The combined effects of enhanced transfer of higher spatial frequencies and the large tolerance for defocusing make quadrupole illumination a preferred method for high-quality lithographic projection systems. The optimum effects are observed for periodic structures with their frequency vector oriented along the y_0- or x_0-axis.

Another illumination method for the enhancement of higher spatial frequencies is annular illumination, see Fig. 10.17b. The effects are less pronounced than for the quadrupole illumination but in the region $1.5 < u_1' < 2$ an enhancement of a factor of 2 is realised when an annulus width w_a of 0.2 is chosen. Both in the case of quadrupole and annular illumination, extra requirements are put on the illumination system that preferably concentrates all source energy in the prescribed regions of the exit pupil.

10.2.4 The Incoherent Transfer Function of Aberrated Systems

When aberrations are present and the pupil function of Eq. (10.2.19) is complex, the incoherent OTF cannot simply be set equal to the geometric overlap area of two shifted pupil functions and the source area. The interference patterns which are formed in the image plane by the diffraction orders belonging to each source point have a nonzero phase difference which is caused by the spatially varying argument of the complex exit pupil function. The periodic pattern produced by each source point is then subjected to a lateral shift in the image plane, parallel to the grating vector. Due to these aberration-induced shifts, the summation of the intensity patterns produced by all source points will lead to a decrease in modulation depth of the periodic image and, for larger aberrations, the pattern may be wiped out or even re-appear with an inversion of the original contrast due to a phase shift of the cumulative intensity pattern by an odd number of π radians. Semi-analytic expressions have recently been given by Janssen **[163]** to compute the incoherent OTF for a general pupil function and source position, based on Hopkins' diffraction theory of optical imaging **[142]**. For more general pupil and source geometries one has recourse to numerical integration of the integral expressions in Eq. (10.2.33) to obtain the value of the OTF for a particular pair of frequency components $(u'_{x,1}, u'_{y,1})$. For the incoherent OTF it is possible to establish a general relation between the loss in frequency transfer and the magnitude of the aberration of the imaging system.

We will consider the case of aberrations that are small with respect to the wavelength of the light (diffraction-limited systems) and the case of large aberrations for which the ray-optics approximation of wave propagation can be used.

10.2.4.1 Weakly Aberrated Imaging Systems

The OTF_W of an aberrated system can be written as the sum of the aberration-free OTF and an increment according to

$$OTF_W(u'_{x,1}, u'_{y,1}) = OTF(u'_{x,1}, u'_{y,1}) + \Delta OTF(u'_{x,1}, u'_{y,1}) , \tag{10.2.60}$$

where ΔOTF has a small value with respect to the aberration-free OTF value if the imaging system is weakly aberrated. To establish a relationship between the aberrated OTF_W and the point-spread image we perform an inverse Fourier transformation (see Eq. (A.8.6)) of OTF_W with the aid of Eq. (10.2.52),

$$\iint \left[OTF(u'_{x,1}, u'_{y,1}) + \Delta OTF(u'_{x,1}, u'_{y,1})\right] \exp\left\{2\pi i(u'_{x,1}x'_1 + u'_{y,1}y'_1)\right\} du'_{x,1}du'_{y,1} = I_0(x'_1, y'_1) + \Delta I(x'_1, y'_1) , \tag{10.2.61}$$

where we have used normalised coordinates in both the exit pupil and the image plane. I_0 is the (normalised) intensity of the non-aberrated diffraction image and ΔI is the change of the normalised intensity as a result of the presence of aberration. For the centre of the diffraction image the expression of Eq. (10.2.61) yields

$$\iint \Delta OTF(u'_{x,1}, u'_{y,1})\, du'_{x,1}du'_{y,1} = \Delta I(0,0) . \tag{10.2.62}$$

An expression for $\Delta I(0,0)$ can be derived in terms of the variance of the wavefront aberration **[259]**, without a need to specify the exact type of aberration **[236]**. The relationship between the (normalised) Strehl intensity I_S, given by $I_S = 1 - \Delta I(0,0)$, and the wavefront variance was given in Eqs. (9.3.21) and (9.3.22). With the aid of Eq. (10.2.62) above, a connection has been established between the change in value of the OTF, averaged over all spatial frequencies, the Strehl intensity I_S **[331]** of the diffraction image and the rms value of the wavefront aberration. Even though Eq. (10.2.62) has general validity, it only makes sense to use this expression for sufficiently small values of $\Delta I(0,0)$, e.g. ≤ 0.20. The corresponding rms wave-front error is then $0.071\lambda_0$ at the most, the peak-to-valley wave-front error on the exit pupil sphere in image space is of the order of $\lambda_0/4$, and the imaging system is said to be *just diffraction-limited* (see Subsection 9.3.1.2, for the definition of 'just' diffraction-limited imaging quality). It will be seen that the decrease of the OTF is not equal for all frequencies. As the spatial frequency increases, the aberration of the system will become more pronounced in the phase difference between the image-forming diffraction orders. On the other hand, the overlap area from which contributions to the image intensity are accumulated is smaller and the difference in shift between the contributing interference patterns from the overlap area becomes smaller.

10.2.4.2 The *OTF* in the Presence of Defocusing

The degradation of the OTF under the influence of an imaging defect is illustrated by means of a defocusing of the image plane. For a defocused system the (normalised) aberration function W' in normalised pupil coordinates is given by

$$W'(X'_1, Y'_1) = \alpha_{20}\,(X'^2_1 + Y'^2_1) , \tag{10.2.63}$$

where α_{20} equals the wavefront deviation of the defocused wavefront at the rim of the exit pupil in units of the wavelength of the light. If the frequency vector along the X'_1-axis has the value $(u'_{x,1}, 0)$ the maximum width w of the overlap area is given by $2 - u'_{x,1}$ (see Fig. 10.19a). For this frequency vector we calculate the phase difference $\Delta\Phi = \Phi_{-1} - \Phi_0$ (see Fig. 10.9) between the -1st-order diffracted wave of a source point and the 0th-order diffracted wave,

$$\Delta\Phi = \Phi_{-1} - \Phi_0 = 2\pi\,\alpha_{20}\, u'_{x,1}(u'_{x,1} - 2X'_1) . \tag{10.2.64}$$

The maximum value of $\Delta\Phi$ is found at the extreme points A and B of the overlap region in the exit pupil (see Fig. 10.19a),

$$\Delta\Phi_A = -\Delta\Phi_B = 2\pi\,\alpha_{20}\, u'_{x,1}(2 - u'_{x,1}) . \tag{10.2.65}$$

$|\Delta\Phi_A|$ and $|\Delta\Phi_B|$ are maximum if the spatial frequency equals

$$u'_{x,1} = 1 , \quad \text{or,} \quad u_{x,1} = \frac{NA}{\lambda_0} . \tag{10.2.66}$$

This result corresponds to the well-known phenomenon that the so-called medium frequencies are most affected by defocusing and, more generally, by aberration of the imaging system. In the vicinity of the frequency 0 and the cutoff frequency $2NA/\lambda_0$ the influence of aberration is negligible in first order as only the curvature of the OTF function is altered in those regions.

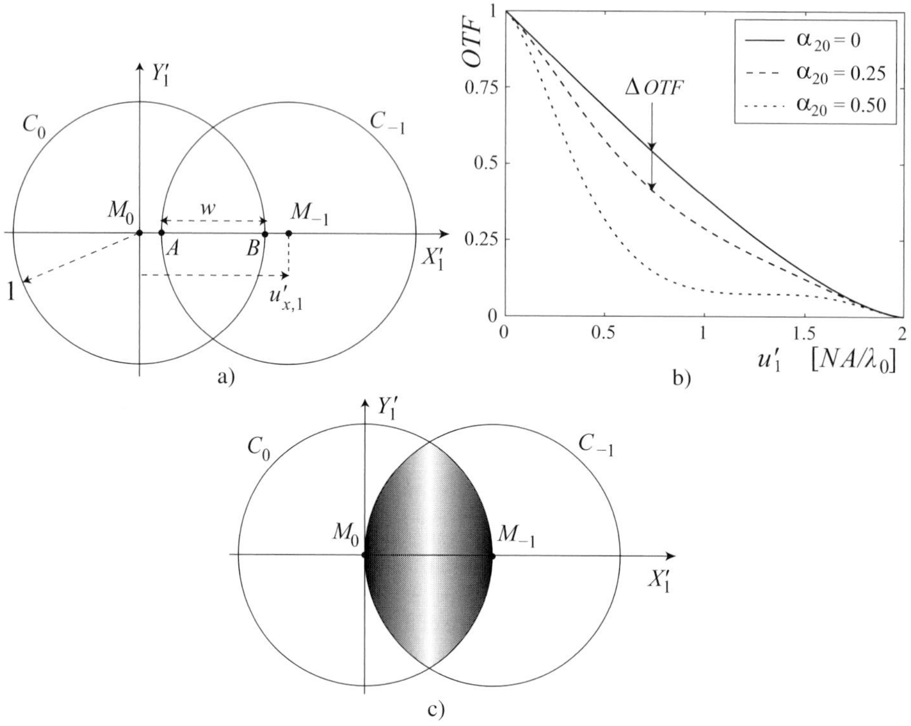

Figure 10.19: a) The overlap area of the zeroth-order disc with rim C_0 and the shifted disc with rim C_{-1} when imaging a periodic object with normalised spatial frequency vector $(u'_{x,1}, 0)$.
b) The modulus of the *OTF* of a perfect system in the optimum focal plane (solid curve) and in two defocused planes with a defocusing of one and two Rayleigh focal depths (dashed and dotted curve, respectively).
c) A plot of the product function $\Re\{f(X_1', Y_1') f^*(X_1' - u'_{x,1}, Y_1')\}$ in the overlap region of two shifted pupil functions. The pupil function f is real and equal to unity; $\alpha_{20} = 0.50$ and $u'_{x,1} = 1$. The grey-shading in the overlap region ranges from black (-1) to white $(+1)$.

The preceding discussion shows that the reduction of the *OTF* value at medium frequencies must be larger than predicted on the basis of the average decrease given by Eq. (10.2.62). Moreover, the reduction is expressed in absolute value and, therefore, is relatively more important at the medium frequencies than at the low spatial frequencies. In Fig. 10.19b three incoherent transfer functions are shown. They are the aberration-free *OTF* in optimum focus and the moduli of the *OTF*'s for the case of an image-plane defocusing over a distance of one and two Rayleigh focal depths, respectively. The defocusing is given by the (normalised) defocus parameter α_{20}. Its values are 1/4 and 1/2 and the wavefront defect at the rim of the exit pupil then equals $\alpha_{20}\lambda_0$. If $\alpha_{20} = 0.25$, the frequency transfer is still 'relatively' good. The defocusing distance δz in image space that corresponds to $\alpha_{20} = 0.25$ is given by Eq. (5.2.28) and this leads us, using frequency transfer analysis, to the definition of the focal depth of a diffraction-limited system,

$$\delta z_f \approx \frac{\lambda_0}{2n_1 \sin^2 \alpha_m} = \frac{n_1 \lambda_0}{2(NA)^2} , \tag{10.2.67}$$

where α_m is the semi-opening angle of the imaging pencil and where we have included the case that the refractive index n_1 of image space is not equal to unity. We conclude that the *perfect* diffraction-limited imaging system exhibits a *just* diffraction-limited quality by the defocusing of the image plane over a distance of one Rayleigh focal depth. Figure 10.19b also shows that a defocusing of two focal depths leads to a transfer function of practically zero at frequencies close to NA/λ_0, half of the incoherent cutoff frequency of the imaging system. If we apply, for instance, a 50% modulation depth criterion for a transmitted frequency, we observe that the transmitted frequency band is more than halved by a defocusing of two Rayleigh focal depths. Figure 10.19c shows the value of the real part of the pupil product function $f(X_1', Y_1') f^*(X_1' - 1, Y_1')$ in the overlap region for the case that $u'_{x,1} = 1$ and $\alpha_{20} = 0.50$.

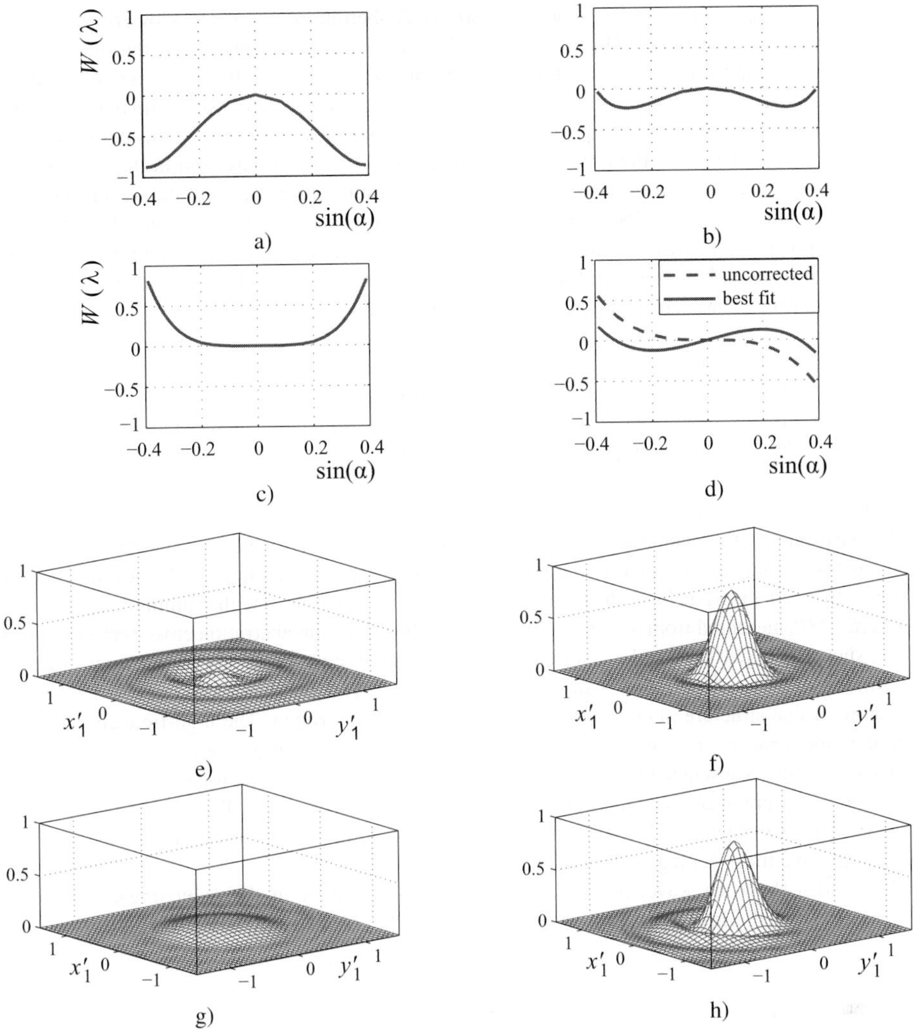

Figure 10.20: a) to c) The wavefront cross-section in the case of spherical aberration for a focus-setting on a) marginal focus, b) the 'best' diffraction focus, and c) the paraxial focus.
d) The meridional cross-section ($X_1' = 0$) of a comatic wavefront.
e) to h) The intensity point-spread functions associated with the wavefronts and focus-settings of a) to d), respectively.
The Seidel spherical aberration and coma both have a magnitude corresponding to the *just* diffraction-limited case ($OPD_{\mathrm{rms}} = 0.071\,\lambda_0$). The lateral coordinates x_1' and y_1' have been expressed in the diffraction unit λ_0/NA in the image plane.

10.2.4.3 The *OTF* for a Specific Type of Aberration

In Fig. 10.20 we show some typical Seidel aberration curves (spherical aberration and coma) and the corresponding point-spread functions in the image plane. Figure 10.20a–c represent cross-sections of wavefronts suffering from spherical aberration for different focus-settings of the image plane. The magnitude of the spherical aberration has been chosen such that the imaging system has 'just' diffraction-limited quality. This amounts to an rms wavefront aberration over the exit pupil of $0.071\,\lambda_0$. In terms of the Zernike polynomial $R_4^0(\rho)$ for third-order or Seidel-type spherical aberration, the coefficient c_4^0 follows from Eq. (9.3.22) and Eq. (D.2.22) of Appendix D, yielding

$$\frac{(c_4^0)^2}{n+1} = \frac{1}{20\pi^2}\,, \quad \text{or,} \quad c_4^0 = \frac{1}{2\pi}\,. \tag{10.2.68}$$

The Zernike polynomial $R_4^0(\rho) = 6\rho^4 - 6\rho^2 + 1$ for Seidel spherical aberration contains two lower-order wavefront terms, defocusing and piston, such that the OPD_{rms} is minimised in a certain image plane and that the Strehl intensity is maximised (for relatively small values of the coefficient c_4^0, definitely smaller than the wavelength of the light). From the first two coefficients in $R_4^0(\rho)$ we see that the fourth-order spherical term and the defocusing term have equal amplitude but opposite sign at the rim of the exit pupil. The position of the image plane corresponding to this defocusing term is called the 'best' diffraction image plane. For the paraxial image plane there is no defocus term in the wavefront polynomial. The defocusing that corresponds to the focus of the marginal rays requires a zero value of the wavefront *derivative* at the pupil rim where $\rho = 1$. There is also the focus-setting, relevant for larger wavefront aberration values, that minimises the ray blur. The Lukosz polynomial $L_4^0(\rho)$ provides us with the defocusing term $(-8\rho^2)$ (see Eq. (5.9.29)). We thus have the following table of wavefront polynomials for the various focus-settings, each of them provided with an extra constant term to obtain zero value of the wavefront aberration at the centre of the pupil,

$$
\begin{aligned}
W_P(\rho) &= 6\rho^4 \,, && \text{P, paraxial focus,} \\
W_D(\rho) &= 6\rho^4 - 6\rho^2 \,, && \text{D, 'best' diffraction focus,} \\
W_B(\rho) &= 6\rho^4 - 8\rho^2 \,, && \text{B, focus with minimum geometrical blur,} \\
W_M(\rho) &= 6\rho^4 - 12\rho^2 \,, && \text{M, marginal focus .}
\end{aligned}
\qquad (10.2.69)
$$

The axial positions of the image plane corresponding to these particular focusing conditions are illustrated in Fig. 10.21. We have plotted the geometrical rays of an imaging pencil with spherical aberration. The magnitude of the aberration has been highly exaggerated in the figure. For Seidel spherical aberration, the best diffraction focus D is exactly halfway between the marginal and the paraxial focus. The focus-setting B with minimum ray blur is found at one third of the distance MP, measured from the marginal focus M. The wavefront aberration cross-sections corresponding to W_M and W_P are shown in Fig. 10.20a and c, respectively. The intensity of the point-spread functions associated with the three focus-settings for spherical aberration is shown in Fig. 10.20e–g. The marginal focus in e) has a very low maximum intensity and shows a fine structure (a central maximum plus diffraction rings). The high-frequency fine structure can be ascribed to the large propagation angles with respect to the axis of the beam in the marginal region of the imaging pencil. Such a fine structure is absent in the image intensity distribution in the paraxial focus. It has a transverse dimension which is given by the aperture over which the wavefront is approximately flat, see c). The best diffraction focus shows the maximum intensity of 0.80 which could be expected from the value of the wavefront variance of the aberration function $W_D(\rho)$ corresponding to a focus-setting in the axial point D.

Figure 10.20d and h show the wavefront cross-section and the intensity of the point-spread function in the case of comatic aberration, for the *just* diffraction-limited situation. The Zernike polynomials $Z_3^{\pm1}(\rho,\theta) = R_3^{|1|}(\rho)\exp(\pm i\theta)$

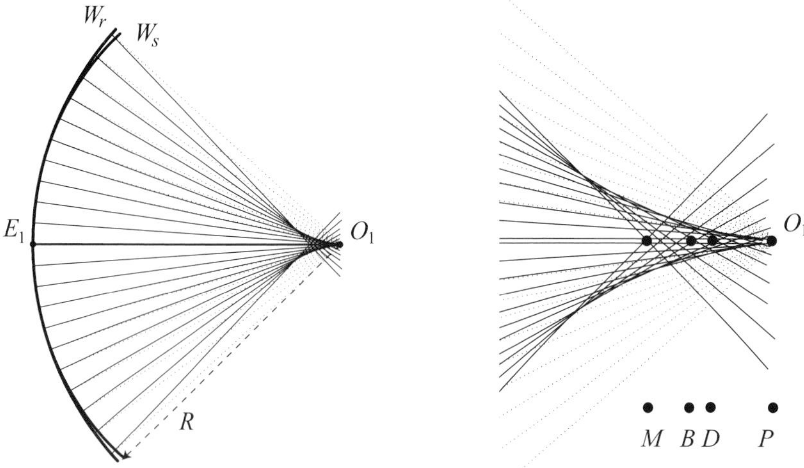

Figure 10.21: Spherical aberration and the focus positions for the marginal rays (M) and the paraxial rays (P). Between these outer focus positions, there is B, the minimum blur position, and D, the 'best' diffraction focus for largest Strehl intensity.

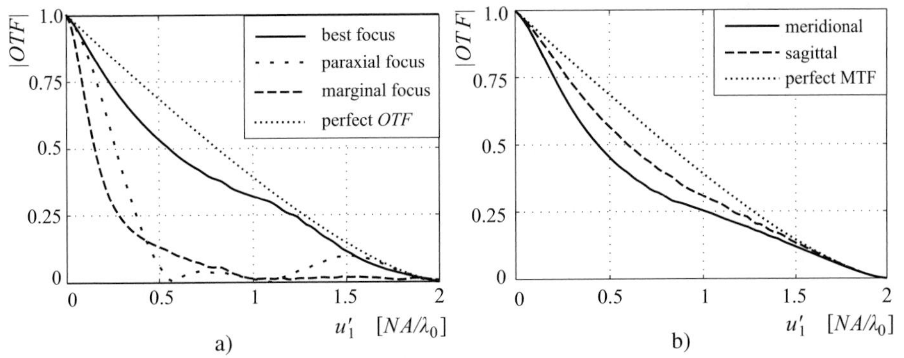

Figure 10.22: a) The *OTF* curves associated with an imaging system suffering from spherical aberration. The parameter of each curve is the position of the image plane in image space.
b) *OTF* of a system with comatic aberration.

have coefficients $c_3^{+1} = -0.1i$ and $c_3^{-1} = +0.1i$ in the case that the comatic wavefront function is given by $W_C(\rho, \theta) = 0.2\, R_3^1(\rho) \sin\theta$ where $R_3^1(\rho) = 3\rho^3 - 2\rho$. The variance of this wavefront amounts to $0.2/(4\pi^2)$, corresponding to the 'just' diffraction-limited case. Figure 10.20d produces a wavefront cross-section with the third-order term only ('uncorrected') and the cross-section with a linear term subtracted according to the radial Zernike polynomial $R_3^1(\rho)$, the 'best fit' wavefront. The intensity point-spread function of Fig. 10.20h shows a peak intensity of 0.8 in accordance with the just diffraction-limited aberration magnitude. The intensity distribution is symmetric with respect to the plane $x'_1 = 0$. The first diffraction ring of the point spread function shows an intensity variation with one maximum and one minimum when the azimuth in the image plane rotates over a full range of 2π.

In Fig. 10.22 we show *OTF* curves for two Seidel aberrations, spherical aberration and coma. In a), corresponding to spherical aberration, the modulus of the *OTF*, also called the *MTF*, has been plotted for the best-focus position of the image plane. The wavefront aberration for this focus-setting is shown in Fig. 10.20b. The decrease in *MTF* value with respect to the perfect system is comparable to that in Fig. 10.19b where the decrease was due to a defocusing only. The two *MTF* curves for spherical aberration with the paraxial or the marginal focus-setting show how sensitive the imaging system is to a defocusing beyond the Rayleigh focal depth. The effective frequency transfer is strongly reduced. From a frequency as low as $0.1\ NA/\lambda_0$, the paraxial *MTF* curve shows a reduced transfer with a gradual reduction to zero for larger frequencies. Zero crossings occur beyond the frequency $u'_1 = 0.50$. The *OTF* for the marginal focus-setting shows the fastest decrease at low spatial frequencies. In comparison with the paraxial *OTF*, however, its first zero crossing is found at almost twice as large a frequency.

The *OTF* graph of Fig. 10.22b is related to comatic aberration. The meridional wavefront cross-section ($Y'_1 = 0$) is shown in Fig. 10.20d, the sagittal cross-section is flat. The *MTF* curve associated with spatial frequency vectors $(u'_{x,1}, 0)$ is most affected by the aberration since the largest gradients in the wavefront are found in the cross-section $Y'_1 = 0$. A reduction of almost 40% with respect to the perfect *MTF* is observed, typically at a frequency of $0.7NA/\lambda_0$. The *MTF* for frequency vectors $(0, u'_{x,1})$ is less sensitive to the aberration; the decrease of the sagittal *MTF* is approximately half that of the meridional one.

10.2.4.4 Severely Aberrated Systems and the Geometrical Optics Transfer Function

When the aberrations of the imaging system are large with respect to the wavelength of the light, the optical transfer function is mainly determined by these geometrical aberrations and much less by diffraction unsharpness. Instead of rigorous calculation of the optical transfer function by means of a convolution of the heavily aberrated pupil function, a computationally more efficient approach can be chosen. For the case of incoherent illumination of the object, the convolution operation is replaced by a Taylor expansion for spatial frequencies that are sufficiently small with respect to the maximum frequency, $2NA/\lambda_0$, that are transferred by the optical system. Before carrying out this approximate calculation, we revisit the relationship between the wavefront aberration $W(X_1, Y_1)$ in the exit pupil and the transverse aberration components $(\delta x_1, \delta y_1)$ in the image plane, previously studied in Subsection 4.9.2.

In Fig. 10.23 we show the wavefront of an aberrated pencil in the exit pupil of an optical system. The origin of the pencil is an off-axis point Q_0 in the object plane, not shown in the figure. The centre of the exit pupil is at E_1 and a

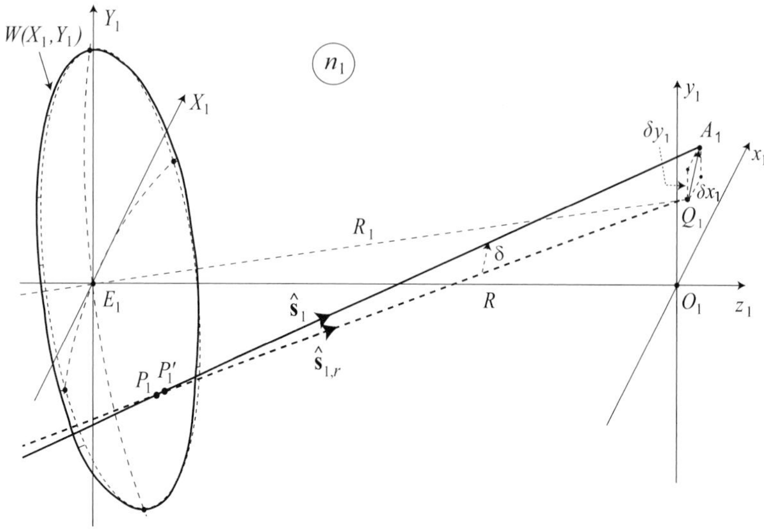

Figure 10.23: An aberrated wavefront surface $W(X_1, Y_1)$ in the exit pupil with the origin at E_1. The aberration is referred to the spherical reference sphere through E_1 which has the reference point Q_1 in the image plane as centre of curvature. A general aperture ray $\hat{\mathbf{s}}_1$ through P_1 on the reference sphere intersects the image plane in A_1. The transverse aberration components of ray $\hat{\mathbf{s}}_1$ in the image plane are $(\delta x_1, \delta y_1)$ The refractive index of the image space is n_1.

reference sphere has been drawn (dashed in the figure) with its centre of curvature in a chosen reference point Q_1 in the image plane. Q_1 can be the paraxial image point of the object point Q_0 or the intersection point of the principal ray with the image plane or any other judiciously chosen reference point. An aperture ray with unit vector $\hat{\mathbf{s}}_1$ intersects the reference sphere in P_1 and the wavefront through E_1 in the point P_1'. The ray through P_1 intersects the image plane in the point A_1 and the transverse aberration components of this particular ray are given by $(\delta x_1, \delta y_1)$. An ideal reference ray, unit vector $\hat{\mathbf{s}}_{1,r}$, intersects the reference sphere in the point P_1 and the image plane in Q_1. The distance $P_1'P_1$ is the wavefront aberration in the image space with refractive index n_1. In the figure, the wavefront aberration function $W(X_{P_1}, Y_{P_1})$ in the point P_1 is negative according to the convention of Chapter 4; the distance $P_1'P_1$ is negative measured along the propagation direction of the light.

The calculation of the transverse aberration components is carried out along the same lines as in Subsection 4.9.2. The only difference from the analysis in Subsection 4.9.2 is that an off-axis pencil is considered and we choose the paraxial point Q_1 with coordinates (x_{Q_1}, y_{Q_1}) as reference point. The axial distance E_1O_1 is R and the radius of the reference sphere is given by $R_1 = (R^2 + x_{Q_1}^2 + y_{Q_1}^2)^{1/2}$. For simplicity of notation we do not use in this context the quantities $R^{(s)}$ and $R_1^{(s)}$ which follow from Eq. (10.2.15) and allow the correct treatment of the sign of the radius of the reference sphere for arbitrary combinations of image and pupil magnification. Using the quantity R_1 which has a positive value in Fig. 10.23 and the characteristic functions employed in Subsection 4.9.2, we find for a general ray of the off-axis pencil through the point P_1 on the exit pupil sphere with coordinates (X_1, Y_1) the following transverse ray aberration components,

$$x_{A_1} - x_{Q_1} = \delta x_1 = \frac{(P_1A_1)}{n_1} \frac{\partial W(X_1, Y_1)}{\partial X_1} , \qquad y_{A_1} - y_{Q_1} = \delta y_1 = \frac{(P_1A_1)}{n_1} \frac{\partial W(X_1, Y_1)}{\partial Y_1} . \tag{10.2.70}$$

For a sufficiently small and slowly varying aberration function $W(X_1, Y_1)$ and for angular aberration values δ of the order of a mrad or less, we can safely replace the distance P_1A_1 by R_1, the radius of the reference sphere.

10.2.4.5 Low-frequency Expansion of the Optical Transfer Function

In Eq. (10.2.33), an expression for the two-dimensional incoherent optical transfer function was derived,

$$OTF(u_{x,1}, u_{y,1}) = \frac{\iint_S f(X_1, Y_1) f^*(X_1 + \lambda_1 R_1 u_{x,1}, Y_1 + \lambda_1 R_1 u_{y,1})\, dX_1 dY_1}{\iint_S |f(X_1, Y_1)|^2\, dX_1 dY_1} , \tag{10.2.71}$$

where S is the integration area in the exit pupil, in general the entire pupil area as we consider the case of the incoherent OTF. In Eq. (10.2.71) the refractive index of the image space is n_1 and does not need to be identical to the index n_0 of the object space. The complex pupil function $f(X_1, Y)$ is written as

$$f(X_1, Y_1) = A(X_1, Y_1)\exp\{i\,k_1\,W(X_1, Y_1)\}\,, \tag{10.2.72}$$

where $k_1 = 2\pi/\lambda_1$ is the wavenumber of the light in the image space with refractive index n_1 and W is the wavefront aberration as a function of the real-space coordinates (X_1, Y_1) in the exit pupil.

For a sufficiently small length of the spatial frequency vector $\mathbf{u}_1 = (u_{x,1}, u_{y,1})$, the second factor of the integrand function in the numerator of Eq. (10.2.71) is written as a Taylor series up to the second order,

$$\begin{aligned} f(X_1 + \lambda_1 R_1 u_{x,1}, Y_1 + \lambda_1 R_1 u_{y,1}) &= f(X_1, Y_1) + \lambda_1 R_1 u_{x,1}\frac{\partial f}{\partial X_1} + \lambda_1 R_1 u_{y,1}\frac{\partial f}{\partial Y_1} \\ &+ \frac{1}{2}\lambda_1^2 R_1^2 u_{x,1}^2 \frac{\partial^2 f}{\partial X_1^2} + \lambda_1^2 R_1^2 u_{x,1} u_{y,1}\frac{\partial^2 f}{\partial X_1 \partial Y_1} + \frac{1}{2}\lambda_1^2 R_1^2 u_{y,1}^2\frac{\partial^2 f}{\partial Y_1^2}\,. \end{aligned} \tag{10.2.73}$$

In Eq. (10.2.73) the following expressions for the derivative functions have to be used,

$$\begin{aligned} \frac{\partial f}{\partial X_1} &= \left\{\frac{\partial A}{\partial X_1} + ik_1 A\frac{\partial W}{\partial X_1}\right\} e^{ik_1 W}\,, \qquad \frac{\partial f}{\partial Y_1} = \left\{\frac{\partial A}{\partial Y_1} + ik_1 A\frac{\partial W}{\partial Y_1}\right\} e^{ik_1 W}\,, \\ \frac{\partial^2 f}{\partial X_1^2} &= \left\{\frac{\partial^2 A}{\partial X_1^2} - k_1^2 A\left(\frac{\partial W}{\partial X_1}\right)^2 + ik_1\left(2\frac{\partial A}{\partial X_1}\frac{\partial W}{\partial X_1} + A\frac{\partial^2 W}{\partial X_1^2}\right)\right\} e^{ik_1 W}\,, \\ \frac{\partial^2 f}{\partial X_1 \partial Y_1} &= \left\{\frac{\partial^2 A}{\partial X_1 \partial Y_1} - k_1^2 A\frac{\partial W}{\partial X_1}\frac{\partial W}{\partial Y_1} + ik_1\left(\frac{\partial A}{\partial X_1}\frac{\partial W}{\partial Y_1} + \frac{\partial A}{\partial Y_1}\frac{\partial W}{\partial X_1} + A\frac{\partial^2 W}{\partial X_1 \partial Y_1}\right)\right\} e^{ik_1 W}\,, \\ \frac{\partial^2 f}{\partial Y_1^2} &= \left\{\frac{\partial^2 A}{\partial Y_1^2} - k_1^2 A\left(\frac{\partial W}{\partial Y_1}\right)^2 + ik_1\left(2\frac{\partial A}{\partial Y_1}\frac{\partial W}{\partial Y_1} + A\frac{\partial^2 W}{\partial Y_1^2}\right)\right\} e^{ik_1 W}\,. \end{aligned} \tag{10.2.74}$$

In many instances it is justified to introduce further approximations. The first- and second-order derivatives of the pupil transmission function $A(X_1, Y_1)$ are put equal to zero, implying slow variations in the intensity part of the pupil function. We also assume that the second derivatives of the wavefront function W are much smaller than the quantities $k_1(\partial W/\partial X_1)^2$ and $k_1(\partial W/\partial Y_1)^2$. Using these approximations and employing the result of Eq. (10.2.70), we obtain for the optical transfer function in the geometrical optics approximation,

$$\begin{aligned} OTF_g(u_{x,1}, u_{y,1}) &= 1 - i2\pi n_1 \frac{\iint_S A^2(X_1, Y_1)\left\{u_{x,1}[\delta x_1(X_1, Y_1)] + u_{y,1}[\delta y_1(X_1, Y_1)]\right\} dX_1 dY_1}{\iint_S A^2(X_1, Y_1) dX_1 dY_1} \\ &- 2\pi^2 n_1^2 \frac{\iint_S A^2(X_1, Y_1)\left\{u_{x,1}[\delta x_1(X_1, Y_1)] + u_{y,1}[\delta y_1(X_1, Y_1)]\right\}^2 dX_1 dY_1}{\iint_S A^2(X_1, Y_1) dX_1 dY_1}\,. \end{aligned} \tag{10.2.75}$$

The expression for the transfer function is weighted by the intensity function $A^2(X_1, Y_1)$ in the exit pupil which means that each ray aberration value in the integrand function in the numerator is weighted by the intensity of that ray. In the particular case that the function $A(X_1, Y_1)$ is uniform, the transfer function reduces to

$$\begin{aligned} OTF_g(u_{x,1}, u_{y,1}) &= 1 - i2\pi n_1 \iint_S \left\{u_{x,1}[\delta x_1(X_1, Y_1)] + u_{y,1}[\delta y_1(X_1, Y_1)]\right\} dX_1 dY_1 \\ &- 2\pi^2 n_1^2 \iint_S \left\{u_{x,1}[\delta x_1(X_1, Y_1)] + u_{y,1}[\delta y_1(X_1, Y_1)]\right\}^2 dX_1 dY_1\,. \end{aligned} \tag{10.2.76}$$

From the OTF function we derive the modulation transfer function $MTF(u_{x,1}, u_{y,1})$ and the phase transfer function $PTF(u_{x,1}, u_{y,1})$ by calculating the modulus and the argument of the OTF.

10.2.4.6 Further Approximation of the Low-frequency Geometrical MTF

In the special case that $u_{x,1}$ (or, equivalently, $u_{y,1}$) equals zero, the modulus of the optical transfer function is conveniently written as

$$\begin{aligned} MTF_g^2(u_{x,1}, 0) &= \left|OTF_g(u_{x,1}, 0)\right|^2 \approx 1 - 4\pi^2(u_{x,1})^2\,\overline{\{\delta x_1(X_1, Y_1)\}^2} + 4\pi^2(u_{x,1})^2\left\{\overline{\delta x_1(X_1, Y_1)}\right\}^2 \\ &= 1 - 4\pi^2(u_{x,1})^2\,\{\delta x_1(X_1, Y_1)\}_{\text{rms}}^2\,, \end{aligned} \tag{10.2.77}$$

where higher than second powers of the spatial frequency components have been omitted. For sufficiently small values of the product $u_{x,1}\,\delta x_1$ we use the approximation

$$MTF_g(u_{x,1},0) \approx 1 - 2\pi^2 (u_{x,1})^2 \, \{\delta x_1(X_1, Y_1)\}^2_{\rm rms} \; . \tag{10.2.78}$$

The quantity $\{\delta x_1(X_1, Y_1)\}_{\rm rms}$ is the x-averaged geometrical blur in the image plane. It can be used to obtain an estimate of the spatial frequency $|\mathbf{u}_1|_{80}$ at which the geometrical transfer function has a value of 80%. From Eq. (10.2.78) we obtain

$$1/5 = 2\pi^2(|\mathbf{u}_1|_{80})^2\,(\delta_{80})^2,$$

or

$$\delta_{80} = \frac{0.10}{|\mathbf{u}_1|_{80}} \, , \tag{10.2.79}$$

where δ_{80} is the rms blur that yields 80% modulation depth at the desired spatial frequency $|\mathbf{u}_1|_{80}$. A further quadratic extrapolation of the shape of the geometrical *MTF* curve yields the less reliable rule of thumb for the rms blur that will yield at least a 50% modulation depth at spatial frequency $|\mathbf{u}_1|_{50}$, viz. $\delta_{50} = 0.16/|\mathbf{u}_1|_{50}$.

In Fig. 10.24 some geometrical and diffraction *MTF* curves have been plotted. The wavefront aberration is coma, the value of $OPD_{\rm rms}$ is $0.71\lambda_0$ and $\lambda_0 = 0.50$ μm. The rms wavefront aberration is ten times the diffraction limit of $0.071\lambda_0$, which means that in the image plane the blur due to the geometrical aberration is much larger than the diffraction unsharpness of typically $\lambda_0/2NA$. In Fig. 10.24a the geometrical optics *MTF* has been plotted. The solid curve corresponds to frequency vectors that are parallel to the cross-section containing the asymmetric comatic wavefront profile. The dashed curve corresponds to frequency vectors that are parallel to the symmetric cross-section of the wavefront (sagittal *MTF*). The dotted curve is the theoretical diffraction-based *MTF* which corresponds to a spherical wavefront in the exit pupil. In b), the same comatic wavefront aberration is present in the exit pupil and the diffraction-based *MTF* curves have been plotted according to Eq. (10.2.71). Using the values for *NA* and λ_0 according to the legend of Fig. 10.24, the cut-off frequency $2NA/\lambda_0$ of the diffraction-based *MTF* is given by 200 mm^{-1}. It is seen that the geometrical and diffraction *MTF* are largely similar in the lower range of spatial frequencies where the approximations that led to the geometrical *MTF* are justified. The computational advantage of the geometrical *MTF* is important because the interpolation and convolution calculations pertaining to a diffraction *MTF* are replaced by a simple multiplication of tabulated transverse aberration components and frequency component values. In Fig. 10.24b we observe that the diffraction *MTF* shows some numerical noise, especially at the mid-spatial frequencies. The reason is that the numerical integration of the *OTF* function has been done with the sampling density of the pupil function that is adequate for a weakly aberrated imaging system. Sampling inaccuracy occurs when using this relatively low sampling density for the severely aberrated imaging system in this example.

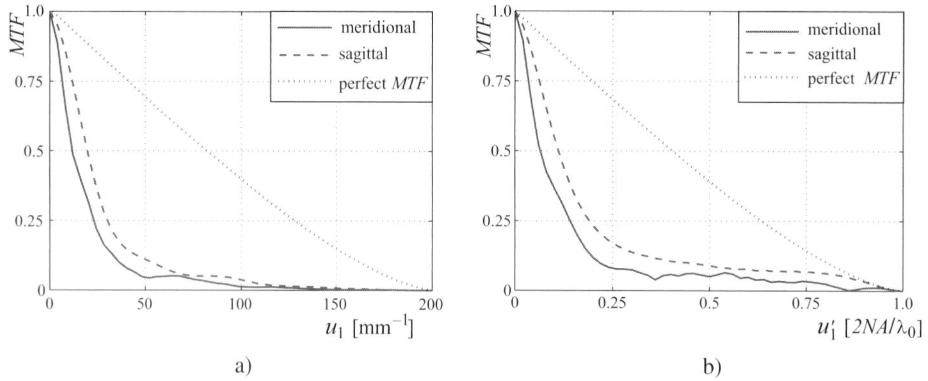

Figure 10.24: Image formation with the following imaging parameters: $NA = 0.05$, $\lambda_0 = 0.5$ μm, $n_1 = 1$, $OPD_{rms} = 0.71\,\lambda_0$ (third-order coma).
a) Geometrical *MTF*.
b) Diffraction *MTF*.

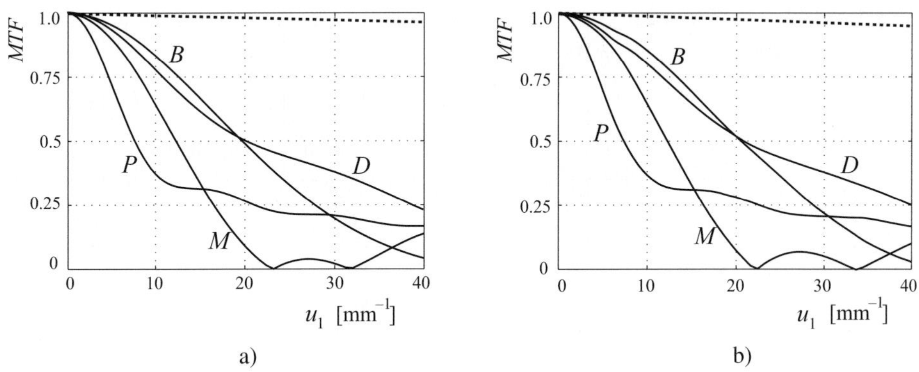

Figure 10.25: a) Modulus of the geometrical *OTF* (*MTF*) for the case of an imaging system with a large amount of spherical aberration. The Seidel coefficient α_{40} equals $10\,\lambda_0$, the corresponding Zernike coefficient $c_4^0 = 1.67\,\lambda_0$.
b) The diffraction *MTF* for the same aberration.

In Fig. 10.25 we show a final example of a highly aberrated imaging system suffering from Seidel spherical aberration. The spherical aberration is 10 times larger than in Fig. 10.22a. The Seidel coefficient α_{40} equals 10, normalised with respect to the wavelength of the light. The corresponding Zernike coefficient c_4^0 of the normalised wavefront aberration function W' equals 1.67. In both Fig. 10.25a and b we have plotted five *MTF* curves. The *MTF* function of a perfect imaging system according to Eq. (10.2.49) is represented by the dotted curve. Only the low spatial frequency part of the *MTF* is shown, the theoretical cut-off frequency in this case being 1000 mm^{-1}. The other four transfer functions have been labelled P (paraxial focus), D (best diffraction focus), B (minimum blur focus) and M (marginal focus), in accordance with Eq. (10.2.69) and Fig. 10.21. The geometrical *MTF* curves of a) and the diffraction-based *MTF*s in b) are very similar, which proves again the validity of the geometrical *OTF* approximation for large values of the aberration coefficients. The paraxial and marginal *MTF* are inferior to the best diffraction focus and the minimum blur focus-setting. The minimum blur curve B has the best performance for the lowest spatial frequencies. For the larger spatial frequencies, the 'best' diffraction focus remains the optimum choice.

10.2.5 The Spatial Coherence Function and the van Cittert–Zernike Theorem

The discussion on the optical transfer function revealed that the maximum frequency (cut-off frequency) which can be transferred by an imaging system is determined by the ratio of the diameter of the source image in the exit pupil and the diameter of the exit pupil itself. Equation (10.2.50) shows that for centred sources the largest cut-off frequency is obtained for a source image that, at least, covers the entire exit pupil area. The illumination of the object by a source with such a minimum size was called incoherent. The other extreme case, illumination by a point-like source, was called coherent. The influence of a general shape of the illuminating source on the image formation and on the transfer of spatial frequencies is discussed in this section with the aid of the so-called van Cittert–Zernike theorem **[70]**,**[386]**. This theorem was developed in the 1930s and put an end to a long discussion on the optimum design of illumination systems, for instance, for a high-resolution microscope. The theorem applies to the scalar model of light and neglects any influence of the state of polarisation of light. It typically applies to 'natural' light, emitted for instance by an incandescent lamp (black-body radiator) with a rapidly changing state of polarisation. With respect to the temporal coherence, quasi-monochromatic light is assumed to be used by means of a narrow spectral filter, although it will be seen that this requirement can be relaxed to broader light spectra and even to white light.

An important development in the field of coherence function analysis was the formulation of the propagation law for the mixed temporal and spatial coherence function. This law was put forward by Wolf **[37]**,**[371]**,**[372]** in the 1950s. More recent developments in optical coherence theory go beyond the limitations of the original van Cittert–Zernike theorem. The theory has been extended to cover sources with a particular temporal spectrum and a specific state of polarisation. Moreover, the light propagation is treated vectorially and the media in which the source light propagates do not need to be homogeneous and isotropic but can be of a more general nature.

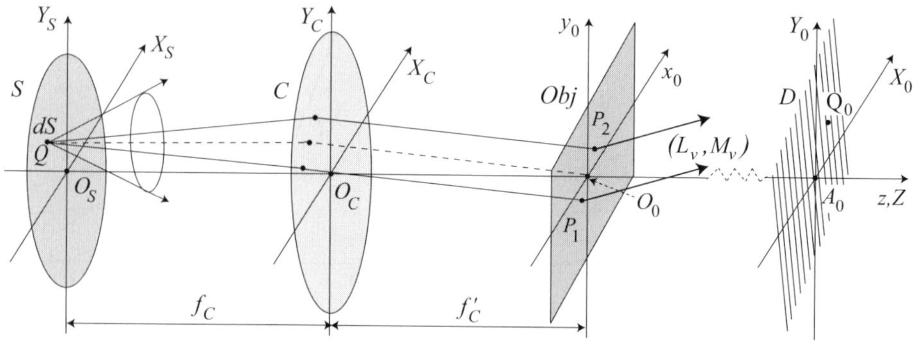

Figure 10.26: The degree of coherence between two points P_1 and P_2 in the plane *Obj*, produced by an incoherent planar source S which is projected to infinity by the condenser lens C having an image-side focal distance f'_C. The degree of spatial coherence is measured in the infinitely distant plane D.

10.2.6 Derivation of the van Cittert–Zernike Theorem

The van Cittert–Zernike theorem is based on the classical definition of spatial coherence, namely the ability of two illuminated neighbouring point sources in a screen to produce an interference pattern with a certain modulation depth on a receiving planar surface at a certain distance. The geometry of the interference set-up to measure the coherence in a receiving plane is shown in Fig. 10.26. The planar radiating surface S has its origin at O_S and a general point of the source has coordinates (X_S, Y_S). The source is relayed to infinity by the condenser lens C and the screen at the plane *Obj* collects the radiation from S. Its origin is at O_0 and the coordinates in this plane are (x_0, y_0). In an optical imaging system *Obj* will be used as the object plane. If the source S is said to be 'fully incoherent' we assume that the source surface contains independent, infinitely small radiation sources, e.g. the atoms in a phosphor layer or in a hot filament, which transmit in a completely chaotic and uncorrelated manner wave packages with finite length in the direction of the plane *Obj*. We consider the general point $Q(X_S, Y_S)$ which emits radiation. Far away from Q, the emitted radiation is assumed to have the shape of a spherical wave with a certain state of polarisation. According to the laws of geometrical optics this wave is transformed by the condenser lens C into a plane wave with unit wave vector

$$\hat{\mathbf{s}} = (L_C, M_C, N_C) = \left(-\frac{X_S}{f'_C}, -\frac{Y_S}{f'_C}, \sqrt{1-(X_S^2+Y_S^2)/f_C'^2}\right) , \tag{10.2.80}$$

and a complex amplitude in the plane *Obj* which is proportional to

$$U(x_0, y_0) = \sqrt{I(X_S, Y_S)}\, \exp\{ik(L_C x_0 + M_C y_0)\} , \tag{10.2.81}$$

where k is the wavenumber of the light in the medium in which source, condenser lens and receiving screen are located. The complex amplitude at the location of the two pinholes P_1 and P_2 with coordinates $-(\Delta_x, \Delta_y)/2$ and $(\Delta_x, \Delta_y)/2$, respectively, is given by

$$\begin{aligned} dU_{P_1} &= \sqrt{I_S(X_S, Y_S)}\, \exp\left\{+i\frac{k}{2f'_C}(X_S\Delta_x + Y_S\Delta_y)\right\} dX_S dY_S , \\ dU_{P_2} &= \sqrt{I_S(X_S, Y_S)}\, \exp\left\{-i\frac{k}{2f'_C}(X_S\Delta_x + Y_S\Delta_y)\right\} dX_S dY_S . \end{aligned} \tag{10.2.82}$$

The two pinholes in the screen *Obj* serve as secondary sources and are capable of producing an interference pattern in the space behind the screen. At a very large distance, the intensity of the interference pattern becomes a function of the viewing angle, defined by the (X, Y) direction cosines (L_v, M_v). The geometric pathlength difference between the pinholes P_1 and P_2 as a function of the viewing angle is given by

$$l_{P_1} - l_{P_2} = \Delta_x L_v + \Delta_y M_v . \tag{10.2.83}$$

The infinitesimally small amplitude in the viewing angle defined by (L_v, M_v) which originates from the source area dS centred at the point Q is given by

$$dU_Q = \sqrt{I_S(X_S, Y_S)}\left[\exp\left\{+i\frac{k}{2}\left[\Delta_x\left(\frac{X_S}{f'_C}+L_v\right)+\Delta_y\left(\frac{Y_S}{f'_C}+M_v\right)\right]\right\} + \exp\left\{-i\frac{k}{2}\left[\Delta_x\left(\frac{X_S}{f'_C}+L_v\right)+\Delta_y\left(\frac{Y_S}{f'_C}+M_v\right)\right]\right\}\right] dX_S dY_S\,. \tag{10.2.84}$$

To obtain the intensity in the far field behind the screen Obj we calculate $|dU_Q|^2$ for each elementary source point Q and add the intensity contributions by integration over the source area. This yields the expression

$$I(L_v, M_v) = \iint_S I_S(X_S, Y_S)\left[2 + \Re\left\{\exp\left[ik\left(\Delta_x L_v + \Delta_y M_v\right)\right]\exp\left[i\frac{k}{f'_C}\left(\Delta_x X_S + \Delta_y Y_S\right)\right] + \exp\left[-ik\left(\Delta_x L_v + \Delta_y M_v\right)\right]\exp\left[-i\frac{k}{f'_C}\left(\Delta_x X_S + \Delta_y Y_S\right)\right]\right\}\right] dX_S dY_S\,. \tag{10.2.85}$$

We introduce a coordinate scaling by using normalised coordinates in the source plane and in the object plane, namely $(X'_S, Y'_S) = (X_S, Y_S)/\rho_S$ and $(\Delta'_x, \Delta'_y) = (\Delta_x, \Delta_y)\rho_S/(\lambda f'_C) = (\Delta_x, \Delta_y)NA_S/\lambda$. The diameter of the source disc is $2\rho_S$ and this disc contains all contributing source elements. $NA_S = n\rho_S/f'_C$ is the aperture of the source (n is the refractive index of the medium). The direction cosines (L_v, M_v) are normalised with the aid of the numerical aperture NA_S of the source, given by the ratio ρ_S/f'_C, yielding the primed normalised direction cosines $(L'_v, M'_v) = (L_v, M_v)f'_C/\rho_S$. Using these normalised source and object plane coordinates we observe that the integral over the source area has a three-term integrand with the form of a 'backward' Fourier transform according to Eq. (A.8.6). The Fourier transformation is performed from the normalised source coordinates (X'_S, Y'_S) to the normalised object plane coordinates $(0, 0)$ and $\pm(\Delta'_x, \Delta'_y)$, respectively. Denoting the Fourier transform operator by $\mathcal{F}$ and omitting non-essential prefactors, we find the expression

$$I(L'_v, M'_v; \Delta'_x, \Delta'_y) = \mathcal{F}\{I_S(X'_S, Y'_S)\}_{0,0} + \frac{1}{2}\Re\left\{\exp\left[+2\pi i\left(\Delta'_x L'_v + \Delta'_y M'_v\right)\right] \times \mathcal{F}\{I_S(X'_S, Y'_S)\}_{-\Delta'_x,-\Delta'_y} + \exp\left[-2\pi i\left(\Delta'_x L'_v + \Delta'_y M'_v\right)\right]\mathcal{F}\{I_S(X'_S, Y'_S)\}_{+\Delta'_x,+\Delta'_y}\right\}. \tag{10.2.86}$$

Since the intensity function $I(X'_S, Y'_S)$ is real it is permissible to write Eq. (10.2.86) as

$$\begin{aligned} I(L'_v, M'_v; \Delta'_x, \Delta'_y) &= \tilde{I}_S(0,0) + \Re\left\{\exp\left[-2\pi i(\Delta'_x L'_v + \Delta'_y M'_v)\right]\tilde{I}_S(\Delta'_x, \Delta'_y)\right\} \\ &= \tilde{I}_S(0,0) + \cos\left[2\pi\left(\Delta'_x L'_v + \Delta'_y M'_v\right)\right]\Re\left\{\tilde{I}_S(\Delta'_x, \Delta'_y)\right\} \\ &\quad + \sin\left[2\pi\left(\Delta'_x L'_v + \Delta'_y M'_v\right)\right]\Im\left\{\tilde{I}_S(\Delta'_x, \Delta'_y)\right\} \\ &= \tilde{I}_S(0,0)\left[1 + \frac{\left(\left[\Re\{\tilde{I}_S(\Delta'_x, \Delta'_y)\}\right]^2 + \left[\Im\{\tilde{I}_S(\Delta'_x, \Delta'_y)\}\right]^2\right)^{1/2}}{\tilde{I}_S(0,0)} \right. \\ &\quad \left. \times \cos\left[2\pi\left(\Delta'_x L'_v + \Delta'_y M'_v\right) - \arctan\left(\frac{\Im\{\tilde{I}_S(\Delta'_x, \Delta'_y)\}}{\Re\{\tilde{I}_S(\Delta'_x, \Delta'_y)\}}\right)\right]\right], \end{aligned} \tag{10.2.87}$$

where we have used the abbreviated form $\tilde{I}_S(\Delta'_x, \Delta'_y)$ for the Fourier transform of the source function $I_S(X', Y')$ and have exploited the fact that $\tilde{I}_S(\Delta'_x, \Delta'_y) = \tilde{I}^*_S(-\Delta'_x, -\Delta'_y)$. The expression for the far-field intensity I is reduced to the two terms on the second line on the right-hand side of Eq. (10.2.87) when the source intensity function is point-symmetric with respect to the origin O_S.

The mutual coherence of the two illuminated pinholes P_1 and P_2 at the screen Obj is defined as the (complex) modulation depth of the far-field intensity pattern produced by the radiation emanating from these pinholes. In experimental terms, the modulus of the mutual coherence equals the visibility of the fringe pattern in the far field. The argument of the complex mutual coherence function is derived from the shift of the interference pattern in the far field divided by its period and then multiplied by 2π. Using the expression on the first line of Eq. (10.2.87) we can write

$$I(L'_v, M'_v; \Delta'_x, \Delta'_y) = \tilde{I}_S(0,0)\left[1 + |\mu(\Delta'_x, \Delta'_y)|\,\Re\left\{\exp\left[i\left\{-2\pi(\Delta'_x L'_v + \Delta'_y M'_v) + \psi(\Delta'_x, \Delta'_y)\right\}\right]\right\}\right]. \tag{10.2.88}$$

As a function of the separation (Δ'_x, Δ'_y) of the two pinholes the complex mutual coherence function μ is then given by

$$\mu(\Delta'_x, \Delta'_y) = \frac{\tilde{I}_S(\Delta'_x, \Delta'_y)}{\tilde{I}_S(0,0)} = \frac{\sqrt{\left[\Re\left\{\tilde{I}_S(\Delta'_x, \Delta'_y)\right\}\right]^2 + \left[\Im\left\{\tilde{I}_S(\Delta'_x, \Delta'_y)\right\}\right]^2}}{\tilde{I}_S(0,0)} \exp[i\psi(\Delta'_x, \Delta'_y)] , \tag{10.2.89}$$

where the argument ψ of the exponential function is given by

$$\psi(\Delta'_x, \Delta'_y) = \arctan\left(\frac{\Im\left\{\tilde{I}_S(\Delta'_x, \Delta'_y)\right\}}{\Re\left\{\tilde{I}_S(\Delta'_x, \Delta'_y)\right\}}\right) . \tag{10.2.90}$$

For the case of a centred source having point-symmetry with respect to the origin, the function $\Im\left\{\tilde{I}_S(u_{x,1}, u_{y,1})\right\} = 0$, as well as $\psi(u_{x,1}, u_{y,1})$. For the frequently occurring case of a uniform circular source with radius ρ_S, the mutual coherence function of the points P_1 and P_2 has the simple form

$$\mu(\Delta'_r) = \frac{\Re\left\{\tilde{I}_S(\Delta'_r)\right\}}{\tilde{I}_S(0)} , \qquad \Delta'_r = \sqrt{(\Delta'_x)^2 + (\Delta'_y)^2}, \tag{10.2.91}$$

where $\mu(\Delta'_r)$ follows from the Fourier transform in polar coordinates

$$\mu(\Delta'_r) = \frac{1}{\pi} \int_0^1 \int_0^{2\pi} \exp\{2\pi i \rho \Delta'_r \cos(\theta - \alpha)\} \rho \, d\rho d\theta . \tag{10.2.92}$$

The coordinate ρ is the radial source coordinate, normalised with respect to half the source diameter $(2\rho_S)$, and α equals $\arctan\left(\Delta'_y/\Delta'_x\right)$.

To compute this integral we use the following properties of Bessel functions and integrals of Bessel functions **[10]**,

$$J_0(x) = \frac{1}{2\pi} \int_0^{2\pi} \exp(ix\cos\theta) \, d\theta , \qquad \int_0^p x J_0(x) dx = p J_1(p) . \tag{10.2.93}$$

Applied to the integral of Eq. (10.2.92), we obtain an expression which has a close relationship with the expression for the image plane *amplitude* of the point-spread function associated with a circular aperture,

$$\mu(\Delta'_r) = \frac{J_1(2\pi\Delta'_r)}{\pi\Delta'_r} . \tag{10.2.94}$$

The essence of the van Cittert–Zernike theorem is *that it relates the mutual coherence function μ of two points on a distant screen to the complex amplitude of the diffraction pattern that would be produced by the source (intensity) function on the same screen.* In Zernike's own words, *for any extended light source the coherence function is found to be equal to the amplitude in a certain diffraction image* **[386]**. Equation (10.2.94) gives this result for a circular and centred incoherent source. From the expression for μ in this special case we find that the mutual coherence becomes zero for a real-space separation Δ given by

$$\Delta_r = \frac{3.83}{2\pi} \frac{\lambda f'_C}{\rho_S} = 0.61 \frac{\lambda_0}{NA_S} , \tag{10.2.95}$$

where we have used use the source aperture $NA_S = n\rho_S/f'_C$, as viewed from the illuminated screen *Obj* in the direction of the source S ($\lambda = \lambda_0/n$ and n is the index of the medium). Figure 10.27 shows a surface plot of the mutual coherence function $\mu(\Delta'_x, \Delta'_y)$ for a centred circular source with uniform source intensity. The coordinates (Δ'_x, Δ'_y) are in units of λ_0/NA_S. Outside the central area with source separation $\Delta = 0.61\,\lambda_0/NA_S$, the function μ first becomes negative which means that the contrast in the interference pattern is inverted. For still larger separations, further oscillations in contrast are observed and the mutual coherence effectively becomes zero. One frequently requires the 80% level as the criterion for a high level of coherence and this yields a maximum separation in the object plane of $\Delta = 0.21\,\lambda_0/NA_S$.

Another important conclusion, stated explicitly by Zernike in his publication, is the irrelevance of the aberration correction of the condenser lens C. In all practical situations, the numerical aperture NA_S of the source will be appreciable, for instance 0.01 or larger. For classical sources, a high irradiance level in the object plane of the imaging system, for instance in a microscope, requires a substantial value of NA_S. In such a case the separation Δ for a high level of mutual coherence is of the order of the wavelength of the light or, at the most, one order of magnitude larger. With reference to Fig. 10.28, one concludes that the footprint over which the condenser lens should be isoplanatic (see Section 4.8) is of that same order of magnitude, typically the distance P_1P_2 over which coherence is present in the object plane. Outside

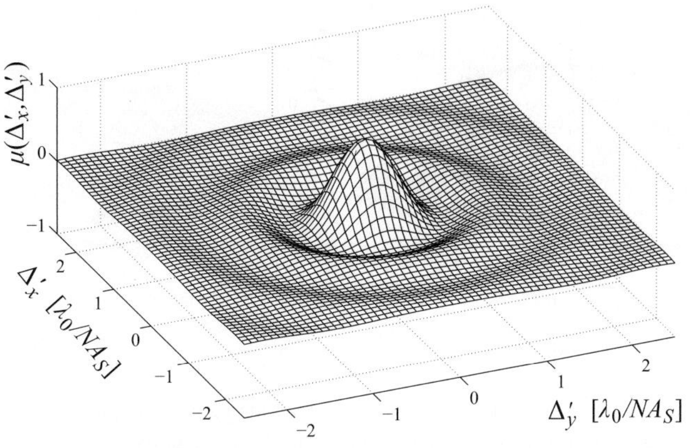

Figure 10.27: Mutual coherence function of a centred and uniform circular source with aperture $NA_S = n\rho_S/f'_C$.

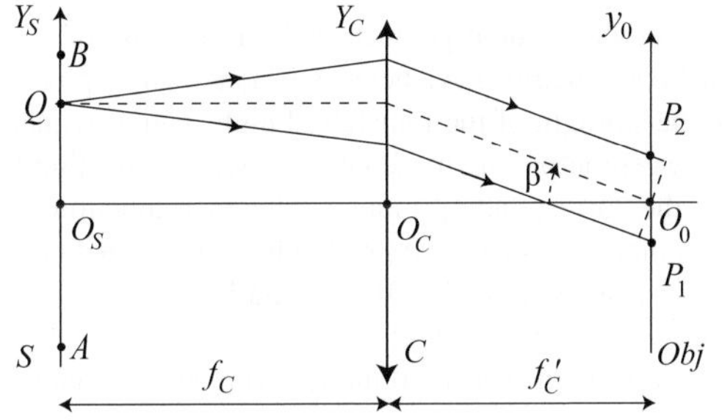

Figure 10.28: Calculation of the mutual coherence μ of P_1 and P_2.

this footprint, the exact phase of the optical waves is irrelevant because they will randomly oscillate without a stable phase relation. For condenser lenses, with lateral dimensions of centimetres or larger, the aberrational change over a lateral distance of some 10 μm or less is negligible. This reasoning led to the insight that the state of aberration of the condenser system does not effectively influence the coherence function in the object plane of, for instance, a microscope. The aberrations of the condenser are important for the correction of pupil aberrations of the entire optical system (see Section 7.7 and the subsection on condenser optics of the microscope).

Figure 10.28 is also useful for obtaining a quick estimate of the coherence area in the plane *Obj*. For an off-axis source point Q, the phase difference between the illuminated points P_1 and P_2 equals $k\Delta \sin\beta$ where Δ is the lateral distance P_1P_2. Let us assume that this quantity varies from $+\pi$ to $-\pi$ when the source point Q moves from the border point A to the opposite border point B. The averaged contributions from the source sections O_SA and O_SB to the interference pattern behind the plane *Obj* are in phase opposition and no modulation is observed. We have that $\sin\beta_A = -\sin\beta_B = \rho_S/f'_C$ and this yields $2k\Delta_0\rho_S/f'_C = 2\pi$, where Δ_0 is the separation value P_1P_2 that corresponds to the first zero in the mutual coherence function. We find that $\Delta_0 = \lambda_0/2NA_S$, for a one-dimensional homogeneous source. A still smaller coherence area can be created when the central source area is obstructed. In the one-dimensional picture of Fig. 10.28, we would have a source geometry that is limited to the two extreme points A and B. It follows immediately that phase opposition is achieved when the pathlength difference for these extreme source points is π and the points P_1 and P_2 are thus considered to be incoherent. The value of Δ_0 then amounts to $\lambda_0/4NA_S$. We note that for this special case with two point sources the coherence function becomes periodic with a period $\lambda_0/2NA_S$.

10.2.7 The Mutual Coherence Function and the Optical Transfer Function

In this subsection we discuss some special shapes of the mutual coherence function which allow enhanced imaging of periodic objects close to the cut-off frequency of an imaging system. In the preceding subsection it was shown

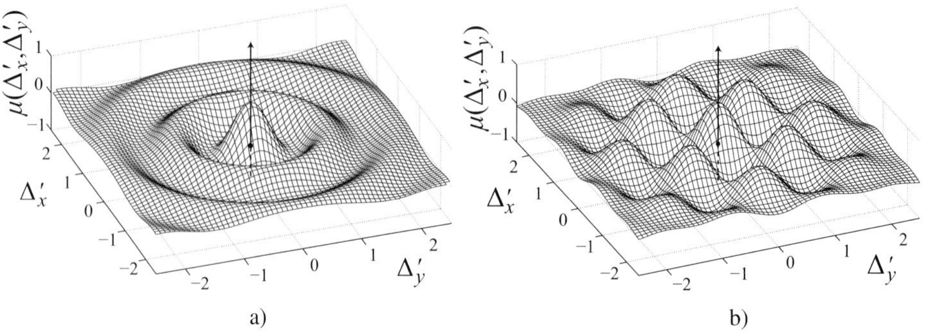

Figure 10.29: a) The mutual coherence function $\mu(\Delta'_x, \Delta'_y)$ at the receiving plane *Obj* in the case of an annular illuminating source (relative width w of the annulus is $0.2\rho_S$). The coordinates (Δ'_x, Δ'_y) are in units λ_0/NA_S where NA_S is associated with the source disc with radius ρ_S.
b) Quadrupole illumination, sub-source diameter $2s = 0.4\rho_S$.

that such an enhancement exists for certain source geometries by inspection of the optical transfer function for these source geometries. Two special source geometries have been considered which produce enhanced imaging of fine spatial features, or, equivalently, increase the strength of the transferred high spatial frequencies close to the cut-off frequency $2NA/\lambda_0$ of the imaging system. These geometries are the annular source and the illumination with quadrupole sub-sources (see Fig. 10.17). We first calculate the corresponding mutual coherence functions produced by these two illumination methods and then give a physical explanation of the contrast enhancement of fine imaging details. The corresponding coherence functions in the object are displayed in Fig. 10.29a and b. The realisation of these illumination geometries without loss of optical energy requires a special design of the condenser system. High-quality imaging systems are often provided with the possibility to switch from one special illumination method to another, depending on the type of pattern to be imaged.

- *Annular illumination*
  The mutual coherence function of the annular source follows from the basic result of Eq. (10.2.92) by calculating the difference of the coherence functions associated with a circular sources with normalised radius 1 and a second source with radius $1 - w$. After normalisation with respect to the total source intensity we find the expression

$$\mu(\Delta') = \frac{1}{w(2-w)}\left[\frac{2J_1(2\pi\Delta')}{2\pi\Delta'} - (1-w)^2\,\frac{2J_1(2\pi(1-w)\Delta')}{2\pi(1-w)\Delta'}\right], \qquad (10.2.96)$$

  where, as defined previously, $\Delta' = \Delta\,(NA_S/\lambda_0)$ is the normalised separation in the radial direction in the object plane *Obj*. Figure 10.29a shows this mutual coherence function for $w = 0.2$. In comparison with the coherence function of a full circular source (see Fig. 10.27), we observe a narrowed central peak and enhanced negative and positive ring sections.
- *Quadrupole illumination*
  The evaluation of the Fourier transform of the intensity pattern of the projected source of Fig. 10.18 with sub-source diameters equal to $2s$ and with positions of the source centres on the diagonals of the Cartesian quadrants yields the expression

$$\mu(\Delta'_x, \Delta'_y) = \frac{2J_1\left[2\pi s\,(\Delta'^2_x + \Delta'^2_y)^{1/2}\right]}{2\pi s\,(\Delta'^2_x + \Delta'^2_y)^{1/2}}\,\cos\left\{\pi\sqrt{2}(1-s)\,\Delta'_x\right\}\,\cos\left\{\pi\sqrt{2}(1-s)\,\Delta'_y\right\}. \qquad (10.2.97)$$

  In Fig. 10.29b we have plotted the mutual coherence function for quadrupole illumination as a function of the normalised distances (Δ'_x, Δ'_y) between two points P_1 and P_2 in the receiving plane *Obj*.

In Fig. 10.30 cross-sections of the mutual coherence functions are given for $\Delta'_y = 0$. With respect to the coherence function of a uniform circular source with disc diameter $2\rho_S$ we observe a narrower central lobe for the annular and quadrupole illumination (dashed and dotted curves in the figure, respectively) and much higher amplitudes of the surrounding minima and maxima. We also note that the quadrupole coherence function is definite positive along the

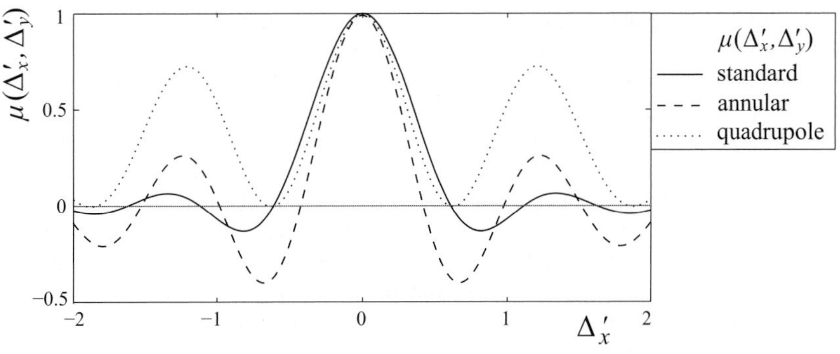

Figure 10.30: Cross-sections of the mutual coherence function $\mu(\Delta'_x, \Delta'_y)$ along the quadrant diagonals ($\Delta'_x = \pm\Delta'_y$) in the case of standard illumination (solid curve), annular illumination (dashed curve) and quadrupole illumination (dotted curve).

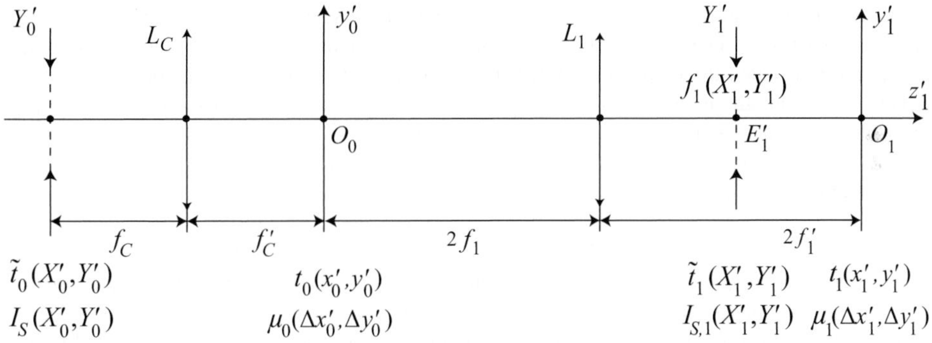

Figure 10.31: Definition of (normalised) coordinates in the optical 'near' and 'far' field that are needed for calculation of the image plane intensity in a partially coherent imaging system. The calculations preferably use the normalised far-field coordinates (X'_1, Y'_1) in the back-focal plane of the imaging objective lens L_1 and the normalised image-plane coordinates (x'_1, y'_1). The focal lengths of condenser and imaging lens, f'_C and f'_1, respectively, have equal value.

quadrant diagonals defined by $\Delta'_x = \pm\Delta'_y$. A larger coherence area with respect to that of the standard illumination suggests that a larger modulation depth in the image plane can be obtained. This speculation is supported by the maximum modulation depth of unity that was obtained for a fully coherent point source illumination, up to the coherent cut-off frequency of NA/λ_0. For the mutual coherence functions with extended coherence area, a larger modulation depth can also be expected in the frequency region $NA/\lambda_0 < |u'_1| < 2NA/\lambda_0$.

10.2.8 Partially Coherent Imaging of a General Object

In this subsection we calculate the image intensity in the case of partially coherent illumination of the object plane. The complex object transmission function $t(x_0, y_0)$ is of general nature. In a later phase we assume that the function t is periodic, in a first instance having a weak modulation so that we can calculate an optical transfer function. Before we address calculation of the intensity in the image plane of the imaging system, we study the schematic layout of an imaging system according to Fig. 10.31. In principle, the layout is that of a microscope equipped with a Köhler illumination system. The source with intensity distribution $I_S(X'_0, Y'_0)$ is located in the front focal plane of the condenser lens L_C and imaged in the back focal plane of the imaging lens L_1. The exit pupil sphere through E'_1 has its centre of curvature at central image point O_1. The coordinates (X'_1, Y'_1) on the spherical exit pupil surface are normalised with respect to the half diameter ρ_1 of the exit pupil which determines the aperture angle of an imaging pencil. The coordinates (X'_0, Y'_0) in the source plane have also been normalised with respect to $\rho_1/m_{T,p}$, where $m_{T,p}$ is the pupil magnification for imaging through the lenses L_C and L_1. For equal focal distances of condenser and imaging lens we have that $m_{T,p} = -1$. The object with transmission function $t_0(x'_0, y'_0)$ is located in the object plane through O_1. As usual, the coordinates (x'_0, y'_0) have been normalised with respect to the diffraction unit λ_0/NA in the object plane. When we consider the imaging of the object, the far-field function of t_0, commonly denoted by $\tilde{t}_1(X'_0, Y'_0)$, is found at the exit pupil of the imaging lens L_1 and has the character of a Fourier transform of t. Alternatively, its spectrum can be projected back to the source plane where it is denoted by $t_0(X'_0, Y'_0)$. The mutual coherence function in the object plane is denoted $\mu(\Delta'_{x_0}, \Delta'_{y_0})$ and, according to the van Cittert–Zernike theorem, it is the diffraction figure produced by the intensity distribution I_S of the source. The far field of the mutual coherence function $\mu(\Delta'_{x_0}, \Delta'_{y_0})$ is the source $I_S(X'_0, Y'_0)$; it is equally found at the exit pupil of lens L_1 through E'_1 where we denote it by $I_{S,1}(X'_1, Y'_1)$. The lateral magnification between the object and image plane is m_T. By using the normalised coordinate pairs (x'_0, y'_0) and (x'_1, y'_1) we effectively use a unit magnification system which can be modified to a real-world system with the general magnification m_T by going back to the real-space physical coordinates. The sign of the image magnification and the sign of the pupil magnification are taken into account by means of the quantities s_T and $s_{T,p}$ which equal ± 1 depending on the sign of the magnification.

We calculate the image intensity at a general point $\boldsymbol{P}'_1$ (the image of the point $\boldsymbol{P}_1$ in the object plane) by partially coherent addition at $\boldsymbol{P}'_1$ of the interfering complex amplitudes produced by general points $P_2(x'_{0,2}, y'_{0,2})$ and $P_3(x'_{0,3}, y'_{0,3})$ in the object plane. The interference of contributions from two neighbouring object points P_2 and P_3 in the image plane depends on their mutual coherence which is given by the value of the mutual coherence function $\mu_0(x'_{0,2} - x'_{0,3}, y'_{0,2} - y'_{0,3})$ in the object plane. It is generally assumed that this function only depends on the separation of two points in the object plane and not on the (average) position of the points in the object plane (shift-invariant property of the mutual coherence function). Shift invariance can be violated if the apparent source size changes over the object plane, for instance if the projected source size changes substantially in the object plane when large field angles occur in a wide-angle imaging system.

If the object function is a spatial δ-function at a general position (x'_0, y'_0), the image plane amplitude is given by the impulse response of the imaging system. According to a basic result of coherent Fourier optics, the image-plane amplitude A_1 is given by (see **[113]**, p. 112)

$$A_1(x'_1, y'_1; x'_0, y'_0) = t_0(x'_0, y'_0)\, A(x'_1 - s_T x'_0, y'_1 - s_T y'_0)\,, \tag{10.2.98}$$

where $t_0(x'_0, y'_0)$ is the object transmission at the location of the pinhole and $A(x'_1, y'_1)$ is the impulse response or 'diffraction pattern' of lens L_1 in the image plane. For a perfect lens the function $A(x'_1, y'_1)$ corresponds to the Airy (amplitude) pattern. In Fig. 10.32 we illustrate how the intensity at the point $\boldsymbol{P}'_1$ in the image plane is constructed from elementary contributions, namely the partially interfering amplitude impulse responses created by pairs of points (P_2, P_3). The total intensity is the sum of all these contributions when P_2 and P_3 occupy all possible positions in the object plane. The image intensity $I(x'_1, y'_1)$ is then given by

$$\begin{aligned} I(x'_1, y'_1) = \iiiint_{-\infty}^{+\infty} & \mu_0(x'_{0,2} - x'_{0,3}, y'_{0,2} - y'_{0,3})\, t_0(x'_{0,2}, y'_{0,2})\, A(x'_1 - s_T x'_{0,2}, y'_1 - s_T y'_{0,2}) \\ & \times t_0^*(x'_{0,3}, y'_{0,3})\, A^*(x'_1 - s_T x'_{0,3}, y'_1 - s_T y'_{0,3})\, dx'_{0,2}\, dy'_{0,2}\, dx'_{0,3}\, dy'_{0,3}\,. \end{aligned} \tag{10.2.99}$$

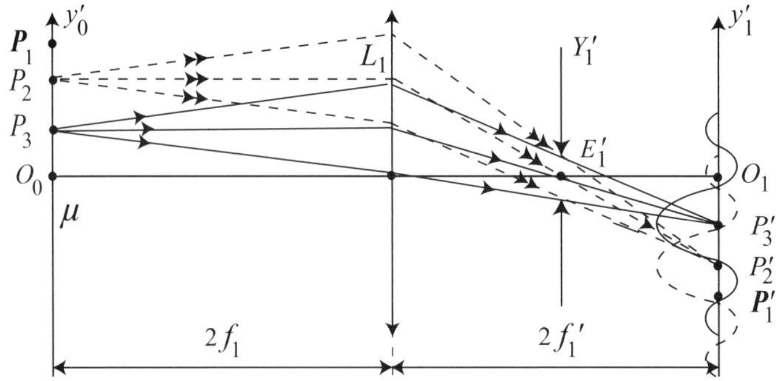

Figure 10.32: The intensity at the point $\boldsymbol{P}'_1$ in the image plane obtained from the partially coherent superposition of complex amplitudes corresponding to displaced coherent point-spread functions produced by the general points P_2 and P_3 (one-dimensional picture of the two-dimensional geometry).

Equation (10.2.99) can be evaluated for an arbitrary object transmission function t and coherence function μ_0. Analytic solutions of the fourfold integral are not available in the general case. The amount of numerical work is important since both the impulse response function A and the coherence function μ_0 show an oscillatory behaviour with slow convergence to zero for large values of their arguments.For this reason we examine an alternative formulation of the image plane intensity in the frequency domain which is more attractive with respect to numerical evaluation.

10.2.8.1 The Spatial Spectrum of the Image Intensity Function

Instead of calculating the image intensity function according to Eq. (10.2.99), the Fourier transform of the image intensity pattern is evaluated. We denote the (forward) Fourier transform of the image intensity by $\tilde{I}(X_1', Y_1')$ and it is given by (see Appendix A),

$$\tilde{I}(X_1', Y_1') = \iint_{-\infty}^{+\infty} I(x_1', y_1') \exp\{-2\pi i(x_1'X_1' + y_1'Y_1')\} \, dx_1' dy_1' \, . \tag{10.2.100}$$

When transforming the function $I(x_1', y_1')$ to the frequency domain, we preferably introduce the frequency counterparts of the near-field functions μ_0, t_0 and A in Eq. (10.2.99).

We use the following expressions for these object- and image-plane-related functions,

$$\mu_0(x_0', y_0') = \iint_{-\infty}^{+\infty} I_S(X_0', Y_0') \exp\{-2\pi i(x_0'X_0' + y_0'Y_0')\} \, dX_0' dY_0' \, , \tag{10.2.101}$$

$$t_0(x_0', y_0') = \iint_{-\infty}^{+\infty} \tilde{t}_0(X_0', Y_0') \exp\{-2\pi i(x_0'X_0' + y_0'Y_0')\} \, dX_0' dY_0' \, , \tag{10.2.102}$$

$$A(x_1', y_1') = \iint_{-\infty}^{+\infty} f_1(X_1', Y_1') \exp\{-2\pi i(x_1'X_1' + y_1'Y_1')\} \, dX_1' dY_1' \, , \tag{10.2.103}$$

where the functions $I_S(X_0', Y_0')$, $\tilde{t}_0(X_0', Y_0')$ and $f_1(X_1', Y_1')$ are associated with the source plane and the exit pupil of the lens L_1, respectively. With respect to the coherence function μ_0, the van Cittert–Zernike theorem shows that the mutual coherence function $\mu_0(x', y')$ is the Fourier transform of the source intensity distribution $I_S(X', Y')$ (see Eqs. (10.2.86)–(10.2.89)). As was concluded previously, only the aperture of the condenser lens L_C influences the function μ_0, the aberration of the condenser lens is irrelevant. The integrals in Eqs. (10.2.101)–(10.2.103) comprise exponential functions which originate from a pathlength difference expression in a diffraction integral of the form of Eq. (9.3.4). We refer to Section A.9 of Appendix A for the various sign options for the argument of the Fourier exponential function associated with forward and backward propagating waves. Non-essential proportionality factors have been omitted from Eqs. (10.2.101)–(10.2.103). According to the definition of the spatial Fourier transform (see Eqs. (A.8.5) and (A.8.6) of Appendix A), the three functions μ_0, t_0 and A are the Fourier transforms of the far-field functions $I_S(X_0', Y_0')$, $\tilde{t}_0(X_0', Y_0')$ and $f_1(X_1', Y_1')$ with respect to negative near-field coordinates.

We insert the expressions for μ_0, t_0 and A in Eq. (10.2.99) and then perform the Fourier transformation given by Eq. (10.2.100). After some lengthy manipulation and performing the integration with respect to the variable pairs (x_1', y_1'), $(x_{0,2}', y_{0,2}')$ and $(x_{0,3}', y_{0,3}')$ we obtain the expression

$$\begin{aligned}\tilde{I}(X_1', Y_1') = \int_{-\infty}^{+\infty} \cdots \int_{-\infty}^{+\infty} & I_S(X_{0,2}', Y_{0,2}') \, \tilde{t}_0(X_{0,3}', Y_{0,3}') \, \tilde{t}_0^*(X_{0,4}', Y_{0,4}') f_1(X_{1,2}', Y_{1,2}') f_1^*(X_{1,3}', Y_{1,3}') \\ & \times \delta(-X_{0,2}' - X_{0,3}' + s_T X_{1,2}', -Y_{0,2}' - Y_{0,3}' + s_T Y_{1,2}') \, \delta(X_{0,2}' + X_{0,4}' - s_T X_{1,3}', Y_{0,2}' + Y_{0,4}' - s_T Y_{1,3}') \\ & \times \delta(-X_{1,2}' + X_{1,3}' - X_1', -Y_{1,2}' + Y_{1,3}' - Y_1') \, dX_{0,2}' dY_{0,2}' dX_{0,3}' dY_{0,3}' dX_{0,4}' dY_{0,4}' dX_{1,2}' dY_{1,2}' dX_{1,3}' dY_{1,3}' \, , \end{aligned} \tag{10.2.104}$$

where the variables $X_{02}', \cdots, Y_{1,3}'$ are auxiliary integration variables. The integrations with respect to the variable pairs $(X_{0,2}', Y_{0,2}')$, $(X_{0,4}', Y_{0,4}')$ and $(X_{1,3}', Y_{1,3}')$ are performed using the sifting property of the δ-functions, yielding the expression

$$\begin{aligned}\tilde{I}(X_1', Y_1') = \iiiint_{-\infty}^{+\infty} & I_S(-X_2' + s_T X_3', -Y_2' + s_T Y_3') \, \tilde{t}_0(X_2', Y_2') \, \tilde{t}_0^*(X_2' + s_T X_1', Y_2' + s_T Y_1') \\ & \times f_1(X_3', Y_3') f_1^*(X_3' + X_1', Y_3' + Y_1') \, dX_2' dY_2' dX_3' dY_3' \, . \end{aligned} \tag{10.2.105}$$

To simplify the analysis, we image the far-field patterns in the source plane onto the exit pupil of lens L_1, such that the patterns associated with the source intensity function I_S, the spectrum $\tilde{t}_0$ of the object transmission function and the lens function f_1 can be directly compared in the same coordinate system (X'_1, Y'_1). For the imaging operation from the source plane to the exit pupil of L_1 we just require that the Abbe sine condition is fulfilled by the lens combination L_C–L_1. In that case, the normalised coordinate pairs (X'_0, Y'_0) and (X'_1, Y'_1) are identical when imaging the pupils (see Eq. (4.9.28)). The images of the far-field object-side functions I_S and $\tilde{t}_0$ are defined by

$$I_{S,1}(X'_1, Y'_1) = I_S[s_{T,p}(X'_0, Y'_0)] , \qquad \tilde{t}_1(X'_1, Y'_1) = \tilde{t}_0[s_{T,p}(X'_0, Y'_0)] , \tag{10.2.106}$$

where the factor $s_{T,p}$ is the sign of the normalised pupil magnification between the source plane and the exit pupil of lens L_1. We thus have the following expression for $\tilde{I}(X'_1, Y'_1)$, using (u_1, v_1) and (u_2, v_2) as integration variables,

$$\begin{aligned}\tilde{I}(X'_1, Y'_1) = \iiiint_{-\infty}^{+\infty} & I_{S,1}[s_{T,p}(-u_1+s_T u_2, -v_1+s_T v_2)]\, \tilde{t}_1[s_{T,p}(u_1, v_1)] \\ & \times \tilde{t}_1^*[s_{T,p}(u_1+s_T X'_1, v_1+s_T Y'_1)] f_1(u_2, v_2) f_1^*(u_2+X'_1, v_2+Y'_1)\, du_1 dv_1 du_2 dv_2 .\end{aligned} \tag{10.2.107}$$

For the frequently occurring case of a real object and image and a Köhler illumination system ($s_T = s_{T,p} = -1$), the expression for $\tilde{I}(X'_1, Y'_1)$ reads

$$\begin{aligned}\tilde{I}(X'_1, Y'_1) = \iiiint_{-\infty}^{+\infty} & I_{S,1}(u_1 + u_2, v_1 + v_2)\, \tilde{t}_1(-u_1, -v_1)\, \tilde{t}_1^*(-u_1 + X'_1, -v_1 + Y'_1)] \\ & \times f_1(u_2, v_2) f_1^*(u_2 + X'_1, v_2 + Y'_1)\, du_1 dv_1 du_2 dv_2 .\end{aligned} \tag{10.2.108}$$

By the change of variables $(u_1, v_1) \rightarrow (-u_1, -v_1)$ and by virtue of the infinite integration intervals, the expression for $\tilde{I}(X'_1, Y'_1)$ can also be written as

$$\begin{aligned}\tilde{I}(X'_1, Y'_1) = \iiiint_{-\infty}^{+\infty} & I_{S,1}(u_2 - u_1, v_2 - v_1)\, \tilde{t}_1(u_1, v_1)\, \tilde{t}_1^*(u_1 + X'_1, v_1 + Y'_1) \\ & \times f_1(u_2, v_2) f_1^*(u_2 + X'_1, v_2 + Y'_1)\, du_1 dv_1 du_2 dv_2 .\end{aligned} \tag{10.2.109}$$

The image spectrum $\tilde{I}(X'_1, Y'_1)$ has the character of a double correlation of the image-side spectrum $\tilde{t}_1$ and the pupil function f_1 with the image-side source intensity function $I_{S,1}$. The latter function plays the role of weighting function for both correlation integrals. The calculation of the image intensity spectrum requires evaluation of a fourfold integral. An advantage of the integral in the spectral domain with respect to the fourfold integral of Eq. (10.2.99) in the spatial domain is that the spectral functions $\tilde{t}_1, f_1$ and $I_{S,1}$ are only nonzero on a finite interval, as compared to the corresponding spatial functions which, in principle, have an infinite extent. For that reason, the spectral domain is generally preferred for image evaluation. However, the integral above requires extensive numerical labour when the continuous Fourier transformed functions $\tilde{t}_1, f_1$ and $I_{S,1}$ are used. This is especially the case if high spatial frequency content is present in $\tilde{t}_1$, if the lens function f_1 suffers from large aberration or when the source intensity function I_S is not a simple disc-like function. In such cases, it is recommended to switch from an analogous Fourier transform to a discrete Fourier series of the object function t_0. In the following subsection we discuss such a discrete Fourier series approach in more detail and show how the harmonic components in the image intensity spectrum can be obtained in a numerically efficient way.

The general expression for the image intensity spectrum of Eq. (10.2.109) allows for some physical interpretation, especially regarding the role of the source intensity function $I_{S,1}$. Often, the source function is a centred disc with a radius ρ_S. We define an incoherence factor σ for the object illumination by

$$\sigma = \frac{\rho_S}{\rho_0} . \tag{10.2.110}$$

The term 'incoherence factor' is in conflict with many literature sources where the factor σ is called the 'coherence' factor. We prefer to use *incoherence* factor for σ to avoid the situation where perfect coherence in the object plane would correspond to a 'coherence' factor of zero. Two extreme cases for object coherence are considered.

a) *Incoherence factor* $\sigma = 0$
The illumination is perfectly coherent and the source intensity function $I_{S,1}$ is approximated by a δ-function at the origin. The fourfold integral in Eq. (10.2.109) is reduced to a double integral according to

$$\tilde{I}(X_1', Y_1') = \iint_{-\infty}^{+\infty} \tilde{t}_1(u,v)f_1(u,v)\left\{\tilde{t}_1(u+X_1', v+Y_1')f_1(u+X_1', v+Y_1')\right\}^* du dv \,. \tag{10.2.111}$$

The integral is the autocorrelation of the spectral product function $\tilde{t}_1(u,v)f_1(u,v)$. Transforming back this function to the spatial domain yields the modulus squared of the convolution of the spatial functions t_1 and A. Inspection of Eq. (10.2.99) indeed shows that in the coherent illumination limit with $\mu_0 \equiv 1$ the fourfold integral is reduced to

$$I(x_1', y_1') = \left| \iint_{-\infty}^{+\infty} t_1(x', y') A(x_1' - x', y_1' - y')\, dx' dy' \right|^2 . \tag{10.2.112}$$

b) *Incoherence factor* $\sigma \to \infty$
For the case of a source diameter that is much larger than the pupil diameter, the incoherent illumination condition is effectively satisfied. We note that the maximum value of the incoherence factor σ of a classical illumination system cannot exceed $(\sin\alpha)^{-1}$ where $n\sin\alpha$ is the object-side numerical aperture of the imaging system, n being the refractive index in which the object is immersed. Supposing that we can effectively reach a quasi infinitely large value for σ for a uniform radiating source, the source intensity function $I_{S,1}$ in Eq. (10.2.109) equals unity over the total integration range and can be taken out of the integrand. The integral is then split into two separate integrals,

$$\tilde{I}(X_1', Y_1') = \iint_{-\infty}^{+\infty} \tilde{t}_1(u_1,v_1)\,\tilde{t}_1^*(u_1+X_1', v_1+Y_1')\, du_1 dv_1 \iint_{-\infty}^{+\infty} f_1(u_2,v_2)f_1^*(u_2+X_1', v_2+Y_1')\, du_2 dv_2 \,. \tag{10.2.113}$$

The inverse Fourier transform of $\tilde{I}(X_1', Y_1')$ is the transform of a product of two autocorrelation integrals and this yields the convolution of the moduli squared of the Fourier transforms of $\tilde{t}_1$ and f_1, thus of $|t_1|^2$ and $|A|^2$. This result also follows from Eq. (10.2.99) when the function μ_0 is put equal to a spatial δ-function, yielding the 'incoherent' result

$$I(x_1', y_1') = \iint_{-\infty}^{+\infty} \left|t(x', y')\right|^2 \left|A(x_1' - x', y_1' - y')\right|^2 dx' dy' \,. \tag{10.2.114}$$

The image intensity is obtained by a convolution of the intensity transmission function and the intensity of the point-spread function. In practice, the condition $\sigma = 1$ is often considered to be equivalent to 'incoherent' illumination because this is the ultimate value which can be reached in a high-aperture system. In principle, however, this statement is incorrect.

10.2.8.2 Graphical Illustration of Partially Coherent Imaging

In this subsection we consider partially coherent imaging and focus on the expression for the image intensity spectrum of Eq. (10.2.109). In particular, this expression is graphically illustrated with the aid of Figs. 10.33a and b. In these diagrams we have sketched how a particular (normalised) frequency pair (X_1', Y_1') of the image intensity spectrum is formed and how its strength is found. The object transmission function is $t_0(x_0', y_0')$. In the graphical constructions, we use the frequency spectrum $\tilde{t}_1$ at the image side of lens L_1. We assume that, in principle, the object spectrum has an infinitely large extent.

- *Coherent imaging*
  In Fig. 10.33a the illumination is coherent with a centred source which yields a plane wave with transverse wave vector components $(0,0)$ in object space. From the (continuous) object spectrum created by the incident wave parallel to the optical axis we have selected two diffracted plane waves, labelled 0 and 1, with strengths $\tilde{t}_1(u,v)$ and $\tilde{t}_1(u+X_1', v+Y_1')$, respectively. They have a frequency difference of (X_1', Y_1') and, if they were the only diffracted components in image space, they would produce an interference pattern with a periodic intensity and a complex interference factor which is proportional to $a(u,v;X_1',Y_1') = \tilde{t}_1^*(u+X_1', v+Y_1')f_1^*(u+X_1', v+Y_1')\,\tilde{t}_1(u,v)f_1(u,v)$. The shift of the periodic intensity pattern depends on the argument of the complex factor a. The summation over all diffracted beam pairs with frequency difference (X_1', Y_1') yields a final periodic pattern whose amplitude and phase are determined by the 'coherent' summation as a function of u and v of the factors $a(u,v;X_1',Y_1')$,

$$a(X_1', Y_1') = I_{S,1}(0,0)\; \overline{\tilde{t}_1^*(u+X_1', v+Y_1')f_1^*(u+X_1', v+Y_1')\,\tilde{t}_1(u,v)f_1(u,v)}\,\Big|_{(u,v)}, \tag{10.2.115}$$

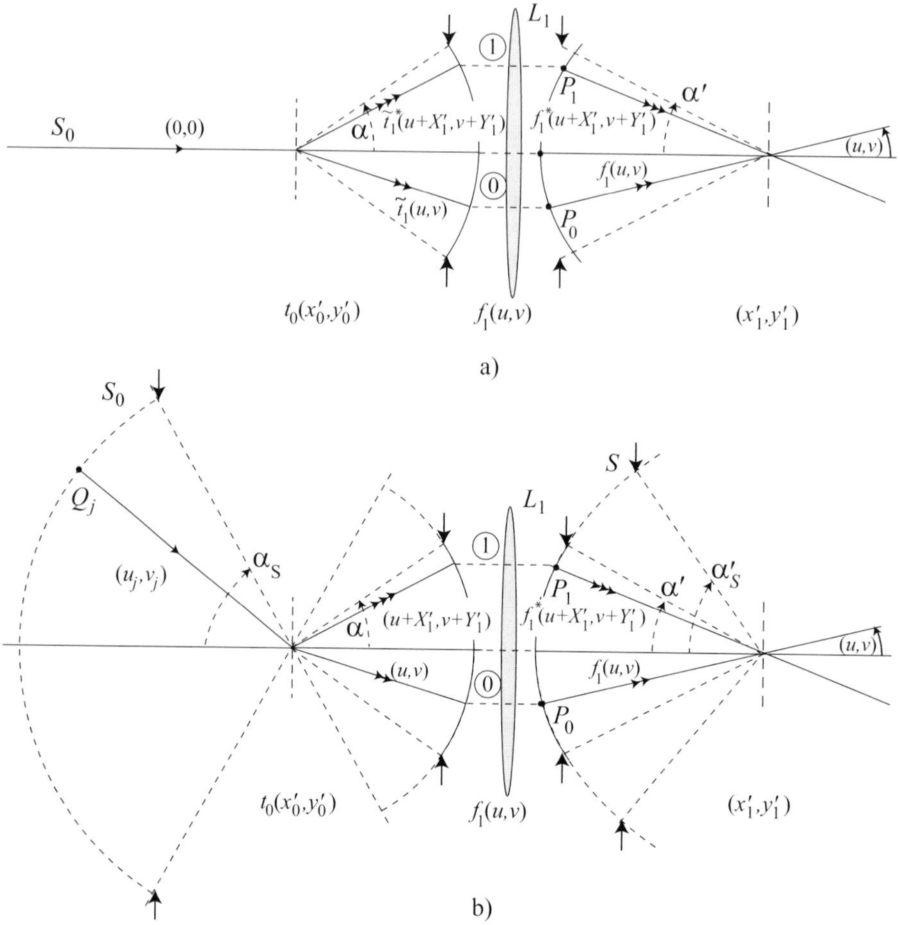

Figure 10.33: The coherent a), and incoherent b), calculation scheme to obtain the spectrum $\tilde{I}(X_1', Y_1')$ of the image intensity.

where $I_{S,1}(0,0)$ is the intensity of the axial source point and the bar denotes averaging over the object frequencies (u, v). It is clear from the diagram that the highest possible spatial frequency that is found in the image plane is given by $2 \sin\alpha'/\lambda$, or, in normalised frequency units, $(X_1'^2 + Y_1'^2)^{1/2} = 2$. This maximum spatial frequency is produced by the interference of diffracted beams with lateral frequency components $\pm \sin\alpha'/\lambda$ in image space, or $\pm$ 1 in normalised frequency units in object or image space. The presence of such a doubled frequency in the image plane is due to the nonlinearity of the image intensity formation. When the contrast in the object transmission function decreases, the nonlinear term, proportional to the square of this contrast, becomes imperceptible.

- *Partially coherent imaging*
  In Fig. 10.33b an extended incoherent source S_0 is shown with an angular size of α_S in object space. A microscope illumination system of the Köhler-type projects the source to infinity such that an incoherent source point Q_j sends a plane wave with lateral frequency components (u_j, v_j) to the object plane. In the object space of the imaging system, we consider again two propagation directions labelled 0 and 1, with the lateral plane wave components (u, v) and $(u + X_1', v + Y_1')$, respectively. For the general source point Q_j, the frequency components $(u - u_j, v - v_j)$ and $(u - u_j + X_1', v - v_j + Y_1')$ in the spectrum $\tilde{t}_1$ give rise in the diagram to the two diffracted beams with directions labelled 0 and 1, respectively. We note that within the scalar approximation, the frequency spectrum $\tilde{t}_1$ is assumed to be independent of the angle of incidence of the plane wave on the object and is uniquely determined by the difference in lateral frequency of the diffracted and incident beams. It is clear that this approximation becomes highly questionable for high-numerical-aperture systems with large diffraction angles. For the complex factor a which was discussed previously for

the single centred source point, we now write $a(u, v; u_j, v_j; X'_1, Y'_1) = I_{S,1}(u_j, v_j)\, \tilde{t}_1(u - u_j, v - v_j)\, \tilde{t}_1^*(u - u_j + X'_1, v - v_j + Y'_1) f_1(u, v) f_1^*(u + X'_1, v + Y'_1)$. The averaging of the factor $a(u, v; u_j, v_j; X'_1, Y'_1)$ is now to be performed over the relevant ranges of (u_j, v_j) in the source and (u, v) in the pupil function f. By a shift of coordinates, we are allowed to average over the quantities $(u_1, v_1) = (u - u_j, v - v_j)$ and $(u_2, v_2) = (u, v)$, respectively. The intensity modulation factor for the frequency (X'_1, Y'_1) is thus given by

$$a(X'_1, Y'_1) = \overline{I_{S,1}(u_2 - u_1, v_2 - v_1)\, \tilde{t}_1(u_1, v_1)\, \tilde{t}_1^*(u_1 + X'_1, v_1 + Y'_1)\, f_1(u_2, v_2) f_1^*(u_2 + X'_1, v_2 + Y'_1)}\,, \tag{10.2.116}$$

where an averaging over the frequency pairs (u_1, v_1) and (u_2, v_2) has to be performed. The averaging operations for the partially coherent imaging case are equivalent to the fourfold integral in the analytic expression for the spectrum of the image plane intensity function of Eq. (10.2.109).

- *Incoherent imaging*

A special case arises if the size of the source is such that the range of nonzero frequencies (u_1, v_1) is much larger than that of the frequencies (u_2, v_2). It is then permissible to write $I_{S,1}(u_2 - u_1, v_2 - v_1) \approx I_{S,1}(-u_1, -v_1)$ and the fourfold averaging operation according to Eq. (10.2.116) can be split into two independent averaging expressions,

$$a(X'_1, Y'_1) = \left.\overline{I_{S,1}(-u_1, -v_1)\, \tilde{t}_1(u_1, v_1)\, \tilde{t}_1^*(u_1 + X'_1, v_1 + Y'_1)}\,\right|_{(u_1, v_1)}$$
$$\times \left.\overline{f_1(u_2, v_2) f_1^*(u_2 + X'_1, v_2 + Y'_1)}\,\right|_{(u_2, v_2)} . \tag{10.2.117}$$

If the source has uniform intensity, this qualitative result corresponds to Eq. (10.2.113) where it was shown that the image intensity spectrum can be written as the product of two independent correlation integrals of the complex object spectrum $\tilde{t}_1$ and the complex pupil function f, respectively.

- *Partially coherent imaging of a knife-edge object*

The effect of the illumination mode on the image intensity is now further illustrated by means of the response of an imaging system to a one-dimensional step-function (knife edge) in the object plane. The amplitude transmission of the knife edge is defined by $H(-x'_0)$ where $H(x)$ is the 'step' or Heaviside function (see Appendix A, Table A.3). The knife edge is an interesting object since its spectrum has nonzero values up to infinity. When the edge is exactly at the position $x'_0 = 0$, the normalised on-axis intensity in the image plane equals 1/4 for the coherent case with $\sigma = 0$ and 1/2 for the fully incoherent case ($\sigma \rightarrow \infty$). The two curves in Fig 10.34 have been calculated with the aid of Eq. (10.2.112) and Eq. (10.2.114), for the coherent and incoherent case, respectively. The convolution of the ideal point-spread function PSF, equal to the function $A(x'_1, y'_1)$ in both equations, and of the knife edge function as the transmission function t, leads to an integral with oscillating integrand, especially for the case of coherent illumination. The amplitude oscillations in the first rings of the coherent PSF (Airy disc) reach a maximum value of approximately 0.15. The oscillations of the intensity PSF are smaller by an order of magnitude. A strong oscillatory behaviour is

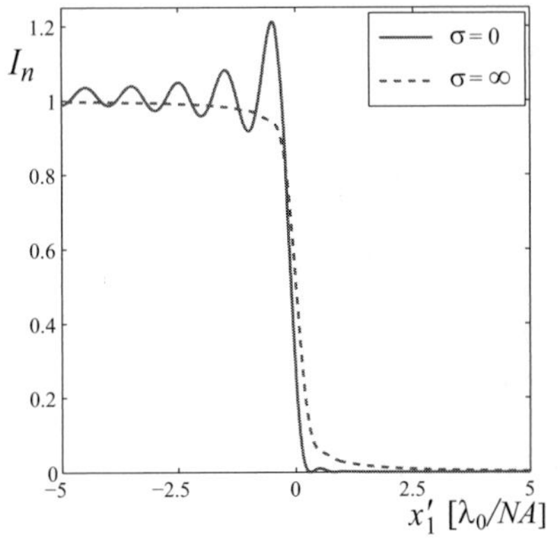

Figure 10.34: The normalised intensity I_n in the image plane as a function of the lateral position x'_1 in units λ_0/NA of a knife edge (amplitude step function) in the object plane. The incoherence factor σ of the illumination is either 0 (coherent illumination) or ∞, the limiting case of incoherent illumination.

thus observed for the coherent knife edge response, with an overshoot in the bright region of the order of 20%. On the contrary, the incoherently imaged knife edge shows a monotonically decreasing intensity profile when proceeding from the transmitting side to the shadow side of the knife edge. The steep slope of the intensity profile in the image plane of the coherently illuminated knife edge might be considered as an advantage. A serious problem, however, of the coherent edge response is the offset between the geometrical position of the knife-edge transition in the object plane and the measured 50% intensity level in the image plane. When using the 50% intensity as the decision level for the geometrical position of the edge, the real position of the knife edge is found a distance of $\approx 0.1\ \lambda_0/NA$ inside the geometrical shadow region of the knife edge image. This shift could be accounted for if it were a reproducible feature. But this shift of $0.1\ \lambda_0/NA$ is influenced by the presence of other neighbouring transients in the object. For that reason, coherent optical metrology needs an inversion procedure in which the presence of all the transients in the object is taken into account.

Measurement of the position of object transients is more reliable when using incoherent object illumination. Unfortunately, high-resolution metrology requires a high image-side numerical aperture and this requirement is incompatible with (fully) incoherent illumination. In Fig. 10.35 we sketch the experimental conditions that would approach as closely as possible the fully coherent, a), and fully incoherent imaging condition, b). In a) the sine of the source aperture angle α_0 is very small. In the case of laser illumination, the divergence angle of the beam would be a good measure of the illumination aperture that determines the mutual coherence function μ. The solid angle into which light coherently scattered by the knife edge is collected is given by $\sin\alpha_I$. In a) this quantity is close to the theoretical limit of unity. The size of the illumination point-spread function in the object plane is inversely proportional to $\sin\alpha_0$. A relatively large coherence is obtained when the coherence area is many image-side point-spread functions wide. In that case, the incoherence factor $\sigma = \sin\alpha_0/\sin\alpha_I$ becomes very small. For a non-lasing source with a much lower brilliance, the value of α_0 has to be increased for reasons of light efficiency and the coherence in the object illumination is reduced, leading to a larger

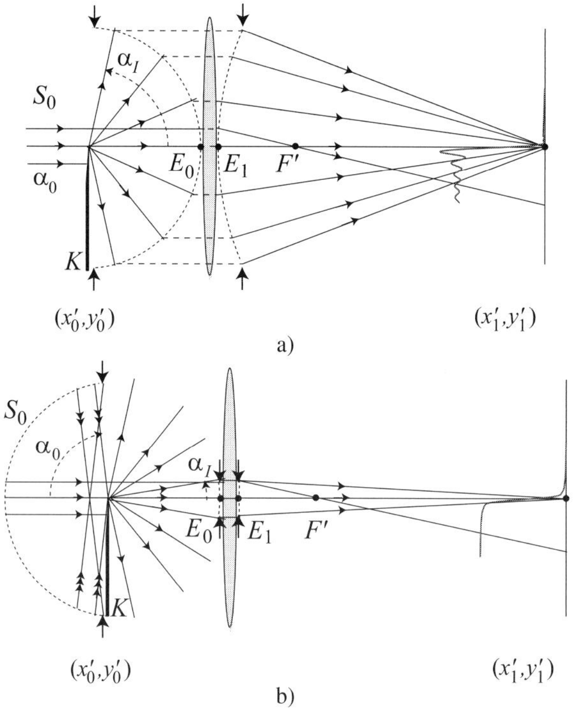

Figure 10.35: Imaging of a knife-edge object K. The source aperture is given by $\sin\alpha_0$, the imaging aperture by $\sin\alpha_I$. E_0 and E_1 are the centres of entrance and exit pupil, respectively, F' is the image-side focal point.
a) Maximum coherence is achieved in the imaging process.
b) The incoherence of the imaging is maximum.

σ value. We conclude that a small σ value is best achieved when the imaging aperture, for instance of a microscope objective, is made as high as possible. For the incoherent illumination situation which is depicted in Fig. 10.35b, the setting is reversed. We require as large as possible a ratio $\sin\alpha_0/\sin\alpha_I$. This is achieved by having a high-numerical-aperture condenser and a low-aperture imaging system. Unfortunately, this leads to a low-resolution extraction of the object features. In general this is not acceptable for high-resolution optical metrology.

- *Intermediate values of the incoherence factor* $(0 < \sigma < \infty)$
  We conclude that for high-resolution imaging an incoherence factor σ of approximately unity corresponds to the largest achievable incoherence in the object illumination. In that case the simplified formulas of Eqs. (10.2.112) and (10.2.114) cannot be applied because the spatial functions μ and A have a comparable spatial extent. For this intermediate case the general theory of partially coherent imaging should be used. Preferably, one switches to the frequency domain and calculates the spectrum of the image intensity according to Eq. (10.2.109). This approach is discussed in more detail in Subsection 10.2.10 where the imaging of strongly modulated objects is treated.

10.2.9 A Weakly Modulated Periodic Object in Partially Coherent Illumination

In this subsection we calculate the image plane intensity for the case of a weakly modulated object using the double convolution integral in the spatial domain given by Eq. (10.2.99). We apply this expression to the simple case of an object containing a single frequency. The amplitude transmission function of the object is given by

$$t(x'_1, y'_1) = 1 + m\cos[2\pi(x'_1 u'_{x,1} + y'_1 u'_{y,1})]\,, \tag{10.2.118}$$

where we have omitted a trivial phase factor in the argument of the cosine function associated with a spatial shift of the object from its zero reference position. As before, we use normalised coordinates in object and image space such that the coordinates can be interchanged at will. We limit m to values $\ll 1$ to avoid noticeable quadratic terms with an m^2 prefactor in the intensity transmission function $T = t\,t^*$. The modulation depth of the function T is then equal to $2m$. The substitution of the periodic transmission function t and the Fourier transformed functions for μ and A according to Eqs. (10.2.101)–(10.2.103) leads to a rather unwieldy expression. With the cosine function written in exponential form we obtain

$$\begin{aligned} I(x'_1, y'_1) = &\iint\cdots\iint \left\{1 + \frac{m}{2}\exp[2\pi i(x'_2 u'_{x,1} + y'_2 u'_{y,1})] + \frac{m}{2}\exp[-2\pi i(x'_2 u'_{x,1} + y'_2 u'_{y,1})]\right\} \\ &\times \left\{1 + \frac{m}{2}\exp[-2\pi i(x'_3 u'_{x,1} + y'_3 u'_{y,1})] + \frac{m}{2}\exp[+2\pi i(x'_3 u'_{x,1} + y'_3 u'_{y,1})]\right\} \\ &\times I_{S,1}(X'_1, Y'_1)\, f(X'_2, Y'_2)\, f^*(X'_3, Y'_3)\, \exp\{+2\pi i[x'_1(X'_3 - X'_2) + y'_1(Y'_3 - Y'_2)]\} \\ &\times \exp\{2\pi i[x'_2(X'_2 - X'_1) + y'_2(Y'_2 - Y'_1)]\}\, \exp\{2\pi i[x'_3(X'_1 - X'_3) + y'_3(Y'_1 - Y'_3)]\} \\ &\times dx'_2 dy'_2 dx'_3 dy'_3 dX'_1 dY'_1 dX'_2 dY'_2 dX'_3 dY'_3\,. \end{aligned} \tag{10.2.119}$$

The collection of the exponential functions that depend on the coordinate pairs (x'_2, y'_2) and (x'_3, y'_3) leads to integrals which, after evaluation, yield δ-functions with arguments $\delta(X'_2 - X'_1 \pm u'_{x,1}, Y'_2 - X'_1 \pm u'_{y,1})$ and $\delta(X'_1 - X_3 \pm u'_{x,1}, Y'_1 - Y'_3 \pm u'_{y,1})$.

We neglect the terms in Eq. (10.2.119) that are quadratic in the modulation factor m and perform the integrations over the variable pairs (X'_2, Y'_2) and (X'_3, Y'_3), resulting in the integral expression

$$\begin{aligned} I(x'_1, y'_1) = \iint I_{S,1}(X'_1, Y'_1) \Big[& f(X'_1, Y'_1) f^*(X'_1, Y'_1) + \frac{m}{2}\left\{ f(X'_1, Y'_1) f^*(X'_1 + u'_{x,1}, Y'_1 + u'_{y,1}) + f^*(X'_1, Y'_1) f(X'_1 - u'_{x,1}, Y'_1 - u'_{y,1})\right\} \\ & \times \exp\{+2\pi i(x'_1 u'_{x,1} + y'_1 u'_{y,1})\} + \frac{m}{2}\left\{ f(X'_1, Y'_1) f^*(X'_1 - u'_{x,1}, Y'_1 - u'_{y,1}) + f^*(X'_1, Y'_1) f(X'_1 + u'_{x,1}, Y'_1 + u'_{y,1})\right\} \\ & \times \exp\{-2\pi i(x'_1 u'_{x,1} + y'_1 u'_{y,1})\} \Big]\, dX'_1 dY'_1\,. \end{aligned} \tag{10.2.120}$$

In terms of the *OTF* function of Eq. (10.2.33), the normalised expression for $I(x'_1, y'_1)$ is given by

$$\begin{aligned} I_n(x'_1, y'_1) = 1 + \frac{m}{2}\Big[& \left\{ OTF_p(u'_{x,1}, u'_{y,1}) + OTF^*_p(-u'_{x,1}, -u'_{y,1})\right\} \exp\{+2\pi i(x'_1 u'_{x,1} + y'_1 u'_{y,1})\} \\ & + \left\{ OTF_p(-u'_{x,1}, -u'_{y,1}) + OTF^*_p(u'_{x,1}, u'_{y,1})\right\} \exp\{-2\pi i(x'_1 u'_{x,1} + y'_1 u'_{y,1})\} \Big]\,, \end{aligned} \tag{10.2.121}$$

where we have written everywhere OTF_p to emphasise the fact that the partially coherent OTF function has to be used, which takes into account the influence of the source shape on the transfer of an optical frequency. The expression for the real function $I_n(x'_1, y'_1)$ can also be written with the aid of the real and imaginary parts of the optical transfer function, yielding

$$I_n(x'_1, y'_1) = 1 + m\,\Re\left\{OTF_p(u'_{x,1}, u'_{y,1}) + OTF_p(-u'_{x,1}, -u'_{y,1})\right\}\cos\{2\pi(x'_1 u'_{x,1} + y'_1 u'_{y,1})\}$$
$$- m\,\Im\left\{OTF_p(u'_{x,1}, u'_{y,1}) - OTF_p(-u'_{x,1}, -u'_{y,1})\right\}\sin\{2\pi(x'_1 u'_{x,1} + y'_1 u'_{y,1})\}\,. \qquad (10.2.122)$$

If the pupil and source function are symmetric with respect to the origin, the function OTF_p is real and we have that $OTF_p(u'_{x,1}, u'_{y,1}) = OTF_p(-u'_{x,1}, -u'_{y,1})$. The expression for $I_n(x'_1, y'_1)$ then reduces to

$$I_n(x'_1, y'_1) = 1 + 2m\,OTF_p(u'_{x,1}, u'_{y,1})\,\cos\{2\pi(x'_1 u'_{x,1} + y'_1 u'_{y,1})\}\,. \qquad (10.2.123)$$

The modulation depth of the intensity pattern is thus given by $2m\,OTF_p(u'_{x,1}, u'_{y,1})$ whilst the initial modulation depth of the intensity transmission function in the object plane is $2m$ for sufficiently small values of m.

It has been shown in this subsection that, in the presence of a weakly modulated periodic object function, the expression for the image plane intensity of Eq. (10.2.99) can be simplified. It turns out that the modulation depth of the periodic intensity function in the image plane is equal to the product of the modulation depth $2m$ of the intensity transmission of the object structure and the partially coherent transfer function of the imaging system.

10.2.9.1 Image Intensity Calculation for Special Mutual Coherence Functions

In the preceding subsections we have derived expressions in the spatial and frequency domains for computation of the modulation depth of the image intensity corresponding to a periodic object (see Eqs. (10.2.99) and (10.2.120)). We now illustrate the spatial domain approach for a relatively simple case where the imaging system is aberration free and has a uniform transmission ($f(X'_1, Y'_1) \equiv 1$). We also assume that the source intensity function is point-symmetric and is zero outside the unit disc (in normalised coordinates). Inspection of Eq. (10.2.120) for this particular case shows that the product of $f(X'_1, Y'_1)\,I_{S,1}(X'_1, Y'_1)$ can be replaced by $I_{S,1}(X'_1, Y'_1)$ only. The inverse Fourier transformation of the corresponding simplified version of Eq. (10.2.120) yields the following expression for the modulation depth m_I of the intensity pattern in the image plane,

$$m_I(u'_{x,1}, u'_{y,1}) = \frac{\iint_{-\infty}^{+\infty} \mu(x'_1, y'_1)\,A^*(x'_1, y'_1)\cos\left\{2\pi(u'_{x,1}x'_1 + u'_{y,1}y'_1)\right\}\,dx'_1\,dy'1}{\iint_{-\infty}^{+\infty} \mu(x'_1, y'_1)\,A^*(x'_1, y'_1)\,dx'_1\,dy'_1}\,. \qquad (10.2.124)$$

We note that only a two-dimensional integration is required instead of the original fourfold integral in the spatial domain. We have calculated the modulation depth m_I in the image plane for two special mutual coherence functions, the first given by Eq. (10.2.96) for annular illumination and the second given by Eq. (10.2.97) for quadrupole illumination. The complex amplitude of the point-spread function of a point object in O_O is given by an Airy disc, centred in O_I,

$$A(x'_1, y'_1) = \frac{J_1\{2\pi(x_1'^2 + y_1'^2)^{1/2}\}}{\pi(x_1'^2 + y_1'^2)^{1/2}}\,. \qquad (10.2.125)$$

In the absence of aberrations, the amplitude $A(x'_1, y'_1)$ of the point-spread function is real and shows circular symmetry.

In Fig. 10.36 we have plotted the modulation depth m_I according to Eq. (10.2.124) for a periodic intensity pattern in the object plane. The three mutual coherence functions that were discussed previously (see Fig. 10.30), have been used in the calculations. The frequencies were limited to the positive frequency interval $0 \le u'_1 \le 2$. A comparison with Figs. 10.15, 10.17c and 10.17d immediately shows that, not surprisingly, the curves in Fig. 10.36 are identical to the optical transfer functions for the standard, annular and quadrupole illumination. The modulation enhancement for higher spatial frequencies by a special illumination method is illustrated in Fig. 10.30 where we have shown the mutual coherence functions for the three illumination modes. The extended coherence regions introduced by annular or quadrupole illumination have virtually equal-sized coherence 'satellites'. These extra coherence regions produce a modulation enhancement when their width and phase are well synchronised with the periodic amplitude transmission function of the object. For a quadrupole illumination with the sub-sources centred on the diagonals of the four source quadrants, the modulation enhancement is more pronounced for frequency vectors **u** along the x'_1- or y'_1-axis and strongest in the frequency interval $1 < u'_{x,1} < 1.5$. Annular illumination produces better modulation in the interval $1.2 < u'_{x,1} < 2$.

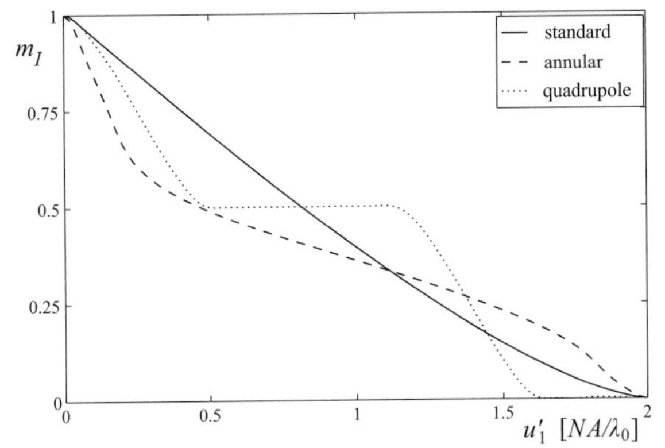

Figure 10.36: The modulation depth m_I of the intensity distribution in the image plane as a function of the spatial frequency vector $(u'_1, 0)$ for a periodic object. Calculation in the spatial domain according to Eq. (10.2.124).

The enhancement of the higher frequencies is at the cost of a reduced modulation transfer in the interval $0 < u'_{x,1} < 1$. This effect follows from Fig. 10.30 where at low values of the frequency $u'_{x,1}$ the object transmission function shows only a modest change in the region of mutual coherence. The strong coherence sidelobes with possibly opposite phase in the case of annular and quadrupole illumination give rise to partial cancellation of complex amplitude in a general receiving point $\boldsymbol{P}'_1$ in the image plane. The consequence is a drop in modulation depth of the intensity pattern for lower spatial frequencies.

The modulation depth m_I of Eq. (10.2.124) has been calculated by integration in the spatial domain while the previous *OTF* calculations were based on integrations in the spatial frequency domain. It should be emphasised that the integrations in the frequency domain are advantageous with respect to calculation time because the integrations extend over well-defined limited domains (pupil cross-section of a lens or a source area). Moreover, it has turned out to be possible to obtain semi-analytic expressions for the general *OTF* of an aberrated imaging system **[163]**. Contrarily, in the spatial domain, we deal with oscillating impulse response functions that are due to the diffraction at sharp edges in the system like the rim of the system stop (diaphragm). There is also the important fact that, apart from dealing with well-behaving functions, the frequency analysis requires the evaluation of only a double integral instead of the fourfold integral of Eq. (10.2.99). For these reasons, the frequency domain analysis of optical systems is generally preferred. When more complicated non-periodic objects with a small lateral extent have to be imaged, the spatial domain analysis can be numerically more advantageous.

10.2.10 Imaging of a Strong Object with Partially Coherent Illumination

The imaging of an arbitrary object implies that the modulation depth in the object transmission function may have large values. Moreover, an entire spectrum of spatial frequencies is present. Most conveniently, using a fast Fourier transform calculation, the object amplitude transmission function t is represented by a two-dimensional Fourier series expansion,

$$t(x'_1, y'_1) = \sum_n \sum_m c_{n,m} \exp\{i2\pi[n\, u'_{x,1} x'_1 + m\, u'_{y,1} y'_1]\} , \tag{10.2.126}$$

where, as before, we describe the object function t as a function of the (normalised) image plane coordinates taking into account the magnification of the imaging system. The object function is defined on a generally rectangular spatial window with area $L_x L_y$ and the frequencies in the Fourier series are multiples of the inverse distances $1/L_x$ and $1/L_y$ in the two orthogonal directions. The complex valued $c_{n,m}$ are the Fourier coefficients of the periodic object function t. To calculate the image plane intensity according to Eq. (10.2.119), we need the product function $t(x'_2, y'_2)\, t^*(x'_3, y'_3)$ which is written as

$$t(x'_2, y'_2)\, t^*(x'_3, y'_3) = \sum_n \sum_{n'} \sum_m \sum_{m'} c_{n,m}\, c^*_{n',m'} \exp\{i2\pi[u'_{x,1}(nx'_2 - n'x'_3) + u'_{y,1}(my'_2 - m'y'_3)]\} . \tag{10.2.127}$$

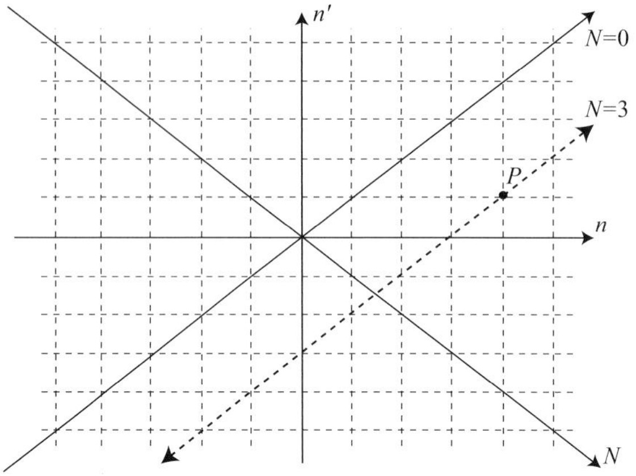

Figure 10.37: Combination of the interfering diffracted beams in image space according to a square (n, n') or a diagonal (N, N') summation scheme. The point P represents a general diffracted beam with indices $(n, n') = (4, 1)$ and is located on a diagonal with index $N = 3$.

Each term of the quadruple sum in Eq. (10.2.127) can be considered as the product of the complex amplitudes of two planar waves in image space which were diffracted by the object structure. The complex amplitudes are measured in the points with coordinates (x'_2, y'_2) and (x'_3, y'_3) and are associated with plane waves having the lateral wave vector components $(nu'_{x,1}, mu'_{y,1})$ and $(n'u'_{x,1}, m'u'_{y,1})$. Periodic features in the image plane intensity pattern depend on the difference between the lateral wave vector components of a pair of diffracted beams. For that reason, it is advisable to carry out the summation in Eq. (10.2.127) in such a way that terms are collected with identical values of the lateral wave vector difference. In Fig. 10.37 such a scheme has been sketched. By means of diagonal summation, all diffracted beams associated with an (off)-diagonal track of the rectangular grid of diffracted beams have equal difference in lateral wave vector component and are added to yield the power of a certain harmonic component in the image plane intensity pattern. We exemplify the diagonal summation for a two-dimensional function $g(x'_2, x'_3) = f(x'_2)f^*(x'_3)$ with a Fourier expansion of $f(x)$ given by $f(x) = \sum_n a_n \exp(i2\pi nu'_{x,1}x')$. For the function g we write

$$g(x'_2, x'_3) = \sum_{n=-n_m}^{+n_m} \sum_{n'=-n_m}^{+n_m} b_n c^*_{n'} \quad \text{where} \quad \begin{cases} b_n = a_n \exp(i2\pi nu'_{x,1}x'_2) \,, \\ c_{n'} = a_{n'} \exp(i2\pi n' u'_{x,1}x'_3) \,, \end{cases} \tag{10.2.128}$$

where n_m is the highest harmonic number that is included in the Fourier expansion. Diagonal summation yields the expression

$$\begin{aligned} g(x'_2, x'_3) &= \sum_{n=-n_m}^{+n_m} b_n c^*_n + \sum_{N=1}^{2n_m} \left\{ \sum_{n=-n_m+N}^{n_m} b_n c^*_{n-N} + \sum_{n=-n_m}^{n_m-N} b_n c^*_{n+N} \right\} \\ &= \sum_{n=-n_m}^{+n_m} b_n c^*_n + \sum_{N=1}^{2n_m} \sum_{n=-n_m+N}^{n_m} (b_n c^*_{n-N} + b_{n-N} c^*_n) \\ &= \sum_{n=-n_m}^{+n_m} |a_n|^2 \exp\{i2\pi u'_{x,1} n(x'_2 - x'_3)\} + \sum_{N=1}^{2n_m} \sum_{n=-n_m+N}^{n_m} \Big[a_n a^*_{n-N} \\ &\quad \times \exp\{i2\pi u'_{x,1}[nx'_2 - (n-N)x'_3]\} + a_{n-N} a^*_n \exp\{i2\pi u'_{x,1}[(n-N)x'_2 - nx'_3]\} \Big] \,, \end{aligned} \tag{10.2.129}$$

where the quantity $2\pi N u'_{x,1}$ denotes the difference of the transverse wave vector components of the diffracted beams in the summation. Extension to the two-dimensional transmission function $t(x, y)$ with its Fourier expansion given by Eq. (10.2.126) is straightforward and the product function $t(x'_2, y'_2)\, t^*(x'_3, y'_3)$ is given by four terms $T(N, M)$,

$$T(0,0) = \sum_{n=-n_m}^{+n_m} \sum_{m=-m_m}^{+m_m} |c_{n,m}|^2 \exp\{i2\pi[nu'_{x,1}(x'_2 - x'_3) + mu'_{y,1}(y'_2 - y'_3)]\}\,, \tag{10.2.130}$$

$$T(N,0) = \sum_{N=1}^{2n_m} \sum_{n=-n_m+N}^{+n_m} \sum_{m=-m_m}^{+m_m} \Big[c_{n,m}c^*_{n-N,m} \exp\{i2\pi[u'_{x,1}(nx'_2 - (n-N)x'_3) + u'_{y,1}m(y'_2 - y'_3)]\} + c^*_{n,m}c_{n-N,m} \exp\{i2\pi[u'_{x,1}((n-N)x'_2 - nx'_3) + u'_{y,1}m(y'_2 - y'_3)]\}\Big]\,, \tag{10.2.131}$$

$$T(0,M) = \sum_{M=1}^{2m_m} \sum_{n=-n_m}^{+n_m} \sum_{m=-m_m+M}^{+m_m} \Big[c_{n,m}c^*_{n,m-M} \exp\{i2\pi[u'_{x,1}n(x'_2 - x'_3) + u'_{y,1}(m(y'_2 - (m-M)y'_3]\} + c^*_{n,m}c_{n,m-M} \exp\{i2\pi[u'_{x,1}n(x'_2 - x'_3) + u'_{y,1}((m-M)y'_2 - my'_3)]\}\Big]\,, \tag{10.2.132}$$

$$T(N,M) = \sum_{N=1}^{2n_m} \sum_{M=1}^{2m_m} \sum_{n=-n_m+N}^{+n_m} \sum_{m=-m_m+M}^{+m_m} \Big[c_{n,m}c^*_{n-N,m-M} \exp\{i2\pi[u'_{x,1}(nx'_2 - (n-N)x'_3) + u'_{y,1}(my'_2 - (m-M)y'_3)]\}$$
$$+ c^*_{n,m}c_{n-N,m-M} \exp\{i2\pi[u'_{x,1}(nx'_2 - (n-N)x'_3) + u'_{y,1}((m-M)y'_2 - my'_3)]\}$$
$$+ c^*_{n,m}c_{n-N,m-M} \exp\{i2\pi[u'_{x,1}((n-N)x'_2 - nx'_3) + u'_{y,1}(my'_2 - (m-M)y'_3)]\}$$
$$+ c_{n,m}c^*_{n-N,m-M} \exp\{i2\pi[u'_{x,1}((n-N)x'_2 - nx'_3) + u'_{y,1}((m-M)y'_2 - my'_3)]\}\Big]\,, \tag{10.2.133}$$

where the integers n_m and m_m determine the harmonic components in the Fourier expansion with the highest frequency along the x- and y-direction, respectively.

The expressions for the quantities T are substituted in the general expression for the image plane intensity $I(x'_1, y'_1)$ of Eq. (10.2.119). We consider a general exponential function $\exp\{i2\pi[u'_{x,1}(n_2x'_2 - n_3x'_3) + u'_{y,1}(m_2y'_2 - m_3y'_3)]\}$ and evaluate the integral of Eq. (10.2.119) with respect to the variable pairs (x'_2, y'_2) and (x'_3, y'_3) which yields two two-dimensional δ-functions. Omitting the factor with the product of Fourier coefficients $c_{n,m}$ we obtain the following contribution $\Delta I(x'_1, y'_1)$ to the image plane intensity,

$$\Delta I_{n_2,m_2;n_3,m_3}(x'_1, y'_1) = \exp\{i2\pi[(n_2 - n_3)u'_{x,1}x'_1 + (m_2 - m_3)u'_{y,1}y'_1]\}$$
$$\times \iint I_{S,1}(X'_1, Y'_1) f(X'_1 + n_2u'_{x,1}, Y'_1 + m_2u'_{y,1}) f^*(X'_1 + n_3u'_{x,1}, Y'_1 + m_3u'_{y,1})\, dX'_1 dY'_1$$
$$= \exp\{i2\pi[(n_2 - n_3)u'_{x,1}x'_1 + (m_2 - m_3)u'_{y,1}y'_1]\}\, D(u'_{x,1}, u'_{y,1}; n_2, m_2; n_3, m_3)\,. \tag{10.2.134}$$

The integrand of the integral in Eq. (10.2.134) is the product of two displaced pupil functions and the source function projected onto the exit pupil. The integral over the exit pupil area has been denoted by $D(u'_{x,1}, u'_{y,1}; n_2, m_2; n_3, m_3)$ and plays a central role in the transfer of spatial frequencies associated with strongly modulated objects. D is commonly given the name of *cross-correlation coefficient* of two frequency pairs, $(n_2u'_{x,1}, m_2u'_{y,1})$ and $(n_3u'_{x,1}, m_3u'_{y,1})$ in the case at hand. Figure 10.38b shows the geometry in the exit pupil which is associated with a general cross-correlation coefficient D for strongly modulated objects. For comparison the geometry which is related to the optical transfer function $OTF(u'_{x,1}, u'_{y,1})$ for weakly modulated objects is shown in Fig. 10.38a. The final expression for the image plane intensity $I(x'_1, y'_1)$ is obtained by summing all contributions $\Delta I_{n_2,m_2;n_3,m_3}$ coming from the four terms T of Eqs. (10.2.130)–(10.2.133), yielding

$$I(x'_1, y'_1) = \sum_{n=-n_m}^{+n_m} \sum_{m=-m_m}^{+m_m} |c_{n,m}|^2 D(u'_{x,1}, u'_{y,1}; n, m; n, m)$$
$$+ \sum_{N=1}^{2n_m} \sum_{n=-n_m+N}^{+n_m} \sum_{m=-m_m}^{+m_m} \Big[c_{n,m}c^*_{n-N,m} \exp\{i2\pi Nu'_{x,1}x'_1\}\, D(u'_{x,1}, u'_{y,1}; n, m; n-N, m)$$
$$+ c^*_{n,m}c_{n-N,m} \exp\{-i2\pi Nu'_{x,1}x'_1\}\, D(u'_{x,1}, u'_{y,1}; n-N, m; n, m)\Big]$$

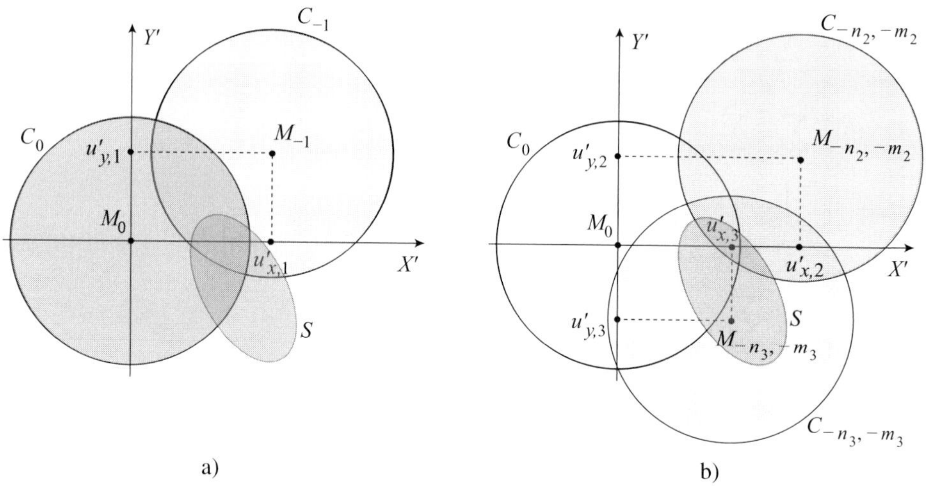

Figure 10.38: a) Calculation of the classical *OTF* function (S delimits the region where the source intensity function $I_{S,1}$ is nonzero).
b) The cross-correlation coefficient $D(u'_{x,1}, u'_{y,1}; n_2, m_2; n_3, m_3)$.
In b) the frequencies $(u'_{x,2}, u'_{y,2})$ and $(u'_{x,3}, u'_{y,3})$ which determine the coordinates of the centres of the circles $C_{-n_2,-m_2}$ and $C_{-n_3,-m_3}$ are given by the harmonics $(n_2u'_{x,1}, m_2u'_{y,1})$ and $(n_3u'_{x,1}, m_3u'_{y,1})$ of a two-dimensional frequency pair $(u'_{x,1}, u'_{y,1})$.

$$
\begin{aligned}
&+\sum_{M=1}^{2m_m}\sum_{n=-n_m}^{+n_m}\sum_{m=-m_m+M}^{+m_m}\Big[c_{n,m}c^*_{n,m-M}\exp\{i2\pi Mu'_{y,1}y'_1\}\,D(u'_{x,1},u'_{y,1};n,m;n,m-M)\\
&+c^*_{n,m}c_{n,m-M}\exp\{-i2\pi Mu'_{y,1}y'_1\}\,D(u'_{x,1},u'_{y,1};n,m-M;n,m)\Big]\\
&+\sum_{N=1}^{2n_m}\sum_{M=1}^{2m_m}\sum_{n=-n_m+N}^{+n_m}\sum_{m=-m_m+M}^{+m_m}\\
&\Big[c_{n,m}c^*_{n-N,m-M}\exp\{i2\pi[Nu'_{x,1}x'_1+Mu'_{y,1}y'_1]\}\,D(u'_{x,1},u'_{y,1};n,m;n-N,m-M)\\
&+c_{n,m-M}c^*_{n-N,m}\exp\{i2\pi[Nu'_{x,1}x'_1-Mu'_{y,1}y'_1]\}\,D(u'_{x,1},u'_{y,1};n,m-M;n-N,m)\\
&+c_{n-N,m}c^*_{n,m-M}\exp\{i2\pi[-Nu'_{x,1}x'_1+Mu'_{y,1}y'_1]\}\,D(u'_{x,1},u'_{y,1};n-N,m;n,m-M)\\
&+c_{n-N,m-M}c^*_{n,m}\exp\{-i2\pi[Nu'_{x,1}x'_1+Mu'_{y,1}y'_1]\}\,D(u'_{x,1},u'_{y,1};n-N,m-M;n,m)\Big]. \qquad (10.2.135)
\end{aligned}
$$

Using the property that $D(u'_{x,1}, u'_{y,1}; n, m; n', m') = D^*(u'_{x,1}, u'_{y,1}; n', m'; n, m)$, pairs of complex conjugate terms in Eq. (10.2.135) are combined and one obtains the expression

$$
\begin{aligned}
I(x'_1,y'_1) &= \sum_{n=-n_m}^{+n_m}\sum_{m=-m_m}^{+m_m}|c_{n,m}|^2\,D(u'_{x,1},u'_{y,1};n,m;n,m)\\
&+2\Re\left\{\sum_{N=1}^{2n_m}\exp\{i2\pi Nu'_{x,1}x'_1\}\sum_{n=-n_m+N}^{+n_m}\sum_{m=-m_m}^{+m_m}c_{n,m}c^*_{n-N,m}\,D(u'_{x,1},u'_{y,1};n,m;n-N,m)\right\}\\
&+2\Re\left\{\sum_{M=1}^{2m_m}\exp\{i2\pi Mu'_{y,1}y'_1\}\sum_{n=-n_m}^{+n_m}\sum_{m=-m_m+M}^{+m_m}c_{n,m}c^*_{n,m-M}\,D(u'_{x,1},u'_{y,1};n,m;n,m-M)\right\}\\
&+2\Re\sum_{N=1}^{2n_m}\sum_{M=1}^{2m_m}\Bigg\{\exp\{i2\pi[Nu'_{x,1}x'_1+Mu'_{y,1}y'_1]\}\\
&\qquad\times\left[\sum_{n=-n_m+N}^{+n_m}\sum_{m=-m_m+M}^{+m_m}c_{n,m}c^*_{n-N,m-M}\,D(u'_{x,1},u'_{y,1};n,m;n-N,m-M)\right]
\end{aligned}
$$

$$+\exp\{i2\pi[Nu'_{x,1}x'_1 - Mu'_{y,1}y'_1]\}$$
$$\times\left[\sum_{n=-n_m}^{+n_m}\sum_{m=-m_m+M}^{+m_m} c_{n,m-M}c^*_{n-N,m}D(u'_{x,1},u'_{y,1};n,m-M;n-N,m)\right]\bigg\}. \quad (10.2.136)$$

The expression for the image intensity $I(x'_1, y'_1)$ is now given as a sum of a constant term, two terms with harmonic components of the frequencies $u'_{x,1}$ and $u'_{y,1}$ in the x'_1- and y'_1- directions and a term with mixed harmonics of the two basic frequencies. It is straightforward to decompose the exponential terms in Eq. (10.2.136) into cosine and sine harmonic components by using the real and imaginary parts of the Fourier components $c_{n,m}$ and the 'four-circle' cross-correlation coefficients $D(u'_{x,1}, u'_{y,1}; n, m, n', m')$.

An interesting feature of the description of the object in the Fourier domain, more specifically by means of a Fourier *series*, is that the final expression for the image intensity according to Eq. (10.2.136) contains distinct factors that are associated with only the object or only the imaging system. In many experimental situations, the imaging system will have several fixed parameters for instance: the wavelength of the light, the source and imaging aperture, the pupil function including transmission function and aberration and a possible defocus with respect to the Gaussian image plane. With these fixed parameters, the functions $D(u'_{x,1}, u'_{y,1}; n, m, n', m')$ are calculated for all frequency pair combinations $(nu'_{x,1}, mu'_{x,1})$ and $(n'u'_{x,1}, m'u'_{y,1})$ which yield a nonzero contribution of the fundamental frequency $(u'_{x,1}, u'_{y,1})$ in the object spectrum. Different object transmission or reflection functions can be devised and when the fundamental frequencies are kept unchanged, they just produce a different set of Fourier coefficients $c_{n,m}$ in the expression of Eq. (10.2.136). Fortunately, calculation of a new set of object coefficients $c_{n,m}$ can be done at low computational cost by performing a fast Fourier transform. Calculation of the coefficients D is much more onerous. Fortunately, these coefficients need to be calculated only once and they are then stored in memory for later use.

We illustrate the calculation of the image intensity associated with a strong object by choosing a binary knife-edge object. Previously, the image intensity distribution of a knife-edge object, commonly called the *edge response* of an imaging system, was calculated for the two extreme cases of an incoherence factor $\sigma = 0$ and $\sigma = \infty$ (see Fig. 10.34). The calculation was based on the convolution of the object function and the point-spread function in the spatial domain. For the quasi-incoherent case with $\sigma = 1$, an interesting analytic result was given by Zernike **[385]** which showed that the normalised intensity in the image plane at the geometric position of the edge is equal to 1/3. For more general conditions of partially coherent illumination, we use Eq. (10.2.136) to calculate the edge response. To this purpose, we construct a double-periodic object that consists of a two-dimensional sequence of individual 'knife edges' that are truncated by a window function with length $p'_{x,1}$ and width $p'_{y,1}$. Both distances are expressed in diffraction units, in this example related to the image plane unit. In Fig. 10.39a the composition of an infinitely large periodic object and a typical point-spread function $A(x'_1, y'_1)$ have been drawn. More flexibility for the construction of general objects is obtained by allowing a horizontal shift s'_1 between successive basic cells (see Fig 10.39b). By adopting slanted features in an individual cell, the object function can be adapted to periodic structures with oblique axes. In the example of Fig. 10.39a, a sequence of knife edge sections is created with the succession of edges well separated at the scale of the point-spread function of the imaging system. In the calculated example, two edges were separated by a distance of 15 diffraction units which makes the optical cross-talk between adjacent edges negligible, even in the case of almost coherent illumination.

In Fig. 10.40 we have plotted the edge response in the image plane according to Eq. (10.2.136). When the incoherence factor σ is increased, the steepness of the edge response decreases and the normalised intensity I_n at the geometrical image position of the knife-edge increases. The data on the slope of the edge response and the intensity I_n at the edge location (units NA/λ_0) are given in the table below,

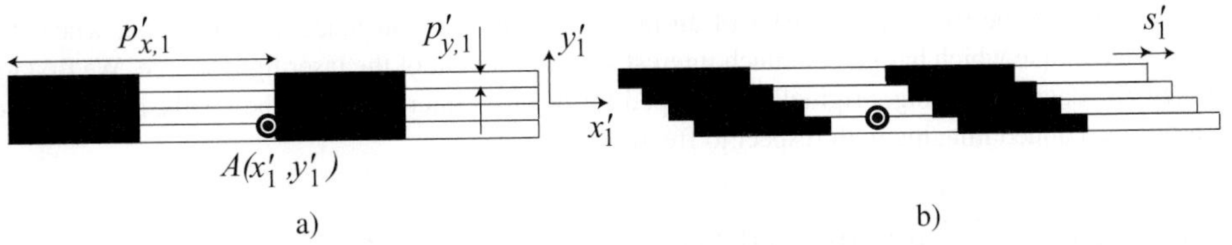

Figure 10.39: The construction of a double-periodic planar object pattern from elementary cells with basic lengths $p'_{x,1}$ and $p'_{y,1}$, as measured in the image plane in units λ_0/NA. An extra degree of freedom in the construction of the pattern is the shift s'_1 of the elementary cell in the x'_1-direction. $A(x'_1, y'_1)$ symbolises the point spread function.

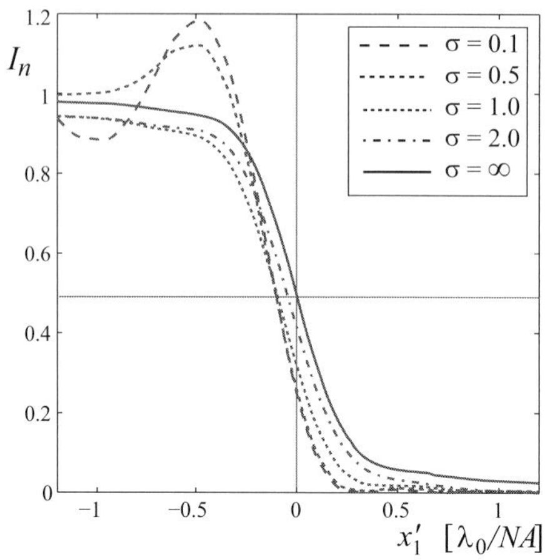

Figure 10.40: The normalised intensity I_n in the image plane as a function of the lateral position x_1' in diffraction units of λ_0/NA. The object is a periodic sequence of knife-edge transitions. The coherence of the illumination varies from virtually coherent ($\sigma = 0.1$) to strongly incoherent ($\sigma = 2.0$).

σ	Slope (NA/λ_0)	I_n
0.0	2.49	0.250
0.1	2.47	0.252
0.5	2.33	0.258
1.0	1.60	0.333
2.0	1.53	0.445
∞	1.50	0.500

The table shows that beyond the value $\sigma = 1.0$ the slope of the edge response hardly changes, the main difference being a change in the value of I_n. This can be explained with reference to Fig. 10.38b where the pupil geometry that leads to the cross-correlation coefficient D has been sketched. For $\sigma > 1$ the source function I_S extends beyond the lens aperture with rim C_0. From the source part outside the back-projected lens aperture, plane waves are injected in the object space at large angles. After being diffracted back at large angles by the knife edge, the incident plane waves produce diffracted waves that still (partly) enter the lens aperture with rim C_0. Some weak interference terms enhance the modulated intensity but most of this large-angle diffracted light simply contributes to the background intensity. This explains why the edge response for $\sigma = 2$ resembles a 'lifted' version of the edge response for $\sigma = 1$. We observe that an apparent shift of the order of 0.10 λ_0/NA occurs in the position of the knife-edge transient in the object plane when we switch from coherent to fully incoherent illumination. Such an effect must be taken into account when precise optical metrology has to be performed on strongly modulated object structures. In practice, each choice of illumination and numerical aperture requires its own image intensity decision level for the correct localisation of object features.

10.3 Frequency Transfer by a Scanning Imaging System

In this section we discuss the frequency transfer of an imaging system, for instance a microscope, when it is used in scanning mode, a technique which has gained much interest with the advent of the laser light source. We first give a short historic overview of scanned imaging, originally developed for electron microscopy in the 1930s, and then focus on the special properties of scanned imaging with respect to frequency transfer.

10.3.1 The Origin of Conventional Scanning Microscopy

Scanned optical imaging using a focused light spot on the object was first used at a larger scale in the 1950s and 60s. In these days there was an urgent need to convert film material into electronic form to make it available to the recently

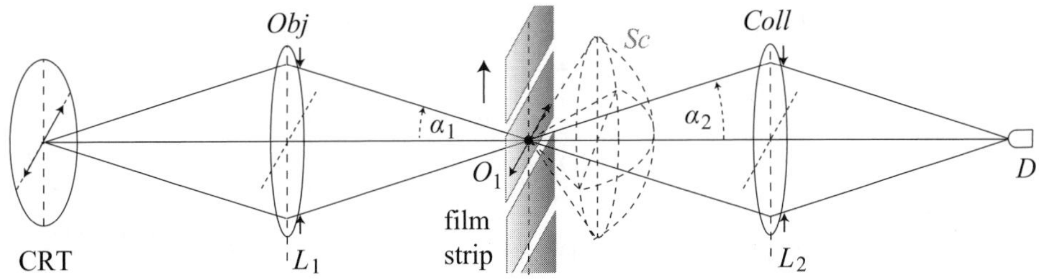

Figure 10.41: A film-scanning apparatus using as source the flying spot on the phosphor screen of a cathode ray tube (CRT). O_1 is the origin of the object plane containing the film strip to be scanned. The lens L_1 serves as the objective lens and produces a sharply focused image of the flying spot on the film strip. L_2 is the optically less-demanding collector lens. The (extended) detector D produces a time-dependent electrical signal proportional to the optical power passing at each moment through the aperture of the collector lens L_2.

developed magnetic storage media (originally the magnetic wire, later magnetic tape). The purpose was to display the recorded electronic film material on a television set. Figure 10.41 shows a schematic drawing of the experimental set-up that was used for this conversion from spatial information on a film transparency into a time-sequential electronic signal. A scanning pointlike source is obtained from the phosphor screen of a cathode ray tube (CRT), preferably having a uniform colour spectrum (white light). The point source executes a line scan and its image is formed on the film strip by a first lens L_1. The film transparency is moved uniformly, perpendicular to the fast line-scanning direction, at the typical film transport speed of 24 frames per second. The light scattered by the film transparency, labelled *Sc* in the drawing, is partly collected by a second lens L_2 and then focused on a detector D which delivers the desired electronic signal. Without major problems the set-up can be extended to produce colour signals also by having colour-selective multiple detector paths or by creating multiple flying spots with different colours. Important optical parameters of the set-up according to Fig. 10.41 are the focusing aperture $\sin\alpha_1$ of lens L_1 and the collecting aperture $\sin\alpha_2$ of lens L_2. In the original film-scanning apparatus, the role of the second lens was to collect as much as possible of the scattered light intensity onto a small detector. The detector surface had to be small to reduce its capacity, such that an electronic detection passband of typically 5 MHz for a standard TV signal could be accommodated. The second lens was not meant to achieve a nearly perfect image of the scanning spot on the detector. The relatively simple apparatus worked in a satisfactory way and the experimenters optimised the various parameters to combine fast-scanning with optimum resolution. High-aperture beams were not allowed as the unflatness and the guiding mechanism of a typical movie filmstrip required a relatively large focal depth of the scanning spot.

The successful scanning (or 'flying spot') devices for film conversion attracted the attention of the microscopy community. A first sign of this interest is the patent filed by Minsky **[247]**. Again with reference to Fig. 10.41, a fixed source point is projected at the point O_1 in the object plane of a microscope. The object is then submitted to a two-dimensional mechanical scanning. In one of the patent claims the resulting time-dependent detector signal is used as the steering signal for the beam intensity of a cathode ray tube of which the scanning spot executes a scanning pattern which is identical to that of the object, apart from a magnification factor. By virtue of the coupled mechanical scanning of the microscopic object and the geometrically amplified scanning movement on the CRT screen, a magnified image of the object is visible on the CRT. We remark that the filed patent explicitly requires a pointlike detector, although it is not explained why this would be an essential feature of the invention. Scanning microscopy of a different nature has been and indeed is currently used in optical metrology. The precise position of transients, for instance the marks on a ruler, is derived from the detection of the light scattered by such a mark from a finely focused incident light spot **[361]**.

The invention of the laser has made available a high-brilliance pointlike light source which can be advantageously used in a scanning system. Semi-conductor lasers enable high-frequency modulation of the source and the integration of such a source with compact semiconductor technology. These developments have paved the way for the development of scanning optical disc systems. The earliest, still primitive version of an optical disc system used a classical light source, viz. a high-pressure mercury lamp **[291]**. Subsequent versions used a laser source and profited from essential system improvements (active tracking of the protected phase-encoded information) paving the way to the CD, DVD and Blu-ray systems for optical storage and recording (see, for instance, **[74]**, **[41]**, **[5]**). The mass-production of optical disc players has enabled the development of (cheap) scanning microscopy techniques.

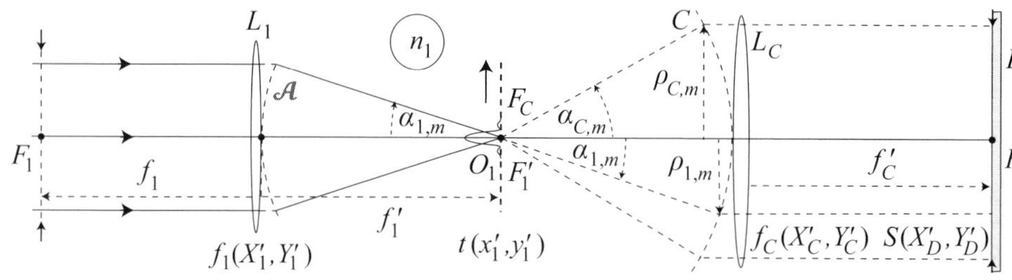

Figure 10.42: Scanning microscopy using a focused light spot O_1 produced by objective lens L_1. D is an extended detector in the back-focal plane F_C' of collector lens L_C and coincides in position with the exit pupil of the collector lens ('Type I' scanning microscope). The object is scanned mechanically or by optical means, for instance with the aid of a pivoting scanning mirror in the front focal plane F_1 of the objective lens L_1.

10.3.2 The Detector Signal of a Conventional Scanning Imaging System (Type I)

To analyse the detector signal of a standard scanning microscope we use the schematic drawing of Fig. 10.42. According to the nomenclature established in the authoritative textbook on optical scanning microscopy **[369]**, this scanning set-up is called a Type I microscope. We assume that the focused light spot at the object plane through the point O_1 is subjected to a translation in time, for instance a fast line scan in the x-direction, accompanied by slow position increments in the perpendicular direction. As usual, we express distances and positions with the aid of normalised coordinate values. In this case, the normalising distance is the diffraction unit λ_0/NA defined by the focusing objective lens L_1 ($n_1 \sin\alpha_{1,m} = NA$). The pupil coordinates (X_1', Y_1') of objective lens L_1 are normalised with respect to the half-diameter $\rho_{1,m}$ of the exit pupil of this lens. As a consequence of the translation of the scanning light spot, its complex amplitude is written as

$$A_s(x_1', y_1'; x_s', y_s') = A_1(x_1' - x_s', y_1' - y_s') \,. \tag{10.3.1}$$

For the case of a line scan, the shift distance x_s' equals $v_x\, T_s$ where v_x is the translation speed and T_s the elapsed scanning time. y_s' is a displacement perpendicular to the scanning direction, for instance a multiple of the line scanning pitch d_y'. The complex amplitude function $A_1(x_1', y_1')$ in the nominal image plane through the origin O_1 of the object plane is given by means of a diffraction integral according to Eq. (9.3.4), written in Cartesian (normalised) coordinates as

$$A_1(x_1', y_1') \propto \iint_{-\infty}^{+\infty} f_1(X_1', Y_1') \exp\{-i2\pi(x_1'X_1' + y_1'Y_1')\}\, dX_1' dY_1' \,, \tag{10.3.2}$$

where $\mathcal{A}$, the open pupil area of lens L_1, is the effective integration area. We have omitted the prefactors that co-determine the amplitude and phase in the focal plane of the lens. The negative sign of the argument of the exponential function in Eq. (10.3.2) stems from the argument $[\theta - (\phi - \pi)]$ of the cosine-factor of the complex exponential function in the integrand of Eq. (9.3.4). Therefore, according to our sign conventions for Fourier transforms of Appendix A, Eqs. (A.8.5) and (A.8.6), the function A_1 is proportional to the inverse Fourier transform of the function $f_1(X_1', Y_1')$ with respect to the negative arguments $(-x_1', -y_1')$. We further note that the exponential factor in Eq. (9.3.4) that is quadratic in the lateral coordinate $r_1' = (x_1'^2 + y_1'^2)^{1/2}$ equals unity everywhere in the object plane if the exit pupil is at a very large distance (telecentric use of the objective lens L_1), or, equivalently, if the entrance pupil coincides with the front focal plane of the imaging lens. A possible defocusing and aberration of the objective lens are included in the complex lens functions $f_1'(X_1', Y_1')$. The complex amplitude that is transmitted by the scanned object at each moment in time is given by

$$A_t(x_1', y_1'; x_s', y_s') \approx t(x_1', y_1')\, A_1(x_1' - x_s', y_1' - y_s') \,. \tag{10.3.3}$$

This approximation is justified if the object is very thin and if the scalar diffraction model can be applied to the problem at hand.

The light diffracted by the object propagates towards a collector lens L_C and the collected light intensity is detected by the (extended) detector D, located in the back-focal plane through F_C' of L_C. All the light incident on the detector within the area defined by the diaphragm rim (represented by the two arrows in Fig. 10.42) is detected. The diffraction of the light from the front focal plane through F_C to the back focal plane through F_C' is represented by a (forward) Fourier transformation. We assume that the collector lens satisfies the Abbe sine condition, such that the far-field coordinates

(X'_C, Y'_C) on the dashed spherical surface C equal the lateral coordinates (X'_D, Y'_D) on the planar detector surface D. The complex amplitude on the detector surface is thus given by

$$B(X'_D, Y'_D; x'_s, y'_s) \propto \iint_{-\infty}^{\infty} t(x'_1, y'_1) A_1(x'_1 - x'_s, y'_1 - y'_s) \exp\left\{-i2\pi(x'_1 X'_D + y'_1 Y'_D)\right\} dx'_1 dy'_1 . \tag{10.3.4}$$

The light intensity at a particular point on the detector surface is proportional to the squared modulus of $B(X'_D, Y'_D; x'_s, y'_s)$. The detector signal I_D follows from an integral of the detector intensity over the detector area, multiplied by the detector sensitivity function S_D, which is allowed to be a function of position,

$$I_D(x'_s, y'_s) \propto \iiiint\!\!\iint_{-\infty}^{+\infty} S_D(X'_D, Y'_D) A_1(x'_1 - x'_s, y'_1 - y'_s) A_1^*(x'_2 - x'_s, y'_2 - y'_s) t(x'_1, y'_1)$$
$$\times t^*(x'_2, y'_2) \exp\left\{-i2\pi[(x'_1 - x'_2) X'_D + (y'_1 - y'_2) Y'_D]\right\} dx'_1 dy'_1 dx'_2 dy'_2 dX'_D dY'_D . \tag{10.3.5}$$

The expression for the detector signal contains an integral that has the structure of a diffraction integral of the detector sensitivity function $S_D(X'_D, Y'_D)$. We compare this integral with Eq. (10.2.101) and, using an analogous notation, we can write

$$I_D(x'_s, y'_s) \propto \iiiint_{-\infty}^{+\infty} \mu_D(x'_1 - x'_2, y'_1 - y'_2) t(x'_1, y'_1) A_1(x'_1 - x'_s, y'_1 - y'_s)$$
$$\times t^*(x'_2, y'_2) A_1^*(x'_2 - x'_s, y'_2 - y'_s) dx'_1 dy'_1 dx'_2 dy'_2 . \tag{10.3.6}$$

In this expression we have introduced the function μ_D which plays the same role as the mutual coherence function $\mu_0(x', y')$ in Eq. (10.2.99) for the image plane intensity of a classical wide-field microscope. The index D was chosen to emphasise that the function μ_D depends on the detector sensitivity function S_D and is the result of an inverse Fourier transform of the detector sensitivity function S_D from the frequency plane to the object plane.

10.3.3 Equivalence of a Classical and a Type I Scanning Imaging Systems

We compare the expression of (Eq. (10.3.6) for the detector signal, $I_D(x'_s, y'_s)$, of a conventional scanning imaging system (Type I) with the expression for the image intensity $I(x'_1, y'_1)$ of Eq. (10.2.99) for a classical wide-field imaging system equipped with a Köhler illumination system. The only difference is the sign of the arguments of the point-spread functions A in Eq. (10.2.99) and A_1 in Eq. (10.3.6). We note that the point-spread function A in the classical microscope is the image of the point-spread function in the object plane. The point spread function A_1 in the scanning microscope is associated with the object plane. The relation between the object-plane and the image-plane point-spread functions in normalised coordinates is governed by the sign s_T of the image magnification. In the layout of Fig. 10.31, the factor $s_T = -1$ and this causes the sign difference between the arguments of A and A_1. We thus conclude that the imaging by the classical microscope and by the 'Type I' scanning microscope is described by identical formulas for the image intensity distribution and the detector signal.

We further observe that the (back-projected) diffraction pattern of the detector sensitivity function $S_D(X'_D, Y'_D)$ plays the same role as the mutual coherence function in the object plane of the classical wide-field microscope. This latter function, as stated by the van Cittert–Zernike theorem, is the diffraction image (or Fourier transform) of the intensity function $I_S(X_S, Y_S)$ of the incoherent illumination source. This means that the detector sensitivity function S_D in a scanning system plays the same role as the source intensity function in a standard system.

The equivalence of the expressions for the detected intensity in a classical wide-field system and in a conventional scanning *optical* imaging system was first stipulated by Welford **[361]** and further elaborated in **[363]**. He studied the light intensity diffracted by a knife-edge transition when illuminated by a well-focused scanning spot. The imaging was described within the framework of scalar wave propagation. His argument for the equivalence of classical wide-field imaging and conventional scanning imaging (Type I) was the similarity of the resulting equations for the image-plane intensity in classical wide-field imaging and the detected intensity in a conventional scanning system, in line with the findings of the preceding subsection. The equivalence of classical wide-field imaging and point-by-point object scanning is further illustrated in Fig. 10.43. The drawings a) and b) of this figure show the schematic layout of both imaging systems. The classical imaging system has been equipped with the Köhler illumination system and this scheme has also been applied in the detection section of the scanning system. Not shown in the drawings is the light diffracted by the

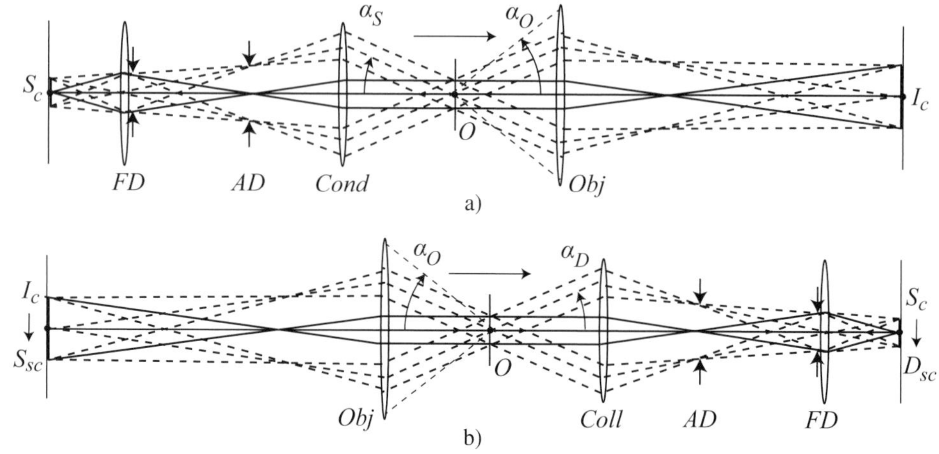

Figure 10.43: Illustrating the equivalence of a) a classical wide-field imaging system, and b) a conventional scanning imaging system of Type I.

object in the plane through O. For the classical wide-field system we consider the local intensity I_c detected by e.g. the grain of a photographic emulsion or the individual pixel of an image sensor. The incoherent points of an extended illuminating source S_c all contribute to the light flux that reaches a specific image point. In the scanning system we select a specific point of the extended source S_{sc} which is focused in the object plane O. The light diffracted by an object point in O is collected with a certain detector aperture angle α_D on the detector with extent D_{sc}. Welford established the equivalent role of the object aperture $\sin\alpha_O$ in both systems and the identical role played by condenser and collector aperture in the partial coherence of the imaging process. A further consequence of the equivalence of both systems is the relaxed imaging quality of the collector lens.

The equivalence of both imaging modes can be proven by means of Helmholtz's reciprocity theorem, even including non-scalar vector-imaging systems (see **[180]**). With respect to Fig. 10.43a, we first consider an incoherent source with intensity I_S and surface S_c that emits radiation to the Köhler condenser system, interacts with the planar object at O and reaches the image plane where it produces a certain intensity I_c over the image region (determined by the field diaphragm *FD*). Helmholtz's theorem states that, if we place in the image plane an incoherent source with the size of the image region and the intensity I_c that radiates back to objective *Obj*, through the object O and to the Köhler condenser system, the measured intensity in a receiving area with the size of S_c is equal to the source intensity I_S used in the *forward* system. If we rotate the system of Fig. 10.43a through an angle of 180°, we obtain the forward light path of Fig. 10.43b with the typical layout of a scanning imaging system. The object scanning is performed by modulating the elementary source points in time such that the contributions from each source point can be discriminated on the detector. The only difference between Figs. 10.43a and b is the inversion of the light propagation direction through the object at O. Based on Helmholtz's reciprocity theorem, perfect equivalence between the classical imaging system in a) and the conventional scanning system in b) exists only when the object structure equally obeys the reciprocity theorem. A well-known exception to reciprocity in light propagation is the presence of Faraday (or Kerr) optical activity in a medium (see Appendix E).

The practical operation of a scanning imaging system can be performed in three ways:

- Time-sequential scanning of an extended source area S_c, for instance with the aid of a spatial light modulator,
- Use of a pointlike source (or a diffraction-limited laser source) which is scanned in position or in angular direction,
- Use of a fixed source point and two-dimensional scanning of the object at O.

10.3.4 The Detector Signal of a Confocal Scanning Imaging System (Type II)

The layout of a scanning microscope of Type II (*confocal* microscope) is shown in Fig. 10.44. In contrast to the Type I scanning microscope, the Type II version does not have its equivalent in classical microscopy. The extension of the

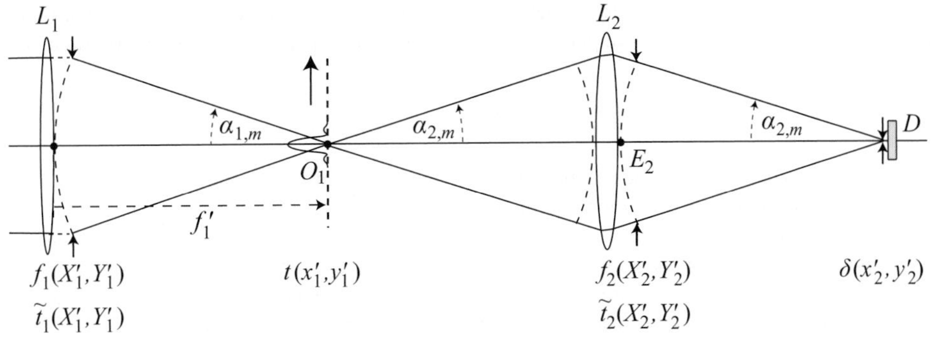

Figure 10.44: Scanning microscopy using a focused point source at O_1 produced by the objective lens L_1. The transmitted scanning light spot at O_1 is re-imaged on a point-detector by means of the collector lens L_2 ('Type II' microscope).

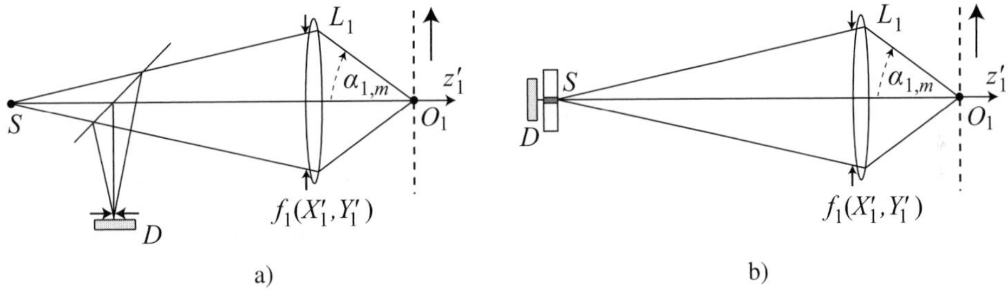

Figure 10.45: a) A confocal microscope used in reflection. The reflected light is guided towards the pointlike detector D with the aid of a semitransparent mirror.
b) A compact semiconductor laser source S functions as a small detector using the (incoherent) intensity modulation effect induced by the beam reflected at O_1. The coherence length of the source S should be significantly smaller than the total optical path length from S to O_1 and back.

Type II microscope with respect to the Type I version is a re-imaging by collector lens L_2 of the scanning light spot at O_1 onto a 'pointlike' detector. It is possible to focus the light on a tiny pinhole that selects the portion of the light to be detected by an extended detector D. Instead of using the object in transmission, it is equally possible to use a Type II microscope in reflection (see Fig. 10.45a). Another 'point'-detection method in reflective mode exploits the response of an optical cavity or of a laser light source to the reflected signal. The laser source is simultaneously source and detector by measuring the laser intensity at the back of the cavity or by monitoring the laser current of a semiconductor laser. Such a detection method was first used in the framework of optical disc read-out (see Fig. 10.45b). The light generated by a specific spatial mode of a laser source is guided back into the lasing cavity after *reflection* at the information structures of an optical disc. The laser intensity is then modulated by scanning the reflecting information structure on the disc and the laser cavity behaves as a (nonlinear) detector which is small with respect to the diffraction image produced by the laser source **[34]**. The principle was developed further by using a compact gallium-arsenide laser which allows for a direct feedback of the reflected light into the tiny cavity of the semiconductor laser **[132]**. The modulation due to the reflected light is optically measured with a detector D at the back of the laser cavity or by measuring the modulation of the laser current.

10.3.4.1 Calculation of the Detector Signal of a Confocal Microscope

Although we introduced a mathematical δ-function for the size of the detector in Fig. 10.44, in practice the physical pinhole will have a finite size, albeit substantially smaller than the half-width of the Airy disc of the lens point-spread function in the detection plane at D. A typical pinhole diameter is 0.2 to 0.4 $\lambda/\sin\alpha_{2,m}$ where $\alpha_{2,m}$ is the half-aperture angle of the focusing beam at D. To calculate the detector signal we assume that the object is scanned in the x- and y-direction with respect to a static and axially aligned system. The instantaneous light distribution, transmitted or reflected by the object, is denoted by

$$A_t = t(x'_1 - x'_s, y'_1 - y'_s)\, A_1(x'_1, y'_1)\,, \qquad (10.3.7)$$

where, as before, A_1 is the complex amplitude of the light generated by a point-source and focused by lens L_1 in the object plane through O_1. The function A_1 is called the amplitude impulse response of the lens L_1. The complex amplitude in the detector plane is given by a convolution of A_t and the impulse response A_2 of lens L_2 in the plane where the detection pinhole has been positioned. In Appendix A, Section A.9, we have derived the expression for the amplitude in the detector plane and it is given by the convolution integral of Eq. (A.9.13). The expression assumes a lens image that is real and has a negative magnification. Using normalised coordinates in both object and detection plane we have

$$A_D(x'_2, y'_2; x'_s, y'_s) = \iint_{-\infty}^{+\infty} A_t(x'_1, y'_1; x'_s, y'_s)\, A_2(x'_2 + x'_1, y'_2 + y'_1)\, dx'_1 dy'_1\,. \qquad (10.3.8)$$

To find the impulse response of a Type II microscope we choose an object with transmission function $t(x'_1, y'_1) = \delta(x'_1 - x'_0, y'_1 - y'_0)$, yielding the result

$$A_D(x'_2, y'_2; x'_s, y'_s) = A_1(x'_s + x'_0, y'_s + y'_0)\, A_2(x'_2 + x'_s + x'_0, y'_2 + y'_s + y'_0)\,. \qquad (10.3.9)$$

The intensity detected after transmission through the (mathematically) small pinhole in front of D is then given by

$$A_D(0, 0; x'_s, y'_s) = A_1(x'_s + x'_0, y'_s + y'_0)\, A_2(x'_s + x'_0, y'_s + y'_0)\,. \qquad (10.3.10)$$

We note that the diffracted light distributions A_1 and A_2 have the same physical dimension if the aperture angles $\alpha_{1,m}$ and $\alpha_{2,m}$ are equal. For a transmissive system this is not necessarily true; $\alpha_{1,m} = \alpha_{2,m}$ is a natural choice in a reflective Type II microscope. The intensity of the light detected by D is given by

$$I_D(0, 0; x'_s, y'_s) \propto I_1(x'_s + x'_0, y'_s + y'_0)\, I_2(x'_s + x'_0, y'_s + y'_0)\,. \qquad (10.3.11)$$

Some interesting conclusions can be drawn about the intensity impulse response according to Eq. (10.3.11). To simplify the discussion we assume equal apertures of lenses 1 and 2. The intensity impulse response is then given by

$$I_D(0, 0; x'_s, y'_s) \propto I_1^2(x'_s + x'_0, y'_s + y'_0)\,. \qquad (10.3.12)$$

10.3.4.2 Impulse Response of a Conventional and a Confocal Microscope

In this subsection we compare the impulse response of the Type II microscope with the impulse response of a coherent and an incoherent classical microscope as represented in Fig. 10.14b and 10.14d. In Fig. 10.46a we show a cross-section of the impulse response of a classical coherent and incoherent imaging system (labelled A and I, respectively) and the impulse response of the Type II microscope (labellel I^2). The full-width at half-maximum of the impulse response in diffraction units of λ_0/NA is quite different for the three curves. The FWHM data are as follows:

Microscope	Coherent (classical)	Incoherent (classical)	Type II (scanning)
FWHM	0.705	0.514	0.369
%	100	73	52

The table above shows that the Type II microscope has a significantly narrower impulse response than the classical wide-field microscope. Especially for isolated objects, this property is advantageous, for instance for the precise localisation of the centre of an object. In Fig. 10.46b we show another advantage of the Type II microscope, the almost complete absence of sidelobes. On the logarithmic scale we observe that the maximum of the first 'light ring' of the impulse response is of the order of 10^{-4}. This means that cross-talk from neighbouring object points when imaging a central object point will be much smaller than with classical wide-field microscopes or the conventional scanning microscope of Type I.

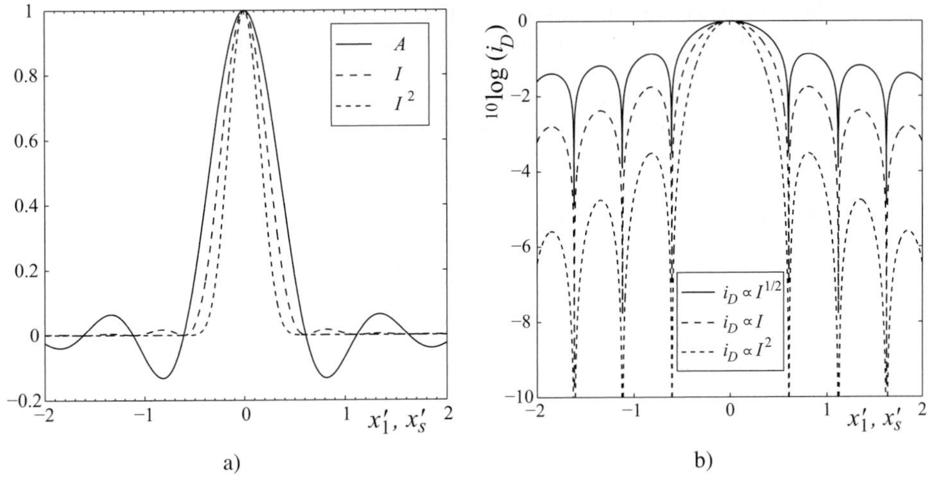

Figure 10.46: The detector plane amplitude A, intensity I and its squared value I^2, incident on a point-detector of a Type II scanning microscope (aberration-free case, perfect focusing, normalised coordinates in units λ_0/NA).
a) Linear scale representation.
b) The $^{10}\log$-values of the detector current i_D, proportional to I^2, together with hypothetical detector currents proportional to I and $|A|$.

10.3.5 Frequency Transfer by a Scanning Imaging System

For both types of scanning microscopes, Type I and Type II, we first calculate the (spatial) spectrum of the detector signal. In a second stage, we calculate the response of the microscope to a periodic function in the object plane, such that the value of the optical frequency transfer function OTF is found.

10.3.5.1 Conventional Scanning Microscope (Type I)

To obtain the frequency transfer function of a scanning microscope, we Fourier transform the detector signal as a function of the (normalised) scanning coordinates (x_s', y_s'). The expression for the detector signal of the Type I microscope is given by Eq. (10.3.6) and is identical to the intensity distribution in the image plane of a classical microscope. For the frequency transfer it is thus permissible to use in a straightforward manner the expression for the frequency response to a particular input frequency of the classical microscope, as given by Eqs. (10.2.120) and (10.2.121).

For strongly modulated objects with a more general spectrum, we use a Fourier series representation of the object by infinitely repeating the region of interest in the object. We then apply the analysis of Section 10.2.10. The final equation (10.2.136) for the image plane intensity in this section is based on such a frequency analysis of a limited portion of the object structure. The spatial object spectrum and the imaging properties of the optical system are well separated in Eq. (10.2.136) and they can be independently calculated. The imaging properties of a Type I scanning system are taken into account by the cross-correlation coefficients D and their calculation is illustrated in Fig. 10.38.

10.3.5.2 Frequency Spectrum of the Signal of a Confocal Scanning Microscope

The detector signal of the confocal microscope is proportional to the squared modulus of the axial light amplitude incident on the detector. Using Eq. (10.3.8) we find the expression

$$I_D(0,0;x_s',y_s') = \iiiint_{-\infty}^{+\infty} A_1(x_1',y_1')\, A_1^*(x_3',y_3')\, A_2(x_1',y_1')\, A_2^*(x_3',y_3')$$
$$\times\, t(x_1'-x_s',y_1'-y_s')\, t^*(x_3'-x_s',y_3'-y_s')\; dx_1'\, dy_1'\, dx_3'\, dy_3' \,, \qquad (10.3.13)$$

where we have used the auxiliary integration variables (x_3', y_3') in the object plane to avoid any confusion with the image plane coordinates (x_2', y_2').

The spectrum of the detector intensity follows from a Fourier transformation of I_D with respect to the spatial coordinates (x'_s, y'_s) to a pair of frequency (or far-field) coordinates (X', Y'). With reference to Fig. 10.44 we note that positive coordinates (X'_1, Y'_1) in the exit pupil of lens L_1 correspond to plane waves with negative transverse wave vector components. Propagation of the plane waves from the exit pupil of L_1 to the entrance pupil of L_2 connects the same (negative) wave vector components to negative pupil coordinates (X'_2, Y'_2). It is straightforward to write the Fourier transform of $I_D(0, 0; x'_s, y'_s)$ as

$$\tilde{I}_D(0,0;X',Y') = \iiiint\!\!\iint_{-\infty}^{+\infty} A_1(x'_1, y'_1)\, A_1^*(x'_3, y'_3)\, A_2(x'_1, y'_1)\, A_2^*(x'_3, y'_3)$$
$$\times\, t(x'_1 - x'_s, y'_1 - y'_s)\, t^*(x'_3 - x'_s, y'_3 - y'_s)\, \exp\{-2\pi i (x'_s X' + y'_s Y')\}\, dx'_1\, dy'_1\, dx'_3\, dy'_3\, dx'_s\, dy'_s\,. \qquad (10.3.14)$$

We then use Eqs. (10.2.102) and (10.2.103) to replace the transmission function t and the point-spread functions A by their diffraction integrals containing the corresponding 'far-field' functions $\tilde{t}$ and f. Given the negative sign of the exponential function in Eq. (10.2.102), the forward propagation of the function $\tilde{t}_1$ yields the object transmission function t. In Fig. 10.44 the function $\tilde{t}_1$ is thus found in the exit pupil of lens L_1. The same holds for the transformation of the lens pupil function f which yields the point-spread function A. The lens function f_1 thus yields the point spread function A_1 at O_1 and, in a similar way, the lens function f_2 yields the point-spread function A_2 at the detector plane D_2. In Appendix A, Fig. A.8 illustrates how the near-field and far-field functions in an optical system are transformed under the influence of a forward or backward propagation through (identical) imaging lenses. Using this scheme, we can easily transport a far-field function from one pupil plane to another with due attention to the inversion of the function. With respect to the spectrum of the object function t, it is thus permitted to use either $\tilde{t}_1(X'_1, Y'_1)$ or $\tilde{t}_2(X'_2, Y'_2)$, using the property $\tilde{t}_2(-X'_2, -Y'_2) = \tilde{t}_1(X'_1, Y'_1)$. We remark that this property is valid if the lenses L_1 and L_2 have equal imaging apertures $(\alpha_{1,m} = \alpha_{2,m})$. If this is not the case, the angular magnification of the image in the detector plane has to be taken into account, yielding

$$\tilde{t}_2(X'_2, Y'_2) = \tilde{t}_1\left[(X'_1, Y'_1)/m_A\right] = \tilde{t}_1\left[m_T(X'_1, Y'_1)\right]\,, \qquad (10.3.15)$$

where m_T and m_A are the transverse and angular magnification of the image of the object plane at O_1, formed by lens L_2 in the detector plane D. In what follows we use $m_T = m_A = -1$: for a reflective system this is the standard situation.

After the introduction of the transformed functions f_1, f_2 and $\tilde{t}_2$ in Eq. (10.3.14) we perform the integrations with respect to the variables (dx'_1, dy'_1), (dx'_2, dy'_2) and (dx'_s, dy'_s) and obtain spatial δ-functions. With the aid of these δ-functions the integrations with respect to the auxiliary frequency variables (X'_i, Y'_i) are easily performed and after some manipulation the expression for $\tilde{I}_D$ takes the form

$$\tilde{I}_D(0,0;X',Y') = \iiiint\!\!\iint_{-\infty}^{+\infty} f_1(X'_3, Y'_3)\, f_1^*(X'_4, Y'_4)\, f_2(X'_5, Y'_5)$$
$$\times f_2^*(X'_3 - X'_4 + X'_5 + X', Y'_3 - Y'_4 + Y'_5 + Y')\, \tilde{t}_2(X'_3 + X'_5, Y'_3 + Y'_5)$$
$$\times\, \tilde{t}_2^*(X'_3 + X'_5 + X', Y'_3 + Y'_5 + Y')\, dX'_3\, dY'_3\, dX'_4\, dY'_4 dX'_5\, dY'_5\,, \qquad (10.3.16)$$

where the variables (X'_3, Y'_3), (X'_4, Y'_4) and (X'_5, Y'_5) are auxiliary integration variables associated with the normalised pupil coordinates of lens L_1 and L_2. We introduce the change of variables $(X'_3, X'_4, X'_3 + X'_5) \rightarrow (X''_3, X''_4, X''_5)$ and a similar change for the corresponding Y' variables. Using the $\star$ sign for convolution (see Appendix A), we find the expression

$$\tilde{I}_D(0,0;X',Y') = \iint_{-\infty}^{+\infty} (f_1 \star f_2)_{(X'',Y'')}\, (f_1 \star f_2)^*_{(X''+X',\,Y''+Y')}$$
$$\times\, \tilde{t}_2(X'', Y'')\, \tilde{t}_2^*(X'' + X', Y'' + Y')\, dX''\, dY''\,, \qquad (10.3.17)$$

where we have replaced the auxiliary integration variables (X''_5, Y''_5) by (X'', Y''). According to Table A.2 of Appendix A we denote the autocorrelation of a function g by $R_{gg}(x, y)$ and write Eq. (10.3.17) as

$$\tilde{I}_D(0,0;X',Y') = R_{gg}(-X', -Y')\,, \qquad (10.3.18)$$

where the function g is given by

$$g(X'', Y'') = \tilde{t}_2(X'', Y'')\, (f_1 \star f_2)_{X'',Y''}\,. \qquad (10.3.19)$$

In the case of unit magnification by lens L_2 it is permissible to simply replace the spectral function $\tilde{t}_1(-X'', -Y'')$ by $\tilde{t}_2(X'', Y'')$. For a general magnification factor m_T we use the result of Eq. (10.3.15). Up to this point we have also assumed that the lenses L_1 and L_2 have not only equal focal distances but also equal apertures. If this not the case, the formulas are easily adapted by replacing the lens function $f_2(X', Y')$ by $f_2[\sin\alpha_{1,m}(X', Y')/\sin\alpha_{2,m}]$, where $\alpha_{1,m}$ and $\alpha_{2,m}$ have been depicted in Fig. 10.44.

10.3.5.3 Frequency Transfer by a Type II Scanning Microscope (Periodic Object)

The frequency response for a single-frequency object (e.g. a grating) can be derived from the general expression for the intensity spectrum of Eqs. (10.3.17) and (10.3.18). We use the periodic object of Eq. (10.2.118) for the object function t and calculate the Fourier transform towards the entrance pupil of lens L_2, yielding the function $\tilde{t}_2(X_2', Y_2')$. We obtain the expression

$$\tilde{t}_2(X_2', Y_2') = \delta(X_2', Y_2') + \frac{m}{2}\,\delta(X_2' - u'_{x,1}, Y_2' - u'_{y,1}) + \frac{m}{2}\,\delta(X_2' + u'_{x,1}, Y_2' + u'_{y,1})\,. \tag{10.3.20}$$

Using the expression for $\tilde{t}_2$ in Eq. (10.3.17) we find for the image plane spectrum

$$\begin{aligned}\tilde{I}_D(0,0;X',Y') = \sum_{p_1=-1}^{+1}\sum_{p_2=-1}^{+1} a_{p_1}a_{p_2} \iiiiiint_{-\infty}^{+\infty} & f_1(X_3', Y_3') f_2(X_5' - X_3', Y_5' - Y_3') f_1^*(X_4', Y_4') \\ & \times f_2^*(X_5' + X' - X_4', Y_5' + Y' - Y_4')\, \delta(X_5' - p_1 u'_{x,1}, Y_5' - p_1 u'_{y,1}) \\ & \times \delta(X_5' + X' - p_2 u'_{x,1}, Y_5' + Y' - p_2 u'_{y,1})\, dX_3'\, dY_3'\, dX_4'\, dY_4'\, dX_5'\, dY_5'\,. \end{aligned} \tag{10.3.21}$$

In this expression we have used the quantity a_p which, for integer p, equals 1 ($p = 0$) or $m/2$ ($|p| = 1$). We integrate Eq. (10.3.21) with respect to the variables (X_5', Y_5') and obtain

$$\begin{aligned}\tilde{I}_D(0,0;X',Y') = \sum_{p_1=-1}^{+1}\sum_{p_2=-1}^{+1} & a_{p_1}a_{p_2}\, D(u'_{x,1}, u'_{y,1}; p_1, p_1; p_2, p_2) \\ & \times \delta\left[X' - (p_2 - p_1)u'_{x,1}, Y' - (p_2 - p_1)u'_{y,1}\right]\,. \end{aligned} \tag{10.3.22}$$

The coefficient $D(u'_{x,1}, u'_{y,1}; p_1, p_1; p_2, p_2)$ in the summation is given by

$$\begin{aligned} D(u'_{x,1}, u'_{y,1}; p_1, p_1; p_2, p_2) = & \iint_{-\infty}^{+\infty} f_1(u'_x, u'_y) f_2(p_1 u'_{x,1} - u'_x, p_1 u'_{y,1} - u'_y)\, du'_x du'_y \\ & \times \iint_{-\infty}^{+\infty} \left[f_1(u'_x, u'_y) f_2(p_2 u'_{x,1} - u'_x, p_2 u'_{y,1} - u'_y)\right]^* du'_x du'_y\,, \end{aligned} \tag{10.3.23}$$

where (u'_x, u'_y) are auxiliary normalised far-field integration variables. The frequency spectrum of the detector signal consists of a certain number of spikes, represented by mathematical δ-functions. In the case of a finite-sized object, the δ-functions become real-world finite-sized diffraction peaks. The width of the peaks is given by the diffraction pattern of the circular disc or rectangular area of the object function t.

The coefficients $D(u'_{x,1}, u'_{y,1}; p_1, p_1; p_2, p_2)$ can be identified with the more general cross-correlation coefficients that were previously defined with the aid of Eq. (10.2.134). In the same way as for the classical microscope or the Type I scanning microscope, we define coefficients $D(u'_{x,1}, u'_{y,1}; n_2, m_2; n_3, m_3)$ for the Type II microscope, given by

$$\begin{aligned} D(u'_{x,1}, u'_{y,1}; n_2, m_2; n_3, m_3) = & \iint_{-\infty}^{+\infty} f_1(u'_x, u'_y) f_2(n_2 u'_{x,1} - u'_x, m_2 u'_{y,1} - u'_y)\, du'_x du'_y \\ & \times \iint_{-\infty}^{+\infty} \left[f_1(u'_x, u'_y) f_2(n_3 u'_{x,1} - u'_x, m_3 u'_{y,1} - u'_y)\right]^* du'_x du'_y\,. \end{aligned} \tag{10.3.24}$$

In the presence of a strongly modulated periodic object, Eq. (10.2.136) is then used for the calculation of the light intensity after the pinhole detector D.

For a weakly modulated object the frequency transfer is given by Eq. (10.3.22) when the terms in the summation are limited to those with at least one p-value equal to zero. After normalisation the corresponding coefficient D then plays the role of optical transfer function, *OTF*, yielding

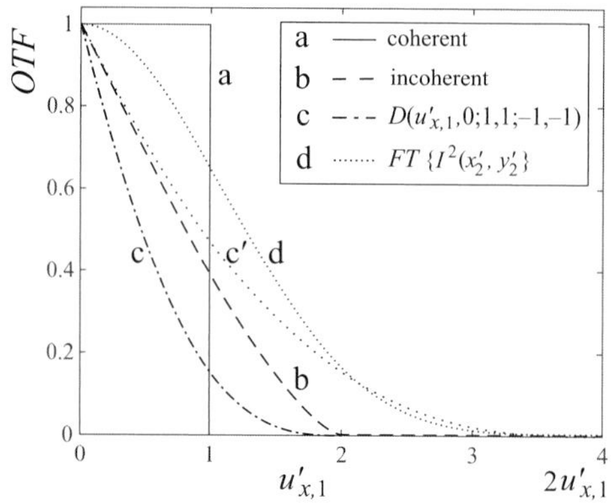

Figure 10.47: Normalised frequency transfer (first and second harmonic of the object frequency). The various curves represent the transfer of a coherent classical and scanning microscope (Type I), curve a) and that of its incoherent analogon, curve b). The curves c), c′) and d) are associated with the Type II microscope. Curve c) displays the cross-correlation coefficient D for indices p_1 and p_2 with opposite sign. Curve c′) has been derived from c) by stretching c) horizontally by a factor of 2, up to the maximum frequency $2u'_{x,1} = 4$. Curve d) is the Fourier transform of the impulse response I^2 of a Type II microscope.

$$OTF(u'_{x,1}, u'_{y,1}) = \frac{D(u'_{x,1}, u'_{y,1}; 0, 0; 1, 1)}{D(u'_{x,1}, u'_{y,1}; 0, 0; 0, 0)}$$

$$= \frac{\iint_{-\infty}^{+\infty} f_1(u'_x, u'_y) f_2(-u'_x, -u'_y)\, du'_x du'_y \iint_{-\infty}^{+\infty} \left[f_1(u'_x, u'_y) f_2(u'_{x,1} - u'_x, u'_{y,1} - u'_y)\right]^* du'_x du'_y}{\left(\iint_{-\infty}^{+\infty} \left|f_1(u'_x, u'_y)\right| \left|f_2(-u'_x, -u'_y)\right| du'_x du'_y\right)^2}. \tag{10.3.25}$$

In Fig. 10.47 we have plotted various optical transfer functions. The curves a) and b) correspond to the coherent and incoherent OTF functions of the classical wide-field microscope and the Type I scanning microscope. For an aberration-free optical system and a weakly modulated object, curve b) also represents the OTF of linearly transferred frequency components $(u'_{x,1}, u'_{y,1})$ in the Type II scanning microscope. These frequencies are the result of interference in the detector plane of a zeroth-order and a first-order diffracted beam by the object. Although the Type II microscope operates with coherent light, we observe that its (linear) transfer of frequencies is identical to the incoherent frequency transfer in the classical microscope.

A special case arises when two first-order beams interfere in the Type II microscope. For the transfer of the second harmonic $2(u'_{x,1}, u'_{y,1})$ in the Type II microscope we have the squared value of curve b) and in the figure this transfer is represented by curve c), in accordance with Eqs. (10.3.22) and (10.3.23), for $p_1 = -p_2$. The second harmonic is transferred provided the first harmonic fits within the incoherent passband (curve b), $|u'_{x,1}| \leq 2$ or $u_{x,1} \leq 2NA/\lambda_0$). The maximum frequency transferred by the Type II microscope is thus given by $4NA/\lambda_0$. In Fig. 10.47 we have shown the transfer of this second harmonic frequency $2u'_{x,1}$ by means of the curve c′). This curve has been derived from curve c) by stretching it horizontally by a factor of two. It is interesting to compare curve c′) with curve d). The latter curve has been calculated by Fourier transforming the expression of Eq. (10.3.12), assuming that the Fourier transform of the impulse response $I_D(0, 0; x'_s, y'_s)$ yields the optical transfer function of the Type II microscope. The relation between impulse response and frequency transfer function has been confirmed for the coherent and incoherent microscope as these are linear imaging systems in amplitude and intensity, respectively. Curve d) of Fig. 10.47 shows that this relation is not valid for the Type II microscope below the classical cut-off frequency of $u'_{x,1} = 2$. The strength of the second harmonic, however, is well approximated by curve d) as the difference between curve c′) and curve d) is small in the region $2 \leq 2u'_{x,1} \leq 4$.

The appearance of the second-harmonic frequency in the Type II microscope is illustrated in Fig. 10.48. From the coherent optical field in the exit pupil of lens L_1 we select the outer portions, denoted by the (normalised) wave vectors $\hat{\mathbf{k}}_0$ and $\hat{\mathbf{k}}_1$. The wave associated with the vector $\hat{\mathbf{k}}_0$ is diffracted by the object. The object frequency $(0, u'_{y,1})$ has been chosen such that the diffraction angle equals $+2\alpha_{1,m}$ and the corresponding wave associated with the diffracted wave vector $\hat{\mathbf{k}}_{0,+1}$ is just accepted by the entrance pupil of lens L_2. It can easily be verified with the aid of a Fourier transformation of Eq. (10.2.118) that the plus first order has a complex amplitude of $(m/2)\exp[-i2\pi(u'_{x,1}x'_s + u'_{y,1}y'_s)]$ if the centre of the

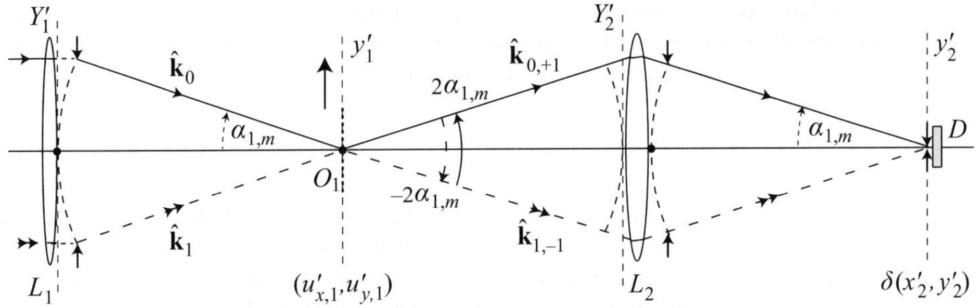

Figure 10.48: Generation of the frequency-doubled signal (frequency $(2u'_{x,1}, 2u'_{y,1})$) at the pinhole detector D of a Type II scanning microscope. Maximum transmitted spatial frequency is the second harmonic frequency, $4\,NA/\lambda_0$. The apertures and focal distances of lenses L_1 and L_2 are equal.

object has been shifted to a point with coordinates (x'_s, y'_s). The phase difference between the two interfering beams with wave vectors $\hat{\mathbf{k}}_{0,+1}$ and $\hat{\mathbf{k}}_{1,-1}$ thus amounts to $-4\pi(u'_{x,1}x'_s + u'_{y,1}y'_s)$. For the one-dimensional frequency $\mathbf{u}_1 = (0, u_{y,1})$, the two 'marginal' waves produce a periodic interference pattern with a frequency $u_{y,1} = 2\sin(\alpha_{1,m})/\lambda_0$ in the detector plane, in accordance with the unit paraxial magnification by lens L_2. The corresponding period $\lambda_0/(2\sin(\alpha_{1,m})$ corresponds to the finest detail that can be produced in the image plane of lens L_2. This interference pattern, however, changes its phase by an amount 2π if the object shift equals half the object period $1/u_{y,1}$. The intensity variations on the pinhole detector thus have a time frequency which corresponds to a spatial frequency $2u_{y,1} = 4\sin(\alpha_{2,m})/\lambda_0$. It turns out that the Type II microscope is capable of a sharpening of the image of an isolated object point, thanks to the presence of transferred frequencies beyond the classical cut-off frequency. However, for an object that is composed of dense lines and spacings, the maximum frequency in the object plane that is transferred to image space is still limited by the cut-off frequency of the classical incoherent microscope.

10.3.5.4 Special Imaging Properties of the Type II Confocal Microscope

The special properties of a confocal imaging system are further illustrated in this subsection with the aid of Figs. 10.45 and 10.48.

- *Defocusing (transmissive mode).*
  An axial shift of the object plane with respect to the point O_1 produces a defocusing for the imaging by both lens L_1 and lens L_2. With the aid of the axial diffraction unit $\lambda/\sin^2\alpha_{1,m}$ we can use the normalised axial coordinate z'_n (see Eq. (5.2.20)). According to Eq. (5.2.26) the defocusing factor of the lens transmission function f_1 is then given by $\exp[-i\pi(\delta z'_n)(X_1'^2 + Y_1'^2)]$, where $\delta z'_n$ is the focal shift of the object in the positive z-direction. The defocusing factor for the second lens has equal size but opposite sign. If the object transmission is uniform, we conclude that the expression for $OTF(0,0)$ is unity. The average signal on the pinhole detector D is thus unaffected by defocusing. The expression for the OTF shows that defocusing has an effect on the signal transfer if $|\mathbf{u}'_1| \neq 0$. For the frequencies that are transferred in a linear way, the effect of defocusing is the same as for the incoherent frequency transfer in a classical microscope (see Fig. 10.19b).
- *Defocusing (reflective mode): depth discrimination.*
  Figure 10.45 shows the reflective set-up of a confocal microscope. In the expressions for the frequency transfer we have to replace the lens function $f_2(X'_2, Y'_2)$ by $f_1(X'_1, Y'_1)$. With respect to defocusing, we observe that the defocusing is the same for lens L_1 and L_2. The cross-correlation coefficient $D(u'_{x,1}, u'_{y,1}; 0, 0; 1, 1)$ of Eq. (10.3.23) is thus given by $(u'_{x,1} = 0)$:

$$D(0, u'_{y,1}; 0, 0; 1, 1) = \iint_{-\infty}^{+\infty} |f_1(u'_x, u'_y)|\,|f_1(-u'_x, -u'_y)| \exp\left\{-i2\pi(\delta z'_n)(u_x'^2 + u_y'^2)\right\} du'_x du'_y$$
$$\times \iint_{-\infty}^{+\infty} |f_1(u'_x, u'_y)|\,|f_1(-u'_x, u'_{y,1} - u'_y)| \exp\left\{+i2\pi(\delta z'_n)(u_x'^2 + u_y'^2 + \frac{u_{y,1}'^2}{2} - u'_{y,1}u'_y)\right\} du'_x du'_y\,. \tag{10.3.26}$$

In this expression we have assumed aberration-free lenses of which the imaging is only influenced by object defocusing. Furthermore, if we assume that the amplitude transmission of the lenses is unity everywhere, we have after proper normalisation the following expression for the first integral,

$$\frac{1}{\pi}\int_0^{2\pi}\int_0^1 \exp\left\{-i2\pi(\delta z'_n)\rho^2\right\}\rho d\rho d\theta = \exp\{-i\pi(\delta z'_n)\}\ \mathrm{sinc}\{\pi(\delta z'_n)\}\ , \tag{10.3.27}$$

where ρ is the normalised radial coordinate on the exit pupil surface of lens L_1. It follows that the linear frequency transfer is exactly zero if the defocusing $\delta z'_n = 1$. This corresponds to a defocusing of the object over a distance of two Rayleigh focal depths. The total wavefront deviation as seen from the detector corresponds to one wavelength at the pupil rim. This means that two Fresnel zones can be constructed over the pupil cross-section and these two zones exactly cancel their contribution at the location of the pinhole in front of the detector D.

We note that the nonlinear frequency transfer is less affected by defocusing. This follows from Fig. 10.48, used in reflection. The diffraction orders related to the non-linearly transferred frequencies $2(u'_{x,1}, u'_{y,1}) \rightarrow 4NA/\lambda_0$ follow an almost symmetric path through the microscope and they acquire a virtually equal wavefront offset in the case of object defocusing. Inspection of Eq. (10.3.23) shows that in the case of nonlinear signal components the first integral does not become zero for $\delta z'_n = 1$. For this typical defocusing value, only nonlinear components are present in the detector signal.

The sharp decrease in strength of the linearly transferred signal as a function of defocusing has an interesting effect if the object is not flat but has a height profile. Defocused portions of the object will appear effectively grey to black in the scanned image. By subsequent scanning of such a three-dimensional object with selected defocusing values, we can reconstruct the object with a sequence of sharply imaged object slices. From the defocusing value $\delta z'_n$ during the imaging of each slice, we obtain the height level attached to each slice. This feature of the reflective confocal microscope is called its depth-discrimination capability, or, alternatively, its *optical sectioning* feature. A further interesting property of the confocal reflecting microscope is its relative insensitivity to asymmetric aberration of lens L_1, of the type $W(\rho,\theta) \propto \cos m\theta$, with m being an odd integer. The total wavefront aberration in reflection equals zero, at least for the zeroth-order diffracted light. Contrarily, any aberration of L_1 with an even value of m yields a wavefront with double aberration after reflection.

10.3.6 Optical Disc Systems and Scanning Microscopy

In this subsection we discuss the read-out of an optical disc, for instance a compact disc, with the aid of the CD light path which is typically a Type I scanning microscope. A relatively simple modification of the read-out light path permits the operation of the CD player as a Type II scanning microscope. In Fig. 10.49 we show the typical information pattern on an optical disc. The elementary structure is a depression (or 'pit') in the plastic substrate such that the light reflected at the bottom of the structure has a phase increment with respect to the light reflected from the intermediate non-modulated region ('land' region). The calculation of the field strength components of the reflected light needs an electromagnetic treatment, for instance by using rigorous coupled wave analysis (see Section 1.12). Such a vector analysis of the diffracted light is required since the diffracting structures have sub-wavelength dimensions and the diffracted field is influenced by

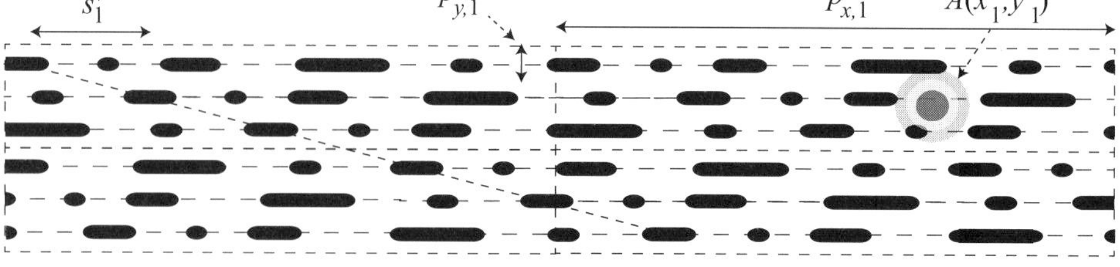

Figure 10.49: Schematic view of an optical disc information layer with repeating sections of length $p'_{x,1}$ in the track direction and $p'_{y,1}$ in the direction perpendicular to the tracks (track pitch). The identical sections along a single track are shifted over a distance $\pm s'_1$ when going to a neighbouring track. This shift of the information simulates the uncorrelated character of the information in adjacent tracks.

the state of polarisation of the incident field. If the scanning system has a sufficiently small numerical aperture, the transverse components of the reflected field are selected and each component of this field is propagated towards the detector region using scalar wave propagation. For the CD system, having a numerical aperture of 0.45, this approach is permitted. Figure 10.49 shows the information pits with varying lengths, obeying a certain modulation rule with lengths ranging from 3 to 11 elementary units, the clock length unit. On the disc, the digital clock period corresponds to the physical clock length l_c, which is determined by the clock frequency and the scanning speed. The probability distribution for the occurrence of each pit length follows from the modulation rule which determines how each digital signal byte is transformed into a sequence of pits and lands ('byte dictionary'). The minimum length of an information pit and the minimum separating distance between two pits is equal to $3l_c$. The clock length unit is a certain fraction of the diffraction unit λ_0/NA on the optical disc. The wavelength of the CD semiconductor laser source is typically 785 nm and the diffraction unit thus equals 1.75 μm. The CD clock length unit l_c equals 0.29 μm and the smallest period $p_{x,1}$ of 1.74 μm is thus equal to one diffraction unit. The track-to-track spacing $p_{y,1}$ is chosen to be 1.6 μm, also very close to the CD diffraction unit. Subsequent optical disc systems such as the DVD and Blu-ray system have a shortest signal period and a track spacing that is substantially smaller than the diffraction unit on the disc surface. The CD light path is a typical Type I reflecting microscope 'avant la lettre' with equal objective and collector aperture. It is a partially coherent imaging device with an incoherence factor σ_D equal to unity. As the optical disc is a strongly modulated object, the calculation of the detector signal follows the scheme given in Subsection 10.2.10. More information on the application of this general calculation scheme to objects with the specific structure of an optical disc can be found in **[144]**.

Figure 10.50a shows the eye-pattern of the optical analogue signal produced by a DVD-disc (λ = 650 nm, NA = 0.60). It is obtained by superposition of the signals associated with a large number of detected sequences of information pits. The signal is obtained by means of a two-dimensional set of Fourier coefficients according to Eq. (10.2.136) and it is evaluated at positions $x_{1,n} = n\, l_c$ along the track, where l_c is the digital clock length unit on the disc and n an integer number. If needed, the signal is further interpolated by the calculation of values in between two clocking positions. In the centre of Fig. 10.50a we have shown a grey-shaded digital eye. Exactly in the middle of each digital eye, a decision is made about the positive or negative value of the analogue signal with respect to the decision level given by the thick dashed line running through the centre of the eye. Although the optical disc information layer was perfectly focused and the read-out system was assumed to be perfect, the eye-height is relatively small and the 'jitter' of the sides of the idealised eye lozenge is large, typically more than 15% of the eye width. The effect of an electronic improvement of the

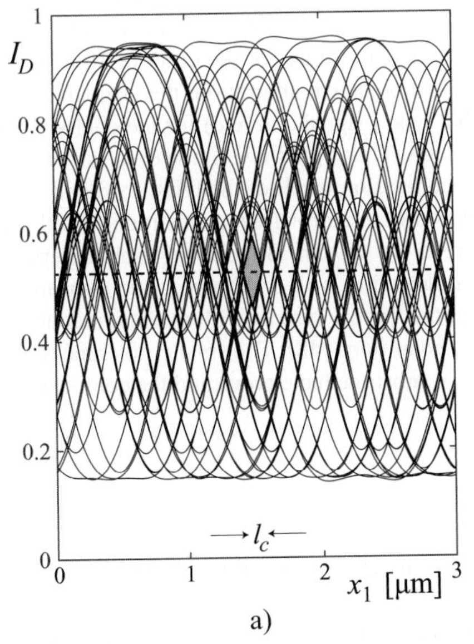

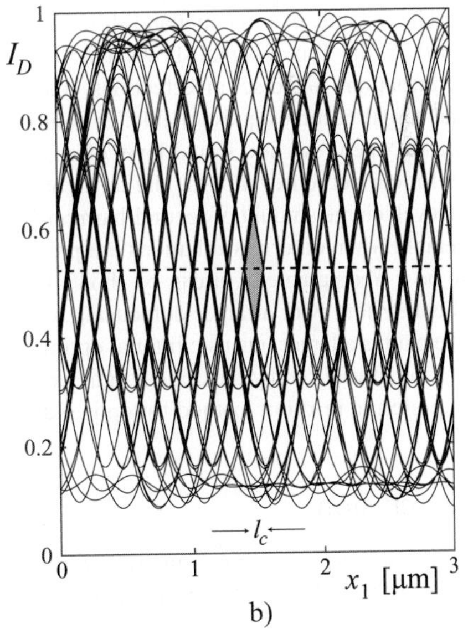

Figure 10.50: The eye-pattern of the analogue detector signal produced by an optical disc system (DVD, standard Type I read-out).
a) bare detector signal. b) electrical signal after appropriate high-pass filtering.

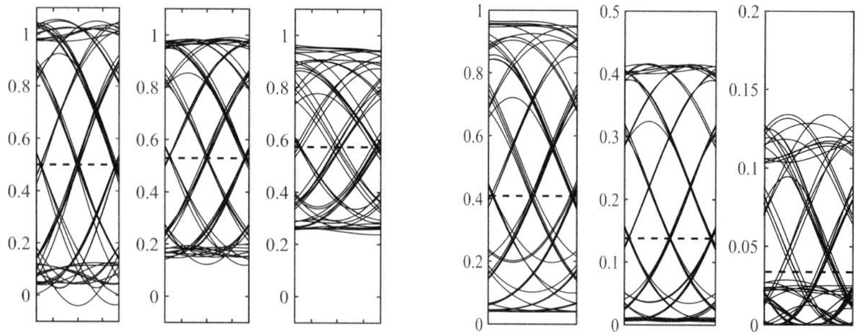

Figure 10.51: The eye-pattern of an information track on an optical compact disc (CD) when imaged with the aid of a classical Type I system (including equalisation) or a Type II confocal read-out system, both used in reflection. a) $\delta z'_n = 0$. b) $\delta z'_n = 0.5$ (one Rayleigh focal depth). c) $\delta z'_n = 0.75$.

detected optical signal is shown in Fig. 10.50b. A high-pass 'equalisation' filter compensates for the reduction in strength of the higher optical frequencies close to the cut-off frequency. An error-free reconstruction of the digital signal is now possible, even in the presence of a weakly aberrated read-out system ($OPD_{\text{rms}} \leq 0.07\,\lambda$) and a small defocusing of the optical disc surface.

In Fig. 10.51 we show portions of digital eye-patterns that have been calculated if the reflecting optical disc is read-out with a standard light path (Type I read-out in reflection) and with a confocal set-up (Type II read-out in reflection). The eye patterns in a), b) and c) correspond to different amounts of defocusing $\delta z'_n$ of the disc information surface. As a function of $\delta z'_n$ the curves a), b) and c) associated with the Type I detection (including electronic equalisation) have an almost constant average value, as does the corresponding decision level (thick dashed horizontal line). The confocal signal has a strongly decreasing average value and optimum decision level as a function of defocusing (note the different vertical scales in the three graphs for Type II detection). If we compare the in-focus signals for both detection modes, we observe a comparable shape of the eye-pattern. For a defocusing $\delta z'_n = 1/2$, equal to one Rayleigh focal depth, the confocal signal has a better eye-opening and shows less jitter in comparison to the Type I signal. Finally, for a defocusing value of $\delta z'_n = 0.75$, the Type I signal shows a digital eye with an increased jitter of the order of 10 to 15%. Combined with other perturbations like disc imperfections, tracking errors and objective aberration, the error-free retrieval of the original digital signal becomes questionable. In contrast with the Type I signal, curve c) for the confocal system shows a well-modulated 'open' eye at a signal level of approximately 0.05. This signal level, however, is *not* the appropriate decision level for correct retrieval of the digital signal. The correct decision level (dashed line) is found one level lower. At this signal level a large amount of second-harmonic distortion is present; error-free reconstruction of the digital signal is quite vulnerable when other perturbations are present in the read-out system. By allowing the numerical aperture of the system to be a free parameter, it can be shown that confocal detection with a slightly lower aperture (minus 10%) yields a read-out system with a larger defocusing tolerance than the Type I system with the nominal aperture of 0.60. The price to be paid is a larger intrinsic jitter at optimum read-out conditions **[48]**. The smaller tolerances with respect to residual aberration introduced by the optical light path and the disc substrate have prevented the confocal detection from becoming the preferred read-out method for optical disc systems.

10.3.7 Conclusion

We conclude this section with some remarks on the dramatically increased versatility of classical imaging methods and microscopy. Before the 1980s, the photographic plate or film was the only surface detector with an adequate resolution over a large area. The possibilities for further image enhancement were limited. Multiple image treatment with scanned images from photographic exposures have been devised to improve image resolution and contrast but such approaches have remained cumbersome and slow. The advent of pixel-based amplitude modulation devices (for instance, the digital mirror device DMD and the liquid-crystal spatial light modulator SLM), high-definition picture memories and the possibility of fast digital storage and handling of image data have given rise to new developments. The application of a DMD device allows a change in the source intensity function on the scale of milliseconds. Captured images with different source settings can be added, subtracted and filtered to achieve image enhancement. A very simple operation is the recording of two images. Each image is made with a complementary part of the total (circular) source. In terms of partially coherent imaging, the subtraction of these two images is equivalent to the imaging of the object with one half of the source provided with a positive intensity, the other half with a 'negative' intensity. The subtracted result indeed yields a differentiated image of the object along the direction perpendicular to the source dividing line. Instead of a simple subtraction, it is also possible to add the two images with a generally complex pre-factor, subtraction being the limiting case with a phase difference of π. Spatial-frequency-dependent image treatment is easily carried out using high-resolution picture memories. An imaging method called *structured illumination* **[256]**,**[122]** uses subsequent source illumination patterns to obtain an improved final image. The use of an SLM in the imaging part of a system enables the correction of imaging aberrations. In its extreme version, the spatial light modulator replaces the imaging system and so-called lensless imaging is achieved **[353]**, for instance of objects that are embedded in a strongly scattering medium. A spatial light modulator also enables the deliberate introduction of an image defocusing or an imaging aberration to introduce so-called phase diversity in object retrieval problems. Digital picture memories can be used to easily switch to the Fourier domain and to perform digital filtering with the result that certain desired features are optimised in the final (digital) image. Since the 1980s, all these practices and possibilities have opened new ways for image formation, image enhancement, image interpretation and object retrieval.

10.4 The Three-dimensional Transfer Function

In the two preceding sections of this chapter we have studied the coherent, incoherent and partially coherent frequency transfer function with the aid of the source intensity function and the complex pupil function of a static optical system. For a Type I scanning imaging system, a comparable analysis was made using the detector sensitivity function. The confocal microscope gave rise to a special treatment with no direct analogon in classical imaging. The transfer function was obtained for the special case of an object and image *plane*, both perpendicular to the optical axis. Moreover, we assumed that the object structure is infinitely thin. When volumetric objects are imaged, for instance biological specimina like living cells or bacteria, it is advantageous to have an expression for the frequency transfer in an arbitrary cross-section of the three-dimensional image of such an object. The frequency transfer in an arbitrary cross-section of a three-dimensional image can be extracted from a general three-dimensional transfer function. A preferred cross-section is the transverse cross-section yielding the frequency transfer for image features that are periodic along the orthogonal x- and y-axes. The three-dimensional transfer function is able to yield the supplementary transfer data for features that are periodic along the optical axis and, in general, for any orientation of a three-dimensional spatial frequency vector. From the elongated shape of the three-dimensional diffraction image, it is evident that the value of the transfer function along the optical axis in a classical imaging system, for instance a low to moderate power optical microscope, is much lower at equal spatial frequency than for the transverse components in the image plane. It is only for high-numerical-aperture systems that the transverse and axial frequency transfer functions have comparable values. A thorough analysis of three-dimensional frequency transfer has been given by Frieden **[296]** and we will use his approach to establish the relationship between the three-dimensional diffraction pattern and the three-dimensional frequency transfer function. The analysis is limited to coherently or incoherently illuminated volumes in object space.

10.4.1 Diffraction by a Three-dimensional Periodic Structure

In this subsection we encounter functions describing complex amplitude distributions, intensity distributions, transmission or reflection functions for amplitude or intensity, all defined in the spatial domain. The functions that are defined in object space are denoted by means of unprimed characters, the same quantities in image space by the corresponding primed characters. The Fourier or inverse Fourier transforms of these functions are denoted by the same function names in object or image space, provided with a tilde sign on top of them. A position in object and image space is denoted by the vectors $\mathbf{r}$ and $\mathbf{r}'$, respectively. The spatial frequencies occurring in the transformed functions in object space are denoted by the angular spatial frequency vector $\boldsymbol{\omega} = (\omega_x, \omega_y, \omega_z)$ or the spatial frequency vector $\boldsymbol{\nu} = (\nu_x, \nu_y, \nu_z) = \boldsymbol{\omega}/(2\pi)$, where the components are generally expressed in mm^{-1}. The angular spatial frequency $\boldsymbol{\omega}$ in calligraphic font is repeatedly used in this chapter to avoid the abundant use of factors of 2π in the equations. The angular spatial frequency $\boldsymbol{\omega}$ should not

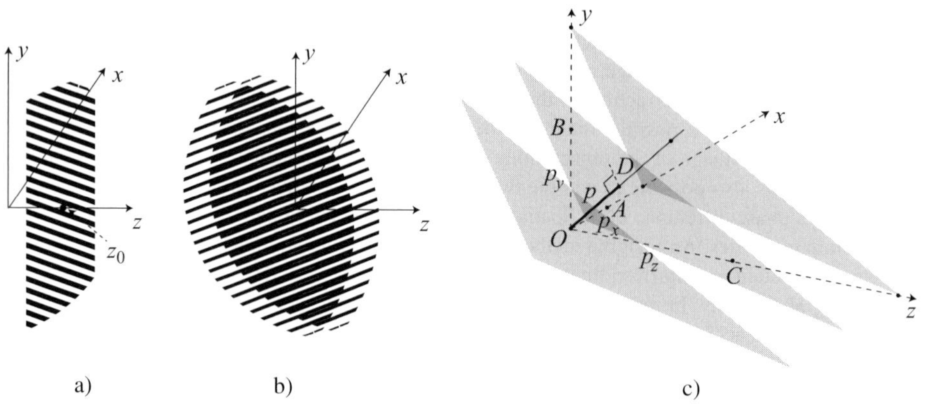

Figure 10.52: a) A thin three-dimensional periodic object. The thin object is perpendicular to the z-axis and located in the plane $z = z_0$.
b) A thick three-dimensional periodic object.
c) The period p of a three-dimensional periodic structure.

be confounded with the temporal frequency ω that is used in other chapters of this book. Likewise, to avoid confusion between the unit frequency vector $\hat{\mathbf{v}}$ and the unit surface normal, we denote the latter vector by $\hat{\mathbf{n}}$ in this section.

Before we study the three-dimensional transfer of spatial frequencies by an optical imaging system, it is necessary to have a closer look at the diffraction of plane waves by periodic structures embedded in a thin or a thick object. Schematic drawings of such three-dimensional objects are given in Fig. 10.52. The period p of a three-dimensional structure, equal to the distance OD in Fig. 10.52c, follows from the periods that can be measured along the orthogonal Cartesian axes (x, y, z), with values $p_x = 1/\nu_x$, $p_y = 1/\nu_y$ and $p_z = 1/\nu_z$, respectively. From elementary vector calculus, we readily obtain that the inverse length $\nu = 1/p$ is given by

$$\nu = [\nu_x^2 + \nu_y^2 + \nu_z^2]^{1/2} . \tag{10.4.1}$$

In the following analysis of diffraction by three-dimensional periodic structures, we consider the two extreme cases of a periodic object that is very thin (with respect to the wavelength of the light) and a periodic object that fills the entire space. In general, for the case of weak or single scattering, we assume that the outgoing field is obtained by a multiplication of the incident field and the transmission function of the three-dimensional object. To obtain the waves that are diffracted by a periodic object when a plane wave is incident on that object, we first apply a Fourier transform to the object transmission function to obtain its spectral representation. The spectrum of plane waves that is the result of diffraction at the periodic structure is the convolution of the Fourier transform of the incident field (a plane wave, represented by a spectral δ-function), and the spatial frequency spectrum of the object. In the analysis that follows, we do not claim reliable results for the strengths of the diffracted plane waves. In many cases, the weak scattering or Born approximation will be insufficient. To obtain the transmitted light amplitude for a certain incident plane wave, we should apply Maxwell's equations to the wave field propagating within the modulated volume and impose the appropriate boundary conditions on the surface that delimits the object. Such an approach yields the correct values of the amplitudes of the transmitted and reflected waves if the modulation inside the object is such that it is not allowed to neglect secondary and higher-order scattering effects (see Section 1.12 for a detailed treatment of the solution of such a boundary-value problem). The analysis we present below uses the weak scattering approximation and, consequently, the values of the complex amplitudes of the diffracted waves are first-order approximations. But the analysis provides the exact direction of the diffracted waves.

10.4.1.1 Harmonically Modulated Thin Planar Object

We consider an object with a (very small) periodic amplitude absorption that is confined to a very thin planar layer. The general equation of a plane is given by $(\mathbf{r} - \mathbf{r}_0) \cdot \hat{\mathbf{n}} = 0$, where $\hat{\mathbf{n}} = (\hat{n}_x, \hat{n}_y, \hat{n}_z)$ is the unit normal vector to the plane. The scalar product $\mathbf{r}_0 \cdot \hat{\mathbf{n}}_0$ is the perpendicular distance from the origin to the plane and is denoted by b. The transmission function of this very thin planar object in space is then given by

$$t(x,y,z) = 1 - a\Big[1 - \cos(\omega_{x,p}x + \omega_{y,p}y + \omega_{z,p}z)\Big]_{\{\mathbf{r}\cdot\hat{\mathbf{n}}-b=0\}} . \qquad (10.4.2)$$

In the spatial frequency domain, the (spectral) object function is given by its Fourier transform, defined by Eq. (A.8.5). To simplify the evalution of the resulting three-dimensional integral we perform a two-stage coordinate transformation. The first is a rotation around the z-axis over an angle ϕ_0, resulting in the (x',y',z')-system. A second rotation around the y'= axis over an angle θ_0 from the positive z'-axis to the x'-axis yields the (x'',y'',z'')-system, such that the z''-axis coincides with the normal $\hat{\mathbf{n}}$ to the modulated planar surface. After some manipulation in the rotated coordinates system, the various terms of the Fourier transform turn out to be given by

$$\begin{aligned}
\tilde{t}(\boldsymbol{\omega}) &= \iiint_{-\infty}^{+\infty} t(x,y,z)\exp[-i(\omega_x x + \omega_y y + \omega_z z)]\,dxdydz \\
&= 8\pi^3\,\delta(\omega_x,\omega_y\,,\,\omega_z) - 4\pi^2\frac{A}{\hat{n}_y}\exp\left\{\frac{-ib\omega_y}{\hat{n}_y}\right\}\delta\left(\omega_x - \frac{\hat{n}_x}{\hat{n}_y}\omega_y\,,\,\omega_z - \frac{\hat{n}_z}{\hat{n}_y}\omega_y\right) \\
&+ 2\pi^2\frac{A}{\hat{n}_y}\exp\left\{\frac{-ib(\omega_y-\omega_{y,p})}{\hat{n}_y}\right\}\delta\left(\omega_x - \omega_{x,p} - \frac{\hat{n}_x}{\hat{n}_y}(\omega_y-\omega_{y,p})\,,\,\omega_z - \omega_{z,p} - \frac{\hat{n}_z}{\hat{n}_y}(\omega_y-\omega_{y,p})\right) \\
&+ 2\pi^2\frac{A}{\hat{n}_y}\exp\left\{\frac{-ib(\omega_y+\omega_{y,p})}{\hat{n}_y}\right\}\delta\left(\omega_x + \omega_{x,p} - \frac{\hat{n}_x}{\hat{n}_y}(\omega_y+\omega_{y,p})\,,\,\omega_z + \omega_{z,p} - \frac{\hat{n}_z}{\hat{n}_y}(\omega_y+\omega_{y,p})\right),
\end{aligned} \qquad (10.4.3)$$

where the factor A is the product of the contrast factor a of the modulated surface and its perpendicular thickness $\Delta z''$. The incident plane wave is represented by a δ-function in the spectral domain. For a plane wave with wave vector $\mathbf{k}$ we have, apart from a complex amplitude factor, the spectral function $\delta(\boldsymbol{\omega}-\mathbf{k})$. The spatial function related to the plane wave is obtained by an inverse Fourier transform and equals $\exp[+i\mathbf{k}\cdot\mathbf{r}]$.

The complex amplitude transmitted by the object, in the small scattering approximation, is given by the product of the three-dimensional transmission function t and the plane wave amplitude distribution in space. The Fourier transform of this product function in the spatial domain corresponds to a convolution of the Fourier transforms in the frequency domain. The resulting spectral function $g(\boldsymbol{\omega})$ is thus given by

$$\begin{aligned}
g(\boldsymbol{\omega}) &= \iiint_{-\infty}^{+\infty}\tilde{t}(\boldsymbol{\omega}')\,\delta(\boldsymbol{\omega}-\mathbf{k}-\boldsymbol{\omega}')\,d\boldsymbol{\omega}' = 8\pi^3\,\delta(\omega_x-k_x,\omega_y-k_y,\omega_z-k_z) \\
&- 4\pi^2\frac{A}{\hat{n}_y}\exp\left\{\frac{-ib(\omega_y-k_y)}{\hat{n}_y}\right\}\delta\left(\omega_x - k_x - \frac{\hat{n}_x}{\hat{n}_y}(\omega_y-k_y)\,,\,\omega_z-k_z-\frac{\hat{n}_z}{\hat{n}_y}(\omega_y-k_y)\right) \\
&+ 2\pi^2\frac{A}{\hat{n}_y}\exp\left\{\frac{-ib(\omega_y-k_y-\omega_{y,p})}{\hat{n}_y}\right\}\delta\left(\omega_x-k_x-\frac{\hat{n}_x}{\hat{n}_y}(\omega_y-k_y)-\omega_{x,p}+\frac{\hat{n}_x}{\hat{n}_y}\omega_{y,p}\,,\right. \\
&\qquad\qquad\left.\omega_z-k_z-\frac{\hat{n}_z}{\hat{n}_y}(\omega_y-k_y)-\omega_{z,p}+\frac{\hat{n}_z}{\hat{n}_y}\omega_{y,p}\right) \\
&+ 2\pi^2\frac{A}{\hat{n}_y}\exp\left\{\frac{-ib(\omega_y-k_y+\omega_{y,p})}{\hat{n}_y}\right\}\delta\left(\omega_x-k_x-\frac{\hat{n}_x}{\hat{n}_y}(\omega_y-k_y)+\omega_{x,p}-\frac{\hat{n}_x}{\hat{n}_y}\omega_{y,p}\,,\right. \\
&\qquad\qquad\left.\omega_z-k_z-\frac{\hat{n}_z}{\hat{n}_y}(\omega_y-k_y)+\omega_{z,p}-\frac{\hat{n}_z}{\hat{n}_y}\omega_{y,p}\right).
\end{aligned} \qquad (10.4.4)$$

The spectral function $g(\omega_x,\omega_y,\omega_z)$ contains four terms which each represent a plane wave by means of the δ-function in the spectral domain. The wave vector of the first term is equal to that of the incident wave and corresponds to the unperturbed wave which is present in all space. The wave vectors of the other terms are not yet fully defined because their δ-functions only determine the ω_x and ω_z components of the wave vector. In each case, the third wave vector component follows from the dispersion relation for a plane wave that was discussed in Chapter 1, see Eqs. (1.6.62) and (1.6.73). In terms of the spatial frequency vector components $(\omega_x,\omega_y,\omega_z)$ of a plane wave we have the property that $\omega_x^2+\omega_y^2+\omega_z^2 = k_x^2+k_y^2+k_z^2$, where $\mathbf{k}$ is the wave vector of the incident wave.

Applying the dispersion relation to the three remaining terms of the function g of Eq. (10.4.4) we have for the wave vector components of the second term the following equations,

$$\begin{aligned}
\omega_x &= \frac{\hat{n}_x}{\hat{n}_y}\omega_y + k_x - \frac{\hat{n}_x}{\hat{n}_y}k_y ,\\
\omega_z &= \frac{\hat{n}_z}{\hat{n}_y}\omega_y + k_z - \frac{\hat{n}_z}{\hat{n}_y}k_y ,\\
\omega_x^2 + \omega_y^2 + \omega_z^2 &= k_x^2 + k_y^2 + k_z^2 = k^2 \, .
\end{aligned} \tag{10.4.5}$$

It is easily verified that a solution for the wave vector components of this plane wave is given by $\boldsymbol{\omega} = \mathbf{k}$ and that this second term thus represents a plane wave which propagates in the direction of the incident wave. This wave corresponds to the zeroth diffraction order of the modulated planar surface with modulation depth a. A less obvious solution is found when solving the quadratic equation in ω_y which follows from Eq. (10.4.5). We obtain

$$\begin{aligned}
\omega_x &= k_x - 2\hat{n}_x(\hat{n}_x k_x + \hat{n}_y k_y + \hat{n}_z k_z) \, ,\\
\omega_y &= k_y - 2\hat{n}_y(\hat{n}_x k_x + \hat{n}_y k_y + \hat{n}_z k_z) \, ,\\
\omega_z &= k_z - 2\hat{n}_z(\hat{n}_x k_x + \hat{n}_y k_y + \hat{n}_z k_z) \, .
\end{aligned} \tag{10.4.6}$$

This wave vector determines the propagation direction of a wave that has been specularly reflected by the modulated surface determined by $\mathbf{r} \cdot \hat{\mathbf{n}} = b$.

The wave vector components related to the third and fourth term follow from a quadratic equation in ω_y; in the case of the third term we obtain,

$$\begin{aligned}
\omega_x &= \frac{\hat{n}_x}{\hat{n}_y}\omega_y + k_x + \omega_{x,p} - \frac{\hat{n}_x}{\hat{n}_y}(k_y + \omega_{y,p}) \, ,\\
\omega_z &= \frac{\hat{n}_z}{\hat{n}_y}\omega_y + k_z + \omega_{z,p} - \frac{\hat{n}_z}{\hat{n}_y}(k_y + \omega_{y,p}) \, ,\\
\omega_x^2 + \omega_y^2 + \omega_z^2 &= k_x^2 + k_y^2 + k_z^2 = k^2 \, .
\end{aligned} \tag{10.4.7}$$

With the short-hand of writing $\omega_x = (\hat{n}_x/\hat{n}_y)\omega_y + c_x$ and $\omega_z = (\hat{n}_z/\hat{n}_y)\omega_y + c_z$, we can readily solve the quadratic equation in ω_y on the last line of Eq. (10.4.7) and obtain the result

$$\omega_y = \hat{n}_y \left\{ -(\hat{n}_x c_x + \hat{n}_z c_z) \pm \sqrt{2\hat{n}_x\hat{n}_z c_x c_z - (\hat{n}_y^2 + \hat{n}_z^2)c_x^2 - (\hat{n}_x^2 + \hat{n}_y^2)c_z^2 + k^2} \right\} . \tag{10.4.8}$$

If $\hat{n}_y$ is not the largest component of the unit normal vector, a new set of equations according to (10.4.6) can be established by cyclic substitution. The quadratic equation in ω_y produces two solutions for the diffracted plane waves. Using the plus sign in Eq. (10.4.8) we obtain from Eq. (10.4.7) the wave vector of a transmitted diffracted wave. The minus sign produces the wave vector of a reflected diffraction wave. In Eqs. (10.4.5)–(10.4.8) the solutions for the diffracted wave directions were found by algebraic means; the modulated surface was free-standing in a medium. In Subsection 5.7.3, comparable calculations were carried out by using ray vector analysis at a modulated portion of a surface that separates two media in space with different refractive indices. In these ray vector calculations we also included the effect of refraction at the interface between the two media.

A frequently occurring geometry is that of a thin modulated surface perpendicular to the z-axis. The z-axis coincides with the optical axis of the imaging system and the axial coordinate z of the planar surface equals z_0. In that case, simple explicit formulas for the diffracted orders are obtained. The following spectral components are found, associated with the presence of the modulated planar surface,

$$\begin{aligned}
g_0(\boldsymbol{\omega}) &= -A \, \exp[-i(\omega_z - k_z)z_0] \; \delta\left\{\frac{1}{2\pi}(\omega_x - k_x, \omega_y - k_y\right\} ,\\
g_{+1}(\boldsymbol{\omega}) &= \frac{A}{2} \, \exp[-i(\omega_z - k_z - \omega_{z,p})z_0] \, \delta\left\{\frac{1}{2\pi}(\omega_x - k_x - \omega_{x,p}, \omega_y - k_y - \omega_{y,p})\right\} ,\\
g_{-1}(\boldsymbol{\omega}) &= \frac{A}{2} \, \exp[-i(\omega_z - k_z + \omega_{z,p})z_0] \, \delta\left\{\frac{1}{2\pi}(\omega_x - k_x + \omega_{x,p}, \omega_y - k_y + \omega_{y,p})\right\} .
\end{aligned} \tag{10.4.9}$$

The functions $g_m(\boldsymbol{\omega})$ with $m = -1, 0, +1$, represent the zeroth and first diffracted orders in the spectral domain. The value of the axial frequency component $\omega_{z,m}$ is obtained by applying the dispersion relation to the spectral components that are described by the two-dimensional δ-functions in Eq. (10.4.4). For a general term g_m we find the expression

$$\omega_{z,m} = \pm \left[k_x^2 + k_y^2 + k_z^2 - (k_x + m\omega_{x,p})^2 - (k_y + m\omega_{y,p})^2\right]^{1/2} . \qquad (10.4.10)$$

The selection of the sign of $\omega_{z,m}$ determines if the transmitted (+ sign) or the reflected set of diffracted orders is chosen. If the argument of the square root itself is negative, the corresponding plane wave is non-periodic in at least one direction and generally damped in this direction. In an amplifying medium or in a so-called metamaterial, the amplitude can increase as a function of distance. The damped evanescent waves have an appreciable amplitude only in the direct vicinity of the periodic object.

10.4.1.2 Harmonically Modulated One-dimensional Line Object

Similarly to the way we defined the transmission function of a two-dimensional planar object by means of Eq. (10.4.2), we can write the transmission function of a one-dimensional periodic line object as

$$t(x, y, z) = 1 - a\left[1 - \cos(\omega_{x,p}x + \omega_{y,p}y + \omega_{z,p}z)\right]_{\{\mathbf{r}\cdot\hat{\mathbf{n}}_1 - b_1 = \mathbf{r}\cdot\hat{\mathbf{n}}_2 - b_2 = 0\}} , \qquad (10.4.11)$$

where the infinitely thin line is defined by the intersection line of two planes with normal unit vectors $\hat{\mathbf{n}}_1$ and $\hat{\mathbf{n}}_2$, respectively, whilst b_1 and b_2 define the perpendicular distance to the origin of these two planes. In a similar way as for the planar modulated surface, we perform a rotation of the coordinate system from (x, y, z) to (x'', y'', z''), so that the rotated z''-axis is parallel to the line object. After some manipulation we find the following expression for the three-dimensional Fourier transform of $t(x, y, z)$,

$$\tilde{t}(\boldsymbol{\omega}) = \iiint_{-\infty}^{+\infty} t(x, y, z) \exp[-i(\omega_x x + \omega_y y + \omega_z z)]\, dxdydz$$

$$= 8\pi^3\, \delta(\omega_x, \omega_y, \omega_z) - 2\pi A_l \exp\left\{i\,(b_{2,x}\omega_x + b_{1,y}\omega_y)\right\} \delta\left\{\omega_z - c_{yz}\omega_x - c_{xz}\omega_y\right\}$$

$$+ \pi A_l \exp\left\{i\left[b_{2,x}(\omega_x - \omega_{x,p}) + b_{1,y}(\omega_y - \omega_{y,p})\right]\right\} \delta\left\{\omega_z - \omega_{z,p} - c_{yz}(\omega_x - \omega_{x,p}) - c_{xz}(\omega_y - \omega_{y,p})\right\}$$

$$+ \pi A_l \exp\left\{i\left[b_{2,x}(\omega_x + \omega_{x,p}) + b_{1,y}(\omega_y + \omega_{y,p})\right]\right\} \delta\left\{\omega_z - \omega_{z,p} - c_{yz}(\omega_x + \omega_{x,p}) - c_{xz}(\omega_y + \omega_{y,p})\right\} , \qquad (10.4.12)$$

where the amplitude factor $A_l = a(\Delta x'')(\Delta y'')\sqrt{1 + c_{xz}^2 + c_{yz}^2}$. The area $\Delta x''\Delta y''$ is the (very small) perpendicular cross-sectional area of the line object. The coefficients c_{yz}, c_{xz}, $b_{2,x}$ and $b_{1,y}$ are given by

$$c_{yz} = \frac{\hat{n}_{2,y}\hat{n}_{1,z} - \hat{n}_{1,y}\hat{n}_{2,z}}{\hat{n}_{1,x}\hat{n}_{2,y} - \hat{n}_{2,x}\hat{n}_{1,y}} , \quad c_{xz} = \frac{\hat{n}_{1,x}\hat{n}_{2,z} - \hat{n}_{2,x}\hat{n}_{1,z}}{\hat{n}_{1,x}\hat{n}_{2,y} - \hat{n}_{2,x}\hat{n}_{1,y}} , \quad b_{2,x} = \frac{-\hat{n}_{2,y}b_1 + \hat{n}_{1,y}b_2}{\hat{n}_{1,x}\hat{n}_{2,y} - \hat{n}_{2,x}\hat{n}_{1,y}} , \quad b_{1,y} = \frac{\hat{n}_{2,x}b_1 - \hat{n}_{1,x}b_2}{\hat{n}_{1,x}\hat{n}_{2,y} - \hat{n}_{2,x}\hat{n}_{1,y}} . \qquad (10.4.13)$$

The 'far-field' spectral function $g(\boldsymbol{\omega})$ created by the periodic line object when a plane wave with wave vector $\boldsymbol{\omega} = \mathbf{k}$ is incident on the object is given by the convolution integral on the first line of Eq. (10.4.4). In order to keep the expressions simple we assume that the values of b_1 and b_2 are zero in Eq. (10.4.12). For this case in which the line object intersects the origin we readily obtain for the function g the following spectral components due to the presence of the line object,

$$g(\boldsymbol{\omega}) = \delta\{(\boldsymbol{\omega} - \mathbf{k})/2\pi\} - A_l\, \delta\left\{\left[\omega_z - k_z - c_{yz}(\omega_x - k_x) - c_{xz}(\omega_y - k_y)\right]/2\pi\right\}$$
$$+ \frac{A_l}{2}\, \delta\left\{\left[\omega_z - k_z - \omega_{z,p} - c_{yz}(\omega_x - k_x - \omega_{x,p}) - c_{xz}(\omega_y - k_y - \omega_{y,p})\right]/2\pi\right\}$$
$$+ \frac{A_l}{2}\, \delta\left\{\left[\omega_z - k_z + \omega_{z,p} - c_{yz}(\omega_x - k_x + \omega_{x,p}) - c_{xz}(\omega_y - k_y + \omega_{y,p})\right]/2\pi\right\} . \qquad (10.4.14)$$

In a similar way as for the thin two-dimensional object we identify the two terms on the first line of Eq. (10.4.14) as the incident wave and the zeroth order light of the line object spread out on a conic surface. The following two terms correspond to the plus and minus first diffracted orders of the periodic object. The diffracted wave corresponding to the plus first-order light on the second line of Eq. (10.4.14) has a z-component which is given by

$$\omega_z = c_{yz}\omega_x + c_{xz}\omega_y + k_z + \omega_{z,p} - c_{yz}(k_x + \omega_{x,p}) - c_{xz}(k_y + \omega_{y,p}) \,. \tag{10.4.15}$$

We use the dispersion relation on the last line of Eq. (10.4.5) and write for the lateral vector components

$$\omega_x^2 + \omega_y^2 + \left(c_{yz}\omega_x + c_{xz}\omega_y + a^{+}_{k,\omega}\right)^2 = k^2 \,, \tag{10.4.16}$$

where, in general, the constant $a^{\pm}_{k,\omega}$ for the plus and minus first order is given by $k_z \pm \omega_{z,p} - c_{yz}(k_x \pm \omega_{x,p}) - c_{xz}(k_y \pm \omega_{y,p})$ and depends only on the direction of the incident wave vector and the spatial frequency vector $\boldsymbol{\omega}_p$ associated with the periodic modulation of the line object.

The quadratic equation in ω_x and ω_y corresponds to an elliptical curve in the two-dimensional (ω_x, ω_y) frequency domain. If the incident field is a plane wave, the first-order light diffracted by the line object is found on a conical surface in space with an elliptical cross-section. The allowed values of the lateral components (ω_x, ω_y) follow in a standard way from the solution of a quadratic equation $f(\omega_x, \omega_y) = 0$ derived from Eq. (10.4.16) and which is given by

$$\begin{aligned} f(\omega_x, \omega_y) &= \omega_x^2 + \omega_y^2 + \left(c_{yz}\omega_x + c_{xz}\omega_y + a^{\pm}_{k,\omega}\right)^2 - k^2 \\ &= a\,\omega_x^2 + b\,\omega_y^2 + c\,\omega_x\omega_y + d\,\omega_x + e\,\omega_y + f_c = 0 \,, \end{aligned} \tag{10.4.17}$$

where the constant factors a, b, c, d, e and f_c are expressed in terms of the fixed quantities c_{yz}, c_{xz} and the components of the wave vector $\mathbf{k}$ and the frequency vector $\boldsymbol{\omega}_p$. The quadratic form is reduced to its principal axis by means of a rotation by an angle γ in the (ω_x, ω_y) frequency plane where $\tan(2\gamma) = c/(a-b)$. The rotation angle γ is counted positive when using the right-hand rule in the $\boldsymbol{\omega}$-coordinate system. After an additional coordinate translation in the rotated (ω_x, ω_y) frequency plane we write the expression for the quadratic surface $f(\omega_x, \omega_y) = 0$ in terms of the rotated and shifted frequency coordinates $(\omega_{xrt}, \omega_{yrt})$ as

$$\begin{aligned} f(\omega_{xrt}, \omega_{yrt}) = &\frac{1}{2}\left[a + b + \sqrt{c^2 + (a-b)^2}\right]\omega_{xrt}^2 + \frac{1}{2}\left[a + b - \sqrt{c^2 + (a-b)^2}\right]\omega_{yrt}^2 \\ &- \frac{(ae^2 + bd^2 - cde)}{4ab - c^2} + f_c = 0 \,. \end{aligned} \tag{10.4.18}$$

The transformed coordinates $(\omega_{xrt}, \omega_{yrt})$ follow from

$$\begin{aligned} \omega_{xrt} &= \omega_x \cos\gamma + \omega_y \sin\gamma + (d\cos\gamma + e\sin\gamma)\Big/\left(a + b + \sqrt{c^2 + (a-b)^2}\right) , \\ \omega_{yrt} &= -\,\omega_x \sin\gamma + \omega_y \cos\gamma + (-d\sin\gamma + e\cos\gamma)\Big/\left(a + b - \sqrt{c^2 + (a-b)^2}\right) , \end{aligned} \tag{10.4.19}$$

the back transformation is given by

$$\begin{aligned} \omega_x &= \omega_{xrt}\cos\gamma - \omega_{yrt}\sin\gamma + \frac{ce - 2bd}{4ab - c^2} \,, \\ \omega_y &= \omega_{xrt}\sin\gamma + \omega_{yrt}\cos\gamma + \frac{cd - 2ae}{4ab - c^2} \,. \end{aligned} \tag{10.4.20}$$

The general expression (10.4.17) for the propagation direction of light diffracted by a line object is used in Subsection 10.4.14 to study the one-dimensional frequency transfer function when imaging an axially modulated object.

10.4.1.3 Thick Object and the Bragg Condition

For a thick object which, in principle, fills the entire space, we have the transmission function of Eq. (10.4.2) in which the function defining the planar surface has been suppressed. For the amplitudes of the spectral components, we have the straightforward result

$$\begin{aligned} g_0(\boldsymbol{\omega}) &= (1-a)\,\delta\left[(\omega_x - k_x, \omega_y - k_y, \omega_z - k_z)/(2\pi)\right] , \\ g_{+1}(\boldsymbol{\omega}) &= \frac{a}{2}\,\delta\left[(\omega_x - k_x - \omega_{x,p}, \omega_y - k_y - \omega_{y,p}, \omega_z - k_z - \omega_{z,p})/(2\pi)\right] , \\ g_{-1}(\boldsymbol{\omega}) &= \frac{a}{2}\,\delta\left[(\omega_x - k_x + \omega_{x,p}, \omega_y - k_y + \omega_{y,p}, \omega_z - k_z + \omega_{z,p})/(2\pi)\right] . \end{aligned} \tag{10.4.21}$$

For the zeroth-order wave we have the wave vector $\boldsymbol{\omega}_0 = \mathbf{k}$, for the first orders we find the wave vectors $\boldsymbol{\omega}_m = \mathbf{k} + m\boldsymbol{\omega}_p$. For a thick object, the vector $\mathbf{k}$ is determined by the condition that the dispersion relation in the medium has to be fulfilled for diffracted waves with order number ± 1. It readily follows that the first-order diffracted waves satisfy the dispersion relation when it holds that

$$\boldsymbol{\omega}_p \cdot \mathbf{k} + m\,\frac{\omega_p^2}{2} = 0\,, \qquad m = \pm\, 1. \tag{10.4.22}$$

If we denote the angle between the vectors $\boldsymbol{\omega}_p$ and $\mathbf{k}$ by $\pi/2 - \theta$, we have the relation

$$\sin\theta = -m\frac{\omega_p}{2k} = -m\frac{\lambda}{2p}\,. \tag{10.4.23}$$

This expression is known as the Bragg *condition* for diffraction of radiation by three-dimensional periodic structures, or, more briefly, as the Bragg reflection condition.

Solutions to Eq. (10.4.23) have symmetry of revolution with respect to an axis determined by the grating vector $\boldsymbol{\omega}_p$. The symmetry property and the construction of the solution vectors that follow from Eq. (10.4.22) are illustrated in Fig. 10.53. Two possible plane wave combinations are shown in each drawing. If $\vec{OA}$ is the incident wave vector, labelled $\boldsymbol{\omega}_0$, the Bragg condition creates the diffracted order labelled $\boldsymbol{\omega}_{-1}$ in the direction given by $\vec{OB}$. Conversely, an incident wave labelled $\boldsymbol{\omega}_0$ with $\vec{OB}$ as propagation direction produces a diffracted wave labelled $\boldsymbol{\omega}_{+1}$ in the direction specified by $\vec{OA}$. The figure shows that for both cases the endpoints of a Bragg diffraction vector pair on the so-called Ewald sphere are joined by the grating vector $\boldsymbol{\omega}_p$. Figure 10.53a applies to a low value of the spatial frequency $\boldsymbol{\omega}_0$ as compared to the value of k in the medium. In Fig. 10.53b a situation is shown where both the incident and diffracted wave are at a grazing angle with the diffracting planes of constant phase and the two beams propagate in almost opposite directions. Freely-propagating wave solutions are found as long as the value of $\sin\theta$ in Eq. (10.4.23) is real. The period of the Bragg grating should be larger than $\lambda/2$ in that case. A smaller period gives rise to an evanescent Bragg wave.

The name Bragg *reflection* condition instead of Bragg *diffraction* condition is inspired by the physical picture in which the spatial modulation in each period of a Bragg grating is replaced by a hypothetical surface that is perpendicular to the grating vector and weakly reflects the incident radiation. The direction of the diffraction order then follows from a symmetric construction of the incident and diffracted wave according to Fig. 10.54. The diffraction angle 2θ follows from the pathlength increment condition per period,

$$[Q_0PQ_1] = 2np\sin\theta = \lambda_0\,. \tag{10.4.24}$$

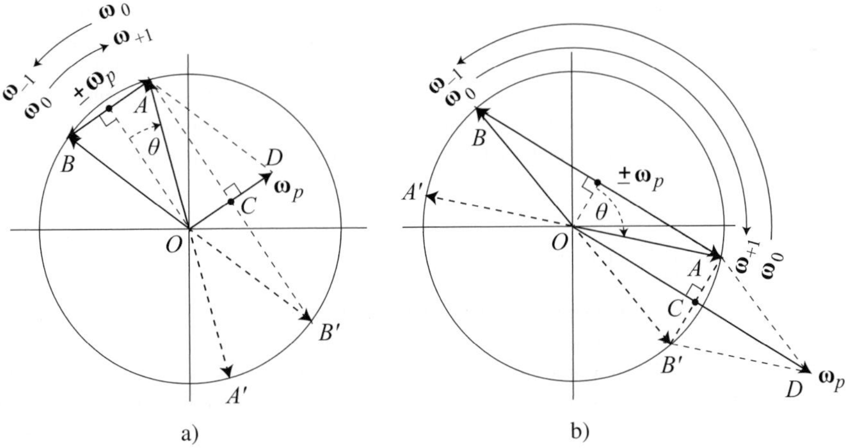

Figure 10.53: Examples of a geometrical construction for obtaining the direction of the incident wave and the diffracted order with order number $m \pm 1$ for a thick three-dimensional grating structure with grating vector $\boldsymbol{\omega}_p$ (Bragg diffraction). The circle through the vector endpoints A, B, A' and B' is a cross-section of the Ewald sphere with diameter $2|\mathbf{k}|$.
a) $|\boldsymbol{\omega}_p| \ll |\mathbf{k}|$.
b) $|\boldsymbol{\omega}_p| \approx |\mathbf{k}|$.

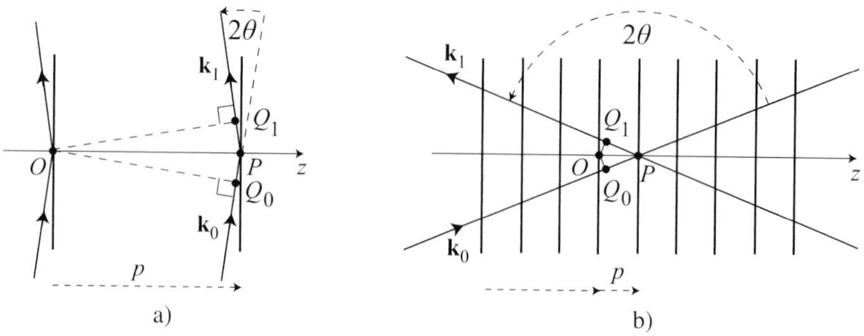

Figure 10.54: Geometrical construction of a Bragg reflected wave using the optical pathlength condition $[Q_0PQ_1] = \lambda_0$.
a) A low-frequency grating.
b) A high-frequency grating with $p \approx \lambda/2$ (λ is the wavelength in the medium).

The diffracted wave returns in the opposite direction to the incident wave if the period p equals exactly half the wavelength in the medium. Such a special condition is used, for instance, in optical fibres which have a Bragg reflecting grating with period p printed on the surrounding surface of the core or on the (thin) cladding of the fibre. Such a grating behaves as a spectral filter by reflecting in the backward direction a certain spectral band that obeys the Bragg condition. It can also be used as a temperature or a pressure sensor when the period p changes under external influence. A spectral analysis of the back-propagating light yields information on the change δp of the Bragg grating period.

The solutions given above for a very thin or for an infinitely extending three-dimensional periodic object are special cases of a more general solution. For instance, if a finite thickness slab-like object is present, the function defining the planar surface in Eq. (10.4.2) is replaced by the function rect$[(z - z_0)/d_s]$, where d_s is the slab thickness. The δ-functions in the Fourier transform $\tilde{t}(\boldsymbol{\omega})$ will be replaced by sinc-functions of the spatial frequency $\boldsymbol{\omega}$. In the expressions for the orders of Eq. (10.4.21), the infinitely sharp Bragg reflection peaks are replaced by a broadened central peak at the Bragg positions with diffraction sidelobes around it. Another approximation, the one about the weak scattering in the periodic structure, is not generally fulfilled in practice. In practical media, on propagation through the medium, the first order accumulates an increasingly large amplitude up to a certain position where all the incident power has been transferred to the first order. For instance, with reference to Fig. 10.53, the power of the wave with propagation direction $\vec{OA}$ is fully transferred after a certain propagation distance to the wave with direction $\vec{OB}$, the minus first order. The typical propagation distance needed for a complete transfer of optical power between the Bragg pair in the medium is called the Bragg distance d_B. Propagation over a larger distance than d_B shows the inverse process and the power is gradually transferred back to the original direction of incidence given by $\vec{OA}$. The optical power as a function of the propagation distance z in the medium in the direction of incidence $\vec{OA}$ is proportional to $\cos^2(\pi z/2d_B)$. In a lossless medium the power in the first-order direction varies as $\sin^2(\pi z/2d_B)$. In an acousto-optic crystal used as a high-frequency modulator, the modulation of the refractive index caused by the elastic wave excited by the acousto-optic actuator is tuned to a level such that, after propagation over a distance equal to the crystal thickness, the incident power is completely transferred to the first-order wave.

10.4.2 Paraxial Imaging in Three Dimensions

Paraxial imaging between optically conjugate planes was studied in Section 4.10. In this subsection we use the results of this section together with the axial magnification to obtain the paraxial change in shape if a three-dimensional volume in object space is imaged and occupies a modified volume in image space. To evaluate the frequency transfer in three dimensions we have to supplement the paraxial results with the geometrical aberrations of the imaging system and the intrinsic diffraction effects.

In Fig. 10.55 we show an optical system which is paraxially represented by four cardinal points. The selected cardinal points are the object and image-side principal points, P_0 and P_0', and the object and image-side focal points, F and F', respectively. The origins for the Cartesian coordinate systems (x, y, z) and (x', y', z') in object and image space are the principal points P_0 and P_0'. The z- and z'-axis are coincident with the optical axis of the optical system. Two examples of conjugate object and image volumes are shown in Fig. 10.55, with volumes (V_0, V_0') and (V_1, V_1'), respectively. The object-side volumes are chosen to be cubic, but in principle, they can have any shape. The shape of an imaged volume V'

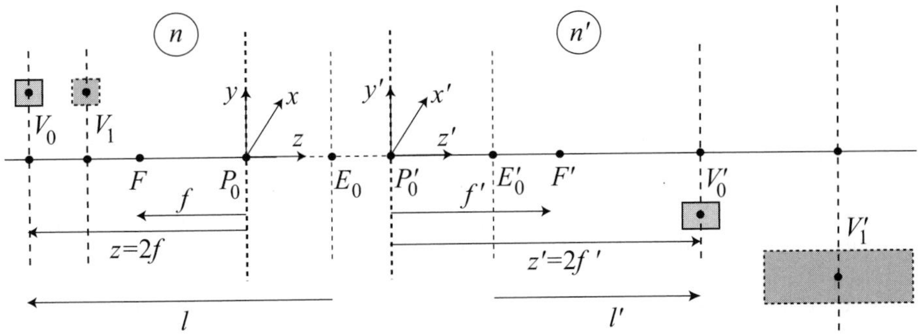

Figure 10.55: Volumetric imaging according to the laws of paraxial optics. The principal planes of the optical system have been denoted by P_0 and P_0', the focal points by F and F'. Two sets of conjugate planes have been shown with $z = 2f$ and $z = 3f/2$, respectively. The objects V_0 and V_1 are cubes and their corresponding paraxial images are shown with volumes denoted by V_0' and V_1', respectively. The refractive indices in object and image space are given by n and n'.

is determined by the paraxial transverse magnification m_T, the ratio z'/z and the axial magnification m_z. These quantities follow from Eq. (4.10.43) and are given by

$$
\begin{aligned}
m_T &= \frac{y'}{y} = \frac{x'}{x} = \frac{nz'}{n'z} = \frac{n}{n+Az}\,,\\
\frac{z'}{z} &= \frac{n'}{n+Az}\,,\\
m_z &= \frac{dz'}{dz} = \frac{n'n}{(n+Az)^2} = \frac{n'}{n} m_T^2\,,
\end{aligned}
\tag{10.4.25}
$$

where the quantity $A = -n/f = n'/f'$ denotes the optical power of the imaging system. For the special case that the refractive indices in object and image space are equal, the expressions can be combined in vector form, yielding

$$\mathbf{r}_p' = m_T \mathbf{r}_p\,, \tag{10.4.26}$$

where the paraxial object and image position vectors are given by $\mathbf{r}_p = (x, y, z)$ and $\mathbf{r}_p' = (x', y', z')$, respectively.

A diffraction-based analysis needs to take into account the position of entrance and exit pupil. In Fig. 10.55 the centres of these pupil surfaces have been denoted by E_0 and E_0', respectively. The paraxial position of the pupil surfaces is uniquely determined by their transverse magnification $m_{T,p}$. The imaging equation for an object plane at V_0 with distance l from the entrance pupil in E_0 is given by Eq. (4.10.38),

$$\frac{l'}{n'\,m_{T,p}} = \frac{m_{T,p}\,l}{n} - \frac{l'}{n'}\frac{l}{n}A\,,$$

or,

$$l' = \frac{n'\,m_{T,p}^2\,l}{n + m_{T,p}\,l\,A}\,. \tag{10.4.27}$$

The other paraxial quantities that are needed in the three-dimensional frequency analysis are the transverse, angular and axial magnification between two conjugate planes V and V' and these follow from Eqs. (4.10.39) and (4.10.42),

$$m_{T,V} = \frac{n\,m_{T,p}}{n + m_{T,p}\,l\,A}\,, \qquad m_{A,V} = \frac{n + m_{T,p}\,l\,A}{n'\,m_{T,p}}\,, \qquad m_{z,V} = \frac{n'}{n} m_{T,V}^2\,. \tag{10.4.28}$$

For the relationship between a paraxial image vector $\mathbf{r}_E' = (x', y', l')$ and the corresponding object vector $\mathbf{r}_E = (x, y, l)$ we thus find

$$\mathbf{r}'_E = \mathbf{M}\,\mathbf{r}_E\,, \qquad \mathbf{M} = \frac{n\,m_{T,p}}{n + m_{T,p}\,l\,A}\begin{pmatrix} 1 & 0 & 0 \\ 0 & 1 & 0 \\ 0 & 0 & n' m_{T,p}/n \end{pmatrix}. \qquad (10.4.29)$$

If the indices in object and image space are equal to unity ($n = n' = 1$) and the paraxial pupil plane positions are coincident with the principal planes P_0 and P'_0, the matrix $\mathbf{M}$ is the identity matrix multiplied by a factor $(1 + l\,A)^{-1}$. The analysis of three-dimensional frequency transfer given in **[296]** has been restricted to this particular case.

The expression $\mathbf{r}'_E = \mathbf{M}\,\mathbf{r}_E$ of Eq. (10.4.29) can be applied to any geometry. It should be noted, however, that this expression is nonlinear in the axial distance l from the centre of the entrance pupil to a point inside the object V. For relatively small changes in axial position within the object volume, it is advisable to use the local (paraxial) magnification factors in transverse and axial directions as given by Eq. (10.4.28). In doing so we obtain a linear relationship between the object and image space coordinates which is a good approximation in a restricted region of space. We then denote the axial distance z_E to a general point in the object volume or on the object surface by $z_0 + q_Z$ where z_0 is adequately chosen in the interior of the object volume. For instance, z_0 can be made equal to the average axial position obtained from all object points inside the object volume under consideration. The transverse and axial magnifications inside the object are then equated to those corresponding to the paraxial object plane through z_0. The matrix equation in (10.4.29) is subsequently split up in two parts,

$$\mathbf{r}'_0 = \begin{pmatrix} x'_0 \\ y'_0 \\ z'_0 \end{pmatrix} = m_{T,z_0}\begin{pmatrix} x_0 \\ y_0 \\ (n' z_0/n)\, m_{T,p} \end{pmatrix}, \qquad \mathbf{q}' = \begin{pmatrix} q'_x \\ q'_y \\ q'_z \end{pmatrix} = m_{T,z_0}\begin{pmatrix} q_x \\ q_y \\ (n' m_{T,z_0}/n)\, q_z \end{pmatrix}, \qquad (10.4.30)$$

where the constant m_{T,z_0} is given by $n\,m_{T,p}/(n + m_{T,p}Az_0)$ which is the lateral magnification between the conjugate planes at distances z_0 and z'_0 from the paraxial pupil planes. The object and image space coordinates of two conjugate points are then related by the equation

$$\mathbf{r}'_E = \mathbf{r}'_0 + \mathbf{q}' = \mathbf{M}_1\mathbf{r}_0 + \mathbf{M}_2\mathbf{q}\,, \qquad (10.4.31)$$

and the matrices $\mathbf{M}_1$ and $\mathbf{M}_2$ are given by

$$\mathbf{M}_1 = m_{T,z_0}\begin{pmatrix} 1 & 0 & 0 \\ 0 & 1 & 0 \\ 0 & 0 & (n'/n) m_{T,p} \end{pmatrix}, \qquad \mathbf{M}_2 = m_{T,z_0}\begin{pmatrix} 1 & 0 & 0 \\ 0 & 1 & 0 \\ 0 & 0 & (n'/n)\, m_{T,z_0} \end{pmatrix}. \qquad (10.4.32)$$

10.4.3 Stationarity of Aberration in Three Dimensions

The application of a three-dimensional frequency transfer function requires the invariance of aberration over the image volume under consideration. If this condition cannot be fulfilled over the specified volume or surface, the frequency transfer is limited to the image volume in which the aberration value is virtually constant. The frequency transfer over a larger volume is then obtained by adding together the results over various partial volumes with constant but different aberration value. The (curved) surfaces or volumes over which aberration can be assumed to be constant have been given the name 'isotomes' in **[296]**.

For the large majority of (flat-field) imaging systems the Abbe sine condition has to be respected as these imaging systems need good imaging quality over an extended planar image field perpendicular to the optical axis. It was demonstrated in Subsection 4.8.2 that compliance with the Abbe sine condition in the design of an optical system leads to a conflict with the Herschel condition for absence (or constancy) of spherical aberration as a function of the axial position of the conjugate planes. Especially for high-numerical-aperture imaging systems, the number represented by the quantity OHC (Offence against the Herschel Condition) takes on a large value. According to Eq. (4.8.28) the wavefront deviation dW_S at the rim of the pupil, consisting of spherical aberration, is given by ($m_T = 0$, object plane at infinity),

$$dW_S = 2n'(dz')\left[\frac{\sin^2(\alpha')}{4} - \sin^2(\alpha'/2)\right], \qquad (10.4.33)$$

where $n'\sin(\alpha')$ is the image-side numerical aperture. If we permit a value for dW_S of $\lambda/4$, we have listed in the table below the values $2\delta z'_H$ of the axial range of an isotomic volume as a function of the imaging aperture in air, together with the corresponding (double) Rayleigh focal depth $2\delta z'_f$:

$n' = 1$	NA	$2\delta z'_f$ (λ)	$2\delta z'_H$ (λ)
	0.1	99.6	40000
	0.2	24.8	2500
	0.4	6.0	144
	0.9	0.8	3.2

The table clearly shows that for high-numerical-aperture systems the intrinsic change in aberration with axial position strongly reduces the allowed axial thickness of an isotomic volume. When the sine condition is obeyed, a high-NA isotomic volume takes on the shape of a pancake having a thickness to width ratio of the order of 1% ($\lambda \approx 500$ nm). The table also shows that for a moderate power microscope objective ($NA = 0.2$) and a useful field diameter of typically 1 mm, the isotomic volume approaches the spherical shape.

In some special cases, the optical design may have been optimised to deliver a good imaging quality over a large axial range. Such systems, which have to obey the Herschel condition, will show an offence against the sine condition (OSC), which was also discussed and quantified in Subsection 4.8.1 (see Eq. (4.8.6)). For the case that the Herschel condition is respected, the comatic aberration value dW_C at the rim of the pupil and the half diameter $\delta y'$ of the transverse image field over which the aberration remains virtually constant are given by

$$dW_C = n'(dy')\sin\alpha'\left[1 - \frac{2\sin(\alpha'/2)}{\sin\alpha'}\right],$$
$$\delta y' = \frac{\lambda}{4n'\left[\sin\alpha' - 2\sin(\alpha'/2)\right]}, \tag{10.4.34}$$

where, somewhat arbitrarily, we have adopted a maximum comatic aberration dW_C of $\lambda/4$ to determine the half field diameter with stationary aberration. With n'=1, we obtain the following table for $2\delta y'$ (the lateral image field diameter with stationary aberration) and the lateral halfwidth $\delta y_d = 0.5\lambda_0/NA$ of the ideal diffraction image, both as a function of the image-side aperture NA:

$n' = 1$	NA	$\delta y'_d$ (in λ_0)	$2\delta y'$ (in λ_0)
	0.1	5.00	4000
	0.2	2.50	490
	0.4	1.25	60
	0.9	0.55	3

The table shows that a high-NA imaging system that satisfies the Herschel condition has a lateral field diameter with stationary aberration that comprises only five halfwidths of the ideal diffraction image. The isotomic volume has a strongly elongated tubular shape. The conclusion from the two tables is that the isotomic volumes contract very strongly at high NA, either in the lateral or in the axial direction. For the establishment of the frequency transfer over a larger volume, many isotomic sub-volumes would be needed and this is an unattractive feature. For this reason we encounter a practical limitation of three-dimensional frequency analysis for imaging systems with a typical maximum numerical aperture value of 0.5.

10.4.4 The Convolution Integral for Three-dimensional-image Intensity

To obtain the image-side intensity distribution $I'(\mathbf{r}')$ in the presence of an incoherent object radiance distribution on a surface S or inside a volume V, denoted by $O_S(\mathbf{r}_E)$ and $O_V(\mathbf{r}_E)$, respectively, we have to calculate a three-dimensional convolution integral. The integral is evaluated over the object surface S or the object volume V, according to

$$I'(\mathbf{r}') = \int_{S,V} \begin{Bmatrix} O_S \\ O_V \end{Bmatrix}(\mathbf{r}_E)\, I'_d(\mathbf{r}' - \mathbf{r}'_E)\, d\mathbf{r}_E\,, \tag{10.4.35}$$

where $I'_d(\mathbf{r}' - \mathbf{r}'_E)$ is the intensity point-spread function due to a point source at Q with position vector $\mathbf{r}_E$, measured from the centre of the entrance pupil E_0. The position vector $\mathbf{r}'_E$ points from the centre E'_0 of the exit pupil to Q', the paraxial image point of Q. The convolution property of this integral follows from the relation between the paraxial image vector

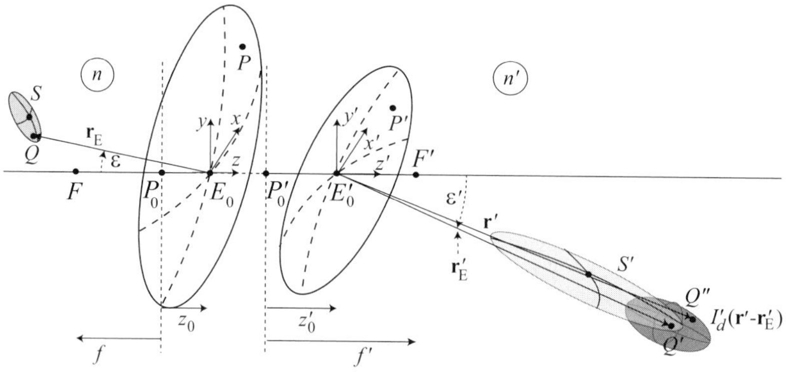

Figure 10.56: Surface (S) or volume imaging described by means of a convolution of the paraxial object radiation function $O(\mathbf{r}_E)$ and the three-dimensional image intensity distribution $I'_d(\mathbf{r}' - \mathbf{r}'_E)$ of the diffraction image of a point source. $\mathbf{r}'_E$ is the position vector of the paraxial image point Q' of a general object point Q on the object surface S or within the object volume V. $\mathbf{r}'$ is the general position vector E'_0Q'' in image space. The exit pupil sphere has its centre at the paraxial image point Q' and intersects the optical axis at E'_0. P' is a general point on the pupil (reference) sphere (reproduced from **[58]** with the permission of the Optical Society of America).

$\mathbf{r}'_E$ and the object vector $\mathbf{r}_E$ where $\mathbf{r}'_E = \mathbf{M}\,\mathbf{r}_E$, according to Eq. (10.4.29). We thus have, continuing with the case of a three-dimensional-surface object,

$$I'(\mathbf{r}') = \int_S O_S(\mathbf{r}_E)\, I'_d(\mathbf{r}' - \mathbf{M}\,\mathbf{r}_E)\, d\mathbf{r}_E \,. \tag{10.4.36}$$

The origins of the object and image space coordinate systems are located at the centres E_0 and E'_0 of the entrance and exit pupil of the imaging system. According to Fig. 10.56, the lateral pupil magnification is given by $m_{T,p} = nz'_0/n'z_0$. The angular pupil magnification $m_{A,p} = z_0/z'_0$ determines the ratio of the angles ϵ and ϵ' which the principal rays E_0Q and E'_0Q' make with the optical axis in object and image space. The image-side angle ϵ' determines the orientation of the long axis of a diffraction intensity pattern given by $I'_d(\mathbf{r}' - \mathbf{r}'_E)$.

The three-dimensional image spectrum is obtained by performing a three-dimensional Fourier transform of the image intensity function $I'(\mathbf{r}')$, yielding,

$$\tilde{I}'(\boldsymbol{\omega}) = \iiint_{-\infty}^{+\infty} I'(\mathbf{r}') \exp(-i\boldsymbol{\omega}' \cdot \mathbf{r}')\, d\mathbf{r}' \,, \tag{10.4.37}$$

where the dot-operator denotes the scalar product of the vectors $\boldsymbol{\omega}' = (\omega'_x, \omega'_y, \omega'_z)$ and $\mathbf{r}'$ ($\boldsymbol{\omega}'$ is the spatial frequency vector in image space).

The value of the Fourier transform at the origin ($\boldsymbol{\omega} = \mathbf{0}$) is a reference value which is evaluated as follows. In an optical system that is free of absorption and in a dielectric image space, equally free of optical loss, the conservation of energy of a wave field on propagation requires that there holds

$$\iint_{-\infty}^{+\infty} I'(x', y', z')\, dx'dy' = P_r \,, \tag{10.4.38}$$

where P_r is the total radiating power flowing through any plane with constant value of z'. For the frequency values $\omega'_x = \omega'_y = 0$, the transformed function $\tilde{I}'(\boldsymbol{\omega}')$ can be written as

$$\tilde{I}'(0, 0, \omega'_z) = \iiint_{-\infty}^{+\infty} I'(\mathbf{r}')\, \exp(-i\omega'_z z')\, d\mathbf{r}' = \int_{-\infty}^{+\infty} P_r \exp(-i\omega'_z z')\, dz' = 2\pi\, P_r\, \delta(\omega'_z) \,. \tag{10.4.39}$$

In physical terms, the axial spectrum of the average value of the lateral intensity distribution is limited to frequency zero. The appearance of the δ-function accounts for the fact that the spatial integration over z' from $-\infty$ to $+\infty$ includes all the energy delivered by the source from $t_0 = -z'/c$ to $t_1 = +z'/c$ with z' tending to infinity and yielding $\int_{t_0}^{t_1} P_r\, dt \rightarrow \infty$.

From the three-dimensional spectral function $\tilde{I}'(\boldsymbol{\omega}')$ we return to the three-dimensional image intensity distribution by means of an inverse Fourier transformation

$$I'(\mathbf{r}') = \frac{1}{(2\pi)^3} \iiint_{-\infty}^{+\infty} \tilde{I}'(\boldsymbol{\omega}) \exp(+i\boldsymbol{\omega}' \cdot \mathbf{r}') \, d\boldsymbol{\omega}' . \tag{10.4.40}$$

10.4.5 Frequency Transfer for an Object Surface or Volume Intensity Function

To calculate the transfer of spatial frequencies through an optical system associated with a certain object intensity function we first calculate the spectrum of the intensity distribution and then calculate the transfer of each component of this spectrum by the three-dimensional imaging operation of the optical system. We focus on the transfer of a volume intensity distribution in which multiple scattering can be neglected. This is certainly the case if the object intensity distribution was itself the result of an imaging operation in a non-scattering medium. If the volume object is self-luminescent, the scattering of the radiation by other parts of the object should be negligible. The same analysis can then be applied to a surface intensity distribution. We require that no multiple reflections of the emitted radiation occur between different parts of the same surface or obstruction.

We start with the basic convolution integral for a volume object given by Eq. (10.4.35). We consider two options for the coordinate systems in object and image space. The first option is to have the origin at the centre of the pupil, both in object and image space. The second option is to use a local origin, centred on the object in object space and on its paraxial image in image space.

10.4.5.1 Centres of the Entrance and Exit Pupil as Coordinate Origins

Referring to Fig. 10.56, we choose the points E_0 and E_0' as origins for the coordinate systems in object and image space. In this case, the three-dimensional Fourier transform of the image space intensity distribution is written as,

$$\tilde{I}'(\boldsymbol{\omega}') = \iiint_V O_V(\mathbf{r}_E) \iiint_{-\infty}^{+\infty} I_d'(\mathbf{r}' - \mathbf{r}_E') \exp(-i\boldsymbol{\omega}' \cdot \mathbf{r}') \, d\mathbf{r}' \, d\mathbf{r}_E, \tag{10.4.41}$$

where the integration variable $\mathbf{r}_E$ explores the object volume. We introduce the change of variables $\mathbf{p}' = \mathbf{r}' - \mathbf{r}_E' = (p_x', p_y', p_z')$ where $\mathbf{r}_E' = \mathbf{M}\,\mathbf{r}_E$. The matrix $\mathbf{M}$ is given by Eq. (10.4.29) and we thus obtain the expression,

$$\tilde{I}'(\boldsymbol{\omega}') = \iiint_V O_V(\mathbf{r}_E) \exp\{-i\boldsymbol{\omega}' \cdot (\mathbf{M}\,\mathbf{r}_E)\} \, d\mathbf{r}_E \iiint_{-\infty}^{+\infty} I_d'(\mathbf{p}') \exp(-i\boldsymbol{\omega}' \cdot \mathbf{p}') \, d\mathbf{p}' . \tag{10.4.42}$$

This expression is conveniently written as

$$\tilde{I}'(\boldsymbol{\omega}') = \tilde{O}_V(\boldsymbol{\omega}') \, \tilde{I}_d'(\boldsymbol{\omega}') , \tag{10.4.43}$$

where $\tilde{O}_V(\boldsymbol{\omega}')$ and $\tilde{I}_d'(\boldsymbol{\omega}')$ are given by

$$\tilde{O}_V(\boldsymbol{\omega}') = \iiint_V O_V(\mathbf{r}_E) \exp\{-i\boldsymbol{\omega}' \cdot (\mathbf{M}\,\mathbf{r}_E)\} \, d\mathbf{r}_E ,$$

$$\tilde{I}_d'(\boldsymbol{\omega}') = \iiint_{-\infty}^{+\infty} I_d'(\mathbf{p}') \exp(-i\boldsymbol{\omega}' \cdot \mathbf{p}') \, d\mathbf{p}' . \tag{10.4.44}$$

If blocking of radiation and multiple scattering effects can be neglected we find for an object that is confined to a certain surface S the expression

$$\tilde{O}_S(\boldsymbol{\omega}') = \iint_S O_S(\mathbf{r}_E) \exp\{-i\boldsymbol{\omega}' \cdot (\mathbf{M}\,\mathbf{r}_E)\} \, d\mathbf{r}_E , \tag{10.4.45}$$

which is an integral that extends over the entire source surface S.

The transforms $\tilde{O}_V$ and $\tilde{O}_S$ are not standard Fourier transforms because of the nonlinear z'-dependence of the exponential term with the matrix $\mathbf{M}$. At each imaging step, the three-dimensional object spectrum manifests itself in image space subject to z-dependent scaling factors in the axial and transverse directions. In analogy with the terminology in the time-frequency domain, we have spatial 'chirping' of the signal. For that reason, to obtain the image spectrum after some subsequent imaging steps, it is not possible to simply multiply the lens transfer spectra $\tilde{I}'_i$ of each lens i. We have to calculate each quantity $\tilde{I}'(\boldsymbol{\omega}')_i = \tilde{O}(\boldsymbol{\omega}')_i\, \tilde{I}'_d(\boldsymbol{\omega}')_i$ for an ith imaging step. The resulting spectrum $\tilde{I}'(\boldsymbol{\omega})_i$ serves as input for the next imaging step by transforming it back to the spatial domain and we then use this spatial function as input function $O(\mathbf{r}_E)_{i+1}$ for the next imaging step.

10.4.5.2 Centres of Object and Image Volume/Surface as Coordinate Origins

The centre of the object volume or surface can be defined in various ways. It can be chosen as the geometrical centre $\mathbf{r}_0$ of the object or as the radiant centre $\mathbf{r}_{0,I}$ by using the local intensity of the object as a weighting function. We have the following expressions,

$$\bar{x}_0 = \frac{1}{V} \iiint_V x\, dV\,, \qquad \bar{x}_{0,I} = \iiint_V x\, I(x,y,z)\, dV \Big/ \iiint_V I(x,y,z)\, dV\,, \tag{10.4.46}$$

with analogous expressions for the y- and z-coordinates of the centre position in the object volume. The coordinate values of the object centre and the image centre are denoted in what follows by the vectors $\mathbf{r}_0$ and $\mathbf{r}'_0$. For the relationship between the coordinates in object and image space we have, with the aid of Eq. (10.4.31), the following expressions,

$$\mathbf{r}_E = \mathbf{r}_0 + \mathbf{q}\,, \quad \mathbf{r}'_E = \mathbf{r}'_0 + \mathbf{q}' = \mathbf{M}_1\mathbf{r}_0 + \mathbf{M}_2\mathbf{q}\,, \quad \mathbf{r} = \mathbf{r}_0 + \mathbf{p}\,, \quad \mathbf{r}' = \mathbf{r}'_0 + \mathbf{p}'\,. \tag{10.4.47}$$

The vectors $\mathbf{r}_E = E_0Q$ and $\mathbf{r}'_E = E'_0Q'$ point to the conjugate points Q and Q' in object and image space, respectively. The vectors $\mathbf{q}$ and $\mathbf{q}'$ have their origin in the object and image centre, respectively, and connect a pair of conjugate points in object and image space. Finally, the vectors $\mathbf{p}$ and $\mathbf{p}'$ define general points in object and image space; they also have their origin in the object and image centres given by the vectors $\mathbf{r}_0$ and $\mathbf{r}'_0$, respectively.

In what follows we use the coordinate relationships provided by paraxial optics by means of the matrices $\mathbf{M}_1$ and $\mathbf{M}_2$ in a limited volume of space. In doing so we avoid the nonlinear exponential function in the expression for $\tilde{O}_V(\boldsymbol{\omega}')$ of Eq. (10.4.44). We thus insert the paraxial relationship between object and image space coordinates in the basic convolution integral of Eq. (10.4.41) and obtain for the image spectrum:

$$\begin{aligned}
\tilde{I}'(\boldsymbol{\omega}') &= \iiint_{-\infty}^{+\infty} O_V(\mathbf{r}_E) \iiint_{-\infty}^{+\infty} I'_d(\mathbf{r}' - \mathbf{r}'_E)\, \exp(-i\boldsymbol{\omega}' \cdot \mathbf{p}')\, d\mathbf{r}_E\, d\mathbf{p}' \\
&= \iiint_{-\infty}^{+\infty} O_V(\mathbf{r}_E)\, \exp(-i\boldsymbol{\omega}' \cdot \mathbf{q}') \iiint_{-\infty}^{+\infty} I'_d(\mathbf{p}' - \mathbf{q}')\, \exp[-i\boldsymbol{\omega}' \cdot (\mathbf{p}' - \mathbf{q}')]\, d\mathbf{r}_E\, d\mathbf{p}' \\
&= \tilde{I}'_d(\boldsymbol{\omega}') \iiint_{-\infty}^{+\infty} O_V(\mathbf{r}_0 + \mathbf{q})\, \exp(-i\boldsymbol{\omega}' \cdot \mathbf{M}_2\mathbf{q})\, d\mathbf{q} \\
&= \tilde{I}'_d(\boldsymbol{\omega}') \iiint_{-\infty}^{+\infty} O_{V,c}(\mathbf{q})\, \exp(-i\boldsymbol{\omega}' \cdot \mathbf{M}_2\mathbf{q})\, d\mathbf{q} \\
&= \tilde{I}'_d(\boldsymbol{\omega}')\, \tilde{O}_{V,c}(\mathbf{M}_2\, \boldsymbol{\omega}')\,,
\end{aligned} \tag{10.4.48}$$

where we have used that $\mathbf{M}_1\mathbf{r}_0 = \mathbf{r}'_0$. The function $O_{V,c}$ is the object intensity distribution measured in a coordinate system with its origin in $\mathbf{r}_E = \mathbf{r}_0$ and having the vector coordinate $\mathbf{q}$. The frequency vector $\mathbf{M}_2\, \boldsymbol{\omega}'$ is given by $(M_{2,xx}\omega'_x\,, M_{2,yy}\omega'_y, M_{2,zz}\omega'_z)$ where the coefficients $M_{2,xx}$, $M_{2,yy}$ and $M_{2,zz}$ are the diagonal elements of the matrix $\mathbf{M}_2$ of Eq. (10.4.32).

In practice, both Eqs. (10.4.42) and (10.4.48) can be used, provided that the extent of the object volume remains small. Equation (10.4.48) is an exact Fourier transformation with a linearised relation between the object and image space coordinates. The expression of Eq. (10.4.42) has the correct nonlinear paraxial relationship between object and image space coordinates with the result that the transformation expression is not that of a linear Fourier transform. If the relative axial extent of the object volume is sufficiently small, for instance smaller than 1%, both formulas can be used without introducing significant errors. If the axial extent becomes larger, the best approach is to use Eq. (10.4.48) sequentially, each time for a different axial section of the object and with an adjustment of the paraxial magnification factors.

10.4.6 Connection between a Three- and a Two-dimensional Transfer Function

The frequency analysis in the Sections 10.2 and 10.3 of this chapter was two-dimensional and dealt with the transfer of a frequency pair (ν_x, ν_y) by a lens system from an object plane, perpendicular to the optical axis, to a receiving plane in image space, equally perpendicular to the optical axis. The connection between the three-dimensional frequency transfer function of a volume or surface object function and the two-dimensional frequency transfer between, for instance, the object and image plane of a lens system is found by taking a two-dimensional Fourier transform of the three-dimensional image-space intensity function $I'(\mathbf{r}')$ of Eq. (10.4.35). For the coordinate systems in object and image space, we choose the transformation defined by Eqs. (10.4.31) and (10.4.32) of the preceding subsection with position vectors $\mathbf{p}$ and $\mathbf{p}'$, measured with respect to the object and image centres which are defined by the coordinates $\bar{r}_0$ and $\bar{r}'_0$, respectively. After some manipulation we obtain for the transverse Fourier transform $\tilde{I}'_t$ the expression

$$\tilde{I}'_t(\omega'_x, \omega'_y; p'_z) = \int\!\!\int\!\!\int\!\!\int\!\!\int_{-\infty}^{+\infty} O_V(\mathbf{r}_0 + \mathbf{q})\, I'_d(\mathbf{r}' - \mathbf{r}'_E)\, \exp[-i(\omega'_x p'_x + \omega'_y p'_y)]\, dp'_x dp'_y\, d\mathbf{r}'_E$$
$$= \int_{-\infty}^{+\infty} \tilde{O}_{V,c}(M_{2,xx}\,\omega'_x, M_{2,yy}\,\omega'_y; q_z)\; \tilde{I}'_{d,t}(\omega'_x, \omega'_y; p'_z - M_{2,zz}\, q_z)\, dq_z\,, \qquad (10.4.49)$$

where the diagonal matrix elements of $\mathbf{M}_2$ in Eq. (10.4.32) have been used, related to the imaging from a paraxial object plane at $z = \overline{z_0}$ to its conjugate image plane in $z' = \overline{z'_0}$.

In the special case that the object is confined to a (single-valued) surface in object space with intensity function $O_{S,c}\left[q_x, q_y, q_{z,S}(q_x, q_y)\right]$ with $q_{z,S}$ the z-coordinate of the object surface, the transverse two-dimensional-transform is given by

$$\tilde{I}'_t(\omega'_x, \omega'_y; p'_z) = \int\!\!\int_{-\infty}^{+\infty} O_{S,c}\left[q_x, q_y, q_{z,S}(q_x, q_y)\right] \tilde{I}_{d,t}(\omega'_x, \omega'_y; p'_z - q'_{z,S}) \exp\left\{-i[\omega'_x q'_x + \omega'_y q'_y]\right\} dq'_x dq'_y\,. \qquad (10.4.50)$$

The two-dimensional 'transverse' transforms of the object intensity function O and the image diffraction point-spread function I_d are given by

$$\tilde{O}_t(M_{2,xx}\omega'_x, M_{2,yy}\omega'_y; p_z) = \int\!\!\int_{-\infty}^{+\infty} O_{V,c}(p_x, p_y, p_z)\, \exp\{-i(M_{2,xx}\omega'_x p_x + M_{2,yy}\omega'_y\, p_y)\}\, dp_x dp_y\,,$$
$$\tilde{I}'_{d,t}(\omega'_x, \omega'_y; p'_z - M_{2,zz}p_z) = \int\!\!\int_{-\infty}^{+\infty} I'_d(p'_x, p'_y; p'_z - M_{2,zz}p_z)\, \exp\{-i(\omega'_x p'_x + \omega'_y\, p'_y)\}\, dp'_x dp'_y\,. \qquad (10.4.51)$$

In an imaging system with circular symmetry, the values of $M_{2,xx}$ and $M_{2,yy}$ are both equal to m_T and the factor $M_{2,zz}$ equals the axial magnification m_z of the imaging system for a pair of conjugate planes.

From the transverse spectrum $\tilde{I}'_{d,t}$ we construct the three-dimensional transfer function by reconsidering the expression for $\tilde{I}'(\boldsymbol{\omega}')$ of Eq. (10.4.44) and writing this expression in the form

$$\tilde{I}'(\boldsymbol{\omega}') = \int_{-\infty}^{+\infty}\left(\int\!\!\int_{-\infty}^{+\infty} I'_d(p'_x, p'_y; p'_z)\, \exp\{-i(\omega'_x p'_x + \omega'_y p'_y)\}\, dp'_x dp'_y\right) \exp(-i\,\omega'_z p'_z)\, dp'_z$$
$$= \int_{-\infty}^{+\infty} \tilde{I}'_{d,t}(\omega'_x, \omega'_y; p'_z)\, \exp(-i\,\omega'_z p'_z)\, dp'_z\,. \qquad (10.4.52)$$

We conclude that the three-dimensional transform is obtained by a one-dimensional Fourier transform of the transverse spectral function $\tilde{I}'_{d,t}$. It is also possible to transform back the three-dimensional transform with respect to the ω'_z frequency component, yielding the two-dimensional transformed function

$$\tilde{I}'_{d,t}(\omega'_x, \omega'_y, p'_z) = \frac{1}{2\pi}\int_{-\infty}^{+\infty} \tilde{I}'(\boldsymbol{\omega}')\, \exp(+i\,\omega'_z p'_z)\, d\omega'_z\,. \qquad (10.4.53)$$

10.4.7 Transfer Functions and the Fourier Slice Theorem

A special situation arises when one of the frequency components of the three-dimensional transfer function is put equal to zero, for instance, ω'_z. The resulting function $\tilde{I}'_s(\omega'_x, \omega'_y)$ is called a slice through the ω'_z-axis of the three-dimensional

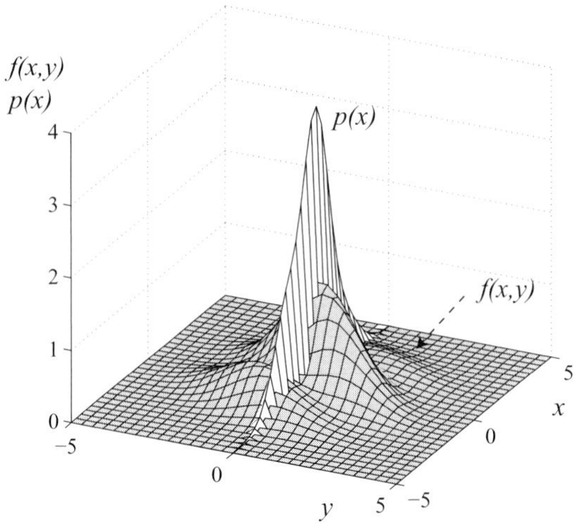

Figure 10.57: The projection $p(x)$ of a two-dimensional function $f(x,y)$ onto the x-axis.

function $\tilde{I}'(\omega'_x, \omega'_y, \omega'_z)$. The following relation exists between the three-dimensional intensity function $I'_d(p'_x, p'_y, p'_z)$ and the sliced transform $\tilde{I}'_s(\omega'_x, \omega'_y)$,

$$
\begin{aligned}
\tilde{I}'_s(\omega'_x, \omega'_y) &= \iint_{-\infty}^{+\infty} \left\{ \int_{-\infty}^{+\infty} I'_d(p'_x, p'_y; p'_z)\, dp'_z \right\} \exp\{-i\,(\omega'_x p'_x + \omega'_y p'_y)\}\, dp'_x dp'_y \\
&= \iint_{-\infty}^{+\infty} I'_{d,p}(p'_x, p'_y)\, \exp\{-i\,(\omega'_x p'_x + \omega'_y p'_y)\}\, dp'_x dp'_y \,,
\end{aligned}
\qquad (10.4.54)
$$

where the function $I'_{d,p}(p'_x, p'_y)$ is the *projection* of I'_d onto the (x', y')-plane. In Fig. 10.57 we have plotted an example of a general function $f(x,y)$ and its projection $p(x)$ onto the x-axis. We thus have that the two-dimensional Fourier transform of the projected function $I'_{d,p}$ equals the slice with $\omega'_z = 0$ of the three-dimensional Fourier transform of I'_d itself. This so-called *slice theorem* of multi-dimensional Fourier transforms is valid for the three- and two-dimensional transfer functions discussed in this paragraph. For instance, putting $\omega'_z = 0$ in the first line of Eq. (10.4.52) shows that the projection of I'_d on the (x', y') plane gives rise to the slice with $\omega'_z = 0$ of the three-dimensional transfer function. From the second line of Eq. (10.4.52) we deduce that the summation of all two-dimensional transverse transfer function values in axial planes with shift p'_z yields the slice for $\omega'_z = 0$ of the three-dimensional transfer function. The slice theorem can also be applied to the original three-dimensional transfer function and to a one-dimensional function derived from it. In this case, we have to take the Fourier transform of the projection of the three-dimensional transfer function onto a line instead of onto a plane. For checking the relationships above, it is important to keep track of all leading constant factors that accompany the proper one- or two-dimensional transfer functions that are derived from the original three-dimensional function. When using the transfer functions separately, it is justified to use their normalised versions and to neglect any constant factors.

10.4.8 The Three-dimensional Transfer Derived from the Intensity Point-spread Function

The three-dimensional point-spread function $I'_d(\mathbf{r}' - \mathbf{r}'_E) = I'_d(p'_x, p'_y, p'_z)$ follows from a diffraction integral over the exit pupil surface. In this subsection we limit ourselves to a scalar diffraction integral. The approximations when using such a scalar integral expression are twofold. First, the polarisation of the light is neglected and, secondly, for the calculation of optical pathlengths from the exit pupil to the image space, we apply a small angle quadratic approximation. This does not mean that we apply a small angle paraxial approach for the light propagation in the imaging system itself. Imaging aberrations may occur in the imaging system. But the light propagation from the exit pupil to the image volume is limited to small aperture angles of the imaging beams.

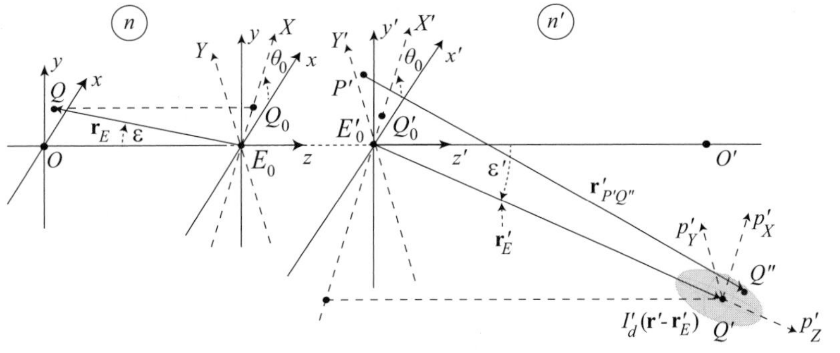

Figure 10.58: Coordinate systems (paraxial imaging) for the calculation of $I_d(\mathbf{r}' - \mathbf{r}'_E)$ by means of the classical Fresnel diffraction integral over the exit pupil surface.

The geometry for the calculation of the diffraction integral is outlined in Fig. 10.58. Instead of the coordinate systems (x, y, z) and (x', y', z') referred to the centre points E_0 and E'_0 of the entrance and exit pupil, respectively, we use tilted coordinate systems to facilitate the calculation of the pathlengths occurring in the Fresnel diffraction integral. We use the small field approximation, which implies that in Fig. 10.58 the field angle $\epsilon' = (x'^2_{Q'} + y'^2_{Q'})^{1/2}/z'_{Q'} \ll 1$. The exit pupil reference sphere has a radius given by $r'_E = |\mathbf{r}'_E|$ and its centre is at Q'. The coordinate system (X', Y', Z') follows from (x', y', z') in two steps. First we rotate the (x', y', z') system by an angle θ_0 around the z'-axis, yielding the (X', Y', z') system. The angle θ_0 follows from the position of the object point Q and its projection Q_0 in the entrance pupil. A second rotation by an angle ϵ' around the Y'-axis yields the new coordinate system (X', Y', Z'). The local coordinate axes (p'_X, p'_Y, p'_Z) at the point Q' are parallel to the coordinate axes E'_0X', E'_0Y' and E'_0Z', respectively. P' is a general point on the exit pupil sphere. The paraxial image point of a point source Q in the object plane has been denoted by Q'; it forms the origin of the local coordinate system (p'_X, p'_Y, p'_Z). Q'' is a general point in the volume where the three-dimensional point-spread function is formed, its local coordinates are given by means of the coordinate system (p'_X, p'_Y, p'_Z).

To obtain the expression for the Fresnel diffraction integral corresponding to a general (defocused) position of the image point, we calculate the pathlength $\Delta = P'Q''$ in the rotated coordinate systems using the classical power series expansion of this distance (see Section 8.3 for a more detailed discussion on the calculation of this pathlength difference). The point P' is a general point on the exit pupil sphere and $Q'' = (p'_X, p'_Y, p'_Z)$ is the general point at which the intensity of the point-spread function needs to be calculated. Using the common approximation, the pathlength difference Δ is evaluated up to second order as a function of the lateral pupil coordinates (X', Y'). The expression for the image-side diffracted amplitude distribution U'_d is given by

$$
\begin{aligned}
U'_d(p'_X, p'_Y, p'_Z) &= \frac{-i}{\lambda\, r'_E} \iint_{-\infty}^{+\infty} U(X', Y') \exp(ik\Delta)\, dX'dY' \\
&= \frac{-ik}{2\pi\, r'_E} \exp\left\{ik\left(r'_E + p'_Z + \frac{p'^2_X + p'^2_Y}{2r'_E}\right)\right\} \\
&\quad \times \iint_{-\infty}^{+\infty} U(X', Y') \exp\left\{-ik\left[\left(\frac{X'^2 + Y'^2}{2r'^2_E}\right)p'_Z + \frac{X'p'_X + Y'p'_Y}{r'_E}\right]\right\} dX'dY', \qquad (10.4.55)
\end{aligned}
$$

where $U(X', Y')$ is the complex amplitude on the exit pupil sphere of the wave which originated from a point source at Q in the object plane. Equation (10.4.55) is the Cartesian equivalent of Eq. (8.4.36), apart from the constant pathlength term $\exp(ikr'_E)$. In line with Section 8.4, we omit the generally irrelevant constant pathlength term. If certain conditions are satisfied, we will also omit the pathlength terms that are quadratic in p'_X, p'_Y.

In Fig. 10.59 we have plotted the normalised intensity distribution $I'_{d,n}$, proportional to $|U'_d|^2$ as given by Eq. (10.4.55). The optical system is aberration free and has uniform transmittance ($U(X', Y') \equiv 1$) within the physical aperture of the system. The physical aperture is determined by the distance $2\rho_0$, corresponding to the diameter of the circular rim of the exit pupil of the imaging system. The intensity at the centre Q' of the diffraction image ($\mathbf{r}' = \mathbf{r}'_E$) has been normalised to

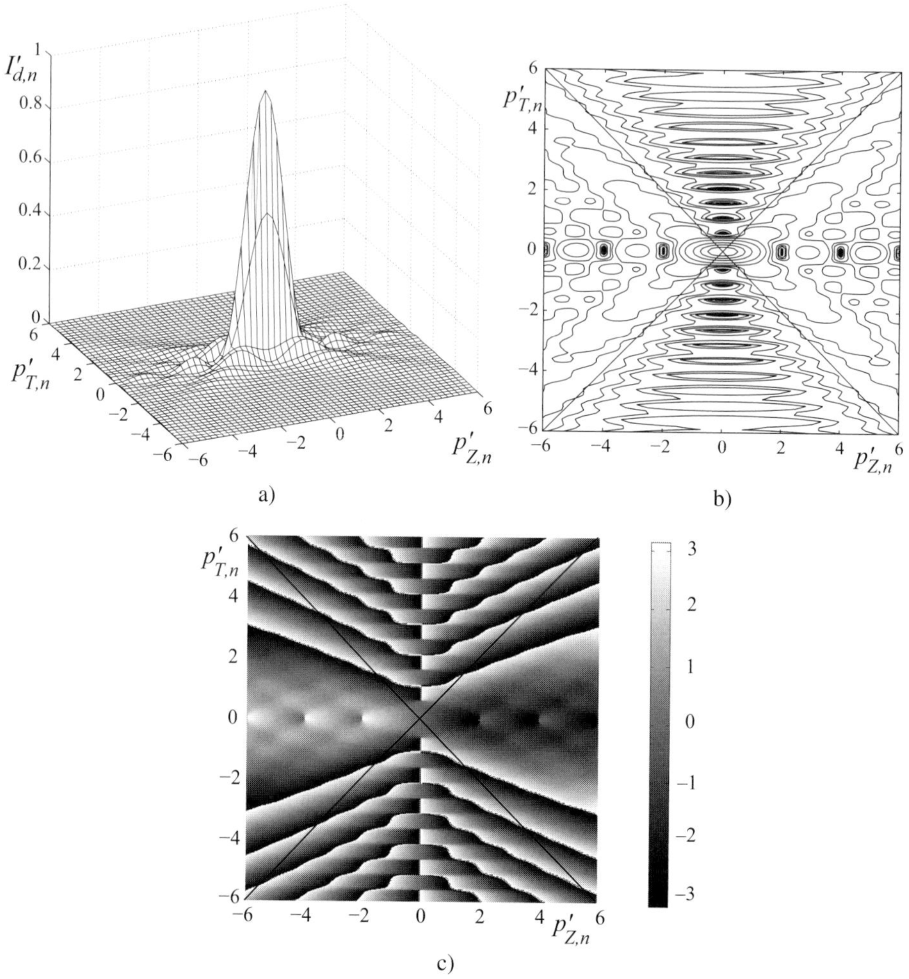

Figure 10.59: The normalised intensity $I'_{d,n}$ in the image region of a perfect imaging system with $U(X', Y') = 1$.
a) A three-dimensional plot of the intensity function $I'_{d,n}$ in a plane through the nominal image point, perpendicular to the z'-axis. The axial and transverse normalised coordinates are $p'_{Z,n}$ and $p'_{T,n}$, respectively.
b) Contour plot of the same intensity pattern.
c) The phase of the through-focus complex amplitude distribution.

unity. The coordinates $p'_{Z,n}$ along the axis E'_0Q' and $p'_{T,n}$ in a plane normal to this axis have been normalised with respect to the axial and lateral diffraction units, resulting in

$$p'_{Z,n} = \frac{p'_Z}{\lambda/\sin^2\alpha'_m}\,, \qquad p'_{T,n} = \frac{p'_T}{\lambda/\sin\alpha'_m}\,. \qquad (10.4.56)$$

The aperture angle α'_m corresponds to the angle between a marginal ray and the central ray of the imaging pencil propagating from the exit pupil to the image region. It is the three-dimensional scalar *intensity* distribution $|U'_d|^2$, without normalisation to unity, which will serve as the intensity impulse response function of an imaging system according to Eq. (10.4.44). Using the normalised axial and transverse coordinates, we have that the geometrical shadow of the focused beam is given by a cone with a half apex angle of $\pi/4$ as shown by the solid lines through the image point in Fig. 10.59b and c. This result for the low-aperture scalar approximation of the diffraction integral is derived with the aid of Fig. 10.60. The wavefront surface W of the imaging pencil with aperture angle α'_m intersects the centre of the exit pupil in E'_0; its geometrical focal point is in F'. A general point A on the boundary between the illuminating pencil and the geometrical shadow area S has real-space coordinates (p'_Z, p'_T), measured with respect to the focal point F'. In the small angle approximation we write

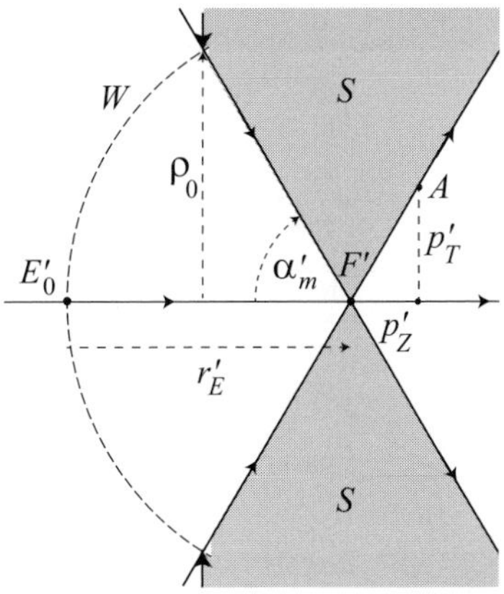

Figure 10.60: A general point A with real-space coordinates (p'_T, p'_Z) at the boundary of the geometrical shadow S of an imaging beam.

$$\sin\alpha'_m = \frac{\rho_0}{r'_E} \approx \tan\alpha'_m = \frac{p'_T}{p'_Z} = \frac{p'_{T,n}}{p'_{Z,n}} \sin\alpha'_m \,, \tag{10.4.57}$$

and observe that, in the normalised coordinate space, the direction tangent of the geometrical shadow boundary $F'A$ equals unity. For finite angles α'_m we obtain,

$$\tan\alpha'_m = \frac{\rho_0}{r'_E \cos\alpha'_m} = \frac{p'_T}{p'_Z} = \frac{p'_{T,n}}{p'_{Z,n}} \sin\alpha'_m \,, \tag{10.4.58}$$

with the result that the ratio $p'_{T,n}/p'_{Z,n} = 1/\cos\alpha'_m$. This expression yields the correct angle for the geometrical shadow boundary at large opening angles α'_m of the focusing pencil.

We now return to Fig. 10.58 and Eq. (10.4.55) where it is permitted, within the small field angle approximation ($\epsilon' \ll 1$), to neglect the rotation of the Z'-axis with respect to the z'-axis. The only difference in orientation between the (x', y', z') and (X', Y', Z') system is then the rotation by an angle θ_0 around the z'- (or Z')-axis. For the spatial frequencies associated with the rotated (X', Y', Z') and (p'_X, p'_Y, p'_Z) coordinate systems we now use the notation $(\omega'_X, \omega'_Y, \omega'_Z)$ with capital subscripts, with the tacit assumption that in the small-field-angle approximation the quantities ω'_z and ω'_Z are identical. The transverse transfer function $\tilde{I}'_{d,t}(\omega'_X, \omega'_Y; p'_Z)$ is then given by

$$\tilde{I}'_{d,t}(\omega'_X, \omega'_Y; p'_Z) = \iint_{\text{pupil}} U'_d(p'_X, p'_Y, p'_Z)\, U'^{*}_d(p'_X, p'_Y, p'_Z) \exp\{-i(\omega'_X p'_X + \omega'_Y p'_Y)\}\, dp'_X dp'_Y \,. \tag{10.4.59}$$

Insertion of the Fresnel integral expression for U'_d yields the expression

$$\begin{aligned}\tilde{I}'_{d,t}(\omega'_X, \omega'_Y; p'_Z) =& \frac{k^2}{4\pi^2\, r'^2_E} \iiiint\!\!\iint U(X'_1, Y'_1)\, U^*(X'_2, Y'_2) \\ &\times \exp\left\{-i\left(p'_X\left[\omega'_X + \frac{k(X'_1 - X'_2)}{r'_E}\right] + p'_Y\left[\omega'_Y + \frac{k(Y'_1 - Y'_2)}{r'_E}\right]\right)\right\} \\ &\times \exp\left\{-ikp'_Z\left[\frac{X'^2_1 - X'^2_2 + Y'^2_1 - Y'^2_2}{2r'^2_E}\right]\right\} dX'_1 dX'_2 dY'_1 dY'_2 dp'_X dp'_Y \,.\end{aligned} \tag{10.4.60}$$

Performing the integration over p'_X and p'_Y we find two delta-functions, $2\pi\delta[\omega'_X + k(X'_1 - X'_2)/r'_E]$ and $2\pi\delta[\omega'_Y + k(Y'_1 - Y'_2)/r'_E]$, respectively. The integrations over $X'_1, \cdots, Y'_2$ then simply yield the expression

$$\begin{aligned}\tilde{I}'_{d,t}(\omega'_X, \omega'_Y; p'_Z) =& \iint_{-\infty}^{+\infty} U(X', Y')\, U^*\left(X' + \frac{\omega'_X r'_E}{k}, Y' + \frac{\omega'_Y r'_E}{k}\right) \\ &\times \exp\left\{+i\left[\frac{\omega'_X X' + \omega'_Y Y'}{r'_E} + \frac{\omega'^2_X + \omega'^2_Y}{2k}\right] p'_Z\right\} dX' dY' \,.\end{aligned} \tag{10.4.61}$$

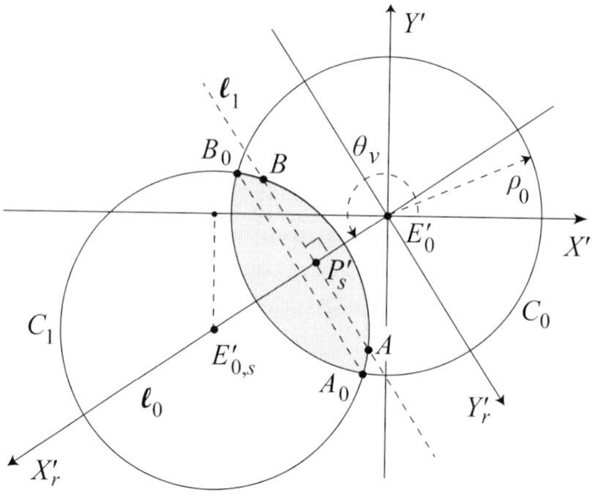

Figure 10.61: The pupil circle C_0 and its shifted copy C_1 in the plane $Z' = 0$, with their centres in E'_0 and $E'_{0,s}$, respectively. The line l_1 determines the integration path for determining the three-dimensional transfer function $\tilde{I}'(\omega'_X, \omega'_Y, \omega'_Z)$. The integration limits are given by the intersection points A and B of l_1 with the border of the overlap area (shaded) of C_0 and C_1. P'_s is the intersection point of l_1 and the line l_0 that joins the pupil centres E'_0 and $E'_{0,s}$.

To obtain the three-dimensional transfer function we evaluate the expression of Eq. (10.4.52) with the aid of Eq. (10.4.61),

$$\tilde{I}'(\omega'_X, \omega'_Y, \omega'_Z) = \iiint_{-\infty}^{+\infty} U(X', Y')\, U^*\left(X' + \frac{\omega'_X r'_E}{k}, Y' + \frac{\omega'_Y r'_E}{k}\right) \times \exp\left\{+i\left[\frac{\omega'_X X' + \omega'_Y Y'}{r'_E} + \frac{\omega_X'^2 + \omega_Y'^2}{2k} - \omega'_Z\right] p'_Z\right\} dX'\,dY'\,dp'_Z\,. \tag{10.4.62}$$

Evaluation of the integral over p'_Z yields the expression

$$\tilde{I}'(\omega'_X, \omega'_Y, \omega'_Z) = 2\pi \iint_{-\infty}^{+\infty} U(X', Y')\, U^*\left(X' + \frac{\omega'_X r'_E}{k}, Y' + \frac{\omega'_Y r'_E}{k}\right) \times \delta\left\{\frac{\omega'_X X' + \omega'_Y Y'}{r'_E} + \frac{\omega_X'^2 + \omega_Y'^2}{2k} - \omega'_Z\right\} dX'\,dY'. \tag{10.4.63}$$

It is useful at this point to make a change to rotated coordinate axes (X'_r, Y'_r) for the transverse coordinates (X', Y'). We introduce the rotation angle θ_v given by $\cos\theta_v = -\omega'_X/\omega'_T$ and $\sin\theta_v = -\omega'_Y/\omega'_T$, where the transverse radial frequency ω'_T is given by $(\omega_X'^2 + \omega_Y'^2)^{1/2}$ (see Fig. 10.61). According to Eq. (10.4.63), the centre $E'_{0,s}$ is then found on the X'_r-axis with coordinate $(-\omega'_X r'_E/k)/\cos\theta_v = (-\omega'_T r'_E/k)$. In a next step, we normalise the rotated pupil coordinates by division through $\rho_0 = r'_E \sin\alpha'_m$, half the transverse diameter of the exit pupil ($n' \sin\alpha'_m$ is the numerical aperture of the imaging system). This normalisation yields the coordinate pair $(X'_{r,n}, Y'_{r,n})$. In the normalised rotated coordinate system $(X'_{r,n}, Y'_{r,n})$, the pupil function is given by $U[\rho_0(X'_r \cos\theta_v - Y'_r \sin\theta_v), \rho_0(X'_r \sin\theta_v + Y'_r \cos\theta_v)]$. In shorthand we denote this expression of the pupil function in rotated coordinates by $U_0(\rho_0 X'_{r,n}, \rho_0 Y'_{r,n})$, where the index zero indicates that no physical rotation of the pupil function has taken place. In the case that the pupil function is represented by an expansion with the aid of Zernike polynomials, the pupil function is simply obtained by using the azimuthal angle $\theta = \theta_r + \theta_v$ when evaluating the Zernike polynomials, where θ_r is the azimuth angle in the rotated coordinate system $(X'_{r,n}, Y'_{r,n})$.

Using the definitions of the normalised original and the normalised rotated pupil coordinate systems we obtain the following expression for the intensity spectrum:

$$\tilde{I}'(\omega'_X, \omega'_Y, \omega'_Z) = 2\pi r_E'^2 \sin^2\alpha'_m \iint U_0(\rho_0 X'_{r,n}, \rho_0 Y'_{r,n})\, U_0^*\left(\rho_0\left[X'_{r,n} - \frac{\omega'_T}{k \sin\alpha'_m}\right], \rho_0 Y'_{r,n}\right) \times \delta\left\{-\omega'_T \sin\alpha'_m X'_{r,n} + \frac{\omega_T'^2}{2k} - \omega'_Z\right\} dX'_{r,n}\,dY'_{r,n}. \tag{10.4.64}$$

Integration over $X'_{r,n}$ yields a one-dimensional 'line' integral in the pupil, perpendicular to the $X'_{r,n}$ coordinate axis. Using $\delta(ax) = \delta(x)/|a|$, the frequency transfer function is thus given by the one-dimensional integral

$$\tilde{I}'(\omega'_X, \omega'_Y, \omega'_Z) = \frac{2\pi r_E'^2 \sin\alpha'_m}{\omega'_T} \int_A^B U_0\left(\rho_0\left[\frac{\omega'_T}{2k\sin\alpha'_m} - \frac{\omega'_Z}{\omega'_T\sin\alpha'_m}\right], \rho_0 Y'_{r,n}\right) \times U_0^*\left(\rho_0\left[-\frac{\omega'_T}{2k\sin\alpha'_m} - \frac{\omega'_Z}{\omega'_T\sin\alpha'_m}\right], \rho_0 Y'_{r,n}\right) dY'_{r,n}. \quad (10.4.65)$$

In Fig. 10.61 we have plotted the line l_1 which follows from the value of X' that is imposed by the δ-function in Eq. (10.4.63). In the original pupil coordinate system the equation of l_1 is given by

$$X' = -\frac{\omega'_Y}{\omega'_X}Y' - \frac{r'_E}{\omega'_X}\left(\frac{\omega_T'^2}{2k} - \omega'_Z\right). \quad (10.4.66)$$

In the rotated and normalised coordinate system $(X'_{r,n}, Y'_{r,n})$ we find

$$X'_{r,n} = \frac{1}{\sin\alpha'_m}\left(\frac{\omega'_T}{2k} - \frac{\omega'_Z}{\omega'_T}\right). \quad (10.4.67)$$

The line l_0 joins the centre E'_0 of the pupil C_0 and the centre $E'_{0,s}$ of its displaced version C_1. The lines l_0 and l_1 are orthogonal. The point P'_s at which these two lines intersect has the coordinates

$$X'_{P'_s} = \omega'_X r'_E\left(-\frac{1}{2k} + \frac{\omega'_Z}{\omega_T'^2}\right), \qquad Y'_{P'_s} = \omega'_Y r'_E\left(-\frac{1}{2k} + \frac{\omega'_Z}{\omega_T'^2}\right). \quad (10.4.68)$$

The integration limits for the integral of Eq. (10.4.65) are given by the points A and B on the rim of the overlap region between the two circular pupils. In terms of the normalised coordinates $(X'_{r,n}, Y'_{r,n})$ we have for the integration limits A and B on the line l_1,

$$Y'_{r,n}\,|_B = -Y'_{r,n}\,|_A = \sqrt{1 - \frac{1}{\sin^2\alpha'_m}\left[\frac{\omega'_T}{2k} - \frac{\omega'_Z}{\omega'_T}\right]^2}. \quad (10.4.69)$$

The distance $d = E'_0P'_s$ from the origin E'_0 to the intersection point of l_0 and l_1 is given by

$$d = r'_E\left(\frac{\omega'_T}{2k} - \frac{\omega'_Z}{\omega'_T}\right). \quad (10.4.70)$$

If $\nu_Z = 0$ we have the special case that $d = (1/2)E'_0E'_{0,s} = \nu_T\lambda_0 r'_E/2n'$, where we have used that $k = n'k_0$ ($k_0 = 2\pi/\lambda_0$, λ_0 is the vacuum wavelength).

We conclude this subsection by giving the expression for the three-dimensional transfer function in terms of spatial frequencies ν instead of the angular frequencies ω. Moreover, we use normalised spatial frequencies $(\nu_{X,n}, \nu_{Y,n}, \nu_{Z,n})$ and the normalised transverse frequency $\nu_{T,n} = (\nu_{X,n}^2 + \nu_{Y,n}^2)^{1/2}$. These are obtained by division of the radial frequency components of the vector $\boldsymbol{\omega}'$ by the factor $NA/\lambda_0 = \sin\alpha'_m/\lambda$,

$$(\nu_{X,n}, \nu_{Y,n}, \nu_{Z,n}) = \frac{1}{k\sin\alpha'_m}(\omega'_X, \omega'_Y, \omega'_Z). \quad (10.4.71)$$

We note that although the normalised components on the left-hand side of the equation are non-primed quantities, they apply nevertheless to spatial frequencies in the image space of the imaging system. Using the rotated normalised pupil coordinate system $(X'_{r,n}, Y'_{r,n})$ the frequency transfer function is then given by

$$\tilde{I}'_{\text{inc}}(\nu_{X,n}, \nu_{Y,n}, \nu_{Z,n}) = \frac{\lambda\, r_E'^2}{\nu_{T,n}} \int_A^B U_0\left(\rho_0\left[\frac{\nu_{T,n}}{2} - \frac{\nu_{Z,n}}{\nu_{T,n}\sin\alpha'_m}\right], \rho_0 Y'_{r,n}\right) \times U_0^*\left(\rho_0\left[-\frac{\nu_{T,n}}{2} - \frac{\nu_{Z,n}}{\nu_{T,n}\sin\alpha'_m}\right], \rho_0 Y'_{r,n}\right) dY'_{r,n}, \quad (10.4.72)$$

where we added the subscript $_{\text{inc}}$ to $\tilde{I}$ to emphasise the fact that this transfer function applies to an incoherently illuminated object surface or object volume.

10.4.9 The 3D Frequency Transfer Function for an Ideal Imaging System

In this subsection we analyse some properties of the transfer function $\tilde{I}'_{inc}(\boldsymbol{\nu})$ for a perfect optical imaging system with incoherent illumination of the three-dimensional object. We first establish the ranges of transverse frequency $\nu_T = (\nu_X, \nu_Y)$ and axial frequency ν_Z for which $\tilde{I}'_{inc}$ is nonzero. For $\tilde{I}'_{inc}$ to be nonzero, we require that P'_s in Fig. 10.61 lies in the overlap area of the two pupil discs with circular rims C_0 and C_1, respectively, such that the section AB of the line l_1 has finite length. We obtain the following condition for the frequency ν_Z,

$$-\nu_T \sin\alpha'_m + \frac{\lambda_0 \nu_T^2}{2n'} < \nu_Z < \nu_T \sin\alpha'_m - \frac{\lambda_0 \nu_T^2}{2n'} , \tag{10.4.73}$$

where $n' \sin\alpha'_m = NA$ is the image-side numerical aperture. We normalise both the lateral and the axial spatial frequencies with the aid of the quantity NA/λ_0 and obtain

$$\left|\nu_{Z,n}\right| < \nu_{T,n} \sin\alpha'_m \left(1 - \frac{\nu_{T,n}}{2}\right) , \tag{10.4.74}$$

where the subscript n refers to the normalised frequencies of Eq. (10.4.71). Within the frequency region where $\tilde{I}'_{inc}$ is nonzero, we normalise the transfer function by means of its limiting value when the spatial frequencies ν_T and ν_Z approach zero. The line segment AB in that case equals $2\rho_0 = 2r'_E \sin\alpha'_m$.

The representation of the three-dimensional transfer function of Eq. (10.4.72) suffers from the pole in the origin that is caused by the factor $1/\nu_{T,n}$. To avoid the singularity in the origin, we plot the function $\tilde{I}'_{n,inc}$ which is given by

$$\begin{aligned}\tilde{I}'_{n,inc}(\nu_{T,n}, \nu_{Z,n}) &= \frac{\nu_{T,n}}{2\lambda r_E'^2}\, \tilde{I}'_{inc}(\nu_{T,n}, \nu_{Z,n}) \\ &= \frac{1}{2} \int_A^B U_0\left(\rho_0 \left[\frac{\nu_{T,n}}{2} - \frac{\nu_{Z,n}}{\nu_{T,n} \sin\alpha'_m}\right], \rho_0 Y'_{r,n}\right) U_0^*\left(\rho_0 \left[-\frac{\nu_{T,n}}{2} - \frac{\nu_{Z,n}}{\nu_{T,n} \sin\alpha'_m}\right], \rho_0 Y'_{r,n}\right) dY'_{r,n}.\end{aligned} \tag{10.4.75}$$

For an ideal imaging system, the incoherent three-dimensional transfer function and its normalised version are thus given by

$$\begin{aligned}\tilde{I}'_{inc}(\nu'_{T,n}, \nu'_{Z,n}) &= \frac{2\lambda r_E'^2}{\nu'_{T,n}} \sqrt{1 - \left(\frac{\nu'_{T,n}}{2} + \frac{|\nu'_{Z,n}|}{\nu'_{T,n}}\right)^2} , \\ \tilde{I}'_{n,inc}(\nu'_{T,n}, \nu'_{Z,n}) &= \sqrt{1 - \left(\frac{\nu'_{T,n}}{2} + \frac{|\nu'_{Z,n}|}{\nu'_{T,n}}\right)^2} .\end{aligned} \tag{10.4.76}$$

In this expression we use primed normalised frequency coordinates, defined by

$$\left(\nu'_{X,n}, \nu'_{Y,n}, \nu'_{Z,n}\right) = \left(\nu_{X,n}, \nu_{Y,n}, \frac{\nu_{Z,n}}{\sin\alpha'_m}\right) ,$$

or, in terms of the real-space spatial frequency components (ν_X, ν_Y, ν_Z),

$$\left(\nu'_{X,n}, \nu'_{Y,n}\right) = \frac{\lambda}{\sin\alpha'_m} (\nu_X, \nu_Y) = \frac{\lambda_0}{NA} (\nu_X, \nu_Y) , \qquad \nu'_{Z,n} = \frac{\lambda}{\sin^2\alpha'_m} \nu_Z = \frac{n'\lambda_0}{(NA)^2} \nu_Z . \tag{10.4.77}$$

The only difference between the primed and unprimed normalised spatial frequencies is the normalisation of the axial frequency component ν_Z with respect to the inverse of the axial diffraction unit, $\lambda/(\sin^2\alpha'_m)$.

The normalised transfer function $\tilde{I}'_{n,inc}$ is given by the ratio AB/A_0B_0 in Fig. 10.61. The modulus of $\nu'_{Z,n}$ is used in Eq. (10.4.76) to simultaneously cover positive and negative values of $\nu'_{Z,n}$ which, at equal absolute value, yield the same value of the transfer function in the aberration-free case. In Fig. 10.62a we have plotted the frequency transfer function $\tilde{I}'_{inc}$ according to Eq. (10.4.72). The factor $\lambda r_E'^2$ in front of the integral has been omitted. The function $\tilde{I}'_{inc}$ has been truncated to a value of 10 close to the origin where the singularity occurs. The value of $NA = n' \sin\alpha'_m$ equals 0.40. In b) we show, for an ideal imaging system, the three-dimensional surface determined by the equation $\tilde{I}'_{n,inc} = 0$ where $\tilde{I}'_{n,inc}$ is given by Eq. (10.4.75). Inspection of this surface shows that the frequency transfer value given by both $\tilde{I}'_{inc}$ and $\tilde{I}'_{n,inc}$ is zero outside the volume in space that is delimited by the doughnut-shaped surface of Fig. 10.62b. With respect to the nonzero transfer volume in space, the difference between $\tilde{I}'_{inc}$ and $\tilde{I}'_{n,inc}$ is, apart from a trivial constant factor, caused by the factor $1/\nu_{T,n}$. This factor strongly influences the transfer at low frequency values, but does not change the volume

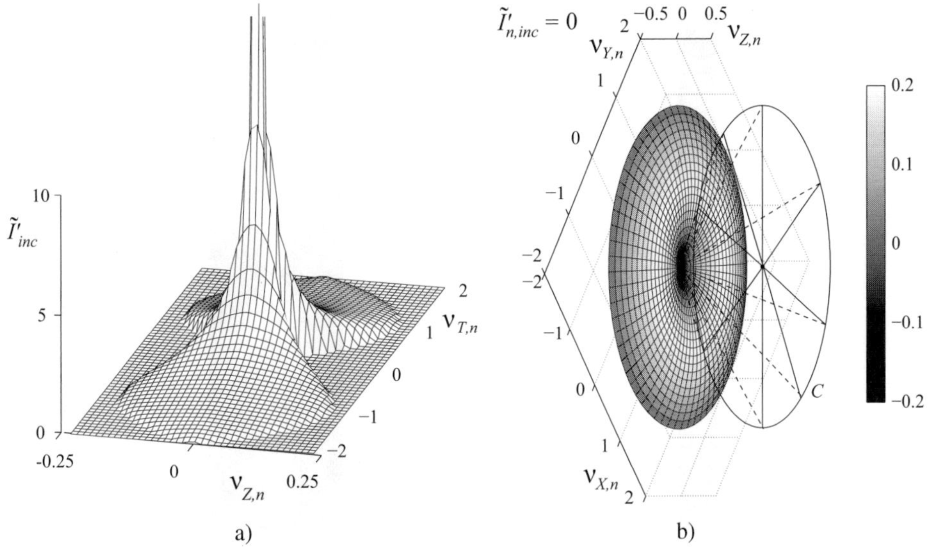

Figure 10.62: a) A plot of the three-dimensional frequency transfer function $\tilde{I}'_{\text{inc}}(\nu_{T,n}, \nu_{Z,n})$ of an ideal imaging system according to Eq. (10.4.72); $NA = 0.40$.
b) A three-dimensional graph of the surface $\tilde{I}'_{n,\text{inc}}(\nu_{X,n}, \nu_{Y,n}, \nu_{Z,n}) = 0$, also for $NA = 0.40$. The grey-shading of the surface is proportional to the value of $\nu_{Z,n}$. The interior volume of the conical surface C represents the so-called 'missing cone' of spatial frequencies.

in space where $\tilde{I}'_{\text{inc}}$ has nonzero transfer values. It is common practice to further normalise the *axial* spatial frequency $\nu_{Z,n}$ by division through $\sin\alpha'_m$, yielding the primed set of transverse and axial normalised frequencies of Eq. (10.4.77). The advantage of using the primed normalised spatial frequencies is that the dependence on the imaging aperture $\sin\alpha'_m$ disappears from both Eq. (10.4.72) and (10.4.75). Within the scalar approximation of wave propagation, the expressions of Eq. (10.4.76) cover all aperture values and a single graph suffices to show the three-dimensional frequency transfer by an imaging system.

In Fig. 10.63 we have plotted two cross-sections of $\tilde{I}'_{n,\text{inc}}$. The value of $\tilde{I}'_{n,\text{inc}}$ has been visualised by grey-shading each mesh in accordance with the function value at the centre of the mesh. Equation (10.4.76) was applied, corresponding to an aberration-free imaging system, without vignetting effects and with uniform transmission of the optics. Figure 10.63a shows a cross-section through the $\nu_{Z,n}$-axis for an imaging aperture of 0.10. It follows from Eq. (10.4.74) that the maximum value of $\nu_{Z,n}$ is attained for $\nu_{T,n} = 1$ and equals $\sin\alpha'_m/2 = 0.05$. Figure 10.63b shows that this value has been increased to 0.20 for an aperture in air of 0.40. The application of the normalised frequencies $(\nu'_{T,n}, \nu'_{Z,n})$ would have yielded a single unique graph. The plotting of two separate graphs for a low and a higher aperture with identical normalisation of the transverse and axial spatial frequencies emphasises the strong difference between transverse and axial frequency transfer that arises at low aperture. The ratio between the maximum transmitted axial and transverse frequency is found for $\nu'_{Z,n} = 1/2$, yielding

$$\nu_{Z,\max} = \frac{n' \sin^2\alpha'_m}{2\lambda_0} = \frac{(NA)^2}{2n'\lambda_0} \,, \qquad \nu_{T,\max} = \frac{2NA}{\lambda_0} \,,$$

or,

$$\frac{\nu_{Z,\max}}{\nu_{T,\max}} = \frac{NA}{4n'} \,. \tag{10.4.78}$$

10.4.10 Incoherent Transfer Function and the Missing Cone of Spatial Frequencies

In this subsection we discuss the 'missing cone' of spatial frequencies exhibited by the incoherent transfer function $\tilde{I}'_{\text{inc}}$ (see Fig. 10.62a and b). The half apex angle of the cone C in this figure follows from the slope $\partial\nu_{Z,n}/\partial\nu_{T,n}$ of the surface $\tilde{I}'_{n,\text{inc}}$ in the origin. From Eq. (10.4.76) we derive that the cone surface is described by the equation

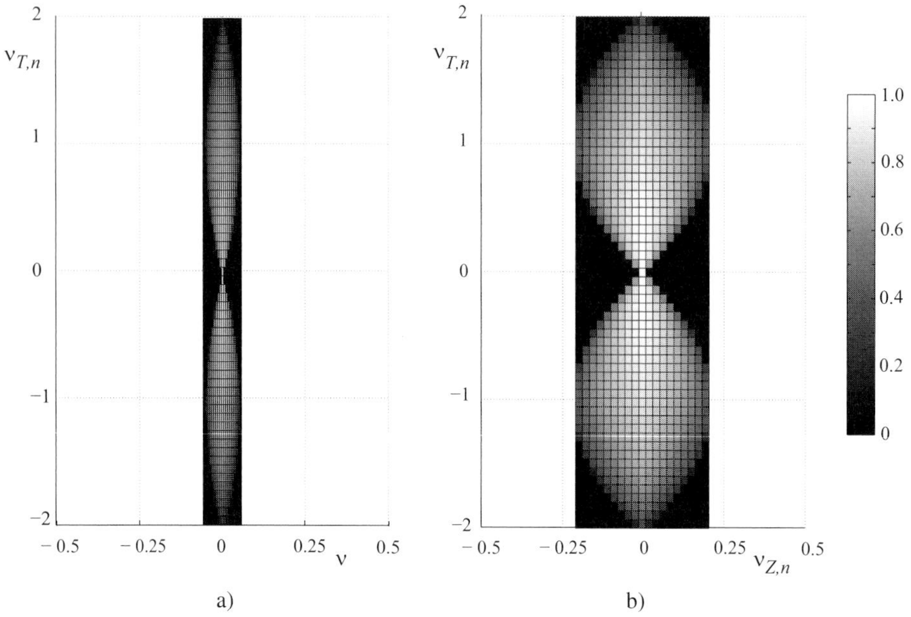

Figure 10.63: a) A cross-section comprising the $\nu_{Z,n}$-axis of the three-dimensional frequency transfer function $\tilde{I}'_{n,\mathrm{inc}}(\nu_{T,n},\nu_{Z,n})$ of Eq. (10.4.75). $\nu_{Z,n}$ and $\nu_{T,n}$ are normalised spatial frequencies; $NA = 0.10$. The grey-shading of the cross-section represents the value of $\tilde{I}'_{n,\mathrm{inc}}(\nu_{T,n},\nu_{Z,n})$.
b) Same legend as a), now for an aperture value $NA = 0.40$.

$$\nu'_{Z,n} = \pm\, \nu'_{T,n}\,, \quad \text{or,} \qquad \nu_{Z,n} = \pm\, \nu_{T,n}\, \sin\alpha'_m\,. \tag{10.4.79}$$

This expression implies that the cone has a half apex angle of $\pi/4$ when using the primed normalised coordinates. In terms of the spatial periods q'_z and q'_x in image space, associated with these coordinates, we have that $q'_x/q'_z = \sin\alpha'_m$. In the scalar approximation with $\sin\alpha'_m \approx \alpha'_m$, this means that the frequency vectors associated with the conic surface of the focusing beam cannot be transferred to image space. The same holds for all frequency vectors that are located inside the conical surface. The conclusion from Fig. 10.62 is that only frequency vectors that point into the geometrical shadow of the focused beam can be transferred to image space. And this conclusion is valid even when the spatial frequencies under consideration are very small.

With the aid of Fig. 10.64 we can qualitatively explain these, at a first sight, surprising results for three-dimensional frequency transfer by an imaging system. To discuss the transfer at very low spatial frequencies, we can approximate the focusing beam by its sharply delimited geometrical optics equivalent without taking into account the fine structure due to diffraction effects. The rays of the beam propagate in the positive z'-direction and all of them pass through the geometrical focal point F'. In Fig. 10.64a the cone of rays through F' (white in the figure) and a periodic pattern are sketched. The equiphase planar surfaces of the periodic pattern are shown, each time with a phase increment of 2π between two neighbouring planes. The equiphase planes are perpendicular to the normalised grating vector $\mathbf{v}'_n$. The sheet S_r of rays through F' in a plane perpendicular to the plane of the drawing is chosen to be coplanar with the equiphase plane labelled ϕ_0 of the periodic pattern. Since the absorption of the periodic pattern varies sinusoidally, the intensity of the rays in the sheet S_r experiences a maximum modulation depth when the periodic pattern is displaced in the direction of the grating vector $\mathbf{v}'_n$. The relative strength of the modulation is given by the ratio of the intensity in the infinitely thin sheet of rays S_r and the total beam intensity. For a finite frequency $\mathbf{v}'_n$, this ratio approaches zero. If, contrarily, we normalise the frequency transfer value for a nonzero frequency to the intensity in the sheet S_r, the transfer of the zero spatial frequency would become infinitely large since all rays of the illuminating cone of light will contribute to the light modulation in image space. In Fig. 10.64b the frequency vector $\mathbf{v}'_n$ is found inside the cone of propagating rays with focal point F'. An equiphase plane S_r through F' with phase ϕ_1 is contained in the geometrical shadow of the focused beam. No sheet of rays can be found in the focused beam that is coincident with S_r. Therefore, the frequency transfer is

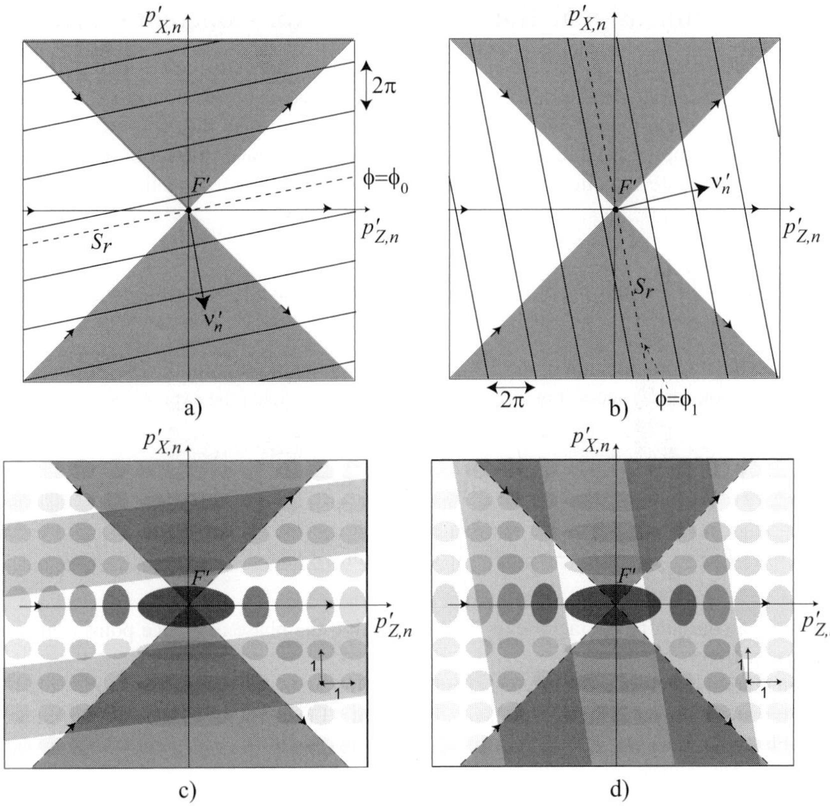

Figure 10.64: Low-frequency three-dimensional transfer by multiplication of the geometrical optics light cone intensity and the periodic object pattern, projected to image space. The cross-section in image space shown is perpendicular to the y'-axis.
a) $|\nu'_{Z,n}| < |\nu'_{X,n}|$, b) $|\nu'_{Z,n}| > |\nu'_{X,n}|$.
c) and d) Imaging cone of light with the diffraction fine structure schematically shown in the focal region. In c) the grey-shaded periodic pattern has frequency $(\nu'_{X,n}, 0, \nu'_{Z,n})$ with $|\nu'_{X,n}| > |\nu'_{Z,n}|$, yielding a nonzero frequency transfer value. In d) we have that $|\nu'_{Z,n}| > |\nu'_{X,n}|$ and frequency transfer is not possible.

rigorously zero for this geometry of focused beam and periodic structure. The conclusion is that the transfer function $\tilde{I}'_{\text{inc}}$ is only nonzero when the three-dimensional grating vector points into the geometrical shadow of the illuminating beam. The solid angle that corresponds to the cone of light of the illuminating focused beam forms the so-called *missing cone* of spatial frequency vectors that cannot be transferred to image space by the optical system.

In Fig. 10.64c and d the focusing beam has been, very schematically, provided with the diffraction fine structure represented in the intensity domain. In terms of the normalised axial and transverse coordinates, the axial central lobe of the intensity pattern has a width w'_Z of 4. This value w'_Z is the distance between the two central neighbouring zeros on the optical axis, closest to the focal point F'. In the transverse direction, the central lobe has a typical width w'_T of the order of 1. For the maximum frequencies that can be transmitted, we use the rule of thumb that $\nu'_{n,\max} = 2/w'$. The maximum frequencies which can be transferred are then given by

$$\begin{cases} \nu'_{Z,n,\max} = 1/2 \,, & \nu_{Z,\max} = 1/2 \; \sin^2\alpha'_m/\lambda, \\ \nu'_{T,n,\max} = 2 \,, & \nu_{T,\max} = 2 \; \sin\alpha'_m/\lambda \,. \end{cases} \tag{10.4.80}$$

These axial and lateral cut-off frequencies also follow from a more rigorous analysis of three-dimensional frequency transfer and were given in Eq. (10.4.78) in terms of the real-space frequencies $\nu_{Z,\max}$ and $\nu_{T,\max}$.

10.4.11 The Three-dimensional Transfer Function for Coherent Object Illumination

To obtain the coherent transfer function, we proceed in a comparable way as for the transfer in the case of an incoherently illuminated object. Referring to Eq. (10.4.42), we replace the object radiance function $O(\mathbf{r}_E)$ and the intensity point-spread function $I_d(\mathbf{r}' - \mathbf{r}'_E)$ by the corresponding complex functions that represent the complex transmission of the three-dimensional object and the complex amplitude $U'_d(\mathbf{p}')$ of the three-dimensional diffraction point-spread function. The three-dimensional coherent transfer function then follows from Eq. (10.4.44), second line, with I'_d replaced by U'_d of Eq. (10.4.55),

$$\tilde{I}'_c(\omega') = \iiint_{-\infty}^{+\infty} U'_d(p'_X, p'_Y, p'_Z) \exp\{-i(\omega'_X p'_X + \omega'_Y p'_Y)\} \exp(-i\omega'_Z p'_Z)\, dp'_X dp'_Y dp'_Z \,. \tag{10.4.81}$$

Within the small-angle scalar approximation of the diffraction problem in the exit pupil, it is permitted to use the Fresnel diffraction integral expression for U'_d according to Eq. (10.4.55) and this yields the expression

$$\begin{aligned}\tilde{I}'_c(\omega') = &\left(\frac{-ik}{2\pi r'_E}\right) \iiiiint_{-\infty}^{+\infty} U(X', Y') \exp\left\{+ik\left[\left(1 - \frac{\omega'_Z}{k} - \frac{X'^2 + Y'^2}{2{r'_E}^2}\right) p'_Z\right]\right\} \\ &\times \exp\left[ik\left(\frac{{p'_X}^2 + {p'_Y}^2}{2r'_E} - \frac{X' p'_X + Y' p'_Y}{r'_E}\right)\right] \exp\left[-i(\omega'_X p'_X + \omega'_Y p'_Y)\right] dX' dY' dp'_X dp'_Y dp'_Z \,,\end{aligned} \tag{10.4.82}$$

where we have omitted a constant phase factor $\exp(ikr'_E)$ by shifting the phase reference point from E'_0 to the paraxial image point Q' (see Fig. 10.58).

Performing the integration with respect to p'_Z yields a delta-function with the argument $\nu_Z - \left\{1 - (X'^2 + Y'^2)/2{r'_E}^2\right\}/\lambda$. The integrations with respect to p'_X and p'_Y require the evaluation of a Fresnel-type integral. For instance, with respect to the integration variable p'_X we have the integral

$$\begin{aligned}&\int_{-\infty}^{+\infty} \exp\left\{+ik\left[\frac{{p'_X}^2}{2r'_E} - \left(\lambda \nu_X + \frac{X'}{r'_E}\right) p'_X\right]\right\} dp'_X \\ &= \sqrt{\lambda r'_E}\, \exp(-i\pi/4) \exp\left\{-\pi i \lambda r'_E \left[\nu_X + \frac{X'}{\lambda r'_E}\right]^2\right\} \,,\end{aligned} \tag{10.4.83}$$

where we have used the expression for the Fourier transform of a complex Gaussian function, evaluated at zero frequency (see Appendix A, Table A.3). Using one of the numerous limit representations of the Dirac delta-function,

$$\delta(x) = \lim_{\epsilon \to 0} \frac{1}{\sqrt{2i\epsilon}} \exp\left\{i\frac{\pi x^2}{2\epsilon}\right\} \,, \tag{10.4.84}$$

we conclude that the integral of Eq. (10.4.83), provided that ${p'_X}^2 \ll \lambda r'_E$, converges to a delta-function with argument $\nu_X + X'/(\lambda r'_E)$. If $r'_E \rightarrow \infty$, the integral effectively yields the delta-function solution. We use a $\pi/2$ criterion for the combined phase factor $ik({p'_X}^2 + {p'_Y}^2)/r'_E$ in the kernel of the integral in Eq. (10.4.82) to justify the choice of the delta-function solution. Using the normalised radial transverse coordinate $p'_{T,n}$, we then readily find the condition

$${p'_{X,n}}^2 + {p'_{Y,n}}^2 = {p'_{T,n}}^2 \leq \frac{N_F}{2} \,, \tag{10.4.85}$$

where $N_F = r'_E \sin^2\alpha'_m/\lambda$ is the Fresnel number of the focused beam. The lateral extent of the intensity distribution of the point-spread function of a well-corrected imaging system is of the order of unity when expressed in diffraction units. We thus require that $N_F \geq 2$ to safely use the delta-function approximation. For a typical imaging lens with $\sin\alpha'_m = 0.1$ and $r'_E = 100$ mm, we have that $N_F = 2000$ ($\lambda = 500$ nm), which shows that the quadratic phase factor can be safely omitted from the calculation up to values $p'_{T,n}$ of the order of 30.

We use the delta-functions that result from the integrations over the coordinates p'_X, p'_Y and p'_Z in Eq. (10.4.82) and obtain the following expression for the transfer function

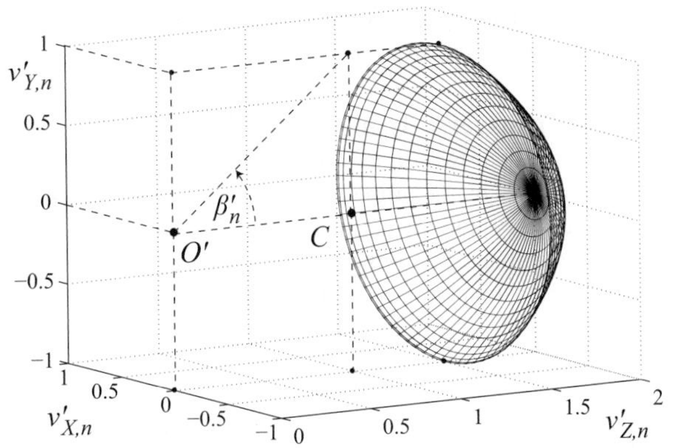

Figure 10.65: The surface in frequency space on which the coherent three-dimensional transfer function *CTF* is nonzero ($\sin\alpha'_m = 0.70$). The coordinate system employs the primed normalised frequencies $(\nu'_{X,n}, \nu'_{Y,n}, \nu'_{Z,n})$. $O'$is the origin of the frequency coordinate system, C is the centre of (paraxial) curvature of the *CTF* surface in frequency space.

$$
\begin{aligned}
\tilde{I}'_c(\mathbf{v}) &= \left(\frac{-ik}{2\pi r'_E}\right) \iint_{-\infty}^{+\infty} U(X',Y')\,\delta\left(\nu_X + \frac{X'}{\lambda r'_E}\right)\delta\left(\nu_Y + \frac{Y'}{\lambda r'_E}\right) \\
&\quad \times \delta\left(\nu_Z - \frac{1}{\lambda}\left[1 - \frac{\lambda^2(\nu_X^2+\nu_Y^2)}{2}\right]\right)\, dX'\,dY' \\
&= -\,i\lambda r'_E\,\delta\left(\nu_Z - \frac{1}{\lambda} + \frac{\lambda(\nu_X^2+\nu_Y^2)}{2}\right)\, U(-\lambda r'_E\nu_X, -\lambda r'_E\nu_Y)\,.
\end{aligned}
\tag{10.4.86}
$$

In terms of the primed normalised coordinates of Eq. (10.4.77) we have the final result

$$
\tilde{I}'_c(\mathbf{v}'_n) = \frac{-i\lambda^2 r'_E}{\sin^2\alpha'_m}\,\delta\left(\nu'_{Z,n} - \frac{1}{\sin^2\alpha'_m} + \frac{\nu'^{\,2}_{X,n} + \nu'^{\,2}_{Y,n}}{2}\right) U(-\rho_0\nu'_{X,n}, -\rho_0\nu'_{Y,n})\,. \tag{10.4.87}
$$

The surface in space where the transfer function is not equal to zero has been plotted in Fig. 10.65. The value of *NA* has been taken relatively large, well outside the domain of scalar diffraction, to keep the surface close to the origin O' of the spatial frequency coordinate system. Using the scalar approximation, the surface is part of a paraboloid. The paraxial radius of the paraboloid is unity and its centre of curvature is found at the axial position $\nu'_{Z,n} = \cot^2(\alpha'_m)$. The rim of the parabolic surface is limited by the condition that $\nu'^{\,2}_{X,n} + \nu'^{\,2}_{Y,n} \leq 1$. We replace the parabolic surface which followed from the scalar approximation by a spherical surface which has the same radius of curvature and also intersects the $\nu'_{Z,n}$-axis at the point $(0, 0, 1/\sin^2\alpha'_m)$. In the small-angle approximation, the opening angle β'_n of this surface in terms of the primed normalised spatial frequencies is given by $\sin\beta'_n \approx \sin^2\alpha'_m$. In terms of the real-space spatial frequencies, the opening angle $\beta' \approx \nu_T/\nu_Z$ of the three-dimensional transfer function surface is equal to the aperture angle α'_m. This means that the *CTF* is such that plane waves are transmitted by the imaging system with wave vectors that are contained within a spherical cap on the Ewald sphere, corresponding to lens aperture.

10.4.12 Two-dimensional Transfer Functions Derived from the Three-dimensional Transfer Function

In this subsection we show the usefulness of the three-dimensional transfer function as a master function from which various transfer functions of lower dimensionality can be derived. We do this for both the incoherent *OTF* and the coherent *CTF*. We first focus on the transverse two-dimensional transfer function that was derived in a more dedicated way in Section 10.2 of this chapter. The derivation of the two-dimensional transfer function from the three-dimensional master function immediately shows the general nature of such a derivation. It yields the two-dimensional transfer function for any defocused plane. The other example applies to the two-dimensional transfer function for objects that contain a mixture of axial and transverse frequencies. This two-dimensional result shows the strong difference between transverse and axial resolution in optical imaging. It also makes visible the forbidden region for frequency transfer, a cross-section of the 'missing cone' of the three-dimensional transfer function.

10.4.12.1 The Two-dimensional Transverse Transfer Function for Incoherent Object Illumination

The starting point for obtaining a z-dependent two-dimensional transverse transfer function is Eq. (10.4.72), where we replace the unprimed normalised frequencies by the primed ones. The two-dimensional transfer function then follows by transforming back the function $\tilde{I}'_{\text{inc}}$ with respect to the normalised axial frequency component $\omega'_Z = 2\pi \sin^2\alpha'_m \nu'_{Z,n}/\lambda$. We also introduce the normalised spatial coordinate $p'_{Z,n}$ in the axial direction where $p'_{Z,n}$ is given by $p'_Z \sin^2\alpha'_m/\lambda$. We obtain the following expression for the two-dimensional transverse function, denoted by OTF_t,

$$OTF_t(\nu'_{X,n}, \nu'_{Y,n}; p'_{Z,n}) = \frac{2\rho_0^2}{\nu'_{T,n}} \int_{-\infty}^{+\infty} \int_A^B U_0 \left\{\rho_0 \left[\frac{\nu'_{T,n}}{2} - \frac{\nu'_{Z,n}}{\nu'_{T,n}}\right], \rho_0 Y'_{r,n}\right\}$$
$$\times U_0^* \left\{\rho_0 \left[-\frac{\nu'_{T,n}}{2} - \frac{\nu'_{Z,n}}{\nu'_{T,n}}\right], \rho_0 Y'_{r,n}\right\} \exp(+i2\pi \nu'_{Z,n} p'_{Z,n})\, dY'_{r,n} d\nu'_{Z,n}, \qquad (10.4.88)$$

where the integration limits for the variable $Y'_{r,n}$ are given by (see Fig. 10.61)

$$Y'_{r,n}\Big|_{A,B} = \pm \sqrt{1 - \left(\frac{\nu'_{T,n}}{2} + \frac{|\nu'_{Z,n}|}{\nu'_{T,n}}\right)^2}, \qquad (10.4.89)$$

and where we have used that the exit pupil radius ρ_0 equals $r'_E \sin\alpha'_m$.

For a general optical system, the complex pupil function U can be expanded with the aid of Zernike polynomials with complex coefficients, and semi-analytic expressions can be obtained for the line integral from A to B over the rotated and normalised pupil coordinate $Y'_{r,n}$. In this subsection we limit ourselves to an *ideal* optical system in which case the integral over $Y'_{r,n}$ is simply given by the length of the line segment AB in Fig. 10.61. It equals twice the square root expression of Eq. (10.4.89). Because of the presence of $|\nu'_{Z,n}|$, the integral should be split in two parts, one for positive values of $\nu'_{Z,n}$ and one for the negative values where $|\nu'_{Z,n}|$ equals $-\nu'_{Z,n}$. When performing the integrations over the positive and negative values of $\nu'_{Z,n}$, it is seen that the total integral equals twice the real part of the integral over the positive $\nu'_{Z,n}$ values. In this latter integral we also introduce a change of integration variable according to $\nu'_s = \nu'_{T,n}/2 + \nu'_{Z,n}/\nu'_{T,n}$ and obtain the expression

$$OTF_t(\nu'_{X,n}, \nu'_{Y,n}; p'_{Z,n}) = 4\rho_0^2\, \Re\left\{\exp(-i\pi \nu'^2_{T,n} p'_{Z,n}) \int_{\nu'_{T,n}/2}^{+1} \sqrt{1 - \nu'^2_s}\, \exp(+i2\pi \nu'_{T,n} p'_{Z,n} \nu'_s)\, d\nu'_s\right\}. \qquad (10.4.90)$$

We substitute $\nu'_s = \cos\phi$ and, for a circularly symmetric optical system, we restrict ourselves to the transverse frequency $\nu'_{T,n}$,

$$OTF_t(\nu'_{T,n}; p'_{Z,n}) = 4\rho_0^2\, \Re\left\{\exp(-i\pi \nu'^2_{T,n} p'_{Z,n}) \int_0^{\phi_0} \sin^2\phi\, \exp(+i2\pi \nu'_{T,n} p'_{Z,n} \cos\phi)\, d\phi\right\}, \qquad (10.4.91)$$

where $\phi_0 = \arccos(\nu'_{T,n}/2)$. For the in-focus transfer function we have

$$OTF_t(\nu'_{T,n}; 0) = 4\rho_0^2 \int_0^{\phi_0} \sin^2\phi\, d\phi = 2\rho_0^2 \left[\phi - \frac{\sin 2\phi}{2}\right]_0^{\phi_0}$$
$$= \pi\rho_0^2 \left\{\frac{2}{\pi}\left[\arccos(\nu'_{T,n}/2) - (\nu'_{T,n}/2)\sqrt{1 - (\nu'_{T,n}/2)^2}\right]\right\}. \qquad (10.4.92)$$

The expression between curly braces on the second line of Eq. (10.4.92) is the normalised two-dimensional incoherent transfer function of an aberration-free optical system which was derived in Section 10.2 of this chapter (see Eq. (10.2.49)).

For the general case of a defocused system with $p'_{Z,n} \neq 0$ we expand the exponential in the integrand of Eq. (10.4.91) with the aid of Eq. (10.2.8). After some manipulation we obtain for the defocused transfer function

$$OTF_t(\nu'_{T,n}; p'_{Z,n}) = 2\rho_0^2\, \Re\left\{\exp(-i\pi \nu'^2_{T,n} p'_{Z,n}) \sum_{m=-\infty}^{+\infty} i^m J_m(2\pi \nu'_{T,n} p'_{Z,n}) \int_0^{\phi_0} \sin^2\phi\, \exp(im\phi)\, d\phi\right\}. \qquad (10.4.93)$$

Having performed the integration over ϕ we obtain the result

$$
\begin{aligned}
OTF_t(v'_{T,n};p'_{Z,n}) &= \pi\rho_0^2\,\Re\Bigg\{\frac{2}{\pi}\,\exp(-i\pi v'^{\,2}_{T,n}p'_{Z,n})\,\phi_0\sum_{m=-\infty}^{+\infty}(i)^m J_m(2\pi v'_{T,n}p'_{Z,n}) \\
&\quad\times\Bigg[2\exp\left(\frac{im\phi_0}{2}\right)\operatorname{sinc}\left(\frac{m\phi_0}{2}\right)-\exp\left(\frac{i(m+2)\phi_0}{2}\right)\operatorname{sinc}\left(\frac{(m+2)\phi_0}{2}\right) \\
&\quad-\exp\left(\frac{i(m-2)\phi_0}{2}\right)\operatorname{sinc}\left(\frac{(m-2)\phi_0}{2}\right)\Bigg]\Bigg\} \\
&= \pi\rho_0^2\,\Re\left\{G(v'_{T,n};p'_{Z,n})\right\},
\end{aligned}
\tag{10.4.94}
$$

where ϕ_0, as defined previously, is given by $\arccos(v'_{T,n}/2)$.

The real part of the complex function $G(v'_{T,n};p'_{Z,n})$ yields the normalised optical transfer function OTF_t. This result, in a somewhat different form, was derived from the expression for the scalar two-dimensional classical *OTF* by Hopkins **[143]**. We have plotted the modulus and the binary phase of the complex function $G(v'_{T,n};p'_{Z,n})$ in Fig. 10.66a and b. The role of modulus and phase of the OTF_t for the image contrast and image position of a periodic object follows from Eq. (10.2.36). This equation for the image intensity applies to the case of incoherent illumination of the periodic object.

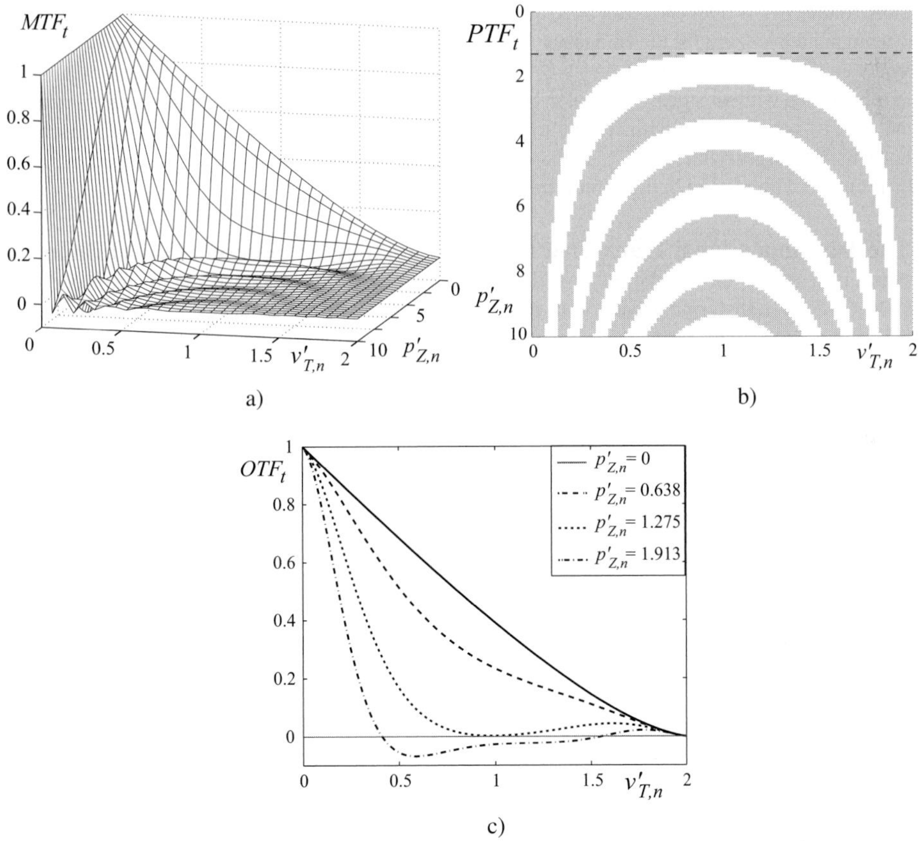

Figure 10.66: The complex transverse transfer function $OTF_t(v'_{T,n};p'_{Z,n})$ as a function of the axial defocus position (perfect imaging system).
a) MTF_t, the modulus of the transfer function.
b) PTF_t, the argument of the complex function G having discrete values of 0 (grey) and π (white) for the aberration-free case.
c) Plot of $OTF_t(v'_{T,n})$ for four specific axial positions $p'_{Z,n}$ = 0, 0.638, 1.275 and 1.913.

The relative modulation depth in the image pattern is directly determined by the modulus of OTF_t. There is no lateral shift of the pattern because the imaginary part of OTF_t is identically zero. The value of the transfer function PTF_t takes on the discrete values of 0 or π, with a shift of π each time the real part of OTF_t shows a zero-crossing. When PTF_t equals π, the periodic pattern appears with inverted contrast.

In Fig. 10.66c we have plotted the OTF_t for some discrete axial positions $p'_{Z,n}$. For small values of $p'_{Z,n}$ the decrease in modulation depth is strongest for the mid-spatial frequencies where $\nu'_{T,n}$ is close to unity. For a defocus value $p'_{Z,n} = 1.283$, the first zero crossing occurs at a spatial frequency $\nu'_{T,n} \approx 0.97$, see also the horizontal dashed line in Fig. 10.66b. In the small-aperture approximation, the phase departure at the rim of the exit pupil due to defocusing is given by $1.283\,\pi$. For larger defocus values, the effective bandwidth of the image is limited by the first zero-crossing of the OTF_t function. In the spectral data of such a defocused image, at least one specific frequency is entirely missing and one or several frequency bands are transmitted with a phase offset of π. As a measure for the reliable frequency bandwidth transmitted to a defocused image plane, one commonly takes as limiting frequency the one that still presents 50% of its original modulation depth.

With respect to the zero crossings of the OTF_t function at the high-frequency side, close to the cut-off frequency $\nu'_{T,n} = 2$, we refer to Eq. (10.2.64). The phase departure $\Phi_A - \Phi_B$ over the common area of the two shifted pupil functions is identical for the spatial frequencies $\nu'_{T,n}$ and $2 - \nu'_{T,n}$. This phase departure is the main factor that determines the appearance of a zero-value of OTF_t, provided it amounts to a value slightly in excess of 2π (2.56π for the example above where we considered a spatial frequency $\nu'_{T,n}$ close to unity). Minor deviations from this value of typically 2π are due to the different shape of the overlapping region between the two pupil circles when going from low frequencies to frequencies close to the cut-off frequency. But as a rule of thumb we can maintain that a first zero $\nu'_{T,n} = \nu'_0$ in the transfer function is accompanied by a zero for the frequency $\nu'_{T,n} \approx 2 - \nu'_0$. These two zeros of the function OTF_t delimit a lower and upper frequency band with intervals $0 \leq \nu'_{T,n} \leq \nu'_0$ and $2 - \nu'_0 \leq \nu'_{T,n} \leq 2$. Within these frequency bands, the spatial information is transmitted without a phase shift of π. Figure 10.66b illustrates this virtually perfect symmetric appearance of the first and the last zero-crossing with respect to the central frequency $\nu'_{T,n} = 1$ in a defocused function OTF_t. The graph also shows that, at larger values of $p'_{Z,n}$, additional zero-crossings appear in the OTF_t, arranged in pairs and almost symmetrically arranged with respect to the central frequency $\nu'_{T,n} = 1$.

10.4.12.2 The Two-dimensional Transverse Transfer Function for Coherent Object Illumination

Along the same lines as for the incoherent two-dimensional transfer function, we obtain the coherent transverse transfer function CTF_t by transforming back with respect to the normalised frequency $\nu'_{Z,n}$ the expression for $\tilde{I}_c(\mathbf{v}'_n)$ of Eq. (10.4.87). If we also use the normalised spatial coordinate $p'_{Z,n} = p'_Z \sin^2\alpha'_m/\lambda$, we have the result

$$CTF_t(\nu'_{X,n}, \nu'_{Y,n}; p'_{Z,n}) = -\,i\lambda r'_E \int_{-\infty}^{+\infty} \delta\left(\nu'_{Z,n} - \frac{1}{\sin^2\alpha'_m} + \frac{\nu'^{\,2}_{X,n} + \nu'^{\,2}_{Y,n}}{2}\right)$$
$$\times\, U_0(-\rho_0\nu'_{X,n}, -\rho_0\nu'_{Y,n})\, \exp(+i2\pi\nu'_{Z,n}p'_{Z,n})\, d\nu'_{Z,n}\,. \qquad (10.4.95)$$

The integration over $\nu'_{Z,n}$ yields the following two expressions, in either the normalised or the real-space coordinates and frequencies,

$$CTF_t(\nu'_{X,n}, \nu'_{Y,n}; p'_{Z,n}) = -\,i\lambda r'_E \exp\left\{i\pi\left[\frac{2}{\sin^2\alpha'_m} - (\nu'^{\,2}_{X,n} + \nu'^{\,2}_{Y,n})\right]p'_{Z,n}\right\} U_0(-\rho_0\nu'_{X,n}, -\rho_0\nu'_{Y,n})\,, \qquad (10.4.96)$$

$$CTF_t(\nu_X, \nu_Y; p'_Z) = -\,i\lambda r'_E \exp(ikp'_Z) \exp\{-i\pi\lambda(\nu_X^2 + \nu_Y^2)p'_Z\}\, U_0\{-\lambda r'_E(\nu_X, \nu_Y)\}\,. \qquad (10.4.97)$$

As a function of the axial defocusing p'_Z, the transfer function accumulates a linear phase shift and a phase term which has a quadratic dependence on the lateral spatial frequencies (ν'_X, ν'_Y). This latter phase shift is added to the phase departures that are related to the aberration of the pupil transmission function $U(X', Y')$. Although the modulus of the CTF_t is not affected by the z-dependent phase departures, the imaging in a defocused plane will be degraded by the unequal phase defects for the various spatial frequencies which are present in a general extended object. Even a purely periodic pattern will give rise to a decrease in the modulation depth of the image intensity pattern when the image plane is shifted away from the optimum focal plane with axial coordinate $p'_Z = 0$. To study this change in modulation depth as a function of defocus, we use Eq. (10.2.41) which gives the intensity distribution in a receiving plane when there is interference between a zeroth-order wave and the (weak) first-order diffracted waves which originate at a periodic object. We assume that the pupil function has unit amplitude transmittance and a phase dependence given by the second exponential term of Eq. (10.4.97). For a zeroth-order incident wave which is parallel to the z-axis, the intensity pattern in a general receiving plane with coordinate p'_Z is given by

$$
\begin{aligned}
I'(p'_X, p'_Y; p'_Z) &\propto 1 + \frac{b}{2}\left\{\exp[-i\pi\lambda(\nu_X^2+\nu_Y^2)p'_Z] + \exp[+i\pi\lambda(\nu_X^2+\nu_Y^2)p'_Z]\right\}\cos[2\pi(\nu_X p'_X + \nu_Y p'_Y)] \\
&= 1 + b\,\cos[\pi\lambda(\nu_X^2+\nu_Y^2)p'_Z]\,\cos[2\pi(\nu_X p'_X + \nu_Y p'_Y)]\,. \qquad (10.4.98)
\end{aligned}
$$

The modulation depth b of the periodic image pattern is affected by the p'_z-dependent cosine factor in Eq. (10.4.98). Two interference patterns are produced in the receiving image plane. One originates from interference by the plus first and the zeroth order, the other from the interference between zeroth and minus first order. The defocus phase factors are written as $\pm\pi\lambda\nu_T^2 p'_Z$ where the transverse frequency ν_T is given by $(\nu_X^2+\nu_Y^2)^{1/2}$. The phase factors in each interference pattern produce a total lateral shift Δ_s of the two periodic interference patterns with period $1/\nu_T$ which is given by $\Delta_s = \nu_T\lambda p'_Z$. If the value of Δ_s equals half a period, $1/(2\nu_T)$, the modulation in the image intensity disappears and this occurs at a defocusing distance of

$$
p'_Z = \frac{1}{2\lambda\,\nu_T^2}\,. \qquad (10.4.99)
$$

A maximum defocus distance can be defined which still produces acceptable imaging quality in the receiving plane. We require that the modulation in an interference pattern attains zero for the maximum frequency $\nu_{T,\max} = \sin\alpha'_m/\lambda$ which is transmitted by the imaging system. It then follows that

$$
|p'_Z| < \frac{\lambda}{2\sin^2\alpha'_m}\,. \qquad (10.4.100)
$$

This criterion for the axial range of good imaging coincides with the earlier defined axial range based on the Rayleigh depth of focus. We recollect that the Rayleigh criterion for defocusing requires that the peak intensity of the perturbed diffraction point-spread function is at least 80% of its theoretical maximum value.

10.4.13 The One-dimensional Axial Transfer Function

For calculation of the transfer of a purely axial frequency component under coherent or incoherent object illumination conditions we transform back the expression for $\tilde{I}'(\nu'_{X,n}, \nu'_{Y,n}, \nu'_{Z,n})$ with respect to the normalised frequency components $\nu'_{X,n}$ and $\nu'_{Y,n}$. In general, numerical evaluation is necessary when the pupil function U has a general amplitude transmission and phase profile. In the case of a perfect imaging system we use the (normalised) result of Eq. (10.4.76) for the case of incoherent illumination of the object. For the case of coherent object illumination we use Eq. (10.4.87). For both illumination types, it turns out that an analytic solution is possible.

10.4.13.1 One-dimensional Frequency Transfer for an Incoherently Illuminated Object

For the backward Fourier transform of the aberration-free three-dimensional transfer function of Eq. (10.4.72) we can write in terms of the normalised spatial coordinates and frequency coordinates,

$$
\begin{aligned}
OTF_z(p'_{X,n}, p'_{Y,n}; \nu'_{Z,n}) &= \frac{2\rho_0^2}{\lambda}\iint_{-\infty}^{+\infty}\frac{1}{\nu'_{T,n}}\sqrt{1-\left(\frac{\nu'_{T,n}}{2}+\frac{|\nu'_{Z,n}|}{\nu'_{T,n}}\right)^2} \\
&\quad\times\exp\left\{+i2\pi(\nu'_{X,n}p'_{X,n}+\nu'_{Y,n}p'_{Y,n})\right\}\,d\nu'_{X,n}d\nu'_{Y,n}\,, \qquad (10.4.101)
\end{aligned}
$$

where $\nu'_{T,n} = \sqrt{\nu'^{\,2}_{X,n} + \nu'^{\,2}_{Y,n}}$. For an imaging system with a circular pupil rim, we change to polar coordinates, both on the pupil sphere and in the area of interest in the image space where the p'-coordinate system was defined. The normalised version of these polar coordinates is given by

$$
\begin{cases} \nu'_{X,n} = \nu'_{T,n}\cos\theta\,, \\ \nu'_{Y,n} = \nu'_{T,n}\sin\theta\,, \end{cases} \quad \text{and,} \quad \begin{cases} p'_{X,n} = p'_{T,n}\cos\gamma\,, \\ p'_{Y,n} = p'_{T,n}\sin\gamma\,. \end{cases} \qquad (10.4.102)
$$

The expression for the *OTF* is then given by

$$
\begin{aligned}
OTF_z(p'_{T,n}; \nu'_{Z,n}) &= \frac{2\rho_0^2}{\lambda}\int_0^{2\pi}\int_0^{\infty}\sqrt{1-\left(\frac{\nu'_{T,n}}{2}+\frac{|\nu'_{Z,n}|}{\nu'_{T,n}}\right)^2} \\
&\quad\times\exp\left\{+i2\pi\nu'_{T,n}p'_{T,n}\cos(\theta-\gamma)\right\}\,d\nu'_{T,n}d\theta\,. \qquad (10.4.103)
\end{aligned}
$$

The integral with respect to θ yields $2\pi J_0(2\pi v'_{T,n} p'_{T,n})$ and the OTF is then given by

$$OTF_z(p'_{T,n}; v'_{Z,n}) = 2k\rho_0^2 \int_0^\infty \sqrt{1 - \left(\frac{v'_{T,n}}{2} + \frac{|v'_{Z,n}|}{v'_{T,n}}\right)^2} \, J_0(2\pi v'_{T,n} p'_{T,n}) \, dv'_{T,n} \, . \qquad (10.4.104)$$

An analytic solution for OTF_z with nonzero values of $p'_{T,n}$, is not available.

For the on-axis transfer function ($p'_{T,n} = 0$) an analytic treatment of the integral is possible. We are indebted to Dr. A.J.E.M. Janssen who pointed out the solution given in the footnote below.[2] Using the expression of Eq. (10.4.105) for the on-axis integral we find the on-axis transfer function

$$OTF_z(0; v'_{Z,n}) = \frac{\pi\rho_0^2}{\lambda} \left(1 - 2|v'_{Z,n}|\right) , \quad |v'_{Z,n}| \le 1/2 \, . \qquad (10.4.106)$$

This triangular transfer function for the aberration-free case follows from the three-dimensional amplitude point-spread function according to scalar diffraction, as given in Eq. (8.4.38). Evaluating this equation for the on-axis amplitude distribution ($\rho_p = 0$), we obtain after some rearrangement,

$$U'(0,0,p'_{Z,n}) = -\frac{iUkr'_E \sin^2\alpha'_m}{2} \exp\left\{i2\pi\left(\sin^{-2}\alpha'_m - 1/4\right)p'_{Z,n}\right\} \operatorname{sinc}\left(\frac{\pi p'_{Z,n}}{2}\right) . \qquad (10.4.107)$$

Here, we have substituted $r'_E = f$, $\sin\alpha'_m = \rho_0/r'_E$ and $p'_{Z,n} = z_p \sin^2\alpha'_m/\lambda$. The quantity U is the constant amplitude on the exit pupil sphere. For the on-axis intensity, we obtain the expression

$$I'(0,0,p'_{Z,n}) = \frac{k^2 r'^2_E \sin^4\alpha'_m}{4} |U|^2 \operatorname{sinc}^2\left(\frac{\pi p'_{Z,n}}{2}\right) . \qquad (10.4.108)$$

The Fourier transformation with respect to $p'_{Z,n}$ transforms the sinc^2-function in the spatial domain into a triangular function in the (axial) frequency domain. The result within the scalar approximation for $p'_{T,n} = 0$ is then given by Eq. (10.4.106). We emphasise that the two-dimensional inverse Fourier transformation which was performed by means of Eq. (10.4.101) to obtain OTF_z assumes an axially modulated line-shaped object, which is incoherently illuminated. For an aberration-free imaging system with circular symmetry, the value of the lateral coordinate $p'_{T,n}$ in the image plane

[2] The values of the transverse normalised frequency $v'_{T,n}$ are limited by the requirement that the square root argument in Eq. (10.4.104) is non-negative; this implies that

$$1 - \sqrt{1 - 2|v'_{T,n}|} \le v'_{T,n} \le 1 + \sqrt{1 - 2|v'_{T,n}|} \, .$$

We introduce the change of variable

$$y = v'_{T,n}/2 + |v'_{Z,n}|/v'_{T,n} \, ,$$

and obtain, with due attention to the signs, two separate inverse relations on the total interval for $v'_{T,n}$,

$$\begin{aligned}
&1 - \sqrt{1 - 2|v'_{T,n}|} \le v'_{T,n} \le \sqrt{2|v'_{Z,n}|} \, , && \sqrt{2|v'_{Z,n}|} \le v'_{T,n} \le 1 + \sqrt{1 - 2|v'_{T,n}|} \, , \\
&1 \le y \le \sqrt{2|v'_{Z,n}|} \, , && \sqrt{2|v'_{Z,n}|} \le y \le 1 \, , \\
&v'_{T,n} = y - \sqrt{y^2 - 2|v'_{Z,n}|} \, , && v'_{T,n} = y + \sqrt{y^2 - 2|v'_{Z,n}|} \, , \\
&dv'_{T,n} = \left\{1 - y\left(y^2 - 2|v'_{Z,n}|\right)^{-1/2}\right\} dy \, , && dv'_{T,n} = \left\{1 + y\left(y^2 - 2|v'_{Z,n}|\right)^{-1/2}\right\} dy \, .
\end{aligned}$$

The integral for OTF_z with the y-integration variable is split in two integrals over each of the two sub-intervals,

$$\int_1^{(2|v'_{Z,n}|)^{1/2}} (1-y^2)^{1/2}\left\{1 - y\left(y^2 - 2|v'_{Z,n}|\right)^{-1/2}\right\} dy + \int_{(2|v'_{Z,n}|)^{1/2}}^1 (1-y^2)^{1/2}\left\{1 + y\left(y^2 - 2|v'_{Z,n}|\right)^{-1/2}\right\} dy$$

$$= 2\int_{(2|v'_{Z,n}|)^{1/2}}^1 (1-y^2)^{1/2}\left(y^2 - 2|v'_{Z,n}|\right)^{-1/2} y\,dy = \int_{2|v'_{Z,n}|}^1 (1-t)^{1/2}(t - 2|v'_{Z,n}|)^{-1/2}\, dt \, .$$

The latter integral can be transformed into a *Beta*-integral, see **[266]**, Chapter 5, to yield the result,

$$(1 - 2|v'_{Z,n}|)\int_0^1 (1-p)^{1/2} p^{-1/2} dp = (1 - 2|v'_{Z,n}|)\,\frac{\Gamma(1/2)\,\Gamma(3/2)}{\Gamma(2)} = \frac{\pi}{2}(1 - 2|v'_{Z,n}|) \, . \qquad (10.4.105)$$

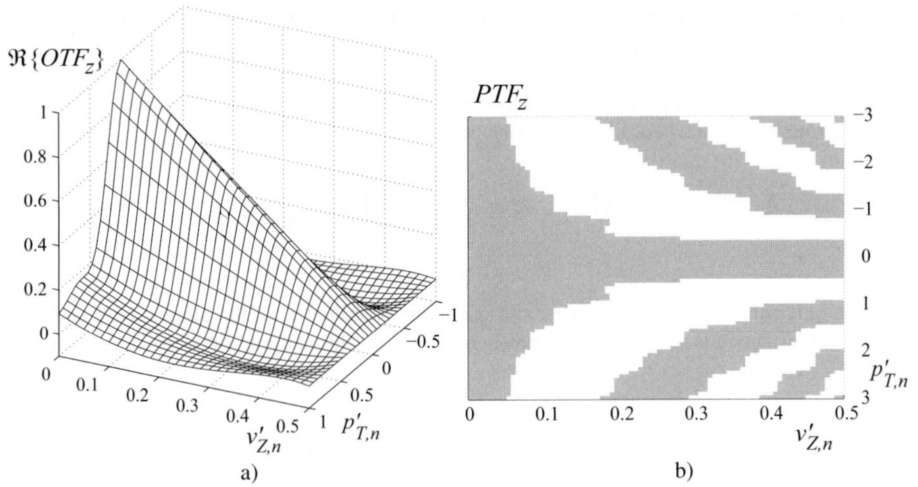

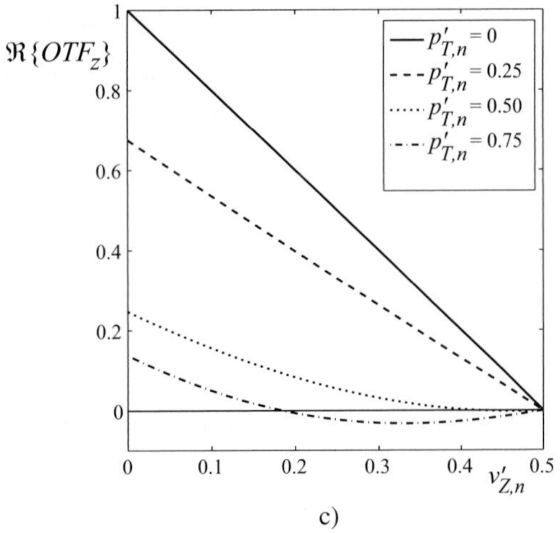

Figure 10.67: The normalised axial frequency transfer function $OTF_z\,(p'_{T,n}; v'_{Z,n})$ for incoherent illumination as a function of the lateral position $p'_{T,n}$ (perfect imaging system).
a) $\Re\,\{OTF_z\}$.
b) PTF_z having discrete values of 0 (grey) and π (white).
c) $\Re\left\{OTF_z\,(v'_{Z,n})\right\}$ at four lateral positions $p'_{T,n} = 0,\ 0.25, 0.50, 0.75$.

(see Eq. (10.4.104)) allows us to study the change in frequency transfer when a lateral offset from the nominal image position of the line object is introduced in image space.

In Fig. 10.67a and b, we have plotted the real part and the phase of the axial transfer function $OTF_z\,(p'_{T,n}; v'_{Z,n})$. The graphs show the degradation of the transfer function if the line image is observed in an off-axis position with normalised lateral coordinate $p'_{T,n}$. Figure 10.67b shows for which combinations of $v'_{Z,n}$ and $p'_{T,n}$ the first changes in sign occur when these quantities assume nonzero values. In c) we have plotted the value of $\Re\left\{OTF_z\,(p'_{T,n}; v'_{Z,n})\right\}$ for some specific off-axis distances $p'_{T,n}$.

It depends on the application how large the effective bandwidth of an optical system should be. Frequency transfer values of, for instance, 80% or 50% have been required for (high-quality) microscopic imaging. More relaxed requirements are possible if some nonlinear post-processing step is part of the image formation. In this context one can think of a photographic development process with nonlinear contrast transfer or of a digital post-processing including a modulation and encoding/decoding step before the final image is produced. In the latter case, transfer values as low as 10% may be acceptable, depending on the signal to noise ratio of the image.

10.4.13.2 One-dimensional Frequency Transfer for a Coherently Illuminated Object

To obtain the one-dimensional transfer function for a modulated line object which is coincident with the z-axis(transfer of 'axial frequencies' for coherent object illumination), we use the function $\tilde{I}'_c(\mathbf{v}'_n)$ of Eq. (10.4.87) and transform it back with respect to the real space x- and y-frequency components. We transform the integral variables from the real space coordinates to the normalised spatial coordinates and obtain

$$\tilde{I}'_c(p'_{X,n}, p'_{Y,n}; \nu'_{Z,n}) = -\, i r'_E \iint_{-\infty}^{\infty} U_0(-\rho_0 \nu'_{X,n}, -\rho_0 \nu'_{Y,n})\, \delta\left(\nu'_{Z,n} - \frac{1}{\sin^2\alpha'_m} + \frac{\nu'^{\,2}_{X,n} + \nu'^{\,2}_{Y,n}}{2}\right) \times \exp\{+2\pi i(\nu'_{X,n} p'_{X,n} + \nu'_{Y,n} p'_{Y,n})\}\, d\nu'_{X,n} d\nu'_{Y,n} \,. \tag{10.4.109}$$

We assume that the exit pupil has a circular shape and use polar coordinates on the pupil sphere and in image space (see Eq. (10.4.102)). We substitute the polar coordinates in Eq. (10.4.109) and introduce the coordinate transformation $\nu'^{\,2}_{T,n} = u$, yielding the expression

$$\tilde{I}'_c(p'_{T,n}, \gamma; \nu'_{Z,n}) = -\, i r'_E \int_0^{2\pi}\int_0^{\infty} U_0(\rho_0 \sqrt{u}, \theta + \pi)\, \delta\left(u + 2\nu'_{Z,n} - \frac{2}{\sin^2\alpha'_m}\right) \times \exp\{+2\pi i(\sqrt{u}\, p'_{T,n} \cos(\theta - \gamma)\}\, du d\theta \,. \tag{10.4.110}$$

In this equation for $\tilde{I}'_c(p'_{T,n}, \gamma; \nu'_{Z,n})$ we have written the pupil function U_0 with the aid of polar coordinates which are related to the Cartesian coordinates $(X'_{r,n}, Y'_{r,n})$ through the expression $X'_{r,n} + iY'_{r,n} = \rho \exp(i\theta)$. We first perform the integration with respect to u and use the Bessel expansion of Eq. (10.2.8) for the exponential function in the integrand,

$$\tilde{I}'_c(p'_{T,n}, \gamma; \nu'_{Z,n}) = -\, i r'_E \sum_{m=-\infty}^{\infty} i^m J_m\left(2\pi p'_{T,n} \sqrt{2(\sin^{-2}\alpha'_m - \nu'_{Z,n})}\right) \times \int_0^{2\pi} U_0(\rho_0 \sqrt{2(\sin^{-2}\alpha'_m - \nu'_{Z,n})}, \theta + \pi)\, \exp\{im(\theta - \gamma)\}\, d\theta \,. \tag{10.4.111}$$

For the case of an aberrated imaging system, without circularly symmetric amplitude and phase functions in the exit pupil, the evaluation of the integral is best performed with the aid of an expansion of the complex pupil function U_0 into Zernike polynomials with complex expansion coefficients given by

$$U_0(\rho, \theta) = \sum_{n=0}^{\infty} \sum_{m=-\infty}^{+\infty} \beta^m_n\, R^{|m|}_n(\rho)\, \exp(im\theta) \,, \tag{10.4.112}$$

where $(n - |m|)$ is even and non-negative. The pupil function U_0 of Eq. (10.4.111) may assume nonzero values in a circular region with radius ρ_0. For the normalised coordinate ρ of a radial Zernike polynomial in Eq. (10.4.112) we then have the value $\sqrt{2(\sin^{-2}\alpha'_m - \nu'_{Z,n})}$. Evaluation of the integral over the azimuthal coordinate θ yields the expression

$$\tilde{I}'_c(p'_{T,n}, \gamma; \nu'_{Z,n}) = -i2\pi r'_E \sum_{n=0}^{\infty} \sum_{m=-\infty}^{+\infty} \exp\left[-im\left(\gamma + \frac{\pi}{2}\right)\right] \beta^m_n R^{|m|}_n(\nu'_a)\, J_m(2\pi p'_{T,n} \nu'_a) \,, \tag{10.4.113}$$

where the value of ν'_a is given by $[2(\sin^{-2}\alpha'_m - \nu'_{Z,n})]^{1/2}$. For the case of an aberration-free imaging system with uniform transmission and a circular pupil rim, it is allowable to put $U_0(\rho_0 \nu', \theta + \pi) = \text{circ}(\nu')$, equal to the radial Zernike polynomial $R^0_0(\nu'_a)$. The integral over θ in Eq. (10.4.111) yields a nonzero value only if $m = 0$. After normalisation of β^0_0 to unity we obtain

$$\tilde{I}'_c(p'_{T,n}, \gamma; \nu'_{Z,n}) = -i2\pi r'_E\, J_0\left(2\pi p'_{T,n} \nu'_a\right)\, \text{circ}\,(\nu'_a) \,. \tag{10.4.114}$$

To have nonzero values of $\tilde{I}'_c(p'_{T,n}, \gamma; \nu'_{Z,n})$ in expressions (10.4.113) and (10.4.114) it is required that the argument of the radial Zernike polynomial $R^{|m|}_n(\rho)$ or of the circ-function is found in the interval $[0, 1]$ and this yields the condition

$$-\frac{1}{2} + \frac{1}{\sin^2\alpha'_m} \leq \nu'_{Z,n} \leq \frac{1}{\sin^2\alpha'_m} \,. \tag{10.4.115}$$

In the aberration-free case we then have the solution

$$\tilde{I}'_c(p'_{T,n};v'_{Z,n}) = -i2\pi r'_E\, J_0\left(2\pi p'_{T,n}\sqrt{2(\sin^{-2}\alpha'_m - v'_{Z,n})}\right) . \qquad (10.4.116)$$

If we omit the constant factor $-i2\pi r'_E$ we have unit frequency transfer for the two frequency intervals

$$\tilde{I}'_c(p'_{T,n};v'_{Z,n}) = 1\,, \text{ if } \begin{cases} a) \; -\frac{1}{2}+\sin^{-2}\alpha'_m \le v'_{Z,n} \le \sin^{-2}\alpha'_m\,, & p'_{T,n}=0\,, \\ b) \qquad v'_{Z,n} = \sin^{-2}\alpha'_m\,, & -\infty < p'_{T,n} < +\infty\,. \end{cases} \qquad (10.4.117)$$

The real part of the axial coherent transfer function CTF_z and the corresponding phase transfer function $CPTF_z$ have been plotted in Fig. 10.68a and b, for the range of spatial frequencies $v'_{Z,n}$ that are present in a focused beam with an aperture of $\sin\alpha'_m = \sqrt{2}/2$. The special cases a) and b) presented in Eq. (10.4.117) are clearly visible in graph a). The special case labelled a) in Eq. (10.4.117) can be understood by inspection of Eq. (10.4.107) for the three-dimensional amplitude point-spread function according to scalar diffraction. The Fourier transform with respect to $p'_{Z,n}$ of this amplitude point-spread function transforms the sinc-function in the spatial domain into a rectangular function in the frequency domain. The result within the scalar approximation for $p'_T = 0$ is thus given by

$$\tilde{I}'(0;v'_{Z,n}) = -\, i2\pi r'_E\, \mathrm{rect}\left[2(v'_{Z,n} - \sin^{-2}\alpha'_m + 1/4)\right] . \qquad (10.4.118)$$

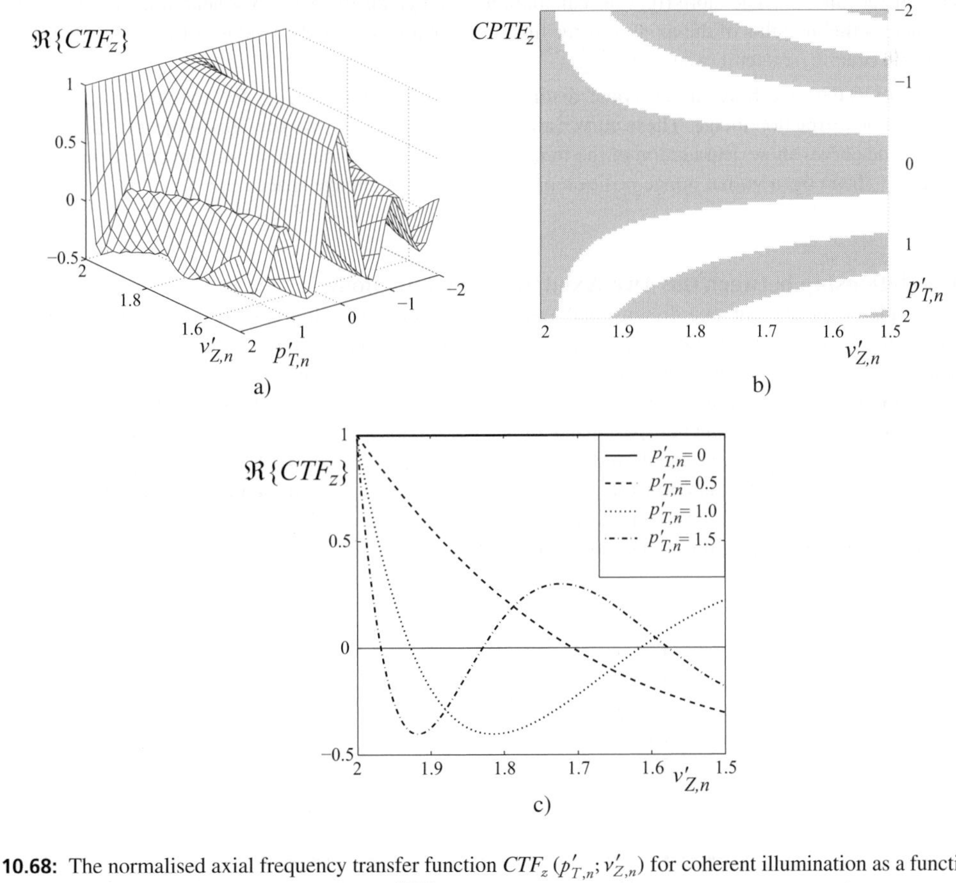

Figure 10.68: The normalised axial frequency transfer function $CTF_z\,(p'_{T,n};v'_{Z,n})$ for coherent illumination as a function of the lateral position $p'_{T,n}$ (perfect imaging system, $NA=\sqrt{1/2}$).

a) $\Re\left\{CTF_z\,(p'_{T,n},v'_{Z,n})\right\}$.

b) $CPTF_z$ with discrete values of 0 (grey) and π (white).

c) $\Re\left\{CTF_z\,(p'_{T,n},v'_{Z,n})\right\}$ at four lateral positions $p'_{T,n} = 0,\ 0.50, 1.00, 1.50$.

This result confirms the unit CTF_z value for $p'_{T,n} = 0$ in the $\nu'_{Z,n}$ interval from $-1/2 + \sin^{-2}\alpha'_m$ to $\sin^{-2}\alpha'_m$ (see Fig. 10.68a). In terms of the z-component of the wave vector of a propagating plane wave, all plane waves with $k\cos\alpha'_m \leq k_z \leq k$ are accepted by the lens pupil and transmitted with unit strength by the lens. In the scalar small-angle approximation, we replace $k\cos\alpha'_m$ by $k\{1 - (\sin^2\alpha'_m)/2\}$ and we arrive, after switching to normalised frequency and wave vector values, at the same result as on the first line of Eq. (10.4.117).

The special case b) of Eq. (10.4.117) refers to the real-space spatial frequency $\nu_Z = 1/\lambda$ and corresponds to a plane wave propagating along the z-axis. This case is best explained by invoking the scalar Lommel diffraction formula of Eq. (8.4.38) in which we equate the amplitude of the incident wave, denoted by U_0 in (8.4.38), to unity. We then calculate the Fourier transform with respect to the axial coordinate p'_Z and obtain in terms of the normalised space and frequency coordinates,

$$\tilde{I}'(p'_{T,n};\nu'_{Z,n}) = -\,i2\pi r'_E \,\exp\left\{\frac{i\pi\lambda p'^{\,2}_{T,n}}{r'_E\sin^2\alpha'_m}\right\} J_0\left\{2\pi p'_{T,n}\sqrt{2(\sin^{-2}\alpha'_m - \nu'_{Z,n})}\right\} . \qquad (10.4.119)$$

The expression of Eq. (10.4.119) is in accordance with Eq. (10.4.116), apart from the extra exponential factor depending on $p'^{\,2}_{T,n}$. The difference between Eqs. (10.4.116) and (10.4.119) follows from the discussion just below Eq. (10.4.85), where this exponential factor was omitted for physically relevant reasons in the calculation of the coherent three-dimensional transfer function. The more rigorous calculation which yields the extra exponential factor of Eq. (10.4.119) shows that, whereas the modulus of the axial transfer function remains unity for all values of $p'_{T,n}$, the $CPTF_z(p'_{T,n}, 0)$ is affected in second order by a lateral excursion $p'_{T,n}$.

Finally, in Fig. 10.68c, we have drawn some distinct curves of $\Re\left\{CTF_z(\nu'_{Z,n})\right\}$ for specific values of the lateral coordinate $p'_{T,n}$ in the diffraction image. These curves are relevant when a very 'narrow' line-object is imaged, coincident with the z-axis. The curves show degradation of the frequency transfer when the line object image is detected at a certain lateral distance $p'_{T,n}$ from the nominal image position of the line object in image space.

10.4.13.3 Relationship between the Two Axial Transfer Functions

The coherent and the incoherent one-dimensional transfer functions according to Eqs. (10.4.109) and (10.4.106), respectively, are the Fourier transform of a three-dimensional complex amplitude function and of the modulus squared value of that same function. It thus follows from Fourier theory that the incoherent transfer function equals the autocorrelation of the coherent transfer function. In Fig. 10.69 the two transfer functions have been depicted for a perfect imaging system. The frequency pass-band of the rectangular coherent transfer function CTF_z is given by $k\cos\alpha'_m \leq \omega_z \leq k$, or, in the paraxial small-angle approximation, by $k[1 - (\sin^2\alpha'_m)/2] \leq \omega_z \leq k$. The autocorrelation function of CTF_z is the triangular function OTF_z, centred at zero frequency and having a bandwidth of $\sin^2\alpha'_m$. This bandwidth is twice the bandwidth of the coherent transfer function, a result which was previously found for the transverse incoherent transfer function OTF_t and its coherent counterpart CTF_t.

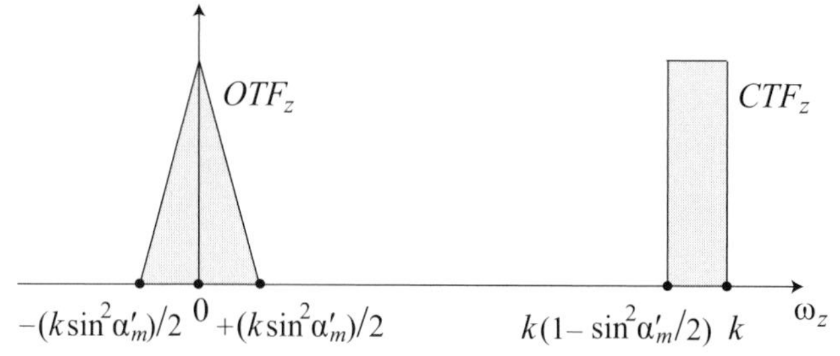

Figure 10.69: The one-dimensional axial transfer functions CTF_z and OTF_z as a function of the real-space axial frequency ω_z. Perfect imaging system, the maximum aperture angle of the marginal ray has been denoted by α'_m.

10.4.14 Imaging of Axially Modulated Periodic Objects

In this subsection we illustrate by means of a physical diffraction picture the mathematical result of the previous subsection about the axial transfer function associated with an axially modulated and *incoherently* illuminated object. With respect to the standard frequency analysis of two-dimensional periodic objects, confined to a plane perpendicular to the optical axis of the imaging system, the three-dimensional frequency transfer studied in this chapter opens the possibility to analyse the 'in-depth' imaging of a three-dimensional object. Living cell imaging in medicine and biology provides us with a typical example of non-planar objects where the features in each dimension have equal importance. For the imaging of objects with axial modulation, we make a distinction between axial line objects (single scattering) and thick objects in which the diffraction of light is governed by the Bragg condition.

10.4.14.1 Diffracted Light Associated with an Axially Modulated Line Object

Light diffraction by a line object with general orientation in (object) space was treated in Subsection 10.4.1 of this chapter. Before using Eq. (10.4.17), which yields the wave vector components of the light diffracted by a line object, we illustrate the diffraction of light by a line object with the aid of Fig. 10.70. We calculate the phase differences which ariseswhen a plane wave is diffracted by the diffracting objects on a line parallel to the x-axis (see a)), or parallel to the z-axis (see b)). The first case has been included as its geometry is comparable to that of a line grating structure with the grating lines perpendicular to the plane of the drawing. In the first case, the phase increment $\Delta\phi$ per period p along the incident and diffracted ray is given by

$$\Delta\phi = \phi_P - \phi_A + (\phi_B - \phi_P) = (\phi_P - \phi_O)_{\mathbf{k}_0} - (\phi_P - \phi_O)_{\mathbf{k}_1} = p_t\,\hat{\mathbf{u}}_t \cdot (\mathbf{k}_0 - \mathbf{k}_1)\,, \qquad (10.4.120)$$

where the transverse grating vector $\mathbf{u}_t$ is directed along the x-axis, parallel to the vector OP ($|\mathbf{u}| = 1/p_t$). The subscripts of k_0 and k_1 in the equation refer to the wave propagation directions which have to be used for the calculation of the phase difference between the points P and O. In b) the phase increment along the paths between two adjacent diffracting points O and P at an axial distance p of the periodic line object is given by a similar expression,

$$\Delta\phi = \phi_P - \phi_A - (\phi_B - \phi_O) = (\phi_P - \phi_O)_{\mathbf{k}_0} - (\phi_P - \phi_O)_{\mathbf{k}_1} = p_z\,\hat{\mathbf{u}}_z \cdot (\mathbf{k}_0 - \mathbf{k}_1)\,, \qquad (10.4.121)$$

where the grating vector $\mathbf{u}_z$ is parallel to the z-axis. The appearance of a diffraction order with order number m requires constructive interference of the diffracted waves originating at each period such that $\Delta\phi = m2\pi$.

For the imaging of the periodic structure in object space it is necessary that the inclination angles of the diffracted waves with the z-axis are such that the propagating plane waves associated with the diffraction orders are not obstructed by the physical stop of the imaging system. The diffraction angles α_0 and α_m of two transmitted waves are given by

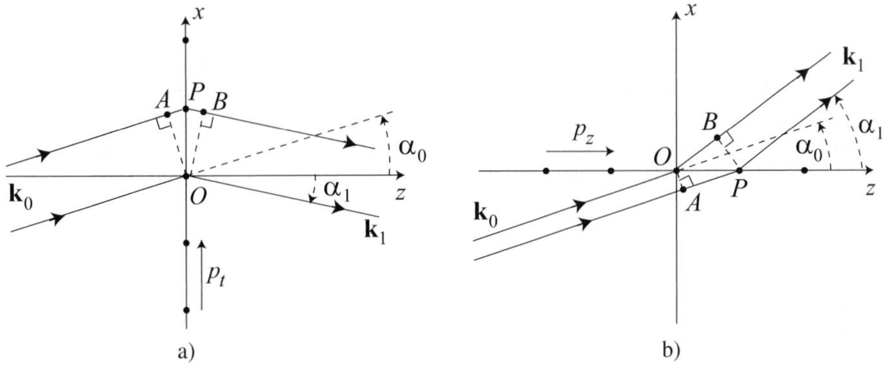

Figure 10.70: The diffraction of an incident plane wave with wave vector $\mathbf{k}_0$ by a line object (planar cross-section $y = 0$).
a) The wave vector $\mathbf{k}_1$ is located in the cross-section $y = 0$ of light which has been diffracted by a periodic line object with period $p_t = OP$ along the x-axis.
b) The wave vector $\mathbf{k}_1$ is located in the cross-section $y = 0$ of light which is diffracted by a periodic line object along the z-axis with identical period $p_z = p_t = OP$.

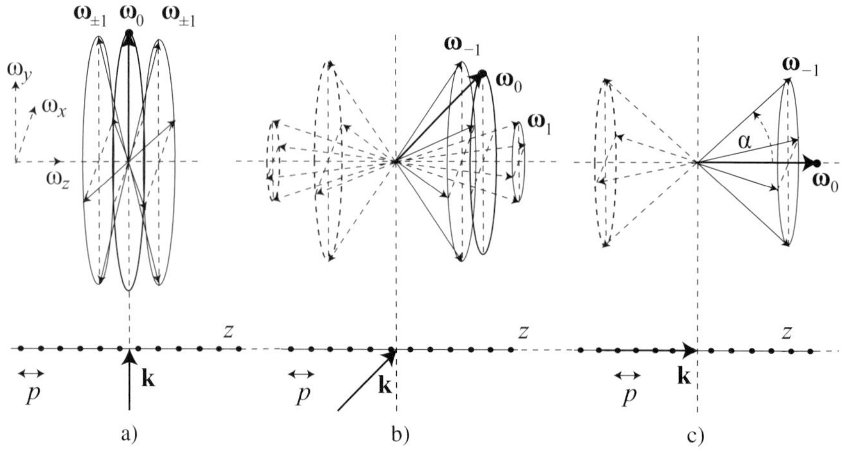

Figure 10.71: Diffraction of light by a periodic line object oriented along the z-axis. The incident wave vector is $\mathbf{k}$, the grating vector $\boldsymbol{\omega}_p = (0, 0, k/4)$. The 'global' zeroth-order wave vector is $\boldsymbol{\omega}_0 = \mathbf{k}$ and has been provided with a bullet-point in the graph. The zeroth-order light associated with the line object is represented by the assembly of wave vectors whose endpoints lie on the cone which also contains the global zeroth-order vector $\boldsymbol{\omega}_0$. The diffracted first-order wave vectors $\boldsymbol{\omega}_{\pm 1}$ are found in the far-field on two cones with a circular cross-section denoted by a solid curve (forward scattering) and a dashed curve (backward scattering).
a) $\mathbf{k}$ is perpendicular to the z-axis.
b) $\mathbf{k}$ is at an angle of 45° to the z-axis.
c) $\mathbf{k} = (0, 0, k)$ is parallel to the z-axis.

$$\sin\alpha_0 - \sin\alpha_m = m\frac{\lambda}{p_t} \,, \qquad \text{(Fig. 10.70a)}$$

$$\cos\alpha_0 - \cos\alpha_m = m\frac{\lambda}{p_z} \,, \qquad \text{(Fig. 10.70b)} \qquad (10.4.122)$$

where λ is the wavelength of the light in the medium surrounding the periodic line object and m the order of the diffracted wave. We remark that in the presence of a line object the diffracted light amplitude shows circular symmetry with respect to the line object. Instead of the discrete diffraction orders associated with the diffraction of a plane wave by a planar object, the line object produces (hollow) cones of diffracted light associated with diffracted light of order m. The equations in (10.4.122) are related to two simple geometries of light diffraction taken from the general three-dimensional geometry of an incident plane wave and line object orientation which was algebraically analysed in Subsection 10.4.1.

Diffraction by a line object is now illustrated by means of Fig. 10.71 and the diffracted light directions are calculated using Eqs. (10.4.14)–(10.4.17). The periodic line object is coincident with the z-axis and the period equals 4λ for the three cases presented in Fig. 10.71 ($\boldsymbol{\omega}_0 = k/4$). Using this geometry Eq. (10.4.16) is greatly simplified since the line object is parallel to one of the Cartesian coordinate axes. The transverse frequency components (ω_x, ω_y) of the diffracted light are given by the expression

$$\omega_x^2 + \omega_y^2 + (k_z \pm \omega_{z,p})^2 = k^2 \,, \qquad (10.4.123)$$

where the plus or minus sign determines if the plus first or minus first diffracted light is considered. The following three geometries are shown:

- $\mathbf{k} = (0, 1, 0)\, k$
  In Fig. 10.71a we have shown the (hollow) cones of diffracted light which are generated by the periodic line object when the incident wave vector $\mathbf{k} = (0, k, 0)$ of the (infinitely) extending plane wave is perpendicular to the line object along the z-axis. Two first-order cones of diffracted light are created, identical for the plus and minus first diffracted light. The zeroth-order cone of diffracted light associated with the line object obeys the equation $\omega_x^2 + \omega_y^2 = k^2$. The complex amplitude of the frequency components on the conical surface depends on the nature of the diffracting

elements which form the periodic line object. The conical surfaces only indicate where in frequency space a nonzero amplitude of diffracted light can be expected.

- $\mathbf{k} = (0, 1, 1)\, k/\sqrt{2}$
  In Fig. 10.71b the incident plane wave is at an angle of 45° degrees to the line object. Distinct cones of diffracted light associated with the minus first, the zeroth and the plus first-order diffracted light are seen. We note that the potential diffracted light directions show circular symmetry with respect to the line object despite the specific direction of the incident wave vector $\mathbf{k} = k(0, 1, 1)/\sqrt{2}$ which is confined to the plane $x = 0$. The incident wave vector (thick arrow with bullet) determines the phase gradient of the incident wave along the line object and this phase gradient is proportional to the scalar product $(\mathbf{k} \cdot \boldsymbol{\omega}_p)$ of the incident wave vector and the frequency vector of the periodic line object.
- $\mathbf{k} = (0, 0, 1)\, k$
  Figure 10.71c shows the special case for which the incident wave vector and the frequency vector are parallel. The plus first-order light corresponds to waves which are evanescent in the lateral direction,

$$\omega_T = \sqrt{\omega_x^2 + \omega_y^2} = i3k/4\,, \qquad \omega_z = 5k/4\,. \tag{10.4.124}$$

The minus first order corresponds to a conical spectrum of freely propagating waves of which the lateral frequency component is given by $k\sqrt{7}/4$. The cone opening angle 2α has the value $2 \times 41.4°$. For this particular case the zeroth-order cone of diffracted light has been reduced to a single vector with the direction of the incident wave vector $\mathbf{k}$. For Fig. 10.71b and c we have also shown the solutions which correspond to backward diffracted waves. In practice, the backward diffracted waves tend to have a (much) smaller amplitude if weakly scattering line objects are considered.

10.4.14.2 Imaging of a Coherently Illuminated Axial Line Object

For the case of coherent illumination of the line object with a plane wave with wave vector $\mathbf{k} = (k_x, k_y, k_z)$, the diffracted light directions should be contained within the coherent passband for ω_z in order to be able to produce an image of the line object. At least the zeroth order and one of the first-order diffracted fields should be transferred to image space. Figure 10.69 shows the coherent passband and, consequently, the diffracted fields with a z-component of the wave vector given by $\omega_z = k_z + m\omega_{z,p}$ should satisfy the condition

$$k\left(1 - \frac{\sin^2\alpha_m}{2}\right) \le k_z + m\omega_{z,p} \le k\,. \tag{10.4.125}$$

The small-angle approximation for k_z reads $k[1 - (\sin^2\alpha)/2]$ and we find for the range of transferable axial frequencies $\omega_{z,p}$,

$$\frac{k}{2}\left(\sin^2\alpha - \sin^2\alpha_m\right) \le m\omega_{z,p} \le \frac{k}{2}\sin^2\alpha\,. \tag{10.4.126}$$

The frequency ranges that are transferred by means of the plus first- and the minus first-order light are given by

$$\begin{aligned} m &= +1\,, & (\sin^2\alpha - \sin^2\alpha_m)/2 &\le \omega_{z,p}/k \le (\sin^2\alpha)/2\,,\\ m &= -1\,, & -(\sin^2\alpha)/2 &\le \omega_{z,p}/k \le (-\sin^2\alpha + \sin^2\alpha_m)/2\,, \end{aligned} \tag{10.4.127}$$

where it is also required that $\sin^2\alpha \le \sin^2\alpha_m$ for the zeroth-order light to reach the image space. The standard imaging method uses central illumination ($\alpha = 0$ or $k_z = k$) and the frequency intervals then equal

$$\begin{aligned} m &= +1\,, & -(\sin^2\alpha_m)/2 &\le \omega_{z,p}/k \le 0\,,\\ m &= -1\,, & 0 &\le \omega_{z,p}/k \le (\sin^2\alpha_m)/2\,. \end{aligned} \tag{10.4.128}$$

We observe that for the case of central illumination, according to our sign convention, the plus first-order diffracted light allows 'single-sideband' imaging with the aid of the negative axial frequencies and the minus first-order light enables imaging with the aid of the positive frequencies. For maximum off-axis illumination with $\alpha = \alpha_m$, the zeroth-order light annulus touches the rim of the lens aperture and we find the following complementary frequency ranges for $\omega_{z,p}$, as compared to the central illumination case,

$$\begin{aligned} m &= +1\,, & 0 &\le \omega_{z,p}/k \le (\sin^2\alpha_m)/2\,,\\ m &= -1\,, & -(\sin^2\alpha_m)/2 &\le \omega_{z,p}/k \le 0\,. \end{aligned} \tag{10.4.129}$$

We conclude that for any illumination angle $\alpha \le \alpha_m$, the bandwidth per order equals $k\sin^2\alpha_m/2$. The total bandwidth thus amounts to $k\sin^2\alpha_m$ when both plus and minus first-order light can be used for imaging.

The conclusions about axial frequency transfer for a coherently illuminated line object can also be reached by using Eq. (10.4.123) which determines the squared value of the lateral frequency $\omega_T = (\omega_x^2 + \omega_y^2)^{1/2}$ associated with diffracted light of order m. We require again that at least (parts of) the zeroth-order and first-order diffracted light reach the image space. The zeroth-order light is transmitted if the condition

$$\omega_x^2 + \omega_y^2 = k_x^2 + k_y^2 \le k^2 \sin^2\alpha_m = k^2 - k_{z,a}^2 \qquad (10.4.130)$$

is satisfied, where $k_{z,a}$ is the value of the z-component of the wave vector of a freely propagating wave which is just accepted by the imaging system. In a similar way we find for first-order diffracted light the 'acceptance' condition

$$\omega_x^2 + \omega_y^2 = k^2 - (k_z + m\omega_{z,p})^2 \le k^2 \sin^2\alpha_m \,. \qquad (10.4.131)$$

With the aid of Eqs. (10.4.130) and (10.4.131) the presence of the zeroth-order and first-order diffracted light is determined by the inequalities related to the propagation angles in object space,

$$\begin{cases} 0 \le \sin^2\alpha \le \sin^2\alpha_m \,, \\ 0 \le \sin^2\alpha - m^2\omega_{z,p}^2/k^2 - 2m\cos\alpha\,(\omega_{z,p}/k) \le \sin^2\alpha_m \,. \end{cases} \qquad (10.4.132)$$

The function $f(\omega_{z,p}/k) = \sin^2\alpha - m^2(\omega_{z,p}/k)^2 - 2m\cos\alpha(\omega_{z,p}/k)$ has been plotted in Fig. 10.72 for two values of the angle α of the incident plane wave. It follows from Eq. (10.4.132) that the same passbands for $\omega_{z,p}/k$ are found as those defined by Eqs. (10.4.128) and (10.4.129). Their widths equal $(1 - \cos\alpha_m)$, the finite value corresponding to the previously found small-angle value $\sin^2\alpha_m/2$. The figure shows the frequency intervals for the object frequency $\omega_{z,p}/k$ which can be imaged by the optical system. For the minus first-order light they are determined by the abscissas of the points A–B and the points C–D, respectively, for the case that the incident beam is parallel to the z-axis ($\alpha = 0$). Similarly, the abscissae of the points A'–B' and C'–D' yield the two frequency intervals of $\omega_{z,p}/k$ for the case that the wave vector of the incident plane wave is parallel to the direction of a marginal ray ($\alpha = \alpha_m$) of the imaging system. The frequency intervals determined by the points C–D and C'–D' in the neighbourhood of the normalised frequency value of 2 have no direct practical value. Such spatial frequency components $\omega_{z,p}$ create backward diffracted waves. A standard imaging system is not capable of capturing these waves, unless the system is meant to be operated in reflection. From the local axial magnification m_z in image space we find the image-side spatial frequency band for $\omega'_{z,p}$ to be $\omega_{z,p}/m_z$.

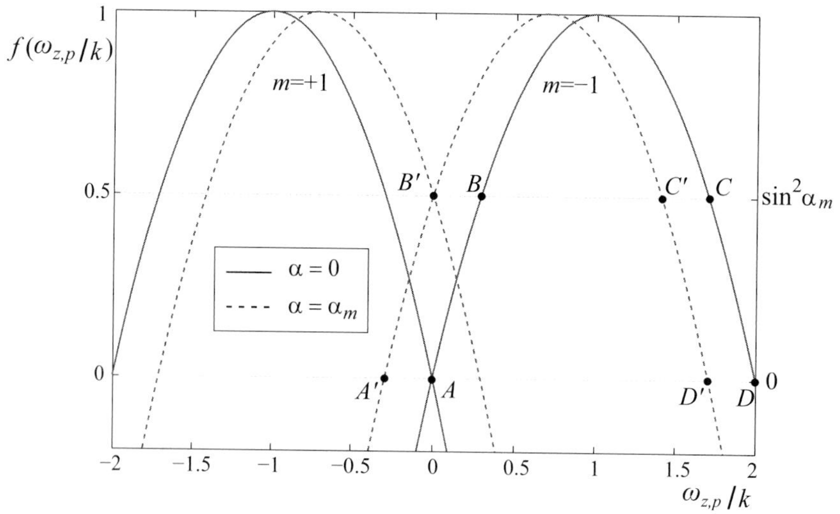

Figure 10.72: The function $f(\omega_{z,p}/k)$ for diffracted light of orders $m = \pm 1$ and its 'acceptance' bounds 0 and $\sin^2\alpha_m$. Parameter α is the angle in object space between the incident wave vector $\mathbf{k}$ and the z-axis. The aperture angle α_m at the object side has been given the value of $\pi/4$ in the figure.

10.4.14.3 Imaging of an Incoherently Illuminated Axial Line Object

The image of an incoherently illuminated axial line object follows from the superposition of the intensities of the images created by each coherent wave which emanates from the extended source which replaces the single coherent point source used in the previous subsection. Incoherent illumination of an object is commonly achieved by using an incoherent source of which the image fills at least the aperture of the imaging system. However, since we have a line object, incoherent illumination can also be achieved by using a line-shaped source of which the image in the entrance pupil of the imaging system intersects the object axis (the z-axis in our special geometry). The incoherent axial transfer function depends on the source shape and we briefly discuss here the calculation of the transfer function for a uniform disc-shaped source and for a uniform line source. We consider the minus first-order diffracted light and the coherent rectangular frequency transfer of the axial frequency $\omega_{z,p}/k$ on the interval as given by the second line of Eq. (10.4.127). For ease of notation we multiply this inequality by $\lambda/\sin^2\alpha_m$ and find, in terms of the normalised axial frequency component $\nu_{Z,n,p}$ in object space, the inequality

$$-\frac{\rho^2}{2} \le \nu_{Z,n,p} \le \frac{1}{2}(1-\rho^2)\,, \qquad (m=-1)\,, \qquad (10.4.133)$$

where ρ is the ratio $\sin\alpha/\sin\alpha_m$ $(0 \le \rho \le 1)$.

For the case of a uniform source image that fills the entire aperture, we find the incoherent transfer function by integration of the rectangular coherent transfer function associated with each source point over the unit circle. The normalised integral over the disc-shaped source is given by

$$OTF_{z,d} = \frac{1}{\pi}\int_0^{2\pi}\int_0^1 \mathrm{rect}\left[\frac{\nu_{Z,n,p} - 1/4 + \rho^2/2}{1/2}\right]\rho\, d\rho\, d\phi\,. \qquad (10.4.134)$$

We make the substitution $y = \rho^2 + 2\nu_{Z,n,p} - 1/2$ and, for positive values of $\nu_{Z,n,p}$, find the result

$$OTF_{z,d} = \int_{2\nu_{Z,n,p}-1/2}^{1/2} \mathrm{rect}\,(y)\, dy = 1 - 2\nu_{Z,n,p}\,. \qquad (10.4.135)$$

For negative values of $\nu_{Z,n,p}$ the integration limits are $-1/2$ and $1/2 + \nu_{Z,n,p}$ and the value of $OTF_{z,d}$ is $1 + 2\nu_{Z,n,p}$. Combining both results we write $OTF_{z,d} = 1 - 2|\nu_{Z,n,p}|$, where $|\nu_{Z,n,p}| \le 1/2$. This expression has previously been given in Eq. (10.4.106) after performing the integration given in Eq. (10.4.105) of footnote 2.

The derivation of the normalised transfer function $OTF_{z,l}$ for a line source which covers a full diameter (normalised length is 2) in the system aperture is similar. The integral expression for a line source image in the plane $y = 0$ over the interval $[0 \le x \le 1]$ is given by

$$OTF_{z,l} = \int_0^1 \mathrm{rect}\left[\frac{\nu_{Z,n,p} - 1/4 + x^2/2}{1/2}\right] dx\,. \qquad (10.4.136)$$

The integral over the negative x-values yields an identical expression. The substitution of $y = x^2 + 2\nu_{Z,n,p} - 1/2$ yields for positive values of $\nu_{Z,n,p}$ the integral expression

$$OTF_{z,l} = \frac{1}{2}\int_{2\nu_{Z,n,p}-1/2}^{1/2} \frac{\mathrm{rect}\,(y)}{\sqrt{y - 2\nu_{Z,n,p} + 1/2}}\, dy = \sqrt{1 - 2\nu_{Z,n,p}}\,. \qquad (10.4.137)$$

Including the OTF expression for negative values of $\nu_{Z,n,p}$ we find for the line source illumination case,

$$OTF_{z,l} = \sqrt{1 - 2|\nu_{Z,n,p}|}\,, \qquad -1/2 \le \nu_{Z,n,p} \le +1/2\,. \qquad (10.4.138)$$

The axial transfer functions $OTF_{z,d}$ and $OTF_{z,l}$ are given as a function of the object-side normalised spatial frequency $\nu_{Z,n,p} = 1/p_{Z,n,p}$. The frequency $\nu'_{Z,n,p}$ of the image of the line object depends on the mapping of the wave vectors of the diffracted orders from object to image space. We consider the zeroth and plus first order of a line object along the z-axis in object space with real-space period p_z. The incident plane wave is parallel to the line object. With reference to Fig. 10.70 we write for the change in phase difference $\Delta\phi_{0,1}$ per period between a zeroth and a first diffracted order

$$\Delta\phi_{0,1} = kp_z(1-\cos\alpha)\,, \qquad (10.4.139)$$

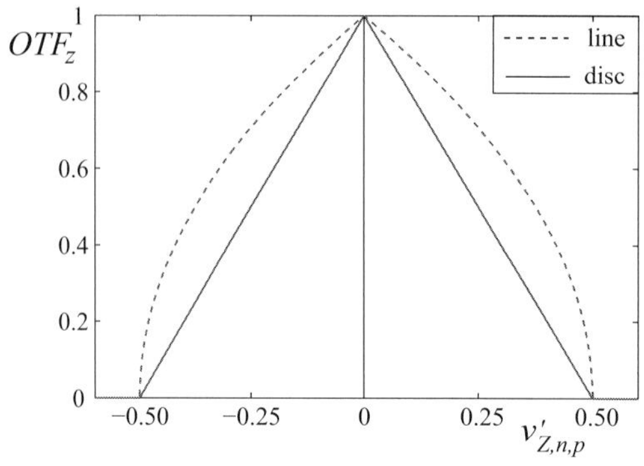

Figure 10.73: The incoherent axial transfer function OTF_z as a function of the image-side normalised spatial frequency $\nu'_{Z,n,p}$ of an axial line object. Solid curve: disc-shaped source, dashed curve: line source.

where α is the diffraction angle of the first-order light in object space associated with the period p_z. In image space the zeroth and first-order interfere and produce an axial period pattern of which the real-space period p'_z is given by

$$p'_z = \frac{\Delta\phi_{0,1}}{k(1-\cos\alpha')} = p_z\left(\frac{1-\cos\alpha}{1-\cos\alpha'}\right) , \tag{10.4.140}$$

where α' is the angle of the diffracted light with the z-axis which is determined by the design of the optical system. Undistorted imaging of a general line object is achieved if the ratio p'_z/p_z has the same value irrespective of the diffraction angle α in object space. Its value for very small diffraction angles is given by paraxial optics and equals m_z, the axial magnification. Using Eq. (10.4.140) we thus impose the condition

$$m_z = \frac{p'_z}{p_z} = \frac{1-\cos\alpha}{1-\cos\alpha'} = \frac{\sin^2(\alpha/2)}{\sin^2(\alpha'/2)} . \tag{10.4.141}$$

Using the relation $m_z = 1/m_A^2$ between the paraxial values of the axial and angular magnification for the case of equal indices in object and image space, we note that, not surprisingly, Eq. (10.4.141) is identical with the ray mapping condition between object and image space as first given by Herschel (see Eq. (4.8.20)). If Eq. (10.4.141) is satisfied by the imaging system it is permissible to write in terms of the image-side spatial frequency,

$$OTF_{z,l}(\nu'_{Z,p,n}) = \sqrt{1-2|\nu'_{Z,n,p}|} , \qquad -1/2 \leq \nu'_{Z,n,p} \leq +1/2 . \tag{10.4.142}$$

Figure 10.73 shows the incoherent transfer function as a function of the normalised frequency $\nu'_{Z,n,p}$ for a system of which the design satisfies the Herschel condition. The parameter is the source shape which is disc-shaped for the solid curve and line-shaped for the dashed curve in the figure.

10.4.14.4 Axial Modulation, Thick Object

For the imaging of a thick object with frequency vector $(0, 0, \omega_{z,p})$ we have to apply the Bragg condition of Eq. (10.4.22) or (10.4.23). An extra restriction on the direction of incident and diffracted beams is imposed. Only two pairs of plane waves can simultaneously propagate through the thick medium, whilst in a thin object an infinite number of incident plane waves is allowed, each accompanied by their diffracted satellite waves. The severe restriction on wave propagation direction in a Bragg medium has drastic consequences for the frequency transfer by such a medium if it is the object in an imaging system. Two extreme cases of Bragg reflection have been visualised in Fig. 10.74a and b. In a) the incident and diffracted wave vectors that satisfy the Bragg condition are almost perpendicular to the z-axis. Both wave vectors are at a small angle θ with the plane through O perpendicular to the z-axis. The angle θ follows from the Bragg condition $\sin\theta = \omega_{z,p}/2k$. In a standard imaging system, the illuminating beam should be at a very large angle $(\pi/2-\theta)$ the z-axis. The corresponding diffraction order is found at the rim of the backward-oriented half space. A standard imaging system, even with a limiting aperture of unity, is not able to bring the two diffracted beams to interference and, consequently, no imaging of the frequency $\omega_{z,p}$ is possible. The solution of this problem is to rotate the axis of the imaging system over an angle of 90° and apply the transfer of a transverse spatial frequency component to obtain the intensity modulation in the (rotated) image plane.

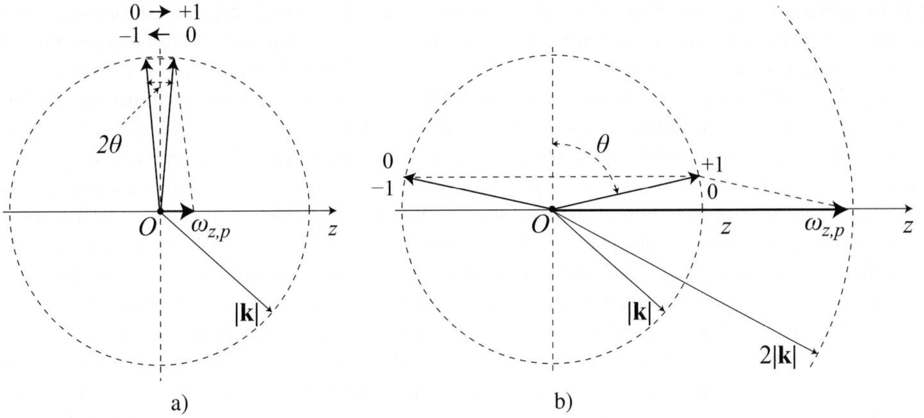

Figure 10.74: Bragg reflection at a periodic structure with frequency vector $(0, 0, \omega_{z,p})$.
a) Coarse structure with $\omega_{z,p} \ll k$.
b) Fine repetitive pattern with $\omega_{z,p}$ close to the value $2k$.

In the geometry of Fig. 10.74b the first and zero-order pairs travel in almost opposite directions and cannot be brought to interference in the image plane. In practice, it would be possible to inject an incident beam in the backward direction (labelled 0 in Fig. 10.74b) and use a partial direct reflection of this beam to interfere with the first-order beam. The reflecting surface should then be perpendicular to the z-axis. The two beams are collinear in image space and the position of the z-periodic structure could be deduced from the phase difference between these two parallel beams. But this phase detection method offers only an indirect means to obtain information about the z-grating position. Direct imaging of the structure can only be achieved by rotation of the imaging system which should have a large numerical aperture to capture both Bragg beams and produce an intensity modulated image.

The discussion above concerned coherent imaging. Incoherent imaging cannot be achieved with an object of which the diffracted beams obey the Bragg condition. Any incident beam directions that do not satisfy the Bragg condition are reflected back in object space by the thick structure. The result is that the illuminating cone of light, emitted by an extended incoherent light source, is largely reflected and only two diffracted beams are transmitted by the object. Such an illumination gives rise to a coherent transfer of spatial frequencies.

10.4.15 Conclusion

The three-dimensional frequency analysis of images formed by an optical system is a useful extension of the classical two-dimensional frequency analysis of planar images as produced on photographic plates or image sensors. The characterisation of objects immersed in media like air, transparent liquids or low scattering medical and biological substances profits from the *scalar* three-dimensional frequency analysis which is presented in this section. The three-dimensional frequency analysis is limited to the two extreme cases of fully coherent or fully incoherent illumination of the object. Further extensions of the theory, not explicitly discussed in this section, are partially coherent object illumination and the influence of imaging aberrations **[308]**,**[58]** and high-aperture imaging **[277]**,**[307]**. Special shapes of the limiting aperture, for instance an annular shape, have been discussed in **[120]**,**[159]**. Image formation in high-aperture microscopy requires a vectorial treatment of the light propagation, including the state of polarisation of the source. The corresponding three-dimensional vector transfer function is treated in **[309]**,**[299]**,**[11]**.

10.5 Light Scattering and Frequency Transfer

The propagation of an electromagnetic wave through a medium or its reflection or refraction at a boundary is generally described under idealised conditions. It is assumed that the medium is homogeneous and that a boundary region between two subsequent media is infinitely thin and perfectly smooth. At short wavelengths, it is the atomic structure of the medium itself which causes scattering of the light as any interface between two media is corrugated at the atomic scale.

For instance, the scattering of X-radiation at an interface or in a medium yields precious information on the atomic structure. At larger wavelengths, for instance in the visible domain, the scattering of radiation in a medium and the light scattering when a wave encounters an interface between two media are strongly reduced. The scattering of light can be produced by partial absorption and by scattering at individual atoms or sub-micron clusters of material. It can also be the result of density variations in a medium (inhomogeneity) or residual roughness of an interface. Such a situation occurs when an optical surface has been ground with a coarse grinding powder or when it has been imperfectly polished.

If a medium shows density variations, the initially flat wavefront of a plane wave is perturbed during propagation and for an observer the rectilinear propagation of the light wave is lost. For example, in the case of stellar observation, the stochastic density fluctuations in the atmosphere lead to a constantly varying local slope of the wavefront from a star at the location of the eye pupil. The point image of the star on the retina is shifted continuously over the detection cone(s) of the retina, giving rise to the 'twinkling' effect of the star (atmospheric 'seeing'). When light strikes an interface that is not perfectly (locally) flat at the sub-micron scale, we observe light scattering at refraction or reflection by such an interface. The surface deviations in the direction perpendicular to the plane of the interface are usually described by means of some stochastic model. In the plane of the surface one defines a correlation distance τ over which the surface deviation remains correlated, for instance above an expectation level of $1/e$. With the aid of such a stochastic model, both the light scattering from a rough surface and the light propagation through a weak random medium such as the Earth's atmosphere can be adequately described. For the case of random density fluctuations in the atmosphere, their integrated effect over the full height of the atmosphere can be thought to be concentrated in one or several phase screens, located at appropriate heights in the atmosphere. An important part of the atmospheric 'seeing' effects can be accounted for by the introduction of a single phase screen at ground level. This screen is able to account for the seeing effects that are common to all viewing directions of the astronomical telescope.

In this section we discuss a scalar, thin-layer theory of light scattering at a rough surface, either in transmission or in reflection **[112]**. The application of such a theory to light scattering from a rough surface limits its validity to lateral correlation lengths τ that are substantially larger than the wavelength of the light. The vertical surface deviations should be such that the surface slope remains small, for instance, below 0.1 radian. Consequently, the angular range in which light is scattered away from the specular direction remains equally small. A scalar scattering theory of this type can also be applied to observation through the fluctuating atmosphere. The statistical parameters of the density variations follow from measurements of light propagation through the atmosphere and these parameters are then used to define the equivalent phase screen(s) through which stellar radiation reaches the aperture of a telescope. The measurement of the wavefront perturbation is preferably based on a wavefront expansion using orthogonal functions over the telescope aperture. In this respect we mention the Zernike polynomials with an adjustment to accommodate for the presence of the central obstruction in a telescope (annular Zernike polynomials, see **[230]**), the disc-harmonic functions (**[355]**, see also Appendix D) and other ad hoc wavefront representations. In the latter case one could think of a set of neighbouring wavefront sub-areas on which the local wavefront perturbation is represented by means of the coefficients of a cubic spline associated with that sub-area. If the perturbation can be measured in real-time and the phase screen can be actively modulated to compensate for the atmospheric perturbation (adaptive optics), it is in principle possible to observe celestial objects without aberration.

10.5.1 Specular and Diffuse Reflection or Transmission at a Rough Surface

In Fig. 10.75a we show part of a rough interface which separates two media with refractive indices n_1 and n_2. The average height level of the rough surface coincides with the plane $z = 0$. The stochastic height variations of the interface are measured along the z-axis which is parallel to the unit normal vector $\hat{\mathbf{n}}$ of the perfectly smooth surface. We assume that the height variations have a normal probability distribution given by the function

$$p(z) = \frac{1}{\sigma\sqrt{2\pi}} \exp\left(-\frac{z^2}{2\sigma^2}\right) , \tag{10.5.1}$$

where σ is the standard deviation of the height variations of the surface. A height deviation of the surface with respect to the reference level is modelled by the insertion of a locally flat thin layer. The surface is shifted in the z-direction to the dashed height level over a distance represented by the vector $\boldsymbol{\delta}$. Such a local deformation approach is permitted if the resulting surface slope remains small, which is certainly the case if the maximum spatial frequency present in the surface deformation $z(x, y)$ has a low value, or, equivalently, if the correlation length of the surface deformation is large. The local path length change Δ of an incident plane wave follows from Fig. 10.75b. We take the origin O as the reference point and

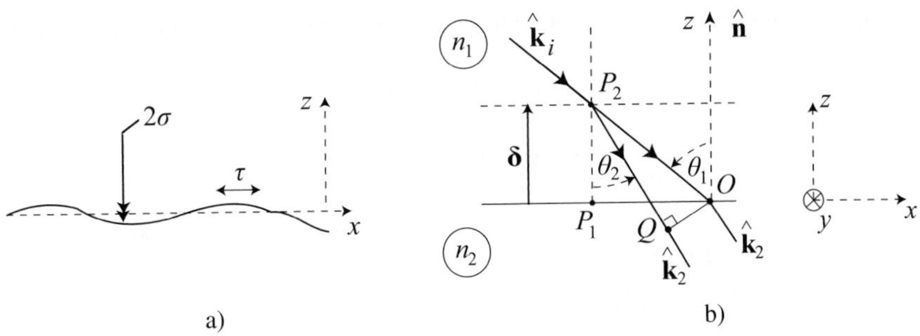

Figure 10.75: a) The height variation z (solid curve) in a cross-section with $y = 0$, with respect to the average level ($z = 0$). τ is the lateral correlation length of the rough surface.
b) Optical pathlength difference Δ_t at transmission through an interface with random height variation given by a function $z(x, y)$.

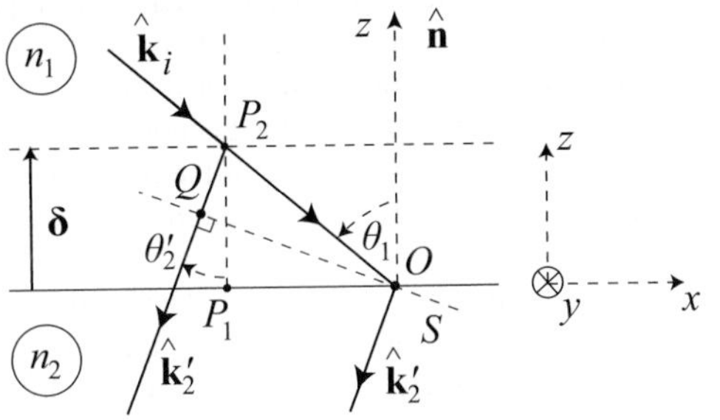

Figure 10.76: The pathlength difference for a general direction of the scattered wave (wave vector $\mathbf{k}'_2$) in the presence of a surface displacement $\boldsymbol{\delta}$. The wave vector of the incident wave is $\mathbf{k}_i$.

define Δ as the distance $[OP_2Q]$ where the brackets denote as usual that the optical distance along the path OP_2Q must be taken. The pathlength changes for the transmissive and the reflective case are given by

$$\Delta_t(x, y) = \boldsymbol{\delta}(x, y) \cdot [n_1\hat{\mathbf{k}}_i - n_2\hat{\mathbf{k}}_2] = z(x, y)\,[n_1 N_i - n_2 N_2]\,, \qquad \text{(transmission)}, \tag{10.5.2}$$

$$\Delta_r(x, y) = 2\, z(x, y)\, n_1 N_i\,, \qquad \text{(reflection)}, \tag{10.5.3}$$

where $\boldsymbol{\delta}(x, y) = z(x, y)\,\hat{\mathbf{n}}$ and where we have used Snell's law to find the direction of the transmitted or reflected ray. The two equations can be combined in a single equation if we require for transmission that $N_i N_2 \geq 0$ and for the reflective case $N_i N_1 < 0$, where $N_1 = -N_i$ for the direction of specular reflection.

For the case of a rough surface with high-frequency height variations we have transmitted and reflected wave directions that do not satisfy Snell's law as is the case for the virtually flat surface. The spatial frequency content of the height modulation of the rough surface produces stochastically diffracted light at larger angles, more commonly denoted by scattered light. In the thin-layer approximation, the amplitude of the scattered wave that is created by an incident planar wave depends on the optical pathlength difference associated with a particular incident wave vector and a scattered wave vector, characterised by angles θ_1 and θ'_2, respectively. Figure 10.76 shows the geometry of the incident wave and a scattered wave. We note that the scattered wave vector $\hat{\mathbf{k}}'_2$ is not restricted to lie in the plane of incidence determined by the vectors $\hat{\mathbf{k}}_i$ and $\hat{\mathbf{n}}$. For the thin-layer model to be valid for large scattering angles we require again that the surface slope of the rough surface remains small. For a roughness profile with high spatial frequencies, this implies that the vertical amplitude is limited to the nanometre scale. To obtain the optical pathlength difference $\Delta_s = [OP_2Q]$ we calculate the distances OP_2 and P_2Q in Fig. 10.76, in a three-dimensional geometry. The incident and scattered unit wave vectors are given by (spherical coordinates)

$$\hat{\mathbf{k}}_i = (\sin\theta_1,\ 0,\ -\cos\theta_1)\,,$$

$$\hat{\mathbf{k}}'_2 = (\sin\theta'_2\cos\phi'_2,\ \sin\theta'_2\sin\phi'_2,\ -\cos\theta'_2)\,, \tag{10.5.4}$$

where we have restricted the incident wave vector $\hat{\mathbf{k}}_i$ to lie in the plane $y = 0$. The optical pathlength Δ_s is given by

$$\begin{aligned}\Delta_s &= \frac{\delta}{\cos\theta_1}\,\{-n_1 + n_2(\sin\theta_1 \sin\theta_2' \cos\phi_2' + \cos\theta_1 \cos\theta_2')\}\\ &= \frac{n_1\,\delta}{\cos\theta_1}\left\{\frac{n_2}{n_1}\,\hat{\mathbf{k}}_i \cdot \hat{\mathbf{k}}_2' - 1\right\}.\end{aligned} \qquad (10.5.5)$$

If we apply Snell's law to the expression for Δ_s and choose the in-plane geometry ($\phi_2' = 0$) we recover the expression for the pathlength difference Δ_t of Eq. (10.5.2) for the transmissive case. After some trigonometry applied to Eq. (10.5.5) the special case $\Delta_s = 0$ leads to a scattering angle θ_2' given by

$$\theta_2' = \arcsin\left(\frac{n_1}{n_2\left(1-\sin^2\theta_1\,\sin^2\phi_2'\right)^{1/2}}\right) - \arccos\left(\frac{\sin\theta_1\,\cos\phi_2'}{\left(1-\sin^2\theta_1\,\sin^2\phi_2'\right)^{1/2}}\right). \qquad (10.5.6)$$

If $\Delta_s = 0$, Eq. (10.5.5) shows that the angle between a scattered wave vector $\hat{\mathbf{k}}_2'$ and the incident wave vector $\hat{\mathbf{k}}_i$ has a fixed value. The scalar product $\hat{\mathbf{k}}_i \cdot \hat{\mathbf{k}}_2'$ is then given by n_1/n_2. The scattered waves that satisfy Eq. (10.5.6) correspond to waves that do not carry energy away from the surface according to the scalar wave propagation model that was used in this subsection. A more realistic model should take into account the state of polarisation of the light and the electromagnetic boundary conditions that have to be applied at the stochastically modulated interface between the two media. The rigorous treatment of electromagnetic boundary value problems, discussed in Section 1.12, provides such a realistic model for the calculation of scattered wave directions and the expectation value of the intensity carried away in a specific scattering direction. The scalar approach used in this section simply gives an estimate for the scattering directions with scattered intensity close to zero.

In Fig. 10.77 we have plotted a set of scattering angles with zero intensity, defined by the wave vector components $(\hat{k}_{2,x}', \hat{k}_{2,y}', \hat{k}_{2,z}')$. The scattering angles correspond to an incident wave vector at an angle of $\pi/4$ to the surface normal (thick arrow in each graph). The total 'umbrella' of scattered waves according to Eq. (10.5.5) has been truncated by the condition $\hat{k}_{2,z}' \leq 0$ for transmission into the second medium.

10.5.1.1 Reflection and Transmission at a Rough Surface in the Specular Direction

In this subsection we first calculate the relative change in specular reflection and transmission of a surface in the presence of stochastic height variations given by the probability function $z(x,y)$ of Eq. (10.5.1). The incident, reflected, transmitted and scattered wave vectors have been depicted in Fig. 10.78. The light amplitude U that is scattered in a general direction given by the direction cosines (L_2', M_2'), is calculated in the far field using the Fraunhofer diffraction integral. The 'in-focus' Fraunhofer integral follows from Eq. (8.4.36) when we insert $z_p = 0$. For the present case we are interested in the far-field pattern which is obtained when the focal plane coordinates (x_p, y_p) in Eq. (8.4.36) are replaced by the direction cosines $(L_2', M_2') = (x_p, y_p)/f$. The far-field pattern of the scattered light is measured on a distant sphere with radius of curvature R. The far-field amplitude as a function of the direction cosines (L_2', M_2') is given by the expression

$$\begin{aligned}U(L_2', M_2') &= -i\,\frac{n_2 k_0 \exp(i n_2 k_0 R)}{2\pi R}\left(\frac{N_2 + N_2'}{2}\right)\\ &\times \iint_S U_2(x,y) \exp\{-i n_2 k_0\,[(L_2' - L_2)x + (M_2' - M_2)y]\}\,dxdy.\end{aligned} \qquad (10.5.7)$$

In the expression for $U(L_2', M_2')$ the wavenumber in vacuum of the radiation has been denoted by k_0, S is the integration area on the rough surface, a disc with radius ρ_0. $U_2(x,y)$ is the complex amplitude function in a plane perpendicular to the z-axis, immediately below (transmission) or above (reflection) the modulated interface region. The phase part of $U_2(x,y)$ follows from the stochastic height variations of the surface. The phase of the incident plane wave is given by the exponential function $\exp\{ik_0 n_1(L_i x + M_i y)\}$. Using Snell's law, this expression can be replaced by $\exp\{ik_0 n_2(L_2 x + M_2 y)\}$, where (L_2, M_2, N_2) is the unit wave vector of the specularly transmitted wave. Any changes in the modulus part of the transmitted field with respect to the value given by the Fresnel transmission coefficients are neglected in this analysis. This means that the scattering angles should remain small. Only the obliquity factor $(N_2 + N_2')/2$, absent in Fraunhofer's basic integral, has been included in Eq. (10.5.7). Since we are only interested in the relative change in transmission or reflection we conveniently equate the transmission and reflection coefficients to unity.

The expectation value of the amplitude function $U(L_2', M_2')$ of Eq. (10.5.7) is obtained by first computing the average value of the z-dependent integrand function $U_2(x,y) = U_0 \exp\{ik_0 z(n_1 N_i - n_2 N_2')\}$, and then inserting this result in

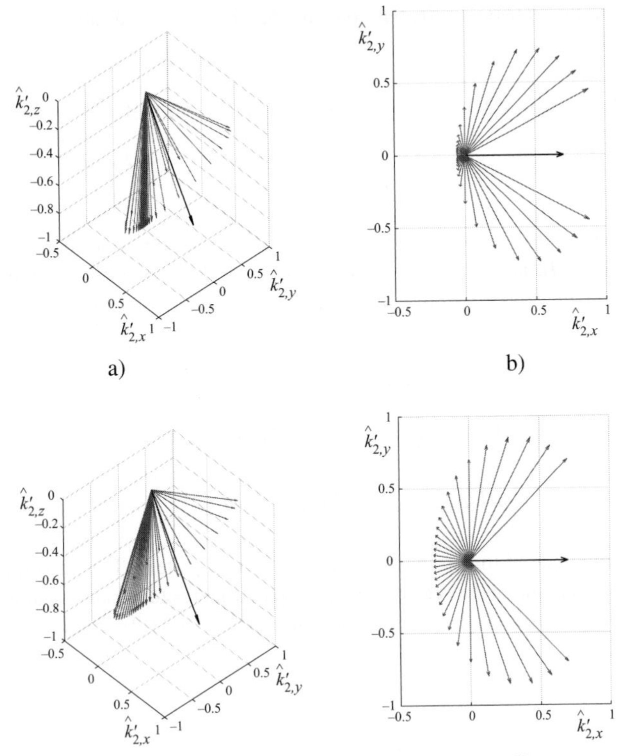

Figure 10.77: Scattered wave directions (unit wave vector $\hat{\mathbf{k}}_2'$) having zero amplitude in the presence of stochastic surface roughness. The thick arrow represents the incident wave vector in the plane $y = 0$, at 45° to the rough surface.
a) and b) The optical contrast $n_1/n_2 = 2/3$.
c) and d) The optical contrast equals 1/2.

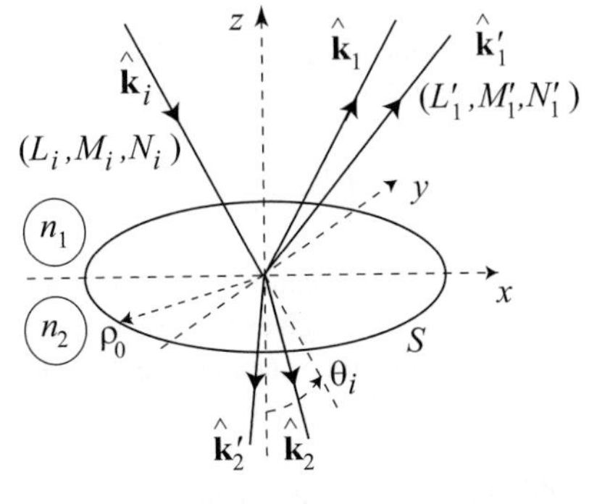

Figure 10.78: Reflection and transmission at a rough surface with stochastic height variations in the z-direction. An incident unit wave vector $\hat{\mathbf{k}}_i$, a transmitted vector $\hat{\mathbf{k}}_2$ and a reflected vector $\hat{\mathbf{k}}_1$ have been plotted. A scattered transmitted wave vector has been denoted by $\hat{\mathbf{k}}_2'$, a reflected one by $\hat{\mathbf{k}}_1'$.

Eq. (10.5.7). This procedure is equivalent to a change in the order of integration. In a first step we integrate the stochastic function $U_2(x, y)p(z)$ where $p(z)$ is the Gaussian probability function. In the following step we perform the integration over the spatial coordinates (x, y). Such an inversion of the order of integration is allowed if the stochastic process is ergodic. For the case of a rough surface, this means that the statistics of the surface height values $z(x_0, y_0)$ on various surface samples is identical to the statistical distribution of the height values over all points on a single sample. For the average value of U_2 we obtain,

$$\langle U_0 \exp\{ik_0 z(n_1 N_i - n_2 N_2')\}\rangle = U_0 \int_{-\infty}^{+\infty} \exp\{ik_0 z(n_1 N_i - n_2 N_2')\} \frac{1}{\sigma\sqrt{2\pi}} \exp\left(-\frac{z^2}{2\sigma^2}\right) dz$$

$$= U_0 \exp\left\{-\frac{[k_0\sigma(n_1 N_i - n_2 N_2')]^2}{2}\right\}, \qquad (10.5.8)$$

where only the even cosine part of the complex exponential function needs to be considered when evaluating the integral (see Appendix A, Table A.3 or **[4]**, page 302, Eq. (7.4.6)). Inserting the expression for the averaged amplitude U_2 in Eq. (10.5.7), we find the following expression for the average amplitude $\langle U(L_2', M_2')\rangle$ of the transmitted field,

$$\langle U(L_2', M_2')\rangle = -i\pi\rho_0^2\, U_0 \frac{n_2 k_0 \exp(i n_2 k_0 R)}{2\pi R}\left(\frac{N_2 + N_2'}{2}\right)\exp\left\{-\frac{[k_0\sigma(n_1 N_i - n_2 N_2')]^2}{2}\right\}$$

$$\times \frac{2J_1\left(n_2 k_0 \rho_0 \sqrt{(L_2' - L_2)^2 + (M_2' - M_2)^2}\right)}{\left(n_2 k_0 \rho_0 \sqrt{(L_2' - L_2)^2 + (M_2' - M_2)^2}\right)}, \qquad (10.5.9)$$

where $2J(r)/r$ is the scalar amplitude of the normalised Airy diffraction pattern. The maximum of the expectation value of the amplitude in the far field is found in the centre of the Airy pattern (specular transmission direction) with a reduction in amplitude given by the real exponential factor on the first line of Eq. (10.5.9). The intensity transmission coefficient T_d in the specular direction is given by

$$T_d = \exp\left\{-[k_0\sigma(n_1 N_i - n_2 N_2)]^2\right\}, \qquad (10.5.10)$$

where the value of unity applies to a perfectly smooth surface.

The transmission coefficient T_d can be measured by detecting the optical flux within a certain solid angle which contains a small fraction of the central lobe of the far-field Airy diffraction pattern. The reference measurement corresponds to the same solid angle and to an interface that has been polished to a high degree of flatness and smoothness such that its σ value is virtually negligible with respect to the wavelength of the light.

The analogous expressions for the case of reflection are obtained by replacing $n_2 N_2$ and $n_2 N_2'$ by $n_1 N_1$ and $n_1 N_1'$, respectively. For instance, the (normalised) reflection coefficient in the specular direction of a diffusing surface is given by

$$R_d = \exp\left\{-(2 n_1 k_0 \sigma N_i)^2\right\}, \qquad (10.5.11)$$

where we have used that $N_1 = -N_i$ for the specular direction of reflection. For well-polished reflecting surfaces the exponential function can be expanded up to the first Taylor term according to

$$R_d = 1 - \left(\frac{4\pi n_1 \cos\theta_i \sigma}{\lambda_0}\right)^2. \qquad (10.5.12)$$

Equation (10.5.12) shows that a high specular reflectivity can be obtained if the value of $\sigma\cos\theta_i$ is much smaller than λ_0. The second term at the right-hand side of Eq. (10.5.12) gives the loss in specular reflection and this quantity is found as scattered intensity in all scattering directions around the specular direction. For this reason, the second term has been given the name 'total integrated scatter', commonly abbreviated as TIS.

In Table 10.1 we give some examples of TIS values (reflective case) for various levels of polishing quality and wavelength values. It follows from the table that a polishing quality with a residual roughness characterised by $\sigma = 3$ nm is sufficient for mirror surfaces. For a surface used in transmission, for instance an air–glass transition, the optical contrast factor $(n-1)/2$ to the second power is introduced and this means that the TIS value is reduced by a factor of 16 (n = 1.5). For this case, even if the optical system has many lens elements, the fine-polishing quality of $\sigma = 3$ nm will be sufficient. For short wavelengths, for example $\lambda_0 = 193$ nm, currently used in optical lithography, the polishing requirements for a complicated projection lens are much more severe and the rms surface roughness value should be smaller than 0.5 nm. The next example in the table applies to mirror surfaces used at a wavelength in the *extreme UV* region ($\lambda_0 = 13$ nm). It is seen that an extremely stringent polishing quality is needed (σ <0.1 nm). Surface undulations that exceed the atomic scale cannot be tolerated. For still smaller wavelengths, down to the hard X-ray spectral band,

Table 10.1. Total integrated scatter (TIS, in %) as a function of rms surface roughness σ, angle of incidence θ_i and wavelength λ_0 (reflection $n_1 = 1$).

λ_0 (nm)	θ_i (degrees)	N_i	σ (nm)	TIS (%)
550.0	0.00	1.000	10.0	5.2
550.0	0.00	1.000	3.0	0.5
550.0	0.00	1.000	1.0	0.1
193.0	0.00	1.000	3.0	4.2
193.0	0.00	1.000	1.0	0.4
13.4	0.00	1.000	0.3	7.9
13.4	0.00	1.000	0.1	0.9
0.1	89.43	0.010	0.3	14.0
0.1	89.43	0.010	0.1	1.6
0.1	89.94	0.001	1.0	1.6
0.1	89.94	0.001	0.3	0.2

the grazing incidence solution has to be used to reduce the effective roughness down to the wavelength size. A surface with 'fine polishing' quality ($\sigma = 1$ nm) requires an angle of incidence which is typically $4'$ away from grazing incidence for a wavelength of 1 Å. Comparable wavelengths are encountered in the case of thermal neutrons, moving at a speed of typically 2 km s^{-1} (the de Broglie wavelength of such a neutron beam is 2 Å). Without significant scattering losses, neutron particle streams can be guided and focused with the aid of well-polished 'optical' surfaces used at grazing angles of a few arcminutes.

10.5.1.2 Angular Distribution of Light Scattered by a Weakly Diffusing Surface

In this subsection we calculate the intensity scattered by a weakly modulated rough surface as a function of the scattering angle. The adverb 'weakly' means that the surface roughness is small with respect to the wavelength of the light, $k_0\sigma(n_1N_i - n_2N_2') \ll 1$. We also assume that the reflected or transmitted light has been scattered a single time by the surface. This assumption is justified if the (average) surface slope of the rough surface is small. In this regime of surface roughness, the intensity of the specularly transmitted or reflected light amplitude gradually decreases if the surface roughness increases. In this subsection we exclude strongly diffusing surfaces such that the specular component becomes negligibly small and all light is found outside the specular direction on one or both sides of the interface. This type of rough surfaces is discussed in the next subsection.

In general, surface roughness leads to a stochastic amplitude variation superimposed on the (narrow) specular component, and to a (weak) background in the directions where the amplitude of the specularly refracted or reflected component has become negligibly small. To find the intensity of the scattered light we calculate the quantity $U_2(L_2', M_2') - \langle U_2(L_2', M_2') \rangle$. Far enough outside the central lobe of the Airy diffraction pattern the value of $|\langle U_2(L_2', M_2') \rangle|^2$ becomes negligibly small and we effectively calculate the stochastic amplitude of the scattered light itself. We obtain the expression

$$\begin{aligned} U_2(L_2', M_2') - \langle U_2(L_2', M_2') \rangle = &-i\, U_0\, \frac{n_2 k_0 \exp(i n_2 k_0 R)}{2\pi R} \left(\frac{N_2 + N_2'}{2}\right) \\ &\times \iint_S \left[\exp\{i k_0 z(n_1 N_i - n_2 N_2')\} - \exp\left\{-\frac{[k_0\sigma(n_1 N_i - n_2 N_2')]^2}{2}\right\}\right] \\ &\times \exp\{-i n_2 k_0 [(L_2' - L_2)x + (M_2' - M_2)y]\}\, dxdy . \end{aligned} \qquad (10.5.13)$$

Since σ is small with respect to the wavelength λ_0/n_2 in the second medium, we can expand the exponentials between brackets in the integrand of Eq. (10.5.13) up to the first order, neglect the term proportional to $(k_0\sigma)^2$ and obtain the expression

$$U_2(L_2', M_2') - \langle U_2(L_2', M_2') \rangle = U_0 \,\frac{n_2\, k_0^2 \exp(in_2 k_0 R)}{2\pi R} \left(\frac{N_2 + N_2'}{2}\right) (n_1 N_i - n_2 N_2')$$
$$\times \iint_S z(x,y) \,\exp\{-in_2 k_0\, [(L_2' - L_2)x + (M_2' - M_2)y]\}\, dxdy\,. \qquad (10.5.14)$$

Taking the squared modulus of the amplitude function $U_2(L_2', M_2') - \langle U_2(L_2', M_2') \rangle$ we obtain for the scattered intensity,

$$I(L_2', M_2') = |U_0|^2 \,\frac{n_2^2\, k_0^4\, (N_2 + N_2')^2\, (n_1 N_i - n_2 N_2')^2}{16\pi^2 R^2} \iiiint_S z(x,y)\, z(x',y')$$
$$\times \exp\{-in_2 k_0\, [(L_2' - L_2)(x - x') + (M_2' - M_2)(y - y')]\}\, dxdydx'dy'\,. \qquad (10.5.15)$$

The expectation value of the scattered intensity then follows from

$$\langle I(L_2', M_2') \rangle = |U_0|^2 \,\frac{n_2^2\, k_0^4\, (N_2 + N_2')^2\, (n_1 N_i - n_2 N_2')^2}{16\pi^2 R^2} \iiiint_S \langle z(x,y)\, z(x',y') \rangle$$
$$\times \exp\{-in_2 k_0\, [(L_2' - L_2)(x - x') + (M_2' - M_2)(y - y')]\}\, dxdydx'dy'\,. \qquad (10.5.16)$$

The autocorrelation function $\langle z(x,y) z(x',y') \rangle$ of the surface height variation appears in the integral expression of Eq. (10.5.16). If the stochastic height variations of the surface are stationary, this function can be written as a function of the coordinate differences $\Delta_x = x - x'$, $\Delta_y = y - y'$ only,

$$\langle z(x,y)\, z(x',y') \rangle = R_{zz}(\Delta_x, \Delta_y) = \sigma^2 \exp\left\{-\,\frac{\Delta_x^2 + \Delta_y^2}{\tau^2}\right\}, \qquad (10.5.17)$$

where the correlation function R_{zz} has a width of 2τ at the $1/e$ correlation level (τ is called the rms correlation length) and (Δ_x, Δ_y) is the two-dimensional translation argument of the autocorrelation function of the spatially stationary height variation $z(x,y)$.

The integral of Eq. (10.5.16) is evaluated with the aid of the relation given in Appendix A, Table A.3,

$$\int_{-\infty}^{+\infty} \exp[-\pi(x/b)^2] \exp(-i\omega x)\, dx = b \exp[-(\omega b)^2/4\pi]\,, \qquad (10.5.18)$$

where we assume that $\rho_0 \gg \tau, \lambda_0$ such that the integration limits can be effectively extended to infinity. We introduce the change of variables $(x, x') \rightarrow (x - x', x') = (\Delta_x, x')$ and a similar change for the (y, y') pair. The integration with respect to the variables (Δ_x, Δ_y) uses the result of Eq. (10.5.18), the integration over the variables (x', y') yields the integration area $\pi\rho^2$ on the rough surface. We thus have

$$\langle I(L_2', M_2') \rangle = |U_0|^2 \,\frac{\pi^4 n_2^2\, \sigma^2 \tau^2 \rho_0^2\, (N_2 + N_2')^2\, (n_1 N_i - n_2 N_2')^2}{R^2\, \lambda_0^4}$$
$$\times \exp\left\{-\,\frac{\pi^2 n_2^2\, \tau^2}{\lambda_0^2} \left[(L_2' - L_2)^2 + (M_2' - M_2)^2\right]\right\}. \qquad (10.5.19)$$

The scattering angles corresponding to a reduction by $1/e$ in the intensity of the scattered light are defined by

$$\sqrt{(L_2' - L_2)^2 + (M_2' - M_2)^2} = \frac{\lambda_0}{\pi n_2 \tau}\,. \qquad (10.5.20)$$

In a good approximation, for small scattering angles and isotropic roughness, the scattered intensity shows circular symmetry with respect to the specular transmission direction. The $1/e$ intensity level corresponds to frequency components in the spatial spectrum with magnitude $\nu_{sp} = 1/(\pi\tau)$ and arbitrary orientation. The wavelength in the second medium equals λ_0/n_2. The corresponding scattering angles can be visualised with the aid of the Ewald sphere in Fig. 5.65, for $m = 1$. The endpoint of the vector **a** with length $(\pi\tau)^{-1}$ sweeps a full circular arc in the plane of the interface. In a straightforward way, we obtain for the scattered light intensity in reflection,

$$\langle I(L_1', M_1') \rangle = |U_0|^2 \,\frac{4\pi^4 n_1^4\, \sigma^2 \tau^2 \rho_0^2\, N_i^2\, (N_i - N_1')^2}{R^2\, \lambda_0^4}$$
$$\times \exp\left\{-\,\frac{\pi^2 n_1^2\, \tau^2}{\lambda_0^2} \left[(L_1' - L_1)^2 + (M_1' - M_1)^2\right]\right\}. \qquad (10.5.21)$$

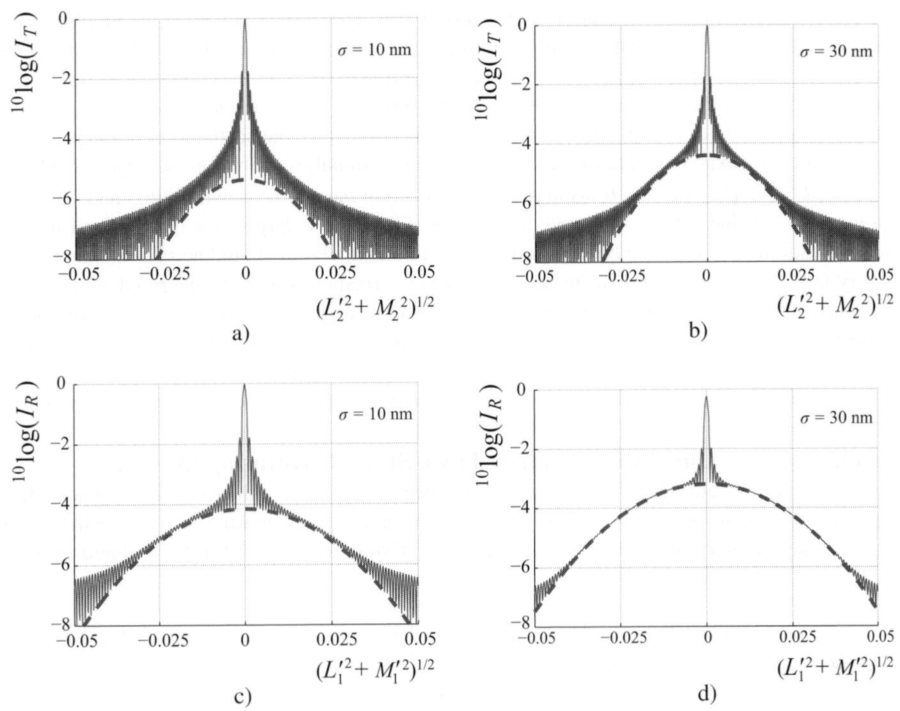

Figure 10.79: Angular scattering in transmission and reflection at a rough surface with stochastic height variations in the z-direction. The refractive indices are $n_1 = 1$, $n_2 = 1.5$, $\lambda_0 = 500$ nm. The beam cross-section $2\rho_0 = 0.6$ mm, the correlation length τ is 10 μm and σ is the rms surface roughness.
a) Transmission, $\sigma = 10$ nm.
b) Transmission, $\sigma = 30$ nm.
c) Reflection, $\sigma = 10$ nm.
d) Reflection, $\sigma = 30$ nm.

To obtain the absolute values of the scattered light according to Eqs. (10.5.19) and (10.5.21), these expressions should be multiplied by the specular intensity reflection coefficient or the specular intensity transmission coefficient of a perfectly flat interface.

In Fig. 10.79 we have plotted the specular and scattered light intensity for two levels of low roughness, in transmission (a) and b)) and in reflection (c) and d)), all at normal incidence. The plotted function for the case of transmission is obtained by the addition of the specular component $|\langle U_2(L_2', M_2') \rangle|^2$ according to Eq. (10.5.9) and the scattered intensity given by Eq. (10.5.19) and then normalising this function with respect to the peak intensity of the Airy pattern of a perfectly flat surface. The expected values for the intensity transmission and reflection factors are given by

$$\langle T(L_2', M_2') \rangle = \frac{4\pi^2 (n_1 - n_2)^2 \sigma^2 \tau^2}{\lambda_0^2 \rho_0^2} \exp\left\{ -\left(\frac{\pi n_2 \tau}{\lambda_0} \right)^2 (L_2'^2 + M_2'^2) \right\} , \tag{10.5.22}$$

$$\langle R(L_1', M_1') \rangle = \frac{16\pi^2 n_1^2 \sigma^2 \tau^2}{\lambda_0^2 \rho_0^2} \exp\left\{ -\left(\frac{\pi n_1 \tau}{\lambda_0} \right)^2 (L_1'^2 + M_1'^2) \right\} , \tag{10.5.23}$$

where we have assumed small scattering angles and normal incidence, such that all z-direction cosines can be put equal to unity. In Fig. 10.79a and c the rms roughness σ amounts to 10 nm ($\lambda_0 = 500$ nm). The diameter of the illuminated surface is 600 μm and the autocorrelation length of the rough surface is 10 μm. The refractive indices equal $n_1 = 1$, $n_2 = 1.5$. In the transmission and reflection graph we observe the peak of the Airy intensity pattern, amounting to 99.6 and 93.9%, respectively. At a logarithmic scale, the patterns clearly show the secondary maxima and minima of the Airy pattern, more densely packed in the transmitted far-field radiation pattern than in the reflected pattern. The cones of scattered light on both sides of the interface have an apex angle ratio of $1/n$. The dashed curves in the graphs correspond to the scattered intensity. In the transmitted pattern, this scattered intensity does not yet affect the heights of the secondary maxima. In the reflected far-field pattern, some 15 secondary maxima away from the central peak, the scattered intensity reaches a value equal to that of the Airy far-field pattern. In the graphs b) and d), the same phenomena are observed, now for a surface roughness parameter σ, equal to 30 nm. In the reflection case, the scattered light intensity gives rises to a dominating background with exponential fall-off as a function of the (radial) scattering angle, represented by the direction cosine value $(L_1'^2 + M_1'^2)^{1/2}$. The specular transmission and reflection have gone down to 96.5 and 56.6%, respectively. In general, for customary values of the refractive indices, the TIS value is largest for the reflected light beam. In the reflection case, the scattering background has reached a level of almost 10^{-3}. It should be said that the roughness level of 30 nm rms is very large for polished optical surfaces. This level corresponds more to the roughness produced by a fine grinding process in optical manufacturing. In the production process of optical surfaces, the measurement of scattered intensity is a useful tool to characterise the surface structure with respect to the surface parameters σ and τ. For very small roughness values, below 1 nm rms, scatterometry becomes rather insensitive and has to be supplemented by direct surface profilometry, for instance by interference microscopy, atomic force microscopy or tunnelling microscopy.

10.5.1.3 Angular Distribution of Light Scattered by a Strongly Diffusing Interface

The average amplitude component $\langle U_2(L_2', M_2') \rangle$ has become negligibly small in the case of strong scattering by a rough surface. For instance, the argument $(n_1 N_i - n_2 N_2) k_0 \sigma$ of the exponential factor in Eq. (10.5.10) for the specular intensity transmission is much larger than unity and T_d is effectively zero. Using Eq. (10.5.7) for the scattered complex amplitude, we find for the scattered intensity,

$$I(L_2', M_2') = |U_0|^2 \frac{n_2^2\, k_0^2\, (N_2 + N_2')^2}{16\pi^2 R^2} \iiiint_S \exp\left\{ik_0(n_1 N_i - n_2 N_2')\,[z(x,y) - z(x',y')]\right\}$$
$$\times \exp\{-in_2 k_0\,[(L_2' - L_2)(x - x') + (M_2' - M_2)(y - y')]\}\, dxdydx'dy' . \qquad (10.5.24)$$

As before, we assume that the surface irregularity is stationary, such that the value of $z(x,y) - z(x',y')$ only depends on the coordinate differences and not on the position on the surface. Therefore, we may write that

$$z(x,y) - z(x',y') = z(x,y) - z(x - \Delta_x, y - \Delta_y) = \frac{\partial z(x,y)}{\partial x}\Delta_x + \frac{\partial z(x,y)}{\partial y}\Delta_y + \cdots \qquad (10.5.25)$$

The relevant values of (Δ_x, Δ_y) to calculate the average of the first exponential function in the integrand of Eq. (10.5.24) will be smaller than the typical autocorrelation distance of the surface profile if the factor $k_0(n_1 N_i - n_2 N_2')[z(x,y) - z(x',y')]$ is substantially larger than 2π. Only a very restricted range of the coordinate differences $\Delta_x = x - x'$ and $\Delta_y = y - y'$ will contribute to the average value and we can limit the expansion of $z(x,y) - z(x',y')$ as a function of (Δ_x, Δ_y) to the first two terms given in the right-hand side of Eq. (10.5.25). We consider $\beta_x = \partial z/\partial x$ and $\beta_y = \partial z/\partial y$ as our new stochastic variables associated with the slope of the surface height deviations $z(x,y)$. These new variables have zero average and their variances are denoted by $\sigma_{\beta,x}^2$ and $\sigma_{\beta,y}^2$, respectively. Using these stochastic quantities the expression for the scattered intensity is given by

$$I(L_2', M_2') = |U_0|^2 \frac{n_2^2\, k_0^2\, (N_2 + N_2')^2}{16\pi^2 R^2} \iiiint_S \exp\left\{ik_0(n_1 N_i - n_2 N_2')(\beta_x \Delta_x + \beta_y \Delta_y)\right\}$$
$$\times \exp\{-in_2 k_0\,[(L_2' - L_2)u + (M_2' - M_2)v]\}\, d\Delta_x d\Delta_y dxdy . \qquad (10.5.26)$$

By a change of variables from (Δ_x, Δ_y) to $(\Delta_x', \Delta_y') = k_0(\Delta_x, \Delta_y)$, we can obtain an expression for $I(L_2', M_2')$ from which the wavelength of the light has disappeared. This is a consequence of the fact that we have represented the surface profile by a stochastic distribution of facets with a two-dimensional slope defined by (β_x, β_y) which determines the normal to each facet. The refraction or reflection of light by such a faceted surface is locally described by the laws of geometrical optics in which the wavelength of the light is absent.

To obtain the average value of $I(L_2', M_2')$ we initially calculate the average of the first exponential function in the integral of Eq. (10.5.26) with respect to the stochastic variable β_x, yielding

$$\frac{1}{\sigma_{\beta,x}\sqrt{2\pi}} \int_{-\infty}^{+\infty} \exp\{ik_0(n_1 N_i - n_2 N_2')\beta_x \Delta_x\} \exp\left\{-\beta_x^2/2\sigma_{\beta,x}^2\right\} d\beta_x$$
$$= \exp\left\{-\frac{[k_0(n_1 N_i - n_2 N_2')\Delta_x \sigma_{\beta,x}]^2}{2}\right\} , \qquad (10.5.27)$$

where $\sigma_{\beta,x}$ is the rms value of the projection of the slope angles of the surface facets onto the plane $y = 0$. A comparable expression is valid for the stochastic variable β_y. We use the result of Eq. (10.5.27) in Eq. (10.5.26) and, performing the integration over (x, y), we find

$$\begin{aligned}\langle I(L_2', M_2') \rangle = |U_0|^2 \frac{n_2^2\, k_0^2\, (N_2 + N_2')^2\, \pi \rho_0^2}{16\pi^2 R^2} \\ \times \iint_S \exp\left\{ - \frac{\left[k_0^2 (n_1 N_i - n_2 N_2')^2 \left\{(\Delta_x \sigma_{\beta,x})^2 + (\Delta_y \sigma_{\beta,y})^2\right\}\right]}{2} \right\} \\ \times \exp\left\{ -i n_2 k_0 \left[(L_2' - L_2)\Delta_x + (M_2' - M_2)\Delta_y \right] \right\} d\Delta_x d\Delta_y \,.\end{aligned} \tag{10.5.28}$$

The integral with respect to (Δ_x, Δ_y) is evaluated with the aid of Eq. (10.5.18), yielding,

$$\begin{aligned}\langle I(L_2', M_2') \rangle = |U_0|^2 \frac{n_2^2\, \rho_0^2\, (N_2 + N_2')^2}{8R^2 (n_1 N_i - n_2 N_2')^2 \sigma_{\beta,x} \sigma_{\beta,y}} \\ \times \exp\left\{ - \frac{n_2^2}{2(n_1 N_i - n_2 N_2')^2} \left[\frac{(L_2' - L_2)^2}{\sigma_{\beta,x}^2} + \frac{(M_2' - M_2)^2}{\sigma_{\beta,y}^2} \right] \right\} .\end{aligned} \tag{10.5.29}$$

In the frequently occurring case of isotropic, uncorrelated roughness we have $\sigma_{\beta,x} = \sigma_{\beta,y} = \sigma_\beta$. The expression of Eq. (10.5.29) is easily adapted to the case of reflection at an isotropic rough surface with the result

$$\langle I(L_1', M_1') \rangle = |U_0|^2 \frac{\rho_0^2}{8\sigma_\beta^2 R^2} \exp\left\{ - \frac{(L_1' - L_i)^2 + (M_1' - M_i)^2}{2\sigma_\beta^2 (N_i - N_1')^2} \right\} , \tag{10.5.30}$$

where we have replaced the obliquity factor $N_2 + N_2'$ in the numerator of Eq. (10.5.29) by $N_i - N_1'$ as the z-direction cosine changes sign in the reflective case.

In Fig. 10.80 we have plotted the intensity scattered by a strongly diffusing rough surface, both for the transmission and the reflection case. After normalisation with respect to the peak intensity of the Airy pattern produced by a perfectly flat interface, we obtain for the transmitted and reflected intensity (normal incidence and small scattering angles),

$$\langle T(L_2', M_2') \rangle = \frac{\lambda_0^2}{2\pi^2 (n_1 - n_2)^2 \rho_0^2 \sigma_\beta^2} \exp\left\{ - \frac{1}{2\sigma_\beta^2} \left(\frac{n_2}{n_1 - n_2} \right)^2 (L_2'^2 + M_2'^2) \right\} , \tag{10.5.31}$$

$$\langle R(L_1', M_1') \rangle = \frac{\lambda_0^2}{8\pi^2 n_1^2 \rho_0^2 \sigma_\beta^2} \exp\left\{ - \frac{1}{8\sigma_\beta^2} (L_1'^2 + M_1'^2) \right\} . \tag{10.5.32}$$

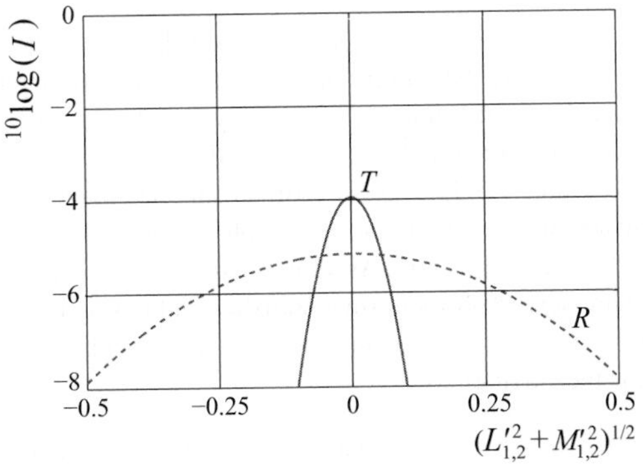

Figure 10.80: Intensity reflection (dashed curve) and transmission (solid curve) at a strongly diffusing surface as a function of the scattering angle. The rms surface roughness σ is 1 μm, the autocorrelation length τ is 20 μm, yielding an rms slope value of ≈0.07 rad. The refractive index values are $n_1 = 1$ and $n_2 = 1.5$, $\lambda_0 = 500$ nm, $2\rho_0 = 0.6$ mm.

The light scattering shows a circularly symmetric behaviour with respect to the direction of incidence (L_i, M_i) of the light. In the case of reflection, for a general angle of incidence, the $1/e^2$ scattered intensity is found in the direction

$$\left\{(L_1' - L_i)^2 + (M_1' - M_i)^2\right\}^{1/2}\Big|_{1/e^2} = 2\sigma_\beta\,|N_i - N_1'| \approx 4\sigma_\beta\,, \tag{10.5.33}$$

where the value $4\sigma_\beta$ applies to the normal incidence case. For the case of transmission, the $1/e^2$ value of the intensity is found at a scattering angle with direction cosine $(L_2'^2 + M_2'^2)^{1/2} = 2\,|n_2 - n_1|\,\sigma_\beta/n_2$. It is seen again that the scattering angles in transmission and reflection are determined by the laws of geometrical optics applied to the surface facets with slope β. In Fig. 10.80 where we have used $n_2/n_1 = 1.5$, the ratio between the scattering angles in reflection and in transmission equals 6.

10.5.1.4 Relationship between the Surface Height and Surface Slope Variables

A relationship can be established between the stochastic height variable z of a rough surface and the stochastic slope variables (β_x, β_y) of the same surface. The autocorrelation function $R_{zz}(\Delta_x, \Delta_y)$ plays a central role in the derivation of such a relationship between z and β. The following property exists between the autocorrelation function of a stochastic process itself and of its derivative function **[269]**,

$$\left\langle \frac{\partial z(x,y)}{\partial x}\,\frac{\partial z(x+\Delta_x, y)}{\partial x} \right\rangle = R_{z_x' z_x'}(\Delta_x, 0) = -\frac{\partial^2 R_{zz}(\Delta_x, 0)}{\partial \Delta_x^2}\,, \tag{10.5.34}$$

where the prime in the subscript z_x' of the autocorrelation function $R(\Delta_x, \Delta_y)$ indicates differentiation of this function with respect to x. If Δ_y is zero we find for the variance of the surface slope in the x-direction

$$E\left\{\left(\frac{\partial z(x,y)}{\partial x}\right)^2\right\} = -\frac{\partial^2 R_{zz}(\Delta_x, \Delta_y)}{\partial \Delta_x^2}\Bigg|_{\Delta_x=\Delta_y=0}\,, \tag{10.5.35}$$

where E is the symbol used for the expectation value of the statistical quantity between braces. We apply the result of Eq. (10.5.35) to the autocorrelation function which is given by Eq. (10.5.17) and find for the variance of the slope angle β_x on the surface the expression

$$\sigma_{\beta,x}^2 = \langle \beta_x^2 \rangle = \frac{2\sigma^2}{\tau^2}\,, \tag{10.5.36}$$

where we have assumed that $\langle \beta_x \rangle = 0$. If the height variations are isotropic, the rms slope angle σ_β equals $\sigma\sqrt{2}/\tau$ for any azimuthal direction on the surface.

10.5.2 Frequency Transfer in the Presence of a Random Phase Screen

In this subsection we analyse the degradation of the frequency transfer by an incoherently illuminated imaging system when a random phase screen is present in the diaphragm of the system. Instead of physically installing a phase screen in the stop of the system, we project the joint influence of (weakly) scattering surfaces in the imaging system into a hypothetical single random phase screen in the diaphragm. For well-polished optical surfaces and a compact imaging system, this simplification is allowed and yields a good estimate of the image quality degradation due to residual scattering in the system. The phase variance of the hypothetical phase screen is put equal to the sum of the stochastic phase variances of the individual surfaces in the system. As a further simplification of the model, we assume that the phase function $\phi(X_1, Y_1)$ of the random phase screen obeys normal statistics and has an isotropic autocorrelation function with a correlation length given by τ. The complex transmission function of the imaging system, projected onto the diaphragm D and including scattering effects, is then written as

$$f_D(X_1, Y_1) = A(X_1, Y_1)\exp\{ik_0 W(X_1, Y_1)\}\, t_s(X_1, Y_1)\,, \tag{10.5.37}$$

where $t_s(X_1, Y_1)$ is the stochastic complex amplitude transmission function. We write the stochastic function t_s as the product of a stochastic amplitude function and a stochastic phase function such that

$$t_s(X_1, Y_1) = A_s(X_1, Y_1)\exp\{i\phi_s(X_1, Y_1)\}\,, \tag{10.5.38}$$

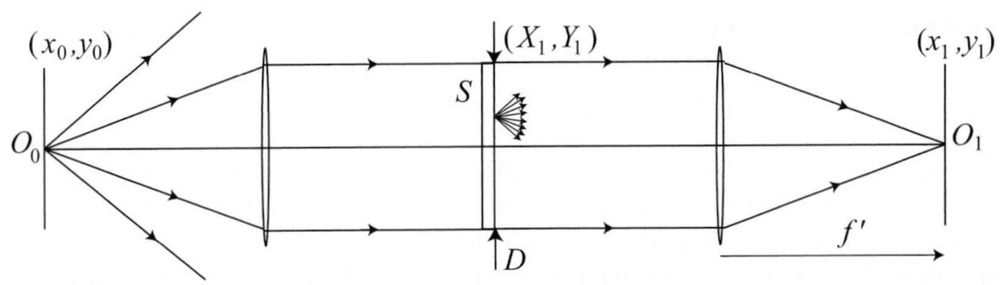

Figure 10.81: A $4f$ imaging system with a random phase screen in the plane of the diaphragm D. The scattering by the random screen is symbolically represented by the scattered rays from some arbitrary point on the screen.

where A_s and ϕ_s are stochastic functions of the lateral coordinates (X_1, Y_1) in the diaphragm. For the case of the random *phase* screen, $A_s(X_1, Y_1)$ is assumed to have unit value over the entire open area S of the diaphragm.

For calculation of the value of the incoherent optical transfer function at a specific spatial frequency with components (ν_x, ν_y) in the image plane, we evaluate the expression of Eq. (10.2.33) in which we replace R_1 by the image-side focal distance f' associated with the $4f$ imaging system of Fig. 10.81,

$$OTF(\nu_x, \nu_y) = \iint_S f(X_1, Y_1) f^*(X_1 + \lambda f'\nu_x, Y_1 + \lambda f'\nu_y) \times \exp\left\{i\left[\phi_s(X_1, Y_1) - \phi_s(X_1 + \lambda f'\nu_x, Y_1 + \lambda f'\nu_y)\right]\right\} dX_1 dY_1 \Big/ \iint_S |f(X_1, Y_1)|^2 \, dX_1 dY_1 , \qquad (10.5.39)$$

where $\lambda = \lambda_0/n$ (n is the refractive index of the image space). In the presence of the random phase-only screen, the average value of the *OTF* is determined by the average value of the numerator because the value of the denominator remains unaffected by the screen. To evaluate the value of $\langle OTF \rangle$, we invert the order of integration in the numerator and first evaluate the average value of the exponential function in the integrand. This inversion is allowed if the moments of the phase difference function $f_s(X_1, Y_1; X_1 + X_s, Y_1 + Y_s) = \phi_s(X_1, Y_1) - \phi_s(X_1 + X_s, Y_1 + Y_s)$ are independent of the position (X_1, Y_1) in the diaphragm (stationary stochastic function). We assume that the random phase function $\phi_s(X_1, Y_1)$ is normally distributed with mean zero and variance σ_ϕ^2. With respect to the difference function f_s, we assume that it is also stationary and that it only depends on the shifted coordinate values (X_s, Y_s). The difference function f_s is also normally distributed, has zero mean and we denote its variance by $\sigma_{f_s}^2$. For the average value of the exponential function in the integrand of the numerator of Eq. (10.5.39) we then find,

$$\langle \exp\{i f_s(X_s, Y_s)\} \rangle = \frac{1}{\sigma_{f_s} \sqrt{2\pi}} \int_{-\infty}^{+\infty} \exp\{i f_s(X_s, Y_s)\} \exp\left\{-\left(\frac{f_s^2}{2\sigma_{f_s}}\right)\right\} df_s = \exp\left\{-\frac{\sigma_{f_s}^2}{2}\right\} , \qquad (10.5.40)$$

a result which is obtained in the same way as the result of Eq. (10.5.8). For the variance σ_{f_s} as a function of the shift (X_s, Y_s) we write

$$\sigma_{f_s}^2 = \langle \{\phi_s(X_1, Y_1) - \phi_s(X_1 + X_s, Y_1 + Y_s)\}^2 \rangle = 2\sigma_\phi^2 - 2R_{\phi\phi}(X_s, Y_s) . \qquad (10.5.41)$$

We assume an exponential profile for the autocorrelation function of the stochastic phase ϕ_s, given by the expression

$$R_{\phi\phi}(X_s, Y_s) = \sigma_\phi^2 \exp\left\{-\frac{X_s^2 + Y_s^2}{\tau^2}\right\} , \qquad (10.5.42)$$

where τ is the (isotropic) correlation length yielding a value of e^{-1} of the autocorrelation function $R_{\phi\phi}$. The variance of the phase difference function f_s is then given by

$$\sigma_{f_s}^2 = 2\sigma_\phi^2 \left[1 - \exp\left\{-\frac{X_s^2 + Y_s^2}{\tau^2}\right\}\right] , \qquad (10.5.43)$$

and the expression of Eq. (10.5.40) is written as

$$\langle \exp\{i f_s(X_s, Y_s)\} \rangle = \exp\{-\sigma_\phi^2\} \exp\left[\sigma_\phi^2 \exp\left\{-\frac{X_s^2 + Y_s^2}{\tau^2}\right\}\right] . \tag{10.5.44}$$

The statistical average of the exponential function $\exp\{i f_s(X_s, Y_s)\}$ is independent of the coordinates (X_1, Y_1) in the plane of the diaphragm and can be taken outside the integral in the numerator of Eq. (10.5.39). This means that the *OTF* can be written as the product of a scattering-induced factor and the *OTF* function that is given by the deterministic part of the optical system without the random phase screen. For the case of a random phase screen of which the autocorrelation function $R_{\phi\phi}$ only depends on the radial distance $\sqrt{X_s^2 + Y_s^2}$ we find the expression

$$\begin{aligned} OTF(\nu_x, \nu_y) &= OTF_s\left(\sqrt{\nu_x^2 + \nu_y^2}\right) OTF_a(\nu_x, \nu_y) \\ &= \exp\{-\sigma_\phi^2\} \exp\left[\sigma_\phi^2 \exp\left\{-\frac{\lambda^2 f'^2 \nu^2}{\tau^2}\right\}\right] OTF_a(\nu_x, \nu_y) \\ &\approx \left\{1 - \sigma_\phi^2\left[1 - \exp\left(-\frac{\lambda^2 f'^2 \nu^2}{\tau^2}\right)\right]\right\} OTF_a(\nu_x, \nu_y) , \end{aligned} \tag{10.5.45}$$

where ν is the length of the frequency vector $\mathbf{\nu}$. OTF_s is the transfer function due to the random phase screen in the diaphragm of the imaging system. OTF_a is the incoherent transfer function, possibly affected by a nonzero aberration function W of the imaging system. To obtain the expression of Eq. (10.5.45) we have used that the diffraction angle θ at a periodic structure (frequency $\mathbf{\nu}$, normal incidence) in a medium with refractive index n is given by

$$\frac{\sqrt{X_s^2 + Y_s^2}}{f'} = \sin\theta = \lambda\nu , \quad \text{or,} \quad \sqrt{X_s^2 + Y_s^2} = \lambda f' \nu . \tag{10.5.46}$$

In the case of a perfect imaging system we obtain

$$\begin{aligned} OTF(\nu_x, \nu_y) &= OTF_s\left(\sqrt{\nu_x^2 + \nu_y^2}\right) OTF_i(\nu_x, \nu_y) \\ &= \exp\left\{-\sigma_\phi^2\left[1 - \exp\left(-\lambda^2 f'^2 \nu^2/\tau^2\right)\right]\right\} \\ &\times \left\{\frac{2}{\pi}\arccos\left[\frac{\lambda\nu}{2\sin\alpha_m}\right] - \frac{\lambda\nu}{\pi\sin\alpha_m}\sqrt{1 - \left(\frac{\lambda\nu}{2\sin\alpha_m}\right)^2}\right\} , \end{aligned} \tag{10.5.47}$$

where OTF_i is the transfer function of a perfect imaging system with a circular diaphragm rim. The quantity $n\sin\alpha_m = n\rho_m/f'$ is the numerical aperture NA of the imaging system ($2\rho_m$ is the diameter of the diaphragm). The diffraction unit in image space is given by $\lambda/NA = \lambda_0/(n\sin\alpha_m)$.

In Fig. 10.82a we have plotted the average of the phase screen scattering function according to Eq. (10.5.44) for several values of the standard deviation σ_ϕ of the random phase function associated with the phase screen at the diaphragm. For a thin random surface used in transmission, the square root of the variance σ_z of the random height function z on the phase screen produces the value of σ_ϕ of the random phase function ϕ_s with the aid of the expression $\sigma_\phi = 2\pi(n_2 - n_1)\sigma_z/\lambda_0$. Each dashed curve of Fig. 10.82a thus corresponds to a different value of the height variance σ_z^2 whilst the optical contrast is given by $(n_2 - n_1) = 0.5$ and the wavelength λ_0 is 500 nm. The values of σ_z are 30, 60, 100 and 200 nm and the corresponding OTF_s curves show the increasing deterioration of the high-frequency transfer when the scattering by the phase screen increases. The correlation length τ of the random phase function ϕ_s of the screen is 100 μm in all graphs. The smaller its value, the broader the central part of the OTF_s with a relatively good transfer of spatial frequency content. In Fig. 10.82b we have plotted the total *OTF* function which is the product of the ideal transfer function $OTF_i(\nu)$ and the scattering transfer function $OTF_s(\nu)$. The dashed curves correspond to the σ_z values of 30, 60, 100 and 200 nm, the solid curve is the scattering-free transfer function. For all curves we have used the values $f' = 20$ mm and $NA = 0.1$. The cut-off frequency ν_c is $2NA/\lambda_0$ and equals 400 mm^{-1}.

A criterion can be set for the *OTF* degradation in the presence of a random phase screen, for example by requiring that $OTF_s(\nu) > 0.99$ for frequency values outside the central lobe of $OTF_s(\nu)$ (see Fig. 10.82a). Equation (10.5.45) shows that, for small values of σ_ϕ, the value of σ_ϕ should be less than 0.1 radian. For an optical contrast of 0.5 in transmission and for the wavelength $\lambda_0 = 500$ nm, we find that the value σ_z of the phase screen profile should be less than 16 nm. For the

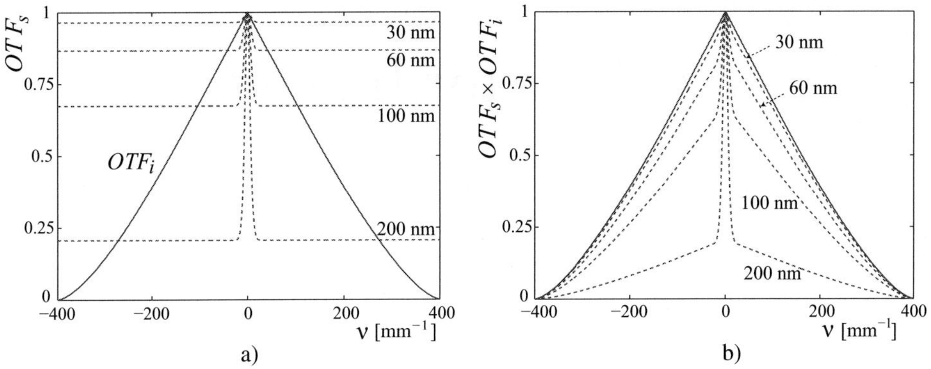

Figure 10.82: Optical transfer functions for various levels of scattering by a random phase screen, positioned in the diaphragm of an incoherently illuminated imaging system. The parameter applying to the various curves is the value σ (in nm) of the surface height variations on the random phase screen.
a) Scattering transfer functions $OTF_s(\nu)$ (dashed curves), and, for comparison, the ideal transfer function $OTF_i(\nu)$ (solid line).
b) Total transfer function obtained as the product of $OTF_s(\nu)$ and $OTF_i(\nu)$.

case of a single, well-polished optical surface having a typical value for σ_z of 3 nm, this criterion is easily met. In the case of a multi-element optical system, the height variances of the residual surface roughness on each surface are added quadratically and the total variance projected into the diaphragm of the system amounts to $N_s\sigma_z^2$ where N_s is the number of scattering glass–air interfaces. It follows that a system with approximately 12 to 13 lenses should be polished down to 3 nm residual roughness in order to meet the criterion of 1% degradation in frequency transfer. In some applications, the 1% criterion is too loose. In an image pattern, this residual value of 1% is found as background illumination, present in the dark portions of the image. For visual observation, the perceived crispness of an image is related to this background level and a more severe criterion than the 1% level has to be applied. Contrarily, if an image is digitised for further treatment, the requirements on the level of the scattered light background can be relaxed because a digital threshold operation can be applied to eliminate the undesired background of an image.

The application of the 1% degradation criterion for the *OTF* to a high-resolution mirror system for the production of microprocessors or memory chips (optical Extreme UV-lithography) leads to much more stringent requirements on the residual stochastic surface errors on the mirrors. A factor of approximately four stems from the change in optical contrast when going from transmission to reflection. Using the extreme UV wavelength of 13 nm and a total of six mirrors in the imaging system we find a maximum permitted surface roughness given by

$$\sigma_z = 0.1 \frac{\lambda}{4\pi\sqrt{6}} = 0.04 \text{ nm}. \tag{10.5.48}$$

This extreme requirement on σ_z implies that the mirror surfaces must be polished with an accuracy down to the atomic level.

11 Theory of Vector Imaging

In this chapter we expand on the ideas presented in Chapter 9 by generalising the Ignatowsky **[153]** and Debye–Wolf theory to describe imaging by high-NA optical systems. En route we develop a matrix formalism built on the use of the generalised Jones matrices, that permits the computation of the electric field of the geometrical optics approximation for the Debye–Wolf integral, presented in Chapter 8. Once in possession of that knowledge we use the developed formalism to calculate the distribution of the time-averaged electric energy density in the focal region of high-NA lenses for linearly polarised illumination and subsequently focusing of complex vector beams also including partially polarised light.

Next, we expand the work to include, by means of the generalised Jones matrices, the description of a high-NA telecentric and afocal optical system. This we do to explain how these optical systems behave when no samples are present. Since scalar paraxial optical systems are seldom analysed without samples this idea might appear strange at first. However, vectorial properties of imaging optics introduce a new degree of freedom, and so understanding polarisation changes induced by the optical system alone requires a different approach. This analysis is followed by a discussion on how high-NA conventional (also called 'incoherent' or Type I) and confocal (also called 'coherent' or Type II) microscopes image point objects. Frequent comparisons are made with results from paraxial (Fourier) theory to highlight how the two descriptions differ. We then look at the properties of one, two and three photon high-NA fluorescent microscopes. In the third part of this chapter we discuss extensions of the imaging theory by generalising the Debye–Wolf theory to include description of focusing and imaging through stratified media – a scenario frequently encountered in practical biological microscopy. Finally, we further generalise our theory which permits the calculation of images formed by high-NA optical systems from samples of almost arbitrary complexity.

Modern microscope design has significantly deviated from the old principles in that microscopes used to be designed such that the high-NA lens formed an intermediate image of the object usually at moderate transverse magnification at around 170 mm from the aperture stop, which was then magnified further by the eyepiece, creating a virtual image of the object at reasonably large transverse magnification. We have discussed the design principles of these types of systems in previous chapters so we shall not elaborate further. The main disadvantage of the traditional design was that it was not possible to freely alter the distance between the high NA lens and the eyepiece to possibly accommodate additional optics, such as beam splitters or dichroic mirrors. In addition, objective lenses with finite conjugates must be designed for specific beam splitter thickness, refractive index and Abbe number so if a different beam splitter, or dichroic mirror is used the performance of the resulting optical system will significantly degrade as compared to the one used with the nominal beam splitter. For these reasons microscope manufacturers modified their design strategy: modern optical microscope objective lenses are designed for infinite conjugates, i.e. the objective lens creates an image of the object at infinity and another lens, the so-called tube lens forms a real image at finite conjugate which is then re-imaged by an eyepiece. We have seen in Chapter 8 that the Luneburg derivation of the Debye–Wolf integral means that when the aperture stop of a high-NA lens is placed in the front focal plane so that the exit pupil is at infinity, as discussed below, geometric optics can be used to determine the electric field incident upon the aperture stop. Because the position of the aperture stop is critical, it is always kept within the mechanical unit of the objective lens. Even though it would be desirable to arrange the tube lens such that it forms an afocal and thus doubly telecentric imaging system with the objective lens,[11.1] in practice

[11.1] The practical advantages to constructing a microscope this way are considerable, but it should be noted that the use of infinity corrected objective lenses does not guarantee arbitrary objective and tube lens distance flexibility. Some manufacturers use afocality most of the time cannot be achieved. Nevertheless for the present analysis we shall assume that the lenses are arranged so for the sake of simpler formulation. Even though our formalism can be adapted to the case when the tube lens is not placed in an afocal arrangement with the objective lens, the practical differences between the original and adapted versions are so small that in the vast majority of cases the effort spent on such extension is not justified.

As discussed, we shall assume that the two lenses are arranged in a $4f$ configuration, that is to say they are placed such that their foci coincide. The object is then placed in the front (first) focal plane of the high-NA lens and the image appears in the back (second) focal plane of the tube lens in a phase correct manner. As discussed above, the aperture stop is placed between the two lenses at the common focus position. This arrangement is therefore afocal and telecentric from both image and object side.

As pointed out in Chapter 8, when diffraction problems are solved using a boundary value at infinity, the boundary value may be specified in terms of geometrical optics. We recognise the following fact: *having a telecentric system from the image side makes the aperture stop appear at infinity. This means that for such an optical system using a boundary value at infinity is perfectly legitimate because the plane on which the specification occurs is the aperture plane which, from both the image and object sides, is situated at infinity.*

Consequently, our task is simply to develop a formalism that is capable of producing the electric and magnetic field vectors at the aperture plane. The solution comes from generalising the Jones matrices.

11.1 Vector Ray Tracing – The Generalised Jones Matrix Formalism

It was described in Chapter 8 that, when the Debye–Wolf integral is used, the boundary value function of the integral is specified by the electric vector of the geometric optics approximation. In this section we introduce the generalised Jones matrix formalism **[343]** that will be used to specify the boundary value function for our diffraction problem. The matrix formalism also includes depolarising effects[11.2] of high-numerical-aperture lenses first explained by Inoué **[156]** and Kubota and Inoué **[198]**. It must be emphasised that our method described in this section does not perform finite ray-tracing as defined in Chapter 4 and neither is it designed to be used with imperfect optical systems.

11.1.1 Preliminary Considerations

There are two classes of optical components that we are concerned with that affect the polarisation of light: one that modifies the direction of propagation of a normally incident ray and one that does not. An example of the former is a lens and an example of the latter is a polariser. Considering lenses first, let us assume an aberration-free optical system of revolution. The axis of revolution (*optic axis*) and the marginal ray define a meridional plane as discussed in Chapter 4. Since the optical system is assumed to be aberration free, the Gaussian image point also lies in the meridional plane. All rays connecting a pair of conjugate on-axis object and image points are non-skew rays, located in the meridional plane. For the moment we shall assume that the object point is located on axis at infinity so all rays in object space are parallel to the optic axis. The aperture stop is located in the front focal plane of the lens and so the principal ray is parallel with the optic axis in image space – the optical system is therefore telecentric from the image side (see Section 4.10.5). Since in our case the point object is at infinity and all rays are parallel to the optic axis, by the Malus–Dupin theorem (see Section 4.2) the wavefronts must be plane, propagating parallel to the optic axis. As described in Chapter 1, the state of polarisation of a fully polarised wave is defined by the time evolution of the end point of the electric vector. Because plane waves are necessarily transverse the electric vector at any instant in time is perpendicular to a ray.

When a ray traverses a lens its direction changes. In practice this occurs at every surface the ray traverses, and so if we have sufficient knowledge of the optical system (angle of incidence and the refractive indices on either side of the surface) it is possible to calculate, using Snell's law, the direction of the refracted ray. With the angle of refraction also known it is a straightforward matter to compute the Fresnel transmission coefficients t_s and t_p (see Eq. (1.8.28)). The incident

off-the-shelf achromats for the tube lens, others design these specifically for their objective lenses often leaving the correction of residual aberrations for the tube lens. Note that changing the distance between the objective lens and tube lens can result in the introduction of off-axis aberrations that depend on the refraction invariant for the principal ray B (Section 5.3.2 – coma, astigmatism and distortion) because with changing lens distances the footprint of the off-axis ray pencil changes on the surface of the tube lens.

11.2 The term 'depolarisation' is discussed in more detail in Section 11.3.

electric field associated with a given ray is then decomposed into s- and p-components, E_s and E_p so that the transmitted electric field is given as $\mathbf{E}_t = t_s E_s \hat{\mathbf{w}}_s + t_p E_p \hat{\mathbf{w}}_p$ with $\hat{\mathbf{w}}_s$ and $\hat{\mathbf{w}}_p$ being the unit vectors in the corresponding directions. This procedure can then be repeated at each surface of the optical system for every ray traversing it. It is important to note that in the present case it is not possible to reproduce an arbitrary non-skew ray from two other non-skew rays, unlike in paraxial systems, because the angle of incidence and therefore the Fresnel coefficients depend on the ray height.

Whilst the procedure outlined above is used in finite ray-tracing by optical design programs, it would be rather difficult to follow it in analytic calculations. Therefore, we elect to follow a different path: as discussed later in this chapter, we replace the effect of refraction at the *individual* surfaces by the refraction at a *single* surface, called the *equivalent refractive surface* or *equivalent refractive locus*. This notion is not unfamiliar in the world of geometrical optics where a paraxial thick lens is represented by two principal planes that are the loci of the refraction of rays. When representing a thick lens by means of its principal planes one does not substitute the physical surfaces of the lens with the principal planes, one merely represents the function of the lens that way (see Section 5.4.1 on the construction of the 'equivalent thin lens' of a thick lens). The principal plane is a special case of the equivalent refractive surface which, upon making the paraxial approximation, becomes a plane. The equivalent refractive surface can be constructed just as the principal plane **[181]**: rays that enter from the object side parallel to the optic axis exit the optical system and converge to the (second) Gaussian focal point; the intersection points produced by extending both an incident ray and its corresponding exit ray define the equivalent refractive surface, which is rotationally symmetric for optical systems of revolution. For example, as discussed above, for an aberration-free optical system satisfying Abbe's sine condition Eq. (4.8.4) the equivalent refractive surface is the spherical exit pupil (see Fig. 9.5).

For aplanatic systems the equivalent refractive surface (*Gaussian reference sphere*) can then be readily modelled by a coordinate transformation that changes the direction of the ray such that, while it remains in the meridional plane, it is directed to the Gaussian focus. The electric field corresponding to the ray is decomposed into two components: one parallel E_p and one perpendicular E_s to the meridional plane. Ignoring for a moment the Fresnel transmission and refraction coefficients, E_s is unaffected by focusing and it will traverse the lens without being changed. On the other hand, E_p needs to remain perpendicular to the ray after the ray changes direction as shown in Fig. 11.1. Because the ray changed direction, E_p, that had no z component before focusing in the x–y–z Cartesian coordinate system, now clearly has a longitudinal (z) component. The resulting electric vector is of course the superposition of E_p and E_s.

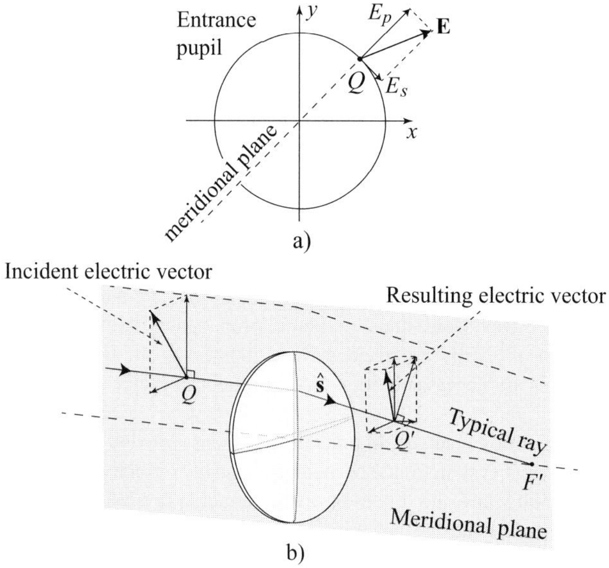

Figure 11.1: Decomposition of the electric field and its transition through a lens. a) The incident electric vector **E** is decomposed to components parallel E_p and perpendicular E_s to the meridional plane.
b) When the ray is refracted by the lens the direction (and magnitude) of the electric vector change so that in the image space a longitudinal component results.

11.1.2 Definition of the Generalised Jones Matrices

In what follows it is convenient to define spherical co-ordinates ϑ and ϕ so that a ray makes an angle ϑ with the negative direction of the optical axis ($-z$) and a meridional plane makes an angle ϕ with the positive x-direction (see Fig. 8.6). As discussed in the previous section, for the treatment of an electric field traversing an optical system we decompose the electric vector into two components: one perpendicular and one parallel to a given meridional plane. This decomposition may be performed by the matrix $\mathbb{R}$:

$$\mathbb{R} = \begin{pmatrix} \cos\phi & \sin\phi & 0 \\ -\sin\phi & \cos\phi & 0 \\ 0 & 0 & 1 \end{pmatrix}, \tag{11.1.1}$$

where ϕ is measured from the $+x$ direction. This is clearly a transformation for rotation around the z axis. The effect of lenses changing the direction of ray propagation can be taken into account by the matrix $\mathbb{L}$:

$$\mathbb{L} = \begin{pmatrix} \cos\vartheta & 0 & -\sin\vartheta \\ 0 & 1 & 0 \\ \sin\vartheta & 0 & \cos\vartheta \end{pmatrix}, \tag{11.1.2}$$

where the angle ϑ is defined with respect to the $-z$ direction as before. Matrices for the interaction of the electric vector with interfaces between two homogeneous and isotropic media may readily be written after considering that when plane waves are reflected or refracted by interfaces, E_s remains unchanged whilst E_p changes direction. The amplitude of both components will change as described by the Fresnel reflection and transmission coefficients. The description for the interaction between the electric vector and surfaces, therefore, requires two matrices:

$$\mathbb{I}_t = \begin{pmatrix} t_p & 0 & 0 \\ 0 & t_s & 0 \\ 0 & 0 & t_p \end{pmatrix}, \tag{11.1.3}$$

where t_p and t_s are the Fresnel transmission coefficients (see Eq. (1.8.28)):

$$t_p = \frac{2\sin\vartheta_t\cos\vartheta_i}{\sin(\vartheta_i+\vartheta_t)\cos(\vartheta_i-\vartheta_t)}, \tag{11.1.4a}$$

$$t_s = \frac{2\sin\vartheta_t\cos\vartheta_i}{\sin(\vartheta_i+\vartheta_t)}, \tag{11.1.4b}$$

where ϑ_i and ϑ_t represent the incident and refracted angles, respectively. Similarly:

$$\mathbb{I}_r = \begin{pmatrix} r_p & 0 & 0 \\ 0 & r_s & 0 \\ 0 & 0 & r_p \end{pmatrix}, \tag{11.1.5}$$

where r_p and r_s are the Fresnel coefficients:

$$r_p = \frac{\tan(\vartheta_i-\vartheta_t)}{\tan(\vartheta_i+\vartheta_t)}, \tag{11.1.6a}$$

$$r_s = -\frac{\sin(\vartheta_i-\vartheta_t)}{\sin(\vartheta_i+\vartheta_t)}. \tag{11.1.6b}$$

It should be noted that the above matrices are only applicable when the electric field is decomposed to s and p components. For more complex interface structures (e.g. a stratified interface structure, composed of a sequence of thin layers) it is possible to create similar matrices. For such a stratified interface we first form the product of the so-called transfer matrices associated with the individual thin layers of the thin-layer stack (see Subsection 1.9.2 for the calculation of the elements of such a transfer matrix). From the matrix elements of the resulting *composite* matrix for the thin-layer stack, we construct the corresponding transmission or reflection matrices, $\mathbb{I}_t$ and $\mathbb{I}_r$ respectively. The procedure to obtain $\mathbb{I}_t$ and $\mathbb{I}_r$ is explained in **[345]** and is discussed later on in this chapter.

In order to be able to apply the matrices to lenses we need to return to the discussion above, when we ignored the Fresnel transmission in computing the electric vector traversing the lens. We said before that for analytic work it is not really feasible to trace Fresnel transmission through individual surfaces of the lens, not the least because microscope manufacturers jealously guard lens prescription data. However, as discussed below, based on experimental results and data fitting **[343]** one can find that a single spherical air/glass surface can model qualitatively the effect of Fresnel transmission through the entire lens. For this reason we shall now incorporate the Fresnel coefficients into the lens model as follows.

11.1.2.1 Transmission through a Lens Including the Fresnel Coefficients

Let a collimated light beam be incident upon a lens. The incident electric vector[11.3] is, therefore, of the form $\mathcal{E}_0 = (\mathcal{E}_x, \mathcal{E}_y, 0)^T$. This vector is now decomposed to $\mathcal{E}_p$ and $\mathcal{E}_s$ components, hence giving the resulting vector:

$$\boldsymbol{\mathcal{E}}_1 = \mathbb{R} \cdot \boldsymbol{\mathcal{E}}_0 = \begin{pmatrix} \mathcal{E}_x \cos\phi + \mathcal{E}_y \sin\phi \\ \mathcal{E}_y \cos\phi - \mathcal{E}_x \sin\phi \\ 0 \end{pmatrix} . \tag{11.1.7}$$

The ideal, high-aperture lens rotates $\mathcal{E}_p$ by an angle ϑ but leaves $\mathcal{E}_s$ unchanged (Fig. 8.6). The rotation due to the lens is expressed by the matrix $\mathbb{L}$ so that the resulting electric vector $\boldsymbol{\mathcal{E}}_2$, still expressed in terms of s and p components, is given by:

$$\boldsymbol{\mathcal{E}}_2 = \mathbb{L} \cdot \boldsymbol{\mathcal{E}}_1 = \begin{pmatrix} \mathcal{E}_x \cos\phi\cos\vartheta + \mathcal{E}_y \sin\phi\cos\vartheta \\ \mathcal{E}_y \cos\phi - \mathcal{E}_x \sin\phi \\ \mathcal{E}_x \cos\phi \sin\vartheta + \mathcal{E}_y \sin\phi\sin\vartheta \end{pmatrix} . \tag{11.1.8}$$

The effect of the Fresnel transmission coefficients is computed by:

$$\boldsymbol{\mathcal{E}}_3 = \mathbb{I}_t \cdot \boldsymbol{\mathcal{E}}_2 = \begin{pmatrix} \mathcal{E}_x t_p \cos\phi\cos\vartheta + \mathcal{E}_y t_p \sin\phi\cos\vartheta \\ \mathcal{E}_y t_s \cos\phi - \mathcal{E}_x t_s \sin\phi \\ \mathcal{E}_x t_p \cos\phi\sin\vartheta + \mathcal{E}_y t_p \sin\phi\sin\vartheta \end{pmatrix} . \tag{11.1.9}$$

Finally, the vector $\boldsymbol{\mathcal{E}}_3$ needs to be transformed back to Cartesian basis:

$$\boldsymbol{\mathcal{E}} = \mathbb{R}^{-1} \cdot \boldsymbol{\mathcal{E}}_3 = \frac{1}{2}\begin{pmatrix} \mathcal{E}_x\left[\left(t_p\cos\vartheta + t_s\right) + \left(t_p\cos\vartheta - t_s\right)\cos 2\phi\right] + \mathcal{E}_y\left(t_p\cos\vartheta - t_s\right)\sin 2\phi \\ \mathcal{E}_y\left[\left(t_p\cos\vartheta + t_s\right) - \left(t_p\cos\vartheta - t_s\right)\cos 2\phi\right] + \mathcal{E}_x\left(t_p\cos\vartheta - t_s\right)\sin 2\phi \\ 2t_p\sin\vartheta\left(\mathcal{E}_x\cos\phi + \mathcal{E}_y\sin\phi\right) \end{pmatrix} . \tag{11.1.10}$$

It should be noted that for an ideal lens $t_s = t_p = 1$, i.e. for perfect anti-reflection coatings, and for an x polarised incident plane wave ($\mathcal{E}_{0,x} = U_0, \mathcal{E}_{0,y} = 0$) hence Eq. (11.1.10) results in a simplified form:

$$\boldsymbol{\mathcal{E}} = \frac{U_0}{2}\begin{pmatrix} (\cos\vartheta + 1) + (\cos\vartheta - 1)\cos 2\phi \\ (\cos\vartheta - 1)\sin 2\phi \\ 2\sin\vartheta\cos\phi \end{pmatrix} . \tag{11.1.11}$$

Matrix operations of the above derivation can be written in a more concise form:

$$\boldsymbol{\mathcal{E}} = \mathbb{R}^{-1} \cdot \mathbb{I}_t \cdot \mathbb{L} \cdot \mathbb{R} \cdot \boldsymbol{\mathcal{E}}_0 . \tag{11.1.12}$$

Equation (11.1.12) constitutes one possible way of using our matrix formalism. The following, second example describes the resulting electric vector when light is focused by a high-aperture lens through a plane interface between two dielectric and isotropic media:

$$\boldsymbol{\mathcal{E}} = \mathbb{R}^{-1} \cdot \mathbb{I}_t^i \cdot \mathbb{P} \cdot \mathbb{I}_t \cdot \mathbb{L} \cdot \mathbb{R} \cdot \boldsymbol{\mathcal{E}}_0 , \tag{11.1.13}$$

where the superscript i signifies that the operation is to be carried out on the interface. Note that the $\mathbb{I}_t \cdot \mathbb{L} \cdot \mathbb{R} \cdot \boldsymbol{\mathcal{E}}_0$ part is of course identical to the case of the lens Eq. (11.1.12). The additional matrix $\mathbb{P}$ rotates the components such that the matrix $\mathbb{I}$ can be applied for the plane interface:

$$\mathbb{P} = \begin{pmatrix} \cos\theta & 0 & \sin\theta \\ 0 & 1 & 0 \\ -\sin\theta & 0 & \cos\theta \end{pmatrix} . \tag{11.1.14}$$

[11.3] We are using here the notation $\mathcal{E}$ to denote the electric vector of geometrical optics approximation.

11.1.2.2 Generalised Jones Matrices for Retarders and Polarisers

It now remains to discuss the last remaining group of optical components: the retarders and polarisers. The Jones matrices are well known **[17]** and readily available for the description of the state of polarisation as light traverses an optical system, but they are inherently not capable of describing non-collimated beams. This is because the Jones matrices are 2×2 in dimension. Matrices of the present formalism are 3×3 in dimension so that they are capable of describing non-collimated beams as well as collimated ones. Within the geometric optics approximation, when a wavefront with oblique propagation direction is incident upon, say, a polariser then its propagation direction does not change after passing through the polariser (whose two surfaces are parallel to each other) apart from a parallel shift with respect to the incident ray.[11.4] By resolving the incident electric vector into parallel and perpendicular components**[343]** with respect to the surface of the polariser it may be stated that the propagation direction can only be maintained if the polariser does not modify the perpendicular component of the incident electric vector. This argument can readily be extended to all polarising devices that consist of two parallel surfaces (such as wave plates, Babinet–Soleil compensators, etc.). In the present formalism, therefore, the Jones matrix of a linear retarder may be written as:

$$\mathbb{BS} = \begin{pmatrix} \cos\frac{\delta}{2} + i\cos 2\beta \sin\frac{\delta}{2} & i\sin 2\beta \sin\frac{\delta}{2} & 0 \\ i\sin 2\beta \sin\frac{\delta}{2} & \cos\frac{\delta}{2} - i\cos 2\beta \sin\frac{\delta}{2} & 0 \\ 0 & 0 & 1 \end{pmatrix}, \tag{11.1.15}$$

where β is the azimuth of the fast axis of the linear retarder, δ is the relative retardation and the unity matrix element $(3,3)$ signifies that the device does not affect the z component of the incident electric vector. It is easy to show from Eq. (11.1.15) that the generalised Jones matrix of a linear quarter-wave plate may be written as:

$$\mathbb{W}_{\lambda/4} = \frac{\sqrt{2}}{2}\begin{pmatrix} 1 + i\cos 2\beta & i\sin 2\beta & 0 \\ i\sin 2\beta & 1 - i\cos 2\beta & 0 \\ 0 & 0 & 1 \end{pmatrix}, \tag{11.1.16}$$

and that for a half-wave plate:

$$\mathbb{W}_{\lambda/2} = \frac{\sqrt{2}}{2}\begin{pmatrix} i\cos 2\beta & i\sin 2\beta & 0 \\ i\sin 2\beta & -i\cos 2\beta & 0 \\ 0 & 0 & 1 \end{pmatrix}. \tag{11.1.17}$$

Finally, the generalised Jones matrix corresponding to an ideal linear polariser is given by:

$$\mathbb{O} = \begin{pmatrix} \cos^2\beta & \sin\beta\cos\beta & 0 \\ \sin\beta\cos\beta & \sin^2\beta & 0 \\ 0 & 0 & 1 \end{pmatrix}. \tag{11.1.18}$$

It is of interest to note that whereas the first three matrices defined above ($\mathbb{BS}$, $\mathbb{W}_{\lambda/4}$ and $\mathbb{W}_{\lambda/2}$) have an inverse, the matrix $\mathbb{O}$ does not. It should also be noted that the above matrices for retarders and the polariser correspond to idealised optical components. It is, however, possible to treat non-ideal devices such as partial polarisers with the same formalism.

11.2 Vectorial Point-spread Function

We are now in possession of all tools to be able to compute the vector PSF using the Debye–Wolf integral and assuming that the Fresnel number of the lens is sufficiently high (see discussion in Section 8.3).

11.2.1 Linearly Polarised Incident Illumination

We start by recalling Eq. (8.3.25):

$$\mathbf{E}\left(\mathbf{r}_p\right) = -\frac{if}{\lambda} \iint\limits_{\hat{s}_x^2+\hat{s}_y^2 \le 1} \frac{\mathcal{E}(\hat{s}_x, \hat{s}_y)}{\hat{s}_z} \exp(ik\hat{\mathbf{s}}\cdot\mathbf{r}_p) d\hat{s}_x d\hat{s}_y, \tag{11.2.1}$$

[11.4] This statement does not, of course, apply to Wollaston prisms and other birefringent components that cause the light ray to change direction.

where $\mathcal{E}$ can be written using the generalised Jones matrices by

$$\mathcal{E} = \mathbb{R}^{-1} \cdot \mathbb{L} \cdot \mathbb{R} \cdot (U_0, 0, 0)^T ,$$

where U_0 is the electric field amplitude, measured in units [V/m], and for simplicity we have assumed that it is x polarised and that Fresnel transmission coefficients at each surface can be set to unity. Of course the choice of any other transverse polarisation would have been possible which will be demonstrated in the next section. From the above equation we have

$$\mathcal{E} = \frac{U_0}{2} \begin{pmatrix} (\cos\vartheta + 1) + (\cos\vartheta - 1)\cos 2\phi \\ (\cos\vartheta - 1)\sin 2\phi \\ 2\sin\vartheta\cos\phi \end{pmatrix} ,$$

which formula is of course identical to Eq. (11.1.11).

Substituting $\hat{\mathbf{s}} = (-\sin\vartheta\cos\phi, -\sin\vartheta\sin\phi, \cos\vartheta)$ for the unit vector in the direction QF (Q is a general point on the exit pupil sphere) and[11.5] $\mathbf{r}_p = (r_t\cos\gamma, r_t\sin\gamma, z_p)$ into Eq. (11.2.1) we have

$$\begin{aligned} E_x(r_t,\gamma,z_p) = -\frac{ifU_0}{2\lambda}\int_0^{\alpha}\int_0^{2\pi} & \sqrt{\cos\vartheta}\,[(\cos\vartheta + 1) + (\cos\vartheta - 1)\cos 2\phi] \\ & \times \exp[-ikr_t\sin\vartheta\cos(\phi-\gamma)]\exp(ikz_p\cos\vartheta)\sin\vartheta d\vartheta d\phi, \end{aligned} \quad (11.2.2a)$$

$$\begin{aligned} E_y(r_t,\gamma,z_p) = -\frac{ifU_0}{2\lambda}\int_0^{\alpha}\int_0^{2\pi} & \sqrt{\cos\vartheta}\,(\cos\vartheta - 1)\sin 2\phi \\ & \times \exp[-ikr_t\sin\vartheta\cos(\phi-\gamma)]\exp(ikz_p\cos\vartheta)\sin\vartheta d\vartheta d\phi, \end{aligned} \quad (11.2.2b)$$

$$\begin{aligned} E_z(r_t,\gamma,z_p) = -\frac{ifU_0}{\lambda}\int_0^{\alpha}\int_0^{2\pi} & \sqrt{\cos\vartheta}\,\sin\vartheta\cos\phi \\ & \times \exp[-ikr_t\sin\vartheta\cos(\phi-\gamma)]\exp(ikz_p\cos\vartheta)\sin\vartheta d\vartheta d\phi , \end{aligned} \quad (11.2.2c)$$

where the term $\sqrt{\cos\vartheta}$ arises from Eq. (9.2.9) and is often referred to as 'apodisation function' or 'radiometric effect', $d\hat{s}_x d\hat{s}_y/\hat{s}_z = \sin\vartheta d\vartheta d\phi$ and α is the angle of the marginal ray in image space. The integration by ϕ can be performed analytically using Eq. (8.4.37) to give

$$\begin{aligned} E_x(r_t,\gamma,z_p) = &-\frac{ikfU_0}{2}\int_0^{\alpha}\sqrt{\cos\vartheta}\,(\cos\vartheta + 1)J_0(ikr_t\sin\vartheta)\exp(ikz_p\cos\vartheta)\sin\vartheta d\vartheta \\ &+ \frac{ikfU_0}{2}\cos 2\gamma\int_0^{\alpha}\sqrt{\cos\vartheta}\,(\cos\vartheta - 1)J_2(ikr_t\sin\vartheta)\exp(ikz_p\cos\vartheta)\sin\vartheta d\vartheta, \end{aligned} \quad (11.2.3a)$$

$$E_y(r_t,\gamma,z_p) = \frac{ikfU_0}{2}\sin 2\gamma\int_0^{\alpha}\sqrt{\cos\vartheta}\,(\cos\vartheta - 1)J_2(ikr_t\sin\vartheta)\exp(ikz_p\cos\vartheta)\sin\vartheta d\vartheta, \quad (11.2.3b)$$

$$E_z(r_t,\gamma,z_p) = -kfU_0\cos\gamma\int_0^{\alpha}\sqrt{\cos\vartheta}\,\sin^2\vartheta J_1(ikr_t\sin\vartheta)\exp(ikz_p\cos\vartheta)d\vartheta . \quad (11.2.3c)$$

These equations can be written in a more concise form:

$$E_x(r_t,\gamma,z_p) = -\frac{ikfU_0}{2}[I_0(r_t,z_p) + I_2(r_t,z_p)\cos 2\gamma], \quad (11.2.4a)$$

$$E_y(r_t,\gamma,z_p) = -\frac{ikfU_0}{2}I_2(r_t,z_p)\sin 2\gamma, \quad (11.2.4b)$$

$$E_z(r_t,\gamma,z_p) = -kfU_0 I_1(r_t,z_p)\cos\gamma , \quad (11.2.4c)$$

where

$$I_0(r,z) = \int_0^{\alpha}\sqrt{\cos\vartheta}\,\sin\vartheta(1 + \cos\vartheta)J_0(kr\sin\vartheta)\exp(ikz\cos\vartheta)d\vartheta, \quad (11.2.4d)$$

$$I_1(r,z) = \int_0^{\alpha}\sqrt{\cos\vartheta}\,\sin^2\vartheta J_1(kr\sin\vartheta)\exp(ikz\cos\vartheta)d\vartheta, \quad (11.2.4e)$$

$$I_2(r,z) = \int_0^{\alpha}\sqrt{\cos\vartheta}\,\sin\vartheta(1 - \cos\vartheta)J_2(kr\sin\vartheta)\exp(ikz\cos\vartheta)d\vartheta , \quad (11.2.4f)$$

[11.5] Note that in our current notation $|\mathbf{r}_p| = \sqrt{r_t^2 + z_p^2}$ so r_t is the length of the vector $\mathbf{r}_p$ projected on the transverse plane.

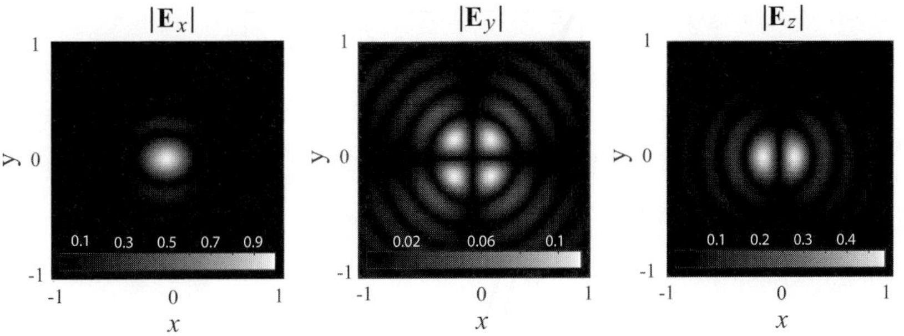

Figure 11.2: The modulus of the Cartesian components of the electric field (given by Eq. (11.2.4)) distribution produced by the Debye–Wolf integral in focal plane for $\lambda = 405$ nm, $\sin\alpha = 1.0$. The plots are normalised to $|\mathbf{E}(0,0)|$. Transverse axes are in μm.

where r and z are dummy variables in the radial and axial directions. For completeness and without giving the derivation we point out the magnetic field $\mathbf{H}$ may also be obtained in terms of the functions I_n. Its Cartesian components are then given by:

$$\begin{aligned} H_x(r_t,\gamma,z_p) &= -\frac{ikfU_{m0}}{2} I_2(r_t,z_p)\sin 2\gamma, \\ H_y(r_t,\gamma,z_p) &= -\frac{ikfU_{m0}}{2}[I_0(r_t,z_p) - I_2(r_t,z_p)\cos 2\gamma], \\ H_z(r_t,\gamma,z_p) &= -kfU_{m0}I_1(r_t,z_p)\sin\gamma\,, \end{aligned} \tag{11.2.5}$$

where U_{m0} [A/m] is the magnetic field strength. Note that these equations can also be written directly from Eq. (9.5.47) where the relationship between the I and the V functions is as follows: I_0 corresponds to $s_1^2 V_{0,0}^{0,E}$, I_1 to $s_1^2 V_{0,1}^{0,E}$, and I_2 to $-s_1^2 V_{0,2}^{0,E}$. Also, as discussed in connection with Eqs. (9.5.46) and (9.5.47), a relationship exists with Ignatowsky's original formulas of Eqs. (G.4.18) and (G.4.20) with the understanding that, apart from Ignatowsky's $\exp(i\omega t)$ convention, $I_0 = I_2$, $I_1 = I_1$ and $I_2 = I_3$ (the first function refers to Eq. (11.2.4) above, the second to Ignatowsky's functions).

The modulus of the Cartesian components of the electric field in the focal plane of a high-NA lens is shown in Fig. 11.2 as calculated from Eq. (11.2.4). The system parameters used to compute the plots were $\lambda = 405$ nm, $\sin\alpha = 1$ and $k = 2\pi/\lambda$. Since these distributions are probably the most studied in the literature of diffraction optics **[225]**, **[292]**, **[293]** we shall not spend a great deal of effort in describing them in minute detail and so we restrict our discussion to the bare essentials. It is interesting to observe that the maximum field strength is in the x component. The maximum field strength in the z and y components is approximately 0.5× and 0.1× that of the x component. This ratio varies as the numerical aperture changes – the general trend is that the relative strength of both the y and the z components decreases with respect to the x component as the numerical aperture decreases. The distribution of the x component lacks rotational symmetry and it leads to the overall distribution that is not rotationally symmetric. In fact, this lack of symmetry, shown in Fig. 11.3, increases with increasing numerical aperture.

When the numerical aperture increases so does the $|\mathbf{E}|^2$ at the Gaussian focal point. Looking at Eq. (11.2.4) it is clear that for $r_t = 0$ and $z_p = 0$ we have

$$\begin{aligned} |\mathbf{E}|^2 &= \frac{k^2 f^2 U_0{}^2}{4}\left|\int_0^\alpha \sqrt{\cos\vartheta}\sin\vartheta(1+\cos\vartheta)d\vartheta\right|^2 \\ &= \frac{k^2 f^2 U_0{}^2}{225}[\cos^{3/2}\alpha\,(5+3\cos\alpha) - 8]^2\,, \end{aligned} \tag{11.2.6}$$

which is plotted in Fig. 11.3b, showing that the peak value of $|\mathbf{E}|^2$ is a nonlinear function of the marginal ray angle.

The electric field along the z axis is seen to be co-polarised with the incident field because from Eq. (11.2.4) for $r_t = 0$ we have $\mathbf{E}(r_t = 0, z_p) = (-ikfU_0 I_0, 0, 0)^T$. The distribution of $|\mathbf{E}(0,z_p)|^2/|\mathbf{E}(0,0)|^2$ is then shown in Fig. 11.4a for marginal ray angles of $\alpha = \pi/2$ and $\alpha = \pi/3$. Apart from the trivial observation, already described in conjunction with Fig. 11.3b, that the field is more concentrated for higher marginal ray angles, it is important to point out that the distribution changes shape for different marginal ray angles. This should be directly contrasted with results from the paraxial theory discussed

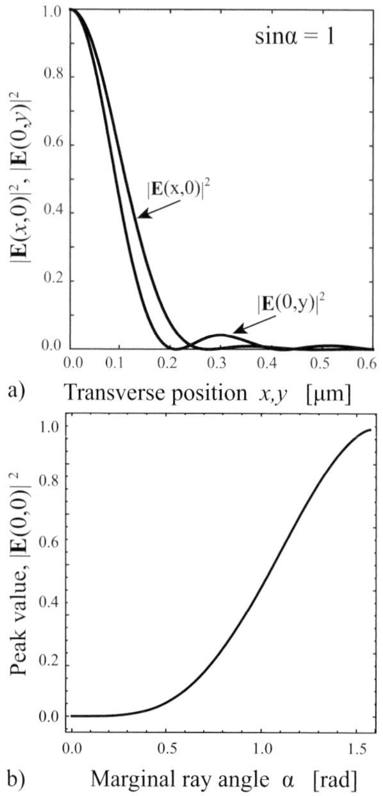

Figure 11.3: a) $|\mathbf{E}|^2$ along the x and y axes for $\alpha = \pi/2$ and $\lambda = 405$nm. The vertical axis is normalised to $|\mathbf{E}(0, 0)|^2$. b) The peak value of $|\mathbf{E}|^2$ as a function of α for $\lambda = 405$ nm. The vertical axis is normalised to $|\mathbf{E}(0, 0)|^2$ at $\alpha = \pi/2$.

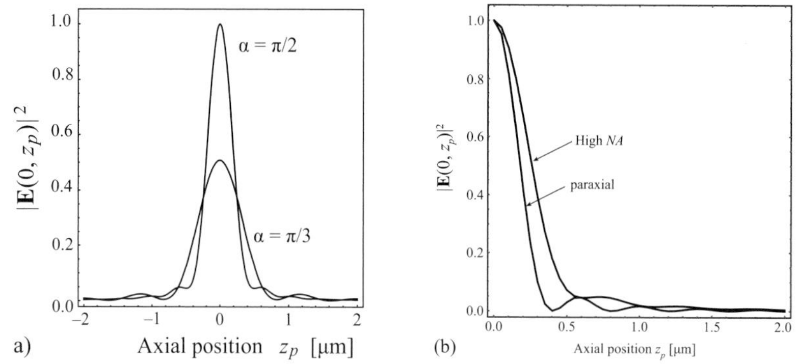

Figure 11.4: a) $|\mathbf{E}|^2$ along the z axis for $\alpha = \pi/2$ and $\pi/3$. The vertical axis is normalised to $|\mathbf{E}(0, 0)|^2$ at $\alpha = \pi/2$. b) Comparison of the axial distributions given by the Debye–Wolf diffraction formula and the paraxial approximation for $\alpha = \pi/2$. The vertical axis is individually normalised to the $|\mathbf{E}|^2$ value at the Gaussian focus. For both plots $\lambda = 405$ nm.

in conjunction with the Lommel integral Eq. (8.4.49) which showed that the distribution of $|U|^2$ does not change shape; it merely scales with the marginal ray angle. As a consequence of this observation it follows that, in our opinion, it does not make much sense to use normalised optical coordinates in the high-numerical-aperture cases, though there are several researchers who have attempted to do so. To appreciate how inaccurate the paraxial theory is in predicting the light distributions for the high-aperture case, we present Fig. 11.4b which shows a direct comparison of the axial $|\mathbf{E}|^2$

distributions for $\alpha = \pi/2$ and $\lambda = 405$ nm for both the high-NA and paraxial cases. Whereas, as discussed in Section 8.4, the paraxial axial distribution possesses real zeros, the corresponding high-numerical-aperture distribution does not. Furthermore, the paraxial theory is shown to underestimate the width of the axial distribution for this case by as much as $\approx 30\%$.

11.2.2 Complex Vector Fields

Since the seminal paper of Quabis *et al.* **[284]** there has been a surge of interest in the literature for vector beams of various sorts. The general discussion of these beams is outside the scope of the current book so it should suffice that they can be produced as linear superposition of the so-called Gauss–Laguerre beams. The general expression for the complex amplitude of these beams as a function of the transverse polar coordinates and the axial coordinate z in the propagation direction is given by Eq. (1.7.65). In the plane of the beam waist ($z = 0$), the complex amplitude is simplified since $|r_L(0)| = 1$ and $w(0) = w_0$ where $2w_0$ is the beam-waist diameter of an elementary Gaussian beam. In the beam waist we also have that, according to Eq. (1.7.16), the factor $1/q(z)$ takes on the value $-i\lambda/(\pi w_0^2)$. Expressed in the pupil plane coordinates (ρ, ϕ), we then have for the scalar amplitude distribution $U_{m,n}$ at the beam waist of a Gauss–Laguerre beam with order numbers (m, n) the expression

$$\begin{aligned} V_{m,n}(\rho,\phi,z=0) &= V_{m,n}(\rho,\phi) \\ &= U_0\sqrt{\frac{2}{\pi}\frac{1}{p_{GL}}}\left(\frac{\rho}{w_0}\right)^{|m|}\exp\left(-\frac{\rho^2}{w_0^2}\right)L_n^{|m|}\left(2\frac{\rho^2}{w_0^2}\right)\exp(im\phi) \\ &= U_{m,n}(\rho)\exp(im\phi)\,, \end{aligned} \qquad (11.2.7)$$

where $L_n^{|m|}(x)$ is the generalised Laguerre polynomial with integer indices n and m ($n \geq 0$). The functions $V_{m,n}$ are a set of scalar functions, derived from the Gauss–Laguerre beam-waist profile, that have been defined in the framework of self-imaging beams **[280]**.

The quantity p_{GL} in Eq. (11.2.7) can be evaluated by equating $V_L(\rho, \alpha, 0)$ of Eq. (1.7.65) and $V_{m,n}(\rho, \phi)$ of Eq. (11.2.7). With the aid of the normalisation constant $c_L(n, |m|)$ for a Gauss–Laguerre beam (see Eq. (1.7.67)), we obtain that

$$p_{GL} = \left\{\frac{(n+|m|)!}{2^{|m|}\; n!}\right\}^{1/2} . \qquad (11.2.8)$$

The dimensionless factor $1/p_{GL}$ in Eq. (11.2.7) quantifies the relative decrease in amplitude of a higher-order Gauss–Laguerre beam with respect to that of the lowest-order Gaussian beam. The factor p_{GL} itself is a measure of the increase in beam-waist size of a higher-order Gauss–Laguerre beam as compared to the value w_0 of the elementary Gaussian beam. To illustrate a particular use of these beams in generating complex vector beams consider the following superposition:

$$\begin{aligned} V_{1,n}(\rho,\phi) + V_{-1,n}(\rho,\phi) &= 2U_{1,n}(\rho)\cos\phi \\ i\left[V_{-1,n}(\rho,\phi) - V_{1,n}(\rho,\phi)\right] &= 2U_{1,n}(\rho)\sin\phi\,, \end{aligned} \qquad (11.2.9)$$

where the equalities follow because $U_{1,n} = U_{-1,n}$. The functional form of these two formulas is shown in Fig. 11.5. We may combine these two beams to produce two different vector beams:

$$\mathbf{E}(\rho,\phi) = (U_{1,n}(\rho)\cos\phi, U_{1,n}(\rho)\sin\phi, 0)^T = U_{1,n}(\rho)(\cos\phi, \sin\phi, 0)^T\,, \qquad (11.2.10a)$$

so that evidently we have a radially polarised electric field because $\hat{\boldsymbol{\rho}} = \cos\phi\hat{\mathbf{x}} + \sin\phi\hat{\mathbf{y}}$ where $\hat{\boldsymbol{\rho}}$ is the unit vector in the (transverse) radial direction. Alternatively we may choose

$$\mathbf{E}(\rho,\phi) = (-U_{1,n}(\rho)\sin\phi, U_{1,n}(\rho)\cos\phi, 0)^T = U_{1,n}(\rho)(-\sin\phi, \cos\phi, 0)^T\,, \qquad (11.2.10b)$$

which is clearly an azimuthally polarised light in anticlockwise direction because $\hat{\boldsymbol{\phi}} = -\sin\phi\hat{\mathbf{x}} + \cos\phi\hat{\mathbf{y}}$ where $\hat{\boldsymbol{\phi}}$ is the unit vector in the anticlockwise azimuthal direction. The experimental realisation of these beams (Fig. 11.6) has been published by several authors using a wide range of methods including interferometry **[339]**, using spatial light modulators **[257]**, laser resonators **[267]** or by specially rubbed liquid crystal converters **[33]**.

The radially and azimuthally polarised beams are just examples of the simplest kind and the complexity of the beams that can be generated is almost limitless. For example by using

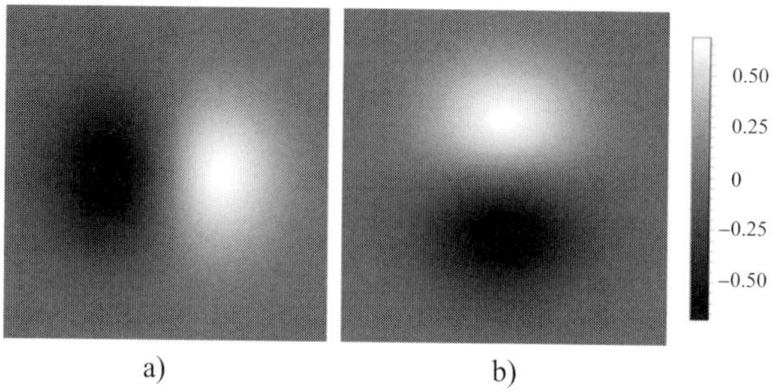

Figure 11.5: Distribution of the scalar field produced by superimposing two Gauss–Laguerre beams as shown in Eq. (11.2.9). a) $U_{1,n}(\rho)\cos\phi$, b) $U_{1,n}(\rho)\sin\phi$, both for $n = 1$.

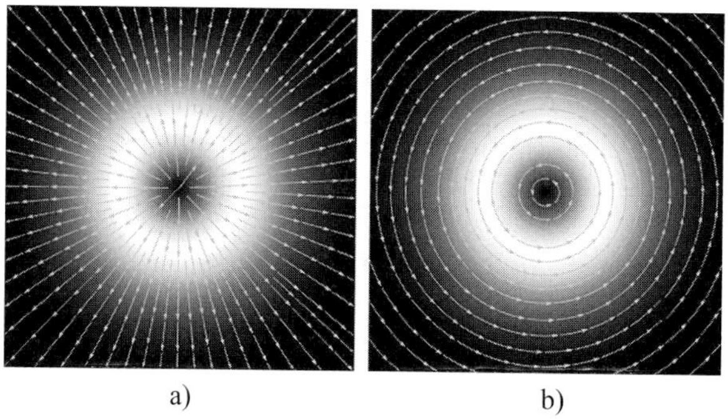

Figure 11.6: a) Radially and b) azimuthally polarised vector beams generated by superposition of Gauss–Laguerre beams given by Eq. (11.2.10). The arrows show the state of polarisation at a given point in the transverse plane and the grey scale of these images is proportional to the modulus of amplitude.

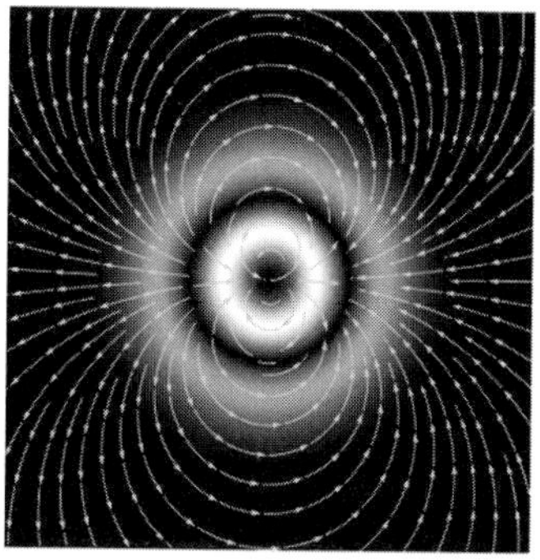

Figure 11.7: Same legend as in Fig. 11.6 but now for an exotic vector beam generated by superposition of Gauss–Laguerre beams as shown by Eq. (11.2.11).

$$\mathbf{E}(\rho,\phi) = \left(\Re\left\{V_{2,4}(\rho,\phi) + V_{-2,4}(\rho,\phi)\right\}, \Im\left\{V_{2,4}(\rho,\phi) - V_{-2,4}(\rho,\phi)\right\}, 0\right)^T , \tag{11.2.11}$$

we can obtain a polarisation distribution similar to that of an electric dipole with a double ring modulus of amplitude distribution as shown in Fig. 11.7.

11.2.3 Focusing of Complex Vector Beams

We now move to consider the complex vector beams that were discussed in the previous subsection focused by a high-NA lens. For simplicity we ignore Fresnel transmission coefficients related to the lens surfaces. Consider the Gaussian reference sphere of radius f centred on the Gaussian focal point and a non-skew ray intersecting the reference sphere at height ρ. After refraction at the equivalent refractive surface (which is the Gaussian reference sphere in this instance) the ray makes an angle ϑ with the optic axis in image space. Then $\rho = f \sin\vartheta$ which means that the ratio ρ/w_0 can be written as

$$\frac{\rho}{w_0} = \frac{f}{w_0}\sin\vartheta = \frac{\sin\vartheta}{\mu} , \tag{11.2.12}$$

where the factor $1/\mu = f/w_0$ serves as a short-hand notation for the ratio $\rho/w_0 = \sin\vartheta/\mu$ in the expressions describing Gaussian beams. We note that the angle α of the marginal ray is determined by the numerical aperture of the lens according to $\sin\alpha = NA/n$. Inevitably, part of the Gauss–Laguerre beam will be truncated by the limited aperture of the lens. The higher the radial and azimuthal order numbers of the Gauss–Laguerre beam, the larger the truncation effect will be. In addition a word of caution needs to be given here because the Gauss–Laguerre beams are only orthogonal for $\rho \in [0, \infty)$ which means that the truncated beams are *not* orthogonal. So it is not a simple matter to represent an arbitrary pupil plane distribution, defined only within the pupil itself, in terms of these beams just by calculating the projection integral of the arbitrary pupil plane distribution on the Gauss–Laguerre modes. A considerable amount of work has been done on how that problem could be solved using a different basis, but this is outside the scope of the book. The curious reader is referred to the papers by Sherif and Török **[314]**, **[313]**.

We can write $\mathcal{E}$ with the help of the generalised Jones matrices as

$$\mathcal{E}^{x}_{m,n} = \mathbb{R}^{-1}\cdot\mathbb{L}\cdot\mathbb{R}\cdot(V_{m,n}, 0, 0)^T . \tag{11.2.13}$$

Alternatively, for circularly polarised illumination we may write

$$\mathcal{E}^{\mathrm{LH}}_{m,n} = \mathbb{R}^{-1}\cdot\mathbb{L}\cdot\mathbb{R}\cdot\mathbb{BS}_{\lambda/4}\cdot(V_{m,n}, 0, 0)^T , \tag{11.2.14}$$

where

$$\mathbb{BS}_{\lambda/4} = \begin{pmatrix} 1 & -i & 0 \\ i & 1 & 0 \\ 0 & 0 & 1 \end{pmatrix}$$

is the generalised Jones matrix corresponding to a left-handed quarter-wave plate (axis set at $\beta = -\pi/4$). For x-polarised incident illumination (Eq. (11.2.13)) the above formulas give the following expression for the electric field $\mathbf{E}^{\mathrm{x}}_{m,n}$ in the vicinity of the focus:

$$E^{\mathrm{x}}_{m,n;x}(\mathbf{r}_p) = -\frac{ikf}{2}\left\{(-i)^m \exp(im\gamma)I_{m,n} - \frac{(-i)^{m+2}}{2}\exp[i(m+2)\gamma]I_{m+2,n} - \frac{(-i)^{m-2}}{2}\exp[i(m-2)\gamma]\,I_{m-2,n}\right\}, \tag{11.2.15a}$$

$$E^{\mathrm{x}}_{m,n;y}(\mathbf{r}_p) = \frac{kf}{2}\left\{\frac{(-i)^{m+2}}{2}\exp[i(m+2)\gamma]\,I_{m+2,n} - \frac{(-i)^{m-2}}{2}\exp[i(m-2)\gamma]I_{m-2,n}\right\}, \tag{11.2.15b}$$

$$E^{\mathrm{x}}_{m,n;z}(\mathbf{r}_p) = -\frac{ikf}{2}\left\{(-i)^{m+1}\exp[i(m+1)\gamma]I_{m+1,n} + (-i)^{m-1}\exp[i(m-1)\gamma]\,I_{m-1,n}\right\} , \tag{11.2.15c}$$

and for left-handed circularly polarised incident illumination (Eq. (11.2.14)):

$$E^{\mathrm{LH}}_{m,n;x}(\mathbf{r}_p) = -\frac{ikf}{2}\left\{(-i)^m \exp(im\gamma)I_{m,n} - (-i)^{m+2}\exp[i(m+2)\gamma]\,I_{m+2,n}\right\}, \tag{11.2.16a}$$

$$E^{\mathrm{LH}}_{m,n;y}(\mathbf{r}_p) = \frac{kf}{2}\left\{(-i)^m \exp(im\gamma)I_{m,n} + (-i)^{m+2}\exp[i(m+2)\gamma]\,I_{m+2,n}\right\}, \tag{11.2.16b}$$

$$E^{\mathrm{LH}}_{m,n;z}(\mathbf{r}_p) = -\,ikf(-i)^{m+1}\exp[i(m+1)\gamma]\,I_{m+1,n} , \tag{11.2.16c}$$

where we have used Eq. (8.4.37) and for brevity omitted the functional dependence of the I functions which are given by

$$I_{m,n}(r,z) = \int_0^{\alpha} U_{m,n}(\vartheta)\sqrt{\cos\vartheta}(1+\cos\vartheta)J_m(kr\sin\vartheta)\exp(ikz\cos\vartheta)\sin\vartheta d\vartheta, \tag{11.2.17a}$$

$$I_{m\pm1,n}(r,z) = \int_0^{\alpha} U_{m\pm1,n}(\vartheta)\sqrt{\cos\vartheta}J_{m\pm1}(kr\sin\vartheta)\exp(ikz\cos\vartheta)\sin^2\vartheta d\vartheta, \tag{11.2.17b}$$

$$I_{m\pm2,n}(r,z) = \int_0^{\alpha} U_{m\pm2,n}(\vartheta)\sqrt{\cos\vartheta}(1-\cos\vartheta)J_{m\pm2}(kr\sin\vartheta) \times \exp(ikz\cos\vartheta)\sin\vartheta d\vartheta\,. \tag{11.2.17c}$$

As a function of the angle ϑ the Gauss–Laguerre-type functions $U_{m,n}$ are given by

$$U_{m,n}(\vartheta) = U_0\sqrt{\frac{2}{\pi p_{GL}^2}}\left(\frac{\sin\vartheta}{\mu}\right)^{|m|}\exp\left[-\frac{\sin^2\vartheta}{\mu^2}\right]L_n^{|m|}\left[2\frac{\sin^2\vartheta}{\mu^2}\right]. \tag{11.2.17d}$$

If we introduce

$$K_{m,n}(r,\gamma,z) = -\frac{ikf}{2}(-i)^m\exp(im\gamma)I_{m,n}(r,z)\,,$$

Eq. (11.2.15) is given by:

$$\begin{aligned} E^{\mathrm{X}}_{m,n;x}(\mathbf{r}_p) &= K_{m,n} - \frac{1}{2}K_{m+2,n} - \frac{1}{2}K_{m-2,n}, \\ E^{\mathrm{X}}_{m,n;y}(\mathbf{r}_p) &= \frac{i}{2}K_{m+2,n} - \frac{i}{2}K_{m-2,n}, \\ E^{\mathrm{X}}_{m,n;z}(\mathbf{r}_p) &= K_{m+1,n} + K_{m-1,n}\,, \end{aligned} \tag{11.2.18}$$

and Eq. (11.2.16) is given by:

$$\begin{aligned} E^{\mathrm{LH}}_{m,n;x}(\mathbf{r}_p) &= K_{m,n} - K_{m+2,n}, \\ E^{\mathrm{LH}}_{m,n;y}(\mathbf{r}_p) &= iK_{m,n} + iK_{m+2,n}, \\ E^{\mathrm{LH}}_{m,n;z}(\mathbf{r}_p) &= 2K_{m+1,n}\,. \end{aligned} \tag{11.2.19}$$

For completeness the right-hand polarised field is given by:

$$\begin{aligned} E^{\mathrm{RH}}_{m,n;x}(\mathbf{r}_p) &= K_{m,n} - K_{m-2,n}, \\ E^{\mathrm{RH}}_{m,n;y}(\mathbf{r}_p) &= -iK_{m,n} - iK_{m-2,n}, \\ E^{\mathrm{RH}}_{m,n;z}(\mathbf{r}_p) &= 2K_{m-1,n}\,. \end{aligned} \tag{11.2.20}$$

The self-consistent nature of these equations can be checked by verifying that $\mathbf{E}^{\mathrm{LH}}_{m,n} + \mathbf{E}^{\mathrm{RH}}_{m,n} = 2\mathbf{E}^{\mathrm{X}}_{m,n}$. These equations reveal the power of the generalised Jones matrix formalism used in conjunction with the Debye–Wolf integral. The focused field for a given incident vector beam can be obtained as a linear superposition of the constituent vector beams as described above.

11.3 Focusing of Partially Coherent, Partially Polarised Light

We have discussed certain aspects of the scalar coherence theory in Chapter 10. There we have introduced $\mu(x,y)$, the mutual coherence function describing second-order spatial correlation, which was defined as the modulation depth produced by the interference of radiation emanating from two pinholes illuminated by a light source. The experimental concept behind this definition is simple: as the distance between the two pinholes is changed, the visibility (modulation) of the interference fringes changes. In general, although not always, the closer the two pinholes are the higher the visibility of the fringes, indicating a higher degree of correlation between the radiation due to the two point sources.

The above concept could be extended to electromagnetic waves in principle, but with the following observation: assume that the radiation reaching the screen having the pinholes is produced by an incandescent light source (i.e. black-body radiation). Let's place a very narrow band spectral filter in front of the source so that only a single wavelength will be selected from the radiation.[11.6] Two pinholes are made on the screen just as above, but now after each pinhole a polariser is placed. In the first instance the two polarisers are co-aligned, i.e. their corresponding polarisation axes are parallel. We then move the two pinholes sufficiently close to each other such that a discernible interference pattern appears on the screen. As discussed above, the modulation depth of the interference pattern corresponds to the mutual coherence function that is supposed to characterise the coherence properties of the source. Let us now gradually turn one of the polarisers from parallel to perpendicular setting. On turning the polariser the modulation of the interference fringes is decreased until the fringes completely disappear at the orthogonal polariser setting. Now, there is a problem, because the zero value of the mutual coherence function implies that there is no phase correlation between the radiation produced by the two pinholes. This is clearly untrue because the source has not changed and therefore visibility of the fringes, according to the above theory, should only depend on the distance between the two pinholes.

This thought experiment suggests that in the electromagnetic case a more subtle approach is needed. Therefore we shall implement a matrix formulation to characterise spatial and temporal coherence, with the second-order properties in the space-time domain described by the mutual coherence matrix **[365, 366, 373]**:

$$\mathbf{\Gamma}(\mathbf{r}_1,\mathbf{r}_2,\tau) = \begin{bmatrix} \langle E_x^*(\mathbf{r}_1,t)E_x(\mathbf{r}_2,t+\tau)\rangle & \langle E_x^*(\mathbf{r}_1,t)E_y(\mathbf{r}_2,t+\tau)\rangle \\ \langle E_y^*(\mathbf{r}_1,t)E_x(\mathbf{r}_2,t+\tau)\rangle & \langle E_y^*(\mathbf{r}_1,t)E_y(\mathbf{r}_2,t+\tau)\rangle \end{bmatrix}, \tag{11.3.1}$$

where E_x and E_y are the two orthogonal Cartesian components of the electric vector, τ is the time difference between two instances of time for the observation of the fields, and, as before, $\langle\ldots\rangle$ denotes the ensemble average over many monochromatic statistical realisations. Clearly, the sum of the diagonal elements is proportional to the intensity, whilst the off-diagonal elements characterise the correlation of E_x and E_y.

An alternative space-frequency domain representation, referred to as the cross-spectral density matrix (CSDM), is also available which is a Fourier transform of the mutual coherence matrix with respect to time **[376]**:

$$\begin{aligned}\mathbb{W}(\mathbf{r}_1,\mathbf{r}_2,\omega) &= \langle \mathbf{E}^*(\mathbf{r}_1,\omega)\mathbf{E}^T(\mathbf{r}_2,\omega)\rangle \\ &= \begin{bmatrix} \langle E_x^*(\mathbf{r}_1,\omega)E_x(\mathbf{r}_2,\omega)\rangle & \langle E_x^*(\mathbf{r}_1,\omega)E_y(\mathbf{r}_2,\omega)\rangle \\ \langle E_y^*(\mathbf{r}_1,\omega)E_x(\mathbf{r}_2,\omega)\rangle & \langle E_y^*(\mathbf{r}_1,\omega)E_y(\mathbf{r}_2,\omega)\rangle \end{bmatrix},\end{aligned} \tag{11.3.2}$$

where * denotes complex conjugation and ^{T} denotes transpose of a matrix. It must be emphasised that not only does the CSDM describe spatial coherence properties of a stochastic field (i.e. correlations between the field at two different points $\mathbf{r}_1$ and $\mathbf{r}_2$ in space), but it is also a measure of the correlation between orthogonal field components. Consequently the theory that we shall develop below using the CSDM-formalism applies to focusing of stochastic electromagnetic fields with arbitrary coherence and polarisation properties and it resolves the problem we encountered with our thought experiment in terms of extending the definition of the mutual coherence to polarised light. It should be noted that it is often said that light is 'partially polarised'. Classically this phrase means that the state of polarisation of the radiation can be written as a superposition of a fully polarised and a fully unpolarised radiation. For the latter the off-diagonal elements of $\mathbb{W}$ are zero.

The *degree of spectral coherence* can be defined **[337]** as

$$\zeta^2(\mathbf{r}_1,\mathbf{r}_2) = \frac{\|\mathbb{W}(\mathbf{r}_1,\mathbf{r}_2)\|_F^2}{\mathrm{tr}[\mathbb{W}(\mathbf{r}_1,\mathbf{r}_1)]\,\mathrm{tr}[\mathbb{W}(\mathbf{r}_2,\mathbf{r}_2)]}, \tag{11.3.3}$$

where $\mathrm{tr}[\cdots]$ and $\|\cdots\|_F$ denote the matrix trace and Frobenius norm respectively.

We have used the phrase 'depolarised' before in connection with polarised light traversing lenses. In order to put this phrase in context with regards to coherence theory, we now clarify that in our use depolarisation does not mean that the correlation between the components of the field is changed, which could only be done by a time varying process, but that transmission through the surfaces of the lens causes the generation of a new vector component that is fully correlated with the original field.

Returning now to discussing partially polarised focusing, from Eq. (11.3.2) we denote the CSDM in the focal region of a lens as

$$\mathbb{W}(\mathbf{r}_{p1},\mathbf{r}_{p2},\omega) \equiv \langle \mathbf{E}^*(\mathbf{r}_{p1},\omega)\mathbf{E}^T(\mathbf{r}_{p2},\omega)\rangle\,. \tag{11.3.4}$$

Given Eq. (11.3.4) and the Debye–Wolf integral Eq. (8.3.25) it is easy to determine the CSDM for light focused by a lens. By substitution we have

[11.6] Such narrow band filters do not exist in practice.

$$\mathbb{W}(\mathbf{r}_{p1}, \mathbf{r}_{p2}) = \frac{f^2}{\lambda^2} \int_0^{2\pi}\int_0^{2\pi}\int_0^{\alpha}\int_0^{\alpha} \langle \boldsymbol{\mathcal{E}}^*(\vartheta_1, \phi_1)\boldsymbol{\mathcal{E}}^T(\vartheta_2, \phi_2)\rangle \exp\left[ik\Delta_{12}\right] \times \exp\left[ik(-z_{p1}\cos\vartheta_1 + z_{p2}\cos\vartheta_2)\right] \sin\vartheta_1 \sin\vartheta_2 d\vartheta_1 d\vartheta_2 d\phi_1 d\phi_2 \,, \tag{11.3.5}$$

where $\Delta_{12} = r_{p1}\sin\vartheta_1\cos(\phi_1 - \gamma_1) - r_{p2}\sin\vartheta_2\cos(\phi_2 - \gamma_2)$ and we have dropped denoting the explicit dependence of $\mathbb{W}$ on ω for brevity but we remind the reader that the frequency dependence is still present in the equations via $k = \omega/c$. Knowledge of the CSDM in a single transverse plane is sufficient to calculate the CSDM on any transverse plane in the focal region via, for example, the Wolf equations **[376]**. Henceforth we shall thus make the simplifying assumption that $z_{p1} = z_{p2} = z_p$, i.e. we restrict attention to a single plane in the focal region. It is convenient to define the CSDM on the Gaussian reference sphere in an analogous way to Eq. (11.3.4) such that

$$\mathrm{w}(\vartheta_1, \phi_1, \vartheta_2, \phi_2) = \langle \boldsymbol{\mathcal{E}}^*(\vartheta_1, \phi_1)\boldsymbol{\mathcal{E}}^T(\vartheta_2, \phi_2)\rangle = \begin{bmatrix} \langle \mathcal{E}_x^*(\vartheta_1, \phi_1)\mathcal{E}_x(\vartheta_2, \phi_2)\rangle & \langle \mathcal{E}_x^*(\vartheta_1, \phi_1)\mathcal{E}_y(\vartheta_2, \phi_2)\rangle \\ \langle \mathcal{E}_y^*(\vartheta_1, \phi_1)\mathcal{E}_x(\vartheta_2, \phi_2)\rangle & \langle \mathcal{E}_y^*(\vartheta_1, \phi_1)\mathcal{E}_y(\vartheta_2, \phi_2)\rangle \end{bmatrix} , \tag{11.3.6}$$

which means that

$$\mathbb{W}(\mathbf{r}_1, \mathbf{r}_2) = \frac{f^2}{\lambda^2} \int_0^{2\pi}\int_0^{2\pi}\int_0^{\alpha}\int_0^{\alpha} \mathrm{w}(\vartheta_1, \phi_1, \vartheta_2, \phi_2) \exp\left[ik\Delta_{12}\right] \exp\left[ikz_p(\cos\vartheta_2 - \cos\vartheta_1)\right] \sin\vartheta_1 \sin\vartheta_2 d\vartheta_1 d\vartheta_2 d\phi_1 d\phi_2 \,. \tag{11.3.7}$$

In some applications it may be more useful to define the focused CSDM in terms of the CSDM in the front-focal plane of the lens, which we denote $\mathrm{w}_i(\vartheta_1, \phi_1, \vartheta_2, \phi_2) = \langle \boldsymbol{\mathcal{E}}_i^*(\vartheta_1, \phi_1)\boldsymbol{\mathcal{E}}_i^T(\vartheta_2, \phi_2)\rangle$. Using the generalised Jones matrices we have

$$\boldsymbol{\mathcal{E}}(\vartheta, \phi) = \sqrt{\cos\vartheta}\, \mathbb{R}^{-1}(\phi) \cdot \mathbb{L}(\vartheta) \cdot \mathbb{R}(\phi) \cdot \boldsymbol{\mathcal{E}}_i(\vartheta, \phi) = \mathbb{P}(\vartheta, \phi) \cdot \boldsymbol{\mathcal{E}}_i(\vartheta, \phi) \,. \tag{11.3.8}$$

Hence

$$\mathrm{w}(\vartheta_1, \phi_1, \vartheta_2, \phi_2) = \mathbb{P}^*(\vartheta_1, \phi_1) \cdot \mathrm{w}_i(\vartheta_1, \phi_1, \vartheta_2, \phi_2) \cdot \mathbb{P}^T(\vartheta_2, \phi_2) \,. \tag{11.3.9}$$

Substituting Eq. (11.3.9) in Eq. (11.3.7) yields

$$\mathbb{W}(\mathbf{r}_{p1}, \mathbf{r}_{p2}) = \frac{f^2}{\lambda^2} \int_0^{2\pi}\int_0^{2\pi}\int_0^{\alpha}\int_0^{\alpha} \mathbb{P}^*(\vartheta_1, \phi_1) \cdot \mathrm{w}_i(\vartheta_1, \phi_1, \vartheta_2, \phi_2) \cdot \mathbb{P}^T(\vartheta_2, \phi_2) \exp\left[ik\Delta_{12}\right] \times \exp\left[ikz_p(\cos\vartheta_2 - \cos\vartheta_1)\right] \sin\vartheta_1 \sin\vartheta_2 d\vartheta_1 d\vartheta_2 d\phi_1 d\phi_2 \,. \tag{11.3.10}$$

Table 11.1 summarises the various notations in this subsection. Figure 11.8 schematically shows a high-numerical-aperture imaging system obeying the Abbe sine condition and illustrates the positions where the CSDM functions of Table 11.1 are found.

Equations (11.3.7) and (11.3.10) are important results but the integrals are difficult to perform. In what follows we consider various cases under which the integrals simplify from the fourfold integrals to separable twofold integrals (Section 11.3.1), or even under certain symmetry assumptions, single integrals (Section 11.3.2).

11.3.1 Coherent Mode Representation

Scalar coherent mode expansions in optical coherence theory (see e.g. **[376]** for a fuller discussion) have seen frequent use in optics **[375]**, **[83]**, **[268]**, **[98]**. Essentially this formalism allows the representation of the (scalar) cross-spectral density function as a coherent superposition of a set of orthonormal functions, the so-called coherent modes. As we

Table 11.1. The various cross-spectral density matrix notations used in this section.

Notation	Definition	Location
$\mathbb{W}(\mathbf{r}_{p1}, \mathbf{r}_{p2})$	$\langle \mathbf{E}^*(\mathbf{r}_{p1})\mathbf{E}^T(\mathbf{r}_{p2})\rangle$	focal region of the lens
$\mathrm{w}(\vartheta_1, \phi_1, \vartheta_2, \phi_2)$	$\langle \boldsymbol{\mathcal{E}}^*(\vartheta_1, \phi_1)\boldsymbol{\mathcal{E}}^T(\vartheta_2, \phi_2)\rangle$	gaussian reference sphere (exit pupil)
$\mathrm{w}_i(\vartheta_1, \phi_1, \vartheta_2, \phi_2)$	$\langle \boldsymbol{\mathcal{E}}_i^*(\vartheta_1, \phi_1)\boldsymbol{\mathcal{E}}_i^T(\vartheta_2, \phi_2)\rangle$	entrance pupil in front focal plane

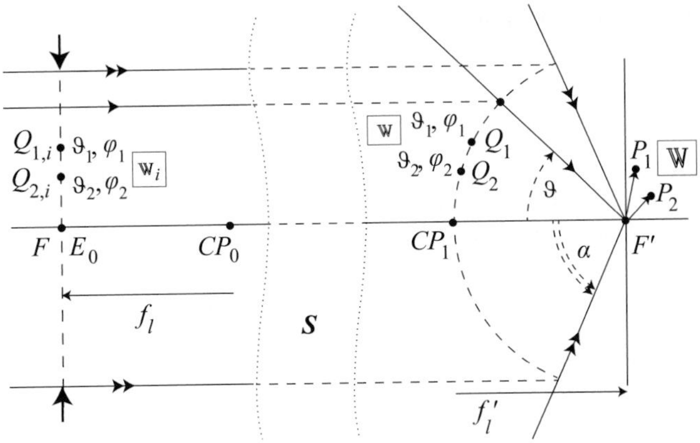

Figure 11.8: The cross-spectral density matrix notation at the planar entrance pupil through the front-focal point F ($\mathbb{W}_i$), on the spherical Gaussian reference sphere through the image-side principal point CP_1 ($\mathbb{W}$) and in the focal region at the back-focal point F' ($\mathbb{W}$). S represents the optical system with image-side focal length f_l' and principal points at CP_0 and CP_1. F is the front-focal point of the system, F' its back-focal point. A single-arrow general aperture ray has been shown, at an angle ϑ with the optic axis in image space. The aperture is given by $\sin\alpha$.

show below this permits separation of the fourfold integral to two double integrals provided the coherent modes can be computed.

All theoretical descriptions originate from the Karhunen–Loève theory **[177]**, **[217]** which has been employed in statistics since the 1940s and it states that given a Hermitian, non-negative definite, square integrable scalar correlation function over a closed domain $\mathcal{D}$, such as a general cross-spectral density function $W(\mathbf{r}_1, \mathbf{r}_2, \omega)$, it is possible to expand it in terms of an infinite, orthonormal set of coherent modes, $\psi_n(\mathbf{r}, \omega)$:

$$W(\mathbf{r}_1, \mathbf{r}_2, \omega) = \sum_{n=0}^{\infty} \lambda_n(\omega)\psi_n^*(\mathbf{r}_1, \omega)\psi_n(\mathbf{r}_2, \omega)\,, \tag{11.3.11}$$

where the coherent modes and associated expansion coefficients $\lambda_n(\omega)$ are found by solution of the Fredholm integral equation

$$\int_{\mathcal{D}} W(\mathbf{r}_1, \mathbf{r}_2, \omega)\psi_n(\mathbf{r}_1, \omega)d\mathbf{r}_1 = \lambda_n(\omega)\psi_n(\mathbf{r}_2, \omega)\,. \tag{11.3.12}$$

In what follows we again drop the explicit dependence of the various quantities on ω. We have a choice to continue our treatment in two possible ways: based on a scalar or a vector formalism. We shall look at both in turn to arrive at the conclusion that the vector-based treatment is more suitable to treat the problem at hand.

The first, scalar-based interpretation applies the scalar formulation described above to each field component individually, hence requiring the solution of two or three *uncoupled* Fredholm integral equations of the form of Eq. (11.3.12), depending on whether the CSDM is specified in the pupil or the focal region. Accordingly the individual elements of a general CSDM are expressed in the form **[376]**

$$W_{ij}(\mathbf{r}_1, \mathbf{r}_2) = \begin{cases} \sum\limits_{n=0}^{\infty} \lambda_n^{(i)} \left(\psi_n^{(i)}(\mathbf{r}_1)\right)^* \psi_n^{(i)}(\mathbf{r}_2) & \text{for } i = j \\ \sum\limits_{n=0}^{\infty}\sum\limits_{m=0}^{\infty} \Lambda_{nm}^{(ij)} \left(\psi_n^{(i)}(\mathbf{r}_1)\right)^* \psi_m^{(j)}(\mathbf{r}_2) & i \neq j \end{cases}, \tag{11.3.13}$$

where $W_{ij}(\mathbf{r}_1, \mathbf{r}_2)$ is the (i,j)th element of $\mathbb{W}(\mathbf{r}_1, \mathbf{r}_2)$ and the expansion coefficients for off-diagonal terms $\Lambda_{nm}^{(ij)}$ are found from

$$\Lambda_{nm}^{(ij)} = \int_{\mathcal{D}}\int_{\mathcal{D}} \psi_n^{(i)}(\mathbf{r}_1) W_{ij}(\mathbf{r}_1, \mathbf{r}_2) \left(\psi_m^{(j)}(\mathbf{r}_2)\right)^* d\mathbf{r}_1 d\mathbf{r}_2\,. \tag{11.3.14}$$

The alternative, vector-based formalism solves the Fredholm integral equation with matrix-valued kernel,

$$\int_{\mathcal{D}} \boldsymbol{\Psi}_n^T(\mathbf{r}_1, \omega)\mathbb{W}(\mathbf{r}_1, \mathbf{r}_2, \omega)d\mathbf{r}_1 = \lambda_n(\omega)\boldsymbol{\Psi}_n^T(\mathbf{r}_2, \omega), \tag{11.3.15}$$

to find vectorial coherent modes **[115]**, **[337]**, such that

$$\mathbb{W}(\mathbf{r}_1, \mathbf{r}_2, \omega) = \sum_{n=0}^{\infty} \lambda_n(\omega) \boldsymbol{\Psi}_n^*(\mathbf{r}_1, \omega) \boldsymbol{\Psi}_n^T(\mathbf{r}_2, \omega) , \tag{11.3.16}$$

where we have re-introduced the explicit dependence on ω for these two equations for the sake of clarity of definition. Although the vector-based approach is more laborious from the mathematics point of view because it requires the solution of two or three *coupled* scalar Fredholm integral equations, depending on whether we want to describe pupil plane or focal region, respectively, it does have the advantage of expressing the off-diagonal elements more concisely. The coherent mode expansions will allow us to simplify our formulas for focusing of partially polarised, partially coherent light. On the way there we shall first consider expansions of the CSDMs $\mathbb{W}(\vartheta_1, \phi_1, \vartheta_2, \phi_2)$ on the Gaussian reference sphere and then that of $\mathbb{W}_i(\vartheta_1, \phi_1, \vartheta_2, \phi_2)$ in the back focal plane.

Consider first the scalar-based expansion of $\mathbb{W}(\vartheta_1, \phi_1, \vartheta_2, \phi_2)$. Using Eq. (11.3.7) we immediately have

$$\begin{aligned} W_{ij}(\mathbf{r}_{p1}, \mathbf{r}_{p2}) = & \frac{f^2}{\lambda^2} \sum_{n=0}^{\infty} \lambda_n^{(i)} \int_0^{2\pi}\int_0^{2\pi}\int_0^{\alpha}\int_0^{\alpha} \left(\psi_n^{(i)}(\vartheta_1, \phi_1)\right)^* \psi_n^{(i)}(\vartheta_2, \phi_2) \exp\left[ik\Delta_{12}\right] \\ & \times \exp\left[ikz_p(\cos\vartheta_2 - \cos\vartheta_1)\right] \sin\vartheta_1 \sin\vartheta_2 d\vartheta_1 d\vartheta_2 d\phi_1 d\phi_2 , \end{aligned} \tag{11.3.17}$$

for $i = j$ and

$$\begin{aligned} W_{ij}(\mathbf{r}_{p1}, \mathbf{r}_{p2}) = & \frac{f^2}{\lambda^2} \sum_{n=0}^{\infty}\sum_{m=0}^{\infty} \Lambda_{nm}^{(ij)} \times \int_0^{2\pi}\int_0^{2\pi}\int_0^{\alpha}\int_0^{\alpha} \left(\psi_n^{(i)}(\vartheta_1, \phi_1)\right)^* \psi_m^{(j)}(\vartheta_2, \phi_2) \exp\left[ik\Delta_{12}\right] \\ & \times \exp\left[ikz_p(\cos\vartheta_2 - \cos\vartheta_1)\right] \sin\vartheta_1 \sin\vartheta_2 d\vartheta_1 d\vartheta_2 d\phi_1 d\phi_2 , \end{aligned} \tag{11.3.18}$$

for $i \neq j$. Separating the integrals yields

$$\begin{aligned} W_{ij}(\mathbf{r}_{p1}, \mathbf{r}_{p2}) = & \left(\frac{f}{\lambda}\right)^2 \sum_{n=0}^{\infty} \lambda_n^{(i)} \\ & \times \int_0^{2\pi}\int_0^{\alpha} \left(\psi_n^{(i)}(\vartheta_1, \phi_1)\right)^* e^{ikr_{p1}\sin\vartheta_1\cos(\phi_1-\gamma_1)} e^{-ikz_{p1}\cos\vartheta_1} \sin\vartheta_1 d\vartheta_1 d\phi_1 \\ & \times \int_0^{2\pi}\int_0^{\alpha} \psi_n^{(i)}(\vartheta_2, \phi_2) e^{-ikr_{p2}\sin\vartheta_2\cos(\phi_2-\gamma_2)} e^{ikz_{p2}\cos\vartheta_2} \sin\vartheta_2 d\vartheta_2 d\phi_2 , \end{aligned} \tag{11.3.19}$$

for $i = j$ and

$$\begin{aligned} W_{ij}(\mathbf{r}_{p1}, \mathbf{r}_{p2}) = & \left(\frac{f}{\lambda}\right)^2 \sum_{n=0}^{\infty}\sum_{m=0}^{\infty} \Lambda_{nm}^{(ij)} \\ & \times \int_0^{2\pi}\int_0^{\alpha} \left(\psi_n^{(i)}(\vartheta_1, \phi_1)\right)^* e^{ikr_{p1}\sin\vartheta_1\cos(\phi_1-\gamma_1)} e^{-ikz_{p1}\cos\vartheta_1} \sin\vartheta_1 d\vartheta_1 d\phi_1 \\ & \times \int_0^{2\pi}\int_0^{\alpha} \psi_m^{(j)}(\vartheta_2, \phi_2) e^{-ikr_{p2}\sin\vartheta_2\cos(\phi_2-\gamma_2)} e^{ikz_{p2}\cos\vartheta_2} \sin\vartheta_2 d\vartheta_2 d\phi_2 , \end{aligned} \tag{11.3.20}$$

for $i \neq j$. Letting

$$\begin{aligned} C_n^{(i)}(\mathbf{r}) &= C_n^{(i)}(r, \gamma, z) \\ &= \frac{-if}{\lambda} \int_0^{2\pi}\int_0^{\alpha} \psi_n^{(i)}(\vartheta, \phi) \exp\left[-ikr\sin\vartheta\cos(\phi - \gamma)\right] \exp(ikz\cos\vartheta) \sin\vartheta d\vartheta d\phi \end{aligned} \tag{11.3.21}$$

gives

$$W_{ij}(\mathbf{r}_{p1}, \mathbf{r}_{p2}) = \begin{cases} \sum\limits_{n=0}^{\infty} \lambda_n^{(i)} \left(C_n^{(i)}(\mathbf{r}_{p1})\right)^* C_n^{(i)}(\mathbf{r}_{p2}) & \text{for } i = j \\ \sum\limits_{n=0}^{\infty}\sum\limits_{m=0}^{\infty} \Lambda_{nm}^{(ij)} \left(C_n^{(i)}(\mathbf{r}_{p1})\right)^* C_m^{(j)}(\mathbf{r}_{p2}) & i \neq j \end{cases} . \tag{11.3.22}$$

Alternatively, using a similar process the vector-based expansion of $\mathbb{W}(\vartheta_1, \phi_1, \vartheta_2, \phi_2)$ in terms of the set of vector coherent modes $\boldsymbol{\Psi}_n(\vartheta, \phi)$ gives

$$\mathbb{W}(\mathbf{r}_{p1}, \mathbf{r}_{p2}) = \left(\frac{f}{\lambda}\right)^2 \sum_{n=0}^{\infty} \lambda_n \times \int_0^{2\pi}\int_0^{2\pi}\int_0^{\alpha}\int_0^{\alpha} \boldsymbol{\Psi}_n^*(\vartheta_1, \phi_1)\boldsymbol{\Psi}_n^T(\vartheta_2, \phi_2) \exp\left[ik\Delta_{12}\right] \exp\left[ik(z_{p2}\cos\vartheta_2 - z_{p1}\cos\vartheta_1)\right] \times \sin\vartheta_1 \sin\vartheta_2 d\vartheta_1 d\vartheta_2 d\phi_1 d\phi_2 \,, \tag{11.3.23}$$

and

$$\mathbb{W}(\mathbf{r}_{p1}, \mathbf{r}_{p2}) = \sum_{n=0}^{\infty} \lambda_n \, \mathbf{C}_n^*(\mathbf{r}_{p1})\mathbf{C}_n^T(\mathbf{r}_{p2}) \,, \tag{11.3.24}$$

where

$$\mathbf{C}_n(\mathbf{r}) = \mathbf{C}_n(r, \gamma, z) = \frac{-if}{\lambda}\int_0^{2\pi}\int_0^{\alpha} \boldsymbol{\Psi}_n(\vartheta, \phi) \times \exp\left[-ikr\sin\vartheta\cos(\phi - \gamma)\right]\exp(ikz\cos\vartheta)\sin\vartheta d\vartheta d\phi \,, \tag{11.3.25}$$

which is of course formally identical to the Debye–Wolf diffraction integral with vector kernel $\boldsymbol{\Psi}_n(\vartheta, \phi)$.

Comparing Eqs. (11.3.22) and (11.3.25) to the definition of the scalar- and vector-based expansions given by Eqs. (11.3.22) and (11.3.24), respectively, it is apparent that the scalar (vector) coherent modes in the focal region can be found by focusing the coherent modes defined on the Gaussian reference sphere by use of the Debye–Wolf integral with scalar (vector) kernel. This result is expected because by construction the modes are fully spatially and temporally coherent in addition to being statistically independent. Consequently each coherent mode can be propagated independently. It should however be noted that

$$\int_0^{2\pi}\int_0^{\infty} \mathbf{C}_m^T(\mathbf{r})\mathbf{C}_n^*(\mathbf{r}) r dr d\gamma = \left(\frac{f}{\lambda}\right)^2 \delta_{nm} \,, \tag{11.3.26}$$

where δ_{nm} is the Kronecker delta meaning that to maintain orthonormality it is necessary to normalise by the factor f/λ, which leads to the alternative but otherwise equivalent expansion

$$\mathbb{W}(\mathbf{r}_{p1}, \mathbf{r}_{p2}) = \sum_{n=0}^{\infty} \lambda_n \left(\frac{f}{\lambda}\right)^2 \hat{\mathbf{C}}_n^*(\mathbf{r}_{p1})\hat{\mathbf{C}}_n^T(\mathbf{r}_{p2}) \,, \tag{11.3.27}$$

where $\hat{\mathbf{C}}_n(\mathbf{r})$ denotes a renormalised coherent mode.

We now look at using the coherent mode expansion of the cross-spectral density written at the entrance pupil, $\mathbb{W}_i(\vartheta_1, \phi_1, \vartheta_2, \phi_2)$. We shall use scalar $\xi_n^{(i)}(\vartheta, \phi)$ and vector $\boldsymbol{\Xi}_n(\vartheta, \phi)$ coherent modes. It is a simple matter to write this problem as we need to relate the coherent modes on the Gaussian reference sphere to those in the entrance pupil. For the vector-based expansion Eq. (11.3.9) gives $\boldsymbol{\Psi}_n(\vartheta, \phi) = \mathbb{P}(\vartheta, \phi) \cdot \boldsymbol{\Xi}_n(\vartheta, \phi)$. For the scalar-based expansion, however, the mixing of the elements of the CSDM caused by the transformation of Eq. (11.3.9) means that the focused CSDM cannot be expressed in the form of Eq. (11.3.22). The lack of a simple, analytic correspondence between the coherent modes in the back focal plane and those in the focal region hence suggests that *a scalar-based coherent mode expansion is unsuitable for focusing in electromagnetic problems*. Consequently we shall consider only vector-based expansions in the derivations of Section 11.3.2.

We now recall Eq. (11.3.3), written for the CSDM in the focal region, to discuss the degree of spectral coherence at various parts of the focusing arrangement. Analogous definitions hold for the light in the front-focal plane, i.e. before focusing. Since the CSDM will in general change upon focusing then so too will the degree of spectral coherence ζ. Numerical examples of this are given in Section 11.3.3, however, it is informative to consider the *effective degree of coherence*, $\bar{\zeta}$, over the domain $\mathcal{D}$ for a general CSDM called $\mathbb{Z}$, as defined in **[337]** by

$$\bar{\zeta}^2 = \frac{\int_{\mathcal{D}}\int_{\mathcal{D}} \|\mathbb{Z}(\mathbf{r}_1, \mathbf{r}_2)\|_F^2 d\mathbf{r}_1 d\mathbf{r}_2}{\int_{\mathcal{D}} \mathrm{tr}[\mathbb{Z}(\mathbf{r}_1, \mathbf{r}_1)] d\mathbf{r}_1 \int_{\mathcal{D}} \mathrm{tr}[\mathbb{Z}(\mathbf{r}_2, \mathbf{r}_2)] d\mathbf{r}_2} \,. \tag{11.3.28}$$

Let us evaluate Eq. (11.3.28). The numerator is given as

$$\int_{\mathcal{D}}\int_{\mathcal{D}} \|\mathbb{Z}(\mathbf{r}_1,\mathbf{r}_2)\|_F^2 d\mathbf{r}_1 d\mathbf{r}_2 = \sum_{ij}\int_{\mathcal{D}}\int_{\mathcal{D}} |Z_{ij}(\mathbf{r}_1,\mathbf{r}_2)|^2 d\mathbf{r}_1 d\mathbf{r}_2 \,. \tag{11.3.29}$$

Using the vector coherent modes we have

$$\mathbb{Z}(\mathbf{r}_1,\mathbf{r}_2) = \sum_n \lambda_n \boldsymbol{\Psi}_n^*(\mathbf{r}_1)\boldsymbol{\Psi}_n^T(\mathbf{r}_2)\,, \tag{11.3.30}$$

or

$$Z_{ij}(\mathbf{r}_1,\mathbf{r}_2) = \sum_n \lambda_n \Psi_{n,i}^*(\mathbf{r}_1)\Psi_{n,j}(\mathbf{r}_2)\,, \tag{11.3.31}$$

which gives for the Frobenius norm

$$\begin{aligned}\|\mathbb{Z}(\mathbf{r}_1,\mathbf{r}_2)\|_F^2 &= \sum_{n,m}\lambda_n\lambda_m \sum_{ij}\Psi_{n,i}^*(\mathbf{r}_1)\Psi_{n,j}(\mathbf{r}_2)\Psi_{m,i}(\mathbf{r}_1)\Psi_{m,j}^*(\mathbf{r}_2)\\ &= \sum_{n,m}\lambda_n\lambda_m\left(\sum_i \Psi_{n,i}^*(\mathbf{r}_1)\Psi_{m,i}(\mathbf{r}_1)\right)\left(\sum_j \Psi_{n,j}(\mathbf{r}_2)\Psi_{m,j}^*(\mathbf{r}_2)\right)\\ &= \sum_{n,m}\lambda_n\lambda_m\left[\boldsymbol{\Psi}_n^*(\mathbf{r}_1)\boldsymbol{\Psi}_m^T(\mathbf{r}_1)\right]\left[\boldsymbol{\Psi}_n^T(\mathbf{r}_2)\boldsymbol{\Psi}_m^*(\mathbf{r}_2)\right]\,.\end{aligned} \tag{11.3.32}$$

Therefore

$$\begin{aligned}\int_{\mathcal{D}}\int_{\mathcal{D}} \|\mathbb{Z}(\mathbf{r}_1,\mathbf{r}_2)\|_F^2 d\mathbf{r}_1 d\mathbf{r}_2 &= \sum_{n,m}\lambda_n\lambda_m \int_{\mathcal{D}} \boldsymbol{\Psi}_n^*(\mathbf{r}_1)\boldsymbol{\Psi}_m^T(\mathbf{r}_1)d\mathbf{r}_1 \int_{\mathcal{D}} \boldsymbol{\Psi}_n^T(\mathbf{r}_2)\boldsymbol{\Psi}_m^*(\mathbf{r}_2)d\mathbf{r}_2\\ &= \sum_{n,m}\lambda_n\lambda_m\delta_{nm} = \sum_n \lambda_n^2\,.\end{aligned} \tag{11.3.33}$$

The first term in the denominator of Eq. (11.3.28) gives

$$\begin{aligned}\int_{\mathcal{D}} \mathrm{tr}[\mathbb{Z}(\mathbf{r}_1,\mathbf{r}_1)]d\mathbf{r}_1 &= \sum_i\sum_n \lambda_n \int_{\mathcal{D}} \Psi_{n,i}^*(\mathbf{r}_1)\Psi_{n,i}(\mathbf{r}_1)d\mathbf{r}_1\\ &= \sum_n \lambda_n \int_{\mathcal{D}} \boldsymbol{\Psi}_n^*(\mathbf{r}_1)\boldsymbol{\Psi}_n^T(\mathbf{r}_1)d\mathbf{r}_1 = \sum_n \lambda_n\,.\end{aligned} \tag{11.3.34}$$

Gathering all equations together, we may write

$$\bar{\zeta}^2 = \frac{\sum\limits_{n=0}^{\infty}\lambda_n^2}{\left[\sum\limits_{n=0}^{\infty}\lambda_n\right]^2}\,. \tag{11.3.35}$$

If we want to apply these results to focusing, the eigenvalues of the above expression should correspond to the coherent mode expansion of $\mathbb{W}_i$ for expressing the CSDM in the front focal plane. For the CSDM in the focal region we need to use the $\lambda_n \to \lambda_n(f/\lambda)^2$ substitution as follows from Eq. (11.3.27). On this occasion therefore we may write

$$\bar{\zeta}^2 = \frac{\sum\limits_{n=0}^{\infty}\left(\lambda_n^2\frac{f^4}{\lambda^4}\right)}{\left[\sum\limits_{n=0}^{\infty}\left(\lambda_n\frac{f^2}{\lambda^2}\right)\right]^2} = \frac{\sum\limits_{n=0}^{\infty}\lambda_n^2}{\left[\sum\limits_{n=0}^{\infty}\lambda_n\right]^2}\,, \tag{11.3.36}$$

which means that it is possible to conclude that *the effective degree of spectral coherence $\bar{\zeta}$ is unchanged upon focusing*.

11.3.2 Harmonic Angular Dependence

The results hitherto obtained have significantly eased the numerical computational burden associated with evaluating a four-dimensional integral. We can simplify the problem further if we assume that $\boldsymbol{\Psi}_n(\vartheta,\phi) = \boldsymbol{\Psi}_n(\vartheta)\exp(im\phi)$ and similarly for $\Xi_n(\vartheta,\phi)$, where $m \in \mathbb{Z}_0^+$. As we show in these cases the azimuthal integration can be evaluated analytically.

For a harmonic angular dependence on the Gaussian reference sphere it is sufficient to consider the $\mathbf{C}_n(\mathbf{r})$ integrals of Eq. (11.3.25) such that

$$\mathbf{C}_n^I(\mathbf{r}_p) = \frac{-if}{\lambda}\int_0^{2\pi}\int_0^{\alpha} \boldsymbol{\Psi}_n(\vartheta)\exp(im\phi)\exp\left[-ikr_t\sin\vartheta\cos(\phi-\gamma)\right]\exp(ikz_p\cos\vartheta)\sin\vartheta d\vartheta d\phi\,. \tag{11.3.37}$$

Using Eq. (8.4.37) we obtain

$$\mathbf{C}_n^I(\mathbf{r}_p) = (-i)^{m+1}kf\exp(im\gamma)\int_0^{\alpha}\boldsymbol{\Psi}_n(\vartheta)J_m(kr_t\sin\vartheta)\exp(ikz_p\cos\vartheta)\sin\vartheta\, d\vartheta\,. \tag{11.3.38}$$

Alternatively, when considering coherent modes on the back focal plane we have

$$\mathbf{C}_n^{II}(\mathbf{r}_p) = \frac{-if}{\lambda}\int_0^{2\pi}\int_0^{\alpha}\mathbb{P}(\vartheta,\phi)\boldsymbol{\Xi}_n(\vartheta)\exp(im\phi)\exp[-ikr_t\sin\vartheta_l\cos(\phi-\gamma)]\exp(ikz_p\cos\vartheta)\sin\vartheta d\vartheta d\phi\,. \tag{11.3.39}$$

Expanding $\mathbb{P}(\vartheta,\phi)$ gives

$$\mathbb{P}(\vartheta,\phi) = \frac{\sqrt{\cos\vartheta}}{2}\begin{pmatrix}(\cos\vartheta+1)+(\cos\vartheta-1)\cos 2\phi & (\cos\vartheta-1)\sin 2\phi & -2\sin\vartheta\cos\phi\\ (\cos\vartheta-1)\sin 2\phi & (\cos\vartheta+1)-(\cos\vartheta-1)\cos 2\phi & -2\cos\vartheta\sin\phi\\ 2\sin\vartheta\cos\phi & \sin\vartheta\sin\phi & 2\cos\vartheta\end{pmatrix}. \tag{11.3.40}$$

Using Eq. (8.4.37) we can perform the integration over ϕ to give

$$\mathbf{C}_n^{II}(\mathbf{r}_p) = \begin{pmatrix} 2\,{}_xK_{m,n} - {}_xK_{m+2,n} - {}_xK_{m-2,n} + i\,{}_yK_{m+2,n} - i\,{}_yK_{m-1,n} - 2\,{}_zK_{m+1,n} - 2\,{}_zK_{m-1,n}\\ 2\,{}_yK_{m,n} + {}_yK_{m+2,n} + {}_yK_{m-2,n} + i\,{}_xK_{m+2,n} - i\,{}_xK_{m-1,n} + 2i\,{}_zK_{m+1,n} - 2i\,{}_zK_{m-1,n}\\ 4\,{}_zK_{m,n} + 2\,{}_xK_{m+1,n} + 2\,{}_xK_{m-1,n} - 2i\,{}_yK_{m+1,n} + 2i\,{}_yK_{m+1,n}\end{pmatrix}, \tag{11.3.41}$$

with

$${}_vK_{m,n} = -\frac{ikf}{4}(-i)^m\exp(im\gamma)\,{}_vI_{m,n}\,, \tag{11.3.42}$$

and

$$\begin{aligned}
{}_{x,y}I_{m,n}(r,z) &= \int_0^{\alpha}\sqrt{\cos\vartheta}\,\Xi_n^{x,y}(\vartheta)(1+\cos\vartheta)\sin\vartheta J_m(kr\sin\vartheta)\exp(ikz\cos\vartheta)d\vartheta,\\
{}_zI_{m,n}(r,z) &= \int_0^{\alpha}\sqrt{\cos\vartheta}\,\Xi_n^{x,y}(\vartheta)\sin\vartheta\cos\vartheta J_m(kr\sin\vartheta)\exp(ikz\cos\vartheta)d\vartheta,\\
{}_vI_{m\pm1,n}(r,z) &= \int_0^{\alpha}\sqrt{\cos\vartheta}\,\Xi_n^{v}(\vartheta)\sin^2\vartheta J_{m\pm1}(kr\sin\vartheta)\exp(ikz\cos\vartheta)d\vartheta,\\
{}_{x,y}I_{m\pm2,n}(r,z) &= \int_0^{\alpha}\sqrt{\cos\vartheta}\,\Xi_n^{x,y}(\vartheta)(1-\cos\vartheta)\sin\vartheta J_{m\pm2}(kr\sin\vartheta)\exp(ikz\cos\vartheta)d\vartheta\,,
\end{aligned} \tag{11.3.43}$$

where Ξ_n^v represents the vth component of $\boldsymbol{\Xi}_n$. Evaluation of the single integrals of Eq. (11.3.43) is all that is necessary to calculate the CSDM of focused, inhomogeneous, partially polarised, partially coherent light for which the coherent modes have a harmonic angular dependence. In general the computation of $\boldsymbol{\Psi}$ is only possible by solving the Fredholm integral equation Eq. (11.3.15). However, in the following subsection we make an assumption as to the exact analytic functional form of $\boldsymbol{\Psi}$. Such functions rarely, if ever, occur naturally and as such the analytic expressions are only used as an example. However, these modes can be generated by using spatial light modulators and as a result it is possible to experimentally produce beams of arbitrary partial coherence properties **[226]**.

In coherence calculations the assumption of a circularly symmetric CSDM is often made (whereby either $\mathbb{W}(\vartheta_1,\phi_1,\vartheta_2,\phi_2)=\mathbb{W}(\vartheta_1,\vartheta_2)$ or $\mathbb{W}_i(\vartheta_1,\phi_1,\vartheta_2,\phi_2)=\mathbb{W}_i(\vartheta_1,\vartheta_2)$) because it allows the dimensionality of analysis to be reduced. Circularly symmetry in the CSDM is inherited by the coherent modes and hence this frequently considered scenario is given as a special case ($m=0$) of the preceding analysis.

11.3.3 Numerical Examples

11.3.3.1 Radially Polarised Gauss–Laguerre Modes

We first consider a focusing of the superposition of mutually uncorrelated, radially polarised Gauss–Laguerre modes (see Eq. (11.2.10a)) located in the back focal plane of a lens. In this scenario the CSDM in the back focal plane is of the form $\mathbb{W}_i(\vartheta_1,\phi_1,\vartheta_2,\phi_2)=\sum_{n=0}^{\infty}\lambda_n\boldsymbol{\Xi}_n^*(\vartheta_1,\phi_1)\,\boldsymbol{\Xi}_n^T(\vartheta_2,\phi_2)$ where

$$\Xi_n(\vartheta,\phi) = U_n(\vartheta)\begin{pmatrix}\cos\phi\\ \sin\phi\\ 0\end{pmatrix}, \tag{11.3.44}$$

and from Eqs. (11.2.7) and (11.2.12) with $m = 0$

$$U_n(\vartheta) = \left(\frac{2}{\pi p_{GL}^2}\right)^{1/2} L_n\left(\frac{2\sin^2\vartheta}{\mu^2}\right)\exp\left(-\frac{\sin^2\vartheta}{\mu^2}\right). \tag{11.3.45}$$

We further consider the case discussed in **[121]** for which $\lambda_n = \pi(1-q^2)q^{2n}/2\mu^2$ for $0 < q < 1$. As we have seen before, the parameter μ is a metric for the beam radius in units of the focal length f, and q determines the effective degree of spectral coherence by $\bar{\zeta}^2 = (1-q^2)/(1+q^2)$, with the limits $q \to 0$ ($q \to 1$) giving a fully spatially (un)correlated source.

Although it is possible to specialise Eq. (11.3.41) to this case, computing $\mathbb{P}(\vartheta,\phi)\Xi_n$ directly results in a particularly simple formula:

$$\mathbb{P}(\vartheta,\phi)\Xi_n(\vartheta,\phi) = U_n(\vartheta)\begin{pmatrix}\cos\vartheta\cos\phi\\ \cos\vartheta\sin\phi\\ \sin\vartheta\end{pmatrix}, \tag{11.3.46}$$

which gives

$$\mathbf{C}_n^{II}(\mathbf{r}_p) = -ifk\begin{pmatrix}-i\,I_{1,n}\cos\gamma\\ -i\,I_{1,n}\sin\gamma\\ I_{0,n}\end{pmatrix}, \tag{11.3.47}$$

where

$$\begin{aligned} I_{1,n}(r,z) &= \int_0^\alpha \sqrt{\cos\vartheta}\, U_n(\vartheta)\sin\vartheta\cos\vartheta J_1(kr\sin\vartheta)\exp(ikz\cos\vartheta)d\vartheta \\ I_{0,n}(r,z) &= \int_0^\alpha \sqrt{\cos\vartheta}\, U_n(\vartheta)\sin^2\vartheta J_0(kr\sin\vartheta)\exp(ikz\cos\vartheta)d\vartheta\,. \end{aligned} \tag{11.3.48}$$

We have computed the trace of the cross spectral density matrix and plotted the resulting distribution in the half-meridional plane as shown in Fig. 11.9 for $\bar{\zeta} = 1$, 2/3 and 1/3. The figure reveals that there is a modest extension in the

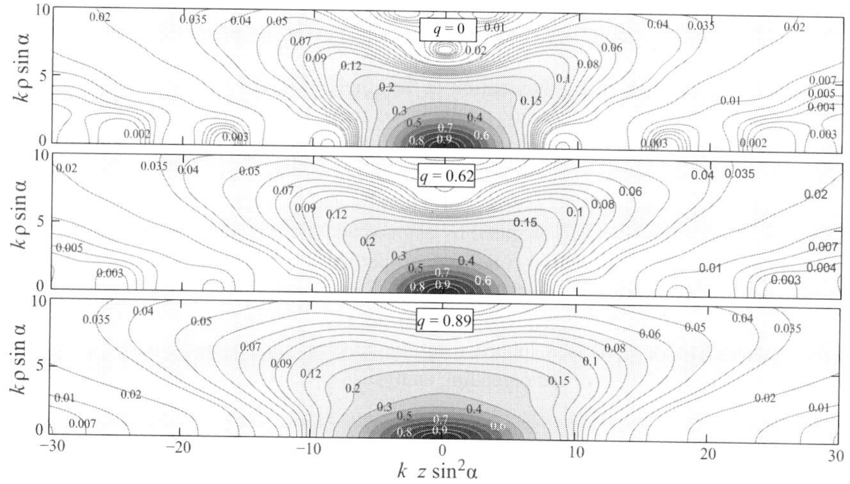

Figure 11.9: Normalised axial focal intensity distributions ($\gamma = 0$) for partially coherent radially polarised collimated sources with differing effective degrees of spectral coherence as specified by a) $\bar{\zeta} = 1$, b) $\bar{\zeta} = 2/3$, and c) $\bar{\zeta} = 1/3$. Other simulation parameters: $NA = 0.97$, $\mu = 1$ and $\lambda = 405$nm.

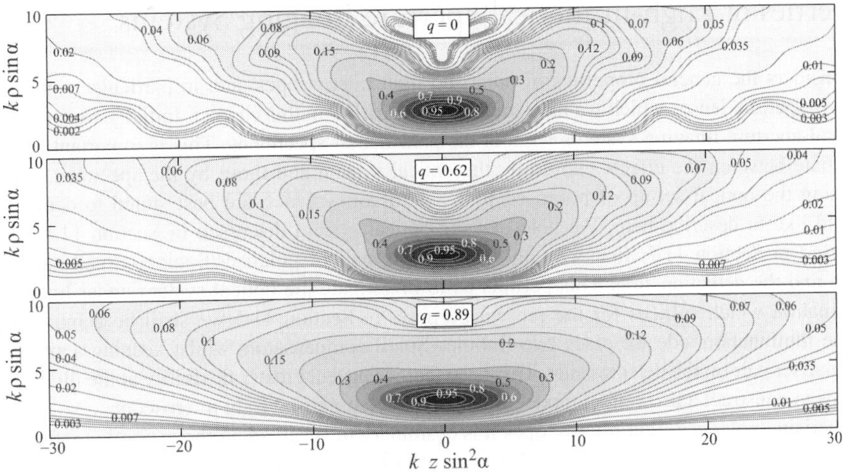

Figure 11.10: Normalised axial focal intensity distributions ($\gamma = 0$) for partially coherent azimuthally polarised collimated sources with differing effective degrees of spectral coherence as specified by a) $\bar{\zeta} = 1$, b) $\bar{\zeta} = 2/3$, and c) $\bar{\zeta} = 1/3$. Other simulation parameters: $NA = 0.97$, $\mu = 1$ and $\lambda = 405$ nm.

depth of field as the source becomes more incoherent. There is also a slight increase in the energy density in the wings of the transverse profile. We note that due to the apodisation[11.7] over the pupil the focal spot is broader than that for uniform intensity since the contribution from the longitudinal field component, responsible for the narrow spot for the unapodised case, is reduced.

11.3.3.2 Azimuthally Polarised Gauss–Laguerre Modes

As a further example we consider an azimuthally polarised beam in the clockwise direction. In this case the vectorial coherent modes are of the form (c.f. Eq. (11.2.10b)),

$$\Xi_n(\vartheta, \phi) = U_n(\vartheta) \begin{pmatrix} \sin\phi \\ -\cos\phi \\ 0 \end{pmatrix}, \tag{11.3.49}$$

which in the focal region yields

$$\mathbf{C}_n^{II}(\mathbf{r}_p) = -ifk \begin{pmatrix} -i\, I_{1,n} \sin\gamma \\ i\, I_{1,n} \cos\gamma \\ 0 \end{pmatrix}, \tag{11.3.50}$$

where

$$I_{1,n}(r, z) = \int_0^\alpha \sqrt{\cos\vartheta}\; U_n(\vartheta) \sin\vartheta J_1(kr \sin\vartheta) \exp(ikz\cos\vartheta) d\vartheta \,.$$

The axial intensity distribution is shown in Fig. 11.10. Similar conclusions to those made for the radially polarised source can be drawn for an azimuthally polarised source, however, the augmentation of the wings of the transverse intensity profile is more pronounced.

With this figure we have completed our initial discussions on how high-numerical-aperture lenses behave in isolation so it is now time to start to build more complex optical systems and use these lenses in conjunction with other lenses and then turn our knowledge to characterising these optical systems for imaging of objects of increasing complexity.

11.7 By apodisation we mean a ray height-dependent amplitude (i.e. real) function that modifies the relative weight of plane waves contributing to the electric field in the focus.

11.4 Properties of High-numerical-aperture Imaging Systems

This section considers the general properties of polarised light microscopes and, in particular, light depolarisation[11.8] produced by high-aperture lenses which we have touched upon earlier in this chapter. Initially we only discuss the optics of high-numerical-aperture imaging systems without the presence of specimens. This is important because we need to clearly differentiate between the effects caused by the optical system and those by the specimen. For this reason we start by discussing the optical set-up of polarised light microscopes which are best suited to demonstrate the subtle details we would like to describe here. We note that the optical system discussed in Section 11.4.1 will be different from that analysed in Section 11.5: first we only use a single pair of polariser/Babinet–Soleil compensator on both the illumination and the detection side. This permits us to investigate the optical system under linearly and circularly polarised illumination which suffices for the present purpose. In Section 11.5 we shall be using a polariser on both sides, but on the illumination side we place two Babinet–Soleil compensators whilst keeping a single Babinet–Soleil compensator/polariser pair on the detector side. This is so that we have a greater flexibility at the illumination side which in turn permits us to construct a theoretical model of an automated polarised light microscope.

When characterising polarised light microscopes it is customary to speak about the extinction ratio (ER). It is usually defined as the ratio of the detected intensities for parallel and cross polarised settings. By parallel and cross polar setting one usually means that the analyser of the microscope set-up (described below in detail) is set to be parallel and perpendicular, respectively, to the polarisation direction of the light illuminating the first optical component of the microscope. Ideally we would like to extinguish all light in cross polar settings so that very small amounts of changes in polarisation can be measured in which case the extinction ratio becomes infinite. In practice sheet polarisers provide ER = 10^2, wire grid polarises ER = 10^3, nanoparticle linear film polarisers ER = 10^3–10^5 and finally calcite-based (Glan–Thompson, Glan–Taylor, Wollaston) polarisers typically have ER = 10^5–10^6. It is of practical importance to note that, especially calcite-based polarisers possess large spatial variations in their polarisation properties and they often depolarise[11.9] light. This is especially true for most polarising beam splitters in the reflected direction. We shall see below that the use of high-numerical-aperture lenses leads to depolarisation that results in much compromised ERs.

We start by considering a pair of lenses. In our derivation below, we first obtain expressions for the field in the back focal plane of the second lens and compare it, in a pictorial format, to those obtained experimentally.

11.4.1 General Properties of the Optical System

We have discussed above that it is not practical to describe high-numerical-aperture lenses by tracing rays through the individual surfaces that make up these often very complex lenses (see Fig. 7.86 for example). We also described how the direction of a light ray is changed, together with the associated electric field vector of the geometrical optics approximation, by a high-numerical-aperture lens. The Debye–Wolf integral (especially in Luneburg's derivation) tells us that lenses, when designed with one infinite conjugate, fundamentally are phase front converters; they are designed to convert either a plane or spherical wavefront to a spherical or plane phase front, respectively. Our present interest is to model a lens designed for focusing so that it converts plane phase fronts into spherical phase fronts. The model for the collimating arrangement follows from the focusing model in a logical manner.

We have discussed above that high-aperture lenses depolarise light, which was first observed by Inoué and co-workers **[156]**, **[157]**, **[198]** who also gave a mathematical explanation for the phenomenon, based on Fresnel transmission coefficients, without the use of diffraction integrals. The work by Hardy and Treves **[125]** brought together the Debye–Wolf formulas with Inoué's formulation to consider a singlet having two surfaces: ellipsoidal on the object side and spherical on the image side. The latter was arranged so that its centre coincided with the Gaussian focal point so that the refraction invariant on the second surface was zero for the marginal ray. Since they considered an on-axis object point located at infinity, the second surface did not introduce depolarisation. Hardy and Treves derived formulas that only differ from the Debye–Wolf integrals by an apodisation term. In this section our approach is based on an idea similar to that of Hardy and Treves: we model the effect of depolarisation through *all* lens surfaces, which are unknown to most experimentalists, by considering Fresnel transmission through a *single* air/glass interface. This is a phenomenological model that is acceptable for objective lenses having both spherical and aspherical surfaces (see Fig. 7.62 and related discussion) but it is expected not to produce results of high accuracy when diffractive optical elements are used. In the

[11.8] The pioneering work on this subject was done by Shinya Inoué **[156]**, **[157]**, **[198]** investigating the basic properties of polarising microscopes.

[11.9] By depolarisation we mean a spatially non-uniform polarisation distribution as discussed in the previous section.

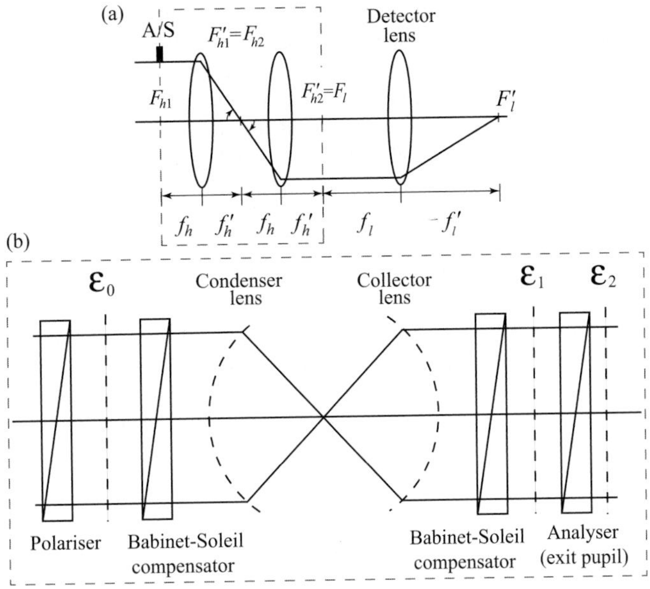

Figure 11.11: Calculating the electric field vector passing through the optical system.
a) The *physical* arrangement of the system with a marginal ray drawn. The incident plane wave is specified at the aperture stop (A/S) that is placed in the front focal plane of the condenser lens (defined by its front focal point F_{h1}). The plane wave is focused into the rear focal point of the condenser lens F'_{h1}, which coincides with the front focal point of the collector lens F_{h2}. The objective lens re-collimates the radiation which then gets focused to F'_l by the detector lens. The detector lens is placed such that its front focal point F_l coincides with the rear focal point F'_{h2} of the collector lens. b) A subset of the full system, as marked by a dashed line box, shows the notation of the model. The different polarising elements (polariser, Babinet–Soleil compensator, analyser), although shown at different axial locations, represent successive applications of the generalised Jones matrices but physically they are not at different locations.

procedure we follow there is no need to make an assumption for the air/glass interface to be spherical as long as it possesses rotational symmetry with respect to the optic axis, and it also remains valid for catadioptric and mixed catadioptric systems (see Chapter 7).

The essence of the procedure we follow is that we set up an optical system so that two identical high-numerical-aperture objective lenses (condenser and collector lenses) are in afocal arrangement (see Fig. 11.11).[11.10] The collector lens is then used in conjunction with a low numerical aperture lens, the so-called detector lens, which is arranged so that its front focal plane coincides with the back focal plane of the collector lens (4-f set-up). For the present purpose we assume that the air/glass interface is spherical and fix the incident polarisation to the x-direction in the laboratory coordinate system:

$$\mathcal{E}_0 = (U_0, 0, 0)^T .$$

Vector ray-tracing is performed as usual using the generalised Jones matrix formalism. The electric field recollimated by the collector lens is given by

$$\begin{aligned}\mathcal{E}_1 &= \mathbb{BS}_2 \cdot \mathbb{R}^{-1} \cdot \mathbb{L}_2^{-1} \cdot \mathbb{I}_2 \cdot \mathbb{R} \cdot \mathbb{R}^{-1} \cdot \mathbb{I}_1 \cdot \mathbb{L}_1 \cdot \mathbb{R} \cdot \mathbb{BS}_1 \cdot \mathcal{E}_0 \\ &= \mathbb{BS}_2 \cdot \mathbb{R}^{-1} \cdot \mathbb{L}^{-1} \cdot \mathbb{I}_2 \cdot \mathbb{I}_1 \cdot \mathbb{L} \cdot \mathbb{R} \cdot \mathbb{BS}_1 \cdot \mathcal{E}_0 ,\end{aligned} \tag{11.4.1}$$

where $\mathbb{R}$, $\mathbb{L}$, $\mathbb{BS}_n$ and $\mathbb{I}$ are given by Eqs. (11.1.1), (11.1.2), (11.1.15) and (11.1.3), respectively, and we have assumed that the two lenses are identical, hence setting $\mathbb{L}_1 = \mathbb{L}_2 = \mathbb{L}$. The subscripts 1 and 2 in $\mathbb{I}$ mean that the matrix is written for an air/glass and a glass/air interface, respectively.

There are two rather subtle but important points that need to be noted at this juncture. First, let's look at how the generalised Jones matrices transmit a transverse electric field through the condenser–collector lens pair. If we take

[11.10] Note that, although the figure shows a transmission set-up it equally applies to reflection (epi) configuration when the condenser lens also acts as the collector lens.

$\mathcal{E} = \mathbb{R}^{-1} \cdot \mathbb{L}^{-1} \cdot \mathbb{L} \cdot \mathbb{R} \cdot (0,1,0)^T$ we get $\mathcal{E} = (0,1,0)^T$ assuming the two lens operators have the same azimuthal angle. This result, however, is incorrect because as Fig. 11.11 shows, the marginal ray that enters through the aperture stop at the top leaves after the second lens at the bottom and so if we had an electric vector at the aperture stop that was perpendicular to the ray and pointing in the $+y$ direction, the electric vector exiting the lens pair should point in the $-y$ direction. The same applies to an incident electric field directed at $+x$ before the condenser lens, which should point in the $-x$ direction upon exiting the collector lens, in order to maintain the right-handed, mutually perpendicular nature of the $(1,0,0)^T$, $(0,1,0)^T$ and $\hat{\mathbf{s}} = (0,0,1)^T$ triplet. So the first conclusion is that the operand $\mathbb{R}^{-1}$ really means $\mathbb{R}^{-1}(\phi) = \mathbb{R}(\pi - \phi)$ because that will permit the right orientation of the vector triplet. The second point is associated with the definition of the $\mathbb{L}$ matrix. As discussed in conjunction with Fig. 8.6, the angle ϑ is defined with respect to the $-z$ direction, so calling $\mathbb{L}(\vartheta)$ causes the ray to be rotated in the meridional plane clockwise by an amount of ϑ. Likewise, the $\mathbb{L}^{-1} = \mathbb{L}(-\vartheta)$ matrix should rotate the ray anticlockwise by an amount of ϑ. However, the effect of the collector lens on the ray should be expressed by means of the azimuthal angle defined with respect to that lens. Looking at Fig. 11.11 we see that the azimuthal angle is defined in clockwise direction and with respect to the $+z$ direction. For this reason, we denote the azimuthal angle corresponding to the collector lens as θ in order to be consistent with our previous notation introduced with Fig. 8.6.

Now returning to the discussion of the optical system, in order to simplify the notation we introduce

$$A_n^{\pm} = \cos\frac{\delta_n}{2} \pm i\cos 2\beta_n \sin\frac{\delta_n}{2}, \tag{11.4.2}$$

$$B_n = i\sin 2\beta_n \sin\frac{\delta_n}{2}\,, \tag{11.4.3}$$

so now

$$\mathbb{BS}_n = \begin{pmatrix} A_n^+ & B_n & 0 \\ B_n & A_n^- & 0 \\ 0 & 0 & 1 \end{pmatrix}, \tag{11.4.4}$$

with $n = 1, 2$. After some straightforward algebraic manipulations the electric field $\mathcal{E}_1$ is obtained as:

$$\mathcal{E}_1 = \frac{U_0}{2}\begin{pmatrix} (A_1^+A_2^+ + B_1B_2)T^+ + T^-\left[(A_1^+A_2^- - B_1B_2)\cos 2\phi + (A_1^+B_2 + A_2^+B_1)\sin 2\phi\right] \\ (A_1^+B_2 + A_2^-B_1)T^+ + T^-\left[(A_1^+B_2 - A_2^-B_1)\cos 2\phi + (A_1^+A_2^- + B_1B_2)\sin 2\phi\right] \\ 0 \end{pmatrix}, \tag{11.4.5}$$

with

$$T^{\pm} = t_{p1}t_{p2} \pm t_{s1}t_{s2}\,, \tag{11.4.6}$$

where $t_{p1,s1}$ is written for an air/glass interface and $t_{p2,s2}$ is written for a glass/air interface. Note that we did not use subscripts in conjunction with the polar angle ϕ because it defines the angle of a meridional plane that does not change as light traverses the optical system. The polarisation of light transmitted by the second Babinet–Soleil compensator is analysed. This means that the resulting field is given by $\mathbb{O} \cdot \mathcal{E}_1$, where $\mathbb{O}$ is given by Eq. (11.1.18), or explicitly:

$$\mathcal{E}_2 = (\mathcal{E}_{1x}\cos^2\beta + \mathcal{E}_{1y}\sin\beta\cos\beta, \mathcal{E}_{1x}\sin\beta\cos\beta + \mathcal{E}_{1y}\sin^2\beta, 0)^T\,, \tag{11.4.7}$$

where β is the orientation of the polariser (via $\mathbb{O}$). Hence the intensity[11.11] is given by:

$$I = \mathcal{E}_2^T\,\mathcal{E}_2^* = |\mathcal{E}_{1x}|^2\cos^2\beta + |\mathcal{E}_{1y}|^2\sin^2\beta + \Re\{\mathcal{E}_{1x}\mathcal{E}_{1y}^*\}\sin 2\beta\,. \tag{11.4.8}$$

Remembering that the vectorial ray-tracing applies to the electric vector of the geometrical optics approximation, it is clear that the above expressions give the field and intensity distribution after the analyser of the optical system depicted in Fig. 11.11 with two arbitrarily aligned Babinet–Soleil compensators.

11.4.1.1 Linearly and Circularly Polarised Illumination

Before we discuss the light distributions at the detector plane we first specialise our results to two cases of particular interest: linearly and circularly polarised illumination. The first case is when the incident light is linearly polarised. We may set $\delta_1 = \delta_2 = 0$ so the field is now given by

[11.11] In this chapter the term 'intensity' is used somewhat liberally. As discussed in Section 4.11, in radiometry the term 'intensity' refers to radiant intensity in W/sr units. The quantity we call intensity in this chapter is closely related to irradiance (measured in W/m^2 units) and it can be converted to irradiance by multiplying our intensity values by $\epsilon_0 c/2$.

$$\mathcal{E}_2 = \frac{U_0}{2} \begin{pmatrix} (T^+ + T^- \cos 2\phi)\cos^2\beta + T^- \sin 2\phi \sin\beta \cos\beta \\ (T^+ + T^- \cos 2\phi)\sin\beta\cos\beta + T^- \sin 2\phi \sin^2\beta \\ 0 \end{pmatrix}, \tag{11.4.9}$$

from which the intensity reads:

$$I = \mathcal{E}_2^T \, \mathcal{E}_2^* = \frac{I_0}{4} |(T^+ \cos\beta + T^- \cos(\beta - 2\phi)|^2 , \tag{11.4.10}$$

where we have introduced $I_0 = |U_0|^2$. Equation (11.4.10) represents the intensity after the analyser when it is set to an angle β with respect to the incident polarisation.

The second important case is when the incident illumination is circularly polarised, i.e. $\delta_1 = \pi/2$, $\delta_2 = -\pi/2$ and $\beta_1 = \beta_2 = \pi/4$. We hence obtain from Eq. (11.4.7):

$$\mathcal{E}_2 = \frac{U_0}{2} \begin{pmatrix} T^+ \cos^2\beta - i\,T^- \exp(2i\phi)\sin\beta\cos\beta \\ T^+ \sin\beta\cos\beta - i\,T^- \exp(2i\phi)\sin^2\beta \\ 0 \end{pmatrix}, \tag{11.4.11}$$

where we have dropped a constant multiplier and so the intensity is given by:

$$I = \frac{I_0}{8} \left\{(T^+)^2 + (T^-)^2 + \left[(T^+)^2 - (T^+)^2\right]\cos 2\beta + 2T^- T^+ \sin 2\beta \sin 2\phi\right\} . \tag{11.4.12}$$

Equation (11.4.12) gives the intensity for a circularly polarised illumination after the analyser when it is set to an angle β with respect to the incident polarisation.

It is interesting to note that for a reflection type microscope the material properties of the reflective specimen can be treated as part of the apodisation function. To show this we write the electric vector $\mathcal{E}_2$ before the analyser for a linearly polarised (i.e. $\delta_1 = \delta_2 = 0$) reflection type microscope as:

$$\mathcal{E}_2 = \mathbb{R}^{-1} \cdot \mathbb{L}^{-1} \cdot \mathbb{I}_2 \cdot \mathbb{P}^{-1} \cdot \mathbb{I}_r \cdot \mathbb{P} \cdot \mathbb{I}_1 \cdot \mathbb{L} \cdot \mathbb{R} \cdot \mathcal{E}_0 , \tag{11.4.13}$$

where $\mathbb{I}_r$ is the matrix representing ray reflection from a surface and is given by Eq. (11.1.5). It is not difficult to show that Eq. (11.4.13) results in:

$$\mathcal{E}_2 = \frac{U_0}{2} \begin{pmatrix} K^- + K^+ \cos 2\phi \\ K^+ \sin 2\phi \\ 0 \end{pmatrix}, \tag{11.4.14}$$

where

$$K^\pm = t_{p1} t_{p2} r_p \pm t_{s1} t_{s2} r_s . \tag{11.4.15}$$

The above equation shows that material-dependent properties of both the lenses and the reflective specimen are included in the functions $K^\pm$ which are independent of any parameter of the microscope.

11.4.1.2 The Field at the Detector

The next step is to determine the field at the detector which is placed at the back focal point F_l' of the detector lens (Fig. 11.12). The numerical aperture of the detector lens is, in practice, very low (< 0.1) in order to achieve high transverse magnification. Vectorial properties of lenses become unimportant with decreasing marginal ray angle, $\alpha \to 0$, because, as discussed in Section 11.2, the image field becomes fully co-polarised with the illumination. As there are no polarisation sensitive optical components beyond the analyser the polarisation of the field in the focal region in terms of the detected intensity is immaterial. Therefore let us denote the field after the analyser by $\mathcal{E}_2 = \mathcal{E}_2 \hat{\boldsymbol{\beta}}$ where $\hat{\boldsymbol{\beta}}$ is the eigenpolarisation of the polariser (unit vector along the axis of transmittance). The optical set-up described above is a true Fourier transforming arrangement (as discussed in Chapter 10 there is a Fourier transform relationship between the aperture plane and the focal

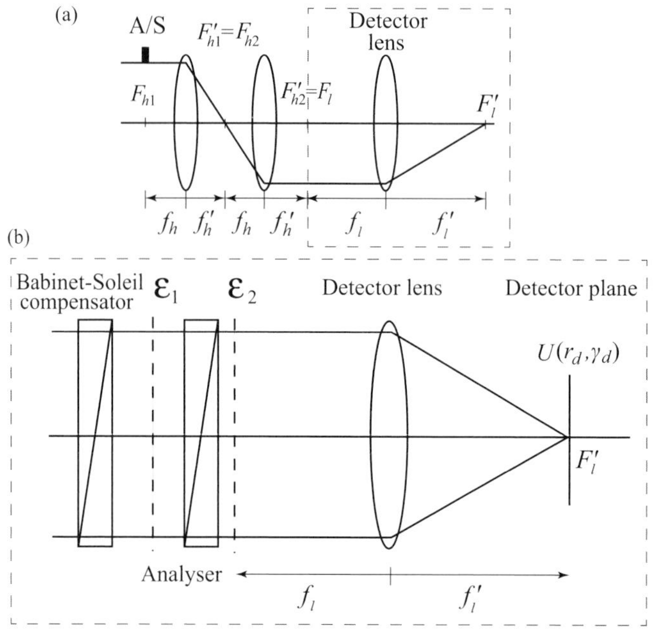

Figure 11.12: The detector lens arrangement.
a) As described in detail in the caption of Fig. 11.11, the entire system as it is physically laid out.
b) A subset of a), as shown by the dashed box, demonstrates the sequential use of the generalised Jones matrices.

plane of a low numerical aperture lens) and hence the scalar field in the focal plane of the detector lens (see Fig. 11.12) is given by the truncated Fourier transform[11.12] of the scalar field $\mathcal{E}_2$ (see Eq. (8.4.23)):

$$U(r_d,\gamma_d) = -\frac{if_l}{2\lambda}\int_0^{\alpha_l}\int_0^{2\pi}\mathcal{E}_2(\theta_h,\phi)\sin 2\vartheta_l\exp[-ikr_d\sin\vartheta_l\cos(\phi-\gamma_d)]d\vartheta_l d\phi\,, \tag{11.4.17}$$

where θ_h and ϑ_l are the azimuthal angles in high- and low-aperture spaces, respectively, the polar coordinates (r_d,γ_d) are defined in the detector plane, α_l is the angle of the marginal ray in detector space and a constant multiplier has been dropped. We note that in Eq. (11.4.17) azimuthal coordinates of both high- and low-aperture spaces are present. However, rather than complicating the expression with the introduction of the relationship $\sin\theta_h/\sin\vartheta_l = f_h/f_l$, f_h being the focal length of the high-aperture lens, we keep the current formalism. It is then easier to use the relationship again, when the final formulas are obtained. We note that for an ideal system, the choice of $\sin\theta_h$ or θ_h as integral variable is relatively irrelevant. However, for imperfect systems one should use $\sin\theta$ as pupil coordinate (see Eq. (9.2.10) and Eq. (9.3.1)). An alternative way of writing Eq. (11.4.17) is by introducing $\rho = f_l\sin\vartheta_l$, which then gives

$$U(r_d,\gamma_d) = -\frac{i}{f_l\lambda}\int_0^{a}\int_0^{2\pi}\mathcal{E}_2(\rho,\phi)\exp\left[-i\frac{k}{f_l}r_d\rho\cos(\phi-\gamma_d)\right]\rho d\rho d\phi\,, \tag{11.4.18}$$

where a is the radius of the exit pupil of the collector lens. In this case we do not need to worry about the high- or low-aperture space since the image of the aperture stop (exit pupil of the collector lens) is the same diameter between the collector and detector lenses.

[11.12] The Fourier transform $F(\rho,\varphi)$ of $f(r,\phi)$, where $F, f, \rho, \varphi, r, \phi$ are all dummy functions/variables with no physical relationship to functions and coordinates used in the book, in cylindrical coordinates is given, after ignoring a constant multiplier, by

$$F(\rho,\varphi) = \int_0^{\infty} rdr\int_0^{2\pi} d\phi f(r,\phi)\exp[-2\pi i\rho r\cos(\varphi-\phi)].$$

If we introduce $r = a\sin\vartheta$ and $dr = a\cos\vartheta d\vartheta$ we have

$$\begin{aligned} F(\rho,\varphi) &= a^2\int_0^{\pi/2}\sin\vartheta\cos\vartheta d\vartheta\int_0^{2\pi} d\phi f(\vartheta,\phi)\exp[-2\pi i a\rho\sin\vartheta\cos(\varphi-\phi)] \\ &= \frac{a^2}{2}\int_0^{\pi/2}\sin 2\vartheta d\vartheta\int_0^{2\pi} d\phi f(\vartheta,\phi)\exp[-2\pi i a\rho\sin\vartheta\cos(\varphi-\phi)]\,. \end{aligned} \tag{11.4.16}$$

Truncation is achieved of course by limiting the radial or polar integration range.

Returning now to the case of the two Babinet–Soleil compensators with arbitrary settings and the analyser, the scalar field $\mathcal{E}_2(\theta_h, \phi)$ after the analyser is given from Eq. (11.4.7) by:

$$\mathcal{E}_2 = \frac{U_0}{2}([M_1 T^+ + T^-(M_2 \cos 2\phi + M_3 \sin 2\phi)] \ , \tag{11.4.19}$$

where we have introduced:

$$\begin{aligned} M_1 &= (A_1^+ A_2^+ + B_1 B_2)\cos\beta + (B_1 A_2^- + A_1^+ B_2)\sin\beta, \\ M_2 &= (A_1^+ A_2^+ - B_1 B_2)\cos\beta - (B_1 A_2^- - A_1^+ B_2)\sin\beta, \\ M_3 &= (B_1 A_2^+ + A_1^+ B_2)\cos\beta + (A_1^+ A_2^- + B_1 B_2)\sin\beta \ . \end{aligned} \tag{11.4.20}$$

We note the functional dependence of

$$M_n = f(\beta_1, \delta_1; \beta_2, \delta_2; \beta) \ ,$$

which means that the effects of all polarising components of the optical system are included in the functions M_n. The detector lens produces the field $U(r_d, \gamma_d)$ at the detector plane which we can calculate from Eq. (11.4.17) by writing:

$$\begin{aligned} U(r_d, \gamma_d) = -\frac{f_l U_0 \pi}{2\lambda} \Big[M_1 \int_0^{\alpha_l} T^+ J_0(kr_d \sin\vartheta_l) \sin 2\vartheta_l d\vartheta_l \\ -(M_2 \cos 2\gamma_d + M_3 \sin 2\gamma_d) \int_0^{\alpha_l} T^- J_2(kr_d \sin\vartheta_l) \sin 2\vartheta_l d\vartheta_l \Big] \ , \end{aligned} \tag{11.4.21}$$

where we have used the identity Eq. (8.4.37). Alternatively, from Eq. (11.4.18) we have

$$U(r_d, \gamma_d) = -\frac{i\pi U_0}{f_l \lambda} \left[M_1 \int_0^a T^+ J_0\left(\frac{kr_d}{f_l}\rho\right)\rho d\rho - (M_2 \cos 2\gamma_d + M_3 \sin 2\gamma_d) \int_0^a T^- J_2\left(\frac{kr_d}{f_l}\rho\right)\rho d\rho \right] . \tag{11.4.22}$$

The corresponding intensity in the detector plane can then be computed as

$$I(r_d, \gamma_d) = U(r_d, \gamma_d) U^*(r_d, \gamma_d) \ . \tag{11.4.23}$$

Equations (11.4.21) and (11.4.22) constitute the general solution for the field in the detector plane of a polarised microscope consisting of two, arbitrarily aligned Babinet–Soleil compensators, two objective lenses and an analyser. The direction of polarisation of this field is $\hat{\beta}$ as discussed before.

11.4.1.3 Amplitude and Intensity Sensitive Detectors

When a detector is placed in the focal plane of the detector lens its output depends linearly on either the intensity or the field present in the focal plane. According to this we distinguish between incoherent and coherent detectors, respectively. Typical examples of intensity sensitive detectors are a photodiode, a camera or a photomultipler. An example of a field-sensitive detector is a *single* mode optical fibre when the fundamental mode of the fibre is excited by the electric field incident upon it. We now discuss both types of detectors.

When an incoherent detector is employed the *integrated intensity*, $\bar{I}$, in the focal plane is given by:

$$\bar{I} = \iint_D |U|^2 S_D \, dD = \int_0^R \int_0^{2\pi} |U(r_d, \gamma_d)|^2 S_D(r_d) r_d dr_d d\gamma_d \ , \tag{11.4.24}$$

where $S_D(r)$ is the (rotationally symmetric) unitless intensity sensitivity function, D is the detector area and R is the radius of the detector. Note that the integrated intensity is related to the optical power P which can be calculated from $\bar{I}$ via $P = \bar{I}\epsilon_0 c/2$. Since $|U|^2 = UU^*$ and noting that upon integration by γ_d the cross-terms all yield zero, we obtain from Eq. (11.4.21)

$$\bar{I} = \frac{I_0 \pi^3}{f_l^2 \lambda^2} \left[2|M_1|^2 \int_0^R |P_0|^2 S_D(r_d)\, r_d dr_d + (|M_2|^2 + |M_3|^2) \int_0^R |P_2|^2 S_D(r_d)\, r_d dr_d \right] , \tag{11.4.25}$$

where we have introduced $I_0 = |U_0|^2$ and

$$\begin{aligned} P_0(r) &= \frac{f_l^2}{2} \int_0^{\alpha_l} T^+ J_0(kr \sin\vartheta_l) \sin 2\vartheta_l d\vartheta_l = \int_0^a T^+ J_0\left(\frac{kr\rho}{f_l}\right)\rho d\rho, \\ P_2(r) &= \frac{f_l^2}{2} \int_0^{\alpha_l} T^- J_2(kr \sin\vartheta_l) \sin 2\vartheta_l d\vartheta_l = \int_0^a T^- J_2\left(\frac{kr\rho}{f_l}\right)\rho d\rho, \end{aligned} \tag{11.4.26}$$

and note that $\Im\{P_n\} = 0$ if $\Im\{T^\pm\} = 0$.

When a coherent detector is employed the detection is sensitive to the field rather than to the intensity:

$$\bar{I} = \left| \iint_D US_C \, dD \right|^2 = \left| \int_0^R \int_0^{2\pi} U(r_d, \gamma_d) S_C(r_d) r_d dr_d d\gamma_d \right|^2 , \tag{11.4.27}$$

where $S_C(r)$, (unit m^{-1}) is the rotationally symmetric amplitude sensitivity function, which gives from Eq. (11.4.21), after integration by γ_d:

$$\bar{I} = \frac{4\pi^4 I_0}{f_l^2 \lambda^2} \left| \int_0^R M_1 P_0(r_d) S_C(r_d) r_d dr_d \right|^2 . \tag{11.4.28}$$

Having obtained expressions for both the field and the intensity in the exit pupil and the detector plane it is instructive to analyse these results. In the following we focus our attention on the two special cases of polarised microscopes, the linearly and the circularly polarised set-ups.

11.4.1.4 Linearly Polarised Illumination

For a linearly polarised microscope ($\delta_1 = \delta_2 = 0$) and for the cross polarised case ($\beta = \pi/2$) the intensity ($|U|^2$) and the field in the front focal plane of the detector lens and in the detector plane are shown in Fig. 11.13. We assume $n_{i1} = 1$, $n_{t1} = 1.5$, $n_{i2} = 1.5$, $n_{t2} = 1$ for the two interfaces. Considering now only the distributions in the detector plane let us imagine that a detector of nonvanishing diameter is placed on axis (centre of the image) as shown by the dashed circle in the figure. For an intensity sensitive detector the detected signal will always be nonvanishing due to the fact that the intensity is only zero on the optic axis. When, however, the on-axis detector is sensitive to the field a vanishing signal results which is a consequence of the symmetry properties of the field in the I–III and II–IV quadrants. It follows directly from what is said above that a detector placed off the optic axis behaves in a markedly different manner than an on-axis detector and in general will yield a nonvanishing signal regardless of whether it is intensity or field sensitive.

For the linearly polarised microscope we may write, from Eqs. (11.4.2) and (11.4.20):

$$\begin{aligned} &A_n^{\pm} = 1 \qquad &&\text{and} \qquad &&B_n = 0, \\ &M_1 = \cos\beta, \qquad &&M_2 = \cos\beta, \qquad &&M_3 = \sin\beta , \end{aligned} \tag{11.4.29}$$

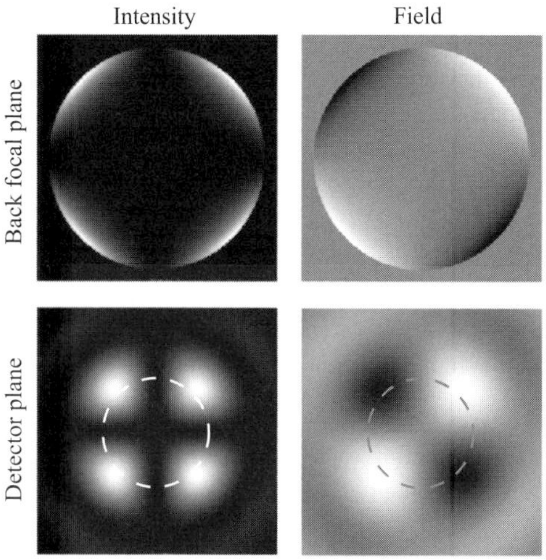

Figure 11.13: Distributions of the intensity and the field for a linearly polarised microscope at the back focal plane of the collector lens (Eq. (11.4.12) and Eq. (11.4.11)) and at the detector plane (Eq. (11.4.23) and Eq. (11.4.21)). The dashed circle in the detector plane images symbolises a detector placed in the detector plane. In the intensity images black level corresponds to zero intensity, whilst white corresponds to maximum intensity. In the field images inside the circular aperture, black corresponds to the minimum (negative) field value, whilst white corresponds to the maximum (positive) field value. Reproduced from **[343]** with permission of Elsevier.

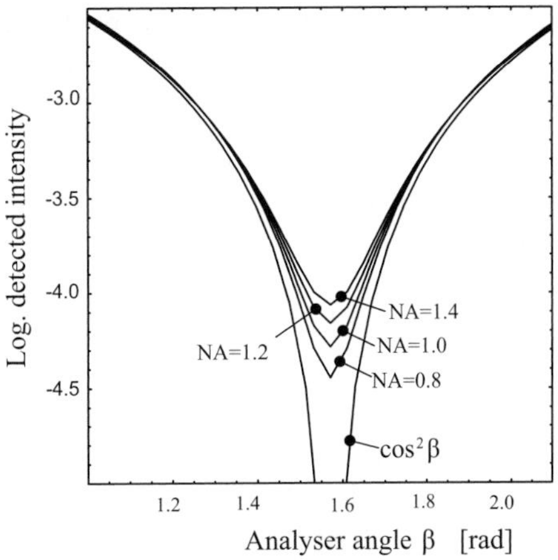

Figure 11.14: Detected intensity as a function of analyser angle β (in radians) for a linearly polarised microscope. Curves correspond from bottom to top to numerical apertures of 0.8, 1.0, 1.2, and 1.4. The lowest curve corresponds to $\cos^2\beta$, Malus' law. We used high values of numerical apertures (defined as $NA = n\sin\alpha$) because such lenses are often used in practical polarised light microscopy. Reproduced from **[343]** with permission of Elsevier.

from which the general expression for the detector signal of a microscope using an on-axis coherent and uniformly sensitive detector $S_C = S_{C0}$ is given from Eq. (11.4.28) by:

$$\bar{I} = C_c\cos^2\beta\left|\frac{f_l^2}{2}\int_0^{\alpha_l} T^+\cos\vartheta_l J_1(kR\sin\vartheta_l)d\vartheta_l\right|^2 = C_c\cos^2\beta\left|\int_0^a T^+J_1\left(\frac{kR\rho}{f_l}\right)d\rho\right|^2, \tag{11.4.30}$$

with $C_c = \pi^2R^2I_0S_{C0}^2/f_l^2$. For a microscope employing an on-axis incoherent and uniformly sensitive detector $S_D = S_{D0}$ the integrated intensity[11.13] is given from Eq. (11.4.25) by:

$$\bar{I} = C_{ic}\left[2\cos^2\beta\int_0^R |P_0|^2\, r_d dr_d + \int_0^R |P_2|^2\, r_d dr_d\right], \tag{11.4.31}$$

with $C_{ic} = I_0S_{D0}^2\pi^3/(\lambda^2f_l^2)$. The above equations permit us to estimate the detector signal for coherent and incoherent detectors for the linearly polarised case. When the polariser and the analyser are set to cross polar position (i.e. when the transmission axes of polariser and analyser are orthogonal) the coherent detector signal is zero and hence the extinction (defined by $I(\beta = 0°)/I(\beta = 90°)$) is infinite. It is important to emphasise that this is only the case when perfect polarisers are used. In practice, as discussed above, good quality polarisers possess an extinction ratio of $\approx 10^5$ which is then the *fundamental limit of linearly polarised microscopes employing coherent detection.* Equation (11.4.31) shows that for an *incoherent detector, and for crossed polars, the detected intensity will always be finite*, except when vanishingly small detectors are used because the second term in Eq. (11.4.31) is independent of β. In this latter case, however, the detection approximates the coherent limit and so zero detected signal is anticipated. Thus, since high-aperture lenses depolarise light and this light makes its way to the detector, measuring the intensity at the detector does not allow us to distinguish between signal due to changes of polarisation by the sample or the depolarisation effect of the objective lenses even if ideal polarisers were used. It also follows that in the absence of lenses an ideal analyser will extinguish all transmitted light in the cross polar setting. By Malus' law[11.14] the transmitted field is proportional to $\cos\beta$ and hence the transmitted intensity is proportional to $\cos^2\beta$. This relationship clearly shows that for $\beta = 90°$ the transmitted intensity would vanish. We may hence conclude that if, in the arrangement shown in Fig. 11.11, for $\beta = 90°$ nonvanishing intensity is detected this must be due to the presence of a high-numerical-aperture lens that depolarises light. To examine the effect of lens numerical aperture on the detected intensity further, we have computed the results shown in Fig. 11.14 from Eq. (11.4.31) using $R = 0.36$ Airy unit (A.u.).[11.15] This figure shows that, as anticipated, the higher the numerical aperture of a lens

[11.13] Note that although it is possible to analytically perform the integral over ρ in Eq. (11.4.31), the resulting formula is long and it actually distracts from understanding the physical meaning of the result so we have omitted its presentation.

[11.14] Malus' law describes the intensity transmitted through a linear analyser. Let us assume an incident electric field $\mathcal{E}_1 = (U_0, 0, 0)^T$. The field transmitted through the analyser is then $\mathcal{E}_2 = \mathbb{O}\cdot\mathcal{E}_1 = U_0(\cos^2\beta, \sin\beta\cos\beta, 0)^T$ and so the intensity is given by $I = \mathcal{E}_2^T\mathcal{E}_2^* = I_0\cos^2\beta$.

[11.15] We define the A.u. as the radial location of the first minimum of the transverse intensity distribution (see Eq. (8.4.47)).

the more it affects the polarisation and hence a higher detected intensity results for analyser angles close to the crossed direction. If we take $R \to 0$ in Eq. (11.4.31) then the second term becomes negligibly small and the first term dominates, resulting in a $\cos^2\beta$ dependence of the detected intensity corresponding to the case of no objective lenses.

To illustrate the dependence of the detected intensity upon the detector radius we present Fig. 11.15 which shows the detected intensity for a linearly polarised microscope with incoherent detection for parallel and crossed polarisers. The figure corresponding to parallel polarisers has an appearance similar to the normalised detected intensity of a paraxial confocal microscope **[80]**, **[288]** given by

$$\frac{\bar{I}(R)}{I_0} = 1 - J_0^2(R) - J_1^2(R) \,. \tag{11.4.32}$$

When this equation is plotted together with the parallel polariser case of Fig. 11.15 we obtain the curves shown in Fig. 11.16. The first position when the rate of change of the function is minimum coincides with first minimum of the intensity distributions in the focal plane. According to this the first position when the rate of change of the function is minimum occurs at $\approx$ 1A.u. and $\approx$ 1.5A.u. pinhole radii for the parallel and crossed polariser cases, respectively, suggesting that by crossing the polarisers the light distribution widens by 50%. This is because as the polariser is crossed

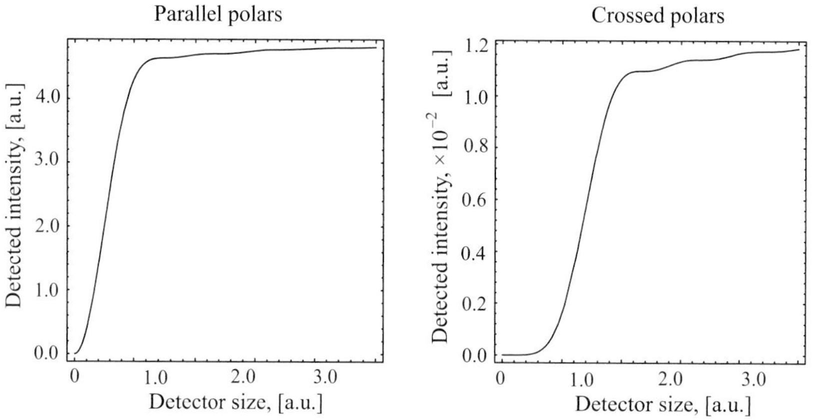

Figure 11.15: Detected intensity in arbitrary units [a.u.] for a linearly polarised microscope with incoherent detection for parallel and crossed polarisers as a function of the detector radius in Airy units [A.u.]. The numerical aperture of the condenser and collector lenses was NA = 1.3. Reproduced from **[343]** with permission of Elsevier.

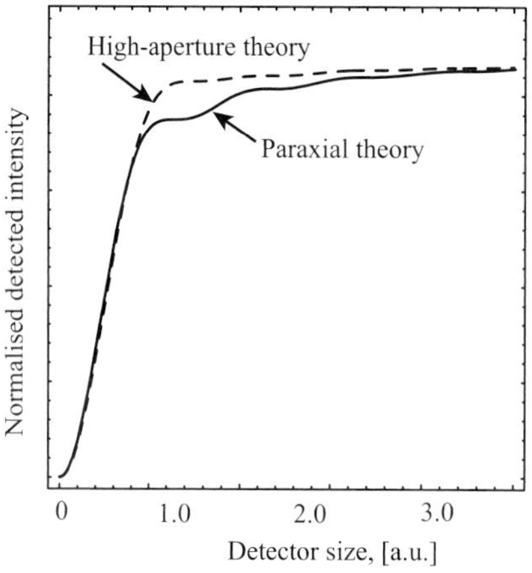

Figure 11.16: Comparison of the normalised detected intensity for a linearly polarised microscope from the paraxial and high-aperture theories as a function of the detector radius in Airy units [A.u.]. The numerical aperture of the condenser and collector lenses was NA = 1.3. The two intensity distributions are normalised to unity at $R \to \infty$.

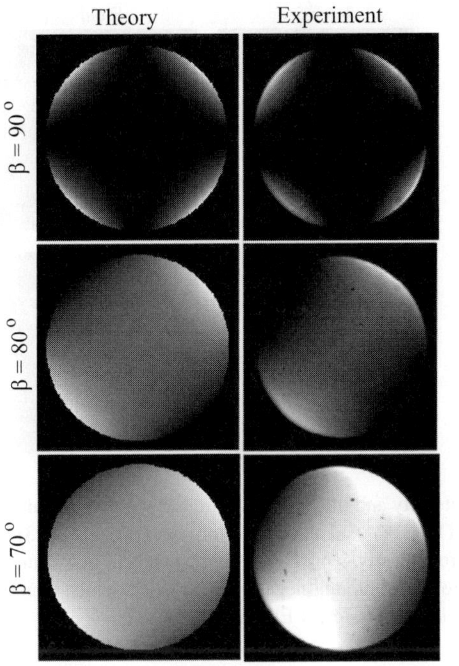

Figure 11.17: Theoretical and experimental exit pupil intensity distributions for a linearly polarised microscope with varying analyser angle, β. Reproduced from **[343]** with permission of Elsevier.

the contribution to the detector plane field distribution from the co-polarised component (E_x in Fig. 11.2) vanishes and the wider cross-polarised component dominates (E_y in Fig. 11.2).

The intensity in the exit pupil of the second objective lens was experimentally measured with a set-up similar to that of a wide-field microscope employing a Bertrand lens. The two objective lenses used in the experiment had a numerical aperture of 1.3 (oil immersion) and were illuminated by light of wavelength $\lambda = 632.8$ nm. The results of this measurement are shown in Fig. 11.17 together with theoretical predictions, as obtained from Eq. (11.4.8). The slight loss of fold symmetry in the experimental images was caused by one of the objective lenses not being centred on the optical axis. The figure shows that the assumption of a spherical interface is highly justified in our theoretical model. We note that our aim was to create a qualitative model to describe the polarisation aberration introduced by high-aperture lenses. A quantitative model would have been mathematically much more involved. For example, birefringence that could be readily present in an objective lens may be caused by residual stress in the glass of the objective lens. It is clear that if we wished to account for such local birefringence it would be impossible to create such a simple practically working model of the lens. Therefore small discrepancies between experimental and theoretical data are anticipated. When more accurate models are needed, for example in the case of lithographic lenses, specialist methods are required. These include the use of Jones pupils **[297]**, orientational Zernike polynomials and Extended Nijboer–Zernike functions discussed in Chapter 9.

11.4.1.5 Circularly Polarised Illumination

For a circularly polarised microscope the field and intensity in the front focal plane of the detector lens (back focal plane of the collector lens) and in the detector plane are shown in Fig. 11.18. This situation emphasises the importance of simultaneous study of the field and the intensity. If, for example, we wanted to determine the intensity in the detector plane, just by measuring the intensity in the back focal plane of the collector lens then a different, non-annular, distribution would result. Because the phase of the field in the back focal plane follows a $2\gamma_d$ dependence (see Eq. (11.4.21)), the resulting distribution in the detector plane is annular. Figure 11.18 also shows that when a coherent detector is employed (i.e. detecting the field) the resulting detector signal is zero. Conversely, for sufficiently large incoherent detectors the

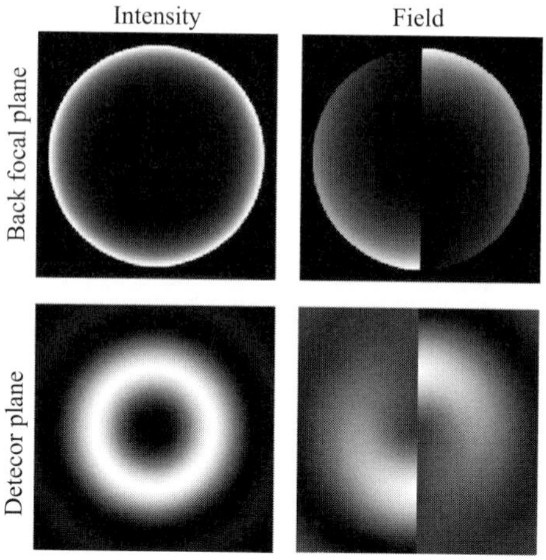

Figure 11.18: Distributions of the intensity and the field for a circularly polarised microscope in the back focal plane of the collector lens (Eq. (11.4.12) and Eq. (11.4.11)) and at the detector plane (Eq. (11.4.23) and Eq. (11.4.21)). In the intensity images black level corresponds to zero intensity, whilst white corresponds to maximum intensity. In the field images inside the circular aperture, black corresponds to the minimum (negative) field value, whilst white corresponds to the maximum (positive) field value. Reproduced from **[343]** with permission of Elsevier.

signal is different from zero. The figure also suggests that we may anticipate zero signal for both coherent and small incoherent detectors.

For the circularly polarised microscope we may write, for left-hand circular polarisation illumination, from Eqs. (11.4.2) and (11.4.20):

$$\begin{aligned} A_n^{\pm} &= \frac{1}{\sqrt{2}}, \qquad & B_1 &= \frac{i}{\sqrt{2}}, \qquad & B_2 &= -\frac{i}{\sqrt{2}}, \\ M_1 &= \cos\beta, \qquad & M_2 &= -i\sin\beta, \qquad & M_3 &= \sin\beta\,. \end{aligned} \tag{11.4.33}$$

It is interesting to note that, as shown by Eqs. (11.4.29) and (11.4.33), parameter M_1 is the same for both the linearly and the circularly polarised microscopes. This means that the detected signal of a coherent microscope and for the circularly polarised case is identical to that obtained for the linearly polarised case (Eq. (11.4.28)).

The incoherent integrated intensity is given from Eq. (11.4.33) by:

$$\bar{I} = 2C_{ic}\left[\cos^2\beta\left(\int_0^R |P_0|^2\, r_d dr_d - \int_0^R |P_2|^2\, r_d dr_d\right) + \int_0^R |P_2|^2\, r_d dr_d\right]. \tag{11.4.34}$$

The above equation shows that the detector signal for a circularly polarised microscope employing an incoherent detector will be finite for $\beta = 90^\circ$ except for, again, the case when small pinholes are used.

In Table 11.2 we compare the linear and circular cases for an incoherent detector to find that the detector signal for a linearly polarised microscope is somewhat higher for $\beta = 0^\circ$ than for a circularly polarised set-up. The detector signals are equal for $\beta = 45^\circ$ and finally when $\beta = 90^\circ$ the circularly polarised signal is twice as high as the detector signal for the linearly polarised microscope.

11.5 High-aperture Scanning Light Microscopes Imaging a Point Object

In this section we discuss the basic, underlying theory for modelling image formation by high-aperture optical microscopes of small, possibly birefringent, scatterers. With reference to Fig. 11.19 we first we compute the field in the focal region for a general incident polarisation using the Debye–Wolf integral Eq. (8.3.25). We then discuss how a birefringent point object scatters light and how the scattered light is collected by a high-aperture lens. As the next step we find expressions for the field distribution in the detector plane. It is not the objective of this section to produce numerical results – these are given in abundance in the following sections.

The optical system under consideration, although rotationally symmetric, is as general as feasibly possible. We discuss two basic geometries: reflection and transmission. The major difference between these two types of arrangements is that

Table 11.2. Special cases for incoherent detector signal as a function of analyser angle β. Constant multipliers have been dropped. Reproduced from **[343]** with permission of Elsevier.

β	Linear	Circular
0°	$\int_0^R (2\lvert P_0\rvert^2 + \lvert P_2\rvert^2)\rho d\rho$	$2\int_0^R \lvert P_0\rvert^2 \rho d\rho$
45°	$\int_0^R (\lvert P_0\rvert^2 + \lvert P_2\rvert^2)\rho d\rho$	$\int_0^R (\lvert P_0\rvert^2 + \lvert P_2\rvert^2)\rho d\rho$
90°	$\int_0^R \lvert P_2\rvert^2 \rho d\rho$	$2\int_0^R \lvert P_2\rvert^2 \rho d\rho$

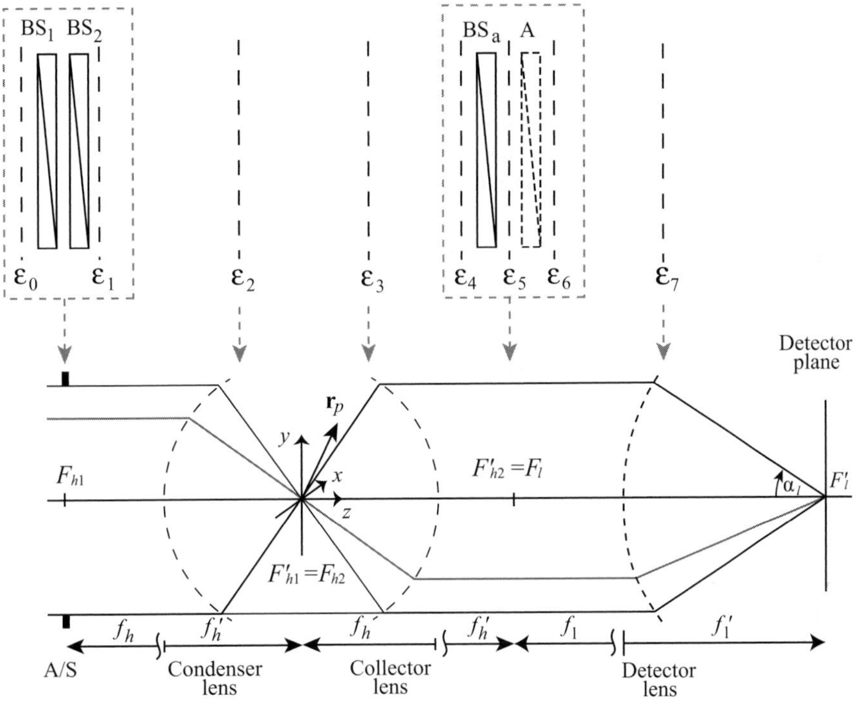

Figure 11.19: The optical system under consideration (lower figure). The point-like, monochromatic source is located on-axis at an infinite distance from the optical system in object (left) space. There is an aperture stop (A/S) placed in the front focal plane (denoted by its front Gaussian focal point F_{h1}) of a high-numerical-aperture (condenser) lens which is assumed to be aplanatic. A second high-numerical-aperture lens (collector lens) is placed such that its front focal point (F_{h2}) coincides with the back focal point (F'_{h1}) of the condenser lens. Another aplanatic lens (detector lens) is then placed so that its front focal point (F_l) coincides with the back focal point of the collector lens (F'_{h2}). The detector is placed in the back focal plane (denoted by its front Gaussian focal point F'_l) of the detector lens. The upper part of the figure shows the definition of the electric field of the geometrical optics approximation at the different locations. The dashed box around the two Babinet–Soleil compensators (BS_1 and BS_2) signifies that from the point of view of the model, the entire assembly is located at the front focal plane of the condenser lens, i.e. diffraction of light via propagation is ignored from $\mathcal{E}_0$ to $\mathcal{E}_1$. The same applies to the dashed box between the collector and detector lenses which includes another Babinet–Soleil compensator (BS_a) and an optional analyser (A) used to analyse the state of polarisation of the light. The fields $\mathcal{E}_2$, $\mathcal{E}_3$ and $\mathcal{E}_7$ are defined over the Gaussian reference spheres of the condenser, collector and detector lenses, respectively. The grey ray is a non-skew ray traversing the system.

for a reflection type microscope only the light scattered off by the specimen reaches the detector. Conversely, in the transmission geometry light detected by the detector is a result of interference between light scattered by the specimen and light that reaches the detector unperturbed, as recognised by Zernike in his Nobel prize winning theory of phase-contrast imaging **[389]**.

Both the transmission and reflection geometries are studied for two cases. First we discuss conventional scanning (Type I) microscopes. In this case, a *high-quality* microscope objective lens is used to illuminate the sample by light from a point-like source and another lens is used to gather light (collector lens). We shall term the first lens 'condenser lens' to adhere to the usual conventions of microscopy. However, as signified by the words emphasised in the previous sentence, we should keep in mind that unlike in conventional wide-field microscopy where the quality of the condenser lens is usually inferior to the collector lens, in scanning microscopy this lens is at least as good as the collector lens. This is because for a Type I system the quality of the condenser lens determines the overall quality of the imaging system. A large area detector that converts the incident electric field to an electric current proportional to the intensity is assumed to be present after the collector lens. Second, we discuss scanning microscopes employing small but finite-sized detectors. In this case another lens is placed after the collector lens (detector lens) which focuses the light onto the detector.

Although we have concluded from Section 11.4 that high-aperture microscope objective lenses depolarise light due to Fresnel transmission coefficients being polarisation-dependent, this finding is ignored in the rest of the chapter. This we mainly do to simplify the mathematics. However, it is also done because this depolarisation merely limits the extinction ratio one can achieve in polarised light microscopy but it does not change the qualitative description of the optical system in any way.

The optical system used in the following subsection is much the same as the one discussed in Section 11.5.6 with the exception that here we introduce an additional Babinet–Soleil compensator just before the condenser lens. It is necessary to consider this component in Section 11.5.6 when modelling the Pol-Scope. We then calculate an image formed by either a conventional or a confocal microscope in the transmission or reflection configuration.

11.5.1 The Field in the Focal region

In Section 11.4 we could readily avoid the direct calculation of the electric field between the condenser and collector lenses. Here it is essential to first calculate the electric field in the focal region and then determine how a point object interacts with this. We have derived the distribution of the electric field in the focal region of a high-numerical-aperture lens illuminated by linearly polarised coherent light in Section 11.2. The illumination we consider here is similar.

The optical system used to illuminate the sample is shown in Fig. 11.20 and is modelled as follows. First, expressions are obtained for the electric field that passes through the two retarders. This field is then focused by a high-aperture lens. Let a collimated electromagnetic beam (see Fig. 11.20) be polarised in the x-direction of a right-handed Cartesian coordinate system. The beam propagates in the $+z$ direction and is incident on two Babinet–Soleil compensators representing the two retarders in the illumination path. The slow axes of the first and second compensators are set to

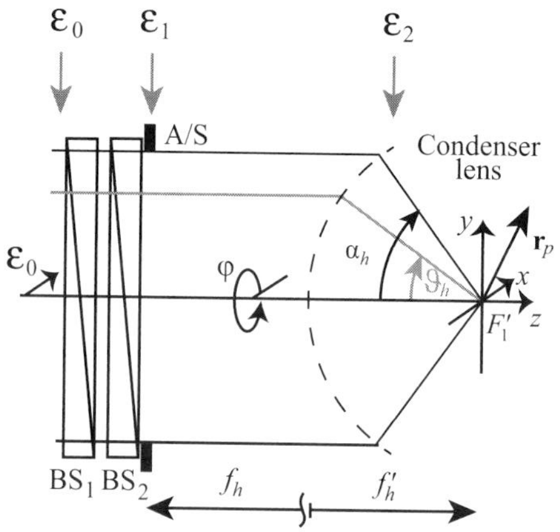

Figure 11.20: The illumination set-up and co-ordinate system notation. The initially x polarised light from a monochromatic point source, placed on axis and at an infinite distance from the system, is passed through two Babinet–Soleil compensators (BS_1 and BS_2) to produce a radiation of spatially homogeneous but otherwise arbitrary state of polarisation. An aperture stop (A/S) is placed in the front focal plane of the (high-numerical-aperture) collector lens. In the vicinity of the back Gaussian focus F_{h1}' a point scatterer is located at $\mathbf{r}_p$ whose dipole moment is induced by the electric field of the collector lens. The marginal ray angle is denoted α_h and ϑ_h is the angle of a typical non-skew ray in the system.

be 45° and 0°, respectively, with respect to the incident polarisation. The combined effect of the two compensators is an arbitrary homogeneous state of polarisation exiting this assembly. Light transmitted through the components is incident on a high-aperture lens that focuses light into a small volume, the so-called probe volume. The electric vector of the geometric optics approximation, after being focused by the high-aperture lens, is given by:

$$\mathcal{E}_2 = \mathbb{R}^{-1} \cdot \mathbb{L} \cdot \mathbb{R} \cdot \mathbb{BS}_{\beta=0,\delta\to\delta_2} \cdot \mathbb{BS}_{\beta=\pi/4,\delta\to\delta_1} \cdot \mathcal{E}_0 \,, \tag{11.5.1}$$

where $\mathbb{R}$, $\mathbb{L}$ and $\mathbb{BS}_n$ are given by equations (11.1.1), (11.1.2) and (11.1.15), respectively. In Eq. (11.5.1) $\mathcal{E}_0 = (U_i, 0, 0)$. Equation (11.5.1) results in:

$$\mathcal{E}_2 = \frac{U_i \exp(-i\delta_2/2)\sqrt{\cos\vartheta_h}}{2} \begin{pmatrix} \mathcal{E}'_{2x} \\ \mathcal{E}'_{2y} \\ \mathcal{E}'_{2z} \end{pmatrix} , \tag{11.5.2}$$

where

$$\begin{aligned} \mathcal{E}'_{2x} &= \cos\frac{\delta_1}{2}\exp(i\delta_2)[(1+\cos\vartheta_h) - (1-\cos\vartheta_h)\cos 2\phi] \\ &\quad - i\sin\frac{\delta_1}{2}(1-\cos\vartheta_h)\sin 2\phi , \\ \mathcal{E}'_{2y} &= -\cos\frac{\delta_1}{2}\exp(i\delta_2)(1-\cos\vartheta_h)\sin 2\phi + i\sin\frac{\delta_1}{2}[(1+\cos\vartheta_h) \\ &\quad + (1-\cos\vartheta_h)\cos 2\phi] , \\ \mathcal{E}'_{2z} &= 2\cos\frac{\delta_1}{2}\exp(i\delta_2)\sin\vartheta_h\cos\phi + 2i\sin\frac{\delta_1}{2}\sin\vartheta_h\sin\phi \,. \end{aligned} \tag{11.5.3}$$

After substituting from Eq. (11.5.2) into Eq. (8.3.25) the integration over ϕ can analytically be carried out to give:

$$\mathbf{E}(r_t, \gamma, z_p) = -iU_i kf \exp\left(-\frac{i\delta_2}{2}\right) \begin{pmatrix} \exp(i\delta_2)\cos\frac{\delta_1}{2}(I_0 + I_2\cos 2\gamma) + i\sin\frac{\delta_1}{2} I_2 \sin 2\gamma \\ \exp(i\delta_2)\cos\frac{\delta_1}{2} I_2 \sin 2\gamma + i\sin\frac{\delta_1}{2}(I_0 - I_2\cos 2\gamma) \\ -2iI_1\left[\exp(i\delta_2)\cos\frac{\delta_1}{2}\cos\gamma + i\sin\frac{\delta_1}{2}\sin\gamma\right] \end{pmatrix} , \tag{11.5.4}$$

with $I_n(r_t, z_p)$ given by Eq. (11.2.4). This equation gives the electric field in the vicinity of the focus of a high-aperture lens, illuminated by a plane wave that has passed through two Babinet–Soleil compensators.

Now we consider a small, possibly birefringent, object placed in the vicinity of the focus of this high-aperture lens. The position of the small scatterer is characterised by cylindrical co-ordinates (r_t, γ, z_p). When the particle is scanned in the vicinity of the focus produced by the high-aperture lens it experiences an electric field given by Eq. (11.5.4) at a position (r_t, γ, z_p). The dimensions of the particle in any direction are required to be much smaller than the wavelength, or $ka \ll 1$ where a is the diameter. If the scatterer is embedded in a medium and an incident electric field is applied the field also polarises the embedding medium. The equation that describes this process is given by (see Eqs. (1.2.12) and (2.2.1)):

$$\mathbf{D} = \epsilon_0 \,\overline{\overline{\epsilon}}_r\, \mathbf{E} = \epsilon_0 \mathbf{E} + \epsilon_0 \,\overline{\overline{\chi}}\, \mathbf{E} = \epsilon_0 \mathbf{E} + \mathbf{P} \,, \tag{11.5.5}$$

where $\mathbf{E}$ is the incident electric field, $\mathbf{D}$ is the induced displacement, and $\mathbf{P}$ is the induced polarisation. In the above equation the permittivity (or dielectric) tensor $\overline{\overline{\epsilon}}_r$ is defined by Eq. (2.2.4) and $\overline{\overline{\chi}}$ is the electric susceptibility tensor defined via $\overline{\overline{\epsilon}}_r = \overline{\overline{I}} + \overline{\overline{\chi}}$ for the general case.

We want to describe the process of light scattering by a birefringent point scatterer so we consider the following model. The birefringent scatterer will have a dipole moment (units Cm) $\mathbf{p}_s$ induced by the incident electric field $\mathbf{E}$ that is given by **[150]**, **[160]**:

$$\mathbf{p}_s = a^3 \epsilon_0 \,\overline{\overline{\chi}}_s\, \mathbf{E}_i = a^3 \epsilon_0 \left(\overline{\overline{\epsilon}}_{r,s} - \overline{\overline{I}}\right) \mathbf{E} \,. \tag{11.5.6}$$

By taking the limit $\overline{\overline{\epsilon}}_{r,s} \to \overline{\overline{\epsilon}}_{r,m}$, where $\overline{\overline{\epsilon}}_{r,m}$ is the permittivity tensor of the embedding medium, we see that the polarisation, and therefore the dipole moment, within the same small region when the scatterer is absent is

$$\mathbf{p}_m = a^3 \epsilon_0 \,\overline{\overline{\chi}}_m\, \mathbf{E} = a^3 \epsilon_0 \left(\overline{\overline{\epsilon}}_{r,m} - \overline{\overline{I}}\right) \mathbf{E} \,. \tag{11.5.7}$$

The difference between the two dipole moments $\Delta\mathbf{p}$ is what causes the image contrast, which for a homogeneous and isotropic embedding medium of relative permittivity $\epsilon_{r,m}$, can be written as:

$$\Delta\mathbf{p} = \mathbf{p}_s - \mathbf{p}_m = a^3\epsilon_0\left(\bar{\bar{\epsilon}}_{r,s} - \epsilon_{r,m}\,\bar{\bar{I}}\right)\mathbf{E} = a^3\epsilon_0\epsilon_{r,m}\left(\frac{\bar{\bar{\epsilon}}_{r,s}}{\epsilon_{r,m}} - \bar{\bar{I}}\right)\mathbf{E}\,. \tag{11.5.8}$$

We shall denote $\Delta\mathbf{p}$ as $\mathbf{p}$ in what follows. Note that the above equation is formally identical to $\mathbf{p} = \bar{\bar{\alpha}}\,\mathbf{E}$, with $\bar{\bar{\alpha}}$ being the polarisability. For the case when the local electric field is identical to the incident electric field we have $\bar{\bar{\chi}} = N\,\bar{\bar{\alpha}}$ where N is the volume density of molecules, which means that our results imply $N = 1$.

The dielectric tensor is in general given by (see Eq. (2.2.4)):

$$\bar{\bar{\epsilon}}_r = \begin{pmatrix} \epsilon_{xx} & \epsilon_{xy} & \epsilon_{xz} \\ \epsilon_{yx} & \epsilon_{yy} & \epsilon_{yz} \\ \epsilon_{zx} & \epsilon_{zy} & \epsilon_{zz} \end{pmatrix}. \tag{11.5.9}$$

For biaxial crystals this tensor can either be diagonalised as shown in Subsection 2.2.1 or we can make the assumption that the principal dielectric axes coincide with the crystallographic axes and the axes of the Cartesian coordinate system, in which case the dielectric tensor is greatly simplified to give:

$$\bar{\bar{\epsilon}}_r = \begin{pmatrix} \epsilon_{xx} & 0 & 0 \\ 0 & \epsilon_{yy} & 0 \\ 0 & 0 & \epsilon_{zz} \end{pmatrix}, \tag{11.5.10}$$

with $\epsilon_{i,j} = \epsilon_0 n_{i,j}^2$ for a nonmagnetic medium. If the birefringent scatterer is rotated to an arbitrary orientation **[195]** we can use Euler rotation matrices:

$$\mathbb{R}_x = \begin{pmatrix} 1 & 0 & 0 \\ 0 & \cos\alpha_x & \sin\alpha_x \\ 0 & -\sin\alpha_y & \cos\alpha_x \end{pmatrix}, \tag{11.5.11}$$

$$\mathbb{R}_y = \begin{pmatrix} \cos\alpha_y & 0 & -\sin\alpha_y \\ 0 & 1 & 0 \\ \sin\alpha_y & 0 & \cos\alpha_y \end{pmatrix}, \tag{11.5.12}$$

and

$$\mathbb{R}_z = \begin{pmatrix} \cos\alpha_z & \sin\alpha_z & 0 \\ -\sin\alpha_z & \cos\alpha_z & 0 \\ 0 & 0 & 1 \end{pmatrix}. \tag{11.5.13}$$

Let the product of these matrices be denoted by $\mathbb{T} = \mathbb{R}_x \cdot \mathbb{R}_y \cdot \mathbb{R}_z$ and so the transformed dielectric tensor by $\bar{\bar{\epsilon}}'$ is given by:

$$\bar{\bar{\epsilon}}' = \mathbb{T}^{-1} \cdot \bar{\bar{\epsilon}} \cdot \mathbb{T}\,,. \tag{11.5.14}$$

It is worth noting here that the above formalism is also suitable to treat small homogeneous scatterers with shape birefringence. In this case the asymmetry in the dielectric tensor occurs due to the deviation in shape from the sphere. The dielectric tensor and the corresponding dipole moment for these cases are given by, for example, van de Hulst **[150]**.

11.5.2 Field Emitted by the Birefringent Point Scatterer and Collimated by a Lens

As we have seen above the small scatterer placed in the focal region of the high-aperture lens 'experiences' an electric field produced by that lens. Hence, if the scatterer was placed off-axis at say $\mathbf{r}_p$ then it would get excited by the electric field $\mathbf{E}(\mathbf{r}_p)$, given by Eq. (11.5.4).

For particles for which $ka \gg \lambda$ the above theory breaks down. The scattered electromagnetic field can however be described by means of Mie theory **[150]**. The solution is obtained in terms of a series expansion with individual terms usually referred to as partial waves. The first-order partial wave in the expansion is the dipole wave. The solution shows that the electric field many wavelengths away from the scatterer that is taken to be much smaller than the wavelength rapidly becomes purely transverse when expressed in terms of its spherical field components, i.e. the dominant term in the expansion is the dipole wave. For this reason one sometimes refers to a small scatterer viewed at large distances as a dipole scatterer. The phase of the scattered field is identical to that of a diverging spherical wave and its polarisation can be shown to be a simple function of the position vector and the dipole moment (see Fig. 1.7).

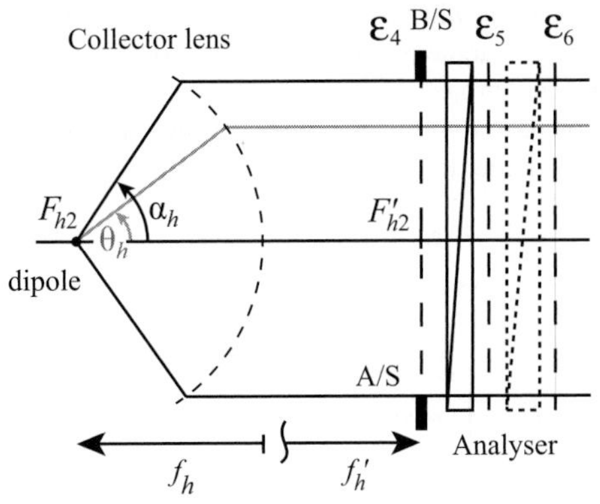

Figure 11.21: Showing how the collector lens recollimates the light emitted by an electric dipole. The electric field $\mathcal{E}_4$ is at the aperture stop (A/S) of the system which is placed in the second focal plane of the collector lens. A Babinet–Soleil (B/S) compensator and a linear analyser are also shown in the collimated path. The coordinate system used is defined in Fig. 11.20, after noting that the focal points F'_{h1} and F_{h2} coincide.

Evidently, if we place such a dipole scatterer into the Gaussian focus of a high-aperture lens the lens collimates the distribution, as follows from the principle of reciprocity, and produces a wave with a plane wavefront propagating parallel to the optic axis. The polarisation of the plane wave can be readily computed using generalised Jones matrices. With reference to Fig. 11.21, because the aperture stop is in the back focal plane of the collector lens, when the dipole is in the front focal plane but off the Gaussian focus the principal ray is parallel with the optic axis in object space of the collector lens. Therefore, in the image space of this lens the plane wave now propagates along the principal ray and at an angle to the optic axis. For small displacements of the scatterer the polarisation of the plane wave does not change and therefore we can model this scenario by adding a linear phase term (tilt phase) to the wavefront corresponding to the on-axis scatterer. When the scatterer is on axis but out of focus, the marginal ray is no longer parallel with the optic axis in image space and so a defocus phase term needs to be added to the phase of the plane wave. Note that strictly speaking any curved wavefront is by definition non-transversal and so in principle the polarisation of the wave in the image space of this lens should be modified. However, in practical cases the radius of the wavefront corresponding to the defocused scatterer is so large that the longitudinal component of the electric vector in image space can safely be ignored. This approach works well when the phase of light scattered by the dipole is a smooth function of either polar or azimuthal angles, i.e. for all practically important cases.

We have discussed in Section 1.6.3 that the electric far field $\mathbf{E}$ of a harmonically oscillating electric dipole is given by Eq. (1.6.48) from which $\mathcal{E}_3$, the electric field of the geometric optics approximation can be obtained by taking the field on the surface of a Gaussian reference sphere (with r = const.). It is thus given by:

$$\mathcal{E}_3 = -\frac{k^2}{4\pi\epsilon_0 r}\frac{1}{r^2}\mathbf{r}\times\mathbf{r}\times\mathbf{p} = -\frac{U_d}{r}\,\hat{\mathbf{r}}\times\hat{\mathbf{r}}\times\hat{\mathbf{p}}\,, \tag{11.5.15}$$

where $U_d = (k^2 p)/(4\pi\epsilon_0)$, as before $\mathbf{r} = r(\sin\theta\cos\phi, \sin\theta\sin\phi, \cos\theta)$ is a position vector and $\mathbf{p} = p\hat{\mathbf{p}}$ is given by Eq. (11.5.8). This field is then collected and recollimated by the collector lens. The collimated electric field $\mathcal{E}_4$ is given by (see Fig. 11.21):

$$\mathcal{E}_4 = \mathbb{R}^{-1}\cdot\mathbb{L}^{-1}\cdot\mathbb{R}\cdot\left(\mathcal{E}_3|_{\theta\to\theta_h, r\to f_h}\right)\,, \tag{11.5.16}$$

or, by substituting from Eqs. (11.1.1), (11.1.2), (11.5.8) and (11.5.15) into Eq. (11.5.16) we obtain:

$$\begin{aligned}
\mathcal{E}_{4x} &= -\frac{U_d}{2f_h\sqrt{\cos\theta_h}}\{\hat{p}_x[(1+\cos\theta_h)-(1-\cos\theta_h)\cos 2\phi]-\hat{p}_y(1-\cos\theta_h)\sin 2\phi \\
&\quad - 2\hat{p}_z\sin\theta_h\cos\phi\} \\
\mathcal{E}_{4y} &= -\frac{U_d}{2f_h\sqrt{\cos\theta_h}}\{-\hat{p}_x(1-\cos\theta_h)\sin 2\phi+\hat{p}_y[(1+\cos\theta_h)+(1-\cos\theta_h)\cos 2\phi] \\
&\quad - 2\hat{p}_z\sin\theta_h\sin\phi\} \\
\mathcal{E}_{4z} &= 0\,.
\end{aligned} \tag{11.5.17}$$

Note the term $\cos^{-1/2}\theta_h$ in the above equation, which has been introduced because the energy projection now occurs from a spherical to a plane surface. One can also argue on the basis of the conservation of energy: when two *identical* lenses are placed in a confocal position, whatever projection factor the first lens introduces the second lens must cancel it. The field collimated by the collector lens is passed through another Babinet–Soleil compensator with its fast axis set at 45° with respect to the $+x$ direction. The field after the compensator is given by:

$$\boldsymbol{\mathcal{E}}_5 = \mathbb{BS}_{\beta=\pi/4,\delta\to\delta_a} \cdot \boldsymbol{\mathcal{E}}_4 \,, \tag{11.5.18}$$

or, explicitly

$$\boldsymbol{\mathcal{E}}_5 = \begin{pmatrix} \mathcal{E}_{4x}\cos(\delta_a/2) + i\mathcal{E}_{4y}\sin(\delta_a/2) \\ \mathcal{E}_{4y}\cos(\delta_a/2) + i\mathcal{E}_{4x}\sin(\delta_a/2) \\ 0 \end{pmatrix} . \tag{11.5.19}$$

We now have to continue our treatment separately for the various types of optical microscopes.

11.5.3 Reflection, Conventional (Type I) Scanning Microscopes

We first discuss the situation when the analyser, shown in Fig. 11.21, normally placed after the Babinet–Soleil compensator, is absent and a large area detector is placed in the collimated optical path. Clearly, in this case, the collector lens merely acts as a light collector and plays no further part in imaging. Assuming that the detector is large enough so that it captures all light incident upon it, the detected integrated intensity is then given by:

$$\bar{I}_r = \frac{f_h^2}{2}\int_0^{\alpha_h}\int_0^{2\pi} |\boldsymbol{\mathcal{E}}_5|^2 \sin 2\theta_h d\theta_h d\phi = \int_0^a \int_0^{2\pi} |\boldsymbol{\mathcal{E}}_5|^2 \rho d\rho d\phi \,. \tag{11.5.20}$$

The above equation can readily be evaluated analytically. The solution is given by:

$$\bar{I}_r = \frac{\pi I_d}{12}\left[16 - 3\left(5|\hat{p}_x|^2 + 5|\hat{p}_y|^2 + 6|\hat{p}_z|^2\right)\cos\alpha_h - \left(|\hat{p}_x|^2 + |\hat{p}_y|^2 - 2|\hat{p}_z|^2\right)\cos 3\alpha_h\right] , \tag{11.5.21}$$

where we have defined $I_d = |U_d|^2$. It is interesting, though altogether not very surprising, to discover that the detected signal of a conventional transmission microscope is independent of the setting of the Babinet–Soleil compensator in the detector path. Note that if we collect light from the 2π solid angle, i.e. set $\alpha_h = \pi/2$, we get

$$I_r = \frac{4\pi I_d}{3} = \frac{4\pi}{3}\frac{k^4p^2}{16\pi^2\epsilon_0^2} = \frac{k^4p^2}{12\pi\epsilon_0^2} \,,$$

from which the total emitted power P_r is exactly half the power given for the 4π solid angle by Eq. (1.6.44).

We now turn to discuss the detected intensity for the cases when a linear polariser is inserted into the collimated optical path just after the Babinet–Soleil compensator. The two important cases are when the polariser is set parallel or perpendicular to the x-axis. The detected integrated intensities are denoted $\bar{I}^{\parallel}$ and $\bar{I}^{\times}$ for parallel and perpendicular polariser settings, respectively. Because we chose the incident polarisation direction to be cross polarised, it is straightforward to write the intensity for parallel polars by simply taking the cross polarised component of the electric field:

$$\bar{I}_r^{\parallel} = \frac{f_h^2}{2}\int_0^{\alpha_h}\int_0^{2\pi} \mathcal{E}_{5x}\mathcal{E}_{5x}^* \sin 2\theta_h d\theta_h d\phi \,, \tag{11.5.22}$$

which, again, has an analytical solution given by:

$$\begin{aligned}\bar{I}_r^{\parallel} = \frac{\pi I_d}{12}\Big[& |\hat{p}_z|^2(8 - 9\cos\alpha_h + \cos 3\alpha_h) + \sin^2\frac{\alpha_h}{2}\left(|\hat{p}_x|^2 + |\hat{p}_y|^2\right)(18 + 4\cos\alpha_h + 2\cos 2\alpha_h) \\ & + \left(|\hat{p}_x|^2 - |\hat{p}_y|^2\right)(15 + 8\cos\alpha_h + \cos 2\alpha_h)\cos\delta_a \\ & + 48\cos^2\frac{\alpha_h}{2}\,\Re\left\{\hat{p}_y\Im\{\hat{p}_x\} - \hat{p}_x\Im\{\hat{p}_y\}\right\}\sin\delta_a\Big] .\end{aligned} \tag{11.5.23}$$

The intensity for a perpendicular polariser can be directly computed from the y component:

$$\bar{I}_r^{\times} = \frac{f_h^2}{2} \int_0^{\alpha_h} \int_0^{2\pi} \mathcal{E}_{5y} \mathcal{E}_{5y}^* \sin 2\theta_h d\theta_h d\phi \,, \tag{11.5.24}$$

or analytically:

$$\begin{aligned} \bar{I}_r^{\times} = \frac{\pi I_d}{12} \Big[|\hat{p}_z|^2 (8 - 9\cos\alpha_h + \cos 3\alpha_h) + \sin^2\frac{\alpha_h}{2} \left(|\hat{p}_x|^2 + |\hat{p}_y|^2\right)(18 + 4\cos\alpha_h + 2\cos 2\alpha_h) \\ - \left(|\hat{p}_x|^2 - |\hat{p}_y|^2\right)(15 + 8\cos\alpha_h + \cos 2\alpha_h)\cos\delta_a \\ -48\cos^2\frac{\alpha_h}{2} \Re\left\{\hat{p}_y \Im\{\hat{p}_x\} - \hat{p}_x \Im\{\hat{p}_y\}\right\} \sin\delta_a \Big] \,. \end{aligned} \tag{11.5.25}$$

The sum of the intensities $\bar{I}_r^{\parallel}$ and $\bar{I}_r^{\times}$ yields $\bar{I}_r$ as it should. With this equation we have successfully obtained all results for a conventional reflection microscope. Now we turn to discussing the situation corresponding to a transmission type set-up.

11.5.4 Transmission, Conventional (Type I) Scanning Microscopes

An essential difference between transmission and reflection type microscopes arises from the fact that in transmission microscopes light not scattered by the object may also reach the detector. This is because part of the light is transmitted through the optical system. When, for example, the ideal Babinet–Soleil compensator and the linear polariser in the detection path are set in such a way that they fully extinguish the transmitted light, only that part of the scattered light which changed polarisation as a result of the scattering process can reach the detector. When, however, the Babinet–Soleil compensator and the linear polariser do not fully extinguish the transmitted light then the portion of light that traversed the two optical components will reach the detector. We, therefore have to consider the transmission arrangement separately.

First we determine the electric field of the light transmitted through the optical system for an arbitrary compensator setting. Let the transmitted field, just after the last Babinet–Soleil compensator, be denoted by $\boldsymbol{\mathcal{E}}_t$. We may write

$$\boldsymbol{\mathcal{E}}_t = \mathbb{BS}_{\beta=\pi/4,\delta\to\delta_a} \cdot \mathbb{R}^{-1} \cdot \mathbb{L}^{-1} \cdot \mathbb{L} \cdot \mathbb{R} \cdot \boldsymbol{\mathcal{E}}_1 = \mathbb{BS}_{\beta=\pi/4,\delta\to\delta_a} \cdot \mathbb{BS}_{\beta=0,\delta\to\delta_2} \cdot \mathbb{BS}_{\beta=\pi/4,\delta\to\delta_1} \cdot \boldsymbol{\mathcal{E}}_0 \,, \tag{11.5.26}$$

or explicitly:

$$\boldsymbol{\mathcal{E}}_t = U_0 \begin{pmatrix} \exp\left(i\frac{\delta_2}{2}\right)\cos\frac{\delta_a}{2}\cos\frac{\delta_1}{2} - \exp\left(-i\frac{\delta_2}{2}\right)\sin\frac{\delta_a}{2}\sin\frac{\delta_1}{2} \\ i\exp\left(i\frac{\delta_2}{2}\right)\sin\frac{\delta_a}{2}\cos\frac{\delta_1}{2} + i\exp\left(-i\frac{\delta_2}{2}\right)\cos\frac{\delta_a}{2}\sin\frac{\delta_1}{2} \\ 0 \end{pmatrix} . \tag{11.5.27}$$

Based on the results of the previous subsection we may write for the detected intensity of a transmission conventional microscope employing no analyser:

$$\begin{aligned} \bar{I}_t &= \frac{f_h^2}{2} \int_0^{\alpha_h} \int_0^{2\pi} \left|\boldsymbol{\mathcal{E}}_5 + \boldsymbol{\mathcal{E}}_t\right|^2 \sin 2\theta_h d\theta_h d\phi \\ &= \frac{f_h^2}{2} \int_0^{\alpha_h} \int_0^{2\pi} \left(|\boldsymbol{\mathcal{E}}_5|^2 + |\boldsymbol{\mathcal{E}}_t|^2 + 2\Re\left\{\boldsymbol{\mathcal{E}}_5^T \cdot \boldsymbol{\mathcal{E}}_t^*\right\}\right) \sin 2\theta_h d\theta_h d\phi \,, \end{aligned} \tag{11.5.28}$$

or explicitly:

$$\begin{aligned} \bar{I}_t = \bar{I}_r + I_0 f_h^2 \pi \sin^2\alpha_h - \frac{4\pi U_d U_0 f_h}{15}\left[8 - \cos^{3/2}\alpha_h(5 + 3\cos\alpha_h)\right] \\ \times \left[\cos\frac{\delta_2}{2}\left(\cos\frac{\delta_1}{2}\Re\{\hat{p}_x\} + \sin\frac{\delta_1}{2}\Im\{\hat{p}_y\}\right) + \sin\frac{\delta_2}{2}\left(\cos\frac{\delta_1}{2}\Im\{\hat{p}_x\} + \sin\frac{\delta_1}{2}\Re\{\hat{p}_y\}\right)\right] . \end{aligned} \tag{11.5.29}$$

Equation (11.5.28) means that the scattered and transmitted light interfere at *the detector*. We note that the first α_h independent term in Eq. (11.5.29) means that there will be a constant background intensity present in the detector signal.

When a linear analyser is inserted in the detector path the detected intensity for a parallel polariser is given by:

$$\bar{I}_t^{\parallel} = \frac{f_b^2}{2}\int_0^{\alpha_b}\int_0^{2\pi}\left(\left|\mathcal{E}_{5x}\right|^2 + \left|\mathcal{E}_{tx}\right|^2 + 2\Re\left\{\mathcal{E}_{5x}\mathcal{E}_{tx}^*\right\}\right)\sin 2\theta_b d\theta_b d\phi\ . \tag{11.5.30}$$

Equation (11.5.30) has an analytical solution that is given by:

$$\begin{aligned}\bar{I}_t^{\parallel} = \bar{I}_r^{\parallel} &+ \frac{I_0\pi f_b^2}{2}\sin^2\alpha_b(1+\cos\delta_1\cos\delta_a - \sin\delta_1\cos\delta_2\sin\delta_a)\\ &- \frac{4\pi U_d U_0 f_b}{15}\left[8-\cos^{3/2}\alpha_b(5+3\cos\alpha_b)\right]\\ &\times\left\{\sin\left(\frac{\delta_2}{2}\right)\cos\left(\frac{\delta_1-\delta_a}{2}\right)\left[\Im\{\hat{p}_x\}\cos\left(\frac{\delta_a}{2}\right)+\Re\{\hat{p}_y\}\sin\left(\frac{\delta_a}{2}\right)\right]\right.\\ &\left.\quad+\cos\left(\frac{\delta_2}{2}\right)\cos\left(\frac{\delta_1+\delta_a}{2}\right)\left[\Re\{\hat{p}_x\}\cos\left(\frac{\delta_a}{2}\right)-\Im\{\hat{p}_y\}\sin\left(\frac{\delta_a}{2}\right)\right]\right\}\ .\end{aligned} \tag{11.5.31}$$

The detected intensity for a transmission microscope with crossed linear polarisers is given by:

$$\bar{I}_t^{\times} = \frac{f_b^2}{2}\int_0^{\alpha_b}\int_0^{2\pi}\left(\left|\mathcal{E}_{5y}\right|^2 + \left|\mathcal{E}_{ty}\right|^2 + 2\Re\left\{\mathcal{E}_{5y}\mathcal{E}_{ty}^*\right\}\right)\sin 2\theta_b d\theta_b d\phi\ . \tag{11.5.32}$$

The analytical solution is given by:

$$\begin{aligned}\bar{I}_t^{\times} = \bar{I}_r^{\times} &+ \frac{I_0\pi f_b^2}{2}\sin^2\alpha_b(1-\cos\delta_1\cos\delta_a + \sin\delta_1\cos\delta_2\sin\delta_a)\\ &+ \frac{4\pi U_d U_0 f_b}{15}\left[8-\cos^{3/2}\alpha_b(5+3\cos\alpha_b)\right]\\ &\times\left\{\sin\left(\frac{\delta_2}{2}\right)\sin\left(\frac{\delta_1-\delta_a}{2}\right)\left[\Im\{\hat{p}_x\}\sin\left(\frac{\delta_a}{2}\right)-\Re\{\hat{p}_y\}\cos\left(\frac{\delta_a}{2}\right)\right]\right.\\ &\left.\quad-\cos\left(\frac{\delta_2}{2}\right)\sin\left(\frac{\delta_1+\delta_a}{2}\right)\left[\Re\{\hat{p}_x\}\sin\left(\frac{\delta_a}{2}\right)+\Im\{\hat{p}_y\}\cos\left(\frac{\delta_a}{2}\right)\right]\right\}\ .\end{aligned} \tag{11.5.33}$$

It is easy to give a physical interpretation of $\bar{I}_t$, $\bar{I}_t^{\parallel}$ and $\bar{I}_t^{\times}$. As Eq. (11.5.29), (11.5.31) and (11.5.33) show the detected intensity contains three terms: one due to the scatterer present in the focal region, one due to the unperturbed transmitted light, and an interference term. When the microscope is set to operate with linear polarisation ($\delta_1 = 0$, $\delta_2 = 0$) the transmitted intensity with a parallel analyser setting is given from Eq. (11.5.31), showing that all three terms are present. On the other hand, when the linear analyser is crossed, as expected from Eq. (11.5.32), we have $\bar{I}_t^{\times} = \bar{I}_r^{\times}$ because the other components are blocked by the analyser. Conversely, when we use circularly polarised incident light ($\delta_1 = \pi/2$, $\delta_2 = 0$) and an analyser wave plate set to opposite handedness ($\delta_a = -\pi/2$) the application of a co-polarised linear analyser yields an intensity with all three intensity terms present, and in the cross-polarised direction $\bar{I}_t^{\times} = \bar{I}_r^{\times}$ as expected.

Let us consider now Eqs. (11.5.20) and (11.5.23)), (11.5.25), (11.5.29), (11.5.31) and (11.5.33). In these equations the only terms that depend on the spatial coordinates are the Cartesian components of the dipole moment of the scatterer **p**, which enter into the expressions weighted by functions determined by both the numerical aperture of the collector lens and the retardance of the wave plates. Consequently, it is possible to conclude that the optical resolution of both reflection and transmission conventional (scanning) microscopes is determined only by the optical properties of the lens illuminating the sample. The numerical aperture of the collector lens merely determines the magnitude of the detected intensity. The numerical aperture of the collector lens also determines the effective spatial coherence between adjacent scatterers, as was shown in Ch. 10 for the low-aperture scalar read-out of an object. The coherence properties associated with the vectorial detection are not discussed in this subsection.

11.5.5 Confocal (Type II) Microscopes

Now we turn to discuss imaging by confocal microscopes. The part of the optical system we consider is shown in Fig. 11.22. Our current system is different from what we have analysed in the previous two sections in that now we need to consider light propagating through, and hence focused by, the detector lens. Before we start the discussion it is worth stopping here to point out the role of the detector lens.

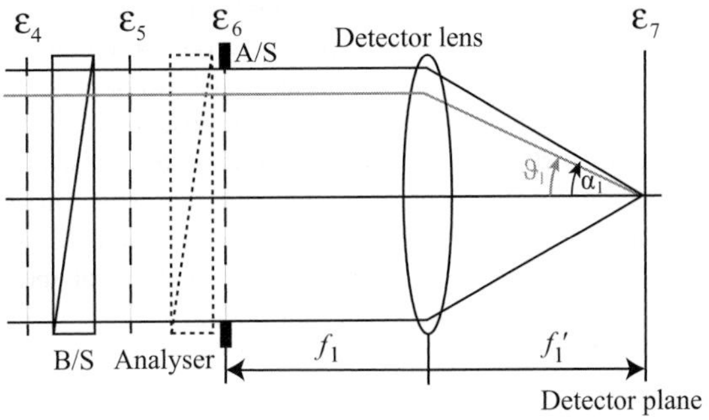

Figure 11.22: The first part of the detection system and corresponding notations. The detector lens having a focal length of f_l is set to focus light on the detector plane. The aperture stop (A/S) is at the front focal plane of the detector lens. A Babinet–Soleil compensator (B/S) together with a linear polariser (shown as Analyser) can together act as an analyser for elliptically polarised light. The angle α_l corresponds to the marginal ray and the angle ϑ_l corresponds to a non-skew ray.

The primary role of the detector is to focus the collimated light produced by the high-aperture collector lens onto the detector. We know from the paraxial theory of confocal microscopy that the properties of such an imaging system are partly determined by the relative size of the detector with respect to the focused distribution produced by the detector lens, or more practically, the size of the small aperture (the *pinhole*) normally placed in front of the detector. The size of the pinhole, together with the wavelength of light and the numerical aperture of the lenses used, determine both the transverse and axial resolution and of course profoundly affect signal levels in the system. The NA, and therefore the focal length, of the high-aperture collector lens are usually determined by the resolution required from the confocal microscope. The physical size of the light distribution at the detector is set by the ratio of the focal lengths of the collector and detector lenses, i.e. the transverse magnification of collector–detector lens assembly that is arranged in a 4-f configuration with the aperture stop placed in the common focus. Because this imaging system is telecentric from both object and image space, the principal ray is parallel to the optical axis in both object and image spaces. Because the vertex of the Gaussian reference sphere always intersects the principal ray and its centre of curvature is the Gaussian image point, it is clear that this, aberration-free, optical system can be modelled for off-axis objects just by translating the Gaussian reference sphere in the transverse direction.

The choice of the focal length of the detector lens is entirely governed by convenience. Since the transverse diameter of the intensity distribution in the focal plane of this lens is determined by the focal length of this lens, and if limiting the overall optical path length is not an objective, often it is easiest to choose a long focal length so as not to make the required pinhole diameter too small. In fact one of the early commercial confocal microscopes on the market used such a long focal length detector lens that the optical path needed to be folded several times by a number of mirrors. Nowadays the focal length of the detector lens is usually on the order of 200 mm so as to permit the use of 10–20 μm pinhole diameters. When the pinhole size needed to be adjusted, early commercial confocal microscopes used to employ mechanically adjustable pinholes. Nowadays, high end systems use a zoom lens that results in a varying transverse magnification of the collector–detector lens optical system hence permitting the use of a fixed pinhole size.

From a wave optics point of view, it is clear from the discussion in Chapter 8 and also known from Fourier optics (see **[113]** and Appendix A.9) that a phase-correct[11.16] relationship exists between the field in the focal region and in the exit pupil plane. Therefore when the two lenses are arranged in a true 4-f configuration, the conjugates are phase-correct, scaled copies of one another. When, on the other hand, the distance between the lenses is not exactly the sum of the focal lengths of the collection and detector lenses, free space propagation of the wavefront from the exit pupil of one lens to the entrance pupil of the other lens, results in an additional quadratic phase term. This phase term in turn causes the Fresnel number of the detector lens to drop, hence resulting in low Fresnel number effects (see Section 8.3) of the intensity distribution in the focal region. Fortunately this effect is almost always negligible so that experimentally it would be difficult to notice.

When the numerical aperture of a lens approaches zero the polarisation-dependent effects become negligible. The distribution in the focal plane can hence be expressed by either a Fourier or a Fourier–Bessel transform. Furthermore,

[11.16] The image *field* is a scaled copy of the object *field* in a transverse plane.

when a low-aperture lens is illuminated by linearly polarised light, the polarisation direction in the vicinity of the focus is identical to that of the incident illumination. It would, therefore, be tempting to suggest that the distribution of light in the detector plane can be calculated as a simple Fourier transform of the distribution in the back focal plane. Indeed, this method of calculation is used when discussing confocal microscopes employing a linear polariser to simplify mathematics. However, we shall use the method of vectorial ray-tracing and subsequently the Debye–Wolf integral to obtain the full electric field in the vicinity of the focus of the low-aperture lens. This we do because in the case when no polariser is present in the detector path, symmetry properties of the electric field make the mathematics particularly simple. Our results can also be used to show that as a result of the low numerical aperture of the detector lens, certain components can be ignored and hence the formulas are much simplified.

11.5.5.1 Reflection Confocal Microscopes

We now continue with the mathematical development for a reflection confocal microscope in the absence of the linear polariser (also referred to as analyser). First we use the generalised Jones matrices to obtain the electric field vector $\boldsymbol{\mathcal{E}}_7$ from Eq. (11.5.19). It is given by:

$$\boldsymbol{\mathcal{E}}_7 = \mathbb{R}^{-1} \cdot \mathbb{L}_{\vartheta \to \vartheta_l} \cdot \mathbb{R} \cdot \boldsymbol{\mathcal{E}}_5 \,, \tag{11.5.34}$$

where generalised Jones matrices of $\mathbb{R}^{-1}$, $\mathbb{L}$ and $\mathbb{R}$ are given by Eq. (11.1.1) and Eq. (11.1.2). After performing the matrix operation it is not difficult to obtain $\boldsymbol{\mathcal{E}}_7$ analytically:

$$\begin{aligned}
\mathcal{E}_{7x} = \frac{p_x}{4}\Bigg[& 2\cos\frac{\delta_a}{2}(1+\cos\vartheta_l\cos\theta_h) - 2i\sin\frac{\delta_a}{2}(1-\cos\vartheta_l\cos\theta_h)\sin 2\phi \\
& - 2\cos\frac{\delta_a}{2}(1-\cos\vartheta_l\cos\theta_h)\cos 2\phi + i\sin\frac{\delta_a}{2}(1+\cos\vartheta_l\cos\theta_h)\sin 4\phi \\
& - i\sin\frac{\delta_a}{2}(\cos\theta_h+\cos\vartheta_l)\sin 4\phi\Bigg] + \frac{p_y}{4}\Bigg[i\sin\frac{\delta_a}{2}(1+\cos\vartheta_l\cos\theta_h) \\
& + i\sin\frac{\delta_a}{2}(\cos\theta_h+\cos\vartheta_l) - 2\cos\frac{\delta_a}{2}(1-\cos\vartheta_l\cos\theta_h)\sin 2\phi \\
& - 2i\sin\frac{\delta_a}{2}(\cos\theta_h-\cos\vartheta_l)\cos 2\phi - i\sin\frac{\delta_a}{2}(1+\cos\vartheta_l\cos\theta_h)\cos 4\phi \\
& + i\sin\frac{\delta_a}{2}(\cos\theta_h+\cos\vartheta_l)\cos 4\phi\Bigg] - \frac{p_z}{2}\Bigg[i\sin\frac{\delta_a}{2}(1+\cos\vartheta_l)\sin\theta_h\sin\phi \\
& + 2\cos\frac{\delta_a}{2}\cos\vartheta_l\sin\theta_h\cos\phi - i\sin\frac{\delta_a}{2}(1-\cos\vartheta_l)\sin\theta_h\sin 3\phi\Bigg] \,,
\end{aligned} \tag{11.5.35a}$$

for the y component:

$$\begin{aligned}
\mathcal{E}_{7y} = \frac{p_x}{4}\Bigg[& i\sin\frac{\delta_a}{2}(1+\cos\vartheta_l\cos\theta_h) + i\sin\frac{\delta_a}{2}(\cos\theta_h+\cos\vartheta_l) \\
& - 2\cos\frac{\delta_a}{2}(1-\cos\vartheta_l\cos\theta_h)\sin 2\phi + 2i\sin\frac{\delta_a}{2}(\cos\theta_h-\cos\vartheta_l)\cos 2\phi \\
& - i\sin\frac{\delta_a}{2}(1+\cos\vartheta_l\cos\theta_h)\cos 4\phi + i\sin\frac{\delta_a}{2}(\cos\theta_h+\cos\vartheta_l)\cos 4\phi\Bigg] \\
& + \frac{p_y}{4}\Bigg[2\cos\frac{\delta_a}{2}(1+\cos\vartheta_l\cos\theta_h) - 2i\sin\frac{\delta_a}{2}(1-\cos\vartheta_l\cos\theta_h)\sin 2\phi \\
& + 2\cos\frac{\delta_a}{2}(1-\cos\vartheta_l\cos\theta_h)\cos 2\phi - i\sin\frac{\delta_a}{2}(1+\cos\vartheta_l\cos\theta_h)\sin 4\phi \\
& + i\sin\frac{\delta_a}{2}(\cos\theta_h+\cos\vartheta_l)\sin 4\phi\Bigg] - \frac{p_z}{2}\Bigg[i\sin\frac{\delta_a}{2}(1+\cos\vartheta_l)\sin\theta_h\cos\phi \\
& + 2\cos\frac{\delta_a}{2}\cos\vartheta_l\sin\theta_h\sin\phi + i\sin\frac{\delta_a}{2}(1-\cos\vartheta_l)\sin\theta_h\cos 3\phi\Bigg] \,,
\end{aligned} \tag{11.5.35b}$$

and finally for the z component:

$$\mathcal{E}_{7z} = \frac{p_x}{2}\Bigg[i\sin\frac{\delta_a}{2}(1+\cos\theta_h)\sin\vartheta_l\sin\phi + 2\cos\frac{\delta_a}{2}\cos\theta_h\sin\vartheta_l\cos\phi$$

$$- i \sin \frac{\delta_a}{2}(1 - \cos\theta_h) \sin\vartheta_l \sin 3\phi\Big] + \frac{p_y}{2}\Big[i \sin\frac{\delta_a}{2}(1 + \cos\theta_h)\sin\vartheta_l\cos\phi$$

$$+ 2\cos\frac{\delta_a}{2}\cos\theta_h\sin\vartheta_l\sin\phi + i\sin\frac{\delta_a}{2}(1-\cos\theta_h)\sin\vartheta_l\cos 3\phi\Big]$$

$$- p_z\left[\cos\frac{\delta_a}{2}\sin\theta_h\sin\vartheta_l + i\sin\frac{\delta_a}{2}\sin\theta_h\sin\vartheta_l\sin 2\phi\right]. \tag{11.5.35c}$$

Note that there is a multiplier term of $\sqrt{\cos\vartheta_l/\cos\theta_h}$ in front of all vector components that we have not written for clarity. Likewise, the $1/f_h$ constant multiplier term is omitted above, but that too re-appears in the constant A below.

The electric field at the detector is given, after substitution in the Debye–Wolf integral Eq. (8.3.25), by: 10.8

$$E_{7x}(r_d,\gamma_d,z_d;z_p) = \frac{Ap_x}{2}\Big[2\cos\frac{\delta_a}{2}I_0^a + 2I_2^a\left(\cos\frac{\delta_a}{2}\cos 2\gamma_d + i\sin\frac{\delta_a}{2}\sin 2\gamma_d\right)$$

$$+ i\sin\frac{\delta_a}{2}\left(I_4^a - I_4^b\right)\sin 4\gamma_d\Big] + \frac{iAp_y}{2}\Big[\sin\frac{\delta_a}{2}\left(I_0^a + I_0^b\right)$$

$$- 2i\cos\frac{\delta_a}{2}I_2^a\sin 2\gamma_d + 2\sin\frac{\delta_a}{2}I_2^b\cos 2\gamma_d + \sin\frac{\delta_a}{2}\left(I_4^b - I_4^a\right)\cos 4\gamma_d\Big]$$

$$+ Ap_z\left[\sin\frac{\delta_a}{2}I_1^a\sin\gamma_d - 2i\cos\frac{\delta_a}{2}I_1^b\cos\gamma_d + \sin\frac{\delta_a}{2}I_3^a\sin 3\gamma_d\right], \tag{11.5.36a}$$

$$E_{7y}(r_d,\gamma_d,z_d;z_p) = \frac{iAp_x}{2}\Big[\sin\frac{\delta_a}{2}\left(I_0^a + I_0^b\right) - 2i\cos\frac{\delta_a}{2}I_2^a\sin 2\gamma_d$$

$$- 2\sin\frac{\delta_a}{2}I_2^b\cos 2\gamma_d + \sin\frac{\delta_a}{2}\left(I_4^b - I_4^a\right)\cos 4\gamma_d\Big] + \frac{Ap_y}{2}\Big[2\cos\frac{\delta_a}{2}I_0^a$$

$$- 2\left(\cos\frac{\delta_a}{2}\cos 2\gamma_d - i\sin\frac{\delta_a}{2}\sin 2\gamma_d\right)I_2^a + i\sin\frac{\delta_a}{2}\left(I_4^a - I_4^b\right)\sin 4\gamma_d\Big]$$

$$- Ap_z\left[\sin\frac{\delta_a}{2}I_1^a\cos\gamma_d - 2i\cos\frac{\delta_a}{2}I_1^b\sin\gamma_d - \sin\frac{\delta_a}{2}I_3^a\cos 3\gamma_d\right], \tag{11.5.36b}$$

and finally the z component is given by:

$$E_{7z}(r_d,\gamma_d,z_d;z_p) = Ap_x\left[\sin\frac{\delta_a}{2}I_1^d\sin\gamma_d - 2i\cos\frac{\delta_a}{2}I_1^e\cos\gamma_d + \sin\frac{\delta_a}{2}I_3^b\sin 3\gamma_d\right]$$

$$+ Ap_y\left[-2i\cos\frac{\delta_a}{2}I_1^e\sin\gamma_d + \sin\frac{\delta_a}{2}I_1^d\cos\gamma_d - \sin\frac{\delta_a}{2}I_3^b\cos 3\gamma_d\right]$$

$$- 2Ap_z\left[\cos\frac{\delta_a}{2}I_0^c - i\sin\frac{\delta_a}{2}I_2^c\sin 2\gamma_d\right], \tag{11.5.36c}$$

with $A = -iU_dkf_l/(2pf_h)$ and

$$I_n^{a,\dots,e}(r,z_d;z_p) = \int_0^{\alpha_l}\sqrt{\frac{\cos\vartheta_l}{\cos\theta_h}}\sin\vartheta_l A_n^{a,\dots,e}(\vartheta_l,\theta_h)J_n(kr\sin\vartheta_l)\exp(ik_hz_p\cos\theta_h)\exp(ikz_d\cos\vartheta_l)d\vartheta_l, \tag{11.5.37a}$$

where

$$A_0^a = A_4^a = 1 + \cos\theta_h\cos\vartheta_l,\ A_0^b = A_4^b = \cos\theta_h + \cos\vartheta_l,\ A_0^c = A_2^c = \sin\theta_h\sin\vartheta_l,$$

$$A_1^a = (1+\cos\vartheta_l)\sin\theta_h,\quad A_1^b = \sin\theta_h\cos\vartheta_l,\quad A_1^c = (1-\cos\vartheta_l)\sin\theta_h,$$
$$A_1^d = (1+\cos\theta_h)\sin\vartheta_l,\quad A_1^e = \cos\theta_h\sin\vartheta_l, \tag{11.5.37b}$$

$$A_2^a = 1 - \cos\theta_h\cos\vartheta_l,\quad A_2^b = \cos\theta_h - \cos\vartheta_l,$$

$$A_3^a = (1-\cos\vartheta_l)\sin\theta_h,\quad A_3^b = (1-\cos\theta_h)\sin\vartheta_l.$$

A few notes are due here. First, the defocus term in Eq. (11.5.37a), $\exp(ik_hz_p\cos\theta_h)$, is included to account for axial position of the dipole in the high-aperture space of the collector lens. This term also includes the wavenumber in that space k_h to account for possible immersion media. There is a second defocus term, $\exp(ikz_d \cos\vartheta_l)$, which models the axial displacement of the detector from its nominal position, at f_l' distance from the detector lens. The position where the field is calculated in detector space is therefore given by (r_d, γ_d, z_d). Furthermore, the Cartesian components of the dipole moment (p_x, p_y, p_z) are no longer normalised in Eq. (11.5.36), which is shown by the definition of the constant A. This is because the magnitude of the dipole moment p is proportional to the magnitude of the electric field that induced that moment, as described by Eq. (11.5.8): this way the dipole 'inherits' the amplitude, phase and orientation of the electric field produced by the condenser lens.

Equation (11.5.36) provides the electric field at the detector space of a microscope that uses the light of a monochromatic (i.e. temporary coherent) point source (i.e. spatially coherent) which is polarised and then passed through two Babinet–Soleil compensators to illuminate the condenser lens. In the high-numerical-aperture space there is a point object that is modelled as an electric dipole, whose moment is induced by the electric field in the focal region of the condenser lens. The light emitted by the dipole is re-collimated by the collector lens. In the collimated path there is a Babinet–Soleil compensator in place whose orientation and retardance can be adjusted at will. The detector lens, that follows the Babinet–Soleil compensator, focuses the light onto the detector.

In what follows we compute the field at the detector in the presence of the analyser (see Fig. 11.22) which is set to either $\beta = 0$ (parallel) or $\beta = \pi/2$ (perpendicular) position in order to simulate confocal polarised microscopes. In order to obtain the field at the detector plane we follow the method used in the previous subsection. If we only consider the cases of a parallel and crossed analyser then the scalar field in the detector plane $E_{d,x}$ and $E_{d,y}$ may be written as a truncated Fourier transform of the field in the front focal plane of the detector lens: $\mathcal{E}_{5x}$ and $\mathcal{E}_{5y}$, respectively:

$$\begin{aligned} E_{d,(x,y)}(r_d, \gamma_d; z_p) &= -\frac{if_l}{2\lambda} \int_0^{\alpha_l} \int_0^{2\pi} \sqrt{\frac{\cos\vartheta_l}{\cos\theta_h}} \mathcal{E}_{5x,y} \sin 2\vartheta_l \exp[-ikr_d \sin\vartheta_l \cos(\phi - \gamma_d)] \exp(ik_h z_p \cos\theta_h) d\phi d\vartheta_l \\ &= \int_0^{a} \int_0^{2\pi} \sqrt{\frac{\cos\vartheta_l}{\cos\theta_h}} \mathcal{E}_{5x,y} \exp\left[-ik\frac{r_d \rho}{f_l} \cos(\phi - \gamma_d)\right] \exp(ik_h z_p \cos\theta_h) \rho d\phi d\rho \,. \end{aligned} \tag{11.5.38}$$

There does not seem to be much utility in making this formula more complicated by substituting $\cos\theta_h = \sqrt{1 - \rho^2/f_h^2}$ and $\cos\theta_l = \sqrt{1 - \rho^2/f_l^2}$. In writing this formula we have not included the defocus term for the detector plane position. If required, this can be done just as in Eq. (11.5.37a). We have from Eqs. (11.5.16) and (11.5.19):

$$\begin{aligned} \mathcal{E}_{5x} = &-(1 + \cos\theta_h)\left(p_x \cos\frac{\delta_a}{2} + ip_y \sin\frac{\delta_a}{2}\right) \\ &+ (1 - \cos\theta_h)\left(p_y \cos\frac{\delta_a}{2} + ip_x \sin\frac{\delta_a}{2}\right) \sin 2\phi \\ &+ (1 - \cos\theta_h)\left(p_x \cos\frac{\delta_a}{2} - ip_y \sin\frac{\delta_a}{2}\right) \cos 2\phi \\ &+ 2p_z \sin\theta_h \left(\cos\frac{\delta_a}{2} \cos\phi + i \sin\frac{\delta_a}{2} \sin\phi\right) , \end{aligned} \tag{11.5.39a}$$

and

$$\begin{aligned} \mathcal{E}_{5y} = &-(1 + \cos\theta_h)\left(p_y \cos\frac{\delta_a}{2} + ip_x \sin\frac{\delta_a}{2}\right) \\ &+ (1 - \cos\theta_h)\left(p_x \cos\frac{\delta_a}{2} + ip_y \sin\frac{\delta_a}{2}\right) \sin 2\phi \\ &- (1 - \cos\theta_h)\left(p_y \cos\frac{\delta_a}{2} - ip_x \sin\frac{\delta_a}{2}\right) \cos 2\phi \\ &+ 2p_z \sin\theta_h \left(\cos\frac{\delta_a}{2} \sin\phi + i \sin\frac{\delta_a}{2} \cos\phi\right) , \end{aligned} \tag{11.5.39b}$$

where we have omitted the $U_d/(2pf_h\sqrt{\cos\theta_h})$ term, which reappears in all expressions below. After substituting from Eq. (11.5.39) into Eq. (11.5.38) we obtain the scalar field corresponding to a parallel and a crossed polariser situation:

$$E_{d,x}(r_d,\gamma_d;z_p) = A\left(p_x \cos\frac{\delta_a}{2} + ip_y \sin\frac{\delta_a}{2}\right)K_0 + A\left[\cos\frac{\delta_a}{2}\left(p_x \cos 2\gamma_d + p_y \sin 2\gamma_d\right) - \right.$$
$$\left. - i\sin\frac{\delta_a}{2}\left(p_y \cos 2\gamma_d - p_x \sin 2\gamma_d\right)\right]K_2$$
$$+ 2iAp_z\left(\cos\frac{\delta_a}{2}\cos\gamma_d + i\sin\frac{\delta_a}{2}\sin\gamma_d\right)K_1\,, \quad (11.5.40a)$$

and

$$E_{d,y}(r_d,\gamma_d;z_p) = A\left(p_y \cos\frac{\delta_a}{2} + ip_x \sin\frac{\delta_a}{2}\right)K_0 + A\left[\cos\frac{\delta_a}{2}\left(p_x \sin 2\gamma_d - p_y \cos 2\gamma_d\right) + \right.$$
$$\left. + i\sin\frac{\delta_a}{2}\left(p_y \sin 2\gamma_d + p_x \cos 2\gamma_d\right)\right]K_2$$
$$+ 2iAp_z\left(\cos\frac{\delta_a}{2}\sin\gamma_d + i\sin\frac{\delta_a}{2}\cos\gamma_d\right)K_1\,. \quad (11.5.40b)$$

Where $A = i\pi f_l U_d/(2pf_h\lambda)$ we have defined:

$$K_0(r,z) = \int_0^{\alpha_l} \sqrt{\frac{\cos\vartheta_l}{\cos\theta_h}}(1+\cos\theta_h)\sin 2\vartheta_l J_0(kr\sin\vartheta_l)\exp(ik_h z\cos\theta_h)d\vartheta_l,$$
$$K_1(r,z) = \int_0^{\alpha_l} \sqrt{\frac{\cos\vartheta_l}{\cos\theta_h}}\sin\theta_h \sin 2\vartheta_l J_1(kr\sin\vartheta_l)\exp(ik_h z\cos\theta_h)d\vartheta_l, \quad (11.5.41)$$
$$K_2(r,z) = \int_0^{\alpha_l} \sqrt{\frac{\cos\vartheta_l}{\cos\theta_h}}(1-\cos\theta_h)\sin 2\vartheta_l J_2(kr\sin\vartheta_l)\exp(ik_h z\cos\theta_h)d\vartheta_l\,.$$

11.5.5.2 Transmission, Confocal Microscopes

As discussed for the case of a transmission conventional microscope, when the analyser in the optical set-up does not exactly extinguish the transmitted light, it produces interference with the light scattered by the object. It would be far too complicated to express the detected intensity of the present case, unlike when the conventional microscope was discussed. Furthermore, the detected intensity clearly depends on the size of the detector too. We can, however, obtain the transmitted field that coherently adds to the light scattered by the object. First, the case of no analyser is discussed. We may write from Eq. (11.5.27):

$$\mathcal{E}_t^c = \mathbb{R}^{-1}\cdot\mathbb{L}_{\vartheta\to\vartheta_l}\cdot\mathbb{R}\cdot\mathcal{E}_t\,, \quad (11.5.42)$$

where we have used the notation $\mathcal{E}_t^c$ for the transmitted field for a confocal microscope to distinguish it from that of a conventional microscope $\mathcal{E}_t$. The confocal transmitted field in the detector plane is then simply obtained by substituting $\mathcal{E}_t^c$ into the Debye–Wolf integral. The result of integration is:

$$\mathbf{E}_t^c(r_d,\gamma_d,z_d) = \frac{if_l k}{2}\begin{pmatrix} \mathcal{E}_{t,x}(I_0 + I_2\cos 2\gamma_d) + \mathcal{E}_{t,y} I_2 \sin 2\gamma_d \\ \mathcal{E}_{t,y}(I_0 - I_2\cos 2\gamma_d) + \mathcal{E}_{t,x} I_2 \sin 2\gamma_d \\ -2iI_1\left(\mathcal{E}_{t,x}\cos\gamma_d + \mathcal{E}_{t,y}\sin\gamma_d\right)\end{pmatrix}, \quad (11.5.43)$$

where $I_{0\ldots2}(r_d,z_d)$ are given by Eq. (11.2.4d) written for the low-aperture azimuthal angle ϑ_l. The total field in the detector plane is hence given as a coherent sum of $\mathbf{E}_t^c$ and $\mathbf{E}_7$ given by Eqs. (11.5.43) and (11.5.36), respectively.

When a linear polariser (analyser) is placed after the Babinet–Soleil compensator BS_a the interference is calculated after the polariser rather than at the detector. However, since the transmitted field and that scattered from the object are added coherently, we can simply calculate the transmitted field and then add the resulting fields. If the transmitted field is analysed by means of a polariser set at an angle β we have

$$\mathcal{E}_{t,a}^c = \mathbb{R}^{-1}\cdot\mathbb{L}_{\vartheta\to\vartheta_l}\cdot\mathbb{R}\cdot\mathbb{O}\cdot\mathcal{E}_t\,, \quad (11.5.44)$$

which gives

$$\mathbf{E}_{t,a}^{c}(r_d, \gamma_d, z_d) = \frac{if_1 k}{2}\left(\cos\beta\mathcal{E}_{t,x} + \sin\beta\mathcal{E}_{t,y}\right)\begin{pmatrix} \cos\beta\left(I_0 + I_2\cos 2\gamma_d\right) + \sin\beta I_2 \sin 2\gamma_d \\ \sin\beta\left(I_0 - I_2\cos 2\gamma_d\right) + \cos\beta I_2 \sin 2\gamma_d \\ -2iI_1\left(\cos\beta\cos\gamma_d + \sin\beta\sin\gamma_d\right) \end{pmatrix}. \tag{11.5.45}$$

The total field can then be calculated by the coherent sum of $\mathbf{E}_{t,a}^{c}$ and $\mathbf{E}_7$ given by Eqs. (11.5.45) and (11.5.36), respectively.

With the above equations we have arrived at the final result of this section. We have obtained expressions for both reflection and transmission conventional and confocal microscopes imaging a birefringent point scatterer. The optical path was taken to be very general, assuming two Babinet–Soleil compensators in the illumination path. These two compensators can produce an arbitrarily polarised plane wave. This plane wave was then incident on a high-aperture lens to produce a focused distribution. The small birefringent object was placed in the vicinity of the focus. The field emitted by the object was recollimated by a second high-aperture lens. In the detector path we have inserted a third Babinet–Soleil compensator and allowed for the possibility of having a linear polariser.

In the following sections we shall first specialise the general solution to describe a commercially available transmission polarised light microscope. We then move to obtain a solution for a reflection polarised light microscope using finite-sized detectors. These cases should be considered as worked application examples illustrating the use of the above general solution.

11.5.6 Theory of Imaging in Transmission Polarised Light Scanning Microscopes

The development of a commercial polarised light microscope, the so-called Pol-Scope **[265]**, has opened new avenues in quantitative light microscopy. The microscope was developed chiefly by Dr. Rudolf Oldenbourg of Molecular Biological Laboratory, Woodshole, US, based on the rich traditions of Dr. Shinya Inoué's work at the same location.

We do not attempt to model here the wide-field set-up of the Pol-Scope. For modelling the wide-field system we would have to know the spatial coherence and polarisation properties of the arc lamp which would then permit calculation of the distribution of radiation in the sample plane using the partially coherent, partially polarised focusing theory described in Section 11.3. The computational overhead of such a calculation is considerable: in reality it would require the numerical evaluation of a sixfold integral. The field scattered by the birefringent point object could be calculated relatively easily given that, whatever the state of spatial coherence is in sample space, sampling that field at a single point 'converts' the light scattering calculations to a fully spatially coherent model. However, the portion of light that does not get scattered on the point object would need to be propagated through the optical system using diffraction theory for partially coherent and partially polarised light. The utility of such a complicated and computationally burdensome model is strongly questionable: researchers usually set up mathematical models to explain salient features of physical systems and predict deviations from expectation. In general, the more analytic the model the simpler it is to understand, and thus predict, these features. We shall see below that the interference of the light scattered by the sample and that transmitted without scattering plays an important part in polarised light imaging. In this respect our model described below will predict the 'best case' behaviour of the system because sources of lower degree of partial polarisation will exhibit less pronounced interference and hence diminished contrast.

Instead, we use some of the optical components of the Pol-Scope to construct an analogous model based on a Type I conventional scanning microscope. We shall then extend this model to include Type II confocal microscopes. In order to understand the operation of the system, consider the optical set-up of the Pol-Scope (Fig. 11.23) as follows: linearly polarised light is incident upon a variable retarder with its principal axis set at 45° with respect to the incident polarisation direction. The principal role of the retarder is to act as a quarter-wave plate with the possibility of introducing small variations in this retardance setting. When it is set to act as a quarter-wave plate the transmitted light becomes circularly polarised. When its retardance is changed to less or more than a quarter wave the transmitted light becomes elliptically polarised. A second variable retarder is placed after the first with its principal axis coinciding with the incident polarisation direction. The main purpose of this retarder is to act as a half-wave plate. It then swaps the sense of orientation of the incident circular polarisation, i.e. if the first retarder produces a right-hand circular polarisation then the second retarder produces a left-handed circular polarisation. As a result of the combined effect of the linear polariser and two retarders any polarisation state can be synthesised. Therefore, the linear polariser and two variable retarders function as a *universal compensator*. In practice it is possible to use, for example, electronically adjustable liquid crystal retarders so that the polarisation state can be switched almost instantaneously, without any mechanical movements. The light

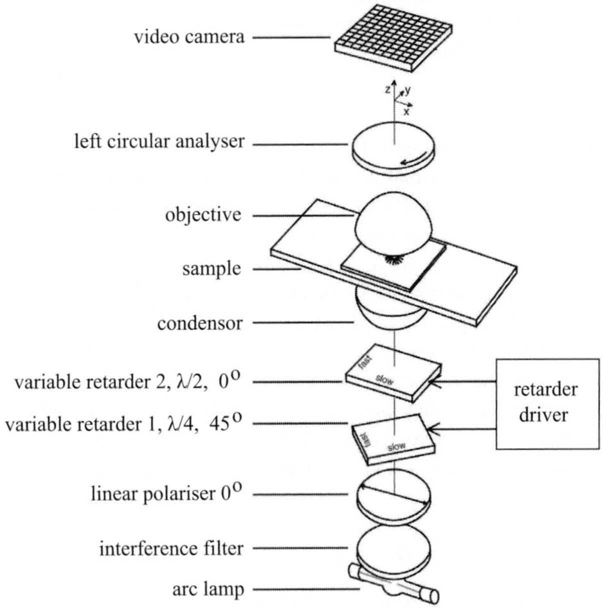

Figure 11.23: The Pol-Scope as a wide-field set-up. The light produced by an arc lamp is passed through an interference filter in order to select a single wavelength. The laboratory coordinate system is established by a linear polariser producing linearly polarised light. The two variable retarders can produce an arbitrary state of polarisation that is incident on the sample via the condenser lens. The objective (collector lens) images the sample onto the video camera via a circular analyser. Note that this is a wide-field system and in this section we analyse a scanning 'Pol-Scope' instead. Image courtesy of Dr. Rudolf Oldenbourg of MBL, Woods Hole, USA. Reproduced from **[341]** with permission of Elsevier.

emerging from the two retarders, represented in our model by the two Babinet–Soleil compensators, is then incident upon a condenser lens followed by the sample.

As usual the light transmitted through the sample is collected and re-collimated by the collector lens. A circular analyser is placed after the collector lens such that it has the opposite handedness of the circular polarisation leaving the universal compensator and is usually constructed from a quarter-wave plate followed by a linear polariser that is rotated by 45° with respect to the principal axis of the quarter-wave plate (in our model these are represented by a Babinet–Soleil compensator and a polariser) in such as way that it is crossed with respect to the incident polarisation emitted by the source (or the linear polariser used after the source). As discussed in the previous section, light scattered off the specimen and that transmitted by a pair of not fully extinguished polariser-analysers will produce interference at the polariser. The structure of the interference pattern carries information concerning the object, as shown below under Numerical Examples. Based on images recorded with the Pol-Scope optical train the retardances and the direction of the birefringence axes of a specimen can be calculated for every image point simultaneously **[265]**. We note that although the Pol-Scope is in practice realised as a wide-field microscope, it is modelled here as a scanning (Type I) microscope which is in accordance with the discussions of Section 10.3.3. As discussed there, since we only consider this microscope imaging a time-stationary (birefringent) point object illuminated by a monochromatic light, approximating the actual optical system this way is, based on the comparison of the theoretical and experimental results, proven to produce results of reasonable accuracy. If a more accurate model became necessary, that would need to be based on the theory of partially polarised, partially coherent focusing as described in Section 11.3. However, the accuracy of that model would depend on how accurately one could measure the cross spectral density matrix of the light source, which is a difficult experimental task. The calculation of extended objects imaged by the Pol-Scope, as represented by the scanning microscope model presented in this section, is possible as long as the time fluctuations of the sample are sufficiently slow so that temporal coherence issues do not affect the outcome. The principles of constructing such a model are discussed in Section 11.8. It is currently not possible to model the imaging of extended objects illuminated by partially polarised, partially coherent light primarily due to computational difficulties.

11.5.6.1 Numerical Examples

The mathematical details of the model describing the Pol-Scope have been developed in the previous sections so we shall only quote the equation numbers and settings that are needed to obtain numerical results. We consider both scanning incoherent (Type I) and scanning coherent (Type II) microscopes. Let us start with scanning incoherent microscopes. The equation that describes the image intensity in this mode is Eq. (11.5.33) where we set for the retardance and angle of the three Babinet–Soleil compensators: $\delta_1 = \pi/2 \pm \Delta\delta, \beta_1 = \pi/4, \delta_2 = \pi, \beta_1 = 0, \delta_a = -\pi/2$ and $\beta_a = \pi/4$. The misalignment of the first compensator $\Delta\delta$ will be used as parameter in the model to simulate imperfect compensator setting. The dipole moment is calculated from Eq. (11.5.8) via Eq. (11.5.4) with the above settings. Note that in Eq. (11.5.33) the Cartesian components of the unit vector of the dipole moment need to be replaced by their un-normalised counterparts in order to be able to transmit the strength of the electric vector illuminating the dipole.

For modelling the Type II, coherent scanning microscope we shall use Eq. (11.5.45) with the same Babinet–Soleil settings as for the incoherent Type I case above and orient the analyser at $\beta = 0$. The total electric field is then given as the coherent sum of $\mathbf{E}^c_{t,a}$ and $\mathbf{E}_7$ given by Eqs. (11.5.45) and (11.5.36), respectively. The integrated intensity is then given by

$$\bar{I}_c = \int_0^R \int_0^{2\pi} (\mathbf{E}^c_{t,a} + \mathbf{E}_7)^T (\{\mathbf{E}^c_{t,a}\}^* + \mathbf{E}^*_7) r_d dr_d d\gamma_d \,,$$

where, as before, R is the radius of the detector. If the detector is a single mode optical fibre, the sum of the electric fields $\mathbf{E}^c_{t,a} + \mathbf{E}_7$ needs to be projected onto the eigenmode of the fibre **[319]**.

It is instructive to study the detected signal for both conventional and confocal microscopes and compare these results directly. All of the simulated images presented here were calculated by scanning the point object located at (r_t, γ, z_p), which has uniaxial birefringence only for which $n_{xx} = n_{yy} = n_o$ is the ordinary refractive index and $n_{zz} = n_e$ is the extraordinary refractive index (see Eq. (11.5.10)).

Figure 11.24 shows the lateral (in the x–y transverse plane) image of a dielectric scatterer ($n_o = n_e = 1.35$) embedded in a homogeneous medium of refractive index $n_m = 1.52$ for both conventional and confocal microscopes for a lens numerical aperture of 0.9 and wavelength $\lambda = 545$ nm. The same parameters were used for numerical computation throughout this section. When the polariser/analyser assembly is perfectly crossed (meaning that the illumination is elliptically polarised and the circular analyser extinguishes all transmitted, i.e. non-scattered, light) the image of the dielectric scatterer assumes a doughnut shape. Clearly, in this situation all detected signal is due to light scattered by the birefringent scatterer. When the first Babinet–Soleil compensator in the illumination path is detuned slightly ($\Delta\delta_1 = \pi/18$) part of the transmitted light reaches the detector and there it interferes with the light scattered by the birefringent scatterer. The image thus exhibits a grey background (due to the transmitted intensity) with dark and bright features. When the dielectric scatterer is defocused, which is done by moving the scatterer with respect to the condenser and collector lenses (whose relative position is maintained), a characteristic spiral pattern appears on both conventional and confocal images. The spiral turns in the opposite direction when a negative defocus is applied.

Figure 11.25 shows the image of a small ($ka \ll \lambda$) KDP crystal ($n_o = 1.5064, n_e = 1.4664$) oriented in such a way that its fast axis coincides with the z-axis of the coordinate system. The crystal is embedded in an isotropic medium of refractive index $n_m = 1.52$. It is interesting to see, when compared to the corresponding figures of Fig. 11.24 that, whereas the confocal cross polarised image of the KDP crystal is virtually indistinguishable from that of the dielectric scatterer, there is a small but discernible difference between the conventional images. On the other hand, when elliptically polarised light is used ($\Delta\delta = \pi/18$) the situation is the opposite. For the defocused images, again, only the confocal images are seen to be slightly different when compared to those of the dielectric scatterer.

It is interesting to examine the image of a birefringent object when it is rotated from its original orientation about the x-axis by $\pi/2$ so that the fast axis now coincides with the y-axis. The result of these computations is shown in Fig. 11.26. It is seen that as a result of the slow axis being parallel to the x-direction and the fast axis parallel to the y-direction, the image for $\Delta\delta = 0$ no longer assumes the doughnut shape but is slightly elliptical. For the elliptical illumination the interference pattern is no longer present. Images become slightly asymmetric, which is further emphasised when the crystal is defocused.

Next, the image of a small calcite crystal ($n_o = 1.66277, n_e = 1.48841$) oriented in such a way that its fast axis coincides with the z-axis of the coordinate system is computed. The crystal is embedded in an isotropic medium of refractive index

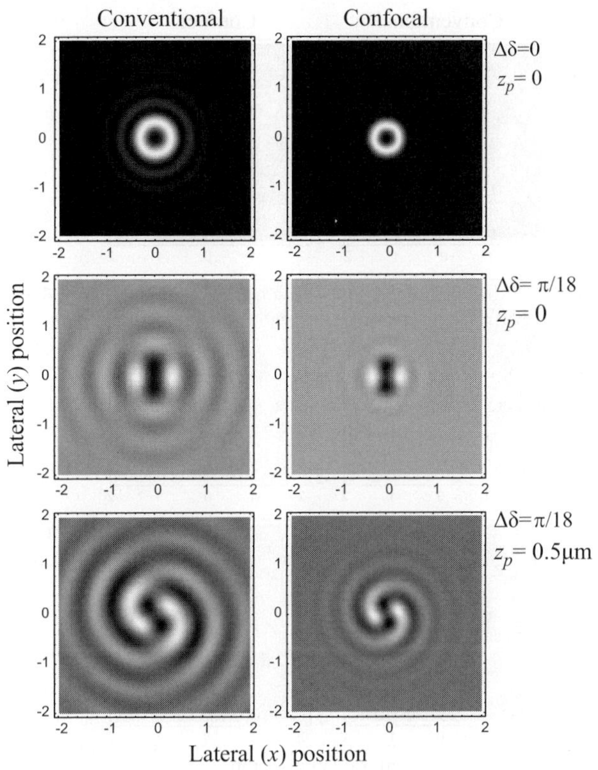

Figure 11.24: Simulated images obtained from a dielectric scatterer ($n_o = n_e = 1.35$ and $n_m = 1.52$). Top row corresponds to a perfectly crossed polariser/analyser set-up ($\Delta\delta = 0$), the second row corresponds to slightly elliptical illumination ($\Delta\delta = \pi/18$) and the third row corresponds to the same retarder setting as the second row but with a small defocus of $z_p = 0.5\mu$m employed. In this and all subsequent images all transverse dimensions are given in μm. The numerical aperture of the condenser and collector lenses was set to 0.9 and a wavelength of $\lambda = 545$ nm was used. The medium embedding the scatterer had a refractive index of $n_m = 1.52$. Images are normalised row-wise to produce maximum contrast and so grey scale levels are not directly comparable between individual rows. Reproduced from **[341]** with permission of Elsevier.

$n_m = 1.52$. The result of this computation is shown in Fig. 11.27. When this figure is compared to Fig. 11.24 it is seen that the images corresponding to the crossed polariser case, either conventional or confocal, are virtually indistinguishable. When elliptical illumination is applied the differences between the corresponding images of either Fig. 11.24 or Fig. 11.25 appear clearly. These are further emphasised when a defocus is also applied. When the small calcite crystal is embedded in a medium that matches the refractive index corresponding to either the extraordinary or ordinary direction, simulated images of the first case ($n_e = n_m$) are shown in Fig. 11.28. It is seen that conventional images of Fig. 11.27 and Fig. 11.28 do not differ. Indeed, according to Eqs. (11.5.8) and (11.5.10), when $n_e = n_m$ then $p_z = 0$ and from Eq. (11.5.33) it is clear that p_z does not contribute to the interference term. Conversely, it is seen that the confocal images of Fig. 11.27 and Fig. 11.28 are significantly different. This is a result of p_z playing a significant role in the detected field (see Eq. (11.5.36)). An even more interesting situation arises when $n_o = n_m$, as shown in Fig. 11.29. The interference pattern due to the non-extinguished transmitted light in the case of an elliptical incident polarisation is completely absent in the conventional images. This is not very surprising when considering that for $n_o = n_m$ Eqs. (11.5.8) and (11.5.10) give $p_x = p_y = 0$ and hence the cross term in Eq. (11.5.33) is zero. On the other hand, for a confocal microscope the electric field due to the scatterer would not vanish and its interference with the transmitted electric field results in the pattern shown in Fig. 11.29.

From these results it is clear that polarised light microscopy brings a new dimension into imaging by revealing details of samples not possible by other imaging methods. Defocus seems to be an important part of observation which, by way of interference, permits a better identification of samples.

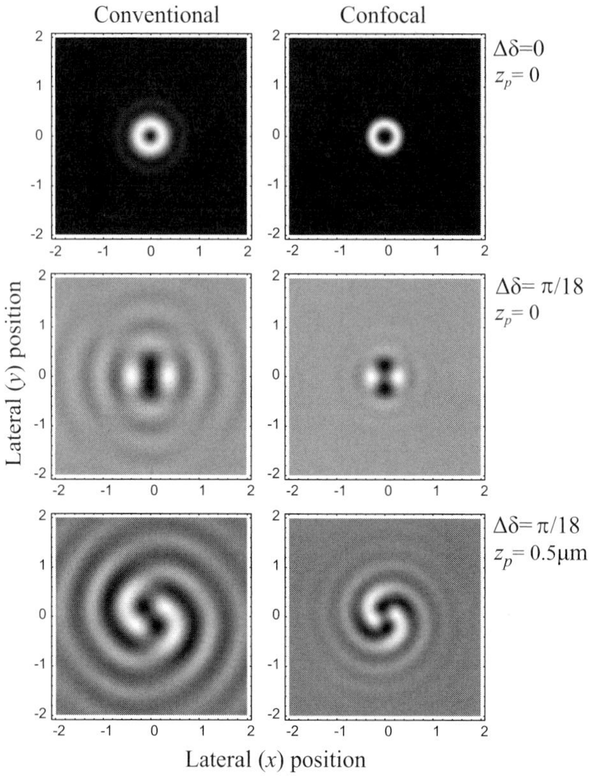

Figure 11.25: Simulated images obtained from a small KDP crystal (n_o = 1.5064, n_e = 1.4664 and n_m = 1.52) with its optic axis oriented in the z direction. All other parameters and simulation settings are the same as in Fig. 11.24. Reproduced from **[341]** with permission of Elsevier.

11.6 Theory of Multiphoton Fluorescence Microscopes

In the previous section we have investigated how microscopes image point scatterers. The fundamental assumption was that the instantaneously induced dipole moment was proportional to the electric field vector the scatterer experienced, which means that the electromagnetic fields 'emitted' by the scatterer and that inducing it are correlated, i.e. the scattered light is fully coherent with respect to the illumination. Consider now the situation when the energy of the incident photon is absorbed by an electron transition in a molecule and the energy gained by this molecule is lost partly via a non-radiative process and partly via the molecule re-emitting a photon of lower energy at some time after the excitation (fluorescence). Clearly, such a process must be described by quantum mechanical means and so the re-emission of photons of lower energy happens at random. In addition, when the excited molecule is not bound, it may rotate in any direction between excitation and emission. Thus, such an interaction with a sample requires a different treatment from that which we have used so far for point scatterers.

In this section we describe the effect of polarisation of the fluorescent emission on the imaging properties of multiphoton microscopes. Numerical examples are given for one-, two- and three-photon excitation and of a particular fluorescent molecule (p-quaterphenyl) for high- and low-aperture objective lenses.

11.6.1 Multiphoton Fluorescence Imaging

As discussed above, a fluorescent molecule can be excited by absorbing the energy of a photon and subsequently radiating another photon as it returns to a lower energy state. The wavelength of the emission for Stoke fluorescence is longer than that of the excitation (the emitted photon possesses less energy than the absorbed photon owing to energy lost between

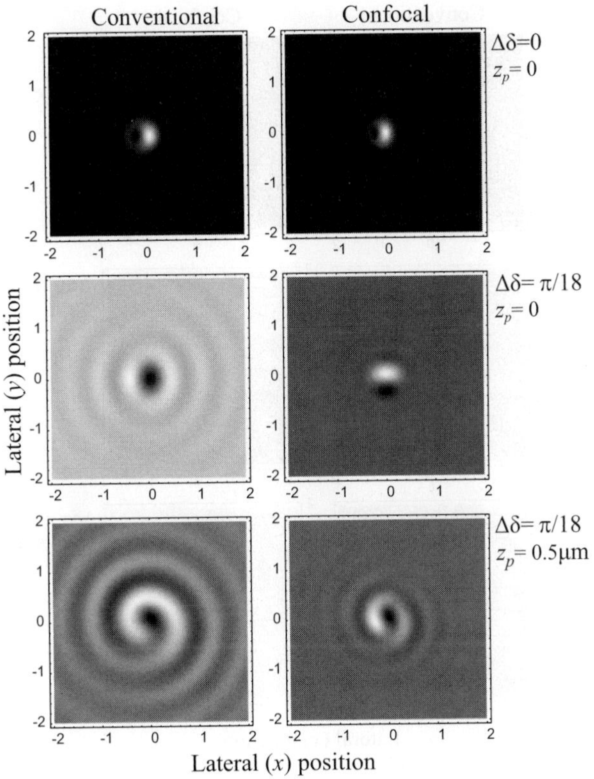

Figure 11.26: Simulated images obtained from a small KDP crystal (n_o = 1.5064, n_e = 1.4664 and n_m = 1.52) with its optic axis oriented in the *y*-direction. All other parameters and simulation settings are the same as in Fig. 11.24. Reproduced from **[341]** with permission of Elsevier.

excitation and emission). Alternatively, the energy of two photons may be absorbed simultaneously by the fluorescent molecule and, again, the energy is irradiated as a single photon. In this case the energy of the emitted photon is limited by the combined energy of the two absorbed photons. Hence the emission wavelength has a lower limit of approximately half the excitation wavelength. In theory, any number of photons, N, may be simultaneously absorbed and a single photon emitted. By similar consideration of the energy involved, the emission wavelength must be greater than $1/N$ of the illumination wavelength. When a single absorbed photon causes the emission of a single photon the process is called one-photon (1–p) fluorescence. When N simultaneously absorbed photons cause the emission of a single photon the process is called N-photon (N–p), or multiphoton, fluorescence. There are two possible ways to model the mechanism of fluorescent emission. The first is to consider the emission as being randomly polarised. Extensive attention has been given in the literature to study of the relationship between absorbed and emitted polarisation states since the theories of Perrin **[275]** and Soleillet **[320]** were presented. In general, fluorescent emission is partially polarised and the number of parameters that determine the state of the polarisation is extremely large **[320]**. The relationship between illumination and emission polarisation direction also depends on a number of parameters such as the fluorescent molecule's rotational mobility **[16]** and its individual characteristics **[118]**. Accounting for the polarisation states in 2-p and 3-p fluorescence is a more involved process than 1-p **[64]**, **[106]**,**[118]** but the conclusion is still that the fluorescent emission is partially polarised. We consider three different models of fluorescent imaging:

- Case I – Incident illumination induces a dipole in the direction of polarisation of illumination at the location of the dipole. The dipole is free to rotate after excitation and, at a random time and orientation, it re-radiates resulting in unpolarised fluorescent emission. The total measured fluorescent intensity is due to many individual dipoles whose re-emission time and orientation are not correlated.

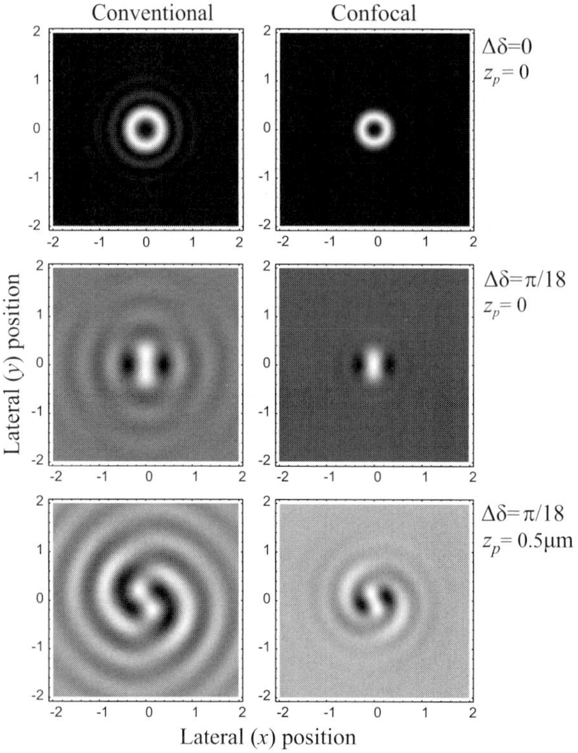

Figure 11.27: Simulated images obtained from a small calcite crystal (n_o = 1.66277, n_e = 1.48841 and n_m = 1.52) with its optic axis oriented in the z direction. All other parameters and simulation settings are the same as in Fig. 11.24. Reproduced from **[341]** with permission of Elsevier.

- Case II – Incident illumination induces a dipole in the direction of preset dipole directions that are fixed in a given orientation, i.e. the dipole is excited only by the projection of the incident electric field on the dipole moment. The dipole is free to rotate after excitation and, at a random time and orientation, it re-radiates which means that the fluorescent emission is unpolarised. The total measured fluorescent intensity is due to many individual dipoles whose re-emission time and orientation are not correlated.
- Case III – Incident illumination induces dipole in the direction of polarisation of illumination at the location of the dipole. The dipole cannot rotate from the preset direction and it re-radiates at a given time instant in the orientation of the excitation. This case is essentially identical to that we have studied in Section 11.5.5.

It should be noted that, while Cases I and II are physically plausible, for example for fluorescent dyes in solution, Case III should be treated with care because it is assumed that there is zero angular displacement between excitation and emission, which is never actually the case but, rather, a limiting model which is a good approximation to a highly polarised emission. A fourth polarised emission case that is not considered here is that of fixed dipoles that are aligned in a given direction. In such a case the direction of the fixed dipole alignment with respect to the incident polarisation direction is the important factor in image formation.

Single-photon fluorescence excitation of a single fluorescent molecule occurs via the relationship:

$$p_i = \sum_{j=x}^{z} \alpha_{ij} E_j; \qquad \mathbf{p} = \overset{=(1)}{\alpha} \mathbf{E}, \tag{11.6.1a}$$

for *two-photon* excitation:

$$p_i = \sum_{j=x}^{z} \sum_{k=x}^{z} \alpha_{ijk} E_j E_k; \qquad \mathbf{p} = \overset{=(2)}{\alpha} \mathbf{E}{:}\mathbf{E}, \tag{11.6.1b}$$

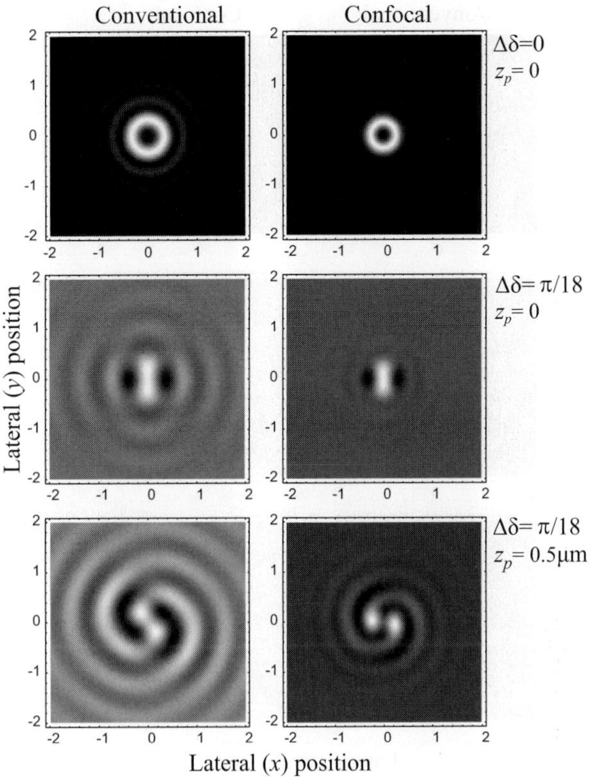

Figure 11.28: Simulated images obtained from a small calcite crystal embedded in a medium such that it is refractive index matched to n_o (n_o = 1.66277, n_e = 1.48841 and $n_m = n_o$) with its optic axis oriented in the z direction. All other parameters and simulation settings are the same as in Fig. 11.24. Reproduced from **[341]** with permission of Elsevier.

and for *three-photon* excitation:

$$p_i = \sum_{j=x}^{z} \sum_{k=x}^{z} \sum_{l=x}^{z} \alpha_{ijkl} E_j E_k E_l; \qquad \mathbf{p} = \overset{=(3)}{\alpha} \mathbf{E}{:}\mathbf{E}{:}\mathbf{E}\ , \tag{11.6.1c}$$

where $\overset{=(1,2,3)}{\alpha}$ are the first-, second- and third-order tensors of the fluorescence process, $i = x, y, z$ and the : denotes the operation expressed by the summation. After the fluorescent molecule is excited it relaxes its energy. The molecule is not usually expected to maintain its direction with respect to the orientation it gained due to the excitation. The ability of the molecule to rotate away from the excited direction depends on many factors which are not discussed here. It is therefore clear that Case I and III are two extreme cases of the same excitation process, i.e. when $\overset{=}{\alpha}$ tensor is a constant times the unit tensor, and Case II is a completely different excitation process.

11.6.2 The Fluorescent Dipole Moment

Practical multiphoton microscopes use either linearly or circularly polarised illumination. Thus we shall use the theory analysing the optical system of Fig. 11.19 and which led to Eq. (11.5.4). We here consider setting of $\delta_2 = 0$ which then results in the electric field at the focus of the condenser lens:

$$\mathbf{E}(r_t, \gamma, z_p) = -iU_i k f_b \begin{pmatrix} \cos\frac{\delta_1}{2} I_0 + I_2\left(\cos\frac{\delta_1}{2}\cos 2\gamma + i\sin\frac{\delta_1}{2}\sin 2\gamma\right) \\ i\sin\frac{\delta_1}{2} I_0 + I_2\left(\cos\frac{\delta_1}{2}\sin 2\gamma - i\sin\frac{\delta_1}{2}\cos 2\gamma\right) \\ 2iI_1\left(\cos\frac{\delta_1}{2}\cos\gamma + i\sin\frac{\delta_1}{2}\sin\gamma\right) \end{pmatrix}, \tag{11.6.2}$$

with $I_{0\ldots 2}(r_t, z_p)$ given by Eq. (11.2.4d).

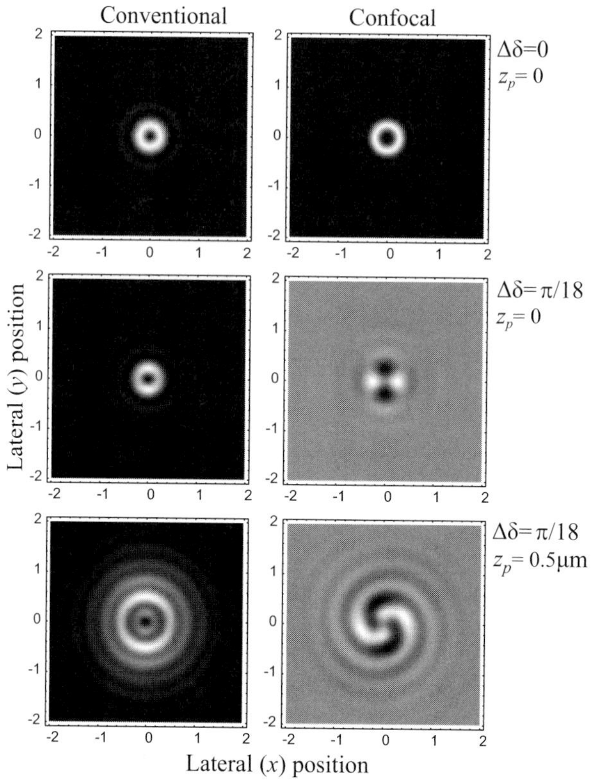

Figure 11.29: Simulated images obtained from a small calcite crystal embedded in a medium such that it is refractive index matched to n_e ($n_o = 1.66277, n_e = 1.48841$ and $n_m = n_e$) with its optic axis oriented in the z direction. All other parameters and simulation settings are the same as in Fig. 11.24. Reproduced from **[341]** with permission of Elsevier.

For Cases I and II the electric dipole moment of the emission is randomly oriented. The unit vector, and hence the dipole moment, oriented at angles θ_{dp} and ϕ_{dp} to the $+z$- and $+x$-axes, respectively, are given by:

$$\hat{\mathbf{p}}^I = \hat{\mathbf{p}}^{II} = \begin{pmatrix} \cos\phi_{dp}\sin\theta_{dp} \\ \sin\phi_{dp}\sin\theta_{dp} \\ \cos\theta_{dp} \end{pmatrix}, \tag{11.6.3}$$

where the angles ϕ_{dp} and θ_{dp} are set at random values uniformly distributed within the interval $[0, 2\pi]$ and $[-\pi/2, \pi/2]$, respectively, for the different individual fluorescent molecules. Integrating the intensity in the detector plane over all dipole orientations (θ_{dp}, ϕ_{dp}) then gives the response of an 'average' dipole moment, corresponding to observation of a random cluster of dye molecules within the point spread function volume.

In Case III the electric dipole moment of the emission, $\hat{\mathbf{p}}^{III}$, is taken to be the unit vector in the direction of the incident electric field at the fluorescent molecule at position $\mathbf{r}_p = (r_t, \gamma, z_p)$:

$$\hat{\mathbf{E}}(\mathbf{r}_p) = \frac{\mathbf{E}(\mathbf{r}_p)}{|\mathbf{E}(\mathbf{r}_p)|}. \tag{11.6.4}$$

The Cartesian components of the unit vector associated with the electric dipole moment are thus given by:

$$\hat{\mathbf{p}}^{III} = \begin{pmatrix} \hat{p}_x \\ \hat{p}_y \\ \hat{p}_z \end{pmatrix} = \hat{\mathbf{E}}(\mathbf{r}_p) = \frac{1}{|\mathbf{E}(\mathbf{r}_p)|}\begin{pmatrix} E_x \\ E_y \\ E_z \end{pmatrix}. \tag{11.6.5}$$

The *normalised* probability of excitation $P_N^{I,II,III}$ is defined such that it is proportional to the modulus squared of the component of the electric field in the direction of the dipole, raised to the power of the multiphoton process order N. The constant of proportionality is chosen to be the maximum value of the electric field in the direction of the dipole raised to the power of the multiphoton process order. This of course is not the true value of the probability as, for an unaberrated optical system $P_N^{I,III}(\mathbf{r}_p = 0) = 1$, but the choice ensures dimensional consistency of our equations and represents real fluorescent systems well as long as it is understood that the un-normalised probability of excitation is a constant times P_N. For Cases I and III, corresponding to an induced dipole, the direction of the dipole is parallel to the incident field, and hence:

$$P_N^I(\mathbf{r}_p) = P_N^{III}(\mathbf{r}_p) = \frac{|\mathbf{E}(\mathbf{r}_p)|^{2N}}{|\mathbf{E}|^{2N}_{\max}}, \tag{11.6.6}$$

where N is the order of the process, i.e. 1-p (N = 1), 2-p (N = 2), etc. From Eq. (11.6.2) it can be seen that

$$|\mathbf{E}|^2 = (U_i k f_b)^2 \left\{|I_0|^2 + 2|I_1|^2 + |I_2|^2 + 2\cos\delta_1 \cos 2\gamma \left(\Re\{I_0 I_2^*\} + |I_1|^2\right)\right\}. \tag{11.6.7}$$

For Case II, the probability of excitation depends only on the component of the incident electric field parallel to the permanent dipole moment and hence:

$$P_N^{II}(\mathbf{r}_p) = \frac{|\hat{\mathbf{p}}^{II}(\mathbf{r}_p)\cdot\mathbf{E}(\mathbf{r}_p)|^{2N}}{|\hat{\mathbf{p}}^{II}\cdot\mathbf{E}|^{2N}_{\max}}. \tag{11.6.8}$$

As before we now consider the detected signal by using a 4-f collector–detector lens assembly to image the dipole onto the detector plane.

11.6.3 The Detected Signal

The light emitted from the fluorescent molecule is collected by a high-aperture collector lens and imaged onto a finite-sized pinhole. The expression for the electric field in the detector plane due to a radiating electric dipole in the front focal plane of the collector lens is given, from Eq. (11.5.36) with the same notation, after setting $\delta_a = 0$, by:

$$\mathbf{E}_7(r_d, \gamma_d; z_p) = A\begin{pmatrix} p_x I_0^a - 2ip_z I_1^b \cos\gamma_d + I_2^a(p_x\cos 2\gamma_d + p_y \sin 2\gamma_d) \\ p_y I_0^a + 2ip_z I_1^b \sin\gamma_d + I_2^a(p_x \sin 2\gamma_d - p_y\cos 2\gamma_d) \\ -2p_z I_0^c - 2iI_1^e(p_x\cos\gamma_d + p_y\sin\gamma_d)\end{pmatrix}. \tag{11.6.9}$$

To obtain the intensity at the detector plane, for Cases I and II, an integration must be performed over all possible dipole orientations, because the dipoles can freely rotate during the fluorescence lifetime, i.e.:

$$I_d^{I,II}(\mathbf{r}_p;\mathbf{r}_d) = \int_0^{\pi}\int_0^{2\pi} P_N^{I,II}(\mathbf{r}_p)\left|\mathbf{E}_7(\mathbf{r}_p; r_d; \theta_{dp}, \phi_{dp})\right|^2 \sin\theta_{dp} d\theta_{dp} d\phi_{dp}, \tag{11.6.10}$$

where we have used $\mathbf{r}_d = (r_d, \gamma_d, z_d)$. For Case III the intensity distribution in the detector plane is given by:

$$I_d^{III}(\mathbf{r}_p;\mathbf{r}_d) = P_N^{III}(\mathbf{r}_p)\left|\mathbf{E}_7(\mathbf{r}_p; r_d; \theta_{dp}, \phi_{dp})\right|^2, \tag{11.6.11}$$

Substituting the probability functions from Eqs. (11.6.6) and (11.6.8) gives:

$$\begin{aligned} I_d^I(\mathbf{r}_p;\mathbf{r}_d) &= \frac{|\mathbf{E}(\mathbf{r}_p)|^{2N}}{|\mathbf{E}|^{2N}_{\max}}\int_0^{\pi}\int_0^{2\pi}\left|\mathbf{E}_7(\mathbf{r}_p; r_d; \theta_{dp}, \phi_{dp})\right|^2 \sin\theta_{dp}\mathrm{d}\theta_{dp}\mathrm{d}\phi_{dp}, \\ I_d^{II}(\mathbf{r}_p;\mathbf{r}_d) &= \int_0^{\pi}\int_0^{2\pi}\frac{|\hat{\mathbf{p}}^{II}(\mathbf{r}_p)\cdot\mathbf{E}(\mathbf{r}_p)|^{2N}}{|\hat{\mathbf{p}}^{II}\cdot\mathbf{E}|^{2N}_{\max}}\left|\mathbf{E}_7(\mathbf{r}_p; r_d; \theta_{dp}, \phi_{dp})\right|^2 \sin\theta_{dp} d\theta_{dp} d\phi_{dp}, \\ I_d^{III}(\mathbf{r}_p;\mathbf{r}_d) &= \frac{|\mathbf{E}(\mathbf{r}_p)|^{2N}}{|\mathbf{E}|^{2N}_{\max}}\left|\mathbf{E}_7(\mathbf{r}_p; r_d; \theta_{dp}, \phi_{dp})\right|^2. \end{aligned} \tag{11.6.12}$$

It is useful to give the analytic form of $|\mathbf{E}_7|^2$ here:

$$\begin{aligned}|\mathbf{E}_7|^2 &= A^2\left(|I_0^a|^2 + |I_1^e|^2 + |I_2^a|^2\right)\left(|p_x|^2 + |p_y|^2\right) + 4A^2\left(|I_0^c|^2 + |I_1^b|^2\right)|p_z|^2 \\ &+ 2A^2\left(|I_1^e|^2 + \Re\{I_0^a I_2^{a*}\}\right)\left[\left(|p_x|^2 - |p_y|^2\right)\cos 2\gamma_d + \left(p_y p_x^* + p_x p_y^*\right)\sin 2\gamma_d\right] \\ &+ 2A^2\Im\left\{\left[2I_1^e I_0^{c*} - (I_0^a + I_2^a)\, I_1^{b*}\right]p_z^*\left(p_x\cos\gamma_d + p_y\sin\gamma_d\right)\right\}. \end{aligned} \tag{11.6.13}$$

For Case I, a substitution for the electric dipole moment into Eq. (11.6.13) and performing the integrations in θ_{dp} and ϕ_{dp}, leads to:

$$I_d^I = \frac{8\pi A^2}{3}|\mathbf{E}|^{2N}\left[|I_0^a|^2 + 2|I_1^b|^2 + |I_2^a|^2 + 2\left(|I_0^c|^2 + |I_1^e|^2\right)\right]. \tag{11.6.14}$$

For Case II it is not possible to obtain a general solution for an arbitrary order of multiphoton process. However, a long analytical calculation leads to the intensity for the first three-order processes:

$$\begin{aligned}I_d^{II} &= \frac{16\pi A^2}{15} f_1^{1p}\left(|I_0^a|^2 + |I_0^c|^2 + |I_1^b|^2 + 2|I_1^e|^2 + |I_2^a|^2\right) \\ &+ \frac{16\pi A^2}{15} f_2^{1p}\left(|I_1^e|^2 + \Re\{I_0^a I_2^{a*}\}\right)\cos 2\gamma_d,\end{aligned} \tag{11.6.15a}$$

with

$$f_1^{1p} = |\mathbf{E}|^2 \tag{11.6.15b}$$

$$f_2^{1p} = |E_x|^2 - |E_y|^2 - |E_z|^2, \tag{11.6.15c}$$

for a 1-photon process. For a 2-photon process the expression becomes:

$$\begin{aligned}I_d^{II} &= \frac{8\pi A^2}{105} f_1^{2p}\left(3|I_0^a|^2 + 2|I_0^c|^2 + 2|I_1^b|^2 + 6|I_1^e|^2 + 3|I_2^a|^2\right) \\ &+ \frac{32\pi A^2}{35} f_2^{2p}\left(|I_1^e|^2 + \Re\{I_0^a I_2^{a*}\}\right)\cos 2\gamma_d,\end{aligned} \tag{11.6.16a}$$

with

$$f_1^{2p} = 3|\mathbf{E}|^4 + 2|E_x|^2\left(|E_y|^2 + |E_z|^2\right) + 6|E_y|^2|E_z|^2 \tag{11.6.16b}$$

$$f_2^{2p} = |E_x|^4 - |E_y|^4 - |E_z|^4 - 2|E_y|^2|E_z|^2, \tag{11.6.16c}$$

and for a 3-photon process:

$$\begin{aligned}I_d^{II} &= \frac{16\pi A^2}{315} f_1^{3p}\left(2|I_0^a|^2 + |I_0^c|^2 + |I_1^b|^2 + 4|I_1^e|^2 + 2|I_2^a|^2\right) \\ &+ \frac{16\pi A^2}{105} f_2^{3p}\left(|I_1^e|^2 + \Re\{I_0^a I_2^{a*}\}\right)\cos 2\gamma_d,\end{aligned} \tag{11.6.17a}$$

with

$$\begin{aligned}f_1^{3p} &= 5|\mathbf{E}|^6 + 3|E_x|^4\left(|E_y|^2 + |E_z|^2\right) + 3|E_x|^2\left(|E_y|^4 + 2|E_y|^2|E_z|^2 + |E_z|^4\right) \\ &+ 15|E_y|^2|E_z|^2\left(|E_y|^2 + |E_z|^2\right),\end{aligned} \tag{11.6.17b}$$

$$\begin{aligned}f_2^{3p} &= 5\left(|E_x|^6 - |E_y|^6 - |E_z|^6\right) + |E_x|^4\left(|E_y|^2 + |E_z|^2\right) \\ &- |E_x|^2\left(|E_y|^4 + 2|E_y|^2|E_z|^2 + |E_z|^4\right) - 15|E_y|^2|E_z|^2\left(|E_y|^2 + |E_z|^2\right).\end{aligned} \tag{11.6.17c}$$

Substituting Eq. (11.6.13) into the last line of Eq. (11.6.12) the intensity for Case III can be computed.

The integrated intensity proportional to the signal detected by a fluorescence microscope employing a pinhole of radius R is given by:

$$\bar{I}^{I,II,III}(\mathbf{r}_p) = \int_0^R \int_0^{2\pi} I_d^{I,II,III}(\mathbf{r}_p, \mathbf{r}_d) r_d dr_d d\gamma_d , \tag{11.6.18}$$

where, as before, (r_d, γ_d) is a coordinate system centred on the pinhole. [11.17]

11.6.4 Numerical Examples

A 3-p fluorescence microscope exhibits better lateral and axial resolution than a 2-p microscope for the same excitation wavelength and for different fluorescence wavelengths **[211]**. Conversely, when the system has different excitation wavelengths and the same fluorescence wavelength the 3-p microscope exhibits worse lateral and axial resolution than a 2-p microscope. This is due to the increase in the spatial distributions of the illuminating field (Eq. (11.6.2)) due to the longer wavelength inherent in a higher-order process. For this reason one has to decide whether Case I and II is computed numerically, although the results can be easily scaled for different wavelength conditions. Multiphoton excitation of p-quaterphenyl (p-QT) dyes has been demonstrated **[119]** in which the fluorescence was measured at different excitation wavelengths for one-, two- and three-photon excitation at 283, 586 and 850 nm, respectively, with the peak emission around 360 nm. These experimental numbers seem to be a good starting point for our numerical work in this section. We shall use absolute dimensions because as we argued before, normalised optical coordinates do not suit high-aperture microscope theory. Pinhole radii will be expressed in terms of Airy units (A.u.) where 1 A.u. is the radial distance of the first zero of the Airy distribution in the *detector plane*[11.18] of the 1.4 NA, 100× microscope objective lens used in our numerical simulations. Thus for our assumed nominal transverse magnification of $m_T = 100$, 1 A.u. corresponds to a pinhole radius of 15.6 μm.

11.6.5 Lateral and Axial Resolution

We begin by considering the two limiting cases of linearly and circularly polarised illumination. It is useful to clarify the physical meaning of the three polarisation-dependent fluorescent emissions (Cases I, II and III) described in Section 11.6.1. Cases I and II, the two limiting models of unpolarised emission, describe much the same fluorescence emission process whatever the polarisation state of the illumination. In Case I the circularly polarised illumination is absorbed and a randomly oriented dipole emits. In Case II the dipole moment of emission is fixed in a random direction and only the component of the illumination polarised parallel to the dipole moment is absorbed. In Cases I and II averaging over all dipole directions leaves the fluorescent emission completely unpolarised. In Case III at any given moment all of the incident illumination is absorbed and the dipole moment of the emission is aligned with the electric vector of the illuminating field. This statement is equally true for circularly as well as linearly polarised illumination. When we say all the illumination is absorbed of course we really mean a constant proportion is absorbed for all three cases, dependent on the excitation cross-section of the dye molecules as discussed in conjunction with Eq. (11.6.19).

11.17 A word of caution needs to be given before one makes a direct comparison between the results corresponding to the different order processes. The time-averaged fluorescent photon flux $\langle F^{(N)}(t)\rangle$ for two- and three-photon processes is proportional to **[67]**

$$\langle F^{(N)}(t)\rangle \propto \frac{1}{N} \frac{\phi\eta\sigma_N C}{(v_R\tau)^{N-1}} \frac{a_N (NA)^{2N-4} \langle P(t)\rangle^N}{\pi^{3-N}\lambda^{2N-3}} , \tag{11.6.19}$$

where $N = 2, 3$ is the order of the fluorescence process, v_R is the repetition rate of the laser usually used for illumination, τ is the pulse duration, λ is the excitation wavelength in vacuum, $P(t)$ is the incident power, ϕ is the collection efficiency, C is the dye concentration, σ_n is the absorption cross-section, η is the fluorescence quantum efficiency, $a_2 = 64$, $a_3 = 28.1$. Instantaneous fluorescent flux is a measure of the number of emitted fluorescent photons per unit area. The instantaneous irradiance (in W/m^2) can be calculated by multiplying the instantaneous fluorescent flux by hc/λ where h is Planck's constant, $h = 6.63 \times 10^{-34}$ Js. This equation shows that the intensity of the florescence signal strongly depends on the experimental conditions, such as the dye, laser and optics used, and so our results should not be used to compare peak intensities of these imaging modes.

11.18 Note that this definition is slightly different from how we used the same unit in Section 11.4.1, where we defined the unit in object (i.e. high NA) space, whereas here we define it in the low-aperture (i.e. detector) space. The two definitions differ from each other by the transverse magnification of the collector–detector lens system only and it is done this way because in the present case the relative size of the pinhole with respect to the image of the fluorescent molecule on the detector plane is shown later to determine the imaging properties of this microscope. In the previous case of the optical system of a polarised light microscope the size of the light distribution produced by the condenser lens has the defining effect on the imaging properties.

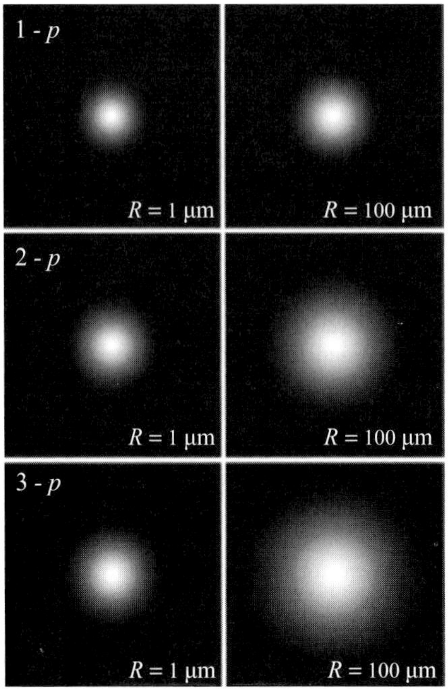

Figure 11.30: Transverse (x–y) image of a fluorescent molecule illuminated with circularly polarised light with unpolarised emission (Case I) for 1-p, 2-p and 3-p and for pinhole radii R of 1 and 100 μm. The image size is 0.8 × 0.8 μm. Individual images are normalised to the peak intensity so as to optimise contrast, and so they cannot be directly compared with regards to the image intensity. The numerical aperture of the condenser/collector lenses was set to 1.4 (oil immersion) with a transverse magnification from object to detector space of 100. The refractive index of the immersion oil was assumed to match perfectly that of the medium the fluorescent molecule was embedded in. The one-, two- and three-photon excitation occurred at 283, 586 and 850 nm, respectively, with the peak emission at 360 nm. Reproduced from **[136]** with permission of John Wiley & Sons, Inc.

In Fig. 11.30 the image of a fluorescent molecule scanned in the focal plane for the specific example of unpolarised emission (Case I) in a microscope employing a 100×, 1.4 NA oil immersion (n_{oil} =1.518) objective lens for 1-p, 2-p and 3-p processes is presented. The right-hand images were computed for a 100 μm (6.4 A.u.) radius pinhole, which is later shown to be close to equivalent to conventional imaging, whilst the left-hand images were calculated for a microscope employing a confocal pinhole of 1 μm (0.064 A.u.) radius. The 100 μm plots give an idea of the relative image size for the first three multiphoton processes in a conventional scanning fluorescence microscope imaging a single molecule. It is seen that the image of the single molecule increases by ≈ 60% (confocal, 1-p to 3-p) and ≈ 215% (R = 100 μm, 1-p to 3-p) in size, and hence the lateral resolution worsens, as the order of the multiphoton process increases. Such a worsening in lateral resolution, caused by the longer excitation wavelength required with higher-order processes, as illustrated in Fig. 11.30, in some cases might be a high price to pay for the advantages that are inherent with the higher-order (2-p and 3-p) processes. However, when the 1 μm pinhole is employed, and hence the imaging is confocal, the process order has a much smaller effect on the lateral resolution, because the predominant factor in resolution is the emission, which is the same wavelength for any order process. Note that as the magnification is specified for a particular focal length of the detector lens, the radius of the aperture stop, and $\sin\alpha_2$, are proportional to NA/m_T. The image size still increases with the number of photons but by a significantly smaller amount (≈ 60%). This is an important result which suggests that, although a confocal pinhole is not required to achieve optical sectioning in 2-p and 3-p fluorescence imaging **[82]**, considerably higher resolution imaging will result if a confocal pinhole is employed. Each of the images in Fig. 11.30 is circularly symmetric. This is due to the fact that the circularly polarised illumination results in a circularly symmetric time-averaged electric field in the focal region when focused by the high-aperture lens and, in Case I, the emission is unpolarised. Cases II and III also result in circularly symmetric images: Case III because the emission, as well as the illumination, is circularly polarised, and Case II for exactly the same reasons as Case I.

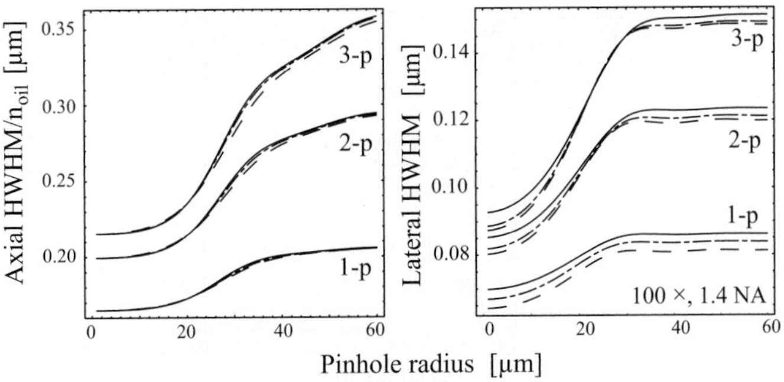

Figure 11.31: Values of the axial and lateral half-width at half-maximum (HWHM) of the image of a fluorescent molecule are shown for 1-p, 2-p and 3-p as a function of pinhole radius for Cases I (continuous line), II (dash-dot line) and III (dashed line) with circularly polarised incident illumination. All simulation parameters were the same as described in Fig. 11.30. Reproduced from **[136]** with permission of John Wiley & Sons, Inc.

To show how the limiting cases of the fluorescent emission affect the resolution properties of the microscope Fig. 11.31 is presented. In this figure the axial and lateral half width at half-maximum (HWHM) curves, as measures of the Rayleigh resolution, see Section 10.1, are plotted as a function of pinhole radius for 1-p, 2-p and 3-p and for Cases I, II and III. Figure 11.31 shows the transition between the two limiting cases of confocal and conventional imaging, which were illustrated in Fig. 11.30, as the pinhole size is increased. The improvement in lateral resolution as the pinhole size is reduced is clearly seen to increase with the order of the process for Cases I, II and III. In fact, the dependence of the lateral resolution on the pinhole radius is similar for Cases I, II and III. So when circularly polarised illumination is employed the state of polarisation of the fluorescent emission does not greatly affect the imaging properties of the microscope. The most significant resolution difference (of less than 7%) is seen for the 1-p process and is between Case III (polarised emission) and Case I (unpolarised with all illumination absorbed). Case III is seen always to exhibit the best lateral resolution and Case I the worst. Hence, the resolution of the microscope can be assumed to fall between these two limits and is determined by the state of polarisation of the emission, which depends on many parameters as discussed in the introduction to this chapter. In general, the higher the degree of polarisation of the fluorescent emission, the better the resolution. Figure 11.31 also shows that, for lateral resolution, pinhole radii greater than 30 μm (1.92 A.u.) correspond to conventional imaging (i.e. to infinite detector radius) whereas pinhole radii smaller than 10 μm (0.64 A.u.) give close to confocal imaging (i.e. to vanishingly small detector radius) for each of the multiphoton processes. It should be noted that these values are applicable only to the current numerical parameters. Although results have also been given for information in normalised units for the current numerical parameters, it is not possible to make general statements that are independent of lens aperture when considering fluorescence processes in high-aperture systems.

A measure of the axial resolution of the multiphoton fluorescence microscope imaging a single fluorescent molecule is also given in Fig. 11.31 in terms of the axial HWHM as it is scanned along the optic axis. It is seen that the polarisation state of the emission has little influence on the axial HWHM and that reducing the size of the pinhole reduces the axial HWHM. This improvement in the axial resolution due to the introduction of a confocal pinhole is seen to increase with process order (as with lateral resolution). In the case of the axial resolution, close to confocal imaging is obtained for pinhole radii less than 20 μm (1.28 A.u.) and conventional imaging for greater than 60 μm (3.84 A.u.). Although it is known that two- and three-photon fluorescence imaging exhibits optical sectioning without the need to use a confocal point detector, it is important, particularly in light of the results presented in Fig. 11.30, to determine whether any advantage results from the use of confocal detection. To investigate this we define the integrated intensity, $\bar{I}_{\text{int}}$, as:

$$\bar{I}_{\text{int}}(z_p) = \int_0^{\infty} \int_0^{2\pi} I_d(r_t = 0; r_d, \gamma_d, z_p) r_d dr_d d\gamma_d \,, \tag{11.6.20}$$

which is proportional to the total energy in the image as a function of defocus of the fluorescent dipole, z_p. In the case of conventional single-photon fluorescence imaging it is well known that this function is constant, whereas for a confocal

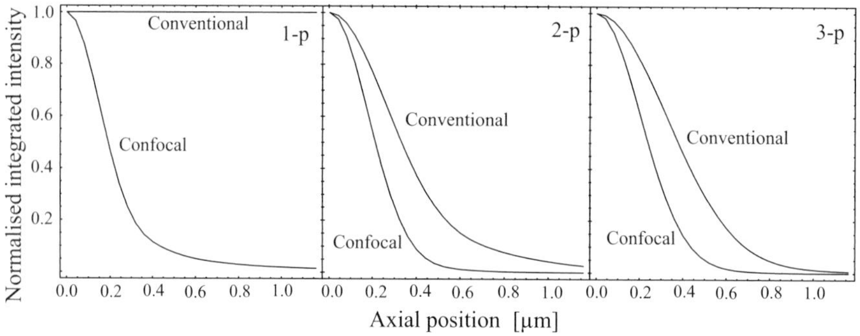

Figure 11.32: Normalised integrated intensity $\bar{I}_{\text{int}}/\bar{I}_{\text{int}}(\mathbf{r}_p = 0)$ (Eq. (11.6.20)) as a function of axial position z_p for 1-p, 2-p and 3-p for Case I and conventional and confocal detection. All simulation parameters were the same as described in Fig. 11.30. Reproduced from **[136]** with permission of John Wiley & Sons, Inc.

geometry it decays, i.e. the system exhibits optical sectioning. This is illustrated in Fig. 11.32 for Case I and circularly polarised illumination.

Similar trends are to be expected for Cases II and III. The 2-p and 3-p cases illustrate the optical sectioning capability of a conventional multiphoton fluorescence microscope. However, as also shown in this figure, the introduction of a confocal pinhole leads to a further improvement in the optical sectioning for both 2-p and 3-p fluorescence scanning microscopes with a decrease of the integrated intensity HWHM of $\approx 30\%$. This result, together with those of Fig. 11.30, suggest that considerable resolution advantage is likely to result from implementing confocal rather than conventional multiphoton fluorescence imaging.

11.6.6 Image Asymmetry

A measure of the image asymmetry is introduced as the ratio of the HWHM of a scan along the x-direction to the HWHM of a scan along the y-direction. For circularly polarised illumination ($\delta_1 = \pi/2$) the asymmetry ratio is unity, whereas if the Babinet–Soleil is set at any relative retardance other than $\pi/2$ ($0 < \delta_1 < \pi$) then the electric field in the focal region becomes asymmetric. In general the incident polarisation is elliptical, and for the special cases of $\delta_1 = 0$ and $\delta_1 = \pi$ the incident illumination is linearly polarised in the x- and y-directions, respectively.

Figure 11.33 shows the asymmetry as a function of retardance setting δ_1 for the case of polarised 1-p, 2-p and 3-p fluorescent emission detected with a 100 μm pinhole. There is a smooth transition in the symmetry ratio with maxima and minima corresponding to the two linear illumination cases as expected. The 3-p process is seen to be the most asymmetric followed by the 2-p and 1-p images.

Figure 11.34 shows the images corresponding to those presented in Fig. 11.30 but for linearly polarised illumination ($\delta_1 = 0$). These lateral images are asymmetric as expected as a consequence of employing a high-aperture microscope objective lens and linear polarisation. Although the form of the image is more complicated for linearly polarised light, the general dependencies of image size on process order and pinhole size are the same as for circular polarisation, i.e. the improvement in lateral resolution due to the confocal pinhole increases with the order of the multiphoton fluorescence process. The symmetry properties in Fig. 11.34 are seen to vary from image to image, with process order and pinhole radius. As the image asymmetry with linearly polarised illumination is of practical importance, Fig. 11.35 is presented. This figure shows the variation in image asymmetry as a function of pinhole radius and it is seen that, in general, the asymmetry is smaller when the pinhole radius is reduced. The images in Fig. 11.34 were plotted for the case of unpolarised fluorescent emission (Case I) and hence the asymmetry is a ‘mapping’ of the asymmetry in the illuminating electric field.

11.6.7 Paraxial Approximation

All of the asymmetry effects are due to the combination of a high-aperture objective (condenser/collector) lens and the linearly polarised illumination used for the preceding numerical results. Figure 11.36 shows axial and lateral HWHM curves calculated for a low-aperture, 10×, 0.15 NA, lens (c.f. Fig. 11.31). The dashed and full lines in the lateral

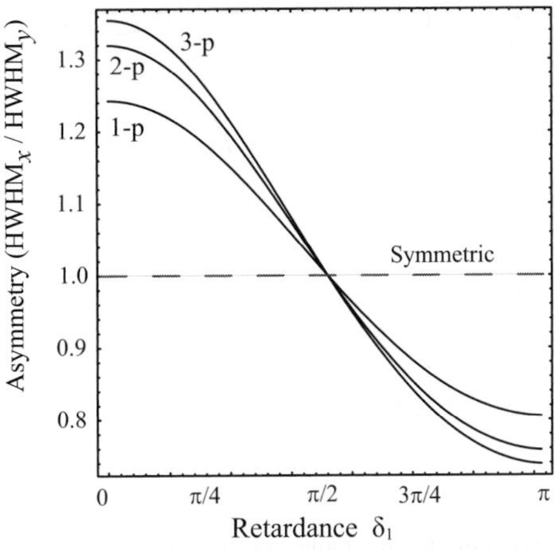

Figure 11.33: Asymmetry in the image of a fluorescent molecule for 1-p, 2-p and 3-p excitation as a function of the retardance setting of the first Babinet–Soleil compensator for R = 100 μm for Case I. All simulation parameters were the same as described in Fig. 11.30. Reproduced from **[136]** with permission of John Wiley & Sons, Inc.

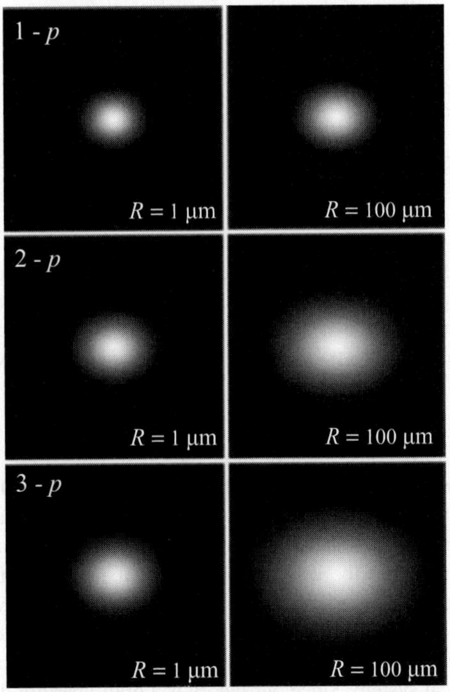

Figure 11.34: Image of a fluorescent molecule illuminated with linearly polarised light (horizontal direction) with unpolarised emission (Case I) for 1-p, 2-p and 3-p for pinhole radii of 1 and 100 μm. The image size is 0.8 × 0.8 μm. All simulation parameters were the same as described in Fig. 11.30. Reproduced from **[136]** with permission of John Wiley & Sons, Inc.

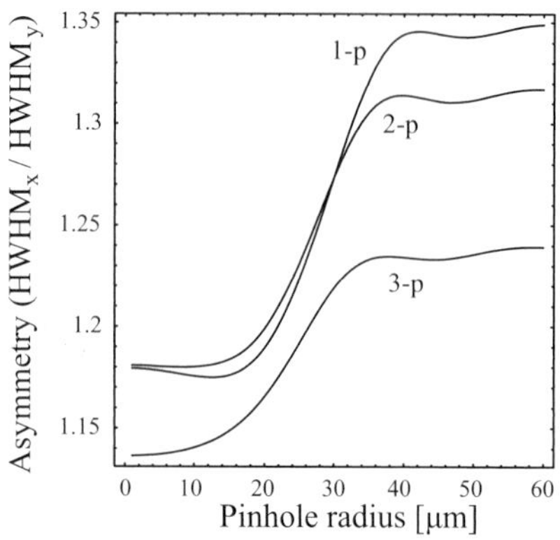

Figure 11.35: Asymmetry in the image of a fluorescent molecule for 1-p, 2-p and 3-p excitation as a function of pinhole size for linearly polarised illumination (Case I). All simulation parameters were the same as described in Fig. 11.30.

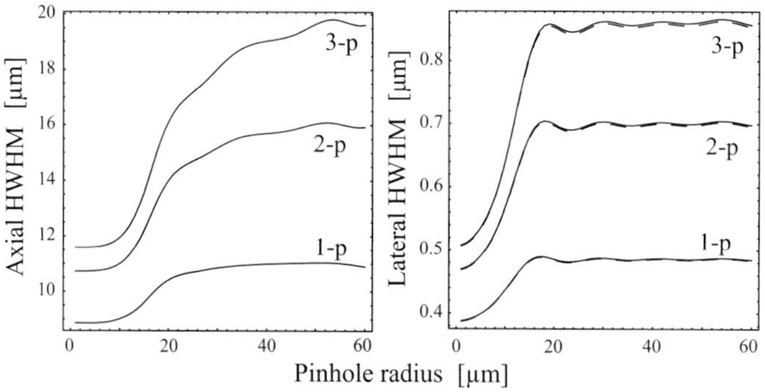

Figure 11.36: Axial and lateral HWHM curves for 1-p, 2-p and 3-p as a function of the pinhole radius for Cases I (continuous line), II (dash-dot line), and III (dashed line) with linearly polarised illumination and for a low NA condenser/collector lens (10×, 0.15 NA). The corresponding curves fully overlap. The dashed curves indicate scans in the direction perpendicular and the full curves parallel to the direction of incident polarisation, (c.f. Fig. 11.31). Reproduced from **[136]** with permission of John Wiley & Sons, Inc.

plot correspond to orthogonal scan directions. Each of the Cases I, II and III are plotted in this figure but cannot be distinguished from one another, indicating that the degree of polarisation does not affect the axial or lateral resolution in low-aperture systems. The asymmetry in the image is also seen to be negligible for each case. It should also be noted that when plotted for circularly polarised illumination almost identical results to those in Fig. 11.36 were obtained.

These results may also be obtained from the less complicated paraxial theory for multiphoton fluorescence as the polarisation state of the emission and high-aperture effects may be ignored. However, in practice, such low numerical aperture lenses are seldom used except in developmental biology where large fields of view are often preferred over high transverse magnification. Nevertheless, the paraxial theory is often used to obtain a rough estimate concerning the performance of a high-aperture optical system. Since our theory is a high-aperture description, it is easy to approximate our formulas paraxially. When such an approximation is made one can compare the results of the approximation to those obtained in the high-aperture case. The result of such a comparison could be used to evaluate the accuracy of the low-aperture approximation.

It is also important to calculate how much the presence of a confocal pinhole would improve on the lateral and axial resolution as compared to the conventional case. The fractional improvement in axial and lateral resolution for unpolarised emission (Case I) from a fluorescent molecule illuminated with circular polarisation between the conventional

Table 11.3. Ratio increase in resolution due to a confocal pinhole over conventional imaging ($HWHM_{conv}/HWHM_{conf}$) for 1-, 2- and 3-photon fluorescence imaging using low- and high-aperture lenses. Reproduced from **[136]** with permission of John Wiley & Sons, Inc.

Lens		Resolution improvement 1-p	2-p	3-p
10×, 0.15	Lateral	1.26	1.52	1.73
	Axial	1.27	1.53	1.74
100×, 1.4	Lateral	1.26	1.49	1.69
	Axial	1.27	1.53	1.77

and confocal imaging limits is given in Table 11.3 for 1-p, 2-p and 3-p processes. Both the lateral and the axial resolution improvement is approximately the same for the 10× and 100× lenses, although there seems to be an increasing difference with increasing process order.

The results for the 10×, 0.15 NA lens are very close to those of the paraxial theory. Since the paraxial theory may be written in normalised coordinates its results are simply scaled according to the numerical aperture of the lens. Therefore, the differences in the presented lateral resolution from the paraxial theory are the same for any lens. Considering the data in Table 11.3, this means that the paraxial theory estimates the *ratio* of the resolution values for the high-aperture lens fairly accurately for the 1-p, 2-p and 3-p processes. On the other hand, we should not forget that the *actual* resolution values are very inaccurate when the paraxial theory is used at large values of numerical apertures.

11.6.8 Signal-to-noise Ratio

In the previous subsections it has been shown that employing a confocal pinhole in 2-p and 3-p fluorescence microscopes leads to great improvements in the lateral and axial resolution. However, it is important to consider that in such microscopes the light budget can be limited. It is therefore the aim in this section to determine an optimum pinhole radius in terms of maximising the signal-to-noise ratio of the microscope. The shot noise on the background is assumed to be proportional to the area of the detector, of radius R, and a measure of the signal-to-noise ratio is given by:

$$\frac{S}{N}(R) = \frac{I(R)}{\sqrt{I(R) + \xi R^2}}, \tag{11.6.21}$$

where the constant ξ represents the noise level. This model is applicable for background arising from autofluorescence of the optical components (cements and coatings), as is important in single molecule fluorescence imaging. Electronic noise arising from amplification/digitisation is not included in this model. A typical example of how the signal-to-noise ratio changes with pinhole radius is given in Fig. 11.37. These curves are plotted for 3-p fluorescence on a log-log scale. Figure. 11.38a shows the optimum pinhole radius plotted as a function of the noise level (ξ) for the two limiting cases of polarised (Case III) and unpolarised (Case I) emission for 1-p, 2-p and 3-p fluorescence in the high-aperture system.

For large values of ξ, when the shot noise dominates the signal, it is seen that the optimum pinhole radius varies between approximately 14 μm and 16 μm (0.89–1.02 A.u.), depending on the degree of polarisation and is independent of the order of the multiphoton process. In the low noise limit the optimum pinhole radius varies between approximately 17 μm and 25 μm (1.09–1.60 A.u.) for unpolarised and polarised emission, respectively. In this case there is seen to be a small variation in the curves corresponding to the 1-p, 2-p and 3-p processes but in practice these differences may be ignored. These optimum pinhole radii correspond to imaging in the regime between the conventional and confocal limits, i.e. employing the optimum pinhole already leads to a distinct improvement in the axial resolution over the conventional imaging case. However, to obtain the maximum possible improvement in resolution an even smaller pinhole needs be employed. Figure 11.38b shows the cost, in terms of the signal-to-noise ratio, of employing a pinhole which leads to

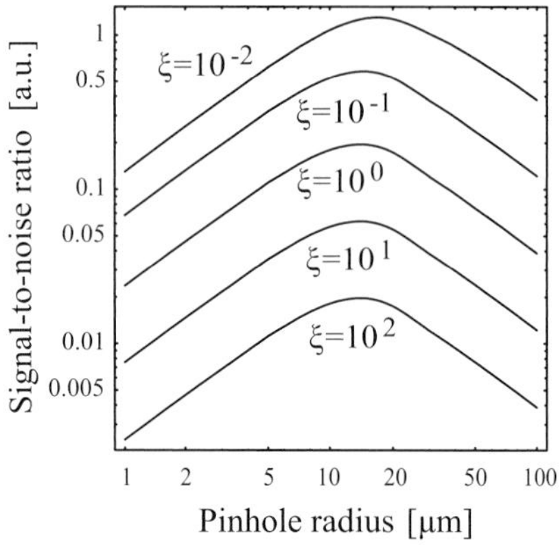

Figure 11.37: Signal-to-noise ratio as a function of pinhole radius for 3-p fluorescence imaging and for Case I. All simulation parameters were the same as described in Fig. 11.30. Reproduced from **[136]** with permission of John Wiley & Sons, Inc.

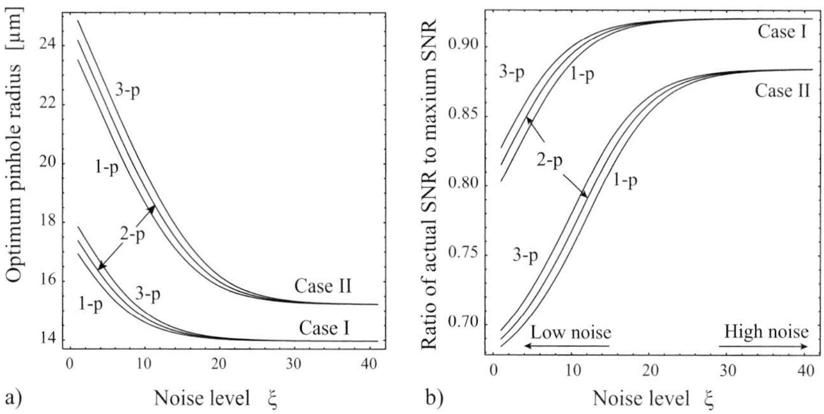

Figure 11.38: a) Pinhole radius corresponding to maximum signal-to-noise ratio (SNR) as a function of noise level for fluorescence imaging for 1-p, 2-p and 3-p excitation and for Cases I and II.
b) SNR for a confocal pinhole (R = 10 μm) as a ratio of the maximum SNR. Plotted as a function of noise level for fluorescence imaging for 1-p, 2-p and 3-p excitation and for Cases I and II. All simulation parameters were the same as described in Fig. 11.30. Reproduced from **[136]** with permission of John Wiley & Sons, Inc.

close to true confocal imaging. For these calculations a pinhole of radius 10 μm (0.64 A.u.) is used. Such a pinhole would lead to resolution properties only slightly worse than an ideal confocal microscope ($R \to 0$). Figure 11.38b shows the signal to noise ratio with the 10 μm pinhole as a percentage of the signal-to-noise ratio obtained by employing the optimum pinhole radius as a function of the noise level.

It is seen that in the shot noise limit the signal-to-noise ratio is reduced from its maximum value by only 8–12% (i.e. $\approx$ 0.5 dB) depending on the state of polarisation of the fluorescent emission. For the case when the detector noise dominates, the decrease in signal-to-noise ratio is more significant at 17–32% of its maximum value (i.e. $\approx$ 1–2 dB). Again, the reduction in signal-to-noise ratio is due mainly to the state of polarisation rather than the order of the multiphoton process. However, it should be noted that 3-p fluorescence, which benefits the most in terms of resolution from confocal imaging, actually suffers the least in terms of signal-to-noise ratio. Ultimately, even the largest loss in the signal-to-noise ratio mentioned above is a relatively small price to pay for the vast improvement in the lateral resolution of the 2-p and 3-p microscopes achieved by employing the confocal pinhole.

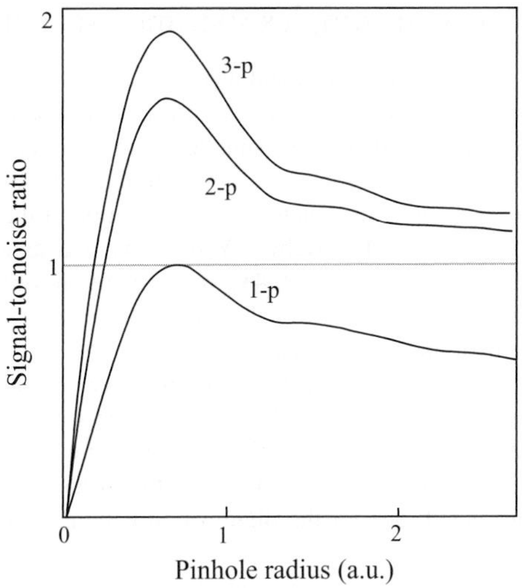

Figure 11.39: The signal to noise ratio in the case when the specimen exhibits strong background fluorescence, as a function of pinhole size. Results are obtained from paraxial theory. The 2-p and the 3-p curves are normalised to a signal-to-noise ratio of unity for large pinholes. The 1-p curve is normalised to a peak signal-to-noise ratio of unity. Reproduced from **[136]** with permission of John Wiley & Sons, Inc.

11.6.9 High Background Noise

In the previous subsection, the signal to noise ratio was investigated based on a model in which noise power is represented by a term proportional to the area of the pinhole, which was shown experimentally **[306]** to be a form of noise present in a confocal fluorescence microscope arising from optical glasses, cementing and coatings on optical components. Another potential source of noise is the background autofluorescence from neighbouring regions of the sample **[107]**, **[298]**. Assuming this to be featureless, it is equivalent to integrating over the overall three-dimensional point-spread function that, in paraxial fluorescence microscopy, is the product of the intensity point-spread functions of the illumination and detection systems. This has been analysed for a 1-photon fluorescence microscope in the paraxial approximation **[107]** and also for 2-photon and 3-photon microscopes **[108]**. It is found that for single-photon fluorescence the noise power increases as the area of the pinhole for small pinhole sizes, but eventually increases linearly with pinhole radius as the noise power becomes proportional to the thickness determined by the depth of focus. A qualitatively similar result is expected for the high-aperture case. The resulting signal-to-noise ratio is shown in Fig. 11.39, where for 1-photon fluorescence the signal-to-noise is normalised to unity at the peak value, and for 2-photon and 3-photon cases it is normalised to unity for large pinholes. Note that in the latter cases the signal-to-noise ratio does not fall off to zero with large pinhole size as the depth of focus is limited by the excitation process. The optimum pinhole sizes are found to be 0.75, 0.79 and 0.80 A.u. for 1-, 2- and 3-photon processes, respectively, i.e. 11.8, 12.3 and 12.5 μm for the 1.4 NA, 100× objective considered here. The improvement in signal-to-noise ratio compared to using a large pinhole is then by a factor of 1.66 and 1.92 for 2-photon and 3-photon, respectively (2.2 and 2.8 dB).

11.7 Extension of the Imaging Theory to More Complicated Optical Systems

In this section we extend some of the methods presented in the previous sections. In particular, we study the case when the dipole scatterer acting as the object is situated inside a stratified medium. Such specimen assemblies frequently occur in the biological sciences. Second, we generalise our results to show how samples of almost arbitrary complexity can be imaged using our model. This requires the use of rigorous electromagnetic solvers, such as finite difference time domain (FDTD) or finite element method (FEM) which, apart from the brief overview given in Chapter 1, are not the subject of this book. Nevertheless we show, even though the output of these computations is numerical, how one can continue using an analytic description.

11.7.1 Extension of the theory to complex specimen structures

The major use of optical microscopy is in the biological and medical sciences, where the sample is mounted on a glass microscope slide and protected from the other side by a cover glass. The entire specimen assembly is then placed under an optical microscope. High spatial resolution is achieved by using high-numerical-aperture oil immersion objective lenses. The refractive index of the immersion oil is usually slightly different from that of the cover slip which is then significantly different from that of the usually watery sample. In Chapter 9 the application of the ENZ theory to multilayer structures was discussed (see discussion from Fig. 9.17). The Debye–Wolf theory was extended to include focusing through a single and multilayer interface in 1994 **[346]** and 1997 **[345]**, respectively. In order to preserve the framework of the discussion presented earlier in this chapter, we now discuss the problem of imaging from a multilayer structure using the Debye–Wolf theory directly without involving the ENZ theory of Chapter 9.

In addition to describing focusing into a multilayer structure **[345]** we also need to consider what happens when our usual dipole representing a point object is embedded in a multilayer structure. Transmission of light into such a structure results in a different amplitude than transmission from the structure to the collector lens because the products of the Fresnel transmission coefficients are not identical for both paths. If an optical microscope is to be modelled the two separate models need to be constructed; one for the illumination path and one for the detection path. We start by discussing the illumination path.

We initially assume that the condenser lens (used in the sense of our discussions in the preceding sections) is perfect, i.e. the spherical wavefront emerging from the exit pupil is centred on the Gaussian image point (focus). We have seen in Section 4.6 and Section 7.7 that, from a geometrical optics point of view, it is not difficult to design lenses such that they are corrected for multilayer image space. High-quality and high-numerical-aperture immersion microscope objective lenses are usually designed using a hyper-hemispheric last lens element (see Fig. 11.40) in such a way that the spherical portion of the surface satisfies the aplanatic condition, with the object placed at B_1, on a flat portion of the surface (Fig. 11.40a). The designer 'replaces' a certain thickness of the glass by the coverslip and immersion liquid, usually assuming that they have the same refractive index as that of the glass of the hemispherical element (Fig. 11.40b). Assuming that the object is not moved from B_1 it is possible to achieve zero spherical aberration, coma and astigmatism due to zero marginal ray height at B_1 and the aplanatic condition being satisfied at the spherical portion of the lens surface.

Since most biological samples are mounted in aqueous media, the refractive index on the left-hand side of the coverslip (Fig. 11.40b) will not match the refractive index of the immersion liquid and if the thickness B_2B_3 is reduced by focusing deeper into the aqueous medium aberrations will result. Therefore it is important to investigate how focusing through a multilayer structure will affect the light distribution at the sample.

11.7.1.1 The Formulation of the Problem

Consider an optical system of revolution with its optic axis z as shown in Fig. 11.41. The origin of a Cartesian coordinate system is positioned at the Gaussian focus F' of the unaberrated rays (dashed line). The electric field is determined at the arbitrary point P in the focal region. The vector $\hat{\mathbf{s}}_j = (\hat{s}_{x,j}, \hat{s}_{y,j}, \hat{s}_{z,j}) = \mathbf{s}_{t,j} + \hat{s}_{z,j}\hat{\mathbf{z}}$ is the unit vector along a typical ray in the jth medium and $\mathbf{r}_p = (x_p, y_p, z_p) = \mathbf{r}_t + z_p\hat{\mathbf{z}}$ is the position vector F' to P. Let the Debye–Wolf integral Eq. (8.3.25) represent the field in the focal region of an aberration-free convergent spherical wave in the homogeneous medium of incidence:

$$\mathbf{E}(\mathbf{r}_p) = -\frac{ifk_{J+1}}{2\pi} \iint_{\Omega_{J+1}} \frac{\mathcal{E}(\hat{s}_{x,J+1}, \hat{s}_{y,J+1})}{\hat{s}_{z,J+1}} \exp(ik_{J+1}\hat{\mathbf{s}}_{J+1}\cdot\mathbf{r}_p)d\hat{s}_{x,J+1}d\hat{s}_{y,J+1} \,, \qquad (11.7.1)$$

where Ω_{J+1} is the solid angle of convergence and k_{J+1} is the wavenumber in the initial medium. We now place a stratified (multilayered) medium in the focal region. The wavenumber corresponding to the jth medium is denoted by k_j. Media are numbered sequentially from left to right as shown in Fig. 11.41, so the wavenumbers in the individual media are $k_{J+1}, k_J, \ldots k_0$. Note that contrary to how we used k_0 before, it now denotes the wavenumber in the last medium and *not* the wavenumber in vacuum. $J+1$ media are separated by J interfaces, which are located at $z = -z_J, -z_{J-1}, \ldots, -z_0$. We reformulate Eq. (11.7.1) as follows. In the first medium and at the first interface ($z = -z_J$), the incident electric field is given by:

$$\mathbf{E}_J(\mathbf{r}_t, -z_J) = -\frac{ifk_{J+1}}{2\pi} \iint_{\Omega_{J+1}} \frac{\mathcal{E}(\hat{s}_{x,J+1}, \hat{s}_{y,J+1})}{\hat{s}_{z,J+1}} \exp[ik_{J+1}(\hat{\mathbf{s}}_{t,J+1}\cdot\mathbf{r}_t - \hat{s}_{z,J+1}z_J)]d\hat{s}_{x,J+1}d\hat{s}_{y,J+1} \,. \qquad (11.7.2)$$

To describe the field in the last (0th) medium we assume that each plane wave component refracting at the interfaces obeys Snell's refraction law and the resulting field in the last medium is constructed as a superposition of multiply

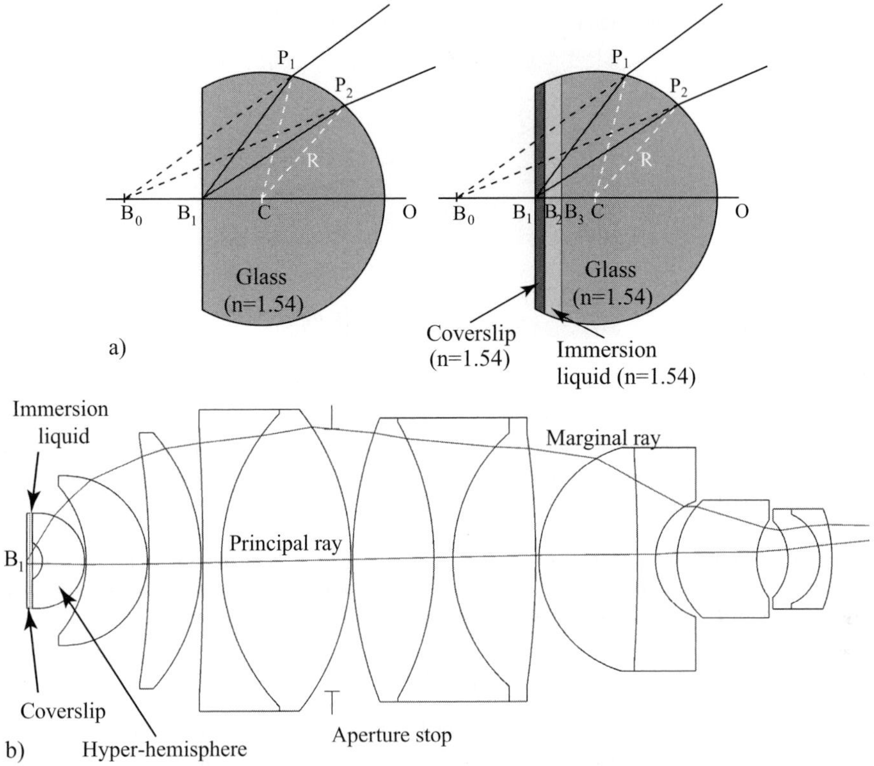

Figure 11.40: a) Arrangement of a hyper-hemispherical lens element, initial design. The object (B_1) is placed on the flat surface, on axis and the image (B_0) is formed on the same side of the lens because the spherical portion of the surface is designed to satisfy the aplanatic condition for stigmatic imaging, i.e. the distances $z_0 = B_0O$ and $z_1 = B_1O$ must satisfy the corresponding relationships given in Table 4.1.
b) When designing an immersion microscope objective lens, the B_1B_3 section is 'cut' away and replaced by the coverslip of thickness B_1B_2, and immersion liquid of thickness B_2B_3. The system remains corrected for spherical aberration, coma and astigmatism as long as the refractive indices of the coverslip, immersion liquid and glass of the first plano-convex lens element shown (shown in the figure as $n = 1.54$ by way of example) remain the same. Note that chromatic aberration might be introduced if the Abbe numbers of the coverslip, immersion liquid and glass differ.
c) A typical high-numerical-aperture immersion lens design (US Patent US6504653, **[242]**) with the marginal and principal ray shown. The object B_1 is located at the left surface of the coverslip. The coverslip, immersion liquid and the composite hyper-hemisphere together make up the original hyper-hemispherical lens of the upper-left drawing.

refracted and reflected plane waves. If the weighted amplitude of the plane waves incident on the interface is given by $\mathbf{W}(\hat{\mathbf{s}}_{J+1}) = \mathcal{E}/\hat{s}_{z,J+1}$, then the weighted amplitude of the transmitted plane waves in the last medium is a linear function of $\mathbf{W}(\hat{\mathbf{s}}_{J+1})$, i.e. $\mathbf{\Lambda W}$ where the elements of $\mathbf{\Lambda}$ are (nonlinear) functions of the angle of incidence, thickness of individual layers and k_{J+1}, $k_J, \ldots$ k_0. The exact form of $\mathbf{\Lambda}$ will be determined indirectly by Eq. (11.7.15) for inward propagation in this subsection, and directly by Eq. (11.7.29) in the next section for out propagation. As opposed to the transfer matrix method presented in Section 9.4, which permits the calculation of the electric field anywhere inside the layers, the current method describes the effect of the multilayer structure as a whole. It therefore cannot be used to calculate the field everywhere inside the layers and so it is limited to yield the electric field only in the very last medium. Hence the method described in Section 9.4 is more computational whereas the present formalism is more suited to analytic work.

The transmitted field in the last medium, just past the last interface is given by

$$\mathbf{E}_0(\mathbf{r}_t, -z_0) = -\frac{ifk_0}{2\pi} \iint_{\Omega_0} \mathbf{\Lambda W}(\hat{\mathbf{s}}_0) \exp[ik_1(\hat{\mathbf{s}}_{t,0}\cdot\mathbf{r}_t - \hat{s}_{z,0}z_0)]d\hat{s}_{x,0}d\hat{s}_{y,0} \,. \tag{11.7.3}$$

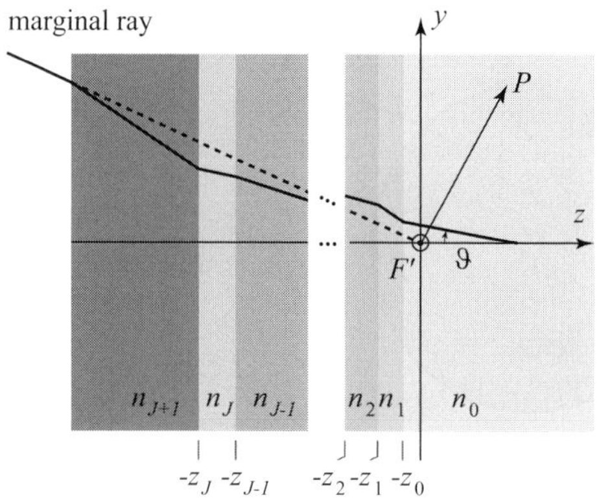

Figure 11.41: The notation used to develop a theory of focusing into multilayered image space. The Gaussian focus F' is set to coincide with the origin of the Cartesian coordinate system and we seek expressions for the electric field at P. The multilayered structure is located at $-z_J, -z_{J-1}, \ldots, -z_0$ for J interfaces.

Note that the phase of this integral is written for $z_p = -z_0$ because the effect of the stratified medium on the phase is carried by $\mathbf{\Lambda}$ in what follows. We compose the field inside the last medium again as a superposition of plane waves. This representation is required for $\mathbf{E}_0$ to be a solution of Maxwell's equations and can be written as:

$$\mathbf{E}_N(x_p, y_p, z_p) = -\frac{ifk_0}{2\pi}\iint_{\Omega_0} \mathcal{F}\exp(ik_0\hat{\mathbf{s}}_{t,0}\cdot\mathbf{r}_p)d\hat{s}_{x,0}d\hat{s}_{y,0}\ . \tag{11.7.4}$$

We have to determine the function $\mathcal{F}$ and for this we shall make use of Eq. (11.7.3), which represents an initial value for Eq. (11.7.4). First, however, we need to establish the relationship between $\hat{\mathbf{s}}_{J+1}$, $\hat{\mathbf{s}}_J$ and, in general, $\hat{\mathbf{s}}_0$. It is evident from the vectorial law of refraction:

$$k_J\hat{\mathbf{s}}_J - k_{J+1}\hat{\mathbf{s}}_{J+1} = (k_J\cos\vartheta_J - k_{J+1}\cos\vartheta_{J+1})\hat{\mathbf{v}}\ , \tag{11.7.5}$$

where $\hat{\mathbf{v}}$ represents the unit normal of the interface, $\vartheta_{J+1} = \cos^{-1}(\hat{\mathbf{s}}_{J+1}, \hat{\mathbf{v}})$ and $\vartheta_J = \cos^{-1}(\hat{\mathbf{s}}_J, \hat{\mathbf{v}})$ (see also Subsection 4.2.5.1), so that:

$$k_0\hat{s}_{x,0} = \ldots = k_J\hat{s}_{x,J} = k_{J+1}\hat{s}_{x,J+1}, \qquad k_0\hat{s}_{y,0} = \ldots = k_J\hat{s}_{y,J} = k_{J+1}\hat{s}_{y,J+1}\ . \tag{11.7.6}$$

On expanding Eq. (11.7.4) at $z_p = -z_0$ and expressing the function $\mathcal{F}$ we obtain the electric field in the last medium:

$$\mathbf{E}_0(x_p, y_p, z_p) = -\frac{ifk_{J+1}}{2\pi}\iint_{\Omega_{J+1}} \mathbf{\Lambda}\mathbf{W}(\hat{\mathbf{s}}_{J+1})\exp[i(k_0z_0\hat{s}_{z,0} - k_{J+1}z_J\hat{s}_{z,J+1})] \\ \times\exp(ik_0\hat{s}_{z,0}z_p)\exp(ik_{J+1}\mathbf{s}_{t,J+1}\cdot\mathbf{r}_t)d\hat{s}_{x,J+1}d\hat{s}_{y,J+1}\ . \tag{11.7.7}$$

From Eq. (11.7.6) it also follows that in Eq. (11.7.7)

$$\hat{s}_{z,j} = \left[1 - \frac{k_{J+1}^2}{k_j^2}(\hat{s}_{x,J+1}^2 + \hat{s}_{y,J+1}^2)\right]^{1/2}\ . \tag{11.7.8}$$

For numerical purposes, however, the above formula cannot be applied directly. This is because the stratified medium can contain interfaces at which refraction is not ordinary (i.e. total reflection or generation of evanescent waves). We shall, therefore, use for numerical purposes a successive computation, i.e. $\hat{s}_{z,J+1} \to \hat{s}_{z,J} \to \ldots \hat{s}_{z,0}$. It is important to emphasise that since both the boundary conditions represented by Eq. (11.7.3) and the integral representation Eq. (11.7.4) are exact solutions of Maxwell's equations, our formula for the electric vector Eq. (11.7.7) in the final medium also satisfies Maxwell's equations.

11.7.1.2 The Electric Strength Vector

Individual layers of the stratified medium are taken to be isotropic and homogeneous and to have an optically smooth planar surface which is perpendicular to the optic axis. For our decomposition, the usual assumptions are made, namely, that the electric vector maintains its direction with respect to a meridional plane and the electric vector remains on the same side of a meridional plane on passing through the system.

For the optical system under consideration, the angle of incidence at the interface is denoted by ϑ_{J+1} and the angle in the last medium by ϑ_0. The unit vector $\hat{\mathbf{s}}_0$ and position vector $\mathbf{r}_p$ are given in spherical polar coordinates by:

$$\hat{\mathbf{s}}_0 = -\sin\vartheta_0 \cos\phi\, \hat{\mathbf{x}} - \sin\vartheta_0 \sin\phi\, \hat{\mathbf{y}} + \cos\vartheta_0\, \hat{\mathbf{z}} \,, \tag{11.7.9}$$

and

$$\mathbf{r}_p = r_t \cos\gamma\, \hat{\mathbf{x}} + r_t \sin\gamma\, \hat{\mathbf{y}} + z_p\, \hat{\mathbf{z}} \,. \tag{11.7.10}$$

The electric strength vector $\mathcal{E}_0$ in the final medium is given by:

$$\mathcal{E}_0 = \mathbb{R}^{-1}\cdot[\mathbb{P}^{(0)}]^{-1}\cdot\mathbb{I}_S\cdot\mathbb{P}^{(J+1)}\cdot\mathbb{L}\cdot\mathbb{R}\cdot\mathcal{E}_{J+1} \,, \tag{11.7.11}$$

where $\mathbb{P}$ is given by Eq. (11.1.14) but also reproduced here:

$$\mathbb{P}^{(j)} = \begin{pmatrix} \cos\vartheta_j & 0 & \sin\vartheta_j \\ 0 & 1 & 0 \\ -\sin\vartheta_j & 0 & \cos\vartheta_j \end{pmatrix} , \tag{11.7.12}$$

and the matrix $\mathbb{I}_S$ describes the Fresnel transmission of the stratified medium:

$$\mathbb{I}_S = \begin{pmatrix} t_p & 0 & 0 \\ 0 & t_s & 0 \\ 0 & 0 & t_p \end{pmatrix} , \tag{11.7.13}$$

specifically, $t_{s,p}$ are the transmission coefficients of the stratified medium describing the s and p polarised light traversing the medium, given by Eqs. (1.10.5) and (1.10.3).

From Eq. (11.7.11) we obtain the components of the electric vector in the last medium after setting $\mathcal{E}_{J+1} = (1, 0, 0)$:

$$\mathcal{E}_0 = \cos^{1/2}\vartheta_{J+1} \begin{pmatrix} t_p \cos\vartheta_0 \cos^2\phi + t_s \sin^2\phi \\ t_p \cos\vartheta_0 \sin\phi\cos\phi - t_s \sin\phi\cos\phi \\ -t_p \sin\vartheta_0 \cos\phi \end{pmatrix} , \tag{11.7.14}$$

where the $\cos^{1/2}\vartheta_{J+1}$ term, again, arises since we consider an aplanatic lens. We then evidently have:

$$\mathbf{\Lambda W}(\hat{\mathbf{s}}_{J+1}) = \mathcal{E}_0 \,. \tag{11.7.15}$$

11.7.1.3 The Electric Field Vector

First we formulate the expressions needed to simplify Eq. (11.7.7). We transform the integral variables $d\hat{s}_{1x}, d\hat{s}_{1y}$ to the spherical coordinate system and define

$$\kappa = -k_{J+1} \sin\vartheta_{J+1} \cos(\phi - \gamma) \,, \tag{11.7.16}$$

and

$$\Psi_i = k_0 z_0 \hat{s}_{z,0} - k_{J+1} z_J \hat{s}_{z,J+1} \,. \tag{11.7.17}$$

Ψ_i is referred to as the *initial aberration function*. With the help of the above results, Eq. (11.7.7) can be rewritten to express the electric $\mathbf{E}_0$ vector in the last medium as:

$$\mathbf{E}_0(\mathbf{r}_p) = -\frac{ifk_{J+1}}{2\pi} \iint_{\Omega_{J+1}} \mathcal{E}_0 \exp(ir_t\kappa) \exp(i\Psi_i) \exp(ik_0 z_p \cos\vartheta_0) \sin\vartheta_{J+1} d\vartheta_{J+1} d\phi \,. \tag{11.7.18}$$

On substituting from Eq. (11.7.15) and Eq. (11.7.14) into Eq. (11.7.18), changing the integration limits, assuming that the system obeys the sine condition, setting α to be the marginal ray angle as usual, and carrying out the integration with respect to ϕ, we obtain the following expression for the electric field components:

$$\begin{aligned} E_{x,0}(r_t, \gamma, z_p) &= -\frac{ifk_{J+1}}{2}(I_0 + I_2 \cos 2\gamma), \\ E_{y,0}(r_t, \gamma, z_p) &= -\frac{ifk_{J+1}}{2} I_2 \sin 2\gamma, \\ E_{z,0}(r_t, \gamma, z_p) &= -fk_{J+1} I_1 \cos\gamma \,, \end{aligned} \tag{11.7.19}$$

and the integrals I_0, I_1 and I_2 are given by:

$$\begin{aligned} I_0 &= \int_0^\alpha \sqrt{\cos\vartheta_{J+1}}\sin\vartheta_{J+1}(t_s + t_p\cos\vartheta_0)J_0(k_{J+1}r_t\sin\vartheta_{J+1}) \\ &\quad\times\exp(i\Psi_i)\exp(ik_0 z_p\cos\vartheta_0)d\vartheta_{J+1}, \\ I_1 &= \int_0^\alpha \sqrt{\cos\vartheta_{J+1}}\sin\vartheta_{J+1}t_p\sin\vartheta_0 J_1(k_{J+1}r_t\sin\vartheta_{J+1}) \\ &\quad\times\exp(i\Psi_i)\exp(ik_0 z_p\cos\vartheta_0)d\vartheta_{J+1}, \\ I_2 &= \int_0^\alpha \sqrt{\cos\vartheta_{J+1}}\sin\vartheta_{J+1}(t_s - t_p\cos\vartheta_0)J_2(k_{J+1}r_t\sin\vartheta_{J+1}) \\ &\quad\times\exp(i\Psi_i)\exp(ik_0 z_p\cos\vartheta_0)d\vartheta_{J+1}\,. \end{aligned} \tag{11.7.20}$$

Equations (11.7.19) and (11.7.20) conclude our solution of the problem. It should be noted that for $J = 1$, $k_{J+1} = k_0 = k$, therefore $\vartheta_{J+1} = \vartheta_0 = \vartheta$, and $t_p = t_s = 1$ such that Eq. (11.7.20) reduces to the Debye–Wolf integral Eq. (8.3.25).

11.7.1.4 Propagation of Electromagnetic Dipole Waves through Stratified Media

Having worked out the field in the multilayer structure, we now turn to discussing how a dipole wave propagates out from inside a stratified medium **[342]**. The electric dipole, representing a fluorescent molecule or a point scatterer, is embedded in a layered medium. It is excited by the illumination produced by the condenser lens focusing through the same layered medium. The radiation emitted by the dipole is collected in the $(J + 1)$th medium by the collector lens. The light so collimated is then focused by the detector lens. Obtaining the image of the dipole inside the multilayered structure permits the construction of the full model of imaging, just as was done in the previous sections.

Consider the geometry shown in Fig. 11.42. The harmonically oscillating dipole is placed at the origin of the Cartesian coordinate system. Just as in the previous subsection the stratified medium contains $J + 1$ media and hence J interfaces and the first, i.e. the one closest to the dipole, dielectric interface is placed perpendicular to the z-axis at $z = -z_0$. Subsequent interfaces are placed at $z = -z_1, \ldots, -z_J$. All other definitions are the same as before. Equation (11.7.1) represents a general *convergent* spherical wave given by $\mathbf{E}$. It is then clear that the electric field of the divergent spherical wave is given by $\mathbf{E}^d = \mathbf{E}^*$:

$$\mathbf{E}^d(\mathbf{r}_p) = -\frac{ifk}{2\pi}\iint_\Omega \frac{\mathcal{E}^*(\hat{\mathbf{s}})}{\hat{s}_z}\exp(-ik\mathbf{r}_p\cdot\hat{\mathbf{s}})\,d\hat{s}_x d\hat{s}_y\,, \tag{11.7.21}$$

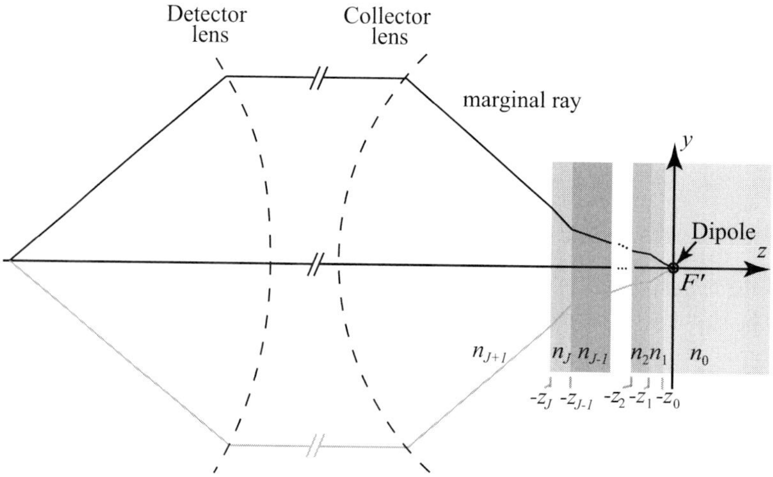

Figure 11.42: The electric dipole, representing a fluorescent molecule or a point scatterer, is embedded in a layered medium. It is excited by the illumination produced by the condenser lens focusing through the same layered medium. The illumination is not shown in the figure. The radiation emitted by the dipole is collected in the $(J + 1)$th medium by the collector lens. The light so collimated is then focused by the detector lens. Note that in this figure, as opposed to the convention, light propagates from right to left. Notation is identical to Fig. 11.41.

where $\mathcal{E}^*$ is the relative strength of the plane wave component outgoing along the direction $\hat{\mathbf{s}}$. The principle of the solution of the present problem is essentially the same as computing the electric field inside the stratified medium, therefore Eq. (11.7.21) gives the field at the first interface to be:

$$\mathbf{E}_0^d(\mathbf{r}_t, -z_0) = -\frac{ifk_0}{2\pi}\iint_{\Omega_0} \frac{\mathcal{E}^*(\hat{\mathbf{s}}_0)}{\hat{s}_{z,0}} \exp(-ik_0\mathbf{r}_t \cdot \hat{\mathbf{s}}_{t,0}) \exp(ik_0\hat{s}_{z,0}z_0)\, d\hat{s}_{x,0} d\hat{s}_{y,0} \,. \tag{11.7.22}$$

We now set, as before, $\mathbf{W}(\hat{\mathbf{s}}_0) = \mathcal{E}^*(\hat{\mathbf{s}}_0)/\hat{s}_{z,0}$. It is a reasonable assumption that the field just after the last interface will be a linear function of $\mathbf{W}$. Hence we may write:

$$\mathbf{E}_0^d(\mathbf{r}_t, -z_0) = -\frac{ifk_0}{2\pi}\iint_{\Omega_0} \mathbf{\Lambda}\mathbf{W} \exp(-ik_0\mathbf{r}_t \cdot \hat{\mathbf{s}}_{t,0}) \exp(ik_0\hat{s}_{z,0}z_0)\, d\hat{s}_{x,0} d\hat{s}_{y,0} \,, \tag{11.7.23}$$

where $\mathbf{\Lambda}$ is some tensor operator, which, just as before, does not need to be computed. The field in the first medium is, in general, represented as:

$$\mathbf{E}_{J+1}^d(\mathbf{r}_t, -z_p) = -\frac{ifk_{J+1}}{2\pi}\iint_{\Omega_{J+1}} \mathcal{F} \exp(-ik_{J+1}\mathbf{r}_t \cdot \hat{\mathbf{s}}_{t,J+1}) \exp(ik_{J+1}\hat{s}_{z,J+1}z_p)\, d\hat{s}_{x,J+1} d\hat{s}_{y,J+1} \,, \tag{11.7.24}$$

and it is noted that $k_{J+1}\mathbf{s}_{t,J+1} = k_J\mathbf{s}_{t,J} = \ldots = k_0\mathbf{s}_{t,0}$ from which the Jacobian of the coordinate transformation from $d\hat{s}_{x,J+1}d\hat{s}_{y,J+1}$ to $d\hat{s}_{x,0}d\hat{s}_{y,0}$ is given by $d\hat{s}_{x,J+1}d\hat{s}_{y,J+1} = (k_0/k_{J+1})^2 d\hat{s}_{x,0}d\hat{s}_{y,0}$. Using this relationship to re-write Eq. (11.7.23) to integrate with respect to $\hat{s}_{x,0}$ and $\hat{s}_{y,0}$, one obtains a boundary value for Eq. (11.7.24) at $z_p = -z_J$:

$$\mathcal{F} = \frac{k_{J+1}}{k_0}\mathbf{\Lambda}\mathbf{W} \exp\left[i(k_0\hat{s}_{z,0}z_0 - k_{J+1}\hat{s}_{z,J+1}z_J)\right] = \frac{k_{J+1}}{k_0}\mathbf{\Lambda}\mathbf{W}\exp(i\Psi_i) \,. \tag{11.7.25}$$

After substituting Eq. (11.7.25) into Eq. (11.7.24) one obtains the general solution of the problem:

$$\mathbf{E}_{J+1}^d(\mathbf{r}_p) = -\frac{ifk_{J+1}^2}{2k_0\pi}\iint_{\Omega_{J+1}} \mathbf{\Lambda}\mathbf{W} \exp(i\Psi_i)\exp(-ik_{J+1}\mathbf{r}_p \cdot \hat{\mathbf{s}}_{J+1})\, d\hat{s}_{x,J+1} d\hat{s}_{y,J+1} \,. \tag{11.7.26}$$

$\mathbf{E}_{J+1}^d$ gives the solution of the electric field on the left-hand side of the left-hand most interface. Note that even though this formula is very similar to Eq. (11.7.7) there are differences. First, the pre-multiplier term is different, which is due to the fact that the integral is expressed over the final rather than the initial medium. Second, the second phase term in the kernel of the integral has the opposite sign. This is due to the fact that we are considering the expansion of a divergent wave.

Clearly, Eq. (11.7.26) provides a solution that can be used with any value of $\mathcal{E}^*$. However, in order to obtain the desired solution for an electric dipole field the following approximations are made. First, the dipole is required to be many wavelengths from the right-hand most interface. Second, it is assumed that the field reflected back from the interface does not re-excite the dipole. Furthermore, in order to keep the formalism simple, evanescent waves shall be excluded from the integration range of Eq. (11.7.26). This latter assumption also means that

$$\hat{s}_{z,j} = \left(1 - \hat{s}_{x,j}^2 - \hat{s}_{y,j}^2\right)^{1/2}, \qquad j = 0, 1, \ldots, J+1 \,, \tag{11.7.27}$$

where $(\hat{s}_{x,j}^2 + \hat{s}_{y,j}^2) \le 1$. Now the specific form of $\mathcal{E}^*$ can be determined. From Eq. (11.5.15) the far electric field $\mathcal{E}$ of an electric dipole is given by

$$\mathcal{E} = -\frac{U_d}{r}\hat{\mathbf{r}} \times (\hat{\mathbf{r}} \times \hat{\mathbf{p}}) \quad \text{and hence} \quad \mathcal{E}^* = -\frac{U_d^*}{r}\hat{\mathbf{r}} \times (\hat{\mathbf{r}} \times \hat{\mathbf{p}}) \,, \tag{11.7.28}$$

where as before $\hat{\mathbf{r}}$ is a unit position vector and $\hat{\mathbf{p}}$ is the unit vector associated with the electric dipole moment. The tensor $\mathbf{\Lambda}$ can be computed with the use of the generalised Jones matrices:

$$\mathbf{\Lambda} = \mathbb{R}^{-1}\cdot[\mathbb{P}^{(J+1)}]^{-1}\cdot\mathbb{I}_{-S}\cdot[\mathbb{P}^{(0)}]\cdot\mathbb{R} \,, \tag{11.7.29}$$

where the subscript $-S$ is used in conjunction with $\mathbb{I}$ to remind us to use the Fresnel transmission coefficients in the right order: $0, 1, \ldots J$ (see Eq. (1.10.13)). It is now easy to calculate $\mathbf{\Lambda}\mathcal{E}^*$, which is given by:

$$\begin{aligned}\mathbf{\Lambda}\mathcal{E}_x^*|_{r=f} = -\frac{U_d^*}{2f}\big[&\hat{p}_x(t_s^- + t_p^- \cos\vartheta_{J+1}\cos\vartheta_0) - 2\hat{p}_z t_p^- \sin\vartheta_0 \cos\vartheta_{J+1}\cos\phi \\ &-(t_s^- - t_p^- \cos\vartheta_{J+1}\cos\vartheta_0)(\hat{p}_x\cos 2\phi + \hat{p}_y \sin 2\phi)\big] \,,\end{aligned} \tag{11.7.30}$$

$$\mathbf{\Lambda}\mathcal{E}^*_y|_{r=f} = -\frac{U_d^*}{2f}\left[\hat{p}_y(t_s^- + t_p^- \cos\vartheta_{J+1}\cos\vartheta_0) - 2\hat{p}_z t_p^- \sin\vartheta_0 \cos\vartheta_{J+1}\sin\phi \right.$$
$$\left. -(t_s^- - t_p^- \cos\vartheta_{J+1}\cos\vartheta_0)(\hat{p}_x \sin 2\phi - \hat{p}_y \cos 2\phi)\right],$$
$$\mathbf{\Lambda}\mathcal{E}^*_z|_{r=f} = -\frac{U_d^*}{f}\left[\hat{p}_z t_p^- \sin\vartheta_{J+1}\sin\vartheta_0 - t_p^- \cos\vartheta_0 \sin\vartheta_{J+1}(\hat{p}_x\cos\phi + \hat{p}_y \sin\phi)\right] .$$

When Eq. (11.7.30) is substituted into Eq. (11.7.26), the following analytic expressions result for the Cartesian components of the electric field of an electric dipole:

$$E^d_{x,J+1}(\mathbf{r}_t, -z_p) = \frac{iU_d^* k_0^2}{2k_{J+1}}\left[\hat{p}_x I_0^{(1)} - 2i\hat{p}_z I_1^{(1)}\cos\gamma + I_2(\hat{p}_x \cos 2\gamma + \hat{p}_y \sin 2\gamma)\right],$$
$$E^d_{y,J+1}(\mathbf{r}_t, -z_p) = \frac{iU_d^* k_0^2}{2k_{J+1}}\left[\hat{p}_y I_0^{(1)} - 2i\hat{p}_z I_1^{(1)}\sin\gamma + I_2(\hat{p}_x \sin 2\gamma - \hat{p}_y \cos 2\gamma)\right],$$
$$E^d_{z,J+1}(\mathbf{r}_t, -z_p) = \frac{iU_d^* k_0^2}{k_{J+1}}\left[\hat{p}_z I_0^{(2)} - iI_1^{(2)}(\hat{p}_x \cos\gamma + \hat{p}_y \sin\gamma)\right] , \tag{11.7.31}$$

with

$$I_0^{(1)} = \int_0^{\pi/2} (t_s^- + t_p^- \cos\vartheta_{J+1}\cos\vartheta_0)\sin\vartheta_{J+1}\frac{\cos\vartheta_{J+1}}{\cos\vartheta_0}$$
$$\times J_0(k_1 r_t \sin\vartheta_{J+1})\exp(ik_{J+1}z_p\cos\vartheta_{J+1})\exp(i\Psi_i)d\vartheta_{J+1},$$
$$I_0^{(2)} = \int_0^{\pi/2} t_p^- \sin^2\vartheta_{J+1}\sin\vartheta_0\frac{\cos\vartheta_{J+1}}{\cos\vartheta_0}$$
$$\times J_0(k_1 r_t \sin\vartheta_{J+1})\exp(ik_{J+1}z_p\cos\vartheta_{J+1})\exp(i\Psi_i)d\vartheta_{J+1},$$
$$I_1^{(1)} = \int_0^{\pi/2} t_p^- \sin\vartheta_0 \cos\vartheta_{J+1}\sin\vartheta_{J+1}\frac{\cos\vartheta_{J+1}}{\cos\vartheta_0} \tag{11.7.32}$$
$$\times J_1(k_1 r_t \sin\vartheta_{J+1})\exp(ik_{J+1}z_p\cos\vartheta_{J+1})\exp(i\Psi_i)d\vartheta_{J+1},$$
$$I_1^{(2)} = \int_0^{\pi/2} t_p^- \sin^2\vartheta_{J+1}\cos\vartheta_{J+1}J_1(k_1 r_t \sin\vartheta_{J+1})\exp(ik_{J+1}z_p\cos\vartheta_{J+1})\exp(i\Psi_i)d\vartheta_{J+1},$$
$$I_2 = \int_0^{\pi/2} (t_s^- - t_p^- \cos\vartheta_{J+1}\cos\vartheta_0)\sin\vartheta_{J+1}\frac{\cos\vartheta_{J+1}}{\cos\vartheta_0}$$
$$\times J_2(k_1 r_t \sin\vartheta_{J+1})\exp(ik_{J+1}z_p\cos\vartheta_{J+1})\exp(i\Psi_i)d\vartheta_{J+1} .$$

The above derivation provides the electric field emerging from the left-hand most interface. We can use a pair of lenses, the collector and detector lenses, to image the dipole onto the detector plane. If these two lenses are placed in a 4-f configuration, the electric field in the vicinity of the focus of the detector lens can be computed. The electric field $\mathcal{E}_d$ of the geometric optics approximation in the aperture stop of the high-aperture and detector lenses is given by:

$$\mathcal{E}_d = \cos^{-1/2}\vartheta_{J+1}\,\mathbb{R}^{-1}\cdot[\mathbb{L}^{(J+1)}]^{-1}\cdot\mathbb{R}\cdot\mathbf{\Lambda}\mathcal{E}^* , \tag{11.7.33}$$

which gives for the electric dipole:

$$\mathcal{E}_{x,d} = -\frac{1}{2\sqrt{\cos\vartheta_{J+1}}}\left[\hat{p}_x(t_s^- + t_p^- \cos\vartheta_0) - 2\hat{p}_z t_p^- \sin\vartheta_0\cos\phi \right.$$
$$\left. -(t_s^- - t_p^- \cos\vartheta_0)(\hat{p}_x\cos 2\phi + \hat{p}_y \sin 2\phi)\right], \tag{11.7.34}$$
$$\mathcal{E}_{y,d} = -\frac{1}{2\sqrt{\cos\vartheta_{J+1}}}\left[\hat{p}_y(t_s^- + t_p^- \cos\vartheta_0) - 2\hat{p}_z t_p^- \sin\vartheta_0\sin\phi \right.$$
$$\left. -(t_s^- - t_p^- \cos\vartheta_0)(\hat{p}_x\sin 2\phi - \hat{p}_y \cos 2\phi)\right],$$
$$\mathcal{E}_{z,d} = 0 .$$

The field in the focal plane of the detector lens can be obtained by substituting this field into the Debye–Wolf integral – a calculation that we have done many times before in this chapter.

11.8 Imaging of Arbitrary Objects

Thus far we have been discussing computational modelling of high-aperture optical systems imaging a point scatterer. The reason for choosing this object was that it helped us to decouple the illumination and detection parts of the optical system. Had we not been able to call upon this trick, we would have ended up with formulas that are far too complicated for analytic work. When imaging extended objects it is usually not possible to make much headway using an analytic approach. Although it is possible to model optical systems fully numerically, commercially available programs use fairly primitive methods of simulation. The main disadvantage of fully numerical approaches, apart from their lack of accuracy and refinement, is that they do not allow understanding of the fine details of the optical system which in turn hinders our ability to understand the underlying physics. There are, however, exceptions which allow for an analytic treatment of an extended object. First, it is possible to treat a self-luminous extended object, such as an incandescent light bulb, as a collection of uncorrelated dipoles emitting in random directions. The image of light emitted by the individual dipoles can be readily calculated using the formalism presented above and the overall image can then be computed by adding the images of individual dipoles incoherently. Another example is that of extended fluorescent objects which are of course not self-luminous. A fluorescent object can be represented as a collection of dipoles that each get excited by the electromagnetic field they experience at their location and re-emit at a different wavelength by one of the three processes (or their combination) discussed above. Since the wavelength of emission is different from that of the excitation, neighbouring dipoles do not re-excite one another, so again the image can be computed by incoherently summing up the intensity images due to the individual dipoles.

The condition that dipoles should not re-excite one another is the key as to whether analytic approaches are possible. For example, light scattered by an extended object can be modelled by means of assuming that the object is made up by a large number of dipole scatterers and taking into account dipole–dipole interactions in computing the overall scattered electromagnetic field. The reason why this approach could not be used in purely analytic calculations is that when we compute the image of an individual dipole we automatically ignore all but the far-field component of the electric field. In fact, as shown in Section 1.6.2, in general the electric field of an electric dipole is not transverse in spherical coordinates. Equation (11.5.15), which we used as the fundamental equation for analytic image computations, does not contain terms corresponding to the near zone which are necessary to describe dipole–dipole interactions.

We therefore require a different approach. As we said before, it is not difficult to compute light scattering by an extended object by means of rigorous electromagnetic methods, (e.g. FEM, FDTD, RCWA, etc., see below). However, the problem is that once such a method is involved, the scattered electromagnetic field is only available numerically which appears to make analytic calculations impossible. In order to solve this seemingly unsurmountable problem, we take the following approach. The object being imaged is delimited by a computational volume. The incident electromagnetic field produced by a high-aperture lens is computed on the surface of this volume. A rigorous electromagnetic solver is then used to compute the scattered field on the surface of the volume from which it propagates off in all space (see Fig. 11.43). In order to compute the field that is relevant for imaging in the given optical system we use the Stratton–Chu

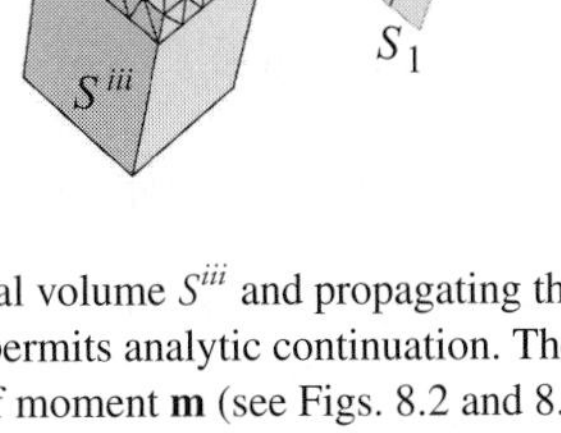

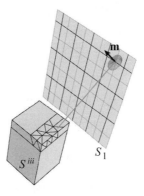

Figure 11.43: Resampling of the computational volume S^{iii} and propagating the field out to the transverse plane S_1 from which the equivalent magnetic dipole method permits analytic continuation. The field on S_1 can be represented as a coherent superposition of equivalent magnetic dipoles of moment $\mathbf{m}$ (see Figs. 8.2 and 8.3 for the role of S^{iii} in a diffraction problem).

theory (Eqs. (8.1.33) and (8.1.34)) where the surface S^{iii} corresponds to that of the computational volume and the field is calculated on a transverse plane S_1 close to the scatterer. We shall shortly show that the field on this plane can be converted by using a special form of the m-theory applied to a collection of dipole fields using the far-field approximation which then permits analytic continuation of the calculations. We now discuss each of these steps individually.

11.8.1 The Equivalent Dipole Method

We start by developing the only missing component needed for this work, which is showing that an arbitrary electric field on the plane S_1 can be represented by the coherent superposition of the electric fields emitted by magnetic dipoles. This follows from an application of the m-theory Eq. (8.1.92). With reference to Fig. 8.4 we have

$$\mathbf{E}(P) = \frac{1}{2\pi}\nabla_p \times \iint_{S_1} (\hat{\mathbf{v}} \times \mathbf{E}(Q))\frac{\exp(ikr)}{r} dS_1 \,, \tag{11.8.1}$$

where $\mathbf{r} = (x_p - x_Q, y_p - y_Q, z_p - z_Q)$ is the vector directed from Q, an arbitrary point on the surface S_1, to the observation point P. Performing the curl operator the kernel of the integral is given by:

$$\begin{aligned}\nabla_p \times \left[\hat{\mathbf{v}} \times \mathbf{E}(Q)\frac{\exp(ikr)}{r}\right] &= \frac{\exp(ikr)}{r}\left(\nabla_p \times (\hat{\mathbf{v}} \times \mathbf{E}(Q))\right) - (\hat{\mathbf{v}} \times \mathbf{E}(Q)) \times \nabla_p\left[\frac{\exp(ikr)}{r}\right] \\ &= -(\hat{\mathbf{v}} \times \mathbf{E}(Q)) \times \left[\frac{\exp(ikr)}{r}\left(\frac{ik}{r} - \frac{1}{r^2}\right)\mathbf{r}\right] .\end{aligned} \tag{11.8.2}$$

Since only the far field is required for imaging, the $1/r^2$ term is omitted from Eq. (11.8.2) yielding the final expression for the electric far field as:

$$\mathbf{E}(P) = \frac{ik}{2\pi}\iint_{S_1} (\hat{\mathbf{r}} \times (\hat{\mathbf{v}} \times \mathbf{E}(Q))\frac{\exp(ikr)}{r} dS_1 \,. \tag{11.8.3}$$

This reveals that each point Q can be considered as the source of a spherical wave with polarisation given by $\hat{\mathbf{r}} \times (\hat{\mathbf{v}} \times \mathbf{E}(Q))$. Note that since $\mathbf{r} \cdot (\mathbf{r} \times (\hat{\mathbf{v}} \times \mathbf{E}(Q))) = 0$, these waves have only field components tangential to the spherical wavefront as expected in the far field. Thus, Eq. (11.8.3) decomposes an arbitrary electric field into a coherent superposition of spherical waves. Interestingly, the spherical wave is equivalent to the electric field due to a harmonically oscillating magnetic dipole of moment $\mathbf{m} = \hat{\mathbf{v}} \times \mathbf{E}$ as shown in Fig. 11.43. Thus, in order to find the image of an arbitrary field we must first calculate the image of an equivalent magnetic-dipole (EMD). Then the image of an arbitrary field may be found by a coherent superposition of the images of EMDs as prescribed by Eq. (11.8.3).

11.8.2 The Image of a Magnetic Dipole

The derivation of the image of a magnetic dipole is identical to that of an electric dipole presented in Section 11.5, except that now $\boldsymbol{\mathcal{E}}_3 = -(U_{dm}/r)\hat{\mathbf{r}} \times \hat{\mathbf{m}}$ in Eq. (11.5.16). Noting $\hat{\mathbf{v}} = (0, 0, 1)$ we therefore have, with identical notation, from Eq. (11.5.34):

$$\begin{aligned}\mathcal{E}_{7,x} &= \frac{\sqrt{\cos\vartheta_l}}{2f_h\sqrt{\cos\theta_h}}\{\hat{m}_x\left[(\cos\vartheta_l + \cos\theta_h) + (\cos\vartheta_l - \cos\theta_h)\cos 2\phi\right] \\ &\qquad + \hat{m}_y(\cos\vartheta_l - \cos\theta_h)\sin 2\phi\}, \\ \mathcal{E}_{7,y} &= \frac{\sqrt{\cos\vartheta_l}}{2f_h\sqrt{\cos\theta_h}}\{\hat{m}_y\left[(\cos\theta_h + \cos\vartheta_l) + (\cos\theta_h - \cos\vartheta_l)\cos 2\phi\right] \\ &\qquad + \hat{m}_x(\cos\vartheta_l - \cos\theta_h)\sin 2\phi\}, \\ \mathcal{E}_{7,z} &= -\frac{\sqrt{\cos\vartheta_l}}{f_h\sqrt{\cos\theta_h}}\sin\vartheta_l\left(\hat{m}_x\cos\phi + \hat{m}_y\sin\phi\right) .\end{aligned} \tag{11.8.4}$$

The electric field $\mathbf{E}_7$ at the detector is therefore given by:

$$E_{7,x} = A\left[\hat{m}_x (I_0 + I_2 \cos 2\gamma) + \hat{m}_y I_2 \sin 2\gamma\right],$$
$$E_{7,y} = A\left[\hat{m}_y (I_0 - I_2 \cos 2\gamma) + \hat{m}_x I_2 \sin 2\gamma\right],$$
$$E_{7,z} = 2Ai\left(\hat{m}_x I_1 \cos\gamma + \hat{m}_y I_1 \sin\gamma\right), \quad (11.8.5)$$

where $A = U_0 k f_l/(2f_h)$ and

$$I_0 = \int_0^{\alpha_l} \sqrt{\frac{\cos\theta_l}{\cos\theta_h}} \sin\theta_l(\cos\theta_h + \cos\theta_l) J_0(kr_l \sin\theta_l) \exp[ik(z_l\cos\theta_l - z_h\cos\theta_h)]d\theta_l,$$

$$I_1 = \int_0^{\alpha_l} \sqrt{\frac{\cos\theta_l}{\cos\theta_h}} \sin^2\theta_l J_1(kr_l \sin\theta_l) \exp[ik(z_l\cos\theta_l - z_h\cos\theta_h)]d\theta_l, \quad (11.8.6)$$

$$I_2 = \int_0^{\alpha_l} \sqrt{\frac{\cos\theta_l}{\cos\theta_h}} \sin\theta_l(\cos\theta_h - \cos\theta_l) J_2(kr_l \sin\theta_l) \exp[ik(z_d\cos\theta_l - z_h\cos\theta_h)]d\theta_l .$$

In the above equations we have added the defocus term $z_h \cos\theta_h$ to account for the axial position of the plane S_1, which for technical reasons cannot be in the focal plane of the collector lens.

11.8.3 Interaction of Light with Specimen

As discussed above, we need to use rigorous numerical methods to solve electromagnetic scattering problems. Examples, some of which have been already discussed briefly in Chapter 1, are the finite element method (FEM) **[170]**, the method of moments (MOM) **[281]** and the FDTD method **[382]**. These methods have been in use since the 1960s and as a result are quite well understood. A number of methods have been developed more recently, including a method based upon the Green's tensor formalism **[240]**, the rigorous coupled-wave method (RCWM) **[250]** and the multiple multipole method (MMM) **[252]**. Lalanne *et al.* **[200]** have performed a detailed comparison of the most prominent numerical techniques in terms of accuracy. It is beyond the scope of this chapter to discuss the strengths and weaknesses of each method; we do, however, note that an advantage of our imaging model is that any rigorous method may be used to calculate the light scattered by the specimen.

We shall now briefly describe as an example the use of the FDTD method in conjunction with the model. As we have seen in Chapter 1, Maxwell's equations result in a set of coupled partial differential equations, and one example from this set, for a source-free region, is:

$$\frac{\partial E_x}{\partial t} = \frac{1}{\epsilon}\left[\frac{\partial H_z}{\partial y} - \frac{\partial H_y}{\partial z} - \sigma E_x\right]. \quad (11.8.7)$$

Most numerical methods require discretisation of field values throughout space. The FDTD method employs a discretisation reported by Yee **[382]**. The field quantities on the Yee cell are described by an indexing system of the form (i, j, k) which corresponds to a position $(i\Delta x, j\Delta y, k\Delta z)$ where Δx, Δy and Δz are the physical dimensions of the Yee cell.

The field values must also be discretised in time and they are calculated at intervals of a specifically chosen time step Δt. The electric and magnetic field values are, however, known half a time step apart. This allows an indexing scheme for time such that a time of index n refers to real time $n\Delta t$. Using this system, Yee showed how each partial differential equation of the form of Eq. (11.8.7) can be approximated to second-order accuracy by a difference equation of the form **[335]**:

$$E_x|^{n+1/2}_{i,j+1/2,k+1/2} = \alpha_{i,j+1/2,k+1/2} E_x|^{n-1/2}_{i,j+1/2,k+1/2} + \beta_{i,j+1/2,k+1/2} \left(\frac{H_z|^n_{i,j+1,k+1/2} - H_z|^n_{i,j,k+1/2}}{\Delta y} - \frac{H_y|^n_{i,j+1/2,k+1} - H_y|^n_{i,j+1/2,k}}{\Delta z}\right), \quad (11.8.8)$$

and similarly for other components of **E** and **H** (see also Subsection 1.12.6.2 for the complete three-dimensional finite-difference scheme for the field components). Note that $\alpha_{i,j+1/2,k+1/2}$ and $\beta_{i,j+1/2,k+1/2}$ are functions of Δt and material properties at location $(i, j + 1/2, k + 1/2)$ and that the superscripts $n + 1/2$, n and $n - 1/2$ are time indices. The set of difference equations allows an incident field to be introduced to the computational grid and the fields leap frogged in time. The incident field is introduced as a pulse with a Gaussian profile to limit its spectral width. The scattered field at the centre wavelength of the pulse may then be found from the time domain data through an application of a discrete Fourier transform. The FDTD algorithm must execute for sufficient iterations so that the scattered field decays to a negligible amplitude.

The incident field is introduced at a plane above the sample. This is an approximation, however, care is taken to ensure that the plane is sufficiently wide and the beam introduced adequately close to its waist such that the incident field is introduced accurately into the computational grid. This is usually described as the 'total-field/scattered-field' technique **[335]**. It would be more accurate to use the 'pure scattered field formulation' **[335]** where the incident field is calculated analytically, everywhere within the computational grid, at each time step.

The FDTD computational domain must be terminated by a boundary condition. A range of conditions has been proposed to model open region scattering problems **[22]**, **[92]**, **[253]**. Berenger's **[25]** perfectly matched layer (PML) is a good method of choice, which is a layer of absorbing material surrounding the scattering region which has a low coefficient of reflection with the interface between it and the scattering region. The low reflection coefficient allows the PML to simulate open region scattering on a finite-sized computational grid.

The primary limitation of the FDTD method is the type of objects which can be accurately modelled. For example, the Yee cell size must be no larger than the smallest feature to be modelled. This can lead to enormous memory requirements when modelling large objects with fine details.

11.8.4 The Use of the Stratton–Chu Integral to Resample the Field

Most rigorous numerical methods for calculating electromagnetic scattering calculate the scattered field on a dense grid. The FDTD method, for example, employs a grid spacing of no larger than $\lambda/20$. This means that the scattered field is calculated only in close proximity to the scattering object. This is because too much computer memory is required to include large quantities of homogeneous space within the computational domain. Our objective is, however, to calculate the image of the scattered light and so it is necessary to resample the numerical data rigorously onto a less dense grid with larger lateral dimensions (see Fig. 11.43).

Since the near-field data are known numerically, the Stratton–Chu integral must be evaluated numerically. The surface of integration and associated complex amplitudes are defined using a mesh of triangles. Such a mesh is represented by a set of vertices $\mathrm{V} = \left\{\mathbf{r}_{s,i} = \left(r_{s,i}^1, r_{s,i}^2, r_{s,i}^3\right) \in \mathbb{R}^3\right\}$, a set of facets $\mathrm{F} = \left\{\left(v_i^1, v_i^2, v_i^3\right) \in \mathbb{N}^3, 1 \le v_i^j \le N_v\right\}$ where N_v is the number of vertices, and two sets of complex amplitudes $\mathrm{E} = \left\{\mathbf{E}_i = \left(e_i^1, e_i^2, e_i^3\right) \in \mathbb{C}^3\right\}$ and $\mathrm{H} = \left\{\mathbf{H}_i = \left(h_i^1, h_i^2, h_i^3\right) \in \mathbb{C}^3\right\}$. Each element of each triplet in F is an index into V, E and H. In this way, each triangle is constructed from three vertices and the field at each vertex is also known. The orientation of the surface is stored according to the order in which the facet indices are stored. Such a representation minimises computer storage and provides an efficient way to traverse the surface. This representation can be used to represent any polyhedral surface and so is very general.

Integration is performed over each facet and the results summed to give the final result. Gaussian quadrature is commonly used for integration over a triangle. In general, a high-order Gaussian quadrature would be employed to improve integration accuracy. However, since the field is known only at the triangle vertices, only first-order Gaussian quadrature may be employed. The first-order scheme provides an exact result if the integral kernel varies no worse than a polynomial of first order **[10]**. This is why it is important to use a fine mesh to represent the closed surface of integration.

By employing first-order Gaussian quadrature integration, the Stratton–Chu integral may be evaluated according to:

$$\mathbf{U}(\mathbf{r}_p) = \sum_{i=1}^{N_{\text{facets}}} \left(\frac{1}{3}\sum_{j=1}^{3} \mathbf{I}\left(\mathbf{r}_p, \hat{\mathbf{v}}_i, \mathbf{r}_{s,v_i^j}, \mathbf{E}_{v_i^j}, \mathbf{H}_{v_i^j}\right)\right)\Delta_i\,, \tag{11.8.9}$$

where $\mathbf{U}$ is the field of interest, N_{facets} is the number of facets, $\mathbf{I}$ is the kernel of the integral being evaluated, $\mathbf{r}_p = (x_p, y_p, z_p)$ is the observation point, $\hat{\mathbf{v}}_i$ is the surface normal of facet i, $\mathbf{r}_{s,v_i^j}$ is the jth vertex of facet i, $\mathbf{E}_{v_i^j}$ is the complex electric field at the jth vertex of facet i, $\mathbf{H}_{v_i^j}$ is the complex magnetic field at the jth vertex of facet i, and Δ_i is the area of facet i.

11.8.5 Application Example

The method described in this section is immensely powerful and can be used in a variety of ways. For simple optical systems, such as the optical path of a Blu-ray disc reader, the model provides a wealth of information as shown in Fig. 11.44. The image on the right was taken as a frame from a long animation that was created to understand certain

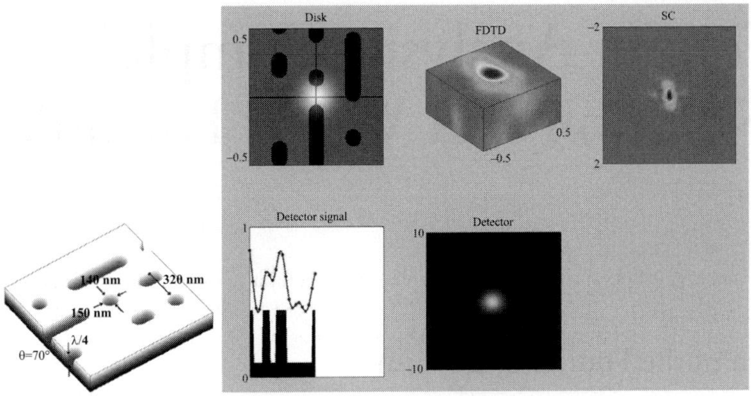

Figure 11.44: Results of simulations of a Blu-ray disc (BD) reader. The image on the left shows the dimensions of BD pits and tracks. The image on the right, which is taken from an animation showing the disc surface moving in front of the focused light distribution, provides details of the model's inner components: top left: the surface of the disk and the focused illumination superimposed over it, top middle: electric field distribution on the surface S^{iii} of the computational volume, top right: electric field distribution on the transverse plane S_1, bottom left: the detector signal as a function of track position and the cross-section of the pits along the vertical central line of the disc surface, bottom middle: the intensity distribution in the detector plane.

features of the Blu-ray disc (BD) detector signal measurements that could not be explained using scalar modelling. In the figure the top left subfigure shows the surface of the disc and the focused illumination superimposed over it, the top middle subfigure shows the electric field distribution on the surface S^{iii} of the computational volume computed using the FDTD method (Section 11.8.3), the top right is electric field distribution on the transverse plane S_1 (Section 11.8.4). The bottom left figure shows the detector signal as a function of track position and the bottom middle figure is the intensity distribution in the detector plane as computed from the equivalent magnetic dipole method (Sections. 11.8.1 and 11.8.2). Computations of this sort can be trivially parallelised and thus on a larger cluster of computers it is not a difficult task to produce long animations.

A Fourier Analysis, Complex Notation and Vector Formulas

A.1 Definition of the Fourier Transform

The complex notation of time-dependent signals is frequently used to convey the amplitude *and* phase of a signal in a convenient way, especially in the case of time-harmonic signals. The complex notation is based on the general Fourier relationship between a time-dependent signal and its frequency spectrum. In this Appendix A we apply the complex notation to a space-time signal $f(\mathbf{r}, t)$ of which the temporal Fourier transform F is given by

$$F(\mathbf{r}, \omega) = \int_{-\infty}^{+\infty} f(\mathbf{r}, t) \exp(-i\omega t)\, dt \,. \tag{A.1.1}$$

The inverse transform is defined by

$$f(\mathbf{r}, t) = \frac{1}{2\pi} \int_{-\infty}^{+\infty} F(\mathbf{r}, \omega) \exp(+i\omega t)\, d\omega \,. \tag{A.1.2}$$

In both equations we have used the circular frequency ω in units of rad s^{-1}. For a real signal $f(\mathbf{r}, t)$ in the time domain, we have the relationship $F(\mathbf{r}, \omega) = F^*(\mathbf{r}, -\omega)$.

Other definitions of the forward and inverse transformation are found in the literature. Differences arise from the treatment of the factor $1/2\pi$ which is needed to recover the initial signal after a forward and an inverse transformation. Another definition than the one used in Eqs. (A.1.1) and (A.1.2) inserts a factor $(2\pi)^{-1/2}$ in both the forward and the inverse transformation. The 2π-ambiguity in the definition can be avoided by choosing the time frequency ν in units of s^{-1} instead of the circular frequency $\omega = 2\pi\nu$. The forward Fourier transformation in terms of the frequency ν is given by

$$F(\mathbf{r}, \nu) = \int_{-\infty}^{+\infty} f(\mathbf{r}, t) \exp(-i2\pi\nu t) dt \,, \tag{A.1.3}$$

the inverse transform is now defined by

$$f(\mathbf{r}, t) = \int_{-\infty}^{+\infty} F(\mathbf{r}, \nu) \exp(+i2\pi\nu t) d\nu \,. \tag{A.1.4}$$

A second difference in definition may arise from the sign of the argument of the complex exponential time function in both transformations. For a plane wave propagating in the positive z-direction it is permissible to write the associated complex exponential function as $\exp\{i(\omega t - kz)\}$ or $\exp\{i(kz - \omega t)\}$. In this book, the second option has been chosen on the basis of our choice for the sign of the spatial part of the exponential propagation function of an elementary plane wave (see also the argument on this subject in Subsection A.8 of this Appendix). The plane wave function is an important basis function in wave analysis. It is used to represent a general wave function by means of an *angular spectrum* of plane waves. The relation between the general wave function and the spectrum of plane waves is a Fourier transformation.

A.2 Periodic Signal

If a spatio-temporal signal $f(\mathbf{r}, t)$ is periodic (period T), we can write the signal as a series with an infinitely large number of terms,

$$f(\mathbf{r}, t) = \sum_{n=-\infty}^{+\infty} f_p(\mathbf{r}, t - nT) , \tag{A.2.1}$$

where $f_p(\mathbf{r}, t - nT)$ is defined on the temporal interval $-T/2 \leq t \leq +T/2$ and is zero elsewhere. We take the Fourier transform of the periodic function according to Eq. (A.1.1) and obtain

$$\begin{aligned} F(\mathbf{r}, \omega) &= \sum_{n=-\infty}^{+\infty} \int_{-\infty}^{+\infty} f_p(\mathbf{r}, t - nT) \exp(-i\omega t)\, dt \\ &= \sum_{n=-\infty}^{+\infty} \int_{-T/2}^{+T/2} \exp(-in\omega T) f_p(\mathbf{r}, t) \exp(-i\omega t)\, dt , \end{aligned} \tag{A.2.2}$$

where the integration limits $[-\infty, +\infty]$ have been replaced by $[-T/2, +T/2]$ by virtue of the limited extent of the function $f_p(\mathbf{r}, t)$ in the time domain. The inverse of $F(\mathbf{r}, \omega)$ is then given by

$$f(\mathbf{r}, t) = \frac{1}{2\pi} \int_{-\infty}^{+\infty} \sum_{n=-\infty}^{+\infty} \int_{-T/2}^{+T/2} \exp(-in\omega T) f_p(\mathbf{r}, t') \exp\{-i\omega(t' - t)\}\, dt'\, d\omega . \tag{A.2.3}$$

The infinite series in the integral above can be written in terms of an array of delta-functions (Dirac's comb) using

$$\sum_{n=-\infty}^{+\infty} \exp\{in(2\pi x)\} = \sum_{n=-\infty}^{+\infty} \delta(x - n) . \tag{A.2.4}$$

In Fig. A.1 we illustrate the creation of the array of delta-functions by plotting the sum of the real part of the exponential functions in Eq. (A.2.4) for maximum values N of the running index n equal to 10 and 100.

Using the scaling property of the δ-function

$$\delta(ax) = \frac{\delta(x)}{|a|} , \tag{A.2.5}$$

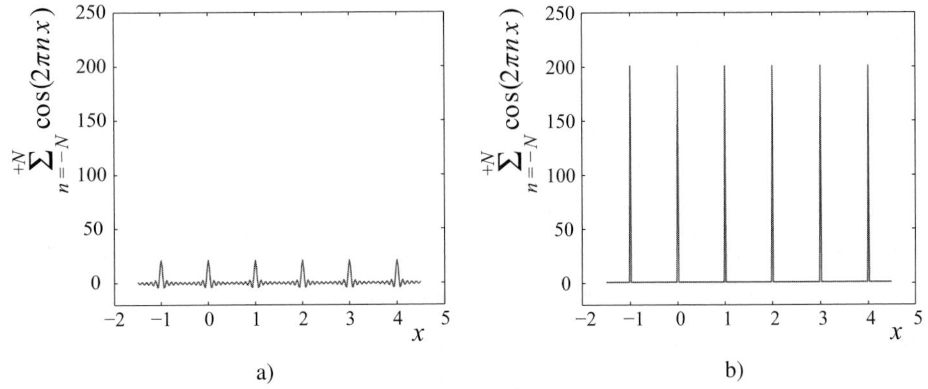

Figure A.1: The function $\sum_{n=-N}^{+N} \cos(2\pi n x)$ and the creation of the Dirac comb for $N \to \infty$.
a) $N = 10$; b) $N = 100$.

we rewrite Eq. (A.2.3) as

$$
\begin{aligned}
f(\mathbf{r},t) &= \frac{1}{2\pi}\int_{-\infty}^{+\infty}\sum_{n=-\infty}^{+\infty}\delta\left(\frac{T}{2\pi}\left[\omega-\frac{2\pi n}{T}\right]\right)\int_{-T/2}^{+T/2} f_p(\mathbf{r},t')\exp\{-i\omega(t'-t)\}\,dt'\,d\omega \\
&= \frac{1}{T}\sum_{n=-\infty}^{+\infty}\exp\left(+i\frac{2\pi nt}{T}\right)\int_{-T/2}^{+T/2} f_p(\mathbf{r},t')\exp\left(-i\frac{2\pi nt'}{T}\right)dt' \\
&= \sum_{n=-\infty}^{+\infty} F_{p,n}\exp\left(+i\frac{2\pi nt}{T}\right), \qquad \text{(A.2.6)}
\end{aligned}
$$

where we have used the sifting property of the delta-function according to

$$\int_{-\infty}^{+\infty} f(x)\,\delta(x-a)dx = f(a)\,. \qquad \text{(A.2.7)}$$

The periodic function $f(\mathbf{r},t)$ of Eq. (A.2.6) is now written as a Fourier series with the Fourier coefficients $F_{p,n}$ given by

$$F_{p,n} = \frac{1}{T}\int_{-T/2}^{+T/2} f_p(\mathbf{r},t)\exp\left(-i\frac{2\pi nt}{T}\right)dt\,. \qquad \text{(A.2.8)}$$

A.3 Harmonic Signal

In the optical domain of the electromagnetic spectrum, the signals often have a relatively narrow bandwidth and a representation of them by means of a harmonic signal is a valid approximation. In this case, the Fourier series of Eq. (A.2.6) is limited to the three terms that satisfy $|n| \leq 1$. We omit the background component with $n = 0$ of the harmonic signal and represent it by the expression $f(\mathbf{r},t) = \mathbf{a}(\mathbf{r})\cos[\phi(\mathbf{r}) - \omega_0 t]$. With the aid of Eqs. (A.2.6) and (A.2.8) we obtain

$$
\begin{aligned}
f(\mathbf{r},t) &= F_{p,-1}\exp(-i\omega_0 t) + F_{p,+1}\exp(+i\omega_0 t) \\
&= \frac{\mathbf{a}(\mathbf{r})}{2}\left(\exp\{+i[\phi(\mathbf{r})-\omega_0 t]\} + \exp\{-i[\phi(\mathbf{r})-\omega_0 t]\}\right) \\
&= \mathbf{a}(\mathbf{r})\cos[\phi(\mathbf{r})-\omega_0 t]\,, \qquad \text{(A.3.1)}
\end{aligned}
$$

where we have defined the basic frequency ω_0 by the quantity $2\pi/T$. We now define the complex amplitude $\mathbf{A}(\mathbf{r})$ of the harmonic signal by

$$\mathbf{A}(\mathbf{r}) = 2F_{p,-1} = 2F^*_{p,+1} = \mathbf{a}(\mathbf{r})\exp[+i\phi(\mathbf{r})]\,. \qquad \text{(A.3.2)}$$

The complex amplitude is a time-independent quantity and, if needed in the calculation, we can recover the harmonic time-dependent signal at circular frequency ω_0 by the operation

$$
\begin{aligned}
f(\mathbf{r},t) &= F_{p,+1}\exp(i\omega_0 t) + F_{p,-1}\exp(-i\omega_0 t) \\
&= F_{p,-1}\exp(-i\omega_0 t) + \left\{F_{p,-1}\exp(-i\omega_0 t)\right\}^* \\
&= \Re\left\{\mathbf{A}(\mathbf{r})\exp(-i\omega_0 t)\right\} = \mathbf{a}(\mathbf{r})\Re\left\{\exp[i(\phi(\mathbf{r})-\omega_0 t]\right\} \\
&= \mathbf{a}(\mathbf{r})\cos[\phi(\mathbf{r})-\omega_0 t]\,. \qquad \text{(A.3.3)}
\end{aligned}
$$

We now consider the expression for the amplitude of a plane wave, having an amplitude $\mathbf{a}$ and propagation vector $\mathbf{k}$,

$$f(\mathbf{r},t) = \mathbf{a}\cos(\mathbf{k}\cdot\mathbf{r} + \phi - \omega_0 t)\,. \qquad \text{(A.3.4)}$$

Following the definition of Eq. (A.3.2), we have the complex amplitude

$$\mathbf{A}(\mathbf{r}) = \mathbf{a}\ \exp[i(\mathbf{k}\cdot\mathbf{r} + \phi)]\,. \qquad \text{(A.3.5)}$$

It is also customary to define a complex time signal according to

$$\mathbf{B}(\mathbf{r},t) = \mathbf{a}\ \exp[i(\mathbf{k}\cdot\mathbf{r} + \phi - \omega_0 t)]\,, \qquad \text{(A.3.6)}$$

where the time-dependent signal is obtained by taking the real part of this expression.

A.4 Discrete Fourier Transform of a Band-limited Signal

In this subsection we describe a numerically stable method of computation for the Fourier transform of a band-limited function. The analysis is based on the treatment given in **[167]**. A band-limited function has a finite number of nonzero Fourier coefficients, up to a maximum coefficient number m. As an example of a band-limited function we consider the product of two higher-order harmonic signals ($f(\theta) = \cos^n\theta \sin^m\theta$). We assume that a function $f(\theta)$ is 2π-periodic and that it is an integrable function of θ, with Fourier series

$$f(\theta) = \sum_{m_2=-\infty}^{+\infty} F_{p,m_2} \exp(+im_2\theta) \,. \tag{A.4.1}$$

Instead of computing the Fourier coefficients of the periodic function f by means of the Fourier integral of Eq. (A.2.8),

$$F_{p,m_1} = \frac{1}{2\pi}\int_0^{2\pi} f(\theta) \exp(-im_1\theta)\, d\theta \,, \quad \text{integer } m_1 \,, \tag{A.4.2}$$

we approximate the integral by a discrete sum of integrand values, sampled at equidistant θ values, and we examine under what circumstances this 'discrete' result yields the exact value of the integral and when it is an approximation of the integral value. For any V=1, 2, ⋯, the Fourier integral of Eq. (A.4.2) in its *discrete* form is given by,

$$\begin{aligned} F_{p,m_1} &= \frac{1}{V}\sum_{v=0}^{V-1} f\left(\frac{2\pi v}{V}\right)\exp\left(-2\pi i m_1 \frac{v}{V}\right) = \frac{1}{V}\sum_{v=0}^{V-1}\sum_{m_2=-\infty}^{+\infty} F_{p,m_2} \exp\left[-2\pi i(m_1 - m_2)\frac{v}{V}\right] \\ &= \sum_{r=-\infty}^{+\infty} F_{p,m_1-rV} \,, \end{aligned} \tag{A.4.3}$$

where we have used that for integer t

$$\sum_{v=0}^{V-1} \exp\left(-2\pi i t\frac{v}{V}\right) = \begin{cases} V \,, & t \text{ is a multiple of } V \\ 0 \,, & \text{otherwise} \,. \end{cases} \tag{A.4.4}$$

It follows from Eq. (A.4.3) that the discrete summation provides the exact value of a general Fourier coefficient F_{p,m_1} if all coefficients F_{p,m_1-rV} in the summation on the right-hand side are zero for $r \neq 0$. We denote the maximum order of the Fourier expansion of $f(\theta)$ by M_1, which means that all Fourier coefficients F_{p,m_1} are zero if the absolute value of the index $|m_1|$ is larger than M_1. The summation over r in Eq. (A.4.3) then yields the value F_{p,m_1} for any $|m_1| \leq M_1$ if we choose V such that $V \geq 2M_1 + 1$. In the case that

$$f(\theta) = \cos^n\theta \sin^m\theta \,, \tag{A.4.5}$$

we have that $F_{p,m_2} = 0$ in Eq. (A.4.1) if $|m_2| > n + m$. Therefore, when m_1 is an integer with $|m_1| \leq n + m$ and $V > 2(n + m)$, the series on the last line of Eq. (A.4.3) has only one nonzero term, viz. the term with $r = 0$. Hence we then have

$$F_{p,m_1} = \frac{1}{V}\sum_{v=0}^{V-1} f\left(2\pi\frac{v}{V}\right)\exp\left(-2\pi i m_1\frac{v}{V}\right) . \tag{A.4.6}$$

Now also note that $f(\theta)$ in Eq. (A.4.5) is real, and so $F_{p,-m_1} = F_{p,m_1}$ for integer m_1.

With reference to the problem described by Eq. (5.9.12), we state that all required numbers $I_{nm}^{m_1}$ that have to be evaluated can be obtained in discrete form with the aid of the expression

$$\begin{aligned} I_{nm}^{m_1} &= \int_0^{2\pi} \cos^n\theta \sin^m\theta \exp(-im_1\theta) d\theta = 2\pi F_{p,m_1} \\ &= \frac{2\pi}{V}\sum_{v=0}^{V-1} \cos^n\left(2\pi\frac{v}{V}\right)\sin^m\left(2\pi\frac{v}{V}\right)\exp\left(-2\pi i m_1\frac{v}{V}\right) , \end{aligned} \tag{A.4.7}$$

for $m_1 = 0, 1, \cdots, n+m, n+m+1, \cdots, V-1$. Equation (A.4.7) has the form of a discrete Fourier transform (DFT) on V points applied to the function $f(\theta)$ of Eq. (A.4.5), sampled at $\theta = 2\pi v/V$, $v = 0, 1, \cdots, V-1$. The DFT formula has a fast implementation (FFT) in which all quantities $I_{nm}^{m_1}$, $0 \leq m_1 \leq n+m$ for a given n and m are computed simultaneously, using only $O[V(\ln V)]$ operations and with very favourable round-off error propagation. The approach of evaluating azimuthal integrals using the *DFT* can also be used for the double integral in Eq. (5.9.14). Nonzero solutions require that $m_1 = -m_2$, yielding the expression

$$2\pi\, \delta_{m_1+m_2} \int_0^{2\pi} \cos^m\theta \exp(-im_1\theta)\, d\theta\,. \tag{A.4.8}$$

The remaining integral is of the form of (A.4.7).

A.5 The Discrete Fourier and Laplace Transforms and the Chirp z-transform

In this section we reintroduce the discrete Fourier transform and the closely related discrete Laplace transform in a slightly broader context, by means of a representation of the frequency variable and the elementary transformation functions in the complex plane. This broader context is then exploited to introduce the discrete Chirp z-transform **[285]**. This transform has a greater flexibility with respect to the choice of the sample points in the time (or spatial) domain and in the frequency domain without losing the FFT computational advantage. In optical diffraction problems, despite the slightly larger computation burden with respect to the standard FFT approach, the Chirp z-transform permits a higher sampling density in the image plane. This is an important advantage since the standard FFT generally provides only a small number of sampling data in the region where the largest intensity in the diffraction image is measured. From the standard FFT, more densely sampled intermediate points have to be obtained by interpolation, which is a tedious operation as the basic interpolating diffraction function is a slowly converging, oscillating function.

A.5.1 The Discrete Fourier and Laplace Transform

To introduce the broader framework for the transforms discussed in this section, we start with the expression for the Fourier coefficients following from the discrete Fourier transform according to Eq. (A.4.6). We formally write this equation as

$$F(z) = \frac{1}{V}\sum_{v=0}^{V-1} f_v\, z^{-v}\,, \tag{A.5.1}$$

where the summation is limited to a finite set V of nonzero sampled values of the function f. The factor $1/V$ has been used here in the 'forward' transformation step. Other conventions attribute this factor to the backward transformation step, or, alternatively, use a factor $V^{-1/2}$ in both steps. We use the factor $1/V$ in the forward transformation, as was also done in the expression for the discrete Fourier transform of the preceding section.

We assume that the sampled values f_v, for instance, taken in the time domain, are equally spaced by a distance T and can be represented by the sequence

$$f(t) = \sum_{v=0}^{V-1} f_v\, \delta(t - vT)\,. \tag{A.5.2}$$

A coefficient F_m following from the (discrete) Laplace transformation of the sequence is then given by

$$F_m = F(z_m) = \frac{1}{V}\sum_{v=0}^{V-1} f_v\, z_m^{-v} = \frac{1}{V}\sum_{v=0}^{V-1} f_v \exp(-vs_m T)\,, \qquad m = 0, 1, \cdots, V-1\,, \tag{A.5.3}$$

where the transforming basic functions are given by

$$z_m = \exp(s_m T)\,. \tag{A.5.4}$$

In Eqs. (A.5.3) and (A.5.4), the quantity s is a complex number defined by $s = a + i\omega$ (s is the complex frequency associated with the Laplace transform).

We represent the (sampled) complex values of s and z in the complex plane. To this purpose we write

$$s_m = a_0 + i\omega_0 + m\,[\Delta a + i\Delta\omega]\,. \tag{A.5.5}$$

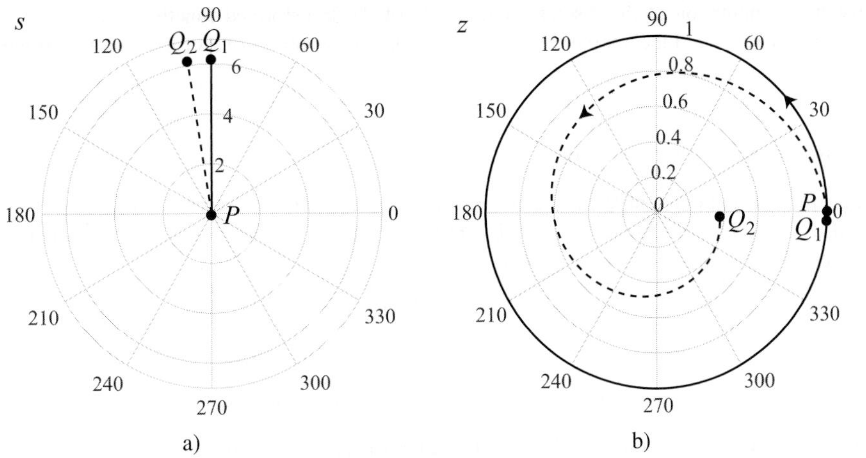

Figure A.2: a) The complex variable $sT = (a + i\omega)T$ in the complex plane.
b) Polar plot of the function $z = \exp(sT)$ in the complex plane.
Solid curves: $a = 0$, dashed curves: $a = -1$. P is the starting point of a contour, Q the endpoint.
$V = 100$, $\Delta\omega = 2\pi/(VT)$, $T = 1$, $a_0 = \omega_0 = 0$.

The contour of s in the complex plane is a straight line. The starting point of the contour is given by (a_0, ω_0) and $\Delta\omega/\Delta a$ determines the slope of the straight line. The sampled values z_m of the function $z = \exp(s_m T)$ are given by

$$z_m = \exp\{(a_0 + i\omega_0)T\}\exp\{m(\Delta a + i\Delta\omega)\ T\}\,. \tag{A.5.6}$$

The increments Δa and $\Delta\omega$ between two neighbouring sample points are chosen such that

$$\Delta a = \frac{1}{V}\frac{a}{T}, \qquad \Delta\omega = \frac{1}{V}\frac{2\pi}{T}\,. \tag{A.5.7}$$

Using these values for the increments of a and ω we write for the z-sampling points

$$z_m = \exp\{(a_0 + i\omega_0)\,T\}\exp\left\{(a + i2\pi)\,\frac{m}{V}\right\}, \quad m = 0, 1, \cdots, V-1\,. \tag{A.5.8}$$

The contours of s and z in the complex plane have been plotted in Fig. A.2 for two cases:

- $a = 0$, $\Delta\omega = 2\pi/T$ $(a_0 = \omega_0 = 0)$.
  The (solid) s-contour coincides with the portion PQ_1 of the imaginary axis. The z-contour is the unit circle, sampled from the starting point $z = 1$ at P to the endpoint Q_1 in the close neighbourhood of P. The equidistant sample points on the s and z contour correspond to those of a discrete Fourier transform and the basic z-function is given by

$$z_m = \exp\left(i2\pi\,\frac{m}{V}\right). \tag{A.5.9}$$

  In line with Eq. (A.5.3), discrete Fourier coefficients F_m of the sampled set of data points f_v are given by

$$F_m = \frac{1}{V}\sum_{v=0}^{V-1} f_v \exp\left(-i2\pi m\,\frac{v}{V}\right), \qquad m = 0, 1, \cdots, V-1\,, \tag{A.5.10}$$

  a result that was given in the previous section on the discrete Fourier transform (see Eq. (A.4.6)).
- $a = -1$, $\omega = 2\pi/T$ $(a_0 = \omega_0 = 0)$.
  The sampled s-points are found on the straight line PQ_2 and the z-sample points on the inward-spiralling contour PQ_2. The function z_m is given by Eq. (A.5.8) and the use of this z-function gives rise to a sampled Laplace transform with coefficients

$$F_m = \frac{1}{V}\sum_{v=0}^{V-1} f_v \exp\left\{-m(-1 + i2\pi)\,\frac{v}{V}\right), \qquad m = 0, 1, \cdots, V-1 \tag{A.5.11}$$

We note that the computation of the discrete coefficients of the transformed function only profits from the speed advantage of an FFT scheme if the number of samples V is an integer power of 2 or if V can be written as the product of two (or more) prime numbers.

A.5.2 The Discrete Chirp z-transform

The Chirp z-transform, commonly abbreviated as CZT, uses a more general transformation from the s- to the z-variable and was described for the first time by Rabiner *et al.* **[285]**. In what follows, we largely use their analysis. The sampled points of the function z are written in a more general manner as

$$z_m = A\,W^{-m}, \qquad m = 0, 1, \cdots, M-1\,, \tag{A.5.12}$$

where M is a positive integer, not necessarily equal to V, and A and W are general complex numbers defined by

$$A = A_0 \exp(i2\pi\theta_0)\,, \qquad W = W_0 \exp(i2\pi\phi_0)\,. \tag{A.5.13}$$

The corresponding s values follow from Eq. (A.5.4). The values of s_0 and of a general point s_m are given by

$$s_m = \frac{\ln A_0 + 2\pi i\theta_0 - m\,(\ln W_0 + 2\pi i\phi_0)}{T}\,, \qquad m = 0, 1, \cdots, M-1. \tag{A.5.14}$$

The contours created in the complex plane by the sampled values s_m of the transformation variable s and the sampled values $z_m = AW^{-m}$ are represented in Fig. A.3. It is important to note that the number of samples amounts to M. The azimuthal increments of the function z are given by $2\pi/V$, so that the total angular range of z amounts to $2\pi(M-1)/V$. This creates an extra degree of freedom for the CZT with respect to the discrete Fourier transform. Before discussing the consequences of this independent choice of M and V, we give the expression for the transformed samples F_m. According to Eq. (A.5.12) we have that

$$F_m = \frac{1}{V}\sum_{v=0}^{V-1} f_v\, z_m^{-v} = \frac{1}{v}\sum_{v=0}^{V-1} f_v\, A^{-v}\, W^{mv}\,, \qquad m = 0, 1, \cdots, M-1\,. \tag{A.5.15}$$

To enhance the speed of calculation of this expression, the power mv of the quantity W is written in a different way,

$$mv = \frac{m^2}{2} + \frac{v^2}{2} - \frac{(m-v)^2}{2}\,, \tag{A.5.16}$$

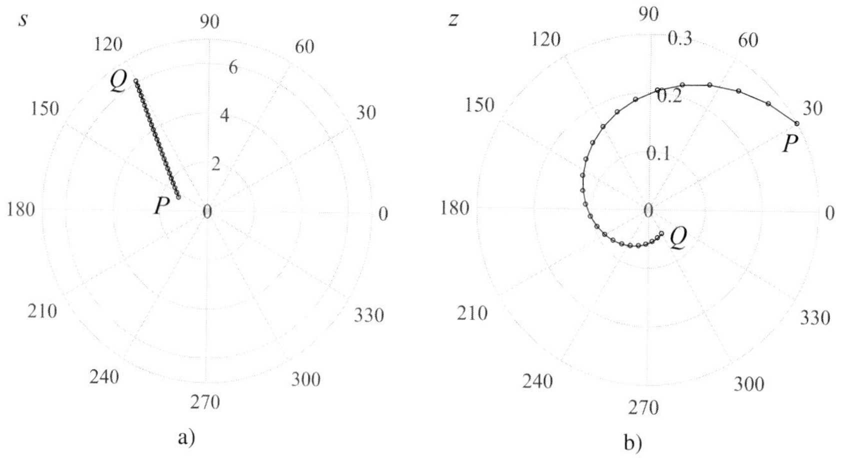

Figure A.3: a) Sampled points s_m of the complex variable s in the complex plane ($m = 0, 1, \cdots, M-1$). P is the starting point of the contour ($m = 0$) and Q the endpoint ($m = M-1$).
b) Polar plot of the sampled points $z_m = AW^{-m}$ in the complex plane.
$V = 32,\ M = 25,\ A_0 = 0.3,\ W_0 = 1.08,\ \theta_0 = 1/12,\ \phi_0 = -1/V, T = 1$.

a suggestion made by Bluestein [28]. Substitution of this expression for the product mv into Eq. (A.5.15) yields the result

$$F_m = \frac{1}{V} W^{m^2/2} \sum_{v=0}^{V-1} f_v A^{-v} W^{v^2/2} W^{-(m-v)^2/2}, \quad m = 0, 1, \cdots, M-1. \tag{A.5.17}$$

At first sight, the expression for F_m of Eq. (A.5.17) seems to be more complicated than the previous one of Eq. (A.5.15). The usefulness of Bluestein's substitution becomes evident if we execute the following substitutions,

- $y_v = f_v A^{-v} W^{v^2/2}$,
- $p_n = W^{-n^2/2}$,

yielding for the coefficients F_m the expression

$$F_m = \frac{1}{V} p_m^{-1} \sum_{v=0}^{V-1} y_v p_{m-v} = \frac{1}{V} p_m^{-1} q_m, \tag{A.5.18}$$

where we have used the shorthand q_m for the convolution-type product

$$q_m = \sum_{v=0}^{V-1} y_v p_{m-v}. \tag{A.5.19}$$

The coefficients q_m are obtained by performing discrete Fourier transformations of the sets of coefficients y and p, multiplying these two sets and then transforming back the result, according to the scheme

$$[q_m] = \mathcal{F}^{-1}\{\mathcal{F}\{[y]\} \cdot \mathcal{F}\{[p]\}\}. \tag{A.5.20}$$

The substitution of Eq. (A.5.16) thus enables us to write the CZT coefficients F_m as the convolution of two sets of coefficients $[y]$ and $[p]$ that are both obtained by means of simple complex multiplications. The convolution is preferably performed by means of numerical operations based on the FFT method. If the numbers of samples V and M satisfy the conditions for an FFT to be applicable, the computation time for the coefficients F_m can become very small, typically three times the speed of a single FFT computation. As the substitution of Eq. (A.5.16) is the essential step for the efficient computation of the Chirp z-transform, the algorithm is also known as 'Bluestein's fast Fourier transform'.

A.5.3 The Chirp z-transform and 'Spectral Zooming'

In this subsection we discuss an application of the CZT in optical diffraction problems which exploits the property of 'spectral zooming'. In the framework of the CZT, spectral zooming refers to the capability of the CZT algorithm to produce a selected set of frequency components that is unrelated to the standard Nyquist frequency set of $2M_1 + 1$ spectral components (maximum frequency is $M_1\omega$), obtained by means of a DFT or FFT from at least V temporal or spatial data points ($V \geq 2M_1 + 1$). The FFT frequency components are separated by the distance $\omega = 2\pi/T$ and cover the interval $[-M_1\omega, +M_1\omega]$.

For the case of a well-corrected (scalar) imaging system, the complex amplitude in the nominal image plane is the (two-dimensional) Fourier transform of the amplitude distribution in the exit pupil. The region in the image plane where the amplitude distribution is appreciable is sampled by only a few frequency components. The continuous amplitude distribution can be obtained by interpolation with the aid of the impulse response function of a uniformly transmitting circular exit pupil (the Jinc_0 function represented in Fig. 9.8). Interpolation with the aid of such an oscillating and slowly converging function is computationally slow. For that reason, it is advisable to increase the lateral extent of the diffraction image by reducing the extent of the exit pupil in the spatial domain. The standard approach is to use 'zero padding' in the spatial domain, such that the nonzero data points associated with the circular exit pupil cover only a minor part of the total rectangular spatial domain. The disadvantage of zero padding is the less precise sampling of the complex amplitude function in the exit pupil.

We now use the 'zooming' property of the CZT to solve the coarse sampling problem of a sharply focused diffraction image. To this purpose we use basic functions $z_m = AW^{-m}$ in which $|A| = 1$ and $|W| = 1$, so that the z-contour is found on the unit circle. The maximum frequency $\omega_{\max}$ on the unit circle is given by $V\omega$. The zoom range is given by the interval $[\omega_1, \omega_2]$ where $\omega_1, \omega_2 < \omega_{\max}$. The quantity W is then given by

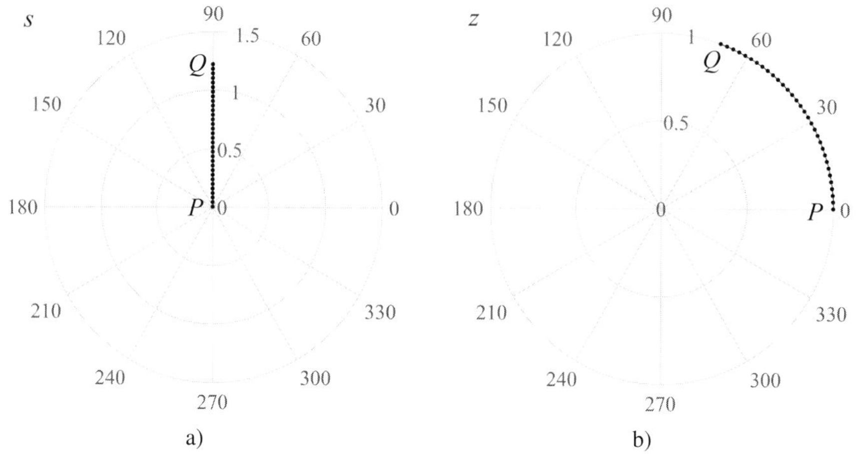

Figure A.4: a) Sampled points s_m of the complex variable s in the complex plane ($m = 0, 1, \cdots, M-1$). P is the starting point of the contour ($m = 0$) and Q the endpoint ($m = M-1$).
b) Polar plot of the sampled points $z_m = AW^{-m}$ on the unit circle.
$V = M = 32,\ T = 1$. W and A are given by Eqs. (A.5.21) and (A.5.22). $\omega_1 = 0$, $\omega_2 = \omega_{max}/5$.

$$W = \exp\left\{\frac{-i2\pi}{V}\frac{(\omega_2-\omega_1)}{\omega_{max}}\right\}, \tag{A.5.21}$$

and the factor A equals

$$A = \exp\left\{i2\pi\frac{\omega_1}{\omega_{max}}\right\}. \tag{A.5.22}$$

The range of frequencies covered by the CZT is given by

$$\omega_m = \omega_1 + \frac{m}{V}(\omega_2-\omega_1), \qquad m = 0, 1, \cdots, V-1. \tag{A.5.23}$$

A specific choice for the s- and z-contour is shown in Fig. A.4. The quantity s is purely imaginary as is the case for a Fourier transform but the s-values do not cover a total range of 2π as, for instance, does the line segment PQ_1 in Fig. A.2. The total range of s is such that the z-contour covers an analogue fraction PQ of the unit circle, $1/5$ of the total circumference of the unit circle in Fig. A.4. The total number of points M on this contour is 32, equal to V itself. Performing the CZT with these data yields a five times higher sampling density in the image plane, up to a five times smaller maximum frequency. By means of the phase θ_0 of the quantity A_0, the CZT can also be given an offset such that a zoomed-in, off-axis set of frequency coefficients F_m is computed.

A.6 Complex Amplitude and Quadratic Quantities

To obtain expressions for energy density, momentum, energy flow, etc., we need the correct expressions for these quantities in terms of the complex notation for the waves that give rise to them. We start by considering the product U of two time-dependent wave amplitudes given by

$$\begin{aligned}
\mathbf{B}_1(\mathbf{r},t) &= \mathbf{a}_1(\mathbf{r})\cos[\phi_1(\mathbf{r})-\omega_0 t] = \Re\{\mathbf{A}_1(\mathbf{r})\exp[-i\omega_0 t]\}\,,\\
\mathbf{B}_2(\mathbf{r},t) &= \mathbf{a}_2(\mathbf{r})\cos[\phi_2(\mathbf{r})-\omega_0 t] = \Re\{\mathbf{A}_2(\mathbf{r})\exp[-i\omega_0 t]\}\,,\\
U &= \mathbf{a}_1(\mathbf{r})\mathbf{a}_2(\mathbf{r})\cos[\phi_1(\mathbf{r})-\omega_0 t]\cos[\phi_2(\mathbf{r})-\omega_0 t]\\
&= \frac{1}{2}\mathbf{a}_1(\mathbf{r})\mathbf{a}_2(\mathbf{r})\{\cos[\phi_1(\mathbf{r})+\phi_2(\mathbf{r})-2\omega_0 t]+\cos[\phi_1(\mathbf{r})-\phi_2(\mathbf{r})]\}\,,
\end{aligned} \tag{A.6.1}$$

where $\mathbf{A}_1(\mathbf{r})$ and $\mathbf{A}_2(\mathbf{r})$ are the complex amplitudes of the waves.

Using the complex notation, the product of the two wave functions is given by

$$U = \mathbf{a}_1(\mathbf{r})\mathbf{a}_2(\mathbf{r})\Re\{\exp[i\{\phi_1(\mathbf{r}) - \omega_0 t\}]\}\Re\{\exp[i\{\phi_2(\mathbf{r}) - \omega_0 t\}]\} . \tag{A.6.2}$$

It is easily shown that the following relationship holds for complex quantities a and b,

$$\Re(a)\Re(b) = \frac{1}{2}\{\Re(ab) + \Re(ab^*)\} , \tag{A.6.3}$$

and the substitution of the complex amplitudes of the two waves in this equation yields the expression for U that was given in Eq. (A.6.1).

In general, at optical frequencies, we are mainly interested in the time-averaged value of the product U and this value is given by the second quantity on the right-hand side of Eq. (A.6.3). It is thus permissible to write

$$\langle U \rangle = \frac{1}{2}\{\mathbf{A}_1(\mathbf{r})\mathbf{A}_2^*(\mathbf{r})\} . \tag{A.6.4}$$

The average electric energy density of a wave is proportional to $< \mathbf{a}^2(\mathbf{r}) > /2$ and it follows that

$$\langle w_e \rangle \propto \frac{1}{4}\mathbf{A}(\mathbf{r})\mathbf{A}^*(\mathbf{r}) = \frac{1}{4}|\mathbf{A}(\mathbf{r})|^2 . \tag{A.6.5}$$

A.7 Operations Using Complex Amplitude

In Table A.1 we give the expressions for frequently occurring operations on an electromagnetic wave function, presented either as a time-dependent signal $\mathbf{E}(\mathbf{r}, t)$ or with the aid of its harmonic complex amplitude $\mathbf{A}(\mathbf{r})$ (time-frequency ω_0). We have:

general wave function,

$$\mathbf{E}(\mathbf{r}, t) = \Re\{\mathbf{A}(\mathbf{r})\exp(-i\omega_0 t)\} , \tag{A.7.1}$$

plane wave function,

$$\mathbf{E}(\mathbf{r}, t) = \Re\{[\mathbf{A}_0\exp(i\mathbf{k}\cdot\mathbf{r})]\exp(-i\omega_0 t)\} . \tag{A.7.2}$$

A.8 Properties of Fourier Transforms and Some Transform Pairs

The definition of the Fourier transform and its inverse according to Eqs. (A.1.1) and (A.1.2) is not the only possible one. The sign of the exponential term in both equations can be changed, the only requirement being that a sequence of forward and inverse transformations reproduces the original signal. In the case of optical signals, the basic function is the time-harmonic plane wave with a temporal frequency ω_0 and a spatial frequency k. In both cases the frequencies are angular frequencies and they appear in the following way in the expression of a plane wave:

$$f(\mathbf{r}, t) = \Re\{\mathbf{A}_0 \exp[i(-\omega_0 t + \mathbf{k}\cdot\mathbf{r})]\} , \tag{A.8.1}$$

where $\mathbf{A}_0$ is the complex amplitude of the plane wave. The plane wave can also be written as

$$f(\mathbf{r}, t) = \Re\{\mathbf{A}_0^* \exp[i(\omega_0 t - \mathbf{k}\cdot\mathbf{r})]\} . \tag{A.8.2}$$

Calculations with complex amplitude quantities based on either Eq. (A.8.1) or Eq. (A.8.2) yield complex conjugate results. Of course, the final time-dependent expressions are independent of the complex notation convention. With respect to the spatial dependence of the plane wave function $f(\mathbf{r}, t)$, we adhere to the notation of Eq. (A.8.1). It has the conceptual advantage that the spatial phase contribution increases in the direction of propagation of the plane wave. For optical signals, we thus define Fourier transform pairs in the temporal and spatial domain in line with the plane wave expression of Eq. (A.8.1), yielding

Table A.1. Frequently occurring operations on a time–space-dependent signal and its corresponding complex amplitude.

	$\mathbf{E}(\mathbf{r}, t)$	$\mathbf{A}(\mathbf{r})$ or $\mathbf{A}_0$
Differentiation (time)	$\frac{\partial \mathbf{E}}{\partial t}$	$-i\omega_0 \mathbf{A}(\mathbf{r})$
Differentiation (time) twice	$\frac{\partial^2 \mathbf{E}}{\partial t^2}$	$-\omega_0^2 \mathbf{A}(\mathbf{r})$
Differentiation (position x)	$\frac{\partial \mathbf{E}}{\partial x}$	$\frac{\partial \mathbf{A}(\mathbf{r})}{\partial x}$
Divergence	$\nabla \cdot \mathbf{E}$	$\nabla \cdot \mathbf{A}(\mathbf{r})$
Curl	$\nabla \times \mathbf{E}$	$\nabla \times \mathbf{A}(\mathbf{r})$
Divergence plane wave	$\nabla \cdot \mathbf{E}$	$i\mathbf{k} \cdot \mathbf{A}_0$
Curl plane wave	$\nabla \times \mathbf{E}$	$i\mathbf{k} \times \mathbf{A}_0$
Product of two wave functions	$\mathbf{E}_1\mathbf{E}_2$	$\frac{1}{2}(\{\mathbf{A}_1\mathbf{A}_2\} + \{\mathbf{A}_1\mathbf{A}_2^*\})$
Time-averaged product	$\langle \mathbf{E}_1\mathbf{E}_2 \rangle$	$\frac{1}{2}\{\mathbf{A}_1\mathbf{A}_2^*\}$
Time-averaged electric density	$\frac{1}{2}\langle (\mathbf{E})^2 \rangle$	$\frac{1}{4}\lvert\mathbf{A}\rvert^2$
Convolution[1] two wave functions	$\mathbf{E}_1 \otimes \mathbf{E}_2$	$\frac{1}{2}\mathbf{A}_1\mathbf{A}_2 \exp(i w_0)$

[1] The convolution $\mathbf{E}_1 \otimes \mathbf{E}_2$ of two real functions is defined by
$\lim_{T\to\infty} \frac{1}{T} \int_{-T/2}^{+T/2} \mathbf{E}_1(\mathbf{r}, t')\mathbf{E}_2(\mathbf{r}, t - t')\, dt'$.

$$F_t(\mathbf{r}, \omega) = \int_{-\infty}^{+\infty} f(\mathbf{r}, t) \exp(-i\omega t)\, dt \, , \tag{A.8.3}$$

$$f(\mathbf{r}, t) = \frac{1}{2\pi} \int_{-\infty}^{+\infty} F_t(\mathbf{r}, \omega) \exp(+i\omega t)\, d\omega \, , \tag{A.8.4}$$

$$F_r(\mathbf{k}, t) = \iiint_{-\infty}^{+\infty} f(\mathbf{r}, t) \exp(-i\mathbf{k} \cdot \mathbf{r})\, d\mathbf{r} \, , \tag{A.8.5}$$

$$f(\mathbf{r}, t) = \frac{1}{(2\pi)^3} \iiint_{-\infty}^{+\infty} F_r(\mathbf{k}, t) \exp(+i\mathbf{k} \cdot \mathbf{r})\, d\mathbf{k} \, . \tag{A.8.6}$$

In the kernel of the integral in Eq. (A.8.6), the spectral function F_r is multiplied with the complex exponential function associated with the elementary plane wave $\exp(+i\mathbf{k} \cdot \mathbf{r})$. This plane wave representation shows an increase in the phase of the wave along the direction of propagation. If the sign of the exponentials is reversed in the definition of the forward and inverse transformation, the relationship between the transforms with plus and minus sign in the exponential part of the kernel, F_r^+ and F_r^-, respectively, is given by $F_r^+(\mathbf{k}, t) = F_r^-(-\mathbf{k}, t)$.

To conclude this section on Fourier transforms and transform pairs we first give some general properties of Fourier transforms in Table A.2. Then follows, in Table A.3, a list of frequently occurring functions and their transforms.

Table A.2. General relationships between a function and its Fourier transform.

	Function	Fourier transform
Nominal functions	$f(t)$	$F(\omega)$
Shift	$f(t-t_0)$	$F(\omega)\exp(-i\omega t_0)$
Exponential term	$\exp(i\omega_0 t)f(t)$	$F(\omega-\omega_0)$
Sideband creation	$f(t)\cos(\omega_0 t+\phi)$	$\frac{1}{2}[F(\omega-\omega_0)\exp(+i\phi)$ $+F(\omega+\omega_0)\exp(-i\phi)]$
Scaling operation	$f(Mt)$	$(1/\lvert M\rvert)F(\omega/M)$
Duality Fourier pair	$F(t)$	$2\pi f(-\omega)$
Differentiation t	$\frac{d^n}{dt^n}f(t)$	$(i\omega)^n F(\omega)$
Differentiation ω	$t^n f(t)$	$i^n \frac{d^n}{d\omega^n}F(\omega)$
Integration	$\int_{-\infty}^{t} f(p)\,dp$	$(i\omega)^{-1}F(\omega)\quad (\int_{-\infty}^{+\infty} f(p)\,dp=0)$
Convolution (f,g)	$f\star g\big\rvert_t=\int_{-\infty}^{\infty} f(\tau)g(t-\tau)\,d\tau$	$F(\omega)\,G(\omega)$
	$f\star g^*\big\rvert_t=\int_{-\infty}^{\infty} f(\tau)g^*(t-\tau)\,d\tau$	$F(\omega)\,G^*(-\omega)$
Autocorrelation (R_{ff})	$f\star f_-^*\big\rvert_t=\int_{-\infty}^{\infty} f(\tau)f^*(\tau-t)\,d\tau$	$\lvert F(\omega)\rvert^2$
Multiplication	$f(t)g(t)$	$F\star G\big\rvert_\omega$

Table A.3. Fourier transform pairs.

	Function	Fourier transform
Delta-function (t)	$\delta(t-t_0)$	$\exp(-i\omega t_0)$
Complex exponential (ω_0)	$\exp(i\omega_0 t)$	$2\pi\delta(\omega-\omega_0)$
Real exponential	$\exp(-\lvert at\rvert)$	$2a/(a^2+\omega^2)$
Gaussian	$\exp[-\pi(t/a)^2]$	$a\exp[-(\omega a)^2/4\pi]$
Complex Gaussian (real $b>0$)	$\exp[\pm i\pi(t/b)^2]$	$b\exp\{\mp i[(\omega b)^2/(4\pi)-\pi/4]\}$
Window ($=0$ for $\lvert t\rvert\geq a/2$)	$\mathrm{rect}(t/a)$	$a\,\mathrm{sinc}(\omega a/2)$
Triangle ($=0$ for $\lvert t\rvert\geq a$)	$\Lambda(t/a)$	$a\,\mathrm{sinc}^2(\omega a/2)$
Periodic function	$\cos(\omega_0 t+\phi)$	$\pi[\delta(\omega-\omega_0)\exp(+i\phi)$ $+\delta(\omega+\omega_0)\exp(-i\phi)]$
Signum function	$\mathrm{sgn}(t)$	$\frac{2}{i\omega}$
Step function (Heaviside)	$H(t)=\frac{1}{2}+\frac{1}{2}\mathrm{sgn}(t)$	$\pi\delta(\omega)+\frac{1}{i\omega}$

We perform the transformation to the time-frequency domain according to Eqs. (A.8.3) and (A.8.4), using the variable pair t and ω.

A.9 Some Elementary Results from Fourier Optics

In this section we discuss use of the spatial Fourier transform in imaging optics. A first example is propagation of a wave in free space. The propagated wave in an axially shifted plane is (partly) described by means of a Fourier transformation of the initial field. The free-space wave propagation is then applied to the imaging by a lens. More specifically, we demonstrate that the imaging of an object can be described in terms of a convolution of the object function and the amplitude impulse response of the imaging system. It is seen that the use of normalised coordinates enhances the clarity of the transformation equations.

A.9.1 Free-space Wave Propagation and the Fourier Transform

In this subsection we first discuss propagation of a plane wave in free space, within the framework of scalar wave propagation. The strength of the Huygens' secondary waves follows from this scalar model for plane wave propagation. The second subject is propagation of a wave through a modulated thin object towards a possibly curved receiving surface.

A.9.1.1 Free-space propagation and the Huygens' Secondary Waves

In Fig. A.5 we illustrate propagation of a plane wave from one planar surface S_0 to a subsequent one, labelled S_1. Using the simplified scalar model for wave propagation based on Huygens' secondary waves, the complex amplitude $U_1(X_1, Y_1)$ in the plane S_1 is given by

$$\begin{aligned} U_1(X_1, Y_1) &= U_0 \iint_{-\infty}^{+\infty} C \exp\{ik_0\, P_0Q_1\}\, dx_0 dy_0 \\ &= \iint_{-\infty}^{+\infty} C \exp\left\{ik_0 \left[d^2 + (X_1 - x_0)^2 + (Y_1 - y_0)^2\right]^{1/2}\right\} dx_0 dy_0 . \end{aligned} \tag{A.9.1}$$

With respect to the coordinate choice, we use in general small lettering for the object-field or 'near-field' coordinates and capital letters for the coordinates in a plane where the diffracted field or 'far field' is found. The wave vector of the incident plane wave is parallel to the z-axis such that the amplitude of the plane wave in the plane S_0 is uniform and equal to U_0. The generally complex quantity C equals the strength of the Huygens' secondary waves. We assume that the secondary waves radiate isotropically and, as a consequence, the 'obliquity factor' is unity. The strength of the secondary waves can now be derived from the argument that the assembly of secondary waves associated with a plane wave should reproduce the plane wave amplitude after propagation over a certain distance d. In Fig. A.5 we show an isolated secondary spherical wave of the incident plane wave with its origin at P_0. The plane wave has a complex amplitude U_0. The amplitude in a distant plane S_1, perpendicular to the z-axis and at a distance d from the plane S_0 is the sum of the complex amplitudes

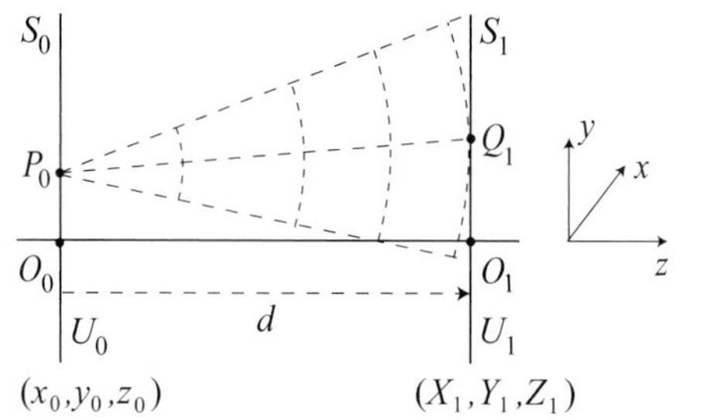

Figure A.5: Propagation of a plane wave in free space between two planes S_0 and S_1. The wave vector of the plane wave is parallel to the z-axis.

produced by the secondary waves at S_0. Written as a diffraction integral we have for the complex amplitude $U_1(X_1, Y_1)$ in the plane S_1 the expression

$$U_1(X_1, Y_1) = U_0 \exp(ik_0 d) \iint_{-\infty}^{+\infty} C \exp\left\{ i \frac{k_0}{2d} \left[(X_1 - x_0)^2 + (Y_1 - y_0)^2 \right] \right\} dx_0 dy_0 \,, \tag{A.9.2}$$

where we have used a first-order expansion of the square root function in the argument of the complex exponential function in Eq. (A.9.1). Such a truncated series expansion can be used if the axial displacement d is large with respect to the lateral dimension of the region in which the incident field has a non-negligible value. The insertion of infinitely large integration limits for x_0 and y_0 in Eqs. (A.9.1) and (A.9.2) is in conflict with the scalar wave propagation model in which propagation angles should remain small. In that sense, the value of $U_1(X_1, Y_1)$ according to Eq. (A.9.2) should be considered as a limit value.

The integrals with respect to x_0 and y_0 can be evaluated with the aid of the Fourier transform of a complex quadratic Gaussian function (see Table A.3). For zero-value of the transformed variable ω we find

$$U_1(X_1, Y_1) = C\, U_0 \exp(ik_0 d) \left\{ \sqrt{\lambda_0 d}\, \exp(+i\pi/4) \right\}^2 . \tag{A.9.3}$$

Equation (A.9.3) shows that the amplitude U_1 in the plane S_1 has a constant value. If we require that the plane wave amplitude U_1 is equal to $U_0 \exp(ik_0 d)$, the constant C is thus given by

$$C = \frac{1}{i \lambda_0 d} \,. \tag{A.9.4}$$

The factor $C = -i/(\lambda_0 d)$ can also be found in the expression of Eq. (8.4.36) for the amplitude of the point-spread function in the focal region of an imaging system (scalar approximation).

A.9.1.2 Free-space Propagation of a General Wave Field and the Fourier Transform

In this subsection we assume that the incident field at S_0 has unit amplitude and traverses an object structure with a transmission function $t_0(x_0, y_0)$. The geometry for the wave propagation is illustrated in Fig. A.6. The spatially modulated light field propagates to the plane S_1 and we apply Huygens' principle to this propagation process. The complex amplitude $U_0(x_0, y_0)$ at the exit side of the 'object' plane S_0 is given by $t_0(x_0, y_0)$. The expression for $U_1(X_1, Y_1)$ follows from Eq. (A.9.2) if we multiply the constant amplitude U_0 by the object function $t_0(x_0, y_0)$,

$$\begin{aligned} U_1(X_1, Y_1) = -\frac{i}{\lambda_0 d} \exp(ik_0 d) \exp\left\{ ik_0 \frac{\left(X_1^2 + Y_1^2\right)}{2d} \right\} \iint_{-\infty}^{+\infty} t_0(x_0, y_0) \\ \times \exp\left\{ ik_0 \frac{\left(x_0^2 + y_0^2\right)}{2d} \right\} \exp\left\{ -ik_0 \frac{(x_0 X_1 + y_0 Y_1)}{d} \right\} dx_0 dy_0 \,. \end{aligned} \tag{A.9.5}$$

The same result for $U_1(X_1, Y_1)$ follows from the Cartesian form of Eq. (8.4.36) when the substitutions $z_p = 0$ and $f = d$ are made. Formally, the integration interval is infinitely large. In practice the interval is limited by the finite extent of the object function t_0.

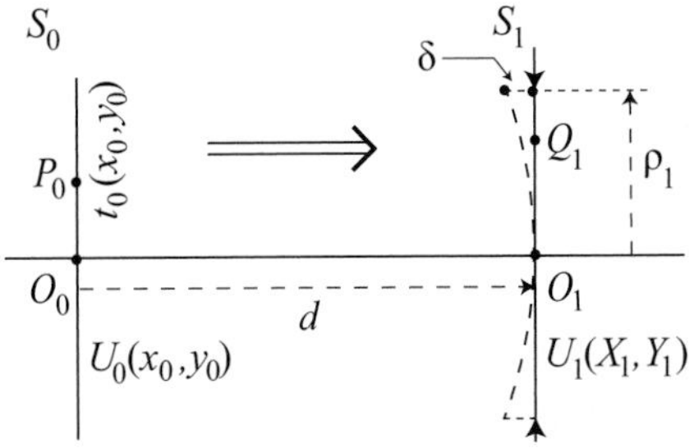

Figure A.6: Propagation of an optical field between two planar surfaces S_0 and S_1 in free space.

For reconstruction of the field U_0 from the propagated field U_1, we note that we use the same free-space propagation formalism with U_0 and U_1 interchanged and with the distance d replaced by $-d$. When we evaluate the integral expression for the 'inverse' propagation from surface S_1 to S_0, we effectively recover the object function $t_0(x_0, y_0)$. If the field at S_1 is reflected by means of a planar mirror and further propagated in the negative z-direction, the diffracted field at S_0 equals that of a forward diffracted field, measured at a distance $2d$.

To simplify the expression of Eq. (A.9.5) it is recommended to use normalised coordinates in the planes S_0 and S_1. To this purpose we have defined in Fig. A.6 a maximum transverse size at the plane S_1, equal to the distance $2\rho_1$. In an optical system this maximum diameter could be the useful diameter of a lens or the dimension of an object frame. With the aid of the lateral distance ρ_1 we define the normalised coordinates (x', y') and (X', Y') in the object or 'near' field and in the field where we measure the diffracted wave ('far' field),

$$(x', y') = \frac{\rho_1}{\lambda_0 d}(x, y), \qquad (X', Y') = \frac{1}{\rho_1}(X, Y). \tag{A.9.6}$$

The quantity $\lambda_0 d/\rho_1$ is called the diffraction unit associated with the object plane S_0. Insertion of the normalised coordinates in Eq. (A.9.5) yields the expression

$$U_1(X_1', Y_1') = -\frac{i}{N_F}\exp(ik_0 d)\exp\left\{+i\pi N_F\left(X_1'^2 + Y_1'^2\right)\right\}\iint_{-\infty}^{+\infty} t_0(x_0', y_0') \times \exp\left\{i\pi\left(x_0'^2 + y_0'^2\right)/N_F\right\}\exp\left\{-i2\pi\left(x_0' X_1' + y_0' Y_1'\right)\right\} dx_0' dy_0'. \tag{A.9.7}$$

In this expression we use the Fresnel number $N_F = \rho_1^2/(\lambda_0 d)$ (see Eq. (8.3.23)). In the problem at hand, N_F equals the number of Fresnel zones that can be constructed from the point O_0 on surface S_1 within a circular area with radius ρ_1. Geometrically, N_F equals the quadratically approximated distance δ in Fig. A.6, divided by half the wavelength λ_0 of the light. In many practical imaging problems, the value of the Fresnel number in object or image space is large, typically $N_F \geq 1000$. For that reason the exponential factor in the integrand depending on $(x_0'^2 + y_0'^2)$ can often be neglected. If this is the case we are left with an integral that is a spatial Fourier transformation (see Section A.8) of the object transmission function $t_0(x_0', y_0')$ with respect to the normalised far-field coordinates (X_1', Y_1'). The exponential factors in front of the integral influence the complex amplitude $U_1(X_1', Y_1')$. If we evaluate the diffracted intensity in the plane S_1, these phase factors are irrelevant.

A.9.2 Imaging by a Lens

In Fig. A.7 we show the subsequent wave propagation steps which are needed to calculate the complex amplitude in the image plane S_2 of a lens. There are three wave-propagation steps, two free-space propagation steps and the lens-traversal step. The free-space propagation steps, as shown in the figure, are from the planar surface S_0 to the spherical surface S_1 centred at O_1 and from a spherical surface S_1' with its centre of curvature at O_2 towards the image plane S_2. It is easily verified that the presence of a spherical surface with its centre at the other surface removes one of the two quadratic phase factors in Eq. (A.9.7). We assume that the planes S_0 and S_2 are paraxially conjugated planes. It is a property of paraxial

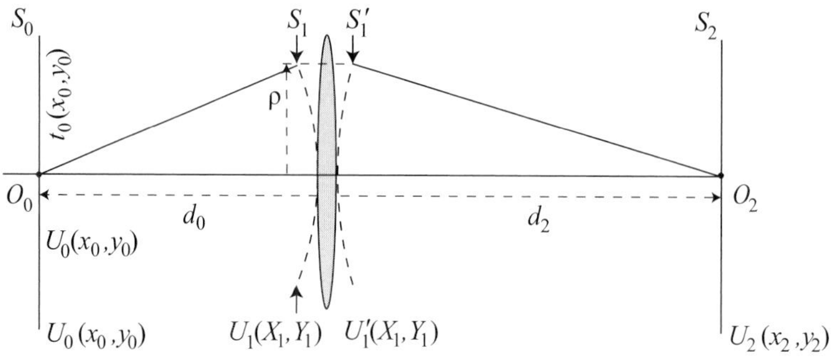

Figure A.7: Imaging of an object by a lens.

optics that, up to the second order in the far-field coordinates (X_1, Y_1), the pathlength from the first spherical surface S_1 to the second one (S_1') has a constant value. This property not only holds for the axial conjugated points but equally well for any pair of off-axis conjugated points. Outside the paraxial domain, the pathlength P_0Q_1 has to be developed up to a higher order and we have to include possible wavefront aberration of the lens. For our analysis up to the second order in the transverse far-field coordinates (X_1, Y_1) it is sufficient to include the spatial limitation of the transmitted beams because of the lens diaphragm. This limitation is taken into account by a (real) lens function $f_1(X_1, Y_1)$. If we suppose that the lens is infinitely thin, the phase difference between the surfaces S_1 and S_1' is zero. In a later phase, lens imperfections (aberrations) can be be included in the analysis by allowing the lens function f_1 to be complex.

The propagation of the incident wave from surface S_0 to the curved surface S_1 yields in normalised coordinates the complex amplitude

$$U_1(X_1', Y_1') = -\frac{i}{N_{F,0}} \exp(ik_0 d_0) \iint_{-\infty}^{+\infty} t_0(x_0', y_0') \exp\left\{-i2\pi\left(x_0' X_1' + y_0' Y_1'\right)\right\} dx_0' dy_0' , \tag{A.9.8}$$

where we have omitted in the integrand the quadratic phase factor in (x_0', y_0'), assuming that the object-side Fresnel number $N_{F,0} = \rho_1^2/(\lambda d_0)$ is large enough. In that case, if the object function $t(x_0', y_0')$ has a sufficiently small extent, it can be guaranteed that the argument of the quadratic phase factor does not exceed the value of, say, $\pi/2$. In a similar manner, the propagation from the spherical surface S_1' to the image plane S_2 is given by a diffraction integral from the far field to the near field, such that with the aid of normalised coordinates the image-plane amplitude can be written as

$$\begin{aligned} U_2(x_2', y_2') = &- i N_{F,2} \exp(ik_0 d_2) \exp\left\{+i\pi\left(x_2'^2 + y_2'^2\right)/N_{F,2}\right\} \\ &\times \iint_{-\infty}^{+\infty} U_1'(X_1', Y_1') \exp\left\{-i2\pi\left(X_1' x_2' + Y_1' y_2'\right)\right\} dX_1' dY_1' . \end{aligned} \tag{A.9.9}$$

When we combine the two free-space propagation steps according to Eqs. (A.9.8) and (A.9.9), the complex amplitude at the spherical surface S_1' is given by $U_1(X_1' Y_1') f_1(X_1' Y_1')$. The image plane amplitude is then given by

$$\begin{aligned} U_2(x_2', y_2') = &-\frac{d_0}{d_2} \exp[ik_0(d_0 + d_2)] \exp\left\{+i\pi\left(x_2'^2 + y_2'^2\right)/N_{F,2}\right\} \\ &\times \iint_{-\infty}^{+\infty} t_0(x_0', y_0') \iint_{-\infty}^{+\infty} f_1(X_1' Y_1') \exp\left\{-i2\pi\left[X_1'(x_2' + x_0') + Y_1'(y_2' + y_0')\right]\right\} dX_1' dY_1' dx_0' dy_0' , \end{aligned} \tag{A.9.10}$$

where we have replaced the quantity $N_{F,2}/N_{F,0}$ by d_0/d_2. In many instances, if the image region of interest is small and the Fresnel number is large, it is permissible to make abstraction of the quadratic phase outside the integral expression. In what follows we also omit the generally irrelevant constant phase $k_0(d_0 + d_2)$ of the field in the image plane.

A.9.2.1 The Amplitude Impulse Response of a Lens

The expression for the image plane amplitude $U_2(x_2', y_2')$ can be further simplified if we consider the integral associated with the far-field coordinates (X_1', Y_1'). In general, if we denote a far-field function by $f(X', Y')$ and a near-field function by $A(x', x')$, both defined with the aid of normalised coordinates, we have a relationship between these two which is given by a diffraction integral of the type (see Eq. (A.8.5)),

$$A(x', y') \propto \iint_{-\infty}^{+\infty} f(X', Y') \exp\{-i2\pi(X'x' + Y'y')\} dx' dy' , \tag{A.9.11}$$

where the light propagation is from the far-field region towards the near field, for instance, from the exit pupil of a lens towards the image plane. Applying this relationship to Eq. (A.9.10) we define

$$A(x', y') = -\frac{d_0}{d_2} \iint_{-\infty}^{+\infty} f_1(X_1', Y_1') \exp\left\{-i2\pi(X_1' x' + Y_1' y')\right\} dX_1' dY_1' . \tag{A.9.12}$$

The function $A(x', y')$ is commonly called the (near-field) impulse response function of an imaging system, since it represents the image plane amplitude if the object contains an isolated pinhole (delta-function). With due attention to signs, we can then write the expression for the image plane amplitude of Eq. (A.9.10) as

$$U_2(x_2', y_2') = \iint_{-\infty}^{+\infty} t_0(x_0', y_0')\, A(x_2' + x_0', y_2' + y_0')\, dx_0' y_0'$$
$$= \iint_{-\infty}^{+\infty} t_0(-x_0', -y_0')\, A(x_2' - x_0', y_2' - y_0')\, dx_0' y_0' \,. \qquad \text{(A.9.13)}$$

On the second line of Eq. (A.9.13) we have written the complex amplitude $U_2(x_2', y_2')$ in the image plane in a different way by a change of coordinates from (x_0', y_0') to $(-x_0', -y_0')$. The expression for the amplitude U_2 is then a convolution of the (inverted) object transmission function and the amplitude impulse response of the imaging system. The convolution integral is evaluated with respect to the normalised coordinate x_2'. The negative sign of the variables of the object function t_0 is there since the chosen imaging geometry (real object to real image) produces an image that is an inverted version of the object.

A.9.2.2 Concatenation of Fourier Transformations in an Optical Imaging System

In Fig. A.8a we have plotted a sequence of far-field and near-field functions, $f_1(X_1', Y_1')$ and $A(x', y')$, respectively, as they would be produced in a periscope-like set-up. The distance between the identical lenses is two times their focal distance. When the light propagates in the positive z-direction, the exponential function in the Fourier-type diffraction integral has a negative sign. If we reconstruct the near-field and far-field functions by an inverse Fourier transformation, we follow the large horizontal arrows in the lower parts of the figure with the positive sign. We note that the positive sign in the exponential function of the diffraction integral does not correspond to free-space wave propagation. It is a mathematical

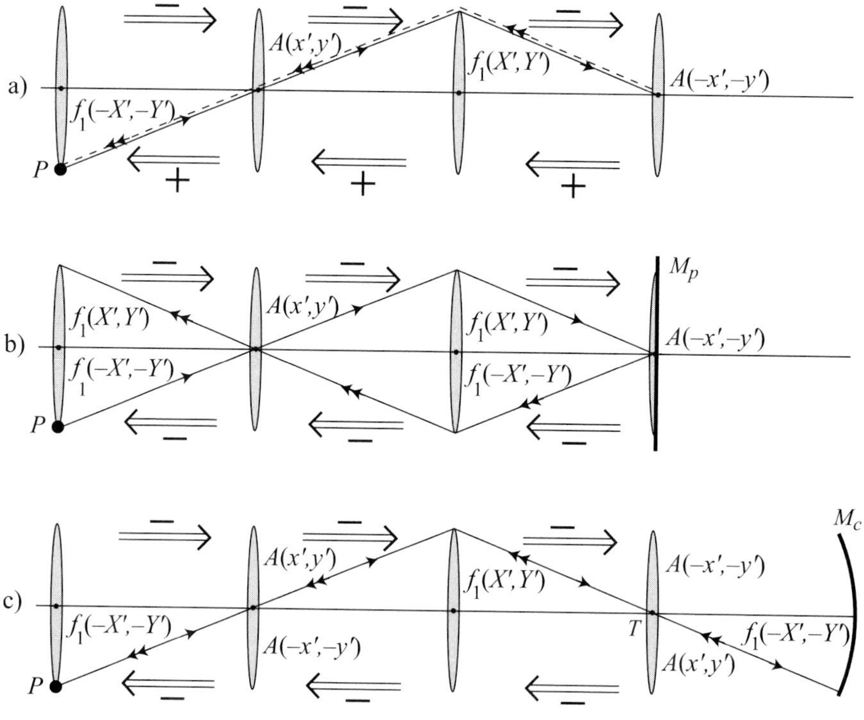

Figure A.8: Graphical illustration of the Fourier transformations performed by a sequence of lenses (periscope layout). The forward rays (starting point is P) have been provided with a single arrow, the real or fictitious rays in the backward direction with two arrows.

a) Forward (negative sign) and inverse (positive sign) Fourier transformation of the far-field pupil function f_1 of a lens.

b) Fourier transformations after reflection at a planar mirror M_p.

c) The same, but now after reflection by a concave mirror M_c.

operation to recover the original functions at each axial position. For that reason, the backward travelling fictitious rays are dashed in Fig. A.8a.

In Fig. A.8b the optical field at the exit of the periscope is reflected by a planar mirror M_p. The mirror introduces a lateral magnification $m_T = +1$ and an angular magnification $m_A = -1$. The exponential function in the diffraction integral associated with the backward propagation has a negative sign. It is seen that the near-field function A is unaltered in the reflected wave, whereas the reflected far field, determined by the angles of real (solid) reflected rays, is inverted. Figure A.8c shows the opposite effect. A concave mirror with its centre of curvature at the centre of the last lens sends back a reflected field with lateral magnification $m_T = -1$ and angular magnification $m_A = +1$. We then observe that the far-field functions f_1 are identical for the forward and backward propagating field whilst the near-field functions A are inverted at each corresponding axial position.

A.10 Some Useful Vector Formulas

In the framework of electromagnetic theory and its application to diffraction problems, frequent use is made of expressions for various products of vectors and gradients of scalar functions, some rather basic but others of a more intricate nature. For that reason we reproduce in this section various vector expressions and identities that are used in this textbook or that are needed by the reader when a subject is further developed. All vectors are in three-dimensional space and are written with bold lower case characters; $\mathbf{r}$ is the three-dimensional position vector and $\hat{\mathbf{r}}$ the unit position vector. For scalar functions which depend on the Cartesian spatial coordinates (x, y, z) we use the symbols f and g.

$$\mathbf{a}\cdot(\mathbf{b}\times\mathbf{c}) = \mathbf{b}\cdot(\mathbf{c}\times\mathbf{a}) = \mathbf{c}\cdot(\mathbf{a}\times\mathbf{b}) \,,$$
$$\mathbf{a}\times(\mathbf{b}\times\mathbf{c}) = (\mathbf{a}\cdot\mathbf{c})\,\mathbf{b} - (\mathbf{a}\cdot\mathbf{b})\,\mathbf{c} \,.$$

If parentheses are omitted in the vector triple product, the product $\mathbf{a}\times\mathbf{b}\times\mathbf{c}$ is expanded

by convention with respect to the central and right-hand vector.

$$(\mathbf{a}\times\mathbf{b})\cdot(\mathbf{c}\times\mathbf{d}) = (\mathbf{a}\cdot\mathbf{c})(\mathbf{b}\cdot\mathbf{d}) - (\mathbf{a}\cdot\mathbf{d})(\mathbf{b}\cdot\mathbf{c}) \,,$$
$$(\mathbf{a}\times\mathbf{b})\times(\mathbf{c}\times\mathbf{d}) = [\mathbf{a}\cdot(\mathbf{b}\times\mathbf{d})]\,\mathbf{c} - [\mathbf{a}\cdot(\mathbf{b}\times\mathbf{c})]\,\mathbf{d} = [\mathbf{a}\cdot(\mathbf{c}\times\mathbf{d})]\,\mathbf{b} - [\mathbf{b}\cdot(\mathbf{c}\times\mathbf{d})]\,\mathbf{a} \,.$$

$$\nabla(fg) = f\,\nabla g + g\nabla f \,, \qquad \nabla(f/g) = (g\nabla f - f\,\nabla g)/g^2 \,,$$
$$\nabla\cdot(f\mathbf{a}) = \nabla f\cdot\mathbf{a} + f\,\nabla\cdot\mathbf{a} \,,$$
$$\nabla\cdot(\nabla\times\mathbf{a}) = 0 \,,$$
$$\nabla\times(\nabla f) = 0 \,,$$
$$\nabla\times(f\mathbf{a}) = \nabla f\times\mathbf{a} + f\,\nabla\times\mathbf{a} \,,$$
$$\nabla^2 f = \nabla\cdot(\nabla f) \,,$$
$$\nabla^2(fg) = f\,\nabla^2 g + g\nabla^2 f + 2(\nabla f\cdot\nabla g) \,,$$
$$\nabla\cdot(\nabla f\times\nabla g) = \nabla g\cdot(\nabla\times\nabla f) - \nabla f\cdot(\nabla\times\nabla g) = 0 \,,$$
$$\nabla\cdot(f\,\nabla g - g\nabla f) = f\,\nabla^2 g + \nabla f\cdot\nabla g - \nabla f\cdot\nabla g - g\nabla^2 f = f\,\nabla^2 g - g\nabla^2 f \,,$$
$$\nabla\times\nabla\times\mathbf{a} = \nabla(\nabla\cdot\mathbf{a}) - \nabla^2\mathbf{a} \,,$$
$$\nabla(\mathbf{a}\cdot\mathbf{b}) = (\mathbf{b}\cdot\nabla)\,\mathbf{a} + (\mathbf{a}\cdot\nabla)\,\mathbf{b} + \mathbf{b}\times(\nabla\times\mathbf{a}) + \mathbf{a}\times(\nabla\times\mathbf{b}) \,,$$
$$\nabla\cdot(\mathbf{a}\times\mathbf{b}) = \mathbf{b}\cdot(\nabla\times\mathbf{a}) - \mathbf{a}\cdot(\nabla\times\mathbf{b}) \,,$$
$$\nabla\times(\mathbf{a}\times\mathbf{b}) = (\mathbf{b}\cdot\nabla)\,\mathbf{a} - (\mathbf{a}\cdot\nabla)\,\mathbf{b} - \mathbf{b}\,(\nabla\cdot\mathbf{a}) + \mathbf{a}\,(\nabla\cdot\mathbf{b}) \,,$$
$$\nabla\{\mathbf{a}\cdot(\mathbf{b}\times\mathbf{r})\} = \mathbf{a}\times\mathbf{b} \,,$$
$$\nabla\cdot f(r)\,\hat{\mathbf{r}} = 2f(r)/r + df(r)/dr \,,$$
$$\nabla\times f(r)\,\hat{\mathbf{r}} = 0 \,,$$
$$\nabla\cdot r\hat{\mathbf{r}} = 3 \,,$$

$$\nabla \times r\hat{\mathbf{r}} = 0 ,$$
$$\mathbf{a} \times \nabla \times \mathbf{a} = (1/2)\nabla(|\mathbf{a}|^2) - (\mathbf{a} \cdot \nabla)\mathbf{a} ,$$
$$\mathbf{a} \times \nabla \times \mathbf{b} = \nabla_b(\mathbf{a} \cdot \mathbf{b}) - (\mathbf{a} \cdot \nabla)\mathbf{b} ,$$

where the subscript b of ∇_b implies that differentiation is performed with respect to the components of the vector $\mathbf{b}$ only.

A.10.1 The Gradient of a Vector

The gradient of a vector, $\nabla\mathbf{a}(\mathbf{r})$, is used in Subsection 4.2.3 and in Eqs. (8.1.51)–(8.1.55), and it needs special attention. The formal definition of the gradient of a vector follows from the definition of the gradient operator ∇. For a Cartesian coordinate system we have the expression,

$$\nabla\mathbf{a} = \begin{pmatrix} \frac{\partial}{\partial x}\hat{\mathbf{x}} \\ \frac{\partial}{\partial y}\hat{\mathbf{y}} \\ \frac{\partial}{\partial z}\hat{\mathbf{z}} \end{pmatrix} \begin{pmatrix} a_x\hat{\mathbf{x}} & a_y\hat{\mathbf{y}} & a_z\hat{\mathbf{z}} \end{pmatrix} = \begin{pmatrix} \frac{\partial a_x}{\partial x}\hat{\mathbf{x}}\hat{\mathbf{x}} & \frac{\partial a_y}{\partial x}\hat{\mathbf{x}}\hat{\mathbf{y}} & \frac{\partial a_z}{\partial x}\hat{\mathbf{x}}\hat{\mathbf{z}} \\ \frac{\partial a_x}{\partial y}\hat{\mathbf{y}}\hat{\mathbf{x}} & \frac{\partial a_y}{\partial y}\hat{\mathbf{y}}\hat{\mathbf{y}} & \frac{\partial a_z}{\partial y}\hat{\mathbf{y}}\hat{\mathbf{z}} \\ \frac{\partial a_x}{\partial z}\hat{\mathbf{z}}\hat{\mathbf{x}} & \frac{\partial a_y}{\partial z}\hat{\mathbf{z}}\hat{\mathbf{y}} & \frac{\partial a_z}{\partial z}\hat{\mathbf{z}}\hat{\mathbf{z}} \end{pmatrix} , \tag{A.10.1}$$

where $(\hat{\mathbf{x}},\hat{\mathbf{y}},\hat{\mathbf{z}})$ are the Cartesian unit vectors. The gradient of the vector $\mathbf{a}$ is a second rank tensor. If we take the scalar product of the tensor $\nabla\mathbf{a}$ and a general vector $\mathbf{v}$, the tensor rank is reduced by one and the resulting quantity is again a vector. The scalar product $(\nabla\mathbf{a})^T \cdot \mathbf{v}$ yields a vector $\mathbf{b}$ which is given by the expression,

$$\mathbf{b} = \begin{pmatrix} \frac{\partial a_x}{\partial x}\hat{\mathbf{x}}\hat{\mathbf{x}} & \frac{\partial a_x}{\partial y}\hat{\mathbf{y}}\hat{\mathbf{x}} & \frac{\partial a_x}{\partial z}\hat{\mathbf{z}}\hat{\mathbf{x}} \\ \frac{\partial a_y}{\partial x}\hat{\mathbf{x}}\hat{\mathbf{y}} & \frac{\partial a_y}{\partial y}\hat{\mathbf{y}}\hat{\mathbf{y}} & \frac{\partial a_y}{\partial z}\hat{\mathbf{z}}\hat{\mathbf{y}} \\ \frac{\partial a_z}{\partial x}\hat{\mathbf{x}}\hat{\mathbf{z}} & \frac{\partial a_z}{\partial y}\hat{\mathbf{y}}\hat{\mathbf{z}} & \frac{\partial a_z}{\partial z}\hat{\mathbf{z}}\hat{\mathbf{z}} \end{pmatrix} \begin{pmatrix} v_x\hat{\mathbf{x}} \\ v_y\hat{\mathbf{y}} \\ v_z\hat{\mathbf{z}} \end{pmatrix} = \begin{pmatrix} \left(\frac{\partial a_x}{\partial x}v_x + \frac{\partial a_x}{\partial y}v_y + \frac{\partial a_x}{\partial z}v_z\right)\hat{\mathbf{x}} \\ \left(\frac{\partial a_y}{\partial x}v_x + \frac{\partial a_y}{\partial y}v_y + \frac{\partial a_y}{\partial z}v_z\right)\hat{\mathbf{y}} \\ \left(\frac{\partial a_z}{\partial x}v_x + \frac{\partial a_z}{\partial y}v_y + \frac{\partial a_z}{\partial z}v_z\right)\hat{\mathbf{z}} \end{pmatrix} . \tag{A.10.2}$$

For the special case that $\mathbf{a} = \mathbf{v}$ and $\mathbf{a}$ is a unit vector, we find that $(\nabla\mathbf{a}) \cdot \mathbf{a} = \mathbf{0}$, in line with Eq. (4.2.39).

In Eq. (8.1.54), associated with the vector Kirchhoff integral, we encounter the gradient of the electric field vector $\mathbf{E}$. The scalar product with the outward normal vector $\hat{\mathbf{v}}$ has to be calculated. For the result to be a column vector we need to evaluate the quantity

$$(\nabla\mathbf{E})^T \cdot \hat{\mathbf{v}} = \{\hat{\mathbf{v}}^T \cdot (\nabla\mathbf{E})\}^T . \tag{A.10.3}$$

With the aid of Eq. (A.10.1) we readily obtain the result

$$(\nabla\mathbf{E})^T \cdot \hat{\mathbf{v}} = \begin{pmatrix} \frac{\partial E_x}{\partial x}\hat{v}_x + \frac{\partial E_x}{\partial y}\hat{v}_y + \frac{\partial E_x}{\partial z}\hat{v}_z \\ \frac{\partial E_y}{\partial x}\hat{v}_x + \frac{\partial E_y}{\partial y}\hat{v}_y + \frac{\partial E_y}{\partial z}\hat{v}_z \\ \frac{\partial E_z}{\partial x}\hat{v}_x + \frac{\partial E_z}{\partial y}\hat{v}_y + \frac{\partial E_z}{\partial z}\hat{v}_z \end{pmatrix} . \tag{A.10.4}$$

Using the property $\hat{v}_x = \partial x/\partial v$ etc., we can find the expression for the required column vectors in the right-hand side of Eqs. (8.1.55)–(8.1.59),

$$(\nabla\mathbf{E})^T \cdot \hat{\mathbf{v}} = \{\hat{\mathbf{v}}^T \cdot (\nabla\mathbf{E})\}^T = \begin{pmatrix} \frac{\partial E_x}{\partial v} \\ \frac{\partial E_y}{\partial v} \\ \frac{\partial E_z}{\partial v} \end{pmatrix} = \frac{\partial \mathbf{E}}{\partial v} . \tag{A.10.5}$$

B Phase and Group Velocity of a Wave Packet

In this appendix we derive the propagation speed v_g of the envelope of a wave packet (group velocity). We consider an assembly of plane waves, propagating in the positive z-direction with a narrow frequency spread around a central temporal frequency ω_0. A spectral component of the wave packet is given by

$$F(z,\omega) = A(\omega)\exp\{ik(\omega)z\} ,$$

where A(ω) is given by

$$A(\omega) = \Delta \exp\left\{-\frac{[(\omega-\omega_0)\Delta]^2}{4\pi}\right\} . \tag{B.1}$$

The Gaussian spectral function $A(\omega)$ follows from a temporal signal, given by the (inverse) Fourier transform $\exp[-\pi(t/\Delta)^2]$ of $F(z,\omega)$, where Δ is the temporal width at $1/e^{\pi}$ level of the Gaussian pulse in the time domain (see Appendix A, Table A.3).

The wavenumber k of a plane wave in a dispersive medium depends on the temporal frequency ω. A quadratic approximation of the generally complex wavenumber in the neighbourhood of the central frequency ω_0 is given by

$$\begin{aligned} k(\omega) &= k^{(0)} + (\omega-\omega_0)\frac{dk}{d\omega}\Big|_{\omega_0} + \frac{1}{2}(\omega-\omega_0)^2\frac{d^2k}{d\omega^2}\Big|_{\omega_0} \\ &= k^{(0)} + k^{(1)}(\omega-\omega_0) + \frac{1}{2}k^{(2)}(\omega-\omega_0)^2 . \end{aligned} \tag{B.2}$$

The inverse Fourier transformation of $F(z,\omega)$ according to Eq. (A.1.2) yields, with integration variable $p = \omega - \omega_0$,

$$\begin{aligned} f(z,t) &= \frac{\Delta}{2\pi}\exp\left\{i[k^{(0)}z-\omega_0 t]\right\}\int_{-\infty}^{+\infty}\exp\left\{-\frac{(p\Delta)^2}{4\pi}\right\}\exp\left\{i\left[k^{(1)}p+\frac{1}{2}k^{(2)}p^2\right]z\right\}\exp(-ipt)\,dp \\ &= \frac{\Delta}{2\pi}\exp\left\{i[k^{(0)}z-\omega_0 t]\right\}\int_{-\infty}^{+\infty}\exp(-ap^2-ibp)dp , \end{aligned} \tag{B.3}$$

where a and b are defined by

$$a = \frac{\Delta^2}{4\pi} - i\frac{k^{(2)}z}{2} , \qquad b = t - k^{(1)}z . \tag{B.4}$$

The value of the integral in the right-hand side of Eq. (B.3) can be found with the aid of **[266]**, (see expression 7.7.3, p. 162),

$$g(z) = \int_0^{\infty}\exp(-ap^2+2izp)dp = \frac{1}{2}\sqrt{\frac{\pi}{a}}\exp\left(-\frac{z^2}{a}\right) + \frac{i}{\sqrt{a}}G\left(\frac{z}{\sqrt{a}}\right) , \tag{B.5}$$

where $G(z) = \exp(-z^2)\int_0^z \exp(t^2)\,dt$ is Dawson's integral (see **[266]**, 7.2.5, p. 160). As the integration interval in Eq. (B.3) is $[-\infty,+\infty]$ and Dawson's integral obeys the relation $G(-z) = -G(z)$, we have the following result for the function $f(z,t)$,

$$f(z,t) = \frac{\Delta}{2\pi}\exp\left\{i[k^{(0)}z-\omega_0 t]\right\}(\pi/a)^{1/2}\exp\left(-b^2/4a\right) . \tag{B.6}$$

The inverse transformation is defined for $b \in C$ and $\Re(a) > 0$. In terms of the dispersion coefficients $k^{(0)}$, $k^{(1)}$ and $k^{(2)}$ the function $f(z,t)$ is given by

$$f(z,t) = \frac{1}{[1 - i2\pi k^{(2)}z/\Delta^2]^{1/2}} \exp\left\{-\pi\left(\frac{k^{(1)}z - t}{\Delta\left[1 - i2\pi k^{(2)}z/\Delta^2\right]^{1/2}}\right)^2\right\} \exp\left\{i\left(k^{(0)}z - \omega_0 t\right)\right\}, \tag{B.7}$$

where the wave amplitude is given by the real part of $f(z,t)$. The condition $\Re(a) > 0$ is equivalent to

$$\Im(k^{(2)}) > -\frac{\Delta^2}{2\pi z}. \tag{B.8}$$

This condition implies that $\Im(k^{(2)})$ should be positive to have a solution $f(z,t)$ that is valid for arbitrary positive values of the coordinate z. Inspection of Eq. (B.7) also shows that convergence of the solution requires

$$\left[\Re(k^{(1)})z - t\right]^2 > [\Im(k^{(1)})z]^2, \tag{B.9}$$

which determines the range of z-values of the wave solution at a certain moment t.

The following conclusions can be drawn about the shape of the propagating wave:

- The second exponential factor in the expression for $f(z,t)$ of Eq. (B.7) contains the high-frequency carrier part of the wave function with a propagation speed v_{ph} given by $\omega_0/k^{(0)}$. In special circumstances, for instance when using metamaterials, v_{ph} can become negative.
- The maximum of the wave envelope (first exponential factor) is found at a position $z = t/k^{(1)}$. This implies that the wave envelope propagates with a speed $v_{gr} = 1/k^{(1)} = d\omega/dk$, where the value of the derivative should be taken at the average frequency, ω_0. In special circumstances, for instance in regions of anomalous dispersion, the group velocity can become negative.
- At $z = 0$, the temporal width of the wave packet equals Δ (at the $e^{-\pi}$ level). The pulse broadens on propagation as it follows from the z-dependent amplitude function in the denominator of the first exponential function (envelope function). Simultaneously, a z-dependent phase shift appears in the envelope function due to a nonzero value of $\Re(k^{(2)})$.
- During propagation, the maximum amplitude of the wave packet decreases, given the z-dependence of the denominator of the function $f(z,t)$. This decrease in amplitude is a consequence of the wave packet broadening on propagation.

In Fig. B.1 we show some typical profiles of wave packets. At $z = 0$ we have the transition from vacuum to a dispersive medium. The transmission loss at the interface has been neglected, as well as the wavelength change at the transition between vacuum and the dispersive medium. In all graphs the dashed line indicates the expected position of the centre of the wave packet when the propagation would take place in a dispersion-free medium. Graph a) shows the initial wave packet, at time $t = 0$. The quantity Δ equals 0.4, resulting in typically 20 oscillations in the pulse. The values of ω_0 and k_0 are $40\pi\,\mathrm{s}^{-1}$ and $40\pi\,\mathrm{m}^{-1}$, respectively. The phase velocity v_{ph} thus equals unity. The value of $k^{(1)}$ is 1.05, resulting in a group velocity of $0.95v_{ph}$. In Fig. B.1b the wave packet is shown after a propagation time of 5 units. Since $\Im(k^{(1)})$ and $k^{(2)}$ are both equal to zero, the pulse shape remains unchanged and we just observe a delay in arrival time of the pulse peak. In c) the influence of a nonzero value of the real value of the group delay dispersion $k^{(2)}$ has been included. The pulse delay is unchanged as compared to b), but the pulse is broadened and the maximum amplitude has been lowered. In d) material absorption has been included by making $\Im(k^{(0)})$ nonzero. The average pulse position after a unit propagation time would have been $z = 1$. Due to the strong absorption, the pulse dies out after a few cycles. In e) we have made the imaginary part of $k^{(0)}$ again zero, but the $k^{(1)}$ value now equals $1.05 + i0.5$. This particular choice of $k^{(1)}$ makes the group velocity larger than the phase velocity. The peak of the wave packet has advanced more than in a dispersion-free medium. Such a phenomenon may occur in a spectral region with anomalous dispersion. In general, the absorption is high in such a region but this effect can be cancelled, for instance, by a population inversion of the atoms at the absorption line.

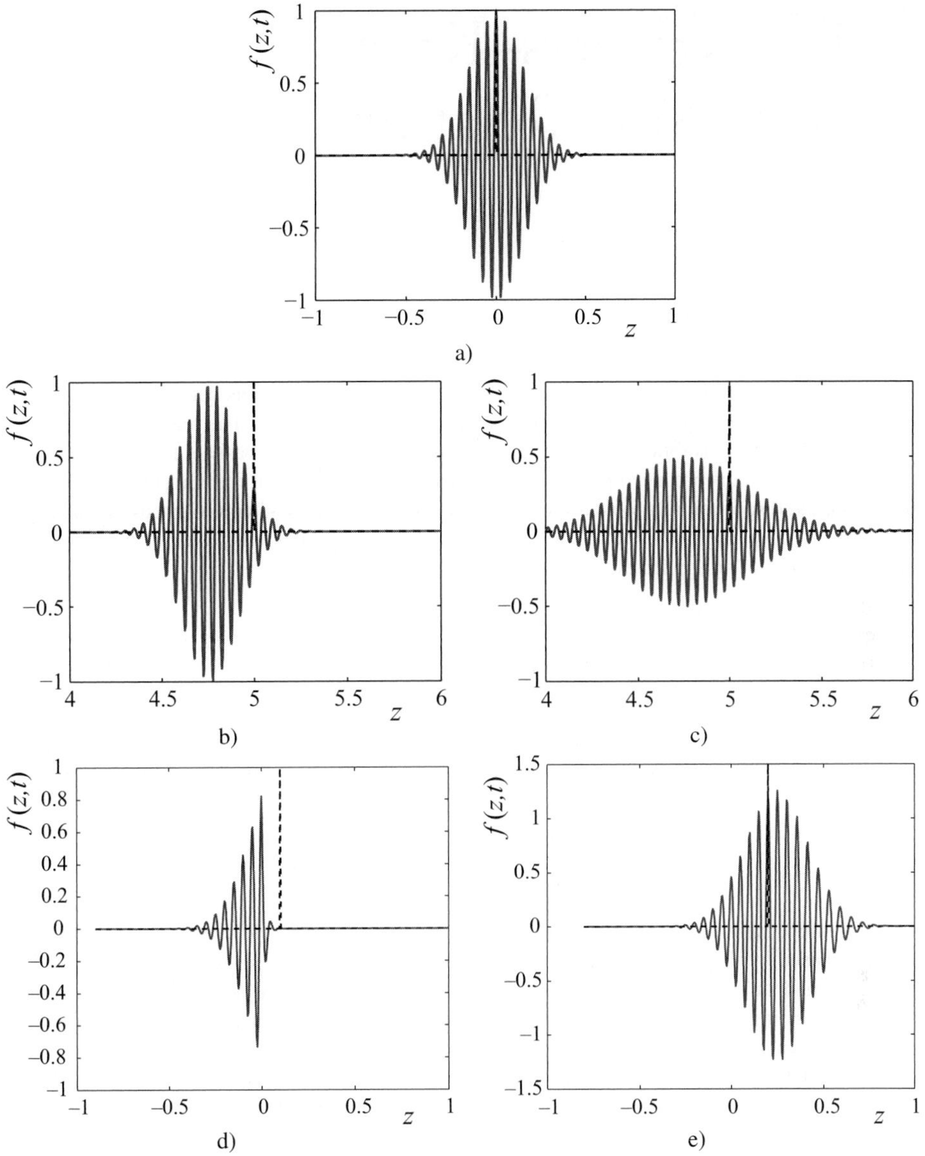

Figure B.1: The propagation of a wave packet in a dispersive medium (wave function $f(z, t)$ at fixed values $t = t_0$). The dashed vertical line indicates the position of the centre of the unperturbed wave packet in a dispersion-free medium.

a) Unperturbed wave packet ($t_0 = 0$),

b) $k^{(1)} = 1.05$, $k^{(2)} = 0$ ($t_0 = 5$),

c) $k^{(1)} = 1.05$, $k^{(2)} = 0.012$ ($t_0 = 5$),

d) $k^{(0)} = 1 + i0.5$, $k^{(1)} = 0$, $k^{(2)} = 0.0$ ($t_0 = 0.1$),

e) $k^{(0)} = 1$, $k^{(1)} = 1.05 + i0.5$, $k^{(2)} = 0.0$ ($t_0 = 0.2$).

C The Kramers–Kronig Dispersion Relations

In this Appendix we derive the Kramers–Kronig dispersion relations, discovered independently by H.A. Kramers and R. de Laer Kronig in 1926 and 1927 **[194]**, **[196]**. The dispersion relations establish a link between the real and the imaginary parts of the electrical permittivity as a function of the frequency of the radiation. The real and imaginary parts govern the propagation speed and the absorption of radiation in a medium. The physical basis of the Kramers–Kronig relations is causality: no physical effect is produced prior to the application of its driving force. Originally derived in the context of X-ray absorption, the dispersion relations also have applications in signal and control theory. In the electromagnetic domain, with respect to the permittivity of a medium, we analyse the creation of a dielectric displacement or electric flux density under the influence of an applied electric field.

C.1 Electric Flux Density and Driving Field

The electric field strength of the driving electromagnetic field in a medium gives rise to a polarisation effect of the charges in the medium. In general, at any position in a homogeneous and isotropic medium, it is permissible to write

$$\mathbf{D}(t) = \epsilon_0 \left\{ \mathbf{E}(t) + \int_{-\infty}^{+\infty} g(t')\mathbf{E}(t-t')\, dt' \right\} , \tag{C.1.1}$$

where $\mathbf{D}$ is the time-dependent electric flux density, $\mathbf{E}(t)$ the exciting electric field vector and $g(t)$ the specific impulse response function caused by the material polarisation. As usual, ϵ_0 denotes the vacuum permittivity. The first term in Eq. (C.1.1) is the immediate reaction of the medium to the electric field as would occur in vacuum. The second term is the extra contribution arising from the presence of the physical medium. As we assume that the medium is homogeneous and isotropic, $g(t)$ is a scalar function.

The time-dependent function $g(t)$ is the Fourier transform of the electric susceptibility, $(\epsilon(\omega)/\epsilon_0) - 1$, commonly denoted by $\chi(\omega)$. In this appendix we define the Fourier transform of $g(t)$ and its inverse transform as

$$\chi(\omega) = \int_{-\infty}^{+\infty} g(t) \exp(i\omega t) dt , \tag{C.1.2}$$

$$g(t) = \frac{1}{2\pi} \int_{-\infty}^{+\infty} \chi(\omega) \exp(-i\omega t) d\omega . \tag{C.1.3}$$

In the frequency domain, employing the frequency-dependent quantity $\chi(\omega)$, we have the basic relation

$$\mathbf{D}_0 = \epsilon(\omega)\mathbf{E}_0 = \epsilon_0 \{(1 + \chi(\omega)\} \mathbf{E}_0 , \tag{C.1.4}$$

where $\mathbf{D}_0$ and $\mathbf{E}_0$ are the complex amplitudes of the flux density and electric field vectors of a harmonic wave with frequency ω in the medium. The dielectric permittivity at a certain frequency ω is generally a complex function, $\epsilon(\omega) = \epsilon'(\omega) + i\epsilon''(\omega)$ or, using the susceptibility function, $\epsilon(\omega) = \epsilon_0\{1 + \chi'(\omega) + i\chi''(\omega)\}$. We use the property that $g(t)$ is a real function of time and find from Eq. (C.1.2) for the real and imaginary parts of $\chi(\omega)$,

$$\chi'(\omega) = \int_{-\infty}^{+\infty} g(t) \cos \omega t \, dt \,, \tag{C.1.5}$$

$$\chi''(\omega) = \int_{-\infty}^{+\infty} g(t) \sin \omega t \, dt \,. \tag{C.1.6}$$

The two equations above show that the real and imaginary parts of the permittivity function are derived from one single function, $g(t)$, and thus cannot have arbitrary, independent values.

The function $g(t)$ is real by definition. Using the real and imaginary parts $\chi'(\omega)$ and $\chi''(\omega)$ of the electric susceptibility, we find the following expression for $g(t)$ with the aid of Eq. (C.1.3),

$$g(t) = \frac{1}{2\pi} \int_{-\infty}^{+\infty} [\chi'(\omega) \cos \omega t + \chi''(\omega) \sin \omega t] \, d\omega \,. \tag{C.1.7}$$

C.2 Causality and the Kramers–Kronig Relations

The essential step in the derivation of the Kramers–Kronig relations is the causality requirement for the action of the field on the medium. For a driving force given by a Dirac delta-function $\mathbf{E}(t) = \delta(t)\mathbf{E}_d$, we have a response function for the electric flux density

$$\mathbf{D}(t) = \epsilon_0 \{\delta(t) + g(t)\} \mathbf{E}_d \,. \tag{C.2.1}$$

The response is the sum of an instantaneous response as in vacuum and a delayed response, $g(t)$, which stems from the presence of the medium. The function $g(t)$ depends on the susceptibility of the medium. For the case of an impulse-like input, causality requires that $g(t) = 0$ for all $t < 0$. We apply this condition to Eq. (C.1.7) and impose for all $t > 0$,

$$g(-t) = \frac{1}{2\pi} \int_{-\infty}^{+\infty} [\chi'(\omega) \cos \omega t - \chi''(\omega) \sin \omega t] \, d\omega = 0 \,. \tag{C.2.2}$$

In terms of the cosine and sine transforms of the $\chi'$and χ'' functions we thus have the causality condition,

$$\frac{1}{2\pi} \int_{-\infty}^{+\infty} \chi'(\omega) \cos \omega t \, d\omega = \frac{1}{2\pi} \int_{-\infty}^{+\infty} \chi''(\omega) \sin \omega t \, d\omega \,, \qquad t \geq 0 \,. \tag{C.2.3}$$

C.2.1 The First Kramers–Kronig Relation

To recover the function $\chi'(\omega)$ we apply an inverse cosine transformation to both sides of Eq. (C.2.3), over the time interval $0 \leq t \leq \infty$ for which the equality of Eq. (C.2.3) has to be respected. Several methods for obtaining the Kramers–Kronig relations have been presented in the literature. In what follows we use the approach that was given in the two original publications on the subject:

- *Cosine transform of the left-hand side of* (C.2.3)
  We need to evaluate the expression

$$\frac{1}{2\pi} \int_{0}^{\infty} \left[\int_{-\infty}^{+\infty} \chi'(\omega') \cos \omega' t \, d\omega' \right] \cos \omega t \, dt \,, \tag{C.2.4}$$

  which contains the improper integral

$$I_l = \int_{0}^{+\infty} \cos \omega' t \, \cos \omega t \, dt \,. \tag{C.2.5}$$

  We evaluate the product of the cosines by means of their complex representations and multiply the integrand with a real exponential factor $\exp(-at)$ where a is a positive real number. The value of the improper integral I is then equated to its limit value when a approaches zero. We have that

$$I_l = \lim_{a\to 0} \frac{1}{4} \left[\int_0^{\infty} \exp[i(\omega' + \omega + ia)\,t] + \exp[-i(\omega' + \omega - ia)\,t] \right.$$
$$\left. + \exp[i(\omega' - \omega + ia)\,t] + \exp[-i(\omega' - \omega - ia)\,t] \right] dt\,. \tag{C.2.6}$$

We perform the integration and observe that it is only the lower integration limit that contributes finite values. We obtain the result

$$I_l = \frac{1}{2} \lim_{a\to 0} \left\{ \frac{a}{(\omega' + \omega)^2 + a^2} + \frac{a}{(\omega' - \omega)^2 + a^2} \right\}. \tag{C.2.7}$$

A so-called 'nascent' definition of the Dirac δ-function is given by

$$\delta(x) = \frac{1}{\pi} \lim_{a\to 0} \frac{a}{x^2 + a^2}\,. \tag{C.2.8}$$

With the aid of this definition we find that

$$I_l = \frac{\pi}{2} \left\{ \delta(\omega' + \omega) + \delta(\omega' - \omega) \right\}. \tag{C.2.9}$$

The transformed left-hand side according to Eq. (C.2.4) is then written as

$$\frac{1}{4} \int_{-\infty}^{+\infty} \chi'(\omega') \left[\delta(\omega' + \omega) + \delta(\omega' - \omega) \right] d\omega' = \frac{1}{2} \chi'(\omega)\,, \tag{C.2.10}$$

where we have used the property that $\chi'(\omega) = \chi'(-\omega)$, according to Eq. (C.1.5).

- *Cosine transform of the right-hand side of* (C.2.3)

The cosine transform of the right-hand side of Eq. (C.2.3) requires evaluation of the integral

$$\frac{1}{2\pi} \int_0^{\infty} \left[\int_{-\infty}^{+\infty} \chi''(\omega') \sin \omega' t \, d\omega' \right] \cos \omega t \, dt\,, \tag{C.2.11}$$

which contains another improper integral with respect to the time variable,

$$I_r = \int_0^{+\infty} \sin \omega' t \, \cos \omega t \, dt\,. \tag{C.2.12}$$

Following the same approach as in Eqs. (C.2.5)–(C.2.7) for the evaluation of I_r we obtain

$$I_r = \frac{1}{2} \lim_{a\to 0} \left\{ \frac{\omega' + \omega}{(\omega' + \omega)^2 + a^2} + \frac{\omega' - \omega}{(\omega' - \omega)^2 + a^2} \right\} = \frac{1}{2} \left\{ \frac{1}{\omega' + \omega} + \frac{1}{\omega' - \omega} \right\}. \tag{C.2.13}$$

Performing the integration with respect to ω' of Eq. (C.2.11) and equating it to $\chi(\omega)/2$ according to Eq. (C.2.10) we obtain

$$\chi'(\omega) = \frac{1}{\pi} \int_{-\infty}^{+\infty} \chi''(\omega') \frac{\omega'}{\omega'^2 - \omega^2} \, d\omega'\,. \tag{C.2.14}$$

From Eq. (C.1.6) it follows that $\chi''(\omega') = -\chi''(-\omega')$ and this leads to the final result

$$\chi'(\omega) = \frac{2}{\pi} \mathcal{P} \int_0^{+\infty} \frac{\omega' \chi''(\omega')}{\omega'^2 - \omega^2} \, d\omega'\,. \tag{C.2.15}$$

We note that evaluation of the integral in Eq. (C.2.15) is hindered by the presence of a singular point of the integrand function at $\omega' = \omega$. The integration interval has to be subdivided and the Cauchy principal part of the integral should be taken. With the result of Eq. (C.2.15) we terminate the derivation of the first of the two Kramers–Kronig relations. The derivation shows that the real part of the susceptibility function can be uniquely obtained from the (measured) values of the imaginary part of the same function.

C.2.2 The Second Kramers–Kronig Relation

To obtain a relation for the imaginary part of $\chi(\omega)$ as a function of its real part, we proceed in a similar manner as before by taking the sine transforms of the left- and right-hand side of Eq. (C.2.3). We leave out the details and just produce the results of the sine transform operations:

- *Sine transform of the left-hand side of* (C.2.3)

$$\frac{1}{2\pi}\int_0^{\infty}\left[\int_{-\infty}^{+\infty}\chi'(\omega')\cos\omega' t\, d\omega'\right]\sin\omega t\, dt = -\frac{\omega}{\pi}\int_0^{+\infty}\frac{\chi'(\omega')}{\omega'^2-\omega^2}\,d\omega', \tag{C.2.16}$$

where we have used that $\chi'(-\omega') = \chi'(\omega')$.
- *Sine transform of the right-hand side of* (C.2.3)

$$\frac{1}{2\pi}\int_0^{\infty}\left[\int_{-\infty}^{+\infty}\chi'(\omega')\sin\omega' t\, d\omega'\right]\sin\omega t\, dt = \frac{\chi''(\omega')}{2}\,. \tag{C.2.17}$$

Combining Eqs. (C.2.16) and (C.2.17) yields the second Kramers–Kronig relation,

$$\chi''(\omega) = -\frac{2\,\omega}{\pi}\mathcal{P}\int_0^{+\infty}\frac{\chi'(\omega')}{\omega'^2-\omega^2}\,d\omega'. \tag{C.2.18}$$

We replace the susceptibility functions χ' and χ'' by the corresponding ϵ-quantities to obtain

$$\epsilon'(\omega) - \epsilon_0 = +\frac{2}{\pi}\mathcal{P}\int_0^{+\infty}\frac{\omega'\,\epsilon''(\omega')}{\omega'^2-\omega^2}\,d\omega', \tag{C.2.19}$$

$$\epsilon''(\omega) = -\frac{2\omega}{\pi}\mathcal{P}\int_0^{+\infty}\frac{\epsilon'(\omega')-\epsilon_0}{\omega'^2-\omega^2}\,d\omega'. \tag{C.2.20}$$

In the literature Kramers–Kronig relations are found that both have an opposite sign. The sign-choice depends on the definition of the imaginary part $\chi''(\omega)$. In this appendix we have made the choice that a positive imaginary part of $\chi(\omega)$ gives rise to damped wave propagation, in accordance with an amplitude decrease at propagation of a plane wave represented by $\exp[i(\tilde{n}\,k_0\hat{\mathbf{k}}\cdot\mathbf{r} - \omega t)]$. The quantity $\tilde{n} = n + i\kappa$ is the complex refractive index and k_0 is the vacuum wavenumber of the radiation. To avoid plane waves with a diverging amplitude, the complex quantity χ is only defined in the upper half of the complex plane. The opposite choice of sign for $\chi''(\omega)$ and the corresponding plane wave representation implies a change of sign in the exponential functions of Eqs. (C.1.2) and (C.1.3) and produces, for instance, the Eqs. (C.2.19) and (C.2.20) with opposite sign.

In Fig. C.1a, we show an example of a time response function $g(t)$. Figure C.1b shows the normalised frequency domain functions $(\epsilon'/\epsilon_0) - 1$ (solid curve) and ϵ''/ϵ_0 (dashed curve).

C.3 Alternative Representation of the Kramers–Kronig Relations

The expressions for the complex susceptibility $\chi(\omega)$ of Eqs. (C.2.15) and (C.2.18) and those for the complex permittivity $\epsilon(\omega)$, Eqs. (C.2.19) and (C.2.20), can be written in a simpler way by the introduction of negative frequencies in the integration interval. Using the property that $\chi(-\omega) = \chi^\star(\omega)$ we readily obtain for the permittivity function,

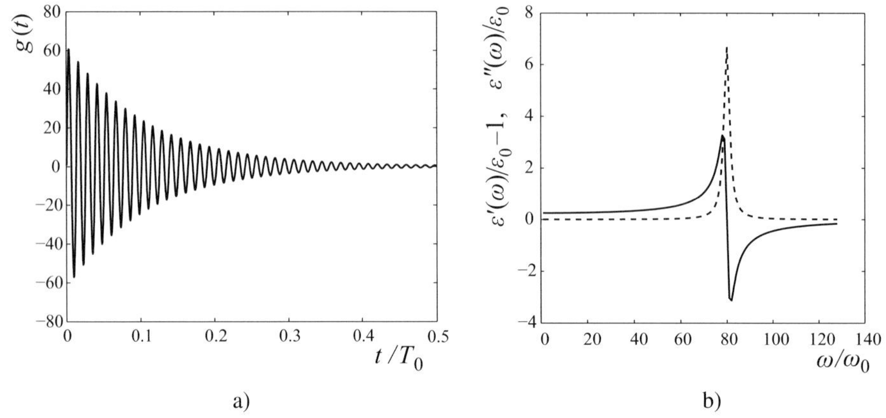

Figure C.1: a) The time response function of a medium.
b): The corresponding functions $(\epsilon'/\epsilon_0) - 1$ (solid curve) and ϵ''/ϵ_0 (dashed curve) in the frequency domain.
The units along the horizontal axes are t/T_0 and ω/ω_0, respectively. T_0 is the time interval length used for the calculation of the Fourier Transform and $\omega_0 = 2\pi/T_0$ is the frequency unit in b).

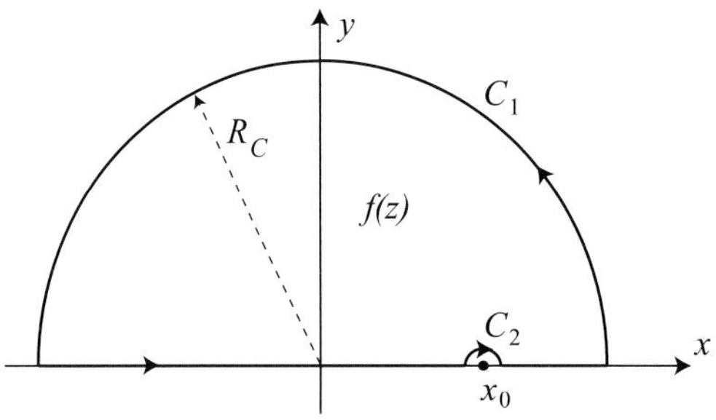

Figure C.2: The integration contour for the complex function $f(z)$, having a singularity at $z = x_0$ on the real axis.

$$\epsilon'(\omega) - \epsilon_0 = + \frac{1}{\pi} \mathcal{P} \int_{-\infty}^{+\infty} \frac{\epsilon''(\omega')}{\omega' - \omega} \, d\omega', \tag{C.3.1}$$

$$\epsilon''(\omega) = - \frac{1}{\pi} \mathcal{P} \int_{-\infty}^{+\infty} \frac{\epsilon'(\omega') - \epsilon_0}{\omega' - \omega} \, d\omega'. \tag{C.3.2}$$

These expressions present a numerical advantage as the singularity in the integrand function is first order whilst the previous expressions show a second-order singularity at $\omega' = \omega$. The real and imaginary permittivity functions of Eqs. (C.3.1) and (C.3.2) are each recognised as a Hilbert transformation. By consequence, the functions $\epsilon'(\omega) - \epsilon_0$ and $\epsilon''(\omega)$ are each other's Hilbert transforms.

The result of Eqs. (C.3.1) and (C.3.2) or the equivalent expressions for $\chi'(\omega)$ and $\chi''(\omega)$ can equally well be obtained by means of contour integration of the (analytic) complex function associated with χ. To this purpose, we use Cauchy's integral formula that, for an analytic complex function of a complex variable, is given by

$$f(z_0) = \frac{1}{i2\pi} \oint_C \frac{f(z)}{z - z_0} dz \,, \tag{C.3.3}$$

where $f(z)$ is the complex function of the variable z and C a closed contour, encircling the point z_0, in a region of the complex plane where $f(z)$ is analytic (see Fig. C.2). For the complex function $f(z)$ we choose $f(z) = \chi'(z) + i\chi''(z)$ and the point z_0 is chosen to be x_0 on the real axis. For this choice of z_0, the value of the contour integral according to Fig. C.2

is zero because the area within the contour does not contain singularities. The integral along the contour consists of four contributions,

$$\oint_C \frac{f(z)}{z-z_0}dz = \frac{1}{2\pi i}\left\{\int_0^{\pi} \frac{f(Re^{i\theta})}{Re^{i\theta}-x_0} iRe^{i\theta}\,d\theta \right.$$
$$\left. + \int_{-R}^{x_0-r_\epsilon} \frac{f(x)}{x-x_0}dx - \pi i f(x_0) + \int_{x_0+r_\epsilon}^{R} \frac{f(x)}{x-x_0}dx \right\} = 0\,, \quad \text{(C.3.4)}$$

where R_C is the radius of the large semi-circle C_1 and r_ϵ the radius of the small semi-circle C_2. The integration over contour C_1 yields zero for $R_C \to \infty$ if the absolute value of $f(z)$ tends to zero for $R \to \infty$ (Jordan's lemma). This is equivalent to the requirement that the optical properties of the medium become those of vacuum for electromagnetic radiation with an extremely high frequency. This is experimentally confirmed for very short wavelength X-ray radiation and for gamma rays. The third term in the right-hand side of Eq. (C.3.4) is the result of the integral over C_2, the half circle with radius r_ϵ, executed in the clockwise direction. For $r_\epsilon \to 0$, this integral over C_2 equals $-i\pi$ times the residue of $f(x)/(x-x_0)$ in the point $x = x_0$, which simply equals $f(x_0)$. We then obtain, if $f(x)$ is replaced by $f(\omega) = \chi'(\omega) + i\chi''(\omega)$,

$$\chi'(\omega) + i\chi''(\omega) = \frac{1}{i\pi}\mathcal{P}\int_{-\infty}^{+\infty} \frac{\chi'(\omega') + i\chi''(\omega')}{\omega'-\omega}d\omega'\,, \quad \text{(C.3.5)}$$

where the Cauchy principal value of the integral above has to be taken because of the singularity of the integrand at $\omega' = \omega$. Separating the real and imaginary parts of Eq. (C.3.5), we immediately obtain the expressions

$$\chi'(\omega) = \frac{1}{\pi}\mathcal{P}\int_{-\infty}^{+\infty} \frac{\chi''(\omega')}{\omega'-\omega}d\omega'\,, \quad \text{(C.3.6)}$$

$$\chi''(\omega) = -\frac{1}{\pi}\mathcal{P}\int_{-\infty}^{+\infty} \frac{\chi(\omega')}{\omega'-\omega}d\omega'\,. \quad \text{(C.3.7)}$$

C.3.1 Kramers–Kronig Relations for the Complex Refractive Index $\tilde{n} = n + i\kappa$

To obtain Kramers–Kronig-type relations between the real and imaginary part of the complex refractive index $\tilde{n} = n + i\kappa = [\epsilon(\omega)/\epsilon_0]^{1/2}$, we assume that the material has a 'weak contrast' with respect to vacuum ('dilute medium' approximation). In many instances, this approximation is valid for gases. We write

$$\tilde{n} = (1 + \chi' + i\chi'')^{1/2} \approx 1 + \frac{1}{2}(\chi' + i\chi'')\,. \quad \text{(C.3.8)}$$

For the case of weak contrast, the relation between the susceptibility χ and the complex refractive index $\tilde{n} \approx 1 + \Delta n + i\Delta\kappa$ is given by

$$\chi' = n^2 - \kappa^2 - 1 \approx 2(\Delta n)\,, \qquad \chi'' = 2n\kappa \approx 2(\Delta\kappa)\,. \quad \text{(C.3.9)}$$

In these approximate expressions we have assumed that Δn and $\Delta\kappa$ are both small with respect to unity and we have omitted second- and higher-order small terms. Using the weak-contrast approximation we find with the aid of Eqs. (C.3.6) and (C.3.7) the expressions

$$\Delta n(\omega) \approx \frac{1}{\pi}\mathcal{P}\int_{0}^{+\infty} \frac{\Delta\kappa(\omega')}{\omega'-\omega}d\omega'\,, \quad \text{(C.3.10)}$$

$$\Delta\kappa(\omega) \approx -\frac{1}{\pi}\mathcal{P}\int_{0}^{+\infty} \frac{\Delta n(\omega')}{\omega'-\omega}d\omega'\,. \quad \text{(C.3.11)}$$

When performing the integration, one should calculate as before the principal value because of the singularity at $\omega = \omega'$.

D Zernike Polynomials

The Zernike polynomials, orthogonal on the unit disc, were introduced in optics by F. Zernike in the context of his work on diffraction theory, the knife-edge test and phase-contrast imaging **[385]**. A set of orthogonal polynomials depends on the weighting function on their domain of existence. For the case of the Zernike polynomials, the weighting function is unity on the unit disc. In principle, the polynomials are a function of the Cartesian coordinates (x, y) on the unit disc. They are, however, commonly expressed in terms of polar coordinates. It turns out that with the aid of these coordinates the Zernike polynomials can be written as the product of two functions, one depending only on the radial coordinate and one depending only the azimuthal coordinate. Using the polar coordinates (ρ, θ), we have the general expression for a Zernike polynomial,

$$Z_n^m(\rho, \theta) = R_n^{|m|}(\rho)\, g_m(\theta)\ . \tag{D.0.1}$$

$Z_n^m(\rho, \theta)$ is the Zernike polynomial with radial degree n and azimuthal order m, $R_n^{|m|}(\rho)$ is the so-called *radial* Zernike polynomial of degree n, associated with the absolute value $|m|$ of the azimuthal order m of the Zernike polynomial. The structure and the exact definition of a general Zernike polynomial $Z_n^m(\rho, \theta)$ and its constituent functions $R_n^{|m|}(\rho)$ and $g_m(\theta)$ are given in the next subsection. We just mention here that the radial Zernike polynomials $R_n^{|m|}(\rho)$ have a weighting function ρ on the existence interval $0 \leq \rho \leq 1$.

The Zernike polynomials play an important role in the systematic representation of the aberration function of an imaging system. In the 'forward' direction, when the light propagates from the exit pupil to the image region, the Zernike polynomials are instrumental for the (analytic) solution of the diffraction integral. For inverse problems in optics, their orthogonality on the exit pupil disc guarantees a stable inversion scheme. For pupil functions with a very broad spatial spectrum, the Zernike polynomials are less apt as an orthogonal basis since they show very rapid oscillations towards the rim of the unit disc when the radial degree assumes a large value. To avoid such rapid oscillations, the so-called 'disc-harmonic' functions are commonly invoked. In a similar way as for the Zernike polynomials, the disc-harmonic functions are expressed in a polar coordinate system and they are orthogonal on the unit disc. They play, on a circular support, the same role as the harmonic exponential functions that are used for the Fourier expansion of a function in Cartesian coordinates.

D.1 Structure of the Zernike Polynomials

The structure of each polynomial from the infinite set should be such that, in the case of a rotation of the coordinate system, the same polynomial will emerge subject to the rotation. This is due to the invariance of the unit disc under rotations about the origin. In **[37]**, App.VII, it is shown that a polynomial function of the Cartesian coordinates $g(x, y)$, written in polar coordinates as $g(x, y) = g(\rho\cos\theta, \rho\sin\theta) = g_1(\rho)\, g_2(\theta)$ satisfies the rotation-invariant requirement, provided that $g_2(\theta)$ is of the form $\exp(im\theta)$, where m is an arbitrary integer, zero included. The complex exponential function $\exp(im\theta)$ allows simultaneous treatment of cosine and sine components in the azimuthal expansion. Subject to a rotation θ_0, a polynomial with azimuthal order m is written in polar coordinates as

$$\begin{aligned} g[x\cos\theta_0 - y\sin\theta_0, x\sin\theta_0 + y\cos\theta_0] =& g[\rho\cos(\theta + \theta_0), \rho\sin(\theta + \theta_0)] \\ =& g_1(\rho)\exp\{im(\theta + \theta_0)\} = \exp(+im\theta_0)\, g_1(\rho)\exp(im\theta)\ . \end{aligned} \tag{D.1.1}$$

The expression shows that we recover the original polynomial, subject to a rotation that for this polynomial is represented by the function $\exp(im\theta_0)$.

The radial functions $g_1(\rho)$) with $0 \le \rho \le 1$ are polynomials with a degree in ρ of n, where n is the largest value of $r+s$ in the Cartesian expression $g(x,y)$ of the Zernike polynomial,

$$g(x,y) = \sum_{r,s \ge 0;\; r+s \le n} a_{rs}\, x^r y^s \,. \tag{D.1.2}$$

The radial Zernike polynomials depend on the azimuthal order m. The radial polynomials associated with an even value of m, including $m = 0$, contain only even powers of ρ. For the polynomials with odd m, only odd powers of ρ occur. Not all even or odd powers of ρ are allowed in the radial polynomial associated with even m or odd m, respectively. To establish the range of powers of ρ, a reasoning which is briefly stipulated in **[37]**, App.VII, Eq. (6), is further developed here. In terms of the polar coordinates (ρ, θ) we can write the function $g(x,y)$ as

$$g(\rho\cos\theta, \rho\sin\theta) = g_1(\rho)\, \exp(il\theta) \,, \tag{D.1.3}$$

where l is an integer which determines the azimuthal order of the Zernike polynomial on the interval $0 \le \theta \le 2\pi$. $g_1(\rho)$ is a polynomial in ρ and we have to establish which general terms $b_k \rho^k$ with integer $k = 0, \cdots, n$ have a nonzero coefficient b_k. Using the summation ranges for (r,s) according to Eq. (D.1.2) and integrating Eq. (D.1.3) with respect to θ over the interval $[0, 2\pi]$, we obtain the expression

$$\begin{aligned} g_1(\rho) &= \frac{1}{2\pi} \int_0^{2\pi} \exp(-il\theta)\, g(\rho\cos\theta, \rho\sin\theta)\, d\theta \\ &= \frac{1}{2\pi} \sum_{r,s \ge 0;\; r+s \le n} a_{rs}\, \rho^{r+s} \int_0^{2\pi} \exp(-il\theta)\, \cos^r\theta \sin^s\theta \, d\theta \,. \end{aligned} \tag{D.1.4}$$

By means of a binomial expansion of the powers of $\cos\theta$ and $\sin\theta$ we see that

$$\cos^r\theta\, \sin^s\theta = \left(\frac{\exp(i\theta) + \exp(-i\theta)}{2} \right)^r \left(\frac{\exp(i\theta) - \exp(-i\theta)}{2i} \right)^s \tag{D.1.5}$$

is a linear combination of powers $\exp(ij\theta)$, $j = r+s,\ r+s-2, \cdots, -(r+s)$. So, the coefficients b_{rsl}, defined by

$$b_{rsl} = \frac{1}{2\pi} \int_0^{2\pi} \exp(-il\theta)\, \cos^r\theta\, \sin^s\theta \, d\theta \,, \tag{D.1.6}$$

vanish unless a power $j = l$ is present in the linear combination of powers $\exp(ij\theta)$ which represents $\cos^r\theta\, \sin^s\theta$. The condition $j = l$ is satisfied provided the integer $r+s \ge |l|$ and $r+s$ and l have equal parity. Hence, denoting $g_1(\rho)$ by $R(\rho)$, the so-called *radial* Zernike polynomial, we have the expression

$$R(\rho) = \sum_{\substack{r,s \,\ge\, 0;\;\; r+s \,\le\, n \\ r+s \,=\, |l|,\, |l|+2,\, \cdots}} a_{rs}\, b_{rsl}\, \rho^{r+s} \,. \tag{D.1.7}$$

The summation range for the powers $r+s$ shows that $R(\rho)$ is a polynomial in ρ of degree $\le n$ containing only the powers $\rho^{|l|}, \rho^{|l|+2}, \cdots$

A general radial Zernike polynomial, named $g_1(\rho)$ up to this point, can now be written in line with standard notation as $R_n^{|m|}(\rho)$. On purpose we write $|m|$ instead of m for the upper index to emphasise that we use the radial polynomials in combination with general complex exponential functions $\exp(im\theta)$. For the index m we permit any integer number including zero, provided that $n - |m|$ is even and ≥ 0. The radial polynomial with degree n and azimuthal order m is of the following form:

$$R_n^{|m|}(\rho) = \sum_{j=0}^{(n-|m|)/2} a_{n,j}^{|m|}\, \rho^{|m|+2j} \,, \tag{D.1.8}$$

where the $a_{n,j}^{|m|}$ are the coefficients to be found. The maximum degree is n, the lowest is $|m|$ and, as was stated above, we have the condition that $(n-|m|)/2$ is a non-negative integer number.

D.2 Construction of the Zernike Polynomials

There are several approaches to constructing the polynomials with the requirements described above. In the original paper by Zernike **[385]**, the polynomials were introduced as the result of the characteristic function of a differential equation. Another possibility for defining a class of polynomials is the use of Rodrigues' formula, provided with the typical functions for that class of polynomials **[4]**. A set of polynomials can also be derived from their generating function. The polynomials $P_n(x)$ of degree n follow from an expression

$$G(x, z) = \sum_{n=0}^{\infty} P_n(x) z^n \,, \tag{D.2.1}$$

where $G(x, z)$ is such a generating function. There is no straightforward procedure to find the generating function belonging to a set of polynomials. Once the polynomials have been shown to belong to a certain class, the previous work on finding the generating function of that class can facilitate the construction of the generating function of a precise set.

D.2.1 Relation between Zernike and Jacobi Polynomials

For the derivation of the radial Zernike polynomials, with their required orthogonality on the finite interval $[0, 1]$, we turn to the Jacobi polynomials as the basic class with a finite existence interval. For the Jacobi polynomials, we use the representation

$$P_k^{(\alpha,\beta)}(t) = \frac{(\beta + k)!}{k!\,(\alpha+\beta+k)!} \sum_{j=0}^{k} (-1)^{k-j} \binom{k}{j} \frac{(\alpha+\beta+k+j)!}{(\beta+j)!} \left(\frac{t+1}{2}\right)^j \,, \tag{D.2.2}$$

that follows from **[4]**, (22.3.2) on p. 775 and (22.4.1) on p. 777. Here k, α and β are non-negative integers. The Jacobi polynomials are orthogonal on the interval $[-1, 1]$ with weight function $(1-t)^\alpha(1+t)^\beta$, so that

$$\int_{-1}^{1} (1-t)^\alpha (1+t)^\beta \, P_k^{(\alpha,\beta)}(t)\, P_{k'}^{(\alpha,\beta)}(t)\, dt = 0. \tag{D.2.3}$$

when $k, k' = 0, 1, \cdots$ and $k \neq k'$.

To connect the radial Zernike polynomials to the appropriate Jacobi polynomials, we observe from Eq. (D.1.8) that, for $m \geq 0$,

$$R_n^m(\rho) = \rho^m \, S_n^m(\rho^2) \,, \tag{D.2.4}$$

where $S_n^m(\rho^2)$ is a polynomial in ρ^2 of degree $\frac{1}{2}(n-m)$. The orthogonality condition for the radial polynomials,

$$\int_0^1 R_n^m(\rho)\, R_{n'}^m(\rho)\, d\rho^2 = 0 \,, \qquad n \neq n' \,, \tag{D.2.5}$$

becomes

$$\begin{aligned} &\int_0^1 \rho^{2m}\, S_n^m(\rho^2)\, S_{n'}^m(\rho^2)\, d\rho^2 \\ &= \frac{1}{2^{m+1}} \int_{-1}^{1} (1+t)^m\, S_n^m\!\left(\frac{1+t}{2}\right) S_{n'}^m\!\left(\frac{1+t}{2}\right) dt = 0 \,, \qquad n \neq n' \,, \end{aligned} \tag{D.2.6}$$

where we have substituted $t = 2\rho^2 - 1 \in [-1, 1]$. Since $(1+t)^m$ is the weight function on $[-1, 1]$ pertaining to the Jacobi polynomials $P_k^{(\alpha,\beta)}(t)$ with $\alpha = 0$ and $\beta = m$, we choose

$$S_n^m\left(\frac{1+t}{2}\right) = P_{(n-m)/2}^{(0,\,m)}(t) \,, \tag{D.2.7}$$

i.e., since $t = 2\rho^2 - 1$ when $\rho^2 = (1+t)/2$,

$$R_n^m(\rho) = \rho^m P_{(n-m)/2}^{(0,\,m)}(2\rho^2 - 1) \,. \tag{D.2.8}$$

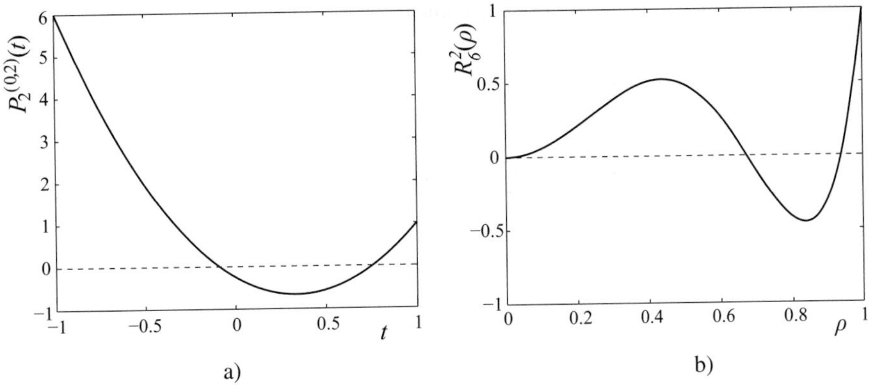

Figure D.1: a) The Jacobi polynomial $P_2^{(0,2)}(t)$.
b) The Zernike radial polynomial $R_6^2(\rho)$ derived from $P_2^{(0,2)}(t)$.

We now use Eq. (D.2.2) with $k = (n-m)/2$, $\alpha = 0, \beta = m$ and $t = 2\rho^2 - 1$ to get

$$R_n^m(\rho) = \frac{1}{k!} \sum_{j=0}^{k} (-1)^{k-j} \binom{k}{j} \frac{(m+k+j)!}{(m+j)!} \rho^{m+2j} . \tag{D.2.9}$$

Letting $s = k - j = 0, 1, \cdots, k$ and noting that $n = m + 2k$, so that $m + k + j = n - s$, $m + j = (n+m)/2 - s$, $m + 2j = n - 2s$, we then finally obtain

$$R_n^m(\rho) = \sum_{s=0}^{(n-m)/2} (-1)^s \frac{(n-s)!}{s!\left(\frac{n-m}{2} - s\right)!\left(\frac{n+m}{2} - s\right)!} \rho^{n-2s} , \tag{D.2.10}$$

which is the standard form for the radial Zernike polynomial of azimuthal order m and degree n. A different notation is sometimes used, reading

$$R_n^m(\rho) = \sum_{s=0}^{(n-m)/2} (-1)^s \binom{n-s}{s} \binom{n-2s}{\frac{n-m}{2} - s} \rho^{n-2s} . \tag{D.2.11}$$

There follows that the coefficients of the power expansion of $R_n^m(\rho)$ are integer numbers. Figure D.1 shows an example of two related Jacobi and Zernike polynomials. Equation (D.2.8) gives their relationship following from the nonlinear variable transformation $t = 2\rho^2 - 1$. At the ρ-origin, for $m \neq 0$, it is the extra factor ρ^m which determines the behaviour of the Zernike polynomial.

In the special case of circularly symmetric Zernike polynomials with upper index $m = 0$, the Jacobi polynomials reduce to the Legendre polynomials $P_n(t)$ and then

$$R_{2n}^0(\rho) = P_n(2\rho^2 - 1) . \tag{D.2.12}$$

Table D.1 shows the expressions for the radial Zernike polynomials of various order numbers n and m.

D.2.1.1 Generating Function and Differential Equation

Now that we have established the relationship between the Jacobi polynomials and the radial Zernike polynomials, it is possible to find the generating function of the Zernike polynomials from that of the Jacobi polynomials. For the latter polynomials we have the generating function (see **[4]**, (22.9.1), p.783)

$$\frac{1}{r(1-z+r)^\alpha(1+z+r)^\beta} = \sum_{j=0}^{\infty} \frac{1}{2^{\alpha+\beta}} P_j^{(\alpha,\beta)}(t)\, z^j , \tag{D.2.13}$$

Table D.1. Radial part of the Zernike polynomials for n and $m \leq 6$.

n→ m↓	0	1	2	3	4	5	6
0	1		$2\rho^2 - 1$		$6\rho^4 - 6\rho^2 + 1$		$20\rho^6 - 30\rho^4 + 12\rho^2 - 1$
1		ρ		$3\rho^3 - 2\rho$		$10\rho^5 - 12\rho^3 + 3\rho$	
2			ρ^2		$4\rho^4 - 3\rho^2$		$15\rho^6 - 20\rho^4 + 6\rho^2$
3				ρ^3		$5\rho^5 - 4\rho^3$	
4					ρ^4		$6\rho^6 - 5\rho^4$
5						ρ^5	
6							ρ^6

where r is given by $\sqrt{1 - 2tz + z^2}$. In the case of the Zernike polynomials, we put $\alpha = 0, \beta = m, t = 2\rho^2 - 1$ and write

$$\frac{(2\rho)^m}{\sqrt{1 - 2(2\rho^2 - 1)z + z^2}\,[1 + z + \sqrt{1 - 2(2\rho^2 - 1)z + z^2}\,]^m} = \sum_{j=0}^{\infty} R^m_{m+2j}(\rho)\, z^j \,. \tag{D.2.14}$$

After some rearrangement this yields the expression

$$\frac{\left[1 + z - \sqrt{1 - 2(2\rho^2 - 1)z + z^2}\right]^m}{(2\rho z)^m \sqrt{1 - 2(2\rho^2 - 1)z + z^2}} = \sum_{j=0}^{\infty} R^m_{m+2j}(\rho)\, z^j \,. \tag{D.2.15}$$

The expression for the generating function can be used to calculate the value of the polynomials at the interval limit $\rho = 1$. For this value, the generating function reduces to $1/(1 - z)$ and, expanding this in powers of z, we have

$$\frac{1}{1 - z} = \sum_{j=0}^{\infty} z^j = \sum_{j=0}^{\infty} R^m_{m+2j}(1)\, z^j \,, \tag{D.2.16}$$

showing that $R^m_n(1) = 1$ for the permitted combinations of n and m. For the value of the radial polynomials at the origin we find from Eqs. (D.2.10)–(D.2.12) and from the property of the Legendre polynomials, $P_n(-x) = (-1)^n P_n(x)$, see **[4]**,

$$\begin{aligned} R^0_{2n}(0) &= (-1)^n \,, & m &= 0 \,, \\ R^m_n(0) &= 0 \,, & m &\neq 0 \,. \end{aligned} \tag{D.2.17}$$

The differential equation that produces the radial Zernike polynomials as characteristic functions is given by

$$\rho(1 - \rho^2)\frac{d^2 R^m_n(\rho)}{d\rho^2} + (1 - 3\rho^2)\frac{dR^m_n(\rho)}{d\rho} + \left\{ n(n + 2)\rho - \frac{m^2}{\rho} \right\} R^m_n(\rho) = 0 \,. \tag{D.2.18}$$

This result is given in **[385]** and **[259]**.

Another method for defining a set of polynomial functions uses Rodrigues' formula. In **[259]**, with the aid of Rodrigues' equation for the Jacobi polynomials, it is shown that there holds

$$R^m_n(\rho) = \frac{\rho^{-m}}{\left(\frac{n-m}{2}\right)!} \left\{ \frac{d}{d(\rho^2)} \right\}^{\frac{n-m}{2}} \left\{ (\rho^2)^{\frac{n+m}{2}} (\rho^2 - 1)^{\frac{n-m}{2}} \right\} \,. \tag{D.2.19}$$

D.2.1.2 Inner Product of Two Zernike Polynomials

We first consider the inner product of two Zernike polynomials with equal upper index m. With the aid of Eq. (D.2.8) we write

$$I^{m,m}_{n,n'} = \frac{1}{2} \int_0^1 \rho^{2m} P^{(0,m)}_{(n-m)/2}(2\rho^2 - 1) P^{(0,m)}_{(n'-m)/2}(2\rho^2 - 1)\, d(2\rho^2 - 1) \,. \tag{D.2.20}$$

For the inner product of two Jacobi polynomials with lower indices k and k' and identical upper indices (α, β) we have from **[4]**,

$$I = \int_{-1}^{+1} (1-t)^{\alpha}(1+t)^{\beta}\, P_k^{(\alpha,\beta)}(t)\, P_{k'}^{(\alpha,\beta)}(t)\, dt = \frac{2^{\alpha+\beta+1}}{(2k+\alpha+\beta+1)} \frac{(k+\alpha)!\,(k+\beta)!}{k!\,(k+\alpha+\beta)!}\, \delta_{kk'}\ . \tag{D.2.21}$$

Applying this result to Eq. (D.2.20) using $\alpha = 0,\ \beta = m,\ t = 2\rho^2 - 1 \in [-1, 1]$, we obtain

$$I_{n,n'}^{m,m} = \begin{cases} \dfrac{1}{n+1}\,, & n = n'\,, \\ 0\,, & n \neq n'\,. \end{cases} \tag{D.2.22}$$

Using the complex azimuthal function $\exp(im\theta)$, we obtain for the inner product of two Zernike polynomials, $Z_n^m(\rho,\theta) = R_n^{|m|}(\rho)\exp(im\theta)$ and $Z_{n'}^{m'}(\rho,\theta) = R_{n'}^{|m'|}(\rho)\exp(im'\theta)$,

$$I_{n,n'}^{m,m'} = \frac{1}{\pi}\int_0^{2\pi}\int_0^1 R_n^{|m|}(\rho) R_{n'}^{|m'|}(\rho) \exp\{i(m-m')\theta\}\,\rho d\rho d\theta = \begin{cases} 1/(n+1)\,, & n = n', m = m'\,, \\ 0\,, & n \neq n'\,, m = m'\,, \\ 0\,, & m \neq m'\,, \end{cases} \tag{D.2.23}$$

or, in more concise notation,

$$I_{n,n'}^{m,m'} = \frac{\delta_{nn'}\,\delta_{mm'}}{n+1}\ . \tag{D.2.24}$$

If the cosine and sine azimuthal functions are used, we find the result

$$\begin{aligned} I &= \frac{1}{\pi}\int_0^{2\pi}\int_0^1 R_n^m(\rho) R_{n'}^{m'}(\rho)\cos(m\theta)\cos(m'\theta)\,\rho d\rho d\theta \\ &= \begin{cases} 1/(n+1)\,, & n = n',\ m = m' = 0\,, \\ 1/[2(n+1)]\,, & n = n',\ m = m' \neq 0\,, \\ 0\,, & n \neq n',\ m = m'\,, \\ 0\,, & m \neq m'\,, \end{cases} \end{aligned} \tag{D.2.25}$$

and comparable expressions apply to the Zernike sine polynomials. The orthogonality properties of the Zernike polynomials yield zero for the inner product of a Zernike cosine and a Zernike sine polynomial.

From the results for the inner products of two Zernike polynomials it follows that the notation using complex exponential functions for the azimuthal part of the Zernike polynomials has clear advantages. No special selection rules have to be applied to assign an inner product value when using the complex exponential functions. A further normalisation of the inner product value has been proposed in the literature, such that the inner product of two (scaled) Zernike polynomials equals unity (*orthonormal* Zernike polynomials). Denoting an orthonormal radial Zernike polynomial by $\overline{R}_n^m$, we have that

$$\overline{R}_n^m(\rho) = \sqrt{n+1}\ R_n^m(\rho)\,, \qquad \overline{c}_n^m = \frac{c_n^m}{\sqrt{n+1}}\,, \tag{D.2.26}$$

where the over-barred radial polynomials $\overline{R}_n^m(\rho)$ and expansion coefficients $\overline{c}_n^m$ are associated with the combination of orthonormal radial polynomials and complex azimuthal functions. As soon as cosine and sine polynomials are used ($m \neq 0$), the prefactors have to be adapted. Although the orthonormal property is an attractive feature, we do not use the normalised radial Zernike polynomials in this book. Their presence in the vast literature on Zernike polynomials has remained rather limited.

D.3 Computation of Zernike Polynomials

The standard representation of Eq. (D.2.10) is most frequently used for computation of the radial part $R_n^{|m|}(\rho)$ of a Zernike polynomial. Its numerical accuracy is low for larger values of n when subsequent powers of ρ with large multiplying integer factors have to be added and subtracted. If a set of measured data has to be represented by means of a Zernike

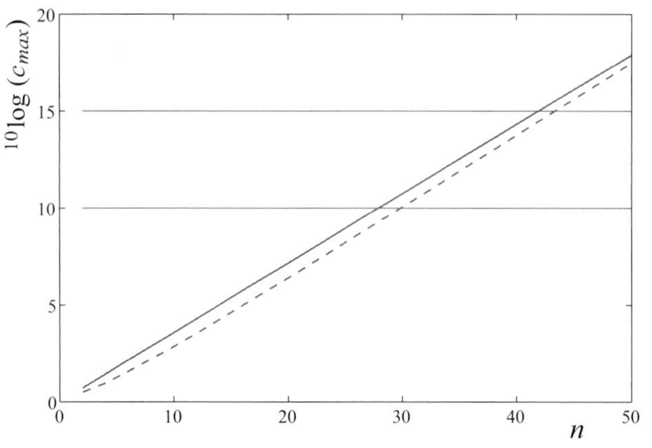

Figure D.2: Maximum value c_{max} of the coefficients of the standard power series of a radial Zernike polynomial (see Eq. (D.2.10)). We assume that $m \ll n$.
Solid line: crude approximation assuming that c_{max} is found for $s_{\text{max}} = n/4$.
Dashed line: refined approximation using $s_{\text{max}} = (2 - \sqrt{2})\, n/4$.

expansion, the Zernike polynomials themselves need to be computed at a general point (ρ, θ) with high accuracy. The standard expression for the radial polynomial $R_n^{|m|}(\rho)$ then fails if individual terms of the power series expansion have a number of significant digits which is equal to that of the machine representation. In practice, for polynomial degrees n beyond a typical value of 30 to 40, the numerical accuracy becomes problematic when double precision computation is used. The accuracy of the computation is most compromised for ρ close to unity. For $m \ll n$ and $\rho = 1$, we can make an estimate of the loss of numerical accuracy:

- A crude estimate assumes, for instance by observation of polynomial coefficients in Table D.1, that the largest coefficient of the power series of Eq. (D.2.10) occurs when the summation index s_{max} equals $n/4$. Assuming that $n/4$ has an integer value, the absolute value of the coefficient $c_{n/4}$ equals

$$c_{n/4} \approx \frac{(3n/4)!}{[(n/4)!\,]^3} \,. \tag{D.3.1}$$

  The approximation sign is used in this equation since s_{max} is an approximated value for the index of the largest coefficient and since we have assumed that $m \ll n$. We then use a rather elementary version of Stirling's formula, $n! = n^n$, to evaluate the logarithm of a factorial number which yields the simple expression $c_{n/4} = 3^{3n/4}$. We calculate the value of $^{10}\log(c_{n/4}) = 0.3578\, n$ and obtain the straight solid line of Fig. D.2. Note that we have also included non-integer values of $n/4$ in this graph. The solid line shows that substantial loss of accuracy occurs, down to 10^{-5}, when $n \geq 28$ and that double precision calculations start to fail for $n \geq 42$.
- A more refined approximation[1] calculates an approximate, analogue value of the index s_{max} by equating two neighbouring coefficients of the power series such that $c_{s+1}/c_s = 1$. Allowing non-integer values in this calculation of the index s_{max}, we find the following expressions for the index s_{max} of the largest coefficient in the series and for the corresponding value c_{max},

$$s_{\text{max}} \approx (2 - \sqrt{2})\,\frac{n}{4}\,, \qquad c_{\text{max}} \approx \frac{\sqrt{2}}{\pi\, n}\left(1 + \sqrt{2}\right)^{n+1}. \tag{D.3.2}$$

  In obtaining this result for c_{max}, we have used the improved value of s_{max} together with a more precise version of Stirling's formula ($n! = n^{n+1/2}\, e^{-n}\, \sqrt{2\pi}$). The approximation sign in the expressions for s_{max} and c_{max} is now there for two reasons, firstly since we have assumed that $m \ll n$ and, secondly, since analogue values of s have been permitted. The values of c_{max} as a function of n according to Eq. (D.3.2) have been plotted as a dashed curve in Fig. D.2. We observe that the range for n is slightly larger now before numerical breakdown occurs for double precision calculations ($n = 43.5$).

[1] Private communication Dr. A.J.E.M. Janssen.

D.3.1 Recurrence Relations for Polynomial Computation

An improved numerical stability for high-order Zernike polynomials is obtained by using a recursive formula. In the case of the radial Zernike polynomials with two indices n and m, the recursion operation can use a shift in the lower index or in the upper index or in both. The basic recursion formula with shifts in the lower index n follows immediately from a recurrence relation for the Jacobi polynomials $P_k^{\{\alpha,\beta\}}(x)$ (**[4]**, (22.7.1), p. 782)

$$P_{k+1}^{\{\alpha,\beta\}}(x) = \frac{1}{2(k+1)(k+\alpha+\beta+1)(2k+\alpha+\beta)} \times \left\{ \left[(2k+\alpha+\beta+1)(\alpha^2-\beta^2) + \Pi_3(2k+\alpha+\beta)\,x\right] P_k^{\{\alpha,\beta\}}(x) - 2(k+\alpha)(k+\beta)(2k+\alpha+\beta+2)\,P_{k-1}^{\{\alpha,\beta\}}(x) \right\}, \qquad \text{(D.3.3)}$$

where Pochhammer's symbol $\Pi_j(a)$ is defined as $\Gamma(a+j)/\Gamma(a)$. Using the substitutions $k \rightarrow (n-m)/2, \alpha = 0, \beta = m$ and $x = 2\rho^2 - 1$, we find the following recurrence relation for the radial Zernike polynomials with equal upper index m,

$$R_{n+2}^m(\rho) = \frac{4(n+1)(n+2)}{(n+2)^2 - m^2} \left\{ \left[\rho^2 - \frac{1}{2} - \frac{m^2}{2n(n+2)}\right] R_n^m(\rho) - \frac{n^2 - m^2}{4n(n+1)} R_{n-2}^m(\rho) \right\}. \qquad \text{(D.3.4)}$$

This expression, together with other recurrence relations that change the upper index, can be found in **[259]**. The expression of Eq. (D.3.4) has also been derived in **[336]** using the Clebsch–Gordan expression for the product of two series.

In Fig. D.3a we show the numerical effect of using the power series expansion for computation of the value of a radial Zernike polynomial. As an example of numerical inaccuracy, the polynomial $R_{44}^0(\rho)$ has been shown in the critical region of ρ close to unity. According to Eq. (D.3.1), the polynomial with such an index is fully in the region of inaccurate computation. For comparison, the exact curve has been plotted in the same graph (dashed curve). In Fig. D.3b a very high-order polynomial, $R_{1000}^0(\rho)$, has been plotted, applying the recursive scheme of Eq. (D.3.4). Numerically stable computation of a polynomial up to an order n as high as 10^4 is feasible with such a recursive scheme. These extremely high orders can be needed if the Zernike polynomials are used for fitting functions with a high-frequency oscillatory behaviour on the unit disc. In optical applications like metrology and aberration characterisation, the polynomial degree rarely exceeds 20.

Another calculation method for a radial Zernike polynomial that has a good numerical behaviour irrespective of the degree of the polynomial relies on the discrete Fourier transform (DFT). The DFT was presented in Appendix A and its application to the calculation of a Zernike polynomial can be found in **[167]**. The Zernike polynomial is written as a series

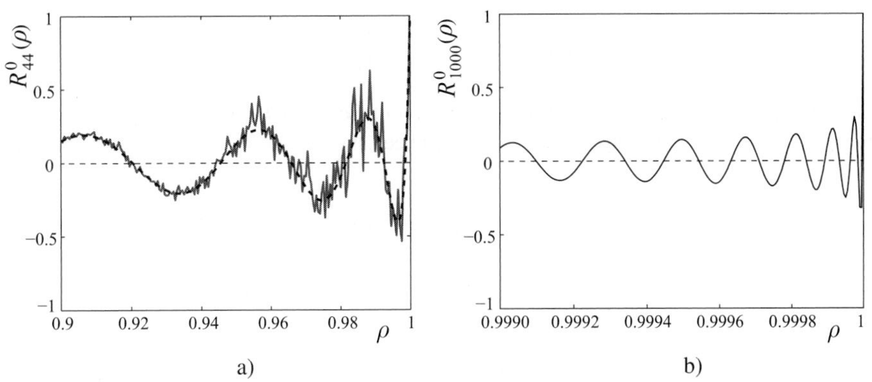

Figure D.3: a) The radial Zernike polynomial $R_{44}^0(\rho)$ on the interval $0.9 \leq \rho \leq 1$, calculated with the aid of Eq. (D.2.9) (solid curve) and using the recursive scheme of Eq. (D.3.4) (dashed curve).
b) The polynomial $R_{1000}^0(\rho)$ on the interval $0.999 \leq \rho \leq 1$, calculated with the aid of the recursive scheme of Eq. (D.3.4).

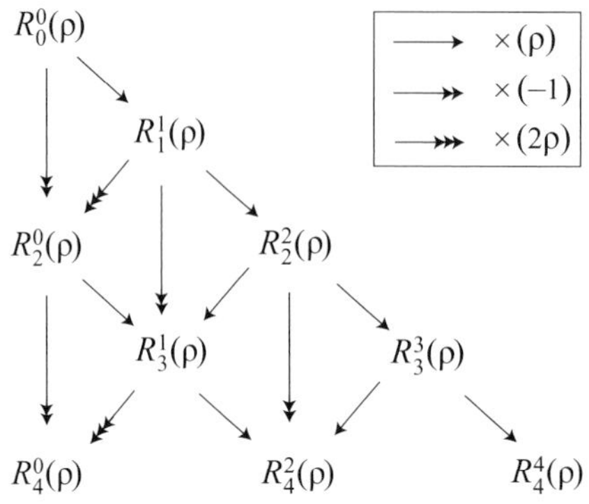

Figure D.4: Flow diagram for recursive calculation **[305]** of all Zernike polynomials up to a maximum degree of n using Eq. (D.3.7).

$$R_n^{|m|}(\rho) = \frac{1}{N}\sum_{k=0}^{N-1} U_n\{\rho\cos(2\pi k/N)\}\cos(2\pi mk/N)\,, \qquad 0 \le \rho \le 1\,, \tag{D.3.5}$$

where N is an integer $> n + |m|$. The function U_n is the Chebyshev polynomial of the second kind and of degree n. This polynomial is easily evaluated using the expression

$$U_n(x) = \frac{\sin(n+1)u}{\sin u}\,, \qquad x = \cos u\,. \tag{D.3.6}$$

To conclude this subsection we mention a global recursion formula that has been published by Shakibaei and Paramesran **[305]**. The adjective 'global' refers to the property that the coefficients in the recursion scheme are independent of the indices (n, m) of the polynomial under consideration. The recursion formula is given by

$$R_n^m(\rho) = \rho\left\{R_{n-1}^{|m-1|}(\rho) + R_{n-1}^{m+1}(\rho)\right\} - R_{n-2}^m(\rho)\,. \tag{D.3.7}$$

Using the initialisation $R_0^0(\rho) = 1$ and the definition $R_n^m(\rho) = 0$ for $n < m$, all radial polynomials up to a maximum degree of n are conveniently calculated with the aid of the single Eq. (D.3.7). The general recursion equation has a simpler form for the polynomials with equal lower and upper index: $R_{n+1}^{n+1}(\rho) = \rho R_n^n(\rho)$. The polynomials with zero upper index satisfy the equation $R_n^0(\rho) = 2\rho R_{n-1}^1(\rho) - R_{n-2}^0(\rho)$. The recursion equation (D.3.7) is illustrated by means of the flow diagram of Fig. D.4 which allows calculation of a full triangle of Zernike polynomials with a maximum degree of 4.

D.3.2 Cosine Representation of the Zernike Polynomials

The numerical instability associated with the power expansion of Eq. (D.2.9) can be avoided by the introduction of a transformation of the variable ρ according to $\cos x = \rho$. It has been shown by Janssen **[164]** that there holds

$$R_n^m(\cos x) = \sum_{j=0}^{\lfloor n/2 \rfloor} a_{n-2j}\cos[(n-2j)x]\,, \tag{D.3.8}$$

where $\lfloor n/2 \rfloor$ denotes the largest integer $\le n/2$. Ref. **[164]** presents an analytic expression for the a_k coefficients which are given by

$$a_k = \epsilon_k \frac{(p_1)!\,(q_1)!}{s!\,t!}\left(\frac{1}{2}\right)^l \left[P_p^{(\gamma,\delta)}(0)\right]^2\,, \tag{D.3.9}$$

where $P_p^{(\gamma,\delta)}$ is the Jacobi polynomial of degree p, ϵ_k is Neumann's symbol ($\epsilon_0 = 1$, $\epsilon_1 = \epsilon_2 = \cdots = 2$) and the various integers and indices are given by

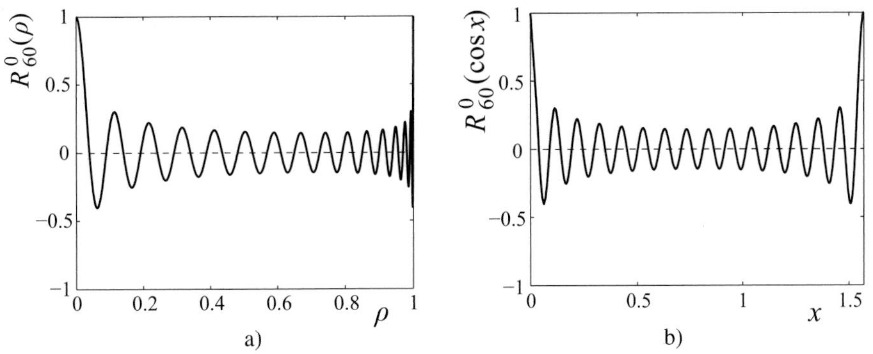

Figure D.5: a) The radial Zernike polynomial $R^0_{60}(\rho)$ on the interval $0 \le \rho \le 1$.
b) The polynomial $R^0_{60}(\cos x)$ on the interval $0 \le x \le \pi/2$.

$$l = \max(m,k)\,, \qquad r = \min(m,k)\,, \qquad p_1 = \frac{n-l}{2}\,, \qquad q_1 = \frac{n+l}{2}\,,$$
$$s = \frac{n-r}{2}\,, \qquad t = \frac{n+r}{2}\,, \qquad \gamma = \frac{l-r}{2}\,, \qquad \delta = \frac{l+r}{2}\,.$$

Substituting $x = 0$ in Eq. (D.3.8) we find that $R^{|m|}_n(1) = 1 = \sum_{j=0}^{\lfloor n/2 \rfloor} a_{n-2j}$. Moreover, from Eq. (D.3.9) we deduce that all coefficients a_k are non-negative. Consequently, each individual coefficient a_k is bound to the interval $[0,1]$.

In Fig. D.5 we have plotted two representations of the Zernike polynomials, as a function of ρ and of x, where $\cos x = \rho$. For high values of n and for $m = 0$, it is seen that the frequency content of the cosine-transformed polynomial gets concentrated around a typical frequency $\nu_x = n/(2\pi)$. In Table D.2 we give the values of the coefficients a_k for the radial Zernike polynomials of order $n \le 6$. The transformation $\rho = \cos x$ has been applied and c_k denotes the harmonic function $\cos(kx)$. The coefficient data have been taken from **[164]**. In Table D.3 we list the names and the expressions for a commonly used set of Zernike polynomials (Zernike *fringe* polynomials) which is extensively used in optical aberration analysis and in optical metrology **[380]**.

D.3.3 Zernike Expansion of a Function on the Unit Disc

The Zernike polynomials are often used with azimuthal cosine and sine functions, leading to an expansion

$$f(\rho,\theta) = \sum_{n,m} R^m_n(\rho)\,\{a^m_n \cos(m\theta) + b^m_n \sin(m\theta)\}\ , \tag{D.3.10}$$

where the coefficients a^m_n and β^m_n are real if the function f itself is real. The indices n and m are both positive with $(n-m) \ge 0$ and even. In the case of a general complex function, the coefficients a and b are also complex. An alternative representation of a function is

$$f(\rho,\theta) = \sum_{n,m} c^m_n R^{|m|}_n(\rho)\exp(im\theta)\ , \tag{D.3.11}$$

where the index m runs from $-m_{\max}$ to $+m_{\max}$ with $n-|m|$ even and ≥ 0. If the function $f(\rho,\theta)$ is real, we find the conditions $\Re\{c^m_n\} = \Re\{c^{-m}_n\}$, $\Im\{c^m_n\} = -\Im\{c^{-m}_n\}$. The Zernike expansion of the function is given by

$$f(\rho,\theta) = \sum_{n,m} R^m_n(\rho)\,\{2\Re\{c^m_n\}\cos(m\theta) - 2\Im\{c^m_n\}\sin(m\theta)\}\ . \tag{D.3.12}$$

For a general complex function $f(\rho,\theta)$, the complex expansion yields the following real and imaginary part in terms of the cosine and sine expansion functions,

$$f(\rho,\theta) = \sum_{n,m} R^{|m|}_n(\rho)\Big\{\Re\{c^m_n + c^{-m}_n\}\cos(m\theta) - \Im\{c^m_n - c^{-m}_n\}\sin(m\theta)$$
$$+\ i\left[\Im\{c^m_n + c^{-m}_n\}\cos(m\theta) + \Re\{c^m_n - c^{-m}_n\}\sin(m\theta)\right]\Big\}\,. \tag{D.3.13}$$

Table D.2. Cosine representation of the radial Zernike polynomials. In the table we use the shorthand notation $c_k = \cos kx$.

n→ ↓ m	0	1	2	3	4	5	6	7	8
0	1 c_0		$2\rho^2 - 1$ c_2		$6\rho^4 - 6\rho^2 + 1$ $\frac{1}{4}c_0 + \frac{3}{4}c_4$		$20\rho^6 - 30\rho^4 + 12\rho^2 - 1$ $\frac{3}{8}c_2 + \frac{5}{8}c_6$		$70\rho^8 - 140\rho^6 + 90\rho^4 - 20\rho^2 + 1$ $\frac{9}{64}c_0 + \frac{5}{16}c_4 + \frac{35}{64}c_8$
1		ρ c_1		$3\rho^3 - 2\rho$ $\frac{1}{4}c_1 + \frac{3}{4}c_3$		$10\rho^5 - 12\rho^3 + 3\rho$ $\frac{1}{4}c_1 + \frac{1}{8}c_3 + \frac{5}{8}c_5$		$35\rho^7 - 60\rho^5 + 30\rho^3 - 4\rho$ $\frac{9}{64}c_1 + \frac{15}{64}c_3 + \frac{5}{64}c_5 + \frac{35}{64}c_7$	
2			ρ^2 $\frac{1}{2}c_0 + \frac{1}{2}c_2$		$4\rho^4 - 3\rho^2$ $\frac{1}{2}c_2 + \frac{1}{2}c_4$		$15\rho^6 - 20\rho^4 + 6\rho^2$ $\frac{3}{16}c_0 + \frac{1}{32}c_2 + \frac{5}{16}c_4 + \frac{15}{32}c_6$		$56\rho^8 - 105\rho^6 + 60\rho^4 - 10\rho^2$ $\frac{9}{32}c_2 + \frac{1}{16}c_4 + \frac{7}{32}c_6 + \frac{7}{16}c_8$
3				ρ^3 $\frac{3}{4}c_1 + \frac{1}{4}c_3$		$5\rho^5 - 4\rho^3$ $\frac{1}{8}c_1 + \frac{9}{16}c_3 + \frac{5}{16}c_5$		$21\rho^7 - 30\rho^5 + 10\rho^3$ $\frac{15}{64}c_1 + \frac{1}{64}c_3 + \frac{27}{64}c_5 + \frac{21}{64}c_7$	
4					ρ^4 $\frac{3}{8}c_0 + \frac{1}{2}c_2 + \frac{1}{8}c_4$		$6\rho^6 - 5\rho^4$ $\frac{5}{16}c_2 + \frac{1}{2}c_4 + \frac{3}{16}c_6$		$28\rho^8 - 42\rho^6 + 15\rho^4$ $\frac{5}{32}c_0 + \frac{1}{16}c_2 + \frac{1}{8}c_4 + \frac{7}{16}c_6 + \frac{7}{32}c_8$
5						ρ^5 $\frac{5}{8}c_1 + \frac{5}{16}c_3 + \frac{1}{16}c_5$		$7\rho^7 - 6\rho^5$ $\frac{5}{64}c_1 + \frac{27}{64}c_3 + \frac{25}{64}c_5 + \frac{7}{64}c_7$	
6							ρ^6 $\frac{5}{16}c_0 + \frac{15}{32}c_2 + \frac{3}{16}c_4 + \frac{1}{32}c_6$		$8\rho^8 - 7\rho^6$ $\frac{7}{32}c_2 + \frac{7}{16}c_4 + \frac{9}{32}c_6 + \frac{1}{16}c_8$
7								ρ^7 $\frac{35}{64}c_1 + \frac{21}{64}c_3 + \frac{7}{64}c_5 + \frac{1}{64}c_7$	
8									ρ^8 $\frac{35}{128}c_0 + \frac{7}{16}c_2 + \frac{7}{32}c_4 + \frac{1}{16}c_6 + \frac{1}{128}c_8$

Table D.3. The 37 *fringe* Zernike polynomials, $R_n^m(\rho)\cos m\theta$ and $R_n^m(\rho)\sin m\theta$, with sequence number N_W, running from 0 to 36, according to **[219]**,**[380]**. There is no consensus in the literature about the nomenclature of the higher-order aberration terms. The aberration names in the table are suggestions.

$\frac{n+m}{2}$	$\frac{n-m}{2}$	n	m	N_W		$R_n^m(\rho)$	Nomenclature
0	0	0	0	0		1	piston
1	0	1	1	1	$\cos\theta$	ρ	tilt
1	0	1	1	2	$\sin\theta$	ρ	
1	1	2	0	3		$2\rho^2-1$	defocus
2	0	2	2	4	$\cos 2\theta$	ρ^2	astigmatism
2	0	2	2	5	$\sin 2\theta$	ρ^2	
2	1	3	1	6	$\cos\theta$	$3\rho^3-2\rho$	coma
2	1	3	1	7	$\sin\theta$	$3\rho^3-2\rho$	
2	2	4	0	8		$6\rho^4-6\rho^2+1$	primary spherical
3	0	3	3	9	$\cos 3\theta$	ρ^3	trefoil
3	0	3	3	10	$\sin 3\theta$	ρ^3	
3	1	4	2	11	$\cos 2\theta$	$4\rho^4-3\rho^2$	secondary
3	1	4	2	12	$\sin 2\theta$	$4\rho^4-3\rho^2$	astigmatism
3	2	5	1	13	$\cos\theta$	$10\rho^5-12\rho^3+3\rho$	secondary
3	2	5	1	14	$\sin\theta$	$10\rho^5-12\rho^3+3\rho$	coma
3	3	6	0	15		$20\rho^6-30\rho^4+12\rho^2-1$	secondary spherical
4	0	4	4	16	$\cos 4\theta$	ρ^4	tetrafoil
4	0	4	4	17	$\sin 4\theta$	ρ^4	
4	1	5	3	18	$\cos 3\theta$	$5\rho^5-4\rho^3$	secondary
4	1	5	3	19	$\sin 3\theta$	$5\rho^5-4\rho^3$	trefoil
4	2	6	2	20	$\cos 2\theta$	$15\rho^6-20\rho^4+6\rho^2$	cubic
4	2	6	2	21	$\sin 2\theta$	$15\rho^6-20\rho^4+6\rho^2$	astigmatism
4	3	7	1	22	$\cos\theta$	$35\rho^7-60\rho^5+30\rho^3-4\rho$	cubic
4	3	7	1	23	$\sin\theta$	$35\rho^7-60\rho^5+30\rho^3-4\rho$	coma
4	4	8	0	24		$70\rho^8-140\rho^6+90\rho^4-20\rho^2+1$	tertiary spherical
5	0	5	5	25	$\cos 5\theta$	ρ^5	pentafoil
5	0	5	5	26	$\sin 5\theta$	ρ^5	
5	1	6	4	27	$\cos 4\theta$	$6\rho^6-5\rho^4$	secondary
5	1	6	4	28	$\sin 4\theta$	$6\rho^6-5\rho^4$	tetrafoil
5	2	7	3	29	$\cos 3\theta$	$21\rho^7-30\rho^5+10\rho^3$	tertiary
5	2	7	3	30	$\sin 3\theta$	$21\rho^7-30\rho^5+10\rho^3$	trefoil
5	3	8	2	31	$\cos 2\theta$	$56\rho^8-105\rho^6+60\rho^4-10\rho^2$	quartic
5	3	8	2	32	$\sin 2\theta$	$56\rho^8-105\rho^6+60\rho^4-10\rho^2$	astigmatism
5	4	9	1	33	$\cos\theta$	$126\rho^9-280\rho^7+210\rho^5-60\rho^3+5\rho$	quartic
5	4	9	1	34	$\sin\theta$	$126\rho^9-280\rho^7+210\rho^5-60\rho^3+5\rho$	coma
5	5	10	0	35		$252\rho^{10}-630\rho^8+560\rho^6-210\rho^4+30\rho^2-1$	quartic spherical
6	6	12	0	36		$924\rho^{12}-2772\rho^{10}+3150\rho^8$ $-1680\rho^6+420\rho^4-42\rho^2+1$	quintic spherical

D.4 Theorems and Expressions Related to Radial Zernike Polynomials

In this section we present some theorems related to radial Zernike polynomials and the corresponding proofs. The theorems play an important role in the chapters on aberration theory and diffraction theory.

D.4.1 The Inner Product of a Monomial and a Radial Zernike Polynomial

In some applications the inner product between the radial part of a Zernike polynomial and the monomial term of a power series expansion is needed. The product is defined by

$$I = \int_0^1 \rho^a R_n^m(\rho)\,\rho d\rho\,, \tag{D.4.1}$$

where m is non-negative and $n-m$ is even and non-negative. To evaluate the product we follow the derivation which is given in **[37]**, **[259]** and, more specifically, in **[57]**. With the aid of Rodrigues' equation for the Jacobi polynomials, a Rodrigues' equation for the radial Zernike polynomials has been derived (see Eq. (D.2.19)), subject to the same restrictions on n and m as given above. Rodrigues' expression is used for calculation of the integral I of Eq. (D.4.1) which is also encountered in Eqs. (5.9.12) and (5.9.16). We use $(n-m)/2 = p$, $(n+m)/2 = q$, insert Rodrigues' expression into Eq. (D.4.1), and, substituting $\rho^2 = x$, we obtain,

$$I = \frac{1}{2(p!)}\int_0^1 x^{(a-m)/2}\left\{\frac{d}{dx}\right\}^p [x^q(x-1)^p]\,dx\,. \tag{D.4.2}$$

By a single integration step by parts, we obtain

$$I = \frac{-1}{2(p!)}\left(\frac{a-m}{2}\right)\int_0^1 x^{(a-m)/2\,-1}\left\{\frac{d}{dx}\right\}^{p-1} [x^q(x-1)^p]\,dx\,. \tag{D.4.3}$$

After p integrations by parts we have, using $q-p=m$,

$$I = \frac{(-1)^p}{2(p!)}\left(\frac{a-m}{2}\right)\cdots\left(\frac{a-m}{2}-p+1\right)\int_0^1 x^{(a+m)/2}(x-1)^p\,dx\,. \tag{D.4.4}$$

The remaining integral over x is equally subjected to p integrations by parts,

$$I = \frac{(-1)^{2p}}{2}\,\frac{\left(\frac{a-m}{2}\right)\cdots\left(\frac{a-m}{2}-p+1\right)}{\left(\frac{a+m}{2}+1\right)\cdots\left(\frac{a+m}{2}+p\right)}\int_0^1 x^{(a+m)/2+p}dx\,. \tag{D.4.5}$$

We then obtain the final result

$$I = \frac{\frac{1}{2}\left(\frac{a-m}{2}\right)\cdots\left(\frac{a-m}{2}-p+1\right)}{\left(\frac{a+m}{2}+1\right)\cdots\left(\frac{a+m}{2}+p+1\right)}\,. \tag{D.4.6}$$

The derivation is valid as long as $(a-m)/2-p>0$. However, both I in Eq. (D.4.1) and the right-hand side of Eq. (D.4.6) depend analytically on a, where a satisfies $\Re(a) > -m-2$. Therefore, the result of Eq. (D.4.6) extends to this range by analyticity. Finally, the result of Eq. (D.4.6) can be written in terms of Γ functions as

$$I = \frac{\frac{1}{2}\Gamma\left(\frac{a-m}{2}+1\right)\Gamma\left(\frac{a+m}{2}+1\right)}{\Gamma\left(\frac{a-m}{2}-p+1\right)\Gamma\left(\frac{a+m}{2}+p+2\right)}\,, \tag{D.4.7}$$

where the right-hand side vanishes when $a = m+2p-2, m+2p-4, \cdots, m$.

A numerically stable evaluation of I in (D.4.6) is based on the product representation

$$I = \frac{1}{a+m+2p+2}\prod_{j=0}^{p-1}\frac{a-m-2j}{a+m+2j+2}\,. \tag{D.4.8}$$

The general factor of the product expression in (D.4.8) is well behaved and ≤ 1 and a numerically accurate calculation of I is possible in standard double precision arithmetic for arbitrary high orders n and a. If $p = 0$ the multiple product is put equal to unity.

D.4.2 Zernike Expansion of the Product of Two Functions

In many physical problems related to functions that are expanded with the aid of Zernike polynomials, the product of two such functions needs to be evaluated. Ad hoc solutions have been proposed for products of two polynomials of low degree, for instance in Nijboer's thesis **[259]**. A systematic solution to the product linearisation problem has been provided by Tango **[336]** in the form

$$Z_{n_1}^{m_1}(\rho,\theta)\, Z_{n_2}^{m_2}(\rho,\theta) = \sum_{n_3,m_3} C_{n_1,n_2,n_3}^{m_1,m_2,m_3}\, Z_{n_3}^{m_3}(\rho,\theta)\,. \tag{D.4.9}$$

The Clebsch–Gordan coefficient C is related to Wigner's $3j$ symbol for angular momentum coupling in quantum mechanics (see **[266]**, §§34.2-3) through the relation

$$C_{n_1,n_2,n_3}^{m_1,m_2,m_3} = (n_3+1)\left|\begin{pmatrix} n_1/2 & n_2/2 & n_3/2 \\ m_1/2 & m_2/2 & -m_3/2 \end{pmatrix}\right|^2 , \tag{D.4.10}$$

where the quantity between curved brackets is the Wigner $3j$ symbol. Since the elements of the $3j$ symbol are related to indices of Zernike polynomials, we require as usual for a general integer index pair (n, m) that $n \geq 0$ and $n - |m|$ is even and non-negative. Many coefficients turn out to be zero since the properties of the Zernike polynomials require for the azimuthal order $|m|$ and the polynomial order n that

$$\begin{aligned} m_3 &= m_1 + m_2, \\ n_3 &= n_1 + n_2, n_1 + n_2 - 2, \cdots, |n_1 - n_2|\,, \end{aligned} \tag{D.4.11}$$

and the number of nonzero coefficients C in the right-hand side of Eq. (D.4.9) would normally be given by $1 + (n_1 + n_2 - m_1 - m_2)/2$. In **[266]**, **[336]** it is shown that, because of the triangle condition that has to be respected by the set of angular momenta, the number of nonzero coefficients is at most $\min\{n_1, n_2\} + 1$.

A general $3j$ symbol is given by **[266]**

$$\begin{aligned} \begin{pmatrix} j_1 & j_2 & j_3 \\ i_1 & i_2 & i_3 \end{pmatrix} &= (-1)^{j_1-j_2-i_3} \left(\frac{(j_1+j_2-j_3)!\,(j_1-j_2+j_3)!\,(-j_1+j_2+j_3)!}{(j_1+j_2+j_3+1)!} \right)^{1/2} \\ &\times \left[(j_1+i_1)!\,(j_1-i_1)!\,(j_2+i_2)!\,(j_2-i_2)!\,(j_3+i_3)!\,(j_3-i_3)!\right]^{1/2} \\ &\times \sum_s \frac{(-1)^s}{s!\,(j_1+j_2-j_3-s)!\,(j_1-i_1-s)!\,(j_2+i_2-s)!\,(j_3-j_2+i_1+s)!\,(j_3-j_1-i_2+s)!}\,. \end{aligned} \tag{D.4.12}$$

The summation over s is applied to all non-negative integers such that the arguments in the factorials are non-negative. The Clebsch–Gordan coefficients have a simpler form if the upper indices m_1 and m_2 of the polynomials to be multiplied are both equal to zero (see **[52]** and **[266]**, (34.3.5). p. 759). For this case we have the expression

$$R_{2n_1}^0(\rho) R_{2n_2}^0(\rho) = \sum_{k=0}^{n_1} \frac{A_{n_1-k} A_k A_{n_2-k}}{A_{n_1+n_2-k}} \left(\frac{2n_1 + 2n_2 - 4k + 1}{2n_1 + 2n_2 - 2k + 1} \right) R_{2n_1+2n_2-4k}^0(\rho)\,. \tag{D.4.13}$$

The expression is valid provided $0 \leq n_1 \leq n_2$ and the coefficients A are then given by

$$A_n = \frac{1}{n!}\, 1 \cdot 3 \cdot \ldots \cdot (2n-1) = \frac{1}{2^n} \binom{2n}{n}\,. \tag{D.4.14}$$

The general function product problem can now be stated as follows. Given two functions

$$f_1 = \sum_{n_1,m_1} \beta_{n_1}^{m_1}(1)\, Z_{n_1}^{m_1} \quad \text{and} \quad f_2 = \sum_{n_2,m_2} \beta_{n_2}^{m_2}(2)\, Z_{n_2}^{m_2}\,,$$

find the expansion of

$$f_1 f_2 = f_3 = \sum_{n_3,m_3} \beta_{n_3}^{m_3}(3)\, Z_{n_3}^{m_3}\,, \tag{D.4.15}$$

where the extra variables 1, 2 and 3 of the β coefficients indicate to which expansion the βs belong. The coefficients $\beta_{n_3}^{m_3}(3)$ follow from

$$\begin{aligned} f_3 &= \sum_{n_1,m_1,n_2,m_2} \beta_{n_1}^{m_1}(1)\beta_{n_2}^{m_2}(2)\, Z_{n_1}^{m_1}\, Z_{n_2}^{m_2} \,, \\ &= \sum_{n_1,m_1,n_2,m_2} \beta_{n_1}^{m_1}(1)\beta_{n_2}^{m_2}(2) \sum_{n_3,m_3} C_{n_1,n_2,n_3}^{m_1,m_2,m_3}\, Z_{n_3}^{m_3} \,, \\ &= \sum_{n_3,m_3} \left(\sum_{n_1,m_1,n_2,m_2} C_{n_1,n_2,n_3}^{m_1,m_2,m_3}\, \beta_{n_1}^{m_1}(1)\beta_{n_2}^{m_2}(2) \right) Z_{n_3}^{m_3} \,, \end{aligned} \tag{D.4.16}$$

with the result that

$$\beta_{n_3}^{m_3}(3) = \sum_{n_1,m_1,n_2,m_2} C_{n_1,n_2,n_3}^{m_1,m_2,m_3}\, \beta_{n_1}^{m_1}(1)\beta_{n_2}^{m_2}(2) \,. \tag{D.4.17}$$

Dr. A.J.E.M. Janssen brought the results of this subsection to the attention of the first author. As has been shown previously in this appendix, a reliable and efficient calculation of coefficients comprising factorials should avoid the explicit calculation of these factorials and, instead, use a stable recursive scheme. A complete recursive scheme[2] for efficient calculation of the coefficients $\beta_{n_3}^{m_3}$ has been constructed. For the special case with $m_1 = 0$ a recursive scheme can be found in **[127]**, (Eq. (151)).

D.4.3 A Theorem Related to the Calculation of Transverse Aberration Components

In Section 5.9.3.1, the expressions on the left-hand side of Eq. (5.9.35) were given as a summation over a set of Zernike polynomials with appropriate coefficients. In this subsection a comprehensive theorem[3] is given **[165]** that produces the results of Eq. (5.9.35) in an unambiguous way, both for positive and negative azimuthal orders m of Zernike polynomials,

$$\left(R_n^{|m|}\right)'(\rho) \pm \frac{m}{\rho} R_n^{|m|}(\rho) = 2 \sum_{k=0}^{(n-|m|)/2} (n-2k) R_{n-2k-1}^{|m\mp 1|}(\rho) \,, \tag{D.4.18}$$

where the prime at the left-hand side of the equation indicates that differentiation with respect to the radial coordinate ρ has to be performed.

For the proof of the theorem, the following integral result for the radial Zernike polynomials is needed,

$$R_n^m(\rho) = (-1)^{(n-m)/2} \int_0^\infty J_{n+1}(r) J_m(\rho r)\, dr \,, \quad 0 < \rho < 1 \,, \tag{D.4.19}$$

where $J_n(r)$ is a Bessel function of the first kind and of order n. This result, holding for integer $n, m \geq 0$ such that $n-m$ is even and non-negative, is often attributed to Noll **[260]**. It is, however, equivalent to the basic integral result of the Zernike–Nijboer diffraction theory, see **[259]** ((Eq. (2.22)), and it is a special case of the Weber–Schaftheitlin integral, see **[4]**, (11.4.33 and 11.4.34). This integral produces a zero result for the right-hand side of Eq. (D.4.19) when $n < m$. By using Eq. (D.4.19) with $|m|$ instead of m and employing $J_{|m|}(z) = (-1)^{(|m|-m)/2} J_m(z)$, one can generalise Eq. (D.4.19) to be valid for all integer n and m with $n - |m|$ even and non-negative as

$$R_n^{|m|}(\rho) = (-1)^{(n-m)/2} \int_0^\infty J_{n+1}(r) J_m(\rho r)\, dr \,, \quad 0 < \rho < 1 \,, \tag{D.4.20}$$

where it is important to note that the right-hand side of Eq. (D.4.20) involves only m and not $|m|$.

To derive the theorem of Eq. (D.4.18), we first consider the case of general integer $m \neq 0$. From **[4]**, (9.1.27), we have

$$J_m'(z) = \pm J_{m\mp 1}(z) \mp \frac{m}{z} J_m(z) \,, \tag{D.4.21}$$

$$r J_{n+1}(r) = -r J_{n-1}(r) + 2n J_n(r) \,, \tag{D.4.22}$$

[2] This scheme was provided by Dr. A.J.E.M. Janssen to the first author in a private communication.

[3] The proof of this theorem has been provided to the first author by Dr. A.J.E.M. Janssen. The proof given here is a careful elaboration of the argument given by Noll to show how it is possible to derive Eq. (13) from Eq. (8) in **[260]** (we note that Eq. (8) of **[260]** is identical to Eq. (D.4.19) here).

for general n and m. We now differentiate Eq. (D.4.20) with respect to ρ and substitute the result in the left-hand side of Eq. (D.4.18) to obtain

$$\begin{aligned}\left(R_n^{|m|}\right)'(\rho) \pm \frac{m}{\rho} R_n^{|m|}(\rho) &= (-1)^{(n-m)/2} \int_0^\infty J_{n+1}(r)\, rJ_m'(\rho r)dr \pm (m/\rho)R_n^{|m|}(\rho) \\ &= (-1)^{(n-m)/2} \int_0^\infty J_{n+1}(r)\, r\left[\pm J_{m\mp1}(\rho r) \mp (m/\rho r)J_m(\rho r)\right] dr \pm (m/\rho)R_n^{|m|}(\rho) \\ &= \pm\,(-1)^{(n-m)/2} \int_0^\infty J_{n+1}(r)\, rJ_{m\mp1}(\rho r)dr\,, \end{aligned} \tag{D.4.23}$$

where the result of Eq. (D.4.21) has been used to proceed from the first to the second line of the equation, and the general property of Eq. (D.4.20) to obtain the last line of Eq. (D.4.23). We then apply Eq. (D.4.22) to obtain

$$\begin{aligned}\left(R_n^{|m|}\right)'(\rho) \pm \frac{m}{\rho} R_n^{|m|}(\rho) &= \pm\,(-1)^{(n-m)/2} \int_0^\infty \left[-J_{n-1}(r)\, r + 2nJ_n(r)\right] J_{m\mp1}(\rho r)dr \\ &= \mp\,(-1)^{(n-m)/2} \int_0^\infty J_{n-1}(r)\, rJ_{m\mp1}(\rho r)dr \pm 2n\,(-1)^{(n-m)/2} \int_0^\infty J_n(r)J_{m\mp1}(\rho r)dr\,. \end{aligned} \tag{D.4.24}$$

In the next step we substitute the result of Eq. (D.4.23), where n is replaced by $n-2$, for the first integral in the last member of Eq. (D.4.24) and the result of Eq. (D.4.20) where (n, m) is replaced by $(n-1, m\mp1)$ for the second integral above, to obtain

$$\left(R_n^{|m|}\right)'(\rho) \pm \frac{m}{\rho} R_n^{|m|}(\rho) = \left(R_{n-2}^{|m|}\right)'(\rho) \pm \frac{m}{\rho} R_{n-2}^{|m|}(\rho) + 2nR_{n-1}^{|m\mp1|}(\rho)\,. \tag{D.4.25}$$

By successive induction steps according to Eq. (D.4.25), each time lowering n by 2 until $|m|$, we get

$$\begin{aligned}\left(R_n^{|m|}\right)'(\rho) \pm \frac{m}{\rho} R_n^{|m|}(\rho) &= 2nR_{n-1}^{|m\mp1|}(\rho) + 2(n-2)R_{n-3}^{|m\mp1|}(\rho) + \cdots + 2|m|\; R_{|m|-1}^{|m\mp1|}(\rho) \\ &= 2 \sum_{k=0}^{(n-|m|)/2} (n-2k)\; R_{n-2k-1}^{|m\mp1|}(\rho)\,, \end{aligned} \tag{D.4.26}$$

so that Eq. (D.4.18) follows. We observe that $R_{|m|-1}^{|m+1|}(\rho) = 0$ when $m > 0$ and that $R_{|m|-1}^{|m-1|}(\rho) = 0$ when $m < 0$.

The identity of Eq. (D.4.18) and its proof for the case that $m = 0$ is entirely similar. One just deletes the terms involving (m/z) and (m/ρ) from Eqs. (D.4.21), (D.4.23) and (D.4.26) to obtain

$$\left(R_n^0\right)'(\rho) = 2\sum_{k=0}^{n/2}(n-2k)\; R_{n-2k-1}^1(\rho)\,, \tag{D.4.27}$$

where $R_{-1}^1(\rho)$ equals zero in the summation.

D.5 Zernike Polynomials and the Disc-harmonic Functions on the Unit Disc

In this last section of the appendix on the efficient and stable representation of wavefront and transverse aberration functions, another approach to wavefront fitting on the unit disc is presented. The Zernike polynomials are frequently used in optics for two reasons. They form an orthogonal set on the unit disc (a laterally scaled version of the physical exit pupil of an imaging system). The consequence of orthogonality is that the inner product of a polynomial with itself provides the variance of the polynomial on the unit disc (the average value of each Zernike polynomial, except Z_0^0, is zero). In the case of small wavefront aberration, the variance of a wavefront is directly related to the central irradiance in the diffraction image of a point source. The second reason why Zernike polynomials are omnipresent in optics is that the coefficients of each polynomial provide the user with the optimum ratios between the aberration coefficients of the classical Seidel and higher-order aberrations of a specific aberration type. This automatic aberration balancing property is very instructive for an optical designer. It should be said that the analytic insight which follows from a Zernike polynomial expansion has gradually become less important for the optical designer. Numerical optimisation based on the minimisation of Zernike coefficients automatically steers the design towards a solution with a balanced aberration distribution.

A practical disadvantage of Zernike polynomials is their strongly varying density of zero crossings, especially towards the rim of the unit disc (see, for instance, Fig. D.5a on this subject). When function fitting with high-order polynomials is needed, the cell dimension of a wavefront sensing device has to be decreased in a disproportionate way. In this respect, spatial sampling on a square or rectangular area with the aid of the basic function of the Fourier transform, the periodic complex exponential function, offers practical advantages when higher frequencies need to be included. The sampling density has to be increased in a linear manner to recover the higher-frequency content of the function to be fitted. A set of orthogonal functions on the unit disc that offers a comparable sampling density requirement to that of the periodic Fourier basic functions is the set of *disc-harmonic* functions. They were proposed a long time ago in Watson's textbook on Bessel functions **[360]**. Recently, they have been further studied in the framework of pattern recognition and adaptive optics **[355]**. In this section we demonstrate how a set of two-dimensional disc-harmonic functions can be constructed on the unit disc and how their zero-crossings are distributed over the unit disc in the radial direction.

D.5.1 Construction of the Disc-harmonic Functions

In line with the definition of Zernike polynomials we write a disc-harmonic function, denoted by $d_n^m(\rho,\theta)$, as the product of a radial function and an azimuthal complex exponential function,

$$d_n^m(\rho,\theta) = J_m(2\pi l_{nm}\rho)\exp(im\theta)\,, \tag{D.5.1}$$

where J_m is the Bessel function of the first kind of order m. The factor l_{nm} determines the lateral scaling of the Bessel function and is used to orthogonalise the radial parts of the disc-harmonic functions. As was the case for the Zernike polynomials, the azimuthal functions $\exp(im\theta)$ take care of the orthogonality of two disc-harmonic functions with different upper indices. The orthogonality of two disc-harmonic functions with equal upper index m requires that the inner product of the radial parts equals zero,

$$\int_0^1 d_n^m(\rho,\theta)\, d_\nu^m(\rho,\theta)\,\rho d\rho = 0\,, \qquad n \neq \nu\,,$$

or,

$$\int_0^1 J_m(2\pi l_{nm}\rho) J_m(2\pi l_{\nu m}\rho)\,\rho d\rho = 0\,. \tag{D.5.2}$$

To determine for which values of the factor l_{nm} in the argument of the Bessel functions J_m the zero inner product condition can be satisfied we use expression (10.22.4) of **[266]** where the following indefinite integral is given,

$$\int \rho J_m(a\rho)J_m(b\rho)\,d\rho = \frac{\rho\,[aJ_{m+1}(a\rho)J_m(b\rho) - bJ_m(a\rho)J_{m+1}(b\rho)]}{a^2-b^2}\,. \tag{D.5.3}$$

Substitution of the lower and upper integration limits in this indefinite integral yields the expression

$$\int_0^1 J_m(a\rho)J_m(b\rho)\,\rho d\rho = \frac{[aJ_{m+1}(a)J_m(b) - bJ_m(a)J_{m+1}(b)]}{a^2-b^2}\,. \tag{D.5.4}$$

We then use **[266]**, (second item in (10.6.2)),

$$J_m'(\rho) = \frac{m}{\rho}J_m(\rho) - J_{m+1}(\rho)\,, \tag{D.5.5}$$

where the prime on the Bessel function indicates that the first derivative with respect to ρ has to be taken. We substitute the result of Eq. (D.5.5) in Eq. (D.5.4) and obtain

$$\begin{aligned}\int_0^1 J_m(a\rho)J_m(b\rho)\,\rho d\rho &= \frac{a\left[\frac{m}{a}J_m(a) - J_m'(a)\right]J_m(b) - bJ_m(a)\left[\frac{m}{b}J_m(b) - J_m'(b)\right]}{a^2-b^2}\\ &= \frac{bJ_m(a)J_m'(b) - aJ_m(b)J_m'(a)}{a^2-b^2}\,.\end{aligned} \tag{D.5.6}$$

The inner product of two disc-harmonic radial functions is made zero by equating to zero the numerator of Eq. (D.5.6), where $a = 2\pi l_{nm}$ and $b = 2\pi l_{\nu m}$ (see Eq. (D.5.2)). We can consider three ways to do this:

I Choose $2\pi l_{nm}$, $n = (0), 1, 2, \cdots$, as the consecutive zeros of $J_m(x)$ with $x > 0$ and $l_{00} = 0$.
II Choose $2\pi l_{nm}$, $n = (0), 1, 2, \cdots$, as the consecutive zeros of $J'_m(x)$ with $x > 0$ and $l_{00} = 0$.
III Let $C \neq 0, \infty$ be a fixed constant and choose $2\pi l_{nm}$, $n = (0), 1, 2, \cdots$, as the consecutive solutions x of the equation

$$\frac{J_m(x)}{x J'_m(x)} = C\,. \tag{D.5.7}$$

The third option is unattractive compared to options I and II since there are well-established results for the consecutive zeros of $J_m(x)$ and $J'_m(x)$, but not for the consecutive solutions of Eq. (D.5.7) for general $C \neq 0, \infty$.

With respect to the disc-harmonic functions of type I, it follows from Eq. (D.5.1) that the radial part $J_m(2\pi l_{nm}\rho)$ of the disc-harmonic function equals zero at the rim of the unit disc. Therefore, these radial disc-harmonic functions are called 'clamped' functions. A similar argument shows that the disc-harmonic functions of type II have a zero-valued first derivative at the rim of the unit disc whilst their function value itself at the rim is not subjected to a specific condition. For this reason the functions of type II are called the 'free' disc-harmonic functions. The most frequently used disc-harmonic functions are those of type II. An important advantage of the 'free' disc-harmonic functions is that not only the functions themselves form an orthogonal set on the unit disc, but also their spatial derivatives (see **[149]** for a proof of this property of the 'free' disc-harmonic functions). A second advantage of the 'free'functions is the decay of the fitted coefficients c_n that appear when a general function is represented by means of a series of disc-harmonic functions. It can be shown[4] that the coefficients $c_{n,I}$ associated with the type I disc-harmonic functions decay according to an $n^{-1/2}$-law, whereas the coefficients $c_{n,II}$ pertaining to the 'free' disc-harmonic functions show a more favourable decay according to $n^{-3/2}$.

The inner product of the radial parts of two disc-harmonic functions is given by **[355]**,

$$\int_0^1 J_m(2\pi l_{nm}\rho) J_m(2\pi l_{n'm}\rho)\,\rho d\rho = \begin{cases} \dfrac{1}{2}, & n = n' = 0\,, \\[2ex] \dfrac{1}{2}\left[\{J'_m(2\pi l_{nm})\}^2 + \left\{1 - \left(\dfrac{m}{2\pi l_{nm}}\right)^2\right\}\{J_m(2\pi l_{nm})\}^2\right], & n = n' \neq 0\,, \\[2ex] 0\,, & n \neq n'\,, \end{cases} \tag{D.5.8}$$

and this expression is valid for all three types of disc-harmonic functions.

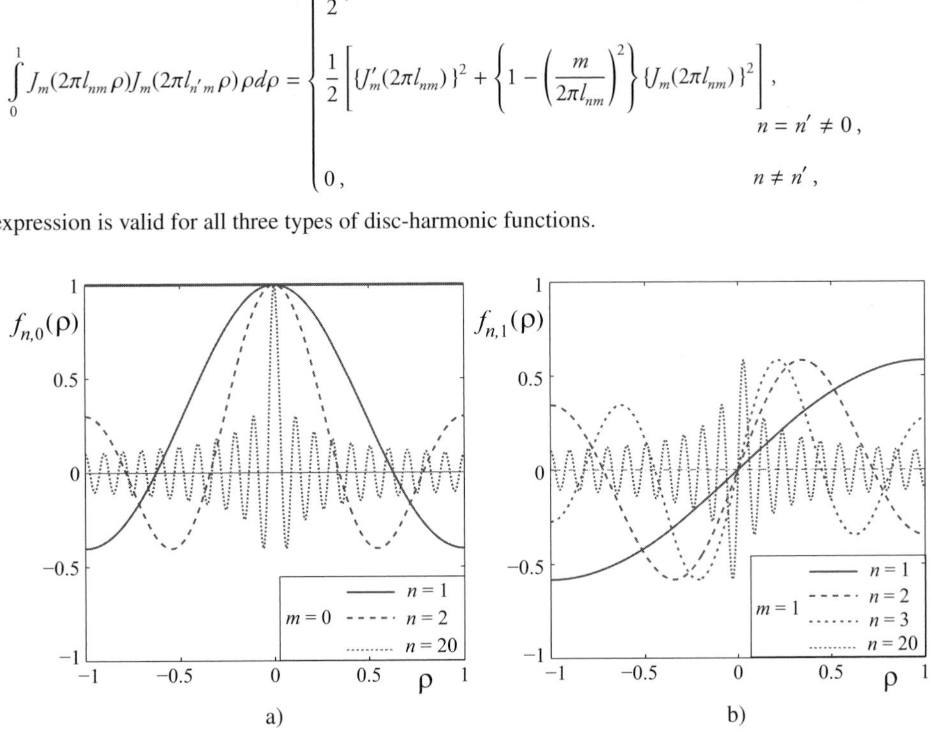

Figure D.6: 'Free' disc-harmonic functions (type II).
a) Radial part of a disc-harmonic function with radial symmetry ($m = 0$).
b) The same, but now for an azimuthal order number $m = 1$.

[4] Private communication Dr. A.J.E.M. Janssen.

We conclude this section on disc-harmonic functions by displaying the radial part of type-II disc-harmonic functions for some combinations of the order numbers n and m. In Fig. D.6a the azimuthal order m is zero and the radial order numbers are 1, 2 and 20. A cross-section through the origin at $\rho = 0$ is shown ($\theta = 0$). The functions show a maximum of unity at the origin and the envelope of the function decreases towards the rim of the unit disc. In contrast to the radial Zernike polynomials, it is seen that the local (spatial) frequency of the disc-harmonic function remains relatively constant over the entire cross-section. In Fig. D.6b, some asymmetric disc-harmonic functions are shown with azimuthal order $m = 1$ ($\theta = 0$). The same remark can be made for the symmetric functions in Fig. D.6a. The oscillations of a specific function with radial order number n show a constant frequency over the entire cross-section of the unit disc. This means that the required sampling density for the fitting of an unknown function has a constant value on the unit disc.

E Magnetically Induced Optical Rotation (Faraday Effect)

In this appendix we first discuss the physical origin of Faraday rotation. The Drude dispersion model is applied to obtain an expression for the medium polarisation in the presence of an external magnetic field. With the aid of the resulting permittivity tensor, we calculate the polarisation eigenstates of the medium and the propagation speed associated with each eigenstate. The strength of the Faraday effect then follows from the medium dispersion, for instance in terms of the Abbe dispersion number V_d. Finally we discuss the non-reciprocity of wave propagation through a 'Faraday medium' and its use as an optical isolator.

E.1 Polarisation of the Medium and the Permittivity Tensor

We use the classical picture of an electron, electric charge $-e$, orbiting around a nucleus, with orbital circular frequency ω_0 (see Fig. E.1). A harmonic electromagnetic wave is incident on the medium, with circular frequency ω and unit wave vector $\hat{\mathbf{k}}$ and a state of polarisation that is characterised by the electric and magnetic field strengths $\mathbf{E}_0$ and $\mathbf{H}_0$. The Coulomb force on the electron applied by the electric field strength $\mathbf{E}_0$ and the Lorentz force due to the external magnetic field cause a perturbation $\mathbf{a}$ of the electron orbit which is synchronised with the frequency ω of the electromagnetic wave,

$$\mathbf{a}(\mathbf{r}, t) = (a_x, a_y, a_x)\exp\{+i(\mathbf{k}\cdot\mathbf{r} - \omega t)\}\,, \tag{E.1.1}$$

where $\mathbf{r}$ is the position vector with respect to the centre of the electron orbit. In almost all instances, we may neglect the effect of the magnetic field $\mathbf{H}_0$ with respect to the equivalent external field $\mathbf{B}_e/\mu_0$, unless the electromagnetic intensity is very large. The dynamic equation for the electron in the orbit follows from

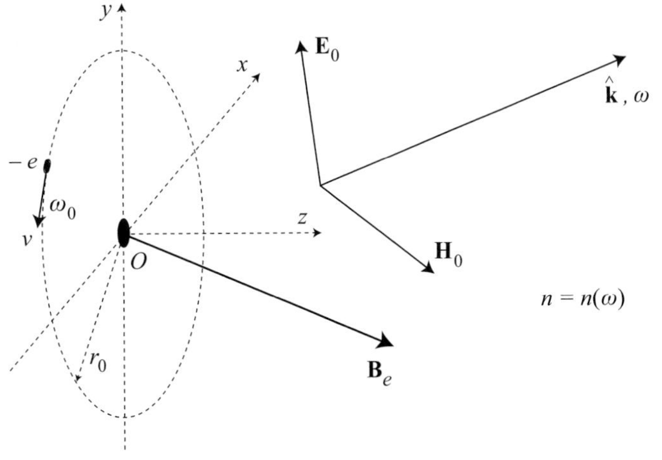

Figure E.1: Classical picture of an orbiting electron (circular frequency ω_0), subjected to the influence of an electromagnetic field (field strength vectors $\mathbf{E}_0$ and $\mathbf{H}_0$, unit wave vector $\hat{\mathbf{k}}$, temporal frequency ω), and a static external magnetic field $\mathbf{B}_e/\mu_0$. The radius of the circular orbit without external perturbation is r_0 and its rotation axis is parallel to the z-direction.

$$m_e \frac{\partial^2 \mathbf{a}}{\partial t^2} = -m_e \omega_0^2 \mathbf{a} - e\mathbf{E}_0 - e\frac{\partial \mathbf{a}}{\partial t} \times \mathbf{B}_e \,, \tag{E.1.2}$$

where m_e is the electron mass. The first term on the right-hand side is the centripetal force, the second the Coulomb force and the third term is the Lorentz force exerted on the moving electron by the external magnetic field. A more complete treatment should also take into account the precession of the electron orbit around the nucleus and the influence of other electrons in different orbits. However, for the basic effect of the external magnetic field on the polarisation of the medium, these details can be omitted.

Performing the differentiations and evaluating the vector product we obtain the following system of equations for the components of the orbit perturbation vector **a**,

$$\begin{pmatrix} m_e(\omega^2-\omega_0^2)/e & +i\omega B_{e,z} & -i\omega B_{e,y} \\ -i\omega B_{e,z} & m_e(\omega^2-\omega_0^2)/e & +i\omega B_{e,x} \\ +i\omega B_{e,y} & -i\omega B_{e,x} & m_e(\omega^2-\omega_0^2)/e \end{pmatrix} \begin{pmatrix} a_x \\ a_y \\ a_z \end{pmatrix} = \begin{pmatrix} E_{0,x} \\ E_{0,y} \\ E_{0,z} \end{pmatrix} . \tag{E.1.3}$$

We use Cramer's rule to solve the system of equations and find the expression

$$\begin{pmatrix} a_x \\ \\ \\ a_y \\ \\ \\ a_z \end{pmatrix} = \frac{e}{A} \begin{pmatrix} \left[m_e^2(\omega^2-\omega_0^2)^2 - \omega^2 e^2 B_{e,x}^2\right] E_{0,x} \\ +\left[-i\omega e B_{e,z} m_e(\omega^2-\omega_0^2) - \omega^2 e^2 B_{e,y} B_{e,z}\right] E_{0,y} \\ +\left[i\omega e B_{e,y} m_e(\omega^2-\omega_0^2) - \omega^2 e^2 B_{e,x} B_{e,z}\right] E_{0,z} \\ \\ \left[i\omega e B_{e,z} m_e(\omega^2-\omega_0^2) - \omega^2 e^2 B_{e,y} B_{e,z}\right] E_{0,x} \\ +\left[m_e^2(\omega^2-\omega_0^2)^2 - \omega^2 e^2 B_{e,y}^2\right] E_{0,y} \\ +\left[-i\omega e B_{e,x} m_e(\omega^2-\omega_0^2) - \omega^2 e^2 B_{e,y} B_{e,z}\right] E_{0,z} \\ \\ -\left[i\omega e B_{e,y} m_e(\omega^2-\omega_0^2) + \omega^2 e^2 B_{e,x} B_{e,z}\right] E_{0,x} \\ +\left[i\omega e B_{e,x} m_e(\omega^2-\omega_0^2) - \omega^2 e^2 B_{e,y} B_{e,z}\right] E_{0,y} \\ +\left[m_e^2(\omega^2-\omega_0^2)^2 - \omega^2 e^2 B_{e,z}^2\right] E_{0,z} \end{pmatrix} , \tag{E.1.4}$$

where the quantity A in the denominator is given by

$$A = m_e(\omega^2-\omega_0^2)\{m_e^2(\omega^2-\omega_0^2)^2 - e^2\omega^2 |B_e|^2\} . \tag{E.1.5}$$

The expression for the perturbation vector **a** is a nonlinear function of the external magnetic induction vector and the resulting Faraday rotation effect should be classified among the nonlinear electromagnetic effects. In practical situations, the quotient $m_e\omega/e$ at optical frequencies is equivalent to an induction value of 20 kT. This allows us to neglect the quantities that are of second order in the components of $\mathbf{B}_e$.

For the case of equally aligned orbiting electrons, the resulting polarisation of the medium is given by

$$\mathbf{P}_0 = -eN\mathbf{a} = -eN\, \overline{\overline{T}}_P\, \mathbf{E}_0 \,, \tag{E.1.6}$$

where N is the number of orbiting electrons per unit volume and the tensor $\overline{\overline{T}}_P$ is given by

$$\overline{\overline{T}}_P = -\frac{e}{m_e(\omega_0^2-\omega^2)} \begin{pmatrix} 1 & +i\dfrac{e\omega B_{e,z}}{m_e(\omega_0^2-\omega^2)} & -i\dfrac{e\omega B_{e,y}}{m_e(\omega_0^2-\omega^2)} \\ -i\dfrac{e\omega B_{e,z}}{m_e(\omega_0^2-\omega^2)} & 1 & +i\dfrac{e\omega B_{e,x}}{m_e(\omega_0^2-\omega^2)} \\ +i\dfrac{e\omega B_{e,y}}{m_e(\omega_0^2-\omega^2)} & -i\dfrac{e\omega B_{e,x}}{m_e(\omega_0^2-\omega^2)} & 1 \end{pmatrix}$$

$$= -\frac{e}{m_e(\omega_0^2-\omega^2)} \overline{\overline{T}}_{P,n} \,, \tag{E.1.7}$$

where $\overline{\overline{T}}_{P,n}$ is the normalised polarisation tensor. The relative permittivity tensor $\overline{\overline{\epsilon}}_r$ of the medium follows from

$$\epsilon_0\, \overline{\overline{\epsilon}}_r\, \mathbf{E}_0 = \epsilon_0 \mathbf{E}_0 + \mathbf{P}_0 \,, \tag{E.1.8}$$

and we obtain the expression

$$\overline{\overline{\epsilon}}_r = \overline{\overline{I}} + \frac{e^2 N}{m_e \epsilon_0 (\omega_0^2 - \omega^2)} \overline{\overline{T}}_{P,n} \ . \tag{E.1.9}$$

As usual we write $e^2 N/(m_e \epsilon_0) = \omega_p^2$ where ω_p is the plasma frequency of the medium. We make the further substitutions

$$n_a^2 = 1 + \frac{\omega_p^2}{\omega_0^2 - \omega^2} \ , \qquad \omega_c = \frac{eB_e}{m_e} \ , \tag{E.1.10}$$

where n_a is the refractive index of the medium and ω_c the cyclotron frequency of the electrons under the influence of the external magnetic field. Differentiating n_a with respect to ω and writing $\omega_c(dn_a/d\omega) = \delta n_a$, we have the following expression for the relative permittivity tensor,

$$\overline{\overline{\epsilon}}_r = \begin{pmatrix} n_a^2 & +in_a(\delta n_a)\hat{B}_{e,z} & -in_a(\delta n_a)\hat{B}_{e,y} \\ -in_a(\delta n_a)\hat{B}_{e,z} & n_a^2 & +in_a(\delta n_a)\hat{B}_{e,x} \\ +in_a(\delta n_a)\hat{B}_{e,y} & -in_a(\delta n_a)\hat{B}_{e,x} & n_a^2 \end{pmatrix} , \tag{E.1.11}$$

where $\hat{\mathbf{B}}_e$ is the unit vector in the direction of the applied external magnetic field.

E.2 Solution of the Wave Equation

To solve the Helmholtz equation for a plane wave in this medium, we construct the appropriate matrix, comparable to the one given in Eq. (2.9.10) and then solve the linear system of equations

$$\begin{pmatrix} \frac{n_a^2}{n^2} - 1 + \hat{k}_x^2 & +i\frac{n_a}{n^2}(\delta n_a)\hat{B}_{e,z} + \hat{k}_x\hat{k}_y & -i\frac{n_a}{n^2}(\delta n_a)\hat{B}_{e,y} + \hat{k}_x\hat{k}_z \\ -i\frac{n_a}{n^2}(\delta n_a)\hat{B}_{e,z} + \hat{k}_x\hat{k}_y & \frac{n_a^2}{n^2} - 1 + \hat{k}_y^2 & +i\frac{n_a}{n^2}(\delta n_a)\hat{B}_{e,x} + \hat{k}_y\hat{k}_z \\ +i\frac{n_a}{n^2}(\delta n_a)\hat{B}_{e,y} + \hat{k}_x\hat{k}_z & -i\frac{n_a}{n^2}(\delta n_a)\hat{B}_{e,x} + \hat{k}_y\hat{k}_z & \frac{n_a^2}{n^2} - 1 + \hat{k}_z^2 \end{pmatrix} \mathbf{E}_0 = M_E \, \mathbf{E}_0 = 0 \ . \tag{E.2.1}$$

After some manipulation, the eigenvalue equation $|M_E| = 0$ reduces to

$$n^4 - \left\{2n_a^2 - (\delta n_a)^2 \sin^2\theta\right\} n^2 + n_a^2 \left\{n_a^2 - (\delta n_a)^2\right\} = 0 \ , \tag{E.2.2}$$

where $\cos\theta = \hat{\mathbf{k}} \cdot \hat{\mathbf{B}}_e$. For sufficiently small values of δn_a with respect to n_a, we obtain the approximate solution

$$n \approx n_a \pm \frac{\delta n_a}{2} \cos\theta \ . \tag{E.2.3}$$

The corresponding eigenstates for the electric field vector are elliptical. It is easily verified that for a propagation direction parallel to the external magnetic field we have circular eigenstates. For instance, if $\hat{\mathbf{k}} = \hat{\mathbf{B}}_e = (0, 0, 1)$, we find the eigenstates

$$\begin{aligned} &\text{RC:} \quad n_{RC} = n_a + \frac{\delta n_a}{2} \ , \qquad \hat{\mathbf{E}}_{RC} = \frac{1}{\sqrt{2}} (1, -i, 0) \ , \\ &\text{LC:} \quad n_{LC} = n_a - \frac{\delta n_a}{2} \ , \qquad \hat{\mathbf{E}}_{LC} = \frac{1}{\sqrt{2}} (1, +i, 0) \ . \end{aligned} \tag{E.2.4}$$

If $\hat{\mathbf{k}}$ and $\hat{\mathbf{B}}_e$ are orthogonal, the eigenstates are linear in the approximation of small $\delta n_a/n_a$. In second order, the states are elliptical and the ratio of the small and long axis is given by $\delta n_a/n_a$.

For the case of interest with $\hat{\mathbf{k}}\cdot\hat{\mathbf{B}}_e = 1$, we have $\delta n_a = \omega_c dn_a/d\omega$, where $dn_a/d\omega$ is a measure for the dispersion of the medium. In Section 5.6, the Abbe V number was introduced to characterise the dispersion of a substance in the optical part of the spectrum. Using the central wavelength λ_d and the two extreme wavelengths λ_F and λ_C in the visual window, we have the following relation between $dn_a/d\omega$ and the Abbe V_d dispersion number,

$$\frac{dn_a}{d\omega} = \frac{(n_d - 1)\lambda_d^2}{2\pi c(\lambda_C - \lambda_F)V_d} \,. \tag{E.2.5}$$

The expression for the index difference between the two circular eigenstates is then given by

$$n_{RC} - n_{LC} = \omega_c \frac{dn_a}{d\omega} = 1.92 \cdot 10^{-4} \, \frac{(n_d - 1)\, B_e}{V_d} \,. \tag{E.2.6}$$

The progression of a linearly polarised wave, parallel to the external magnetic field, shows a rotation of the plane of polarisation. Starting with an x-polarised wave in the plane $z = 0$ that propagates in the positive z-direction, the field strength components in a plane $z = z_0$ are given by,

$$\begin{aligned} E_{0,x} &= (1/2)\left[\exp\{i(n_a + \delta n_a/2)k_d z_0\} + \exp\{i(n_a - \delta n_a/2)k_d z_0\}\right] \\ &= \exp\{i n_a k_d z_0\}\cos[(\delta n_a)k_d z_0/2]\,, \\ E_{0,y} &= (1/2)\left[-i\exp\{i(n_a + \delta n_a/2)k_d z_0\} + i\exp\{i(n_a - \delta n_a/2)k_d z_0\}\right] \\ &= \exp\{i n_a k_d z_0\}\sin[(\delta n_a)k_d z_0/2]\,. \end{aligned} \tag{E.2.7}$$

The rotation angle α of the linear state of polarisation is given by

$$\alpha = k_d z_0(\delta n_a)/2 = 1.92 \cdot 10^{-4} \, \frac{\pi z_0 (n_d - 1)\, B_e}{\lambda_d V_d} \,. \tag{E.2.8}$$

Equation (E.2.8) shows that a positive magnetic induction B_e gives rise to a right-handed rotation of the plane of polarisation when progressing along the positive z-axis.

E.3 Molecular Rotation Versus Faraday Rotation, Reciprocity

An interesting difference is observed when a plane wave is propagated forth and back through a medium that exhibits optical rotation, for instance, by inserting a mirror in the light path. We compare two media with optical rotatory power, one because of its molecular structure and one because of an externally applied magnetic field. For the permittivity tensor of a medium with molecularly introduced rotation, we use Eq. (2.9.9). In the case of a medium with Faraday rotation we have the expression of Eq. (E.1.11). The Helmholtz equations for these two cases follow from Eq. (2.9.11) and Eq. (E.2.1), respectively. We neglect second-order terms in the gyrotropic constants g_0 and g_e and find the matrix equations,

$$\begin{pmatrix} (n_a^2/n^2) - 1 & -2icg_0\hat{k}_z/n & 0 \\ +2icg_0\hat{k}_z/n & (n_a^2/n^2) - 1 & 0 \\ 0 & 0 & (n_a^2/n^2) \end{pmatrix} \mathbf{E}_0 = M_M\,\mathbf{E}_0 = 0\,, \tag{E.3.1}$$

$$\begin{pmatrix} (n_a^2/n^2) - 1 & +in_a(\delta n_a)\hat{B}_{e,z}/n^2 & 0 \\ -in_a(\delta n_a)\hat{B}_{e,z}/n^2 & (n_a^2/n^2) - 1 & 0 \\ 0 & 0 & (n_a^2/n^2) \end{pmatrix} \mathbf{E}_0 = M_F\,\mathbf{E}_0 = 0\,, \tag{E.3.2}$$

where M_M and M_F are the matrices associated with a medium with molecular rotatory power and a medium exhibiting Faraday rotation, respectively. Solving for the eigenvalues of the phase index n yields the solutions and the corresponding eigenstates of the electric field:

Table E.1. Wave propagation speeds and wave polarisation eigenstates for media showing Faraday rotation under the influence of an external magnetic field and for media exhibiting molecular rotatory power.

	Faraday rotation		Molecular rotation	
	$\hat{k}_z = +1$	$\hat{k}_z = -1$	$\hat{k}_z = +1$	$\hat{k}_z = -1$
Index	$n_a + (\delta n)/2$	$n_a + (\delta n)/2$	$n_a + cg_0$	$n_a - cg_0$
Eigenstate	$(1, -i, 0)/\sqrt{2}$ (RC)	$(1, -i, 0)/\sqrt{2}$ (LC)	$(1, +i, 0)/\sqrt{2}$ (LC)	$(1, +i, 0)/\sqrt{2}$ (RC)
Index	$n_a - (\delta n)/2$	$n_a - (\delta n)/2$	$n_a - cg_0$	$n_a + cg_0$
Eigenstate	$(1, +i, 0)/\sqrt{2}$ (LC)	$(1, +i, 0)/\sqrt{2}$ (RC)	$(1, -i, 0)/\sqrt{2}$ (RC)	$(1, -i, 0)/\sqrt{2}$ (LC)

Molecular rotation:

$$
\begin{aligned}
n_1 &= n_a + cg_0\hat{k}_z\,, & \hat{\mathbf{E}}_1 &= \frac{1}{\sqrt{2}}(1, +i, 0)\,,\\
n_2 &= n_a - cg_0\hat{k}_z\,, & \hat{\mathbf{E}}_2 &= \frac{1}{\sqrt{2}}(1, -i, 0)\,,
\end{aligned}
\tag{E.3.3}
$$

Faraday rotation:

$$
\begin{aligned}
n_1 &= n_a + \frac{(\delta n)}{2}\hat{B}_{e,z}\,, & \hat{\mathbf{E}}_1 &= \frac{1}{\sqrt{2}}(1, -i, 0)\,,\\
n_2 &= n_a - \frac{(\delta n)}{2}\hat{B}_{e,z}\,, & \hat{\mathbf{E}}_2 &= \frac{1}{\sqrt{2}}(1, +i, 0)\,.
\end{aligned}
\tag{E.3.4}
$$

In the case of forward and backward propagating waves with $\hat{\mathbf{k}} = (0, 0, \pm 1)$ and in the presence of a magnetic induction vector pointing in the positive z-direction ($\hat{\mathbf{B}}_e = (0, 0, 1)$), we have the following scheme (see Table E.1) for the wave propagation speed and the polarisation eigenstates. For the case of Faraday rotation, δn is given by $\omega_c(dn_a/d\omega)$.

To discuss circular eigenstates in the context of optical rotation, it is useful to characterise a circular eigenstate by its transverse components, $(1, +i)$ and $(1, -i)$, as shown in the table. It was pointed out in Subsection 1.6.6 that a polarisation eigenstate defined in this way avoids the confusion about the handedness of a state of polarisation. For instance, in our convention, the eigenstate $(1, +i)$ corresponds to an LC wave propagating in the positive z-direction and to an RC wave in the backward propagation direction. When a plane wave reflects at a mirror surface, an RC forward state becomes an LC backward propagating state. When the transverse field components are used for defining the eigenstate, an automatic coupling occurs between the two states that are connected at reflection.

In the case of Faraday rotation, the table shows that a plane wave with a circular eigenstate, propagating forth and back through the same medium after reflection by a mirror, experiences the same refractive index during the entire trajectory through the medium. If a plane polarised wave is sent through the Faraday medium and reflected back through it, the total phase difference between the two circular eigenstates is doubled and the rotation of the plane of polarisation is also doubled with respect to a single passage through the medium. The same experiment in a medium with molecular rotation shows a null effect in double passage.

E.3.1 Reciprocity in Wave Propagation

The theorem of reciprocity in wave propagation was first stated by Stokes and Helmholtz and was further developed for acoustic waves by Rayleigh **[286]** and in the electromagnetic domain by Lorentz **[220]**. More general formulations have recently been given by de Hoop **[138]**,**[139]**. In Fig. E.2 we sketch the geometry for demonstrating the reciprocity of wave propagation. A source at the point S emits a polarised ray with its direction given by the unit vector $\hat{\mathbf{s}}_S$. The wave travels through the medium M and passes through a point D, where the ray vector direction is given by the unit vector $\hat{\mathbf{s}}_D$. At the point S, two orthogonal planes S_1 and S_2 have been drawn, defined by the unit vectors $\hat{\mathbf{s}}_1$ and $\hat{\mathbf{s}}_2$, respectively,

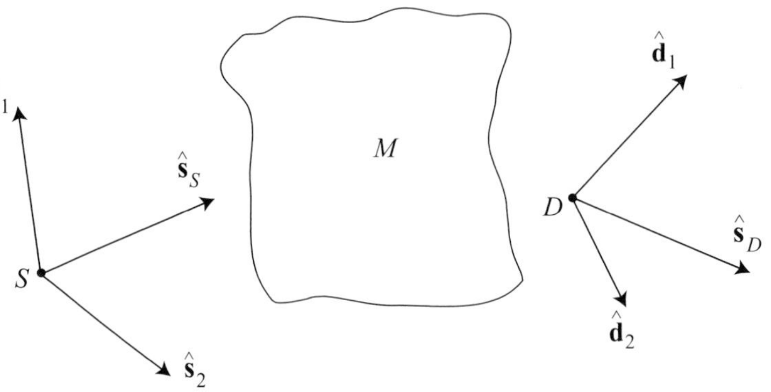

Figure E.2: Helmholtz' formulation of reciprocity in wave propagation. A source S emits a polarised wave of light with its propagation direction given by the ray vector $\hat{\mathbf{s}}_S$ which represents the direction of the Poynting vector of the radiation in the point S. The wave travels through a medium M and produces a disturbance in a detection point D. S_1 and S_2 are two mutually orthogonal planes, defined by the ray vector $\hat{\mathbf{s}}_S$ and the unit vectors $\hat{\mathbf{s}}_1$ and $\hat{\mathbf{s}}_2$, respectively. The planes D_1 and D_2 through the detection point D are orthogonal and defined by the ray vector $\hat{\mathbf{s}}_D$ in D and the unit vectors $\hat{\mathbf{d}}_1$ and $\hat{\mathbf{d}}_2$, respectively.

and by their intersecting line, the common ray vector $\hat{\mathbf{s}}_S$. A comparable set of planes D_1 and D_2 has been constructed in the point D. Light is sent out by the source in S with an intensity Q and with a linear state of polarisation contained in the plane S_1. In the point D, the intensity in the plane of polarisation D_1 is measured and equals pQ. The reciprocity theorem says that if a wave with the same light intensity Q and a state of polarisation defined by the plane D_1 is sent back through the system, the intensity measured in S and belonging to the state of polarisation defined by the plane S_1 will also be given by pQ. Any reflection, refraction or diffraction effect is allowed in the medium M. An exception should be made for materials that change their refractive index as a function of time or as a function of light intensity (nonlinear optical effects). Light propagation under the influence of an external magnetic field (Faraday effect) was also excluded by Helmholtz in his memoir. With respect to optical rotation, molecular rotatory power complies with the reciprocity theorem. Table E.1 shows that forward and backward propagating light with the same handedness travels at the same speed in a medium with molecular rotatory power. For instance, for RC light, the refractive index in both directions is given by $n_a - cg_0$. For the case of Faraday rotation, light with a particular circular state of polarisation changes its speed of propagation when the propagation direction is reversed; consequently, the reciprocity of wave propagation is lost.

F Vector Point-spread Function in a Multilayer Structure

In this appendix we discuss evaluation of the Debye diffraction integral when a vector point-spread function is formed in a multilayer in image space. More particularly, we give the expressions for the Cartesian electric field components in a particular thin layer of the multilayer with index number h which is sufficiently close to the paraxial image plane in the multilayer. As before, the evaluation of the Debye diffraction integral uses a Zernike expansion of the X_1- and Y_1-components of the electric field strength vector on the exit pupil sphere. The Zernike expansions of these field components on the exit pupil sphere for a diffraction problem in homogeneous image space was given by Eqs. (9.5.18) and (9.5.19). It comprised two sets of coefficients, $\beta^m_{n,x}$ and $\beta^m_{n,y}$, each associated with an orthogonal direction on the exit pupil sphere. In the case of diffraction in a multilayer this relatively simple expansion does not suffice. We first have a doubling of the number of coefficients due to the difference in strength and phase of the p- and s-polarisation of a particular plane wave component at the interfaces of the multilayer. A further doubling in the number of coefficients is caused by the presence of forward transmitted and backward reflected waves in each thin layer. The number of sets of β coefficients to be included then amounts to eight in total. The transmitted plane wave components created in a particular thin layer are projected back to the exit pupil sphere and create a vector field there to be used as input field in the integrand of the Debye diffraction integral, just as in the case of a homogeneous image space. The reflected components can be thought of as originating from a mirrored version of the exit pupil that is created by the reflection of the physical pupil surface at the entrance surface through C_s of a particular layer h (see Fig. 9.28). The reflected components propagate back towards layer h in the negative z_1-direction and interfere with the transmitted waves. The reflected waves form a vector field on the mirrored exit pupil to be used in the Debye integral for the 'backward' diffraction problem. To simplify the analysis from the situation with a physical pupil and a mirrored version of it to that with a single pupil, we reflect back the mirrored exit pupil for the backward propagating waves to the real exit pupil of the imaging system as illustrated in Fig. 9.28. We then evaluate the Debye integral comprising the combined forward and backward propagating waves, issued from the real physical exit pupil. The backward propagation is accounted for by the opposite signs of the backward wave vector components with respect to the wave vector components of the forward propagating waves.

F.1 Effective Cartesian Field Components on the Exit Pupil Sphere

The Cartesian components of the electric field strength on the exit pupil sphere which are the addition of field components associated with forward and backward propagating waves follow from the substitution of the p- and s-components in layer h (see Eqs. (9.5.54) and (9.5.55)) into the expressions for the Cartesian components that were given by Eqs. (9.5.56)–(9.5.58). We obtain

$$\begin{aligned}
E_{x,h}(\rho,\theta) = \left|\frac{f_1}{R_p}\right| T_R(\rho)\sqrt{\frac{n_0}{n_1}}\Bigg[\frac{\cos\alpha_h}{\alpha_0}\bigg\{& t_{p,h}(\rho) f_L'(\rho,\theta)\left[E_{x,0}(\rho,\theta)\cos^2\theta + E_{y,0}(\rho,\theta)\cos\theta\sin\theta\right] \\
&- r_{p,h}(\rho) f_L'(\rho,\theta+\pi)\left[E_{x,0}(\rho,\theta+\pi)\cos^2\theta + E_{0,y}(\rho,\theta+\pi)\cos\theta\sin\theta\right]\bigg\} \\
&+ \bigg\{ t_{s,h}(\rho) f_L'(\rho,\theta)\left[E_{x,0}(\rho,\theta)\sin^2\theta - E_{y,0}(\rho,\theta)\cos\theta\sin\theta\right] \\
&+ r_{s,h}(\rho) f_L'(\rho,\theta+\pi)\left[E_{x,0}(\rho,\theta+\pi)\sin^2\theta - E_{y,0}(\rho,\theta+\pi)\cos\theta\sin\theta\right]\bigg\}\Bigg],
\end{aligned} \qquad \text{(F.1.1)}$$

$$E_{y,h}(\rho,\theta) = \left|\frac{f_1}{R_p}\right| T_R(\rho)\sqrt{\frac{n_0}{n_1}}\left[\frac{\cos\alpha_h}{\alpha_0}\left\{t_{p,h}(\rho)f_L'(\rho,\theta)\left[E_{x,0}(\rho,\theta)\cos\theta\sin\theta + E_{y,0}(\rho,\theta)\sin^2\theta\right]\right.\right.$$
$$- r_{p,h}(\rho)f_L'(\rho,\theta+\pi)\left[E_{x,0}(\rho,\theta+\pi)\cos\theta\sin\theta + E_{y,0}(\rho,\theta+\pi)\sin^2\theta\right]\Big\}$$
$$-\Big\{t_{s,h}(\rho)f_L'(\rho,\theta)\left[E_{x,0}(\rho,\theta)\cos\theta\sin\theta - E_{y,0}(\rho,\theta)\cos^2\theta\right]$$
$$\left.\left. + r_{s,h}(\rho)f_L'(\rho,\theta+\pi)\left[E_{x,0}(\rho,\theta+\pi)\cos\theta\sin\theta - E_{y,0}(\rho,\theta+\pi)\cos^2\theta\right]\right\}\right], \qquad \text{(F.1.2)}$$

$$E_{z,h}(\rho,\theta) = \left|\frac{f_1}{R_p}\right| T_R(\rho)\frac{R_1}{|R_1|}\frac{|\sin\alpha_h|}{\cos\alpha_0}\sqrt{\frac{n_0}{n_1}}\Big\{t_{p,h}(\rho)f_L'(\rho,\theta)\left[E_{x,0}(\rho,\theta)\cos\theta + E_{y,0}(\rho,\theta)\sin\theta\right]$$
$$- r_{p,h}(\rho)f_L'(\rho,\theta+\pi)\left[E_{x,0}(\rho,\theta+\pi)\cos\theta + E_{y,0}(\rho,\theta+\pi)\sin\theta\right]\Big\}. \qquad \text{(F.1.3)}$$

The β coefficients as a function of the X_1- and Y_1-directions on the exit pupil are calculated using the product functions $f_L' t_{s,h}, f_L' t_{p,h}, f_L' r_{s,h}$ and $f_L' r_{p,h}$. It is also possible to include birefringence of the optical system by allowing non-identical aberration functions, $f'_{x,L}$ and $f'_{y,L}$, acting on the x- and y-components of the incident field, respectively. The following sets of Zernike coefficients have to be calculated for the forward propagating field,

$$E_{x,0}(\rho,\theta)\, t_{s,h}(\rho)\, f'_{x,L}(\rho,\theta) = \sum_{n,m}\beta^m_{n,x,t_s}R_n^{|m|}(\rho)\exp(im\theta), \qquad \text{(F.1.4)}$$

$$E_{x,0}(\rho,\theta)\, t_{p,h}(\rho)\, f'_{x,L}(\rho,\theta) = \sum_{n,m}\beta^m_{n,x,t_p}R_n^{|m|}(\rho)\exp(im\theta), \qquad \text{(F.1.5)}$$

$$E_{y,0}(\rho,\theta)\, t_{s,h}(\rho)\, f'_{y,L}(\rho,\theta) = \sum_{n,m}\beta^m_{n,y,t_s}R_n^{|m|}(\rho)\exp(im\theta), \qquad \text{(F.1.6)}$$

$$E_{y,0}(\rho,\theta)\, t_{p,h}(\rho)\, f'_{y,L}(\rho,\theta) = \sum_{n,m}\beta^m_{n,y,t_p}R_n^{|m|}(\rho)\exp(im\theta), \qquad \text{(F.1.7)}$$

and a corresponding set for the backward propagating field,

$$E_{x,0}(\rho,\theta)\, r_{s,h}(\rho)\, f'_{x,L}(\rho,\theta) = \sum_{n,m}\beta^m_{n,x,r_s}R_n^{|m|}(\rho)\exp(im\theta), \qquad \text{(F.1.8)}$$

$$E_{x,0}(\rho,\theta)\, r_{p,h}(\rho)\, f'_{x,L}(\rho,\theta) = \sum_{n,m}\beta^m_{n,x,r_p}R_n^{|m|}(\rho)\exp(im\theta), \qquad \text{(F.1.9)}$$

$$E_{y,0}(\rho,\theta)\, r_{s,h}(\rho)\, f'_{y,L}(\rho,\theta) = \sum_{n,m}\beta^m_{n,y,r_s}R_n^{|m|}(\rho)\exp(im\theta), \qquad \text{(F.1.10)}$$

$$E_{y,0}(\rho,\theta)\, r_{p,h}(\rho)\, f'_{y,L}(\rho,\theta) = \sum_{n,m}\beta^m_{n,y,r_p}R_n^{|m|}(\rho)\exp(im\theta). \qquad \text{(F.1.11)}$$

For analysis of the imaging in layer h, we need these eight sets of coefficients. When going to another layer, new sets have to be constructed. Moreover, it turns out that each sublayer also requires new diffraction integrals to be integrated and tabulated ($V(r, f)$ functions). A definite advantage of the ENZ approach is that the field does not have to be evaluated sequentially, layer by layer, in the propagation direction.

With the aid of the coefficients of the Zernike expansions according to Eqs. (F.1.4)–(F.1.7) and (F.1.8)–(F.1.11) the integrand of the Debye integral can be calculated. For reasons of compatibility with the previous analysis for a homogeneous image space, we prefer to replace the eight sets above by their sum and difference sets for the forward and backward diffracted fields,

$$\beta^m_{n,x,t_+} = \frac{\beta^m_{n,x,t_p} + \beta^m_{n,x,t_s}}{2}, \qquad \beta^m_{n,x,t_-} = \frac{\beta^m_{n,x,t_p} - \beta^m_{n,x,t_s}}{2}, \qquad \text{(F.1.12)}$$

$$\beta^m_{n,y,t_+} = \frac{\beta^m_{n,y,t_p} + \beta^m_{n,y,t_s}}{2}, \qquad \beta^m_{n,y,t_-} = \frac{\beta^m_{n,y,t_p} - \beta^m_{n,y,t_s}}{2}, \qquad \text{(F.1.13)}$$

$$\beta^m_{n,x,r_+} = \frac{\beta^m_{n,x,r_p} + \beta^m_{n,x,r_s}}{2}, \qquad \beta^m_{n,x,r_-} = \frac{\beta^m_{n,x,r_p} - \beta^m_{n,x,r_s}}{2}, \qquad \text{(F.1.14)}$$

$$\beta^m_{n,y,r_+} = \frac{\beta^m_{n,y,r_p} + \beta^m_{n,y,r_s}}{2} \, , \qquad \beta^m_{n,y,r_-} = \frac{\beta^m_{n,y,r_p} - \beta^m_{n,y,r_s}}{2} \, . \tag{F.1.15}$$

The β coefficients in the second column with the minus sign are all zero if the complex p- and s-transmission and reflection coefficients in layer h are identical for any propagation direction.

Using the composite Zernike coefficients in the expressions Eqs. (F.1.1)–(F.1.3) for the Cartesian electric field components, we obtain for the forward and backward propagating field components with upper indices t and r, respectively,

$$\begin{aligned} E^t_{x,h}(\rho,\theta) = & \left|\frac{f_1}{2R_p}\right| \frac{1}{\cos\alpha_0} \sqrt{\frac{n_0\cos\alpha_1}{n_1\cos\alpha_0}} \sum_{n,m} R^{|m|}_n(\rho)\exp(im\theta) \\ & \times \Big\{ \beta^m_{n,x,t_+}(\cos\alpha_h + \cos\alpha_0) + \beta^m_{n,x,t_-}(\cos\alpha_h - \cos\alpha_0) \\ & + \left[\beta^m_{n,x,t_+}(\cos\alpha_h - \cos\alpha_0) + \beta^m_{n,x,t_-}(\cos\alpha_h + \cos\alpha_0)\right]\cos 2\theta \\ & + \left[\beta^m_{n,y,t_+}(\cos\alpha_h - \cos\alpha_0) + \beta^m_{n,y,t_-}(\cos\alpha_h + \cos\alpha_0)\right]\sin 2\theta \Big\} \, , \end{aligned} \tag{F.1.16}$$

$$\begin{aligned} E^r_{x,h}(\rho,\theta) = & \left|\frac{f_1}{2R_p}\right| \frac{1}{\cos\alpha_0} \sqrt{\frac{n_0\cos\alpha_1}{n_1\cos\alpha_0}} \sum_{n,m} R^{|m|}_n(\rho)(-1)^m\exp(im\theta) \\ & \times \Big\{ -\beta^m_{n,x,r_+}(\cos\alpha_h - \cos\alpha_0) - \beta^m_{n,x,r_-}(\cos\alpha_h + \cos\alpha_0) \\ & + \left[-\beta^m_{n,x,r_+}(\cos\alpha_h + \cos\alpha_0) - \beta^m_{n,x,r_-}(\cos\alpha_h - \cos\alpha_0)\right]\cos 2\theta \\ & + \left[-\beta^m_{n,y,r_+}(\cos\alpha_h + \cos\alpha_0) - \beta^m_{n,y,r_-}(\cos\alpha_h - \cos\alpha_0)\right]\sin 2\theta \Big\} \, , \end{aligned} \tag{F.1.17}$$

$$\begin{aligned} E^t_{y,h}(\rho,\theta) = & \left|\frac{f_1}{2R_p}\right| \frac{1}{\cos\alpha_0} \sqrt{\frac{n_0\cos\alpha_1}{n_1\cos\alpha_0}} \sum_{n,m} R^{|m|}_n(\rho)\exp(im\theta) \\ & \times \Big\{ \beta^m_{n,y,t_+}(\cos\alpha_h + \cos\alpha_0) + \beta^m_{n,y,t_-}(\cos\alpha_h - \cos\alpha_0) \\ & + \left[-\beta^m_{n,y,t_+}(\cos\alpha_h - \cos\alpha_0) - \beta^m_{n,y,t_-}(\cos\alpha_h + \cos\alpha_0)\right]\cos 2\theta \\ & + \left[\beta^m_{n,x,t_+}(\cos\alpha_h - \cos\alpha_0) + \beta^m_{n,x,t_-}(\cos\alpha_h + \cos\alpha_0)\right]\sin 2\theta \Big\} \, , \end{aligned} \tag{F.1.18}$$

$$\begin{aligned} E^r_{y,h}(\rho,\theta) = & \left|\frac{f_1}{2R_p}\right| \frac{1}{\cos\alpha_0} \sqrt{\frac{n_0\cos\alpha_1}{n_1\cos\alpha_0}} \sum_{n,m} R^{|m|}_n(\rho)(-1)^m\exp(im\theta) \\ & \times \Big\{ -\beta^m_{n,y,r_+}(\cos\alpha_h - \cos\alpha_0) - \beta^m_{n,y,r_-}(\cos\alpha_h + \cos\alpha_0) \\ & + \left[\beta^m_{n,y,r_+}(\cos\alpha_h + \cos\alpha_0) + \beta^m_{n,y,r_-}(\cos\alpha_h - \cos\alpha_0)\right]\cos 2\theta \\ & + \left[-\beta^m_{n,x,r_+}(\cos\alpha_h + \cos\alpha_0) - \beta^m_{n,x,r_-}(\cos\alpha_h - \cos\alpha_0)\right]\sin 2\theta \Big\} \, , \end{aligned} \tag{F.1.19}$$

$$\begin{aligned} E^t_{z,h}(\rho,\theta) = & \left|\frac{f_1}{R_p}\right| \frac{1}{\cos\alpha_0} \sqrt{\frac{n_0\cos\alpha_1}{n_1\cos\alpha_0}} \left(\frac{R_1}{|R_1|}\right) s_h\rho \sum_{n,m} R^{|m|}_n(\rho)\exp(im\theta) \\ & \times \Big\{ \left[\beta^m_{n,x,t_+} + \beta^m_{n,x,t_-}\right]\cos\theta + \left[\beta^m_{n,y,t_+} + \beta^m_{n,y,t_-}\right]\sin\theta \Big\} \, , \end{aligned} \tag{F.1.20}$$

$$\begin{aligned} E^r_{z,h}(\rho,\theta) = & \left|\frac{f_1}{R_p}\right| \frac{1}{\cos\alpha_0} \sqrt{\frac{n_0\cos\alpha_1}{n_1\cos\alpha_0}} \left(\frac{R_1}{|R_1|}\right) s_h\rho \sum_{n,m} R^{|m|}_n(\rho)(-1)^m\exp(im\theta) \\ & \times \Big\{ -\left[\beta^m_{n,x,r_+} + \beta^m_{n,x,r_-}\right]\cos\theta - \left[\beta^m_{n,y,r_+} + \beta^m_{n,y,r_-}\right]\sin\theta \Big\} \, , \end{aligned} \tag{F.1.21}$$

where the values of $\cos\alpha_0$ and $\cos\alpha_h$ are given by Eq. (9.5.6). The exponential function $\exp(im\theta)$ in the expressions for the reflected fields E^r is accompanied by a factor $(-1)^m$ to account for the fact that these fields have been produced by plane waves with wave vectors that have an azimuth $\theta+\pi$ as compared to the transmitted fields E^t corresponding to a plane wave azimuth θ. The reason is the origin $Q_{1,r}$ of a backward propagating wave on the exit pupil that is diametrically opposed to the origin $Q_{1,t}$ of the corresponding collinear transmitted wave (see Fig. 9.28).

F.2 The Electric Field Components in a Thin Layer of the Multilayer Stack

The next step is the insertion of the Cartesian field strength components in the vector Debye diffraction integral. The propagation factor for a transmitted wave is $\exp\{i[k_{x,h}x+k_{y,h}y+k_{z,h}z]\}$; for the back propagating wave we have the exponential $\exp\{-i[k_{x,h}x+k_{y,h}y+k_{z,h}z]\}$. In both cases, the $k_{z,h}$ component is obtained from the dispersion relation in the layer with index n_h. The field components of the forward and back propagating waves are inserted in the Debye integral. For forward propagating waves, the resulting expression has been given in Eq. (9.5.30). The addition of the backward propagating waves with reversed wave vector components yields the following expression for the electric field strength vector E_i in thin layer h:

$$\begin{aligned}\mathbf{E}_i(r_1',\gamma,f_{v,h}') = \frac{-in_1s_1^2|R_1|}{\lambda_0}\Bigg[&\exp\left(\frac{if_{v,h}'}{u_h}\right)\iint_C \frac{\mathbf{E}_h^t(\rho,\theta)}{(1-s_1^2\rho^2)^{\frac{1}{2}}}\exp\left\{\frac{-if_{v,h}'}{u_h}\left[1-(1-s_h^2\rho^2)^{\frac{1}{2}}\right]\right\}\\ &\times\exp\{-i(R_1/|R_1|)\,2\pi r_1'\rho\cos(\theta-\gamma)\}\rho\,d\rho d\theta\\ &+\exp\left(\frac{-if_{v,h}'}{u_h}\right)\iint_C \frac{\mathbf{E}_h^r(\rho,\theta)}{(1-s_1^2\rho^2)^{\frac{1}{2}}}\exp\left\{\frac{if_{v,h}'}{u_h}\left[1-(1-s_h^2\rho^2)^{\frac{1}{2}}\right]\right\}\\ &\times\exp\{i(R_1/|R_1|)\,2\pi r_1'\rho\cos(\theta-\gamma)\}\rho\,d\rho d\theta\Bigg],\end{aligned}\tag{F.2.1}$$

where

$$u_h = 1-\sqrt{1-s_h^2}\,;\qquad f_{v,h}' = \frac{2\pi n_h u_h z_1}{\lambda_0},\tag{F.2.2}$$

and λ_0 is the vacuum wavelength of the radiation. Note the following about the Debye integral above:

- The diffraction integral is thought to be evaluated in a medium with index n_h. This would normally lead to a leading factor that is given by $-in_hs_h^2|R_h|/\lambda_0$ where s_h is the maximum aperture value in the thin layer and R_h the apparent position of the exit pupil as seen from the medium with index n_h. It is easily verified that the quantity $ns^2|R|/\lambda_0$ is a refraction invariant quantity. For this reason we have kept the quantities that are valid for the image space medium with index n_1.
- In the second pathlength exponential we could have used the quantity $r_h' = r_1s_h/\lambda_h$. Also in this case, because of Snell's law or, equivalently, the continuity of the tangential electric field components at an interface between two media, we have that $\lambda_h/s_h = \lambda_1/s_1$ and $r_1' = r_h'$.
- It is possible to combine the factors $|R_1|$ and E_h^t or E_h^r in single quantities $|R_1|E_h^t$ and $|R_1|E_h^r$, respectively. This is advantageous if the exit pupil is at infinity (telecentric imaging), because the combined quantity RE, called the ray invariant field strength, remains finite in that case.

After the substitution of the expressions for E_h^t and E_h^r of Eqs. (F.1.16)–(F.1.21) into the Debye integral above, we perform the integration over θ and, after some manipulation, we obtain for the field strength components of E_i in thin layer h,

$$\begin{aligned}E_i^t(r_1',\gamma,f_{v,h}') = \frac{-i\pi n_1s_1^2}{\lambda_0}&\left|\frac{f_1}{[1+m_Tf_1/R_1]}\right|\sqrt{\frac{n_0}{n_1}}\exp\left(\frac{-if_{v,h}'}{u_h}\right)\sum_{n,m}[-i(R_1/|R_1|)]^m\exp[im\gamma]\\ &\times\left[\beta_{n,x,t_+}^m\begin{pmatrix}V_{n,0,t_+}^m-\frac{1}{2}\left\{V_{n,+2,t_+}^m\exp[+2i\gamma]+V_{n,-2,t_+}^m\exp[-2i\gamma]\right\}\\ +\frac{i}{2}\left\{V_{n,+2,t_+}^m\exp[+2i\gamma]-V_{n,-2,t_+}^m\exp[-2i\gamma]\right\}\\ -i\left\{V_{n,+1,t_+}^m\exp[+i\gamma]-V_{n,-1,t_+}^m\exp[-i\gamma]\right\}\end{pmatrix}\right.\end{aligned}$$

$$
\begin{aligned}
&+\beta^m_{n,x,t_-}\left(\begin{array}{r} V^m_{n,0,t_-} - \frac{1}{2}\left\{V^m_{n,+2,t_-}\exp[+2i\gamma] + V^m_{n,-2,t_-}\exp[-2i\gamma]\right\} \\ +\frac{i}{2}\left\{V^m_{n,+2,t_-}\exp[+2i\gamma] - V^m_{n,-2,t_-}\exp[-2i\gamma]\right\} \\ -i\left\{V^m_{n,+1,t_-}\exp[+i\gamma] - V^m_{n,-1,t_-}\exp[-i\gamma]\right\} \end{array}\right) \\
&+\beta^m_{n,y,t_+}\left(\begin{array}{r} +\frac{i}{2}\left\{V^m_{n,+2,t_+}\exp[+2i\gamma] - V^m_{n,-2,t_+}\exp[-2i\gamma]\right\} \\ V^m_{n,0,t_+} + \frac{1}{2}\left\{V^m_{n,+2,t_+}\exp[+2i\gamma] + V^m_{n,-2,t_+}\exp[-2i\gamma]\right\} \\ -\left\{V^m_{n,+1,t_+}\exp[+i\gamma] + V^m_{n,-1,t_+}\exp[-i\gamma]\right\} \end{array}\right) \\
&\left.+\beta^m_{n,y,t_-}\left(\begin{array}{r} +\frac{i}{2}\left\{V^m_{n,+2,t_-}\exp[+2i\gamma] - V^m_{n,-2,t_-}\exp[-2i\gamma]\right\} \\ V^m_{n,0,t_-} + \frac{1}{2}\left\{V^m_{n,+2,t_-}\exp[+2i\gamma] + V^m_{n,-2,t_-}\exp[-2i\gamma]\right\} \\ -\left\{V^m_{n,+1,t_-}\exp[+i\gamma] + V^m_{n,-1,t_-}\exp[-i\gamma]\right\} \end{array}\right)\right] ,
\end{aligned}
\tag{F.2.3}
$$

for the forward propagating contribution. In a similar fashion we get for the counter propagating contribution

$$
\begin{aligned}
E^r_i(r'_1,\gamma,f'_{v,b}) &= \frac{-i\pi n_1 s_1^2}{\lambda_0}\left|\frac{f_1}{[1+m_T f_1/R_1]}\right|\sqrt{\frac{n_0}{n_1}}\exp\left(\frac{if'_{v,b}}{u_b}\right)\sum_{n,m}[-i(R_1/|R_1|)]^m\exp[im\gamma] \\
&\times\left[\beta^m_{n,x,r_+}\left(\begin{array}{r} -V^m_{n,0,r_-} + \frac{1}{2}\left\{V^m_{n,+2,r_-}\exp[+2i\gamma] + V^m_{n,-2,r_-}\exp[-2i\gamma]\right\} \\ -\frac{i}{2}\left\{V^m_{n,+2,r_-}\exp[+2i\gamma] - V^m_{n,-2,r_-}\exp[-2i\gamma]\right\} \\ -i\left\{V^m_{n,+1,r_-}\exp[+i\gamma] - V^m_{n,-1,r_-}\exp[-i\gamma]\right\} \end{array}\right)\right. \\
&+\beta^m_{n,x,r_-}\left(\begin{array}{r} -V^m_{n,0,r_+} + \frac{1}{2}\left\{V^m_{n,+2,r_+}\exp[+2i\gamma] + V^m_{n,-2,r_+}\exp[-2i\gamma]\right\} \\ -\frac{i}{2}\left\{V^m_{n,+2,r_+}\exp[+2i\gamma] - V^m_{n,-2,r_+}\exp[-2i\gamma]\right\} \\ -i\left\{V^m_{n,+1,r_+}\exp[+i\gamma] - V^m_{n,-1,r_+}\exp[-i\gamma]\right\} \end{array}\right) \\
&+\beta^m_{n,y,r_+}\left(\begin{array}{r} -\frac{i}{2}\left\{V^m_{n,+2,r_-}\exp[+2i\gamma] - V^m_{n,-2,r_-}\exp[-2i\gamma]\right\} \\ -V^m_{n,0,r_-} - \frac{1}{2}\left\{V^m_{n,+2,r_-}\exp[+2i\gamma] + V^m_{n,-2,r_-}\exp[-2i\gamma]\right\} \\ -\left\{V^m_{n,+1,r_-}\exp[+i\gamma] + V^m_{n,-1,r_-}\exp[-i\gamma]\right\} \end{array}\right) \\
&\left.+\beta^m_{n,y,r_-}\left(\begin{array}{r} -\frac{i}{2}\left\{V^m_{n,+2,r_+}\exp[+2i\gamma] - V^m_{n,-2,r_+}\exp[-2i\gamma]\right\} \\ -V^m_{n,0,r_+} - \frac{1}{2}\left\{V^m_{n,+2,r_+}\exp[+2i\gamma] + V^m_{n,-2,r_+}\exp[-2i\gamma]\right\} \\ -\left\{V^m_{n,+1,r_+}\exp[+i\gamma] + V^m_{n,-1,r_+}\exp[-i\gamma]\right\} \end{array}\right)\right] .
\end{aligned}
\tag{F.2.4}
$$

The integrals $V^m_{n,j,t_\pm}(r'_1,f'_{v,b})$ with indices $j = -2,-1,0,+1,+2$, occurring in the expressions above, are given by

$$
V^m_{n,j,t_\pm}(r'_1,f'_{v,b}) = \int_0^1 \rho^{|j|} f^E_{|j|,\pm}(\rho)\exp\left[\frac{-if'_{v,b}}{u_b}\left(1-\sqrt{1-s_b^2\rho^2}\right)\right] R^{|m|}_n(\rho)\, J_{m+j}(2\pi r'_1\rho)\,\rho\, d\rho\, , \tag{F.2.5}
$$

where the functions $f^E_{|j|,\pm}(\rho)$ for $j = 0, 1, 2$ are given by

$$
f^E_{0,\pm}(\rho) = \frac{\left\{\left(1-s_b^2\rho^2\right)^{\frac{1}{2}} \pm \left(1-s_0^2\rho^2\right)^{\frac{1}{2}}\right\}}{\left(1-s_1^2\rho^2\right)^{\frac{1}{4}}\left(1-s_0^2\rho^2\right)^{\frac{3}{4}}}\, , \tag{F.2.6}
$$

$$
f^E_{1,\pm}(\rho) = \frac{s_b}{\left(1-s_1^2\rho^2\right)^{1/4}\left(1-s_0^2\rho^2\right)^{3/4}}\, , \tag{F.2.7}
$$

$$f_{2,\pm}^{E}(\rho) = \frac{s_0^2 - s_h^2}{\left\{\left(1 - s_h^2\rho^2\right)^{1/2} \pm \left(1 - s_0^2\rho^2\right)^{1/2}\right\}\left(1 - s_1^2\rho^2\right)^{1/4}\left(1 - s_0^2\rho^2\right)^{3/4}} \; . \tag{F.2.8}$$

The integrals $V^{m}_{n,j,r_\pm}(r'_1, f'_{v,h})$ which are for the reflected field are obtained from the integrals $V^{m}_{n,j,t_\pm}(r'_1, f'_{v,h})$ for the transmitted field with the aid of the property $f'_{v,h} = -f'_{v,h}$. For absorption-free media we then have that

$$V^{m}_{n,j,r_\pm}(r'_1, f'_{v,h}) = \left(V^{m}_{n,j,t_\pm}(r'_1, -f'_{v,h})\right)^* \; . \tag{F.2.9}$$

An efficient semi-analytic evaluation of the V-integrals above is given in **[127]**. Truncation rules for the infinite series that occur in the expressions for the V integrals can be found in **[128]**.

G V. S. Ignatowsky: Diffraction by a Lens of Arbitrary Aperture

V. S. Ignatowsky, 'Diffraction by a lens of arbitrary aperture,'
Trans. Opt. Inst. **I** (*IV*), 1–36 (1919, Petrograd)

Denoted as reference D by the author

(An exact translation of the original publication in the Russian language.)

G.1 Introduction

The goal of this work is, firstly, to investigate the phenomenon of diffraction at the focus of a lens at any aperture, assuming that, from the point of view of geometrical optics, the lens is fully corrected (i.e. a plane perpendicular to the optical axis has as an image a plane which is also perpendicular to the optical axis), in particular, the sine law is fulfilled precisely. Secondly, we aim to show how the above diffracted field is imaged by a second lens, situated after the first one. Finally, we outline the analogy between our results and Abbe's theory of the microscope.

This work is based on my previous work 'Relation between geometrical and wave optics and diffraction of a homocentric beam,'[1] which we refer to as 'C' in what follows. In that work, on the basis of the generalised sine law that I have derived, formulas are derived to calculate the diffraction of a homocentric beam, whose focus is situated at an arbitrary distance from the optical axis and the aperture is of arbitrary size and shape. The present work considers the case when the focus is on the axis.

In the previous work we have pointed out the drawbacks and mistakes in the theory of microscopy by Abbe and Lummer.[2] In particular, in all places where we have $\sqrt{\cos\alpha}$, Abbe and Lummer's theory would have yielded $\cos\alpha$. In this work we use the same notation as in C. We do not cite other authors who studied lens aberrations, due to the fact that our starting point and the results are completely different, and only partly agree with the results of previous authors (calculation of the energy, though only at small beam apertures; c.f. (G.10.7) §G.10). One particular result of this work needs to be pointed out. The object to be imaged by the second lens is some diffracted field distribution. This object, while satisfying Lambert's law at large distances, emits light rigorously in agreement with Maxwell's laws everywhere, including the close vicinity of the object. Therefore, our object takes the place of Hertz's dipole, which was commonly used to represent a radiating point.[3] Whereas Hertz's dipole, by its nature, is arbitrary and ill-defined, our object is quite natural and well defined. The former is due to the fact that it exists in reality, as it is produced by the first lens; and the latter is due to the fact that it agrees with Maxwell's equations everywhere and absolutely rigorously.

G.2 Starting Expressions

In the following we will deal with an optical system which (in the language of geometrical optics) produces an image of a point O_0 on the optical axis at a point O_1 (Fig. G.1). The amplitude of the electric field at a unit distance from O_0

[1] V. S. Ignatowsky, *Trans. Opt. Inst. Petrograd*, **I**, (*III*), Petrograd (1920).

[2] E. Abbe, Die Lehre von der Lichtentstehung im Mikroskop. Herausgegeben von O. Lummer und F. Reiche. Braunschweig Vieweg & Sohn.

[3] c.f. for example, F. Reiche, *Ann. d. Phys.* **Bd. 29**, p. 65, 401 and **Bd. 30**, p. 182 (1909).

along the ray $\mathfrak{s}_0$ in the object space, is denoted a_0; similarly in the image space, the amplitude of the electric field at a unit distance from O_1 along the ray $\mathfrak{s}_1$ in the object space, is denoted a_1. Then according to Eq. (4.11) of §.4 of C we have[4]

$$a_1 = -\frac{\beta a_0 \sqrt{n_1 \cos\alpha_1}}{\sqrt{n_0 \cos\alpha_0}}, \tag{G.2.1}$$

where β is the magnification and is equal to:

$$\frac{n_0 \sin\alpha_0}{n_1 \sin\alpha_1} = -\beta. \tag{G.2.2}$$

Equations (G.2.1) and (G.2.2) yield:

$$a_1 = a_0 \frac{\sin\alpha_0 \sqrt{n_0}\sqrt{\cos\alpha_1}}{\sin\alpha_1 \sqrt{n_1}\sqrt{\cos\alpha_0}}. \tag{G.2.3}$$

Let us draw a sphere of radius ρ_0 with the centre at O_0. The amplitude at this sphere is equal to $A_0 = a_0/\rho_0$, and therefore we have instead of (G.2.3):

$$a_1 = A_0\rho_0 \frac{\sin\alpha_0 \sqrt{n_0}\sqrt{\cos\alpha_1}}{\sin\alpha_1 \sqrt{n_1}\sqrt{\cos\alpha_0}}. \tag{G.2.4}$$

Let us draw a plane E_0 tangential to the sphere at the point B. Leaving the plane E_0 in place, we will now move that point O_0 away. As ρ_0 gradually increases, the product $\rho_0 \sin\alpha_0$ approaches the value of h_0, the height of the intersection of the ray $\mathfrak{s}_0$ with the plane E_0. The angle α_0 approaches zero. Therefore, at $\rho_0 = \infty$, i.e., when the system is illuminated by a beam, parallel to the axis, we have

$$a_1 = \frac{A_0 h_0 \sqrt{n_0}\sqrt{\cos\alpha_1}}{\sqrt{n_1}\sin\alpha_1}, \tag{G.2.5}$$

and A_0 is therefore the amplitude of the incident electric vector.

This case is the subject of the current study and it is presented in Fig. G.2. The plane E_0 corresponds to the same plane in Fig. G.1, C is the intersection point of the incident ray with the plane E_0 at the distance h_0 from the axis, L is the optical system. The plane E_1 is the image plane. The angle between the meridional plane containing the ray CDO_1, and the plane YO_0X is denoted ϑ; this angle is assumed to be positive for clockwise rotations, when viewed in the direction of the X axis. The unit vector $\mathfrak{c}_0$ is directed along h_0 and points away from the axis. The unit vector $\mathfrak{c}_1$ lies in the meridional plane of the ray, is perpendicular to the ray and points away from the axis. The vector $\mathfrak{P}$ is directed from O_1 to P. $\mathfrak{i}$, $\mathfrak{j}$, $\mathfrak{k}$ are unit vectors along the coordinate axes. Φ_1 is the surface of a sphere of unit radius, i.e. spherical solid angle corresponding to the aperture of the outgoing homocentric beam.

The electric vector $\mathfrak{E}_0$ of the incident plane polarised wave is assumed to be parallel to the Y axis, so we have:

$$\mathfrak{E}_0 = \mathfrak{j} A_0 \exp\left[i\omega\left(t - \frac{xn_0}{c}\right)\right], \tag{G.2.6}$$

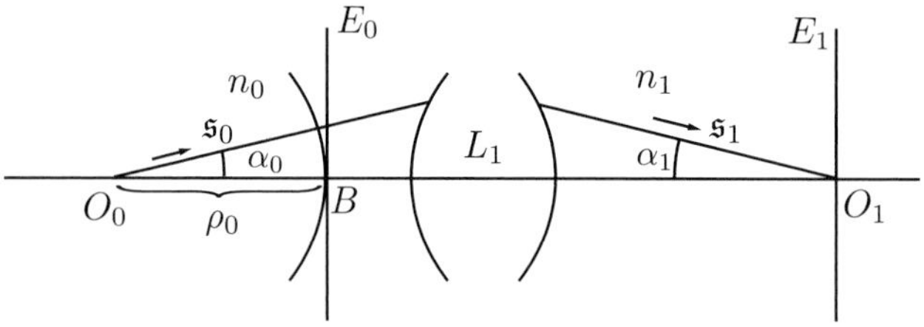

Figure G.1

4 Note translator: a_0 denotes the amplitude on the surface of a unit sphere in object space, centred on the object point. In a similar way, a_1 is defined as the amplitude on a unit sphere in image space.

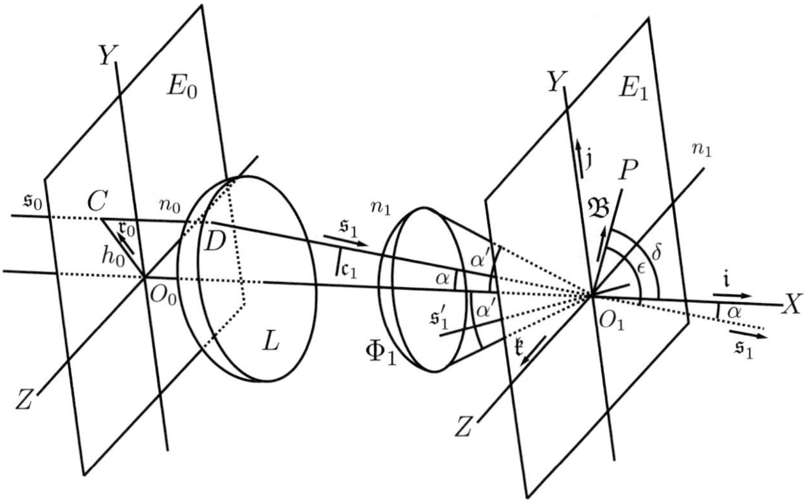

Figure G.2

with the eikonal taken with respect to the plane E_0. For the magnetic field $\mathfrak{H}_0$ of the incident ray we obtain from (G.2.6)

$$\mathfrak{H}_0 = \mathfrak{k}\frac{n_0 A_0}{c}\exp\left[i\omega\left(t - \frac{xn_0}{c}\right)\right]. \tag{G.2.7}$$

Since we will only deal with the angle α_1 (as α_0 has dropped out of (G.2.5)), we will simply write α instead of α_1. The maximum value of α that corresponds to the aperture of the homocentric beam is denoted as α'. Denoting the maximum value of h_0 as h, we obtain, as is well known,

$$\frac{h_0}{\sin\alpha} = \frac{h}{\sin\alpha'} = F_1, \tag{G.2.8}$$

where F_1 is the focal distance in image space. Therefore we obtain for the electric field, $\mathfrak{E}_1$, at the point P in image space, according to Eq. (5.12) of §5 of C and (G.2.5) and (G.2.6):

$$\mathfrak{E}_1 = B\int_{\Phi_1}\mathfrak{a}_1\sqrt{\cos\alpha}e^{-ik\mathfrak{s}_1\cdot\mathfrak{B}}d\varphi_1, \tag{G.2.9}$$

where

$$B = \frac{iA_0F_1\sqrt{n_1n_0}}{\lambda}\exp\left[i\omega\left(t - \frac{E}{c}\right)\right], \tag{G.2.10}$$

and the eikonal is calculated from the plane E_0 to the point O_1. Since we have assumed a circular aperture, we have:

$$\int_{\Phi_1}(\cdots)d\phi_1 = \int_{\alpha=0}^{\alpha=\alpha'}\int_{\vartheta=0}^{\vartheta=2\pi}(\cdots)\sin\alpha d\alpha d\vartheta. \tag{G.2.11}$$

Let $\mathfrak{B}_0$ be the unit vector along $\mathfrak{B}$. Then

$$\mathfrak{B} = b\mathfrak{B}_0 \tag{G.2.12}$$

and

$$\mathfrak{s}_1\mathfrak{B}_0 = \cos\varepsilon. \tag{G.2.13}$$

Introducing the further notation:

$$\kappa b = \frac{\omega n_1 b}{c} = \frac{2\pi n_1}{\lambda}b = u, \tag{G.2.14}$$

we can rewrite (G.2.9) taking (G.2.11) into account:

$$\mathfrak{E}_1 = B\int_{\alpha=0}^{\alpha=\alpha'}\int_{\vartheta=0}^{\vartheta=2\pi}\left(\mathfrak{a}_1\sqrt{\cos\alpha}e^{-iu\cos\varepsilon}\right)\sin\alpha d\alpha d\vartheta. \tag{G.2.15}$$

For the magnetic field, $\mathfrak{H}_1$, at the point P, we have the following expression, according to (2) §2, C:

$$\mathfrak{H}_1 = -\frac{Bn_1}{c}\int_{\alpha=0}^{\alpha=\alpha'}\int_{\vartheta=0}^{\vartheta=2\pi}\left([\mathfrak{a}_1\mathfrak{s}_1]\sqrt{\cos\alpha}e^{-iu\cos\varepsilon}\right)\sin\alpha d\alpha d\vartheta. \tag{G.2.16}$$

Let us draw a sphere of radius ρ_1 with centre at O_1 so that ρ_1 is greater than the radius of the limit sphere. Then at the surface of this sphere, within the homocentric beam, we obtain:

$$\mathfrak{E}_1' = \frac{a_1\mathfrak{a}_1}{\rho_1}e^{i\omega\left(t-\frac{E_1}{c}\right)} = \frac{\mathfrak{a}_1A_0F_1}{\rho_1}\sqrt{\frac{n_0}{n_1}}\sqrt{\cos\alpha}e^{i\omega\left(t-\frac{E_1}{c}\right)}, \tag{G.2.17}$$

where E_1 is the eikonal from the plane E_0 to the sphere. For the magnetic field, $\mathfrak{H}_1$, (G.2.17) yields the following (note that $\nabla\rho_1 = -\mathfrak{s}_1$):

$$\mathfrak{H}_1' = \frac{n_1a_1}{c\rho_1}[\mathfrak{a}_1\mathfrak{s}_1]\left(1+\frac{i\lambda}{2\pi n_1\rho_1}\right)e^{i\omega\left(t-\frac{E_1}{c}\right)}.$$

Or, neglecting the second term in the brackets, since λ_1/ρ_1 is much smaller than one:

$$\mathfrak{H}_1' = \frac{n_1a_1}{c\rho_1}[\mathfrak{a}_1\mathfrak{s}_1]e^{i\omega\left(t-\frac{E_1}{c}\right)} = -\frac{A_0F_1\sqrt{n_1n_0}\sqrt{\cos\alpha}}{c\rho_1}[\mathfrak{a}_1\mathfrak{s}_1]e^{i\omega\left(t-\frac{E_1}{c}\right)}. \tag{G.2.18}$$

The mean value $\overline{\mathfrak{S}}$ of the Poynting vector $\mathfrak{S} = [\mathfrak{E}\mathfrak{H}]/4\pi$ can be obtained by taking real parts of (G.2.6) and (G.2.7), and remembering that $\mathfrak{s}_0 = \mathfrak{i}$ and $\mathfrak{a}_1\mathfrak{s}_1 = 0$,[5]

$$\overline{\mathfrak{S}}_0 = \frac{1}{T}\int_0^T\mathfrak{S}_0dt = \frac{A_0^2n_0}{8\pi c}\mathfrak{s}_0 = w_0\mathfrak{s}_0, \tag{G.2.19}$$

and in image space, at a sphere of the radius ρ_1, according to real parts of (G.2.17) and (G.2.18):

$$\overline{\mathfrak{S}}_1' = \frac{A_0^2F_1^2n_0\cos\alpha}{8\pi c\rho_1^2}\mathfrak{s}_1 = w_1'\mathfrak{s}_1. \tag{G.2.20}$$

Now denote the total mean energy that flows into the system as W_0; this is given by:

$$W_0 = \int_{\vartheta=0}^{\vartheta=2\pi}\int_{h_0=0}^{h_0=h}w_0h_0dh_0d\vartheta = \frac{A_0^2n_0h^2}{8c} = w_0\pi h^2. \tag{G.2.21}$$

For the total energy W_1' that flows through the sphere of radius ρ_1 within the homocentric beam, we obtain from (G.2.20):

$$W_1' = \int_{\alpha=0}^{\alpha=\alpha'}\int_{\vartheta=0}^{\vartheta=2\pi}\rho_1^2w_1'\sin\alpha d\alpha d\vartheta = \frac{A_0^2F_1^2n_0\sin^2\alpha'}{8c} = \frac{A_0^2h^2n_0}{8c} = w_0\pi h^2, \tag{G.2.22}$$

i.e.

$$W_0 = W_1', \tag{G.2.23}$$

as it should be.

Extending the rays to the right of the plane E_1, according to §6 C, we obtain a homocentric beam, which carries away the same energy (G.2.20) at the distance ρ_1 from O_1, i.e., the result would be the same as if there was a surface element at O_1 which emitted according to Lambert's law.

G.3 Some Preliminary Calculations

Let us denote the angle between the meridional plane that contains the point P (Fig. G.2), and the plane YO_1X as γ. The angle γ is positive clockwise, when viewed along the x axis. Then we obtain:

[5] c.f. §3, C.

$$\mathfrak{B}_0 = \mathfrak{i}\cos\delta + \mathfrak{j}\sin\delta\cos\gamma + \mathfrak{k}\sin\delta\sin\gamma. \tag{G.3.1}$$

We also have:

$$\mathfrak{s}_1 = \mathfrak{i}\cos\alpha - \mathfrak{j}\sin\alpha\cos\vartheta - \mathfrak{k}\sin\alpha\sin\vartheta. \tag{G.3.2}$$

Equations (G.3.1) and (G.3.2) yield:

$$\begin{aligned}\mathfrak{s}_1\mathfrak{B}_0 = \cos\varepsilon &= \cos\alpha\cos\delta - \sin\alpha\sin\delta(\cos\vartheta\cos\gamma + \sin\vartheta\sin\gamma)\\ &= \cos\alpha\cos\delta - \sin\alpha\sin\delta\cos(\vartheta-\gamma),\end{aligned} \tag{G.3.3}$$

and, finally

$$\mathfrak{c}_1 = \mathfrak{i}\sin\alpha + \mathfrak{j}\cos\alpha\cos\vartheta + \mathfrak{k}\cos\alpha\sin\vartheta. \tag{G.3.4}$$

We know that

$$\mathfrak{a}_1\mathfrak{s}_1 = 0 \tag{G.3.5}$$

and

$$\mathfrak{a}_1\mathfrak{c}_1 = \mathfrak{j}\mathfrak{c}_0 = \cos\vartheta. \tag{G.3.6}$$

Assuming that:

$$\mathfrak{a}_1 = \mathfrak{i}x_1 + \mathfrak{j}x_2 + \mathfrak{k}x_3, \tag{G.3.7}$$

we can determine x_1, x_2 and x_3, i.e. the vector $\mathfrak{a}_1$,[6] on the basis of (G.3.2) and (G.3.4)–(G.3.6). Equation (G.3.7) yields:

$$\mathfrak{a}_1^2 = x_1^2 + x_2^2 + x_3^2. \tag{G.3.8}$$

It follows from (G.3.2), (G.3.7) and (G.3.5) that

$$x_1\cos\alpha - x_2\sin\alpha\cos\vartheta - x_3\sin\alpha\sin\vartheta = 0, \tag{G.3.9}$$

and (G.3.4), (G.3.6) and (G.3.7) yield

$$x_1\sin\alpha + x_2\cos\alpha\cos\vartheta + x_3\cos\alpha\sin\vartheta = \cos\vartheta. \tag{G.3.10}$$

From this we obtain:

$$x_1 = \cos\vartheta\sin\alpha \tag{G.3.11}$$

and

$$x_2\cos\vartheta + x_3\sin\vartheta = \cos\vartheta\cos\alpha. \tag{G.3.12}$$

Finally, (G.3.8), (G.3.11) and (G.3.12) yield:

$$x_2 = \cos\alpha\cos^2\vartheta + \sin^2\vartheta = \cos\alpha + \sin^2\vartheta(1-\cos\alpha) \tag{G.3.13}$$

and

$$x_3 = \cos\vartheta\sin\vartheta(\cos\alpha - 1). \tag{G.3.14}$$

Hence

$$\begin{aligned}\mathfrak{a}_1 &= \mathfrak{i}\cos\vartheta\sin\alpha + \mathfrak{j}\left\{\cos\alpha + \sin^2\vartheta(1-\cos\alpha)\right\} + \mathfrak{k}\cos\vartheta\sin\vartheta(\cos\alpha-1)\\ &= \mathfrak{i}x_1 + \mathfrak{j}x_2 + \mathfrak{k}x_3,\end{aligned} \tag{G.3.15}$$

and consequently

$$\begin{aligned}[\mathfrak{a}_1\mathfrak{s}_1] &= -\mathfrak{i}\sin\vartheta\sin\alpha + \mathfrak{j}\cos\vartheta\sin\vartheta(1-\cos\alpha) + \mathfrak{k}\left\{-1 + \sin^2\vartheta(1-\cos\alpha)\right\}\\ &= \mathfrak{i}y_1 + \mathfrak{j}y_2 + \mathfrak{k}y_3.\end{aligned} \tag{G.3.16}$$

6 c.f. §6, C.

Now introducing the following notation:

$$\mathfrak{i}\mathfrak{E}_1 = e_1; \quad \mathfrak{j}\mathfrak{E}_1 = e_2; \quad \mathfrak{k}\mathfrak{E}_1 = e_3; \tag{G.3.17}$$

$$\mathfrak{i}\mathfrak{H}_1 = h_1; \quad \mathfrak{j}\mathfrak{H}_1 = h_2; \quad \mathfrak{k}\mathfrak{H}_1 = h_3; \tag{G.3.18}$$

then (G.2.13) §G.2 yields:

$$e_s = B\int_{\alpha=0}^{\alpha=\alpha'}\int_{\vartheta=0}^{\vartheta=2\pi}\left(x_s\sqrt{\cos\alpha}e^{-iu\cos\varepsilon}\right)\sin\alpha d\alpha d\vartheta; \qquad s = 1, 2, 3; \tag{G.3.19}$$

and (G.2.14) §G.2 yields

$$h_s = -\frac{Bn_1}{c}\int_{\alpha=0}^{\alpha=\alpha'}\int_{\vartheta=0}^{\vartheta=2\pi}\left(y_s\sqrt{\cos\alpha}e^{-iu\cos\varepsilon}\right)\sin\alpha d\alpha d\vartheta; \qquad s = 1, 2, 3. \tag{G.3.20}$$

G.4 Calculation of $\mathfrak{E}_1$ and $\mathfrak{H}_1$ using Bessel Functions

It is known that

$$e^{ix\cos(\vartheta-\gamma)} = J_0(x) + 2\sum_{s=1}^{s=\infty} i^s J_s(x)\cos s(\vartheta-\gamma), \tag{G.4.1}$$

where $J_s(x)$ is the Bessel function of order s with the argument x. Further, we have

$$\int_{\vartheta=0}^{\vartheta=2\pi}\cos m\vartheta\cos n\vartheta d\vartheta = \begin{cases} 0 & \text{if } m \neq n, \\ \pi & \text{if } m = n \neq 0, \\ 2\pi & \text{if } m = n = 0, \end{cases} \tag{G.4.2}$$

$$\int_{\vartheta=0}^{\vartheta=2\pi}\cos m\vartheta\sin n\vartheta d\vartheta = 0 \tag{G.4.3}$$

$$\int_{\vartheta=0}^{\vartheta=2\pi}\sin m\vartheta\sin n\vartheta d\vartheta = \begin{cases} 0 \text{ if } m \neq n, \\ \pi \text{ if } m = n \neq 0, \\ 0 \text{ if } m = n = 0, \end{cases} \tag{G.4.4}$$

and

$$\begin{cases} \cos\vartheta\sin\vartheta = \frac{1}{2}\sin 2\vartheta, \\ \sin^2\vartheta = \frac{1}{2}(1-\cos 2\vartheta). \end{cases} \tag{G.4.5}$$

Introducing the notation

$$u\sin\alpha\sin\delta = p, \tag{G.4.6}$$

we obtain, according to (G.4.1)–(G.4.6),

$$\int_{\vartheta=0}^{\vartheta=2\pi} e^{ip\cos(\vartheta-\gamma)}d\vartheta = 2\pi J_{0},(p) \tag{G.4.7}$$

$$\int_{\vartheta=0}^{\vartheta=2\pi}\cos\vartheta e^{ip\cos(\vartheta-\gamma)}d\vartheta = 2i\pi\cos\gamma J_1(p), \tag{G.4.8}$$

$$\int_{\vartheta=0}^{\vartheta=2\pi}\sin\vartheta e^{ip\cos(\vartheta-\gamma)}d\vartheta = 2i\pi\sin\gamma J_1(p), \tag{G.4.9}$$

$$\int_{\vartheta=0}^{\vartheta=2\pi}\sin^2\vartheta e^{ip\cos(\vartheta-\gamma)}d\vartheta = \pi J_0(p) + \pi\cos 2\gamma J_2(p), \tag{G.4.10}$$

$$\int_{\vartheta=0}^{\vartheta=2\pi}\cos\vartheta\sin\vartheta e^{ip\cos(\vartheta-\gamma)}d\vartheta = -\pi\sin 2\gamma J_2(p). \tag{G.4.11}$$

So instead of (G.3.19) and (G.3.20) §G.3, we can write:

$$e_1 = 2i\pi B\cos\gamma I_1, \tag{G.4.12}$$

$$e_2 = \pi BI_2 + \pi B\cos 2\gamma I_3, \tag{G.4.13}$$

$$e_3 = \pi B\sin 2\gamma I_3, \tag{G.4.14}$$

and

$$h_1 = 2i\pi \frac{Bn_1}{c} \sin\gamma I_1, \tag{G.4.15}$$

$$h_2 = \frac{\pi Bn_1}{c} \sin 2\gamma I_3, \tag{G.4.16}$$

$$h_3 = \frac{\pi Bn_1}{c} I_2 - \frac{\pi Bn_1}{c} \cos 2\gamma I_3, \tag{G.4.17}$$

where

$$I_1 = \int_{\alpha=0}^{\alpha=\alpha'} \sin^2\alpha \sqrt{\cos\alpha} J_1(p) e^{-iu\cos\alpha\cos\delta} d\alpha, \tag{G.4.18}$$

$$I_2 = \int_{\alpha=0}^{\alpha=\alpha'} \sin\alpha \sqrt{\cos\alpha}(1+\cos\alpha) J_0(p) e^{-iu\cos\alpha\cos\delta} d\alpha, \tag{G.4.19}$$

$$I_3 = \int_{\alpha=0}^{\alpha=\alpha'} \sin\alpha \sqrt{\cos\alpha}(1-\cos\alpha) J_2(p) e^{-iu\cos\alpha\cos\delta} d\alpha. \tag{G.4.20}$$

Let us assume that the point P (Fig. G.2) lies in the plane E_1, i.e., in the focal plane, such that $\delta = \pi/2$, $\cos\delta = 0$ and $\sin\delta = 1$. In this case the integrals (G.4.18) to (G.4.20) are real and we denote them by an additional index '0'. Furthermore, we also have

$$p = p_0 = u\sin\alpha; \quad \delta = \frac{\pi}{2}. \tag{G.4.21}$$

Denoting the real parts of e_s and h_s as (e_s) and (h_s), respectively, we obtain from (G.4.13), (G.4.14), (G.4.16) and (G.4.17):

$$(e_2) = -\pi B_1 (I_{20} + I_{30}\cos 2\gamma) \sin\omega\left(t - \frac{E}{c}\right), \tag{G.4.22}$$

$$(e_3) = -\pi B_1 I_{30} \sin 2\gamma \sin\omega\left(t - \frac{E}{c}\right), \tag{G.4.23}$$

$$(h_2) = -\frac{\pi B_1 n_1}{c} I_{30} \sin 2\gamma \sin\omega\left(t - \frac{E}{c}\right), \tag{G.4.24}$$

$$(h_3) = -\frac{\pi B_1 n_1}{c} (I_{20} - I_{30}\cos 2\gamma) \sin\omega\left(t - \frac{E}{c}\right), \tag{G.4.25}$$

where

$$B_1 = \frac{A_0 F_1 \sqrt{n_1 n_0}}{\lambda}; \quad \frac{\pi B_1^2 n_1}{8c} = w_0 \frac{F_1^2 n_1^2 \pi^2}{\lambda^2}. \tag{G.4.26}$$

We therefore obtain:

$$\mathrm{i}\mathfrak{S} = \frac{\mathrm{i}[\mathfrak{E}_1 \mathfrak{H}_1]}{4\pi} = \frac{\pi B_1^2 n_1}{4c} \sin^2\omega\left(t - \frac{E}{c}\right)\left\{I_{20}^2 - I_{30}^2\right\}, \tag{G.4.27}$$

and, consequently,

$$\mathrm{i}\overline{\mathfrak{S}} = w_1 = \frac{\pi B_1^2 n_1}{8c}\left\{I_{20}^2 - I_{30}^2\right\} = w_0 \frac{F_1^2 n_1^2 \pi^2}{\lambda^2}\left\{I_{20}^2 - I_{30}^2\right\}. \tag{G.4.28}$$

This value is nothing but the average energy that passes perpendicularly through the plane E_1, i.e., the focal plane, per unit area. It is quite interesting that this energy depends neither on θ nor on γ. In other words, it does not depend on the position of the plane of polarisation, nor on the position of the point P on the focal plane, at constant b. Thus the energy W_1 flowing through the entire focal plane is equal to:

$$W_1 = 2\pi \int_0^\infty w_1 b\,db = \frac{2\pi^2 B_1^2 n_1}{8c} \int_0^\infty \left\{I_{20}^2 - I_{30}^2\right\} b\,db. \tag{G.4.29}$$

We will prove later on that

$$W_0 = W_1' = W_1. \tag{G.4.30}$$

G.5 Evaluation of the Integrals I_{20} and I_{30} in the Case $\alpha' = \pi/2$

We have:

$$J_0(p_0) = J_0(u \sin\alpha) = \sum_{s=0}^{s=\infty} \frac{\cos^{2s}\alpha\,(u/2)^s J_s(u)}{s!}, \tag{G.5.1}$$

$$J_2(p_0) = J_2(u \sin\alpha) = \sin^2\alpha \sum_{s=0}^{s=\infty} \frac{\cos^{2s}\alpha\,(u/2)^s J_{s+2}(u)}{s!}, \tag{G.5.2}$$

and hence

$$I_{20} = \frac{1}{2}\sum_{s=0}^{s=\infty} \frac{(u/2)^s J_s(u)}{s!}\left\{\frac{1}{s+3/4} + \frac{1}{s+5/4}\right\} \tag{G.5.3}$$

and

$$I_{30} = \frac{1}{2}\sum_{s=0}^{s=\infty} \frac{(u/2)^s J_{s+2}(u)}{s!}\left\{\frac{1}{(s+3/4)(s+7/4)} - \frac{1}{(s+5/4)(s+9/4)}\right\}. \tag{G.5.4}$$

This yields[7]

$$\sum_{s=0}^{s=\infty} \frac{(u/2)^s J_s(u)}{s!\,(s+3/4)} = \sum_{s=0}^{s=\infty} \frac{(u/2)^s \Gamma(s+3/4) J_s(u)}{s!\,\Gamma(s+7/4)} = \Gamma\left(\frac{3}{4}\right)\left(\frac{2}{u}\right)^{\frac{3}{4}} J_{\frac{3}{4}}(u). \tag{G.5.5}$$

Taking analogous summations, we finally obtain:

$$I_{20} = \frac{1}{2}\left\{\Gamma\left(\frac{3}{4}\right)\left(\frac{2}{u}\right)^{\frac{3}{4}} J_{\frac{3}{4}}(u) + \Gamma\left(\frac{5}{4}\right)\left(\frac{2}{u}\right)^{\frac{5}{4}} J_{\frac{5}{4}}(u)\right\}. \tag{G.5.6}$$

Let us denote

$$A = \sum_{s=0}^{s=\infty} \frac{(u/2)^s \Gamma(s+3/4) J_{s+2}(u)}{s!\,\Gamma(s+11/4)} \tag{G.5.7}$$

and

$$B = \sum_{s=0}^{s=\infty} \frac{(u/2)^s \Gamma(s+5/4) J_{s+2}(u)}{s!\,\Gamma(s+13/4)}, \tag{G.5.8}$$

thus yielding

$$I_{30} = \frac{A-B}{2}. \tag{G.5.9}$$

Equation (G.5.7) yields:

$$\frac{dA\,(u/2)^2}{du} = \left(\frac{u}{2}\right)^2 \sum_{s=0}^{s=\infty} \frac{(u/2)^s J_{s+1}(u)\Gamma(s+3/4)}{s!\,\Gamma(s+11/4)} = \left(\frac{u}{2}\right)^{\frac{5}{4}} J_{\frac{7}{4}}(u)\Gamma\left(\frac{3}{4}\right), \tag{G.5.10}$$

and hence[8]

$$A = \left(\frac{2}{u}\right)^2 \Gamma\left(\frac{3}{4}\right)\int_0^u \left(\frac{u}{2}\right)^{\frac{5}{4}} J_{\frac{7}{4}}(u)du = \left(\frac{u}{2}\right)^{\frac{1}{4}} \Gamma\left(\frac{3}{4}\right)\sum_{s=0}^{s=\infty} \frac{(u/2)^s J_{s+\frac{7}{4}}(u)}{\Gamma(s+3)},$$

or

[7] N. Nielsen, *Handbuch d. Cylinderfunctionen*, Tuebner, Leipzig, p. 268 (1904).
[8] N. Nielsen, *Handbuch d. Cylinderfunctionen*, Tuebner, Leipzig, p. 97 (1904).

$$A = \left(\frac{u}{2}\right)^{-\frac{7}{4}} \Gamma\left(\frac{3}{4}\right) \sum_{s=2}^{s=\infty} \frac{(u/2)^s J_{s-\frac{1}{4}}(u)}{s!}$$

$$= \left(\frac{2}{u}\right)^{\frac{7}{4}} \Gamma\left(\frac{3}{4}\right) \left\{ -J_{-\frac{1}{4}}(u) - \frac{u}{2} J_{\frac{3}{4}}(u) + \frac{(u/2)^{-\frac{1}{4}}}{\Gamma(3/4)} \right\}. \tag{G.5.11}$$

Making analogous calculations, we obtain

$$B = \left(\frac{2}{u}\right)^{\frac{9}{4}} \Gamma\left(\frac{5}{4}\right) \left\{ -J_{\frac{1}{4}}(u) - \frac{u}{2} J_{\frac{5}{4}}(u) + \frac{(u/2)^{\frac{1}{4}}}{\Gamma(5/4)} \right\}. \tag{G.5.12}$$

Therefore

$$I_{30} = \frac{1}{2} \left\{ \Gamma\left(\frac{5}{4}\right) J_{\frac{1}{4}}(u) \left(\frac{2}{u}\right)^{\frac{9}{4}} + \Gamma\left(\frac{5}{4}\right) J_{\frac{5}{4}}(u) \left(\frac{2}{u}\right)^{\frac{5}{4}} \right.$$

$$\left. - \Gamma\left(\frac{3}{4}\right) J_{-\frac{1}{4}}(u) \left(\frac{2}{u}\right)^{\frac{7}{4}} - \Gamma\left(\frac{3}{4}\right) J_{\frac{3}{4}}(u) \left(\frac{2}{u}\right)^{\frac{3}{4}} \right\}, \tag{G.5.13}$$

and consequently

$$I_{20} + I_{30} = \Gamma\left(\frac{5}{4}\right) J_{\frac{5}{4}}(u) \left(\frac{2}{u}\right)^{\frac{5}{4}} + \frac{\Gamma\left(\frac{5}{4}\right) J_{\frac{1}{4}}(u) 2^{\frac{5}{4}}}{(u)^{\frac{9}{4}}} - \frac{\Gamma\left(\frac{3}{4}\right) J_{-\frac{1}{4}}(u) 2^{\frac{3}{4}}}{(u)^{\frac{7}{4}}} \tag{G.5.14}$$

and

$$I_{20} - I_{30} = \Gamma\left(\frac{3}{4}\right) J_{\frac{3}{4}}(u) \left(\frac{2}{u}\right)^{\frac{3}{4}} - \frac{\Gamma\left(\frac{5}{4}\right) J_{\frac{1}{4}}(u) 2^{\frac{5}{4}}}{(u)^{\frac{9}{4}}} + \frac{\Gamma\left(\frac{3}{4}\right) J_{-\frac{1}{4}}(u) 2^{\frac{3}{4}}}{(u)^{\frac{7}{4}}}. \tag{G.5.15}$$

For $u = 0$ we obtain:

$$I_{20} + I_{30} = \frac{4}{5} + \frac{4}{15} = \frac{16}{15}; \quad I_{20} - I_{30} = \frac{4}{3} - \frac{4}{15} = \frac{16}{15},$$

i.e.,

$$I_{20} = \frac{16}{15}; \quad I_{30} = 0 \quad \text{at } u = 0. \tag{G.5.16}$$

G.6 Evaluation of the Integrals I_{20} and I_{30} for Small Values of α'

Before evaluating the integrals I_{20} and I_{30} for small values of α', we must consider the general case of arbitrary α', and then take the limit of small α'. We introduce the following notation:

$$A_{2\nu+1} = \int_0^{\alpha'} \cos^{2\nu+1} \alpha \sin \alpha J_0(u \sin \alpha) d\alpha, \tag{G.6.1}$$

$$B_{2\nu+1} = \int_0^{\alpha'} \cos^{2\nu+1} \alpha J_1(u \sin \alpha) d\alpha. \tag{G.6.2}$$

Making use of the relation between Bessel's functions:

$$\frac{2(\nu+1) J_{\nu+1}(x)}{x} = J_\nu(x) + J_{\nu+2}(x), \tag{G.6.3}$$

it is easy to show that

$$I_{20} = A_{\frac{1}{2}} + A_{\frac{3}{2}} \tag{G.6.4}$$

and

$$I_{30} = -A_{\frac{1}{2}} + A_{\frac{3}{2}} + \frac{2}{u}\left(B_{\frac{1}{2}} - B_{\frac{3}{2}}\right), \tag{G.6.5}$$

and consequently

$$I_{20} + I_{30} = 2\left\{A_{\frac{3}{2}} + \frac{1}{u}\left(B_{\frac{1}{2}} - B_{\frac{3}{2}}\right)\right\} \tag{G.6.6}$$

and

$$I_{20} - I_{30} = 2\left\{A_{\frac{1}{2}} - \frac{1}{u}\left(B_{\frac{1}{2}} - B_{\frac{3}{2}}\right)\right\}. \tag{G.6.7}$$

Denoting

$$\sin\alpha = z; \quad \sin\alpha' = z_1, \tag{G.6.8}$$

yields

$$A_{2\nu+1} = \int_0^{z_1} (1 - z^2)^\nu z J_0(uz)dz \tag{G.6.9}$$

and

$$B_{2\nu+1} = \int_0^{z_1} (1 - z^2)^\nu J_1(uz)dz, \tag{G.6.10}$$

or

$$A_{2\nu+1} = \frac{1}{u^2}\int_0^{z_1 u}\left(1 - \frac{t^2}{u^2}\right)^\nu tJ_0(t)dt \tag{G.6.11}$$

and

$$B_{2\nu+1} = \frac{1}{u}\int_0^{z_1 u}\left(1 - \frac{t^2}{u^2}\right)^\nu J_1(t)dt. \tag{G.6.12}$$

Since the upper integration limit is equal to $z_1 u$ with $z_1 \leq 1$, we obtain $t/u \leq 1$. Therefore the brackets under the integrals can be expanded in series, so we obtain:

$$\left(1 - \frac{t^2}{u^2}\right)^\nu = \Gamma(\nu + 1)\sum_{s=0}^{\infty} \frac{(-1)^s t^{2s}}{s!\,\Gamma(\nu - s + 1)u^{2s}}. \tag{G.6.13}$$

Thus (G.6.11) and (G.6.12) can be rewritten as:

$$A_{2\nu+1} = \frac{\Gamma(\nu + 1)}{u}\sum_{s=0}^{\infty} \frac{(-1)^s}{s!\,\Gamma(\nu - s + 1)u^{2s+1}}\int_0^{z_1 u} t^{2s+1}J_0(t)dt \tag{G.6.14}$$

and

$$B_{2\nu+1} = \frac{\Gamma(\nu + 1)}{u}\sum_{s=0}^{\infty} \frac{(-1)^s}{s!\,\Gamma(\nu - s + 1)u^{2s}}\int_0^{z_1 u} t^{2s}J_1(t)dt. \tag{G.6.15}$$

We have[9]

$$\int_0^{z_1 u} t^{2s+1}J_0(t)dt = (uz_1)^{2s+1}\Gamma(s + 1)\sum_{n=0}^{n=s} \frac{(-1)^n(2n + 1)J_{2n+1}(uz_1)}{\Gamma(s + 2 + n)\Gamma(s - n + 1)} \tag{G.6.16}$$

and

$$\int_0^{z_1 u} t^{2s}J_1(t)dt = (uz_1)^{2s}\Gamma(s + 1)\Gamma(s)\sum_{n=0}^{n=s-1} \frac{(-1)^n(2n + 2)J_{2n+2}(uz_1)}{\Gamma(s + 2 + n)\Gamma(s - n)}; \tag{G.6.17}$$

where s must be greater than or equal to 1.

For $s = 0$, we obtain immediately from (G.6.15):

$$\frac{1}{u}\int_0^{z_1 u} J_1(t)dt = \frac{1}{u}\{1 - J_0(uz_1)\}. \tag{G.6.18}$$

[9] N. Nielsen, *Handbuch d. Cylinderfunctionen*, Tuebner, Leipzig, p. 98 (1904).

From (G.6.14)–(G.6.18) we obtain

$$A_{2\nu+1} = \frac{z_1 \Gamma(\nu+1)}{u} \sum_{s=0}^{\infty} \frac{(-1)^s z_1^{2s} s!}{\Gamma(\nu - s + 1)} \sum_{n=0}^{n=s} \frac{(-1)^n (2n+1) J_{2n+1}(uz_1)}{\Gamma(s+2+n)\Gamma(s-n+1)} \tag{G.6.19}$$

and

$$\begin{aligned} B_{2\nu+1} =& \frac{1}{u}\{1 - J_0(uz_1)\} \\ & - \frac{z_1^2 \Gamma(\nu+1)}{u} \sum_{s=0}^{\infty} \frac{(-1)^s z_1^{2s} s!}{\Gamma(\nu - s)} \sum_{n=0}^{n=s} \frac{(-1)^n (2n+2) J_{2n+2}(uz_1)}{\Gamma(s+3+n)\Gamma(s-n+1)}. \end{aligned} \tag{G.6.20}$$

At small α', and consequently, small z_1, we cannot nevertheless assume the value of uz_1 to be small, due to the small value of the wavelength λ. Therefore, when expanding as a series we will not consider the product uz_1; this is permitted since $J_\nu(t) \le 1$.

Retaining only the powers of z_1 not higher than the third, we consequently obtain:

$$A_{\frac{1}{2}} = \frac{z_1 J_1(uz_1)}{u}\left(1 + \frac{z_1^3}{8}\right) - \frac{z_1^3 J_3(uz_1)}{8u}, \tag{G.6.21}$$

$$A_{\frac{3}{2}} = \frac{z_1 J_1(uz_1)}{u}\left(1 - \frac{z_1^3}{8}\right) + \frac{z_1^3 J_3(uz_1)}{8u}, \tag{G.6.22}$$

and, using (G.6.3):

$$B_{\frac{1}{2}} = \frac{1}{u}\{1 - J_0(uz_1)\} + \frac{z_1^2 J_2(uz_1)}{4u} = \frac{1}{u}\{1 - J_0(uz_1)\} + \frac{z_1^3}{16}\{J_1(uz_1) + J_3(uz_1)\}, \tag{G.6.23}$$

and similarly:

$$B_{\frac{3}{2}} = \frac{1}{u}\{1 - J_0(uz_1)\} - \frac{z_1^3}{16}\{J_1(uz_1) + J_3(uz_1)\}. \tag{G.6.24}$$

Finally we obtain:

$$I_{20} + I_{30} = \frac{2z_1}{u}\left(J_1(uz_1) - \frac{z_1^2 J_3(uz_1)}{4}\right), \tag{G.6.25}$$

$$I_{20} - I_{30} = \frac{2z_1}{u}\left(J_1(uz_1) + \frac{z_1^2 J_3(uz_1)}{4}\right), \tag{G.6.26}$$

and consequently,

$$I_{20}^2 - I_{30}^2 = \frac{4\sin^2\alpha'}{u^2}\left(J_1^2(u\sin\alpha') - \frac{\sin^4\alpha'}{16} J_3^2(u\sin\alpha')\right). \tag{G.6.27}$$

G.7 Calculation of $\mathfrak{E}_1$, $\mathfrak{H}_1$ and the Energy w_1 at the Focus

In this case we have

$$u = p = 0 \tag{G.7.1}$$

and consequently,

$$I_1 = I_3 = 0 \tag{G.7.2}$$

and

$$I_2 = \int_{z_1}^{1} \sqrt{z}(1+z)dz = \frac{16}{15}\left(1 - \frac{5}{8}\cos^{\frac{3}{2}}\alpha'\left[1 + \frac{3\cos\alpha'}{5}\right]\right), \tag{G.7.3}$$

where

$$z = \cos\alpha'; \quad z_1 = \cos\alpha'. \tag{G.7.4}$$

Therefore we have:

$$e_1 = e_3 = 0, \tag{G.7.5}$$

$$e_2 = \pi B I_2, \tag{G.7.6}$$

$$h_1 = h_2 = 0, \tag{G.7.7}$$

$$h_3 = \frac{\pi B n_1}{c} I_2 \tag{G.7.8}$$

and

$$w_1 = \frac{\pi B_1^2 n_1}{8c} I_2^2 = \frac{w_0 F_1^2 n_1^2 \pi^2}{\lambda^2} I_2^2. \tag{G.7.9}$$

All these expressions are valid for an arbitrary value of α'. When $\alpha' = \pi/2$, $I_2 = 16/15$, in agreement with (G.5.16) §G.5.

G.8 Calculation at Large Distances from the Focus

We know from Section G.2 that the electric and magnetic fields on a sphere of radius ρ_1 greater than the limit sphere radius, within the homocentric beam, on the left of the focal plane (Fig. G.2) are given by:

$$\mathfrak{E}_1' = \frac{\mathfrak{a}_1 A_0 F_1}{\rho_1} \sqrt{\frac{n_0}{n_1}} \sqrt{\cos\alpha}\, e^{i\omega\left(t - \frac{E_1}{c}\right)} \tag{G.8.1}$$

and

$$\mathfrak{H}_1' = -\frac{A_0 F_1 \sqrt{n_1 n_0} \sqrt{\cos\alpha}}{c\rho_1} [\mathfrak{a}_1 \mathfrak{s}_1]\, e^{i\omega\left(t - \frac{E_1}{c}\right)}. \tag{G.8.2}$$

At the same sphere, also within the homocentric beam, but on the left of the focal plane, i.e., after passing through the focus, we have (recalling that the eikonal makes a half-wavelength jump at the focus[10]):

$$\mathfrak{E}_1'' = \frac{\mathfrak{a}_1 A_0 F_1}{\rho_1} \sqrt{\frac{n_0}{n_1}} \sqrt{\cos\alpha}\, e^{i\omega\left(t - \frac{E_1 + 2\rho_1 n_1 + \lambda/2}{c}\right)} \tag{G.8.3}$$

and

$$\mathfrak{H}_1'' = -\frac{A_0 F_1 \sqrt{n_1 n_0} \sqrt{\cos\alpha}}{c\rho_1} [\mathfrak{a}_1 \mathfrak{s}_1]\, e^{i\omega\left(t - \frac{E_1 + 2\rho_1 n_1 + \lambda/2}{c}\right)}. \tag{G.8.4}$$

For the average energy $\overline{\mathfrak{S}}_1'' = w_1'' \mathfrak{s}_1$ flowing out of a unit surface area of this sphere, we obtain an expression identical to (G.2.20), i.e.,

$$\overline{\mathfrak{S}}_1 = w_1 \mathfrak{s}_1 = \frac{A_0^2 F_1^2 n_0 \cos\alpha}{8\pi c \rho_1^2} \mathfrak{s}_1, \tag{G.8.5}$$

and, as we mentioned at the end of Section G.2, the result is the same as if at the focus there was a fictitious element df_1 radiating according to Lambert's law. Denoting the intensity of this element as J_1, we obtain from (G.8.5):

$$J_1 df = \frac{A_0^2 F_1^2 n_0}{8\pi c}. \tag{G.8.6}$$

Of course, we cannot determine J_1 and df_1 separately without making special assumptions. We will come back to (G.8.6) in the following sections.

In the work C, we have derived expressions similar to (G.2.15) and (G.2.16), on the assumption that (G.8.1) and (G.8.2) were valid, and on the basis of the latter we have demonstrated the existence of the eikonal jump at the focus. It is interesting, and necessary for complete clarification of the issue, to show the reverse, i.e., accept (G.2.15) and (G.2.16) as given, and then derive (G.8.1) and (G.8.2) for the case under consideration, and to demonstrate the eikonal jump. This can be done on the basis of the theory of spherical functions.

[10] See §6, C.

Consider a function $\varphi(\alpha, \vartheta)$ of α and ϑ, which equals $f(\alpha, \vartheta)$ on the surface Φ_1 of a sphere of unit radius (Fig. G.2), and equal to zero elsewhere, i.e.,

$$\varphi(\alpha, \vartheta) = \begin{cases} f(\alpha, \vartheta) & \text{on } \Phi_1 \\ 0 & \text{outside } \Phi_1. \end{cases} \tag{G.8.7}$$

Suppose that the length $O_1P = 1$ and introduce the angle $\varepsilon_1 = \pi - \varepsilon$, i.e., ε_1 is the angle between $\mathfrak{B}$ and the negative direction of the ray. Then, as is known from the theory of spherical functions, the function $\varphi(\alpha, \vartheta)$ that meets condition (G.8.7), can be represented as follows:

$$\begin{aligned} \varphi(\alpha, \vartheta) &= \frac{1}{4\pi} \sum_{m=0}^{\infty} (2m+1) \int_{\Phi_1} f(\alpha, \vartheta) P_m(\cos \varepsilon_1) d\varphi \\ &= \frac{1}{4\pi} \sum_{m=0}^{\infty} (2m+1) \int_{\alpha=0}^{\alpha=\alpha'} \int_{\vartheta=0}^{\vartheta=2\pi} f(\alpha, \vartheta) P_m(\cos \varepsilon_1) \sin \alpha \, d\alpha \, d\vartheta, \end{aligned} \tag{G.8.8}$$

i.e., if the point P (the end of the unit vector O_1P) is located on Φ_1, then the left-hand side of (G.8.8) is equal to $f(\alpha, \vartheta)$. When P does not lie on Φ_1, the left-hand side of (G.8.8) yields zero. Here $P_m(x)$ is Legendre's spherical function of order m and argument x. Using the fact that

$$P_m(-x) = (-1)^m P_m(x), \tag{G.8.9}$$

we can rewrite (G.8.8) in terms of ε rather than ε_1:

$$\varphi(\alpha, \vartheta) = \frac{1}{4\pi} \sum_{m=0}^{\infty} (-1)^m (2m+1) \int_{\alpha=0}^{\alpha=\alpha'} \int_{\vartheta=0}^{\vartheta=2\pi} f(\alpha, \vartheta) P_m(\cos \varepsilon) \sin \alpha \, d\alpha \, d\vartheta. \tag{G.8.10}$$

Imagine that to the right of the focal plane E_1 in Fig. G.2 there is a spherical fragment Φ_1' of unit radius, symmetric to Φ_1. Then, according to the above consideration, the function

$$\varphi'(\alpha, \vartheta) = -\frac{1}{4\pi} \sum_{m=0}^{\infty} (2m+1) \int_{\alpha=0}^{\alpha=\alpha'} \int_{\vartheta=0}^{\vartheta=2\pi} f(\alpha, \vartheta) P_m(\cos \varepsilon) \sin \alpha \, d\alpha \, d\vartheta, \tag{G.8.11}$$

meets the following condition:

$$\varphi'(\alpha, \vartheta) = \begin{cases} -f(\alpha, \vartheta) & \text{on } \Phi_1 \\ 0 & \text{outside } \Phi_1. \end{cases} \tag{G.8.12}$$

The integration in (G.8.10) is performed over Φ_1 (not Φ_1'), which should be understood as follows. When the point P coincides with a certain point on Φ_1', the left-hand side of (G.8.12) yields $-f(\alpha, \vartheta)$, where $f(\alpha, \vartheta)$ has the same value as in (G.8.7), and corresponds to the same ray.

Now we construct the functions

$$M_1 = \frac{2\pi}{iq} \varphi(\alpha, \vartheta) e^{\frac{i\omega \rho_1 n_1}{c}}, \tag{G.8.13}$$

$$M_2 = \frac{2\pi}{iq} \varphi'(\alpha, \vartheta) e^{\frac{-i\omega \rho_1 n_1}{c}}, \tag{G.8.14}$$

and

$$M = M_1 + M_2, \tag{G.8.15}$$

where

$$q = \frac{\omega \rho_1 n_1}{c}. \tag{G.8.16}$$

It is clear from the above that, if the point P moves from the surface Φ_1 to the surface Φ_1' while remaining on the same beam, M_1 transforms into M_2 at the same time. According to (G.8.7) and (G.8.12), we therefore obtain:

$$M_2 = M_1 e^{\frac{-i\omega}{c}(2\rho_1 n_1 + \lambda/2)}, \tag{G.8.17}$$

and

$$M = \begin{cases} M_1 & \text{if } P \text{ is on } \Phi_1 \\ M_2 & \text{if } P \text{ is on } \Phi_1' \\ 0 & \text{otherwise.} \end{cases} \tag{G.8.18}$$

It is known that

$$\begin{aligned} e^{-iu\cos\varepsilon} = e^{iu\cos\varepsilon_1} &= \frac{1}{2}\sum_{m=0}^{\infty}(2m+1)i^m\psi_m(u)P_m(\cos\varepsilon_1) \\ &= \frac{1}{2}\sum_{m=0}^{\infty}(-1)^m(2m+1)i^m\psi_m(u)P_m(\cos\varepsilon), \end{aligned} \tag{G.8.19}$$

where

$$\frac{1}{2}\psi_m(u) = \sqrt{\frac{\pi}{2u}}J_{m+\frac{1}{2}}(u). \tag{G.8.20}$$

At large u we can write:

$$J_{m+\frac{1}{2}}(u) = \sqrt{\frac{2}{\pi u}}\sin\left(u - \frac{m\pi}{2}\right) = \sqrt{\frac{2}{\pi u}}\frac{i^m}{2i}\left\{(-1)^m e^{iu} - e^{-iu}\right\}. \tag{G.8.21}$$

Inserting (G.8.19) into (G.2.15) and (G.2.16), we obtain that for large u, according to (G.8.21), these expressions transform into (G.8.1) and (G.8.2) within the homocentric beam to the left of the focal plane, and into (G.8.3) and (G.8.4) to the right of the focal plane. (G.8.17) yields the half-wavelength jump of the eikonal.

Equations (G.2.15) and (G.2.16) represent a set of plane waves,[11] whose envelope is a sphere with the centre at O_1. Therefore these expressions precisely satisfy Maxwell's equations everywhere in space, which explains the previous conclusions. This yields, as well, that the energy W_1 passing through the entire focal plane, is equal to the energy W_1' that flows into the homocentric beam; we will prove this statement directly in the next section.

In conclusion, we point out the work by Debye,[12] who makes the calculations analogous to the calculations of this section, with the difference that he does not introduce an arbitrary function $f(\alpha, \vartheta)$ and defines a function analogous to M from the condition that this function is finite at O_1. We start with equations of the type (G.8.1) and (G.8.2) and derive (G.2.15) and (G.2.16), for waves of an arbitrary polarisation, arbitrary tilt of the homocentric beam to the optical axis, and for arbitrary shape of the solid angle (see the work C). Furthermore, this section yields the value of the radius of the limit sphere. Namely, ρ_1 is the radius for which we can use (G.2.21) instead of (G.2.20).

G.9 Calculation of the Energy W_1 Flowing through the Entire Focal Plane

For this energy we have (G.4.29):

$$W_1 = w_0\frac{F_1^2 n_1^2 2\pi^3}{\lambda^2}\int_0^{\infty}\left\{I_{20}^2 - I_{30}^2\right\} b\,db. \tag{G.9.1}$$

Let us introduce the notation:

$$Q = \int_0^{\infty}\left\{I_{20}^2 - I_{30}^2\right\} b\,db. \tag{G.9.2}$$

We need to evaluate the integral Q for an arbitrary α'. If the equality

$$W_1' = W_1 \tag{G.9.3}$$

is indeed valid, then (G.2.22) yields:

$$Q = \frac{\lambda^2\sin^2\alpha'}{2\pi^2 n_1^2}, \tag{G.9.4}$$

which we will now prove.

[11] See §6, C.
[12] P. Debye, *Ann. d. Phys.* **Bd. 30**, p. 755 (1909).

As follows from the known relations between Bessel's functions:

$$I_{20} - I_{30} = \frac{2J_1(u\sin\alpha')\sin\alpha'}{u\sqrt{\cos\alpha'}} + \frac{1}{u}\int_{\alpha=0}^{\alpha=\alpha'} \frac{2\cos^3\alpha - \cos^2\alpha - 1}{\cos^{\frac{3}{2}}\alpha} J_1(u\sin\alpha)d\alpha, \tag{G.9.5}$$

and

$$\begin{aligned} I_{20} + I_{30} =& \frac{2\sqrt{\cos\alpha'}\sin\alpha' J_1(u\sin\alpha')}{u} \\ &+ \frac{1}{u}\int_{\alpha=0}^{\alpha=\alpha'} \frac{2\cos\alpha + 1 - 3\cos^2\alpha}{\cos^{\frac{1}{2}}\alpha} J_1(u\sin\alpha)d\alpha. \end{aligned} \tag{G.9.6}$$

We have:

$$\int_{b=0}^{b=\infty} (\cdots)bdb = \frac{c^2}{n_1^2\omega^2}\int_{u=0}^{u=\infty} (\cdots)udu, \tag{G.9.7}$$

and therefore:[13]

$$\int_0^\infty \frac{\{J_1(u\sin\alpha)\}^2}{u^2} bdb = \frac{c^2}{n_1^2\omega^2}\int_0^\infty \frac{\{J_1(u\sin\alpha)\}^2}{u} du = \frac{c^2}{2n_1^2\omega^2} = \frac{\lambda^2}{8n_1^2\pi^2}. \tag{G.9.8}$$

In an analogous way we find:

$$\begin{aligned} \int_0^\infty \frac{J_1(\sin\alpha')J_1(u\sin\alpha)}{u^2} bdb &= \frac{c^2}{n_1^2\omega^2}\int_0^\infty \frac{J_1(p_0)J_1(p_0 q)}{p_0} dp_0 \\ &= \frac{c^2\sin^2\alpha}{2n_1^2\omega^2\sin^2\alpha'} = \frac{\lambda^2\sin^2\alpha}{8n_1^2\pi^2\sin^2\alpha'}, \end{aligned} \tag{G.9.9}$$

where

$$q = \frac{\sin\alpha}{\sin\alpha'} \leq 1. \tag{G.9.10}$$

We further find:

$$\begin{aligned} Q_1 &= \int_0^\infty \left\{ \frac{2J_1(u\sin\alpha')\sin\alpha'}{u^2} \int_0^{\alpha'} \frac{2\cos\alpha - 3\cos^2\alpha + 1}{\cos^{\frac{1}{2}}\alpha} J_1(u\sin\alpha)d\alpha \right\} bdb \\ &= \frac{c^2}{n_1^2\omega^2\sqrt{\cos\alpha'}} \int_0^{\alpha'} \frac{2\cos\alpha - 3\cos^2\alpha + 1}{\cos^{\frac{1}{2}}\alpha} d(\cos\alpha) \end{aligned} \tag{G.9.11}$$

and

$$\begin{aligned} Q_2 &= \int_0^\infty \left\{ \frac{2\sqrt{\cos\alpha'}\sin\alpha' J_1(u\sin\alpha')}{u^2} \int_0^{\alpha'} \frac{2\cos^3\alpha - \cos^2\alpha - 1}{\cos^{\frac{3}{2}}\alpha} J_1(u\sin\alpha)d\alpha \right\} bdb \\ &= -\frac{\sqrt{\cos\alpha'}c^2}{n_1^2\omega^2} \int_0^{\alpha'} \frac{2\cos^3\alpha - \cos^2\alpha - 1}{\cos^{\frac{3}{2}}\alpha} d(\cos\alpha). \end{aligned} \tag{G.9.12}$$

With the values of Q_1 and Q_2 calculated, we obtain:

$$\begin{aligned} Q_1 + Q_2 = -\frac{c^2}{n_1^2\omega^2} \Bigg\{ & \frac{4}{5}\cos^3\alpha' - \frac{28}{15}\cos^2\alpha' + \frac{4}{3}\cos\alpha' \\ & - \frac{32}{15}\sqrt{\cos\alpha'} - \frac{32}{15\sqrt{\cos\alpha'}} + 4 \Bigg\}. \end{aligned} \tag{G.9.13}$$

Let us denote

$$\begin{aligned} Q_3 = \int_0^\infty \Bigg\{ & \frac{1}{u^2} \left[\int_0^{\alpha'} \frac{2\cos^3\alpha - \cos^2\alpha - 1}{\cos^{\frac{3}{2}}\alpha} J_1(u\sin\alpha)d\alpha \right] \\ & \times \left[\int_0^{\alpha''} \frac{2\cos\alpha - 3\cos^2\alpha + 1}{\cos^{\frac{1}{2}}\alpha} J_1(u\sin\alpha)d\alpha \right] \Bigg\} bdb, \end{aligned} \tag{G.9.14}$$

[13] Formula (10) of N. Nielsen, *Handbuch d. Cylinderfunctionen*, Tuebner, Leipzig, p. 200 (1904).

such that

$$Q = \frac{\lambda^2 \sin^2 \alpha'}{2\pi^2 n_1^2} + Q_1 + Q_2 + Q_3. \tag{G.9.15}$$

We need to prove therefore that

$$Q_1 + Q_2 + Q_3 = 0, \tag{G.9.16}$$

which requires evaluation of Q_3. For this purpose we take some value of α in the range between 0 and α' and denote this value as α_0. We then have[14]

$$\int_0^\infty \frac{J_1(u\sin\alpha_0)J_1(u\sin\alpha)}{u^2} bdb = \begin{cases} \frac{c^2}{2n_1^2\omega^2}\frac{\sin\alpha}{\sin\alpha_0}, & \alpha_0 > \alpha \\ \frac{c^2}{2n_1^2\omega^2} & \alpha_0 = \alpha \\ \frac{c^2}{2n_1^2\omega^2}\frac{\sin\alpha_0}{\sin\alpha}, & \alpha_0 < \alpha. \end{cases} \tag{G.9.17}$$

Since in evaluating the integral only $\alpha_0 \neq \alpha$ are meaningful, the integral Q_3 splits into two integrals Q_3' and Q_3'', namely:

$$\begin{aligned} Q_3' &= -\frac{c^2}{2n_1^2\omega^2}\int_0^{\alpha'} \frac{2\cos\alpha - 3\cos^2\alpha + 1}{\cos^{\frac{1}{2}}\alpha\sin\alpha}\varphi_1(\cos\alpha)d\alpha \\ &= \frac{c^2}{2n_1^2\omega^2}\int_0^{\alpha'} \frac{(3\cos\alpha + 1)\varphi_1(\cos\alpha)}{\cos^{\frac{1}{2}}\alpha(1+\cos\alpha)}d(\cos\alpha), \end{aligned} \tag{G.9.18}$$

and

$$Q_3'' = -\frac{c^2}{2n_1^2\omega^2}\int_\alpha^{\alpha'} \frac{2\cos\alpha - 3\cos^2\alpha + 1}{\cos^{\frac{1}{2}}\alpha}\varphi_2(\cos\alpha)d(\cos\alpha), \tag{G.9.19}$$

where

$$\varphi_1(\cos\alpha) = \int_0^\alpha \frac{2\cos^3\alpha - \cos^2\alpha - 1}{\cos^{\frac{3}{2}}\alpha}d(\cos\alpha), \tag{G.9.20}$$

and

$$\varphi_2(\cos\alpha) = \int_\alpha^{\alpha'} \frac{2\cos^3\alpha - \cos^2\alpha - 1}{\cos^{\frac{3}{2}}\alpha\sin\alpha}d\alpha = \int_\alpha^{\alpha'} \frac{2\cos^2\alpha + \cos\alpha + 1}{\cos^{\frac{3}{2}}\alpha(1+\cos\alpha)}d(\cos\alpha). \tag{G.9.21}$$

Evaluating these two latter integrals, we obtain:

$$\varphi_1(\cos\alpha) = \frac{4}{5}\cos^{\frac{5}{2}}\alpha - \frac{2}{3}\cos^{\frac{3}{2}}\alpha + \frac{2}{\sqrt{\cos\alpha}} - \frac{32}{15}, \tag{G.9.22}$$

and

$$\begin{aligned} \varphi_2(\cos\alpha) = 4&\left\{\sqrt{\cos\alpha'} - \sqrt{\cos\alpha} - \operatorname{arctg}\sqrt{\cos\alpha'} + \operatorname{arctg}\sqrt{\cos\alpha}\right\} \\ &- 2\left\{\frac{1}{\sqrt{\cos\alpha'}} - \frac{1}{\sqrt{\cos\alpha}}\right\}; \end{aligned} \tag{G.9.23}$$

but

$$\frac{(3\cos\alpha + 1)\,d(\cos\alpha)}{\cos^{\frac{1}{2}}\alpha(1+\cos\alpha)} = 2d\left\{3\sqrt{\cos\alpha} - \operatorname{arctg}\sqrt{\cos\alpha}\right\}, \tag{G.9.24}$$

and

$$\frac{2\cos\alpha - 3\cos^2\alpha + 1}{\cos^{\frac{1}{2}}\alpha}d(\cos\alpha) = d\left\{\frac{4}{3}\cos^{\frac{3}{2}}\alpha - \frac{6}{5}\cos^{\frac{5}{2}}\alpha + 2\cos^{\frac{1}{2}}\alpha\right\}. \tag{G.9.25}$$

[14] Formula (10) of N. Nielsen, *Handbuch d. Cylinderfunctionen*, Tuebner, Leipzig, p. 200 (1904).

Therefore, using partial integration we obtain from (G.9.18) and (G.9.19), as a consequence of (G.9.22) and (G.9.23) and bearing in mind that:

$$\frac{d\varphi_1}{d(\cos\alpha)} = \frac{2\cos^3\alpha - \cos^2\alpha - 1}{\cos^{\frac{3}{2}}\alpha} \tag{G.9.26}$$

and

$$\frac{d\varphi_2}{d(\cos\alpha)} = -\frac{2\cos^{\frac{1}{2}}\alpha}{(1+\cos\alpha)} - \frac{1}{\cos^{\frac{3}{2}}\alpha}, \tag{G.9.27}$$

and that $\varphi_1(1) = 0$ at $\alpha = 0$ and $\varphi_2(\cos\alpha') = 0$ at $\alpha = \alpha'$:

$$\begin{aligned} Q_3' =& \frac{c^2}{n_1^2\omega^2}\left\{\left[3\sqrt{\cos\alpha'} - 2\mathrm{arctg}\sqrt{\cos\alpha'}\right]\left[\frac{4}{5}\cos^{\frac{5}{2}}\alpha' - \frac{2}{3}\cos^{\frac{3}{2}}\alpha' + \frac{2}{\sqrt{\cos\alpha'}} - \frac{32}{15}\right]\right. \\ &\left. - \int_0^{\alpha'} \frac{\left(3\sqrt{\cos\alpha} - 2\mathrm{arctg}\sqrt{\cos\alpha}\right)\left(2\cos^3\alpha - \cos^2\alpha - 1\right)}{\cos^{\frac{3}{2}}\alpha} d(\cos\alpha)\right\}, \end{aligned} \tag{G.9.28}$$

and

$$\begin{aligned} Q_3'' =& -\frac{c^2}{2n_1^2\omega^2}\left\{-\frac{128}{15}\left[\sqrt{\cos\alpha'} - 1 - \mathrm{arctg}\sqrt{\cos\alpha'} + \frac{\pi}{4}\right] + \frac{64}{15}\left[\frac{1}{\sqrt{\cos\alpha'}} - 1\right]\right. \\ &\left. + \int_0^{\alpha'}\left(\frac{4}{3}\cos^{\frac{3}{2}}\alpha - \frac{6}{5}\cos^{\frac{5}{2}}\alpha + 2\cos^{\frac{1}{2}}\alpha\right)\left(\frac{2\cos^{\frac{1}{2}}\alpha}{1+\cos\alpha} + \frac{1}{\cos^{\frac{3}{2}}\alpha}\right)d(\cos\alpha)\right\}, \end{aligned} \tag{G.9.29}$$

or

$$\begin{aligned} Q_3' =& \frac{c^2}{n_1^2\omega^2}\left\{\frac{2}{5}\cos^3\alpha' - \frac{9}{10}\cos^2\alpha' + \frac{22}{15}\cos\alpha' - \frac{96}{15}\sqrt{\cos\alpha'} + 2\lg\sqrt{\cos\alpha'}\right. \\ &\left. + \frac{163}{30} - \frac{16\pi}{15} + \frac{8}{15}\lg(1+\cos\alpha') - \frac{8}{15}\lg 2 + \frac{64}{15}\mathrm{arctg}\sqrt{\cos\alpha'}\right\}, \end{aligned} \tag{G.9.30}$$

and

$$\begin{aligned} Q_3'' =& \frac{c^2}{n_1^2\omega^2}\left\{\frac{2}{5}\cos^3\alpha' - \frac{29}{30}\cos^2\alpha' + \frac{2}{15}\cos\alpha' + \frac{64}{15}\sqrt{\cos\alpha'} - \frac{32}{15\sqrt{\cos\alpha'}} + \frac{16\pi}{15} - \frac{43}{30}\right. \\ &\left. -2\lg\sqrt{\cos\alpha'} - \frac{8}{15}\lg(1+\cos\alpha') + \frac{8}{15}\lg 2 - \frac{64}{15}\mathrm{arctg}\sqrt{\cos\alpha'}\right\}, \end{aligned} \tag{G.9.31}$$

which yields:

$$\begin{aligned} Q_3 &= Q_3' + Q_3'' \\ &= \frac{c^2}{n_1^2\omega^2}\left\{\frac{4}{5}\cos^3\alpha' - \frac{28}{15}\cos^2\alpha' + \frac{4}{3}\cos\alpha' - \frac{32}{15}\sqrt{\cos\alpha'} - \frac{32}{15\sqrt{\cos\alpha'}} + 4\right\}. \end{aligned} \tag{G.9.32}$$

This and (G.9.13) yield (G.9.4). Q.E.D.

G.10 Numerical Calculations

In what follows we present some numerical data, which, together with the graphs, illustrate the previous results more visually. We are primarily interested in the energy passing through the focal plane. We shall start by considering this. With the definition:

$$\frac{\pi B_1^2 n_1}{8c} = \frac{w_0 F_1^2 n_1^2 \pi^2}{\lambda^2} = q, \tag{G.10.1}$$

we obtain from (G.4.28)

$$w_1 = q\left(I_{20}^2 - I_{30}^2\right). \tag{G.10.2}$$

Table G.1. Values of w_1/q at the focus for different α'.

$\sin\alpha'$	w_1/q
0	0.000000
0.1	0.000100
0.2	0.001600
0.4	0.025615
0.6	0.130079
0.8	0.416392
0.9	0.686367
0.979796	0.999768
1.0	$1.137778 = (16/15)^2$

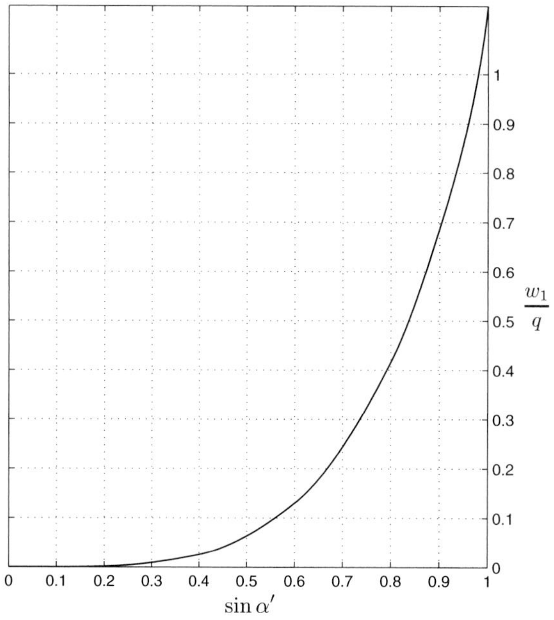

Figure G.3

In the first place we determine the energy w_1 which passes through the focus at different α'. We know that

$$\sin\alpha' = \frac{h}{F_1} = \frac{2d}{F_1}, \tag{G.10.3}$$

where d is the lens diameter. We can therefore construct Table G.1, on the basis of (G.7.3).

When the lens aperture is $F_1 : 5$, $\sin\alpha' = 0.1$ and in this case w_1 is 11378 times smaller than at $\alpha' = \pi/2$. Figure G.3 illustrates the data presented in Table G.1. It follows from (G.7.3) that the tangent to the curve at the points $\alpha' = 0$ and $\alpha' = \pi/2$ is parallel to the horizontal axis.

Calculating the product in (G.5.14) and (G.5.15), we get values of w_1/q as a function of $u = (2\pi n_1 b)/\lambda$ for $\alpha' = \pi/2$, as given in Table G.2. The data in Table G.2 are graphically illustrated in Fig. G.4. The tangent to the curve at $u = 0$ is parallel to the horizontal axis. Curve II shows a part of curve I with the vertical scale expanded tenfold for clarity.

Table G.2. The values of w_1/q for $\alpha' = \pi/2$.

u	w_1/q	u	w_1/q	u	w_1/q	u	w_1/q
0	1.1378						
0.2	1.1278	2.2	0.2610	4.2	−0.0028	6.2	0.0045
0.4	1.0887	2.4	0.1837	4.4	0.0042	6.4	0.0016
0.6	1.0343	2.6	0.1152	4.6	0.0105	6.6	−0.0006
0.8	0.9712	2.8	0.0693	4.8	0.0155	6.8	−0.0017
1.0	0.8697	3.0	0.0331	5.0	0.0181	7.0	−0.0020
1.2	0.7693	3.2	0.0071	5.2	0.0188	7.2	−0.0016
1.4	0.6626	3.4	−0.0077	5.4	0.0177	7.4	−0.0005
1.6	0.5392	3.6	−0.0137	5.6	0.0152	7.6	0.0008
1.8	0.4491	3.8	−0.0119	5.8	0.0118	7.8	0.0023
2.0	0.3541	4.0	−0.0092	6.0	0.0081	8.0	0.0032

Table G.3.

x	$J_1(x)$	$J_1^2(x)$	$J_3(x)$	$J_3^2(x)$
3.80	0.012821	0.000164	0.416973	0.173866
3.81	0.008766	0.000077	0.417809	0.174564
3.82	0.004722	0.000022	0.419576	0.176044
3.83	0.000687	0.000000	0.420324	0.176672
3.831706	0.000000	0.000000	0.420449	0.176777
3.84	−0.003337	0.000011	0.421054	0.177286
3.85	−0.007350	0.000054	0.421768	0.177882

Table G.4.

	$\sin\alpha' = 0.1$	$\sin\alpha' = 0.2$
$u\sin\alpha'$	$t \times 10^4$	$t \times 10^4$
3.80	1.629	1.466
3.81	0.759	0.595
3.82	0.209	0.044
3.83	−0.011	−0.077
3.831706	−0.011	−0.177
3.84	+0.099	−0.067
3.85	+0.529	+0.362

Table G.3 contains auxiliary data needed for calculation of Table G.4. In Table G.4 we use the notation

$$t = J_1^2(u\sin\alpha') - \frac{\sin^4\alpha'}{16}J_3^2(u\sin\alpha') = \frac{w_1 u^2}{4q\sin^2\alpha'}. \tag{G.10.4}$$

Figure G.5 illustrates the data given in Table G.4: curve I represents the value of $10^6 \times J_1^2(u\sin\alpha')$, and curve II shows $t \times 10^6$ for $\sin\alpha' = 0.2$. The curve of $t \times 10^6$ for $\sin\alpha' = 0.1$ is not shown since it is practically coincident with curve I.

From Tables G.2 and G.4 we notice an interesting phenomenon: the energy can take negative values. This fact points to the reverse flow of the energy, which results in widening of the dark rings. Indeed, when observing the energy distribution in the focal plane, we can only see the energy which flows towards us. The energy that flows away from us cannot be seen, i.e., the area of reverse energy flow should be observed as dark. Therefore, this results in widening and greater contrast of the dark rings. Such a reverse energy flow must evidently be a common feature of all interference phenomena; it should be more pronounced for sharper and wider dark stripes. We will come back to this question in another work.

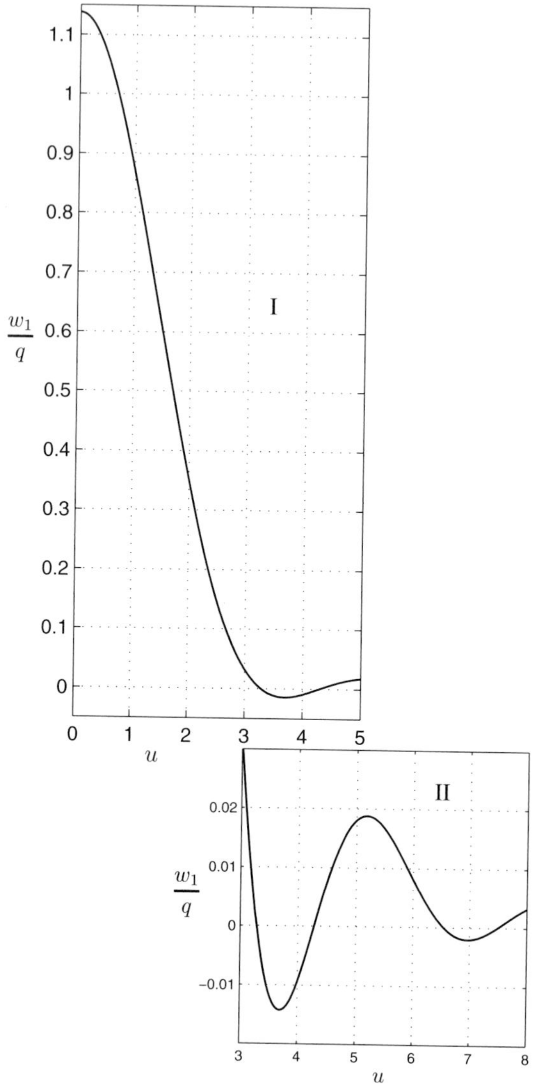

Figure G.4

Let us denote the theoretical width of the dark ring as Δb. From Table G.2 we can see that w_1 is approximately less than zero between $u = 4.3$ and $u = 3.3$, i.e.,

$$\Delta b = \frac{\lambda}{2\pi n_1}; \quad \alpha' = \frac{\pi}{2}. \tag{G.10.5}$$

From Table G.4 and Figure G.5 we can see that according to (G.6.27), $w_1 < 0$ is satisfied when the difference of values of $u \sin \alpha'$ is approximately 0.02 at $\sin \alpha' = 0.2$. Therefore we obtain

$$\Delta b = \frac{0.1\lambda}{2\pi n_1}; \quad \sin \alpha' = 0.2. \tag{G.10.6}$$

In practice, the values (G.10.5) and (G.10.6) are meaningless, since due to the shallowness of the curves around $w_1 = 0$, the rings would appear significantly wider.

Table G.4 was calculated for the values of x close to the root of the equation $J_1(x) = 0$, since it is only there that we can expect negative values of w_1, because the second term in (G.10.4) is small. In general, this term can be neglected and we can write:

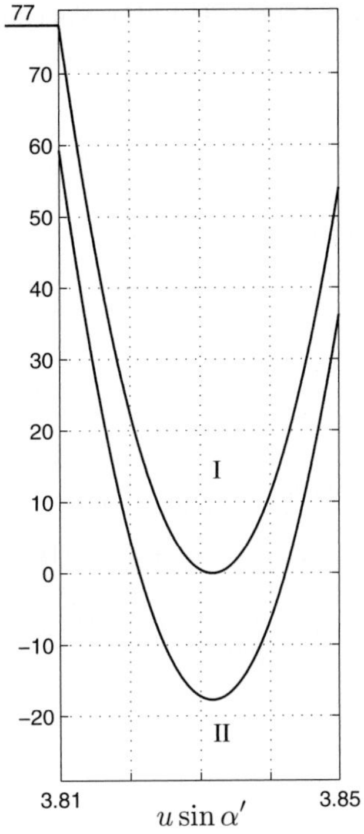

Figure G.5

$$w_1 = \frac{4q \sin^2 \alpha'}{u^2} J_1^2(u \sin \alpha'). \tag{G.10.7}$$

This expression agrees with previous results of other authors.

It is also interesting for us to determine the energy e that flows through a circle with the centre at the focus and of radius b_1, i.e.,

$$e = 2q\pi \int_0^{b_1} (I_{20}^2 - I_{30}^2) b db, \tag{G.10.8}$$

or

$$\frac{e}{W_0} = \frac{2\pi^2 n_1^2}{\lambda^2 \sin^2 \alpha'} \int_0^{b_1} (I_{20}^2 - I_{30}^2) b db, \tag{G.10.9}$$

or, finally:

$$\frac{e}{W_0} = \frac{P}{2 \sin^2 \alpha'}, \tag{G.10.10}$$

where

$$P = \int_0^{u_1} (I_{20}^2 - I_{30}^2) u du. \tag{G.10.11}$$

When $u_1 = \infty$ and $\alpha' = \pi/2$, we obtain, according to (G.9.2), (G.9.14) and (G.9.7), $P = 2$, i.e., $e/W_0 = 1$, as it should be. From (G.10.7), (G.10.10) and (G.10.11) we get:

$$\frac{e}{W_0} = 2 \int_0^{u \sin \alpha'} \frac{J_1^2(x) dx}{x} = 1 - J_0^2(u \sin \alpha') - J_1^2(u \sin \alpha'). \tag{G.10.12}$$

For $\alpha' = \pi/2$ we must insert (G.5.14) and (G.5.15) into (G.10.11). Skipping rather complicated derivation, we give the following result without proof:

$$P = \int_0^{u_1} (I_{20}^2 - I_{30}^2) u du$$

$$= 2 - \frac{3}{u_1^2} + \frac{\pi^2 2^{\frac{3}{2}} J_{-\frac{1}{4}}^2(u_1)}{u_1^{\frac{3}{2}} \Gamma(\frac{1}{4})\Gamma(\frac{1}{4})} + \frac{2^{\frac{3}{2}} \Gamma(\frac{5}{4})\Gamma(\frac{5}{4}) J_{\frac{1}{4}}^2(u_1)}{u_1^{\frac{5}{2}}}$$

$$+ \frac{\pi J_{\frac{1}{4}}(u_1) J_{-\frac{1}{4}}(u_1)}{2^{\frac{1}{2}}} \left\{ \frac{1}{2u_1^2} - 1 \right\} - \frac{2^{\frac{1}{2}} \pi J_{-\frac{1}{4}}(u_1) J_{-\frac{3}{4}}(u_1)}{u_1}$$

$$= 2 - \frac{3}{u_1^2} + \frac{2.12365}{u_1^{\frac{3}{2}}} J_{-\frac{1}{4}}^2(u_1) + \frac{2.32372}{u_1^{\frac{5}{2}}} J_{\frac{1}{4}}^2(u_1)$$

$$+ 2.22139 J_{\frac{1}{4}}(u_1) J_{-\frac{1}{4}}(u_1) \left\{ \frac{1}{2u_1^2} - 1 \right\} - \frac{4.44287}{u_1} J_{-\frac{1}{4}}(u_1) J_{-\frac{3}{4}}(u_1). \quad \text{(G.10.13)}$$

If follows immediately from (G.10.13) that $P = 2$ at $u_1 = \infty$. If u_1 is so large that we can use the approximate formula:

$$J_\nu(u_1) = \sqrt{\frac{2}{\pi u_1}} \sin\left(u_1 - \frac{2\nu - 1}{4}\pi\right), \quad \text{(G.10.14)}$$

then instead of (G.10.13) we obtain, assuming that

$$u_1 = \pi m, \quad \text{(G.10.15)}$$

where m is integer:

$$P = 2 - \frac{0.15916}{m} - \frac{0.54856}{m^2} + \frac{0.065966}{m^{\frac{3}{2}}} + \frac{0.0080629}{m^3} + \frac{0.0039420}{m^{\frac{7}{2}}}. \quad \text{(G.10.16)}$$

Equation (G.10.16) yields:

$$\left\{ \begin{matrix} m = 3, & \frac{e}{W_0} = 0.9495 \\ m = 10, & \frac{e}{W_0} = 0.9894 \end{matrix} \right\} \quad \alpha' = \pi/2. \quad \text{(G.10.17)}$$

We are interested in determining the energy that passes through the central zone within the first ring. We then get from (G.10.12), taking $u \sin \alpha' = 3.831706$ (the first root of $J_1(x) = 0$), and knowing that at this value of x, $J_0(x) = -0.402759$,

$$\frac{e}{W_0} = 0.838; \quad \sin \alpha' \text{ is small.} \quad \text{(G.10.18)}$$

In the case $\alpha' = \pi/2$, we take a value of u in Table G.3, which corresponds approximately to the middle of the first dark ring, namely $u = 4$.[15] In this case we cannot use (G.10.16), but should perform the calculations according to (G.10.13). We then obtain:

$$\frac{e}{W_0} = 0.761; \quad \sin \alpha' = 1. \quad \text{(G.10.19)}$$

Despite the fact that, for example, at $\sin \alpha' = 0.1$ the value of w_1 is 11378 times smaller than at $\sin \alpha' = 1$, the values (G.10.18) and (G.10.19) differ only little, and even (G.10.19) is smaller than (G.10.18). This can be explained partly by the fact that the ring diameter D is much larger for small α' than for $\alpha' = \pi/2$, and partly by the distribution of energy in the focal plane. Assuming that u corresponds to the ring diameter D for $\alpha' = \pi/2$, and $u \sin \alpha' = 3.832$ for small α', we can construct Table G.5 of approximate values of D. In the above calculations we have used for $J_{1/4}(x)$ and $J_{3/4}(x)$ the tables found in Dinnik.[16] For other values we used.[17]

[15] We take this value for ease of calculation.

[16] A. Dinnik, *Archiv d. Math u. Phys.* **21**, p. 324 (1913) (see also Izvestiya Donskogo Politekhnicheskogo Instituta, Novocherkassk (1913)).

[17] A. Gray and G.B. Matthew, *A Treatise on Bessel Functions*, London Macmillan and Co. (1895) and E. Jahnke and F. Emde, *Functionentaflen*, Tuebner, Leipzig (1909).

Table G.5. Approximate values for the first ring diameter D.

$\sin \alpha'$	D
1	$1.22\lambda : n_1$
0.2	$6\lambda : n_1$
0.1	$12\lambda : n_1$

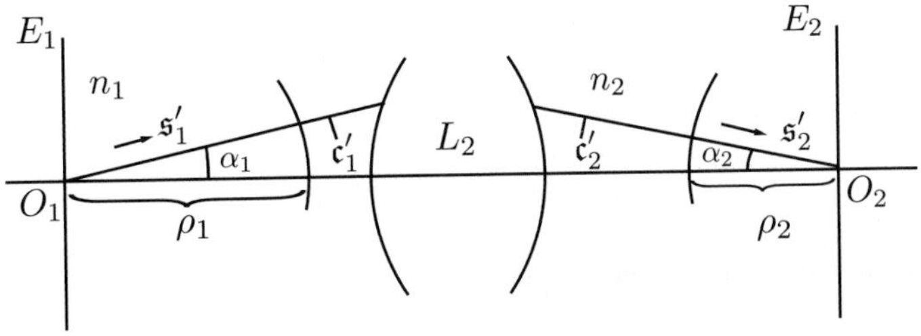

Figure G.6

G.11 The Secondary Image and the Relation to the Microscope Theory

Let us assume that behind the lens L (Fig. G.2) there is another second lens (L_2), which creates a real image O_2 of the point O_1. We assume that the lens L_2 is well corrected, so that the points O_1 and O_2 are aplanatic points for this lens. The question now is what we observe at point O_2 assuming that at O_1 we have a diffraction pattern produced by the lens L_1.

Let us follow a ray when it passes through the lens L_2. For this purpose we take a ray $\mathfrak{s}_1'$ that lies in the same meridional plane as the ray $\mathfrak{s}_1$, has the same angle α, but the angle ϑ is greater by π than the corresponding angle for the ray $\mathfrak{s}_1$. Equation (G.3.2) then yields:

$$\mathfrak{s}_1' = \mathfrak{i} \cos \alpha + \mathfrak{j} \cos \alpha \cos \vartheta + \mathfrak{k} \cos \alpha \sin \vartheta, \tag{G.11.1}$$

and (G.3.2) yields:

$$\mathfrak{c}_1' = \mathfrak{i} \sin \alpha - \mathfrak{j} \cos \alpha \cos \vartheta - \mathfrak{k} \cos \alpha \sin \vartheta, \tag{G.11.2a}$$

where ϑ refers to the ray $\mathfrak{s}_1$.

Let us now follow the ray $\mathfrak{s}_1'$ through the focus and beyond the limit sphere. We take the plane of the drawing (Fig. G.6) to be the meridional plane of the ray $\mathfrak{s}_1'$ and the ray $\mathfrak{s}_2'$, which is a continuation of the ray $\mathfrak{s}_1'$ past the lens L_2. We consider phenomena at the surface of a sphere of the radius ρ_1 greater than the radius of the limit sphere.

Since the vector $\mathfrak{c}_1'$ always points away from the axis and we have gone past the focus, the expression for $\mathfrak{c}_1'$ that corresponds to Fig. G.6 has a different sign to (G.11.2a), i.e.,

$$\mathfrak{c}_1' = -\mathfrak{i} \sin \alpha + \mathfrak{j} \cos \alpha \cos \vartheta + \mathfrak{k} \cos \alpha \sin \vartheta. \tag{G.11.2b}$$

Equation (G.11.1) does not change, but now in (G.11.1) and (G.11.2b) ϑ refers to the ray $\mathfrak{s}_1'$ and is the same as ϑ of the ray $\mathfrak{s}_1$.

Denoting the unit vector along the electric field of the ray $\mathfrak{s}_1'$ as $\mathfrak{a}_1'$, and on the basis of the above formulas and (G.3.15) we get:

$$\mathfrak{a}_1' = -\mathfrak{i} \cos \vartheta \sin \alpha + \mathfrak{j} \left\{ \cos \alpha + \sin^2 \vartheta (1 - \cos \vartheta) \right\} + \mathfrak{k} \cos \vartheta \sin \vartheta (\cos \alpha - 1), \tag{G.11.3}$$

and, as follows from (G.11.2b)

$$\mathfrak{a}_1' \mathfrak{c}_1' = \cos \vartheta. \tag{G.11.4}$$

According to previous sections, we have the electrical vector $\mathfrak{E}_1'$ of the ray $\mathfrak{s}_1'$, at the sphere of radius ρ_1:

$$\mathfrak{E}_1' = \frac{\mathfrak{a}_1 a_1}{\rho_1} e^{i\omega\left(t - \frac{E_1 + 2\rho_1 n_1 + \lambda/2}{c}\right)}, \tag{G.11.5}$$

where

$$a_1 = A_0 F_1 \sqrt{\frac{n_0}{n_1}} \sqrt{\cos\alpha_1}. \tag{G.11.6}$$

Let us draw a sphere with the centre at O_2 and the radius ρ_2 greater than the radius of the limit sphere. At the surface of this sphere we obtain the expression for the electric field $\mathfrak{E}_2'$ of the ray $\mathfrak{s}_2'$:

$$\mathfrak{E}_2' = \frac{\mathfrak{a}_2' a_2}{\rho_2} e^{i\omega\left(t - \frac{E_1 + \Delta + 2\rho_1 n + \lambda/2}{c}\right)}, \tag{G.11.7}$$

where Δ is the eikonal between the spheres ρ_1 and ρ_2. According to Eq. (4.11) of Section 4 in C, we get:

$$a_2 = -\beta \sqrt{\frac{n_2}{n_1}} \sqrt{\frac{\cos\alpha_2}{\cos\alpha_1}} a_1, \tag{G.11.8}$$

where β is the magnification of the system, which is equal to

$$\beta = -\frac{n_1}{n_2} \frac{\sin\alpha_1}{\sin\alpha_2}, \tag{G.11.9}$$

and is constant for all α_1 and α_2. As a consequence of (G.11.8), we can write instead of (G.11.6):

$$a_2 = -\beta A_0 F_1 \frac{\sqrt{n_0 n_2}}{n_1} \sqrt{\cos\alpha_2}, \tag{G.11.10}$$

and consequently

$$\mathfrak{E}_1' = -\frac{\mathfrak{a}_2' \beta A_0 F_1 \sqrt{n_0 n_2}}{\rho_2 n_1} \sqrt{\cos\alpha_2} e^{i\omega\left(t - \frac{E_1 + \Delta + 2\rho_1 n_1 + \lambda/2}{c}\right)} \tag{G.11.11}$$

and

$$\mathfrak{H}_2' = +\frac{\beta A_0 F_1 n_2 \sqrt{n_0 n_2}}{c\rho_2 n_1} \sqrt{\cos\alpha_2}\, [\mathfrak{a}_2' \mathfrak{s}_2']\, e^{i\omega\left(t - \frac{E_1 + \Delta + 2\rho_1 n_1 + \lambda/2}{c}\right)}. \tag{G.11.12}$$

We now determine the vector $\mathfrak{a}_2'$. We know that the following should hold:

$$\mathfrak{a}_2' \mathfrak{s}_2' = 0 \tag{G.11.13}$$

and

$$\mathfrak{a}_2' \mathfrak{c}_2' = \mathfrak{a}_1' \mathfrak{c}_1' = \cos\vartheta. \tag{G.11.14}$$

We further have:

$$\mathfrak{c}_2' = \mathfrak{i}\sin\alpha_2 + \mathfrak{j}\cos\alpha_2\cos\vartheta + \mathfrak{k}\cos\alpha_2\sin\vartheta. \tag{G.11.15}$$

Similarly to Section G.3 we obtain:

$$\mathfrak{a}_2' = -\mathfrak{i}\cos\vartheta\sin\alpha_2 + \mathfrak{j}\left\{\cos\alpha + \sin^2\vartheta(1 - \cos\vartheta)\right\} + \mathfrak{k}\cos\vartheta\sin\vartheta(\cos\alpha - 1). \tag{G.11.16}$$

Now we denote the eikonal of the entire system from the plane E_0 to the sphere ρ_2 as E_2. Then

$$E_2 = E_1 + \Delta + 2\rho_1 n_1. \tag{G.11.17}$$

Furthermore, from geometrical optics, the focus F_2 at the rear of the system is given by:

$$F_2 = -\frac{\beta F_1 n_2}{n_1}, \tag{G.11.18}$$

and consequently, instead of (G.11.11) we have

$$\mathfrak{E}_2' = \frac{\mathfrak{a}_2' A_0 F_2}{\rho_2} \sqrt{\frac{n_0}{n_2}} \sqrt{\cos\alpha_2} e^{i\omega\left(t - \frac{E_2 + \lambda/2}{c}\right)}, \tag{G.11.19}$$

and instead of (G.11.12) we have

$$\mathfrak{H}_2' = -\frac{A_0 F_2 \sqrt{n_0 n_2}}{c\rho_2} \sqrt{\cos\alpha_2}\,[\mathfrak{a}_2'\mathfrak{s}_2']\, e^{i\omega\left(t-\frac{E_2+\lambda/2}{c}\right)}. \tag{G.11.20}$$

Comparing (G.11.19) and (G.11.20) with (G.8.1) and (G.8.2), we see that they are respectively identical, apart from the half-wavelength jump of the eikonal. The latter circumstance has no bearing on the determination of the energy. The phenomenon occurs the same way as if we had a system with focus F_2 instead of the system L_1. Therefore the energy $\overline{\mathfrak{S}}_2'' = w_2''\mathfrak{s}_2'$, going out of the unit surface of the sphere ρ_2 to the right of O_2, is equal, according to (G.8.5) to:

$$w_2''\mathfrak{s}_2' = \frac{A_0^2 F_2^2 n_0 \cos\alpha_2}{8\pi c\rho_2}\mathfrak{s}_2'. \tag{G.11.21}$$

Denoting a fictitious surface element at the focus O_2 as df_2, which emits according to Lambert's law, and J_2 its intensity, we get, recalling (G.11.18):

$$J_2 df_2 = \frac{A_0 F_2^2 n_0}{8\pi c} = \frac{A_0^2 \beta^2 n_0 n_2^2}{n_1^2 8\pi c}, \tag{G.11.22}$$

or, as a result of (G.8.6),

$$\frac{J_2 df_2}{n_2^2} = \frac{J_1 df_1 \beta^2}{n_1^2}. \tag{G.11.23}$$

If we assume that there is a relationship between these fictitious elements df_1 and df_2:

$$df_2 = df_1\beta^2, \tag{G.11.24}$$

as is required for real elements in geometrical optics, then (G.11.23) yields the known result:

$$\frac{J_1}{n_1^2} = \frac{J_2}{n_2^2}. \tag{G.11.25}$$

We have written $\mathfrak{a}_1'$, $\mathfrak{a}_2'$, $\mathfrak{s}_2'$, etc. instead of $\mathfrak{a}_1$, $\mathfrak{a}_2$, $\mathfrak{s}_2$, etc because we were following the ray $\mathfrak{s}_1'$, not $\mathfrak{s}_1$. If we wish to go along $\mathfrak{s}_1$ (which of course does not affect the energy calculations), we only need to replace ϑ by $\vartheta + \pi$ in the respective formulas.

When we go along the ray $\mathfrak{s}_2'$ or $\mathfrak{s}_2$ and get past the focus O_2, the eikonal to the right of the focal point E_2 increases by a further half-wavelength. Since the jump by an integer number of wavelengths does not affect the amplitude, the resulting eikonal is $E_2 + 2\rho_2 n_2$, i.e., when passing through an even number of real foci, the eikonal does not exhibit jumps.

Now, there is an interesting question, as to what is the relationship between the diffraction patterns, or, plainly speaking, whether the rings at O_2 are images of the rings at O_1. On the basis of Table G.4, Fig. G.5 and (G.6.27), we see that for values of $u\sin\alpha'$ that are equal to the roots of $J_1(x) = 0$, w_1 is less than zero., i.e., the corresponding b lies within a dark ring. We notice the same in Table G.2 and Fig. G.4. We therefore can assume that the diameter of the dark rings is given (as we did indeed in Table G.5) by the expression:

$$u\sin\alpha = \text{const} \tag{G.11.26}$$

for an arbitrary α, where α stands for the limiting angle of the system aperture and the constant is equal to one of the roots of the equation $J_1(x) = 0$. Let α_1' and α_2' be the limiting values of α_1 and α_2 in Fig. G.6. Then

$$u_2\sin\alpha_2' = \text{const}. \tag{G.11.27}$$

Imagine that the light were travelling in the reverse direction in Fig. G.7. We would then have rings at O_1, whose diameter D_1' would be given by

$$u_1\sin\alpha_1' = \text{const}. \tag{G.11.28}$$

Equations (G.11.27) and (G.11.28) yield:

$$u_2\sin\alpha_2' = u_1\sin\alpha_1', \tag{G.11.29}$$

or, according to (G.11.9),

$$b_2 = -\beta b_1'. \tag{G.11.30}$$

From (G.11.28) we obtain:

$$D_1' = -2b_1' = \frac{\text{const}\ \lambda}{\pi u_1 \sin \alpha_1'}. \qquad \text{(G.11.31)}$$

Indeed, at O_2 we observe rings of diameter $D_2 = 2b_2$, which are (according to (G.11.30)) the image of the rings of diameter $D_1' = 2b_1'$, virtually positioned at O_1. We say 'virtually' because the maximum angle of the aperture of the lens L_1 can be greater than α_1' and can even be equal to $\pi/2$, i.e., in reality the rings at O_1 would have completely different diameter, $D_1 \neq D_1'$.

In answer to the above question, we should therefore say that the diffraction pattern observed at O_2 depends on the aperture $n_1 \sin \alpha_1'$ of the system L_2, and does not depend on the object (i.e., the rings) that is situated at O_1. At this point we notice a complete similarity with Abbe's theory of the microscope. To demonstrate the numerical similarity as well, let us recall that the theory yields:

$$d = \frac{m\lambda}{n_1 \sin \alpha_1'}, \qquad \text{(G.11.32)}$$

where d is the grating pitch, $n_1 \sin \alpha_1'$ is the aperture and m is the order of the spectrum. To be able to see the correct structure of the grating, one should at least satisfy the condition:

$$n_1 \sin \alpha_1' = \frac{\lambda}{d}. \qquad \text{(G.11.33)}$$

In our case, when the role of d is played by the diameter D_1, in order to have $D_1' = D_1$ it is necessary to have:

$$n_1 \sin \alpha_1' = \frac{\text{const}\ \lambda}{\pi D_1} = \frac{3.83\lambda}{\pi D_1} = \frac{1.22\lambda}{D_1}. \qquad \text{(G.11.34)}$$

References*

[1] E. H. L. Aarts, J. H. M. Korst, and P. J. M. van Laarhoven. *Local Search in Combinatorial Optimisation*, chapter 4, *Simulated Annealing*, pages 91–120. John Wiley, New York, USA, 1997. **409**

[2] E. Abbe. Beiträge zur Theorie des Mikroskops und der mikroskopischen Wahrnehmung. *Arch. mikrosk. Anat.*, **9**(*1*):413–468, 1873. **218**, **660**

[3] E. Abbe. On new methods for improving spherical correction, applied to the construction of wide-angled object-glasses. *J. R. Microsc. Soc.*, **2**:812–824, 1879. **503**

[4] M. Abramowitz and I. A. Stegun. *Handbook of Mathematical Functions*. Dover Publications, New York, USA, 9th edition, 1972. **395**, **600**, **772**, **890**, **891**, **892**, **893**, **895**, **902**

[5] R. Adler. An optical video disc player for NTSC receivers. *IEEE Trans. Broadc. Telev.*, **BTR-20**:230–234, 1974. **711**

[6] I. Agurok. Double expansion of wavefront deformation in Zernike polynomials over the pupil and field-of-view of optical systems: lens design, testing, and alignment. In *Proc. SPIE*, volume 3430, pages 80–87, Bellingham WA, USA, 1998. **360**

[7] Y. Aharonov and D. Bohm. Significance of electromagentic potentials in the quantum theory. *Phys. Rev.*, **115**(*3*):485–491, 1959. **13**

[8] G. B. Airy. On the diffraction of an object-glass with circular aperture. *Trans. Camb. Phil. Soc.*, **5**:283–291, 1835. **582**

[9] T. B. Andersen. Automatic computation of optical aberration coefficients. *Appl. Opt.*, **19**:3800–3816, 1980. **339**, **360**

[10] G. Arfken. *Mathematical Methods for Physicists*. Academic Press, Boston, third edition, 1985. **19**, **169**, **692**, **858**

[11] M. R. Arnison and C. J. R. Sheppard. A 3D vectorial optical transfer function suitable for arbitrary pupil functions. *Opt. Commun.*, **211**:53–63, 2002. **767**

[12] G. Artzner. Aspherical wavefront measurements: Shack-Hartmann numerical and practical experiments. *Pure Appl. Opt.*, **7**:435–48, 1998. **496**

[13] A. Ashkin. Acceleration and trapping of particles by radiation pressure. *Phys. Rev. Lett.*, **24**(*4*):156–159, 1970. **640**

[14] A. Ashkin, J. M. Dziedzic, J. E. Bjorkholm, and S. Chu. Observation of a single-beam gradient force optical trap for dielectric particles. *Opt. Lett.*, **11**(*5*):288–290, 1986. **640**

[15] J. S. Asvestas. Diffraction by a black screen. *J. Opt. Soc. Am.*, **65**(2):111–118, 1975. **561**

[16] D. Axelrod. Carbocyanine dye orientation in red cell membrane studied by microscopic fluorescence polarization. *Biophys. J.*, **26**:557–574, 1979. **833**

[17] R. M. A. Azzam and N. M. Bashara. *Ellipsometry and Polarized Light*. North-Holland, Amsterdam, third edition, 1992. **787**

[18] B. B. Baker and E. T. Copson. *The Mathematical Theory of Huygens' Principle*. Oxford University Press, second edition, 1953. **550**

[19] J. G. Baker. On improving the effectiveness of large telescopes. *IEEE Trans. Aerosp. Elect. Syst.*, **5**(2):261–272, 1969. **488**

[20] M. F. Bal, F. Bociort, and J. J. M. Braat. Analysis, search, and classification for reflective ring-field projection systems. *Appl. Opt.*, **42**:2301–2311, 2003. **537**

* The number(s) in bold type following the year of publication of a bibliographic reference indicate the book pages where the particular reference has been cited.

[21] S. M. Barnett and L. Allen. Orbital angular momentum and nonparaxial light beams. *Opt. Commun.*, **110**(*5*):670–678, 1994. **591**

[22] A. Bayliss and E. Turkel. Radiation boundary conditions for wave-like equations. *Commun. Pure Appl. Math.*, **33**(*6*):707–725, 1980. **858**

[23] C. Benoît. Note sur une méthode de résolution des équations normales provenant de l'application de la méthode des moindres carrés à un système d'équations linéaires en nombre inférieur à celui des inconnues (Procédé du Commandant Cholesky). *B. Geod.*, **2**:67–77, 1924. **402**

[24] M. Berek. *Grundlagen der praktischen Optik*. W. de Gruyter, Berlin, 1930. **309**, **446**, **447**

[25] J.-P. Bérenger. A perfectly matched layer for the absorption of electromagnetic waves. *J. Comput. Phys.*, **114**(*2*):185–200, 1994. **103**, **858**

[26] F. W. Bessel. Untersuchung des Teils der planetarischen Störungen, welcher aus der Bewegung der Sonne entsteht. *Abh. d. K. Akad. Wiss. Berl.*, pages 1–52, 1824. **582**

[27] K. P. Birch and M. J. Downs. An updated Edlén equation for the refractive index of air. *Metrologia*, **30**:155–162, 1993. **325**

[28] L. I. Bluestein. A linear filtering approach to the computation of the discrete Fourier transform. *IEEE Trans. Acoust. Speech*, **18**(*4*):451–455, 1970. **867**

[29] F. Bociort, T. B. Andersen, and L. H. J. F. Beckmann. High-order optical aberration coefficients: extension to finite objects and to telecentricity in object space. *Appl. Opt.*, **47**:5691–5700, 2008. **339**, **360**, **460**

[30] F. Bociort, E. van Driel, and A. Serebriakov. Network structure of the set of local minima in optical system optimization. In *Proc. SPIE*, volume 5174, pages 26–34, Bellingham WA, USA, 2003. **412**

[31] F. Bociort and P. van Grol. Systematics of the design shapes in the optical merit function landscape. In *Proc. SPIE*, volume 7717, pages 77170F–1 –77170F–14, Bellingham WA, USA, 2010. **412**

[32] F. Bociort and M. van Turnhout. Looking for order in the optical design landscape. In P. Z. Mouroulis, W. J. Smith, and R. B. Johnson, editors, *Current Developments in Lens Design and Optical Engineering VII*, volume 6288 of *Proc. SPIE*, page 628806, Bellingham WA, USA, 2006. **412**, **414**

[33] G. C. Boer. LCD convertor. http://www.arcoptix.com/radial_polarization_converter.htm, 2015. **791**

[34] M. Boerner, G. Glasmachers, O. Bernecker, V. Riemann, and S. Maslowski. Signal playback system transducer with optical resonance cavity, 1976. US patent 3,941,945, German application DE 19,722,244,119, September 1973, TELDEC company. **715**

[35] C. F. Bohren and D. R. Huffmann. *Absorption and Scattering of Light by Small Particles*. John Wiley, New York, USA, 1983. **106**

[36] M. Born. *Optik*. Springer, Berlin, 1933. **3**, **108**

[37] M. Born and E. Wolf. *Principles of Optics. Electromagnetic Theory of Propagation, Interference and Diffraction of Light*. Cambridge University Press, Cambridge (UK), seventh edition, 1999. **3**, **46**, **164**, **189**, **191**, **197**, **224**, **231**, **324**, **339**, **384**, **545**, **550**, **553**, **611**, **612**, **689**, **888**, **889**, **900**

[38] P.-P. Borsboom and H. J. Frankena. Field analysis of two-dimensional focusing grating couplers. *J. Opt. Soc. Am. A*, **12**(*5*):1142–1146, 1995. **515**

[39] P.-P. Borsboom and H. J. Frankena. Field analysis of two-dimensional integrated optical gratings. *J. Opt. Soc. Am. A*, **12**(*5*):1134–1141, 1995. **515**

[40] A. van den Bos. *Parameter Estimation for Scientists and Engineers*. John Wiley, New York, USA, 2007. **403**

[41] G. Bouwhuis and P. Burgstede. The optical scanning system of the Philips VLP record player. *Philips Tech. Rev.*, **33**:186–189, 1973. **711**

[42] C. J. Bouwkamp. Diffraction theory. *Rep. Prog. Phys.*, **17**:35–100, 1954. **88**, **545**, **561**

[43] G. D. Boyd and J. P. Gordon. Confocal multimode resonator for millimeter through optical wavelength masers. *Bell Sys. Tech. J.*, **40**:489–508, 1961. **35**

[44] G. D. Boyd and H. Kogelnik. Generalized confocal resonator theory. *Bell Sys. Tech. J.*, **41**:1347–1369, 1962. **35**

[45] S. Boyd and L. Vandenberghe. *Convex Optimization*. Cambridge University Press, Cambridge, UK, 2004. **411**

[46] J. J. M. Braat. Polynomial expansion of severely aberrated wave fronts. *J. Opt. Soc. Am. A*, **4**:643–650, 1987. **366**, **368**

[47] J. J. M. Braat. Quality of microlithographic projection lenses. In *Proc. SPIE*, volume 811, pages 22–30, Bellingham WA, USA, 1987. **531**

[48] J. J. M. Braat. Optics of recording and read-out in optical disc systems. *Jpn. J. Appl. Phys.*, **28**(*3*):103–108, 1989. **724**

[49] J. J. M. Braat. Analytic expressions for the wave-front aberration coefficients of a tilted plane-parallel plate. *Appl. Opt.*, **36**:8459–8467, 1997. **322**

[50] J. J. M. Braat. Mirror projection system for a scanning lithographic projection apparatus, and lithographic apparatus comprising such a system. *US Patent 6396067*, May 28, 2002. **538**

[51] J. J. M. Braat, P. Dirksen, and A. J. E. M. Janssen. Assessment of an extended Nijboer-Zernike approach for the computation of optical point-spread functions. *J. Opt. Soc. Am. A*, **19**:858–870, 2002. **584**

[52] J. J. M. Braat, P. Dirksen, A. J. E. M. Janssen, and A. S. van de Nes. Extended Nijboer-Zernike representation of the vector field in the focal region of an aberrated high-aperture optical system. *J. Opt. Soc. Am. A*, **20**:2281–2292, 2003. **363**, **584**, **627**, **630**, **644**, **901**

[53] J. J. M. Braat and P. F. Greve. Aplanatic optical system containing two aspheric surfaces. *Appl. Opt.*, **18**:2187–2191, 1979. **391**

[54] J. J. M. Braat, S. van Haver, A. J. E. M. Janssen, and P. Dirksen. Energy and momentum flux in a high-numerical-aperture beam using the extended Nijboer-Zernike diffraction formalism. *J. Eur. Opt. Soc.-Rapid*, **2**(*07032*), 2007. **642**, **651**, **653**, **654**, **655**

[55] J. J. M. Braat, S. van Haver, A. J. E. M. Janssen, and P. Dirksen. Assessment of optical systems by means of point-spread functions. In E. Wolf, editor, *Prog. Optics*, volume 51, chapter 6, pages 349–468. Elsevier B.V., 2008. **611**

[56] J. J. M. Braat, S. van Haver, A. J. E. M. Janssen, and S. F. Pereira. Image formation in a multilayer using the Extended Nijboer-Zernike theory. *J. Eur. Opt. Soc.-Rapid*, **4**(*09048*), 2009. **636**, **637**, **638**

[57] J. J. M. Braat and A. J. E. M. Janssen. Double Zernike expansion of the optical aberration function from its power series expansion. *J. Opt. Soc. Am. A*, **30**(*6*):1213–1222, 2013. **361**, **363**, **364**, **900**

[58] J. J. M. Braat and A. J. E. M. Janssen. Derivation of various transfer functions of ideal or aberrated imaging systems from the three-dimensional transfer function. *J. Opt. Soc. Am. A*, **32**:1146–1159, 2015. **736**, **767**

[59] J. J. M. Braat and M. O. E. Laurijs. Geometrical optics design and aberration analysis of a focusing grating coupler. *Opt. Eng.*, **33**(*4*):1037–1043, 1994. **513**

[60] L. Brillouin. *Wave Propagation and Group Velocity*. Academic Press, New York, 1960. **169**

[61] W. Brouwer. *Matrix Methods in Optical Instrument Design*. W. A. Benjamin, New York, USA, 1964. **234**

[62] H. A. Buchdahl. *Aberrations of Optical Systems*. Dover Publications, New York, USA, 1968. **339**, **360**, **395**, **460**

[63] M. Cagnet, M. Françon, and J.-C. Thrierr. *Atlas of Optical Phenomena*. Springer-Verlag, Berlin, 1962. **582**, **583**

[64] P. R. Callis. On the theory of two-photon induced fluorescence anisotropy with application to indoles. *J. Chem. Phys.*, **99**:27–37, 1993. **833**

[65] E. Charney. *The Molecular Basis of Optical Activity*. John Wiley, New York, USA, 1979. **143**

[66] N. P. Chavannes. *Local Mesh Refinement Algorithms for Enhanced Modeling Capabilities in the FDTD Method*. PhD thesis, ETH, Zürich, 2002. **104**

[67] L.-C. Cheng, N. G. Horton, K. Wang, S.-J. Chen, and C. Xu. Measurements of multiphoton action cross-sections for multiphoton microscopy. *Biomed. Opt. Express*, **5**(*10*):3427–3433, 2014. **839**

[68] C.-W. Chong, P. Raveendran, and R. Mukundan. A comparative analysis of algorithms for fast computation of Zernike moments. *Pattern Recogn.*, **36**:731–742, 2003. **610**

[69] P. E. Ciddor. Refractivity index of air: new equations for the visible and near infrared. *Appl. Opt.*, **35**:1566–1573, 1996. **324**, **325**

[70] P. H. van Cittert. Die wahrscheinliche Schwingungsverteilung in einer von einer Lichtquelle direkt oder mittels einer Linse beleuchteten Ebene. *Physica*, **1**:201–210, 1934. **498**, **673**, **689**

[71] R. Clausius. Über die Concentration von Wärme- und Lichtstrahlen und die Grenzen ihrer Wirkung. *Ann. Phys. (Pogg. Ann.)*, **121**(*1*):1–44, 1864. **218**

[72] H. Coddington. *A Treatise on the Reflexion and Refraction of Light, being Part I. of a System of Optics*. Cambridge University, 1829. **350**

[73] S. A. Collins. Lens-system diffraction integral written in terms of matrix optics. *J. Opt. Soc. Am.*, **60**(*9*):1168–1177, 1970. **571**

[74] K. Compaan and P. Kramer. The Philips 'VLP' system. *Philips Tech. Rev.*, **33**:178–180, 1973. **711**

[75] E. U. Condon. Theories of optical rotatory power. *Rev. Mod. Phys.*, **9**:432–457, 1937. **143**

[76] A. E. Conrady. Star discs. *Mon. Not. R. Astr. Soc.*, **79**:575, 1919. **583**

[77] A. E. Conrady. *Applied Optics and Optical Design*. Oxford University Press, London, 1929. **329**, **330**, **454**

[78] A. W. Conway and J. L. Synge. *The Mathematical Papers of Sir W. R. Hamilton*, volume I (Geometrical Optics). Cambridge University Press, Cambridge, 1931. **203**

[79] A. Couder. Sur un type nouveau de télescope photographique. *C. R. Acad. Sci.*, **183**:1276–1279, 1926. **481**

[80] I. J. Cox and C. J. R. Sheppard. Information capacity and resolution in an optical system. *J. Opt. Soc. Am. A*, **3**(*8*):1152–1158, 1986. **812**

[81] A. J. den Dekker and A. van den Bos. Resolution: a survey. *J. Opt. Soc. Am.*, **14**:547–557, 1997. **660**

[82] W. Denk, J. H. Strickler, and W. W. Webb. Two-photon fluorescence scanning microscopy. *Science*, **248**:73–76, 1990. **840**

[83] T. van Dijk, G. Gbur, and T. D. Visser. Shaping the focal intensity distribution using spatial coherence. *J. Opt. Soc. Am. A*, **25**:575–581, 2008. **796**

[84] D. C. Dilworth. Pseudo-second-derivative matrix and its application to automatic lens design. *Appl. Opt.*, **17**:3372–3375, 1978. **406**, **407**

[85] P. A. M. Dirac. Quantised singularities in the electromagnetic field. *Proc. Roy. Soc. Lond. A*, **133**(*821*):60–72, 1931. **5**

[86] A. Dodoc, W. Ulrich, and H.-J. Rostalski. Projection objective for immersion lithography. *US Patent 7969663*, June 28 2011. **533**

[87] J. Dollond. XCVIII. An account of some experiments concerning the different refrangibility of light. By Mr. John Dollond. With a letter from James Short. *Phil. Trans.*, **50**:733–743, 1758. **2**

[88] P. Drude. *Optics*. Longmans, Green and Co., New York, USA, 1901. **143**

[89] P. M. Duffieux. *L'intégrale de Fourier et ses applications à l'optique*. Privately published, Printing house Oberthur, Rennes, 1946. **660**

[90] K. Edlén. The refractive index of air. *Metrologia*, **2**:71–80, 1966. **325**

[91] J. van den Eerenbeemd and S. Stallinga. Compact system description for systems comprising a tilted plane parallel plate. *Appl. Opt.*, **46**:319–326, 2007. **322**

[92] B. Engquist and A. Majda. Absorbing boundary conditions for the numerical simulation of waves. *Math. Comput.*, **31**:629–651, 1977. **858**

[93] J. H. Erkkila and M. E. Rogers. Diffracted fields in the focal volume of a converging wave. *J. Opt. Soc. Am.*, **71**:904–905, 1981. **575**

[94] L. Euler. Recherches sur la courbure des surfaces. *Mém. Acad. Sc. Berlin*, **16**:119–143, 1767. **288**

[95] M. Faraday. On the magnetic affection of light, and on the distinction between the ferromagnetic and diamagnetic conditions of matter. *Philos. Mag.*, **29**(*193*):153–156, 1846. **143**

[96] M. Faraday. On the magnetization of light and the illumination of magnetic lines of force. *Phil. Trans. R. Soc. Lond.*, **136**:1–20, 1846. **143**

[97] D. P. Feder. Automatic optical design. *Appl. Opt.*, **2**:1209–1226, 1963. **400**, **407**

[98] S. Flewett, H. Quiney, C. Tran, and K. Nugent. Extracting coherent modes from partially coherent wavefields. *Opt. Lett.*, **33**:2198–2200, 2009. **796**

[99] A. D. Fokker. Over het anomale phaseverloop bij een brandpunt. *Physica*, **3**:334–337, 1923 (in Dutch). **603**

[100] A. D. Fokker. Hyperbolische zones van Fresnel. *Physica*, **4**:166–172, 1924 (in Dutch). **603**, **604**

[101] G. W. Forbes. Shape specification for axially symmetric optical surfaces. *Opt. Expr.*, **15**:5218–5226, 2007. **343**

[102] A. G. Fox and T. Li. Resonant modes in a maser interferometer. *Bell Sys. Tech. J.*, **40**:453–488, 1961. **35**

[103] A.-J. Fresnel. Mémoire sur la diffraction de la lumiére. *Mem. Acad. Sci. Inst. France*, 1818. **603**

[104] A.-J. Fresnel. Mémoire sur la double réfraction. *Rec. Acad. Sci. Inst. France*, **VII**:45–176, 1824. **2**

[105] A.-J. Fresnel. Mémoire sur les modifications que la réflexion imprime à la lumière polarisée. *Mem. Acad. Sci. Inst. France*, **11**:373–434, 1832. **62**

[106] D. M. Friedrich. Tensor patterns and polarization ratios for three-photon transition in fluid media. *J. Chem. Phys.*, **75**:3258–3268, 1981. **833**

[107] X. S. Gan and C. J. R. Sheppard. Detectability: A new criterion for evaluation of the confocal microscope. *Scanning*, **15**:187–192, 1993. **847**

[108] R. Gauderon and C. J. R. Sheppard. Effect of a finite-size pinhole on noise performance in single-, two-, and three-photon confocal fluorescence microscopy. *Appl. Opt.*, **38**(*16*):3562–3565, 1999. **847**

[109] A. Gerrard and J. M. Burch. *Introduction to Matrix Methods in Optics*. John Wiley, New York, USA, 1974. **235**

[110] E. Glatzel. New lenses for microlithography. In R. E. Fisher, editor, *Proceedings 1980 International Lens Design Conference*, volume 237 of *Proc. SPIE*, pages 310–320, Bellingham WA, USA, 1980. **531**

[111] M. A. Golub and A. A. Friesem. Effective grating theory for resonance domain surface-relief diffraction gratings. *J. Opt. Soc. Am. A*, **22**:1115–1126, 2005. **388**

[112] J. W. Goodman. *Statistical Optics*. John Wiley, New York, USA, 1985. **768**

[113] J. W. Goodman. *Introduction to Fourier Optics*. Roberts & Co. Publishers, Greenwood Village, USA, 2004. **660**, **696**, **823**

[114] F. Goos and H. Haenchen. Ein neuer und fundamentaler Versuch zur Totalreflexion. *Ann. Physik*, **436**:333–346, 1947. **62**

[115] F. Gori, M. Santarsiero, R. Simon, G. Piquero, R. Borghi, and G. Guattari. Coherent-mode decomposition of partially polarized, partially coherent sources. *J. Opt. Soc. Am. A*, **20**(*1*):78–84, 2003. **798**

[116] L. G. Gouy. Sur une propriété nouvelle des ondes lumineuses. *C. R. Hebd. Acad. Sci. Paris*, **110**:1251–1253, 1890. **600**

[117] F. M. Grimaldi. *Physico-mathesis de lumine, coloribus, et iride aliisque adnexis libri duo*. V. Bonati, Bologna, 1665. **1**, **543**

[118] I. Gryczynski, H. Malak, and J. R. Lakowicz. Multiphoton excitation of the DNA stains DAPI and Hoechst. *Bioimaging*, **4**:138–148, 1996. **833**

[119] I. Gryczynski, H. Malak, and J. R. Lakowicz. Three-photon excitation of P-Quaterphenyl with a mode-locked femtosecond Ti:Sapphire laser. *J. Fluoresc.*, **6**:139–145, 1996. **839**

[120] M. Gu and C. J. R. Sheppard. Three-dimensional imaging in confocal fluorescent microscopy with annular lenses. *J. Mod. Opt.*, **38**:2247–2263, 1991. **767**

[121] G. Guattari, C. Palma, and C. Padovani. Cross-spectral densities with axial symmetry. *Opt. Commun.*, **73**:173–178, 1989. **802**

[122] M. G. Gustafsson. Surpassing the lateral resolution limit by a factor of two using structured illumination microscopy. *J. Microsc.*, **198**(*2*):82–87, 2000. **725**

[123] J. Haisma, E. Hugues, and C. Babolat. Realization of a bi-aspherical objective lens for the Philips Video Long Play system. *Opt. Lett.*, **4**(*2*):70–72, 1979. **509**

[124] W. R. Hamilton. On some results of the view of a characteristic function in optics. *Rep. Br. Ass. (Cambridge)*, pages 360–370, 1833. **138**

[125] A. Hardy and D. Treves. Structure of the electromagnetic field near the focus of a stigmatic lens. *J. Opt. Soc. Am.*, **63**(*1*):85–90, 1973. **804**

[126] P. Hariharan. *Optical Interferometry*. Elsevier Science, London, 2nd edition, 2003. **591**

[127] S. van Haver and A. J. E. M. Janssen. Advanced analytic treatment and efficient computation of the diffraction integrals in the Extended Nijboer-Zernike theory. *J. Eur. Opt. Soc.-Rapid*, **8**(*13044*), 2013. **611**, **628**, **902**, **918**

[128] S. van Haver and A. J. E. M. Janssen. Truncation of the series expressions in the advanced ENZ-theory of diffraction integrals. *J. Eur. Opt. Soc.-Rapid*, **9**(*14042*), 2014. **628**, **918**

[129] O. Heaviside. *Electrical Papers, Vol. I and II*. MacMillan & Co., London and New York, 1892. **13**

[130] B. Hecht and L. Novotny. *Principles of Nano-Optics*. Cambridge University Press, Cambridge, UK, 2006. **180**

[131] A. C. S. van Heel. *Inleiding in de optica*. Martinus Nijhoff, The Hague, Netherlands, 1958. **234**

[132] J. P. J. Heemskerk. Noise in a video disc system: experiments with an (AlGa)As laser. *Appl. Opt.*, **17**:2007–2012, 1978. **715**

[133] H. von Helmholtz. Die theoretische Grenze für die Leistungsfähigkeit der Mikroskope. *Ann. Phys. (Pogg. Ann., Jubelband J. C. Poggendorf gewidmet)*, pages 557–584, 1874. **218**

[134] J. F. W. Herschel. On the aberrations of compound lenses and object-glasses. *Phil. Trans. R. Soc. Lond.*, **111**:222–267, 1821. **222**

[135] M. Herzberger and N. R. McClure. The design of superachromatic lenses. *Appl. Opt.*, **2**:553–560, 1963. **334**

[136] P. D. Higdon, P. Török, and T. Wilson. Imaging properties of high aperture multiphoton fluorescence scanning optical microscopes. *J. Microsc.-Oxford*, **193**(2):127–141, 1999. **840**, **841**, **842**, **843**, **844**, **845**, **846**, **847**

[137] G. W. 't Hooft. Comment on 'Negative refraction makes a perfect lens'. *Phys. Rev. Lett.*, **87**:249701, 2001. **178**, **186**

[138] A. T. de Hoop. Reciprocity of the electromagnetic field. *Appl. Sci. Res. B*, **8**:135–139, 1959. **911**

[139] A. T. de Hoop. Time-domain reciprocity theorems for electromagnetic fields in dispersive media. *Radio Sci.*, **22**:1171–1178, 1987. **911**

[140] H. H. Hopkins. *Wave Theory of Aberrations*. Clarendon Press, Oxford, UK, 1950. **361**, **389**, **398**, **425**

[141] H. H. Hopkins. The wave aberration associated with skew rays. *Proc. Phys. Soc. B*, **65**:934–942, 1952. **226**, **339**, **357**

[142] H. H. Hopkins. On the diffraction theory of optical images. *Proc. R. Soc. Lond. Ser. A*, **217**:408–432, 1953. **660**, **680**

[143] H. H. Hopkins. The frequency response of a defocused optical system. *Proc. R. Soc. Lond. Ser. A*, **231**:91–103, 1955. **753**

[144] H. H. Hopkins. Diffraction theory of laser read-out systems for optical video discs. *J. Opt. Soc. Am.*, **69**(*1*):4–24, 1979. **723**

[145] H. H. Hopkins. Canonical and real-space coordinates used in the theory of image formation. In *Applied Optics and Optical Engineering*, volume IX, chapter 8, pages 307–369. Academic Press, New York, USA, 1983. **229**, **230**

[146] H. H. Hopkins and P. M. Barham. The influence of the condenser on microscopic resolution. *Proc. Phys. Soc. B*, **63**:737–744, 1950. **660**

[147] Z. Hou, I. Livshits, and F. Bociort. One-dimensional searches for finding new lens design solutions efficiently. *Appl. Opt.*, **55**:10449–10456, 2016. **412**

[148] W. Hsu and R. Barakat. Stratton-Chu vectorial diffraction of electromagnetic fields by apertures with application to small-Fresnel-number systems. *J. Opt. Soc. Am. A*, **11**(*2*):623–629, 1994. **562**

[149] S. Huang, F. Xi, C. Liu, and Z. Jiang. Eigenfunctions of Laplacian for phase estimation from wavefront gradient or curvature sensing. *Opt. Commun.*, **284**:2781–2783, 2011. **905**

[150] H. C. van de Hulst. *Light Scattering by Small Particles*. Dover Publications, New York, USA, 1981. **817**, **818**

[151] R. G. Hunsperger. *Integrated Optics, Theory and Technology*. Springer, New York, USA, sixth edition, 2009. **168**

[152] C. Huygens. *Treatise on Light*. Dover Publications, New York, USA, 1962. **107**, **267**, **268**, **274**

[153] V. S. Ignatowsky. Diffraction by a lens of arbitrary aperture. *Trans. Opt. Inst. Petrograd*, **I**(*4*):1–36, 1919. **565**, **584**, **622**, **627**, **628**, **782**

[154] V. S. Ignatowsky. Diffraction by a parabolic mirror of arbitrary aperture. *Trans. Opt. Inst. Petrograd*, **1**(*5*):1–30, 1920. **565**, **622**

[155] V. S. Ignatowsky. The relationship between geometrical and wave optics and the diffraction of a homocentric beam. *Trans. Opt. Inst. Petrograd*, **I**(*3*):1–30, 1920. **565**, **622**

[156] S. Inoué. Studies of depolarization of light at microscope lens surfaces. I. The origin of stray light by rotation at the lens surfaces. *Exp. Cell Biol.*, **3**:199–208, 1952. **783**, **804**

[157] S. Inoué and W. L. Hyde. Studies of depolarization of light at microscope lens surfaces. II. The simultaneous realization of high resolution and high sensitivity with the polarizing microscope. *J Biophys. Biochem. Cy.*, **3**:831–838, 1957. **804**

[158] M. Isshiki, H. Ono, K. Hiraga, J. Ishikawai, and S. Nakadate. Lens design: Global optimization with escape function. *Opt. Rev.*, **2**:463–470, 1995. **411**, **412**

[159] D. G. A. Jackson, M. Gu, and C. J. R. Sheppard. Three-dimensional optical transfer function for circular and annular lenses with spherical aberration and defocus. *J. Opt. Soc. Am. A*, **11**:1758–1767, 1994. **767**

[160] J. D. Jackson. *Classical Electrodynamics*. John Wiley, New York, USA, third edition, 1998. **3**, **12**, **180**, **552**, **648**, **817**

[161] K. Jahn and N. Bokor. Vector Slepian basis functions with optimal energy concentration in high numerical aperture focusing. *Opt. Commun.*, **285**:2028–2038, 2012. **614**

[162] A. J. E. M. Janssen. Extended Nijboer-Zernike approach for the computation of optical point-spread functions. *J. Opt. Soc. Am. A*, **19**:849–857, 2002. **584**, **609**, **610**, **627**

[163] A. J. E. M. Janssen. Computation of Hopkins' 3-circle integrals using Zernike expansions. *J. Eur. Opt. Soc.-Rapid*, **6**(*11059*), 2011. **680**, **705**

[164] A. J. E. M. Janssen. New analytic results for the Zernike circle polynomials from a basic result in the Nijboer-Zernike diffraction theory. *J. Eur. Opt. Soc.-Rapid*, **6**(*11028*), 2011. **896**, **897**

[165] A. J. E. M. Janssen. Zernike expansion of derivatives and Laplacians of the Zernike circle polynomials. *arXiv*, math-ph (1404.1766v1), 2014. **902**

[166] A. J. E. M. Janssen, J. J. M. Braat, and P. Dirksen. On the computation of the Nijboer-Zernike aberration integrals at arbitrary defocus. *J. Mod. Opt.*, **51**:687–703, 2004. **627**

[167] A. J. E. M. Janssen and P. Dirksen. Computing Zernike polynomials of arbitrary degree using the discrete Fourier transform. *J. Eur. Opt. Soc.-Rapid*, **2**(*07012*), 2007. **863**, **895**

[168] O. T. A. Janssen, S. van Haver, A. J. E. M. Janssen, J. J. M. Braat, H. P. Urbach, and S. F. Pereira. Extended Nijboer-Zernike (ENZ) based mask imaging: efficient coupling of electromagnetic field solvers and the ENZ imaging algorithm. In *Proc. SPIE, Optical Microlithography XXI*, volume 6924, Bellingham WA, USA, 2008. **418**

[169] T. Jewell, J. M. Rodgers, and K. P. Thompson. Reflective systems design study for soft-X-ray projection lithography. *J. Vac. Technol. B*, **8**:1519–1523, 1990. **536**

[170] J. Jin. *The Finite Element Method of Electromagnetics*. Wiley Interscience, New York, USA, 2002. **857**

[171] R. C. Juškaitis and T. Wilson. The measurement of the amplitude point-spread function of microscope objective lenses. *J. Microsc.*, **189**(*1*):8–11, 1998. **581**

[172] F. M. Kahnert. Numerical methods in electromagnetic scattering theory. *J. Quant. Spectr. Rad. Transf.*, **79–80**:775–824, 2003. **100**

[173] N. G. van Kampen. The method of stationary phase and the method of Fresnel zones. *Physica*, **24**:437–444, 1958. **603**

[174] R. Kant. An analytical solution of vector diffraction for focusing optical systems with Seidel aberrations. I. Spherical aberration, curvature of field and distortion. *J. Mod. Opt.*, **40**(*11*):2293–2310, 1993. **614**

[175] R. Kant. An analytical solution of vector diffraction for focusing optical systems. *J. Mod. Opt.*, **40**(*2*):337–347, 1993. **614**

[176] B. Karczewski and E. Wolf. Comparison of three theories of electromagnetic diffraction at an aperture. Part I: Coherence matrices, Part II: The far field. *J. Opt. Soc. Am.*, **56**:1207–1219, 1966. **557**

[177] K. Karhunen. Über lineare Methoden in der Wahrscheinlichkeitsrechnung. *Ann. Acad. Sci. Fenn. A1*, **A137**:1–79, 1947. **797**

[178] J. B. Keller. Geometrical theory of diffraction. *J. Opt. Soc. Am.*, **52**:116–130, 1962. **197**

[179] M. M. Kelly, J. B. West, and D. E. Lloyd. Reflectance of silicon carbide in the vacuum ultraviolet. *J. Phys. D: Appl. Phys.*, **14**:401–404, 1981. **536**

[180] D. Kermisch. Principle of equivalence between scanning and conventional optical imaging systems. *J. Opt. Soc. Am.*, **67**:1357–1360, 1977. **714**

[181] R. Kingslake. *Lens Design Fundamentals*. Academic Press, New York, USA, 1978. **454**, **461**, **784**

[182] R. Kingslake. Who discovered Coddington's equations? *Opt. Phot. News*, **5**:20–23, 1994. **350**

[183] E. C. Kintner and R. M. Sillitto. A new 'analytic' method for computing the optical transfer function. *Opt. Acta*, **23**:607–619, 1976. **590**, **609**

[184] S. Kirkpatrick, C. D. Gelatt, and M. P. Vecchi. Optimization by simulated annealing. *Science*, **220**(*4598*):671–680, 1983. **409**

[185] B. H. Kleemann, M. Seesselberg, and J. Ruoff. Design concepts for broadband high-efficiency DOEs . *J. Eur. Opt. Soc.-Rapid*, **3**(*08015*), 2008. **391**

[186] M. Kline and I. W. Kay. *Electromagnetic Theory and Geometrical Optics*. Interscience Publishers, New York, USA, 1965. **191**, **197**

[187] K. Knop. Rigorous diffraction theory for transmission phase gratings with deep rectangular grooves. *J. Opt. Soc. Am.*, **68**:1206–1210, 1978. **100**

[188] H. Kogelnik. Coupled wave theory for thick hologram gratings. *Bell Syst. Tech. J.*, **48**:2909–2947, 1969. **100**

[189] H. Kogelnik and T. Li. Laser beams and resonators. *Appl. Opt.*, **5**(*10*):1550–1567, 1966. **35**

[190] A. Köhler. Ein neues Beleuchtungsverfahren für mikrophotographische Zwecke. *Z. Wiss. Mikrosk.*, **10**(*4*):433–440, 1893. **497**

[191] G. A. Korn and T. M. Korn. *Mathematical Handbook for Scientists and Engineers: Definitions, Theorems, and Formulas for Reference and Review*. McGraw-Hill, New York, USA, 1968. **580**

[192] F. Kottler. Diffraction at a black screen I. *Prog. Optics*, **4**:283–314, 1965. **88**, **545**

[193] F. Kottler. Diffraction at a black screen II. *Prog. Optics*, **6**:336–377, 1967. **88**, **545**, **561**

[194] H. A. Kramers. La diffusion de la lumière par les atomes. In *Atti del Congresso Internationale dei Fisici (Transactions of Volta Centenary Congress, Como)*, volume 2, pages 1–13 (545–557), 1927. **882**

[195] E. E. Kriezis, S. Filippov, and S. J. Elston. Light propagation in domain walls in ferroelectric liquid crystal devices by the Finite-Difference Time-Domain method. *J. Opt. A-Pure Appl. Opt.*, **2**:27–33, 2000. **818**

[196] R. de Laer Kronig. On the theory of dispersion of X-rays. *J. Opt. Soc. Am.*, **12**(*6*):547–557, 1926. **882**

[197] J. Kross. Beschreibung, Analyse und Bewertung der Bildfehler optischer Systeme durch interpolierende Darstellungen mit Hilfe von Zernike-Kreispolynomen. *Optik*, **29**:65–80, 1969. **366**

[198] H. Kubota and S. Inoué. Diffraction images in the polarizing microscope. *J. Opt. Soc. Am.*, **49**:191–198, 1959. **783**, **804**

[199] I. W. Kwee and J. J. M. Braat. Double Zernike expansion of the optical aberration function. *Pure Appl. Opt.*, **2**:21–32, 1993. **360**

[200] P. Lalanne, M. Besbes, J. P. Hugonin, S. van Haver, O. T. A. Janssen, A. M. Nugrowati, M. Xu, S. F. Pereira, H. P. Urbach, A. S. van de Nes, P. Bienstman, G. Granet, A. Moreau, S. Helfert, M. Sukharev, T. Seideman, F. Baida, B. Guizal, and D. van Labeke. Numerical analysis of a slit-groove diffraction problem. *J. Eur. Opt. Soc.-Rapid*, **2**(*07022*), 2007. **100, 857**

[201] P. Lalanne and G. Morris. Highly improved convergence of the coupled-wave method for TM-polarization. *J. Opt. Soc. Am. A*, **13**:779–784, 1996. **100**

[202] H. Lamb. On group-velocity. *Proc. London Math. Soc.*, **1**:473–479, 1904. **169**

[203] C. L. Lawson and R. J. Hanson. *Solving Least Squares Problems*. Prentice-Hall, Englewood Cliffs, New Jersey, USA, 1974. **403, 406**

[204] J. Lee, M. van der Aa, C. Verschuren, F. Zijp, and M. van der Mark. Development of an air gap servo system for high data transfer rate near-field optical recording. *Jpn. J. Appl. Phys.*, **44**:3423–3426, 2005. **639**

[205] D. C. van Leijenhorst, C. B. Lucasius, and J. M. Thijssen. Optical design with the aid of a genetic algorithm. *Biosystems*, **37**:177–187, 1996. **409**

[206] N. v. d. W. Lessing. Selection of optical glasses in superachromats. *Appl. Opt.*, **9**:1665–1668, 1970. **334, 335**

[207] K. Levenberg. A method for the solution of certain non-linear problems in least squares. *Quart. Appl. Math.*, **2**:164–168, 1944. **403**

[208] L. Li. Formulation and comparison of two recursive matrix algorithms for modeling layered diffraction gratings. *J. Opt. Soc. Am. A*, **13**:1024–1035, 1996. **86, 98**

[209] L. Li. Use of Fourier series in the analysis of discontinuous periodic structures. *J. Opt. Soc. Am. A*, **13**:1870–1876, 1996. **101**

[210] Y. Li and E. Wolf. Three-dimensional intensity distribution near the focus in systems of different Fresnel numbers. *J. Opt. Soc. Am. A*, **1**:801–808, 1984. **575**

[211] S. Lindek, N. Salmon, C. Cremer, and E. H. K. Stelzer. Theta microscopy allows phase regulation in 4Pi (A)-confocal two-photon fluorescence microscopy. *Optik*, **98**(*1*):15–20, 1994. **839**

[212] M. Linnemann. *Über nicht-sphärische Objektive*. PhD thesis, Universität Göttingen, 1905. **391**

[213] W. P. Linnik. A simple interferometer for the investigation of optical systems. *C. R. Acad. Sci. URSS*, **5**:210, 1933; English translation in Appl. Opt. **18**, pp. 2010-2011 (1979). **600**

[214] J. J. Lister. On the improvement of achromatic compound microscopes. *Phil. Trans. Roy. Soc.*, **120**:187–200, 1830. **500**

[215] H. Lloyd. Further experiments on the phaenomena presented by light in its passage along the axes of biaxal crystals. *Philos. Mag.*, **2**:207–210, 1833. **138**

[216] H. Lloyd. On the phaenomena presented by light in its passage along the axes of biaxal crystals. *Philos. Mag.*, **2**:112–120, 1833. **138**

[217] M. Loève. *Probability Theory*. Springer-Verlag, Berlin, 1978. **797**

[218] E. Lommel. Die Beugungserscheinungen einer kreisrunden Oeffnung und eines kreisrunden Schirmchens. *Abh. Bayer. Akad. Math. Naturwiss. Kl*, **15**:233–328, 1885. **582, 611**

[219] J. Loomis. A computer program for analysis of interferometric data. In *Optical Interferograms - Reduction and Interpretation*. ASTM International, West Conshohocken, Pa., USA, 1978. **899**

[220] H. A. Lorentz. The theorem of Poynting concerning the energy in the electromagnetic field and two general propositions concerning the propagation of light. *Versl. K. Akad. W. Amsterdam*, **4**:176, 1896. **911**

[221] H. A. Lorentz. *Collected Papers*, volume 2-3. Martinus Nijhoff, The Hague, The Netherlands, 1936. **143, 536**

[222] E. Louis, H.-J. Voorma, N. B. Koster, F. Bijkerk, Y. Y. Platonov, S. Y. Zuev, S. S. Andreev, E. A. Shamov, and N. N. Salashchenko. Multilayer coated reflective optics for Extreme UV lithography. *Microelectron. Eng.*, **27**:235–238, 1995. **536**

[223] W. Lukosz. Der Einfluss der Aberrationen auf die optische Übertragungsfunktion bei kleinen Orts-Frequenzen. *Opt. Acta*, **10**:1–19, 1963. **366**

[224] R. K. Luneburg. *Propagation of Electromagnetic Waves (lecture notes)*. New York University, New York, USA, 1947-1948. **191**

[225] R. K. Luneburg. *Mathematical Theory of Optics*. University of California Press, Berkeley and Los Angeles, 1966. **565, 567, 789**

[226] C. Macías-Romero, R. Lim, M. R. Foreman, and P. Török. Synthesis of structured partially spatially coherent beams. *Opt. Lett.*, **36**(*9*):1638–1640, 2011. **801**

[227] T. G. Mackay and A. Lakhtakia. Electromagnetic fields in linear bianisotropic mediums. In E. Wolf, editor, *Prog. Optics*, volume 51, chapter 3, pages 121–209. Elsevier B.V., 2008. **146**

[228] H. A. Macleod. *Thin-film Optical Filters*. CRC Press Taylor and Francis, London, 3rd rev. edition, 2001. **68**

[229] V. N. Mahajan. *Optical Imaging and Aberrations, Part I, Ray Geometrical Optics*. SPIE Press, Bellingham WA, USA, 1998. **295, 361**

[230] V. N. Mahajan. *Optical Imaging and Aberrations, Part II, Diffraction Optics*. SPIE Press, Bellingham WA, USA, 2001. **768**

[231] D. Malacara, J. M. Carpio, and J. J. Sánchez. Wavefront fitting with discrete orthogonal polynomials in a unit radius circle. *Opt. Eng.*, **29**:672–675, 1990. **610**

[232] L. Mandel and E. Wolf. *Optical Coherence and Quantum Optics*. Cambridge University Press, Cambridge, 1995. **26, 545, 570, 581**

[233] H.-J. Mann, G. Seitz, and W. Ulrich. 8-Spiegel-Mikrolithographie-Projektionsobjektiv. *European Patent EP1199590 A1*, April 24, 2002. **542**

[234] H.-J. Mann, W. Ulrich, and R. M. Hudyma. Reflective projection lens for EUV-photolithography. *US Patent 7199922*, April 3, 2007. **538**

[235] D. Marcuse. *Theory of Dielectric Optical Waveguides*. Academic Press, New York, USA, 2nd revised edition, 1991. **168**

[236] A. Maréchal. Étude des effets combinés de la diffraction et des aberrations géometriques sur l'image d'un point lumineux. *Rev. Opt. Theor. Instrum.*, **26**:257–277, 1947. **361, 599, 681**

[237] A. Maréchal. Study of the combined effects of diffraction and geometrical aberrations on the image of a luminous point. *Rev. Opt.*, **26**:257–277, 1947. **583**

[238] A. Maréchal and M. Françon. *Traité d'optique instrumentale; la formation des images. Tome 2. Diffraction, structure des images, influence de la cohérence de la lumière*. Éditions de la 'Revue d'optique théorique et instrumentale', 1960. **660**

[239] D. W. Marquardt. An algorithm for least-squares estimation of nonlinear parameters. *J. Soc. Indust. Appl. Math.*, **11**:431–441, 1963. **403**

[240] O. J. F. Martin, A. Dereux, and C. Girard. Iterative scheme for computing exactly the total field propagating in dielectric structures of arbitrary shape. *J. Opt. Soc. Am. A*, **11**(*3*):1073–1080, 1994. **857**

[241] T. Matsuyama and T. Ujike. Orthogonal aberration functions for microlithographic optics. *Opt. Rev.*, **11**:199–207, 2004. **360**

[242] M. Matthae, L. Schreiber, A. Faulstich, and W. Kleinschmidt. High aperture objective lens, 2003. US Patent 6,504,653, Jan 7 2003. **849**

[243] D. Maystre. Rigorous vector theories of diffraction gratings. In E. Wolf, editor, *Prog. Optics*, volume 21, chapter 1, pages 1–67. Elsevier, Amsterdam, 1984. **89**

[244] S. C. McClain, L. W. Hillman, and R. S. Chipman. Polarization ray tracing in anisotropical optically active media. II. Theory and physics. *J. Opt. Soc. Am. A*, **10**:2383–2393, 1993. **143**

[245] C. W. McCutchen. Generalized aperture and the three-dimensional diffraction image. *J. Opt. Soc. Am.*, **54**(*2*):240–244, 1964. **585**

[246] P. W. Milonni. *Fast Light, Slow Light and Left-Handed Light*. Institute of Physics Publishing, Bristol, UK, 2005. **186**

[247] M. Minsky. Microscopy apparatus, US Patent 3,013,467, filed 19 December 1961. **711**

[248] T. M. Mitchell. *Machine Learning*. McGraw-Hill, New York, USA, 1997. **409**

[249] M. G. Moharam. Coupled-wave analysis of two-dimensional gratings. In I. Cindrich, editor, *Holographic Optics: Design and Applications*, volume 883 of *Proc. SPIE*, pages 8–11, Bellingham WA, USA, 1988. **100**

[250] M. G. Moharam and T. K. Gaylord. Rigorous coupled-wave analysis of planar-grating diffraction. *J. Opt. Soc. Am.*, **71**:811–818, 1981. **91, 100, 857**

[251] M. G. Moharam, E. B. Grann, D. A. Pommet, and T. K. Gaylord. Formulation for stable and efficient implementation of the rigorous coupled-wave analysis of binary gratings. *J. Opt. Soc. Am. A*, **12**:1068–1076, 1995. **100**

[252] E. Moreno, D. Erni, C. Hafner, and R. Vahldieck. Multiple multipole method with automatic multipole setting applied to the simulation of surface plasmons in metallic nanostructures. *J. Opt. Soc. Am. A*, **19**(*1*):101–111, 2002. **857**

[253] G. Mur. Absorbing boundary conditions for the Finite-Difference approximation of the Time-Domain electromagnetic-field equations. *IEEE Trans. Electromagn. C.*, **EMC-23**(*4*):377–382, 1981. **858**

[254] G. Mur and A. T. de Hoop. A finite-element method for computing three-dimensional electromagnetic fields in inhomogeneous media. *IEEE Trans. Magn.*, **MAG-21**(*6*):2188–2191, 1985. **106**

[255] J.-C. Nédélec. *Acoustic and Electromagnetic Equations. Integral Representations for Harmonic Problems*. Springer Verlag, Heidelberg, 2001. **550**

[256] M. A. A. Neil, R. C. Juškaitis, and T. Wilson. Method of obtaining optical sectioning by using structured light in a conventional microscope. *Opt. Lett.*, **22**:1905–1907, 1997. **725**

[257] M. A. A. Neil, F. Massoumian, R. C. Juškaitis, and T. Wilson. Method for the generation of arbitrary complex vector wave fronts. *Opt. Lett.*, **27**(*21*):1929–1931, 2002. **791**

[258] I. Newton. *Opticks: or, a Treatise of the Reflexions, Refractions, Inflexions & Colours of Light*. Dover Publications (fourth edition, 1730), New York, USA, 1979. **1**

[259] B. R. A. Nijboer. *The Diffraction Theory of Aberrations*. PhD thesis, Rijksuniversiteit Groningen, Groningen, The Netherlands, 1942. Downloadable from: http://www.nijboerzernike.nl. **368, 583, 595, 596, 599, 600, 606, 607, 681, 892, 895, 900, 901, 902**

[260] R. J. Noll. Zernike polynomials and atmospheric turbulence. *J. Opt. Soc. Am.*, **66**(*3*):207–211, 1976. **902**

[261] E. Noponen and J. Turunen. Eigenmode method for electromagnetic synthesis of diffractive elements with three-dimensional profiles. *J. Opt. Soc. Am. A*, **11**:2494–2502, 1994. **91, 98**

[262] A. M. Nugrowati. *Vectorial Diffraction of Extreme Ultraviolet Light and Ultrashort Light Pulses*. PhD thesis, Technische Universiteit Delft, 2008. **91, 98, 99, 100**

[263] A. M. Nugrowati, S. F. Pereira, and A. S. van de Nes. Near and intermediate fields of an ultrashort pulse transmitted through Young's double-slit experiment. *Phys. Rev. A*, **77**(*053810*):1–8, 2008. **100**

[264] A. Offner. New concepts in projection mask aligners. *Opt. Eng.*, **14**:130–132, 1975. **527**

[265] R. Oldenbourg and G. Mei. New polarized light microscope with precision universal compensator. *J. Microsc.*, **180**(*2*):140–147, 1995. **828, 829**

[266] F. W. J. Olver, D. W. Lozier, R. F. Boisvert, and C. W. Clark. *Handbook of Mathematical Functions*. Cambridge University Press, Cambridge, UK, 1st edition, 2010. **44, 47, 492, 595, 756, 879, 901, 904**

[267] R. Oron, S. Blit, N. Davidson, A. A. Friesem, Z. Bomzon, and E. Hasman. The formation of laser beams with pure azimuthal or radial polarization. *Appl. Phys. Lett.*, **77**(*21*):3322–3324, 2000. **791**

[268] C. Palma and G. Cincotti. Imaging of J_0 correlated Bessel-Gauss beams. *IEEE J. Quantum Elect.*, **33**(*6*):1032–1040, 1997. **796**

[269] A. Papoulis. *Probability, Random Variables and Stochastic Processes*. McGraw-Hill, New York, USA, 1st edition, 1965. **420, 423, 778**

[270] M. Paul. Systèmes correcteurs pour réflecteurs astronomiques. *Rev. Opt. Theor. Instrum.*, **14**(*5*):169–202, 1935. **488, 490**

[271] E. R. Peck and K. Reeder. Dispersion of air. *J. Opt. Soc. Am.*, **62**:958–962, 1972. **324**

[272] J. B. Pendry. Negative refraction makes a perfect lens. *Phys. Rev. Lett.*, **85**(*18*):3966–3969, 2000. **179**

[273] J. B. Pendry. Negative refraction. *Contemp. Phys.*, **45**(*3*):191–202, 2004. **68, 160, 179**

[274] R. Penrose. A generalized inverse for matrices. *Proc. Camb. Philos. Soc.*, **51**:406–413, 1955. **404**

[275] F. Perrin. La fluorescence des solutions. *Ann. Phys.-Paris*, **12**:169–275, 1929. **833**

[276] R. Petit, editor. *Electromagnetic Theory of Gratings*. Springer Verlag, Berlin, 1980. **89, 100**

[277] J. Philip. Optical transfer function in three dimensions for a large numerical aperture. *J. Mod. Opt.*, **46**:1031–1042, 1999. **767**

[278] J. Picht. Über den Schwingungsvorgang, der einem beliebigen (astigmatischen) Strahlenbündel entspricht. *Ann. Phys.*, **382**:785–882, 1925. **583**

[279] J. Picht. Die Intensitätsverteilung in einem astigmatischen Strahlenbündel in Abhängigkeit von dem Brennlinienabstand und der Öffnung auf Grund der Wellentheorie des Lichtes. *Ann. Phys.*, **385**:491–508, 1926. **583**

[280] R. Piestun, Y. Y. Schechner, and J. Shamir. Propagation-invariant wave fields with finite energy. *J. Opt. Soc. Am. A*, **17**(*2*):294–303, 2000. **791**

[281] H. Pocklington. Electrical oscillations in wires. *Proc. Camb. Philos. Soc.*, **9**:324–332, 1897.**857**

[282] J. H. Poynting. On the transfer of energy in the electromagnetic field. *Phil. Trans.*, **175**:277, 1884. **11**

[283] W. H. Press, S. A. Teukolsky, W. T. Vetterling, and B. P. Flannery. *Numerical Recipes: The Art of Scientific Computing*. Cambridge University Press, New York, USA, 3rd edition, 2007. **393**

[284] S. Quabis, R. Dorn, M. Eberler, O. Glockl, and G. Leuchs. Focusing light to a tighter spot. *Opt. Commun.*, **179**:1–7, 2000. **791**

[285] L. R. Rabiner, R. W. Schafer, and C. M. Rader. The Chirp z-Transform algorithm. *IEEE Trans. Acoust. Speech*, **17**(*2*):86–92, 1969. **864, 866**

[286] Lord Rayleigh. *Treatise on Sound*, volume II. Macmillan, London, 1878. **911**

[287] Lord Rayleigh. Investigations in optics with special reference to the spectroscope. *Philos. Mag.*, **8**(*49*):261–274, 1879. **657**

[288] Lord Rayleigh. On images formed without reflection or refraction. *Philos. Mag.*, **11**(*67*):214–218, 1881. **411, 812**

[289] R. Redheffer. Difference equations and functional equations in transmission-line theory. In E. F. Beckenbach, editor, *Modern Mathematics for the Engineer*. McGraw-Hill, New York, USA, 1961. **82, 85**

[290] S. R. Rengarajan and Y. Rahmat-Samii. The field equivalence principle: Illustration of the establishment of the non-intuitive null fields. *IEEE Antennas Propag. Mag.*, **42**(*4*):122–128, 2000. **555**

[291] P. Rice, A. Macovski, E. Jones, H. Frohbach, R. Crews, and A. Noon. An experimental television recording and playback system using photographic discs. *J. Soc. Motion Pict. T.*, **79**:997–1002, 1970. **711**

[292] B. Richards. Diffraction in systems of high relative aperture. In Z. Kopal, editor, *Astronomical Optics and Related Subjects*, pages 352–359. North Holland Publishing Company, Amsterdam, 1955. **789**

[293] B. Richards and E. Wolf. Electromagnetic diffraction in optical systems. II. Structure of the image field in an aplanatic system. *Proc. Roy. Soc. Lond. A*, **253**:358–379, 1959. **565, 622, 627, 789**

[294] R. Richter. Zur beugungstheoretischen Untersuchung optischer Systeme. *Z. Instrumentenkd.*, **45**:1–15, 1925.**583**

[295] M. von Rohr. *Theorie und Geschichte des photographischen Objektivs*. J. Springer, Berlin, 1899. **447**

[296] B. Roy Frieden. Optical transfer of the three-dimensional object. *J. Opt. Soc. Am.*, **57**(*1*):56–66, 1967. **725, 734**

[297] J. Ruoff and M. Totzeck. Orientation Zernike polynomials: a useful way to describe the polarization effects of optical imaging systems. *J. Microlith. Microfab.*, **8**(*3*):031404, 2009. **813**

[298] D. R. Sandison, R. M. Williams, K. S. Wells, J. Stricker, and W. W. Webb. Quantitative fluorescence confocal laser scanning microscopy (CLSM). In J. B. Pawley, editor, *Handbook of Biological Confocal Microscopy*, pages 39–53. Plenum, New York, USA, 1995. **847**

[299] A. Schoenle and S. W. Hell. Calculation of vectorial three-dimensional transfer functions in large-angle focusing systems. *J. Opt. Soc. Am. A*, **19**:2121–2126, 2002. **767**

[300] K. Schwarzschild. Untersuchungen zur geometrischen Optik. II. Theorie der Spiegeltelescope. *Abh. Königl. Ges. Wiss. Göttingen, Math. Phys. Kl,* Neue Folge Band IV, **2**:1–28, 1905. **391, 479, 480, 482, 484**

[301] L. Seidel. Über die Entwicklung der Glieder 3ter Ordnung welche den Weg eines ausserhalb der Ebene der Axe gelegene Lichtstrahles durch ein System brechender Medien bestimmen. *Astr. Nach.*, **43**:289–304, 1856. **276, 278, 283, 295, 425, 446**

[302] N. Seong, K. Lai, A. E. Rosenbluth, and R. Gallatin. Assessing the impact of intrinsic birefringence on 157nm lithography. In B. W. Smith, editor, *Optical Microlithography XVII*, volume 5377 of *Proc. SPIE*, pages 99–103, Bellingham WA, USA, 2004. **536**

[303] A. Serebriakov, E. Maksimov, F. Bociort, and J. J. M. Braat. Birefringence induced by the spatial dispersion in deep UV lithography: theory and advanced compensation strategy. *Opt. Rev.*, **12**:140–145, 2005. **536**

[304] H. Severin. Zur Theorie der Beugung elektromagnetischer Wellen. *Z. Phys.*, **129**:426–439, 1951. **556**

[305] B. H. Shakibaei and R. Paramesran. Recursive formula to compute Zernike radial polynomials. *Opt. Lett.*, **38**:2487–2489, 2013. **896**

[306] C. J. R. Sheppard. Stray light and noise in confocal microscopy. *Micron*, **22**:239–243, 1991. **847**

[307] C. J. R. Sheppard, M. Gu, Y. Kawata, and S. Kawata. Three-dimensional transfer functions for high- aperture systems. *J. Opt. Soc. Am. A*, **11**:593–598, 1994. **767**

[308] C. J. R. Sheppard and M. Hole. Three-dimensional optical transfer function for weak aberrations. *J. Mod. Opt.*, **42**:1921–1928, 1995. **767**

[309] C. J. R. Sheppard and K. G. Larkin. Vectorial pupil functions and vectorial transfer functions. *Optik*, **107**:79–87, 1997. **767**

[310] C. J. R. Sheppard and P. Török. Efficient calculation of electromagnetic diffraction in optical systems using a multipole expansion. *J. Mod. Opt.*, **44**:803–818, 1997. **614**

[311] C. J. R. Sheppard and P. Török. Effect of Fresnel number in focusing and imaging. In O. Nijhawan, A. K. Gupta, A. K. Musla, and K. Singh, editors, *Optics and Optoelectronics Theory, Devices and Applications*, chapter 106, pages 635–649. Narosa Publishing House, New Delhi, India, 1999. **564**

[312] S. S. Sherif, M. R. Foreman, and P. Török. Eigenfunction expansion of the electric fields in the focal region of a high numerical aperture focusing system. *Opt. Express*, **16**:3397–3407, 2008. **614**

[313] S. S. Sherif and P. Török. Pupil plane masks for super-resolution in high-numerical-aperture focusing. *J. Mod. Opt.*, **51**(*13*):2007–2019, 2004. **793**

[314] S. S. Sherif and P. Török. Eigenfunction representation of the integrals of the Debye-Wolf diffraction formula. *J. Mod. Opt.*, **52**(*6*):857–876, 2005. **793**

[315] A. E. Siegman. *Lasers*. University Science Books, Mill Valley CA, USA, 1986. **591**

[316] M. Singh and J. J. M. Braat. Design of multilayer extreme-ultraviolet mirrors for enhanced reflectivity. *Appl. Opt.*, **39**:2189–2197, 2000. **536**

[317] R. N. Smartt and J. Strong. Point-diffraction interferometer. *J. Opt. Soc. Am.*, **62**:737, 1972. **600**

[318] T. Smith. On multiple reflection within a symmetrical optical instrument. *J. Opt. Soc. Am.*, **18**(*2*):75–81, 1929. **235**

[319] A. W. Snyder and J. D. Love. *Optical Waveguide Theory*. Chapman and Hall, London, 1983. **830**

[320] P. Soleillet. Sur les paramètres caractérisant la polarisation partielle de la lumière dans les phénomènes de fluorescence. *Ann. Phys.-Paris*, **10**(*12*):23–97, 1929. **833**

[321] A. Sommerfeld. Mathematische Theorie der Diffraction. *Math. Ann.*, **47**:317–374, 1896. **88**

[322] A. Sommerfeld. *Optics*, volume IV of *Lectures on Theoretical Physics*. Academic Press, New York, USA, 1954. **570**, **580**, **581**

[323] C. M. Sparrow. The influence of the condenser on microscopic resolution. *Astrophys. J.*, **44**:76–86, 1916. **658**

[324] E. Spiller. Reflective multilayer coatings for the far UV region. *Appl. Opt.*, **15**:2333–2338, 1976. **536**

[325] J. J. Stamnes. *Waves in Focal Regions*. Adam Hilger, Bristol and Boston, 1986. **545**, **557**, **565**

[326] G. C. Steward. Aberration diffraction effects. *Phil. Trans. Roy. Soc. A*, **225**:131–198, 1926. **583**

[327] G. C. Steward. The aberrations of a symmetrical optical system. *Trans. Camb. Phil. Soc.*, **23**:235–263, 1926. **583**

[328] J. A. Stratton. *Electromagnetic Theory*. McGraw-Hill, New York, USA, 1941. **3**, **12**, **546**, **547**, **550**, **552**, **554**, **561**, **648**

[329] J. A. Stratton and L. J. Chu. Diffraction theory of electromagnetic waves. *Phys. Rev.*, **56**:99–107, 1939. **550**

[330] K. Strehl. *Die Theorie des Fernrohrs auf Grund der Beugung des Lichtes. Teil I*. J.A. Barth, Leipzig, 1894. **583**, **599**

[331] K. Strehl. Aplanatische und fehlerhafte Abbildung im Fernrohr. *Z. Instrumentenkd.*, **15**:362–370, 1895. **498**, **681**

[332] J. J. Sylvester. Sur les puissances et les racines de substitutions linéaires. *C. R. Acad. Sci.*, **XCIV**:55–59, 1882. **255**

[333] A. Taflove. *Computational Electrodynamics: the Finite-difference Time-domain Method*. Artech House, Norwood (MA), USA, 2005. **101**, **180**

[334] A. Taflove and M. E. Brodwin. Numerical solution of steady-state electromagnetic scattering problems using the time-dependent Maxwell's equations. *IEEE T. Microw. Theory*, **23**:623–630, 1975. **101**, **102**, **103**

[335] A. Taflove and S. Hagness. *Computational Electrodynamics*. Artech House, Norwood (MA), USA, 2nd edition, 2000. **857**, **858**

[336] W. J. Tango. The circle polynomials of Zernike and their application in optics. *Appl. Phys.*, **13**:327–332, 1977. **598**, **607**, **609**, **895**, **901**

[337] J. Tervo, T. Setälä, and A. T. Friberg. Theory of partially coherent electromagnetic fields in the space-frequency domain. *J. Opt. Soc. Am. A*, **21**(*11*):2205–2215, 2004. **795**, **798**, **799**

[338] K. P. Thompson. *Aberration Fields in Tilted and Decentered Optical Systems*. PhD thesis, The University of Arizona, Tucson, USA, 1980. **361**

[339] S. C. Tidwell, D. H. Ford, and W. D. Kimura. Generating radially polarized beams interferometrically. *Appl. Opt.*, **29**(*15*):2234–2239, 1990. **791**

[340] P. Török. Focusing of electromagnetic waves through a dielectric interface by lenses of finite Fresnel number. *J. Opt. Soc. Am. A*, **15**(*12*):3009–3015, 1998. **562**

[341] P. Török. Imaging of small birefringent objects by polarised light conventional and confocal microscopes. *Opt. Commun.*, **181**(*1*):7–18, 2000. **829**, **831**, **832**, **833**, **834**, **835**, **836**

[342] P. Török. Propagation of electromagnetic dipole waves through dielectric interfaces. *Opt. Lett.*, **25**(*19*):1463–1465, 2000. **852**

[343] P. Török, P. D. Higdon, and T. Wilson. On the general properties of polarised light conventional and confocal microscopes. *Opt. Commun.*, **148**(*4*):300 – 315, 1998. **783**, **786**, **787**, **810**, **811**, **812**, **813**, **814**, **815**

[344] P. Török, P. R. Munro, and E. E. Kriezis. Rigorous near- to far-field transformation for vectorial diffraction calculations and its numerical implementation. *J. Opt. Soc. Am. A*, **23**(*3*):713–722, 2006. **548**, **550**

[345] P. Török and P. Varga. Electromagnetic diffraction of light focused through a stratified medium. *Appl. Opt.*, **36**(*11*):2305–2312, 1997. **785**, **848**

[346] P. Török, P. Varga, Z. Laczik, and G. R. Booker. Electromagnetic diffraction of light focused through a planar interface between materials of mismatched refractive indices: an integral representation. *J. Opt. Soc. Am. A*, **12**(*2*):325–332, 1995. **609**, **848**

[347] M. van Turnhout and F. Bociort. Chaotic behavior in an algorithm to escape from poor local minima in lens design. *Opt. Express*, **17**:6436–6450, 2009. **408**

[348] M. van Turnhout, P. van Grol, F. Bociort, and H. P. Urbach. Obtaining new local minima in lens design by constructing saddle points. *Opt. Express*, **23**:6679–6691, 2015. **412**

[349] W. Ulrich, H.-J. Rostalski, and R. M. Hudyma. The development of dioptric projection lenses for DUV lithography. In P. K. Manhart and J. M. Sasián, editors, *International Optical Design Conference 2002*, volume 4832 of *Proc. SPIE*, pages 158–169, Bellingham WA, USA, 2002. **533**

[350] S. Ura, T. Suhara, H. Nishihara, and J. Koyama. An integrated-optic disk pickup device. *J. Lightwave Technol.*, **LT-4**(*7*):913–918, 1986. **513**

[351] H. P. Urbach and D. A. Bernard. Modeling latent-image formation in photolithography, using the Helmholtz equation. *J. Opt. Soc. Am. A*, **6**:1343–1356, 1989. **106**

[352] E. M. Vaskas. Note on the Wasserman-Wolf method for designing aspheric surfaces. *J. Opt. Soc. Am.*, **47**:669–670, 1957. **391**

[353] I. M. Vellekoop and A. P. Mosk. Focusing coherent light through opaque strongly scattering media. *Opt. Lett.*, **32**(*16*):2309–2311, 2007. **725**

[354] C. H. F. Velzel and J. L. F. de Meijere. Characteristic functions and the aberrations of symmetric optical systems. II. Addition of aberrations. *J. Opt. Soc. Am. A*, **5**:251–256, 1988. **339**

[355] S. C. Verrall and R. Kakarala. Disc-harmonic coefficients for invariant pattern recognition. *J. Opt. Soc. Am. A*, **15**(*2*):389–401, 1998. **768**, **904**, **905**

[356] V. G. Veselago. The electrodynamics of substances with simultaneously negative values of ϵ and μ. *Sov. Phys. Usp.*, **10**(*4*):509–514, 1967 (1968: English translation). **68**, **160**, **169**, **179**, **186**

[357] W. Wang, A. T. Friberg, and E. Wolf. Focusing of partially coherent light in systems of large Fresnel numbers. *J. Opt. Soc. Am. A*, **14**:491–496, 1997. **581**

[358] W. Wang and E. Wolf. Far-zone behavior of focused fields in systems with different Fresnel numbers. *Opt. Commun.*, **119**:453–459, 1995. **581**

[359] G. D. Wassermann and E. Wolf. On the theory of aplanatic aspheric systems. *Proc. Phys. Soc. B*, **62**:752–756, 1949. **391**

[360] G. N. Watson. *A Treatise on the Theory of Bessel Functions*. Cambridge University Press, Cambridge, UK, second edition, 1995. **575**, **904**

[361] W. T. Welford. Length measurement at the optical resolution limit by scanning microscopy. In P. Mollet, editor, *Optics in Metrology*, pages 85–91. Pergamon Press, New York, USA, 1960. **711**, **713**

[362] W. T. Welford. A new total aberration formula. *Opt. Acta*, **19**:719–727, 1972. **358**

[363] W. T. Welford. On the relationship between the modes of image formation in scanning microscopy and conventional microscopy. *J. Microsc.*, **96**(*1*):105–107, 1972. **713**

[364] W. T. Welford. *Aberrations of Optical Systems*. Adam Hilger (IOP Publishing Ltd), Bristol, UK, 1986. **224**, **295**, **297**, **339**

[365] N. Wiener. Coherency matrices and quantum theory. *J. Math. Phys.*, **7**(*1–4*):109–125, 1928. **795**

[366] N. Wiener. Harmonic analysis and the quantum theory. *J. Franklin Inst.*, **207**(*4*):525–534, 1929. **795**

[367] E. Wigner. Einige Folgerungen aus der Schrödingerschen Theorie für die Termstrukturen. *Z. Phys.*, **43**:601–623, 1927. **609**

[368] R. N. Wilson. *Reflecting Telescope Optics I, II*. Springer, Berlin, 1996. **474**, **480**

[369] T. Wilson and C. J. R. Sheppard. *Theory and Practice of Scanning Optical Microscopy*. Academic Press, London, 1984. **712**

[370] E. Wolf. On the designing of aspheric surfaces. *Proc. Phys. Soc.*, **61**:494–503, 1948. **385**

[371] E. Wolf. A macroscopic theory of interference and diffraction of light from finite sources. II. Fields with a spectral range of arbitrary width. *Proc. Roy. Soc. Lond. A*, **230**:246–265, 1955. **689**

[372] E. Wolf. Reciprocity inequalities, coherence time and bandwidth in signal analysis and optics. *Proc. Phys. Soc.*, **71**:257–269, 1958. **689**

[373] E. Wolf. Coherence properties of partially polarized electromagnetic radiation. *Nuovo cimento*, **13**(*6*):1165–1181, 1959. **795**

[374] E. Wolf. Electromagnetic diffraction in optical systems. I. An integral representation of the image field. *Proc. Roy. Soc. Lond. A*, **253**:349–357, 1959. **622**, **627**

[375] E. Wolf. New theory of partial coherence in the space-frequency domain. Part I: spectra and cross spectra of steady-state sources. *J. Opt. Soc. Am.*, **72**:343–351, 1982. **796**

[376] E. Wolf. *Introduction to the Theory of Coherence and Polarization of Light*. Cambridge University Press, Cambridge, UK, 2007. **795**, **796**, **797**

[377] E. Wolf and Y. Li. Conditions for the validity of the Debye integral representation of focused fields. *Opt. Commun.*, **39**:205–210, 1981. **581**

[378] W. Wöltche. Optical systems design with reference to the evolution of the Double-Gauss lens. In R. E. Fisher, editor, *Proceedings 1980 International Lens Design Conference*, volume 237 of *Proc. SPIE*, pages 202–215, Bellingham WA, USA, 1980. **523**

[379] A. Wu and C. Yang. Evolution of the concept of the vector potential in the description of fundamental interactions. *Int. J. Mod. Phys. A*, **21**(*16*):3235–3277, 2006. **13**

[380] J. C. Wyant and K. Creath. Basic wavefront aberration theory for optical metrology. In *Applied Optics and Optical Engineering*, volume XI, chapter 1, pages 27–39. Academic Press, Cambridge MA, USA, 1992. **897**, **899**

[381] C. G. Wynne. Monocentric telescopes for microlithography. *Opt. Eng.*, **26**:300–303, 1987. **527**

[382] K. Yee. Numerical solution of initial boundary value problems involving Maxwell's equations in isotropic media. *IEEE Trans. Antenn. Propag.*, **14**(*3*):302–307, 1966. **101**, **857**

[383] T. Young. On the mechanism of the eye. *Phil. Trans. R. Soc. Lond.*, **91**:23–88, 1801. **350**

[384] A. Zangwill. *Modern Electrodynamics*. Cambridge University Press, Cambridge, U.K., 2013. **12**, **648**

[385] F. Zernike. Beugungstheorie des Schneidenverfahrens und seiner verbesserten Form, der Phasenkontrastmethode. *Physica*, **1**:689–704, 1934. **583**, **595**, **709**, **888**, **890**, **892**

[386] F. Zernike. The concept of degree of coherence and its applications to optical problems. *Physica*, **5**:785–795, 1938. **498**, **673**, **689**, **692**

[387] F. Zernike. Phase contrast, a new method for the microscopic observation of transparent objects, part I. *Physica*, **9**(*7*):686–698, 1942. **665**

[388] F. Zernike. Phase contrast, a new method for the microscopic observation of transparent objects, part II. *Physica*, **9**(*10*):974–980, 1942. **665**

[389] F. Zernike. How I discovered phase contrast. http://www.nobelprize.org/nobel_prizes/physics/laureates/1953/zernike-lecture.html, 1953. **664**, **815**

[390] F. Zernike. Diffraction theory of the knife-edge test and its improved form, the phase-contrast method (English translation of the original publication in the German language in Physica 1, 689–704, 1934). *J. Microlith. Microfab.*, **1**(2):87–94, 2002. **595**

Author Index

Subject Index